HANDBOOK FOR DESIGNING GREEN CEMENT PLANTS

SECOND EDITION

HANDBOOK FOR DESIGNING GREEN CEMENT PLANTS

SECOND EDITION

S.P. Deolalkar
Director
Deolalkar Consultants

A unit of **BSP Books Pvt., Ltd.**

4-4-309/316, Giriraj Lane, Sultan Bazar,
Hyderabad - 500 095

Handbook for Designing Green Cement Plants, *Second Edition* by
S.P. Deolalkar

Deolalkar Consultants
B-32, Shanti Shikhara Apts.,
Raj Bhavan Road, Somajiguda
Hyderabad - 500 082

Published by :

BSP BS Publications
A unit of BSP Books Pvt., Ltd.
4-4-309/316, Giriraj Lane, Sultan Bazar,
Hyderabad - 500 095
Phone : 040 - 23445688, 23445600
e-mail : info@bspbooks.net
www.bspbooks.net

ISBN : 978-93-89974-42-3 (Hardbound)

Photo on the cover shows
Sagar Cements' *3 mtpa Cement Plant* at Mattampally in Telangana, District Suryapet, Telangana, India.

This Book is for

My beloved daughter Anuradha
and
grandsons Pratik and Sanket

FOREWORD

This revised edition of "Handbook for Designing Green Cement Plants" is a major milestone in the history of literature concerning Cement and Cement-making machinery. There is no book anywhere in the world, which covers as many aspects of this industry as this book.

The author, S.P.Deolalkar is eminently qualified to write this book. He has been part of Cement industry since 1956 when he joined Associated Cement Companies. Since then he has operated cement plants, designed cement plants, manufactured cement plants. There is no aspect of the cement industry with which he has not been actively associated. I have known the author since 1970 as his Trainee Engineer and have been a first hand witness to his extraordinary technical brilliance, his immense capacity to put in very hard work, day in and day out and his superb capability as a Teacher and Mentor of young talent.

For a developing country, cement is a very important raw material for the development and maintenance of its infrastructure. By the end of 2020, India's installed capacity to manufacture cement would be app 500 million tons per year. By any standard this is a huge number. Stupendous efforts have gone into installing this capacity. A very important aspect of these efforts has been the art, science and engineering of the sizing, selection, designing of these cement plants. This Handbook teaches the professionals of this industry how to efficiently deal with all aspects of this. The Handbook starts with the basics of cement and through different chapters covers subjects like what is cement, how cement is made, what raw materials are used, what machinery is used for making cement, how this machinery is sized and selected, how the plant layouts are designed and finalized, etc. The Handbook also discusses how over a period of time cement manufacturing processes have become increasingly efficient resulting in reduced power consumption, reduced heat consumption, reduced NO_X emissions. The author also covers the design, manufacture and use of blended cements. All these efforts are contributing to the 'greening' of the cement manufacturing process.

This Handbook is a must for everyone all over the world who is connected with any aspect of cement. All cement manufacturing companies, all cement-making machinery companies, all cement consultants should have this Handbook in their libraries.

My very best wishes are with the Author and with the Publisher for the global success of this revised edition of the Handbook.

Pradeep Kapoor

Former Managing Director and CEO, FLSmidth India.

Former Managing Director and CEO, Trafalgar House Construction India.

PREFACE TO THE SECOND EDITION

The first Edition of this book appeared in 2008 – a reprint in 2009

Under the name 'Handbook for Designing Cement Plants'

In this Second Edition, the title has been slightly modified to 'Handbook for Deigning Green Cement Plants' to include developments in Cement Plants in the interim period.

So much has happened in the last two decades in the ' World of Cement' in terms of size of plant and its contents.

Capacity of kiln has gone beyond 10000 tpd. Dominating process continues to be Dry Process. There have been a great many improvements in designs of kiln, preheaters and grate coolers so that operating efficiencies have reached new high.

Two decades back awareness for 'Sustainable Development' was beginning to catch on. Today it is the basis in the design of new cement plants.

Conscious efforts are being made all over the World to combat 'global warming' by using alternate fuels, renewable sources of energy and making and using blended cements.

Increased size has brought about changes in types and designs of machinery. VR Mills and Roller Presses have replaced Ball mills. 6 stage preheaters have same or even lower pressure drops compared to old 4 stage prheaters. Grate coolers have new designs like static grate, walking floor and a great many others. This has resulted in substantial increase in operational efficiencies.

Automation in process and quality control has increased substantially to ensure quality of cement produced even at increased scale of operations.

As a result of all these developments, Designing of Cement Plants now has new dimensions. They should be brought out for the Cement Industry of today and of tomorrow.

This has been the motivation behind this revised edition of 'Handbook for Designing Cement Plants'- I mean Green Cement Plants.

I have tried to make it as up to date as possible with the continuing help of Designers and Manufacturers of Cement Machinery, and Auxiliaries therein and of course Cement Producers.

Cement Consultants also play a major role in designing new cement plants.

Some items of process and machinery have now become irrelevant because of developments in size and process and in machinery like wet process kilns, planetary and rotary coolers, even semi dry process. This edition makes only a passing reference to them.

While the the first edition had 3000 tpd kiln as the base size, this Second Edition has 10000 tpd kiln as the base size.

The first edition had a 'CD' for basic data and process calculations. This edition has 'web pages' to furnish readers similar and even better information so that he can do his own calculations according to his/her requirements. Readers will find that 'active' web pages are many more in this edition. In addition to web pages in excel format, I have added a number of 'nomograms' which would serve the same purpose.

Since all new cement plants and expansions of existing plants would be 'green' cement plants, a section has been added covering those aspects of green cement in which project engineers would be normally involved. For in depth treatment my book ' Designing Green Cement Plants' is available.

All in all, I have made an honest attempt to present 'state of the art' situation in designs of new cement plants.

I hope that it will be found to be useful to all concerned in the design and even operation of cement plants. That will be my reward.

I would very much appreciate receiving Readers' feedback on it.

-Author

THANKSGIVING

This is a revision of the first edition of the 'Handbook'. The structure remains the same. I am therefore grateful to all those Individuals and Institutions who helped me bring out that edition.

New Cement Plants are going to be 'Green Cement' Plants. Even existing ones will be moving in that direction. This revision includes aspects of making Green Cement Plants. I am thankful to all those who helped me to bring out my book 'Designing Green Cement Plants'.

I am thankful to Cement Machinery Manufacturers FLSmidth, KHD, Thyssen Krupp, Fives, IKN, Loesche, Pfeiffer, Hazemag BEML Claudius Peters, Siemens for furnishing latest Brochures and literature. They generously allowed me to include photographic plates of their respective products and pertinent text from their brochures and literature.

There are a host of manufactures of supporting machinery like materials handling, dust collectors, fans, compressors and blowers who also helped me in this respect.

In Section 10 titled 'Sources' I have acknowledged the sources from which various details like data, figures and plates were obtained. I am thankful to all of them.

I am thankful to Mr.V.K. Batra M.D. and Mr. Jagdeep Vermsa Sr,. Executive of Holtec Consulting Engineers for helping with information on Capital Costs and Costs of Production.

Mr. Pradeep Kapoor, ex-M.D. and CEO of FL Smidth India and Trafalgar House Construction India Ltd., who has been associated with the Cement Industry for half a century, has been helping me right from the beginning in every possible way. He kindly agreed to write the 'Foreword'. I am thankful to him for his kind words.

I thank Mr. R. Vasudevan, Mr. P. Biksha Reddy, Mr. A.D. Deshpande, Mr. Prajapati, Mr. V.B.Raju for their help and guidance. Mr. P. Rajesh helped me with the drawings.

I found feasibility study by Neco Industries very useful.

I collected material also by visiting 'Google'. In Section 11, I have listed out Companies and Institutions for data on cement plants, capacity and production as also feasibility studies. I am thankful to them for the information obtained from their publications.

Mr. Anil Shah, of BSP Books suggested to me that it was time to revise the 'Handbook'. But for this prompting and encouragement, I might not have taken up this work. I thank him.

I' m also thankful to their executives Mr. Nikunjesh Shah and Mr. Naresh Davergave for their efforts in bringing out this book maintaining their high quality standards in a short time.

Thanksgiving

Photograph on the cover of the book is that of the 3 mtpa capacity of Cement Plant of Sagar Cements Ltd, located at Mattampally, District Suryapet, in Telangana. I am thankful to its President, Projects - Mr. K.Ganesh.

In a book of this kind, it was necessary to refer to a great many Sources. I had to go through articles published in various technical magazines. I have sincerely tried to include as many as I could in Sections 10 and 11. I may have inadvertently left out some sources. I apologize for unintentional omissions.

Last but not the least, my grandson Sanket Gupte helped me with computer related problems all along. My daughter Doctor Anuradha has also been a great support.

I certainly miss my wife Urmila. I dedicate this edition to her memory.

-Author

CONTENTS

Foreword **(vii)**

Preface to the Second Edition **(ix)**

Thanksgiving **(xi)**

Abbreviations **(xix)**

SECTION - 1 BASICS

Chapter 1	Overview of Manufacture of Cement	1-10
Chapter 2	Types of Cements – Their Properties and Uses	11-17
Chapter 3	Processes of Manufacturing of Cement	18-23
Chapter 4	Selection of Process and Machinery	24-29
Chapter 5	Raw Materials, Additives and Fuels	30-37
Chapter 6	Taking up the Design of a Cement Plant	38-46
Chapter 7	Technical Concepts – Process Flow Charts	47-58
Chapter 8	Technical Concepts – Conversion Factors	59-65
Chapter 9	Technical Concepts – Design Margins to Arrive at Capacities	66-68
Chapter 10	Technical Concepts – Running Hours in Various Sections	69-74
Chapter 11	Technical Concepts – Sectional and Individual Capacities	75-86
Chapter 12	Process Gases	87-89
Chapter 13	Storages	90-95
Chapter 14	Machinery Schedules	96-99
Chapter 15	Drawing Specifications for Inviting Offers	100-101
Chapter 16	Properties of Materials that Affect Sizing and Selection of Machinery	102-103
Chapter 17	Expansion	104-111

SECTION - 2 MACHINERY USED IN MAKING CEMENT

Chapter 1	Quarrying Operations	1-6
Chapter 2	Crushers	7-14
Chapter 3	Stacker Reclaimer Systems	15-20

Chapter 4	Grinding Mills	21-29
Chapter 4a	Ball Mills	30-36
Chapter 4b	Vertical Mills	37-46
Chapter 4c	Roller Press	47-52
Chapter 5	Screens, Separators and Classifiers	53-66
Chapter 6	Blending Systems	67-75
Chapter 7	Metering Systems for Raw Meal and Coal	76-82
Chapter 8	Preheaters	83-93
Chapter 9	Calciners	94-107
Chapter 10	Kilns	108-117
Chapter 11	Clinker Coolers	118-127
Chapter 12	Pollution Control	128-144
Chapter 13	Cement Mills and Cement Storage	145-152
Chapter 14	Packing of Cement	153-158
Chapter 15	Despatches of Cement, Bagged and in Bulk and Despatches of Clinker	159-166
Chapter 16	Material Handling Systems	167-181
Chapter 17	Fans	182-192
Chapter 18	Compressors and Blowers	193-199
Chapter 19	Motors	200-202
Chapter 20	Gearboxes	203-205

SECTION - 3 TECHNOECONOMIC FEASIBILITY STUDIES

Chapter 1	Investment Decision	1-2
Chapter 2	Setting up a Cement Plant Techno-Economic Feasibility Studies	3-14
Chapter 3	What should the T.E.F.S Contain	15-16
Chapter 4	Demand Forecast and Market	17-20
Chapter 5	Limestone Deposits	21-26
Chapter 6	Environment	27-31
Chapter 7	Infrastructural Facilities	32-37
Chapter 8	Manpower Requirements	38-44
Chapter 9	Implementation Schedule	45-52
Chapter 10	Erection	53-56
Chapter 11	Commissioning	57-61
Chapter 12	Capital Costs of the Project	62-75
Chapter 13	Costs of Production	76-86
Chapter 14	Profitability and Investment Analysis	87-89

Chapter 15 Executive Summary and Conclusion 90-90
Chapter 16 Factors Governing the Size of a Cement Plant 91-93

SECTION - 4 CIVIL DESIGN AND CONSTRUCTION

Chapter 1 Civil Design and Civil Engineering 1-11
Chapter 2 Non-Factory Buildings 12-13
Chapter 3 Design Aspects of some of the Non-Factory Building 14-19
Chapter 4 Colony, Guest Houses and Clubs 20-24
Chapter 5 Storages, Silos, Bins and Hoppers 25-30
Chapter 6 Water Supply 31-38

SECTION - 5 ELECTRICALS AND INSTRUMENTATION

Chapter 1 Electrical Systems, General 1-15
Chapter 2 Power Distribution and Cables 16-32
Chapter 3 Power Consumption and Its Calculation 33-40
Chapter 4 Instrumentation and Process Control 41-64
Chapter 5 Lighting 65-66
Chapter 6 Communications 67 68
Chapter 7 Quality Control and Laboratory 69-71
Chapter 8 Computer in Cement Industry 72-74

SECTION - 6 LAYOUTS AND DETAILED ENGINEERING

Chapter 1 Cement Plants General Layouts 1-43
Chapter 2 Departmental Layouts 44-46
Chapter 3 Layouts of Crushing Plants 47-59
Chapter 4 Auxiliaries in Crushing Section 60-62
Chapter 5 Storage of Crushed Lime Stone and Stacker Reclaimer Systems 63-72
Chapter 6 Layouts of Raw Mills and Separators 73-99
Chapter 7 Maintenance of Mills 100-106
Chapter 8 Layouts of Coal Mills 107-112
Chapter 9 Conveying of Raw Meal / Pulversied Coal / Cement 113-118
Chapter 10 Batch Blending and Storage Systems for Raw Meal 119-124
Chapter 11 Continuous Blending Systems 125-131
Chapter 12 Kiln Feed Systems 132-136
Chapter 13 Pyroprocessing Section 137-138
Chapter 14 Preheaters Lepol Grate 139-141
Chapter 15 Cyclone Preheaters 142-166
Chapter 16 Preheater Fan 167-174
Chapter 17 Calciners 175-186

Chapter 18	Tertiary Air Duct, Dust Chamber and Connecting Duct	187-196
Chapter 19	Rotary Kiln	197-226
Chapter 19A	Coal and Oil Metering and Firing Systems	227-236
Chapter 20	Clinker Coolers - Planetary Coolers	237-239
Chapter 21	Grate Coolers	240-251
Chapter 22	Clinker Coolers, Collecting Spillage and Product	252-261
Chapter 23	Cooling Air Fans	262-268
Chapter 24	Clinker Coolers - Cooler Vent and Dust Collectors	269-277
Chapter 25	Other Clinker Coolers	278-280
Chapter 26	Clinker Conveying and Storage	281-299
Chapter 27	Handling Coal and Gypsum	300-305
Chapter 28	Handling Slag to make Slag Cements	306-313
Chapter 29	Handling Fly Ash to Make Pozzolana Portland Cement	314-321
Chapter 30	Cement Mills	322-334
Chapter 31	Cement Storage	335-338
Chapter 32	Cement Storage and Packing	339-344
Chapter 33	Bags Godown	345-348
Chapter 34	Layout of Packing Plant and Dispatches	349-357
Chapter 35	Simultaneous Loading of Trucks and Wagons	358-363
Chapter 36	Bulk Loading by Road and by Rail	364-367
Chapter 37	Bulk Loading of Clinker	368-376
Chapter 38	Exporting Cement	377-383
Chapter 39	Import of Coal	384-384
Chapter 40	Railway Siding	385-390
Chapter 41	Venting	391-404
Chapter 42	Chutes	405-415
Chapter 43	Ducts	416-453
Chapter 44	Chimneys	454-458
Chapter 45	Pipelines and Supports for Compressed Air and Pneumatic Conveying Pipe Lines	459-464
Chapter 46	Oil Lubrication and Cooling Water Systems	465-471
Chapter 47	Bed Plates	472-476
Chapter 48	Foundations and Load Data	477-482
Chapter 49	Civil Drawings and Design of Departmental Buildings	483-484
Chapter 50	Working Platforms	485-488
Chapter 51	Conveyor Gantries	489-494
Chapter 52	Lifts, Hoists and Lifting Tackles	495-499
Chapter 53	Safety	500-508
Chapter 54	Sampling	509-511

Chapter 55	Refractories	512-526
Chapter 56	Insulation and Lagging	527-528
Chapter 57	Fans	529-531
Chapter 58	Compressors and Blowers	532-533
Chapter 59	Good Housekeeping	534-535
Chapter 60	Co-generation	536-538

SECTION - 7 SELECTING AND ORDERING MACHINERY

Chapter 1	Ordering Machinery	1-2
Chapter 2	Formulating Enquiries	3-6
	ENQS	7-13
Chapter 3	Technical Data Sheets	14-15
	TDS 1-8	16-29
Chapter 4	Evaluation	30-32
Chapter 5	Crtical Operational and Design Parameters	33-35
Chapter 6	Time Factor	36-36

SECTION - 8 SUSTAINABLE DEVELOPMENT

Chapter 1	Sustainable Development	1-3
Chapter 2	Blended Cements & Designing Cement Plants to make Blended Cements	4-9
Chapter 3	Alternate Fuels and Raw Materials	10-18
Chapter 4	Co-Generation	19-22

SECTION - 9 WEB PAGES

Supplementary materials for this Section - 9 is a available in http://www.bspublications.net/9789389354829/

1. General
2. Psychrometry
3. Properties of Materials
4. Specific Heats
5. Hardness and Grindability
6. Temperatures and Pressures
7. Slopes and Gradients
8. Comparison of Sieves Blaines and Wagner
9. Fuels Calorific Values
10. Combustion and Products of Combustion
11. Kiln Gases
12. Drying

13. Quality Control Formulae
14. Log Mean Temperature Difference
15. Heat Balances
16. Quarry Equipment
17. Crushers, Screens
18. Grinding Media
19. Ball Mills
20. Vertical Mills
21. Separators
22. Fan Laws
23. Preheaters
24. Kilns
25. Burners
26. Grate Coolers
27. Materials Handling Mechanical Conveyors
28. Pneumatic Conveyors
29. Vent Volumes and Dust Burdens
30. Log Log, Semi Log Sheets
31. Motors
32. Gearboxes

(Web Page 2)

CALCULATIONS

W2.C1 Workout for Arriving at Capacity of AF Firing System
W2.C2 Workout for Calculating Nett Heat Available from Waste Gases
W2.C3 Heat Content of Waste Gases at Different Inlet and Outlet Temperatures
W2.C4 Design of Coal Firing Pipe with Ventury
W2.C5 Calculation of Pressure Drop in a Branched Circuit
W2.C6 Balancing in a Simple Parallel Circuit
W2.C7 Sizing Silo for Clinker
W2.C8 Sizing Continuous Blending Silo
W2.C9 Sizing Storage Silo for Cement
W2.C10 Calculation of Dimensions of Clinker Storage
W2.C11 Sp. Gas Volume from Preheater
W2.C12 Vent Air from Clinker Cooler in nm^3/kg Clinker

FIGURES

W2.F.1	Nomogram Showing Relation between Carbonates and Loss of Ignition
W2.F2.1	Nomogram for Calculating various Moduli
W2.F3.1	Pertinent Capacities in TPD OPC, PPC, BFSC etc
W2.F3.2	Pertinent Capacities in TPD of Gypsum, Iron ore, Clay
W2.F.4	Capacity of a Calciner Kiln for ……
W2.F.5	Nomogran for Calculating Capacity of Linear Stock Pile
W2.F.6	Nomogram for Calculating Capacity of Linear Stock Pile for Coal
W2.F.7	Hourly Capacities

TABLES

W2.T.1	Cement to Clinker Ratios
W2.T.2	Capacity in MTPA from Capacity in TPD for OPC and Blended Cements
W2.T.3	Factors for Blended Cements
W2.T.4	Overall Cement Clinker Ratios for Different Proportions of Blended Cements
W2.T.5	Conversion and Multiplying Factors
W2.T.6	Running Hours in Various Sections
W2.T.7	Working out Sectional Capacities
W2.T.8	Capacities of Cement Mills making Blended Cements
W2.T.9	Desirable Storages in Cement Plants
W2.T.10	Storages in a Plant of 10000 TPD Capacity

SECTION 10 - SOURCES

	Sources	1-1
1.	Reference Books	2-2
2.	Technical Periodicals	3-3
3.	Reference Manuals	4-4
4.	Sources of Plates	5-8
5.	Sources of Layouts	9-9
6.	Sources of Text and Figures	10-15

SECTION 11 - RECOMMENDED READING

1.	Recommended Reading Books/Reports/Articles	1-1
2.	Recommended Sources of Data on Cement Production and Capacity	2-2

ABBREVIATIONS

Abbreviation	Full form
A / amp	amperes
a.c.	air conditioner
AC/ a.c.	alternating current
AF	alumina ratio
Al_2O_3	alumina
AM	alumina modulus
Annex.	Annexure
Atmos.	atmospheric pr.
b.c.	belt conveyor
B.D. / b.d.	bulk density
B.F.	burnability factor
B.G./ b.g.	broad gauge
B.I.	burnability index
BE	back end (of kiln)
B.F.S.C./ b.f.s.c.	Blast Furnace Slag Cement
B.H.N.	brinell hardness number
B.I.S.	Bureau of Indian Standards
B.T.U. / b.t.u.	British thermal unit
BZ	burning zone (of kiln)
C	carbon
C.E.O.	Chief Executive Officer
c.i.	cast iron
C.I.F./ c.i.f.	cost insurance and freight
C.S.T.	central sales tax
C_2S	Di calcium silicate

(xxi) *Abbreviations*

C_3S	Tri calcium silicate
C_3A	Tricalcium aluminate
C_4AF	Tetra calcium aluminoferrite
$Ca\ CO_3$	calcium carbonate
Ca	calcium
CaO	calcium oxide
CCS	cold crushing strength
CCTV	closed circuit television
CH_4	methane
Cl_2	chlorine
cm	centimeter
cm^2	square centimeter
cm^3 / c.c.	cubic centimeter
CMA	Cement Manufacturers' Association
CO	carbon monoxide
CO_2	carbon dioxide
D.B.C. / d.b.c.	deep bucket conveyor
d.g.set	diesel generating set
D.O.L.	direct on line
D/E ratio	debt / equity ratio
D.C. / d.c.	direct current
DDC	direct digital control
dia.	Diameter
E.O.T	electric overhead traveling
ENQS	enquiry specifications
E.S.I.S.	Employees' State Insurance Scheme
ESP / esp	electrostatic precipitator
f.a.d.	free air delivered at 20 °C
F.I.	financial institution
f.k.pump	fuller kinyon pump
F.O.B./f.o.b.	free on board
F.O.R. / f.o.r.	free on rail
Fe_2O_3	iron oxide
Fig.	figure
FLS	F.L.Smidth
G.M.	general manager
GCT / gct	gas conditioning tower
gm	gramme
gms/cc	grammes per cubic centimeter

h.d.p.e.	high density polyethylene
H.O.	head office
H.P.	Himchal Pradesh
h.s.d.	high speed diesel oil
H.T. / h.t.	high tension
H_2SO_4	sulphuric acid
HHV/hhv	high heat value
HM	hydraulic modulus
HP / hp	horse power
hr	hour
ht	height
I & P C	instrumentation and process control
I.C.I.C.I.	Industrial Credit and Investment Corporation
I.D.B.I.	Industrial Development Bank of India
I.F.C.I.	Industrial Financial Corporation of India
I.F.M.R.	Institute of Financial Management Resources
I.R.R.	internal rate of return
I.T.I.	Industrial Training Institute
ILC	in line calciner
IR	insoluble residue
ISO	International Standards Organisation
Jr.	junior
K_2O	potassium oxide
KCAL / kcal	Kilo calorie
Kcal /kg	kilo calorie per kilogram
KG / kg	Kilogramme
kg/kg	kilogramme/kilogramme
Kg/cm^2	load in kilogrammes per sq. centimetre
kg/m^3	kilogrammes per cubic metre
km	kilometer
KV	kilovolts
KVA	kilovolt amperes
KW /kw	kilowatt
KWH /kwh	kilowatt hour
kwh/ton	kilowatt hours/ton
L.O.I./ l.o.i.	loss on ignition
L.S.F.	lime saturation factor
l.s.h.s.	low sulphur high sp. gravity
L.T. / l.t.	low tension
l/d ratio	length to diameter ratio

(xxiii) ***Abbreviations***

LHC	low heat cement
LHV/lhv	low heat value
M / m	metre
M.C.C.	motor control center
M.D.	Managing Director
M.G. / mg	metre gauge
M.I.S.	management information system
M.T.P.A. / m.t.p.a.	million tons per annum
m^2	square metre
m^3	cubic metre - actual volume
max	maximum
MD	maximum demand
mgm	milligramme
mgm/nm^3	milligram per normal cubic metre
mgm/m^3	milligram per cubic metre
MgO	magnesium oxide
min.	minute / minimum
mm	milli metre
mmwg	pressure /draft in mms of water above /below atmos.pressure
MVA	mega volt amps = 1000 * kva
N_2	nitrogcn
Na_2O	sodium oxide
NCBM	National Council of Cement and Building Materials
nm^3	normal cubic metre
NO	nitrous oxide
No. /no.	number
NO_2	nitrogcn dioxide
NO_x	nitrogen oxides
$^o c$ / $^o C$	degrees centigrade
$^o f$ / $^O F$	degrees Fahrenheit
O_2	oxygen
OPC / opc	ordinary portland cement
P & I	process and instrumentation
P.C.	personal computer
P.C.B.	Pollution Control Board
P.D.S.	power distribution system
p.f.	power factor

P.L.C.	programmable logic controller
P.O.	post office
P_2O_5	phospherous pentoxide
Para	paragraph
PPC / ppc	Portland Pozzolana Cement
RDSO	Research Design & Standards Organisation
RITES	Rail India Technical & Economic Services
R.M.C.	ready mix concrete
r.m.s.	root mean square
r.p.m.	revolutions per minute
RBI	Reserve Bank of India
RCC	reinforced cement concrete
RHC	rapid hardening cement
RSP	reinforced suspension preheater
S	sulphur
s.c.	screw conveyor
S.M.	silica modulus
S.R.	silica ratio
S.T.	sales tax
S/D	star delta
SC/ sc	squirrel cage
SCA / sca	specific collection area
sec.	Second
SiO_2	silica / silicon dioxide
SLC	separate line calciner
SO_2	sulphur dioxide
Sp. gr.	Specific gravity
Sq.cm/gm	square centimeter per gramme
Sq.cm	Square centimeter
Sr	senior
SR/ sr	slip ring
SRC	sulphate resistant cement
SRR	stator rotor resistance
STD	straight trunk dialing
T.E.F.R.	techno economic feasibility report
T.E.F.S.	techno economic feasibility studies
T.P.A. / t.p.a.	Tons per annum
T.P.D. / t.p.d.	Tons per day
T.P.H. / t.p.h.	Tons per hour
T.P.S.	thermal power station
T/m^3	tons per cubic meter

(xxv) *Abbreviations*

TDS	technical data sheet
Temp.	temperature
TiO_2	titanium dioxide
Transfr	transformer
TV	television
U.P.S.	uninterrupted power service
V	volts
v.r.m./V.R.Mill	vertical roller mill
V.S.Kiln	vertical shaft kiln
VAT	value added tax
VDU	visible display unit
v.i.p.	very important person
v.v.i.p.	very very important person
w.r.t.	with respect to
WPC	white cement
wt	weight
yr	year

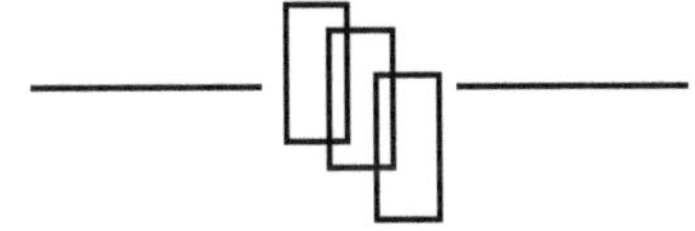

SECTION - 1

Basics

Contents

Chapter 1	Overview of Manufacture of Cement	1-10
Chapter 2	Types of Cements – Their Properties and Uses	11-17
Chapter 3	Processes of Manufacturing of Cement	18-23
Chapter 4	Selection of Process and Machinery	24-29
Chapter 5	Raw Materials, Additives and Fuels	30-37
Chapter 6	Taking up the Design of a Cement Plant	38-46
Chapter 7	Technical Concepts Process Flow Charts	47-58
Chapter 8	Technical Concepts Conversion Factors	59-65
Chapter 9	Technical Concepts Design Margins to Arrive at Capacities	66-68
Chapter 10	Technical Concepts Running Hours in Various Sections	69-74
Chapter 11	Technical Concepts Sectional and Individual Capacities	75-86
Chapter 12	Process Gases	87-89
Chapter 13	Storages	90-95
Chapter 14	Machinery Schedules	96-99
Chapter 15	Drawing Specifications for Inviting Offers	100-101
Chapter 16	Properties of Materials that Affect Sizing and Selection of Machinery	102-103
Chapter 17	Expansion	104-111

CHAPTER 1

OVERVIEW OF MANUFACTURE OF CEMENT

1.1 Beginning of Cement

"Cement" as Portland Cement was first made in a shaft kiln using dry process and later in rotary kilns.

That "Slaked" lime hardens with water was well known and was used as "Mortar" with sand in construction industry before the advent of cement.

Some "Natural rocks" contained all ingredients like CaO, SiO_2, Al_2O_3 and Fe_2O_3 in approximately right proportions so that they did not need any additions and when ground calcined and sintered in a kiln produced clinker which when ground with 5% gypsum produced what has come to be known as "Portland Cement".

Cement has hydraulic properties like slaked lime and hardens when mixed with water. Compressive strength increases in time and reaches its practical top limit after 28 days.

Mixing crushed stone, sand, cement and water makes "Concrete". When hardened it is like rock and hence is called "Synthetic rock". It has similar properties of high compressive but low tensile strengths.

When concrete is poured around steel it becomes Reinforced Cement Concrete - popularly known as RCC - and has high tensile strength also.

RCC has revolutionized construction industry and it is well nigh unimaginable to construct roads, dams, skyscrapers and silos and many other large and heavy buildings for residential or for industrial purpose without RCC.

Cement the main strength giving and binding ingredients is thus an all-important part of RCC and thus plays a vital role in the progress and development of a nation.

At present there is no substitute for cement. Hence, it will continue to play an all-important role in construction industry.

Yardsticks like inter alia per capita consumption of steel, power and cement are used to indicate state of development of countries. Advanced and developed countries have per capita consumption of cement of 400-500 kgs. As against it, in India per capita consumption of cement is only about 195/200 kgs.

1.2 Making Cement

Cement Industry started in a very small way, first as shaft kilns using dry process.

When it was found that the proper composition of raw mix required to making good quality clinker almost always needed, additions or correcting materials to compensate for constituents like Silica, Alumina and Iron Oxides, it became necessary to "Blend" the constituents after "Grinding".

Blending was then more convenient in wet stage in the form of slurry.

By this time Rotary Kiln had come to be used to make Cement. It would conveniently receive slurry as well. Thus process of cement making changed from Dry to Wet.

1.3 Dry to Wet to Dry Process of Manufacture

Wet Cement plants continued to grow in number and size and wet process was the predominant process of manufacture of cement till 1950 or so. It continued to be the dominant process in India for another two decades.

Wet process was simple and required less process control and instrumentation and manpower. But it consumed a large quantity of water and also heat energy in drying the slurry.

As fuel costs rose, alternative processes were investigated to reduce water content of slurry and thereby fuel consumption.

1.4 Semi-dry Process

Thus came into use 'semi-dry' process which needed only 8-10% water compared to 35-36% for wet slurry. Raw materials were ground and blended dry. Water was added to dry raw mix in a revolving pan to make nodules. The nodules were dried on a travelling grate preheater before feeding them to a rotary kiln or to a shaft kiln.

1.4.1 V. S. Kilns

Vertical Shaft Kilns also came to be developed for capacities up to 300 tpd (in Europe). They needed low volatile fuels like coke breeze.

1.5 Dry Process

When wet process plants had reached their peak capacity of 750-1000 tpd, developments in processes and machinery took place that once again changed the course of cement making.

1.6 Suspension Preheaters

In early 50s of the 20th Century, an epoch making concept was developed - that of Suspension Preheater.

The suspension preheater, with rotary kiln and grate cooler formed the heart of the cement plant and 'dry process' came to be adopted fast and number of wet process plants and semi dry process plants declined.

Even in India, which had predominance in wet process plants in numbers and capacity as late as sixties, the percentage has decreased to less than 3 % now.

1.7 Vertical Mills and Calciners

Other epoch making developments took place in 70s. They were using Vertical mills for grinding raw materials and coal and development of calciners which calcined raw meal before it entered the kiln. With a calciner, output of the same kiln could be increased by about two and a half times.

Vertical mills have also came to be used for grinding cement clinker and slag.

Roller Presses are also now part of grinding syatems

These two developments gave tremendous boost to the size of the plant and also to economies in power consumption.

1.8 Benchmarks in Manufacture of Cement

Progress in making cement described above has been shown in **Table 1.1.1** and pictorially in **Figs. 1.1.1(a), 1.1.1(b) and 1.1.2**

1.9 Fuel and Power Consumption

Fuel consumption was also steadily brought down from 1500 Kcal/kg clinker for wet process kilns to 800 Kcal/kg clinker for dry process kilns with 4 stage preheater and grate cooler. This has further come down to less than 700 Kcal/kg by using 6 stage preheaters.

Power Consumption has come down to 85-90 Kwh/Ton.

1.10 Green Cement

Cement industry is now conscious of the emission of green house gases emitted in the process of making cement. Green house gases cause global warming.

Cement Industry is therefore taking steps to reduce these emissions. Therefore it makes ' blended cements' like Pozzolana cement (PPC) and slag cement (BFSC) wherever possible. It also uses Alternate fuels to reduce emissions arising out of combustion.

Cement Industry in countries, like India which were short of power, had begun to install captive power plants to ensure continuity of power. This is taken one step further by using waste heat in exit gases from kiln and cooler to generate power.

These new developments have been dealt with in this new edition to the extent cement plant engineers are commonly required to deal with them.

1.11 Differences in Processes

Differences in various processes of making cement and equipment used therein have been brought out in **Tables 1.1.2 and 1.1.3 and in Fig. 1.1.2.**

1.12 Size of the Plant

From beginning with a 20-30 t.p.d. capacity, individual kilns have gone up in size upto 7500-10,000 tpd capacity. Plants of 3-4 mtpa capacity in one place have also become common.

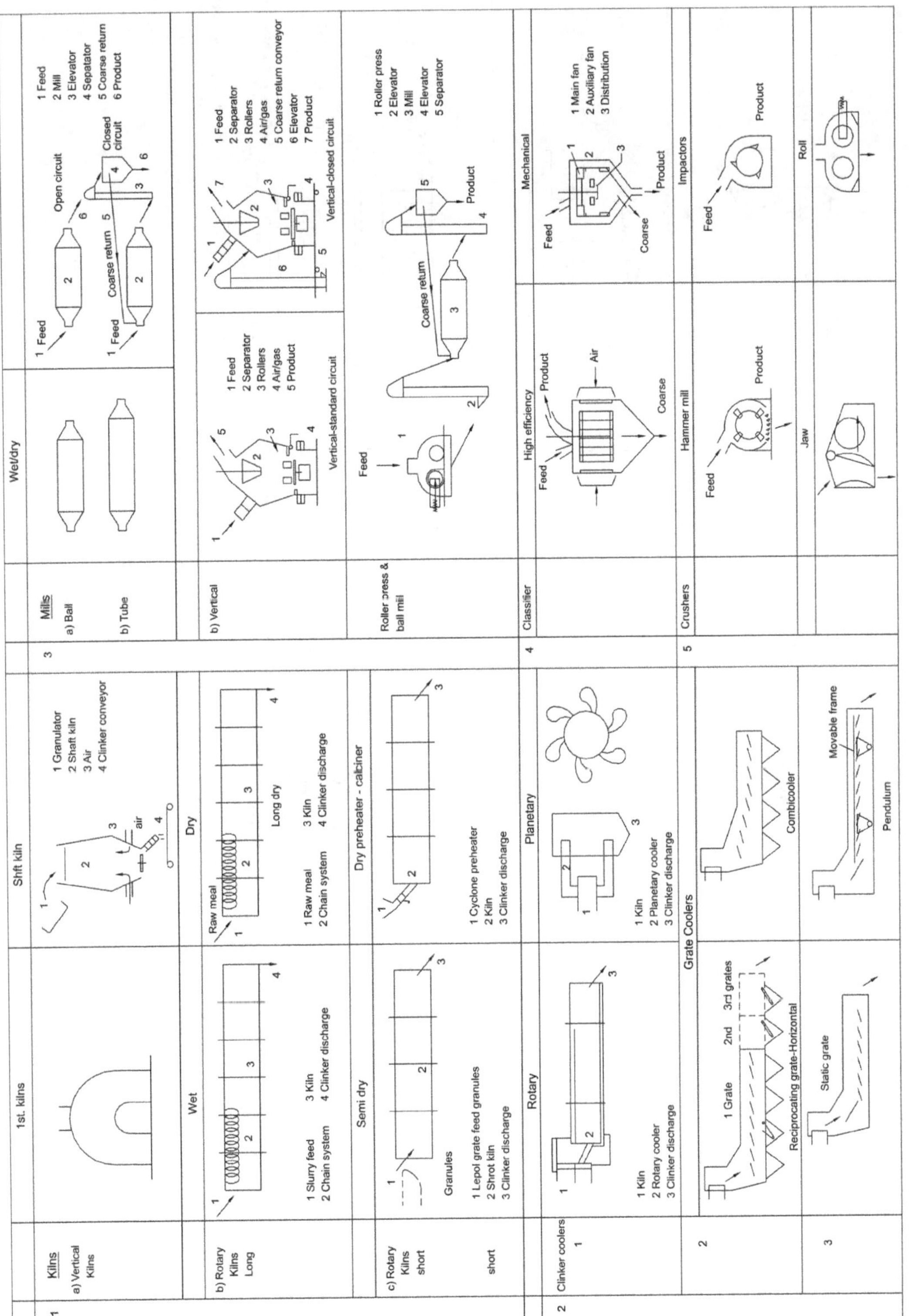

Fig. 1.1.1 (a) Benchmarks in manufacture of cement.

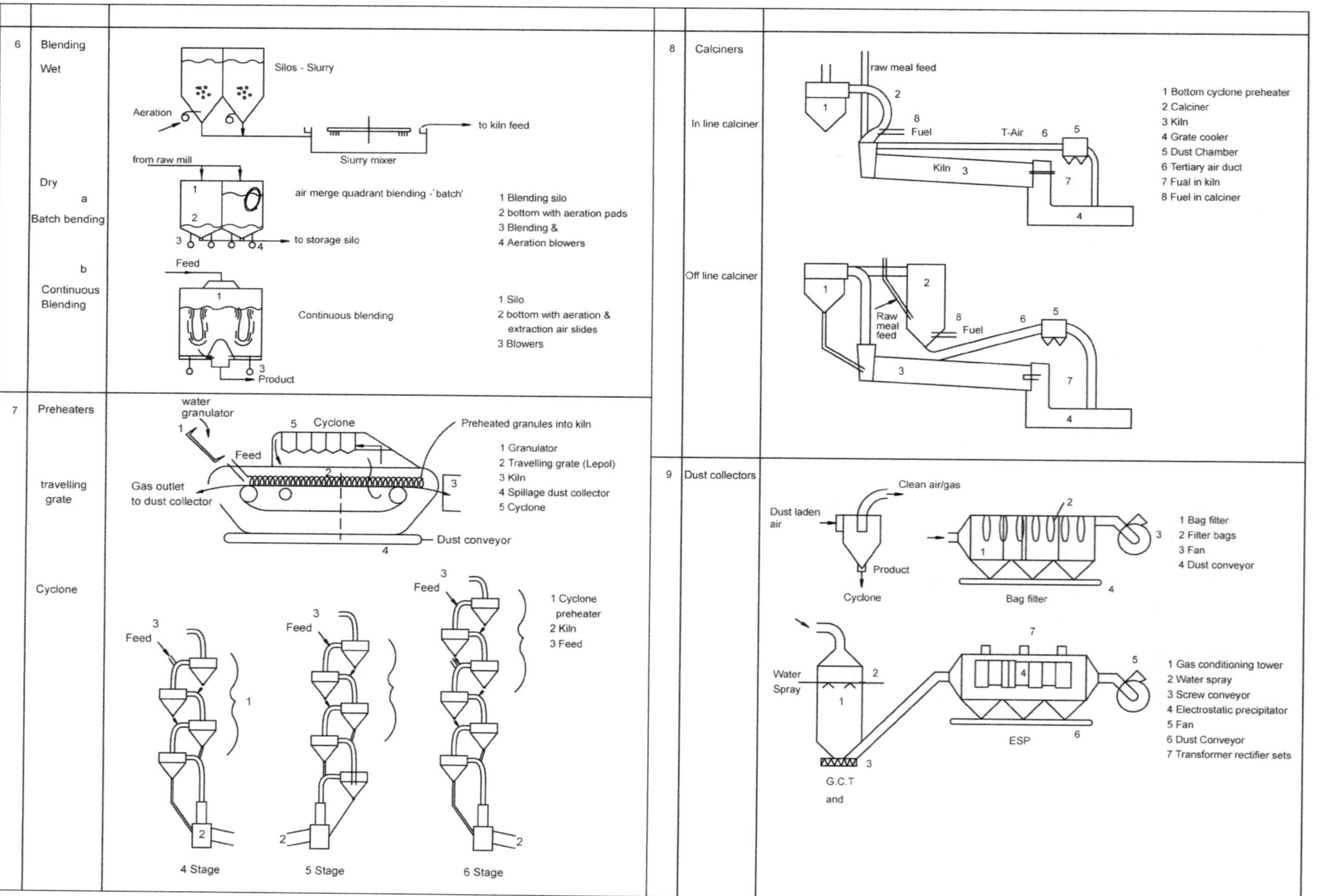

Fig. 1.1.1 (b) Benchmarks in manufacture of cement.

Table 1.1.1 Benchmarks in manufacture of cement.

Benchmarks	
Kilns	Shaft kiln-Rotary Kiln - Wet - Long dry - Short - Semi dry and Dry preheater kilns
Coolers	Rotary Cooler – planetary cooler; Traveling and Reciprocating grate coolers; Static grate, pendulum coolers and cross bar coolers
Preheaters	Calciners in wet kilns, traveling grate preheaters; Suspension preheaters 4 – 6 stages.
Mills	Ball and tube mills – mills with slide shoe bearings; Vertical ring and roller mills; Roller press and ball mill combinations; Horizontal Roller mill
Blending	Slurry blending – slurry mixer; Air merge blending – batch and continuous
Classifiers	Wet classifiers; Dry classifiers- grit separators, mechanical air separators; High efficiency separators
Crushers	Jaw, Hammer, Roll and Impact, mobile and semi mobile crushers
Packing machines	Stationary – rotary – rotary electronic
Calciners	In and off line; spouted bed, fluidized bed and many others
Despatches	Mechanized-automated loading of bagged cement; Bulk cement by road and rail and sea
Dust collectors	Cyclones; poly and multiclones; Bag filters- glass bag filters; Gravel Bed filters; Electrostatic Precipitators

The trend even in India is to go in for large plants to avail of economies of scale which can be achieved by using machineries (Vertical Mills, ESPS, etc.) which are suitable for and affordable by large plants.

The days of small plants and particularly those of VSKs are over and they would not be considered for any future cement projects.

1.13 Sizes of Cement Plants in India

Even in conventional plants using rotary kilns, there was a distinction.

1. Mini Plants of capacity - 300 tpd - to start with- but expanded to 1000-1200 tpd capacity by installing calciners
2. Large Plants of capacity - 600 tpd and above earlier, 3000 tpd and now to 10000 tpd above

Thus as far as process and types of machinery used are concerned, the distinctions between large and small plants have almost blurred. New projects are almost invariably for dry process plants of +1.0 MTPA capacity.

1.13.1 Large Cement Plants

There are already about 183 cement plants of +1 MTPA capacity. Average capacity per plant is 1.7 mtpa. Total installed capacity is 480 mtpa.

Table 1.1.2 Basic Differences between Various Processes of Manufacture of Cement.

Sr. No.	Section	Wet	Semi Wet	Semi Dry	Dry	Dry with calciner
(A)	**Raw material preparation**					
1.	Crushing	Common to all processes;machinery selected depends on size of plant and properties of stone				
2.	Grinding	Wet grinding in ball mills sometimes in closed circuit; product slurry with 35-40% water		Dry grinding; drying during grinding; mostly in closed circuit; using kiln gases for drying. ball mills, vertical mills and roller press in hybrid grinding product dry raw meal with less than 1% moisture		
3.	Homogenising	wet, air and mechanical agitation product blended slurry wth 35 % water		fluidisation techinques for blending dry pulverised raw meal batch or continuous blending;continuous blending will be preceded by prehomogenising in stock piles; product dry blended raw mix		
4.	Kiln feed	slurry with ~ 34 % moisture synchronised with kiln	either extruded pellets with 15% moisture or, dry raw meal dried in flash dryer	pellets with 8-10% % moisture	dry pulverised raw meal fed through volumetric or gravimetic feeders	

Table 1.1.2 *Contd....*

Sr. No.	Section	Wet	Semi Wet	Semi Dry	Dry	Dry with calciner
	(B) Pyroprocessing					
1. 2.	drying preheating	in kiln in kiln	drying of pellets in travelling grate preheater preheating in preheater	drying of nodules	preheating in kiln for long dry kiln	preheating outside kiln in dry preheater kilns; mostly cyclone preheaters
3.	calcining	in kiln	dissociation of CO_2 beginning at ~ 600 °C and completing at ~ 950 °C in kiln	largely in kiln	partly in preheater balance in kiln	almost totally out of kiln in calciner
4.	sintering		formation of clinker at 1250-1450 °C in kiln			
5.	clinker cooling		cooling of clinker to 65- 150 °C common to all processes			
			now mostly reciprocating grate coolers with variations like static grate, controlled flow and pendulum type are usec			

Table 1.1.3 Different types of Equipment used in various Processes.
also see Fig. 1.1.2

Sr. No.	Process	Wet	Semi Wet	Semi Dry	Dry	Dry with Calciner
1.	crushing	common to all two stage crushing for ball mills; single stage crushing for Vertical mills and Roller Presses Jaw crusher – Hammer crusher combination for two stage crushing Impactors – single or two stage for single stage crushing semi mobile, mobile crushers for large plants				
2.	grinding	ball mill open or closed circuit autogenous mill		ball mills-air-swept or bucket elevator in closed cir cuit with conventional or high efficiency separators vertical roller mills - vrms with external circuit and high efficiency separators roller press and ball mill in various combinations		
3.	prehomogenising			stacker reclaimer systems used for prehomogenising of limestones and coal during building up of and extraction from stock piles		
4.	homogenising	pneumatic and mechanical agitation slurry mixers		blending systems basedon 'fluidising' techniques batch and continuous blending systems		
5.	kiln feed	metering of slurry synchronised with kiln	filter press and disagglomerator and flash dryer or filter press and extruder	noduliser with variable speed	weigh feeder or solids flow meters with prefeeders	

Table 1.1.3 *Contd....*

Sr. No.	Process	Wet	Semi Wet	Semi Dry	Dry	Dry with Calciner
6.	preheater	calcinator	travelling grate preheater or suspension preheater	travelling grate preheater or suspension preheater or shaft kiln	suspension preheater or shaft kiln	
7.	calcining					calciners in or off line
8.	clinkering	rotary long kiln	rotary short kiln	rotary short kiln shaft kiln	rotary short kiln rotary long kiln	rotary short kiln
9.	clinker cooling	common to all; rotary and planetary coolers for small plants – now almost discarded; reciprocating grate coolers of various designs like static grate, controlled flow grate pendulum cooler cross bar cooler etc.				
		Following processes are common to cement plants of all types differing in scale according to size of plant				
10.	coal grinding	ball or vertical mills with drying facilities				
11.	cement grinding	ball mills in closed circuit anc high efficiency separators vertical mills with external circuit and high efficiency separators roller press and ball mill and high efficiency separators in a number of ways				
12.	cement packing loose cement	rotary or stationary packers to pack cement in jute/paper bags cement sent in bulk in bulk carriers by road byships etc.				
13.	cement despatches	by road or rail of bagged cement using semi or fully mechanised loading machines also by ship loads for export				

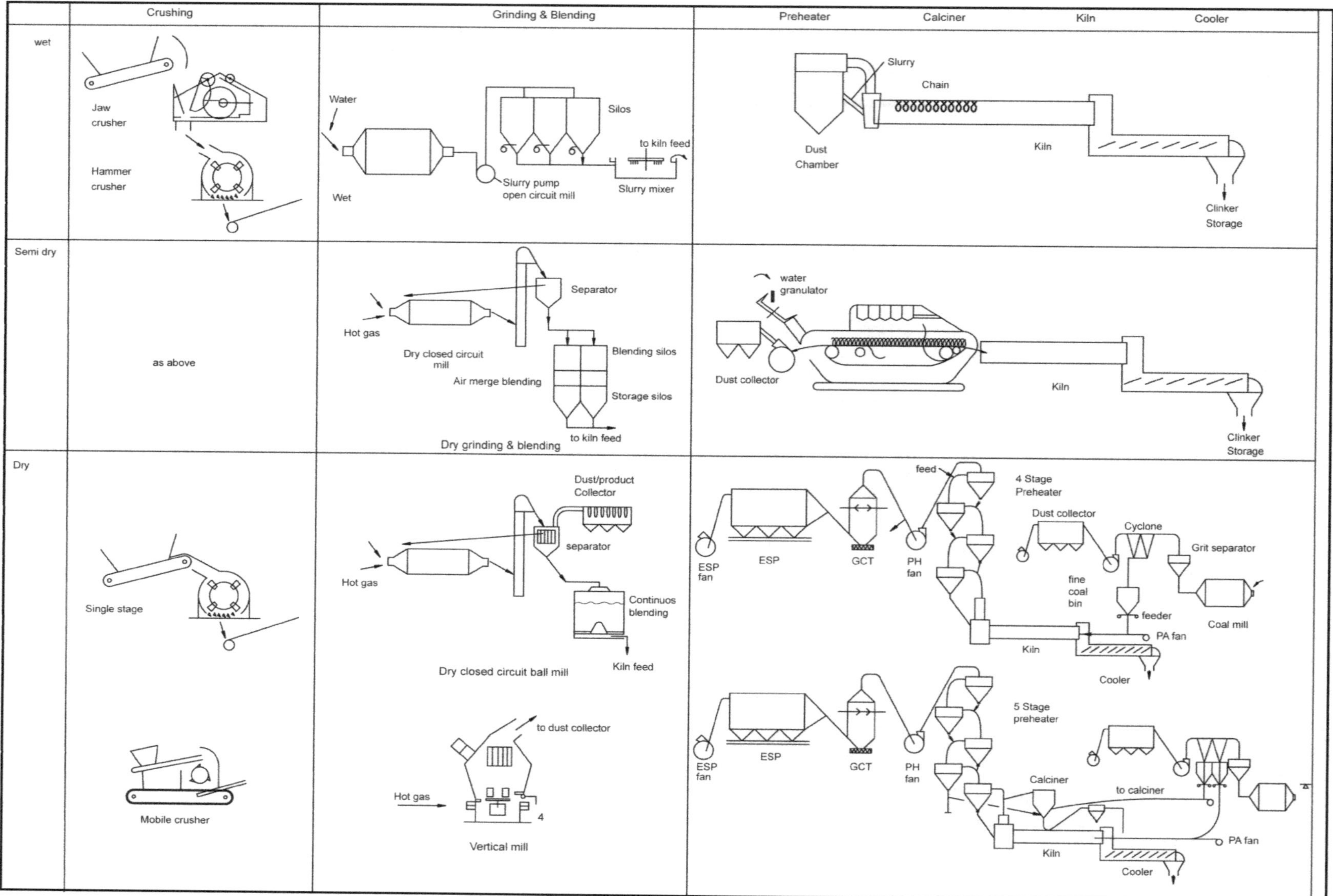

Fig. 1.1.2 Equipment used in manufacture of cement.

CHAPTER 2

TYPES OF CEMENTS - THEIR PROPERTIES AND USES

2.1 Cements

"Cement" is a general nomenclature and is known all the world over as Ordinary Portland Cement (OPC). Over a period of time, Special Cements have been developed to suit specific needs of construction. Further for specialized heavy duty applications cements that develop higher strengths in shorter time and with low evolution of heat have also been developed.

OPCs

In "OPC" itself there are three grades now in the Indian context

1. OPC - 33 Grade - BIS Ref. - IS 269 2013
2. OPC - 43 Grade - BIS Ref. - IS 8112 2013
3. OPC - 53 Grade - BIS Ref. - IS 12269 2013

Broadly speaking 33, 43 and 53 grade cements develop 330, 430 and 530 Kg/cm^2 compressive strengths after 28 days.

2.2 Other OPCs

Other types of OPCs are :	Ref.	
1. Rapid hardening cement	IS - 8041	1990
2. Oil well cement	IS - 8229 E	1986
3. Low heat cement	IS - 269	1989
4. White cement	IS - 8042	1989
5. High Alumina cement	IS - 6952	1989

2.3 Blended Cements

Apart from the OPC's - "blended" cements have also been developed. Blended cements serve two purposes :

1. Conserve clinker and hence limestone deposits of cement grade.
2. Use "waste materials" of other industries and thus help in keeping environment cleaner.

Principal blended cements are :

1. Blast Furnace Slag Cement (BFSC) IS – 455 1989
2. Portland Pozzolana (Fly Ash) Cement (PPC) IS – 1489 1991

Blended Cements can meet requirements of OPC but have an additional advantage that they give out 'less heat' during hydration and hence are well suited for "mass constructions" like dams.

There were some prejudices about "blended cements" the principal being that they were 'adulterated' hence "inferior" compared to pure OPC. If properly made by choosing blending materials of right qualities and used in specified proportions, they are not "adulterated" cements.

These prejudices have disappeared and bulk of cements produced in India are blended cements.

2.4 Detailed Specifications of Various Types of Cements

Detailed of specifications of these various types of cements are covered in respective Standards mentioned above. These standards specify physical and chemical properties that the respective cements should possess.

Further each Country has its own Standards. It is useful to know corresponding Standards for given type/s of cement/s of different Countries.

Each of these types of cement meets specific needs and should really be used for the purposes for which they are developed.

2.5 Special Cements

Special cements like 'Oil well cement' and 'White Cement' have limited demand and hence the plants producing these would necessarily be smaller in size in comparison to OPC Plants. The price that special cements fetch can still make them viable. But obviously plants producing them will come up at irregular intervals.

2.6 Making Different Types of Cement

All these types of cements with the exception of 'White Cement' require basically same raw materials and use the same manufacturing process. Hence any cement plant in principle can produce all types of cements with existing facilities but at somewhat different capacities.

However, there would be small differences in qualities of materials. 43 and 53 grade cements require comparatively higher grades of limestone and are generally ground finer. So the cost of production is slightly higher.

2.7 White Cement

'White Cement' requires special lime stones with negligible iron content, which causes pigmentation or colouring. Till a few years back, white cement was produced by using oil as fuel so that ash from coal would not discolour it. With reduction in fuel consumption, by using multistage preheaters, and quenching coolers, coal can now be used as fuel. Clinker is rapidly quenched in reducing atmosphere to avoid oxidation and hence coloring. Special silica stone liners and grinding media are used in grinding mills to maintain 'whiteness'.

See Fig. 1.2.1.

2.8 Slag Cement - BFSC

Blast Furnace Slag, a waste product of Steel Industry has 'hydraulic properties' of hardening like Cement clinker and hence serves as an ideal 'blending' material so much so that slag and clinker can be mixed in proportions of 60 : 40 and ground with gypsum to make slag cement. Thus a million ton slag cement plant needs to install a kiln of only 1500 tpd capacity.

2.9 Pozzolana Cement – PPC

Fly Ash' collected from exhausts of power plants using high ash fuels is another material which serves as a 'blending material' but to a smaller extent of 25-30 % only.

There are other natural pozzolanic materials and clays burnt bricks among them, which also serve as blending materials for producing Portland Pozzalona Cements.

2.10 Tables of Specifications and Properties

Tables 1.2.1 to 1.2.5 furnish specifications and properties of different types of cement, raw materials and blending materials

Table 1.2.1	Principal Types of Cements.
Table 1.2.2	Properties of different types of Cement.
Table 1.2.3	Typical compositions of different Cements.
Table 1.2.4	Typical compositions of slag and fly ash.
Table 1.2.5	Typical chemical compositions of lime stones and correcting materials.

2.11

The entrepreneurs would make up their mind as to what cements they would like to produce taking into account, various technical and commercial factors.

It stands to reason though that wherever there is a possibility to make blended cements due proximity to steel plants or large thermal power plants, they should design the plant for making blended cements as capital cost of investment per ton of installed capacity would be substantially less compared to OPC of the same capacity.

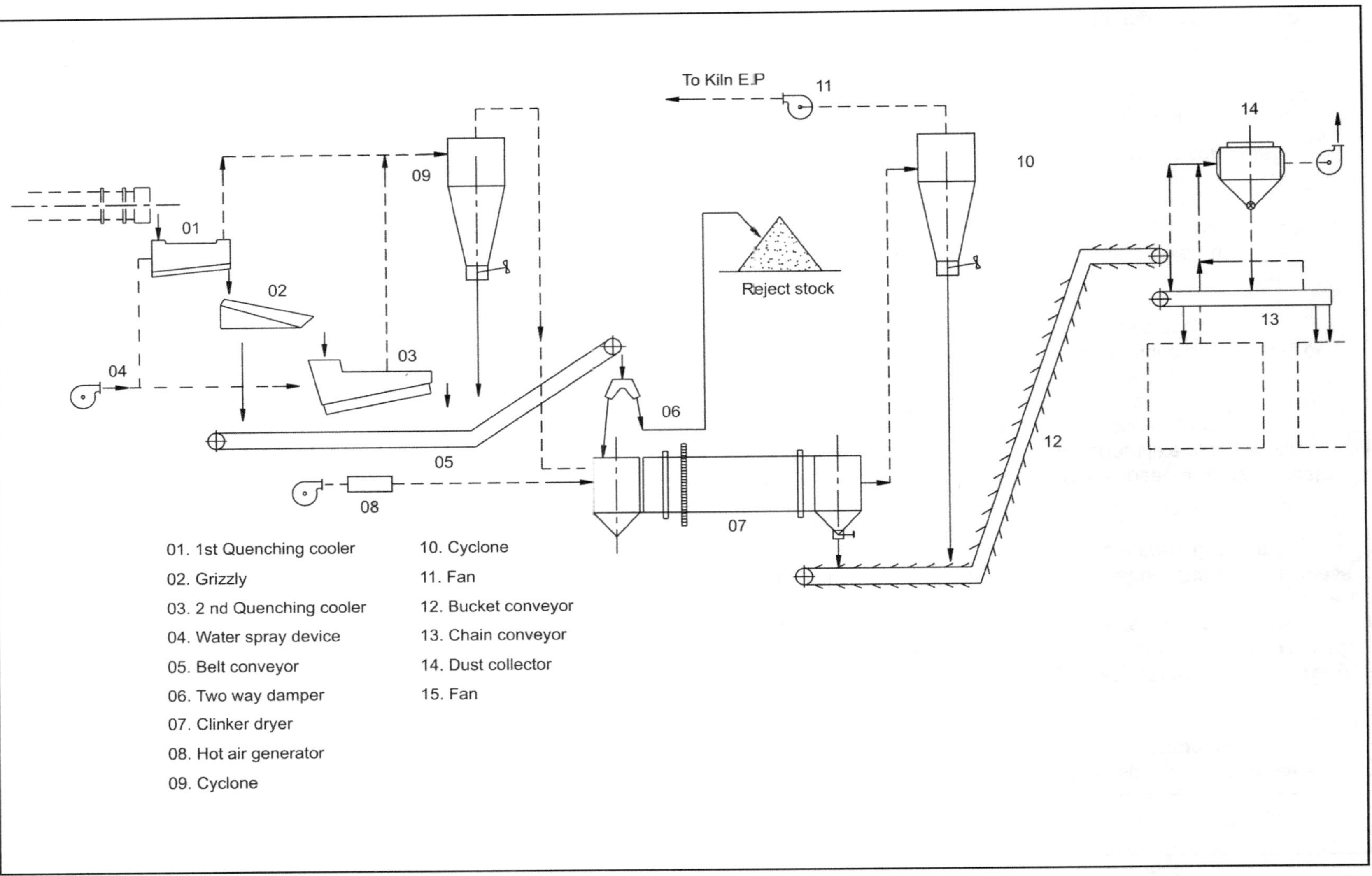

Fig. 1.2.1 White cement plant - special air free quenching of clinker to prevent oxidation.

Annexe 1.2.1

Principal Types of Cements and their uses.

Sr.No.	Type of cement	Short Name	Salient Properties	Uses
1	rapid hardening cement	RHC	similar to OPC; ground finer; develops strengths more rapidly	where speed of construction is important
2	sulphate resistant cement	SRC	composition adjusted to give increased resistance to sulphate bearing waters	for lining of canals, culverts and where construction is exposed severe alkali conditions.
3	low heat cement	LHC	composition reduces heat of hydration; strengths developed more slowly	for constructions involving mass concrete like dams
4	high alumina cement		made by sintering limstone and bauxite;develops strengths rapidly approaches final strength in 24 hours	where speed of work is most important like roads in cities and casting piles
5	oil well cement		cement of special chemical composition; rather coaresly ground	for cementing steel casing of walls of bores
6	white cement	WPC	very low < 0.4 % iron oxide	white in color; used for decorative purposes, mosaic tiles; cement paints
7	pozzolana cement	PPC	has low heat of hydration	for mass concrete
8	blast furnace slag cement	BFSC	low heat of hydration; is also more sulphate resistant	used for dams and other water retaining structures
9	early strength cement			used in constructions where shuttering is required to be removed early; in construction of silos with slip form shuttering
10	railway sleepers			for making sleepers for railways

Annexe 1.2.2

Detailed properties of different types of cements.

Sr.No.	Type of Cement	Short Name	B.I.S. Standard No. / IS	Year	Modulii	Chemical Requirements % max. MgO	SO_3	I.R.	L.O.I.	Sulphur	Fineness Blaine sq.cm/gm min.	Compressive Strength days kg per sq.cm, min. 3	7	28
1	Ordinary Portland Cement 33 Grade	O.P.C. 33	269	2013	LSF 0.66-1.02 A/F > 0.66	6		2	5	<2.75	2250	160	220	330
	43 Grade	O.P.C. 43	8112	2013	LSF 0.66-1.02 A/F > 0.66	6	2.5-3	2	5		2250	230	330	430
	53 Grade	O.P.C. 53	12209	2013	L.S.F 0.8-1.02 A/F > 0.66	6	2.5-3	2	4		2250	270	370	530
2	Low heat cement	LHC	12600	1989	LSF 0.66-1.02 A/F > 0.66						3200	100	160	350
3	Rapid hardening cement	RHC	8041	1990		6	2.5-3	4	5		3250	275		
4	Sulphate resistant cement	SRC	DOC/ BDC2 4335			6	2.5	4	5		2250	102	163	337
5	High alumina cement		6452	1989							2250	350		
6	Oil well cement		8229 E	1986		5	2.5	0.75	3					
7	White cement	WPC	8042	1989		6	2.75-3	2			2250	144	198	297
8	Puzzolana portland cement fly ash not < 10, nor > than 25 %	PPC	1489	1991		6	3				3000	160	220	330
9	Blast furnace slag cement slag not < 25,nor > 65 than 65 %	BFSC	455	1995		8	3	2.5	4	1.5		160	220	

Note : Though fineness has been mentioned as 2250 blaine, minimum, in many cases, very rarely cements are ground to less than 2700 blaine; ppc is ground generally to 3200 blaine and slag cement to 3400 blaine.

Annexe 1.2.3

Typical chemical compositions of different types of cements.

Sr.No.	Compound	OPC	Slag Cement	High Alumina Cement	White Cement
		<——		—%—	——>
1	SiO_2	19-24	24-30	7-17,	23-25
2	Al_2O_3	4-9,	7-16,	35-50	4.5-7
3	Fe_2O_3	1.6-6	1-3,	6-12,	0.2-0.5
4	CaO	60-67	43-35	36-47	65-67
5	MgO	not > 5	not > 6	not > 2	0.8-1.5
6	SO_3	not > 3	not > 4	not > 0.5	0.5-1.0

Annexe 1.2.4

Typical compositions of Blast Furnace Slag and Fly Ash used to make Slag and Pozzolana Cements respectively.

Sr. No.	Compound	Slag	Fly Ash
		<——— —%	———>
1	SiO_2	30-40	45-55
2	Al_2O_3	8-18	25-30
3	Fe_2O_3	0.5-1	5-10
4	CaO	30-40	2-6
5	MgO	0-8	2-3
6	SO_3	0.2,	0.5-3

Annexe 1.2.5

Typical chemical compositions of Lime stones and correcting materials.

Sr. No.	Description	Lime-stone	Sea shells	Clay	Shale	Bauxite	Laterite	Iron ore	Sand	Coal ash
		←----------	----------	----------	----------	---------- %	----------	----------	----------	----------→
1	SiO_2	8-13	0.5-2	40-70	40-80	5-15	10-30	5-10	55-95	50-70
2	Al_2O_3	1-4	0.5-1	15-30	15-30	40-55	20-40	2-5	2-5	15-30
3	Fe_2O_3	0.2-5	0.1-0.5	3-10	3-10	2-10	20-40	85-95	1-3	5-10
4	CaO	40-50	53-55	1-10	1-10	2-4	2-4	Trace	1-3	2-5
5	MgO	0.2-4	0.5-1	1-5	1-5	1-2	1-2	Trace	1-3	1-3
6	Alkalies	0.2-1	0.3-0.5	1-4	1-4	Trace	Trace	Trace	1-2	2-4
7	Chlorides	0-0.1	0.2-0.5	Traces						
8	SO_3	0-1	0.1-0.3	Traces						1-3
9	L.O.I.	35-40	42-45	5-15	2-5	20-30	15-25	0-5	2-5	0-2

CHAPTER 3

PROCESSES OF MANUFACTURING OF CEMENT

3.1 Process of Making Cement

Cement is a good example of developments in manufacturing processes used to make a product, which by itself has not changed much.

Developments in processes of manufacture of cement are reflections of the needs of the periods such as scale of production, product quality and specific consumption of thermal and electrical energy to produce it.

3.2 Predominance of Wet Process

A developed country like America persisted for a long time (compared to Europe) with the wet process of manufacture of cement because of :

1. Alkaline aggregates that were used in Construction Industry in America.
2. Higher requirements of manpower.

Till about middle of 20th century, power and fuel were cheap. There was no oil crisis. America used oil as fuel even in wet process plants, which required 1300-1500 Kcal/Kg clinker, fuel consumption.

Labour costs used to be higher than power costs. Therefore 'simple' wet process, which did not have too many 'auxiliaries', was preferred.

Another fact on account of which 'wet process' had a long innings was the ease with which constituents of raw mix feed to kiln could be blended and homogenized in a 'slurry' form.

3.3 Marginal grade Limestone and Froth Flotation

Decades ago even when cement plants were few and high quality limestones were readily available, marginal grade limestones was sought to be used to make cement by 'enriching' them. The 'froth flotation' process used to enrich limestone, removed silica from limestone and thus increased the carbonate content. This process required 'slurry' with moisture content of 36-38%. The slurry was agitated and aerated in cells with additions of doses of chemicals that removed 'silica' in a 'froth' form.

Naturally therefore where it is proposed to use limestone of marginal grade 60-70% CaCO3, wet process of grinding and formation of slurry was a necessity. Therefore in the past, wet process kilns were used to make cement from marginal grade limestones.

3.4 Dry and Semi Dry Processes

Dry process has evolved with time. The first cement was produced by dry process. White cements were produced by dry process. Initially dry process kilns were long kilns with chain systems like wet process kilns. Chains had to be of heat resisting steels.

However as sizes of plants increased due to growth in market and importance of blending to mix uniform raw mix was realized, the wet process took over as the main process used for manufacture of cement and continued to do so for almost half a century.

3.4.1 Semi Dry Process

Semi dry process was developed to make cement in early 40's of the last century. It took two routes

(i) Development of shaft kiln that could use low volatile fuels.

(ii) Lepol Grate Preheater followed by a rotary kiln.

3.5 Dry Grinding and Blending

As a corollary, two developments became necessary :

1. Drying and Grinding of Raw Materials in 'Dry Grinding'. This was not difficult as coal mills and cement mills were already dry grinding mills. Hence, same ball mills could be used to grind materials dry.
2. Blending Dry - As a result of dry grinding it became necessary to blend dry ground fine powders. Development of 'fluidization' techniques resulted in 'Airmerge blending' systems - first 'batch' systems and then 'continuous' systems. Air was used as medium which 'fluidized' or aerated raw meal and created different densities cyclically within the sections of silos. This resulted in internal flows and circuits within the silo that thoroughly mixed the raw meal. In continuous blending, it helped to break and mix layers of raw meal in the silo. Filling, mixing and extracting operations took place simultaneously.

Dry process or semi dry process became a reality after dry blending and pneumatic conveying was developed.

3.6 Semi Wet Process

Many Paper and Fertilizer plants produced calcium carbonate sludge as byproduct. It could be used to make cement with small corrections for composition and fineness.

In early days, wet process was used to make cement from such sludge. Filter presses reduced the moisture in sludge which was subsequently treated in a long wet kiln.

Now 'Semi wet' process would be used to produce cement. In this process preparation of materials is wet just as in regular wet process. Slurry produced is passed through filter presses to reduce moisture to 15-17 %. Cake so produced would be passed through either :

1. an extrusion press and made into briquettes which would be dried and preheated in a 'Lepol' preheater as in semi dry process.

or

2. a disagglomerator and flash drier and would be preheated in a three stage preheater as in case of dry kilns with cyclone preheaters.

3.7 Preheaters

The next stages in development were :

1. Traveling grate preheaters.
2. Suspension Preheaters

The long dry kilns - performed all functions or processes like preheating, calcining and sintering within themselves. Though shorter than long wet kilns which had a length to diameter ratio (l/d) of 30-35 : 1, long dry kilns still had l/d ratios of 25-30 : 1.

3.7.1 Lepol Grate Preheater

"Lepol Grate' was a traveling grate preheater which was used to dry granules formed in a granulator by adding 8-10% water to dry pulverised raw meal.

Lepol grate preheater took away the function of 'preheater' out of the kiln. Kiln thus became much shorter - l/d s were 18 :1, or less.

3.7.2 Suspension Preheater

Suspension Preheater - 4 stages to start with - used kiln exit gases to 'preheat' dry raw meal, when suspended in hot gases. Cyclones arranged one over the other received raw meal in connecting ducts. It was carried by kiln gases coming from lower stage to the upper stage; in transit, raw meal received heat while gases lost it. In cyclones, raw meal was separated from gases, then this meal entered the duct of the lower stage. Thus in stages, kiln exit gases lost temperature from 1000 °C to 340 °C and raw meal got preheated from 60-80 °C to 800 °C. In the process, raw meal entered the kiln partially calcined (25-30%). Thus kiln became ever shorter as it had to do only balance calcining and sintering. Kiln lengths were reduced to 14:1 to 16:1.

Suspension preheater was an epoch making development in making of cement and all plants which can use dry process, now use suspension preheaters.

Plant sizes jumped from 300 tpd to 1200 tpd. In wet process, sizes of kiln almost peaked at 1000 tpd though there were some exceptions.

Dry Process today is exclusively based on suspension prehaters.

They evolved in number of stages - from 4 to 6 as developments in design of cyclones reduced pressure drop across them without sacrificing efficiency.

Presently pressure drop in a 6 stage preheater is the same as that in a 4 stage preheater of older design

Capacities - of using cyclones in single and multiple streams gave a boost to the capacity of kiln. Presently single sream prehater have gone up to 4000 - 6000 tpd capacity.

Calciners further accelerated the growth of capacity of kiln.

3.8 Clinker Coolers

Clinker coming out of the kiln had to be cooled to temperatures at which it could be handled by the then available conveying equipment like pan, drag chain conveyors and by even belt conveyors

Like preheaters, clinker coolers have also seen great developments so that are able to cope with increasd capacities of kilns and contribute to increased operational efficiencies.

3.8.1 Rotary and Planetary Coolers

In early stages rotary coolers were used to cool clinker. Cooling air was heated to temperatures of about 600 °C for burning as secondary air in the kiln.

To avoid a separate rotating cylinder, a ring of concentric tubes attached to the kiln itself at discharge end served as coolers called Planetary Coolers. In this type too, cooling air served as secondary air for combustion'. However they are a thing of the past now.

3.8.2 Grate Coolers

In grate cooler, which consisted of one stationary grate and another gate 'reciprocating' over it, it it was possible to add more cooling air to cool clinker quickly and to temperatures lower than 100 °C.

Only air required for combustion entered the kiln; the balance was vented out.

Reciprocating Grate cooler was another break-through which helped to reduce fuel consumption of clinkering process because of better recuperation or recovery of heat from hot clinker.

Lepol grate used as preheater was also used for cooling clinker but did not become very popular.

3.9 Increase in Size of the Cement Plant – Large Kilns

1950s and 60s saw great strides in sizes and capacities of individual kilns. In Europe and Japan kilns of 2000-3000 tpd came to be installed. In Japan this limit was crossed and kilns as large as 7000-7500 tpd with diameters of +5 meters were operating.

At this point it was realized that large diameter kilns had shorter brick life and gain in capacity tended to be lost due to lesser availability

3.10 Calciners

At this juncture another epoch making development - that of 'calciner' - took place. Calciner carried out the process of calcination outside the kiln almost up to 90%. The kiln was thus liberated from the task of calcination and hence the same kiln could achieve 2-2 ½ times more production. The total fuel was divided between kiln and calciner; kiln receiving only 40-50% and calciner receiving 60-50%.

As a result large capacities of 10000 tpd and above were possible with much smaller kilns.

Presently kiln capacities of 10000 tons and above have become a norm.

The same kiln thus jumped in capacity and hence many existing preheater kilns were converted into calciner kilns. Particular beneficiaries were small plants of 300 tpd in this country. After installing calciner, their capacities have been raised to 800 t.p.d. and above.

Interestingly beneficiaries of this development were small cement plants also. In India many small plants have increased their capacity from 300 tpd to 800 tpd by installing calciners with additional stream of preheater.

Since calcination was taken out of the kiln, kilns could be much shorter-with R/d ratios as small as 12:1. Two support kilns were feasible for large capacities.

3.11 Technical Collaborations

India had Technical Collaborations with World's leading Process designers and Machinery Manufacturers of Cement Plants since early Sixties. But rate of transfer of 'Know how' was rather slow in the decades of sixties to eighties. After that due to policies of liberalization, the rate of transfer of technology accelerated greatly. New developments in Process and Machinery were soon available either by way of importing machinery or by way of fresh Collaborations. Thus Calciners of various designs came to be installed in the country from early eighties. Plants with Vertical Roller Mills and Roller Presses also came up in the country since late eighties. Cement Plants became modern in every sense and plants that have come up since nineties can compete with the best plants in developed countries.

Presently India is number two in the World in production of cement. Because of the size of the Indian market, World's leading Cement Machinery Maufacturers like FLSmidth, Polysius and Humboldts have established facilities for manufacturing their products in India with Indian franchises

On account the size of the market and future prospects, many foreign cement machinery manufacturers began operations from India also

3.12 Various Processes, Machinery and Size of Cement Plant

In a way process and machinery to be selected were related to the size of the plant. Equipment like Vertical Roller mill and process like Continuous blending could be afforded by large plants. In each type of machinery also there were several options: for example in vertical mills there were options like external circuit and mills fitted with high efficiency separators. Such diverse possibilities have been summarized briefly in **Annexe 1.3.1.**

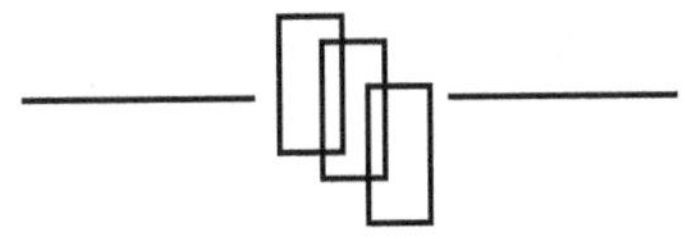

Annexe 1.3.1

Summary of Developments.

Section	1	2	3
Quarrying Operations	Manual	Semi-Mechanised	Mechanised
	Hand drilling and breaking of stone, manual loading of stone	Pneumatic & hand drilling; stone size still small – 300 mm- loaded in trucks manually	Pneumatic drilling; shovels and dumpers used to load stone; capacity of shovels gone up to 4--5 m3 ; and that of dumpers to 50-100 tons
Crushing	Single stage Hammer; double rotor hammer crusher; impactors for vertical mills	Two stage Jaw crusher; gyratory/cone crushers; Hammer; Impactor; (for ball mills)	Three stage jaw crusher gyratory Hammer/ impact; tertiary crusher; (for ball mills)
Raw Mills	Ball mill wet grinding	Vertical roller mill drying & grinding with built in separator	
	Ball Mill dry grinding closed circuit with separator	Vertical roller mill with external circuit built in separator	
	Ball Mill Dry Grinding Closed Circuit High efficiency Separator	Vertical roller mill with external circuit and High efficiency separator	
	Roller press with ball mill And high efficiency separator		Roller press and disagglomerator and High efficiency separator Horizontal roller mill And high efficiency separator
Blending and storage	Wet grinding slurry Aeration in silos	Dry Grinding Raw meal Air merge blending system Batch blending	Continuous blending

Annexe 1.3.1 *Contd....*

Section	1	2	3
Silos	Slurry Slurry Silos Mixer	Double decked Blending and storage silos	Storage cum blending Silos large capacity 15000 t onwards
Kilns	Wet - Long	Dry	Semi-dry
	Wet – Calcinator kilns Semi wet process	Long Kilns Short kiln and preheater	Shaft kilns
preheaters	Wet short kiln with Filter press and	Suspension preheater	Lepol grate
calciners	Preheater and calciner	Short kiln with Preheater and calciner	
Coolers	Rotary	Planetary Old & New	Grate-travelling Grate- reciprocating Conventional Static grate Walking floor, Cross bar cooler, Poly track and other types
Coal Grinding	Ball Mill Air Swept drying and grinding cyclone separator	Vertical Mill Ring ball Mill Vertical Mill roller Built in separator Bag filter / esp to collect product	
Cement Grinding	Open circuit Long Cement Mill	Closed Circuit Cement Mill Conventional separator / High efficiency Separator Roller press with ball mill and High efficiency separator	Vertical mill High efficiency Separator Mill with external circuit
Cement Storage	Silos	Multi compartment Silos	
Packing	Stationary packer	Rotary packer 1 or 2 loading points	Rotary packer Electronic weighing
Loading and Dispatches of bagged cement	Manual loading	Mechanized loading of trucks and wagons	Automatic feeding of Packing machines and Mechanized loading of Trucks and wagons
Bulk handling of cement		Loading of bulk carriers on road and rail	Bulk dispatches by ship

CHAPTER 4

SELECTION OF PROCESS AND MACHINERY

4.1 Wet Process Unlikely to be Selected in Future

Unadulterated wet process would hardly be used in future.

Wet process by itself is unlikely to be used hereafter even where raw materials are available in a slurry form like waste 'by products' of fertilizers plants and paper mills. Even for such situations, as mentioned in chapter 3, 'Semi Wet' route would be used in future.

As high grade limestones are used up it may become necessary to use low grade and hence in future we may have to go back to 'semi-wet' process.

On the other hand 'semi dry' process is not at all likely to be used in future to make use of low grade / marginal limestones because 'enrichment' is presently not known in dry process.

4.2 Using Sweeteners

If a quarry contains deposits with marked variations in $CaCO_3$ content, 'blending' of high and low grade limestones would be resorted to get raw mix of correct composition.

If the 'high grade' available, is not sufficiently high then 'sweetener' would be added. This may be required to be brought in from distant sources.

In normal course, limestones containing say + 80 % Carbonates when mixed with clays containing 20 % carbonates in proportion of 94 : 6 would yield raw mix containing 76 % Carbonates.

If the deposits have substantial quantities of limestone with 75% carbonates, attempt should be made to use it by blending the two grades with a 'sweetener' of say 90 % carbonate brought from outside. In this instance, 75 % and 90 % grades blended in the ratio of 2 : 1 would yield a blend containing 80 % carbonates.

4.3 Froth Flotation Process for Enriching Limestone

However there is a limit to the minimum grade of limestone, which can be used in this fashion. But by using 'froth flotation' process it would be possible to use limestone containing 65-70 % $CaCO_3$. In such circumstances, wet process would have to be selected for grinding and 'semi wet' for pyroprocessing.

4.4 Alternative Raw Materials

Sludges from fertilizer and paper plants that are wastes for them could be and have been used for making cement. ACC have been making cement from the sludge of a fertilizer plant in Sindri for about five decades. This aspect has gained significance in the present times from the angle of Sustainable Development. Cement Plants are therefore looking for possibilities of using alternative raw materials that would replace limestone to some extent

4.5 Dry Process

Dry process is suitable for almost all limestones except when they contain alkalies higher than 3-3.5% and chlorides higher than 0.03%.

This is because in a preheater kiln, in the riser duct between kiln and bottom cyclone and in the bottom two cyclones, alkalies and chlorides go into cyclic evaporation and condensation. They need to be absorbed by sulphur in coal to form sulphates.

If the alkalies continue to condense, they form coatings on the walls of preheater cyclones and in the riser duct between kiln and in the duct between two bottom cyclones and thereby restrict passage of gases through the system.

The coatings soon assume massive proportions and literally choke or throttle the preheater kiln system.

See Fig. 1.4.1

4.6 Alkali Bypass

If a quantity of gases at kiln exit is 'bypassed' that is removed from the system, then the rate of formation of coating is greatly reduced and it is then possible to run the kiln on a continuous basis.

With Precalciner this problem is greatly reduced as only 40-50% of gases pass through kiln and balance through precalciner - where temperatures are lower than at kiln inlet.

See a note on 'Alkali Bypass' in **Annexe 1.4.1.**

4.7 Kilns with Calciner

Thus a calciner makes it more feasible to use difficult raw materials containing high alkalies and chlorides. Much smaller quantity of gases is required to be 'bypassed' and cost of by passing, cooling and passing through an ESP is much reduced. Increase in fuel consumption is also much less.

Grate cooler helps in supplying air for combustion separately to calciner whether 'in line' or 'off line' type according to the quantity required for combustion in it.

For all these reasons the predominant process of manufacture of cement in future would continue to be dry process using 'preheater, calciner, kiln and grate cooler combination'.

4.8 Kilns with Planetary Coolers

It is possible to a limited extent to draw air for combustion that would be required to burn fuel to be fired in calciner, through the kiln, but this quantity is limited to about 30% only. Beyond that, it cools down combustion zone and increases velocity of gases through kiln requiring a bigger kiln for the same output. Therefore kilns with planetary coolers and 'air through' calciners would also be seldom used or selected particularly for new installations.

4.9 Equipment and Processes for the Future

Annexe 1.4.2, shows fuel and power consumption for different processes and for different types of mills used for grinding. It brings out the differences and it would be seen that :

1. mechanized mining
2. single stage crushing and stacker reclaimer systems
3. vertical mills for drying and grinding all three materials viz. raw materials, coal and clinker
4. 6 stage preheater, kiln, calciner and grate coolers with static grates, and various forms of movement of clinker in cooler like moving floor, cross bar et al
5. continuous blending systems
6. mechanized loading of bagged cements and more and more dispatches of bulk cement

would be the most preferred processes and systems in cement plants of the future.

However, size of the plant which in turn depends on availability of funds, may tilt the balance in favour of ball mills which are much cheaper in first costs than vertical mills for plants upto 1000 tpd capacity. Even small plants are installing vertical mills for making slag cement.

For higher capacities, vertical mills will be used almost always for grinding raw materials and coal and increasingly for cement grinding.

Older large plants have conventional machinery; newer plants have the latest energy saving equipment like vertical mills for cement and static grate coolers. **Annexe 1.4.2** furnishes salient features of large and small plants including operational parameters.

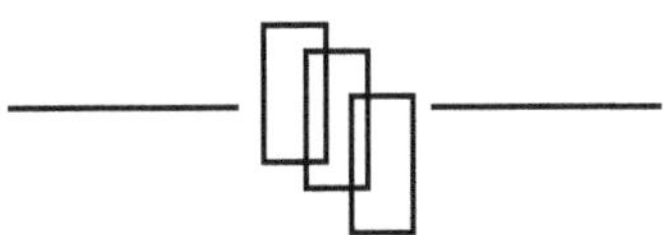

Annexe 1.4.1

Note on Alkalies and Chlorides and Bypass.

1.0 Alkalies and Chlorides

Minor components in raw mix like alkaies and chlorides evaporate in burning zone and condense when cooled at kiln inlet and in riser pipe from kiln. Repeated cycles of evaporation and condensation result in increasing internal circulation till equilibrium is reached.

Compounds involved are mainly of Potassium, Sodium, Chlorides and Sulphur

All compounds, not removed, end in clinker and affect its properties.

When escaping through gases there are problems of disposal of dust. Operational problems occur when circulating compounds reach high concentrations in the system.

1.1 Evaporation of Alkalies and Chlorides

Chlorides evaporate almost totally and they facilitate evaporation of alkalies; Sulphur retards evaporation of alkalies.

Practical limits of alkalies and chlorides in raw mix feed to preheater kiln is 1.5% Alkalies and 0.03 % chlorides. If alkalies do not combine with SO_2 coming in with fuel and raw mix, heavy condensations take place chiefly in the riser duct from the kiln and in between the last two cyclones. They increase to such proportions as to choke flow of gases and affect kiln production.

See Fig. 1.4.1.

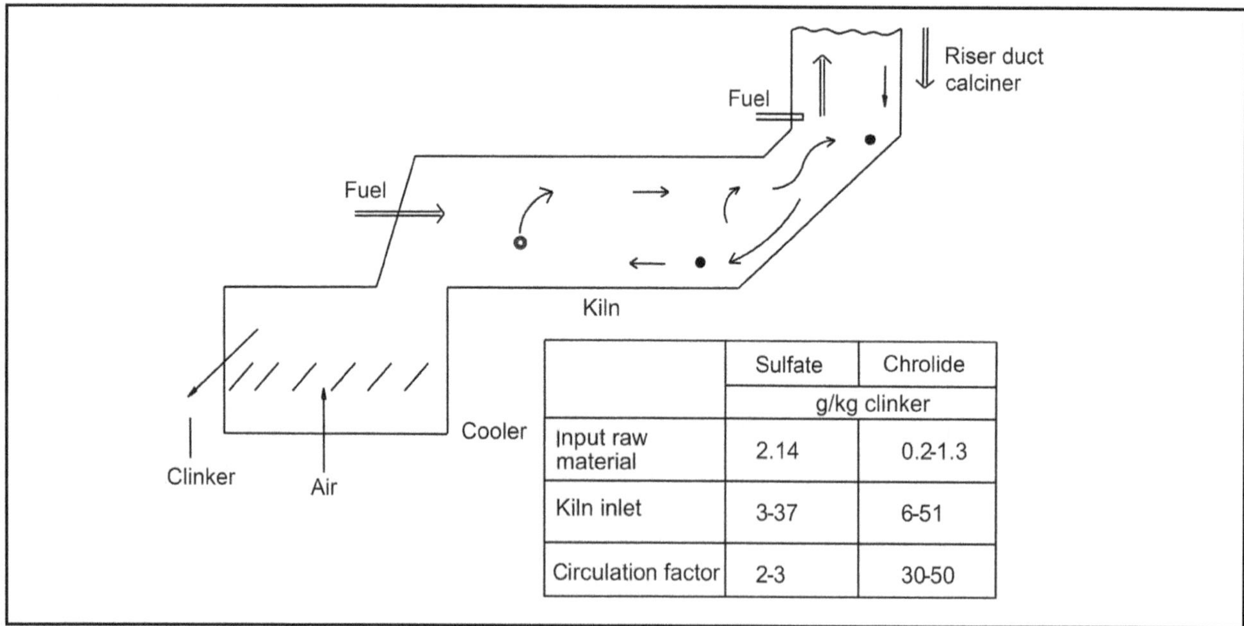

	Sulfate	Chrolide
	g/kg clinker	
Input raw material	2.14	0.2-1.3
Kiln inlet	3-37	6-51
Circulation factor	2-3	30-50

Fig. 1.4.1 Cycles of alkali sulphates and alkali chloride between Kiln and preheater.

1.2 Overcoming Effect of Alkali Cycles

One way to overcome this is to by pass part of the kiln gases at kiln inlet. However bypass increases sp. fuel consumption. Practically it is feasible to remove about 25 % gases maximum. Such a large quantity of bypass will increase fuel consumption by 100 to 125 kcal/kg of clinker.

1.3 Alkali Cycles in Calciner Kilns

With precalciner kilns only 40 % gases pass through kiln and k-line preheater. Therefore problem is very much reduced and hence it is now possible to use materials with high alkalies and chlorides in preheater kilns. If 10 % by pass was necessary as a preheater kiln, only 4 % would be necessary as a calciner kiln and increase in sp. fuel consumption would be limited to 20 kcal/kg of clinker.

1.4 Provision for ByPass

Bypass is taken at kiln inlet. Temperature of gases here is 1000 °C. Gases need to be cooled either by air or in a gas cooling tower before they can be handled in a dust collector. An ESP would be used to collect the dust contained in them.

However it is possible to precipitate dust in separate cyclones and take gases back to preheater stream in between first and second stages.

1.5 Return of Collected Dust

When dust is collected in a separate ESP, it may be possible to return it to the system; however, if the concentrations of chlorides are very high, then it would have to be thrown out.

1.6 Choosing Process to Suit Raw Materials

Thus raw materials and fuel both need to be carefully assessed at planning stage and when selecting the process. Some limestones - those that are sedimentary deposits in or near the sea will naturally have more chlorides. In some situations they must be used. Then the kiln must be a precalciner kiln restricting flow of gases through kiln to 40 %.

Lay out must be designed to allow for above mentioned auxiliaries when a by pass has to be installed **See Flow Charts 1.4.1** and **1.4.2** showing arrangements for providing alkali by pass.

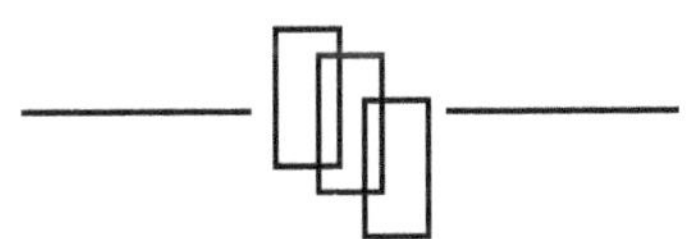

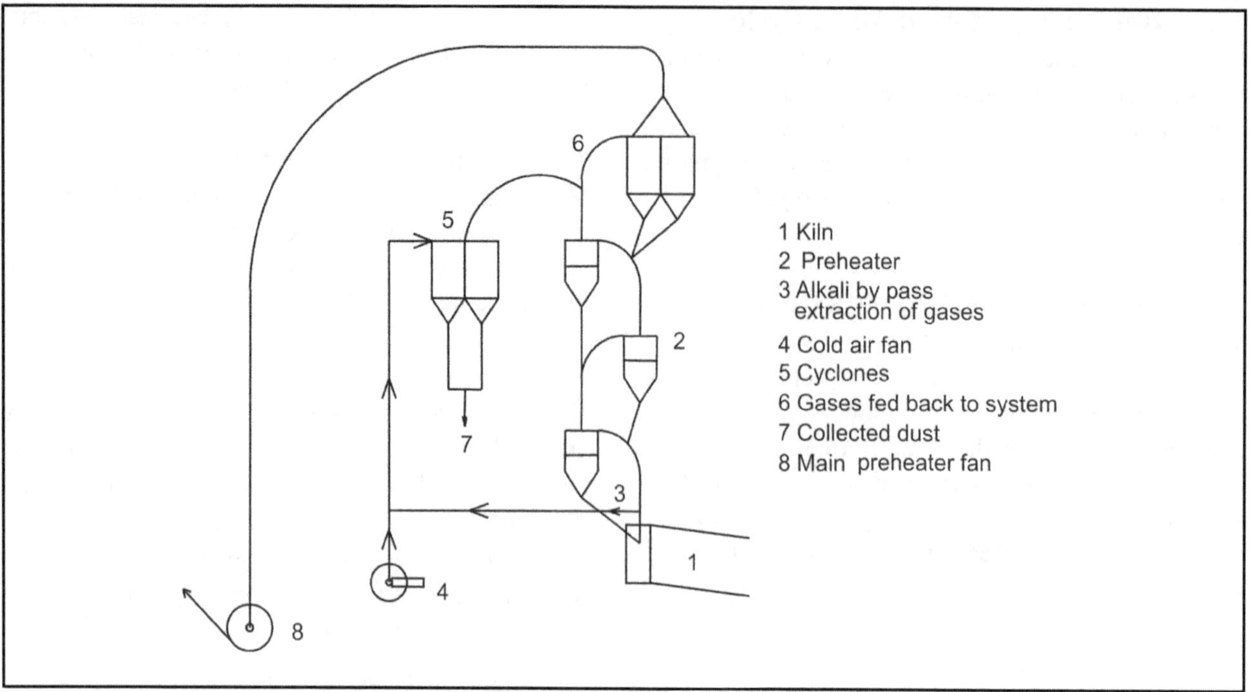

Flow chart. 1.4.1 Alkali by pass arrangement - 1 Gases withdrawn are fed back to the system.

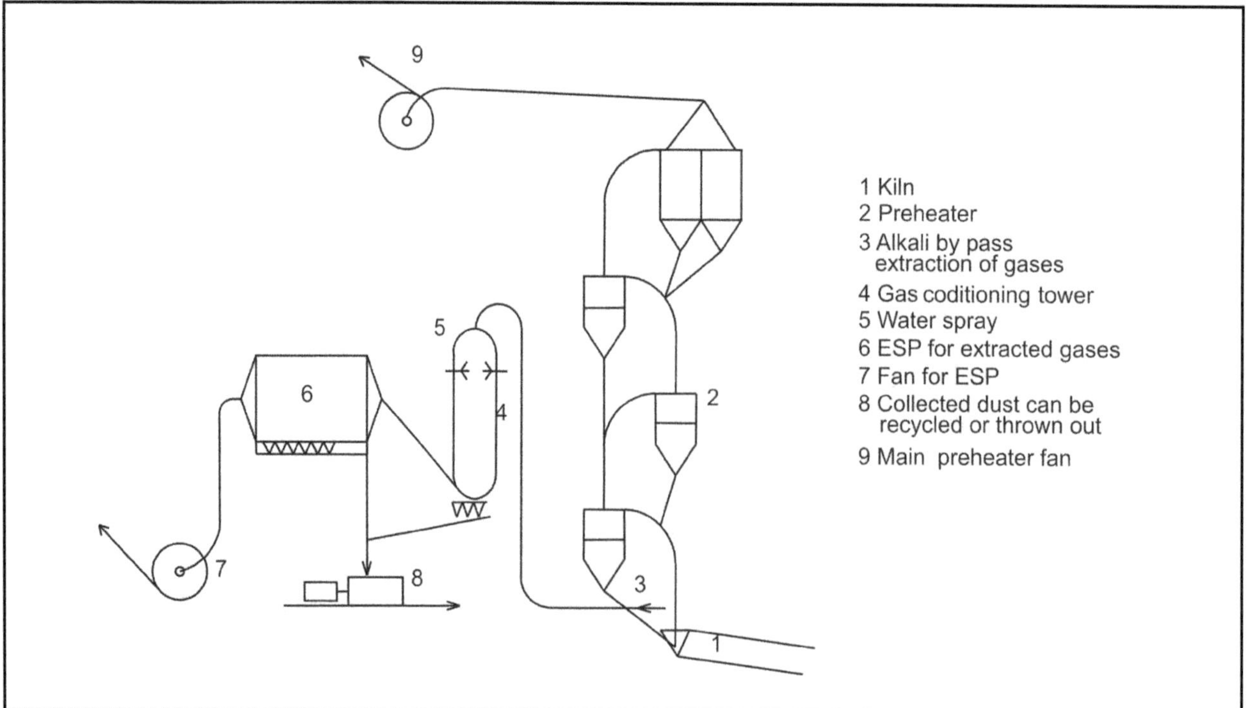

Flow chart. 1.4.2 Alkali by pass arrangement - 2 Gases withdrawn are cooled & passed through separate ESP.

Annexe 1.4.2

Fuel and power Comsumption in Large and Small Plants.

Sr. No.	Section	(1) 1 small plant		(2) 2 large plant		(3) 3 large plant	
		Sp. power consumption kwh/ton cement					
1	Crushing						
	Single stage			yes	2.5	yes	2.0
	Two stage	yes	3.5				
2	Raw mill						
	Ball mill	yes	25				
	Vertical mill			yes	20	yes	16-18
3	Batch / continuous Blending and Kiln feed	yes	5 batch blending	yes	4 cont. blending	yes	3 cont. blending
4	Pyroprocessing						
	5stage preheater conventional cooler	yes	25				
	6stage preheater static grate cooler			yes	25	yes	23
5	Coal crushing Grinding Vertical mill	yes	6	yes	6	yes	5
6	Cement mill						
	Ball mill	yes	30				
	Roller press and ball mill			yes	26		
	Vertical mill					yes	22-23
7	Packing despatches	yes	1.5	yes	1.5	yes	1.0
8	Miscellany	yes	5	yes	5	yes	4
	Total		101		90		75- 80
		Sp. fuel consumption kcal/kg clinker					
Pyroprocessing section							
5stage preheater conventional cooler		yes	750				
6stage preheater conventional grate cooler				yes	720		
6 stage preheater static grate cooler						yes	700-650

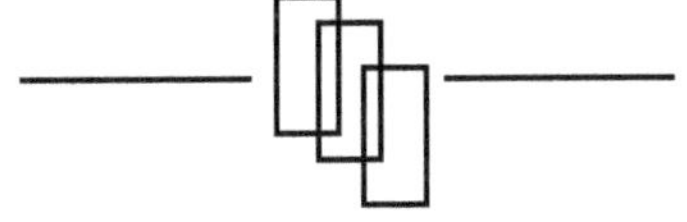

CHAPTER 5

RAW MATERIALS, ADDITIVES AND FUELS

5.1 Raw Materials for Making Cement

In **Chapter 2,** different types of Cements and their properties were seen. Cement concrete is 'synthetic rock' made up by processing raw materials like limestone, clayey materials and ferruginous materials in proportions that would yield clinker of suitable quality- after going through processes like calcining and sintering.

Very few materials will be 'natural rocks' which would have same chemical composition as that of clinker after processing.

Therefore it is necessary to make a 'raw mix' of suitable composition by 'blending and processing' two or more suitable materials in right proportions.

5.2 Constituents of Raw Mix and their Proportioning

As mentioned above, limestone, and clay, sand and iron ore or laterite are usual components of raw materials which when added in suitable proportions and ground produce 'raw mix'.

5.2.1 Limestone

Quarries hardly ever contain limestone of uniform grade. Different sectors contain carbonates in different proportions. Sometimes the difference is marked; so as to call them 'high' and 'low' grade limestones. To conserve high grade limestone it can be blended with low grade. **See para 4.2 in Chapter 4**

Low grade limestones can be 'enriched' by 'froth floatation process'

Alternate Raw Materials like sludges of Fertiliser and Paper Plants can also be used along with the main source of limestone.

Thus limestone, the principal constituent itself can be of more than one grade

5.2.2 Silica and Alumina

Other major constituents are silica and alumina which can be added by way of sand or clayey material and bauxite. Most limestones contain both silica and alumina. The 'additives' as these 'correcting' materials are called are added to limestone/s in suitable proportions to produce 'raw mix' of the desired composition.

5.2.3 Iron Oxide

Iron oxide is obtained by adding iron ore or laterite in small quantities usually 1-2 %.

5.3 Proportioning Constituents

How to arrive at these proportions is decided by following certain norms and applying certain yard-sticks. These yardsticks are :

Hydraulic Modulus or Lime Saturation Factor
Silica ratio
Alumina ratio.

5.3.1 Correcting Materials

Depending on the representative composition of principal raw material that is limestone, suitable correcting materials are obtained. Normally there would be 3 materials so blended viz. limestone for CaO, clays for Silica and Al_2O_3; and iron ore/ laterite for Fe_2O_3.

When individual compositions of the constituents are available, their proportions with one another can be calculated.

5.4 Composition of Clinker

Clinker has complex compounds, that are formed during the process of sintering, in certain proportions to obtain desired strengths in cement; they are C_3S and C_2S and C_4AF.

Free lime in clinker should also be limited to less than 1.5 %. Burnability factor is another yardstick which furnishes information on temperatures to be maintained in the burning zone to limit free lime to desired values.

5.5 Quality Control

Quality Control in manufacture of cement is a highly specialized field and is not the subject of this book. It is touched upon to highlight their impact on design of Plant and machinery in a cement plant.

Annexe 1.5.1, lists the various pertinent quality control aspects with their definitions and the ranges in which they should be maintained to produce clinker and cement of good quality.

For a plant designer it is sufficient at design stage to know the representative Chemical compositions of lime stones and correcting materials to be used and their proportions; the coal proposed to be used and its ash content, calorific value and composition of coal ash; Gypsum proposed to be used and its composition.

5.5.1 Commonly Found Proportions

Quarries supply limestone/s. Often overburden itself serves as an additive to correct silica and alumina. Interstitial material found alongside the deposits also serves as correcting material. Otherwise clays and sand are obtained from nearby sources like sand from river beds.

Iron ore or laterite is obtained from nearest sources of iron deposits.

Commonly found proportions would be :

Limestones	clay	iron ore
←-----------------	%	--------------→
80 – 85	10-15	1-2

5.6 Blended Cements

When making blended cements compositions of fly ash and slag should be known so that in what proportions they could be added to clinker could be worked out to produce Pozzolana and Slag Cements respectively. As of today up to 65% slag can be added to make slag cement; up to 30 % fly ash can be added to make pozzolana cement.

5.7 Fuels

Fuels supply the heat energy required to make clinker from raw mix. Solid fuels like coals which have ash in proportions varying from 10 to 40 % also influence the composition of clinker.

5.8 Fuels for Rotary Kilns

Rotary kilns can use all types of fuels – solid i.e., coals, lignite and petcoke; liquid i.e., oil and also gas.

Solid fuels are fired in pulverized form. Liquid fuels are atomized and fired through oil burners. They need to be heated to obtain correct viscosity to facilitate atomization. Gas – commonly natural gas is used – can be fired most easily.

Liquid and gaseous fuels are uniform in physical and chemical properties. Ash content is negligible and calorific value is consistent varying only slightly in similar fuels. From this point it is advantageous to use them as fuels. Clinker of uniform quality can be produced consistently.

5.9 Coals as Fuel

Coals vary in principal properties like calorific value and ash content from place to place or even in the same place. They are more difficult to burn and quality of clinker produced needs to be watched closely as ash in coal gets almost wholly absorbed in clinker formed affecting its composition thereby. As a corollary, to maintain uniform quality of clinker, raw meal composition has to be modified to counteract effect of coal ash.

Oils and gas can be transported and handled easily as compared to coal.

5.9.1 Preparation of Coal for Firing

Because coal is needed to be pulverized for firing in kilns, preparation required for it is maximum – involving crushing, drying and grinding. Metering it while firing is also not easy.

5.9.2 Preparation of Oil for Firing

Oils need only heating and atomization which can be done by a self contained heating and pumping unit. Atomisation can be done by pressure or by compressed air. Firing and metering is easy and accurate.

5.10 Alternate Fuels

Presently, Alternate Fuels like Municipal Solid Wastes and Biological Wastes are being used on an increasing scale as fuels in kilns and also in calciner/ Besides conserving fossil fuels, this also reduces emission of greenhouse gases.

There are problems about consistency in quality. Like coal they require lot of preparation to make them suitable for firing in kiln and caliner.

5.11 Costs of Fuels

Cost wise oils are more expensive than coal. Let cost of oil be ~rs 75 rupees per litre or 0.9 kgs. Useful calorific value of oil is 9600 kcal/kg. Therefore cost of 1 kcal is 0.8 paise.

Let cost of coal be 4000 Rs/ton. Useful calorific value is 4500 kcal/kg. Therefore cost of 1kcal is 0.09 paise. Costs of coal as fired would be say 0.1 paise per kcal. Even then it is considerably cheaper than oil.

Therefore presently coal is the dominant fuel even in developed countries and even in countries that have substantial oil reserves of their own.

5.12 Coals

Composition of coal is commonly expressed in two ways:

1. proximate analysis which is sort of a ready reckoner.

   It furnishes : carbon, volatiles, moisture, ash and calorific value. ← →

2. ultimate analysis which furnishes : carbon, hydrogen, oxygen, nitrogen, sulphur, moisture, ash and calorific value.

5.12.1 Typical Compositions of Coal and Coke are

	Coals	coke
	-----%---	
carbon	50-90	80-90
hydrogen	1-5	0.4-2
oxygen	2-14	-
nitrogen	0.3-2	-
sulphur	0.5-4	0.4-1
ash	35-50	8-14
moisture	2-15	1-1.5

5.12.2 Typical Composition of Coal Ash is

	%
Al_2O_3	15-20
SiO_2	25-40
Fe_2O_3	20-45
CaO	1-5
MgO	0.5-1
SO_3	2-8

Coals normally contain 8-10 % moisture. It can go up to 15 % in wet season. Coals are therefore required to be dried during grinding, reducing moisture to 1.5- 2 %. It is not desirable to dry coals completely.

5.12.3 Volatiles

Loss in weight as a result of carbonization of coal under exclusion of air is the measure of volatile mater. Volatiles promote combustion in pulverized coals. Therefore Indian coals though they contain high ash - more than 30 %, are not difficult to burn as the volatiles are between 28-30 %.

5.13 Calorific Value of Fuels

All fuels yield heat when burnt. Calorific value is a measure of heat thus available per unit weight of fuel (volume in case of gaseous fuels).

There are several definitions of calorific values ranging from 'gross on dry' to 'net on wet'. From practical point, the useful calorific value is the most important.

Theoretically calorific value can be obtained from ultimate analysis. It is readily calculated from empirical formulae.

Lower heat value (LHV) = HHV (higher heat value) - (VW + FW) × R

where VW is water generated during combustion

FW is moisture in fuel

R is latent heat of evaporation of water-539 kcal/kg

Various formulae have been furnished in web page 1.28 in Section 9.

Annexe 1.5.2 furnishes empirical formulae for calculating calorific values of solid, liquid and gaseous fuels.

5.14 Oils

Significant fuel oils are:

Mineral fuel oils

Coal tar oils

Bituminous coal tar oil.

Typically fuel oils contain

Carbon	85-90 %
Hydrogen	5-10 %
Oxygen, Nitrogen, Sulphur	together 3-4 %

In some fuels sulphur may go up to 3%

Calorific value of oils range between 8500-10000 kcal/kg.

Most important property of oil is its viscosity. It is measured on different scales in different countries. Oils are classified as light medium and heavy.

5.15 Gas as Fuel

Mostly natural gas containing ethane C_2H_6 and Methane CH_4 in varying proportions is used as fuel. Heating value is in the range of 8000-10000 kcal/m^3 at $\simeq$ 16.5 °C.

Typical properties are given in **Table 1.5.1.**

As mentioned earlier gas does not need any preparation for firing in the kiln.

5.16 Specific Fuel Consumption Obtainable using Different Fuels

Relative values of sp. fuel consumption achievable with the three types of fuel from actual installations indicate that sp. fuel consumption with coal as fuel is lower than that for oil or gas. For the same plant Sp. fuel consumption obtainable with coal and oil would be:

be	coal	oil	gas
Kcal/kg clinker	700	740	780

Flame temperature in case of natural gas is 70-150 °C lower than that of coal. When using natural gas, secondary temperature must be high and excess air must be minimum.

5.17 Requirements of Raw Materials and Fuel

With these details, requirements of individual materials and fuels, their rates of feed and capacities of conveyors and feeders needed at different stages of manufacture can be worked out.

Table 1.5.1 Typical properties of natural gas by volume.

Sr.No.	CO_2	N_2	% CH_4	C_2H_6	sp. gr. Air :1	Calorific value kcal/m^3 Higher	Lower
1.	0.8	3.2	96	-	0.57	8600	7766
2.	-	1.3	80.5	18.2	0.65	10000	9118
3.	-	1.1	67.5	31.1	0.71	10930	9900

(Ref: Duda-Cemtent Data book)

Annexe 1.5.1

Quality Control Formulae.

Sr No	Item	symbol	unit	formula	normal range
1.01	loss of iginition	L.O.I	%	$0.44CaCO_3 + 0.524MgCO_3$	37-39
	$CaCO_3$		%		
	$MgCO_3$		%		
		L.O.I.			
1.02	silica ratio (modulus)	SR	ratio	$SR = SiO_2 / (Al_2O_3 + Fe_2O_3)$	1.2-2.4 ave. 2.4-2.7
	SiO_2	%			
	Al_2O_3	%			
	Fe_2O_3	%			
1.03	alumina ratio	A/F	ratio	$A/F = Al_2O_3 / Fe_2O_3$	1.2-4.0 ave 1.5-2.5
	Al_2O_3	%			
	Fe_2O_3	%			
1.04	lime saturation factor	LSF	ratio		
a	A/F > 0.64			$LSF = CaO/(2.8 \times SiO_2 + 1.65 \times Al_2O_3 + 0.35 \times Fe_2O_3)$	0.80-0.95
	CaO	%			
	SiO_2	%			
	Al_2O_3	%			
	Fe_2O_3	%			

Annexe 1.5.1 *Contd....*

Sr No	Item	symbol	unit	formula	normal range
b	A/F < 0.64			$LSF=CaO/(2.8\times SiO_2+1.1\times Al_2O_3+0.7\times Fe_2O_3)$	90-95 for OPC
	CaO		%		
	SiO_2		%		95-98 for high alumina cement
	Al_2O_3		%		
	Fe_2O_3		%		
1.05	hydraulic modulus	HM	ratio	$HM=CaO/SiO_2+Al_2O_3+Fe_2O_3$	1.7-2.2 for OPC 0.5- 0.55 for high alumina cement
	CaO		%		
	SiO_2		%		
	Al_2O_3		%		
	Fe_2O_3		%		
1.06	percent liquid	L	%	$L=1.13C_3A+1.35C_4AF+MgO+Alkalies$	
	C_3A			$C_3A=2.65\times Al_2O_3-1.692\times Fe_2O_3$	
	C_4AF			$C_4AF=3.043\times Fe_2O_3$	
	A/F=>0.64				
	C_3S			$C_3S=4.071CaO-(7.602\times SiO_2+6.718\times Al_2O_3+1.43\times Fe_2O_3+2.852\times SO_3)$	45-55
	C_2S			$C_2S=2.867\times SiO_2-0.7544\times C_3S$	25-35

Annexe 1.5.1 *Contd....*

Sr No	Item	symbol	unit	formula	normal range
A/F = < 0.64				$C_3S=4.071\times CaO-(7.602\times SiO_2+4.479\times Al_2O_3+2.859\times Fe_2O_3+2.852\times SO_3)$	
				$C_2S=2.867\times SiO_2-0.7544\times C_3S$	
				$C_3A=0$	
1.07	burnability index	BI		$BI=C_3S/(C_4AF+C_3A)$	2.8-3.5
1.08	burnability factor	BF		$BF=LSF+10\times SR-3\times(MgO+Alkalies)$	100-110

source : cement managers handbook

Annexe 1.5.2

Empirical Formula for Calorific values of Fuels.

Sr. No.		
	Unit of calorific value	kcal/kg for liquid and solid fuels
		kcal/nm^3 for gaseous fuels
	Gross calorific value Hh	Cal. Value measured in bomb calorimeter
	Useful or effective value H1	steam in combustion is discharged outside the system without being utilised
A	*Empirical formulae for Solid and Liquid fuels*	
	$Hh = 8100 \times C + 34200 \times (H - O/8) + 2500 \times S$	kcal/kg
	$H1 = 8100 \times C + 29000 \times (H - O/8) + 2500 \times S - 600W$	kcal/kg
	where, C=Carbon; H=Hydrogen; S=Sulphur; O=Oxygen; W=Water are quantities in kg of fuel.	
B	*Formulae for gaseous fuels*	
	$Hh = 3035 \times (CO) + 3055 \times (H_2) - 9530 \times (CH_4) + 14900 \times (C_2H_4)$	kcal/nm^3
	$H1 = 3035 \times (CO) - 2570 \times (H_2) - 8570 \times (CH_4) + 13940 \times (C_2H_4)$	kcal/nm^3

Source : Onoda Handbook

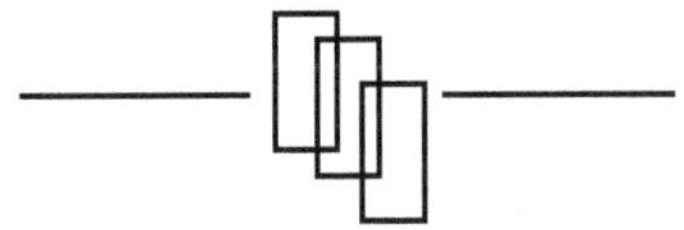

CHAPTER 6

TAKING UP THE DESIGN OF A CEMENT PLANT

6.1 Techno Economic Feasibility Studies

Previous five chapters would have served to give a fairly good idea of the present status of the processes and machinery used to make cements.

Taking up the design of a cement plant presupposes that an Investment Decision has been taken to install a cement plant.

The next logical step would be to take up 'Techno Economic Feasibility Studies' (TEFS) to examine all aspects – technical and commercial and financial of taking up the Project. Since this is a comprehensive and specialized activity it is best to entrust it to specialist Consultants. However Companies already in the field can do these studies themselves.

Because of its great importance, the various aspects of TEFS have been dealt with in **Section 3.**

In the process of carrying out TEFS such aspects as availability of suitable Deposits and their probable locations and markets to be targeted would be examined. This serves as basis for working out the design of the plant and even the types of cements to be made, their mode of dispatches. Location of plant with respect to the deposits may also be decided.

6.2 TEFS and Design of Cement Plant

TEFS uses this data to work out the basic design of the plant – such as sizes and capacities of main sections and major machinery and auxiliaries. Thus TEFS contain all the steps needed to be taken to design a plant. The only difference would be the details and the depth in which various aspects would be gone into.

Having carried out the TEFS, and having arranged for the finances, Detailed Engineering of the Plant would be taken up. This activity may also be entrusted to experienced Consultants.

In this chapter the various steps in taking up the design have been explained. The steps have been further dealt within subsequent chapters.

6.3 Basic Approach in the Design of a Cement Plant

In designing a cement plant it is necessary:

1. to select a process that would be most suitable for raw materials and fuels proposed to be used,
2. to use technology and designs and machinery that would require minimum inputs,
3. to take into account infrastructural facilities in designing the layout and related aspects,
4. to take into account requirements of the market not only for the present but also for the future and to incorporate facilities to make different cements and different modes of dispatching them,
5. to design process control and instrumentation and quality control facilities that would ensure production of high quality continuously at optimum costs,
6. to leave adequate room for expansion,
7. to keep investment costs within the limits of the resources of the entrepreneur.

6.3.1 Basis of Design to be Finalised

To achieve these objectives it is very essential that the Entrepreneur and his Consultants are on the same wavelength in the various aspects of design.

It is necessary to agree upon, margins and factors to be considered in arriving at sectional capacities, sizes of storages, locations of plant, crusher, and mines; types of cements to be made and their proportions;

modes of dispatches of cement – in bagged or in bulk – by rail or road; provisions to be kept for expansion in future.

6.4 Questionnaires

A systematic beginning can be made by devising 'Questionnaires' and to try to answer them in as much depth as possible. Alternatively the Consultant would furnish such Questionnaires and ask the Entrepreneur to fill them.

Questionnaires would have to be specially designed for various situations like :

1. a new plant at a greenfield site,
2. expansion of an existing plant,
3. upgrading or modernization of an existing plant accompanied by increase in capacity.

6.5 Specimen Questionnaries

Specimen Questionnaires of the three types are furnished at the end of the Chapter as **Annexes 1.6.1 to 1.6.3.**

Details of Infrastructural facilities are required for TEFS. They are required to finalise several aspects of technical design and also other aspects like size of colony and social amenities to be provided for the Workers.

Hence a Questionnaire on Infrastructural facilities is also necessary. Such Questionnaire is furnished as **Annexe 1.6.4.**

6.6 Minimum Agreement

Entrepreneurs and Consultants should mutually agree upon at least on the following points at the time of taking up the design:

1. rated clinkering capacity in tpd / tpa,
2. design margin,
3. types of cements to be made and their proportions,
4. gypsum whether natural or synthetic to be used – its source,
5. coals to be used – their average useful calorific values and ash content,
6. representative physical and chemical properties of limestones,
7. sources of correcting materials and their chemical and physical properties,
8. sources of slag / fly ash when blended cements are to be made,
9. modes of receiving crushed limestone, correcting materials, coals, gypsum and blending materials,
10. modes of dispatches of cement – bagged and or in bulk – their proportions; by rail and or road and their proportions,
11. source and availability of power – voltage of transmission,
12. source of water,
13. provisions to be kept for expansion in view of availability of deposits and size of markets,
14. Use of Alternate Fuels & Raw Materials should also be examined,
15. Possibility of waste heat recovery should be looked into when deciding upon captive power plant.

6.7 System Flow Charts

In **Chapter 7**, necessity of finalizing 'flow charts' has been explained. In **Chapters 8 to 10** procedures of arriving at requirements of materials and sectional capacities have been outlined. In **Chapter 11**, sectional capacities have been arrived at using these procedures.

As an illustration, sectional capacities for a 10,000 tpd plant have also been worked out. **See Annexe 1.11.1.**

6.8 Visits to Cement Plants

It would be a very good idea for a new entrepreneur and even for an old one to visit a few cement plants with capacities corresponding to his objectives and using state of the art processes and machineries and that have very good operational efficiencies.

Such plants can now be seen in India also. Visits to plants overseas will be a 'bonus point'. Such visits would give a much better idea to an entrepreneur as to what he can expect and what he can achieve. They also help to crystallize ideas about machinery to be installed and plant layouts.

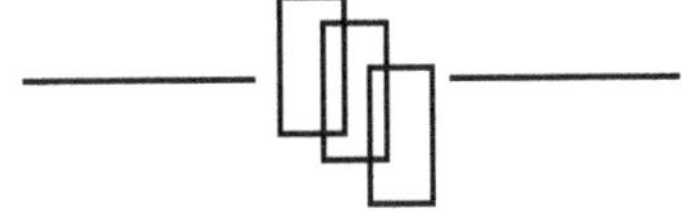

Annexe 1.6.1

Questionnaire For Setting Up A New Cement Plant At A Green Field Site.

1 Name of Company
2 Location / address
3 Phones/ fax/ e-mail details
4 Capacity of proposed cement plant tpd / tpa
5 Types of cements proposed to be made and rough proportions thereof
  1 O.PC.s – 43 and 53 grades
  2 Blast furnace slag cement
  3 Portland Pozzolana Cement
6 Proposed location of the plant
  1 village, mandal, district and state
  2 nearest town, city
  3 nearest railhead- gauge of rail link
  4 nearest highway
7 Data on deposits proposed to be used
  1 exact location
  2 location with respect to proposed location of plant
  3 representative quality of limestones found and estimated quantities in million tones
    chemical analysis of limestone
    high grade / low grade

SiO_2	Al_2O_3	CaO	MgO	Fe_2O_3	Alkalies	Chlorides	L.O.I.

←——————————%——————————→

8 Fuel coal / oil
8 Sources of fuel
9 Coal proximate and ultimate analysis
  Ash content min., max., and average
  Moisture in coal – min., max., and average
  Calorific value min, max and average—also useful heat value
10 Distance over which coal is to be brought and how
11 Similar information when oil is to be used
12 Likely sources of correcting materials (clay, sand, iron ore etc.), gypsum, slag and fly ash (whichever blended cement is proposed to be made)
13 Distances from proposed plant site and modes of transport
14 Source of power and voltage of transmission availability of power
  Will it meet full requirements all round the year
  Will it also meet future requirements after expansion
15 Provision for diesel generator set / thermal power station

Annexe 1.6.1 *Contd....*

16 Source of water supply – perennial- distance from plant approximate capacity in litres / day

17 Conversion factors and margins to be used for arriving at sectional capacities
design margin
safety factor for fuel consumption and power
conversion ratios raw meal / clinker; cement / clinker
for blended cements clinker / slag and clinker / fly ash
margins for feeders and conveyors
running hours in various sections in hrs/shift; nos of shifts/day; hrs/day; days/week; days per year etc.

18 Markets for different types of cement and seasonal demands for them

19 Proportions of cements bagged and loose

20 Proportions of dispatches by road and by rail

21 Storages to be maintained for:
i. limestone crushed and correcting materials
ii. coal crushed
iii. gypsum
iv. slag / fly ash
v. cements of different types
types of storages proposed for them

22 Stacker reclaimer cum preblending systems to be included for limestone and for coal

23 Extent of process control and instrumentation and automation and extent of sampling and quality control systems – x ray analyzer – to be included

24 Targeted operational norms like
sp. fuel consumption
sp. power consumption
clean gas dust burdente
man hours per ton

Note: This is a typical Questionnaire. A specific one can be developed for any particular project to take into its specific needs like terrain and soil Conditions; climatic extremes and such.

Entrepreneur should indicate and discuss his preferences if any with his Consultants for machinery such as vertical mills / roller press for grinding clinker; bag filter or electrostatic precipitator for venting kiln exhaust gases for example.

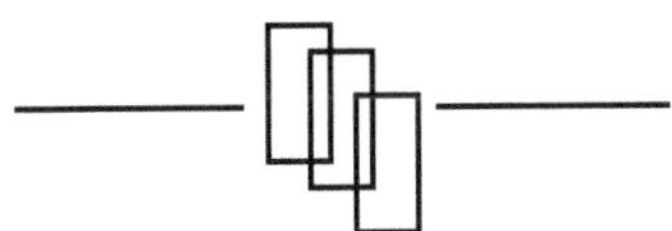

Annexe 1.6.2

Basic Data Normally Required To Take Up Any Assignment On Design Of Cement Plants.

1 Exact location of Plant
2 Altitude : meters
3 Ave. ambient temperature : °C
4 Average annual rainfall : cms
5 Detailed chemical analysis of limestones/ correcting materials, Clinker, gypsum, blast furnace slag, fly ash, coal ash etc.

	%								
	SiO_2	Al_2O_3	Fe_2O_3	CaO	MgO	Alkalies	Cl_2	L.O.I.	moisture
Limestone Grade1									
Limestone Grade2									
Clay									
Sand									
Bauxite									
Iron ore									
Hematite									
Gypsum									
Clinker									
b. f .slag									
fly ash									
coal ash									

6 Proportions of limestones and additives in raw meal feed to kiln
7 Common quality control ratios

	Hydraulic modulus	Silica ratio	iron modulus	C_3S	C_2S
Raw meal					
Feed to kiln					
Clinker					

8 Properties of cement produced

Type	Grade	Fineness Blaine	strength kg/cm²			SO_3
			-------- days--------			
			3	7	21	

9 Sources of fuel and details (assuming coal as fuel)
proximate and ultimate analysis of coals used, at least
fixed carbon volatiles ash sulphur
moistures max. min & average
calorific values -- gross and useful to be considered for design

Annexe 1.6.2 *Contd....*

10 Flow charts of various sections and general plant layout for existing plants

11 Machinery schedules of various sections describing size, capacity, nos and drive details a typical format would be

item	nos	size	capacity tph	motor kw

12 For proposed new plants and expansions and modernization etc.
intended norms for the following should be advised
i. design margins above rated capacities
ii. running hours in various sections days/year
days/week
shifts/day
hours/shift
hours/day
iii. storages of various materials in various stages of production
in terms of requirement for number of days
limestone/clay/iron ore/gypsum
raw meal
clinker
fly ash
b. f. slag
cements of different types
coal/s
iv. margins in feeders and conveyors
v. plans for expansion should be outlined
vi. data on mines distance from factory
representative quality of deposits
inferred / indicated / proved quantities of reserves

13 When TEFS to be made involves, market studies, then details of markets proposed to be served such as :
distance/s of market/s from proposed location
size of market/s in terms of cement consumed in tpa
growth potential—estimates of rate of growth
cement plants already existing or proposed to be added in the vicinity to serve the market/s
whether market is saturated
how is the market being served and share that can be expected in near / distant future

14 Transport and communications facilities
Rail links
Highway links

15 Power distribution and availability
Location of substation
Capacity
Voltage of transmission
Surplus capacity as of today
Proposed addition to demand in near future

16 Size of own power generating facility in terms of % of power requirements

Annexe 1.6.3

Questionnaire On Increase In Capacity Of Clinkering Section.

1 Present Capacity – tpd

2 Proposed capacity after expansion – tpd

3 Logic behind the proposed expansion

(i) limestone reserves permit / do not permit expansion with calciner

(ii) funds do not permit large scale investment

(iii) propose to make use of surplus capacity in other sections

(iv) propose to improve operational efficiency – increase in production to match improvement in efficiency

(v) propose to install grate type clinker cooler and hence propose to realize maximum kiln apacity without calciner

(vi) propose to install a calciner and a 2nd stream preheater to get maximum output from existing kiln

Alternatives (i) to (iv) above can be achieved in several ways such as

(a) modifying and enlarging cyclones of preheater

(b) installing smaller capacity calciner

Alternative (v) can be combined with (i) to (iv); Alternative (vi) would cost maximum but increasein capacity would be 2.5 times and hence returns would be higher.

Alternatives (i) to (v) would involve upgrading of mills to cope with enhanced clinker production;whereas in alternative (vi) above, almost all sections would need enhancement of capacity.

4 On learning more details and logic behind the proposals, Consultants can formulate specific questionnaire to collect all the relevant information and data on plant and machinery as also full operational data

5 Full details such as chemical and physical properties of raw materials, correcting materials and fuel/s used.

6 Types of cements produced and their quantities.

7 Furnish sizes and capacities of major machinery and auxiliaries and rated and achieved capacities of various sections starting from quarries and ending with cement mills and packing.

8 Running hours in each section.

9 Also furnish general plant layout.

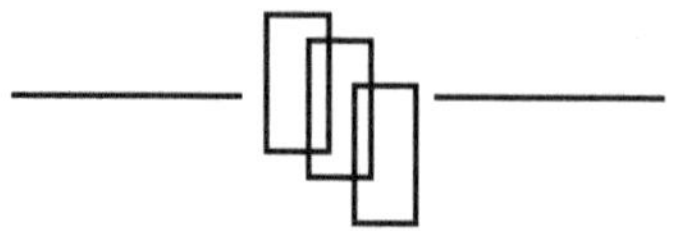

Annexe 1.6.4

Questionnaire for Infrastructure.

A Physical Location of site / sites

1 village
2 mandal
3 taluk
4 district
5 State
6 Latitude Longitude
7 Altitude m above sea level

Attach map

B Data on deposits to be exploited to make cement

1 Name
2 Location village / mandal / district / State as above
3 nearest village / town /
4 estimated area hectares
5 between longitudes and
between latitudes and
6 Is it in area marked as forests or sanctuary
7 broad nature of deposits
a plain
b hilly
c outcrops
d on fallow or cultivated land
e state or private land
f access road / tracks to deposits
g estimated reserves : inferred / indicated / proved in million tons
h other details like over burden, interstitial clays,
j clay and other correcting materials – nearby sources

C Details about proposed location of factory site

1 availability of land
2 private / state owned
3 cultivated / fallow
4 is it in area earmarked as forests / sanctuary ?
5 is land level or needs development
6 contour map of area showing streams, rivulets and other natural drainage
7 climatic details for the last ten years like temperatures mean, max.and min; year round rain fall for average and minimum and maximum
8 ground water level in different parts of the year min, max and average
9 depth of rock bed from surface / load bearing capacity of soil
10 seismic zone of area
11 wind velocities at different heights and wind rose

Annexe 1.6.4 *Contd....*

D Distance between proposed plant site and deposits min and maximum;
Is terrain plain or uneven ?

E Proposed site and its linkages with the outside world

1 road links between site and nearest village / town / city
existing / under construction / planned
pucca / tar / concrete
distance in kms

2 rail links
nearest railway station kms
single or double line
gauge
if rail link does not exist, are there plans for future

3 Nearest grid substation
Distance
Capacity
Voltage of transmission
Routes of transmission lines around proposed site
Distance from site
Is surplus capacity available in substation and in transmission line

4 Perennial sources of water in the vicinity
River / stream
Estimated quantity available in different parts of the year

F General nature
size of nearest town /city and facilities availale there by way of

1 labour force for construction and operation of plant
2 housing facilities
3 markets for provisions , milk, vegetables etc.
4 educational facilities – primary secondary schools and colleges, Industrial Training Institute etc.
5 Health and medical care like hospitals and dispensaries – ESIS Establishments etc.
6 commercial establishments like Banks
7 Government facilities like Post and Telegraph offices, courts and other government offices
8 Entertainment - Cinema, Cable TV etc.
9 Sizes and numbers of cities around the proposed location as possible sources of MSW which can be used as AF.
10 availability of blending materials - fly ash and slag should be obtained with a view to make blended cement/s

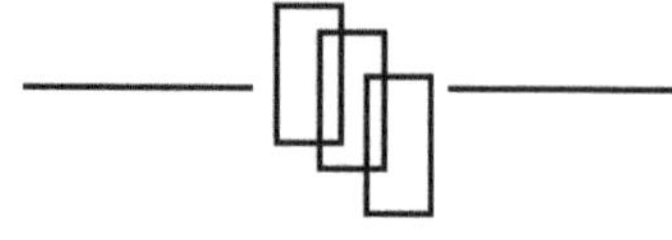

CHAPTER 7

TECHNICAL CONCEPTS
PROCESS FLOW CHARTS

7.1 Process Flow Charts

After selecting the process of manufacture and types of cement to be made, the next steps would be :

1. to select process flow charts of various sections,
2. to decide on sectional capacities.

Selecting process flow charts presupposes that some basic decisions have already been taken regarding selection of major machinery. For example:

1. single stage crushing
2. vertical roller mills for raw materials and coal
3. off line calciner - two stream preheater - grate cooler with static grate

Once these decisions are taken, it follows to fit them in a system along with auxiliaries so that as a group they produce or process what is expected of them in a given section.

For example : crushing section should produce crushed stone suitable for the subsequent grinding section and so on.

7.2 What Flow Charts Show

Process flow charts show pictorially how various machinery are arranged in a section. They try to show sequence or direction of material and gas flows in the section and help to arrive at a clear understanding of what happens in the section.

Process flows can be depicted in several ways. One way is to just show blocks for Major machinery and auxiliary and draw lines to connect them in proper sequence.

The other, more prevalent one is to show machinery and auxiliaries in outline for ease of identification.

7.3 Different Systems and Flow Charts

Illustrations of two types are shown in **Flow charts 1.7.1** and **1.7.2.**

Same type of machinery and auxiliaries can be arranged in more than one way to arrive at the same end result.

In two stage crushing, both crushers can be in open circuit, or the secondary crusher can be in closed circuit to ensure a definite product size.

See Flow charts 1.7.3.1 and **1.7.3.2.**

In raw material grinding, drying can be done either in mill or in separator;

Fresh feed can be taken either to mill first or to separator first.

See Flow charts 1.7.4 a to c.

In vertical roller mill, mill shall be either in conventional circuit with conventional separator, or it can be in external circuit with high efficiency separator built in. Such alternatives result in changes in layouts as also in gas and material flows.

See Flow chart 1.7.5.

Each section of a cement plant can thus have a 'family' of commonly used flow charts.

7.4 Uses of Flow Charts

Process flow charts can also be used to :

1. Build up machinery schedules for sections, indicating numbers, capacity, size and drive details. From such a schedule specifications of major machinery and auxiliaries can be built up for procurement.
2. to show material and gas flows in the section. These are used to size materials handling

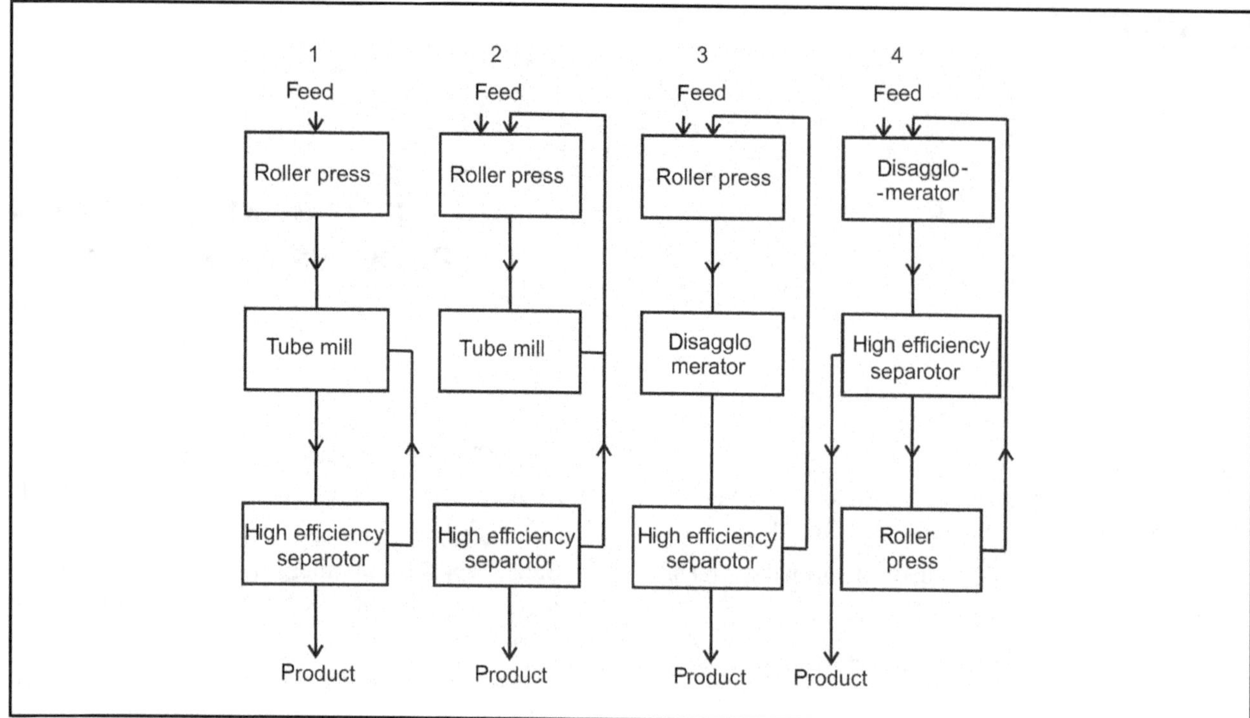

Flow chart. 1.7.1 Flow chart (block type) showing various possibilities of using roller press.

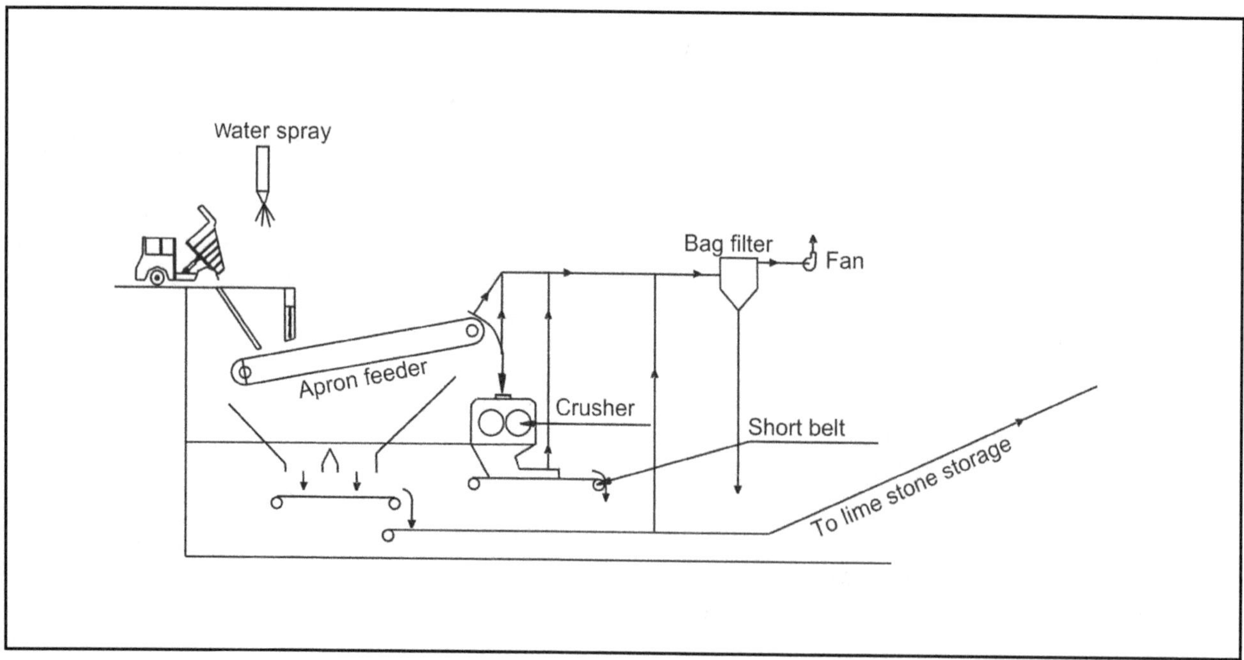

Flow chart. 1.7.2 Flow chart of single stage crushing.

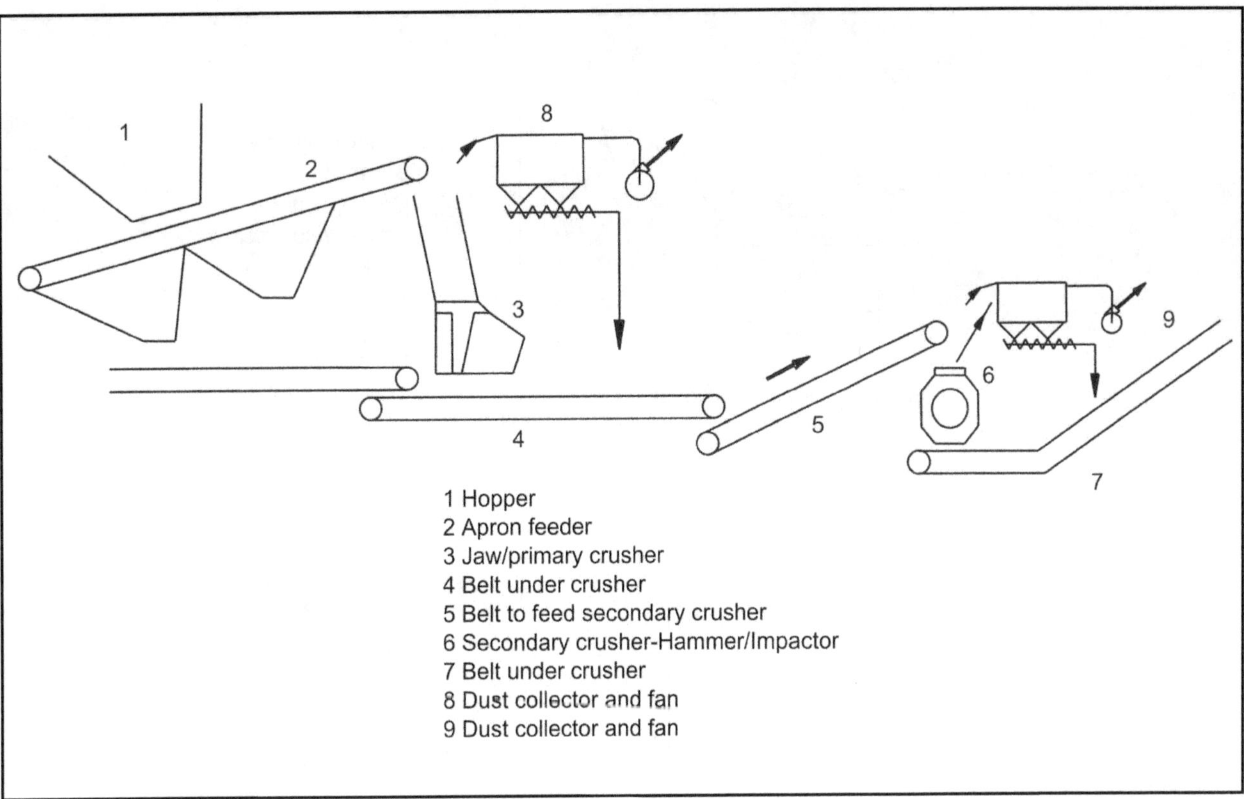

Flow chart. 1.7.3.1 Flow chart of two stage crushing.

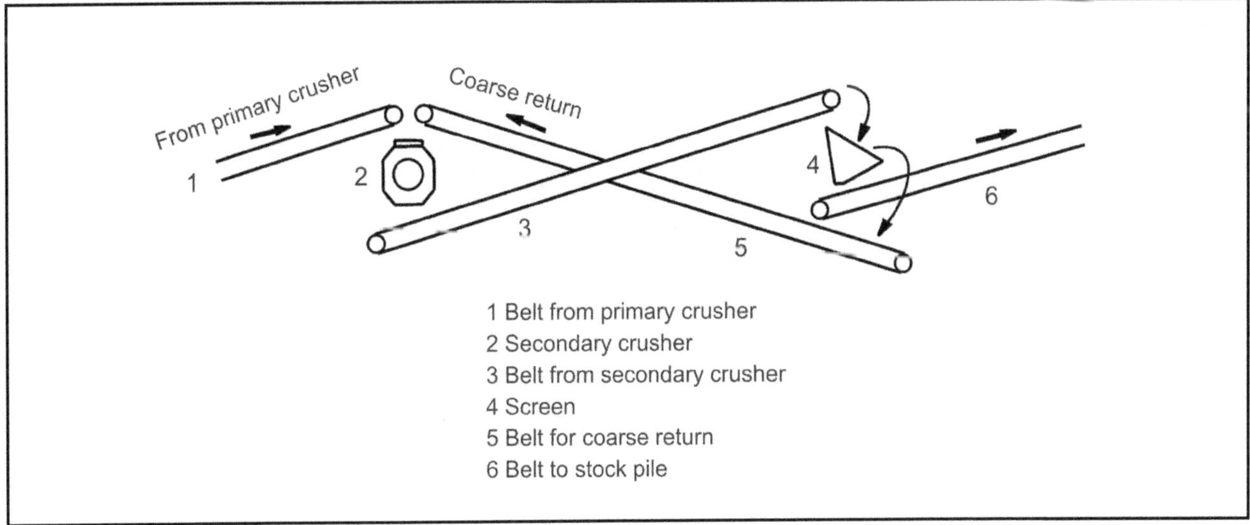

Flow chart. 1.7.3.2 Flow chart of two stage crushing, wih secondary crusher in closed circuit.

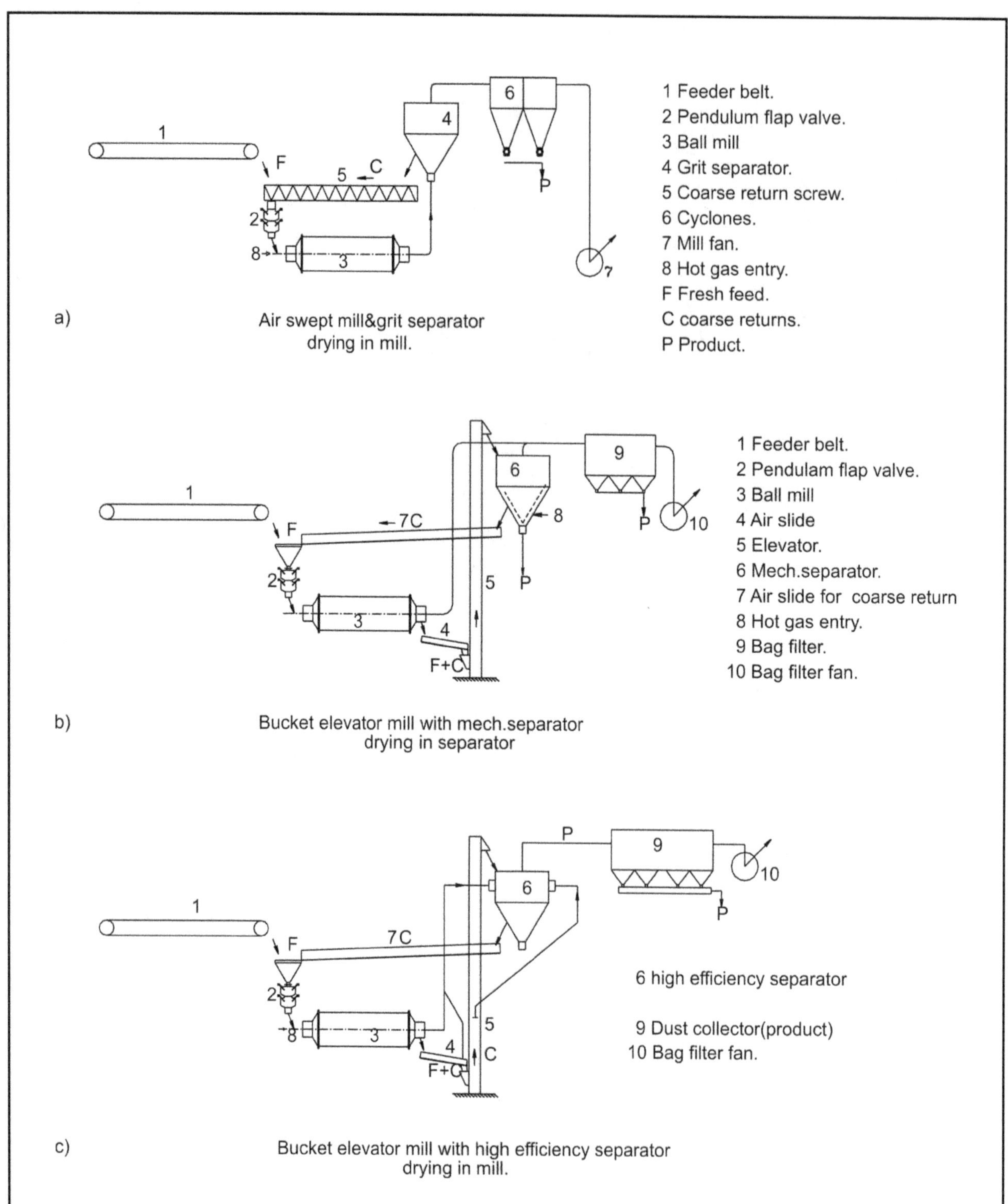

Flow chart. 1.7.4 Flow charts of drying and grinding ball mill systems.

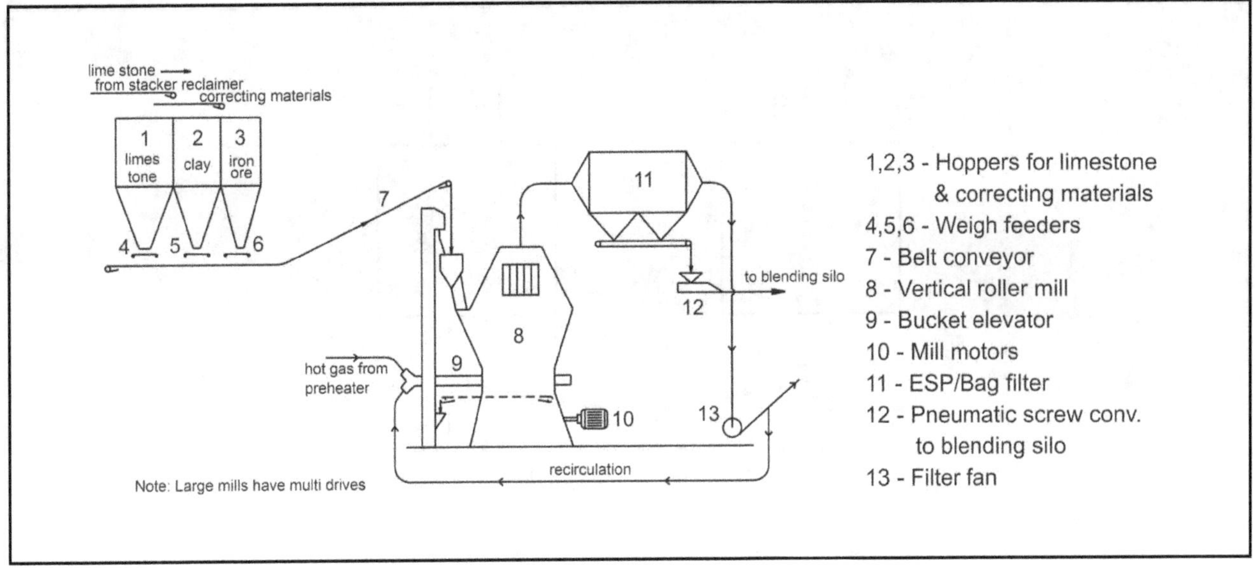

Flow chart. 1.7.5 Flow chart of Vertical roller mill basic system with external circuit.

equipment between processing machines and also to arrive at capacities of fans to handle required gas flows and to work out sizes of ducts and chutes,

3. to indicate temperature and pressure / draught profiles in the section as also important parameters like Oxygen content, CO content etc. They can thus be used to work out details of process control and instrumentation for the section including locations of field sensors.

7.5 Typical Flow Charts

Possibilities of various arrangements in various sections of a cement plant are far too numerous to be put together. More commonly found flow charts for the various sections of a cement plant beginning with crushing and ending with dispatches are enclosed. **See Flow Charts 1.7.2 to 1.7.13.**

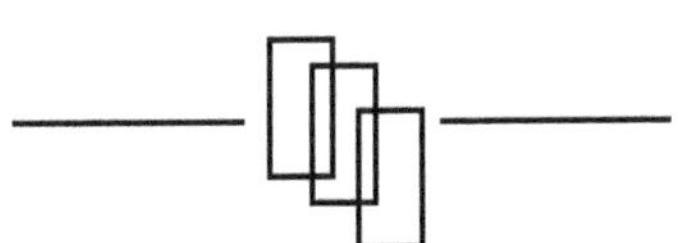

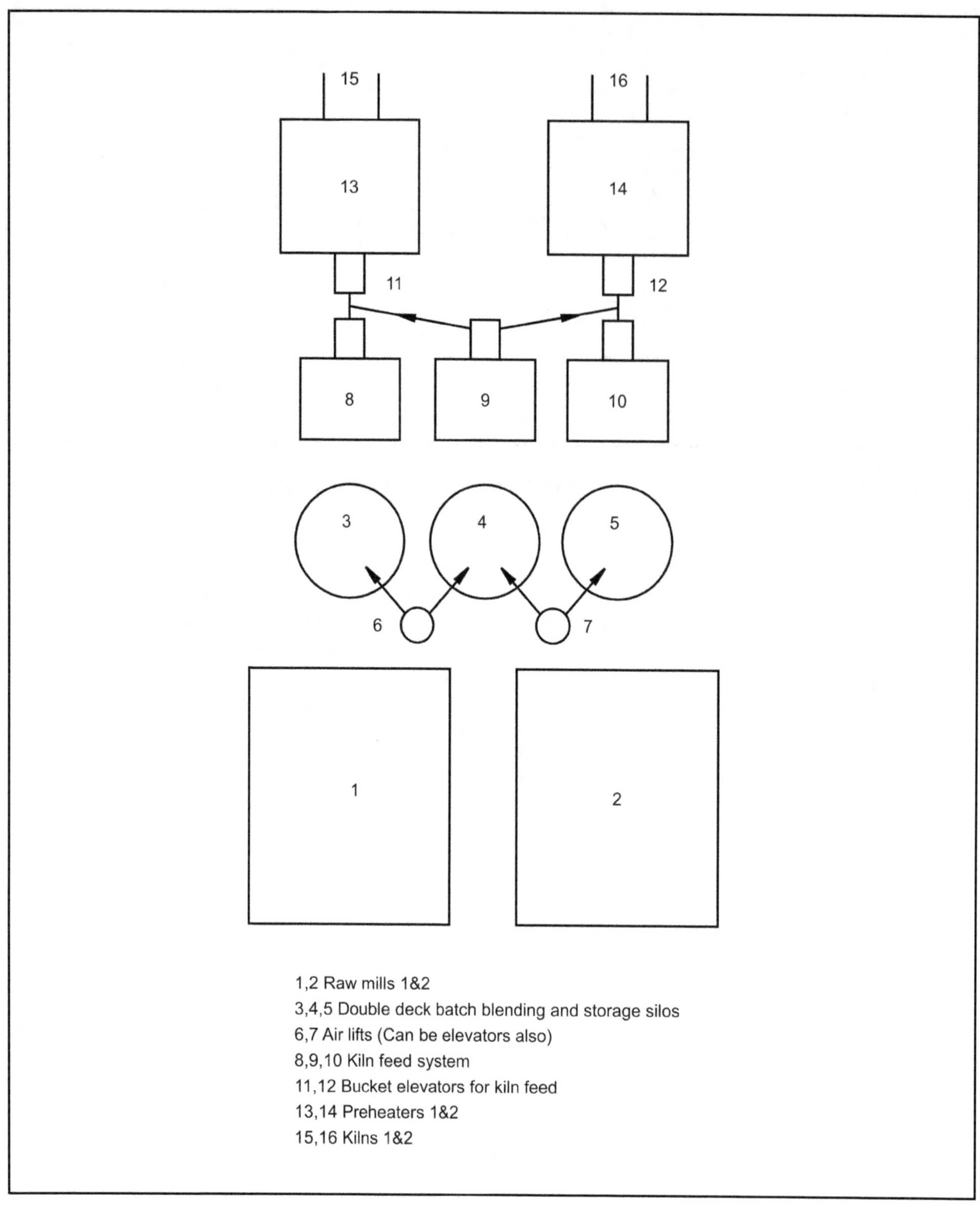

Flow chart. 1.7.6 Flow chart for two production lines with batch blending system.

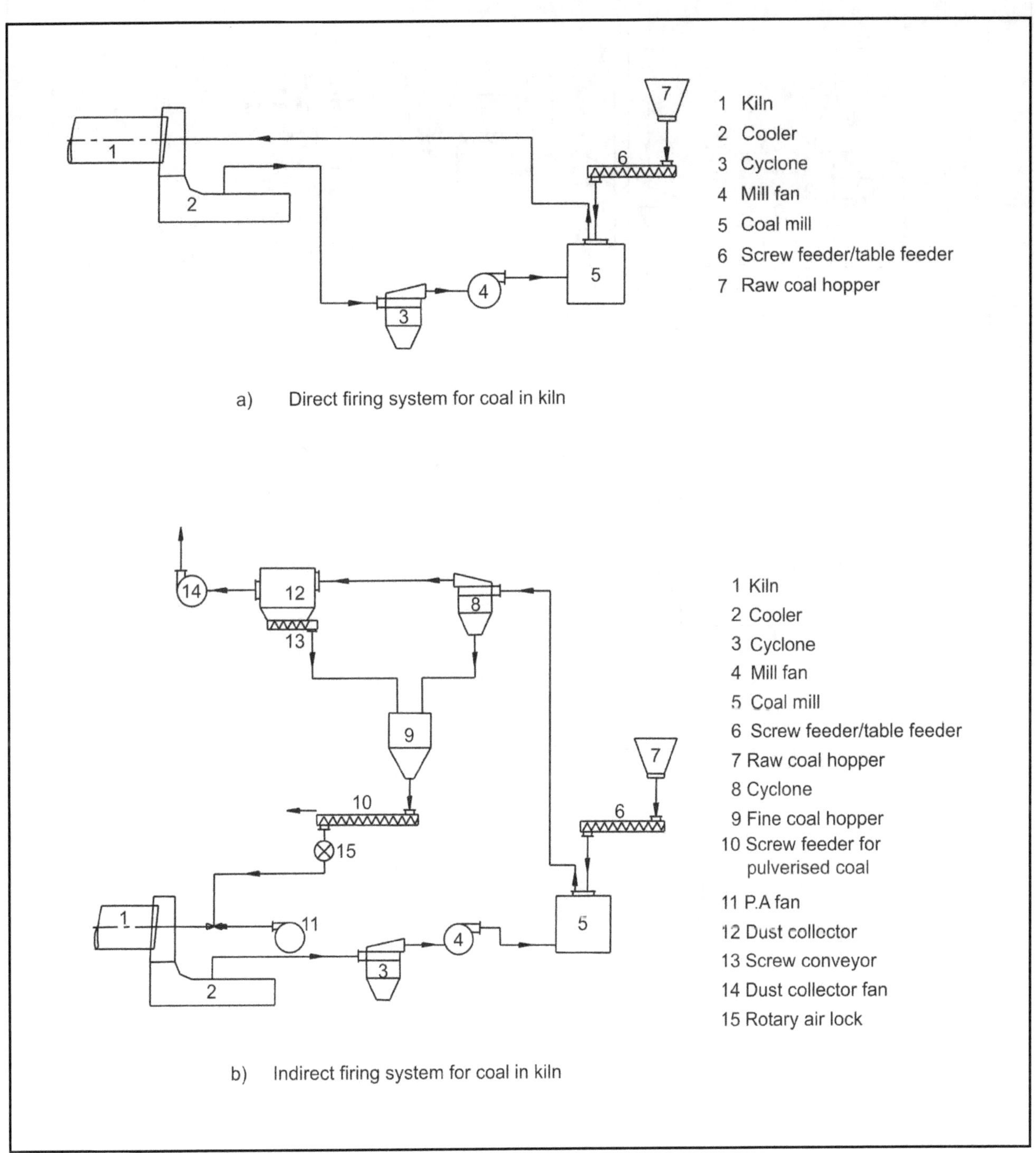

Flow chart. 1.7.7 Flow charts of coal firing systems for cement kiln.

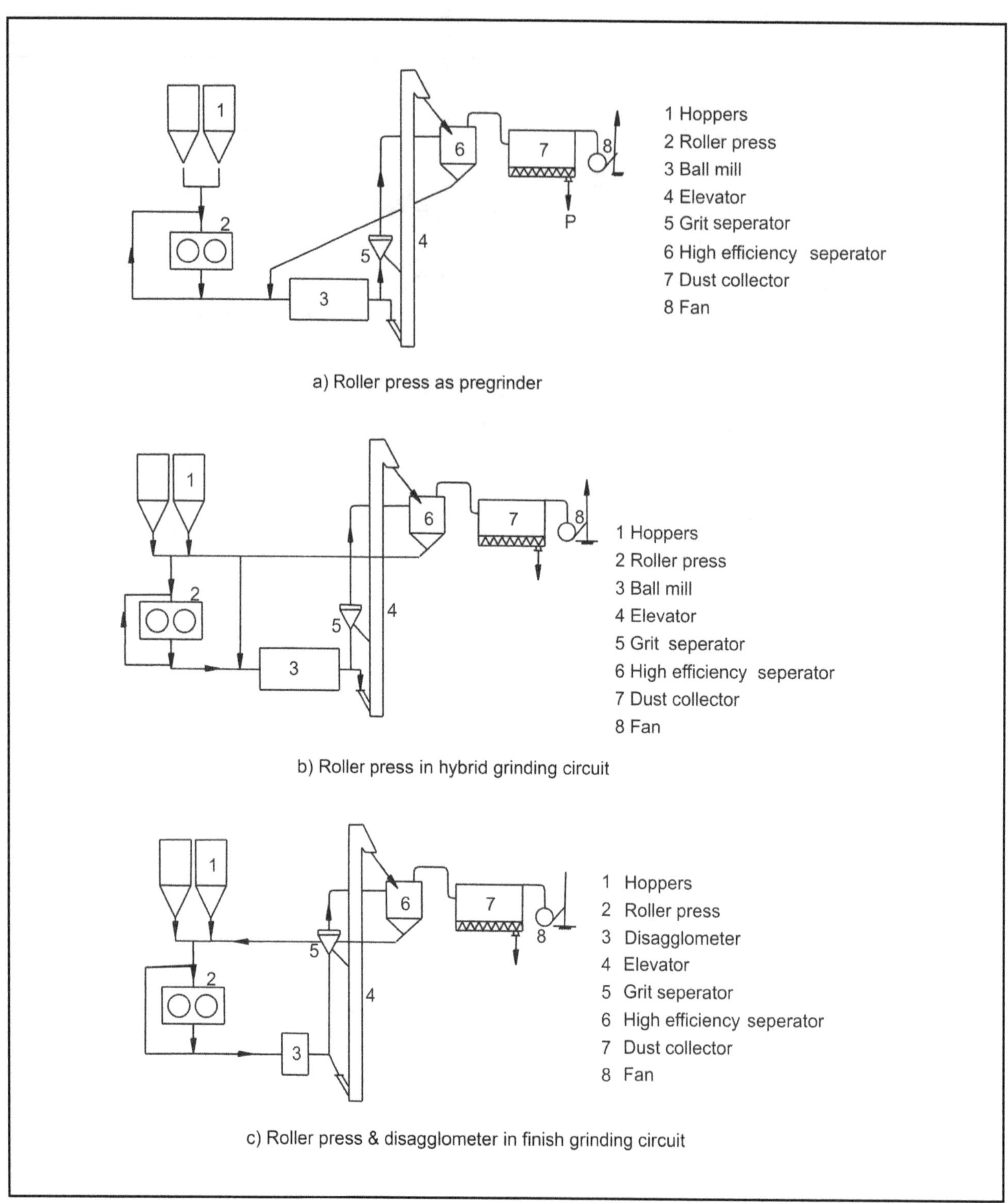

Flow chart. 1.7.8 Flow charts of Roller Press and ball mill in different combinations.

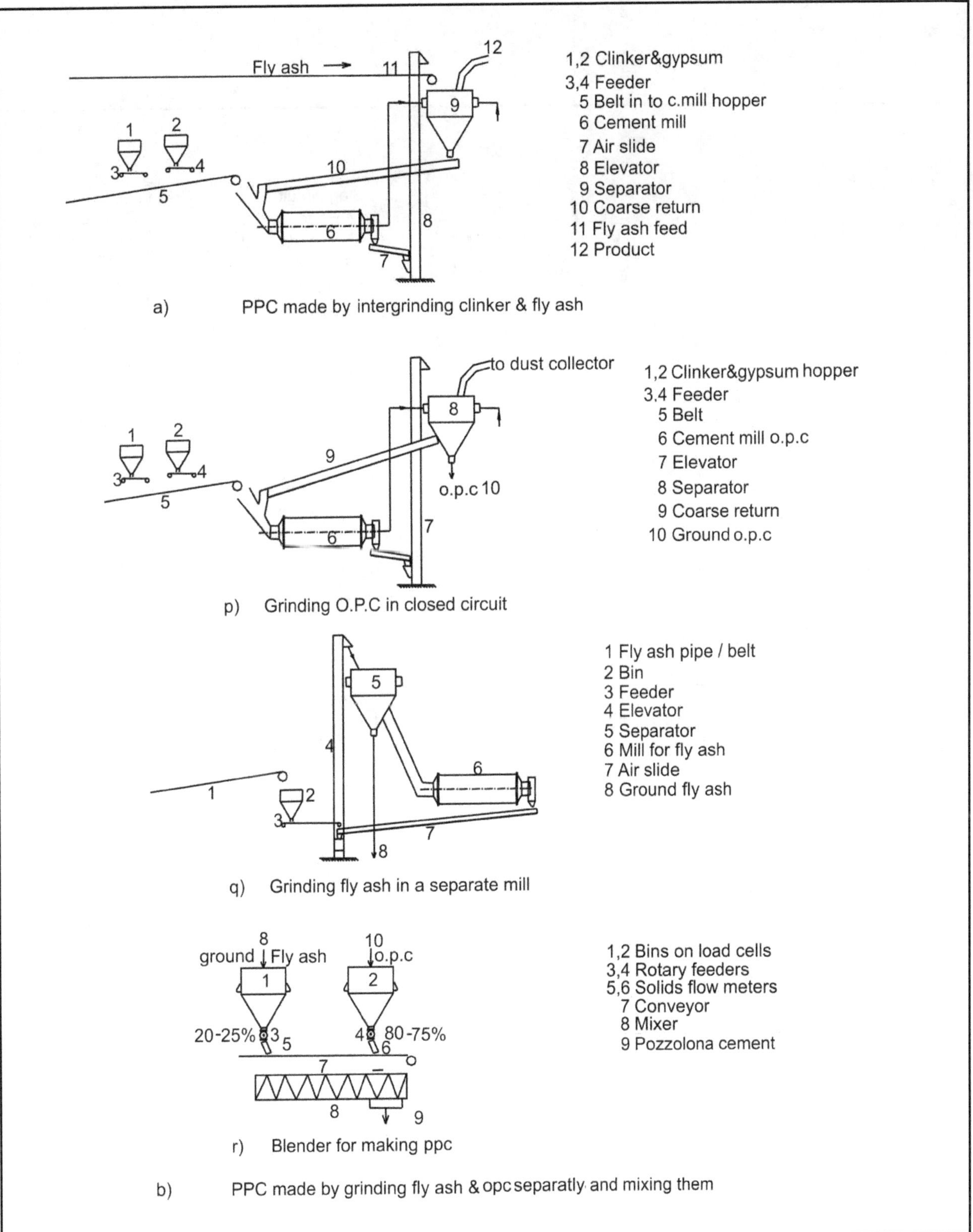

Flow chart. 1.7.9 Flow charts of different systems to make PPC.

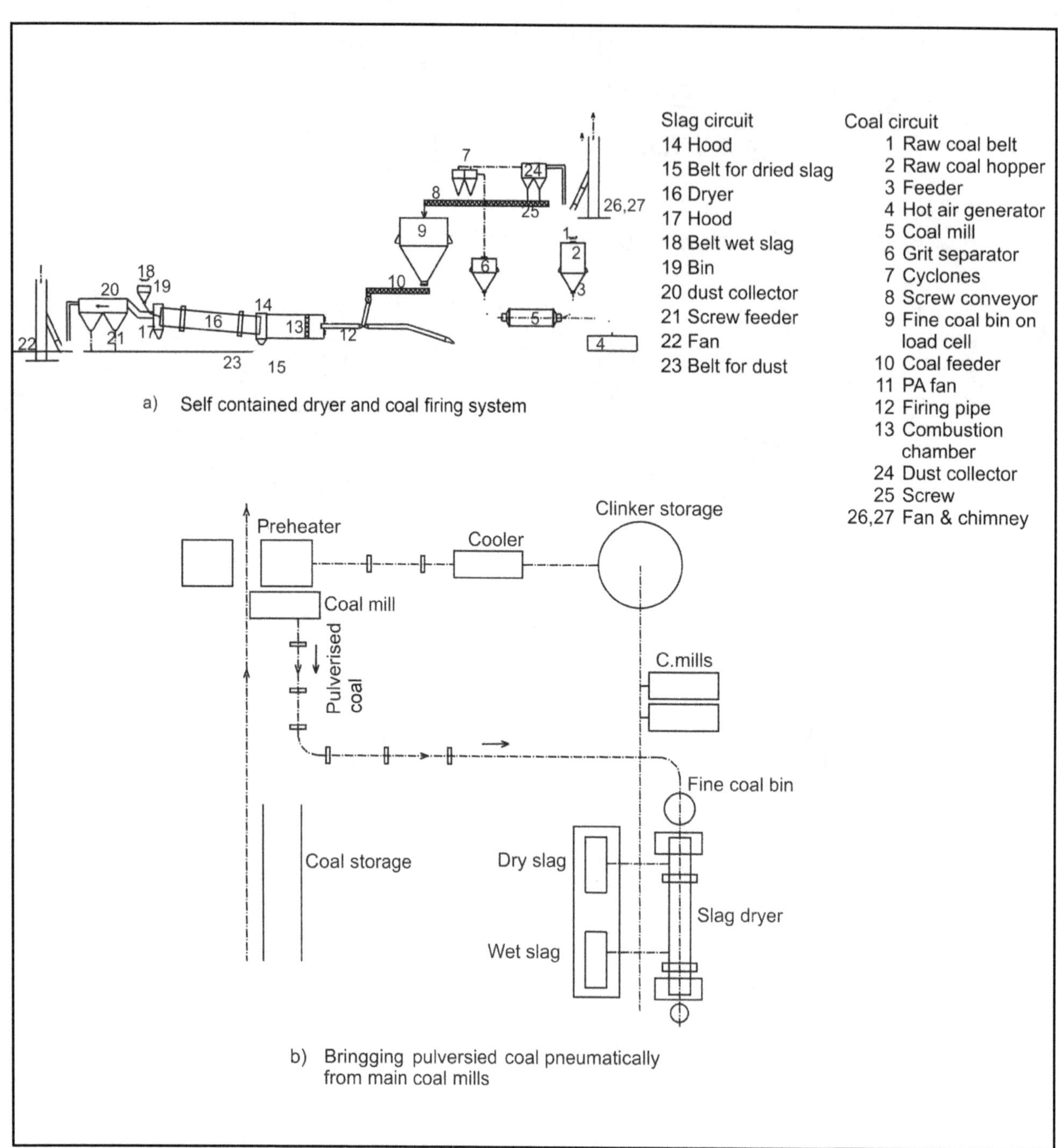

Flow chart. 1.7.10 Flow charts of coal fired rotary dryer for slag.

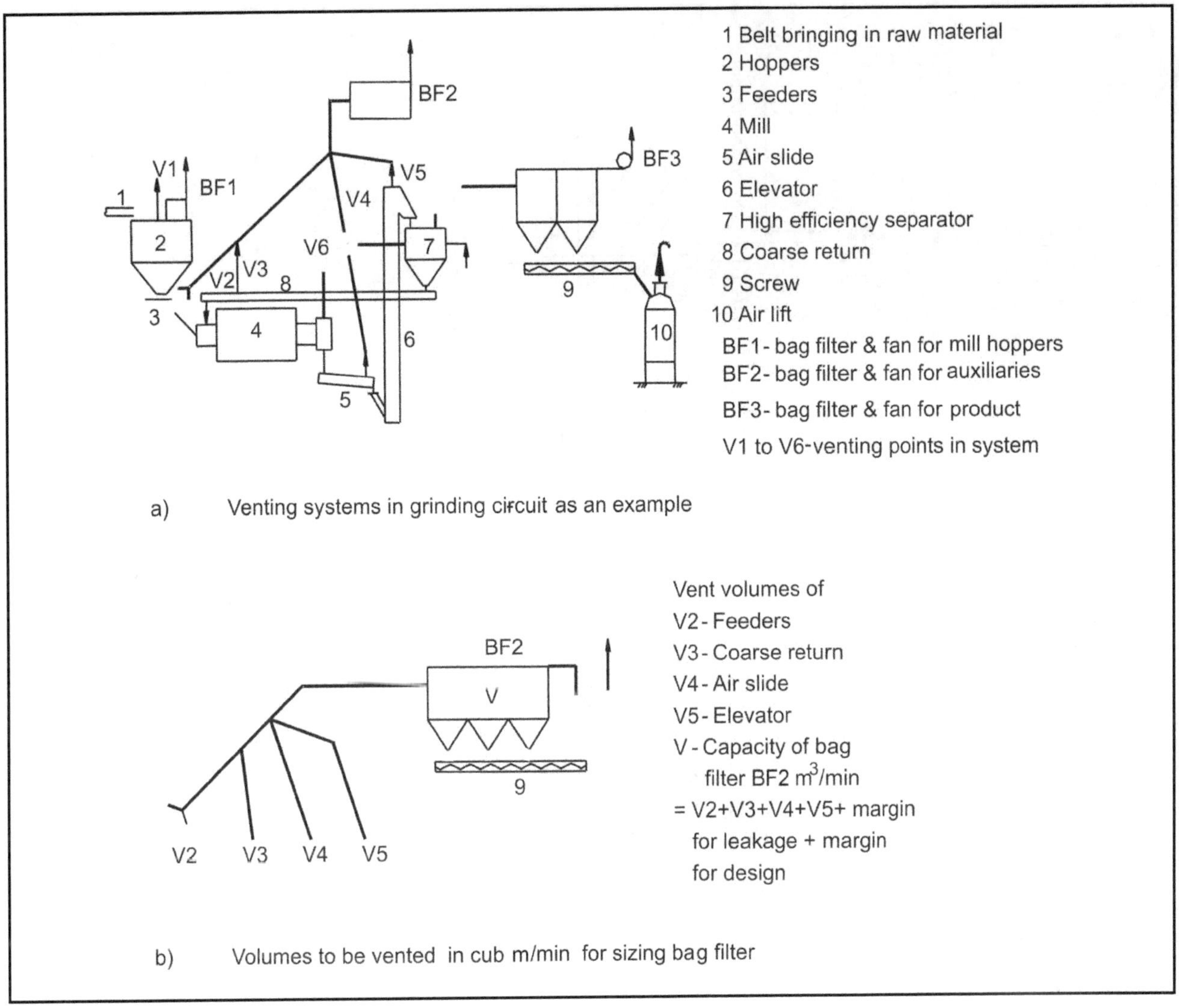

BF2
V1
BF1
1
2
V3
V2
V4
V5
V6
7
8
3
4
6
5
BF3
9
10
1 Belt bringing in raw material
2 Hoppers
3 Feeders
4 Mill
5 Air slide
6 Elevator
7 High efficiency separator
8 Coarse return
9 Screw
10 Air lift
BF1- bag filter & fan for mill hoppers
BF2- bag filter & fan for auxiliaries
BF3- bag filter & fan for product
V1 to V6-venting points in system
a) Venting systems in grinding circuit as an example
BF2
V
9
V2 V3 V4 V5
Vent volumes of
V2- Feeders
V3- Coarse return
V4- Air slide
V5- Elevator
V - Capacity of bag filter BF2 m³/min
= V2+V3+V4+V5+ margin for leakage + margin for design
b) Volumes to be vented in cub m/min for sizing bag filter

Flow chart. 1.7.11 Flow chart of vent system in a ball mill circuit.

1
2
3
4
14
5
6
7
13
11
12
8
9
10
1 Cement silo
2 Cement extraction
3 Air slide
4 Bucket elevator
5 Screen
6 Packing machine
7 Slat conveyor
8 Loading belt
9 Loading belt
10 Truck
11 Dust collector
12 Fan
13 Spillage collection
14 Spillage screw

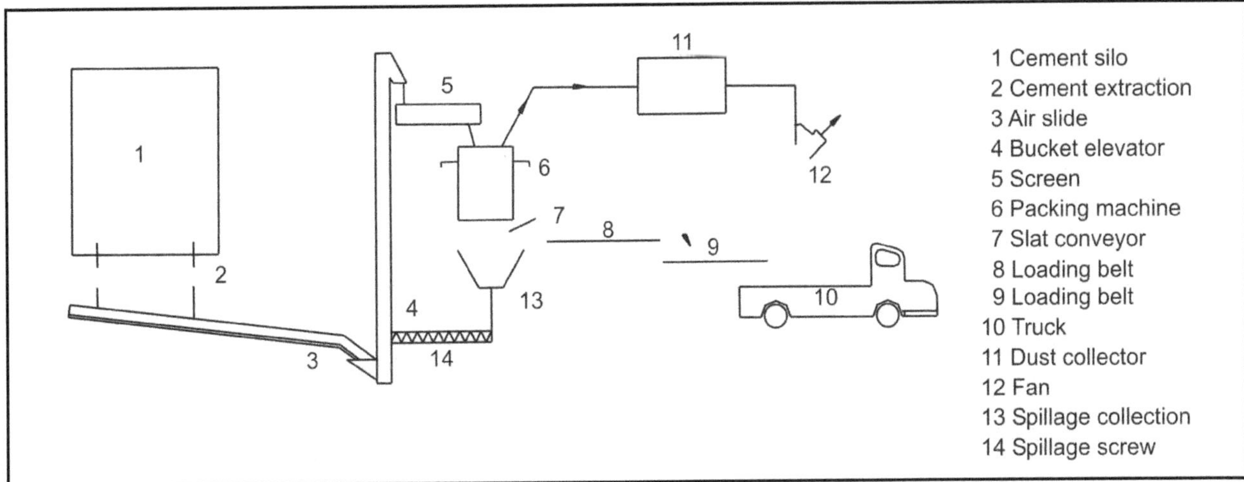

Flow chart. 1.7.12 Flow chart of cement packing and despatches.

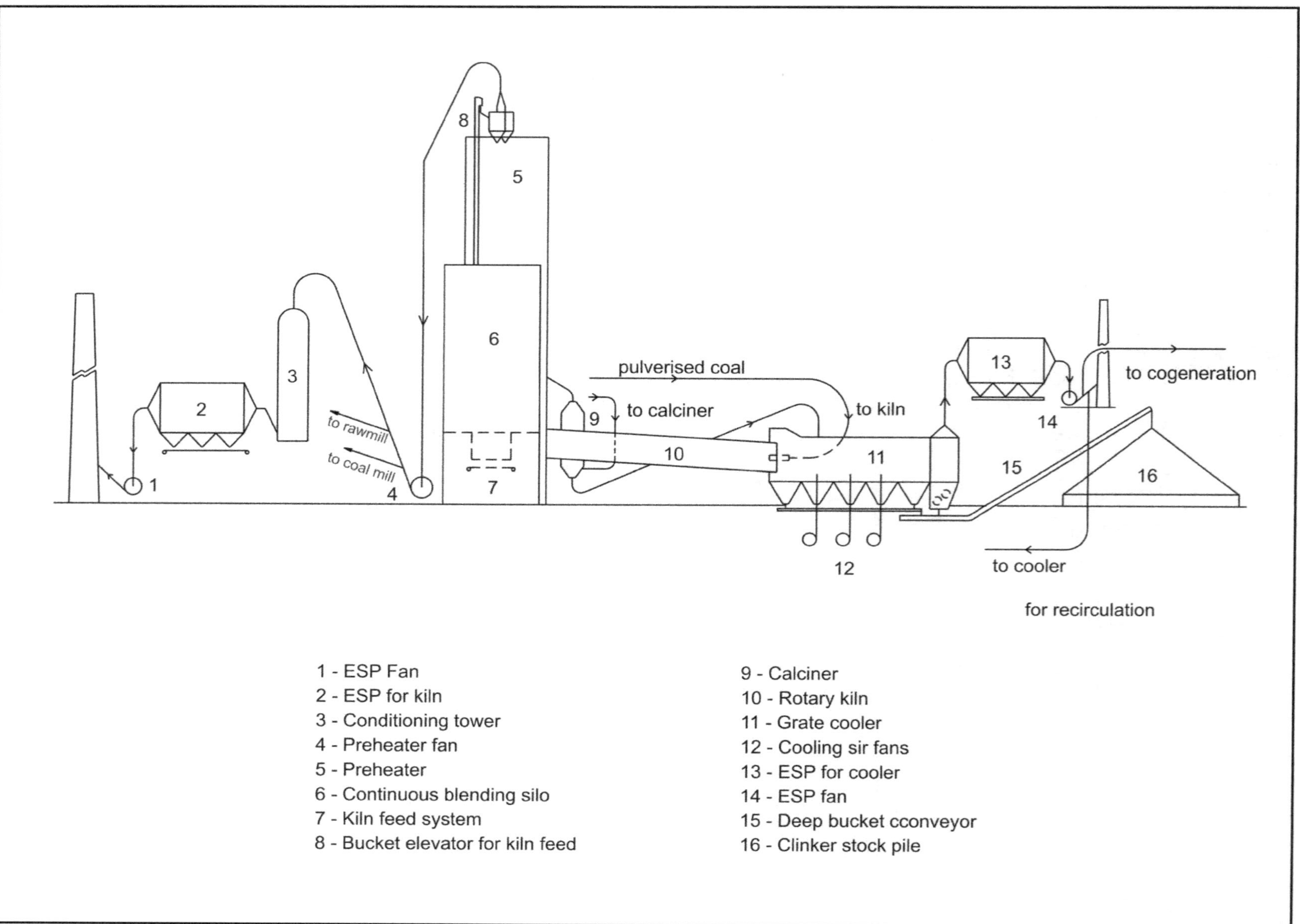

Flow chart. 1.7.13 Flow chart for pyro processing section.

CHAPTER 8

TECHNICAL CONCEPTS CONVERSION FACTORS

8.1 Conversion Factors

As mentioned earlier, Cement is made by processing minerals like limestone, clay, bauxite, sand, iron ore, in suitable proportions to prepare a 'raw mix' for sintering it to make cement clinker. Proportioning is done to achieve various 'modulii' as explained in chapter 5.

Conversion factors could be defined as factors that are dependent on chemical composition of constituent materials individually and collectively to make unit quantity of clinker or cement as the case may be.

8.2 Conversion Factors for Raw Materials

When Carbonates in 'raw mix' feed to kiln are 'calcined' to turn them into oxides Carbon Dioxide is given out and is lost with exhaust gases.

$Ca\ CO_3 = CaO + CO_2$

If raw mix consists of 100 % Ca CO3, to make 1kg of CaO would require 100/56 = 1.79 kg of Carbonate. The loss due to dissociation is 44/100 or 44 % expressed in terms of carbonates.

If limestone has 85 % carbonates then loss on ignition (L.O.I.) would be $0.85 \times 44 = 37.4$ %. Therefore limestone required to make 1 kg clinker would be 1/(1-0.374) = 1.60. This is the conversion factor. It is used to find quantity of limestone required to make 1 kg of clinker.

8.2.1 Raw Meal to Clinker Ratio (Consumption)

Raw mix is made up of limestone and additives like clay or sand or bauxite and Iron ore or laterite in various proportions to maintain the various modulii. These proportions decide quantities required to make unit quantity of clinker.

If efficiency of top cyclone is 93 %, then 7 % of raw meal is lost with kiln exhaust gases. Therefore to maintain required conversion factor of 1.59, raw meal feed to preheater would have to be 1.59/0.93 = 1.7 kg/kg clinker. The meal lost in exhaustgases is recovered in the ESP and returned to the system. Actual consumption of raw meal would then be 1.59 kg/kg clinker as mentioned earlier.

Let raw mix be made from 85% limestone, 13 % clay and 2 % iron ore.

Then to make 1 kg clinker, quantities of these constituents needed would be

$1.59 \times 0.85 = 1.35$ kg limestone

$1.59 \times 0.13 = 0.21$ kg clay and

$1.59 \times 0.02 = 0.03$ kg iron ore.

Total conversion factor for raw mix would be 1.59 and individual conversion factors for limestone, clay and iron ore would be 1.35, 0.21 and 0.03 respectively.

They are needed to work out quantities of these materials and also to calculate capacities of feeders and conveyors.

8.2.2 Raw Meal to Clinker Ratio (Kiln Feed)

Preheater gases carry with them quantity of dust depending upon the efficiency of the cyclones which is generally between 0.92 - 0.94. Thus to produce 1 kg Clinker, it is required to feed to preheater-kiln raw meal factor above / efficiency of top cyclones. Thus if efficiency is 93 %, and raw meal factor 1.59, raw meal to clinker ratio for kiln feed would be 1.59 /0.93 = 1.71.

8.3 Conversion Factors for Fuel

A quantity of heat input is required representative of the process used to make unit quantity of clinker. This heat input measured in kilocalories per kg of clinker is known as 'sp. heat consumption' or 'fuel efficiency'. 'Useful' heat value is heat released by unit quantity of fuel 'as fired' in a kiln. In case of solid fuels it will contain 1-2 % residual moisture. Sp. fuel consumption is often also expressed as % of coal (fuel) required to produce unit weight of clinker.

Thus if sp. heat consumption for a dry process kiln with 6 stage preheater is 700 kcal/kg, and useful calorific value of coal used is 4500 kcal/kg, then coal consumption is 15.55 %.

This conversion factor can be used to find out quantity of coal to be obtained and processed.

8.4 Conversion Factors for Blending Materials

1. Conversion factor of cement to clinker

   In OPC 1kg clinker + 0.05 kg gypsum makes 1.05 kg cement.

   If plant capacity is expressed in terms of kiln capacity of clinker produced, then to arrive at capacities of cement mills this factor would be taken into account.

   Conversion factor of cement to clinker = 1.05

2. *conversion factors for slag cement*

   Slag cement is made by blending blast furnace slag, clinker and gypsum. A maximum of 65 % of slag can be used to make slag cement.

   slag cement could be made from say 35 % clinker, 60 % slag and 5 % gypsum. Conversion factors with reference to clinker would be :

   cement /clinker = 100/35 = 2.86

   slag / clinker = 60/35 = 1.71

   slag /cement = 60/100 = 0.6

   Actual factors can be worked out by using actual proportions

   These factors would be useful to find out quantities of slag needed and total cement produced etc.

3. *conversion factors for Pozzolana cement*

   Pozzolana cement is made by blending fly ash, clinker and gypsum. A maximum of 35 % fly ash can be used. Pozzolana cement may be made from say 65 % clinker, 30 % fly ash and 5 % gypsum. Conversion factors with reference to clinker would be :

   cement/ clinker = 100/65 = 1.54
   fly ash / clinker = 30/65 = 0.4

   fly ash / cement = 30/100 = 0.3

   like slag above, these factors would needed to work out quantities of fly ash and capacities of respective feeders.

8.5 Factors for Moisture in Materials as Received

When reference in quantities of raw mix or proportions of constituents is made, it is with respect to 'dry' materials. Most materials like clays, iron ore, coal, gypsum and slag are received wet with moisture content varying between 5 - 20 % and with seasonal variation therein during dry and wet seasons. It is therefore necessary to know how much wet or as available material is to be received and processed to obtain desired dry quantities.

In most locations in India, limestones do not contain more than 2-3 % moisture during dry months and seldom more than 5 % even during wet months. Therefore moisture in limestone is ignored in process calculations except where very wet materials like sea shells are used.

Coals habitually contain moistures ranging between 8 to 15 %. Granulated slag would also often contain 10-15 % water.

Quantity of wet material = quantity of dry material + moisture

Thus if,

moisture	factor for allowing it
10 %	(100 + 10) / 100 = 1.1
20 %	(100 + 20) / 100 = 1.2

Quantity of wet material to be obtained and processed would be multiplied by factors as explained above.

8.6 Loss in Transit

Many a times materials like coal, gypsum etc., are procured from long distances and their receipt at site from the place of loading involves a number of transshipments and shifting from railway wagons to trucks and vice versa. A certain amount of loss in transit is therefore inevitable. As a convention it is being taken as 3 %. Thus to obtain at site 100 tons wet material, 103 tons would have to be procured and paid for. This has a direct effect on variable costs of materials. Therefore where necessary it is allowed for.

8.7 Total Multiplying (Conversion) Factor

Various conversion factors mentioned above are applied where needed for accurate estimation of quantities and capacities of feeders and conveyors.

Over all multiplying factor is arrived at by multiplying various factors mentioned above as applicable in each case. It is shown in **Table 1.8.1.**

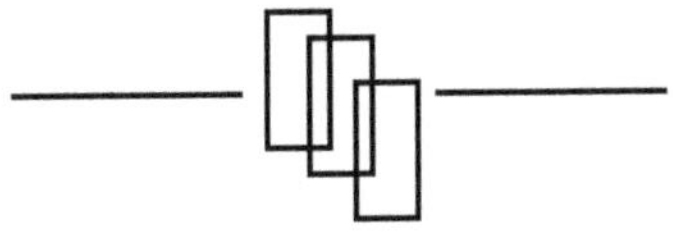

Table 1.8.1

Conversion & Multiplying factors.

Sr. No.	Item	Unit	Design Margin	Conversion Ratio	Proportion	Overburden	Moisture	Loss in Transit	Margin on design	Days per week	Total Multiplying Factor
1	quarry raising	tpd	1.1	1.6	1	1.1	1.05	1	1	1.17	2.4
2	Limestone wet, crushed,	tpd	1.1	1.6	0.9	1	1.03	1	1	1.17	1.91
3	clay,wet	tpd	1.1	1.6	0.1	1	1.2	1	1	1	0.21
4	sand,wet	tpd	1.1	1.6	0.06	1	1.1	1	1	1	0.12
5	ironore/ laterite wet	tpd	1.1	1.6	0.02	1	1.15	1	1	1	0.040

Table 1. 8.1 ***Contd....***

Sr. No.	Item	Unit	Design Margin	Conversion Ratio	Proportion	Over burden	Moisture	Loss in Transit	Margin on design	Days per week	Total Multiplying Factor
							coal in kg/kg clinker				
6	coal as fired	cal. Value		kcl/kg			4000	4500	5000	6000	
		sp.fuel con.		kcal/kg	650		0.16	0.14	0.13	0.11	
					700		0.18	**0.16**	0.14	0.12	
					750		0.19	0.17	0.15	0.13	
			1.1	**0.16**	1	1	1.1	1.05	1	1	0.20
7	coal,wet	tpd			select factor as applicable h30 to k32						
8	coke breeze, wet	tpd	1.1	0.17	1	1	1.1	1	1	1	0.21
		tpd	1.1	0.18	1	1	1.1	1	1	1	0.22
			1.1	0.19	1	1	1.1	1	1	1	0.23
			1.1	0.15	1	1	1.1	1	1	1	0.18
			1.1	0.14	1	1	1.1	1	1	1	0.17
9	clinker	tpd	1.1	1	1	1	1	1	1	1	1.100
10	gypsum, wet	tpd	1.1	0.053	1	1	1.15	1.05	1	1	0.070
11	fly ash	tpd	1.1	0.2	1	1	1	1.03	1.1	1	0.25
			1.1	0.3	1	1	1	1.03	1.1	1	0.37
			1.1	0.35	1	1	1	1.03	1.1	1	0.44

Table 1.8.1 *Contd....*

Sr. No.	Item	Unit	Design Margin	Conversion Ratio	Proportion	Over burden	Moisture	Loss in Transit	Margin on design	Days per week	Total Multiplying Factor
12	bf slag,wet	tpd	1.1	0.4	1	1	1.1	1.03	1.1	1	0.55
			1.1	0.5	1	1	1.1	1.03	1.1	1	0.69
			1.1	0.6	1	1	1.1	1.03	1.1	1	0.82
			1.1	0.65	1	1	1.1	1.03	1.1	1	0.89
13	cement opc	tpd	1.1	1.053	1	1	1	1	1.1	1	1.27
14	raw meal	tpd	1.1	1.55	1	1	1.01	1	1	1	1.722
			1.1	1.75	1	1	1.01	1	1	1	1.944
15	kiln feed	tpd	1.1	1.65	1	1	1.01	1	1.1	1	2.016
16	pulverised coal	tpd	1.1	0.14	1	1	1.02	1	1.1	1	0.173
			1.1	0.15	1	1	1.02	1	1.1	1	0.185
			1.1	0.16	1	1	1.02	1	1.1	1	0.197
			1.1	0.17	1	1	1.02	1	1.1	1	0.210
			1.1	0.18	1	1	1.02	1	1.1	1	0.222
			1.1	0.19	1	1	1.02	1	1.1	1	0.234

Table 1.8.1 *Contd....*

Sr. No.	Item	Unit	Design Margin	Conversion Ratio	Proportion	Over burden	Moisture	Loss in Transit	Margin on design	Days per week	Total Multiplying Factor
17	blended										
	cements										
	ppc										
	fly ash 20 %		1.1	1.33	1	1	1	1	1.1	1	1.61
	30%		1.1	1.54	1	1	1	1	1.1	1	1.86
	35%		1.1	1.67	1	1	1	1	1.1	1	2.02
18	slag cements										
	slag 30%		1.1	1.54	1	1	1.1	1	1.1	1	2.05
	40%		1.1	1.82	1	1	1.1	1	1.1	1	2.42
	50%		1.1	2.22	1	1	1.1	1	1.1	1	2.95
	60%		1.1	2.86	1	1	1.1	1	1.1	1	3.81
	65%		1.1	3.33	1	1	1.1	1	1.1	1	4.43
19	oil		1.1	0.08	1	1	1	1	1.1	1	0.10

CHAPTER 9

TECHNICAL CONCEPTS
DESIGN MARGINS TO ARRIVE AT CAPACITIES

9.1 Sectional Capacities

Sectional capacities depend on 'running hours' in each section and also on factors or margins to be provided for to arrive at capacities of major machinery and auxiliaries in the section.

Knowing the total tonnage of materials to be processed in the section and the number of running hours, basic capacity can be arrived at. This capacity multiplied by margins mentioned above gives the 'design capacity' of a given machine for procurement and sizing.

Running hours in various sections of the plant will be dealt within the next Chapter. In this Chapter the rationale behind various factors and margins commonly used for arriving at capacities and sizing is explained.

9.2 Design Margin

It is a margin over the rated capacity more or less for all sections of the plant. Usually it is 10 %. This margin ensures that rated capacity will be achieved in spite of fluctuations normally expected to occur in quality of raw materials and fuel.

It takes care of some unforeseen stoppages or interruptions.

On the positive side, it helps in higher production under favourable operating conditions

9.3 Safety Margin

Even when there is absolute certainty about the efficacy of the process and machinery selected, it is prudent to play safe as regards sp. fuel consumption and sp. power consumption. Therefore a 'safety margin' is added to arrive at capacities of all entities dependent on them.

For example a margin of 5 % on fuel consumption would directly govern size of coal grinding and coal firing sections and also gas flow in kiln and preheater sections.

Safety factor on power consumption would influence even maximum demand and ratings of transformers etc.

9.4 Margin for Feeders and Conveyors

Material handling equipment of any kind should never be a bottleneck to achieve the production over margins mentioned above in major machinery and auxiliaries. Therefore a margin of 20 to 25 % is allowed for over design capacity in sizing material handling equipment and in feeders.

9.5 Margin for upset Conditions

Machinery like kiln can run under upset conditions by as much as 30 %. Such conditions can last for several hours at a time. During this time, clinker conveyors handle under burnt clinker at high temperatures and high rates. Therefore conveyors for spillage and main product for clinker going right up to clinker storage are sized with a margin for 30 % for this condition.

9.6 Margins for Fans

A margin of 10 to 15 % is allowed in arriving at capacities of fans over volumes arrived at after applying design and safety margins mentioned above. Since system resistance is proportional to square of velocity, a 10 % increase in volume results in an increase of 20 % in system resistance. Therefore a margin of 20-30 % is necessarily added on static pressures of fans.

9.7 Margin for Crushers

Because of irregularities in mining operations, a margin of 20 % is added over design margin, conversion factor and working hours to arrive at crusher capacity.

However primary crushers are sized to suit bucket capacities of shovels selected. It may so happen that crusher so selected would render it unnecessary to use above irregularity factor separately.

9.8 Cement Mills

A margin of at least 10 % is usually added on design capacity in sizing cement mills. It takes care of spurts in demand and also helps in clearing accumulated stocks of clinker.

9.9 Packing and Dispatches

In actual practice, proportions of bagged and bulk cement- dispatches by road and rail could be quite different from those assumed at project stage - based on data collected.

Even if 30 % dispatches should be in bulk it could so happen that there would be periods when all of dispatches are for bagged cement. Therefore capacities of packing machines should be such as to be able to cope for such a contingency. Same logic applies to designing facilities such as mechanized loading for dispatches by road or rail.

In short it should be assumed that

100 % cement would be bagged or in bulk

100 % dispatches could be either by road or rail

Table 1.9.1 summarises various margins mentioned above.

Table 1.9.1 Margins and multiplying factors.

Sr. No.	Item	Margin	Multiplying factor	Remarks
1	Design Margin	10 %	1.1	In special cases this could be higher
2	Safety margins for fuel and power consumption	5 %	1.1 × 1.05	
	For sizing substation capacity, maximum demand	25%	1.1 × 1.25	
3	Margin for feeders and conveyors	15-20 %	1.1 × (1.15-1.2)	
4	Upset condition of kiln	30 %	1.1 × 1.3	
5	Fans			
	On volume	10-15 %	1.1 × (1.1-1.15)	
	On pressure	20-30 %	1.2 –1.3	
6	Crusher	20 %	1.1 × 1.2	
7	Cement mills	10%	1.1 × 1.1	
8	Packing and dispatches			
	Cement bagged and bulk	100 %	1.1 × 1.1 × 1	
	Cement by road / rail	100 %	1.1 × 1.1 × 1	

Above factors are to be used in conjunction with conversion factors explained in **Chapter 8 (See Table 1.8.1).**

9.10 Margin for power cuts

When power cuts are severe, hours available for running the plant are curtailed. Design margin of 10 % cannot compensate for production lost. One way of course is to include a power plant in capital costs. Another is to allow for reduced available number of hours of running in various sections so that machinery selected has a larger capacity and produces required rated capacities even in less available time.

Scheduled or unscheduled power cuts result in a far greater loss in production because it takes some time in each department to reach normal production levels after an interruption.

Some entrepreneurs prefer to oversize machinery to meet this contingency.

Factor to be allowed for will vary from location to location. It could be between 20 to 30 %.

It is assumed that in normal circumstances such a margin is not required. Hence it is not included in above **Table 1.9.1.**

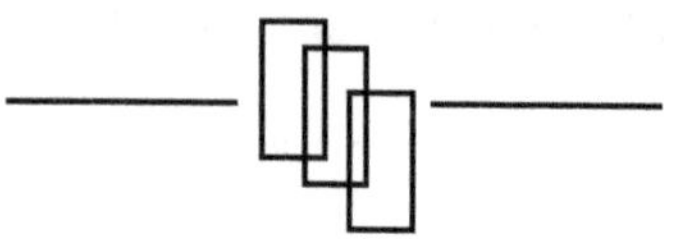

CHAPTER 10

TECHNICAL CONCEPTS
RUNNING HOURS IN VARIOUS SECTIONS

10.1 Running Hours

Running hours in various sections per day, per week and per year decide the capacity of the section required to achieve the yearly production targets.

These capacities in turn influence directly the sizes and drive ratings of the major machineries and auxiliaries in respective sections.

Over a period of time and for practical reasons, some conventions have already been established in the cement industry in this regard. These are described below.

10.2 Clinkering Operations

Clinkering being a continuous process, the operation of kiln and connected sections and auxiliaries must go round the clock for days on end.

For kiln section, conventionally, 330 working days are assumed per year of 365 days. This gives about 10 % of time for 2-3 stoppages in a year for 'brick lining' and maintenance and upkeep of the various machineries in kiln section itself and in sections immediately upstream and down stream of it.

For all practical purposes, the upstream section starts with extraction from raw meal storage silos and includes metering of kiln feed and its feed to the preheater/s and calciner.

Down streamside ends with clinker conveyor discharging into clinker storage.

Fuel metering and firing section must also work continuously with kiln; so also dust collecting equipment for kiln, cooler and kiln feed sections.

Design running hours of a pyroprocessing section per year are therefore $24 \times 330 = 7920$.

Before the advent of calciners in case of large kilns (5 m and above in dia) running days used to be taken as 300 only as life of brick lining in large kilns used to be shorter. With calciner kilns this is no longer necessary.

10.3 Preparation of Raw Materials

10.3.1 Quarrying and Crushing

In the past quarries worked mostly in day time in one shift to start with and in two shifts later.

With plant capacities now reaching 3000-5000 tpd even in India, this is hardly practical any longer.

Crushing section would normally be designed for 2 shifts working to start with and then 3 shifts working after expansion (which in normal course would be assumed as duplication or doubling of capacity)

The crusher capacity would therefore be so selected as to supply requirements of crushed stone for double capacity in 3 shifts.

Quarries likewise would work initially in 2 shifts and later in 3 shifts.

For quarrying and crushing operations, effective hours per shift are not all of 8 hours of the shift as time is lost because :

(a) at beginning of first shift in starting mining machinery, in directing them to faces in operation; dumpers would normally return to the garage. They too would have to be fuelled before going to the loading points.

(b) There is a lead of say 1-3.5 kms between loading face and crusher hopper. While it is planned to have a continuous supply of stone to crusher to keep it filled, this may not always be so. Some 'idle hours' are to be expected.

(c) It is a good idea to empty the hopper of crusher at the end of the day so that when restarted feeder / crusher do not start on load.

For these reasons, the effective time of crushing and quarrying operations is taken as 6 hours / shift in first two shifts and 5 hours in third or night shift.

Quarry and crusher sections work for 6 days of the week. Hence are required to supply 7 days' requirement of crushed limestone in 6 days.

Thus number of hours working / week in quarrying and crushing :

	2 Shifts	3 Shifts
Per Day	$6 \times 2 = 12$	$2 \times 6 + 1 \times 5 = 17$
Per Week	72 hours	102 hours
Per year	3744 hours	5304 hours

10.3.2 Grinding Operations

In size reduction crushing is followed by grinding including drying where required.

Grinding mills prepare raw materials, coal for further processing and convert clinker into cement.

Conventionally, for all the 3 grinding mills - raw, coal and cement,

Running hours are taken as 20 per day.

Running days per year are taken as 360 per year.

Total running hours per year are 7200.

10.4 Raw Material Grinding

This operation grinds crushed stone and additives to a fine powder, drying out moisture in its constituents during grinding by using hot gases from kiln.

Convention of 20 running hours per day allows for 4 hours /day for maintenance and upkeep. It is not as if mill would be stopped after 20 hours. It will be allowed to run continuously as long as possible. These hours are of course insufficient for major grading and loading of grinding media in a ball mill or change of liners of rollers, table etc., in a vertical mill. These operations would take anything between 3-6 shifts. A long stoppage will be taken for such jobs taking advantage of the margins available because of 20 design running hours per day.

In case of large ball mills, material handing equipment is installed over the mill for loading the mill, taking out lining plates etc.

Short portable conveyors are also used to load the mill with grinding media.

It is also necessary to shift quickly unloaded grinding media, to keep the area clean.

Vertical roller mill installation has to be provided with cranes and gantries to take out liners of rollers and tables and for jacking up the mill to remove gear box. The equipment becomes an integral part of the mill and the layout.

As major jobs mentioned above would take longer than 4 hours, it is customary to synchronise them with stoppage of kiln for brick lining. In doing so care is taken to fill in the blending / storage silos of raw meal before stopping the mill.

10.5 Coal Mills

Coal Mills are similarly assumed to be working for 20 hours/day. They use either cooler exhaust gases or kiln exhaust gases for drying. Moisture in coal to be ground varies considerably in dry and wet seasons but would seldom be less than 8 %. Output of a mill particularly ball mill is sensitive to moisture in coal to be ground dropping by as much as 50 % in wet season. Therefore in specific locations coal mills may be sized on the basis of 16 to 18 running hours per d ay. 18 hours would require operation in third shift also and hence 16 would be preferred.

Advantage of long kiln stoppage is taken to attend to maintenance in other departments like crushing and grinding

10.6 Stacker and Reclaimer Systems

Stackers work together with crushers which supply crushed limestone and coal for stockpiling to them. Reclaimers work with grinding mills.

Therefore hours of working of stacker would correspond to working hours of crusher feeding them.

In case of limestone, this would be 2 or 3 shift working of total 12 or 17 hours.

Mills have hoppers before them. Hopper capacities can be between 4 to 8 hrs for limestone in case of raw mills and coal mills. As mentioned above, mills work for 20 hours per day. Without hoppers in between, reclaimer would have to work also for 20 hours. However with hoppers it is possible to reduce number of working hours of reclaimer by increasing capacity of reclaiming. Normally reclaimer would work for half the hours - their job being to keep hoppers full. Running hours will double if same reclaimer supplies stone / coal to two mills after duplication.

10.7 Cement Packing and Despatches

Packing plants normally operate in all three shifts so that wagons or trucks do not have to wait. This is also true of dispatches in bulk.

When cement is dispatched in bags, packing and loading operations are synchronized, that is, filled bags are taken straight to trucks or wagons - without any intermediate storage.

Till now, loading of trucks and wagons were labour intensive operations necessitating short breaks at regular intervals to rest loading gangs. Therefore number of effective hours per shift for these operations were taken as 5 and total per day of 3 shifts 15. Like mills packing operations will be for 360 days per year i.e., 5400 hours per year.

With increasing use of mechanized wagon and truck loading machines, these running hours could be increased to say 6 / shift.

10.8

Annexe 1.10.1 presents running hours for main plant and duplication in various sections of the plant.

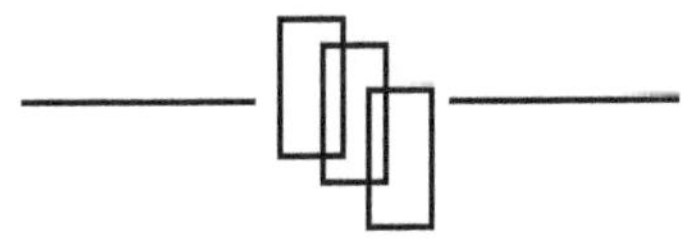

Annexe 1.10.1

Working Hours in Various Sections of a Cement Plant.

A Main Plant

Sr No	Section	hrs/shift	shifts/day	hrs/day	day/week	hrs/week	day/year	hrs/year	Remarks
1	Qyarrying	6	2	12	6	72	312	3744	
2	Crushing	6	2	12	6	72	312	3744	
3	Stacker	6	2	12	6	72	312	3744	
4	Reclaimer	5	2	10	7	70	360	3600	assuming supplies day's requirement of mill in 10 hours
5	Raw mill			20	7		360	7200	
6	Blending			20	7		360	7200	works with raw mill
7	Extraction & kiln feed	24			7		330	7920	works with kiln
8	Kiln	24					330	7920	
9	Coal stacker					to suit capacity of wagon tippler			

Annexe 1 *Contd....*

Sr No	Section	hrs/shift	shifts/day	hrs/day	day/week	hrs/week	day/year	hrs/year	Remarks
10	reclaimer			10	7		360	3600	
11	coal mill	20			7		360	7200	
12	cement mill	20			7		360	7200	
13	packing and despatches	5	3	15	7		360	5400	

B Duplication

Sr No	Section	hrs/shift	shifts/day	hrs/day	day/week	hrs/week	day/year	hrs/year	Remarks
1	Quarrying	6	3	17	6	102	312	5304	same crusher in 3 shifts
		6	4	24	6	144	312	7488	2 crushers in 2 shifts
2	crushing	6	3	17	6	102	312	5304	same crusher in 3 shifts
		6	4	24	6	144	312	7488	2 crushers in 2 shifts
3	stacker	6	3	17	6	102	312	5304	same stacker
		6	4	24	6	144	312	7488	2nd stacker
4	reclaimer	5	2	10	7	70	360	3600	same reclaimer
		5	4	20			360	7200	second reclaimer
5	raw mill	20		40			360	14400	2 raw mills
6	blending	20		40			360	14400	2 blending silos

Annexe 1.10.1 *Contd....*

Sr No	Section	hrs/shift	shifts/day	hrs/day	day/week	hrs/week	day/year	hrs/year	Remarks
7	extraction, kiln feed			24			660	15840	2 production lines
8	kiln			24			660	15840	2 kilns
9	coal stacker					to suit capacity of wagon tippler			
10	reclaimer			20	7		360	7200	same reclaimer
11	coal mill	20		40			360	14400	2 coal mills
12	cement mill	20		40			360	14400	2 cement mills
13	packing and despatches	5	6	30			360	10800	2 packing machines

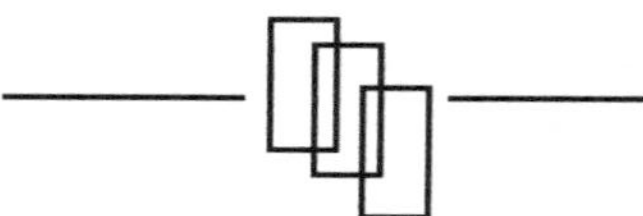

CHAPTER 11

TECHNICAL CONCEPTS SECTIONAL AND INDIVIDUAL CAPACITIES

11.1 Sections of Cement Plant

A Cement Plant can be broadly divided into following sections:

1. Quarrying
2. Crushing
3. Storage of limestone and correcting materials
4. Raw materials grinding
5. Blending and storage of raw meal
6. Pyro processing beginning with kiln feed, preheaters and ending with clinker coolers and coal / fuel firing equipment
7. Clinker storage
8. Coal crushing and storage
9. Coal mills
10. Cement mills
11. Cement storage
12. Cement packing and dispatches.

If the plant is making blended cements, then it will have additional sections for handling blending materials like slag and fly ash.

11.2 Capacity of a Cement Plant

The capacity of a cement plant is commonly expressed in tons per year (TPA or tpa) or Million tons per year (MTPA or mtpa). It then refers to cement/s produced.

It is also quite often expressed as tons per day (TPD or tpd). It then refers to the clinkering capacity.

Thus capacity of kiln is universally expressed in tons per day (TPD or tpd).

Cement capacity is obtained from clinker capacity by increasing it by amount of gypsum added to it.

When making blended cements conversion factors as applicable will be used to convert clinkering capacity into cement capacity. See **Chapter 8, Table 1.8.1. of this Section.**

11.3 Sectional Capacities

Capacities of other sections are commonly expressed in tons per hour (TPH or tph).

For example, for crushers it is stone crushed in tph; for mills it is materials ground in tph; for packing it would be cement packing in tph.

Despatches would commonly be expressed in tpd.

A Plant designer has to break up Overall Plant Capacity into capacities of various sections mentioned above.

11.4 Arriving at Sectional Capacities

These are calculated by using (a) conversion factors, (b) margins and (c) running hours in respective sections as explained in **Chapters 8 to 10.**

Simply put,

$$\text{sectional capacity} = \frac{\text{Rated capacity} \times \text{margins} \times \text{conversion factors}}{\text{running hours}}$$

in the pertinent section.

This is illustrated by the following :

1. Design capacity 10 %, factor 1.1
2. Let Rated capacity be 1 m.t.p.a. cement

   Clinkering capacity m.t.p.a. including design capacity = 1.1 × 1/1.04 (4 % gypsum) = 1.06 mtpa.
3. design capacity of kiln = 1.06 mtpa/330

   = 3200 tpd.

   rated capacity = 2900 tpd

Capacities of various sections listed above and major machineries in them can be worked out in this manner.

11.5 Overall Multiplying Factors

Annexure 1 shows calculated sectional and individual capacities for a cement plant of 10000 tpd clinkering capacity.

11.6 Capacities of Major Auxiliaries

Capacities of major auxiliaries like feeders and conveyors are worked out by using pertinent margins explained in **Chapter 9, Table 1.9.1.**

Capacities of various fans are dependent on process requirements and are worked out using pertinent process calculations. After arriving at base requirement in terms of volume and static pressure, at rated capacities, specifications of fans are worked out by applying margins on volume and pressure as explained in **Chapter 9, Table 1.9.1.**

11.7 Storages

Storages are expressed as stocks in tons in number of days' requirement and consumption.

To arrive day's requirements, the procedure is the same as explained above and in **Chapters 8 to 10.**

However to arrive at space requirement of quantities to be stored, two other factors viz, moisture content and bulk density have to be taken into account.

Storages would be dealt with in greater detail in **Chapter 13.**

11.8 Nomograms

Pl. also see nomgrams **Figs. 1.11.1 to 1.11.3** to arrive at secional capacities.

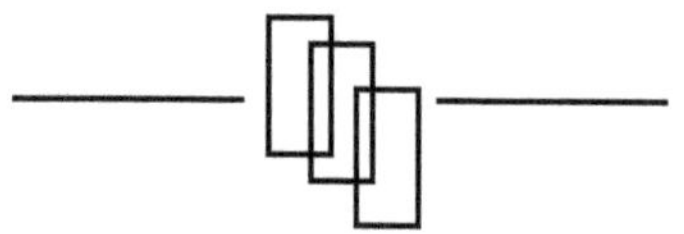

Annexe 1.11.1

Conversion & Multiplying Factors.

Sr. no.	Item	Unit	Design margin	Conversion ratio	Proportion	Over burden	Moisture	Loss in transit	Margin on design	Days per week	Total multiplying factor
1	quarry raising	tpd	1.1	1.6	1	1.1	1.05	1	1	1.17	2.4
2	Limestone wet, crushed	tpd	1.1	1.6	0.9	1	1.03	1	1	1.17	1.91
3	clay, wet	tpd	1.1	1.6	0.1	1	1.2	1	1	1	0.21
4	sand, wet	tpd	1.1	1.6	0.06	1	1.1	1	1	1	0.12
5	Iron ore / laterite wet	tpd	1.1	1.6	0.02	1	1.15	1	1	1	0.040

Annexe 1.11.1 ***Contd....***

Sr. no.	Item	Unit	Design margin	Conversion ratio	Proportion	Over burden	Moisture	Loss in transit	Margin on design	Days per week	Total multiplying factor
							coal in kg/kg clinker				
6	coal as fired	cal. Value		kcl/kg			4000	4500	5000	6000	
		sp.fuel con.		kcal/kg	650		0.16	0.14	0.13	0.11	
					700		0.18	**0.16**	0.14	0.12	
					750		0.19	0.17	0.15	0.13	
				select factor as applicable from above table							
7	coal,wet	tpd	1.1	**0.14**	1	1	1.1	1.05	1	1	**0.18**
	coke breeze,wet	tpd	1.1	0.15	1	1	1.1	1.05	1	1	0.19
		tpd	1.1	**0.16**	1	1	1.1	1.05	1	1	**0.20**
			1.1	0.17	1	1	1.1	1.05	1	1	0.22
			1.1	0.18	1	1	1.1	1.05	1	1	0.23
			1.1	0.19	1	1	1.1	1.05	1	1	0.24
8	clinker	tpd	1.1	1	1	1	1	1	1	1	1.10
9	gypsum, wet	tpd	1.1	0.053	1	1	1.15	1.05	1	1	0.07

Annexe 1.11.1 *Contd....*

Sr. no.	Item	Unit	Design margin	Conversion ratio	Proportion	Over burden	Moisture	Loss in transit	Margin on design	Days per week	Total multiplying factor
10	fly ash	tpd	1.1	0.2	1	1	1	1.03	1.1	1	0.25
			1.1	0.3	1	1	1	1.03	1.1	1	0.00
			1.1	0.35	1	1	1	1.03	1.1	1	0.44
11	Blast furnace	tpd	1.1	0.4	1	1	1.1	1.03	1.1	1	0.55
	slag, wet		1.1	0.5	1	1	1.1	1.03	1.1	1	0.69
			1.1	0.6	1	1	1.1	1.03	1.1	1	0.82
			1.1	0.65	1	1	1.1	1.03	1.1	1	0.89
12	cement opc	tpd	1.1	1.053	1	1	1	1	1.1	1	1.27
13	raw meal	tpd	1.1	1.55	1	1	1.01	1	1.1	1	1.894
14	kiln feed	tpd	1.1	1.55	1	1	1.01	1	1.1	1	1.894

Annexe 1.11.1 ***Contd....***

Sr. no.	Item	Unit	Design margin	Conversion ratio	Proportion	Over burden	Moisture	Loss in transit	Margin on design	Days per week	Total multiplying factor
15	pulverised coal										
			1.1	**0.14**	1	1	1.02	1	1.1	1	**0.173**
		tpd	1.1	0.15	1	1	1.02	1	1.1	1	0.185
			1.1	**0.16**	1	1	1.02	1	1.1	1	**0.197**
			1.1	0.17	1	1	1.02	1	1.1	1	0.210
			1.1	0.18	1	1	1.02	1	1.1	1	0.222
			1.1	0.19	1	1	1.02	1	1.1	1	0.234
16	blended										
	cements										
	1) ppc										
	flyash										
	20%		1.1	1.32	1	1	1	1	1.1	1	1.60
	30%		1.1	**1.54**	1	1	1	1	1.1	1	1.86
	35%		1.1	1.62	1	1	1	1	1.1	1	1.96

Annexe 1.11.1 ***Contd....***

Sr. no.	Item	Unit	Design margin	Conversion ratio	Proportion	Over burden	Moisture	Loss in transit	Margin on design	Days per week	Total multiplying factor
	2) slag cement										
	slag										
	30%		1.1	1.5	1	1	1	1	1.1	1	1.82
	40%		1.1	1.75	1	1	1	1	1.1	1	2.12
	50%		1.1	2.11	1	1	1	1	1.1	1	2.55
	60%		1.1	**2.86**	1	1	1	1	1.1	1	**3.46**
	65%		1.1	3	1	1	1	1	1.1	1	3.63
17	oil		1.1	0.08	1	1	1	1	1.1	1	0.10

Annexe 1.11.2

Overall multiplying factors and sectional capacities.

Sr. no.	Section	Item	Multiplying factor	Hours per day	Factor for hourly capacity	Hourly capacity
1	Quarries	stone raised	2.4	12	0.200	2000
2	Crushing	limestone to be crushed	1.91	12	0.159	1592
		crusher cpacity	2.3	12	0.192	1917
3	Stcker Reclaimer	Stacker (20% above crusher cap.	2.8	12	0.233	2333
		reclaimer	1.76	10	0.176	1760
4	Raw Material Grinding	raw mill	1.76	20	0.088	880
5	Blending	feed rate	1.76	20	0.088	880
		extraction rate	1.91	24	0.080	796
	kiln	feed rate kiln, preheater	1.91	24	0.080	796
		cooler	1.1	24	0.046	458
6		clinker conveyor	1.72	24	0.072	717
		spillage	0.52	24	0.022	217
	Coal Grinding	coal mill	0.2	20	0.010	100
			0.21	20	0.011	105
			0.22	20	0.011	110
			0.23	20	0.012	115
7		coal firing	0.2	24	0.008	83
			0.21	24	0.009	88
			0.22	24	0.009	92
			0.23	24	0.010	96

Annexe 1.11.2 ***Contd....***

Sr. no.	Section	Item	Multiplying factor	Hours per day	Factor for hourly capacity	Hourly Capacity
8	Cement Grinding	cement mill				
		OPC	1.27	20	0.064	635
		PPC 30% ash	1.82	20	0.091	910
		BFSC 60 %	3.15	20	0.158	1575
		slag				
9	Cement Packing	bulk	1.27	6	0.212	2117
		PPC bagged	1.82	15	0.121	1213
		bulk	1.82	6	0.303	3033
		BFSC bagged	3.15	15	0.210	2100
		bulk	3.15	6	0.525	5250

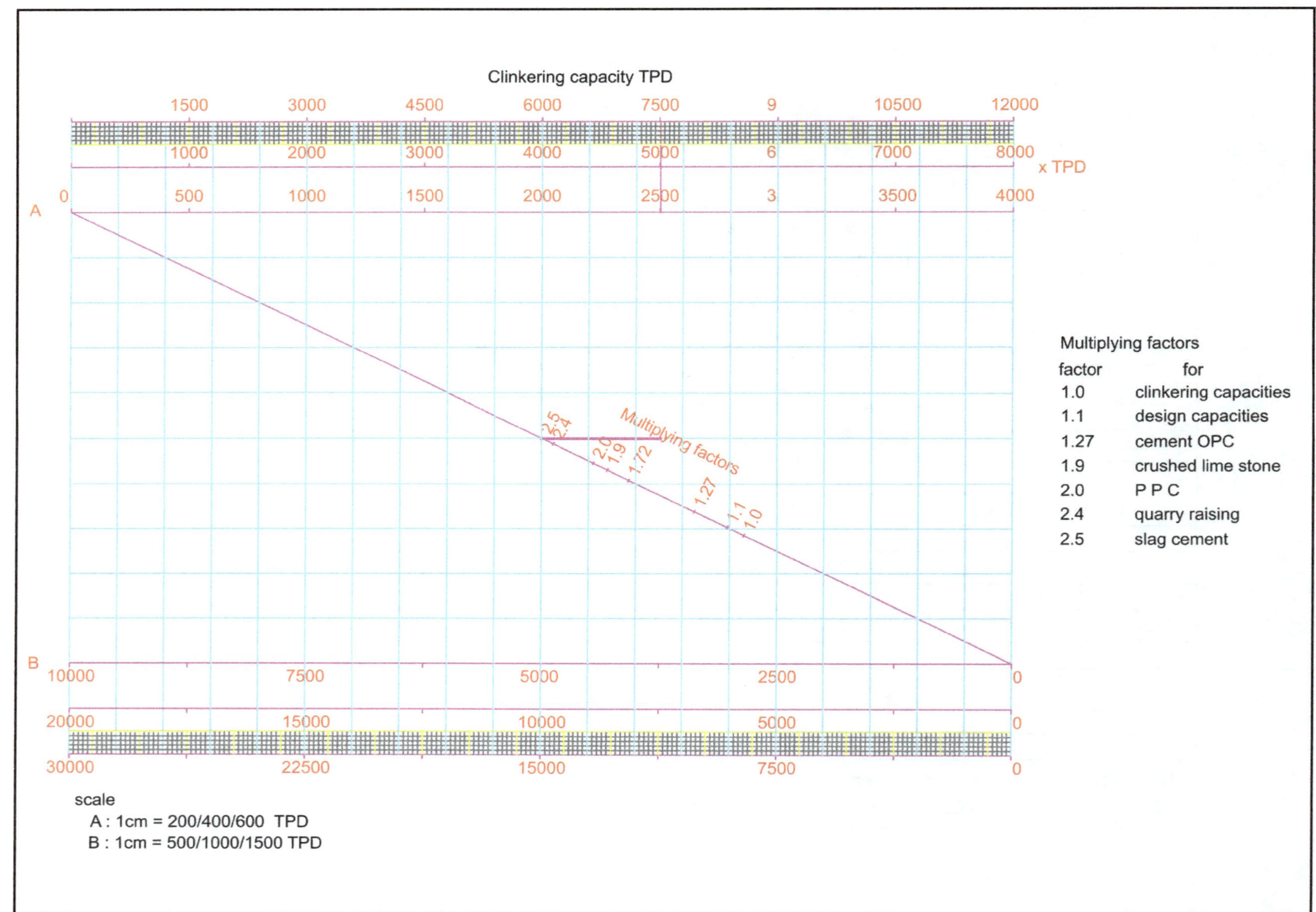

Fig. 1.11.1 Pertinent capacities TPD for design capacities cement mill.

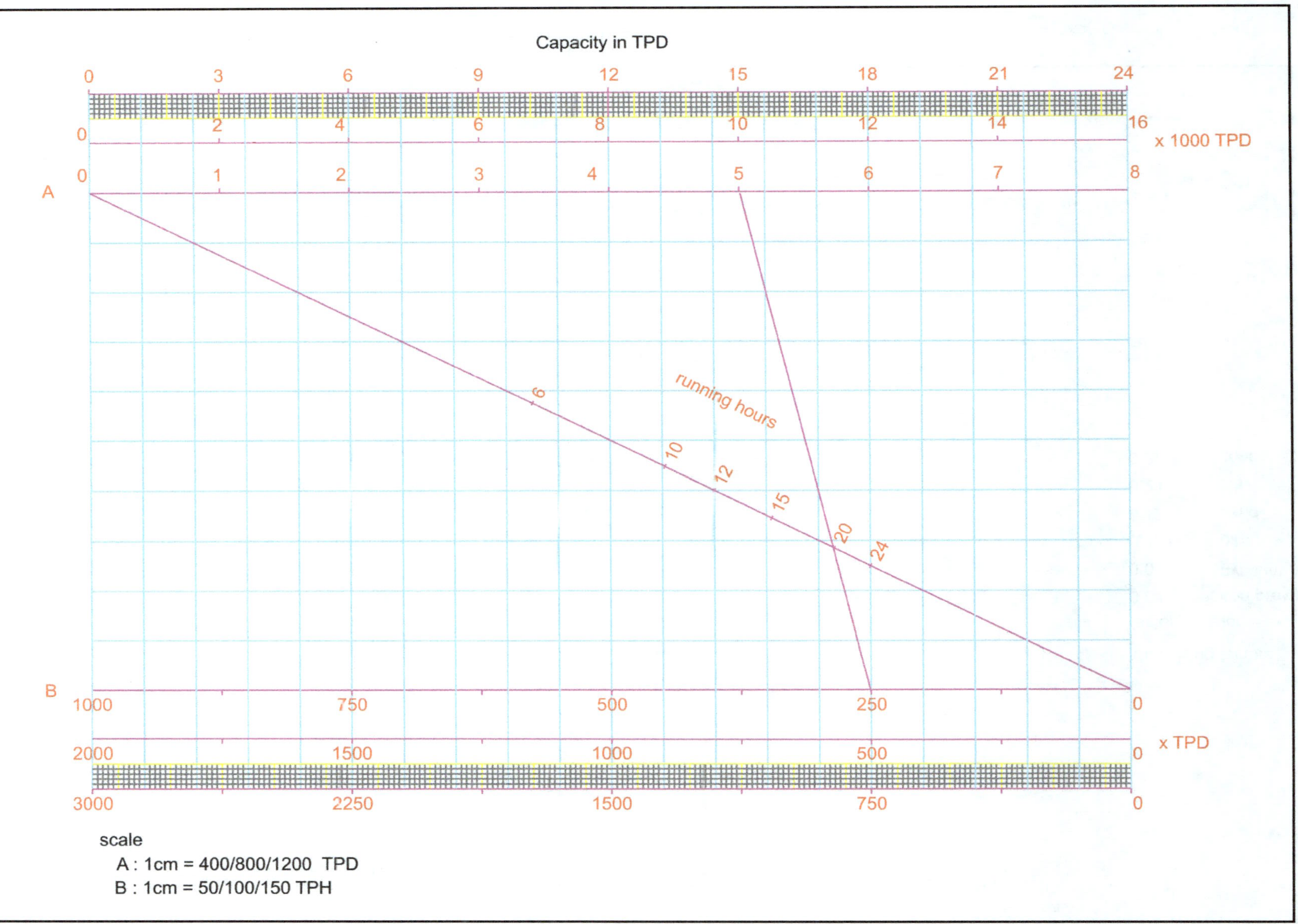

Fig. 1.11.2 Hourly capacities.

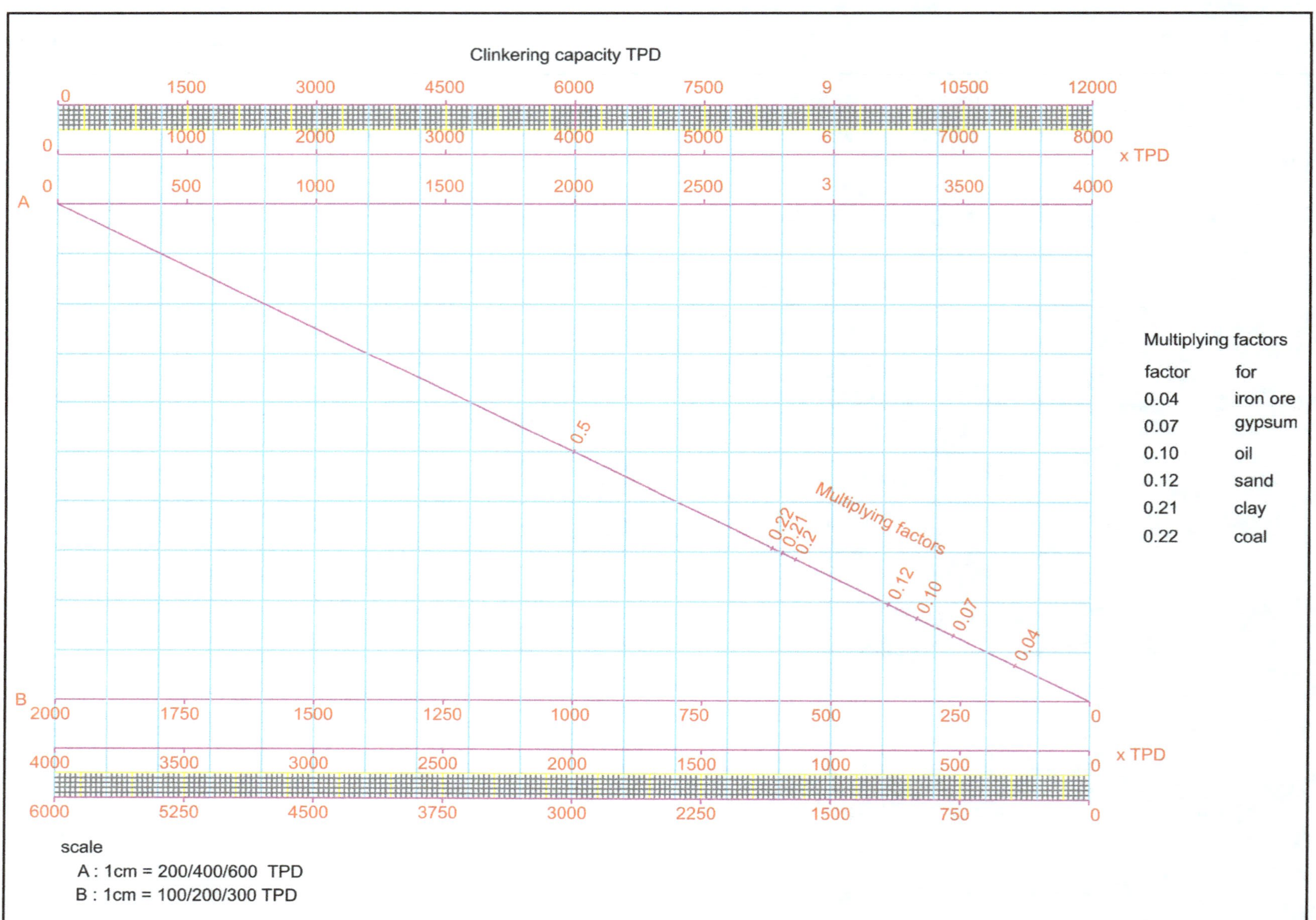

Fig. 1.11.3 Pertinent capacities TPD for gypsum, iron ore, clay, sand etc.

CHAPTER 12

PROCESS GASES

12.1 Making Clinker

In Cement Plants the main process and operation is 'clinkerization'. It takes place in the 'Pyroprocessing' section containing Rotary Kiln at its heart, flanked by Preheater and Calciner on one side and Cooler on the other side. It produces clinker from raw meal fed to it in preheater.

12.2 Clinkerization Involves

1. Drying in case of wet kilns only.
2. Preheating.
3. Calcination i.e., dissociation of CO_2 from carbonates in raw meal feed.
4. Sintering and formation of clinker.
5. Cooling of clinker

It requires heat to be supplied to raise temperatures to appropriate levels for processes of preheating, calcining and sintering. Heat is supplied to calciner and in kiln by firing fuel in them.

System exit gases from this section consist of:

(i) Combustion gases from fuel fired.
(ii) CO_2 released from carbonates in feed.
(iii) Excess air admitted in the system to ensure combustion.
(iv) False air entering system by way of in leakages.
(v) Vapor from water in feed and coal.
(vi) Air used to convey raw meal into preheater.

12.3 Kiln Exit Gases

Gases from combustion and dissociation can be calculated quickly by using empirical formulae. Let coal be the fuel fired.

Let

H = useful calorific value of coal in kcal /kg coal

A = theoretical air for combustion in nm^3/kg

= $1.01 \times H/1000 + 0.5$ nm^3/kg coal

m = excess air in %

G = products of combustion in nm^3/kg coal

= $0.89 \times H/1000 + 1.65$ nm^3/kg coal

q = sp. heat consumption in kcal/kg clinker, then

V = exhaust gases leaving system

= $q/H \times (G + m/100 \times A)$ nm^3/kg clinker

In this formula m represents excess combustion air and also false air.

In klin preheater system, excess air is taken as 10% and total false air as 15%. Therefore m = 25%.

Example : H = 4500, q = 750 and m = 25

Then V = 1.15 nm^3/kg clinker

CO_2 released can be calculated from LOI of kiln feed. For quick calculation it is taken as 0.3 nm^3/kg clinker

Conveying Air

Conveying air is 0.1 nm^3/kg if air lift is used to lift raw meal into preheater and 0 if bucket elevator is used.

12.4 Total Gases or Sp. Gas Volumes

Therefore total gases leaving per Kg clinker also called Sp. gas volume would be

A With air lift = 1.15 + 0.3 + 0.1 = 1.55 nm^3/kg clinker

B With elevator = 1.15 + 0.3 = 1.45 nm^3/kg

Rounding off they would be 1.6 and 1.5 nm^3/kg respectively. This is Sp. gas volume decreases with increase in calorific value of coal and with decrease in in fuel consumption as following **Table 1.12.1** would shoe (when air lift is used for conveying).

Table 1.12.1

Calorific value of coal	Sp. gas volume in Nm^3/kg clinker		
Kcal/kg	Sp. heat consumption Kcal/kg clinker		
	700	750	800
4000	1.51	1.59	1.67
4500	1.48	1.56	1.63
5000	1.45	1.53	1.60

As explained in Chapter 9, at design stage it would be prudent to round off and keep a margin of 5 % to calculate gas flows and fan capacities.

12.5 Gases used for drying

Kiln and preheater gases are used for drying in mills-always in raw and coal mills.

Earlier cooler gases were used to dry coal in coal mills. Now preheater gases are preferred because they are inert.

Cooler gases are also used to dry in cement mills if required.

Heat content of gases depends on their temperature, and quantity of gases required, on the quantity of water to be evaporated. Roughly sp. heat of air / preheater gases could be taken as 0.3 kcal / nm^3.

12.6 Heat Required for Drying

Quantity of heat required to evaporate 1 kg water depends on the initial moisture content and on material to be dried. **Table 1.12.2** shows approximate values for limestone and coal.

Table 1.12.2 Heat required to evaporate 1kg of water in kcal.

Material	Initial moisture %			
	4	6	10	12
limestone	1600	1300	1200	1150
coal	1800	1500	1300	1250

Roughly it could be taken as 1200 kcal/ kg both for limestone and coal because of the difference in their moisture contents.

12.7 Quantity of Water Evaporated

Quantity of water evaporated is calculated as follows:

mw = % moisture in feed; mp = % moisture in product

P = Output of mill with mp % moisture in kg/hr

Water evaporated W kg/hr = P × (mw-mp)/(100-mw)

12.8 Quantity of Gases Required for Drying

Let T be temp. of gas available for drying; 's' its sp.heat at that temp.; and A, ambient temp.

Then heat content H of gas would be

$$H = 1 \times (T\text{-}A) \times s \quad \text{kcal / nm}^3$$

Quantity V of gas to be brought for drying would be V = W × 1200 / H nm^3/hr if heat of grinding is ignored.

Approximately 80 % of power drawn in case of ball mill and 60 % in case of v.r. mill is converted into heat which is available for drying.

If p = Power drawn by mill in kw then heat of grinding useful for drying = 0.8 × p × 860 kcals/hr for ball mill and 0.6 × p × 860 kcal/hr for v.r. mill are available for drying. To that extent quantity of gases to be drawn from preheater would be reduced.

12.8.1 Gases Leaving Mill Systems

Mills, particularly airswept ball mills and vertical roller mills without external circuit require gases /air through them to lift material ground out of the mill. This quantity V is around 1.5 to 2.0 nm^3 / kg of mill capacity. This quatity may be sufficient to dry material also if feed moisture is low. else, quantity leaving mill will be quantity require for drying.

Gas leaving mill system would be

V + W/0.805 + leakage in system in nm^3/hr

Leakage is taken as 10 % of gas brought in

It is then converted into actual volume at actual temperature. With number of stages of preheater going up to 6, temperature of gases leaving preheater has come down from ~ 350 °C for a 4 stage preheater to 270 °C for a 6 stage preheater.

Total quantity of gases in raw and coal mill systems at inlet and outlet can be calculated in this fashion.

System resistance is calculated to arrive at specifications of fans.

12.9 Vent Gases

12.9.1 Cooler Vent Gases

Approximately 2.2 to 2.5 nm^3/kg of clinker air is admitted into cooler for cooling. About 0.8 to 0.9 nm^3/kg is used up as air for combustion. Balance is required to be vented. This air is hot - temperatures ranging from 150 to 300 °C and contains clinker dust. It is vented through dust collectors - mostly esps.

12.9.2 Venting of Material Handling Systems

In material handling systems mechanical and pneumatic, there is displacement of air which needs to be vented through dust collectors.

12.9.3 Venting Main Machines

Main machines like mills, crushers, separators also need venting.

Suppliers of machines furnish data on vent volumes and dust contents there of. This data is used to calculate total quantity of gases to be vented in each section and to select numbers and sizes of dust collectors in it as also specifications of fans.

In **Section 6** on detailing, these matters will be dealt with further.

They have been mentioned here so that a total picture of contents of various sections of a cement plant would be clear.

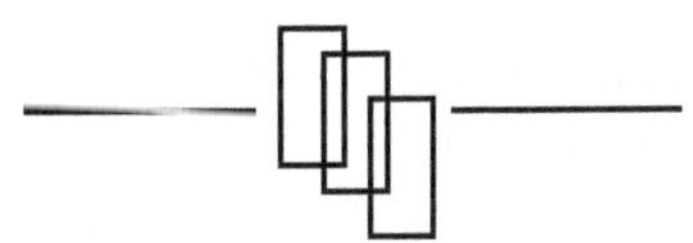

CHAPTER 13

STORAGES

13.1 Storages

Storages are stocks of materials to be maintained in between one part of the process and another to maintain continuity of operation in the event of a breakdown in a section and to tide over fluctuations in supply and demand of material brought in and sold out.

Stocks are maintained of:

1. Raw materials i.e., limestones and additives like clay, iron ore, gypsum and blending materials fly ash and slag.
2. Fuel - Coal /oil
3. Semi Finished product - Clinker
4. Finished Products - Cements

13.2 Daily Requirements of Various Materials

In **Chapter 9** Design Margins and Multiplying Factors to arrive at Capacities have been shown in **Table 1.9.1** and in Chapter 11 overall multiplying factors to be applied have been shown in **Annexe 1.11.1**

Applying those factors daily consumption/production of various materials have been shown in **Table 1.13.1** for a 10,000 tpd capacity cement plant.

13.3 Conventions in Storing Various Materials

Conventionally, storages, expressed in terms of daily consumption, are maintained as shown in **Table 1.13.2.**

Table 1.13.1 Daily requirements of a 10, 000 TPD cement plant.

	Factor	Tons
Quarried stone	2.4	24000
Crushed stone wet	1.9	19000
Raw meal consumption	1.72	17200
Raw meal kiln feed	1.87	18700
Coal wet	0.21	2100
Additive clay wet	0.7	2700
Iron ore wet	0.04	400
Gypsum	0.07	700
Cement o.p.c.	1.27	12700
Fly ash	0.41	4100
Slag	1.87	18700
Cement p.p.c.	1.69	16900
Cement slag	3.21	32100

13.4 Factors Governing Storages

Factors governing storages to be maintained for materials bought from outside would be :

1. The distance over which these items are brought, time taken in processing the order and actual receipt of material at site have to be taken into account.
2. Continuity factor, reliability of source/s.
3. Climate or weather factor, like differences in moisture content in dry and wet seasons and necessity to maintain stocks dry.

Table 1.13.2 Storages to be maintained.

Material	1st Unit	After duplication
Uncrushed limestone	Nil	Except in special cases
	1st unit	after duplication
Crushed limestone	6-7 days	5 days
Additive clay	30-45 days	30 days
Iron ore	30 days	30 days
Gypsum	30 days	30 days
Coal	20 days	14 days
Clinker	14 days	14 days
Cement	7 days	5 days
Raw meal	$2^{1/2}$ days	2
Pulverised coal	8 hours	8 hours
Fly ash	3 days	3 days
Slag	2 days	2 days
Cement PPC/BFSC	3 days	3 days
Oil	8 days	6 days

13.5 Storages of Semi Finished and Finished Products

Factors governing storages of semi-finished products and finished products within the plant are:

1. Frequency and duration of breakdowns in respective sections.
2. Process requirements like blending operations, which influence quantity to be stored.
3. Interruptions in production for long periods for brick lining of kilns changing and grading of grinding media and lining plates, roller liners etc., for ball and vertical mills.

13.6 Storages in Various Sections

13.6.1 Quarry

Breakdown of shovel / dumper can cause disruption in supply of stone to crusher; but at least 3 shifts are available per week for maintenance; normally standby units are also provided. Therefore no stock of uncrushed stone is normally maintained.

13.6.2 Crusher

Crusher breakdown will stop quarrying operations; will disrupt milling operations unless there is a stock of crushed stone. Often crushers are far from the plant. Maintenance jobs like changing hammers, liners etc., can take longer than time available per week. Therefore 7 days' stocks are normally maintained between crusher and raw mill.

13.6.3 Stacker Reclaimer

Stacker reclaimer system design depends on degree of blending required. It depends on number of layers in the stock pile, which in turn are decided by quantity of stone in stockpiles.

13.6.4 Raw Mill and Blending

Blending silo does not receive raw meal when raw mill stops. Kiln stops when raw meal is exhausted.

Major stoppages of raw mill for change of lining plates, rollers, grading of grinding media can take - 2-3 days or 6-9 shifts (normally such jobs will be'planned' in advance to coincide with stoppage of kiln). However for maintaining running of kiln, raw meal stocks of 2-2½ days are maintained.

Continuous blending like stacker reclaimer depends for blending effect on no. of layers formed and broken during extraction.

A continuous blending silo, is never emptied to contain less than 40% of its capacity. This is also a factor which decides size and capacity of raw meal storage.

13.6.5 Kiln feed

It is prudent to have a standby kiln feed system to maintain continuity of kiln operation rather than create a large intermediate storage immediately after the raw meal storage. A small bin on load cells holding 15 to 20 minutes' requirements of raw meal is provided.

13.6.6 Preheater – Calciner – Kiln and Cooler

As the material is continuously processed from the time it enters preheater till it leaves clinker cooler there cannot be any intermediate storage.

When cooler stops, kiln stops, coal firing stops; coal Mill stops (when bin is filled up).

Kiln is stopped only in emergencies as sintering is a continuous process. Kiln stoppages are planned for brick lining. To allow for maintenance, as mentioned earlier, number of working days of kiln are taken as 330 only.

Clinker stocks serve two purposes. First they allow kiln to run when cement grinding has stopped for some reason. Secondly they allow mills to run continuously when there is a spurt in demand for cement.

Kiln is sized for working for 330 days i.e., a million ton cement plant, with 360 working days per year would require a kiln of $1,000,000/(1.04 \times 360) = 2670$ tpd rated capacity.; but because it is sized for 330 days, the rated capacity would be 2910 tpd. This difference makes it possible to build up clinker stocks.

13.6.7 Cement

Cement storage would depend on continuity to be kept in supplies of cement to meet demand at all times promptly.

This is in turn is governed by average radius of the market and modes of transport and time taken for cement to reach market from the date of receipt of order.

Further, demands fluctuate. There could be sudden bunching of orders as well as lean periods.

Cement mill is sized to allow for bunching of orders; storage is sized to allow for slack periods in orders so that mills do not have to stop.

If a cement plant already exists, then it is possible to do in depth studying of market and its fluctuations to arrive at a minimum safe quantity of cement that needs to be stored.

Only as a thumb rule, about 6-7 days stocks are maintained. Normally, in duplication, number of days stock could be reduced.

	Capacity tpd	
	Tons	Tons
	10000	20000
7 days' stock	70000	140000
5 days' stock	50000	100000

Thus, if 2 silos are constructed in 1st phase, only one more silo would be added in the 2nd phase.

1st phase – 2 × 35000 T capacity silos

2nd phase – 3 × 35000 T capacity silos

13.7 Storages after Expansion

This logic can be applied to other storages also to reduce costs of storages as shown in **Table 1.13.3.**

Table 1.13.3 Storage of crushed limestone.

	1st Phase	**2nd Phase**
Limestone stock pile	1.55: 1 conversion ratio 85% limestone	
Plant capacity	10000 tpd	20000 tpd
With design margin	11000 tpd	22000 tpd
Crushed limestone/ day	14500 Tons	29000 Tons
7 day's stock in both phases	101500 Tons	203000 Tons
2 Stock Piles	Each ~51000 Tons	Each 102000 Tons
5 days' stock in 2nd phase		145000 Tons
2 stock piles		Each ~ 72000 Tons

Capacity could be increased by extending existing stock piles by 50%.

13.8 Storage of Coal

Let specific fuel consumption be 750 kcal/kg and useful calorific value of coal be 4500 kcal/kg. Then coal required per day for Ist unit and after duplication would be as shown in **Table 1.13.4.**

Table 1.13.4 Requirements of coal.

	1st Phase	**2nd Phase**
	Tons	**Tons**
Plant capacity TPD	10000	20000
With design margin	11000 Tons	22000
Coal @ designed capacity	1850	3700
Coal stocks raw coal with 10% moisture 3% loss in transit	2100	4200
20 days' storage	42000	84000
14 days' storage	29500	59000

It would be prudent to go in for 20 days stocks to start with, create storage facilities for same and reduce stocks, in the light of experience during expansion, to 14/15 days.

Preblending facilities are even more advantageous in case of storage of coal as compared to limestone.

Hence stacker reclaimer systems for coal are becoming more popular in case of large plants.

They can be circular or linear but on a smaller scale in size and capacity as compared to limestone.

13.9 Storage of Clinker

Clinker will normally be stored under a covered shed. It is a semi-finished product. Weathering (without getting wet) improves grindability.

It would be advisable to have separate stockpiles of clinker for each kiln. There can however be a common stock pile for under burnt clinker and spill over.

It is customary to keep about 14 days' stock of clinker.

Plant capacity	- tpd	- 10000 Tons	20000 Tons
designed margin	- tpd	- 11000 Tons	22000 Tons
14 days' stock	- Tons	- 154000 Tons	308000 Tons

Inter changeability should be provided for the transfer clinker from kiln 1 to stock 2 and vice versa.

In clinker storages, facilities are always provided for storing extra quantity and also under burnt clinker and for recovering them.

13.10 Storages in a 10000 tpd Plant

Table 1.13.5 shows starages in a 10000 tpd clinkering capacity plant for first phase and after duplication

13.11 Space for Storages

Creating storages requires heavy investments also considerable space. In a way it is un-productive investment. But it helps ensuring production and sales but indirectly.

Therefore it is desirable for keeping down the investment, to keep storages to a minimum.

However, it is also not advisable to cut things too fine. Stoppage in production due to lack of storage facilities would cause direct loss of revenue.

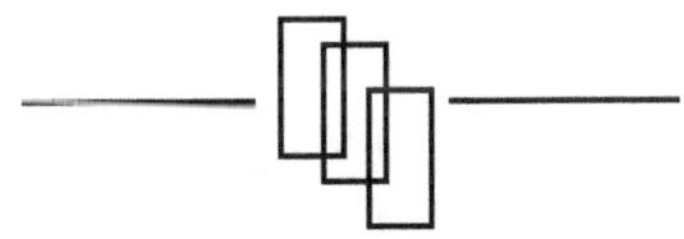

Table 1.13.5 Desirable Storages of Varrious materials in Cement Plants of different capacities.

Sr. No.	Material	Unit		1st unit clinkering capacity tpd 10000	After duplication clinkering capacity tpd 20000
1	Crushed limestone	Factor daily reqmt no. of day's stock storage	tons tons	1.91 19100 7 133700	1.91 38200 5 **191000**
2	Clay	Factor daily reqmt no. of day's stock storage	tons tons	0.21 2100 30 63000	0.21 4200 20 **84000**
3	Sand	Factor daily reqmt no. of day's stock storage	tons tons	0.12 1200 30 36000	0.12 2400 20 **48000**
4	Gypsum	Factor daily reqmt no. of day's stock storage	tons tons	0.07 700 30 21000	0.07 1400 20 **28000**
5	Fly ash	Factor daily reqmt no. of day's stock storage	tons tons	0.4 4000 3 12000	0.4 8000 3 **24000**
6	Slag	Factor daily reqmt no. of day's stock storage	tons tons	0.85 8500 2 17000	0.85 17000 2 **34000**
7	Cement OPC	Factor daily reqmt no. of day's stock storage	tons tons	1.27 12700 7 88900	1.27 25400 5 **127000**
8	Cement PPC	Factor daily reqmt no. of day's stock storage	tons tons	1.86 18600 3 55800	1.86 37200 3 **111600**
9	Cement BFSC	Factor daily reqmt no. of day's stock storage	tons tons	3.46 34600 3 103800	3.46 69200 3 **207600**

Table 1.13.5 *Contd....*

Sr. No.	Material	Unit		1st unit clinkering capacity tpd 10000	After duplication clinkering capacity tpd 20000
10	Blended raw meal	Factor daily reqmt no. of day's stock storage	tons tons	1.82 18200 2.5 45500	1.82 36400 2.5 **91000**
11	Pulverized coal	Factor daily reqmt no. of day's stock storage	tons tons	0.2 2000 0.3 600	0.2 4000 0.3 **1200**
12	Oil	Factor daily reqmt no. of day's stock storage	tons tons	0.1 1000 8 8000	0.1 2000 6 **12000**

CHAPTER 14

MACHINERY SCHEDULES

14.1 Machinery Schedules

Flow Charts are a pictorial presentation of machinery and auxiliaries contained in a section. A Machinery Schedule lists the machinery in it with details like numbers, size and capacity and drive detail with brief description.

It is a very useful document for all activities to be taken in developing the design of a cement plant.

14.1.1 Building of Machinery Schedules

Chapters 8 to 13 of this **Section** progressively explained how step by step numbers and capacities of machinery and auxiliaries; capacities of storages and volumes of gases to be handled could be worked out- albeit briefly at this stage.

Sizes of main machinery could be obtained from Suppliers; or Consultants can use their own experience in the matter.

Drive details are normally furnished by Suppliers.

Standard Catalogues for commonly used material handling machinery are useful in sizing them and in selecting drives.

In early stages of the Project the Schedules so built up need not be and will not be final.

14.1.2 Lists of Motors and Gearboxes

From machinery schedules, lists of motors and gear boxes can be built up.

Lists of motors are useful to design power distribution system beginning with various sections of the plant and going up to Substation.

14.2 A Machinery Sehedule Shows:

1. name and brief description of a machine
2. numbers there of
3. size in form typically used for specific machines for example:
   size of ball mill is indicated as 4 m dia. × 8 m long
   size of kiln is indicated as 4 m dia × 60 m long
   for belt conveyor
   800 mm wide- 40 metre centres
4. in brief description would be included for mill dry, closed circuit, bucket elevator type for kiln dry, preheater kiln with calciner for cooler grate, horizontal, 3 grates - 1 static for belt conveyor
   horizontal/inclined; troughed-angle of trough; angle of inclination
5. Weight
6. Drive detail indicating rating of motor and its speed
7. capacities and numbers of hoppers bins and silos in section

   for example

   for mill

   hopper for limestone, correcting material - numbers; capacity in tons
8. refractories when needed in tons, type and location.
9. fabricated items like chutes and ducts and working platforms

In departments like mills and kilns these are considerable in size and in total tonnage.

14.3 Other Uses of Machinery Schedules

Machinery schedule will also be the basic document for building up Capital Costs of machinery and hence has great importance.

Just as it is used to make lists of motors and gear boxes, it is also used to list civil construction work in a department including main building, hoppers and bins with approximate floor area and number of storeys. It is thus useful to make estimates of Capital Costs of civil construction.

Enclosed Annexes 1.14.1 to 1.14.4 furnish typical schedules for just one of the plant - crushing section.

14.4 Annexes

1.14.1 machinery schedule for crushing section,
1.14.2 list of gearboxes,
1.14.3 list of motors,
1.14.4 list of civil construction work in section.

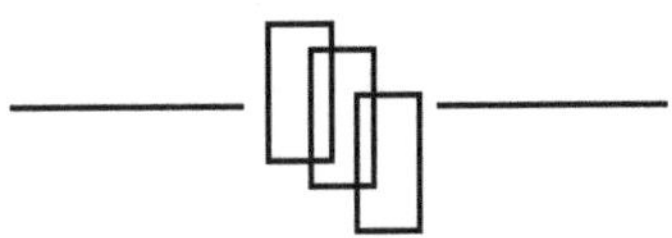

Annexe 1.14.1

1500 TPD Cemeent Plant Project
of
XXXXXXXXXXXXXX
Section : Single Stage Crushing
Flow Chart drawing no: 1.7.2

Machinery Schedule For Crushing Section.

Sr. No	Description	Nos	Capacity tph	Size mms	Drive kw	Remarks
1	Ramp	1	-	-	-	civil
2	Hopper above crusher	1	100 tons	-	-	civil
3	Apron feeder	1	600	1.8 m wide * 6 m long	20	Variable speed
4	Single Stage Impact Crusher	1	500	1200 mm dia* 1800mm	600	h.t. motor
5	Short belt under crusher	1	750	1500* 5 m	10	
6	Long belt conveyor to Stacker reclaimer	1	750	1000mm*100m	50	
7	Bag filter	1	45000 m^3/hr			
8	Fan for bag filter	1	50000m^3/hr @ 300 mm wg		50	
9	Hopper under Apron Feeder for spillage	1				
10	Belt conveyor for spillage	1	50 tph	600 mm wide × 10 m long	5	

Annexe 1.14.2

List of Gearboxes.

Sr. No	Gear Box For	Nos	Rating KW	Speed ratio Input/out put r.p.m.	Type
1	Apron Feeder	1	20	1000/ 96	Helical
2	Short Belt	1	10	1500/37.5	Worm
3	Long Belt	1	50	1500/37.5	Worm
4	Belt under apron feeder	1	5	1500/30	worm

Annexe 1.14.3

1500 TPD Cemeent Plant Project
of
XXXXXXXX
Section : Single Stage Crushing
Flow Chart drawing no: 1.7.2

List of Motors.

Sr. No	Motor for	Nos	Rating kw	Speed r.p.m.	Voltage volts	Type
1	Apron feeder	1	20	1000	415	d.c.vvariable speed
2	Crusher	1	600	1000	6600	Slip ring
3	Short belt	1	10	1440	415	s.c.
4	Long belt	1	20	1440	415	s.c.
5	Fan for bag filter	1	50	1440	415	s.r.
6	Compressor for Filter	1	3	1440	415	s.c.
7	Belt under apron feeder	1	5	1440	415	sc

Annexe 1.14.4

Section : Single Stage Crushing
Flow Chart drawing no: 1.7.2
Civil Construction Work.

Sr. No	Description	Nos	Area m^2	Remarks
1	Ramp to reach crusher	1		10 m wide * 200m long
2	Hopperr for crusher	1	60 m^2	10 m* 6 m
3	Crusher House	1	200 m^2	2 storeys above ground
4	Foundations for crusher etc.	1 Lot		
5	Pit for belts	1		3 m below ground level

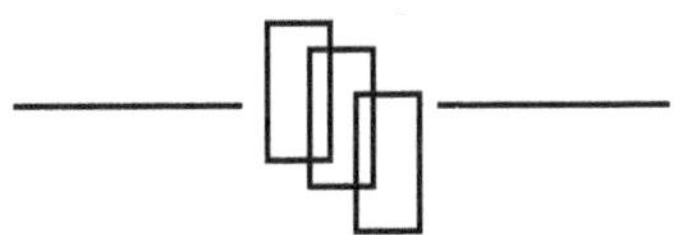

CHAPTER 15

DRAWING SPECIFICATIONS FOR INVITING OFFERS

15.1 Plant and Machinery to be Procured

Based on the findings of TEFS and the size of the plant proposed to be installed, the Consultants will carry out basic engineering and in dialogue with the Entrepreneur, fix the basic skeleton of the plant and equipment to be ordered.

15.2 Enquiry Specifications and Technical Data Sheets

It is necessary to invite offers for the Major plant and Machinery and auxiliaries. Consultants would draw out 'Enquiry Specifications' for machinery furnishing all pertinent details like:

1. Raw materials to be used with detailed properties - physical and chemical - thereof; fuel and its properties.
2. Duty requirements which will stipulate working conditions and also indicate properties to be achieved during processing like drying to be done, fineness to which it is required to be ground.
3. Desired operational norms to be satisfied like sp. fuel consumption and sp. power consumption.
4. Capacity and various margins to be kept.
5. Properties which affect sizing, like grindability.

15.2.1 Technical Data Sheets

To bring uniformity in the various offers received, Consultants will prepare 'Technical Data Sheets' and Vendors will complete them and return them with their offers. If Vendors cooperate, evaluation of offers becomes very easy.

15.2.2

Bureau of Indian Standards have published a great many standard enquiry specifications and technical data sheets for various machines used in cement plants which could be used in formulating enquiries and in evaluation of offers.

15.3 Representative Samples

Entrepreneur will also make available representative samples of raw materials to help Vendors in sizing the machinery.

15.4 Floating Enquiries

Either the Entrepreneur or the Consultant on his behalf will float enquiries inviting offers from reputed Vendors who have the expertise in design and manufacturing experience in the Machinery.

To save time, Vendors may be asked to submit offers on a specific basis like fixed prices f.o.r.; to indicate duties and taxes separately. It then becomes easier to understand differences between different offers. The scope of supply should also be clearly defined.

15.5 Evaluation of Offers

Entrepreneur and Consultant must agree upon the process of evaluation of various offers. One way is to give weightages to specific properties and add them up to arrive at rankings of various vendors.

IIt is customary to 'short list' 2 or preferably 3 vendors and carry on further detailed negotiations with them. Orders would be finalized on those whose offers are better in long term operations and who can offer better 'after sales service'.

15.6 Responsibility for Performance

There are other practical considerations such as pin pointing responsibilities for performance. From this point it is often better to order a whole section from one vendor. For example, kiln, preheater, cooler and calciner - though it is feasible to order them from different vendors.

15.7 Finalising Orders

When Consultants submit their recommendations in a systematic way, it becomes easy for the entrepreneur to decide on the selection. Finalising orders is a long drawn out process and sufficient time must be given for it. **See Chapter 9 of Section 3.**

If Vendors follow the guidelines mentioned above, offers would be received in similar formats with all technical and commercial details and drawings. Less time will then be required in discussions and negotiations with them.

15.8 Consultants' Responsibility

It is Consultants' responsibility to present evaluation of all shortlisted offers in a fair and uniform manner and to give them weightages using the same yardsticks. Final decision would be left with the Company.

This matter being of great importance, has been dealt with in greater detail in Section 7. Typical Enquiry Specifications (ENQS) and Technical Data Sheets (TDS) for principal machinery have been included therein. Typical **Evaluation** has also been shown.

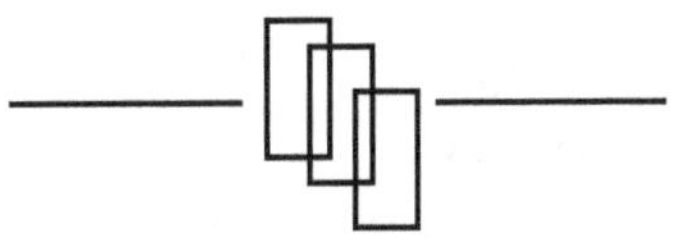

CHAPTER 16

PROPERTIES OF MATERIALS THAT AFFECT SIZING AND SELECTION OF MACHINERY

16.1 Properties of Materials that Affect Sizing of Machinery

In **Chapter 11** the procedure of arriving at Sectional and Individual Capacities has been explained. Preceding **Chapters 7 to 10** showed the various steps leading to the procedure in **Chapter 11.**

For the same duty requirements, different machinery would be selected taking into account properties of material to be processed.

For the same capacity, sizes of machinery would differ for different materials according to their own physical properties and the fineness to which the materials are to be ground for example.

It is therefore necessary to know the impact of various such properties of materials being processed on the sizing of machinery.

16.2 Crushing

Selection of primary crusher is dependent on the size of the shovel used as explained in **Chapter 2** of **Section 2** and **Chapter 3** of **Section 6.**

Power required for the crusher is dependent on the crushing strength of the stone being crushed and the size reduction obtained in crushing.

Sizing could also be affected if a crusher is required to work in closed circuit as it would receive fresh feed plus circulating load.

Moisture may not affect sizing directly, but there could be buildups inside the crusher and time could be lost. In crushers required to handle wet stone, breaker plates could be rotating so that build ups could be removed without requiring to stop crusher.

16.3 Grinding Mills

Grindability is the most important property of a material. It is expressed as power required to grind materials in standardized conditions like for example, Bond's grindability test to establish Bond's Work Index (BWI) and also Hard Grove Index (HGI)

Sp. power required is directly proportional to BWI. It is inversely proportional to Hard Grove Index.

It is inversely proportional to square root of fineness of product expressed in microns.

For example if sp. power consumption is 16 kwh/ton when grinding to a fineness of 80 % residue on 90 microns, then when ground to a fineness of 80 % residue on 53 microns, sp. power consumption will increase by $(90/53)^{0.5} = 1.3$ times; i.e., it will be 21 kwh/ton. Therefore for the same capacity, mill will be so much bigger.

Multiplying factors are also used when material is ground finer than say 90 microns over and above those used on account of grindability.

See Table 1.16.1.

Table 1.16.1

Blaine surface	80% passing size microns	Fineness factor
2520	62.4	1.018
2700	53.6	1.04
2880	45.7	1.07
3060	40.7	1.094
3150	37.6	1.11
3600	28.2	1.19

Ready reckoner curves are available to arrive at sp. power consumption for different grindabilities and for different finenesses.

Moisture content in feed also affects sizing in case of ball mills.

Above conditions are also valid for grinding coal and cement in ball mills because in case of ball mills BWI method has been universally accepted.

In case of cement mills, sp. power also depends on C_3S and C_2S contents of clinker.

See Table 1.16.2.

Table 1.16.2

% C_2S	Multiplier for output
5	1.1
10	1.05
15	1.0
20	0.95
25	0.88
30	0.82
35	0.72

In case of vertical mills, Hard Grove Index is used for sizing E Mills. In case of vertical roller mills each manufacturer claims to have his own special index for sizing purpose.

Moisture in raw materials or coal decides the quantity of hot gases to be brought in the circuit. Output can fall if moisture in feed is higher than the design value..

16.4 Kilns, Preheaters and Calciners

Kilns, Preheaters calciners are volumetric machines and sizing is mainly dependent on capacity. Properties like burnability affect temperatures to be maintained in burning zone.

Because they are volumetric, following affect sizing:

1. Altitude, particularly > 300 m above sea level. This is because gas volumes are higher on account of lower density,
2. sp. fuel consumption. Lower the sp. fuel consumption, lower the gas volumes and for same capacity, sizes will be smaller,
3. division of fuel between kiln and calciner. Increase in percentage of fuel in calciner will reduce kiln size but will increase calciner size.

16.5 Dust Collectors

Bag Filters

Sizing is dependent on the parameter 'air to cloth ratio'. Different values are used for different applications and for different materials of bags.

Glass bags can withstand temperatures up to 240 °C and hence dilution air required is much less as compared to polyester bags which can withstand temperatures up to 120 °C only.

ESPS

Sizing depends very much on the specific resistivity of dust in the gases. Preheater exhaust gases are cooled in a gas conditioning tower to improve performance of ESP

Humidity lowers resistivity. Therefore kiln exhaust gases are passed through a conditioning tower before they enter an ESP

Coal has low sp. resistivity as compared to raw meal; clinker is worse.

When clean gas dust burden required is less than 110 mg/nm^3, then it is required to use low migration velocities which pushes up sizes of esps disproportionately to increase in efficiency.

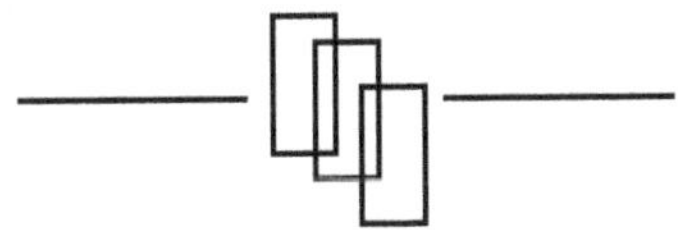

CHAPTER 17

EXPANSION

17.1 Expansion now and in the Past

About three decades back, rate of growth in cement industry in terms of capacity was small. Developments in processes and equipment were also slow. In that period expansion generally meant duplication- that is installing the same line of production.

Now the situation is very different. Cement Industry is growing very fast in terms of capacity and contents both. With advent of calciners expansion has taken many dimensions. The capacity of a kiln increased two and a half times, requiring corresponding increases in capacities of all departments.

Side by side, operational efficiencies also improved with new equipment like roller mills – roller press, 6 stage low pressure cyclones for preheaters, and a great many developments in grate cookers like static grate, cross bar, polytrack coolers. Thus there was additional advantage in improvements in operating norms like reduced power, fuel consumption, gas flows et al. It became imperative to take cognisance of these developments whatever the difference in capacity of the first line and the new line.

17.2 Major Developments in Design

Calcineres are of a great many designs, online, off-line, those that can fire Alternate Fuels, low Nox design and so on.

Preheaters have high efficiency low pressure drop cyclones, capacity can be increased by adding a second stream of preheater,, operational efficiencies were improved by increasing number of stages.

Grate Coolers have seen a great many changes These have been brought out in the pertinent chapters of Section 2 and also in Authors book 'Designing Green Cement Plants'.

With these kind of developments, single kilns of + 10000 tpd capacity became possible thereby reducing capital costs substantially for the same increase in capacity.

Earlier thermal load in a kiln mattered and that had impact on the choice of refractory. In calciner kilns, only 40 % fuel is fired in the kiln and temperature proile in the calciner is much lower. Therefore this bottleneck has also been removed.

Kiln diameter for same capacity is much smaller, requiring a narrow and longer cooler. However with coolers of high specific loading in terms of tpd/m^2, this problem has also been solved.

Now cyclones or multclones are used only as primary dust collectors. ESP is universally used for exhaust gases of coolers and coal mills also.

17.3 Selection of Equipment

Equipment selected /to be selected often decides the choice of preceding and following machinery

(i) Crushing and Raw mill

For example, if Raw Mill is a ball mill, feed to it would have to be fine, less than 25 mm in size.

Crushing will be in two stage- 1st stage a jaw crusher and second stage a hammer or impactor. Vertical roller mills can take much larger feed and crushing can be in one stage -impactor.

In case of VRMs (vertical mill) separator is part of the mill itself.

(ii) Stacker Reclaimer and Continuous Bending

If a stacker recclaimer system is to be installed after expansion because of tonnages to be handled, advantage is taken of preblending during stacking operations. It is then possible to go in for Continuous blending. Installing X ray analyser maintaains quality within required parameters.

17.4 Changes due to 'Green Cement'

New dimensions have been added by the concept of 'Green Cement'. Blended cements – PPC and BFSC have become 'norms' requiring much less clinker for the same quantity of cement. The direct impact is on the size and capacity of production line from crushing to coal mill. Of course cement grinding capacity has to be increased.

Using Alternate Fuels requires new fuels to be brought in, processed and fired- albeit in smaller quantities.

One more impact of Green Cement is Co generation that is generating power by using waste heat in kiln and cooler gases.

Instrumentation has improved and there is substantial automation. Continuous monitoring of quality of raw meal, by X ray and gas analysers have become common. Computer, internet and mobile phones have become integral part of the production line.

17.5 Impact on Machinery and Layouts

These developments have impact on additional plant and machinery for expansion and also on General Layouts of the Plant for accommodating them.

1. Because of blending materials like fly ash and slag which result in greater quantities of cements produced, Railway siding would generally be bigger and would need more careful planning right from acquiring a bigger plot of land. It is also needed to provide space for them though stocks in number of days would be small.
2. Stacker Reclaimer Systems for lime stone also take considerable space.
3. Installing Co generation would also require space.

17.6

In Section 1, various parameters conventionally used for designing and sizing of machinery for main plant and expansion / duplication have been given in Chapters 8 to 13.

As explained above, presently there can be considerable difference in the capacities of first and second production lines. Therefore a few Tables have been added for design parameters of 1st production line of 3000 tpd and second production line of 10000 tpd capacity to high light the points made above.

See **Tables Nos 1.17. 1 to 1.17.4.**

Table 1.17.1

Running Hours in Various Sections of a Cement Plant.

A Main Plant									
S. No.	**section**	**hrs/shift**	**shifts/day**	**hrs/day**	**day/week**	**hrs/week**	**day/year**	**hrs/year**	**remarks**
1	Qyarrying	6	2	12	6	72	312	3744	
2	Crushing	6	2	12	6	72	312	3744	
3	Stacker	6	2	12	6	72	312	3744	
4	Reclaimer	5	2	10	7	70	360	3600	assuming supplies day's requirement of mill in 10 hours
5	Raw mill			20	7		360	7200	
6	Blending			20	7		360	7200	works with raw mill
7	Extraction & kiln feed	24			7		330	7920	works with kiln
8	Kiln	24					330	7920	
9	Coal stacker				7				to suit capacity of wagon tippler
10	Reclaimer			10	7		360	3600	stacker reclaimer may have to work for thermal
11	Coal mill	20			7		360	7200	power statoion also may consider same stacker reclaimer for main plant & 2nd unit
12	Cement mill	20			7		360	7200	
13	Packing and despatches	5	3	15	7		360	5400	

Table 1.17.1 *Contd...*

B Duplication / Expansion									
S. No.	section	hrs/shift	shifts/day	hrs/day	day/week	hrs/week	day/year	hrs/year	remarks
1	Quarrying	6	3	17	6	102	312	5304	same crusher in 3 shifts
		6	4	24	6	144	312	7488	2 crushers in 2 shifts
2	Crushing	6	3	17	6	102	312	5304	same crusher in 3 shifts
		6	4	24	6	144	312	7488	2 crushers in 2 shifts
3	Stacker	6	3	17	6	102	312	5304	same stacker
		6	4	24	6	144	312	7488	2nd stacker
4	Reclaimer	5	2	10	7	70	360	3600	same reclaimer
		5	4	20			360	7200	second reclaimer
5	Raw mill	20		40			360	14400	2 raw mills
6	Blending	20		40			360	14400	2 blending silos
7	Extraction, kiln feed			24				15840	2 production lines
8	Kiln			24			660	15840	2 kilns
9	Coal stacker				7				tosuit capacity of wagon tippler see remark above
10	Reclaimer			10	7		360	7200	
11	Coal mill	20		40			360	14400	2 coal mills
12	Cement mill	20		40			360	14400	2 cement mills
13	Packing and despatches	5	6	30			360	10800	2 packing machines

*Quantities can be modified in the light of operational experience

Table 1.17.2

Conventions in Storing Various Materials
Conventionally, storages expressed in terms of daily consumption are maintained as shown.

Material	1st Unit	after duplication/expansion
Uncrushed Limestone	Nil	Except in Special cases
Crushed Limestone	6 – 7 days	5 days
Additive Clay	30-45 days	30 days
Iron Ore	30 days	30 days
Gypsum	30 days	30 days
Coal	20 days	14 days
Clinker	14 days	14 days
Cement	7 days	5 days
Raw Meal	2 ½ days	2
Pulverised Coal Fly Ash* Slag* Cement PPC/BFSC* Oil	8 hours 3 days 2 days 3 days 8 days	8 hours 3 days* 2 days* 3 days* 6 days

*Quantities can be modified in the light of operational experience.

Table 1.17.3

Desirable Storages of Varrious materials in Cement Plants of different capacities.

S. No.	Material		Unit	1st unit clinkering capacity tpd 10000	After duplication clinkering capacity tpd 20000
1	Crushed limestone	factor		1.91	1.91
		daily reqmt	tons	19100	38200
		no. of day's stock		7	5
		storage	tons	**133700**	**191000**
2	clay	factor		0.21	0.21
		daily reqmt	tons	2100	4200
		no. of day's stock		30	20
		storage	tons	**63000**	**84000**
3	sand	factor		0.12	0.12
		daily reqmt	tons	1200	2400
		no. of day's stock		30	20
		storage	tons	**36000**	**48000**

Table 1.17.3 *Contd...*

S. No.	Material		unit	1st unit clinkering capacity tpd 10000	After duplication clinkering capacity tpd 20000
4	gypsum	factor		0.07	0.07
		daily reqmt	tons	700	1400
		no. of day's stock		30	20
		storage	tons	**21000**	**28000**
5	fly ash	factor		0.4	0.4
		daily reqmt	tons	4000	8000
		no. of day's stock		3	3
		storage	tons	**12000**	**24000**
6	slag	factor		0.85	0.85
		daily reqmt	tons	8500	17000
		no. of day's stock		2	2
		storage	tons	**17000**	**34000**
7	cement OPC	factor		1.27	1.27
		daily reqmt	tons	12700	25400
		no. of day's stock		7	5
		storage	tons	**88900**	**127000**
8	cement PPC	fator		1.86	1.86
		daily reqmt	tons	18600	37200
		no. of day's stock		3	3
		storage	tons	**55800**	**111600**
9	cement BFSC	factor		3.46	3.46
		daily reqmt	tons	34600	69200
		no. of day's stock	3	3	3
		storage	tons	**103800**	**207600**
10	blended raw meal	fctor		1.82	1.82
		daily reqmt	tons	18200	36400
		no. of day's stock		2.5	2.5
		storage	tons	**45500**	**91000**
11	pulverised coal	fadtor		0.2	0.2
		daily reqmt	tons	2000	4000
		no. of day's stock		0.3	0.3
		storage	tons	**600**	**1200**
12	oil	factor		0.1	0.1
		daily reqmt	tons	1000	2000
		no. of day's stock	8	8	6
		storage	tons	8000	12000

Note: Quantities for duplication can be modified in the light of operational experience.

Table 1.17.4

Working out Sectional Capacities for cement plans of (kiln) of 3000 & 10000 tpd Capacity.

S. No.	Section	Item	Multiplying Factor	Hours per day	Factor for hourly Capacity	hourly capacity for	
						3000 tpd	10000 tpd *
1	Quarries	stone raised	2.4	12	0.200	600	2000
2	Crushing	limestone to be crushed	1.91	12	0.159	477.5	1592
		crusher capacity	2.3	12	0.192	575	1917
3	Stcker Reclaimer	stacker (20% above crusher cap.	2.8	12	0.233	700	2333
		reclaimer	1.76	10	0.176	528	1760
4	Raw Material Grinding	raw mill	1.76	20	0.088	264	880
5	Blending	feed rate	1.76	20	0.088	264	880
		extraction rate	1.91	24	0.080	238.75	796
6	kiln	feed rate	1.91	24	0.080	238.75	796
		kiln, preheater cooler	1.1	24	0.046	137.5	458
		clinker conveyor	1.72	24	0.072	215	717
		spillage	0.52	24	0.022	65	217
7	Coal Grinding	coal mill	0.2	20	0.010	30	100
			0.21	20	0.011	31.5	105
			0.22	20	0.011	33	110
			0.23	20	0.012	34.5	115
		coal firing	0.2	24	0.008	25	83
			0.21	24	0.009	26.25	88
			0.22	24	0.009	27.5	92
			0.23	24	0.010	28.75	96

Table 1.17.4 ***Contd...***

S. No.	Section	Item	Multiplying Factor	Hours per day	Factor for hourly Capacity	hourly capacity for	
						3000 tpd	10000 tpd *
8	Cement Grinding	cement mill					
		OPC	1.27	20	0.064	190.5	635
		PPC 30% ash	1.82	20	0.091	273	910
		BFSC 60 % slag	3.15	20	0.158	472.5	1575
9	Cement packing	bulk	1.27	6	0.212	635	2117
		PPC bagged	1.82	15	0.121	364	1213
		bulk	1.82	6	0.303	910	3033
		BFSC bagged	3.15	15	0.210	630	2100
		bulk	3.15	6	0.525	1575	5250

Note: * Quantities can be modified in the light of operational experience

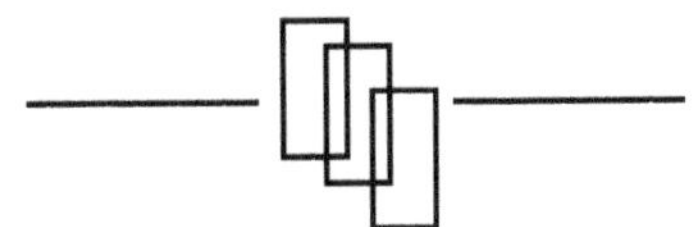

Section - 2

Machinery used in Making Cement

Contents

Chapter 1	Quarrying Operations	1-6
Chapter 2	Crushers	7-14
Chapter 3	Stacker Reclaimer Systems	15-20
Chapter 4	Grinding Mills	21-29
Chapter 4a	Ball Mills	30-36
Chapter 4b	Vertical Mills	37-46
Chapter 4c	Roller Press	47-52
Chapter 5	Screens, Separators and Classifiers	53-66
Chapter 6	Blending Systems	67-75
Chapter 7	Metering Systems for Raw Meal and Coal	76-82
Chapter 8	Preheaters	83-93
Chapter 9	Calciners	94-106
Chapter 10	Kilns	107-116
Chapter 11	Clinker Coolers	117-126
Chapter 12	Pollution Controls	127-143
Chapter 13	Cement Mills and Cement Storage	144-151
Chapter 14	Packing of Cement	152-157
Chapter 15	Despatches of Cement, Bagged and in Bulk and Despatches of Clinker	158-165
Chapter 16	Material Handling Systems	166-180
Chapter 17	Fans	181-191
Chapter 18	Compressors and Blowers	192-198
Chapter 19	Motors	199-201
Chapter 20	Gearboxes	202-204

CHAPTER 1

QUARRYING OPERATIONS

1.1 Quarrying Limestone

Process of making cement begins with quarrying operations in which lime stone in mines is quarried and brought to crusher for crushing.

Limestone quarries are opencast mines. Often limestone appears as an outcrop right on the surface. At other times it is under an overburden that can vary in depth but is generally less than 3 metres thick.

Limestone deposits occur in a variety of ways. Some are in the form of a hill. Some are plain thick slabs; some are fragmented with substantial interstitial impurities.

Deposits are won by mining them. The first step is to drill holes, fill them with explosives and blast them so as to loosen the rock.

Loosened rock is collected by shovels and loaded into dumpers in mechanized mining. In manual mining stone only up to 300 mm in size is loaded into trucks / wagons. Therefore stone is subjected to secondary blasting to break big boulders.

1.2 Wagon Drills and Portable Compressors

Equipment most commonly used for drilling holes in rock are truck mounted, compressed air operated wagon drills. Size of holes and type of explosive to be used would be decided by mining experts to suit the deposits. Spacing of holes and their depth will also be decided by them according to depth of deposits and height of bench.

See plate 2.1.1.

Plate 2.1.1 Wagon Drill.

Since power will not be available in mines in early stages, portable diesel engine operated compressors supplying compressed air at around 7 kg/cm^2 would be used. Compressors will be shifted from place to place by attaching them to tractors. Wagon drills can be either truck mounted or pulled by other vehicles and hence can be shifted from place to place. If compressors are water cooled, provision will have to be made for its supply.

Since quarrying is a continuing operation, cement plants would prefer to have their own drills and compressors along with other quarrying equipment.

1.3 Shovels, Dumpers and Bulldozers

These are the most commonly used earth moving equipment used in quarrying operations. They come in specific sizes and capacities and are selected according to tonnages to be handled per hour or per shift. Numbers will depend among other things on:

1. Number of faces in operation simultaneously,
2. distance between quarries and crusher and time taken for a round trip,
3. provision for standby for ensuring uninterrupted work.

1.4 Shovels / Wheel Loaders / Front End Loaders

Shovels can be electrically operated but more commonly they are diesel operated. They may have 'track' to negotiate rough and uneven terrain in quarries or will have 'tyres'.

They will be moving along the 'face' and from 'face to face'.

Size is designated by the capacity of bucket expressed in cubic yards or cubic metres. Size of bucket also decides the maximum size of stone the bucket can hold. **Table 2.1.1** furnishes sizes of shovels, width of buckets and maximum size of stone they can deliver. This in turn decides the sizing of crusher as crusher must be big enough to receive and crush stone delivered by the shovel.

Generally crusher opening would be 125 % of the max. size of stone.

Table 2.1.2 furnishes sizes of Jaw crushers for some sizes of shovels.

Capacities of various sizes of shovels are made available by the Manufacturers in terms tons loaded per hour. Assumptions made would be number of swings per minute, bulk density of stone and efficiency of filling which is taken on an average as 70 %.

Please **See W1.51** in **Web Page** 1of **Section 9**

See plate 2.1.2.

1.5 Dumpers

Commonly used vehicles for transporting 'run off mine' stone picked by shovels to crushing plant would be :

1. Self tipping trucks when mining is manual,
2. rear end dumpers,
3. tipping railway wagons when stone is transported by company's own railway to crushing plant.

The most common among them are rear end dumpers.

Shovels load dumpers. Dumpers are heavy duty diesel vehicles.

Open heavy duty body which holds stone is tilted end to end hydraulically to unload stone into crusher hopper.

Plate 2.1.2 Wheel Loader.

Dumpers come in standard sizes specified by carrying capacity expressed in cubic metres or in tons. Dumpers are selected to match loading capacity of shovels. Numbers are dependent on the time taken for a round trip to crusher. Generally for each shovel there would be three dumpers.

See plate 2.1.3.

Plate 2.1.3 Rear End Dumper.

Table 2.1.3 furnishes typical sizes and capacities of dumpers, and also minimum radius of turning and optimum gradient for hauling. They are useful in designing ramps and roads on which Dumpers would ply.

1.6 Bulldozers

Bulldozers are required to loosen over burden and to level ground. They carry out several tasks with attachments and are an essential part of earth moving machinery. **Table 2.1.4** furnishes data on Bull dozers. **See plate 2.1.4.**

1.7 Rippers

If type of deposits permits it, blasting is avoided and stone is 'ripped' up from the bed. This avoids vibrations caused by blasting.

Geologists, Mining Engineers and Manufacturers of earth moving equipment should be consulted in formulating mining plans and in selection of equipment. They can advise on costs of operations and help in optimizing capital and operational costs.

1.8 Earth Moving Machinery for large Cement Plants

For a 10000 tpd capacity single kiln, stone to be raised in quarries would be of the order of 2000 tph based on 2 shift working for 6 days a week.

This would require various quarry machinery as shown in **Annexe 2.1.1**

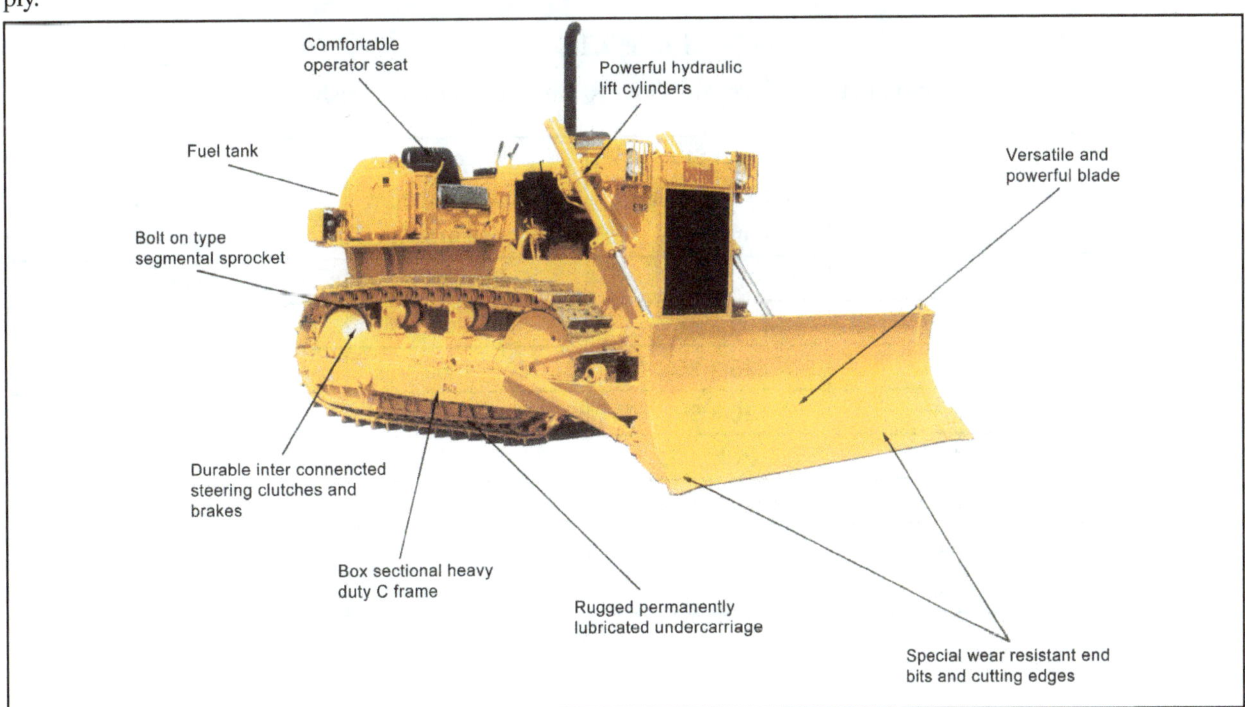

Plate 2.1.4 Bulldozer.

Table 2.1.1

Shovels, their Sizes and Capacities.

Excavator Shovels

Sr. No.	Nominal size	Bucket capacity	Hourly capacity	Operative weight	Width of bucket	Max. size of stone delivered
		m^3	tph	tons	mm	mm
1	100	0.5	154	12	1120	740
2	220	1	239	21	1250	825
3	300	1.8	387	31	1660	1095
4	650	3.7	663	65	2070	1366
5	1000	5.5	910	95	1940	1280
6	1600	11	1578	160	2710	1788

Loading shovels

Sr. No.	Nominal size	Bucket capacity	Hourly capacity	Operative weight	Width of bucket	Max. size of stone delivered
		m^3	tph	tons	mm	mm
1	650	4.5	807	67	2320	1530
2	1000	7	1159	98	2650	1750
3	1600	13	1865	162	3800	2500

Table 2.1.2

Relation between Sizes of Shovels, Buckets and Jaw Crushers.

Sr No.	Size of Shovel m^3	Size of Stone passing bucket mm × mm	Size of Jaw crusher depth × breadth	Size of Gyratory crusher
			mm*mm	mm
1.	0.57	800 × 875	600 × 900	400
2.	0.77	825 × 950	750 × 1050	500
3.	1.15	750 × 900	900 × 1200	520
4.	1.53	825 × 1125	900 × 1200	520
5.	2.29	1000 × 1200	1200 × 1500	1200
6.	3.06	1200 × 1425	1400 × 1800	1500
7.	3.83	1200 × 1500	1650 × 2150	1800

Source : Based on Handbook of M/S Hammermills, U.S.A.

Table 2.1.3

Rear end Dumpers Sizes and capacities.

Sr. No.	Nominal size	payload tons	heaped capacity m^3	max. speed kmph	min. radius of turning m
1	35	32	22.5	47	7.16
2	40	36.5	24	66	10.5
3	50	54	34	57	9.8
4	60	55	36.5	62	9.9
5	70	64	42	65	9.9
6	85	78	52	65	9.9
7	90	85	58	65	9.9
8	100	91.5	61	60	9.9

Maximum and optimum gradient for hauling in quarries and for crusher ramp is **one in sixteen.**

Table 2.1.4

Sizes of Bulldozers.

Sr. No.	Nominal size	Blade dimensions metres	Operating weight tons	Drawbar pull tons
1	31	2.42 × 0.75	6.2	7
2	50	3.35 × 0.85	11	10.3
3	65	4 × 1.15	16.7	27
4	80	4.3 × 1.1	21.5	19.6
5	155	4.13 × 1.6	36	66
6	230	4.3 × 1.13	23	40
7	355	4.32 × 1.9	44	90
8	475	5.3 × 2.6	95	150

Information in Tables 2.1.1 to 2.1.4 based on data furnished by Bharat Earth Movers Ltd.

Annexe 2.1.1

Equipment in Quarries for a10000 tpd capacity clinkering plant.

1. Stone to be raised in quarries 24000 tons per day (multiplying factor 2.4)
2. Assume two faces will be worked simultaneously to maintain continuity of supply an keeping in mind possibility of blending.
3. Crusher will be 5 kms from the quarries – mean distance. It will be a single stage crusher as raw mill proposed would be VR Mill
4. Quarries and Crusher will work for 6 days in a week in 2 shifts of 6 useful hours each
5. Daily raisings 24000 tons. Assuming depth of deposit to be 20 metres average and a bulk density of $2t/m^3$, area covered in mines would be 600 m^2 total in two faces perday. One wagon drill for each face should suffice. One standly
6. Crusher

   Crusher would have a capacity of 2000 tph to be able to supply 24000 tons of stone in 12 hours.
7. Shovels

   To maintain continuity of supply there would two 2 shovels per face. Capacity of each shovel will be 500 trips to supply crusher stone at 200 tph. One stand by
8. Dumpers.

   Each dumper @ 30km per hour will make the return to crusher in 20 minutes. Taking into account at require for waiting, unloading one return trip would take 30 minute.

   One dumper can make 2 trips per hour. A 200 ton capacity dumper can supply can supply 400 tons per hour. With two faces working simultaneously, toal 5 dumpers should suffice between two faces. Standby can be one or two

   To summaries:

1. Wagon Drills	2	1 standby
2. Shovels 200 tph	4	1 standby
3. Dumpers 200 ton capacity	5	1 or 2 standby
4. Bulldozers	2	1 in Quaary, 1 in Plant

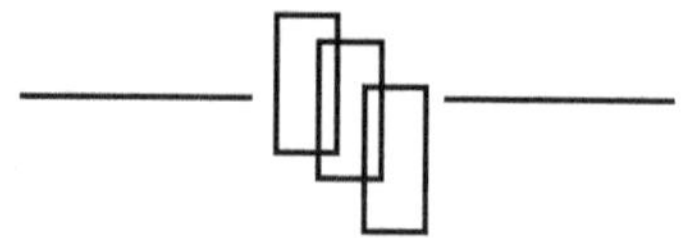

CHAPTER 2

CRUSHERS

2.1 Crushing

The second step in making cement begins with crushing of limestone as mined and received. In turn crushing operation prepares stone for the subsequent operation of grinding or pulverizing.

The extent of size reduction in crushing operation depends on one hand the 'run off mine' stone as received from quarries and the size of stone as required by the mills selected for grinding on the other.

Size of 'run off mine' stone depends on whether mining is 'manual' or 'mechanised'.

Manual mining is relevant only to small plants. Mined stone pieces are carried as head loads and size is limited to - 300 mms.

Mechanised mining uses shovels and dumpers for bringing stone to crusher. Therefore size of stone is directly related to size of shovel used, that is the size of its bucket as has been brought out in Chapter 1.

Ball mills require for grinding size of feed generally less than 25 mm.; smaller the better.

Vertical Roller Mills, Roller Presses and Horo mills can receive stone between 75 to 100 mm size.

2.2 Ratio of Reduction

Ratio of reduction decides the number of stages of crushing required and the type of crushers to be used. **Table 2.2.1** shows ratios of reduction for various possibilities of mining and grinding mentioned above and number of stages of crushing required.

Table 2.2.1

Sr no	Description	Manual mining	Mechanised Mining				
			Size of shovel m^3 bucket capacity				
			1.5	2	2.5	3	4
1.	Size of stone mms	- 300	800×1150	900×1200	1000×1200	1200×1400	1200×1500
2.	Type of mill	Ball mill					
3.	Size of feed	- 25 mm					
4.	Reduction ratio	12	46	48	48	56	60
5.	No. of stages of crushing	1	←.......... 2 ....→ Single stage hammer crusher for manual mining Jaw and hammer crusher for two stage crushing				
6.	Type of mill		V.R.Mill or Roller Press				
7.	Size of feed		- 75 mm				
8.	Reduction ratio		15	16	16	18	20
9.	No. of stages of crushing		1 Stage Hammer or Impactor				

2.3 Types of Crushers

Main types of crushers used in Cement Industry to crush limestone are:

1. Jaw Crushers
   - single or double toggle
2. Hammer crushers
   - unidirectional or reversible
   - full or partial grate bars
   - single or double rotor
3. Impactors
   Single or double rotor
4. Gyratory and cone crushers

Jaw crushers and gyratory crushers are used as Primary crushers in two stage crushing. Size reduction is limited to 5 to 6 : 1. Size of crusher designated by feed opening is dependent on size of shovel used.

Hammer crushers and Impactors can be used both as Primary crusher in single stage crushing and as Secondary crusher in two stage crushing.

2.4 Jaw Crushers

In jaw crushers as the name suggests crushing is done by exerting crushing force between a fixed jaw and a moving jaw which moves back and forth like a human jaw. Reciprocating motion is imparted by an eccentric and a toggle.

See plate 2.2.1.

Jaws can be straight or curved, plain or corrugated. Toggle can be single or double.

Angle between the jaws grips the stone and crushes it and pushes it down for further reduction in the next squeezing stroke

Because of the 'throw' of the moving jaw there is always about 15 % oversize in the product of a jaw crusher. Production charts give outputs at different sizes of closed side settings. Hence if size of product is to be restricted, capacity at a setting smaller by the throw should be considered

Jaw crushers are not run in closed circuit to control size of product for the same reason.

Jaw crushers are slow moving machines. Crushing faces are generally made of manganese steel. They are replaced when worn out. A toggle is used to adjust / maintain closed side setting to suit size of product desired and to allow for wear on the jaws.

Jaws need to be taken out of crusher for replacement. They may be in one piece or in two pieces.

A lifting tackle above the jaw would be provided for this purpose.

A reciprocating feeder is likely to result in rush of stone in the mouth of the crusher andjam it. For large capacities Apron Feeders are preferred.

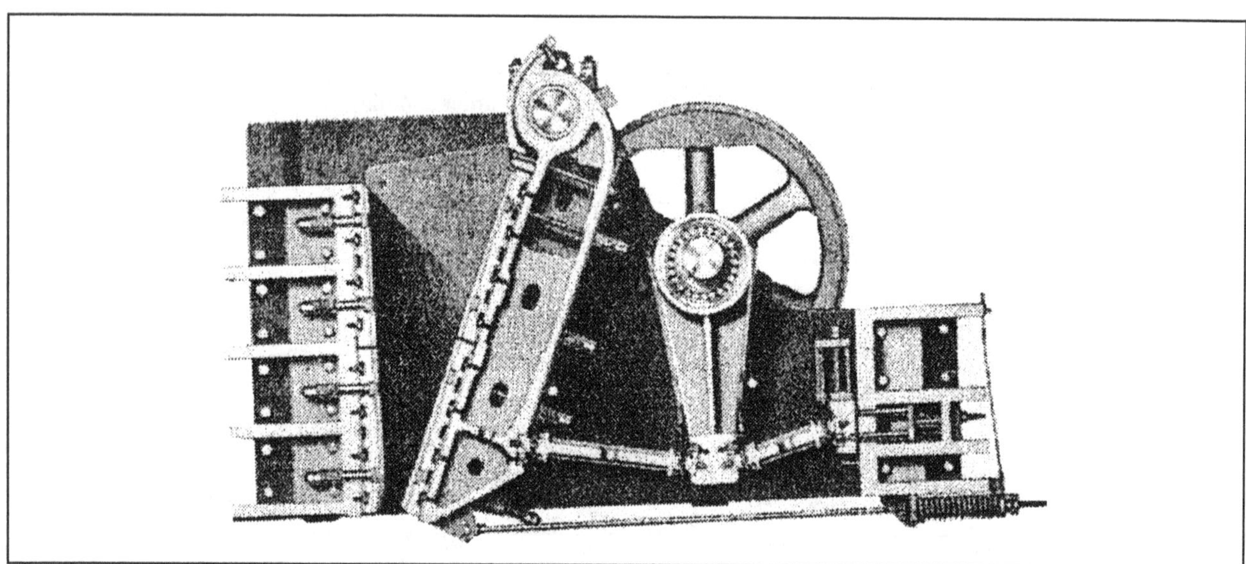

Plate 2.2.1 Double Toggle Jaw Crusher.

Jaws can be straight or curved, plain or corrugated. Toggle can be single or double.

Angle between the jaws grips the stone and crushes it and pushes it down for further reduction in the next squeezing stroke.

Because of the 'throw' of the moving jaw there is always about 15 % oversize in the product of a jaw crusher. Production charts give outputs at different sizes of closed side settings. Hence if size of product is to be restricted, capacity at a setting smaller by the throw should be considered Jaw crushers are not run in closed circuit to control size of product for the same reason.

Jaw crushers are slow moving machines. Crushing faces are generally made of manganese steel. They are replaced when worn out. A toggle is used to adjust / maintain closed side setting to suit size of product desired and to allow for wear on the jaws.

Jaws need to be taken out of crusher for replacement. They may be in one piece or in two pieces.A lifting tackle above the jaw would be provided for this purpose.

A reciprocating feeder is likely to result in rush of stone in the mouth of the crusher and jam it. For large capacities Apron Feeders are preferred.

2.5 Hammer Crushers

As the name suggests rotating hammers mounted on discs pound the stone fed into the crusher against breaking plates. Stone is broken into smaller pieces and gets further reduced because of reducing clearance between tips of hammers and breaking plates.

The bottom is fitted with grate bars fully or partially through which stone crushed smaller than the spacing in the grate, passes.

Over size is subject to repeated pounding till it is finished to size.

When Hammer crushers are unidirectional, only one face of the hammers is used in crushing and is worn out. Hammers are reversed to bring second face in use. Feed must enter the crusher so that rotating hammers catch it at a point where impact is maximum.

Hammer crushers can also be reversible when feed falls right on top and crusher can run in either direction. It has the advantage that both faces of hammers are used for breaking stone without requiring to take a stoppage and reverse hammers.

Grate bars are fixed in cages which are hinged on one side for access and replacement.

See plates 2.2.2 and 2.2.3.

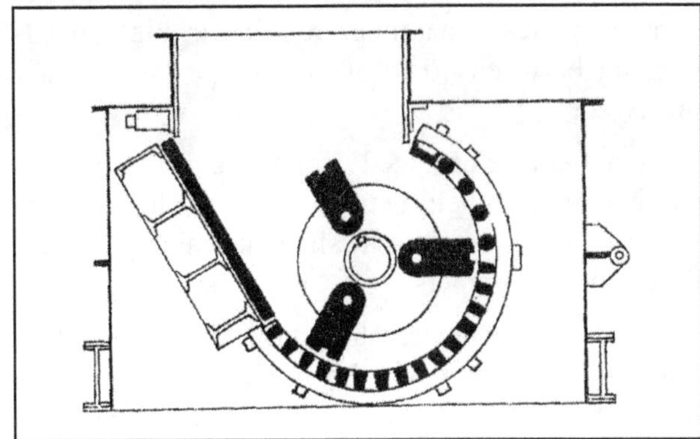

Plate 2.2.2 Single Rotor Hammer Crusher.

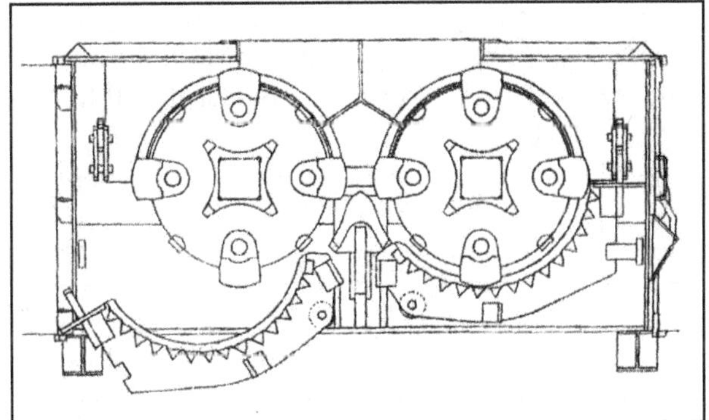

Plate 2.2.3 Double Rotor Hammer Crusher.

2.5.1 Control of Product Size

Hammers and grate bars are made of manganese steel. Hammers are built up by hard facing. Normally one set of hammers will be kept ready to replace the worn out. In installing built up set, care is taken to 'match' the hammers so that rotor is not unbalanced.

Some designs have only partial grate bars.

Size control is thus two ways - one by adjusting breaker plates and two - by adjusting spacing between grate bars. Therefore hammer crusher is well suited for running in closed circuit to maintain a product of fixed size. As a secondary crusher it will often be run in closed circuit and particularly when subsequent grinding mill is a ball mill.

It is not necessary to run it in closed circuit when the mill is a V.R.Mill or a Horo mill or when there is a Roller Press.

2.5.2 Various Designs of Hammer Crushers

In some designs there are two rotors in the same casing turning in opposite directions.

See plate 2.2.3.

There are a great many designs of hammer crushers and hammers themselves. Presently a smaller number of heavy hammers running at high speeds ranging between 40 to 50 metres per second are preferred.

Hammers are arranged 3 or 4 to a round and in a number of rows held between pairs of discs. There are spacers between discs to slide discs along the shaft for installing hammers.

As in case of Jaw crushers, an overhead traveling crane with a hoist will be installed to facilitate maintenance of hammers. For large crushers a hydraulic puller will be installed to take out shaft from the side.

When working as a secondary crusher, it will receive feed directly by a belt conveyor from the primary crusher. As a primary crusher it will have a hopper above it.

2.6 Impactors

Impactors have 3 or 4 more blow bars fitted on a rotor along its width. They throw stone on impact breaker plates and in a series of impacts between plates and blow bars stone is fragmented to desired size.

There are no grate bars for controlling the size. Hence impactors are better suited when grinding mills are V.R.Mills or when a roller press is used as pregrinder.

Impactors are more prone to wear when stone is hard and has high silica in it. Therefore while selecting between a hammer mill and an impactor, hardness of rock and its silica content should be looked into.

Blow bars will be made of manganese steel. They can be of reversible design so that both edges can be used for breaking stone. Rotor will be cast steel mounted on a shaft of forged steel.

See plates 2.2.4 and 2.2.5.

2.6.1 Two stage impactor

In some designs stone is broken to required size in two stages in the same casing. Product of stage one is broken in stage two.

See plate 2.2.6.

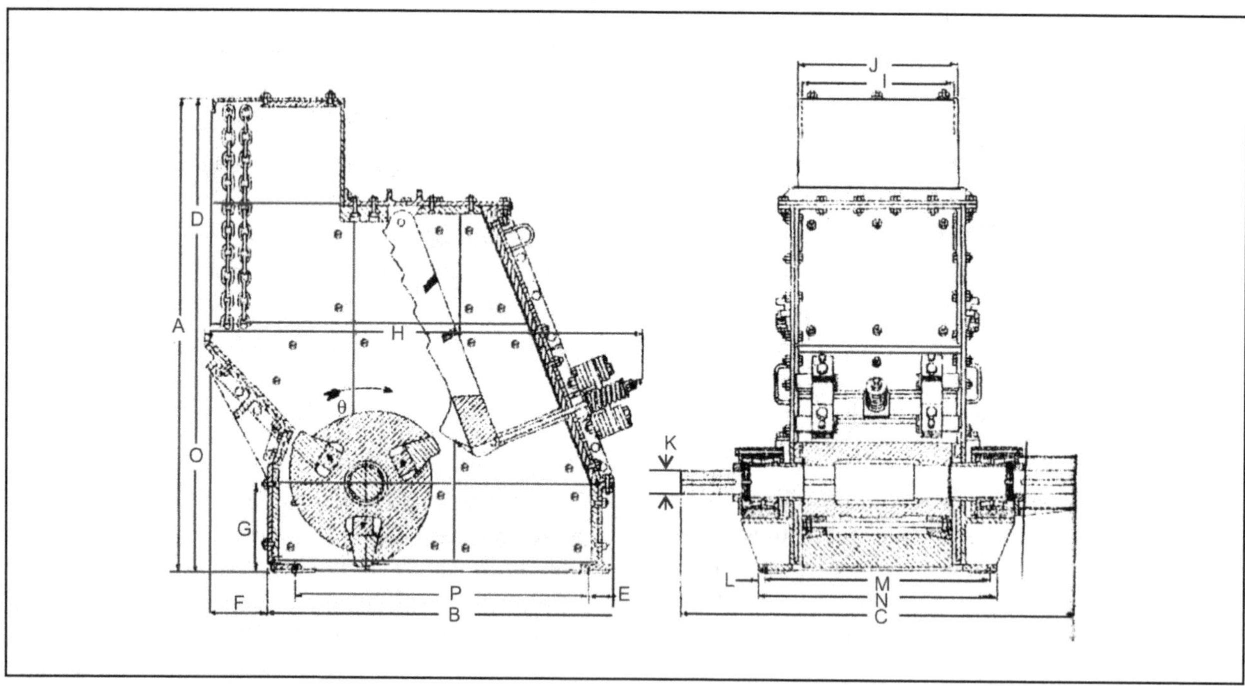

Plate 2.2.4 Impact Crusher.

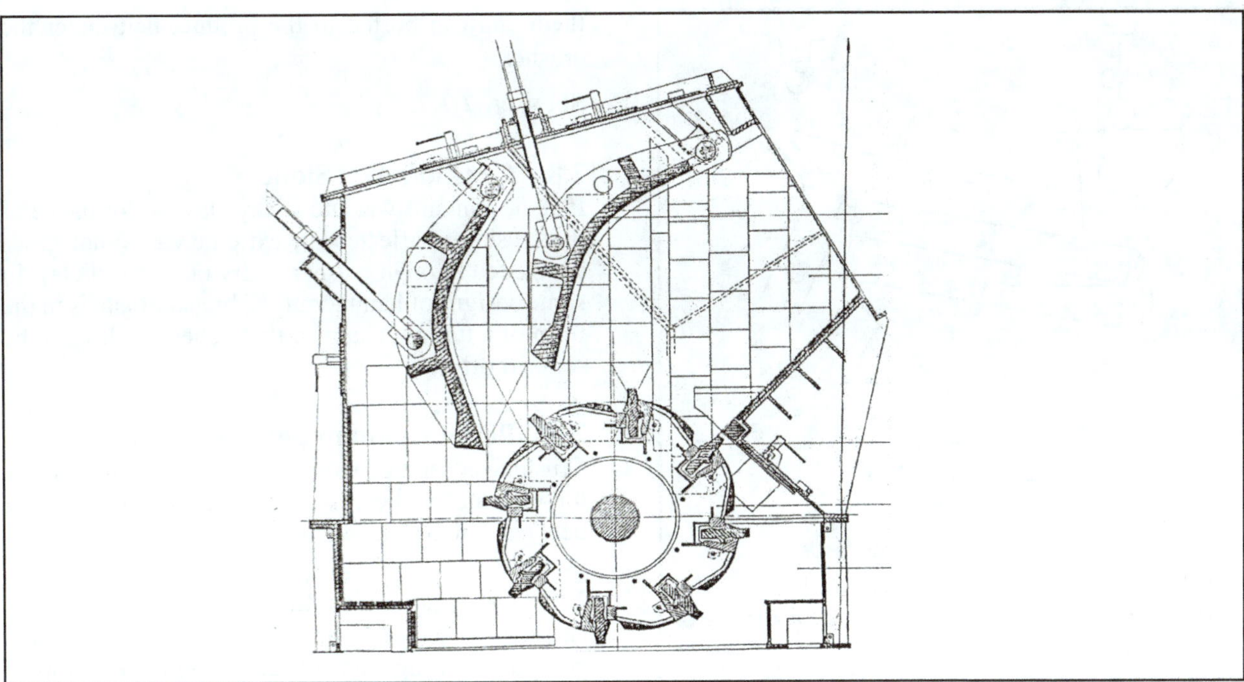

Plate 2.2.5 Impact Crusher with Push-in type Reversible Blow-bars.

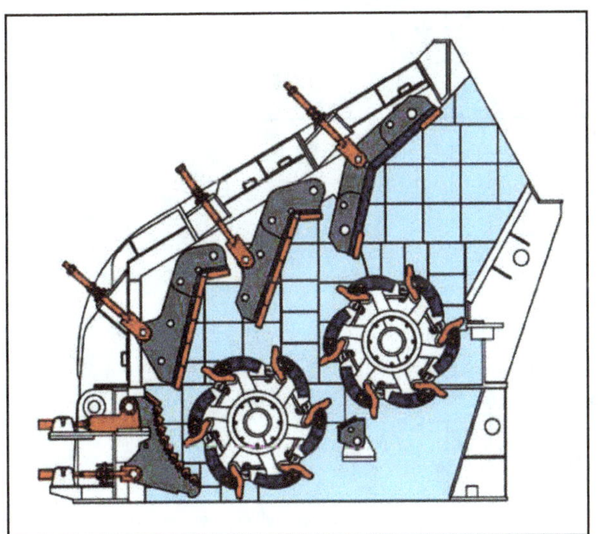

Plate 2.2.6 Hazemag AP -CM Compound Primary Crusher.

2.7 Feeders

2.7.1 Apron Feeder

Apron feeder is the preferred feeder for crushers particularly for large capacity single stage crushers receiving run off mine stone.

Because of the feeder's upward slope, stone does not rush into crusher. It also helps to reduce depths of pit in the crusher. It is rugged in construction and is slow moving. Therefore widths are more. Its width should match the width of crusher. Feeder will have a variable speed drive to regulate feed.

Apron feeder carries on its under side lot of fines causing spillage.

This is collected in hoppers and conveyed by a belt onto the main belt.

See Fig. 2.2.1.

2.7.2 Reciprocating Feeder

Reciprocating feeders are better suited for small capacity crushing installations and where mining is manual or semi mechanized.

In reciprocating feeders stone travels along the downward inclined and reciprocating carriage. Rotary motion of motor is converted into reciprocating by an eccentric and a connecting rod.

Because of its slope, stone tends to accelerate into crusher mouth.

In some designs, part of carriage has spaced bars so that stone that does not need crushing falls through

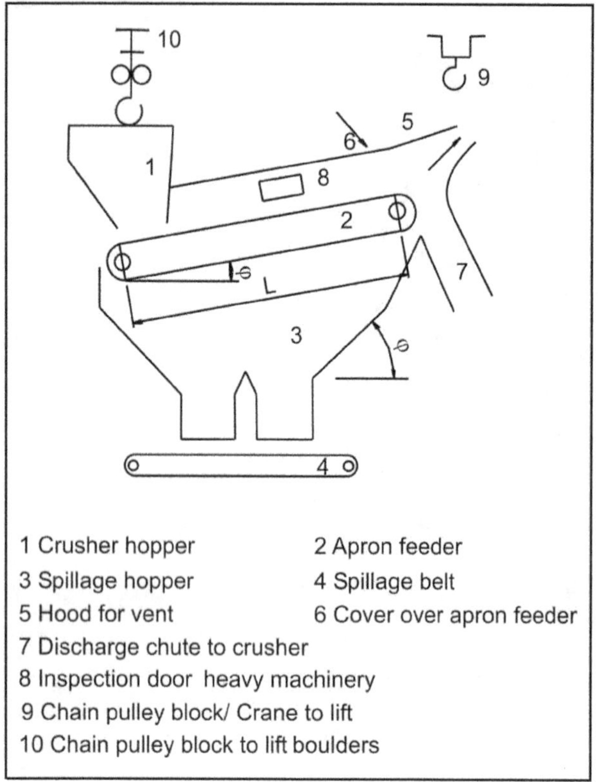

Fig. 2.2.1 Apron feeder for crusher.

them and can be fed to the product belt under the crusher.

See Fig. 2.2.2.

2.8 Crushing Wet Stone

If stone contains wet and sticky clay, performance of the crusher is affected. In extreme cases, hot gases are passed through crusher to dry out the wet clay. In some designs of hammer mills, breaker plate is in the form of a turning chain so that adhering clay can be scraped off it.

2.9 Power Consumption in Crushing

Crushing consumes about 1.5 to 2 units of power per ton of stone crushed. In terms of cement it would be between 2 to 3 units per ton.

2.10 Venting of Crushers and Feeders

Crushers and feeders are vented by collecting vent gases and passing them through a bag filter. Rotors of hammer mills and impactors behave like impellers of fans and displace considerable quantities of air that needs to be vented.

See Fig. 2.2.3.

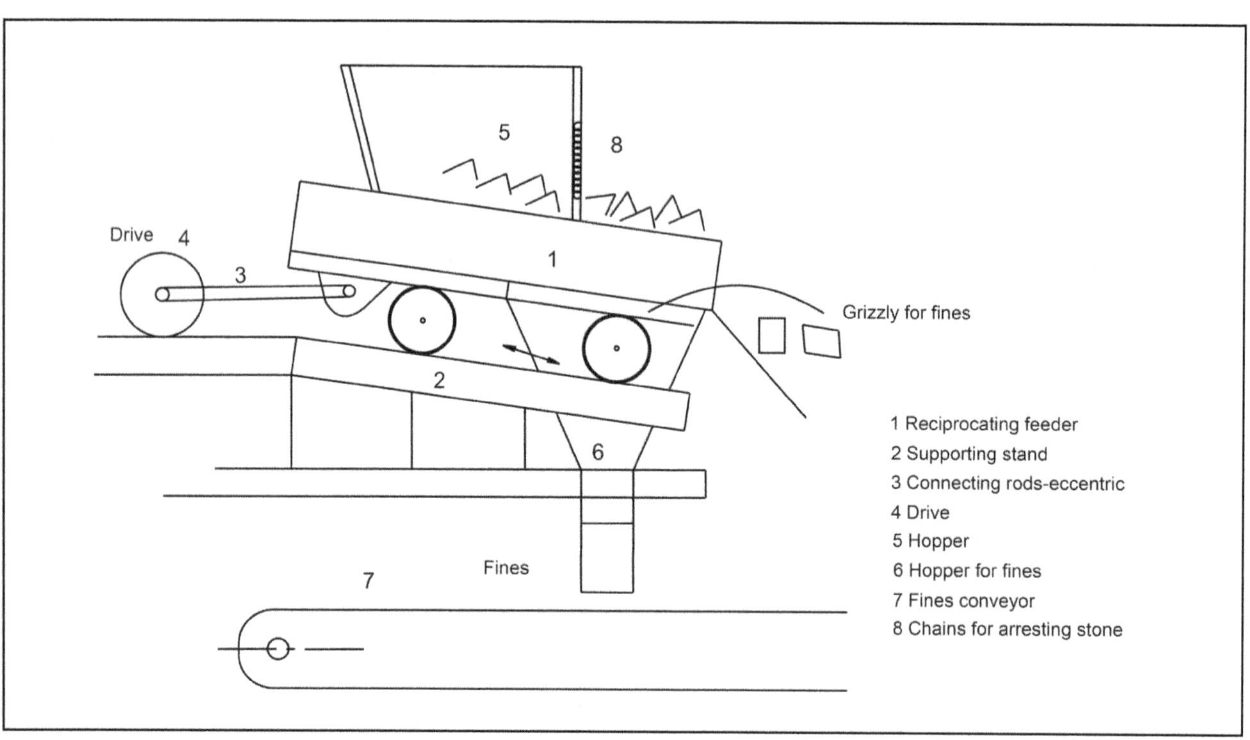

Fig. 2.2.2 Reciprocating feeder for crusher.

2.11 Mobile / Semi Mobile Crusher

In case of large cement plants mines would be spread over hundreds of hectares. Leads from working faces to the crusher would be short initially but would increase by several kilometers in course of time. Operational costs of earth moving machinery -dependent as they are on cost of diesel oil- are high and they can be kept down by not allowing the lead to increase. This means crushing unit should move as faces recede.

Mobile or semi mobile crushers make this possible. The complete crushing unit consisting of hopper, feeder, crusher and a short discharge conveyor is either mounted on a carriage- crawler track mounted or is tyre mounted. When the unit is self propelled it is a 'mobile' unit. When it is required to be towed it is a 'semi mobile' unit.

Crushed stone would be conveyed to the plant by a system of belt conveyors whose total length would increase as crusher is moved further into quarries.

See plates 2.2.7 to 2.2.9.

2.12 Automatic Sampler

An automatic sampler is installed to collect samples of crushed stone before it is fed to the stock piles. This helps in monitoring quarrying operations and also helps in ascertaining efficiency of 'preblending' achieved in following 'Stacker Reclaimer' systems.

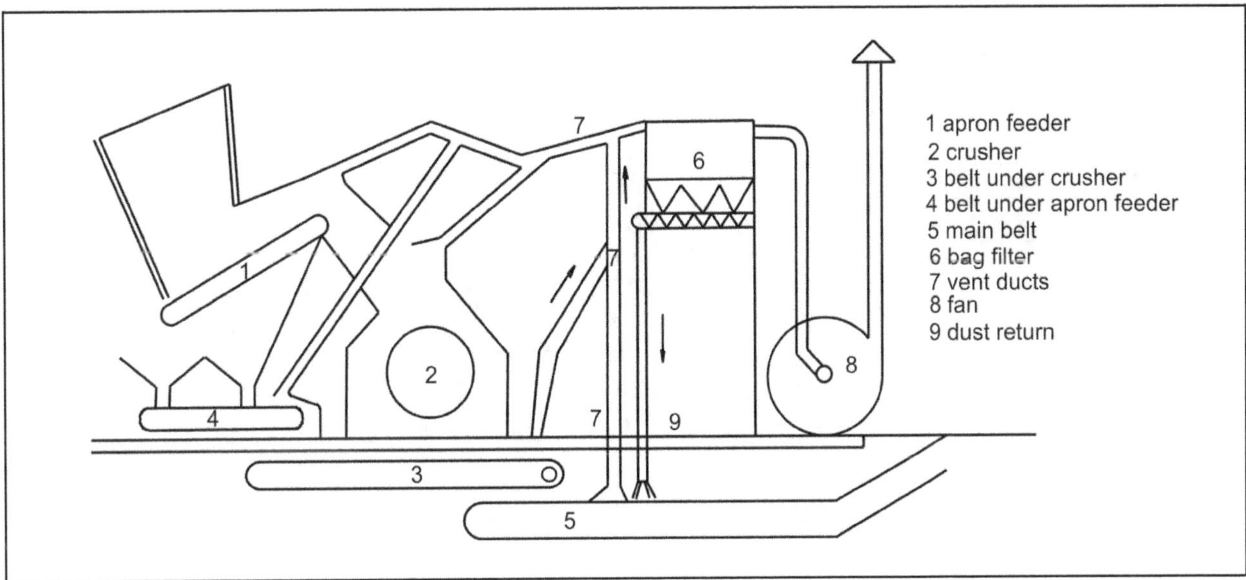

Fig. 2.2.3 Bag filter in crusher circuit.

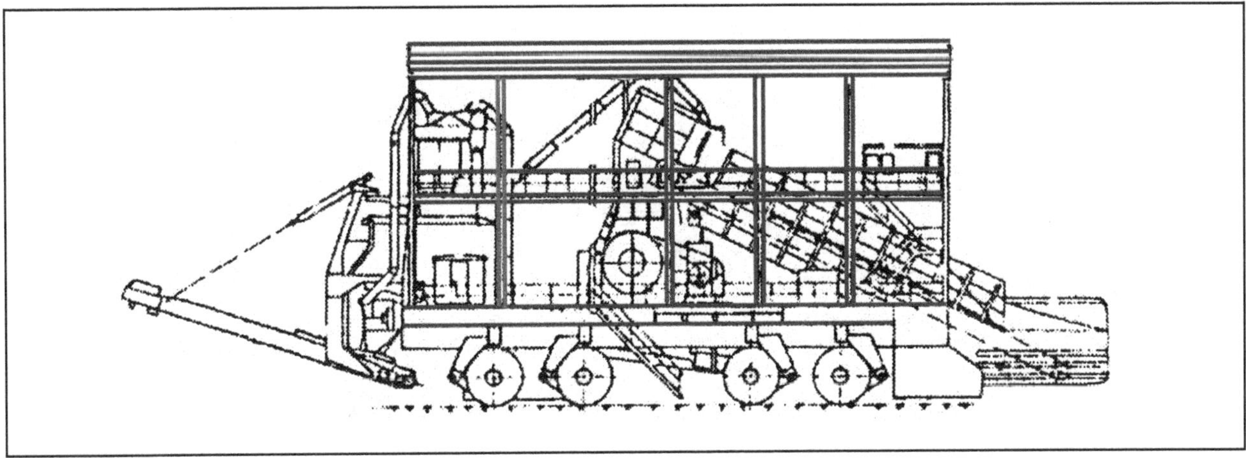

Plate 2.2.7 Mobile crushing unit – wheel mounted.

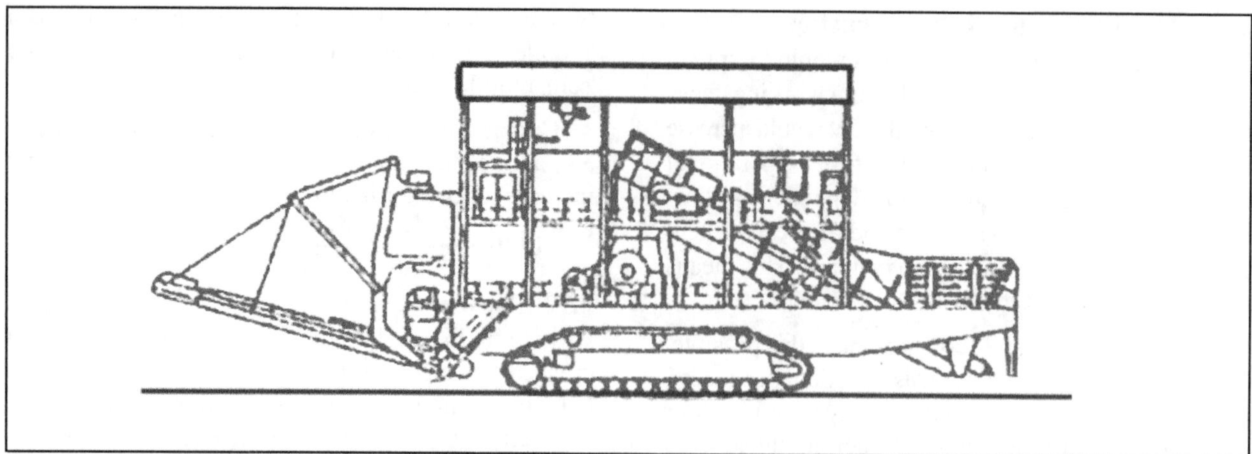

Plate 2.2.8 Mobile Crushing Unit with crawler track.

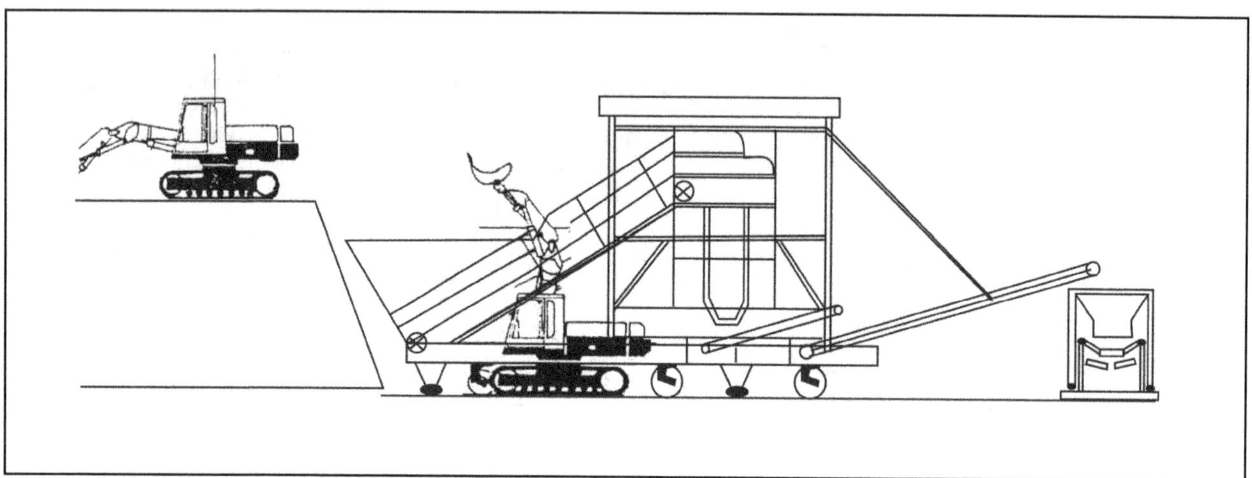

Plate 2.2.9 Semi Mobile Crushing Unit.

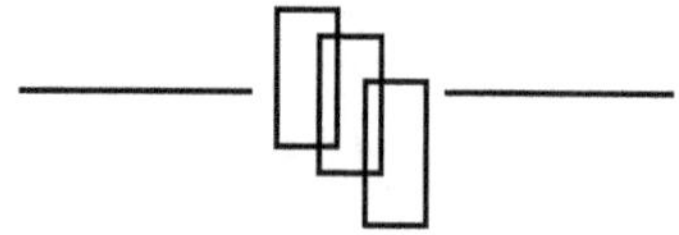

CHAPTER 3

STACKER RECLAIMER SYSTEMS

3.1 Stacker Reclaimer Systems

Stacker Reclaimer Systems are storage and extraction systems for crushed and granular materials like crushed limestone and coal.

They can handle material at high rates of feed and extraction because conveyors in the systems are belt conveyors.

Materials are stored as triangular stock piles. Quantity to be stored is determined by the stocks to be maintained in number of days' consumption.

3.2 Types of Stock Piles

There are either two linear stock piles- arranged side by side or in line; or a circular stock pile.

In case of linear piles when one pile is built up, material is extracted from the second pile. Thus two piles supply material to the plant alternately.

See Figs. 2.3.1 and 2.3.2.

In case of a circular stock pile, pile is formed to occupy three fourths of the circle. Ple is built up from one end and extracted from the other end.

Circular piles are generally used for coal because capacities and handling rates are small.

See plate 2.3.1 and Fig 2.3.3.

3.3 Stacker

It receives crushed stone / coal from the crusher. It consists of a movable carriage which moves on rails along the proposed stock piles. Length of travel is equal to length of two piles when they are lineally arranged

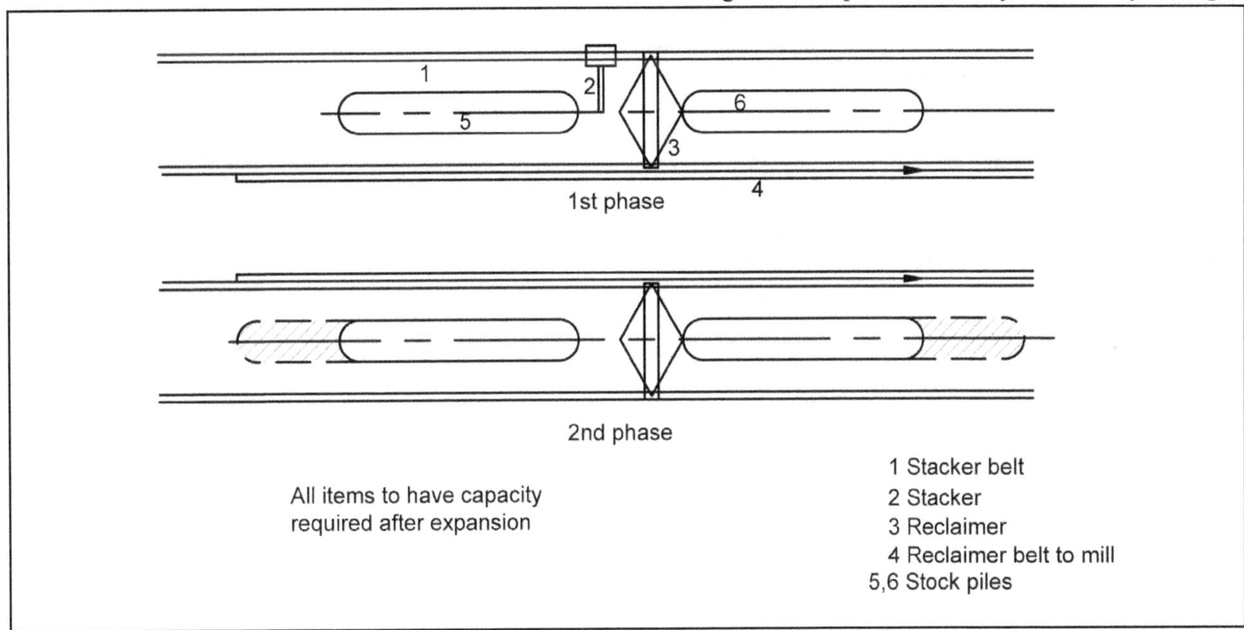

Fig. 2.3.1 Linearly arranged stock piles extend lengths of stock pile.

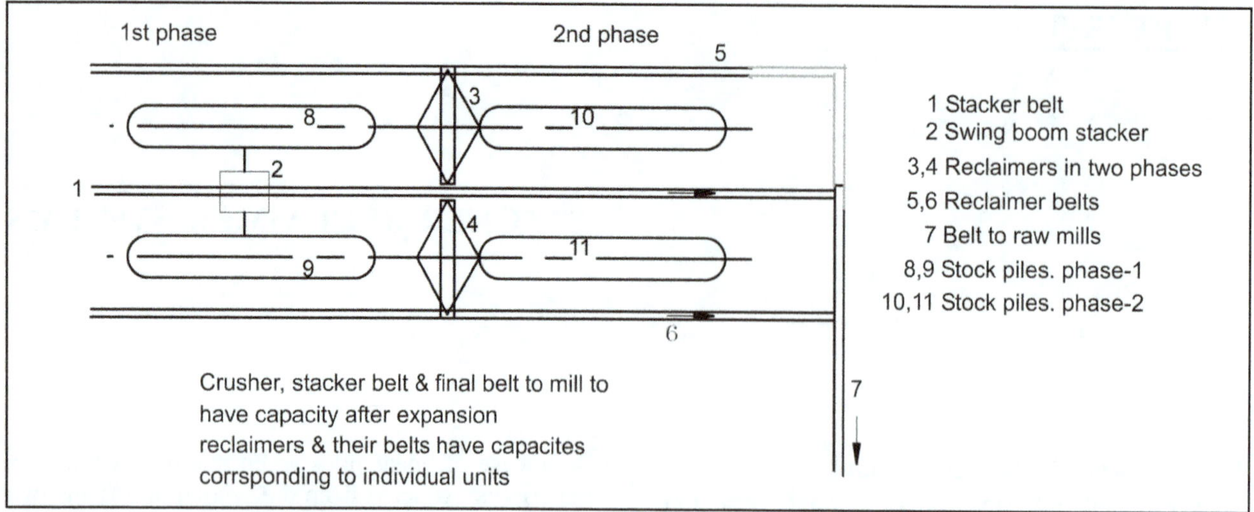

Fig. 2.3.2 Piles arranged side by side same crusher-swing boom stakcer two reclaimers.

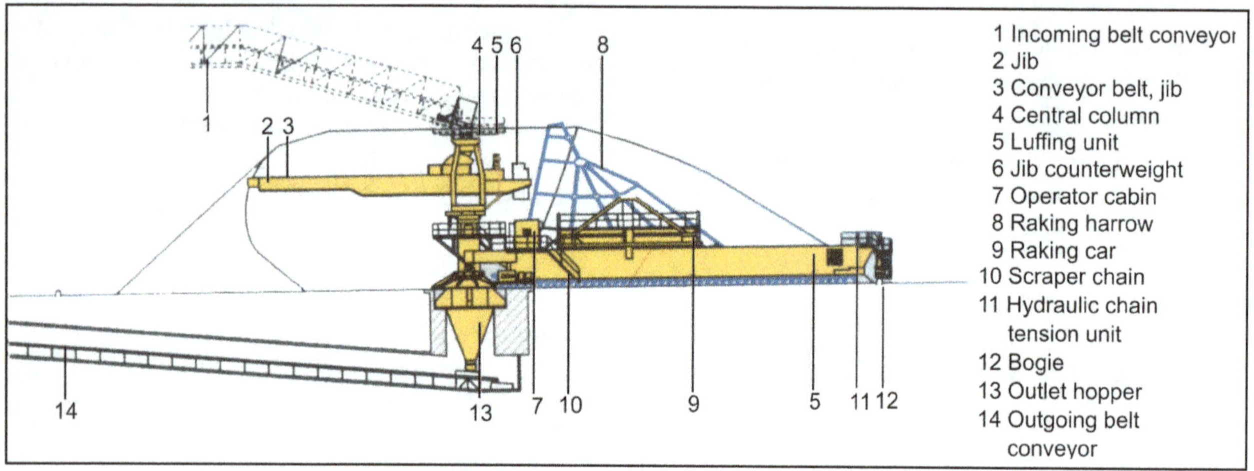

Plate 2.3.1 Circular stacker reclaimer system.

plus some more. Stacker belt goes over this movable carriage and drops material it is carrying on a cross belt which reaches up to the center of the stock pile and which can slew upward and downwards pivoted at the feeding end. Material is dropped from this belt as the main belt moves at a predetermined rate along the pile. When cross belt reaches the end of its travel, carriage starts in reverse and deposits another layer on the pile. Pile thus gets built up in layers.

See plate 2.3.2 and plate 2.3.3

3.3.1 Building up of Stock Piles

There are many ways in which layers can be spread and piles built up; the most common being 'windrows' and' chevron'.

See Figs 2.3.4 and 2.3.5

3.3.2 Stacker for Coal

For building up piles of crushed coal different methods are used so as to form trapezoidal piles because piles are to be restricted to a height of 3 metres.

See Fig. 2.3.6 and 2.3.7

3.4 Reclaimer

Reclaimer consists of a hoe or a rake which dislodges material in layers in small thicknesses from the entire cross section of the pile.

Dislodged material falls on a pan conveyor running across the width of the pile and drops it on a belt to take it to hoppers of mills. Thus the hoe and pan conveyor also move on rails for the total length of the pile. Speed of travel of carriage is variable. Angle of hoe may also be adjustable to suit angle of repose of the material.

See Plate 2.3.4

Plate 2.3.2 Stacker.

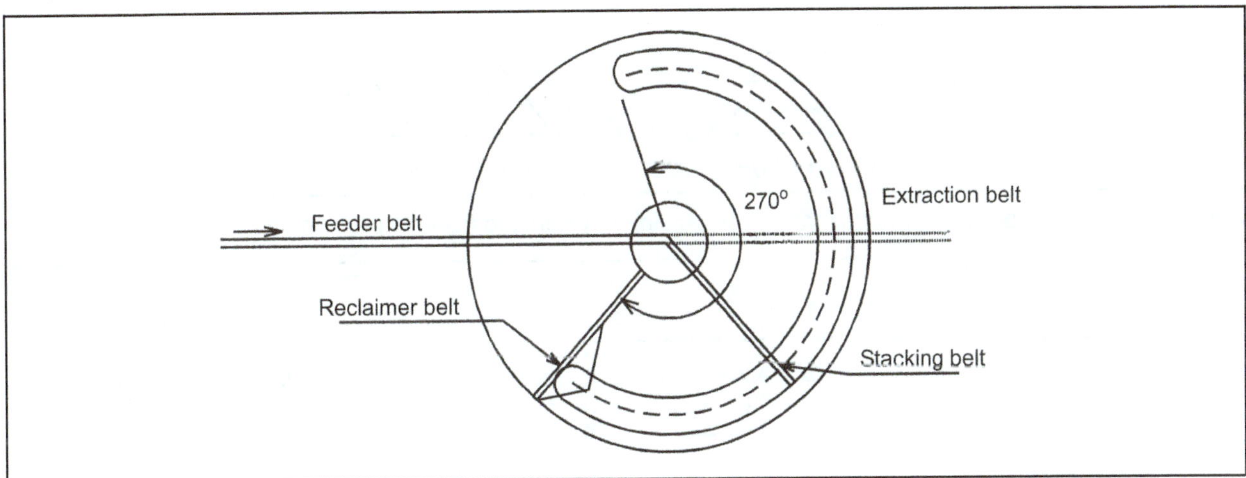

Fig. 2.3.3 Circular stacker reclaimer system for coal.

Plate 2.3.3 Swing boom stacker and two reclaimers for two production lines.

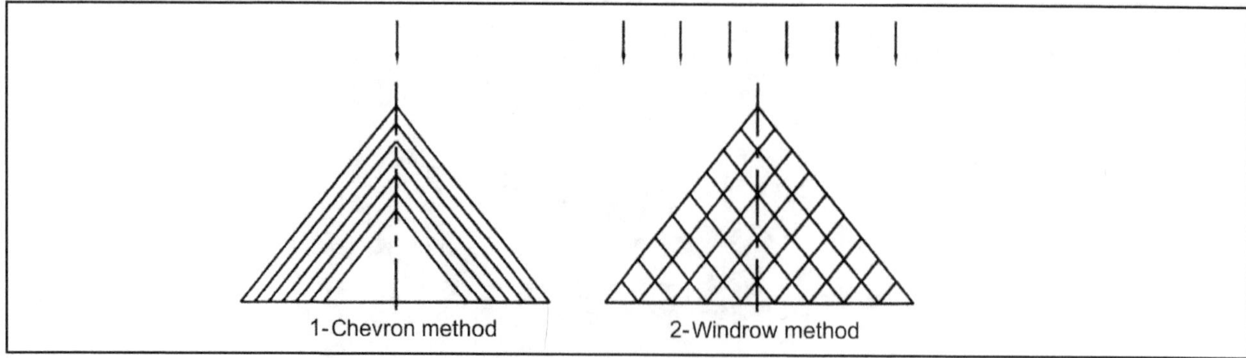

Fig. 2.3.4 Various stacking methods for stacker reclaimer systems.

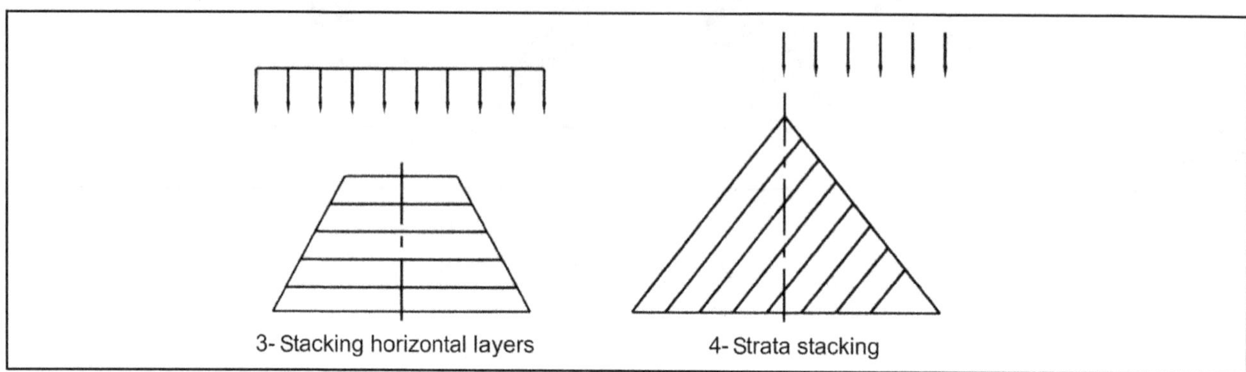

Fig. 2.3.5 Various stacking methods for stacker reclaimer systems to achieve preblending.

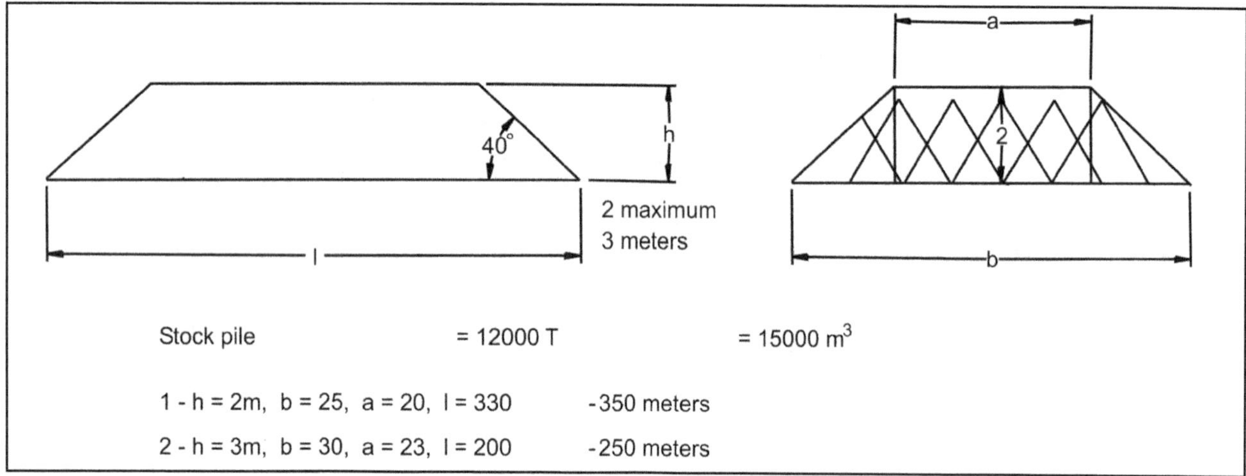

Fig. 2.3.6 Stock piles for coal.

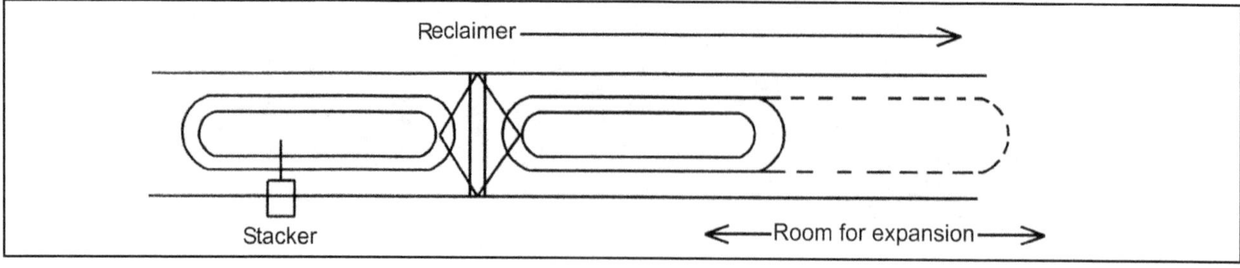

Fig. 2.3.7 Stacker reclaimer system for coal.

Plate 2.3.4 Reclaimer.

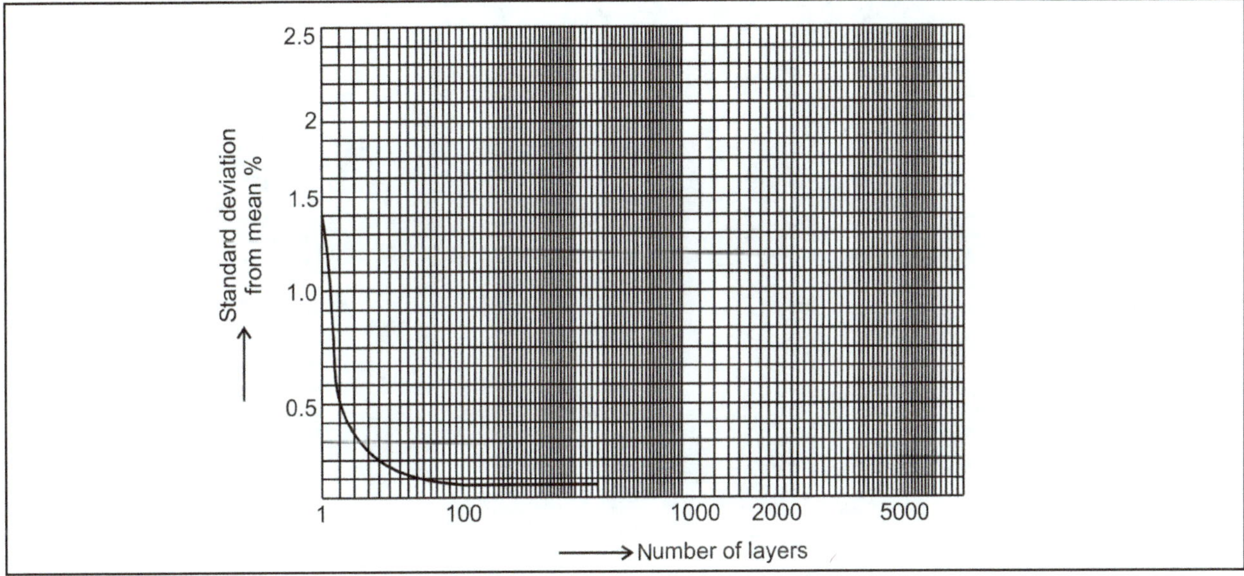

Fig. 2.3.8 Blending effect as a function of number of stacked layers of material.

3.5 Preblending While Stacking

Besides serving as storage and extraction system, this arrangement of stacking and reclaiming serves one more important purpose, that of 'preblending'. Because hoe dislodges material across the total cross section of the pile, it contains material from all the layers of the pile thereby mixing them. This is a great advantage for both limestone and coal.

3.6 Space Required

Stacker reclaimer systems take a considerable amount of space.

Piles can be fully in the open or partially or fully covered. Coal piles will be mostly covered.

3.7 Blending Effect Achieved

Knowing the quantity to be stored and the feeding and extraction rates, period for building up piles and number of layers in forming full pile can be worked out. Extent of preblending that can be achieved can also be influenced. In general degree of blending achieved is 5-6 :1. Blending effect as a function of the number of layers is shown in **Fig. 2.3.8.**

With the help of X-Ray analyser it should be possible to monitor composition of limestone extracted by feeding stone from different sectors of mines to crusher.

3.8 Several Designs of Stacker Reclainer System

There are several designs of the Stacker - Reclaimer systems and they can be 'tailor made' to suit specific needs.

Capacities of system also depend on allowance to be made for expansion. There are many options. Crusher may be so selected that it would work in 3 shifts after expansion.

Alternatively a second crusher could be added during expansion.

Consequently there could be one stacker reclaimer system even after expansion or there could be two.

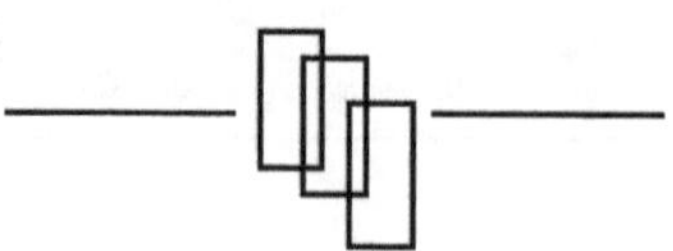

CHAPTER 4

GRINDING MILLS

4.1 Grinding Mills

Grinding mills are used in cement plants to grind :

1. Limestone and additives to produce raw meal.
2. Coal (or other solid fuels) for firing it in kiln and calciner.
3. Clinker and blending materials to produce O.PC. and blended cements like P.P.C. and B.F.S. Cement.

4.2 Classification of Mills

Mills can be classified in several ways. For example:

Wet grinding or dry grinding mill

Open circuit or closed circuit

Ball or tube mill or vertical mill

Air swept or bucket elevator mill

Drying and grinding or only grinding

Autogenous mills

Trunnion bearings or slide shoe bearings

Mill with wrap around motor

Annexe 2.4.1 shows broad types of mills and duties for which they have been commonly used in Cement Industry.

Same type of mill can have different constructional features according to its size and application. For example , ball mills have trunnion bearings on both sides or one trunnion and one slide shoe or both slide shoe bearings. **Annexe 2.4.2** shows broad differences of this type in ball mills and vertical mills.

Mills will be selected according to type most suitable for the capacity and duty required and for overall power consumption.

It would be seen that the most significant types used in grinding operations are:

1. ball and tube mills
2. vertical mills - ball and roller
3. horizontal roller mill
4. roller press with ball mill

Ball mills have been dealt with in **Chapter 4a;** vertical mills in **Chapter 4b** and roller press in **Chapter 4c** Horo mills have also been briefly dealt with.

4.3 Grindability and Power Consumption

There are several theories of communition. Most commonly accepted theory which is relcvant to ball / tube mills is Bond's.

Simply put, this theory states that power input in comminuting process is proportional to the surface generated in the process and the grindability of the material.

Bond developed 'Bond's Work Index' (BWI) to measure grindabilities of various material and also a 'test mill' and a testing procedure. Using this it was possible to work out power required to grind a material from a given feed size to a product of given fineness. Power drawn can be expressed as an exponential function of the diameter of a ball mill.

See Fig. 2.4.1

Table 2.4.1 shows relation between sp.power consumption and grindability (Bond's Work Index) on fineness of material ground (Blaine) on the other in case of ball mills.

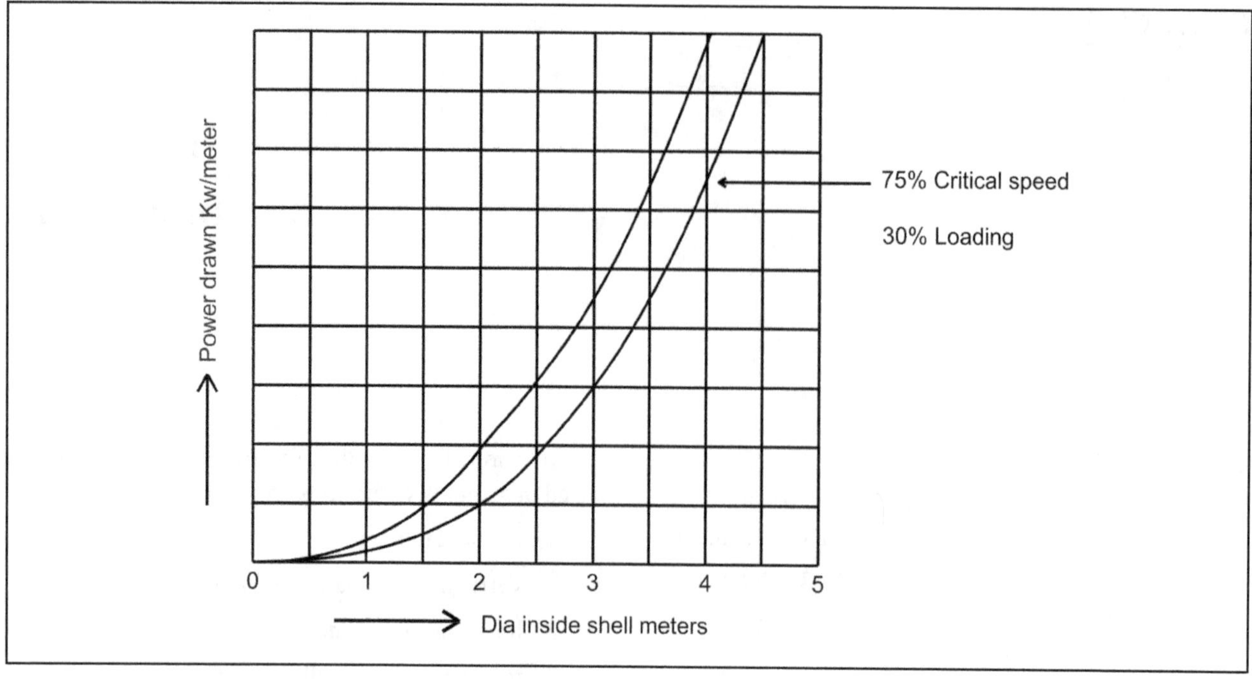

Fig. 2.4.1 Ball Mill Power Chart.

Table 2.4.1 Relative Sp.Power Consumption Of ball and Vertical mills.

1. Grinding OPC

Sp. Power Consumption in kwh /ton

Blaine	3000		3300	
Cm^2/gm	VR Mill	Tube mill	VR Mill	Tube mill
mill	20	33	22.5	36.5
Fan etc.	6	5	6.5	5.5
total	26	38	29	42

2. Grinding Slag

Sp. Power Consumption in kwh/ton

Blaine	4000		4400	
Cm^2/gm	VR Mill	Tube mill	VR Mill	Tube mill
mill	27	52.5	29.5	58
Fan etc.	8	6.5	8.5	7
total	35	59	38	65

4.3.1 Hardgrove Index

A similar index known as 'Hardgrove Index" was developed. The type of test mill developed to establish Hardgrove Index is very similar in action to a vertical ball mill. Hardgrove index is universally used to work out power requirements of vertical coal mills.

Plate 2.4.1 shows relation between HG Index and sp. Power when grinding coal. Finer the product, higher the sp. Power consumption.

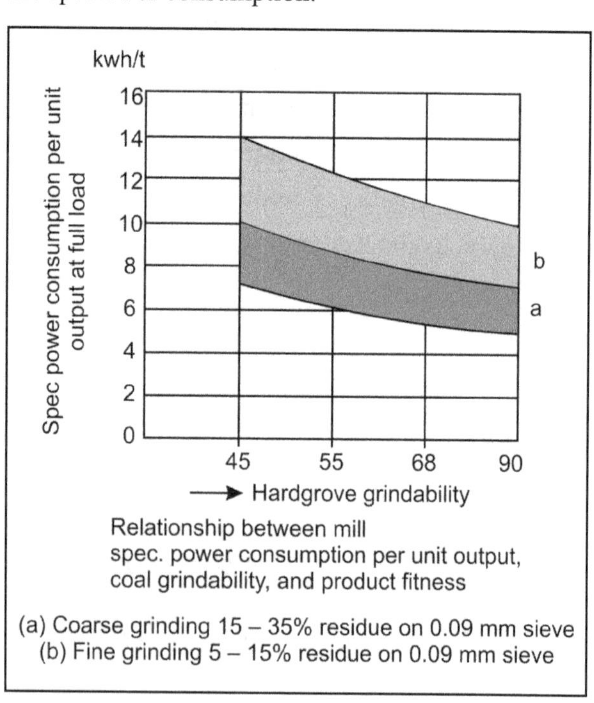

Plate 2.4.1 Sp. power in vertical mill for grinding coals of different Hardgrove Indices to different finenesses.

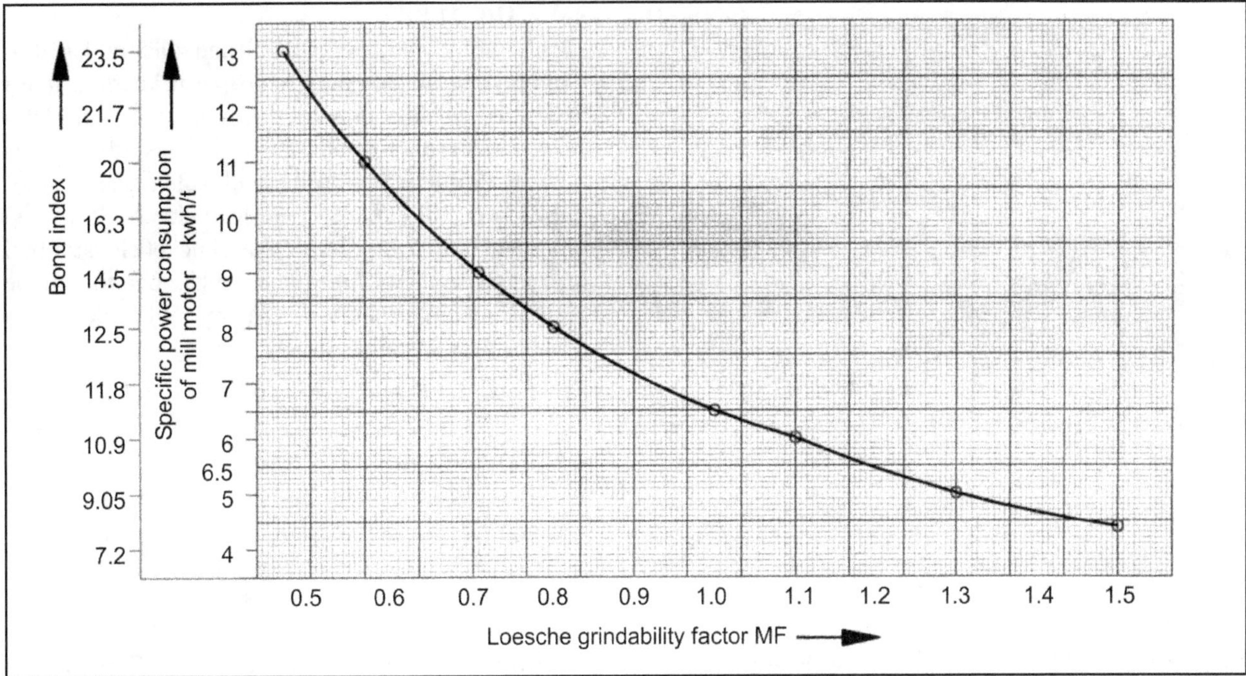

Fig. 2.4.2 Specific power consumption - of Loesche roller mill as a function of Loesche Grindability Index.

Relation between BWI and Hardgrove Index can be described by the following equation.

$$B.W.I. = 435/(H.G.I.)^{0.91}$$

However Hardgrove index is seldom used to size vertical roller mills. Each manufacturer has developed his own index for grindability and hence only he can size a vertical roller mill for any specific application.

See Fig. 2.4.2, which expresses relation between sp. power consumption of the v.r.m. for different values of 'Loesche' grindability indices. This relation is not universal but is specifically for 'Loesche' mills.

An important difference between a ball mill and a v.r.mill is that for a given load, a ball mill draws practically same power regardless of rate of feed. On the other hand in case of a v.r. mill, power drawn is directly proportional to the throughput.

See plate 2.4.2

Unlike ball or tube mills, roller mills are made in standard sizes. Same mill gives widely different outputs depending on grindability of material. Size of mill that comes nearest to the required capacity is selected. Motor installed could be smaller or bigger.

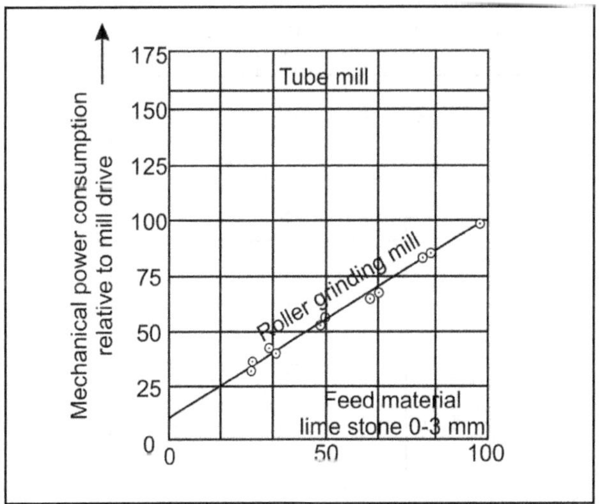

Plate 2.4.2 In a V.R. Mill, actual power consumption varies with thruput.

4.3.2 Grindability and Roller Press

Grindability of the material pressed is improved by one third, when it is passed through a Roller Press.

See plate 2.4.3.

Pressing operation also produces 30 % fines, that is finished product.

Sp. power for ball mill following the Roller Press can be worked out on above basis.

See Fig. 2.4.3 and 2.4.4.

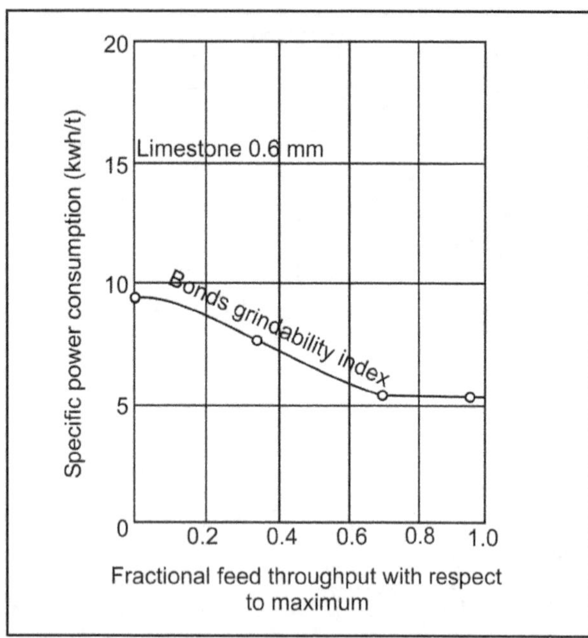

Plate 2.4.3 Roller press reduces sp. power (improves grindability).

Thus a buyer has to depend on the Supplier to size either a v.r.m. or a roller press.

4.4 Horo mill

This is a combination of ball mill and roller mill. Mill is a cylinder turning on horizontal axis in bearings like a ball mill. Instead of charge of grinding media, mill has a cylindrical roller turning inside it. This roller grinds much the same way as roller and table in case of v. r. mill. Horo mill can also take feed size of 75 to 100 mms. Mill has a gas circuit like a bucket elevator ball mill when drying is done in the mill. Separator will be high efficiency separator as in any mill system. The mill has been introduced only of late.

See Figs. 2.4.5 and 2.4.6 and plate 2.4.4

4.5 Drying and Grinding

When grinding raw materials and coal, it is also necessary to dry moisture in them. Normally this is done by passing through the grinding system hot gases from preheater. However when starting from cold, such gases are not available. Therefore Hot gas Generators are provided in Raw Mill and Coal Mill Departments to provide hot gases of requisite capacity and temperature. As they are used only occasionally, they would be oil fired.

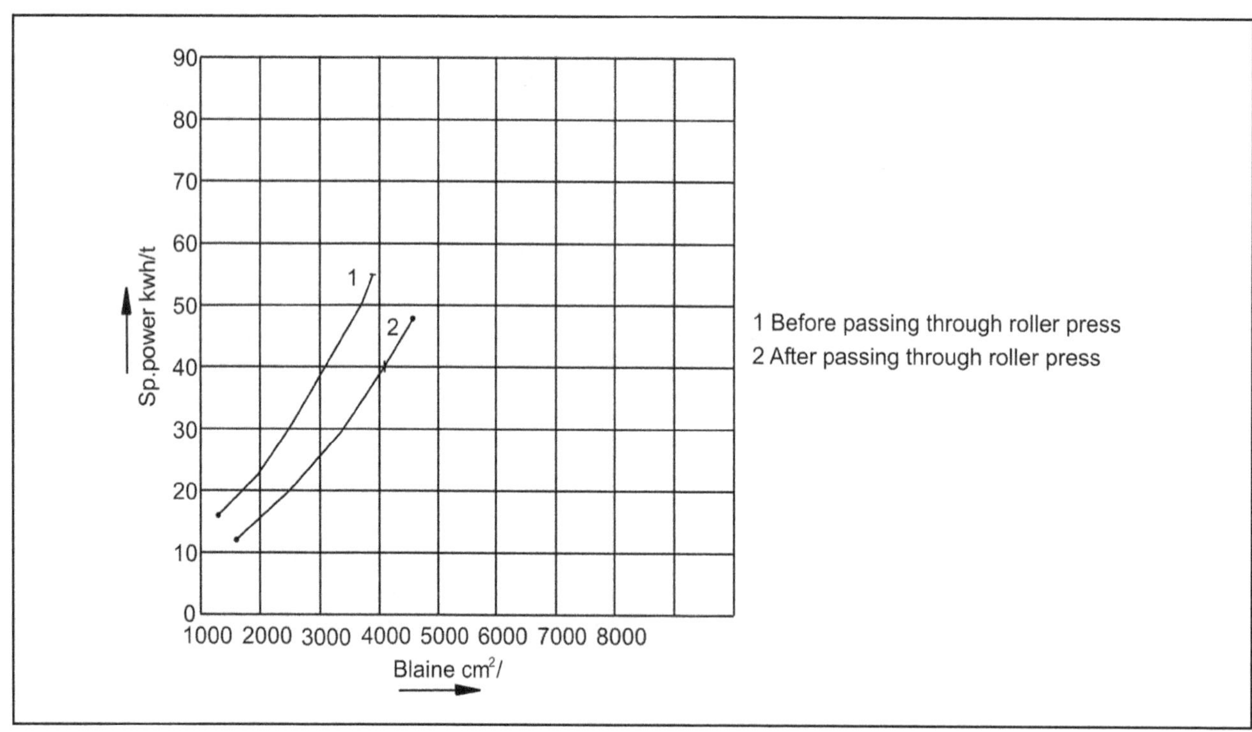

Fig. 2.4.3 Roller press Sp. power consumption before and after roller press.

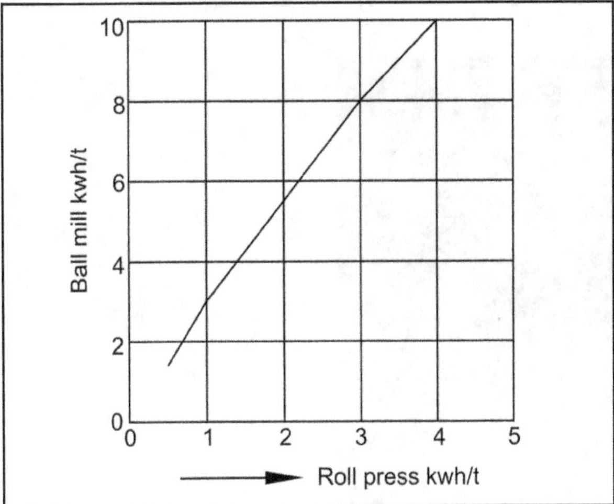

Fig. 2.4.4 Power consumption roll press v/s ball mill.

In case of coal mills, there is a risk of fire and or explosion if hot air is used in the system. Therefore to protect the machinery, provision is made to inject inert gas in the system or to introduce CO_2 from a battery of CO_2 cylinders.

Alternatively, coal mills are installed at preheater end and inert preheater gases are used for drying coal. This is more common now.

Plates 2.4.5 and 2.4.6 shows Hot Gas Generator and Inert Gas Generator respectively.

When Coal mill is located at Preheater end and inert kiln exhaust gases ae used for drying it is not necessary to provide for inert hot gas generator.

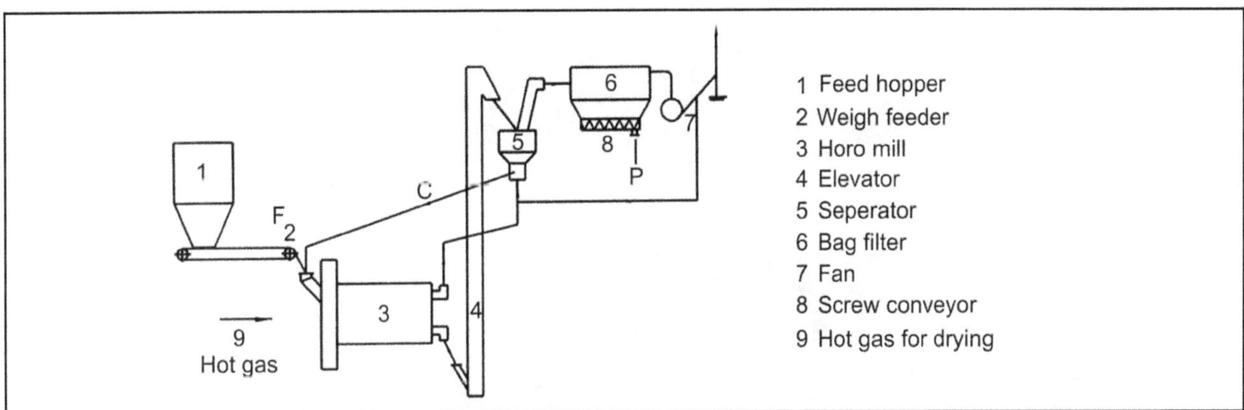

Fig. 2.4.5 Grinding circuit of horo mill.

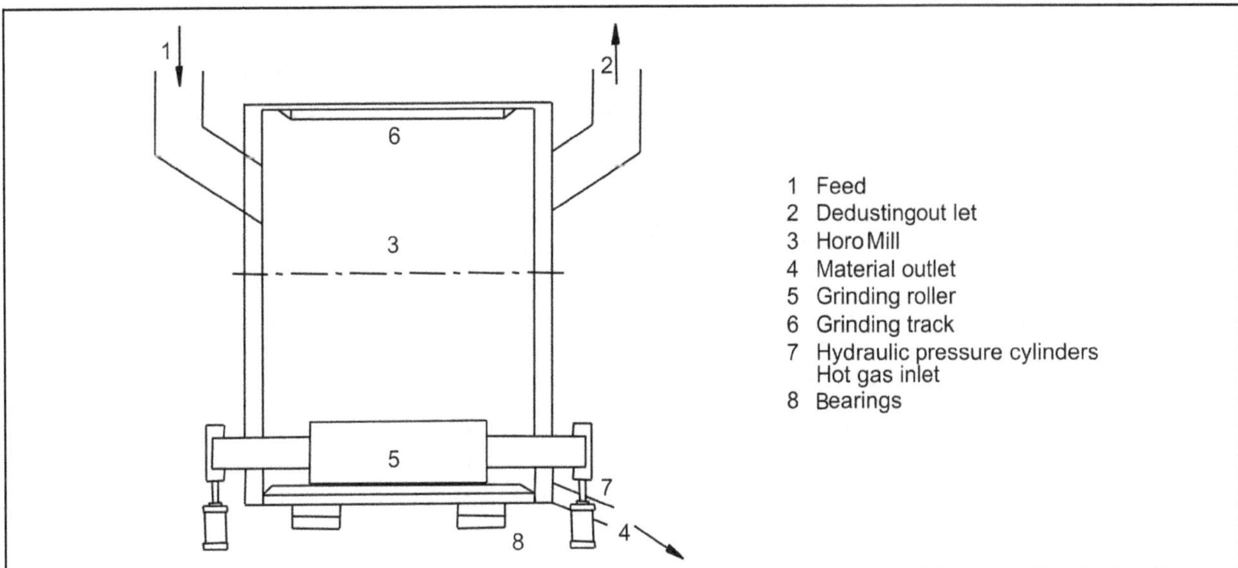

Fig. 2.4.6 Horo mill – sectional view.

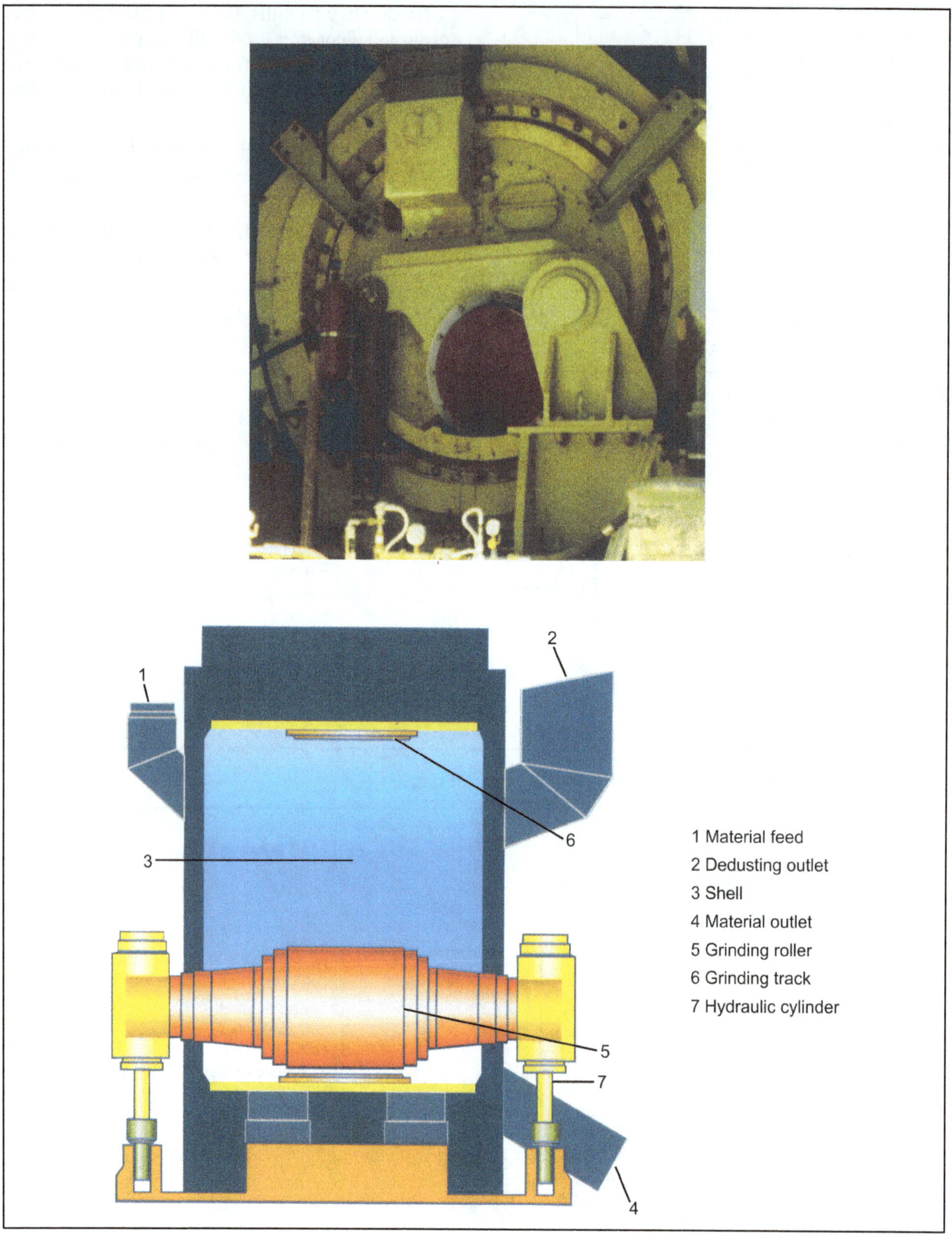

Plate 2.4.4 Horo mill and its section.

Plate 2.4.5 Hot gas generator.

Plate 2.4.6 Inert gas generator.

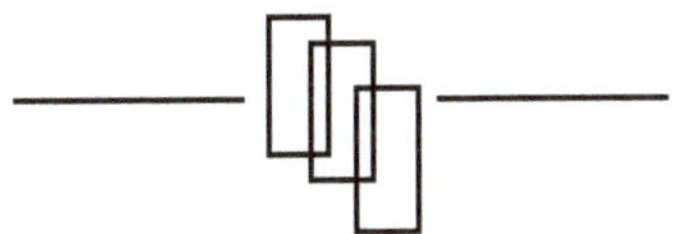

Annexe 2.4.1 Application of various types of mills in a cement plant.

Sr.No.	Application	mill				
		ball	**vertical**	**Roller press & ball mill**	**Roller press**	
	grinding					
	primary	✓	✓		✓	
	finish	✓	✓	✓	✓	
	process					
	Wet grinding Open/ closed circuit	✓				
	Dry grinding					
	open	✓				
	closed	✓	✓	✓	✓	
	Drying& grinding	✓	✓	✓		
	Materials to Be ground					
	Raw materials	✓	✓	✓	✓	
	coal	✓	✓			
	clinker	✓	✓	✓	✓	
	slag	✓	✓	✓	✓	
	pozzolana	✓				
	Type of mill					
	Air swept	✓	✓			
	Bucket elevator	✓		✓		
	External circuit		✓			
	Type of separator					
	grit	✓	✓			
	Mechanical	✓	✓	✓	✓	
	High efficiency	✓	✓	✓	✓	
	disagglomerator			✓	✓	

Annexe 2.4.2 Broad constructional differences in ball mills and vertical mills.

1 Ball Mills

Sr.No.	item	options	
		1	2
1	shell	End discharge	Central discharge
	length	short	long
2	flange	external	internal
3	mill heads	Cast steel	fabricated
4	trunnion	Integral	bolted
5	bearings	Trunnion, 1 or 2	Slide shoe, 1or 2
6	girth gear	Spur	helical
7	drive	Single	dual
8		Side	Central No girth gear
9	Motor	Normal with gear box	Wrap around No gearbox
10	diaphragms	double	Double with flow control

2 Verical mills

Sr.No.	item	Ring ball			roller		
		options			options		
		1	2	3	1	2	3
1	Rollers shape	balls			cylindrical	taper	Shaped
2	numbers	To suit size of mill			2	3	4
3					single	double	Master/ slave
4	Table	grooved			plain	profiled	
5	Liners rollers				Single piece	sections	
6	Throat gap	fixed			fixed	adjustable	
7	separator	grit			static	Whizzer	High efficiency
8	maintenance	Top ring to be removed			Rollers swing out	Rollers turn around	Crane in mill
9	drive	Single/multiple			Single/ multiple		

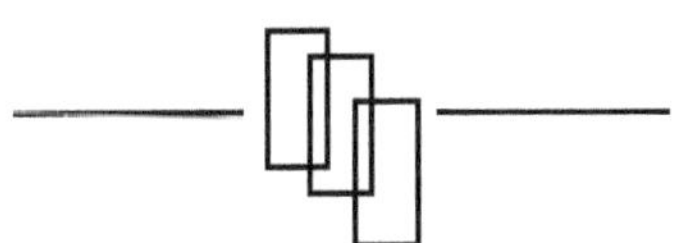

CHAPTER 4a

BALL MILLS

4a.1 Ball Mills

Ball and tube mills have been the 'work horse' of the cement plants all these years. They are truly versatile in that they have been suitable for all applications and have grown and integrated in their circuits new auxiliaries like high efficiency separators. However as cement plants grew bigger and bigger and power became costlier, alternatives had to be found for mills of high capacity and low power consumption. In this ball mills have lost to vertical mills.

4a.2 Wet Grinding

When wet process was used to make cement, limestone and correcting materials were ground wet by adding about 36 % water to it. Ground raw meal came out as a slurry. Mostly ball mills were used for this purpose. Ball mill is a cylinder turning on two trunnions in white metal bearings. Cylinder is lined with lining plates to protect it from wear. In case of very large mills either one or both the trunnions (with mill heads) will be replaced by slide shoe bearings.

See plate 2.4a.1 for general construction of ball mills.and **plate 2.4a.2** for mill with slide shoe bearings.

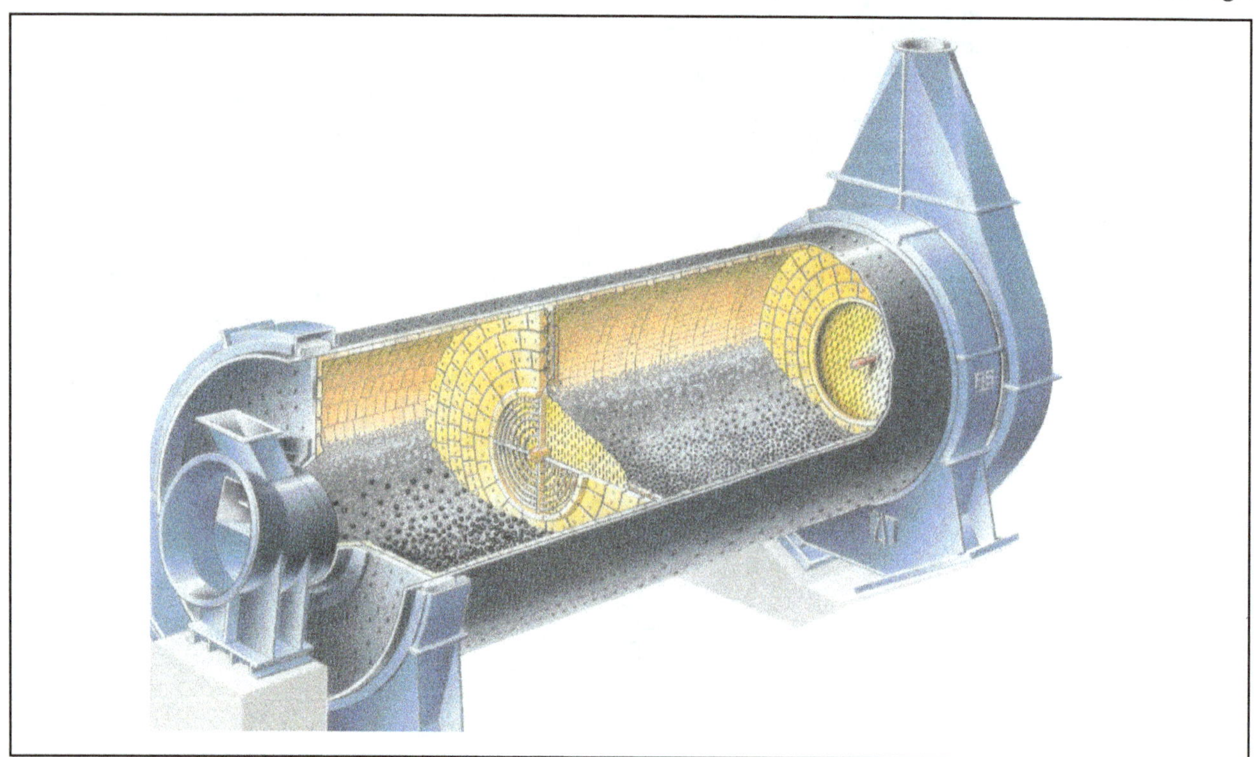

Plate 2.4a.1 Sectional view of a Ball Mill.

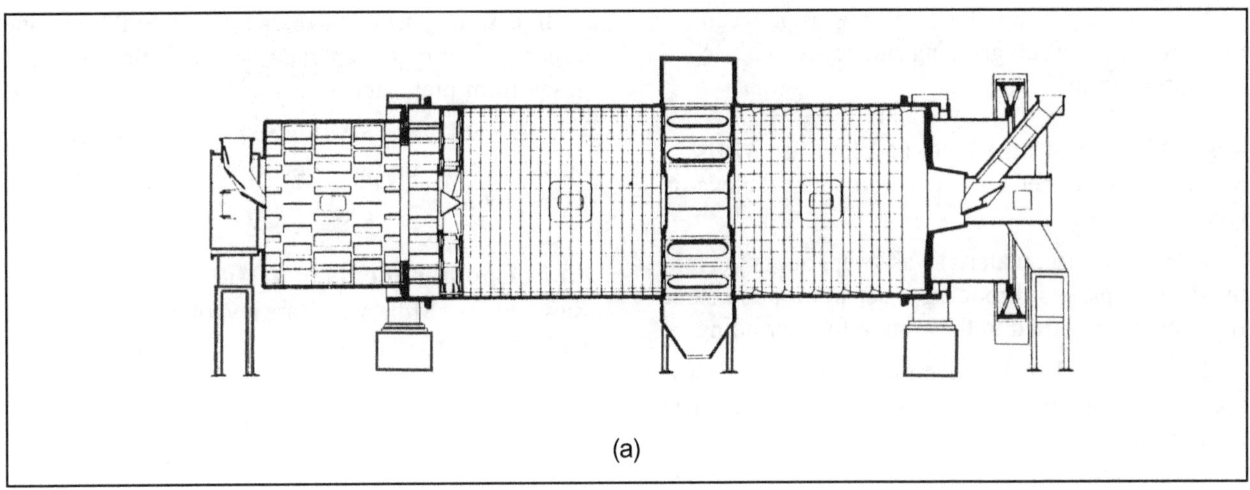

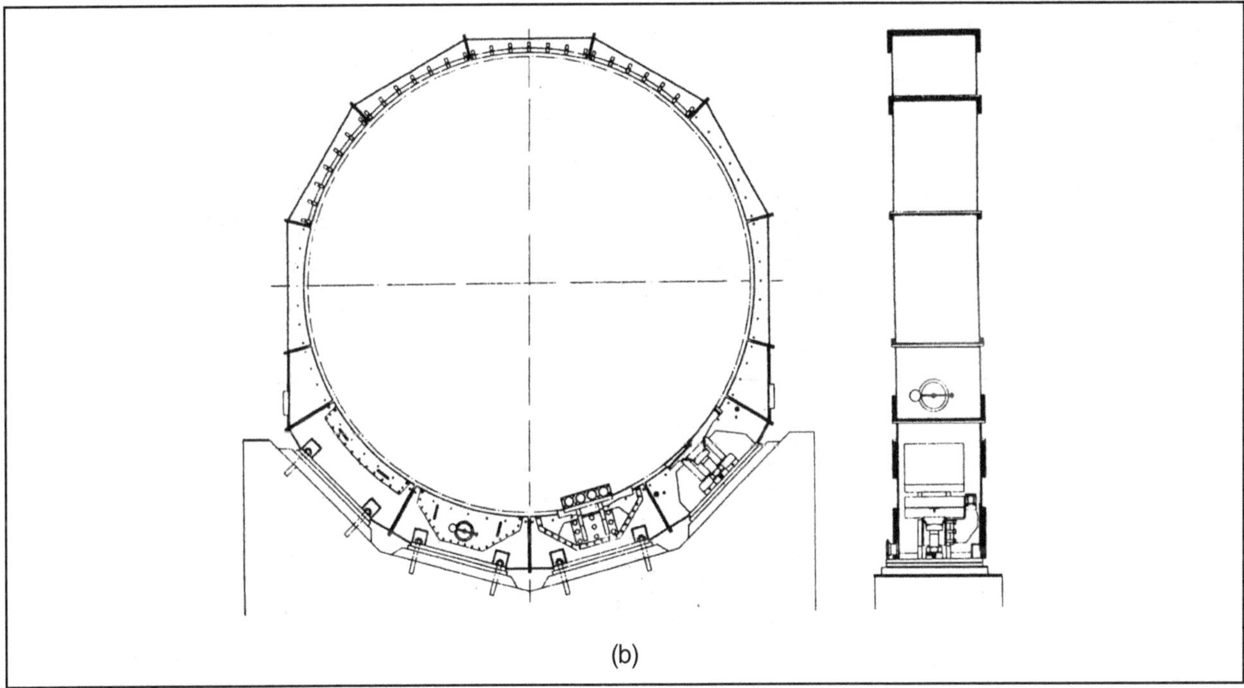

Plate 2.4a.2 (a) Central discharge and external drying chamber. (b) Slide shoe bearings.

Wet grinding was mostly in open circuit but sometimes also in closed circuit in which wet classifiers returned coarse fraction to mill for regrinding.

Mill was divided into 2 or more compartments. Grinding media was steel balls and cylpebs (truncated taper cylinders).

Wear rates were high as slurry had to be ground fine and acted like a grinding paste.

In wet grinding rubber lining plates could also be used with advantage. They have better wearing properties and are much lighter in weight.

4a.3 Dry Grinding

When dry process came to be adopted, it became necessary to grind dry, limestone and correcting materials.

Same ball mills were used. Dry grinding was already being done in case of coal and clinker in ball and tube mills respectively,

Length of mill depended on the application. For raw materials grinding, which was almost always in closed circuit, ratio of shell length to its diameter was between 1.5 and 2.

For cement grinding the l/d ratio was between 3.5 to 4.5 for open circuit grinding and between 2.5 to 3.5 for closed circuit.

4.a.4 Air Swept and Bucket Elevator Mills

Mills could be either 'air swept' or of 'bucket elevator' type.

In the first case, material was swept out of the mill by air /gas, passed through grit separator and coarse fraction was returned to the system for regrinding.

In case of bucket elevator mills, mill discharge was taken by conveyors to a separator. Coarse was returned to the mill.

4a.5 Drying and Grinding

In case of grinding raw materials, it was necessary to dry the materials so that product was dry ground raw meal with less than 0.5 % moisture. In case of air swept mills drying was done in mill by passing hot gases from preheater through the mill. Depending on quantum to be dried, mill would have a drying chamber or an external overhung drying drum attached to the inlet trunnion. Drying chamber or drum would have lifters in it to throw material in the path of the gases.

See plate 2.4a.2.

In case of bucket elevator mills it could be done either in mill or in separator by using the same hot gases from preheater.

Presently when mostly vertical mills are used for large capacities drying is done in mills taking hot gases from preheater.

4a.6 Compartments in a Mill

Raw and Coal Mills, which are invariably closed circuit mills, have two compartments filled with grinding balls. Compartments are divided by a double diaphragm.

Cement mills used to be open circuit. They have three compartments. When in closed circuit, they have only two compartments.

4a.6.1 Double Diaphragm with Flow Control

To improve performance further, double diaphragms with flow control have been developed. In ordinary double diaphragms, level of material dropped sharply near the diaphragm; it is maintained in case of flow control diaphragms thereby improving outputs.

See Figs. 2.4a.1 and 2.4a.2.

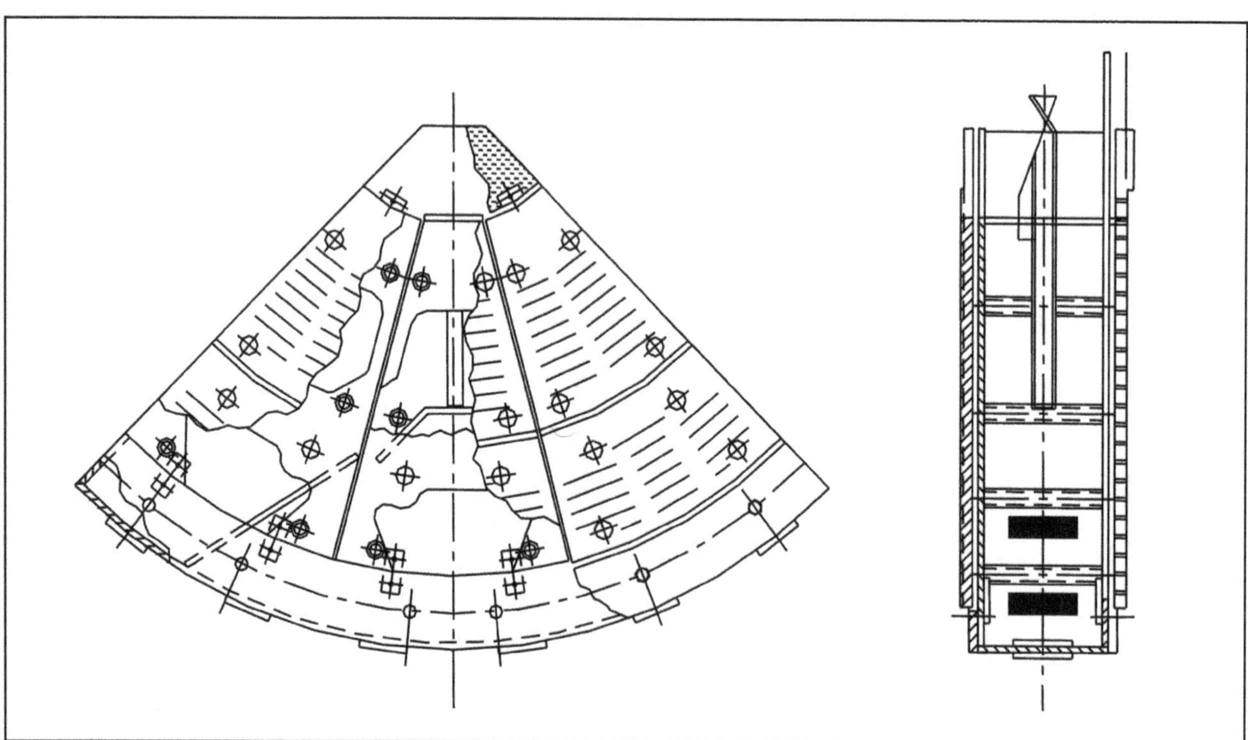

Fig. 2.4a.1 Flow control diaphragm.

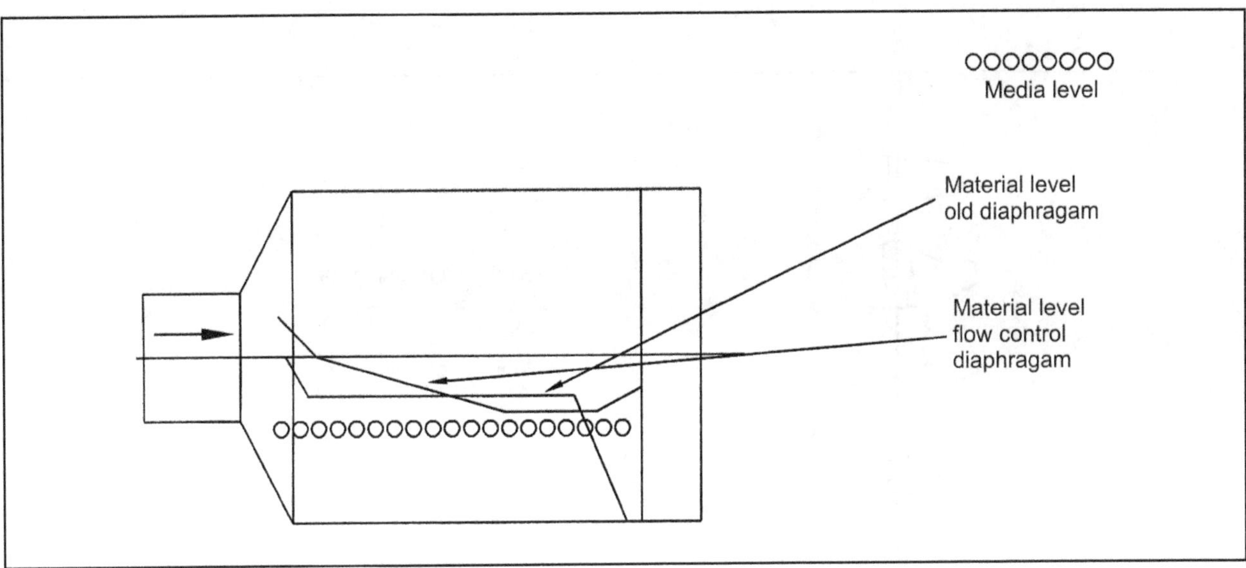

Fig. 2.4a.2 Flow control diaphragm retains material longer.

4a.7 Charge of Grinding Media

Grinding balls in a compartment arc not of one size but contain balls of 3 to 4 sizes in different proportions. About 27 to 35 % volume of mill is filled with grinding media. Heavier and larger balls are towards feed end where material is fed. Sizes reduce progressively as material becomes finer. 'Equilibrium charge' is that charge where compensation for wear can be done by balls of one size only usually the largest size in the compartment.

Grinding media could be made of forged steel, cast steel or even cast iron. Cylpebs were mostly used in the last compartment and were generally made of c.i. They are seldom used now. Presently grinding media used are high chrome steel balls.

4a.8 Lining Plates

Shell is fitted with lining plates made of manganese steel. Now high chrome liners are used which last much longer. Lining plates in the first compartment were 'lifting' type. Grinding action in first compartment was largely by impact. Second compartment was generally longer. Lining plates in this compartment were of the 'classifying' type in that they prevented movement of media along the length, kept larger sizes near first diaphragm and finer towards discharge end. Design of lining plates has considerable influence on grinding efficiency and hence on specific power consumption. For small mills wedged lining plates were used.
See Figs. 2.4a.3 to 2.4a.5.

For small mills . wedged lining plartes were used. See **Figs. 2.4a.6 and 2.4a.7.**

To provide more surface for grinding by attrition, grooved liners have also been developed.
See Fig. 2.4a.8.

4a.9 Power Consumption in Grinding

Almost 80 % of electrical power consumption is in grinding. Typical power consumption figures are shown in **Table 2.4a.1.**

4a.10 Various Designs of Mills

Different machinery manufacturers have developed their own designs for components of the mill like shell, mill heads, bearings, girth gears, diaphragms and lining plates.

Mention has been made of 'slide shoe bearings' in which a tyre or riding ring is mounted on mill shell. This is supported on a pair of slide shoes and rotates on them much like a kiln tyre. Mill head is thus not required to support the weight of the shell and hence its design is much simplified. Slide shoe bearings may be at one end only or at both ends.
See plate 2.4a.2.

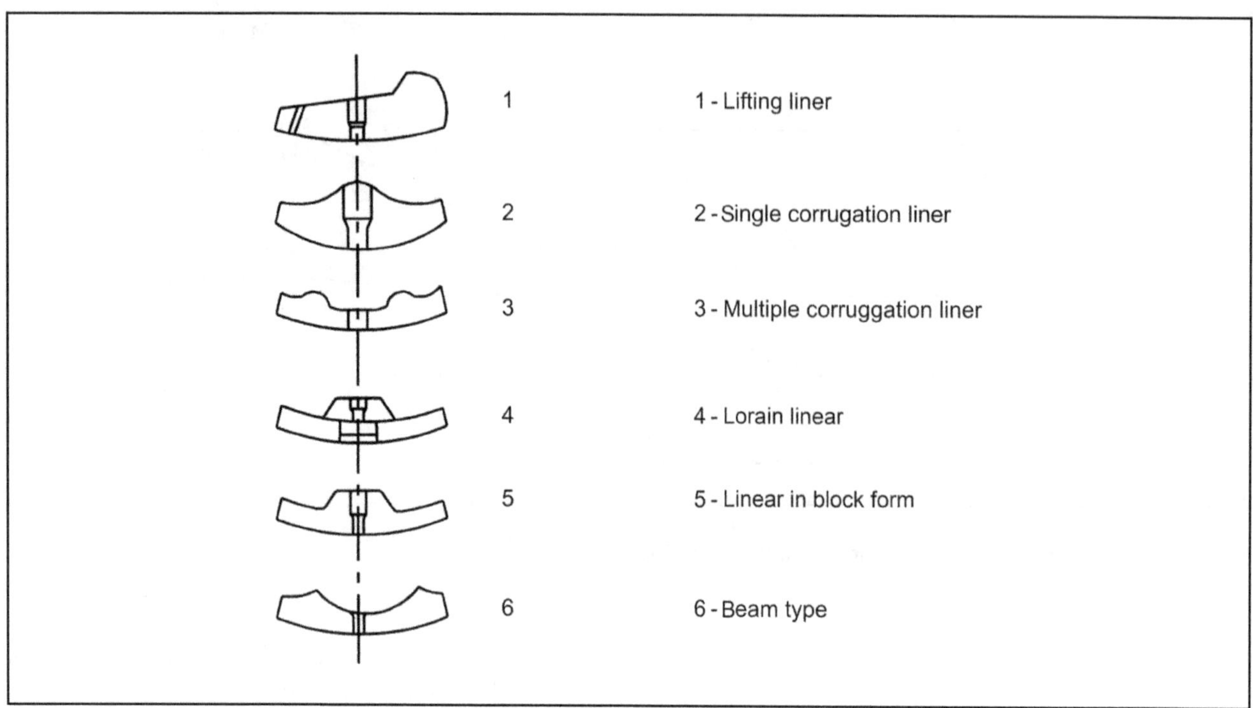

Fig. 2.4a.3 Different designs of lining plates for 1st compartment of ball/tube mills.

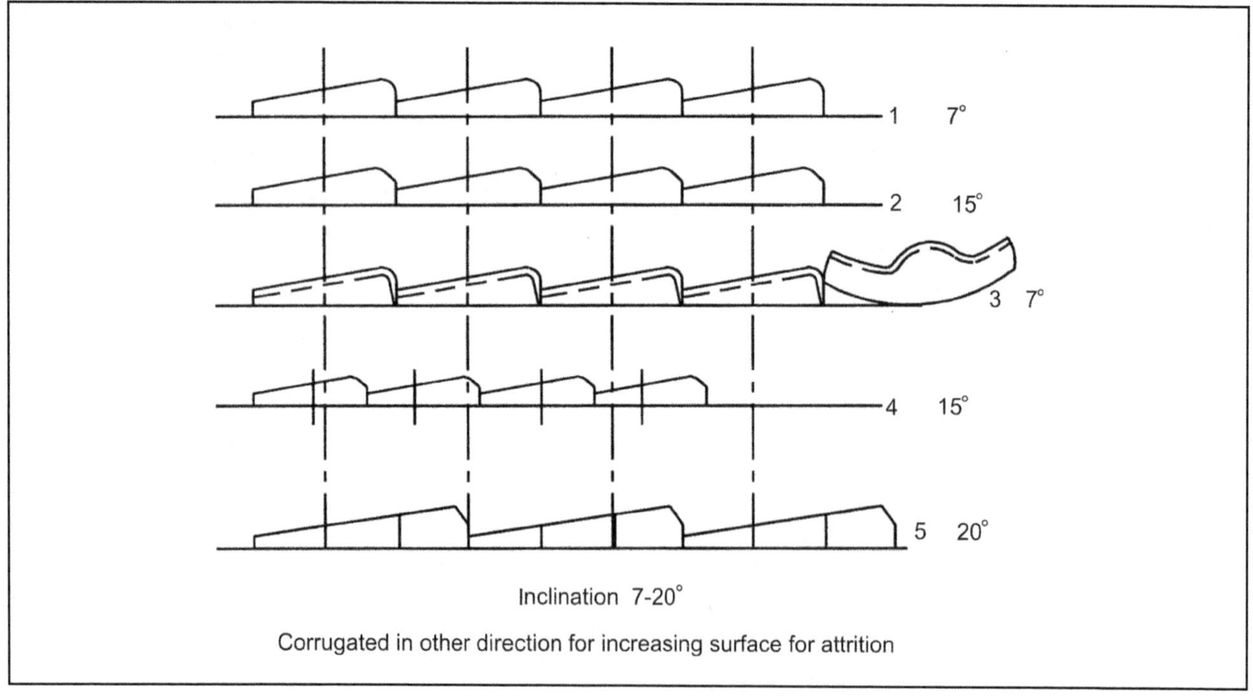

Fig. 2.4a.4 Different designs of lining plates classifying type for 2nd compartment of ball/tube mills.

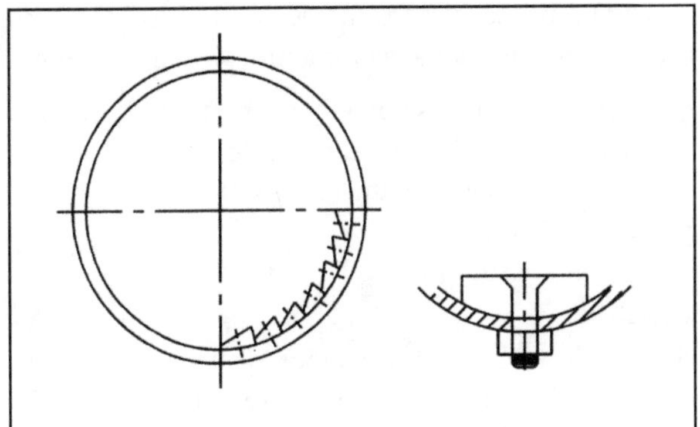

Fig. 2.4a.5 Bolted lining plates.

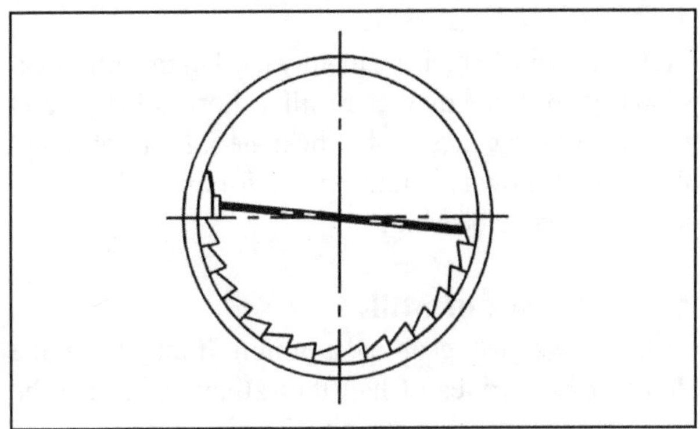

Fig. 2.4a.6 Wedged lining plates.

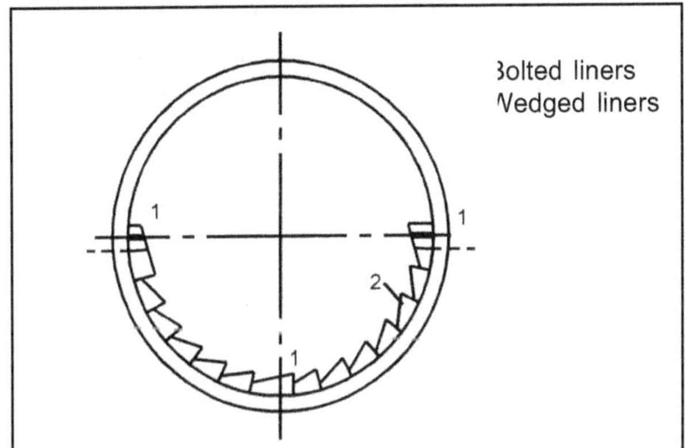

Fig. 2.4a.7 Partially wedged and bolted liners.

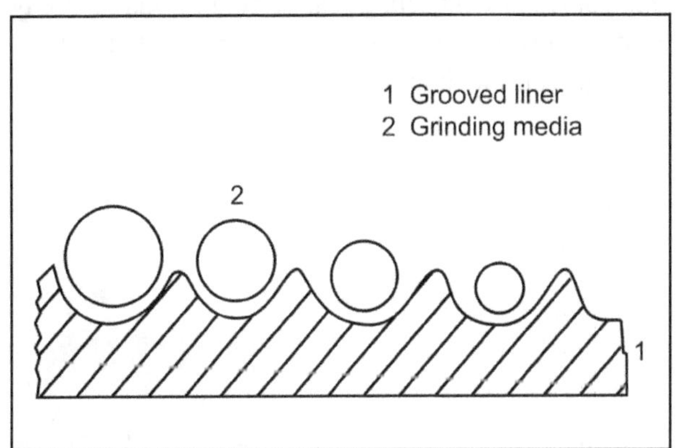

Fig. 2.4a.8 Grooved liners for ball mills.

Table 2.4a.1

Ball mills				
Sr.No.		**Raw materials**	**Coal**	**Cement**
		Kwh/ton material		
1	Wet grinding	12		
2	Dry grinding	16	24	
		Kwh/ton clinker		
3	Wet grinding	19		
4	Dry grinding	25	For wet process 10 For dry Process 6	
		Kwh/ton cement		
	Wet grinding	18		
	Dry grinding	23	For wet process 9.6 for dry process 5.75	For open circuit 35 for closed circuit 30

Note: Based on coal of calorific value 4500 kcal/kg and sp. fuel consumption of 1500 kcal.kg for wet process and 700 kcal.kg for dry process. Power consumption for cement is for OPC.

4a.11 Aero Fall Mills

'Aero fall' mills which are mainly used in metallurgical industry are large diameter mills, short in length and mostly use large pieces of same material for grinding - this is also known as 'autogenous grinding'. They will have slide shoe bearings.

4a.12 Drives of Mills

Mills have a girth gear and a pinion. It may have one drive or two drives of half the ratings. Gear can be spur or helical. There is a gear box in between motor and pinion shaft. Some times the girth gear and pinion are avoided by connecting gear box output shaft directly to mill. This is known as 'central drive' and requires a large gearbox with high ratio of reduction.

Dual drive has the advantage that width of the girth gear remains the same as for a mill with half rating.

In very large mills not only girth gear but gear box is also eliminated by installing a slow variable speed 'wrap around' motor on the mill itself.
See plate 2.4a.3.

4a.13

Ball mills work with all types of separators- grit, mechanical and high efficiency. Presently mill circuits will incorporate high efficiency separators. More on separators in **Chapter 5**.

Developments for ball mills have more or less been put on shelf after Vertical mills have come into use for grinding all the three materials - limestone, coal, and clinker for large capacities.

Plate 2.4a.3 Wrap around or ring motor.

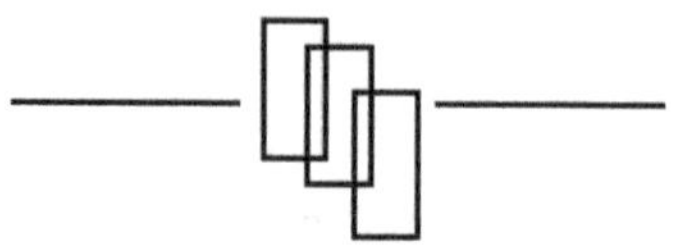

CHAPTER 4b

VERTICAL MILLS

4b.1 Vertical Mills

Ball and tube mills turn on a horizontal axis. Vertical mills turn on a vertical axis. There are two variants of the vertical mills- ring ball mills and vertical roller mills.

4b.2 E Mills

'ring ball mills' better known as 'E' mills were developed by Claudius Peters. The mill is like a thrust ball bearing. Cast steel or nihard balls turn in grooves of a pair of rings. Bottom ring is mounted on table bolted to the gear box. Top ring is fixed and presses on the balls by means of springs or hydraulic cylinders. Balls turn in groove and also on their axes. Material to be ground is admitted by a chute on the center of bottom ring and is forced by centrifugal forces into groove to be ground under the balls. Hot gas /air is admitted through an annular ring at high velocity. It lifts partially ground material as it passes through the grit separator above the grinding table. Coarser fractions fall back for regrinding before gases enter the separator. Coarse fraction from separator also falls on to grinding ring for regrinding. Control of fineness is by changing settings of vanes of grit separators.

There are no bearings inside the mill. Hence maintenance is minimum. When balls wear down one extra ball can be accommodated in the ring. Rings wear out and are replaced. There are no lining plates to be replaced.

See Fig. A, Plate 2.4b.1.

Plate 2.4b.5 shows sizes and capacities of E Mills when grinding coal

4b.2.1 E Mills in Thermal Power Stations

'E mills' have been universally used in thermal power stations to grind coal to be fired in boilers. A bank of mills supplies ground coal to a boiler; firing is 'direct' in that ground coal goes straight to burners without intermediate storage.

E mills were not developed seriously for grinding other materials like limestone and hence size and capacity wise their development was stunted. They have been used to grind coal in cement plants. When so used, 'indirect' firing system is adopted to reduce wear on primary air fan and in this application can compete with vertical roller mills.

Main advantage of E mills is that they do not have any liners to be replaced or any rotating parts that need lubrication inside the mill.

They work under positive as well as negative pressure. If under positive pressure, like when they are used to fire coal in boilers, a seal air fan is provided to protect the gearbox.

New designs have built in dynamic separator in the mill housing.

4b.3 Vertical Roller Mills

The second variant is Vertical Roller mill . Principle of operation is the same. 2 , 3 or 4 rollers turning on their axles press on a turning table mounted on the yoke of a gear box. Pressure is exerted hydraulically. This mill has a built in separator above the rollers. It also has an annular ring which admits air at velocity as high as 70 m /sec. Material is admitted through a feed chute on the center of the table and passes under rollers by centrifugal action. Material gets partially ground and as it falls over the edge of the table, it is picked up by the air /gas. Gas and material circuit in mill and in separator is similar to E mill.

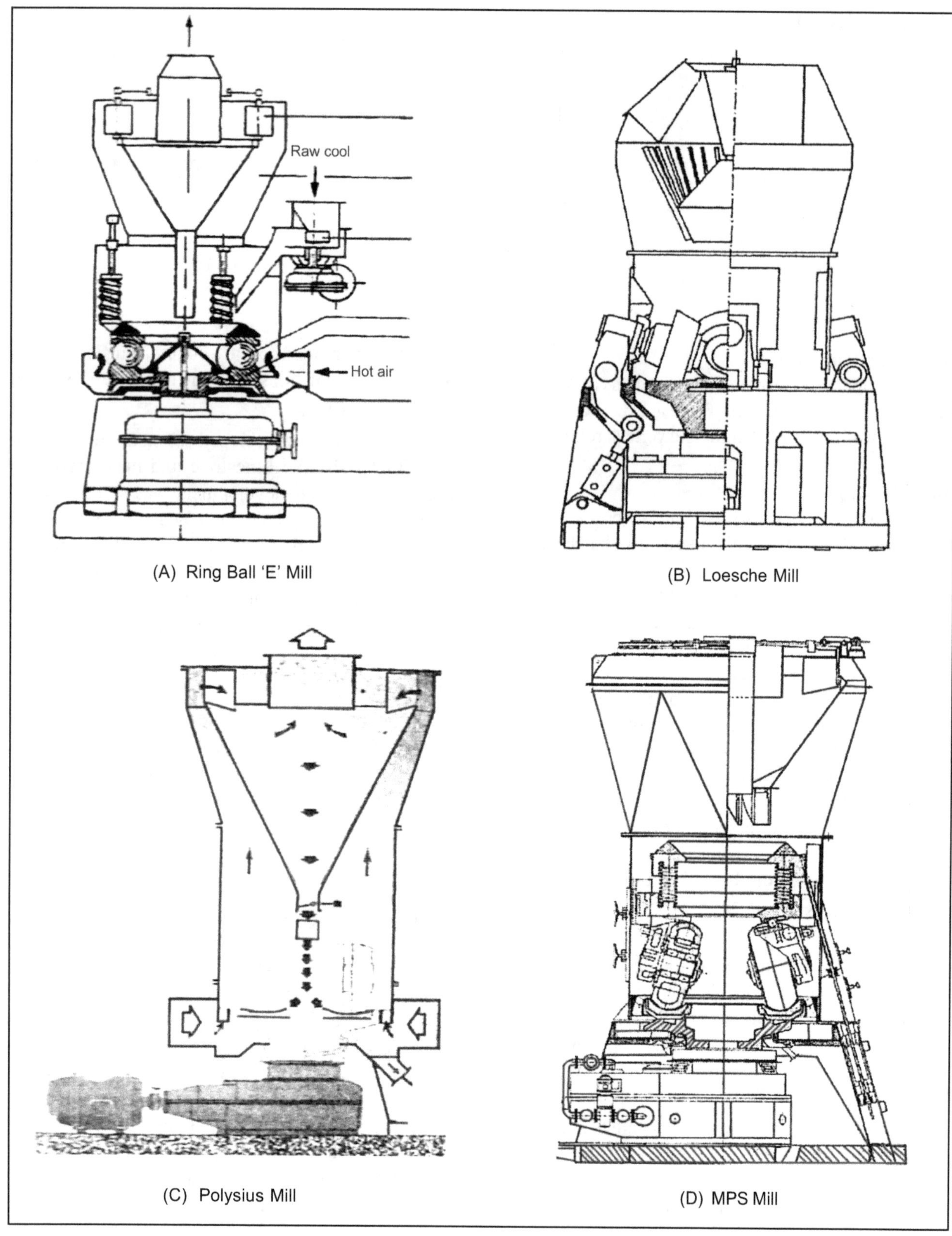

Plate 2.4b.1 Some well known vertical mills.

Separators are either grit or stationary separators or 'whizzers'. When rotating, separators have variable speed drives to change speed of rotor to control fineness of product. Cross sectional views of more popular v.r.ms have been shown in **Plate 2.4b.1 items B to D and Plate 2.4b.2.**

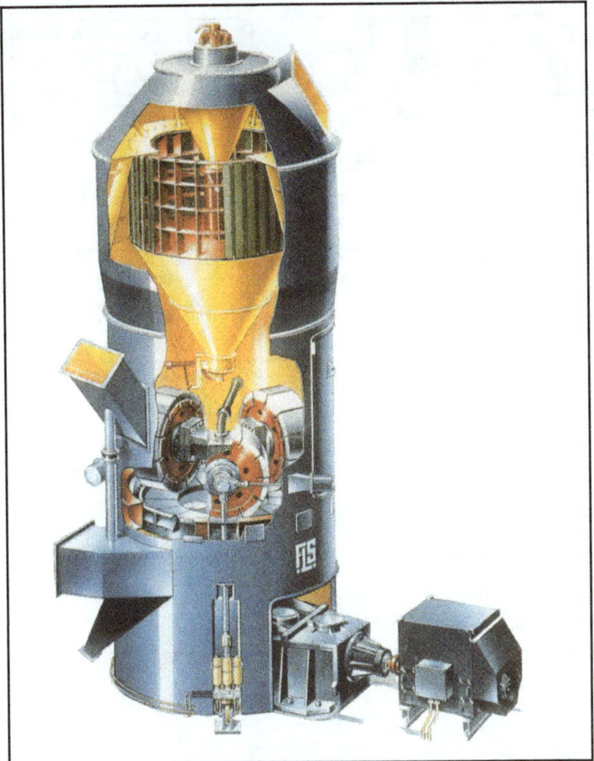

Plate 2.4b.2 Atox vertical roller mill.

Finished product is drawn out of the mill by a fan. Gases leaving mill and carrying the product pass either through a bank of cyclones first and then enter the dust collector or go to the dust collector directly.

Table is lined with replaceable liners of wear resisting material like manganese steel. Rollers are also fitted with replaceable liners which can be in one piece or in segments.

4b.3.1 Many Designs of VR Mills

There are a great many designs of rollers and table developed by different manufacturers. A Comparison of salient features is shown in **Fig. 2.4b.1.**

4b.4 Salient Features of Vertical Mills

These mills are dry grinding and operate in closed circuit. They are air swept mills. Because large quantities of air are admitted to mill for air sweeping, mill has a higher drying capacity. Hence mills are used as drying and grinding mills for raw materials and coal.

Because separator is part of the mill itself, number of auxiliaries are less. Floor space taken by a vertical mill is much less than a closed circuit ball mill of similar capacity.

See Fig. 2.4b.2

As in case of E mills, Rollers exert a direct vertical force on table which is transmitted to the gearbox. Hence gear box for a vertical mill is of special design to be able to withstand the thrust load; also input shaft is horizontal.

In case of roller mills , rollers turn on shafts inside a very dusty and hostile surroundings. Bearings are to be specially designed and lubricated with forced lubrication systems. Pressurised seals protect bearings from dust.

4b.4.1 Pressure Drop in VR Mills

Because of high velocity of gases at annular ring and high circulating loads inside it, pressure drop in mill and classifier is high ranging between 600 to 800 mms wg. Fan power in mill circuit is therefore high sometimes as high as mill itself. This disadvantage has been overcome by installing mills with external circuit. **See Plate 2.4b.3.**

4b.4.2 Power Consumption in VR Mills

Vertical mills are more efficient in grinding for two reasons

1. There is no unbalanced mass like that in ball mills to be lifted.
2. Grinding action in ball mills is kind of hit and miss. Grinding action in v. r. mills is more positive.

Compared to ball mills therefore v.r. mill itself will consume about 50 % less power in grinding the same material.

Total power consumption for grinding circuit as a whole is also less in spite of higher fan power.

4b.4.3 Feed Size

Because of large diameter of rollers, feed size can be as large as 75 to 100 mm. Size of stone that can be 'pinched' by a roller and drawn under it for grinding is roughly 5 % of its diameter.

Type	Roller	Roller	Roller	Roller	Ring ball
make	Loesche	Polysius	Pfiffer(mps)	FLS	Babcock/Peters
Shape of table					
Liners of table	Segmental	in sections	Segmental	Segmental	
Shape of rollers					
Liners of rollers	Single piece	Single piece	Segmental	Segmental	
Nos of rollers	2,3, or 4	2 pairs of rollers	3@ 120° spacing	3@ 120° spacing	6 to 11 balls
Lubrication	Required for antifriction bearings of rollers		Required for antifriction bearings of rollers		Not required
Application pressure or load	Hydro-pneumatic individual to each roller	Hydro-pneumatic individual to each pair	Hydro-pneumatic to the group of rollers	Hydraulic to the group of rollers	Hydraulic
Materials for liners	Generally Ni - hard for wearing liners				Hollow cast steel or alloy iron balls
Type of classifier	rotating variable speed	stationary - adjustable * vanes	rotating variable speed	stationary *	stationary - adjustable vanes
Pressure drop across the mill	Approx.	800 mm WG.	For raw materials	Approx. 600 mm wg for coal	

* rotating high efficiency classifier on new designs

Fig. 2.4b.1 Vertical grinding mills.

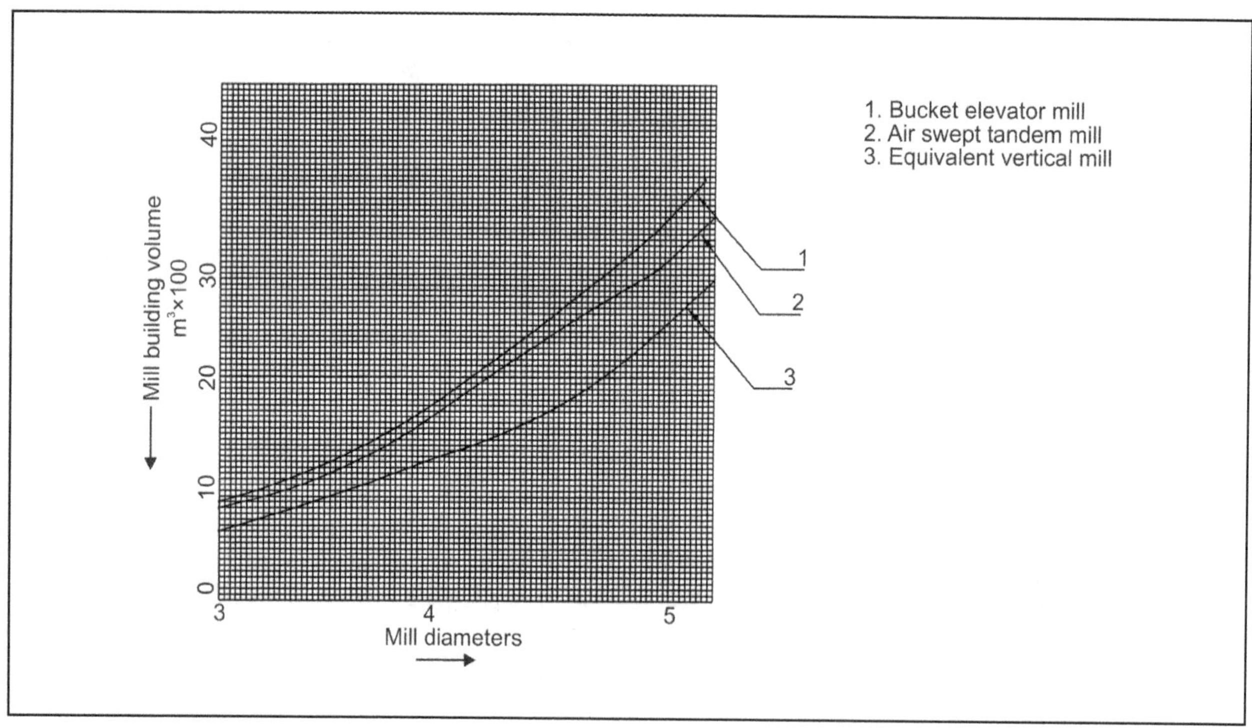

Fig. 2.4b.2 Building volume required for equivalent output for different mill systems.

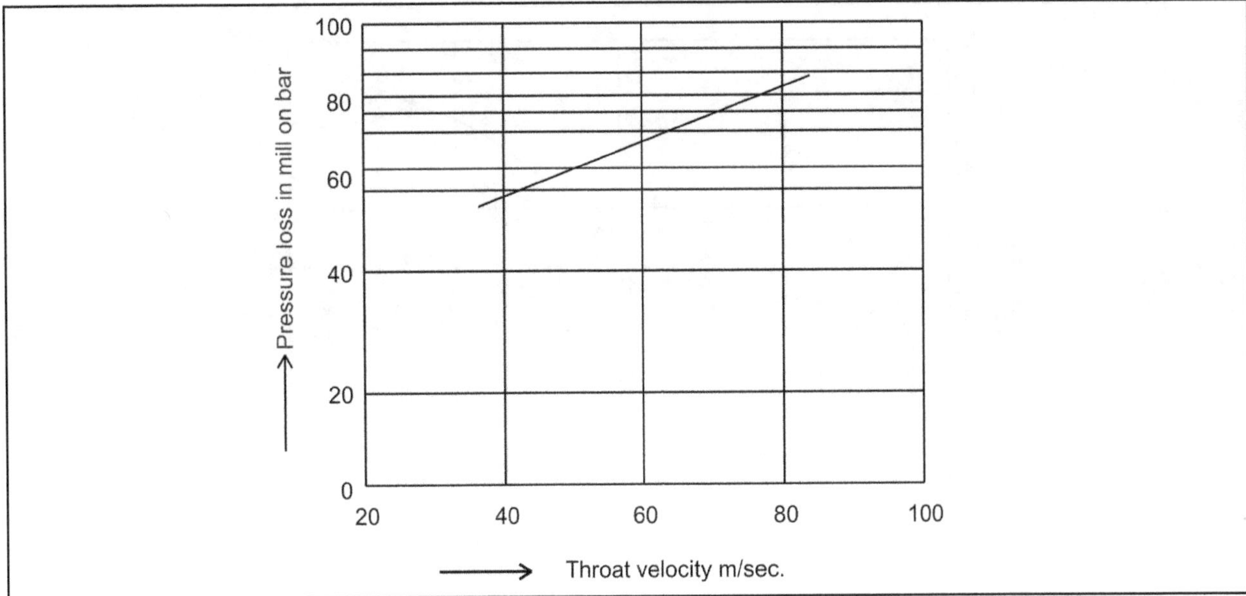

Fig. 2.4b.3 Vertical Roller Mill throat velocity & pressure drop.

4b.4.4 Vibrations and Wear

Vertical mills are subject to vibrations if material is too dry. Hence if moisture in raw materials is too low, provision is made to spray water inside the mill.

Wear rates are high if limestone has coarse silica. In some designs provision is made to remove coarse fraction from inside the mill to reduce silica content.

4b.5 Mill with External Circuit

High pressure drop in the mill is largely at the throat because velocity of gas entering the mill is as high as 70 m/esc.

See Fig. 2.4b.3.

Velocity can be greatly reduced if pick up size of stone over the ring is allowed to be less than 10 mms. Larger quantity and coarser fractions fall through the ring and are collected and fed back to the mill at the top along with fresh feed. **See Fig. 2.4b.4**

Plate 2.4b.3, showing v.r.m. in external circuit.

Velocity at throat (annular ring) is reduced to around 40 m / sec. This reduces pressure drop to about 400 mm reducing fan power by a corresponding amount. **See Fig. 2.4b.5.**

Table 2.4b.1 shows power consumption of roller mill with external circuit and compares it with ball mills.

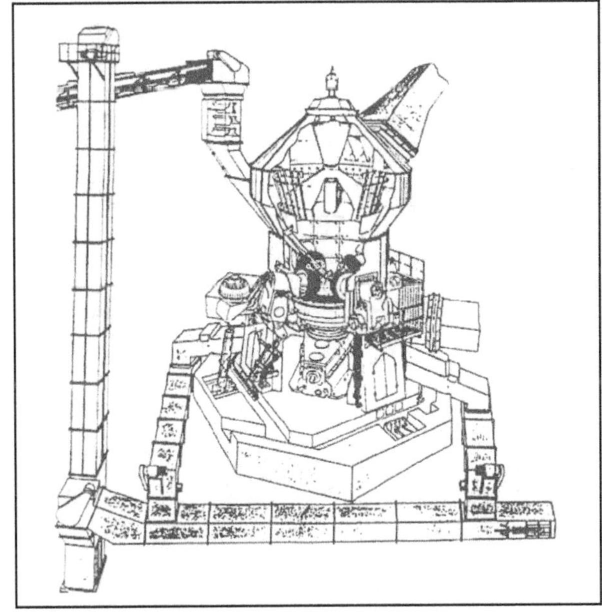

Plate 2.4b.3 Vertical mill with external circuit.

4b.6 High Efficiency Separators within the Mill

Another development relates to incorporating 'high efficiency separators'within the casing of the mill to replace old static or whizzer separators. This needed some modifications to mill housing but almost all manufacturers of v.r.mills have adopted these two changes with benefit to user in terms of reduced power consumption.

See Table 2.4b.1 and Plate 2.4b.2

100
50
40
30
20
10
5
4
3
2
1
0.5
0.4
0.3
0.2
0.1
Terminal velocity (m/s)
0.01
0.05
0.1
1
5
10
20
50
100
Particle size (mm)

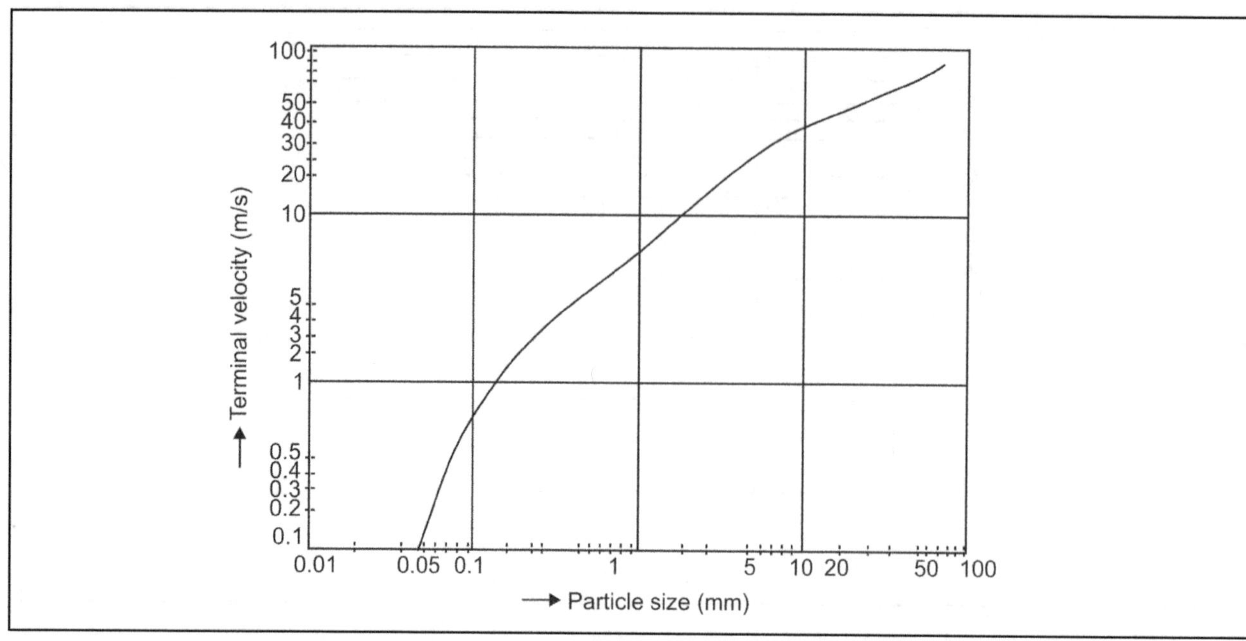

Fig. 2.4b.4 Terminal velocity vs particle sizes.

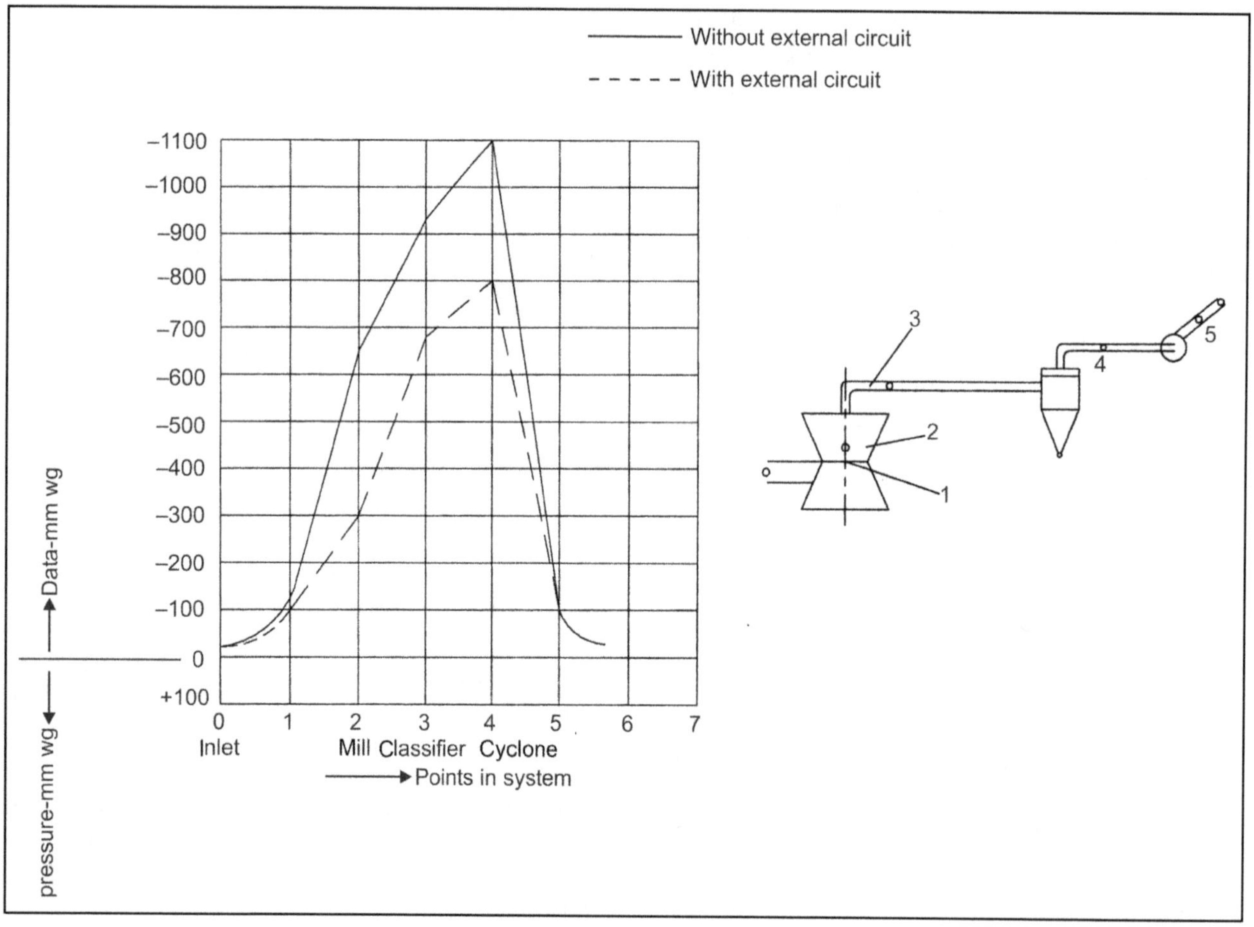

Without external circuit
With external circuit
–1100
–1000
–900
–800
–700
–600
–500
–400
–300
–200
–100
0
+100
Data-mm wg
pressure-mm wg
0
Inlet
1
2
Mill
3
Classifier
4
Cyclone
5
6
7
Points in system
1
2
3
4
5

Fig. 2.4b.5 Pressure distribution in VR Mll without external circuit.

Table 2.4b.1

Power consumption in mill systems							
Sr no.	**Item**	**unit**	**Ball mills**		**v.r.mills**		
			Air swept	**Bucket elevator**	**Standard circuit**	**External circuit**	**External circuit And high efficiency separator**
1	Sp.power mill only	Kwh/tom	12	12	6	6	6
2	Pressure drop in mill	mmwg	200-400	200	1100	650	750
3	Sp. power Fan	Kwh/ton	6	5	8	4	5
4	auxiliaries	Kwh/ton	0.3	0.5	0.5	0.8	0.8
5	Total system	Kwh/ton	18.3	17.5	14.5	10.8	11.3

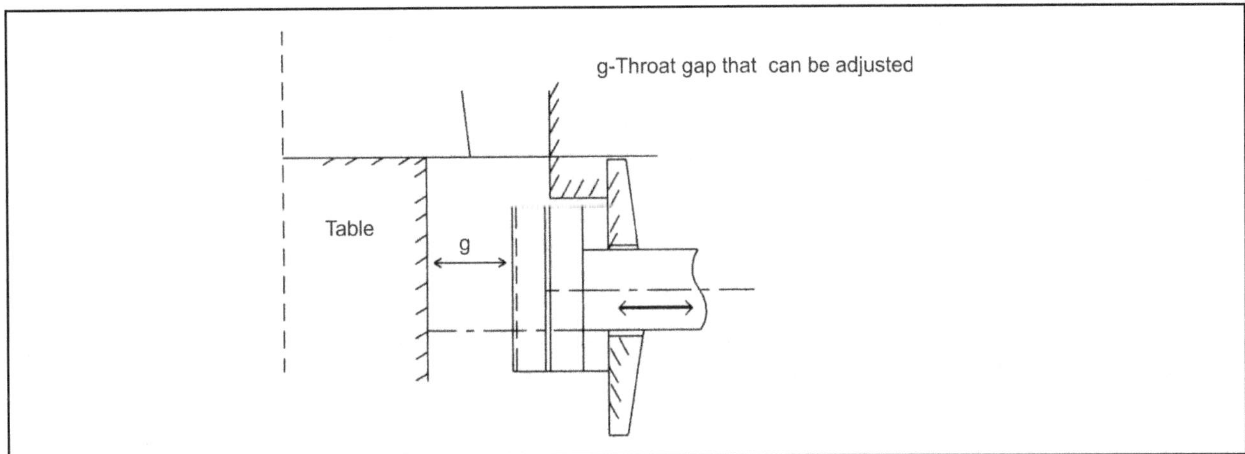

Fig. 2.4b.6 Adjustable throat gap in vertical roller mill.

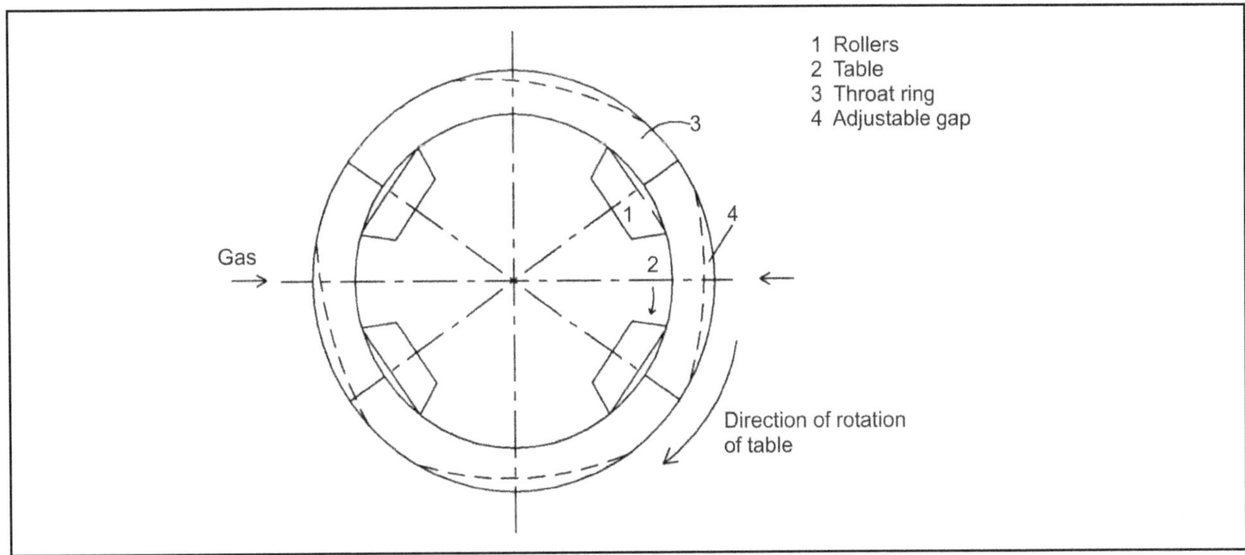

Fig. 2.4b.7 Adjustable thraot gap to suit flow pattern of material.

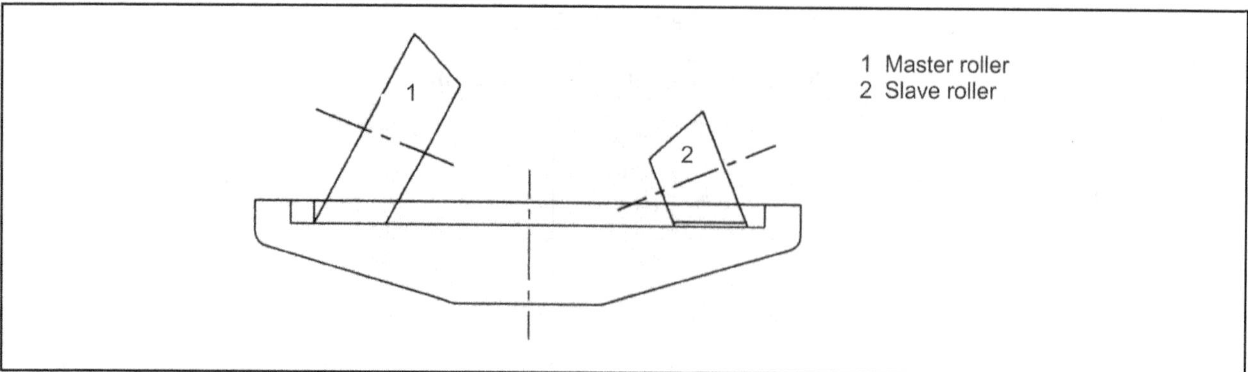

Fig. 2.4b.8 Slave and master rollers of a V.R. mill.

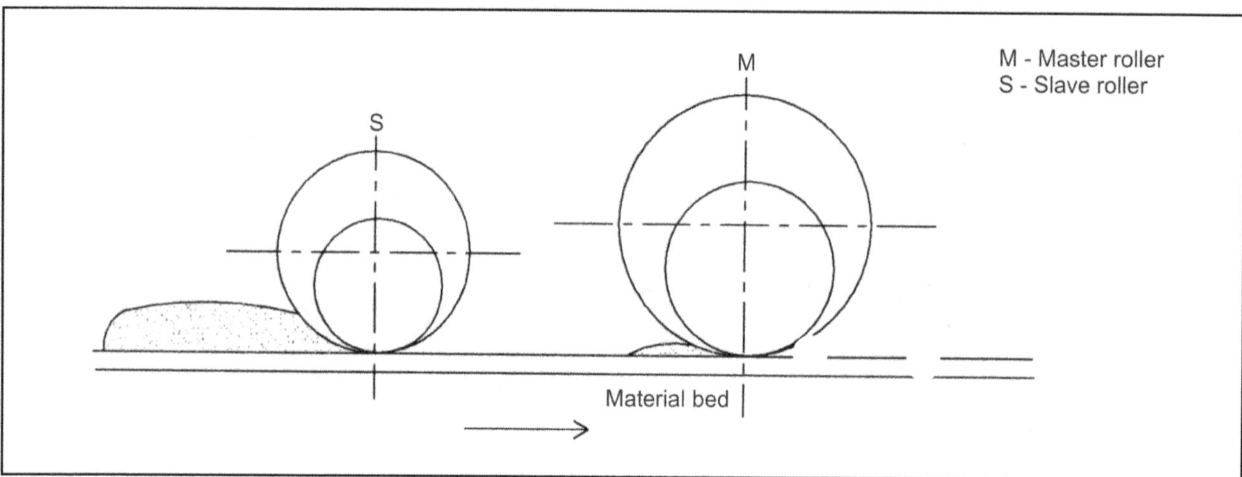

Fig. 2.4b.9 Action of slave and master rollers.

4b.7 Master and Slave Rollers and adjustable throat Gaps

For better and more even distribition of material around the throat gap, some designs have:

1. adjustable throat gaps.
   See Figs. 2.4b.6 and 2.4b.7 and or
2. pairs of large and small diameter rollers working as master and slave.
   See Figs. 2.4b.8 and 2.4b.9.

4b.8 Vertical Roller Mills for grinding Raw Materials and Clinker and slag.

Presently Vertical Roller Mills (VRMs) have replaced Ball and Tube Mills for Grinding both Raw materials and clinker in cement plants large and small. Roller Press alone or in combination with ball mills for secondary grinding is another preferred choice.

It thus implies that VRMs or RPs (Roller Presses) should be available in size and capacity to meet the requirements of large plants.

Table 2.4b.2 shows capacities of raw mills for a cement mills for cement plant of 10000 tpd clinker. Capacities for cement mills vary widely depending on the type of cement/s being made. (for a 10000 tpd clinker capacity, cement grinding capacities would be ~ 640 tph for OPC, 900 tph for PPC and 1590 tph for BFSC).

In case of Raw mills capacities would range between 450 to 900 tph for plant capcities of 5000 to 10000 tpd. Quite a few manufactures have designs for cpapacities ranging between 700 - 900 tph.

Reputed leading manufactures like Loesche and Pfeiffer have designs and have supplied VRMS for grinding ra material, coal and cement of capacities required for a 10000 tpd clinkering plant as mentioned in **Table 2.4b.2**

Table 2.4b.2 Working out Sectional Capacities offor a cement plant (kiln) of 10000 tpd Capacity.

Sr. no.	Section	Item	multiplying factor	hours per day	factor for hourly capacity	plant capacity tpd 10000
1	Quarries	stone raised	2.4	12	0.200	2000
2	Crushing	limestone to be crushed	1.91	12	0.159	1592
		crusher cpacity	2.3	12	0.192	1917
3	Stcker Reclaimer	stacker (20% above crusher cap.	2.8	12	0.233	2333
		reclaimer	1.76	10	0.176	1760
4	Raw Material Grinding	raw mill	1.76	20	0.088	**880**
5	Blending	feed rate	1.76	20	0.088	880
		extraction rate	1.91	24	0.080	796
6	kiln	feed rate	1.91	24	0.080	796
		kiln, preheater cooler	1.1	24	0.046	458
		clinker conveyor	1.72	24	0.072	717
		spillage	0.52	24	0.022	217
7	Coal Grinding	coal mill	0.2	20	0.010	100
			0.21	20	0.011	105
			0.22	20	0.011	110
			0.23	20	0.012	115
		coal firing	0.2	24	0.008	83
			0.21	24	0.009	88
			0.22	24	0.009	92
			0.23	24	0.010	96
8	Cement Grinding	cement mill				
		OPC	1.27	20	0.064	**635**
		PPC 30% ash	1.82	20	0.091	**910**
		BFSC 60% slag	3.15	20	0.158	**1575**
9	Packing	OPC bagged	1.27	15	0.085	847
		bulk	1.27	6	0.212	2117
		PPC bagged	1.82	15	0.121	1213
		bulk	1.82	6	0.303	3033
		BFSC bagged	3.15	15	0.210	2100
		bulk	3.15	6	0.525	5250

2.4b.9 Multi Drive Units

Multi drive units are therefore coming into vogue up where connected power is supplied through two or more sets of motors and gearboxes. Therefore it should be possible to use a single mill requiring power as high12000kw. **See Plate 2.4b.4**

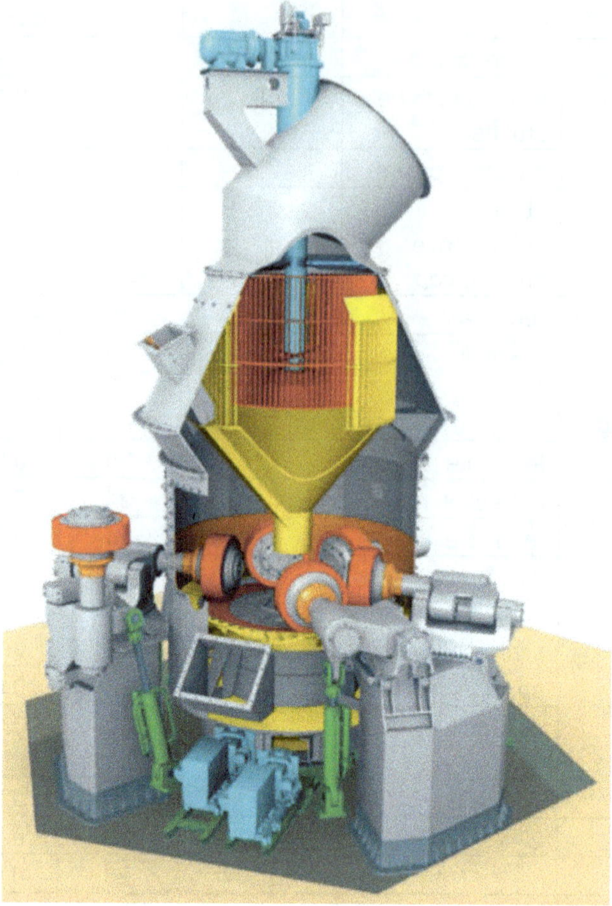

Source: Pfeiffer Brochure

Plate 2.4b.4 GBR Pfeiffer MVR Vertical Roller Mill.

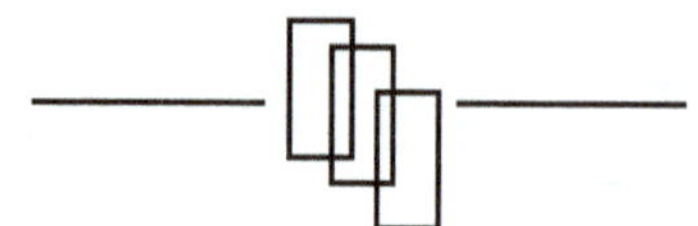

CHAPTER 4c

ROLLER PRESS

4c.1 Roller Press

A roller press is like a roll crusher. However, pressure exerted between rollers is very high - of the order of 400 kg/cm2 as compared to common roll crushers. This high pressure brings about structural changes in the material and it becomes easier to grind; about 30 % material gets pulverized to the desired fineness and less even if gap between rollers is 6 mm.

See Figs. 2.4c.1, 2.4c.2 and 2.4c.3.

As mentioned in para 4.3.2 in chapter 4, roller press reduces power consumption improved grindability. **See plate 2.4.3 & Fig. 2.4.3.**

Fig. 2.4c.2 furnishes comparison between a roll crusher and a roller press.

One roller is fixed and the other movable to exert pressure which is applied hydraulically. Feed is fed on the rollers by a central chute over the total width of the roller. Pressed material comes out as cake. This cake is divided into middle fraction and end fractions. End fractions are sent back to the press for recrushing.

See Figs. 2.4c.4 and 2.4c.5 Also Plate 2.4c.1.

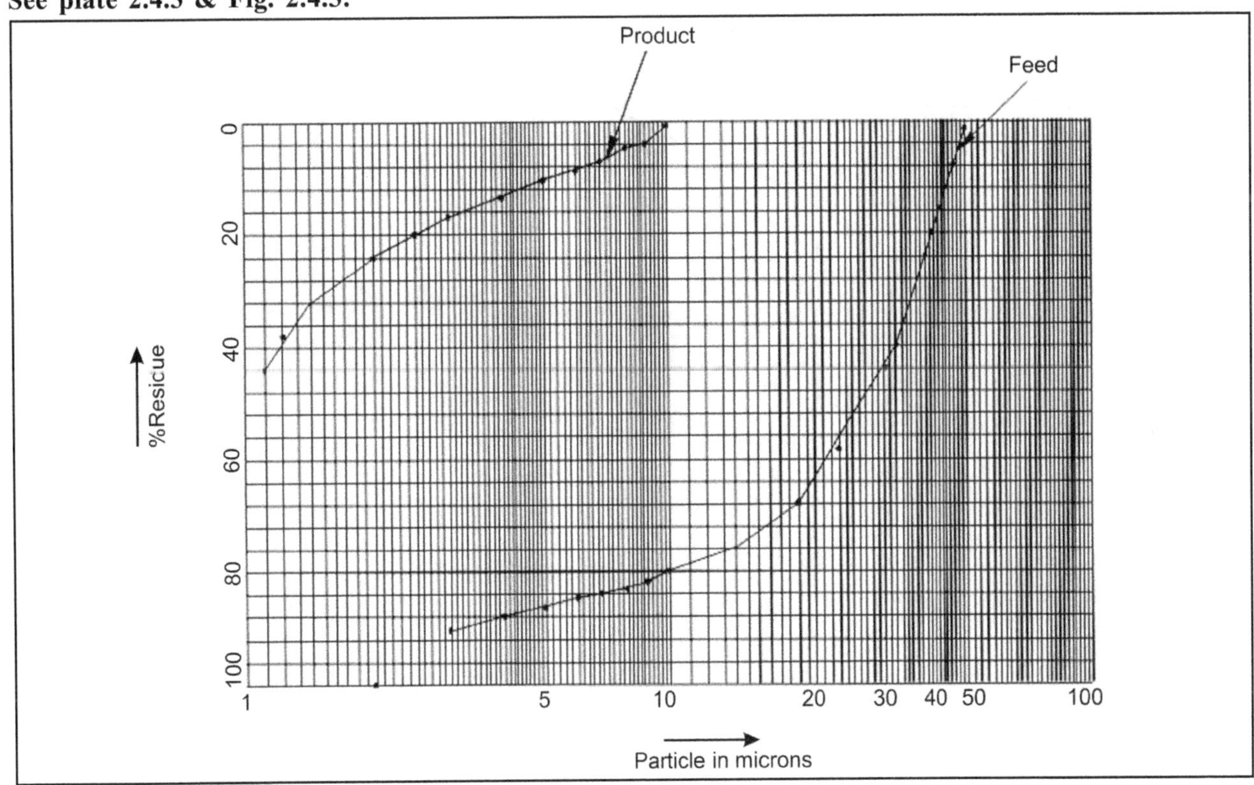

Fig. 2.4c.1 Roller press - feed and product.

1 b
1 a
2 b
2 a
1 a Feed to roll crusher
1 b Product of roll crusher
2 a Feed to roller press
2 b Product of roll press
%residue
.2
.4
.6
.8
1.0
10
20
30
40
Particle size microns

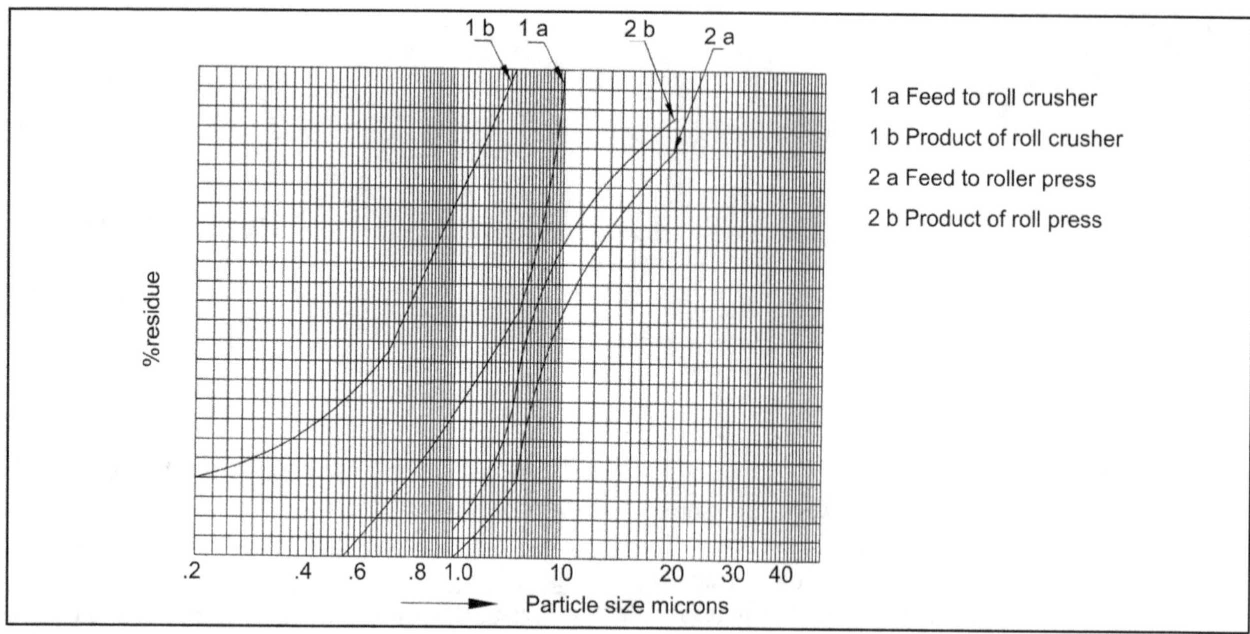

Fig. 2.4c.2 Particle size distribution of product from roll crusher and roller press.

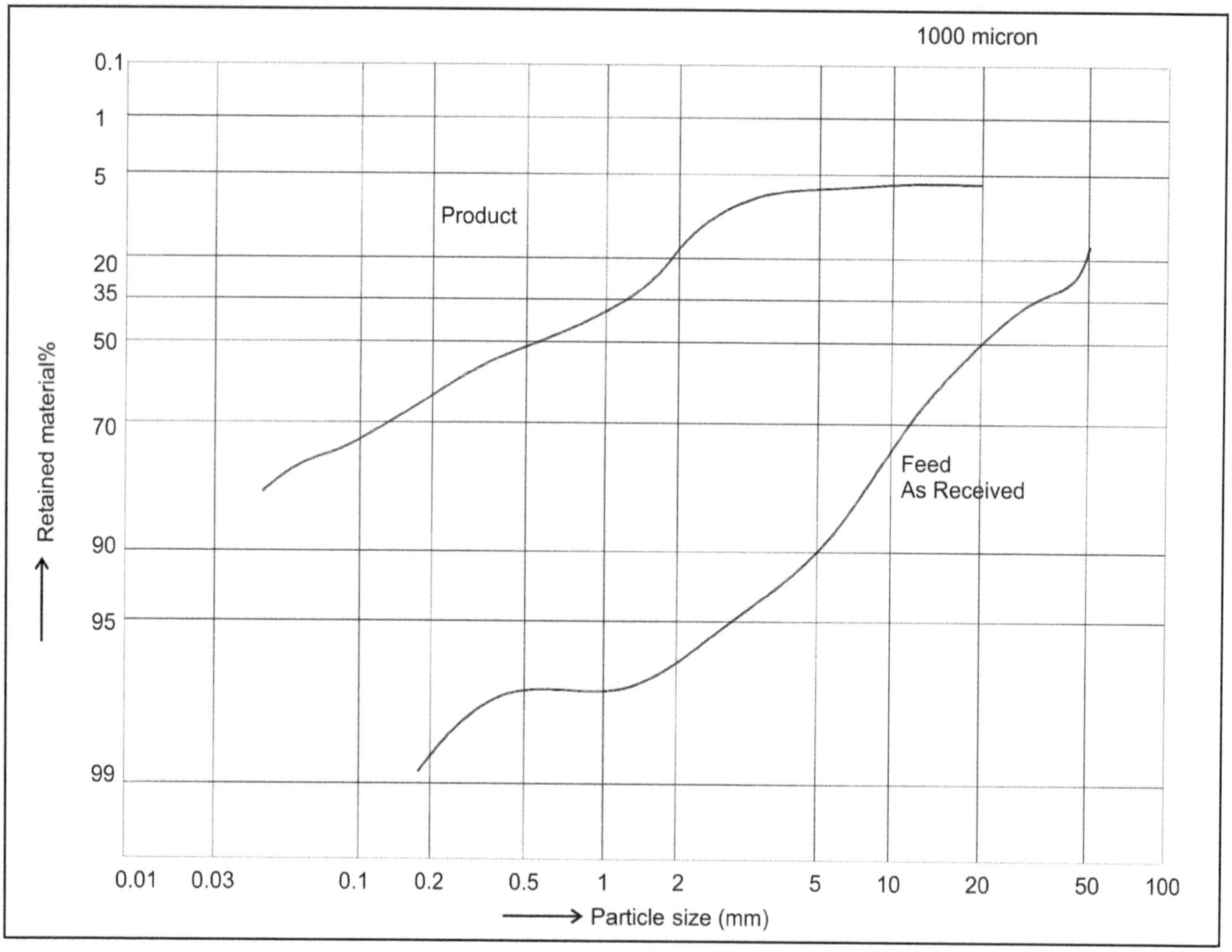

1000 micron
0.1
1
5
20
35
50
70
90
95
99
Retained material%
Product
Feed
As Received
0.01
0.03
0.1
0.2
0.5
1
2
5
10
20
50
100
Particle size (mm)

Fig. 2.4c.3 Particle size distribution of a clinker sample before and after the roller press.

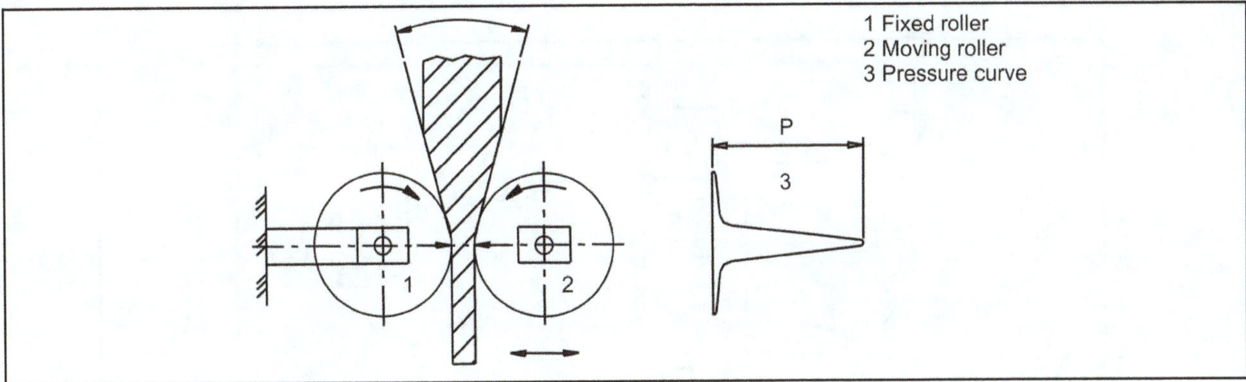

Fig. 2.4c.4 Action of roller press.

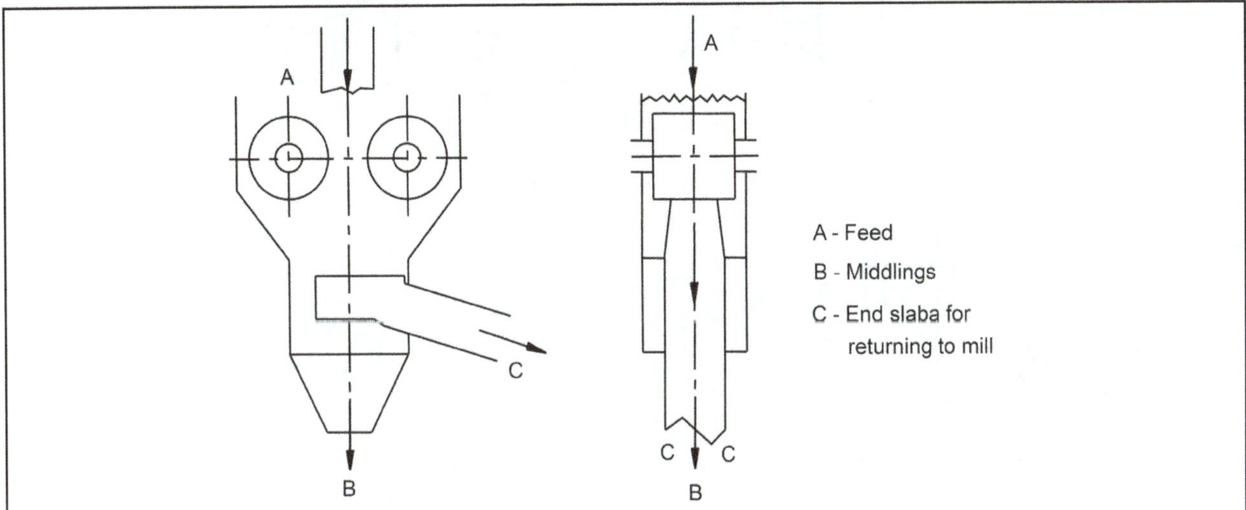

Fig. 2.4c.5 Roller press–cake at edges collected separately for recirculation.

Plate 2.4c.1 Roller press.

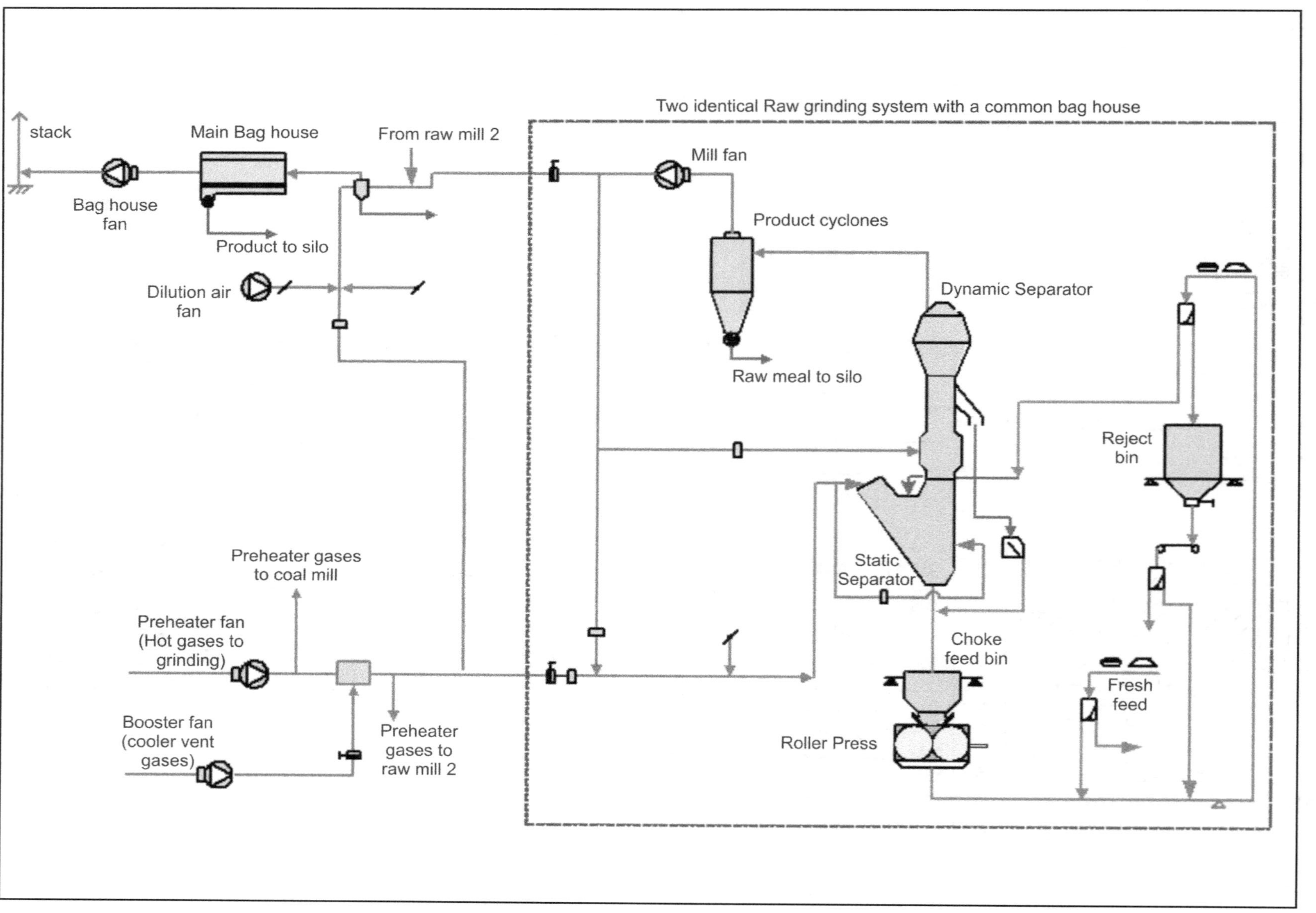

Plate 2.4c.2 Roller press circuit.

4c.2 Different Ways of Using Roller Press

There are various ways in which the roller press can be used in the system.

See Flow Chart 1.7.8 in Chapter 1 of Section 1.

1. As a precrusher in open circuit. Cake emerging is sent to a ball mill operating in closed circuit with a high efficiency separator. Hot gases are passed through mill for drying. Mill separator circuit is like the usual dry closed circuit grinding system. However mill is much smaller in size as it has to do less work.

   See Flow Chart 1.7.8 a

2. In 'hybrid' grinding circuit. In it both roller press and mill operate in closed circuit. Cake from the roller press is sent to a classifier and coarse fraction returned to it. Often the fractions pressed by edges of rollers are found to be coarser than the middle fraction. The edge fractions are collected separately and returned to press. The middle fraction only is sent to mill for grinding.

   See Flow Chart 1.7.8 b.

3. Using press only as pressing and grinding unit. In this circuit, ball mill is eliminated. Middle fraction is sent to a disagglomerator through which hot gases are passed for drying. Disagglomerated material is passed on to separator.

   See Flow Chart 1.7.8 c Also see Plate 2.4c.2

4c.3 Power Consumption with Roller Press

Table 2.4c.1 shows relative power consumption when using roller press in comparison with ball mill.

Rollers are plain or corrugated. They are built up 'in situ' when worn out.

4c.4 Roller Press Circuit with Ball Mill

Roller press - ball mill system has a large number of auxiliaries as compared to a vertical roller mill even with external circuit. Further when the press is out, ball mill system though it is self contained is not of use as it is too small and grinding media pattern is selected to suit size fractions in the cake from the press.

Therefore this system has not become very popular. Advantages in power savings are more pronounced when material is hard like clinker. Therefore roller press ball mill system has found greater acceptance in cement mill systems. These have been dealt with in **Chapter 13.**

4c.5 V Separator

This is a static separator used in grinding circuit with roller press that separates coarse fraction before sending the rest to the high efficiency separator in the circuit. **See Plate 2.4c.3.**

Table 2.4c.1

Power consumption with roller press						
Sr.No.	**item**	**unit**	**Bucket elevator Ball mill**	**Roller press and ball mill**	**Roller press in hybrid grinding**	**Roller press only**
1	Power for mill	Kwh/ton	12	8.0	7.0	
2	Power for press	Kwh/ton	-	2.0	2.0	6.5
3	Power for auxiliaries	Kwh/ton	6.5	4.5	4	4.5
4	total	Kwh/ton	18.5	14.5	13	11

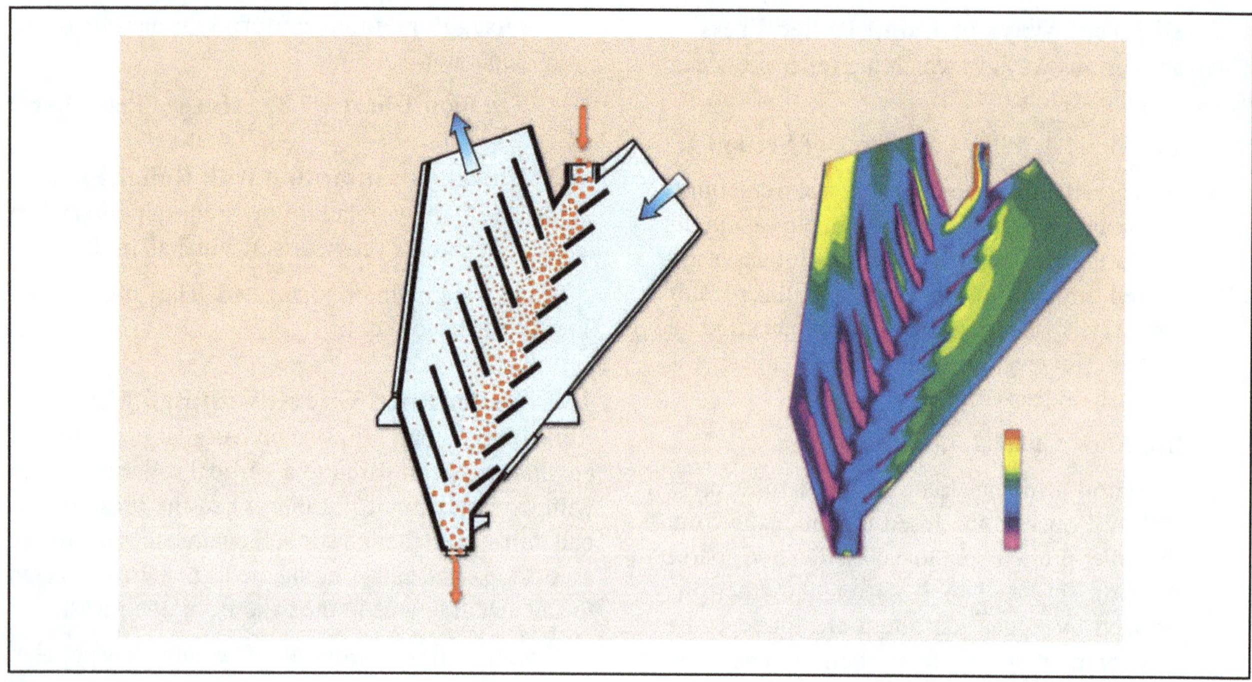

Plate. 2.4c.3 V Separator.

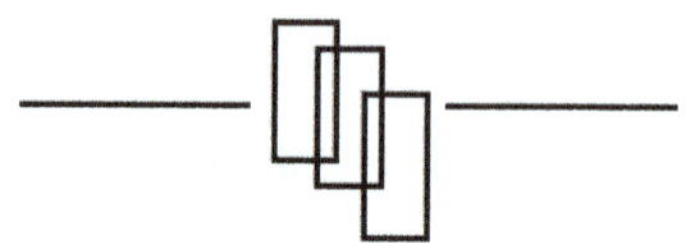

CHAPTER 5

SCREENS, SEPARATORS AND CLASSIFIERS

5.1 Screens, Classifiers and Separators

Screens and classifiers are devices which separate material into fractions size wise. Separators are devices which separate air or gas borne solids from air or gas.

Screens are devices used for crushed and granulated materials. Examples are rotating and vibrating screens.

Classifiers are used to separate and classify ground materials. Classifiers separate the fractions into coarse and fines according to requirements, from feed received in suspension in gases / air. Examples are grit separator, mechanical air separator and high efficiency separator.

Separators separate ground pulverized material from a stream of gases. Well known example is cyclones of Preheaters.

5.2 Screens

Screens are used in crushing circuits and in preparation of aggregates in construction business.

Their purpose is to divide the feed in two or more fractions passing the screens or meshes as required.

In a single deck screen, feed will be divided into two fractions-oversize and undersize. Undersize is finished product. Oversize is returned to crusher for re-crushing.

In case of a double deck screen, there will be three fractions.

Screens are thus used to obtain a product of specific size. It is thus an important part in 'closed circuit' operation of the system.

5.2.1 Vibrating Screens

Most commonly used screens are vibrating screens. They can be horizontal or inclined. Feed is spread at the receiving end evenly across the width of the screen and progresses along its length.

Screens are designated by size of aperture. A 6 mm screen will have 6 mm square clear aperture. Sometimes apertures are round also.

Material below the size of aperture falls through the screen and is collected as finished product.

Screens have an efficiency ranging between 50 to 70 %. That is 50 to 30 % finished product is not separated and goes back unnecessarily for re-crushing. Quantity that goes back is known as 'circulating load'.

Thus both crusher and screen should be capable of processing fresh feed plus circulating load.

See Fig. 1.7.3.2 in Chapter 7 of Section 1

Vibrating motion is imparted by an eccentric or by an electromagnet.

See plate 2.5.1

5.3 Classifiers

Classifiers serve the same purpose as that of a screen for pulverized materials in grinding systems. There are three fractions viz. fresh feed, coarse return and fines or product. Coarse fraction goes back to the mill for regrinding. It is also the circulating load. Therefore mill receives fresh feed plus circulating load so also classifier.

Simplest classifier used in mill systems is the'grit separator'.

See Fig. 2.5.1

Plate 2.5.1 Inclined and horizontal vibrating screens.

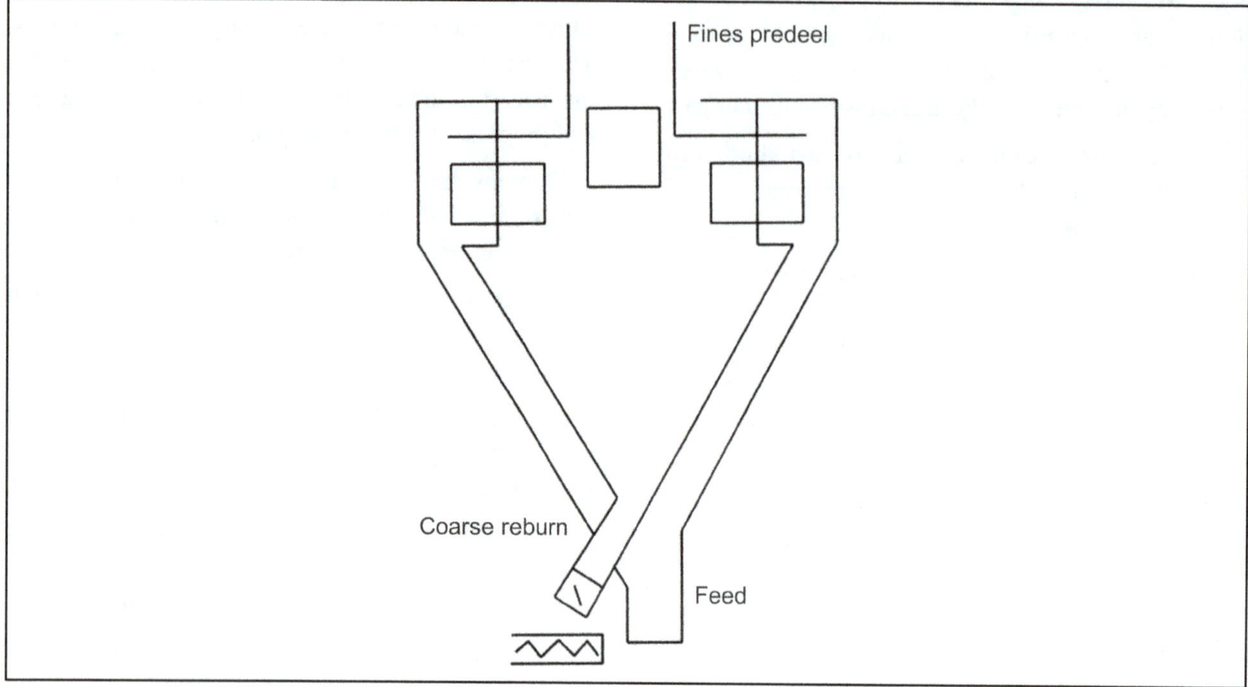

Fig. 2.5.1 Grit separator

It is used in air swept mills. It is installed in the outlet duct from the mill. It consists of a double cone and a ring of adjustable vanes installed at top of inner cone and a central outlet of adjustable height projecting inside the separator.

Fineness is adjusted by setting angle of vanes. Coarse fraction is separated in the inner cone and is collected and returned to mill by a screw conveyor. Fines are carried by air stream out of the separator.

5.3.1 Mechanical Sparators

A 'Mechanical separator' also used in mill systems depends on action of centrifugal forces on particles suspended in air stream.

See Fig. 2.5.2

Feed is received on a distribution plate and thrown by centrifugal force in an air circuit created by the main fan at the top. An auxiliary fan suppresses rising of particles with the air drawn by the main fan. Heavier that is coarser particles drop out and are collected in an inner cone and taken to the mill by an air slide. Fines are separated from the air stream by cyclonic action in the outer casing and are collected and sent out as finish product. Fineness of product is adjusted by changing blades in number or shape or radius of the auxiliary fan. In cement mill systems fineness is adjusted by varying speeds of distribution plate and auxiliary fan.

These separators have an efficiency of 50 to 55 %.

In mill systems circulating load is kept intentionally high to prevent over grinding. Length of the mill is important from this point. Shorter the mill higher the circulating load. For raw mills, l/d ratio is kept between 1.5 and 2. For cement mills it is between 3 to 3.5. For coal mills it is 1.25.

In production of cement not only its fineness expressed in Blaine surface (cm2 /gm) is important but also particle size distribution.

These separators have a shallow selectivity curve and range of particle sizes is very wide. High efficiency

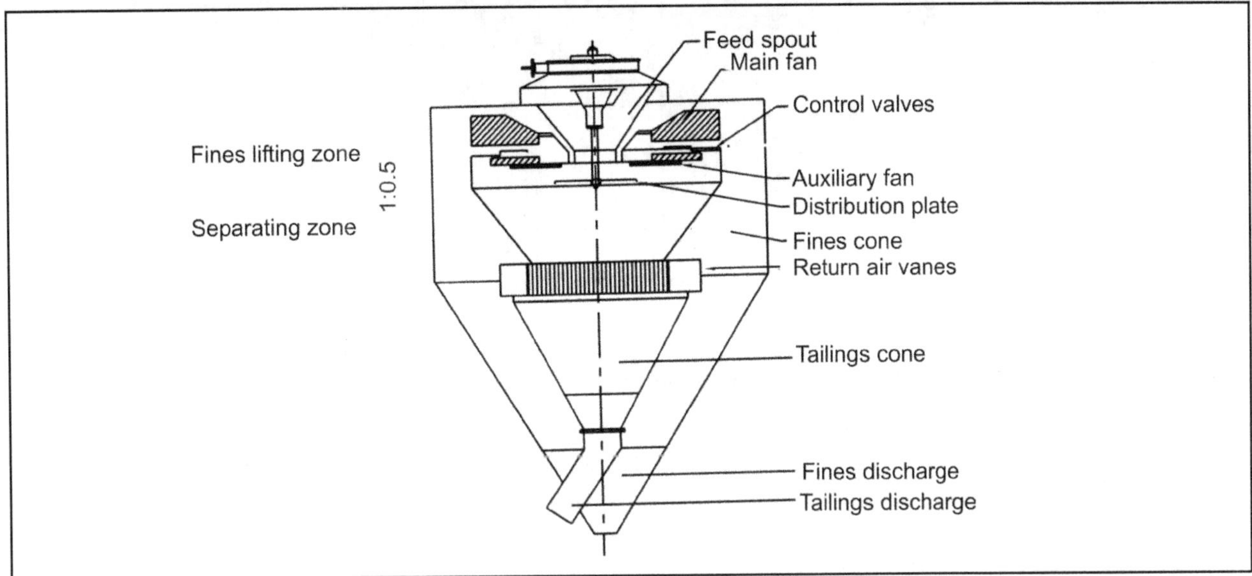

Fig. 2.5.2 Schematic of a mechanical air separator.

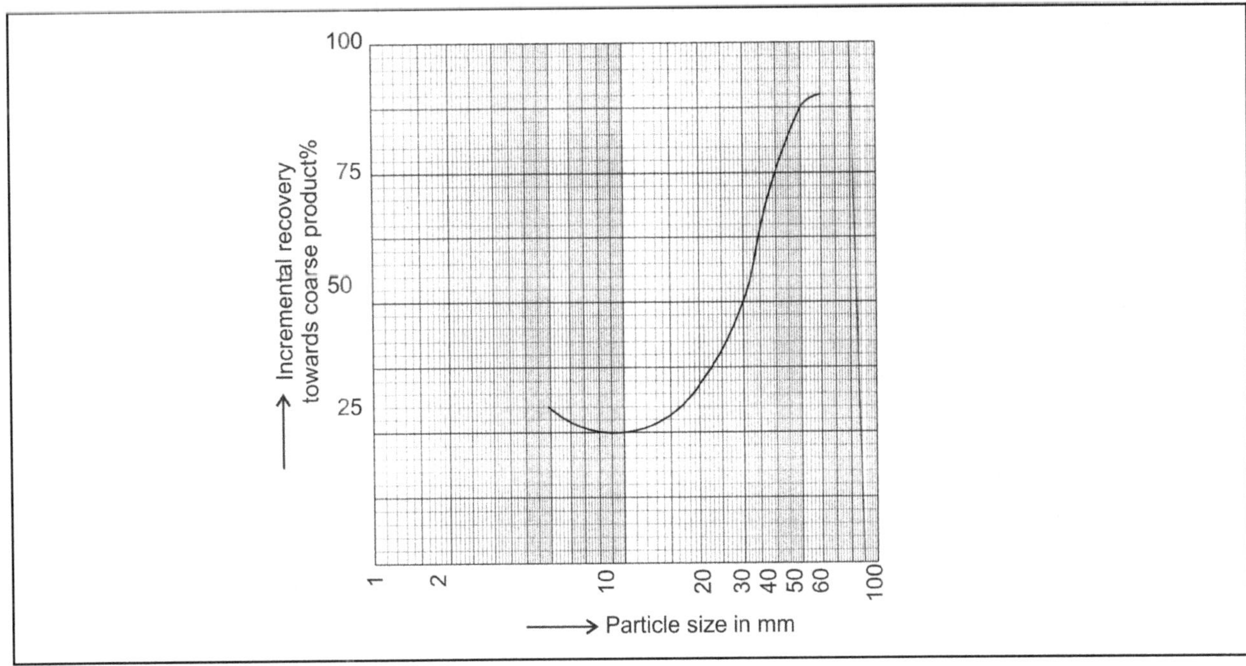

Fig. 2.5.3 Selectivity curve of conventional separator.

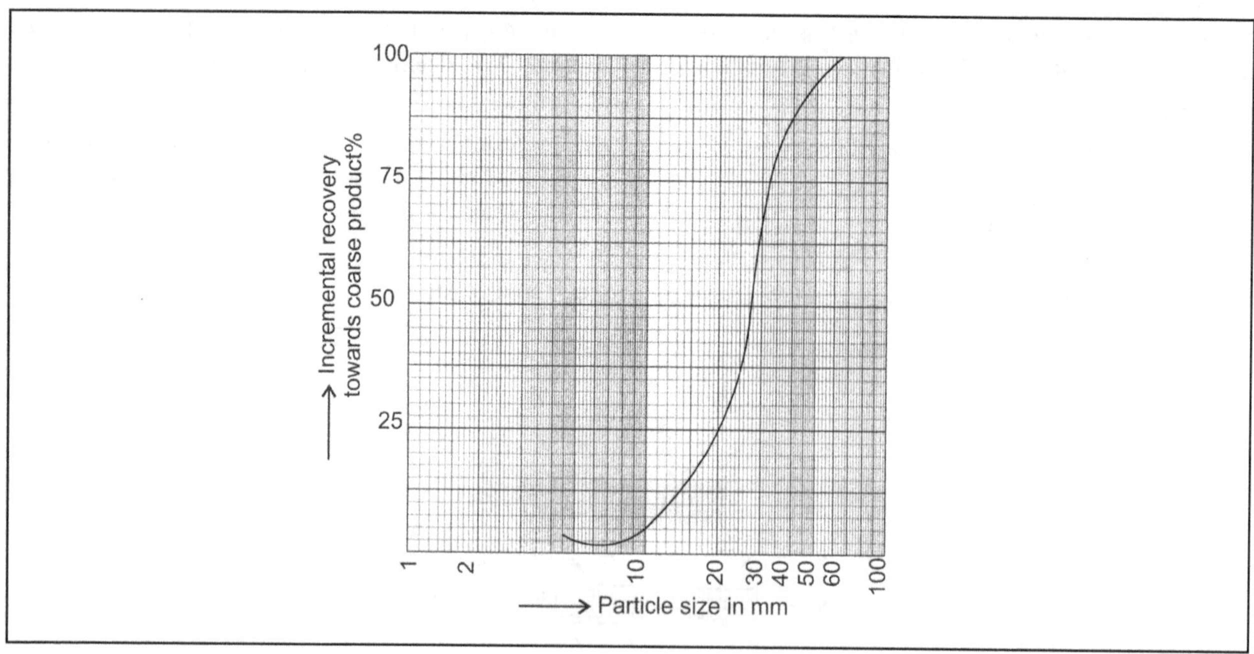

Fig. 2.5.4 Selectivity curve of high efficiency separator.

separators were developed to overcome this disadvantage.

See Figs. 2.5.3 and 2.5.4.

5.4 High Efficiency Separators

High efficiency separator consists of a rotating cage of considerable depth. Fresh feed falls over the side of the cage. It is subjected to action of air stream which enters the casing through an annular ring of vanes. Coarse fractions are collected from the bottom cone and fines are taken out to a dust collector for collecting the product.

Speed of the rotor is varied by a d.c. motor. Quantity of air flow is also adjustable. In some designs, angle of stationary vanes is adjustable.

As compared to mechanical separators, in high efficiency separators fines are collected from the air / gas stream in dust collectors of high efficiency outside the separator itself.

5.5 Different Designs of H.E.S

There are a great many designs of high efficiency separators.

See Figs. 2.5.5 and 2.5.6.

Plates 2.5.2 to 2.5.6 show some of the different types of high efficiency separators. Efficiency of these separators is between 70 to 80 %. Therefore circulating loads are less. Selectivity curves are near vertical straight lines. Majority of particles fall within range of 3 and 30 microns.

See Figs. 2.5.4, 2.5.7 and also 2.5.8

Figs 2.5.9 and 2.2.5.10 show particle size distribution and Blaine surface for three fractions viz. feed, coarse and fines for conventional and high efficiency separators.

5.5.1 Power Consumption - Saving in Power

Fig. 2.5.11 shows saving in power relation to efficiency.

Fig.2.5.12 Power requirement of the rotor at the production of PZ 35F / PZ 45F.

Today high efficiency separators have become an integral part of grinding systems so much so that even vertical roller mills which have a separator built in the mill housing are now fitted with them.

5.6

Grinding systems for grinding raw materials and coal are similar in that it is necessary to dry raw materials and coal while grinding. In cement mills this operation is not necessary unless slag or fly ash are wet when making blended cements.

Slno	Trade name	Typical sketch	Feed dispersion	Main classification	Other features
Notation		(F = Feed , P = Prod , R = Rejects , A = Air)			
1	O - sepa (fuller)	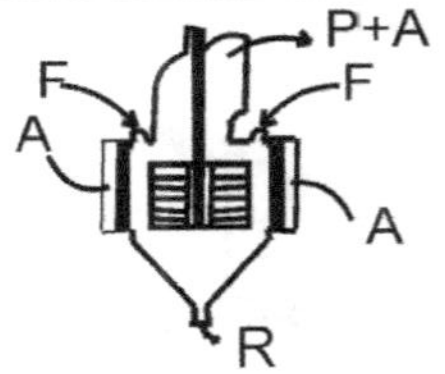 P+A F F A A R	on top angular plate of rotating cage two feed paints	Rotor with vertical blades guide vanes for air flow	Tangential air entry (double) ceramic lining on high wear areas
2	Sepol (polsius)	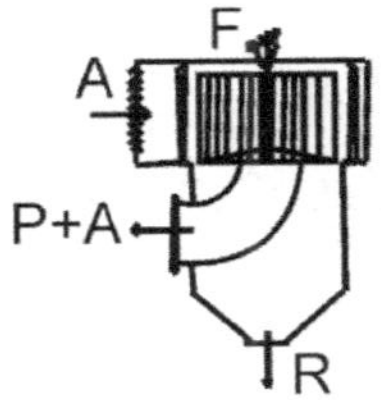 F A P+A R	on top cover plate of rotating cage central feeding	- do -	spiral air entry damper control of air flow separating air fines draw downwards
3	Sepax (fls)	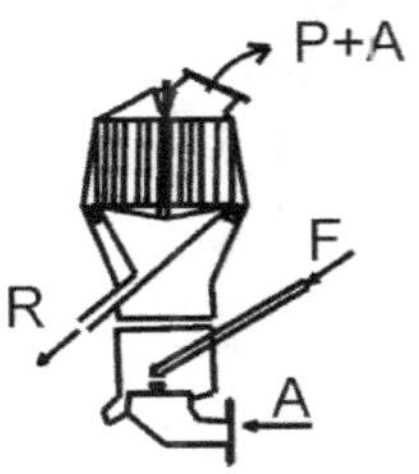 P+A F R A	on impact cone and in air stream feed dispersed in air stream	- do -	circumferential air entry. Efficient drying and cooling easy to incorporate in air swept mills

Fig.2. 5.5 High efficiency separators of different designs.

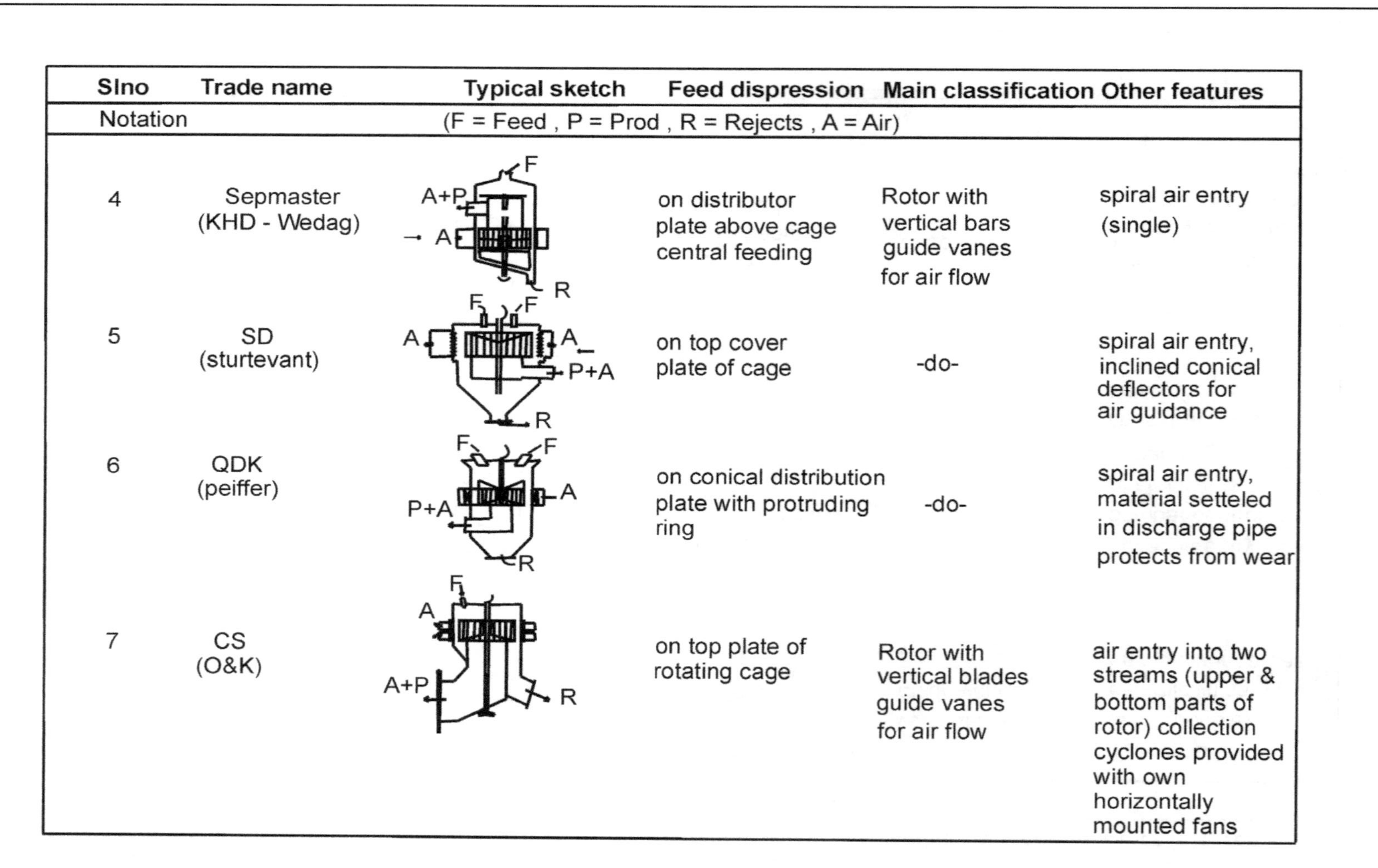

Slno	Trade name	Typical sketch	Feed dispression	Main classification	Other features
Notation		(F = Feed , P = Prod , R = Rejects , A = Air)			
4	Sepmaster (KHD - Wedag)	F, A+P, A, R	on distributor plate above cage central feeding	Rotor with vertical bars guide vanes for air flow	spiral air entry (single)
5	SD (sturtevant)	F, F, A, A, P+A, R	on top cover plate of cage	-do-	spiral air entry, inclined conical deflectors for air guidance
6	QDK (peiffer)	F, F, A, P+A, R	on conical distribution plate with protruding ring	-do-	spiral air entry, material setteled in discharge pipe protects from wear
7	CS (O&K)	F, A, A+P, R	on top plate of rotating cage	Rotor with vertical blades guide vanes for air flow	air entry into two streams (upper & bottom parts of rotor) collection cyclones provided with own horizontally mounted fans

Fig. 2.5.6 High efficiency separators of different designs.

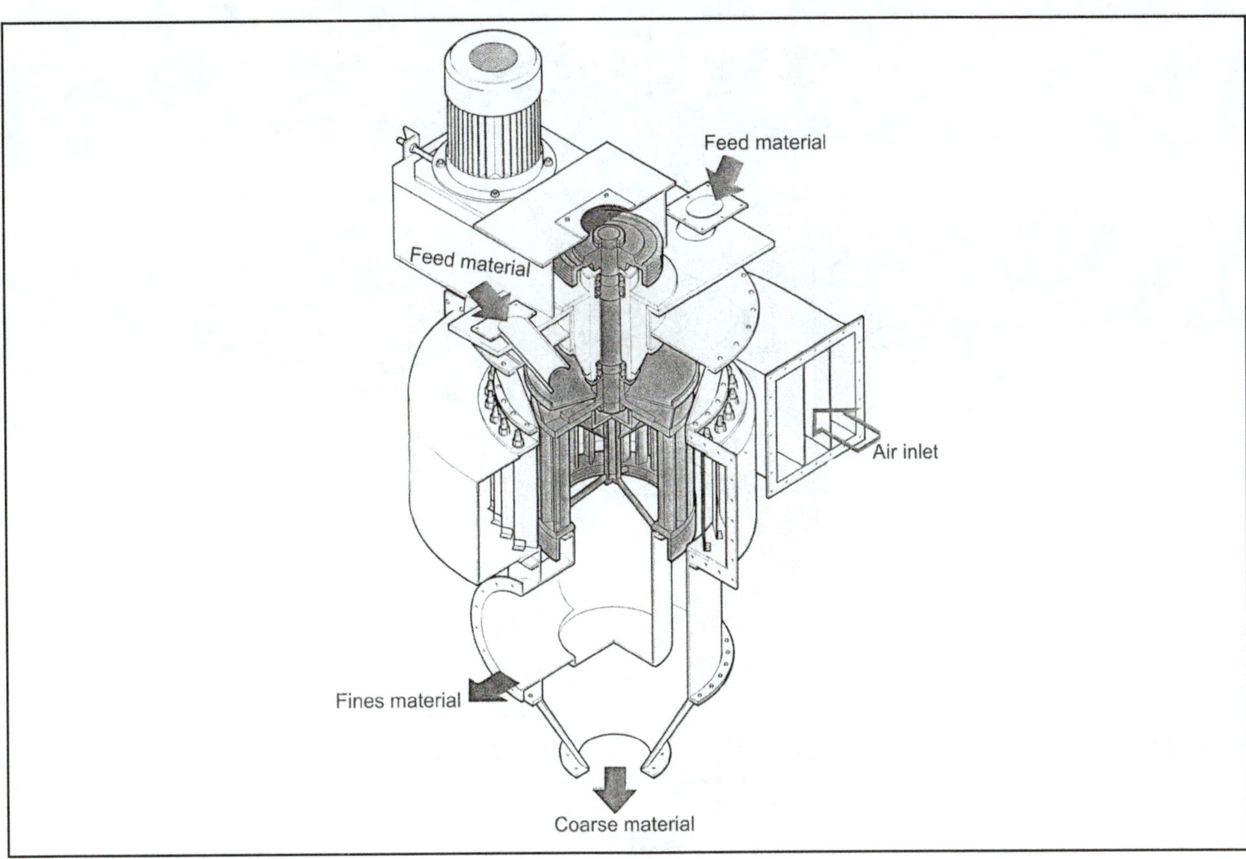

Plate 2.5.2 QDK Pfeiffer high efficiency separator.

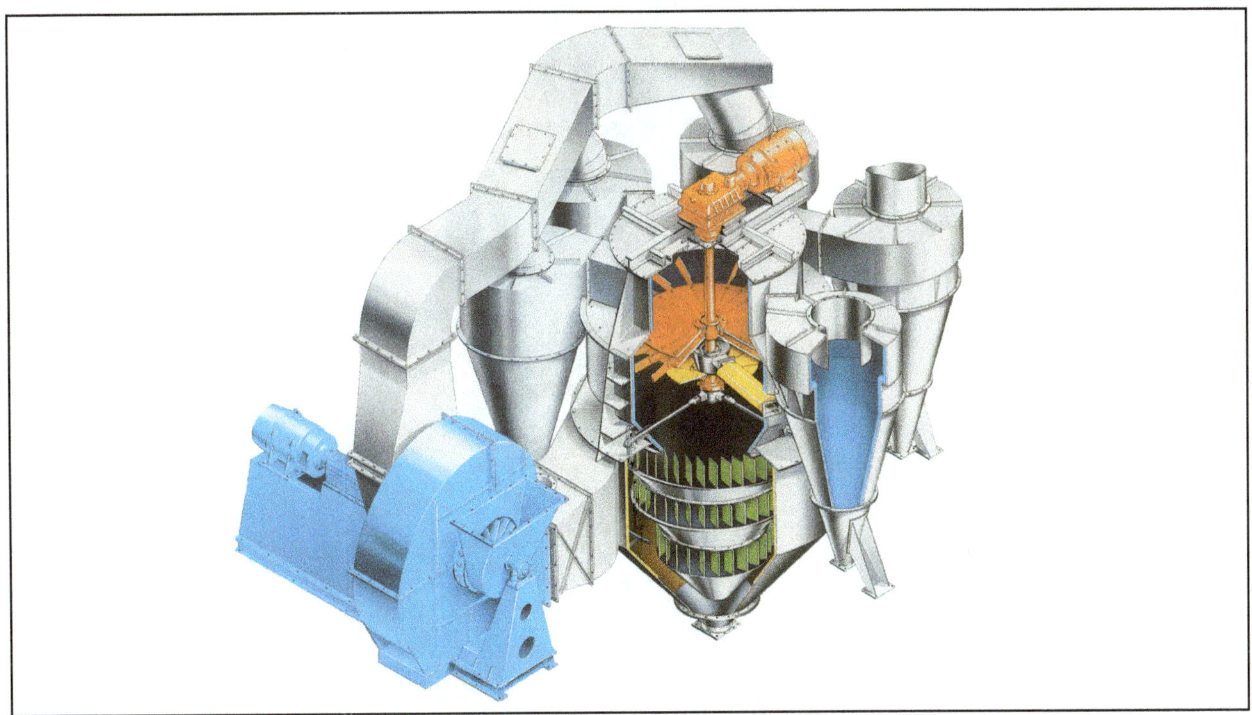

Plate 2.5.3 High efficiency separator in closed circuit with external cyclones to separate product.

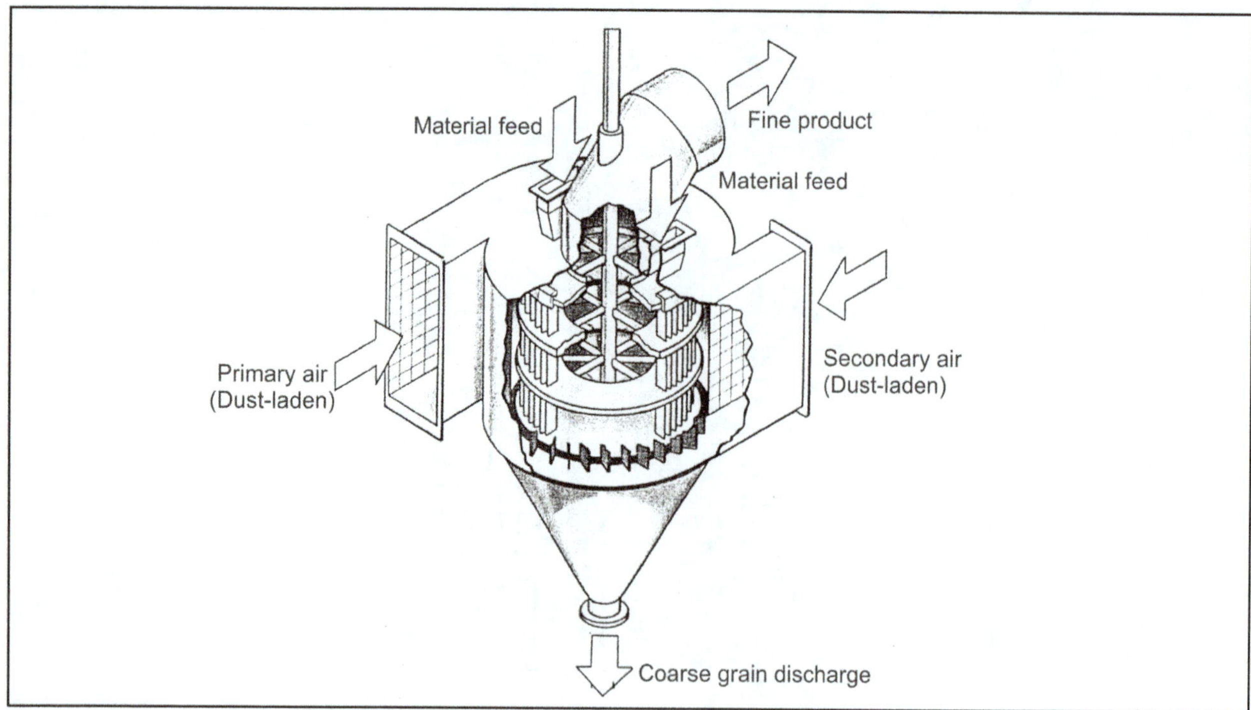

Plate 2.5.4 O-Sepa high efficiency separator.

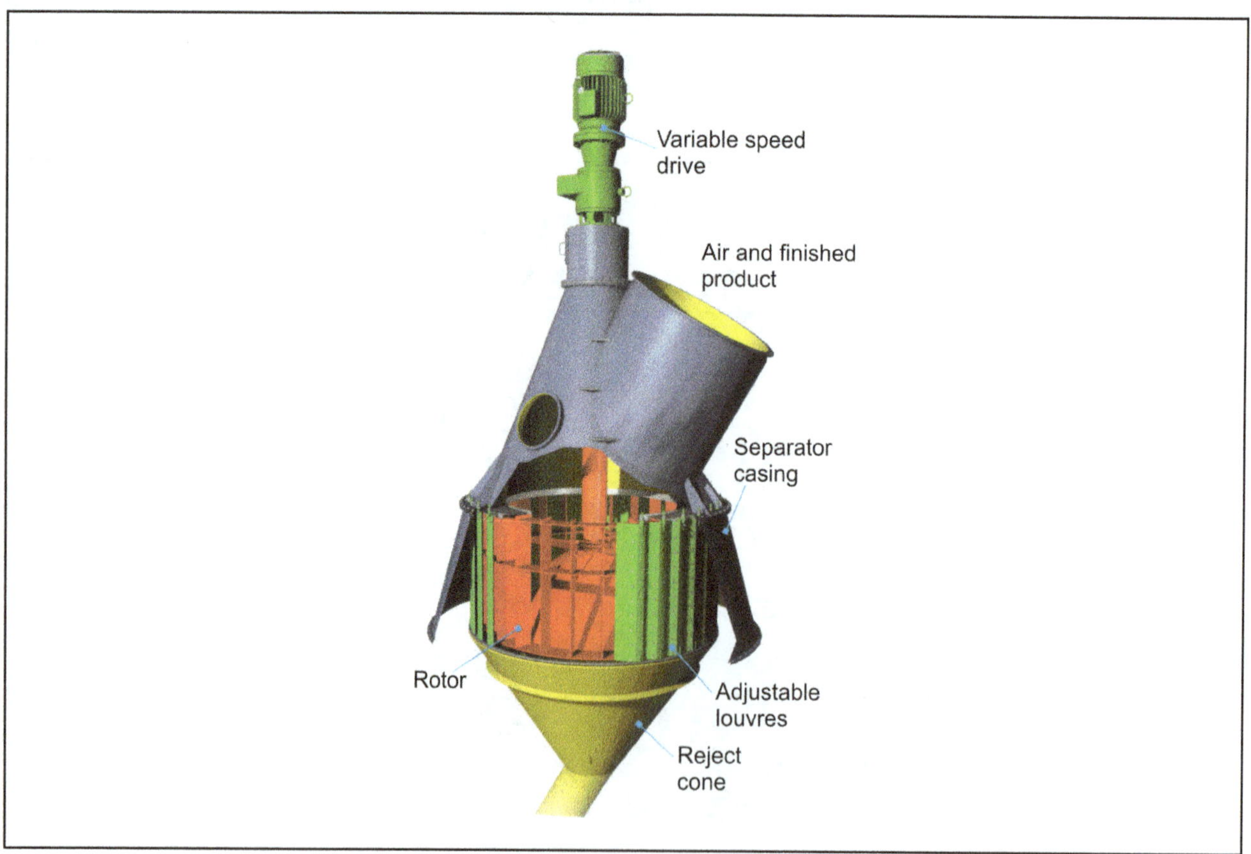

Plate 2.5.5 RTKM high efficiency separator for air swept ball mills.

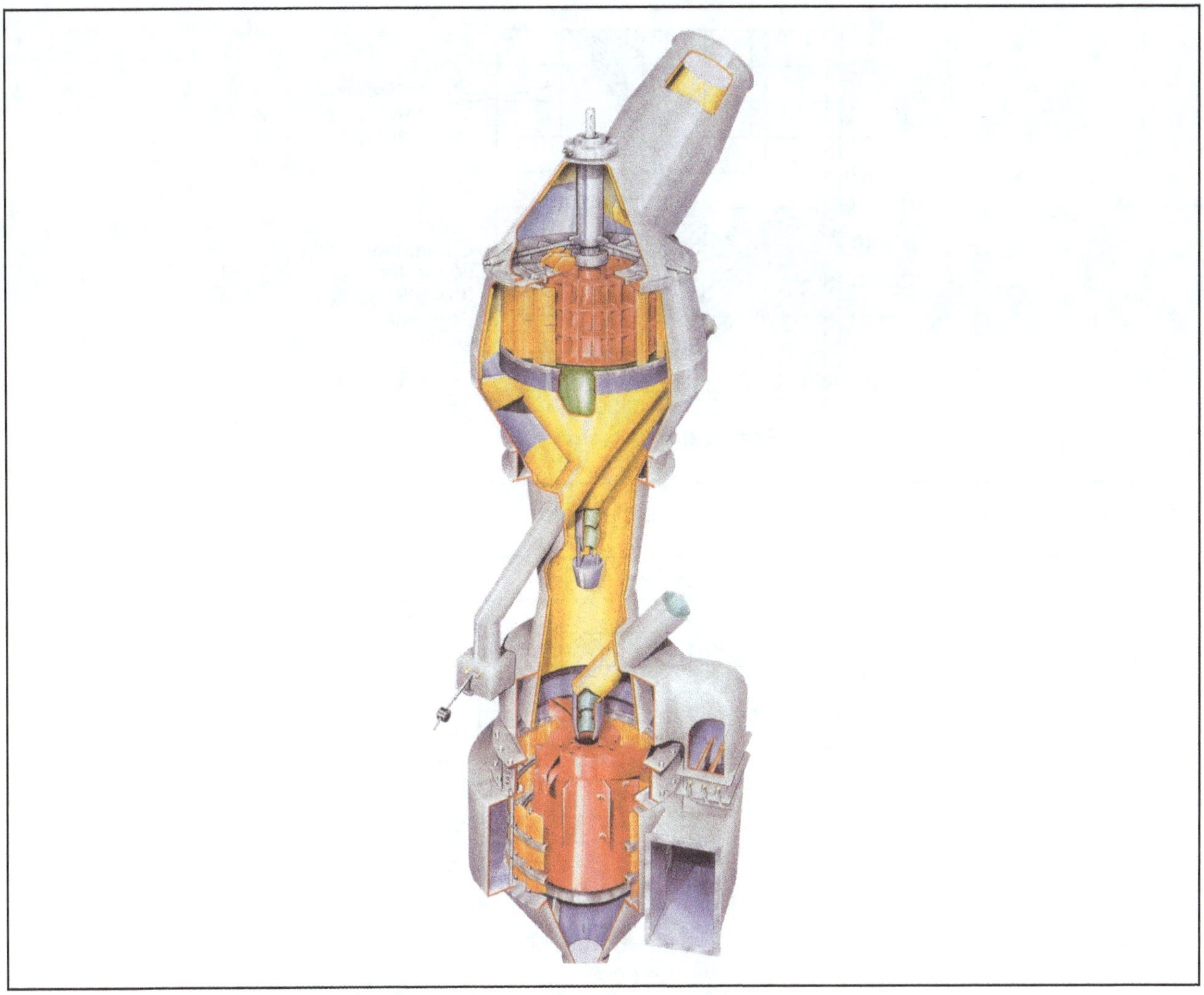

Plate 2.5.6 Sepax separator

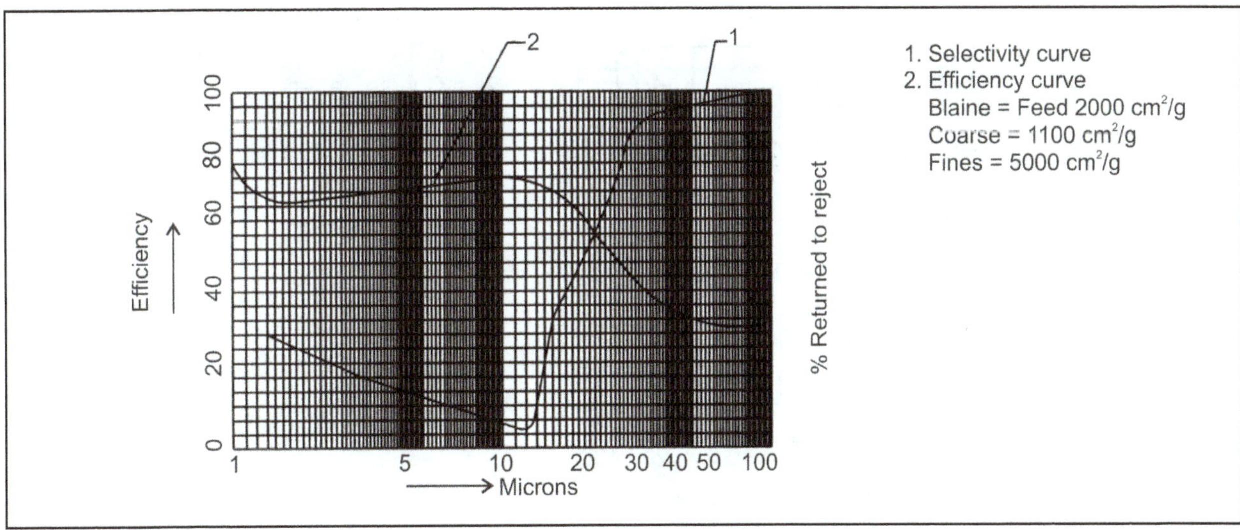

Fig. 2.5.7 Selectivity curve and efficiency of high efficiency separator.

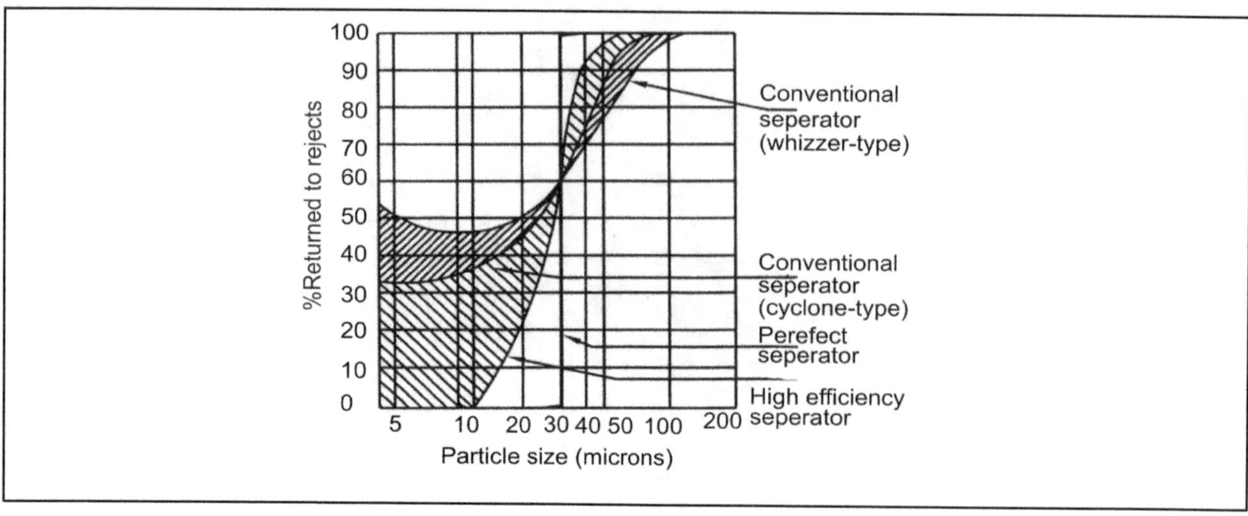

Fig. 2.5.8 Selectivity curves of different types of seperators.

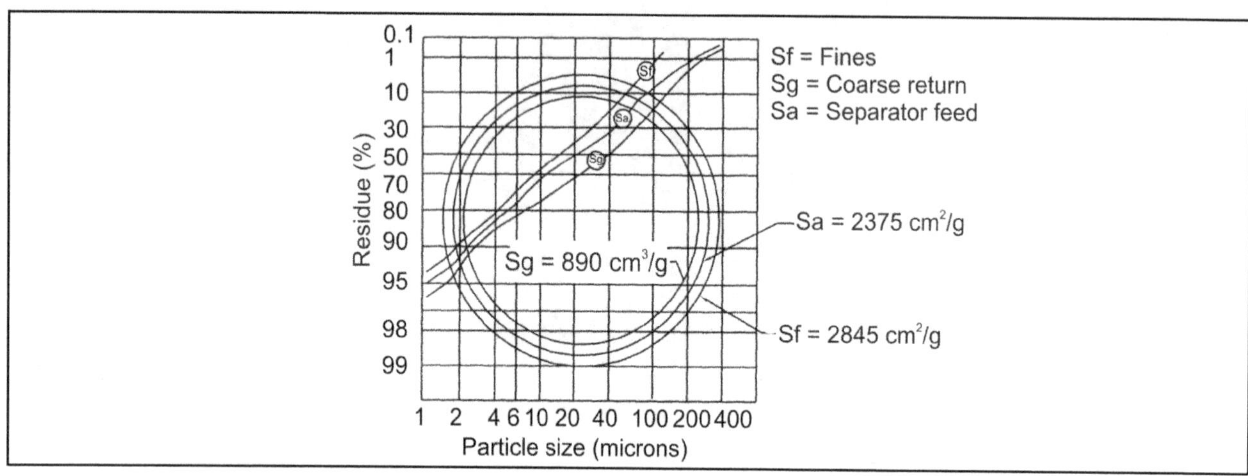

Fig. 2.5.9 Particle size distribution of feed, coarse return fines for conventional seperators.

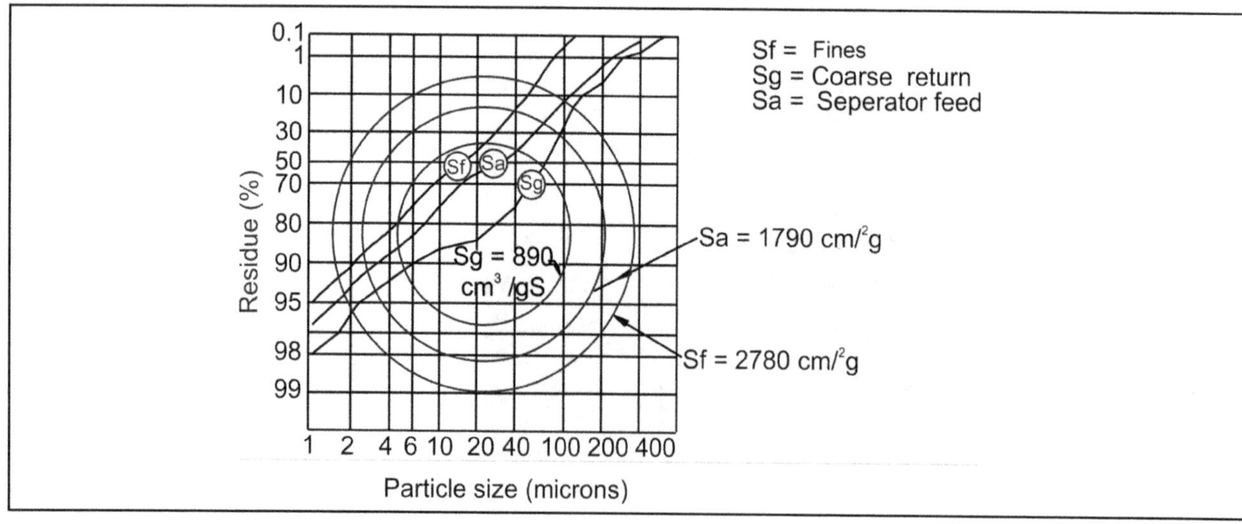

Fig. 2.5.10 Particle size distribution of feed, coarse return and fines for high efficiency seperators.

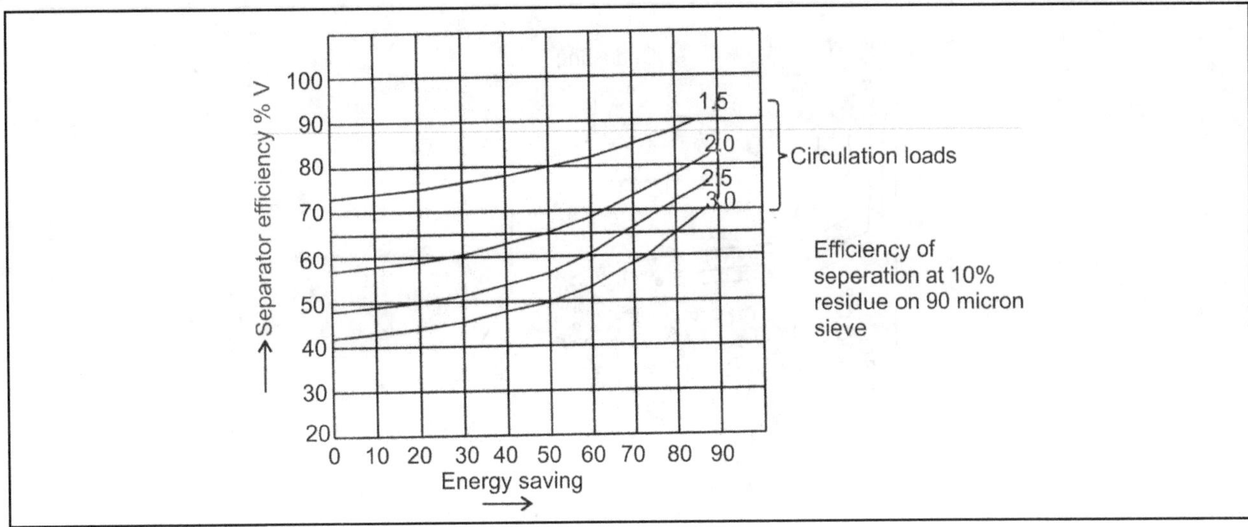

Fig. 2.5.11 High efficiency seperators efficiency of seperation and power saving in %.

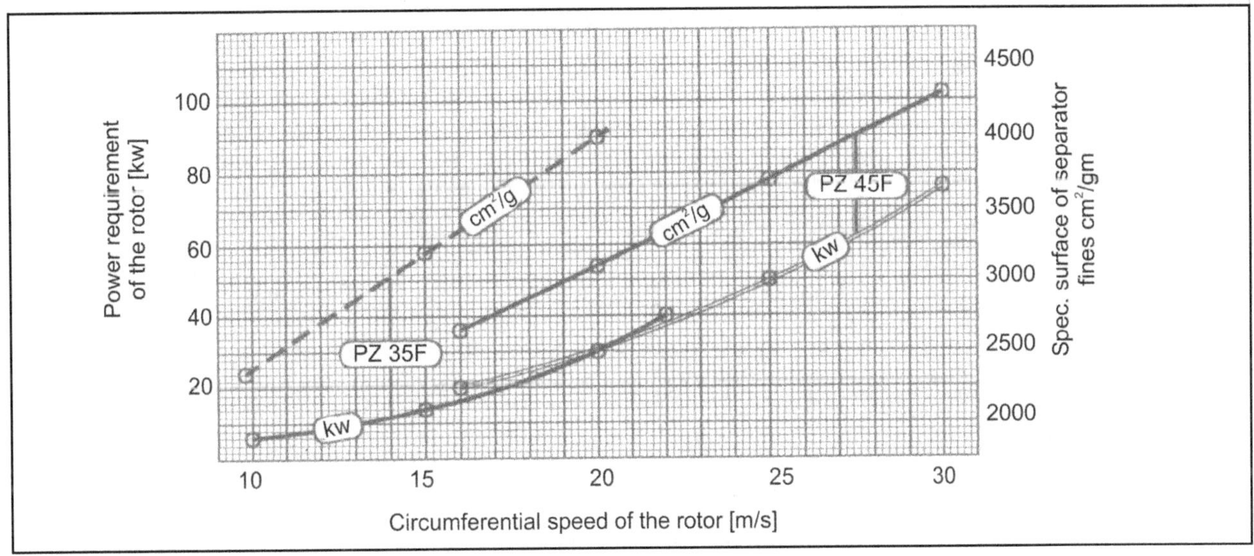

Fig. 2.5.12 Power requirement of the rotor at the production of PZ 35F / PZ 45F.

Construction wise there is no difference in separators of whatever type used for grinding raw materials or cement. Though HESs were used mainly in raw and cement mills, they are now also used in coal grinding systems.

5.7 Dust Collector for Venting HESs

HESs require a considerable quantity of air passing through them and hence require large dust collectors to separate the product from the air streams. Pressure drop in the separator is high compared to mechanical separator. Fan power is therefore higher. It is compensated by the higher efficiency of separation.

In general HESs are smaller in size than mechanical separators and it is often possible to replace mechanical separators by HESs.

5.8 Fluid Bed Separator

Fluid bed separators have been developed as preliminary separators in mill systems which use a pre grinding mill to increase capacity.

See Figs. 2.5.13 and 2.5.14

5.8.1 V Separator

This is a static separator used in grinding circuits with Roller Press to separate coarse fraction to return it to Roller press **See Fig. 2.4c.3** in this section.

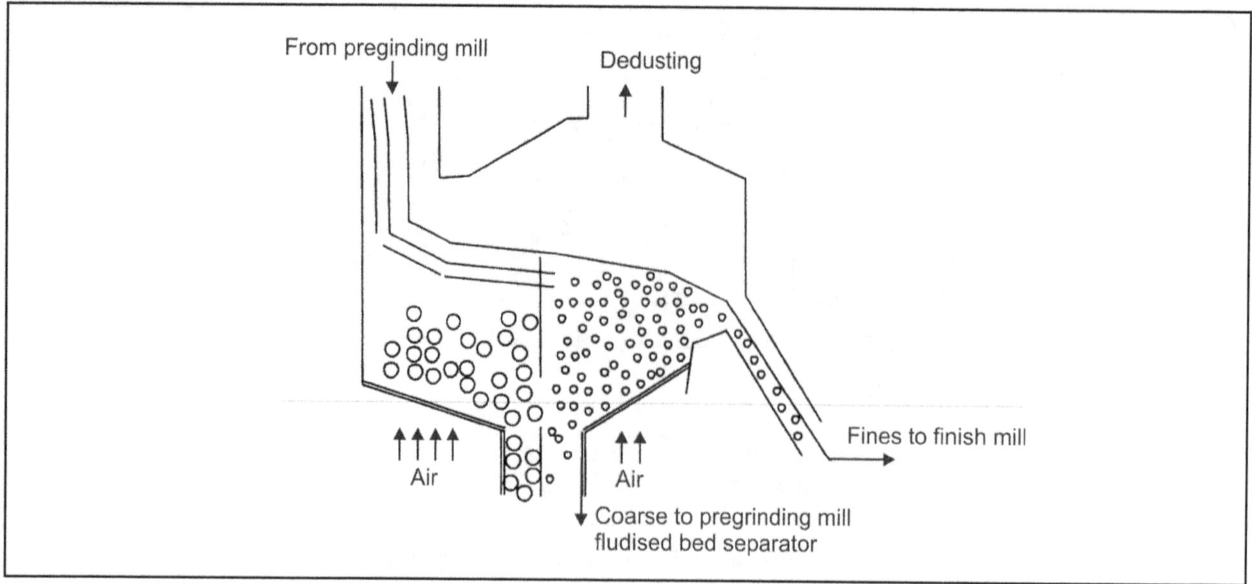

Fig. 2.5.13 Fluidised bed separator in conjuction with pregrinding mill.

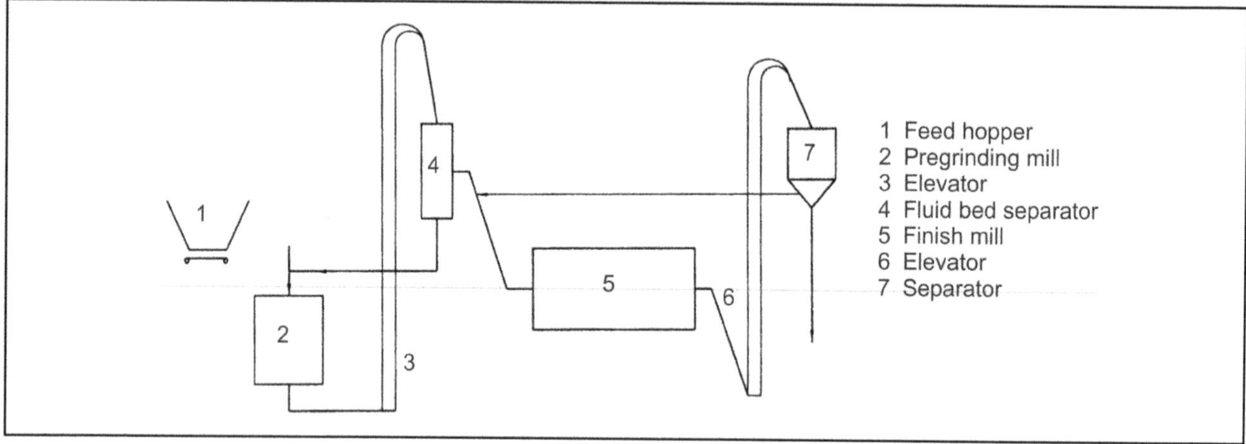

Fig. 2.5.14 Mill circuit with pregrinding mill; fluidized bed separator and finish mill.

5.9 Cyclones

Cyclones are merely separators. They separate pulverized material in a gas or air stream. Collected material is either a product or it is passed on to another cyclone for further processing.

Earlier, cyclones were used to collect pulverised material in mill systems. Now because of stringent pollution norms cyclones are used only as intermediate collectors.

Efficiency of cyclones is dependent on particle size distribution in the material. It is almost 100 % for + 40 micron size particles but drops sharply for small particles.

See Fig. 2.5.15

Cyclones in a preheater system is a good example of cyclones used to separate material from carrying gases. Separated material is passed on to the stage below.

Smaller cyclones have a higher efficiency than large cyclones. Therefore often a bank or group of cyclones is used to dedust large volumes of gases.

See Figs. 2.5.16 and 2.5.17

This logic was carried forward to design multiclones which are cyclones of ~ 150 mm dia arranged in groups. These were commonly used to clean vent gases from clinker cooler.

See Fig. 2.5.18

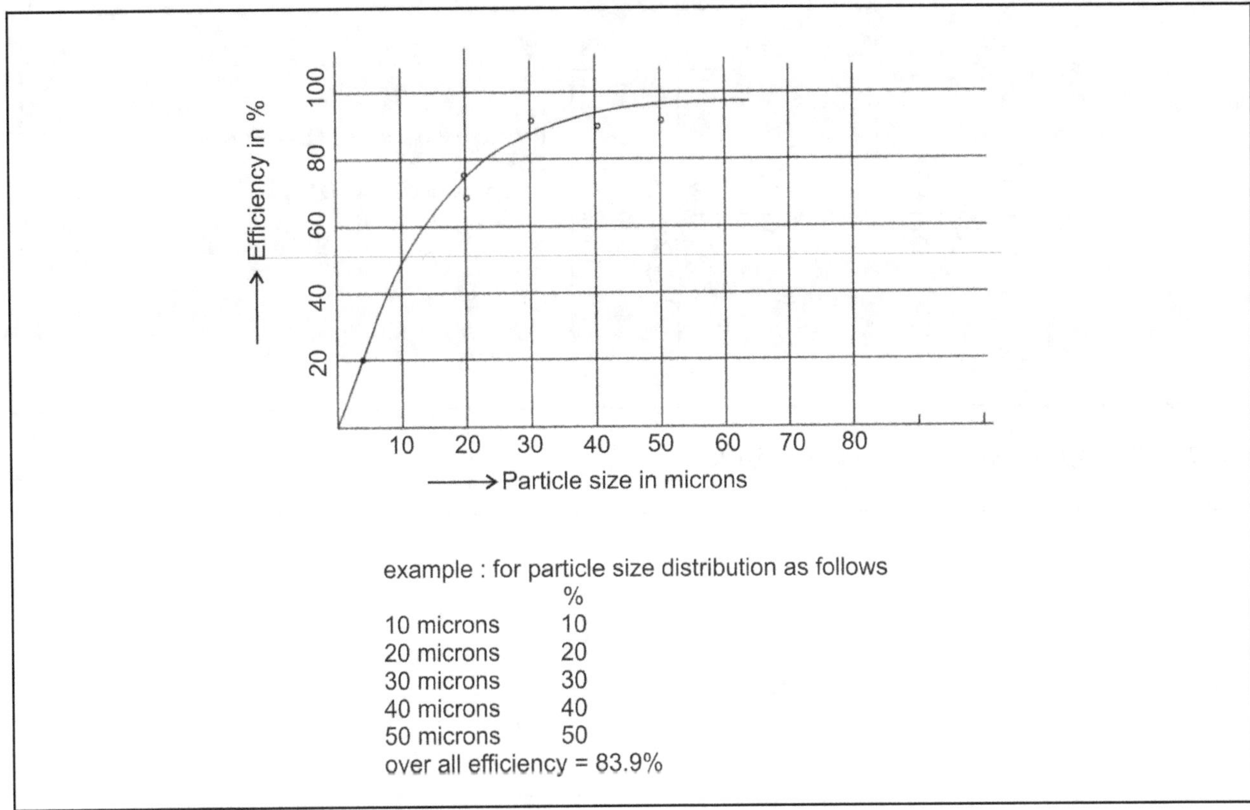

Fig. 2.5.15 Efficiency of cyclone dust collector depends on size distribution of dust particles.

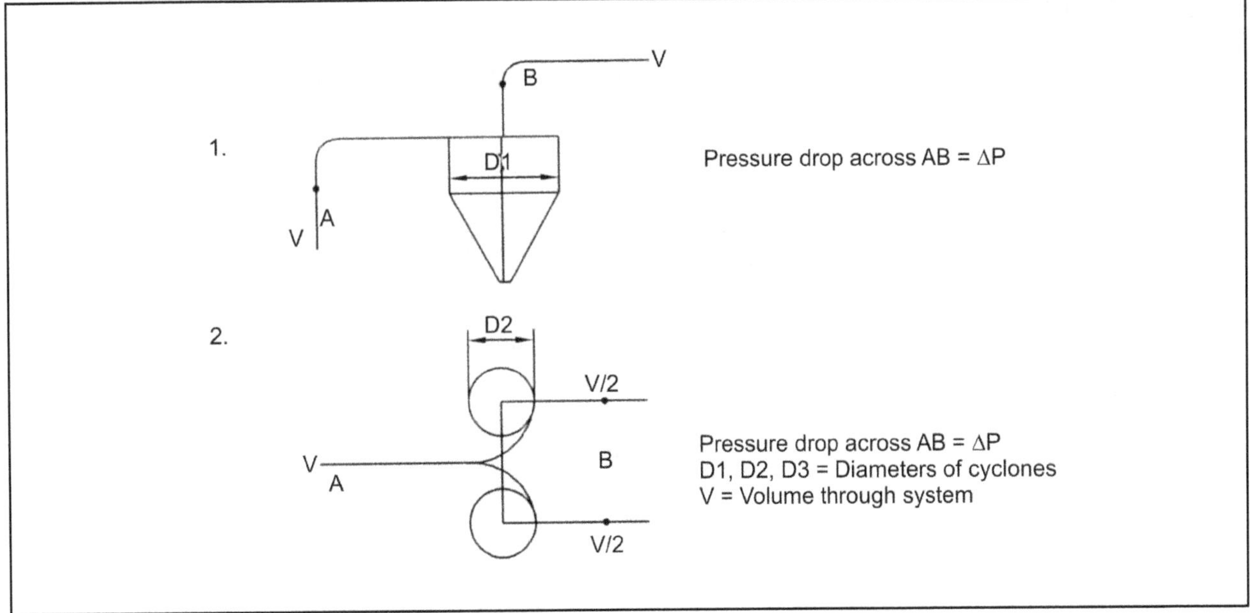

Fig. 2.5.16 Cyclone dust collector single and in group of two.

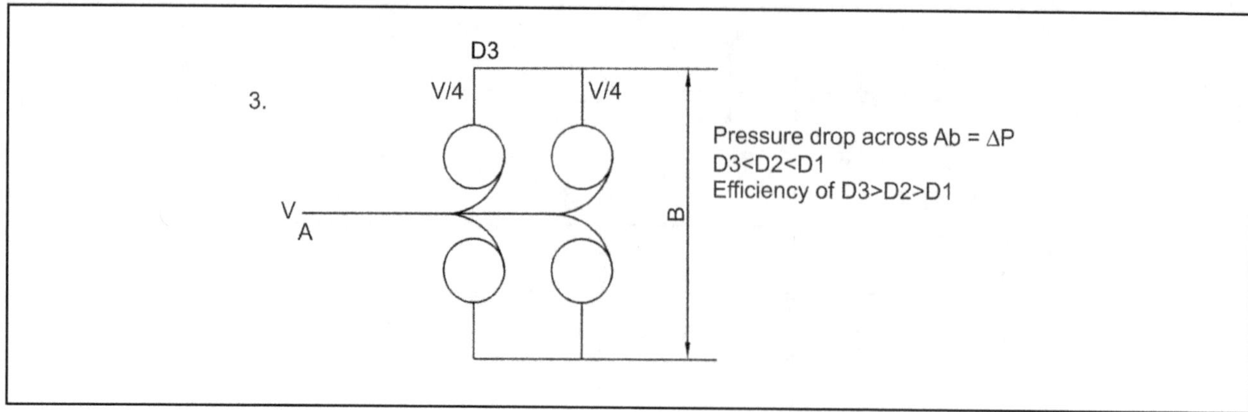

Fig. 2.5.17 Cyclone dust collector in group of four.

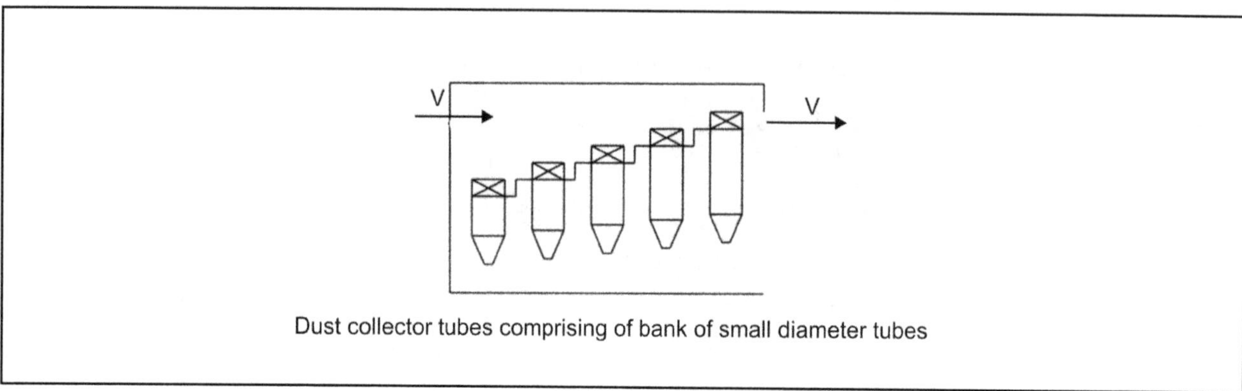

Fig. 2.5.18 Multiclone.

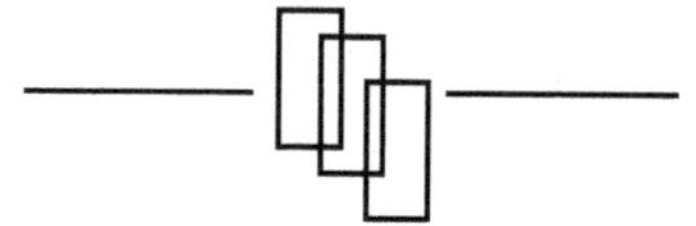

CHAPTER 6

BLENDING SYSTEMS

6.1 Blending Operation

In preparing raw meal, limestone is mixed with additives or correcting materials to produce clinker of right composition. Materials limestone, clay, iron ore are proportioned while being fed to raw mill. While some mixing may take place during grinding it is not enough to produce clinker of uniform quality.

A specific blending operation is therefore necessary to produce raw meal of quality that is consistently uniform in chemical composition and also particle size distribution.

6.2 Blending in Wet Process

In wet grinding, product from mill was in a slurry form. Liquids were easy to stir and blend. Slurry was stored in a silo. Simply admitting compressed air at the bottom of the silo did blending. Air bubbled through the height of raw meal stored and blended it.

However raw meal tended to settle down. Therefore it was necessary to keep blended slurry in agitation. This was done in a 'slurry mixer'. Slurry mixer was a shallow tank, 20 to 30 m in diameter. A slowly rotating stirrer turning on a vertical axis in the center kept slurry in agitation and from settling down.

Slurry mixer was also used to regulate water content in the slurry by skimming off water from the top. Thus if water content in slurry from raw mill was say 36 %, in slurry fed to kiln it would be around 34 %.

Some times clay was ground separately. A separate silo for clay slurry would be maintained. Slurries of limestone and clay would be transferred to a third silo in required proportions and blended in it by aeration as before.

As a matter of fact the ease of this blending is what gave such a long life to wet process.

6.3 Blending in Dry Process

6.3.1 Preblending

In dry process plants, 'blending' is taken to include Preblending also. Preblending is achieved in 'Stacker Reclaimer' systems as described in **Chapter 3**.

The degree of blending that can be achieved is 5-6 : 1.

That is if $CaCO_3$ varies between + 5% in feed, the variation in material extracted from piles would be + 1%.

To obtain uniform quality of clinker it is necessary to keep the variation in carbonates in kiln feed to + 0.2 % measured as a standard deviation.

If in preblending degree of variation has been brought down to + 1 %, then the blending to be achieved in homogenising silos is restricted to 1/0.2 = 5:1

6.3.2 Blending of raw meal

In dry process, product from raw mill is a dry pulverized powder.

Techniques used for aeration of slurry were not suitable for blending dry raw meal.

Fluidisation techniques developed in forties of the last century, made blending of dry raw meal possible. In fluidisation, air at comparatively low pressures is passed through the raw meal stored in a flat bottom silo. Silo was fitted with porous tiles overall of its area.

Air emerged through the tiles and fluidized raw meal over them. It became lighter and tended to rise in volume due to reduction in its bulk density. Such fluidized raw meal behaved like a liquid and tended to flow if aerating surface was at a slope. That was the beginning of 'air slides' and also of 'Batch blending'.

6.4 Air Merge Batch Blending Systems

In 'batch blending' systems, a blending silo was designed to receive raw meal ground by the raw mill in 8-10 hours. This was the size of the 'batch'. This batch was then blended by 'air merge blending'. Each manufacturer has its own designs and equipment for air merge blending but basic principle is the same that is to 'fluidise' raw meal, create differential densities within the silo by supplying air at different pressures and in different quantities to sectors of the silo so that circular movements of raw meal take place in vertical planes, thereby homogenising raw meal.

See plates 2.6.1, 2.6.2 and 2.6.4

Blending effect (also known as blending efficiency) achieved is between 8:1 to 10:1. That is if standard deviation in raw meal feed to silo is + 2 %, then standard deviation of blended raw meal would be + 0.2 %. Since a deviation of 0.2 to 0.3 % is tolerated, 'preblending' would generally not be required.

Operation of the system is such that X-Ray analyzer would also not be required. Therefore batch blending system was eminently suitable for small plants.

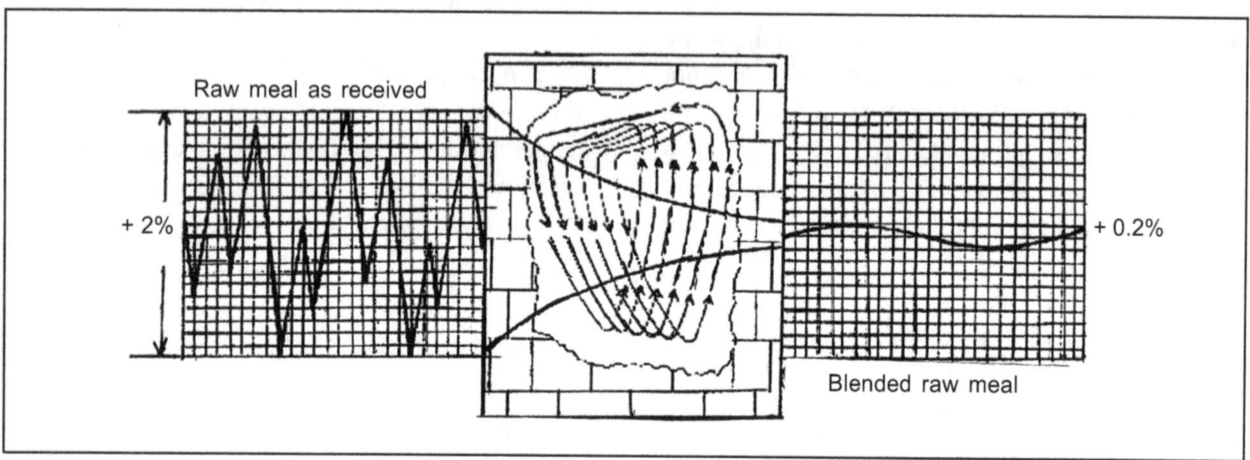

Plate 2.6.1 Blending effect in a quadrant blending system.

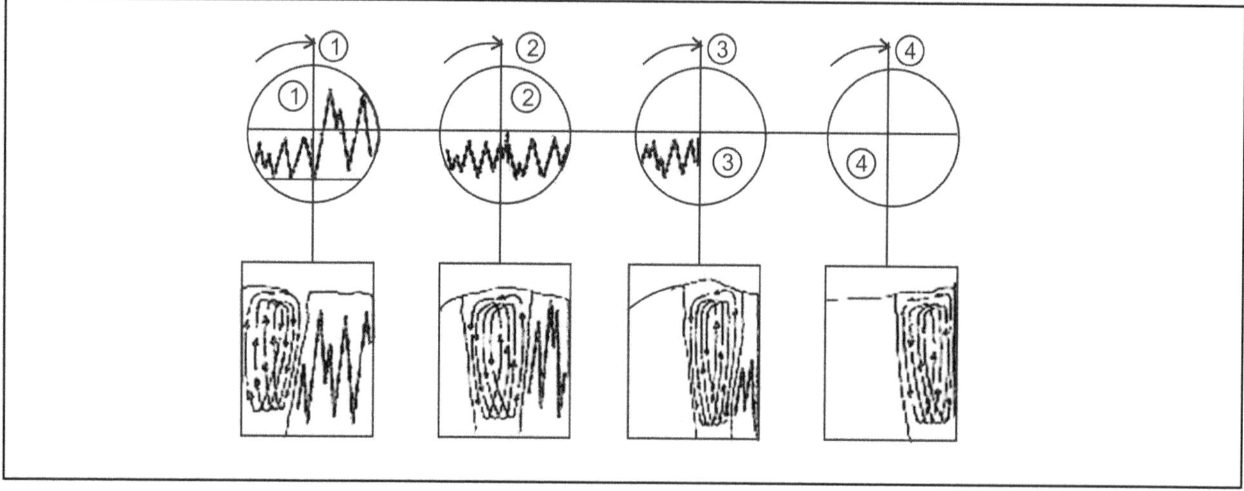

Plate 2.6.2 Blending cycle and progress of blending in a quadrant blending system.

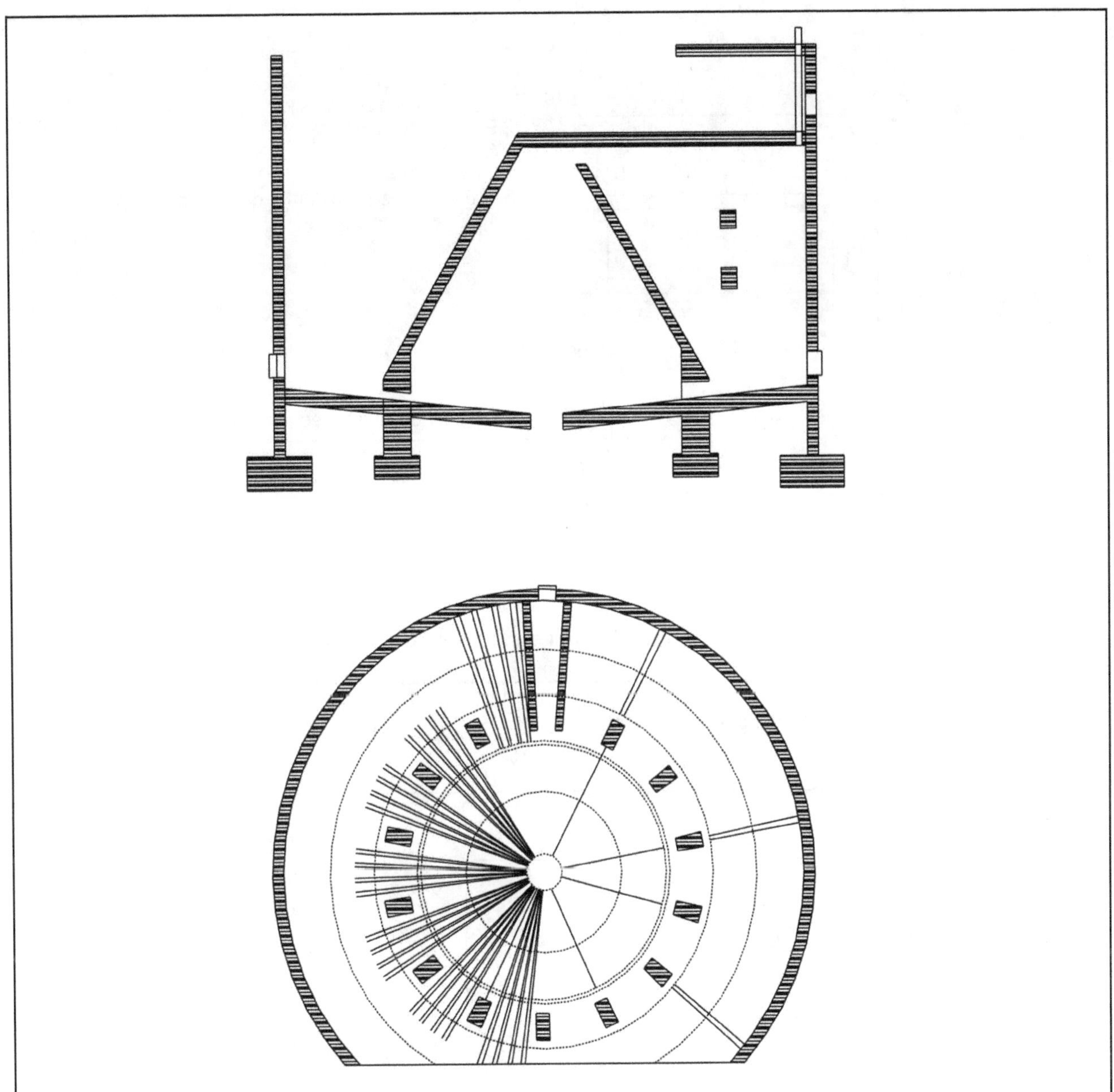

Fig. 2.6.1 Bottom of continuous blending silo - open airslides for aeration.

In subsequent developments in the batch blending system, blending and storage silos were arranged in 'double deck systems' that is blending silo on top and storage silo under it. It eliminated one pumping operation and saved power.

See plate 2.6.3.

6.5 Continuous Blending

As plant sizes grew raw mill capacities also grew. A 10,000 tpd capacity plant required a raw mill of 800 tph capacity. A 10 hours batch would be 8000 tons requiring large silos.

Large kilns also required stocks of 2 - 2 ½ days of raw meal as a buffer stock between grinding and kiln sections. A 10,000 tpd kiln would require a storage of 37500 tons of raw meal. 'Batch blending system' requiring 4 silos in double deck system would be quite expensive.

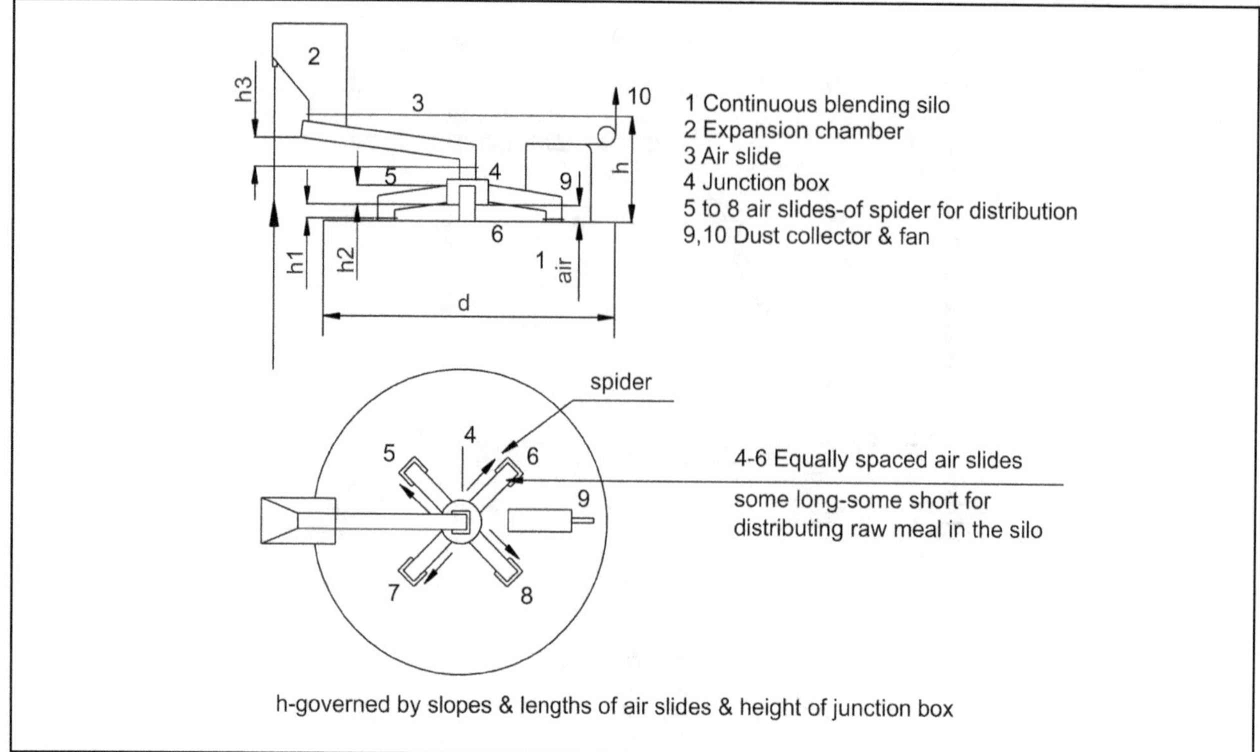

Fig. 2.6.2 Continuous blending feeding arrangement.

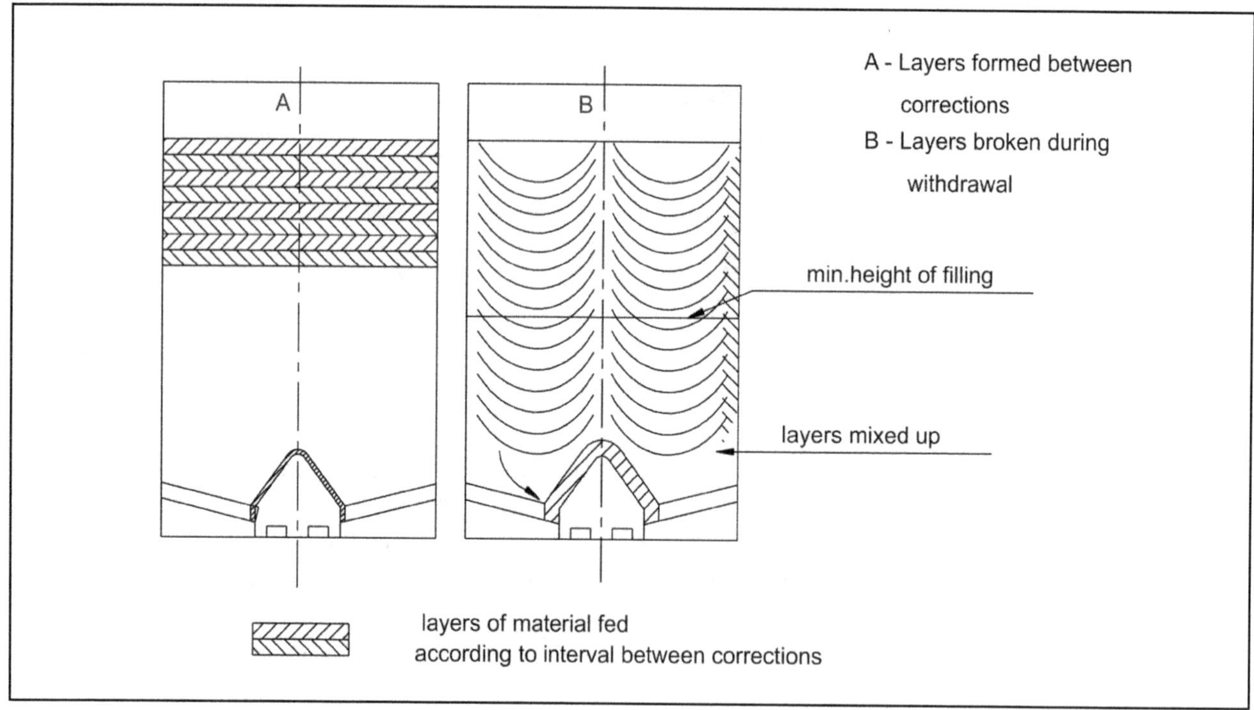

Fig. 2.6.3 Bending action in continuous blending.

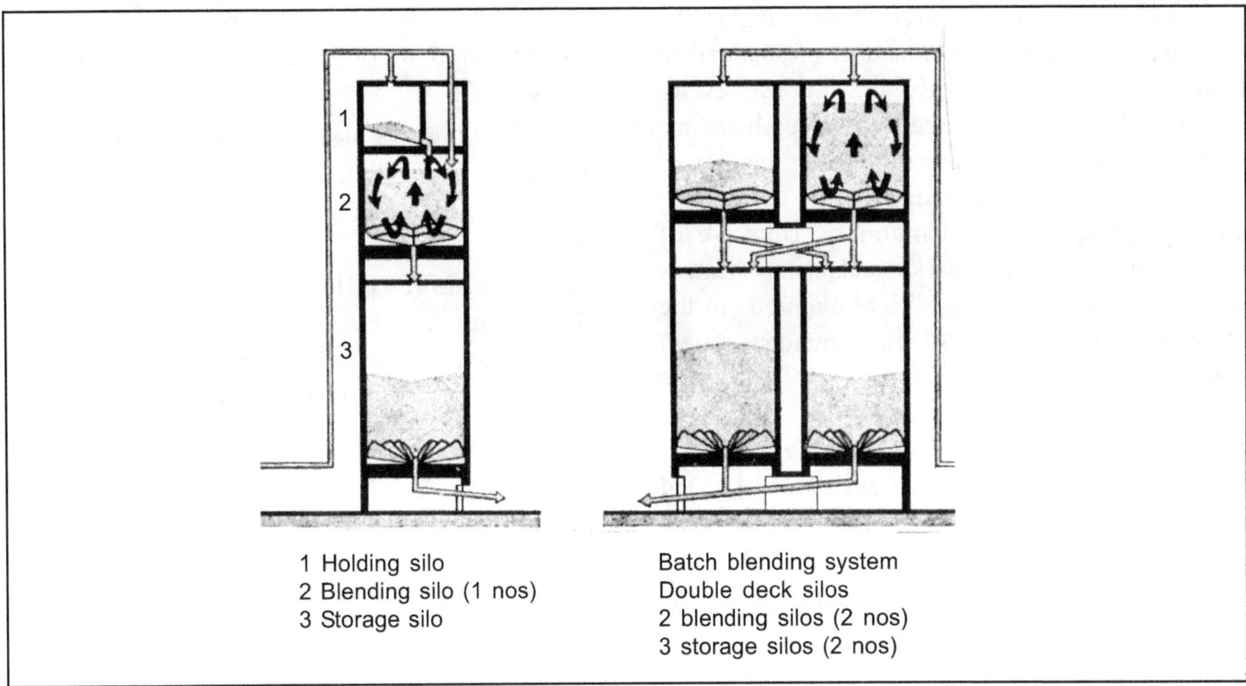

Plate 2.6.3 Arrangements of double deck batch blending systems.

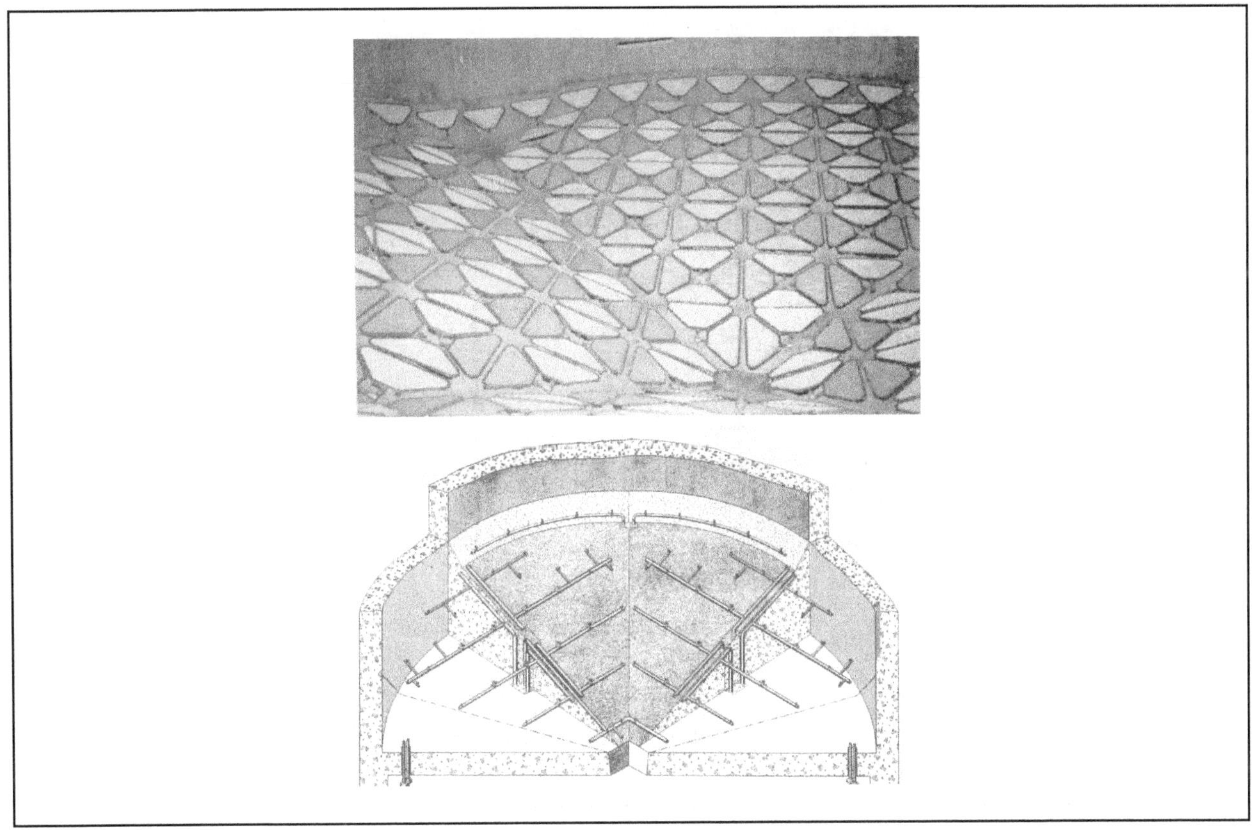

Plate 2.6.4 Piping and aeration media for bottom of the blending silo in a batch blending system.

It was more economical to construct a single large diameter silo - rather than a number of small silos. However this silo would have to achieve necessary degree of blending. Thus the 'continuous blending' system was evolved. As the name suggests, operations of filling, blending and extraction are carried out simultaneously and continuously. Because of 'preblending' achieved in stacker reclaimer system, as mentioned earlier blending effect required in the blending silo was only 5:1-6:1. The continuous system was able to achieve it.

6.5.1 Operation of Continuous Blending System

In continuous blending horizontal layers are broken and extracted across the cross section of the silo like scraper of the reclaimer.

Because of continuous and simultaneous operations of filling, blending and extracting, it was no longer necessary to have separate silos for blending and storage purposes. Same silo serves for both blending and storage.

See plate 2.6.5.

A typical graph showing blending effect acheived as compared to variation in feed is **shown in plate 2.6.7.**

Fig. 2.6.3 shows blending action in the continuous blending silo and **Plates 2.6.5 and 2.6.6** show scheme of aeration and extraction for it.

6.6 Continuous Blending Silo

Thus continuous blending silos of ~ 40,000 tons capacity with diameters 23-25 meters came to be used for cement plants of 10,000 tpd capacity.

Aeration of silo bottom for blending and aeration is shown in **Fig. 2.6.1.** Because of the large diameter of the silo aeration would be done sector wise thereby reducing capacities of blowers / compressors required.

All new plants would be with continuous blending system with 1 large silo of capacity of 2-2 ½ days requirement of raw meal.

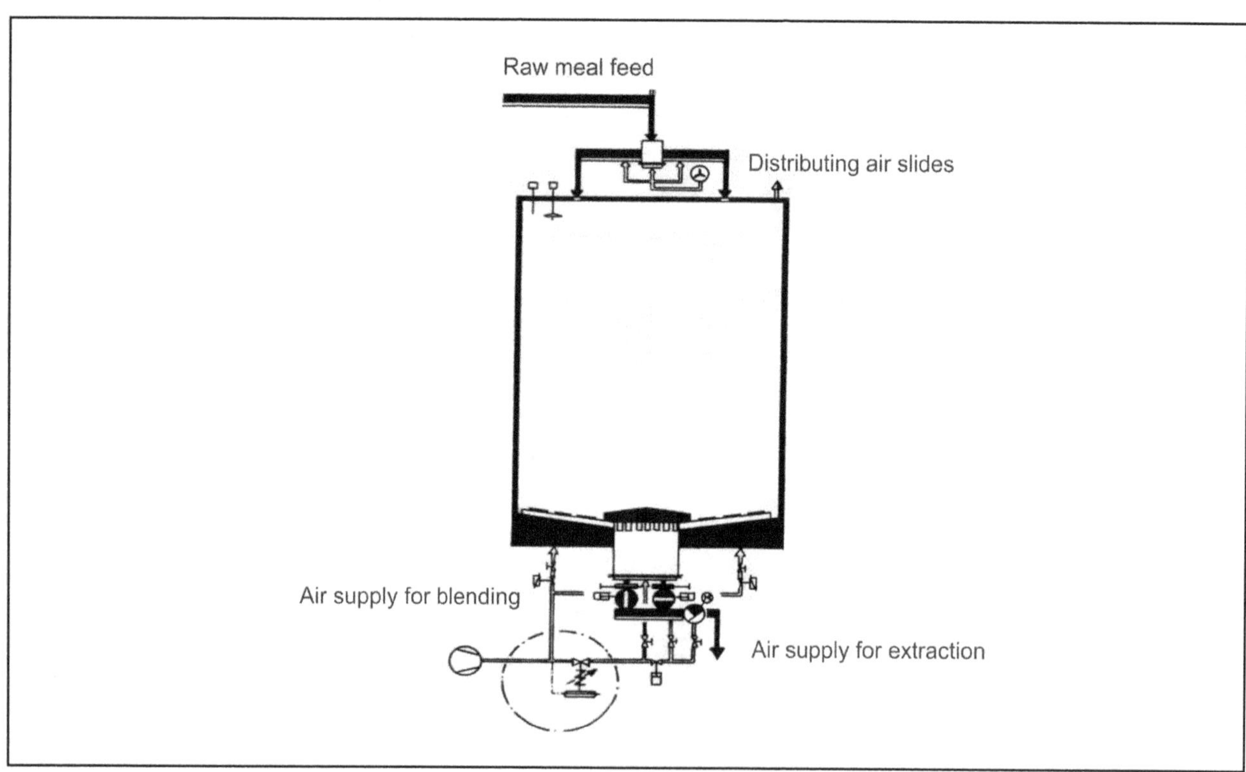

Plate 2.6.5 Continuous blending system.

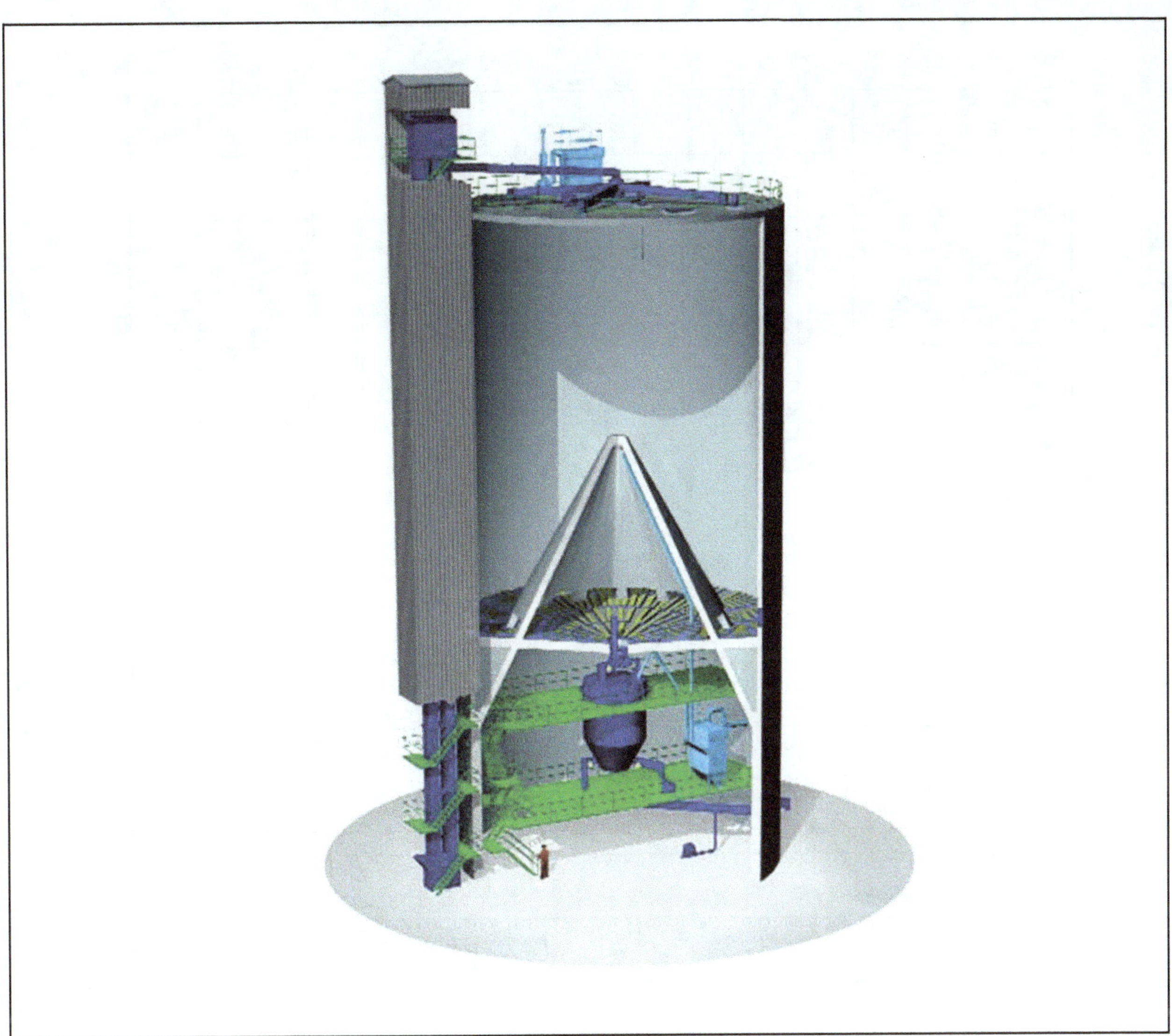

Plate 2.6.6 Continuous Blending System.

In exceptional cases, where the degree of blending achieved in continuous blending is not sufficient to maintain a standard deviation of + 0.2 %, there may be two silos, working in parallel and from which feed would drawn simultaneously for achieving further blending during extraction.

Preblending and Continuous blending are complementary to one another. Either system by itself may not be adequate to achieve the degree of blending required. These systems also require that quality is checked and measured continuously while building up stock piles, while extracting from stock piles, while feeding to the continuous blending silo and while extracting from it for kiln feed.

6.7 X-Ray Analyser

Continuous monitoring is done during grinding of limestones and correcting materials in ball / vertical mills with the help of computer attached to the 'X-Ray analyser' which sends signals automatically to change proportions of limestone (high and low grade), clay - iron ore etc., to achieve a raw mix blend which when mixed with raw meal already in the silo would maintain raw mix quality of kiln feed within the set norms of + 0.2% standard deviation.

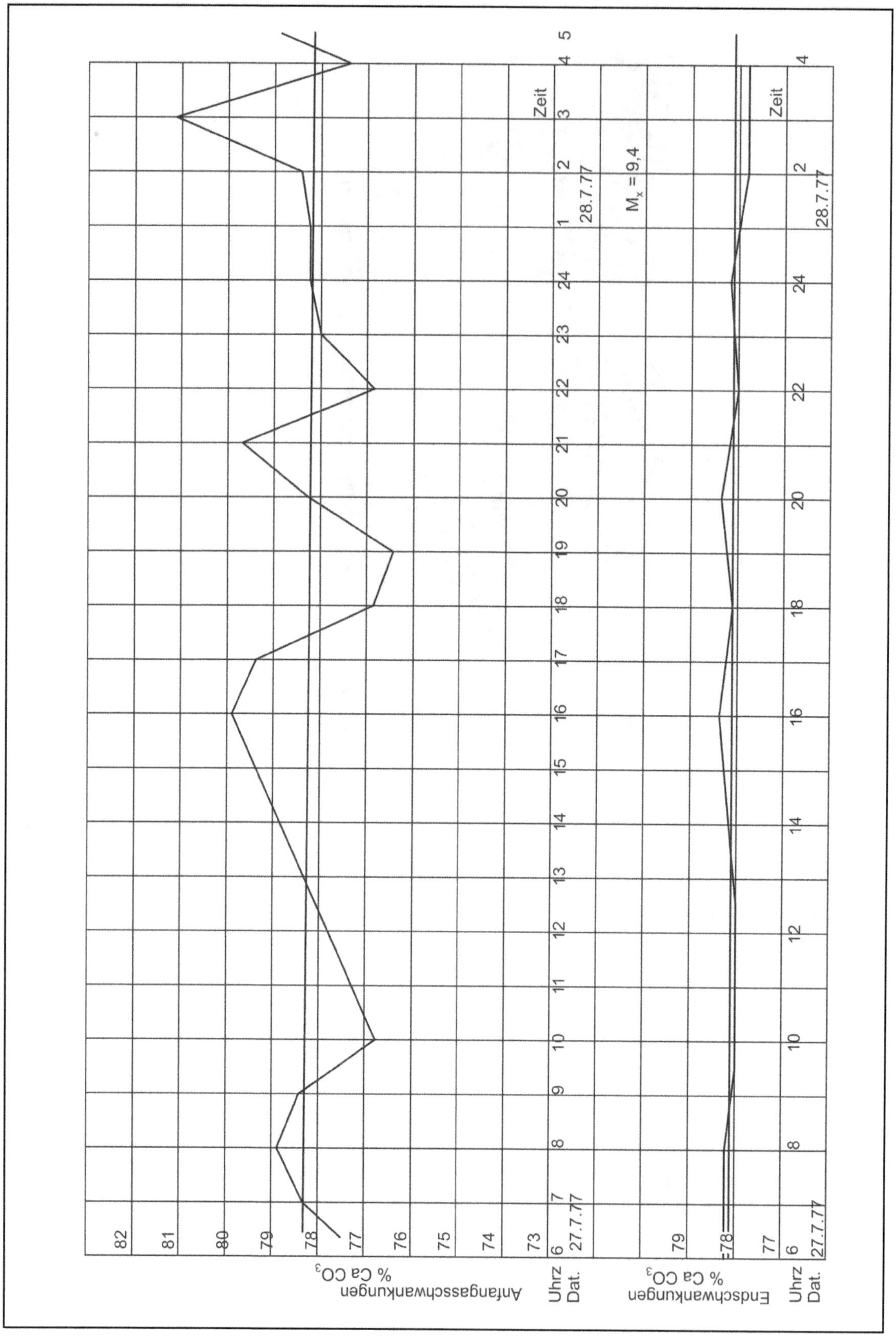

Anfangasschwankungen % Ca CO_3

82
81
80
79
78
77
76
75
74
73

Zeit

Uhrz 6 7 8 9 10 11 12 13 14 15 16 17 18 19 20 21 22 23 24 1 2 3 4 5
Dat. 27.7.77 28.7.77

M_x = 9,4

Endschwankungen % Ca CO_3

79
78
77

Zeit

Uhrz 6 8 10 12 14 16 18 20 22 24 2 4
Dat. 27.7.77 28.7.77

Plate 2.6.7 Blending effect in a Continuous Blending Silo.

Source: CPAG manual.

This will not be possible without an 'X-ray analyzer' (or equivalent instant analysis facilities). It would be too risky to use stacker reclaimer system and continuous blending without an X-ray analyzer. X-ray analysers are expensive, as also are automatic sampling systems and can be afforded by large plants only. However one X-ray analyzer can serve for the whole plant to test clinker and cement as well.

Small low budget plants would therefore be adopting conventional batch blending systems.

6.8 Power Comsumption in Continuous Blending

Power consumption wise continuous blending would require less power because it is not necessary for air admitted for blending to force its way through the entire height of the raw meal in the silo.

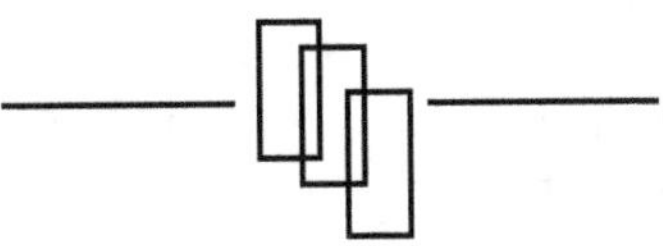

CHAPTER 7

METERING SYSTEMS FOR RAW MEAL AND COAL

7.1 Kiln Feed Systems

Kiln, the main processing unit of a cement plant needs to be fed with blended raw meal in the form of either slurry or powder or granules to produce clinker.

The main purpose of a kiln feed system is to feed the kiln at any desired rate according to the rate of production and to be able to maintain it with a close degree accuracy throughout the range. It should also be possible to monitor the rate according the operational conditions to produce clinker of consistently uniform quality.

7.2 Types of Kiln Feed Systems

There are three main types of kiln feed systems. They are :

1. Volumetric systems where rate of feed is assessed more by volume of slurry or raw meal delivered than its actual weight.
2. Gravimetric systems in which weight of raw meal is actually measured as it is fed to kiln.
3. Indirect systems in which rate of feed is assessed by measuring an apparently unrelated quantity.

+ 3 %. Lesser the variation the better. From this point, gravimetric systems are more accurate with a variation of not more than + 1%.

In dry kilns long or preheater kilns or in kilns with lepol grate, measurement is of dry pulverized raw meal extracted from the storage silo. This measurement could be by volumetric or gravimetric measurements or by indirect methods mentioned above.

7.3 Dry kilns

In dry kilns long or preheater kilns or in kilns with Lepol grate, measurement is of dry pulverized raw meal extracted from the storage silo. This measurement could be by volumetric or gravimetric measurements or by indirect methods mentioned above.

7.4 Volumetric Systems

In this system screw conveyors and rotary vane feeders with close tolerances between casing and screw or rotor are used. Speeds of screw or feeder are variable. Most often d.c. motors would be used. Screws are mounted directly under the serving bin. Head of raw meal inside bin is maintained constant by using an overflow device or level indicators. This ensures that there are no fluctuations in rate of discharge due to change of head of material.

A refinement can be by installing the bin on load cells so that weight delivered can be obtained directly. **See Fig.2.7.1.**

7.5 Gravimetric Systems

In these systems a load cell measures the weight of material passing over a belt, which is translated into feed in tons per hour by measuring belt speed. Belt speed is variable. Gravimetric systems best operate with a pre feeder. Pre feeder extracts material from a bin. If the bin is on load cell it helps to calibrate the system. **Plate 2.7.1 shows weigh feeder**

Pre feeder also has a variable speed drive to change rate of extraction. Pre feeder and weighing belt or weigh feeder work as a pair. **See Fig. 2.7.2**

7.6 Indirect Systems

Most commonly used indirect systems measure impact of falling material on a curved plate which is translated into feed rate in tph.

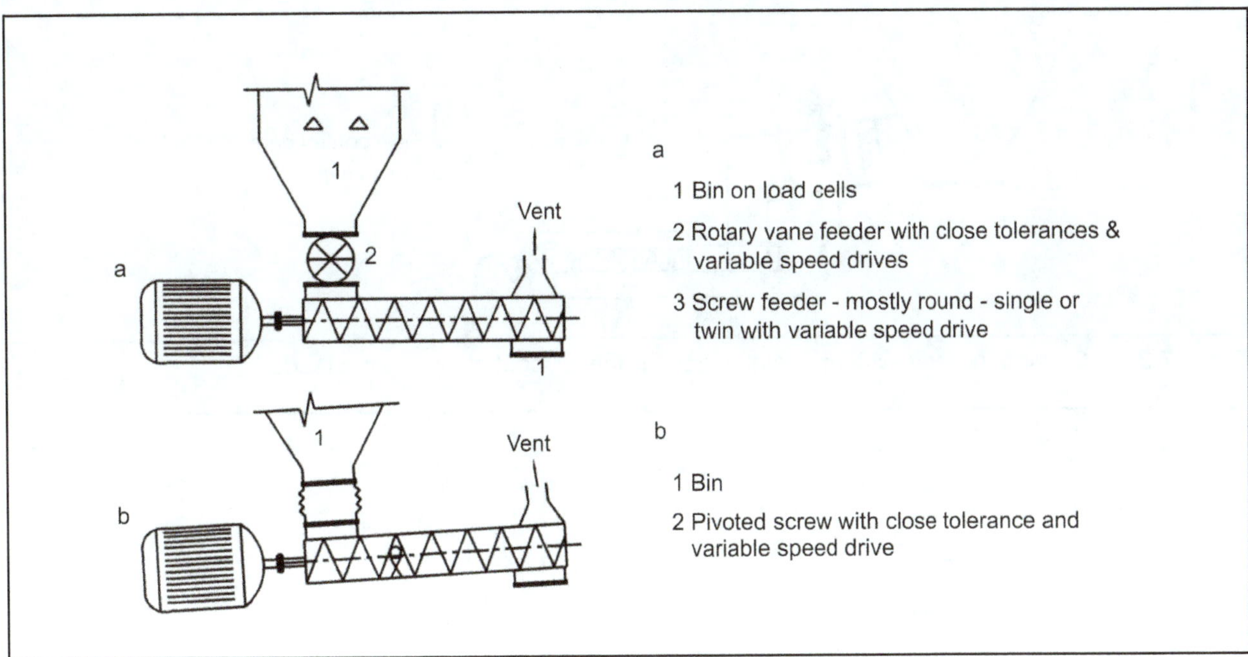

Fig. 2.7.1 Volumeteric kiln feed system – commonly used firing coal in kiln and calciner.

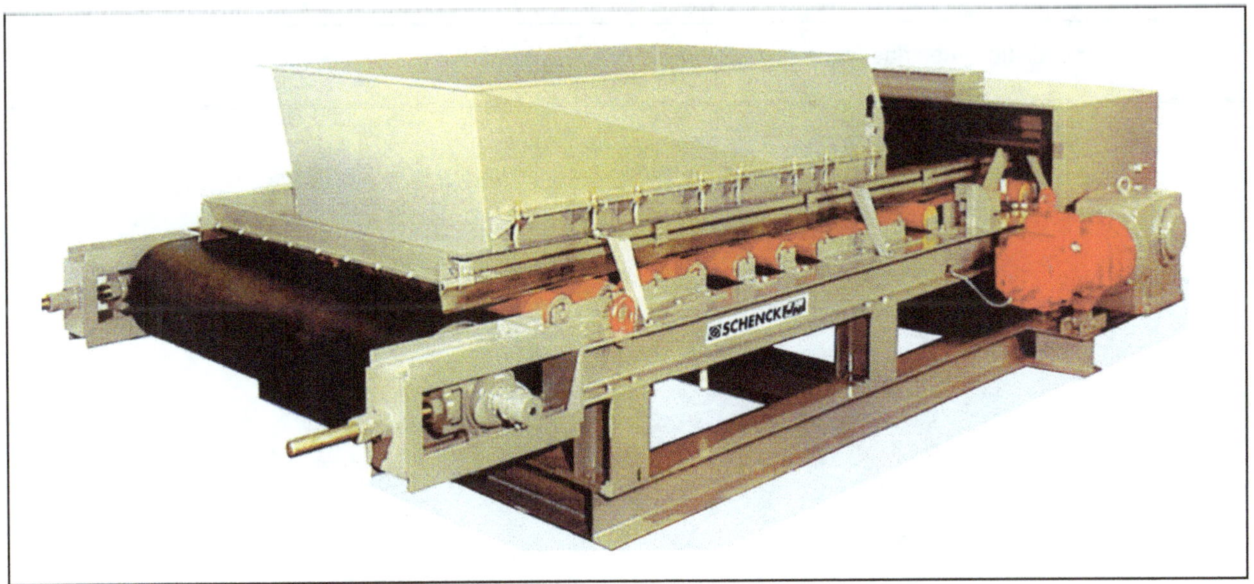

Plate 2.7.1 Weigh feeder in gravimetric system for kiln feed.

This system is known as 'solids flow meter'. See **Fig. 2.7.3 and plate 2.2.7.2.**

In another indirect system, pressure of air in the air chamber of 'air lift' is used to monitor rate of feed with the help of a 'flow control valve'. Other things remaining same, air pressure is a measure of rate of feed.
See Fig. 2.7.4.

In yet another/indirect system, gamma rays are passed thru material -coal- passing over a belt and loss in energy is used as an indication of rate of feed.

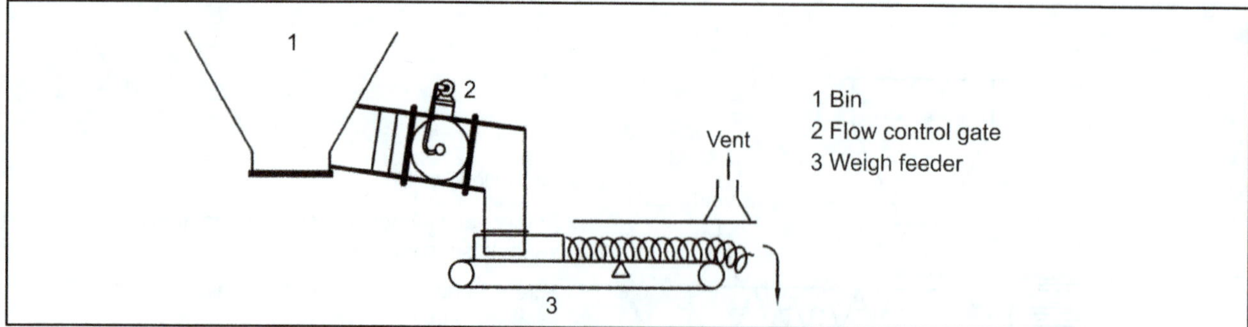

Fig. 2.7.2 Gravimetric feed system - weigh feeder as feeder, control gate as pre feeder.

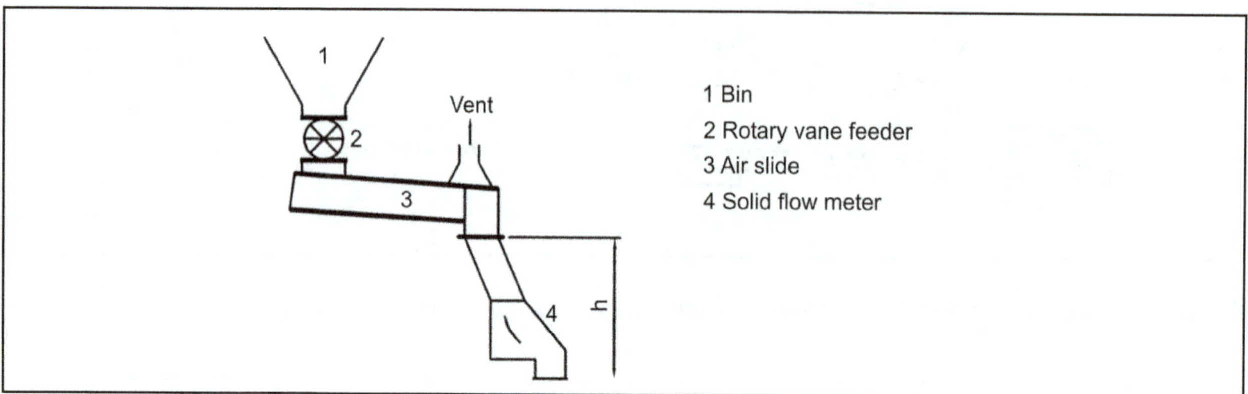

Fig. 2.7.3 Solids flow meter as feeder; rotary vane feeder as prefeeder.

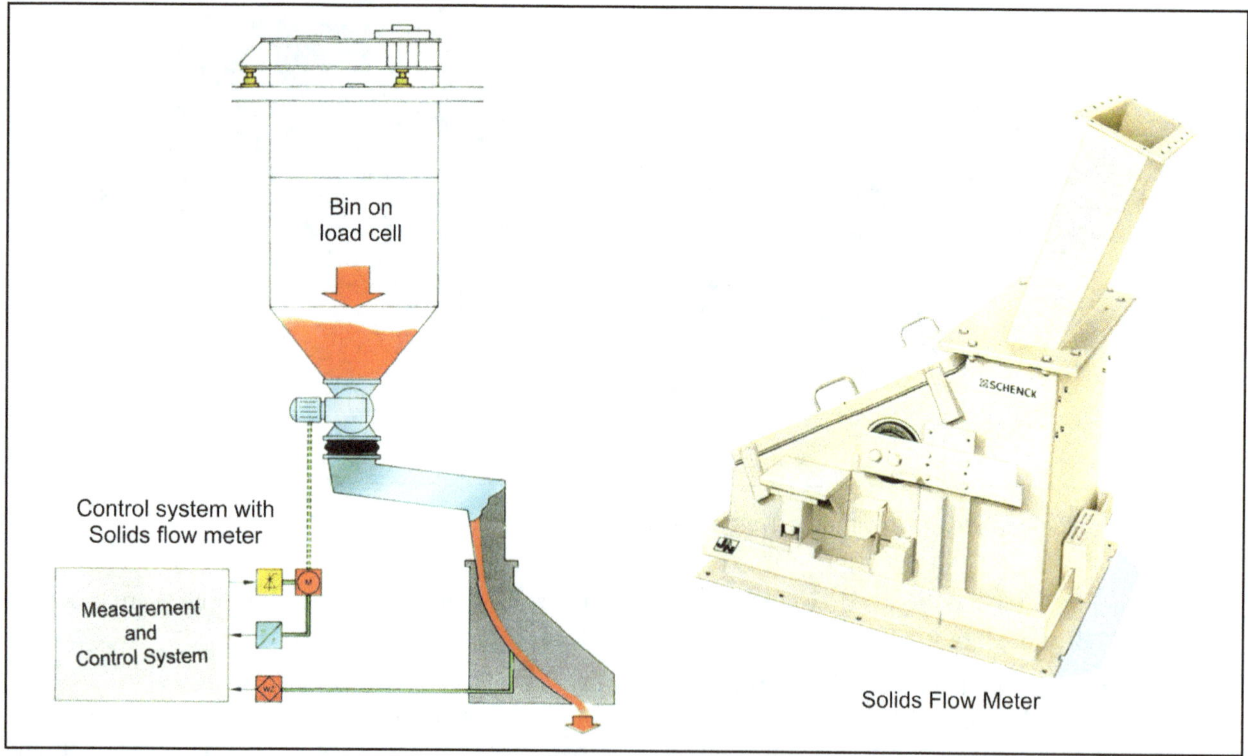

Plate 2.7.2 Kiln feed system using solids flow meter.

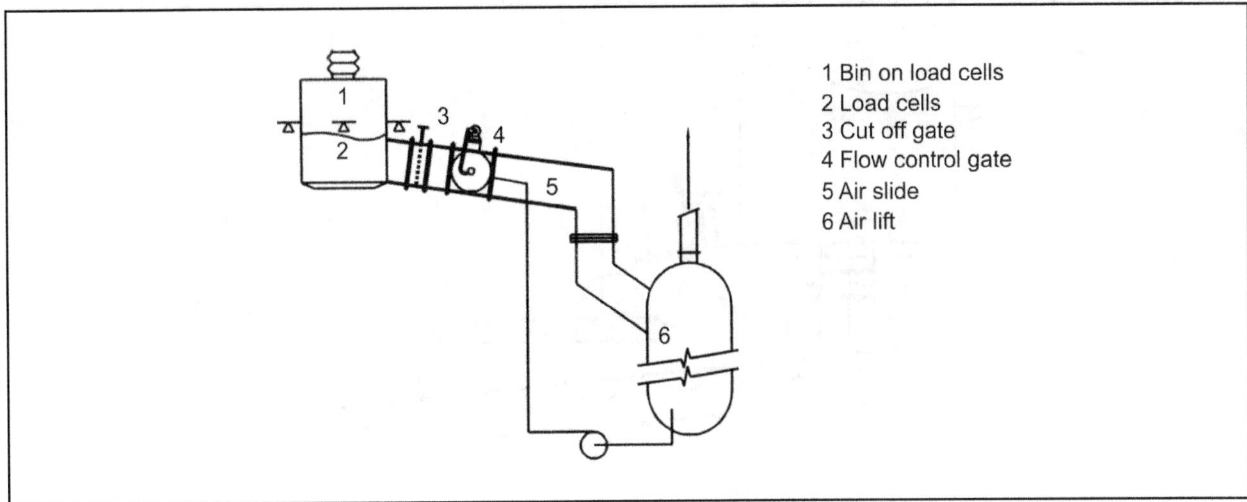

Fig. 2.7.4 Flow control gate alone used for kiln feed : feed regulated by air pressure of air lift.

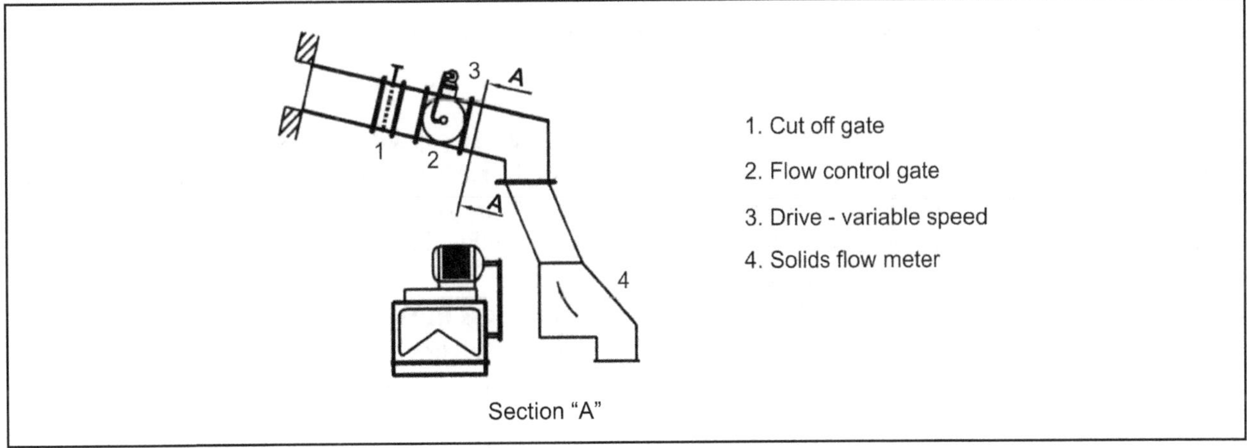

Fig. 2.7.5 Flow control gate for regulating feed – used as prefeeder, solids flow meter as feeder.

Indirect system used more commonly with coal is 'Loss in weight system'. In it the time of emptying a given quantity of material is used as an indication of rate of feed.

7.7

Most commonly used feeders for kiln feed for large cement plants are 'weigh feeders' and 'solids flow meters'. Rotary vane feeders with close tolerances and 'flow control gates' are the commonly used pre feeders.

See Figs. 2.7.2, 2.7.3, 2.7.4 and 2.7.5.

Sometimes a flow control gate may be used by itself as a feeder.

See Fig. 2.7.4.

7.8 Standby for the System

A kiln feed system needs a 'standby'. Weigh feeder serves as the main feeder and flow control gate as the standby.

See Fig. 2.7.6.

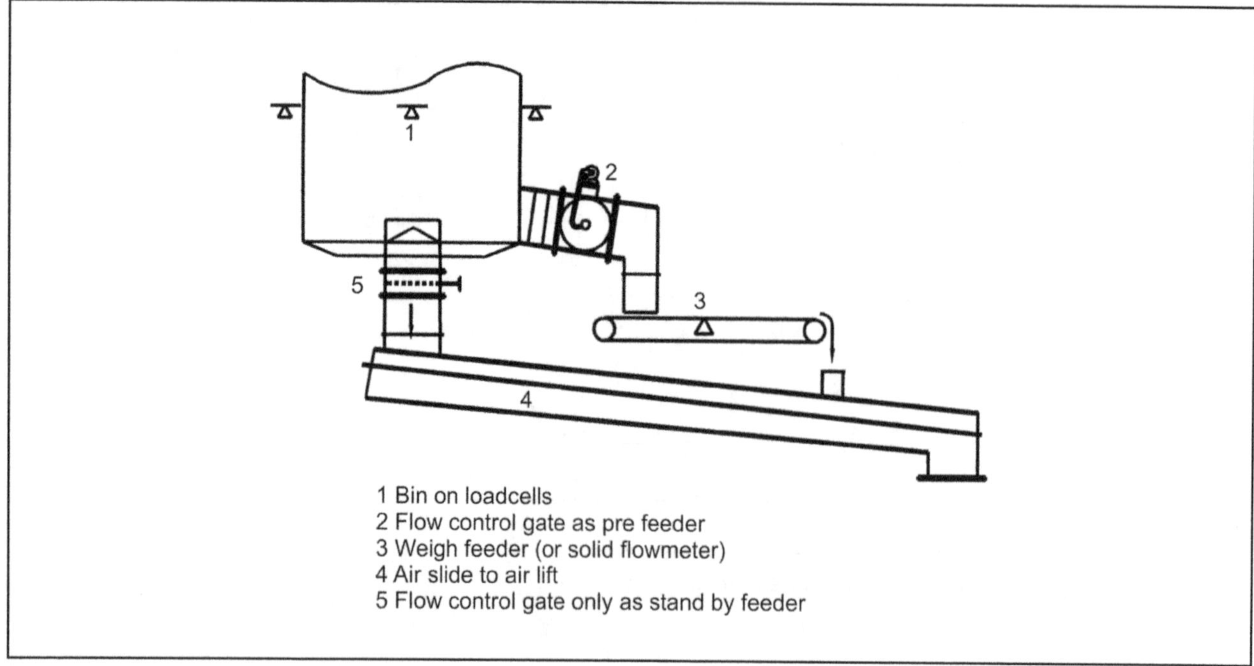

Fig. 2.7.6 Flow control gate as prefeeder and standby feeder for kiln feed.

7.9

Kiln feed system taken as whole should include the final 'conveyor' to lift raw meal to the preheater. An air lift was the most common pneumatic conveyor till recently. However it has now been replaced by a belt bucket elevator. Standby is then provided by an air lift.

7.10

There are many variations of the basic systems and they can be fitted into a plant layout in several ways. These will be dealt with in chapters on layouts.

A typical kiln feed system installation located under a continuous blending silo is shown in **Plate 2.7.3.**

7.11 Metering Pulverized Coal for Kiln and Calciner

Just as raw meal is fed to kiln at preheater end after metering, coal is fired into kiln and calciner after metering. Coal metering systems can be volumetric or gravimetric same as for raw meal.
See Figs. 2.7.1 and 2.7.2.

Because capacities are small, round twin screws with decreasing pitch are commonly used.They are mounted directly under the bin without a prefeeder as shown in Fig. 2.7.1.

7.12 Loss in Weight System

Loss in weight system is also often used to meter and fire coal in kiln and calciner. In it, rate of emptying is monitored with the help of a computer and is used to control the feed rate. **See Plate 2.7.4**

7.13 Measurement of Feed Rates to Mills

Measurement of feed rates are also necessary for all the mills. In case of Raw mills not only rates of feed of constituents of raw mix but their continuous proportioning is also required. This is done with help of X-Ray analyzer. So is the case with cement mills. In making blended cements slag and fly ash have to be proportioned.

For these applications belt weigh feeders are used. They are installed directly under hoppers. Belt feeders will have variable speed drives.

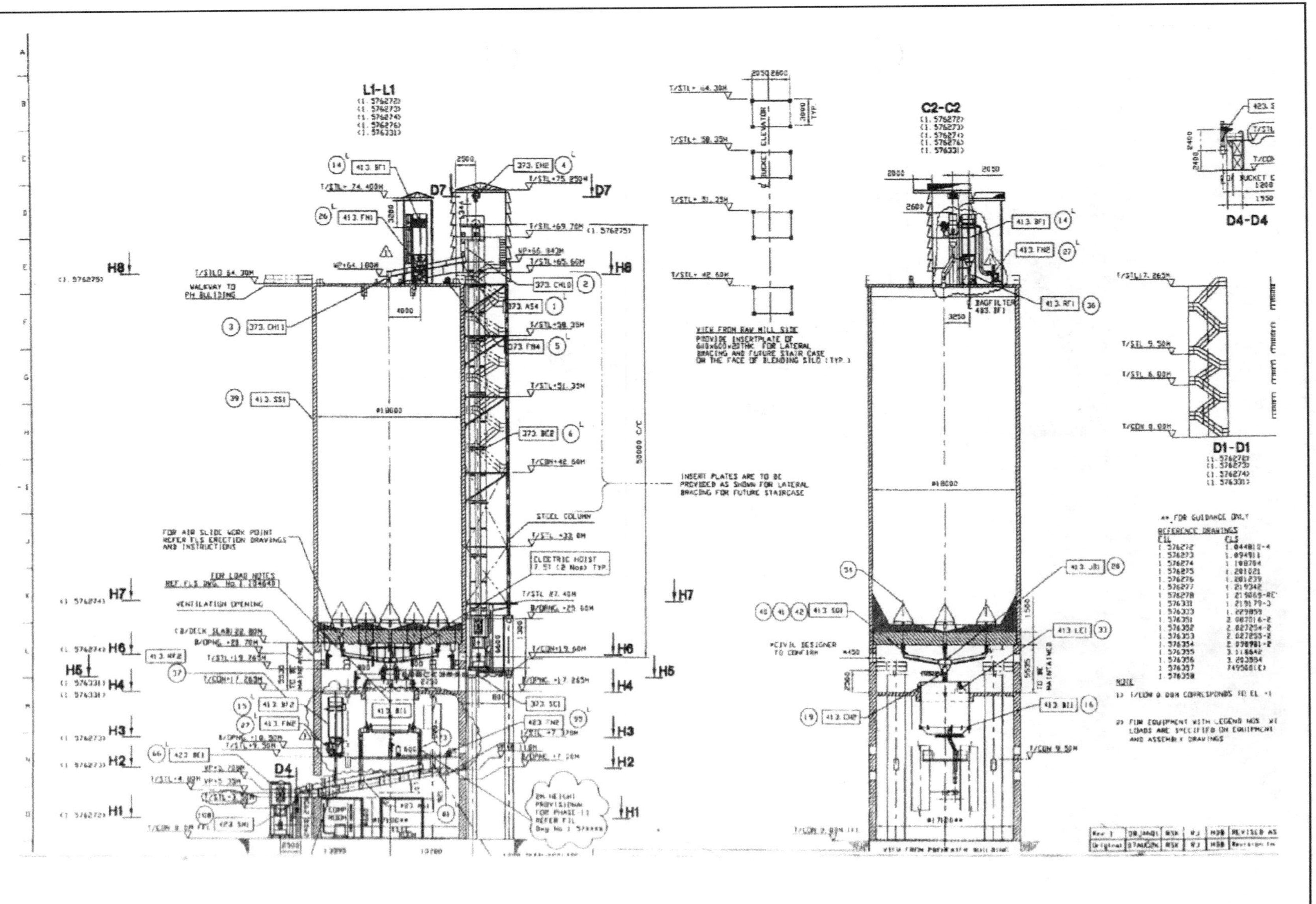

Plate 2.7.3 Continuous blending system with kiln feed system under the silo.

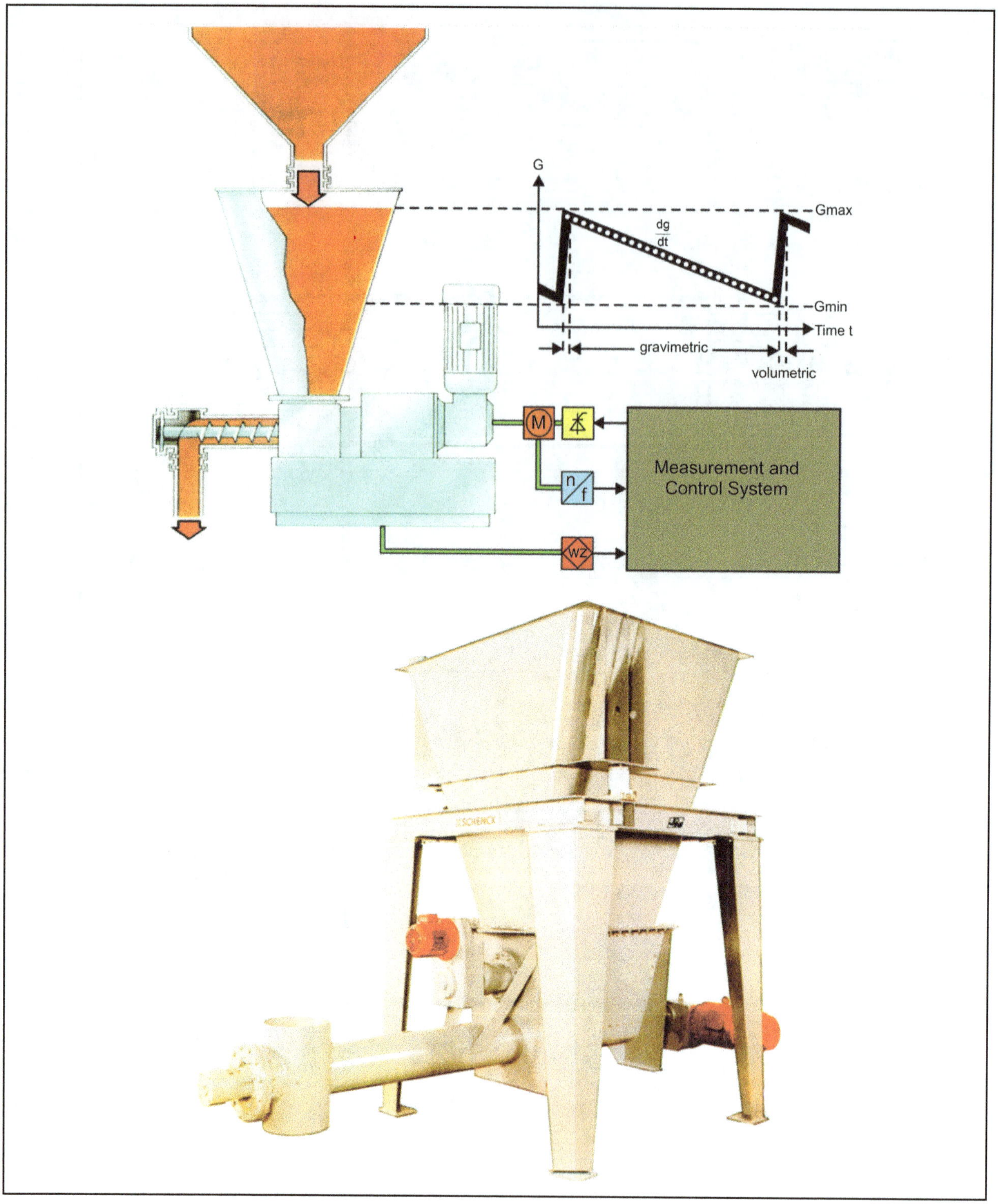

Plate 2.7.4 Loss in weight system generally used for metering pulverised coal.

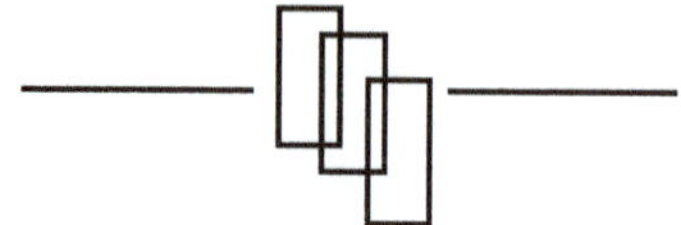

CHAPTER 8

PREHEATERS

8.1 Preheaters

Preheaters were developed to take the process of heating raw meal to a temperature where calcinations or dissociation of CO_2 begins, outside the kiln.

This function was done inside the kiln in long kilns. Length of the kiln shortened as a consequence of taking the function of preheater outside the kiln.

8.2 Lepol Grate Preheater

In case of dry process or to be exact semi dry process, the first attempt to take preheating outside the kiln resulted in the development of Lepol Grate Pre heater. Lepol grate consists of a traveling grate to which nodules of raw mix are fed and travel slowly with it. Hot gases from kiln at temperatures are passed through the bed of granules more than once. Gases entering grate pre heater at 1000 °C lose their heat to granules and heat them to temperatures of 800 °C. Gases leave grate at 180 °C. Grate consists of a chain turning on a number of shafts inside a refractory lined casing. On the chain are fitted grate plates with slots for passage of gases. Smaller granules and dust fall through the slots and are collected in a chamber with hoppers under the gate. The spillage so collected is hot and is conveyed by a Redler conveyor to an elevator which lifts it and feeds it to kiln. Both Redler conveyor and elevator are made of special heat resisting material to withstand high temperatures. Chain and grate plates are also subject to high temperatures at least for a part of their travel and are also required to be made of heat resisting material.

Besides reduction in fuel consumption as compared to long dry kilns, Lepol grate preheater in the initial years when pollution control laws were not so severe, also had the advantage that exhaust gases from grate could be let into the atmosphere straight.

Main disadvantages were high maintenance costs and difficulty to maintain sealing in long casing through which a large number of shafts projected.

Control was by changing speed of grate, by changing speed of granulator and by flow of gases.

Lepol grates were developed up to 2000 tpd capacity. Their development stopped after Cyclone pre heaters were established.

8.3 Cyclone Preheaters

Cyclone preheaters are a series of cyclones arranged one over the other. Gases from the kiln enter the bottom cyclone through a riser duct. In this duct raw meal collected in the stage above is introduced. Raw meal is picked up by kiln gases and carried into the cyclone. In this journey gases transfer their heat to raw meal which gets heated further. When gases with raw meal in suspension enters the cyclone, raw meal is separated from gases by cyclonic action and gases go to the cyclone above it. Raw meal on the other hand enters the kiln. Each cyclone, connecting duct and raw meal pipe thus form a 'stage'. The process is repeated as many times as there are stages. In the final stage or top cyclone, raw meal from kiln feed is introduced from the kiln feed system and gases go to preheater fan which draws them through the system.

In each stage heat transfer takes place when raw meal is in suspension in gases which is a very favourable condition for transfer of heat. Gases from kiln enter bottom cyclone at about 1000 °C. Raw meal enters preheater in the top stage at about 60 °C. Gases leave prheater at 350 °C for in a 4 stage preheater, at ~270 °c in a 6 stage prheater and raw meal enters kiln at about 800-850 °C.

8.3.1 Pressure drop in Preheater

Initially number of stages used to be 4 only. Pressure drop across a 4 stage preheater used to be 500 to 550 mmwg. Any increase in number of stages resulted in additional pressure drop which unfavorably offset the gain in fuel efficiency. Therefore cement industry continued for quite some time with 4 stage preheaters till cyclones of low pressure design were developed.

Attention was paid by Designers of preheaters to the geometry of cyclones and their internals to reduce pressure drop. As a result in present designs, pressure drop is ~ 400-450 only even in a 6 stage preheater. Total pressure drop including that across calciner is around550 mm **See Figs. 2.8.1 and 2.8.2.**

Plate 2.8.1 shows a 5 stage preheater of cyclones of new design with low pressure drop.

Presently pressure drops in 6 stage preheater are comparable to or even less than that in a 4 stage preheater of old designs.

SeeTable 2.8.1

See Fig 2.8.7 shows temperature draft profile in a 6 stage preheater

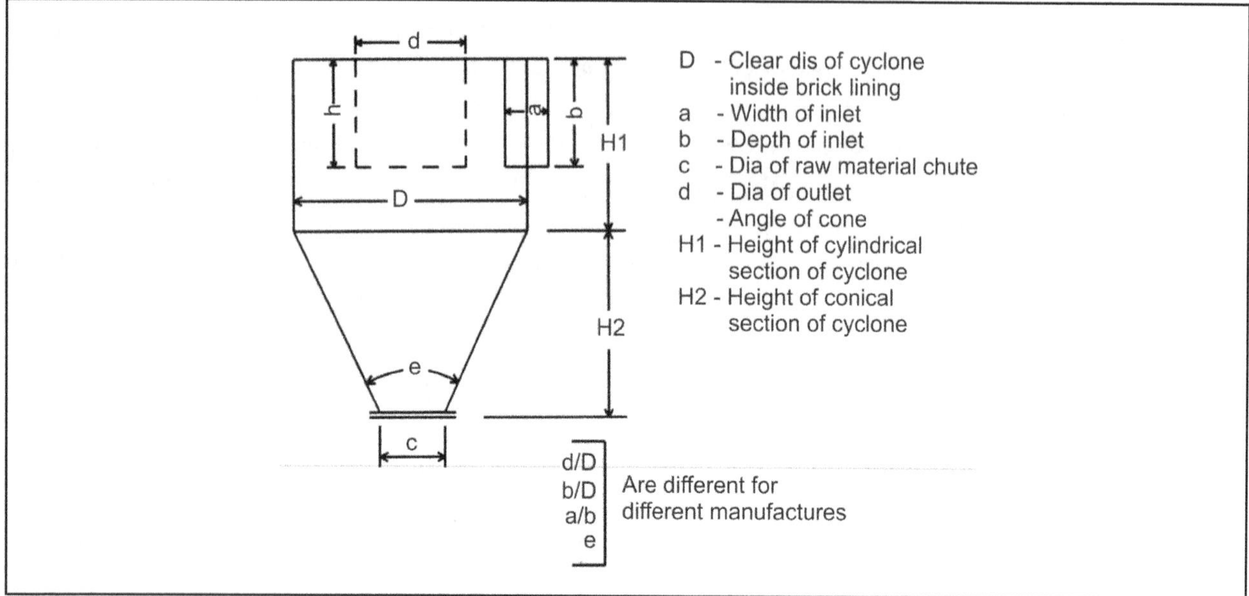

Fig. 2.8.1 Proportions of preheater cyclone.

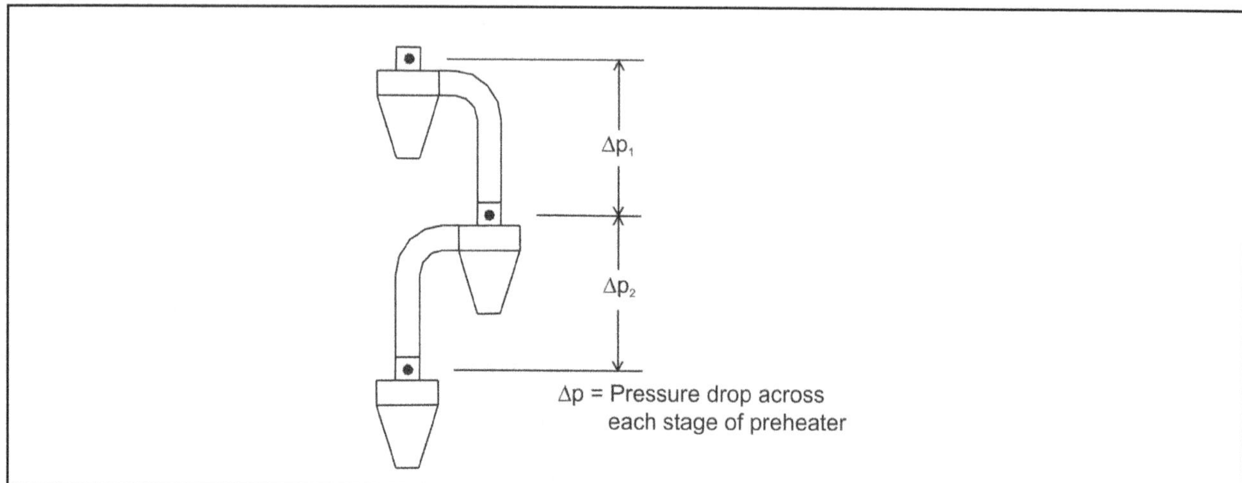

Fig. 2.8.2 Pressure drop across one stage of cyclone.

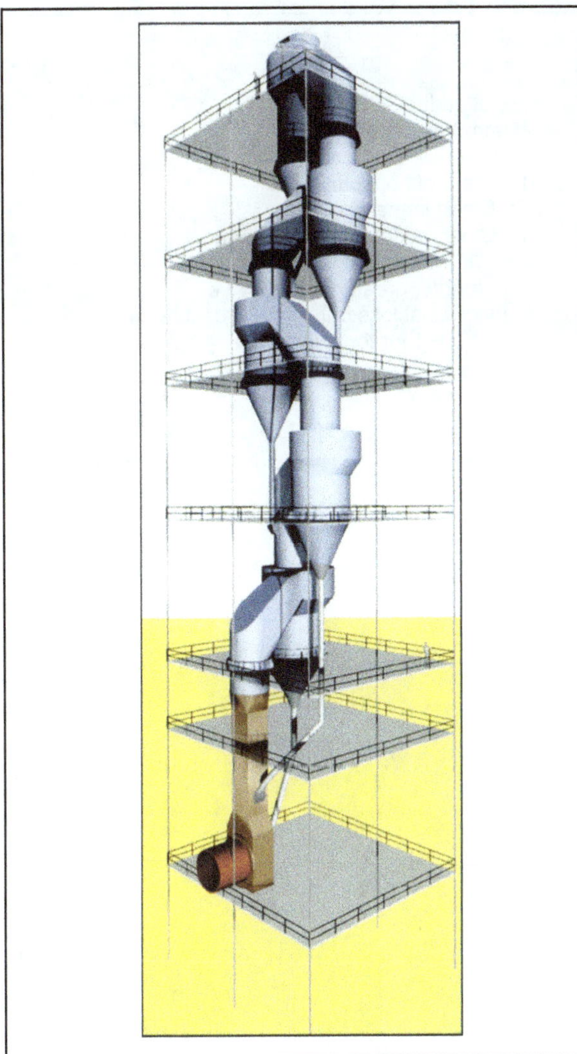

Plate 2.8.1 5 Stage Suspension Preheater with low pressure drop cyclones.

Table 2.8.1

No. of stages	Pr. Drop across Preheater mmwg	temp. of gas °C at exit
4	280 - 300	350
5	320 - 370	300
6	400 - 450	270

As a result, fuel consumption has steadily gone down and specific gas volume expressed in nm3 / kg clinker has also come down.

8.3.2 Sp. Power Consumption

With reduced pressure drop, temperature and reduced gas volume specific power for preheater fan expressed in kwh/ton of clinker has also come down.

It is now even less when kiln feed is lifted by a bucket elevator and not by airlift. **Table 2.8.2** shows typical values for sp. fuel consumption, sp. gas volume and sp. power for preheater fan.

Table 2.8.2 Operational norms in preheaters

Operational norms in preheaters				
Using air lift				
Sr. No.	No. of stages	Sp.fuel consumption	Sp. gas volume	Sp. power Preheater fan
		Kcal kg	Nm^3/kg	Kwh /ton
1	4	800	1.65	5.75
2	5	750	1.55	6.05
3	6	700	1.45	6.3
Using elevator				
1	4		1.55	5.4
2	5		1.45	5.65
3	6		1.35	5.87

It would be seen that development of low pressure drop cyclones has resulted in bringing close specific power consumption of preheaters of 4 to 6 stages.

Earlier with cyclones of older designs when for 4 stage preheater, sp. fuel consumption used to be 850 kcal/kg and temperature at exit used to be370 oC, sp. power consumption for preheater fan used to be + 10 kwh/ton .

Thus with this development, it has been possible to increase number of stages from 4 to 6 without paying penalty in terms of power consumption.

8.4 Developments in Design of Cyclones

2.Changes in design of cyclones are related to :

1. Cylindrical height in relation to clear diameter of cyclone. Different ratios are used in different stages
2. relative dimensions of width and height of duct entering cyclones. Now in almost all designs ratio of width / height is 1 : 2.
3. Velocity at entrance of cyclones is very much reduced. It is now 20-22 m /sec for bottom cyclones reducing to 14-16 m /sec for top cyclones.
4. Velocities at exit of each stage are much less than velocities at inlet. They are between 10 - 15 m sec.
5. Inlet duct is arranged at a slope so that there is no accumulation of dust in the duct.
6. Volute joins the cylindrical portion of the cyclone at a slope so that accumulation of dust is avoided.

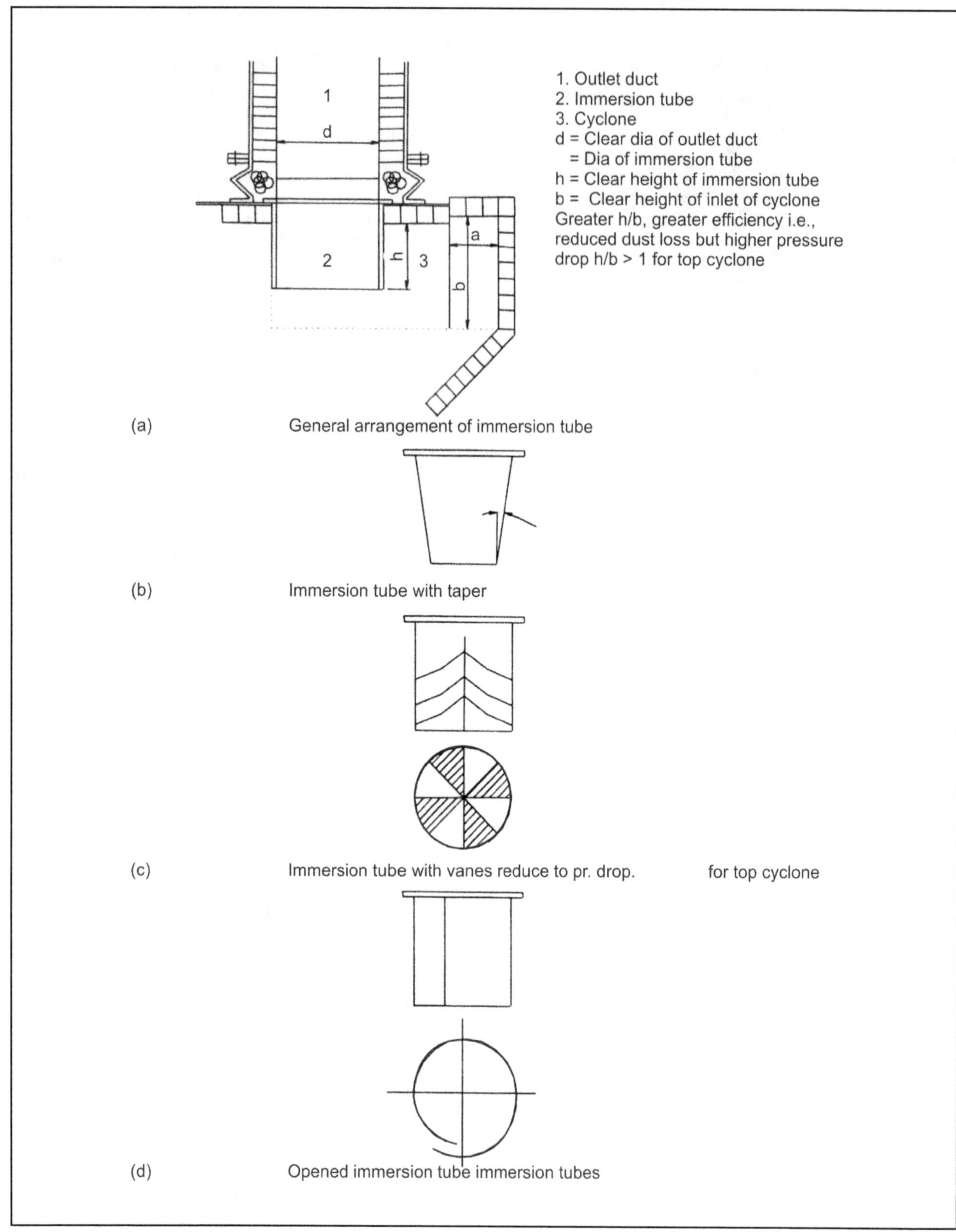

Fig. 2.8.3 Design of immersion tubes.

7. Bottom two stages have double cones to prevent accumulation of material inside cone. Angle of cone is also more steep.
8. Design of immersion tubes has seen significant changes from point of view reducing pressure drop without sacrificing efficiency.

   See Fig. 2.8.3.
9. Leading Companies like FL Smidth use Computer assisted CFD Analysis to design ducts and cyclones in such as way as to improve heat transfer and also to reduce pressure drop in the system.

   See Plate 2.8.5
10. **Down draft cyclone.**

    FLSmidth have developed a 'downdraft' cyclone for the top stage. **See Plate 2.8.4**

    It avoids the big bend at the top and its support.

Single streams are now possible for kiln capacities upto 6000 tpd (cyclone dia ~ 9m). Beyond this two streams are used, beyond that there will be three streams. following **Table 2.8.3** furnishes relevant data on Kiln capacities and corresponding number of streams presently possible.

Beyond 11000 tpd, there will be three streams, **Figure 2.8.6** shows a flow chart of three stream preheater and two clciners and a single kiln.

See plates 2.9.1 to 2.9.6 in Chapter 9.

8.5 Preheater in Calciner Kilns

Because of higher quantum of fuel being fired in calciner, preheater stream for calciner is correspondingly larger than stream on kiln. This results in two dissimilar towers side by side with floors at different spacing and different total heights.

In some designs gases from two streams are mixed and divided equally so that the two streams are identical.

Two stream preheaters can be arranged in a plane perpendicular to kiln or in line with kiln.

Top cyclones used to be twin cyclones. It is not uncommon to find single cyclones in the top stage when preheater is a two stream preheater.

8.6 Feed to Preheater

Raw meal used to be fed by FK Pumps or air lifts to duct between 1st and 2nd stages. Now bucket elevators are used. To reduce heights of elevators, 'enmass conveyors' are used.

8.7 Brick Lining of Cyclones

To reduce radiation losses and weight and size of cyclones, insulating bricks or blocks are used between brick and steel walls of cyclones in all stages.

8.8 Various Designs of Preheater

Reputed designers and manufacturers of Cement Machinery like FLS, Polysius, Humboldt, Mitsubishi and others have their own designs of preheaters and calciers. A comparative statement of proportions of preheater cyclones of their respective designs prevalent in 2000- 2010 has been presented in **Annexe 2.8.1.** Performance wise these designs were close to one another

It is not valid for present designs. But the fact that various new designs also have similar sizes, features and similar pressure drops indicates that the comparability in performance is continuing.

8.9

Plates 2.8.2 and 2.8.3 show preheater towers and preheaters from actual installations.

Table 2.8.3 Kiln capacity and number of preheater streams

Capacity tpd	Preheater streams	Fls system *	Kiln support	Calciners
4000	1	sp	3	1
5000	1	ilc-e	3	1
6000	1	ilc	2/3,	1
7000	2	ilc/slc	2/3,	1
8000	2	Ilc,slc,slc-l, slc-d	2/3,	1
9000	2	ilc,slc,slc-l,slc-d	2/3,	1
10000	2	ilc,slc,slc-l,slc-d	3	1
11000	2 3	ilc,slc,slc-l,slc-d	3	1 2

*nomenclature from FLS Brochure Dry Process Kiln Systems'

Source: above mentioned FLS Brochure

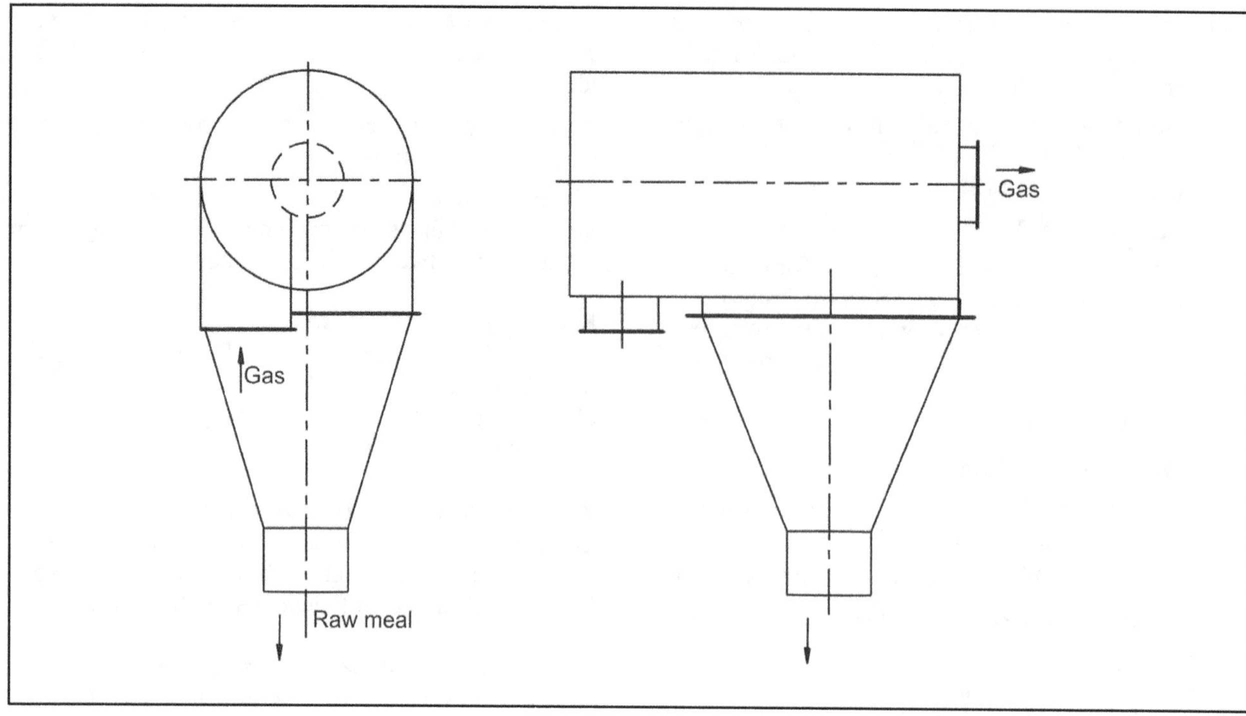

Fig. 2.8.4 Horizontal cyclone for preheater.

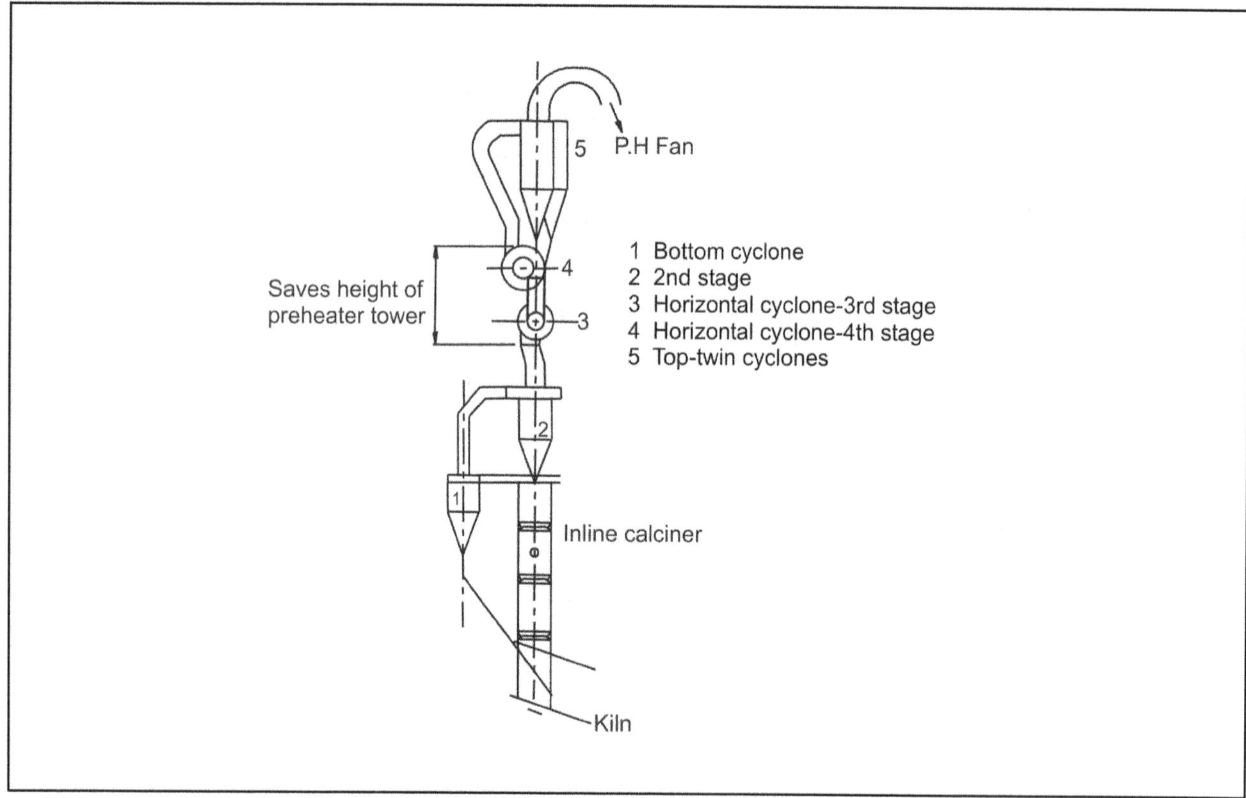

Fig. 2.8.5 Incorporating horizontal cyclones in a preheater system.

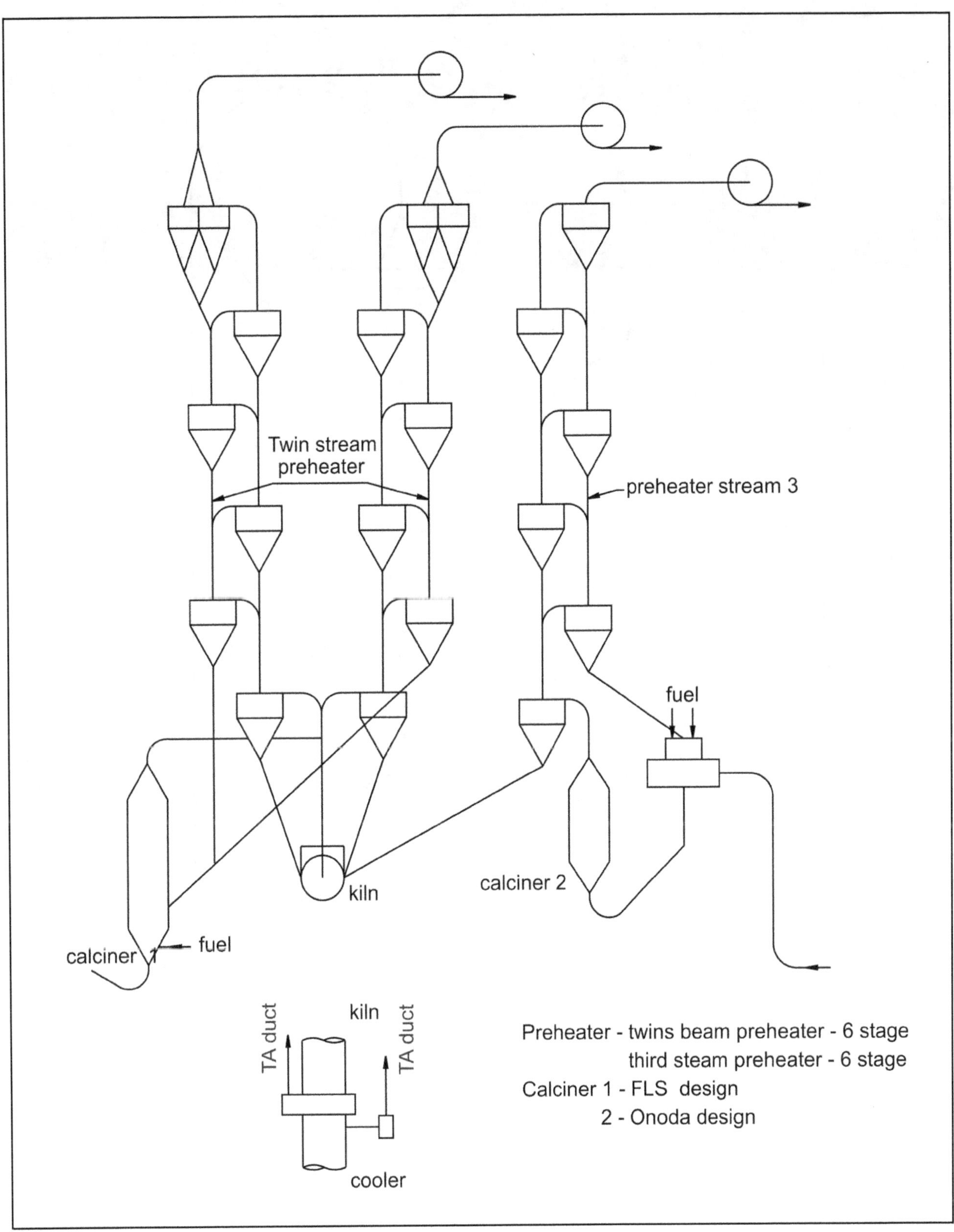

Fig. 2.8.6 Flow chart for kiln with three streams of preheater two calciners for illustration only.

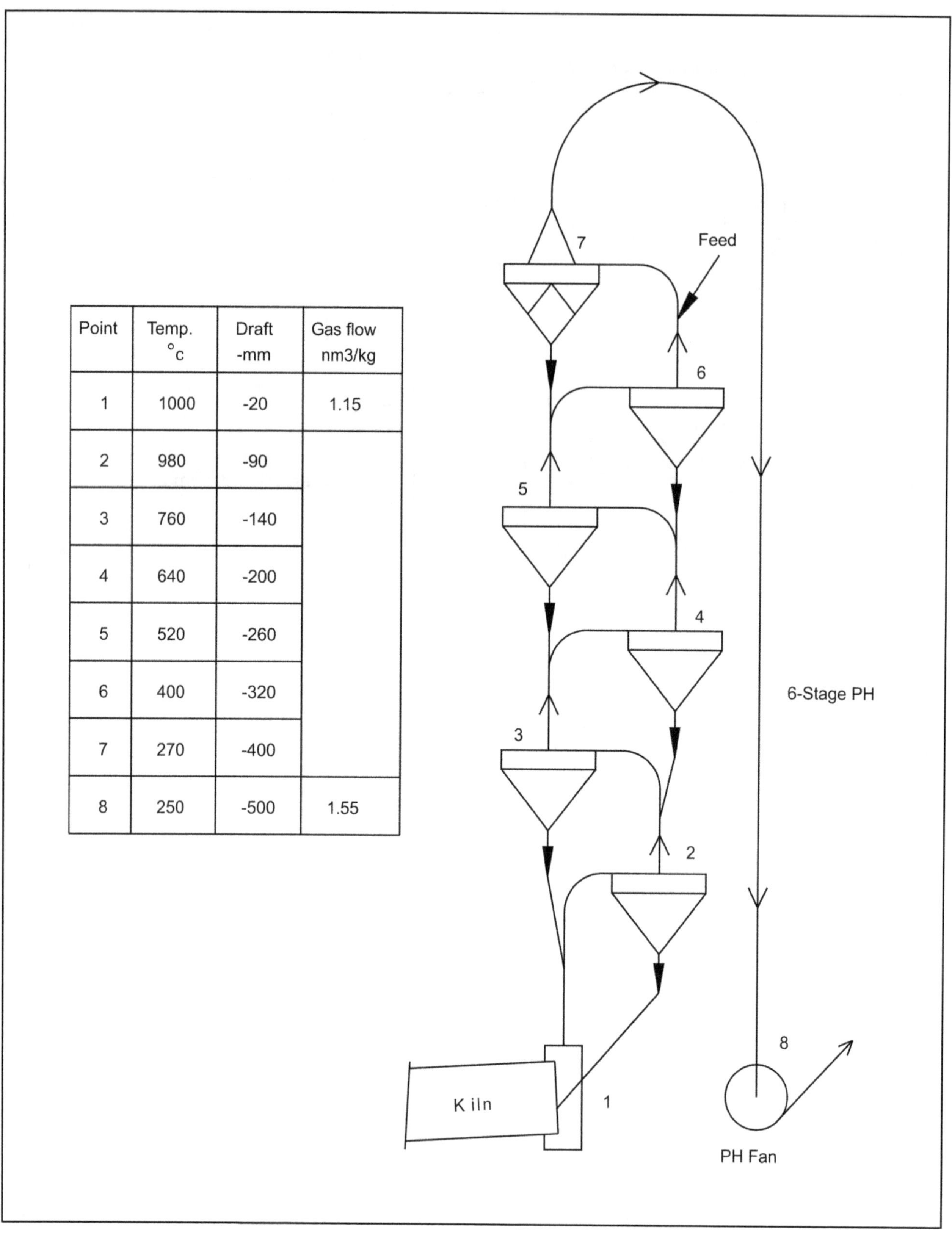

Point	Temp. °c	Draft -mm	Gas flow nm3/kg
1	1000	-20	1.15
2	980	-90	
3	760	-140	
4	640	-200	
5	520	-260	
6	400	-320	
7	270	-400	
8	250	-500	1.55

Fig. 2.8.7 Temperature & draft profile in a 6-stage preheater.

Plate 2.8.2 Preheater Tower.

Plate 2.8.3 Preheater Tower.

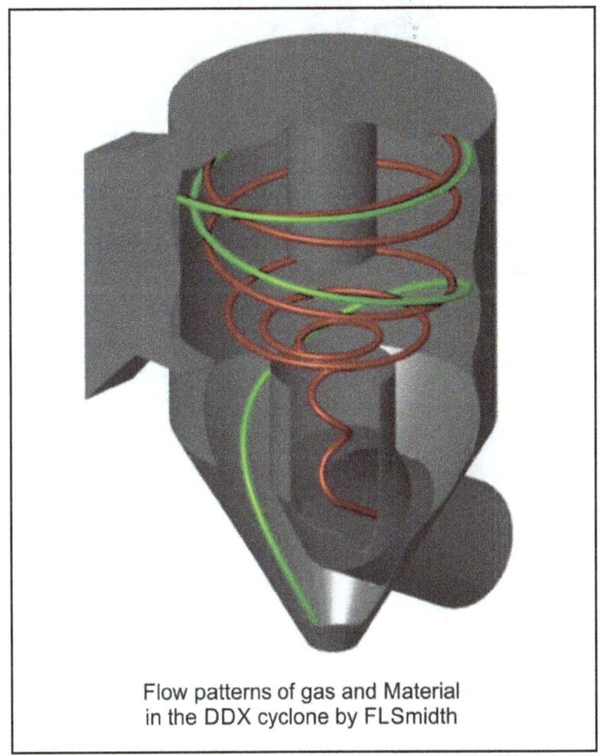

Flow patterns of gas and Material
in the DDX cyclone by FLSmidth

Plate 2.8.4 Preheater Down draft Cyclone.

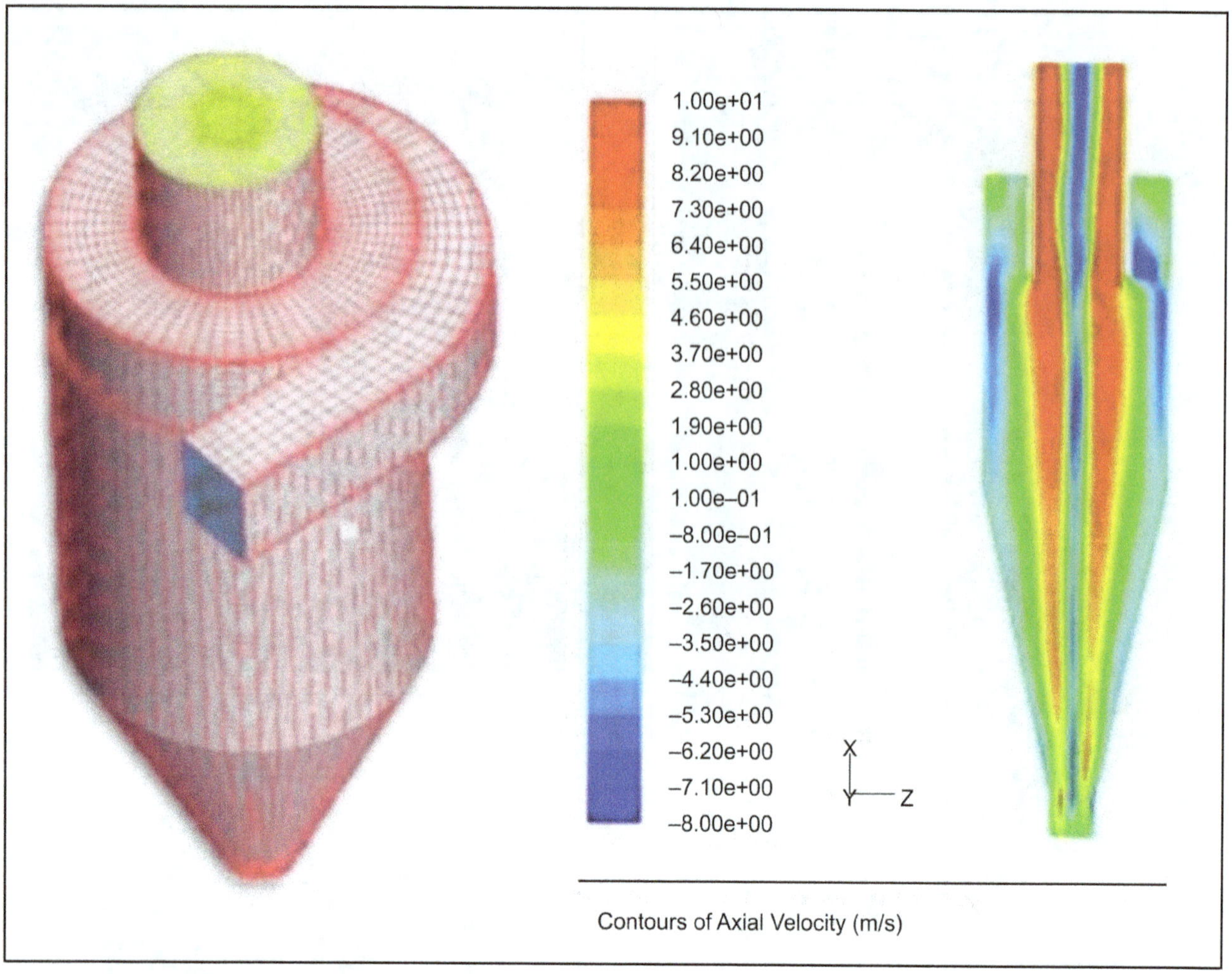

1.00e+01
9.10e+00
8.20e+00
7.30e+00
6.40e+00
5.50e+00
4.60e+00
3.70e+00
2.80e+00
1.90e+00
1.00e+00
1.00e–01
–8.00e–01
–1.70e+00
–2.60e+00
–3.50e+00
–4.40e+00
–5.30e+00
–6.20e+00
–7.10e+00
–8.00e+00

X
Y Z

Contours of Axial Velocity (m/s)

Plate 2.8.5 Computer assisted design of cycbine for improving efficiency

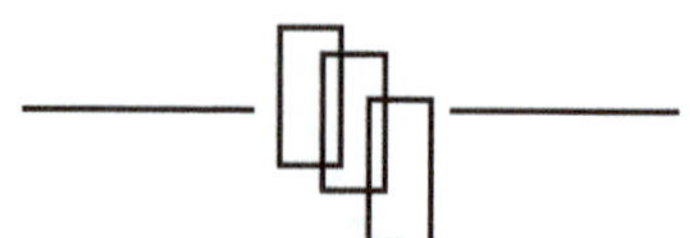

Annexe 2.8.1

sr no	item	unit	MMCC				Onoda				Humboldt				FLS			
			stages				stages				stages				stages			
			1	2	3	4	1	2	3	4	1	2	3	4	1	2	3	4
			top 2				top 2				top 2				1cyclone			
	d		d	d	d	d	d	d	d	d	d	d	d	d	d	d	d	d
	a		0.25d	0.275d	0.275d	0.275d					0.275d	0.4d	0.4d	0.4d	0.326	0.390d	0.404	0.404
	b		0.5d	0.55d	0.55d	0.55d					0.46d	0.4d	0.4d	0.4d	0.69d	0.702d	0.701	0.793d
	a/b		0.5	0.5	0.5	0.5	0.5	0.5	0.5	0.5	0.597	1	1	1	0.47	0.55	0.575	0.08
	a*b	kd2	0.125d2	0.151d2	0.151d2	0.151d2					0.127d2	0.16d2	0.16d2	0.16d2	0.22d2	0.27d2	0.283d2	0.32d2
	e		0.5d	0.5d	0.5d	0.5d					0.512d	0.467d	0.467d	0.467d	0.598d	0.648d	0.65	0.65d
	ea	m2	0.196d2	0.196d2	0.196d2	0.196d2					0.205d2	0.171d2	0.171d2	0.171d2	0.281d2	0.323d2	0.331	0.323d2
	c	mm	300	300	300	300	0.09d	0.095d	0.095d	0.095d	600	600	600	600				
	H		1.75d	1.2d	1.2d	1.2d	1.8d	0.9d	0.9d	1.5d	1.75d	0.838d	0.838d	0.838d	2.2d	1.46d	1.49d	1.58
	L						0.98d	0.6d	0.6d	0.65d	0.79d	0.895	0.895	0.895	1.1d	1.06d	1.06d	1.16d
	o	degree	60	60	60													
	o1					50					50	50	50	50	49	51	50	46
	o2					30												
			MMCC				Onoda				Humboldt				FLS			
			1	2	3	4	1	2	3	4	1	2	3	4	1	2	3	4
			top 2				top 2				top 2				1cyclone			
	immersion tube																	
	e		0.5d	0.5d	0.5d	0.5d					0.5d	0.394d	0.747d	0.747d				
	h		0.5d	0.5d	0.5d	0.25d					0.87d	0.5d	0.15d	0.15d				
	Vi		18	18	18	22	17	16	16	17								
	Vo		11.5	14	14	17	17	15	15	17								
	velocity in cyclone	m/sec.							2.8	3	3.4							

Table: Proportions of cyclones of leading preheater designs

CHAPTER 9

CALCINERS

9.1 Claciners

Just as development of Preheaters took the process of preheating outside the kiln so also Cacliners have taken the process of calcination or dissociation of CO_2 from carbonates in raw mix feed outside the kiln.

In Cyclone Suspension Heaters, retention time of gases is about 30 to 40 seconds. In this time raw meal gets preheated from a temperature of 60 °C to about 800-850 °C. Calcination process begins at 600 °C. Therefore raw meal enters the kiln partially calcined say 20 to 35 %. Balance of calcinations is completed inside the kiln.

9.2 Process of Calcination

Calcination is an endothermic process during which temperature of raw mix remains constant. Heat is required to be supplied for calcinations as per chemical reaction shown below:

$$CaCO_3 = CaO + CO_2 - 420 \text{ kcals.}$$

In calciners , raw meal is taken from the last but one stage of preheater and taken to a vessel where heat for calcination is supplied by firing fuel in it. Preheated air for combustion can come from kiln or from grate cooler. Preheater fan draws products of combustion and CO_2 dissociated through calciner and through cyclones of preheater. Raw meal is separated from gases and is fed to kiln. Degree of calcination achieved is directly related to the amount of fuel fired in the calciner. As a thumb rule, when 60 % fuel is fired in calciner, degree of calcinations achieved is 90 to 95 %. Temperature of raw mix begins to rise when calcination is complete changing its characteristics of flow. Therefore calcination to be achieved is intentionally limited to 90 %.

See Fig. 2.9.1

9.2.1 Output of a Calciner Kiln

As a direct result of carrying out process of calcination outside the kiln, capacity of the kiln increases in proportion.

Ratio of output of a calciner kiln to a preheater kiln = 100/(100-% fuel in calciner).

Relations between fuel fired in calciner and kiln output are given in **Table 2.9.1. and also Fig. 2.9.1**

Table 2.9.1

%fuel in calciner, calcination and kiln output			
Sr.No.	Fuel in calciner %	Degree of Calcinations %	Kiln output Preheater kiln = 1
1	0	30	1
2	10	40	1.11
3	20	50	1.25
4	30	60	1.43
5	40	70	1.66
6	50	80	2.0
7	60	90	2.5

Up to 20 % fuel can be fired in kiln riser duct itself with minor modifications. Air from combustion comes through kiln.

For higher quantities a proper calciner needs to be installed.

9.3 Types of Calciners

Calciners are classified into :

1. ***'In line' or 'off line' calciners.***

   ''In line' calciners are installed between kiln and bottom cyclone. They can be in the form of large diameter ducts with constrictions or rectangular ducts bent to increase retention time.

Air for combustion can come through kiln or through an external duct from grate cooler.

'Off line' calciners are independent vessels installed within or outside preheater tower. Air for combustion comes from grate cooler.

Plates 2.9.1 to 2.9.6 show arrangements of 'in line' and 'off line' calciners with preheater cyclones for one and two stream preheaters. It would be seen that there are a great many possibilities to choose from.

Plate 2.9.6 shows inline calciner in the form of a duct.

2. *Air through kiln or air from cooler.*

   Air through kiln calciners are suitable for kilns with planetary coolers. Quantity of fuel that can be fired in calciner is however limited to about 30 %.

3. *Short or long retention times.*

   In most calciner designs retention time is between 4 to 6 seconds to complete burning of coal fired and to complete calcination. This requires that both coal and raw meal are ground fine - 10 to 12 % residue on 90 micron sieve. In fluidised bed calciners retention time is 30 seconds. Therefore coarsely ground / high ash coals can be burned completely and even raw mix can be ground to a fineness of 16-20 % residue on 90 micron commonly found in preheater kilns.

   In some designs of calciners even crushed coal and fuels like rice husk and shredded tyres can be burnt. **See Plate 2.9.8.**

9.4 Firing Fuels in Calciner

All calciners need fuel firing arrangements and also for feeding raw meal. Hot air is brought in a duct from grate cooler. Air for combustion in calciner is known as 'tertiary air' and duct bringing it as tertiary air duct.

9.4.1 Alternate Fuels in Calciners

Now Calciners are being designed to receive and burn Alternate Fuels. This causes combustion gases contain NOx in undesirable proportions. Therefore Low NOx caliners with multiple entry points for fuel and raw meal have been designed. **See Fig 2.9.10 and Plate 2.9.8.**

For example rise husk.

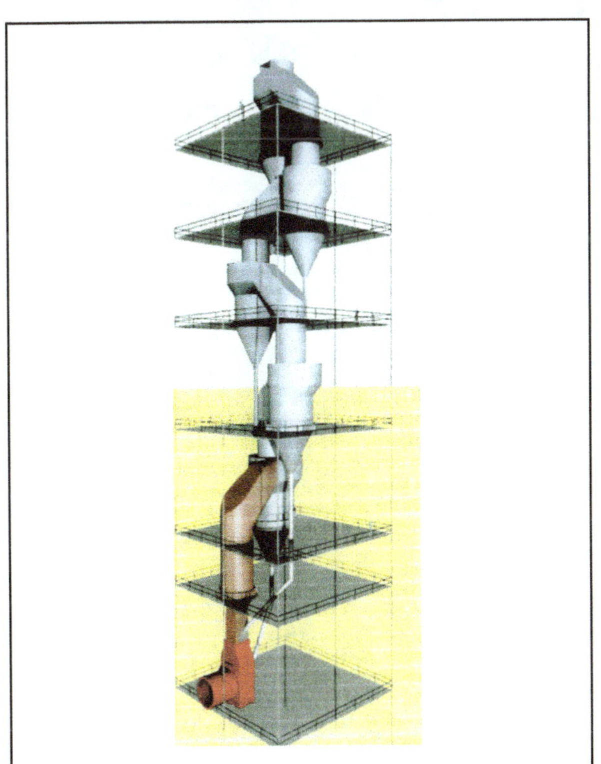

Plate 2.9.1 In line calciner with air through kiln.

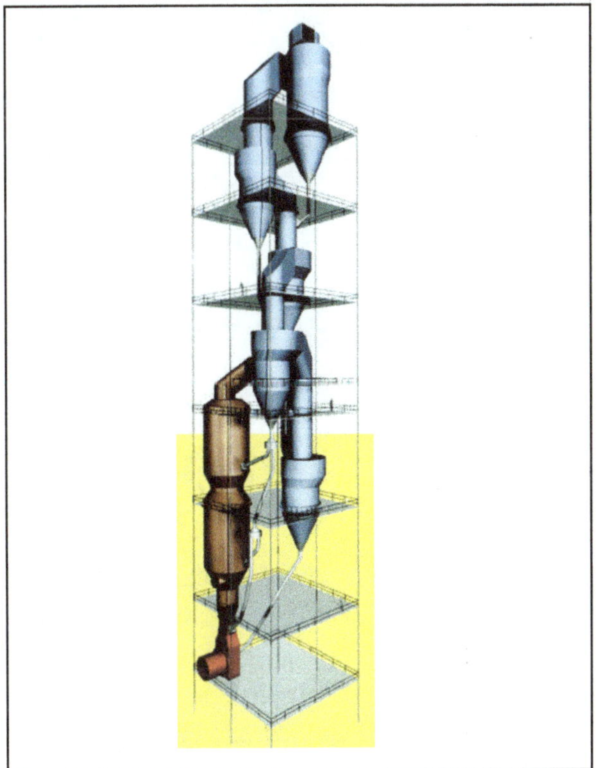

Plate 2.9.2 In line calciner with air from cooler.

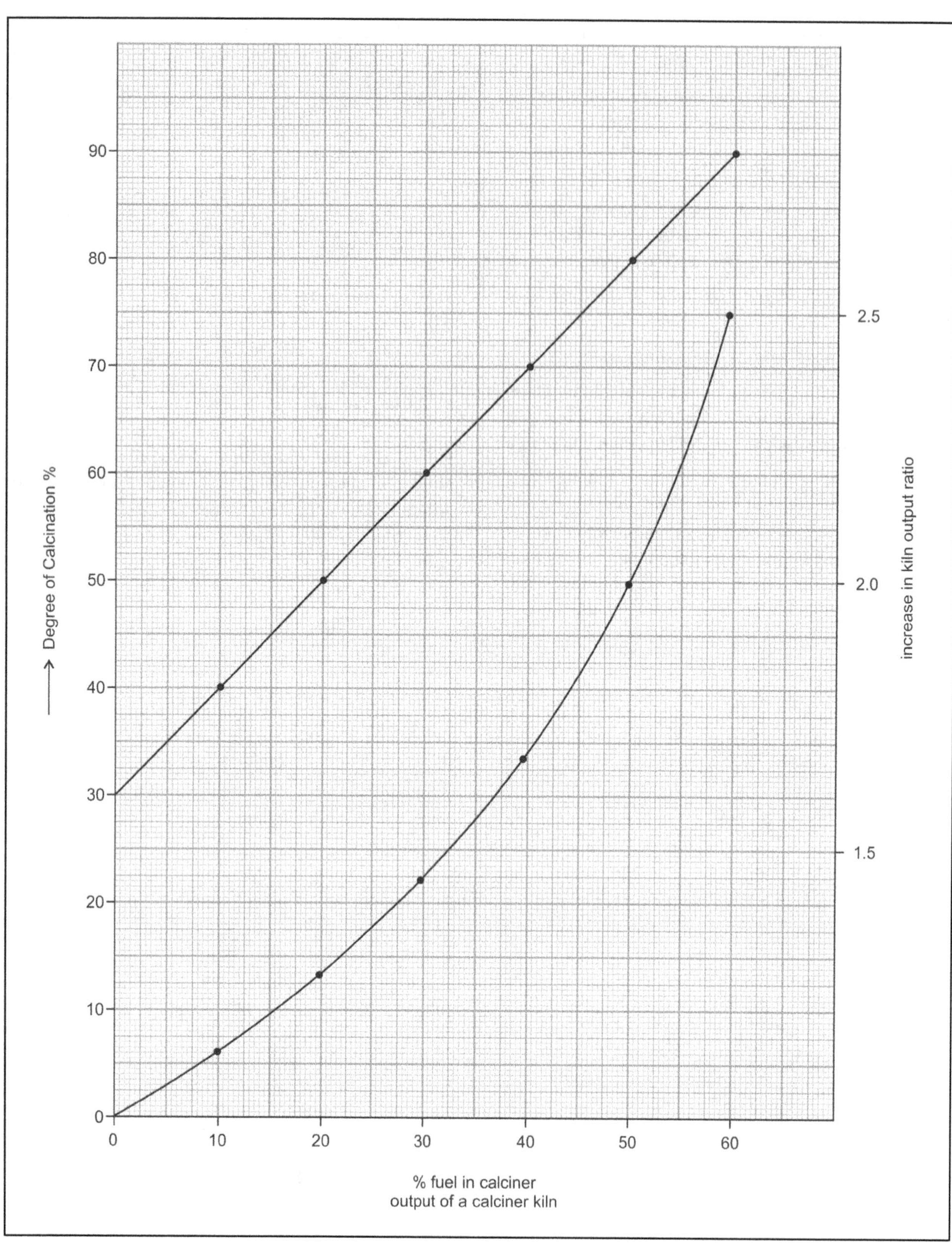

Fig. 2.9.1 In line calciner with air from cooler.

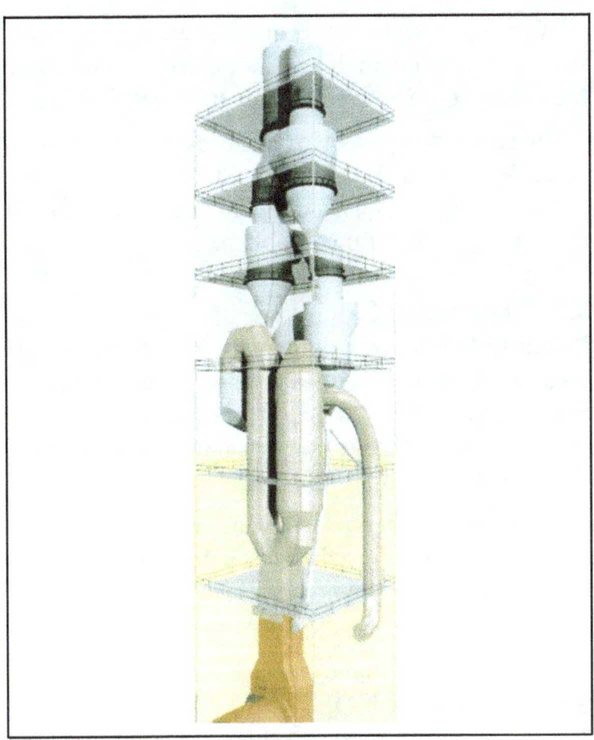

Plate 2.9.3 Off line down draft calciner.

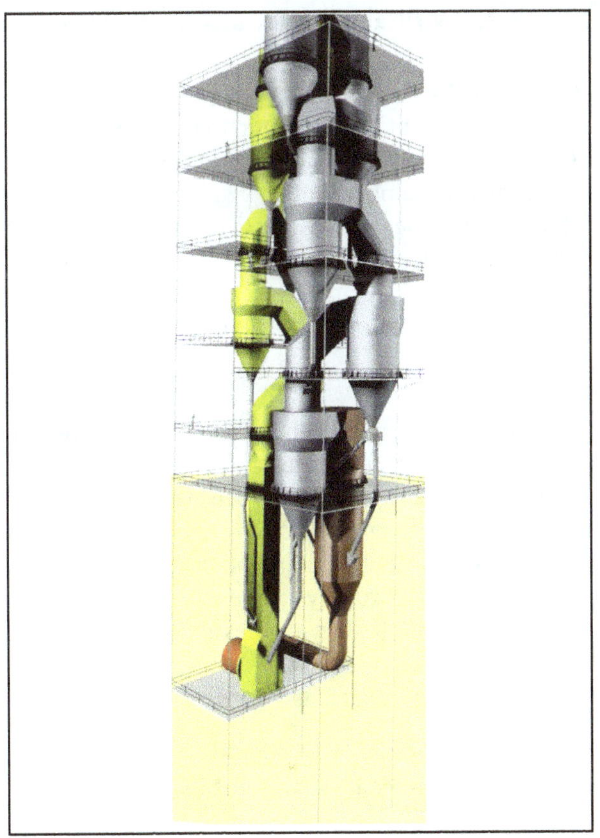

Plate 2.9.4 Off line calciner with its own preheater.

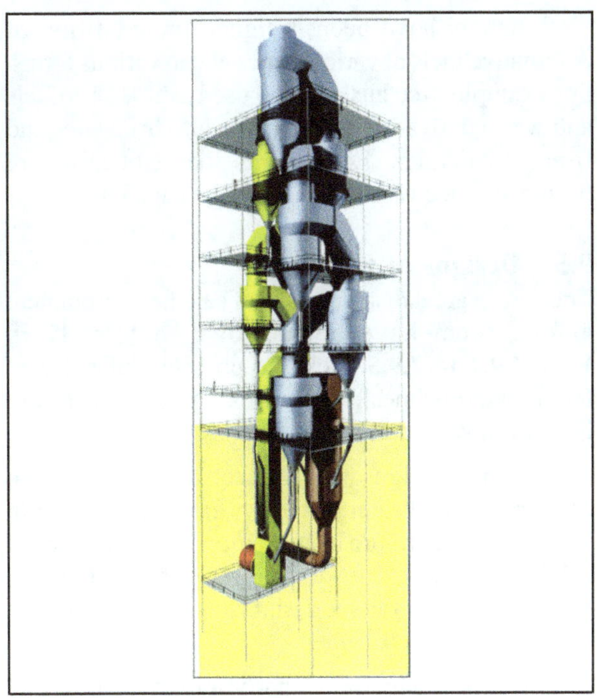

Plate 2.9.5 Two stream preheater with calciner in each stream.

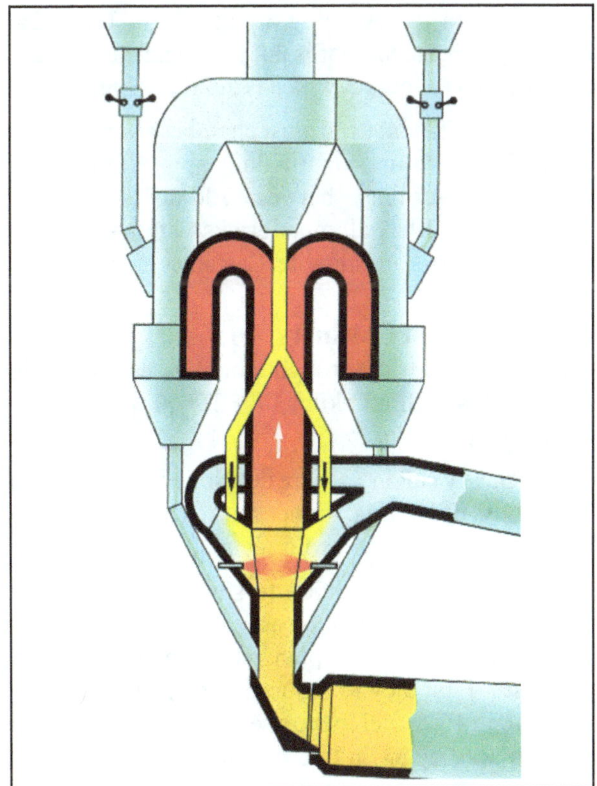

Plate 2.9.6 In line bent duct calciner drawing air from cooler.

Calciners have been designed to suit firing of Alternative fuels of various types and in various forms. For example, rice husk can be used as it is. Biofuels and waste derived fuels are made into briquettes and fired in calciners. Equipments like 'Hot Disc' are designed to feed Alternative fuels in calciners.

9.5 Designs of Calciners

There are a great many designs of calciners. Prominent among them have been shown pictorially in **Figs. 2.9.2 to 2.9.5.** Relative positions of entry of tertiary air, raw meal, fuel and exit gases in each case have been shown in these sketches.

In some designs the connecting duct between calciner and preheater also plays an important part in the process. Therefore in evaluation, often volumes of calciner and connecting duct are to be taken together. **Fig. 2.9.6** shows for six designs relative volumes of calciner and duct together.

Per se, a calciner does not improve specific fuel consumption of an installation as compared to operation as a preheater kiln. Calciner and T.A. duct which have large surfaces increase radiation losses considerably. But radiation losses being a small % of total heat balance, increase thereof is not significant as they are now divided for output which is 2.5 times higher.

Calciner, TA duct and dust chamber are refractory lined. Temperatures in t.a.duct and calciner and connecting duct range between 700 oC to 1000 oC. Therefore 35-40 % alumina bricks with backing of insulating bricks are suitable. Castables are also used.

9.6 Impact of Calciners

Calciners have truly brought about a sea change in cement plants. Single kilns of + 10000 tpd capacity have been possible.

Originally calciners were developed to keep down kiln size as life of brick lining shortened considerably when diameters exceeded 5 metres.

They have been a boon to small plants as well. They have been able to increase capacity of their kilns 2.5 times. Many a small plant has installed 'calciners' in phases. First they introduced 'secondary firing' by firing 10 to 15 % fuel in the riser duct. Later riser duct was replaced by an 'in line' calciner.

Increase in kiln and plant capacities triggered growth of capacities and developments of grate coolers and grinding mills as well.

Various designs have been operating in India successfully using high ash low calorific value coals. Most designs do not pose any operational problems.

In some designs of off line calciners there is occasionally a small amount of spillage. This being very hot dust needs careful handling. In line and down draft calciners are free of this problem.

9.7 Two Stage Calcining

Recently two stage calcining has been introduced consisting of a primary and a secondary calciner. **See Fig. 2.9.7.** Operational profiles of conventional calciner and two stage calciner have been shown in **Fig. 2.9.8.**

In another design of two stage calcining, a small fluid bed caciner is installed just before the inlet of the bottom cyclone of the preheater.

See Fig. 2.9.9.

9.8 Control of NO_x

To reduce NO_x fuel is introduced at more than one level in a calciner.

See Figs. 2.9.10 and 2.9.11. also plate 2.9.7.

9.9 Alkalies and Chlorides

It has been mentioned earlier that if alkalies and chlorides in raw materials are high and are not absorbed chemically by sulphur in fuel, alkali cycles are set up which lead to eventual clogging of riser duct and bottom cyclones. In calciner kilns only about 40 % fuel is fired in the kiln The problem is therefore less acute. Much smaller quantities of gases need to be bypassed.

9.10

Calciners have thus been a great boon in more than one way to Cement Industry.

Burning is more uniform and hence clinker produced is also uniform in quality.

Raw meal need not be ground as fine.

Fuels poorer in quality can be burnt.

Raw materials high in alkalies and chlorides could also be used.

Capacity of kiln increases 2.5 times.

9.11

Table 2.9.2 shows data on various FLS calciners and preheaters.

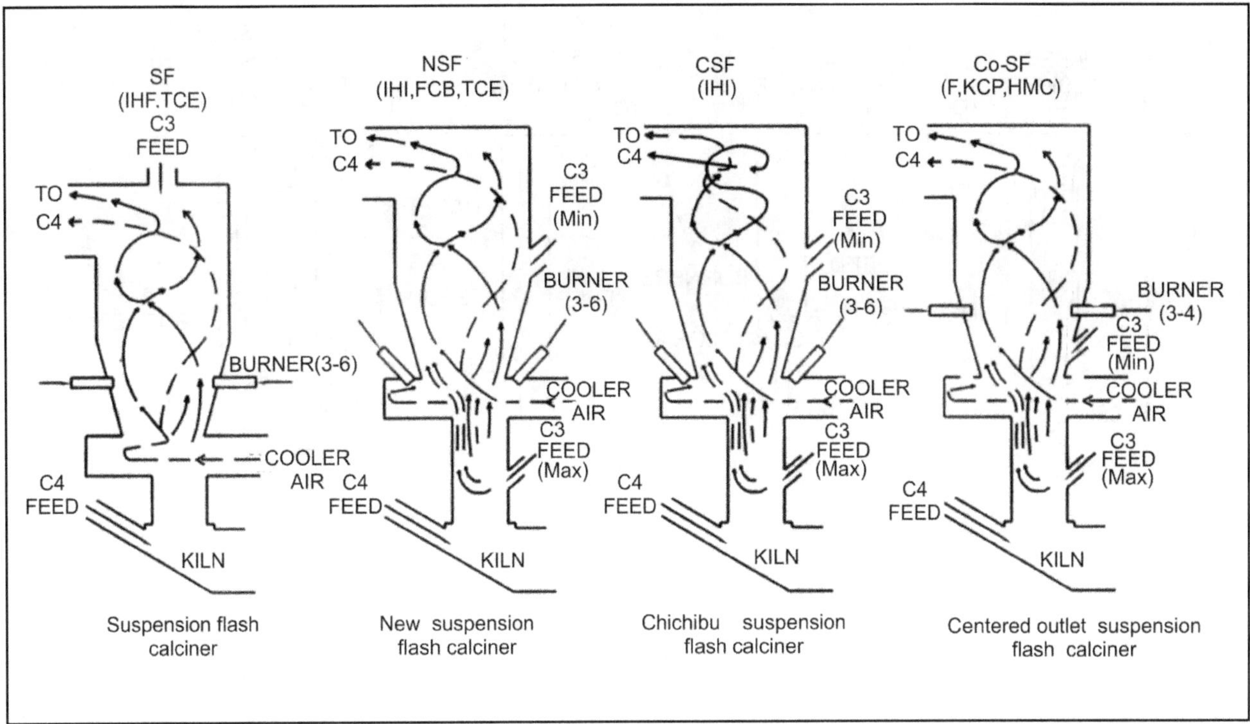

Fig. 2.9.2 Calciners of different designs.

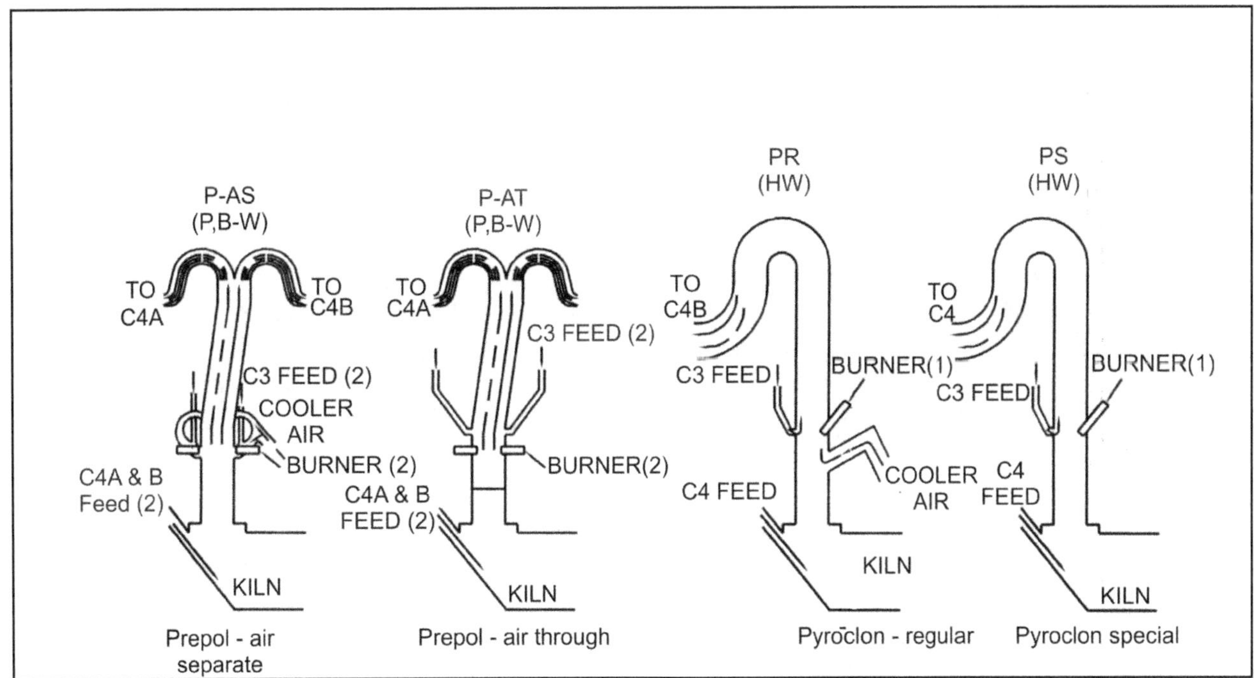

Fig. 2.9.3 Calciners of different designs.

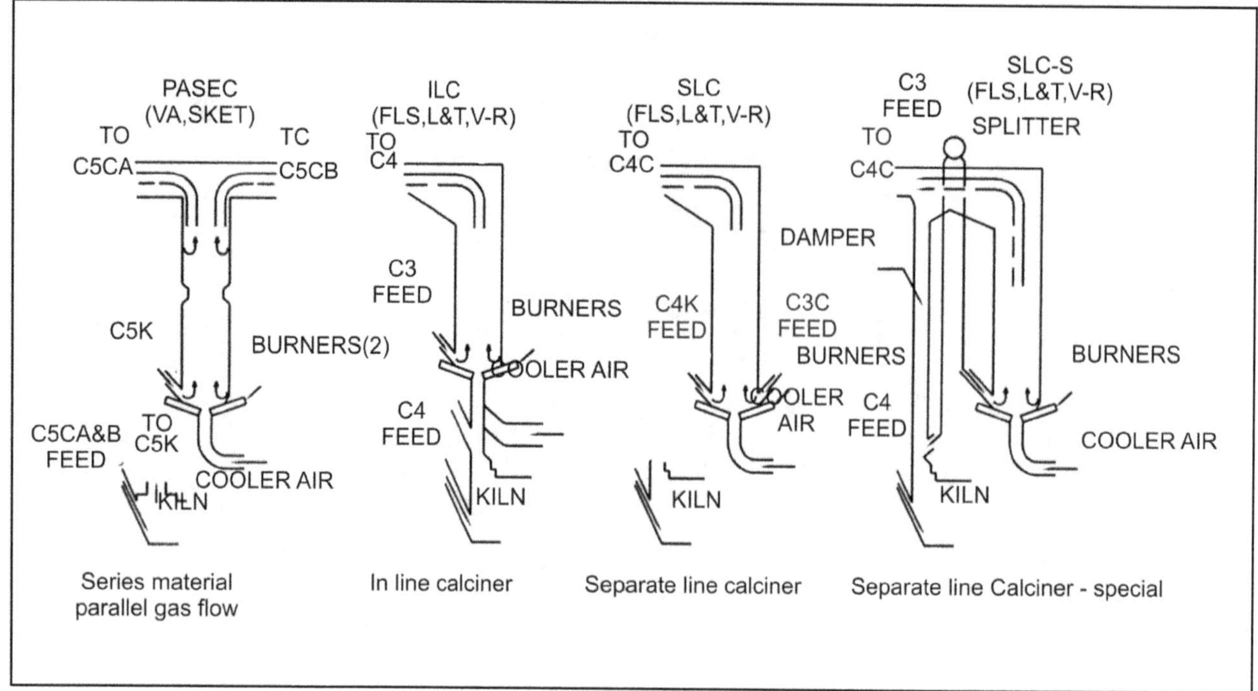

Fig. 2.9.4 Calciners of different designs.

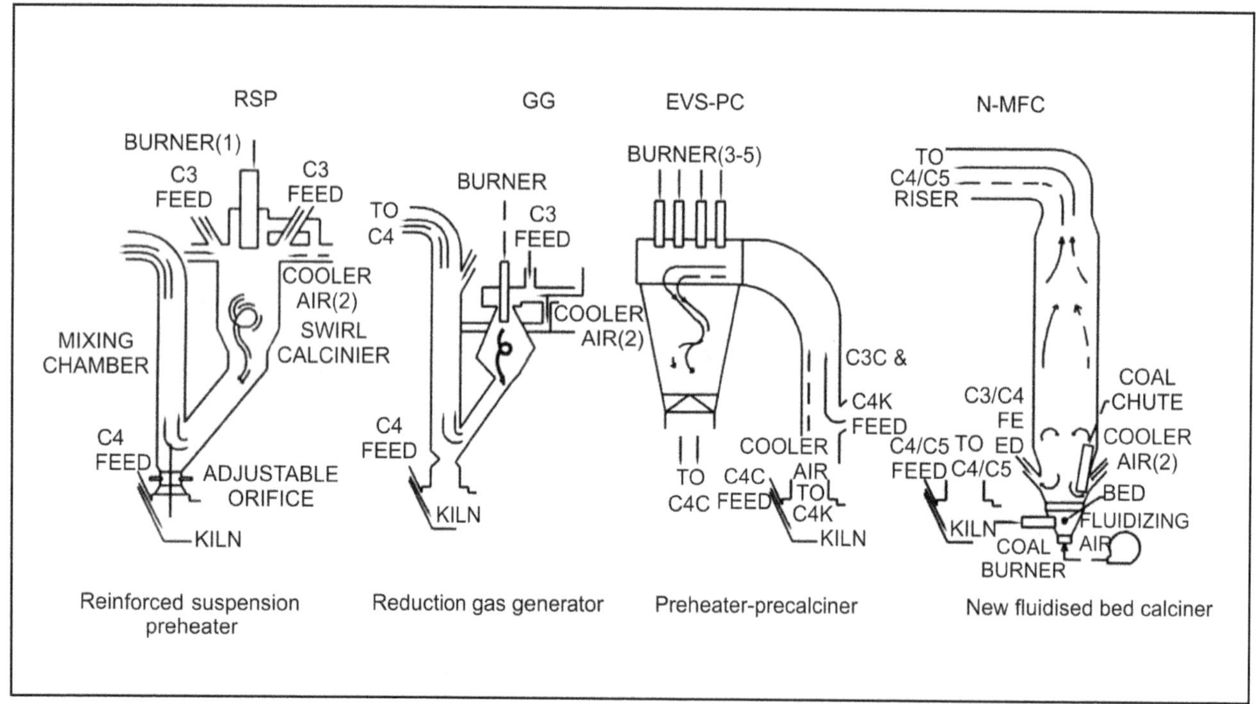

Fig. 2.9.5 Calciners of different designs.

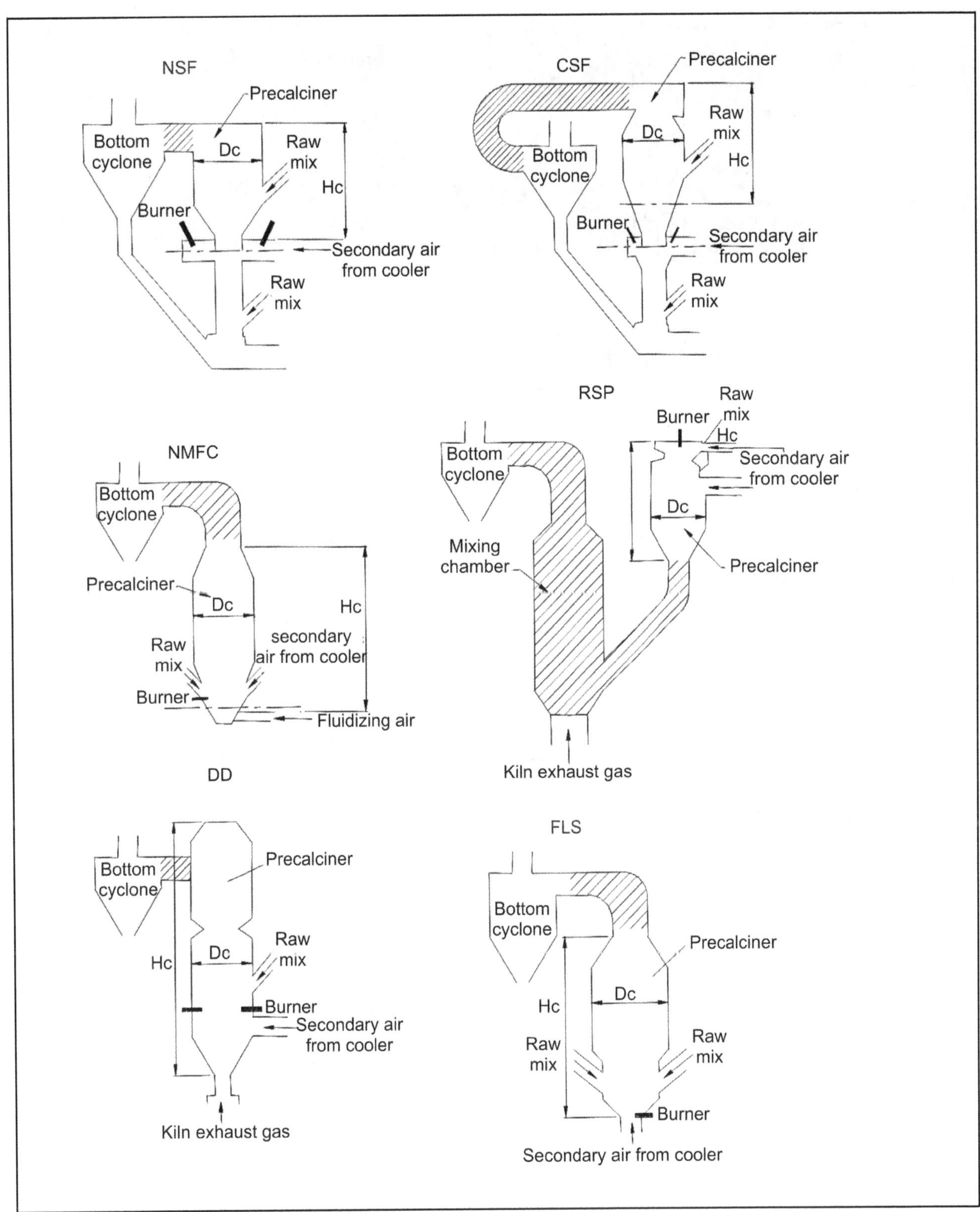

Fig. 2.9.6 Calciners and connecting ducts – rough relative proportions in different designs.

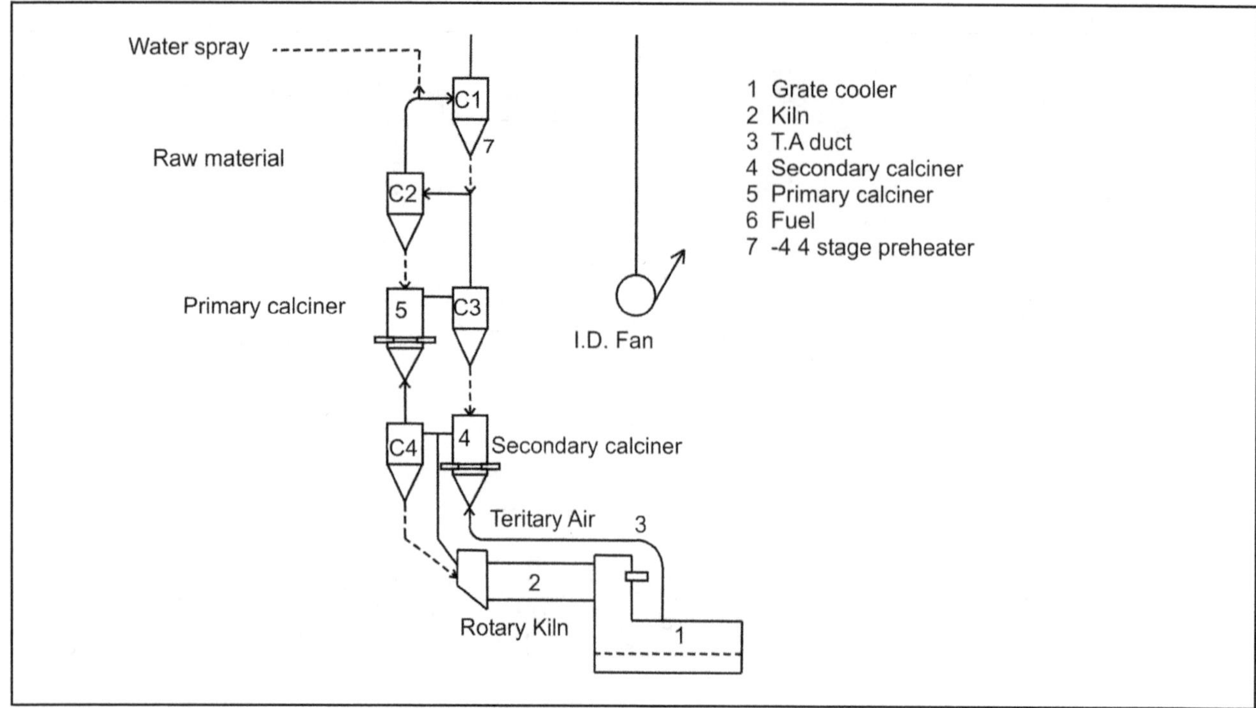

Fig. 2.9.7 Two stage calcining system.

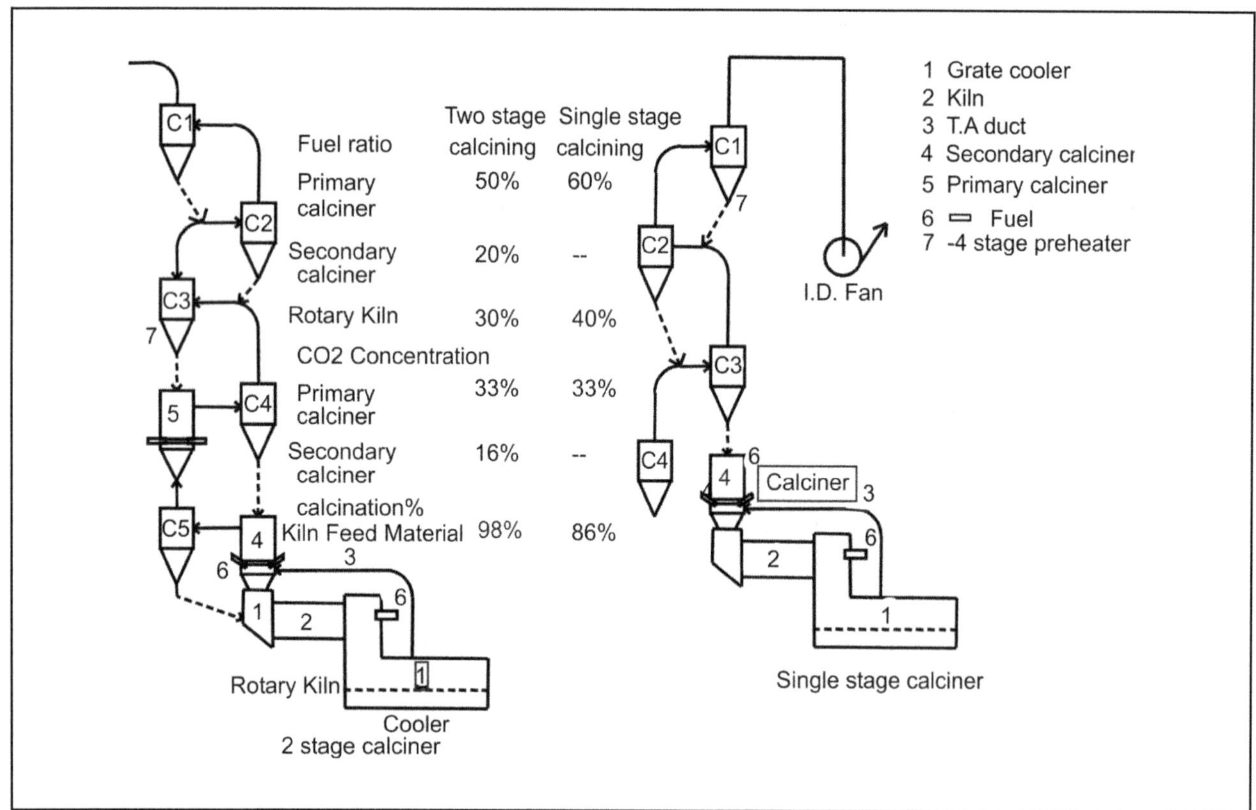

Fig. 2.9.8 Operational parameters of one and two stage calciners.

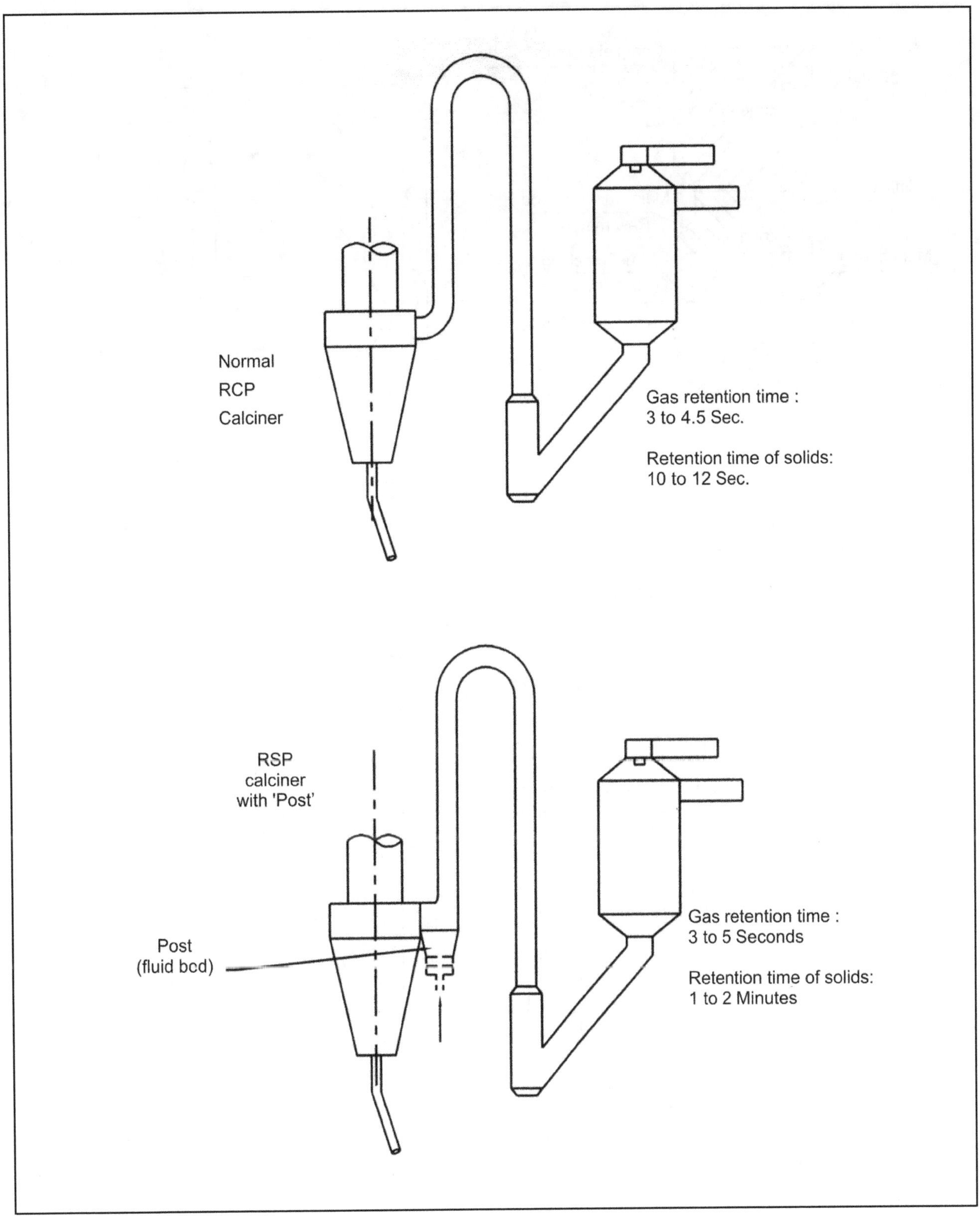

Fig. 2.9.9 RSP calciner – increase in retention time by 'Post'.

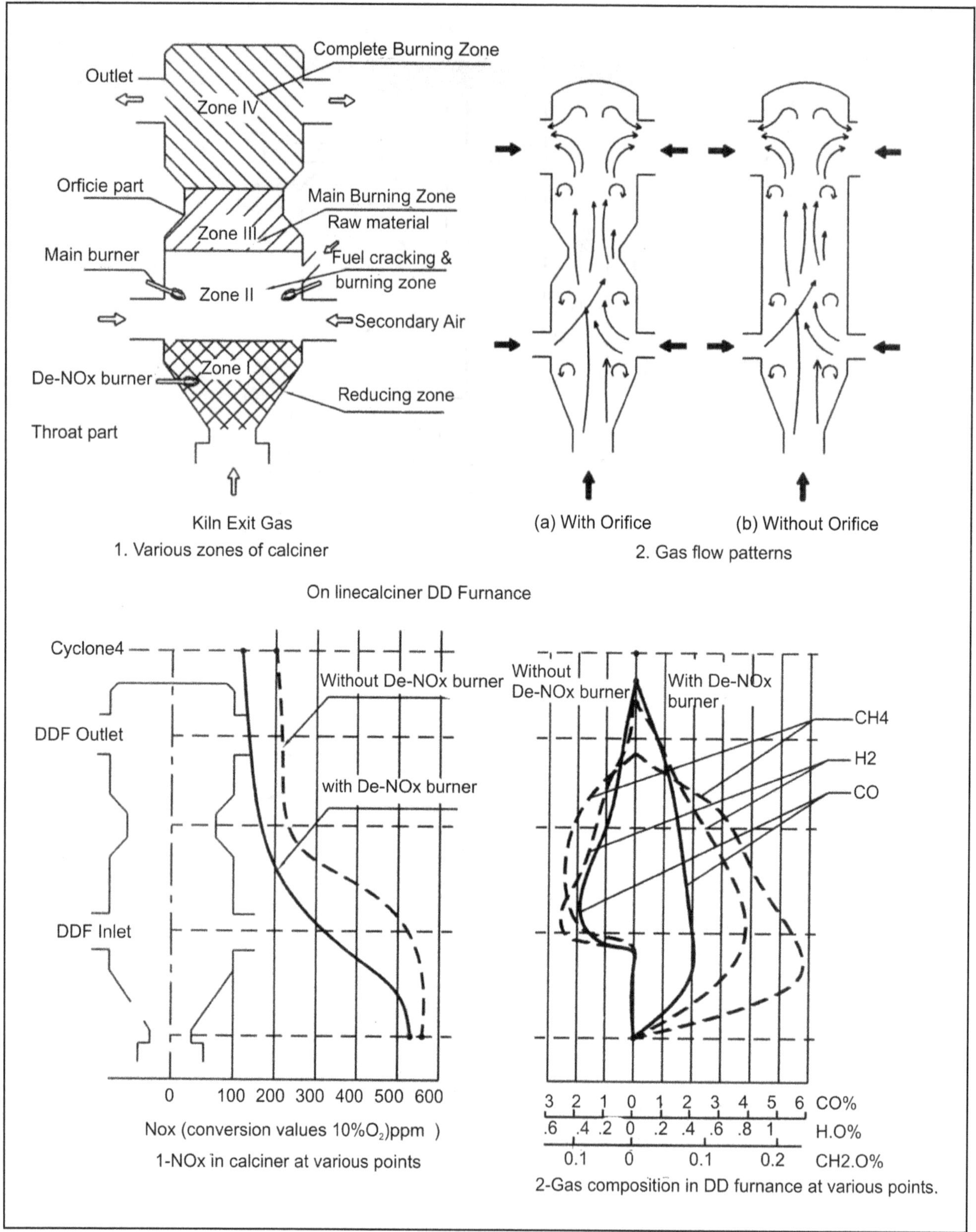

Fig. 2.9.10 Bringing down Nox with De-NOx burner in DD furnance.

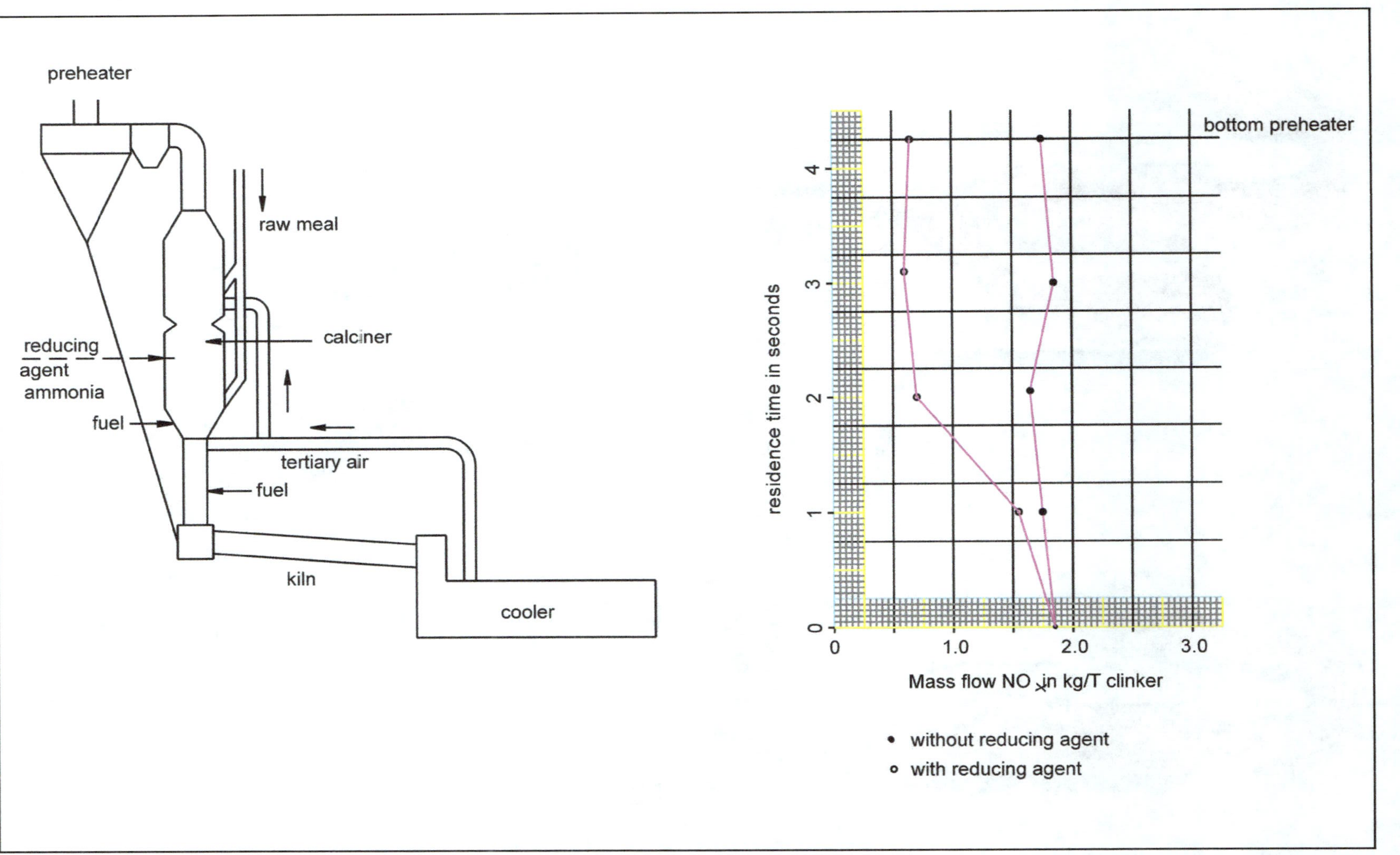

Source: VDZ Activity report 2003-2005

Fig. 2.9.11 Control of NO_x by multiple feed of fuel & raw meal and injecting Reducing agent ammonia.

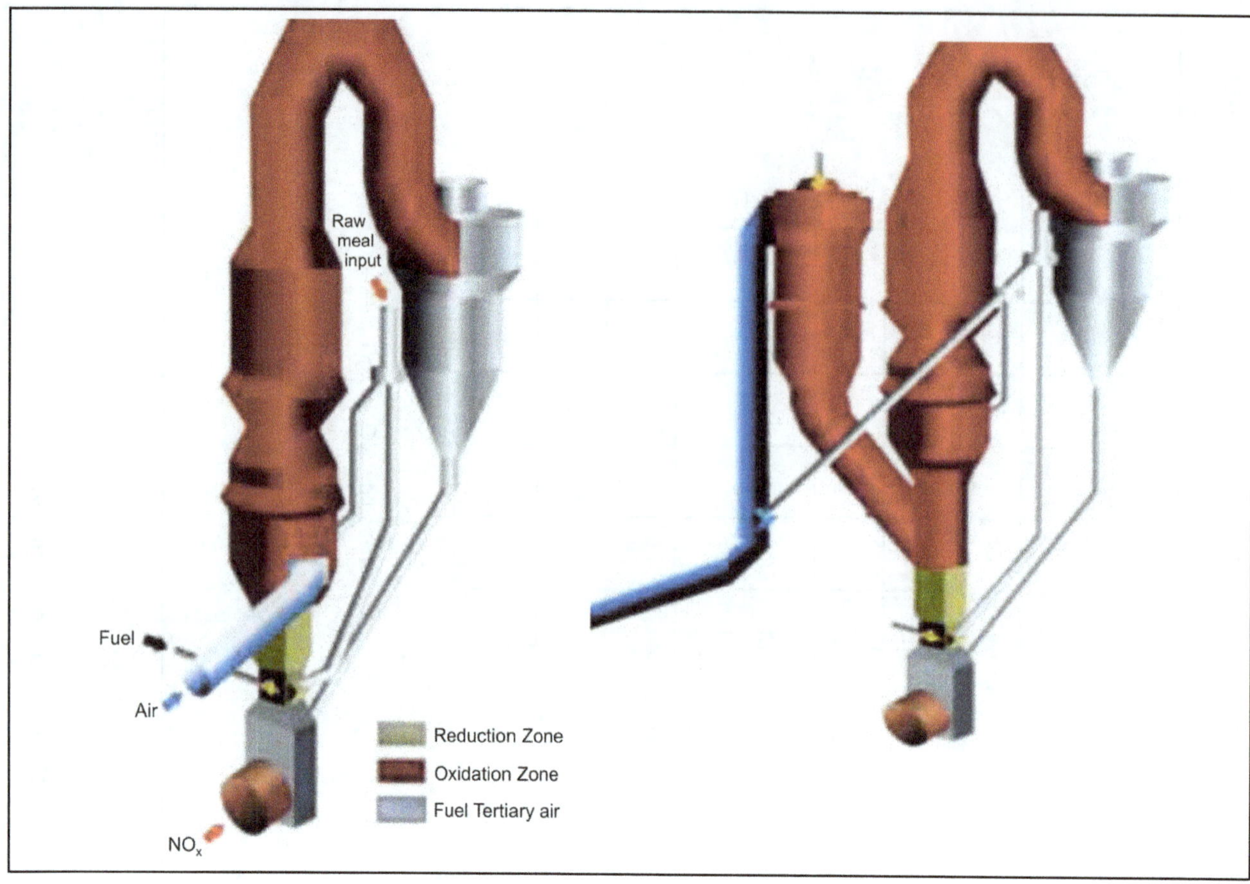

Plate 2.9.7 Low NOx Calciner.

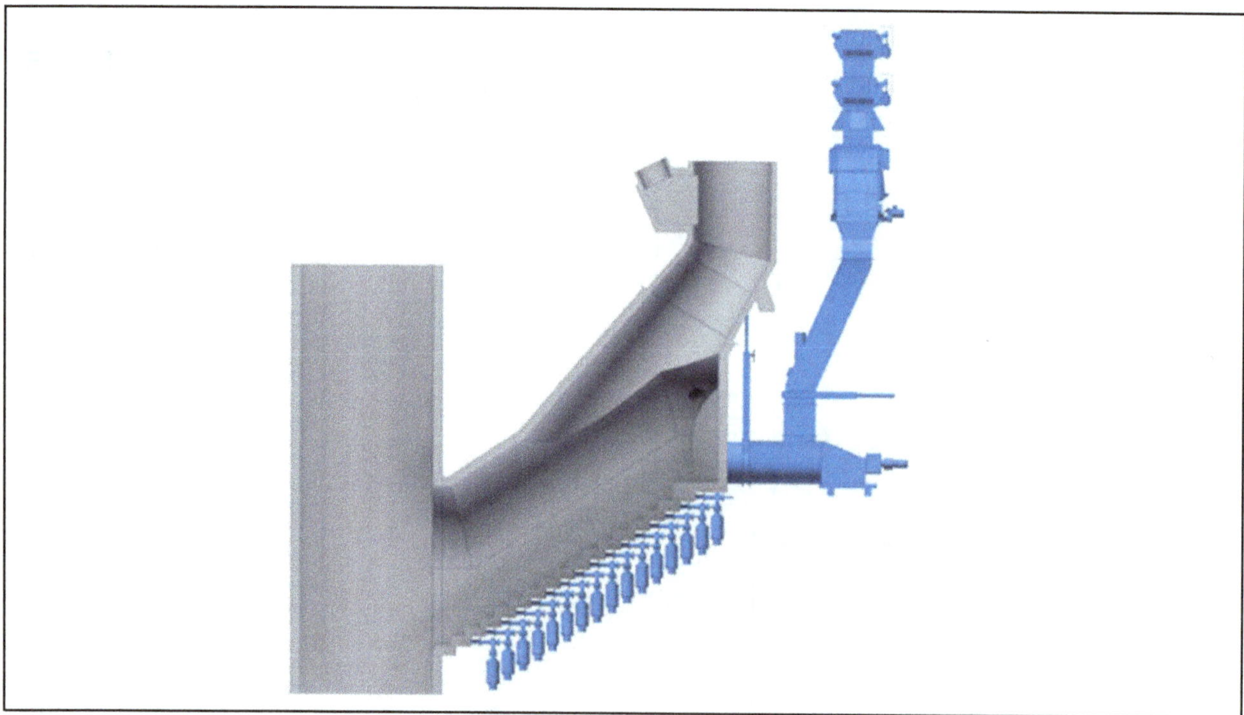

Plate 2.9.8 Thysson Krupp.

Table 2.9.2 FLS kiln, preheater and caciner systems.

Sr. No.	Pyroprocesing system	SP	ILC-E	ILC	SLC-S	SLC
		Air thru kiln		Air thru t.a.duct		
	Max. for					
1	Normal preheater	3000				3000
2	Preheater & Sec. firing		3700			
3	Preheater & calciner			4500	4500	4500
4	Normal layout					
	Preheater streams	1	1	1	1	2
	calciner		1[a]	1	1	1
	Special layouts					
	Preheater streams	(2)	(2)	2 2	2 2	3
	calciner	-	-	1 2	1 2	2
5	Total cap. Normal layout tpd	3000	3700	4500	4500	12000
6	Specific output tpd/m^3 of kiln	1.8	2.5	4.8	4.8	4.8
7	Coal fired calciner	-	-	yes	yes	yes
8	Low No_x calciner	-	-	yes	yes	yes
9	Max. firing in sec. firing %	15	25			
	In calciner %			55-65	55-65	55-65
10	Cooler grate	Yes	Yes	Yes	Yes	yes
	planetary	yes	yes	no	no	no
11	Max. bypass for Alkali chlorides %	30	25	100	60	100

Notes : Special layouts () used for expansion of existing plants

[a] riser duct modified for sec. Firing

SP (preheater kiln)

ILC (in line calciner)

SLC (separate line calciner)

Source : FLS Catalogues and literature

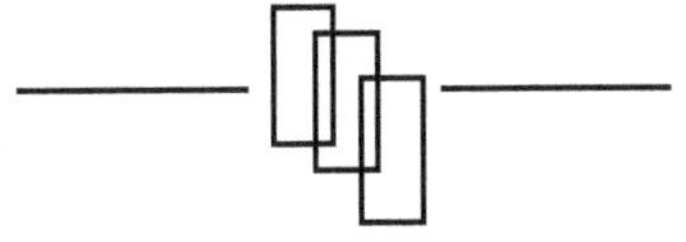

CHAPTER 10

KILNS

10.1 Rotary Kilns

Rotary kilns carry out a number of operations within them. They are suitable for all processes of manufacture of cement viz. wet, semi wet, semi dry and dry.

Length of the kiln or more precisely, ratio of length to diameter depends on the number of operations carried out in the kiln.
See Fig. 2.10.1.

Naturally wet process kilns were the longest because they carried out maximum number of operations starting from drying to sintering.

Dry process kilns with precalciners would be the shortest because in it are carried out only about ten percent calcining and the sintering processes.

All kilns have a short cooling zone at the discharge end. In kilns with planetary coolers, clinker enters the cooler tubes attached to the kiln shell. **See Plate 2.10.1.**

Length of the kiln decides the number of supporting stations required. Wet and long dry kilns had maximum number of supports 5 or more. Dry precalciner kilns can have only two which of course is the minimum number.Till now number of supports used to be three. of late however, even large kilns are being designed with two support.

Rotary kiln is a rotating cylinder, installed at a slope of 2 to 4 degrees. Each supporting station has 2 rollers and 4 bearings.

All are mounted on one fabricated bedplate. On the kiln are mounted tyres which are usually floating. Tyre. Of late tyres of splined design have come into use.**See plate 2.10.6** of a Thyssen Krupp kiln.

Tyre rests on rollers which have an angle of about 30 degrees at the center of the kiln.

Self aligning supporting stations are availale. Even drive to the kiln can be thru supporting rollers, **See Plates 2.10.5 and 2.10.10.**

Kiln is lined with refractories 150-250 mm thick. In burning Zone they are 70 % alumina or magnesite bricks. Rest of the kiln is lined with $\simeq$ 35 % alumina bricks.
See Plate 2.10.3.

Wet and long dry kilns had chains for about one third of the length from inlet end to improve heat transfer between kiln gases and kiln feed-(slurry in case of wet kilns and raw meal. In case of dry kilns). Chains in dry kilns had to be of heat resisting steels.

Though such kilns are now history, chain systems in their days needed and received considerable attention to improve speed of transfer of heat. Chains were curtain and garland chains and their 'density' was a parameter of the efficacy of heat transfer.

10.2 Sizing of Kilns

Sizing of kilns has been largely empirical. Factors influencing sizing are :

1. Retention time of material inside kiln from inlet end till it comes out at discharge end as clinker. It is a function of completion of the various operations mentioned above that take place inside the kiln.

   It is shortest for dry preheater- preclaciner kilns where a retention time of 27-30 minutes is adequate.

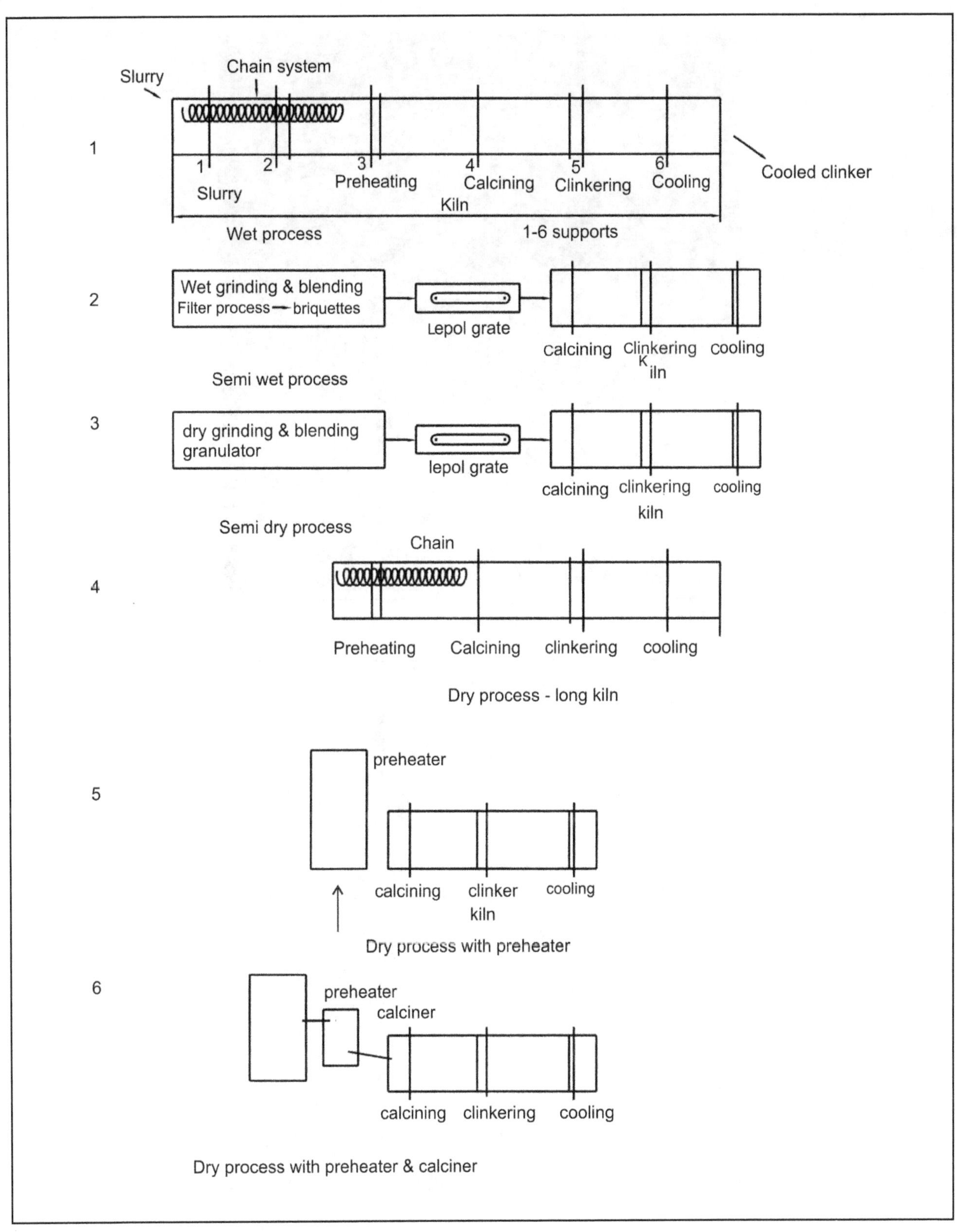

Fig. 2.10.1 Processes in rotary kilns – producing cement clinker.

Plate 2.10.1 Rotary cement kiln with planetary coolers.

Plate 2.10.2 Dry process roatry kiln with preheater and grate cooler.

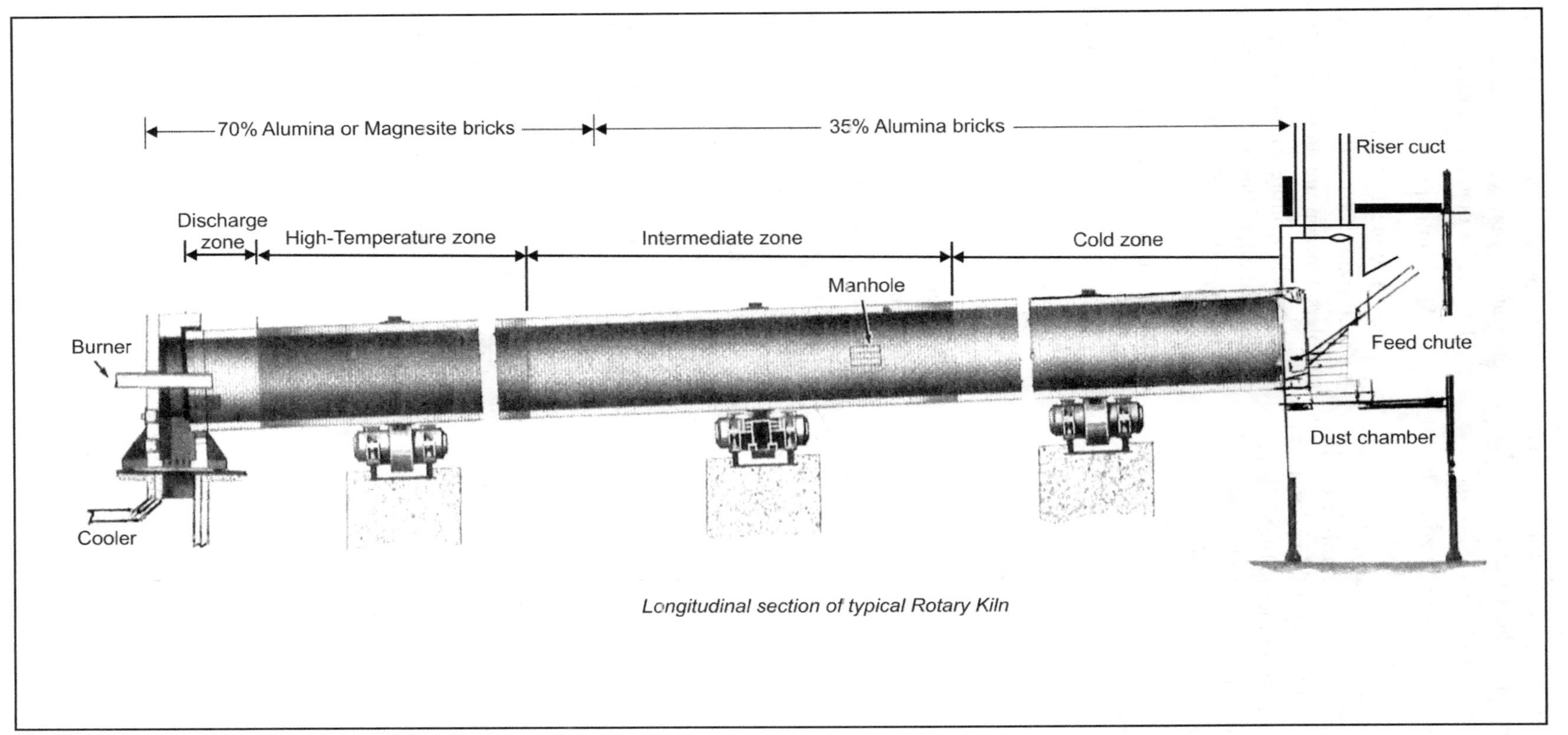

Plate 2.10.3 Refractory lining of a cement kiln.

2. Degree of filling of charge in the kiln. This is important from point of heat transfer to the charge . Usually degree of filling is 5-8 %.
3. Thermal loading in the burning zone. It is measured as heat released in kilocalories per m^2 of clear cross section of kiln per hour. Thermal loading affects the life of refractory in the burning zone. Quality of bricks also influences the permissible thermal load. With basic or magnesite bricks, thermal loads of 4.5 million kcals/m^2/hr are permissible.

   Thermal load has become less relevant in case of calciner kilns where only 40 % of total heat is released in the kiln.

10.2.1 Sp. Outputs and l/d Ratios

Empirically kilns were sized volumetrically on the basis of otput in tpd /m^3 of clear volume of the kiln (inside brick lining). Typical values for various types of kiln are :

Type of kiln	Sp. output in tpd /m^3
Wet	1.5 - 2
Dry long	2 - 2.5
Dry preheater	3 - 4
Dry calciner	5 - 6

Degree of filling is about the same for all processes. Earlier kilns were run at linear speeds of 25-35 cms/sec. Now they run at almost double the speeds.

Once the volume required for a given capacity is known, its Diameter and length can be worked out using length to dia.ratio appropriate to the process. Typical l/d ratios are:

Type of kiln	l/d ratio
Wet	35-40
Long dry	30-35
Preheater	14-16
Calciner	10-12

L/D ratio is also dependent on the fuel used. In case of coals where calorific values fluctuate and also in case of high ash coals it is prudent to opt for longer kilns to ensure complete combustion of fuel.

One of the commonly used formulae for sizing rotary preheater dry process kilns is $C = 3 \times d^{3.283} \times l^{0.337}$.

Here 'd' is diameter of kiln inside shell.

'l' is length of the kiln. And 'C' is capacity in tpd.

Specific fuel consumption and altitude at site also affect sizing. With 6 stage preheaters and better coolers, fuel consumption of < 700 kcal/kg is possible. Kilns have therefore shrunk in size. This is particularly so for calciner kilns whose output increase 2.5 times as compared to preheater kilns.

10.3 3 m dia × 40 m Long Small Kiln

An interesting example is that of a 3 m dia × 40 m long kiln which was once considered as standard for a 300 tpd preheater kiln.

Kiln	3 × 40 m
Clear dia	2.6 m
Clear volume	212 m^3

Capacity as preheater kiln 300 tpd

Sp. fuel consumption 900 kcal /kg clinker.

Sp. output = 300 / 212 = 1.42 tpd/m^3

Thermal load 2.12 million kcal/hr /m^2

This kiln is today operating at many places at capacities of 1500 tpd as calciner kiln with 6 stage preheater. About 45 % fuel was fired in the kiln.

Thus 6 stage preheaters and calciners have reduced the sizes of kilns dramatically. Small diameters are good for life of refractory also.

10.4 Kiln Components

Kiln design has seen many changes particularly as regards seals at inlet and outlet. Shell is cooled at discharge end to prolong life of refractory.

Seals

Kiln design has seen many changes particularly as regards seals at inlet and outlet. **See plate 2.10.8 for kiln inlet seal.**

Shell Cooling

Shell is cooled at discharge end to prolong life of refractory.

Hydraulic thrusters

Hydraulic thrusters are used to prevent slipping of kiln down hill.

Various Designers and Manufacturers have their own designs of the components of the kiln like riding rings, rollers, roller bearings, girth gear and its mounting on shell etc.

10.5 Kiln Drive

Kilns generally have variable speed DC Motors to drive them.

Because of higher running speeds gear box required is much smaller.

Increase in drive rating could be coped by installing a dual drive thereby retaining the same gear.

10.6 Mechanical Design

Mechanical design even for the largest calciner kiln of 10000 tpd was not a problem because increase in capacity was obtained by increasing speed. While power increased, torque remained the same. Hence cross sections corresponding to a 1200 tpd kiln were adequate for a 3000 tpd kiln. However as mentioned above linear speeds increased. Hence rate of wear increasd. Materials of construction for tyres and rollers and gear and pinion of calciner kilns should have better wear resistant properties

10.7 Burners

Pulverised coal or oil is fired into the kiln through a burner inserted into it from the kiln hood.

Coal burners

Earlier coal burner consisted of a refractory lined pipe with a tapered cone at the end to obtain desired velocity at the tip. Pulverised coal was carried into kiln by a primary air fan.

Multi channel burners

To ensure complete combustion of coal with minimum primary air, multi channel burners were developed. Total air is about 12.5 to 15 % out of which about 2.5 % is used to convey coal.

Burners are thus very much a part of the kiln.

A typical multi channel burner for coal has been shown in **plate 2.10.10.** There are designs in which oil can also be fired for preheating of refractory when starting from cold.

Oil Burners

Oil burners fire oil which is atomized either by compressed air or by mechanical pressure. To do proper atomization, oil is required to be preheated. Oil burners come with heating and pumping units.

10.8

In **Annexe 2.10.1** typical data on radiation losses in kiln cooler calciner system is furnished.

In **Annexe 2.10.2** typical performance data on dry processes kilns with 5 and 6 stage preheaters, planetary and grate cooler is furnished.

Plate 2.10.4 kiln support station.

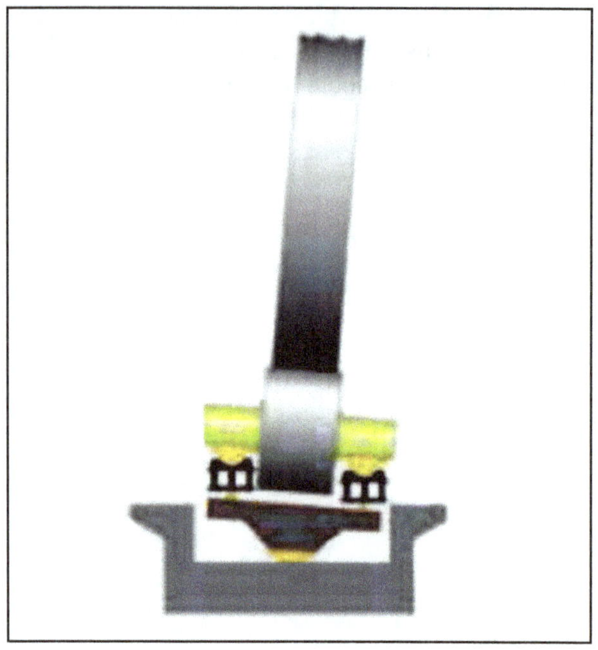

Plate 2.10.5 Self aligning roller support for kiln.

Plate 2.10.6 Preheater – Inline Calciner -Kiln-Cooler Thysson Krupp.

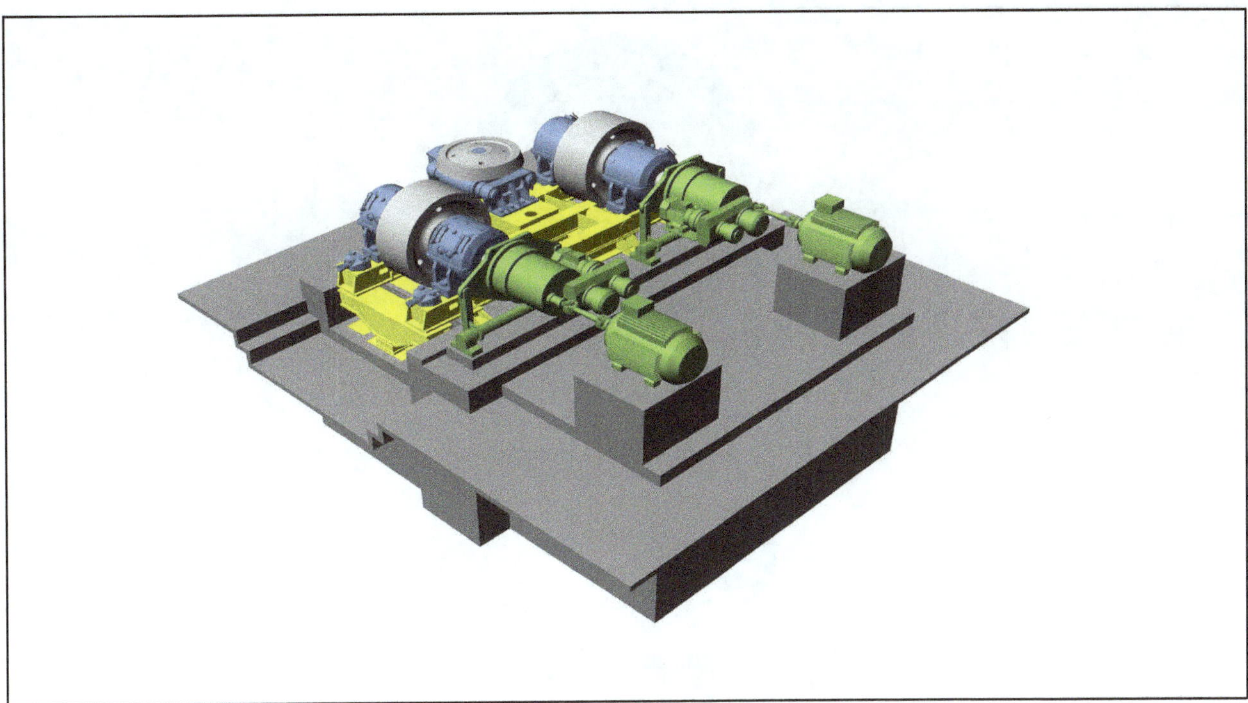

Plate 2.10.7 Friction drive of kiln thru supporting rollers.

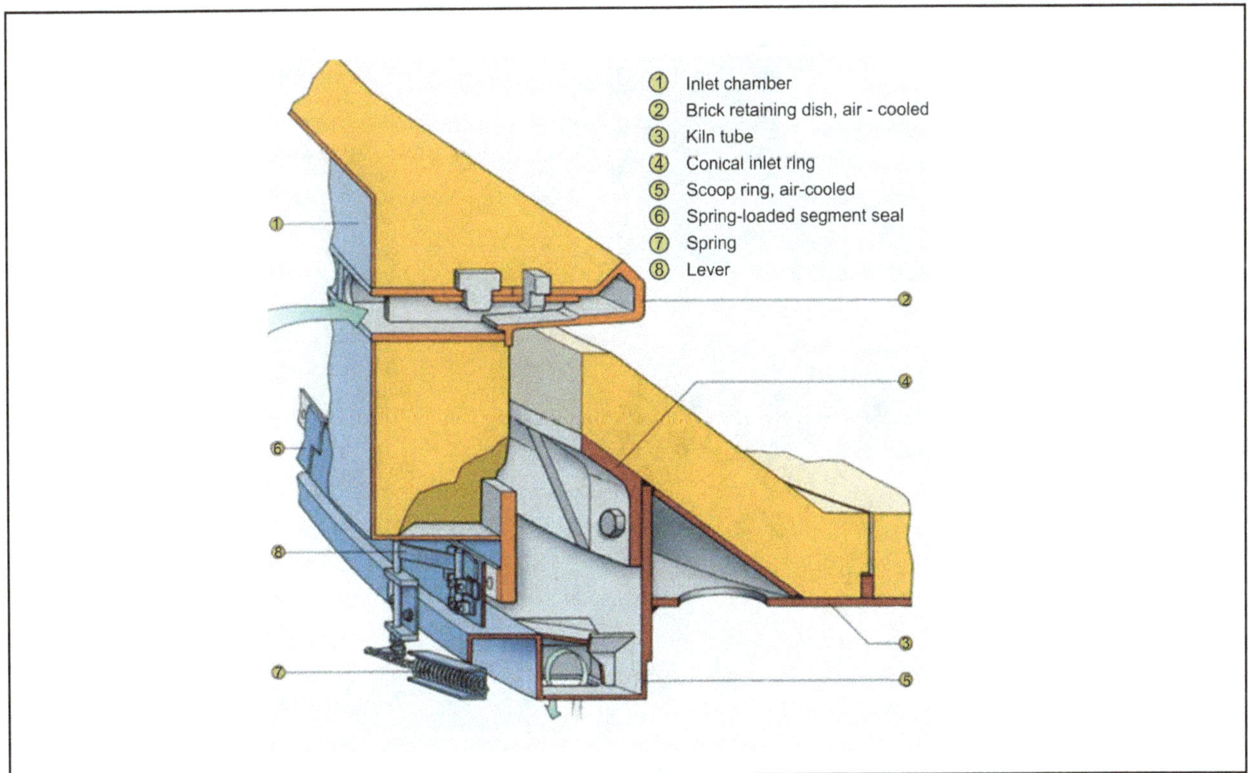

Plate 2.10.8 Kiln inlet seal - KHD.

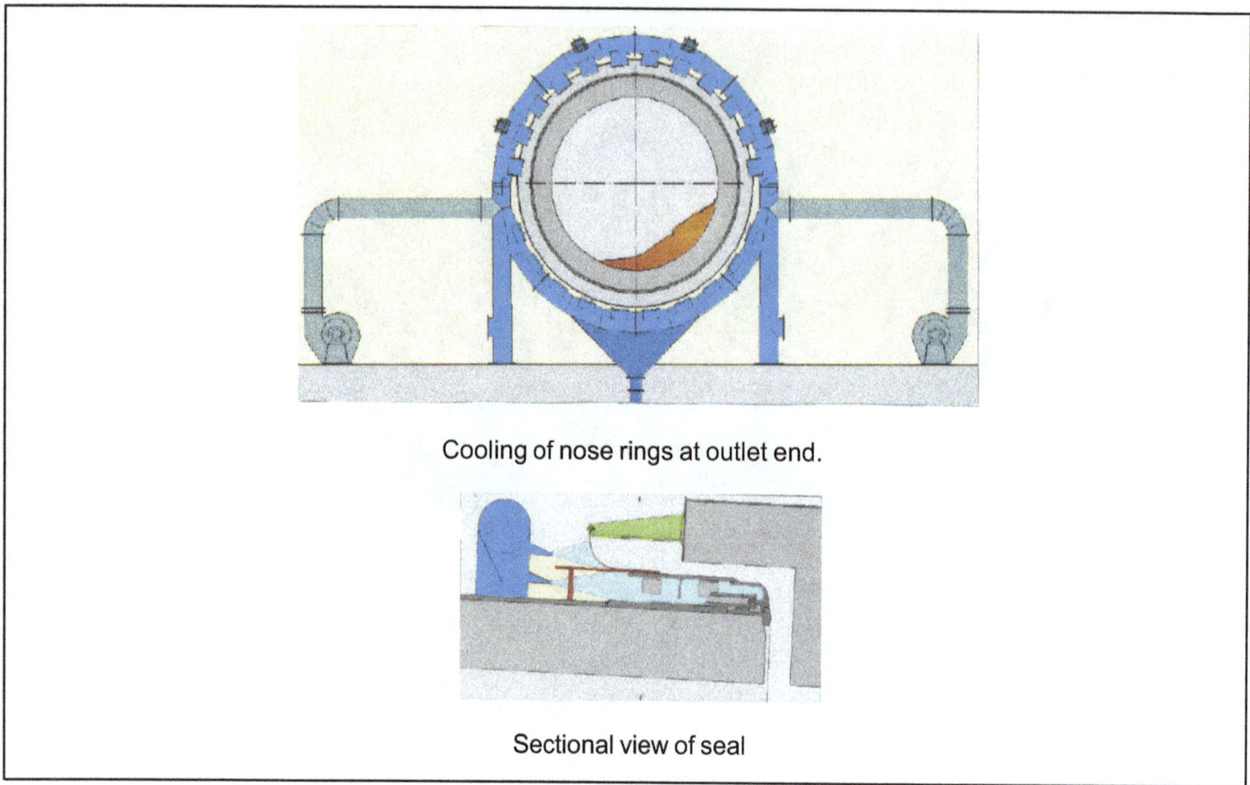

Cooling of nose rings at outlet end.

Sectional view of seal

Plate 2.10.9 Cooling of nose rings and seal at outlet end.

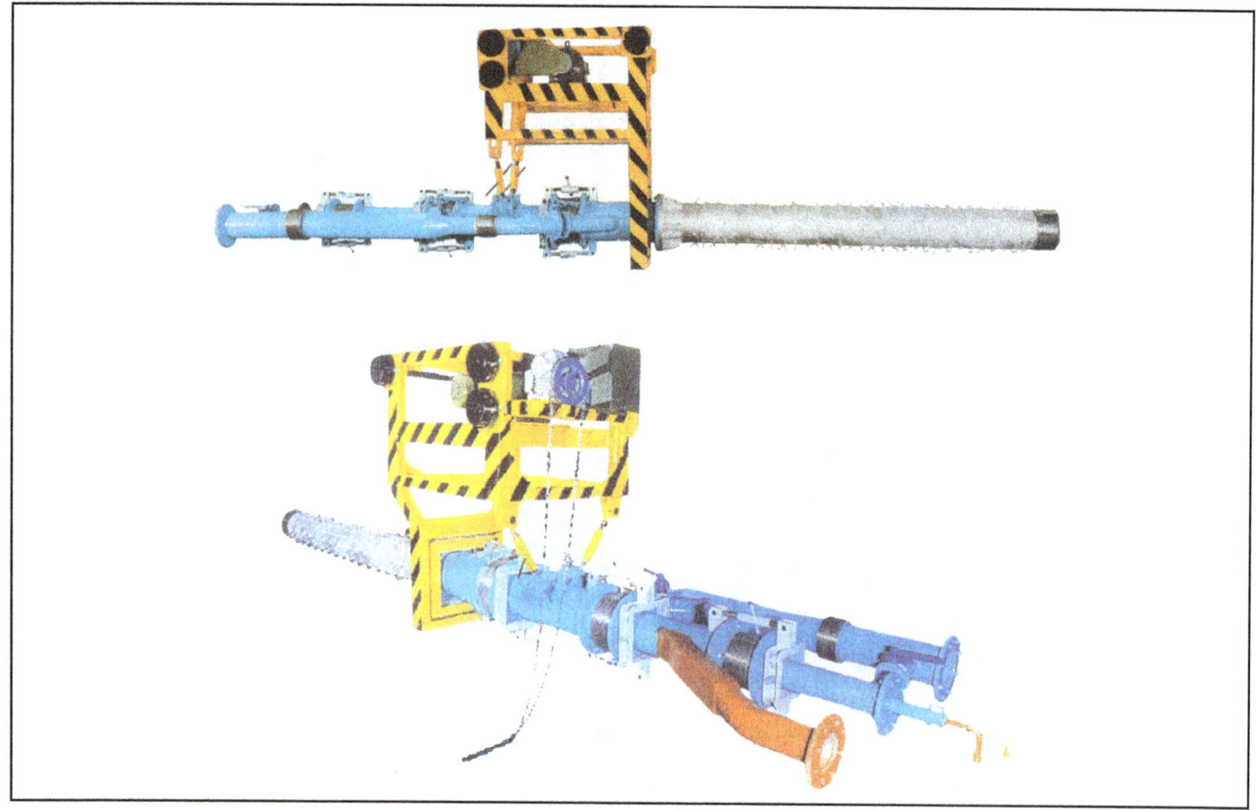

Plate 2.10.10 Multi Channel Burner for Kiln.

Annexe 2.10.1

Typical Radiation Losses in kiln systems
Radiation losses in kcal/kg clinker.

Sr.No.	Kiln and 4 stage preheater			Kiln with 5 stage preheater
Radiation loss in	**Grate cooler**	**Planetary cooler**	**Calciner with t.a.duct**	**Grate cooler and Calciner and t.a.duct**
preheater	14	14	18	17.5
kiln	55	49.5	35	31
cooler	1.8	63	1.4	2
Calciner	-	-	*	*
t.a. duct	-	-	6.5	7
total	$\simeq 71$	126.5	$\simeq 61$	$\simeq 58$

Annexe 2.10.2

Typical performance data on dry process kilns
with preheaters, calciners, grate and planetary coolers.

Sr.No.	Item	Preheater kiln		Calciner kiln	
		Planetary cooler		**Grate cooler**	
	Preheater	**5 stage**	**6 stage**	**5 stage**	**6 stage**
1	Heat balance Kcal/kg clinker				
	Heat in exhaust gas	157	144	160	145
	Radiation loss from Kiln and preheater	65	67	54	56
	Heat of reaction	385	385	385	385
	Cooler loss	141	143	116	116
	Heat in clinker at ambient temp.	3	3	3	3
	Heat in raw meal, fuel, air	34	34	33	33
	Nett sp. heat consumption	717	708	685	672
2	Exhaust gas temp. °C	295	272	311	283
3	Pr. loss in preheater mmwg	280	328	279	325
	Pr. loss in rest of system mmwg	71	71	140	140
	Total pr. Loss	351	399	419	465
4	Power consumption kwh/ton clinker				
	Preheater fan	4.6	4.9	5.6	5.9
	Cooler drive and fans	-	-	5.1	5.1
	Kiln	3.1	3.1	1.7	1.7
	p.a.fan	1.5	1.5	1	1
	Dust transport	1	1	1	1
	Total sp. power consumption	10.2	10.5	14.4	14.7

Power consumption for different kiln feed systems

Power consumpton for	elevator	Air lift	Screw pump
Kwh/ton clinker	0.5-0.6	1-1.5	2.0-2.5

CHAPTER 11

CLINKER COOLERS

11.1 Cooling Clinker

Clinker comes out of the kiln at around 1350 oC. It needs to be cooled down to temperatures where it can be handled by conveyors available like belt, chain, elevators that is below 100-200 °C.

Cooler was a part of the shaft kiln. Clinker left shaft kiln at about 150 °C.

11.2 Coolers External to the Kiln

Coolers external to the Kiln are of 5 types

1. Shaft
2. Rotary
3. Planetary.
4. Traveling grate.
5. Reciprocating grate.

11.3 Shaft, Rotary and Planetary Coolers

In shaft, rotary and planetary coolers all of cooling air is drawn into the kiln for combustion as secondary air. In passing over clinker and cooling it, air itself gets preheated and thereby assists combustion of fuel greatly and hence improves fuel efficiency

However, much larger quantum of cooling air is required to bring down temperatures below 200 °C. Therefore there is a limit to cooling in these three types of coolers.

11.3.1 Rotary cooler

A rotary cooler is like a rotary dryer or a small kiln with internal fittings to promote heat transfer between air and clinker. It needs a separate drive. Rotary cooler does not pose any problems in construction.
See **Fig. 2.11.1.**

Rotary coolers were replaced by planetary coolers.

11.3.2 Planetary coolers

In case of planetary coolers, instead of one cooling tube, 8 to 10 small diameter tubes are mounted on the kiln itself and rotate with kiln.
See Plate 2.10.1 in Chapter 10.
Clinker enters tubes through ports and travels in them. Tubes are fitted with refractory and with liners and lifters made of heat resisting materials. Clinker comes out at around 150 °C. Sometimes tubes are cooled externally with water or air to achieve better cooling.

But planetary coolers which form an overhanging weight on the end section of the kiln requires that section of shell, tyre and last supporting station is made of heavier construction.

In older designs, l/d ratio of cooler tubes was about 5-6/1. In new designs it is around 10/1. Thus coolers are much heavier and overhang is longer. In large kilns it requires an additional supporting station at discharge end to take weight of cooler.
See Figs. 2.11.2 and 2.11.3

Major disadvantage of the planetary coolers was that air could not be drawn from it for precalciner except in small amounts.

An advantage though in their case is that venting is not required because there is no excess air. They were suitable for small kilns but not for kilns with calciners.

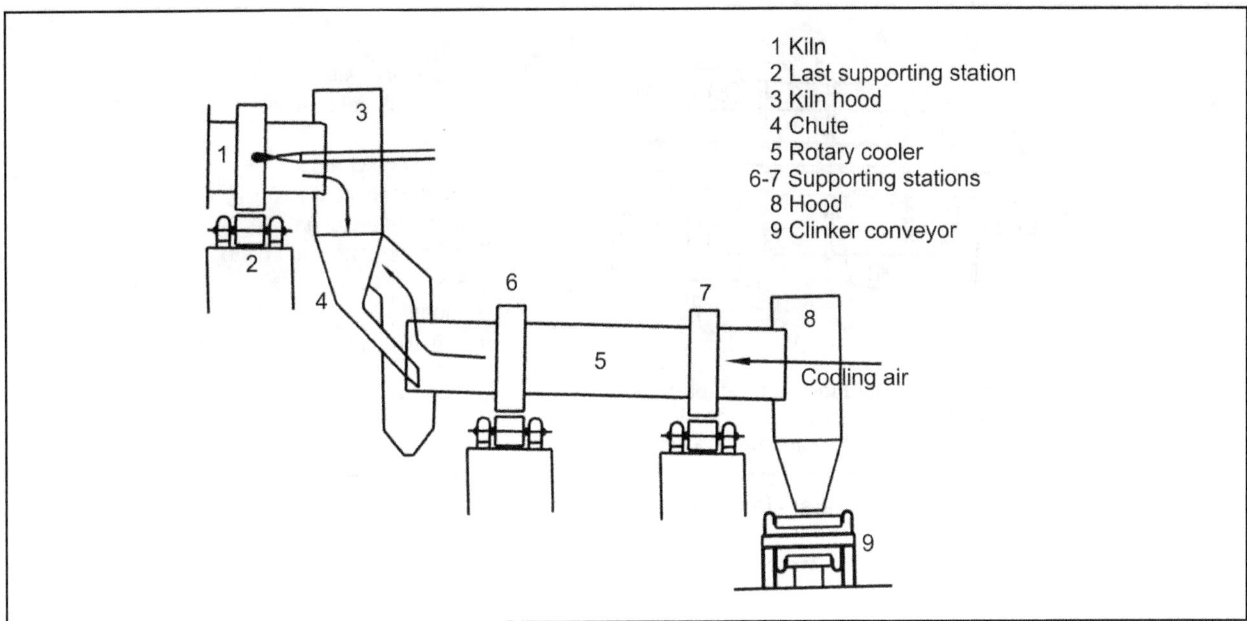

Fig. 2.11.1 Rotary cooler.

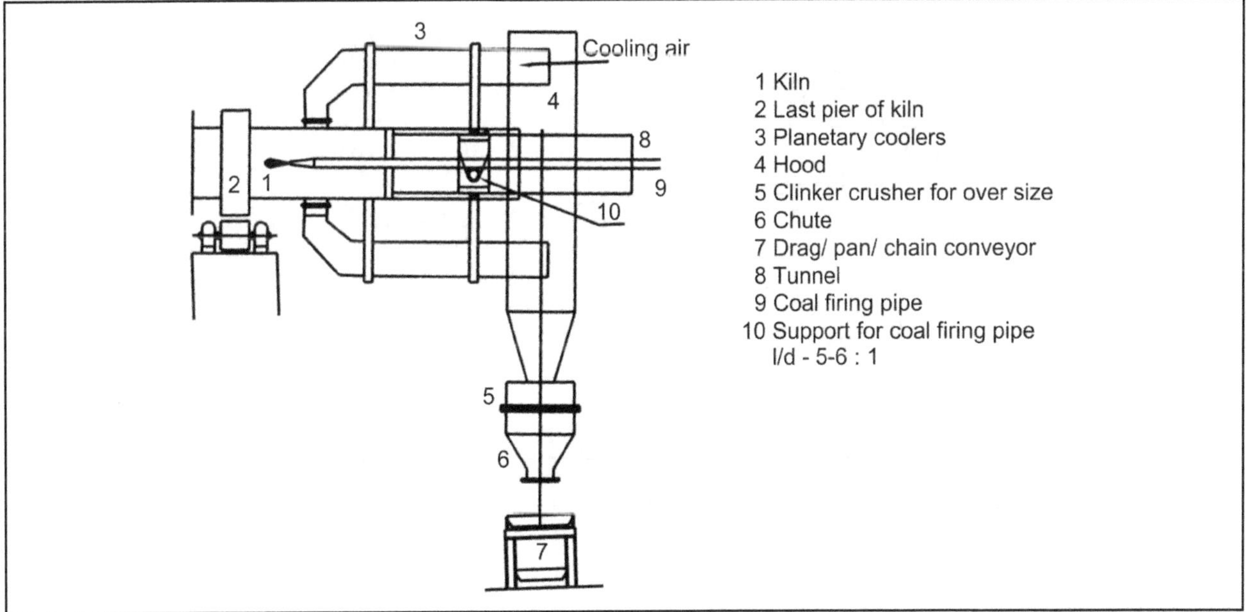

Fig. 2.11.2 Kiln with planetary coolers – cooler tubes 10 to 12 nos. l/d = > 5-6 : 1.

11.4 Traveling Grate Cooler

Traveling grate coolers were 'Lepol grates' used to cool clinker. In construction and in operation they were similar to Lepol preheaters. They did not become popular for similar reasons viz a great many moving parts of special heat resistant materials and high maintenance requirements.

See Fig. 2.11.4

11.5 Reciprocating Grate Coolers

Reciprocating grate cooler has become the dominant clinker cooler and has seen many developments.

Construction wise two sets of grate plates with small holes to pass cooling air are mounted on two frames- one frame is stationary and the other reciprocates on it.. Reciprocating action of one frame over the other, moves clinker slowly along the length of the cooler

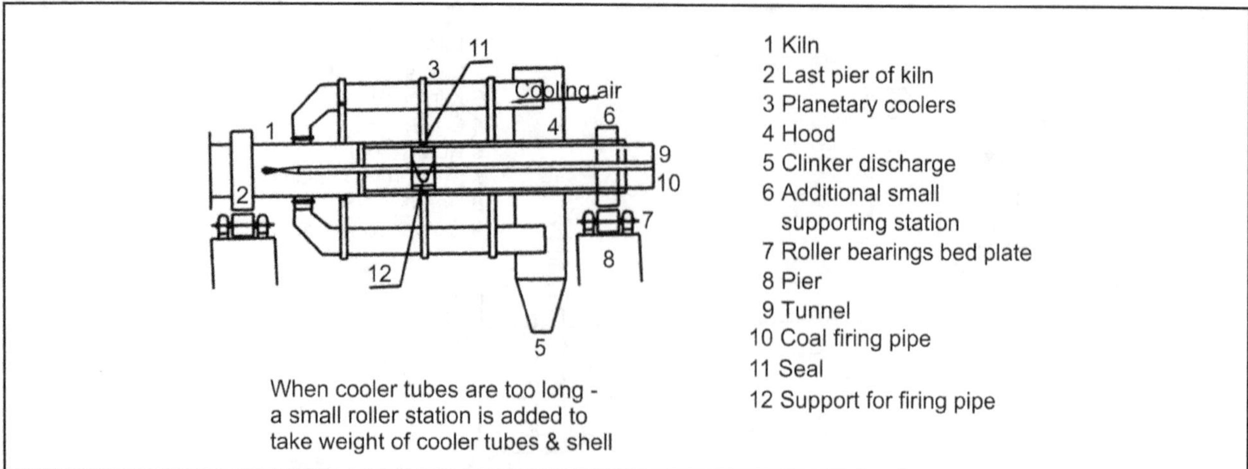

Fig. 2.11.3 Kiln with planetary coolers – long cooler tubes l/d = > 10 : 1.

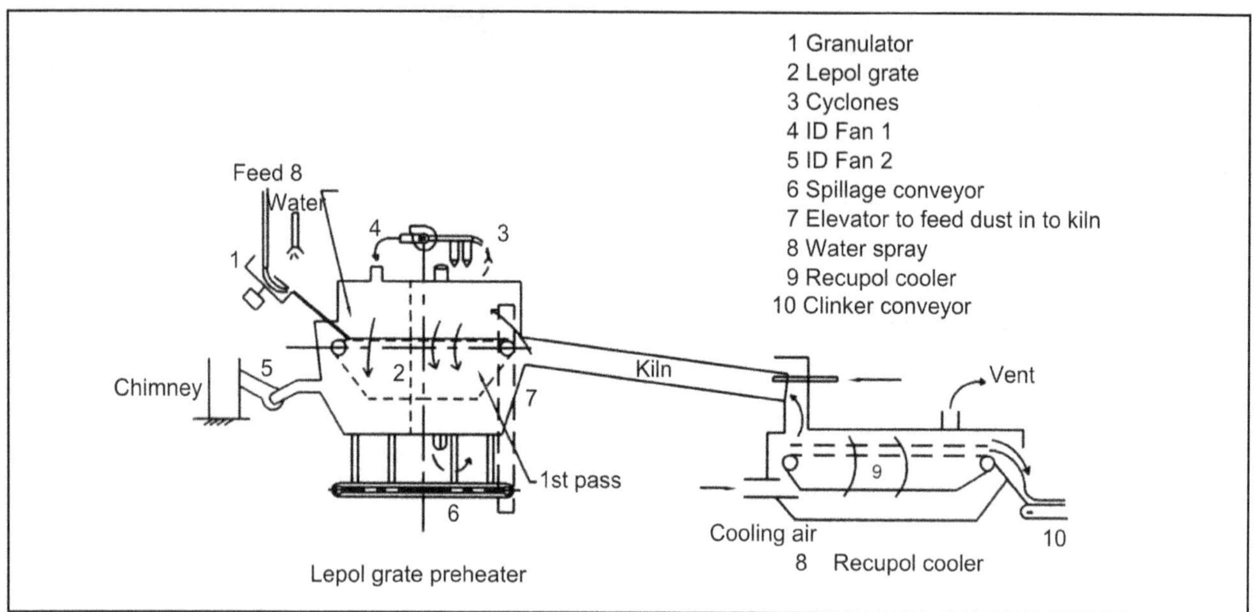

Fig. 2.11.4 Double pass lepol grate and recupol clinker cooler.

from feed end to the discharge end. A set of connecting rods and eccentrics convert rotating motion of motor into reciprocating motion of the grate plates. Frames are within a casing. Bottom part of the casing receives and distributes cooling air. It penetrates through clinker bed and cools it and gets heated in turn. The top part of cooler above the plates is therefore lined with refractory.

See Figs. 2.11.5 and 2.11.6 and Plate 2.11.1.

Cooling air to be admitted depends on the temperature to which it is intended to cool clinker. Excess of cooling air over the air for combustion is drawn off by an induced draught or vent fan and vented through a dust collector.

To make best use of cooling air, bottom half of cooler is divided into a number of compartments - each compartment gets cooling air from its own fan.

Thus a reciprocating cooler has a number of drives for fans for cooling and vent air and for clinker breaker and for spillage drag chain under the air compartments in addition to those for reciprocating grates.

See Plate 2.11.1.

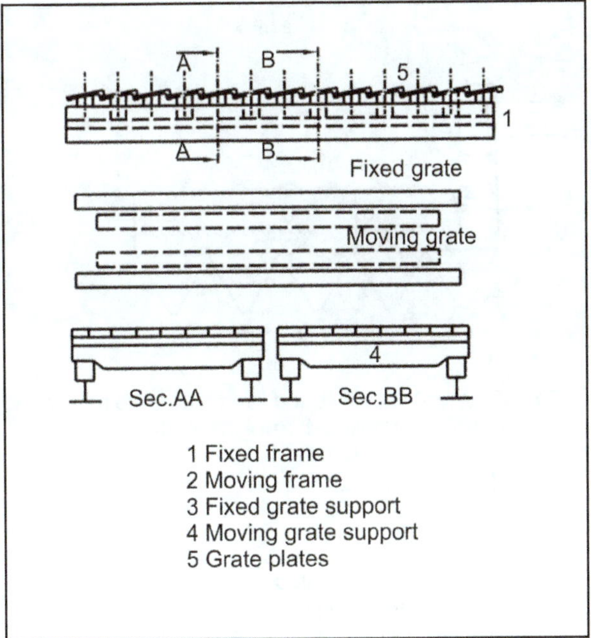

Fig. 2.11.5 Reciprocating grate cooler basic features.

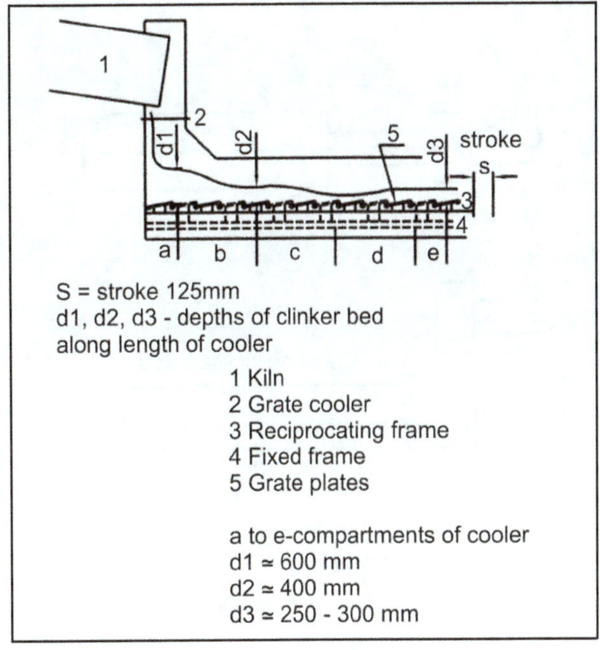

Fig. 2.11.6 Bed thickness of clinker in various sections of cooler.

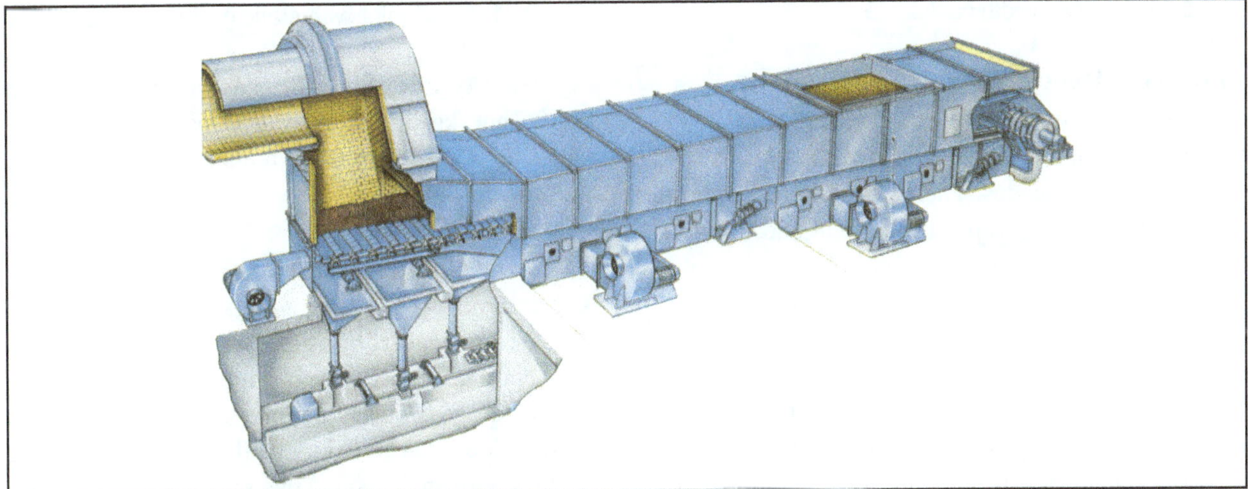

Plate 2.11.1 Conventional reciprocating grate cooler.

Power consumption is high; but cooling efficiency is also very high around 70 % + and radiation losses are negligible.

Cooler has a hot zone and a cold zone. Heat resisting materials are used for grate plates and side castings only in hot zone. Because of temperature profile of clinker inside the cooler, drop in pressure across a clinker bed of same thickness reduces progressively from hot end to cold end. Fans of appropriate static pressures can be selected for various compartments to reduce sp. power consumption.

Capacity of a given cooler can be increased by merely extending it. Thus large kilns may have three travelling grates each with its own drive.

11.5.1 Various Arrangements of Grate Coolers

Coolers are inclined that is grates are at a slope or horizontal or combi i.e., first grate is inclined and subsequent is horizontal. Now mostly horizontal coolers are used. Grates can be of different widths to reduce over all length.

See Fig. 2.11.7

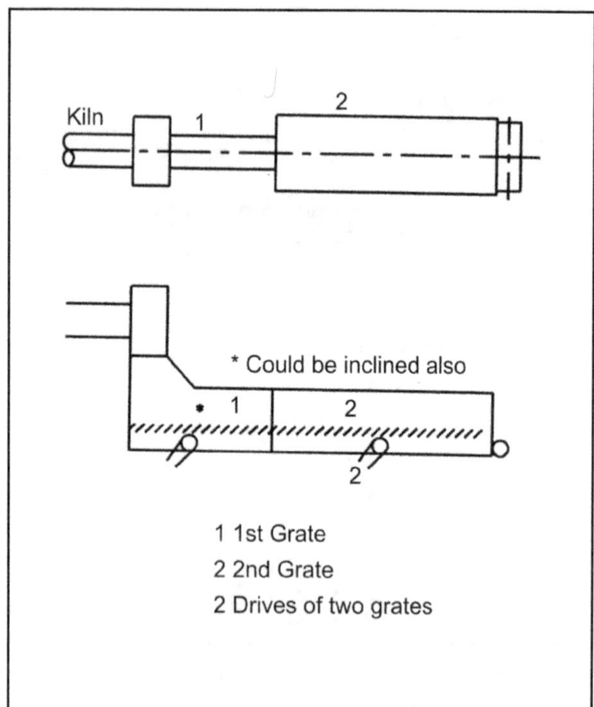

Fig. 2.11.7 Two width cooler.

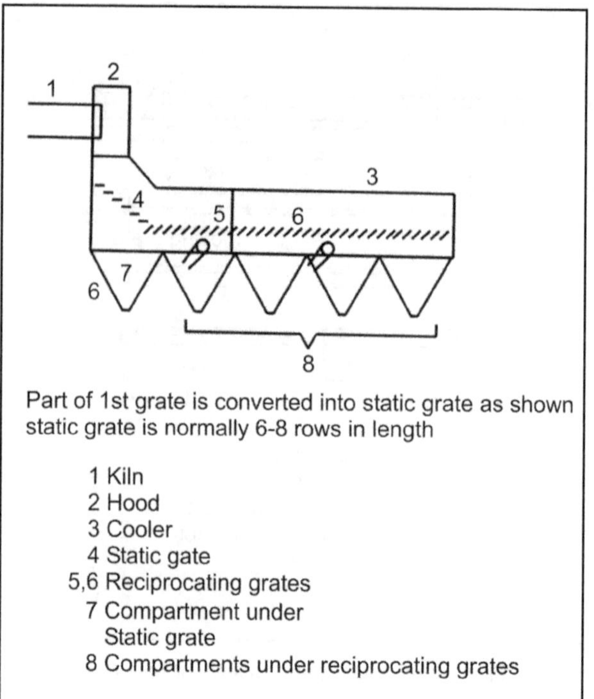

Fig. 2.11.8 Cooler with static grate.

11.6 New Dvevelopments in Design of Grate Coolers

Grate Cooler design has seen a great many changes in the last two decades. Their objective has been to cope with increased capacities of kiln and improve recuperation efficiency of coolers

11.6.1 Controlled Flow of Cooling Air through Grate Plates

Air flow through individual or group of grate plates is controlled by using grate plates that have their own air compartments. Air flow in right and left half of cooler can also be adjusted as required.

11.6.2 Static Grate

First 7 to 8 rows are 'static'. They are installed at a greater degree of inclination to promote flow of clinker over them. Pressure of fan in compartment with static grate is much higher at about 1000 mmwg.

See Fig. 2.11.8 and Plate 2.11.2.

Spillage in coolers with static grates or with controlled flow grates is much less than in case of conventional designs.

Quantity of cooling air has come down from 2.5-2.7 nm^3/kg to 2 –2.2 nm^3/kg. Consequently quantity of air to be vented has also reduced from 1.8 nm^3 to 1.3-1.4 nm^3/kg.

11.7 Developments in Constructional Design

Other developments are as regards suspension and imparting reciprocating motion to grates. In conventional designs each grate had its own drive located at the center of the grate. Though motor was on one side, an even motion to the whole width of the grate was given by having two sets of connecting rods and eccentrics one on each side of the cooler. Even then it was possible that some kind of ‘skewing' of moving grate could take place of grate if it was too wide or too long.

1 Cross bar Cooler : see Plate 2.11.3

In this reciprocating grates are replaced by a series of cross bars moving in a casing over stationary grate plates with their own air compartments. Flow to each grate can be controlled according resistance of clinker flowing over it.

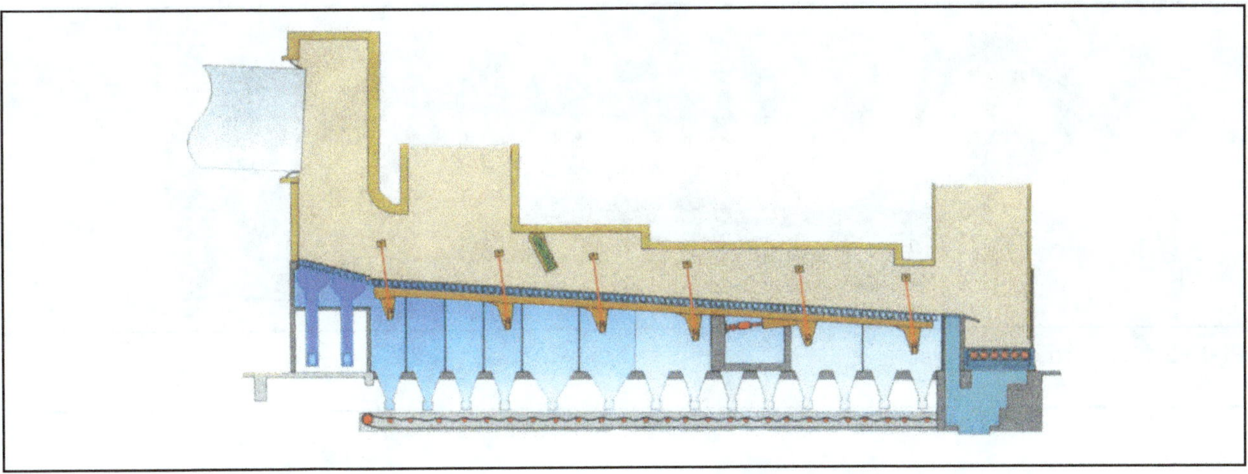

Plate 2.11.2 KIDS reciprocating grate cooler with static grate at inlet.

2 Walking Floor : see Plate 2.11.4 to insert plate
In it, the width is made up of a number of plates. While all move together in forward motion, they return at different speeds and the sequence of return is also different along the length

3 Poly track cooler see Plate 2.11.5

4 Coolax Cooler
Yet another variation is the Coolax Cooler.
See Plate 2.11.6
These developments have resulted in changes in layouts of cooler itself as also with respect to kiln.

11.8 Operational Parameters of New Coolers to check and para nos also

Parameters for sizing coolers have been empirical. Most commonly used parameter is capacity expressed as tpd of clinker cooled/m2 of Grate area. Its value has progressively increased from 25 when bed thickness of clinker used to be 200 mm to 40-45 when bed thickness is 600-800 mm. It has further increased to 50-55 with static grates and controlled flow grate coolers.

Cooling air has come down from 2.7 to 2-2.2 nm^3 / kg clinker.

Different manufacturers have their own practices of compartment lengths and air loading. Commonly used air loadings are 120 m^3 /min/ m^2 for first compartment to 50 m^3 /min / m^2 for the last compartment.

As mentioned, static pressures of fans reduce progressively from hot end to cold end but at same location they are directly proportional to bed thickness.

Cooler efficiency is measured as heat recovered from heat in clinker at inlet in preheating secondary and tertiary airs going to kiln and calciner respectively.

See Figs 2.11.10 and 2.11.11.

As fuel consumption has progressively come down from 900 kcal/kg to < 700 kcal/kg, quantity of air for combustion has come down. Drop in quantity is compensated by increase in temperatures to which air for combustion is preheated. Secondary air temperatures as high as 1000 °C are common.

Recuperation efficiency is about 70 %.

11.9 G Cooler

G Cooler needs to be mentioned though it has not caught on. It has considerable potential of providing a solution for increasing capacity of an existing kiln. Cooler can be installed in line or at right angles and even at some distance from the existing cooler.

Addition does not require any dust collecting equipment as cooling air does not come in contact with clinker.
See Plate 2.11.7 and 2.11.8.

11.10 Conveying Spillage

In new designs, quantum of spillage is much less. - in some, it is totally eliminated.

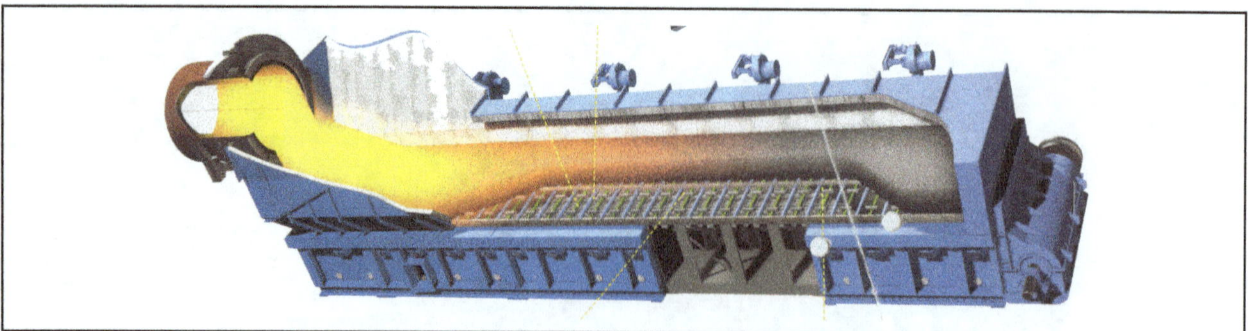

Plate 2.11.3 Cross bar cooler.

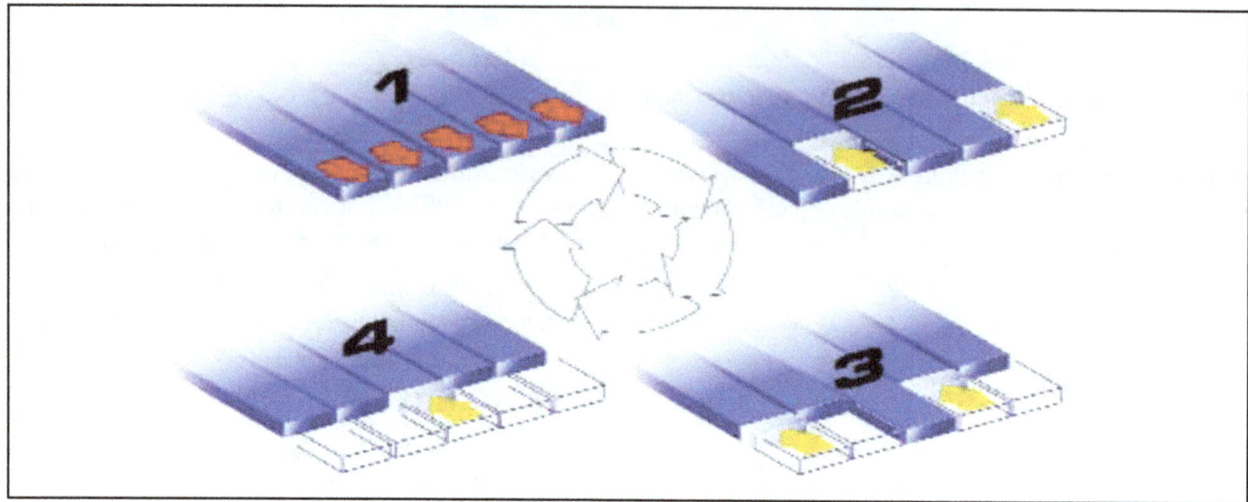

Plate 2.11.4 Principle of walking floor of N cooler.

Source Claudius Peters AG Brochure

Plate 2.11.5 Poly track cooler.

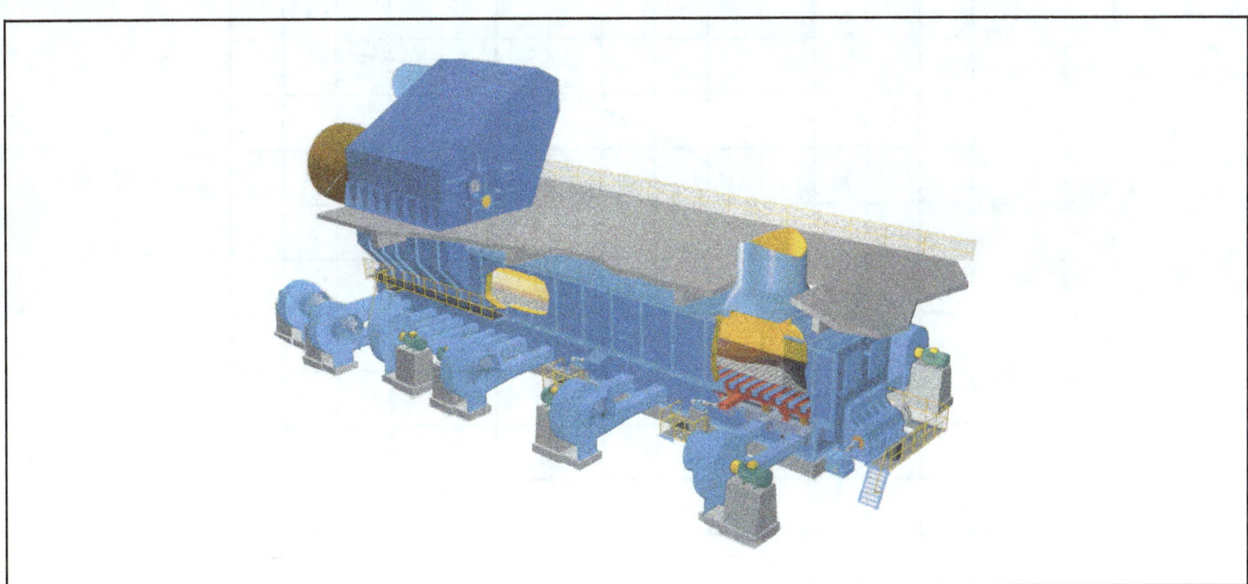

Plate 2.11.6 Coolax grate cooler.

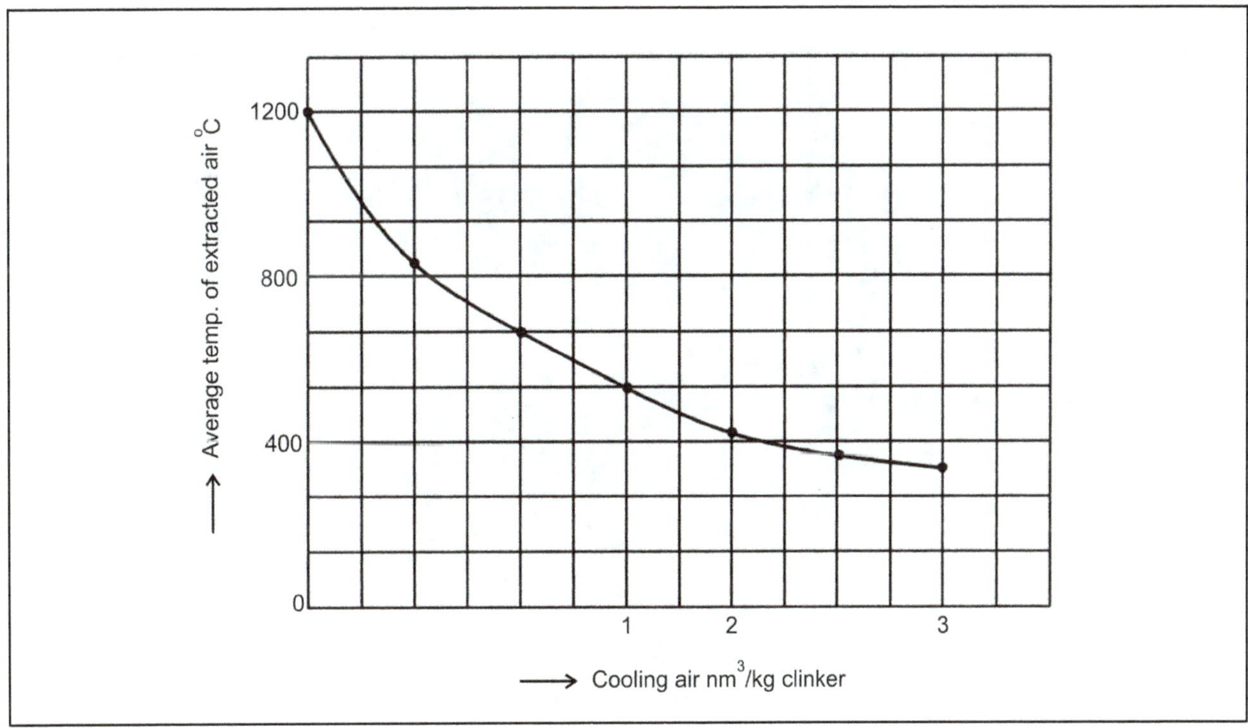

Fig. 2.11.9 Average temperature air extracted from cooler.

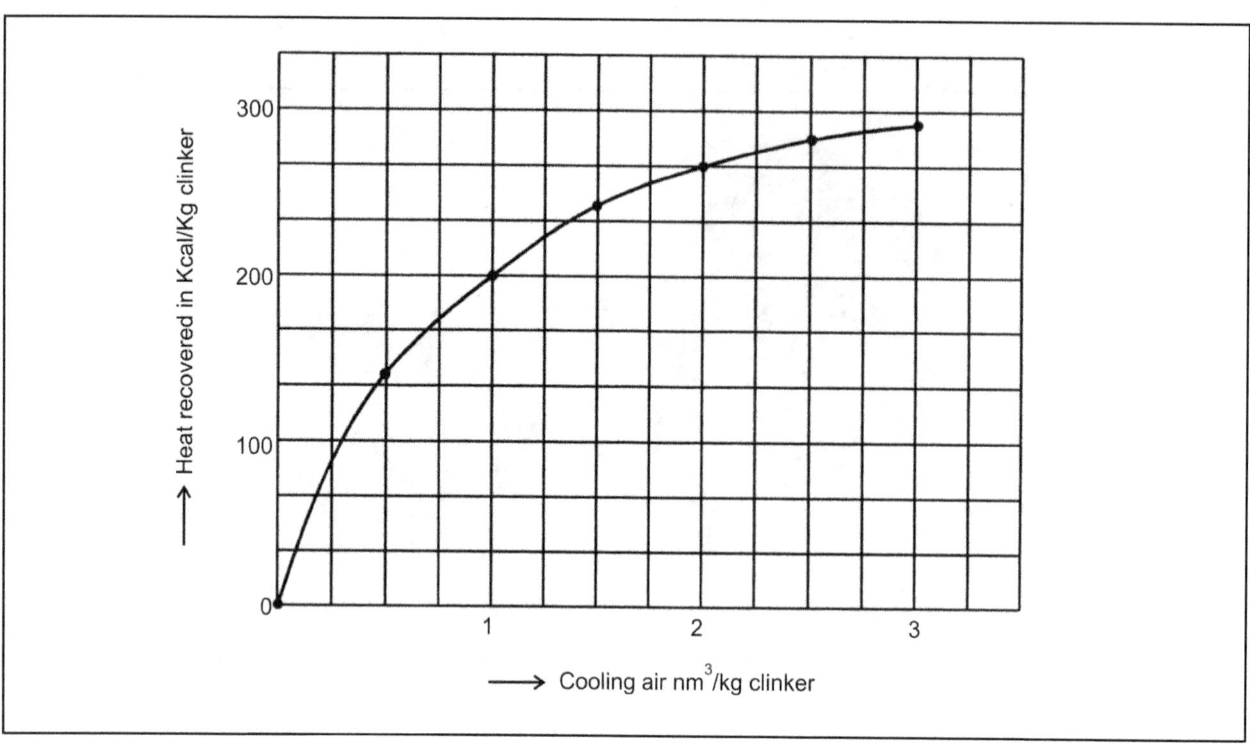

Fig. 2.11.10 Heat recovered by extracted air.

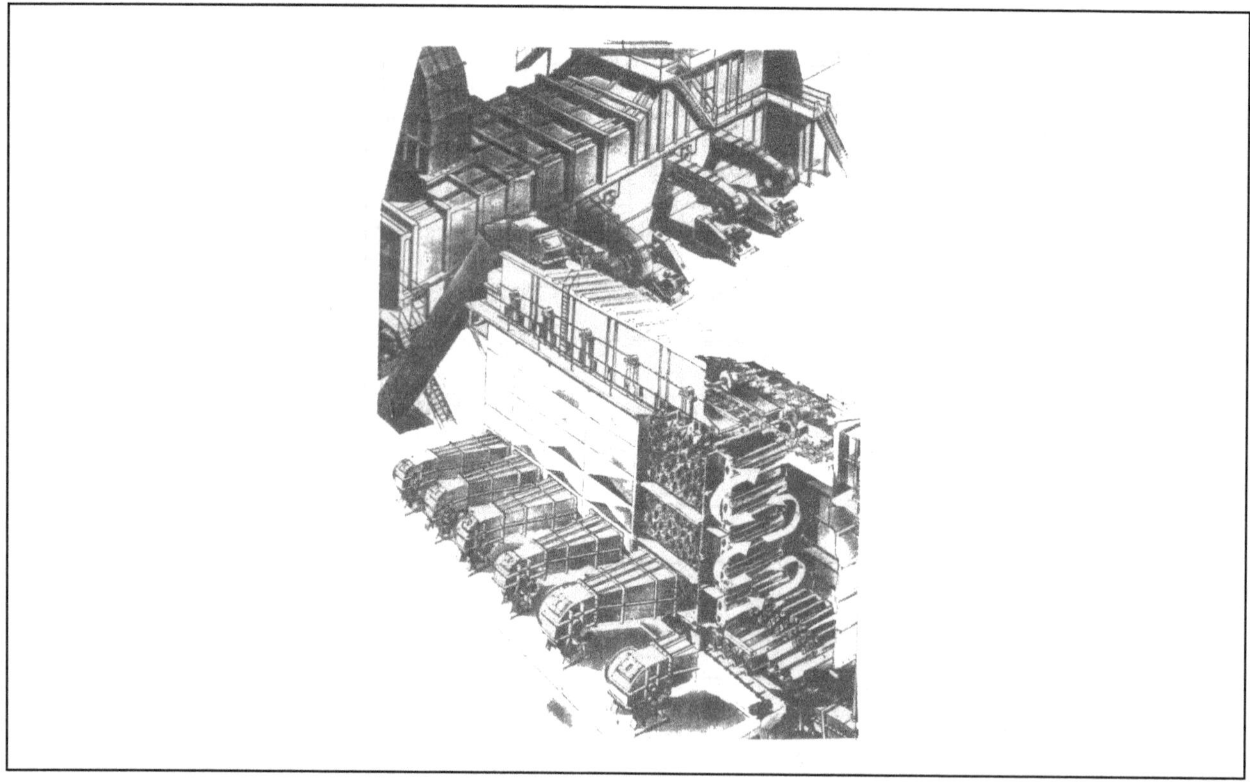

Plate 2.11.7 G cooler installed as extension of conventional grate cooler to increase its capacity.

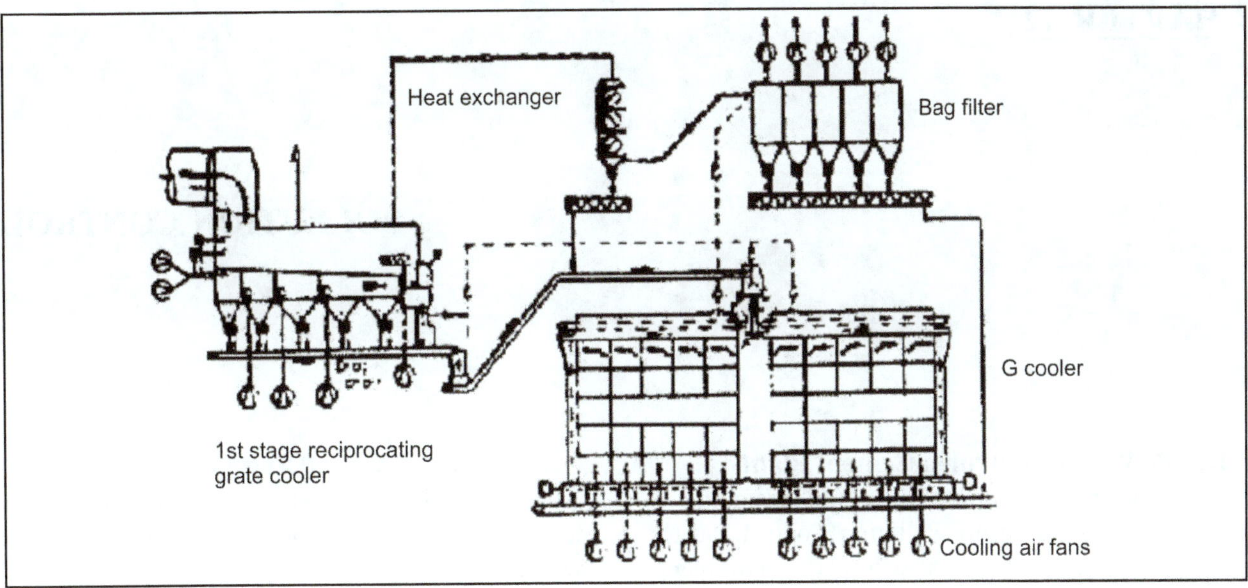

Plate 2.11.8 Indirect or G-Cooler.

Spillage is sometimes conveyed pneumatically from collecting hoppers on continuous basis doing away with a number of pairs of pendulum flap valves. See **Fig. 2.11.11.**

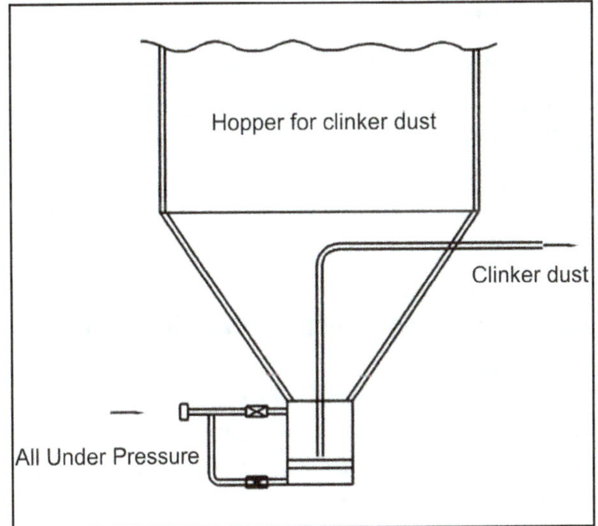

Fig. 2.11.11 Continuous pneumatic conveying of clinker dust to main clinker conveyor.

11.11 Hydraulic Drive

Conventional drive consisting of motor-gearbox-chain and a pair of connecting rods to give reciprocating motion are all replaced by hydraulic drives which impart reciprocating movement directly.

11.12 Roll Crusher

Hammer type clinker breaker in conventional designs is now replaced by one or more pairs of full width roll crushers. These may be installed at the end of the last grate or in between two grates.
See Plate 2.11.9.

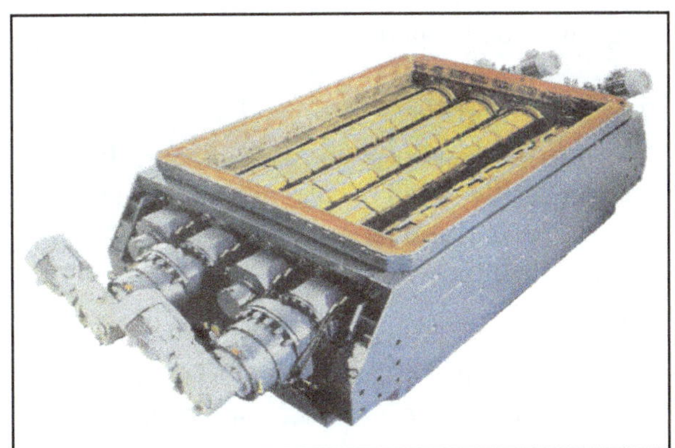

Plate 2.11.9 Full width Roll Crusher replaces conventional clinker breaker.

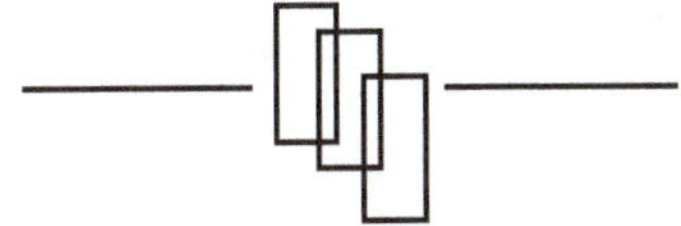

CHAPTER 12

POLLUTION CONTROL

12.1 Pollution Control in Cement Plants

One of the most important aspects of designing a cement plant is to ensure Pollution Control and to provide clean ambient air around the plant for the plant itself and for the community.

Cement Plants cause air and noise pollution.

12.2 Noise Pollution

Most cement plants are in isolated areas remote from towns and hence noise pollution has not yet merited the attention of the public. Further, with the use of vertical mills for grinding coal,raw materials and even clinker,where noise level is much less than that of ball mills, the problem has resolved by itself to a great extent. Cement mills to some extent crushers are the major sources of noise pollution still.

Even if Industry continued with ball mills, it would take care as is taken in developed countries to enclose the mill and to install soundproof doors and walls to bring down noise pollution.

12.3 Air Pollution

Major cause of worry is the air pollution. In a dry process plant the major sources of air pollution are:

12.3.1 Process Gases

1. Kiln & Cooler

   Total volume of gases emitted = 1.7 + 1.7 = 3.4 nm^3/kg clinker or 3.25 nm^3kg cement

2. Grinding Mills
   1. Raw Mill = 1.6 nm^3/kg raw meal drawn from kiln
   2. Coal Mill = 1.6 nm3/kg coal drawn from kiln
   3. Cement Mill = 0.5 nm3/kg cement

Hence Total volume to be vented from mills would be about 2.5 nm^3/kg cement depending on moisture to be divied and capacities of mills.

12.3.2 Venting in departments like

1. Crushing
2. Blending - Continuous blending
3. Kiln Feed.
4. Cement Packing.
5. Pneumatic conveying and other material handling Systems

All above add up to 1 nm3/kg cement

Total volume of gases to be vented, could be taken as 7-7.5 nm^3/kg of Cement.

There are intermittent polluting factors like fugitive dusts - crusher hoppers when dumpers are unloaded and building up of stockpiles.

12.4 Permissible Dust Burdens in Exhaust Gases

Pollution Control Boards- State and Central- lay down norms for permissible dust burdens in exhaust gases for various Industries from time to time. For Cement Industry present norm is 110 mgm / nm^3. It is applicable to all exhaust chimneys large and small individually.

In designing dust collectors it is necessary to plan for expansion right from the beginning.

This is because Pollution Control Authorities now insist that total dust emitted in **'tons per day'** at the initial plant capacity should also remain the same after expansion.

See Table 2.12.1.

Table 2.12.1

		Clinkering Capacity	
	unit	**3000 tpd**	**10000 tpd**
Capacity	Kg/hr clinker	125000	416700
Gases cleaned	nm^3/hr	750000	2500000
Dust emitted at 110 mg/nm^3	tph	0.0825	0.275
	Tons per day	≃2	≃6.6

Taking say 6 nm3/kg as normal volume of gases emitted in a cement plant:

To maintain rate of emission at 0.0825 tph at 10000 tpd capacity, would require that dust emission norms should be reduced to 33 mg/nm3 for the first plant itself.

This can be very expensive, as the efficiency would have to be increased from 99.89 % to 99.95 %. Though the increment looks small - only 0.06 %, repercussions on size and hence cost of dust collector are very high, more than 10 %. Further this increase in cost is to be incurred in the first phase itself.

Wherever possible, cement plant should be located at least 5 Kms from population centres, preferably outside the 'wind rose' and main chimneys should be tall so that they can spread dust over a wide area and dust concentration can come down.

See Annexe 2.12.1.

Because of stringent norms laid down by Pollution Control Authorities, only bag filters and esps can be used as final dust collectors.

12.5 Cyclones for Collecting Dust

Cyclones are still used in grinding circuits but as 'intermediate collectors', which clean mill gases to the extent of 90 -92% of its dust content before they enter esp / bag filters.

The use of cyclones strictly speaking is not warranted as pre collectors before esps / bag filters because both of them are capable of handling full load of dust content, which may be as much 500-600 gm/ nm3 in mill systems.

In kiln system, gases in direct operation, would have dust burden of 5-7 gm or less assuming 93% efficiency of top cyclones of preheater.

But cyclones help in emergencies to bypass the dust collector. By using gas tight dampers, it would be possible to attend to the maintenance and repairs of the dust collector without requiring to stop the plant.

Cyclones however, add pressure drop of the order of 100-150mm and hence require more power.

Let us say gas flow is 100 nm^3/min

Pressure drop in system without cyclone - 150 mm.

Pressure drop in system with cyclone would be - 250 mm.

Power in 1st case = $K \times 100 \times 150$ kw

Power in 2nd case = $K \times 100 \times 250$ kw

By themselves cyclones would hardly be used now. This is because efficiency of a cyclone drops sharply as particle size reduces as would be seen from the graph. **See Fig. 2.12.1.**

To be able to predict the efficiency of a cyclone, it would be desirable to carry out a sieve analysis of a sample of dust in microns fractions as shown in **Table 2.12.2.**

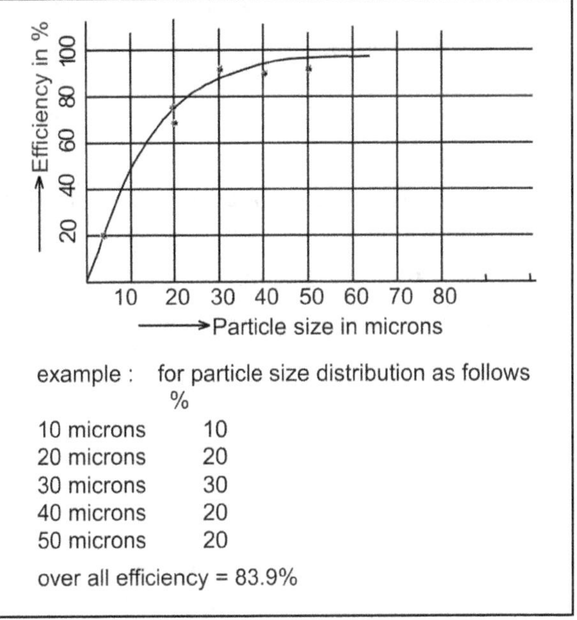

Fig. 2.12.1 Efficiency of cyclone dust collector depends on size of dust particles.

Table 2.12.2

Particle size microns	Quantity %	Efficiency %	Calculation of overall efficiency	
100	5%	99	5 x 99	495
50	15%	99	25 x 99	1485
40	30%	98	30 x 98	2940
30	20%	60	20 x 60	1200
20	15%	50	15 x 50	750
10	10%	30	10 x 30	300
5	5%	20	5 x 20	100
0				
				7270

Overall efficiency would be 72.7 %.

such low efficiencies are not tolerated at all these days.

Typical contents of fines in vent gases from various sources have been shown in **Annexe 2.12.2.**

12.5.1 Design of Cyclones

Design of cyclone, its sizing and pressure drop across it are still governed by empirical rules.

Each manufacturer has its own geometrical configuration of a cyclone. They establish the efficiency achieved or that can be achieved and the pressure drop across it for a given gas volume (at a given temperature and density).

The entrance velocity to a cyclone ranges between 14-16 m/sec (lower end) to 22-24 m/sec (upper end).

The geometry of cyclone is also fixed in relation to its diameter.

See Fig. 2.8.1 in Chapter 8.

To be able to pass pollution norms which lay down 55-110 mg/nm^3 as dust content of clean air, efficiencies required are of the order of + 99.5%. These can be achieved only by using bag filters or esps.

12.6 Polyclones and Multiclones

Reference has already been made to use of cyclones or banks of cyclones called Polyclones to collect material from streams of gases, in **Chapter 5.** Multiclones were a further development of this concept. Though efficiency improved and pressure drops were maintained between 100 to 200 mmwg levels, they were still not adequate to serve as final dust collectors. **See Figs. 2.5.16 to 2.5.18 in Chapter 5.**

Even if multi clones have an efficiency of 95%, it would still not achieve 110 mg / nm3, clean gas dust burden.

12.7 Bag Filters

Bag filters are commonly used for large and small volumes of gases, for individual locations, for material handling equipment, for silos, hoppers and bins where temperatures of gases would generally be 50-100 °C.

See Plate 2.12.1 and 2.12.2.

Bag filters, use different materials for bags for different temperatures of gases.

See Annexe 2.12.4 to 2.12.6 for properties of various materials used to make filter bags.

From ordinary cloth bags to glass bags, temperatures of gases that could be handled would range between 80-90 °C to 240 °C.

In compound operation, temperature of gases to be handled would be between 100-120 °C. Thus polyester bags, which can withstand temperature of up to 140 °C, could be used.

But for design purposes it is to be assumed that for some part of the day (mill runs 20-22 hours), gases would go directly to the filter.

Therefore, the bag filter system should be designed to cool gases to the temperature, bags can withstand viz., 240 °C for glass bags and 120 °C for polyester bags.

12.7.1 Replacement Costs of Bags

Costs of bags vary considerably according to material used. Further they are to be maintained in good condition. Development of holes in bags would adversely affect dust emission. They also have a life. Hence, a whole set of bags needs to be replaced at periodic intervals of say 2-4 years.

12.7.2 Higher Power Consumption

Pressure drop across bags is also high at 150 mm as against a combined pressure drop of 60 mm or so in GCT and ESP.

Therefore, bag filter would consume more power as compared to ESP.

12.7.3 Diluting Air

Cooling gases to 120 °C from 350 °C or 280 °C (depending on number of stages of preheater) requires large quantity of ambient air to be brought in.

Plate 2.12.1 Bag filter.

Plate 2.12.2 Bag filters to vent bins and silos.

This results in requiring very large dust collectors, which for large plants can be as big as sizeable godowns. Layouts must be planned to accommodate them.

There are many permutations and combinations between gases going from kiln to exhaust and to mill according to the moisture to be dried, capacity of mill and water to be added in mill to avoid vibration in case of roller mills.

12.7.4 Sizing of Bag Filters

For good consistent performance, bag filters are sized on basis of a parameter known as $m^3/min/m^2$ or air to cloth ratio. It is actually velocity of gas through the filter cloth.

Recommended air to cloth ratios for different applications have been furnished in **Annexe 2.12.7.**

When bag filters are large, they are divided into several small compartments or modules.

See Plates 2.12.1 and 2.12.2.

For example, a filter requiring 240 bags would be divided into 4 compartments of 60 bags each.

Compartments go through cleaning cycles to dislodge cake of material collected on surface of bags. Airflow is reversed in the compartments under going cleaning. Thus out of 'n' compartments, filter bags in (n-1) compartments would be actually effective for cleaning.

In a 4 compartment dust collector, effective cloth area is 75 %.

More the number of compartments (larger the filter) the effective area in proportion to total area increases.

Sizing the dust collector would be done on effective cloth area.

Air cloth ratios, depend upon material of cloth, dust concentration and characteristics of dust such as its abrasiveness. For abrasive dusts and for coal low air to cloth ratios would be selected.

Designers of bag filters should be consulted for recommended air to cloth ratios.

12.7.5 Example of Sizing a Bag Filter

Kiln capacity 3000 tpd or 125000 kgs/hr Sp. gas volume 1.6 nm^3/kg

At 300 °C; ambient temperature 30 °C.

(i) Bag filter with glass bags

Direct operation

Gas volume from preheater = 200000 nm^3/hr at 300 °C.

Therefore diluting air to bring down temp. to 230 °C,

required = 70000 nm^3/hr or 35 %

If air to cloth ratio = 1.6 for glass bags, effective cloth area of bag

Filter would be 5200 m^2

(ii) Bag filter with polyester bags

Direct operation

Gases to be cooled to 120 °C. Therefore diluting air required would be 400000 nm^3/hr or 200 %.

For same air to cloth ratio (1.6), effective cloth area of bag filter would be 8700 m^2.

The replacement costs should also be taken into account. Thus if a glass bag costs 2 times of polyester bag, the glass bag filter would be costlier in first costs. But if life is twice as long, replacement costs / ton of material would still be the same.

12.8 Heat Exchanger and Bag Filter

It has been mentioned above that temperature of vent gases was brought down by adding ambient air to vent gases. Another way of achieving the same objective would be to cool vent gases through an 'air to air' heat exchanger. They would work together just as gct and esp work together.

Because of high cost of an esp many a small plant had installed such a combination of heat exchanger + bag filter to clean preheater exhaust gases.

Today this combination has found favour in cleaning cooler vent gases. This application requires that in designing heat exchanger , volume and temperature under upset condition of kiln has to be taken into account.

Plate 2.12.3 showing the heat exchanger.

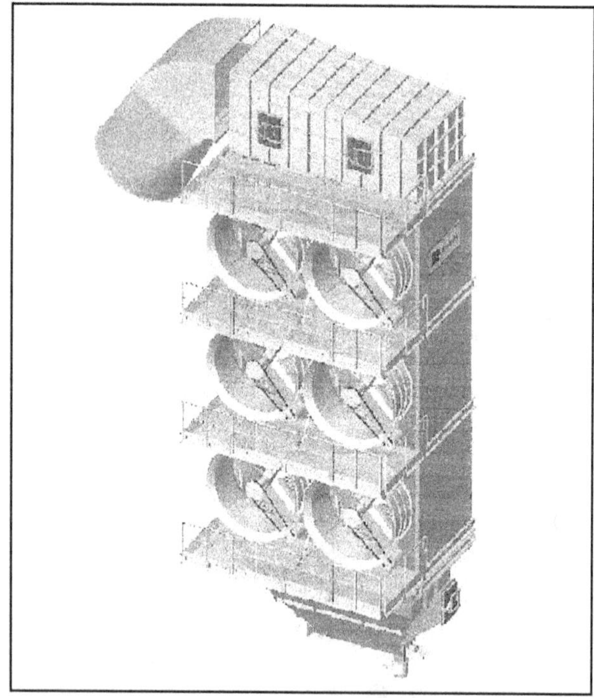

Plate 2.12.3 Air to air heat exchanger for cooling vent gases of cooler before entering bag filter.

12.9 ESP & GCT

GCT that is Gas Conditioning Tower and ESP that is Electostatic Precipitator together form a dust collecting unit for Kiln and Raw Mill gases.

12.10 GCT

GCT is used to bring down the temperature from 350-280 °C to about 140 °C. This also improves dew point and thereby improves resistivity of gases. The dew point is critical for the performance of the ESP.

Quantity of water to be added depends on the initial temperature of gases.

Dew point improves from 35 °C to start with to 52-55 °C after water spray.

From practical considerations, 140-120 °C is the lower limit of cooling. This is because, lower temperatures can create 'wet bottom' conditions.

See Fig. 2.12.2 and 2.12.3.

Normally, an oversized screw / chain conveyor is installed at the bottom with a reversible drive.

If dry dust is collected, it is conveyed to the dust collected in ESP and then two together are conveyed either to blending silo or fed to kiln either independently or after proportioning it with fresh feed.

At both ends of screw are installed pendulum flap valves which prevent in leakage of air.

See Fig. 2.12.4.

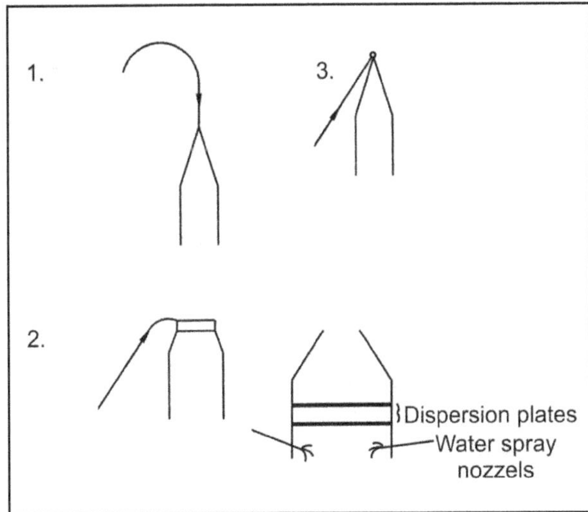

Fig. 2.12.2 Gas entries in gas conditioning towers.

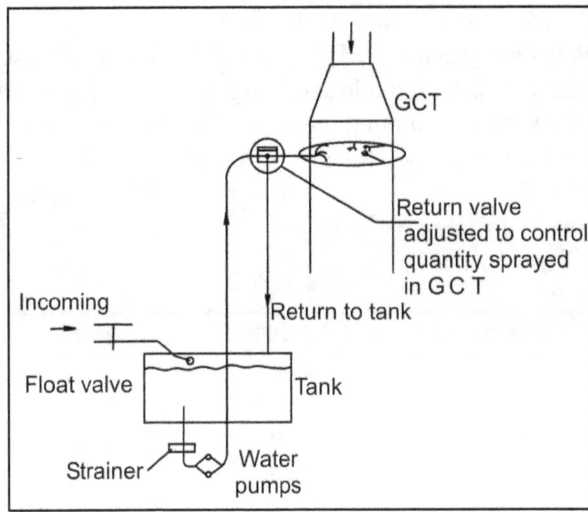

Fig. 2.12.3 Water spray circuit in gas condition tower.

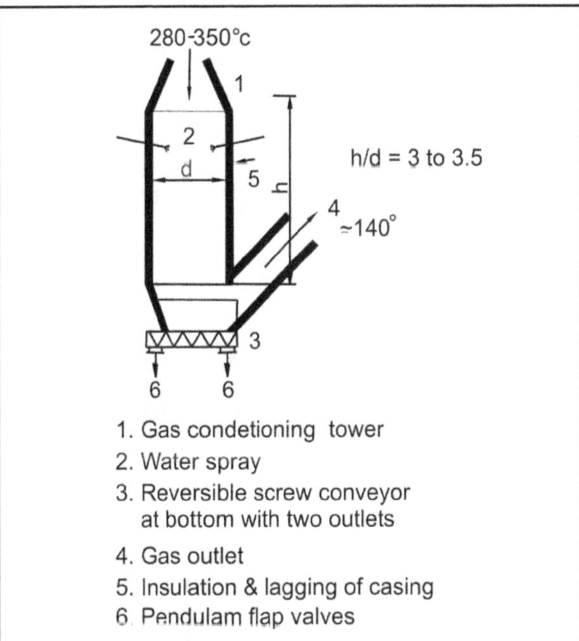

Fig. 2.12.4 Gas cooling tower.

However, there is bound to be some in leakage. For design purposes of gas system, it is taken as about 10%.

In indirect operation, gases will enter ESP at 90-100 °C because of drying in raw mill; dew point increases further.

Quantity of gases, leaving the GCT thus consists of:

(i) In coming volume.

(ii) Vapour from water sprayed.

(iii) In leakage at say 10%.

12.10.1 Water Injected in GCT

Water sprayed in GCT is lost as it escapes with exhaust gases. As a thumb rule assuming an inlet temperature of 300 °C and outlet temperature of 140 °C, quantity of water required for GCT for plants of different capacities would be as shown in **Table 2.12.3. Also See Web page W1.81**

Table 2.12.3

Capacity tpd	Litres/Hr.	Tons/Hr.
1000	4000	4
2000	8000	8
3000	12000	12
4000	16000	16
10000	40000	40

For this reason where there is scarcity of water, ESP would not be chosen. In that case Bag filters would be selected to clean exhaust gases.

12.10.2

Broad design criteria of GCT are :

(i) average velocity of gases inside GCT should be 2m/sec.

(ii) retention time in GCT should be ~ 10-12 seconds.

Equivalent to ~ 20-24 Meters height

(iii) L/D ratio is generally between 3: 1 to 3.5:1

Larger the GCT the better it is. It will ensure maintaining temperature of gases at outlet at 140 °C without wet bottom.

GCTs are always insulated and lagged to maintain skin temperatures of walls constant.

Water quantity sprayed in gct is regulated by measuring either humidity or temperature of gases at its outlet.

The pump draws water from a tank and injects it through nozzles at high pressure (35-40 Kg/cm2) for good atomisation.**See Fig. 2.12.3.**

The regulation is done by adjusting quantity of water returned to the tank. Balance is sprayed through the nozzles.

The number of nozzles and their spacing is governed by the diameter of gct. Finer atomisation will result in smaller droplets of water and hence faster evaporation. Numbers are so decided that the overlapping of sprays is minimum.

See Fig. 2.12.5.

In some designs, the gct has a double casing with hot gases going round the gct to maintain uniform temperature of casing of gct.

See Fig. 2.12.6.

It is necessary to maintain negative pressure in the gct to be able to take out nozzles and clean them in running.

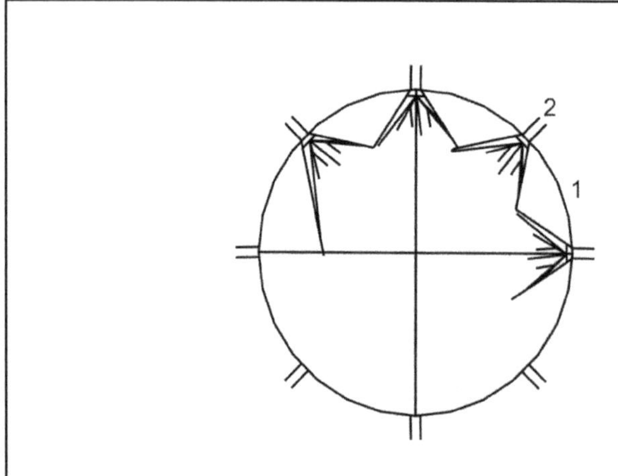

Fig. 2.12.5 Water spray in gas conditioning tower.

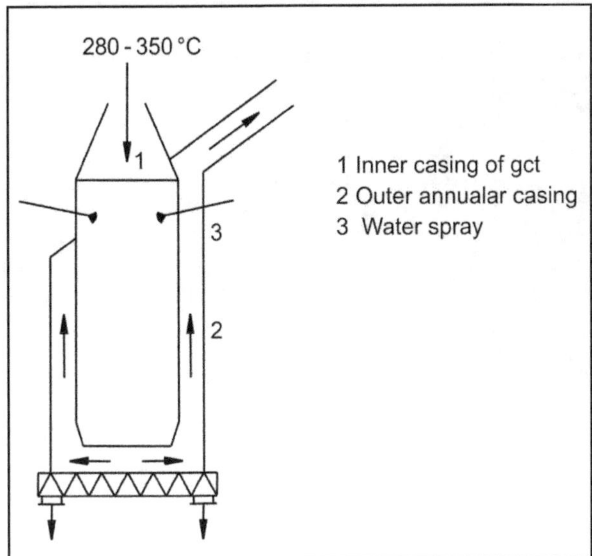

Fig. 2.12.6 Gct with annular casing.

12.11 ESPs

Selecting an esp and its design is also governed by certain factors. Only factors that are common to all designs will be considered here.

See Fig. 2.12.7 and Plate 2.12.4 which shows cross section of an esp.

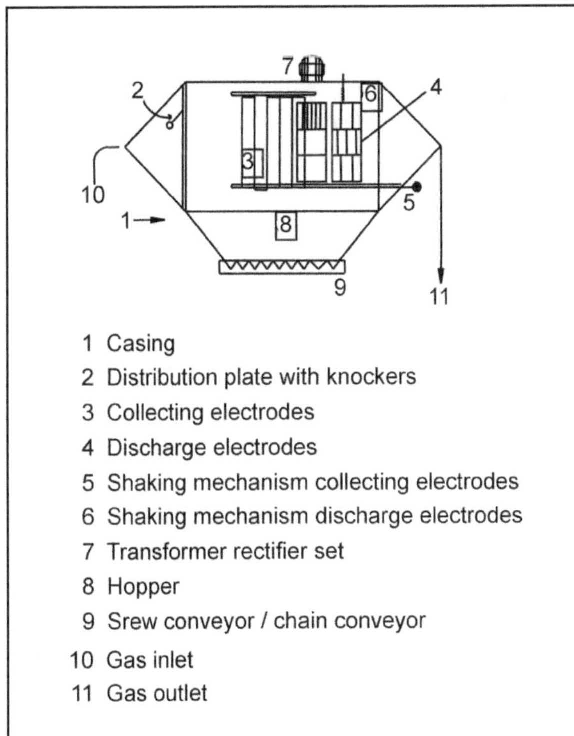

Fig. 2.12.7 Sectorial view.

An esp is a box as shown with inlet, outlet and hoppers. In it are suspended collecting and discharge electrodes. High voltage applied to discharge electrodes ionizes gases and dusts passing through esp. Ionized particles move towards grounded collecting electrodes and lose their change. Dust particles are thus deposited on collecting electrodes. At set intervals, electrodes are shaken to dislodge the dust particles, which fall in the hopper below. A screw conveyor installed under the hoppers carries the dust to the dust conveying system, mechanical or pneumatic, to be treated further.

Plate 2.12.5 showing installations of esps for kiln and cooler.

The collecting electrodes and discharge electrodes form alternate passages as shown.

See Fig. 2.12.8.

Discharge electrodes are generally tensioned wires or strips or tubes with points or spikes for helping ionization.

See Fig. 2.12.9.

Discharge electrodes are connected to transformer rectifier sets, which supply unidirectional voltage at 70-120 KV depending on spacing between collecting electrodes. They are insulated from the casing through high-tension insulators.

See Fig. 2.12.7.

Collecting electrodes are made of thin plates 1.2 mm thick, about 500 mm wide and are suspended, interlocked to form a continuous surface for deposition. **See Fig. 2.12.8**

An esp is divided into a number of fields each being supplied with its own H.T. Recitifier set.

Each field has a number of collecting electrodes. Electrodes are made in single lengths of 3 to 10 metres. To prevent bending, warping etc., they are corrugated as shown. Each manufacturer has its own design. Some time they are welded with round tubes.

12.11.1 Design Criteria for ESPs

The criteria for design of ESP is its collecting surface which is measured in m^2. Since both sides of the electrode are used, the collecting surface of one set in each field is arrived at as follows:

N = no.of collecting electrodes in each field

W = width of each electrode

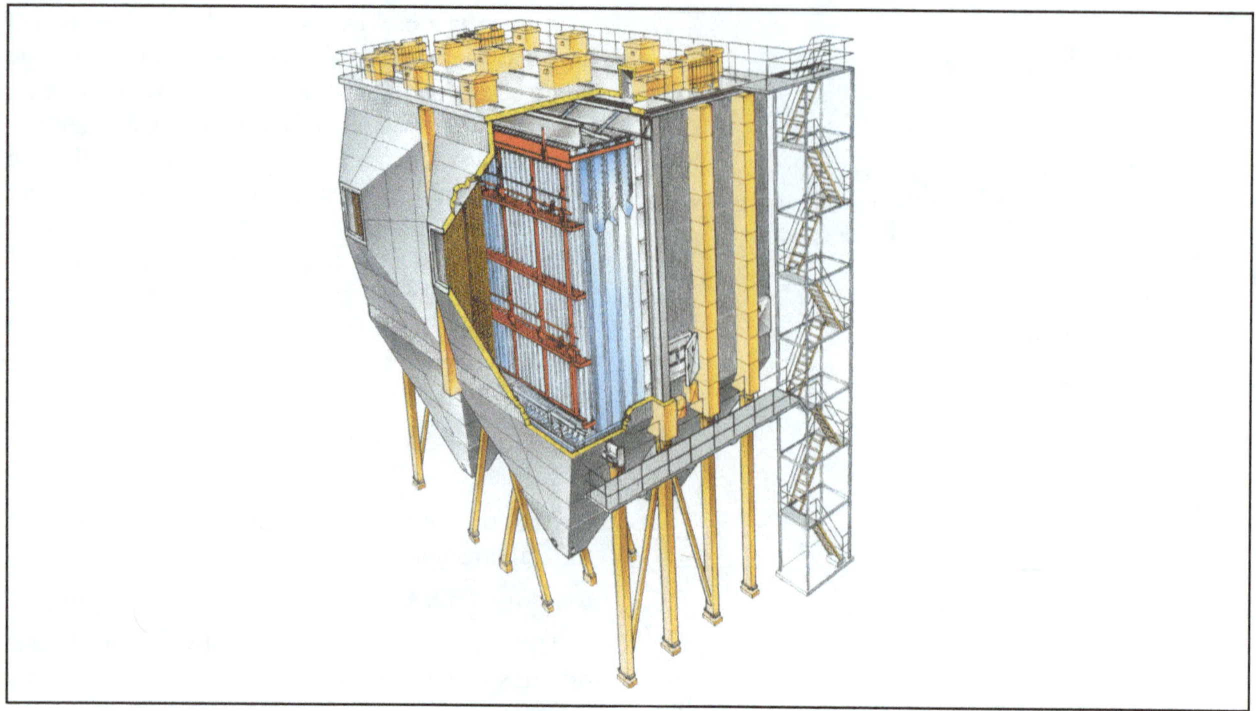

Plate 2.12.4 Cross section of a static precipitator.

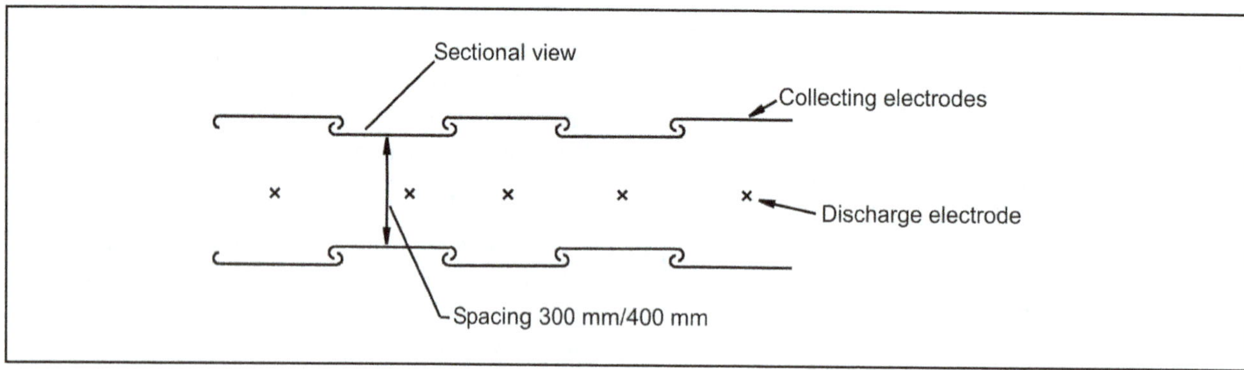

Fig. 2.12.8 Assembly of electrodes.

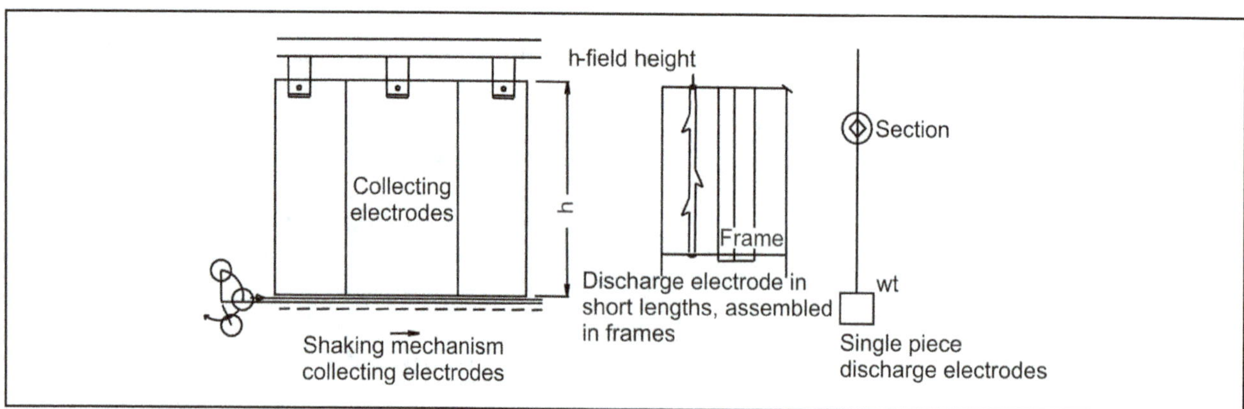

Fig. 2.12.9 ESP – basics.

ESP for Cement Kiln

ESP for Clinker Cooler

Plate 2.12.5

H = height of electrodes (also known as height of field)

S = number of passages

Therefore collecting area of one field is

$$= 2 \times S \times N \times W \times H \text{ m}^2$$

Total collecting surface, F if number of fields is F

$$= 2 \times S \times N \times W \times H \times F \text{ m}^2$$

Number of fields can be 1 in a very small esp to 6 or 7 (or even more) for very large esps like those which handle boiler flue gases in thermal power stations.

Other criteria are:

(i) Velocity of gas through ESP = 1-1.2 m/sec

(ii) Retention time in ESP total = 10-12 sec

(iii) Aspects ratio b / a = 1.00

See Fig. 2.12.10.

It is therefore first necessary to workout the collection area required.

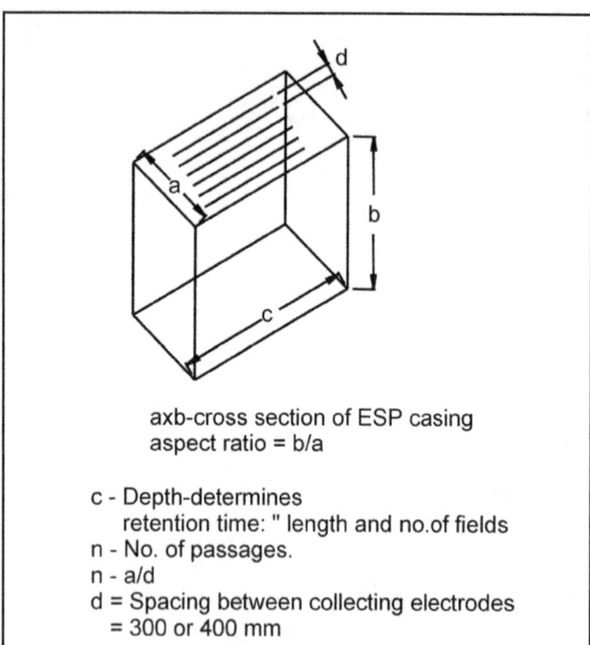

Fig. 2.12.10 ESPs basics.

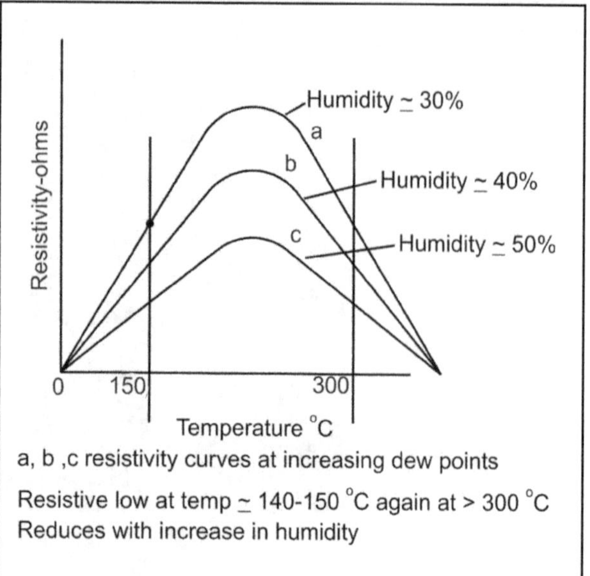

Fig. 2.12.11 Resistivity curves.

12.11.2 Migration velocity and Efficiency

This is related to:

(i) Efficiency of dust collection to be achieved,

(ii) 'migraton velocity' of ionized particles.

Migration velocity is a 'conceptual velocity' of ion particles in their journey between discharge and collecting electrodes.

It is in turn dependent on :

(i) efficiency desired,

(ii) temperature of gases treated,

(iii) their humidity,

(iv) resistivity of dust contained in gases,

(v) spacing between collecting electrodes.

The resistivity curves of dusts of different materials vary but follow a similar shape. It is low at temperature 0-150 °C and again after 300 °C. Peak values decrease with humidity.

See Fig.2.12.11.

Migration velocities vary between 3 to 12 cms / sec for different materials. They would vary for same materials according to gas conditions and desired efficiency.

Thus for efficiencies

between	Spacing between electrodes	
	300 mm	400 mm
	migration	velocity in cms / sec
90 to 99 %	w = 10	13.3
99 to 99.5 %	w = 9	12
99.5 to 99.9 %	w = 8	10.6

Deutsch formula is the most common formula used to arrive at 'specific collection area' from a given design migration velocity.

SCA = -{100 / w} × In (1-n)

SCA = Collection area in m^2 / volume through esp in m^3/sec.

n = efficiency = (dust at inlet – dust at outlet) / dust at inlet

Presently, dust at outlet has been fixed at 55 mg / nm^3 for new plants.

Specific collection area will increase in proportion to inlet dust burden if w migration velocity were to remain constant. But as mentioned earlier migration velocities would drop for higher efficiencies requiring even higher SCAs.

The impact is more severe when outlet gas burden is decreased, inlet dust burden remaining the same.

Thus impact of reducing clean gas dust burden is substantial on size of esp and hence investments are also substantially high.

Recommended migration velocities for esps for different applications have been shown in **Annexe 2.12.3.**

12.11.3 Factors Govering Performance

Performance of ESP also varies with other factors like dew point, temperature of gases etc.

12.12 CO in Gases Through ESP

Incidence of CO is dangerous for an esp. CO + O_2 burn to make CO_2 under right conditions. ESP performance is related to sparking which causes ionization. Therefore, CO in presence of O_2 is a potentially dangerous mixture and can result in explosions inside ESP.

O_2 content should not exceed 6% (even after allowing normal leakage). Kiln exit gas can be called 'inert' when O_2 is limited to 8%. It is attempted to keep the CO content less than 0.03%.

When CO exceeds this level, power to ESP, HT rectifier is tripped so that sparks stop.

It is therefore necessary to monitor CO content in ESP continuously. When CO is indicated in an increasingly trend curve, the ESP must be tripped. In case of sporadic and momentary peaks, it need not be tripped.

See Figs. 2.12.12 and 2.12.13.

12.13 Constructional Features of ESP

Depending on size of ESP, the hopper can be trough or pyramid type. In the 1st Case, screw or chain conveyor is attached to hopper bottom directly.

The conveyor is sized generously much in excess of rate of deposition of the dust, so that it is never over loaded.

Even then, it must operate continuously with the kiln and cannot be maintained or attended to when kiln is in operation. This is a serious disadvantage of this arrangement.

Hopers must have steep angles; valley angles should not be less than 50-55 °C.

Hoppers should hold 6 to 8 hours' deposition in the event of subsequent conveying equipment breaking down.

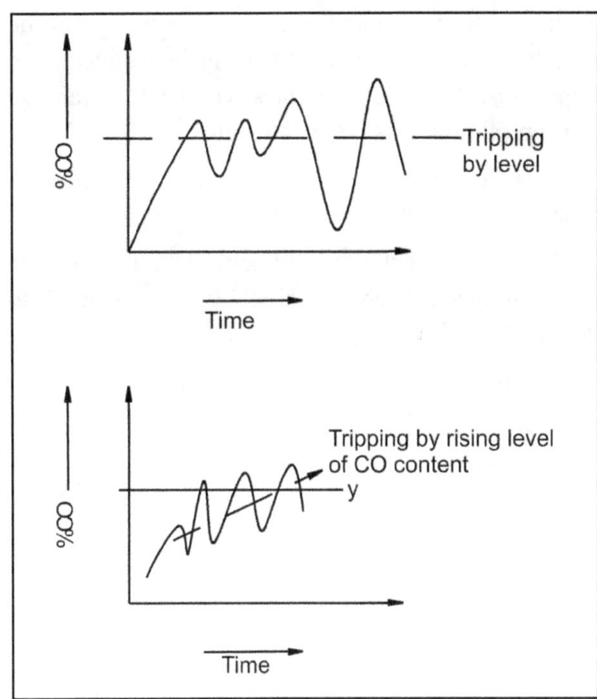

Fig. 2.12.12 Monitoring CO in exhaust gas.

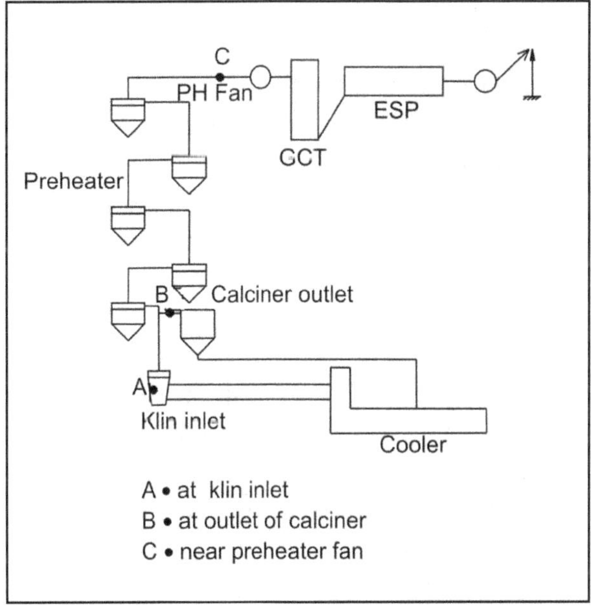

Fig. 2.12.13 Points of measurement of CO in kiln with calciner.

Very long screw conveyors would need to have intermediate bearings. It may not be possible to attend to the bearings without requiring to detach or lower the screw conveyor in an arrangement as shown. **See Fig. 2.12.7.**

In pyramid arrangement, screw conveyor can be readily and easily maintained. It can be isolated from hoppers by dampers. It has however the disadvantage of promoting in leakage in the system.

A redler type chain conveyor may be preferred as it has no intermediate bearings.

The dust collected in ESP being fine, it cannot be easily fluidized. Therefore, airslides are not suitable to convey ESP dusts.

Chain and screw conveyors are commonly used to convey esp dusts.

Number of Fields :

Number of fields would be governed by the gas volume to be handled and desired clean gas dust burden.

Generally speaking there should be minimum 2 fields so that in the event of failure of 1 field, ESP does not stop together.

For cement plants, to ensure good performance 3 fields, would be ideal; mostly through 2 fields would be used.

12.14 ESP for Coal Mill

Coal has much better resistivity than raw meal, clinker and cement.

Gases are at low temperature (< 80 °C) and dew point is high because of high moisture in coal (8-10%). Therefore ESP is an ideal dust collector for venting coal mill gases. It is also possible to use a single field ESP.

However, in coal mill circuit, there is always a possibility of incidence of CO in gases entering ESP due to slow combustion of coal; or burning coal entering the mill for grinding from stockpile.

Coal dust and air mixtures, if volatiles in coal are high, are potentially explosive at certain concentrations. However, it is difficult to predict the dangerous conditions with any degree of accuracy.

Coal mill ESPs are therefore more prone to explosions. ESPs are therefore fitted with explosion flaps on roof and explosion doors on walls to protect them from damage in the event of actual explosions. Such devices installed to protect ESP can also, ironically be the cause of further damage as release of pressure within may also bring in oxygen rich atmospheric air into esp.

12.14.2 Inert Gases Through ESP

The best way is therefore to use inert gases in the mill circuit for drying so as to eliminate all possibilities of explosion.

Preheater exhaust gases at 250-280 °C and with a dew point of 35°C and O_2 content of 6-8% (maximum) are ideal for drying in the coal mill system. However, these gases may contain CO from time to time. Hence, it is necessary to monitor CO at both inlet and outlet of esp of the coal mill.

Provision should be made for injecting CO_2 in the system and also in the coal hopper. In extreme cases spraying water in the ESP is also provided for. Alternatively inert gas generator could be installed. **See Plates 2.4.5 and 2.4.6 in Chapter 4 of this section.**

12.15 Bag Filter for Coal Mill

Bag filter is the other solution and has also been used extensively in coal mill systems.

It requires using bags with metal threads to make them antistatic. Besides detecting CO at inlet and outlet, rise in temperature between inlet and outlet should also be continuously measured as it would be an indication of slow combustion of coal inside the filter.

The filter should be constructed so that there are no corners or ledges, which would permit accumulation of coal dust.

For coal mill circuits, it is best to use preheater gases for drying whatever the filter. It ensures safety of equipment and men.

12.16 Gravel Bed Filter

Gravel bed filters were developed to clean cooler vent gases. Besides the fact that temperature of these gases under normal running conditions is high at 150-200 °C, and they contain coarse and abrasive clinker dust, the gases are also liable to increase in quantity and temperature during upset condition of the kiln. Bag filter was not then a good choice. ESP had not been tried out. In gravel bed filter, cooler vent gases passed through a bank of cyclones and are then passed through beds of gravel in shallow containers. Gravel was periodically stirred and clean air flow was passed through gravel bed to take out the precipitated clinker dust.

See Plate 2.12.6.

These filters were successful in that they cleaned gases as well as bag filters did, but pressure drop was high because of drop in cyclones and beds.

After ESP was successfully used for this application, development of gravel bed filters was not pursued.

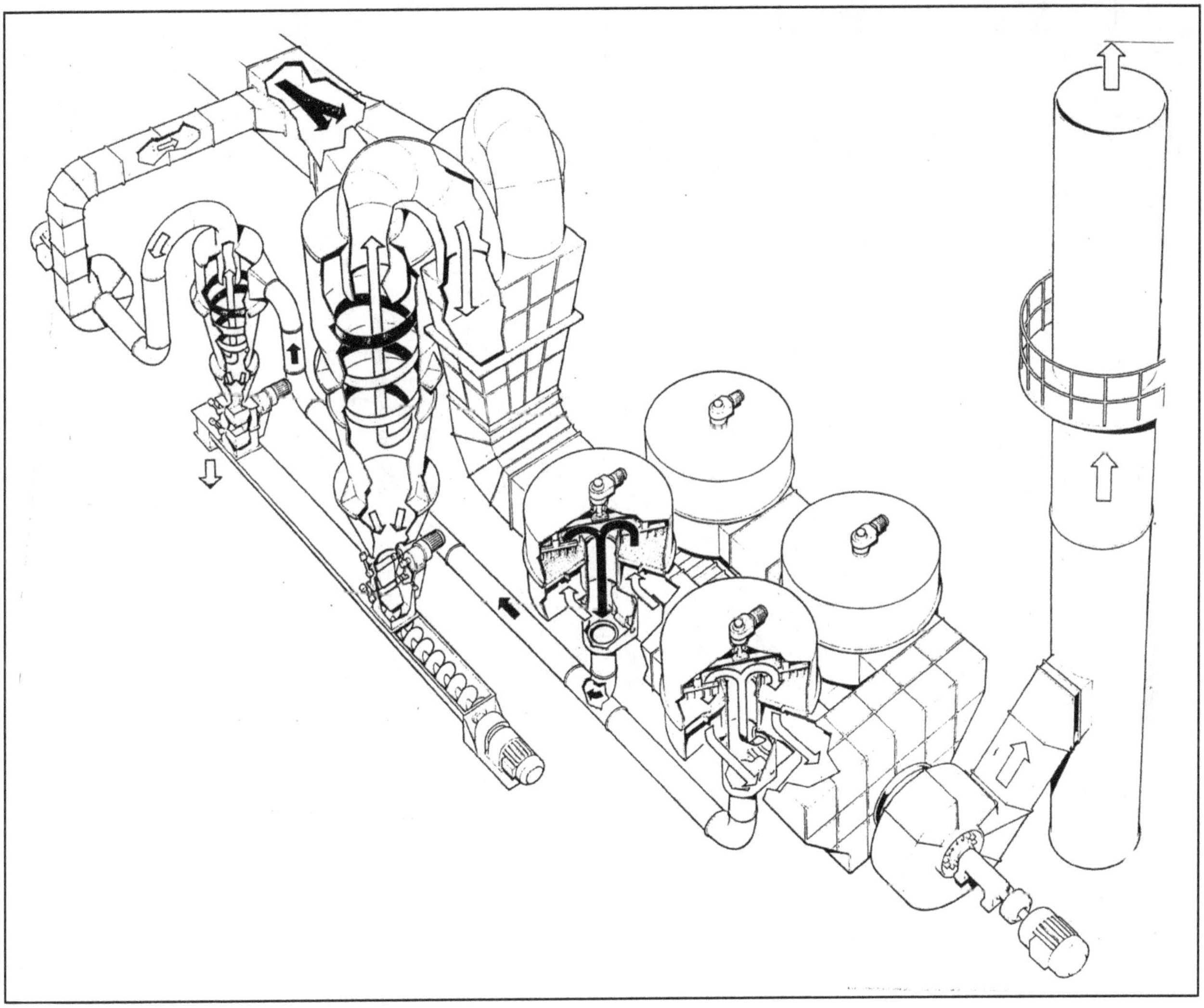

Plate 2.12.6 Gravel bed filter to clean cooler vent gases.

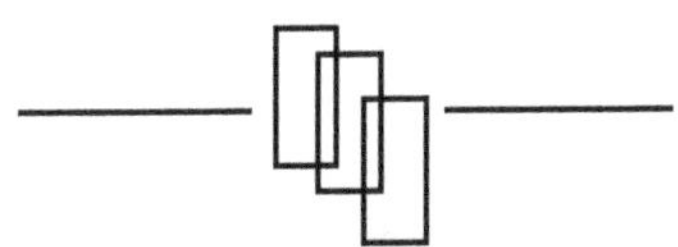

Annexe 2.12.1

Reduction of concentration of air borne dust with distance.

Sr.No.	Height of Stack meters	Vel. of wind kmph	Distance from Stack in meters								
			200	400	600	800	1000	1200	1400	1600	1800
			dust concentration in ppcc								
1	42		106	178	252	429	655	525	140	108	74
2	48		213	193	140	140	133	100	100	93	80
3	53	10	250	353	410	450	500	445	280	165	160
4	60		120	92	80	105	40	40	33	27	19

Annexe 2.12.2

Quantum of fines in vent gases of various machinery.

Sr.No.	Item	% of -10 micron particles
1	Crushers	5-20
2	Rotary dryer	40-70
3	Drying and grinding plant	40-90
4	Rotary kiln with cyclone preheater	85-100
5	Grate cooler	0-15
6	Tube mills Raw, coal and cement	40-80
7	Material handling systems Vents of packing machines	10-50

Annexe 2.12.3

Electrostatic precipitators, migration velocity and gas velocity in case of some common applications in cement plants.

Sr.No	Application	Mean gas Velocity m/sec.	Migration Velocity cm/sec.
1	Wet, rotary kiln	0.9-1.3	10-13
2	Kiln with lepol preheater	0.8-1.2	8-12
3	Dry kiln with Preheater and gct	0.8-1.2	8-12
4	Shaft kiln	0.6-1.0	6-10
5	Grate cooler	0.5-0.8	4-8
6	Dryer, drying and grinding plant	0.6-1.2	6-12
7	Cement mill with water injection	0.6-1.0	6-10

Above migration velocities are for electrode spacing of 300 mm, for spacing of 400 mm, multiply by 1.33.

Annexe 2.12.4

Data on filter fabrics.

Sr.No.	Type of Fibre	unit	Natural Fibres		Man made Fibres					
1	name		wool	cotton	nylon	nomex	dralon T	polyester	teflon	glass
2	density	gm/c.c.	1.32	1.54	1.15	1.38	1.15	1.38	2.1	2.5-2.7
3	tearing resistance	g/den	2.5-5	1-2,	4-6.	5.5	3.1-3.5	4-7.	1.6	7-12.
4	elongation	%	25-35	7-10.	25-45	20-25	15-30	10-20.	18-75	5-Feb
5	moisture absorption	%	10-15.	8-9.	4-4.5	2.5-5	1-1.5	0.3-0.4	0	0
6	water retention cap.	%	50-70	50-80	15-Oct	0	8-12.	2-5.	0	0
7	melting point	°C	>130	> 200	250/215	375	>300	250-260	>275	>800
8	temp. resistance									
	of filter material	°C	80-90	75-85	75-85	190-200	125-135	130-140	200-220	250
9	continuous max.	°C	100	95	95	220	140	160	250	310

Annexe 2.12.5

Properties of fabrics for filter bags.

Sr.No.	fabric trade name	colour	sp. density gm/c.c.	tenacity CN/TEX	continuous operating temp. °C	max. short duration temp. °C	chemical resistance
1	polyester						
	dacron	white	1.38	60	150	170	fair-good
2	polyamide						
	nylon	white	1.14	60	110	120	fair-good
3	polyramid						
	nomex	cream	1.38	33	220	250	fair-good
4	acrylic						
	dralon T	cream	1.15	35	135	150	good

Annexe 2.12.6

Short and long period temperatures withstood by different filter fabrics.

Sr.No.	trade name	chemical classification	maximum operating temp. continuous °C	maximum operating temp. short time °C
1	cotton	cellulose	70	120
2	dacron	polyester	135	175
3	glass	glass	275-285	340
4	nomex	polyamide	230	257
5	nylon	polyamide	90	120
6	teflon	tetrafluorothylene	230	285
7	wool	protein	90	120

Annexe 2.12.7

Commonly used air to cloth ratios.

Sr.No.	Application	Type of dust	Temp. °C	Air to cloth ratio $m^3/m^2/hr$
1	Raw materials Handling, storage	Limestone clay	ambient	180-200
2	Drying and drying Plants raw meal and coal	Raw meal Coal dust	70-130	120-140
3	Conveying and storage Raw meal and pulverized coal	Raw meal Coal dust	< 50	150-180
4	Grate coolers	clinker	~140-150	120-140
7	Handling and storage of clinker	clinker	~70	120-140
8	Cement mills	cement	80-100	120-140
9	Handling and storage of cement	cement	80-100	120-140
10	Packing plant and bulk dispatches	cement	80-100	150-180

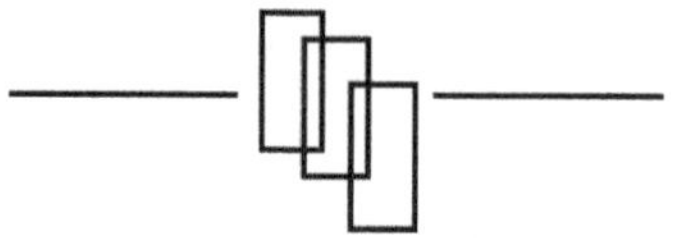

CHAPTER 13

CEMENT MILLS AND CEMENT STORAGE

13.1 Cement Mills

Ball / Tube mills are also suitable for grinding clinker into cement. For decades only tube mills have been used to make cement.

Till about 1960s cement mills used to be 'open circuit' mills. Most Cement Plants made only one grade of cement and changing fineness of cement produced was hardly ever required. Even when some plants made slag cements or pozzolana cements fineness was more or less fixed. Open circuit mills served the purpose of the cement industry for a good many years.

Open circuit mills were longer with length to diameter ratios of + 4:1. Hence they were also called 'tube mills' in comparison with short raw and coal mills.

It was realized though that 'closed circuit' mills were a far better option when grinding to higher finenesses as compared to fineness of plain opc.

They were energy efficient and it was also very convenient to change fineness as and when required. Further many cement plants started to make more than one grade of cement. Therefore cement came to be made in shorter mills with l/d ratios of 3 to 3.5 operating in 'closed circuit' with mechanical separators with dual drives. Dual drive made it convenient to change fineness of product in running.

13.2 High Efficiency Separators

When 'high efficiency separators' were developed they came to be used in cement mill systems and thus further enhanced operating efficiency reducing power consumption.

Therefore where balls mills were used to grind cement, the system is now invariably, closed circuit with high efficiency separators.

13.3 Vertical Mills to Make Cement

Now Vertical Roller Mills are increasingly used to make cement. This has been possible for two reasons.

1. Development of better wear resistant materials for liners of rollers and table.
2. Using external circuit where velocities at air ring are very much on lower side thereby reducing wear further.

Vertical mills are now fitted with High Efficiency Separators improving their performance further. Fan power of the system is also reduced..

Large mills now have 'multi drives'.This has helped increase in size of mill. It has also increased availability.

Mills can now run on partial loads with lesser number of roller pairs in operation.

13.4 Roller Press and Ball Mill Combination

Roller Presses are used as 'precrushers' before the ball mill in closed circuit. As has been mentioned earlier, Roller Presses used very high pressure between Rolls (as against simple roll crushers). Material passing through rolls undergoes constructional changes which makes it far more easier to grind. Also about 30 % material is pressed to final size.

Therefore ball mill following in the circuit has to do much less work and is much smaller. Total power consumption is also less by about 30 %.

Roller Press can be used in 'hybrid grinding' with mill or by itself with a disagglomerator and high efficiency separator.

See para 4c.2 in Chapter 4c.
See Figs. 2.13.1 and 2.13.2.
See Plate 2.4c.1 in **Chapter 4c**.

13.5 Precrushers for Clinker

Cheaper pre crushers were available by way of vertical impact crushers. They worked in closed circuit.

See Fig. 2.13.3 and Plate 2.13.1.

A Vertical Roller mill has also been developed to work as a pregrinder or primary mill. One such mill could serve for two or more secondary or finish mills.

See Plate 2.13.2 and Fig. 2.13.4.

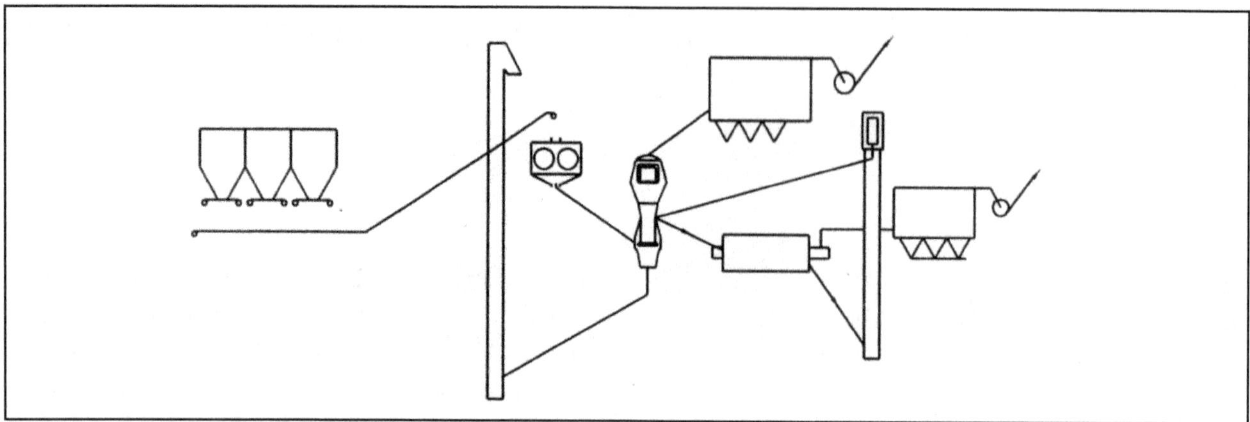

Fig. 2.13.1 Flow chart – roller press and ball mill for cement grinding.

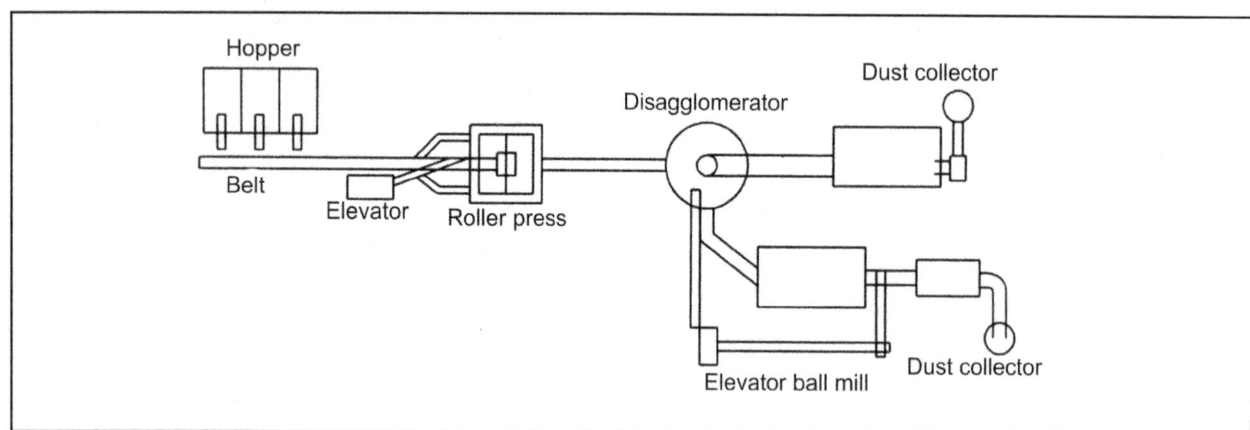

Fig. 2.13.2 Layout of roller press and ball mill and disagglomerator.

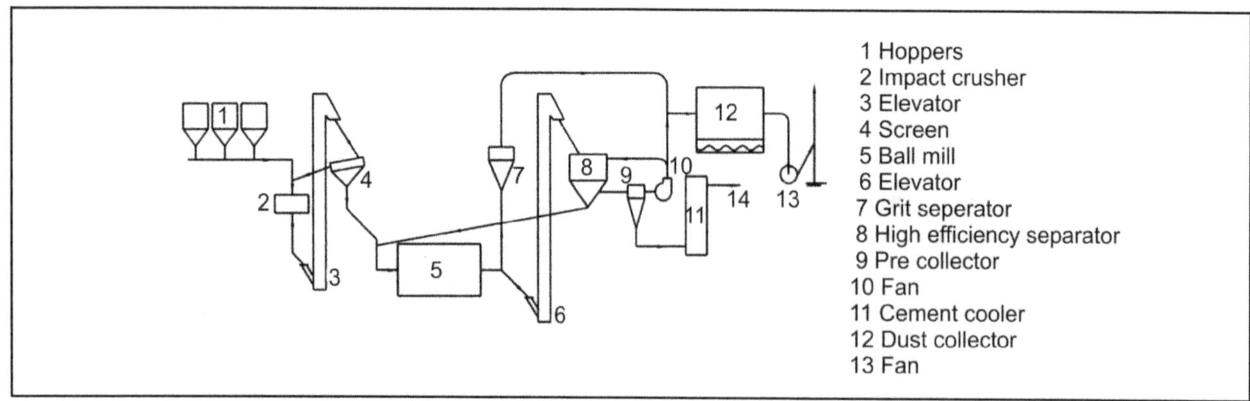

Fig. 2.13.3 Grinding circuit with precrusher and external cement cooler.

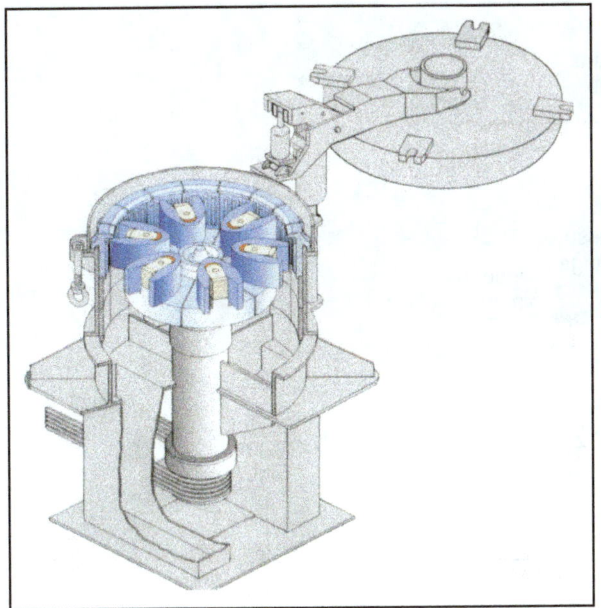

Plate 2.13.1 Pre crusher or clinker.

Plate 2.13.2 Vertical roller mill as a pregrinder.

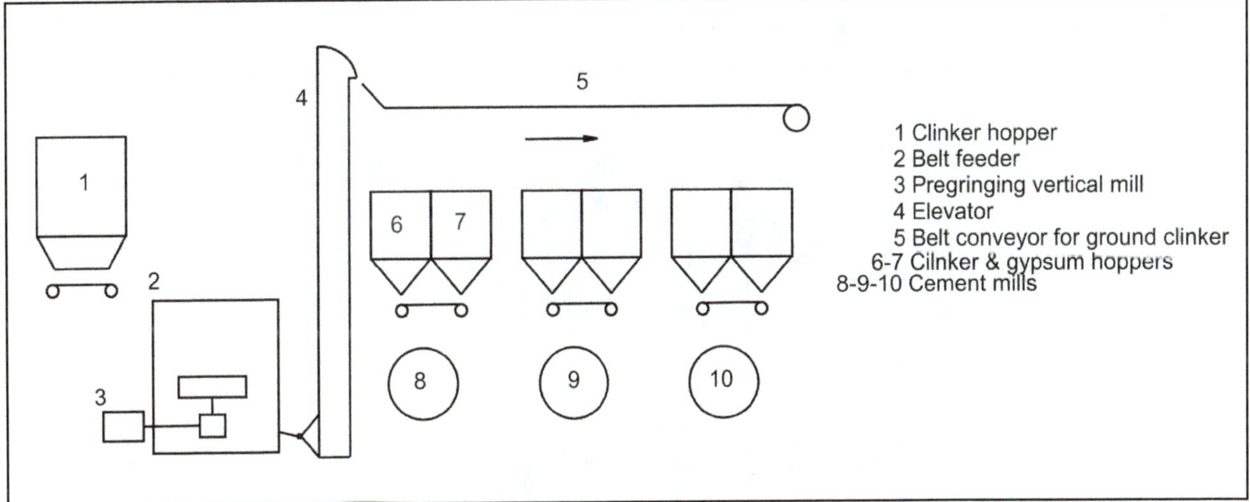

Fig. 2.13.4 Using pregriding vertical mill for a batch of finish mills.

13. 6 Cooling of Cement

Cement has to be kept at temperatures less than 110 oC. Earlier, mills had internal water spray systems to achieve this objective. Even external water cooled cement coolers were developed. **See Fig. 2.13.5** showing internal water spraying system to cool cement and **See Fig. 2.13.6** showing external cement cooler in system.

With closed circuit grinding where high efficiency separators are used, cement comes in contact with large quantities of air. Therefore separate cooling systems are not necessary.

13.7 Sizing Cement Mills

As explained earlier, in working out capacities of cement mills, it is practical to add a further margin of at least 10 % over and above design margin.

For a 10000 tpd kiln, cement mill required on this basis would be ~ 580-600 tph for making OPC. It would probably require two large ball mills with slide

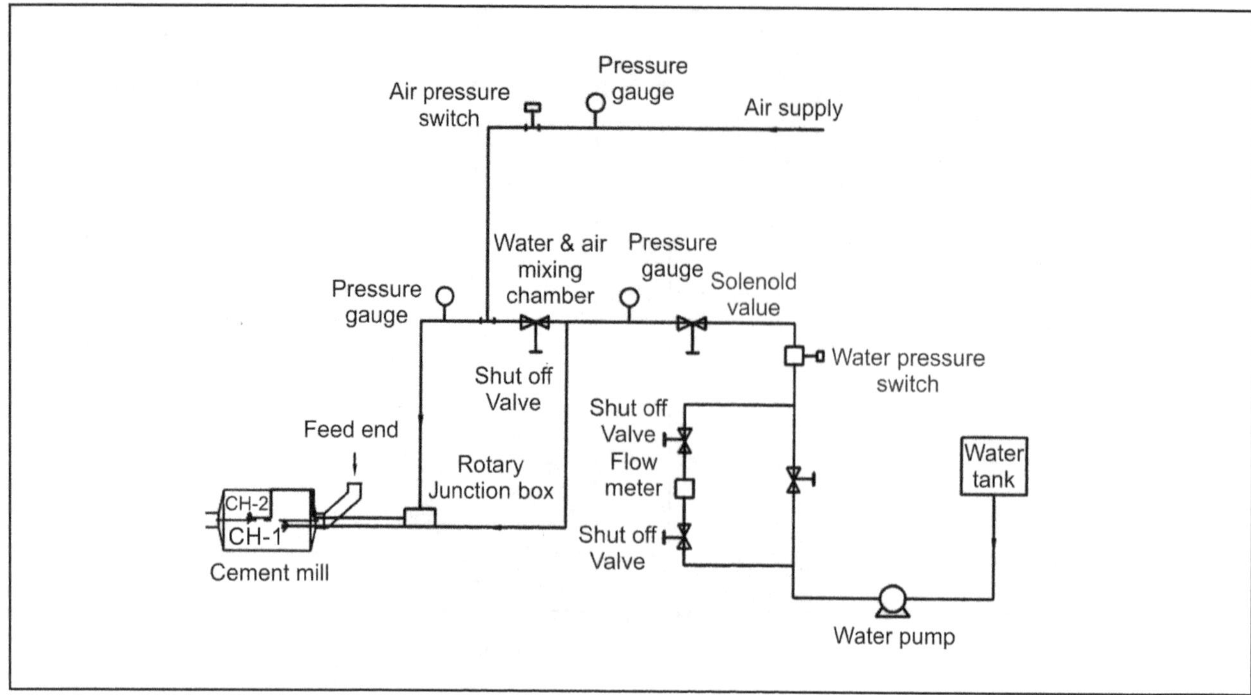

Fig. 2.13.5 Internal water cooling system for a cement mill.

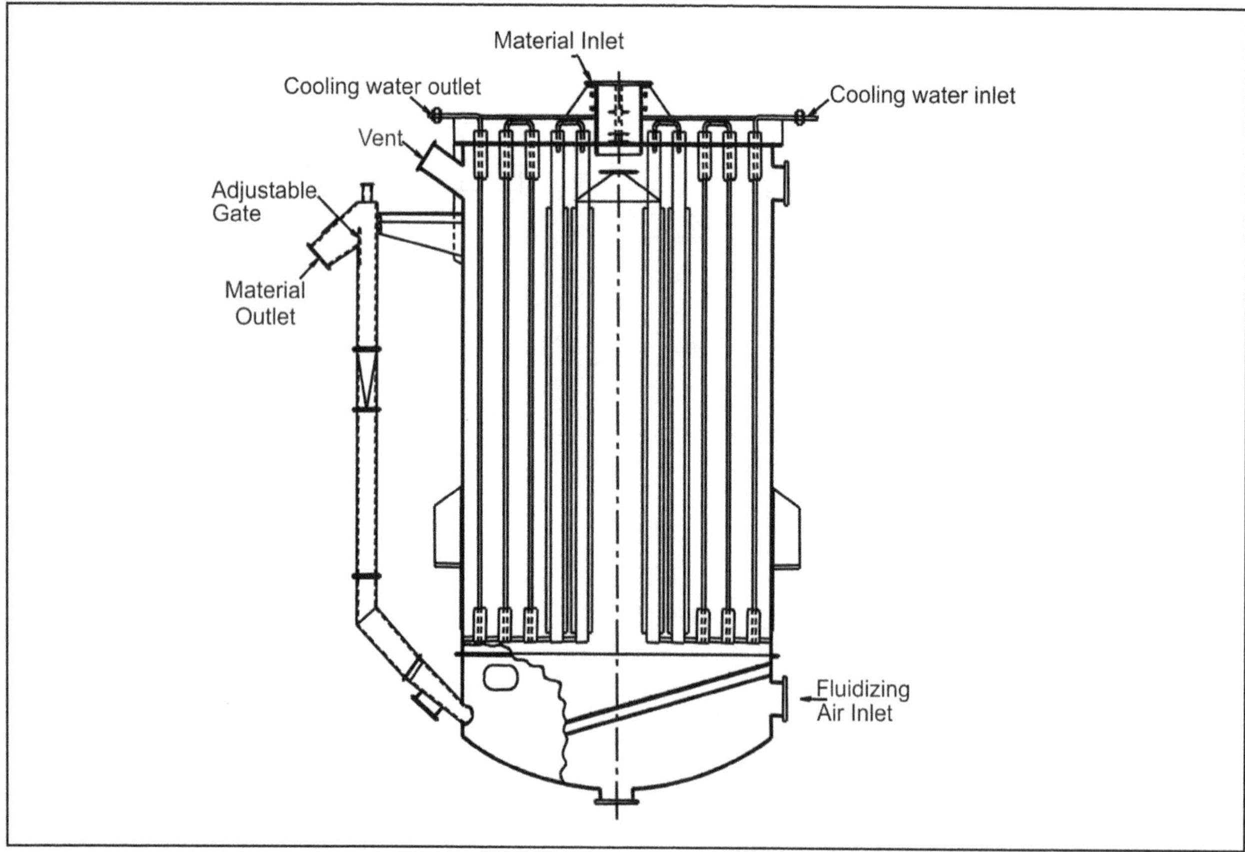

Fig. 2.13.6 External cement cooler.

shoe bearings and dual drive or a wrap around motor. **See Plates 2.4a.2 and 2.4a.3 in Chapter 4.**

Even a vertical mill would need a multidrive arrangement.

When making blended cements, mills of even larger capacities would be required. because of differences in cement to clinker ratio

Cement	cement to clinker ratio
OPC	1.05
PPC (30% fly ash)	1.54
BFSC(60 % slag)	2.86

Thus for same clinkering capacity cement mill capcities would be higher in proportion to respective cement clinker ratios.

Mill sizes and power consumption would be governed by composition of clinker (C2S), fineness of cement produced and composite grindanility.

When producing PPC, same mill gives higher output but will give much lower otput when producing BFSC

Very large mills > 6000 kw would have slide shoe bearing/s and dual drive or a 'wrap around motor,.

Both are very expensive alternatives and hence it would be more common to find that instead of one, two or more mills are preferred in large cement plants.

It would be prudent to install 2 mills rather than 1 so that in the event of something going wrong, one mill would always be working and production would not be stopped altogether.

Another reason for going in for two mills is that 2 types of cement can be ground simultaneously one say OPC and the other a blended cement like Pozzalona / Slag Cement.

13.7.1 Procedure for Sizing

Procedure for sizing the mill (ball mill) would be the same as for mill to be sized for grinding raw materials. Two additional factors, though would have to be taken into account.

1. Fineness factor.

   When cement is required to be ground finer than 3000 Blaine, a multiplying factor is used to arrive at sp. power consumption.

2. C_2S factor.

   When C_2S in cement is more than 15-20 %, it requires more power.

   See Tables 2.13.1 and 2.13.2 and also Figs. 2.13.7 and 2.13.8

13.8 Cement Mills to Make Blended Cements

It would be economical for a plant to make blended cements like

Pozzolana Cement (PPC) and Slag Cement (BFSC) when blending materials fly ash and slag are available nearby and when market accepts blended cements.

Slag is a waste product of steel plants. Hence if a steel plant is in the vicinity of say (100-200 km) it would definitely be advantageous to produce slag cement, there by more than doubling the production of cement, kiln capacity remaining the same.

This is equally true of another blending material - fly ash. Up to 30% Fly ash can be added to make PPC. Fly ash is also a waste product of Thermal Power Stations.

However when there are possibilities of making blended cements then the planning of cement grinding has to be done carefully. It would be necessary to identify demands of blended cements, assess quantities of blending material available on a continuous basis and to arrive at proportions in which OPC and blended cements would be produced.

13.8.1 Capacities When Making Blended Cements

Mill capacities tend to increase when producing fly ash cement as compared to OPC, because fly ash is already a very fine material and is required to be blended rather than ground with clinker.

It is therefore possible to increase capacity of mill by 15-20 % when making PPC.

Fly ash can be either inter ground with clinker or ground separately in a smaller mill and blended with OPC produced separately in a bigger mill. Even when inter grinding, fly ash may be first added to the separator rather than the mill so that only coarse fraction will be ground with clinker. The mill capacity can then be increased further.

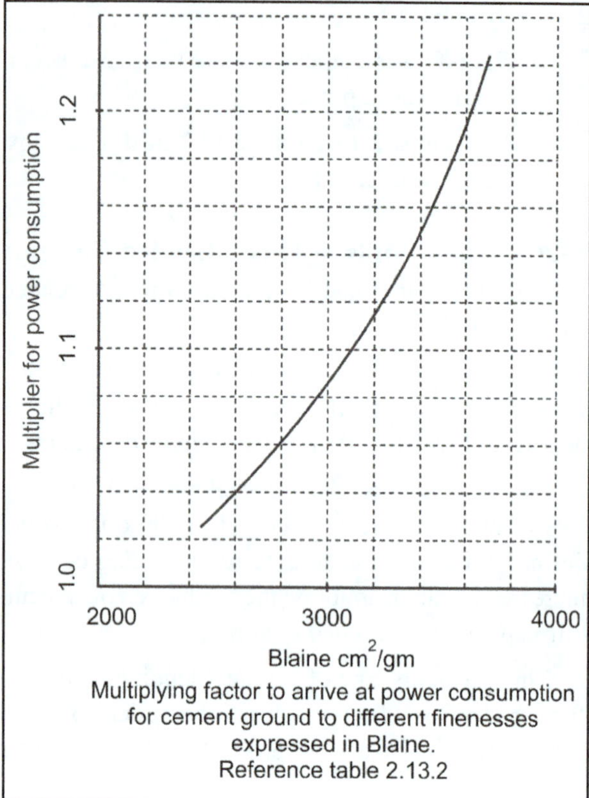

Multiplying factor to arrive at power consumption for cement ground to different finenesses expressed in Blaine.
Reference table 2.13.2

Fig. 2.13.7

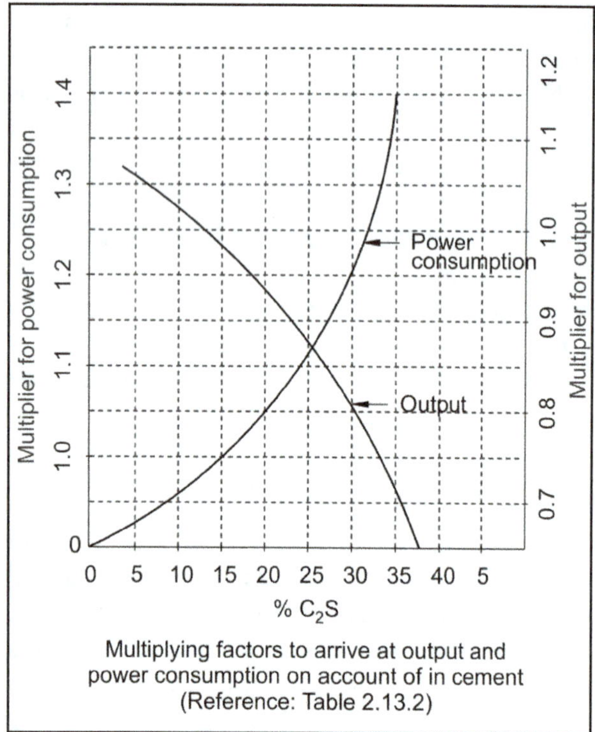

Multiplying factors to arrive at output and power consumption on account of in cement
(Reference: Table 2.13.2)

Fig. 2.13.8

Thus the plant has to size the mill first on basis of OPC and allow for material handling equipment and cement storages on basis of PPC.

See Fig. 6.29.3 in Chapter 29 of Section 6.

The case is exactly opposite when producing slag cement. Same mill when grinding BFSC will produce only at 60 % capacity as compared to OPC. Thus mills will have to be sized for BFSC and material handling equipment on basis of OPC.

13.9 Storing Different Cements

Producing Blended cements as well as OPC simultaneously requires cements to be stored separately. Now a days almost all plants will be producing more than one type of cement. Even in OPC there would be 43 & 53 grades, which, strictly speaking, should be stored separately. From time to time the company my produce special cements in small quantities against specific orders. Thus in a plant facilities are required to be provided to store separately different types of cement.

A typical cement storage silo has been shown in **Plate 2.13.3.**

This can be done in two ways:

(i) Construct different silos for each type of cement. Sizes and capacities of silos would be in accordance with demand for each type.

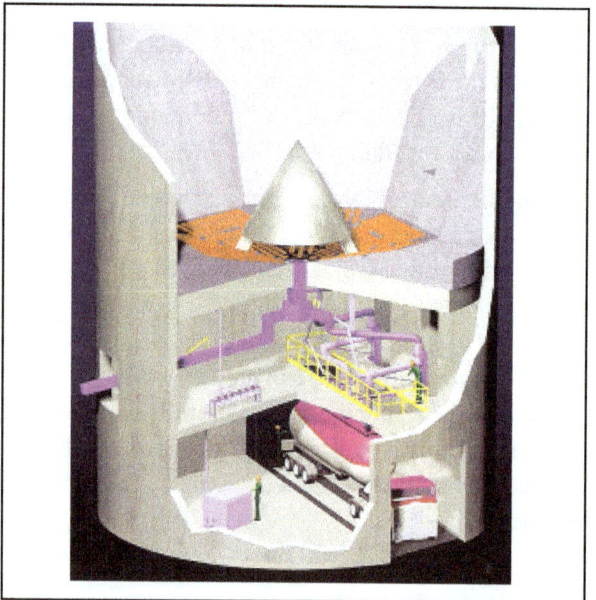

Plate 2.13.3 Cement storage silo with aerated bottom and with facility for bulk loading under it.

Thus if demand for PPC is say 20% then silos for OPC would be for 80% and silos for PPC 20% of production capacity.

In case of slag cement may be all silos would be identical in size and capacity, and cements could be stored in any one that is empty.

(ii) Some times demand for special cements may not be sizeable. In such a case a silo is compartmentalized so that cements of different type can be stored separately within the same silo.

(a) Rings

(b) Sectors and Ring as shown

See Figs. 6.32.2 to 6.32.7 and plate 6.32.1 in Chapter 32 of Section 6.

13.9.1 Extraction from Multi Compartment Silos

With different type of cement stored in the same silo it also becomes necessary to provide for separate extraction systems for each type leading to packing machines.

See Plate 2.13.4 in which cement is extracted from a sectionalized silo at more than one level.

Packing machine and dispatches also would be so laid out that two types of cement can be extracted simultaneously and packed and dispatched simultaneously.

This layout has to be planned carefully. It should also have flexibility in it that any one packing machine or system of conveyors and loaders will handle all types of cements made.

13.10 Handling Bulk Cement

In bulk cement, this becomes easier as cement of a type can be taken to a small bin /silo used for bulk loading and bulk carriers loaded from it. It is also possible for bulk carriers for road transport to stand directly under the silo and get loaded with the type of cement that is to be dispatched.

See Chapter 36 in Section 6.

Layout aspects of cement mill systems and cement storage have been dealt with in greater detail in **Chapters 30 to 32 of Section 6.**

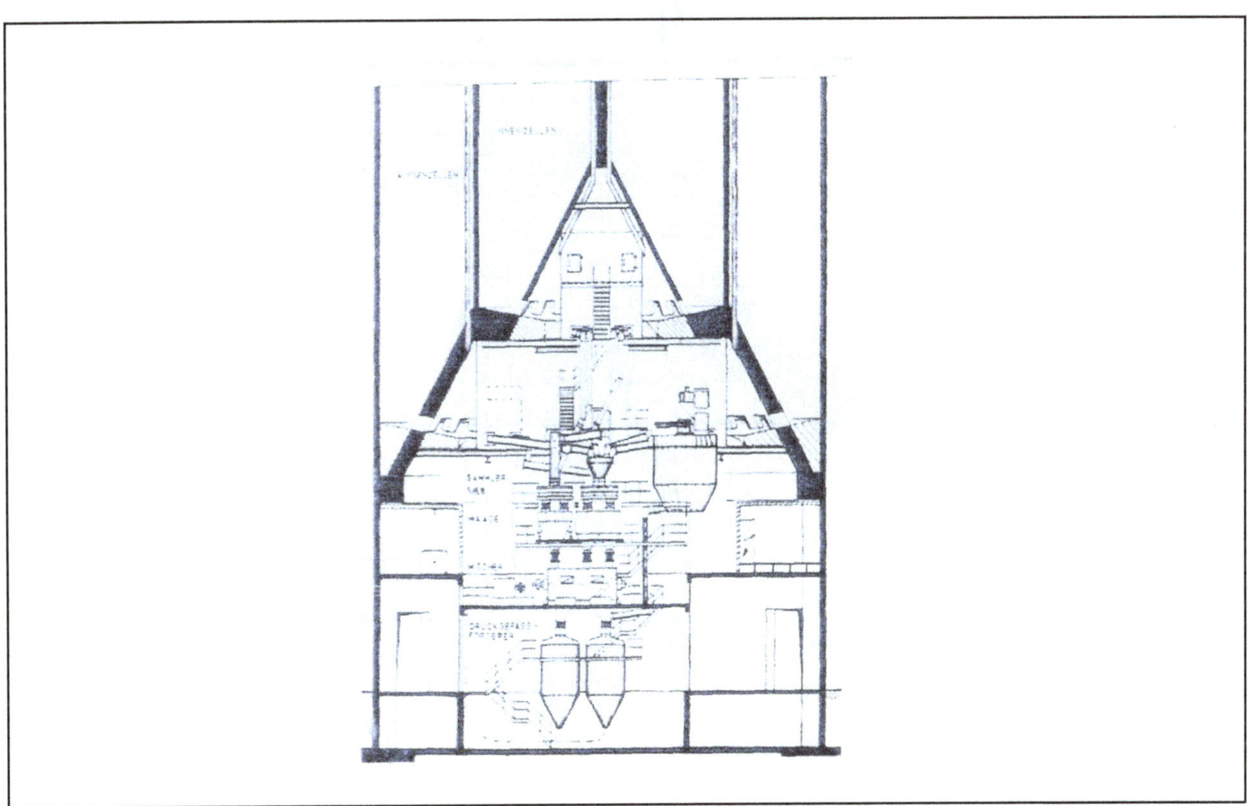

Plate 2.13.4 Multi compartment cement silo extraction from different levels.

Table 2.13.1 Multiplying factors to arrive at power consumption for different fineness of cement.

Blaine surface	80% passing size microns	Fineness factor
2520	62.4	1.018
2700	53.6	1.04
2880	45.7	1.07
3060	40.7	1.094
3150	37.6	1.11
3600	28.2	1.19

Table 2.13.2 Multiplying factors for output and power consumption on account of C_2S content in cement.

% C_2S	Multiplier for output	Multiplier for Power consumption
5	1.1	0.91
10	1.05	0.95
15	1.0	1.0
20	0.95	1.05
25	0.88	1.14
30	0.82	1.22
35	0.72	1.4

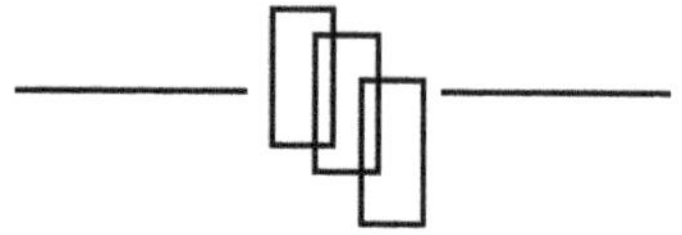

CHAPTER 14

PACKING OF CEMENT

14.1 Packing Cement for Despatches

Cement is produced in bulk and stored in cement silos of large capacities. It is sent to market in general and to individual customers in bags of specific weight filled in the cement plant.

Use of and dispatches of cement in bulk has been common in developed countries but in India it is catching up only now.

Filling cement in bags of specific weight required packing or bagging machines which could pack large quantities accurately. Packing machines were developed to satisfy this need. In developed countries paper bags are being used. In India cheapest and durable and reusable material was jute. Hence till now cement used to be packed in jute bags. Jute has now been replaced by high density polyethylene (h.d.p.e.)

Loading of bags had been manual to start with. Therefore bags were sized to hold 50 kgs of cement. For a large number of bags to be packed and got ready for dispatch it was not practical to fill and weigh and sew and seal each bag individually.

14.2 Packing Machines

Packing machines were developed to carry out the operation of filling each bag with specific weight with a tolerance of + 0. 5%

Necessity to sew a bag was eliminated by making bags with a self closing valve. Bag to be filled was slipped on a spout of the packing machine. It rested on supports which tilted when bag was full and allowed to slip it off the spout. Pressure of cement closed the valve.

See Fig. 2.14.1.

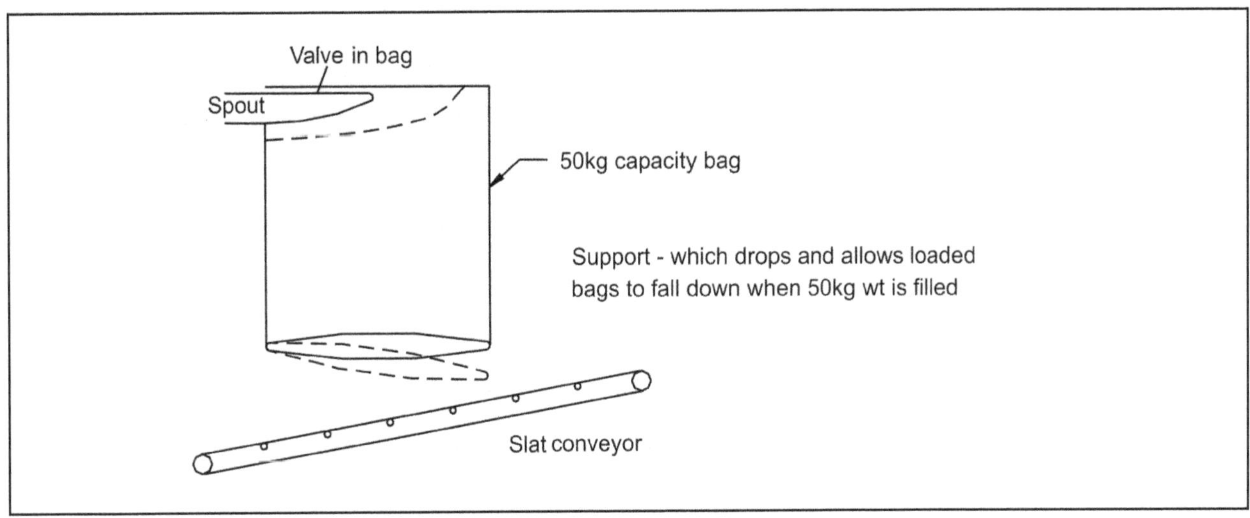

Fig. 2.14.1 Filling of cement in bags-automatic discharge of bags.

Bags were filled by feeders fixed in the hopper of the machine.

A weighing mechanism similar to a weighing scale with a fixed weight of 50 kgs on one side and the bag being filled on the other side of the arm was used for many years to ensure correct weighing of bags. This mechanism is now replaced by electronic weigh scales. **See Fig. 2.14.2 to 2.14.4.**

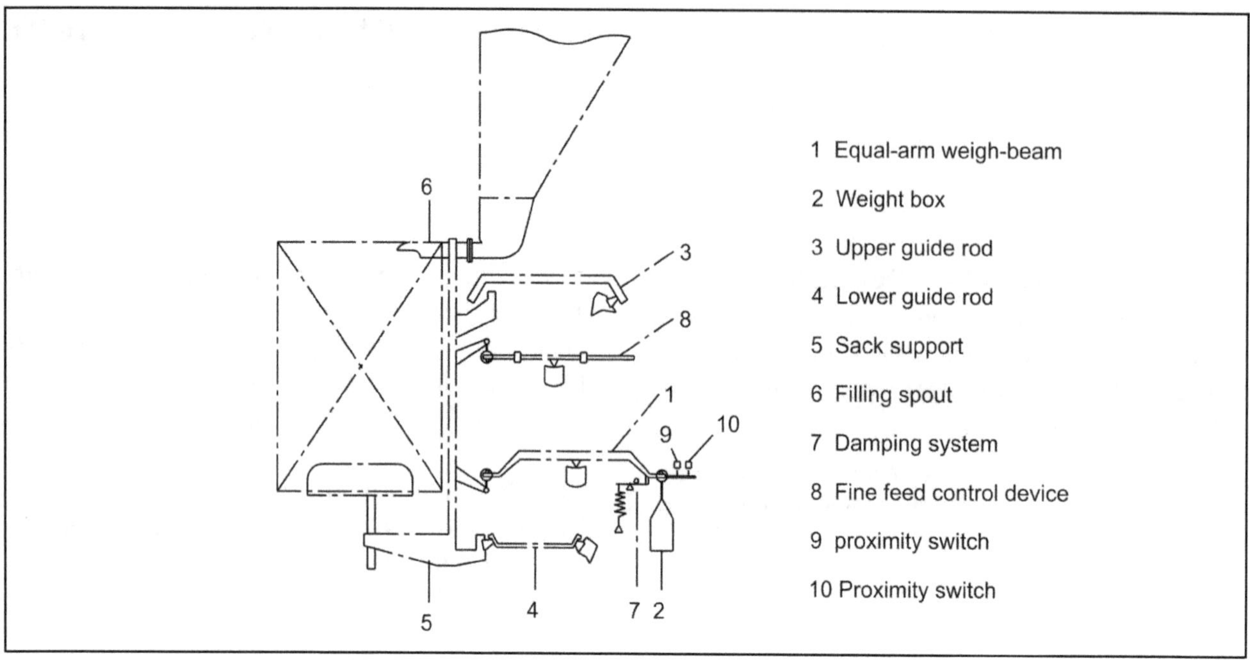

Fig. 2.14.2 Stationary packing machine equal-arm weighing mechanism.

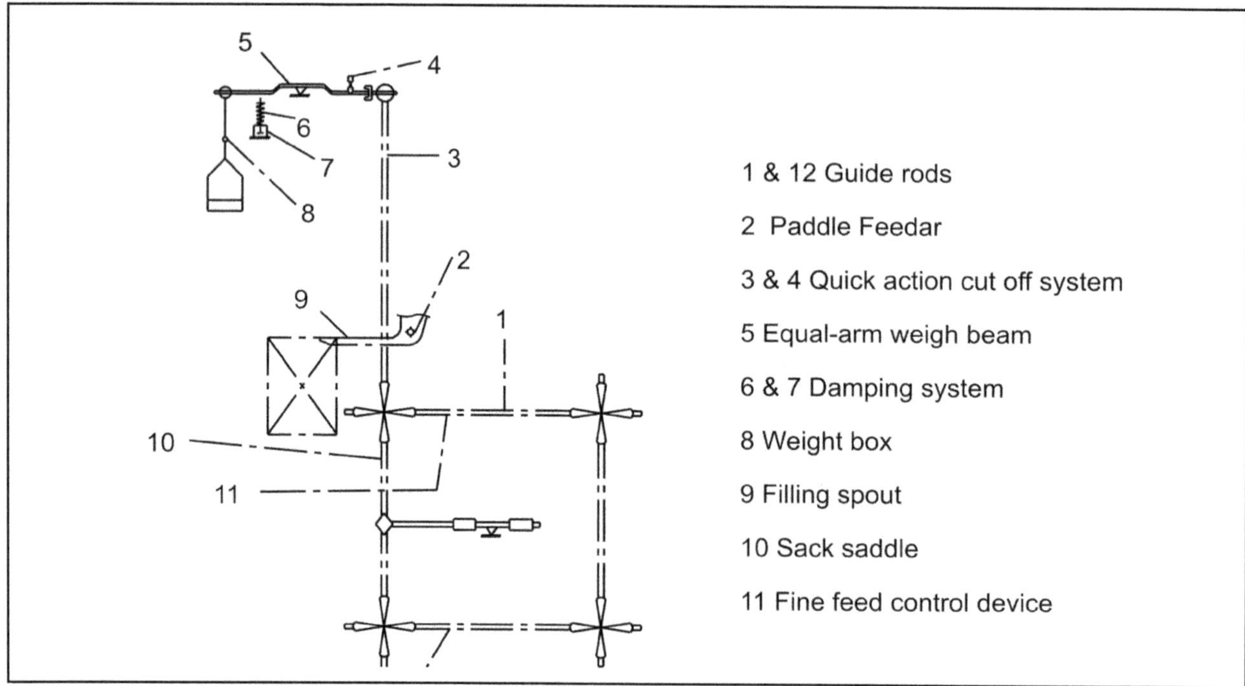

Fig. 2.14.3 Rotary packing machine equal-arm weighing mechanism.

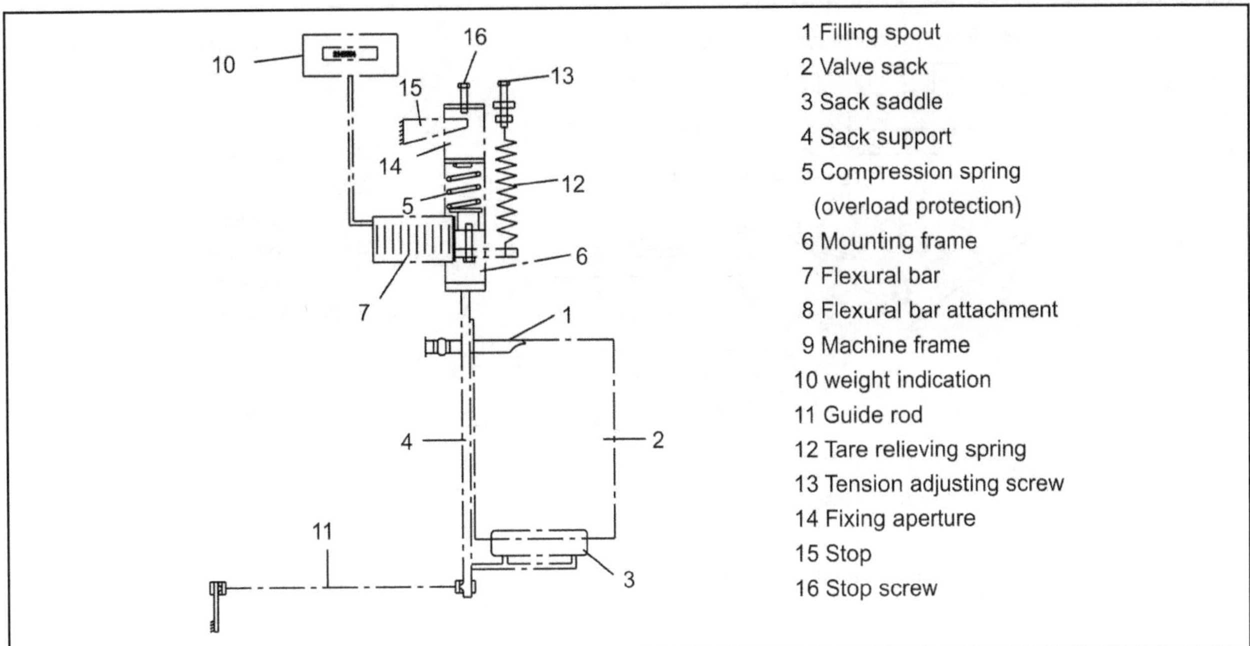

Fig. 2.14.4 Electronic weighing system.

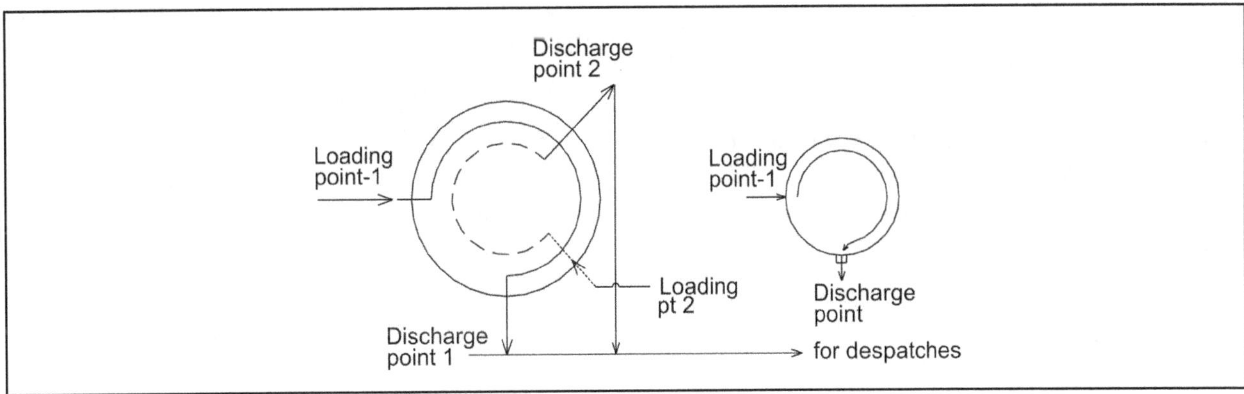

Fig. 2.14.5 Bagging of cement.

It was necessary to feed cement free of foreign bodies to the packing machine. A screen either vibrating or rotary was installed above the hopper of the packing machine to remove foreign bodies if any.

See Fig. 6.32.1 and 6.32.8 in Chapter 32 of Section 6.

14.2.1 Stationary Packing Machines

Stationary packing machines consisted of a number of spouts arranged in a row. Bags were slipped on the spouts and when filled were dropped on a belt conveyor below for conveying to trucks or wagons for loading. For practical reasons number of spouts in a row was 4, max. 6. Therefore capcity of a stationary packing machine was limited to 60 to 90 tph.

14.2.2 Rotary Packing Machines

In rotary packing machines, spouts were arranged on a rotating bin complete with rests for the bags. Operator slipped bags on successive spouts as they were positioned in front of him for the purpose. Bags got filled as machine turned and dropped at another point. One operator was sufficient for capacities up to 100 tph.

For higher capacities rotating machines with two loading and two dropping off points were developed. **See Fig. 2.14.5.**

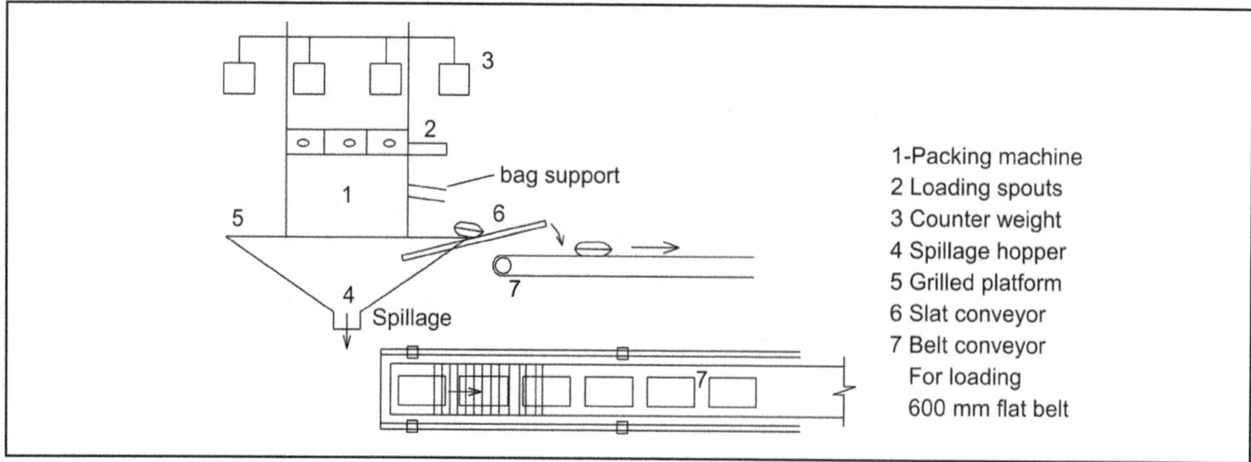

Fig. 2.14.6 Packing of Cement in bags.

Packing machines come fitted with a hopper under it to collect spillage. A short slat conveyor leads bags on to loading conveyor. Spilled cement is collected and returned to elevator
See Fig. 2.14.6.

For higher capacities rotating machines with two loading and two dropping off points were developed.
See Fig. 2.14.5.

Packing machines come fitted with a hopper under it to collect spillage. A short slat conveyor leads bags on to loading conveyor. Spilled cement is collected and returned to elevator
See Fig. 2.14.6.

Packing machine has a hood around it to vent it to avoid dust nuisance.

14.3 Paper and HDPE / Jute Bags

There is considerable difference in handling paper bags and jute or polythene bags. Handling is easier and faster with paper bags. There is no spillage. Jute bags are more prone to get stuck and spillage is more.

Automatic bags feeding machines have been developed for paper bags. Either rolls of paper bags or bags in a stack are fed to the packing machine by mechanical pneumatic conveyors.

See Figs. 6.38.1 and 6.38.2 in Chapter 38 of Section 6.

Packing and dispatching operations with paper bags have been fully automated in developed countries.

Lay out aspects of a packing plant have been dealt within **Chapters 32 and 34 of Section 6.**

14.4

Plate 2.14.1 shows layout of a packing plant designed for truck and wagon loading.

Plate 2.14.2 shows a Rotary Packing Machine.

Plate 2.14.3 shows a truck loading machine.

Plate 2.14.4 shows a wagon loading machine.

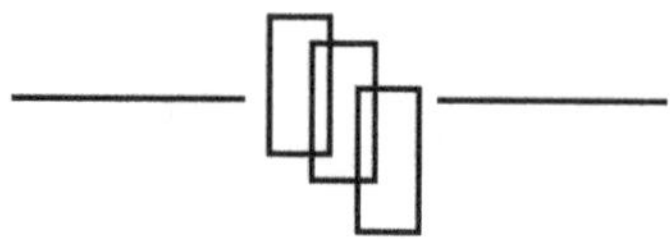

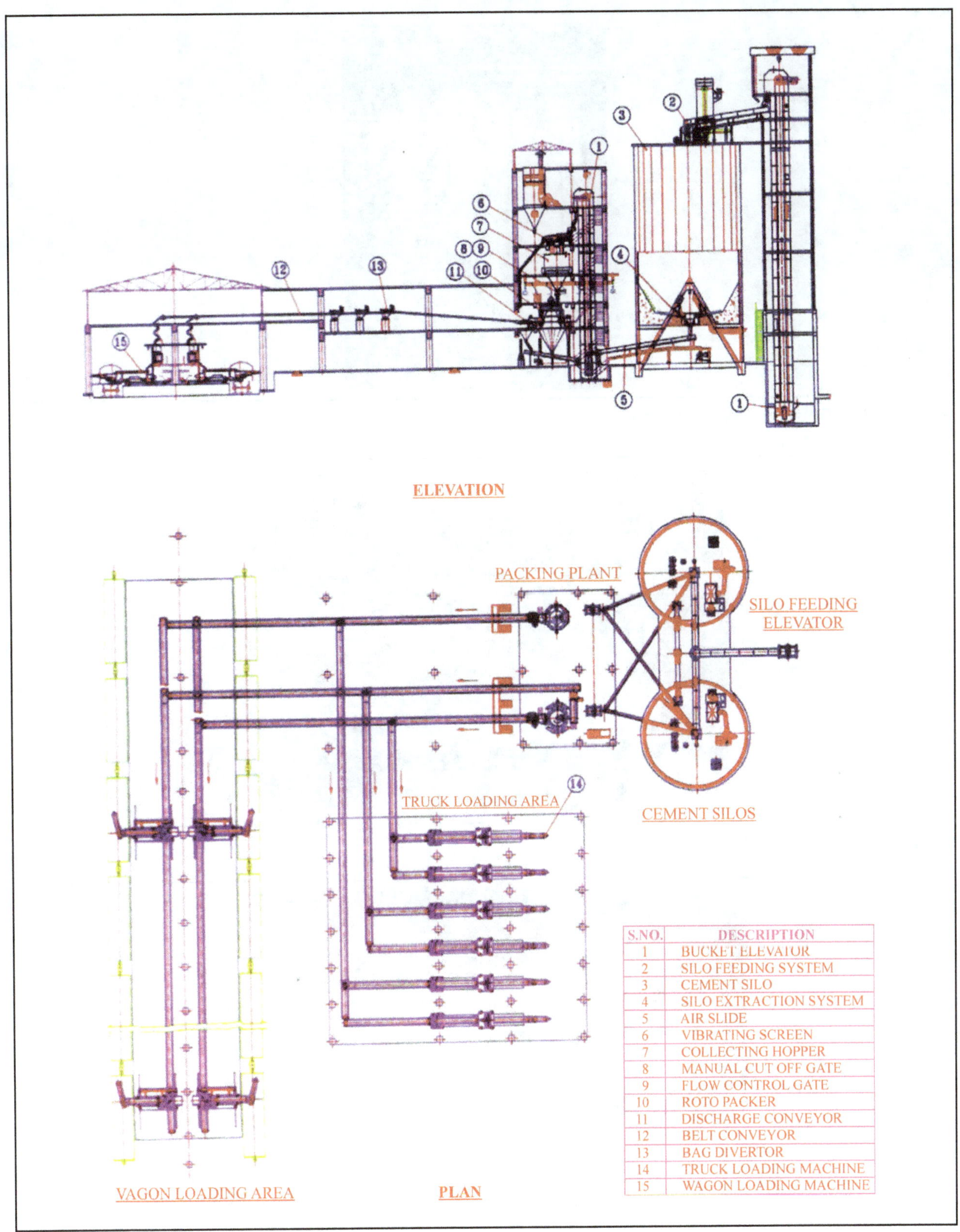

Plate 2.14.1 Layout of a cement packing and despatching plant.

Plate 2.14.2 Rotary packing machine.

Plate 2.14.3 Truck loading machine.

Plate 2.14.4 Wagon loading machine.

CHAPTER 15

DESPATCHES OF CEMENT, BAGGED AND IN BULK AND DESPATCHES OF CLINKER

15.1 Cement Despatches

1. Cement Dispatches can be in bulk and or in bags exclusively or in various proportions.
2. Dispatches can be by rail or by road, again exclusively or in various proportions.

Aspects of layout for arranging facilities in such situations will be dealt within Section 6 dealing with layouts and detailed engineering. Here equipment used in dispatches of bagged and bulk cement and clinker would be dealt with in brief.

15.2 Dispatches of Bagged Cement

(A) By Rail

In most cases bagged cement would be sent in closed wagons. Wagons were till recently loaded manually. Loading platform would have a flat belt with diverters. Wagons would be positioned on either side of the platform. Bags would be lifted off manually from the belt and stacked manually in wagons.

However this is not practical for large sized plants. Therefore 'wagon loading machines' are now commonly used to dispatch bagged cement in wagons by rail. Wagon loading machine is a portable unit with 3 belts which can be manipulated with respect to one another. It can reach all corners of the wagon and can also stack bags. Loading is speeded up and manpower is considerably reduced.

See Plate 2.14.4 in Chapter14.

(B) By Road

As in case of dispatches by rail 'truck loading machines' are used to load cement bags in trucks. They are conceptually similar to wagon loading machines.

See Plate 2.14.3 in Chapter 14.

Trucks mostly used, are 10 t capacity standard trucks; occasionally bags are also loaded on long trailers requiring loading machines with longer reach.

15.3 Handling of Cement Bags

When cement is exported, bags can be loaded on pallets and pallets would be loaded in containers.

Bags are also wrapped in plastic sheets to eliminate pallets. Weights commonly handled are half a ton or one ton.

Cement for export will be packed in paper bags which are easier to handle and handling is free of dust.

Bagged cement docs not nced 'weighing' facilities as each bag is weighed on the packing machine

15.4 Dispatches in Bulk

(A) By Rail

Cement can be dispatched in bulk in closed wagons. It would be necessary to weigh wagons before and after loading. If loaded in open wagons, care would have to be taken to cover the wagon with tarpaulins /plastic sheets to prevent ingress of water.

At unloading end wagons will be emptied manually or by using 'cement unloaders'.

Now 'special self unloading' tank like wagons are available for transporting cement by rail in bulk. Cement plants can have their own wagons for this purpose.

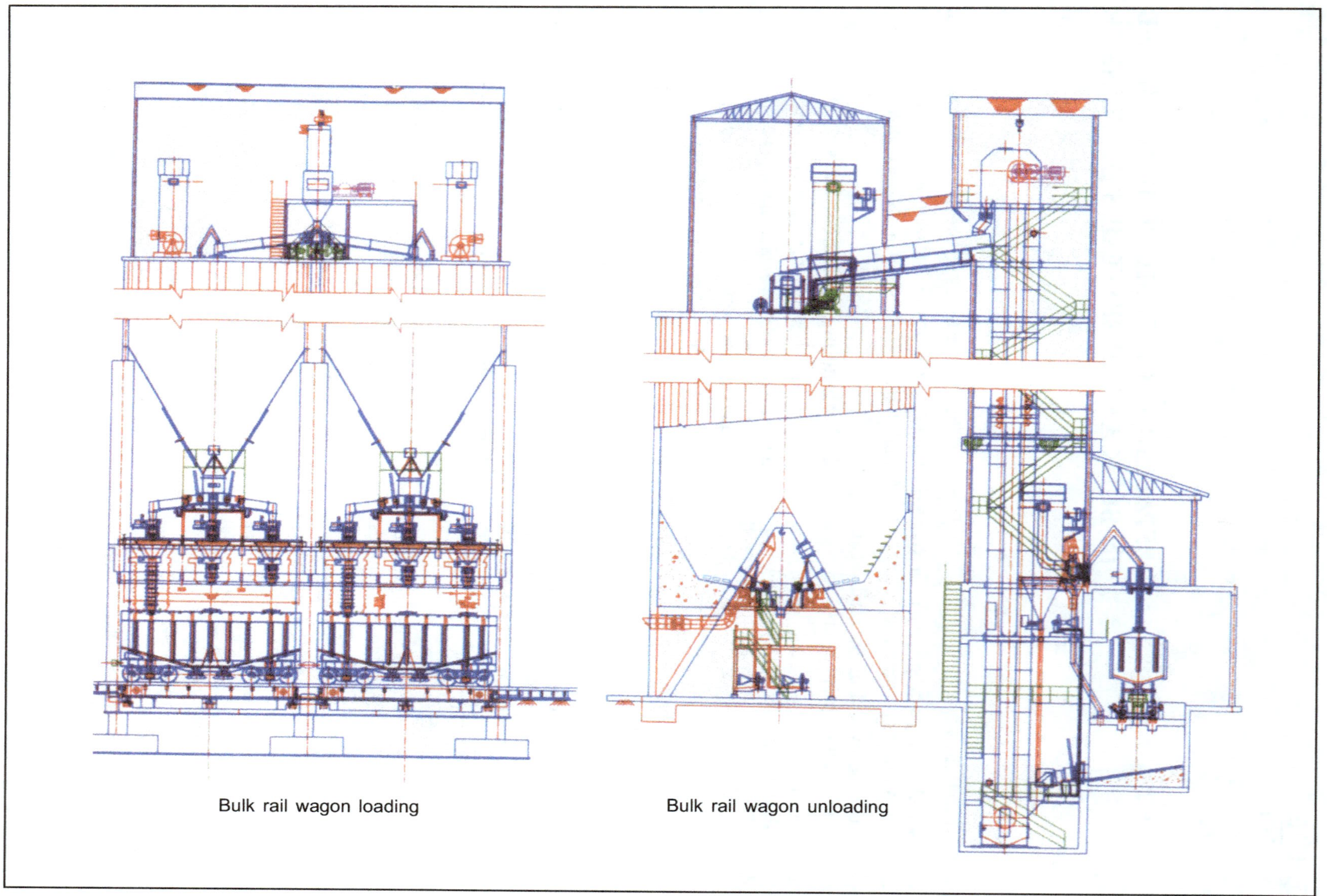

Plate 2.15.1 Loading and unloading bulk cement in wagons.

Wagons are unloaded pneumatically. See **Plate 2.15.1** which shows wagon unloading with the help of a bucket elevator

(B) By Road

Special bulk carriers of 10-15 ton capacity are available to carry cement in bulk by road. These carriers are unloaded pneumatically at the destination. **See Plate 2.15.2**

Plate 2.15.2 Bulk carrler for road transport.

15.4.1 Loading of Cement in Bulk carriers or Wagons

Cement can be directly loaded in wagons or carriers from silos or bins. A weigh bridge is required to know weight of cement loaded.

Alternatively small bins on load cells could be used to load wagons or carriers.

Bulk handling of cement requires dedusting arrangements to clean air being displaced while loading.

15.5 Selecting Plant Machinery

Sizing of machines, their capacities and numbers depend on volume to be handled, and time available for loading and unloading. This is particularly relevant for loading wagons where railways and ships allow a specific time for loading or unloading.

It is also dependent on proportions of cement to be sent from a plant in bulk and bagged and by rail and by road. In a majority of cases, all these four variables could be taking place simultaneously.

15.6 Export of Cement

Cement may be exported to neighboring countries. It will be sent by road or rail as above.

It could also be sent to distant countries. While a small portion would be sent in paper bags, most of it would be sent in bulk, both by ship.

Cement sent in bags and by ship would be sent in containers to the port of shipment. At the port containers will be loaded on the ship

If the plant exporting cement is far from a port, it would prefer to send it in bulk. It would create storing facility at Port. Cement would be sent to silos on port by bulk carriers by road or rail. Cement would be unloaded from it directly into ship's holds pneumatically.

15.6.1 Shipping Bulk Cement

Exporting in ship loads would require very high material handling rates because like railway wagons, ships would be docked for limited hours.

Assume a ship which carries 50,000 T of cement docks into port. Maximum time available for loading this quantity is say 7 days.

Every day 50,000 / 7 = 7150 tons of cement is to be loaded. Working round the clock loading rate would have to be 300 tph or more.

15.6.2 Facilities for handling Bulk Cement for Export

Cement in silos will be first fluidized by admitting air at bottom. Cement will be extracted from silos and metered and conveyed pneumatically by a system of air slides and dosing valves and pneumatic conveying systems for long distances such as, FK Pump, Airveyor- Air stream and Fluxo systems to ship's holds.

15.6.3 Pneumatic and Mechanical Conveying Systems

Pneumatic conveying systems have the advantage that they convey cement when air borne in pipelines and hence it is easily possible to change direction, elevation and so on. Also dust nuisance is nil during transport itself. Conveying air has to be vented at the end, which can easily be done through a bag filter.

For shore based plants, cement could be conveyed pneumatically from the plant itself to storage silos.

For bulk loading of cement at high rates, pneumatic aeration and handling and conveying is the most convenient method.

If the following are compared, the efficacy of pneumatic conveying for high conveying rates would be evident.

(i) Pipeline size for 300 tph for pneumatic conveying.

(ii) Belt conveyor, screw conveyor and Elevator sizes to convey at the rate of 300 tph.

The mechanical conveying systems besides themselves being large in size also require cumbersome and heavy supporting gantries.

Any change of direction and level means another machinery unit, another drive and also dust nuisance during transfer from one unit to another.

Lastly when dropping in the ships' holds, telescopic chutes would be used. These chutes would also have a suction system to arrest escaping air, thereby minimizing dust nuisance. Alternatively if the holds are generally used to handle cement and similar materials a bag filter can be permanently provided on ships hold to arrest fugitive dusts.

Plate 2.15.3 Ship loading and unloading machines for bulk cement.

Plate 2.15.4 Unloading into bulk carriers straight from ship.

Plate 2.15.5 Unloading and loading facilities installed on ships – bulk cement.

Plate 2.15.6 Loading and unloading terminal for cement at port.

15.6.4 Layouts for Despatches

Therefore, the layout of a cement storage and dispatches section is one that requires meticulous planning and it should be designed not only for the present but for the future also.

See Plates 2.15.3 to 2.15.6.

15.7 Bulk Handling of Clinker for Export

Bulk handling of clinker is an altogether different proposition. Clinker would be best conveyed by mechanical conveyors like belt, screw, elevator etc., because it is granular and also abrasive.

In most cases unidirectional mechanical handling systems would suffice for dispatching clinker for export.

See Plates 215.7 and 2.15.8.

A bi-directional system would be used when it is required to receive another granular bulk material like coal and dispatch clinker from the same port.

When handling such materials on a regular basis, ships would be fitted with extractors or machinery that would draw clinker / coal from the hold and bring it to a point outside it; from there the mechanical handling system installed on the port would convey it either to storage silos or open stock piles.

Like handling cement, loading and unloading rates for clinker too would also be high viz + 250 to 300 tph.

Cement companies may come together to create a common handling facility at the port / dock for loading ships with clinker and cement. It would be a good example of co-operation within the Industry. It distributes burden of capital expenditure.

See Plates 2.15.3 to 2.15.6.

15.8 Split Location

Some Cement Companies have split location. In split location clinker is made in one place and cement is ground in another place. Examples of split locations are now many.

Cement plants that are located along the coastal line can conveniently transport clinker in barges or ships. Water transport is the cheapest transport system for bulk materials. The grinding units at split locations would also be located along the coast. These companies have created their own jetties and handling facilities at port or jetties.

Designing of such facilities is a specialist's job and he has to take into account, tides, winds, monsoons, and also maximum and minimum drafts available that would limit size of the ship / barge that can be docked.

15.9 Transporting Clinker in Bulk by Rail or Road

When Cement Companies make blended cements, like Slag Cement or Pozzolana Cement, often movement is in two directions. Blending materials are brought in and clinker is sent out in bulk by rail or by road.

Plant needs to arrange for loading facilities like hoppers on load cells and feeders to load trucks or wagons.

Trucks used would mostly be standard self tipping trucks which would unload clinker in storage facilities provided at the destination.

Plate 2.15.7 Sending clinker in bulk by road.

Plate 2.15.8 Loading station for bulk dispatches of clinker.

Plate 2.15.9 Loading bulk carriers for rail transport of cement/fly ash.

Wagons will be open wagons which would be unloaded manually.

However if the volumes to be handled are large, wagon tipplers could be installed to unload rake loads of clinker received.

Dispatches of cement and clinker have been dealt with in greater detail in particular with reference to layout aspects in **Chapter 37 of Section 6.**

15.10 Handling Fly ash and Slag

Cement companies are now requied to handle fly ash and slag when making Blended Cements. Fly ash is fire and would be mostly dry. It can be transported in bulk carriers similar to those for cement.

would be mostly dry. It can be transported in bulk carriers similar to those for cement. It can be unloaded and transported pneumatically. Cement Plants would have to install facilities to suit. **Plate 2.15.9** shows handling of fly ash by rail.

Slag is granular and can be wet.It is best handled by mechanical conveyors. Quantities to be handled would be double that required for fly ash.

15.11 Wagon Tipplers

Wagon tipplers are necessary to handle large volumes in a limited time. They would be installed in cement plants to unload rake loads of coal, slag and clinker as mentioned above. Railways give limited time to unload rake loads which may bring in 4000 tons of coal or slag or clinker at a time.

15. 12 Import of coal

Many Cement Companies now import coal. Trading Companies import coal anticipating demand from indigenous consumers. They would have created unloading, storing and loading facilities at ports for this purpose.

Coal so received will be sold to individual companies according to their needs and at times it is needed.

Imported coal may be brought into plant by trucks or by rail and will be handled as they would handle indigenously procured coal.

15.13 Handling Cement in Bulk Internally

There is growing demand for cement in bulk in large consumer centers as it is cheaper and now there is no fear of adulteration. Therefore Cement Companies are creating facilities in big cities by constructing cement silos and supplying cement in bulk from there.

See Plate 2.15.9.

'Ready mix concrete' (RMC) is more and more in demand. Many Cement Companies themselves have entered into business of RMC. RMC also requires handling cement in bulk.

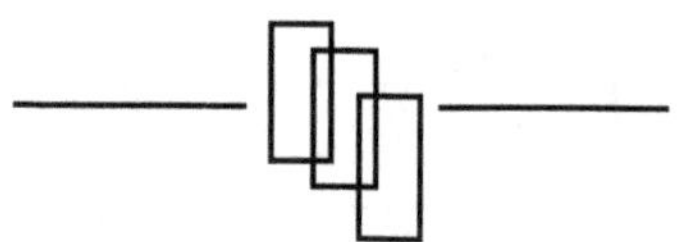

CHAPTER 16

MATERIAL HANDLING SYSTEMS

16.1 Bulk Materials to be Handled in a Cement Plant

In a cement plant, a great variety of materials is required to be handled in bulk and in large quantities. Starting with quarries, materials to be handled are :

1. Over burden which is clayey soil with moisture, sticky and non abrasive.
2. Uncrushed Limestone :
   With interstitial material, size varying from large boulders to dust; sometimes slabby, some times fragmented, moisture limited to 2-3 % for greater part of the year; free flowing when without clayey material.
3. Crushed Limestone :
   -100 mm to -20 mm size, depending on mill used, free flowing when without interstitial clayey material and moisture less than 5 % generally non abrasive but sharp corners can damage belts.
4. Correcting Materials - Clay, Sand and Bauxite:
   Lumps to fines; moisture 10-15 %; sticky, non-abrasive except for sand which because of silica can be abrasive; high moisture content causes problems of flow. Covered storage may be necessary if moisture is high all round the year.
5. Raw Meal:
   Powdered, dry (< 1% moisture) free flowing, non abrasive can be fluidized; entrains air in the process.
6. Clinker:
   Granular, less than 25 mm size, dry, hot (100-140 oC), abrasive; large amount of fines would also be present.
7. Coal:
   (a) Uncrushed as received : Wet 8-15 % moisture; lumps -300 mm; not free flowing not abrasive.
   (b) Crushed : Wet ; -50 to -20 mm depending on mill used; considerable amount of fine dust makes it sticky and difficult to flow in hoppers and bins.
   (c) Pulverized : Dry, free flowing when hot; if dew point temperature too close, can become sticky and difficult to flow; non abrasive - can be fluidized when moisture less than 2 %.

   Coal can catch fire in storage in hot months, particularly uncrushed coal. Even pulverized coal can catch fire.
8. Gypsum:
   , if natural, 8-10 % moisture, very wet (15-20 % moisture) if synthetic; natural gypsum would need crushing; synthetic gypsum is fine, non-abrasive, difficult to flow in hoppers / bins.
9. Fly Ash:
   Very fine, requires drying if collected wet in power plant (moisture 10-15%); dry, if collected dry - very dusty difficult to handle because of fineness, free flowing mildly abrasive.
10. Slag :
    Granulated blast furnace slag, -6mm to dust; wet 10-15 % moisture; very abrasive, not free flowing; free flowing when dried.
11. Cement :
    Very fine; free flowing, abrasive.
12. Cement Bags.

16.2 Selection of Machinery for Handling Various Materials

Material handling systems are selected to suit materials and operating conditions and quantities to be handled. **See Table 2.16.1.**

1. For Uncrushed Stone & Over burden :

   Shovels and dumpers are used as a combination almost invariably. Lead may be long or short. Sizes selected according to quantities be handled and fragmentation properties of stone. Bull dozers are used to level area and to shift overburden.

2. For Crushed Stone :

   Possibilities are :

   (i) Belt Conveyor - From crusher to stacker reclaimer system.

   (ii) Overland belt conveyor - From crusher to stacker reclaimer if crusher is in quarries.

   (iii) Ropeway - if crusher is in quarries and terrain is difficult.

   (iv) Self tipping railway wagons and trucks.

3. For clinker,

   suitable conveyors are:

   (i) Drag chain conveyor,

   (ii) deep bucket conveyor,

   (iii) when clinker is not hot, and clinker is to be transported over long distances, belt conveyors would be used.

4. For coal

   For uncrushed and also crushed coal belt conveyors are the most commonly used conveyors.

5. For slag

   For slag also belt conveyor would be the most suitable conveyor.

16.3 Belt Conveyors

It would thus be seen that for crushed and granular materials belt conveyors are very well suited. A cement plant may have several kilo meters of belt conveyors to handle the various materials mentioned above.

Belt Conveyors have low power consumption and if maintained, have long life. Damage occurs at feeding points where sharp corners of stone hit belt when falling from crusher or from belt to belt. Closely spaced impact idlers are used to take impact of falling stone.

When belt conveyors are several kms long they would be covered to protect them and to prevent damage from wind, rain, etc.

There have been considerable developments in design of belt conveyors and belting.

(i) A troughed belt has a much higher capacity than a flat belt. Now 30o troughs are quite common as against 20o used earlier.

(ii) High belt speeds of 2.0 m/sec and more are used as against speeds of 1.0 m/esc of old. This has been possible because of use of nylon as construction material for belts.

(iii) closed belt conveyors in the form of a pipe havecome into use for specific applications

16.3.1 Capapcity and Size Belt Conveyor to be Installed

Thus in case of existing belts, it is possible to increase the capacity of the same belt substantially by changing angle of trough and belt speeds.

In case of long conveyors, it would be expensive to install a second conveyor along side when plant capacity is duplicated. The conveyor would therefore be sized from the beginning for expanded capacity.

However, when the plant capacity is likely to be doubled even trebled, it would be essential to have a standby so that production is not interrupted to attend to belt maintenance / breakdown. Hence, at some stage the plant should consider installing a standby line. The same logic will also apply to belt conveyors of stacker reclaimer systems, such as :

(i) use existing stacker and reclaimer after expansion,

(ii) use existing stacker but add a second reclaimer during expansion,

(iii) add both stacker and reclaimer during expansion.

Capacities and sizes of belt conveyors can be selected accordingly.

In extreme weather conditions, stock piles would be covered, at least partially, otherwise, they would be open. Belt conveyors would be covered to the extent possible.

Belt conveyors are used to convey cement from cement mills to silos.

Table 2.16.1 Materials handling systems in cement plants.

Materials handling Systems in Cement Plants														
sr no.	conveying systems	Materials to behandled					Material to be handled							
		uncrushed stone	crushed stone	pulverised raw meal	dust	clinker	uncrushed coal	crushed coal	pulverised coal	gypsum	cement	fly ash	slag wet /dry	correcting materials
1	shovels - dumpers	*												
2	wagons	*	*				*			*	*	*	*	
3	trucks						*			*	*		*	*
4	rope way		*				*							*
5	belt conveyors		*			*	*	*		*	*		*	*
6	screw conveyors			*	*				*		*			
7	bucket elevators			*	*	*		*	*		*			
8	chain conveyors				*	*								
9	drag chains					*								
10	deep bucket conveyors					*								
11	air slides			*	*						*			
12	p. s. pumps			*	*				*		*	*		
13	air lifts			*	*						*			
14	other pneumatic systems			*	*				*		*	*		

Plate 2.16.1 Belt conveyors.

Belt conveyors are truly the work horse in cement plants.

See Plate 2.16.1.

16.4 Bucket Elevators

Bucket elevators lift materials only vertically. They are suitable for granular free flowing materials and also for pulverized materials.

Elevators are either central discharge or centrifugal discharge.

They are either chain bucket or belt bucket elevators. Earlier elevators were limited to 30 meters height of centers.

Today belt bucket elevators are designed for > 80 metres height of centers and for high capacities. Their reliability has increased so much that they are used for the most rigorous of the application viz., feeding raw meal to preheater. A standby is necessary for this application.

See Plate 2.16.2.

Plate 2.16.2 Bucket elevator.

16.5 Screw Conveyors

Screw conveyors are used for pulverized and granular materials for short lengths. For long lengths they need intermediate hangar bearings. They are used to handle coal, cement and raw meal.

16.6 Chain Conveyors

They are also largely horizontal conveyors used for short distances particularly under dust collectors and for hot materials.

16.7 Drag Chain Conveyors

These are used almost exclusively to handle hot clinker discharged from grate coolers. They may be installed up to an inclination of 12 degrees. They are open trough conveyors.

16.8 Deep Bucket Conveyors / Pan Coveyors

They are slow moving chain bucket conveyors which run horizontally and also at steep angles of inclination up to 40o to horizontal. They are almost exclusively used to handle hot clinker with temperatures as high as 200 oC to take it from coolers to clinker storages. This application requires availability of 24 hours per day, months on end. A standby is most certainly necessary.

Further, interchangeability is desirable between two clinker storages. For this purpose pan conveyors would be used.

See Plates 2.16.3 and 2.16.4.

Plate 2.16.3 Deep bucket conveyor.

Plate 2.16.4 Pan conveyor.

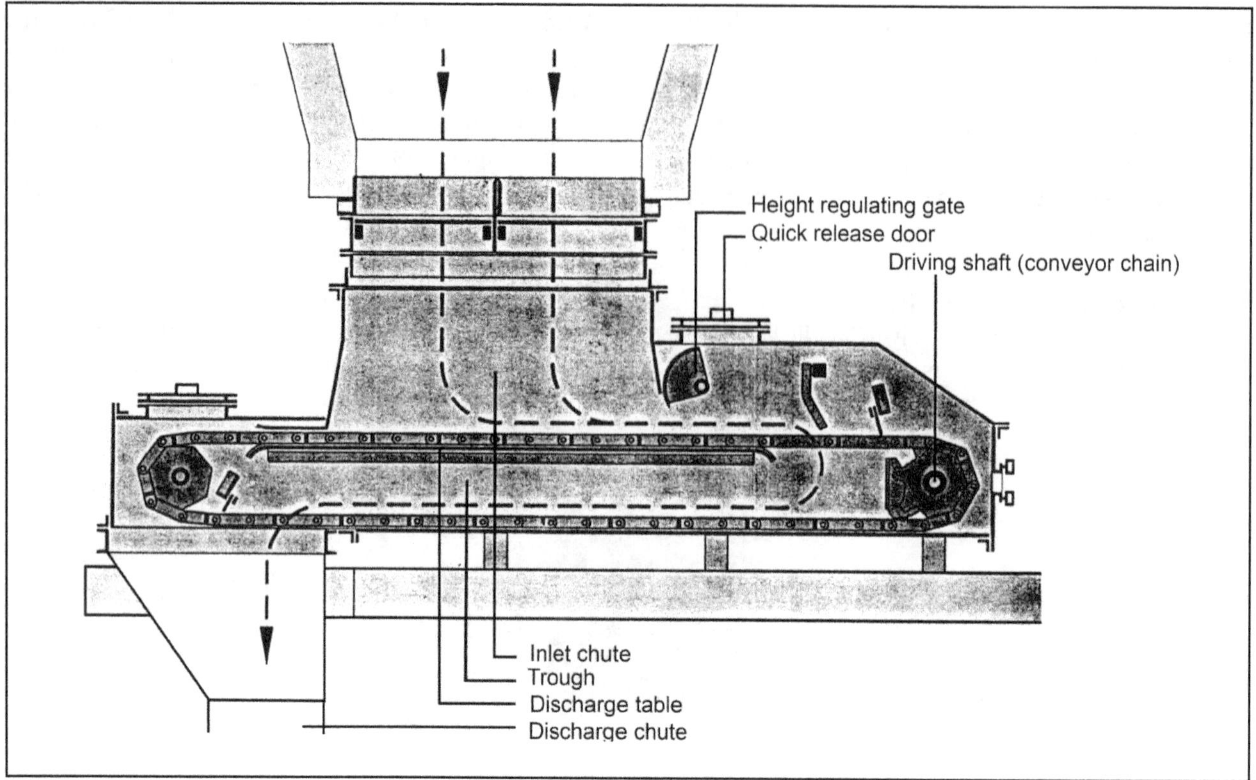

Plate 2.16.5 Drag link chain feeder/conveyor.

Pivoted pan conveyers were used for discharging clinker at more than one point like in case of G-Cooler or in silos.

See Plate 16.6.

16.9

The mechanical conveyors mentioned above were replaced in the sixties for such applications as conveying pulverized raw meal, coal and cement by pneumatic conveyors. However they have made a comeback with improved designs due to their low power consumption.

16.10 Pneumatic Conveyors

Principal among pneumatic conveyors that are mainly suitable to convey pulverized dry products in a cement plant are:

1. Air slides

See Plate 2.16.7.

2. F.K.Pumps

See Plate 2.16.8.

3. Air Lifts

See Plate 2.16.9.

4. Dense / lean phase conveying systems

See Plate 2.16.10.

16.10.1

Pulverized raw meal is conveyed in three stages.

(i) From raw mill to storage (Blending) silo,

(ii) From blending silo to metering device / kiln feed,

(iii) From kiln feed to preheater.

For (i) above, 'air lifts' are used, which lift material directly vertically into a blending silo. A system of air sides in raw mill department collects finished product and brings it to a point where air lift would be installed.

In the mill system itself, bucket elevator would be used to lift mill discharge to separator. These are large elevators as mills have large circulating loads.

For (ii) above, a system of air slides and bucket elevator would be used.

For (iii) above, till now air lift was the most common. But to minimize power consumption, bucket elevators

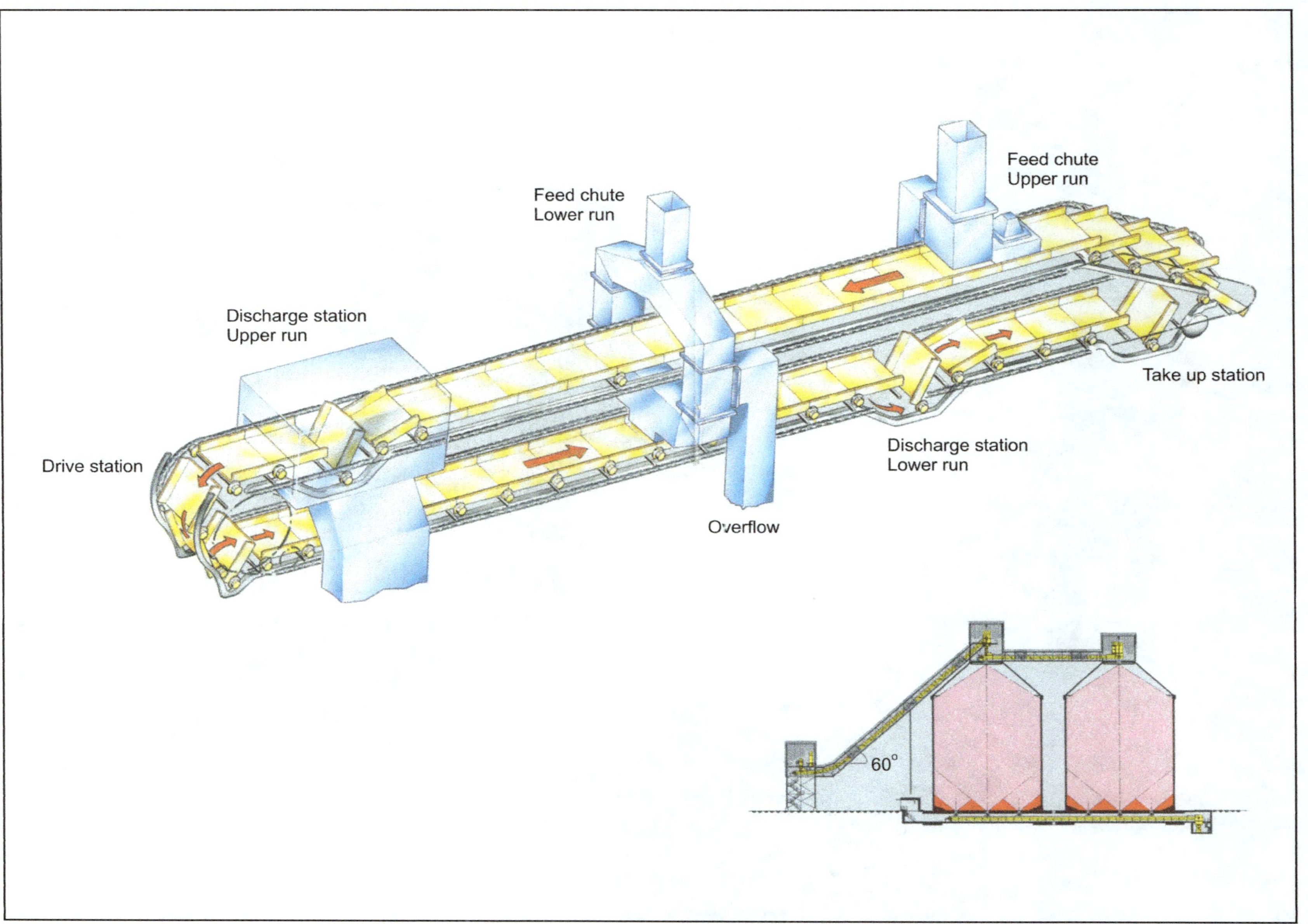

Plate 2.16.6 Pivoted pan conveyor for clinker.

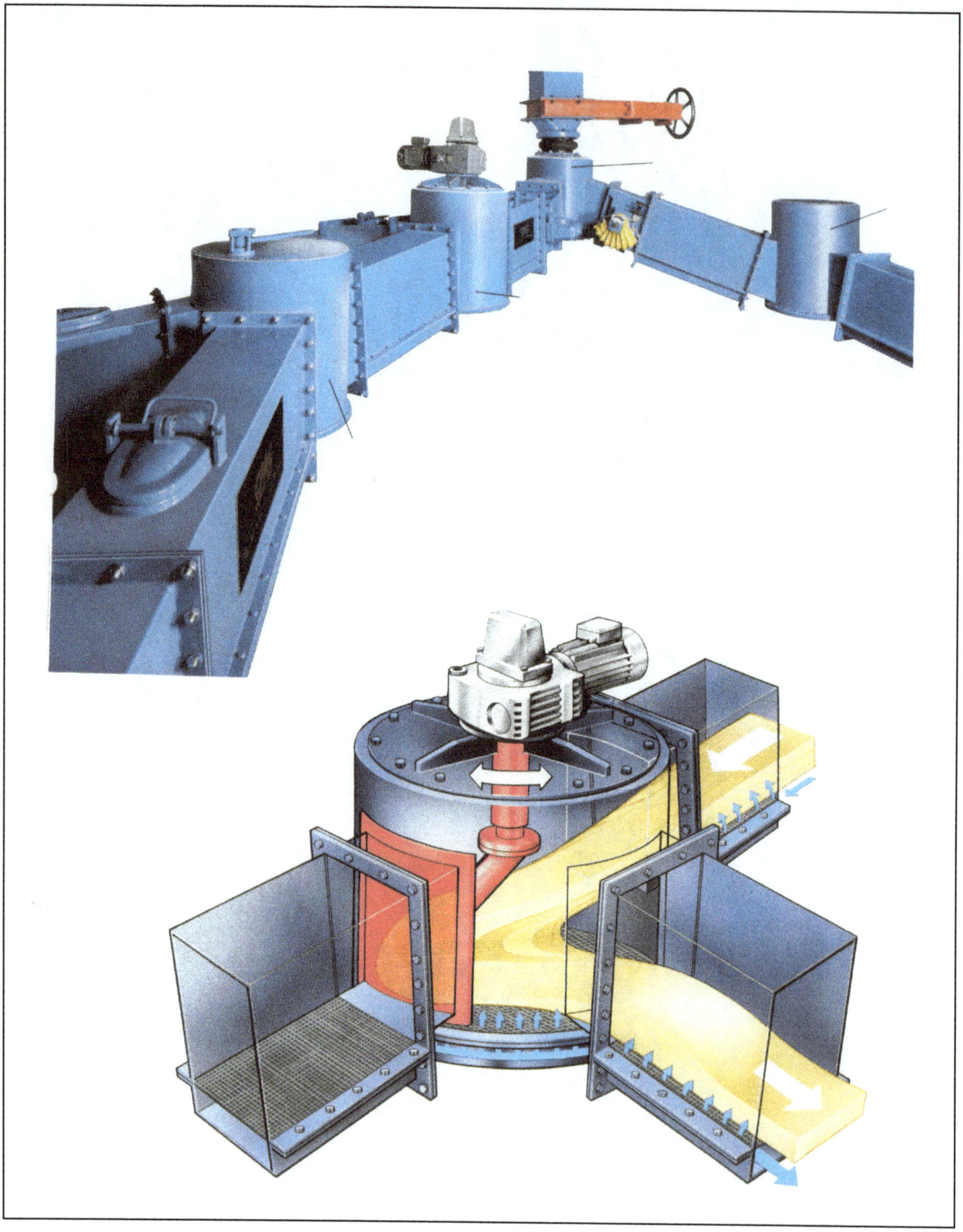

Plate 2.16.7 Air slides for conveying, diverting and dividing flows.

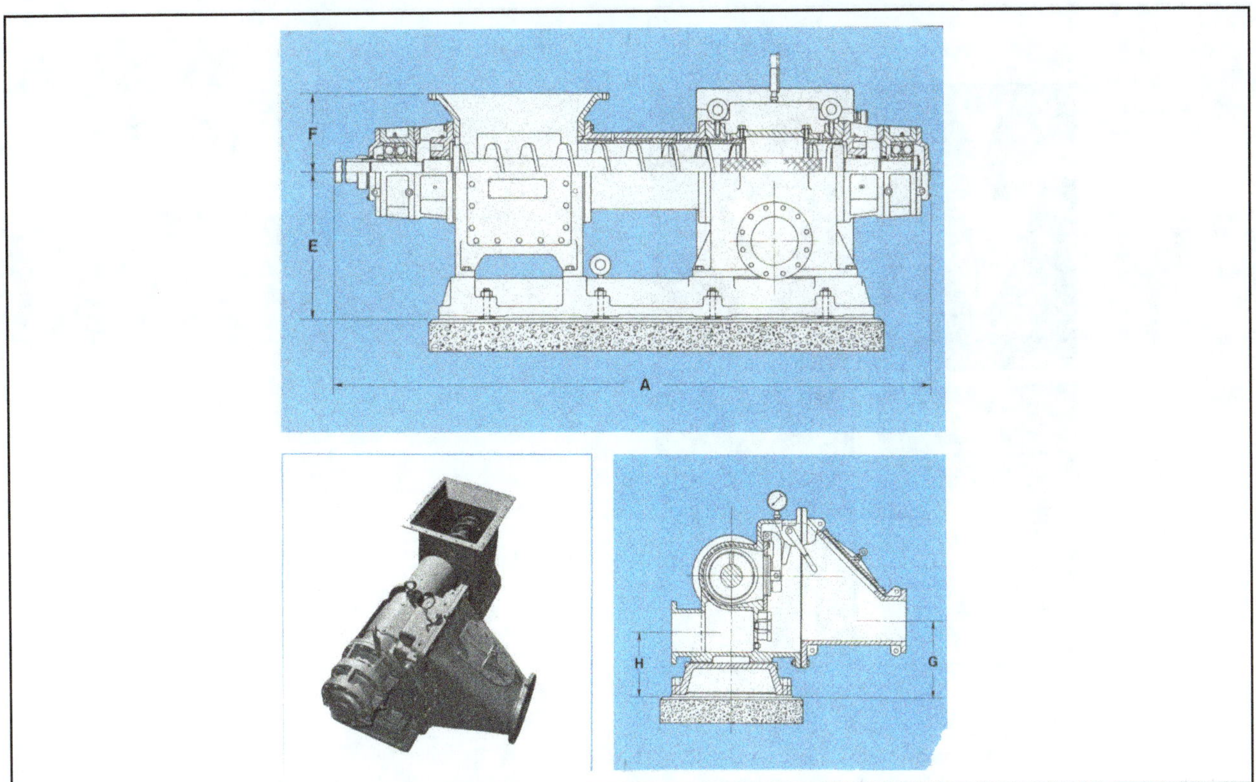

Plate 2.16.8 Fuller kinyon pump M type.

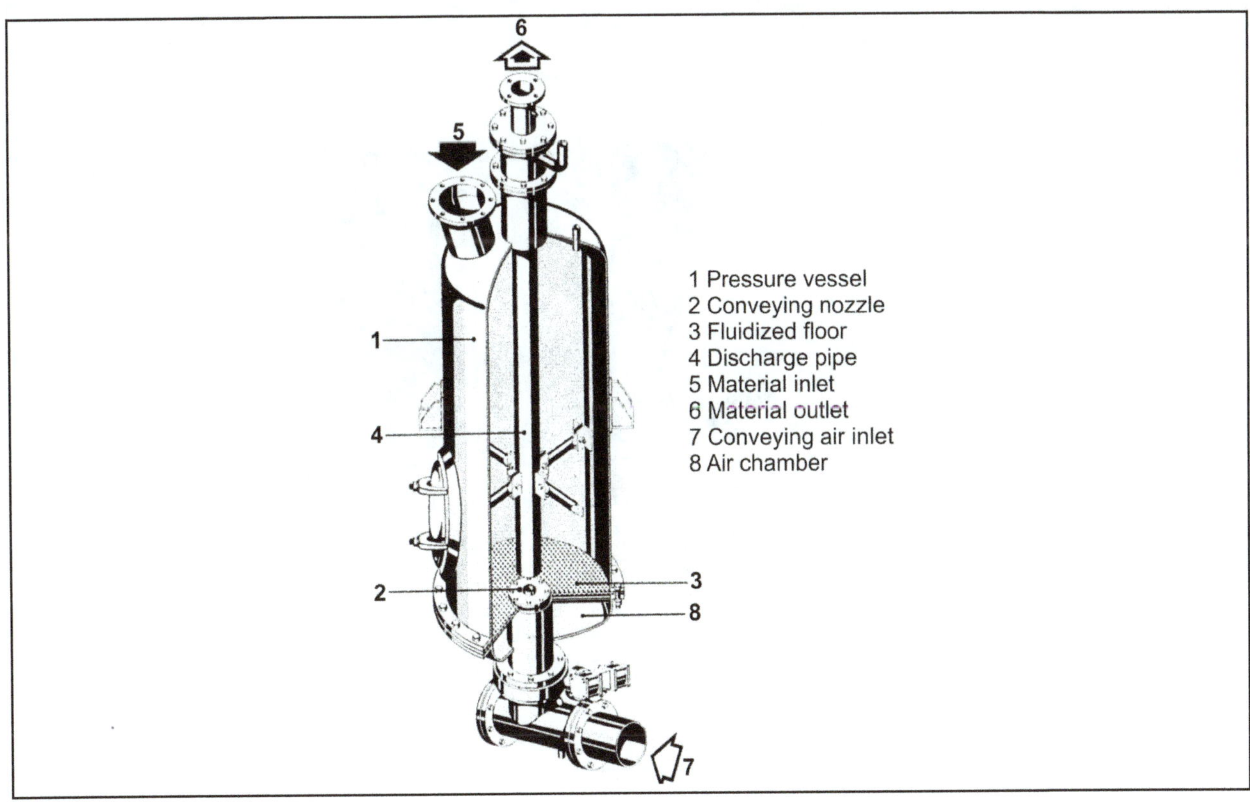

Plate 2.16.9 Air lift pump.

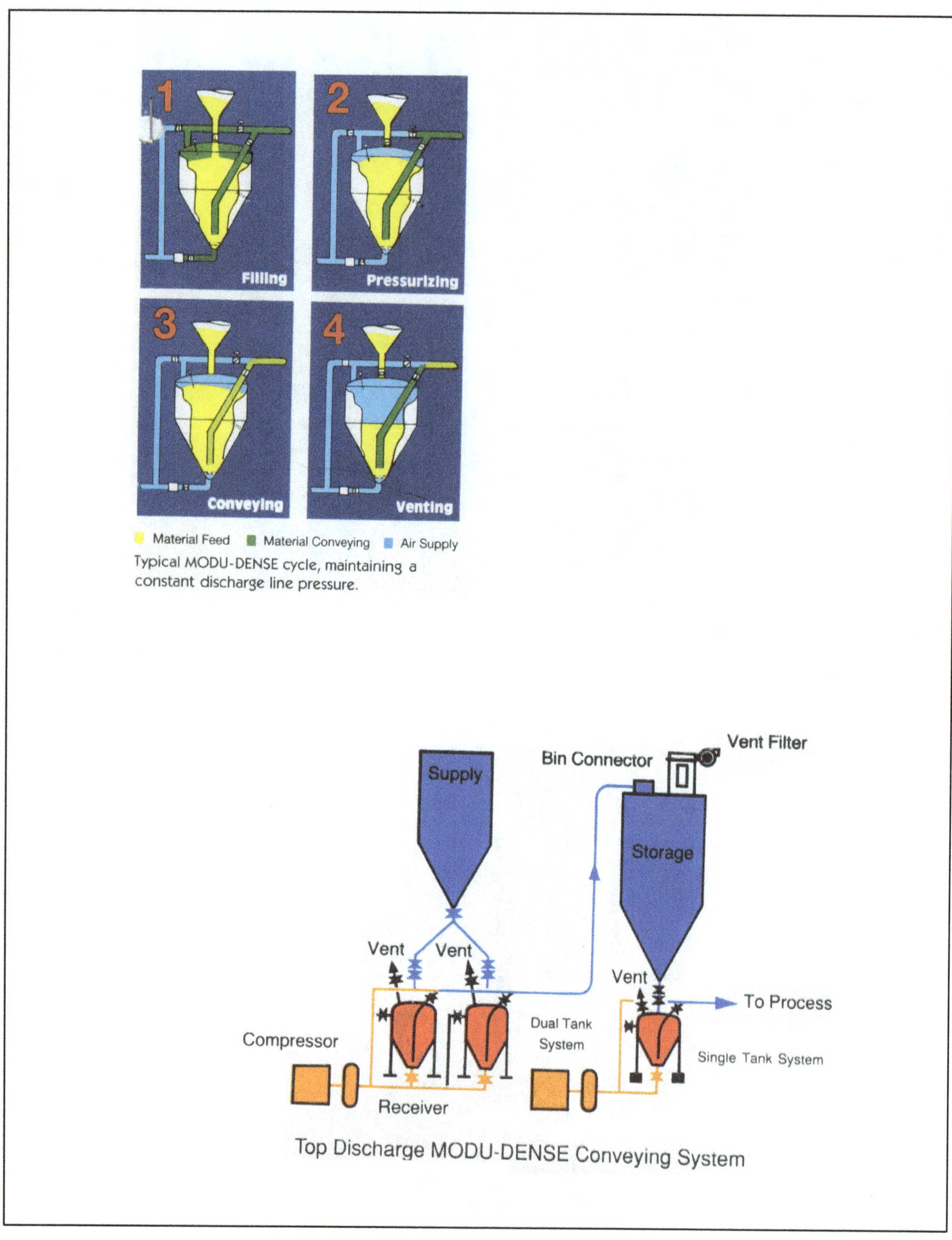

Plate 2.16.10 Dense phase conveying system arragement and action.

are used even for 6 stage preheaters and also because conveying air introduced false air in the system.

If kiln exhaust gas volume is 1.5 nm3/kg, the exhaust gas to be handled by preheater fan would be 1.6 nm3/kg, for air lift and 1.55 nm3/kg for PS Pump; increases of 6 and 3 % respectively.

16.10.2 Features of Pneumatic Conveying Systems

Principal advantage of pneumatic conveyors is that they could change direction. Air slides carry materials only down stream.

Others carry vertically and horizontally. They are flexible and dust free. Conveying air can be and is dedusted in bag filters.

They were very popular in the sixties to eighties to convey raw meal, pulverized coal, and cement. Today their use is limited to air slides and dense /lean phase conveying systems to convey pulverized coal for firing it in kilns and calciners.

Air lifts were almost exclusively used till recently for kiln feed. They have been replaced by bucket elevator on account of power.

This has been possible owning to developments of chains and belting of superior quality. They have made elevators of 100 metre height in single lengths possible. Sometimes elevators are used in tandem to reach the required heights.

An elevator would be much more expensive than an air lift and its compressor for air supply. But it is now universally adopted to save power.

A standby system should be installed else kiln production could come to a stop. The standby would be an 'air lift'.

The only area where pneumatic conveying would be continued to be used would be: firing coal in kiln and calciner.

Even here with multi channel burners, quantum of primary air is now less than 12% compared to 15-20% of the past.

Out of the total primary air, only 2.5-3 % would be used to convey coal dust into the burner. It will be mixed with clean (some times preheated) primary air in the burner.

In calciners, coal is conveyed pneumatically into the calciner, via two or more burners. There are calciners which can burn coarse (grits) fuel. In these cases, coal is fed into the calciner through a chute.

16.10.3 Power Consumption in Pneumatic Systems

Impact of pneumatic conveying on specific power consumption per ton of cement could be seen.

Assume for convenience that the same pneumatic conveying system is used for different applications listed below and that it required 1 kwh / ton of material conveyed.

(i) Raw meal conversion ratio

1.55 : 1 on consumption.

1.65 : 1 on Kiln feed

(ii) coal consumption 16 %.

(iii) gypsum added 4 %.

Power consumption converted into kwh / ton of cement would be as shown in **Table 2.16.2.**

Thus impact of conveying coal pneumatically is 1/10th of that of conveying raw meal.

16.11 Feeders

All feeders are conveyors of one type or the other with an added feature of a variable speed drive so that the rate of feed can be regulated to suit operational requirements.

Rate of feed can also be changed in some cases mechanically by:

1. Changing stroke in reciprocating feeder and in grate cooler.

Table 2.16.2

Sr.No.	Section	Power consumption per ton of material	Per ton of Cement	
1	Raw Mill	1 Unit	1.55 / 1.04 : 1	1.49
2	Kiln Feed	1 Unit	1.65 / 1.04	1.59
3	Coal	1 Unit	0.16 / 1.04	0.15
4	Cement	1 Unit	1.0	1

2. Changing gear ratio between motor and the driven machine through a chain drive, or a belt drive.

However such a change requires stoppage of the plant to change chain sprockets, gears, and belt pulleys and belts. This option is thus not very suitable when speeds are to be changed frequently.

Now change of speed is obtained almost always by using variable speed motors. Variable speed motors can be A.C. or D.C.

The type of drive to be selected would depend on :

(i) Application.
(ii) Ratio of speed variation required.
(iii) Whether speed changes could be stepped or continuous.
(iv) Frequency of changing and intervals between changes.
(v) Remote operation.

In the present context of automation, remote control and operation would be a must for all options.

16.12 Type of Feeders

Feeders like conveyors are selected to suit materials being conveyed. As seen earlier materials range in size from boulders to pulverized very fine materials like cement and ash.

16.12.1

Most commonly used feeders in cement plants are :

(A) For boulders of uncrushed limestone and coal

Apron feeders.

Reciprocating feeders.

Wobbler feeders.

See Figs. 2.2.1 and 2.2.2 in Chapter 2.

(B) For crushed and granular materials like crushed stone, crushed coal, clinker, slag and correcting materials.

Reciprocating feeders.

Vibro feeders.

Belt feeders.

Weigh feeders.

(C) *For pulverized materials like raw meal, pulverized coal, cement and flyash :*

Weigh feeders.

Solids flow meters.

Screw feeders.

Rotary vane feeders.

Flow Control gates.

16.12.2 Feeders for Crushers

They are required to handle large boulders -600 × 600; 800 × 800 and 1000 × 1000 mm in size depending on size of shovel. Commonly used types are:

(i) Wobblers feeders when clayey material is to be scraped off stone before feeding crusher.
(ii) Reciprocating feeder.
(iii) Apron feeder.

Out of these, apron feeders, though expensive are preferred because they prevent stone from rushing into crusher mouth because of their upwards inclination.

Indirectly they also reduce the depth of pit for conveyor underneath the crusher.

This is of particular importance when underground water table is high.

For small plants Reciprocating Feeders are common. They are installed sloping downward and hence are more suitable for stone received from manual mining. Stone pieces tend to rush into crusher and jam it.

Wobbler feeders are installed horizontally. Stone moves over a series of rotating shafts with elliptical projections. Clayey material adhering to stone gets scraped and clean stone enters crusher.

Of necessity the feeders for crusher need to be of rugged construction capable of withstanding impact of falling boulders.

They need to have variable speed drive to change rates of feed.

16.13 Feeders for Grinding Mills

For raw mills

They handle crushed lumpy materials like limestone and clay and iron ore. Clay and iron ore are wet and are not free flowing. Feed size for ball mills would be -25 mms and for vertical mills -75 mms.

Table feeders and Belt feeders directly installed under the hoppers are most commonly used feeders.

For coal mills

They handle wet crushed coal -25 mm for ball mills and -50 mms for vertical mills.

Table feeders and belt feeders and drag link feeders are commonly used feeders

See Plate 2.16.5.

For Cement mills

They handle granular clinker -25 mm in size, wet natural gypsum-crushed also to -25 mm or very wet fine synthetic gypsum; also dried slag.

Table feeders and belt feeders are the most common feeders.

Belt Feeders would be actually weigh feeders in that they would weigh, indicate and rates of feed.

16.13.1 Drives for Feeders

Feeders for grinding mills also need variable speed drives to change rates of feed but in case of raw and cement mills it should also be possible to maintain or change proportions of material ratios of limestone/ clay/iron ore in case of raw mills; and ratios of clinker/ slag/gypsum in case of cement mills..

Individual feeders would be selected so that changes both in proportioning and in feed rates could be coped with. A margin of 20 % on design capacity would be added as explained earlier.

It is preferable to choose slow moving feeder conveyors, which would be belt conveyors. This would result in wider belts and hence first costs would be high.

Feeders are selected also on basis of accuracy in terms of changes in feed rate and in rate of feed. Accuracy should be high for the entire range of speed variation.

Feeders for mills will be directly installed under hoppers with rod gate in between. The feeders are generally 2 to 3 m long and discharge on to a common conveyor which will take feed to mill.

Feeders for mills meter feed rates and also totalize tons fed.

16.13.2 Prefeeders

Feeders used for metering raw meal feed to kiln and coal firing in kiln and calciners are used in conjunction with 'prefeeders'. The two act as a pair.

See Figs. 2.7.2 to 2.7.6 in Chapter 7 of this Section

Prefeeders feed the main feeder at predetermined rates; feeders weigh/ meter the rate and any variations detected are corrected by changing speeds of prefeeders. Rates of feed corresponding to different outputs are taken care of by running feeder faster or slower.

Accuracy of weigh feeders is + 1% and that of solids flow meters + 2%.

16.13.3 Metering on Feeders

'Metering' can be direct or indirect. Indirect metering arrives at quantity fed or rate of feed by measurement of cross section of material and rate at which it flows and taking into account bulk density of material.

By totalling number of revolutions in the period tons fed in that period are found.

Indirect method is suitable for crushed / granular materials flowing out of silos / hoppers without aeration.

Where material is aerated like raw meal, bulk density fluctuations can be continuous and wide. Therefore this indirect method of metering is not so suitable for them.

A long, slow moving belt helps in 'de aeration' of the raw meal.

Another indirect method is to measure the impact of flowing material on a curved plate. This impact is through a given height of fall and hence is in a way independent of bulk density. But if particles tend to agglomerate, or tend to stick together the readings are vitiated.

It is mostly for dry free flowing powdered material like raw meal. Such indirect meters are called 'solids flow meters'.

See Plate 2.7.2 in chpter 7 of this Section

The direct method of course consists of weighing the material passing over a load cell and thus directly measuring rate of flow in kg or t/hr.

Direct system consists of weigh feeder and its pre-feeder.

16.13.4 Indirect Metering Systems

Indirect system is volumetric using volume as measurement like rotary vane feeders or screw feeders with close tolerances and variable speed drives; like disc feeders which also work on similar principle.

The 'loss in weight system' fills up a hopper above the feeder in a very short time and the rate of its emptying is monitored to arrive at the rate of flow. This system, requires beside quick mass filling of a hopper on load cell but also a computer to monitor the rate of emptying. This makes it an expensive system. **See Plate 2.7.4 in Chapter 7 of this Section**

Metering of both raw meal feed to kiln and coal fired in kiln and calciner are both crucial measurements and enough care in selecting feeders must be taken. One measurement indicates the kiln output the other coal consumption and hence specific fuel consumption.

16.13.5 Calibration and Drop Test Facilities

In large plants (in some old plants also there is provision to measure clinker produced. Batch weighing facilities are provided for weighing clinker produced, by 'drop tests'.

Weigh feeders therefore generally come with self-calibration devices. These must be included in the order for weigh feeders.

A simple way of 'self-calibration' is to have the bin the the feeder fitted with load cells. It is filled up and time taken to empty it can be measured at any given setting thereof (speed of feeder).

16.14 Ropeways

Shovels and Dumpers have been dealt with in **Chapter 1** on quarrying operations. Various other conveyors have been dealt with in above paragraphs. Ropeways have been mentioned as being suitable for conveying crushed material. However they need special mention.

Ropeways have been used to carry crushed and granular materials like crushed limestone, coal, sand and ores.

They have been particularly useful when terrain was difficult and was not easily negotiable by road or rail.

Ropeways several kilometer long have been in use for decades to carry coal from collieries to rail heads. Many a cement plant has its mines in forest and hilly areas. Factory cannot be located there. Stone is then brought by ropeway to the plant.

Ropeways are also useful when it becomes necessary to cross rivers or intervening hills.

As the name suggests a ropeway no consists of a pair of ropes strung between trestles. Buckets filled with material run over them. In bi-cable ropeways which are more common for heavy duty, there are carrying ropes and also a hauling rope to which buckets are attached, and which pulls buckets over the carrying ropes. Hauling cable is a closed loop. Buckets are clamped to it at loading station and are detached at unloading stations. It is pulled by an electric drive that may have facility to generate power. One of the carrying cables carries loaded buckets and the other empties.

Advantages of a ropeway is that mines and plant are connected without requiring heavy investment in acquiring land, construction of road or railway, bridges and culverts.

Ropeway trestles can be constructed at suitable spacing and cost thereof would be much less than that of a continuous road or railway. Land required for trestle footings would be minimum.

Since contours are not a problem, a ropeway would be the shortest distance between the two points.

Power required by a ropeway is small. Power can be regenerated while ropeway is running taking advantage of downwards stretches.

Ropeways can be wholly automated requiring minimum manpower.

Principal disadvantage of a ropeway is that its capacity cannot be increased substantially.

Thus if plant capacity is likely to be doubled, it is necessary to install a ropeway of higher capacity from the beginning.

Present cement plants of +1mtpa capacity would require a ropeway of 600 tph capacity for the first phase itself. Ropeways of such large capacities have not been installed / developed.

This has been primarily because of development of overland conveyors which follow the contour of the terrain. Such conveyors can carry material for long lengths running into several kilometers. Change of direction is possible by installing two or more shorter belts.

Overland conveyors cannot go over hills. In such situations it has been possible to dig tunnels in the hills and install conveyors in it

Maintenance costs of ropeway increase with years. Ropes- carrying and hauling need to be replaced. Buckets need to be repaired. In comparison, maintenance costs of overland conveyor remain at a lower level. It is true that belt is required to be taken care off at points of feed.

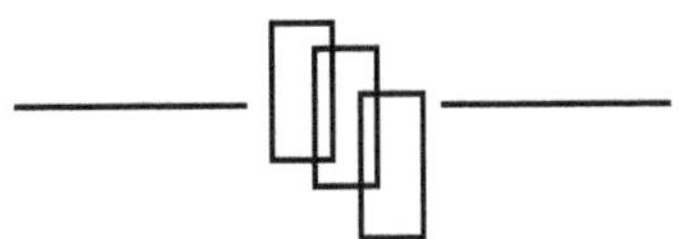

CHAPTER 17

FANS

17.1 Fans

Fans come next to grinding mills in consumption of electric power in a cement plant.

There are two basic types of fans. They are axial flow fans and centrifugal fans.

In axial flow fans, inlets and outlets are in the same direction. These fans are mainly used for applications which require small pressures and large volumes.

Centrifugal fans are the most commonly found fans. The capacity range is very wide -from less than 100 m3/min to several thousand m3/min. They can develop pressures as high as 1000 mmwg.

In cement plants axial fans are used for cooling kiln shell in the burning zone. They are also used in air to gas/air heat exchangers to cool hot vent gases for further treatment in bag filters.

In Centrifugal fans air /gas enters the impeller at center and leaves from the periphery. Pressure build up is due to involute shape of casing.

17.2 Costuction of Fans and Impellers

Impellers are designed to suit duty requirements. There are three principal types of impellers.

1. Radial bladed,
2. backward curved bladed,
3. forward curved bladed.

Of these radial bladed are suitable for handling dust laden gases, but their efficiency is low.

Backward curved bladed impellers have 'limit load' characteristic and hence are preferred to forward curved bladed fans.

Construction of fans varies with its size. In small fans, impeller would be mounted on motor shaft. In larger fans it could be of overhung type. In very large fans it will be supported on both sides.

Entry could be single or double depending on the capacity. Bed plate could be common to hold motor and fan or separate.

Drive could be direct or through 'v - belts' or sometimes even through a gear box. Many applications in a cement plant require variable speed drive.

When fans are required to handle dust laden gases impellers and casing could be subject to wear and may have to be lined with wear resistant liners.

In such a situation fans would be subject to vibrations and may have to be mounted on vibration dampening pads.

Thus there is a great variety in constructional features of fans.

The most important consideration in selection would be the operating efficiency. Radial bladed fans have efficiencies of 45-55 %. Therefore presently mostly backward curved bladed fans would be selected as their efficiencies range between 70-80 %.

17.3 Fans in Cement Plants

In cement plants fans are used for handling:

1. *Process Gases:* In kiln, cooler, preheater, calciner and in grinding systems for drying and venting.
2. *Dedusting & Venting:* During material handling, during filling of silos, etc., air gets displaced. Displaced air carries dust with it. Bag filters are used to dedust air. Fans are used to draw air to be vented through the system.

These are more or less well defined applications and for each application fans will differ only in capacity.

For example, when plant has a 6 stage preheater, preheater fan will be required to handle gases at 270 °C and with a dust burden of 70 gms /nm^3; static pressure would be ~ 450 mm for a preheater kiln and 550 mm for a calciner kiln regardless of the capacity.

Similarly cooling air fans for grate coolers would be handling ambient air ; pressures developed would depend on the thickness of clinker and temperature of clinker depending on its location.

17.4 Dedusting

Volume of gases to be handled depends on the sizes and capacities of various machines in the system to be dedusted.

For example : volume to be vented from an elevator would depend on the size of its casing (hence capacity) and its height. For more than 20 metre height, elevator would have to be vented at top and bottom - both.

In long air slides, air is admitted in air trough at more than one point and vented also from more than one point.

Powdered materials like raw meal, coal, cement can be conveyed pneumatically, using PS pumps, Fluxo Pumps, Air lifts, Airveyor Airstream and other dense / lean phase conveying systems.

Conveying air is to be vented after material is discharged in silos, bins, etc., after passing through cyclones / bag filters.

Depending on distance to be conveyed and type of system to be used for the purpose, fans, blowers or compressors would be required to be used.

Coal is fired into kiln and calciner through fan / blower / screw pumps.

17.5 Process Gases

1. Fuel is fired in kiln and calciner resulting in combustion.
2. In Kiln and Preheater and calciner, CO_2 is dissociated.
3. In Cooler - Cooling air is introduced to cool clinker and to supply heated secondary and tertiary air to kiln and calciner for combustion. Fans are used to draw gases through these systems.

Quantum of gases depends on:

(i) Capacity,

(ii) fuel Consumption,

(iii) moisture content,

(iv) temperature to which clinker is to be cooled.

Flow of gases in these systems (kiln, preheater, calciner and bed of clinker in cooler) and in ductings causes loss of pressure which depends on:

(i) Volume,

(ii) velocity of gases,

(iii) length of travel,

(iv) dust content,

(v) density and temperature,

(vi) resistance offered by the system.

For example, greater the thickness of bed of clinker, greater the pressure drop.

Pressure drop in preheater is a function of design of cyclones and number of stages of the prehrater in addition to other factors mentioned above.

17.6 Fan Power

Fans have thus a very important part to play and power consumption due to fans is next only to grinding mills; their proper selection is therefore very important.

17.7 Capacities and Specifications of Fans

Like conveying and feeding systems, fans should never become bottlenecks in achieving production and should therefore be designed with sufficient margin both in terms of pressure and volume.

According to 'fans laws', volume drawn by a fan is inversely proportional to pressure drop or resistance in the system.

The relation between volume and pressure is shown by 'fan curves', which are characteristic of type of impeller used. A 'fan curve' is typical of a family of fans.

A series of fan curves are constructed for different speeds and damper settings for a given fan.

The operating point of a fan is the crossing of fan and system resistance curves.

See Fig. 2.17.1.

Pressure drop in a system depends on 'velocity pressure' which in turn is proportional to square of velocity and to density of gases.

Thus in the same system of ducting or cyclones, if the flow increases 10%, pressure drop would increase by (1.1)2 = 1.2 times, other things remaining same.

Therefore when providing margin on capacity i.e., volume (which is directly proportional to production rate - other things remaining same), it is also necessary to provide adequate margin in the static pressure of fan.

17.7.1 Fan Curves and Operating point

For a given fan, operating point would shift from P1 to P2 (fan operates at crossing point of fan curve and system resistance curve)

See Fig. 2.17.1.

Therefore to increase output from V1 to V2, the fan should be capable of developing static pressure P2, to be able to pass vol. V2 through the system.

Ratio (P2/P1) = (V2/V1)2

In practice this is achieved by changing the speed of the fan.

Again as per fan laws,

V is proportional to N - speed of fan

P is proportional to N^2

And power drawn is proportional to N^3

It is however advisable not to keep too much margin above the design operating point because fan efficiency is different at different operating points.

Using damper to control volume / resistance causes high loss of power and power consumption measured in units / ton increases

See Fig. 2.17.2.

It is better to control by changing speed.

See Fig. 2.17.3.

Therefore for all these reasons a margin of only about 10-15% is provided in volume over and above worst operating conditions at design capacity. Square of this margin is allowed for in the static pressure.

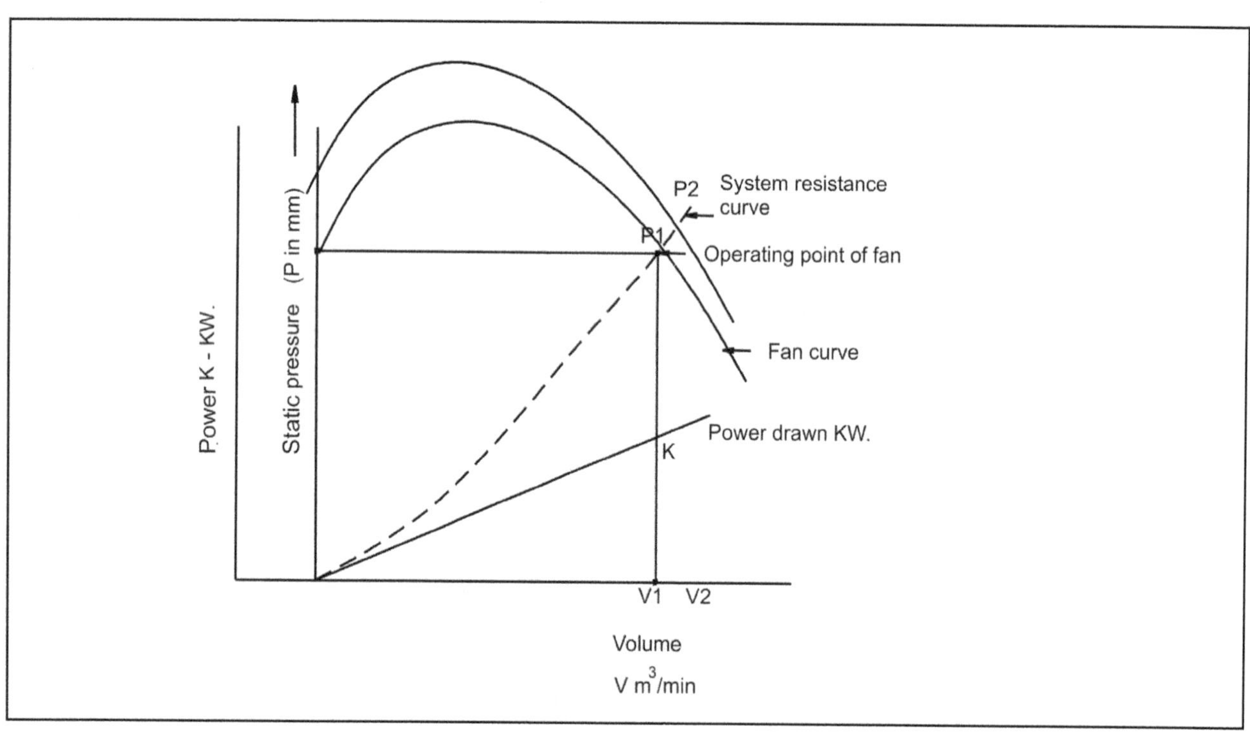

Fig. 2.17.1 Fan curve and system resistance curve.

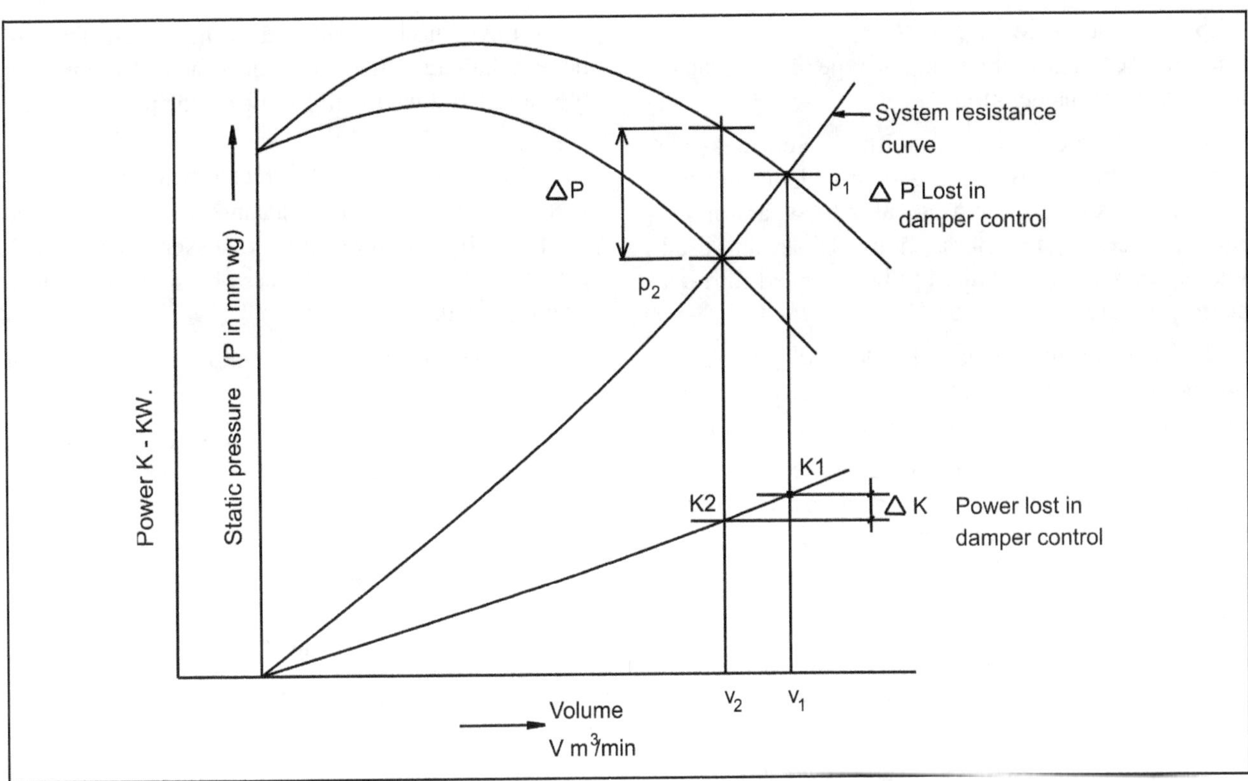

Fig. 2.17.2 Control by damper.

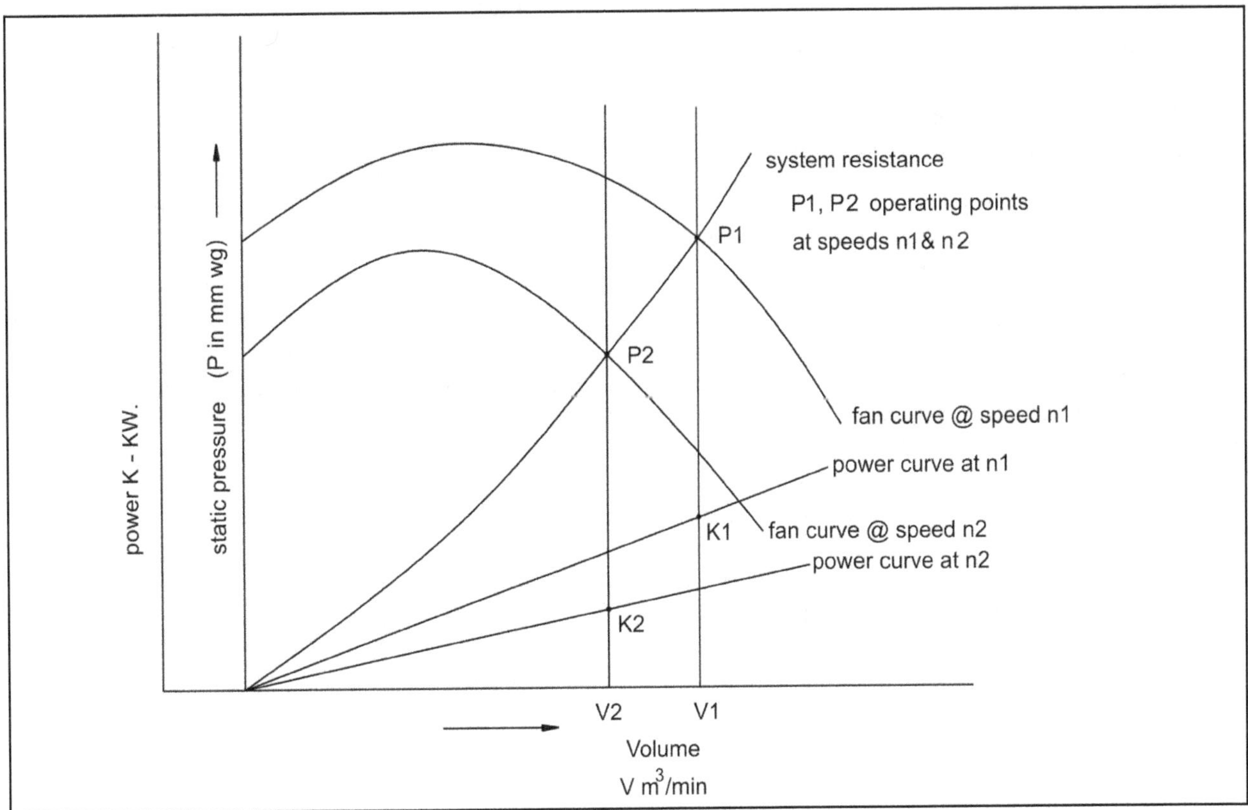

Fig. 2.17.3 Control by speed – no loss in power.

17.8 Fans in Grinding Systems

For fans used in grinding systems, type of mill used is an important consideration.

In bucket elevator mills with conventional air separators, material is discharged through the mill and is lifted by air slides and elevator to separator for further processing i.e., classification. Hence fans used with system are not required to be designed for high static pressures.

In air swept mills, ball and roller mills, the entire product is lifted out of the mill by air. Thus the pressure drop in mill, ducting and separator together is very high.

Bag filter/ esp may be installed before or after the fan. If before the fan, pressure drop across bag filter / esp has also to be allowed for.

In air swept mills it is important to maintain a minimum ratio of weight of air to weight of material.

For example in ball mills, 1.5 kg of air is required to lift 1 kg of product or 1.2 nm^3/kg. In vertical mills it is 1.6 nm^3/kg.

Moisture in the material to be ground and dried also decides volume of hot gases to be drawn in and volume of gases leaving.

When quantum of water to be dried is low it may be possible to recirculate part of exhaust gases to meet requirements of air sweeping mentioned above.

In sizing the fan therefore this situation would have to be taken into account.

Once the operating condition for which fan is to be designed is decided, margins for fixing capacity of fan could be taken as before.

17.9 High Efficiency Separators

In ball mill systems using high efficiency separators, large volumes of air are passed through air separators. Pressure drop in air separators are high as much as 200-300 mmwg. This is to be taken into account in working out static pressure of fan in the system. Pressure drop in dust collector 150-200 mm if bag type should be included when fan is after the dust filter.

In vertical mills, separators are housed in the mill itself. Pressure drop across the mill is thus inclusive of that in separator. Pressure drop across vertical mill is much less if the mill has an external circuit.

Dust load in gases leaving mill can be very high 600 gms/nm^3.

If a cyclone is introduced in the system fan will have to handle 60 gms/nm^3 dust load. If fan is after the bag filter, then it will handle clean air.

17.10 Dust Content of Gases Handled by Fans

It is important to specify dust content of gases handled by fans. It is commonly expressed as gm/nm^3. Characteristic of dust, whether abrasive should also be mentioned.

In preheater circuit, dust burden would be about 70 gms / nm^3.

In mill systems it could be between 200 to even 600 gm / nm^3 depending on whether fan is before or after the dust collector.

17.11 Cooling Air Fans for Grate Cooler

In grate coolers, cooling air is forced through bed of clinker thicker the bed, higher the pressure drop.

To draw out maximum heat from clinker, thicker beds are used.

In reciprocating grate cooler, clinker beds vary from 600 mm in 1st grate to 400 mm in 2nd Grate and 300mm in 3rd grate.

In static grates, bed thicknesses are between 800 to 1000 mm.

Fans have to be designed to force the designed quantity of cooling air through these beds progressively to achieve the desired clinker temperature at discharge.

Cooler is divided into a number of compartments so that there is not too much difference between condition at the beginning and at the end of the compartment. It also helps passing through required quantity of cooling air where it helps most. Each compartment has with own fan for this purpose

Supplier of the cooler will recommend the fans to be used.

17.12 Cooler Vent Air

Air admitted into cooler is used as secondary and tertiary air into kiln and calciner. Two together are directly related to specific fuel consumption. Balance quantity is vented. Vent air fans are sized on this basis.

Under upset conditions, kiln and cooler system is required to handle under burnt clinker at a much higher rate - generally taken as 30 % over the normal rate. It results in cooling air measured in nm^3/kg getting reduced correspondingly.

Clinker is discharged at high temperatures and vent temperatures also rise considerably. Therefore vent fan is to be designed for upset conditions. This is a special requirement of vent air fan.

By running a heat balance, temperature of clinker at outlet end and temperature of vent air can be calculated.

Vent air fan can be used to control flow of combustion air into kiln and calciner on a continuous basis. Therefore it should has a variable speed drive.

17.13 Fans for Venting

A great majority of fans are used in various departments for venting gases through dust collectors. Principle of sizing these fans are the same as outlined above.

Total quantity of air to be vented from the system is established; maximum system resistance is worked out; to it is added pressure drop in the dust collector to arrive at static pressure of fan.

Margins on capacity and pressure will be added as explained.

17.14 Primary Air Fan

This is a special application for conveying coal for firing into kiln.

These are low volume high static pressure fans. When ambient air is used as primary air, they would be handling clean air at ambient temperatures. When however, primary is drawn from cooler, it would be at about 300 °C and would carry abrasive clinker dust. In such cases a cyclone will first clean the air before it enters the fan. Even then dust burden could be as high as 750 to 1000 mgms / nm^3. Impellers of such fans would be lined with special wear resistant materials.

17.15 Dampers

Fans are normally started on no load. Dampers are used to isolate fans for this purpose. They can be:

1. Butterfly.
2. Louvre.
3. Radial.

If used only for this purpose, the loss in the damper itself is immaterial. They will be capable of remote operation.

As mentioned earlier, dampers can also be used to regulate flow in the system but this method wastes energy. **See Fig. 2.17.2.** However if a damper must remain in circuit during operation, it must be so selected that there would be minimum loss of pressure in the damper itself. From this point, louvre and radial bladed dampers are better than butterfly dampers.

Effectiveness of a damper to regulate depends on its design: Louvre dampers can be effective for a much wider range of regulation than butterfly dampers.

See Figs. 2.17.4 to 2.17.6.

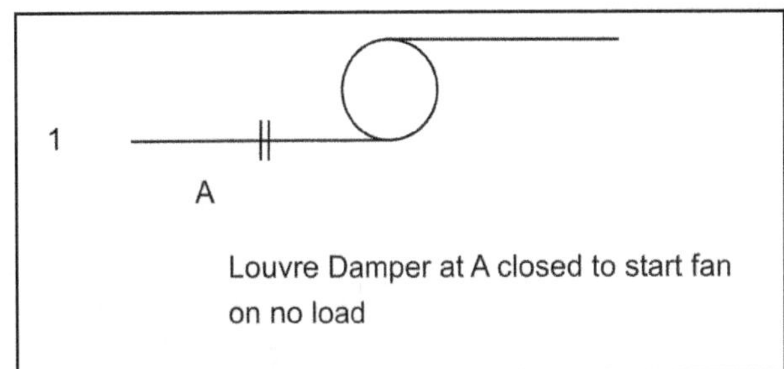

Fig. 2.17.4 Fans started on no load.

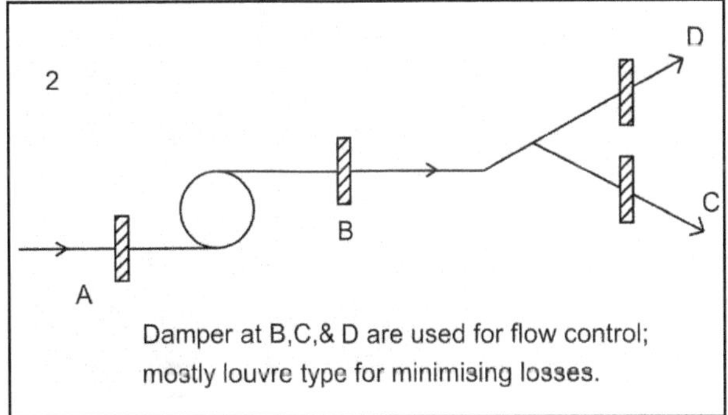

Fig. 2.17.5

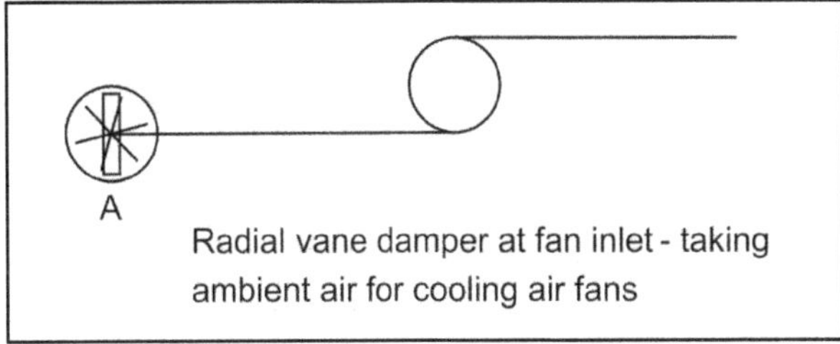

Fig. 2.17.6 Dampers for fans.

Dampers are sometimes used to isolate bag filters and electrostatic separators from the circuit. For this application dampers need to be gas tight.

See Plate 2.17.4.

17.16

It will be thus seen that fans are used in almost all departments of the cement plant.

They vary in size from a few hundred m^3/min to tens of thousands of m^3/min. Static pressure also varies from 100 mmwg for axial fans for cooling of kiln shell to + 1000 mmwg for primary air fans.

Speeds at which fans run depends on pressure that they need to develop. Fans handling hot air need to be run much faster as compared to fans handling ambient air.

Fans should be balanced statically as well as dynamically to avoid vibrations in running.

17.17 Constructional and Mounting Features and Drives

Plate 2.17.1 shows typical assembly of a fan.

Plate 2.17.2 shows fans with different orientations of inlets and outlets, mounting arrangements of fans and also different orientations of inlet and outlets to suit layouts.

Plate 2.17.3 shows different constructions and mountings of fans.

Figs. 2.17.7 and 2.17.8 show typical fan and its drive and typical arrangements of inlet and outlet ducts.

Plate 2.17.1 Fan – general arrangement.

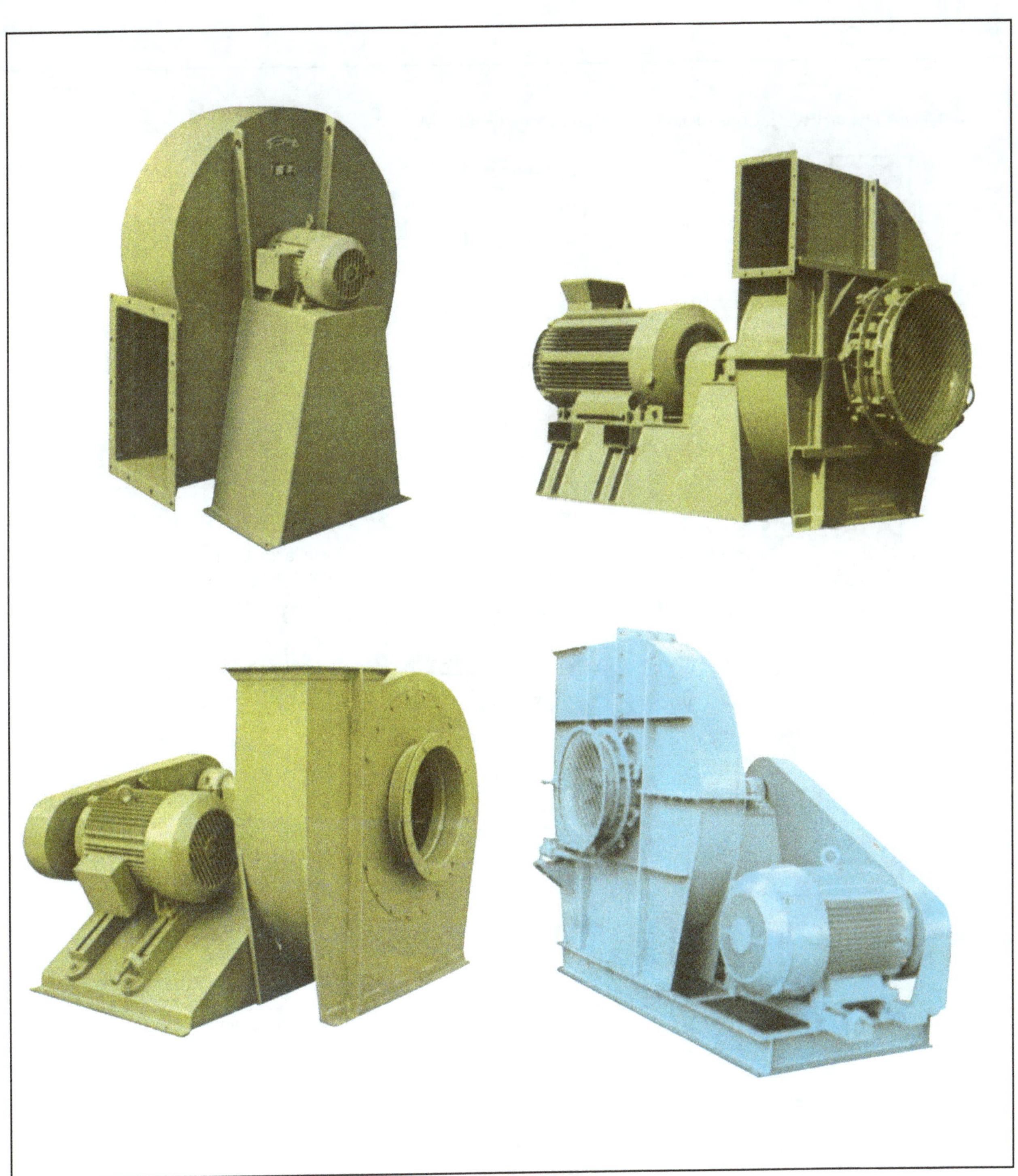

Plate 2.17.2 Fans – different orientations of inlet and outlet.

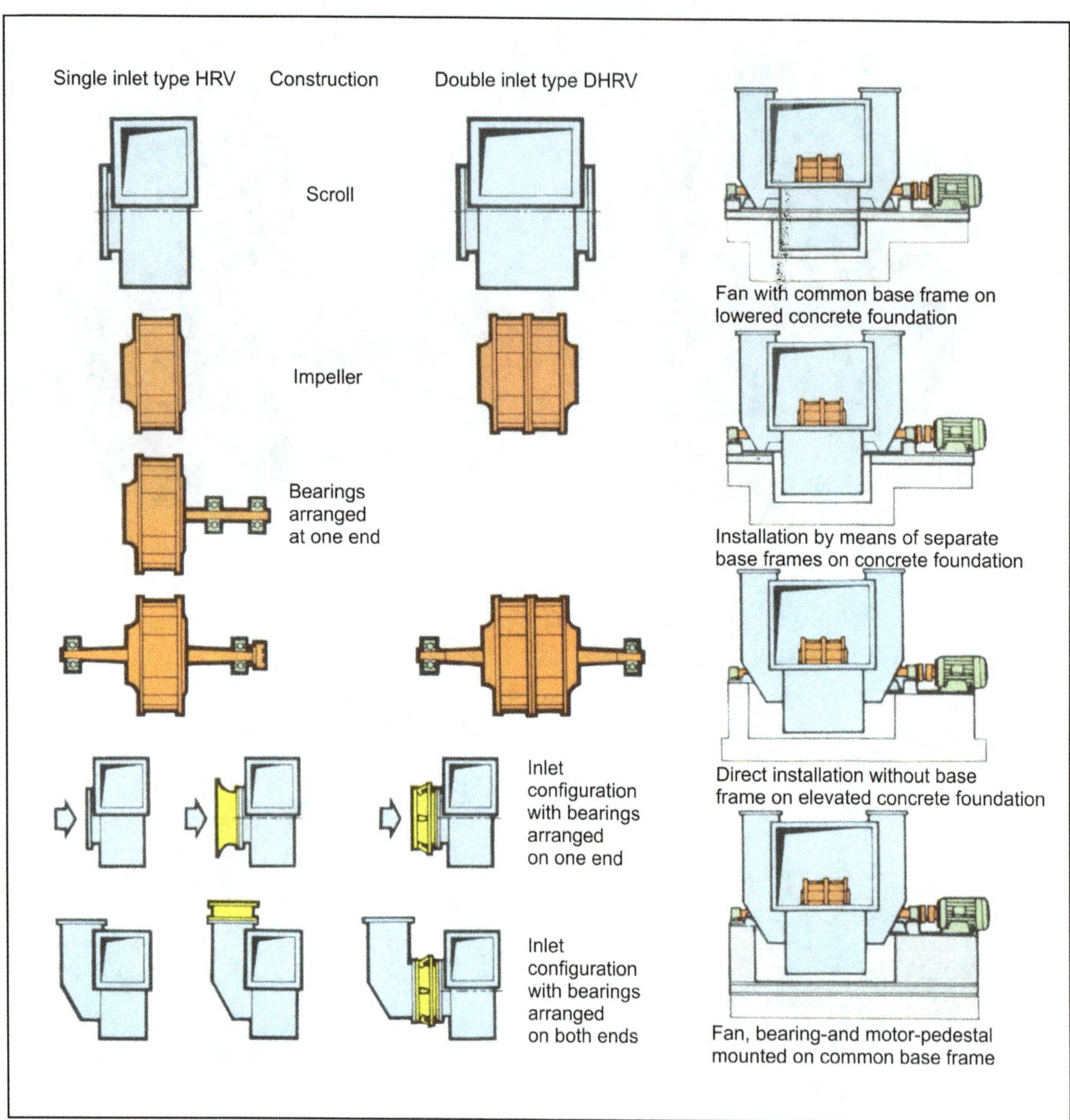

Plate 2.17.3 Fans – different constructions and mountings.

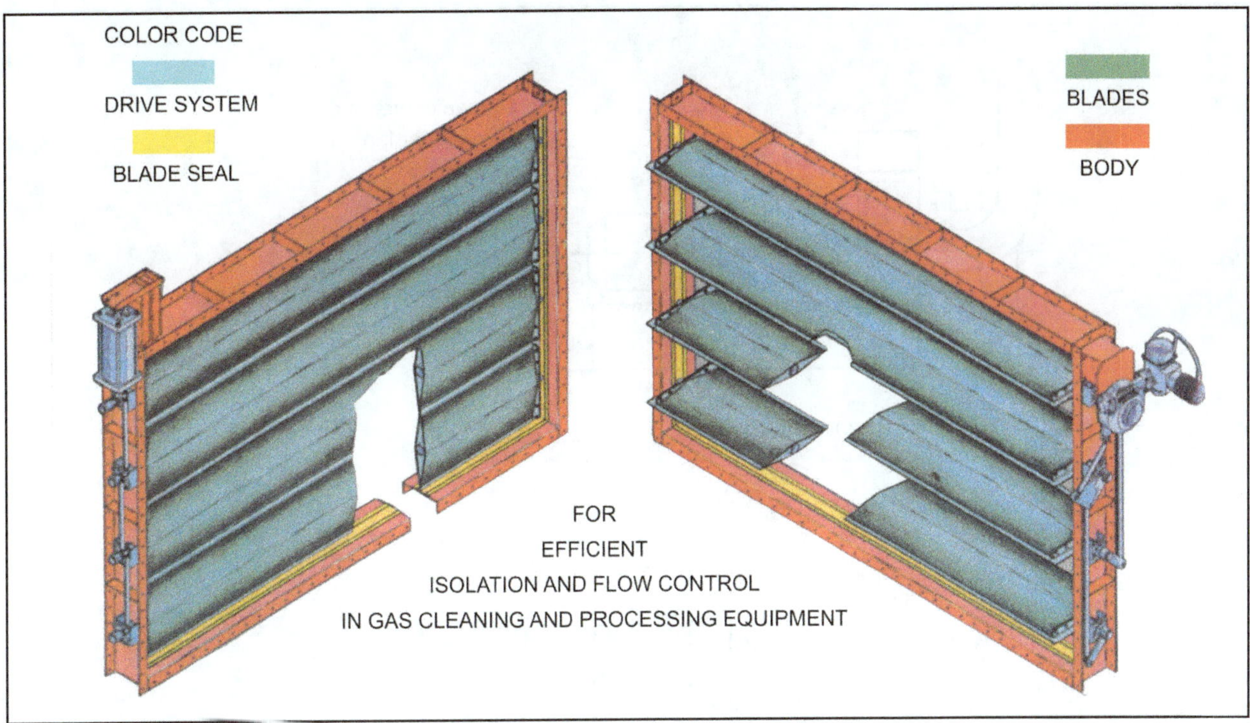

Plate 2.17.4 fGas tight isolation dampers.

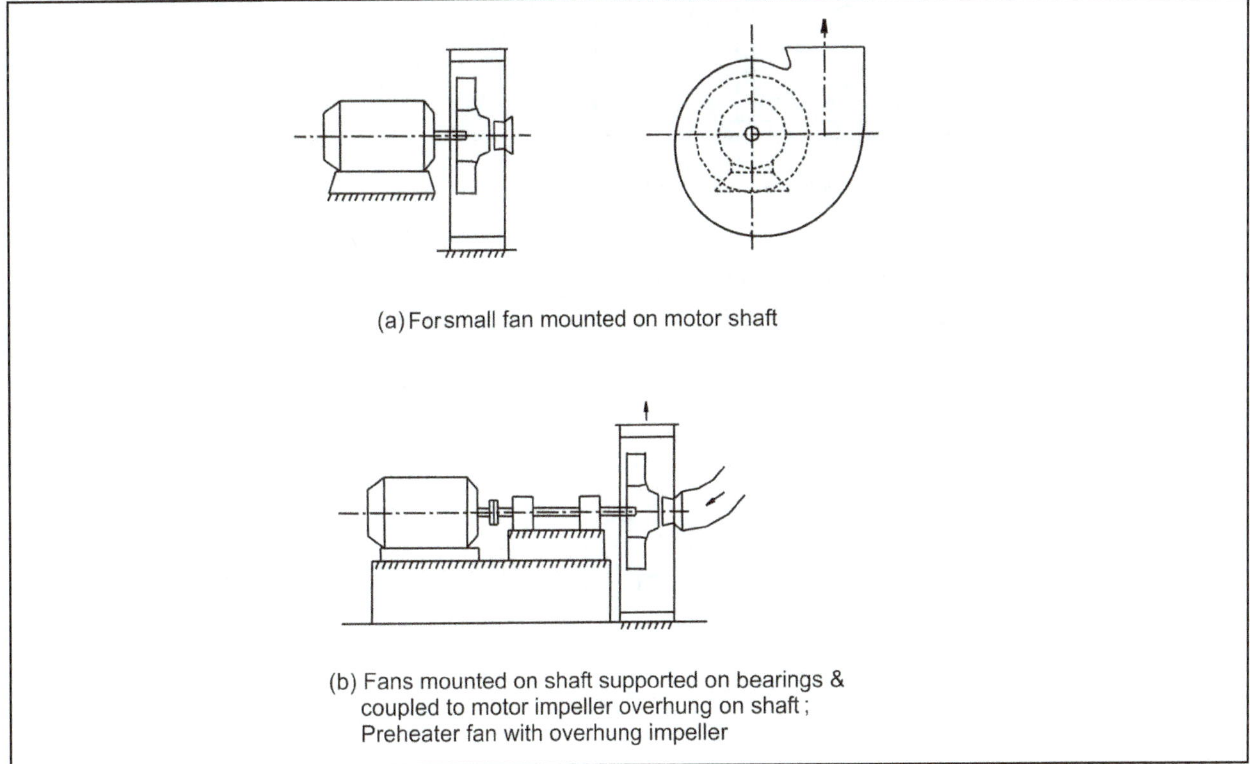

Fig. 2.17.7 Preheater fans with overhung impeller.

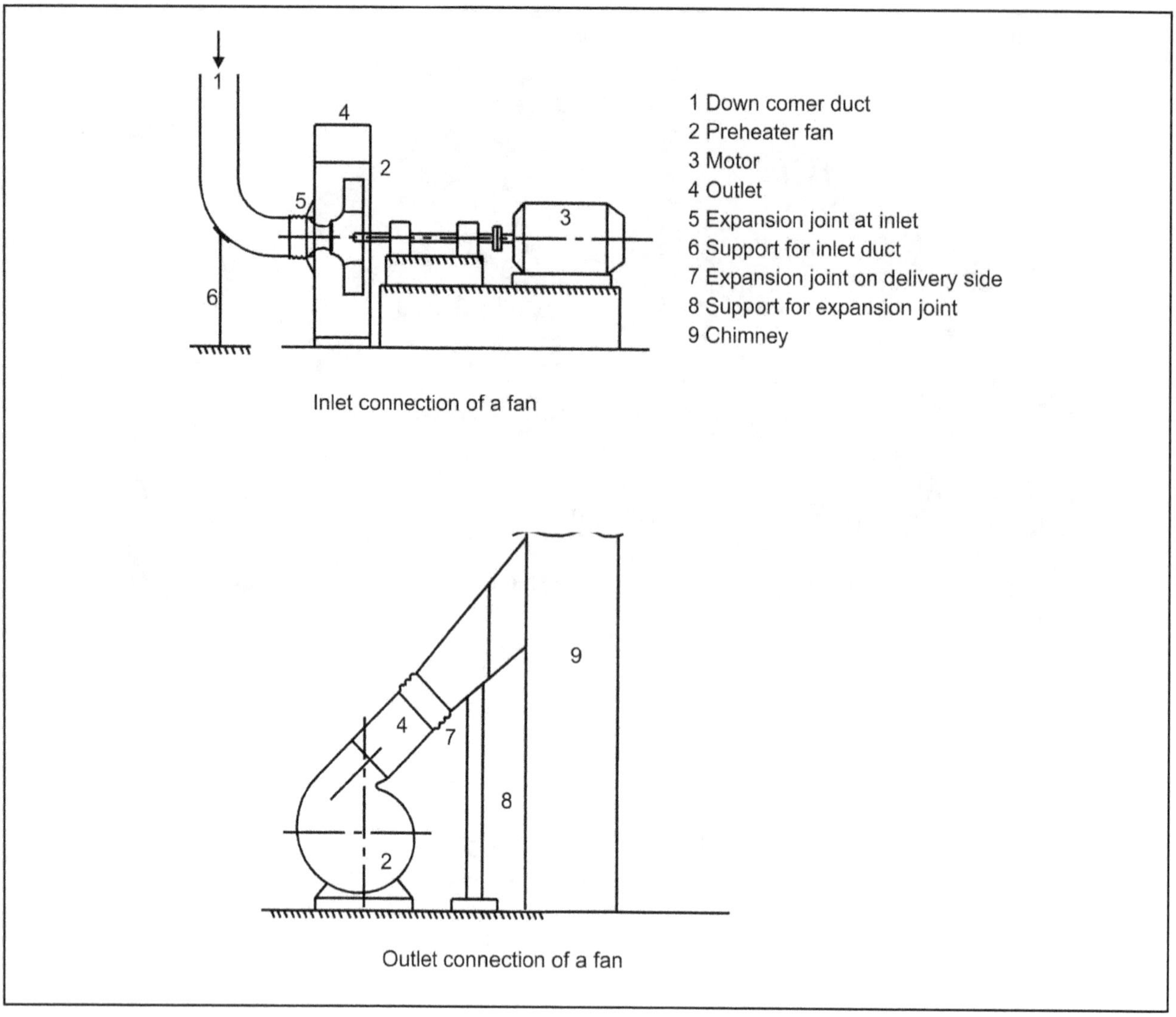

1 Down comer duct
2 Preheater fan
3 Motor
4 Outlet
5 Expansion joint at inlet
6 Support for inlet duct
7 Expansion joint on delivery side
8 Support for expansion joint
9 Chimney

Inlet connection of a fan

Outlet connection of a fan

Fig. 2.17.8 Typical arrangements of inlet and oulet ducts.

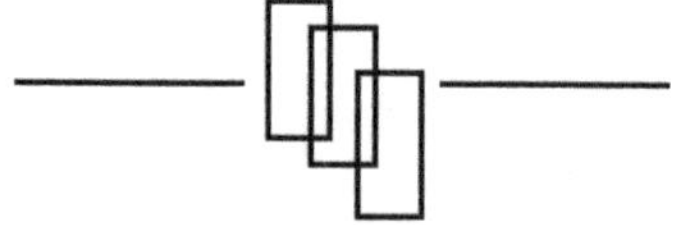

CHAPTER 18

COMPRESSORS AND BLOWERS

18.1 Compressors and Blowers in a Cement Plant

A Cement plant uses compressors and blowers for :

1. Blending and Aeration in silos / bins.
2. Pneumatic Conveying.
3. Drilling by wagon drills in quarries.

Application of drilling holes in mines for blasting has already been dealt with in Chapter 1.

See also Fig.2.18.4

18.2 Types of Compressors

Compressors can be Rotary or Reciprocating.

In rotary compressors an eccentric rotor turns in a circular casing. Compressed air is delivered at a uniform pressure.

In conventional reciprocating compressors, 1 out of 4 strokes in a cycle is a pressure stroke. Therefore pressure fluctuates. Installing an air receiver after the compressor dampens fluctuations. Reciprocating compressors are therefore not suited for applications which require a steady and uniform pressure like conveying by F.K. Pumps.

To overcome the disadvantage 'opposed piston' reciprocating compressors were developed. They had two pressures strokes in a cycle.

Reciprocating compressors are generally suited for large volumes at high pressures.

See Plate 2.18.3.

Screw compressors

Screw compressors are used mostly for portable compressors used in mines for drilling holes where pressures required to be developed are high.

See Plate 2.18.4.

18.3 Blowers

Blowers have come to mean Roots blowers with mostly two elliptically shaped Rotors rotating in an oblong casing. There are also blowers with three lobes. Roots blowers can be multi stage also to build up high pressures; but normally in single stage they would supply compressed air up to 2 kg/cm2. Roots blowers supply air free from water and oil.

See Plates 2.18.1 and 2.18.2

As a matter of fact centrifugal fans are also used as blowers in cases where air pressures do not exceed 1000 nmwg as in case of primary air fans.

Choice of compressor or blower depends on specific application.

18.4 Application of Blending and Aeration

In blending operation, material in the silo is fluidized. In a **'batch blending'** operation, differential is created in densities within the silo resulting in vertical movement of raw meal within it. It therefore requires vertical movement of raw meal within it. It therefore requires low pressure, high volume compressors for aeration and high pressure, low volume compressors for blending. Pressure of blending compressor must be high enough to force blending air through the height of raw meal.

There are many variations of basic 'quadrant' air merge blending systems. Suppliers will furnish specifications of the compressors or blowers to be provided.

Blending and aeration cycles will be automated. Timing and frequency of cycles could be variable.

Plate 2.18.1 Twin lobe roots blower.

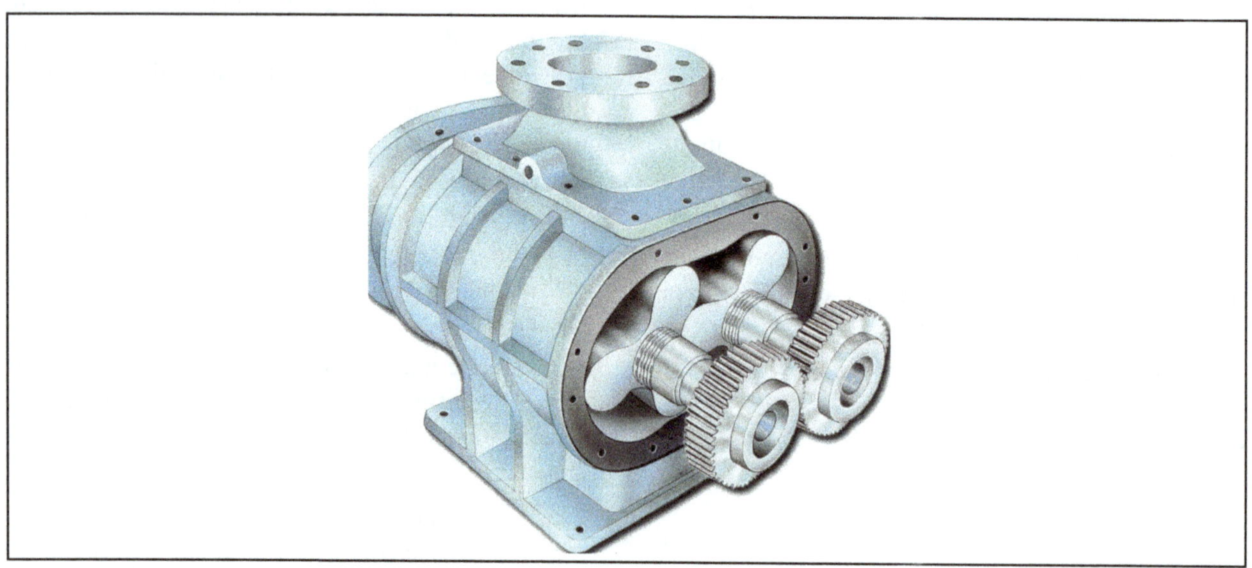

Plate 2.18.2 Three lobe roots blower.

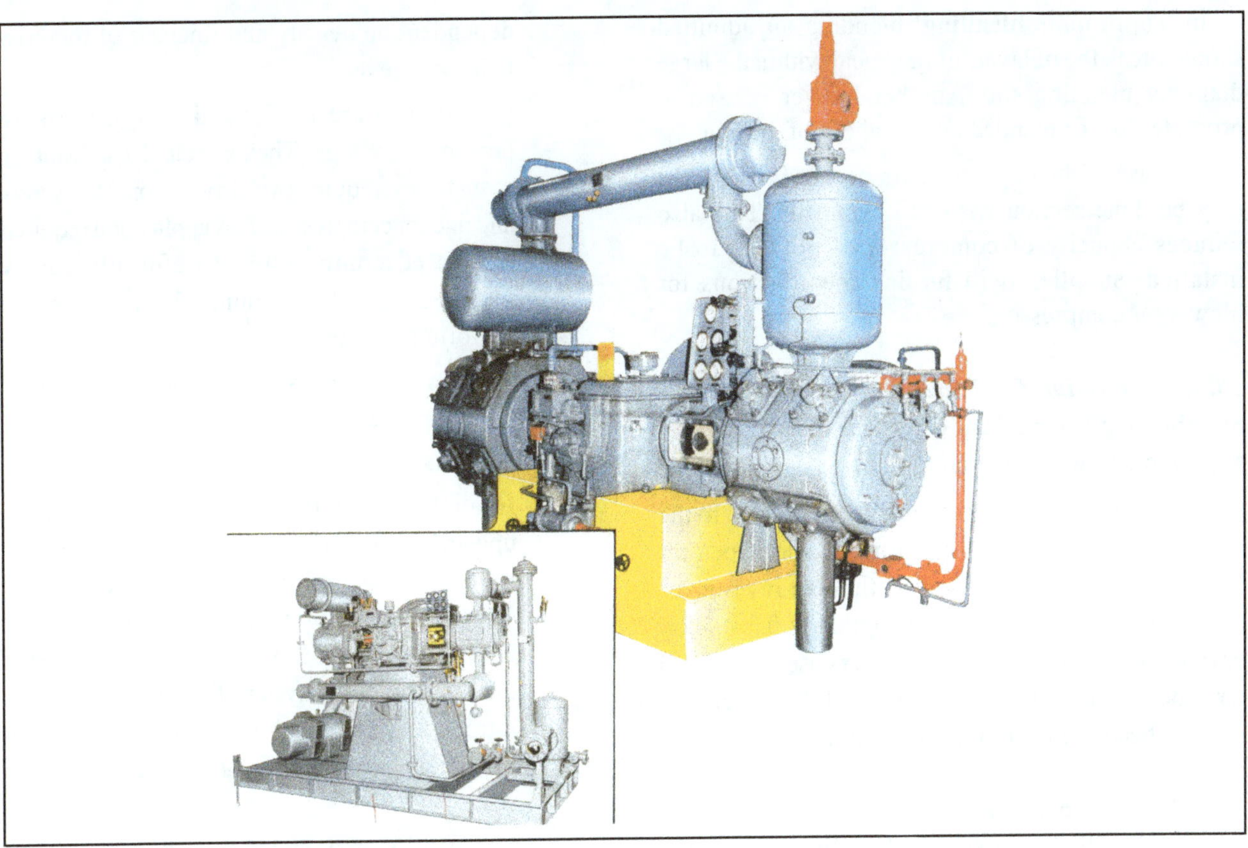

Plate 2.18.3 Opposed piston reciprocating compressor.

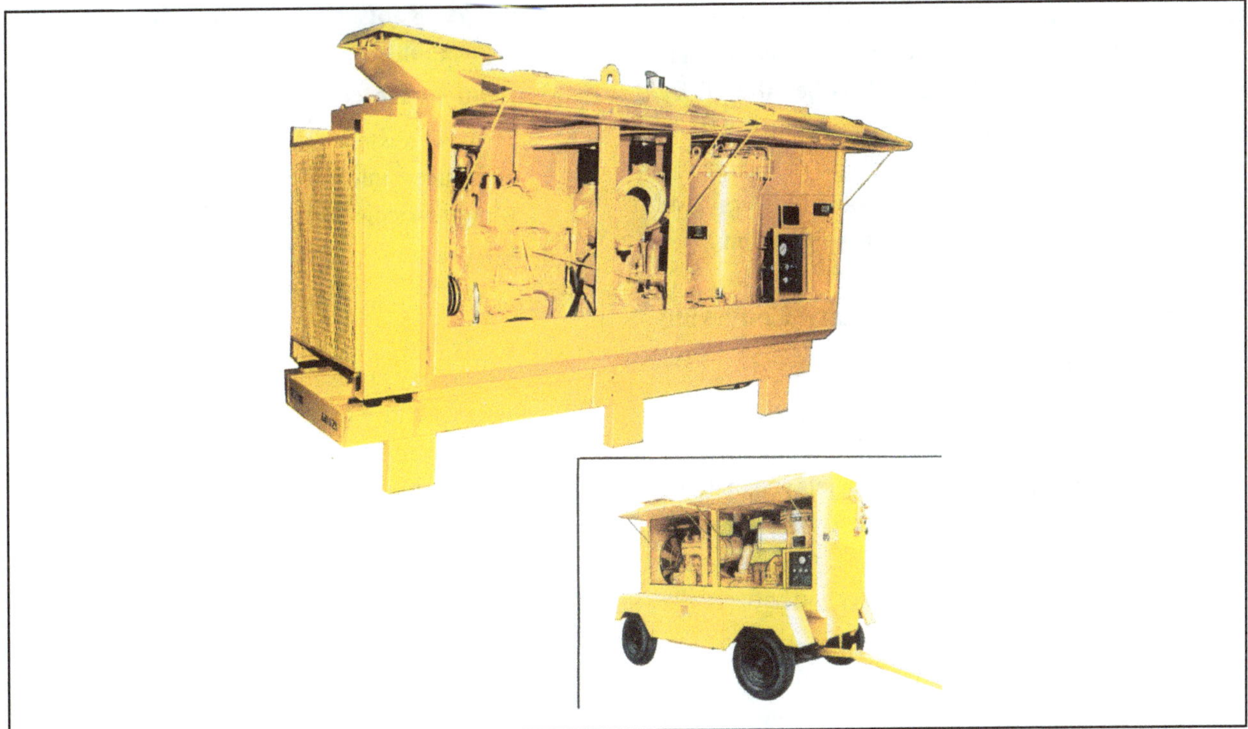

Plate 2.18.4 Screw compressor.

In **'continuous blending'** blending air admitted should break the of layers of raw meal within the large diameter blending silo. Another blower is used to promote flow of blended raw meal out of the silo.

Because of the large diameter of the silo, aeration may be done section wise and cyclically. This also reduces capacity of compressors / blowers to be installed. Suppliers will furnish specifications for blowers / compressors.

18.4.1 Aeration for Extraction

Silos holding raw meal and cement also need 'aeration' to promote flow of material.

Quantum of air admitted is directly related to the area of silo so fitted with aerating media. Open air slides in bins and silos used to fluidize raw meal / cement etc., require air to be supplied at 0.3-0.5 kg/ cm^2 pressure. Hence fan type blowers used for closed air slides are not suitable for open air slides which also support head of material in silo or bin.

18.4.2 Quality of Air

Air to be supplied to the silos for blending and aeration has to be free of oil and water. Pressures required ranged between 1 to 3 kg/cm^2. Steady pressure was desirable. Rotary compressors were suitable. But air delivered contained oil and had to be filtered by installing oil and water filters. Reciprocating compressors were not suitable because of fluctuation of pressure for this application. Roots blowers have been found to be best suited for this application.

18.5 Compressed Air for Pneumatic Conveying

1. **F.K.Pumps**

   In this system, fast running screws drop compressed slugs of material in an air chamber which receives conveying air in jets and material is conveyed through pipe line to silo, preheater etc. FK Pump systems are used to convey raw meal, coal and cement vertically and horizontally over long distances.

   Lay out of pipe line, its total length and number of bends etc., determine quantum of conveying air and its pressure. It is also dependent on density and fineness of material being conveyed.

   F.K. Pumps are low quantity high pressure pneumatic systems. Their principal disadvantage is that they require two drives - one for screw and one for compressor. This application required dedicated compressor - preferably rotary installed close to the pump delivering air at uniform pressure.

   However mostly 'opposed piston' reciprocating compressors are used.

2. **Air lifts**

   In air lifts, material is lifted only vertically upwards and hence air lift vessels are located right at the foot of the silo or preheater tower. They require almost twice the quantity of air as F.K. pumps but at lower pressures. They were the most favoured system for kiln feed i.e., for introducing raw meal into preheaters; and for lifting raw meal and cement into their respective silos.

   Quantity of air and its pressure is related to the 'lift', size of conveying pipe line among other things. Roots blowers are suitable for this application also.

3. **Lean phase conveying systems**

   In these systems material to be conveyed is introduced into the pipe line with the help of venturies or ejectors; Roots blowers are suitable.

4. **Fluxos**

   In this system usually there are two tanks. While one is being filled, the other is being emptied. This system is useful for transporting material pneumatically over long distances. Air at high pressure of 7 kg/cm^2 is required. Reciprocating compressors are suitable.

18.6 Compressors for Bag filters

Compressors are also required for cleaning dust collector bags. Pressure is 7 kg/cm^2 but quantities are very small only 1-2% of air volume handled by the bag filter. It is best to install these compressors close to the bag filters they serve.

There is a temptation to use a common large capacity compressor to supply air to a group of bag filters. There is a possibility however, that a particular bag filter would not receive the cleaning air when required in correct quantity and correct period because the air supply is already going to another dust collector. Therefore this should be avoided.

18.7 Air Blasters for Preheaters

Compressors are also used for 'air blasters' used in preheater system to clear blockages due to collections of material in cyclones and ducts. These compressors 1 working, 1 standby should preferably be installed in the preheater building and should be dedicated. Capacity of such a compressor should be arrived at in consultation with suppliers of air blasters and prheater.

Air blasters can be used on 'as and when required' basis - if blocks are occasional. Else they can be used at set intervals on a regular basis.

18.8 Housing Compressors Centrally in a 'Compressor House'

In the early days it was not difficult to do it because all mills, raw and cement used to be installed side by side. However, this is no longer the case and mills are often installed far apart.

Even then, cement plants prefer to locate compressors and blowers together under / near the blending silo which is close to both mill department and kiln feed sections and can thus serve all three departments needing compressed air.

Compressors / blowers used for cement mill and cement silos and packing are installed under the cement silo or close to it.

18.8.1 A Central Compressor Station has its Merits

(i) There is interchangeability, flexibility and continuity,

(ii) 'Standby' to be provided are reduced,

(iii) number of sizes can be reduced,

(iv) maintenance is centralized,

(v) elaborate filtration of air can be provided for the total group.

If decided to go in for central compressor station, allowance has to be made for loss in pressure in the pipe line. Pipe line also acts as a receiver and hence allowance has to be made for capacity also.

Another advantage of installing compressors in a separate location is that they could be protected better against dust nuisance and hence filtering requirements would not be very stringent.

The performance and availability of a compressor or a blower would very much depend on the care taken to filter in take air. Dry washable generously sized filters are preferred. If the compressors are located in a room, the doors and windows of the room are also fitted with filter panels. These are regularly cleaned.

See Fig. 2.18.1 and 2.18.2.

Thus air entering the compressor filter itself is first filtered. This simple precaution pays handsome dividends.

When compressors and blowers are grouped together, the number of standby can be reduced and supervision of maintenance and operation is easier.

18.9 Specifications of Compressor and Blowers

Compressors / blowers would be procured according to Vendor's specifications. Vendor should indicate the distance of compressor from point of use assumed by him in arriving at the specifications.

When compressors / blowers need water cooling, water pipe line should be arranged in a header format to deliver water to compressors/ blowers, to after coolers, etc., and then to return circuit to cooling pond.

See Fig. 2.18.1.

When air is required to be cleaned of oil and water, filters should be installed near the point of use also.

See Fig. 2.18.3.

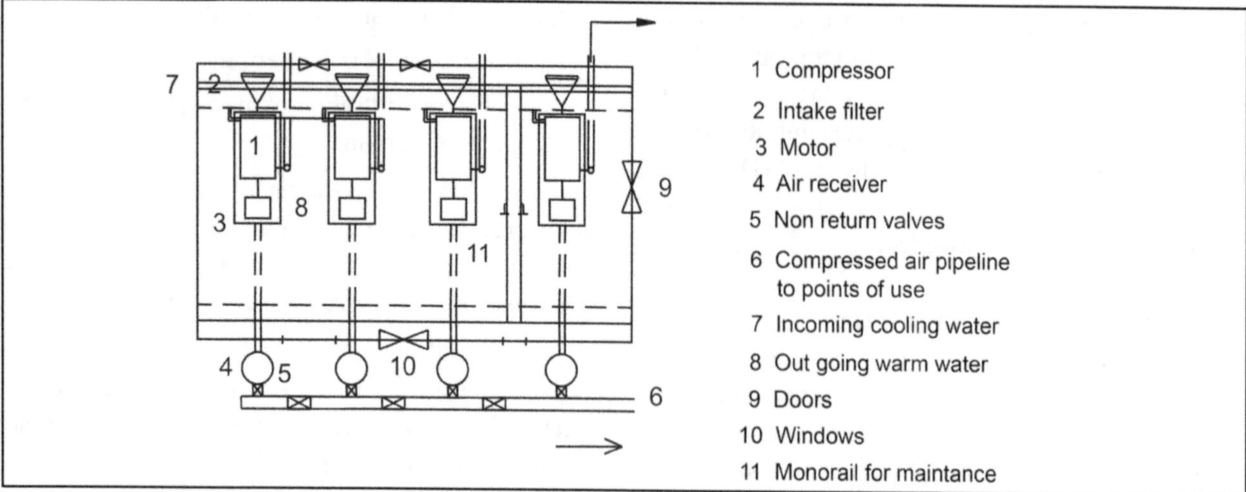

Fig. 2.18.1 Layout of compressor house.

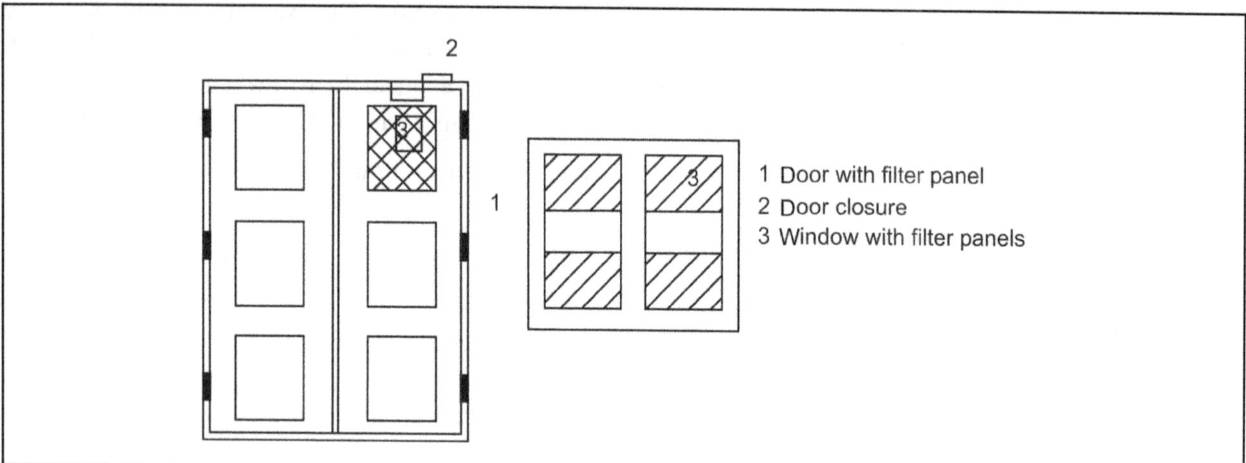

Fig. 2.18.2 Filter panels on doors and windows to keep dust out.

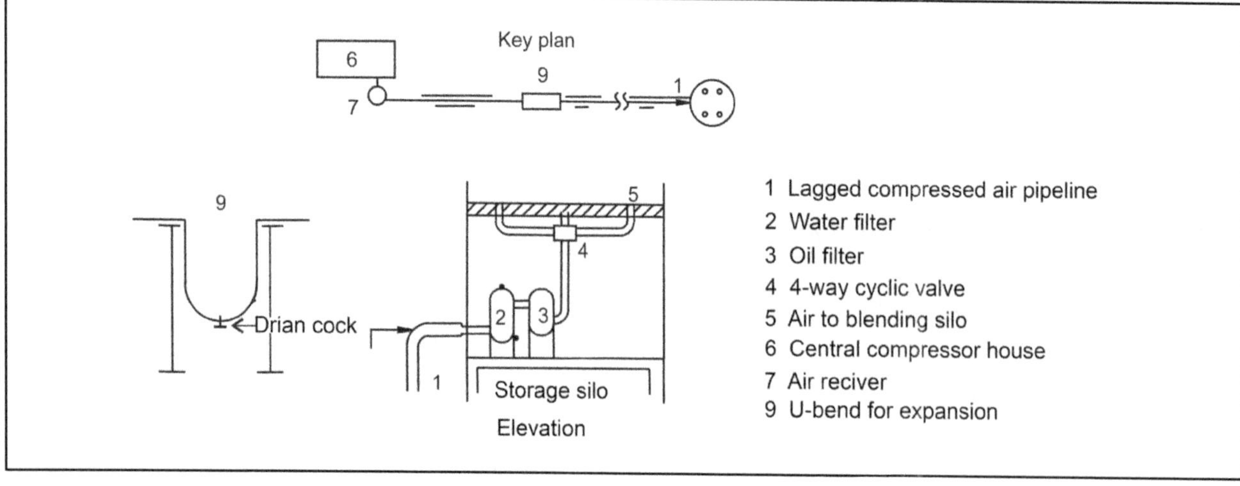

Fig. 2.18.3 Cleaning compressd air of oil and water at point of use.

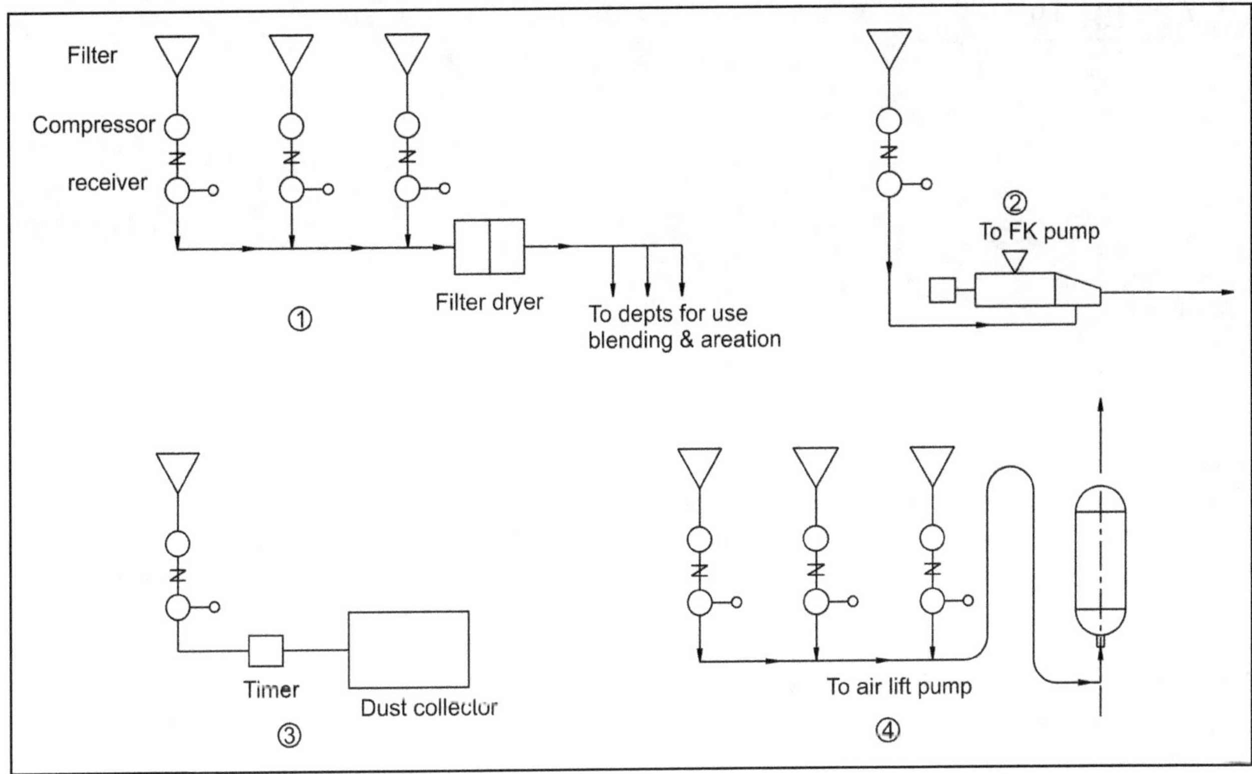

Fig. 2.18.4 Compressd air for different applications.

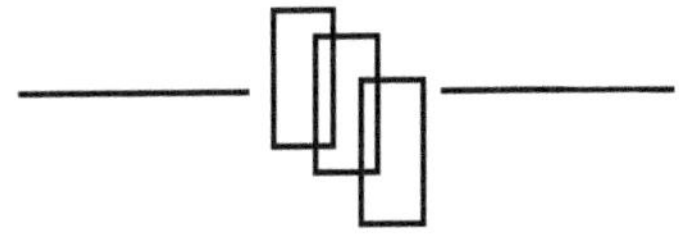

CHAPTER 19

MOTORS

19.1

Electric Motors drive the machinery in cement plants. Sizes range from fractional kws to several thousand kws.

19.2

Motors used can be classified in several ways.

1. H.T. or L.T. according to voltage.
2. Fixed or Variable speeds.
3. Unidirectional or reversible.
4. A.C. or D.C. (alternating or direct current).
5. According to type like slip ring, squirrel cage, synchronous, or synchronous induction.

19.3

Most common are A.C. motors slip ring and squirrel cage. Beyond certain ratings (say 200 kw and above) H.T. motors are used. Though insulation is costlier, currents are small. H.T. Voltage to be selected is a matter of choice though limited. Commonly used voltages are 3.3, 6.6 and 11 KV.

19.4 Voltage

Voltage to be selected in a way depends on incoming voltage of transmission.

Incoming voltage	Voltages of motors	
KV	H.T. (KV)	L.T (V)
132	11/6.6/3.3	415
33	11/6.6/3.3	415
11	11/6.6//3.3	415

Generally only two steps in stepping down would be kept.

1. From incoming to H.T. transmission within the plant.This would then also be voltage for H.T. motors.
2. From H.T. transmission to L.T. i.e., 415 /440 volts.

19.4.1

If H.T. transmission selected is 6.6 kv, then H.T. motors would also be 6.6 kv.

19.4.2

For small plants, motors could be 3.3 or 6.6 kv. For large plants they would be 11 or 6.6 kv.

19.5 Frequency

Frequency of generation decides 'synchronous speeds'. In India frequency is 50 cycles per second; in America it is 60 cycles.

Therefore synchronous speeds are

No. of poles	50 cycles	60 cycles
	<—— r.p.m.	——>
2	3000	3600
4	1500	1800
6	1000	1200
8	750	900

19.5.1

Synchronous motors run at synchronous speeds. Induction motors run at slightly lower speeds. Difference is known as 'slip'. It is between 3 to 5 %.

Therefore actual running speeds would be:

Synchronous speed rpm	Actual running speed rpm
3000	2950
1500	1440
1000	960
750	730

19.6 A.C. or D.C

D.C. motors would be used when it is necessary to change speeds. There are several advantages like:

1. Speeds can be changed smoothly and not in steps,
2. they can be changed over almost the entire range,
3. there is little loss of energy.

19.6.1

Therefore d.c. motors would be used for drives of :

(i) Apron feeders,

(ii) belt feeders,

(iii) weigh feeders and prefeeders,

(iv) kilns,

(v) grate coolers,

(vi) fans.

Ratings would range between fractional kws to several hundred kws.

19.6.2

One disadvantage of using d.c. motors is that it is necessary to convert incoming a.c. into d.c. by installing converters.

19.6.3 Variable speed A.C. motors.

It is also possible to change speeds of a.c. motors by:

(i) Using rotor resistance starters for slip ring motors.

(ii) changing frequency of supply by frequency converters.

(iii) using slip recovery system which uses difference between synchronous speed and actual running speed to change the speed.

This system was profitable when a machine was required to run much below synchronous speed for long periods.

19.7

Most commonly used motors are a.c. induction motors. There are two principal types :

1. squirrel cage,
2. slip ring.

Squirrel cage motors have high starting torque; starting current is 4 to 5 times full load current. Therefore these motors are used up to ratings of 100-150 kw.

Two types of starters are used :

(i) direct on line,

(ii) star delta.

Direct on line starters have a high starting current and therefore beyond certain ratings star delta starters are used.

19.7.1

Slip ring motors use rotor resistance starters in which full speed is reached by cutting out resistance in rotor circuit in a number of steps to limit the starting current. Number of steps can be as many as 10 to 15 also. A small drive can cutout this resistance so that motors can be started remotely.

19.7.2

When this resistance is left in circuit motor runs at correspondingly lower speed. Thus a stepped speed variation is available. It is generally limited to reducing speed by one third or maximum by half. Further resistance remaining in circuit causes loss in energy and hence this is a wasteful means of changing speed and is seldom used now.

19.8 Synchronous Induction Motors

Synchronous motors cannot be started on their own. They are started as induction motors and then run as synchronous motors when speed is close to synchronous speed. They are commonly used to drive grinding mills. Principal advantage of synchronous motors is that they can be used to correct the power factor of the entire plant. **They can run on leading power factor and thus correct lagging power factor of the rest of the electric load.**

19.8.1

Now capacitors are used to correct power factors. Hence induction motors are also used to drive mills.

19.9

Motors are made to various National and International Standards. These must be referred to in selecting motors. Insulation used has a direct bearing on permissible rise in temperature of windings. Ambient temperatures also play an important part in selection of motors. The three are in a way related. Hence it is good to specify the desired values at the time of inviting offers.

For cement plant application, generally speaking, motors of Totally Enclosed Fan Cooled (TEFC) construction, Class F Insulation and IP 54/IP55 Protection are preferred.

Motors are now made in standard International Frame sizes. This is an advantage as a frame size furnishes all pertinent details of dimensions of a motor.

19.10 Starting Torque and Moment of Inertia

Different machines require different starting torques. For example, a ball mill has a load of grinding media. It is to be started from rest by overcoming this unbalanced dead load. Similarly a kiln has also to start on load. Therefore it is necessary to ascertain the starting torque required for a particular machine and also Moment of Inertia it must overcome from makers of the machinery. Motors must be selected on this basis. This precaution should also be taken when specifying motors for fans.

19.11 Standardisation

In a cement plant many motors of similar ratings and speeds are required. This is so particularly for material handling systems. It pays to look into the possibility of 'standardizing' on ratings and frame sizes to reduce inventory of spare motors to be maintained.

Even large mill motors which may have ratings between 1000 to 5000 kw or more, can be standardized so that only one motor can be kept as a standby for two or more mill motors.

19.12

Specifications of motors are also standardized. In selecting motors it should be seen that they conform to pertinent standards. For example IS 325 of Bureau of Indian Standards (BIS) is for Induction motors. BIS have published Standards for other aspects like for testing of motors; for determining their efficiency; for improving their power factor.

19.12.1

Relevant Standards should be studied to serve as guidelines in formulating enquiries for procurement. It saves time all round and ensures quality if Suppliers follow prevalent National or International Standards.

19.13

Efficiency of motor is not the same at all loads. Therefore over sizing is not good. Further power factor also reduces at lower operating loads. From this point also motor should not be oversized. It is common to have a margin of 10 to 20 % in motor over machine brake horse power.

19.13.1

A 'Star Rating' System is now practiced. Number of stars are indicative of efficiency.This should be used in selecting machibery motors etc.

Except for applications like winding drums of cranes and hoists, most drives are unidirectional. If specified accordingly there is a small gain in efficiency.

19.14

See web page no. W1-97 in Web Page 1of Section 9 for more information on protection of motors and factors to be considered in selection of motors.

19.15

Availability of motors for 10000 tpd capacity cement plants 10000 TPD Capacity cement plants would require motors for raw and cement mills in the range of 15000-16000 kw depending on hardness of limestone and type of cement to be produced.

Synchronous Induction motors are preferred for mills because they can correct power factor for the total plant. It may happen motors in this range are not available.

For Large Ball mills dual drives can be used. Multidrive Vertical Roller mills are avilable for large capacity. Another option would be to use two mills.

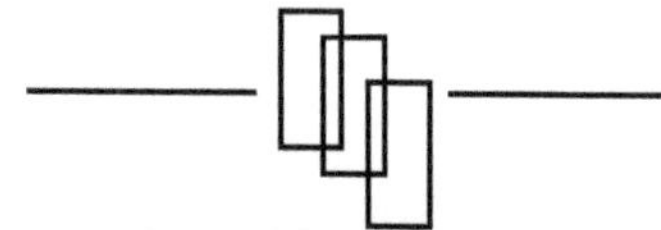

CHAPTER 20

GEARBOXES

20.0 Gearboxes

Gearboxes are connecting links between motors and driven machines. Their function is to reduce speed of motor to speed of machine to be driven.

Often gearbox will be coupled straight to the machine. In some cases though speed is reduced further by a 'chain and sprocket' between gearbox and machine like in case of grate coolers; or an open gear train is introduced between gearbox and machine like girth gear and pinion in case of mills and kilns.

The total ratio of reduction in speeds of motor and machine can be achieved in a number of ways. For example, let machine speed be 20 rpm.

It can be achieved by:

1. motor of 1500 rpm and a gear box with a reduction ratio of 75:1,
2. motor of 1000 rpm and a gear box with a reduction ratio of 50:1.

Gearbox of ratio 75: 1 will be costlier than gearbox of ratio 50:1.

But a 1500 rpm motor will be cheaper than a 1000 rpm motor.

A balance is to be achieved to minimize total cost of drive.

20.1 Open Gear Train or Chain Drive

An open gear train or a chain drive will be cheaper than a gearbox.

For example a drive consisting of a gearbox of 1000/120 rpm ratio and an open gear train of 120/20 rpm would be cheaper than a gearbox of 1000/20 rpm.

In taking a decision increase in maintenance costs due to open gear train would also have to be taken into account.

20.2 Selecting Gearboxes

Gearboxes like motors come in standard sizes. As torque reduces with increase in speed to transmit same power, a given gearbox will have different ratings for different input speeds. For example

input speed rpm	ratio	output speed rpm	rating kw
1000	50:1	20	200
1500	50:1	30	300

This is useful in changing drive of kiln when introducing a calciner to increase its capacity.

Helical gearboxes are selected for better efficiency and for drives of high ratings. Reduction ratio can be increased by increasing number of stages.

See Plate 2.20.2.

Worm reduction gearboxes are used where efficiency is not critical and for drives of small ratings and high reduction ratios like for drives of belt conveyors and screw conveyors etc.

See Plate 2.20.1.

20.3

Mostly both input and output shafts are horizontal. They are also parallel for spur and helical gearboxes. In worm gearboxes they are at right angles **Plates 2.20.2 and 2.20.1**

Plate 2.20.1 Worm reduction gearbox.

Plate 2.20.2 Helical gearbox.

20.4 Gearboxes for Vertical Mills

In case of Vertical Mills though, input shaft is horizontal and output shaft is vertical. It requires a bevel gear train. Moreover in vertical mills table is mounted directly on the gearbox requiring gearbox to withstand high thrust loads. Gearboxes for v. r. mills are therefore specially designed.

Very large Vertical mills now have **'multidrive'** that is a number of sets of motor and gearbox supply the (Total power like dual drive in ball mills). In this arrangement, bevel gear train is eliminated and gear box does not have to take the load of the mill. **See Plate 2.4b.4 in Chapter 4b of this section.**

20.5 Couplings

Couplings between motor and gearbox and gearbox and machine are designed to be able to take a certain amount of misalignment of machine. They can be standard flexible couplings with rubber bushes in most cases. For drives of large ratings like mill and kiln drives, gear couplings are used.

It is also common to use a 'torsion shaft' between machine and output shaft of gearbox to take not only misalignment but to serve also as a safety device.

20.6 Inching Drives

Machines like ball mills and kilns need an 'inching drive' to turn mill slowly and stop it in a particular position to load or unload grinding media and for laying refractory. Inching drive is fitted on the input shaft of the gearbox, which is extended on the other side for this purpose.

Kiln gearbox is installed at an angle corresponding to slope of kiln.

Gearbox is selected according to its application. Nominal power required is found from catalogues to suit, type of machine to be driven, number of hours in use, number of starts per hour etc. Thermal capacity of the gearbox and starting torque it has to withstand are also ascertained.

20.7 Lubrication and Cooling

Some gearboxes would need external 'forced lubrication system' for cooling oil in the gearbox. Otherwise a larger gearbox would be required. For large gearboxes this should be insisted upon.

Some would need a bedplate. Smaller gearboxes would have a common bedplate with motor. Large gearboxes may or may not have bedplates.

It is useful to take help of Suppliers in selecting gearboxes.

20.8

Gearboxes also come in standard sizes like motors and conform to respective National and International standards. It is therefore useful to be familiar with the standards in formulating specifications and to see to it that gearboxes supplied conform to them.

See **web page W1.98 WEB PAGE 1 in Section 9** about selection of gear boxes.

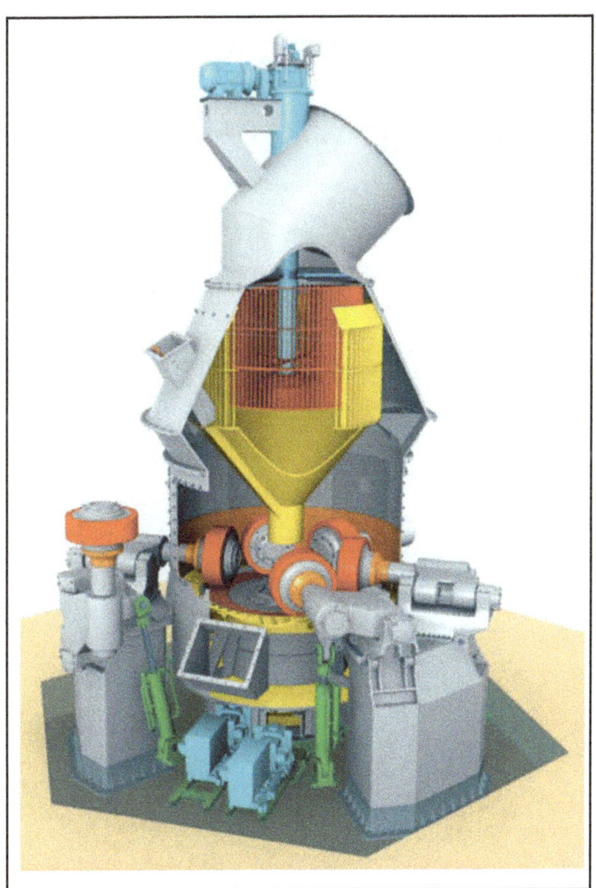

Plate 2.20.3 GBR Pfeiffer MVR Vertical Roller Mill.

Source: Pfeiffer Brochure.

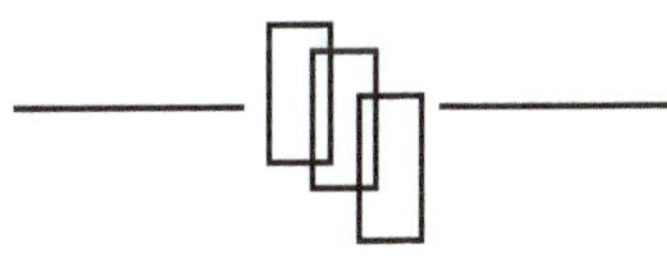

SECTION - 3

Technoeconomic Feasibility Studies

CONTENTS

Chapter 1	Investment Decision	1-2
Chapter 2	Setting up a Cement Plant Techno-Economic Feasibility Studies	3-14
Chapter 3	What should the T.E.F.S Contain	15-16
Chapter 4	Demand Forecast and Market	17-20
Chapter 5	Limestone Deposits	21-26
Chapter 6	Environment	27-31
Chapter 7	Infrastructural Facilities	32-37
Chapter 8	Manpower Requirements	38-44
Chapter 9	Implementation Schedule	45-52
Chapter 10	Erection	53-56
Chapter 11	Commissioning	57-61
Chapter 12	Capital Costs of the Project	62-75
Chapter 13	Costs of Production	76-85
Chapter 14	Profitability and Investment Analysis	86-89
Chapter 15	Executive Summary and Conclusion	90-90
Chapter 16	Factors Governing the Size of a Cement Plant	91-93

CHAPTER 1

INVESTMENT DECISION

1.1 Preliminary Investigations

Shrewd entrepreneurs investigate several options available / open to them for investment. These options are evaluated and compared against yardsticks like:

1. quantum of investment,
2. gestation period,
3. rate of return,
4. continuity of future prospects.

The foremost factors that should be investigated would be :

(i) Prospects - present and future.

(ii) 'Demand and supply' situation and size of market.

1.2 Prospects

Prospects are important. An industry may look prosperous and profit-making today, but if substitutes have already made their appearance then the prospects could diminish or even vanish altogether. In such a case it would not be prudent to venture to invest in such an Industry.

The industry must have good prospects till foreseeable future.

1.3 Market

Size of the market is also an important consideration for investment. Size of the market and its rate of growth and location are important factors.

The share of the market that the new entrant can hope to have would depend on the difference between 'demand and supply' and the competition.

It is therefore necessary to know the movement of demand and supply for the previous ten years and projections for the next ten years also.

1.4 Price

The product may come under 'free market' conditions or under 'controlled or regulated' price conditions.

Gap between demand and supply would ensure a good price. Surplus supply invites cut throat competition and affects prices and profitability.

The price structure consists of elements of :

1. Cost of Sales.
2. Cost of production including costs of material, fuel, power, packing, salaries & wages.
3. Taxes and Duties to be paid to State & Central Governments.
4. Cost of freight.
5. Overheads.
6. Profits

This structure should be known in great detail so as to anticipate movement in costs of productions.

1.5 Profits

Profit is the difference between net sales realization and costs of production.

Market factors govern the sales price and are generally beyond the control of the Entrepreneur.

1.6 To Earn Profits - He Has To

1. Choose unsaturated segments of the market.
2. Locate the unit conveniently in relation to market and raw materials, power, fuel, etc.

3. Select 'State of the Art' process of manufacture and plant and machinery that would ensure lowest cost of production and highest returns.

1.7 Viability

Over a period of time, the size of the unit of production that would become viable increases with increase in costs of production.

The viability of the project being the most important consideration, a new entrant has to set up a much larger unit requiring higher quantum of investment as compared to say that was setup 5 years back.

Therefore, only those entrepreneurs who have their own funds or who can raise funds by borrowing or by public issue can take up new projects. Others would seek opportunities in other fields.

1.8 Investment Decision

The prospective entrepreneurs must therefore obtain factual information on:

1. Market - its size and location.
2. Demand and its rate of growth.
3. Supply and its rate of growth.
4. Price structure prevailing and its trend.
5. Special consideration for locating the plant, with respect to market like availability of raw materials, power, fuel, etc.
6. The profitability of units, already manufacturing the product.
7. The size of viable unit and funds required to set it up.
8. Investment yardsticks in relation to capacity.
9. Sources of financing and costs of raising capital and gestation period.
10. Process and technical know-how and availability of machinery and other items.
11. Consultancy services available in various fields required for taking up the project.

1.9 Sources of Information

Possible resources for collecting the information with special reference to Cement Industry would be:

1. Cement Manufacturers Association.
2. Government agencies like Ministry of Industry
3. Reputed Consultants in Cement field
4. Financial Institutions like, IDBI, IFCI, ICICI, etc.
5. Reviews of Cement Industry brought out by various agencies like: Hindu,

   Cement Industry Review.
6. Report of the Development Council for Cement Industry (1994).

   (This report contains a detailed bibliography on various aspects related to Cement Industry).
7. VDZ Verein Deutscher Zementwerke Germany
8. PCA Portland Cement Association USA
9. Ntional Council for Cement and Building Materials

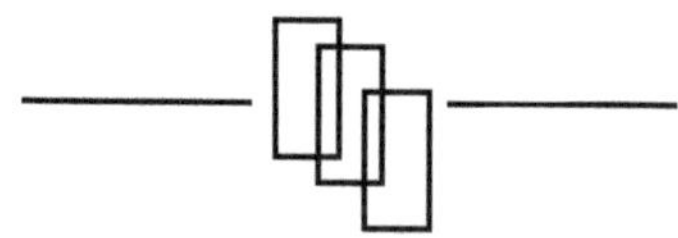

CHAPTER 2

SETTING UP A CEMENT PLANT TECHNO-ECONOMIC FEASIBILITY STUDIES

2.1 First Steps in Setting up a Cement Plant

Having taken a decision in principle to invest in Cement Industry and to install a cement plant, the first steps to be taken are:

1. Identify likely sites for setting up of the plant with respect to market and limestone deposits of cement grades.
2. Take up Techno Economc Feasibility Studies (TEFS) to establish viability.

Though specific reference has been made to India in the following paragraphs, it is by way of explaining the procedures and steps to be taken. Each Country will have Institutions and Organisations similar to those mentioned. Assistance should be taken from such Institutions.

New Entrants should engage experienced Consultants in the field. Cement Companies wishing to expand should also take up TEFS. They may be able to do so with their inhouse strengths.

2.1.1 Identify Possible Sites

It is therefore necessary to identify sites that have sufficient limestone deposits of cement grade that would sustain the cement plant of viable size for a minimum of 30 years- possibly - 40 years.

2.1.2 Preliminary Investigation of Deposits - Availability of Deposits

Preliminary investigations are needed to be undertaken at this stage, with the help of Geologists, to locate such deposits as are still available for exploitation.

Assistance of Central and State Geological and Mining Departments of a country can be taken in the search of such deposits.

Speaking about India by way of illustration, National Council for Cement and Building Materials (NCBM) have brought out a very useful publication on limestone deposits of cement grade in India titled 'Comprehensive Appraisal of Cement Grade Limestone Deposits of India'. It could be a good starting point for entrepreneurs in India.

Preliminary geological investigations should be carried out for the prospective locations. Quality and Quantity of limestone deposits could be established with the help of:

(i) Surface samples.

(ii) Exposed pits or faces.

Quality can be estimated at this stage for indicated reserves by using data available with State Mining and Geological Department. It would have preliminary data on important minerals found in the State. They would be estimating the quantity of reserves using statistical methods.

2.2 Statistics on Production and Consumption of Cement

Preliminary Investigations would also have shown state wise cement production and consumption in a country.

Country wise there are Institutions like PCA in America, VDZ in Germany who regularly make available vital statistics on Cement Industry in their respective countries.

Earlier, Cement Manufacturers Association -CMA- provided relevant Data/statistics on capacity, production, types of cement wise, data on cement companies and so on vide their annual publication – Cement Statistics. This has been discontinued.

In India presently following Institutions serve that purpose.

1. Care Ratings
2. India Brand Equity Foundation (IBEF)
3. ICRA Ltd
4. Industrial Bureau of Mines- Govt. of India
5. Ministry of Commerce and Industry Govt. of India through their periodic publications.

See Section 11 Annexe 3.2.1 furnishes list of coment placed in India in 2011

2.2.1 Important Statistics on Indian Cement Industry

Obtained from perusal of Reports and Publications of above mentioned Institutions, important statistics on Indian Cement Industry in 2017- 18 is

India ranks no 2 in the World in production of cement

1. Installed Capacity	:	430 million tons per annum (mtpa)
2. Production	:	~ 300 mtpa
3. Capacity Utilisation	:	~ 70 %
4. Large plants	:	~210
5. Types of cements made	:	OPC – 25-30 % PPC – 60 – 65 % BFSC – 10 %
6. Per capita consumption of cement		~ 200 kgs

2.2.2 Operational Efficiemcies

Indian Cement Industry compares very well with the World's best in

Operational Efficiencies

1. Sp. fuel consumption : 700-710 kcal/kg clinker
2. Sp. power consumption : 75- 80 kwh/ton

2.3 Super Impose Markets and available Sites

Super imposition of markets and limestone sites would help in short listing likely locations that could be taken up for in depth studies.

Short listed likely locations could be super imposed on a map showing railway lines and road routes to check the transport and communication facilities with respect to them. Similarly locations of collieries and power stations with respect to them could also be examined.

2.4 Selection of Probable Sites

Selection of the site /s for locating the proposed Cement plant.

Thus pertinent data on short listed sites could be readily and visibly seen by placing transparencies of locations of promising sites on maps of identical size showing:

1. Production and consumption– **See Fig. 3.2.1**, areas of circles show relative volumes.
2. **Fig. 3.2.2** shows locations of large cement plants including Grinding units.
3. **Fig. 3.2.3** shows limestone and coal deposits in various parts of the Country.
4. **Fig. 3.2.4** shows networks of Railways and Highways.

   These figures are not uptodate but serve to bring out the logic behind the selectiof site for a cement plant.
5. **Fig. 3.2.5** identifies the 'clusters' of factories is for the year Year 2011.

Annexe 3.2.1 is a detailed statement of Company wise position of large cement plants in the Year 2011.

From Fig. 3.2.1 and Table 3.2.3 it could be readily ascertained which States are surplus and which were deficit in the yar 2003-04

This type of data needs to be updated when taking up setting up of a new plant.

Alternately a table can be made by listing various factors and by giving them weightages. It can be used to take a decision as shown in **Table 3.2.1.**

2.5 Techno Economic Feasibility Studies

Techno Economic Feasibility Studies are 'in depth' studies of an Industrial Project that take into account all Technical and Economic aspects of the Project and work out its viability and profitability.

2.6 Why is Techno Economic Study Necessary

1. It is a scientific and systematic investigation of all aspects that influence profitability of the venture.

2. It ensures that all major and minor factors are looked into in the correct manner.
3. It ensures that projections of profitability are made on factual basis.
4. It studies market trends, demand, supply and price trends.
5. It studies infrastructural facilities that are necessary to put up the plant including :
   (a) Limestone deposits.
   (b) Land for mining lease and factory.
   (c) Transport and road communications.
   (d) Power.
   (e) Water.
   (f) Social structures including availability of skilled and unskilled labour.
6. It also goes into the aspects of sizes, capacities and machineries most suited, taking into account 'State of the Art' conditions of different types of processes and machinery available so as to obtain optimum operating conditions at low operating costs.
7. It translates all these into capital costs, financial costs of barrowings and also costs of production, works out profitability, payback period, and other relevant, financial ratios commonly used as yardsticks by financial institutions.
8. Last but not the least it prepares solid groundwork for applying for loans from financial institutions.

2.7 TEFS for One Site Should Suffice

Techno Economic feasibility need not be carried out for each of the 2-3 shortlisted sites. It should be carried out for the most promising site.

Major factors of various sites can then be brought out for comparison.

The main report can be then modified for other sites to allow for these similarities and differences. **See Table 3.2.2.**

These differences can then be used to allow for, railway link, roads, ropeway as the case may be and in selection of machinery appropriate to quality of limestone / fuel.

Thus one TEFS would serve for 2-3 sites.

The site that would give maximum returns at least operating costs would then be selected for installation of the plant.

2.8 Requirements of Financial Institutions

Financial Institutions like Industrial Development Bank of India (IDBI) would have their own formats in form of 'Question and Answer' which would have to be filled in by the Entrepreneurs seeking loans from them.

An already made TEFS makes it very easy to submit applications for loan to FIs in the form required by them.

2.9

Figs. 3.2.1 to 3.2.5 are with particular reference to India by way of illustration. However procedure would remain same for all Countries. Similar maps will be drawn for the Country in which the Cement Plant is to be established. Various steps mentioned above would also be the same. There could be minor variations in details peculiar to each Country.

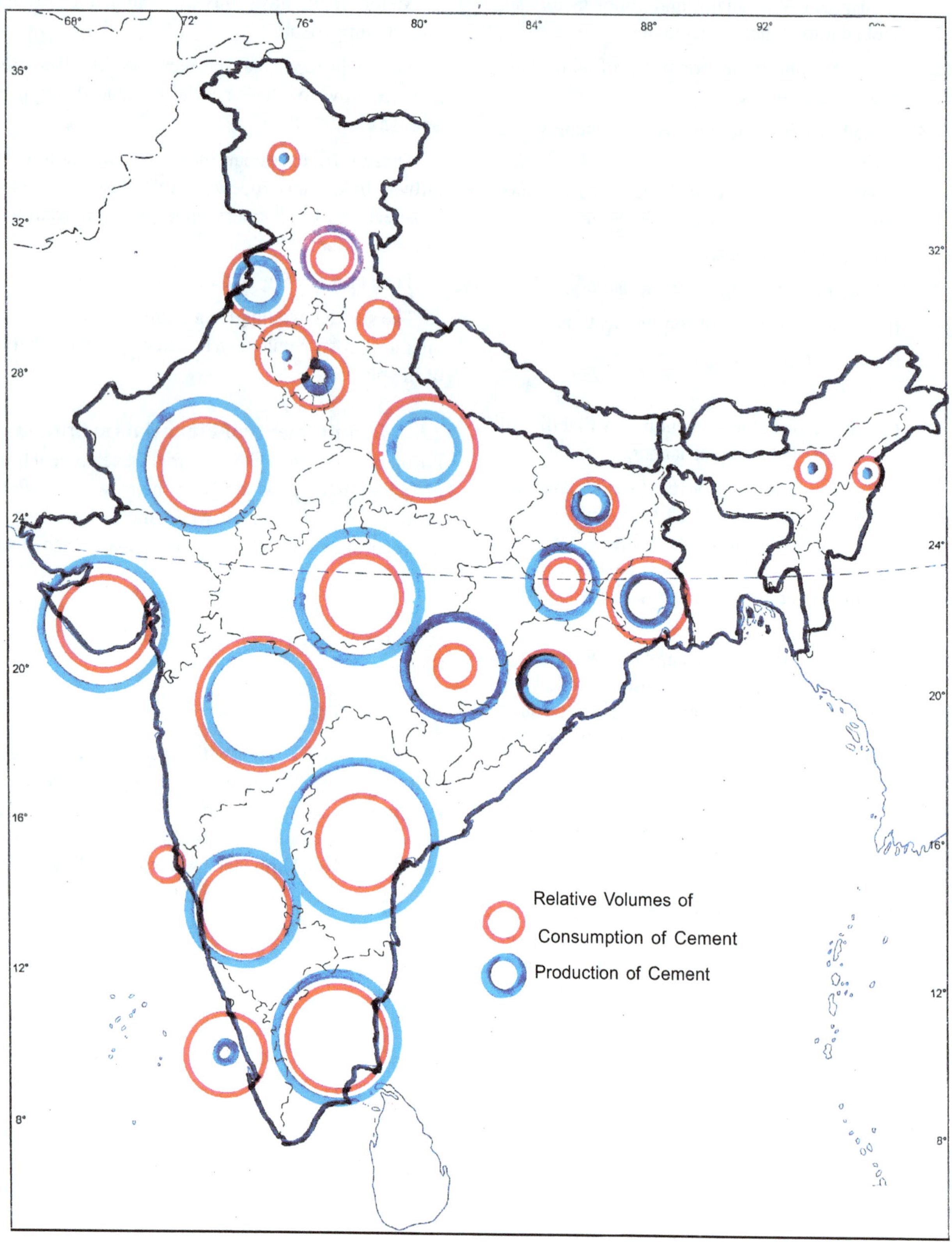

Fig. 3.2.1 Volumes of consumption and production of cement in various states of india.

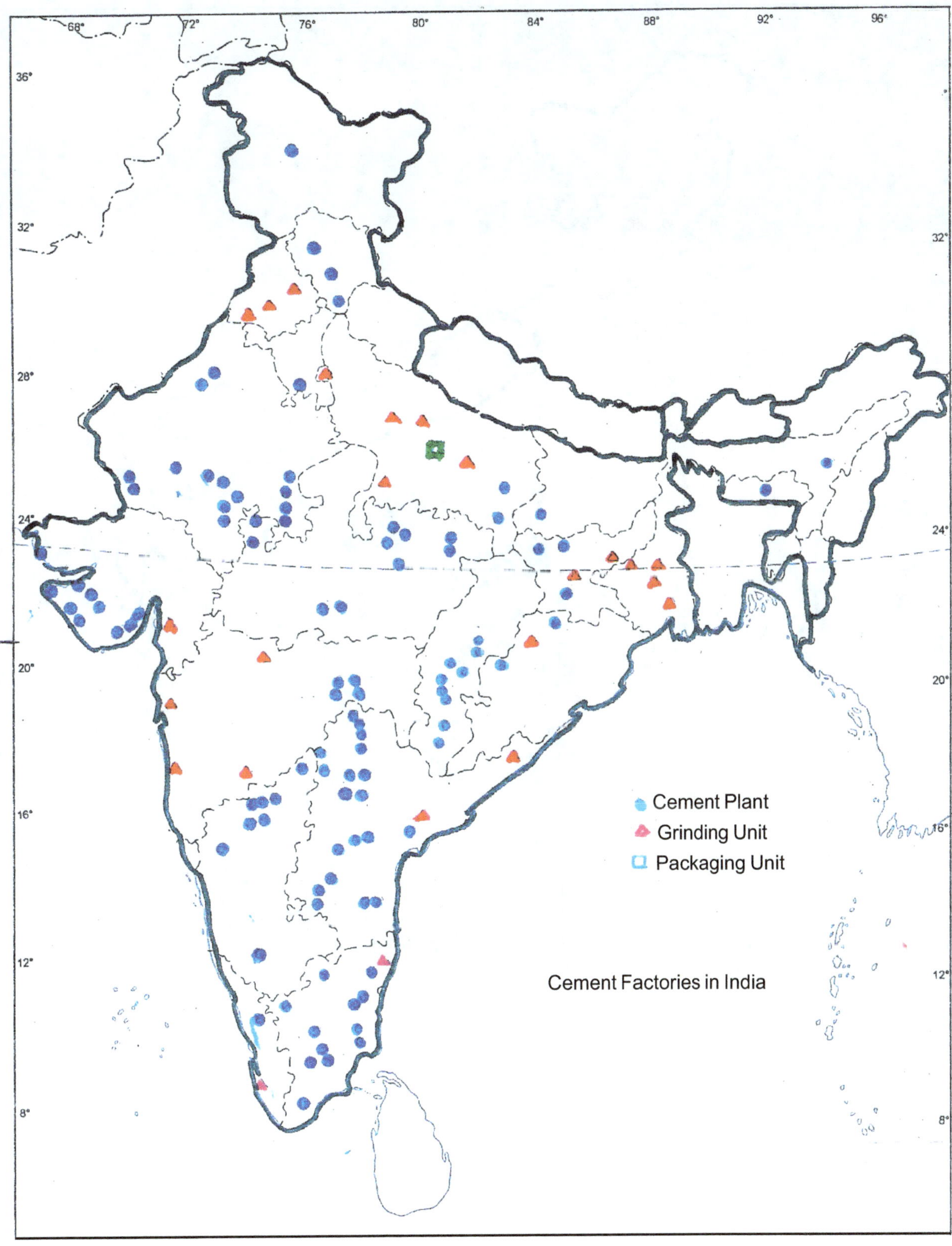

Fig. 3.2.2 Locations of existing cement factories in india.

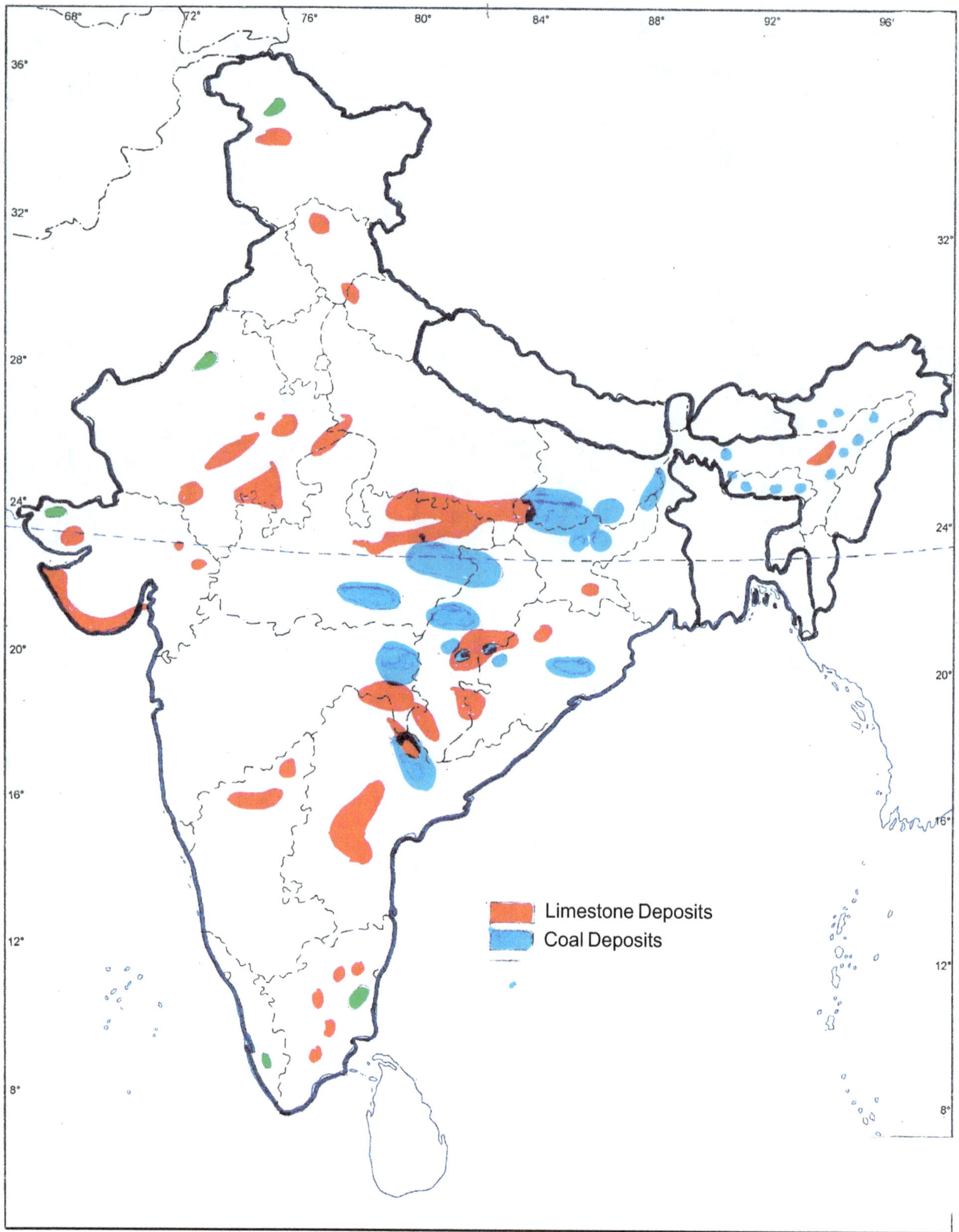

Fig. 3.2.3 Locations of limistone and coal deposits in india.

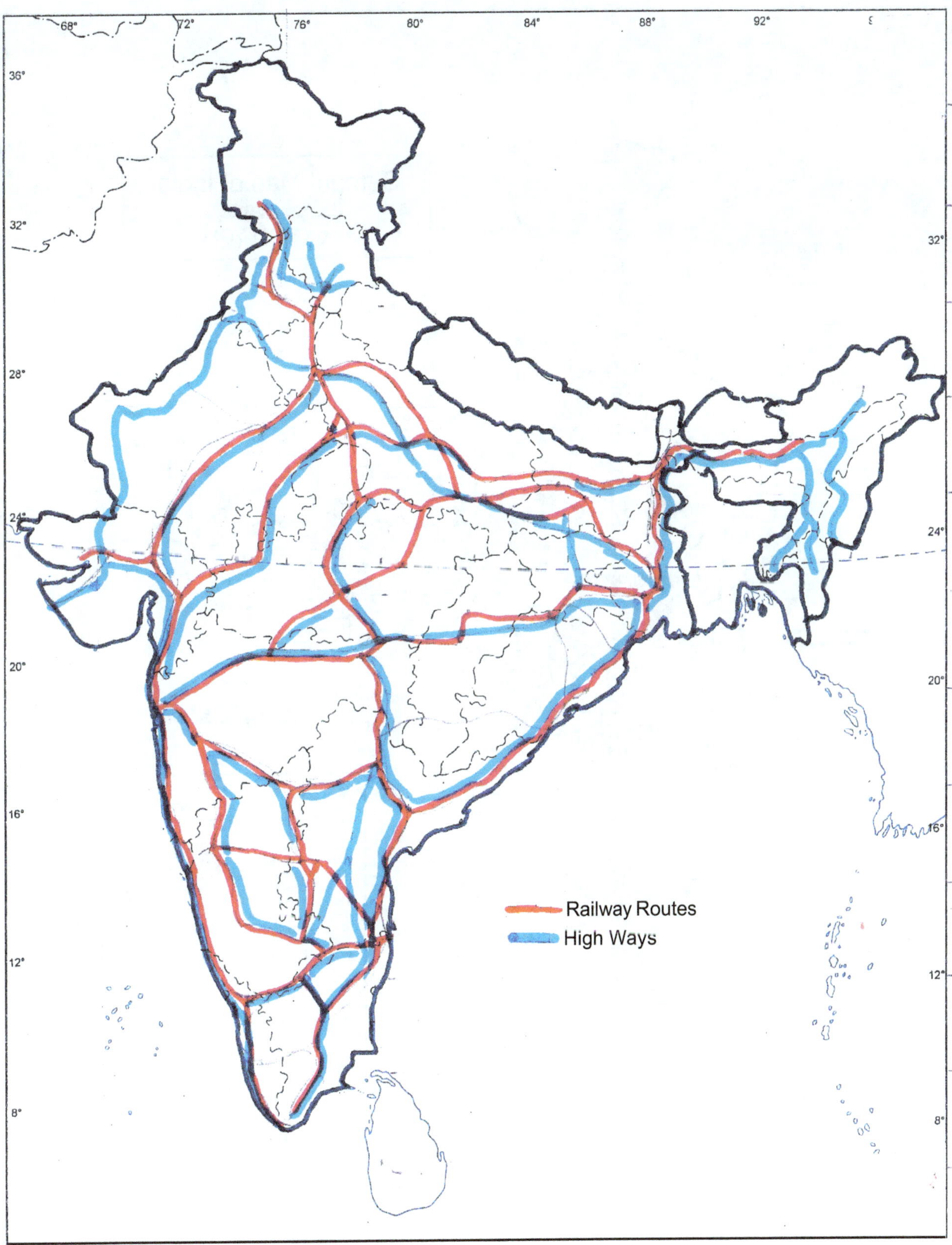

Fig. 3.2.4 Networks of railways and highways in india.

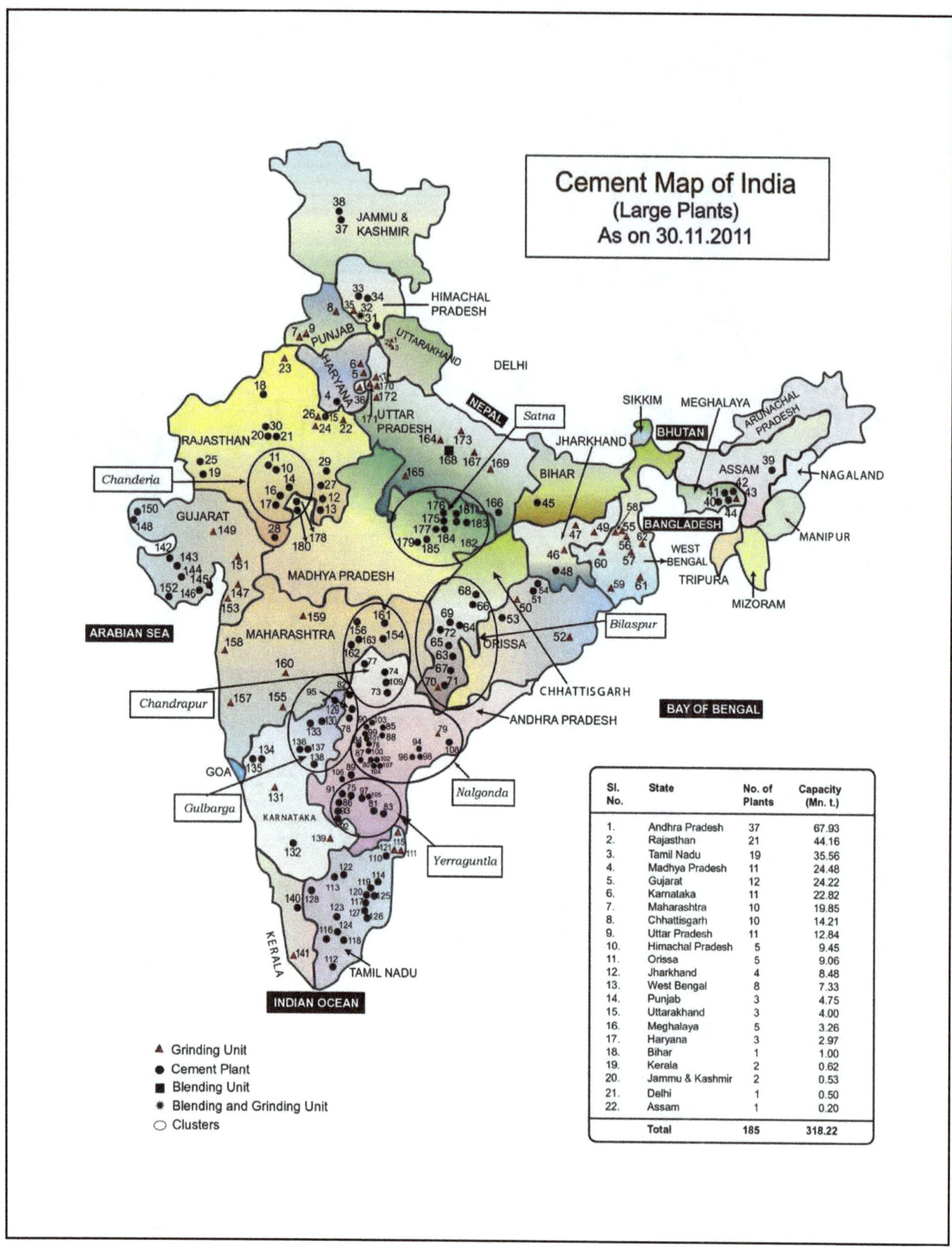

Sl. No.	State	No. of Plants	Capacity (Mn. t.)
1.	Andhra Pradesh	37	67.93
2.	Rajasthan	21	44.16
3.	Tamil Nadu	19	35.56
4.	Madhya Pradesh	11	24.48
5.	Gujarat	12	24.22
6.	Karnataka	11	22.82
7.	Maharashtra	10	19.85
8.	Chhattisgarh	10	14.21
9.	Uttar Pradesh	11	12.84
10.	Himachal Pradesh	5	9.45
11.	Orissa	5	9.06
12.	Jharkhand	4	8.48
13.	West Bengal	8	7.33
14.	Punjab	3	4.75
15.	Uttarakhand	3	4.00
16.	Meghalaya	5	3.26
17.	Haryana	3	2.97
18.	Bihar	1	1.00
19.	Kerala	2	0.62
20.	Jammu & Kashmir	2	0.53
21.	Delhi	1	0.50
22.	Assam	1	0.20
	Total	185	318.22

Fig. 3.2.5 Clusters of cement plants in india.

Table 3.2.1 Evaluation of Seclected sites.

Sr.No.		unit	weightage	Site 1		Site 2		Site 3	
			%		%		%		%
1	Name & location								
2	District, State								
3	Size of Market	mtpa							
3a	Distance of Market from site	km							
4	Likely share of Market	mtpa							
5	Size of deposits	Million tons							
6	Distance of deposit from site	km							
7	Railway Link	km							
8	Highway	km							
9	Collieries	km							
10	Power Grid	km							
11	Substation distance	km							
Total weightage 100									

Table 3.2.2 Comparison of sites.

Item	Unit	Site – A	Site – B	Site – C
reserves	**Million tons**	**60**	**70**	**120**
grade	**High, med, low**	**High**	**High**	**high**
Distance from factory	km	3	6	5
Market size	mtpa	10	8	O 10 P 5
Distance	km	300	200	O 400 P 200
Railway Link, distance	km	Yes 5	No	Yes 5
Highway, distance	km	2	2	3
Substation	km	132kv 10	132kv 15	132kv 7
Mode of transport of stone		dumpers	Belt conveyor	Belt conveyor
Collieries distance	km	100	200	200
Ave. quality of coal	Kcal/kg	4000	4500	5000

High grade-85-90 % ; medium grade- 80-85 % ; low grade- 75-80 % calcium carbonate

O = OPC; P = PPC

Table 3.2.3 Cement production and consumption in various States of India in 2003-04
In million tons per annum.

Sr.No.	State	Production	Consumption	Surplus	Deficit
	North Zone				
1	Haryana	0.2	3.7		3.5
2	Punjab	3.0	5.4		2.4
3	Rajasthan	18.5	6.6	11.9	
4	Himachal Pradesh	4.1	1.4	2.7	
5	Uttaranchal	-	1.4		1.4
6	Delhi	0.5	3.4		2.9
7	Jammu & Kashmir	0.2	0.8		0.6
8	Chandigarh	-	0.2		0.2
	Total North Zone	**26.5**	**22.9**	**14.6**	**11**
	nett			**3.6**	
	East Zone				
1	Assam	0.2	1.0		0.8
2	Meghalaya	0.2	*		
3	Behar	1.0	3.1		2.1
4	Jharkhand	4.6	2	2.6	
5	Orissa	3.0	3.4		0.4
6	West Bengal	3.1	5.8		2.7
7	Chattisgarh	10.2	1.4	8.8	
8	Other N.E. States	-	0.7* include Meghalaya		0.5
	Total East Zone	**22.3**	**17.4**	**11.4**	**6.5**
	nett			**4.9**	
	South Zone				
1	Andhra Pradesh	21.1	8	13.1	
2	Tamilnadu	14.6	9.1	5.5	
3	Karnataka	10.1	8.2	1.9	
4	Kerala	0.6	5.9		5.3
	Total South Zone	**46.4**	**31.2**	**20.5**	**5.3**
	nett			**15.2**	
	West Zone				
1	Gujerat	17.6	7.8	9.8	
2	Maharashtra	11.7	14.7		3
	Total West Zone	**29.3**	**22.5**	**9.8**	**3**
	nett			**6.8**	
	Central Zone				
1	Uttar Pradesh	5.7	13.4		7.7
2	Madhya Pradesh	16.2	5.6	10.6	
	Total Central Zone	**21.9**	**19**	**10.6**	**7.7**
	nett			**2.9**	
	Grand Total	**146.4**	**113**	**33.4**	

Overall surplus of Production over Consumption
33.4 million tons per annum

Source : "Cement statistics 2004" by Cement Manufacturer's Association, India.

Annexe 3.2.1

CEMENT CAPACITY (Large Plants)
(As on 30-11-2011)

Sl. No.	Region/State	Location	Installed Capacity (Mn. T.)
	NORTHERN REGION		**66.36**
	UTTARAKHAND		**4.00**
1	Jaypee Cement (G)	Roorkee	1.20
2	Shree Cement (G)	Roorkee	1.80
3	Ambuja Cements Ltd. (G)	Roorkee	1.00
	HARYANA		**2.97**
4	CCI Ltd.	Charkhi-Dadri	0.17
5	Jaypee Cement (G)	Panipat	1.50
6	UltraTech Cement Ltd. (G)	Panipat	1.30
	PUNJAB		**4.75**
7	UltraTech Cement Ltd. (G)	Bhatinda	1.75
8	Ambuja Cements Ltd. (G)	Ropar	2.50
9	Ambuja Cements Ltd. (G)	Bhatinda	0.50
	RAJASTHAN		**44.16**
10	Birla Cement }	Chittorgarh	2.00
11	Chanderia Cement }		
12	Mangalam Cement }	Morak	2.00
13	Neer Shree Cement }		
14	UltraTech Cement Ltd.-Aditya	Shambhupura	5.00
15	UltraTech Cement Ltd.	Kotputli	3.10
16	J.K. Cement Ltd.	Nimbahera	3.30
17	J.K. Cement Ltd.	Mangrol	0.75
18	J.K. Cement Ltd.	Gotan	0.47
19	JK Lakshmi Cement Ltd.	Sirohi Road	4.20
20	Shree Cement	Beawar	3.00
21	Shree Cement	Ras	3.00
22	Shree Cement (G)	Khushkhera	3.00
23	Shree Cement (G)	Suratgarh	1.20
24	Shree Cement (G)	Jaipur	1.39
25	Binani Cement	Sirohi	4.85
26	Binani Cement-Sikar (G)	Neem Ka Thana	1.40
27	Shriram Cement	Kota	0.40
28	India Cements-Trinetra	Banswara	1.80
29	ACC Ltd.	Lakheri	1.50
30	Ambuja Cmt.-Rabriyawas	Pali	1.80
	HIMACHAL PRADESH		**9.45**
31	CCI Ltd.	Rajban	0.20
32	Jaypee Cement (B & G)	Bagheri	1.75
33	ACC Ltd.	Gagal	4.40
34	Ambuja Cmts. Ltd.	Suli	1.60
35	Ambuja Cmts. Ltd. (G)	Nalagarh	1.50
	DELHI		**0.50**
36	CCI Ltd. (G)	Tughalakabad	0.50
	JAMMU & KASHMIR		**0.53**
37	J & K Cements	Khrew	0.20
38	Khyber Indus. (P) Ltd.	Khyber	0.33
	EASTERN REGION		**43.54**
	ASSAM		**0.20**
39	CCI Ltd.	Bokajan	0.20
	MEGHALAYA		**3.26**
40	Mawmluh Cherra	Cherrapunji	0.20
41	Meghalaya Cement Ltd.	Lumshnong	0.30
42	Cement Manu. Co. Ltd.	Lumshnong	0.59
43	Megha T & E (P)Ltd (G)	Lumshnong	0.67
44	Adhunik	Jaintia Hills	1.50
	BIHAR		**1.00**
45	Kalyanpur Cement	Banjari	1.00
	JHARKHAND		**8.48**
46	Lafarge India (G)-Jojobera	Singhbhum	4.60
47	Bokaro Jaypee (G)	Bokaro	2.10
48	ACC Ltd.	Chaibasa	0.87
49	ACC Ltd. (G)	Sindri	0.91
	ORISSA		**9.06**
50	UltraTech Cement Ltd. (G)-JCW	Jharsuguda	1.00
51	OCL India Ltd.	Rajgangpur	4.00
52	OCL India Ltd. (G)	Kapilas	1.35
53	Bargarg Cement Works	Bargarh	2.11
54	Shiva Cement	Kutra(SunderGarh)	0.60
	WEST BENGAL		**7.33**
55	Birla Corpn. Ltd. (G)	Durgapur	0.60
56	Birla Corpn. - Durga Hitech (G)	Durgapur	1.00
57	UltraTech Cmt. Ltd.-WBCW (G)	Durgapur	1.20
58	Lafarge India (P) Ltd. (G)	Mejia	1.00
59	Madras Cements (G)	Kolaghat	1.00
60	Damodhar Cmt. Works (G)	Purulia	0.53
61	Ambuja Cmts. Ltd. (G)	Sankrail	1.00
62	Ambuja Cmts. Ltd. (G)	Farakka	1.00
	CHHATTISGARH		**14.21**
63	Century Cement	Tilda	2.10
64	UltraTech Cement Ltd.-Rawan	Raipur	2.50
65	UltraTech Cement Ltd.-HCW	Hirmi	1.90
66	CCI Ltd.	Akaltara	0.40
67	CCI Ltd.	Mandhar	0.38
68	Lafarge India (P) Ltd.-Arasmeta	Bilaspur	1.60
69	Lafarge India (P) Ltd.	Sonadih	0.55
70	Bhilai Jaypee (G)	Bhilai	2.20
71	ACC Ltd.	Jamul	1.58
72	Ambuja Cmt-Bhatapara	Bhatapara	1.00
	SOUTHERN REGION		**126.93**
	ANDHRA PRADESH		**67.93**
73	Kesoram Cement	Ramagundam	1.50
74	Orient Cement	Devapur	3.00
75	Zuari Cement	Krishna Nagar	2.20
76	Sri Vishnu Cement	Sitapuram	1.20
77	CCI Ltd.	Adilabad	0.40
78	CCI Ltd.	Tandur	1.00
79	Andhra Cements (G)	Vishakhapatnam	0.62
80	Andhra Cements	Nadikude	0.80
81	India Cements	Chilamkur	1.46
82	India Cements-Visaka	Tandur	2.40
83	India Cements	Yerraguntla	0.73
84	India Cements-Rassi	Wadapally	2.50
85	Madras Cements	Jaggayyapet	3.60
86	UltraTech Cement Ltd. - APCW	Tadipatri	5.60
87	K.C.P Ltd.-Macherla	Macherla	0.83
88	K.C.P Ltd.-Kuktyala	Jaggayyapet	1.52
89	Panyam Cements	Bugganipalle	0.53
90	Rain Cements Unit - I	Ramapuram	1.40
91	Rain Cements Unit - II	Racherla	2.60
92	Penna Cement	Tadipatri	1.50
93	Penna Cement	Boyareddypalli	2.00
94	Penna Cement	Ganeshpahad	1.00
95	Penna Cement	Tandur	2.00
96	My Home Indus. Ltd.	Mellacheruvu Mandal	3.20
97	Dalmia Cement - Kadapa	Jammalamadugu	2.50
98	Anjani Portland Cement Ltd.	Mellacheruvu Mandal	1.16
99	Deccan Cements	Huzurnagar	2.20
100	NCL Indus.	Matampally	2.00
101	Sagar Cements	Huzurnagar	2.50
102	Sri Chakra Cements	Gurajala	0.75
103	Keerthi Indus.	Kodad	0.62
104	Parasakti Cmt.	Rentachintala	1.68
105	Bharathi Cement	Kamalapuram	5.00
106	Jayajothi Cement	Banaganapalle	3.20
107	Bhavya Cement	Dachepally	1.20
108	My Home - Vizag	Visakapatnam	1.20
109	MCC (Mancherial)	Mancherial	0.33

Annexe 3.2.1 *Contd...*

CEMENT CAPACITY (Large Plants)
(As on 30-11-2011)

Sl. No.	Region/State	Location	Installed Capacity (Mn. T.)	
	TAMIL NADU			35.56
110	UltraTech Cement Ltd.	Reddipalayam	1.40	
111	UltraTech Cmt. (G)-ARCW	Arakonam	1.10	
112	India Cements-Sankarnagar	Tulaiyuth	2.05	
113	India Cements	Sankandurg	0.86	
114	India Cements-Dalavoi	Trichy	1.85	
115	India Cements (G)	Vallur	1.10	
116	Tamil Nadu Cements	Alangulam	0.40	
117	Tamil Nadu Cements	Ariyalur	0.50	
118	Madras Cements	R.S. Raja Nagar	1.80	
119	Madras Cements	Alathiyur	3.12	
120	Madras Cements	Ariyalur	2.00	
121	Madras Cements (G)	Uthiramerur	0.60	
122	Madras Cements (G)-Salem	Valapadi	0.60	
123	Chettinad Cement	Karur	1.60	
124	Chettinad Cement I & II	Karikalli	4.30	
125	Chettinad Cement	Ariyalur	4.60	
126	Dalmia Cement	Dalmiapuram	4.00	
127	Dalmia Cement - Ariyalur	Thamaraikulam	2.50	
128	ACC Ltd.	Madukkarai	1.18	
	KARNATAKA			22.82
129	Vasavadatta Cement	Sedam	5.75	
130	UltraTech Cmt. -Rajashree	Malkhed	3.20	
131	UltraTech Cement Ltd (G)	Ginigera	1.30	
132	Heidelberg Cement(I) Ltd.	Ammasandra	0.57	
133	CCI Ltd.	Kurkunta	0.20	
134	Bagalkot Udyog Ltd.	Bagalkot	0.30	
135	J.K. Cement Ltd.-Muddapur	Mudhol	3.00	
136	ACC Ltd.	Wadi	2.60	
137	ACC Ltd.-New	Wadi	3.20	
138	ACC Ltd.	Bellary	1.10	
139	ACC Ltd. (G)	Kolar	1.60	
	KERALA			0.62
140	Malabar Cements	Palghat	0.42	
141	Malabar Cements (G)	Alappuzha	0.20	
	WESTERN REGION			44.07
	GUJARAT			24.22
142	Shree Digvijay-Sikka	Sikka	1.07	
143	Saurashtra Cement	Ranavav	1.50	
144	Gujarat Sidhee Cement	Veraval	1.20	
145	UltraTech Cement - GCW	Pipavav	5.80	
146	UltraTech Cement Ltd.	Jafrabad	0.50	
147	UltraTech Cement Ltd.- (G)	Magdalla	0.70	
148	Sanghi Indus. Ltd.	Abdasa	2.60	
149	JK Lakshmi Cement Ltd. (G)	Kalol	0.55	
150	Jaypee Cement-Kutch I & II	Sewagram	2.40	
151	Jaypee Cmt-Wanakbori (G)	Sonipur	2.40	
152	Ambuja Cement	Kodinar	4.50	
153	Ambuja Cmt.-Magdalla (G)	Magdalla	1.00	
	MAHARASHTRA			19.85
154	Manikgarh Cement	Manikgarh	1.90	
155	UltraTech Cement Ltd. (G)	Hotgi	1.80	
156	UltraTech Cement Ltd. - ACW	Chandrapur	3.60	
157	UltraTech Cement Ltd.- (G)	Ratnagiri	0.40	
158	Heidelberg Cmt(I) Ltd.-Dolvi (G)	Raigad	1.00	
159	Orient Cement (G)	Jalgaon	2.00	
160	India Cements (G)	Parli	1.10	
161	ACC Ltd.	Chanda	2.65	
162	Maratha Cement	Chandrapur	2.40	
163	Murali Agro	Chandrapur	3.00	
	CENTRAL REGION			37.32
	UTTAR PRADESH			12.84
164	Birla Cement (G)	Raebareli	0.63	
165	Heidelberg Cement(I) Ltd. (G)	Jhansi	0.50	
166	Jaypee Cement	Dalla	0.50	
167	Jaypee Cement (G)	Chunar	2.50	
168	Jaypee Cement (B)	Sadva Khurd	0.60	
169	Jaypee Cement-Ayodhya (G)	Tanda	1.00	
170	Jaypee Cement-Sikandrabad (G)	Sikandrabad	1.00	
171	UltraTech Cement Ltd. (G)	Dadri	1.30	
172	UltraTech Cmt. Ltd.-Aligarh (G)	Koil	1.30	
173	ACC Ltd. (G)	Tikaria	2.31	
174	Ambuja Cmts.Ltd. (G)	Dadri	1.20	
	MADHYA PRADESH			24.48
175	Birla Vikas }	Satna	1.55	
176	Satna Cement }			
177	Maihar Cement	Maihar	3.80	
178	UltraTech Cement Ltd.-Vikram	Jawad Road	3.00	
179	Heidelberg Cement(I) Ltd.	Damoh	1.03	
180	CCI Ltd.	Neemuch	0.40	
181	Jaypee Cement	Rewa	3.00	
182	Jaypee Cement	Bela	2.40	
183	Jaypee Cement	Sidhi	1.50	
184	Prism Cement-I&II	Satna	5.60	
185	ACC Ltd.	Kymore	2.20	
	GRAND TOTAL	**185 Plants**		**318.22**

(G) : Grinding Unit
(B) : Blending Unit
(B & G) : Blending & Grinding Unit

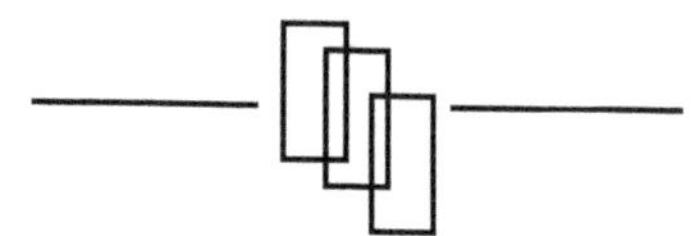

CHAPTER 3

WHAT SHOULD THE T.E.F.S. CONTAIN

3.1 Contents of Techno Economic Feasibility Studies (TEFS)

To be useful to serve as a tool in the hands of the Management to take Investment decisions and for obtaining loans from Financial Institutions (FIs), the TEFS should contain among other things :

1. Future of cement as the principal building material, substitutes available or likely to be available.
2. Types of Cement and their proportions in use.
3. Demand forecast which will include updated data of cement production, growth trends in capacity and demand, movement of Cement within the country and quantity exported, price trends and markets that have potential of growth and are yet unsaturated.
4. Lime stone Deposits vis-à-vis market : The most important aspect after 'markets' is the location of deposits suitable in quality and quantity for making cement. Preliminary assessment of deposits of short listed sites, their quality and quantity and the size of plant / plants they can sustain now and in future.

   If deposits are located in 'clusters' then likelihood of competition from others seeking same location.
5. Broad outline of Processes and Machineries to be used for making cement in an economical way.
6. Environmental aspects, that should take into account, such factors as:

   Reserved Forest areas, sanctuaries, population centers and other inhibiting factors and special cares required to be taken for satisfying fully, on a continuing basis, pollution control norms laid down by competent Authorities.
7. Infrastructure facilities required to support the cement plant like :

land, water, power fuel, communication, transport, labour, and social amenities like schools, hospitals etc., for the size of the plant.

3.2 Technical Aspects, the Report Should Deal With

1. Availability of various processes of manufacture of cement and selection from amongst them giving reasons.
2. Operative norms to be aimed at like :

   (i) Specific fuel and power consumption.

   (ii) Man hours / ton.

   (iii) Dust emissions in mgm/nm^3.
3. Types of machinery available and reasons for selection of specific machinery.
4. Build up of flow charts and machinery schedules for the plant starting with quarries and ending with cement dispatches.
5. Requirements of power and how they would be met from grid power and from captive power.
6. Process and Quality Control aspects and Instrumentation and Automation to be adopted to achieve them.
7. Civil engineering aspects, which include special factors due to location, like seismic zones, load bearing capacity of soil, under ground water tables, wind velocities etc.
8. Water resources.

3.3 Man Power Requirements For

1. Managerial functions,
2. operational functions,
3. skilled and unskilled jobs,
4. marketing, accounting, financial and personnel functions.

3.4 Capital Costs of the Project

After covering the various Technical aspects mentioned above, TEFS should then work out cost estimates for the project including inter alia, cost of acquisition of land and construction of railway siding, roads, power transmission line and all those factors touched upon earlier.

3.5 Costs of Production

TEFS should work out costs of production consisting of material costs, fuel and power costs, and stores and spares and maintenance costs, that together could be termed 'variable costs'.

TEFS should also work out 'fixed costs' such as salaries, wages and overheads.

The two together can then be converted into annual expenses or costs for a given volume of production.

3.6 Realisation from Sales

The TEFS should contain work outs of revenues from sales for different prices and for different shares of the market - on 'optimistic' and 'pessimistic' basis.

Net Sales realizations are worked out after deducting duties, taxes, freights, transport, packing, dealers' commission, costs of sales such as expenditure on advertising, maintenance of depots and dealer network.

Surplus of net sales realization over total costs of production less interest and depreciation would yield gross profit. Net profit is arrived at by deducting taxes.

3.7 Finances for the Project

TEFS also deals with aspects of raising finances and hence cost of capital. Finance raised could be :

1. Totally from own resources.
2. Partially own, partially public.
3. Loans from financial institutions.

Normally, it would be a combination of all three.

3.8 Working out Probitability

The costs of capital are worked out on basis of current interest rates, preliminary and pre-operative expenses.

Financial Institutions (F.I.s) have their own norms for financing new projects and expansions and entrepreneurs approaching them should acquaint themselves with these norms and should be able to satisfy FIs that these norms would be met.

TEFS then works out profitability and prepares various Financial Statements and establishes financial yardsticks used by FIs to bring out viability and soundness of the proposed investment in the project.

Yardsticks like 'debt servicing ratio', 'internal revenue returns' and 'pay back period' are very useful.

Financial Projections are made for 10 to 15 years as required by FIs.

3.9 Conclusions

TEFS must make its recommendation in the light of all this study as regards location, market, size of plant, raising of finances etc. The recommendation should be fair and unbiased.

3.10 Executive Summary

An executive summary is a vital part of the TEFS specially prepared for the Boards of Directors and Financial Institutions.

3.11 Format of TEFS

The same TEFS becomes a part of the application for loan to Financial Institutions and helps to fill in the application for loan and expedites its sanctions.

As a matter of fact, TEFS can be fashioned content wise to contain details in the sequence and format in the Q&A devised by Financial Institutions.

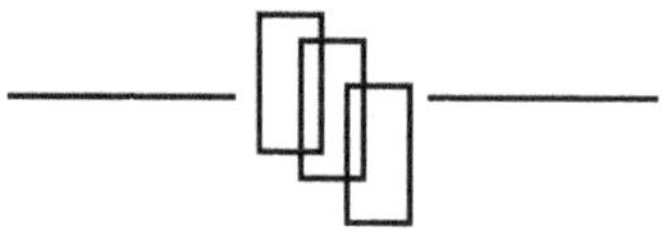

CHAPTER 4

DEMAND FORECAST AND MARKET

4.1 Demand Forecast and Markets

One of the most important chapters in TEFS is the study of Demand Forecast and Market conditions. This study has to be done at 'micro' as well as 'macro' levels.

Firstly, the situation prevailing in the country as a whole in terms of installed capacity, production and consumption has to be studied and taken into account.

It will bring out rates of growth of demand, consumption and supply - (increase in installed capacity and production) - in various regions of the country.

4.2 Demand Forecast Studies: How they Help

In India for example, Cement Manufacturers' Association and Several Government and Quasi Governmental agencies specially appointed by the Government bodies have carried out studies of Demand Forecast from time to time. One in depth study was commissioned in 1994 by the Development Council for Cement Industry.

The agencies use different 'models' for projecting demand. Normally, they come out with

1. Pessimistic }
2. Normal } Projections
3. Optimistic }

These help in assessing the worst and best scenarios and the gap between them.

The trends in growth of manufacturing facilities (installed capacity) also show if the gap is positive or negative between demand and supply and whether it is closing or widening.

If 'supply' is already in excess of demand, the trends would show when demand would overtake and exceed supply (rate of growth of which will slow down in saturated conditions) and when gap would be large enough to warrant further investment in generation of fresh capacity.

A cement plant takes about 2 to 2½ years to commence production from the time orders are placed. Another one year may be allowed for it to reach full capacity. Thus timing of deciding to invest and of going on stream is very important. The first ones to deliver cement when gap appears between demand and supply are the ones who obtain maximum benefits.

4.2.1 Studies at Macro Level

The macro or national level studies also bring out:

(i) Zones which have unsaturated demand,

(ii) zones where demand is growing at a much faster rate than others,

(iii)zones that are ill served because of their location, lack of communications and hence are always 'starved'.

4.2.2 Movement of Cement within the Country

The 'unevenness' of demand and uneven distribution of production facilities give rise to movement of cement across States.

Speaking about India, because of abundance of limestone reserves in them, States like Telangana, Gujarat, M.P. and Rajasthan produce cement far in excess of their own needs. These States 'export' cement to deficient Zones like the East.

Movement of cement is 'mixed' in that same States will 'export' and 'import' cement simultaneously. This is particularly true of adjoining States.

4.3 Clusters

Unevenness of distribution of limestone deposits in the Country has caused formation of 'Clusters' that is grouping of plants installed in close vicinity of one another. Movement of cement from the clusters to other areas is heavy and density of traffic of cement on railways is high in these areas.

See Fig. 3.2.5 in Chapter 2.

While such locations offer advantages of developed infrastructural facilities, a new entrant has to compete so much harder to win a share of the market and may have to go out further to sell cement.

4.4 Proximity of Markets

Proximity of the market is one of the most important considerations in locating a cement plant today. In 'free market' conditions in which Cement Industry operates, the costs of transport erode the profitability of the plant as the price in the market would be same for all whether cement comes to it from near or far.

See Figs.3. 4.1 and3.4.2.

The demand forecast studies coupled with study of limestone reserves in the country thus helps to locate:

1. Markets.
2. Limestone reserves that are close to the market and thus to shortlist areas where a new plant could be located.

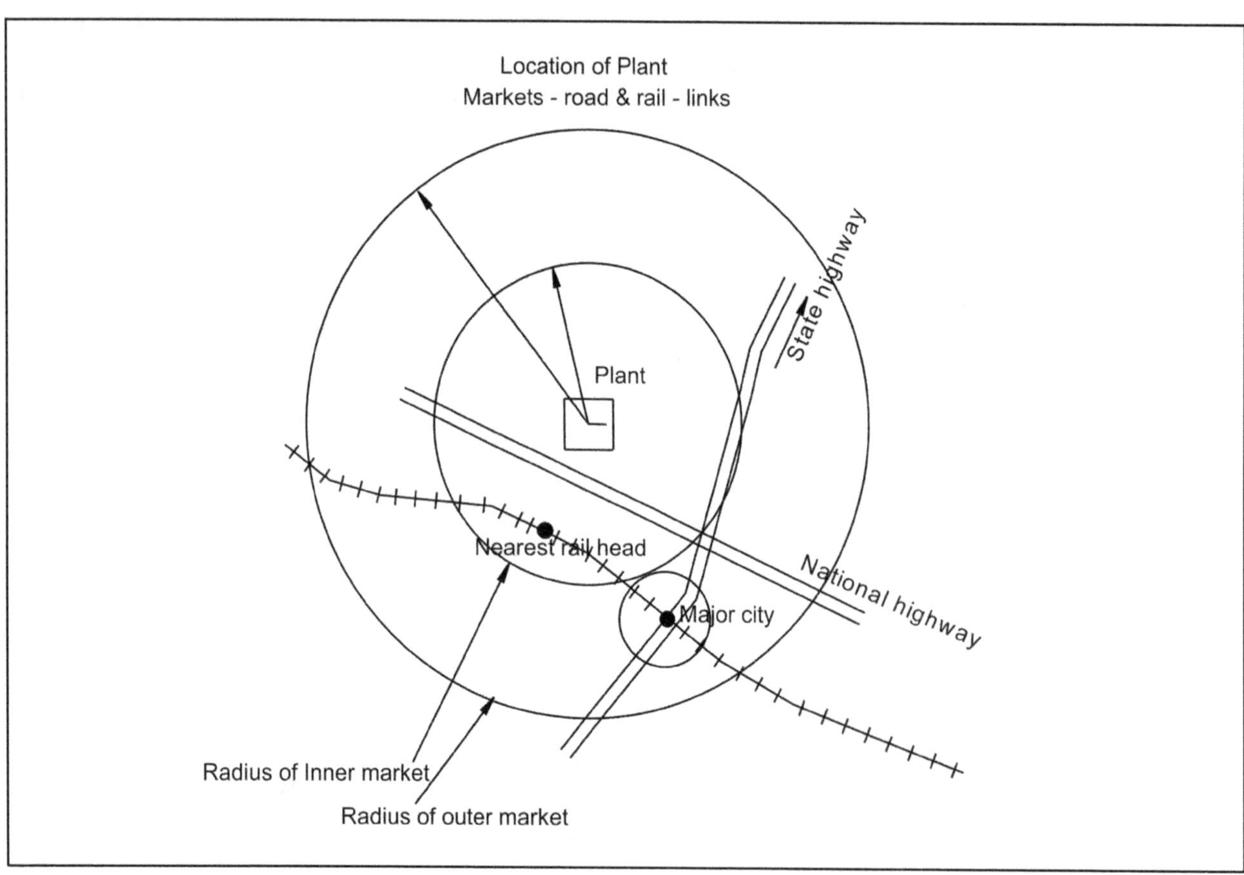

Fig. 3.4.1 Dispatches – inner market – by road; outer market – by rail/by road.

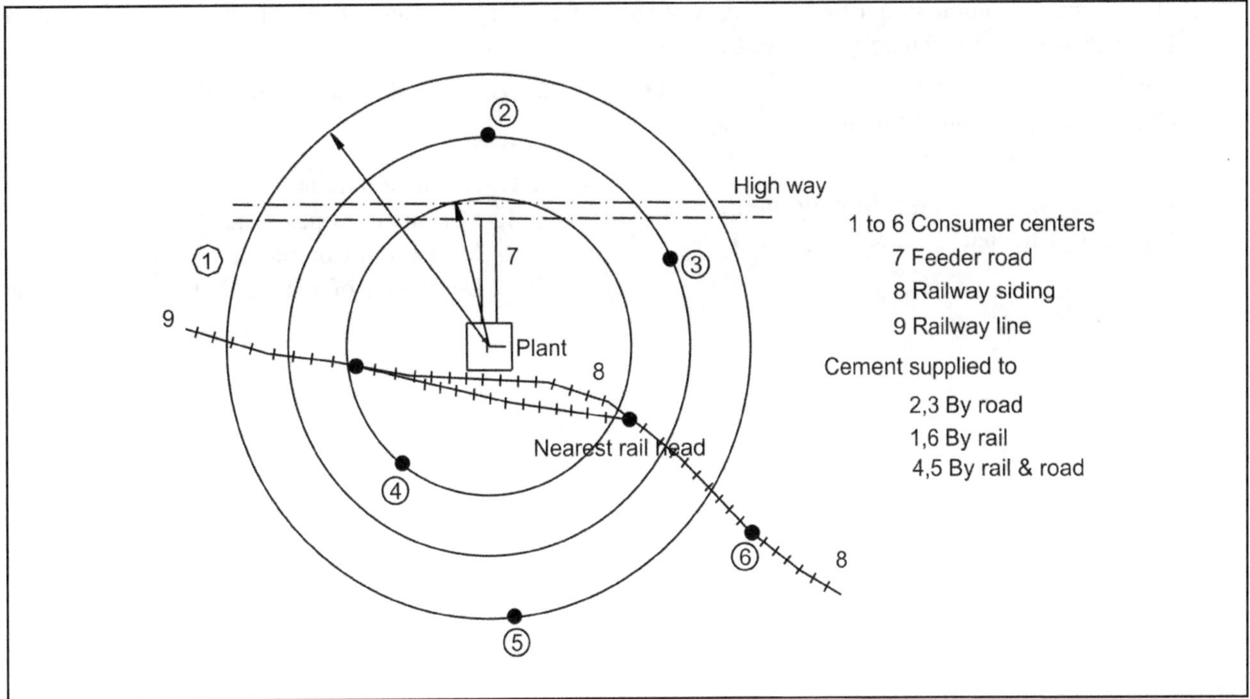

Fig. 3.4.2 Reaching consumer centres – by road and rail.

4.5 Price Movements

The study of price movements in the last 5 years or so, study of profitability of prominent / established cement companies across the country give a fairly good idea as to whether it would be worth while to invest in a new cement plant / project.

4.6

The demand for cement is both in 'public' and 'private' sectors. In private sector it is largely for housing and industry.

In public sector it is for projects like dams, roads, bridges and public sector industrial enterprises.

4.7 Factors Favourable for Growth of Cement Industry

Two factors that are still favourable to Cement Industry are:

1. While a few substitutes like Calix,Novacem, Aether have made their appearance, they are still in developmental stage and offer no real competition to cement presently.
2. Growth of population has not slowed down appreciably. That means private sector demand for housing will continue to grow (though per se housing consumes only small quantities of cement).

The Government of India has launched a massive programme of developing 'infrastructure' in the country like building roads that will give an appreciable boost to demand for cement.

Country has been exporting both clinker and cement for the last several years. In the year 2003-04 India exported ~ 3.4 million tons of cement and 5.6 million tons of clinker. Prospects of export should therefore also be taken into account while studying 'demand forecast and market'.

4.8 References in Indian Context would be:

1. Bureau of Industrial Costs and Prices' Study on Cement Industry.
2. Cement Industry : Annual Reviews by several Specialised Publications.

3. Development Commissioner for Cement, New Delhi 'Prospective Technology Plan for Cement Industry'.
4. Consindia, New Delhi 'Demand for Cement in the Indian Economy, 1988'.
5. Institute of Financial Management and Research, Chennai.
6. RITES, New Delhi 'Studies on movement of cement'.
7. IBEF – India Brand Equity Foundation
8. ICRA Ltd
9. Government o India
   1. Ministry of Mines – Indian Bureau of Mines
   2. Ministry of Commerce and Industry – Department of Indusrial Policy and Promotion

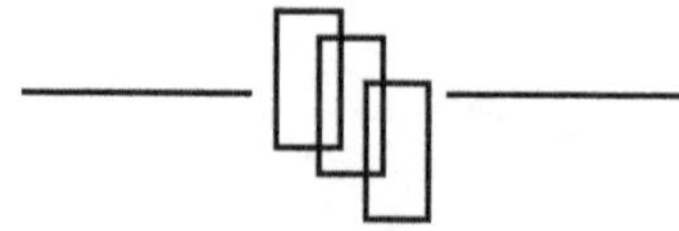

CHAPTER 5

LIMESTONE DEPOSITS

5.1 Limestones of Cement Grade

Limestone is the predominant constituent of raw materials used to make cement. It is to be blended with correcting materials or additives to make up for other constituents like Silica, Alumina and Iron Oxides. For design purposes it can be assumed that it takes about 1.6 tons of limestone to make one ton of cement clinker.

Those limestones that have +80% $CaCO_3$ are suitable for making cement and are called 'cement grade'. When blended with additives, the carbonate content comes down.

All Limestones have a small percentage of $MgCO_3$. Those that have more than 5 % $MgCO_3$, are required to be carefully blended with other limestones and correcting materials so that MgO or $MgCO_3$ in total raw mix is kept within limits to prevent 'expansion' of cement which gives rise to cracking.

5.2 Assesment of Deposits

The first step in assessing a deposit is to collect as much information on it as is available with State Geological and Mining Departments (Central also). These departments have records of not only limestone but also of other commercially valuable minerals in the State.

They would have carried out tests on surface samples or on faces wherever such faces are available to determine chemical analysis of the stone in terms of $CaCO_3$, $MgCO_3$, SiO_2, Al_2O_3, etc.

If found, promising for use of cement, chemical and steel industries, the departments would have carried out random drilling to take out core samples to determine quality of deposits and also its extent and depth so that the quantity of reserves could be estimated on a rough basis.

The area is fixed and demarcated on maps. Using statistical methods the quantity of deposits is estimated.

This data with pertinent macro and micro maps of the likely deposits under consideration becomes the starting point.

A good beginning could be made by referring to the book brought out by NCBM titled 'Comprehensive Appraisal of Cement Grade Limestone Deposits of India'. **Fig. 3. 2.3 of Chapter 2** shows distribution of limestone deposits in India.

5.3 Investigations in Phases

The investigation of deposits would generally be carried out in 3 phases.

1. 'Preliminary' investigations to establish that sufficient deposits of suitable grade are available.
2. More detailed investigations of a promising site are carried out to establish quality and quantity more accurately by drilling a few more bore holes and by digging pits and trenches.

This investigation supplies more detailed information on deposits in terms of :

- (i) Variation in quality (measured as percentage of $CaCO_3$ content).
- (ii) Extent of contaminants including, $MgCO_3$, clays, dolomitic limestone bands etc.
- (iii) Over burden and the depth to which limestone can be commercially mined.

(iv) The area over which cement grade limestone reserves are spread.

5.3.1 Testing Core Samples

Core samples from bore holes are brought out and tested, washed and unwashed, to ascertain quality of 'Clean limestone' and quality of commercially mined limestone. If there is too much difference between the two, mining could be a costly proposition.

Similarly, if there is too much variation in quality then blending while mining itself would become necessary. More than one face would have to be worked upon simultaneously.

5.3.2 Detailed investigation

The third phase would be the 'detailed investigation' of deposits that would be actually exploited to make cement and for which mining lease would have been applied and obtained.

It would be carried out for the area that is finally selected for putting up the cement plant after earlier investigation have shown conclusively that sufficient reserves of suitable grade are available in this specific area.

It would be carried out simultaneously with other activities like acquiring land for factory, colony and carrying out basic design of the plant.

The feed back from these investigations if available in time is of great help in selecting the type of major machinery and their sizing.

These investigations can also be used to formulate mining plans, which need to be filed with Geological Departments and Environmental Authorities.

At the TEFS stage, the entrepreneur must have sufficient information to take a decision based on deposits in shortlisted locations for investment.

5.4 Quantity of Deposits

As a thumb rule, a cement plant is designed for a life of 30-40 years.

To allow for wastages, over burdens and losses and inaccuracies in preliminary investigations, it is desirable to apply a factor of 2:1 instead of 1.5:1 in arriving at the required quantity of reserves. Thus a 1.0 mtpa capacity OPC plant, ignoring 4.5% gypsum, would need to have reserves as shown in **Table 3.5.1.**

Table 3.5.1 Reserves required in million tons for a 3 mtpa capacity cement plant.

Qty. of Reserves	30 Years	35 Years	40 Years
Making100% OPC	180	210	240
Making 100% ppc	130	150	170
Making 100% BFSC	70	80	90

Normally a design margin of 10% is added on the rated capacity.

If the location is very promising from marketing point of view, the expansion should also be allowed for. If a simple duplication is to be allowed for, then the required reserves would be doubled.. See new table at the end of chapter

5.5 Area of Mining Lease

Assuming an in situ density of 2t/m^3 (Specific gravity of limestone is 2.7) and an average depth of say 20 meters to which the deposits could be commercially mined, then one hectare would contain $2 \times 20 \times 10000$ = 4.0 lakhs tons of limestone.. A 3mtpa cement plant making OPC only would require

	For 30	35	40	years' life
reserves of	180	210	240	million tons
area of mining lease required would be	450	525	600	hectares
and for double the capacity,	900	1050	1200	hectares

The area required is obviously a function of depth of deposits and its variations. Naturally some allowances have to be made for boundaries where deposits would be inferior. Hence mining lease to be applied for would be say 10% extra.

~ 1000 1200 1300 hectares

Quantity of reserves required for making blended cements as shown in Table 3.5.1 above. Area for mining lease required would also reduce correspondingly

This exercise, clearly brings out advantages of making blended cements by way requirements of limestone and reduction in requirements of area to be obtained on lease for mining.

There would be more choice in selection of deposits and costs of obtaining leases would also be considerably reduced.

5.5.1 Types of Deposits

Deposits occur in a myriad ways. Some are solid blocks without interstitial materials, some have faults in them in that the deposits take sharp humps and dips with contaminants/ soil in troughs. Some are fragmented with lots of interstitial materials and some are slabby. Others have bands of calcite, dolomite and some have other minerals.

5.6 Categorisation of Reserves

Deposits have been therefore categorized into:

1. Measured reserves.
2. Indicated reserves.
3. Inferred reserves.

Measured reserves are those which would be exploited and supplied to the plant according to its capacities.

For example, assume a raw mix : clinker ratio of 1.55 : 1, and 85% limestone in raw mix, 1 ton of clinker would require 1.32 tons of limestone as mined and supplied to the plant. If it has a moisture of 5%, limestone mined and supplied would be = 1.4 tons per ton clinker. For a design capacity of 1.1 mtpa, stone to be supplied = 1.54 mtpa.

To be able to supply this quantity of stone, it is first necessary to remove over burden and allow for mining losses, etc. On an average 30% is allowed for this purpose. Therefore 'indicated reserves' should be = 2.0 mtpa as mentioned earlier.

The 'inferred reserves' are the roughest estimates of reserves based on very preliminary data. In practice say 25% would take care of inaccuracies inherent in estimating inferred reserves. That is for indicated reserves of 2.0 mt inferred reserves would be 2.5 mt.

Since it would be most difficult to acquire new mining leases later in the same area, it would be prudent to apply for mining lease for at least one duplication in the first phase itself.

Therefore for a 3 million ton plant the mining lease should be applied for ~1200 hectares on the basis of inferred reserves.

Alternatively, plant should switch over to making Blended Cements OPC or BFSC.

5.7 Forests and Sanctuaries

In selecting the deposits, there are other factors, which have become now very important because of the impact of exploiting the deposits on environment.

Some deposits could be in a cultivated area and land would be privately owned, such deposits are likely to be more easy to exploit but it would be difficult to acquire land for mining purpose. Others would be under fallow government lands.

Some reserves are likely to happen in areas designated as 'forests'. In such cases, the grant of mining lease would no longer be in the hands of State Governments. Central Government Agencies only could clear the area for mining lease. It would take a lot of time to obtain clearances from the concerned Environment monitoring bodies.

One of the stipulations made by the Authorities could be to create new forest on an equivalent area that would be made available by the Government for the purpose of planting new trees.

5.7.1 Sancturaries

Another sensitive factor is a 'sanctuary'. Some areas in consultation with naturalists have been categorized as 'Sanctuaries for birds and animals'. It is almost impossible to get a mining lease for such areas even if limestone is excellent in grade and all other factors are favourable.

In selection of deposits therefore at the project stage itself it would be better to avoid such deposits as occur in areas categorized as forests and sanctuaries. Such areas should only be selected as a last resort.

5.8 Preserving Trees During Mining

Trees serve as dust and sound barriers. The mining operations should therefore be planned taking into account:

(a) Creation of new plantings and walls of trees.

(b) Disposal of unusable wastes and collection of and use of water in mines.

5.9 Landscaping

Landscaping of mines has become increasingly popular and serves many purposes like giving an aesthetic look, beautifies the surroundings, provides recreation spots, including swimming and boating facilities.

While all this expenditure is not to be incurred in the project stage there should be definite plans for them at this stage. This can in many cases be simply provided for by acquiring more area than is strictly necessary for mining purposes.

Water found in mines could be profitably collected and used not only for quarrying operations but for purposes of landscaping mentioned above.

5.10 Representative Quality of Limestone Deposits

It is necessary to establish 'representative' samples of the deposits. In nature the quality would scarcely be uniform along the length, breadth and depth of the deposits in terms of chemical and in terms of physical properties like hardness, crushing strength, grindability, etc.

It is very necessary to assess :

1. Correcting materials that would have to be obtained for mixing with limestone.
2. To select and size the machinery.

After knowing the type / types of cements that would be made in the plant. viz., OPC - 43 / 53 Grades, Blended Cements - Slag and or Pozzolona, the quality of clinker and hence 'raw mix design' becomes known. From this the chemical analysis of limestone (which is available) and the desired analysis of correcting materials and their properties can be ascertained so that all criteria that go into making good clinker like :

(i) Hydraulic modulus or lime saturation factor.

(ii) Silica modulus.

(iii) Iron modulus.

(iv) C_3S and C_2S contents.

are met.

Against this the analysis of various samples of limestone from different areas are to be compared and deviations noted.

From the core samples, variations depth wise, would be known and from surface and pit samples, variations over the area could be known. The mines could then be broadly be divided into Zones - A,B,C & D etc., with rough estimates of reserves in each zone and their qualities.

The representative analysis can then be obtained by - physically collecting and testing several samples over the entire area and at various depths using statistical methods.

The samples can be then mixed and quartered several times to obtain representative samples. These should be available in quantities suitable for sizing and for analysis. For instance, 10-15 kg samples are sufficient for chemical analysis and grindability tests like Bonds. For determining crushing strengths, samples of 150 mm cube size would be necessary.

For vertical mills, for testing grindability and for sizing of mill, samples of ½ to 1 ton would be necessary.

It is the responsibility of the entrepreneur to make representative samples available to suppliers and consultants for the process design, basic engineering and for sizing of machinery.

Therefore in his own interest, he should take all necessary steps and make necessary efforts to see that samples called representative are truly so.

5.11 Correcting Materials

Once limestone quality is known - correcting materials should be looked into.

In most, or at least in a large number of cases, over burdens or top soil in the surrounding area serves as one of the correcting materials to balance the required SiO_2 and $Al_2 O_3$ contents.

If this is so, it is very fortunate in that stone as mined could be fed to stockpiles and crushers. There is no wastage and there is no need to arrange for disposal of wastes.

If this is not so, it is necessary not only to separate stone from wastes but also to look for correcting materials of suitable composition elsewhere. Further unusable overburden has to be disposed of.

Generally speaking, clays would be added to extent of 10-15%.

Thus for a 1 mtpa plant with 1.55 ratio, annual requirement of clay would be 0.16 to 0.23 million tons on dry basis.

Iron ore requirement would be about 1-2% or 0.02-0.04 million tons. Generally, it would be obtained from ferruginous deposits nearest to the plant.

5.12 Locations of Crushing Plant and Factory

The investigation of deposits in sufficient detail is useful even at TEFS stage. It helps in arriving at location of factory, crushing and mining operations with respect to one another.

At TEFS stage itself decisions can be taken about:

1. Location of crushing vis-a-vis mines.
2. Location of factory vis-à-vis crusher.

There can be many possibilities here. But three most common options are :

1. *Mines adjacent to factory (0 - 3 Kms)*

   Crushing would be located in the plant, limestone would be brought in dumpers / trucks to plant.

   If mobile / semi mobile crushing is preferred because of the size of plant, crushed stone would be brought to the plant by overland belt conveyors.

2. *Mines located between 5-10 kms from the plant*

   This may be necessary for reasons of communication and transport of cement such as railway link, and road network.

   It may also be necessary for reasons of non-availability of land near mines.

   Crushing would be located in mines / near mines and crushed stone taken to plant either by over land conveyor, or by, Ropeway - depending on the terrain.

3. *Mines located more than 10 kms from the plant*

   This situation would come up in inaccessible areas with poor infrastructural facilities for locating the plant near the mines.

   Crushing would be invariably located near mines and crushed stone taken to factory by belt conveyors / ropeways.

5.13 Deposits Come in All Shapes and Sizes

1. Deposits with out crops of cement grade limestone in plane land - fallow or cultivated.
2. Deposits with 1-2 meter over burden of soil / clay in fallow / cultivated or even in forest lands.
3. Deposits traversed by running streams and with uneven contours.
4. Deposits in the form of hills.
5. Deposits submerged under water / sea.

5.14 Machinery for Mining

Each one of these would require careful selection of mining machinery consisting of:

1. *Drills and Compressors*

   Pattern of drilling, size and spacing and depth of holes will depend on whether deposits are solid, fragmented and or with interstitial materials.

2. *Shovels and Dumpers*

   These are the basic units for recovery of limestone loosened from working faces by blasting and for conveying it to the crushing plant.

3. *Rippers*

   For some deposits, 'rippers' which literally rip stone from its bed could be used. This is possible for sedimentary deposits near sea shores. Ripping can save expenditure on drilling and blasting. It can also avoid vibrations in sensitive areas.

Sizing would be governed by the capacities of the plant and number of shifts and hours in operation. Numbers would depend on lead from faces to crushing plant.

As operational and maintenance costs are high for earth moving machinery, the leads are kept as short as practicable under the given circumstances.

Sizing of shovel would also decide size of primary crusher to be installed.

This equipment would be included in capital costs to be built up for the project.

For these reasons preliminary detailed investigation should precede placing of orders of plant and machinery.

5.15 Availability of Deposits

Another very important factor in case of deposits is their availability.

While there are extensive deposits of cement grade limestone in the country (they are not evenly spread), some locations are sought after by more than one entrepreneur for obtaining mining lease and for the exclusive right of exploiting them.

Mining leases would be granted by State Departments on 'first come first served' basis.

There is therefore always a possibility that deposits that appear most promising are not available as some one has already acquired mining lease for them.

Therefore in the early stages of the project it is necessary to ascertain which deposits in a given area are still free to be exploited. Clear demarcation of such deposits is essential.

In the past mining leases used to be granted for long periods - 20 years or more. There is many a deposit that has simply remained unutilized.

Mining leases are now given for short periods of 6 to 10 years (even shorter for mines in forests) and are therefore to be renewed again and again. This is cumbersome procedure for a running plant.

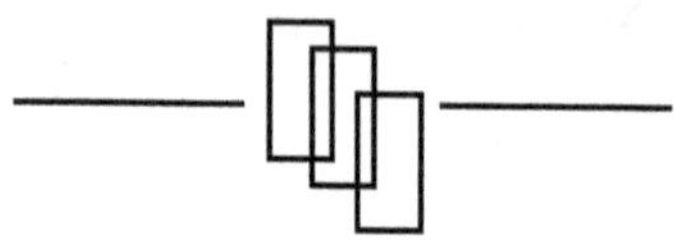

CHAPTER 6

ENVIRONMENT

6.1 Pollution Control Norms

State and Central Pollution Control Boards (PCB) lay down 'norms' to be satisfied by the entrepreneur to ensure 'clean ambient air' in the area surrounding the cement or any industrial plant.

An entrepreneur desirous of putting up a cement plant has to satisfy the Authorities of concerned PCBs that he has taken all necessary steps to ensure clean ambient air as per norms laid down not only for the present but also for future when the plant will be expanded.

Presently it is necessary to take into account emission of 'greenhouse gases' as they warm the environment.Objective would be to minimize their emission to the extent by choosing to

1. make blended cements
2. use Alternate Fuels
3. recover waste heat of Kiln an Cooler gases to generate power
4. use renewable sources of energy like wind and solar power.

6.2 Environmental Aspects

In setting up a cement plant, environmental aspects other than greenhouse gases mentioned above that need to be considered are :

1. Felling of trees / bushes in mines and for construction of plant and colony, roads, railway siding, etc.
2. Loss of land for agriculture and grazing.
3. Creation of ugly pits as a result of mining operation and accumulations of heaps of waste material.
4. Emissions of dust in exhaust gases in various sections of the plant like crushing, grinding, kiln and cooler, etc.
5. Fugitive dusts in material handling operations.
6. Dust nuisance caused by traffic of trucks and dumpers in the plant and on the roads.
7. Noise of working machines like crushers and grinding mills.

An entrepreneur is required to take steps to ensure that the plant installed is harmoniously blended in the surrounding in such a way that it only enriches 'quality of life'.

6.3 Environmental Planning

Thus the entrepreneur has to plan for :

1. Planting of trees to replace those that need to be cut down.
2. Create screens or walls of trees that would serve as 'dust and sound' barriers. This calls for creation of 'green belts' around the plant and colony and even sections of mines.
3. Landscaping so that the waste material from quarries is judiciously spread over and made use of in creating green slopes with trees, flowering plants, bushes, etc.
4. Install the 'state of the art' dust collecting equipment at all vulnerable points that would satisfy not only the present norms viz., 110 mgm/ nm^3 (or whatever is prevailing), but would also go beyond that so as to satisfy the norms of maintaining the same 'quantity' of dust emission even after expansion.
5. Take practical steps to keep nuisance from fugitive dusts down to a minimum.

6. Make asphalt / concrete roads to minimize dust nuisance from plying of trucks.
7. Minimize noise levels.

6.4 Location of the Plant

The important point here would be the location of the plant vis-a-vis centres of population.

If a minimum safe distance is maintained between plant and the populated area the nature itself will absorb dust emissions to a great extent and clean ambient air could be maintained at less cost.

6.5 Tall Chimneys Spread Dust

Similarly tall chimneys spread dust over wider areas and bring down dust concentration. Hence it is advisable to install chimneys much taller than those stipulated.

6.6

A good clean plant is an asset for the plant itself. It promotes good atmosphere, keeps high the morale of the workers who feel proud of and are happy to work in clean surroundings. A clean and dust free plant also keeps down maintenance costs of the plant.

6.7 Choice of Dust Collectors

The choice for selection of dust collection is between Bag Filters and ESPS. Only these two types can cope with dust emission norms of 110 mgm / nm^3 or 50-60 mg / nm^3.

Out of the two, ESPs are expensive for small gas volumes and the smallest application would be cleaning vent gases from cement mills and coal mills.

They are compatible with bag filters for large gas volumes like kiln and raw mill exhausts and cooler exhausts.

6.8 ESPS

ESPs are a little bit 'inconsistent' in dust emissions during 'start' and 'stop' periods of the plant. They also need continuous operation for reaching the designed dust emission levels.

6.9 Performance of ESP

Performance of ESP is dependent on :

1. humidity of gases,
2. resistivity of dust in the gases and
3. temperature of gases.

Humidity greatly improves the performance. Therefore kiln exhaust / raw mill system gases are 'conditioned' to increase the dew point to 50-55 °C by spraying water in a 'conditioning tower' through which gases are passed before entering ESP. Moisture in raw materials increases dew point. Therefore in 'combined operation' i.e., when gases enter ESP after wholly or partially passing through raw mill, quantity of water required to be sprayed would be much less.

In vertical mills, if moisture in raw materials is less, 2-3 % only, water is sprayed into mill to stabilize the mill operation (mill is likely to vibrate when grinding very dry materials).

Water thus introduced is totally lost and it is a sizeable quantity for direct operation - about 0.12 litres/sec/nm3 for a temperature difference of 200 °C.

Therefore for a 10000 tpd plant – which produces 417000 kg clinker/hr, quantity of water required for conditioning would be ~ $85m^3$/hr.

Thus if the plant is located in areas of scanty rainfall and poor water resources, ESP for kilns would be avoided.

6.9.1 ESPS for Coal Mills

ESPs are also used for cleaning vent volumes of coal and cement mill systems.

Coal dust has low resistivity and because coals generally have high moisture (8-10% on average) the humidity of exhaust gases is high. Therefore ESP is ideally suited for venting coal mill gases except for possibility of explosion. Explosions can take place if CO is generated due to slow combustion of coal. Coal dust and air also become explosive mixtures; when and at what temperatures cannot be predicted accurately. It depends on concentration of coal in air, temperature, and volatiles in coal among other things.

Hence ESPS used in coal mill applications would generally be designed for withstanding high pressure (internal and external) and would be fitted with explosion flaps or membranes which would let out safely, high pressures built up within. To avoid the possibility of explosion, at the root itself, hot inert gases from kiln exhaust with oxygen content of 6-8% are preferred for drying in coal mills.

6.9.2 ESPS for Cement Mills

Cement dusts have unfavourable resistivity and are too fine. Cement mills are sometimes equipped with 'internal water cooling' in the mill to keep down temperature of cement to < 110 °C. This increases humidity and hence situation becomes more favourable for application of the ESP.

Cement being the end product the higher the efficiency the better.

To maintain uniform temperature of gases inside the ESP, all ESPs are insulated and lagged. This prevents fluctuation of skin temperatures between day and night due to changes in ambient temperatures.

6.9.3 Cleaning Cooler Vent Gases

There can be 2/3 choices for dust collector to be used for cleaning cooler exhaust gases.

Till recently multiclones were used. Clinker dust is coarse and multiclones have high efficiency of +95% for coarse dusts. Even then, with say 10 gm/nm^3 as dust in raw gases, dust emitted for an efficiency of 95% would be 500 mgm/nm^3, which is too high by today's norms. PCBs have applied the norm of '110 mm/nm3' to cooler exhaust also. Therefore, multiclones are no longer acceptable.

The next choice is a bag filter. Cooler exhaust gases have temperatures between 180 to 250 °C. The volume and temperature increase considerably when kiln runs under upset conditions. Clinker dust is also abrasive. For using bag filters, cooler gases are first cooled down in 'air to air heat exchangers'.

ESPs are extensively used now for cooler vent gases.

However, because of poor resistivity of clinker and because of fluctuation of temperature and volume of gases, cooler ESPs are sized with low design migration velocities and for volumes under upset condition. This results in a very large ESP.

At 1.7 nm^3/kg clinker, in case of a 10000 t.p.d. kiln, vent gases from cooler would be 710000 nm^3/hr and assuming raw gas dust burden of 10 gm/nm^3, dust cntent would be 7.1 tph or 170 ton/day or annually 56000 tons; a gain of less than 2 % in the annual revenue. The ESP could cost as much as 120- 150 millions of Rupees.

The costs of investment for installing an ESP on a cooler are very high to attain a clean gas content of even 110 mgm/nm^3 and much higher if clean gas dust burden is brought down to 55 mg/nm^3.

The investment costs could be brought down appreciably if PCBs would permit a dust burden of around 200 mgm/nm^3 in cooler exhaust. Since cooler dust is much coarser than raw meal and kiln or coal and cement dust, it tends to settle within a short distance of the plant. A tall chimney would also help.

6.10 Bag Filters

6.10.1 Bag Filter in Kiln Preheater System

Bag filters are another alternative for dust collectors. Bag filters are either pulse jet or reverse jet type.

Performance of a bag filter depends on the fabric selected. The selection is governed by the temperature of the gases. Synthetic fabrics like nylon etc., can withstand temperatures of 120-140 °C maximum. Glass bag filters can stand temperatures up to 240 °C.

Kiln gases even with 6 stage preheater would leave kiln at 260-280 °C. The gases would have to be diluted with ambient air to bring down temperature to 120 °C when polyester bags are used. A glass bag filter could be used with much less dilution air to bring down temperature to 220 °C (to be certain).

1 nm^3 at 260 °C to be diluted to 120 °C would require, 1.65 nm^3 of ambient air.

Therefore the bag dust collector would have to be sized for a volume of 2.65 nm^3.

If all kiln gases are taken to raw mill for drying moisture and emerge at a temperature of 90-100 °C, there is no need to dilute them in 'combined' operation.

However there would be periods when mill would be stopped in which cases, dilution, would be necessary.

The volume and size of bag filter could be reduced if kiln gases are taken to coal mill for drying. But volume taken to the coal mill is much smaller than the volume taken to the raw mill. Therefore even if it is possible that raw mill and coal mills do not stop at the same time, it would be prudent to size the bag filter ignoring the coal mill. Over sizing of bag filter in the system is therefore a must. It, however, increases capital costs.

Compared to ESPs, bag filters cost, much less for the same volume, but the operational costs are high because of higher pressure drop across the dust collector. Compared to a pressure drop of 25-50 mm in an ESP, pressure drop across a bag filter could be 150-180 mms. (fabrics have been developed which would cause a pressure drop of only 100-110 mms; but such fabrics are very expensive in first costs).

Let kiln exhaust be 1 nm3 and pressure drop across an ESP be 25 mm. Therefore power of fan after ESP would be K × 1 × 25 kw.

For direct operation, bag filter would be sized for 2.65 nm^3 and pressure drop across it would be 150 mm requiring a power of K × 2.65 × 150 kw.

Thus power costs of bag filter in direct operation would be much higher.

Actually it may be necessary to pass larger quantum of gases through mill as it is an air swept mill. But then temperature of gases emerging from mill would be higher than 90-100 °C. To bring down the temperature to 100 °C, additional water would have to be sprayed into mill; or gas after the dust collector would be recirculated.

In combined operation, dilution with ambient air would not be required. Nevertheless it is necessarily required to oversize a bag filter to cope with conditions under direct operation.

6.10.2 Bag Filters for Coal Mills

In application for venting coal mill gases, bag filter had certain limitations but they have been successfully overcome.

The gases in the mill system are to be kept preferably between 80-85 °C. In actual practice they hardly reach 60-65 °C probably because of fear of explosions and fire hazards in the bag filter.

Moisture in coal being generally between 8-10 % (or more in wet season) dew point is high. If sufficient difference between gas temperature and dew point temperature is not maintained (min. 30 °C) then condensation can take place and 'caking' can take place on the fabric which immediately affects the performance of the bag filter.

Another worry is about bags catching fire either due to explosion in or before filter or duct leading to filter due to burning of coal and generation of CO in the filter itself. This is not an uncommon occurrence in cement industry. As in case of ESP, an inert gas helps in minimizing the risk.

It is customary to install CO_2 injection system in bag filter (or even ESPs) if presence of CO is detected and reaches dangerous level.

For this reason, coal mills are now being located near prheater end and inert preheater gases are used for drying coal.

6.10.3 Efficiency of a Bag Filter

The cleaning efficiency of a well maintained bag filter is consistent. It is not affected by changes in composition of gases, temperature, humidity or dust resistivity.

However, its performance is as good as the condition of the bags. Hence bags must be regularly inspected for their condition. This is not a small job. A bag filter for kiln and mill gases could be as big as a small godown with thousands of bags. Also bags have a limited life and hence need to be replaced periodically.

6.10.4 Bag Filter for Vent Gases

Bag filters are eminently suitable for small applications like venting of silos, bins, pneumatic conveying systems, crushers and so on. The temperatures are around 50-60 °C and dust burdens are not very high. Specially, designed bag filters are also available for installing on individual belt conveyors to arrest dust from vent gases at discharge points. For these applications therefore only bag filters could be considered.

Thus bag filters are expensive in operational costs (power) and also in maintenance. But they do not need water and hence are preferred in dry areas with scarcity of water.

6.11 Fugitive Dusts

PCBs now a days are concerned about fugitive dusts also. These arise in loading of hoppers and bins with stone, coal, etc., and in building up of stockpiles of stone, clinker, coal, etc.

Coal being wet does not cause much nuisance; clinker would generally be stored within a totally or a partially enclosed building. A bag filter would be installed at discharge point of the clinker conveyor bringing in clinker from cooler. A shaft with opening at different levels ensures that free fall is limited and dust nuisance is minimum.

In linear stockpiles telescopic chutes are used to maintain short free fall as piles are built up.

The linear stock piles of stacker reclaimer system for limestone are generally in the open except in locations which have extreme weather conditions. Stacker booms can be lowered to ensure a short fall. Thus even for building up stockpiles, problem of fugitive dust has been contained.

Water / Chemicals are sprayed in crusher hoppers where dust nuisance is inevitable as stone is dumped from rear dumpers into hoppers.

6.12 Green Belts

'Green belts' around the plant, even inside the plant and in colony provide not only barriers against nuisance of sound and dust but have a very pleasing aesthetic and beautifying effect. They also help to bring down the temperature by a few degrees.

To find beautifully landscaped gardens in a cement plant is becoming common. It has such a pleasing effect. Cement plants are tending to blend beautifully with the surroundings.

6.13 Noise

Principal sources of noise in a cement plant are Crushers and Grinding Mills. Crusher buildings can be enclosed but noise arising out of stone falling from dumpers into crushers cannot be eliminated altogether.

Vertical mills make much less noise. Cement Industry commonly uses vertical mills for grinding raw materials and coal. Even for grinding clinker and slag vertical mills are now preferred because of low power consumption. Therefore magnitude of 'noise pollution' has diminished considerably. Mill buildings can be enclosed to reduce it further.

Where ball mills are used and where the plant is close to population centers such enclosing of mill buildings would be essential

6.14 Emission of Greenhouse Gases

A new dimension has been added – that of 'Greenhouse Gas Emissions'

Carbon Dioxide is emitted in kiln exhaust gases as a result of calcination of limestone and burning of fuel. It is responsible for warming of the climate.

It is therefore now attempted at the design stage of the plant to adopt ways and means that would reduce ghg emissions- that is primarily emission of CO_2. Making blended cements reduces clinker to cement ratio and hence quantum of CO_2 released per ton of cement. Reducing fuel consumption would also do that.

Most new cement plants would be therefore deigned to make 'green cement'. Besides making blended cements, they would be replacing fossil fuel to some extent by carbon neutral alternative fuel, use, waste heat of kiln and cooler gases to generate power.

Kiln exhaust gases would be continuously monitored for CO_2 and Other constituents like NOX dioxin et al. **See Section 8** which deals with these aspects.

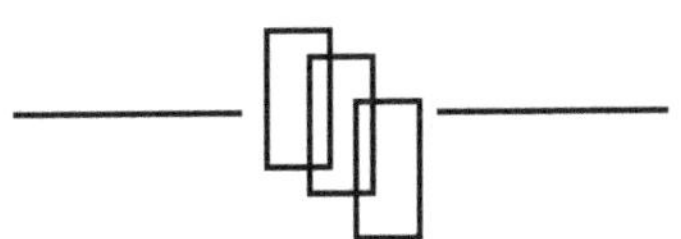

CHAPTER 7

INFRASTRUCTURAL FACILITIES

7.1 Infrastructural Facilities for a Cement Plant

Setting up a Cement Plant in a given location and running it there requires many infra structural facilities such as transport and communication, power, water, labour force etc.

Therefore all other factors being comparable, the site which has better infra structural facilities, would be preferred.

There would hardly be a place where all necessary facilities would be available in quantities required. In almost all cases the plant would have to add to and create some facilities that are lacking.

Availability of infra structural facilities like, a railway link or a highway or a sizeable substation in the vicinity can reduce capital costs substantially and also running costs.

A sizeable township like a district / taluq head quarters would almost certainly have basic amenities like primary and secondary schools, hospital / dispensary, market for daily needs of employees and facilities for housing. To that extent the company would have to spend less on creating facilities for colony, schools, hospitals and shopping.

7.2 Rail Links

Vicinity of the Market for cement is a highly desirable feature for any cement plant of any size and located anywhere.

However, "vicinity" has different connotations for small and large plants. For small plants, Markets may fall within a 200 Kms radius and for large plants, the radius could be 500 Kms or longer.

As time goes and competition grows, cement plants would have to go further and further to sell their product.

Therefore a railway line or a highway becomes a very vital link between the factory and the market.

Though in the early days many a cement plant was located right on the railway line, it is unlikely that this could happen in future.

A railway siding or a feeder road would be necessary to connect the plant to the main line or highway.

7.3 Road Links

As of now ~ 35 % cement is dispatched by rail in wagons and

~ 65 % by road . The share of road transport will continue to increase as more and more not so easily **accessible** deposits get exploited.

In the last two decades, quite a number of plants have come up without rail links. These plants depend entirely on road for receipt of materials and for dispatch of cement. Road Links would therefore assume even greater importance in future.

Feeder roads must be good, all weather "pucca" roads i.e. either tar or concrete as they have to carry heavy truck traffic that would only grow with passage of time.

In some locations like HP and Meghalaya, there are hardly any rail links. Hence a good network of asphalted / concrete roads is a great asset for such States.

Table 3.7.1

Capacity of plant	Cement Despatched daily	No. of Trucks 10 T/Load	density of traffic in Trucks/Hr.*	Coal brought in per day	Truck Loads	Trucks per hour*	Total Trucks per hour
t.p.d.	tons	numbers		tons	numbers		
3000	3200	320	22	540	54	6	56
5000	5300	530	35	900	90	10	90
10000	10600	1060	71	1800	180	20	182

* Assuming all dispatches of cement by road and all coal brought in trucks by road

The proposed location would have to be examined from this point of view. The available links should have sufficient capacity to cope with addition to traffic of the new plant and likely additions in future.

Assuming that the highway is about 15 Kms from the plant, a feeder road of 15 Kms would be necessary for trucks to ply on it to dispatch cement.

Truck traffic per day for cement plants with road link only would be as shown in **Table 3.7.1**.

Number of hours of dispatches are 15 and hours of receipts are 9 per day. Total trucks per day includes return journey.

For OPC

Coal consumption assumed to be 0.18kg per kg clinker on wet basis

It would therefore be seen that as plant capacity increases the density of truck traffic measured in trucks per hour increases.

It would require a good heavy duty all whether minimum 2-lane road to bear this traffic for a plant of 3000 t.p.d. capacity alone and wider for plants of 10000 t.p.d. capacities, solely dependent on road traffic

Further if the feeder road passes through a town or a sizable village, road within the township literally falls in no man's land. The township is expected to construct and look after roads in its area. In most cases they have no funds and hence roads within city / township area are largely in bad condition and too narrow to handle the heavy truck traffic.

Further passing through a populated area slows down movement and increases risks of accidents to human beings, animals, etc.

Hence it would be most desirable to "bypass" such towns. The imposes additional burden on project costs but it is worthwhile.

This burden may be shared with Roads and Buildings Departments. It may also be possible for Cement Companies who use the feeder road to come together to share costs of such a 'bypass' road.

7.4 Railway Siding

Railways may invite project entrepreneurs to finance costs of laying railway siding from the take off point on the main line to the plant. If the take off point is 2/3 kms from the plant, it would mean financing of about 6-8 kms of railway siding. If there are numerous culverts/ small bridges enroute, the cost goes up further.

If all of the cement produced was to be dispatched by rail, the traffic to be handled would be as shown in **Table 3.7.2**.

The railway traffic i.e, number wagons to be received and dispatched per day can be estimated and converted into rake loads as shown above.

The railways permit a fixed time for wagons to remain inside the plant and charge heavy demurrage if this time is exceeded.

Railway siding and loading and shunting facilities are to be designed so as not to incur demurrage.

7.4.1 Lengths of Railway siding

Length of railway siding can be worked out from lengths of rake Loads. See **Table 3.7.3**.

Table 3.7.2 Railway traffic to be handled at rated capacities.

Plant Capacity TPD	Cement Dispatched daily	40 Tons Wagons Loaded Daily	Rake loads of 50 Wagons each per day	Coal required @ 18%	Wagons	Rake Load of Coal Every
	Tons	No.s	No.s	Tons	No.s	Days
3000	3200	80	1.6 rakes	540	14	3-4
5000	5300	133	2.7 rakes	900	23	2
10000	10600	265	5.3	1800	45	1

Table 3.7.3 Rake lengths.

Wagons type	Railways designation	Capacity tons	Numbers Per rake	Capacity Of rake tons	Buffer to buffer length In metres	Length of rake* meres
Closed 4 wheeler	CRT	27.5	65	1800	8.82	610
Closed 8 wheeler	BCX	52.7	35	1850	15.75	587
Closed 8 wheeler	BCX	52.8	58	3060	15.75	970
Open 8 wheeler	BOX N	58.8	58	3400	10.7	667

* includes engine and guards cabin
Above data on wagons is for Indian Railways.
Source: Pawan Machinery Ltd, 'Railway Track Materials'

7.5 Power

All pertinent information on Generating Stations, major substations and transmission lines, their routes and capacities should be obtained from State Electricity Boards and their agencies. The Central Grid should also be looked into.

The surplus capacity available, should also be ascertained as also plans for addition to capacity in the pipeline and on drawing board should also be ascertained.

Quite a few large capacity Generating Plants have come up in the last decade in Private Sector. Details thereof should also be obtained..

7.5.1 Requirements of Power

As a thumb rule, (though in actual practice the plant would consume less than 75 Kwh in producing 1 ton of cement), it may be assumed that specific power consumption for large + 2 MTPA size plants would be80 kwh/ton of cement.

Thus a million ton plant or a 3000 tpd capacity plant making OPC would require

$$(3000 \times 1.05) / 24 * 80 = 10500 \text{ KWs}$$

or ~ 12000 KWs with a design margin of 10%

Considering a power factor of 0.95 and a load factor of 0.80, power required in KVA rating would be 16000 KVA

Allowing a capacity margin of 25%, the KVA capacity of the substation required would be _ 20000 KVA or 20 MVA

A 10000 tpd plant making OPC at 80 kwh/ton of cement would require(10000x1.05) / 24 *80 =35000 KWs or~ 38500 KWs with a design margin of 10 %

Allowing similar factors as mentioned above,capacity of the substation would be 63 MVA

Cement plants install captive power plants of between 30 –40 % of their requirements. This should be ignored while contracting power from the grid.

7.5.2 Voltage of Transmission of Power

For such capacity, the power would be delivered almost certainly at 132 KV in grid transmission lines and hence routes of such lines as also locations of major substations and their capacities should also be found out.

It would be perhaps necessary to string a new transmission line of 132 KV of sufficient capacity especially for the new plant from an existing substation.

In such cases it would be prudent to assume that plant capacity would be doubled and hence both substation and line should be designed for this capacity i.e., 40 MVA for a million ton plant.

The Electricity Boards invite financing of stringing transmission lines and installing substation. The costs incurred are deducted from monthly energy bills. This imposes additional burden and increases initial project costs but it definitely speeds up completion of the project.

7.6 Water

It is necessary to identify a perennial source of water like a river or a stream for the plant and the colony for its operation and for human consumption.

Wells would be dug in river/ stream bed and water pumped from them to the plant where it can be stored in ground level or overhead water tanks.

The requirement of water depends on the process and machinery employed and size of the plant.

Per se "Dry process" does not require water for the process of making cement. It requires water for cooling of bearings; compressors, blowers, D G Sets and such machinery. Cooling water can be recycled. It is collected and led to a cooling pond or cooling towers and then pumped back for cooling again. However, such 'recycling' requires that water is soft and free of suspended matter. If water available is naturally hard, a water treatment plant would be a must.

If the plant proposes to use ESP for cleaning exhaust gases of raw mill and kiln system, it would require substantial quantities treated water to condition the exhaust gases to bring down their temperature and to improve their humidity. This water is totally lost

7.6.1 Quantity of water required and its sources

The total requirement of water in a cement plant can be estimated as shown in **Table 3.7.4** by way of illustration.

Table 3.7.4 Requirement of water for a 3 MTPA capacity dry process cement plant.

sr no	section	water for cooling	make up	water lost	for human consumption	total quantity to beprocured
			20% in GCT			
		Quantity in m^3/day				
			a	b	c	a+b+c
1	Quarries and Crushing	2400	480			
2	Grinding Mills	4000	800			
3	Pyroprocessing	2000	400			
4	ESP			1000		
5	Comptrssrs	2000	400			
6	Miscelleneous	400	80			
7	D.G,Set	1500	300			
8	Colony	200			200	
9	leakage @ 2%	250			250	
	total	12750	2460	1000	450	16660
			~2500			~16700
	water cosumed		2500	1000	450	3950
	system to be designed with 20 % margin					20000

Actual requirements will vary from case to case depending on the contents. Mills of different types require different quantities. If a bag filter is proposed in place of ESP, requirement will reduce substantially. Similarly if a thermal power plant is installed in place of D.G. set mentioned in the table, requirement of water to be procured will be different.

It is in principle feasible to collect waste water from houses and treat it and re-use it. But this is hardly done on account of "prejudices". But it could be at least used for gardening and other non-potable use

Waste Heat Recovery

As and when a Waste Heat Recovery System is installed, requirements of of water would be increased in proportion to the capacity of Captive Power Plant.

7.6.2 Quarries as a source of water

Quarries can also be a source of water. It depends on the depth of under ground water table. As mining operations go deeper, it would be necessary to pump out water collected in pits. It could be used for cooling machines used in mining operations, like drills, compressors, etc. If available in copious quantities, it could be collected in disused or exhausted pits of the quarry and pumped to the plant for use in dry months or even all the year round.

7.6.3 Water from Borewells

Another source of water if there are no nearby perennial rivers or stream would be to tap underground water by drilling bore wells. The higher the underground water table, the surer this source of water would be. Depending on the size of the plant and its spread, a number of bore wells spread over the area would supply water either to a central tank / or to several tanks from which it could be distributed.

7.6.4 Rain Water Harvesting

However, the underground water table needs to be replenished. This normally happens in rainy season. Hence when source of water is significantly, bore wells the plant should make specific arrangements for **'rain water harvesting'** that is collecting rainwater and feeding it to specially dug pits to return it to soil

A water treatment plant to soften water is a necessity where ESPs and captive power plants are installed.

7.7 Manpower

One of the most important factors of production is human resources.

The productivity of Cement Plants has steadily increased and man hours per ton of cement produced have steadily gone down. Though we are still using more man hours compared to plants in advanced countries, the new plants coming up in the country are nearing these norms.

Chapter 8 deals with various aspects of manpower in greater detail.

7.7.1 Reduced Man Power in New Plants

Main reasons for reduction in manpower requirements are:

(i) Mechanized operation of quarries.

(ii) Mechanized material handling within the plant.

(iii) Semi mechanized operation of wagon and truck loading in case of bagged cement – mechanized or even automated "bulk loading".

(iv) Automated and microprocessor based process control in which the machines can be started / stopped from a central location and from where performance can be watched and monitored. Therefore, machinery 'attendants' are now conspicuous by their absence.

(v) Automated sampling and testing techniques in laboratory.

(vi) Computerized accounting and recording systems.

Thus even a 3 million t.p.a. capacity plant would now require a work force of about 480.

Presently Human Productivity expressed in manhours per ton cement produced is 0.31. **See Chapter 8** for further details.

7.7.2 Qualifications of Workforce

These changes have however necessitated higher educational qualifications and training of personnel required to run the plant.

Unskilled workers are now needed only for unloading of coal wagons, etc. and for menial work like house keeping. Even "Semi-skilled" workers like machine attendants samplers, 'oilers' 'greasers' are required fewer and fewer.

The number of "skilled and trained" workers, who "understand" process and are "computer literate", has increased. This means a minimum school / college level education / ITI training / to be supplemented by "Specific on job training" as per the need of the job is necessary at all levels.

7.7.3 Sources of Manpower

It is a good idea to obtain the operating personnel at junior levels from the surrounding area. They have their roots and are likely to be more permanent. The turnover could be less. This also solves "language problems", which is an important factor in a large multi lingual country like India.

The personnel in senior levels could be drawn / recruited from the Industry by providing better remuneration and perks and of course housing.

It is sometimes very difficult for a cement plant to recruit skilled personnel from outside if infra structural facilities like educational facilities and social amenities are not available locally.

The vicinity of a town / village can be a blessing in more then one-way.

(i) It provides personnel with basic training and education.

(ii) It may not be necessary to provide housing facilities for all.

(iii) It may not be necessary to provide for "schools and hospitals' facilities except by way of minimum.

However vicinity of a town / city / village can be a 'nuisance' also. Deviations in clean air norms would be magnified out of proportions and pressures brought on the plant to keep the ambient air clean. Sometimes, such pressures turn into ugly agitations and in interrupted production.

7.8 Social Amenities

Among the social amenities would come market facilities for daily needs of the staff. A large market for vegetables, provisions, milk dairy products, cloth and consumer goods would go a long way for happiness of the employees. Otherwise responsibility falls on the plant to provide for such facilities.

Some plants do run co-operative dairies and stores to meet daily needs of its employees.

"Hospital" facilities in the vicinity could be a great blessing, as they would come in handy for emergencies till better treatment can be arranged for.

Posts and Telegraphs and Banks are other necessary social amenities. Mobile phones have brought in a sea change in communications.

Internet has also a great impact on communications.

Satellite TV has greatly made up for "entertainment". Almost all plants now install dish antenna and provide Cable TV for the inmates of the colony. TV also has the virtue of making people feel part of the society at large and not cut off or isolated.

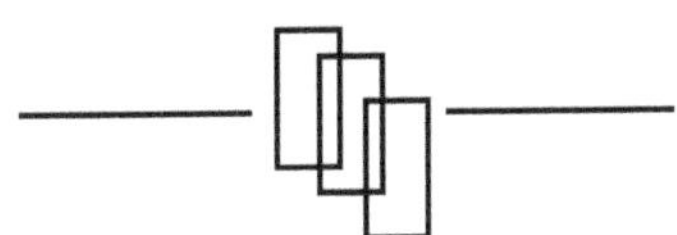

CHAPTER 8

MANPOWER REQUIREMENTS

8.1 Organisation for the Proposed Cement Plant

It is required to work out the Organization of the proposed cement plant in the early stages for several reasons.

1. Firstly, to define numbers and qualifications of workers to be recruited for the running of the plant,
2. secondly, for working out housing and other facilities to be provided for them,
3. thirdly, for working out salaries and perquisites, which are part of 'fixed costs' and
4. to assess requirements of training.

Important aspects that need to be given careful thought at project stage are:

(i) Qualifications and experience.

(ii) Numbers.

(iii) Emoluments and Perks.

8.2 Manpower Required for Different Functions

The total operations of the cement plant requiring manpower has

A. Technical & Operational side beginning with Quarrying operations and ending with dispatches of cement.

Alongside is required work force to ensure quality of cement through :

(i) Quality control.

(ii) Process control and Automation.

B. Non technical side which includes supporting activities like:

(i) Buying inputs like, power, fuel, additives, gypsum, etc., stores and spares.

(ii) Marketing and selling cement produced, and services associated with it.

(iii) Finance and Accounting – including all financial and accounting matters including inter alia, payment of salaries, wages and perks; payments to vendors; receipts from cement sold; working out all financial statements and profitability; payment of taxes and duties.

(iv) Personnel and administration including welfare and other matters related to the work force.

(v) Interaction with Government and other agencies.

8.3 Policy Making Body

Over and above all this organizational structure is the policy making Board of Directors –generally with a three/four tier structure:

1. Chairman.
2. Full time Directors.
3. Managing Director / CEO / President.
4. Working Directors who are Executives of the Company.

CEO / MD is the king pin who first brings into reality and later runs the project within the frame work of policies laid down.

To assist him are working Directors who are experts in their own disciplines like:

(a) Plant- operation and engineering,

(b) Marketing,

(c) Finance,

(d) Administration.

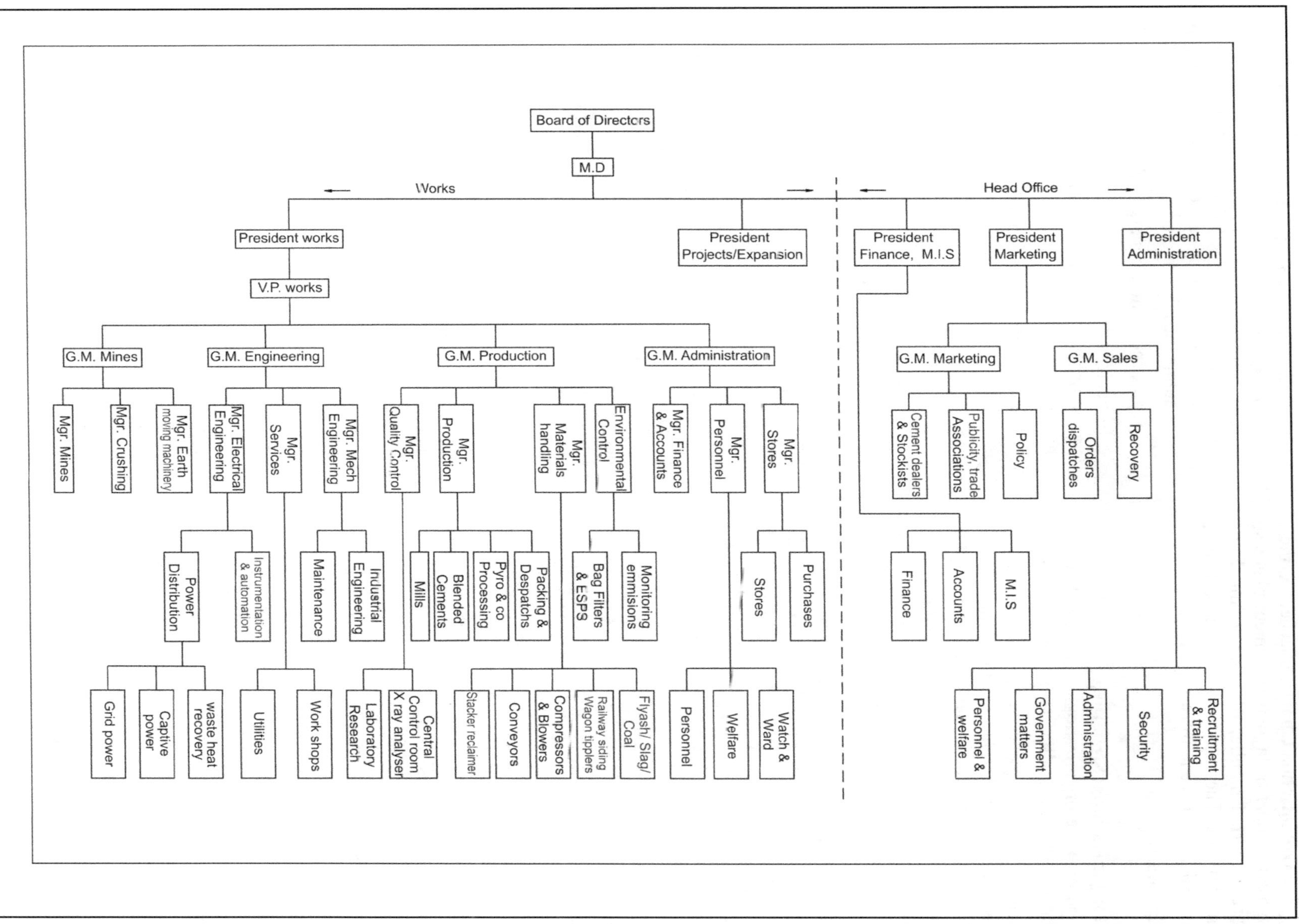

Fig. 3.8.1 Typical organizational chart - for a single unit / large cement plant.

A typical **Organisation Chart** is attached. As there are a myriad number of layouts of cement plants so there are a myriad number of organizational and manning patterns. Therefore chart presented here is to be treated as typical and illustrative only.

See Fig. 3.8.1.

8.4 Organisational Structure

Each one of these organizational sections shown in the chart has several smaller and related sections.

When each separate section is divided into sub-sections, levels of executives and their numbers and qualifications required and experience come to light.

In this manner the total requirement of personnel beginning either from top downwards or from bottom working upwards can be built up.

8.5 Productivity of Manpower

Among the various 'yardsticks' used to judge the operation of a cement plant, one yardstick is for measuring 'productivity of man power'; it measures cement produced in tons per man hour or man hours needed to make one ton of cement to judge its efficiency.

A cement plant of 3.5 mtpa capacity produces 3 million tons of cement / year (@ 85 % capacity).

The plant employs say 480 personnel excluding Head Office.

480 people – say each putting 7 effective hours / day for 283 working days in one year. (arrived at by deducting Sundays, paid holidays, earned leave etc.)

Therefore, man hours in put per year by one person = 283 x 7 = 1981; and by 480 persons = 950880

Cement produced in ton / man hour in 3 mtpa plant making OPC

= 3000000/950880 = ~ 3.2 ton

Man hours / ton of cement = ~ 0.31

They can be lower with fully mechanized loading and bulk dispatches and a greater degree of automation.

The organizational built up should be measured against such a yardstick.

8.6 New Concepts in Manpower Planning

Old concepts in manning have become outdated like :

1. Machinery attendants and machine operators (With motors and groups of motors being started from central control station, they are no longer necessary).
2. Oilers & Greasers – are new required only in a few locations because of centralized and automatic lubrication.
3. Sweepers and cleaners for house keeping.

In quarries, with mechanized operations, staff is required only to operate the machines like, drills, shovels and dumpers.

They are the few machines, which still require personal drivers.

In packing section

(i) Bulk loading – has reduced manpower needs substantially.

(ii) Where paper bags are used, packing machines are fed automatically. But in India, packing machines still require operators to feed jute / hdpe bags.

(iii) Loading operations of packed cement bags in trucks – trailers and wagons have also been automated / mechanized to a large extent.

The maximum number of personnel used in a cement plant was in quarries and in packing departments. Their number has shrunk substantially because of mechanized mining and mechanized loading operations.

8.7 Manpower in Quality Control

In 'Quality Control' and Laboratory, manpower was required for collecting samples from sampling points; in preparation of samples and in carrying out tests on them..

With 'X – ray analyzers' and automatic sampling and conveying systems such a force of semi-skilled and unskilled workers is no longer required or only a fraction there of is required.

8.8 Maintenance and Repairs

With preventive maintenance (and also predictive maintenance), with all the historical / statistical back up computers can now give readily by way of history cards, it is now mostly planned maintenance and servicing; refractory re-lining; periodical replacing of parts like hammers and liners; regarding and changing grinding media in ball mills and rollers and liners in vertical mills.

As machine sizes have gone up with plant capacities, many machines come equipped with their own lifting tackles and, removing equipment like hydraulic jacks, cranes, etc.

For example in case of ball mills, over head mechanised cranes facilitate lifting of lining plates. Portable conveyor belts speed up loading / unloading of mills.

In roller mills, rollers can be swung out and mills have systems of hoists and jacks to remove rollers and liners.

All this reduces number of maintenance workers in the class known as Khalasis greatly.

8.8.1 Replacing Sub-assemblies Reduces Manpower

Practice to remove and replace sub-assemblies and assemblies to save time, rather than carry out repairs of faulty parts is growing. This saves considerable down time. Removed sub assemblies are then attended to and reconditioned and kept ready for replacement. Sub-assemblies are kept in stock and replaced when used.

Where hammers / impact bars are to be built up, it is necessary to keep 3 sets, one working; one built up and ready to be installed and one in shops being built up.

8.8.2 Plannned Shutdown

That means when a planned shutdown is taken in any section, all or almost all parts that need to be replaced / adjusted are all available and maintenance jobs are carried out as planned.

Manpower required for the planned maintenance can also be planned in detail.

8.9 Outsourcing Maintenance and Repairs

These days it is not always necessary for the Works to keep its own staff for maintenance and repairs. Suppliers also take responsibility of maintenance against 'annual service contracts'. This is particularly true of machines like compressors, blowers, air conditioners and computers.

It is also true of 'process control and automation'. A new class of experts is available like 'hardware and software' engineers who take annual contracts and reduce Works' own permanent staff.

8.10 Knowledge Workers

'Knowledge workers' is how the operating staff in today's cement plant would be described.

Computers can serve as diagnostic tools. They can, not only analyse what is wrong and where and how much (deviations from normal conditions) but also help in arriving at a solution.

'Simulation programmes' are good training tools for the staff to get acquainted with the process and operating conditions and to judge actual conditions by comparing them with optimum / ideal conditions.

A plant does not require 'millers' and 'burners' now. It needs 'process experts' in grinding and in burning and combustion processes.

It needs 'chemists' who are experts not only in analysing samples of limestone, clinker, cement etc., but also in understanding the influence of various parameters like hydraulic modulii, silica and iron ratios; influence of C_3S, C_2S, etc., on quality of cement. With this knowledge they can better control the constituents and blending operations.

The 'know why' of the process and operation makes for better understanding of operation and hence it is easy to run a plant at optimum levels with minimum manpower.

8.11 Marketing

Marketing has assumed the greatest importance in present days of 'Free competition'. Decisions are required to be taken in :

1. Selecting / targeting markets.
2. Assessing likely share of that market.
3. Steps to be taken to achieve the targets such as:
   - (i) Advertising.
   - (ii) Appointing Dealers and Stockists.
   - (iii) Selling directly.or
   - (iv) all of these.

Selling costs are incurred for all these activities and take away profits.

Decisions are required to be taken as regards 'Distribution net work', and modes of reaching market- by rail / road.

Whether to have one's own fleet of trucks / special railway wagons or hire them is to be decided.

Quantum of commission to dealers, and credit period to be allowed are to be decided.

Communication net work is to be established, so that company's presence is felt all over in the market and potential customers can be easily contacted and orders processed and executed promptly.

In the situation where Supply exceeds Demand, crux is to supply quality cement at a given time and place in required quantities at a competitive price before any body else.

If a company can also furnish 'after sales services' so much the better.

Computers have proved to be a very useful tool in the marketing organization and have brought down the sales force.

8.12 Financing and Accounting Services

Financing and Accounting, has also been greatly assisted by computers and computer systems. Thus a plant should be making maximum use of computers and Information Technologies and advances in communications to reduce manpower.

8.13 Casual labour / Contract Labour

Large cement plants are subject to jurisdiction of 'wage tribunals' and are required to pay minimum wages and perks according to the stipulations that are in force at any given time.

With growing reliability of plant and machinery, with the developments in monitoring the health of the plant and above all with the reliability and accuracy and maintainability of the process control and automation, the need to depend on 'variable and error prone' human judgment is reduced greatly.

Even then workers are still required now and then to do jobs (though largely assisted by machines) like loading / unloading of wagons and trucks; brick laying; grinding media loading / grading; for house keeping and for shifting and lifting heavy machinery. Periodic housekeeping over and above regular one also requires labour force.

Managements may use either 'contract labour' or 'casual labour' for these purposes, remaining within the framework of laws laid down by concerned State and Central departments/ Tribunals.

8.13.1 Casual labour

Casual Labour as the name suggests is 'need-based' requirements of additional work force for mostly unskilled work. Casual labour is hired on 'daily wages' basis by recruiting at the gate from among the job seekers who gather at the gate every day.

When hired, these labourers are directed to the supervisor who has filled in the requisition for work on that day. Even hiring casual labour should be 'planned' for according to the maintenance and operational schedules prepared in advance. There has to be a 'protocol' laying down procedure for engaging the labour.

Over a period of time, daily requirement of casual labour becomes evident and casual labour can then be ordered on that basis.

However, when the situation indicates that casual labour is regularly required in certain departments, then the manning in that department should be scrutinized once again and revised.

8.13.2 Contract Labour

Where as casual labour could be recruited directly, Contract labour would be supplied by a 'labour contractor' on some long-term understanding. In this category would fall semi-skilled labour used to load trucks, trailers, wagons with cement bags (with and without assistance of truck and wagon loading machines) and unloading of wagons.

The request of contract labour for this purposes reduces by half when truck and wagon loading machines are used. If a wagon tippler is installed, labour for unloading wagons would not be required.

8.13.3 Labour for Housekeeping

One regularly irregular requirement is for labour for 'house keeping'.

The plant buildings, control room, offices, etc., can be kept clean by using industrial vacuum cleaners. They are particularly useful to collect cement spilled on loading platforms. Cement so collected can be returned to silo for reuse.

It is more difficult to deal with spillages of granular materials, crushed stone etc., spillages occur at transfer points of conveying systems and if not cleared regularly result in build up of piles that may reach the belt itself.

'Chronic' location of spillages must of course be examined from engineering point and corrective action taken. Spillage is also required to be cleared when jams are cleared from bins and hoppers. As has been mentioned earlier, some of this material can be very hot and dangerous to handle.

8.14 Contract Jobs

Some jobs like brick lining in kiln, cooler, preheater, have to be done periodically. In earlier days, brick laying was carried out under the supervision of the plant's civil engineer and was mainly manual.

With kilns becoming large, (albeit not so large because of calciner), fork lift trucks carrying brick loads can go right into the kiln and can also bring out removed bricks.

However, the total brick-laying job can be given on contract to outside parties with the plant personnel requiring only to supervise the quality of workmanship.

Similarly, repairs and overhaul of diesel engines, compressors etc., can be given on contract.

8.14.1

Thus, the plant may not be using much labour force directly, but it adds to salaries and wages when such indirect use of labour is also resorted to.

Managements should then look into the various aspects mentioned above and reassess the requirements of direct and indirect labour force taking into account facilities available at the site of the plant.

8.15 Manpower for a 3 MTPA Cement Plant

For illustrative purposes requirement of manpower for a 3 m.t.p.a. capacity Cement Plant has been shown in **Table 3.8.1.** Workforce has been divided into 6 categories :

1. Managerial (senior executives).
2. Senior Officers.
3. Middle level officers.
4. Junior Officers / Skilled Workers.
5. Semi Skilled Workers.
6. Unskilled Workers.

Table 3.8.1 shows Workforce for Mines, Plant operation, Maintenance, Administration, and Staff in Head Office. It is a summary of detailed break up department wise and category wise. The Organisational Chart presented in **Fig. 3.8.1** would be the basis of such an exercise.

This summary could be used to arrive at monthly or annual bill of Salaries and Wages which form part of Fixed Costs of Production. It will also be useful in deciding allocation of housing facilities in Company's Housing Colony.

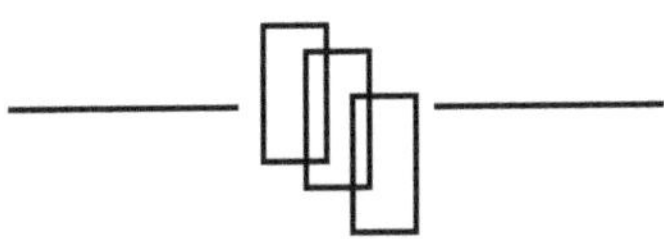

Table 3.8.1 Summary of estimated manpower department and category wise for a 1 mtpa capacity cement plant.

Category	Mines	Plant Operation	Plant Maint.	Quality Laboratory Sampling	Electrical Instrument Aytomation	Power Plant	Finance Admin.	Security	Marketing	Head Office	Total
Executives	1	1	1	1	1	1	1	1	1	1	10
Senior Officers	2	7	3	4	4	2	3	2	4	3	34
Middle level Officers	4	10	4	6	8	6	6	8	8	6	66
Junior Officers/ Technicians	8	20	12	10	10	6	10	16	12	12	116
semiskilled	30	40	20	20	20	10	10		10		160
Unskilled workers	20	20	30	10	5	5	2	2	2	5	101
Total	**65**	**98**	**70**	**51**	**48**	**30**	**32**	**29**	**37**	**27**	**487**

say 490

CHAPTER 9

IMPLEMENTATION SCHEDULE

9.1 Implementation Schedule

It is necessary to work out first a broad 'implementation schedule' at TEFS stage based on the historical data like:

1. Delivery periods of major machinery.
2. Civil construction period.
3. Erection and Commissioning periods.

To this also needs to be added period beginning with the TEFS and ending with placing orders for main machinery and auxiliaries. This period itself can be sometimes as long as 6-8 months.

9.2 Project Monitoring

As the project progresses and detailing is taken up, activities are broken down in small steps and time factors for each are assessed.

Many activities 'dovetail' into each other so that the time for completion is not an arithmetic addition of time taken by individual activities.

'Project Monitoring' is a multi-dimensional activity and a 'live' one. It is of great importance to :

1. Complete the project on time.
2. To monitor costs of project.
3. To monitor availability of funds for the project.

Machinery should be ordered taking into account, the delivery periods and knowing when it would be required at site.

For example – Motors and Gearboxes are required to be installed after the machinery is erected. Their periods of delivery are however short.

Hence even when the specifications and numbers have been frozen, smaller motors and gearboxes could be ordered much later thereby avoiding locking of funds.

See Fig. 3.9.3.

9.3 Tools for Monitoring

Monitoring can be done by the Management with the help of computers. They may also engage Consultants who have devised special software for the purpose. In the past Managements used :

1. Bar charts.
2. Critical path .methods.

Now, computers can be used to monitor each and every activity and present to the Management update of progress and also sound alarms for actions not taken at specified dates and point out likely delays in the completion of the specific jobs. Re-scheduling can also be done with ease.

Works and Head Office can send and receive files and bar charts on computer and can co-ordinate with Consultants and Contractors for ensuring smooth progress.

They can co-ordinate with Finance department to whom they have given requirements of 'cash flow' and ensure that funds are available for paying advances, progress payments etc., as agreed upon with respective vendors.

9.4 Project Implementation Schedules

Project Implementation Schedules are required to be obtained from:

1. Main Consultants for
   (a) TEFS
   (b) Detailed engineering – for inquiry

specifications and for floating inquiries, evaluating offers and for placing orders.

(c) Flow sheet and General Layout drawings.

(d) Departmental drawings, fabrication and utilities drawings

2. Suppliers for

(a) General arrangement drawings of machines.

(b) Foundations drawings and load data.

(c) Departmental drawings.

(d) Specifications of bought outs in their departments.

(e) Sizes and layout of ductings, chutes, pipes, hoppers and bins.

(f) refractories and insulation and lagging.

3. Civil Designers & Consultants for

(a) Civil design drawings for construction.

(b) Floating enquiries for appointing Civil Contractors.

(c) Appointing Contractors.

(d) Load bearing capacity of soil and other tests.

4. Civil Contractors for Construction schedule.

5. Erection Contractors for Erection schedule.

6. Geological & Mining Consultants for

(a) Preliminary investigations and for detailed investigation.

(b) Mining Plan.

(c) For making available representative samples for carrying out various tests on them for sizing and selection of machinery.

7. Commissioning Contractors / Suppliers for

(a) Carrying out no load runs and dry runs.

(b) Running up to capacity.

(c) performance trial/s.

9.5 Integrating Different Schedules

A concise bar chart is prepared at TEFS stage showing above mentioned activities in brief. This chart is then converted into a number of separate charts for individual activities when the project is taken up for execution.

See **specimen** bar charts attached.

Figs. 3.9.1 to 3.9.4.

Same charts can be used for regular follow up if put into computer. In **Figs. 3.9.1 to 3.9.4,** bottom blank bars are meant to be used to mark actual progress.

Computer printouts can be designed to high light deviations and also predict delays in project completion on their account.

It is a fast tool in the hands of the project team to monitor a project. Weekly / Monthly summaries with deviations and recommendations for actions to be taken can be submitted to the Management to enable them to take up the necessary corrective steps and to give advance notice for funds.

Implementation Schedules in **Figs 3.9.1 to 3.9.4**.are illustrative only

9.5.1 Project Completion Time

This can be reckoned from

1. the date of taking up the Techno Economic Feasibility Studies

   or from

2. the date of Applying for Financial Loan

   Or from

3. the date of beginning to place orders of machinery

A new Project Completion Schedule is attached in **Fig. 3.9.5** which shows completion times reckoned as mentioned above.

Presently it appears that a large green field cement can be completed

In ~ 40 months reckoned from the date of taking up T.E.F.S.

Or in~ 24 months reckoned from the date of ordering Machinery.

9.6 Funds Management

Funds Management will be a very important aspect of project implementation schedule. It takes into account:

1. Estimated project costs.
2. Funds classification – own, public issue and loans from Financial Institutions.

Table 3.9.1

Month	Funds - Required Projected Rs million	
January	50	own
February	50	own
March	100	equity
April	400	Loan in three
May	600	installments
June	500	
July	500	Own
August	500	equity
September	300	
	3000	

Interest rates may be fixed or floating; interest charges should be kept to a minimum, and this can be done by drawing funds as and when required according to the progress of the project rather than as per any fixed schedule. Even Financial Institutions would ask for progress reports while advancing funds. Working out 'Cash Flow' would be a necessary part of funds Management.

A typical statement of requirements and funds and possible sources is shown in **Table 3.9.1**.

debt / equity ratio - 1 : 1

Therefore, own funds and

equity = Rs 1500 million

Loan = Rs 1500 million

The loan could be arranged to be available half way through the project in 3 installments as shown there in. Loan repayment schedule has also to be drawn in consultation with Financial Institutions and adhered to.

9.6.1 Working Capital

Plant needs 'Working Capital' to start and operate the plant till such time as it generates its own funds from proceeds of Sales and becomes self sufficient.

This should also be programmed to be available for the start of the plant in right amounts. 'Working Capital' is worked out in the manner explained in Chapter 12 on Capital Costs of the Project. Project implementation schedule should see to it that it is available according to the programme for starting of the plant.

9.7 Managing Cash Flow

Government Institutions like Electricity Board, Railways would have to be paid large funds for transmission lines, railway siding, etc. 'Cash flow' statements should include such funds also and ensure that they are available when required.

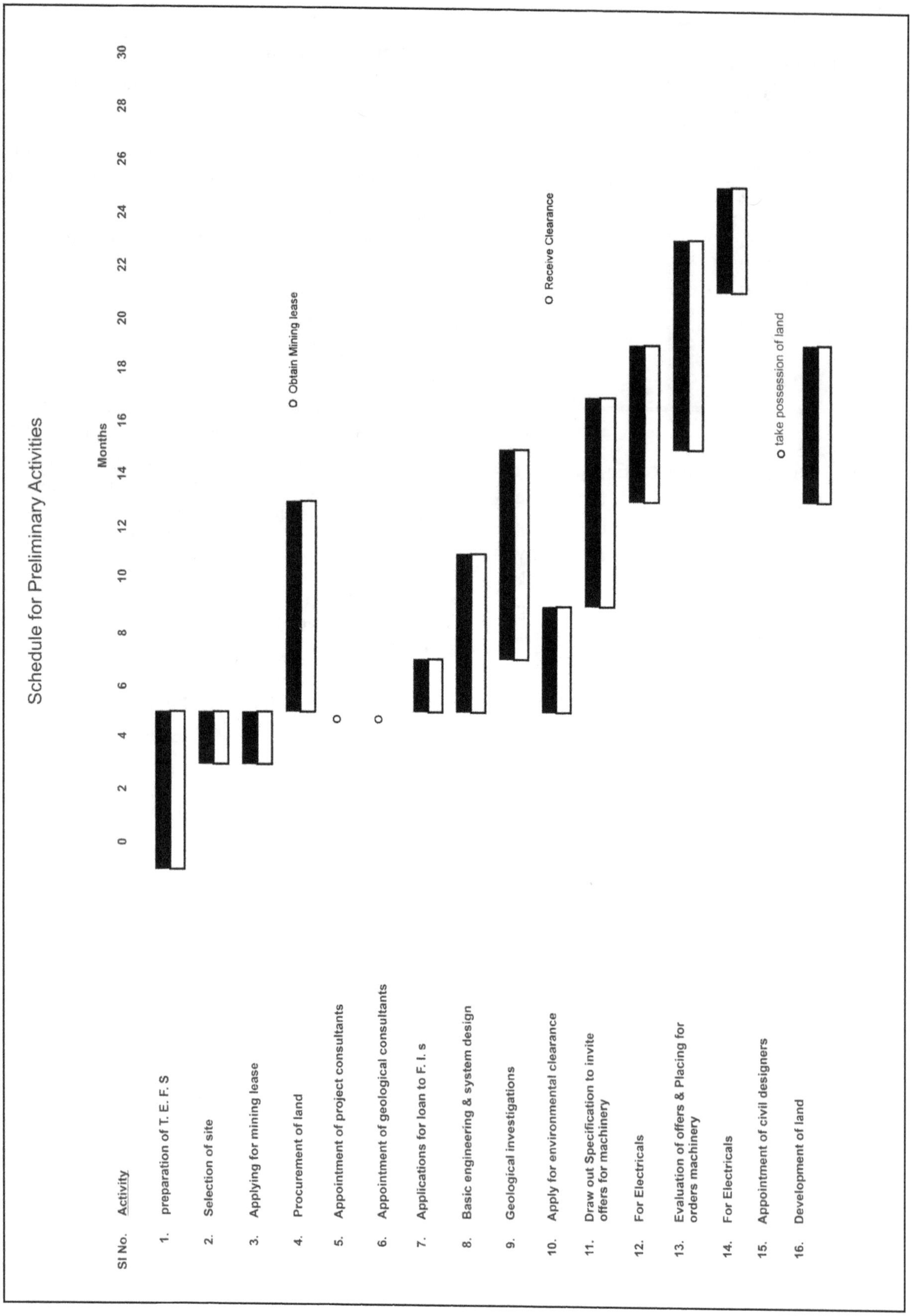

Fig. 3.9.1 Part 1-Preliminary Activities.

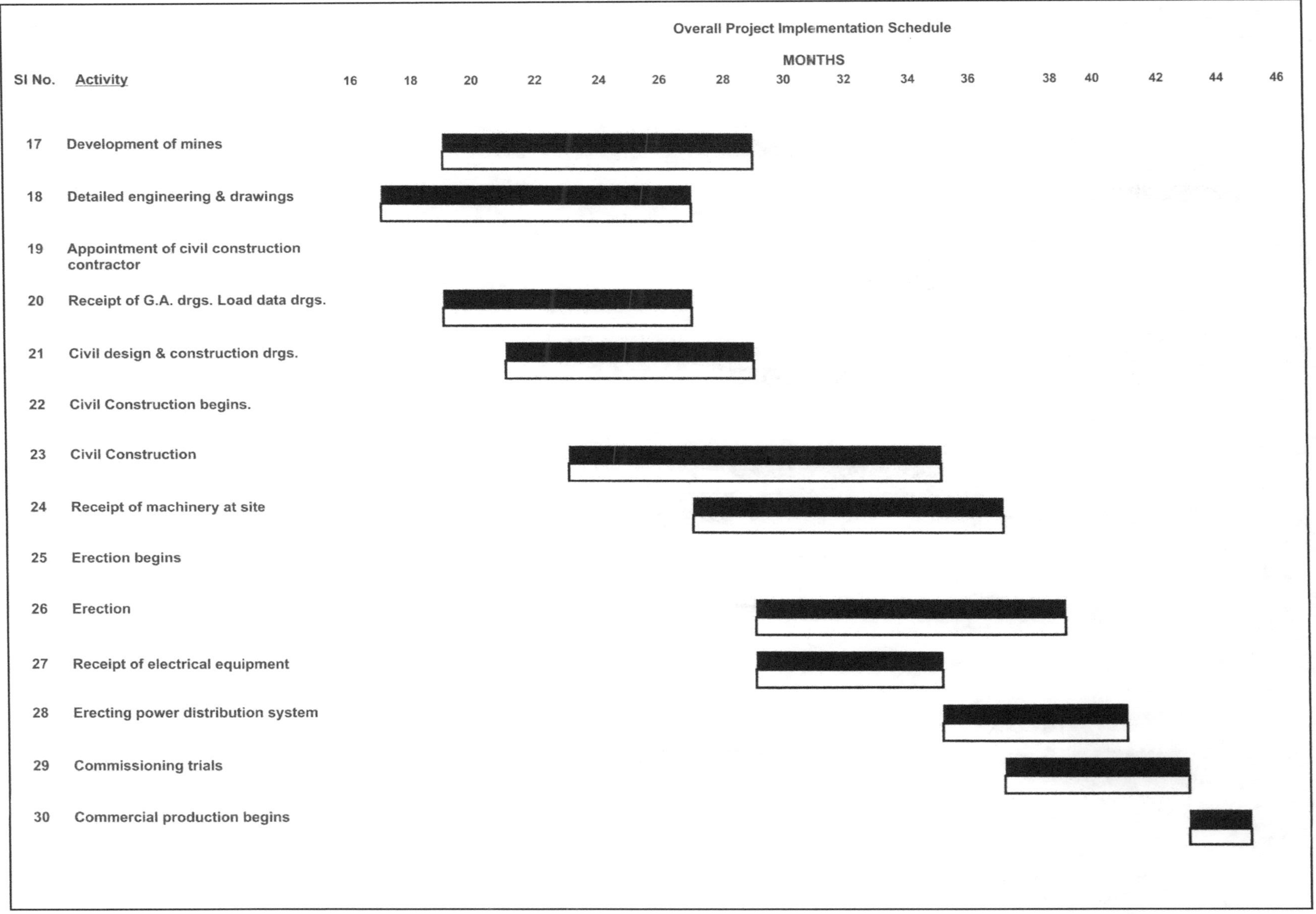

Fig. 3.9.2 Part 2 - Execution.

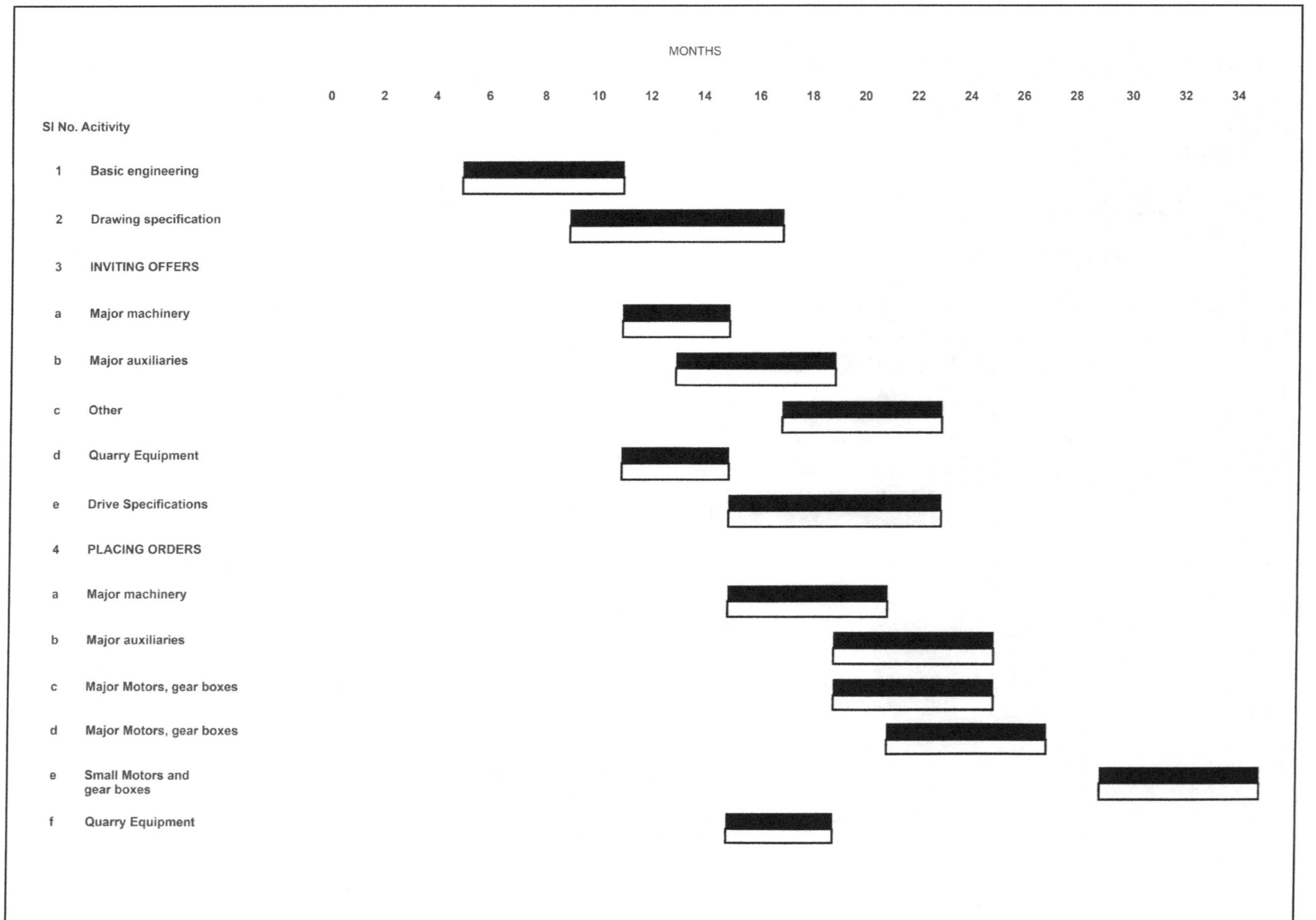

Fig. 3.9.3 Schedule for ordering machinery.

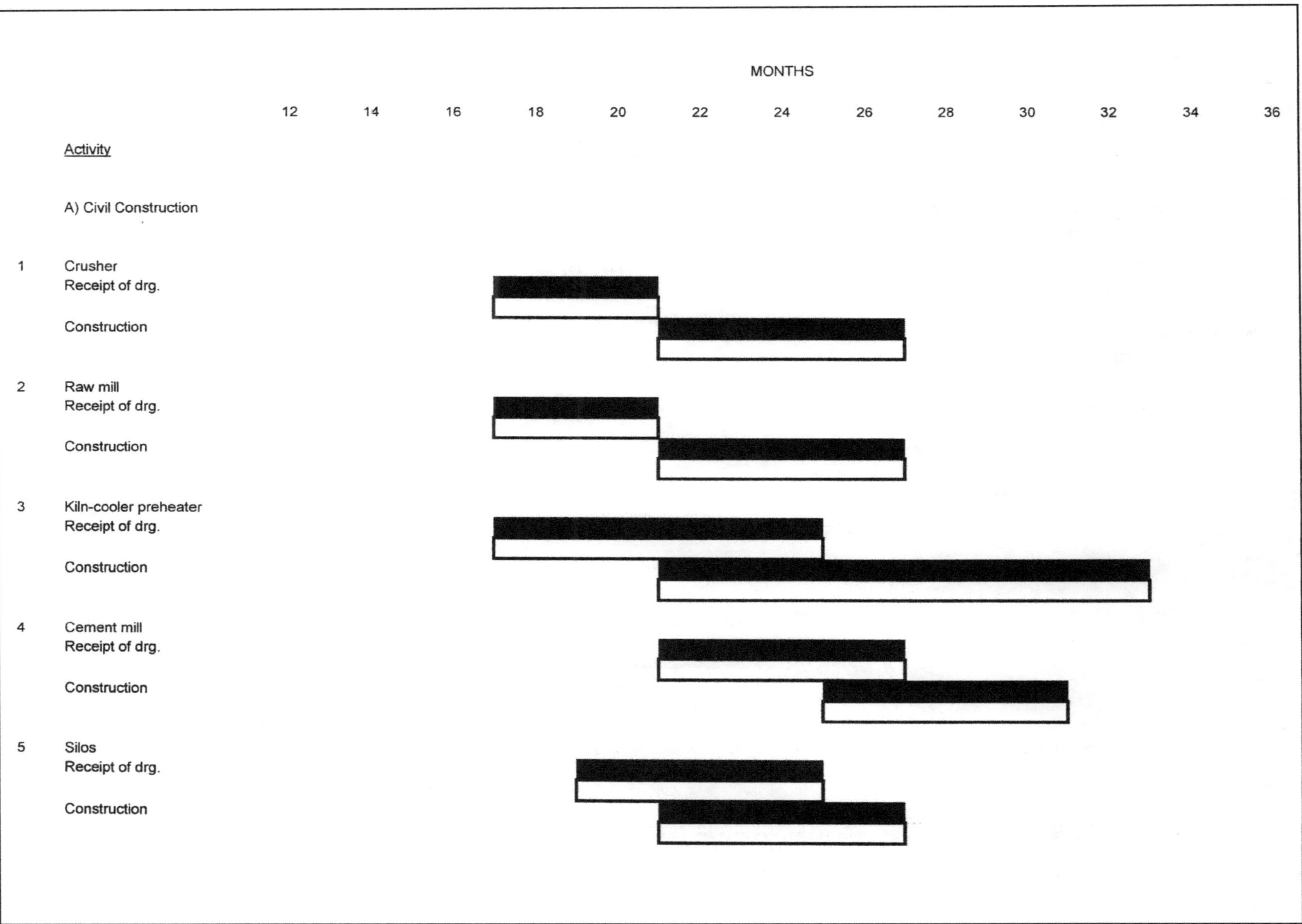

Fig. 3.9.4 Implementation schedule for civil design and construction.

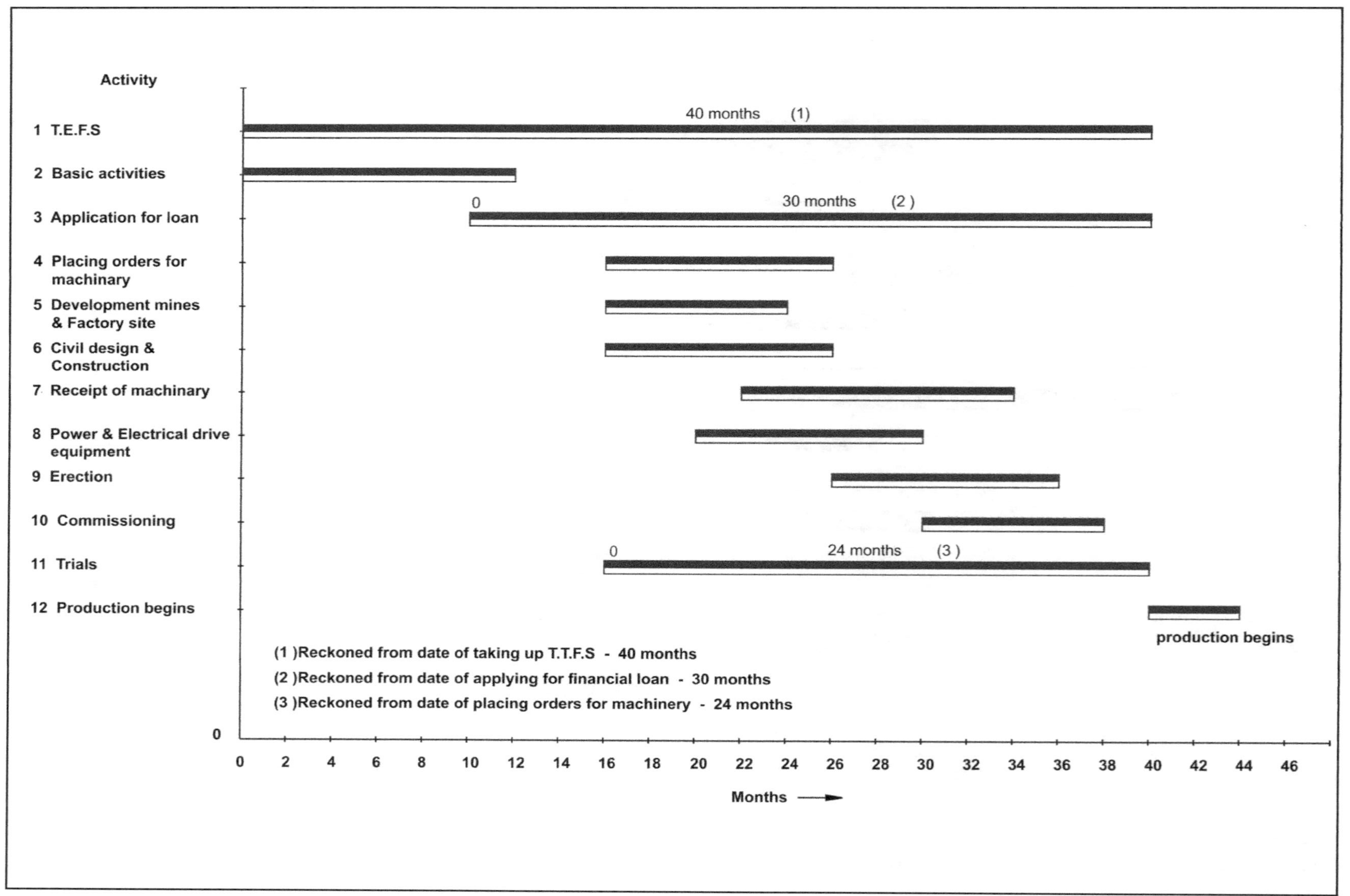

Fig. 3.9.5 Project completion schedule.

CHAPTER 10

ERECTION

10.1 Erection of Plant and Machinery

Erection of machinery is an important activity prior to the plant commencing production.

Erection signifies all those activities, which conclude in placing and fixing various machines and their drives on their foundations / base plates and connecting them up with one another through chutes, ducts and to hoppers. It includes where required insulation and lagging and brick lining. It includes connecting machines with power so that they are ready to run.

At the end of erection, machinery is ready for going on stream and for production.

10.1.1 Erection of Electrical and Power Equipment

In case of 'Electricals', it includes beginning with grid substation and ending with drives and instrumentations, - all activities, such as installation of Transformers, bus bars, laying of cables, HT, LT Boards, MCCS, starters and connecting them to respective drives through cables / wires of appropriate sizes so that when switched on, various drives are ready to drive machinery connected to them.

It includes installing field sensors for Instrumentations and Control, Sampling Systems, sample conveying systems, laying power and instrumentation cables to respective panels and instrumentation and control panels, program logic controllers and PCs.

It includes installation of X-ray analyzer and PCs in different sections as per design so that when switched on, it is ready to monitor the operation.

10.2

In short 'Erection' is an all-encompassing multidisciplinary activity, which results in the plant being got ready for starting production.

There are three distinct disciplines :

1. Erection of machinery and connecting them with one another to be able to process materials.
2. Erection of drives and power distribution system.
3. Instrumentation and process and quality control.

10.3 Organisation for Erection

Generally speaking –

1. 'Erection' is given on contract to experienced and well equipped erection contractors in these three disciplines.
2. Suppliers of machinery and equipment provide personnel who undertake 'supervision' of erection of their machinery and ensure that it is done according to the specifications laid down in respective erection manuals.

Most certainly, suppliers would insist that erection should be carried out under their Supervision and should be certified by them.

The Company provides co-ordination between Suppliers' supervisors and Erection Contractors and sees to it that erection goes according to plan and is completed on schedule.

Company's Executives also do overall supervision to see that erection is completed to theirs and Suppliers' satisfaction.

In case of Instrumentation and Process Control, including PLCs, PCs, many Suppliers undertake to do erection, wiring, connections, etc., themselves thus a separate agency for supervising erection may not be required.

10.4 Erection on Contract

When given on 'Contract', Company takes responsibility to see that :

1. Various civil structures and foundations are ready for installing machinery with suitable markings of centre lines and levels.
2. Machineries are available and are complete and that erection can be carried out in correct sequence.
3. Fabrication contractors provide, chutes, ducts, hoppers, platforms, etc., in sequence well before they are required for erection.
4. Stores and consumables and basic facilities like, access roads, and electric power are available.
5. Facilities for handling lifting and shifting heavy machinery are adequately available as and when required (when it is not the responsibility of erection contractors).

10.4.1 Contractor's Responsibilities

'Contractors' would take responsibility for :

(i) Providing skilled / semi-skilled and unskilled labour force for handling and shifting and lifting and erecting machinery.

(ii) Providing tools and tackles, like cranes, chain blocks, jacks and all and sundries like wire ropes as also sleepers, trailers and trolleys required for erection. They may also bring in their own heavy duty cranes and hoists.

They must co-ordinate with Suppliers and find out the heaviest weight and the largest piece of machinery that they need to handle and equip themselves with the tools / cranes hoists of adequate capacity as and when required.

10.4.2 Tools and Tackles for Erection

Erection contractors must also equip themselves with tools and instruments required for alignment, like piano wires, plumbs, spirit levels of different accuracies, theodolites, laser beams; dial gauge indicators, feeler gauges, calipers, micrometers, and so on and also copper shims and machined plates and wedges of various thickness for alignment. They would also require welding sets, gas-cutting equipments at site.

In case of 'overhead' cabling, contractors may undertake to make and install trestles and cable platforms as per drawings or alternately these could be made by a separate fabricator contractor and installed for laying cables.

10.5 Erection is a Coordinated Activity

Thus considerable co-ordination is required so that various pieces of equipment are available and are inspected;

basic work like cable trenches or trestles, cable gantries and cable trays are available :

all stores like materials for cable jointing, earthing, splicing, and insulation are available.

10.5.1 Company's Role

The company would probably give the responsibility from its side to one of its experienced Senior Executives say a General or a Deputy General Manager. Several Managers in disciplines like machinery, fabrication, electrical and instrumentation in turn would assist him.

These broad disciplines would then be sub-divided into sections like:

(i) Crusher and Auxiliaries.

(ii) Raw Mill and Auxiliaries.

(iii) Kiln, Preheater, Cooler and Auxiliaries.

(iv) Coal Mill and Auxiliaries.

(v) Cement Mill.

A sub-section may also be worthwhile for material handling equipment like belt conveyors, chain, and deep bucket conveyor and for Dust Collectors and ESPs.

The activities and responsibilities of various agencies need to be 'dovetailed' so that total time of erection is kept to a minimum by carrying out to the extent possible activities in different disciplines, simultaneously.

10.6 Erection Schedule

An 'erection schedule' is prepared which would give :

Overall erection programme showing beginning and completion of erection.

Section-wise detailed programmes.

Synchronization with civil and other activities is necessary so that their activities are completed in a sequence that will ensure taking up of erection work.

10.7 Progress of Erection

Erection will seldom wait till civil work of the building is completed. For example, in a preheater tower, the erection of cyclones would commence when 2 / 3 floors are completed and then completed progressively as construction of tower is completed.

Bed plates can be raised and placed on kiln piers as soon as each pier is ready.

10.8 Erection Drawings and Data

Suppliers must furnish 'erection drawings' which will furnish full data on centre lines, heights of centres, foundation bolts, grouting margins and 'erection and expansion' tolerances.

Generally, machined plates are embedded in faces of foundations to install bed plates or jacking devices for leveling.

First, half grouting is done when machines are positioned and aligned.

Fine tuning is done and grouting is completed using special shrink proof compounds.

Each machine would have its special requirements for alignment, leveling, expansion, locating, and for facial and diameter 'outs'. Suppliers must make these available to the erection contractor and erection finally checked with them after completion.

10.9 Fabrication and Erection of Chutes and Ducts

Once main machines are in their place, connecting dampers, gates, chutes and platforms can be erected and linked.

Large ducts may need sizeable supporting structures. Expansion joints would be required and should be suitably supported.

Fabricator contractors must manufacture these items according to drawings so that they are available well before they are to be taken up for erection.

Supporting structures and platforms must be ready and erected to receive ducts, dampers, expansion joints, etc.

Long lengths of ducts are made in straight pieces of convenient lengths. Flanges are made in pairs and generally supplied one fixed and one loose.

Flange joints and ducting must be carefully aligned so that there is no leakage at them, asbestos packing is used to prevent leakages.

Expansion joints and flanges there of are also to be aligned carefully so that there is no stress on the joints that would result in their cracking.

Joints must be carefully supported so that there is no hindrance for designed expansion nor do the joints take load.

10.10 Completion of Erection

Erection is complete when machinery rotates freely and moves freely by hand and later by power and that tolerances are within limits in all positions of the machine.

'Dry runs' is an important part of erection in which machinery erected is turned on power empty with of course necessary lubricants.

'Dry runs' will bring out :

1. Misalignments.
2. Vibrations.
3. Tight or loose bolts.
4. Over heating of bearings.
5. Leakages in oil and water pipe lines and even air leakages when fans are running.

These defects would be corrected before machines are fed with load or material.

10.11 Brick Lining and Insulation and Lagging

Brick lining is a major activity in the erection of kiln, cooler, calciner, preheater and connecting ducts which carry hot gases.

Brick lining will include :

1. Refractories – basic and alumina bricks.
2. Insulating bricks / blocks.
3. Castables.
4. Suitable mortars.
5. Retaining rings to hold bricks; these should be welded in place as shown in drawings given by Suppliers in kiln, cyclones, ducting, etc.

Insulating bricks and blocks require bolts to hold them in place. These may be MS or Stainless Steel according to temperatures they are subject to. 'Castables' need specially designed anchors to hold the castable.

Anchors are welded to casings ducts, shells of kiln. Castables, need great care in pouring, vibrating for consolidation and in curing. Heat generated during is to be taken away by using chilled water.

While main Suppliers will indicate quality and thickness of lining, details may be worked out either by Suppliers of bricks or Consultants who would make 'brick drawings' for procurement.

For large plants, brick lining is a massive job, requiring handling of thousands of tons of bricks, lifting them to various levels, laying them in position. It would require scaffolding to be made for supporting men and materials inside cyclones, ducts, etc.

In case of kiln, kiln needs to be rotated and bricks laid in position with the help of jacks, etc. Bricks are generally supplied from burners' platform. Thus for kilns of 60 m length, bricks have to be taken inside over a distance of 60 m. For large kilns, fork lift trucks are used to take brick loads in the kiln; other devices like portable belts could also be used.

Hoist and may be lift (if ready) will be used to lift refractories in a preheater tower.

Now a days ISO Standards are followed in dimensioning bricks. Even then, in a kiln, 2 types of bricks are used with different angles for better grip and stability.

All hot ducts and bag filters and esps and also gas conditioning tower need to be insulated and lagged for conserving heat in ducts and to maintain skin temperature of casings of bag filters and esps and gct above dew point.

This also is a sizeable job and erection is complete in true sense when it is carried out. Insulation and aluminium cladding is generally given on contract.

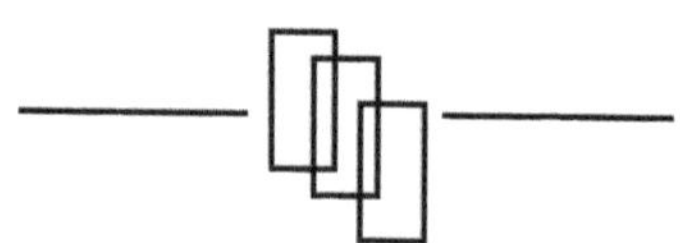

CHAPTER 11

COMMISSIONING

11.1 Commissioning

This is the last phase of the Cement Plant Project. It begins with 'dry runs' of individual and groups of machineries and ends with the stabilization of the plant and achieving desired / rated / guaranteed outputs and other parameters like specific fuel consumption, specific power consumption, etc.

Erection ends when 'dry runs' or no load trials are successful. Commissioning begins there.

Dry individual runs are followed by 'no load' trials of groups of machines in a section or in a department.

11.2 No Load Trials

'No load' trials are preparatory to feeding the machines with materials and gases they are supposed to handle and process.

They are essential in that they bring out faults in alignment and erection. They give an opportunity for testing machines for several hours to see if there are any unusual noises and vibrations, whether bearings are running hot, whether cooling water is adequate and the degree to which it is heated.

They also bring out faults in alignments of belt conveyors, whether belting tends to run on one side – belt tensions, V belt tensions and alignments. In short they help in getting ready the machinery in a section to receive load for processing.

Advantage should be taken of 'no load' trials to note and record 'no load currents' of major drives like mills, kilns, fans, etc.

In 'no load' trials mills and kilns are checked for alignments in all directions – radial, facial, levels, etc. The gears are 'run in' and impressions are taken of gear / pinion contact by applying colour / ink on teeth and taking impressions on paper, cello tape, etc.

If the contact is not adequate, then the gears are 'run in' by applying 'emery paste and oil' on gear / pinion teeth, so that in running high spots are rubbed out. This is true also of kiln gear.

11.3 Running in of Mills

The mills are 'loaded' gradually. If the design load at rated capacity is 100%, then mill is run with say 15%, 25%, 50%, 75%, 90% loads of grinding media with corresponding, material and gas flows in the system. This step by step 'running in' procedure helps in getting good 'seating' of main bearings and gear drive and must be followed though number of steps could be reduced.

11.4 Sequence of Commissioning

When a mill is fed with materials, it is also necessary to see that the material is carried off to its storage like :

Raw Meal from raw mill to Blending Silo.

Fine Coal from coal mill to fine coal Hopper.

Cement from Cement mill to Cement Storage Silos.

It is therefore necessary to get ready the following section first. That is for running the raw mill, the pneumatic conveying equipment and the blending / storages silos should be ready in all respects to receive the ground raw meal and blend it. This rule will apply to all sections.

11.4.1 Sequence in Crushing

In crushing and stock piling sections, the sequence would be as shown in **Fig. 3.11.1.**

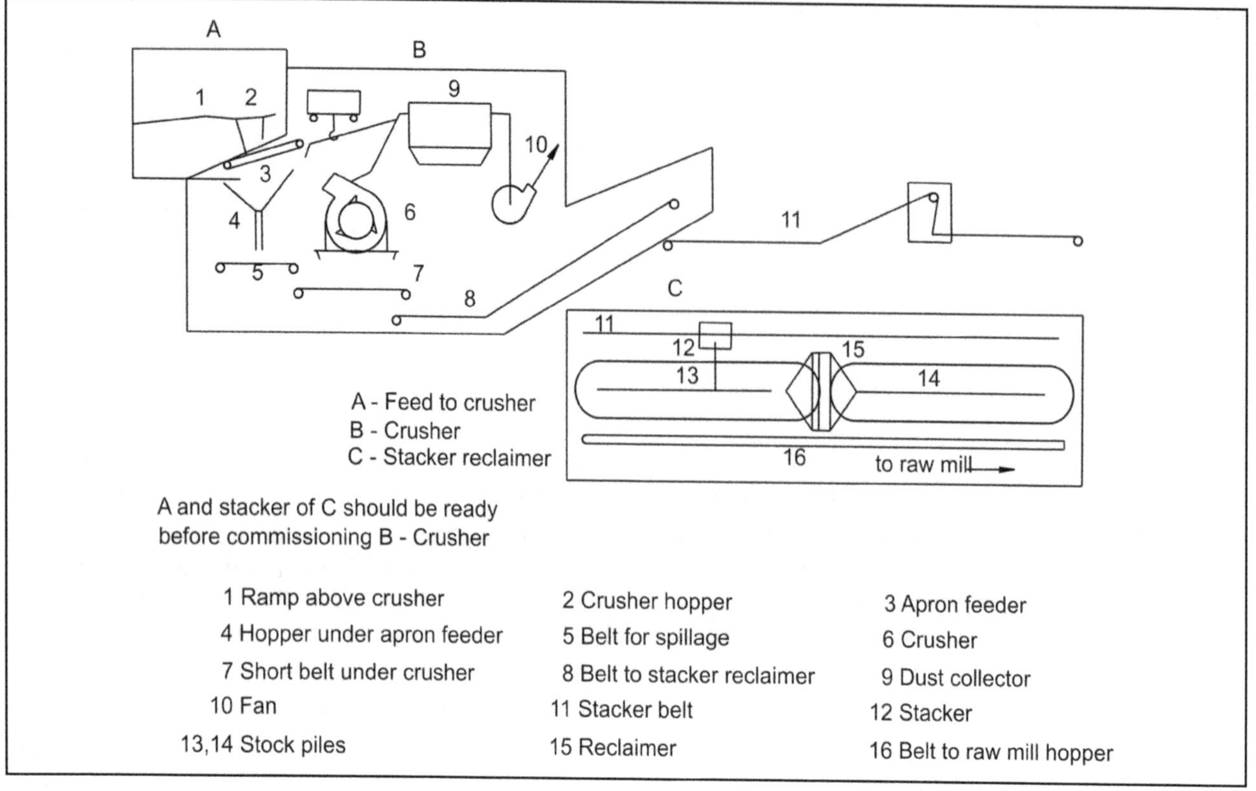

Fig. 3.11.1 Commissioning of crushing and stock pilling sections ; single stage crusher and stacker reclaimer system.

Stacker system (C) should be got ready to receive crushed stone from the crusher. Crusher ramp and hopper and apron feeder (A) should be got ready to feed crusher. In turn quarry face from which run off mine stone would be brought to crusher should be in a position to supply stone. In crusher house (B), belts after crusher taking stone to stacker belt should be ready.

Venting system to clean gases should also be ready before stone is fed to crusher.

Reclaimer system should then be got ready to send stone to mill hoppers

11.4.2 Sequence in Raw Mill

In a raw mill section various groups of machinery would be commissioned in the sequence shown.

See Fig. 3.11.2.

A 'hot gas generator' would be necessary when starting from cold when hot gases from kiln are not available for drying. Such a generator would have been included in the project. A similar (smaller) generator would also be required for Coal Mill.

11.4.3 Sequence in Kiln

In pyroprocessing or clinkering section, preceding sections like blending and kiln feed and dust collection system should be ready ahead of commissioning of kiln; following sections like clinker conveyor and clinker storage should also be ready simultaneously.

See Fig. 3.11.3.

11.5 Checks After Partial Load Trials

It would be a good idea to inspect the machinery after partial load trials are taken to make sure that :

1. Seatings of bearings, gears are as they should be.
2. Lubricant supply and oil scrapers working satisfactorily.
3. There are no vibrations.
4. Bearings are not running hot.
5. There are no unusual noises.
6. Alignments of mill surfaces like head flanges, gears are within limits in all positions.
7. Kiln keeps contact with all rollers at all times.

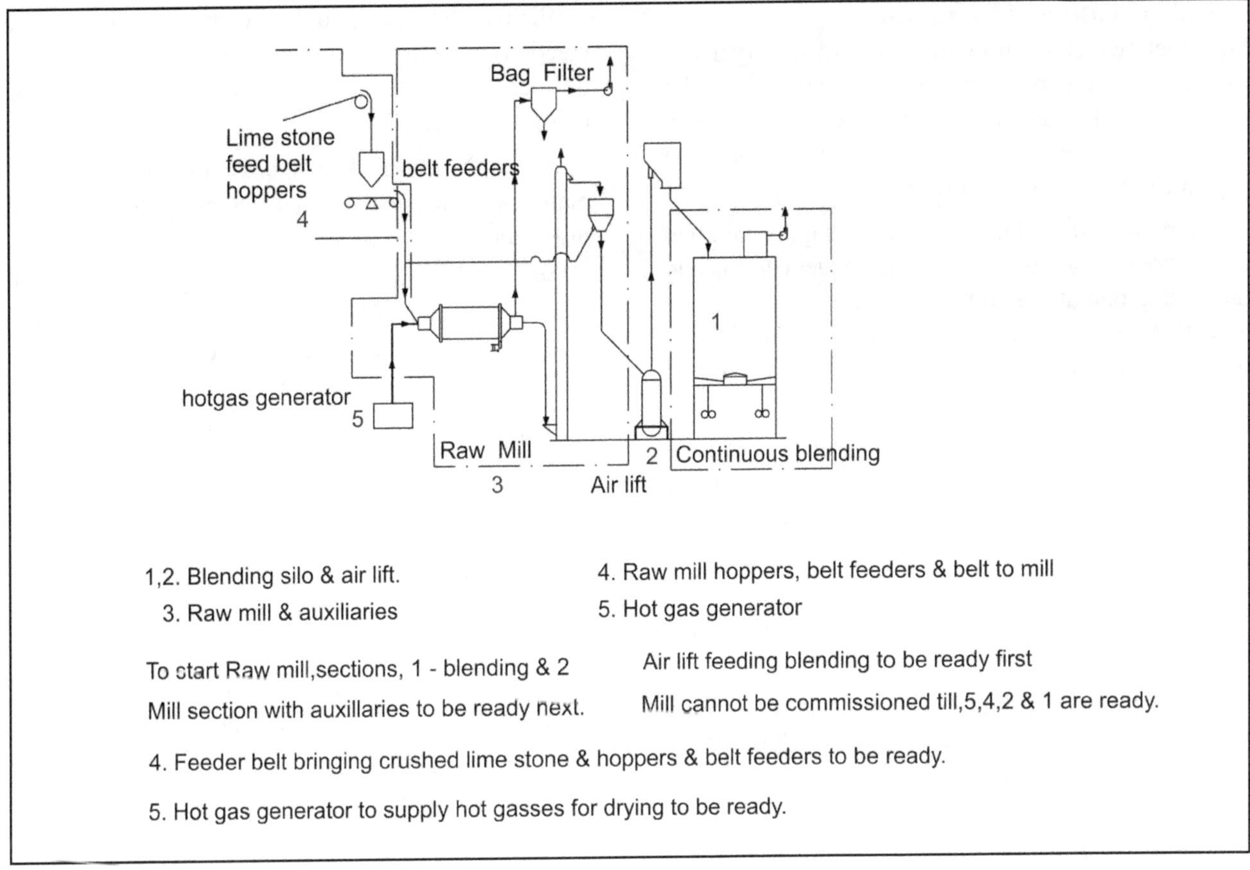

Fig. 3.11.2 Commissioning – raw mill section.

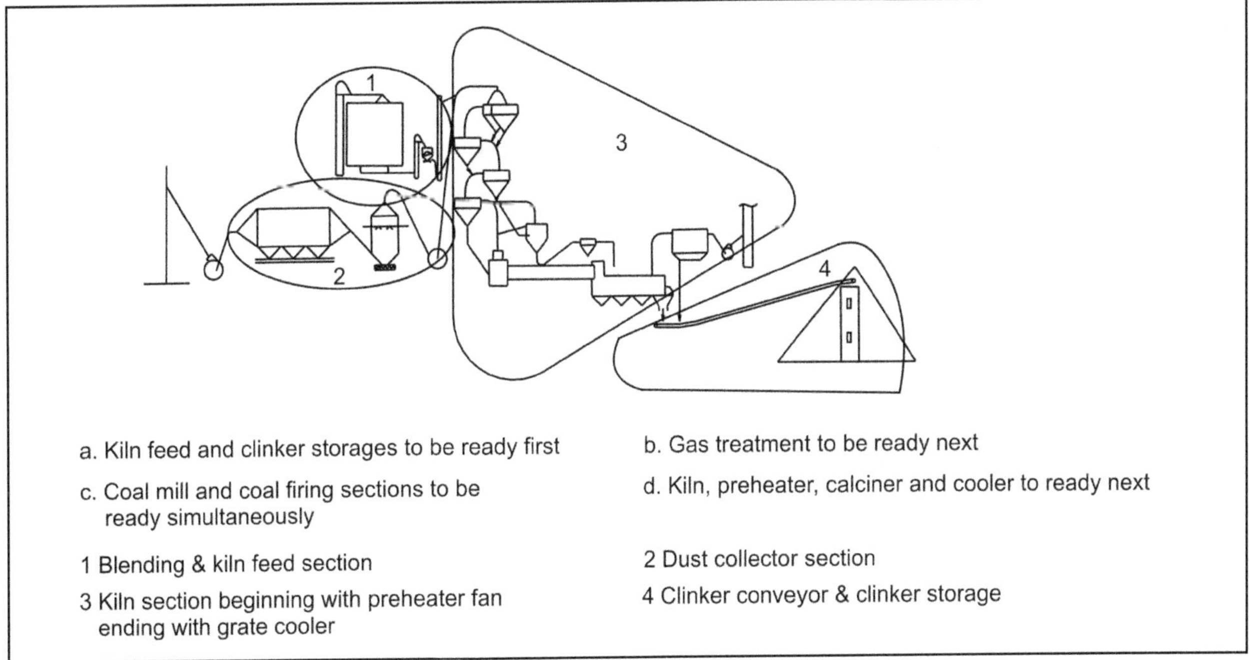

Fig. 3.11.3 Commissioning of kiln section.

11.6 Preheating of Refractory

Machinery which has been lined with refractory needs to follow a step by step procedure of drying the refractory and preheating it slowly to get it ready to withstand the temperatures to which it would be exposed during actual operation.

Refractory manufacturers would supply detailed instructions on laying as well as subsequent drying and preheating operations. It may take 2-3 days or longer to do this for the first time as temperatures inside kiln, preheater etc., are raised gradually to avoid thermal shocks.

It is also desirable that in the process of drying and preheating, power supply is continuous so that there are no long interruptions in the process and kiln runs continuously and bricks are heated evenly.

In this period burners are run in 'turned down' position. Burners should therefore be selected to have a good 'turn down ratio'.

Initial drying and preheating can also be done more conveniently with the help of oil burners. Hence in the project on oil burner of requisite capacity should be included for the purpose.

Temperature can be raised after initial preheating by starting coal firing again following Suppliers' instructions step by step. This procedure is of great importance for first lighting up when bricks / castables are freshly laid.

Subsequently, if there are long shutdowns, refractories should be again preheated slowly in a similar fashion but then it can take a much shorter time.

11.7 Commissioning Manuals

Before taking up commission it is necessary to collect Suppliers' manuals for commissioning. These should be studied and machinery should be commissioned following instructions given there in.

Notes on observations, abnormalities, etc., during commissioning should be made and brought to the attention of Suppliers' representatives.

11.8 Supplier's Supervision

If Suppliers have taken responsibilities for performance (capacity, power, fuel, etc.) they would insist on machinery being erected and commissioned under their supervision.

Plant authorities should provide all facilities to the erection / commissioning personnel. These may include besides, electrical power and fuel, raw materials and manpower also.

Some Suppliers take responsibility for training the plant operating personnel at different levels in different disciplines.

Plant authorities should plan for such training well in advance so that 'trained' personnel of requisite educational and working experience are available for the commissioning of the plant.

They should actively associate themselves with the actual commissioning so that they get 'hands on' job training during commissioning.

11.9 Bringing Plant up to Rated Capacity

The various sections of the plant should be gradually brought up to their respective rated capacities.

It means that the plant has planned in advance and has arranged for uninterrupted supply of :

1. Raw materials of stipulated quality and physical and chemical properties and in requisite quantities.
2. Fuel of quality, assumed in the design.
3. Power supply, including captive power plant.
4. Manpower.

In the running of the plant to rated capacity record must be kept of operational parameters and they should be checked and compared with designed parameters for the respective rates of feed.

Significant deviations should be noted and brought to be attention of Suppliers.

On their part Suppliers should also point out deviations in properties of materials, fuel actually supplied as compared to the properties on which design performance was based.

11.10 Performance Trials

Once a section has been stabilized and has been running continuously, a performance trial can be taken on that section to prove its performance according to guarantees taken by Suppliers.

Primarily it is the responsibility of Plant Authorities to ensure that raw materials and fuel as per design specifications are available for proving the performance. The deviation in performance if any, cannot then be attributed to the raw materials and fuel.

11.10.1 Protocol for Conducting Trials

A 'protocol' or procedure for conducting the trials should be worked out jointly by Plant Authorities and Suppliers. This procedure will include inter alia:

For raw mills and coal mills

(i) Size of feed.

(ii) Product fineness.

(iii) Moisture to be dried.

(iv) Grindability of materials.

(v) Capacity guaranted in tph at specified fineness.

(vi) Specific power consumption.

(vii) Specific gas volume.

(viii) Ash content and its particle size.

(ix) Silica content and its particle size.

For kiln section

Raw mix feed to the kiln should bc as per representative samples tested and raw mix arrived at that would satisfy norms like LSF, SM AM etc. Clinker to be produced, litre weight, free lime, C_3S, C_2S etc.

Fuel to be used - its useful calorific value/ moisture/ fineness and ash content.

The most important points on which agreement should be reached are with respect to how to establish outputs and fuel rates.

It should be spelt out if installed feeders can be used to establish feed rates of raw meal and coal.

Whether clinker output could be established by drop test/ or by using conversion factors or by silo measuments. There should be no ground for disputes.

All pertinent factors should also be listed such as specific fuel consumption and specific power consumption. Facilities should be lined up to establish them.

Results of performance trial/s should be recorded and signed by all the concerned parties. If trials are successful it is to the credit of all. Should there be a shortfall, concerned parties should get together and agree upon further cource of action.

The earlier the performance is proved the better it is for all concerned.

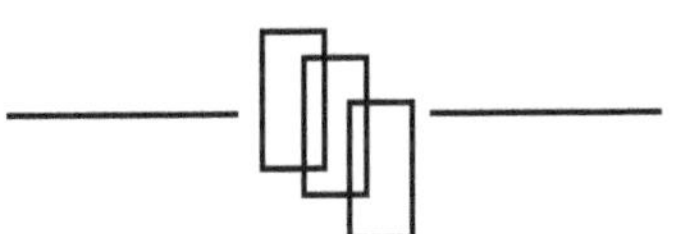

CHAPTER 12

CAPITAL COSTS OF THE PROJECT

12.1 Capital Costs of the Project

Techno-Economic Feasibility Study – furnishes Budgetary Capital Costs of the project and investment decisions are taken there upon after working out profitability, pay back period, and other financial ratios.

The capital costs therefore should include all items large and small that make up the project costs.

The prices should be either updated from earlier cost estimates or worked out especially for the project.

It is best to leave the cost estimates to Consultants, who work regularly on Capital Costs Estimates for different projects of their clients.

They have thus a 'data bank' from which pertinent information can be collected and collated intra extrapolated to build up the estimates.

12.1.1 Suppliers Furnish Budgetary Prices

Many machinery suppliers also furnish budgetary costs of plants of their design and supply when requested. In this they may or may not include 'bought out items'

But if specifically requested, they would include these along with machinery costs. Capital costs thus received would of course be pertinent to the plant of their design and sizing and the process flow charts normally followed by them.

Machinery schedules corresponding to process flow charts of respective Sections should also be submitted so that the picture is complete and there is less chance of missing any items.

12.1.2 Buildingup of Project Costs by the Cement Company

If Project Management of a Cement Company wish to build up the project costs themselves, they would have to take first all the steps described earlier leading to selection of Process, types of machinery, capacities etc., at the selected site.

They should generally approach three machinery suppliers to obtain competitive budgetary prices.

The Suppliers may be asked to indicate broad specifications of bought outs and enquiries would have to be floated to the suppliers of bought outs to obtain budgetary prices from reputed suppliers. A list of motors and gearboxes can be made and enquiries floated to suppliers for obtaining these prices.

All this work can take 2-3 months' time or even longer. Consultants may be able to work out budgetary prices in a shorter time because of their 'data banks' which they keep periodically updated.

12.2 Civil Construction Costs

One of the major item of costs are 'Civil Construction Costs'. Once the project machinery schedule has been built-up and capacities of machinery and their sizes are known, civil costs can be built up for an 'average location' with average load bearing capacity of soil and using known seismic and wind load factors for the region in which the project would be located.

The costs can then be revised to allow for specific site conditions like:

1. Low load bearing capacity – (Piling if required).
2. Depth of foundations to reach rock surface.
3. For large storages like blending and cement silos – raft areas to sustain loads of filled silos; for kiln piers total cubic content of piers and the loads coming on them- dead as well as operating.

For cement machineries like ball mills, crushers, vertical mills and kilns – factors for static and dynamic loads are known and are more or less similar for all machinery manufacturers. These factors together with sizes of base plates are useful in arriving at size of foundations and their weights.

12.2.1 Empirical Methods of Working Out Civil Construction Costs

Costs of building for preheater towers, mill houses, packing plants can be judged from the total area of the floors of the respective buildings.

From such broad preliminary data, civil costs can be estimated.

Data Bank of old and completed projects are a very useful source both for consultants and managements. RBI indices for labour costs and data on prices of cement, steel for reinforcement and structural purposes come handy in working out changes in costs in the data bank and the date of the project.

Costs of foundations are worked out separately and are included in the costs of Plant and Machinery (to avail of higher depreciation).

12.3 Build up of Total Civil Construction Costs

Table 3.12.1 illustrates build up of Civil construction costs for various factory and nonfactory buildings. The Overall costs are presented by regrouping as shown in **Table 3.12.2.** This format is approved by F.I.s. (Financial Institutions).

Table 3.12.1 Illustrative work out of costs of Civil Construction.

Sr.No.	Description of building	Type of construction	Area m^2	Rate Rs/m^2	Estimated cost Rs million
1	Crusher ramp m^3	rubble	1120	$550/m^3$	0.62
2	Crusher house	rcc	2500	500	1.25
3	Limestone storage	rcc	1250	2200	2.75
4	Raw mill house	rcc	600	3300	1.98
5	Blending and storage Silos storage cap. 3100 t	rcc		1000/ton	3.10
6	Kiln feed	rcc	70	3300	0.23
7	Preheaer tower	rcc	750	4000	3.0
8	Kiln piers	rcc	800	3000	2.4
9	Grate cooler	rcc	600	3300	1.98
10	Coal mill building	rcc	400	3300	1.32
11	Clinker conveyor and Clinker storage	Steel and rcc	1600	2200	3.52
12	Cement mill house	rcc	750	3300	2.48
13	Cement silos 4000 t	rcc		1000/ton	4.0
14	Packing plant	Steel and rcc	400	2500	1.0
15	Storages for clay, coal, gypsum	Steel and rcc	2600	1300	3.38
16	Compressor house	rcc brick	350	2600	0.91
17	Substation, h.t.and l.t. board Rooms, m.c.c. s, control room etc.		600	3000	1.8
18	Stores, workshops, garages Time office, laboratory, bags godown etc.	rcc and brick	1800	2200	3.96
19	Admin. Office, guest house, Rest hall etc.	rcc and brick	900	2200	2.0
20	Water tanks over head and underground	rcc	650	2300	1.5
21	Miscellaneous				2.5
	total				45.68
	say				46.0

Table 3.12.2 Civil construction costs are presented to F.I.s in the following form.

Sr.No.	Item	Cost Rs million
1	Factory buildings	18.15
2	Silos and storages	15.35
3	Godowns stores etc	3.96
4	Admin, buildings	2.0
5	Non factory buildings	3,8
6	Miscellaneous	3.88
7	Colony	10.0
8	Architect's fees	2.0
9	Railway siding*	5.0
	total	62.67
	Rounding off	63.0

*when required.

12.4 Costs of Electrical Power Distribution System

Electrical Consultants can similarly furnish costs of power distribution systems and process and automation and quality control equipment.

12.5 Mining Equipment

Costs of few machines involved in quarries like drills, compressors, shovels, dumpers, explosive, van, jeep, maintenance, tools, etc., can be obtained from respective suppliers of these machines, once sizes and capacities and numbers of various types of equipment is fixed taking into account lead from quarry faces to crusher.

12.6 Major Machinery and Auxiliaries

1. ***Crushing – Transport of crushed stone***

   While crusher and immediate preceding and following auxiliaries could be the same for plants of a given size, differences would arise in equipment for transport of crushed stone depending on location of crusher vis-à-vis works and mode of transport. These would be peculiar to a project and should be allowed for specifically.

2. ***Raw Mill***

   Given type and size of mill and flow sheet, costs of mill and auxiliaries can be estimated with the help of Suppliers.

   Since this is quite costly, it would be a good idea to obtain the 'updated' costs from the suppliers. Even Consultants should do this, rather than depend on data bank.

3. ***Blending & Kiln Feed***

   Flow charts would show whether blending silos will be at ground level or raised to house aeration equipment and kiln feed system under them. Choice of kiln feed system, particularly weigh feeder / solid flow meter and airlift / bucket elevator would decide cost of this section.

4. ***Preheater – Calciner – Kiln – Cooler***

   Preheater **costs would depend on:**

   Number of streams (One or Two) and number of stages 5 or 6, apart from sizes of cyclones.

   Calciner costs are related to and depend on whether calciner is in line or off line. It is a fabricated volumetric machine. Knowing the volume of the calciner its size and cost be roughly estimated.

   In calciner costs should also be included, take off point from kiln /cooler dedusting chamber and TA duct and coal firing system for calciner.

   Kiln – knowing dimensions of kiln and number of supports, its weight can be roughly gauged and costing can be carried out. However it would be best to obtain price from Suppliers. Hydraulic thruster should be included for large kilns and also seals of latest design and throttle valve at inlet for one stream preheater.

Cooler – Decision will have been taken about the design from among the several available designs.

Their prices must be ascertained from vendors.

There is also substantial difference in specifications of fans for cooling and vent air and should be finalized with the Supplier of the cooler.

It is advantageous to have spillage conveyor for the length of the cooler and a full product conveyor subsequently.

5. ***Clinker storage and extraction***

   Once type of storage and system of extraction are finalized, capital costs can be worked out. Provision must be kept for standby for clinker conveyor - better still, it should be included in capital costs.

   When plant is expanded there should be interchangeability between storage and conveyor systems of the two kilns and provisions would have to be made in capital costs for the same.

   If there is a possibility of selling /exporting clinker, provision should be made in capital costs for the necessary equipment.

6. ***Cement Mills***

   Like raw mill, capital costs can be estimated once circuit, size and capacity are fixed.

   Cement Mills will be in closed circuit and with high efficiency separator, whether ball mill or vertical roller mill.

   Combination of Roller Press and Ball mill with high efficiency separator could also be considered

7. ***Packing Plant***

   High capacity – 100 / 200 tph capacity, electronic packing machine with two loading points would be most certainly used to suit the layout.

   Loading – Truck and Wagon loading machines would be used. Whole layout and number of machines would depend on proportions of rail, truck loading in bagged and bulk dispatches of cement.

12.7 Data Banks of Prices of Machinery and Auxiliaries

Consultants can and do build up a data bank of costs / prices of individual machines and auxiliaries of various capacities from various machinery manufactures and update them regularly. and update them progressively. Their help should be taken. It will save lot of time and effort.

It can be a beginning of building up of Company's own data bank.

12.7.1 Computer Helps in Creating Data Banks

With computer it is easy to add and update and also delete obsolete data.

Data Bank must show – date, year – for which the price is shown.

In times of urgency, the prices can be updated by a factor, taking into account rates of inflations for the last 5 years or so.

For example, if the last data on prices was 3 years old and rates of inflation in these three years were 5,4 and 3%, projected price in 2003 based on these rates of inflation would be:

$1.05 \times 1.04 \times 1.03 = 1.12$ times, i.e., 12 % increase over the price in year 2000.

12.8 Taxes, Duties and Insurance

In working out project costs and price buildup for plant and machinery, it is necessary to include the following costs also:

1. Packing, forwarding.
2. Freight and Insurance.
3. Sales Tax – Central / State.
4. Excise Duty.
5. Unloading at site and octroi duty if any.
6. VAT or any other taxes of individual States.

It is not practical to do this exercise individually for each item of machinery. It is therefore done on an overall basis like adding a percentage to ex-works / f.o.r. price quoted by the vendor (If price quoted is f.o.r., then packing and forwarding need not be included).

Local fabrication work does not attract tax in many States; Some States charge 'works contract' tax on such work. A Consultant has to acquaint himself with all such matters.

12.9 Imported Machinery

For imported machinery, prices quoted would be generally on FOB basis. To this are to be added ocean freight and insurance to arrive at its c.i.f. price. This price is to be paid in foreign currencies and should be recorded separately as such to arrive at the total foreign exchange required for the project.

All imported prices would be subject to fluctuations in 'rates of exchange' up to a certain mutually agreed date between buyer and supplier. Project costs should provide insurance against such escalation separately.

Customs duties on Imports vary from time to time and are different for different machinery.

A group of machines imported for a 'Project" would normally be charged duties at lower rates. Consultants or entrepreneurs should obtain updated rates of customs duties under various categories.

12.10 Fluctuations in Rates of Taxes and Duties

The taxes and duties, and also freight rates vary from time to time. Some items of machinery attract special taxes and duties and should be segregated so that this tax is not omitted.

Special rate of 5% import duty is applicable under EPCG against export of the product manufactured by the Company within a specified time.

12.11 Machinery Wise and Department Wise Buildup of Total Taxes and Duties

Hence two statements would be necessary in building up capital costs.

First showing for different categories of machinery taxes and duties they attract as shown in **Table 3.12.3.**

The second statement would be to workout average factor for a section of Cement plant as shown in **Table 3.12.4**.

Of course it would be best to treat all imported machinery separately irrespective of the department to which it belongs. Average mentioned above can be worked out for the rest of the machinery in the section.

Excise Duty varies from product to product and for big or small manufacturer. This should also be included in Data Bank. State Sales tax rates vary from State to State. There is also Central Sales Tax. It is customary

Table 3.12.3

Description	Price ex works	Packing & forwarding	Freight & insurance	Excise duty	Sales tax	Octroi / Unloading	Customs duty	others	Total
	In million Rs								
		%	%	%	%	%	%	%	%
Earth moving									
Roller Mill									
Crusher									
Conveyor									
Elevators									
Packing machine									
Compressor									
Fan									
Motors									
Transformers									
Grinding media									

Items included above are only indicative.
This statement indicates prevalent rates that could be used to build up project costs.

Table 3.12.4

Description	AVERAGE - % SECTION					
	crushing	Raw mill	kiln	Coal mill	Cement mill	packing
Packing & Forwarding	3					
Insu. % Freight	3					
Sales Tax	4					
VAT	-					
Excise Duty	10					
Octroi	1					
Unloading	1					
Custom Duty	-					
Port clearing						
Local transport						
Charges for over Sized machinery	2					
total						
Values shown are indicative rates						

Table 3.12.5 Capital Costs of Crushing Section.

Sr.No.	Item of machinery	Nos off	Unit F.O.R. Price Rs.	Total F.O.R. price Rs.
1	feeder	1	10,00,000	
2	crusher	1	75,00,000	
3	Short belt under crusher	1	3,00,000	
4	Long belt to stacker	1	10,00,000	
5	Dust collector and fan	1	5,00,000	
6	crane	1	2,00,000	
	total			105,00,000

to include Sales Tax at 4 % under 'C' form, that is Central rate, as majority of the items would be coming from other States.

12.12 Building up Capital Costs of Plant and Machinery

A good way to build up capital costs is to take the help of flow charts and Machinery schedules for various sections.

Against each item of equipment in the schedule the cost structure is written to arrive at cost of the item at site. If the various elements of duties and taxes are similar, a short cut can be taken by adding F.O.R. prices for the section and to add up taxes and duties for the whole section.

As an example see price of Crushing Section, in **Table 3.12.5** below.

Table 3.12.6 Computation of a section with duties and taxes.

Item	Rate %	Price Rs
F.O.R. Price of crushing section from Table 12.9 (a)		105,00,000
Insurance and freight (b)	4	4,20,000
Excise duty on (a+b)	16.64	17,47,200
Sales tax on (a+b+c) or VAT	4.83	6,11,800
Unloading, octroi	1	1,05,000
VAT		
Customs duty	-	
Total price of machinery at site		133,84,000
Items made at site like chutes and ducts		5,00,000
Total for the section Rounding off		138, 84,000 139,00,000

Table 3.12.7 Capital cost of plant and machinery.

Sr.No.	Section	Price in millions Rs
1	Quarry equipment	10.0
2	Crushing	13.9
3	Stacker reclaimer	60.0
4	Raw mill	120.0
5	Blending and kiln feed	10.0
6	Kiln	300.0
7	Coal stacker	20.0
8	Coal mill	30.0
9	Clinker storage	15.0
10	Cement mill	150.0
11	Cement storage and packing	30.0
12	Motors and control gear	50.0
13	Gear boxes	20.0
14	Instrumentation	20.0
15	Total plant and machinery say	848.9 850

(These are illustrative only)

12.12.1 Capital Cost of a Section with Duties and Taxes

Assume that rates of duties and taxes are similar for the various items listed. Then next step of adding them could be taken for the whole Section.

See Table 3.12.6.

12.12.2 Total Capital Costs for the Plant and Machinery

Next step would be to add each section to arrive at total for the whole plant and Machinery by adding all sections.

See Table 3.12.7.

Table 3.12.8

	Rs millions
1. Land & Site Development	2000
2. Civil Buildings and Structures	8500
3. Plant and Machinery	11000
4. Technical Knowhow	300
5. Miscellaneous Fixed Assets	4700
6. Preoperative & IDC	200
7. Contingency	4500
8. Margin Money and Working Capital	1000
Total	**32200**

In Rs / ton installed capacity rs ~ 9200

12.12.3 Present day Capital Costs of a 10000 tpd Cement Plant

Following **Table 3.12.8** shows present day typical CAPEX buildup of a 10000 tpd clinkering capacity integrated plant making OPC 3.5 mtpa capacity plant

12.13 Elements of Capital Costs

The major elements of capital costs of the project would be:

1. Land and Site development – Mines development.
2. Railway siding and approach and internal Roads (generally included in Misc. fixed assets).
3. Civil construction costs consisting of :
   - (i) Factory buildings.
   - (ii) Non-Factory buildings.
   - (iii) Storages, Silos, Godowns and Sheds.
   - (iv) Water supply and storage.
   - (v) Drains.
   - (vi) Colony.
   - (vii) Administrative buildings.
   - (viii) Foundations, Soil testing, Gates, Fencing & Boundary walls.
4. Plant & Machinery & Electricals consisting of :
   - (i) Imported machinery.
   - (ii) Indigenous machinery.
   - (iii) Motors and control gear, gear boxes.
   - (iv) Process and Quality control and Instrumentation.
   - (v) Taxes and Duties as described above.
   - (vi) Machinery Foundations.
5. Spares and Stores.
6. Erection and Commissioning.
7. Consultancy Fees of :
   - (i) Mining and Geological consultants.
   - (ii) Consultants for :
     - (a) TEFS, Basic design and Engineering.
     - (b) Detailed Engineering.
     - (c) Electrical Power distribution.
     - (d) Process control.
     - (e) Quality control and laboratory.
     - (f) Quality control and laboratory.
   - (iii) Civil Design consultants and architects.
   - (iv) Special Consultants for locating water resources, pile foundations, vastu, load bearing capacity of soil etc.
8. Miscellaneous Fixed Assets

   This is a medley which clubs together all other items which do not belong to any section of the Project.

   Conventionally therefore they have come to include inter alia :
   - (i) Power distribution system costs including transmission line, substations, transformers, cables, switch boards, cables, lighting etc.
   - (ii) D.G. Sets or Thermal Power Station.
   - (iii) Railway siding and roads.
   - (iv) Water supply system.
   - (v) Furniture and Office Equipment.

(vi) Tools and Tackles for erection.

(vii) Cars, Trucks & Jeeps.

(viii) Laboratory Equipment.

(ix) Equipment for Mechanical and Electrical Workshops and Garages.

(x) Fire fighting.

(xi) Weigh bridges.

(xii) Effluent treatment.

(xiii) Rope way when one is required to be included.

8.1. Taxes and duties, costs of spares and erection and commissioning of items of machinery included in Miscellaneous Fixed Assets are included here only.

8.2. Logic of clubbing these items in this fashion is not clear but this has become the convention and what is more pertinent, is that this convention has been approved by Financial Institutions.

What is important is that every attempt should be made to include all the items large or small that to go to make the project complete, taking help of flow charts and machinery schedules.

8.3 In this category are to be included payments to be made for transmission lines to Electricity Boards; for railway siding to Railways, though these costs are later deducted from bills.

12.14 Contingencies

An important part of build up of capital cost is 'contingencies' to be allowed for 'omissions and commissions' and for unforeseen circumstances.

The more detailed the work done in building up various machinery schedules, civil costs, electricals, etc., less would be the need to provide for 'omissions and commissions' and unforeseens.

Contingencies vary from 5% to 15%. To be prudent, 7.5% would be a practical value. Financial Institutions however, advise contingencies at 10 percent.

12.15 Preliminary and Preoperative Expenses

Two other kinds of expenses are incurred for new projects and expansions particularly where capital is raised by equity. They are :

1. Preliminary expenses.
2. Pre-operative expenses.

Preliminary expenses include:

Fees payable to the Registrar of Companies for Authorised Capital.

Capital Issue expenses.

Techno Economic Feasibility Studies; Geological Investigations etc.

***Pre-operative expenses include*:**

Establishment expenses during construction.

Rents, Rates and taxes.

Traveling expenses.

Insurance during construction.

Start up expenses.

Commitment Charges.

Interest and commitment charges on loan during construction.

Deposits if any to agencies like Electricity Boards.

Miscellaneous items not covered in any of above.

These may either be worked out or allowed on a percentage basis in the light of past experience.

12.16 Margin Money

'Margin money' or 'working capital' is funds required to start the plant and to bring it to a position till it steadily starts earning revenues of its own to sustain continuous operations.

There are norms for working these out that have the approval of Financial Institutions (F.I).

Generally speaking, Funds for margin money are not provided by Financial Institutions like, IDBI, ICICI or IFCI but by Banks whose terms could be different than those of F.I.s.

Items normally included in margin money have been shown in **Table 3.12.9** The amounts are worked out on basis of number of days' consumption and prevalent costs of the items.

See Table 3.12.9.

Table 3.12.9 Basis of calculation of margin money.

Item	No. of days' consumption	Cost or rate Rs / ton	Total amount in Rs
Limestone	30 - 40		
Clay	30 - 40		
Additives	30-40		
Gypsum	50 - 70		
Iron ore	30 - 40		
Coal	60		
Power	30		
Consumables	270		
Salaries & wages	30		
Clinker	15		
Cement	15		
Receivables	30		
Packing materials	15		

Actucal values would vary

Thus margin money would vary according to size of plant and costs of individual items

Out of the total requirement of margin money or working capital, Banks provide 75% by way of loan and 25% is to be arranged by the Promoters.

12.17 Interest Charges

This is a substantial burden on the project costs and has to be included in it.

Project Costs for Cement Plants can be assumed to start with by using yardsticks like investment costs in rupees per ton. For example, if prevalent investment costs are Rs 9000/ ton, Project Costs for a 10000 tpd or 3.5 million ton per annum capacity plant would be Rs 31500 millions

F.I.s lay down norms for financing a project from time to time. One of the norms is debt/equity ratio. For different debt/equity ratios loans likely to be available for a given project costs would be :

Total project cost for a million ton per annum capacity plant is Rs 9000 millions

D/E ratio	loan available from F.I. in Rs. millions
2:1	10500
1.5:1	12600
1:1	15750

The balance in each case would have to be arranged by Promoters either by themselves or through friends or by 'equity' that is Public Issue.

Thus term loans that could be available from F.I.s would be roughly known at the beginning of the project. When Capital Costs are built up step by step in the manner explained above, loan to be applied for can be estimated with greater degree of accuracy – taking into account the funds – promoters themselves can raise.

Rate of interest to be paid on term loans, whether fixed or floating, should also be ascertained..

12.18 Cash Flow Statements - Borrowing of Funds

The 'cash flow statements' prepared would indicate when to borrow and what amounts.

To keep down the interest charges and also to begin the plant execution on time, it is best to use 'own' finances in initial stages up to the point say of ordering main machinery.

The project completion schedule together with cash flow should furnish information on the quantity of funds required and their timing.

It takes considerable time in obtaining loans from Financial Institutions. Major steps being preparations of:

1. Techno Economic Feasibility Studies.

2. Project Report to be submitted to Financial Institution for approval in the format prepared by them.
3. Project appraisal and approval of the project by teams of executives of Financial Institutions.
4. Signing of agreements and other formalities.

All this can sometimes take more than a year.

12.18.1 Initial Activities with Own Funds

If the company can arrange for funds for initial activities like:

(i) Mining and Geological investigations of deposits.
(ii) Techno Economic Feasibility Studies.
(iii) Load bearing tests of soil / wind rose.
(iv) Basic Engineering and detailed engineering.
(v) Initial work at site like, land acquisition (which can cost a lot).
(vi) Mining lease and land for mining.
(vii) Transmission lines and railway siding.
(viii) Civil design consultants.
(ix) Environmental clearances.

costs of which can be roughly assessed, the Project can progress without having to wait for sanction of term loans. Real requirements of funds start with ordering of machinery.

12.18.2 Finances for Plant and Machinery

Plant and Machinery (excluding erection and commissioning, duties and taxes) would account for as much as 40-50% of project costs. Roughly, 10%-15% advance is required to be paid with the order (may be in more than one installment).

Banks would sanction loans on 75 % of working capital. Promoter has to arrange for the balance.

Civil engineering would start say in 15th month and complete in 21st month. Funds for civil engineering would be required in three installments. required in 3 installments.

In this fashion requirement of funds can be worked out.

Financial loan may therefore taken in say three to four installments to keep down interest charges.

Interest charges. See **Table 3.12.10**

Table 3.12.10

Say total loan rs 10000 million		
Installments	Amounts	
	Rs millions	
1st	1500	
2nd	1500	
3rd	5000	
4th	2000	
	10000	

Interest charges would then be calculated according to period of each loan, in this way interest charges can be saved.

Different configurations would give different values for interest charges.

Loan for margin money should also be drawn when required, first using company's own funds to start the project on time and to complete it on time.

12.19 Equity

Out of Rs.1500 million to be arranged by the company for a debt/equity ratio of 1:1 minimum 35% (or prevailing norms) are required to be raised by the Promoters (their friends, relatives etc.) i.e., ~ Rs. 5500 million.

Balance Rs.10250 million are to be raised by Equity.

To raise funds by 'public issue' involves many factors. Whether the company would be able to raise funds by equity or public issue would depend not only on its past, performance and reputation, but also on external factors like :

1. General economic situation and investment climate.
2. Health of the industry – Cement.
3. Market conditions and prospects for cement prices in future.
4. Profitability of leading cement companies in the last 2-3 years.

TEFS would have brought out most of these factors in its chapter on market.

A calculated risk would have to be taken as regards timing of bringing out the public issue and quantum thereof.

Another approach is to float public issue when the project is nearing completion so as to create confidence in the minds of investing public.

However it would mean a total review of financing of the project as the Promoter would have to find bulk of funds himself to be able to start and complete the project on time.

Various options in this regard should be considered by Management and their Financial Consultants and the choice made should be advised to Consultants preparing TEFS.

12.20 Yardsticks

A commonly used yardstick for a green field site cement plant is investment costs in Rs/Ton of annual capacity.

Assuming that annual capacity is 3.5 mtpa cement, and capital costs worked out are Rs 32200 million.

Then investment cost in Rs 9200/- per ton.

Investment costs vary slightly according to size of plant, site conditions and according to special features like, railway siding, ropeway, location of mines etc.

But often it is a good starting point and a check on the exercise of estimating capital costs. If the built up capital costs vary by more than + 15% from the prevailing value of the yardstick, then it would be desirable to look into the matter closely once again.

12.21 Capital Costs Based on New and Bought Machinery

The above exercise of building up of project costs is based on the assumptions that :

1. All machinery is new and has been procured from reputed manufacturers.
2. While some fabrication would have to be done by the company at site like chutes, ducts, hoppers, bins, trestles, etc., it is assumed that machinery itself would to be bought out.

Some savings can be made by the Company undertaking to fabricate items like cyclones of preheaters and casings of dust collectors by obtaining manufacturing drawings for them from respective suppliers. Company can also assemble belt conveyors by buying separately, idlers, belting etc. Savings by such means can at best be ~ 20 % of the cost if bought out fully. It will however require the Company to make their own arrangements for fabricating on a large scale and also build up its strengths in engineering.

Table 3.12.11 Costs in % of total costs.

Department	This Project	Earlier Projects of similar capacity
Civil	12.00	13.00
Machinery	40.00	42.00
Erection / Commission	3.00	3.00
Misc. fixed assets	17.00	15.00
Consultants	1.50	2.00
Preliminary & Preoperative Expenses	3.00	3.00
Margin Money	3.50	3.00
Contingencies	10.00	7.50
Interest Charges	10.00	11.50
Total	100.00	100.00

12.22 Cross Checking Build up of Costs

The build up of project costs should also be checked for being 'in line' by comparing percentage costs of various sections of the plant with an earlier project of similar size as shown in **Table 3.12.11.**

A quick comparison like this will show if the build up is on right lines. Significant deviations should be looked into and it should be possible to account for them.

12.22.1 Approximations in the Exercise

Some items of cost are taken on percentage basis without going into minute details such as :

(i) Packing & Forwarding.
(ii) Freight and Insurance.
(iii) Duties and Taxes.
(iv) Erection and Commissioning.
(v) Stores & Spares.
(vi) Contingencies.
(vii) Basis for Margin Money.

The rates assumed should be checked with earlier rates taken for a completed project (Prevailing at that time). The project costs can once again be compared.

12.22.2 Differences in Capital Costs

Major costs which would differ from project to project would be :

(i) Railway siding.
(ii) Lengths of roads.
(iii) Cost of land for plant, colony and mining.
(iv) Soil conditions influencing foundation costs like, depths of foundations, raft foundations, piling, etc.
(v) Conveying of crushed stone, type of conveying system and its length.
(vi) Capacity of captive power plant proposed to be installed.

12.22.3 Core Costs

The 'core' costs should compare well with earlier project costs after allowing for :

(i) Size and Capacity.
(ii) Type of Machinery, Roller Mills / Ball Mills.
(iii) Inflation.

It is possible that two plants, which have same nominal capacity, would have widely different capital costs because of type, sizes and capacities of major machineries selected and also because of specific items listed above.

12.22.4 Check and Cross Check Basis

Thus every exercise of project costs build up should be checked and rechecked to make sure that it is done on the right lines, with right kinds of assumptions and includes all items, reducing 'omissions and commissions' to absolute minimum.

12.23 Contingencies

Strictly speaking when such care is taken in building capital costs, the 'contingencies' should be reduced to say 5% only. For a Rs 30000 million project, the differences, in amount of contingencies at different rates would be:

Contingencies at	5%	7.5%	10%
Amount in million Rs	1500	2250	3000

Thus contingencies are not small amounts. But in this respect one has to be guided by norms laid down by the Financial Institutions approached for loans.

12.24 Discounts

The pattern of 'discounts' given by machinery suppliers is generally known at the project stage.

These discounts vary widely between 5% (perhaps minimum) to 30% (perhaps maximum).

Question is whether the discounts should be taken into account in building up of capital costs. For example:

Rs. 32000 million total project.

Plant and Machinery including Electricals – say Rs.11000 million if the average discount is 5 %, saving in overall price thereof would be rs 550 millions.

There would be further savings in duties and taxes and in cost of spares.

If it is possible to arrive at an understanding in this respect then only discounts should be taken into account. Otherwise not.

However, to guard against the possibility of expected discounts not materializing, it would be better to increase contingencies.

A better way would be to reassess the project costs after negotiating orders for main machinery and reduce quantum of loan to be taken from F.I.s and thereby interest charges.

12.25 Works Contract

Slightly different procedure would have to be adopted if a Project is to be treated like a 'works contract' where a Vendor takes order on a 'lump sum' price basis. It has some advantages for the Buyer. However this procedure is not very common and hence has not been dealt with here.

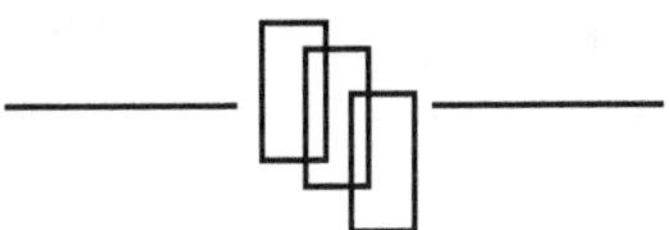

CHAPTER 13

COSTS OF PRODUCTION

13.1 Working Out Costs of Production

First step would be to decide on types of Cements to be made and their proportions. Costs of production should be worked out for all types.

Ascertain all costs involved in the production of each type including inter alia :

1. Royalties, cess etc., on limestone and other correcting materials.
2. Pit head costs of coal, iron ore, gypsum, etc., including governmental taxes and duties and royalties to be paid on them.
3. cost of blending materials like fly ash and blast furnace slag at points of collection.
4. Allowance to be made for moisture if it exceeds 3 to 5 % and for loss in transit.
5. Freight costs by rail and or road to be ascertained for all materials to be brought into the works.
6. Limestone raising costs to be worked out comprising of;
   (i) Explosives consumed.
   (ii) Operation and Maintenance cost of drills, shovels, dumpers compressors etc.
   (iii) Limestone transport costs to crusher.
7. Cost of Electricity comprising of :
   (i) Unit cost in Rs / kwh for power from grid.
   (ii) Unit cost for power generated in captive power plant.
   (iii) Maximum demand charges – in Rs per KVA / Month.
   (iv) Other Levies.
8. Labour costs, derived from salary structure of wages and salaries and perks.
9. Stores and Spares consumed – These may have to be judged from operation of existing plants – either Company's own or from other plants.
10. Overheads – of Factory and Head Office.
11. Marketing and Selling Costs including advertising costs.

13.2 Fixed and Variable Costs

Costs are broadly classified as :

A. Variable costs – which are directly related to production.

B. Fixed costs – which are to be incurred regardless of production.

As their names suggest variable costs are incurred when plant produces cement. Fixed costs are always incurred whether plant is in production or is idle.

In fixed costs the following are included :

(i) Salaries and Wages and perks of employees on permanent roll of the company, including Executives, Directors, Managing Director, etc.

(ii) Plant and Head Office overheads.

(iii) Marketing expenses, advertising, Sales network etc. Costs of maintaining depots and godowns.

Variable costs are worked out in terms of Rs./Ton of cement produced.

Fixed costs are worked out for the year and then converted into costs in Rs / ton. As the following exercise will show.

Plant Capacity – 10000 tpd or 3.5 mtpa OPC

Cement produced in a year ~ 3 mtpa (at 85% capacity utilization).

Variable costs for OPC are say 1963 Rs.2175/- per ton. **See Table 3.13.5**

13.3 Total Costs of Production

Cement produced in the year		3 million tons
Variable costs of production at Rs.2125/Ton (See Table 3.13.5)	= Rs	6375 million
Fixed cost of production in same period	= Rs	1470 million
Total	= Rs	7845 million
Therefore overall cost of cement produced	= Rs	2620/ton

Proportions of costs

Variable costs 81%

Fixed costs 19%

The proportions of fixed and variable costs should be checked with data available from running plants.

13.4 Variable Costs are Worked in Two Steps

A. Cost of Clinker.

B. Cost of Cement.

The elements of variable costs of production for naked, ex-works cement are :

(i) Raw Materials.

(ii) Fuel.

(iii) Power.

(iv) Stores, Spares and Maintenance.

(v) Contribution to research.

Following **Table 3.13.1** lists various elements in the two steps mentioned above.

13.5 Working out Costs of Materials and Fuel

Thus first, costs of raw materials and fuel (coal and oil) and gypsum and blending materials like fly ash and slag (if used) should be worked out.

1. It is necessary to know raw meal / clinker ratio for consumption.
2. Proportions of limestone and other correctives, clay and iron ore.

Table 3.13.1

S. No.	1st Step	2nd Step
1	Cost of Clinker comprising of	Cost of Cement
	Raw material costs	
	Cost of Limestone	Cost of clinker from 1st step
	Cost of Additive 1	Cost of gypsum
	Cost of Additive 2	*
2	Cost of Coal burnt in Kiln and Calciner	Fuel (sometimes for drying slag/ flyash).
3	Cost of power -upto and including clinkering stage	Cost of power for Cement mills Paking & loading of cement Other departments not covered in step1
4	Stores and Spares, consumables and Maintenance upto clinkering stages.	Stores and Spares and Maintenance for cement mill and packing etc.
5	Total variable cost of clinker in Rs. / Ton of Clinker	Statutory costs for research,
6		Variable cost of cement in rs/ton

* when blended cements are made to add cost of blending materials like slag and fly ash.

3. The proportions in raw meal are on dry basis with moisture less than 1% in raw meal.
4. quantity of coal used in making clinker expressed in %. It is arrived at from calorific value of coal and specific fuel consumption.
5. proportions of gypsum, fly ash and slag in respective cements.

13.5.1 Costs of Raw Materials

Assume raw meal / Clinker = 1.55 : 1

Let the proportions of limestone and additives be

Limestone	-	85%
Clay	-	13%
Iron Ore	-	2%
	=	100%

Therefore to make 1 kg clinker it would require on dry basis

Limestone	1.318 kg
Clay	0.202 kg
Iron ore	0.03 kg

13.5.2 Allowances for Transit Loss, Moisture

Allowances are to be made for moisture in material and losses in transit if any to arrive at realistic costs of these materials. This can be done either by converting costs of materials as received into costs of dry materials or convert quantities used in production into wet materials and apply costs of materials as received.

Let us say moisture in clay and iron ore are 8 and 12 % respectively

And that loss in transit be 3 % for both.

Thus wet clay and iron ore to be bought to make 1 kg clinker would be

Clay $= 1.08 \times 1.03 \times 0.202 = 0.225$ kg

Iron ore $= 1.12 \times 1.03 \times 0.03 = 0.035$ kg

Moisture in limestone can be ignored as it is generally less than 5% even in wet seasons.

13.5.3 Royalties to be Paid

In raising limestone – over burden is also to be removed and shifted.

Royalties are paid on stone removed from the quarries. Let us assume that over burden is left in quarries and only stone is brought into the plant.

Therefore limestone required at 85 % $= 1.55 \times 0.85$

$= 1.318$ kg.

If over burden is 20%, stone to be raised to make 1 kg clinker = 1.58 kg

Costs of raising limestone will be applied to 1.58 kg/kg clinker but

Royalties would be calculated for 1.318 kgs.

Royalties including cess are levied at the rate of rs 107 /ton.

13.5.4 Cost of Raising Limestone

Costs of raising limestone would include operating costs of, drills, shovels and dumpers, for transport, compressors, mainly diesel oil and lubricants, and other consumables. At current cost of diesel oil and lubricants, spares they work out to about rs 113 /ton

Royalties including cess are	rs 107/ton
Total	rs 220/ton
Cost of limestone /ton clinker	rs 290/ton

(Royalties are revised for time to time by central Government /State Government may also import some 'cess'. The prevalent Rates are to be included)

13.5.5 Cost of Correcting Materials

Iron ore, clay, gypsum, coal, fly ash and slag are generally bought and their costs 'as received' are worked out as flows

Pithead cost + allowance for moisture and transit loss + freight payable .as mentioned in para **13.5.2**.

See **Table 3.13.2**

13.6

When crusher is in quarries and stone is transported by belt conveyor to the plant the cost of raising limestone would come down appreciably as dumpers would ply over very short distances.

Operating costs of earth moving machinery could be obtained from suppliers of such machinery.

Costs incurred in crushing and transport of crushed limestone would be included in power costs and maintenance costs up to clinkering section.

Table 3.13.2 Costs of correcting materials.

item	Quantity kg/kg clinker	Price rs/ton	Freight rs/ton	Cost as received rs/ton	Cost rs/ton clinker
clay	0.225	50	20	70	16
iron ore	0.035	800	300	1100	38.5

It is a convention that for working out all costs, current prevailing costs are taken into account and not projected costs when the plant would go into production. Profitability is projected for 10 –15 year period using the current costs.

13.7 Coal / Fuel

As in 99% of cases coal would be used as fuel, we shall be assuming coal to be the fuel used. Coal may be brought from nearest Collieries. Sometimes plants used imported coal also in varying proportions to improve quality of coal fired in the kiln.

Cost structures could vary widely for the two alternatives.

13.7.1 Indigenous Coal

It may be bought from the nearest Collieries or from allocated Collieries in quantities allocated.

A mix may be desirable to ensure continuity of supply and for blending purposes.

Let us say desired calorific value of coal as fired should be maintained at 4500 Kcal/Kg. Let coal have a moisture of 10 %

1 kg wet coal = 0.9 kg dry coal and 0.92 kg coal as fired with 2% moisture.

Coals have been graded A to D. D being the worst with very high ash content. Cement Industry is generally allocated Grade C coal with ash content of 30 –35% and useful calorific value 3800-4300 kcal/kg Thus it would be not only desirable but also necessary to obtain some high-grade coal and blend it with regularly available coal from nearest collieries.

The cost of coal received has the same elements- viz pithead cost plus moisture plus loss in transit plus freight.

Cost of coal as received is therefore worked out in the manner explained above.

13.7.2 Selecting coals to reduce costs.

In principle, it is possible to obtain coal from different sources or even to import it.

An exercise can be done to workout which combination gives lowest cost on as fired basis as shown in the following table based on sp. fuel consumption of 700 kcal/kg clinker. See **Table 3.13.3**

For a sp. fuel consumption of 700 kcal/kg, coals required to be procured would be

Category

A (700/4100)*1.03*1.08 = 0.19 ton/ton clinler

B (700/4500)*1,03*1.08 = 0.17 ton/ton clinker

C (700/4700)*1.03*1.06 = 0.16 ton /ton clinker

Costs/ton clinker would be

Category	rs
A	1007
B	1134
C	1176

An exercise of this kind would help to arrive most economical combination without sacrificing fuel efficiency.

13.8 Cost of Power

Cost of Power is split into two :

1. All production sections.
2. Balance like:
   - (i) Lighting.
   - (ii) Utilities.
   - (iii) Losses.

As mentioned earlier, power cost is in 2/3 tiers;

- (i) Cost of energy consumed in Rs/Kwh.
- (ii) Any surcharges, etc., as imposed from time to time.
- (iii) Maximum demand costs in Rs/ KVA/ month.

Table 3.13.3 Cost of coal per kg of clinker.

Sr. no	basis : 700 kcal/kg clinker					
	source	useful calorific value kcal/kg	moisture %	loss in transit %	qty to be procured kg	qty for 700 kcal kg/ kg clinker
			quantity /kg clinker			
	coal A	4100	10	3	1.13	0.19
	coal B	4500	10	3	1.13	0.18
	coal C	4700	8	3	1.1	0.16
			costs			
	source	pithead cost rs/t	freight rs/t	cost as received rs/t	per ton clinker rs/t	
	coal A	5000	300	5300	1007	
	coal B	6000	300	6300	1134	
	coal C	7000	350	7350	1176	

13.8.1 Example of Working out Power Costs

Thus if for a 10000 TPD or 3.5MTPA plant, specific power consumption of **90 kwh/ton** of cement, power requirement in kws ~ 440*90= 39600 say 40000 kw per hour

With design margin of 10 % = 44000 kw

Using power factor of 0.95 and load factor of 0.8 ,

Power required in kva = 58000

Allowing 25% margin for immediate additions or to cover omissions,

Maximum demand would be ~72500 kva

Annual cement produced @ 85% capacity utilization = 3000000 tons

Units consumed = 90 × 3000000 = 270 million kwh

Let cost of energy per unit including surcharges be Rs 4/kwh

Therefore energy costs = 4 × 270 = 1080 milion rupees.

Let maximum demand charges Rs. 250 / KVA / month = 3000 / KVA / year.

Maximum demand charges in one year = 72500 *3000 = Rs 217.5 million

Total power charges = Rs 1297.5 million

Units consumed = 270 million

Therefore overall power cost in Rs 4.8 /kwh

13.8.2 Power Costs with Captive Power Plant

If D.G. Set power is used, its cost of generation would be approximately (3.5 units/litre at Rs 80/litre of fuel oil) would be rs 22.8/kwh

Say in months from

April to July – power cut was	60 %.
August to January	0 %
February to March	30 %

To be made up by captive power. Out of total power consumption of 270 million units, power generated by captive power plant would be 25 % or 67.5 million units and balance bought from grid.

Overall power costs : grid power-maximum demand charges remaining same

	rs
grid power @ 4 rs/unit for 202.5 million units	810 million
captive power @ rs 22.8/unit for 67.5 million unis	1530 million
max. demand charges	217.5 million
Total	2566.5

Total units consumed = 270 million units

Overall energy costs in Rs/kwh = 9.5 rs

During months of power cut, maximum demand would be appreciably less. But it would not be practical to reduce contracted demand for 4 months of the year. The maximum demand charges should be paid for whole of contracted demand for the whole of the year.

13.8.3 Splitting of total power consumption

To work out cost of production in two steps as mentioned earlier, power Consumption in sections up to clinkering and balance would be clubbed As shown in **Table 3.13.4**

Table 3.13.4 Split up of power consumption.

	Section	kwh/t material	kwh/t clinker	kwh/t cement
1	Quarry, crusher, Stacker Reclaimer	2	3.1	3
2	Raw mil	12	18.6	18
3	Blending	2	3.1	3
4	Kiln preheater cooler	25	25	24
5	coal mill	30	5.1	5
	total upto clinkering		55	53
6	cement mill			30
7	packing			1.5
8	water supply			1
9	lighting			1
10	miscellany			1
11	losses			2.5
	total for 6 onwards			37
	grand total			**90**

13.9 Stores and Spares and Maintenance Costs

These include consumables and spares used in a given period of 1 year.

For a running plant, data would be available on actual consumption.

For a new green field site plant, the consumption would have to be taken from data banks or by obtaining data on costing from running cement plants.

Let the total consumption of stores and spares be Rs.300 million / annum for a 3.5 million ton plant. Let it be divided in proportions 70: 30 between sections upto clinkering and the rest.

At a production of 3 milion tons, for 85% utilization of capacity,

Expenditure on stores and spares	=	rs 100/ton
expenditure in sections upto clinkering	=	Rs 70 /ton cement
or		Rs 73.5 / ton clinker
expenditure for cement and other sections	=	Rs 30 / ton of cement

13.10 Exercise of Arriving at Variable Costs

Variable costs of cement can now be worked out in two steps

As follows : See **Table 3.13.5**

Table 3.13.5 Variable costs of clinker and cement.

		1 Variable costs of clinker		
	Description	Quantity kg/kg kwh/ton	Costs rs/ton rs/kwh	Cost rs/ton clinker
1	limestone	1.318	220	290
2	clay	0.225	70	16
3	iron ore	0.035	1100	39
4	fuel	0.16	7350	1176
5	power	53	4.8	254
6	stores/spares			75
	total			**1850**
		2 variable costs cement		
1	Clinker	0.95	1850	1760
2	gypsum	0.05	3100	155
3	power	37	4.8	178
4	srtores / spares			30
	total			**2123**

13.10.1 Cost of Bagged Cement

Above is the Cost of Cement ex-works

To derive cost of bagged cement delivered to site, it is required to add to it :

(i) packing and loading charges,

(ii) freight to market,

(iii) prevailing taxes and duties.

Packing charges are cost of bags for bagged cement. For bulk cement, there are no packing costs only very small power and loading charges.

13.11 Fixed Costs

As explained earlier, these costs are incurred whether the plant is running or is stopped. Main elements of the fixed costs are, salaries and wages, overheads and part of marketing and selling costs like advertising, maintaining godowns etc.

They add up to ~ 400 - 450 R/T **See Table 3.13.9.**

13.11.1 Salaries Wages

Salaries and wages are worked out from systematic build up of manpowerneeds and prevailing salary structure. See **Table 3.13.6.**

Strength of staff is as per **Table 3.8.1** in **Chapter 8.**

Total annual wage bill as shown in **Table 3.13.6** RS 365 million

Rated capacity of plant is 3.5 million tons per annum. At rated capacity cost of salaries amount to ~ rs 122/ton of cement.

13.11.2 Over heads

Factory and Head office overheads are also worked out.

Marketing executives can advise on, and work out fixed portion of marketing and selling costs.

Total fixed costs are then added to the total variable costs for the year under consideration to arrive at total production costs expressed as Rs /ton as explained earlier.

Financial and profitability statements can then be worked out either on total yearly turn over or on Rs/ ton basis which ever is convenient or what is more relevant, which is preferred by Financial Institutions.

Presently all overheads – salaries, factory and administrative and marketing together add up to ~ rs 350-400/ton at rated capacity.

13.12 Price of Cement and its Elements

There was a time when price of cement to a consumer was the same anywhere in India, regardless of who had made it and how much distance it had traveled to reach the ultimate consumer.

Now cement industry operates in 'free market' conditions. Competition dictates market prices. Quality has come to the fore. Already in OPC itself there are 3 grades 33, 43 and 53. All cement companies claim to be making 43 and 53 grade cements.

Marketing and selling strategies have also undergone considerable change. 'Brand Loyalty' is a thing of the past. Many a time 'dealers' dictate terms to manufacturers unless manufacturers can sell directly through their own networks of outlets. Some manufacturers have their own transport fleets to transport cement. Under these conditions, profitability is to be closely watched.

Table 3.13.6 Annual Bill of Wages and Salaries.

Sr. No.	Category	nos	salary per person		total for
			per month rs	per year rs	all numbers rs
1	Executives	10	480000	5760000	57600000
2	Senior Officers	34	240000	2880000	97920000
3	Middle level Officers	66	100000	1200000	79200000
3	Junior Co-officers / Technicians	116	50000	600000	69600000
4	Semiskilled Workers	160	24000	288000	46080000
5	Unskilled Workers	101	12000	144000	14544000
6	Total	487			364944000
				say	365000000
	plant capacity 10000 tpd clinker- 3.5 million tons per annum salaries expressed in Rs/ton at 85 % of rated capacity works out to ~ 122				

13.12.1 Elements of Cost of Cement

The elements of build up of costs at consumer's door would be:

(i) Cost of naked cement ex-works Rs/ton.

(ii) Cost of bagging / loading.

(iii) Freight to a given market by road and or rail converted into Rs/ton according to distance of transport.

(iv) G.S.T. (which has replaced Excise duty) at 28% of Invoice value

(v) Dealers' commission.

13.13 Gross Profit available is

Price ruling in the market < sum total of above costs

Price ruling varies from State to State; city to city and varies in the same place from season to season. In 'free market' economy, 'demand' and 'supply' decide the prevailing sales prices. When supply exceeds demand, prices take a beating. In free market conditions a plant can sell cement anywhere regardless of distance if the prices are right.

13.14 Bagged versus Bulk Cement

Bagging Costs

Cement is packed now almost exclusively in h.d.p.e. woven bags. Jute is hardly used though Government did try to protect Jute Industry by insisting that a percentage of bags used should be jute.

Bagging cost work out to between Rs.11 – 16/- per bag or Rs. 220 – Rs. 320 per ton depending on material used.

This will bring out the advantage of bulk cement dispatches.

Bulk cement

Bulk cement is now readily accepted. Even 'ready mix concrete' is accepted in many major cities

Major players in the Cement Industry have their own silos for storing bulk cement in major cities / centres of market. They are also now suppliers and promoters of ready mix concrete.

While a certain amount of double handling is inevitable in reaching small consumers, large consumers can now get bulk cement 'ready to use' at their doorstep.

The costs of handling of bulk cement would be a fraction of the cost of bagging –

Say Rs.50/- per ton.

13.14.1 Bulk Carriers

One element of 'bulk cement handling' is the need to have special wagons on rail and special bulk carriers on road with unloading facilities either mounted on wagons / carriers themselves or located at point of discharge. Thus cement can be unloaded straight from wagon / carriers into storage.

Now railways are encouraging cement producers to have their own specially designed self unloading bulk wagons and agree to ply them on railways' network.

Truck mounted bulk carriers up to 40 tons capacity are now available and serve those plants which do not have rail head facility. They are also used to transport fly ash.

13.15 Increase in Surplus due to Bulk Cement

Costs can thus come down appreciably by Rs. 500 to Rs. 220 per ton depending on the proportion of bagged and bulk cements.

Thus if a plant sold one Million Ton Cement and dispatched 30% of it in bulk, and its cost of naked ex-works cement was say Rs.1000/- per ton then its profits would increase as shown in **Table 3.13.7.**

Table 3.13.7

	Bagged Cement 70%	Bulk Cement 30%
	Rs. / Ton	
Naked Ex-works	2120.00	2120.00
Bagging / Loading	220.00	50.00
Freight Costs	300.00	300.00
Gst @ 28 %	740.00	690.00
Dealers Commission	500.00	400.00
	3880.00	3560.00
	say 3900.00	say 3600.00
Sales Price at rs 275/ bag	5500.00	5500.00
Gross profit	1600	1900

70% × 1600 + 30% ×1900 = 1690

Gross profit would thus go up by Rs.90 per ton or rs 2700 Million / Annum if 30% of production is sold in bulk.

13.16 Freight

This is a cost which increases every year with the budget. Cement companies have no control over it. On the other hand, competition forces them to go further and further to sell their cement.

Those plants that have no rail head, have to send all their cement in can truckloads or in bulk carriers. Those that have rail head, send it wholly by rail or partly by road and partly by rail. Rail freights are cheaper for long distances. Road transport is cheap for short distances. Companies may therefore work out a mix of desptches which will reduce overall freight costs.

Let the conveying distances and proportion of cement sold for a plant be as shown in **Table 3.13.8.**

Also **see Figs. 3.13.1 and 3.13.2.**

Table 3.13.8

Distances in kms

	< 100	< 200	< 300	< 500	> 500
% Cement	10	30	40	10	10
Transported by	road	Rail or Road		Rail	Rail

Say about 70% of cement can be dispatched either by road or by rail.

Freights incurred for various proportions of rail and road can be worked out.

In **Fig. 3.13.2**, 1 to 6 are consumer centers. Quantities sold at these centres would be known.

It would generally be possible to send cement direct by road to the market centers 2 and 3 and to that extent, save costs of double handling, warehousing, etc.

For market – 5 rail transport could be cheaper but as the center is not on rail head, a combination of rail + road transport would be necessary.

A mix of road and rail transport can thus be worked out to reduce freight costs.

Company would have no control on other costs like dealers' commission, Sales tax and excise duty.

13.17 Current Costs of Production

Table 3.13.9 shows current level of Costs of Production.

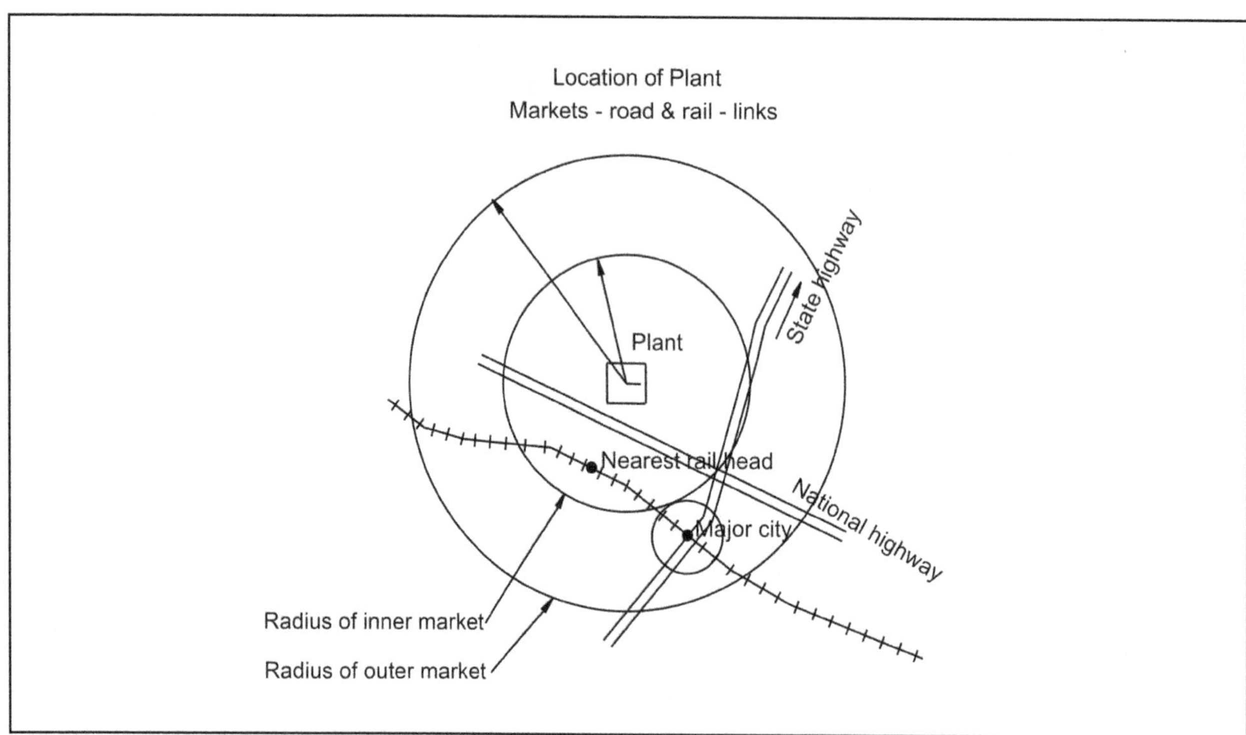

Fig. 3.13.1 Dispatches – inner market – by road; outer market – by rail/by road.

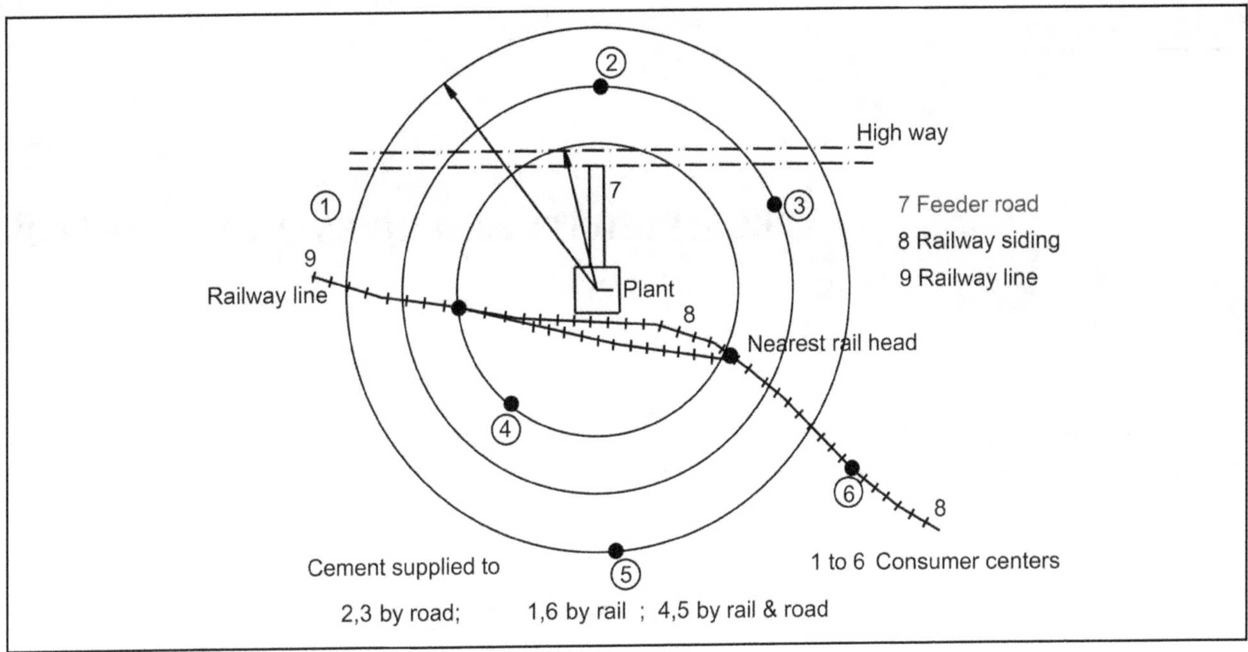

Fig. 3.13.2 Reaching consumer centres – by road and rail.

Table 3.13.9 Current Costs of Production.

Item	unit cost	OPC		PPC	
	Rs/t	Qty/uni	Cost Rs/t	Qty/unit	Cost Rs/t
Raw Materials and Conumables					
limestone	210	1.33	278	0.91	190
correctives	810	0.15	119	0.1	82
gyosum	3100	5%	155	5%	155
fly ash	550	0	0	30%	165
consumble			55		55
			607		**647**
Utilities					
power	4.2/kw	85	357	75	315
fuel	6000	0.2	1200	0.14	840
packing			210		210
Fixed cash			**1767**		**1365**
Salaries and wages			110		110
Factory Overheads			100		100
Admin Overhead			50		50
Seiling & Distribution			150		150
			410		**410**
Unit cost of Production Rs/t			**2784**		**2422**
for this work out fuel cots have been based on coal of 4500 kcal/kg and sp. fuel consumption of 700 kcal/kg has been assumed **It is based on data provided by Holtec Consultants**					

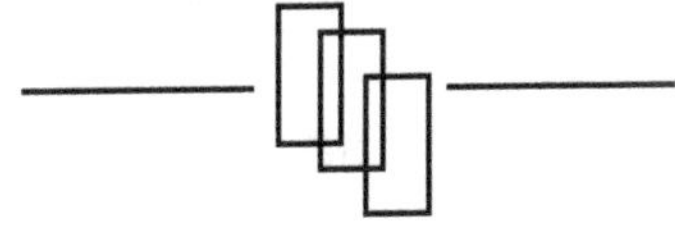

CHAPTER 14

PROFITABILITY AND INVESTMENT ANALYSIS

14.1 Working out Profitability

Profitability is then worked out after taking into account yearly costs of :

1. Interest.
2. Depreciation.
3. Tax.

For example **See Table 3.14.1.**

Table 3.14.1 Working out profitability.

1 Plant capacity : 10000 tpd clinker
2 Cement capacity OPC = 3465000 tons/year
3 Cement produced at 85% capacity = 2945,250 tons
Say 3000000 tons
4 Selling price of cement – rs 5500/ton or rs 275/ bag
5 Average at rs 600/ton
6 Dealer's commission = 500/ton
7 G.S.T. at 28%

	Item	Quantity	Value Rs
1	**Gross sales**	**1 ton**	**5500**
	LESS		
	Freight at ave 10%		550
	G.S.T.		750
	Dealers commission		500
	Total deduction		**1800**
2	**Nett sales realization / Ton**		**3700**
3	**Variable costs of production see Table 3.13.5**		**2125**
4	**Fixed costs @ 85% production**		**480**
5	**Total costs of production @ 85% capacity**		**2605**
6	**Profit on net sales realization**		**1095**
7	**Less depreciation @ 10%**		**110**
8	**Nett profit**		**985**
9	**Nett profit as % of sale alue**		**17.9%**

Note : Interest payable on loan not included.

14.2 Projections of Profitability

Profitability is projected for a period of 10- 15 Years with assumptions like :

1. Capacity Utilization - 80% 1st Year.
   - 90% 2nd Year.
   95% 3rd Year.
2. Selling price of cement remains the same throughout the period. Profitability is however worked out for different sales prices on either side of the current selling price.
3. Freight, Bagging, Sales Tax and Excise Duty remain same for the period.
4. In fixed costs, labour costs are increased by 5% or so every year.
5. Variable costs remain same that is cost of fuel, power and raw materials remain same.
6. Interest charges are worked out on loans taken and repayment period and moratorium if any agreed upon with financial institutions.
7. Depreciation charges are worked out according to well-established methods of calculating them for different items like :
   (i) Civil structures.
   (ii) Plant and Machinery.
   (iii) Pollution control equipment.

   According to the proportions of these in total costs and rates of depreciation for example 10% or 25% st. line, etc.

14.3 Profitability – Financial Ratios

Profitability is worked out in the manner described for a period of 10-15 years.

However, each Industry has its own norms for judging its adequacy or otherwise, to be worthwhile for investment.

Therefore certain yardsticks are worked out to judge and compare the profitability with that of similar plants in the Industry or with other Industries to take an investment decision.

Financial Institutions are also very much interested in some of these yardsticks, which they use to evaluate viability of the project and to assess if loans given would be repaid with interest with certainty within the prescribed time.

Some of these yardsticks are :

1. Payback period.
2. Break even point.
3. Debt servicing ratio.
4. Internal rate of return. (IRR)

14.3.1 Concessions Available in First years

Cognizance has to be taken of facilities and concessions provided by Financial Institutions and State and other Government Agencies.

Some State Governments give Sales Tax exemptions for 5 years from the date of commissioning of the plant. Electricity Boards give concessional tariff for fixed periods.

Money paid to them for laying transmission line / etc., can be deducted from bills.

14.3.2 Moratorium on Interest Payable

There is a 'Moratorium' for beginning to pay interest charges to Financial Institutions. This could be between 2-5 years.

There could be moratorium on Income Tax also. All these should be taken into account and incorporated in computer programmes to work out Internal Rate of Return on investment.

This is a statement of cash inflows against capital expenditure and working capital. Thus for a 15 years period as an example,

Cash inflow = Rs 700 million

And capital expenditure and working capital

= Rs 572.5 million

Net. Cash flow = Rs 127.5 million

I R R = 18.2 %

IRR is calculated on equity also.

Payback periods and cost benefit ratios are calculated with and without discount.

Packages of software programmes for working out profitability and various financial ratios are now readily available.

Financial Consultants can furnish all the financial statements and financial ratios and also sensitivity analysis as required by Financial Institutions.

14.4 Sensitivity Analysis

It is not enough to carry out profitability for one set of parameters. Impact of factors like variations in capital costs, gestation period, costs of production and sales price on various yardsticks mentioned above viz.

I.R.R.

Break Even Point.

Debt Servicing Ratio.

Payback period.

should also be worked out.

Typical proforma is attached in **Annexe 3.14.1.**

Such Investment and Sensitivity Analysis based on increase in cost of raw materials, in power costs, reduction in sales price of cement, lower production of cement, increase in power costs gives a clearer picture about the Viability and profitability of the Project and helps in taking decisions about going ahead with it or to postpone it.

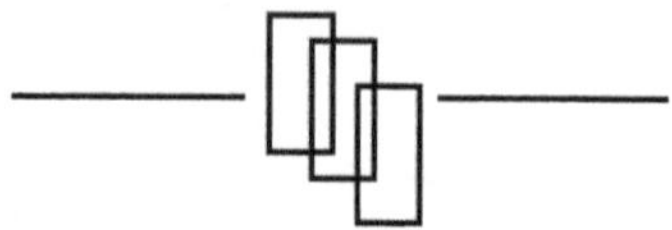

Annexe 3.14.1

1. Investment Analysis

From the financial statements following are worked out.

			Case 1	Case 2
1.	Sales price	Rs/ton		
2.	Total capital costs	Rs million		
3.	I.R.R. (post tax)			
	(a) on investment	%		
	(b) on equity	%		
4.	Break Even Point	%		
	(a) in 3rd year			
	(b) in 10 th year			
5.	Debt Servicing Ratio average	ratio		
6.	Pay Back Period	years		
	(a) without discount			
	(b) with discount			
7.	Capital employed / Sales	ratio		
8.	Raw materials / output	ratio		
9.	Wages / output	ratio		
10.	Interest / output	ratio		
11.	Gross profit / capital			
12.	Return on Investment	%		
13.	Return on Equity	%		

2. Sensitivity Analysis

1.	I.R.R. varies by due to variation of	%		
	(a) percent in capital cost			
	(b) gestation period in –months			
	(c) rupee /ton in variable cost			
	(d) rupee /ton in sales realization			
2.	Break even point varies by due to variation of	%		
	(a) rupee /ton in variable cost			
	(b) rupee /ton in sales price	%		
	(c) percent variation in capital costs			

CHAPTER 15

EXECUTIVE SUMMARY AND CONCLUSION

15.1 Recommendations

Techno Economic Feasibility Studies should end with clear and unambiguous conclusions and recommendations so that the Entrepreneur can take the right decisions about investment.

15.2 Scope of Recommendation

Recommendations should be as regards :

1. site and location,
2. size of plant,
3. types of cements and their proportions,
4. specific markets and their share to be aimed at,
5. selection of process and machinery,
6. operating parameters,
7. man power requirements,
8. suggestions about raising finances.

TEFS should clearly bring out viability or otherwise of the Project so that the clients' investment is fruitful and he has a clear idea of the returns he can expect.

If required additional investigations should be carried out to clear doubts if any.

15.3 Executive Summary

TEFS is a voluminous document. Hence for the benefit of the Top Directors or the Board of a Company it is useful to prepare an Executive summary so that they can grasp the Essence of the Project quickly and easily without leaving out vital aspects thereof.

An Executive Summary is thus a very important part of the TEFS.

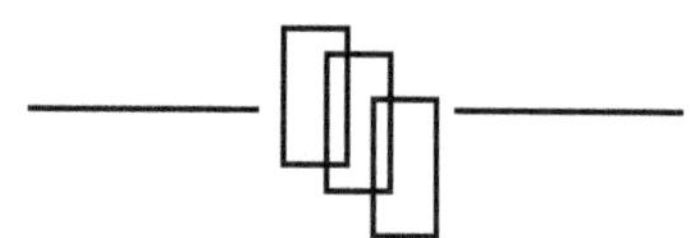

CHAPTER 16

FACTORS GOVERNING THE SIZE OF A CEMENT PLANT

16.1 Factors Governing Size of a Cement Plant

Size of the Cement Plant is governed largely by :

1. State of development of technology.
2. Size of market.

The two are related and influence each other. An assured and expanding market spurs development of technology so as to meet demands of market.

A large size plant would generally mean a more efficient plant in terms of fuel consumption, power consumption and manpower requirements.

When a product is not required to be tailor made and when market is large, like cement, large sized plants would afford economies of scale. Advantages of increased sophistication would be better realized in large plants.

16.2 Total Capacity in one Place and Size of Single Productive Unit

Size of a plant in terms of total capacity in one place and size of a single productive unit have regional and geographical overtones in addition to advances in technology. For instance, USA and European Countries have been always in the forefront of cement technology, and yet the size of a single unit has not exceeded 5000 tpd capacity even with precalciners. On the other hand, in Japan, 3000 tpd capacity would perhaps be the smallest. A great many plants have single units between 8000 to 10000 tpd capacity.

USA is a vast country with extensive deposits and extensive markets. To have a few very large plants would mean very long hauls for transporting cement by road and rail to reach markets. It is advantageous to have a large number of smaller plants spread over the country.

In Japan on the other hand, there are narrow islands with short overland transport. Transport is convenient and economical by sea. Hence a small number of large plants are more suited for that country.

16.3 Optimum Size of a Plant

Optimum size of a plant is not a fixed entity. It will vary with time, as it is influenced by costs of raw materials, fuel, power and labour. For the same country optimum size would tend to go up with passage of time. At the same point of time, different countries would have different optimum sizes.

Specific cost of production tends to decrease with increase in size of plant. The relation becomes asymptotic in that there is a limiting cost even for very large plants.

16.4 Mathematical Model

A mathematical model can be worked out to arrive at an optimum size of a plant based on certain assumptions and taking into account cost factors influencing it.

As far back as 1979 one Mr. Rouss had published an article in Zement Kalk Gyps on the subject. The model was expressed as an equation:

$$(K_P - k_k)\, P^m \simeq D$$

where $K_p = f(P)$ = specific manufacturing cost which is a hyperbolic function

k_k = asymptotic constant of the curve,

P = annual production of cement works

m = exponent of curve

D = constant of curve

k_k depends on costs, particularly costs of raw materials and fuel;

m and D depend on factors like amortisation and operating expenses, process employed, equipment selected and also on standard of development of the country of the cement works. With progress of industrialization and rising standard of living, 'optimum' shifts towards higher outputs.

It will be interesting to note the 'optimum' sizes worked out from the mathematical model above for some countries in 1979.

		Plant capacity	single unit
		mtpa	tpd
1	China	1.7	2500
2	USA	2.7	6700
3	Japan	6.25	19000
4	Switzerland	4.8	14000

One single plant of 14000 tpd capacity would have met requirements of a small country like Switzerland; 12 to 14 plants of 19000 tpd would have met requirements of Japan. Difference in optimum sizes for Japan and USA may also be noted.

At that point of time in India capacity of single unit had not reached 2000 tpd. Today India has quite a few + 10000 tpd capacity clinkering units.The installed capacity has crossed 475 mtpa

16.5 Size of Plant and Limestone Deposits

Apart from the above, size of plant at a given site would also depend on size of deposits and their accessibility.

Life of a cement plant is generally taken to be 30 years. Therefore large plants would need correspondingly large reserves. In any country there would be areas with 'massive' deposits that could support large plants. But there would also be 'small' deposits that would be suitable for small plants only.

16.5.1 Blended Cements

Blended Cements go a long way in saving limestone and it is easier to install large capacity plants with much smaller clinkering capacities.

16.6 Cement Plants for Specific Purposes

Accessibility of deposits also governs size of the plant. Deposits in hilly and inaccessible areas are difficult to exploit. Reaching markets is also difficult in such locations.

Sometimes cement plants are set up for a specific purpose like for supplying cement to a large dam project or for roads in border areas to meet needs of defense. Such plants will have 'short' lives and hence they would have small capacity.

16.7 Availability of Finances

Another important factor that governs the size of the plant is the availability of finances. Capital costs for installing a cement plant are often expressed in Rs per ton of annual capacity. They tend to increase with passage of time.

Cappital Costs of Cement Plants of various capacities would vary according to prevailing Costs expressed in Rs/ton of annual capacity.

Capital costs for plants of different capacities for a capital cost Of Rs 9000/ton of annual capacity

For OPC would be

capacity mtpa	capital costs million Rs.
2.0	18000
3.0	27000
5.0	45000

Capital costs would be less on rs/ton basis for blended cements but higher on production from the same clinkering capacity.

Economic climate dictates availability of finances of this magnitude. Finances could be raised as loans from Financial Institutions and or by equity. Buoyancy of economy will decide the chances of raising money as equity. As the size of plant goes up chances of small entrepreneurs decrease. Only large Industrial Houses can raise the kind of finances required for large plants. Thus growth in Cement Industry will come by way of a few large or very large plants and number of Companies will become less and less.

This can be seen in the present situation in India where a small number of Industrial Houses have a substantial share in the total capacity.

16.8 Economies of Scale

Because of economies of scale, installation costs tend to come down as size of plant increase and cost of production also tend to go down, raw material costs remaining about the same. Labour costs go down because of increased automation. Thus relative proportions of elements of costs also change with size of plant.

The net realization is the difference between sales price and various costs. Artificial regulation of sales prices (as it happens in controlled economy) pushes up the size of the viable plant. In free economy small plants can also be viable but only in specific cases.

Since sales price is the same in a given market at a given point of time for all plants, large or small, it stands to reason that large plants would earn more profits. Hence for this reason also trend will be to install larger and larger plants.

16.9 Acquisitions and Mergers

There would be acquisitions and mergers. Even foreign Companies could be involved. This has happened in developed countries like America where European Cement Companies have acquired interest in and control of American Cement Companies. It is beginning to happen in India also where till now Cement Industry was wholly indigenous. World's leading Cement Producers like Holcim anf Lafarge have Acquired large well established Indian Cement Companies.

SECTION - 4

Civil Design and Construction

Contents

Chapter 1	Civil Design and Civil Engineering	1-11
Chapter 2	Non-Factory Buildings	12-13
Chapter 3	Design Aspects of some of the Non-Factory Building	14-19
Chapter 4	Colony, Guest Houses and Clubs	20-24
Chapter 5	Storages, Silos, Bins and Hoppers	25-30
Chapter 6	Water Supply	31-38

CHAPTER 1

CIVIL DESIGN AND CIVIL ENGINEERING

1.1 Civil Design Engineering and Construction

The Civil Design and Civil Engineering of the Cement Plant should be entrusted to a reputed company specialized in the design of structures, and buildings commonly found in cement plants.

Apart from main factory buildings and storages, civil construction also includes :

1. development of land - leveling thereof for plant, railway siding and colony,
2. planning and laying of roads in plant and colony,
3. drainage in plant and colony,
4. periodical laying of refractory in kiln,
5. design and installation of main water supply scheme, underground and overhead water tank and water pipe lines in plant and colony,
6. developing of green belts in and around plant, quarries and colony,
7. development of gardens and parks,
8. non factory buildings, administrative offices and colony.

1.2 Factory Buildings and Machinery Foundations

Cement machinery in general is among the heaviest machinery in Industry with rotary kiln being the heaviest and dimension wise longest single piece of machinery. With increase in capacity of plant in terms of single kilns - sizes of machinery have increased corresponding to capacities required.

It may be necessary to install two numbers to obtain the desired capacity like two cement mills for the main unit itself.

Cement plants have perhaps the tallest structure in Preheater towers in any Industry apart from chimneys. For large capacities, 2 streams preheater are used, which reduce height somewhat. But with number of stages increasing to 6 now, preheater towers as tall as 80-90 meters are to be commonly found.

Storage silos for blending (Blending cum storage) are designed for 20000 30000 tons capacity and diameters ranging between 20 - 25meters and cement silos of similar capacities and sizes are other large civil structures.

Clinker can also be stored in silos. More commonly it is stored conical sheds with a central shaft; sheds are 80 to 100 metres in diameter and are totally covered.

1.3 Preliminary Investigations

Among the first things to be ascertained / obtained at the proposed location of the plant are :

1. Map of Contours of the plot showing levels, natural drains, streams etc.
2. Load bearing capacity of the soil.
3. Depth of rock on which foundations can rest.
4. Wind velocities at different heights in the area.

   These are important for tall structures like preheater towers and tall chimneys.
5. Seismic zone of the area in which the plant is located and hence the factors to be allowed for in the design of various structures.
6. Underground water tables in dry and wet seasons.

1.4 Functional Design

The civil design of a structure has to be functional in the first place. It must be designed to house the machinery in correct sequence and at different levels as required for flow of material and gases being processed in the section, with ease and without any impediment.

It must provide well-designed stable foundations for machinery, motors, gearboxes, etc.

Foundations must withstand static or dead loads and also dynamic loads arising out of running of machines and vibrations arising therefrom.

1.4.1 Rotary Kiln

The rotary kiln may be supported on 2 or more supporting stations with each tyre resting on two rollers. Kiln is an inclined rotating cylinder which tends to slide down, which expands in hot conditions and which flexes at every revolution under its own weight.

Loads on rollers may not always be uniform. Kiln may leave contact with one roller altogether (for 2 support) or with more than one roller for 3 or more support kilns, thus subjecting other rollers to very high loads albeit for short periods in each revolution. Because of inclination, there are thrust loads along the axis of the kiln.

It is therefore necessary that detailed information on loads under conditions - normal and abnormal - is obtained from respective machinery manufacturers and passed on to civil designers for designing kiln foundations.

1.4.2 Vertical Mills

In vertical roller mils, dead weight of mill itself is very large. In addition there is a direct thrust load on the foundation. During grinding mill also vibrates.

Foundations for vertical roller mill must be designed to take into account all these factors. To spread the heavy loads on a wide area, the foundations assume such proportions that their own weight is several hundred tons, which also must be added to the load on the soil.

Load bearing capacity of soil is measured in T/cm2. The base of foundation should be so designed that the loads transmitted to soil in T / cm2 are well within the load bearing capacity of soil (weight should include weight of civil structures).

1.5 Building Design

Floor levels in a building are arranged to house machinery and drives at different levels.

Column spacings are arranged to facilitate taking the machinery into the building and for its erection consistent with their impact on depths of beams. Column spacing has a direct influence on depths of beams.

Depths of beams and sizes of columns determine clear space for machinery.

The different floors of the building can have their own net work of secondary beams to support the machinery so that the load is in turn transferred to main beams from them to main columns and down to the foundations of column footings.

Column footings are sized so that loads transmitted to soil are within permissible load bearing capacity of the soil.

See Fig. 4.1.1.

1.6 Data Needed for Civil Design

Thus a civil designer needs to be furnished with a departmental layout drawing which furnishes :

1. Dimensions of the buildings. It is a universal convention that building dimension is given in terms of centers of columns and heights are reckoned from tops of floor levels.
2. Locations of machinery with their outlines; their relative centerlines in vertical and horizontal planes.
3. Over all sizes and space occupied by machinery and connecting chutes and ducts in different views - plan, front and side elevations.
4. Dead loads and dynamic loads.
5. Expansion tolerances.
6. Entry and discharge points.
7. Layout of chutes and ducts in the departments with their dimensions and orientation.
8. Openings in floor with sizes and locations.
9. Access platforms to be provided for.
10. Staircases and ladders and landings for access between floors.
11. Maintenance requirements like :
    (a) Storing and handling refractory.
    (b) Lifting Heavy weights (rollers of mills).

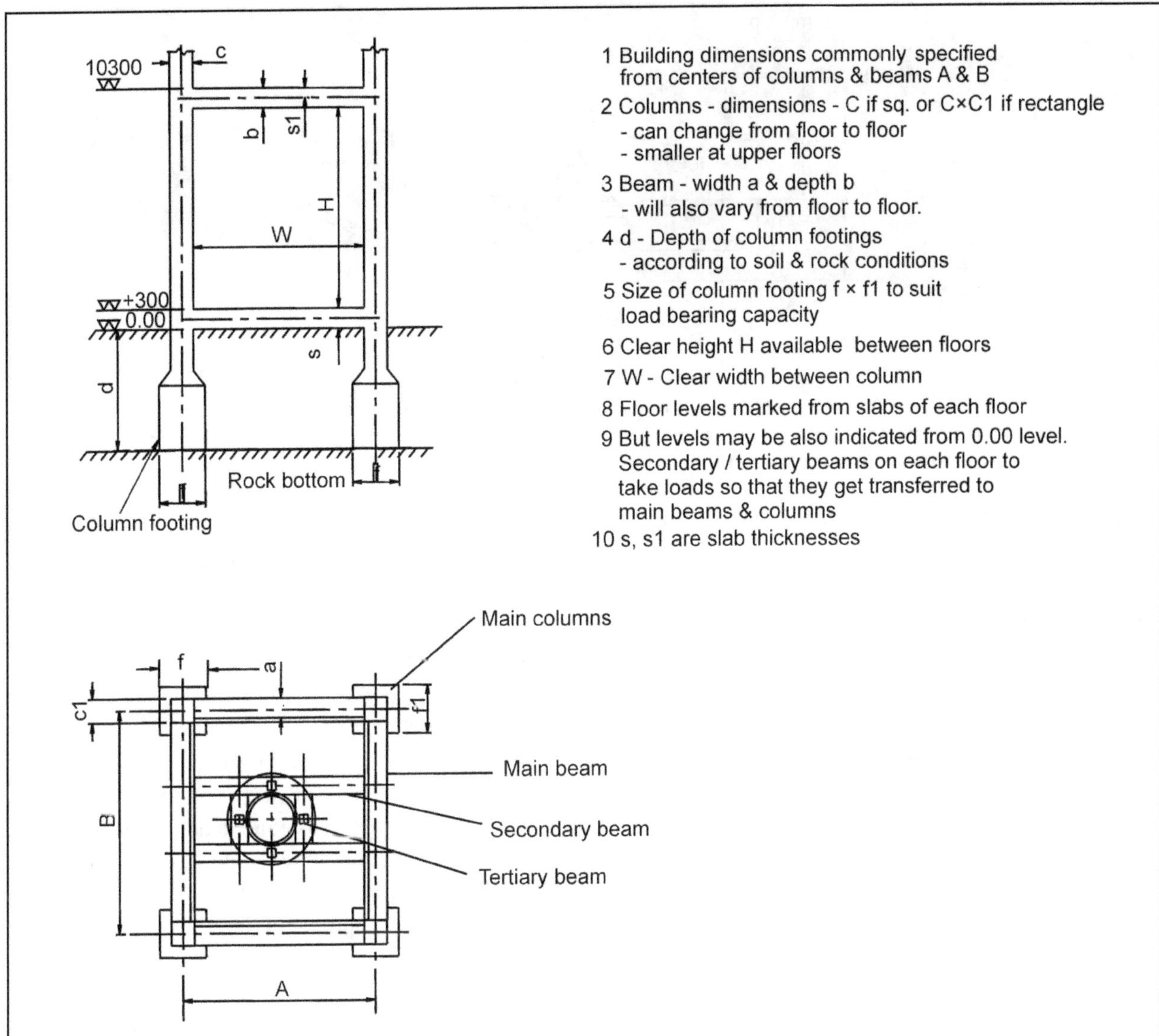

Fig. 4.1.1 Civil design aspects.

(c) Clear space required for maintenance i.e., for removal of shafts, etc.

12. Location for lift when provided.
13. Walkways for inspection and corresponding ladders and platforms.

1.7 Foundations

Major foundations should be marked with benchmarks from which levels can be measured.

Foundations will have marked on them, Centre lines of machines to be installed on them - center lines will be marked in horizontal plane and if need be also in vertical plane.

Foundations will have:

1. Level surfaces which can be used for positioning and leveling bed plates.
2. Marked routes / channels for installing oil lubrication and water-cooling pipelines.
3. Pockets for foundation bolts of adequate size and depth.

Each building should have a 'benchmark' for reference from which heights and distances within

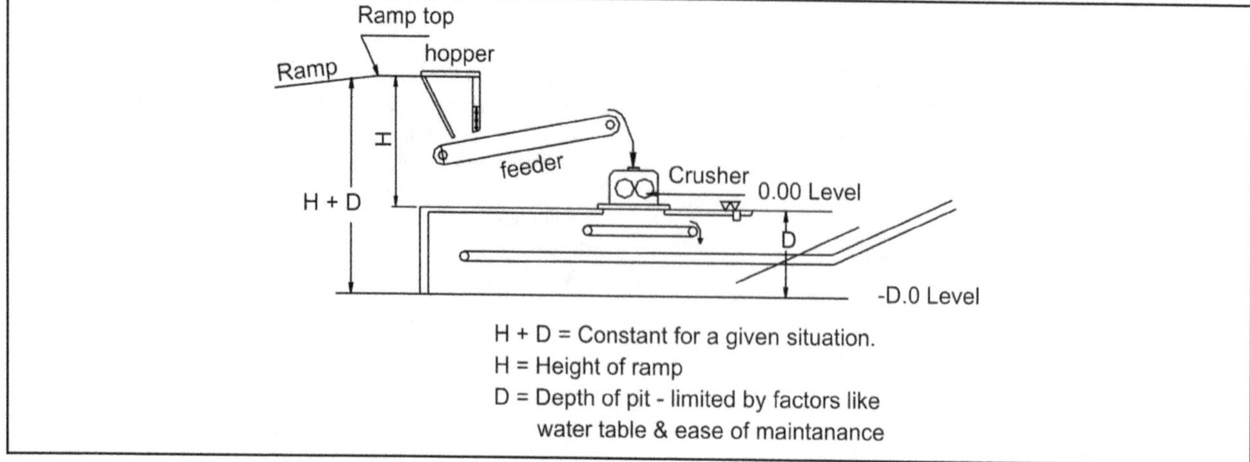

Fig. 4.1.2 Crusher – height of ramp and depth of pit.

building and with respect to other buildings could be measured.

1.8 Utilities

Civil designers should also be informed in advance about :

1. Routing of cables, within the building, whether they are to be under ground / overhead; whether they would be concealed or exposed.
2. Piping layout for utilities.
3. Lighting within and outside building.
4. Sides to be covered for protecting machinery against wind / rain.

1.9 Underground Water Table

Underground water table is important as it limits the depths to which pits should be designed for. Else all pits face risk of seepage and have to be treated with waterproofing.

In crusher for example, choice is between height of ramp and depth of pit. An optimum solution is to be found by comparing cost of ramp with depth of pit and its waterproofing.

See Fig. 4.1.2.

1.10

It is always understood that building and foundations are for the machinery housed in them and not the other way round.

Civil expenditure is a 'one time' expenditure and erring on the generous side is better than shortsighted economizing.

1.11 Design of Silos

In designing silos, among other factors to be considered are :

1. If material is to be fluidized like in blending silos and in cement silos, then the walls of the silo would be subject to 'fluid' loads exerted by fluids like water and hence would be designed from this angle.
2. If insides should be dry at all times as the seepage of moisture, rainwater, through cracks, etc., would result in setting of the material (cement) in hard mass and in formation of lumps which would make it difficult to extract it; it could also affect fluidisation adversely and hence blending and extraction.

Silos should also be designed to withstand pressures from within. Safety hatches should be provided on top to release build up of pressures.

For silos located at a distance from a building it is necessary to install a structure to go to the top where machinery like dust collectors would be installed. Such structure should preferably be independent of silo walls as otherwise danger of seepage increases.

See Fig. 4.1.3.

Some silos rest on the ground; others are raised and house machinery underneath them. These silos would have ring beams and columns of sufficient cross

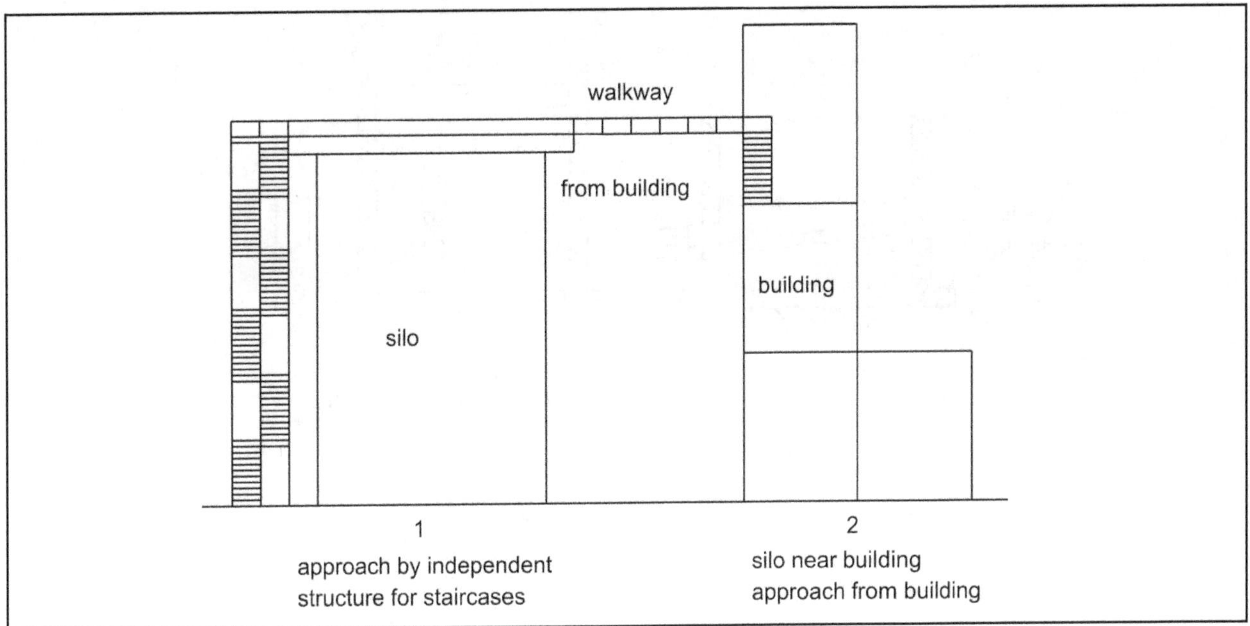

Fig. 4.1.3 Silos - approach to top.

section, which support the load of the silo and also that of material in it.

See Fig. 4.1.4.

Such silos are therefore expensive but they save cost of another building structure like kiln feed building; sometimes even a separate packing plant or more often, bulk loading facility.

It is always desirable to install airlift vessels at ground level (or in pits of not more than 1 metre depth). But airlift vessels are 5-6 m in height and feed points to them being at top, increases the height of building from which they are fed.

1.12 Contours

A contour map is very useful in ascertaining levels in different parts of the plot and to ascertain natural drainages in the plot. These could be used for draining the plant of rainwater etc. There could be natural seasonal streams flowing through the plot. They could be made use of to drain rainwater so that the plant does not get flooded.

See Fig. 6.1.4 of Chapter 1 in **section 6.**

In a flat contour, it is possible that a plant will get flooded in rainy season, if water has no where to go. Planning good drainage in all sections of the plant is therefore very important.

Pits which house machinery and floors that go below ground level should also be protected against flooding.

Contours could also be made use of in the planning of the layout of the plant. A sloping contour can be made use of to install kiln and cooler.

There is considerable difference between levels at A & B. Civil costs could be considerably reduced if the natural slopes are made use of.

In level contours, kiln pier heights would increase from discharge end to feed end. If ground was sloping, they could be of the same height as shown reducing civil costs.

See Fig. 4.1.5.

As mentioned earlier, belts under crusher, part of deep bucket conveyor and conveyors under clinker storage are generally housed in pits; but pits cause problems of :

1. Water seepage.
2. Dust nuisance.
3. Difficulty in maintenance.

If the ground is rocky excavating pits to install these conveyors can also be expensive. Now the trend is to install drag chain under the cooler at ground level.

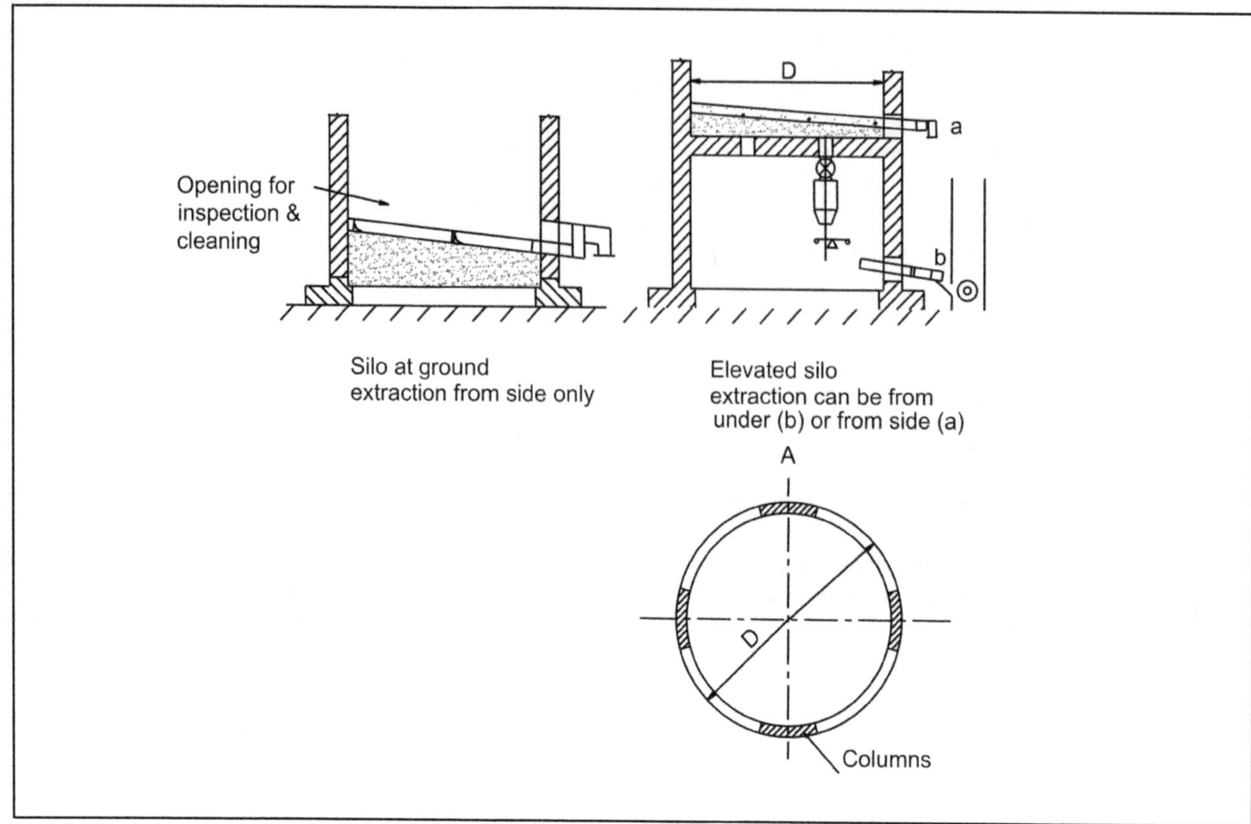

Fig. 4.1.4 Large diameter, flat bottom silos.

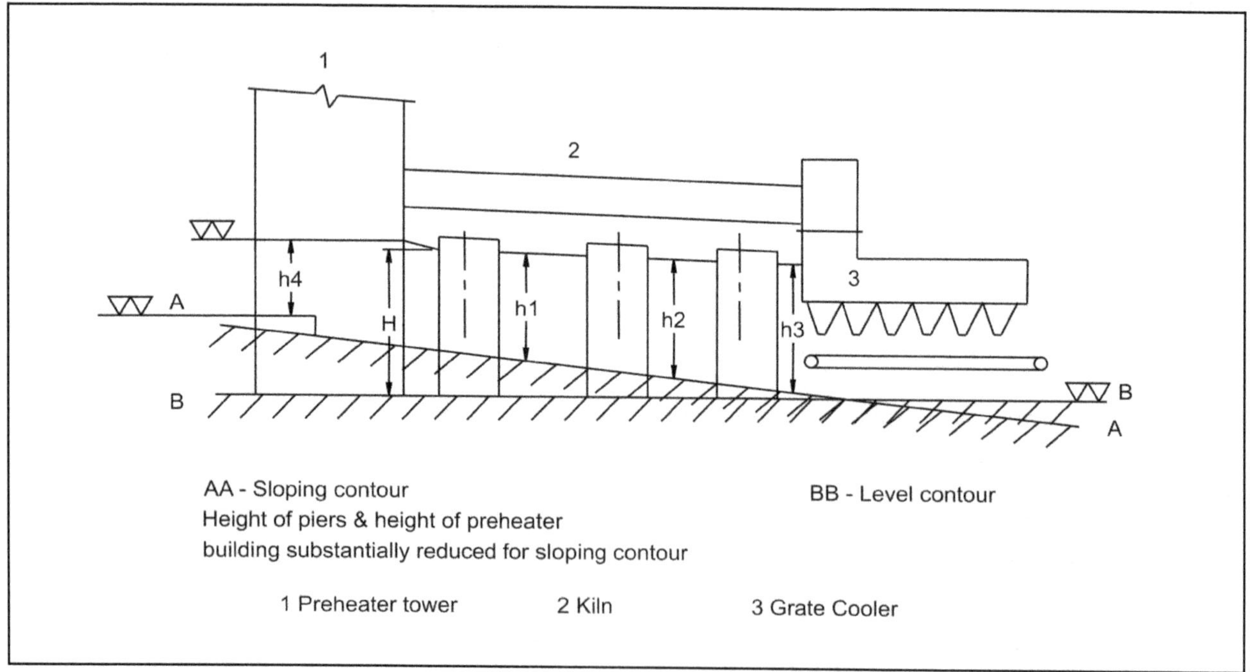

Fig. 4.1.5 Taking advantage of contours.

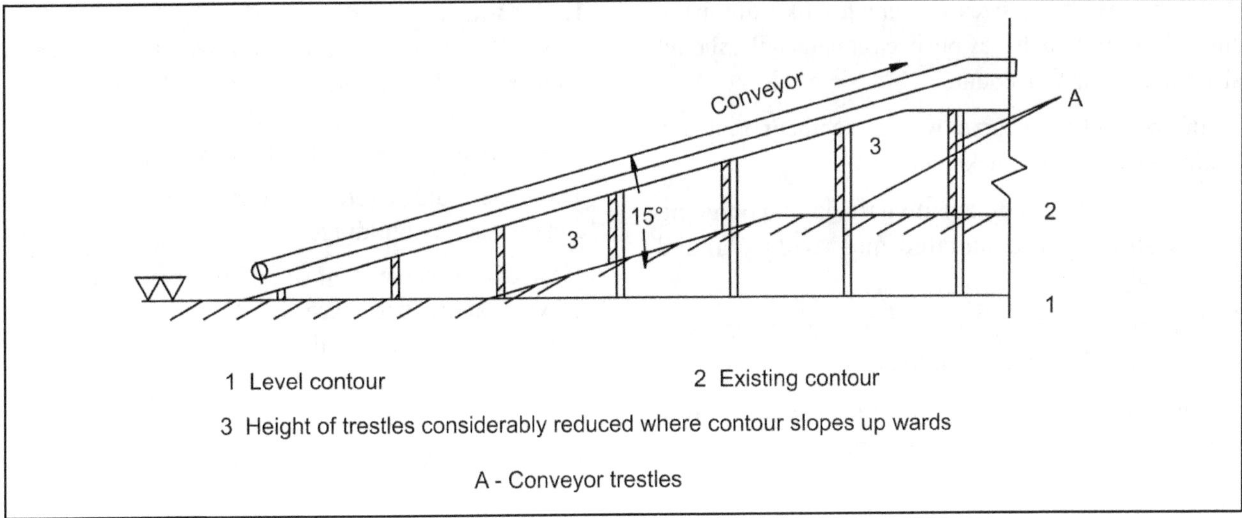

Fig. 4.1.6 Taking advantage of contours.

This no doubt increases height of preheater building by several meters, but is a convenient arrangement to reduce maintenance costs of cooler.

Contours can be put to good use for laying out trestles of conveyors.

See Fig. 4.1.6.

All departments need not have the same floor level within the plant. Small difference of 1.00 meter could be permissible. This would save considerable earth work.

1.13 Roads

Roads provide vital access between :

1. The plant and the rest of the world.
2. Various sections of the plant.
3. Quarries from the plant.
4. Crushers from quarries.

Roads in quarries, between quarry and crusher carry heavy traffic- sometimes all round the clock. As working faces develop, limestone is to be hauled up over greater depths from pit level to reach the crushing plant.

These roads should be wide enough to allow the large dumpers to pass each other and the slopes should be minimum (practical) to save fuel consumption.

This also applies to the ramp along which dumpers ply to reach the crusher. The ramp should have minimum two lanes and a good level area at the top for dumpers to turn around.

See Fig. 4.1.7.

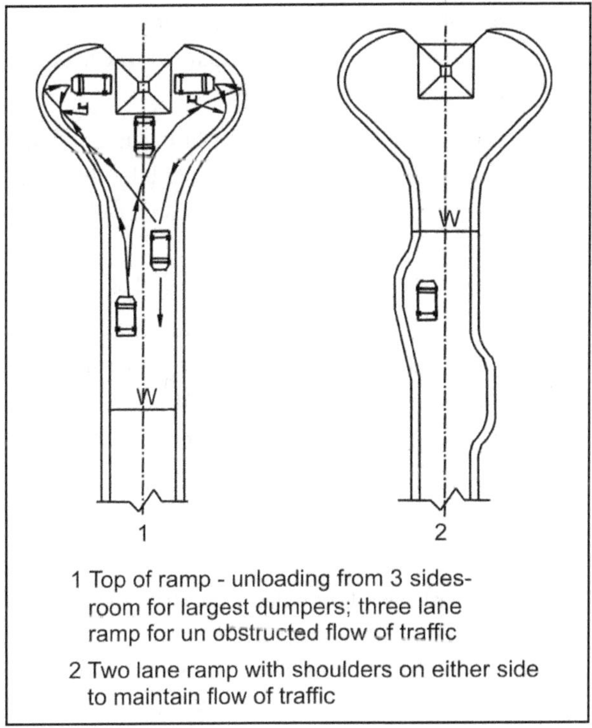

Fig. 4.1.7 Design of ramp for crusher.

Shoulders should be provided at convenient points for stopping the vehicle on side so that the regular traffic is not blocked.

Roads should be sized according to :

(a) Density of traffic.

(b) Sizes of vehicles plying on them.

In this exercise, sizes of vehicles like dumpers, cranes, dozers, which may ply occasionally only, should also be taken into account.

In a cement plant part or whole of cement produced could be sent out in truck loads.

Volume of traffic of incoming trucks and outgoing trucks should be calculated and roads planned accordingly.

It is also necessary to provide for parking space for trucks that have to await loading.

This external traffic should be kept separate from the other internal traffic of the plant.

If possible, separate pathways could be provided for two wheelers and also parking for them, so that they do not get parked on road and obstruct traffic.

In present times planning of parking spaces has also assumed importance because of increase in vehicles owned by the work force.

All roads within the plant do not carry same volume of traffic. Hence widths of roads should be designed to suit density of traffic and sizes of vehicles plying on them.

However, it would be a good idea to provide a 2-way ring road in the plant to provide access to all sections for maintenance. In doing so plans for expansion should also be kept in mind.

Concrete roads would be ideal. They reduce dust nuisance and save fuel particularly on gradients. Therefore roads leading to crusher from quarries in particular should be concreted.

Roads inside plant and also road from highway to the plant may not be concreted to start with. They should be concreted after the machinery is brought into the plant and is erected. Otherwise, roads would get damaged.

Concrete roads have long life and require less maintenance. They also reduce complaints from community about dust nuisance. At project stage, it would be wise to provide for concrete roads. Actual construction can be taken after erection is over.

1.14 Building Code

Green Business Centre of Confederation of Indian Industries- CII have brought out a code for buildings with a view to conserve natural resources and to make maximum use of natural light and ventilation.

This code could be referred to while designing factory and non factory buildings.

One of the aspects that should be kept in mind is to provide facilities for access and maneuverability like ramps for specially abled people,

Extracts from it are attached in **Annexe 4.1.1.**

1.15 New Features of Modern Large Cement Plants

Modern Large Cements have special features.

1 Storages are scattered. Stacker reclaimer systems are used for storing and preblendiing limestone and coal. They require considerable open space for the first unit and for subsequent expansion.

2 Blended cements are commonly made. Space is required for storing and processing fly ash and slag.

3 In quantity slag is more than double that of clinker produced. To handle it requires extended railway siding and wagon tippler. It needs to be dreid also.

4 Plants now use Alternate fuels to substitute coal on an increasing scale.

5 Plants now install Waste Heat Recovery Systems and generate power in addition to Captive Power Plants.

The plant layout must have space for all these features and buildings concerned need to be designed to meet their respective requirements.

6 Today's large 3mtpa capacity plants require a large area as large as

300 hectares. It is unlikely that it will be an even plot. Datum levels and contours can be different over the area of the proposed plant site.

This should be kept in mind along with features mentioned above.

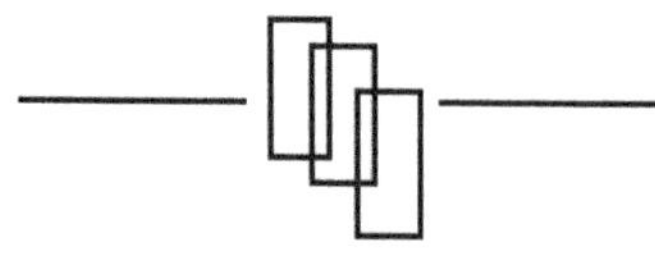

Annex 4.1.1

Norms for Green Buildings for factories

As devised by Confederation of Indian Industries (CII)

1. Rating system is voluntary. It is based on currently available materials and technologies. Objective is to develop eco friendly factory buildings –that would energy and water efficient and healthy and productive
2. Selection of site

   building to comply with local statutory and regulatory options

   control soil erosion – National Building Code of India
3. Accessibility to public transport

   important from point of air pollution due to personal automobiles
4. Basic amenities within the building to include – first aid centre, creche, lockers and showers, canteen, guest house, cyber café- internet facility 1 for 10 persons
5. Topography and landscape

   avoid site disturbance by retaining natural topography
6. Heat island effect

   Thermal gradient between developed and un developed areas

   cover 50 % of roof by vegetation -75 % area open parking by trees- roof over parking or parking in basement
7. Differently abled people

   building should be user friendly to differently abled people
8. Water conservation

   mandatory 50 % of roof and non roof run off should be under RWH. Increase ground water table or reduce water consumption by effective rain water management- rain water storage for 2-3 days of rainfall

   install low flow water fixtures

   plant drought tolerant trees/ shrubs

   non process water- treatment either individual or common effluent treatment plant (ETP)

   ETP. Treated water should meet standards of quality laid down by CPCB
9. Energy conservation

   To avoid use of CFC based heating ventilating and air conditioning HVAC equipment

   Minimum energy performance – prescribe

   SHGC for glazing areas

   Lighting power design

   Chiller efficiency and air conditioned building

   Avoid use of HCFC based refrigerants

   Optimize energy performance

   Metering to be provided to monitor energy consumption

10. On and off site promotion of Renewable Energy

    Minimum 5 % energy used in building excluding process energy to be from renewable source green power- promote investment in renewable energy 50 to 100 %

    Eco friendly generation of captive power by using low emission bio fuels

    Use dg sets certified by CPCB and ISI rated generators

11. Eoo friendly captive power generation

    low CO_2 emitting fuels to be used in CPP-Bio fuels. Use dg sets certified by CPCB and ISI rated generators.

12. Material conservation

    To ensure effective management of non process generated wastes and arrange for

    Their segregation- organic-plastic-glass- e-wastes etc.

    Waste reduction during construction so minimum required to be sent for land filling

    Use materials with recycled content

    To encourage products with recycled content

    Encourage use of local materials

    Minimize use of virgin wood

    Use forest dept certified wood.

13. Indoor environment and occupational health

    Tobacco smoke

    To protect non smokers from passive smoking

    Minimum fresh air requirements

    To install fresh air delivery system for better ventilation

    Minimum air changes for building with forced ventilation

    Avoid use of asbestos

    Day lighting

    Provide good day lighting- day light fctor of min 2 % for min 50 % of total floor area

    Day light design criteria fig and table

    Use paints with Low Volatile Organic Compounds

    Use eco friendly house cleaning materials

Green Homes

1. Benefits of Green Homes

   Most tangible benefits are savings in energy (20-30 %) and in water consumption (30-50 %). Enhanced air quality and better day lighting

2. Water Efficiency

   same as for factories – increase ground water table/reduce usage of water thru effective water management turf design

   to limit such landscape as consumes large quantities of water. Turf area to be < 20 -40 % of total plot area to plant drought tolerant species of trees at least 50 % of landscape planting beds must have drip irrigation

   2.1 Rain water harvesting

   same as for factories

   survey water table in area. Factors to be considered are weathering, fractures and joints for rocky sites and thickness of aquifer in sedimentary rocks

   2.2 Grey water treatment

   grey water is all that waste water resulting from bathrooms , kitchens washing and so on. To provide facility for trating at least 50 % of grey water to supply it for gardening etc.

   reduce demand for fresh water by using grey water for landscaping.

3. Energy Efficiency

   mostly same as for green factories

   install energy meters wherever possible for monitoring consumption

   refrigerators – min 3 star rating by BEE

   Install solar water heating system

   3.1 Captive power generation same as factories

   On site renewable energy – same as factories

   3.2 Energy efficient lighting system and light power density

   Lighting luminaries min 3 star BEE rating

   Power density ~ 20 % less than base line values

   3.3 Energy efficient appliances

   To select efficient motors and pumps and lifts etc

   To install sensor s to control lighting in common areas

4. Materials etc

   same as for factories

   organic waste segregation ;install on site waste treatment technology to prevent such waste being sent to land fills

   use of recycled materials, certified wood erc. Same as for factories

5. Indoor Environment and Occupational Health same as for Factories

   minimum fresh air requirement- same as for factories

   exhaust system

   kitchen and bathrooms to be better ventilated

   enhanced fresh air ventilation – same as for factories

   air conditioners COP/EER equivalent to 3 star BEE rating

CHAPTER 2

NON-FACTORY BUILDINGS

2.1 Non-Factory Buildings

Non-Factory Buildings are facilities which indirectly help in production and which are necessary for the smooth operation of the plant on a continuous basis.

They include facilities like :

A

1. Workshops for machinery and electrical repairs.
2. Stores for spares and consumables, bags, explosives and refractories.
3. Laboratory.
4. Engineering offices within the plant.
5. Civil Engineering offices.
6. Water treatment plant.
7. Pumping stations, etc.
8. Grid and Plant substations, H.T. and L.T. control rooms and panels, room for MCCs.
9. Captive Power Station and Waste Heat Recovery System

B

1. Watch and ward - security office.
2. Time office.
3. Administrative office.
4. Garages, maintenance facilities for vehicles.
5. Petrol / Diesel pumps.
6. Parking facilities.
7. Public Conveniences.

C

1. Guest house.
2. Colony and recreational facilities.
3. Shops, Post Office, Bank & other commercial facilities.

2.2 Requisites for Non-Factory Building

First requisite for all these is that they should be functional and convenient.

Mere opulence and lavish expenditure on décor and materials used do not make a building suitable for the purpose it was meant.

Therefore each building is to be planned in consultation with people, who are going to use it and after ascertaining their present and future needs.

Locations of buildings should also be decided from the point of convenience and accessibility.

Load centers for different categories of buildings should be found out (which sections need / use what most) and building located accordingly.

Facilities like ramps should be provided for access and movement of specially abled people.

2.2.1 Locations of Non-Factory Buildings

Out of necessity some buildings can be located only at specific points in the plant like:

1. Main security office and time office at the main entrance of the plant.
2. Water treatment plant and water pumping station near the incoming point of water supply into the plant.
3. Central control station cum laboratory near main production lines.
4. Bags godown near packing plant.
5. Electrical substation and Control Rooms near load centers.
6. Quarry office and Garages for maintenance of earth moving machinery near quarries;

explosives magazine at a secluded spot near quarries.

7. Administrative office should be easily accessible to outside parties and so located that outside parties do not have to enter the plant.

   Hence it would normally get located just outside the plant but with access to it from the rear for plant personnel.
8. Parking lots for trucks and conveniences for drivers, also just outside the plant.
9. Canteen and rest room for workers at a central point within the plant or just outside the plant
10. Captive Power Station and WHR System at convenient locations from operational angles.

2.3 Drinking Water

All buildings should have facility for drinking water. If water coolers are installed, so much the better. Disposable cups can be used. But these would have to be replenished regularly.

2.4 Fans

In India, temperatures are high almost all months of the year except two, three winter months. In monsoon humidity is high.

Therefore providing adequate cross ventilation in designing buildings is a must. Even then fans and coolers would be found to be necessary for 6 to 9 months of the year.

Working places like offices most of course have fans at close spacing but fans must not disrupt working by blowing away papers.

2.5 Air-Conditioning

Important offices will be air-conditioned. Air conditioning can be central for a multi storied building housing Central Control Room, X-ray analyzer, important sections of laboratory, engineering offices, etc. Offices of senior executives would have room air conditioners.

For rooms of large sizes, Industrial size air conditioners can be used.

Room coolers are an inexpensive way of keeping temperatures down but they increase humidity and also need water supply, there can be spillage also.

Providing comfortable working conditions by way of fans, coolers, air-conditioning should be treated as a necessity and not a luxury. For computer systems and control panels, they are a must as they keep dust out besides maintaining comfortable temperatures and humidity.

2.6 Toilets

Toilets should be located in every building conveniently located and adequately sized for both gents and ladies. This simple but common need for plant personnel whether for officers, executives or workers has been long neglected. A few blocks located at remote places was the practice a few decades back. Toilets must have continuous and adequate water supply.

2.7 Sizing and Layout of Buildings

Sizing and layout of buildings should be arrived at with the help of architects to whom must be furnished:

1. Primary and secondary functions and activities to be carried out.
2. Number of personnel - using the facility always.
3. Transiting personnel.
4. Need of lighting general and focused at work places.
5. Need for ventilation and air conditioning.
6. Locations of lifts and stairs.

Adequate room should be kept for expansion also.

Architects should be selected on basis of their past experience and expertise in this type of work.

Recommendations made in the Building Code mentioned in Chapter 1 should be kept in mind.

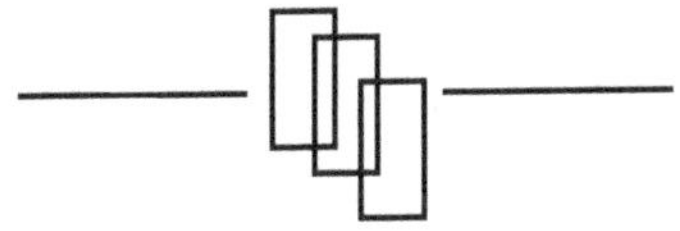

CHAPTER 3

DESIGN ASPECTS OF SOME OF THE NON-FACTORY BUILDING

3.1 Design Norms for Non-Factory Building

Basic design norms to be followed for various non factory buildings have been explained in **Chapter 2.**

Now specific requirements of some of the important buildings are indicated to help in their design and hence cost estimates.

3.2 Stores

These are generally 'central' for cement plants except perhaps for quarries. Stores contain 'consumables' and also 'spares'. Stores should be planned from point of ease of accessibility.

It should be easily accessible for receiving diverse materials, refractories, grinding media, castings, lubricants and so on. If a railway line can be brought up to it, it can be a great convenience.

Facilities should be provided to unload trucks and wagons and to shift unloaded material either into yard or into the stores.

An open store yard is a must to store bulk materials like refractories, grinding media, heavy castings like rollers, gears, lining plates, cables etc., among other things.

These could be handled by fork lift trucks or movable cranes as and when required. While castings could be stored in the open there should be roof over items like cables and refractories. Finish machined emergency spares like riding rings, rollers, gears etc., should be stored with protective covering over them.

Oils and Greases, Cotton waste, are consumables, which are also fire hazards. Such materials, which are required regularly, should be stored under cover but 'separately' with adequate protection against fire and to prevent fire spreading to other parts of the store.

A separate delivery or issue counter is arranged for daily issues. Further, they should be so stored that dispensation is easy and free from dust.

Stores also need facilities to weigh incoming and outgoing goods. The scales installed should be adequate for this purpose. It may have a portable scale also.

Considerable advance has been made in design of racks and bins and in equipment for handling goods within the store. Aisles and passages should be designed so that such modern equipment could be used.

These aspects should be closely gone into and expert advice taken in planning a store and its layout.

Heavier items should be stored on ground floor and lighter items on mezzanine. There should also be a closed room with cabinets for storing high cost items.

With computers it is now easy to maintain inventory of stocks and their values can be obtained almost instantly.

Computers can also provide information on stocks, consumption, orders in transit etc., so that ordering procedure can be stream lined in such a way that stores do not run out of stock of any materials, stores / spares at any time nor do they overstock.

Protection against fire is a specially important aspect of stores layout and design.

Another important aspect is precautions against theft and pilferage.

Open store yard and stores should not be accessible to all and sundry. Issue counters should be cordoned off and personnel from other sections of the plant who have come to draw goods from stores should wait there.

A typical layout of General Stores is shown in **Fig. 4.3.1.**

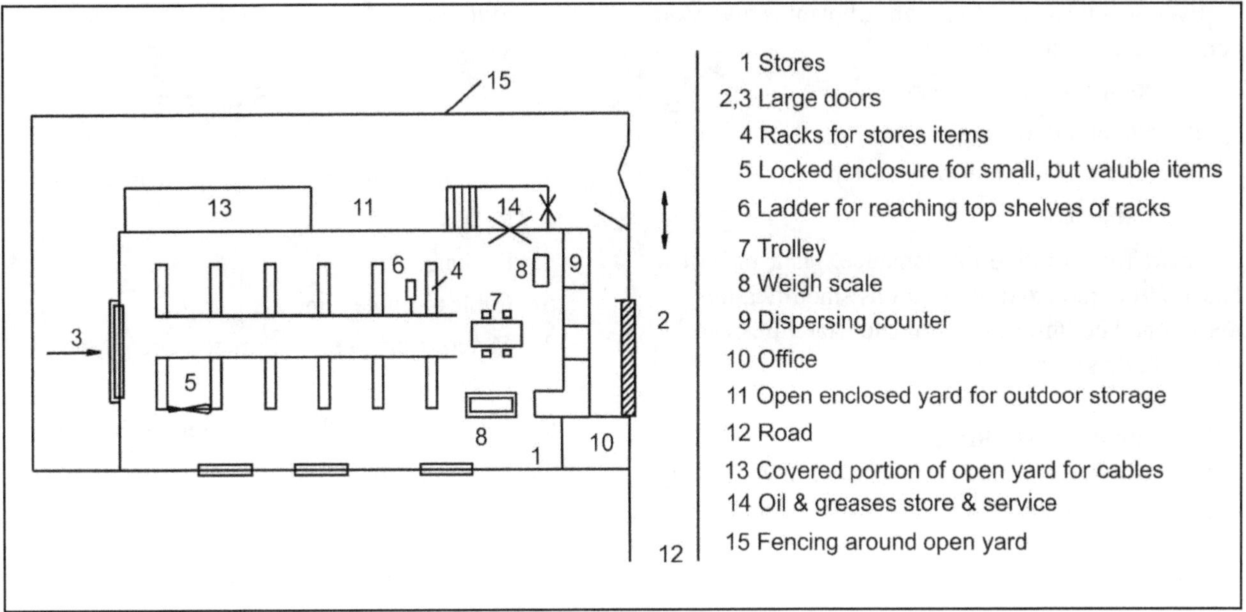

Fig. 4.3.1 Layout of general stores.

3.3 Workshops

The extent to which a cement plant should equip its workshop to carry out repairs / machining of damaged parts, would depend on the location of the plant and the infrastructure facilities available in the neighborhood.

Remotely located plants would have to spend considerable time in procuring parts from cities / suppliers located far away and hence should have better equipped workshops.

3.3.1 Basic Facilities in Workshops

In any case some of the 'basic' facilities should always be included in planning a workshop.

(i)Turning and machining facilities like lathes; Lathe may also have a grinding attachment.

(ii) Milling and planing facility including a milling cum boring machine and a shaping machine.

(iii) Drilling facility to drill holes; a radial drill is a great help.

(iv) Cutting facilities.

(v) Welding facilities for electric and gas welding for thin sheets and also for plates up to 25 / 50 mm thick.

(vi) Forging facilities to forge small components.

(vii) An open yard for marking development of plates for facilitating fabrication of chutes, ducts and hoppers.

(viii) Plant may also have a cupola for casting small items.

(ix) Plate bending machine for rolling shells and pipes.

(x) Balancing machine for small impellers and rotors.

(xi) If the plant is doing its own hard facing of hammers, rollers of vertical mills and roller presses, facility for welding in shops and in situ with fixtures using requisite electrodes should be available.

3.3.2 Selection of Machinery in Workshops

Sizes and capacities of respective machines would depend on size of the plant and individual largest components required to be made, hard faced, machined and ground.

3.3.3 Outsourcing

It is common these days to get maintenance and repair jobs done outside.

Even maintenance of sections of plant is given 'on contract' like maintenance of :

(i) compressors / blowers.

(ii) air conditioners.

(iii) laying of refractories - castables.

(iv) computers.

Apart from routine maintenance, plant may also take up from time to time additions / modifications in a section. The 'in house' workshop facilities can be handy on these occasions.

3.4 Workshop Building

Size of workshop building can be planned after deciding on the various facilities to be provided and largest size/ weight of machines / components that may be required to be handled in it.

Individual machines would be selected taking these into account.

While gears / pinions often need to be reconditioned, there is no point in installing a gear hobbing machine for the purpose. They are best treated as spares and stocked and replenished from time to time.

Handling of heavy and large sized individual pieces of machinery, placing them on machine table / beds etc., should be done with the help of overhead traveling crane, which should cover entire area of the workshop.

While main building would contain machine tools of various types, annexes would include welding and cutting and forging facilities. An adjacent open yard with smooth concrete floor would provide facility for marking templates.

The plate bending machine should also be installed by the side with overhead facility for holding plates being bent.

Heat treatment facilities are required many a time to relieve stresses, to obtain desired hardness. Most common is the tempering facility. Also needed are facilities for heating pinion shaft assemblies to take out and fit on pinions and rollers on to them. Oil / gas fired furnaces with quenching tanks should be available for this purpose.

3.5 Electrical Repair Shops

Electrical repair shop should have facilities for attending to problems related to:

1. Starters - DOL, S/D, SRR.
2. Motors - H.T & L.T
S.C., S.R.
D.C.
3. Switch fuse units - Circuit breaker – oil – air
4. Relays.
5. Busbars.
6. Cables – cable joints.
7. Meters – ammeter, voltmeter, energy meter.
8. Lighting.
9. Transformers – Power, Current & Voltage.
10. Earthing / Lightning.

3.5.1 Planning of Electrical Repair Shops

Planning of Electrical Shops should be done in the light of :

(i) Facilities available in the neighborhood for servicing and repairs.

(ii) Possibilities of getting jobs done through suppliers / their service stations.

(iii) Spares that can be maintained to tide over difficulty including spare starters, circuit breakers and motors.

Even then the shop should be equipped to carry out some basic jobs:

(i) Megger tests to check insulation.

(ii) Drying of windings when megger tests show need to do it.

(iii) Cable jointing.

(iv) Periodical inspection and checks of relays, safety tripping devices, and calibration of meters.

(v) filtering and changing of transformer oil.

(vi) Winding of motors.

The electrical shop should have winding facility for repairing windings of motors, for repairing insulation, for cleaning commutators and so on.

Often bearings of motors need to be replaced. These would almost always be antifriction bearings. A list of bearings in use should be made with the help of suppliers and spare bearings should be maintained in stock. The workshop should have hydraulic gear to remove old bearings and to fit new ones.

3.5.2 Motors

The motors list could be regrouped according to type, kw rating, speed, voltage, etc.

Conveying equipment, blowers, compressors feeders, air locks, remotely operated dampers etc., may have motors of similar specifications. Thus if 10 numbers 7.5 kw, 1440 rpm, 440 V Squirrel Cage motors are in use in the plant, then it would be prudent to order one spare motor of this rating.

At the other end when Raw mills and Cement mills are both ball mills, they could be having similar ratings. A plant may thus have 2 to 4 motors of similar ratings like of say 2000 KW, 6.6 kv, 750 rpm, slip ring induction type. One spare motor even for only 4 motors would be justified as loss of production would be much more costly.

This logic should be followed to list out spare starters, switchgear, cables, relays, motors, etc., and they should be ordered and so identified.

Whenever occasion arises, the faulty unit is removed and spare available put in its place. The faulty unit is then repaired either in company's workshop or sent for repairs to a properly equipped service shop. When it comes back is put in the 'spare' lot to serve as spare again. The 'spare lot' should always remain at the level planned.

3.6 Lighting

In lighting it would be prudent to have similar kinds of fittings, tubes, bulbs, sodium vapour lamps and so on, so that a minimum quantity needs to be maintained as spares.

The electrical workshop would need a 'movable' trolley or van with a platform, which can be elevated to the height of light fittings for street lights.

LED bulbs and tubes save a lot of power. Solar lights can also be provided selectively.

Locating electrical shop near mechanical workshop is advantageous. However, some would prefer to locate it near HT LT control panel rooms.

3.7 Facilities for Maintenance of Instrumentation

Maintenance facilities for central control panel and instrumentation, which are largely electronic and digital would largely call for 'replacement', as such components can hardly be repaired.

Precaution is the best safeguard for long service life of instruments by way of providing dust free, air conditioned surroundings of right humidity. Power supply of constant voltage and frequency also lengthens life of instruments, computers etc.

Such shops would be located next to central control rooms.

As in case of machinery, maintenance and servicing of specific electrical equipment can also be given on 'contract'. It saves recruiting of personnel with requisite expertise on permanent basis.

3.8 Testing Laboratories

Tests of different kinds are to be conducted on various materials and results recorded at various stages of cement production.

Some tests are 'statutory' laid down by Bureau of Indian Standards and are to be recorded and results submitted to BIS for approval so that cement sold in market can be stated to be BIS approved or as per specific grade of BIS.

Tests start with basic raw material - Limestone. Primary tests, which need to be regularly carried out, are chemical analysis, in particular oxides - CaO, MgO, SiO2, Al2O3 , Fe2O3 and other constituents like alkalies and chlorides and with various statutory tests on cements produced.

Earlier all these tests- destructive, non destructive were carried out in the Laboratory.

Now with use of automatic sampling and 'X-Ray analyzer, most of the analytical tests are carried out in air conditioned rooms adjacent to Central Control Room. For convenience sake the Laboratory is now located in the same building to carry out other tests like testing strengths of cement, testing grindability, calorific value etc.

Laboratory is to be designed to include machinery for all these tests and laid out to carry them out in an efficient manner.

The laboratory should also be equipped to carry out physical tests to establish crushing strengths, grindability etc. These tests are useful to provide a pointer as regards power consumption and production from mills.

It is also useful to carry out tests to determine calorific value, moisture and ash in coal as received and as fired on a regular basis.

For cement, fineness tests to establish 'blaine surface' and sieve analysis down to smallest micron sizes for various fractions in grinding circuit are important and are regularly made; as also strengths of cements made at regular intervals. Same 'X-Ray analyzer' would carry out analytical tests on clinker and cement.

If it is proposed to use Alternate Fuels, additional equirement would have to be installed for testing

Like-Brook field viscometer

Ion Analyser

CHNS Analyser

Absrption Spectrophotometer

It may also become necessary to monitor emission of greenhouse gases on continuous basis. Provision

Should be made for that also.

The whole purpose of tests apart from statutory tests, is to ensure that quality product is consistently produced with minimum in puts. Hence the laboratory must feedback instructions for appropriate corrections, regularly / continuously and they must work closely with production departments.

It should be equipped and laid out to carry out these Quality Control Functions. Its planning is now integrated with planning of Central Control Rooms.

3.9 Offices

Offices of junior and senior executives and Management personnel should also be planned in such a way that offices of persons interacting most with one another are located nearby so that time is not wasted in going from one place to the other.

A conference or meeting room on each floor for holding internal meetings or meeting visitors is very useful. The conference room would have facilities like black boards and over head slide projectors and microphone.

There are at least two distinct schools of thought in planning of offices.

1. One believes in 'open' layouts which do not have cabins but provide enough privacy for the officers when they want to concentrate on their work.
2. The other believes in closed offices like cabins and closed halls for their executives and staff.

   It needs a great deal of thought to plan offices.

(i) They should be planned for present and future needs.

(ii) Should take into account currently available facilities for communication, data processing, personal computers, etc.

The company may even have a server station and several outlets working in parallel. The office layout should facilitate installation of PCs, with cabling properly laid out.

Lighting in the office should be adequate for the purpose; indirect soft lighting for general area and table lamps on tables would be ideal.

Even when offices may be centrally air conditioned or with room AC units, fans would still be required.

The 'offices' can become status symbol and a prestige issue on account of:

(a) Area.

(b) Carpeting.

(c) Decor & quality of furniture.

The differences should be kept to a minimum so that there are no heart burns affecting morale.

3.10 Head Office

Most Cement Companies would have their Head Office in a city or in a metropolis distant from the plant.

Plant and Head Office can be linked by latest means of communications now available, including 'tele-conferences' so that it is not necessary for Head Office staff to visit works to clarify a few points and for works executives to visit H.O. to obtain sanctions, etc.

Head Office may also be the center for 'Finance and Accounts' functions for complying with statutory requirements like audit, board meetings, etc.

It would also be the hub from which the Sales and Marketing functions would be directed.

The Head Office must therefore have:

1. Central communications and data centers computerized and linked with plant and other offices.

2. Board room for Board Meetings.
3. Finance, Accounts and Statistics and Audit Sections for monitoring Company's financial health; interest to be paid on loans; collection of dues and so on.
4. Sales and Marketing sections which would direct and execute policies of the Company in promoting sales.

The Head Office staff would thus have very senior executives like:

(a) M.D. and CEO - a full time executive.

(b) Marketing Director.

(c) Finance Director.

(d) Director M.I.S.

Location of Head Office in a city is important. It should be convenient for visitors; should be easily reached and should have sufficient parking place around it. Size of head office would depend on the scale of operations of a Company and hence the staff to be accommodated in the H.O.

In big cities parking space is becoming critical and if Head Office can be planned and constructed due attention should be given to provide for this facility.

Head Office should also have central / room air conditions or a combination of two. (The building itself may be air conditioned if taken on rent).

UPS facilities for power supply to be maintained in times of scheduled and unscheduled power cuts particularly for computer systems, lifts, lights and fans are necessary.

A DG set may also be installed if office is sufficiently large.

3.11 Parking

Parking facilities are now required on a much larger scale in the plant because of steep increase in number of two and four wheel, powered vehicles like scooters, mopeds and cars.

Thus adequate space should be planned in the layout at several locations for this purpose. It is best to have a parking space around each building and office.

3.12 Garages

Garages will house company's vehicles when not in use and will also have facilities for normal servicing and maintenance. These should naturally be installed to suit both large vehicles like trucks and for cars.

Garages for earth moving equipment like dumpers, shovels, dozers, cranes, etc., would need different kind of facilities and space required would be larger. If vehicles are to be lifted then platform with higher capacity hydraulic lift would be required.

There should be two lifting platforms one for normal servicing and the other for repairs as vehicles under repairs will block the lifting facility for a longer time.

Company will also have to plan and provide for 'Service Stations' to dispense petrol and diesel for its vehicles and also for vehicles of staff.

Quarries will most certainly have its own service station for its earth moving machinery.

3.13 Facilities for Workers

Facilities connected with labour force such as:

1. Canteens and Rest room.
2. Creches.
3. Time office.
4. Washing and cleaning facilities.
5. Drinking water- water coolers.
6. Public conveniences.
7. Parking.

Should be planned with great care so that these do not become cause of discontent. They should be designed and planned for, looking at them from workers' angle or point of view.

Facilities for handicapped workers /officers like ramps for access and special toilets shouldalso be provided

3.14 Solar Power

Cement Company should install roof top solar power facilities on office buildings.On a bigger scale, solar power facility can be installed on Company's land surrounding the plant and colony. This can save power bills considerably.

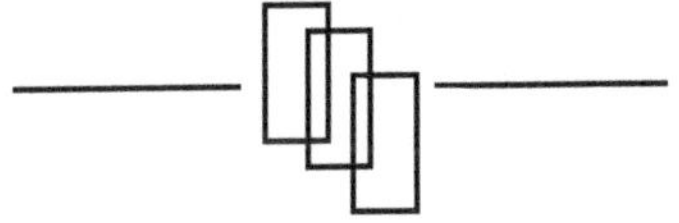

CHAPTER 4

COLONY, GUEST HOUSES, CLUBS

4.1 Colony

Company would be making organizational charts listing Executives, Officers and Staff (skilled and semi-skilled) that would be working in the plant.

It would also list out sections working in :

1. General shift,
2. Two Shifts,
3. Three Shifts.

and identify the staff in various sections according to their status and skills.

Company should also list unskilled workers and casual labour that would be needed from time to time.

The company has then to decide for whom they would have to arrange for quarters from the point of operational convenience including maintenance and repairs.

In taking these decisions company will take into account locations of nearby township / village etc., and availability of residential accommodation there.

Recruiting 'local' people of requisite skills is an advantage in that they have 'roots' there and they are unlikely to leave their jobs. This is particularly true of 'unskilled' and casual labour.

The Company would need in this age of computerization and Information Technology, more skilled and semi-skilled workers and 'knowledge workers' as explained in **Section 3, Chapter 8**

4.1.1 Allocation of Housing

Staff working in sections that work round the clock like mills, kilns, packing plant, control rooms, (sometime even quarries and crusher); staff of maintenance and repair section; staff of utilities like power and water supply sections would most probably be housed by the Company regardless of their qualifications or status in hierarchy.

4.2 Designs of Accommodation to be Provided

Having thus arrived at a total number of housing facilities, the next point to be settled is size and type of houses. In old times, land was cheap and hence individual houses (though small) were common even at lowest level with even a small compound for workers to do their gardening etc.

With land costs being what they are it is now customary to build multistoried apartments, which can be categorized according to sizes of flats.

1. One room, Kitchen 2. Two Room, Kitchen 3. Three Room, Kitchen	According to the hierarchical status of the worker / member of staff

Broad policy decisions would have to be taken to classify workers and the type of quarter they would be entitled to as shown in **Table 4.4.1**

It is best to keep down the varieties of types of houses / flats.

In a colony, a 'house' is a 'status' symbol and announces salary and status of the person. Therefore there is a lot of heartburn in this matter, which cannot be avoided altogether.

4.3 Planning the Housing Colony

After this preliminary work, planning the layout of the colony can be

Table 4.4.1

	Designation	Salary range	Type of House	Area sq. metres
A	General Manager/ Deputy gen. manager			
B	Manager / Dy. Managers			
C	Foremen / skilled workers			
D	Semi-skilled workers			
E	Un-skilled workers			

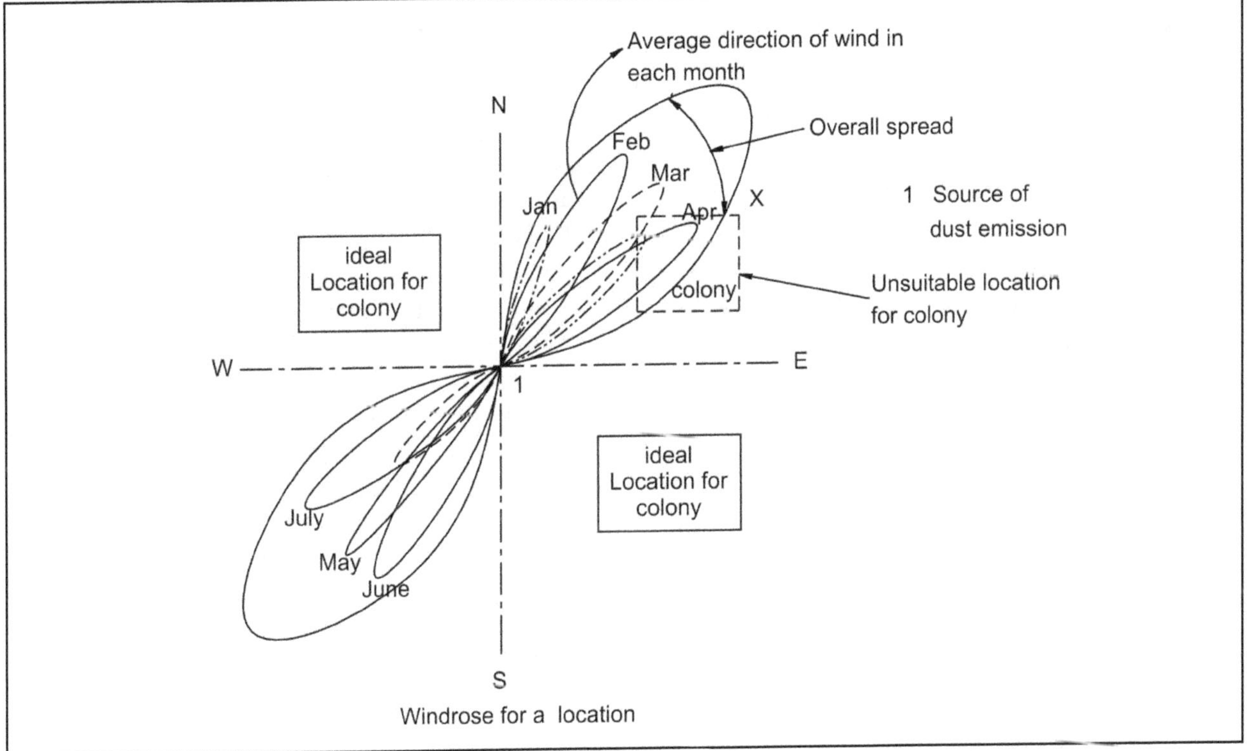

Fig. 4.4.1 Overall spread of directions of wind; scale can indicate average velocities also.

Taken in hand.

First its location :

In old times colony was next to the plant. Care was generally taken to locate it outside the 'wind rose' so that, exhausted dust will not be settling in it. **See Fig. 4.4.1**

Other factors affecting location would be proximity to highway, village / town, etc.

So location nearer to highway, village etc., would be selected assuming that, land was available.

Now, almost all workers have either cycles or two wheelers (a great may officers have cars also). Therefore, colonies can be located at a distance from the plant. This reduces noise level and also dust nuisance.

It is possible to create green belts around the plant and colony.

If the plant is at a distance of > 2 - 5 Kms from the town / village, the Company would have to provide amenities like, supply of milk; provisions; vegetables; primary school and dispensary for out patients as also a Post Office and a Bank

Apart from 'wind rose' considerations mentioned earlier, it should not be so located that workers and officers have to cross highways / railway lines. It should not be too near nor too far from the plant and the town if close by and highway.

4.3.1 Facilities in the Colony

Colony will also have recreational facilities for workers like :

(i) Satellite transmission station, dished antenna for TV.

(ii) Club with facilities for sports like, badminton, table tennis, cards, etc.

(iii) Gardens with play corners for children and for old people.

(iv) Playground for sports like foot ball, cricket, etc.

Sometimes managements also construct a temple / a mosque on premises.

Facilities to be provided for all residents would include:

(a) Water supply – 24 hours or regulated.

(b) Electricity - 24 hours.

(c) Conservation and sanitation.

(d) Depending on climate and location, fans, etc.

It is for each Company to decide whether to charge for water (mostly free) and electricity- some times free.

Load of lighting and electrical gadgets commonly used in houses should also be taken into account.

A civil engineer supported by staff like plumbers, electricians, masons, etc., should be available for maintenance

4.4 Layout of the Colony

The layout of a colony has to follow principles of 'town planning' in almost all respects.

The layout should provide roads designed according to traffic, metalled or concrete.

Trees, and gardens and parks should also be an integral part of the design of a colony.

A typical layout of a Colony for a Cement Plant is attached.

See Fig. 4.4.2.

4.5 Cost of the Colony

A Colony adds to capital costs of the project. It is possible to obtain loans for housing from Government agencies in some states. These could be made use of.

Cost of colony can be estimated with the help of architects by submitting to them the preliminary design of Bungalows, houses, duplex houses, apartments, etc. It may be a good idea to appoint Architects to design the colony who have specialized in such work.

4.6 Green Belts

A green belt is now statutory between high way and the plant. It serves as absorber of noise and dust. Though a good many workers could be having two wheelers now, it is best to arrange the colony in such a way that workers' houses are close to the plant and officers' after that. **See Fig. 4.4.2.**

Companies now spend large amounts on gardening in the plant and in colony and in landscaping in plant and in quarry. They encourage staff to have their own private gardens. A nursery as well as a horticulturist is now a part and parcel of a cement plant.

The excavations and over burdens from quarries provide almost free fertile soil for gardening and landscaping. Landscaping can be extended to quarries also. Excavated and abandoned pits can be turned into lakes for recreation and sports. However, they should be well protected from operations going on in adjacent working sectors of the quarry. Wastewater treatment plant can save substantial quantities of water particularly for non-drinking uses like gardening.

4.7 Medical Centre, Shopping centre etc.

The medical center and shopping facilities and community center should be more or less centrally located within the colony.

When workers are covered by ESIS, Companies do not have to provide hospitalisation facilities. They can have arrangement with ESIS hospitals or with private hospitals for 'beds' reserved for their staff and workers.

The purpose of listing these things out is to provide for them in capital costs and in plant layouts and layouts of colonies.

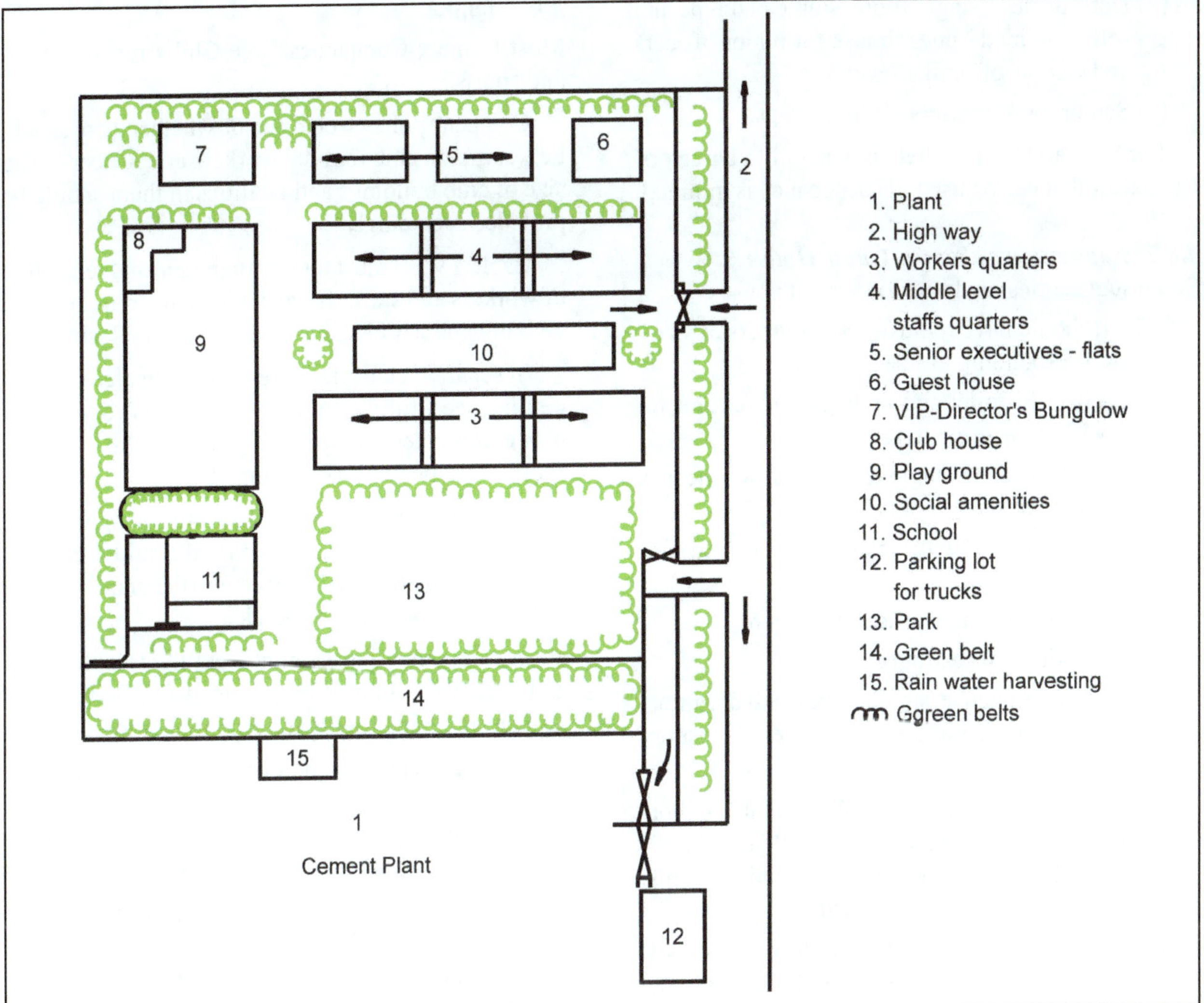

Fig. 4.4.2 Typical layout of a colony for a cement plant.

4.8 Guest Houses

Cement plants being located in remote places are often required to provide boarding and lodging facilities for visitors on long and short term basis.

Most cement plants would therefore have guest house/s for visitors, located either near the plant or in the colony.

Visitors can be diverse in status, according to their businesses with the Company, for example,

(a) Vendors' representatives at junior and senior levels.

(b) Erection and Commissioning, Supervisory personnel.

(c) Government officials at various levels.

(d) Dealers

Company's own junior bachelor staff like trainees would also need guest house or dormitory like facilities.

4.8.1 Types of Guest Houses

Guest houses thus fall into various categories:

(i) Dormitories.

(ii) Junior guest house.

(iii) Middle / senior level executives.

(iv) Very senior executives and Directors.

Only large plants would afford to have separate guest houses for each of these categories. It means duplication of services and personnel.

To save on capital costs and yet maintain standards and quality, it would be best to have,

(a) Dormitories - single room, single bed type and self contained - guest house for junior officers and visitors of similar level.

(b) Senior guest house/s for higher ups.

For VIPs and VVIPs, there is generally a Directors' Bungalow that can be used. Its occupancy is minimal.

4.8.2 Suggestions about a Guest House

Some suggestions on design of a guest house :

(i) It should have room air conditioners of size corresponding to size of room.

(ii) Attached toilet and bath (No tub except for VVIPs) with geysers.

(iii) Single beds in single rooms and twin beds in double rooms.

(iv) Room should have a small desk, chair and table lamp for working.

(v) A common lounge and dining room.

(vi) Parking space around.

(vii) The guest house should have small dormitory like rooms with toilets for drivers of visitors' cars

Distance of guest house from plant should be short, say a walking distance except for VVIP guest house. This way company does not have to provide transport for visitors to take them to the plant.

Company's policies would decide whether only vegetarian meals would be served and whether liquor could be consumed in guest houses.

Facilities in the kitchen should be up to date including, gas burners, ovens, refrigerators large and small- with deep freeze facilities, ventilators for kitchen to take away fumes, water coolers and so on.

Some companies have introduced self service in canteens. It is a good idea when large number of people are to be served in a short time.

It should also have recreation facilities like card room, table tennis and TV and News papers and Magazines.

4.9 Clubs

Most Cement Companies have Clubs for recreation and sports.

Company policy would decide whether there should be a common club for both workers and officers. The size of club buildings and facilities in them should be provided accordingly.

Cable TV is a must for entertainment and relaxation of workers and staff. It links works in a remote place with outside world and workers do not feel isolated.

In separate clubs, facilities get duplicated without getting used fully. In sports activities, borderline between workers and officers becomes thin. Hence, sports activities could be common

A large ground, on which outdoor sports like Hockey, Foot ball and Cricket as also athletic sports like races can be played, adjacent to the main building of the club would be a good idea.

Indoor games like badminton, billiards table tennis and carom would also be provided for. Some clubs also have reading rooms and lending libraries.

Clubs do have a value of their own. With active support and participation of Management they are a good meeting ground between workers and officers and can bring about harmony in relations.

If the officers, which are few in number must have their own club, it should be sort of annexure to the senior guest house so that even visitors can use the club's facilities.

Some companies depending on preferences of promoters also have :

1. Swimming pools.
2. Auditorium for cinema and stage shows with microphone and loudspeakers.
3. Restaurant / bar

These facilities can be added later when a plant starts making profits.The use of club facilities should be subsidized but not free.

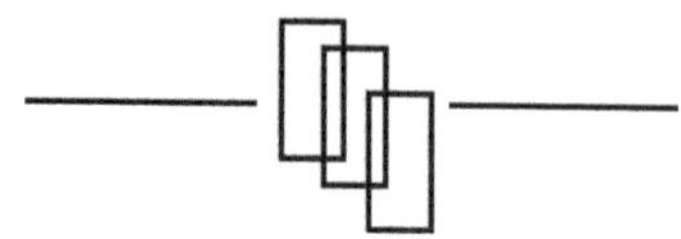

CHAPTER 5

STORAGES, SILOS, BINS AND HOPPERS

5.1 Storages of Bulk Materials

Bulk materials are stored in silos, bins and hoppers for processing.

Crushed limestone and coal are stored in 'stock piles' either covered or uncovered which is a different form of storage. Clinker is stored in linear or circular stock piles or even in silos

During building up stock piles in 'stacker reclaimer' systems 'preblending' is done. Thus processing is also done in these storages.

5.2 Silos

Granular and powdered free flowing materials are stored in silos.

Commonly in cement plants, raw meal and cement - both are pulverized, dry and free flowing materials, are stored in silos. Clinker can also be stored in silos.

Silos generally denote large storage capacity and for economical reasons are generally made in RCC. Capacity could be 2-3 days' requirements like say raw meal or 7 days' for cement.

Raw meal and Cement are invariably stored in silos. With increase in sizes of plants, capacities to be stored increase and hence sizes of silos also increase.

As capacity of a silo increases in proportion to square of the diameter and in proportion to its height, tendency would be to increase diameter to obtain higher capacities. Therefore present day silos are of 16- 20 meters in diameters as against 7-10 m diameter silos of the past.

Proportioning of dimensions of a silo would be governed by costs of construction which in turn would depend on soil conditions, its load bearing capacity, depths of foundation and constructional costs all specific to a particular location.

5.2.1 Flat and Conical Bottom Silos

Small diameter silos (up to say 10 m) had conical bottoms, with single or multiple cones to facilitate free flow of material and to achieve complete emptying of the silo. A few open air slides fitted in the cone facilitated flow of material.

See Figs. 4.5.1 and 4.5.5.

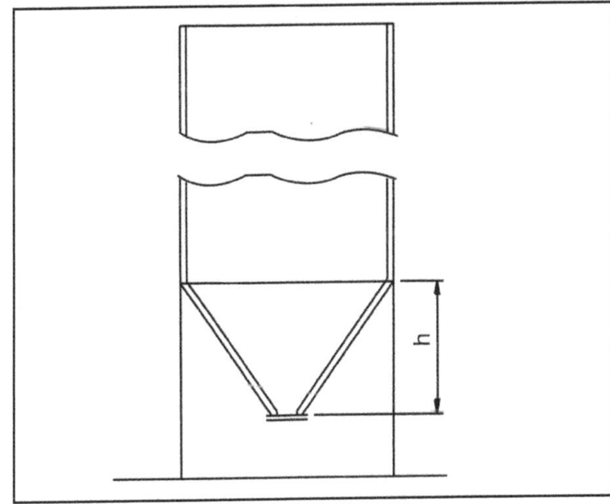

Fig. 4.5.1 Small diameter silo with conical bottom.

However as diameters became larger and larger the height of cone increased in proportion and this solution was no longer feasible. Therefore flat bottom silos had to be adopted.

Flat bottom silos had to be 'aerated' to fluidize material in it to make it flow - like it does in an air slide. Flat bottom silos can have two or more outlets.

These could be at bottom or on side depending on layout or from other considerations.

See Figs. 4.5.2 to 4.5.4; 4.5.6,4.5.7 and fig. 4.1.4 in Chapter 1.

Sometimes the flat bottom silo rests on the ground, in which case, material is extracted from the side. The silo bottom would be sloping on 3 sides as shown to facilitate flow. The bottom is sometimes made with ridges as shown with air slides at bottom so that the material flows freely. This arrangement requires much less air for aeration. The construction results in a slight loss in capacity but ensures smooth flow.

See Fig. 4.5.7.

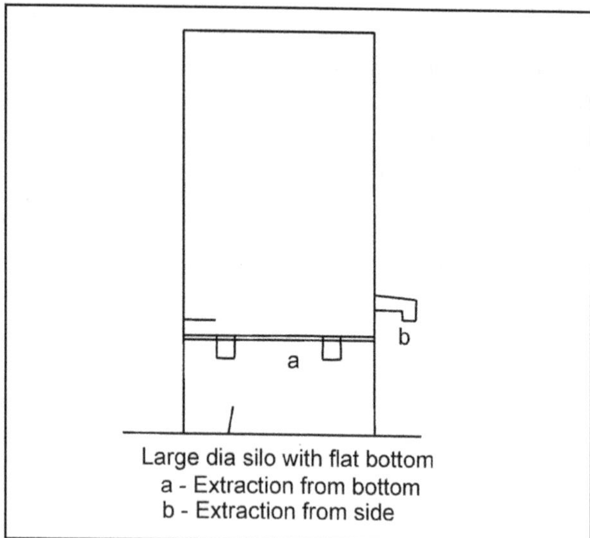

Fig. 4.5.2 Large diameter silo with flat bottom.

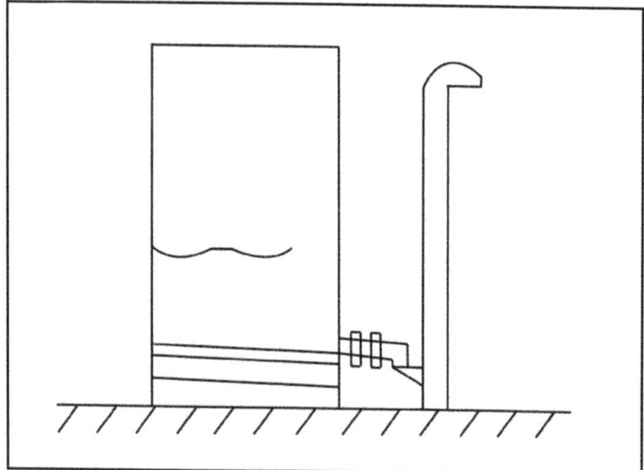

Fig. 4.5.3 Flat bottom silo on ground; extraction from side.

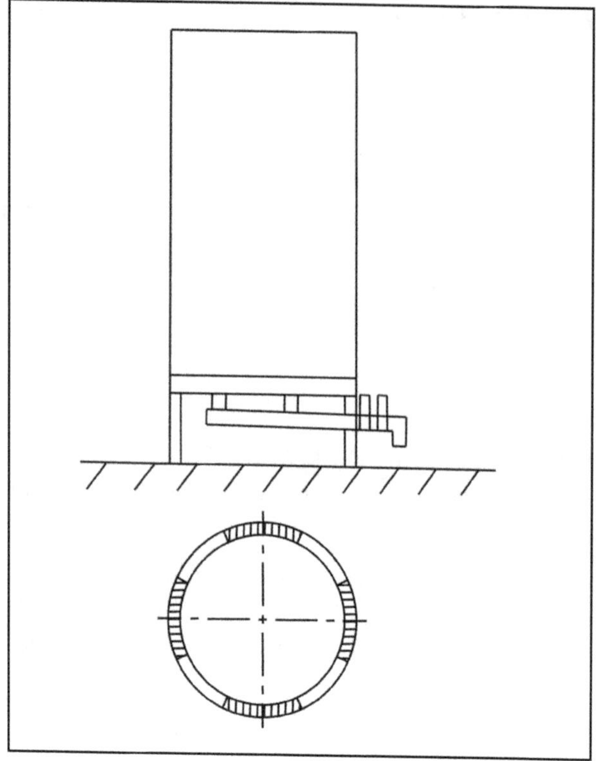

Fig. 4.5.4 Flat bottom silo raised on stilts extraction from bottom.

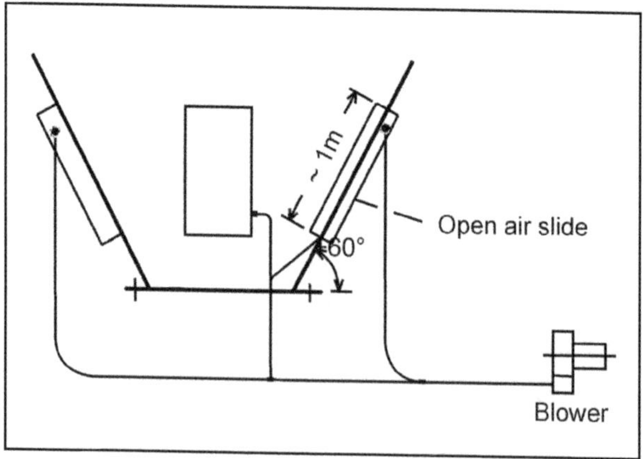

Fig. 4.5.5 Aeration of conical bottom.

This arrangement requires equipment for processing the material further to be installed outside in a separate building. The extracted material is required to be lifted to feed bins and metering devices.

To avoid this, though at the cost of higher construction costs, silo is raised by 6 - 10 meters above

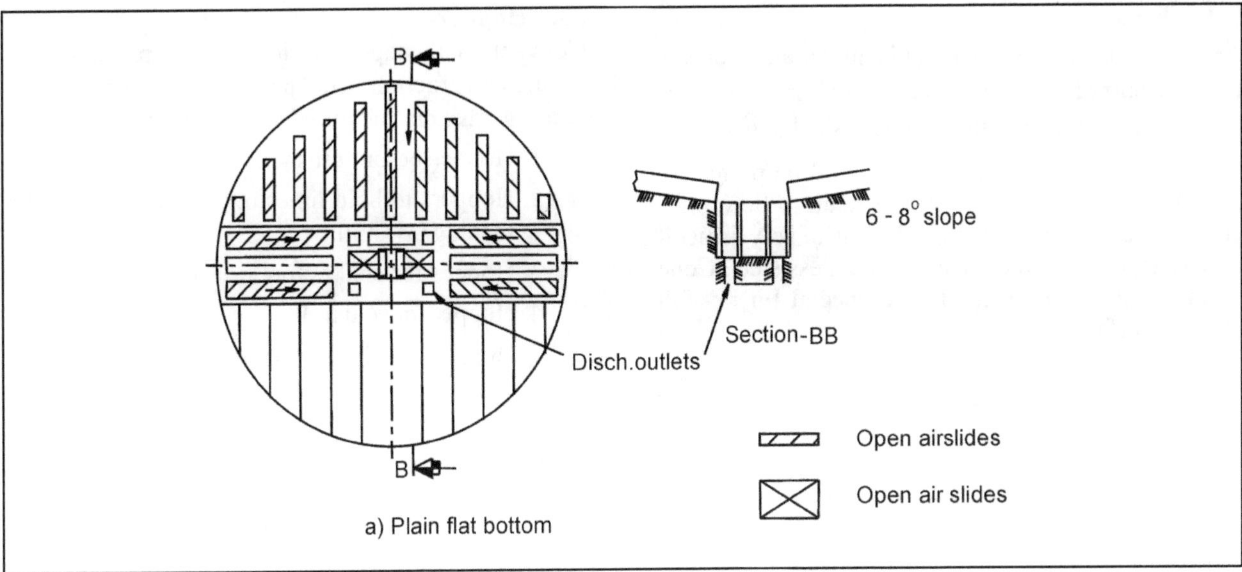

Fig. 4.5.6 Area covered by airsides - 12-15 %.

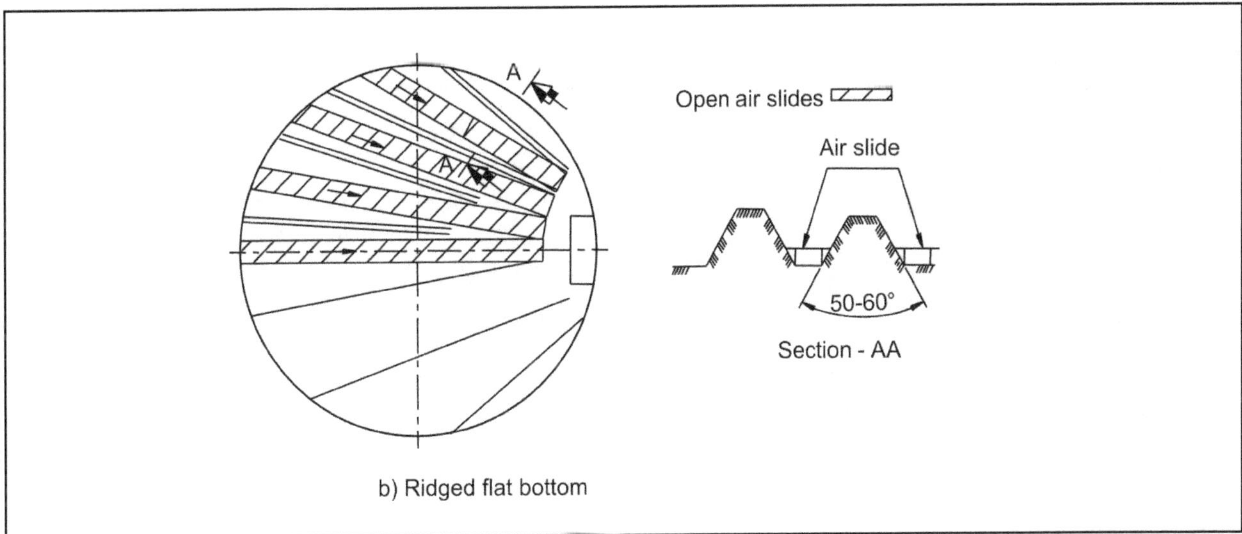

Fig. 4.5.7 Area covered by air slides $\simeq$ 8 to 10% silos with aerated flat bottom.

ground so that the metering devices and conveying devices are installed along with blowers and compressors right under the silo.

See Figs. 4.5.2, 4.5.4 and Fig. 4.1.4 in Chapter 1.

A 18 m dia silo for example, would have approximately 200 m2 of free area underneath it. It saves costs of a separate building and double handling of raw meal.

This concept is stretched to install even packing machines under the silos in case of cement silos.

In bulk handling of course, as has been mentioned earlier, silos are raised so high that bulk carriers can stand under the silos. Even weigh bridges are installed under the silo so that bulk carriers are filled with known weight and billed.

See Plate 6.32.1 in chapter 32 and Fig. 6.36.6 in chapter 36 of Section 6.

5.3 Bins

A bin is a circular silo much smaller in size and capacity. Bins are sort of 'service' containers holding a day's or a shift's requirements. Bins can be in steel or in R.C.C.

If capacities are small, steel would be preferred. Also for wet materials that are difficult to flow, steel cones at steep angles help flow. In case of very difficult materials, cones can be made in stainless steel. Cone surfaces can also be lined with special liners with smooth surface.

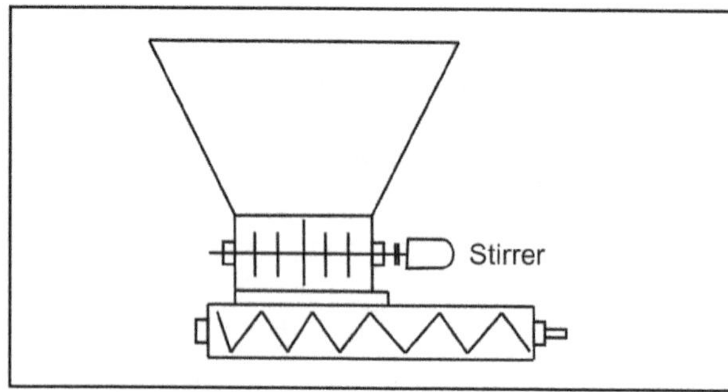

Fig. 4.5.8 Stirrer to prevent arching.

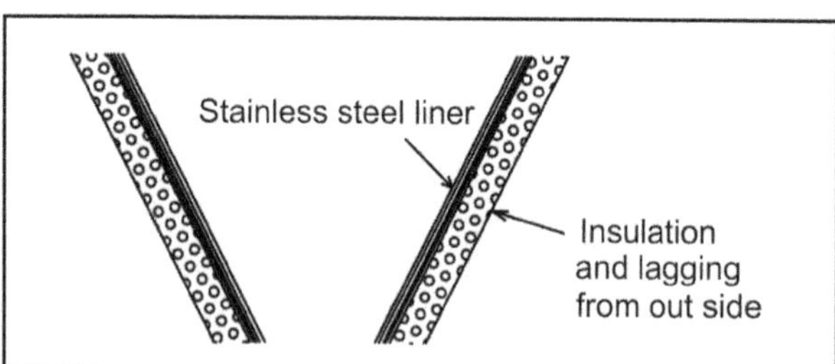

Fig. 4.5.9 Stainless steel or plastic gur liner to help smooth flow.
Also insulation to prevent condensation.

Bins for pulvarised coal to be used for firing in rotary kilns or calciners are sometimes fitted with stirrers to prevent formation of arches.

See Figs. 4.5.8 and 4.5.9.

For wet materials cones at outlet are heated with heating pads in more difficult cases and are fitted with vibrators in others.

See Figs. 4.5.10 and 4.5.11.

In any case even for free flowing materials, the angle of cone should never be less than 50 degrees to horizontal. For coals this would go up to 70 degrees. Principal advantage of a cone is that there are no valley angles and discharge is hundred percent.

5.4 Hoppers

Conventionally hoppers bring to mind a rectangular or square construction with open tops, They are used for feeding crushers and mills, for example:

1. Hopper before crusher.
2. Hoppers to hold limestone/ clay/ iron ore for feeding raw mill.
3. Hopper for coal before crusher.
4. Hopper for coal mill.
5. Hoppers for clinker, gypsum, fly ash, slag before cement mill.

Whereas bins and silos would generally be fully covered at top to prevent ingress of water, hoppers could be either open or closed fully or partially. Crusher hoppers which receive materials through rear dumpers / trucks have to be necessarily open and therefore cause lot of dust nuisance.

In case of dry material like limestone, water may have to be sprayed when material is being dumped into the hopper to suppress dust.

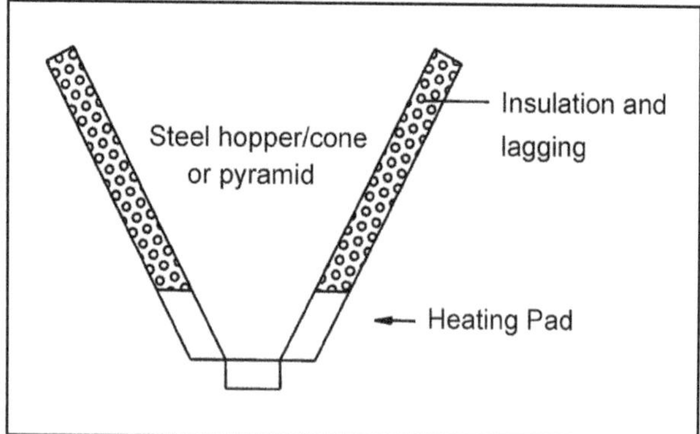

Fig. 4.5.10 Heating outlet for smooth flow.

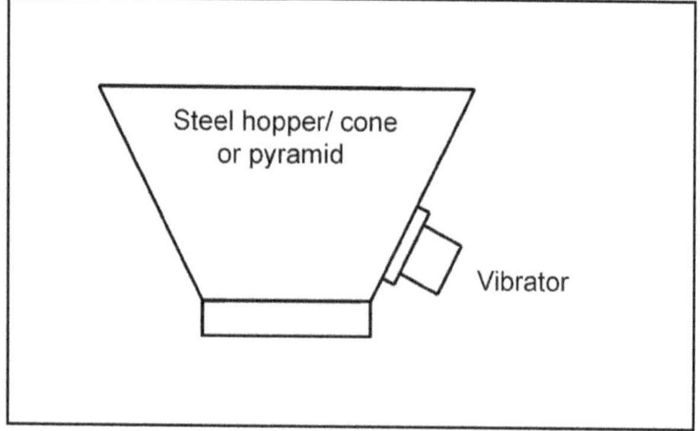

Fig. 4.5.11 Vibrators for breaking arches.

Coal is generally wet and no such measures need to be taken for it.

5.4.1 Shapes of Hoppers

Hoppers particularly for crushers, would be rectangular / square as dumpers come up to them along a ramp. The layout is so arranged that a hopper can receive material from 2 dumpers simultaneously using two sides of the hopper.

See Fig. 4.1.7 in Chapter 1.

Hoppers for crushed limestone can either be rectangular or circular. A circular cone is preferred as it eliminates valley angles.

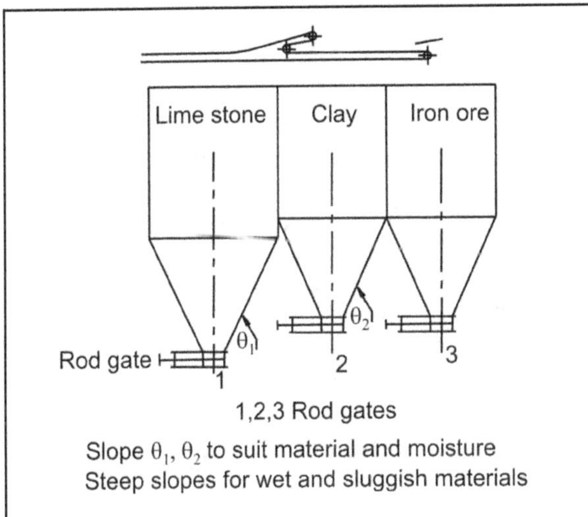

Fig. 4.5.12 Square/rectangular hoppers arrangements - 1.

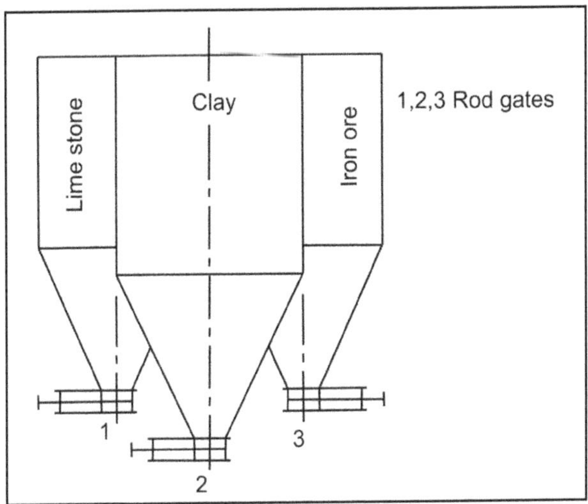

Fig. 4.5.13 Rectangular hoppers arrangement - 2.

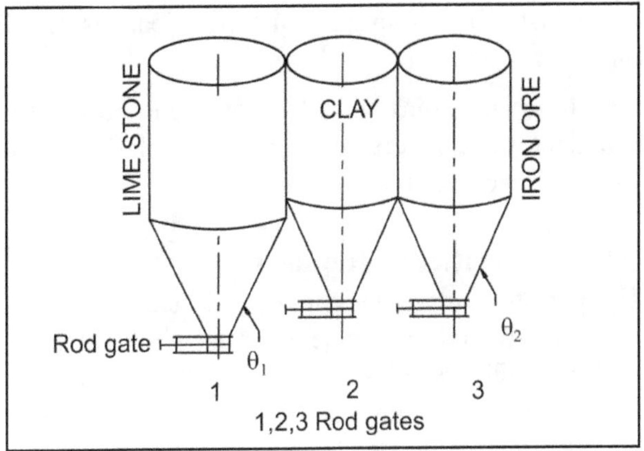

Fig. 4.5.14 Round hoppers and conical outlets.

But changing from rectangular to square shape at top to conical at discharge creates shelves at junction, where material can accumulate.

See Figs. 4.5.12 to 4.5.14.

This may cause only problems of flow for limestone, clay etc., but in case of coal it is very undesirable as accumulated coal and dust can catch fire. Therefore, corners angles ledges, projection should be avoided for bins and hoppers for coal.

5.4.2 Protection Against Wear

Hoppers for a primary crusher receive boulders. In case of limestone these can be be as big as 800 × 800 mm depending on the size of bucket of loading shovels. A single piece can be very heavy and its impact while falling from dumper can damage the hopper in no time. Therefore, such hoppers are invariably lined with steel rails projecting over concrete face. Boulders and rocks slide over the rails and the space between rails is filled with dust and small pieces and thus protects the concrete.

See Fig. 6.3.7 and 6.3.8 in Chapter 3 of Section 6.

For abrasive granular materials like clinker and slag wear is prevented by welding rings of angles on conical face, so that material slides over material rather than steel.

See Fig. 6.42.10 in Chapter 42 of Section 6.

5.4.3 Venting

Air is displaced when material is fed to a hopper. This escaping air carries dust with it. Presently, therefore silos, bins and hoppers for mills would be totally covered and a dust collector installed on top to clean displaced air.

All hoppers, bins and silos, require to be vented. For open top hoppers venting is difficult. The dust escaping becomes fugitive dust.

5.5 Capacities of Hoppers

Hoppers for limestone, clinker and slag used to be designed to hold requirements of one shift. However with raw mill capacities reaching 250-400 tph and even cement mill capacities reaching 100 tph and above, it is no longer practical to do so. Further with reclaimer belts of stacker reclaimer system running round the clock it is no longer necessary to do so. Thus hoppers may be designed to hold only 4 hours' requirements.

On the other hand requirements of clay, iron ore in relation to lime stone and of gypsum in relation to clinker are so small that hoppers to hold 8 hours' needs are too small. Therefore for practical reasons hoppers for them would be designed to hold about 24 hours' needs; but seldom more because of sluggish flow of these materials in hoppers.

5.6 Stock Piles

Limestone, coal correcting materials, clinker and gypsum are stored in stock piles.

Piles are mostly triangular and linear for all materials. Clinker is stored in circular stock piles of large diameter. Coal will also be sometimes stored in circular piles but of a different kind.

Limstone stock piles could be open or only partly covered. When in open, not much civil engineering work is involved except leveling the ground and provide for drainage. Footings for stacker and reclaimer belts are to be provided alongside as per Vendor's drawings.

See Figs. 6.5.1 to 6.5.4 in Chapter 5 of Section 6.

When covered, footings and columns for shed over the stock pile with roof and side cladding would have to be provided in addition.

Because of the costs involved, limestone piles will be in the open. Coal piles (sometimes circular) will be covered with a roof and side claddings.

Correcting materials like clay and iron ore and also gypsum would be stored in covered sheds.

See Figs. 6.5.10 and 6.5.11 in Chapter 5 of Section 6.

Slag will be stored in a covered gantry with an E.O.T. Crane.

5.6.1 Storing Clinker

Clinker is stored in a covered circular conical shed with diameters as large as 60 to 80 metres. To increase storage capacity a retaining wall of 2 to 3 metres high will be constructed all round.

Roof is supported on columns raised from the wall.

At the center is a hollow shaft with slots for discharge of clinker received by the d.b.c. from grate cooler.

Clinker is extracted by pan and belt conveyors at the bottom. Therefore there are a number of tunnels at two levels underground. The floor area being as large as 5000 m2, great care is to be taken in civil design and construction of clinker storage.

See Figs. 6.26.5, 6.26.14 and6. 26.16 in Chapter 26 of Section 6.

5.7 Estimates of Costs of Civil Construction

Estimates of civil construction costs are obtained from experienced civil designers who would have worked out procedures for it.

For silos and clinker storage civil designers would have to take into account soil conditions and load bearing capacity at site among other things.

While silos will be all R.C.C. sheds and clinker storage will be mix of R.C.C. and fabricated construction and of galvanized iron corrugated sheets.

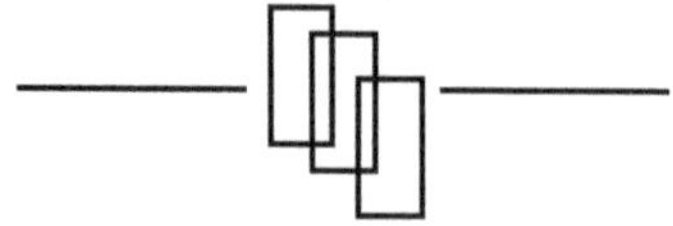

CHAPTER 6

WATER SUPPLY

6.1 Utilities - Water

Amongst the 'utilities' most important could be water. It is used in the plant as well as for drinking and household uses by workers and in departments and in colony.

Water may come from bore wells or from perennial rivers or streams or lakes or canals.

Overheads tanks are necessary in the plant premises and in colony, crusher and quarries to provide uninterrupted supply of water.

While process water like that used in conditioning tower in ESP is lost, cooling water can be re-circulated. It therefore needs cooling ponds or cooling towers.

6.2 Design of Water Supply System

Water supply system needs to be carefully planned as water is a scarce resource and is becoming scarcer.

It may not be possible to plan a water system in minute detail at the project stage but requirements and how they can be met and capital costs thereof should known in sufficient depth so that errors of omission and commission would be minimum. It would be better to err on safer side as it would be expensive to upgrade the system or enhance its capacity later.

For example, if the requirement of water is 5 lakhs litres per day for the main plant and say 8 lakhs litres per day after expansion, it would be prudent to install wells in the river bed that would supply 8 lakhs litres per day and also arrange for pumps and pipe lines that would pump at that rate in the first phase only. Initially, pumps would work for smaller number of hours and after expansion for longer number of hours.

Overhead tank to be provided should also have capacity required after expansion.

Separate overhead tanks for crusher and quarries and colony would be desirable even if quarries were close to the plant.

6.3 Water for Green Belts, Gardens

In present times, there is considerable emphasis on 'green belts' and gardens and parks inside plant and outside and in colony. These require a good amount of water.

Chemically treated sewage water could be used even for drinking purposes but that may not be possible due to sentimental objections to such schemes. However it can certainly be used for non drinking purposes like gardening mentioned above. It would be necessary to install a water purifying plant and separate distribution system to implement the scheme.

6.4 Treatment of Water

Water needs to be softened for long life of nozzles of conditioning tower and tubes of after coolers of compressors, oil coolers and other water-cooled machinery. Hence, a water treatment plant would be an integral part of the water distribution system.

6.5 Water from Bore wells

When source of water has to be 'bore wells' it would be necessary to establish ground water level in different parts of the year and in different parts of the plant and colony. Scientific assessment of availability of ground water should be done for locating bore holes.

There should always be more than one bore well that too located in different parts of the plot available for plant and colony, so that all do not dry up at the same time and continuity of water supply is maintained.

6.6 Study of Resources for Water

Study of rainfall, maximum and minimum levels of water in river bed, under ground water tables - minimum and maximum - for at least 15 years should be carried out. It would help to arrive at decision of how best to procure water.

6.7 Water for Captive Power Plants

Among the major consumers of water would be diesel generating sets or thermal power stations for captive power stations. Their requirements should be assessed and included in planning procurement and distribution of water.

When plants install Waste Heat Recovery System, that would also need water. Water in power plants is recirculated after condensing it, like water required in cooling systems.

It may be possible to designing a common cooling / condensing and recirculating system for all these.

Their requirements should be assessed and included in planning procurement and distribution of water.

Monsoons are unpredictable and hence to depend only on water from river, or stream would not be wise even if these sources were available. It would be better to have a standby source in bore wells also.

6.8 Rain Water Harvesting

Water is a scarce resource. It gets replenished in every monsoon. 'Water harvesting' is a must for all future plants so as to ensure that water tables do not dip too low in dry months.

Quarry pits could be used to store water. Quarries could even be a source of water when water tables are high and working faces go deeper and deeper.

6.9 Typical Water Distribution System

A typical water distribution system drawing water from wells in a river bed is enclosed. **See Fig. 4.6.1.**

Quantity of water required is estimated as shown in **Tables 4.6.1**

6.9.1 Drinking Water

Drinking water can be taken as 500 litres per day per employee.

Minimum quantity recommended is 500 litres / hr.

An overhead tank of say 25000 litres should be included.

6.9.2 Properties of Raw Water

Table 4.6.2 furnishes desirable properties of raw water

6.10

A note on Conservation of water and Rain Water Harvesting together with drawings of scheme of garland canals rainwater harvesting is attached.

See **Figs. 4.6.2 and 4.6.3.**

Guide Lines for Estimating Requirements of Water

1. *Service (cooling) water and process water*

Table 4.6.1 Requirement of water for a 3 MTPA Capaciry Dry process Cement Plant.

sr no	section	water for cooling	make up	water lost	for human consumption	total quantity to beprocured
			20% in GCT			
		Quantity in m^3 /day				
			a	b	c	a+b+c
1	Quarries and Crushing	2400	480			
2	Grinding Mills	4000	800			
3	Pyroprocessing	2000	400			
4	ESP			1000		
5	Comptrssrs	2000	400			
6	Miscelleneous	400	80			
7	D.G,Set	1500	300			
8	Colony	200			200	
9	leakage @ 2%	250			250	
	total	12750	2460	1000	450	16660
			~2500			~16700
	water cosumed		2500	1000	450	3950
	system to be designed with 20 % margin					20000

2. *Drinking water*

   Drinking water could be taken as 0.5 m^3 per day per employee.

   Its consumption will not be uniform. Minimum quantity recommended is 5 m^3 /hr.

   An overhead storage tank of minimum capacity 25000 litres is necessary

3. Please see **Table 3.7.4 in Chapter 7 of Section 3 also.**

Desired Properties of Raw Water

1. Water circulating in cooling systems should have following properties :
   1. It should not have aggressive or corroding effect on metal and concrete parts of the system
   2. It should not promote growth of organisms in system
   3. It should not contain oil or grease
   4. It should be cold enough for effective heat exchange
2. It should conform to following limiting values

Table 4.6.2

item	unit	Limiting values
pH value Carbonate hardness	 °n mval/l	7-9 4-14 1.4-5
Total salt content	mg/l	3000
chlorides	mg/l	500
sulphates	mg/l	500
Total chlorides and sulphates	mg/l	500
iron	mg/l	1
manganese	mg/l	0.15
magnesium	mg/l	60
Suspended Solids ≃ 0.05 mm	mg/l	40
Temp. at entry to circuit, max	°C	30
Ave. rise in temp.	°C	25-30

Source: Otto Labahn – Cement Engineers Handbook

3. Even water to be sprayed in Gas conditioning tower also needs to be soft and with minimum solid impurities for maintaining effectiveness of nozzles

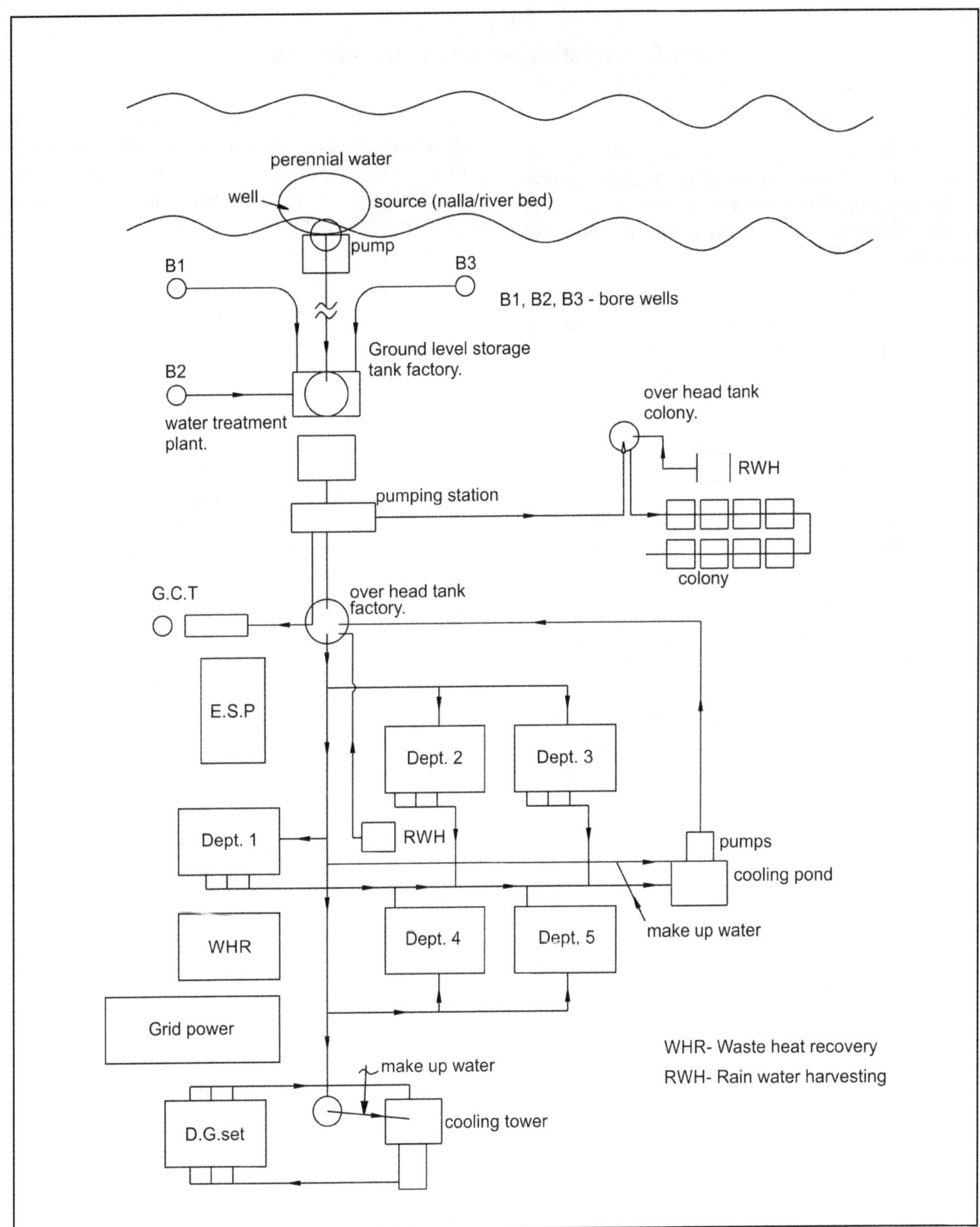

Fig. 4.6.1 Basic Water Supply Scheme for Plant and Colony.

Note on
Water Conservation and Rain Water Harvesting

Water Conservation

Cement Plants have been conserving water in their plants from the beginning as most cement plants have had to make their own arrangements to obtain water required for the plant and for drinking and for household purposes.

Cement plants procure water from nearest perennial sources of water like rivers and streams by digging wells in their beds and pumping it and storing it in the plant / quarries/ housing colony.

Plenty of water is required even in dry processcement plants for cooling bearings, compressors, afteroolers, gearboxes and for conditioning towers preceding ESPs. All water used for cooling is invariably collected and taken to a cooling pond and recirculated in the system. Only 10 to 15 % water is added as make up water to allow for loss by evaporation.

Process water per say is not required in a dry process cement plant. However if an ESP is used to clean prheater exhaust gases, a cooling tower is necessarily installed to bring down the temperature to about 140 ^{o} c. Gases are cooled by spraying water on the gases in the cooling tower. Water gets evaporated and is consequently lost. This is a significant quantity.

Loss of water on this account can be avoided if ESP is replaced by a bag filter. However penalty is required to be paid on account of higher pressure drop in the bag filter and on account of necessity to cool down gases to ~ 120-140 ^{o}c to suit materials of bags. If glass bags are used which can stand a temp. Of 400 ^{o}c (?) this dilution can be avoided. Generally speaking esp would be avoided at the design stage itself if the plant is located in an area of scanty rains and water scarcity.

Thermal Power Station / Waste Heat Recovery

When a cement plant installs a Thermal Power Stations (TPS) to supplement or to ensure availability of power on a continuous basis, it would require large quantities of water for generation of steam.

The steam is condensed in condensers and returned to the circuit. Water used to condense steam is itself cooled in cooling towers operating in closed circuit – that water is used again by recirculation. Here also therefore only make up water is required. Same is true of waste heat recovery boilers. Even where DG Sets are used to generate power, diesel engines are cooled by water which in turn is cooled in cooling towers and returned to the circuit.

Air cooled cooling towers and condensers are now used to minimize use of water.

Water for Laboratories, Offices and for Domestic use

This water is generally wasted after use. Though sewage water can be used after treatment for non drinking purposes like gardening . As a matter of fact Authorities who sanction a cement plant project stipulate that a Sewage Treatment Plant has to be installed in the plant. There should be ZERO EFFLUENT DISHARGE from the plant.

Mines as a Source of Water

Often as mines get developed underground resources of water become available and actually supplement the main source of water. Pits in excavated / exhausted mines can be used to serve as reservoirs of water. It can be readily available around the whole year for mining machinery and crushing plant when it is located in mines.

Above mentioned authorities also stipulate that planning of mines should include:

Creation of reservoirs which can recharge ground water and increase height of water table.

Many a times now used mines are consciously developed and landscaped to serve as recreation or picnic spots. Reservoirs in mines can thus serve a dual purpose as a source of water and as lakes. When treated it could also be used in swimming pools.

Presently there is great emphasis on 'greening' of the plant and its surroundings including housing colony. Green belts are created around the plant and colony to serve as dust and sound barriers. It is mandatory to create such belts between plant and highway /township. Water in mines can thus be put to many uses thereby saving water from the main and perennial source of water.

Rain Water Harvesting

In the context of cement plants, Rain Water Harvesting (RWH) has many dimensions.

1. rain water is collected and stored in natural/ artificial ponds or lakes to counter the salination of ground water in coastal areas.

   For this purpose check dams are constructed across streams /rivulets.

2. A system called 'garland canals' is constructed to collect off the ground water and lead it to eservoirs in quarries or reservoirs created by check dams.

   This water can be used in the cement plant for manufacturing, in captive power plants and for domestic use in colony.

   As a matter of fact cement companies are supplying water for drinking purposes and for agricultural purposes to neighbouring communities on an increasing scale.

   The Authorities sometimes stipulate that the cement company should not draw water from adjoining river / stream.

3. RWH is used to recharge borewells within the plant's own area and colony.

   Water is collected from roof tops and led thru pipes to collection pits near the borewells so as to recharge the borewells. Water is then available round the year – even in summer months.

   Extracts from a typical letter of consent to take up a cement plant project at a green field site or to take up an expansion shows the emphasis the Authorities are laying on water conservation. Of necessity therefore the cement plants of the future would have to be green. See **Annexure 1**

A typical check dam and system of garland canal is shown in **Fig no. 4.6.2.**

A typical system to collect rain water from roofs is shown in **Fig. no. 4.6.3.**

Capital and Maintenance Costs of Rain Water Harvesting

Considering the total costs incurred by a cement plant in designing and installing the main plant the costs incurred on RWH / Water conservation systems mentioned above are insignificant. Returns in terms of good will of neighbouring communities far outweigh the expenditure.

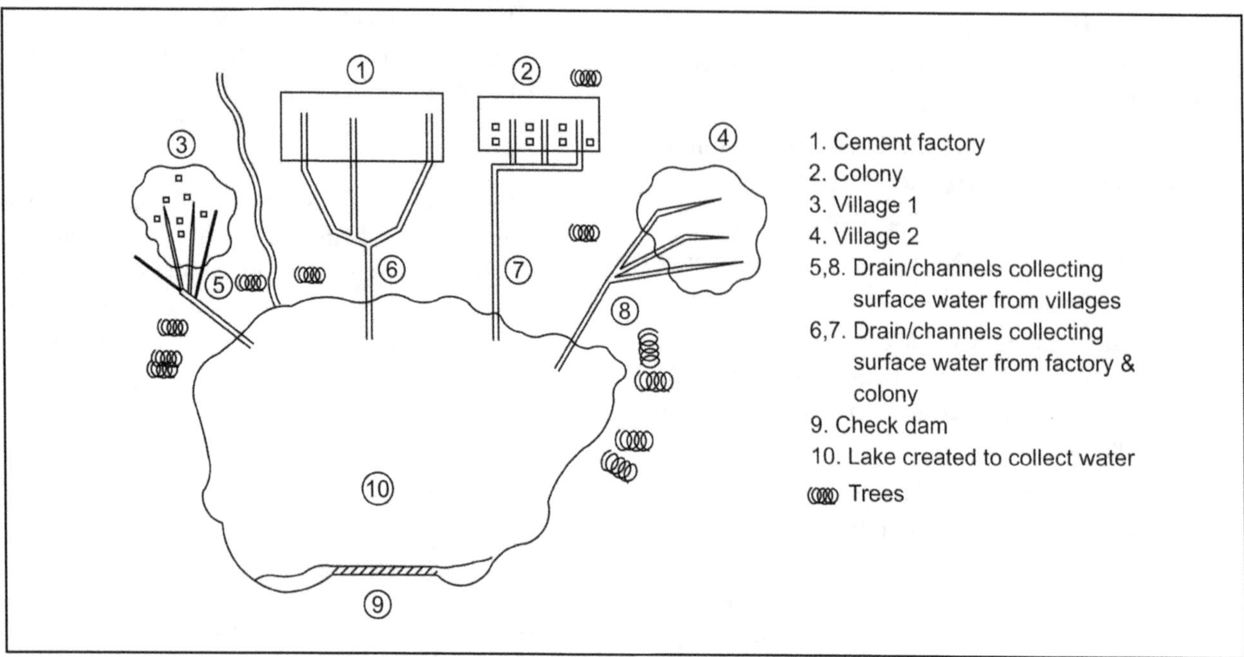

Fig. 4.6.2 Scheme of garland canals and check dam to collect rain water for use community and agriculture.

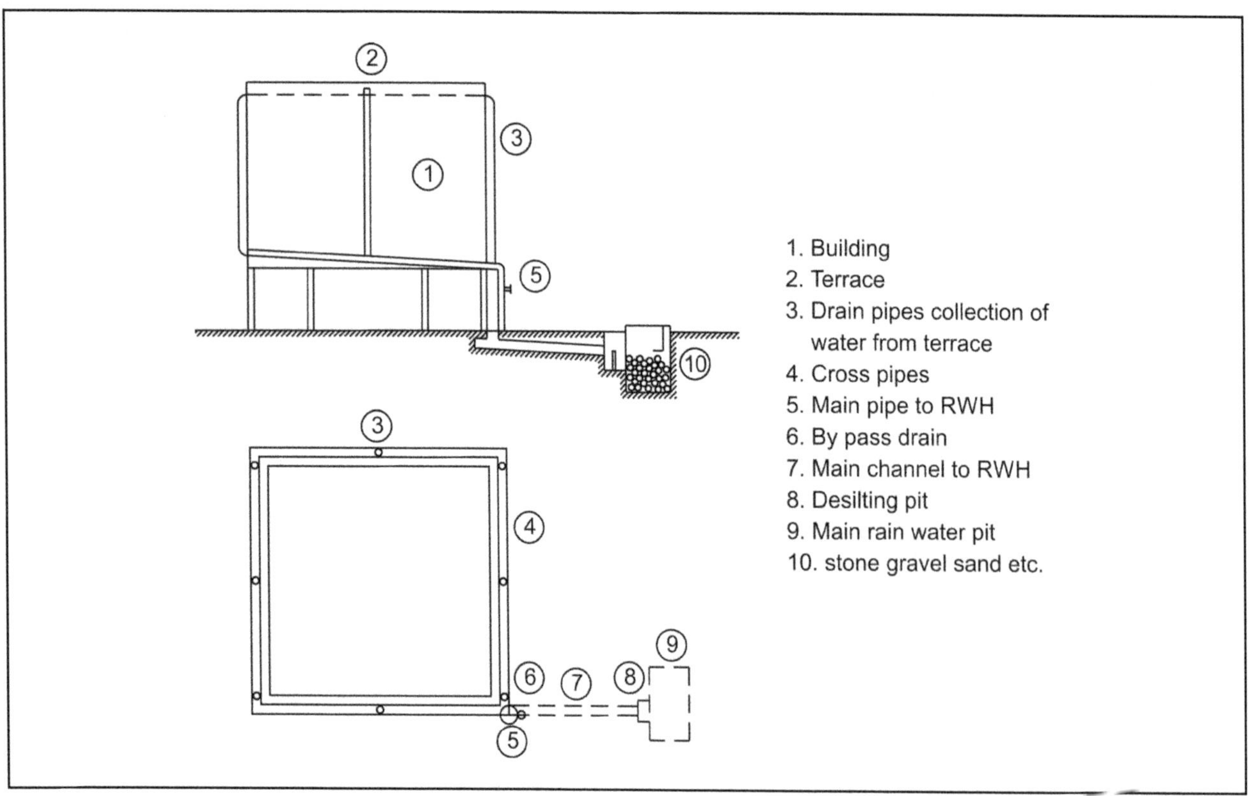

Fig. 4.6.3 Rain water harvesting scheme for a residential building

SECTION - 5

Electricals and Instrumentation

Contents

Chapter 1	Electrical Systems General	1-15
Chapter 2	Power Distribution and Cables	16-32
Chapter 3	Power Consumption and Its Calculation	33-40
Chapter 4	Instrumentation and Process Control	41-64
Chapter 5	Lighting	65-66
Chapter 6	Communications	67-68
Chapter 7	Quality Control and Laboratory	69-71
Chapter 8	Computer in Cement Industry	72-74

CHAPTER 1

ELECTRICAL SYSTEMS, GENERAL

1.1 Power Supply and Distribution in Cement Plants

All machines are driven by electric motors. Majority of the motors are 400- 440 volts. A selected few motors of higher ratings are HT motors with 3300, or 6600 or 11000 volts.

Most motors are fixed speed and unidirectional motors.

Motors for kilns and coolers and feeders and some fans are variable speed motors.

Variable speed motors can be either AC or DC with ratings ranging from fractional kws, for motors for dampers, valves etc., to several hundred kws for motors for fans, kilns and coolers.

All these motors are to be supplied with electric power at voltage and frequency and type of circuit for which they are designed.

1.2 Voltage and Frequency

Power Supply is A.C. in generation and distribution.

In India, frequency of power supply is 50 cycles. Voltage of transmission could be as high as 222 KV or 132 KV for large capacities and 66 KV, 33 KV or 11 KV down the line depending on MVA capacity of the substation.

The voltage of generation itself would be say 6.6 KV. It is stepped up for transmission; longer the distance over which power is to be transmitted higher the voltage of transmission to minimize losses in transit.

1.3 Power Distribution Grids - Macro Scale

The whole system of distribution of power can be shown pictorially as a 'grid' linking the several generating stations in a State. Grid ensures that power is assured to all consumers, even to remotest customers at all times.

See Fig. 5.1.1.

Thus customers 'a' and 'b' can draw power from generating station 1, 2 or 3.

Grid within one State is connected to grids in another State and also to 'national grid'.

1.4 Generation of Power

Bulk of the Power is generated in States by Statc Electricity Boards. It can be thermal power using coal as fuel; steam is generated in boilers and drives turbines which in turn drive alternators or generators.

Power is also generated in 'hydel' stations by using water power. A balanced proportioning of generation between thermal and hydel power ensures availability of power in all circumstances in all seasons of the year.

Hydel power is obtained from water, which in turn depends on seasonal rain fall. When rainfall is scanty, quantity of water collected is affected and power generation is reduced.

Hydel power generation is thus close to the source or storage of water.

1.4.1 Thermal Power Stations

Thermal power stations based on generating steam in boilers by burning 1coal depend on regular supply of coal. This coal can come from near or from far. Its supply may be affected by strikes in collieries or by bottlenecks in transport system, and sometimes also by fluctuations in quality of coal supplied.

Monopoly of State Electricity Boards to generate power has been diluted of late by Central and State Governments permitting private parties to set up large and small power plants.

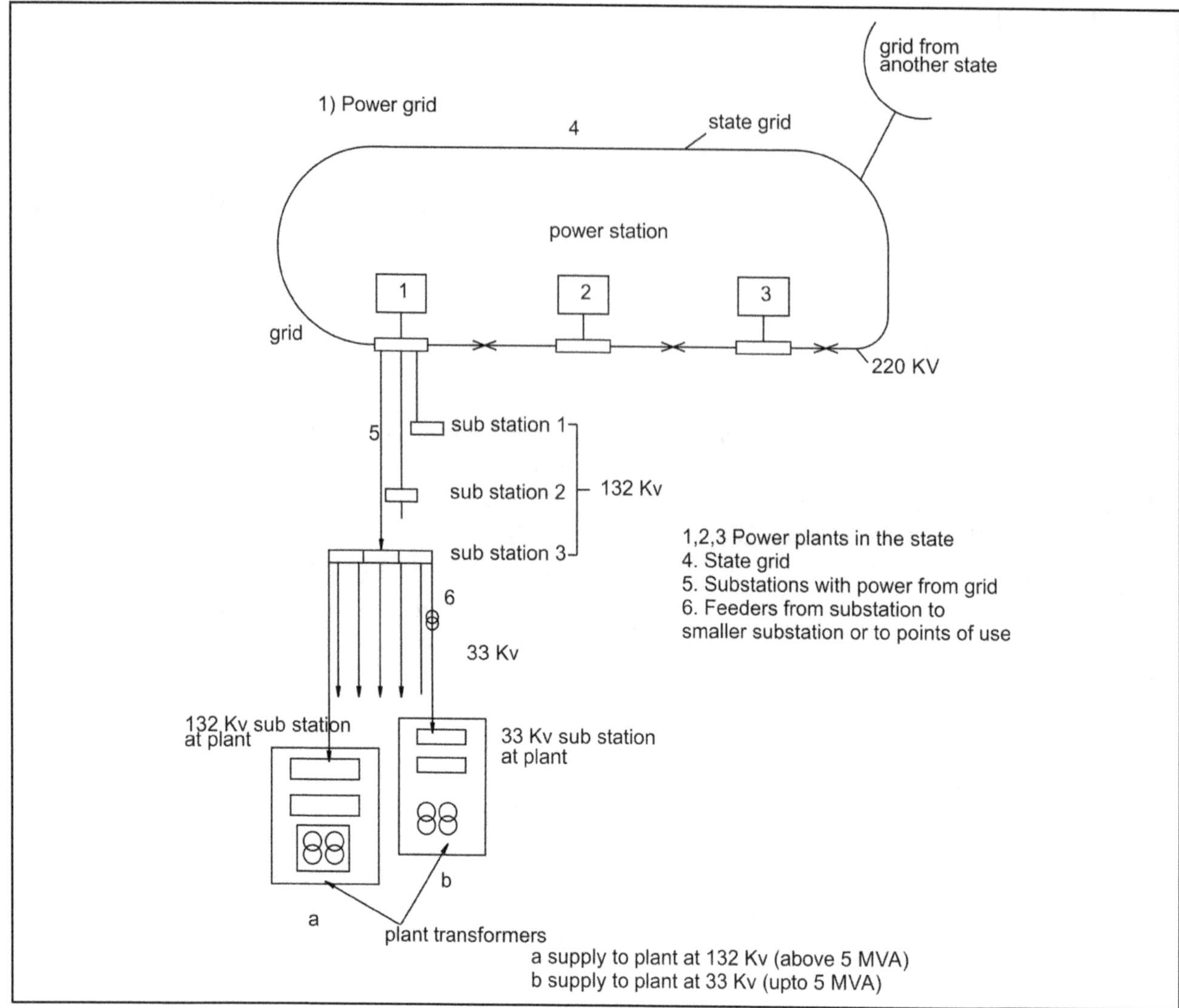

Fig. 5.1.1 Electrical power distribution system - macroscale.

1.4.2 Captive Power Plants

Power situation had become so critical in the last 2-3 decades that Financial Institutions insisted that Industry should install its own Power Plant to meet at least 40% of their requirements.

A large number of captive power plants installed by cement industry have been d.g.sets; many large cement plants have however opted for coal based thermal power plants.

1.4.3 Co generation

There is an increasing trend now to install Waste Heat Recovery Systems to generate power using waste heat of kiln and cooler exhaust gases that is still available after using it for drying coal and raw materials. It has an additional advantage that it reduces quantum of emission of green house gases.

Technologies like Organic ranking Cycle, Kailna process are now available to make this kind of cogeneration of power possible.

1.4.4 Renewable Energy

Captive Power can also be obtained from renewable sources like solar and wind energies. There is no running costs of fuel from these sources. Solar and wind power plantscan be located at convenient places - even away from the main cement plant.

Their design would be done by specialist consultants and companies. See Author's Book ' Designing Green Cement Plants'.

1.5 Distribution System Within the Plant

In a cement plant also there is a 'mini grid' receiving power from main electricity grid and also from plant's captive power plant and from co- generation by Waste Heat Recovery System.

They have to work simultaneously and hence should be 'synchronized' when working in parallel, i.e., the voltage, frequency and phase must match.

Alternatively, captive power is used only to supply power to specific sections like kiln and one of the two major mills.

However, the present trend is to size captive power station large enough and to synchronize captive power with grid power.

1.5.1 Ring Mains Within the Plant

The plant has what is known as a 'ring main' which supplies grid power to load centers within the plant. If the captive power plant is large enough there can be two ring mains in parallel.

Internal Power grid in the plant now has feeders from three sources, Main grid, captive power and power from co generation. It is necessary to synchronise them to ensure availability of power all the year round and round the clock.

1.5.2 H.T. Drives

Normally, drives of following machinery would be HT motors.

(i) Main crusher.

(ii) Raw Mill.

Table 5.1.1 Variable Speed Drives.

Sr. No.	Motor for	DC	AC
1.	Crusher Feeder	✓	✓
	Reclaimer drives		✓
2.	Raw Mill Feeders	✓	✓
3.	Raw Mill air separator	✓	
4.	Preheater Fan	✓	✓
5.	Prefeeders and Feeders for kiln Feed	✓	
6.	Kiln	✓	
7.	Grate Cooler grates	✓	
8.	Cooler vent fan	✓	✓
9.	Cooling air fans for at least first two compartments	✓	✓
10.	Coal mill feeder	✓	
11.	Feeders and prefeeders for fine coal for kiln and calciner	✓	
12.	Feeders for cement mill	✓	✓
13.	Separator for cement mill	✓	
14.	Packing machine	✓	

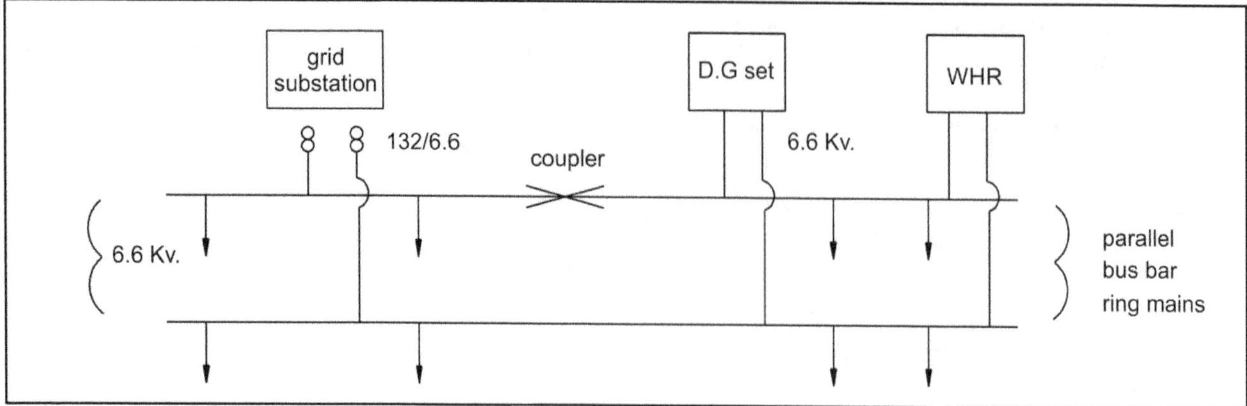

Fig. 5.1.2 Power distribution - with grid and D.G set in parallel.

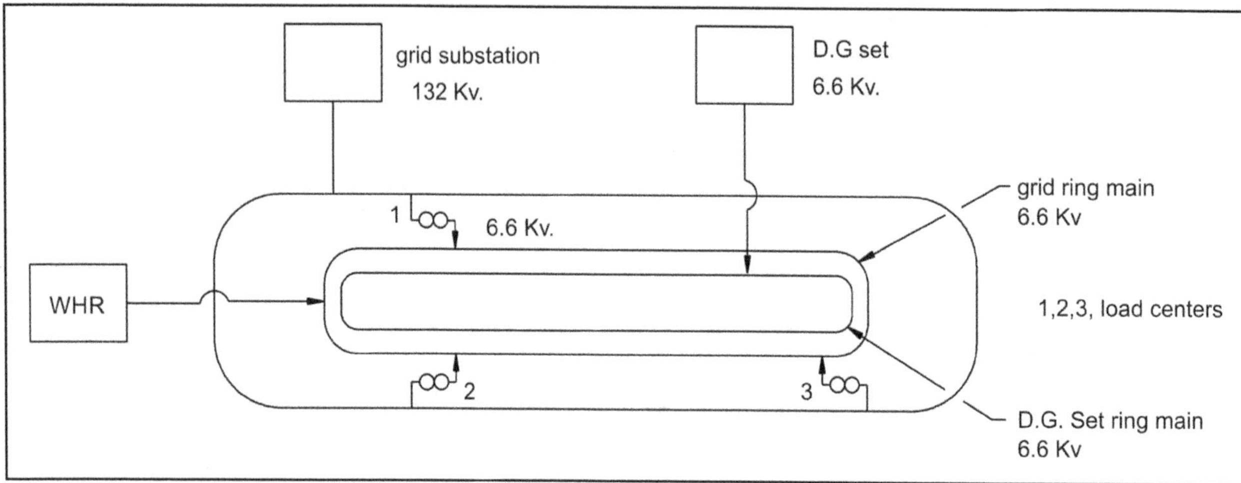

Fig. 5.1.3 132/6.6 Kv transformers near load centers.

(iii) Preheater fan.
(iv) Coal Mill.
(v) Cement Mill.

1.5.3 Variable Speed Drives
Generally speaking, variable speed motors would be used for Motors listed in **Table 5.1.1.**

1.6 H.T. and LT Drives in a Section

Thus some sections will have only LT supply and others both HT and LT supply and also d.c. drives.

One option is to step down voltage from grid voltage for the total plant in one place and lead HT lines from it to respective sections in which HT motors are used. **See Fig. 5.1.2.**

This arrangement required long 6.6 KV cables to various sections.

Another option is to locate transformers near load centers i.e., near mills and kiln.

See Figs. 5.1.3 and 5.1.4.

Both arrangements have their own positive and negative points. Main objective should be to achieve continuity of supply of power of right quality (fluctuations in voltage and frequency and phase should be within permissible limits) and flexibility (supply can be either from grid or captive power) without interruption.

1.7 Synchronization of Grid and Captive Power

Even when parallel bus bar system or ring system is not used, the grid supply and captive supply must synchronize and work in parallel or individually without disrupting power supply to any section of the plant.

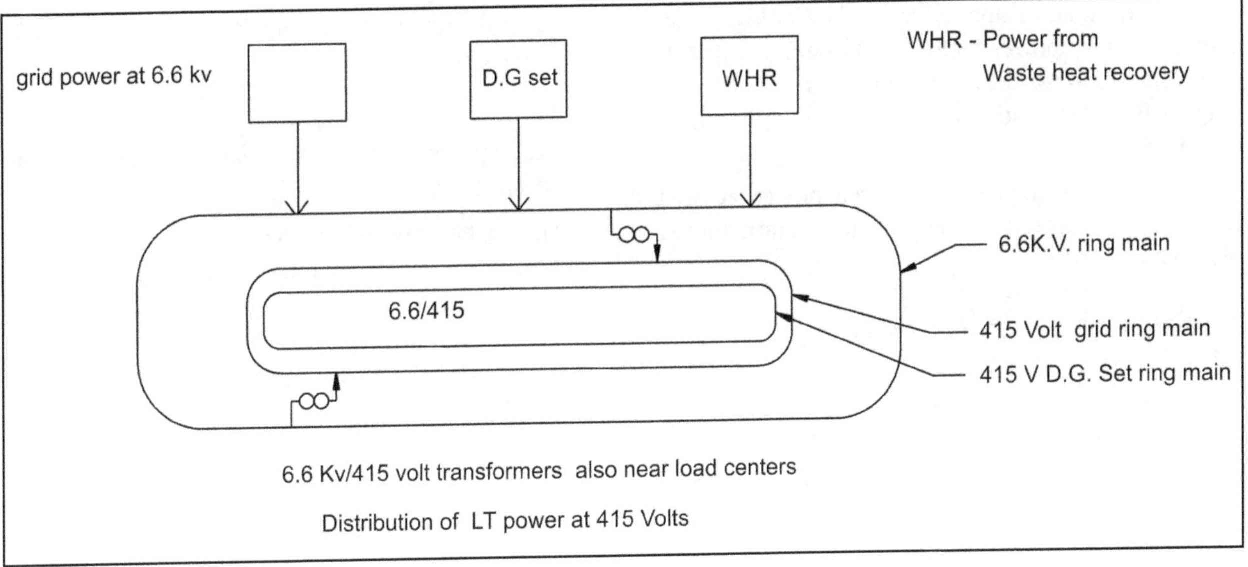

Fig. 5.1.4 Power Distribution of Plant.

Table 5.1.2 Ratings of motors.

Power factor	10 kw	15 kw	25 kw	50 kw	100 kw
	Current in amps				
1	13.9	20.9	34.8	69.5	139
0.9	15.4	23.1	38.5	77	154
0.8	17.4	26.1	43.5	87	174

(all motors 3 phase, 415 volts)

1.8 Power Factor

Another important aspect is that of power factor.

1. Induction motors work at varying lagging power factors depending on load.
2. D.C. motors have unity p.f. on d.c. side.
3. In case of synchronous induction motors, power factors can be altered between leading and lagging.

Power factor as close to unity between 0.95 to 0.90 lagging is desirable to get maximum from power purchased from the grid station. Lower the power factor, less is the useful energy available and greater the losses.

When power factor is unity.

$$V \times A \times 1/1000 = KVA = KWs$$

When power factor is say 0.80.

Useful power available is

$$V \times A \times 0.8/1000 = 0.8 \text{ VA KW}$$

Power bills are based on KWHs consumed and charges for maximum demand are on KVAs.

1.8.1 Current Carried and Power Factor

Current required to be carried by cables varies according to power factor.

See Table 5.1.2.

Say total load of above motors is 200 kws; total current at unity power factor would be 278 amps. If power factor were 0.85, the current would be 328 amps. All cables would draw correspondingly higher current right up to the substation.

It is therefore, important to correct power factor at levels of individual drives and also main power distribution panels.

1.8.2 Correction of Power Factor

It is done by:

(i) Using individual capacitors for motors and by using 'capacitor banks' in total sections.

(ii) Using synchronous induction motors to correct overall power factor by running them on leading power factor.

Let total load be 10000 KWs and let there be 2 synchronous induction motors each of 2000 KW rating; Other load is then 6000 KW

Total load consists of

6000 KW of 0.8 lagging p.f.

4000 KW of 0.95 leading p.f.

overall power factor can be improved to 0.96 lagging. Current will reduce by 17 %.

See Fig. 5.1.5.

Total power factor for the plant can be worked out in this fashion.

If synchronous induction motors are not available, capacitor banks of suitable ratings and voltage are added in the circuit to obtain desired improved p.f.

1.9 Designing Power Distribution System

One of the first things to be done in designing the distribution system is to work out the power to be drawn from the grid for the plant.

This can be drawn on a broad basis starting from overall power consumption of similar cement plants. For example if similar plants are consuming 95-100 KWH/ton of cement, it may be assumed that the new plant will also consume the same amount and total power to be drawn calculated as shown in subsequent paragraphs.

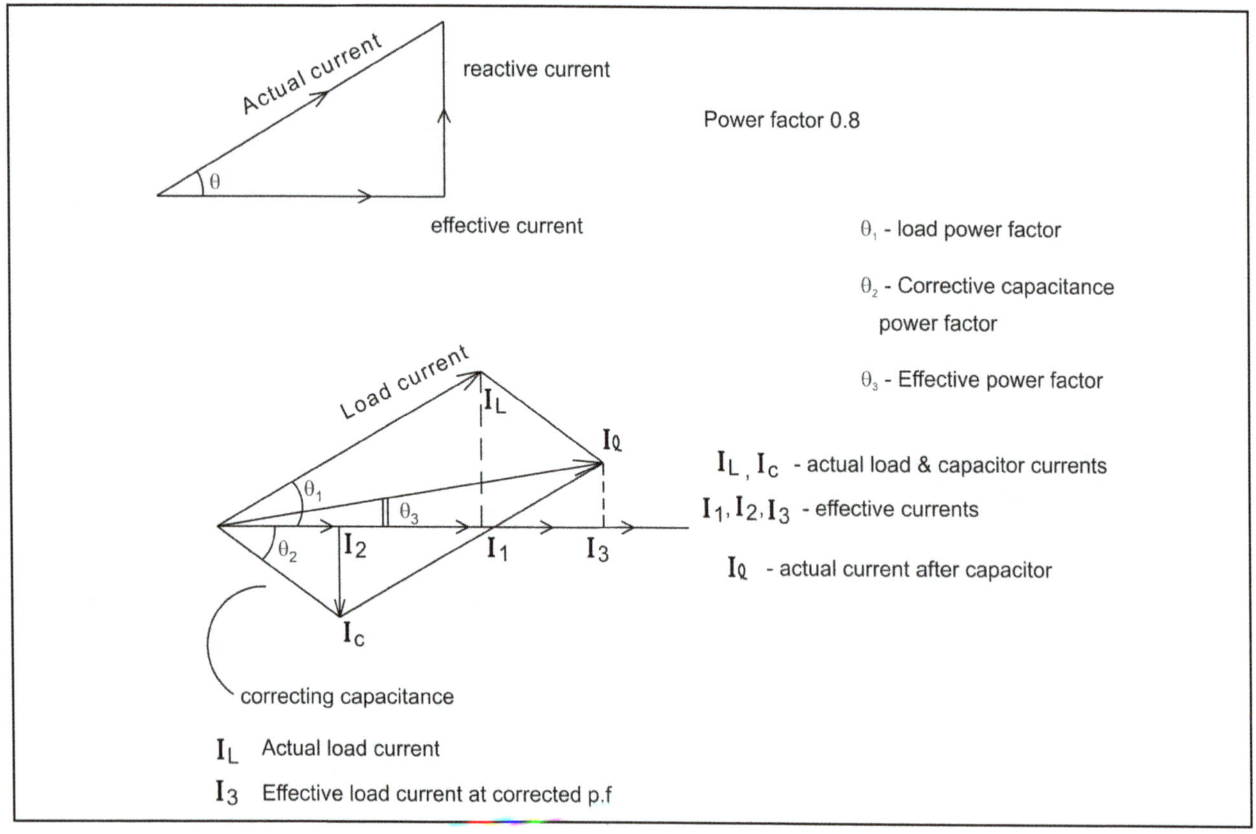

Fig. 5.1.5 Effect of power factor.

1.9.1 Working Requirements of Power Step by Step

Alternately the exercise may be done step by step.

For this the first step is to arrive at departmental power consumption of each section in terms of power / ton of material handled. Then 'convert' it into clinker and then into power per ton of cement. Take example of Raw mill Department.

List all drives with their ratings.

List power drawn at rated capacity.

Work out for total section.

See Table 5.1.3.

Total connected load is 2979 say 2980 KW

Let Material ground be 4000 tons in 20 hours.

Therefore Tons / hour = 4000 / 20 = 200 tph.

Average Kwhr/ton of material ground = 2608/200 = 13KWH

Power consumption per ton of clinker with a conversion ratio for consumption 1.55 : 1

=13 × 1.55 = 20 KWH /ton

Using 5% gypsum - 1 ton clinker = 1.05 tons of cement

Therefore power consumption per ton of cement = 19 KWH

In this manner list all sections of the plant.

See Table 5.1.4. (revised at the end)

Presently, there would be additional machinery in some sections like

cement mill section would include blended cements,

kiln section would include alternate fuels

captive power will include waste heat recovery

1.9.2 Margins to be Provided

For working out requirements of power to be drawn, it is desirable to add to 80 units **(Tabe 5.1.4 above)**

(i) 5-10% on actual consumption.

Assume 10 %

(ii) Design margin of 10% on rated capacity

Therefore, specific power consumption

= 96 KWH/ton.

Let rated capacity of plant = 100000 tpd

Capacity in cement producing OPC = 10500 tpd ; hourly capacity = 438 tph

with design margin = 480

Therefore kw = 96*480

= 46000 KWs

This needs to be converted into KVAs.

1.9.3 Load Factor

For conversion power factor and load factors are to be taken into account.

Table 5.1.3

Motors numbers	Drive rating kw	Power drawn as % of drive rating	Total actual power drawn in kw
4 x	1	0.8	3.2
3 x	5	0.8	12
1 x	60	0.8	48
1 x	2000	0.9	1800
1 x	500	0.85	425
1 x	200	0.8	160
10 x	20	0.8	160
Total	2980		2608

Table 5.1.4 Power consumption for whole plant.

Sl. No.		Unit / Tons Of material	Material Conversion factor	Units/ton clinker	Units/ton cement
		KWH		KWH	KWH
	Quarrying	0.5	2	1	1.0
	Crushing	1.5	1.5	2.25	2.2
	Raw Material Grind	13	1.55	20	19.3
	Blending	2	1.55	3.1	3.0
	Kiln feed	0.5	1.65	0.825	0.8
	Kiln, calciner				
	Preheater Cooler	20	1	20	19
	Coal Mill	26	0.17	3.9	4.2
	Cement Mill	23	1	-	23
	Packing	2	1	-	2
	Water supply				
	Lighting - Factory &Colony				4 all together
	Other Utilities & Losses				
				total	78.5
				Say	80.0

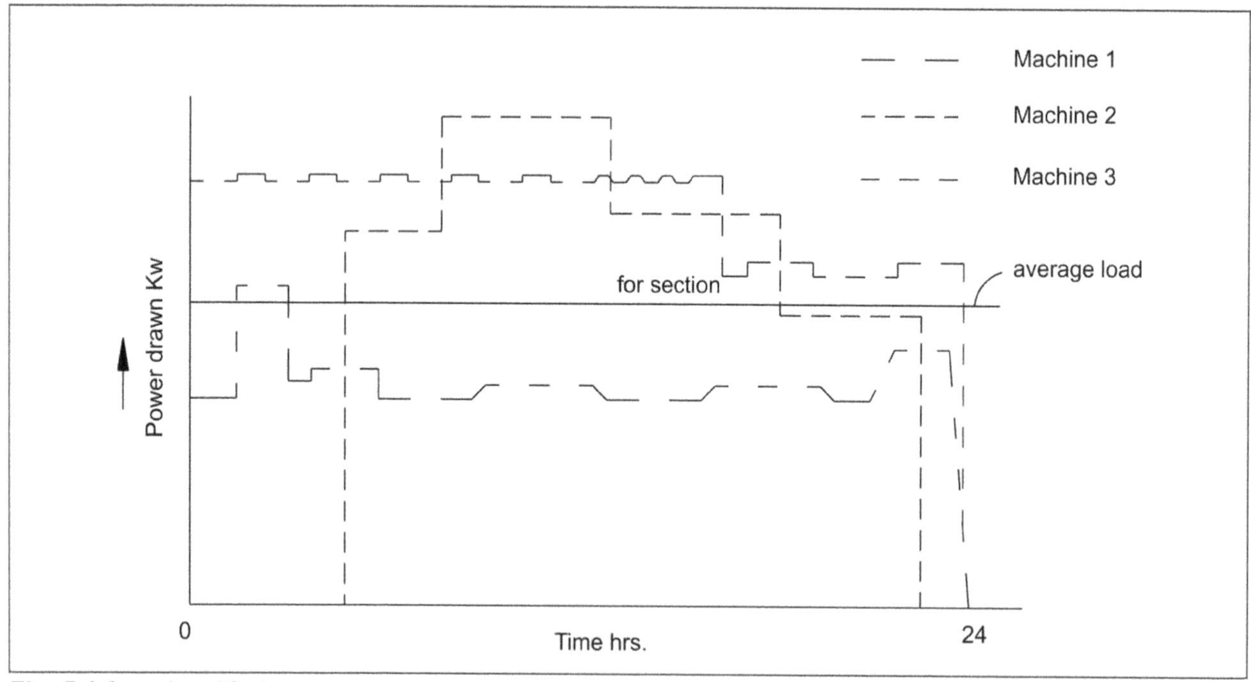

Fig. 5.1.6 Load factor.

Power factor has been dealt with above. Though it would be maintained at 0.95, for working out power to be procured, it may be taken as 0.9.

Load factor is a usage factor; all machines do not run for all the 24 hours and at a constant load. Peak loads occur at different times in different sections. To obtain required capacity in spite of fluctuations, higher capacity is needed to be installed.

See Fig. 5.1.6.

For cement plants where most sections work in three shifts, it is customary to take load factor as 80 %.

Thus if 46000 KW are required and load factor is 80%, rating to be installed would be 46000/0.8 = 57500 KW.

Power required to be obtained in KVA at 0.9 p.f. = 57500/0.9 = ~ 64000 KVA

1.9.4 Margins for Substation Capacity and Expansion

The substation capacity should allow for some margin for additions etc., that are not foreseen today. Therefore for the first phase, electrical demand would be fixed at with say a 25% margin.

= 64000 × 1.25 = 80000 KVA say 80 MVA

If duplication is imminent, substation should be designed for double capacity, i.e., incoming line should have capacity to deliver 160 MVA.

In substation, transformers can be added phase-wise.

1.9.5 Selecting Transformers

In selecting transformers, it is a good idea to select two transformers in parallel rather than one.

Transformers are generally loaded up to 70 to 80 % of their rating.

In above case total load = 80 MVA. Total rating of transformer would be 80/0.7 = 114 MVA. It would be preferable to select two Transformers of 57 MVA each or even 60 MVA each.

If a ring system is used, there will be more than two transformers located near load centers. Each of these transformers should also be loaded to 70% capacity in actual operation.

1.9.6 Standards of Design and Manufacture of Electrical Equipment

All electrical equipment required in distribution of power starting with transformers, LT and HT switch boards, Motor control centers, cables etc., are governed by specifications and standards laid down by Bureau of Indian Standards and Electricity Rules for India and by similar standards in other countries. In selecting various electrical equipment, pertinent standards should be referred to and followed.

Standards however specify 'minimum' requirements and hence it is prudent to err on safer side in design.

All vendors must be asked to furnish detailed specifications of their equipment and standards to which they conform.

Presently a 'star rating' system is in vogue which is indicative of operational effiecency of the equipment. Higher the number of stars, higher the operational efficiency.

Entrepreneurs should furnish 'duty requirements' and 'technical data sheets' to vendors who must return them duly filled. This helps evaluation of various offers.

1.10 Electrical Power Tariff

Electrical tariff is in two / three parts :

1. Energy consumption in Rs./KWH.
2. Maximum demand costs in Rs./KVA/month.
3. Any other duties as may be in force in each State from time to time.

Power is measured by energy meters installed by Grid / Electricity Board in their Substation. Maximum demand is also recorded there.

Energy meters must of course be calibrated periodically.

1.11 Maximum Demand

Many a time sudden over loads or simultaneous working of all sections can push up the maximum demand and power drawn will exceed contracted maximum demand. In such a case, fine is to be paid.

In times of power cuts, contracted demand would not be utilized. In general the plant would opt out to be penalized for not using the contracted demand rather than reduce it.

In times of scarcity of power, Electircity Boards come down heavily on unutilized maximum demand and may reduce the contracted demand.

The impact of tariff on account of maximum demand however is generally about 10% of direct costs of energy.

1.12 Cost of Captive Power

The cost of captive power would be different from that of grid power.

Separate records need to be kept for the units generated and consumed in the year from captive power plant.

The cost of generation can be worked out from (for a d.g. set) :

1. Consumption of fuel (efficiency of d.g. set can also be expressed in terms of KWH/litre of fuel oil consumed, which can be converted into Rs/KWH,
2. spares and maintenance,
3. consumables,
4. manpower,
5. depreciation, etc.

All added together would give overall cost of captive power in Rs/KWH

1.12.1 Captive Power by Waste Heat Recovery.

If the Cement Plant installs a Power Plant generating power by using the Waste Heat in exhaust gases of the kiln and cooler, then its generating costs would be much less than those of captive power plant as there are no fuel costs.

If the existing captive power plant is thermal, then it is possible to augment its capacity by the exhaust gases of kiln and cooler. Further, emission of greenhouse gases per ton of cement would also be reduced.

1.13 Overall Power Costs

Let total units consumed be 100 million and 30 millions are furnished by a DG Set and costs per kwh for grid power and captive power be as shown in **Table 5.1.5.**

Maximum demand 26 MVA

Let maximum demand charges be Rs 100/KVA/month

Table 5.1.5

Grid Power	Captive Power
70 x 10^6 Units Rate Rs 3.0/KWH	30 x 10^6 Units Rate Rs 6.0/KWH

They would be payable even if contracted max. demand is not fully utilized.

Therefore total power cost

Grid power direct = 210 + 31.2 = Rs 241.2 million

Cost of captive power used = Rs 180 million

Total = Rs 421.2 million

The overall power cost would thus be Rs 4.21/kwh as against Rs 3.31 for grid power.

Cost of captive power when it is generated by d.g.set would be directly proportional to cost of fuel used for generation.

Cost of diesel oil fuel was regulated by the government till recently. It used to be subsidized. Of late, diesel / petrol costs are governed by market forces and fluctuate frequently.

Every now and then fuel costs and freight rates are revised and its impact is to be borne by the user.

A calculated decision may be taken as regards maximum demand to be contracted.

In general, it will be contracted without taking cognizance of captive power.

But, if it is cheaper to use d.g. set and if it is used almost continuously, then it may be beneficial to reduce level of contracted maximum demand.

1.13.1

With Co-generation becoming part and parcel of power system in cement plants, it is necessary to assess operational costs of three systems and arrive at use thereof to minimize unit cost of power in production of cement.

On the face of it, Co generation which does not need any fuel looks like the first choice. However, its quantum and availability round the clock will also have to be taken into account.

Using captive power reduces overall loss in transmission –(grid power suffers from this)- but then its operational costs are higher. A careful analytical study and observation of actual operation is necessary.

1.14 Thermal or Hydel Power Plants

While in any given year, generation of power for coal based power plant could be maintained at its rated capacity, generation of power from Hydel power plants is very much subject to good or bad monsoon and levels of water in lakes which supply water to Hydel power projects. It is therefore good to have a balanced mix between thermal and hydel power.

1.14.1 Capacity of Captive Power Plant

All States in India suffer from chronic power cuts; in fact have been suffering so for more than two decades. In some States the problem is so severe that for new projects, or for expansion, Financial Institutions themselves insist that new cement plant projects should incorporate captive power generation to the extent of 40% of plants' requirements and the viability of the project should be examined on this basis

Capacity of the captive power plant to be installed would be different for plants of same capacity located in different parts of the country.

Some States are chronically so short of power that power cuts begin within 2-3 months of end of monsoon and cannot be restored till after 1 month of onset of monsoon.

A typical pattern of availability from grid is would be as shown in **Table 5.1.6.**

This pattern would vary from State to State, depending on the proportion of hydel power generated in it.

If a State has for example 60% thermal power and 40% hydel power, it can supply grid power as shown in **Table 5.1.7.**

Table 5.1.6 Availability of grid power.

Month	Power supply available (%)
January	80
February	80
March	70
April	70
May	70
June	60
July	100
August	100
September,	100
October	100
November	90
December.	90
Overall average	83.3

Table 5.1.7 Spilt up of thermal and Hydel power from Grid.

Month	Thermal Power %	Hydel Power %	Thermal	Hydel	Total (%)
January	100	80	100 x 60	+ 40 x 80	92
February	100	80	100 x 60	+ 40 x 80	92
March	100	70	100 x 60	+ 40 x 70	88
April	100	70	100 x 60	+ 40 x 70	88
May	**90**	**60**	**90 x 60**	**+ 40 x 60**	**78**
June	**90**	**60**	**90 x 60**	**+ 40 x 60**	**78**
July	100	100	100 x 60	+ 40 x 100	100
August	100	100	100 x 60	+ 40 x 100	100
September	100	100	100 x 60	+ 40 x 100	100
October	100	100	100 x 60	+ 40 x 100	100
November	100	90	100 x 60	+ 40 x 90	96
December.	100	90	90 x 60	+ 40 x 90	90
Average	97.5	83			91.8

Thus in dry months maximum of 78 % power could be available.

For a period of about 3 months, actually available power could only be about 70 %.

From such analysis, it can be seen that the captive power plant to be installed should have a capacity of at least 30% of requirement of power.

Table 5.1.8

Department	units/ton cement
Crusher + Raw mill	26
Blending and kiln feed	3
Kiln and coal mill	30
Total	59 say 60

1.14.2 Rating of d.g. Sets

However to arrive at the capacity of d.g. set as captive powe rplant be installed, other aspects also need to be looked into.

Clinker is a half finished product that can be converted into cement at convenience if properly stored.

It would be the objective of all entrepreneurs to produce clinker continuously in spite of power cuts.

Therefore, in arriving at capacity of c.p.p set it is customary to include requirements of power for sections from raw mill to clinker storage and load required to run these sections.

See Table 5.1.8.

In the example given earlier,

At rated capacity of 416 tph for a 10000 tpd plant, kw required

$= 50 \times 416 = 20800$ kw

Raw Mill and its fan together would have a rating of (roller mill) of about 10000 kw

For starting up the mill, the d.g. set should have a rating of at least $1.7 \times 100000 = 17000$ kw.

D.G. sets are seldom loaded beyond 75 - 80 % of their capacity.

Therefore d.g. set to be installed should have a rating of $(20800 - 10000)/0.8 = 15400$

Total $17000 + 15400 = 32400$ KW

Applying a load factor of 0.8 and p.f. of 0.95, its capacity would be 43 MVA.This is about 53 % of the 80 MVA grid power to be contracted arrived at earlier.

Many a plant install (large or small) 40-50% captive generation capacity.

1.14.3 D.G. Sets and Their Utilisation

Maintenance of d.g. sets is the plant's responsibility.

For large capacities it would be advisable to have at least 2 d.g. sets, so that they are fully utilized according to requirement without interruption in generation of power.

In above case, 1×43 MVA capacity set will be very much under utilized when power cut is less. It would be preferable to install say 3×14 MVA d.g. sets, so that according to requirements 1, 2 or 3 could sets run at a time. Any one set could be easily taken out for overhauling without affecting production.

1.14.4 Maintenance of D G Sets

Normal d.g. set maintenance consists largely of cleaning and supplying clean air to the diesel engine. No efforts to clean air of even the smallest dust particles (0-5 microns) should be spared.

For this purpose various cleaning devices are used:

(i) Dry filters.

(ii) Oil bath filters.

(iii) Water curtains, for cleaning and cooling air.

Cooling the air at inlet, the greater weight and hence better the output.

1.14.5 Location of D. G. Sets

D G Set should be always located at a distance from known sources of dust within the plant.

The room containing D. G. Set is generally pressurized to prevent ingress of dust from outside.

A slow speed engine would be larger to deliver same power but would be an insurance against wear and will have long life.

Exhaust gases leave D. G. Set at high temperatures. For large capacity d.g. sets these could be used to generate power by 'cogeneration'. About 5% power can be generated from exhausts of d.g. sets.

1.15 Installation for Supply of Oil for D. G. Set

An integral part of D. G. Set installation is the receipt, storage and handling of oil used.

Since, oil is combustible and inflammable, precautions are to be taken in storing it. All statutory and constructional norms for oil storage tanks should be followed; like constructing an earthen bund around the oil tanks.

Heavy oils need preheating for extraction and they are to be heated to achieve correct viscosity for atomisation in the engines.

Therefore tanks are required be fitted with heating coils (steam or electric); oil pipe lines are not only lagged but are also heated by winding electric tape around them to keep oil at correct temperature.

1.16 Layout of D G Set Installation

Layout must allow for all these facilities.

Large d. g. set engines are water cooled. Therefore there is a water circuit also. Most often, it will have its own independent water circuit consisting of pumps, cooling tower, piping, etc. Water should be softened to minimize incrustation inside the engine.

Plant must make arrangements to bring in required quantities of oil used in d.g. sets and for cooling water. Area required for d.g. set should also be added to plant's requirements.

A typical layout of a d.g. set is attached and also a plan showing overall plan of power plant and oil storage.

See Figs. 5.1.7 and 5.1.8.

1.17 Thermal Power Plant for Captive Power

When captive power plant is a coal based thermal station, its capacity will be arrived at in a similar manner. In a thermal power plant, turbines will convert power from steam into electricity by running generators.

A very different layout involving coal handling and coal preparation plant and water supply system and boilers to generate steam at desired temp. and pressure and also condensers and recovery system for collecting water after steam has passed through turbine and also cooling towers and other auxiliaries would be required.

This is a specialized job and should be entrusted to Experts.

Thermal plant should also be integrated in the main plant augmenting needs of land, quantities of coal and water, railway siding etc., for the purpose.

1.17.1 Captive Power by Waste Heat Recovery

If the Cement Plant installs a Power Plant generating power by using the Waste Heat in exhaust gases of the kiln and cooler, then its generating costs would be much less than those of captive power plant as there are no fuel costs.

If the existing captive power plant is thermal, then it is possible to augment its capacity by the exhaust gases of kiln and cooler. Further, emission of greenhouse gases per ton of cement would also be reduced.

1.18 Locations of Captive Power Plant and Waste Heat Recovery Plant

Captive power plant can be either thermal or d.g.set.

If thermal it would be located from the point of ease of handling coal for cement plant.

Power plant by co generation is required to be located necessarily with respect to kiln exhaust gases and cooler vent gases.

The various possibilities in this respect should be examined at planning stage for green field cement plants.

General layout drawing **figure 6.1.23** shows locations of captive plant and waste heat recovery plant.

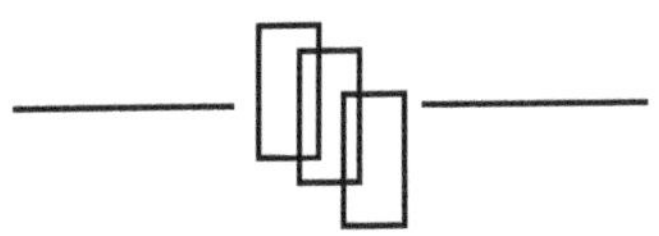

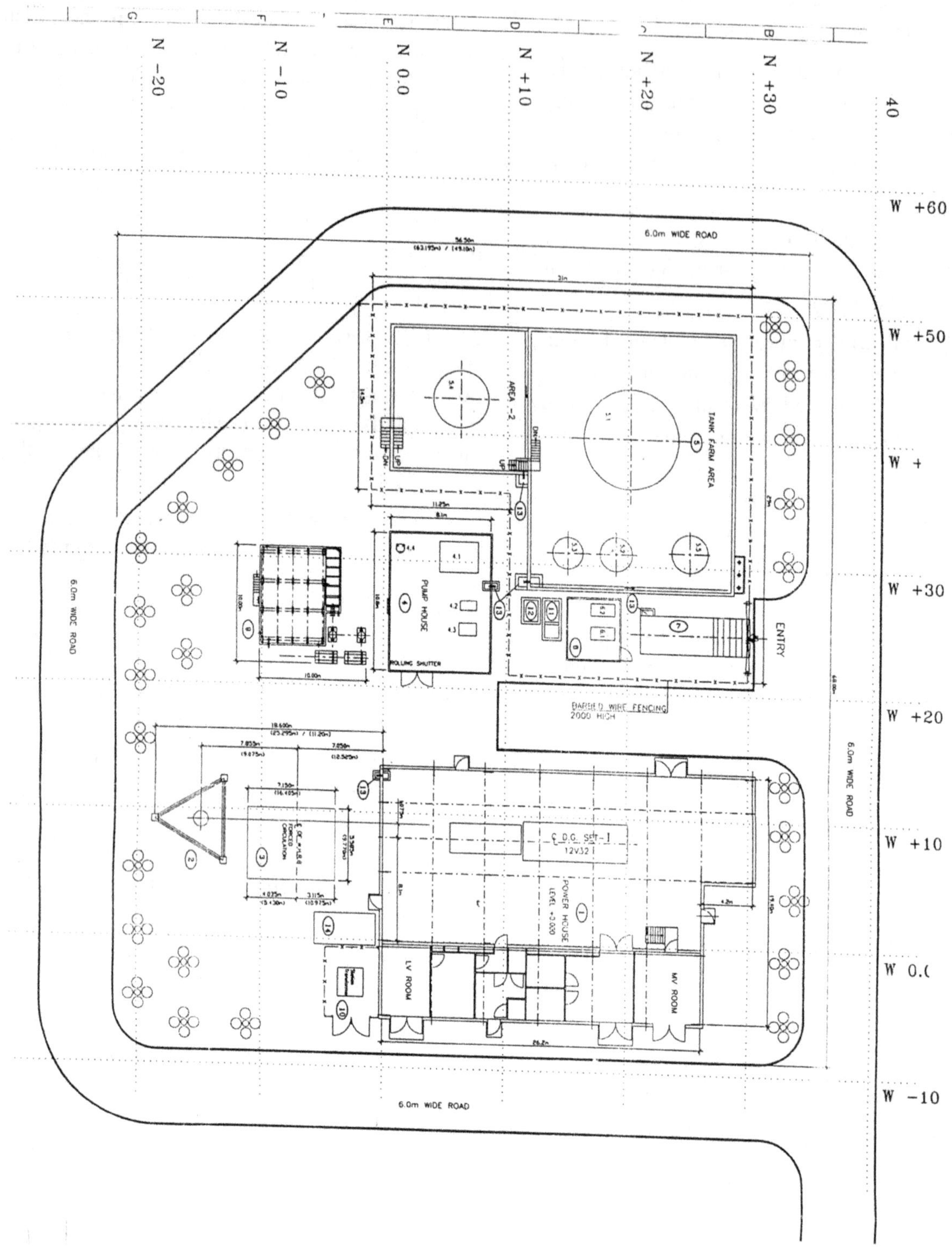

Fig. 5.1.7 Typical layout of a D.G. set installation and oil storage.

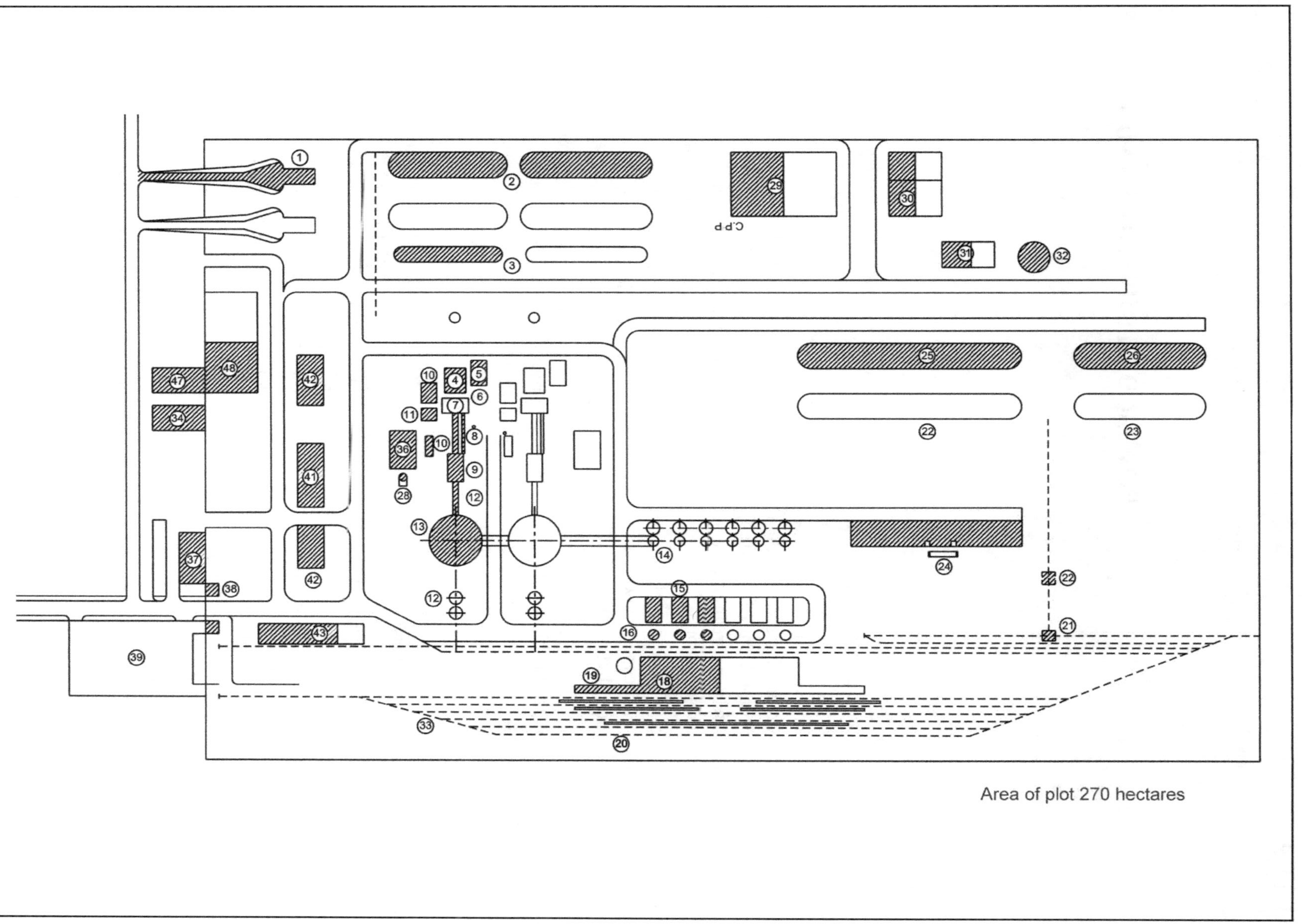

Fig. 5.1.8 General plant layout for a 10,000 TPD Clinkering unit making 6.5 MTPA OPC & BFSC.

CHAPTER 2

POWER DISTRIBUTION AND CABLES

2.1 Power Distribution System in the Cement Plant

Power Distribution System (PDS) of a Cement Plant begins with the Substation of the grid where power is received and ends with individual drives and points of usage. It is a large network consisting of elements like Transformers, H.T. and L.T. Control Panels, individual Distribution Boards and Motor Control Centres; necessary switch gear for safety and regulation and metering of power used at various points.

Power Distribution System within the plant now has several feeders, 1 main grid, 2 captive power plant, 3- power from co generation that is Waste Heat Recovery System

Motors and their control gear; cables connecting various units from beginning to individual points of usage; lighting; earthing are other components of the system.

Design of the P.D.S. is a specialised job and should therefore be entrusted to the Specialists.

In this section only the various elements of the PDS will be touched upon.

Almost all components of the PDS are governed by Standards laid down by Bureau of Indian Standards or respective National Standards of other countries. The prevailing Electricity Rules also lay down guide lines for selecting and installing the equipment.

All such Standards should be meticulously followed for ensuring safety of machinery and men.

2.2 Main Components of the PDS

Starting from grid substation sequence of various components would be :

1. Main transformers reducing voltage of transmission to voltage of h.t. motors in the plant.
2. Transformers to reduce voltage further to 415 / 440 volts which would be the voltage of majority of motors.
3. H.T. Control panel to which main transformers would be connected and from which feeders would go to h.t. motors in various sections and to various sections of the plant.
4. L.T. Control panels which would be as many as the major sections in the plant. They will receive h.t. voltage in the plant; transformers reducing voltage to 415 v will also be connected to them. From these panels will go feeders to motor control centers in various sections; there can be more than one motor control center (MCC) in a section.
5. MCCs are distributed conveniently at various points in a department to make operation and maintenance easy and also from point of view of simplifying cable layout and reducing cable costs and for calculation of energy consumption.
6. Motor control centers will also include starters and voltmeters and ammeters and energy meters and relevant safety equipment like circuit breakers, relays etc.
7. Presently almost always motors and groups of motors would be started from central control room through Programmable Logic Controllers (PLC s).
8. System of cables H.T. and L.T. as required to

connect various components with one another to ensure continuity of supply of Power.

9. Motors themselves to suit each application.

As mentioned earlier, each one of these components is governed by the standards laid down by the Bureau of Indian Standards.

2.3 Motors List

The starting point for designing the power distribution system would be the detailed Lists of Motors in each section along with details like type, numbers, rating, voltage and speed and type of starter. A simplified list was **shown in Table 5.1.3 of Chapter 1.**

Their locations would be shown on pertinent departmental drawings.

A typical list is **shown in Annexe 5.2.1** for Cement mills department.

2.4 Locations and Numbers of MCCS

Next step would be to decide on locations and numbers of MCCs to which these drives would be connected.

In doing so, it should be kept in mind that the computation of sectional power consumption is facilitated.

In this way, number of drives, their ratings and spare panels to be kept in each MCC and capacities of incoming and outgoingFeeders are worked out.

2.5 Panels in HT and LT Boards

Working in this fashion, c0apacity and numbers of panels in sectional L.T.Boards can be worked out. And proceeding in the same manner, numbers of H.T.Boards and capacities and numbers of panels in each can be arrived at.

2.6 Typical PD Schemes

Procedure to work out capacities of main transformers has already been explained in **Chapter 1.**

The entire power distribution scheme can be shown in a diagram/diagrams

See Figs. 5.2.1 to 5.2.3.

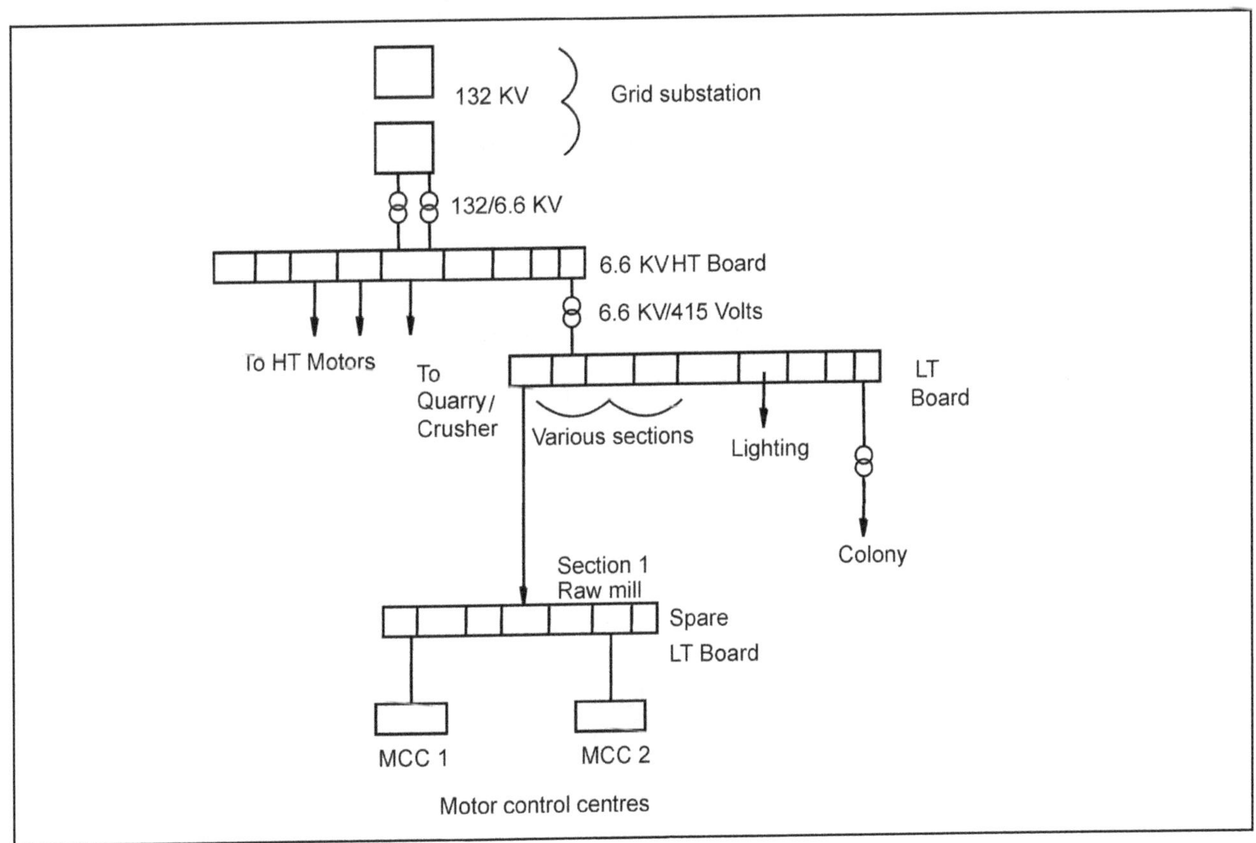

Fig. 5.2.1 Only grid power.

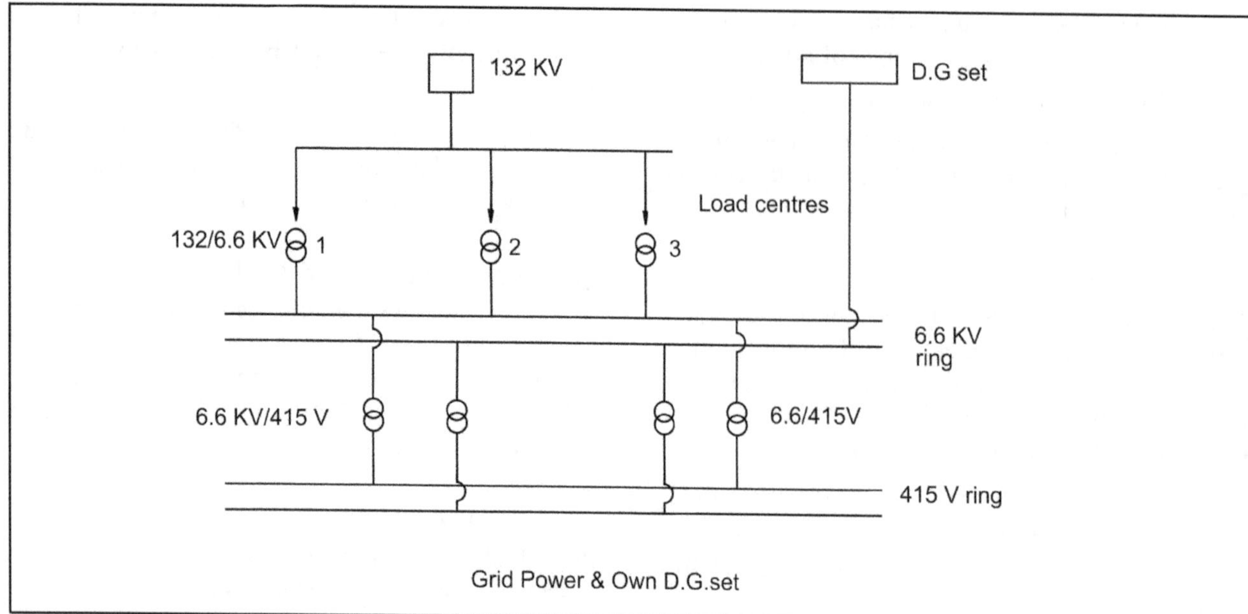

Fig. 5.2.2 Power distribution schemes in the plant.

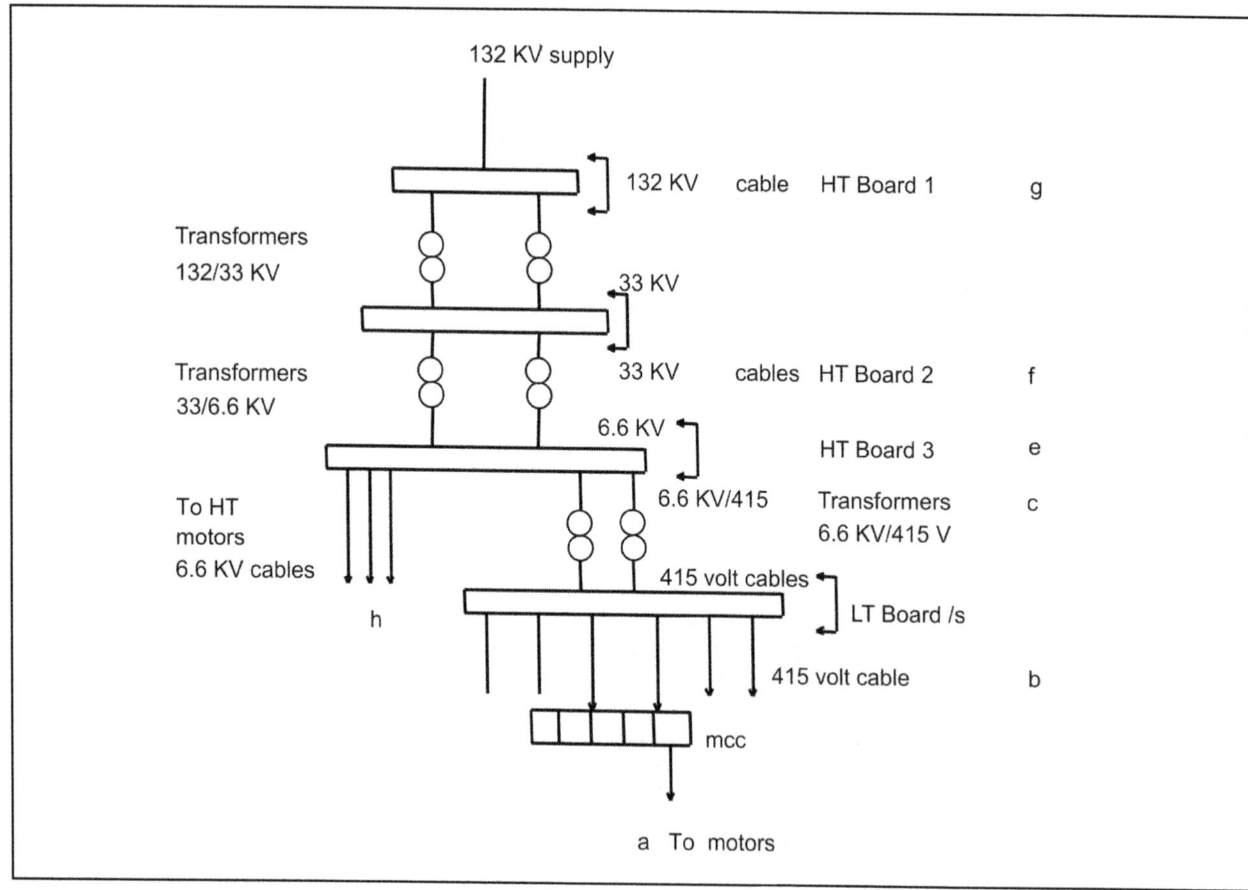

Fig. 5.2.3 Power distribution scheme.

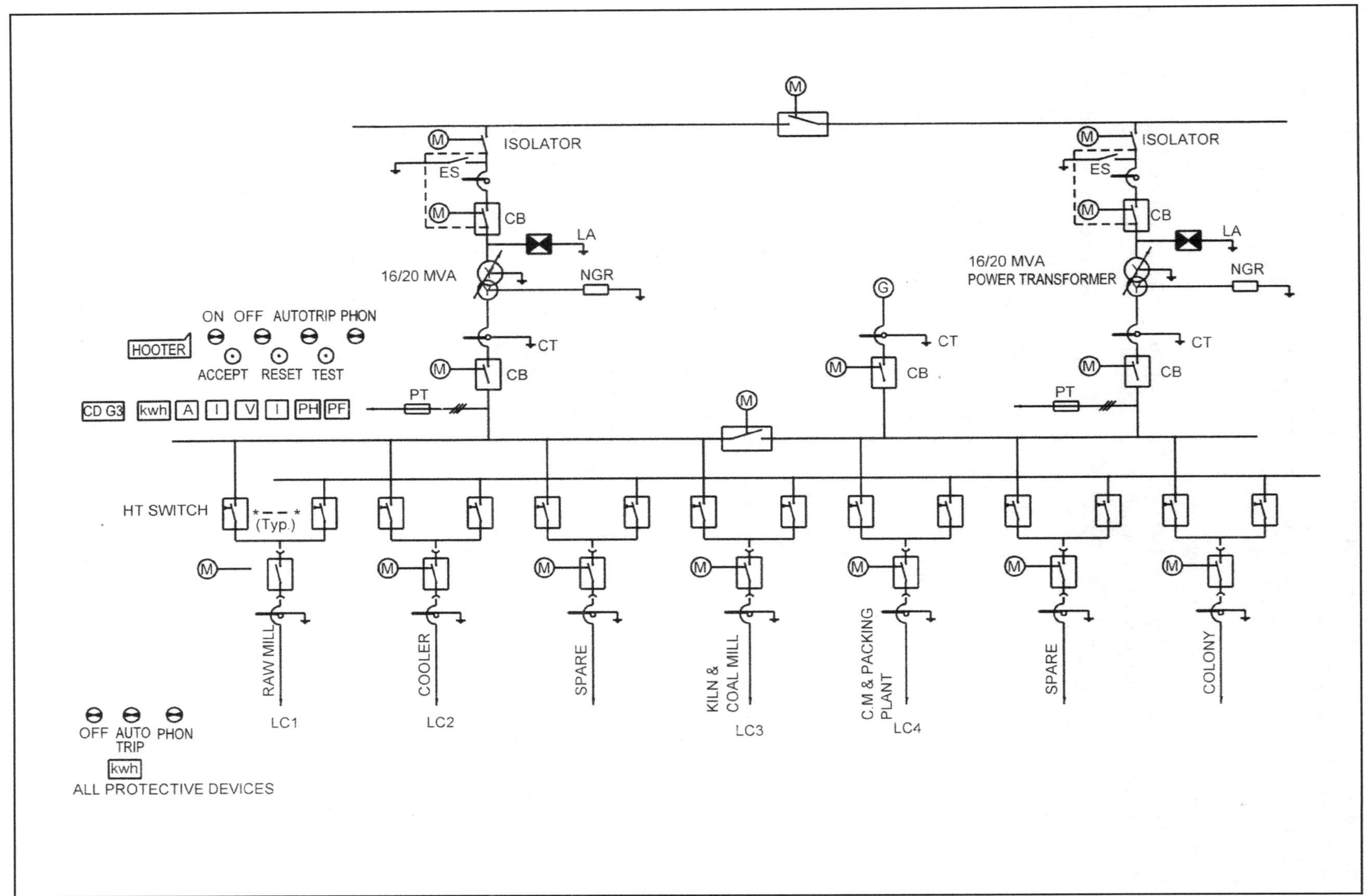

Fig. 5.2.4 Typical power distribution – double bus bars scheme.

Also attached are 3 drawings of typical power distribution schemes in greater detail.

See Figs. 5.2.4 to 5.2.6.

2.7 Cables

Cables convey power to the point of use. There are HT and LT Cables according to the voltage of power supplied. Cables will be insulated and protected according to voltage. Most drives would be three phase motors. Majority of cables will also therefore be suitable for three phase.

Power = V × I × cos single phase and

Power = 1.73 × V × I × cos for 3 phase supply.

Therefore when voltage is high, I is small, losses are small. It is therefore economical to transmit power at high voltage up to the point of use. However, HT cables need high degree of insulation and hence are costly.

Size and type of cables, depends on the amount of current they carry and the voltage.

2.7.1 Cables in Various Locations

There are thus individual cables from motor starters to motors.

There are group cables between :

(i) MCCS and LT Boards.
(ii) LT Boards and HT Boards.
(iii) Transformers and HT / LT Boards.
See Figs. 5.2.3 and 5.2.7.
In **Fig. 5.2.3** are

(a) cables to individual L.T. motors of 415/ 440 V.
(b) LT cables to MCCS from LT boards.
(c) LT cables from transformer 6.6 KV / 415 V to LT board.
(d) H T cables from HT board to 6.6kV/415 transformer.
(e) HT cables 6.6 KV from 33/6.6 KV transformer to 6.6 KV HT Board.
(f) HT cables 33 KV to HT Board for transformer.
(g) HT cables 33 KV from 132/33 KV transformer to HT Board of 33 KV .
(h) cables for individual HT motors.

2.8 Cable Lengths

As a Cement plant is spread over a wide area of approximately 200-300 hectares or more, total length of cables adds up to substantial capital costs.

Costs of cables vary according to their current capacity and voltage because of insulation.

Thus it is most important to locate the substation and the distribution net work in such a way that cables costs are less.

2.8.1 Reducing Quantities of Cables

Cables lengths of individual drives can also be reduced if MCCS are split and located at different levels in a department.

Smaller MCCS are installed at different floors. These are fed with single cables from main MCC on the ground floor. There is thus considerable reduction in overall lengths and costs of cables.

See Fig. 5.2.7

Once locations of machines and drives are fixed, the MCC can be located and routing of cables and hence their lengths can be worked out.

It is best to keep a MCC as near its load center as possible.

2.9 Selection of Cable Sizes, Types

Once the motor ratings are known, it is easy to size individual cables. Allowances are to be made in sizing cables for several factors like grouping, laying over-ground or underground, properties of soil and depth at which cables are laid in the ground etc. Suppliers / manufacturers of cables or Consultants for PDS should be consulted in this regard.

However, in sizing group cables, i.e., a cable which is feeding an MCC and thus feeds a number of drives, it is necessary to keep margins for additions that may take place in future.

It is also necessary to keep spare panels for say welding sets and additions of drives, utilities and so on in all MCCs and distribution boards.

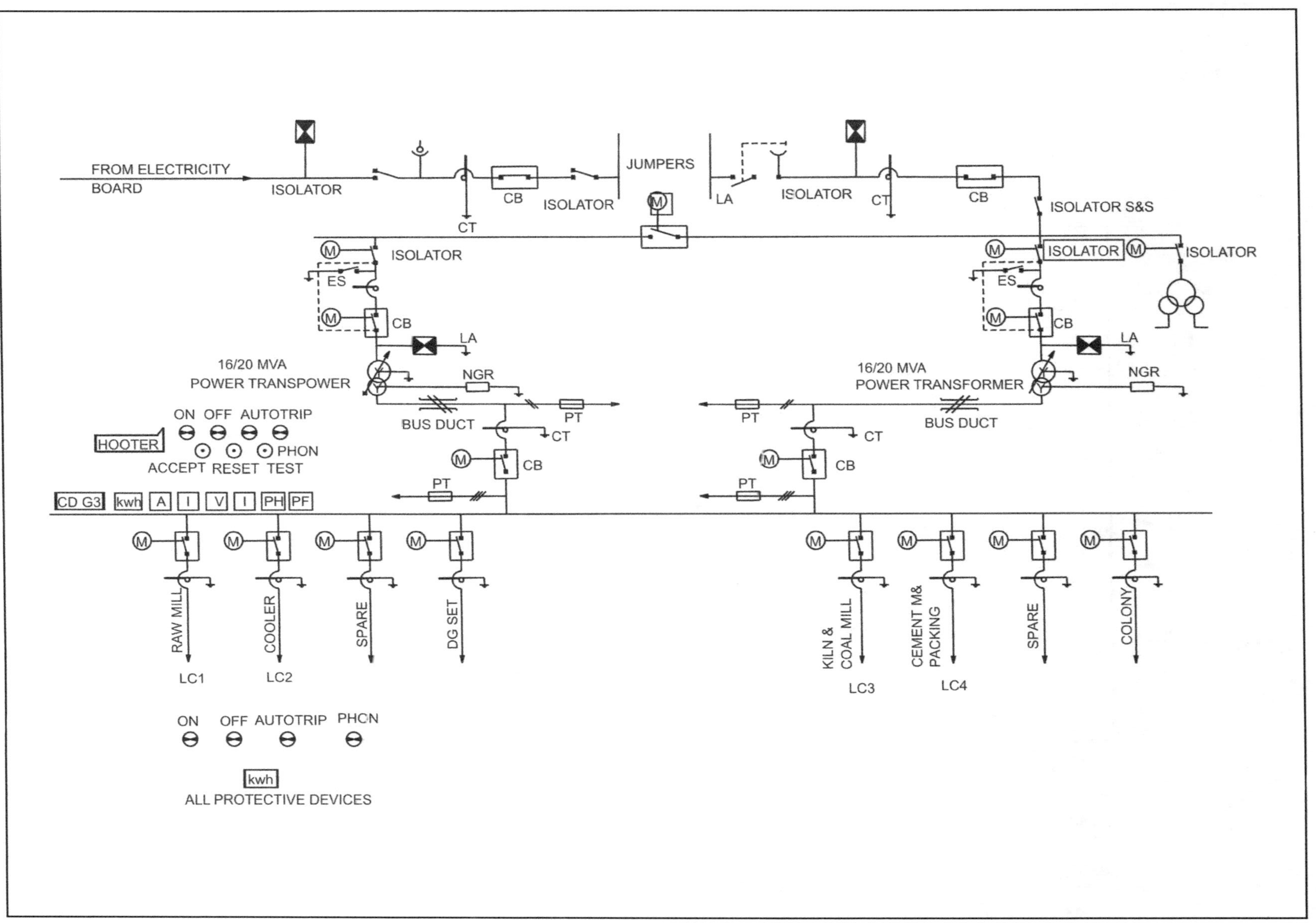

Fig. 5.2.5 Typical power distribution scheme.

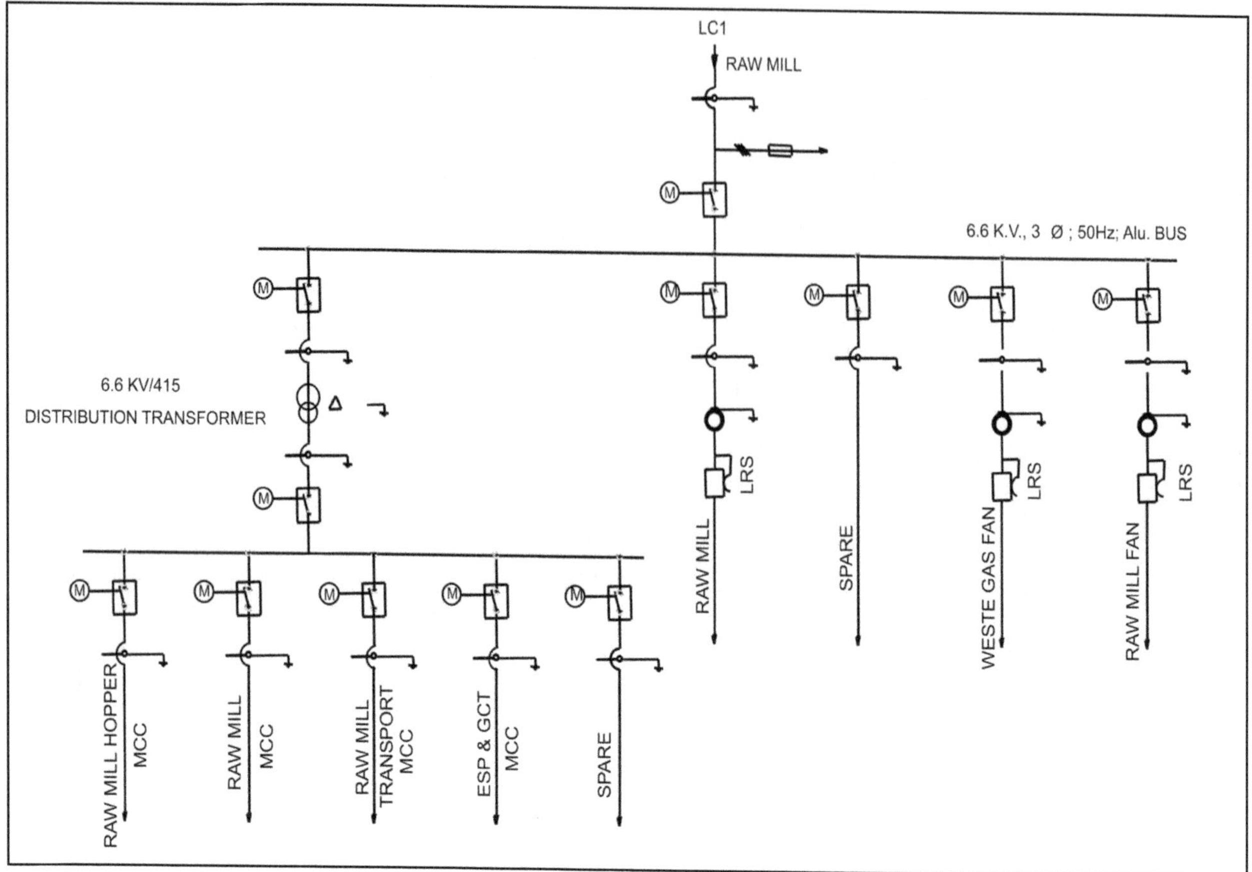

Fig. 5.2.6 Typical power distribution schme – raw mill section.

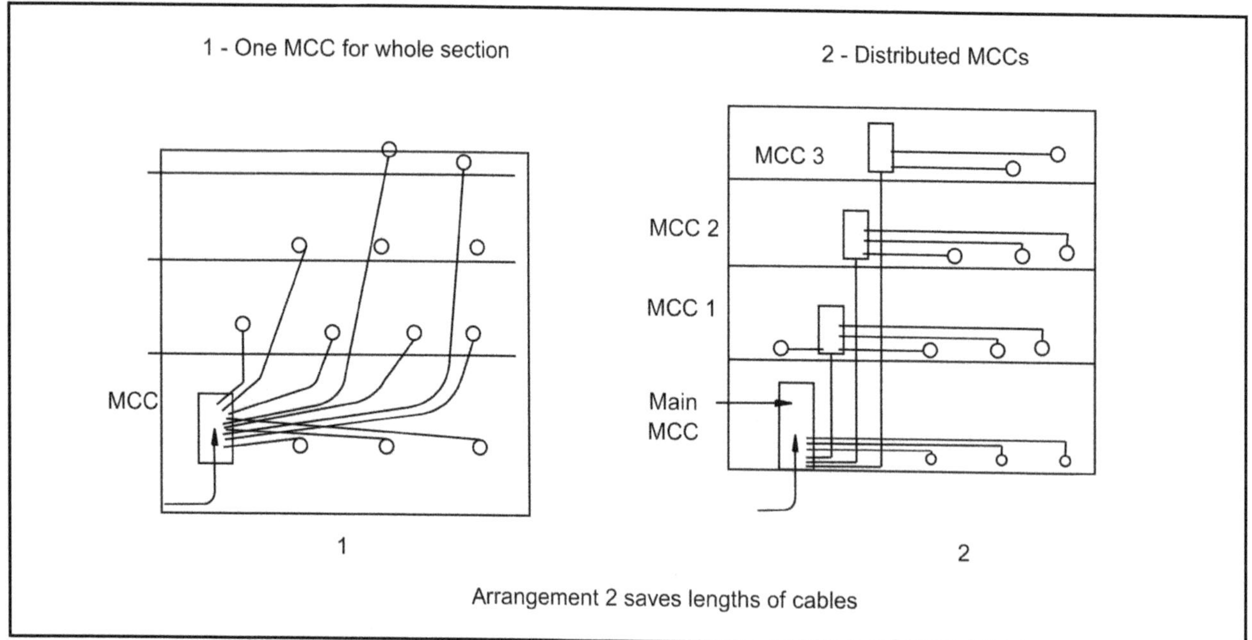

Fig. 5.2.7 Power distribution in a section.

Therefore group cables b,c,d,e,f,g, in **Fig. 5.2.3** should be sized generously, so that they do not get over loaded and are not damaged on account of short circuit currents.

2.9.1 Identifying Cables

A cable is designated by number of cores and number of strands, and cross sections of strands and total area conductors in it. It can also be designated according to its current carrying capacity. Important features of a cable would be :

(i) Voltage of supply.
(ii) Current to be carried.
(iii) No. of cores / No. of strands / core / sizes / area of strands.
(iv) Total cross sectional area of conductors in the cable.
(v) Material of construction copper / aluminium.
(vi) Insulation.

Now a days it will almost always be aluminium cables.

Cables joining is important and is a specialized job and needs attention now and then. Cable lengths should be so selected as to have minimum number of joints. The joints should be easily accessible for maintenance.

2.10 Cable Layout

This is also a very important aspect of engineering and requires considerable thought and planning.

There are 2 systems of cable layouts.

1. 'Overhead', where cables are supported on steel racks supported on trestles and columns. They are overhead and hence are easily accessible for maintenance.
2. 'Underground', where cables are laid in underground trenches up to point of usage like from HT and LT boards to LT boards and MCCs.

   See Figs. 5.2.8 to 5.2.12

In planning, room should be kept for expansion so that it is not necessary to install a completely new set of cable supports or dig new trenches, even when the entire lot of cables will almost be duplicated.

Cables spacing on the racks is also important and minimum spacing between cables should be adhered to.

Cable radii - depending on size of cable and insulation used, cables require a minimum radius of bends. Bends should be generously sized. Bends add to cable lengths and hence number of bends and their radius have to be taken into account in working out total lengths.

Cable laid between A to B underground has changes in levels and direction.

See Fig. 5.2.8.

Total length = L1+L2+L3+L4+L5+L6 = 100 m

Additions due to bends : 4 bends each 90o at radius of say 1 m, extra length due to bends = 6.25 m or 6 %.

If there are joints enroute, allowance has to be made for the same.

Cable lengths are to be worked out in this fashion. A cable layout is thus a must.

2.10.1 Overhead and Underground Layouts of Cables

In overhead layout of cables, the major consideration is the head room to be allowed for when cables cross roads used in the plant for movement of cars, trucks, sometimes even dumpers. Occasionally cranes used for construction and maintenance also have to go along these roads.

See Fig. 5.2.10 and 5.2.11.

Underground cables, to avoid digging too may trenches would be laid along the main drains taking care that cables do not get submerged.

See Fig. 5.2.12.

2.10.2 Using Plant Layouts for Cable Layout

Thus several editions of the general layout the plant have to be made for various purposes like, roads, cables, water drains, cooling water, return system to cooling ponds, etc.

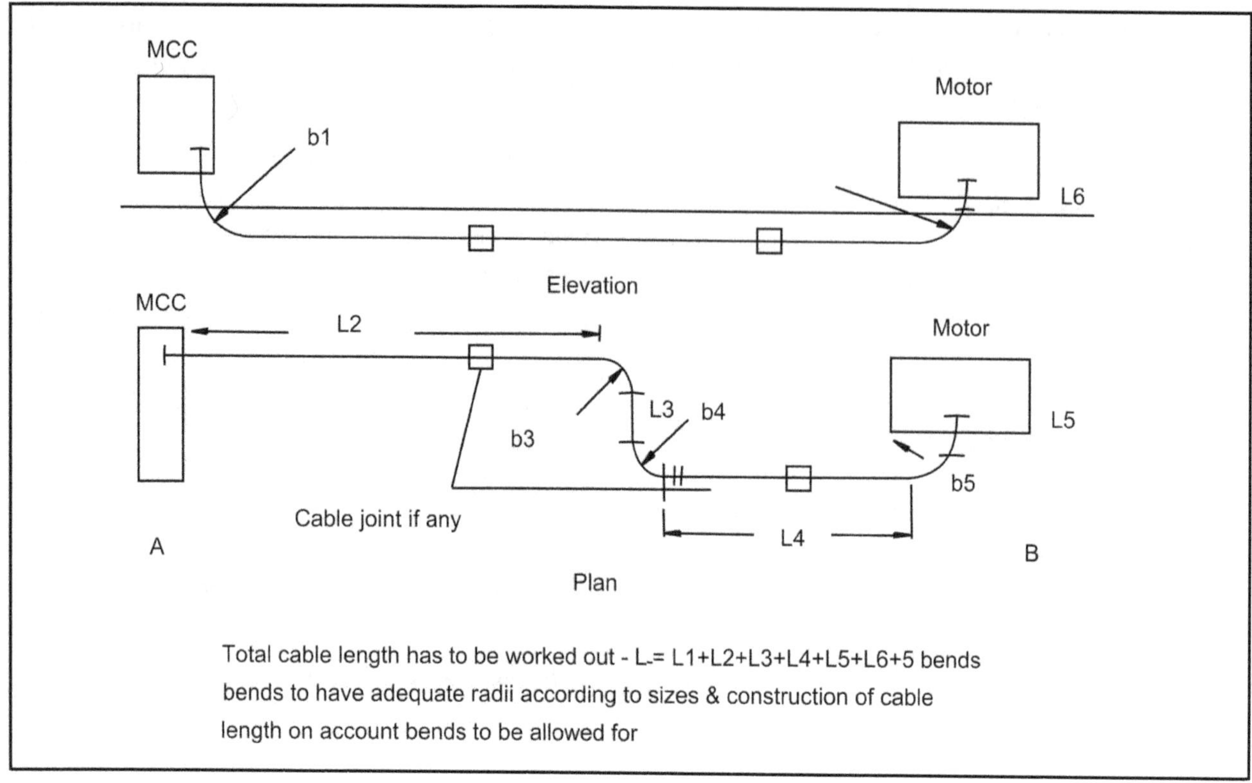

Fig. 5.2.8 Cable layout – from MCC to motor power distribution in a section.

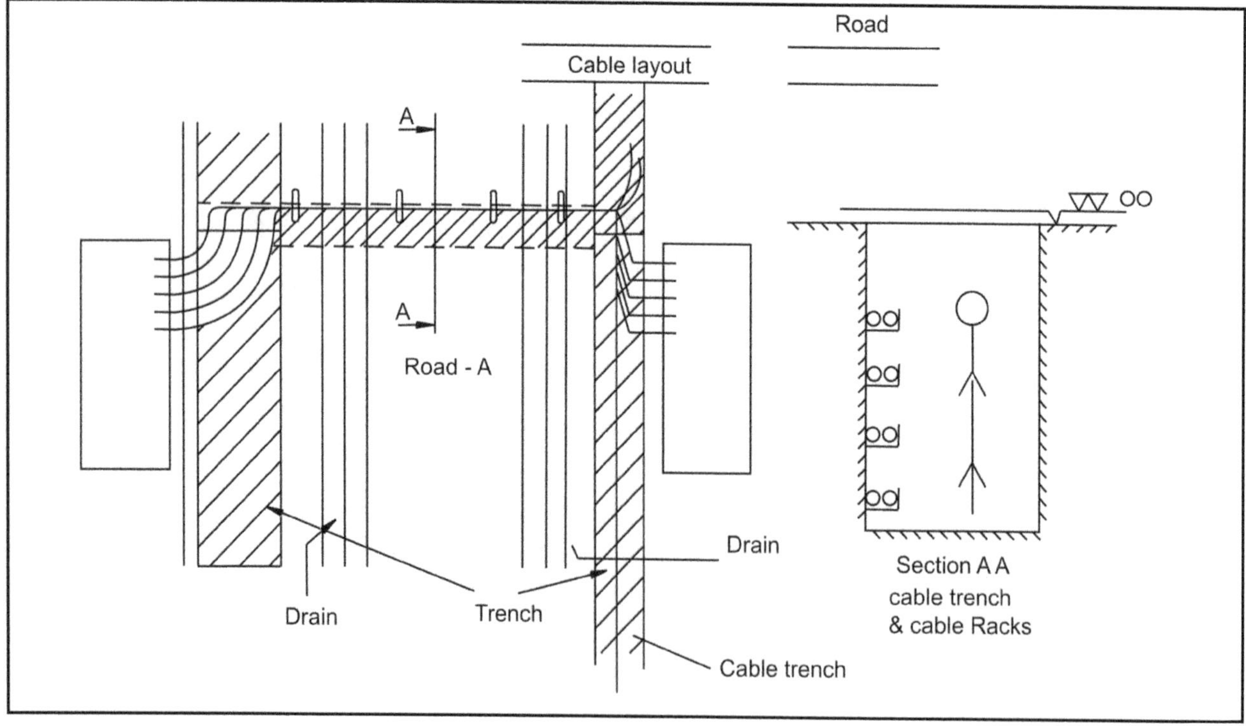

Fig. 5.2.9 Cable layout in trenches.

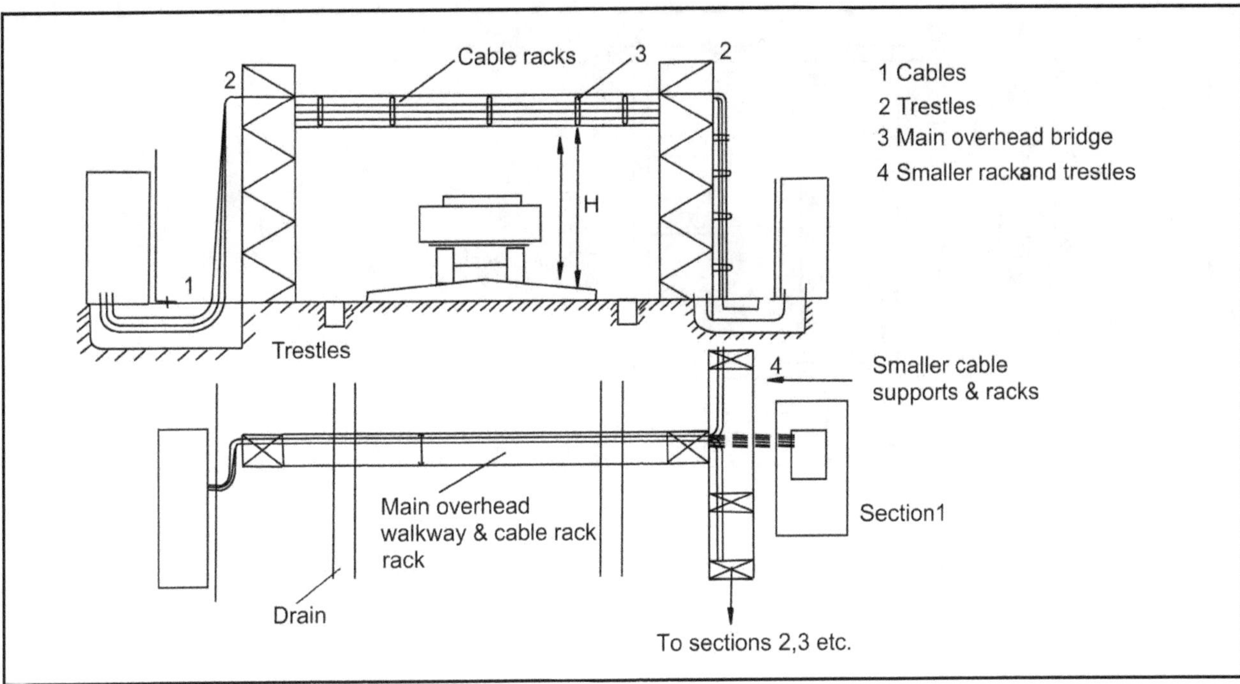

Fig. 5.2.10 Overhead cable routing.

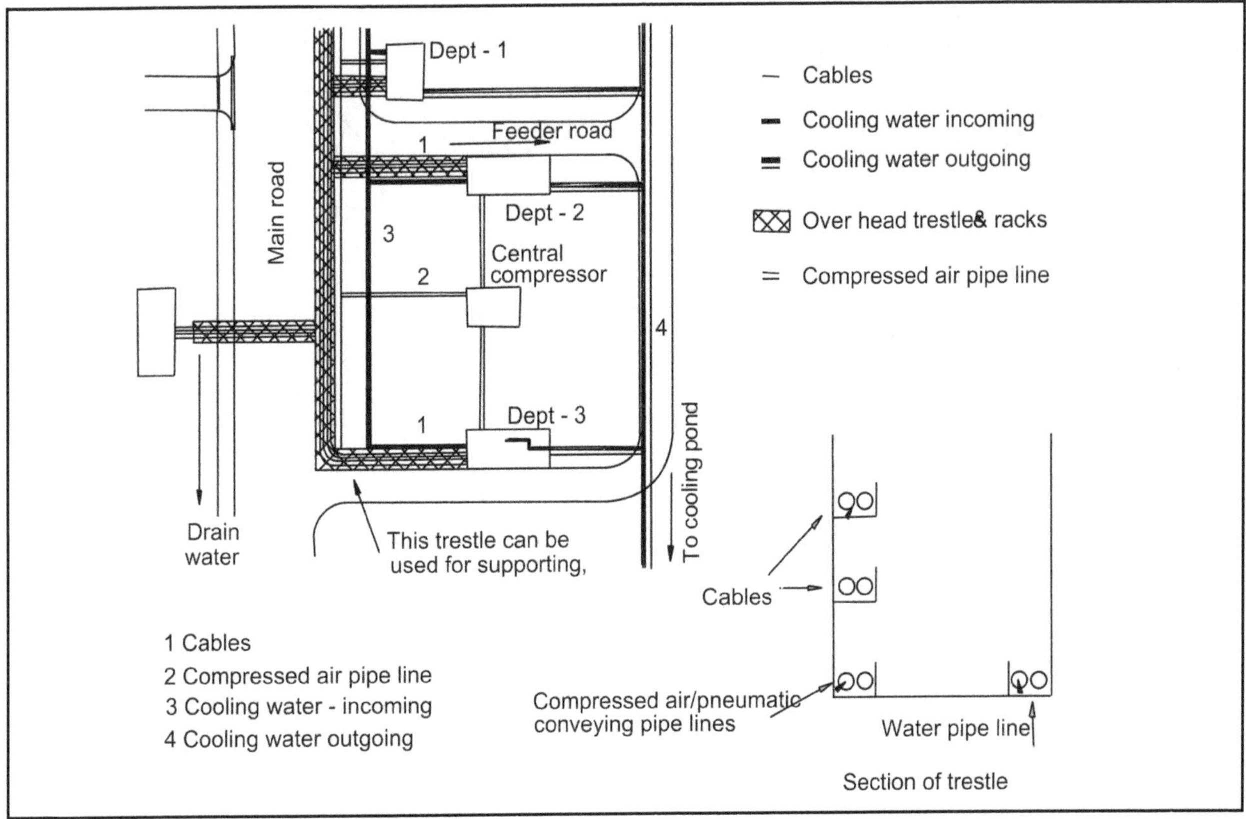

Fig. 5.2.11 Overhead cables.

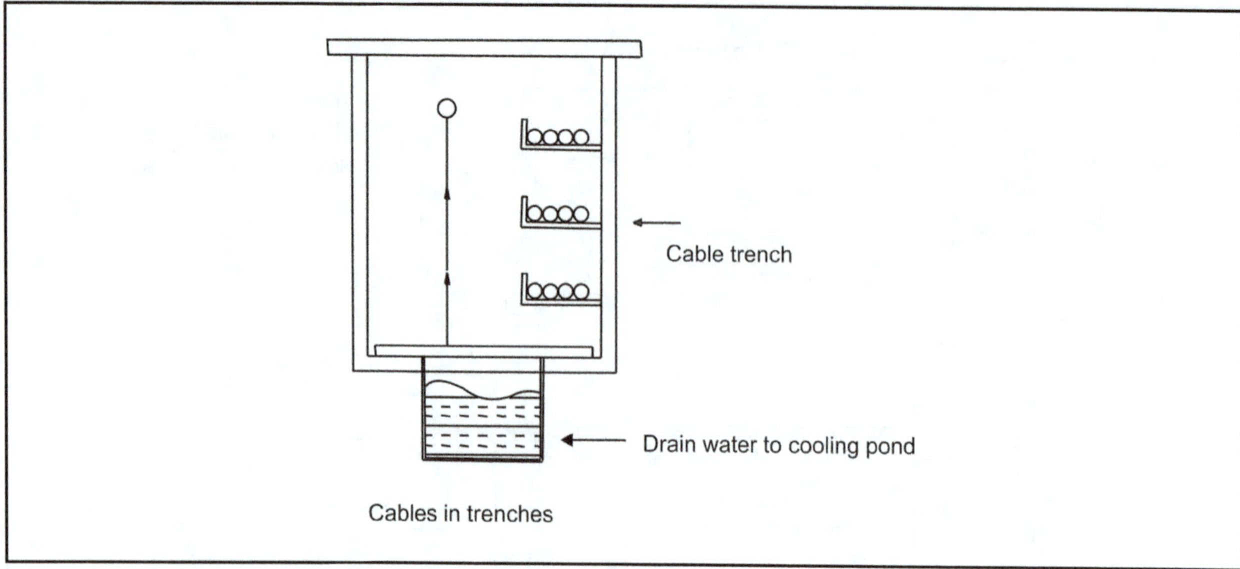

Fig. 5.2.12 Cable layouts to be integrated with other utilities.

2.10.3 Layout within Buildings

Cables and wires are to be taken to each drive from motor control centers in each section. Layouts of cables in various buildings should also be planned and made available to Civil Designers before the buildings are constructed.

Lengths of cables going from ground floor to other floors will be taken along column faces. Some prefer concealed wiring as is done in housing.

If this be the case, channels must be left (with accesses for maintenance) in columns or beams through which wires and cables can be drawn.

Another way to avoid chipping and making holes in slabs, which look ugly and give an impression of 'after thought' is to grout steel plates at regular intervals in faces of columns and beams, which would later on carry cables. This should perhaps be done whether cables are laid or not. Cables trays can then be fixed on these without requiring chipping and drill holes etc.

See Figs. 5.2.13 to 5.2.16.

Attention to small details pays dividends as work can be done fast without loosing time.

2.11 Motor Foundations

Motor and Gearbox foundations also in a way form part of the planning of power distribution

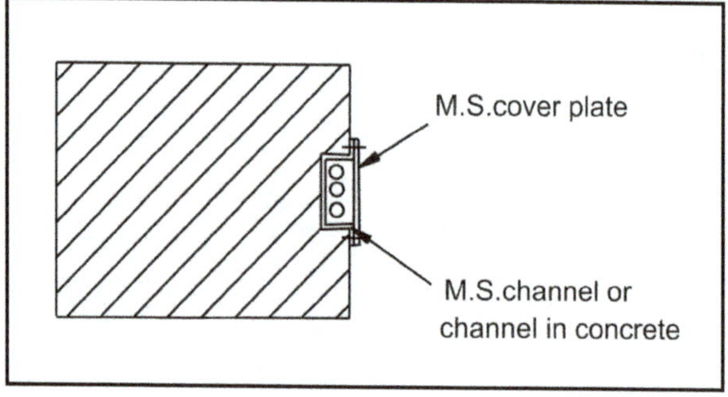

Fig. 5.2.13 Along columns.

Large motors will have their own foundations. For small and medium size motors, a common base frame for motor and gearbox would be used and grouted in a foundation.

Designing base frame for motor and gearbox would need drawings for both with details of bolt holes and their spacing, height of center etc.

The cables box on the motor must come to the right side.

2.12 Motors

An important aspect is direction of rotation. With a few exceptions all motors would be running in one direction all the time. If 'Unidirectional' motors are specified, there is a small but significant gain in operational efficiency of motors.

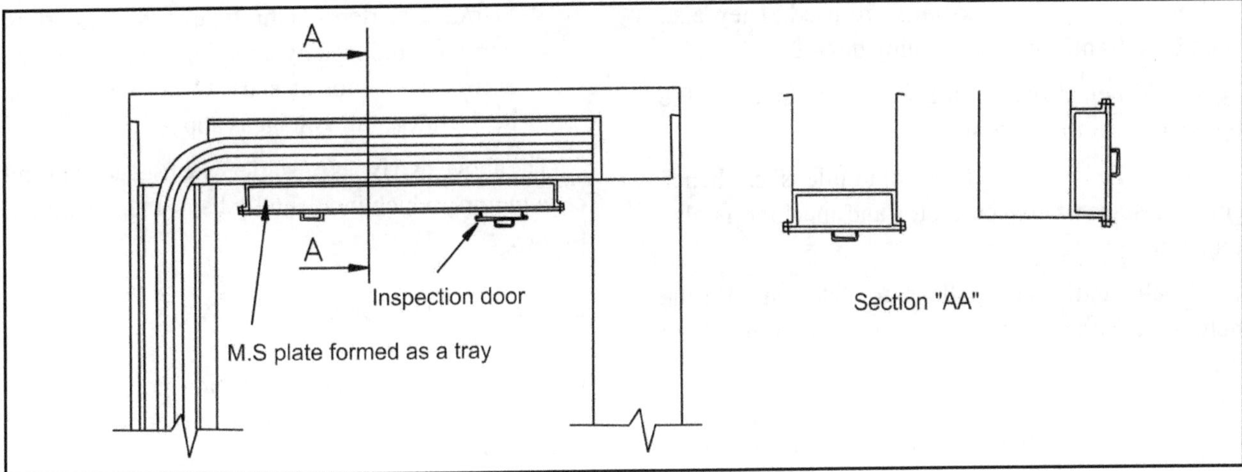

Fig. 5.2.14 Along beams.

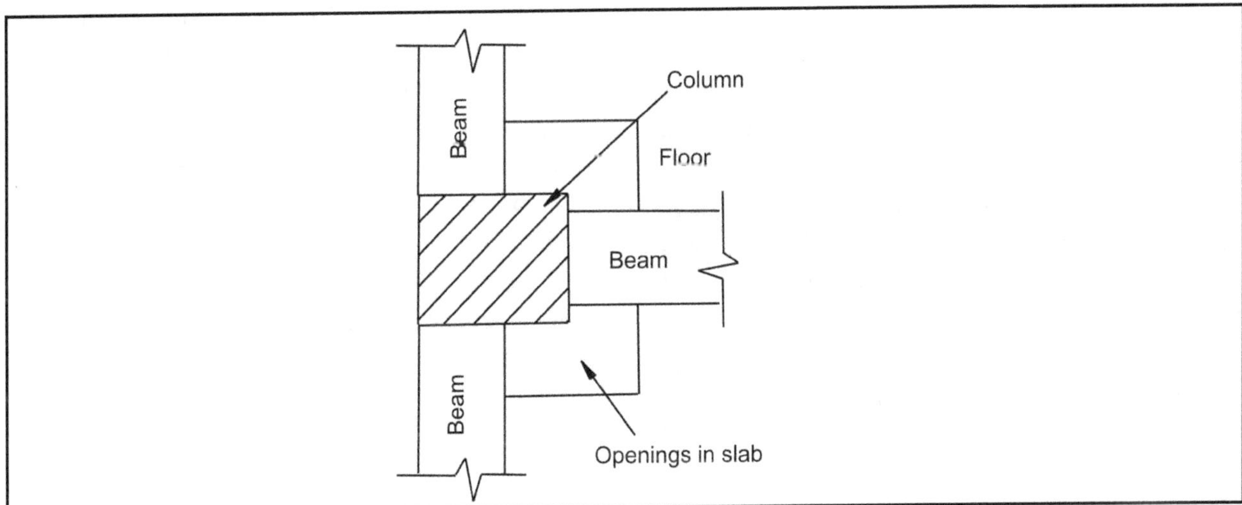

Fig. 5.2.15 Through floors.

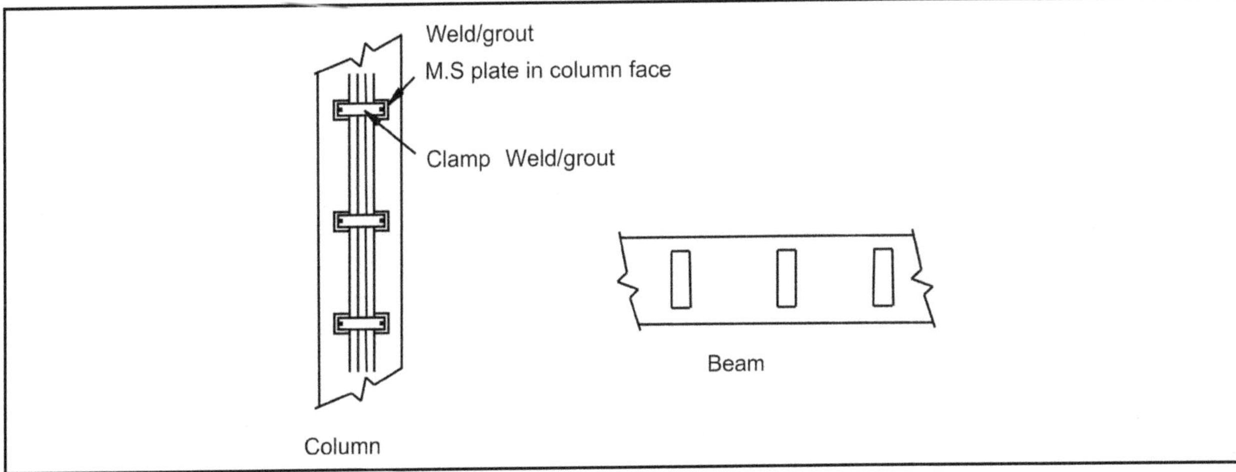

Fig. 5.2.16 Cables in a building.

For large drives, jackshafts are used. They also need base frames with tensioning device.

The V belt drive should be so arranged that the tight side is at the bottom.

The angles of contact for both pulleys and hence centre distance between motor and machine is also important.

V belts and V belt pulleys are selected with the help of the suppliers for the given drive rating of the motor.

There would be occasions like sudden power tripping when machines like conveyors and even kiln, cooler, etc., would have to be started on load.

For such locations motor ratings should be chosen allowing for this fact in consultation with Suppliers. For kiln for instance motor rating would normally be 1.6 times the full load power requirement.

Apart from this, for such applications high torque motors are selected.

In case of motors for kilns, mills and fans it is also necessary to know Moment of Inertia of driven machine and time available to bring it up to full speed from rest.

Motors are selected on the basis of enclosures, cooling, permissible rise of temperature of the winding above ambient. Specifying ambient temperature and also this permissible rise indicates class of insulation of the windings of the motors.

Most common enclosures are TEFC (Totally enclosed fan cooled), common classes of insulation are class B and F.

See Web page W1.99 in Section 9 on selection of motors.

2.13 Starters

Start the motors from zero speed to full speed. Starting currents are large.

There are different types of starters.

1. Direct on line starters (DOL) are for small motors. Starting current is large and hence switch gear, cables and relays are selected for short circuit conditions pertaining to starting current.
2. Star/delta starters - in this type of starter, starting current is reduced by applying smaller voltage across the motor at start. Once the motor is accelerated, full voltage is applied.

The above two types of starters are used for squirrel cage motors, which incidentally have a high starting, torque.

2.13.1 Starters for Slip Ring (SR) Induction Motor

For large motors that is slip ring motors, rotor resistance starters are used. Resistance is introduced which limits starting current, resistance is cut gradually in steps and motor is accelerated slowly to reach full speed.

It takes time to bring motor to full speed. Mostly the starters would have their own drive which would cut out the resistance in a predetermined manner speeding the motor.

When the resistance or part there of is left in circuit, motor runs at slower speed. Thus rotor resistance can also be used to vary the speed of the slip ring induction motor.

Variable speed induction motors, therefore have rotor resistance starters. However, when resistance is left in circuit there is loss of power. When motors are required to run for long periods at low speeds, such starters are not the ideal choice.

2.13.2 Slip

When no resistance is left in circuit, motor runs at full speed, which is synchronous speed, less slip. Normally, slip would be 3-5%. Synchronous speeds are a direct relation of number of pole and frequency, which is 50 cycles per second in India.
See Table 5.2.1.

2.14 Inertia of Load

Motors should be able to overcome, inertia of the driven machine and bring it up to speed. Therefore, for large motors and large fan drives it is necessary to know the moment of inertia of the driven machine converted to motor speed.

If this is not taken into account, it would take inordinately long to start a machine and to bring it to speed.

Table 5.2.1

No. of Poles	Synchronous Speed in r.p.m.	Running Speed in r.p.m.	Slip (%)
2	3000	2900	3.3
4	1500	1440	4
6	1000	960	4
8	750	730	3

2.15 Earthing

This is an important aspect and should not be neglected. Many accidents can be avoided if earthing is done properly.

All drives and electrical equipment like transformers and switch gear and also distribution panels and boards need to be adequately earthed as per practice approved by Electricity Boards and other statutory authorities.

See Figs. 5.2.17 and **5.2.18.**

It should be done properly, using wires / strips of correct cross sections and materials. Grounding is done by digging pits of prescribed dimensions and depth in the soil.

2.16 Lightning arresters

Preheater tower, silos and chimneys are tall structures and need to be adequately protected against lightning.

Structures standing by themselves distant from taller structures should also be provided with lightning arresters as precautionary measure.

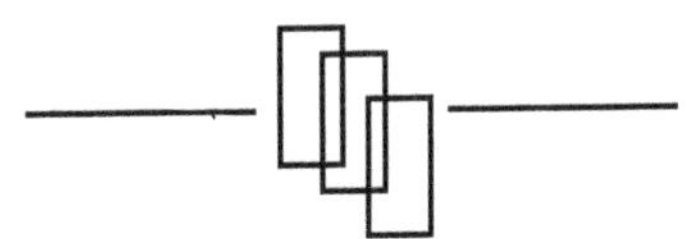

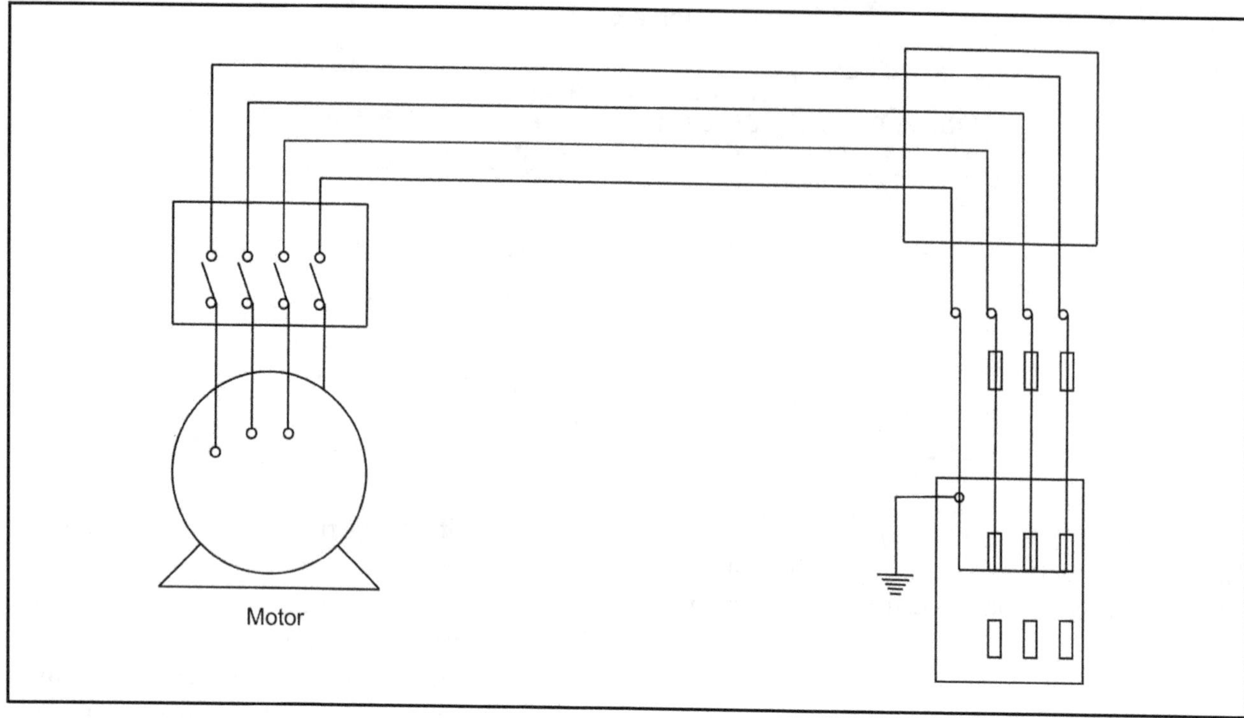

Fig. 5.2.17 Four wire earthed neutral system of substation to motor control.

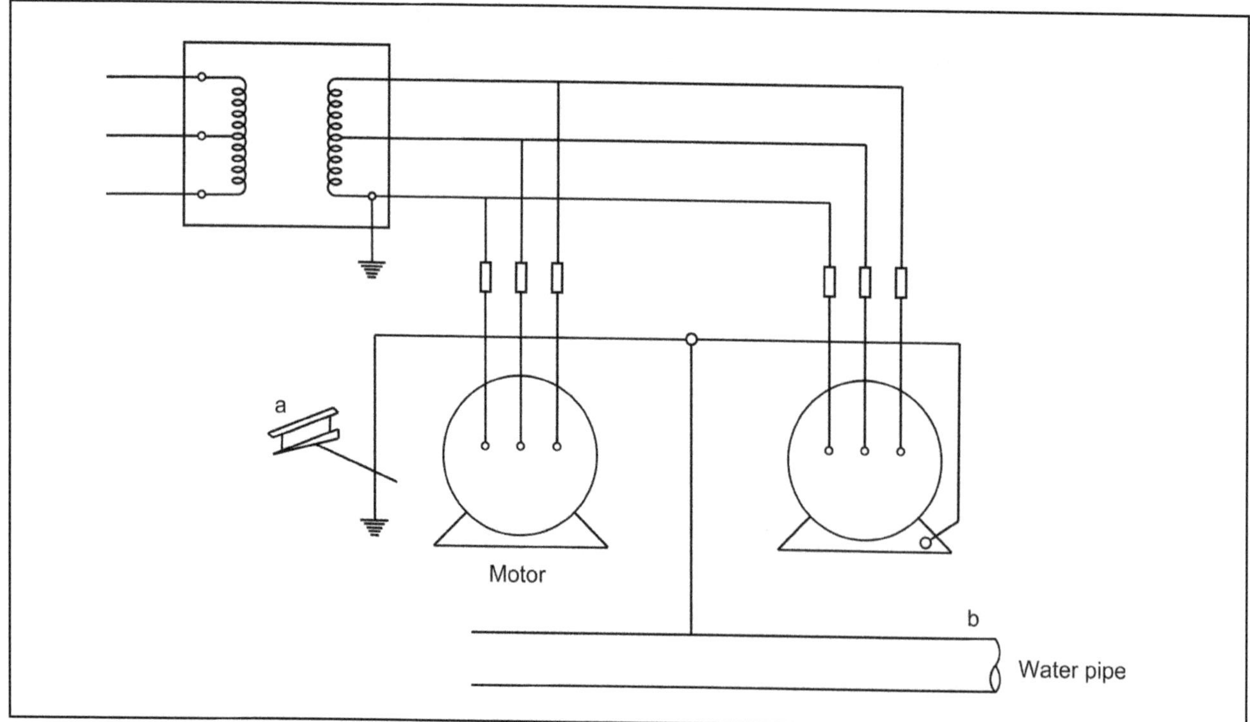

Fig. 5.2.18 Protection conductor system.
(a) Section of steel frame building.
(b) Water pipe.

Annexe 5.2.1

List of Motors for Cement Mill Department.

sr no	section	motor for	nos	rating kw	speed r.p.m.	voltage volts	type	starter	notes
1.	materal handling	reciprocating feeder	1	10	1500	415	variable speed		high torque
2.		belt conveycr	1	25	1440	415	s.c.	d.o.l.	high torque
3.		tripper on belt	1	2	1440	415	s.c.	d.o.l.	
4.		belt conveyor	1	15	1440	415	s.c.	d.o.l.	high torque
5.		fan for bag filter	1	2	1440	415	s.c.	d.o.l.	
6.	cement mill	weigh feeder for clinker	1	2	150	415	variable sped d.c.		
7.		weigh feeder for gypsum	1	0.5	1500	415	variable speed d.c.		
8.		belt conveyor	1	7.5	1440	415	variable speed d.c.		
9.		roll crusher	2	60	1440	415	s.r.	s.r.s.	
10.		belt conveyor	1	7.5	1440	415	s.c.	d.o.l.	high torque
11.		bucket elevator	1	20	1440	415	s.c.	d.o.l.	
12.		cement mill	1	1500	960	6600	s.r./ syn. Ind.	l.r.s.	power factor correction
13.		auxiliary drive	1	30	1440	415	s.c.	d.o.l.	
14.		blower for air slide	1	2	1440	415	s.c.	d.o.l.	
15.		pump for water	2	1	1440	415	s.c.	d.o.l.	
16.		pump for oil	2	1	1440	415	s.c.	d.o.l.	
17.		high efficiency separator	1	100	1500	415	variable speed d.c.		

Contd...

sr no	section	motor for	nos	rating kw	speed r.p.m.	voltage volts	type	starter	notes
18.		screw conveyor under dust collector	1	5	1440	415	s.c.	d.o.l.	
19.		fan for dust collector	1	15	1440	415	s.c.	d.o.l.	
20.		fan for airslide	1	5	1440	415	s.c.	d.o.l.	
21.		bucket elevator	1	10	1440	415	s.c.	d.o.l.	
22.		fan for airslide on cement silo	1	5	1400	415	s.c.	d.o.l	
23.		auxiliary drive	1	30	1400	415	s.c	d.o.l	

CHAPTER 3

POWER CONSUMPTION AND ITS CALCULATION

3.1 Power Consumption

A very significant operational and efficiency parameter is power consumption.

The electrical power distribution system and the distribution of meters of different sections should be organized to facilitate measurement of 'sectional' power consumption in terms of materials treated in them.

Location of every drive and the Motor Control Center (MCC) from which it is fed should be clearly known.

MCCs should therefore be so divided to facilitate calculation of power, Section wise.

3.2 Power Consumption in Crushing Section

For example, in crushing section, MCCs should have energy meters to record units of energy consumed in 24 hour periods. Knowing stone crushed in that period, it should be possible to calculate units/ton of material and also per ton of clinker and cement.

Stacker Reclaimer system would have two rates :

One for stacking which is equal to crushing rate or crusher output.

Another for reclaimer which feeds mill hoppers.

It should be possible to know :

1. Stone crushed.
2. Units consumed by stacker belts and arrive at specific power (1) for stacking in KWH/ton of limestone.
3. Stone reclaimed.
4. Units consumed by reclaimer and belts and arrive at specific power (2) for reclaimer in KWH/ton of limestone.

Total for stacker and reclaimer = (1 + 2) units/ton of limestone.

Alternately, stacker units could be added to crusher and reclaimer units to raw mill.

3.3 Raw mill and Blending Sections

Similar logic would apply to raw meal conveyed to blending silo and raw meal extracted from blending silo because rates of feed and extraction and hours of working are similarly different.

The confusion can arise in defining where raw mill section ends and where blending section begins.

Location of airlift would suggest that the blower for it should be located in blending section - normally it would be located under the blending silo.

Thus even if physically located in another section, the drive should be clubbed in appropriate section for calculating sectional power consumption.

To do this, it would be necessary to define the limits of the section and list drives included in it with ratings.

The location of MCCs of each drive may also be listed to facilitate clubbing. Flow charts can become a handy tool for identifying drives. **See Figs 5.3.1, 5.3.2 and 5.3.3.**

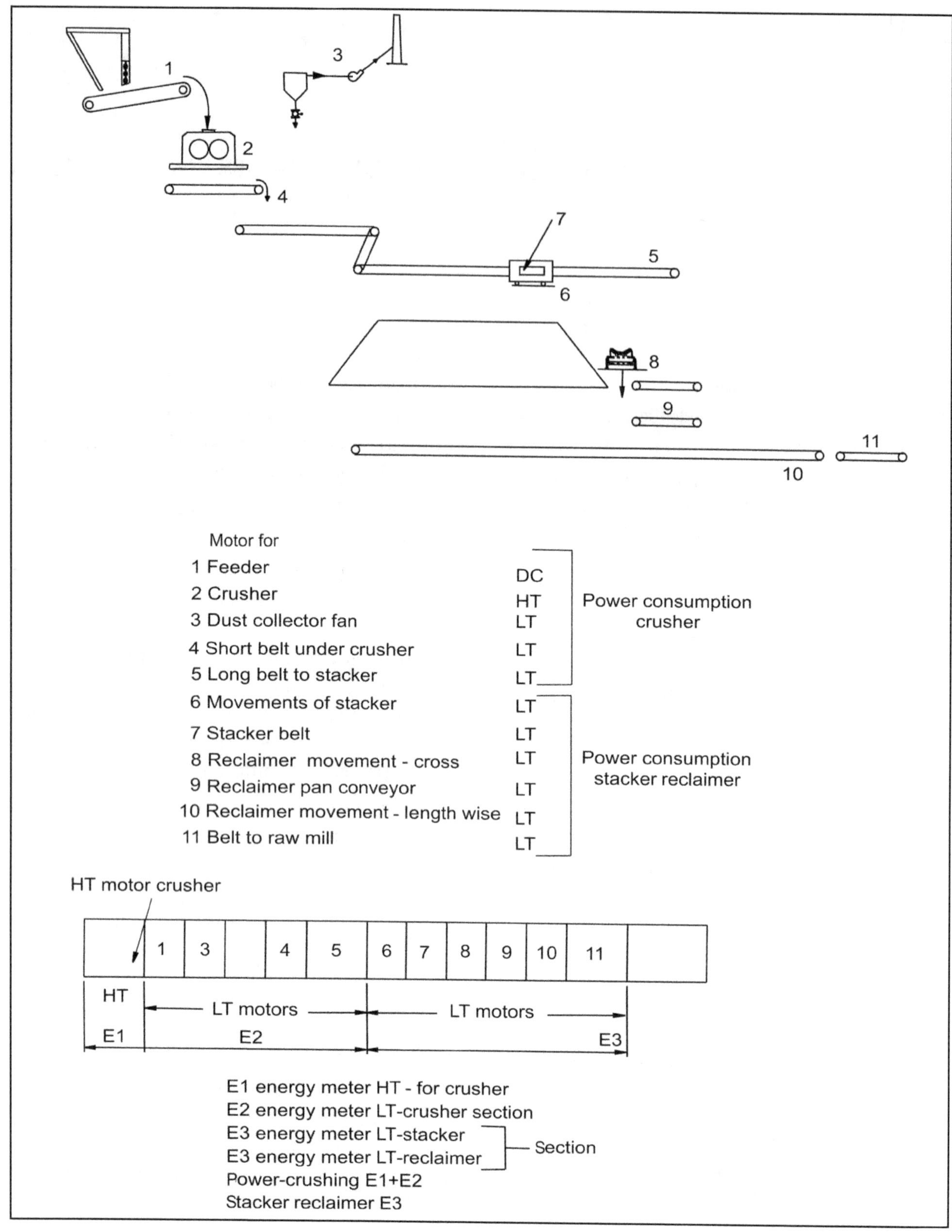

Fig. 5.3.1 Power consumption of crusher and stacker reclaimer.

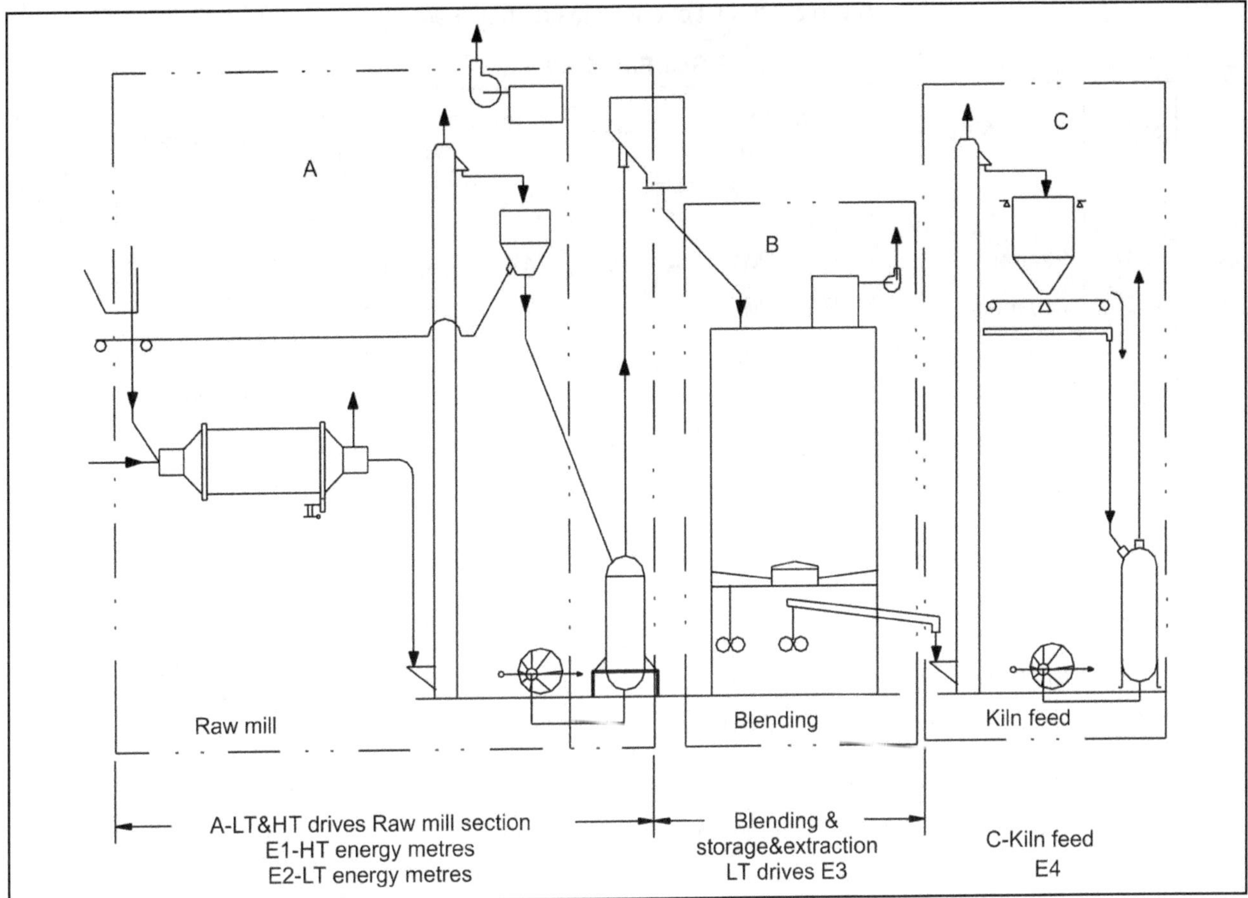

Fig. 5.3.2 Raw mill - blending - kiln feed.

3.4 Procedure for Calculating Power Consumption

In **Chapter 1** the procedure for calculating section wise power consumption and converting it into sp. power consumption for clinker and cement has already been outlined in **para 1.9.1.** It need not therefore be repeated.

In any section after defining the limits as explained above, and using Energy meters, first total units consumed in a given period say one day are found. Number of hours of operation in that section and tons of material processed in that period are ascertained from log sheets.

As an example calculation for crushing plant is shown in **Table 5.3.1.** Limits - as shown in **Fig. 5.3.1.**

In some cases only 1 MCC will be installed in the crusher section. It will be necessary to isolate drives if sectional power consumption is to be known separately for crusher and stacker reclaimer in that MCC.

Confusion generally arises with conveying equipment and storages between departments.

3.4.1 Coal Mill and Coal Firing

Confusion arises as regards where to include power for coal fired, in Coal mill section or in kiln section. Logically it should be in kiln section though it has been Shown in Fig. 5.3.3 along with coal mill.

See Fig. 5.3.3.

(i) 'A' shows limits of Coal Storage and Crushing and Coal belt to mill.

(ii) 'B' shows limits of Coal Mill section.

(iii) 'C' shows limits of Coal firing in kiln and calciner.

Physically MCCs for A & C would be located in Coal Mill building / burners platform. If only 1 MCC is installed, then drives for mill and blowers and fans and

Table 5.3.1 List of drives in crushing section.

See Fig. 5.3.1

Sr.No.	Motor	Identification number	Nos	Rated kw	MCC No.	Location of MCC	Remarks
1	Feeder	M1	1	10	1	Crusher	Club for specific power for crusher
2	Crusher	M2	1	250	1		
3	Lub. Equipment	M3	1	1	1		
4	Belt Under Crusher	M4	1	5	1		
5	Fan for Bag Filter	M5	1	10	1		
6	Compressor for Bag Filter	M6	1	2			
7	Belt to Stacker	M7	1	15	2	Stacker -reclaimer	Club for specific power for stacker reclaimer
8	Stacker	M8	1	5	2		
9	Reclaimer	M9	1	10	2		
10	Belt to Raw Mill	M10	1	15	2		

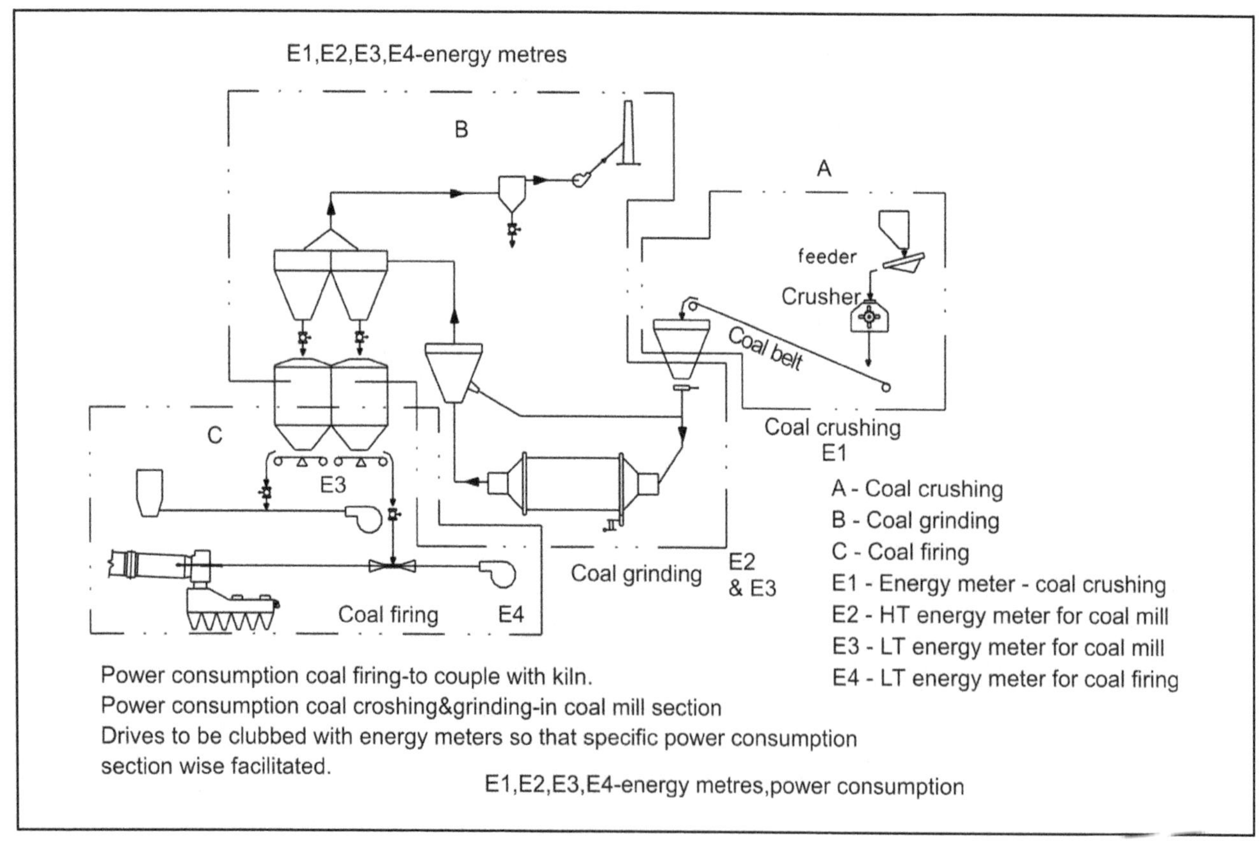

Fig. 5.3.3 Coal crushing - grinding and firing.

feeders for coal firing would also be included in the same MCC. To get a clear picture it would be better to have a separate MCC for Coal firing. MCC for coal mill and auxiliaries could be located on ground floor and MCC for coal firing on burners platform. With this arrangement it would be easy to calculate power for coal firing and add it to kiln section for computing sp. power consumption for kiln.

Table 5.3.2 , brings out the drives 2 sections shown in **Fig. 5.3.3.** It would be ideal to have 3 MCCs for 3 sections with their own energy meters. If MCCs A & C are to be combined because of locations, it should have 2 separate energy meters.

Power consumption calculated from meters in sections of coal mill and coal crushing will go to coal mill and power consumption calculated from coal firing will go to kiln section.

3.4.2 Power Consumption in Kiln Section

The total power consumption of Kiln section could include various sections as shown in **Fig 5.3.4** and in **Table 5.3.3**

3.4.3 Power Consumption on Account of Common Dust Collector for Raw Mill and Kiln

Yet another area which can cause confusion is where to include drives for GCT / ESP / ESP Fan and or conveyors under dust collector or Bag Filter, its compressor which would be common to raw mill and kiln sections.

To make kiln section complete these should be included in kiln section regardless of locations of MCCs. In **Table 5.3.3**, these are included in kiln section.

In such a case it would be clearly understood that specific power calculated for raw mill ends up at raw mill fan only.

It may be better to have different MCCs with their own energy meters for:

(i) Kiln up to and including preheater fans, GCT pumps and Screws conveyors- A.

(ii) Mill, with Feeders, Mill, Mill fans - B.

(iii) ESP, HT rectifier, shaking mechanisms, screws locks, ESP Fan - C.

(iv) ESP Dust Conveyor and compressor / blower for airlift - D.

Power consumption may then be calculated as follows :

(i) Only kiln works, specific power kiln

= A + C + D

(ii) Kiln and Mill both working

For Kiln = A + C + D

For Raw Mill = B

As explained above it is decided to add power for dust collector to kiln section even when dust collector is common to both sections.

Table 5.3.4 illustrates this point.

When dust collector is to be clubbed with kiln,

Sp. power consumption with

Only kiln running : for kiln 30 kwh/ton cement

Kiln and raw mill both running : for kiln 30 kwh/ton cement + for raw mill 18.5 kwh/ton of cement.

= 48.5 kwh/ton cement

In this fashion, power consumption and sp. power consumption for each Section can be worked out and converted into sp. power con. /ton of Cement.

Table 5.3.4

section	Sp.power consumption kwh/ton			
	Per ton material	Per ton clinker	Per ton cement	
Kiln feed	2	3.4	3.25	
Kiln, preheater, Calciner, and coal firing		13	12.5	
Grate cooler		10	9.6	
Raw mill /kiln esp	3	4.8	4.6	
Raw mill excluding esp	12	19.2	18.5	

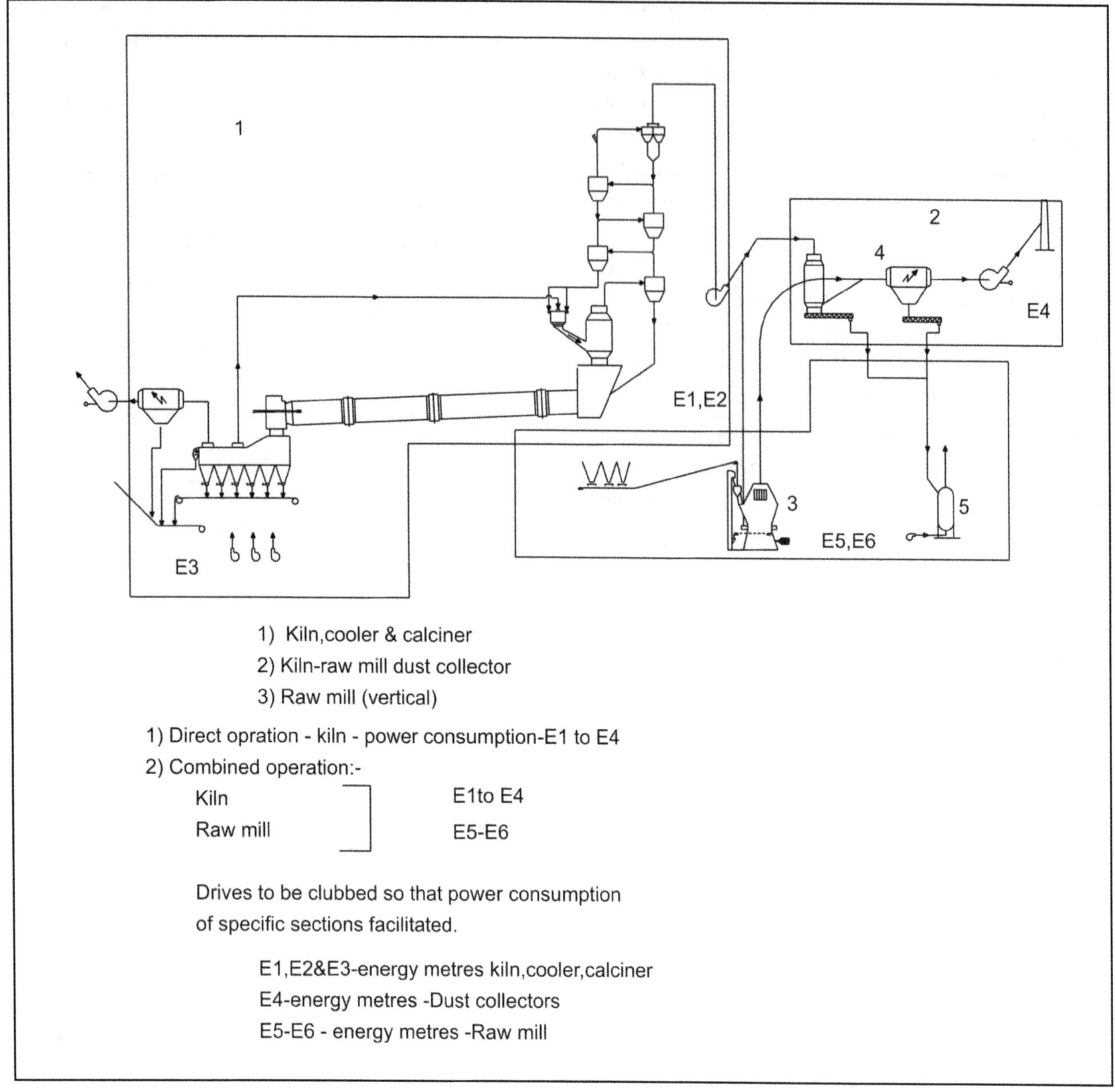

Fig. 5.3.4 Power consumption kiln, cooler, calciner, dust collector and raw mill.

3.4.4 Power consumption for 'co processing'

When 'co processing' is used, drives for feeding AF will have to be allowed for in appropriate section kiln or coal mill depending on location of facility for its preparation.

Table 5.3.2 Table showing drives in coal mill and coal crushing.

Sl. No.	Description	No	Rated KW	Location of mcc	Energy Meter
	A. Coal Crushing			mcc no 1	
	1. Feeder for uncrushed coal	1	35	5 near coal stacker	1
	2. Crusher for coal belt conveyor for coal	1	5		
	B. Coal Mill			mcc no 2	
	1. Feeder for raw coal	1		3 ground floor	2 nos 1 for coal mill 1 for other drives
	2. Mill	1	150		
	3. Mill gearbox	1	150		
	4. Rotary lock, cyclone	1	1		
	5. Circulating air/bag filter fan	1	80		
	6. Screw conceyor under filter	1	3		
	7. Rotary air lock for bag filter	1	1		
	8. Compressor for bag filter	1	2		

coal firing, through shown in Fig. 5.3.3
drives are included in kiln section Fig. 5.3.4

Table 5.3.3 Drives in kiln section.

Sl. No.		Limits	Materials treated	Sp. power/ton material	Sp. power/ton clinker	Sp. power/ton cemnet
1	Kiln Feed Section	Extraction from Blending Silo Ending with Raw Meal Airlift/elevator for Kiln pH fan, Air	Raw meal			Fig. 5.3.4
2	Preheater-Calciner-Kiln	Blasters, dampers, kiln drive, Hydraulic thruster, fans for nose and shell cooling	Clinker Clinker			
3	Grate Cooler	Cooler grates, fans & dampers, spillage drag chain, main deep bucket conveyor IF Fan and damper HT recetifier for cooler ESP, Drives for shaking mechanisms Screw/chain conveyors Under ESp	Clinker			
4	Coal Firing	Weigh feeder blowers, fans, rotary lock as shown above	Coal -			Fig. 5.3.3
5	Dust Collection equipment for Kiln and raw mill Exhaust Gases	A) ESP 1. Water pumps for GST 2. HT Rectifier sets 3. Shaking mechanisms 4. Screws conveyors under ESP GCT valves 5. ESP fan Or B) BAG FILTERS 1. Fan for dilution air 2. compressor for shaking 3. Fan for bag filter	Raw Meal Dust			Fig. 5.3.4

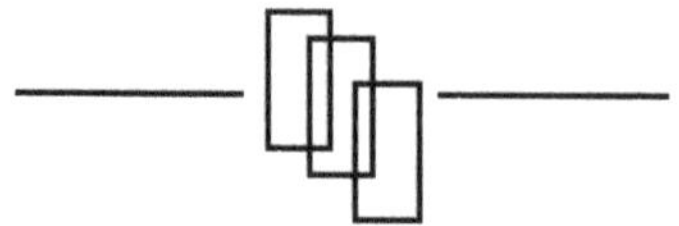

CHAPTER 4

INSTRUMENTATION AND PROCESS CONTROL

4.1 Process Control

No other section of a cement plant has made so much progress as this. With Process Computers (PCs) acquiring the capabilities of old time mainframe computers, it has become not only feasible to install more comprehensive instrumentation and process control but it has become cheap also and has come within the reach of even small plants. As the subject is vast, we will deal with the basics only.

In the operation of any process plant, which produces products useful for the community, materials are bought in and worked upon physically, chemically and in other ways to bring about the product, useful to the individuals, to society and to Industry.

The process of production can generally be divided into sections and sub-sections with materials handling and storages in between.

For smooth, continuous operation, the processes in each section must be carried out efficiently and according to the standards laid down for them. Finished product of one section is the raw material of another section. Therefore, each section must receive raw materials of given specifications and process it to another set of given specifications and pass it on the next section.

The transition from raw materials to finished product is not always a strictly linear sequence of operation. There are inputs from auxiliary or supporting processes, which are used to make finished products in that section -like preparation of fuel for instance.

All this processing and sub-processing require continuous monitoring to maintain quality and parameters of efficient operation.

To do that it is necessary to measure them and record them continuously. It is also necessary to be able to make changes in case of deviations as and when required so that the deviations are within the range that can be tolerated.

Thus, it is first necessary to list the parameters that need to be measured and recorded and those that also need to be monitored and regulated.

4.2 Parameters for Measurement and Control

Parameters may be categorized as:

1. Primary parame1ters like:
   (i) Rates of feed and fuel & product.
   (ii) Gas Flows.
   (iii) Temperatures and drafts.
   (iv) Chemical composition and related ratios to maintain quality.
2. Secondary parameters like:
   (i) Actual speeds of machines.
   (ii) In leakage in systems.
   (iii) Introduction of other gases / materials, water in the process.
3. Tertiary parameters like:
   Dust losses.
4. Efficiency parameters like:
   (i) sp. gas flows,
   (ii) sp. power consumption,
   (iii) sp. fuel consumption,
   (iv) conversion ratios.
5. Quality parameters.

4.3 Instrumentation and Monitoring

Instrumentation is designed to measure the various parameters using appropriate sensing devices and converting them into parameters being measured.

Recorders are used to record continuously the quantity measured so that they can present a trend or a pattern and are available for reference when required.

Instrumentation can be local or central or both.

4.3.1 Monitoring of Parameters

Once the parameters are measured like for instance:

(i) Temperatures by thermometers.

(ii) Pressure drop by pressure / draught gauges.

(iii) O_2, CO, CO_2 NOX by gas analysers.

(iv) Rates of feed by speeds and weight measurements.

it is necessary to monitor them continuously so that they remain within narrow limits of preset values.

For this it is necessary first to set the standards of operation, like for example:

(a) Raw Meal / Clinker conversion ratios.

(b) Gas flows in terms of nm3 /Kg of product.

(c) Temperatures and draught / pressure profiles.

(d) O_2, CO, CO_2 contents.

(e) Sp. fuel consumption.

(f) Sp. power consumption.

(g) Chemical composition and moduli which will ensure quality of clinker.

The actual values measured are then compared with the set or ideal values. Deviations thus brought out are corrected by changing settings of rates of gas flow, material flow, speeds, etc., in relation to one another maintaining desired quality of clinker or cement at all times.

These changes used to be done manually by expert / trained operators called 'millers' and 'burners' by observing the values on the control panel. They would 'judge' conditions of burning zone in the kiln and flow of feed inside the kiln, fuel rates, by observing burning zone at frequent intervals and by colour and shape of flame and then do adjustments to say speed of kiln, speed of coal feed, raw meal feed, damper of preheater fan, etc.

4.4 Automation

Closed Control Loops

When the operational parameters are changed automatically by linking 'cause and effect' relationships, then it is called 'automation'. 'Closed control loops' are an example of automation.

Even before microprocessors came, kiln sections used to be operated with 'closed control loops' like :

1. single control loops:
   (i) fuel rate adjusted as temperature of burning zone increases / decreases,
   (ii) kiln speed adjusted as feed rate increases / decreases,
   (iii) open close damper of cooler I.D. Fan to maintain draft in kiln hood,
   (iv) increase/decrease speed of cooler grates to maintain bed thick ness,
   (v) adjust throttle in kiln riser duct to maintain desired flow rates in kiln and calciner in case of single stream preheater,
   (vi) flow of cooling air in 1st two compartments of a grate cooler is maintained by measuring air flow and by changing speed of fan.
2. interlock loops
   (i) switch off esp if gas analyzer shows high CO for longer than set period,
   (ii) turn on inert gas if temperature at coal mill outlet exceeds set limits.

4.5 Instrumentation and Control Schemes (I & PC)

Typical Instrumentation and Control Schemes have been shown in **Figs. 5.4.1 to 5.4.8**.

Fig. 5.4.9 shows scanner for monitoring shell temperature of kiln. It has proved to be a very useful tool to monitor condition of refractory in the kiln particularly in the burning zone.

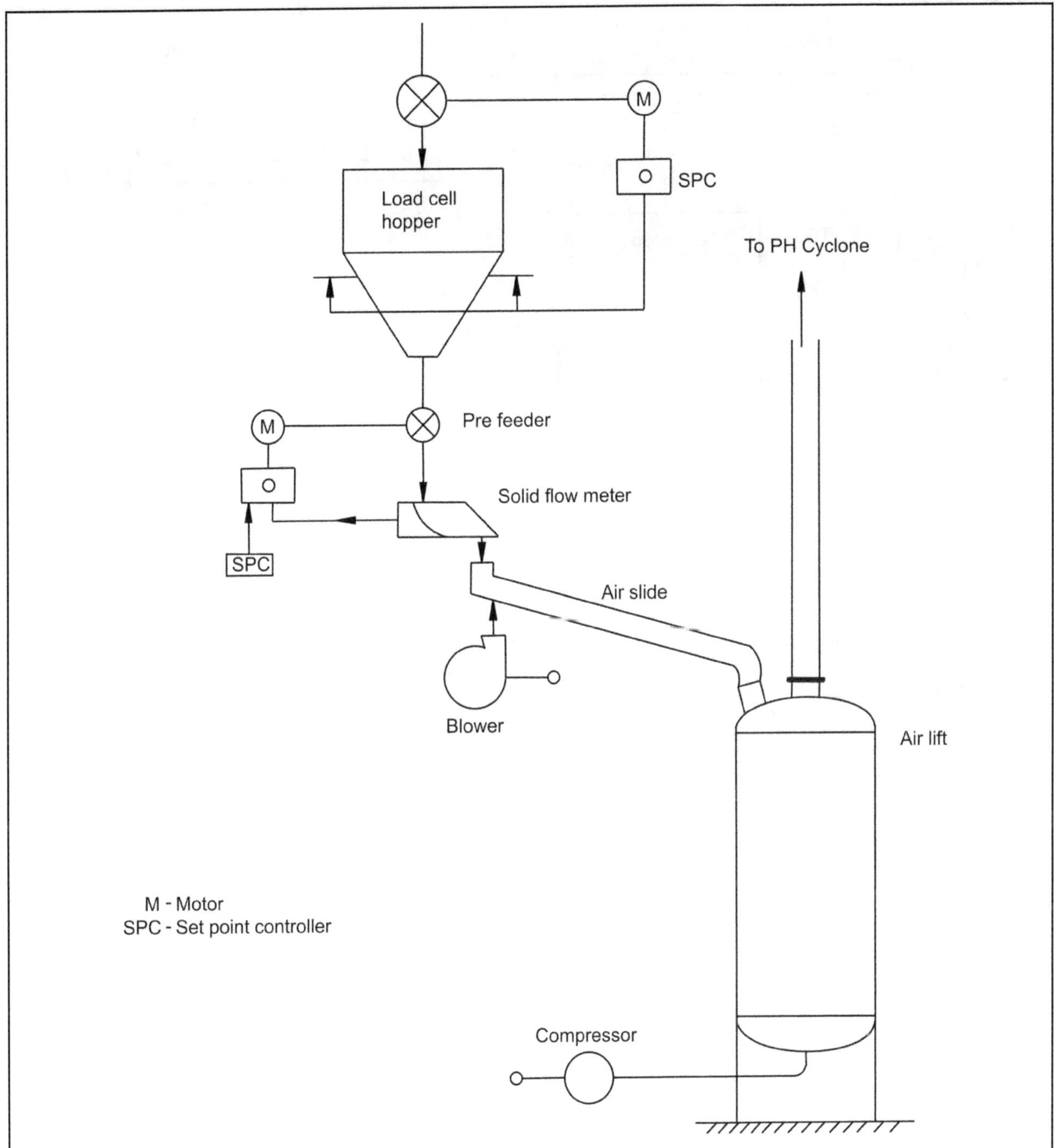

Fig. 5.4.1 Kiln feed control through solid flow meter.

4.5.1 Cascade Control

Now with help of micro processors and computers it is also possible to introduce 'cascade control'.

When cause and effect are complex - for example, burning zone temperature could be affected not only by fuel rate but also by other parameters like :

(i) oxygen in exit gas and its temperature,

(ii) speed of preheater fan,

(iii) speed of kiln.

In cascade control, analysis of various possibilities and selection of appropriate cause is done by the processor. It then selects course of action.

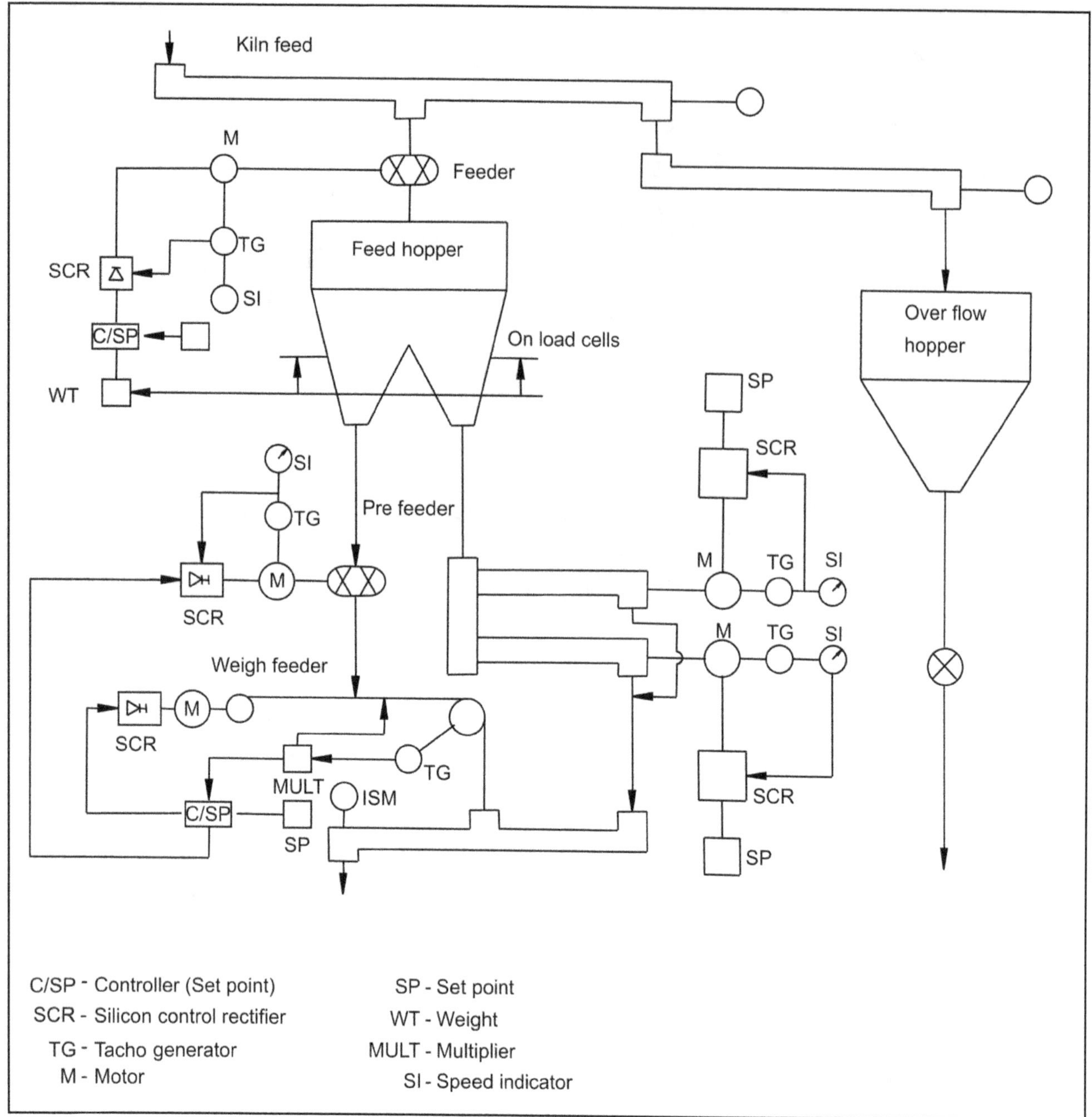

Fig. 5.4.2 Kiln control with weigh feeder system.

4.6 Process Control Scheme

Designers and Suppliers of Machinery would submit their recommendations for I & PC. These would be in the form of P& I Diagrams. Specifications of Instruments and their ranges and other details will also be furnished along with. These should be taken into account. Consultants would help in putting together P&I Diagrams of various sections together to prepare an integrated comprehensive scheme. This would then be presented to the Designers of the I & PC Scheme.

In 'automation' it is implied that the deviation is measured and a correcting signal matching with the deviation is sent to correct it. It also implies that where required, measurements are 'memorized' to store the result after each correction.

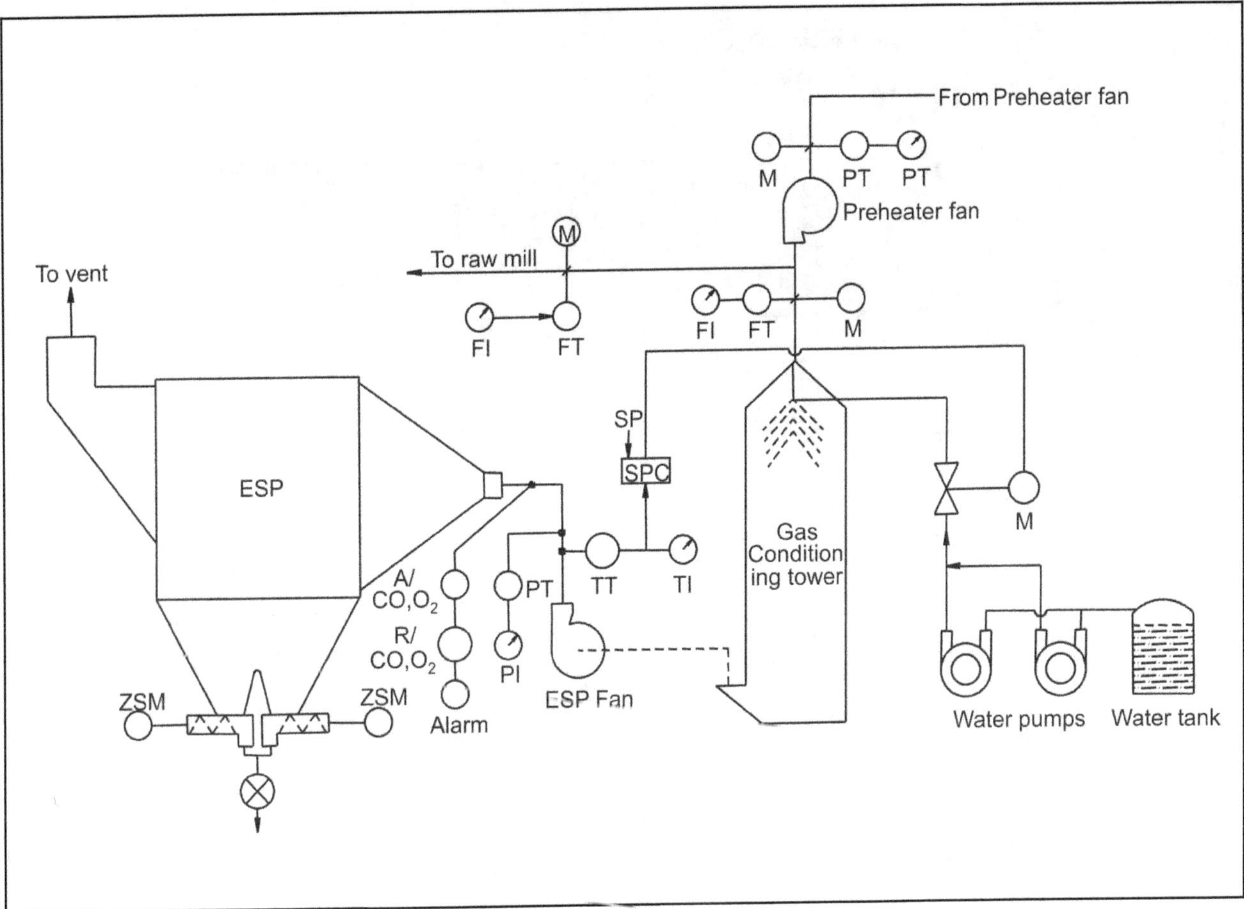

Fig. 5.4.3 Control scheme for ESP and GCT.

4.7 Computers for Process Control

Computers are best suited to do the tasks of :

1. Continuous monitoring.
2. Comparing signals.
3. Reiterations.
4. Calculating results.
5. Storing in memory.

Thus today instrumentation and process control without computers is unthinkable.

4.8 Programmable Logic Controllers

Another important development in operation is starting and stopping groups of drives in proper sequence and at correct speeds to avoid starving or jamming, burnouts etc. in short to protect plant and machinery and to ensure its availability all the time.

This is achieved by 'Programmable Logic Controllers' (PLC) that ensure sequences of starting and stopping and which can be easily changed as required. They also have built in 'safety and protection' programmes.

Motors can be started singly or in groups from the Central Control Station which supervises operations.

PLCs also note interruptions in flow, stoppages of drives, and send audio and visual alarms for operators to take corrective action when deviations exceed danger limits.

The erstwhile 'machine operators' who used to start various drives locally are no longer required, thereby reducing manpower considerably.

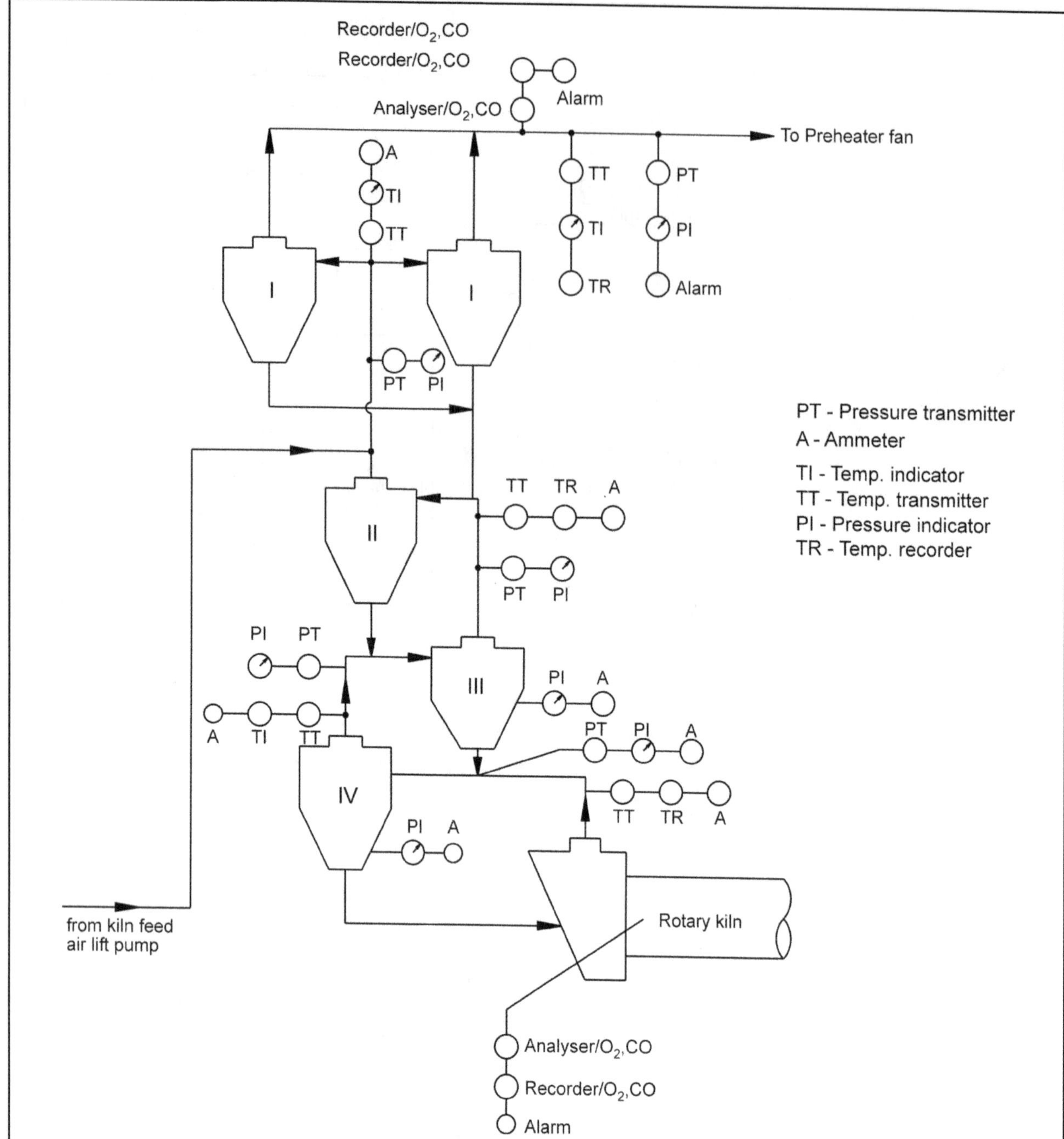

Fig. 5.4.4 Typical instrumentation scheme for preheater and kiln inlet (4 stage preheater shown).

4.9 Instrumentation and Process Control : (I & PC)

Instrumentation and Process Control thus :

1. Measures,
2. monitors,
3. starts, stops in sequence,
4. protects and sounds alarms-even stops machines if needs be,
5. eliminates– operators- reduces, skills and experience required of operators.

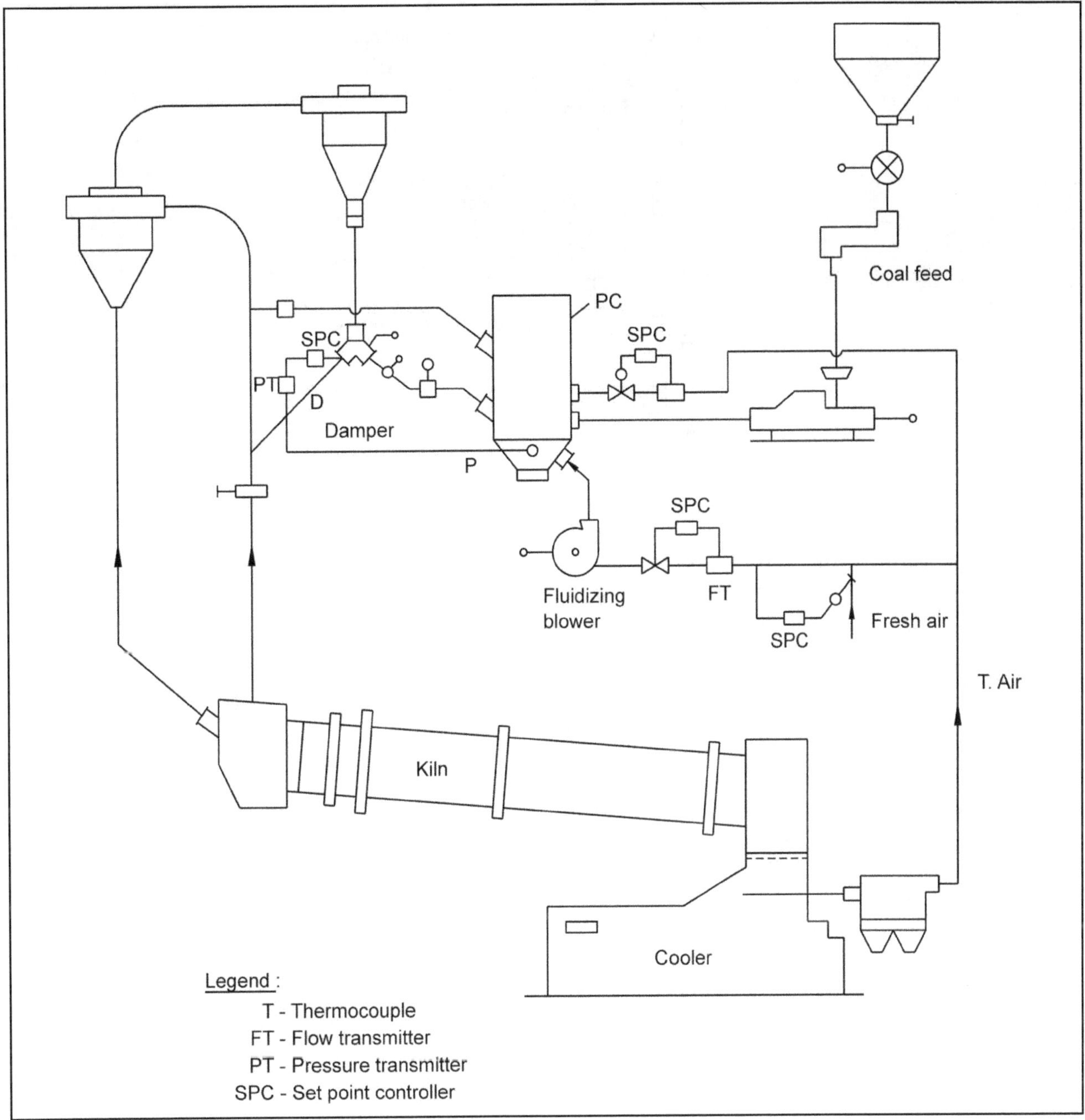

Fig. 5.4.5 Typical instrumentation and control scheme for a precalcinator.

While 'minimum' requirements in process control can be defined, there is no 'maximum' as such.

There is no end to the 'sophistication' or as some may care to put it to 'complications' in the process control.

I & PC as designed should be a tool which can be actually used.

It should not be merely so much data, computer pages and diagrams and so on.

4.10 Trend Curves

An important tool for analysis is the historical movement of a given parameter or 'trend curves'.

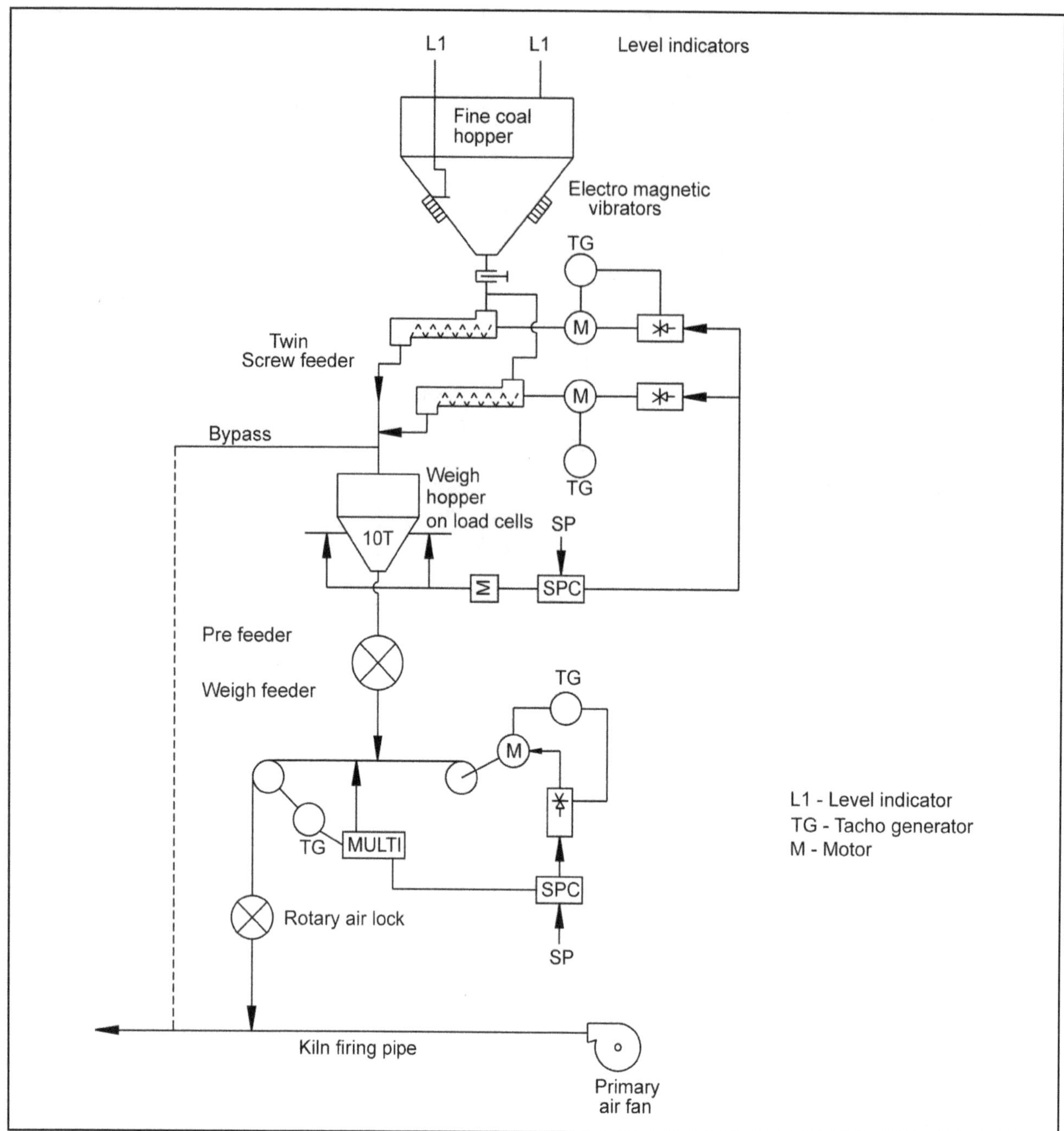

Fig. 5.4.6 Coal feed control system.

These trend curves indicate changes in a parameter on a time scale and are thus important tools in analysis and in solving problems.

Important efficiency parameters, when incorporated in the system of I & P C also help in monitoring operation to keep it at optimum level. These would include among other things:

1. Measurements of percentages of O_2 - CO and sometimes CO_2 present in exhaust gases of kiln and mill systems.
2. Direct specific fuel consumption in kcal/kg of clinker.
3. Actual outputs – conversion ratios of raw meal / clinker.

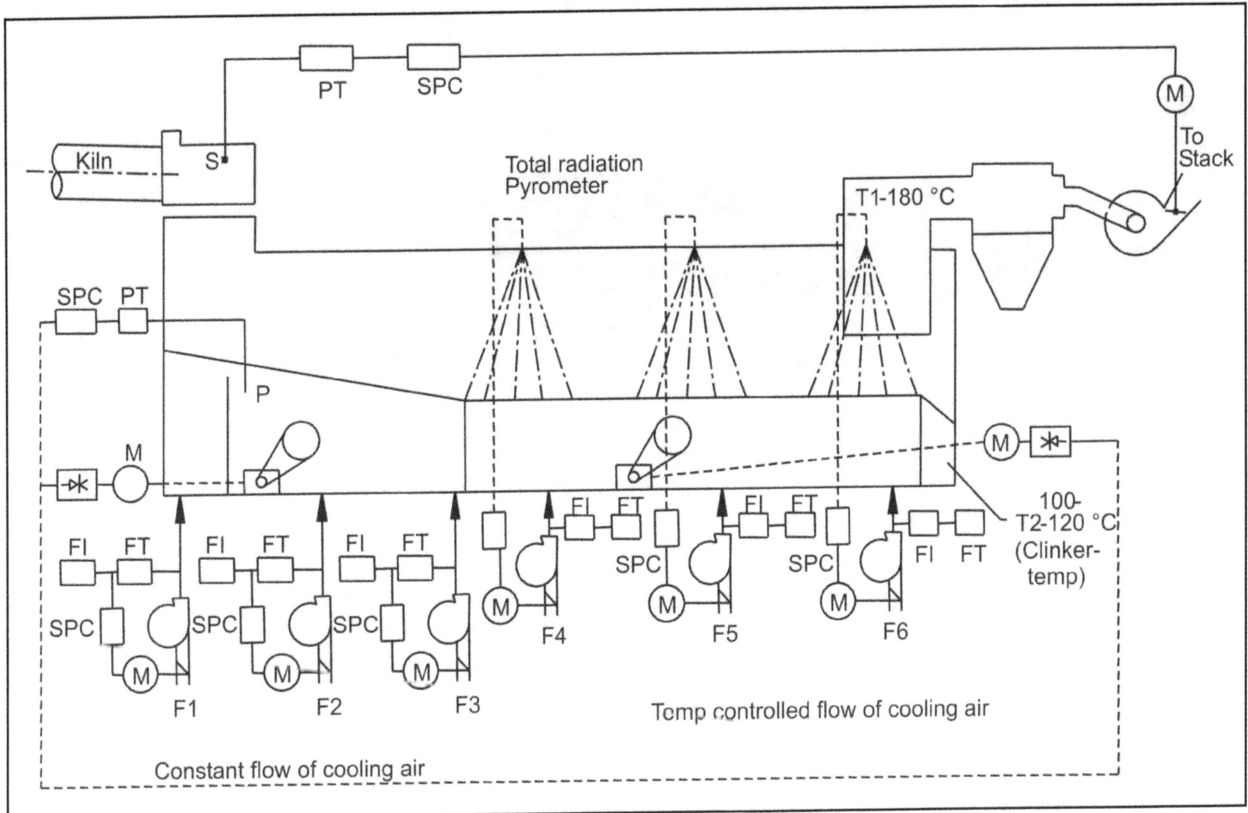

Fig. 5.4.7 Grate cooler control.

4. Specific Gas volume.
5. Quality deviation in raw mix feed and clinker.
6. Dust loss.
7. Sp. power consumption.
8. Heat balance- and also material and gas balances. Monitoring temperature of kiln shell and assess condition of Refractories in kiln- particularly in the burning zone.

It should be possible to select the parameters, type them together in the log sheet against design and actual values.

4.11 Operators' Responsibilities

Job of today's 'burners' is not so much as to look into the kiln but to understand the data presented on screen, the trend curves and the ways in which deviations are to be brought under control speedily.

'Alarm Bells' should ring if O_2 levels go up by more than 2% at kiln inlet. Such a deviation, other things remaining the same should be accompanied by changes in gas volume often due to inleakages.

The burner should therefore start looking for accompanying changes in other parameters, which could have caused this deviation.

4.12 Direct Reading of Secondary Parameters

It should be possible to read directly :

1. Speeds of important machines like kiln, cooler grates, fans and feeders.
2. Rates of feed like raw meal feed to kiln, coal feed to kiln and calciner; rates of feed of limestone and correcting materials; coal; clinker, gypsum, slag or fly ash to their respective grinding mills in tons per hour.

   Unfortunately, more often only speeds of weighing machines are noted like 600 rpm, 800 rpm and so on which need to be converted into feed rates in tons per hour.

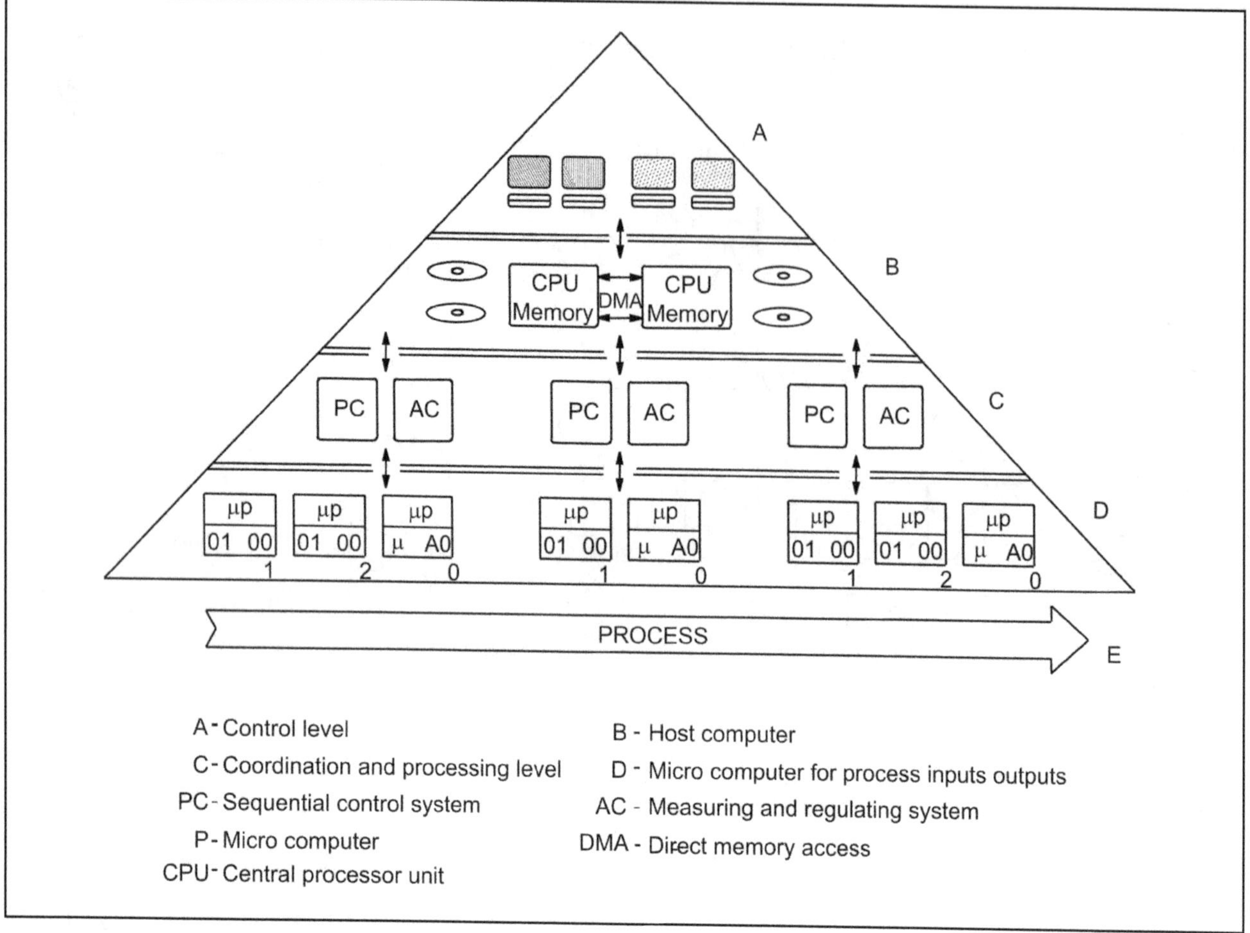

Fig. 5.4.8 Control technique with hierarchical system configuration and distributed intelligence.

3. Gas flows in nm^3, particularly in kilns and mills sections.
4. Power consumption section-wise and total in kwh/t of material and cement.
5. Heat balances of pyroprocessing and mills sections.

4.13 Calibration

It should be possible to calibrate the measuring devices without requiring to stop the pertinent sections of the plant.

4.14 Fuzzy Logic

'Fuzzy Logic' is now used to operate the kiln at optimum level, it imparts analytical and deductive dimensions to measurements. It should be applicable to grinding sections also.

Many a time, operation of Ball mills is a 'black box' because their response time is too long.

4.15 Quality Control

Quality and Process Controls are interlinked. To obtain good quality product it is necessary to measure and monitor :

1. Chemical analysis of Raw material and additives fed to mills, moisture and fineness of product. It is useful to derive ratios like lime standard or hydraulic modulus; silica modulus; alumina modulus etc.
2. Uniformity of raw mix feed to kiln in terms of chemical composition and also fineness.
3. Useful calorific value and ash content and moisture in coal as fired.

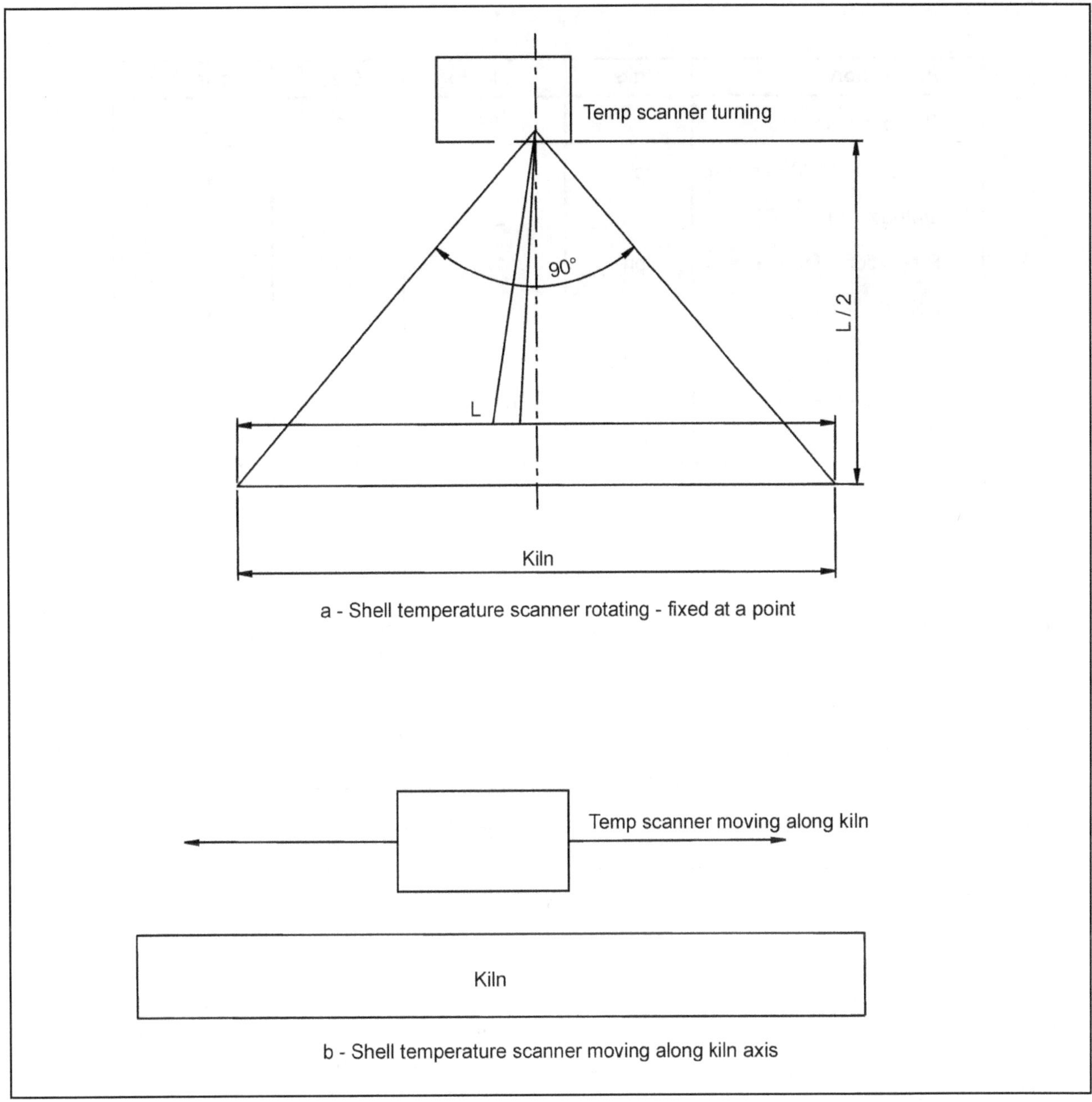

Fig. 5.4.9 Shell temperature scanners.

4. Chemical analysis of clinker- its various modulii like in case of raw mix; free lime; litre weight; MgO; sulphur; and also constituents like C_3S, C_2S etc.

Quality Control Section or Production Departments fix up the parameters to be maintained in operation like hydraulic modulus, silica and iron ratios burnability and so on and arrive at proportioning of raw materials and additives.

In Raw Mill Section, weigh feeders are run at speeds designed to obtain and maintain these proportions at levels that would yield desired output rates.

See Table 5.4.1.

4.15.1 Quality of Raw Mix in Batch Blending

Earlier in batch blending systems, it was sufficient to take hourly samples and test them for chemical composition and make hourly corrections. The average

Table 5.4.1

Proportion	Units	Limestone	Clay	Iron Ore
Proportion desired	%	85	13	2
Capacity at 100 % speed	tph	250	50	10
Output 1	tph	200		
Proportional feed rates	tph	170	26	4
Speed - % full speed	%	68	52	40
Output 2	tph	225		
Propotional feed rates	tph	191.25	29.25	4.5
Speeds	%	76.8	58.5	45

Table 5.4.2

Hourly samples	Limstone tons	Clay tons	Iron Ore tons	Quantity Added tons	Proportions	Carbonate %
1	170	26	4	200	85 : 13 : :2	76
2	174	22	4	200	87 : 11 : 2	78
3			and so on			
4						
5						
6						

analysis of raw mix after blending the 'batch' was arrived at by testing the blended raw meal.

These corrections were made to remain close to the desired composition.

Considering carbonates only, if the desired value was 77 % and one sample showed a value of 76 %, then to restore the value back to 77, feeders would be adjusted to yield a carbonate of 78 % till it was time for another sample and so on till the 'batch' was completed.
See Table 5.4.2.

Average proportions of limestone , clay and iron ore in % in the 'batch' are calculated from above and chemical composition of 'raw meal' calculated from chemical compositions of these materials.

Chemical composition of blended raw meal is also found out by testing samples thereof and also samples of kiln feed. From the composition various modulii are derived to check that they too are in line.

The ideal is to maintain deviation of blended raw meal within + 0.2 % in terms of standard deviation.

4.15.2 Continuous Blending

However when blending process charged to 'continuous blending', hourly corrections were no longer sufficient.

In continuous blending the quality of raw meal entering the silo itself should be as close to the set value as possible always.

This brought about a sea change (revolution) in sampling and testing and monitoring methods.

It became necessary to draw samples, prepare them and analyse them quickly sometimes even continuously and to correct continuously to keep within the narrow range.

4.16 X-ray Analyser

The 'X-ray analyser' and automatic sampling and preparation of samples have become part and parcel of present quality control system and also I & PC.

The same 'X-ray analyser' can be used to test clinker and cement samples also.

Computers analyze samples, arrive at constituents used for monitoring various modulii, add results to previous results, calculate values of total quantity and indicate revised proportions as required. Signals are sent to weigh feeders of raw mill, which now feed raw materials and additives in new proportions till a new signal arrives.

Thus like I & PC, Quality Control System (QCS) has also become automated. There is no going back.

For smaller plants instead of elaborate X-ray analyzer and fully automated (untouched by hand) methods of collecting samples, conveying, and preparing samples, semi automatic methods are used; 'on line' analyzers which can analyze a smaller number of elements are used. But when continuous blending systems are used, some kind of continuous analysis and monitoring is a must.

4.17 Log Sheets or 'Hard Copies'

Design of a 'log sheet' is very important. It is a document, which should provide record of all important and relevant parameters, and help to diagnose operation and to bring it to optimum level. It is of course not possible to record in log sheets every entity that is measured. Therefore, selection is very important and considerable thought must be given to designing it.

'Logs' should also record parameters 'directly' i.e., feed rates, speeds, gas flows, etc., rates rather than in percentage of full speeds, flows, etc.

Kiln speed 0-4 rpm should not be recorded as 0-100%. Feed rates should not be recorded in speeds of weigh feeders but directly in tons per hour.

With computers, it is easy to make changes and hence changes should be made to include new items found relevant and omit others in 'hard copies'.

A log sheet with a 'key flowchart' would be the ideal thing as abbreviations, code numbers, machine numbers, are not known to all - particularly to outsiders like Consultants, etc.

It is good to know not only Feed rate in tph but also output in one shift / in one day of clinker and coal consumption and specific fuel consumption.

'Weekly' log sheets should be printed with 'trend curves' programmed.

With computers, operators do not have to make entries in the log sheets and hence if the instruments were reading correctly entries would be correct. It also implies that instruments must be regularly calibrated to give correct readings. Control Room staff should however be trained to spot the deviations and to go to the root causes thereof.

4.18 Inter Related Operational Parameters

Several parameters in operation are inter related and hence should match i.e., value of specific fuel consumption should match with gas flow at exit of preheater. There is a direct relation between these two.

Quantity of coal consumed and specific fuel consumption should match.

Raw meal consumed should match conversion ratios.

Log sheet should be designed to bring together the related entities.

4.19 Simulators

'Simulation programmes' or 'Simulators' are now available to train control room staff to operate the plant with computers. They are good tools to acquaint the staff with the process and operation. It should be possible to use Simulation programmes to find out 'cause and effect' relations between different parameters at least broadly.

For example,

Increase rate of feed to kiln – keeping other parameters same and see differences in temperature profile across preheater.

Watch effect of change of speed of kiln.

Change of speed of cooler grates.

Such studies also help in arriving at the exact magnitude of correction needed in relation to magnitude of deviation.

They can thus give 'hands on' training to control room staff who become better equipped to cope with abnormal and critical situations.

4.20 Process Control Computers

Process Control Computer will display on its Monitor, Flow sheets or pages of each section of the plant like:

<table>
<tr><td>Crushing</td><td rowspan="11">Each section will have one or more pages. It is good not to clutter up the page with too much data.
It may be a good idea to have different pages for –
1. Temperature, Draughts, Gas Flows and Feed Rates,
2. Drive- amps, KWs, etc.</td></tr>
<tr><td>Stracker
Reclaimer</td></tr>
<tr><td>Raw Mill</td></tr>
<tr><td>Blending</td></tr>
<tr><td>Kiln Feed</td></tr>
<tr><td>Kiln Preheater
Calciner</td></tr>
<tr><td>Cooler</td></tr>
<tr><td>Clinker Transport</td></tr>
<tr><td>Cement Mill</td></tr>
<tr><td>Cement Storage</td></tr>
<tr><td>Packing</td></tr>
</table>

Now, it is also possible to 'enlarge' a flow chart or a section thereof.

4.20.1 Windows

Concept of 'Windows' has now been widely adopted in PCs used for process control. Some programmes like Windows NT and Windows XP are specially designed for processes.

Like in Windows, it is easy to go from Folder, Sub-folder to file, etc. Thus any operation can be studied progressively in depth.

4.21 Today's Control Desk

Earlier, the Central Control Panel consisted of sections beginning with Raw Mill and grinding with cement mills. Now, all sections of the plant are brought on the panel. A 'mimic' was used as a standby. This is no longer required.

Design of a 'control desk' was a challenge. Desk used to contain hundreds of buttons for remote starting, measuring instruments, recorders and so on. It used to take lot of space. Plant flow chart was displayed with the help of a live mimic. All these are things of the past. Today's Control Room contains just 2 or 3 monitors, CPUs, keyboards, mice and printers.

So much sophistication and automation is now not only possible but is within reach of all.

4.22 Using Computerised Automation

All the fears regarding using sophisticated delicate and sensitive instrumentation and sensors and microprocessors have disappeared.

Operators all over the country including those in small plants have taken to operate cement plants with the help of microprocessors like 'duck takes to water'.

Emphasis on containing pollution within the plant has also created clean and dust free environment that helps to obtain good and reliable service from instrumentation on continuous basis.

Computers are now designed to work in hot humid conditions generally found in India. Even air conditioning of Control Room and back up panels is no longer a must though it is certainly desirable in the long run.

4.23 Central Control Station

Because of remote control of all drives and because all parameters are now on panel, location of the Control Room is no longer 'chained' to the burners platform; it need not be near kiln even. It can be located any where in the plant.

The Control Room building would now house 'X-Ray analyser' mentioned above. It would have facilities for maintenance of instrumentation.

Staff of Engineering and Quality Control and Production departments - sometimes even the Laboratory where the tests would be carried out is also located in the same building. Back up panels of weigh feeders and other variable speed drives would also be located in adjoining rooms.

4.24 Calibration

Plant must have facilities to calibrate 'on line' various feeders and meters in the plant. It makes sure that the readings are correct and could be used to compute outputs, fuel consumption etc.

If the Laboratory regularly establishes various conversion ratios mentioned earlier, then when weigh feeders are regularly calibrated, output and fuel efficiency can be established with certainty.

Small bins on load cells facilitate calibration

Weigh feeders under hoppers holding large quantities like , clinker, coal, etc., should be calibrated by filling the hoppers and emptying them.

Apart from field sensors installed, for purpose of verification and for taking measurements at other points for study of operations, the plant should have thermometers and draft / pressure gauges (inclined manometer) and Orsat's apparatus with them for physically checking readings.

4.25 Article by Mr. P.S. Sharma

An article written by Late Mr. P.S. Sharma ex G.M. A.C.C. on the subject of Instrumentation and Process Control is attached as **Annexe 5.4.1**. Basic principles remain the same though there has been lot of sophistication since the article was written. I & PC Schemes shown in **Figs. 5.4.1 to 5.4.8** were also suggested by him. They may appear too simple for the same reason.

4.26 Wider Scope of Computer Control

Computers now do not concern themselves only with Process and Quality Control. Comprehensive programmes have been developed for the total management of an Industrial Unit like a cement plant.

As mentioned earlier, dispatches of cement in bulk have already been automated. Where paper bags are used even packing and dipstching of cement in bags has also been greatly automated. **See plate 5.4.1**

Specially developed programmes like SAP and ERP deal with Marketing and Finances; in fact with every aspect of the Plant's operation.

4.26.1 Environmetal Control

This refers to particulate emissions from vent stacks in the plant and also to noise levels.

It is statutorily binding to comply with standards for particulate emissions as laid down by the concerned pollution control authority in each country. Therefore it is now necessary to monitor

Continuously particualate emissions.

4.26.2 Emission of Green house gases.

Cement Industry is conscious of the fact that in operation it lets out into atmosphere huge quantities of green house gases particularly CO_2 . It is therefore taking steps like making blended cements, using alternate fuels, co generation of power to reduce their quantum. For that it has become necessary to monitor emission of green house gases alsocontinuously.

4.26.3 Key Performance Indicators (KPIs)

World Business Council for Sustainable Development - WBCSD- has Suggested Key Performance Indicators KPIs for cement Industry. They have become benchmarks for assessment of performance of existing plants and guidelines for design of new plants. Cement companies have begum to develop their own KPIs and include them in 'mobile/internet' assisted control systems. this is the new mobile dashboard control system.

4.26.4 Advent of Internet and Mobile Phones

With advances in personal computers (PCs) and advent of Internet, it has become possible to make available all data to all concerned executives of the company any where at any time. Laptops become monitors in the hands of executives responsible for process and quality control.

Mobile phones further advance mobility of processed data. It is now available at the touch of a finger anywhere any time and instructions can be passed oninstantaneously.

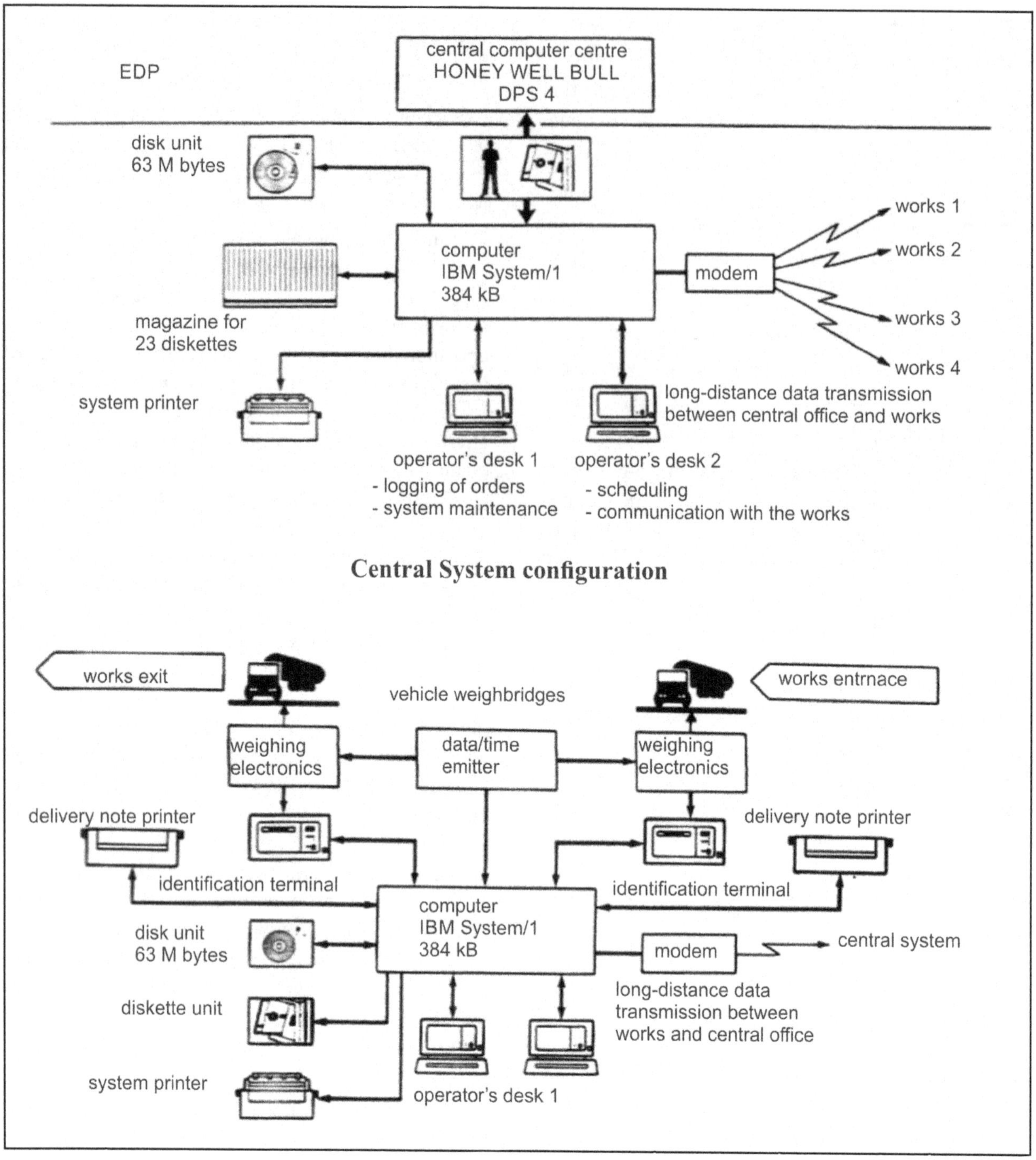

EDP

central computer centre
HONEY WELL BULL
DPS 4

disk unit
63 M bytes

computer
IBM System/1
384 kB

modem

works 1
works 2
works 3
works 4

magazine for
23 diskettes

system printer

long-distance data transmission
between central office and works

operator's desk 1
- logging of orders
- system maintenance

operator's desk 2
- scheduling
- communication with the works

Central System configuration

works exit

vehicle weighbridges

works entrnace

weighing
electronics

data/time
emitter

weighing
electronics

delivery note printer

delivery note printer

identification terminal

identification terminal

computer
IBM System/1
384 kB

disk unit
63 M bytes

modem

central system

long-distance data
transmission between
works and central office

diskette unit

system printer

operator's desk 1

Plate 5.4.1 Centralised dispatch automation in Cement Works.

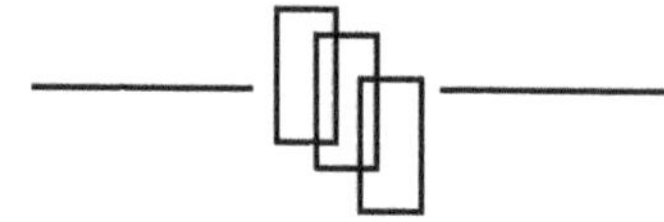

Annexe 5.4.1

AUTOMATION IN CEMENT INDUSTRY

Article by
Late Mr. P.S. Sharma, ex G.M. A.C.C.

In the recent past every segment of society has felt the impact of the oil crisis, drastically altering economic parameters. Energy bills have increased, raw material costs have gone up and there has been a gradual depletion of quality limestone, specially around old cement plants. The quality of coal available has varying and high ash content and there are also constraints on supply which rule out the possibility of blending to obtain at least uniform quality of coal. Statutory regulations on environmental pollution have become more stringent. The cement industry reckons automation with instrumentation as a key factor to meet the above challenges and dictate the tempo of technological progress.

For the manufacture of cement, the basic raw material is lime stone and coal is used as fuel; various process available have been tabulated in **Fig. 1.**

Irrespective of the process employed, the plant can be divided into some of the following interdependent, (but largely independent units) which can be operated independent of each other for a considerable time depending on the storage facilities built into the system.

1. Raw Meal
2. Quarrying, crushing, grinding, blending and homogenizing.
3. Coal Pulverisation.
4. Burning and cooling.
5. Cement grinding.
6. Packing and dispatch.

1. Raw Meal

The production of cement requires the chemical composition of the raw meal to be kept constantly, uniform. This is achieved by blending and bedding. To further even out fluctuations in the quality of raw meal, limestone from different batches are mixed in or on-line weighing and proportioning system. X-ray analysers are also used for quick analysis and corrective action.

The raw meal quality is determined by a co-efficient known as Lime Saturation Factor (LSF), Silica Modulus (SM) and Alumina Modulus (AM). These are derived from the concentration of Calcium Oxide (CaO), Silicon Dioxide (SiO_2), Alumina (Al_2O3) and ferric Oxide (Fe_2 O3) denoted by C.S.A & F. respectively.

$$LSF \quad \frac{100\ C}{2.8\ S + 1.1\ A + 0.78\ F}$$

$$SM = S/(A + F)$$

$$AM = A/F$$

For automatic control, software packages are available for on-line sampling with an X-ray fluorescence analyzer with processor, to give actual values of different coefficients. These are compared with set values and the error is fed to the weighing and proportioning device to obtain desired values. For further refinement in consistency, the raw meal is homogenized.

For open circuit mill an electronic ear (microphone) to control the mill loading is provided, and for closed circuit mill the input from the bucket elevator drive is also additionally monitored for effective control.

2. Coal Grinding

Simultaneous drying and grinding of coal is more popular due to ease of operation, but installation of separate drying and grinding units are more logical from the point of fuel efficiency.

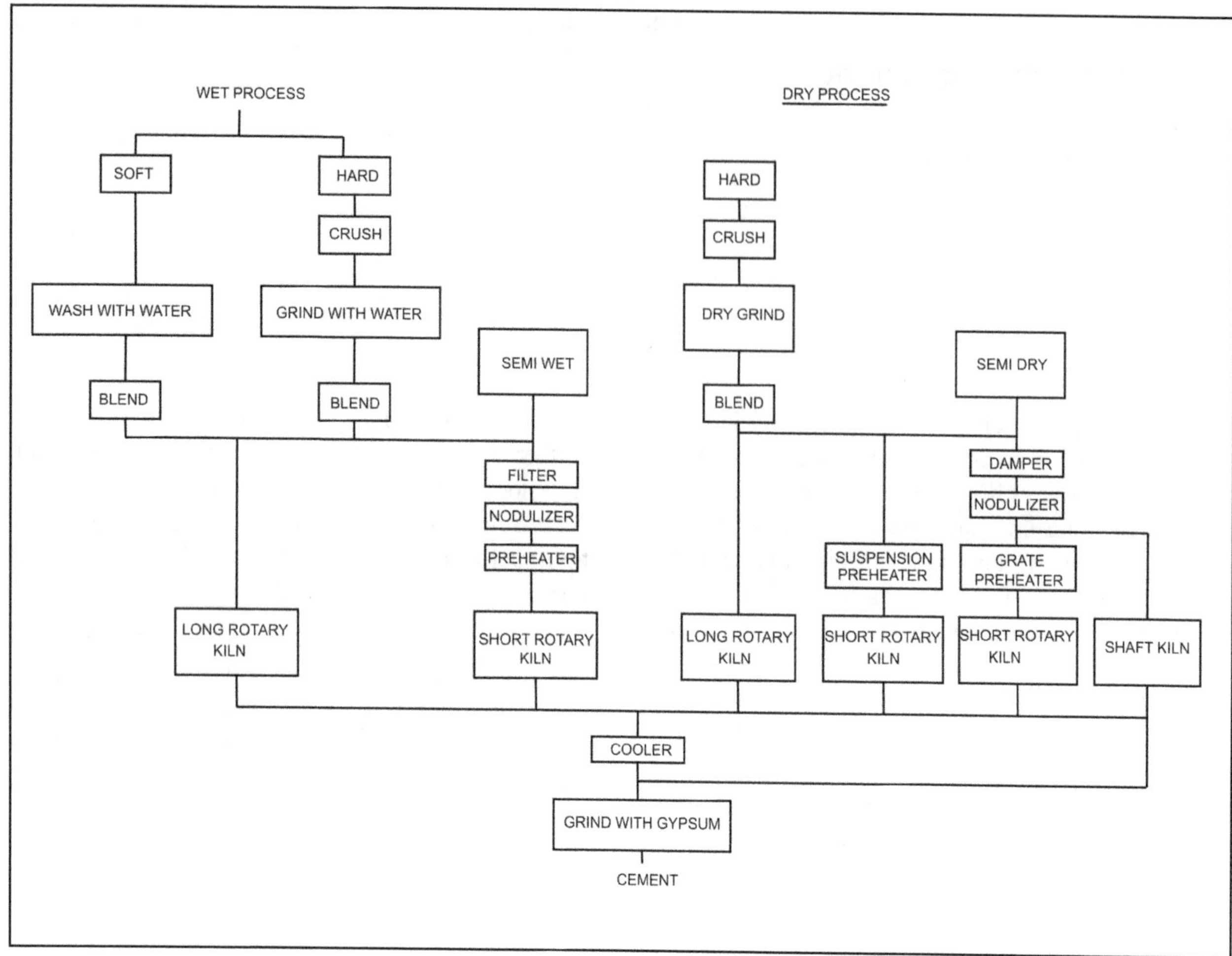

Fig. 1 Processes of manufacture of cement.

The coal mill motors are provided with winding temperature and bearing (temperature) sensors; the bearings of gear boxes and bearing oil are also provided with temperature sensors. All theses temperatures are continuously scanned and an alarm signal vitiated in the event of temperature rise beyond the preset value.

In the Coal Mill, temperature sensors are provided at inlet and outlet. In the fine coal hopper and feeders temperature sensors are provided to detect glow nests so that the possibility of explosion is minimized. It is desirable to monitor CO in the fine coal hopper to detect any incipient fire (in the hopper).

Pressure sensors are provided at the mill inlet and outlet, circulation fan inlet, and differential pressure is determined between against any jamming in the mill.

On line weighing devices are incorporated to determine the coal feed and fine coal output from the mill. In case of the primary air fan for coal transport, speed indicator in case of variable speed drive, or damper position indicator in fixed speed drive are provided.

3. Burning and Cooling

(Kiln & Cooler): A Kiln is a counter current heat exchanger in which heat is transferred from combustion fuel to the raw meal. A typical temperature profile of raw meal and gasses is shown in **Fig. 2.** This profile would vary for gas, oil and coal fired kilns. The performance of a kiln is determined by t/m3 of kiln volume, t/m^2 of burning zone area, t/m2 of cooler area and, KCal/Kg of clinker

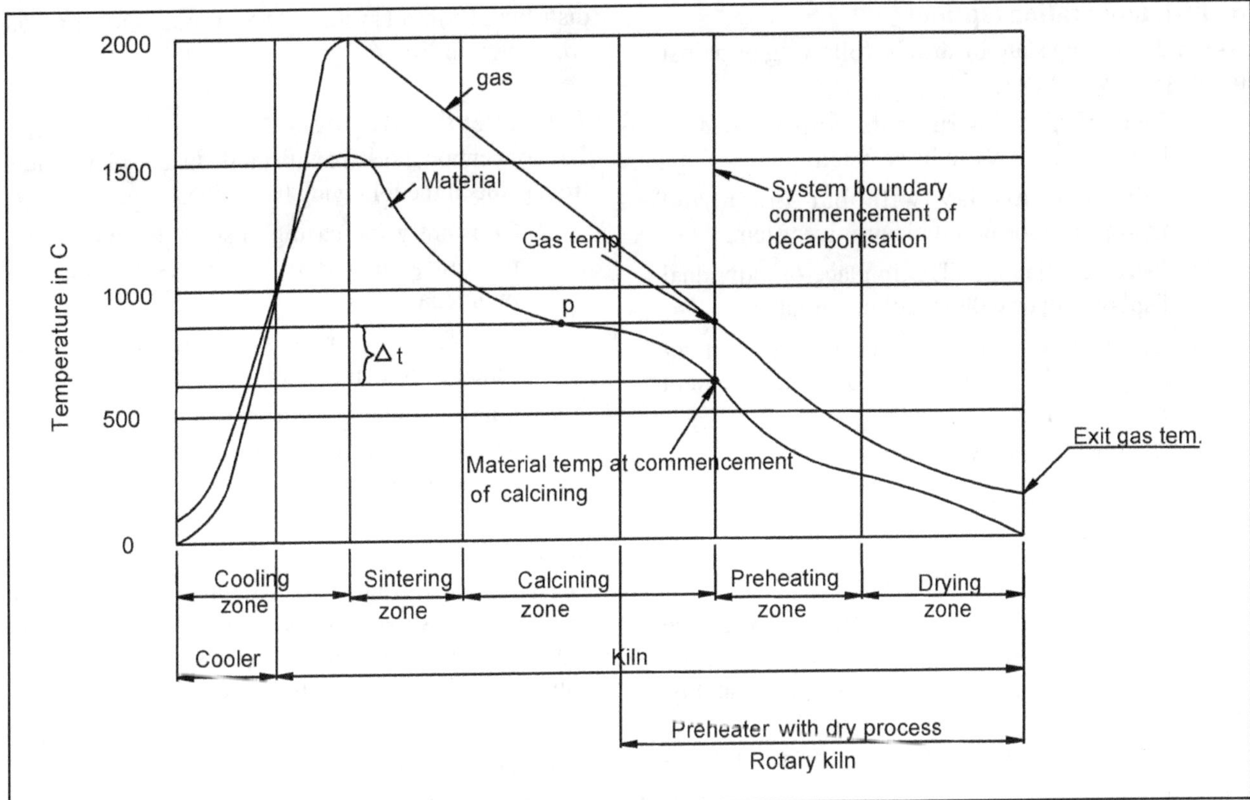

Fig. 2 Gas and material temperature in a rotary cement kiln.

For steady state operation of kiln the most important parameters are:

1. Back end temperature (BE).
2. Burning Zone Temperature (BZ).
3. Oxygen in exhaust gases (O_2).

The above mentioned parameters can be controlled through:

(i) Kiln speed.

(ii) Feed rate.

(iii) Fuel rate.

(iv) Air (primary air from P.A. fan and secondary air from cooler).

In order to evaluate the kiln's as well as burner's performance, the continuous recording of the following parameters is recommended.

BE temp., BE draft, hood draft, O_2, kiln speed, kiln torque, secondary air temp. and coal feed rate.

4. Cement Grinding

Grinding of clinker consumes the highest electrical energy of all the cement plant units. Pre crushing of clinker gives definite reduction in power consumption. These are generally closed circuit mills and all the instrumentation are identical to raw mill: additional control of temperature and humidity through water spray in the Mill is provided for the proper functioning of the electrostatic precipitator. Gypsum and clinker are ground together to produce cement. Blended Portland cements are produced with blast furnace slag and pozzolana.

5. Planning for Instrumentation

Basic units and instruments requirements for each unit should be listed and classified with regard to indicating, recording, alarm (Visuals & Audio) totalizer and close loop control requirement. A separate list with respect to location such as field, local, remote or central control room should be prepared.

6. Instrumentation Options

As per the present state-of-art, the following principal options are available:

1. Convention al instruments with individual display. (not likely to be selected).
2. Individual display with bar graph and microprocessor based electronic system.
3. Shared display on VDU in place of individual display with or without colour graphics.
4. Same as above but with a back up mini or micro-computer for optimization and management information system.
5. Computer control with:
   (a) data logging,
   (b) static mode for operator guidance,
   (c) on line computer control.

Besides the above some more options are available with direct digital control with single or back to back double computer with redundancy for higher reliability. The reliability of direct digital control, even with redundancy, has not proved to be as reliable as distributed control system. The various configuration are shown in **Fig. 4.**

7. Conventional System

The conventional system with individual display suffers from some of the following limitations:

1. Constantly increasing cost of device wiring, planning, assembly, commissioning and maintenance.
2. Poor display of process dynamics due to low function content.
3. Large number of instruments and information beyond human capability to observe and use meaningfully.

8. Bar Graph

The bar graph instrument with plasma and fluorescence display is a positive step forward in the development of front and panel instruments. These instruments are completely 'solid state' and are free of complexities.

A ready reckoner for kiln control in the event of kiln upset is given in **Fig. 3**.

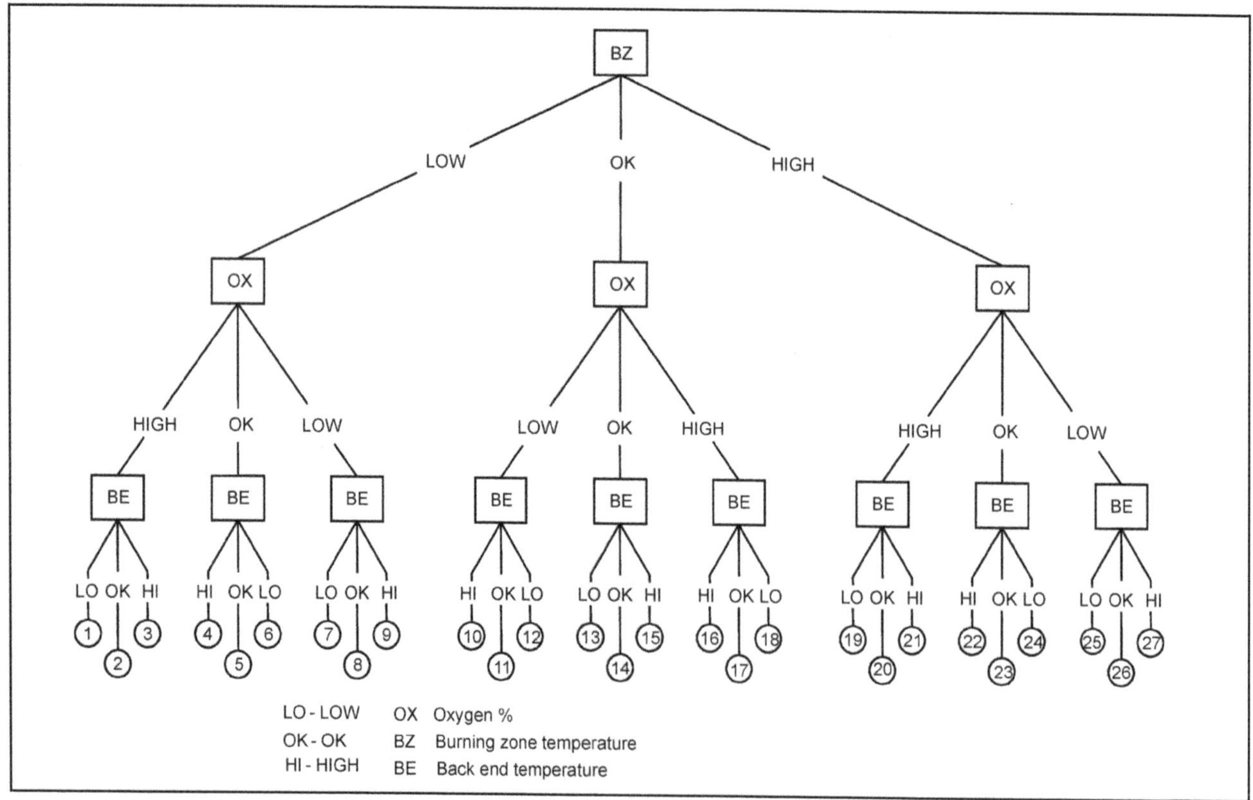

Fig. 3 Evaluation of the three basic variables, burning-zone temperature (BZ), percentage of oxygen in the exit gas (OX), and back-end temperature (BE) for control of kiln operation.

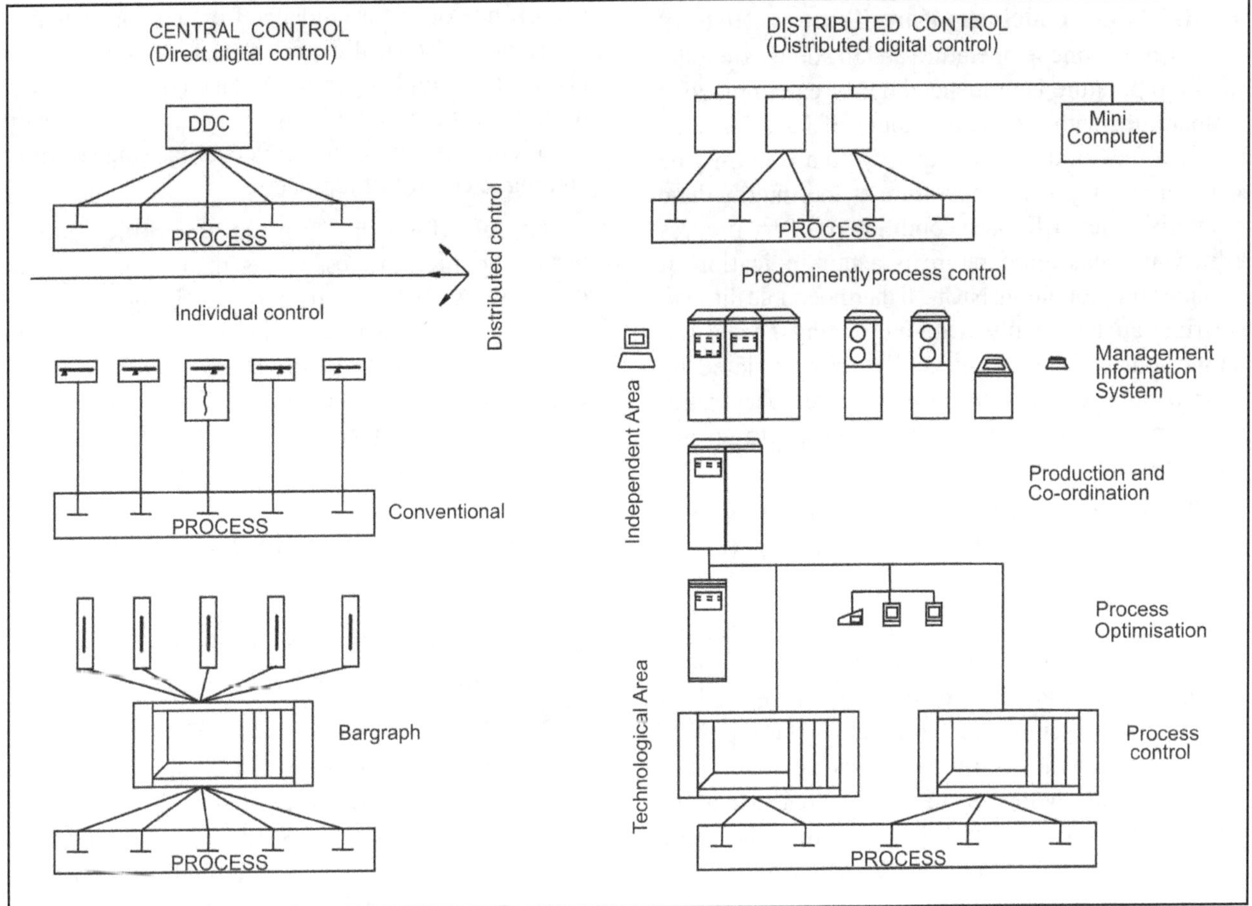

Fig. 4 Central and distributed digital control.

The evaluation of burning zone temperature is possible through:

1. Visual observation.
2. Temperature through two colour pyrometers.
3. Drive motor torque.
4. Nitric Oxide (NOx).

TThe kiln drive motor torque indicates the length and status of the Burning Zone. Recent trials with the measurement of NOx have proved it to be an effective index of burning zone condition.

Post mortem information on burning zone can be obtained from:

(i) Clinker litre weight

(ii) Analysis of free lime in clinker.

For kiln performance, flame temperature is one of the most important parameters and it depends on the following:

(i) Temperature of combustion air.

(ii) Quantity of excess air.

(iii) Fuel/Air ratio.

(iv) Coal Particle size.

(v) Coal moisture content.

(vi) Coal ash content.

Even ten percent excess air can result in 80-100 °C fall in flame temperature, and 100 oC temperature rise in secondary air can give 50 to 70 oC rise in flame temperature.

Shell temperature scanners are provided to warn against hot spots due to refractory wear or damage. Kilns are also provided with CCTV with water cooling and scavenging air to keep the optical system free of dust. In the event of water or air failure the system retrieves automatically. Under Indian conditions, both for Pyrometers as well as for CCTV, besides the automatic retrieval, an automatic closing diaphragm should also be provided.

9. Mid Kiln / Calcining Zone Temperature

The burning zone temperature and to some extent the shell temperature is influenced by the calcining zone temperature with a certain amount of lag. This also has an effect on the clinker quality. In a wet kiln, the kiln being long, the temperature at the intermediate point gives better furnace condition. In a dry process kiln, waste gas temperature is a fair indication of burning zone condition. NiCr-Ni thermocouple through slip rings are commonly used for detecting the mid kiln temperature. Contactless oscillators modulated by thermocouple output have been tried with success. For our ambient conditions only slip rings are recommended.

The cooler air temperature influences the state of the burning zone in the kiln; it is also indicative of cooler efficiency.

The air flow in Fuller cooler from the individual fans is continuously monitored to avoid damage due to burning of the grate plates owing to undue increase in the clinker bed height and thus blocking the supply of air, associated with electro-mechanical devices. These instruments are being used for individual display of parameters and structurally suffer from the same limitation as that of conventional instruments.

10. Share Display

The limitation with regard to information of individual display systems are overcome by the share display system. In this case the information is logically split and only the relevant portion of it is presented to the operator through the Visual Display Unit (VDU) in the form of multi-coloured alpha-numerical display with an overall selective view of the process with real time information on some of the process parameters. Mini or microcomputer backup is sometimes provided for management information and process and or energy optimization. The structuring of the system could be radial or bus system based on function monitoring, equipment or sub-units in the plants. For cement industry, bus system with grouping according to units which can be independently operated have been found to be an effective system as a process control strategy.

11. Computer Control

Computer control can be provided as a direct on-line control or started with data logging and retrieval system and off line operator guidance from computer which stores both historical experiences as well as process algorithm to guide operator, to take corrective action to improve the process as well as handle emergency situations. On line computer is the final stage of fully automatic control of a process.

The proliferation of micro-electronics and in particular microprocessor has made it possible to dispense with hardwired logic. Programmable controllers (PCs) which can modify and enhance the programme without making changes in the external circuit are replacing the hardwire logic which could not possibly provide this flexibility. Some of the advantages of the microprocessor based instruments are:

1. High function content at low hardware cost.
2. Higher reliability.
3. Lesser design and manufacturing lead time.
4. Adaptive, being software based.
5. Can provide enhanced system performance.

Distributed Control System with shared display on VDU with colour graphics is emerging as the most favoured concept in instrument system design. VDUs can provide an overall selective view of the process with alphanumerical display of process parameters in different colour and flashing signals to indicate deviations. Alarms in their sequence and process deviations are stored and can be retrieved for corrective action. Besides process monitoring, trend curves, tabulated information on the process can be displayed.

The distributed control with share display provides some of the following advantages:

(i) Information is processed and logically split and only the relevant portion is displayed or passed on to other levels.

(ii) Failure is restricted to the sub unit unlike the direct digital control (DDC).

(iii) Computer can be subsequently added.

(iv) Task at every level is well defined and hence easier to design, install, commission, operate and maintain.

(v) Provides simple user oriented solutions even for complex control tasks by using higher algorithms.

12. Noise Immunity

(Spurious electrical signals) Separate mention is being made of this concept because in the enthusiasm to introduce high level of technology this aspect is generally being lost sight of by designers. The fusion of electronics and electrical power equipment in process industry has given rise to some new problems. The modern 'chip' has low thermal mass, fast response and low power consumption. These qualities which are otherwise desirable render these devices susceptible to damage as well as malfunction in electrically hostile environment. The qualitative requirement of power supplies for microprocessors are also stringent when the system does not even meet the requirements of Indian Electricity Rules (+ 6% voltage, + 3% frequency). The frequency tolerance for microprocessor/ computer may be 0.5 to 0.75 Hz; for disc storage: not greater than 0.1Hz; the wave form distortion due to harmonies: not greater than 5% or even more stringent requirements. The system should be able to withstand transients surges, black out for ½ Hz, brown out for 1 Hz on 50% voltage. For reliable and effective operation of micro-processor based equipment uninterrupted power supply (UPS) is a must in view of the poor quality of power supply. The equipment should be designed and tested for immunity against noise and appropriate distance of sensitive cables to be kept from power cables. It should also be flicker free to avoid undue strain to the eyes.

13. Energy Management

Energy cost (fuel + power) is the largest component in the cost structure of cement industry in India. Energy flow diagram in a typical dry process cement plant is shown in Fig. 5. The specific energy consumption of kiln reduces with the increase in size of kiln. The fuel efficiency can also be improved by converting wet process to dry process. Energy optimization in the kiln is possible by controlling the quality of raw meal and fuel; by blending with energy saving materials like blast furnace slag, flyash, etc., with Portland cement.

The electric power consumption is about 63% in raw meal preparation and cement grinding, 23% for burning and cooling and 14% for remaining operations. From the foregoing, it is evident that most of the power is consumed by the mills. Ring motors (slow speed) constructed around the mill housing have been successfully used eliminating gear box and hence gear losses. Cost effective converters/ inverters to produce 50 Hz at efficiency as high as 91-93% have been developed which offset the cost by saving in the losses in gear box as well as saving in space.

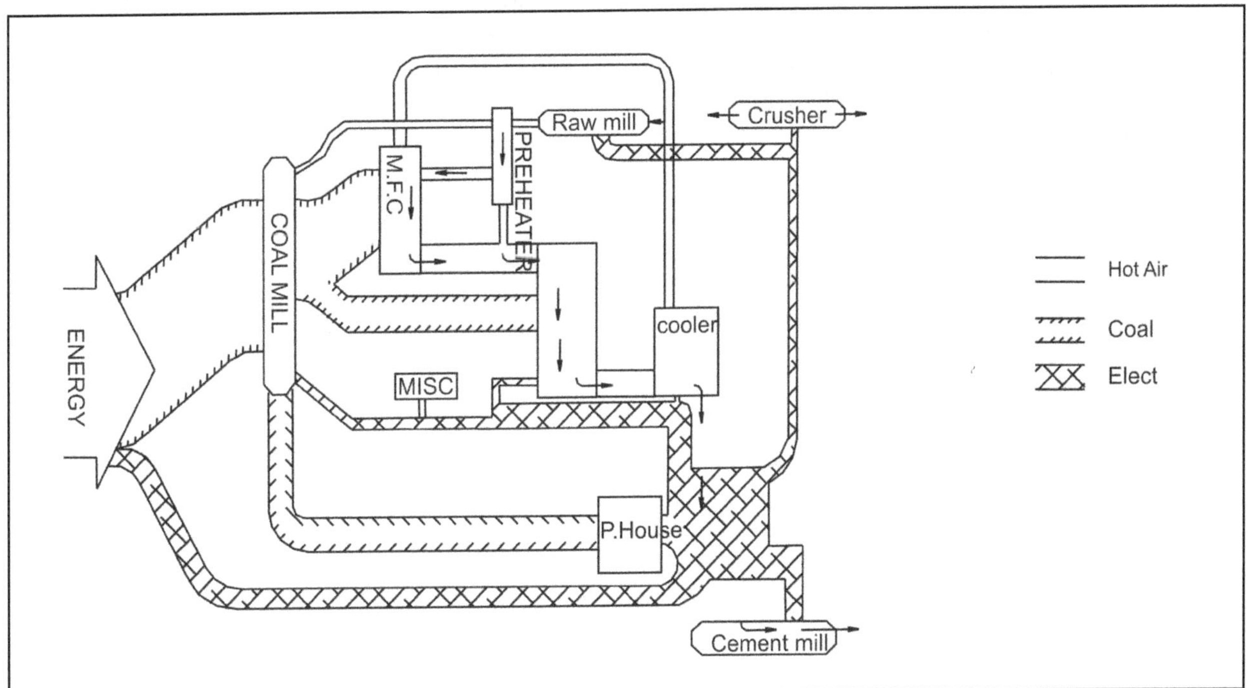

Fig. 5 Energy flow in a dry process cement plant.

Energy saving can be achieved by using variable speed drives for blowers and pumps required to deliver variable quantity instead of control through dampers. For lighting, high efficiency fluorescent tubes and HP sodium vapour lamps could be considered. Control of electrostatic precipitator (ESP) could be though pulse energisation for energy saving. Incidentally this also affects reduction in size of precipitators. In certain cases even conditioning tower can be dispensed with by using this technique.

14. Pollution Control

The growing awareness of the pollution problem and worldwide concern for a cleaner and safer environment, government has identified eighteen industries of which cement is one making it obligatory for anti-pollution steps as a precondition for industrial licenses. This measure has been taken to check and prevent air, water and soil pollution. Instrumentation have made valuable contribution to the detection and control of pollutants. For conditioning tower, simple controller like step-controller in place of modulated controller is recommended. The design should be such that in the event of failure of automatic control, manual control should be possible. About ten percent of cement is lost by way of emission, collected and recirculated into the system; the pay back period on investment of electrostatic precipitators would be less than a year. For coal, ESP has been tried with success. CO is continuously monitored in the ESP outlet to reduce the possibility of explosion. To meet emergency situation, installation of wet hydrants and twin CO2 batteries (one standby) are recommended.

15. Conslusion

The proliferation of micro-electronics and in particular micro-processors has lent a new dimension to the development of instrumentation. The instruments were at one time eyes and ears of the process and now they have also become the brain of the process. In the enthusiasm to use higher technology, the problem of noise and poor quality of power should not be ignored.

Proven and conventional instruments would continue to be used for critical parameters even with shared display. The major share of growth will be taken by microprocessor based instruments with intelligent terminals.

Programmable Controllers(PCs) and other functional blocks would be available as a standard hardware off the shelf resulting in shorter delivery and design lead time. For well established processes, proven software (firmware) shall be available. The man machine interface shall be through Visual Display Unit (VDU) terminal to provide multi-coloured alpha-numerical and graphic display in place of large mimic panels and control room full of instrument panels and control desks.

For variable speed drives and control of electrostatic precipitators, variable voltage inputs would be through thyristors and triacs. Even AC induction motor shall be progressively controlled through variable voltage and variable frequency inverters. The control of flow through dampers and valves shall be replaced by variable speed AC drives. With the increase in the cost of electrical energy variable speed drives would provide reliable and cost effective solution for flow control.

In spite of all these developments, man would always remain in the center of activities still he has no equal at taking decisions and estimate probabilities in the even of incomplete information, a situation which can arise any time.

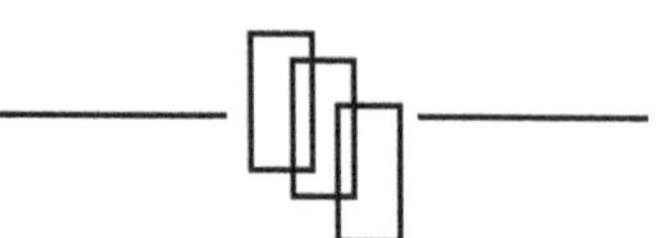

CHAPTER 5

LIGHTING

5.1 Lighting

Adequate lighting needs to be provided in all sections of the plant; in all factory and non factory buildings; in quarries; on roads and in storage yards.

Lighting should be provided as per norms laid down for :

1. Reading.
2. General illumination.
3. Special illumination in specific areas.
4. Standby in case of failures.

5.2 Basic Principles

1. Lifts will be provided with emergency lighting and power.
2. All staircases and landings shall have lighting also from emergency power system (also called UPS).
3. On each floor of the building, emergency light points shall be provided at suitable locations to avoid accidents.
4. In places where readings are to be taken from time to time, in night shifts and so on, additional localized illumination with switch to be provided to turn it on when required (Cooler - inside lights turn on when inspection door is opened).
5. Spare 3 pins, 2 pin sockets should be provided on each floor. All lighting and power points should be earthed.
6. Safety lamps - 110 Volts supply should be available when lights are to be taken inside machinery for inspection of mills, cooler, preheater, kiln, etc. For such purposes, such locations should have 110 V supply-alternately adapters that convert 220 V into 110 V.
7. Corners of buildings and edges and openings in floors should be clearly illuminated, landings and wells should also be illuminated even if hand railings and grills and floorings are provided.
8. Where readings are to be taken from totalizers of weigh feeders or where calibration is to be done, better lighting to facilitate readings at site is to be provided.
9. Lighting in Laboratories / control rooms and offices should be as per norms laid down for Industrial lighting. Besides general lighting, work desks and tables should have local lighting for reading and taking measurements.

5.3 Street Lighting

Roads inside the plant should be well lit so that it is not necessary to drive with headlights on. In fact, driving with headlights on should be discouraged as this can cause accidents.

Lighting can be overdone. It must be functional. It consumes power. Lights are left burning in broad daylight. To avoid this, it should be possible to turn on and off lights, actuated by light meters, photo electric cells, etc., rather than time switches.

5.4 Choice of Lights

Longer lasting, fluorescent lights and sodium vapors lumps and new energy saving tube lights (CFL) should be used. LED bulbs and tubes are also available.Though high in first costs, power consumption is greatly reduced.

Tall buildings like preheater tower, silos, tall chimneys, etc., should be provided with, blinking red lights as per statutory requirements.

5.5 Quarries

When quarries are worked in shifts, working faces in quarries need to be illuminated by powerful floodlights.

These need to be portable as faces shift from time to time (Cabins of shovels and dumpers, cranes, etc., would have their own lighting systems).

Quarry roads, and conveyor system for crushed stone also need to be properly lighted.

5.6 Workshops and Garages

In additionto general lighting, Workshops and Garages would need to have flexible light points according to work being carried out- often under the vehicles. They would need safety hand lamps also. Machines like lathes, milling and grinding machines would necessarily have local light points focusing on the job being machined.

5.7 Walkways Along Conveyors and Tunnels

All walkways along conveyors and tunnels also need to be lighted. There is lot of dust in clinker tunnels (Cooler) and clinker storage - lighting should be selected from this point.

5.8 Stores and Storages

Stores should be well lit for easy identification of labels. Some areas would need lighting for 24 hours. Open store yards must be well lit in night to prevent pilfering.

5.9 Installation of Lighting System

When departmental and building drawings are ready, a 'Lighting' contractor should be selected in the light of his experience in Industrial Lighting. When civil drawings for buildings and departments are ready, same should also be made available to the lighting contractor stressing areas which need to be given special attention. Special requirements of illumination should also be informed to him.

Decisions as regards type of wiring to be selected and also routing of cables need to be taken before construction actually starts. This is particularly so when 'concealed' wiring is opted for. Its routes in column and beams should be carefully planned and civil designers and contractors should be well acquainted with them. Inspection covers for pulling out wires during actual wiring should be clearly marked and provided for.

For large plants with a large area to be covered, separate cables for lighting and lighting transformers at strategic points should be provided for to minimize ill effects of voltage drop and fluctuations.

Voltage stabilizers and UPS systems are essential not only for computers but also for many household gadgets like refrigerators, air conditioners etc. They receive power from the general power distribution system.

Therefore there should be provision for voltage stabilizers / UPS individually or group wise.

6.0 Uninterrupted Power Supply for Lightin

Arrangements should be made for power supply from Captive Power Plant to be available automatically in the event of failure of grid power.

7.0 Roof Top Soalar Power

Roof top Solar power with inverters can be used for street lighting and lighting of common areas of buildings

8.0

Large plants may go in for solar power on a large scale if geographical and meteorological conditions are right by acquiring a plot of land near the cement plant.

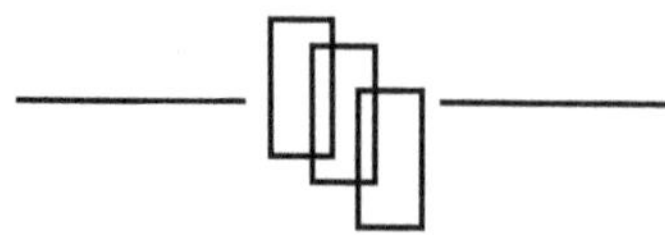

CHAPTER 6

COMMUNICATIONS

6.1 Communications

Communications is assuming great importance as plant sizes are getting bigger and bigger and they are spread over wider areas with simultaneous reduction of manpower. Fortunately communication facilities have also developed fast in the last few years.

6.1.1 Communications in a Cement Plant

Communications for a cement plant can be described in 3 levels..

(i) within the plant itself.

(ii) between plant and Head Office.

(iii) with the World at large.

Operations and maintenance teams and other sections like quarries need to be in touch with one another and with control rooms to pass on information and to receive instructions.

6.2 Means of Communications

Internal communications can be by :

1. Telephones located strategically.
2. Pager system with loud speakers and announcers.
3. Walkie-talkies.
4. Mobile phones.

6.2.1 Telephone Exchange

The size of the internal telephone exchange system depends on size of the plant and number of telephones planned.

It is basically meant for departmental heads and executives to communicate with one another as long as they are near the telephone.

6.2.2 Communications in Sections of the Plant

Communications are also required within a section of the plant. The departmental head may like to give instructions to workers inside the section to make some adjustments, alternatively workers may like to bring to the attention of the head anomalies / problems observed. This can be done through:

(i) Pager systems - with loud speakers installed at / selected floors asking concerned worker operators to go to nearest phone and talk.

(ii) Mobile Phones

6.2.3 Mobile Phones

(iii) Mobile Phones - advent of mobile phones has drastically changed the situation as regards communication - long or short distance. Now almost everybody carries mobile phone - certainly at the work place.

What needs to provided for is the facility for charging.

6.3 Internal Telephone System

Internal telephone system would provide telephones at residences of executives and important operating personnel and also in guest house, hospital, clubs et al.

6.4 Public Telephone System

An internal system however cannot communicate with outside world, not even with Head Office. Officers and Executives of the plant are required to contact Head Office or other parties in the outside world. A plant would therefore also have a public telephone service - government like BSNL or private service providers like Tatas, Reliance etc.

These telephones would be installed in strategic locations like offices of all executives, administrative office, time office, accounts section, watch & ward, colony, guest house, quarry, hospital and so on.

Company would also provide outside telephones in residences of senior and junior executives.

6.5 Internet

Like mobiles, Internet has also changed/ facilitated communications greatly. E -Mails have replaced telegrams. With Computers, even Video Conferences are possible.

6.6 Postal Service.

This is a service run by the Government. Cement Plan can avail of it. In India there is no Telegraph Service any longer.

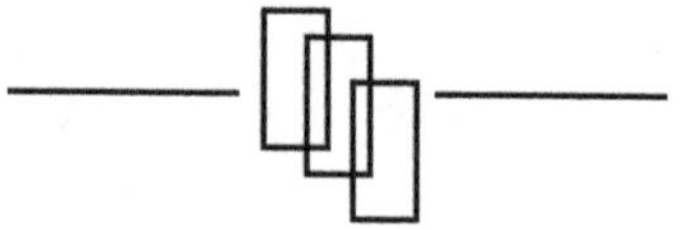

CHAPTER 7

QUALITY CONTROL AND LABORATORY

7.1 Quality Control

'Quality Control' is of as much importance to producing good quality cement as 'Instrumentation and Process Control' is to the operation of the plant.

'X-Ray Analyser' has linked the two as, has been explained in the **Chapter 4** on Instrumentation & Process Control. Therefore presently 'Laboratory' which is the prime set up for 'quality control' may be housed together with I & PC in the same building commonly known as 'Central Control Station'.

7.2 Laboratory

Laboratory in a cement plant has to carry out repeatedly and regularly a number of tests.

Some tests are statutory in that their results are to be submitted to statutory authorities at prescribed intervals.

Others are carried out by the plant to maintain quality of product consistently at various stages of production

7.2.1 Types of Tests Carried out

Tests can be divided / categorized into different types like :

(i) Destructive and non-destructive.

(ii) Physical.

(iii) Chemical.

In chemical tests are included detailed chemical analyses of

(a) Limestones / additives - clays, sand, iron ore.

(b) Raw Mix for kiln feed/ dust precipitated.

(c) Coal.

(d) Clinker - gypsum.

(e) Cement.

The tests should also help in determining the various ratios and modulii for raw meal and clinker mentioned earlier. They should also determine in clinker :

(i) Free lime.

(ii) Burnability.

(iii) C3S, C2S and other constituents.

7.2.2 Statutory Tests

All statutory tests as laid down by the Bureau of Indian Standards should be carried out exactly in the manners prescribed by them using procedures and equipment and materials defined by them.

They should be carried out at frequencies recommended by BIS. Records should be kept as prescribed, identifying the samples, their date and other pertinent details.

7.3 Physical and Destructive Tests

In case of destructive tests, a part of the sample should be preserved to carry out the test again if required.

Physical Tests include - among other things :

1. Crushing and compressive strengths. Compressive strengths of cement at intervals of 3, 7 and 28 days have special significance.
2. Grindability tests.
3. Moisture measurement, absorption of moisture.
4. Sieve analysis for crushed materials, clinker, coal, slag, fly ash etc.

5. Particle size distribution for powdered ground raw meal, coal, and cement in micron sizes.
6. Blaine, surface for cement.

Other tests, would include determination of calorific value of coals and their proximate and ultimate analyses.

They should be measured at regular intervals for raw coal coming from different sources and also for coal as fired in the kiln.

7.4 Equipping the Laboratory

Laboratories should be equipped to carry out as many of these tests as possible.

Testing equipment and procedure for conducting non statutory tests are also prescribed and tests should be carried out exactly in the prescribed manner. Sieving for example should be wet sieving as required.

Whereas gas analyzers would be installed in the plant to measure, CO, O_2, NO_2, etc., it should also be possible to carry out spot checks, if in doubt, by using Orsat's apparatus in additional places.

Dust emission is also required to be measured at stacks and quality of ambient air is to be measured at selected spots around the plant. For continuous monitoring, the plant must have necessary equipment like opacity meter etc., for the purpose.

Some tests are carried out for control of operation like very detailed sieve analysis in micron sizes going down to 0-5 microns. It could be used to establish 'selectivity' curve of the separators etc. Such tests are carried out only to check and hence such tests could also be got done in outside laborarories.

7.4.1 Special Instrumentation, Equipment when using AF

When the plant is using Alternate Fuels and Raw Materials, it is necessary to add

1. Brook field viscometer
2. Ion analysre
3. CHNS analyser
4. Absorption spectrophotometer

7.5 Operational Parameters

Laboratory should regularly establish :

1. Raw meal / clinker ratio for consumption.
2. Raw meal / clinker ratio for kiln feed.
3. Dust collected as % of raw meal.
4. Coal consumption on as fired basis.

Then when weigh feeders / solids flow meter are calibrated - clinker output can be directly calculated from readings and totalisers of raw mix feeder to kiln.

Specific fuel consumption can be calculated as percentage of clinker produced from readings of feeder or as Kcal/kg by multiplying it with average calorific value of coal for the period.

7.6 Continuous Monitoring

Because of need to do continuous monitoring, X-Ray Analysers have brcome 'a must ' in Laboratories.

Many Cemeent Plants now continuously monitor 'greenhouse gas emissions' and also NOX dioxins

If they are using Alternate Fuels.

They equip themselves with

Brookfield Viscometer

Ion Analyser

CHNS Analuser

Absorption spectrophotometer

7.7 Design aspects of a Laboratory

Design aspects of Laboratory building have been brought out in **para 3.8 of Chapter 3 of Section 4.**

7.8 Collecting Samples for Testing

An important part of carrying out tests is to collect samples from field of right type, size and at required frequencies and to send them to the Laboratory for testing.

There would be samples from mines, from stock piles, from raw mill from blending system; samples of clinker and cement and coal as received and as fired. These are to be collected in a proper manner so that they are truly representative of the respective materials at respective locations. Collecting samples and sending them to laboratory required a sizeable amount of work force. Presently 'sampling' has been automated in various degrees as will been seen in **Chapter 54 of Section 6.**

When using Alternate fuels, it becomes necessary to test samples at source. It would then be necessary to have vans fitted with necessary testing apparatus.

7.9 Special Requirements of Modern Times

Presently Laboratories are required to monitor

1 Greenhouse Gas emissions

2 NOx Fluorins and other compounds as a result of using Alternate Fuels

Key Performanc Indicators (KPI) which each plant may choose to monitor its performance.

Central Control Stations and Laboratories have to equip themselves for these functions.

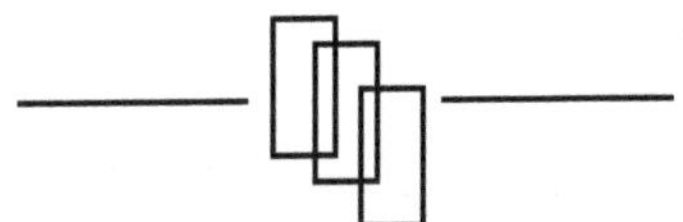

CHAPTER 8

COMPUTER IN CEMENT INDUSTRY

8.1 Computer in Cement Industry

Computer today is playing an all pervading role in each and every aspect of Cement Industry. Some like its roles in Instrumentation and Process Control are well known. Others are less known.

This chapter briefly brings out the role computer plays in lesser known areas.

Anything which requires repetition, which must select from various possibilities, where there are a great many variables resulting in permutations and combinations, which needs data to be stored, processed and retrieved, is best handled by computers.

8.2 Limestone Deposits

Overall quality and quantity of available limestones deposits can be established by using statistical methods and computers.

Representative quality of limestone in various sectors of quarries can be established with the help of computers.

Mining plans short term and long term can be developed by computers.

Correcting materials required can be worked out by the computer in terms of quality and quantities.

8.3 Proportioning of Raw Materials

Computers can continuously yield data on chemical analysis of limestone in the stock pile of a stacker reclaimer system.

With the help of X-Ray analyzer and its computer, quality of raw meal can be monitored and kept close to the desired quality by continuously adjusting proportions of limestones and correcting materials.

8.4 Quality of Clinker

X-Ray analyzer is also used to monitor quality of clinker taking into account quantity and quality of ash absorbed and to suit types of cements to be produced.

8.5 Project Engineering

For a given size of a plant, sectional capacities and capacities of individual machinery can be worked out readily with the help of a computer.

See Section 1.

Capacities of storages and their sizes can also be worked out similarly.

Preliminary data to develop a General Layout (**see Chapter 1 of Section 6**) can be worked out using a computer.

Process gases in various sections at various stages of production can be calculated with the help of a computer knowing targets of performance.

See pertinent **Web pages in Section 9**

8.6 Sizing of Machinery

Main machinery and major auxiliaries will be sized for a given performance using specified raw materials and fuel with the help of computers.

8.7 Design of Machinery and Layouts

Designs of machinery are computerized so that manufacturing drawings can be made quickly. Since computerization requires standardisation of data and procedures and selection of norms, once a program is developed, it is available for use any number of times till it is revised. Anybody can use it.

Computer programmes are developed to furnish stress profiles, bending moment diagrams and similar information to convince potential buyers of the soundness of designs.

Flow patterns in pipes and bends help in designs to reduce pressure drop. **(See Chapter 8 in Section 2, Plate 2.8.5).**

8.8 Drawings

General Layout and Departmental drawings are now made using Autocad and other programmes at a much faster speed. Errors due to human element are reduced to a minimum.

8.9 Civil Design

Civil design of structures like preheater tower and storage silos and kiln piers are computerized. Hence design and drawings for new capacities and sizes can be readily developed and drawings and bills of materials are produced at a faster rate.

8.10 Scheduling and Followup of Sequences of Events

Schedules for placing orders, follow up of deliveries, progress of erection and commissioning are now readily and continuously done by computers.

Monitoring cash flows and receipts and repayment of loans and also overall management of funds can be easily done with the help of computers.

Computers are used in 'perpetual inventory' of stores and spares.

8.11 Operational Control

Role of computer in this field has already been brought out in **Chapter 4.**

Computers furnish diagnostic tools to analyase deviations from norms laid down and to bring back operation on the right track and to optimize it.

8.12 Automation of Dispatches

Computers help automation of all aspects- beginning with receipt of orders, scheduling them, actual delivery and billing or raising of invoices.

Bulk loading greatly facilitates automation.

See Chapters 36 to 38 in Section 6.

8.13 Administrative Aspects

Computers are also used to stream line activities like:

Office work, time keeping,

Inventory control,

Maintenance.

See Plates 5.8.1, 5.8.2 and 5.8.3.

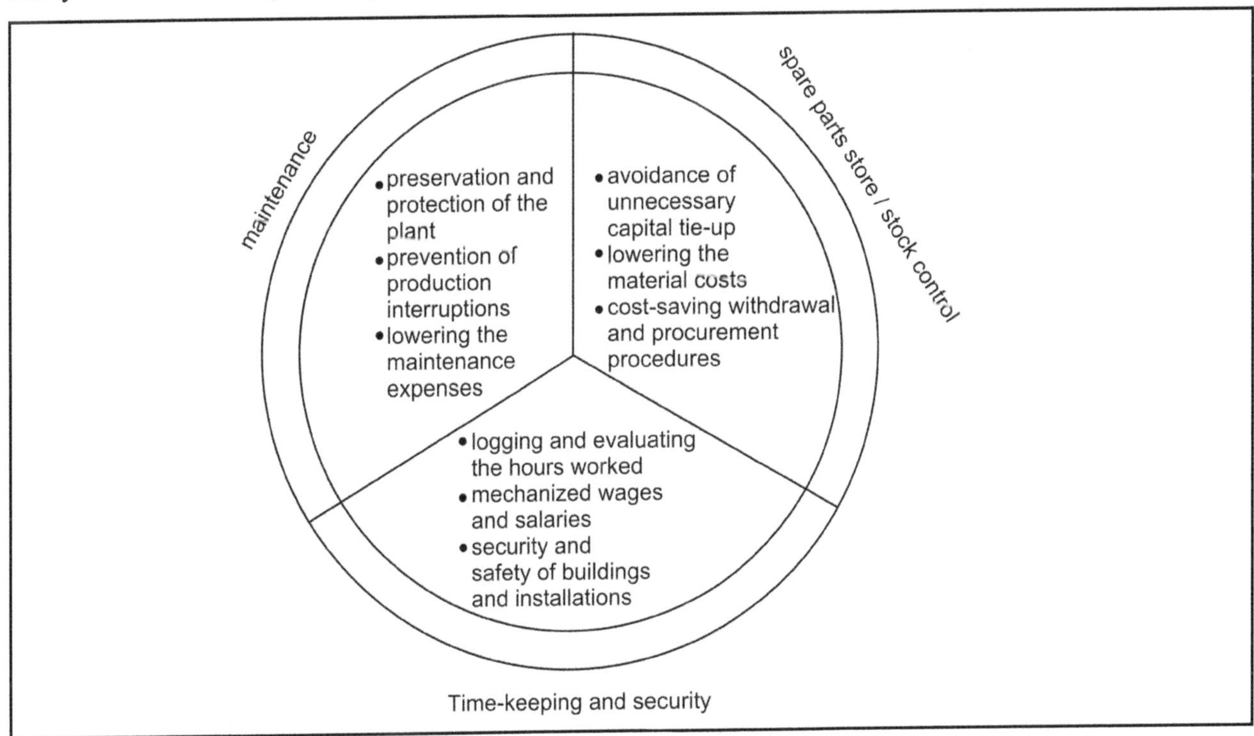

Plate 5.8.1 Computerised administration systems in cement industry.

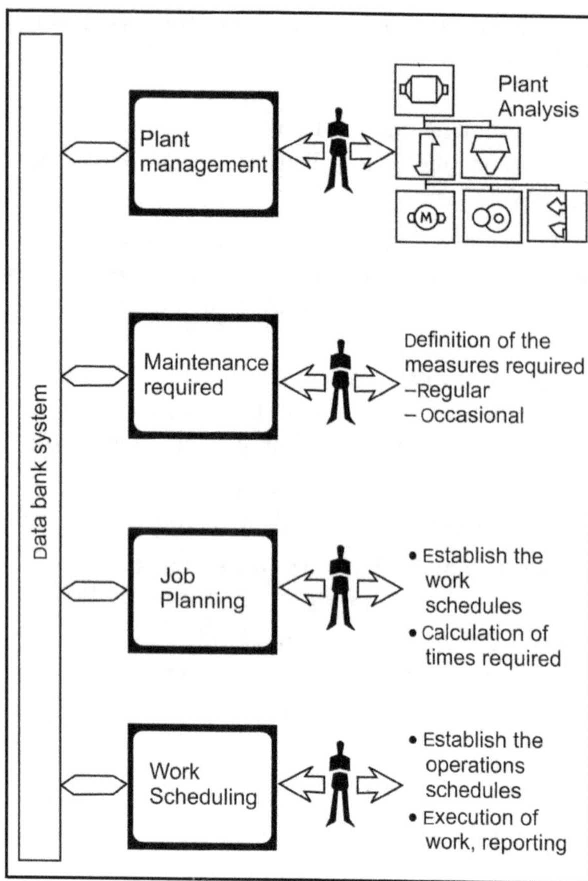

Plate 5.8.2 Scope of maintenance systems.

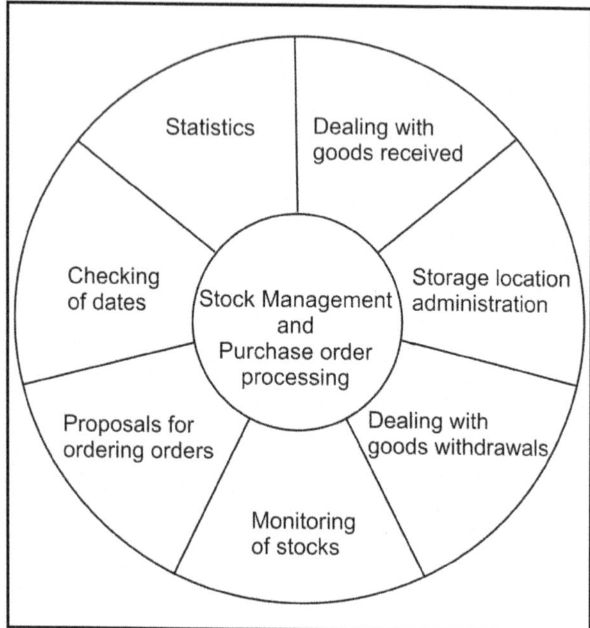

Plate 5.8.3 Stock management and purchase order processing.

8.14 Marketing

Market trends in terms of demand and prices can be studied and analysed using computer programmes. They are a great tool for planning of production and expansion

8.15 Management of Finances

Proven and tailor made computer programmes are available for working out feasibility of schemes; to decide between various options for raising funds. These programmes also carry out 'sensitivity analysis'.

8.16 Computers, Internet and Mobile

There is hardly any activity in the design and operation of Cement Plants which does not use computer. As a matter of fact comprehensive programmes like SAP and ERP are now available to cement industry.

Computer linked to Internet is a mighty tool in the hands of Executives and Operators. It is no longer necessary to control the operation from the 'Central Control Room'. Mobile Phones give additional flexibility. Three together make control of operation of modern large cement plant consistent and accurate.

8.17 Users of Computers

Designers of Machinery developed computer programmes for their own use. It is true today also. So is true of programmes for design of civil structures. Consultants use programmes for project engineering and process calculations. They were used to work on the assignments received by them.

Today however a large number of programmes are available 'generally'. They could be bought and used by anyone. Financial packages would fall in this category. Programmes like Mathcad and Mathcem are available to any one interested in them. This brought the tool of computer in the hands of a far greater number of people.

Mathcem was developed by Late Mr. S.Pal of SoftIdeas Pvt. Ltd. It is a pioneering work of its kind. Even in this Book, **Section 9** contains a good many programmes developed on 'Excel'by the Author which are 'live.' He has been a pioneer in using computer for design of cement machinery and for Project Engineering thereof. Today personal computers have same capabilities of old main frame computers

Cement Plant -large or small- would be unthinkable without computers, internet and mobiles.

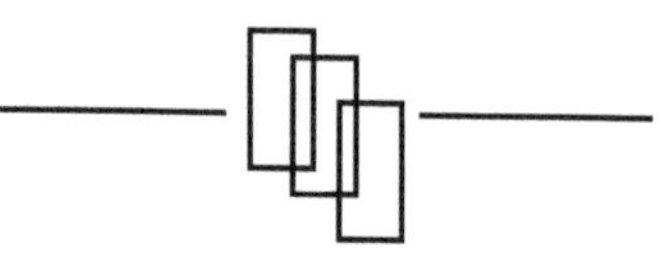

SECTION - 6

Layouts and Detailed Engineering

CONTENTS

Chapter 1	Cement Plants General Layouts	1-43
Chapter 2	Departmental Layouts	44-46
Chapter 3	Layouts of Crushing Plants	47-59
Chapter 4	Auxiliaries in Crushing Section	60-62
Chapter 5	Storage of Crushed Lime stone and Stacker Reclaimer Systems	63-72
Chapter 6	Layouts of Raw Mills and Separators	73-99
Chapter 7	Maintenance of Mills	100-106
Chapter 8	Layouts of Coal Mills	107-112
Chapter 9	Conveying of Raw Meal / Pulversied Coal / Cement	113-118
Chapter 10	Batch Blending and Storage Systems for Raw Meal	119-124
Chapter 11	Continuous Blending Systems	125-131
Chapter 12	Kiln Feed Systems	132-136
Chapter 13	Pyroprocessing Section	137-138
Chapter 14	Preheaters Lepol Grate	139-141
Chapter 15	Cyclone Preheaters	142-166
Chapter 16	Preheater Fan	167-174
Chapter 17	Calciners	175-186
Chapter 18	Tertiary Air Duct, Dust Chamber and Connecting Duct	187-196
Chapter 19	Rotary Kiln	197-226
Chapter 19A	Coal and Oil Metering and Firing Systems	227-236
Chapter 20	Clinker Coolers ; Planetary Coolers	237-239
Chapter 21	Grate Coolers	240-251
Chapter 22	Clinker Coolers, Collecting Spillage and Product	252-261
Chapter 23	Cooling Air Fans	262-268
Chapter 24	Clinker Coolers - Cooler Vent and Dust Collectors	269-277
Chapter 25	Other Clinker Coolers	278-280
Chapter 26	Clinker Conveying and Storage	281-299
Chapter 27	Handling Coal and Gypsum	300-305
Chapter 28	Handling Slag to make Slag Cements	306-313
Chapter 29	Handling Fly Ash to Make Pozzolana Portland Cement	314-321

Chapter 30	Cement Mills	322-334
Chapter 31	Cement Storage	335-338
Chapter 32	Cement Storage and Packing	339-344
Chapter 33	Bags Godown	345-348
Chapter 34	Layout of Packing Plant and Dispatches	349-357
Chapter 35	Simultaneous Loading of Trucks and Wagons	358-363
Chapter 36	Bulk Loading by Road and by Rail	364-367
Chapter 37	Bulk Loading of Clinker	368-376
Chapter 38	Exporting Cement	377-383
Chapter 39	Import of Coal	384-384
Chapter 40	Railway Siding	385-390
Chapter 41	Venting	391-404
Chapter 42	Chutes	405-415
Chapter 43	Ducts	416-453
Chapter 44	Chimneys	454-458
Chapter 45	Pipelines and Supports for Compressed Air and Pneumatic Conveying Pipe Lines	459-464
Chapter 46	Oil Lubrication and Cooling Water Systems	465-471
Chapter 47	Bed Plates	472-476
Chapter 48	Foundations and Load Data	477-482
Chapter 49	Civil Drawings and Design of Departmental Buildings	483-484
Chapter 50	Working Platforms	485-488
Chapter 51	Conveyor Gantries	489-494
Chapter 52	Lifts, Hoists and Lifting Tackles	495-499
Chapter 53	Safety	500-508
Chapter 54	Sampling	509-511
Chapter 55	Refractories	512-526
Chapter 56	Insulation and Lagging	527-528
Chapter 57	Fans	529-531
Chapter 58	Compressors and Blowers	532-533
Chapter 59	Good Housekeeping	534-535
Chapter 60	Co-generation	536-538

CHAPTER 1

CEMENT PLANTS GENERAL LAYOUTS

1.1 Objective

A Layout of an Industrial Plant or Process is a logical and practical arrangement of the various sections and stages of the plant and the processes that would help in the production and disposal of the products and wastes if any from available raw materials and energy resources in an efficient manner using optimum inputs in terms of space, raw materials, energy and manpower.

The general plant layout includes various sections of the plant starting with receipt of raw materials, and ending with dispatches of the product. It also includes receiving of power and fuel and water required for process/cooling/consumption.

While principles of good layout are the same for all processes and plants, there is a large variety of layouts according to the peculiarities of each individual plant and process.

Even for the same Industry layouts vary considerably according to the location, size of plant and process used, storages to be provided for and space available.

Layouts also depend on external factors like size and shape of plot of land available, orientation with roads, river, railway line, power transmission line etc.

This section aims at laying down broad principles and practices and factors governing them for General Layouts of Cement Plants.

1.2 Scope

Cement can be made by using different processes like wet, semi wet, semi dry and dry. The dominant process prevailing is Dry Process. Here primarily dry process is considered.

Cement plants come in all sizes; sizes ranging from 30000 tpa for a mini VSK plant to a + 9 million tpa capacity plant making blended cements.. Cement plants with rotary kiln/s would be considered here.

Fuel can be coal or oil or gas according to economic availability thereof. In India, coal is the fuel used by almost all plants; electric power is bought from the Grid in the vicinity though many plants would have their own captive power plants (coal or oil fired). For this reason the layout would be built around coal as fuel and electricity from grid.

Many plants now use imported coal and some use lignite and petcoke as fuel on regular basis. Layouts need to be slightly modified in such cases.

There are a few new dimensions in recent years like using Alternate Fuels and installing Waste Heat Recovery Plants with a view to conserve exhaustible fossil fuels and to reduce Green house Gases.

1.3 Making of Cement

For better understanding of the design of general layouts of cement plants the process of making cement may be illustrated by block diagrams / flow charts as shown in **Figs. 6.1.1 to 6.1.3.**

Making of cement has already been dealt with in **Section 1** with typical flow charts in **Chapter 7** thereof and hence need not be repeated.

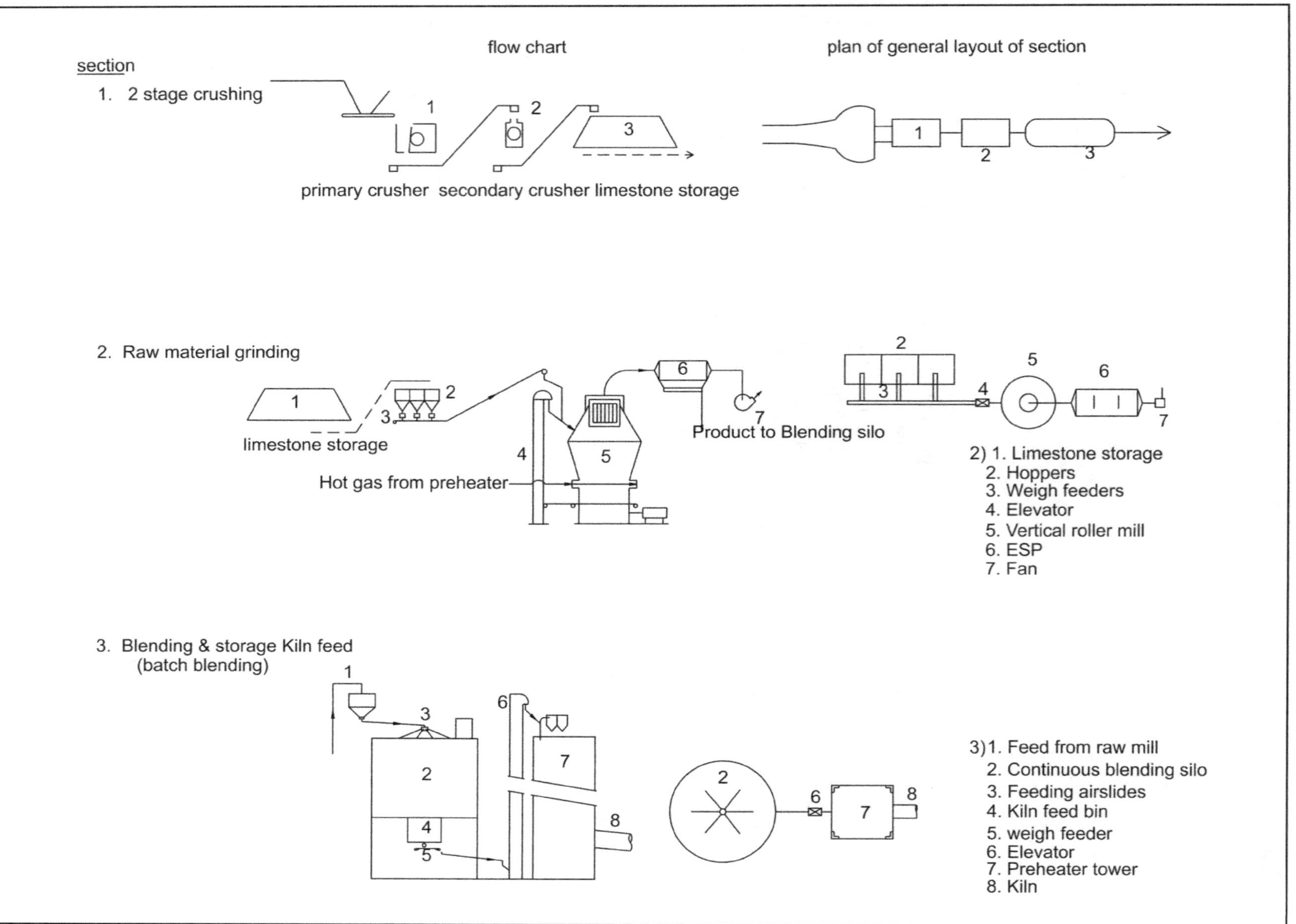

Fig. 6.1.1 Flow charts and plans of sections from crushing to kiln feed.

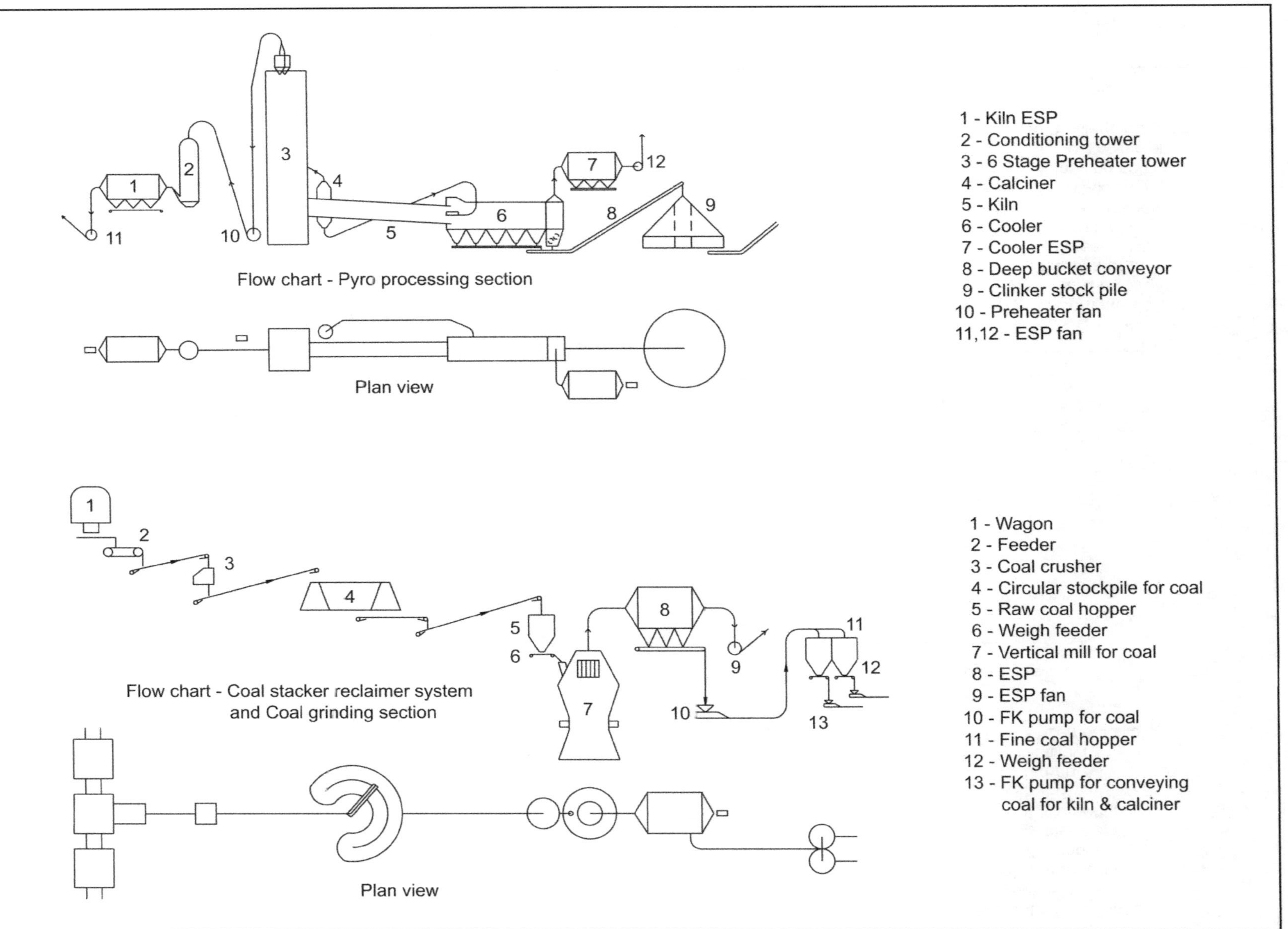

Fig. 6.1.2 Flow charts and plans for clinkering and coal mill sections.

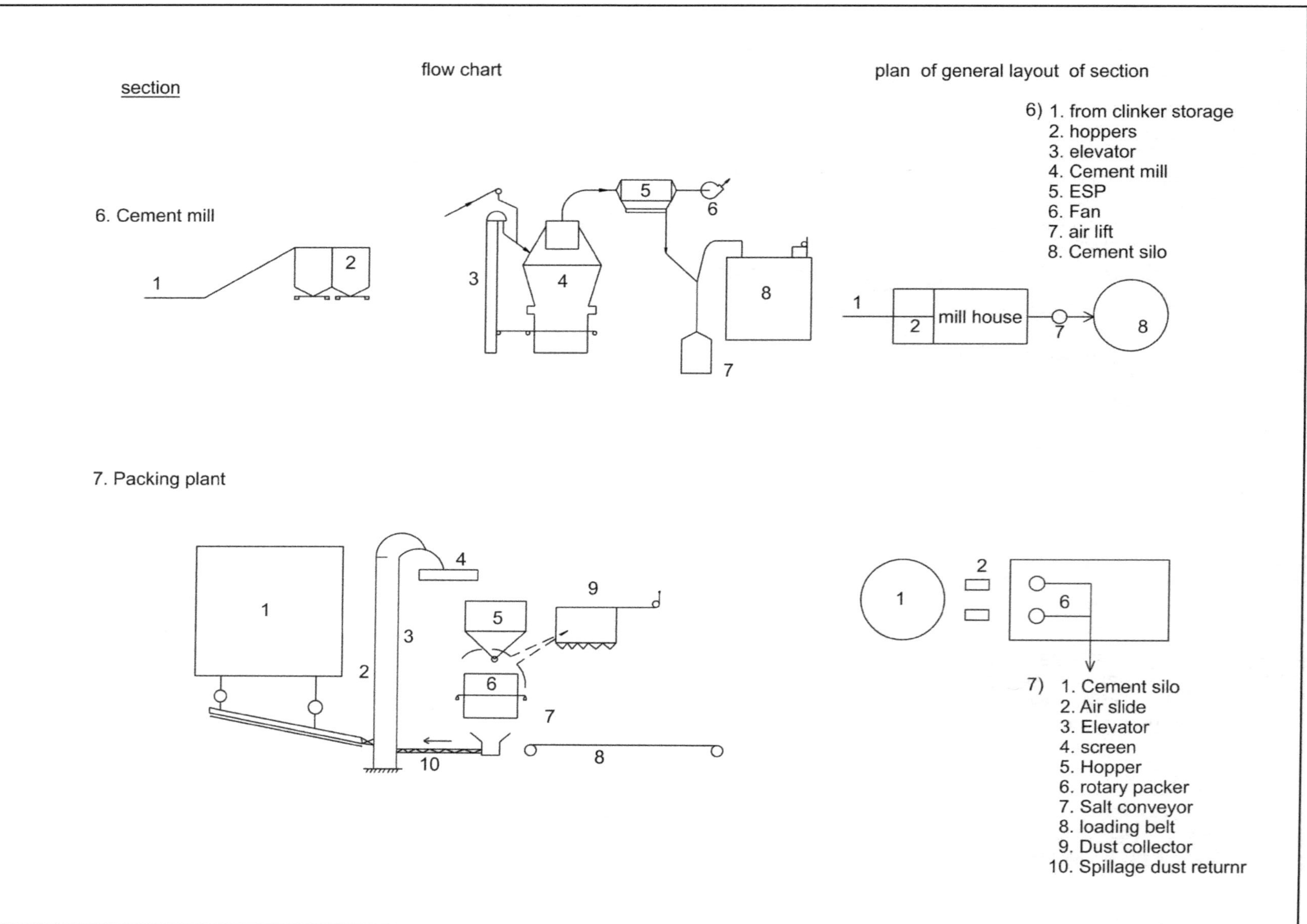

Fig. 6.1.3 Flow charts and plans for cement mill and packing plant sections.

1.4 Initial Steps

Various steps leading to deciding on the size of the plant and its location vis a vis lime stone deposits proposed to be exploited have also been dealt with in **Section 1** on **Basics** and **Section 3** on **T.E.F. Studies.**

Taking all these factors into account, the entrepreneur shortlists two to three sites and investigates them in depth. The primary and most important investigation is of course with respect to the suitability and availability of deposits.

After this, the entrepreneur has to look into the following :

1. obtaining lease for mining for a sufficient area of land,
2. availability of land for factory and colony,
3. relative locations of mines and factory: while the location of quarries is fixed, location of plant would be finalised giving consideration to proximity to roads, rail link, power line etc. As far as possible factory and mines should adjoin but it is not always possible to do so.

Side by side the entrepreneur has a dialogue with Process Technologists, Consultants and Machinery Manufacturers to obtain broad requirements of land for factory, manpower requirements – hence housing to be provided, power fuel and water requirements for the present and for the future. He also prepares preliminary layouts with the help of his Consultants to highlight relative locations of various sections of the plant and storages of various materials to be provided and find out broadly area required for the plant.

Armed with this he can now start developing actual layouts at proposed site/s

1.5 Basic Considerations

Basic considerations in developing good layouts are :

1.5.1 General

Land to be procured should be adequate leaving elbow room even after expansion.

The factory should not be located as far as possible on useable limestone deposits.

Load bearing capacity of soil should be ascertained over a large area and depth to which foundations would have to go should be ascertained for heavy and for tall buildings.

Seismic zone of the region in which factory is located must be ascertained.

Underground water table in different parts of the year and minimum depth below surface should be ascertained. This helps in taking decisions on maximum depths of pits for conveying equipment, crusher etc., in the plant and also with respect to water proofing.

Wind directions and velocities at different heights and in different parts of the year and the wind rose as well as details of precipitation (rain) in different parts of the year over a period of the last 15 years or so should be obtained from the Meteorological Department.

1.5.2 Material Handling

The modes of conveying uncrushed and crushed limestone, inside and outside the plant; conveying ground raw meal and coal and cement should also be selected; transport of clinker from kilns, its storage and transport to cement mills should also be finalised. These decisions help in providing for adequate lengths for conveyors whether belt or deep bucket etc. Presently mechanical conveying is preferred to pneumatic conveying to save power.

1.5.3 Storages

Crushed limestone and correcting materials and coal and clinker and gypsum and also ground raw meal, ground coal and cement are required to be stored in the various stages of the production of cement to ensure uninterrupted supply of materials between stages and for ensuring continuity of operation and dispatches. Quantities to be stored depend on a great many factors which vary from site to site, should be decided upon.

When the plant proposes to make blended cements then, in addition to above it is also required to store blending materials – fly ash and or granulated slag – as the case may be and also blended cements.

Most plants go in for captive power plants using diesel / hsd/ lshs oils, or coal fired depending on the size of the cement plant. Thus the plant has also to provide for storage of oil and if coal fired, then for additional quantities of coal for the thermal power station.

With the introduction of Stacker & Reclaimer systems it is possible to achieve blending of stored materials while reclaiming. This method of storing is now commonly used for crushed lime stones and for coal. But they require considerable amount of space. This is particularly so for coal where height of stock pile is limited to 2.5-3 metres.

Clinker would be generally stored in covered tent like circular covered sheds 60-100 meters in diameter. For small plants it may be stored in silos also.

Raw meal and ground coal and cements would be stored in silos; so also fly ash because of its fineness.

1.5.4 Railway Siding

If the plant can have a rail link a large part of the production would be dispatched in bags or in bulk by rail as this mode of dispatch is cheaper for long distances. Still a substantial part would be dispatched by road. It is now possible to send cement in bulk by road in specially designed self unloading carriers and also by rail in specially designed wagons.

Thus the plant layout must be designed with a railway siding and also for road dispatches. If there is no rail link nor is it likely to be there in the near future, then the layout would be designed only for road dispatches.

A railway siding requires long length to accommodate rakes of box wagons of 55 ton capacity each and capacities of 2200 tons for a rake of 40 wagons and 3300 tons for a rake of 60 wagons. Total rake length will be about 700 metres. Number of lines shall be minimum 3 to allow for incoming and outgoing and for internal placing of wagons for loading.

If coal is also received by rail, additional railway lines or siding may have to be added for unloading of coal (if plant has coal fired power plant quantity of coal to be handled increases substantially).

1.5.5 Location of Crushing Plant

If the quarries are close by it would be advantageous to locate the crushing plant (and subsequent storage also) inside the main plant. If stone is brought in by dumpers as it would be in most cases, the plant would have to have a long ramp for reaching a suitable height for unloading run off mine stone in the crusher hopper. Otherwise crusher would have to be installed in a deep pit. This ramp takes considerable amount of space and the layout has to accommodate it either inside the plant or just outside the plant.

This situation also assumes that the area under mining lease is adjoining the main factory area. Else the plant would have to acquire a narrow strip of land for constructing the road between quarries and plant.

Another possibility, if the plant is very large, is to have a mobile crusher (or a semi mobile). In this case crushed limestone would be brought into the plant by an overland belt conveyor.

1.5.6 Incoming Power Supply

The incoming electric power supply would be, in most cases, at 132 KV or at higher voltages. The Electricity Board would have its own substation.

Generally speaking it would be just outside the main plant. From there, power lines would be taken to the plant's own outdoor substation. Voltage would be reduced to the voltage to be used for HT motors at this point itself.

The location of the grid substation would depend on the orientation of the overhead transmission line in the area vis a vis the factory. If the quarries are far from the plant, or if the crusher is located in the quarries a HT line of smaller capacity would have to be arranged for supplying power to the quarries.

The size of the grid substation and company's own substation must have provision for future expansion.

Location of the outdoor substation within the plant depends on many factors. Primarily it should be at the load centre of the total plant. This location should also be convenient to locate the thermal power station. This would reduce lengths of expensive HT cables and of cables between captive power plant and the main distribution centres of the plant.

1.5.7 Captive Power Plant

Location of captive power plant depends on what fuel is being used. If oil, then its receipt, storage and subsequent handling has to comply with specific norms pertaining to fire hazards which require that oil tanks should be isolated by constructing a wall around them. If the oil used is lshs, then arrangements are required to be made for heating it to make it suitable for use. Naturally therefore, the leads from storage to the diesel generator sets should be as short as possible.

Life of diesel engines used in the generation of power is almost directly proportional to the extent the intake air is dust free. Hence elaborate arrangements are required to be made to clean and wash the intake air.

Even then it helps greatly if the power plant is located in such a way that it is far from sources of generation of dust within the plant.

New Plants also resort to **co – generation** and generate power by using waste heat in kiln and cooler gases. This possibility should also be kept in mind

1.5.8 Water Supply

Plant requires water for;

(a) cooling of machines like bearings, compressors, diesel engines etc.,

(b) process purposes like treating exhaust gases from kiln before the esp and

(c) for consumption as drinking water in the plant and in the colony.

Water required for (b) is totally lost. In some cases water from drains of houses etc., may be collected and treated for non drinking purposes like gardening. Water used for (a) is collected and cooled and recirculated. Only about 20 % is required for makeup purposes. It is required to provide for cooling ponds and for recirculation.

The d.g. sets would also require water to be cooled. It would most probably have its own cooling towers.

If the plant has coal fired power station, boilers would require substantial quantities of water for generation of steam. Steam would be condensed after use and water re used. Thus thermal power stations would need arrangements like cooling towers or ponds. It helps if these cooling ponds are not in the direct path of dust sources.

Water to be used for spraying in conditioning tower and for cooling of compressors and for raising steam needs to be soft and hence water treatment plants are installed.

1.5.9 Contours

Land available for the factory would seldom be level. If cultivated it has a greater chance of being level; but then such land is hard to come by and is very expensive. Fallow land or government lands would be cheaper to get. But such plots of land would need developing for installing the plant. It is therefore necessary to have or draw a contour map of the area. The contours should be used to advantage in locating the various sections of the plant. For example the slope could be used to install the kiln and cooler so that excavations are minimum. This is particularly so if the soil is rocky.

Slopes also help in the drainage of rain water and in flow of water to cooling ponds so that power is not required to be spent for the purpose. Maximum use should be made of natural contours for such purposes. **See Fig. 6.1.4.**

1.5.10 Drainage

Drainage is an important function to be kept in mind. Poor drainage may cause flooding in some sections of the plant in times of heavy rain. There could be streams in the area chosen for the factory. These should be made use of to take the water out of the plant area. If required streams could be diverted so they do not cause flooding.

1.5.11 Green Belts and Landscaping

It is now statutorily required to locate the plant, leaving a certain minimum distance between the plant and the state highway. Further it is required to create 'green belts' between the plant and the highway, between plant and colony or adjoining populated areas, also between plant and mines. The green belts serve as breathing dust filters and also as noise barriers besides beautifying the landscape. In the planing of the plant and colony due attention needs to be paid to this aspect.

In developing quarries to excavate the limestone large quantities of overburden and interstitial clays or shale needs to be handled and set aside if it cannot be used as additives for designing the raw mix.

Over the years this assumes massive proportions and then it is very difficult to dispose it off. The planning of developing quarries should include schemes for storage and disposal of such waste material. As the mines get developed, deep excavations are created over large areas which were level ground to start with. Further as benches get developed the depths increase and it is required to deal with under ground water. Some times and particularly in rainy season this becomes an almost continuous operation necessary to keep mining uninterrupted. The quarry water can be used in the plant for its use.

Fig. 6.1.4 Contours at site.

Landscaping of quarries is now receiving more attention from entrepreneurs to preserve environment and ecology. The excavated material is used for landscaping and to plant trees. The discarded pits are used for making ponds suitable for boating and fishing and discarded quarries become picnic spots.

1.5.12 Pollution Free Environment

The communities all over have become environment conscious. They insist that Industries coming up in their midst do not cause pollution in their surroundings.

The State and Central Governments lay down norms for keeping the environment clean. Cement Industry basically causes air and noise pollution. Over the years air pollution has come down because the Industry now uses electrostatic precipitators (esp) or bag filters to clean exhaust gases.

Noise pollution disturbs only the community in the immediate vicinity. However even here the level has come down considerably by the use of Vertical mills for grinding limestone, coal and increasingly cement also.

Communities act like watchdogs and report to the concerned authorities, violations by the Industry. Industry is therefore required to monitor the levels of emissions almost continuously and is required to keep the community happy.

For all these reasons it is required to give a great deal of thought at the planning and layout stages itself to these matters.

Proximity of the village/s town nearby in relation to the plant should be examined with reference to the proposed plant site. Here the information on wind rose is very useful; the plot should be so selected that dust does not get deposited over the populated areas. The dust nuisance comes down in proportion to the square of the distance of the plant from the target. It would therefore be advisable to keep a safe distance between the plant and the village/s and township.

Green belts mentioned earlier filter dust and noise and hence should be planned with reference to surrounding community. Planting of trees within the factory area around open stock piles also serves the same purpose. But it takes time for trees to grow. Therefore a beginning has to be made in the initial stages of developing the layout.

When dumpers or trucks ply on roads dust is thrown about particularly in dry season. Communities are known to complain about this nuisance also. Hence ways and means are to be found out to keep the community happy.

1.5.13 Process Control and Computerisation

Whether the plant is large or small it will have almost certainly microprocessor based computerised instrumentation and process control. Degree of centralisation and automation and sophistication would vary greatly from case to case.

In a great many plants there are automatic sampling stations; samples are taken pneumatically to the X-ray analyser, prepared for testing and tested in X-ray analysers; results are used to correct proportioning on a continuous basis; even mining operations are guided from the central station.

Operations of the various sections are monitored from the Central Control Panel starting from crushing and ending with dispatches. The Control stations need no longer be in the respective production departments.

The Central Station can be located at any convenient central point in the plant.

PCs have revolutionised and facilitated transfer and storage of data, feedback etc.

The testing laboratory of the cement plant is no longer what it used to be. The production and operation and quality control aspects are now brought together and integrated for optimum results. It has become the show piece of the plant and visiting dignitaries are taken there first. Its design and planning assumes great importance today.

1.6 Developing the Layout

The work on developing the layout begins against this background. It is a joint effort between the Consultant, the Entrepreneur and the Suppliers of machinery.

The entrepreneur has to furnish all pertinent details of site and infrastructural conditions, contour maps, wind rose, Geological Survey of India maps, power line routes, locations of substations, locations of rivers, streams, reservoirs, state roads and highways, rail routes etc., to the Consultant.

1.6.1 Finalising Process and Machinery

The Consultant would be involved in finalisation of processes and types of machinery to be used in the various stages of manufacture; he would also advise on storages to be maintained at various stages and types thereof.

The entrepreneur would have invited offers from the prominent Suppliers of various machinery on the basis of specifications drawn by the Consultant. The Suppliers submit their quotations with system flow charts and typical departmental drawings for buildings housing their machinery. Thus overall areas required for various stages in production like crushing, grinding, kiln-preheater-cooler etc., become known.

See Figs. 6.1.1 to 6.1.3.

These are used by the Consultant to develop the General Plant Layout.

1.6.2 Linking of Sections

An important aspect here is the linking of the stages of production in sequence. It involves transport of finished product of a section to the following section as its raw material and so on till cement is produced, stored and dispatched from the plant. The departments are so arranged with reference to one another that the distances are short and direct, avoiding transfer points and changes of direction etc.

1.6.3 Straight Line Layout

In one approach in developing the plant layout, all departments are arranged in a straight line one after the other. This is known as a 'straight line layout'. This requires a long and a narrow shape of plot to be able to arrange departments in a straight line. When the plant is expanded a second line parallel to the first unit is laid down.

See Fig. 6.1.5a.

1.6.4 Shape of Plot Available

Layout design also depends on the shape of the plot of land available. Departments have to be arranged with respect to one another to suit the shape of the plot. **See Fig. 6.1.6 a and b.**

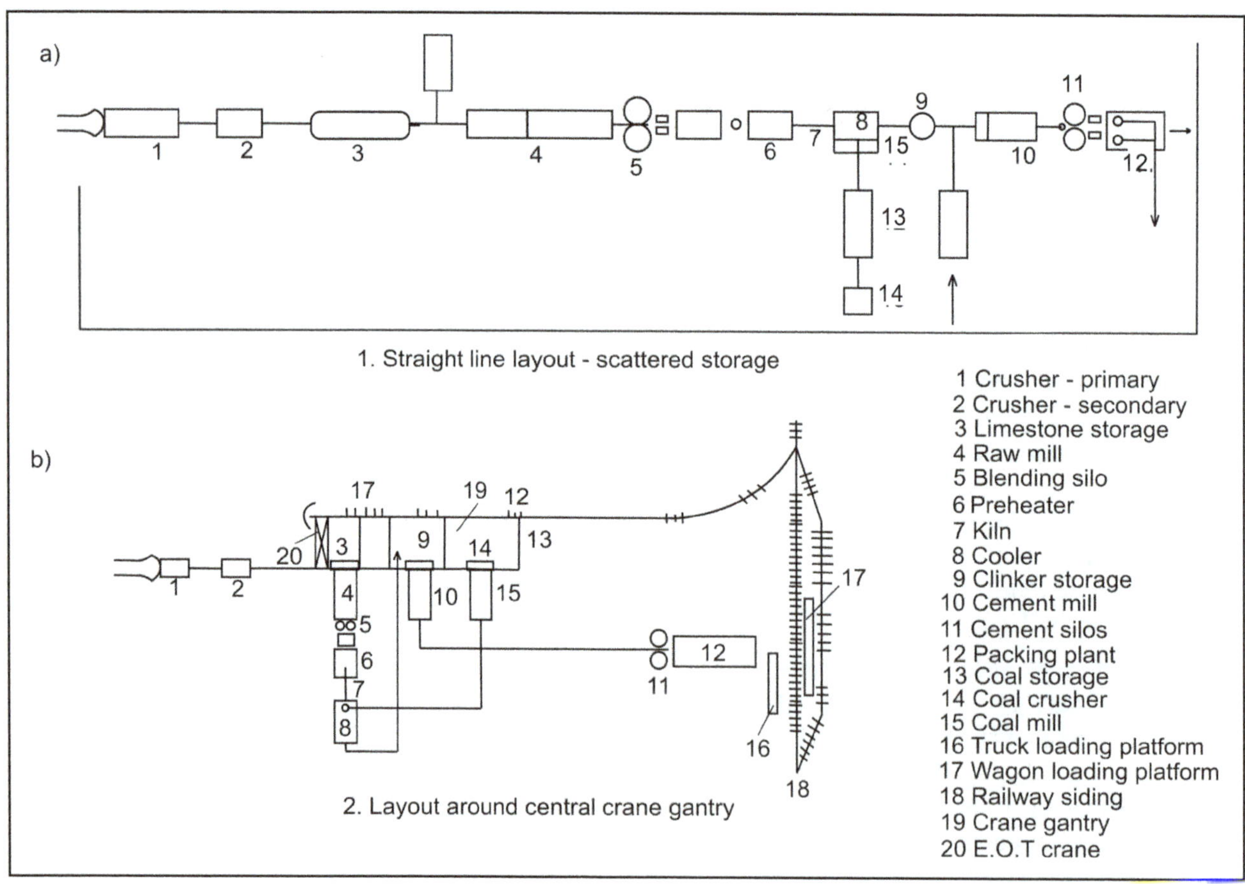

Fig. 6.1.5 Types of layout according to storages – scattered, centralised.

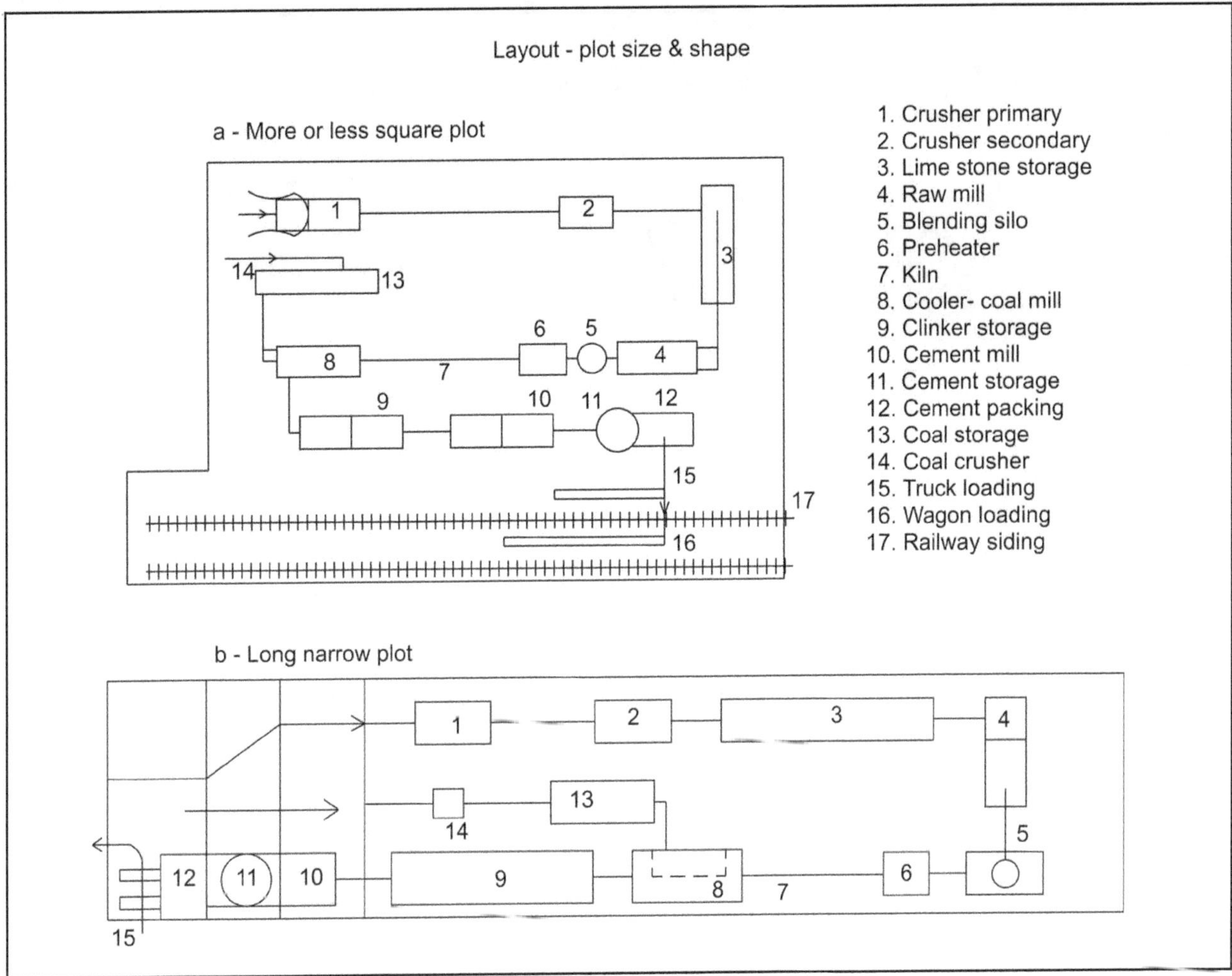

Fig. 6.1.6 Layout arranged to suit shape of plot of land available.

In another approach, facilities of the same kind are grouped for instance blending and storage silos for raw meal, cement silos for the first production line and for subsequent expansions.

Many a time the crushing plant is common for 1st unit and duplication. Capacity of stacker reclaimer system is increased by lengthening the stock piles. Thus crushing plant and stacker reclaimer feeding and extraction systems would be common to both the units.

In any case even where two lines are separate, they are interlinked so that production is maintained in both the lines. Interlinking thus gives flexibility and continuity of operation and is an important feature of planning a layout.

1.6.5 Centralised Storage

Layout is also required to be designed to suit receipts and storages of various materials. In the past layouts used to be built around a central storage like a crane gantry in which were stored various materials like lime stone, clinker, coal and gypsum.

See Fig. 6.1.5 b.

However it is not so now. Storages are scattered all over the plant. Receipts and extraction facilities are designed to suit volumes and rates of consumption. **See Figs. 6.1.5 a and 6.1.6.**

When plant is making blended cements it has to handle receipt and storage of blending materials like slag and fly ash. Layout has to be developed in such a way that all materials are handled efficiently. **See Fig. 6.1.8.**

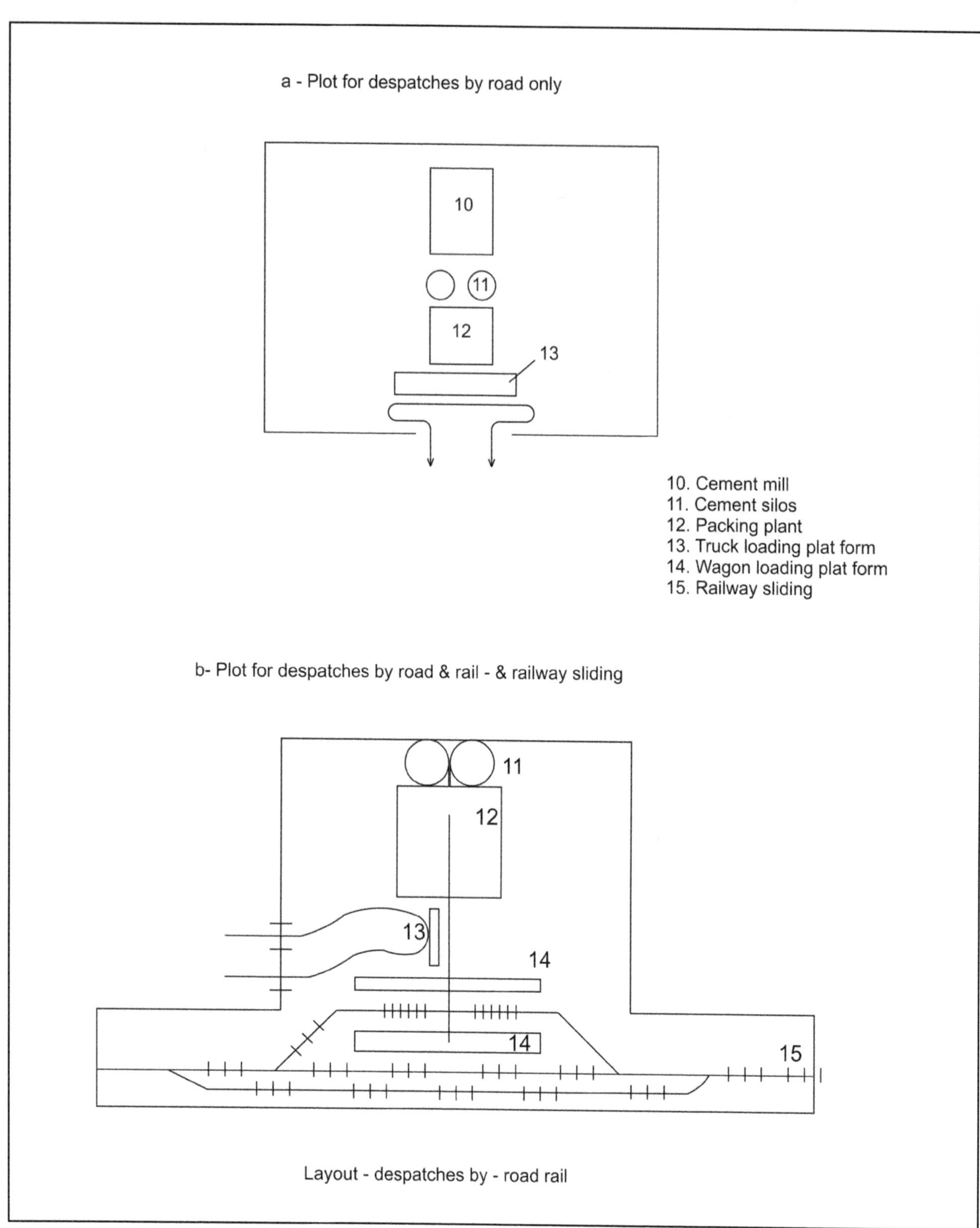

Fig. 6.1.7 Layout - despatches by road, road and rail.

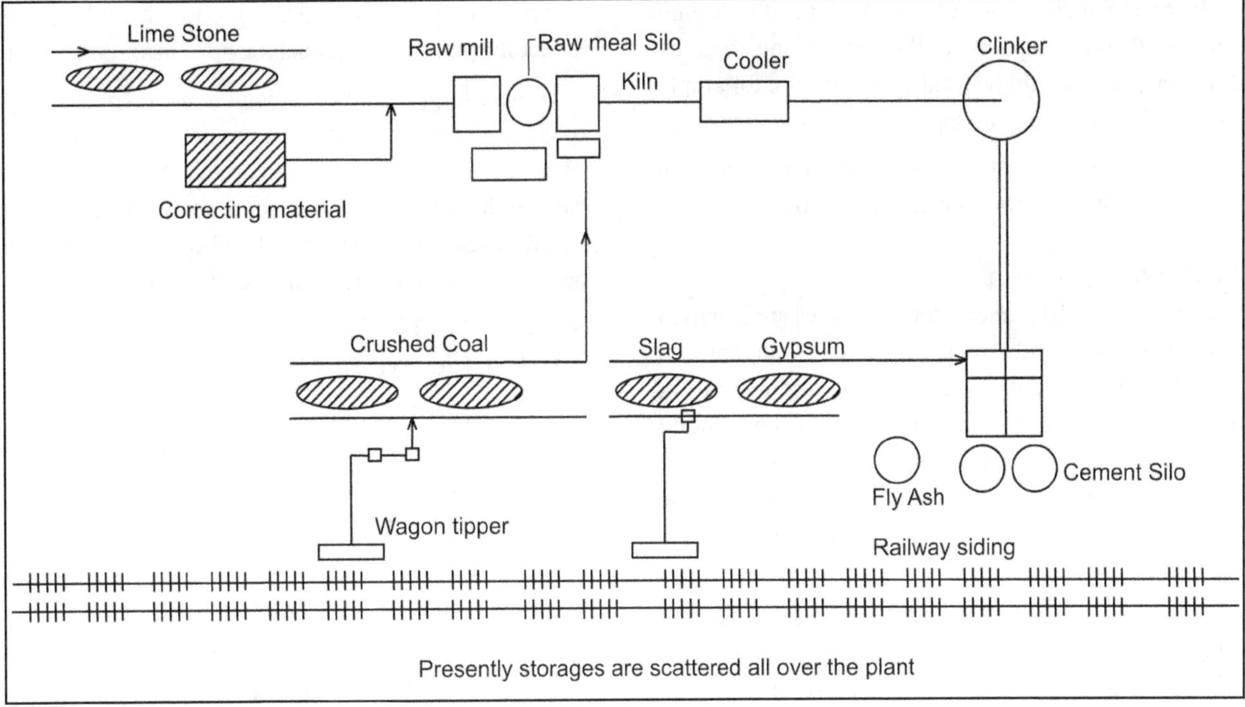

Fig. 6.1.8 Storages in a cement plant.

Where railway siding is concerned it would almost certainly be at right angle to the straight line arrangement. If it has to be in line with the straight line, then straight line would be broken at clinker storage and cement mills and silos and packing plant sections arranged at right angles.

See Figs. 6.1.5 b, 6.1.6 a and 6.1.7 b.

1.6.6 Lay out is a Plan

The General Plant Layout is a plan showing arrangements of various sections. Heights are not shown. Ground levels or datum levels of various sections are shown with reference to a common benchmark.

It is ideal if all ground levels are the same. If the plot of land is fairly level it can be leveled to this end. However if the differences are too high requiring removal or filling up of huge proportions, then different sections of the plants may have different datum levels. If the connecting link is a belt conveyor then its inclination can be reduced to advantage. In case of stock piles for storage, lower datum levels increase stored quantity. Same applies to heights of silos. As mentioned earlier different ground levels between kiln and cooler would reduce excavation and cooler can be installed at ground level.

While developing the general layout it helps to make side by side 'single line' sketches of elevations so that correct distances are maintained between the departments and heights of silos and building are taken into account.

Typical examples of crushing plant and stock pile of crushed lime stone and clinker storage have been shown in **Figs. 6.1.10, 6.1.11 and 6.1.18.**

1.6.7 Expansion

In developing a layout it is most important to provide for expansion. Therefore the entrepreneur must have some idea of the capacity after expansion. This expansion can be 2-3 times the initial capacity if the original kiln is without the calciner.

If the first kiln already has a calciner then the first expansion coming shortly after the first line has gone into operation, would be duplication in almost all sections. This way machinery would be duplicated requiring same spare parts.

If the expansion however takes place after a long interval, then the capacity of the second line could be much larger. It would be prudent to provide for ample space for such an expansion.

The center line distance between two production lines is arrived at from such considerations.

1.6.8 Maintenance

In kiln, cooler, and preheater, refractories are installed in the beginning and then are required to be replaced periodically. The replacement is at regular intervals in the burning zone of the kiln and in the discharge hood and throat of the cooler. This requires provision of quick handling of discarded refractories, bringing in of the new lot and providing room for its storage near or inside the building till they are installed.

From time to time rollers of vertical roller mills are taken out and are replaced or tyres are changed. This needs provision for access of trucks right up to the mill. This is also true of kiln rollers. In case of kilns occasionally shells need to be changed. Kilns used to be erected with the help of a travelling structure straddling the kiln piers with the help of chain blocks hung from it. Now shells and bed plates are lifted up with the help of cranes. This requires access as also head room for the crane to approach the kiln piers.

1.6.9 Roads

Another important consideration is provision of access roads within the plant and around the sections for movement of machinery and materials and also for pedestrian traffic of workers. Free access must be provided at least on two sides of a building even after expansion.

Roads would be designed according to the density of traffic. It is good to provide for a 'ring road' round the plant and in between the units and feeder roads of smaller width for sections within the plant. The main road from the main entrance should be a two way 4 lane road so that traffic flows in both directions without hold-ups.

For raw materials that come in truck loads by road and are stored in different places it would be desirable to provide two way roads with dividers at least for the common length. There are occasional hold–ups at unloading points due to bunching up of incoming trucks.

Where large volumes of cement are dispatched by road, it would be prudent to create a well designed parking place outside the plant premises with toilet and canteen facilities. Only such number of trucks should be admitted inside in a regulated manner as can be handled by the loading facilities provided.

See Figs. 6.1.23 and also **Figs. 6.35.2, 6.35.6 and 6.35.7 in Chapter 35.**

Even where truck loading is mechanised, the trucks may have to reverse for stationing themselves at loading points. A parking and maneuvering place has to be provided for this purpose near the loading station. **See Fig. 6.35.6 in Chapter 35.**

For bulk loading trucks will stand directly under the silos. Bulk loading would be faster.

Care should be taken to see that traffic of cement trucks does not get entangled with trucks of other materials.

Design of roads is thus an important aspect in the plant layout.

1.6.10 Parking and Garages

Space also needs to be provided for parking of Company's own trucks, vans, cars, ambulances and school buses etc. Company would also have to provide for service stations for petrol and diesel and for garages for routine service and maintenance of its vehicles.

The number of two wheelers coming in the plant is increasing though number of workers employed is shrinking. Quite a large number of executives bring in their cars. So parking places on a much larger scale are to be provided all over the plant.

1.6.11 Earth Moving Machinery

Group of Vehicles for mining operations would generally be handled separately. Quarry equipment like shovels, dumpers, drills and compressors etc., need more frequent servicing, repairs and maintenance and hence separate adequate facilities would be provided for them in the quarries itself or near the plant according to the distance of the mines from the plant.

1.6.12 Service Buildings

Besides main factory buildings or production departments there are a good many other service departments – beginning with Time Office and Security. Then there are Main and Sectional Laboratories for testing samples and for carrying out various statutory tests to maintain uniform quality of cements produced.

There are Engineering and Design Offices and offices for executives. Some supervisory staff may have their offices in their respective production / operation sections but for sake of convenience higher grade staff would be sitting in a central office.

1. **Administrative Office**

   Plant would also have an administrative office to house administrative staff, marketing and accounting staff, purchasing section etc. Generally these sections have a large number of visitors from other companies, government departments, vendors etc. It is better not to let them into the plant. Therefore the Administrative offices would be outside the main plant.

2. **Stores**

   Plant has to have a General Store for keeping and issuing maintenance stores and for consumables. This store has to be in a central place for ease of access from various sections. Some items like refractories, lining plates, grinding media etc. take lot of space and would be stored outside in the open in an enclosure. The Store has to receive above mentioned materials. If it is feasible it would be good to take a rail track into the store yard for unloading wagons. There should be good approach road for trucks.

3. **Workshops**

   Plant would have a Workshop for routine maintenance and repairs for the machinery and an Electrical Shop attending to electrical motors switch gear etc. These two would normally be housed in the same building. The Workshop would have commonly required tools like lathe, milling and drilling machines, welding machines etc. The building would be planned to meet all these requirements and would be located near the load centre.

   Location and sizing of these service centres is also an important aspect of plant layout.

1.6.13 Railway Siding

Design of a Railway siding starting from the main line to the Plant is a specialised job and as such should be entrusted to specialists who have been associated with Railways doing such work. Important considerations are volume of traffic to be handled in terms of incoming and outgoing rake loads and retention time permitted.

For large volumes the Company would arrange for its own diesel locomotives for shunting and placing of wagons inside the plant. Railway locomotive will bring in empties for cement, loaded coal, gypsum etc., and take away loaded cement wagons etc., from a given point.

The lengths and numbers of loading platforms shall be in accordance with the number of packing machines, loading points per machine and facilities for wagon loading. If wagon loading is mechanized, the platform has to be designed to suit wagon loading machines. The empties are to be placed in position so that in shifting valuable time is not lost.

Whereas in other sections it would be enough to show a block of the building (because machinery would be housed in it), the truck loading and wagon loading facilities would have to be shown in detail because they are at ground level.

1.6.14 Cables

Cables are laid out from grid substation to Plant's own outdoor substation; from there to H.T. and L.T. Control rooms and to individual MCCs in various departments. An electrical distribution scheme with a diagram is drawn out for this purpose.

The power generated in Company's own captive power plant is also to be connected to the grid distribution system and synrchronised with it so that in case of failure of grid power the selected departments get uninterrupted power supply from the captive power plant. If it is intended to include H.T. motors of large rating also in the scheme, then generation voltage of power plant would be same as that of H.T. motors. Thus a large network of cables, H.T. and L.T., for power and also for Instrumentation and Controls and

Communication is required to be laid down. This is to be integrated in the plant general layout from the beginning.

Cables can be laid under ground in trenches or over head, supported on trestles and gantries. Both systems have their advantages and disadvantages. Plant has to take an early decision in this respect. So that while general layout is being developed, due care and attention is also paid to routes of cables and facilities to be provided for them. If cables are to be underground, many a time cable trenches are combined with water drains. However it would be better to keep them separate. Cable trenches should not be allowed to get flooded with water. There must be facility for inspection and maintenance. For this purpose cables are fixed in cable trays which are supported from brackets fixed in the walls of trenches.

The choice between overhead or underground cables may also be dictated by soil conditions. In case of over head cables, sufficient clearance or head room needs to be provided for passage of dumpers, cranes where cables cross roads. Plans for future expansion should also be kept in mind in designing trenches and gantries.

1.6.15 Sewers and Drains

Sewers for disposal of waste water from bathrooms and toilets etc., and also drains for rain water are to be provided for and integrated in the general plant layout. These are different from trenches required to be provided for taking cooling water to cooling ponds. They may be laid out side by side to follow slopes or contours of the ground.

1.6.16 Gardens

Gardens are becoming an integral part of today's cement plants and colonies attached to them. Well planned and maintained gardens spread out in the plant and around buildings inside it create pleasant surroundings and help to keep morale of the workers high. It also makes them environment conscious. The impact on visitors is great. Blooming gardens in plant premises is a proof of commitment of the plant to keep environment clean.

In some cases the plots selected for factory and colony would have a large number of trees on them. Conscious efforts should be made at the planning stage to preserve as many of these as possible. Today it is possible to transplant even fully grown trees.

Over burden from the mines can many a time provide fertile soil for growing trees and gardens.

Similarly waste water from colony or underground water collected in mines could be used for these purposes.

1.6.17 Plants on Coast

In a great many countries like Japan, Korea, Greece, etc. Cement Plants are located on the water front. These plants have their own particular requirements in designing plant layouts. First and foremost, the land available would be very restricted and very expensive. Almost always they would receive crushed limestone from mines situated far away from the plant. The quantum of storages to be maintained would have to be minimum and because of lack of space they would be vertical that is in silos for limestone and clinker and cement also. If coal is imported, it would come in shiploads and facilities have to be provided for unloading ships at fast rates and for stacking and reclaiming coal in large quantities in the plant or at the port itself. **See Fig. 6.1.25.**

In these locations most of the cement would be shipped in bulk in shiploads. This would require facilities for handling of cement in bulk bringing it from the plant and discharging it into ship's holds. The facilities for handling coal and cement would be quite different from each other because of the difference in characteristics of the two materials.

In India it is unlikely that any cement plant itself would be located on the coast because of the locations of limestone deposits in it, except in Gujarat.

1.6.18 Split Locations

Some plants have however come up with split locations i.e., clinkering facility in one place (near limestone deposits) and cement grinding facility in another place near the markets. In some of these cases, clinker is shipped in barges and unloaded at another coast where the grinding unit is located. Cement can be loaded into ships in bulk and exported.

There will be more instances of split locations as export of cement increases and there is further increase in blended cements.

1.6.19 Export of Cement

In some cases cement is brought in bulk from plants located in the hinterland and a storage facility is created on the coast. Such facility is created in common by a group of Companies joining together. At the depot, facilities are provided to load ships in bulk to export cement. India is also exporting clinker. It would need different kind of facility to handle clinker in bulk.

1.7 Careful Selection of Options

It would thus be seen that there are a great many permutations and combinations of a large number of variables so that there are a myriad possibilities of layouts. Each layout must therefore be designed specifically for its own case taking into account the various factors listed above. It is a challenge to design a good layout. Once it comes into existence very little improvements can be done subsequently to it.

Therefore several possibilities should be carefully examined, their merits and demerits weighed before taking a final decision.

A good well designed layout is a permanent thing and is the very basis of smooth and efficient operation of the plant.

1.8 Preliminary Work on Developing Layouts

As mentioned earlier it is necessary to fix distances between sections by drawing single line sketches of belt conveyors to reach desired elevations of stock piles, silos and hoppers.

1.8.1

To do that it is first necessary to arrive at dimensions of stock piles, silos and hoppers such as their diameters and heights.

1.8.2

For stacker reclaimer systems it is necessary to work out overall lengths and widths of stock piles after expansion.

1.8.3

For developing railway siding it is necessary to know traffic in rake loads and hence number of tracks and their lengths after expansion.

1.8.4

Sketches are developed to generate necessary information.

See Figs. 6.1.10 to 6.1.19

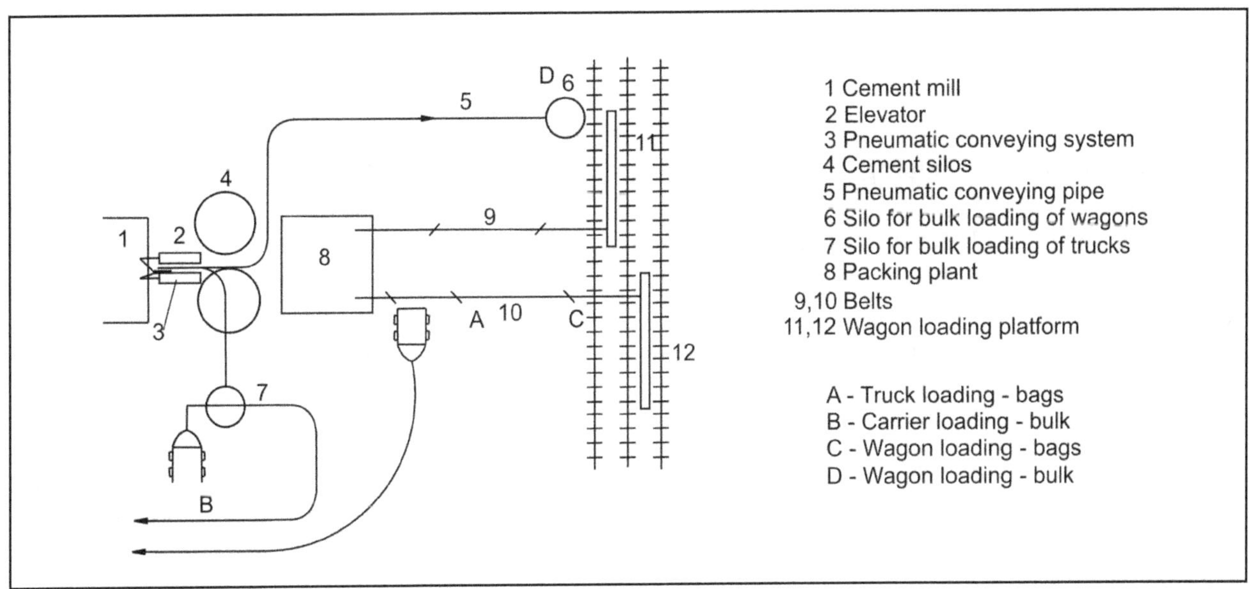

Fig. 6.1.9 Facilities for simutaneous truck and wagon loading of bagged and bulk cement.

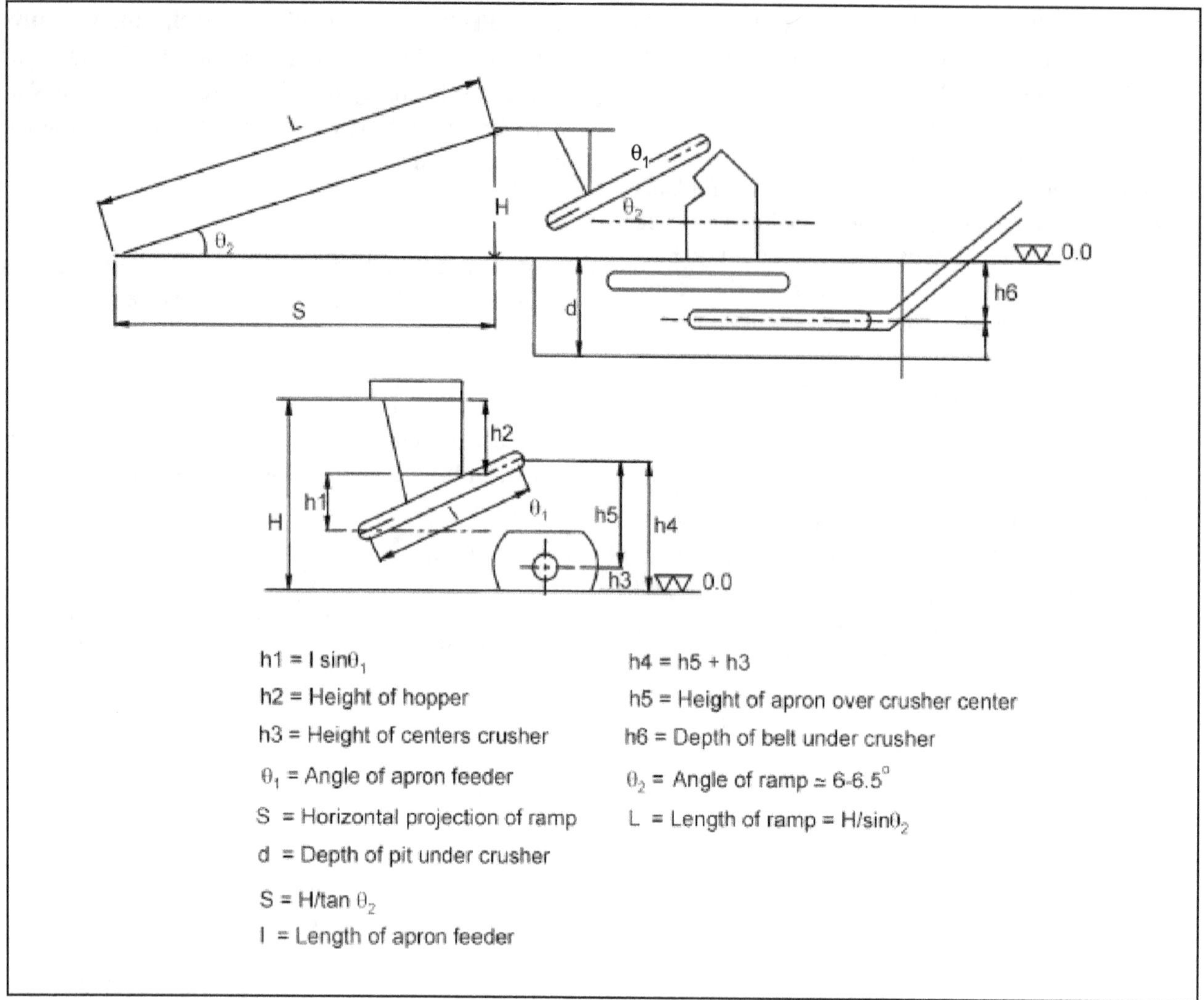

Fig. 6.1.10 Crusher ramp.

1.9 Examples of Developing General Layouts

Making use of the guide lines mentioned above, a General Plant Layout has been developed for: a large plant with railway siding. **See Fig. 6.1.23**

Fig. 6.1.23 shows a general layout of a 10000 TPD clinkering unit making 6.5 mtpa of OPC and BFSC.

In case of the large plant besides the general layout, plans of plant and its mines, orientation and routes of roads, railway line and grid power transmission line have also been shown.

See Figs. 6.1.20 to 6.1.22

1.10 What Should the General Plant Layout Show ?

The general plant layout serves a specific purpose. It is a basic document for the plant and also for various outside agencies for obtaining their approvals in various fields like say from a factory inspector.

It should contain all the pertinent information but should not suffer from cluttering.

General Plant Layout should be either size A1or A0. The scale would have to be so chosen that the plot of land required for the factory is comfortably accommodated in it.

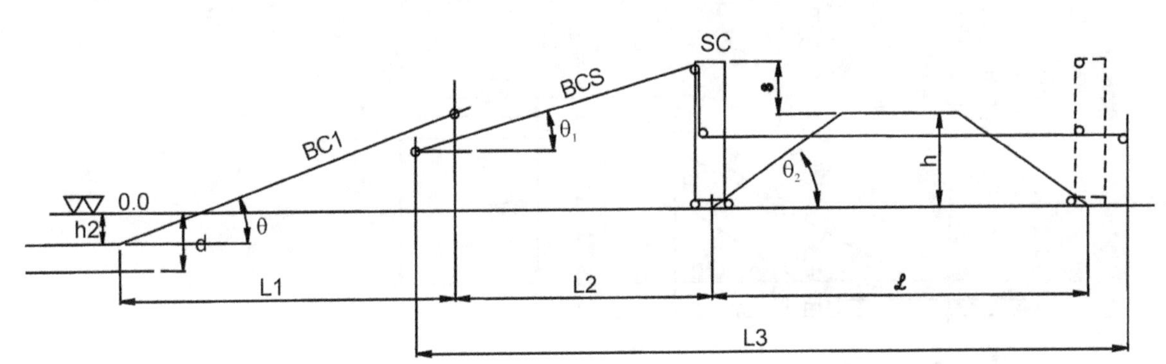

d = Depth of pit
h2 = Depth of conveyor BC1 below ground level
L1 = Horizontal distance of crusher belt to stacker belt.
L2 = Horizontal distance between belt Bc1 and beginning stock pile
ℓ = Length of stock pile
L3 = Horizontal length of stacker belt

BC1 - Belt conveyor from crusher
BCS - Stacker belt
SC - Stacker carriage
BCSL - Stacker luffing belt
θ - Angle BC1
θ_1 - Maximum angle BCS
θ_2 - Angle of repose

Distance between crusher and stock pile

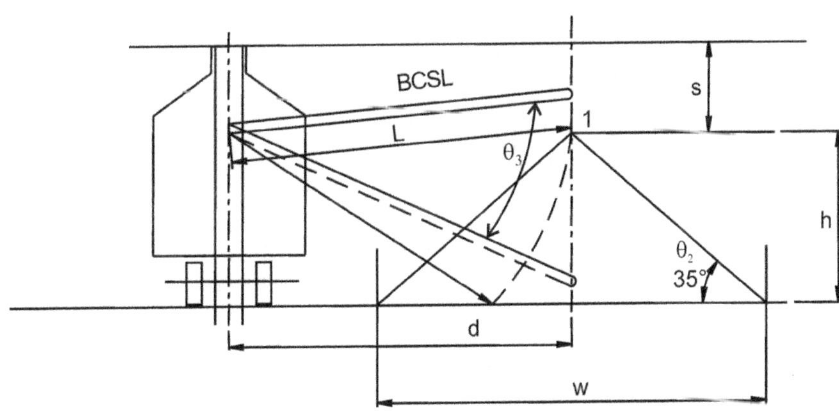

θ_3 = Angle of movement of luffing belt
L = Length stacker luffing belt
d = Distance between stacker belt & center of stock pile
s = Height above stock pile
w = Width of stock pile
h = Height of stock pile
Angle of repose for lime stone 35°

Elevation of stacker and stacker pile

Fig. 6.1.11 Sketches to develop layout of a stacker reclaimer system.

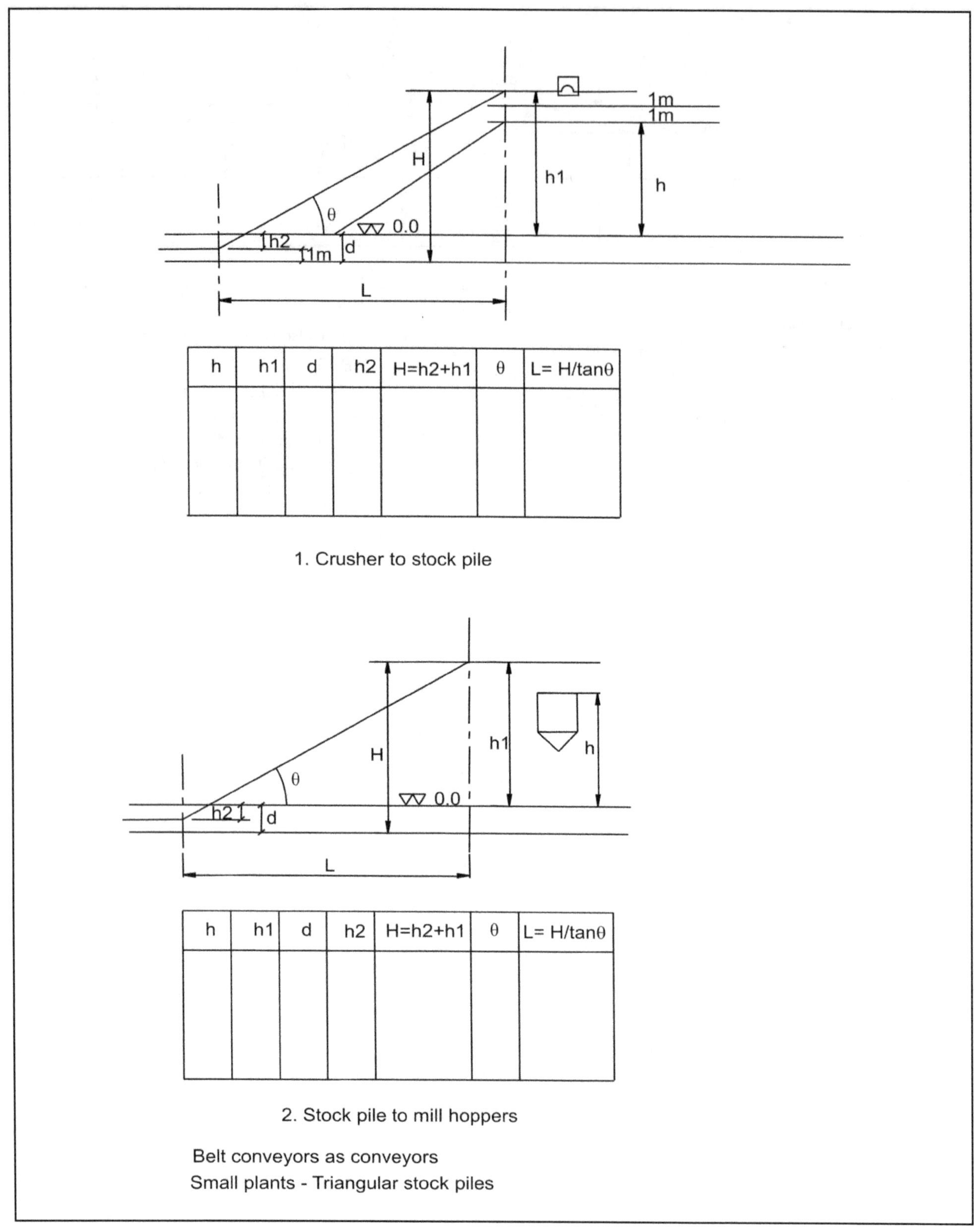

h	h1	d	h2	H=h2+h1	θ	L= H/tanθ

1. Crusher to stock pile

h	h1	d	h2	H=h2+h1	θ	L= H/tanθ

2. Stock pile to mill hoppers

Belt conveyors as conveyors
Small plants - Triangular stock piles

Fig. 6.1.12 Distance between departments and storages; applicable to rawmill, coalmill, cement mill hoppers.

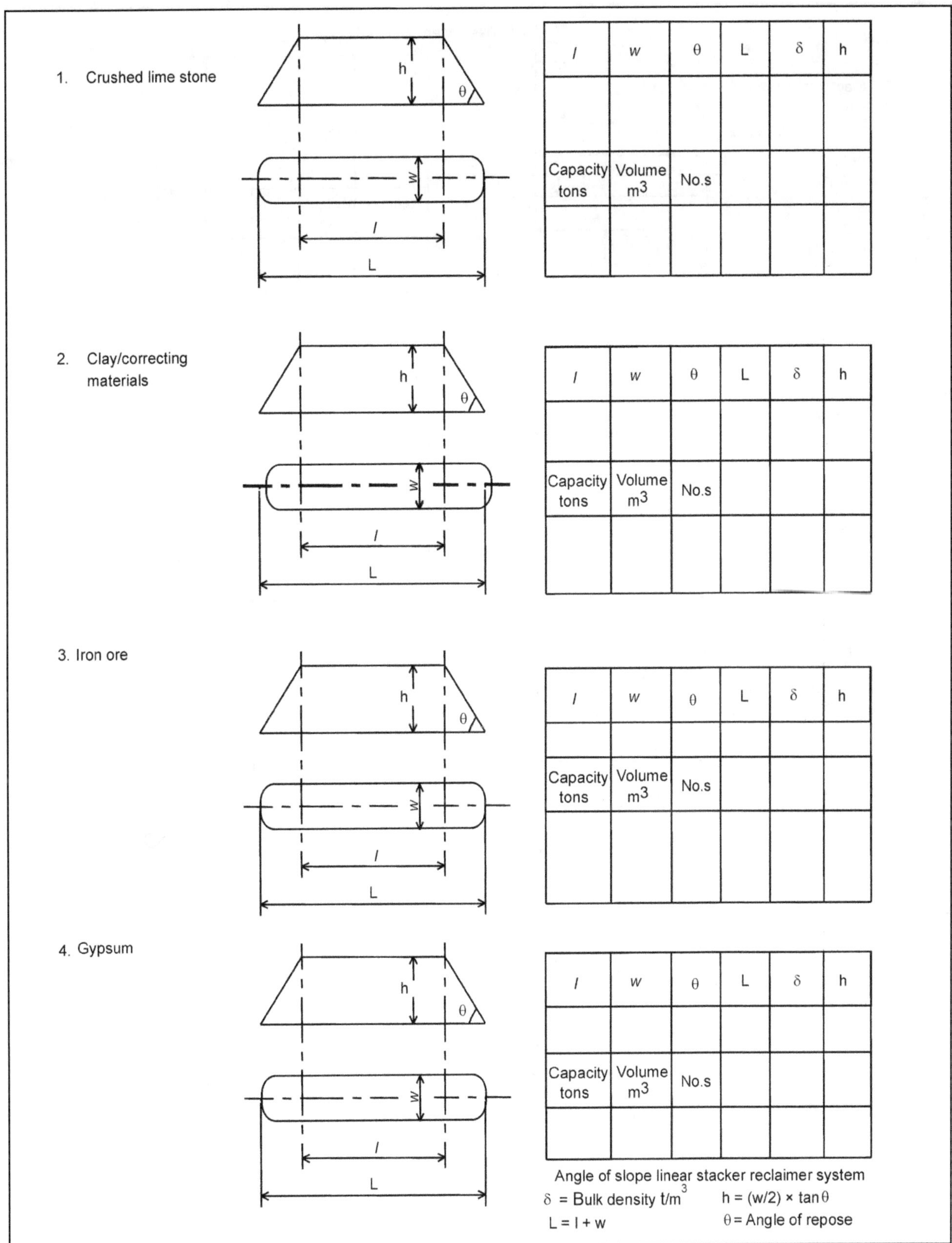

Fig. 6.1.13 Stock piles – estimation of sizes from capacities.

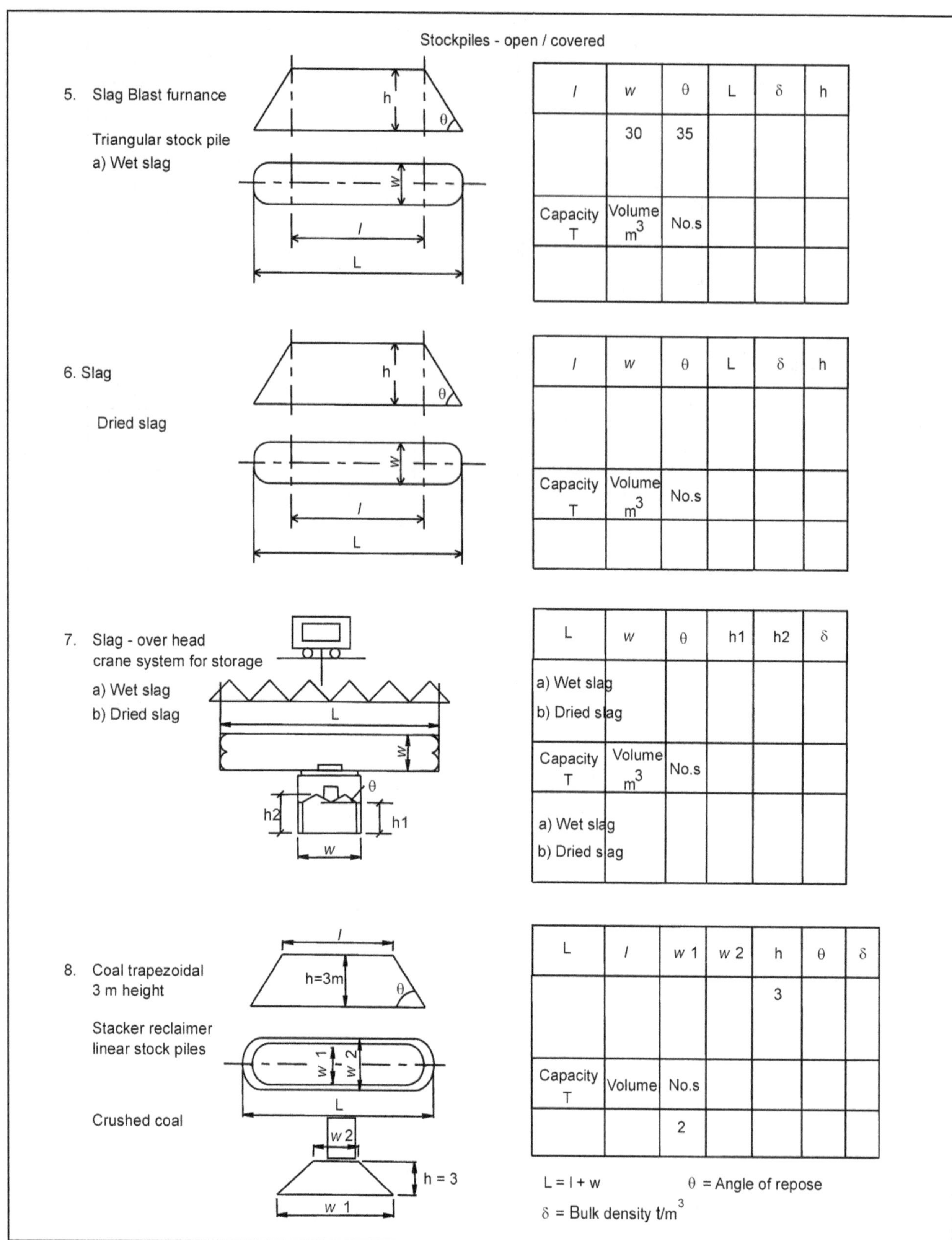

Fig. 6.1.14 Stock pile estimation of sizes.

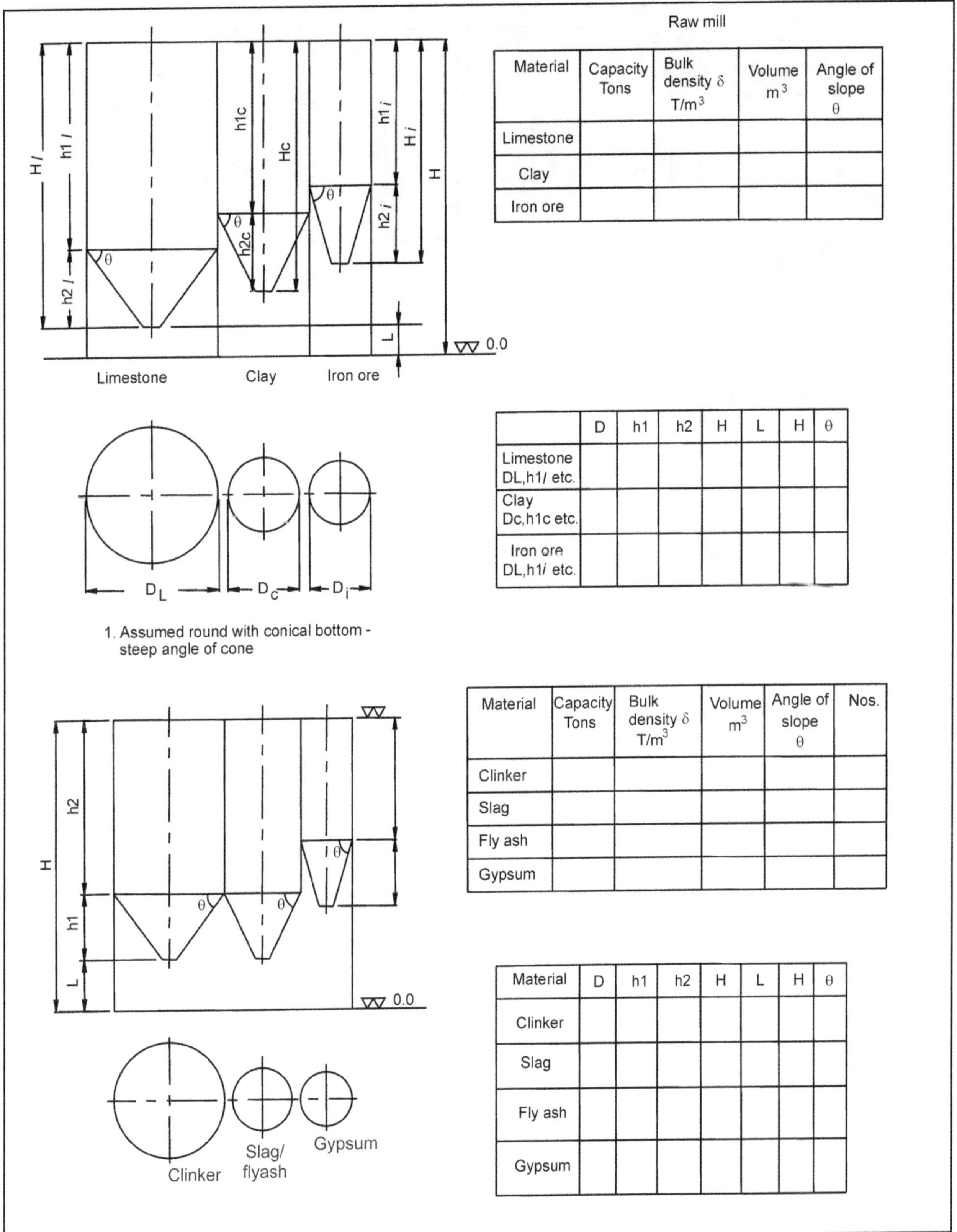

Raw mill

Material	Capacity Tons	Bulk density δ T/m³	Volume m³	Angle of slope θ
Limestone				
Clay				
Iron ore				

	D	h1	h2	H	L	H	θ
Limestone DL,h1/ etc.							
Clay Dc,h1c etc.							
Iron ore DL,h1i etc.							

Material	Capacity Tons	Bulk density δ T/m³	Volume m³	Angle of slope θ	Nos.
Clinker					
Slag					
Fly ash					
Gypsum					

Material	D	h1	h2	H	L	H	θ
Clinker							
Slag							
Fly ash							
Gypsum							

Fig. 6.1.15 Hoppers for mills - sizes and capacities.

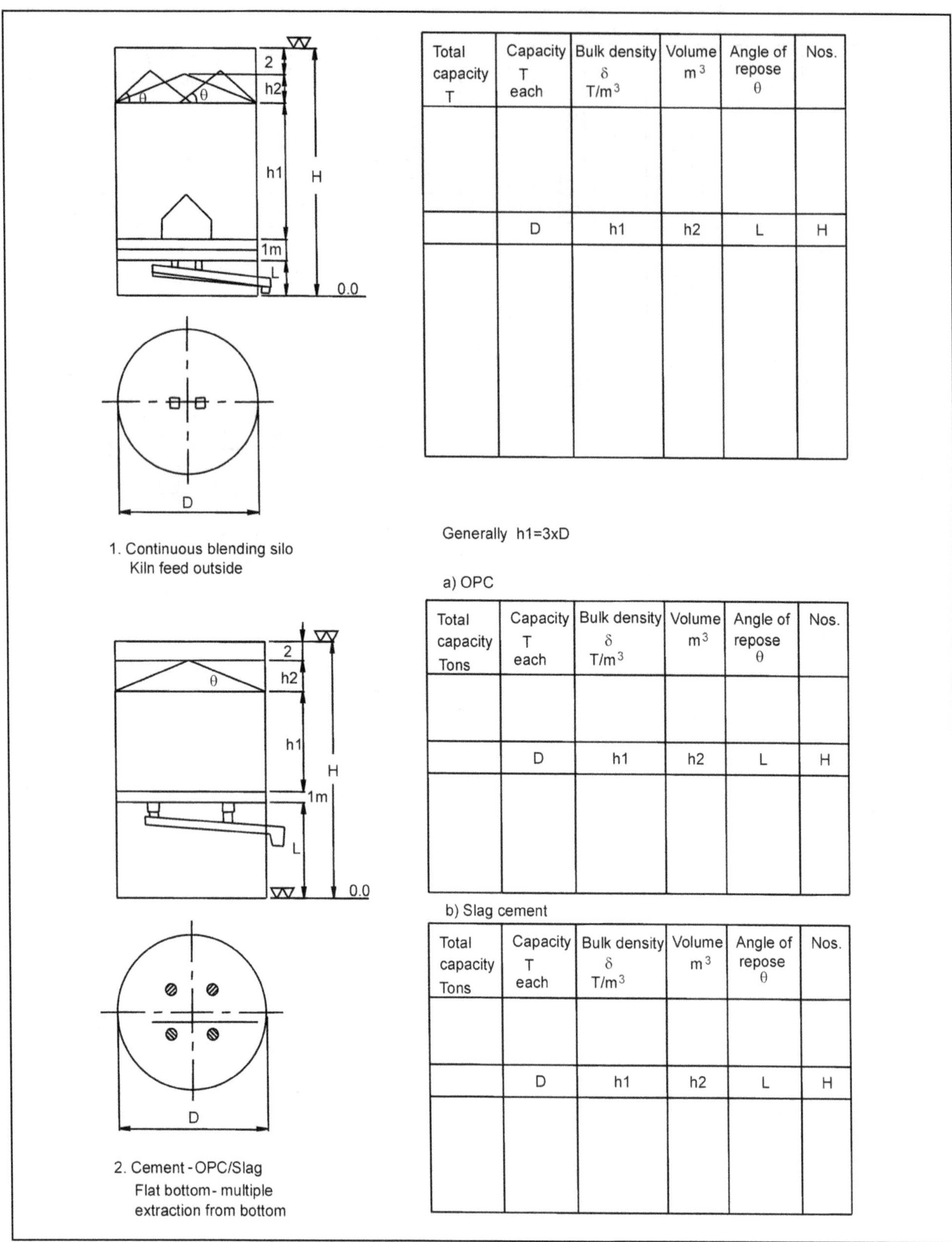

Total capacity T	Capacity T each	Bulk density δ T/m³	Volume m³	Angle of repose θ	Nos.
	D	h1	h2	L	H

Generally h1=3xD

a) OPC

Total capacity Tons	Capacity T each	Bulk density δ T/m³	Volume m³	Angle of repose θ	Nos.
	D	h1	h2	L	H

b) Slag cement

Total capacity Tons	Capacity T each	Bulk density δ T/m³	Volume m³	Angle of repose θ	Nos.
	D	h1	h2	L	H

Fig. 6.1.16 Silos - sizes and capacities.

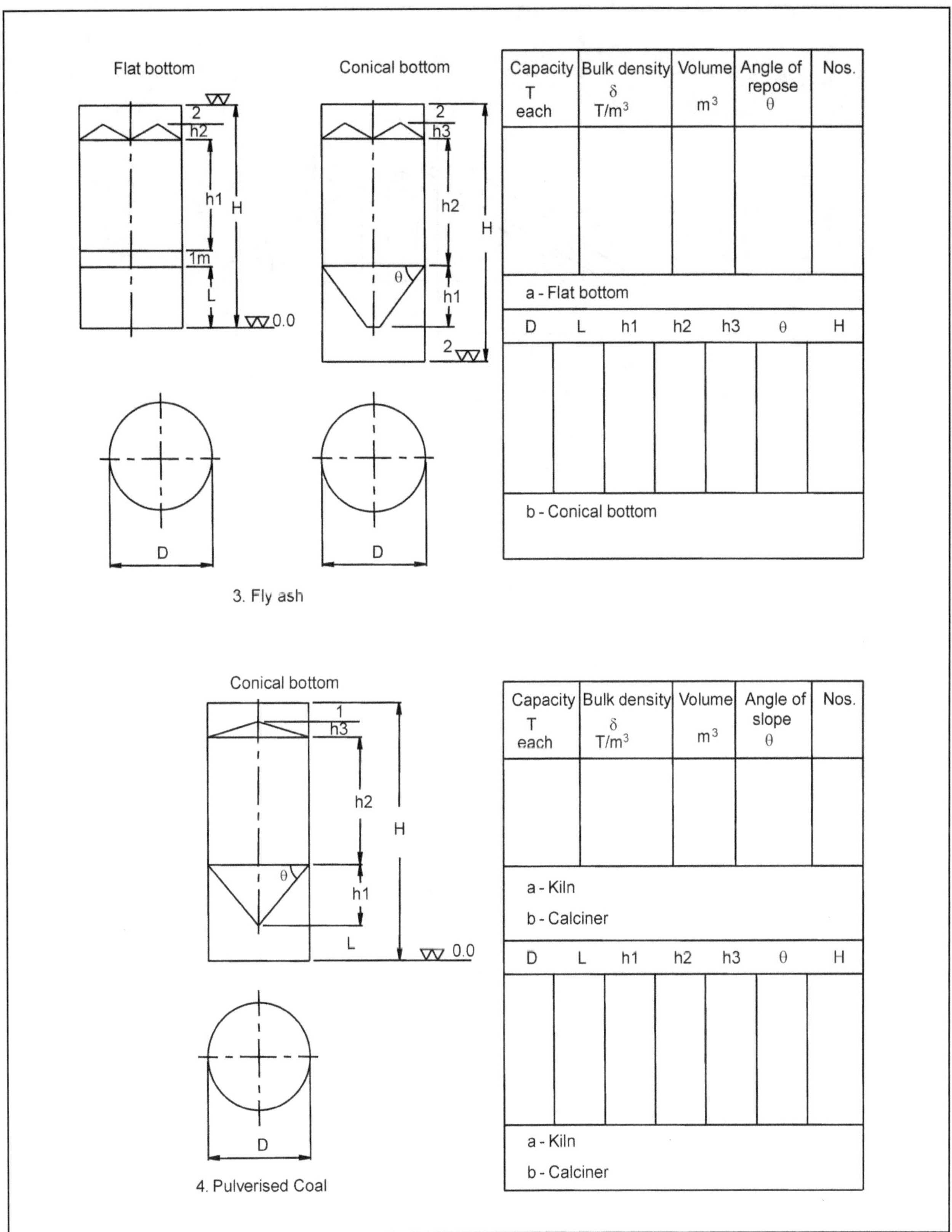

Fig. 6.1.17 Silos.

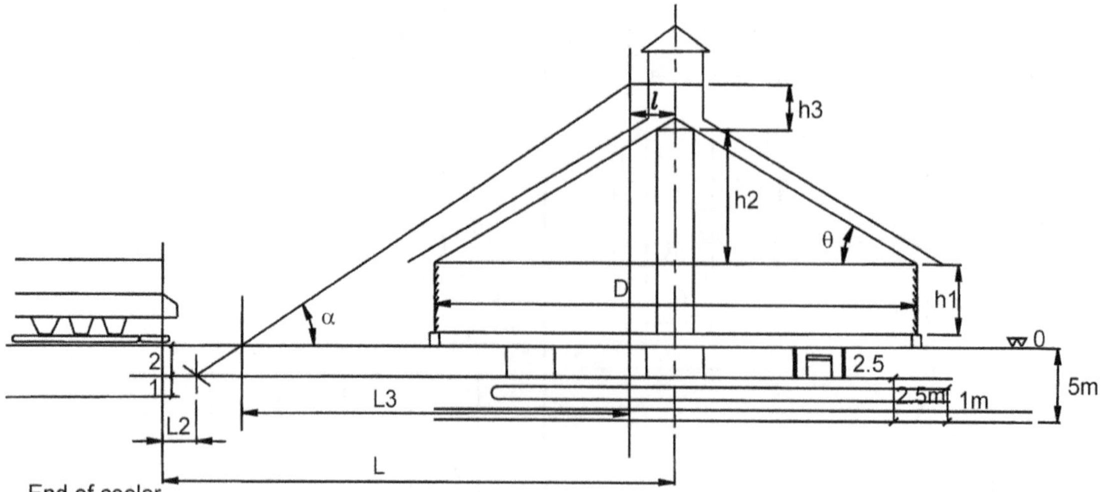

Clinker storage

Capacity Tons	Density kg/m2	Volume m3	Cyclone ht. h1	Angle of repose θ	h2
				35°	

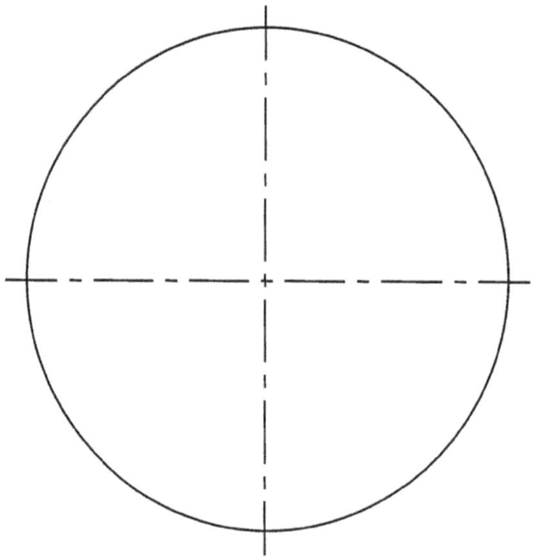

Fig. 6.1.18 Clinker storage.

$L = L1 + L2 + (h1 + h2 + h3) / \tan\alpha$

α = Angle of d.b.c.
$h2 = D/2 \tan\theta$
θ = Angle of repose
$h = 0.35\,D$
h1= Clinker height ≃ 2 to 3 meters

Clinker

D	h1	h2	h3	L1	L2	L3	L

Pit under cooler

Assume spillage drag chain ground level

d = Depth of pit = 3 - 4 m

h3 = 2 - 3 m

L = 60-65 meters for a diameter of 100 meters

L2 = 5 meters approx.

Pit under clinker stock pile

Cross conveyor - pit ≃ 2.5 m

Main conveyor to cement mill pit ≃ 2.5 m

Deepest point below ground level ≃ 1- 5 m maximum

Fig. 6.1.19 Distance of clinker storage from end of cooler.

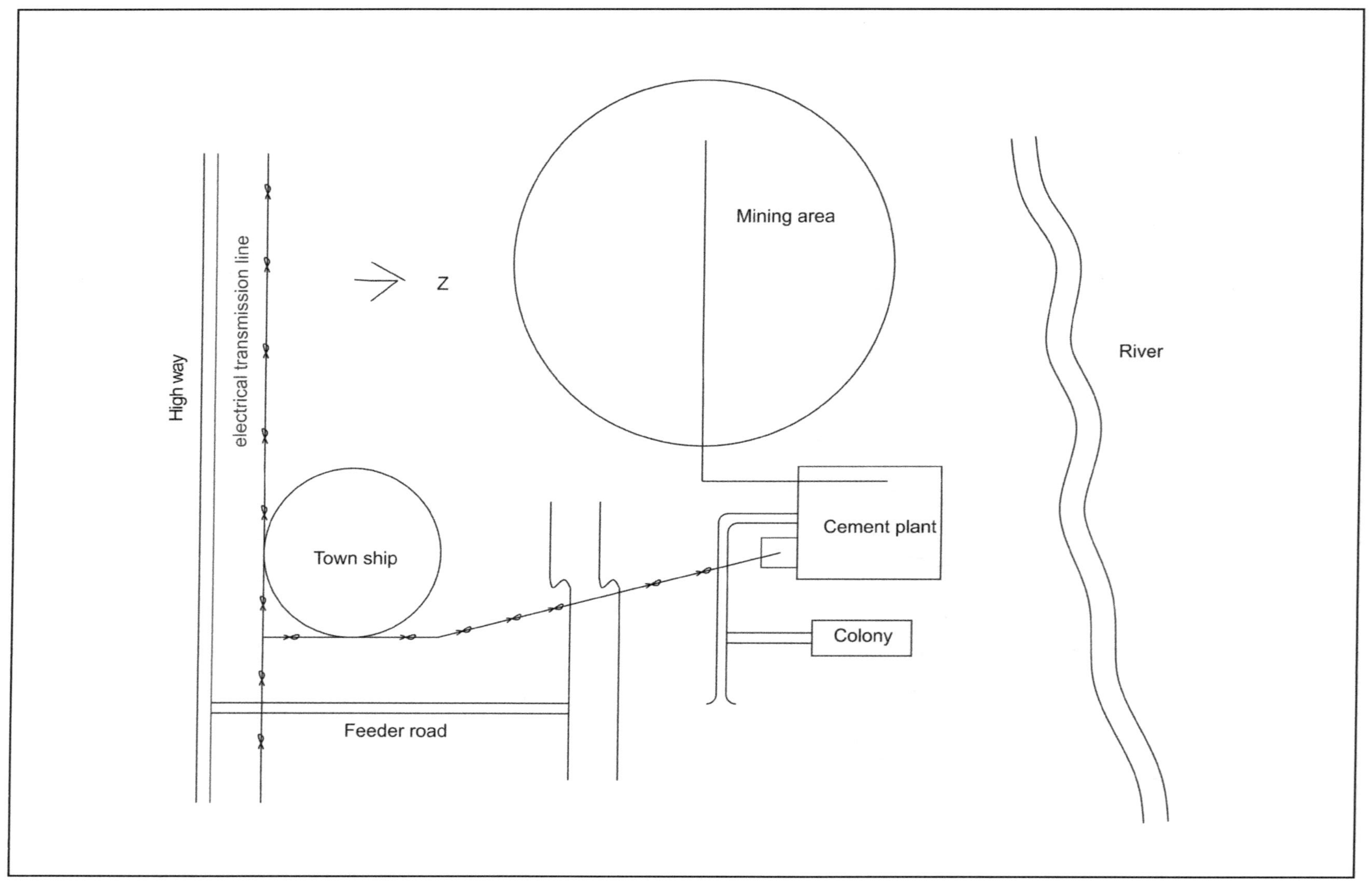

Fig. 6.1.20 Location of proposed cement plant with respect to surrounding.

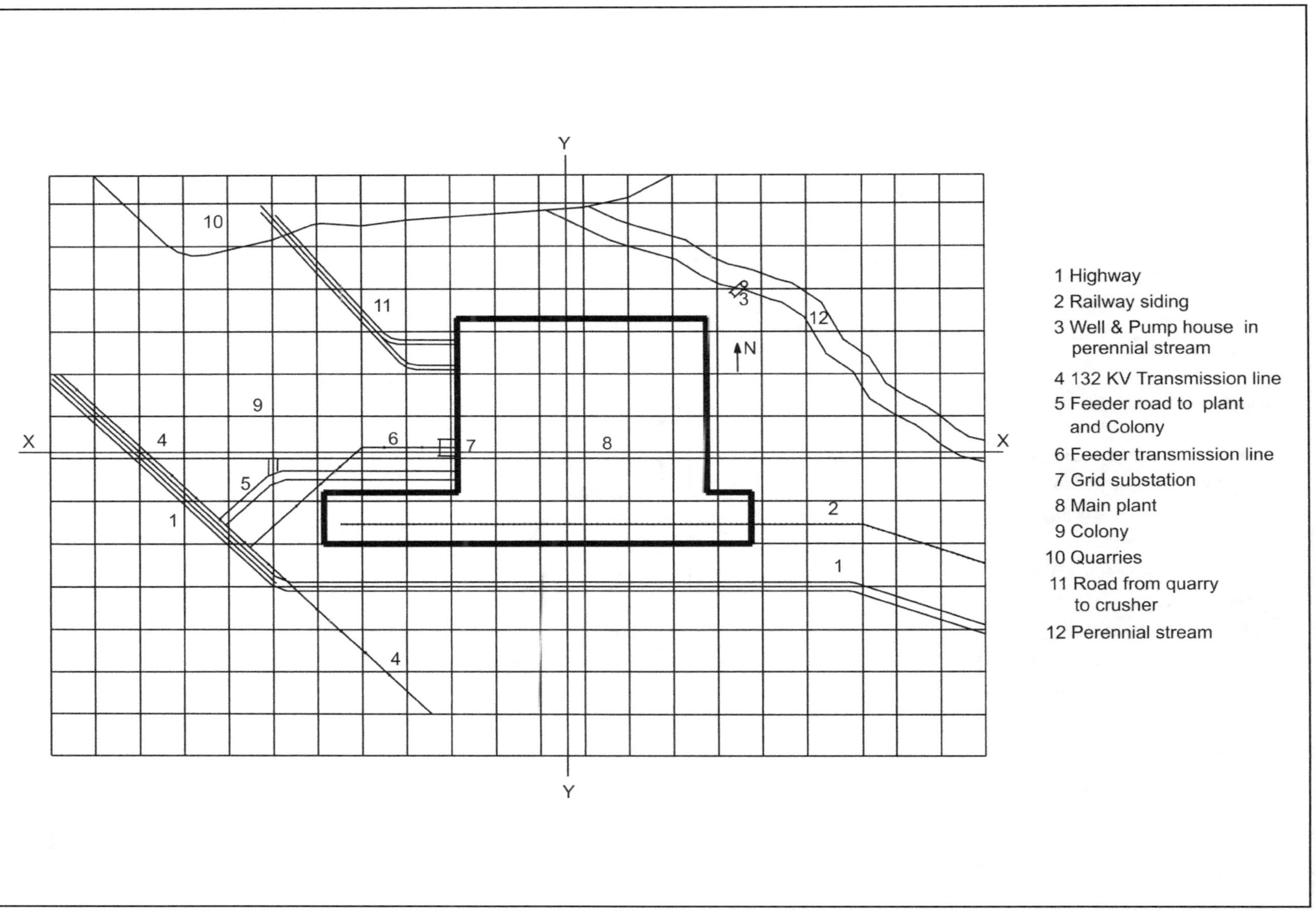

Fig. 6.1.21 Overall layout of cement plant and colony with respect to railway line, highway and quarries and electrical transmission line.

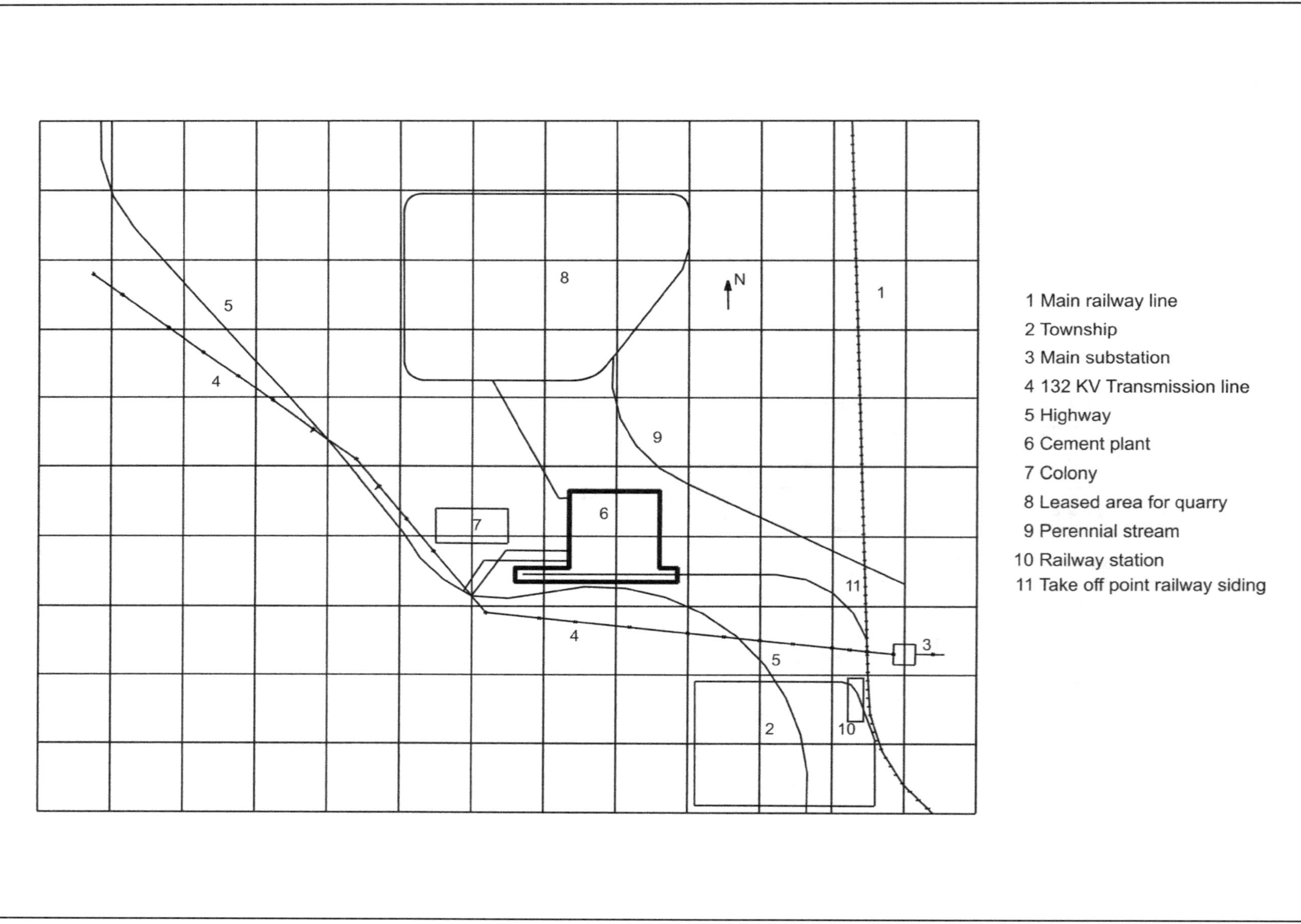

Fig. 6.1.22 Location of the cement plant with reference to surroundings, town ship, quarries etc.

1.10.1 The Name Plate Should Show

(i) Name of the project.

(ii) Name of the Company.

(iii) Size/capacity of the plant.

(iv) Name of the Consultant who is preparing the layout.

(v) Scale.

(vi) North direction (convention is that unless other wise mentioned North would be at the top of the paper).

(vii) Benchmark for the plant which is the reference level for the plant. There may be different levels for different sections; they should all be mentioned with the Main benchmark. Normally the benchmark would be the altitude of the place. This should be clearly mentioned. For convenience it is referred as + 00.

1.10.2

Contours at intervals of 1 or more metres.

1.10.3

Orientations of power line, road, rail link, stream or river, incoming point or entry point of crusher and colony. These are external to the layout proper but are important in the locations of the various departments and sections.

1.10.4 Details Like

(i) Centre lines of various sections with reference to one another. The layout has reference axes x-axis and y-axis with reference to which various sections, storages etc., are located.

(ii) Outlines of buildings with their centre lines and dimensions and also centre lines of conveyors, pipe lines etc., storage areas with overall dimensions of stock piles, diameters of silos together with centre lines and storage capacities.

(iii) Location of all non factory buildings with overall dimensions and centre lines.

(iv) Roads with their centre lines and widths; railway siding with overall length, number of lines and spacing between lines.

(v) Centre lines of cable routes and locations of grid and company's substations, HT and LT control rooms;central process control and Instrumentation station.

(vi) drains and sewer routes and location of cooling ponds and water pipe line routes and overhead storage tanks and pump house.

(vii) In packing plant lengths and numbers of loading platforms and also loading points of trucks.

(viii) Ramp and abutments for dumpers to load crusher hopper with centre line width and approximate length and its alignment in the layout.

(ix) Alignments of incoming power line, water pipe line and approach roads also need to be shown.

(x) In short all that occupies ground space or would do so should be shown so that it should not be difficult at a later to accommodate them.

1.10.5

As mentioned earlier the layout should clearly show space provided for expansion in dotted lines.

Green belts should also be shown to ensure that adequate space is provided for them.

A General Layout can seldom be built up in one go. Additions are made to it from time to time as things get finalised. Therefore there would be several revisions to the layout. They should be marked with date of revision and explanation thereof.

1.11 Supporting Drawings

The Plant Layout is the final picture of the project. It is built up gradually from plans and drawings obtained from several agencies. These supporting drawings and layouts are also important and need to be referred to now and then.

1. The location of the plant and the quarries in the State and the District of the State in which it is located.
2. The Location of the plant in the District and Taluq. This map would show the railway lines, roads major and minor and grid power lines and major and minor rivers and also towns and villages.
3. Geological Survey of India's Plates of specific areas in which factory and mines would be located. The location of the plant and mines would be marked on it. These plates show in greater detail what map in (2) above would show; besides it would also show contours and elevations.
4. Maps from Geological Department of the State to show and to demarcate the area for which mining lease has been granted.

5. More detailed maps of the proposed locations of factory, colony and mines showing the plots of land according to ownership. These are very important for purchasing the land either from the Government or from private owners.
6. Such maps and drawings help in assessment of land required for acquisition and help as guides in the process thereof.
7. A map is then made out showing the precise area acquired, the alignments of power lines, roads etc.
8. In this one drawing are shown relative locations of factory, colony, mines and also near by villages, townships, railway station, road etc.
9. A map showing alignment of railway siding from the railway station, branching point to and inside the plant. This will also show contours and culverts / bridges if any to cross streams, rivulets etc.
10. A similar map showing alignment of overhead transmission line from the substation from which power would be drawn to the Grid substation outside the plant.
11. If water is drawn from a nearby river then the location of the well in the river bed, pumping station and route of the pipe line into the plant to the water treatment plant and overhead storage tanks in the plant.
12. Roads from quarries and in quarries are also to be shown with reference to the mines and the plant.
13. Quarry development plans are special drawings which would be prepared by the mines manager from the drawings obtained from the Geological Department and using the detailed report on the investigations of the deposits that would be used for making cement. A five year mining plan would be prepared showing the areas and their sequence of development; cross sectional drawings would be made to show how benches would be developed and to what depth for economic mining. Roads in the mines get developed as mining progress. Development of mines is required to be taken up ahead (by 6 month or so) of the commissioning of the plant.
14. Then there are layouts and drawings that are required to be made and submitted for approval to different Government agencies like Mines Inspector, Factory Inspector, Electrical Inspector and so on.
15. The General plant lay out is thus not one drawing but several drawings showing information pertinent for specific purpose for which it is used. This way the layout itself does not get cluttered up and become difficult to read.

1.12

A workout for developing a General Laout for a 10000 TPD Cement Plant (clinkering capacity) making blended cements and with facilities for Co processing, and co-genraation is presented in **Annexe 6.1.1** of the Chapter.

General Plant Layouts of a few existing cement plants are also attached.

See Figs. 6.1.24 to 6.1.26.

Fig. 6.1.23 shows the general layout and **Annexe 6.1.2** legend for it.

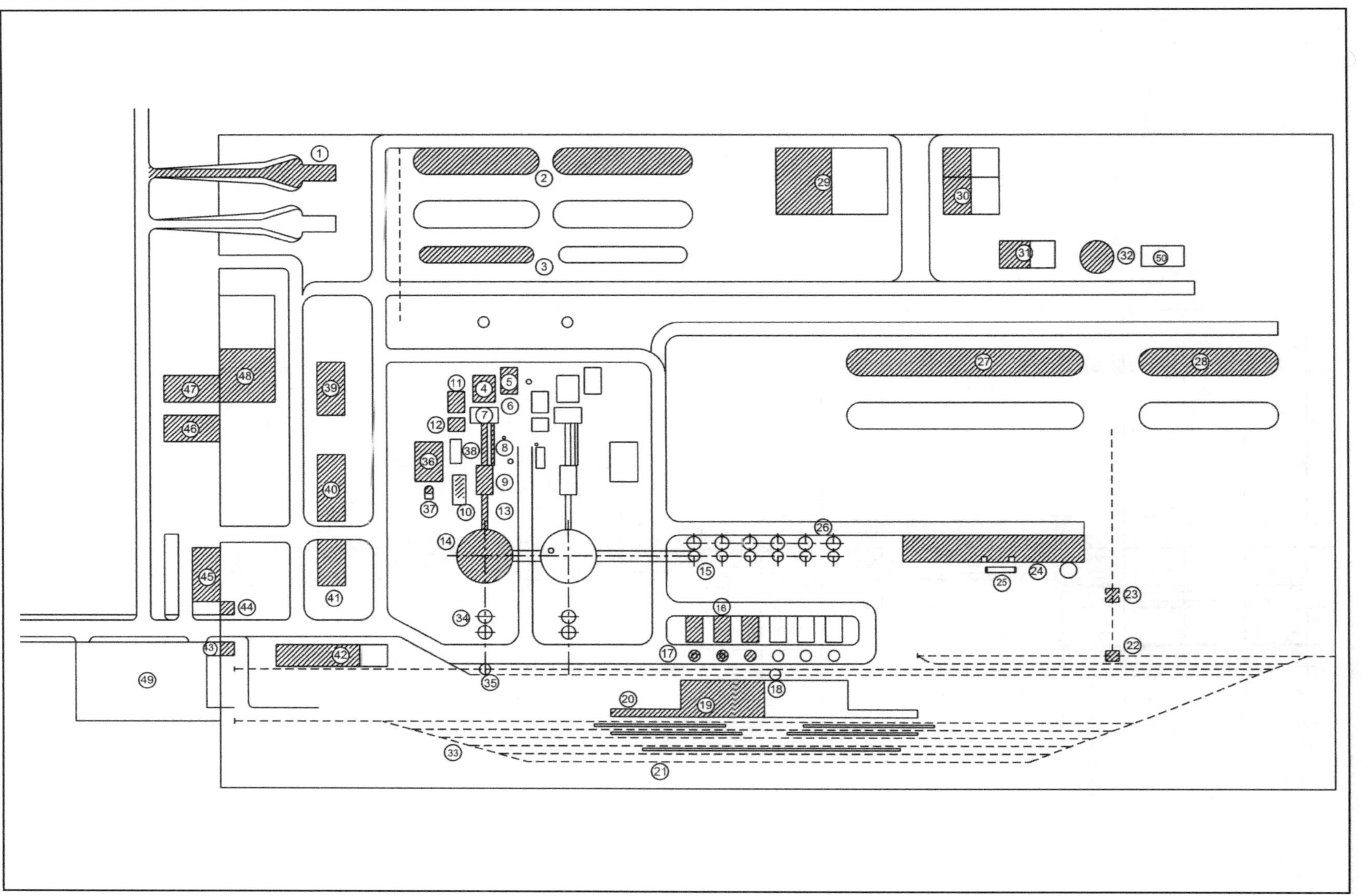

Fig. 6.1.23 General plant layout for a 10,000 tpd clinkering unit making 6.5 mtpa OPC & BFSC.

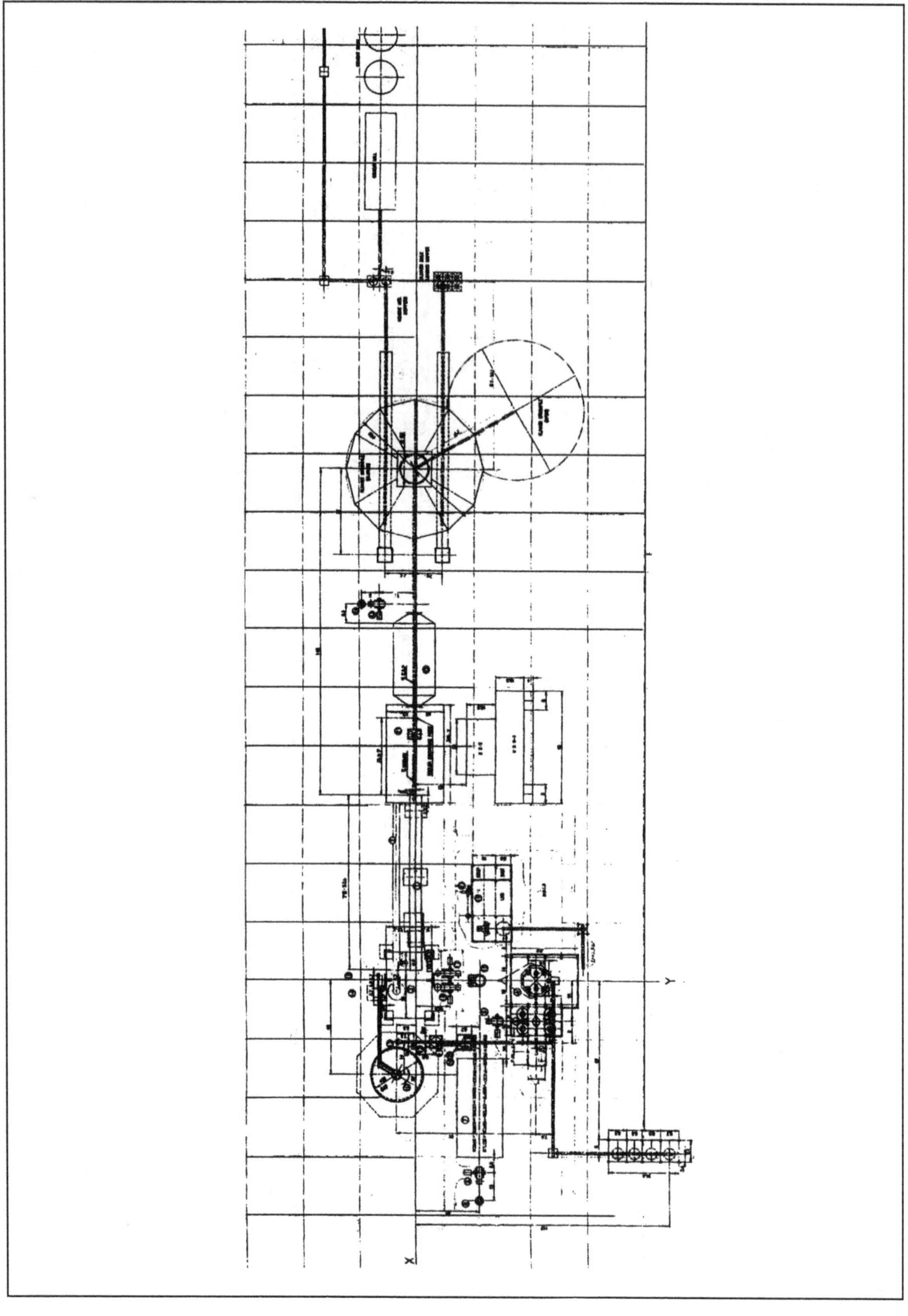

Fig. 6.1.24 General plant layout of a 1.2 MTPA capacity cement plant.

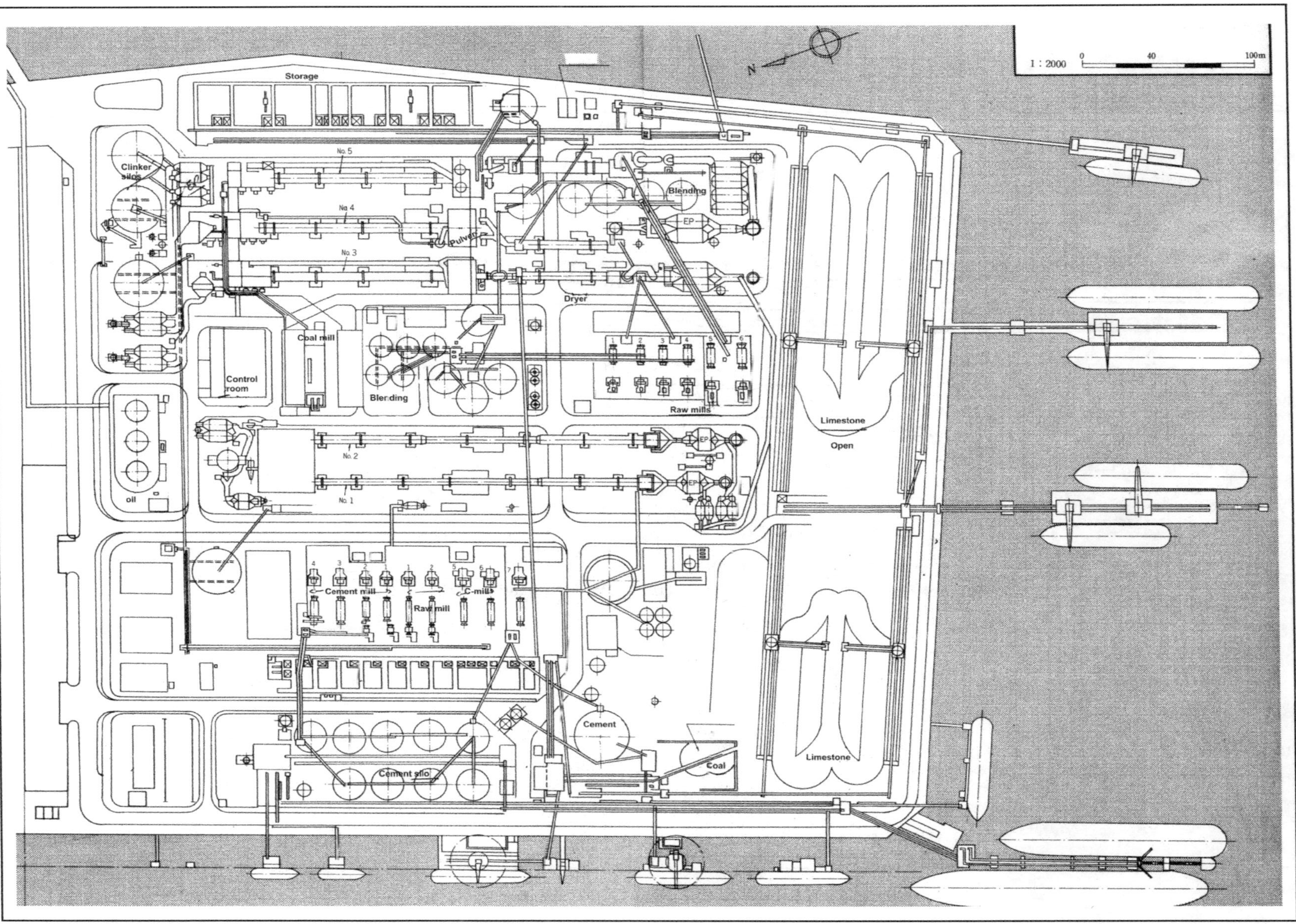

Fig. 6.1.25 General layout of a coastal cement plant in Japan.

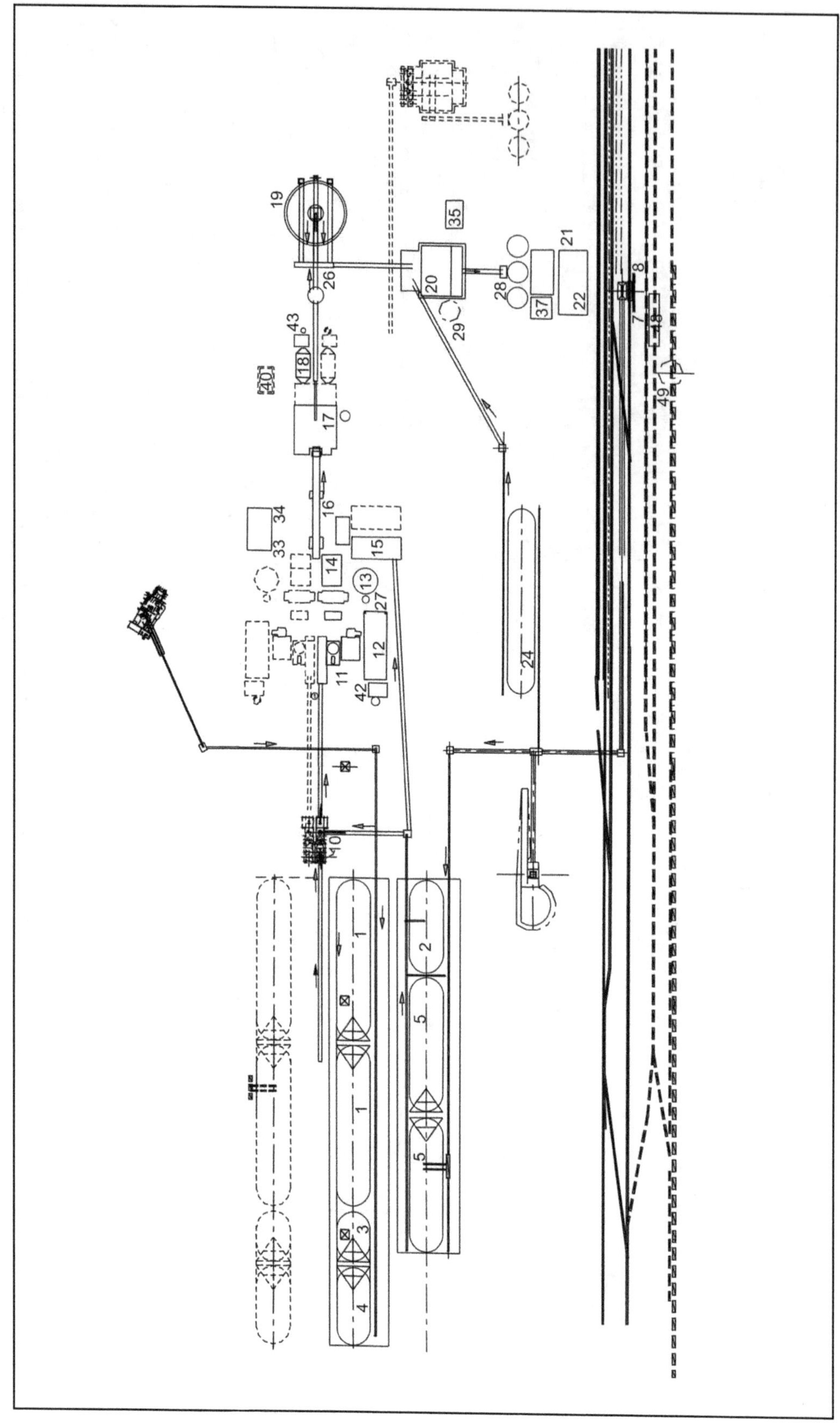

Fig. 6.1.26 General plant layout of a 2.0 mtpa cement plant.

Annexe 6.1.1

Basic Data and workout for the Layout developed for a 10000 tpd clinker capacity Cement Plant making OPC & BFSC

Part 1 Cement mills, silos and packing and despatches

Sr. no.	item	unit	quantity	remarks	
	Capacity	tpd	10000	clinker	1st unit
			10000	clinker	2nd unit
	design margin	%	10		
				for margins and conversion factors see Table 7.1.2	
	cements made				
	OPC	tpd	6930	using 60 % clinker	
	annually	tons	2286900	0	
	BFSC	tpd	11572		
	annually	tons	3818760	62%	
	total		6105660		
		say	6500000	100%	
	cement mill capacities				
	for OPC	tph	380		
	nos		1		
	for BFSC	tph	650		
	number of mills		**2**		
	OPC to be stored	days	7		
		tons	53200		
		say	50000		
	no of OPC silos		1		
	size	dia × ht			
		m	24*82		
	BFSC to be stored	days	7		
		tons	91000		
		say	100000		
	no of BFSC silos		2		
	capacity each	tons	50000		
	size	dia × ht			
		m	24*82		

Part 1 Cement mills, silos and packing and despatches

Sr. no.	item	unit	quantity	remarks
	cement despatches		OPC	BFSC
	per day total	tons	7000	13000
	by road	65%	4550	8450
	by rail	35%	2450	4550
	bagged	50%	3500	6500
	bulk	50%	3500	6500
	hourly bagging rate	tph	233	433
		say	250	450
	packing machines	no	**1**	**2**
	bulk loading rate		OPC	BFSC
	road	tph	152	282
	no of loading points		1	1
		say	150	300
	bulk loading rate	tph	204	379
	rail	say	200	400
	no of loading points		1	2
	rake loads bulk and bagged cement		5	10

Note: main cement silo will be used for bulk loading by road. Cement will extracted and taken to packing machines.

One bulk loading platform for each type of cement will straddle a railway line. It will be a service silo to which cements will be transferred from main cement storage silos as and when required. Its size would be 10 m dia × 30 m high

Part 2 Coal and slag to be received

1 coal to be received for firing in kiln and calciner

capacity of kiln	tpd	10000
wet coal	t/day	2000
days stock		20
stock	tons	40000
size of linear		
stock pile	width × length	**45 × 410**
two stock piles	m	840

Part 2 Coal and slag to be received

2 coal to be received for Captive Power Plant

ave.sp.power consumption	kwh/t cement	77	
hourly production of cement	ton	925	
kwh		71225	
mw		71.2	
after allowing for margins	mw	90	
capacity of CPP at 40 %	mw	36	captive power plant
assume efficiency	%	40	
coal required	tph	22 tph	cal. Value
say		25	3500 kcal/kg
wet coal	tph	29	
per day	tons	700	
stock to be maintained days	day	20	
	tons	14000	
size of stock pile	width × length	**40 × 160**	
two stock piles in line	m	**340**	

Part 3 Stock piles for crushed limestone

crushed stone	tpd	19100
stock	days	7
stock	tons	133700
size stock pile	width × length	**50 × 215**
length 2 piles		450

Part 4 Continuous blending & storage silo

raw meal per day	tons	17050
stock to be maintained	days	2.5
	tons	42625
no of silos		1
height/ dia		3
dia of silo	m	25
height	m	85

Part 5 Clinker storage

clinker	tpd	11000
stock	days	14
	tons	154000
dimensions		
clinker stock pile	dia *	**100dia* 5 ht**

Part 6 Sectional capacities

			Nos
quarrying	tph	2000	
crusher 1 stage	tph	1600	1
raw mill	tph	900	1
kiln	tpd	10000	1
prehater 6 stage		2 stream	1
calciner		off line	1
clinker cooler		10000	1
coal mill	tph	100	1

Part 7 Alternative fuels

calorific value	kcal/kg	4000
sp. Fuel con.	kcal/kg	750
quantity	%	20
to be fired in		calciner
rate	tph	20
af per day	tons	480
	say	500
Stock days		14
	tons	7000

Part 8 Slag for making BFSC

clinker used to make BFSC	tpd	4000
slag requiired	tpd	6000
with margin	tpd	6600
wet slag	tpd	8500
storage wet slag	days	2
	tons	17000
storage dry slag	days	2
	tons	15500

Part 9 Captive Power Plant & Waste Heat Recovery Plant

Plant	Main machinery, auxiliaries and Electrical switch gear, condensers, cooling towers etc, as required
Waste Heat Recovery System	capacity 15 mw operating on Organic Rankine Cycle complete with main machinery and auxiliaries as required for CPP and systems for thermal and working fluids

refer to tables 7.2.1, 7.3.1 & 7.5.1 to 7.5.8 for the basis of quantities, sizes arrived at in this exercise

Annexe 6.1.2

Legend for General Plant Layout 6.1.23
For a
10000 tpd clinkering unit making
OPC & BFSC

Note: Basis and workouts are as per
1st Unit is shown shaded. Duplication is shown blank

Sr. no.	Legend	Remarks
1	Crusher	
2	Stacker Reclaimer System For crushed limestone	two linear piles
3	Storage for additives and correcting Materials	covered shed
4	VRMill for grinding Raw Materials	
5	Dust Collector for raw mill and kiln	
6	Continuous Blending and Storage Silo	
7	Two Stream 6 stage preheater & calciner	
8	Rotary kiln	
9	Clinker Cooler	
10	ESP for cooler	
11	Coal mill	
12	AF firing for calciner	
13	Clinker conveyor	
14	Clinker storage	
15	Hoppers for Cement Mill	
16	Cement Mills VRMs	1 for OPC, 2 for BFSC
17	Cement Silos	1 for OPC,2 for BFSC
18	Silo for bulk loading by rail	1 each forOPC & BFSC
19	Packing machines	1 for OPC, 2 for BFSC
20	Truck loading facility For bagged cement	
21	Wagon loading facility for Bagged cement	3 platforms for 1st unit

Sr. no.	Legend	Remarks
22	Wagon tippler for coal and slag	
23	Coal crusher	
24	Storage for wet and dry slag	
25	Slag dryer	
26	Slag hoppers	
27	Stacker Reclaimer for coal For cement plant	linear piles in line
28	Stacker Reclaimer for coal for Captive Power Plant	linear piles in line
29	Captive Power Plant	
30	Condensers	
31	Cooling Towers	
32	Overhead Water tank	
33	BG Railway siding	
34	Silos for extra clinker	
35	Silo for wagon loading of clinker	
36	WHR System	common for preheater And cooler
37	condenser cooling tower for WHR	
38	Central Control Room	
39	Engineering Offices	
40	Laboratory and Quality control	
41	General Stores	
42	Open Store Yard	
43	Security	
44	Time office	
45	Main Administrative Office	
46	AF Store and Processing	
47	Grid Substation	
48	Main Substation of the plant	
49	Yard for trucks / bulk carriers	
50	Cooling pond	

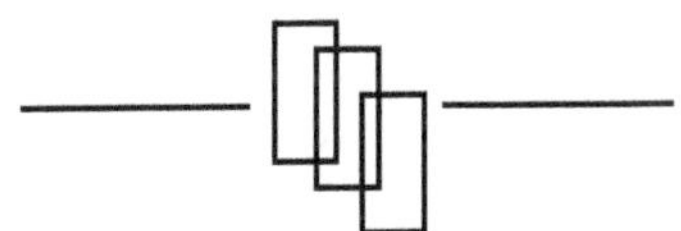

CHAPTER 2

DEPARTMENTAL LAYOUTS

2.1 Departmental Layouts

Along with the General Plant Layout evolved in the manner explained earlier, sectional layouts for various Departments would be developed. They would be progressively interlinked with one another and with storages in between.

Departmental layouts are dictated by the type of major machineries used and by the nature of raw materials processed at the beginning and at the end of operations carried out in the department.

Basic principles, in evolving the departmental layouts are the same as for the general layout. However, more detailed attention is paid in planning them to suit the type and size of machinery, to flow of materials and to flow of gases. Provision of utilities like cooling water, compressed air and supply of electric power and lighting are other considerations.

Departmental layouts show all three views that is plan and two elevations. They also show plans of each floor of the building.

2.2 Development of Departmental Layouts

The Suppliers of the major machinery in respective sections normally furnish Departmental Layouts or at least preliminary layouts

Consultants to whom the task of developing the layouts has been assigned, develop the final Departmental Layouts by incorporating in them data and drawings collected from Suppliers of auxiliaries, motors and gearboxes and also ductings for process and vent gases and chutes for materials.

Each type of processing machine requires a definite orientation with the machine feeding it and the machine to which it discharges. In specific cases it needs gases to be brought in from one section and to be sent them to other sections.

Consultants also have a 'data bank' of layouts from several plants designed by them earlier or collected from vendors, for different types of machines and these can be used as the starting point.

In any given department, even the smallest, machinery would be coming from different sources, for exa1mple :

Core machinery from one source

Auxiliaries like fans, dust collectors, conveying equipment.

Motors and gear boxes

2.2.1

It is necessary to obtain pertinent g.a. drawings of these machines from respective vendors. Vendors will also furnish load data and foundation drawings for machinery of their supply.

Consultants who have been given the responsibility of preparing the departmental drawing will coordinate obtaining these drawings and details from respective vendors.

2.3 A Departmental Drawing Must Contain

1. all machinery in outline with overall dimensions and their feed and discharge points in plan and elevations,
2. centre lines of machines w.r.t. building in which they are housed and also w.r.t. each other both in plan and elevations,

3. dimensions of buildings showing area of each floor, floor heights, total heights, column centers and main and auxiliary beams which will take load of machines,
4. initially column and beam dimensions could be taken on empirical basis. They will be confirmed/ modified by the consultant doing Civil design,
5. stair cases and floor openings,
6. load data and foundation drawing showing spacing of pockets for foundation bolts,
7. space to be kept clear for maintenance,
8. location and capacity of lifting tackles, cranes, hoists etc.,
9. lifts in buildings like preheater tower, silos etc.,
10. chutes and ducts conveying material and gases as the system requires in any section. Layouts should be so developed that the chutes and ducts must be at correct angles and are of right sizes,
11. later on, separate drawings will have to be developed only for Chutes and ducts and their supports to take up their fabrication. Such drawings will show actual lengths and inclinations and not what is shown in standard drawings,
12. arrangement of drives of different types like direct, v-belt, chain, open gears with centerlines with respect to main machine and with arrangement for tensioning as required,
13. location of measuring points for instrumentation and process control and control panels to be installed and also sampling points,
14. if any provision is to be kept to protect machines from rain,
15. separate departmental drawings will be developed to show routes of cooling water and lubrication pipelines and for compressed air.

2.4 Preparation of Drawings – Time Factor

A departmental drawing can seldom be completed in one go. It is customary to receive a preliminary drawing. Even this drawing will be received only after about 4 weeks from the Vendor after he receives a technically and commercially clear order.

It is necessary to agree to a time bound programme with Vendors for departmental drawings that they have to furnish.

Drawings can be submitted progressively and in sequence :

1. preliminary departmental drawings,
2. load data and foundation drawings,
3. chutes and ducts,
4. details of refractory and insulation.

2.5 Departmental Drawings for Civil Design

Since departmental drawings are used for taking up 'civil design' of buildings, it is important that they contain all pertinent details about weights and their distribution, dimensions of building at various levels, floor plans and loads at various levels. Departmental drawings passed on to civil designers should not be 'cluttered' so as to make them confusing.

Often it is advantageous to show more sectional views and enlarged views of parts of elevations and plans so that machinery that could not be shown in conventional views is seen properly.

Besides dead loads of machines and loads arising out of motion (which would be shown in load data) it is assumed that for design purposes, hoppers of bag filters and esps and cones of cyclones are full. Building columns and beams have to be designed to allow for this load.

Floors are designed for distributed load according to norms followed in Industry. They take care of additions in future and also of occasional loads like refractory for laying.

Site conditions like load bearing capacities of soil, wind loads and seismic zone requirements also come into play and must be conveyed to Consultants and Civil designers.

2.6

All departmental layouts should be taken up bearing always in mind that they must be functional. Buildings and structures must be designed to house the machinery for which they are meant and not the other way round.

2.7

In subsequent sections various aspects of departmental layouts and detailed engineering have been dealt with, beginning with Crushing Department.

2.8 Datum Levels

It has been mentioned in the **Chapter** on General Layouts that all departments need not have same datum level. Some times existing contours can be used with advantage. Separate drawings for equipment interconnecting departments like belt conveyors should be prepared. These drawings should show levels in two departments properly in absolute terms if datum of two departments are different.

For example, if departmental drawings are with respect to datum level, and there is a difference of say 1m in datum of the two sections, 00 levels in two will differ by 1m. This could be taken care of in designing drawings of interconnecting conveyor.

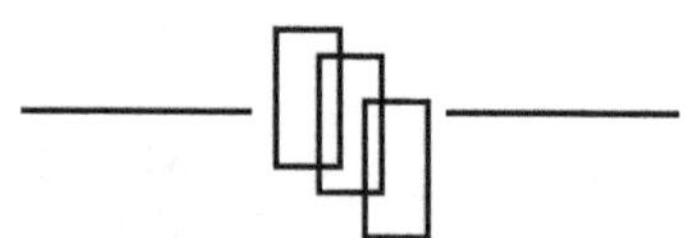

CHAPTER 3

LAYOUTS OF CRUSHING PLANTS

3.1 Layout of Crushing Plants

In **Chapter 2 of Section 2**, the basis of selecting number of stages of crushing and sizing of Primary Crushers have been explained.

3.2 Ratio of Reduction and Number of Stages

Ratio of reduction in crushing is decided on one hand by the size of shovel used in mining operations and on the other by the type of mill used for grinding.

A plant may need 'two stage' crushing if it uses / proposes to use ball mills for grinding and has mechanized mining.

It can have a 'single stage' crushing if the plant uses, 'vertical roller mills' or 'Horo Mills' or a combination of roller press and ball mill or roller press only.

A small plant using manual mining will also have single stage crushing.

See Table 2.2.1 in Chapter 2 Section 2.

Reduction ratios of 20 : 1 are within the range of hammer crushers and impactors. Thus when a V R Mill is used the crushing would be 1 stage and mostly hammer or impactor type crushers would be used.

3.2.1

Choice between the two would be guided by :

(i) Abrasiveness of stone – its Silica content.

(ii) Content of wet / sticky materials like clay in it.

Hammer Crusher with outlet grates (full or partial) would be suitable for dry material. Impactors do not have grate bars and operate in closed circuit to keep down oversize when used with ball mills.

3.2.2 Closed or Open Circuit Crushing

Thus a decision would also have to be taken as regards whether crushing would be open or closed circuit.

Feed size is critical for ball mills, which prefer fines. Hence when ball mills are used, the second stage crushing would generally be in closed circuit.

In impactors, the control on size is by adjusting clearance between rotor and impactor / breaker plates. If material is abrasive, there can be heavy wear both on impact bars and plates resulting in oversize. Normally, however used as a single stage crusher, impactors would work in open circuit.

Closed circuiting requires screening of the product and returning oversize to the crusher for recrushing.

3.3 Apron Feeder

To avoid, 'flushing' of stone or its rolling into the crusher mouth, apron conveyors / feeders inclined upwards are preferred. They help in avoiding stoppage due to overloading of crusher and its getting choked.

3.3.1

An apron feeder also reduces the overall heights in the crushing department.

The level difference between A and B = H is considerable. It would be ideal for A to be at 00 or ground level. That would however push up B requiring a very long ramp and higher abutting walls.

See Fig. 6.3.1.

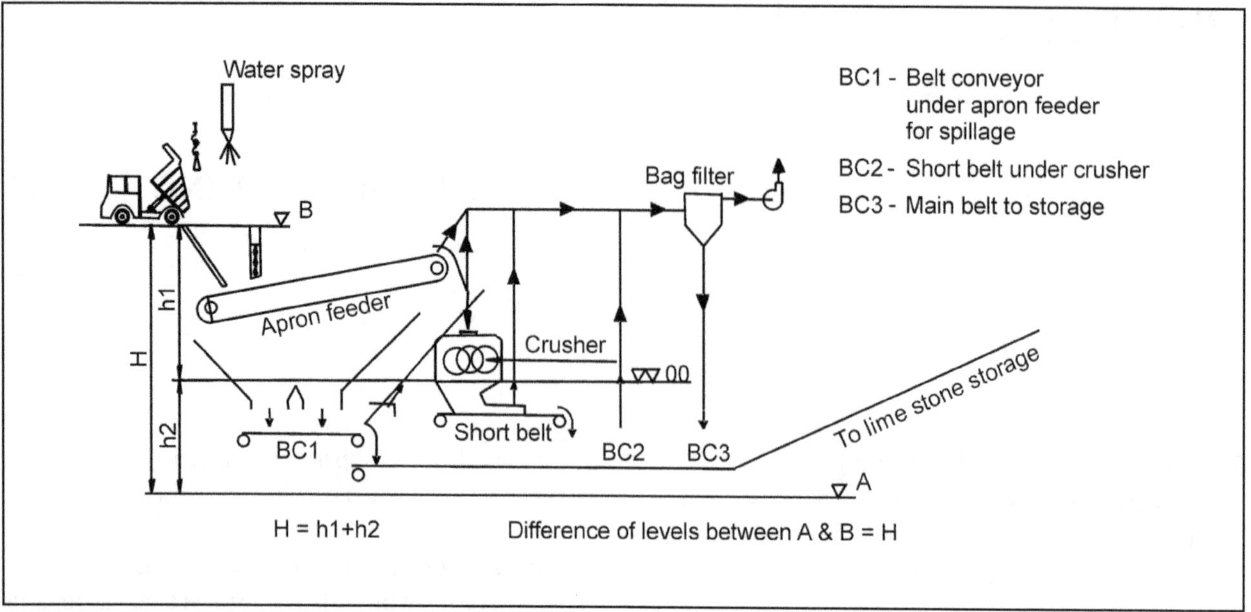

Fig. 6.3.1 Single Stage crushing - Basic layout.

A compromise is generally made to locate belt BC3 below ground level in a tunnel. The depth of tunnel would be guided by the underground water table. It should be restricted to 3 m below ground level. Tunnel should be water proofed in any case.

3.3.2

Apron Feeder should be long enough to decelerate the stone. It should be slow moving and totally covered, with inspection windows at sides. There should be a walkway on both sides (or at least on one side) for access along its length.

The hood at discharge point should have facility to connect it to dust collector through a duct. **See Fig. 6.3.2.**

Apron feeders should be wide enough – almost – as wide as the crusher opening so that stone is fed across the full width of crusher. Apron feeder would have a variable speed drive to regulate feed to crusher.

3.4 Ramp

It should be possible to unload into the hopper from all 3 sides or at least from 2 sides. The ramp should be designed to allow for this. This facility would avoid 'bunching' of dumpers at crusher

There should be room for reversing and placing dumpers for unloading.

See Fig. 6.3.3.

3.4.1

The space to be provided at the top of the ramp is dictated by the radius required by the dumper for reversing and for taking a turn. The Manual of dumpers should be referred (or this information ascertained from Supplier) to plan the ramp.

See Table 2.1.3, Chapter 1 in Section 2.

The space at the top could be reduced if only two sides are to be used for unloading.

3.4.2

It would be best to provide a ramp with 3 lanes width so that dumpers move to and from crusher smoothly even if there is breakdown of a dumper on the ramp occasionally.

See Fig. 6.3.4.

However this would be rather expensive even for large plants. Width of the ramp may be reduced to 2 lanes with shoulders on either side for dumpers to pass one another. Dumper needing attention can be stopped at the shoulder leaving road clear for normal traffic.

See Fig. 6.3.5.

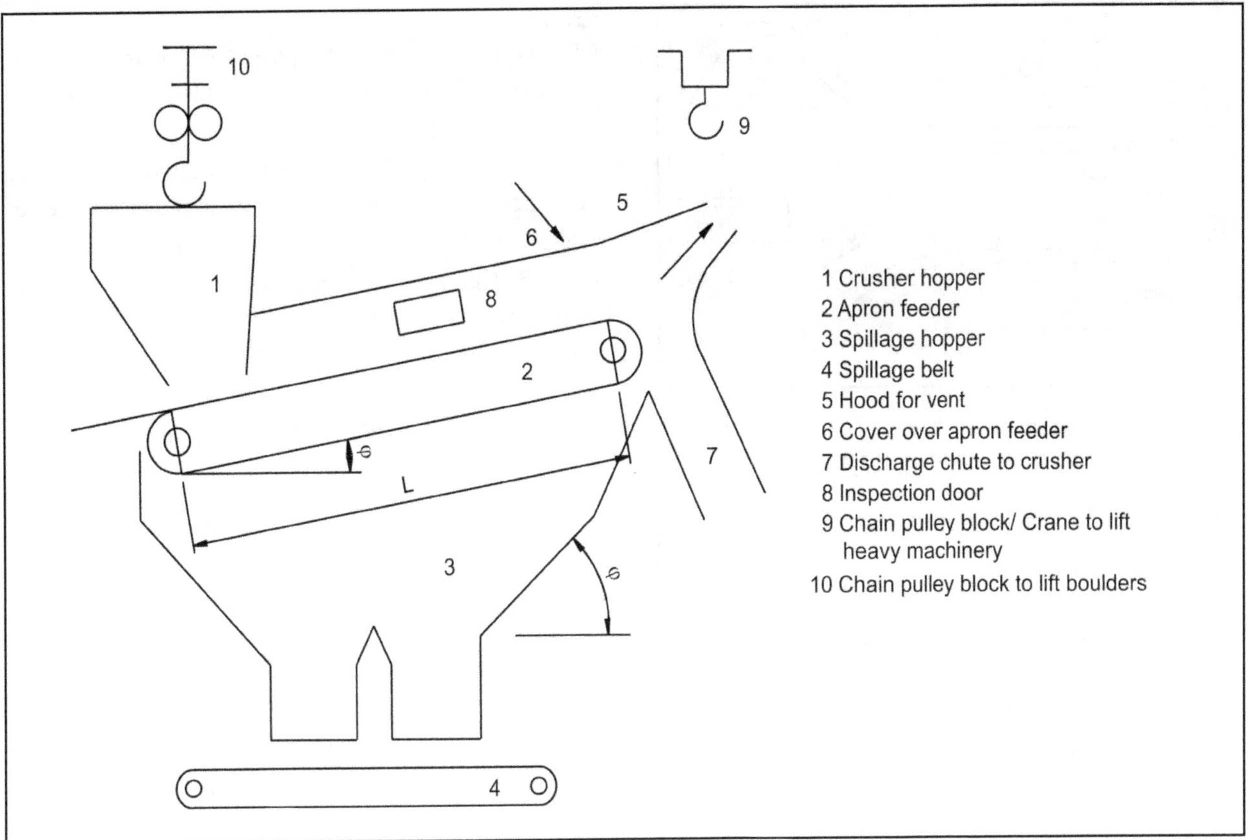

Fig. 6.3.2 Apron feeder for crusher.

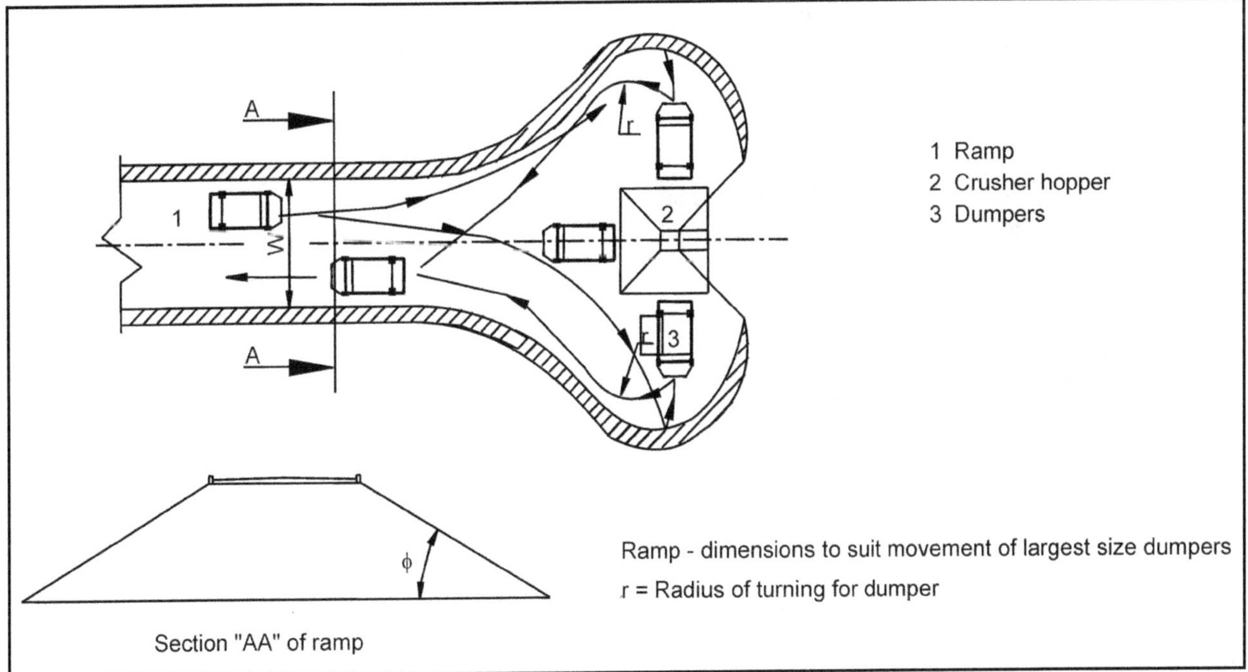

Fig. 6.3.3 Ramp for crusher.

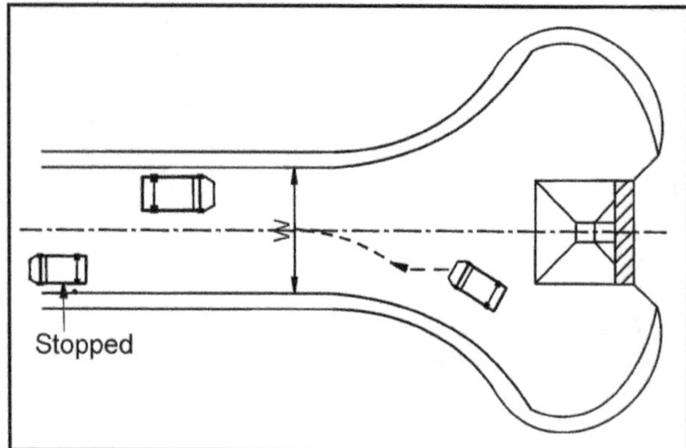

Fig. 6.3.4 Ramp of crusher.
option 1 - Three lane width.

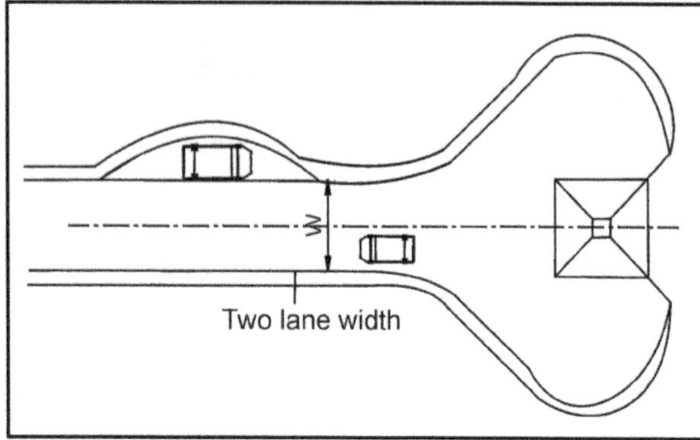

Fig. 6.3.5 Ramp of crusher.
Two lane width with shoulders for dumpers to wait.

3.4.3

The slope of the ramp is important. It should be gentle not exceeding 1 in 16 to save fuel. Surface should be either asphalted or in concrete. This would save fuel consumption in negotiating the climb.

A generously sized ramp is expensive but it is a one time expenditure and it could be justified as it helps in uninterrupted feed to crusher.

3.5 Size and Shape of Hoppers

Hopper should generally be kept full to avoid large rocks hitting the apron feeder . Its capacity is related to the capacity of dumpers unloading into it and the frequency of dumping. If the dumper capacity is 100 tons, hopper should hold minimum 100 tons.

3.5.1

Time required for 'a return trip' from quarries and the number of dumpers available (or planned) would decide the hopper capacity.

If the receipt from quarries is fairly regular i.e., dumpers arrive regularly, then hopper capacity can be maintained at

$2 \times 100 + 1 \times 100 = 300$ tons, even if crusher capacity were as high as 600-1000 tph i.e., a capacity of about 20 minutes.

3.5.2 Crusher Hopper Shape

Slopes of hopper should promote sliding of stone without accelerating the stone too much till it reaches the apron conveyor. Stone with wet clay mixed with it would need steeper slopes.

See Fig. 6.3.6.

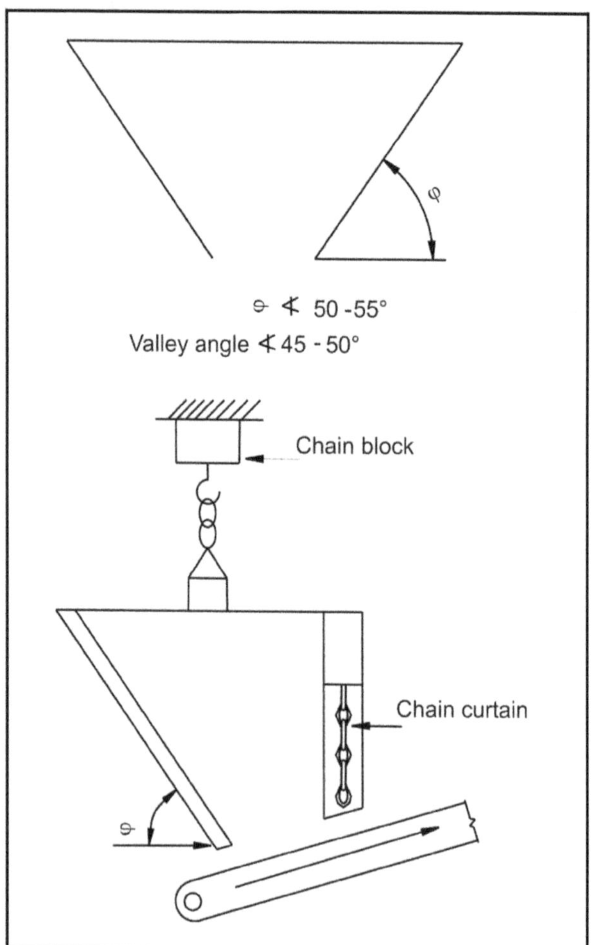

Fig. 6.3.6 Slopes of crusher hopper.

3.5.3

Concrete would not stand the impact of falling stones. Therefore hoppers would be lined with rails (railways rails) – closely placed embedded into the concrete. Dust coming along with the run off mine stone would fill the crevices and protect concrete surface.

See Fig. 6.3.7

The vertical wall of the hopper should also be lined with steel plates.

See Fig. 6.3.8.

3.5.4

The kerbs of the hopper should be protected by wooden sleepers or tyres (old discards from dumpers) so that in the operation of backup, dumpers do not accidentally fall into the hoppers nor do they damage the kerbs. **See Fig. 6.3.9.**

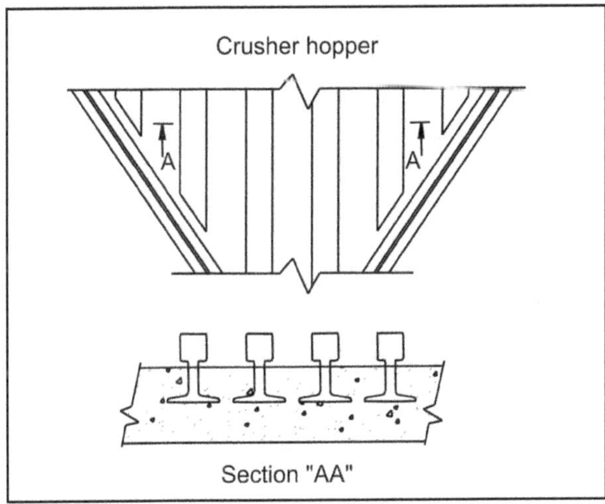

Fig. 6.3.7 Rails embedded concrete to take impact of stone and wear.

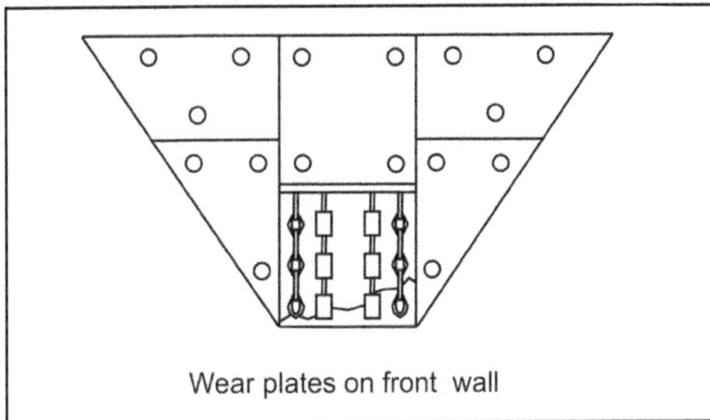

Fig. 6.3.8 Curtain of link chain or rail pieces to decelerate stone.

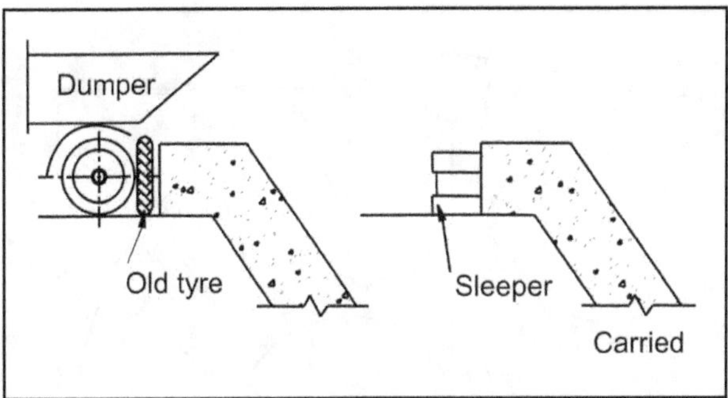

Fig. 6.3.9 Crusher hopper
Protection of kerb and prevention of accidental falling of dumper in hopper.

3.5.5

Run off mine stone when dumped into a large hopper generates lot of dust. This fugitive dust is difficult to catch. It is therefore suppressed by spray of water or water and dust depressant chemicals.

See Fig. 6.3.1.

3.5.6

When feeders are 'reciprocating', stone tends to accelerate into crusher mouth. Its acceleration is arrested by suspending heavy chain links or rail pieces from the vertical wall.

See Fig. 6.3.8.

3.5.7

There should be facility over the crusher hopper to handle / lift boulders out of the hopper.

See Fig. 6.3.6.

3.6 Point of Discharge into Crusher

The point of discharge of stone into the crusher is important in hammer crusher and impactors. To deliver maximum impact, the impact bars and hammers should hit the falling stone at correct height and point.

See Fig. 6.3.10.

If this relation is not maintained, impact will be less effective therefore the relative locations of discharge points of feeder and crusher rotor should be ascertained from the Supplier and adhered to in the layout.

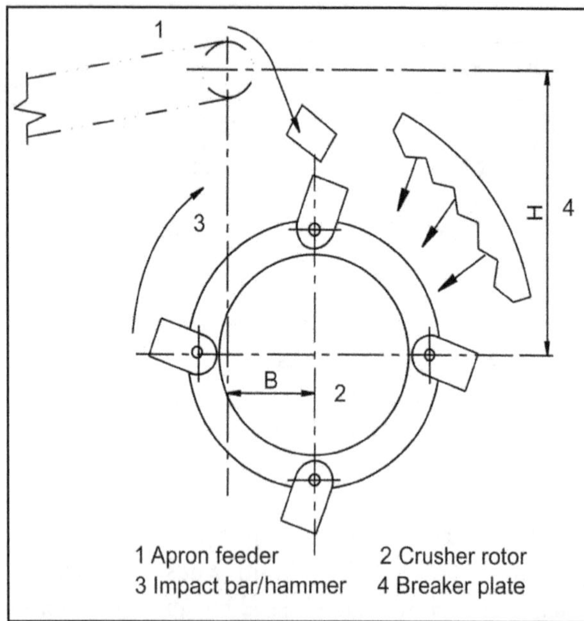

Fig. 6.3.10 Feed into crusher
Dimensions H and B important from point of view of most effective impact and breaking.

3.7 Handling Facilities for Maintenance

Large crushers would need facility for handling replaceable hammers and impact bars. The crusher would therefore need to have an overhead traveling crane for handling them and lifting them out of crusher. The discharge chute from apron feeder to crusher and hood over it should be so designed that access into crusher from top is easy and overhead traveling crane could be used to handle heavy parts of crusher during maintenance.

See Fig. 6.3.11.

When replacing hammers, spacer rings in between are to be removed to slide apart the hammer arms / rotor discs and take out hammers.

See Fig. 6.3.12.

Shaft is removed from end to end. Space and supporting structure needs to be provided for this purpose in the layout along with a hydraulic pusher.

These are repetitive operations and hence should be completed in a planned manner with proper tools and in the shortest possible time.

Large crushers come with hydraulic jacks and pushers to facilitate maintenance.

See Fig. 6.3.13.

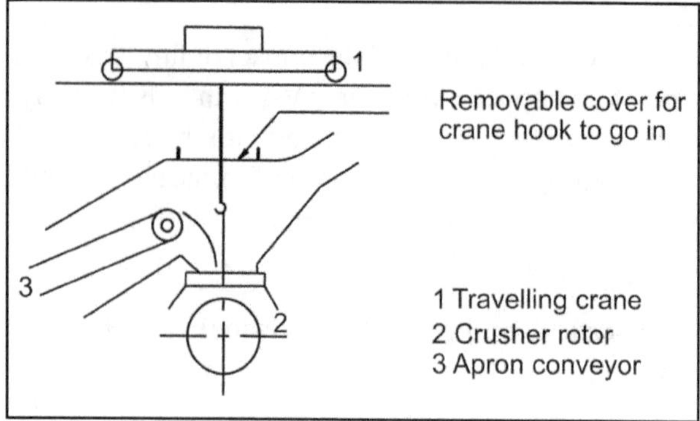

Fig. 6.3.11 Crusher-lifting tackles.

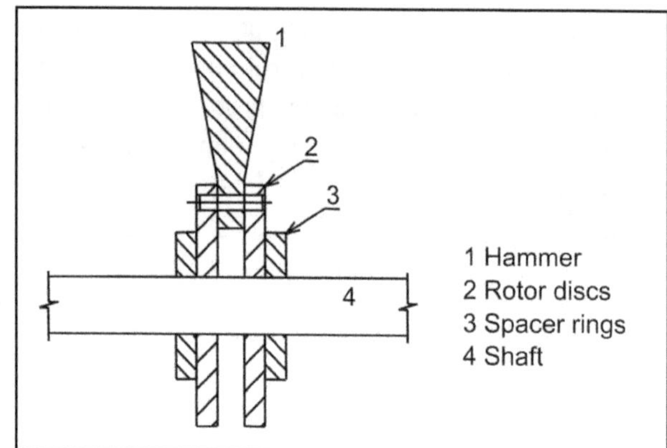

Fig. 6.3.12 Removal of hammers.

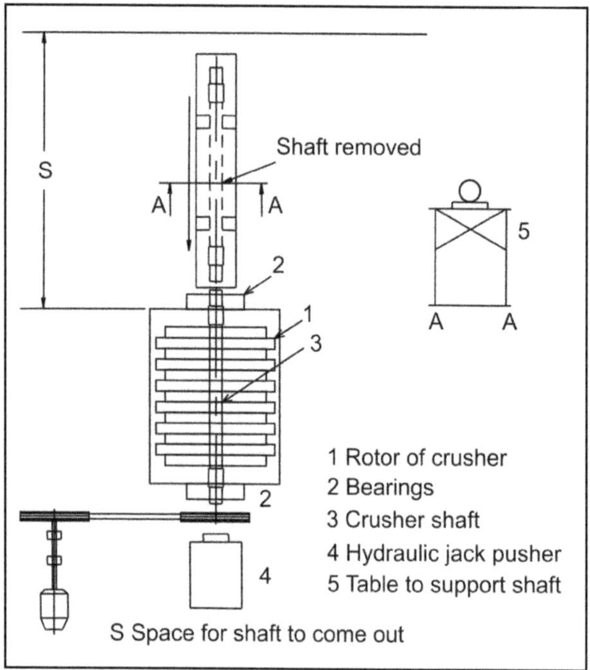

Fig. 6.3.13 Removal of shaft.

3.7.1

Very large crushers would have built-in hydraulic jacks to swing part of crusher top for access into crusher and also for swinging cage bars at the bottom for maintenance.

3.8 Lubrication of Bearings

Crusher bearings would be either bush bearings or roller bearings. There would be a forced lubrication system and perhaps also a cooling system for them. It should be located at a suitable place in the layout preferably in a dust free corner.

3.9 Discharge from Crusher

The discharge of crusher should permit free flow of crushed stone. Hence the opening should not be narrowed down too much for the belt conveyor below it.

See Fig. 6.3.14.

3.10 Belt Conveyors for Crushed Stone

A short belt conveyor with rubber impact idlers, closely spaced, should take the impact of stone falling from the crusher. The belt should extend at least a meter behind the crusher to prevent spillage from near end. It will be longer if it has to receive spillage from the apron /reciprocating feeder.

See Figs. 6.3.14 and 6.3.15.

Belt under the crusher is frequently damaged by sharp edges of the falling stone. It is easier and less expensive to replace a short belt. This belt should feed a long belt of smaller width that would take it to storage of crushed stone.

All belts must have skirt board rubbers and scrapers fitted to prevent spillage. They will be vented.

See Fig. 6.3.15.

3.10.1

The long belt would generally be in a tunnel below ground level. In most cases, it would be a horizontal cum inclined belt. Such a belt is likely to be lifted off idlers when empty, due to tension of belt. Hence enough headroom should be provided for such an eventuality. **See Fig. 6.3.15.**

3.11 Scrapers

A well designed scraper, preferably of rotating type, will clean the belt and prevent accumulation of dust under the belt. It should be possible to adjust it and it should always remain in contact with the belt. An ill designed or badly maintained scraper will permit build up of dust heaps under the belts. As mentioned earlier, water is often sprayed in crusher hopper to suppress dust. Hence dust could be wet and would tend to stick to the belt. In course of time it falls off the belt and heaps of dust begin to form which can soon assume ugly proportions if not attended to in time.

Scrapers are a must for all belts but are particularly important when material has wet dust which tends to stick to the belt.

See Fig. 6.3.16

3.12 Tensioning Devices

All belts need a tensioning device to maintain tension of belt as belt stretches in running. Simple screws are sufficient for short belts. For long belts, a tower for tensioning weights needs to be provided generally towards discharge end where sufficient heights would be available. For a shorter belt a tensioning device of similar counter weight type may be fixed at the rear end of the belt.

See Fig. 6.3.17.

3.13 Selection of Belt Conveyors

Selection of belts for crushed stone should be done carefully taking into account :

(i) Flushing – though minimal with apron feeder.

(ii) Moisture.

(iii) Impact of jagged sharp edged stone.

Belt should be slow enough but not too slow as to require very wide belts. Number of plies and thickness of ducks at top and bottom should be selected with the help of Suppliers.

Belts should have a margin of about 50 % over the capacity of the crusher to take care of rush of material through the crusher.

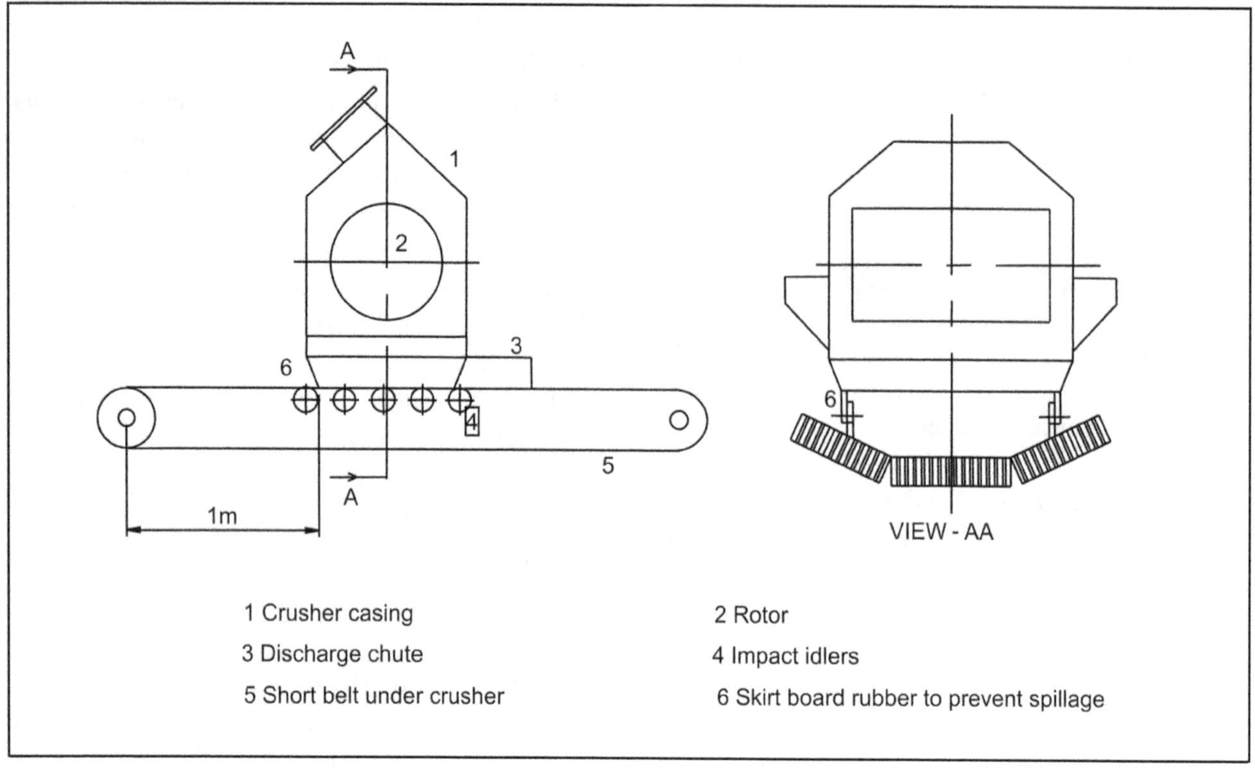

Fig. 6.3.14 Discharge from crusher.

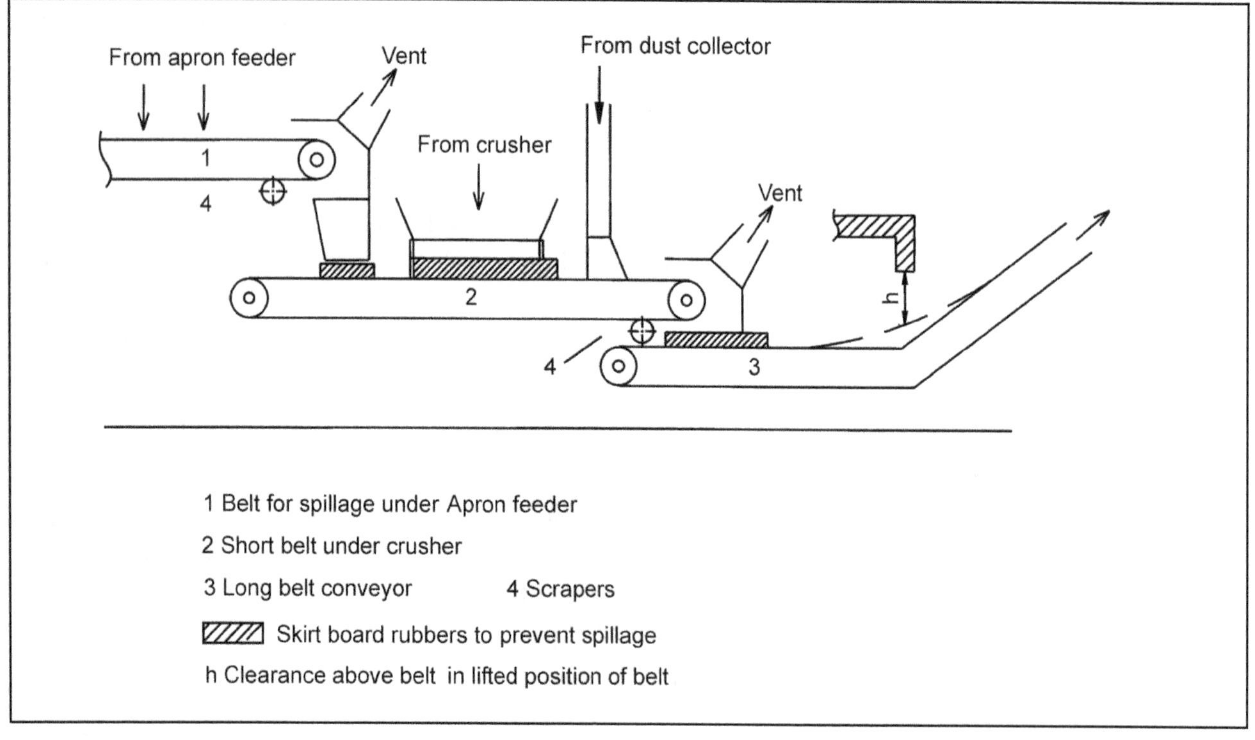

Fig. 6.3.15 Belt conveyors under crusher.

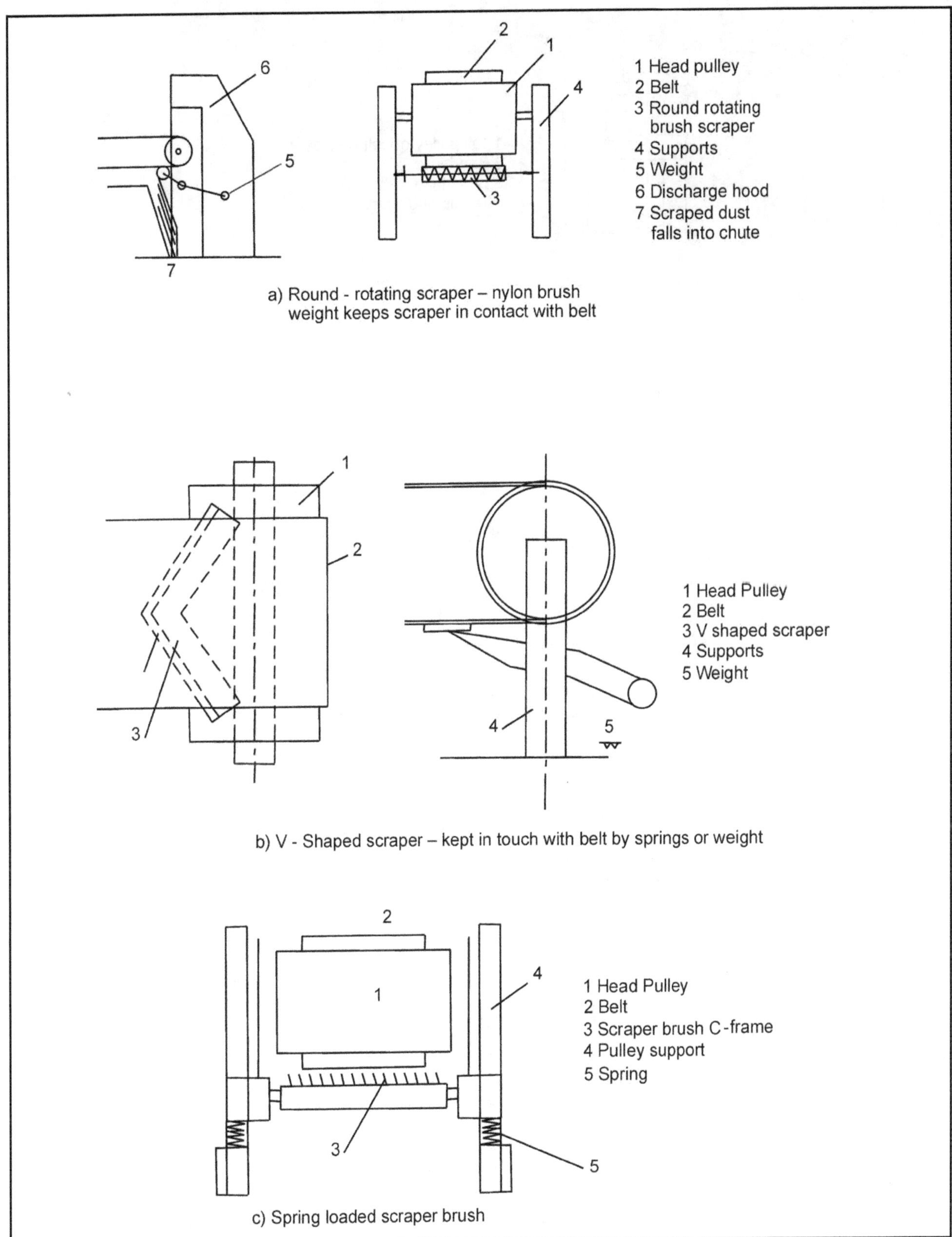

Fig. 6.3.16 Scrapers for Belt Conveyors.

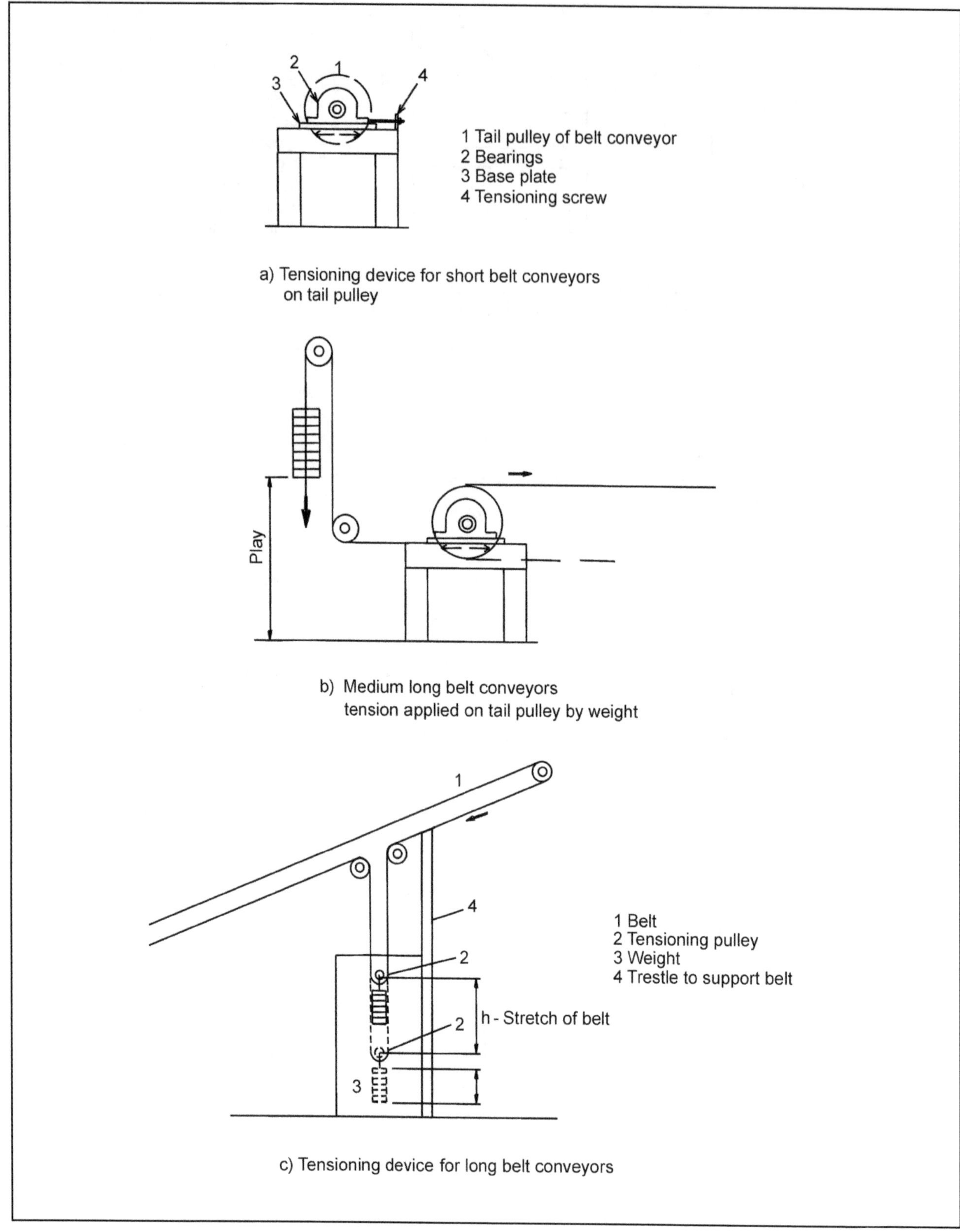

Fig. 6.3.17 Belt conveyors - Tensioning devices.

3.14 Bag Filters

Rotors of hammer crusher and impactor act like fan and circulate dust. Bag filters are used to arrest dust generated at discharge points of feeder, in crusher and as stone falls onto belt conveyors under the crusher. Dust collected is fed back to the long belt.

Bag filter is sized to handle vent volumes of crusher and of feeder and belt conveyors. Respective suppliers should furnish data on vent volumes.

See Fig. 6.3.1

3.15 Two Stage Crushing

When two stage crushing is required. First stage would generally be a jaw crusher and second stage would be a hammer crusher or impactor.

The two crushers would be arranged in a straight line with crushed stone from jaw crusher being generally taken straight to secondary crusher without an intermediate hopper or feeder.

See Fig. 6.3.18

Jaw crushers have a reduction ratio of 5 : 1 to 6 : 1. The jaw crusher would be selected to receive stone delivered by the shovels regardless of capacity required.

Thus a 4 m^3 shovel, stone would deliver stone of size 1200 × 1500 mm and would require jaw crusher of size 1650 × 2150 mm. Capacity of a Jaw crusher of this size would be far in excess of capacity of crusher actually needed for the plant. 15 % stone delivered by a jaw crusher is oversized that is larger than the closed setting, because of movements of the jaw.

Hence the jaws should be set to have closed side setting in such a way that maximum size of crushed stone does not exceed the designed size and yet meets the desired capacity.

3.15.1

The secondary crusher would have to handle stone of 250 × 300 mm size and crush it down to –20 mm. Its capacity should be compatible with the capacity of primary Jaw crusher. Reduction ratio would be 300 / 20 15 or higher.

It could be a hammer crusher with full or partial grate bars, in which case product size is assured and controlled.

If an impactor is installed or a hammer crusher without grate bars, then size of product could become bigger with wearing of impact bars and wear plates / hammers in spite of regularly adjusting the clearance between rotor and impact / breaker plates.

It may also be the case that raw mill shows potential for higher capacity if smaller stone is supplied to it. In such cases it would be a good idea to run secondary crusher in closed circuit.

See Fig. 6.3.19

3.16 Crusher in Closed Circuit

Scondary crusher would then receive feed at a higher rate to include return or circulating load.

Therefore, when it is intended to run a crusher in closed circuit, Suppliers should be consulted as regards its capacity for handling circulating loads.

3.16.1

Screen in the circuit should also be sized on the basis of fresh feed and circulating load and screening efficiency.

The layout should provide for closed circuting. If installed at a later date, things can be complicated when the belt is broken to install the screen. It may not be possible to install the screen right under it.

See Fig. 6.3.20.

In such a case it would be necessary to install the screen at right angles to the belt.

See Fig. 6.3.22.

To install closed circuit, at a later date thus becomes a costly proposition and there should be space in the layout to install screen at a considerable distance

3.16.2

Alternatively, if there is a possibility of closed circuiting the secondary crusher, then the storage should be kept at a longer distance so that angle of inclination of the broken belt q2 still remains within 16 – 18% or so. **See Fig. 6.3.21.**

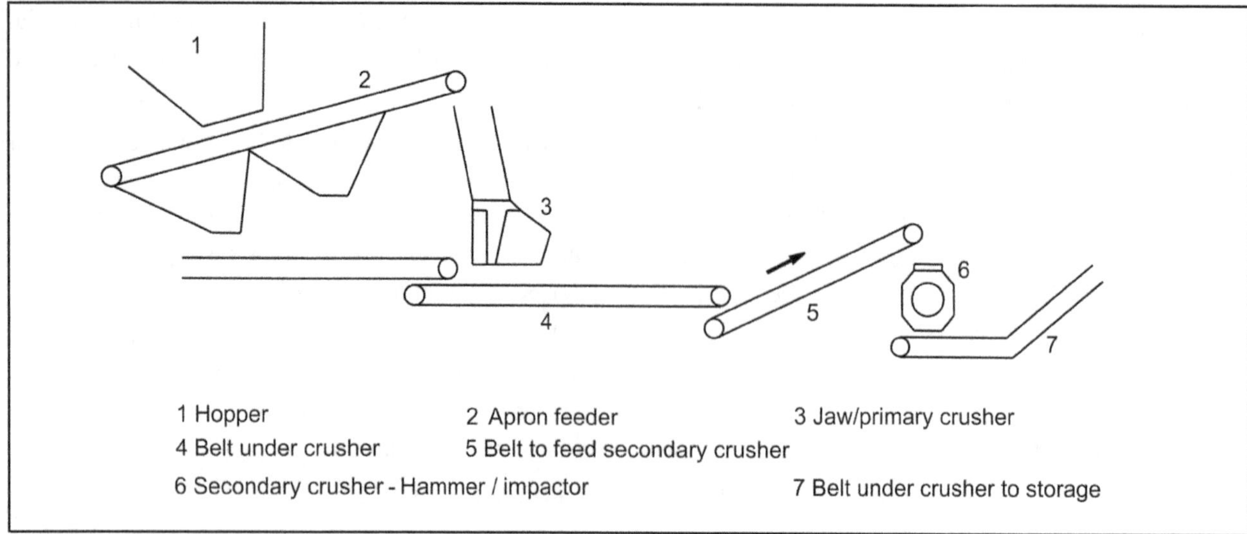

Fig. 6.3.18 Two stage crushing basic flow chart.

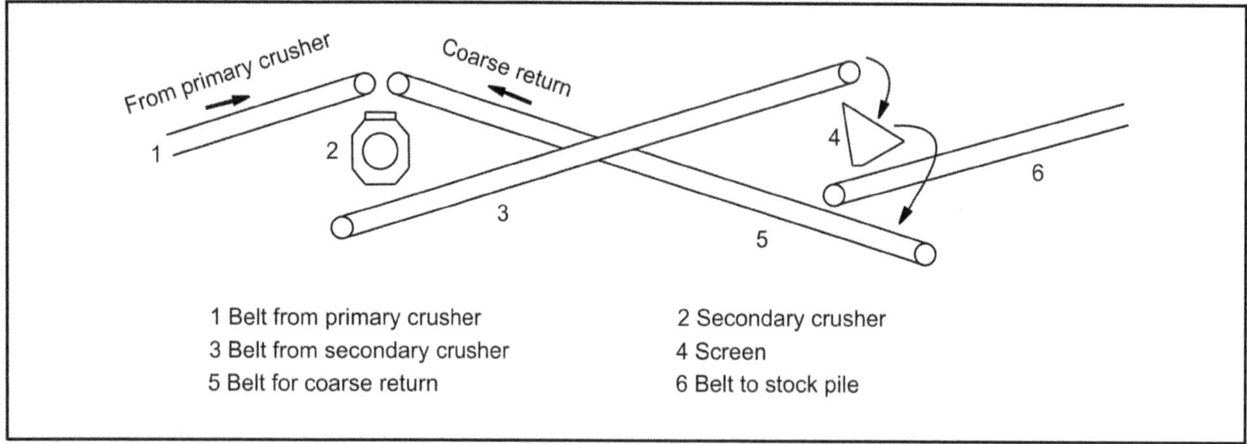

Fig. 6.3.19 Two stage crushing with secondary crusher in closed circuit.

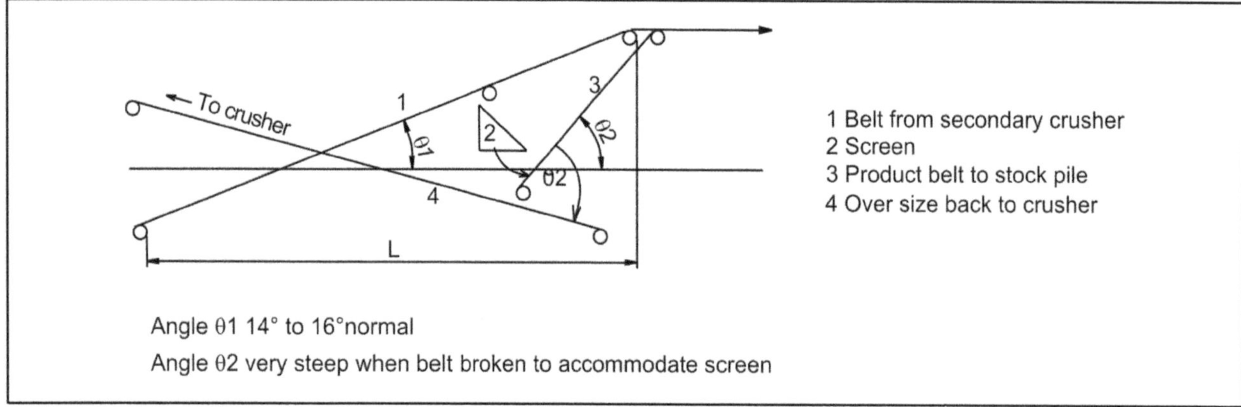

Fig. 6.3.20 Difficult to acccommodate screen if angle of belt-1 is normal – 14 - 16°.

On the whole it would be cheaper to install closed circuit secondary crusher at beginning itself.

See Fig. 6.3.19.

3.17

For secondary crushers, a grill of rails on the hopper would prevent oversize stones from primary crusher over loading it. However, a grill soon gets blocked and needs to be cleaned now and then.

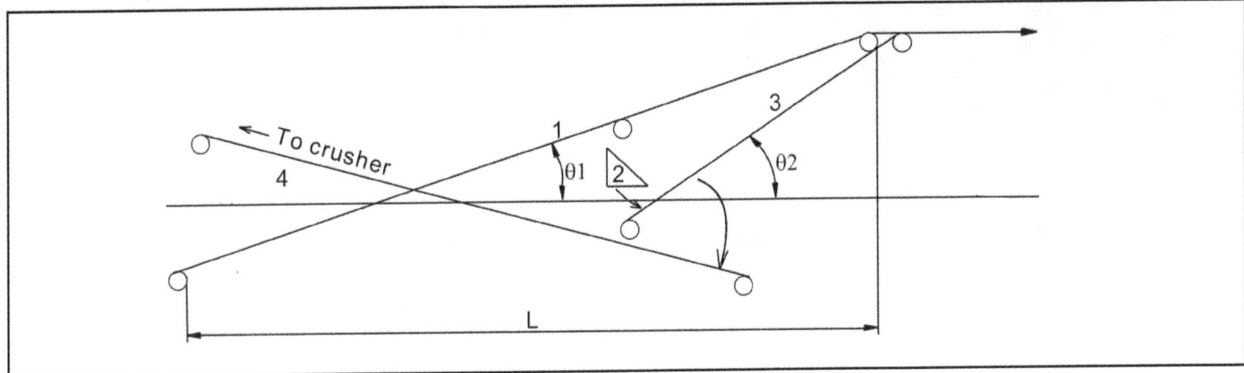

Fig. 6.3.21 For closed circuiting of secondary crusher in future
Install belt 1 at a small angle. L increases correspondingly
θ_2 : 14° - 16° of belt 3 after breaking belt.

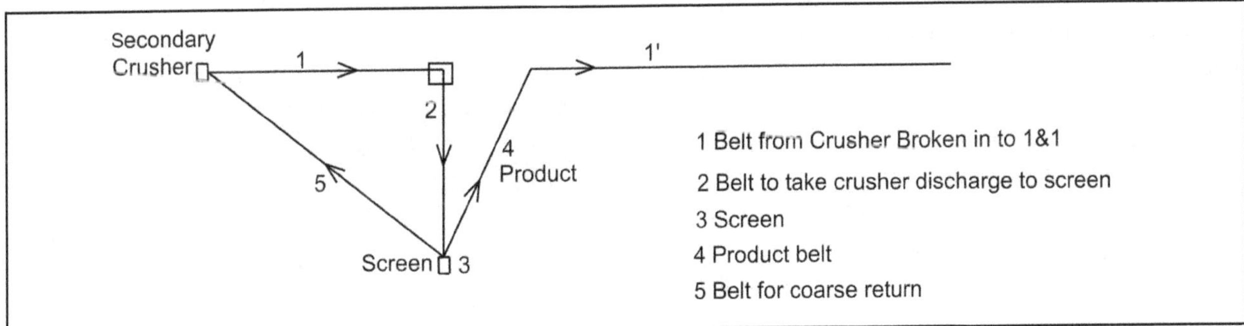

Fig. 6.3.22 Screen on side
Closed circuiting at a later date.
Closed circuiting of secondary crusher.

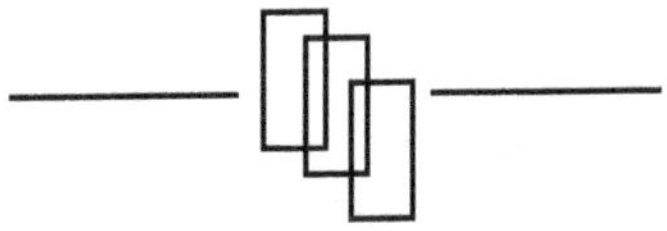

CHAPTER 4

AUXILIARIES IN CRUSHING SECTION

4.1 Feeders and Conveyors in Crushing

All feeders are conveyors but all conveyors are not feeders.

Feeders have variable speed drives and have short center-to-center lengths. It is possible to regulate rate of feed to processing machinery.

In crushing section, 'feeders' used to feed crusher are mostly 'Apron feeders' and sometimes Reciprocating feeders. It is possible to regulate feed but not precisely in terms of tons per hour because of irregular size of stone and its impact on the feeder. Thus feeders for crusher cannot be 'weigh feeders'.

4.1.1 Regulation of Feed

Feeder is 'electrically coupled' with the crusher i.e., as load on the crusher which is a measure of rate of feed, increases or decreases, feeder speed is decreased or increased to maintain it at a set level and not permit its over or under loading.

Of course, a single large boulder can overload the crusher greatly and upset electrical coupling but by and large it should be possible to run the crusher at a steady level of output when so coupled with the feeder.

4.1.2 Rugged Construction

Feeders for Primary crusher need to be very rugged and capable of withstanding impacts of boulders falling from a height.

4.1.3 Safety Device for Feeders

In theory, crusher hopper should not be allowed to get empty to avoid direct impact but sometimes they do. Feeders should have a safety device to protect the drive and also the feeder itself against over loads. It can be in the form of a torque coupling that would slip if torque exceeds a set value or a 'sheer pin/s', which would break and disconnect feeder from its drive.

4.2 Apron Feeders

In **Chapter 3**, Apron feeders and their incorporation in the layout have already been dealt with.

Apron feeders tend to carry back dust on their returning surface. This dust drops off and soon creates heaps of spillage.

Therefore, in a layout, it is necessary to provide for a hopper under the feeder and a short belt to collect dust and feed it to the main belt as shown. This would reduce housekeeping considerably.

See Fig. 6.3.2 in Chapter 3.

It needs the short belt under the crusher to be extended backwards to receive dust.

This belt should have proper hood and skirt boards to prevent spillage and dedusting and also a scraper as explained earlier.

See Fig. 6.6. 3.15.

4.3 Reciprocating Feeders

Reciprocating feeders would be used for smaller outputs and where boulder size is limited; for example, when mining is manual, limiting size of feed to 250 / 300 mm.

It is possible (though not in running) to adjust the output rate of the reciprocating feeder by adjusting its 'stroke' i.e., the length of backward and forward movements. It is desirable therefore to have a variable speed drive.

See Fig. 6.4.1.

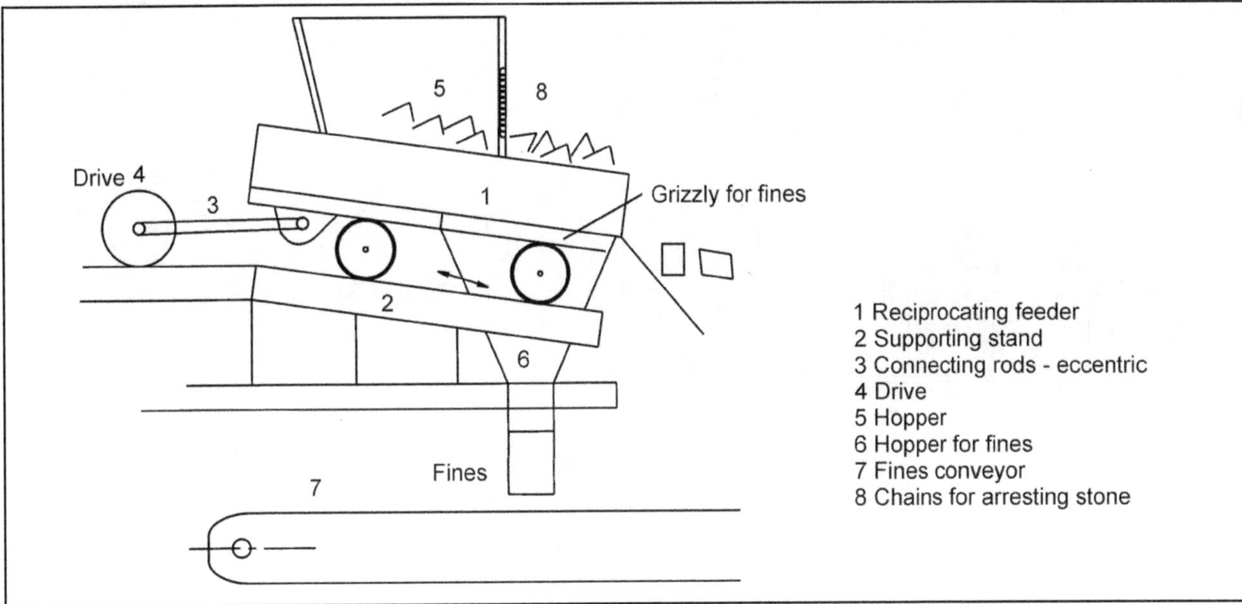

Fig. 6.4.1 Reciprocating feeder for crusher.

4.4 Crusher Drives

Crusher can be either directly coupled to a motor or through a V-belt or through a hydraulic coupling.

See Figs. 6.4.2 to 6.4.4.

A 'hydraulic coupling' would permit speeding up of the motor and then transfer load of crusher to it. Large crushers would have considerable inertia and would have flywheels. Motors should be capable of overcoming the inertia when starting crusher from rest.

When there is a V-belt drive, there would be a jackshaft. Motor and jackshaft would be mounted on a common base frame, which in turn would be fixed on slide rails to adjust belt tension.

See Fig. 6.4.4.

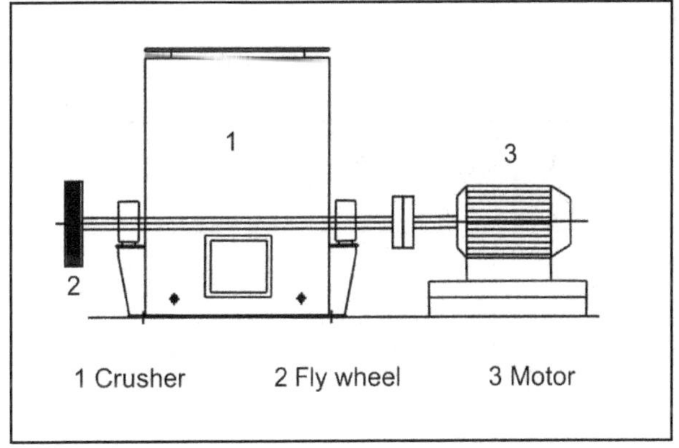

Fig. 6.4.2 Direct drive.

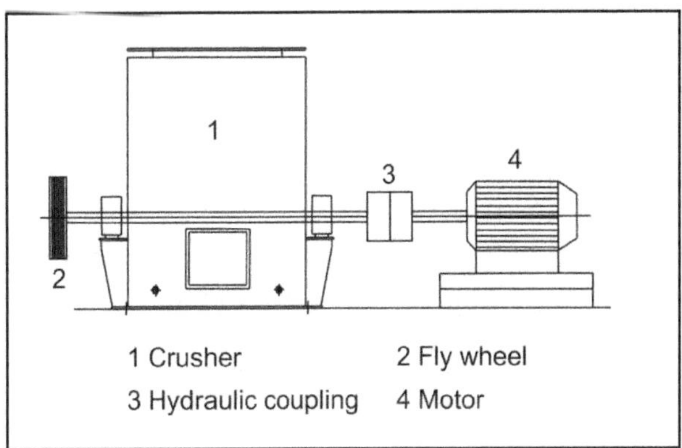

Fig. 6.4.3 Through hydraulic coupling.

4.5 Bag filters

Earlier 'vents' from crushers were cleaned in cyclone collectors.

Now bag filters are used to clean vent gases in the crushing department. Temperature of vent air is 50 °C and dust burden is 60-70 gms / nm^3. Therefore bag filters with cloth or polyester bags are quite suitable. Usually one bag filter vents crusher, feeder and belt conveyors. Collected dust is chuted down on to the main belt conveyor.

See Fig. 6.4.5.

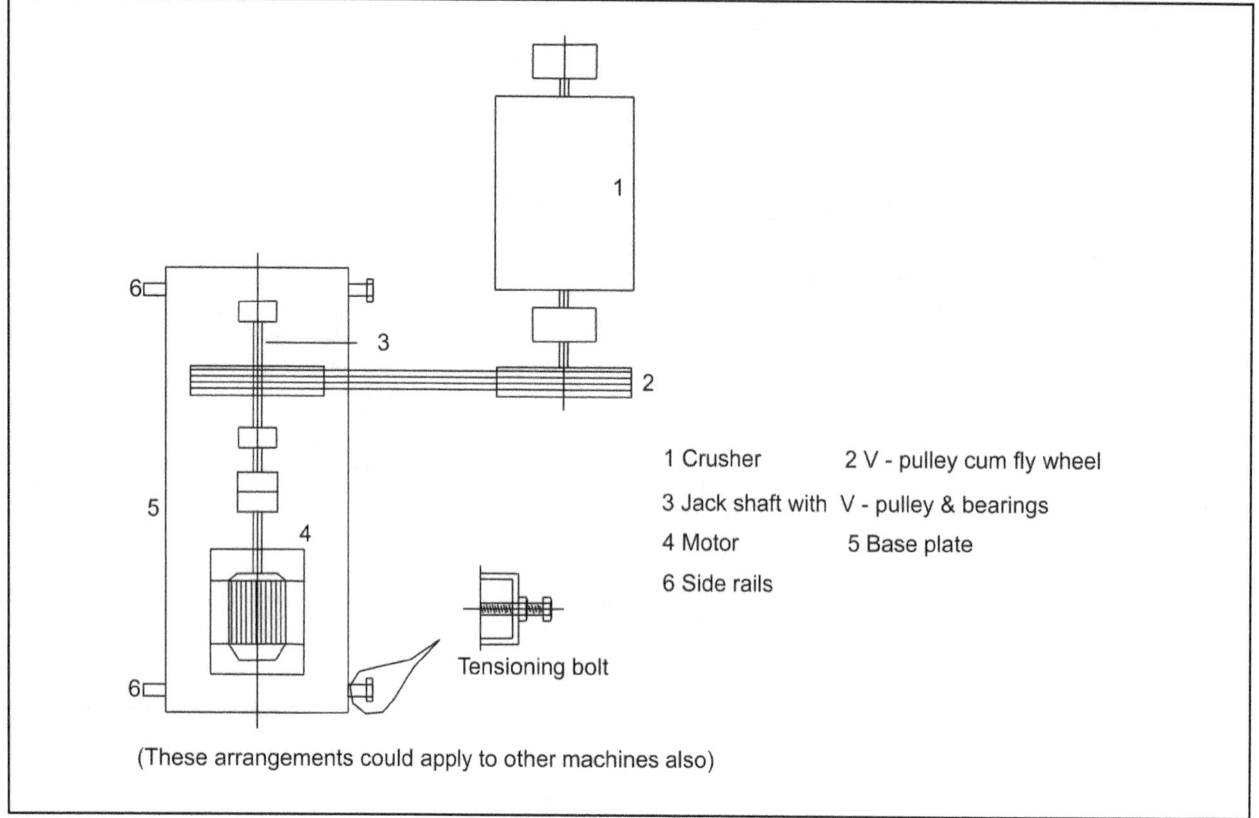

Fig. 6.4.4 Crusher drives.

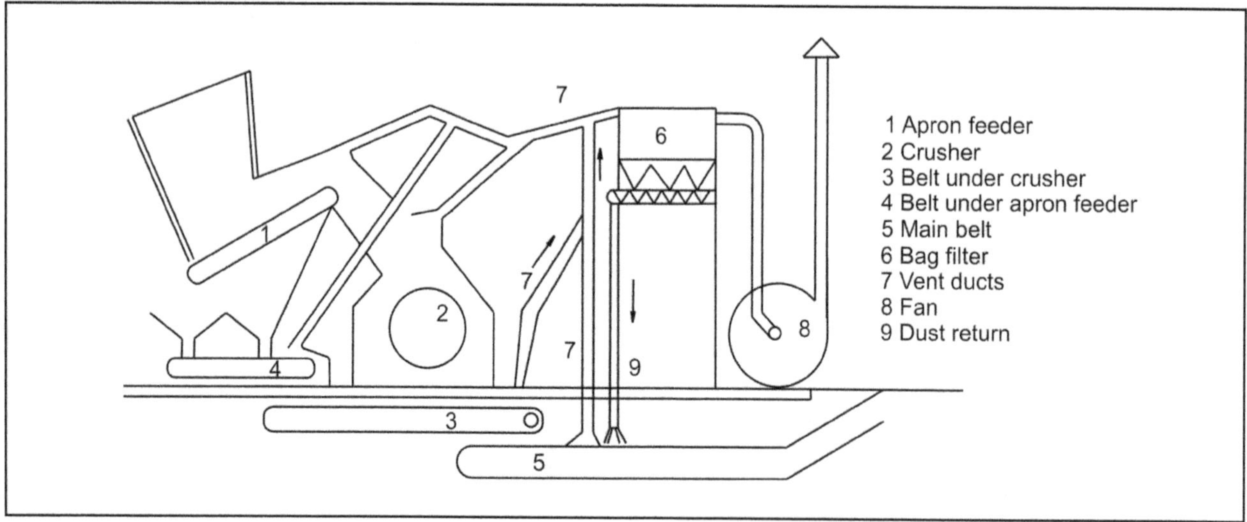

Fig. 6.4.5 Bag filter in crusher circuit.

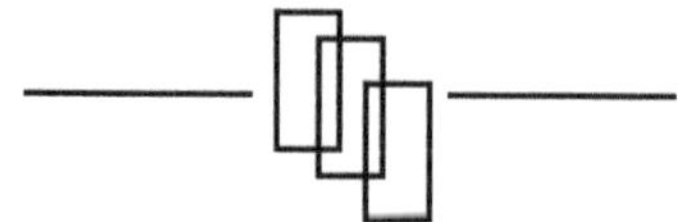

CHAPTER 5

STORAGE OF CRUSHED LIME STONE AND STACKER RECLAIMER SYSTEMS

5.1 Conveying Crushed Lime Stone

Belt Conveyors are almost exclusively used to convey crushed stone to storage. They can be designed for high capacity; can be installed at angles up to 18°; have low power consumption and maintenance costs are low.

Belt conveyors can discharge at more than one point. In lengths they can be as short as 2 metres and also several kilometers long.

When crushing plant is in the plant conveying lengths from crusher to stock pile or stacker reclaimer systems would be short. Whereas if crushing plant is in quarries, the distance could be several kilometers. In such a case, mostly 'overland' conveyors would be used.

When terrain is hilly, a ropeway could be used to convey crushed stone.

5.2 Storages

Earlier crushed lime stone was stored in central crane gantry.

See Fig. 6.1.5 (b) in Chapter 1.

Now in small plants it is stored as triangular stock piles inside a shed or in the open, fed from top and extracted from the bottom.

See Fig. 6.5.1.

In large plants requiring large quantities to be stored, stacker reclaimer systems are installed which besides storing also achieve blending during extraction.

See Figs. 6.5.2 to 6.5.4.

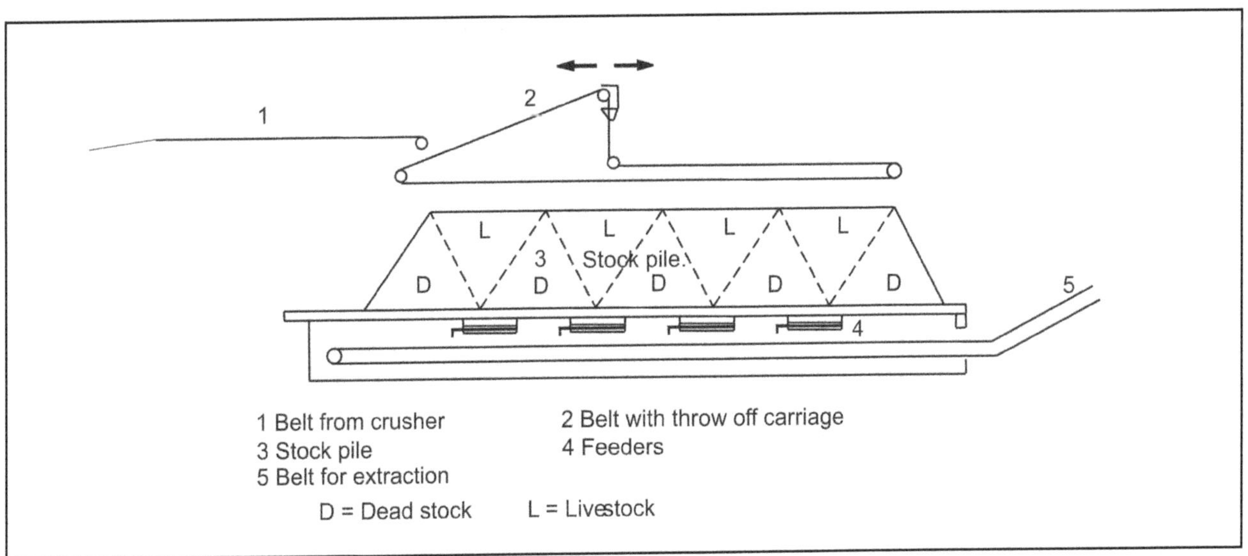

Fig. 6.5.1 Stock pile for a small plant.

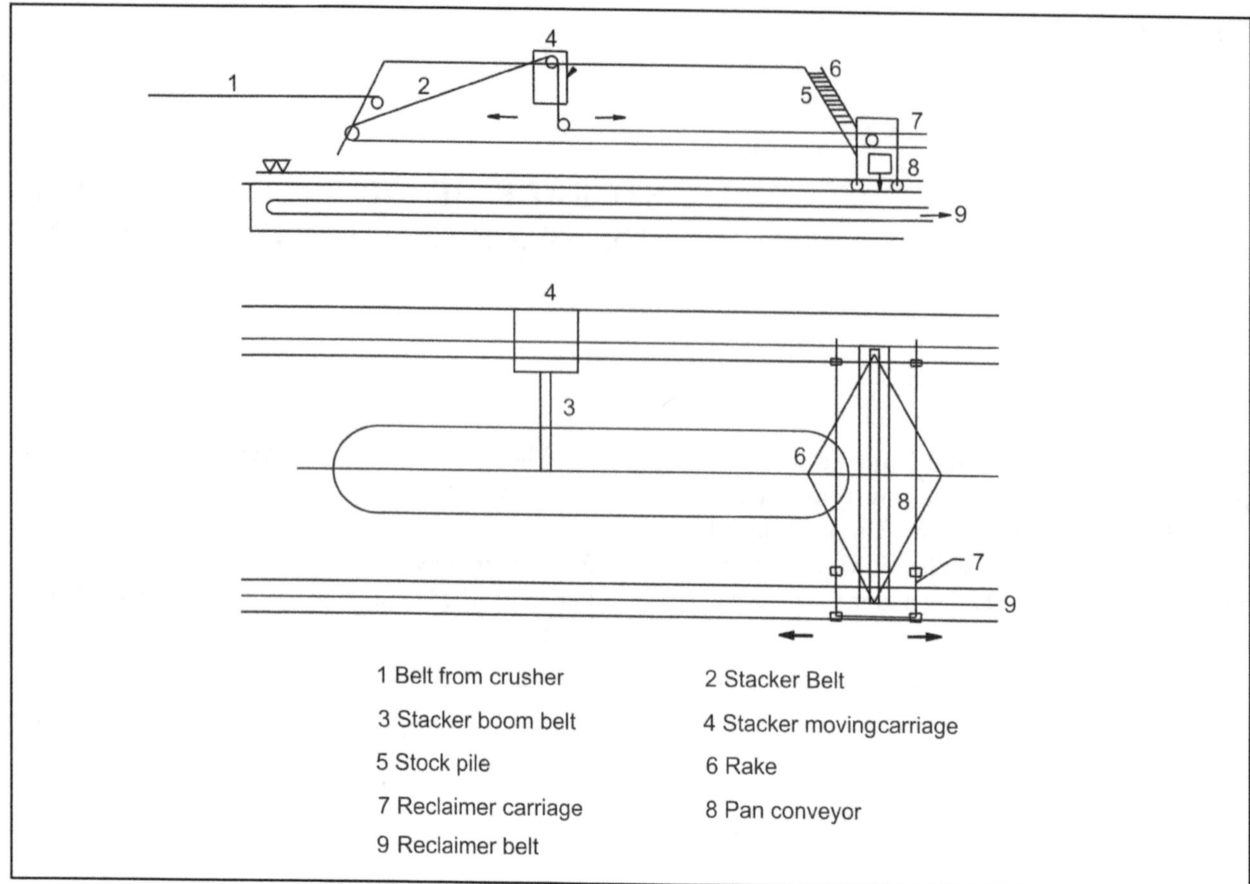

Fig. 6.5.2 Stacker reclaimer system.

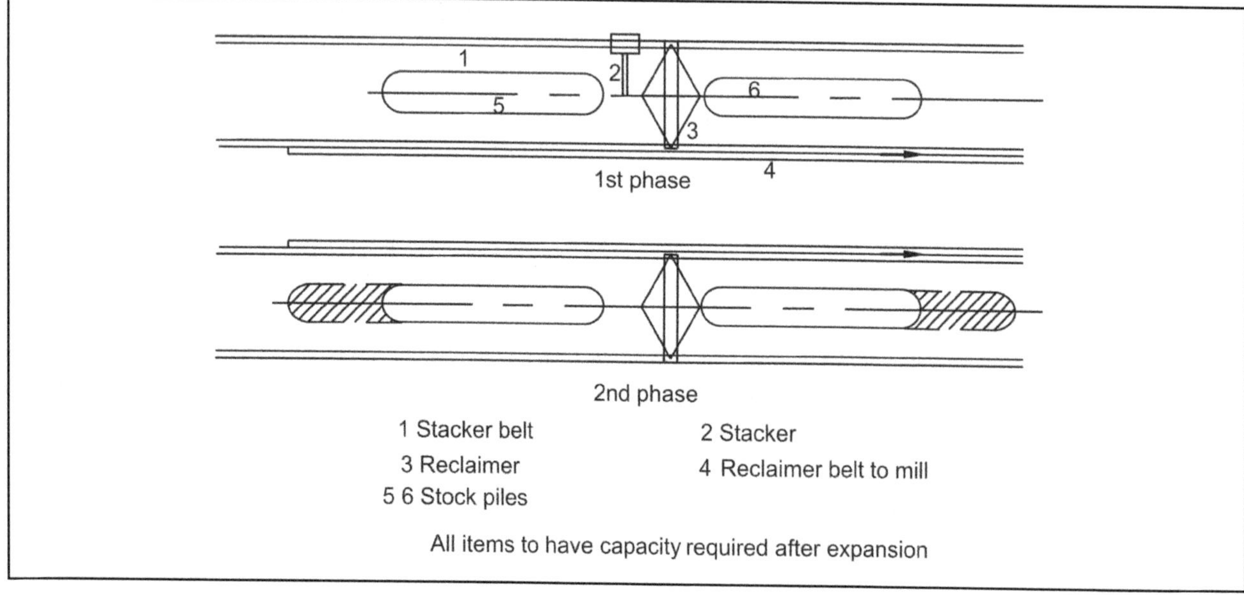

Fig. 6.5.3 Lineally arranged stock piles - extend lengths of stock pile.

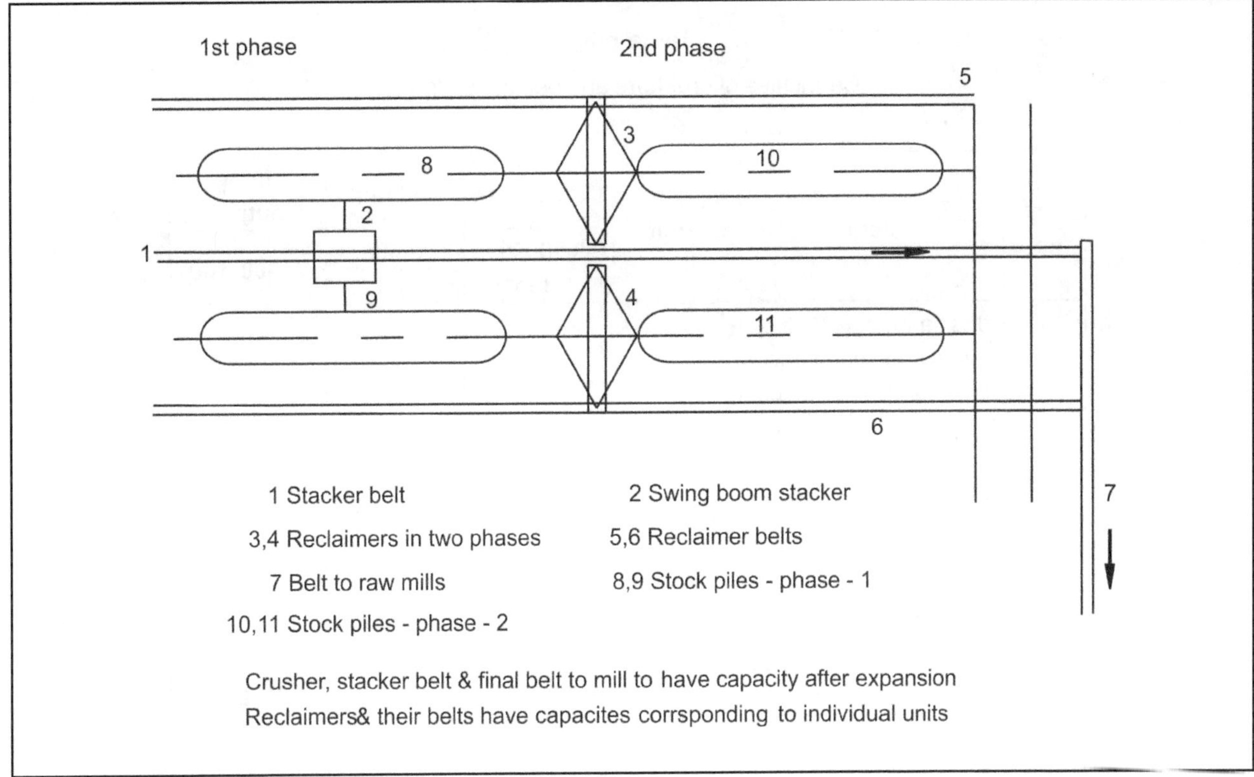

Fig. 6.5.4 Piles arranged side by side same crusher - swing boom stacker two reclaimers.

5.3 Belt Conveyors

In selecting belt conveyors, it would be a good idea to opt for deep troughs. Earlier, 20° angle was common. Now mostly 30° is selected for angle of trough. Narrower belts can be selected for same capacity.

5.3.1 Running Hours

It should be worked out at the planning stage, if crusher would be run for longer hours after expansion to meet the output required or whether a new crusher would be added. If the same crusher were to be used by running it for longer hours, initial capacity of crusher and hence belt conveyor would be higher. See **Chapters 10 and 11 in Section 1**.

5.3.2 Duplication

Secondly whether a new stacker reclaimer would be added or the same would be used by extending lengths of stockpiles.

In case of stacker reclaimer systems initial capacity would be higher if same system were to be used after expansion.

Table 6.5.1 shows requirements of capacities of crusher, belt to storage, and stacker for three alternatives for a plant of initial capacity of 10000 tpd and 20000 tpd after expansion.

It would thus be seen that the belt conveyor capacity after duplication would be increased 1½ times when using same stacker but extending the piles and 2 times if crusher and belt conveyor were to be installed for final capacity from the beginning.

5.3.3 Options Available

Based on **Table 6.5.1**, there are three alternatives/ options available

1. to install belt of 2250 tph one in each phase
2. to install a belt of 4500 tph from the begeiining
3. to install a belt of 3000 tph in the first phase and speed it up to 4500 tph capacity in the 2nd phase

Table 6.5.1

Capacities of crushers and belt conveyors.

Sr. no.	Item	unit	1st Unit	2nd unit	final output
			10000 tpd	10000 tpd	20000 tpd
1	design margin 10 % over limestone to be crushed 1.4 t/t clinker				
2	working hours 6/shift for first two shifts and 5 in 3rd shfit				
3	no of working days per week 6				
4	limestone to be crushed per day	tons	15400	15400	30800
5	limestone required in a week	tons	107800	107800	215600
6	A crusher working 2 shifts 72 hours per week				
	(i) capacity of crusher when each unit has its own crusher	tph	1500	1500	total 3000 tph
	(ii) whencrusher expanded for final capcity from beginning	tph	3000		
	(iii) crusher works in 2 shifts for first unit and 3 after duplication	tph	2000	2000	2000
		Belt Conveyors			
7	Belt conevor from crusehr to				
	#NAME?				
	capacity				
	6 I above	tph	2250	2250	individual stackers
8	6 ii above	tph	4500		one stacker
9	6 iii above	tph	3000		onestacker

5.3.4 Selecting Belts - Belt Capacities

It is also possible to select idlers and gantry for wider belt but install belting for lower capacity of 1000 tph to start with and install wider belting when plant capacity is duplicated.

Capacity tables and charts provided by belt manufacturers can be very useful for this purpose and should be made use of.

Vendors should be consulted and advice taken for best solution.

1. Recommended belt speeds

   Table 6.5.2 below shows recommended speeds for belts of different widths for different materials.

Table 6.5.2 Recommended belt speeds.

Carrying material	Belt width in mm				
	600	750	900	1000	1200
	Belt speed in m/min				
Coal, crushed stone	150	165	180	210	210
Ores, crushed stone	120	135	150	180	180

2. Capacities for various belt widths at different speeds

Table 6.5.3 below shows capacities of belts of different widths at different speeds assuming 20° troughing angle and an angle of repose of 35°.

Table 6.5.3 Capacities of belt conveyors.

Conveyor capacity in m^3 / hr					
Belt width mm	Belt speed in m / min				
	60	90	120	150	180
600	108	160	216	270	325
750	175	250	350	437	525
900	260	390	520	550	780
1000	360	540	720	900	1080
1200	475	710	950	1200	1425

For a trough angle of 30° multiply by 1.2

5.3.5 Drives for Conveyors

The options are :

(i) To install a drive of maximum rating from the beginning.

(ii) To install motor of higher rating later when belt is speeded up.

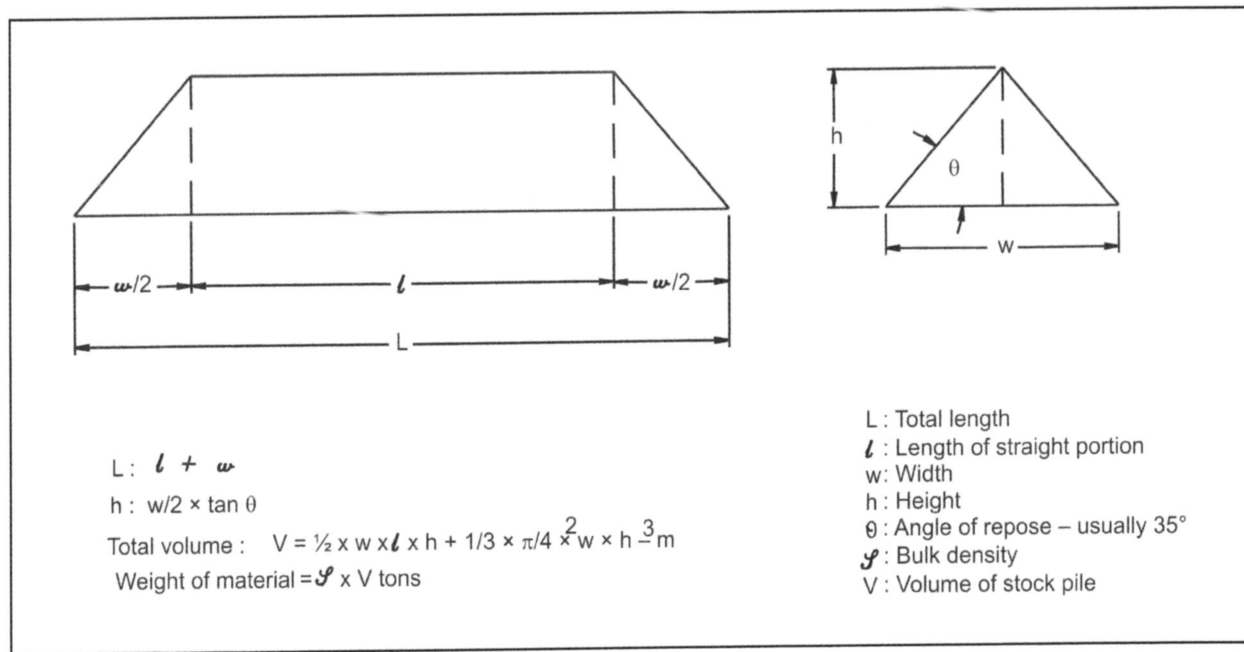

Fig. 6.5.5 Triangular stock pile - most common form of storage.

Table 6.5.4

	1000 tph	1450 tph
Belt speed	1.5 m/sec	2.2 m/sec
Gear box output speed	27 rpm	40.32 rpm
Motor speed	960 rpm	1440 rpm
Gear box ratio	35.7	35.7
Gear box power	50 kw	75 kw

Same gear box could be used as its rating would go up in proportion to ratio of input speeds. **See Table 6.5.4.**

Inclined Belts

Long belts leading to storage of crushed stone would be inclined wholly or partially.

5.4 Stacker Reclaimer Systems

For large plants, the most common systems would be to install 'stacker reclaimers'. When using stacker reclaimer systems there would be 2 stock piles – one being built up and material being withdrawn from the other. The piles, would be arranged either side by side or end to end in line. Capacities would depend on the stocks to be maintained.

7 days stock (only limestone) = 10000 × 1.1 × 1.4 × 7 = 108000 tons. If width of pile is 25 metre, its length would be 200 metres.

See Figs. 6.5.2 to 6.5.5.

5.4.1 Preblending With Stacker Reclaimer System

Stacker-Reclaimer systems are used to 'preblend' limestone (also coal) in stock piles. Degree of homogenization depends on number of layers in a stock piles and also on the way in which pile is built up. Hence the system should be selected to achieve desired blending.

Fig. 6.5.6 shows relation between blending effect and number of layers. **Figs. 6.5.7 and 6.5.8** show some of the common ways of building up stockpiles.

Space would have to be kept in layout for which ever way piles would be arranged – side by side or end to end for 2 piles. In the layout provision should also be kept for expansion choosing from the alternatives explained above.

In the first arrangement, pile lengths have to be increased when plant is duplicated requiring very long linear belts for stacking and for reclaiming. It would need to have stacker belt only on one side.

See Fig. 6.5.3.

In the 2nd arrangement, a twin boom or a swing boom stacker would be required; reclaimer, can be one or two. If one, a transferring facility needs to be provided to shift it from one stock pile to the other. Two reclaimer belts would be required to take reclaimed material to mills.

See Fig. 6.5.4.

5.4.2 Dimensions of Stock Piles

In planning the layout therefore approx. dimensions of stock piles should be worked out as shown in **Fig. 6.5.5.**

Nomgram shown in **Fig. 6.5.12** helps to work out capcities in volume from dimensions of the stock plies of crushed stone with angle or repose of 35°.

Generally speaking, widths would vary between 20-30 metres; most common being 25 metres.

Having arrived at the lengths in this fashion, layout can be planned, starting from crusher to stacker-reclaimer system.

5.4.3 Upgrading of Capacities

The possibility of upgrading capacity of stacker reclaimer system economically (from 900-1350 tph as in above example) should be looked into before taking a decision in their selection and layout can be finalized only after that.

Till then, various options explained above could be worked out on drawing board to arrive at requirements of space.

See Figs. 6.5.3 and 6.5.4.

5.5 Limiting Angles of Inclination for Belt Conveyors

Angles of inclination of belt conveyors are governed by the nature of material carried and size of maximum pieces.

Generally recommended angles fall within 14-16 degrees. Steeper slopes reduce belt lengths but can cause problem like roll back. Capacity is also to be derated. Suppliers' manuals should be used to select angle of inclination.

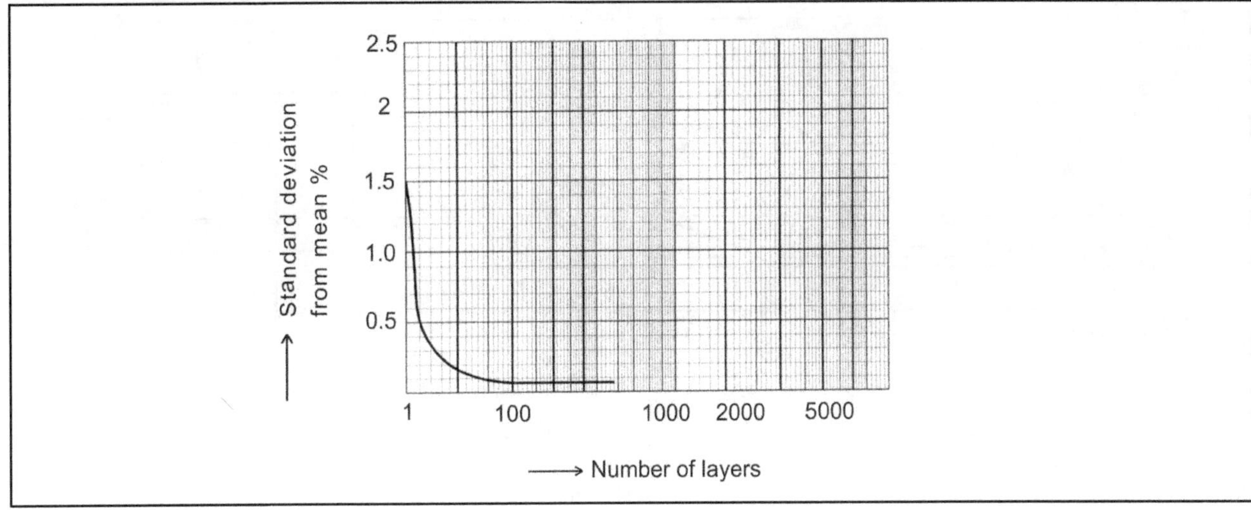

Fig. 6.5.6 Blending effect as a function of number of stacked layers of material.

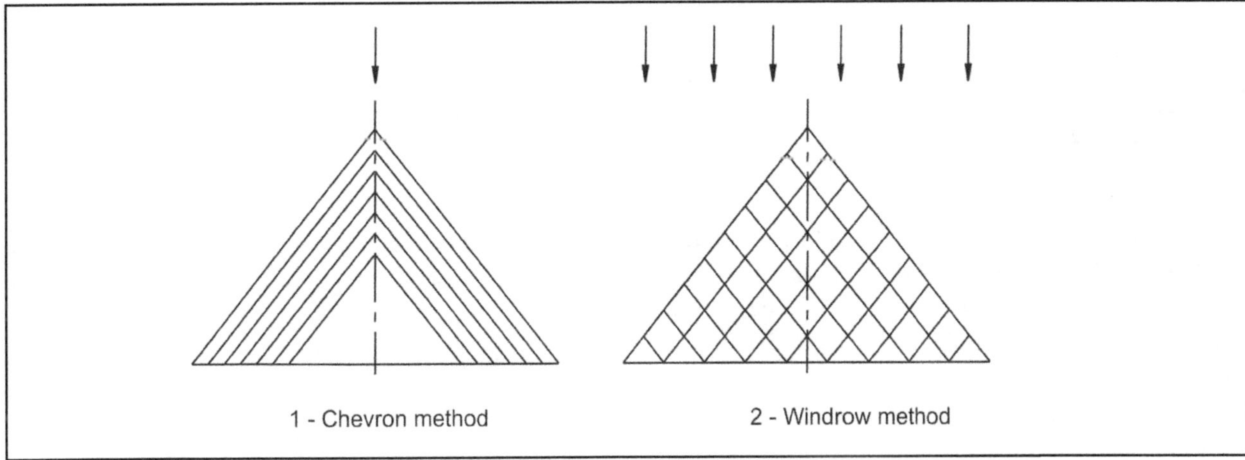

Fig. 6.5.7 Various stacking methods for stacker recalimer systems.

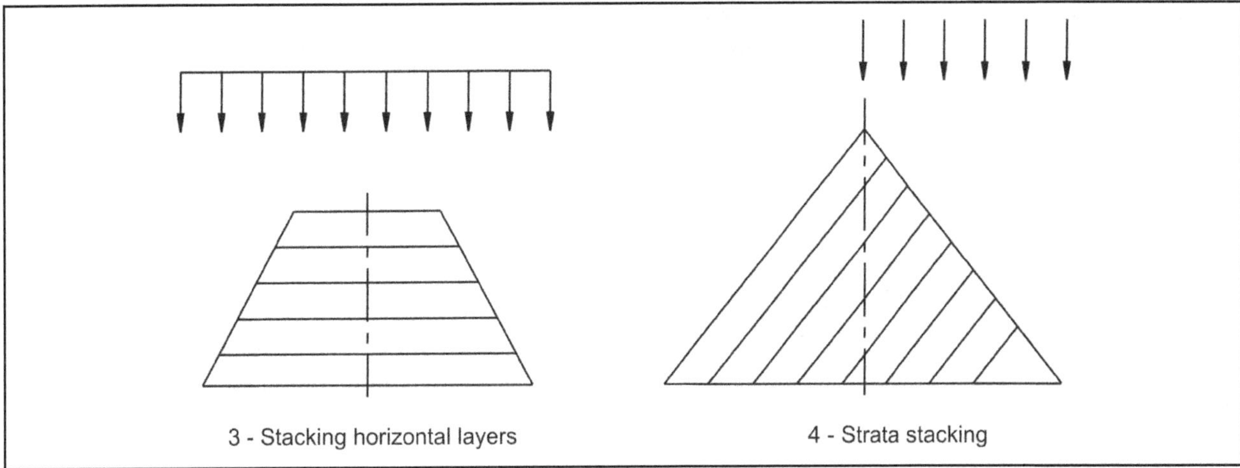

Fig. 6.5.8 Various stacking methods for stacker recalimer systems to achieve prebelnding.

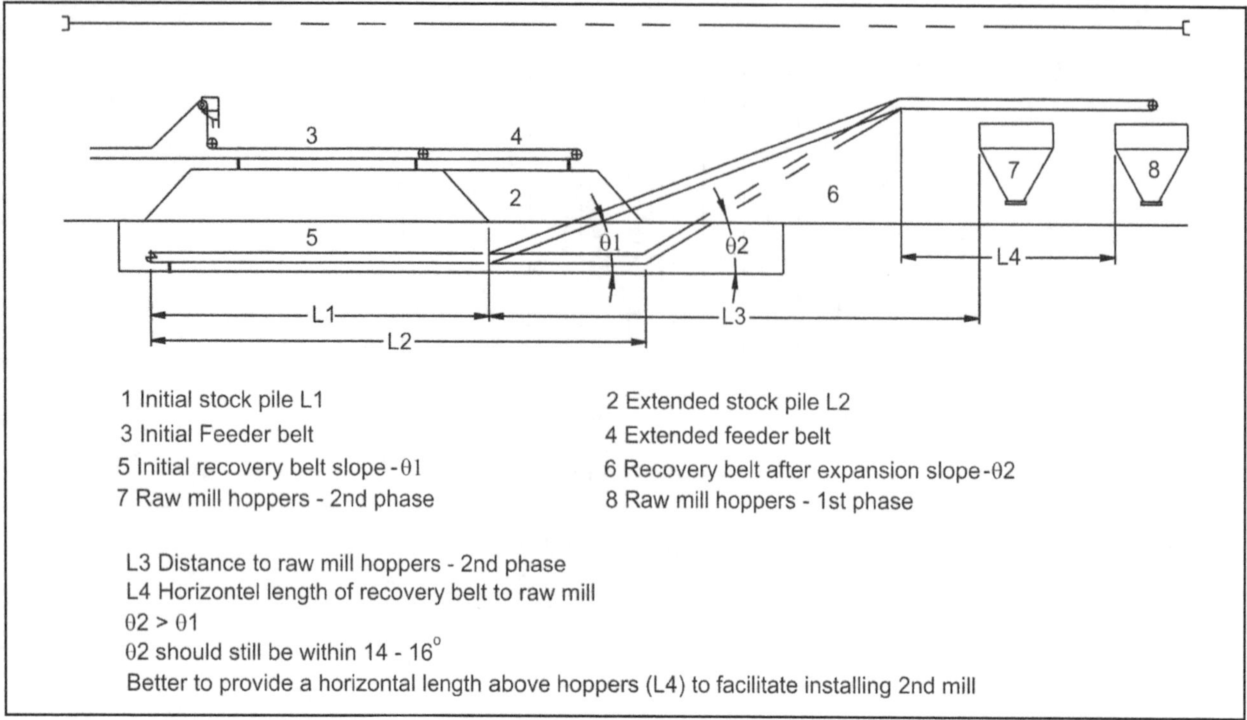

Fig. 6.5.9 Stock pile providing room for expansion.

5.5.1 Splitting of Belt

As mentioned earlier, if there is a possibility of splitting the belt at a later date for close circuiting a crusher, then a smaller angle of 12° or so should be used to start with so that after spillage, the angle of split belt does not exceed 16°.

See Fig. 6.3.21 in Chapter 3 of this Section

Even if the secondary crusher is not to be close circuited, it would be prudent to provide room for expansion of the stockpile and installation of hoppers for a second raw mill. This arrangement may be more expensive in first costs but would save considerable time and money later when expansion materializes. **See Fig. 6.5.9.**

Rated capacity of belts decreases with increasing inclination. Due cognizance should be taken of this in sizing the belts.

The guiding principle for all material handling equipment would be that it should never be a bottleneck at any time.

5.6 Layout of Limestone Storage in Small Plants

In small plants crushed limestone will be stored in triangular stock piles. Belt conveyor from crusher will feed the pile from the top. This belt could be split into two; inclined belt coming up to stock pile and a separate horizontal belt with throw off carriage over the stock pile.

Throw off carriage will help building up of the pile evenly along its length. It will be motorized. Stone will be extracted from the bottom. There will be a number of openings with rod gates and feeders and a belt conveyor in a central longitudinal tunnel for this purpose. **See Fig. 6.5.11.**

5.6.1 Effective Capacity

In this system effective capacity is about 50 % of total capacity of the pile because of the dead stock. Effective capacity can be increased by increasing number of outlets. Often bulldozers are used to push materials over feeding points. **See Fig. 6.5.11.**

	Plant capacity tpd	capacity of stock pile tons	no of days' stock	length of pile metres
1st phase	500	5400	7	37-40
2nd phase	1250	9600	5	66

5.6.2 Increasing Capacity

Storage capacity is increased by extending the length of pile and by extending belt conveyors.

Thus additional 30 metres length should be provided in the layout for expansion. To that extent raw mill department would be shifted further.

An additional horizontal length of belt feeding raw mill hoppers should also be provided so that a second raw mill could be installed.

See Fig. 6.5.9.

5.7 Layout and Capital Costs

It is necessary to keep these things in mind when planning a layout.

If the belt layouts are planned in this fashion, capital costs no doubt would be higher to start with but no modifications are needed to the system at a later date if belts are selected for the final capacity of the plant.

Plant Authorities- Buyers should obtain pertinent catalogues of machinery – crushers, feeders, stacker reclaimer systems and belt conveyors from vendors and use them for developing the layout.

It is best to ask Vendors to develop departmental layouts in consultation with Buyers and their Consultants and submit them for approval. Either Buyer or his Consultant will incorporate them in the General Plant Layout.

5.8 Storage and Extraction of Correcting Materials

Correcting materials clay and iron ore are required in small quantities as compared to limestone. When stacker reclaimer systems are used for limestone, correcting materials would be received and stored separately in covered sheds. More often they would be brought in truck loads but sometimes would be brought in wagons.

5.8.1 Over head Crane

An over head grab bucket crane could be used to feed hopper in the gantry to take clay and iron ore to hoppers before raw mill. Alternatively a payloader could be used to feed materials to hopper.

See Fig. 6.5.10.

5.8.2 Extraction from Bottom

Another option would be to store them in a covered shed same way as lime stone storages of a small plant.

See Fig. 6.5.11.

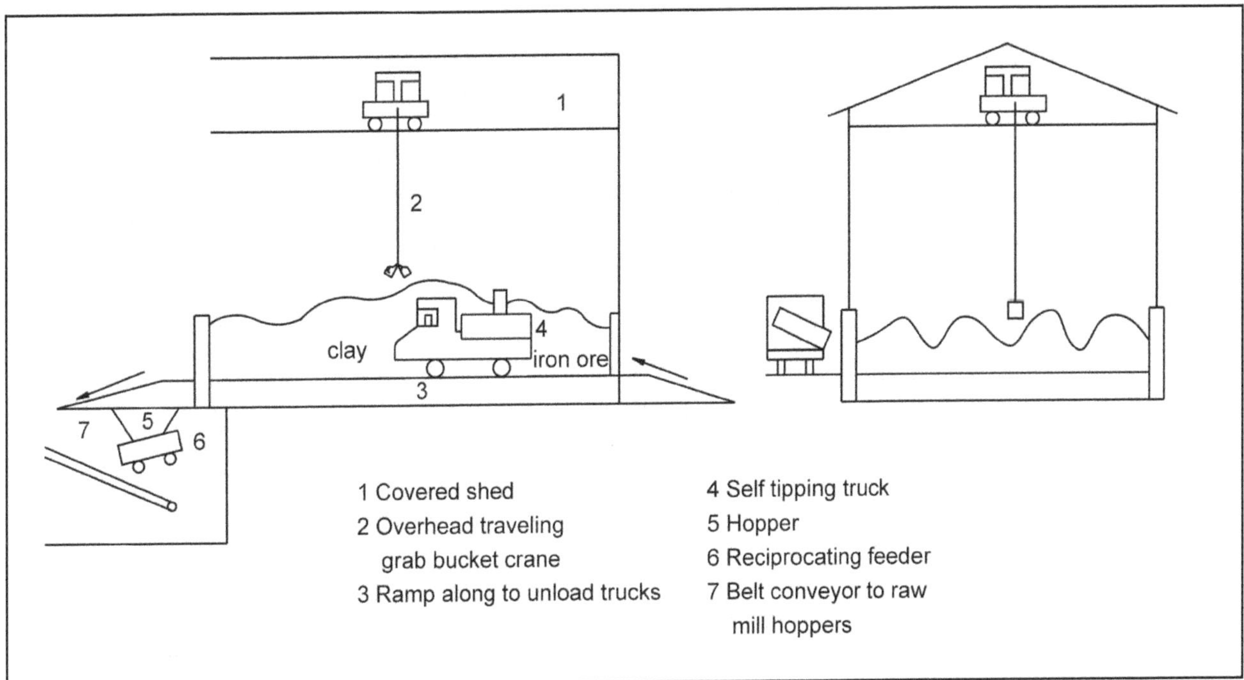

Fig. 6.5.10 Receipt, storage & extraction of correcting materials.

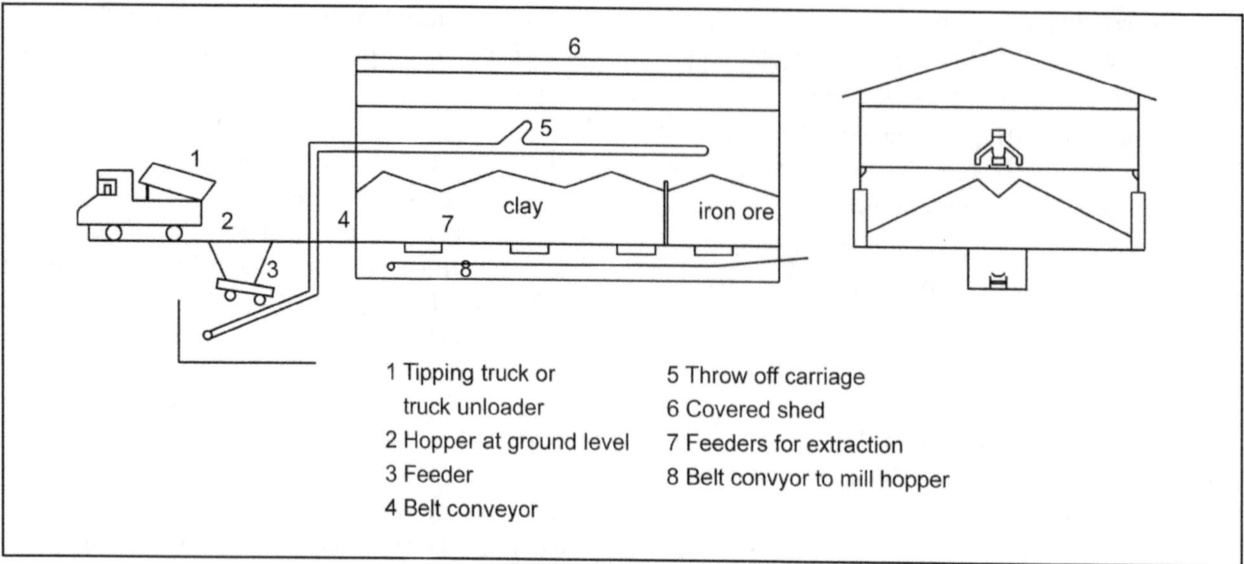

Fig. 6.5.11 Receipt, storage & extraction of correcting materials.

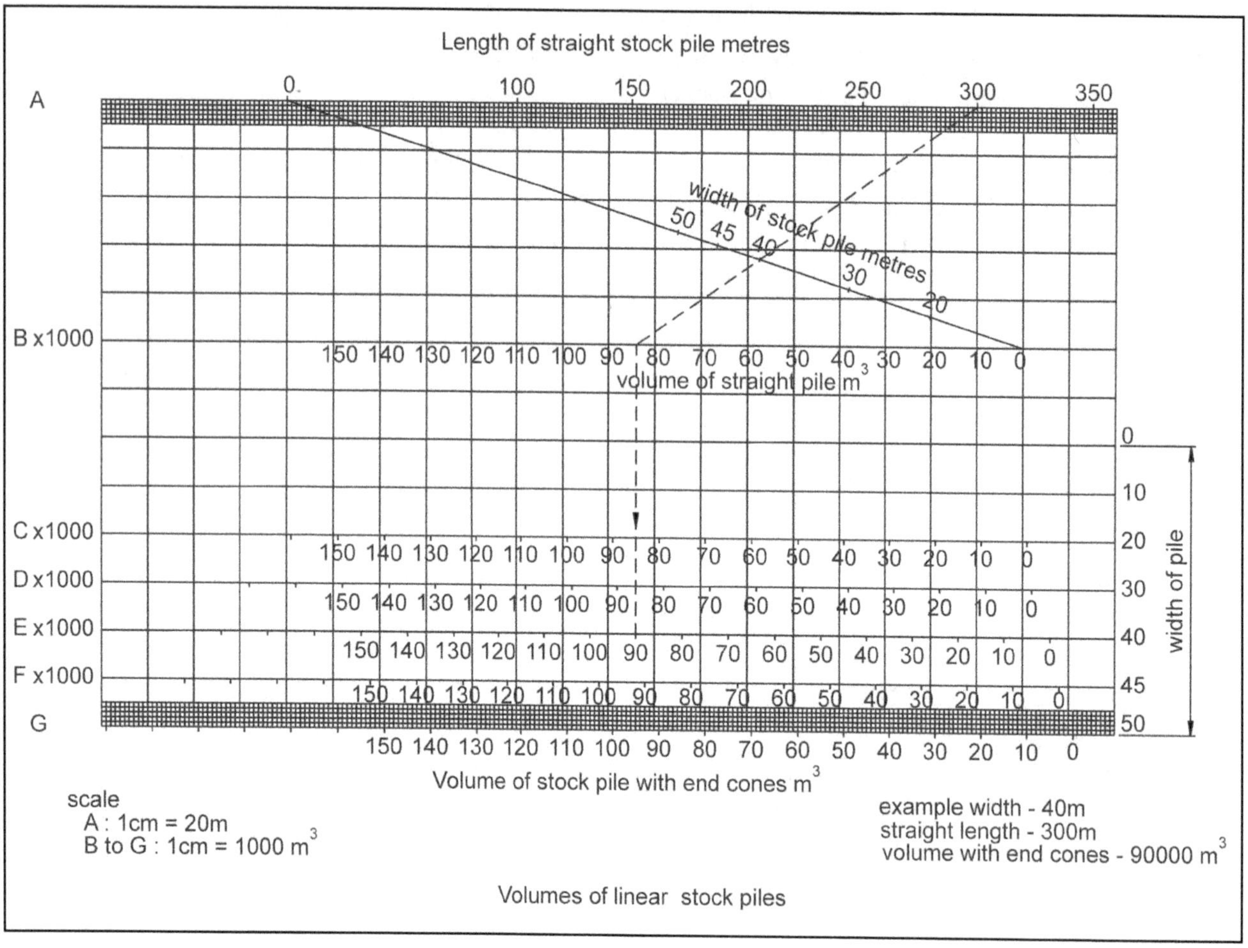

Length of straight stock pile metres

A 0 100 150 200 250 300 350

width of stock pile metres 50 45 40 30 20

B x1000 150 140 130 120 110 100 90 80 70 60 50 40 30 20 10 0

volume of straight pile m^3

C x1000 150 140 130 120 110 100 90 80 70 60 50 40 30 20 10 0

D x1000 150 140 130 120 110 100 90 80 70 60 50 40 30 20 10 0

E x1000 150 140 130 120 110 100 90 80 70 60 50 40 30 20 10 0

F x1000 150 140 130 120 110 100 90 80 70 60 50 40 30 20 10 0

G

150 140 130 120 110 100 90 80 70 60 50 40 30 20 10 0

Volume of stock pile with end cones m^3

width of pile: 0 10 20 30 40 45 50

scale
A : 1cm = 20m
B to G : 1cm = 1000 m^3

example width - 40m
straight length - 300m
volume with end cones - 90000 m^3

Volumes of linear stock piles

Fig. 6.5.12

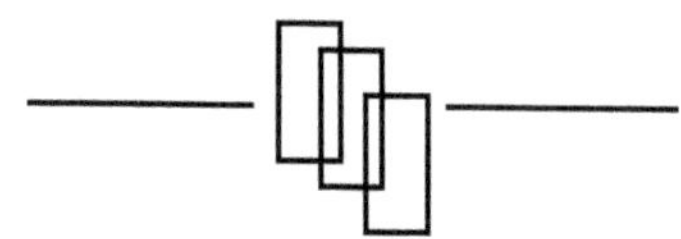

CHAPTER 6

LAYOUTS OF RAW MILLS AND SEPARATORS

6.1 Mills and Separators for Grinding Raw Materials

Mills and Separators work together as a Team. In vertical mills they are actually part of the mill. Hence it would be best to treat them together. First the mills.

6.2 Raw Mills

Two major options for raw mills would be:

1. Vertical roller mills (or ring ball mills) that take a feed size of -75/100 mm.
2. Ball Mills.

6.2.1 Raw Material Grinding Systems in Dry Process Plants

(i) All grinding systems for raw materials work in 'closed circuit'.

(ii) They are required to 'dry' raw materials and correcting materials while grinding.

(iii) Should 'proportion' limestone and correcting materials so that raw meal ground in them is as close to desired raw mix composition as possible.

There are a great many options available in each of these types of mills. Hence a decision on selection of the mill should be taken before developing layouts.

6.3 Options for Ball Mills

1. Air swept mills with grit separators – for small plants- now used more or less for grinding coal only.
2. Bucket elevator mills – for plants of medium size 1000-3000 tpd capacity with either conventional or high efficiency separators.

   See Figs. 6.6.1 to 6.6.3.

   Ball mills prefer fine feed; feed size between 12 and 20 mm will be common (further options are drying can be either in mill or in separator)
3. Horo mills – which is a hybrid of a 'roller' grinding in a ball mill. It can take large feed size like a vertical mill and has the simplicity of a ball mill.

   See Figs. 6.6.4 and 6.6.5.
4. Ball mills working with Roller Presses in a variety of ways.

   See Figs. 6.6.6 to 6.6.8.

Horo mill and Roller press can take a feed size of - 75 mms like vertical mills.

6.4 Options for Vertical Mills

All vertical mills have built in separators. Separators can be static or rotary.

1. **Ring Ball Mills** – E mills – for small plants for capacities of 50-70 tph. These mills had a static separator.Now they too have Mechanical Separators.

   They have been primarily developed for grinding coal. Do not find them as Raw Mills in Large Cement Plants
2. **Roller Mills**

   (i) Conventional separator - rotating or static separator.

   (ii) External circuit mills with conventional separators / high efficiency separators.

   (iii) Existing Mills retrofitted to house high efficiency separator and also converted to external circuit.

   See Figs. 6.6.9 and 6.6.10 and 6.6.48

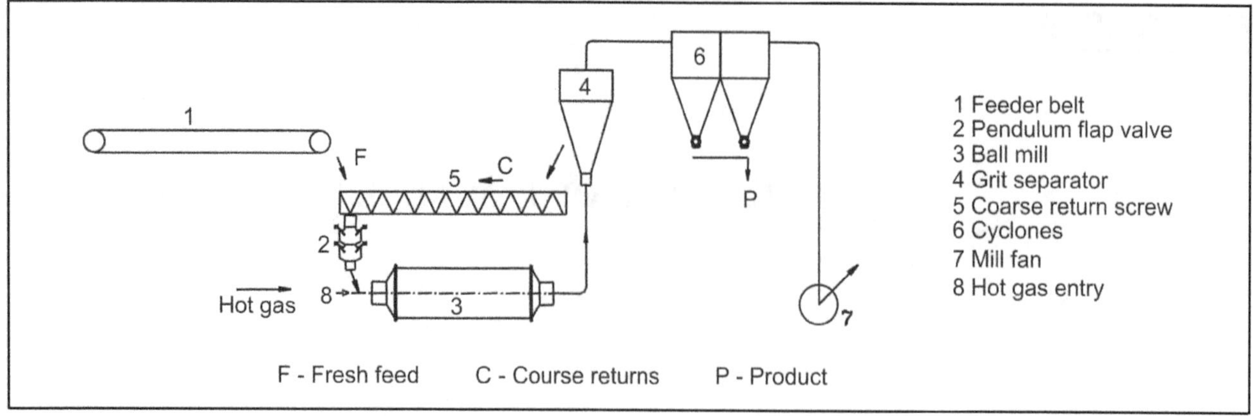

Fig. 6.6.1 Air swept mill & grit separator drying in mill.

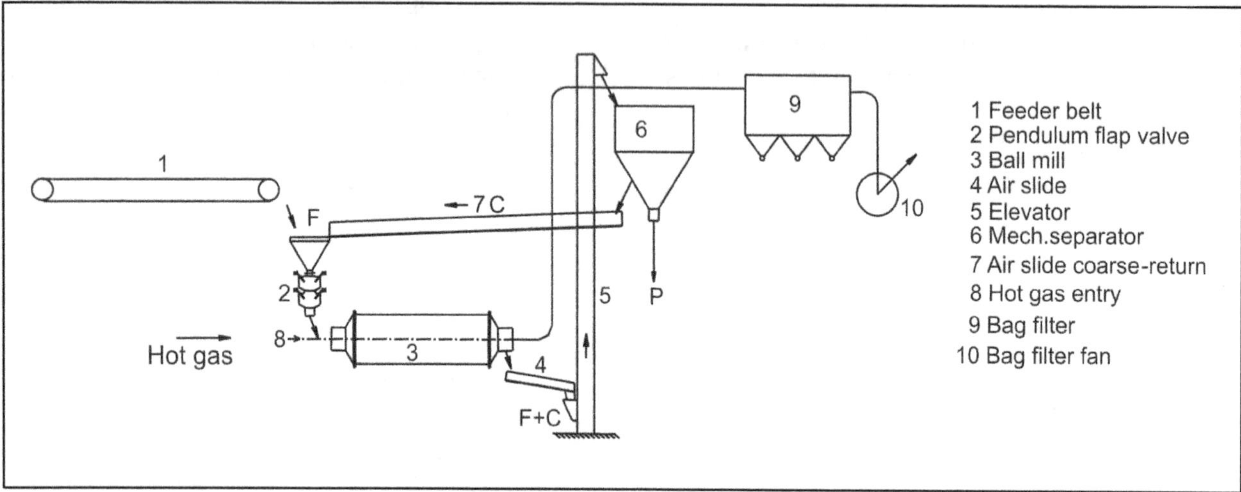

Fig. 6.6.2 Bucket elevator mill with mech. separator drying in mill.

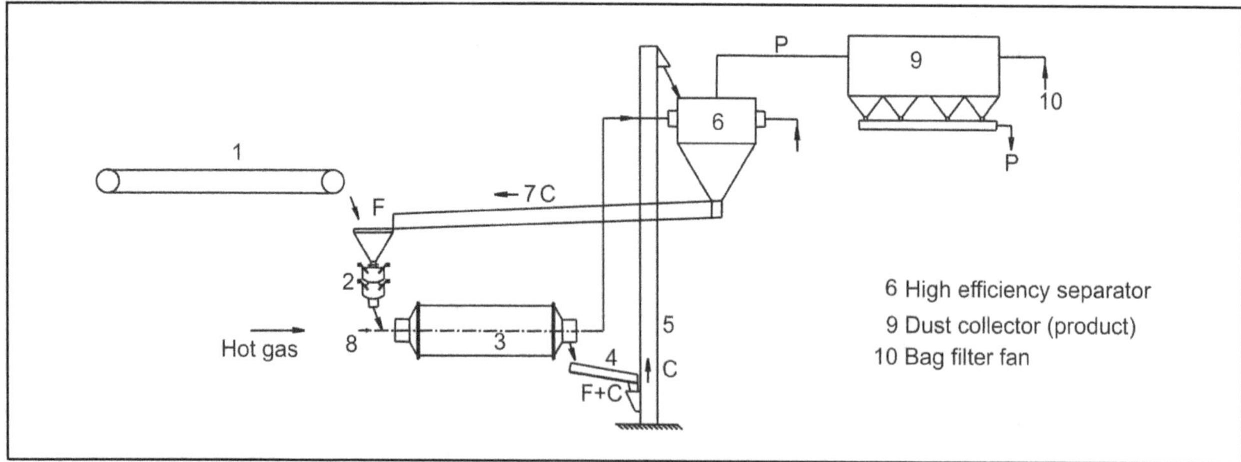

Fig. 6.6.3 Bucket elevator mill with high efficiency separator – drying in mill. Ball mills grinding systems in closed circuit

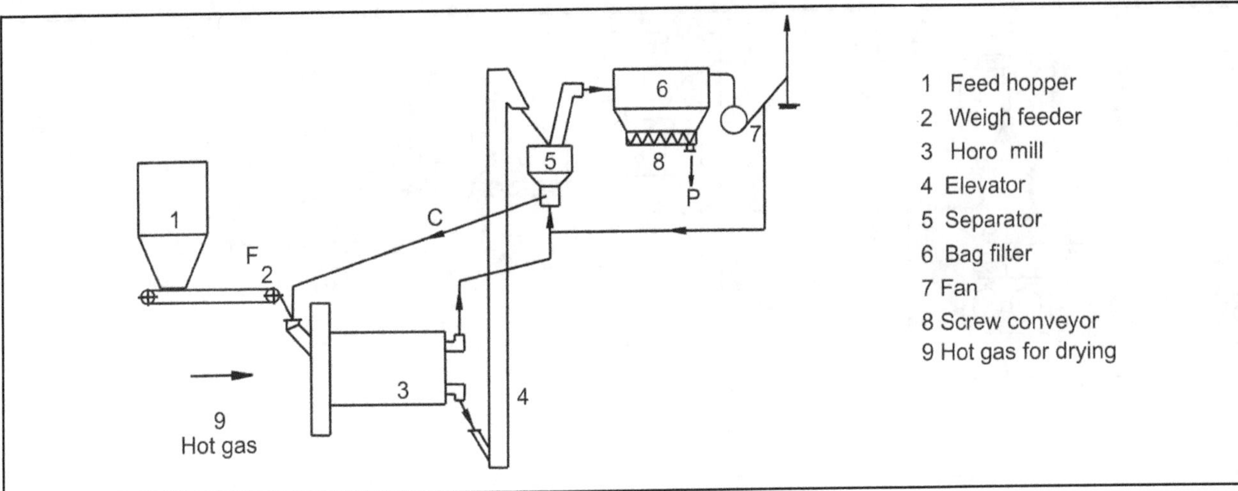

Fig. 6.6.4 Grinding circuit of Horo mill.

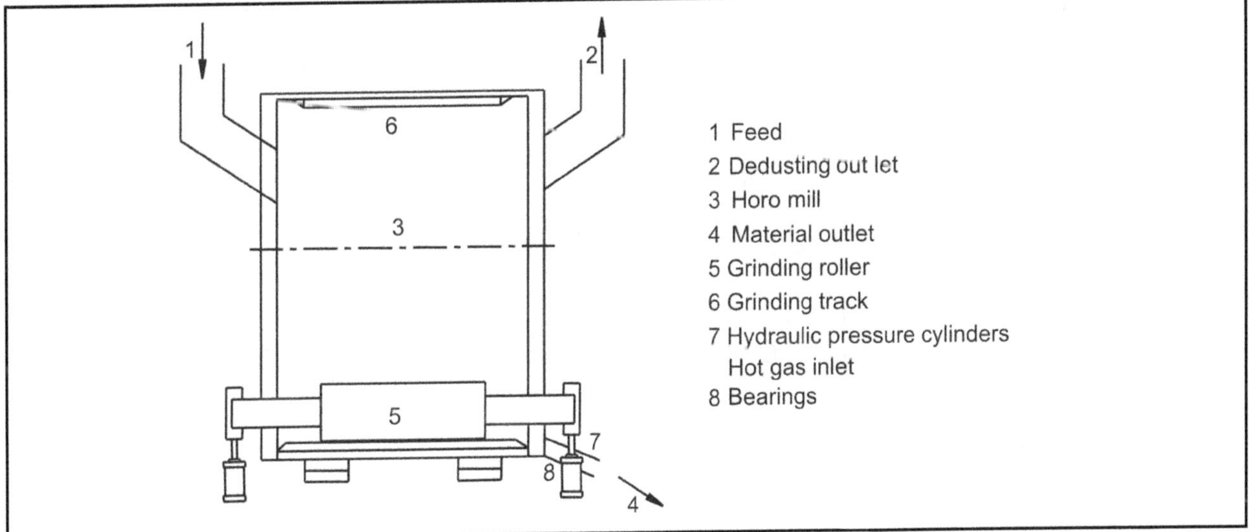

Fig. 6.6.5 Horo mill - sectional view.

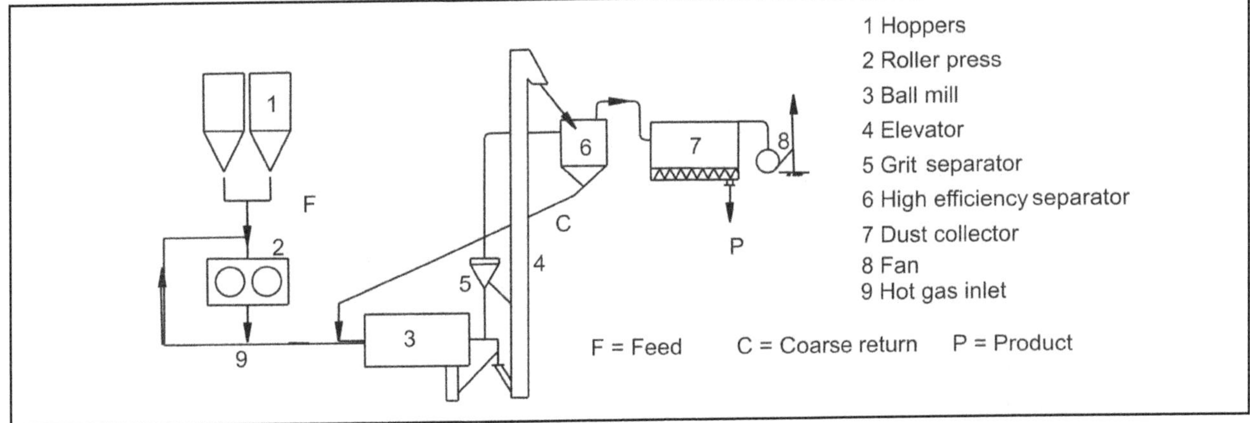

Fig. 6.6.6 Roller press as pregrinder.

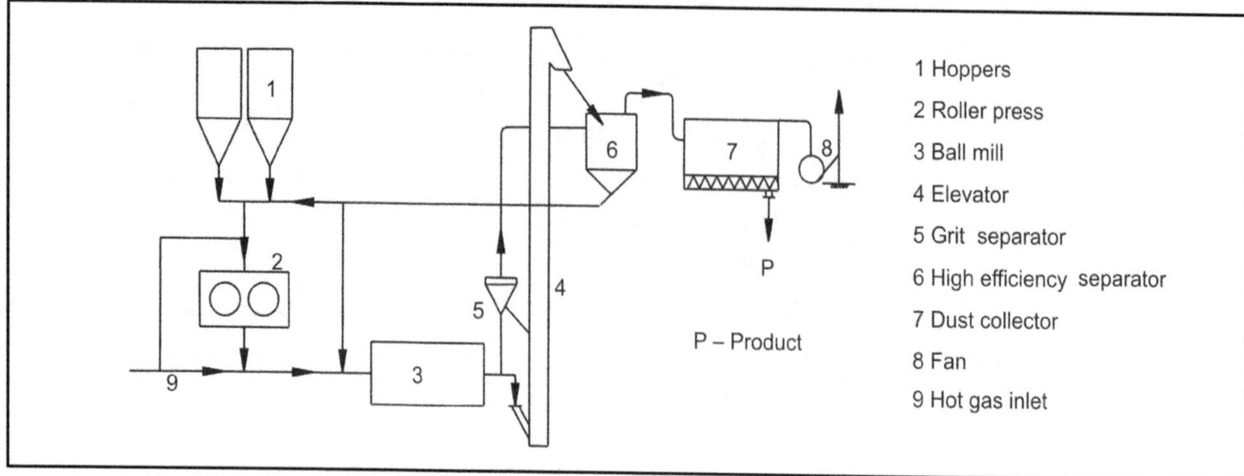

Fig. 6.6.7 Roller press in hybrid grinding circuit.

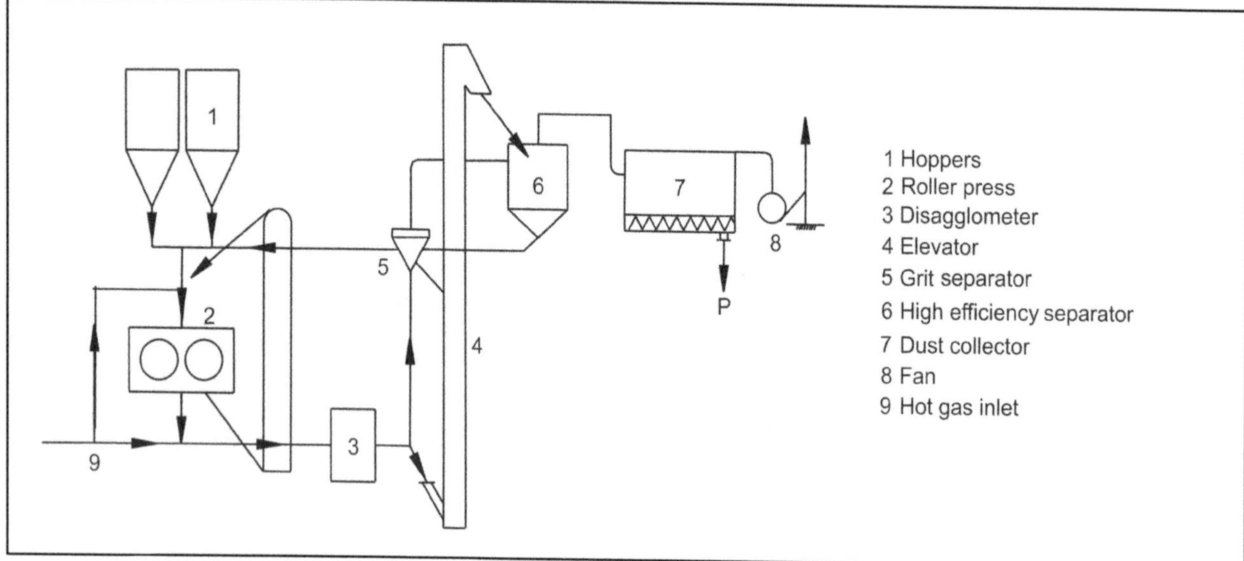

Fig. 6.6.8 Roller press & disagglometer in finish grinding circuit.

Vertical mills systems take smaller floor space compared to ball mills of comparable capacities because :

(a) Gear box is under the mill except for multi drive mills

(b) Separator is built into the mill.

Presently they are the mills commonly used in all sizes of plants - particularly large plants for all applications that is for grinding Raw Materials, Coal and Cements of all types.

6.5 Roller Press and Ball Mill

Because of reduction in grindability index and generation of fines during passage of materials between the rollers, ball mill following it has to do much less work and has to grind a softer material and is therefore much smaller in power and size for a given capacity than when used singly.

There are a great many ways in which a roller press and a ball mill and a separator can work together.

See Figs. 6.6.6 to 6.6.8.

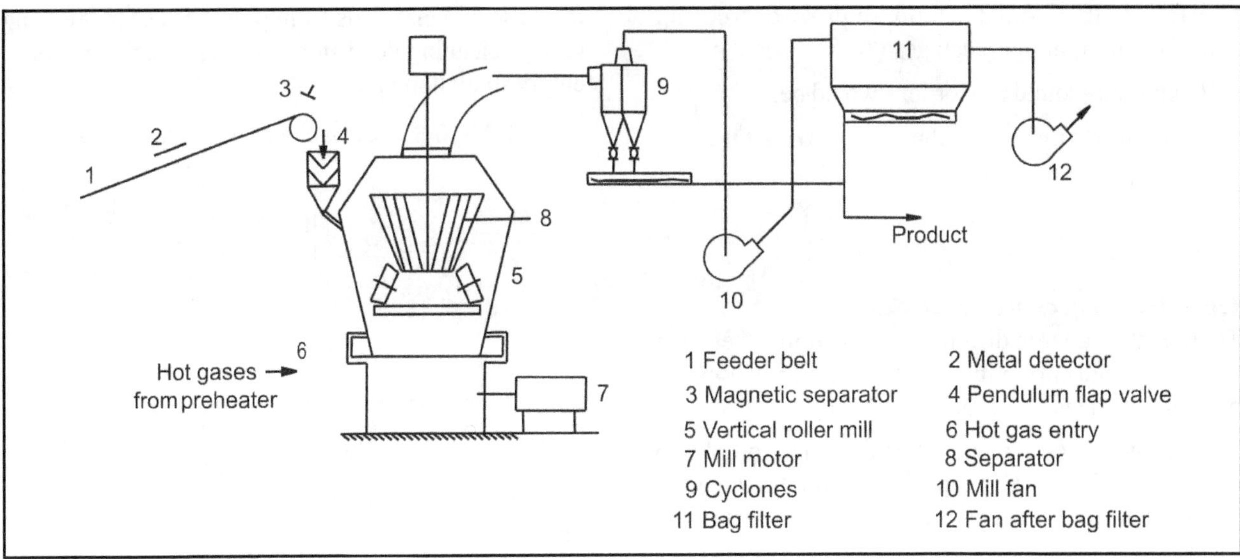

Fig. 6.6.9 Flow chart of vertical roller mill.

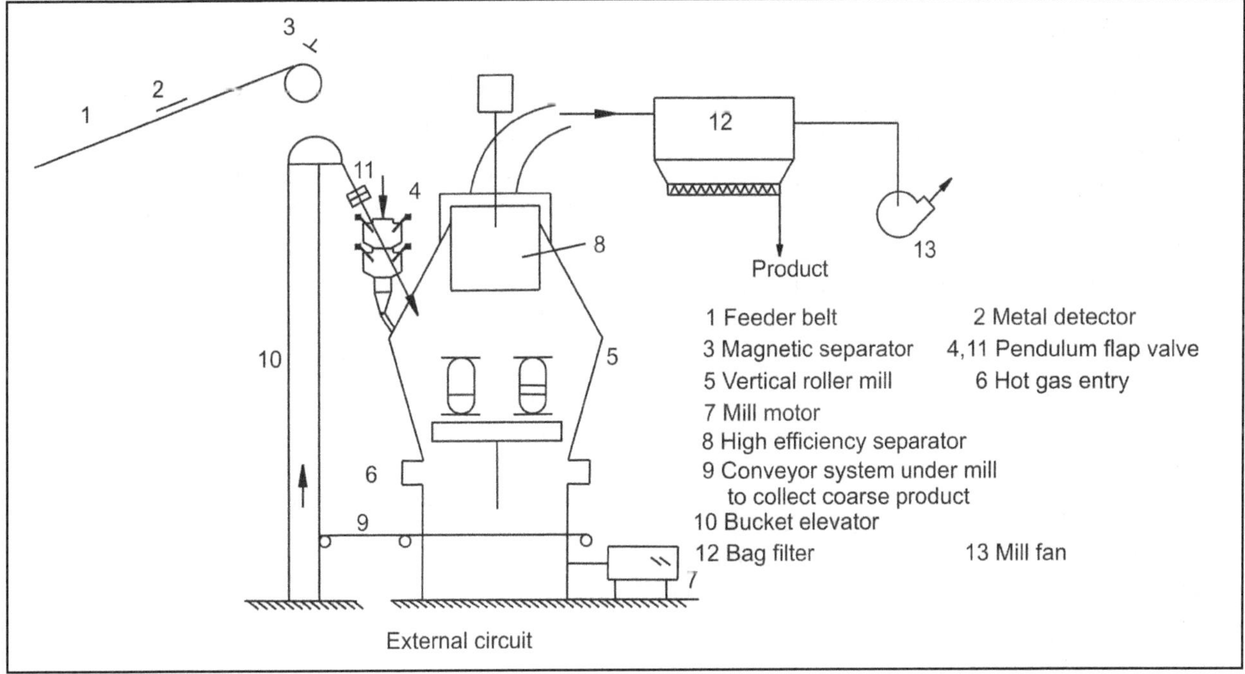

Fig. 6.6.10 Flow chart of vertical roller mill with external circuit.

Ultimately ball mill can be eliminated altogether. A disagglomerator is used to break cake.

See Fig. 6.6.8.

6.6 Mill Hoppers

Raw mix consists of limestone and additives in various proportions.

Limestones could be in two grades – high and low – which could be blended to conserve high grade.

Depending upon the differences in carbonate contents, they could be used in proportions of 50 : 50 to 80 : 20.

Correcting materials are generally clay, sand and bauxite to correct silica and alumina; and iron ore / laterite, murram, etc., to correct iron oxide.

Richer limestones would require correcting materials in higher percentages (%).

Commonly found proportions would be:

Limestone	Clay	Iron Ore
80	18	2
85	12	3
90	9	1

6.6.1 Differences in Capacities

There is thus a large difference in quantities between limestones and correcting materials requiring hoppers of different sizes and capacities.

Limestone hoppers would be designed to hold up to 8 hours of feed. The capacity could be reduced to 2-4 hours if a reclaimer system can feed hoppers round the clock.

For a 10000 tpd plant, capacity of the raw mill, allowing for :

design margin - 10%
conversion ratio - 1.5
running hours/day - 20

raw mill capacity in tph = 850 tph

Therefore capacities of hoppers would be as shown in **Table 6.6.1**

Table 6.6.1 Capacities of hoppers in Tons.

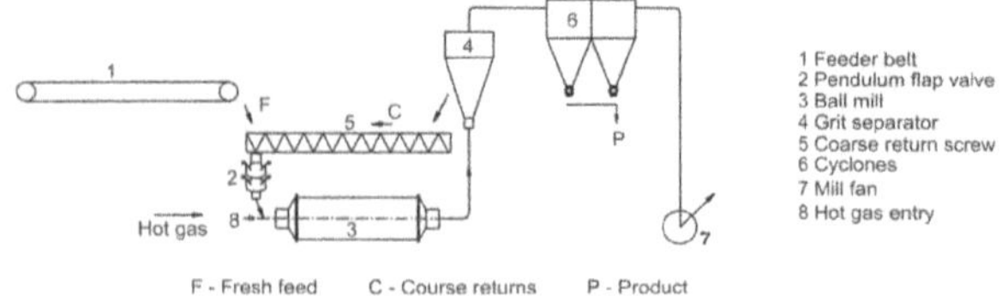

If the plant has a stacker reclaimer system-as a plant of this size is most likely to have-it should be possible to reduce capacities of hoppers to say 2 hours to reduce sizes of hoppers.

6.6.2 Hoppers for Clay and Iron Ore

Thus hoppers for clay and iron ore would be too small in size compared with limestone. From practical considerations, hoppers for clay and iron ore would be designed for 24 hours requirements; sometimes even more. Even then capacities would be small compared to lime stone as would be seen from **Table 6.6.2.**

Problem in having larger storages in clay and iron ore is that both contain high moisture and are not free flowing. It is therefore best to have smaller, storages for these two materials from point of ease in breaking arches, cleaning dead stock etc. It is much simpler to empty small hoppers.

Table 6.6.2 Capacity of hoppers in tons.

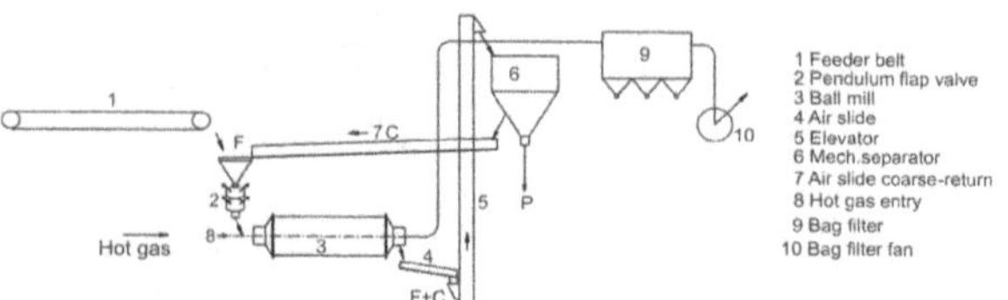

6.6.3 Shape of Hoppers

Construction wise, hoppers can be rectangular / square with pyramid shaped lower portions; or these could be circular in shape with cylindrical height and cone bottom.

See Figs. 6.6.11 to 6.6.13.

The tops of all hoppers are in one level to facilitate installing a belt to feed them.

Because of large differences in capacity, discharge points would be at different levels as shown.

See Fig. 6.6.11.

This arrangement is not so practical, from the point of installing belt feeders or table feeders as platforms needed to have access to them would have to be at different levels. Ideal situation is to have bottoms also at the same level.

See Figs. 6.6.12 and 6.6.13.

This is achieved by arranging hoppers in the fashion shown. Care needs to be taken to see that slopes of side and valley angles are steep enough for smooth flow of wet materials. Wetter the material, steeper the slope.

See Fig. 6.6.12.

For clay and iron ore, valley angles shall not be less than 60°-70°. A valley angle is the slope of the line where two sloping planes meet, it is shallower than the angles of surfaces.

A round cylindrical hopper with a cone is preferred because there are no valley angles or corners where material can accumulate and prevent smooth flow.

See Fig. 6.6.13.

6.6.4 Orientation of Hoppers

Orientation of hoppers with respect to feeding belt and mill is also important.

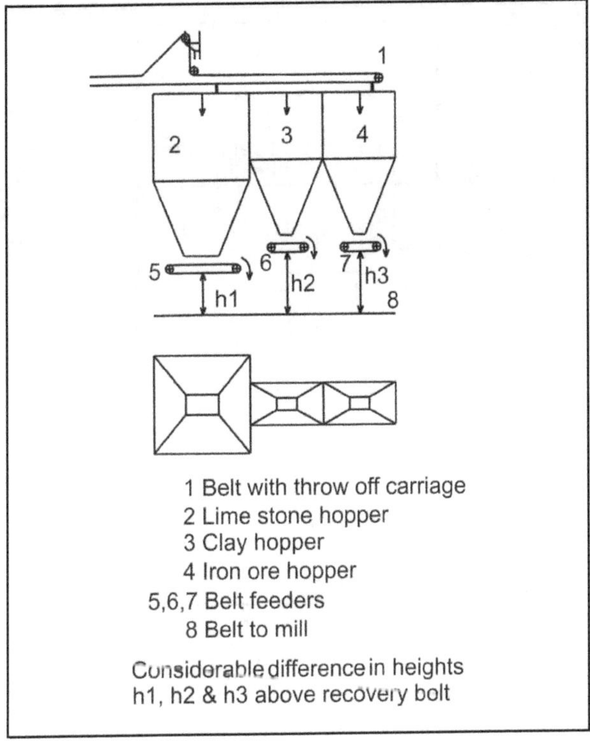

Fig. 6.6.11 Top at same level - bottoms at diferent

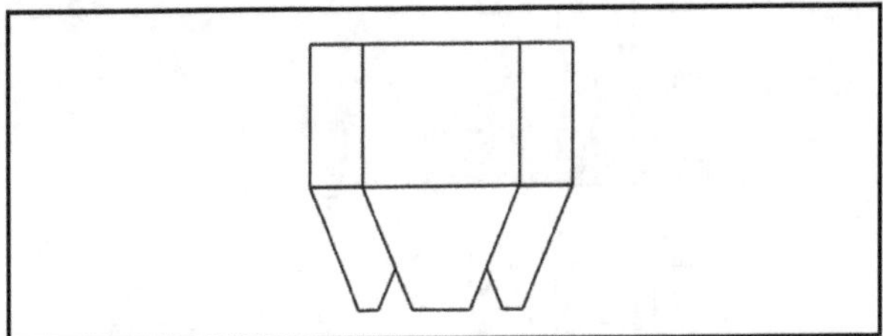

Fig. 6.6.12 Both top & bottom at same levels.

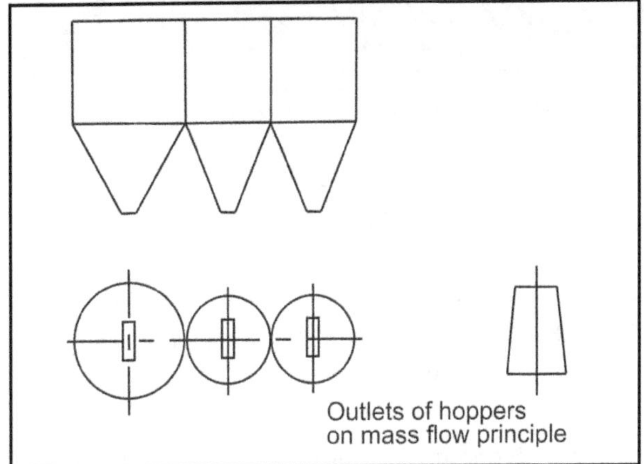

Fig. 6.6.13 Cylindrical hoppers with cone bottoms.

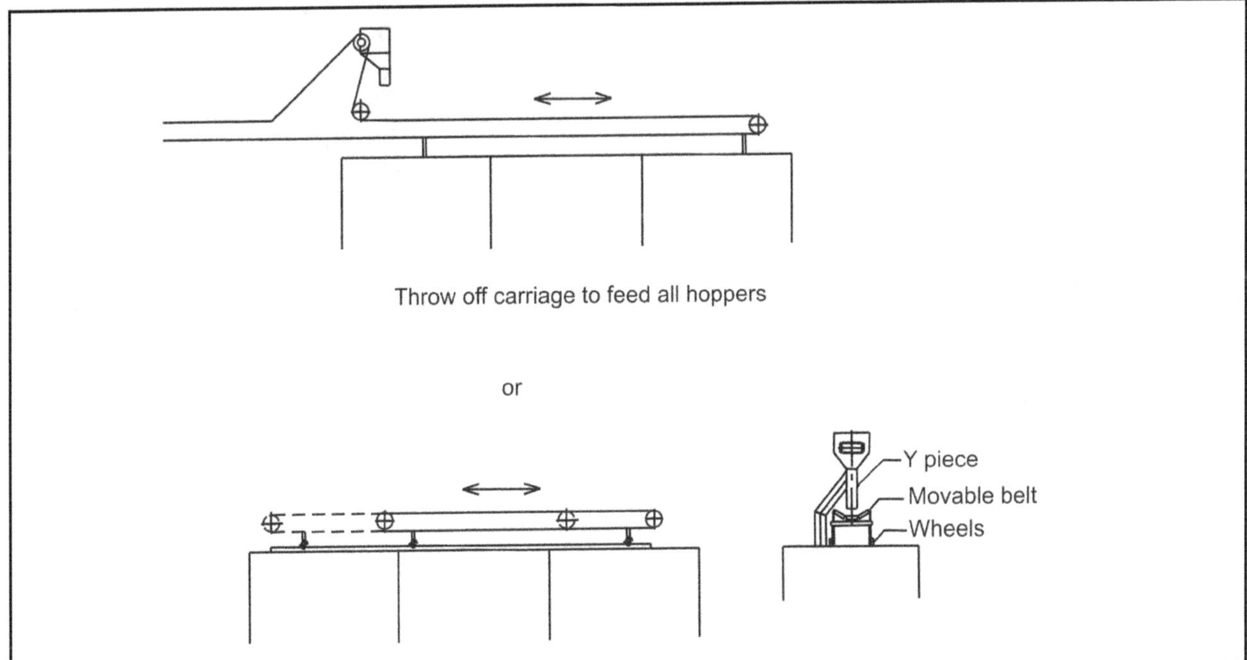

Fig. 6.6.14 Movable short belt at top to feed correcting materials.

If one and the same belt feeds limestone and correcting materials to the hoppers, it is necessary that these are in line.

See Figs. 6.6.11 and 6.6.14.

With this arrangement, weigh feeders at the bottom would be arranged as shown.

See Fig. 6.6.15.

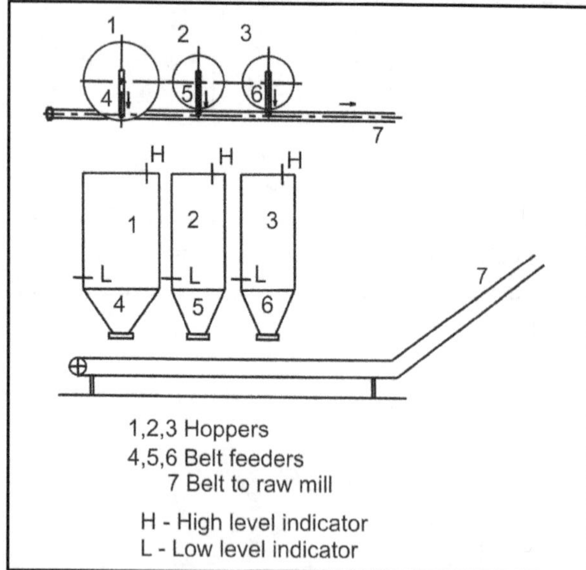

Fig. 6.6.15 Feeding arrangement of hoppers and raw mill.

When a second mill is added a second set of hoppers will be installed by the side.

See Fig. 6.6.16.

Second set of hoppers can be fed either from a common incoming belt or by an independent belt system for the second unit.

See Figs. 6.6.16 and 6.6.17.

6.6.5 Separate Building for Hoppers

Because of large capacity of mills 250 tph and above,, hoppers and proportioning stations are now separate from the mill building itself.

A belt of horizontal and inclined carries proportioned raw mix to the mills.

See Fig. 6.6.15.

If same belt brings in correcting materials also, then hoppers of correcting materials can be filled by a short movable conveyor **See Fig. 6.6.14**

6.6.6 Conveyors for Correcting Materials

If correcting materials are brought by a separate system of belt conveyors, because they are stored at a different place, then the layout must allow for one more

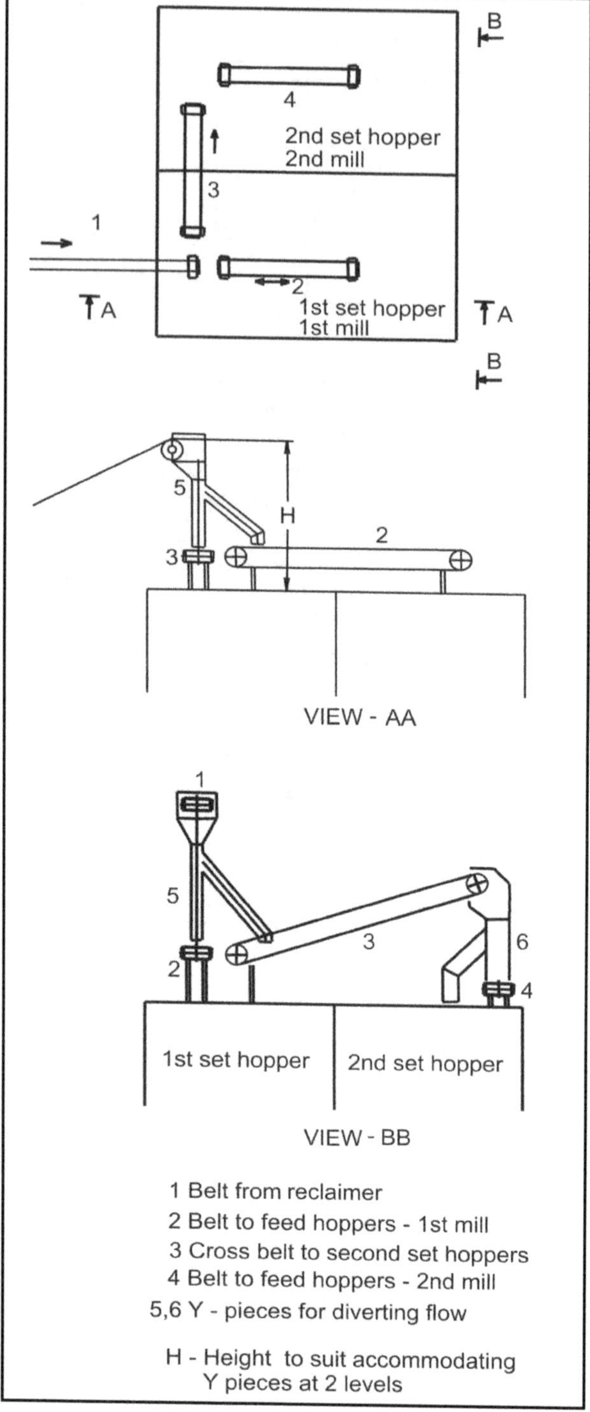

Fig. 6.6.16 Common feeder belt for two mills.

set of belt conveyors on the hoppers except that the movable / reversible belt for filling correcting materials could be eliminated.

See Figs. 6.6.18 to 6.6.20.

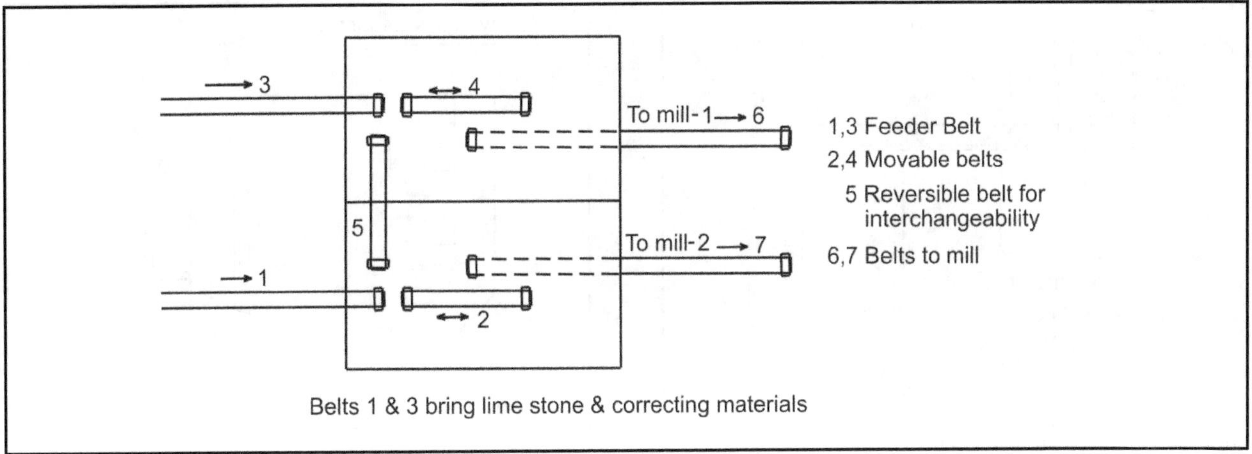

Fig. 6.6.17 Totally independent belt systems for two mills.

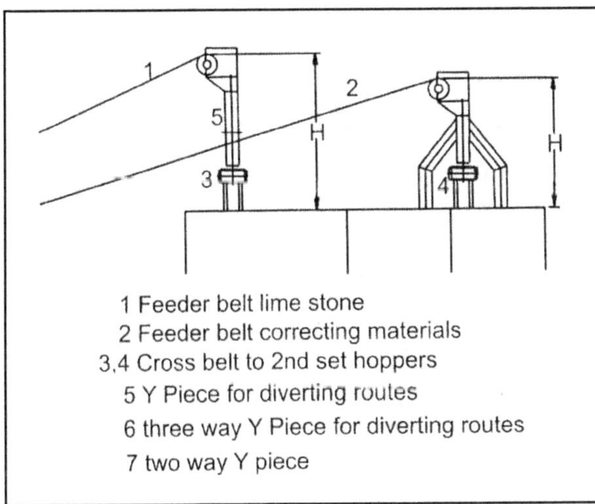

Fig. 6.6.18

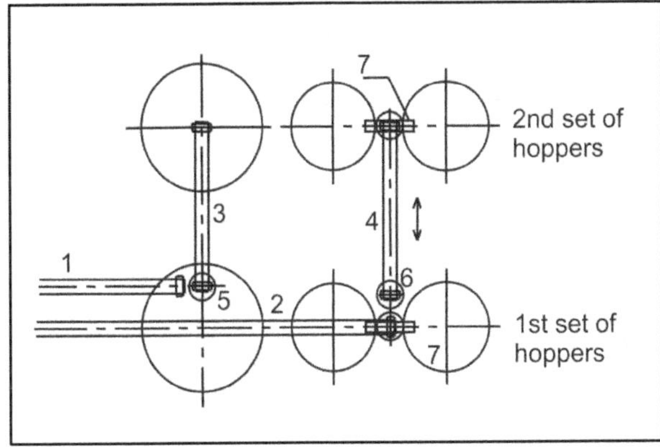

Fig. 6.6.19

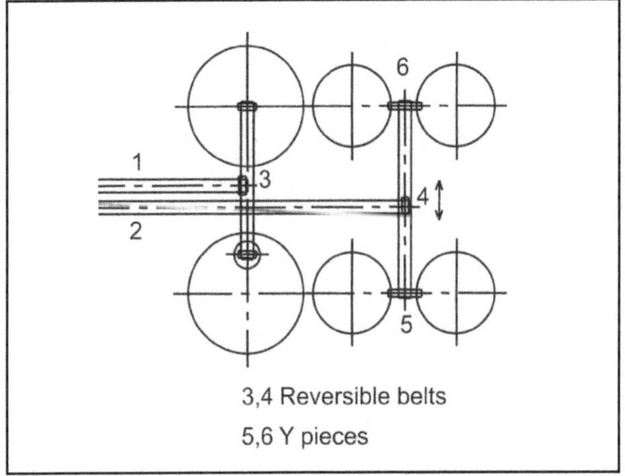

Fig. 6.6.20

6.6.7 Hopper – Construction

All hopper tops must have hand railings throughout. Hoppers should be covered either by a concrete slab or steel plates. There would be a common dust collector to vent the hoppers. It is done by interconnecting ducting between hoppers. The dust collector will also vent discharge hoods of belt conveyors.

See Figs. 6.6.21 and 6.6.22.

There should be provision on hoppers for level indicators, both high and low level.

See Figs. 6.6.15 and 6.6.22.

They should be fitted with rungs on inside walls for going in for cleaning etc.

See Fig. 6.6.23.

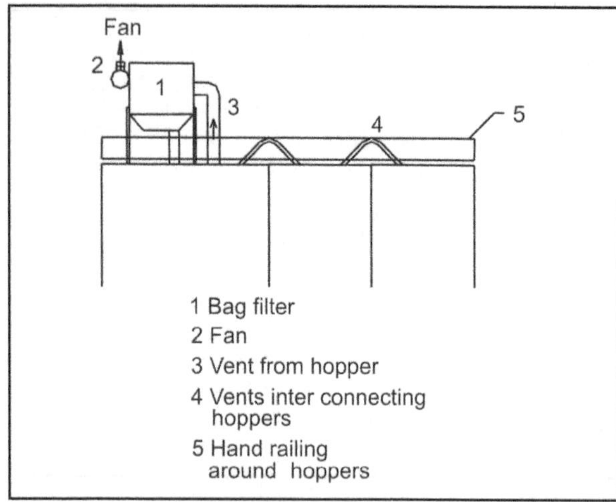

Fig. 6.6.21 Venting of hoppers.

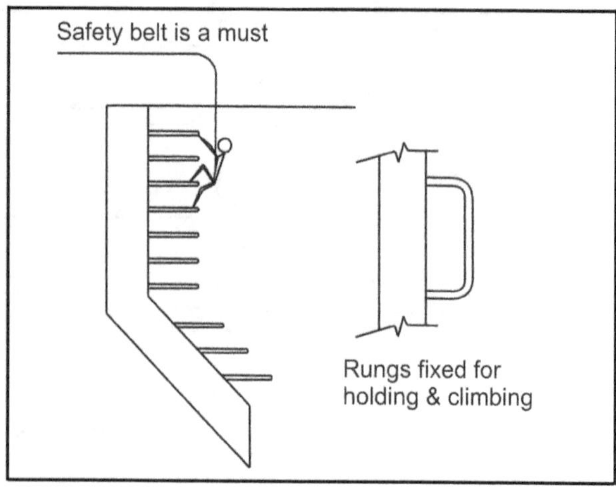

Fig. 6.6.23 Rungs in hoppers & silos for entering them.

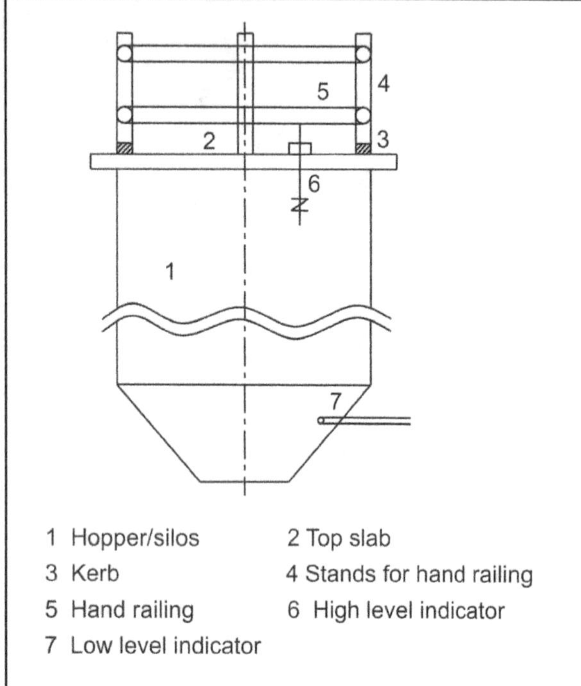

Fig. 6.6.22

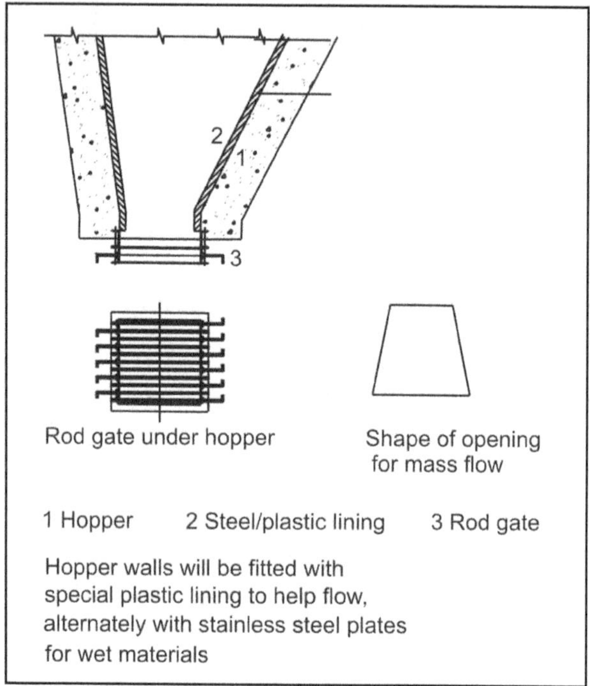

Fig. 6.6.24 Rod gate under hopper.

Sloping walls of hoppers for correcting materials could be lined with stainless steel plates or special plastic, which will permit smooth flow.

See Fig. 6.6.24.

6.7 Hopper Outlets

Openings at the bottom would be rectangular to facilitate discharge on belt feeders.(for large capacities, table feeders are unlikely to be considered). They also promote mass flow.

See Figs. 6.6.13 and 6.6.24.

For small plants, where table feeders are used, the openings could be circular with telescopic chutes.

6.7.1 Gates / Dampers

Dampers / gates are inserted for regulating and cutting off feed.

Sliding plate dampers are not suitable for granular materials. Mostly, therefore 'rod gates' are used to cut off the hopper or to regulate flow on to feeders. **See Fig. 6.6.24.**

The extent to which rods are pulled out, increases flow on to the belt.

The belt feeder should not be starved of materials due to rod gate.

6.7.2 Cleaning or Poking of Hoppers

Inspite of all the precautions taken, arching would sometimes take place and it becomes necessary to break it. Poking holes are therefore provided. However, care should be taken to see that workers engaged in this work do not stand under the hole / openings; a sudden release of arch would result in a rush of material that could injure workers.

See Fig. 6.6.25.

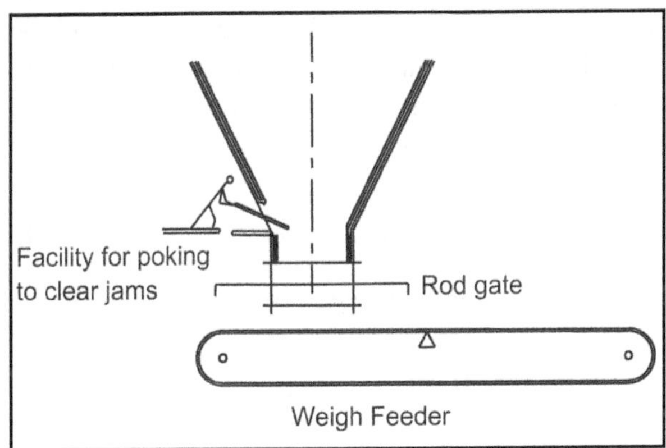

Fig. 6.6.25 Safe arrangement of facility for pocking in a hopper.

Dimensions of openings should match the width of the belt feeder fitted under the hopper.

6.7.3 Devices for Extraction of Wet Materials

For wet granulated materials which do not flow easily, hopper outlet is split so that arches formed are broken. At times a large diamond shaped pendulum is suspended which breaks arches as it moves.

See Fig. 6.6.26.

For wet pulverized coal and synthetic gypsum, a stirrer is installed at the outlet, keeping walls at outlet vertical. Stirrer prevents formation of arches by keeping material in motion.

See Fig. 6.6.27.

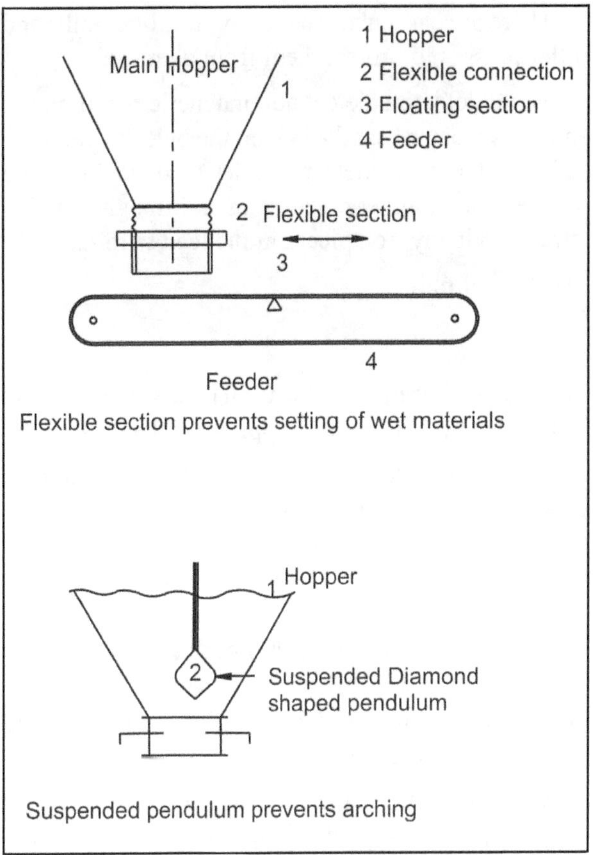

Fig. 6.6.26 Arrangements for wet materials.

6.8 Metal Detector and Magnetic Separator

An important aspect of layout is to install a metal detector and a magnetic separator on belts feeding the mill.

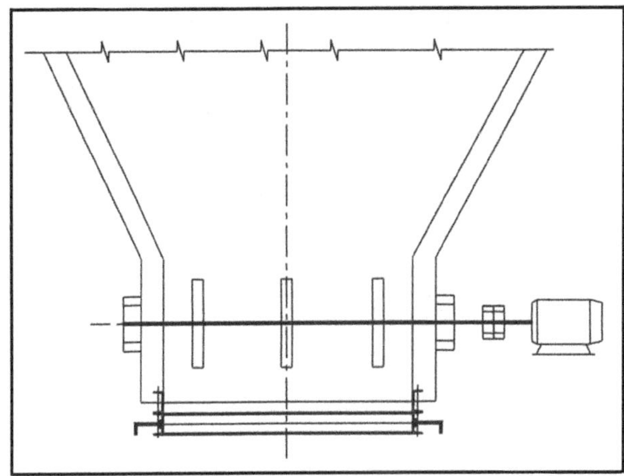

Fig. 6.6.27 Stirrer at bottom to keep pulverised material agitated.

These are particularly necessary for roller mills and roller presses to protect them from damage.

Metal detectors detect non magnetic metal pieces and drop a bag of sand and stop the belt. It should be so located that, the metal piece detected would still be on the belt. Magnetic separator is a magnet which attracts to it any iron pieces in the feed to mills.

See Fig. 6.6.28.

6.9

The layout of the mill itself with all its auxiliaries should be best obtained from the Suppliers of the mill.

It is possible that fans and dust collectors would be obtained from other suppliers. The Consultant preparing overall layouts should obtain drawings there of either through the Buyer or directly from Vendors.

6.10 Feeders and Weigh Feeders

Weigh Feeders and belt feeders are now commonly used to weigh and feed lime stones and correcting materials in desired proportions.

When 'continuous blending' system is used, the 'X-ray analyzer' installed, continuously monitors the proportions between the weigh feeders so that the raw meal ground in the mill always is close to the desired composition of raw mix.

Weigh feeders come with facility for 'calibrating' i.e., the weigh feeders are loaded with chains or similar 'standard' loads and rates indicated are compared and corrected.

However the layout should also provide facility for 'drop tests'. In a drop test, the discharge of feeder is collected in a receptacle for a given time and rate in tph is actually established and compared with that indicated on meters.

When only belt feeders are used, i.e., feeders with variable speed drives but without load cells, carrying out drop tests is essential.

See Fig. 6.6.29.

6.11 Drying

All raw mills systems (and also coal mills) carry out drying while grinding. The drying may be done either in 'mills' or in separators. Now it is more common to do drying in the mills. Mills therefore have a 'drying' chamber for this purpose.

Drying chamber is omitted if the requirements of drying are not severe i.e., the amount of moisture to

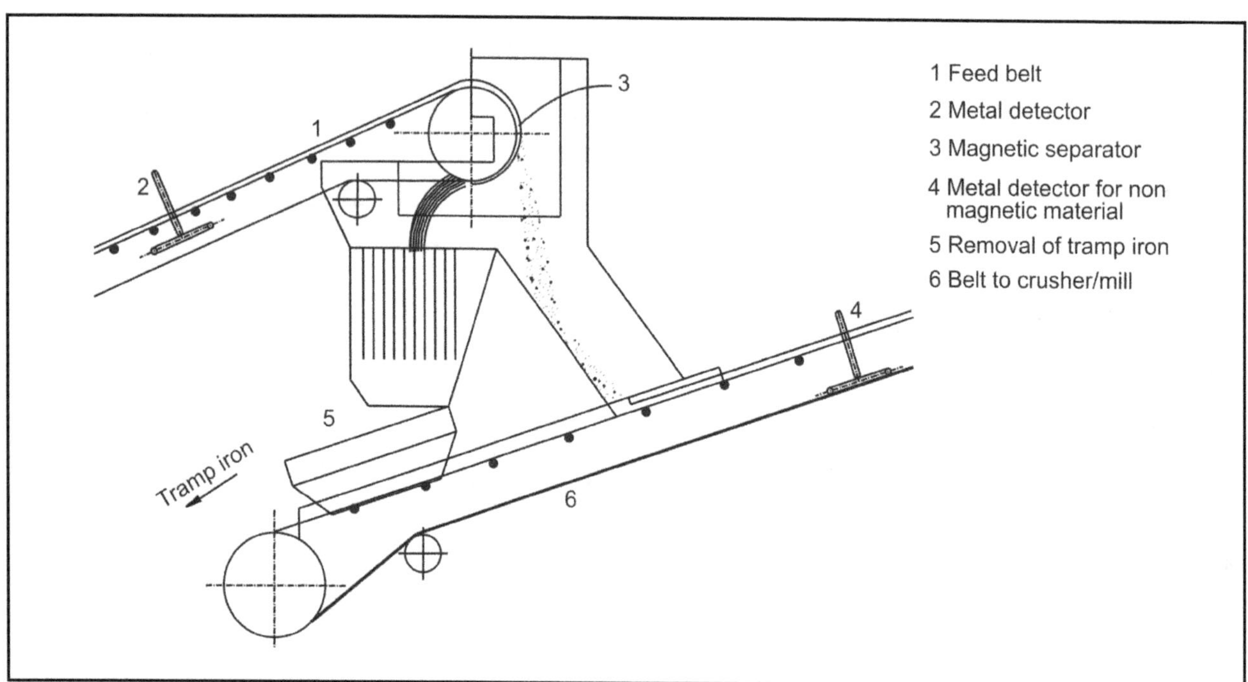

Fig. 6.6.28 Metal detector & magnetic separator to remove magnetic & non magnetic material from feed.

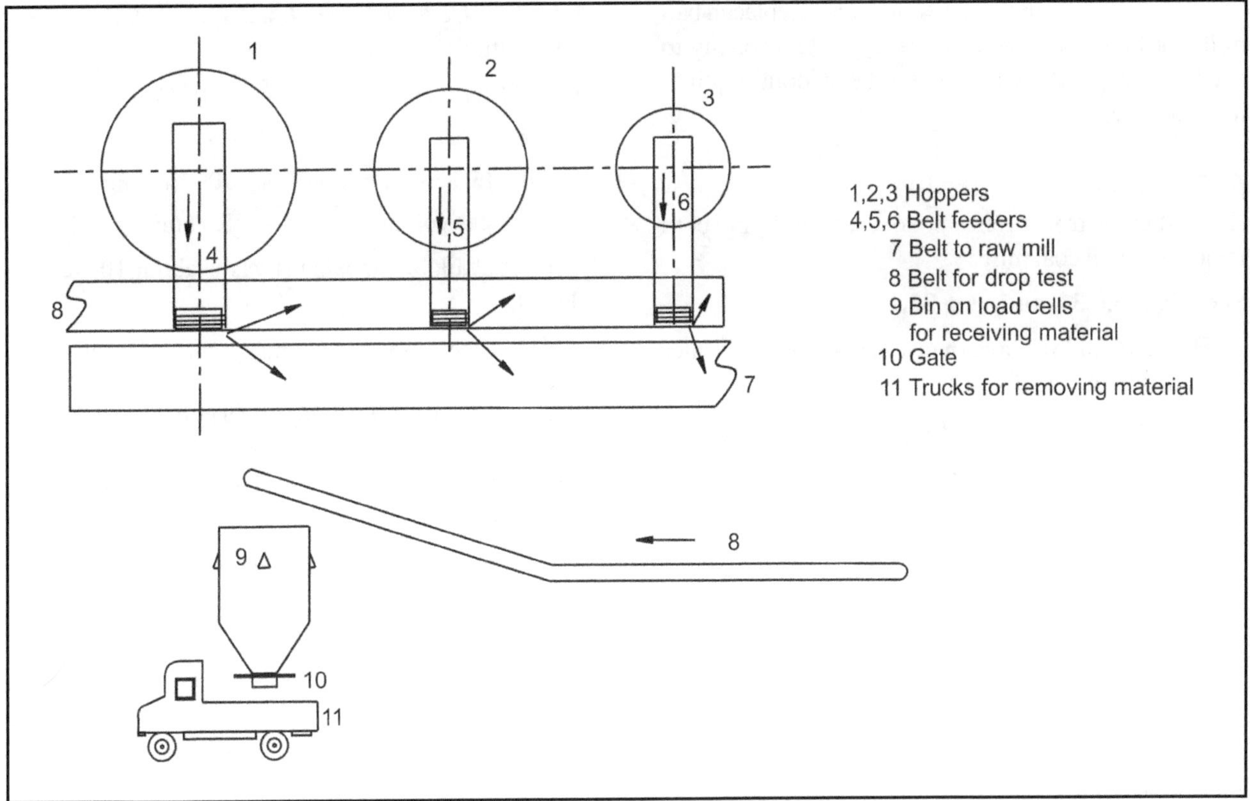

Fig. 6.6.29 Arrangement for drop tests.

be driven out does not exceed 5-6% even in wet seasons. The full length of the mill is then used for grinding.

In 'air swept' mills drying is done in mills only. This is true for both ball mills and vertical mills.

In bucket elevator mills with 'high efficiency' separators drying is done in mills as separator draws gases from mill and also from auxiliaries at low temperatures.

6.11.1 Requirement of Drying

In India, in most parts of the country, moisture in raw materials is high during 'wet' months or in rainy season only.

Exceptions are those parts of the country that get rains from both northeast and southwest monsoons.

In areas where lime stones are close to sea shores, moisture in them will exceed 2 – 3% even in dry season.

Clay and Iron ore contribute more to moisture to be evaporated than limestone in spite of the differences in their quantities in raw mix;

	Proportions		
	Limestone	Clay	Iron ore
	90	9	1
Moisture %	2	10	15

Total moisture to be dried

$$= (90 \times 2 + 9 \times 10 + 1 \times 15) / 100$$
$$= 2.85\%$$

In general, capacity to dry 5-6% moisture is adequate in raw material grinding systems in most parts of India and in most parts of the year. However the situation will vary from country to country and the requirements in specific cases can be worked out in the above manner.

ia and in most parts of the year.

6.11.2 Separate Dryers

Some plants that use sea shells / chalk or limestones close to sea shores install dryers for drying limestone before it is fed to the mill but such installations are exceptional.

Vertical mills which have more or less replaced ball mills for limestone and coal, have greater capacity to dry hence separate dryers would be seldom required in their case.

6.12 Hot Gases for Drying

Exhaust gases from preheater are used for drying both in raw mill and coal mill.

See Figs. 6.6.31 and 6.6.32.

Heat available from exhaust gases from preheater has reduced on account of:

1. Lower fuel consumption (smaller nm^3/Kg.).
2. Lower temperature (more number of stages 5 - 6).

6.12.1 Heat in Gases Used For Drying

Now with 6 stage preheater

Exhaust temperature	-	270 °C
Ambient		30 °C
Specific fuel consumption	-	700 Kcal/kg
Specific gas volume	-	1.4 nm^3/kg

Heat available in exhaust gas is about 105kcal/kg clinker.

This heat is used for drying raw materials and coal.

6.12.2 Heat Required for Drying

Heat required to dry raw materials with 5% moisture is ~ 1500 Kcal/kg and for coal with 10 % moisture it is ~ 1250 kcal/kg of water evaporated.

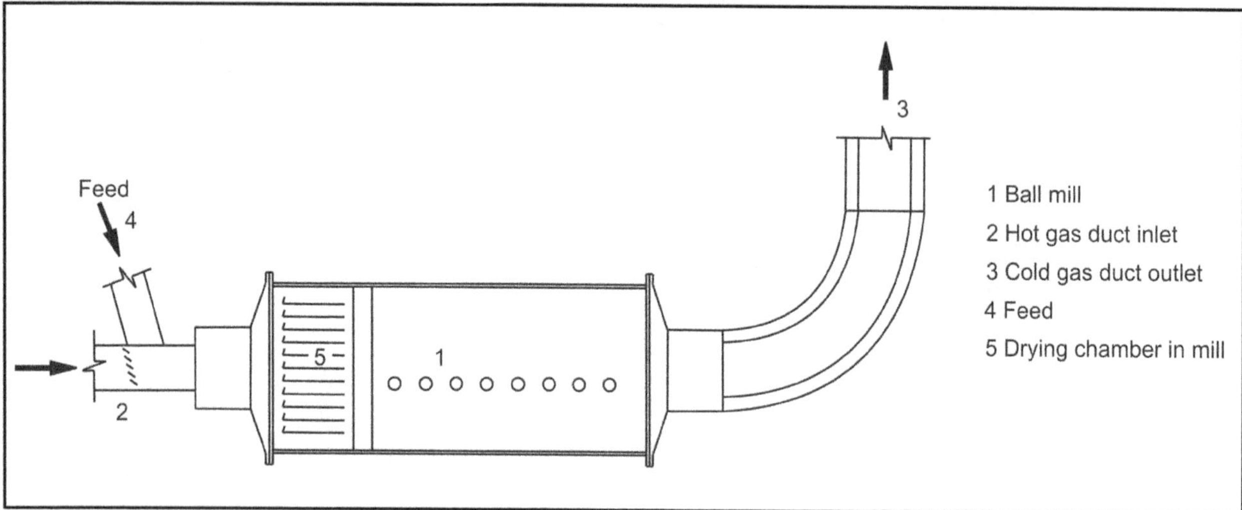

Fig. 6.6.30 Gas ducting at ball mill inlet and outlet.

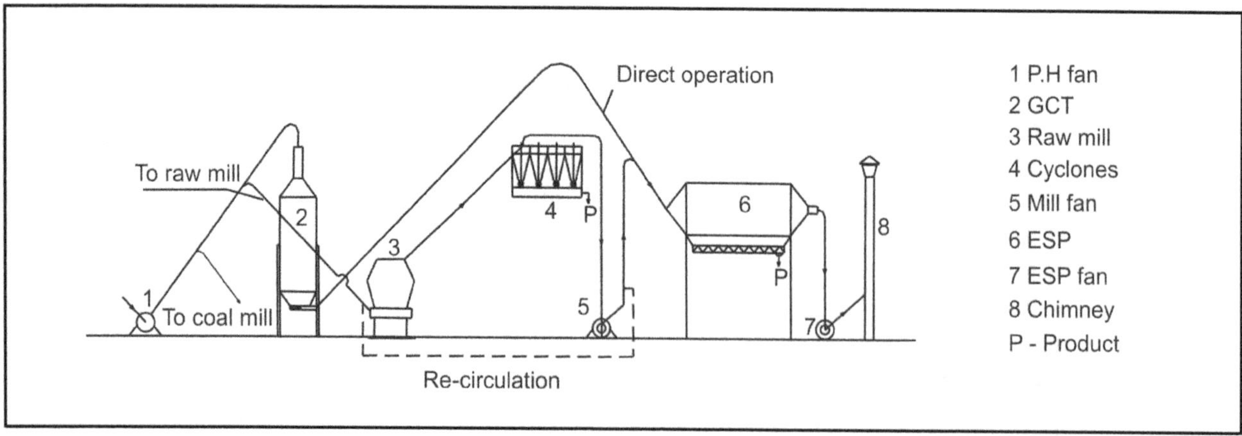

Fig. 6.6.31 Ducting in kiln & raw mill circuit GCT & ESP.

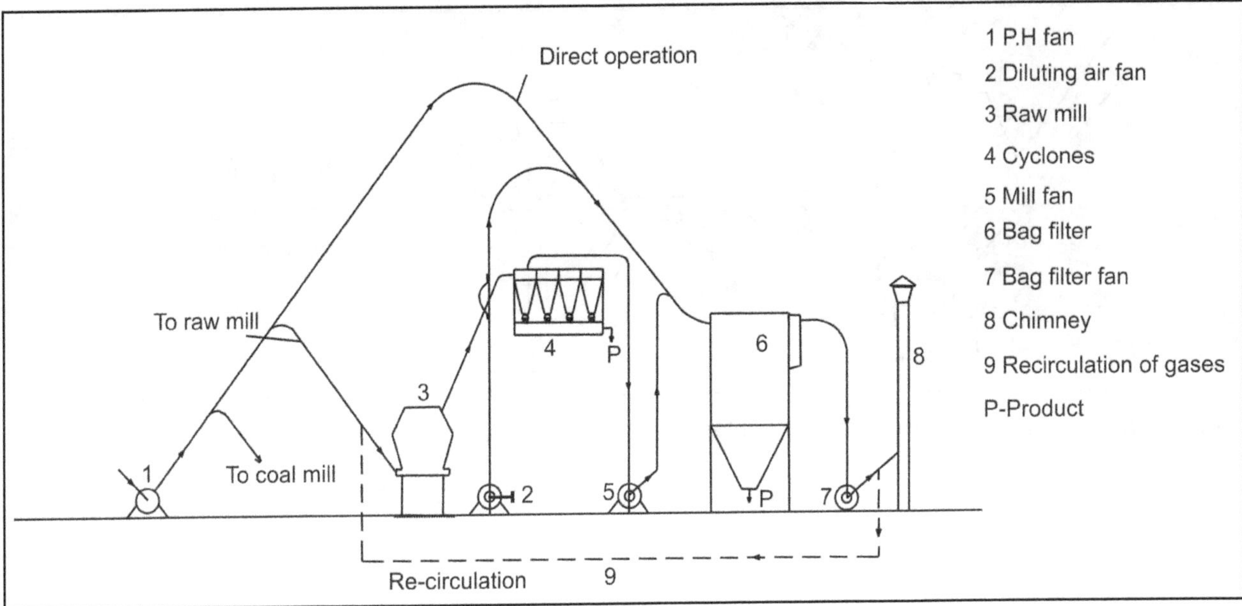

Fig. 6.6.32 Ducting in kiln and raw mill circuit with bag filter.

In terms of raw meal and coal ground it would be for raw meal -dried from 5 to 1% moisture =1.5*0.04*1500 = 90 kcal/kg clinker.

For coal dried from 10 to 2 % moisture and coal consumption of 700/4500 = 016kg coal.kg clinker, it would be 0.16*0.08* 1250 = 16.25 kcal/kg clinker.

Total heat for drying = 90 +16.25 = 106.25 say 107 kcal/kg clinker.

Heat content of preheater exhaust gases at 270 °C and 1.4 nm^3 /kg clinker = 1.4*(270-30)*0.32 = ~ 105 kcal/kg.

That is just sufficient. However when moisture in limestone would go up as it does in wet season, it would not be adequate.

6.13 Preheater and Raw Mill for a 10000 tpd kiln

10000 tpd kiln will have a two stream preheater. Raw material grinding capacity woud be 825 tph and hence raw mills could also be two though VRM of ~1000 tph with multidrive is available.

6.13.1 Layout of Ducting

It would thus be seen that ducting in kiln and raw mill depts. for prheater exhaust gases and for hot gases for drying in raw mills. Needs to be designed very carefully because they are large in size, carry hot gases and have slope and bends and branched circuits.

1. It Should be properly supported with fixed and sliding Supports designed to take the weight of large ducting with insulation cladding.
2. In hot ducting, expansion joints must be provided.
3. Since it contains dust, slopes of ducts should be adequate –minimum 50 ° to horizontal.
4. Horizontal lengths should be avoided.

See Figs. 6.6.31 and 6.6.32 See also Chapter 43 on Ductings.

In a ball mill, there would be one inlet. In a vertical roller mill, the ducting is normally split into two at entry of mill so that gases are evenly spread under the table.

See Fig. 6.6.30 and 6.6.33.

Where horizontal lengths cannot be avoided, arrangement should be made to collect settled dust in a screw conveyor with air locks under the horizontal length.

See Fig. 6.6.34.

Since preheater fans would be installed on ground floor, the ducting has to take a devious route to inlet of mill as shown.

See Figs. 6.6.31, 6.6.32 and 6.6.35

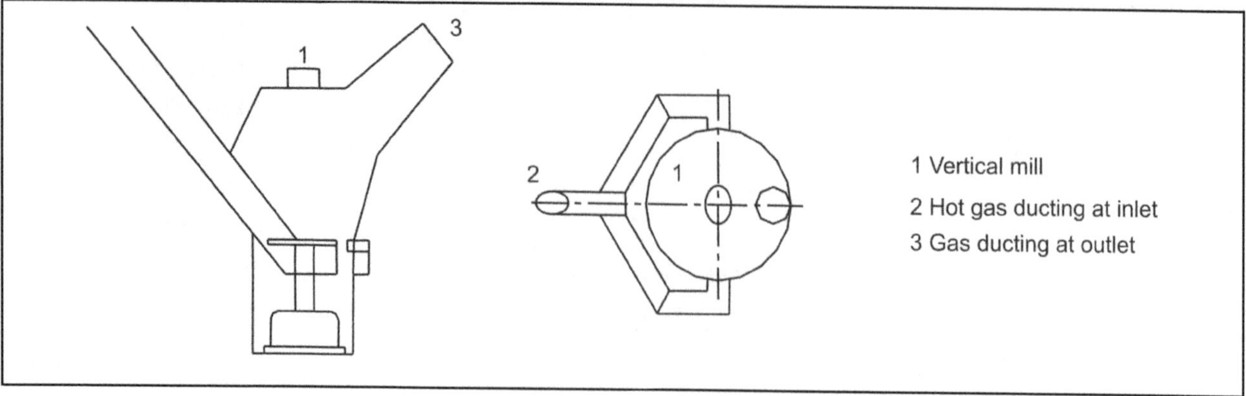

Fig. 6.6.33 Hot gas ducting at vertical mill inlet and outlet. (avoid horizontal length)

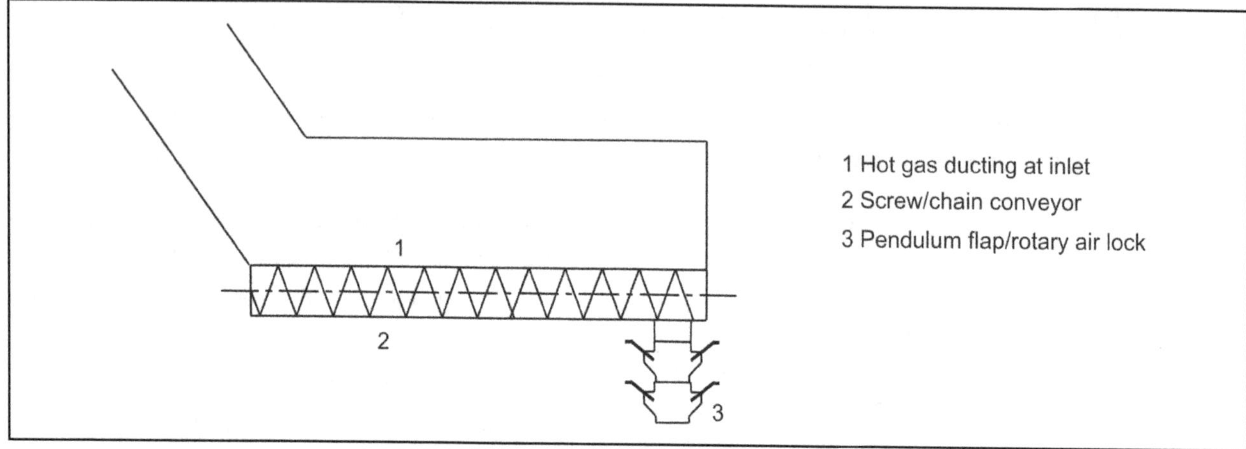

Fig. 6.6.34 Arrangement to collect dust setted in horizontal length of ducting.

6.14 Dust Collectors for Venting

When the mill is not running that is in direct operation, the gases would be taken directly to the dust collector either an ESP or a bag filter.

When ESP, a conditioning tower would be added in the circuit as shown. C.T. will cool gases from kiln at temperatures between 340 and 280 °C to 140 °C.

See Figs. 6.6.31 and 6.6.35.

In case of a bag filter temperature of gases is brought down by diluting Gases with ambient air. Glass bags can stand temperatures of 240 °C hence diluting air will be very small. But when bags are polyester, they withstand temperatures of 120 °C only and hence require large quantities of diluting air. Volume of gases to be handled by the bag filter increases substantially almost doubles.

See Fig. 6.6.32.

Even on 'cold' side ducting is very large. Further all this ducting needs to be insulated and lagged on hot side to preserve temperature.

6.15 Dust Collector in Circuit for Product

There are two schools in the design of the raw mill systems.

1. One system prefers to install a set of cyclones and collect 90% of mill product before passing gases into dust collectors.

   This introduces one more fan in the system and an additional pressure drop of 100-120 mm.
2. The other school, prefers to take entire volume of gases straight to the dust collector whether esp or bag filter. In this case a precollector is introduced before the dust collector. But the loss in precollector is much less than in cyclones. **See Figs. 6.6.31 and 6.35.**

6.16

The whole system of handling preheater gases to the point of cleaning them includes three departments.

1. Raw Mill.
2. Coal Mill.
3. Dust Collector.

There are three operating conditions

(i) Kiln running	-	Mills not running.
(ii) Kiln running	-	One Mill running
	-	Two Mills running
(iii) Kiln not running	-	Mills running.

The third condition arises during start up after long stoppages. To cope with this condition hot gas generators oil / coal fired are provided to supply hot gases to raw mill and coal mill till the kiln gets going and hot preheater gases are available.

The plant layout must therefore include hot gas generators also. The ducting must also be laid out to accommodate ducting from them.

6.17 Balancing of System Under Different Conditions

The system should be balanced in that oversizing or undersizing of units with respect to one another should be avoided.

Let plant capacity be		3000 tpd
With design margin of 10%	=	3300 tpd
Coal consumption	=	18%
Coal mill capacity	=	30 tph
Raw meal	=	1.5 Kg/kg clinker
Raw mill capacity	=	250 tph

Gas volume available from 6 stage Preheater (750 kcal/kg clinker and coal of 4500 kcal/kg)

= 1.6 nm^3/kg at 280°C

= 2,20,000 nm^3/hr.

To dry moisture in coal mill from 10% to 2% would require 34,66,700 Kcal/hr. And gas volume required would be 44,200 nm^3/hr.

To dry moisture in raw mill from 5% to 1 % would require 157,90,000 Kcal/hr, requiring a gas volume of 201,400 nm3/hr from preheater.

Total gas volume for drying = 245, 600 nm3/hr.

Available preheater gases thus are not adequate for drying in the two mill systems under the conditions stipulated; however in dry months when moisture in raw materials would be 3 %, they would be sufficient.

If the mills are oversized, then situation would be worse.

6.18 Gases Required to Lift Material

Another factor is volume of gases required to lift raw materials / coal out of the mill whether ball mill or vertical mill.

In air swept ball mills, volume of gases required for this purpose is 1.2-1.3 nm^3 / kg of material. In vertical roller mill it is 1.6 nm^3 / kg.

For only 'air lifting' coal mill will require

30,000 x 1.6 = 48,000 nm^3 / hr and raw mill would require 2,50,000 x 1.6 = 4,00,000 nm^3/ hr.

Thus to make up for this, a certain amount of recirculation is required in the circuit i.e., to bring back gases to the mill after dust collector. This further complicates ducting of the raw mill system.

See Figs. 6.6.31 and 6.6.32.

6.19 Developing the Layout

The layout therefore has to be developed taking all these factors into account with the help of the Suppliers of machinery and Consultants who have experience of designing the system. In a branched circuit, **see Fig. 6.6.36**, b and d are junction points. Therefore drop in pressure in two branches is the same. Ductings should be sized according to volumes of gases flowing through them and resistance so arranged (if need be artificially) to ensure above condition.

6.19.1 Parameters for Velocities in Ducts

In the layout and design of ducting the principles of sizing ductings, and balancing have to be kept in mind viz :

Velocity in ductings should be commensurate with the dust content.

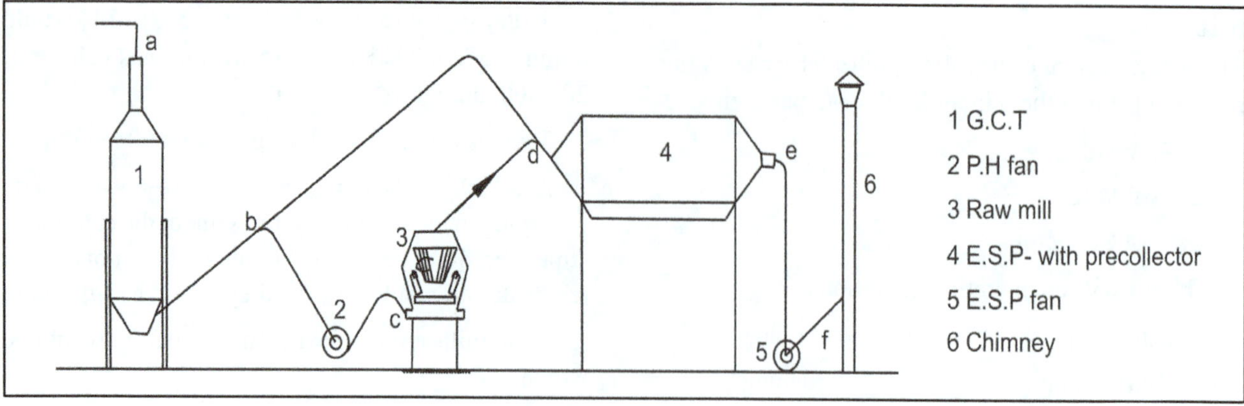

Fig. 6.6.35 Ducting in kiln and raw mill circuit (GCT before PH fan ; raw mill connected to ESP).

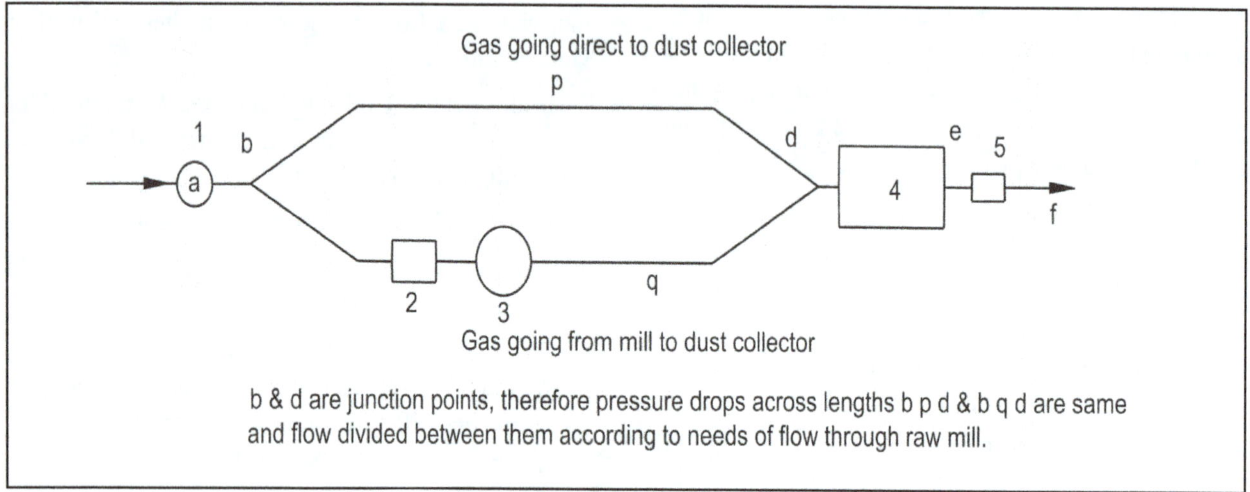

Fig. 6.6.36 Dividing gas flows in a branched circuit.

	designed velocity m/sec
Dust laden gases	20-22
Clean gases	16-18
Dampers	Pressure drop across them to be allowed for
Bends, converging and diverging ductings and different lengths of ductings	Pressure drop in them should also be worked out

Balancing can be achieved, particularly for venting systems, by introducing dampers or orifices. A fan can also be introduced for this purpose.

See Fig. 6.6.36.

6.20 Venting of Auxiliaries

Various auxiliaries in the system also need to be vented. Among them would be :

Belt conveyors, elevators, feeders, pneumatic conveyors, aeration systems, bins, separators and similar others.

The principle is that volume of air corresponding to material being handled gets displaced.

Norms for venting have been established for different types of machines and auxiliaries and should be used for sizing of venting ducting. For details **See Chapter 41** on venting in **this section**.

It would be best to install a separate dust collector for venting of auxiliaries rather than take them to the main dust collector.

The dust so collected is taken to the final product conveyor.

See Fig. 6.41.2 in Chapter 41 in this Section

6.21 Closed Circuit Grinding – Recirculation of Coarse Fraction

In closed circuit grinding systems, material is not ground to final fineness in one pass. The material coming out of the mill is half finished. It is then passed through a separator where coarse fraction is separated and returned to the mill for regrinding. Air Separators or Separators thus help to save energy in grinding operation.

6.21.1 Circulating Load

The return material is called 'circulating load'. In a ball mill, the amount of circulating load is related to the length / diameter ratio of the mill. Longer the mill, smaller the circulating load.

Commonly used l / d ratios for ball mills are:

Coal mill	-	1.25 – 1.5	
Raw mill	-	1.5 – 1.75	Maximum 2
Cement Mill	-	2.5 – 3.5	

6.22 Separators

Separators being the vital part of a mill system, have to be integrated in the system according their own design features and also that of the mill they have to work with. There are three main designs.

6.22.1 Grit Separators

They are static. Dust laden air/gas enters from the bottom and exits from the top after passing through adjustable vanes. These separators can therefore be located in only one position with respect to the mill which is air swept that is in the outlet duct from the mill.

See Fig. 6.6.37.

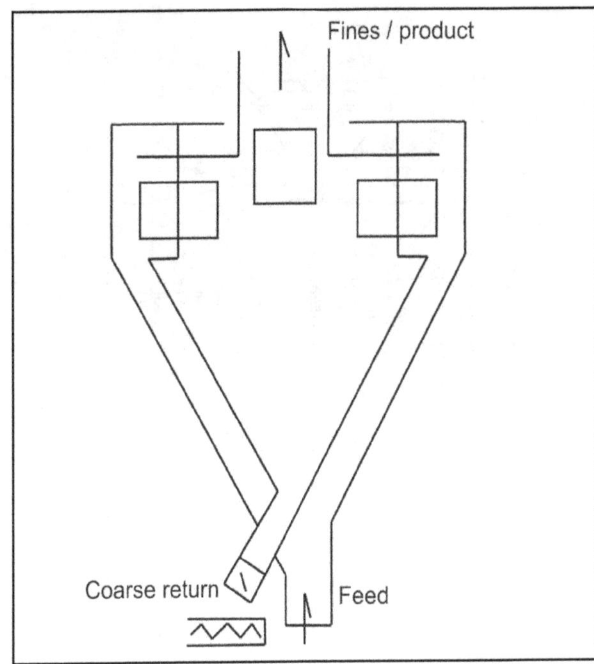

Fig. 6.6.37 Grit separator.

Static separators were part of vertical roller mills like 'E' mills and Polysius Vertical Roller mills also.

6.22.2 Mechanical Separators

In them separation of fines from feed was done by action of centrifugal force on particles. Fines were recovered by cyclonic action within the separator.

Some times hot gas was passed through the separator to dry material. Otherwise these separators needed only a small amount of air /gas flow through them.

Material discharged by the mill was conveyed to an elevator which then lifted it to feed the separator at the top or from the side.

See Fig. 6.6.38.

Thus air slide, elevator and separator formed on integrated unit.

Separator could be located at the feed end or the discharge end of the mill according to the percentage of fines in feed.

When feed contains substantial amount of fines, fresh feed would be taken to separator first; coarse will go to mill.

See Fig. 6.6.39.

1 Feed spout
2 Main fan
3 Control valves
6 Fines cone
5 Distribution plate
4 Auxiliary fan
7 Return air vanes
8 Tailings cone
9 Fines discharge
10 Tailings discharge
A Fines lifting zone
B Separating zone

Fig. 6.6.38 Schematic of a mechanical air separator.

When it does not contain much fines, fresh feed will go to mill and separator will receive mill discharge.

See Fig. 6.6.40.

The two locations decide lengths of airslides for mill discharge and coarse return and also depth of pit for elevator.

See Fig. 6.6.41.

6.23 High Efficiency Separators

They have replaced the two designs mentioned above.

In them air or gas is used as the classifying agent to separate fines from coarse. Fines so recovered were separated from the air/gas stream by passing them through dust collectors like bag filters or esps. Therefore efficiency is very high. Speed of rotor can also be varied to vary fineness of product.

6.23.1 Designs of High Efficiency Separators

There are many designs of high efficiency separators as shown in **Fig. 2.5.5 and 2.5.6 in Chapter 5 of Section 2.**

In most designs feed is from the top. Outlet can be from the top or from the bottom. Mill vent gases are used for classification in the separators. If they are not adequate, part of the gases is recirculated. Drying is almost always done in mills.

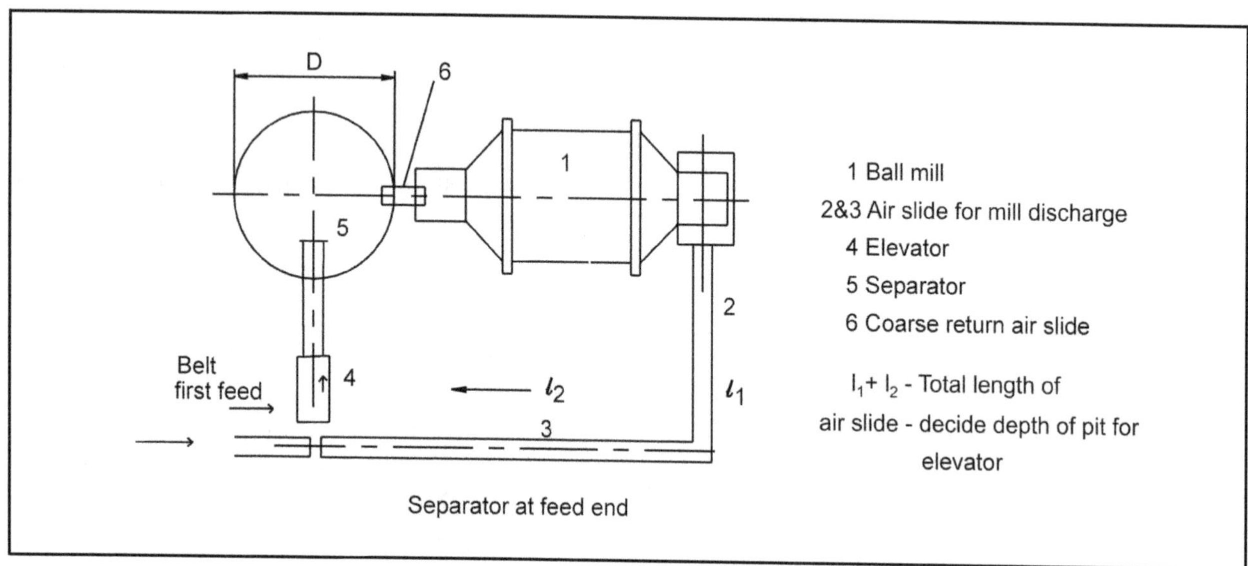

Fig. 6.6.39 Separator at feed end.

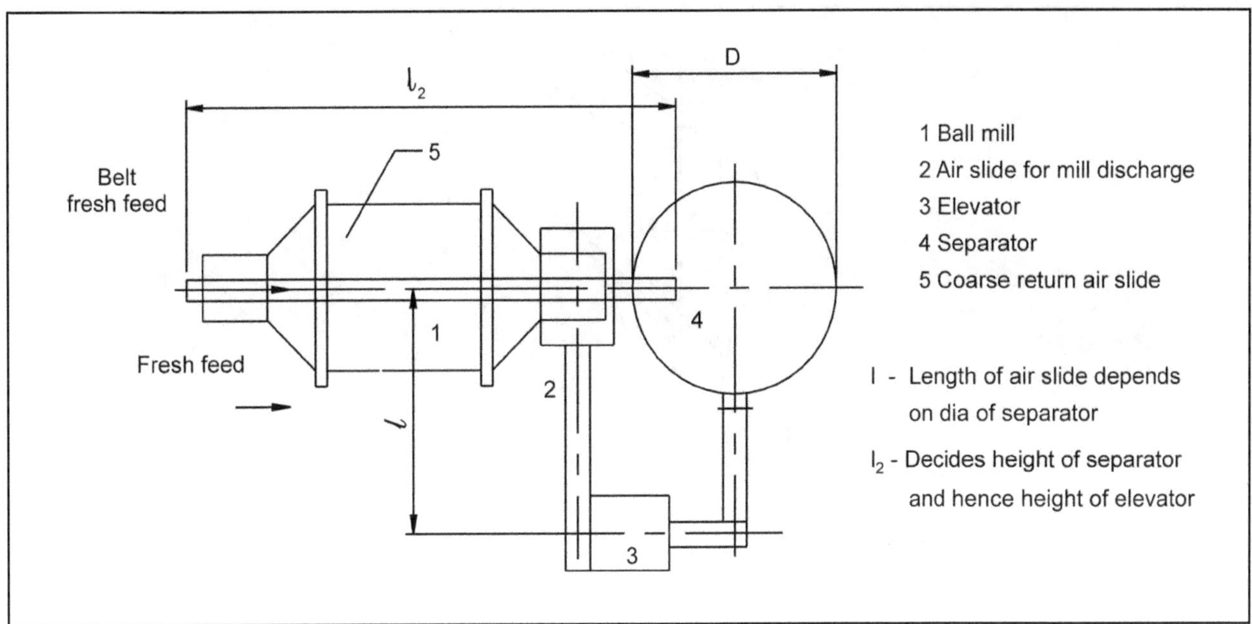

Fig. 6.6.40 Separator at discharge end.

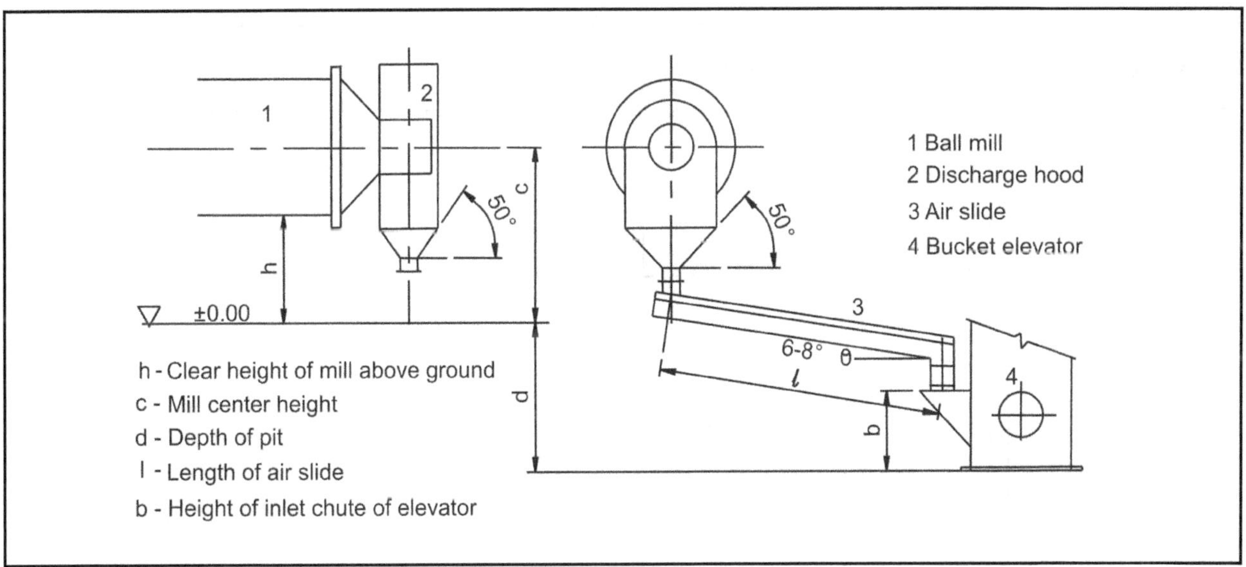

Fig. 6.6.41 Layout of ball mill in closed circuit depth of pit – determined by – clear height h, dia of mill and length of air slide.

These separators are therefore so located at the discharge end of the mill so as to suit its entry into separator as per its design and exit of product to dust collector.

See Fig. 6.6.3 also Fig. 6.6.46 and 6.6.47.

In Sepax separator, which combines grit separator in the lower part, entry of gases is from the bottom. Mill discharge is introduced in the separator at a higher point.

See Fig. 6.6.42.

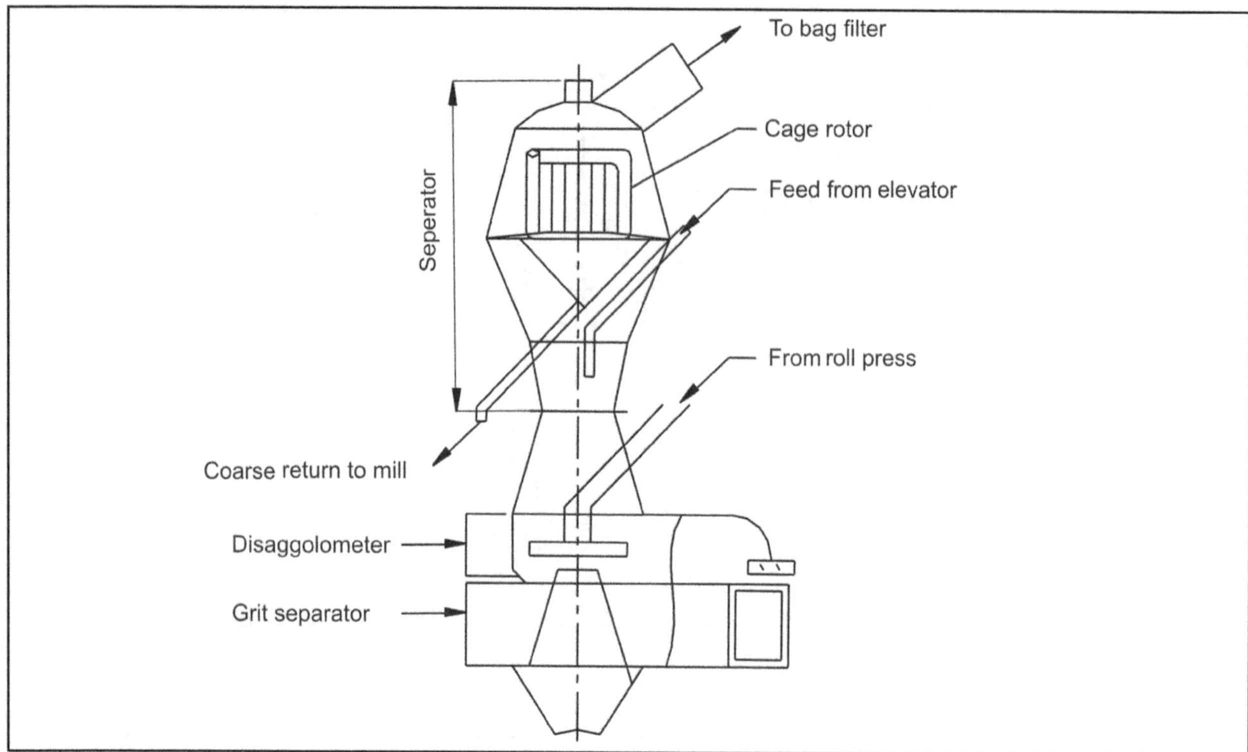

Fig. 6.6.42 Sepax seperator cum disagglomerator.

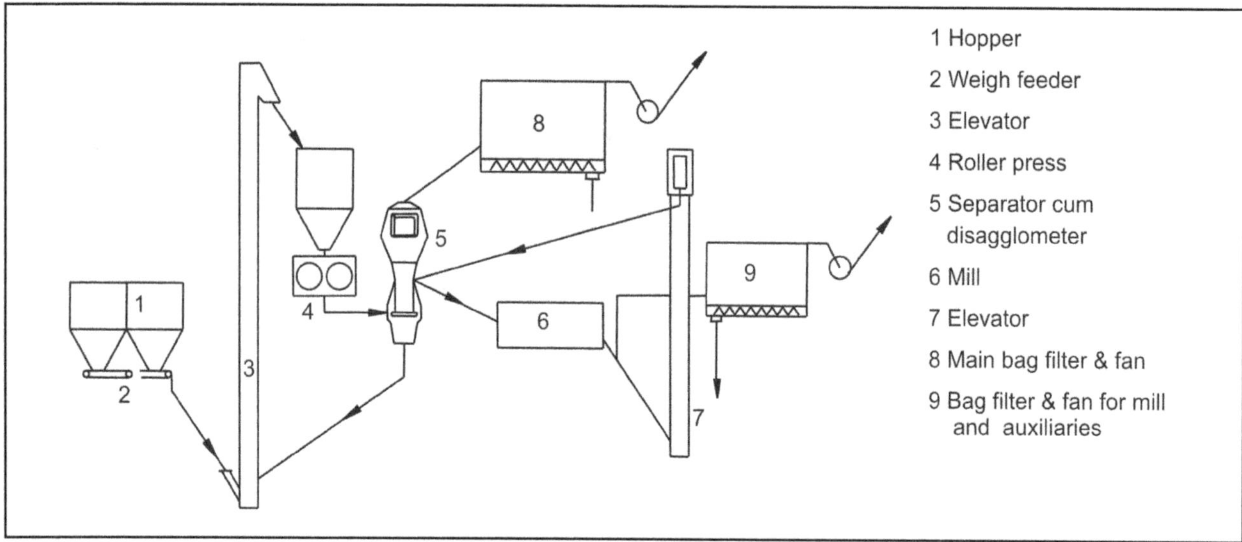

Fig. 6.6.43 Roller press, tube mill, separator and disagglomerator circuit.

When Sepax is used with roller press it also incorporates a disagglomerator to break cake from press and to disperse material in air stream. This locates the separator directly above mill outlet. Three components i.e., grit separator, disagglomerator and separator rotor being in vertical line one above the other requires corresponding head room in mill department and also needs tall elevator. Product is taken out from the top – requiring further head room.

See Fig. 6.6.43

Sepax has been modified to work with air swept ball mlls grinding coal.

6.23.2 High Efficiency Separators in V. R. Mills

High efficiency separators have now been incorporated in V.R.M.s of all designs by modifying the casing. However since they are installed in the casing of the mill itself, no special provision for them in lay out have to be made.

6.23.3 Retrofitting of Separators

Often in old installations of ball mills, performance could be improved by replacing mechanical separators by high efficiency separators.

A high efficiency separator is smaller in size compared to the conventional separator. Therefore it is not difficult to accommodate in the layout. But this change requires a much larger dust collector to be installed. The layout shouldbe examined from this angle.

This is also true of v. r. mills with separators of old design built in. Casing is modified to accommodate new high efficiency separator and may require some minor changes in the layout.

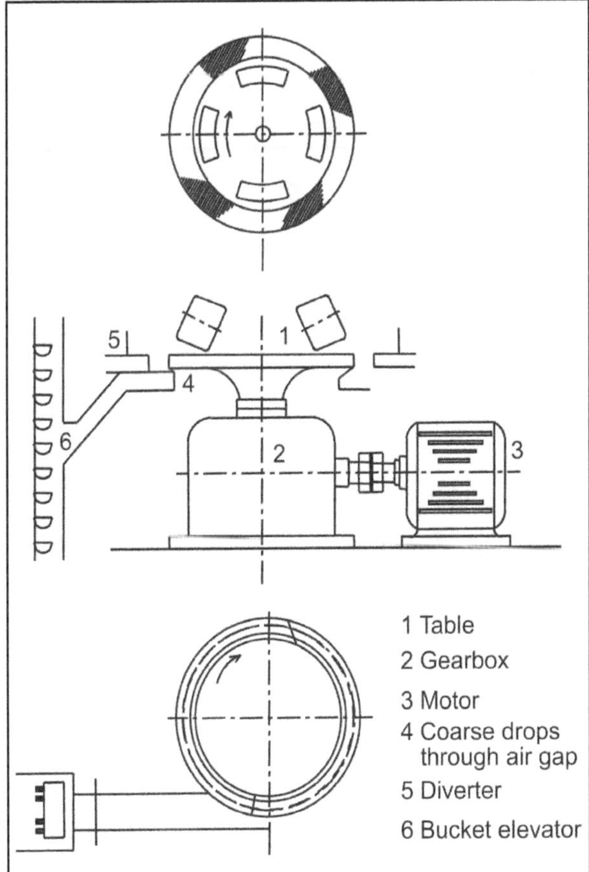

Fig. 6.6.44 External circuit V.R.M. collecting oversize to convey it to mill by an elevator.

6.24 External Circuit of V. R. Mills.

It would be slightly difficult to convert an existing v.r.m. into external circuit mill because there would not be enough head room under the table to install conveyors for collecting coarse dropped from the air ring. An elevator is added to lift coarse so collected to feed it them back into mill.

See Fig. 6.6.44 and 6.6.48.

There would not be any difficulty in case of new installations for adopting external circuit.

6.25 Elevators in Mill System for Closed Circuiting

Bucket elevators in ball mill systems lifting circulating should be generously sized so that they do not become a bottleneck. They carry fresh feed and circulating load. Therefore for a mill with a capacity of 100 tph and a circulating load of 200 %, airslide and elevator would have to carry 300 tph. They should be sized accordingly.

See Fig. 6.6.3.

6.26 Equipment to Convey Finished Product

Layout has also to take into account the equipment selected for conveying finished product that is raw meal to the next section, that is blending silos in case of raw mills.

The feed points on air lift and elevator for example would be at different heights. Air slides from bag filters/ esps collecting the product of the high efficiency separator should be at a suitable height to do this since they are installed only at a downward slope.

This in turn means that dust collectors and separators are in turn located at appropriate height.

Thus layout gets built up backwards from the delivery point of the product.

See Fig. 6.6.45.

6.27

Typical layout drawings of ball and v.r. mills of actual installations have been shown in **Figs. 6.6.46 and 6.6.47.**

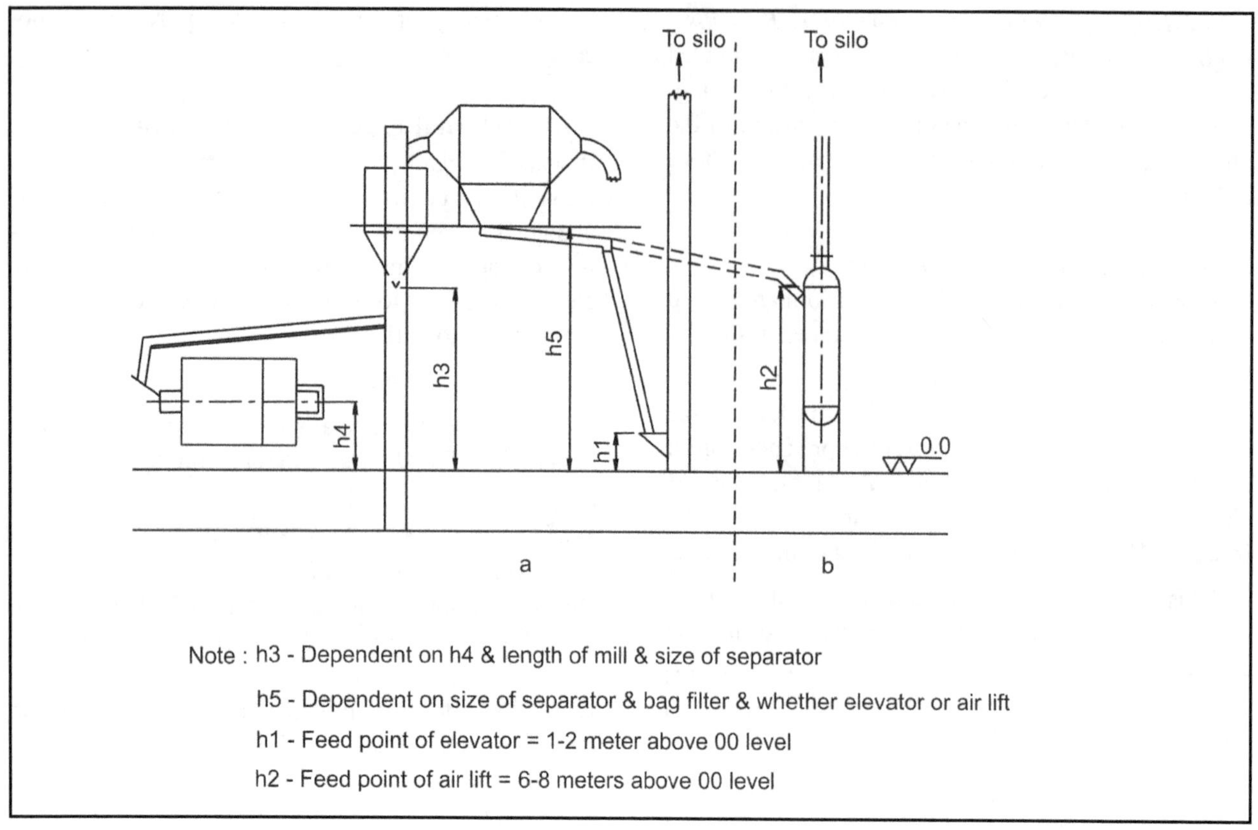

Fig. 6.6.45 Location of dust collector and discharge points.

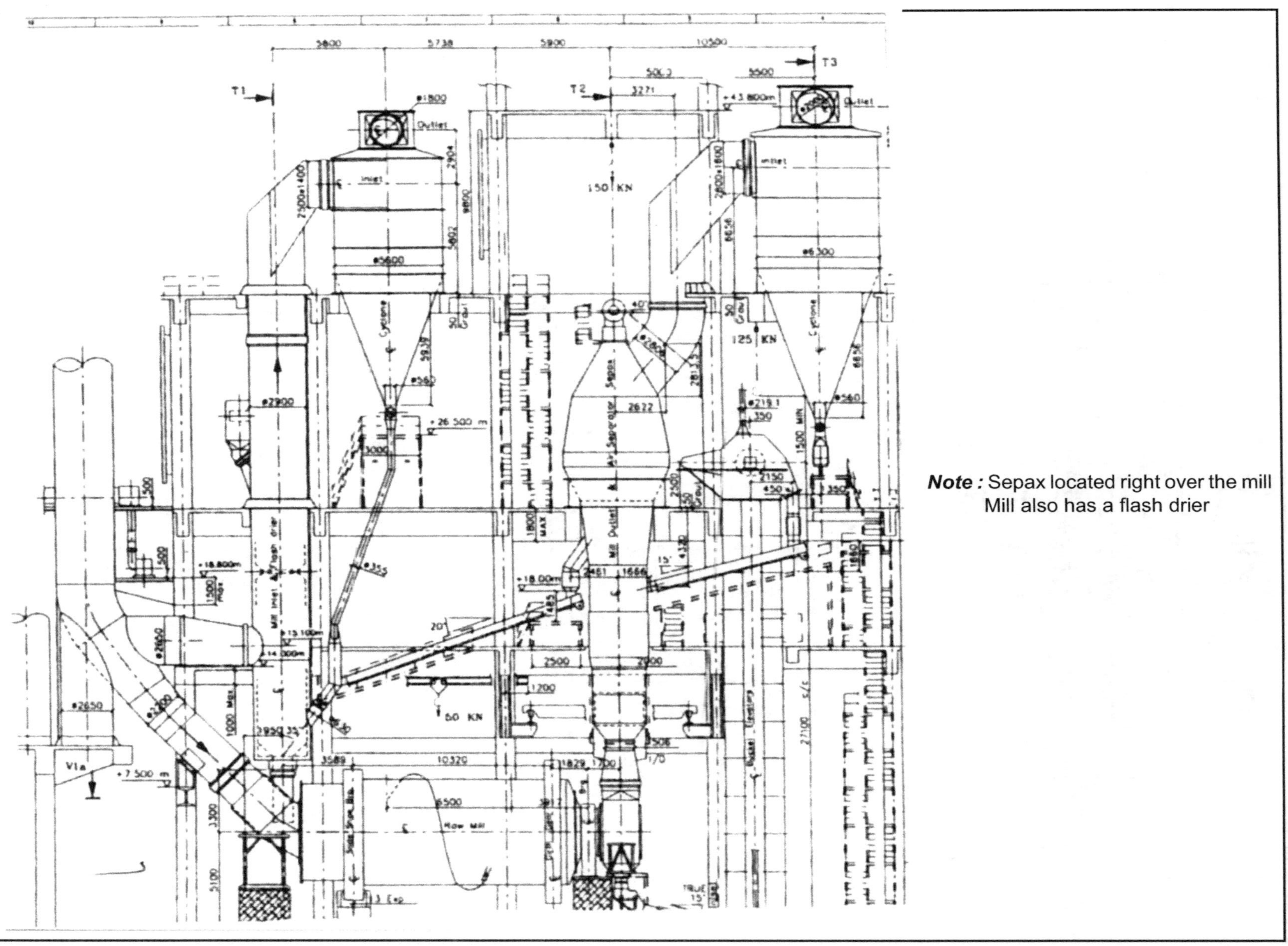

Fig. 6.6.46 Ball mill and sepax separator in raw material grinding system.

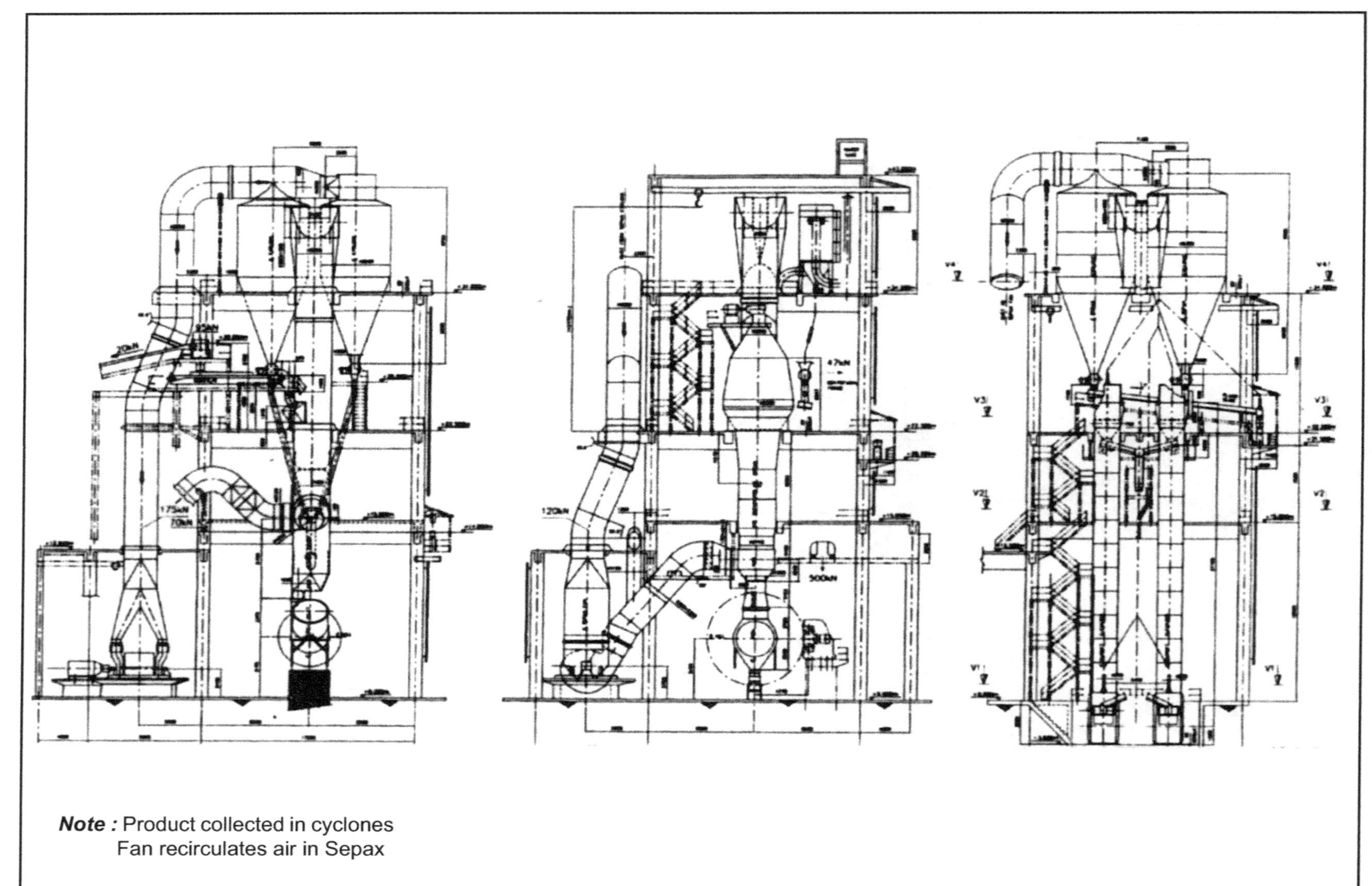

Fig. 6.6.47 Sepax in ball mill system, another view.

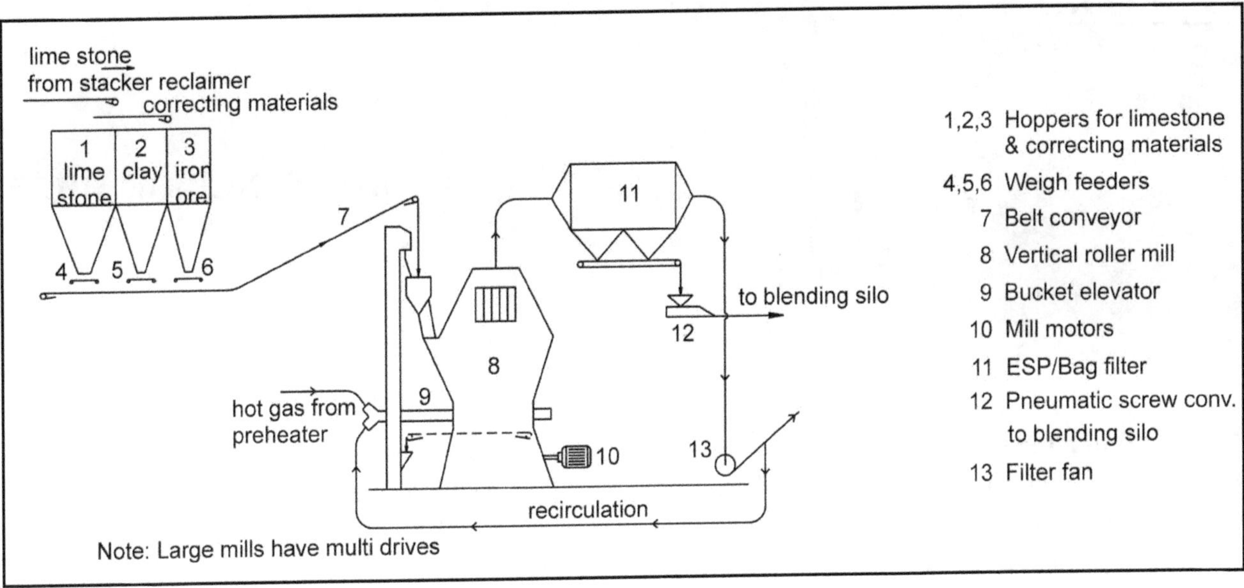
lime stone
from stacker reclaimer
correcting materials
1
lime
stone
2
clay
3
iron
ore
4
5
6
7
8
9
10
11
12
13
to blending silo
hot gas from
preheater
recirculation
Note: Large mills have multi drives
1,2,3 Hoppers for limestone & correcting materials
4,5,6 Weigh feeders
7 Belt conveyor
8 Vertical roller mill
9 Bucket elevator
10 Mill motors
11 ESP/Bag filter
12 Pneumatic screw conv. to blending silo
13 Filter fan

Fig. 6.6.48

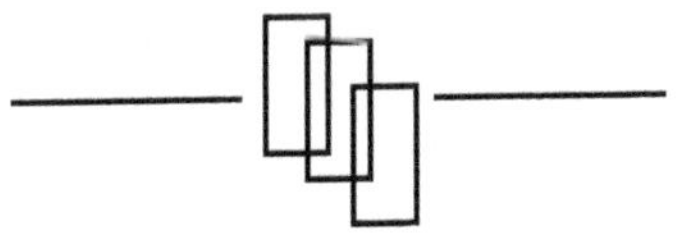

CHAPTER 7

MAINTENANCE OF MILLS

7.1 Maintenance of Ball Mills

In case of ball mills, lining plates are required to be changed now and then so also diaphragm plates.

Grinding media need to be unloaded and regraded as they wear out. If the grading composition is corresponding to 'equilibrium changes', the makeup charge consists of one sized grinding ball usually the largest in the compartment.

Wear is judged by the drop in power drawn (amps) and make up quantity added through chute at inlet or by opening the manhole in 2nd compartment.

7.2 Handling Grinding Media

When there is a sharp drop in the performance of the mill, one of the reasons could be that grinding media have shifted inside the mill and hence are ineffective. Charge requires to be sorted and put back in right place in the compartment.

7.2.1 Regrading

In 'regrading' entire ball charge is dropped out by rotating the mill with man holes open and sorted out size wise, quantity wise and mill is reloaded according to the charge that has given best results.

For large mills, tonnage to be handled could be as large as - 200 tons or more.

7.2.2 Facilities for Handling Grinding Media

The layout of the ball mill should therefore provide facilities like :

(i) Portable belt conveyor to load the mill.

(ii) Screen to sort out media size wise.

(iii) Access for trucks to bring and take away grinding media.

Therefore one side of the mill should be totally free for this purpose.

See Fig. 6.7.1.

7.3 Mill House Constructuion

The column spacing of the mill building at ground floor and also height of first floor should be so selected that mill shell can be easily taken into building between them during erection stages.

See Figs. 6.7.2 and 6.7.3.

If the mill heads are assembled outside, then the mill would have to be raised high enough to clear the bearings and foundations.

Diaphragms are generally inserted from ends. Hence mill heads would generally be fitted after mill has been taken inside and kept in position.

See Fig. 6.7.2.

Unlike kilns which are in the open and hence cranes can be used for lifting, mills are inside a building and hence it is more difficult to use a crane for this purpose.

See Fig. 6.7.4.

7.4 Handling of Lining Plates, Diapharagm Plates

For fitting and removing internals like lining plates / diaphragm plates etc., an overhead monorail with a motorized and remote operated pulley block is used.

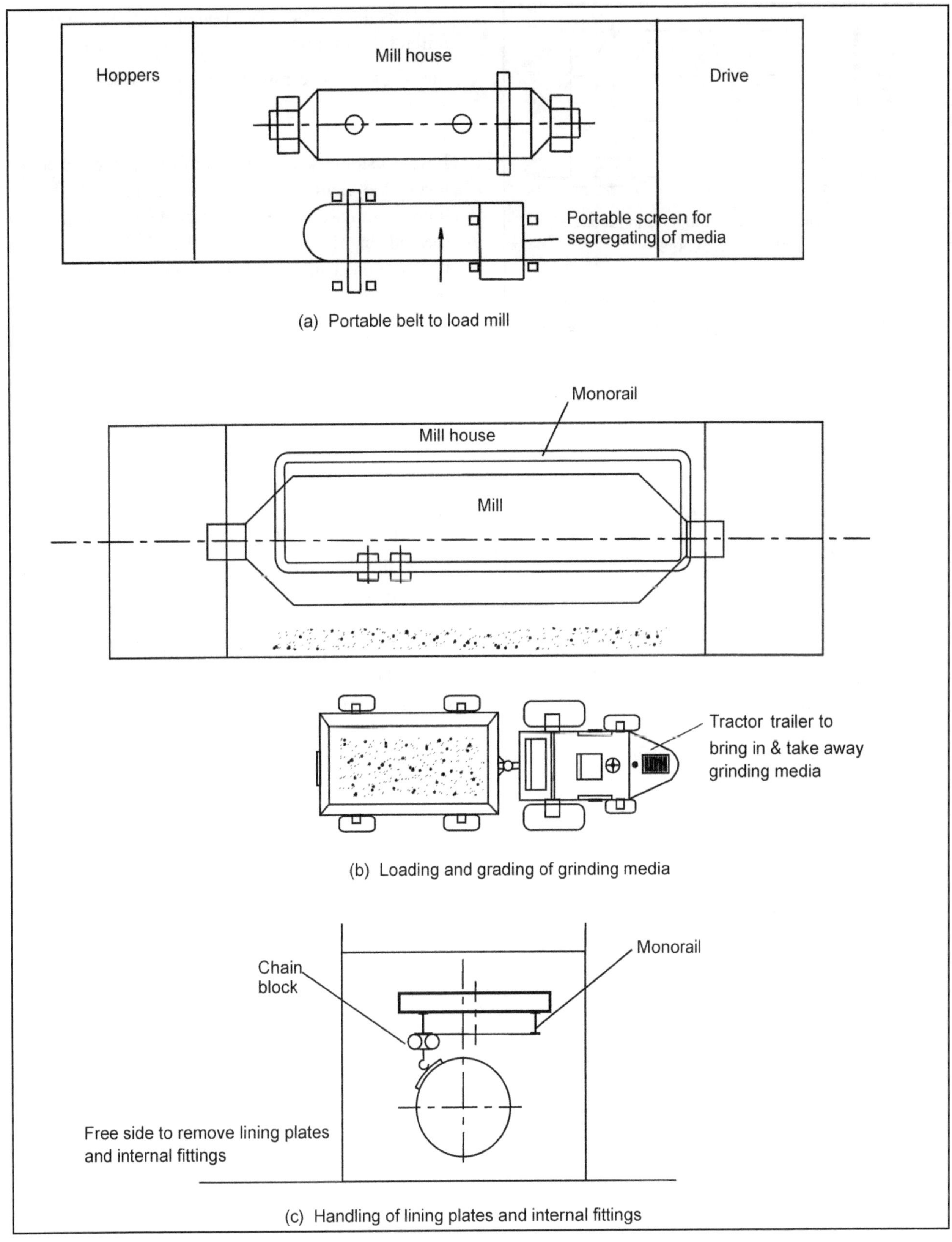

Fig. 6.7.1 Handling of grinding media and internal fittings of ball mill.

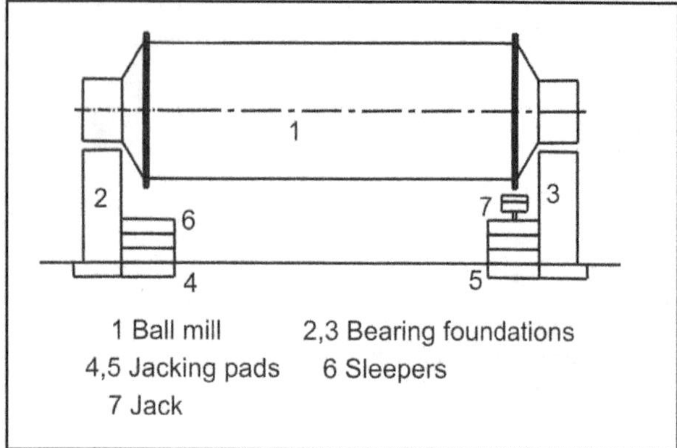

Fig. 6.7.2 Ball mill being placed on foundation.

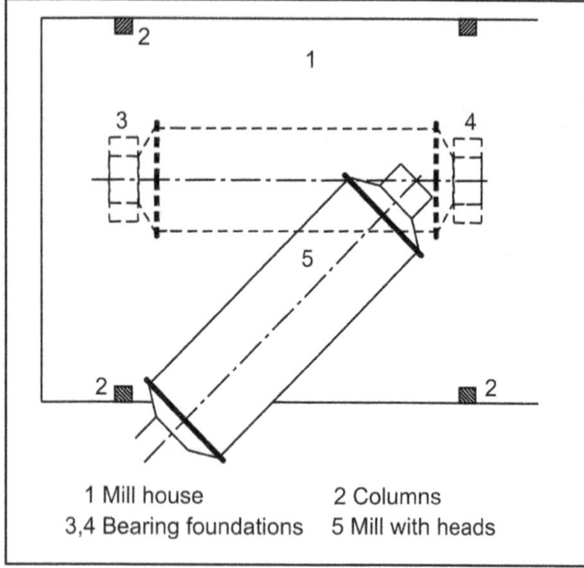

Fig. 6.7.3 Mill assembled with heads being taken in.

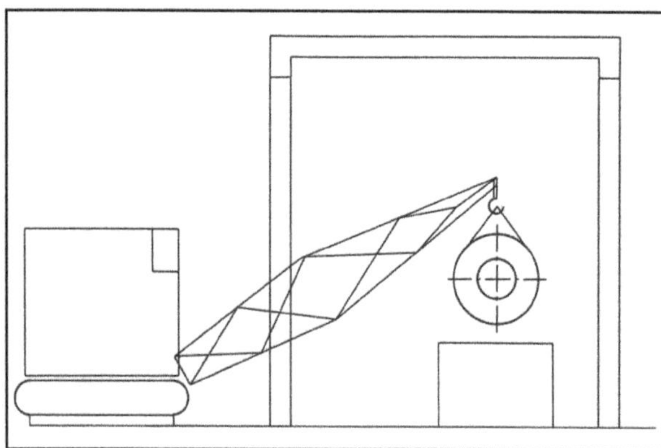

Fig. 6.7.4 Mill assembled on foundation with the help of a crane.

Thus there should be enough clearance above the mill for this purpose.

See Figs. 6.7.1 c and 6.7.5.

7.5

It was customary to locate all mills together when the layout was centered around a crane gantry. This is no longer the case and raw and cement mills are now located far off from one another. Even then, when there is more than one mill, they would for convenience, be installed side by side.

7.6 Motor and Gearbox

Motor and gearbox would be installed in a separate enclosure separated from the mill and auxiliaries as shown.

An overhead traveling crane would be used to lift and remove motors and gearboxes.

See Figs. 6.7.5 and 6.7.6.

On the flat roof of the building could be installed dust collectors.

7.7 Maintenance of Vertical Mills

Vertical rollers mills come in different designs. In one design, rollers can be swung out for maintenance. In another they are lifted and turned around a vertical axis out of the mill.

In others, beams are provided inside the mill housing for facilitating handling. The beams extend outside for keeping rollers on trailers trucks etc.

See Figs. 6.7.7 to 6.7.10.

In smaller mills, with three / two rollers, entire assembly can be taken out of the mill.

These facilities for handling come with the mill; the layout and requirement of free space around the mill will also be suggested by the Supplier and should be followed strictly.

7.7.1 Maintenance Requirements of Rollers

Rollers come in different sizes and shapes. Some have single piece liners; others multiple pieces wedged together.

The maintenance requirements vary with the design. Supplier's instructions should be followed.

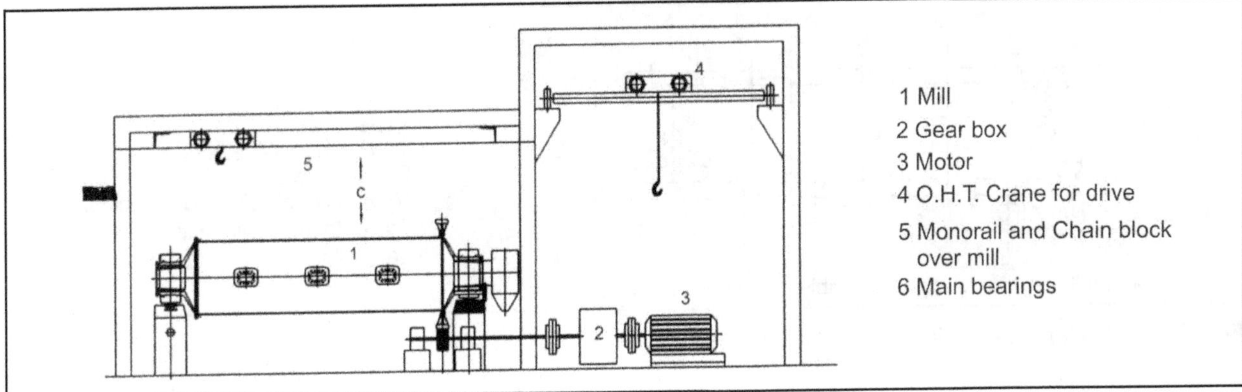

Fig. 6.7.5 Ball mill layout – Elevation.

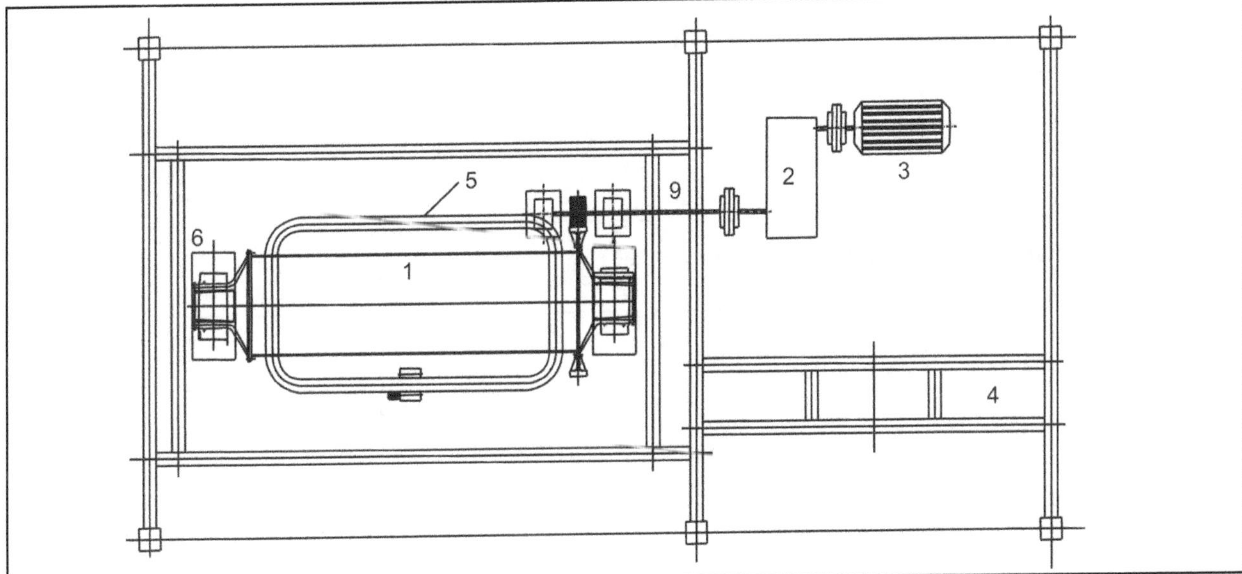

Fig. 6.7.6 Ball mill layout – Plan (material handling equipment for grinding media and lining plate).

The aim in designing layout would be to minimize stoppage time required for changing rollers / replacing liners / including liners of table also.

7.7.2 Table and Gearbox

In a roller mill, the table is required to be jacked up for removing the gearbox. Gearbox is put on a trolley on rails and pulled out from end opposite the motor so that motor is not required to be disturbed.

See Fig. 6.7.11.

In large mills that have 'multi drives this is not necessary. 2,4, gear boxes drive a ringgear under the table. It is also not necessary to have bevel train in the gear box as output shaft is also horizontal **See Plate no. 2.4b.4 in Chapter 4b of Section 2.**

A typical layout of a vertical roller mill from maintenance angle is shown. It should not be necessary to disturb hot air ducting while attending to the internals of the mill.

See Fig. 6.7.12.

7.7.3 Maintenace in Mill with External Circuit

When mill has an external circuit, coarse material falling through the air gap is required to be collected and brought out from under the mill to a single point to take it to the elevator. The group of conveyors under the mill would need maintenance and hence there should be enough head room and elbow room under the mill for this purpose.

See Fig. 6.6.44 in Chapter 6.

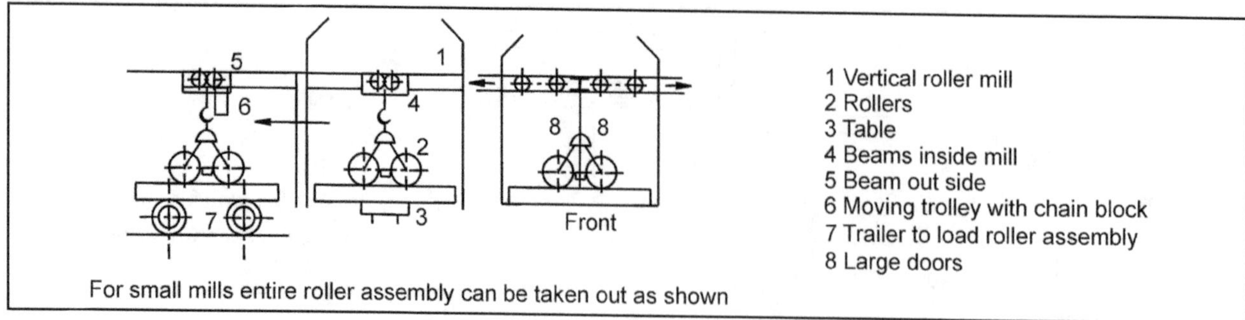

Fig. 6.7.7 Handling facilities for vertical roller mills - 1.

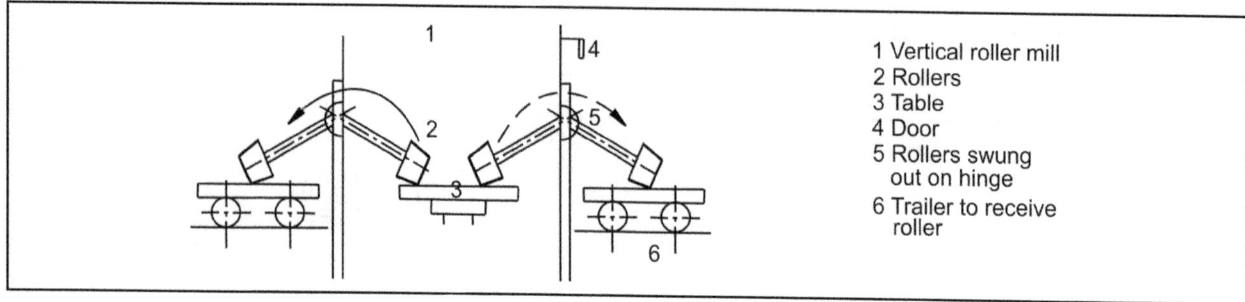

Fig. 6.7.8 Handling facilities for vertical roller mills - 2.

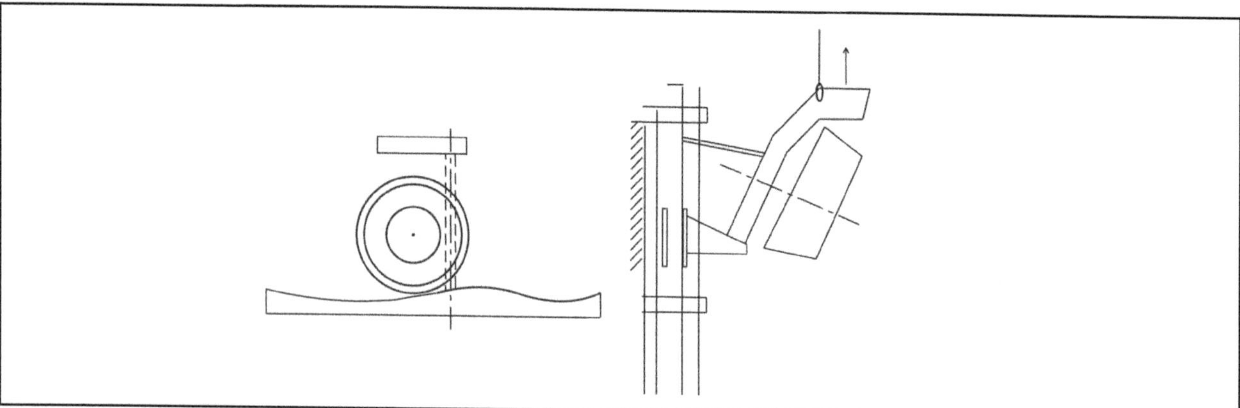

Fig. 6.7.9 Lifting arrangement for roller - elevation.

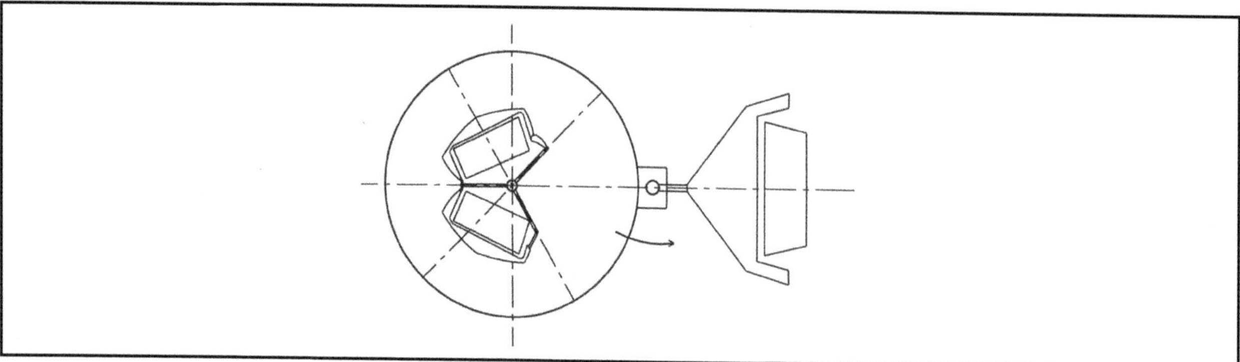

Fig. 6.7.10 Lifting arrangement of rollers - plan.

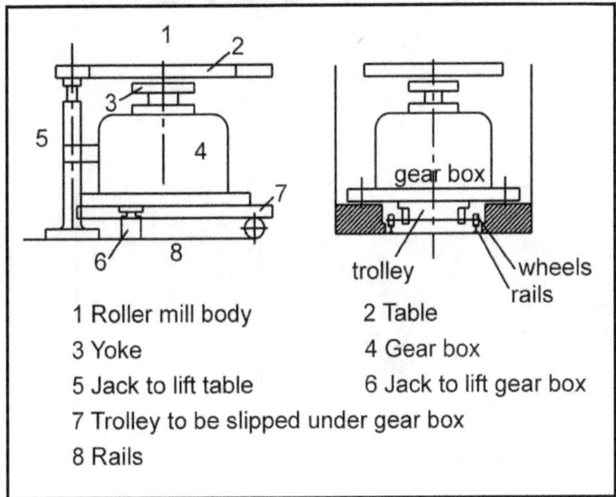

Fig. 6.7.11 Taking gearbox out from under vertical roller mill.

7.8 Maintenance of a Roller Press

Roller press will be used either alone or with ball mill. When ball mill follows a roller press, it will be much smaller in size for reasons explained earlier. Layout considerations for ball mill and auxiliaries would be same as for ball mills in general.

7.8.1 Roller Press

Roller press itself would be housed in a separate building more commonly on the first floor of the building. This building should have facility to hoist either the complete assembly of roller press and place it on foundation or at least the rollers.

Rollers will have to be removed at times for maintenance.

It is usual to build up worn out rollers 'in situ' by fixing a portable welding machine on the press. The inlet chute will be removed for this purpose.

Two rollers will have two drives – motors and gearboxes. An overhead crane will be provided for lifting motors and gear boxes.

As in case of ball mill and v.r.mill, at least one side of the building should be free for access to tractors and trailers to load rollers on them.

See Fig. 6.7.13.

As in case of v.r.mills, Suppliers of roller press should furnish the layout of the roller press showing orientation of lifting gears and also specifications if they are not included in scope of their supply.

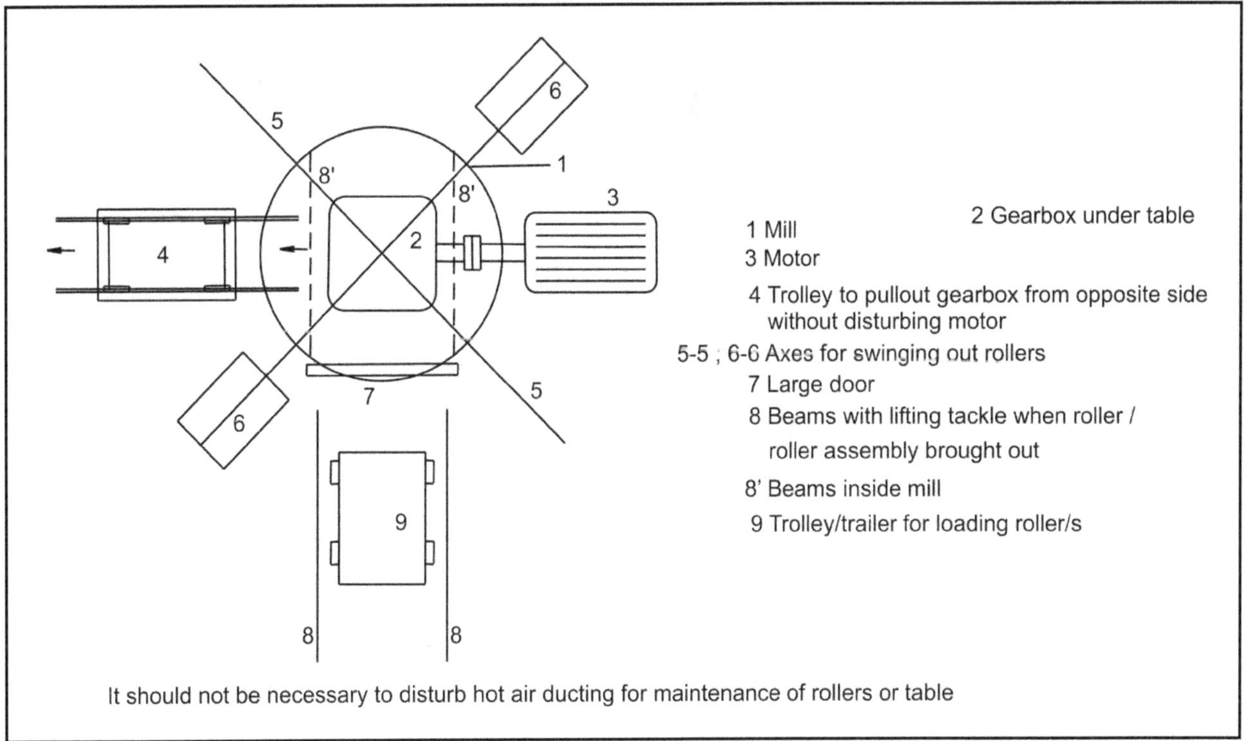

Fig. 6.7.12 Roller mill layout for maintenance.

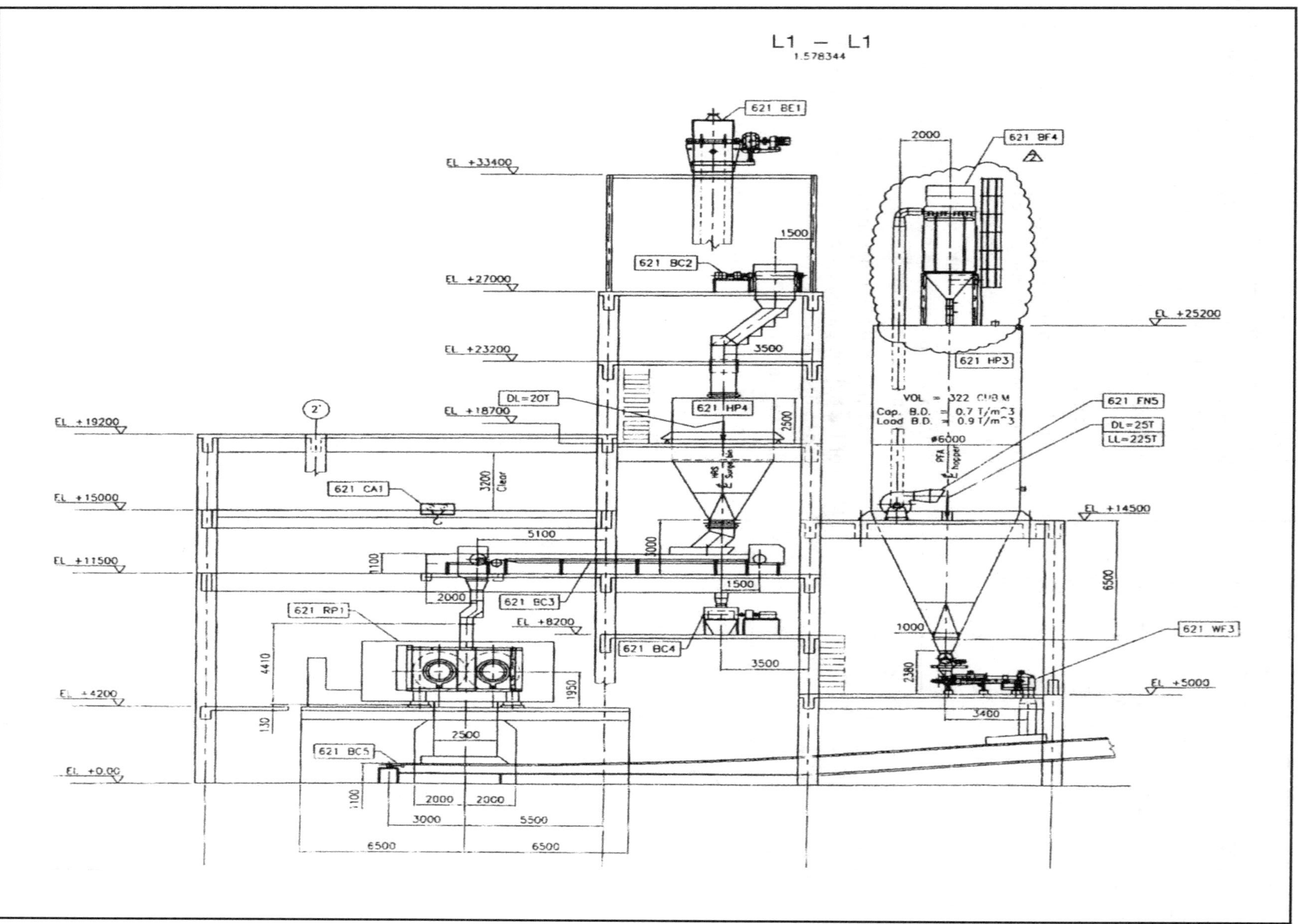

Fig. 6.7.13 Roller press building with hoppers in cement mill circuit.

CHAPTER 8

LAYOUTS OF COAL MILLS

8.1 Coal Mills

Coal mills are either air swept ball mills or vertical roller mills. Therefore all those factors which come into play in developing layouts of raw mills will apply to coal mills also. Coal mills naturally will be much smaller than raw mills. Hence maintenance aspects covered in **Chapter 7** will be on a smaller scale. For example, in most cases entire rollers assembly could be taken out for maintenance.

8.2 Safety Aspects

A very important aspect of layout particular for coal mills is the 'safety' against explosions and fire.

The danger of fire / explosions is more when mill uses hot air from cooler for drying. When 'inert' gases (< 6 % O_2) from preheater exhaust are used for drying, the danger is very much reduced.

See Figs. 6.8.1 to 6.8.3.

Even then it is prudent to provide explosion flaps on ducts, bins and also on dust collectors in the system

Explosion flaps swing out under pressure and snap back in position as they are spring loaded.

See Figs. 6.8.4 to 6.8.6.

8.2.1 Explosion and Rupture Flaps

Rupture foils are thin foils which burst under pressure and have to be replaced.

See Fig. 6.8.7.

Explosion flaps are also provided on ducts so that explosive forces do not reach the dust collector.

See Fig. 6.8.5.

8.2.2 Location of Safety Devices

These safety devices should be so located that there is no possibility of injury to human beings in the event of their coming into action.

See Fig. 6.8.4.

Rupture flaps must be immediately attended to as otherwise with system being under negative pressure, it will admit atmospheric air and situation will worsen.

8.3 Smothering and Extinguishing Fires

There should be a provision to spray limestone powder in bins to smother the fire, should there be one.

8.3.1 Injecting CO_2

Another common practice to cope with fires is to inject CO_2 in the system – in bin and in dust collector. Fire dies for want of oxygen.

A battery of CO_2 cylinders is installed in the department which when actuated injects CO_2 in dust collector and bin and thus extinguishes fire.

See Fig. 6.8.8.

But once discharged the battery needs to be replaced immediately to be available again. It thus requires three sets of cylinders –

(i) One set installed

(ii) One set in store

(iii) One set in transit for filling up.

Occasionally provision is also kept for spraying water in the ESP.

Bag filters and ESPs in coal system are specially designed to withstand high pressure of 500 mmwg.

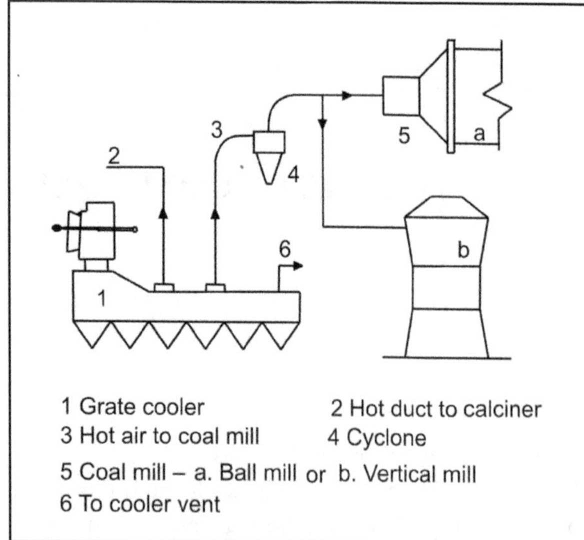

Fig. 6.8.1 Hot air from grate cooler to coal mill – ball, vertical.

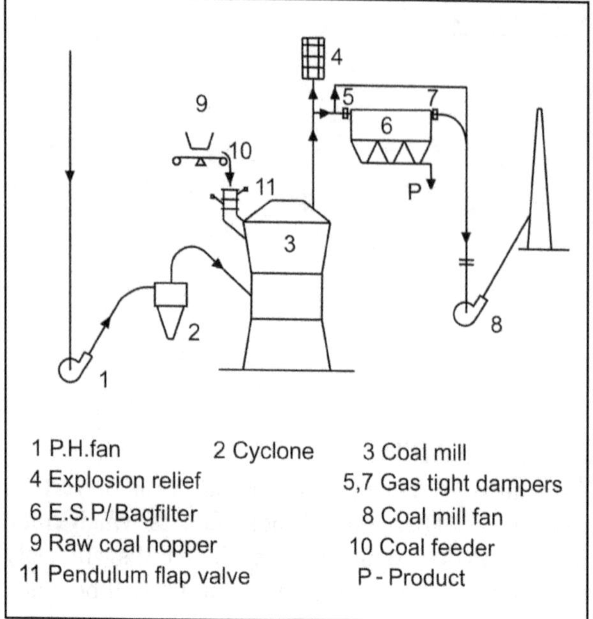

Fig. 6.8.2 Coal mill system vertical mill at preheater end.

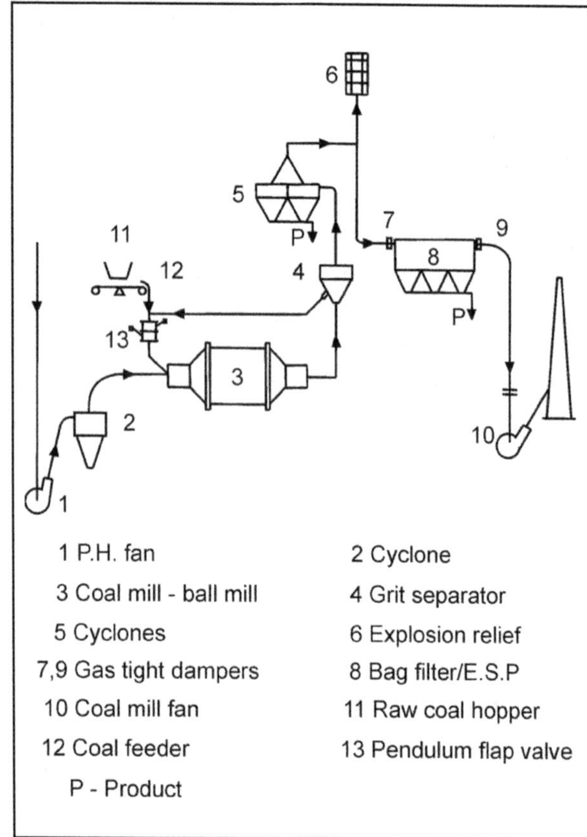

Fig. 6.8.3 Coal mill system – air swept ball mill at preheater end.

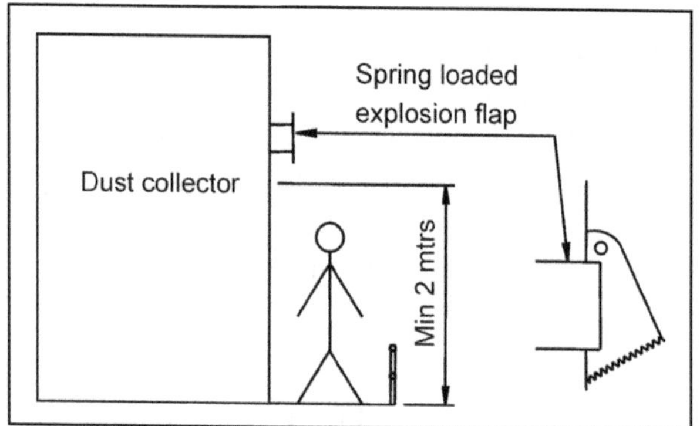

Fig. 6.8.4 Dust collector casing.

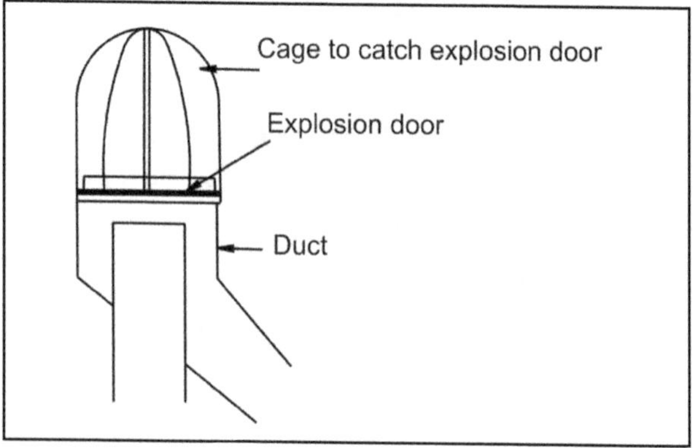

Fig. 6.8.5 Ducting.

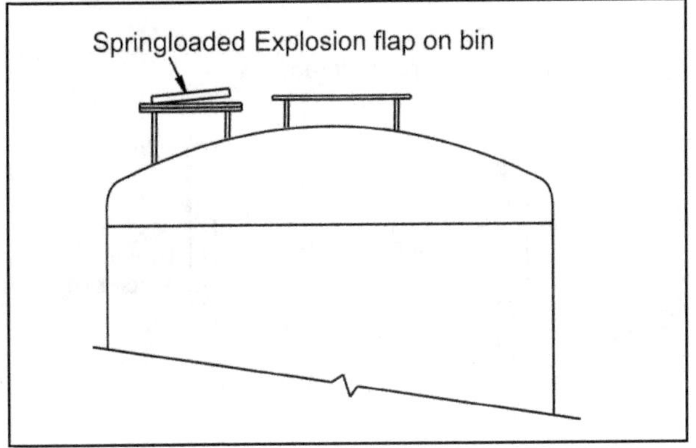

Fig. 6.8.6 Spring loaded explosion flap on bin.

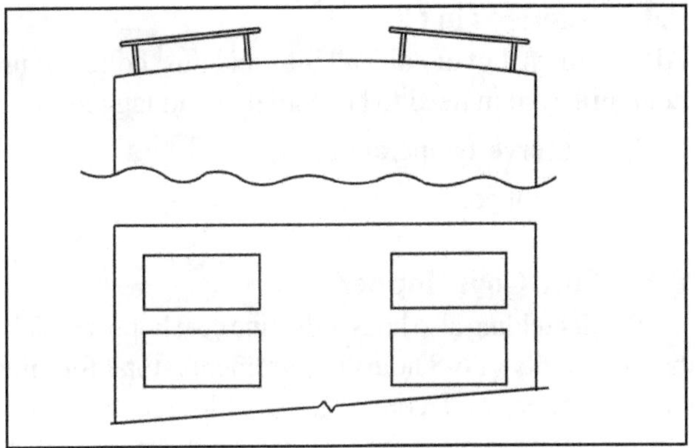

Fig. 6.8.7 Rupture foils on roof of ESP.

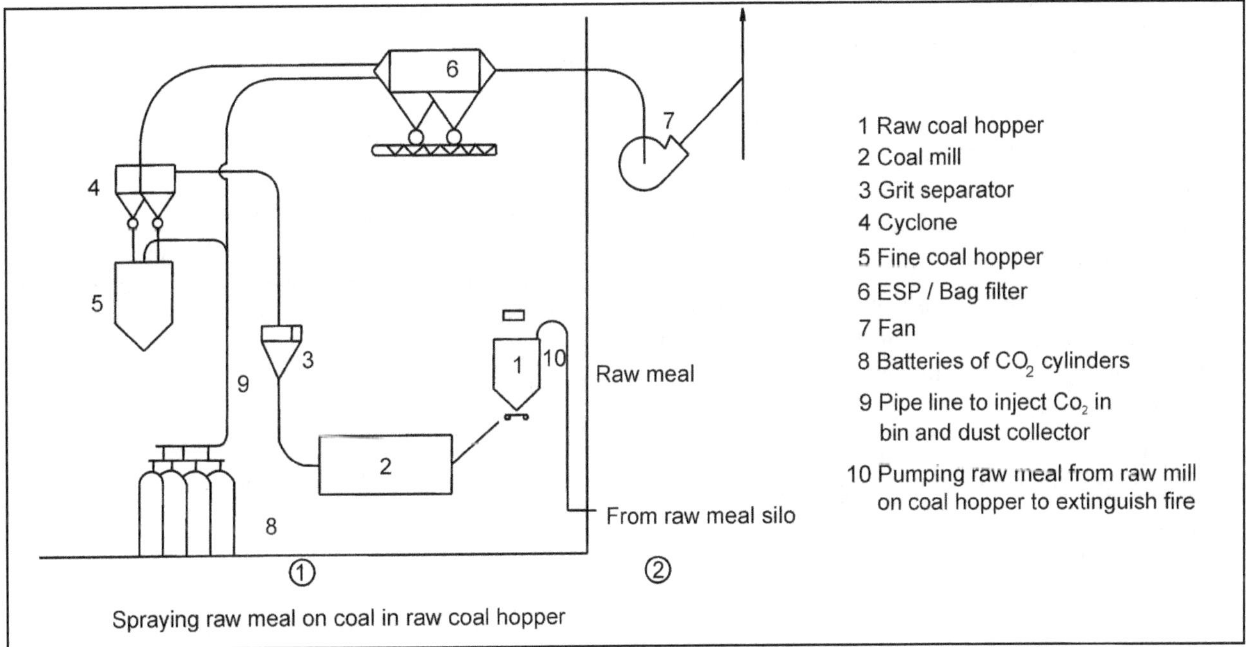

Fig. 6.8.8 1. Injecting CO_2 in fine coal bin and in ESP to extinguish fire.
2. Spraying raw meal in raw coal hopper.

8.3.2 Inert Gas Generator

It is also possible to install an 'inert gas generator' to supply inert gas to extinguish fire. It would eliminate necessity to keep and replenish stocks of CO_2 cylinders.

A large receiver will store inert gas just like receivers of compressors. It will start supplying inert gas automatically when fire or presence of CO is detected. It will also actuate starting of the 'inert gas genearator'.

The screw conveyors at bottom are oversized so as to eliminate possibility of accumulation of coal dust in the hopper and their getting over loaded.

8.3.3 Layout to House Gas Cylinders

The layout thus must provide room, at ground floor for the battery of CO_2 cylinders and its bringing in / taking out for replacement or for an inert gas generator and gas receiver.

The area around bag filter and explosion flaps on ducts should be cordoned off and should not be accessible to men under normal circumstances.

8.4 Ductings in Circuit

All ductings, hot or cold, all bins and dust collector in coal mill system need to be insulated and lagged to :

1. conserve temperature,
2. prevent condensation.

8.5 Fine Coal Hopper

Coal mill building also houses the fine coal hopper which has a capacity of 6-8 hours requirements (total for kiln and calciner) and coal metering device and coal conveying equipment for conveying pulverised coal to kiln and calciner.

There would be two coal bins if coals fired in kiln and calciner differ in fineness.

Coal firing and metering systems are being dealt with separately.

8.6 Location of Coal Mill

When coal mill is located near preheater end to use inert gases, the calciner would be close to it and kiln at a distance from it.

Reverse would be the case when coal mill is at cooler end. In this situation hot air from cooler would be drawn for drying in the mill through a cyclone. **See Fig. 6.8.1**

When coal mill is located at the preheater end, major changes are required in the general layout as compared to the hitherto conventional arrangement of coal mill at cooler end.

See Fig. 6.8.9.

8.6.1

When coal mill is located at the cooler end, it is installed near the cooler on the ground floor. If it is a ball mill, many of its auxiliaries like grit separator and coarse return screw would be installed on the extension of the burners' platform. Coal mill house would be very much attached to the cooler building.

When it is a v.r. mill., dust collector would be installed as shown in **Fig. 6.8.12** next to burners platform.

When mill is installed at preheater end, coal mill building will house, mill all the auxiliaries and also fine coal bins and coal firing systems for kiln and calciner.

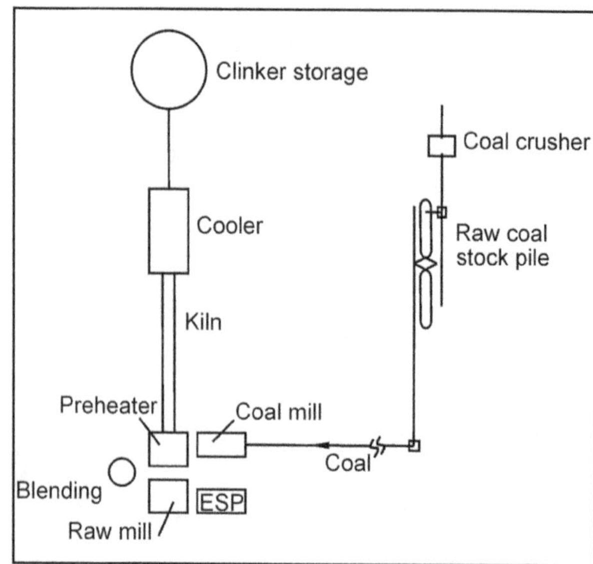

Fig. 6.8.9 Coal mill located at preheater end to use inert gas from preheater.

8.7 Fans in Coal Mill System

Coal mill system may have one fan or two fans to draw gas/air through the system from preheater / cooler and through bag filter/ esp. Though using one fan is a feasible proposition many prefer two fans because of length of ductings.

See Figs. 6.8.10 and 6.8.11.

Figs. 6.8.10 and 6.8.11 show hot gases being drawn from cooler but they could apply to system drawing hot gases from preheater as well.

8.8 Ducting Layout

The comments made earlier about the layout of ducting from preheater for drying in raw mill section also apply to coal mill. In raw mill section, dust contained in preheater gases does not matter.

In taking preheater gases to coal mill this dust needs to be removed as otherwise it would bring down calorific value of coal.

See Figs. 6.8.1 to 6.8.3.

8.9 Gas Tight Dampers

In the coal mill circuit airtight, 'quick closing' dampers are installed to isolate the dust collector in the event of fire / explosion.

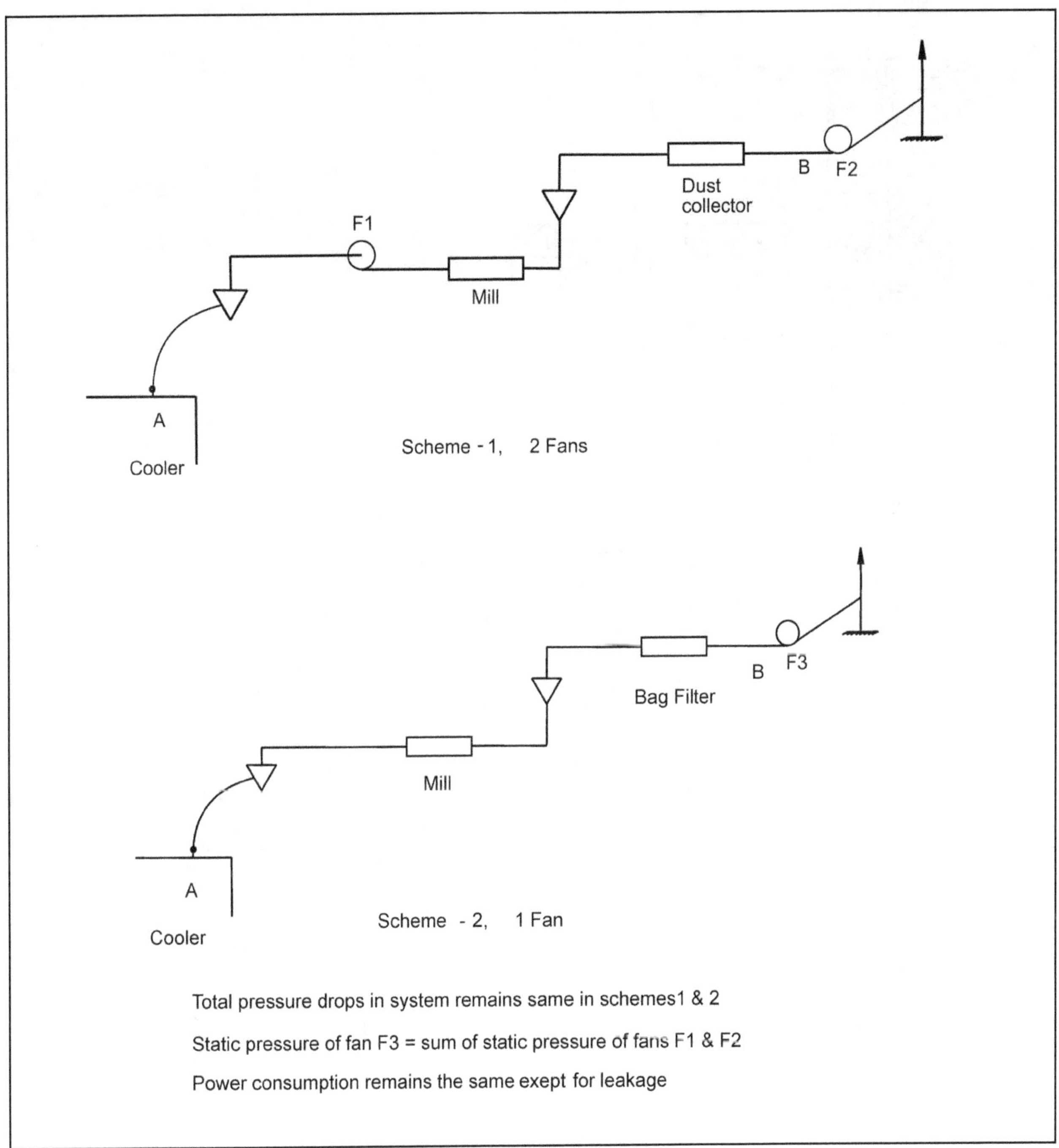

Figs. 6.8.10, 6.8.11 Drying and grinding system for coal mill with 2 fan and 1 fan.

These could also be used for carrying out maintenance if a bypass is provided.

See Fig. 6.43.32 in Chapter 43 on ductings.

8.10

A typical layout of a coal mill department with coal bins and coal metering arrangement is attached.

See Fig. 6.8.12.

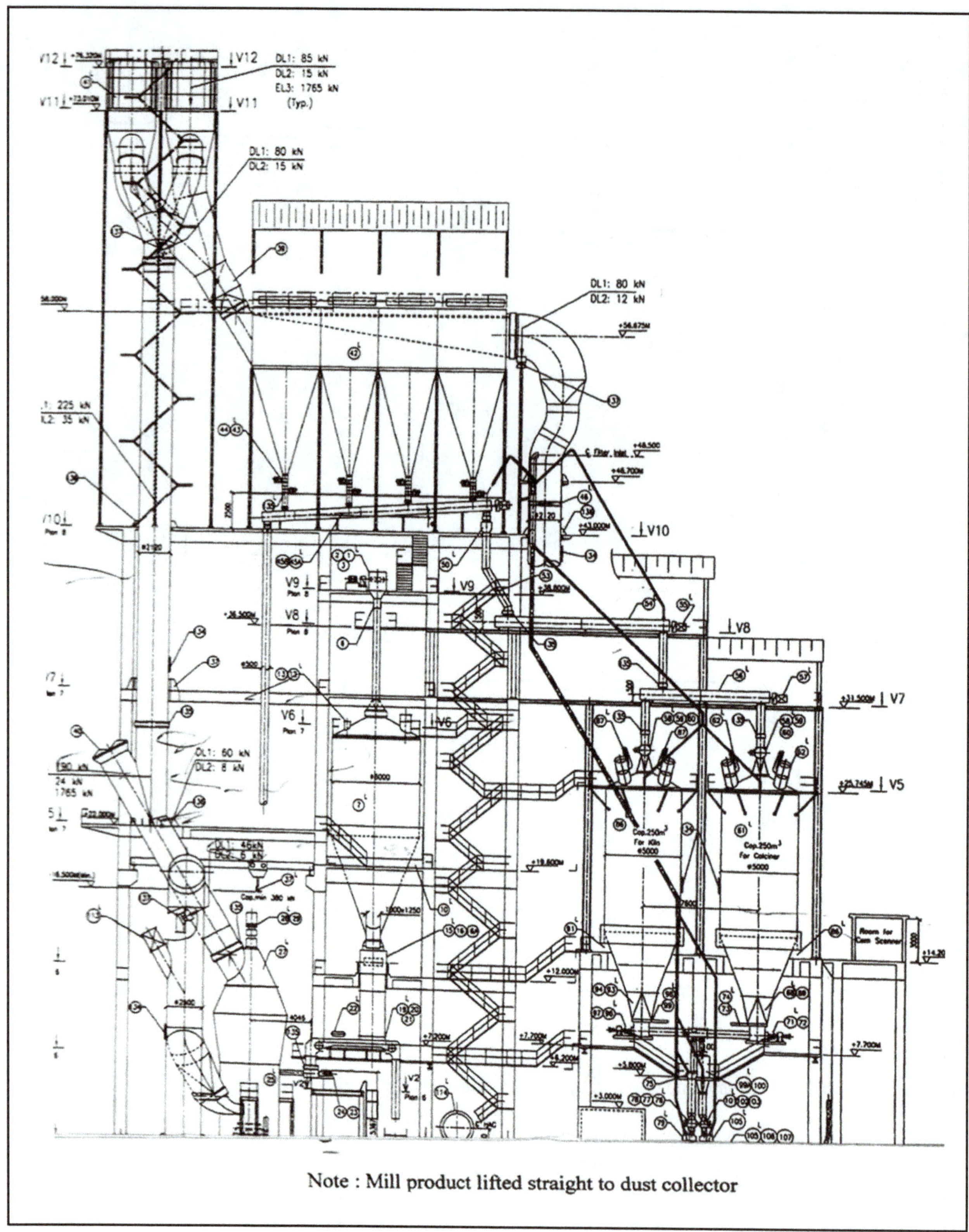

Fig. 6.8.12 Coal grinding and coal metering system with pneumatic conveying of coal to kiln and calciner.

CHAPTER 9

CONVEYING OF RAW MEAL / PULVERISED COAL / CEMENT

9.1 Raw Meal, Pulverised Coal and Cement

Raw Meal i.e., ground raw mix is required to be conveyed first from raw mill to blending silo then from storage silo to kiln feed bin and from kiln feed system to preheater.

Dry pulverised coal is conveyed for firing into kiln and calciner.

Cement is transported from mill to storage silos and from silos to packing machines.

All three materials are finely ground powders and are free flowing when dry. They can be conveyed with ease pneumatically with the help of 'air slides', FK pumps, Fluxos and airlifts; or mechanically by bucket elevators, belt conveyors, screw conveyors and chain conveyors.

9.2 Pneumatic Conveying

9.2.1 Airslides

Airslides are stationary rectangular ducts with an air chamber and a porous medium, installed at a slope of 6-10° and hence can convey fluidized material downwards only. They can be fed at a number of points and can discharge also at a number of points.

See Fig. 6.9.1.

9.2.2 FK Pumps

FK pumps are screw pumps which push material as 'slugs' in a chamber with air nozzles, air borne material is conveyed through a pipe line.

FK Pumps can convey horizontally and or vertically. Material can be discharged at a number of points by using 2 or three way diverting valves in the pipeline.

See Fig. 6.9.2.

It is a clean and flexible system. Conveying air is vented out at discharge through a dust collector. There is thus no dust nuisance.

It however requires two drives – one for the screw and the other for compressor or blower. On the whole power consumption in kwh/ton of material conveyed is high.

9.2.3 Lean and Dense Phase Systems

Other pneumatic conveying systems have done away with the screw and convey pneumatically by admitting material in air stream through an ejector – lean phase system – **see Fig. 6.9.3** or by pushing it out of a vessel by compressed air – dense phase system. Air consumption is high but overall power consumption is less.

Dense phase systems generally require two vessels –one for filling and one for emptying – for continuous conveying. **See Fig. 6.9.13 and Plate 2.16.10 in Chapter 16 of Section 2.**

They are seldom used now for conveying raw meal or coal.

9.2.4 Air Lifts

Air lifts are systems which convey vertically only. They are useful for lifting raw meal / cement etc., to blending silos, cement storage silos and to lift raw meal to preheaters.

Power consumption is much less than that for FK Pumps but, require large volumes of air. They lift material vertically only.

See Fig. 6.9.4.

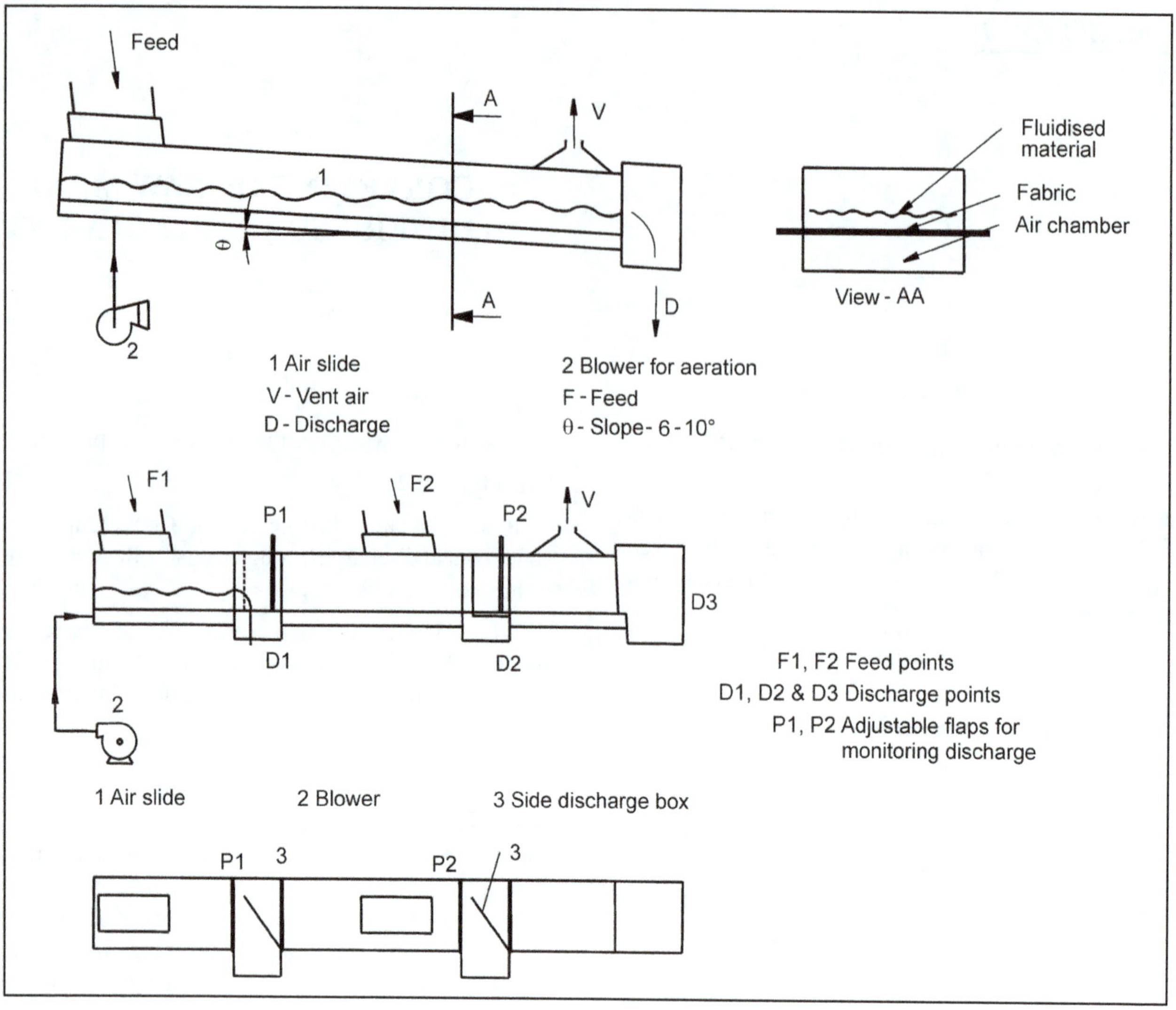

Feed
A
V
1
θ
Fluidised material
Fabric
Air chamber
A
D
View - AA
2
1 Air slide
V - Vent air
D - Discharge
2 Blower for aeration
F - Feed
θ - Slope - 6 - 10°
F1
F2
P1
P2
V
D3
D1
D2
2
F1, F2 Feed points
D1, D2 & D3 Discharge points
P1, P2 Adjustable flaps for monitoring discharge
1 Air slide
2 Blower
3 Side discharge box
P1
3
P2
3

Fig. 6.9.1 Pneumatic conveying of pulverized material by air slide.

3
4
5
6
7
1 FK pump
2 Conveying Pipe line
3,4 Two way diverting valves
5,6,7 Bag filters for venting
8 Compressor for supplying compressed air
Material feed
Vent
Silos
1
2
Diverting valve
8

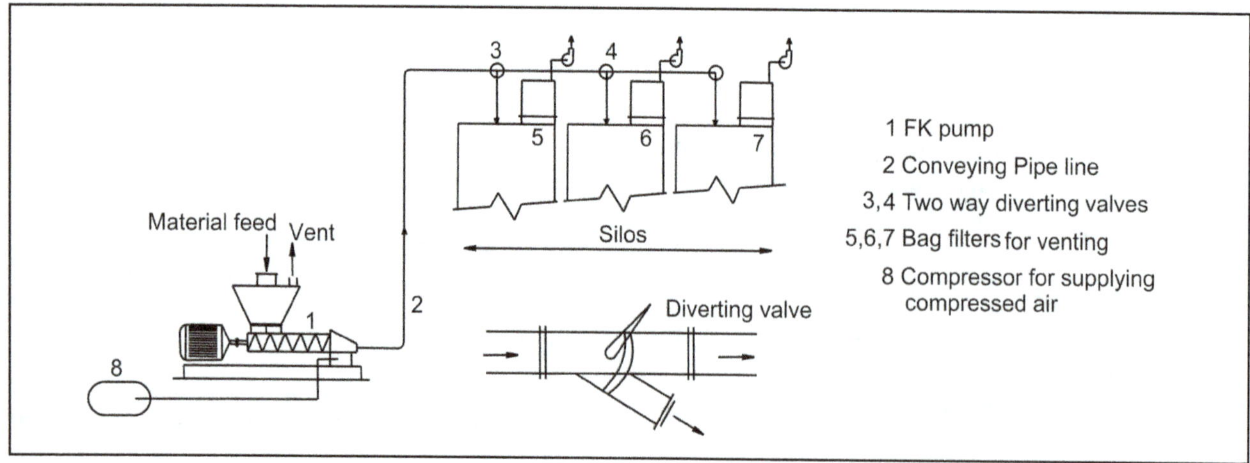

Fig. 6.9.2 FK pump system for pneumatic conveying raw meal / coal / cement.

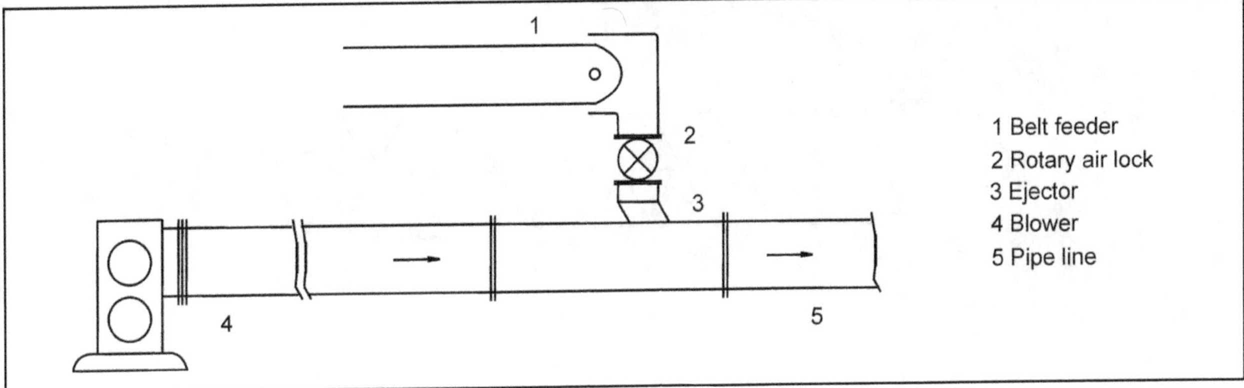

Fig. 6.9.3 Pneumatic conveying system using ejector to feed material in pipe line.

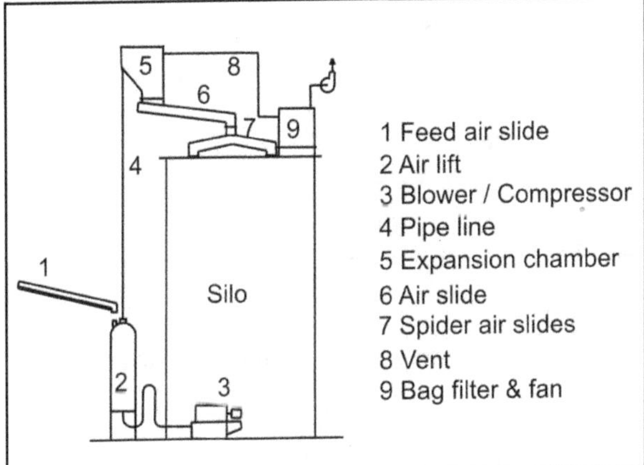

Fig. 6.9.4 Air lift for raw meal, cement, for vertical lift only.

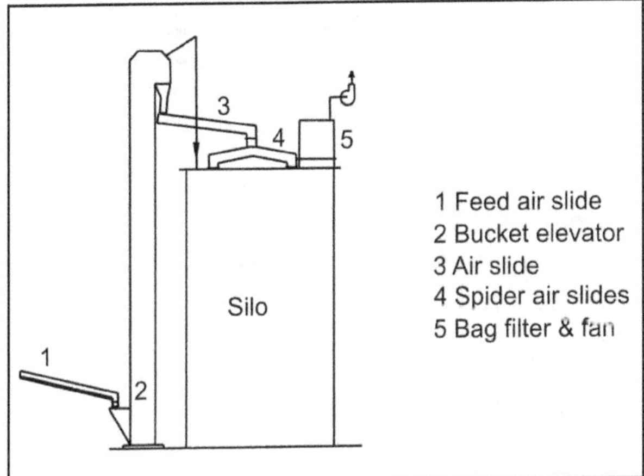

Fig. 6.9.5 Bucket elevator to lift materials pulverised / granular into silos.

9.3 Mechanical Conveyors

Mechnical conveyors like screw, belt, and elevator - require lowest power / ton of material conveyed and of late have become popular again and are preferred to pneumatic conveying systems. They have several limitations though.

1. Size wise they are bulky.
2. Change of direction / elevation is difficult without another drive.
3. Inclined belt conveyors are commonly used to convey and lift materials; but angle of inclination is limited to 20° or less requiring very long lengths of conveyor for same lift.
4. Elevators convey only vertically.
5. Screw and chain conveyors convey only horizontally. They can be installed at an angle up to 12° but capacity gets derated.

See Figs. 6.9.5 and 6.9.6.

9.4 Transport of Raw Meal

Presently the most commonly used modes of transport for raw meal are :

1. Air lift.
2. Bucket elevator.

Since both lift only vertically, they are required to be located almost vertically below the silo and by the side of preheater tower as the case may be.

See Figs. 6.9.4 and 6.9.5.

Ground raw meal collected in cyclones, dust collector and air separator is brought to the air lift / elevator almost always by air slides or by chain conveyors from esps / bag filters.

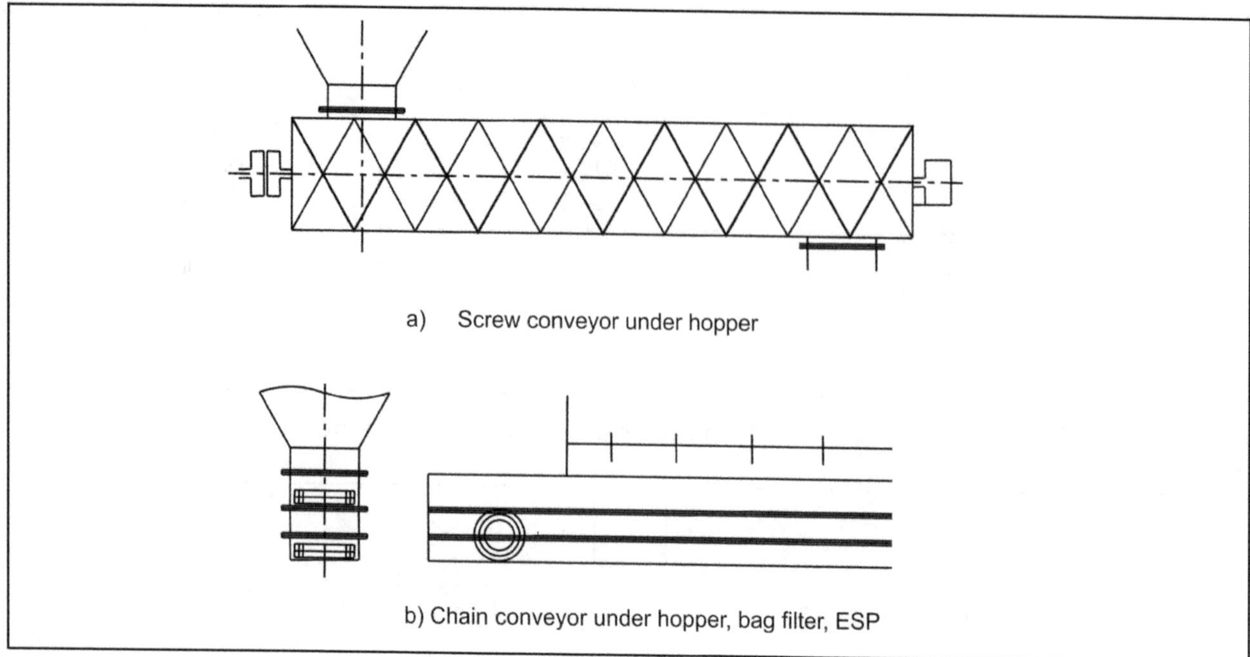

Fig. 6.9.6 Screw and Chain conveyor under hopper.

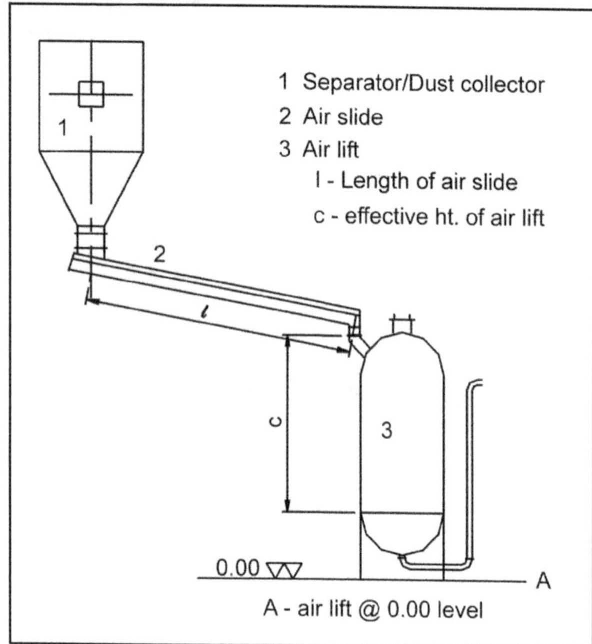

Fig. 6.9.7 Section of mill department - location of air lift at 0.00 level or in pit.

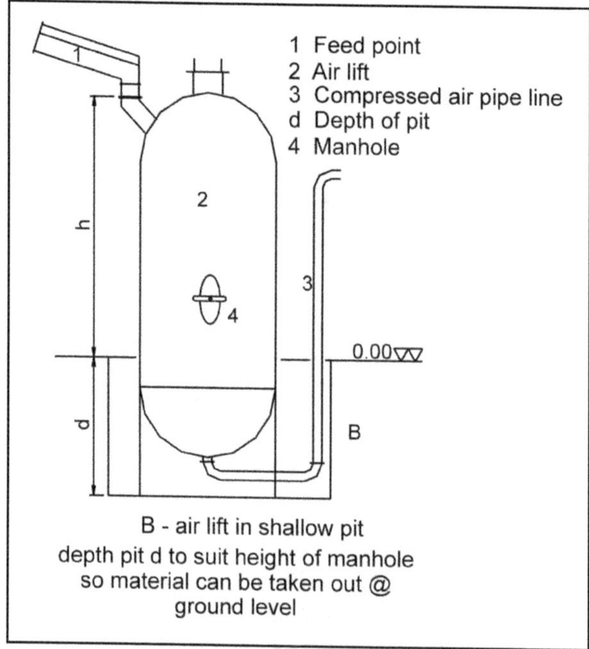

Fig. 6.9.8 Air lift installed in a pit to reduce height h of feed point.

9.4.1 Using Air Lift

Air lift vessels range from 0.8 to 2.2 m in diameter. If installed at ground floor, the feed point an go beyond 10 metres from ground level depending on diameter of vessel and diameter of v conveying pipe. Capacities go up to 1400 tph **See Fig. 6.9.7 and 6.9.8.**

Height of vessel is dependent on diameter of conveying pipe and density of meal laden air.

They consume more power than bucket elevators. Bucket elevators of + 80 m in height (for 6 stage preheaters), being available now, they are preferred over air lifts.

9.5 Depth of pit in Ball Mill Cicuit

In closed circuit mills partially ground material is conveyed out of mill to separators using first an airslide and then a bucket elevator.

See Figs. 6.6.39 to 6.6.41 in chapter 6

For bucket elevators, the feed point would be about 1 - 2 Meters from ground level if elevator is installed at ground level.

Though ideal, it is not practical to install either airlift or elevator at ground level. However pit should be as shallow as possible. The depth should be limited to 3 metres as far as possible.

Depth of pit is dependent on location and size of separator and height of mill above ground level. Pit should be water proofed.

9.5.1 Pit for Air Lift

For airlifts, pit depth should be such that manhole provided for emptying the vessel should be above ground level.

See Figs. 6.9.7 and 6.9.8.

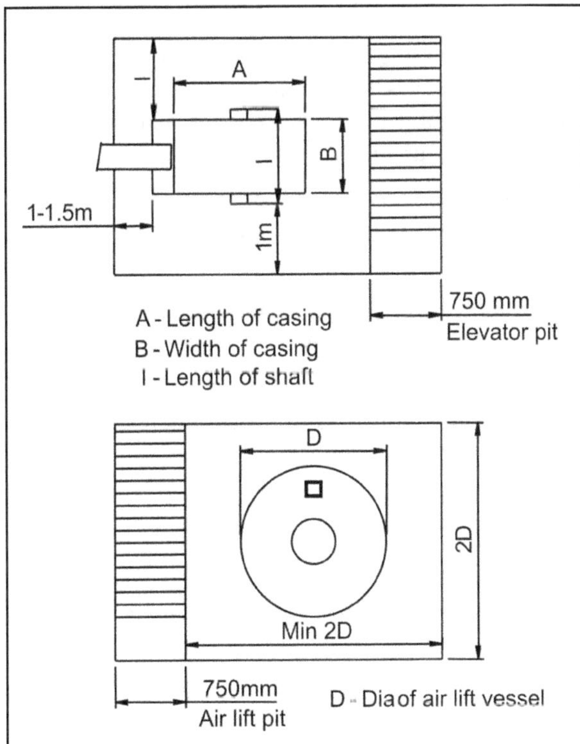

Fig. 6.9.9 Dimensions of elevator and air lift pits.

9.5.2 Dimensions of Pits

A good amount of space needs to be provided around bucket elevator for maintenance and for taking out shaft.

See Fig. 6.9.9.

Air lift itself does not require any maintenance. The blower needs maintenance common to the compressor and blowers.

9.6 Compressors and Blowers

The compressors / blowers can be located either under the silo or in a separate room (compressor house).

See Fig. 6.9.10.

9.6.1 Pressurisation of Compressor House

The entire room whether under silo or a separate one should be pressurized to prevent dust getting into the

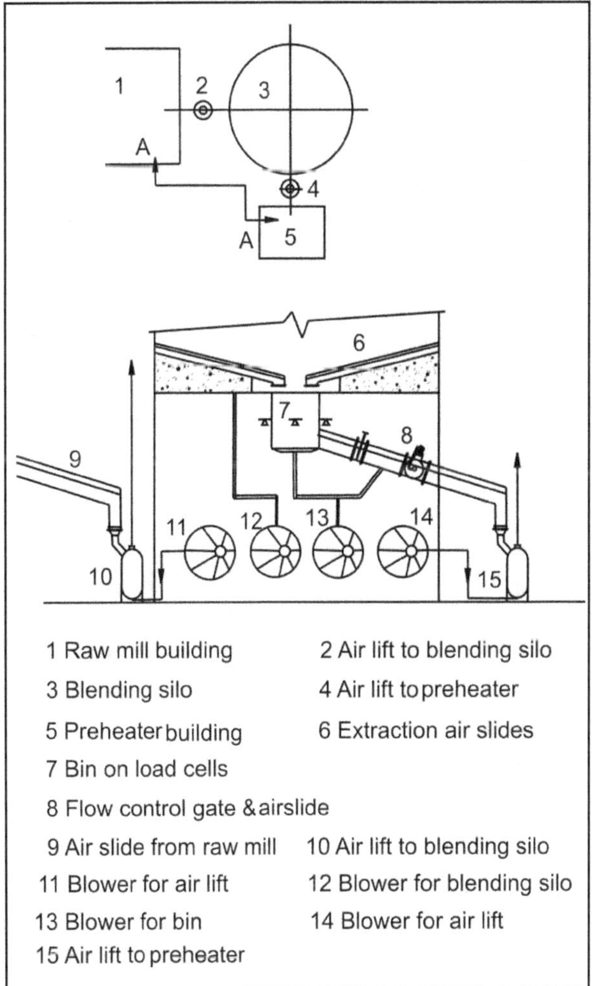

Fig. 6.9.10 Compressors - blowers for air lift mostly located under blending silo.

room. Alternately, doors / openings should be fitted with filters for preventing entry of dust.

Compressors and blowers themselves should have generous dry air filters. It will improve life of compressor / blower.

See Figs. 6.9.11 and 6.9.12.

This is particularly applicable for compressors used for aeration of blending silo which would also be installed in the same place.

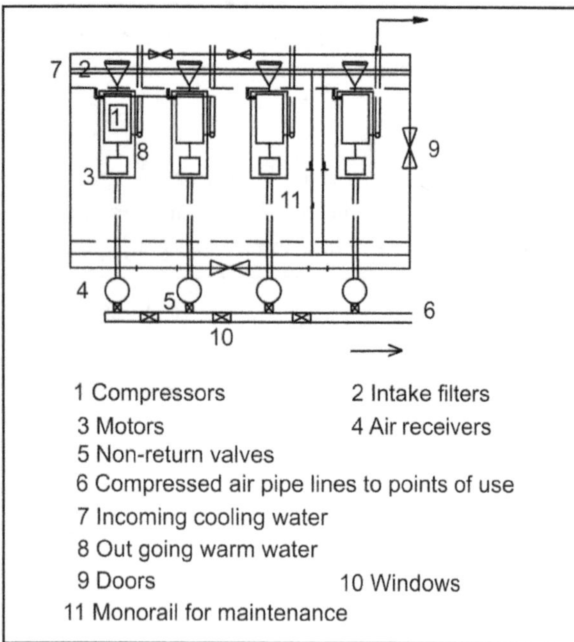

Fig. 6.9.11 Layout of compressor house.

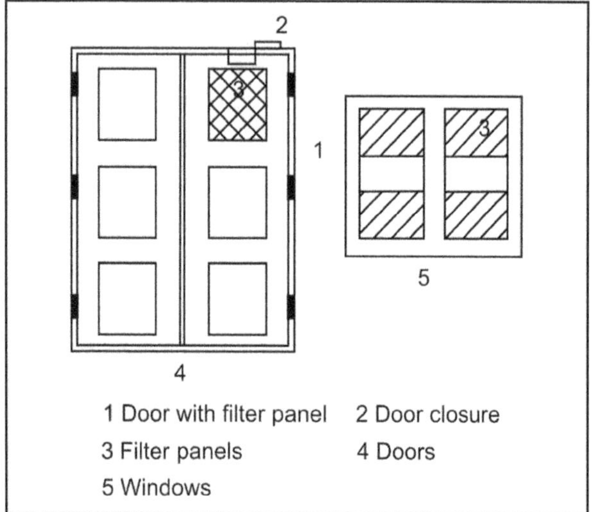

Fig. 6.9.12 Filter panels on doors and windows to keep dust out.

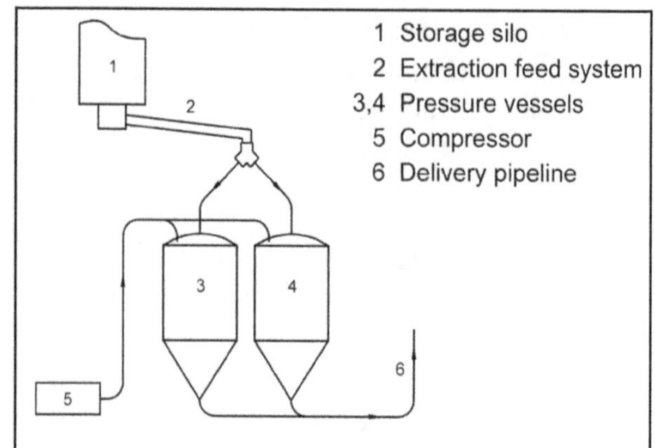

Fig. 6.9.13 Dence phase conveying system.

9.7 Conveying of Cement

Conveying of cement, whether within the cement grinding department or from mill to storage silo is exactly the same as raw meal. There are small differences due to differences in their densities and finenesses. Cement is more abrasive than raw meal.

Layouts of conveying equipment for cement are developed exactly in the same way as for raw meal described above. Even here elevators are preferred to air lifts which are mostly used as standby.

9.8 Conveying of Pulverised Coal

Pulverised coal is extracted from bins and metered and conveyed for firing into kiln and calciner almost always pneumatically.

Pneumatic conveying ensures instant dispersal of pulverized coal in kiln and calciner to achieve complete combustion.

Earlier primary air fans were used to convey coal into kiln. Distances rarely exceeded 30 metres. Both bin and P.A. Fan were installed on burner's platform.

With increasing use of multi-channel burners, blowers are used for the purpose.

FK Pumps or lean phase pneumatic conveying systems are used to convey coal to calciner.

A lot depends on the location of mill. As mentioned earlier in **Chapter 8**, mill could be at cooler end or at preheater end.

Though power consumption wise pneumatic conveying is more expensive its impact is small when conveying coal because consumption of coal is less than 20 % of clinker.

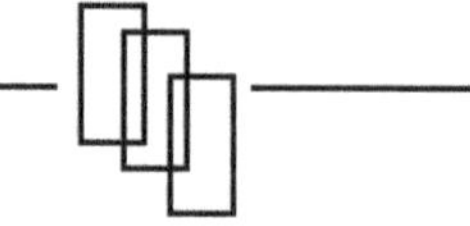

CHAPTER 10

BATCH BLENDING AND STORAGE SYSTEMS FOR RAW MEAL

10.1 Blending Systems

Two basic systems of 'air merge' blending systems that depend on 'fluidisation' of raw meal exist. They are:

1. Batch blending.
2. Continuous blending.

10.2 Batch blending

As the name suggests, in this system, a quantity of - batch – is collected at a time in a silo. It undergoes blending or homogenising operation in it aand is then transferred to storage silo.

A second batch.is then taken into the silo.

A batch is generally 8 t0 10 hours' output of the raw mill.

Thus a minimum of two blending silos are necessary to maintain continuity.

Quantity of storage of blended raw meal is generally equivalent of two days' requirement of the kiln. Blending and storage are generally found to be in a'double decked ' system that is blending silo at top and storage silo below. This way, transfer of blended meal is fast and does not need any power. **See Fig 6.10.1** which shows such a system for a small plant.

10.2.1 Aeration of Silo Bottom

Most commonly used are open air slides or open square boxes fitted with aeration media like fabric or porous tiles fitted on the bottom of the silo. Approximately 60 to 65 % of area of the silo is aerated.

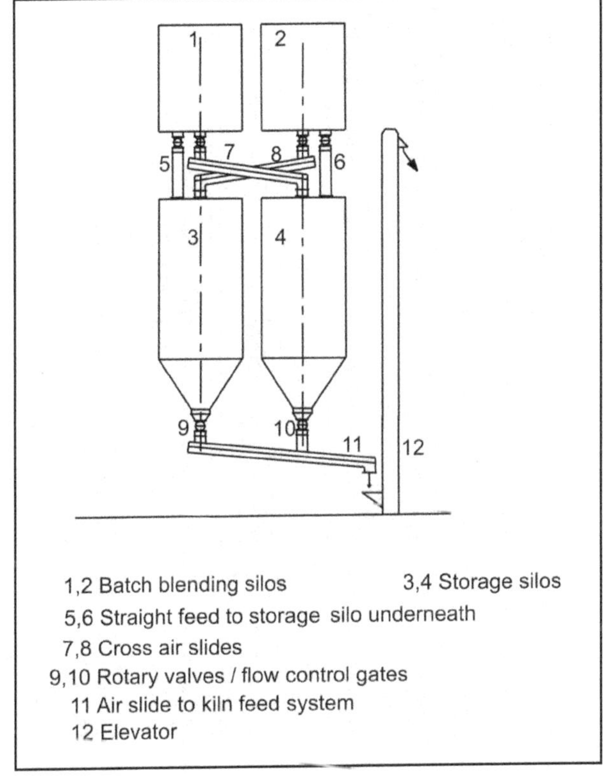

Fig. 6.10.1 Batch blending system - for small plants - double deck blending and storage silos.

10.3 Operation of Batch Blending

When there are two mills, output from both the mills is taken to one blending silo till it is filled and then to the other silo cyclically. For two raw mills, 2 blending silos would suffice if between them they produce sufficient quantity of blended raw meal if both mills are installed together.

A total blended raw meal of 2 ½ days' requirements of kiln is stored between storage and blending silos.

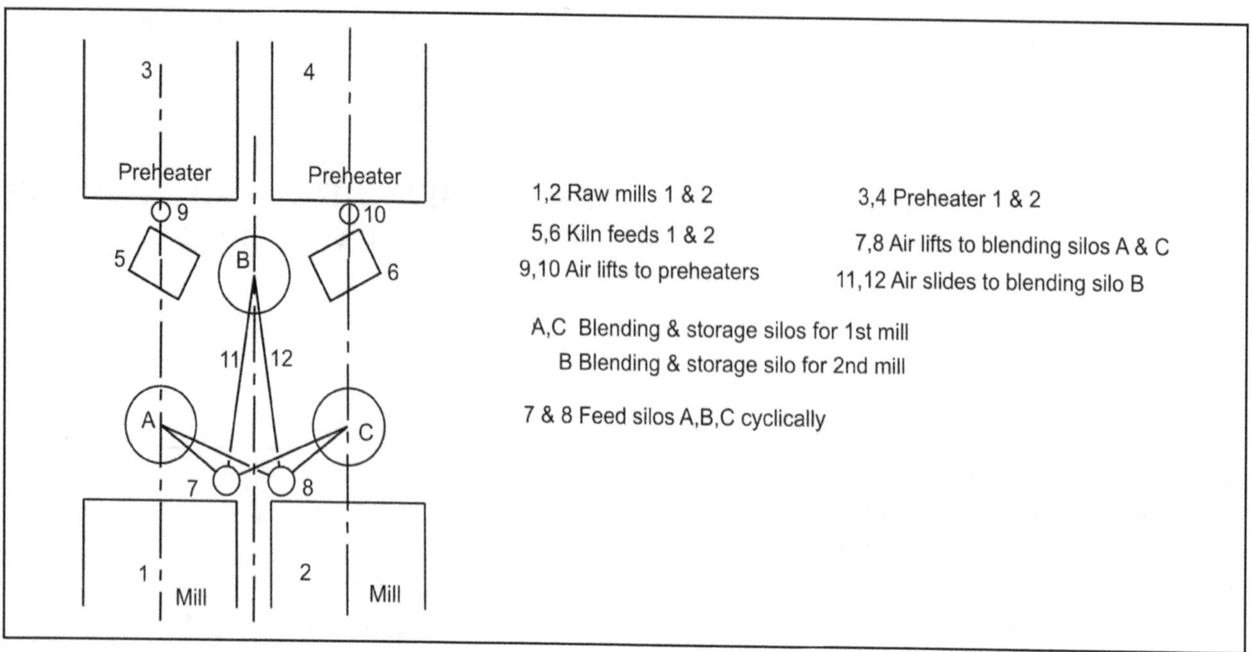

Fig. 6.10.2 Layout – batch blending and storage silos - 2 kilns offers - flexibility and inter changeability.

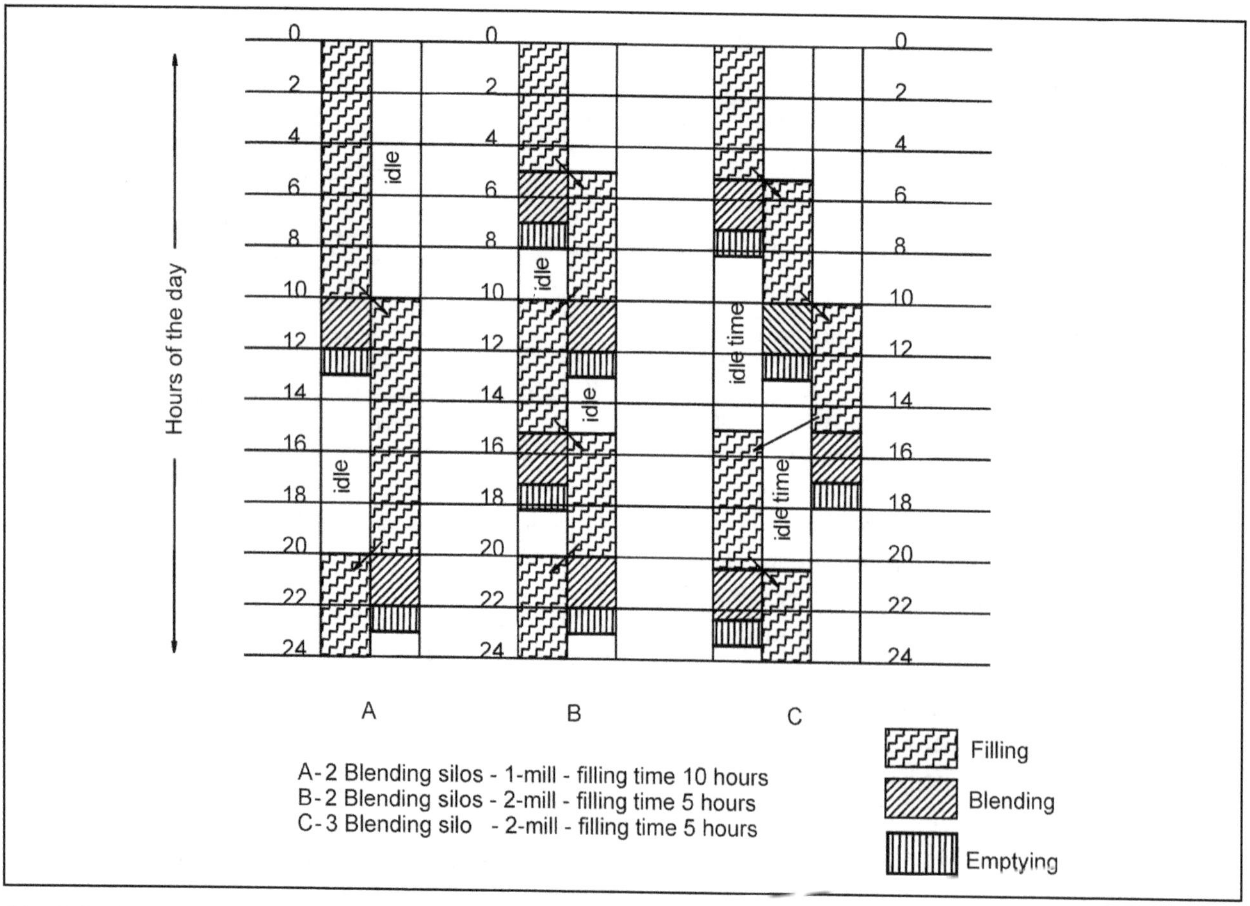

Hours of the day

0 2 4 6 8 10 12 14 16 18 20 22 24

idle

idle time

A B C

A-2 Blending silos - 1-mill - filling time 10 hours
B-2 Blending silos - 2-mill - filling time 5 hours
C-3 Blending silo - 2-mill - filling time 5 hours

Filling
Blending
Emptying

Fig. 6.10.3 Blending in batch blending system.

10.4 Blending Cycles

Blending and emptying and filling cycles for 3 sets of silos is shown in **Fig. 6.10.3.**

Emptying time should be about 1 hour.

When two mills are used to fill one silo, filling time will reduce to half i.e., 5 hours. 4 batches are available corresponding to the output of 2 mills.

10.5 Compressors for Blending Air

Blending and aeration compressors are best located right under the silo to avoid long leads and condensation. Rotary air compressors are best suited to give air supply without fluctuation of pressure but air delivered would be mixed with oil.

Therefore roots blowers are preferred to rotating compressors. Even then the system should include coke filters and moisture traps.

10.5.1 Cycling Valve

A four way cycling valve is installed under the silo to supply cyclically blending air to one quadrant and simultaneously aeration air to the other three quadrants for about 15 minutes and then switches on to the next quadrant. A cycle of 4 quadrants of 15 minutes duration each would take 1 hour. The period of admitting blending air to a quadrant is adjustable. If need the cycle can be repeated. When there are more than 4 sections solenoid operated air valves are used to supply air sequentially same as 4 way cycling valve does for 'quadrant' system.

10.6 Head Room above Storage Silos

There should be sufficient head room between blending and storage silo to install pipe lines and cycling valves and also filters for moisture and oil. Head room is also needed to install air slides between the silos to convey blended raw meal to both the storage silos underneath.
See Fig. 6.10.4

10.7 Storage Silos

Storage silos can be conical bottom for small diameters and flat bottom for large diameters.

10.7.1 Aeration of Storage Silos

Storage silos are also aerated for helping flow. In flat bottom silos, 12-15% of their area is covered with aeration pads / airslides. Bottom can be ridged to reduce aeration area to 8-10 %. Discharge can be from center of bottom or from side.

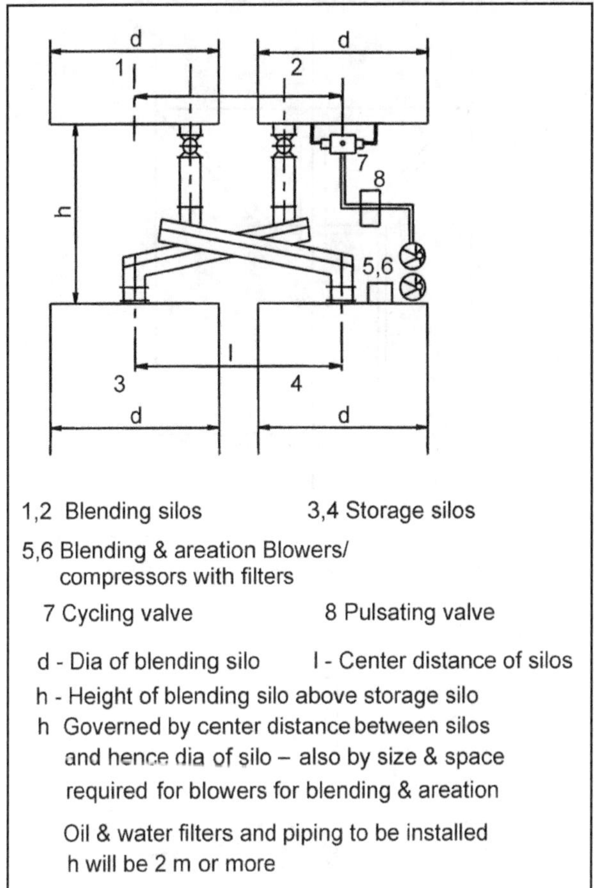

Fig. 6.10.4 Blending and storage silos – height of blending silo above storage.

Storage silo is continuously aerated when material is withdrawn.

See Figs. 6.10.5 to 6.10.9.

Conical bottom silos can be totally emptied without any dead stock. But the cone takes considerable height and head room, minimum angle of cone being 60°. It is also aerated by a few open air slides to help flow.

See Fig. 6.10.5.

10.8 Kiln Feed System

In batch blending systems, kiln feed sections would generally be outside the silos.

See Figs. 6.10.1 and 6.12.8 in Chapter 12.

Silos in double deck batch blending systems are limited to 10/12 metre in diameters.

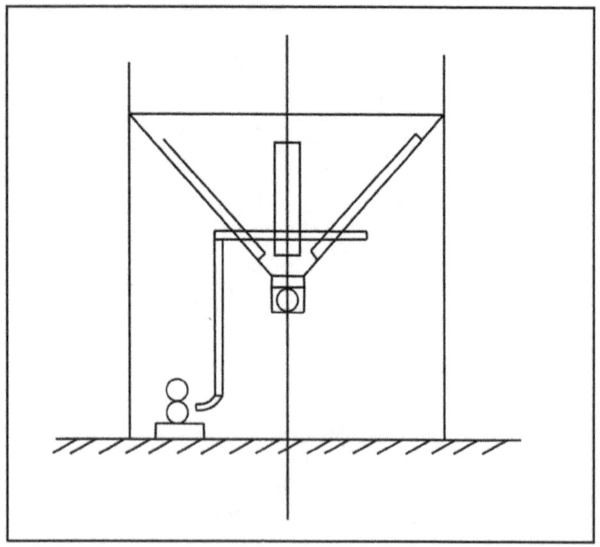

Fig. 6.10.5 Aeration of conical bottom silo.

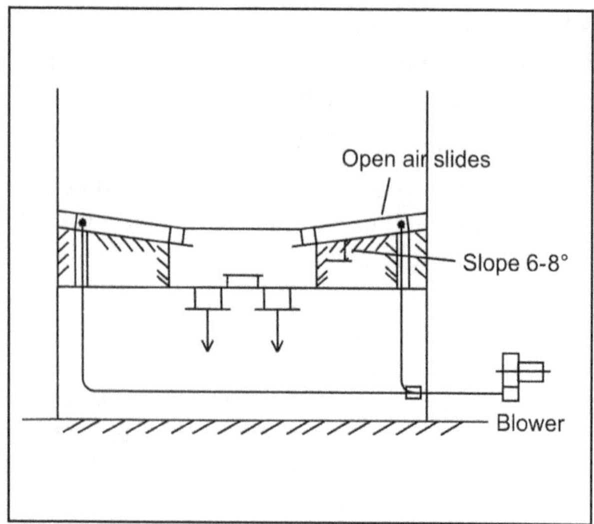

Fig. 6.10.6 Flat bottom – central discharge.

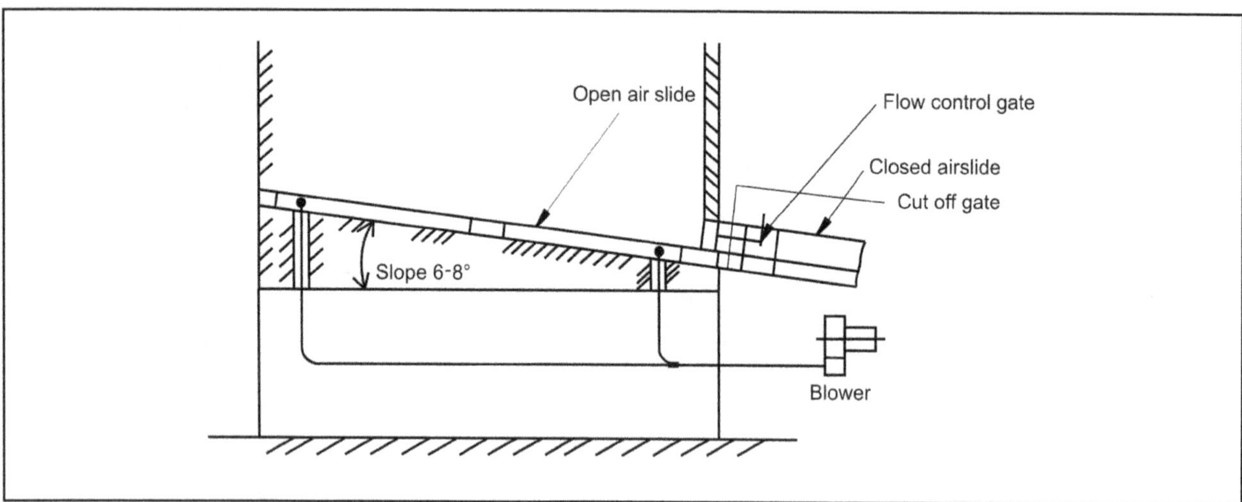

Fig. 6.10.7 Flat bottom – end discharge.

Siols - aerated cones and flat bottoms

10.9 Pressure Release Flaps

Silos are vented and blending silos have pressure releasing flaps at the top.

See Fig. 6.10.10.

10.10 Venting of Blending and Storage Silos

At the end of the pipe line conveying raw meal from air lift to blending silo, there is an expansion chamber which separates raw meal from air. Air is vented into silo and from there to a dust collector. Raw meal is conveyed to centers of silos as needed through air slides with side discharge gates.

See Fig. 6.10.10.

10.11 Dimensioning of Silos

Conventionally in batch blending, height of material filling is limited to the diameter of silo. Height corresponding to about 25 % of the diameter is allowed for the expansion of raw meal during the blending operation. and another 1 meter or so for escaping air.

See Fig. 6.10.11.

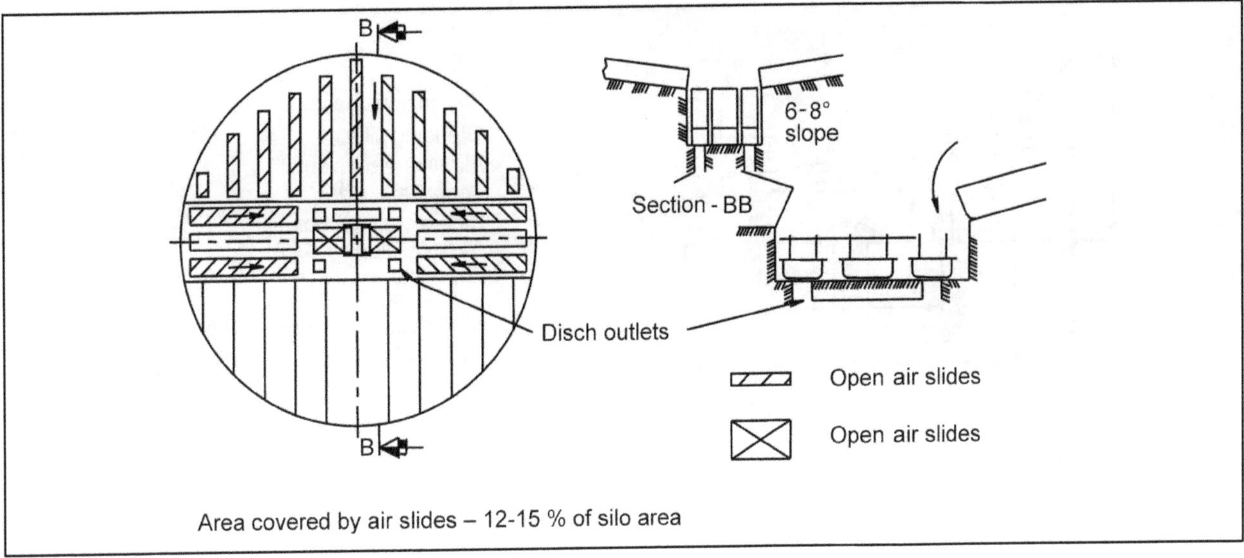

Fig. 6.10.8 Plain flat bottom.

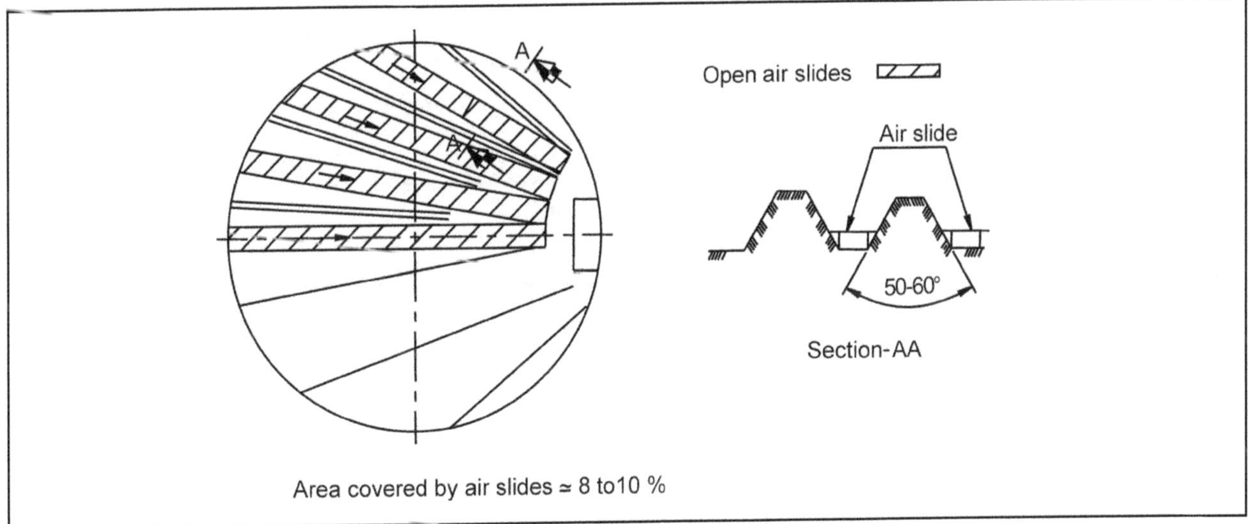

Fig. 6.10.9 Ridged flat bottom.

10.12 Blending Effect Achieved

In batch belnding, blending effect achieved is 8:1 – 10:1. That is if variation in feed is $\pm$ 2% in terms of standard deviation, blended raw meal would have a deviation of $\pm$ 0.25-0.2%.

See Chapter 6 in Section 2.

Better blending is obtained if material is withdrawn simultaneously from 2 storage silos rather than from one only.

10.13

Continuous Blending has replaced batch blending in present times because

1. of 'Preblending' achieved in Stacker Reclaimer Systems which have replaced old crane gantry systems for storing crushed limestone.
2. Blending effect obtainable by continuous blending jointly with preblending is sufficient to to maintain standard deviation of blended raw meal at the desirable level of +/- 0.2 %
3. Size of single kiln has gone beyond 10000 tpd. Hence from costs and operational point of view to have 2 sets of silos as required in batch blending is not a viable proposition.

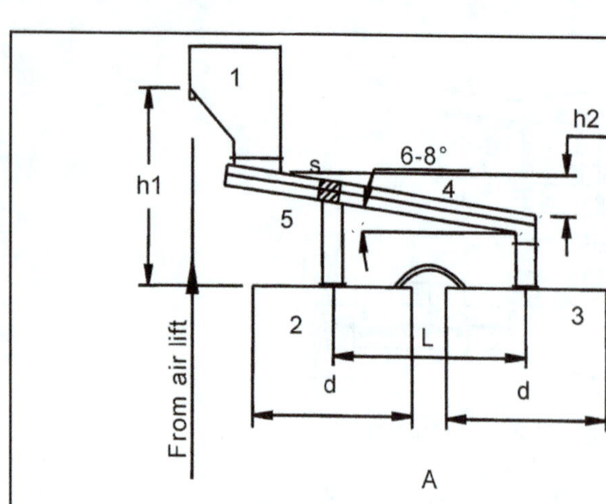

1 Expansion chamber
2,3 Blending silos
4 Air slide = horizontal
Length = L + d/2
5 Side discharge box
Air slide slope = 6 - 8°

Height difference h2 = (l + d/2) tan 6 - 8°
h1 - Depends on size of expansion chamber

6 Dust collector
7 Fan for venting
8 Pressure release flap

B

Fig. 6.10.10 Distribution of raw meal to blending silos and venting.

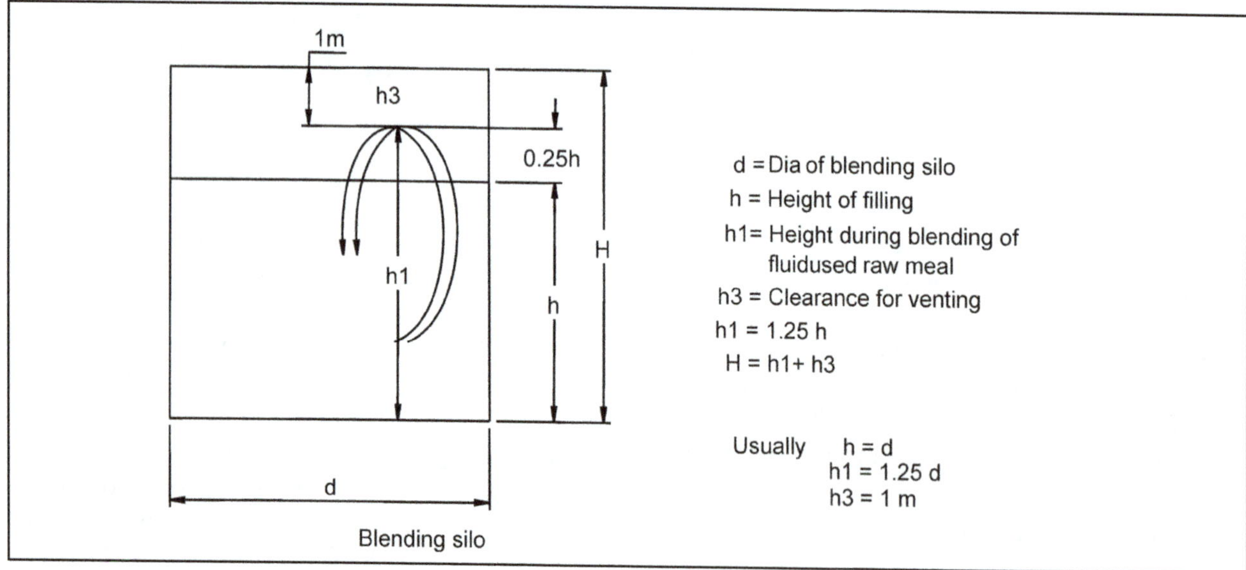

Fig. 6.10.11 Proportions of batch blending silo.

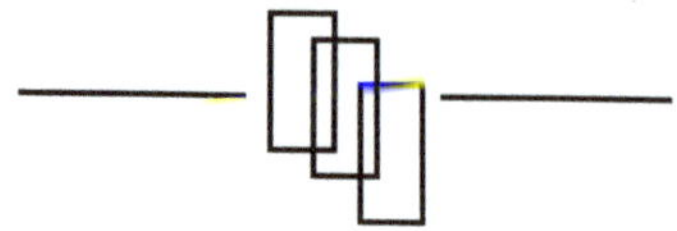

CHAPTER 11

CONTINUOUS BLENDING SYSTEMS

11.1 Switch over from Batch Blending to Continuous Blending

Batch blending system which needs two sets of double deck silos for maintaining continuity of supply of blended raw meal to kiln has been more or less replaced by continuous blending.

In quadrant blending followed in batch blending, it is necessary to have aeration blowers with large pressures as air has to penetrate the toal height of the batch collected in the silo.

Stacker Reclaimer Systems have replaced crane gantry systems of storage of crushed lime stone. They also achieve preblending. Hence blending effect to be achieved in blending silo is reduced to 5-6 /1 which is achievable by continuous blending. This is another reason in favour of continuous blending.

Sizes of plants have also gone up considerably requiring large storage capacity.

For these reasons, continuous blending system was found to be more suitable than batch blending.

11.2 Operation of Continuous Blending System

This system also depends on fluidizing properties of the raw meal to achieve blending. It however operates differently from batch blending system.

As the name suggests raw meal is admitted, blended and extracted continuously, all operations taking place simultaneously. Raw meal delivered by air lift (from raw mill) is spread all over the silo with the help of 4 / 6 radially arranged air slides looking like a spider's legs as shown.

See Fig. 6.11.1.

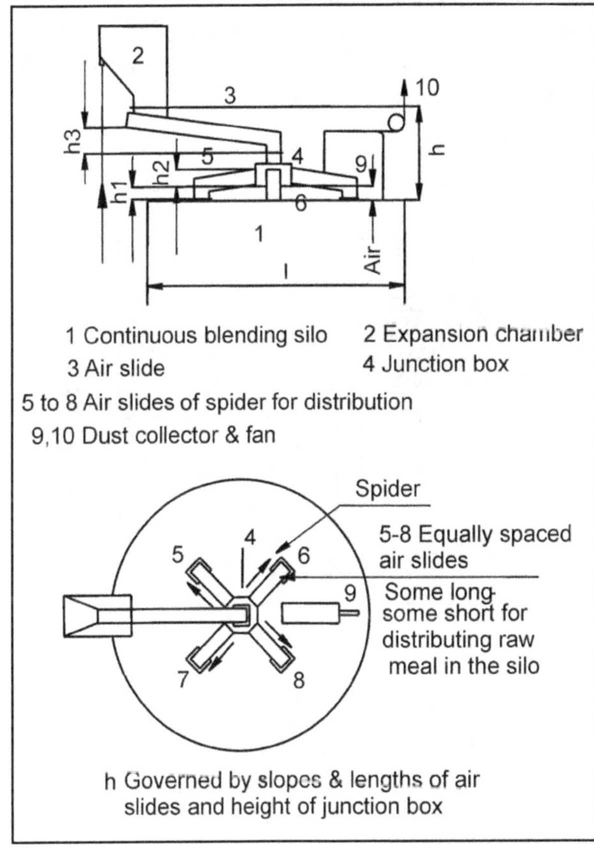

Fig. 6.11.1 Continuous blending feeding arrangement.

Filling rate corresponds to the output of mill and emptying rate corresponds to production rate of kiln. For good efficiency of blending, the silo should always be more than half full. Blending effect achieved is between 5:1 / 6:1.

This coupled with a similar blending effect achieved in stacker reclaimer system earlier results in a total blending effect of 25: 1 / 36 : 1.

An X-ray analyzer is a vital component of the continuous blending system.

11.3 Proportioning of Raw Materials

The proportioning of raw materials is done continuously with the help of X-ray analyzer by monitoring speeds of weigh feeders for limestone and correcting materials feeding the raw mill. This results in raw mix composition being close to the designed value, at all times.

11.4 Construction of Continuous Blending Silo

There is an inverted cone at the center about half the diameter of the silo with 6 or more openings. Annular surface at the bottom is fitted with open air slides to aerate the material in the silo.

See Figs. 6.11.2 and 6.11.3.

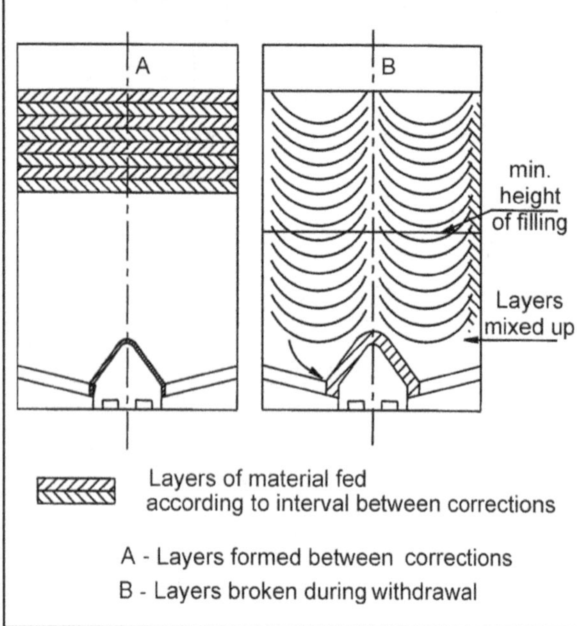

Fig. 6.11.2 Blending action in continuous blending.

The raw meal flows through the openings towards outlet breaking the layers formed during filling as shown. Thus raw meal coming out contains all the layers and is thus blended (like reclaimer cutting across the face of the stock pile).

11.5 Continuous and Quadrant Blending

A further refinement of the system is to use a quadrant blending system in the central portion of the silo. In the floor of the silo a small inverted cylinder is installed concentrically and the bottom is fitted with quadrant blending system described earlier.

See Fig. 6.11.3.

Thus blending takes in two stages :

1. By breaking of layers.
2. By air merge quadrant blending.

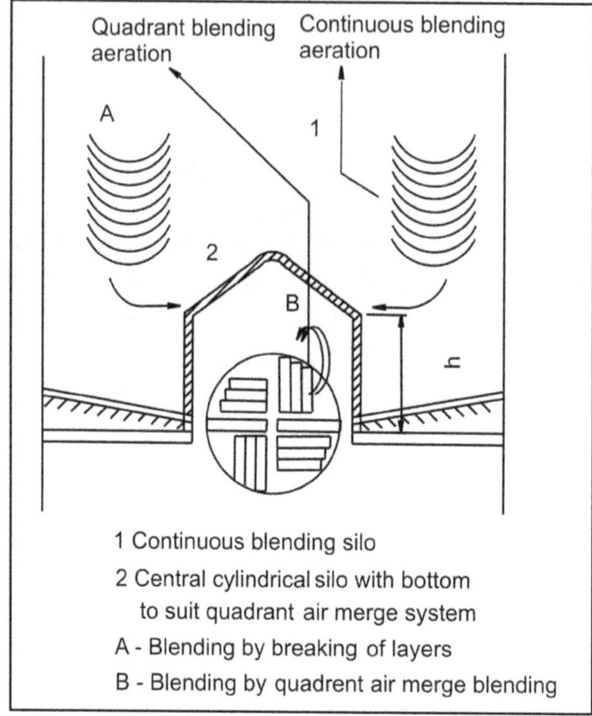

Fig. 6.11.3 Continuous cum quadrant blending.

11.6 Size and Capacity of Continuous Blending Silo

Table 6.11.1 shows workout for sizing a continuous blending silo for a plant of 10000 tpd capacity.

11.7 Aeration of Continuous Blending Silo

Annular space between silo and central cone is aerated by a large number of open air slides that receive air from a number of blowers. Since the area is large, aeration is done cyclically aerating small sectors at a time.

Corresponding air slides in the central cone are aerated at the same time so that flow is maintained right to the discharge points.

See schematic arrangement of aeration blowers for this purpose.

See Fig. 6.11.4.

Table 6.11.1 Sizing Continuous Blending & Storage Silos.

Sr. No.	Item clinkering capacity	Unit ton/day	10000
1	raw meal/clinker	ratio	1.55
	design factor		1.1
	total factor		1.705
2	raw meal per day	tons	17050
3	storage capacity recommended	no of days	2.5
		tons	42625
4	bulk density aerated raw meal	t/m^3	0.96
5	volumetric capacity of blending silo	m^3	44401
6	let h/d ratio be	ratio	4
	volume of silo		$3.14*d^3$
7	volumetric capacity of blending silo	m^3	44401
	let h/d ratio		3
	volume of silo		$2.36*d^3$
	dia of silo	m	24
		say	26
	height		**78**
	allow for air space	m	3
	total height	m	81
	height above ground		5
	total height		**86**

11.8 Discharge of Blended Raw Meal

Continuous blending silo can be discharged from the side or from the bottom.

See Figs. 6.11.5 and 6.11.6.

In the latter case silo will be elevated above ground level and blowers and kiln feed arrangement would be installed under the silo.

See Fig. 6.11.6.

11.9 Venting

Silo is vented through a bag filter mounted on top of the silo. Vent air will include:

1. Conveying air of air lift,
2. air admitted into air slides,
3. air admitted for fluidisation in airslides at the bottom,
4. air admitted for extraction air slides.

11.10 Aeration and Fluidizing Blowers

Unlike in batch blending where blending air is required to break through the entire height of material to reduce its density and start circulation, fluidizing air admitted in continuous blending is required to start the flow and break layers. Therefore pressure of blowers is smaller. Though blowers operate continuously, power consumption is lower than that of batch blending and storage system taken as a whole.

Aeration can be sectionalised-providing aeration cyclically to sections in the silo bottom.

It can be further reduced if an elevator is used to lift raw meal from raw mill department into the silo. If an elevator is used, a standby should be provided. The standby could be an air lift.

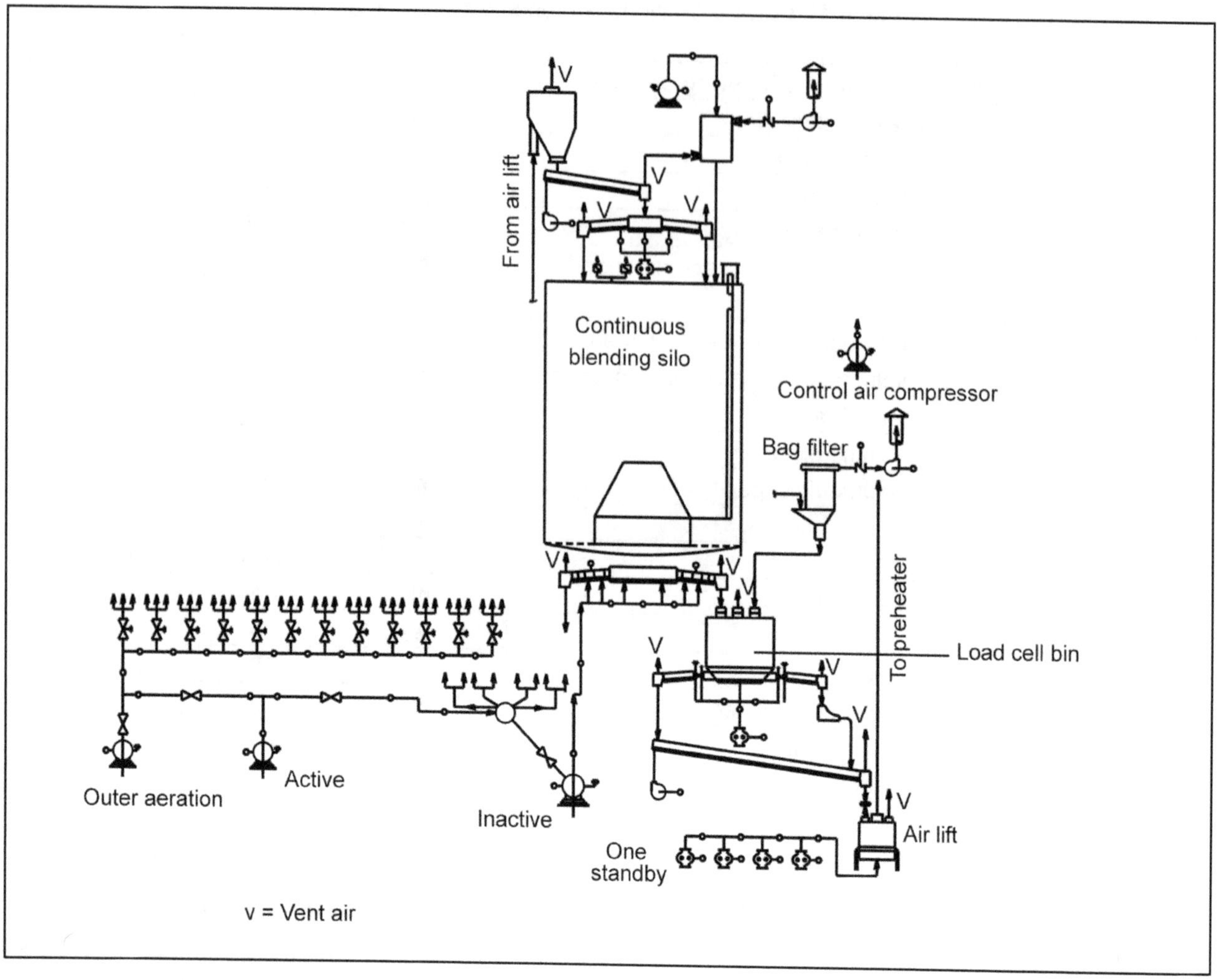

V
From air lift
Continuous
blending silo
Control air compressor
Bag filter
To preheater
Load cell bin
Outer aeration
Active
Inactive
One
standby
Air lift
v = Vent air

Fig. 6.11.4 Continuous blending and kiln feed systems aeration of bottom of silo and for extraction.

1
2
3
4 5
6
+0.00
1 Blending silo
2 Areation air slide
3 Discharge air slide
4 Cut off gate
5 Flow control gate
6 Elevator

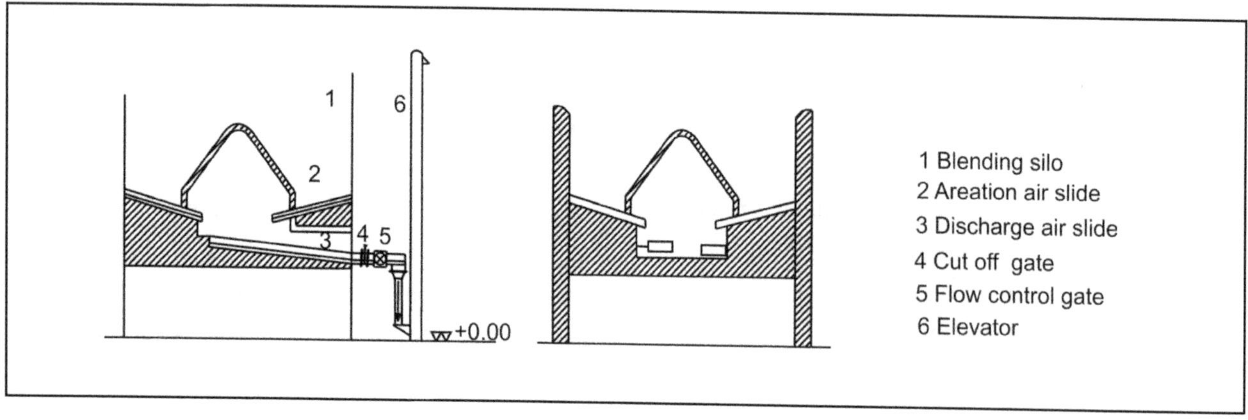

Fig. 6.11.5 Extraction from blending silo and kiln feed – extraction from side – silo at ground level.

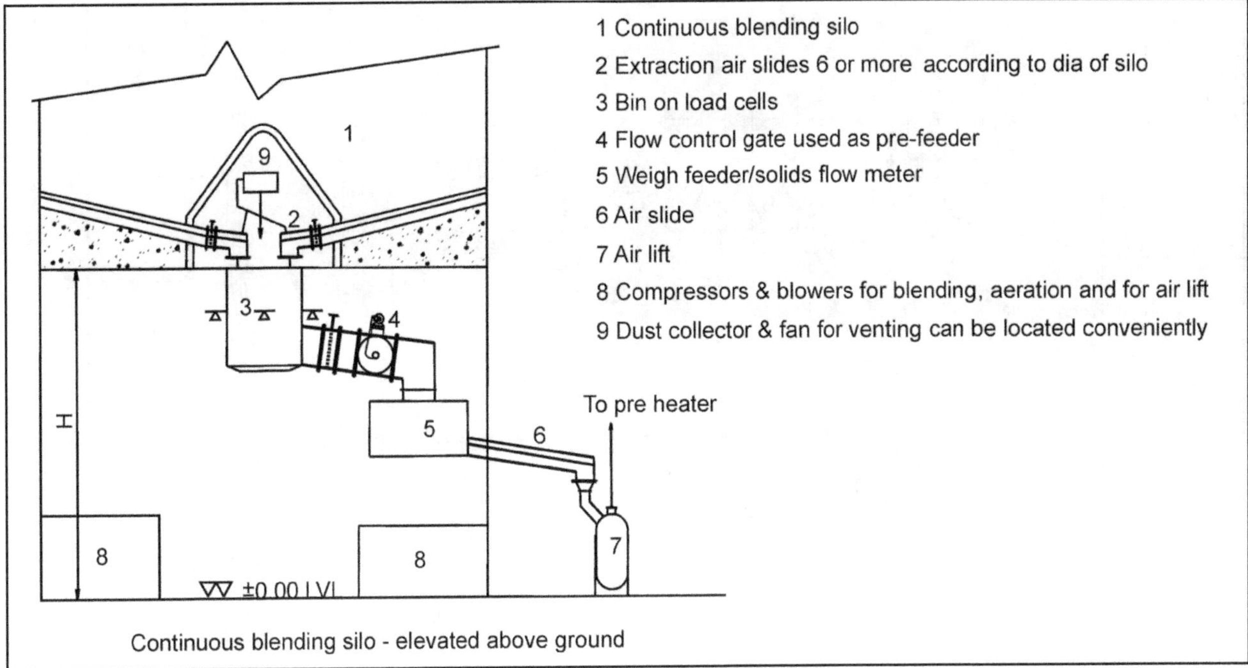

Fig. 6.11.6 Extraction and kiln feed under the continuous blending silo.

11.10.1 Location of Blowers

Blowers for aeration and fluidisation of the blending silo and also blowers for air lift for kiln feed would all be located under the silo. A 16 meter diameter silo would have an area of 200 m^2 under it. Therefore the blowers can be easily accommodated under the silo.

Recommendations for blowers would be furnished by the Suppliers of the Blending equipment.

11.11 Location of Kiln Feed System

Even entire kiln feed system can be installed under the silo as shown in **Fig. 6.11.6** and **Fig 6.11.8** which shows layout of two production lines with continuous blending system, also as described in the next Chapter.

11.12

Layout of a continuous blending system and kiln feed under the silo is shown in **Fig. 6.11.7.**

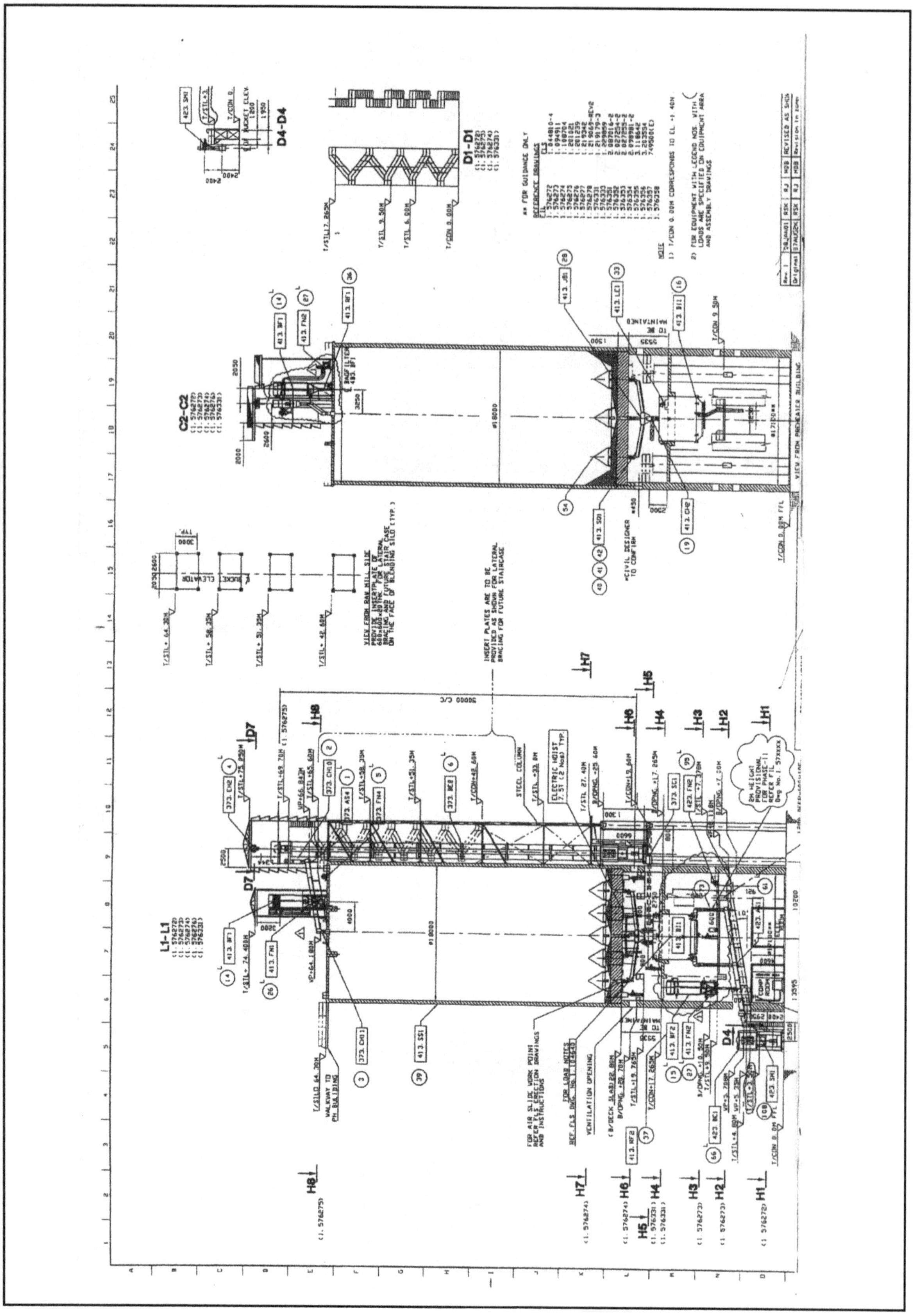

Fig. 6.11.7 Continuous blending system with kiln feed under the silo.

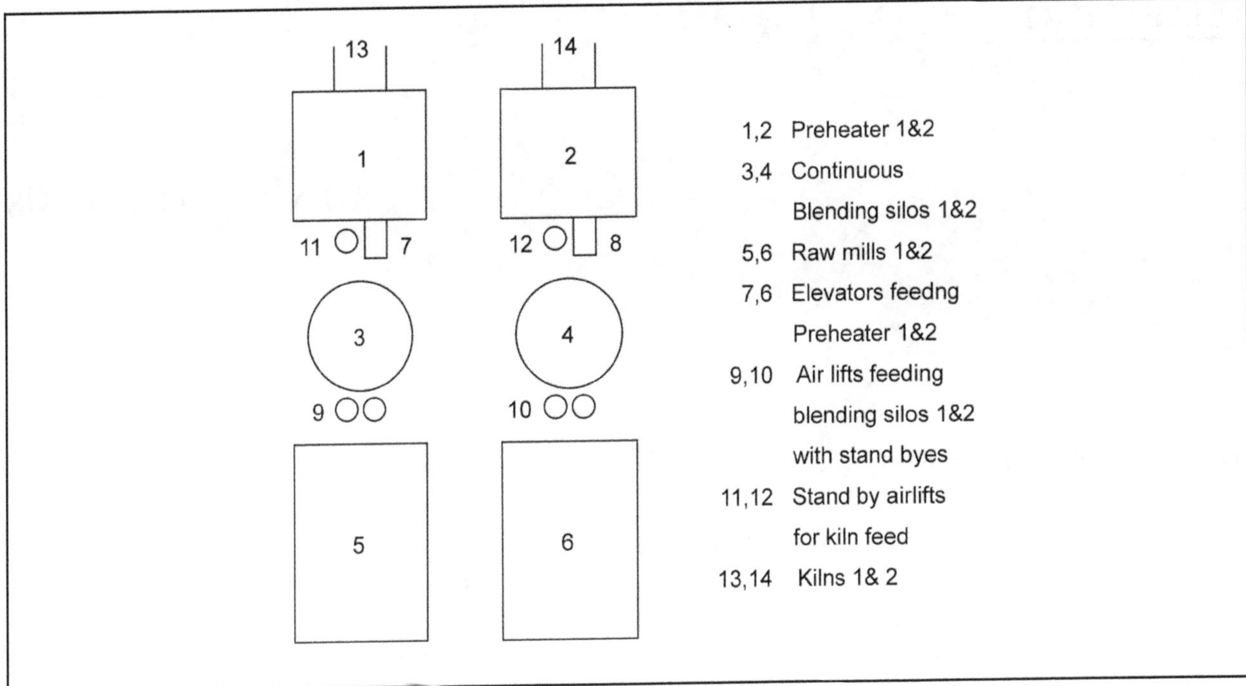

Fig. 6.11.8 Layout of two production lines with continuous blending kiln feeders under blending silos.

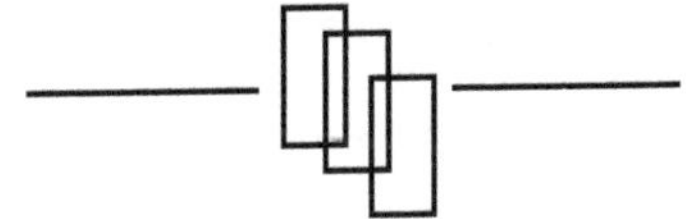

CHAPTER 12

KILN FEED SYSTEMS

12.1 Raw Mix Feed to Kiln

For its operation kiln requires a steady feed of uniform quality, the rate capable of being adjusted to suit production levels.

Kiln feed systems have been evolved from simple volumetric systems to gravimetric systems of a high degree of accuracy.

Blending silos achieve uniform quality in terms of chemical composition and fineness in term of sieve analysis.

Kiln feed systems are designed to achieve uniform steady rate of feed with a close degree of accuracy. The rate is adjustable in accordance with production / operational conditions of the kiln.

12.2 Requisites of Kiln Feed System

The feed system is thus required :

1. To measure – volumetrically or gravimetrically raw mix being fed and to maintain it constant at a given setting.
2. To increase / decrease rate of feed as demanded by operational conditions, the accuracy being maintained at the new setting and throughout the full range.

Degree of accuracy varies with types of feeders used. It would be:

$\pm$ 1% for weigh feeders.

$\pm$ 2% for solids flow meters.

$\pm$ 5% for volumetric feeders like screws.

12.3 Components of a Kiln Feed System

In principle, a kiln feed system consists of :

1. A 'prefeeder' with a variable speed drive, which feeds the feeders. Prefeeder can be either a rotary vane feeder with close tolerances or a flow control gate.

   See Figs. 6.12.1 and 6.12.2

2. Feeder which meters the material passing through / over it and gives signals to the prefeeder to increase or decrease its speed to maintain the rate of flow at the desired setting within $\pm$ 1 or $\pm$ 2 % accuracy according to its design.

   Feeder can be a volumetric feeder (screw) or a gravimetric feeder (weigh feeder) or one which measures flow by impact (solids flow meter).

   See Figs. 6.12.1 to 6.12.3.

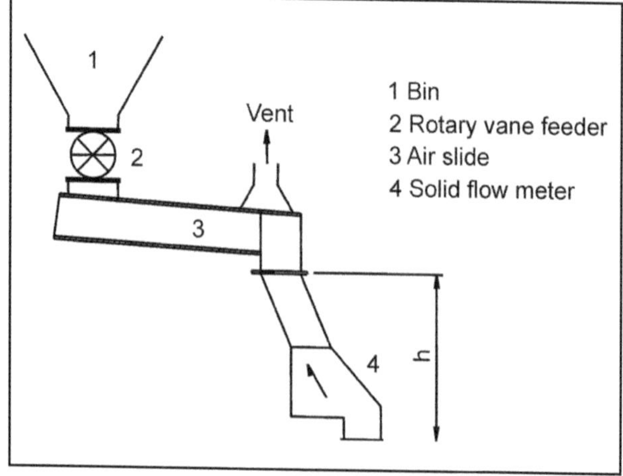

Fig. 6.12.1 Solids flow meter as feeder ; rotary vane feeder as pre feeder.

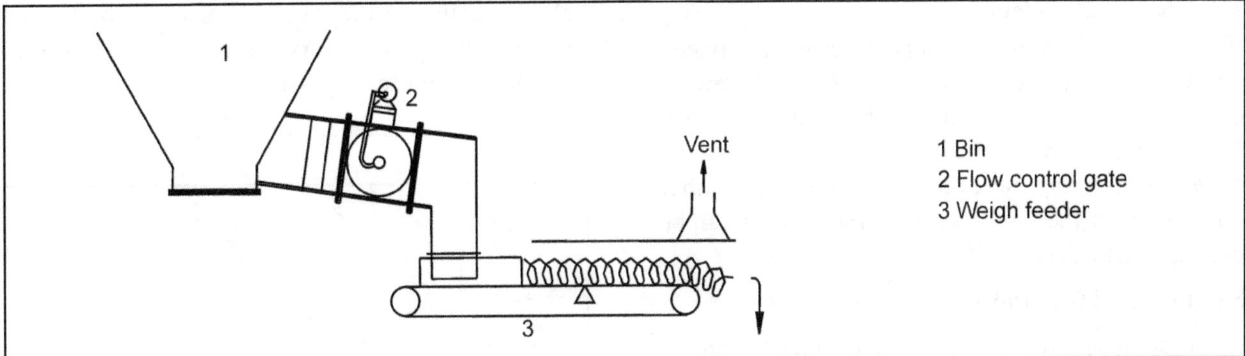

Fig. 6.12.2 Gravimetric feed system – weigh feeder as feeder, control gate as prefeeder.

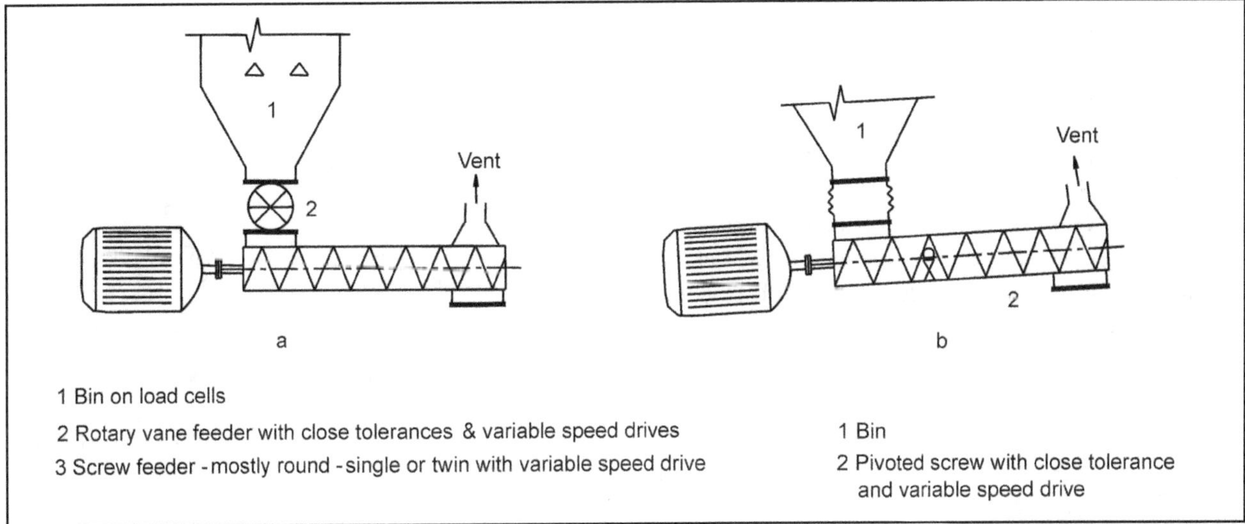

Fig. 6.12.3 Volumetric kiln feed system – commonly used for firing coal in kiln and calciner.

12.3.1 Variation of Rate of Feed

For changing rate of feed, the feeder also must have a variable speed drive. The two are electrically coupled so that the change of setting of feeder also changes settings of prefeeder.

12.3.2 Volumetric Feeding

For volumetric feed, which is not common for large plants, the feeder with variable speed drive would be installed directly under the feeding bin. To maintain consistency of feed, the level of material in the bin is maintained at a constant height.

See Fig. 6.12.3.

12.3.3 Gravimetric Feeding

In case of gravimetric feeders, Weigh feeders are the most common; they actually weigh through a load cell, the weight of material passing over the belt. Wide slow moving belts are preferred to narrow fast moving belts.

Length should be long enough for deaeration and for forming a layer of uniform thickness so that load cell can be accurate and effective.

See Fig. 6.12.2.

12.3.4 Indirect Metering

In case of solids flow meters, impact of material falling over a fixed height on a curved plate is used to determine the rate of flow. The accuracy of solid flow meter will be around $\pm$ 2% as compared to $\pm$ 1% of weigh feeder. It can be affected by agglomeration of particles due to moisture.

See Fig. 6.12.1.

12.4 Screw Feeders

In case of volumetric feeders, tolerance between screw and casing needs to be very close. Screws – mostly twin screws – run full and have variable speed drives. Screws have decreasing pitch to compress and deaerate material. Screws can be installed either directly under the bin or a rotary vane feeder can be used as prefeeder.

See Fig. 6.12.3 a and b.

Sometimes screws are pivoted to gauge the rate of feed and there by give signal to increase or decrease its speed. However when there is no prefeeder, accuracy of screw feeder cannot be certain.

See Fig. 6.12.3 b.

12.5 Flow Control Gate

This is actually a volumetric feeder where rate of flow is controlled by a rotor with a notch as shown.

The flow control gate is electrically coupled to another flow control rate or a slide gate preceding it. It is also often used as a prefeeder to weigh feeders. **See Figs. 6.12.4 to 6.12.6.**

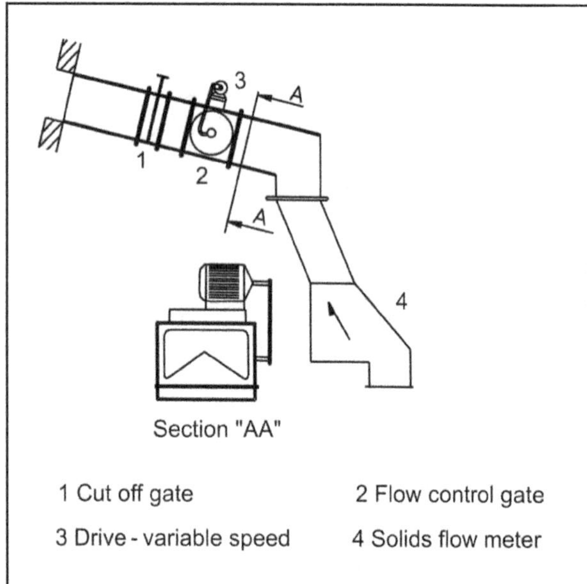

Fig. 6.12.4 Flow control gate for regulating feed - used as prefeeder, solids flow meter as feeder.

12.6 Bin on Load Cell and Calibration

The bin receiving material from silo is invariably on load cells and can be used to calibrate the feeders. It helps to calibrate the weigh feeder. Calibration at regular intervals is very necessary to maintain accuracy of kiln feed systems.

See Figs. 6.12.3, 6.12.5 and 6.12.6.

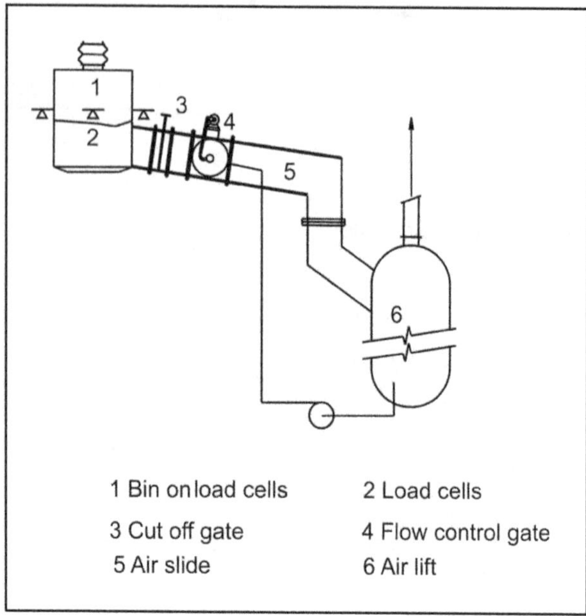

Fig. 6.12.5 Flow control gate alone used for kiln feed; feed regulated by air pressure of air lift.

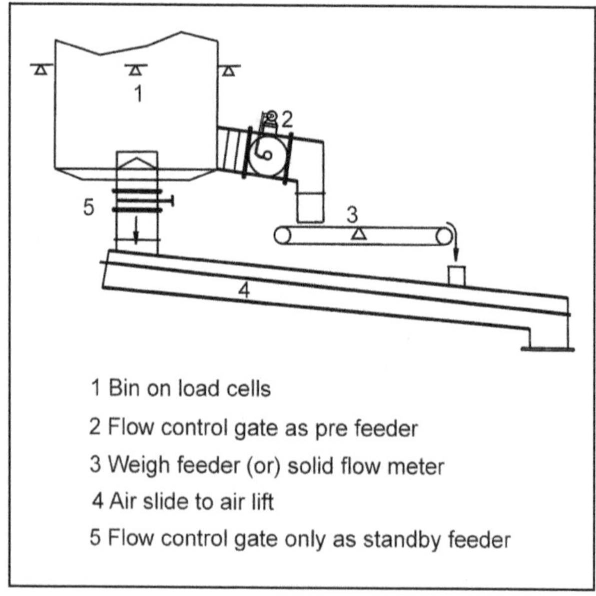

Fig. 6.12.6 Flow control gate as pre feeder and standby feeder for kiln feed.

In one kiln feed system, back pressure of air lift is used to regulate flow from the bin through flow control gate.

See Fig. 6.12.5.

12.7 Standby or Bypass

A 'bypass' or alternative arrangement is also provided to maintain continuity of kiln feed should regular line go out of order.

See Fig. 6.12.6.

12.8 Bucket Elevators in Kiln Feed System

Now belt bucket elevators are universally used to feed even 6 stage prheaters with heights going up to 80-90 metres or more.

A 'standby' for kiln feed elevator is always desirable. However a standby bucket elevator would be very expensive. Many times therefore an airlift would be used for this purpose. This is particularly so when an existing airlift is replaced by an elevator. The layout must therefore be suitable to feed bucket elevator and also standby elevator or airlift.

See Fig. 6.12.7.

12.9 Venting

Venting of various auxiliaries, bins, air slides, elevators, feeders is very important for maintaining steady feed rate.

Dust collector is housed in the kiln feed building whether separate or under the silo.

12.10 Housing Kiln Feed System

Earlier kiln feed systems were housed in a separate building outside the storage silo.

See Fig. 6.12.8.

With continuous blending systems where silos are large in diameter, it was possible to install the entire kiln feed system under the silo by raising it and thus avoid the necessity of a separate building and also eliminate need to have equipment like bucket elevators to feed bins for weigh feeders.

See Fig. 6.11.6 in Chapter 11.

Arrangement of kiln feed system under a continuous blending silo has also been shown in **Fig 6.11.7** in **Chapter 11.**

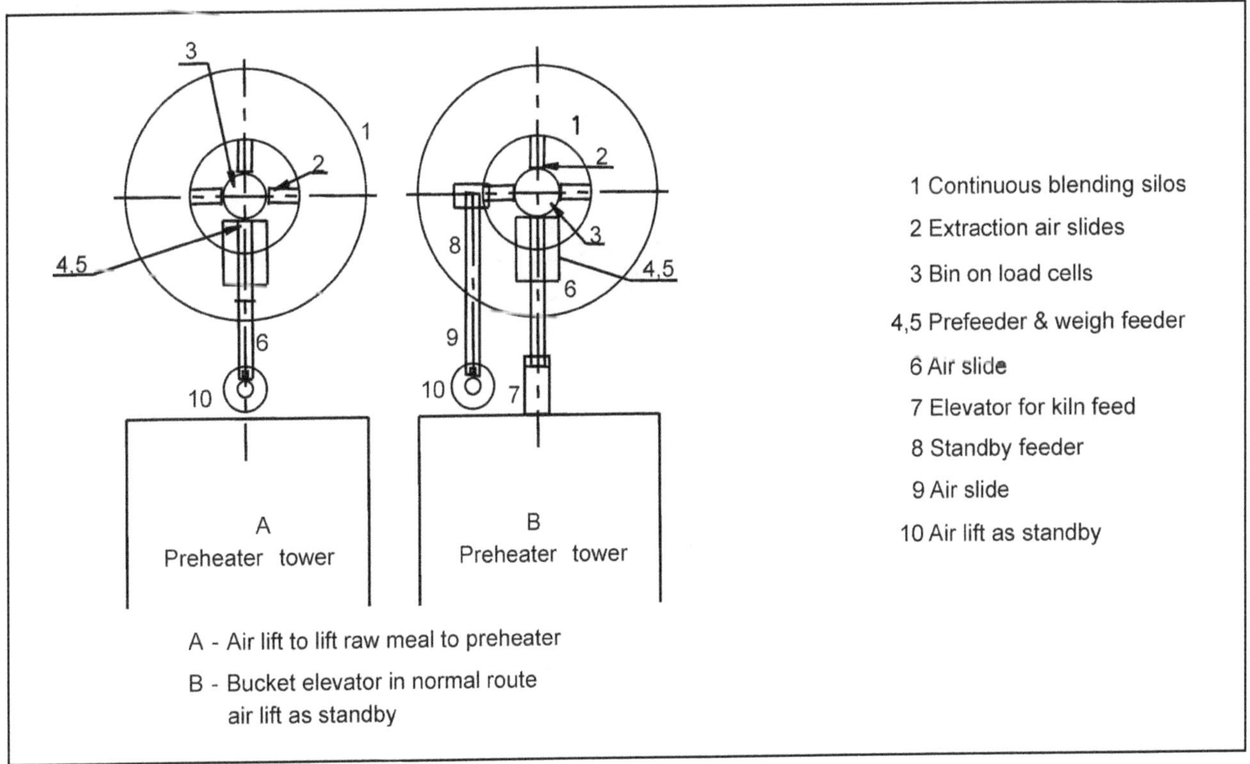

Fig. 6.12.7 Kiln feed systems with air lift and bucket elevator.

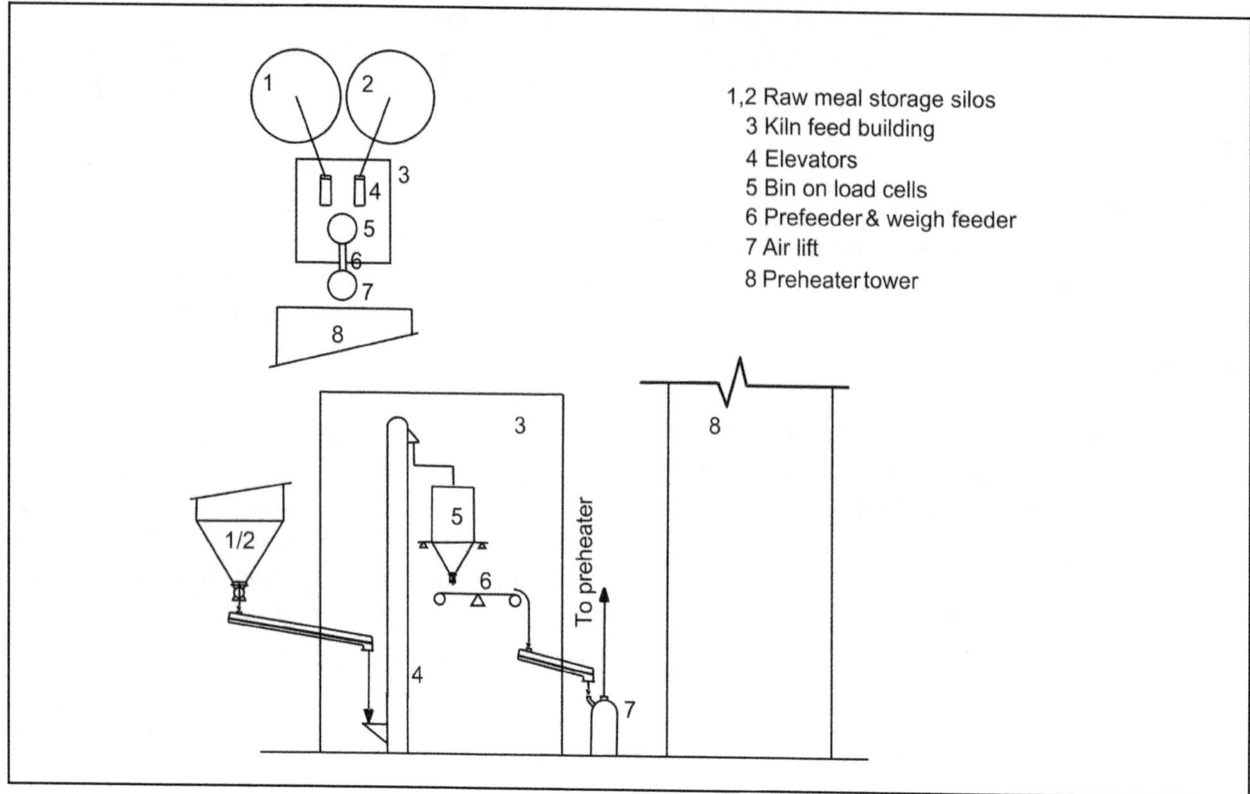

Fig. 6.12.8 Overall kiln feed system for batch blending and storage silo system.

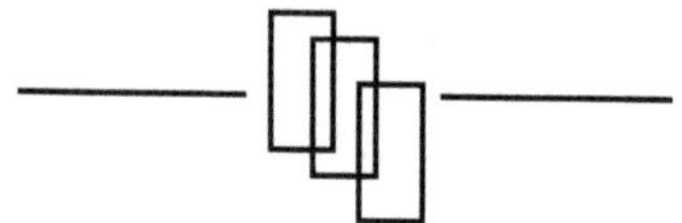

CHAPTER 13

PYROPROCESSING SECTION

13.1 Heart of a Cement Plant

Pyroprocessing section is the heart or core section of a cement plant and includes Preheater, Calciner, Rotary Kiln and Clinker Cooler as the major machinery units in it.

Among the major auxiliaries would include coal firing systems for kiln and calciner; fans; clinker conveying systems and dust collectors.

This section has seen major changes which have brought about sea changes in scales of production and in operational efficiencies.

Kilns have shrunk in size but individual capacities of plus 10000 tpd have been reached.

13.2 Operational Efficiencies

Specific fuel consumption has come down to less than 700 kcal/kg as against 1500 kcal/kg earlier for wet process kilns.

Specific power consumption had initially increased after the introduction of dry process and grate cooler. But due to substantial reduction in volumes of gases to be handled in the section and because of development of low pressure cyclones of the preheater, and because of significant changes in clinker coolers it has come down again.

After introduction of calciners, capacities of kiln have increased two and a half times and more for the same size.

13.3 Physical Changes and Appearances

These developments have brought about changes in the layouts of the pyroprocessing section as a whole.

Conspicuous are very tall preheaters and adjoining calciners with their tertiary air ducts which look like smaller kilns in parallel with the main kilns.

Main kilns are short and very small for their outputs. Clinker coolers have seen great developments. Original grate coolers themselves have seen several changes in design and layout; new designs like static grate, controlled flow grate, pendulum and cross bar coolers have almost replaced conventional reciprocating grate.

Large dust collectors at kiln and cooler ends let out clean exhaust gases. Dust is conspicuous by its absence. It is hard to know if a plant is running or not by watching its chimneys

13.4 Kilns and Grate Coolers

Construction wise kiln itself has not changed much except by way of sophistication of design of seals and use of better materials for tyres and rollers and gears and for heat resisting castings.

Two support kilns are practical because of shortened length.

Supporting rollers of self aligning type, splined kiln tyres and drive thru a roller station are some of the major changes in new kilns.

Burners have also been developed to meet requirements of firing multiple fuels and alternate fuels. Special deigns have been developed like hot disc which can burn shredded alternate fuels in shredded form. Even whole tyres can be introduced from kiln outlet end

Grate coolers perhaps have seen maximum changes. They include a short static grate at inlet, grates themselves seen major changes in construction and distribution of air supply. Moving floor type and cross bars which push clinker rather than reciprocating grate are some of the major changes.

Efficiency of cooling is now + 70 % and cooling air has come down to 2-2.2 nm^3 /kgclinker.

Vent gases from cooler are circulated back to cooler in **duotherm** systems or used for co-generation of power.

In 6 stage preheaters of low pressure drop design, pressure drop across the preheater is same or less that than in 4 stage preheaters of old design. Down draft cyclones have been developed for the top stage.

Calciners have also seen major changes that have enabled them to receive and burn alternate fuels of various types in many forms and to the extent of 20 to 30% of total fuel. Low NO_x designs have been developed to be able to fire AF.

Refractories used in kiln have seen similar developments. Basic and Magnesite bricks have replaced high alumina bricks in burning zones. Insulating bricks are now commonly used to reduce radiation losses. Castables are also widely used. Sp. consumption of refractories in kgs/ton of clinker produced has reduced considerably.

13.5

Against this background layouts of pyroprocessing section would be elaborated and detailed engineering would be explained as in previous Chapters.

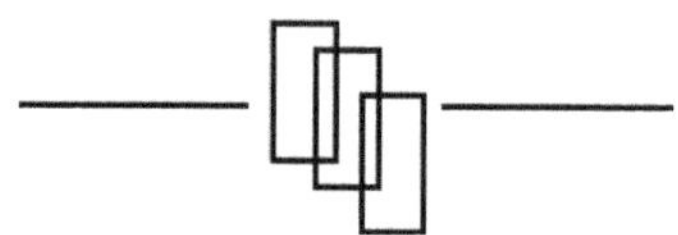

CHAPTER 14

PREHEATERS LEPOL GRATE

14.1 Preheaters in Wet Kilns

Wet kilns had 'calciners' outside the kiln as preheaters. Their function was to reduce moisture in the slurry and achieve lesser fuel consumption.

Wet kilns also had a chain system for almost one third of their lengths from feed end to reduce moisture and dry out the slurry and preheat it to temperatures where calcination could begin. Thus 'chain systems' were the first kind of preheaters which were part of kilns themselves.

Chain systems were designed for efficient heat transfer. They consisted of combinations of garland and curtain chains. 'Density' of the chain system indicating weight of chains per metre length of kiln was a guide to their design and effectiveness. Since wet kilns were predominant for several decades, lot of effort was put in designing chain systems. However it is now history.

14.2 Preheaters in Long Dry Kilns

Initially a long rotary kiln was used to make cement by dry process. For better heat exchange, the kiln was fitted with stainless steel chains / lifters inside the kiln as in case of wet kilns. Only here their function was purely preheating.

The length / diameter ratio of long dry kiln used to be 25-30:1. Specific fuel consumption was 1000-1200 Kcal/Kg. Temperature of gases leaving kiln was around 600-800 °C. Long Kilns had 4-6 supporting stations.

In all respects it was like a wet process kiln except that it did not have to evaporate water and length was shorter to that extent.

14.3 External Preheaters

14.3.1 *Travelling Grate Preheater*

Traveling grate preheaters external to the kiln were developed when cement industry switched over to dry grinding and dry blending operations. They are better known as **'Lepol Grates'**.

The process came to be known as **'semi dry'** process because dry raw meal was converted into granules by adding 8 to 10 % water to it before feeding it to preheater. The Lepol preheater had to drive out this moisture and also raise temperature of granules to 800 °C before feeding them to kiln.

14.3.2 Cyclone Suspension Preheater

Subsequent development was the 'cyclone suspension preheaters' which treated dry pulverised raw mix fed to them and preheated it to 800 to 900 oc in a series of cyclones arranged one above the other.. Thus a full fledged dry process using short kilns came to be developed.

Cyclone type suspension preheater with 6 stages is the most dominant type of preheater used today.

14.4 Lepol Grate Preheater

Fig. 6.14.1 shows arrangement of semi dry process including a granulator, lepol grate preheater and rotary kiln. It also shows same Lepol grate used as cooler for cooling clinker.

Kiln was required to do calcining and sintering only, l / d ratio of kiln came down to 15-18:1.

Hot gases from kiln passed through a layer of granules on a traveling grate fitted with slotted plates. To recover more heat, hot gases were passed again

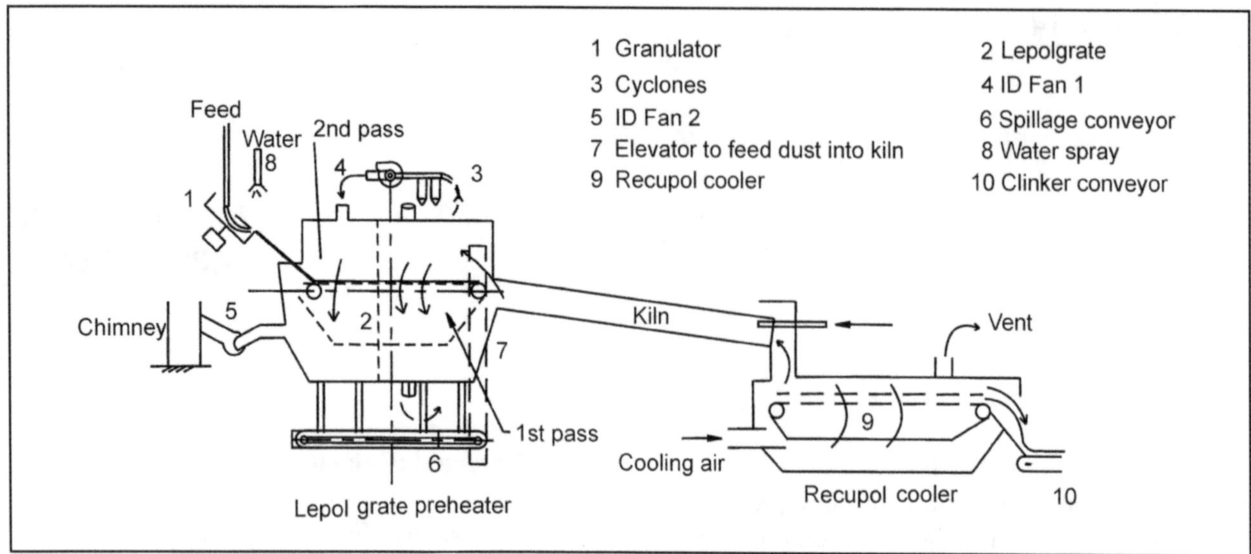

Fig. 6.14.1 Double pass Lepol grate and recupol clinker cooler.

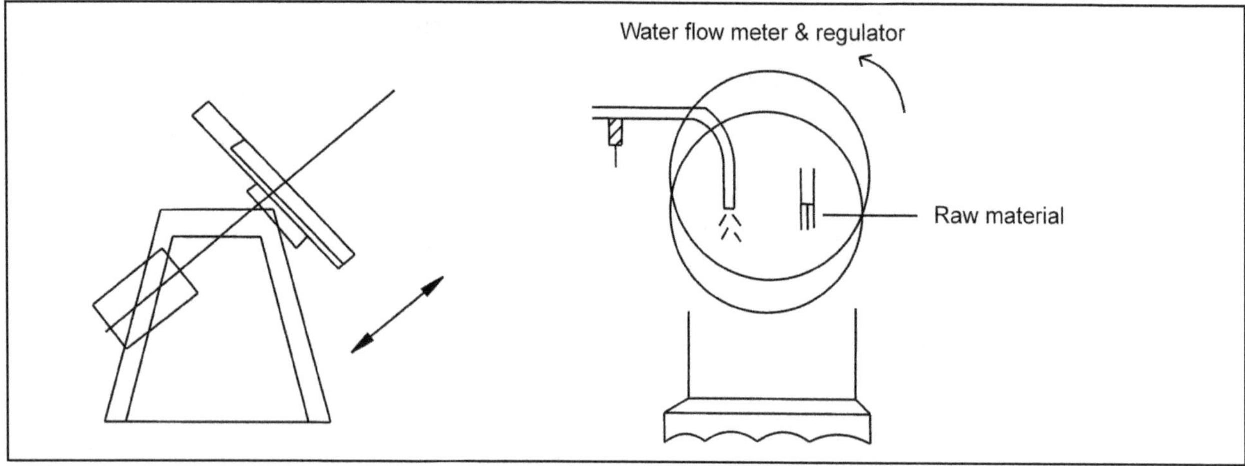

Fig. 6.14.2 Granulator - inclination and speed can be changed.

and again through the grate and (single pass / double pass) emerged cooled down to temperatures of 180-200 °C.

Another advantage of this grate was that the dust content of gases emerging was very low and in 1950s did not require a dust collector after the kiln.

14.4.1 Noduliser

Noduliser used to form nodules of raw mix feed was an inclined rotating pan, which received raw mix as well as 8-10% water spray.

See Fig. 6.14.2.

This nodulizing process required that the correcting materials had properties, which helped forming of nodules, and remaining in nodular form after water was evaporated.

14.4.2 Kiln feed System for Lepol Grate

Kiln feed system required a slightly different arrangement. Raw mix extracted from the storage silo was metered and fed to the noduliser. Speed of noduliser was adjusted to suit rate of feed and to maintain size of granules made. Quantity of water was similarly metered and regulated. Granulator thus had a variable speed drive. Its angle of inclination could also be changed.

Speed of lepol grate was changed to maintain desired thickness of the bed of granules on the traveling grate.

14.5 Fans in Lepol System

Lepol grate needed two fans to handle kiln gases. In double pass grate, Lepol grate was divided into two sections. Hot gases from kiln were passed over granules into first section; taken out from it with the help of a fan. A set of cyclones cleaned gases of their dust content. Cyclones were arranged close to the lepol grate along its length. Fan delivered the gases back to the lepol grate for a second pass for further exchange of heat. Gases were drawn out of lepol grate by the second or i.d. fan and let out to a chimney direct as it used to be the case earlier; now they would be passed through a dust collector.

14.6 Lepol Preheater Structure

Because lepol grate is a horizontal traveling grate and it has to feed the kiln it is installed at a suitable height requiring a tall supporting structure. It needed hoppers at bottom to collect dust falling through the grate plates.

A redler type chain conveyor collected the hot dust and an elevator returned it to the kiln. Elevator had to be specially designed to handle very hot dust with temperatures between 600 to 800 °C.

See Fig. 6.14.1.

14.7 Problems with Lepol Grate

The grate and its chain which had to withstand high temperature + 900 °C, required frequent maintenance and in such a long grate maintaining sealing was a constant problem. These disadvantages went against Lepol grate particularly when Cyclone preheater which had no moving components was developed.

14.8 Operational Efficiency

Fuel consumption came down from 1300–1500 KCal/Kg in wet process to 1000 KCal/Kg.

Sp. power consumption on account of fans to draw gases through kiln and grate was lower than in case of cyclone preheater.

Its main advantage was very low dust loss in exhaust gases due to granular feed. In fifties and also till eighties lepol grates were without any dust collectors.

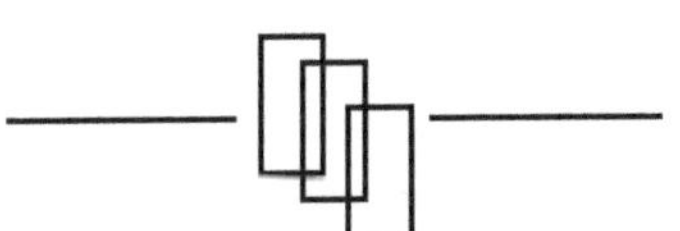

CHAPTER 15

CYCLONE PREHEATERS

15.1 Cyclone Preheaters

Cyclone preheaters, came to be developed by almost all cement machinery manufacturers though pioneers were Humboldts of W. Germany. Polysius also of W. Germany and FLSmidth of Denmark were other prominent designers with KHD and Polysius leading the field till 70s or so.

Their respective licensees – for example Fuller in USA and Mitubishi in Japan – took the development further.

There has been considerable development in the design of cyclone preheaters.

4 stage preheaters are now replaced by 5-6 stage preheaters mainly on account of development of low pressure drop cyclones.

Fuel consumption has come down to 700 KCal/Kg and less. Pressure drop in 6 stages is now less than that in 4 stage preheaters of vintage design.

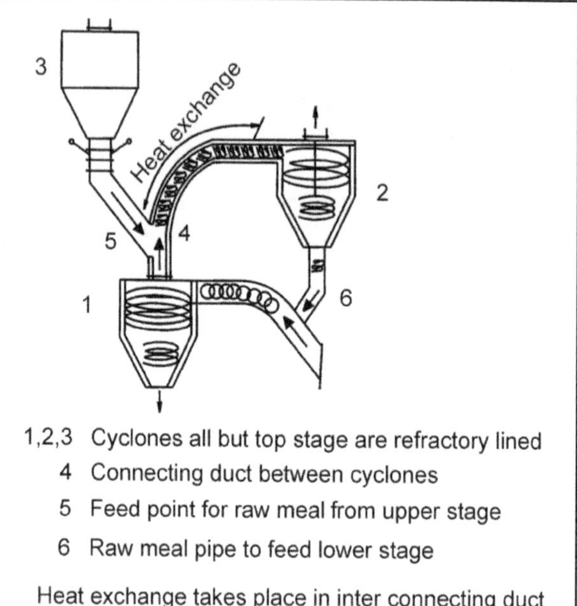

Fig. 6.15.1 Cyclone preheater basic elements.

15.2 Design and Operation of Preheaters

Basically all preheaters operate on the same principle. Heat exchange takes place in each stage when raw meal fed in the connecting duct travels in suspension with hot gases from lower stage to inlet of upper stage.

See Fig. 6.15.1.

Various Cement Machinery manufactures have their own designs of cyclones i.e., geometry of cyclone, raw meal feeding arrangements, immersion tubes, duct geometry etc. However, in course of time there has been a kind of convergence of many designs so that performance wise they have come very close to one another. Among the commonly found design features are :

1. Top Cyclones are taller, cylindrical height / diameter ratio being 1.75-2.
2. Proportions of width / depth of inlet area are generally 1 to 2.
3. Bottom 2-3 stages have double cones for preventing choking.

   See Fig. 6.15.4.
4. Out going ducts are round and meet the rectangular duct from upper cyclone as shown. **See Figs. 6.15.2, 6.15.3.**
5. Slopes are such as to prevent accumulation of dust.

   See Figs. 6.15.3, 6.15.5.

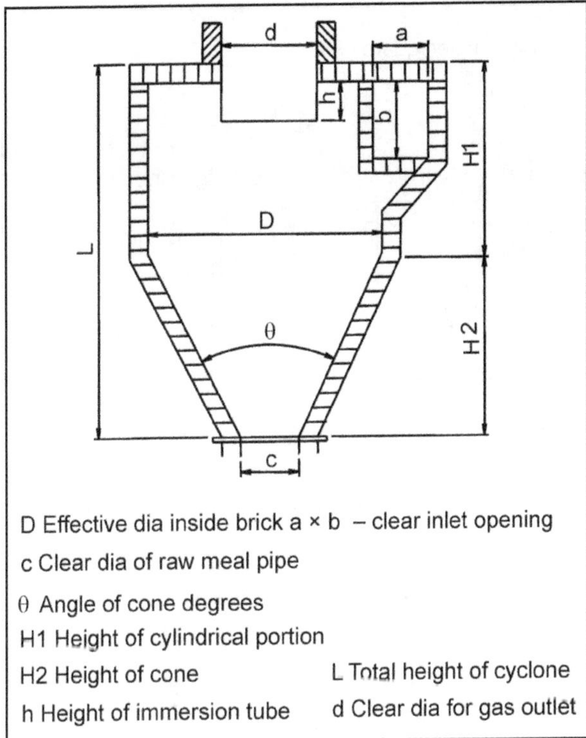

Fig. 6.15.2 Proportioning of cyclones.

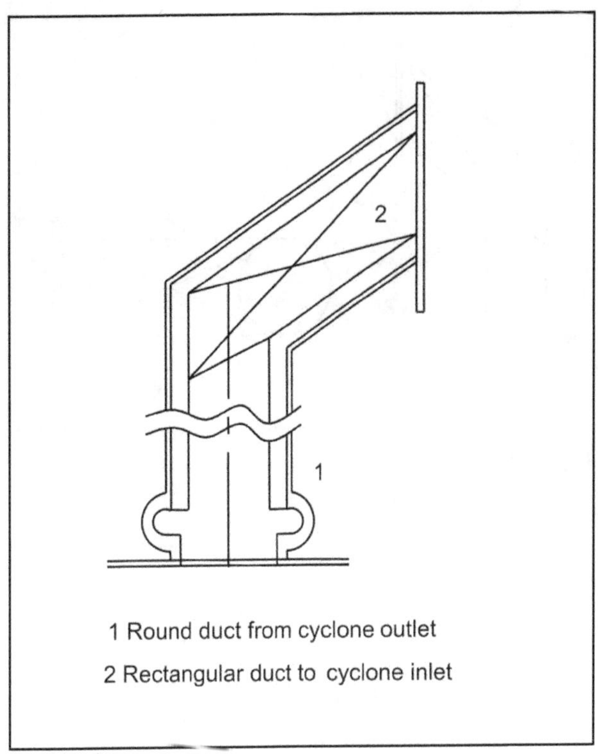

Fig. 6.15.3 Preheater cyclones common features

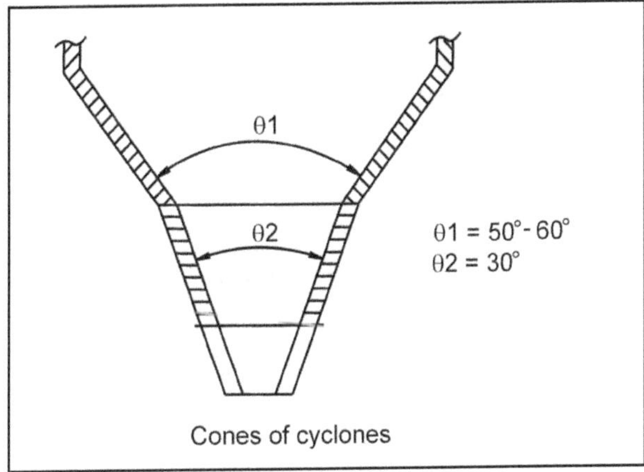

Fig. 6.15.4 Double cone for bottom 2 stages to prevent jamming.

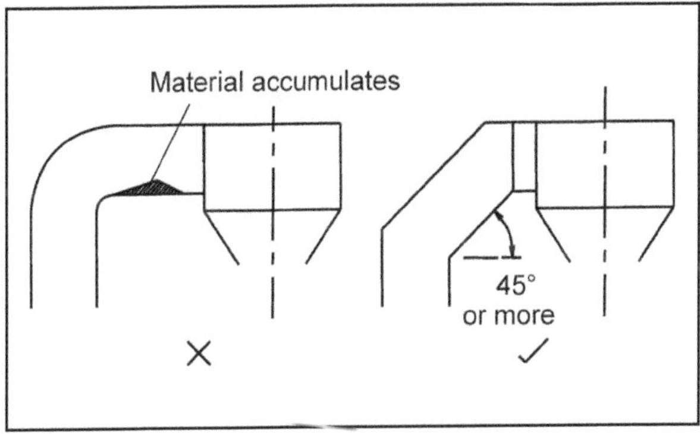

Fig. 6.15.5 Layout of ducts and pipes (aviod horizontal lengths at bends)

6. Even the volute portion is slanted to prevent accumulation of dust.

   See Fig. 6.15.2.
7. Velocities at inlet are between 20-22 m/sec, for bottom cyclones and progressively reduce to 14-16 m/sec, for top cyclones.
8. Velocities at outlet are lower than inlet velocities.
9. Pressure drop in 4, 5 and 6 stage preheaters only would be 350, 400 and 450 mm wg. respectively.

15.2.1 Low Pressure Drop Cyclones

Emphasis on lower pressure drops without sacrificing efficiency have brought about various developments in cyclone design.

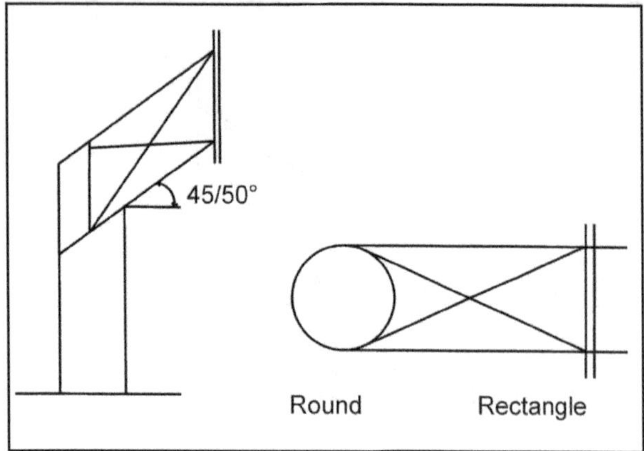

Fig. 6.15.6 Conversion round to rectangle and rectangle to round.

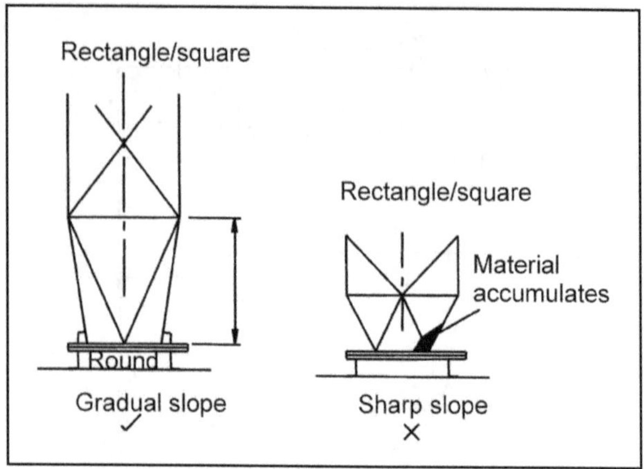

Fig. 6.15.7 Avoid ledges when converting from round to rectangular/square shape.

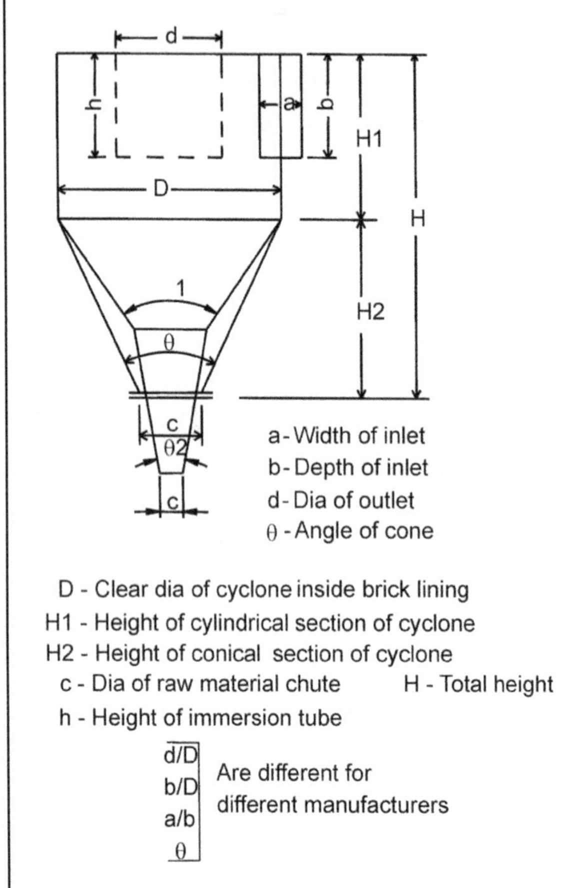

Fig. 6.15.8 Proportions of preheater cyclone.

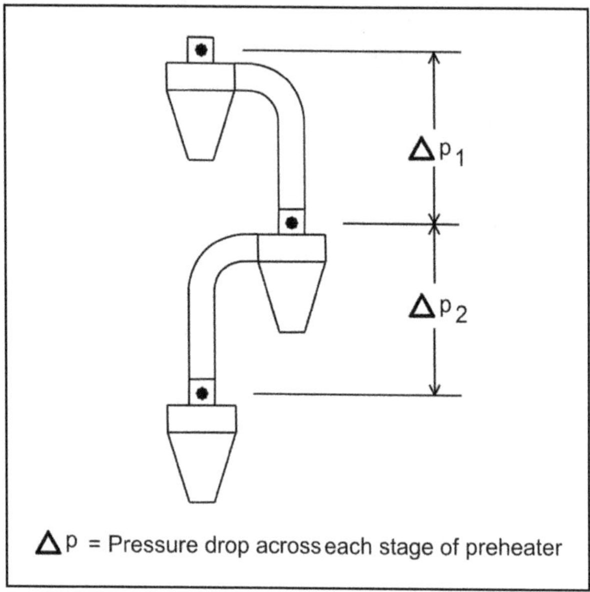

Fig. 6.15.9 pressure drop across one stage of cyclone.

Top and bottom cyclones are required necessarily to be more efficient; bottom cyclone, to reduce circulating load at kiln inlet and top cyclone, to reduce dust loss with exhaust.

Pressure drops across 4, 5 & 6 stages preheaters with cyclones of modern design has been shown in **Table 6.16.2 to in Chapter 16.**

15.3 Immersion Tubes

Immersion tubes influence the dust content in gases leaving the cyclone.

See Figs. 6.15.10 and 6.15.11.

1. Longer tube reduces dust loss but increases pressure drop in the cyclone.

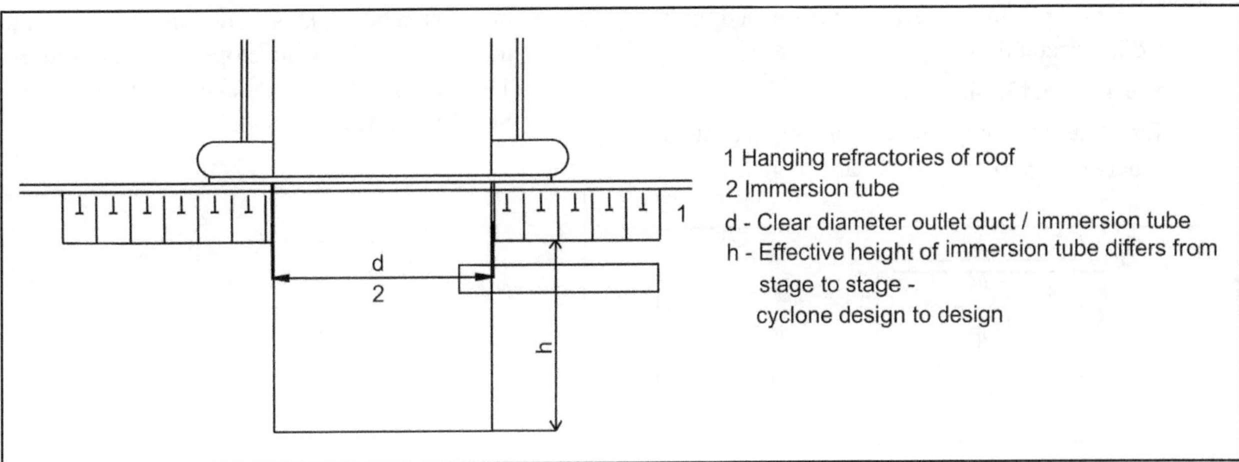

Fig. 6.15.10 Immersion tubes.

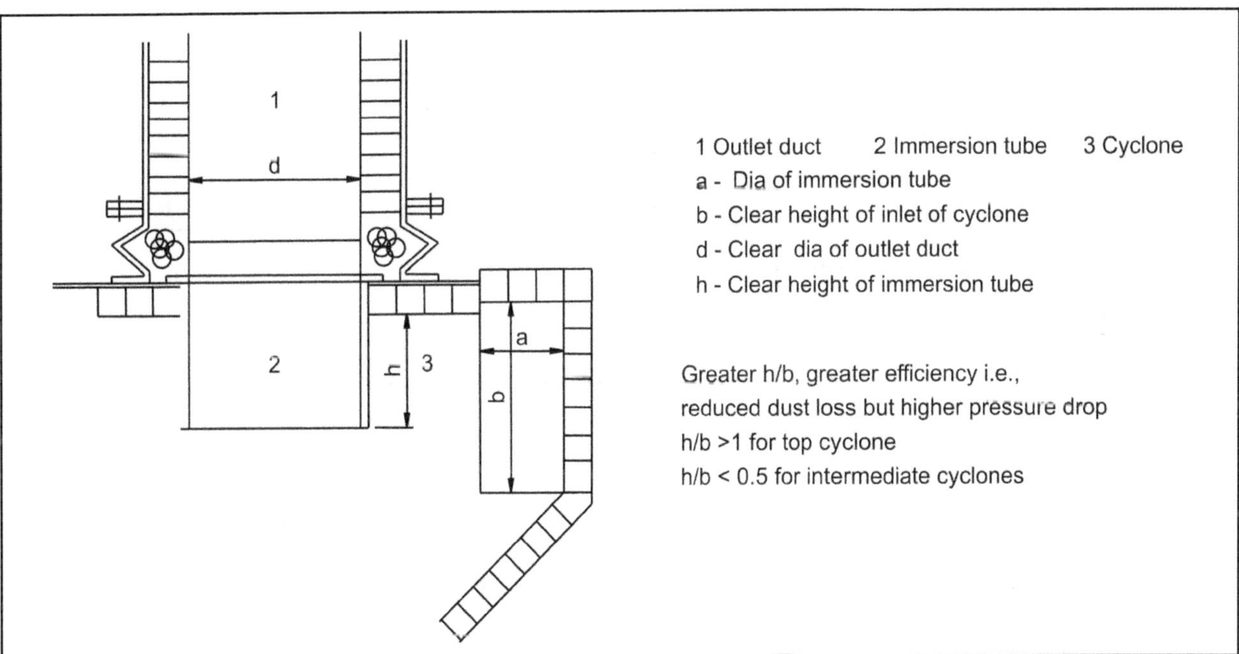

Fig. 6.15.11 General arrangement of immersion tube.

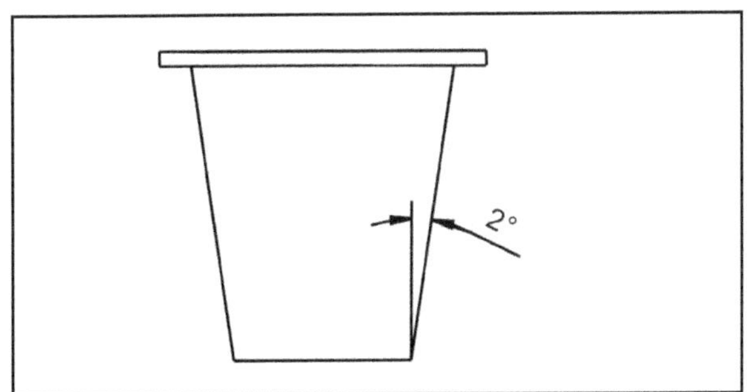

Fig. 6.15.12 Immersion tube with taper.

2. Some designs of immersion tube have a slight taper.

   See Fig. 6.15.12.

3. Immersion tubes of top cyclones are sometimes fitted with vanes to streamline flow out of the cyclone so that dust loss is reduced without increase in loss of pressure.

   See Fig. 6.15.13.

4. In one design immersion tube is opened up to reduce circulation.

   See Fig. 6.15.14.

5. Temperatures are the highest in bottom cyclone – between 850 to 950 °C and hence life of immersion tube is short. Tubes made of ceramic material have been developed for this location. They are made of interlocking ceramic pieces. **See Fig. 6.15.15.**

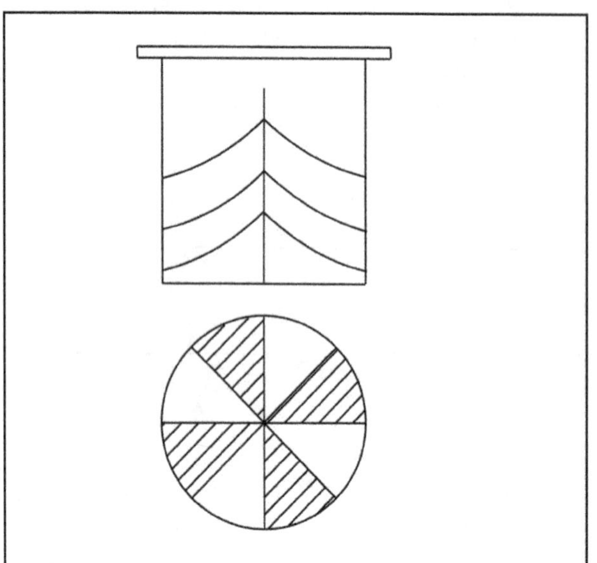

Fig. 6.15.13 Immersion tube with vanes reduce to pressure drop for top cyclone.

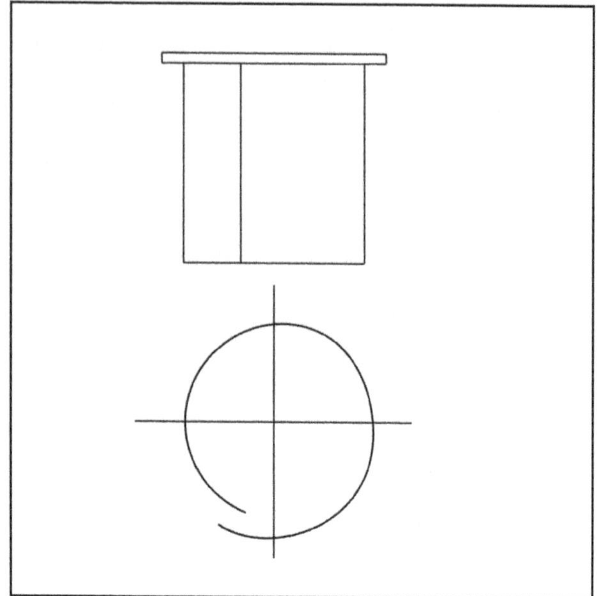

Fig. 6.15.14 Opened immersion tube.

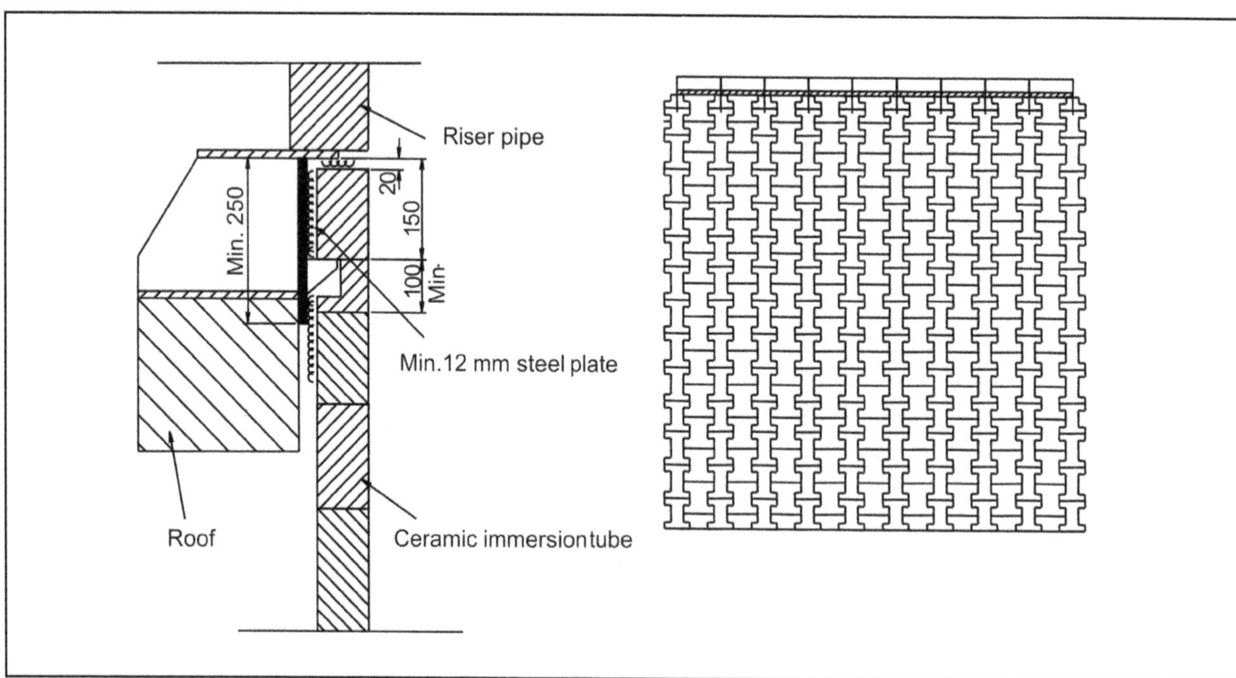

Fig. 6.15.15 Ceramic immersion tube for bottom cyclone of preheater.

15.3.1 Installation of Immersion Tubes

Immersion tubes need to be replaced in all cyclones now and then.

Different manufacturers have different designs to facilitate replacement. In some designs the tube is made in two vertical halves and bolted.

In most designs immersion tubes just rest on the top plate of the cyclone.

See Fig. 6.15.11.

Sealing needs to be provided around it. The cyclone outlet is a round duct and there is an expansion joint immediately after the cyclone. Weight of the duct between two stages of cyclone is taken on the floor between the cyclones. Hence no load comes on the expansion joint.

See Fig. 6.15.44.

15.4 Refractory Lining of Cyclones

All cyclones and interconnecting ductings are lined with refractory, except for the top stages where temperature are less than 350 °C.

In case of 4 stage preheater, temperature of gases leaving preheater would be 350 °C and in case of 6 stage it would be 270 °C.

15.5 Raw Meal Pipes

Material collected in a cyclone is conveyed to the connecting duct from the cyclone below it by a raw meal pipe. These pipes are either made of heat resisting material or are lined with refractory / castable because of high temperature of raw meal. This pipe has an expansion joint and also a single / double pendulum flap valve in it to prevent short circuiting of gas circuit between two cyclone stages. Sharp bends in the pipe should be avoided to ensure smooth flow.

See Figs. 6.15.16 and 6.15.17.

15.5.1 Pendulum Flaps

They are single and double. They are required to with -stand high temperatures and hence are made of heat resisting material. **Fig. 6.15.18** shows typical designs, location and material for single and double pendulum flaps.

15.6 Preheater Tower

The preheater is installed in a structure called preheater tower mentioned earlier. When preheaters had 4 stages, towers used to be 40 meters tall. Now with 6 stages and with capacities of single stream preheaters going up to 4000 tpd, towers have reached heights of 80 meters and higher.

See Fig. 6.15.19.

In the past or early stages of development, single stream preheaters were used upto 2000 tpd capacity only. Beyond that twin stream preheaters were used. Polysius design has been practically a two stream preheater from the beginning. Therefore, it suited well for larger capacities.

15.6.1 Two Stream Preheater

Two stream preheaters are used to reduce height of preheater tower. While floor area would increase height could be reduced considerably.

See Fig. 6.15.20.

15.7 Calciner and Preheater Streams

After the introduction of calciner kiln capacities increased dramatically requiring larger cyclones. Gases could be divided in to two streams one passing through kiln and other through calciner.

About 40 % of total gases passed through kiln stream and 60 % through calciner stream.

Because of the difference in quantities of gases in these two streams preheaters required for them were also dissimilar in size. They required towers of different sizes. Such dissimilar preheaters could also be housed in one tower. Design of such a tower would be difficult as it would have floors and heights at different levels for two streams.

When using two streams of dissimilar preheaters, it is possible to arrange them in line with kiln.

In some designs gases from kiln and calciner were combined and divided equally between cyclones of a twin stream preheater.

See Fig. 6.15.22.

Civil design for such tall structures have to take into account, load bearing capacities of soil, wind loads, earthquake factors – depending on the location of the plant.

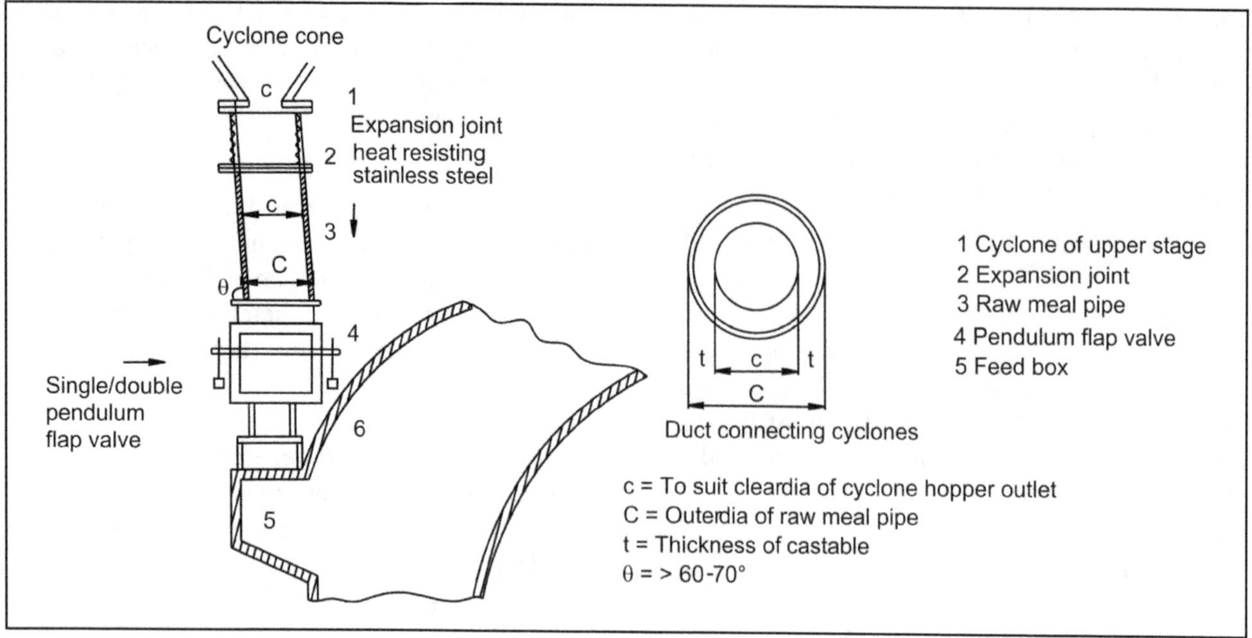

Fig. 6.15.16 Raw meal pipe between cyclones.

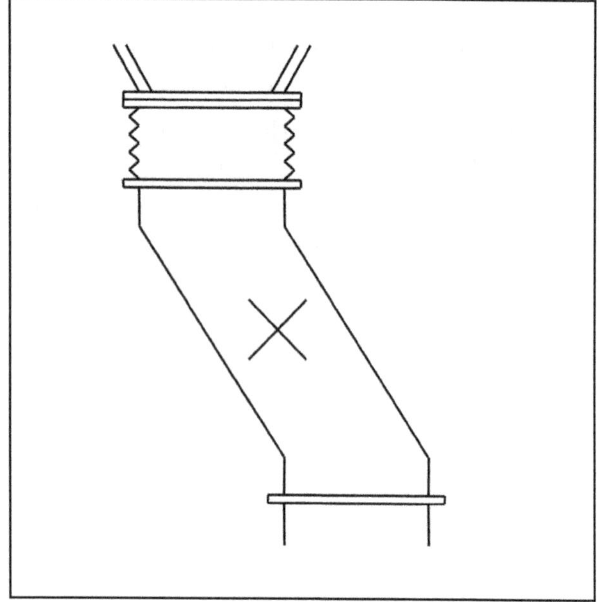

Fig. 6.15.17 Raw meal pipes between preheater cyclones – sharp bends to be avoided.

15.8 Calciners

Calciners are of Two Types:

1. In Line.
2. Off Line.

In line calciner is best suited for a single stream preheater or a twin stream preheater with identical cyclones in both streams or Polysius type preheater.

See Figs. 6.15.21 to 6.15.23.

Off line calciner is suited for all types of preheaters. It is suited for 'twin stream' preheater and it is also suited for preheaters with different sizes of cyclones for kiln and calciner streams.

See Figs. 6.15.24 to 6.15.26.

15.9 Features of Preheater Towers

Preheater tower – layout wise is to be designed to :

1. House and support all cyclones and inter connecting ducting.
2. Support down comer duct.
3. House calciner (sometimes).
4. Provide enough elbow room for taking out and installing refractories.
5. Provide access to tops of cyclones to attend to sensors for temperatures / draughts, to manholes / inspection doors, to expansion joints, raw meal feed pipes and pendulum flap valves.

   This can be by way of steel platforms and ladders or additional floors may be added in concrete structure itself.

sketch	size (mm)	temperature °C	construction	no.s	dimensions		lining
Below twin cyclone	300	350-400	Flaps - 309 SS Housing - MS.	2	300 × 700	Double pendulum	
Below II nd. stage cyclone	300	550-650	Flaps - 309 SS Housing - MS.	1	300 × 700	Double pendulum	Housing is lined with 65 mm thick. Refractory bearings to be antifriction and protected from heat
Below III rd. stage cyclone	300	700-800	Flaps - 309 SS Housing - MS.	1	300 × 700	Double pendulum	Housing is lined with 65 mm thick. Refractory bearings to be antifriction and protected from heat
Below IV th. stage cyclone	500	900-1000	Flaps - 316 SS Housing - MS.	1	500 × 1000	Single pendulum	Housing is lined with 65 mm thick. Refractory bearings to be antifriction and protected from heat

Fig. 6.15.18 Typical single and double pendulum flap valves in preheater.

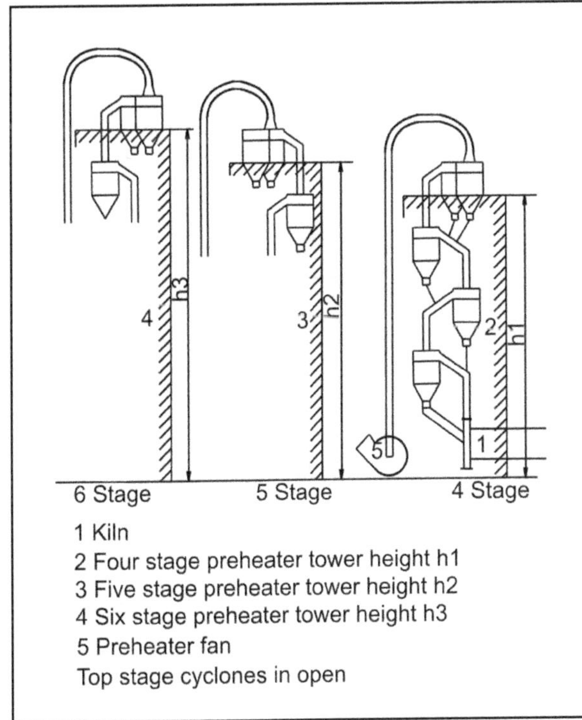

Fig. 6.15.19 Single stream preheater - 4 to 6 stages.

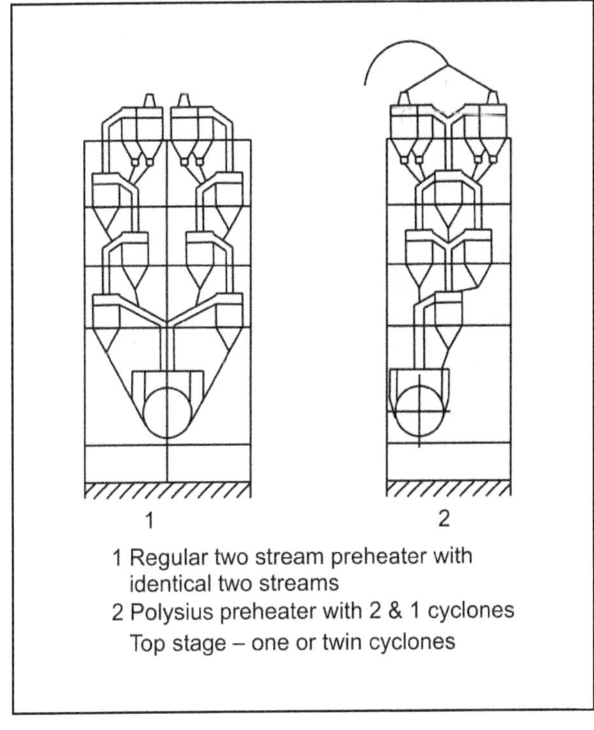

Fig. 15.20 Twin stream preheater.

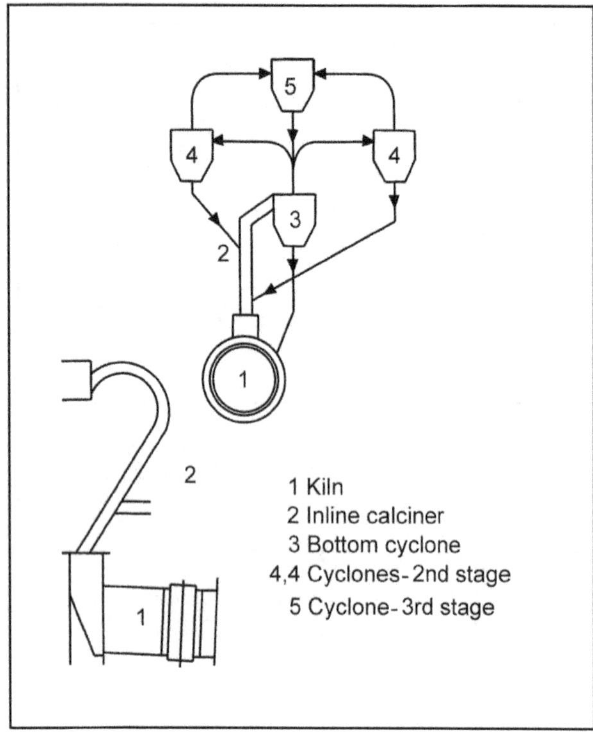

Fig. 6.15.21 In line calciner on Polysius type preheater.

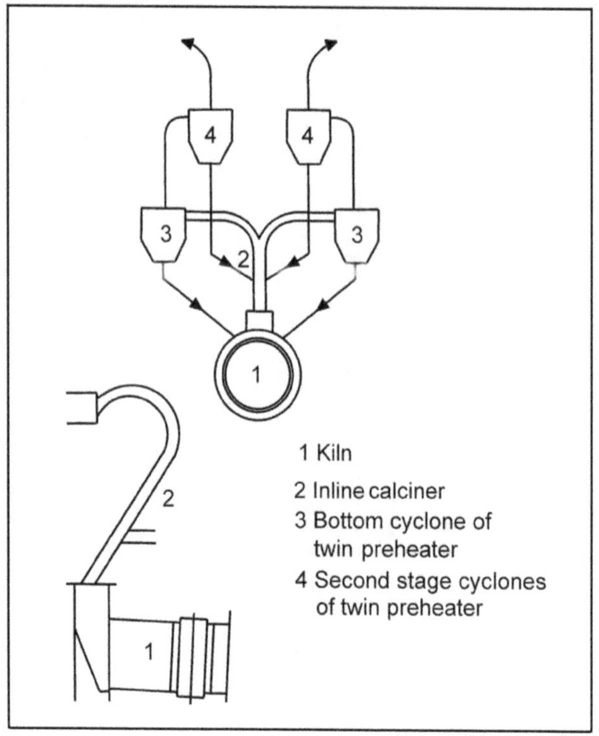

Fig. 6.15.22 In line calciner with twin (identical) stream preheater.

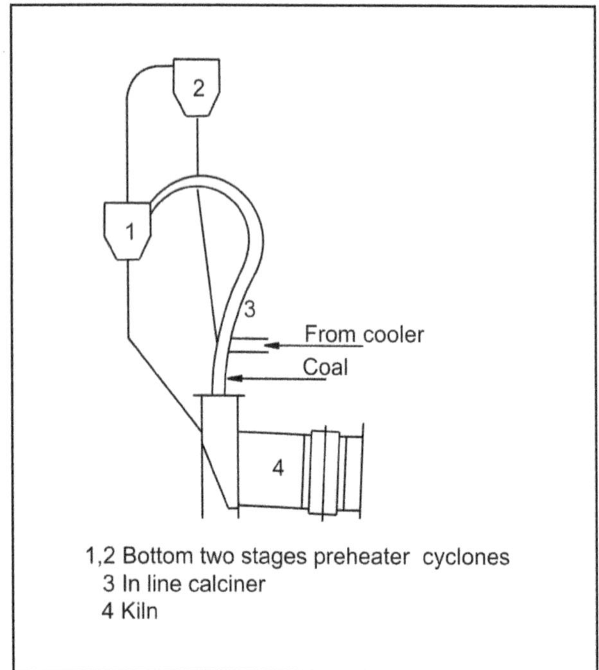

Fig. 6.15.23 In line calciner-single stream preheater.

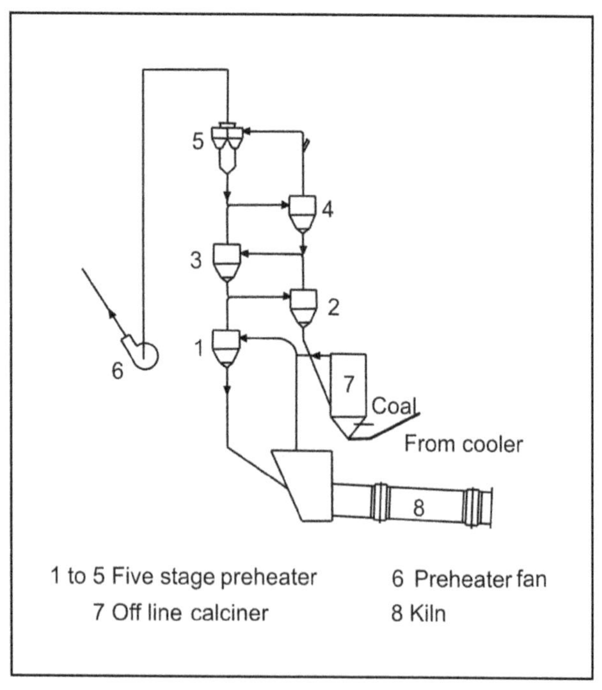

Fig. 6.15.24 Off line calciner 1-stream preheater (shown 5 stage)

6. Stairs with hand railing and landings to go from ground floor to top.
7. 'Lift' to carry workers and materials for maintenance purposes.

   With heights of towers equaling that of 25-30 storied buildings, lift is a necessity.

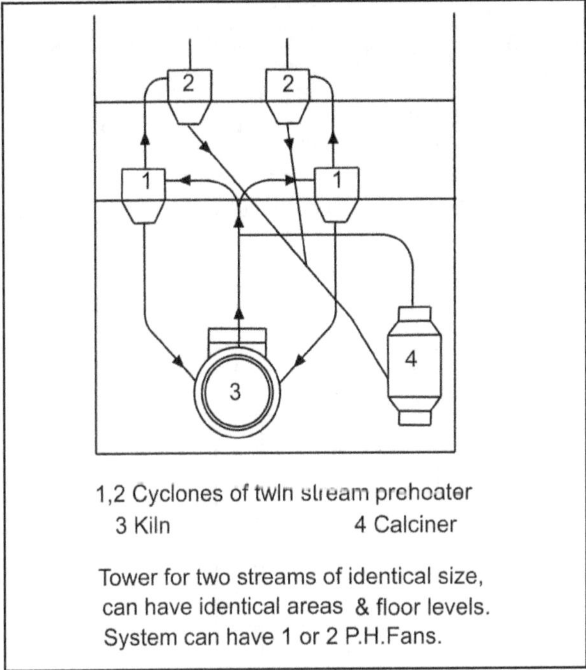

Fig. 6.15.25 offline calciner with a twin stream preheater.

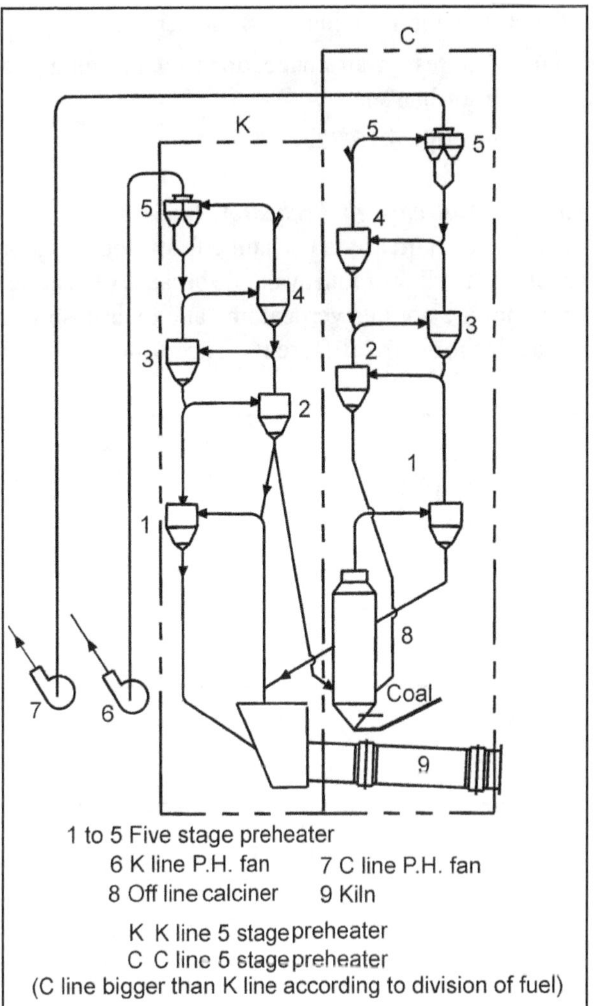

Fig. 6.15.26 Off line calciner on two unequal stream preheater.

15.9.1 Preheater Fan

Preheater fan/s would generally be installed on the ground floor inside the tower. Thereby driving motor is also protected.

See Figs. 6.15.27 and 6.15.28.

15.9.2 En Mass Conveyor

When a bucket elevator is used, to save on height of an already tall elevator, 'en mass' conveyor with rotary air lock is used to feed raw mix to preheater to reduce the height of the elevator.

See Fig. 6.15.29.

15.9.3 Chimney Alongside Preheater Tower

Since chimneys are required to be tall for dissipation of dust over a wide area, it was common practice to support the chimney also along the tower.

See Fig. 6.15.30.

Supported like this the chimney has to be taller than the preheater by 4-5 meters though in principle when an ESP or a bag filter is used, the dust emission is so low that the height arrived at by using formula evolved by Central Pollution Control Board would not exceed 30 meters.

It is not uncommon now to find self-supporting chimneys installed at a distance from the preheater tower. This simplifies ducting layout.

See Fig. 6.15.31.

15.9.4 Provisions in Preheater Towers

Preheater tower should also provide for :

(i) Lifting arrangements for lifting cyclones / bricks etc. (particularly necessary when lift has not been installed).

(ii) Chute for dropping down rubbish.

(iii) Compressed air connections for cleaning and for air blasters.

See Fig. 6.15.32.

15.9.5 Staircase in Preheater Tower

Positions of staircase may change from floor to floor according to elbow room available because of ducting. Cyclones are not in a vertical line and their positions on each floor are different. The dimensions of

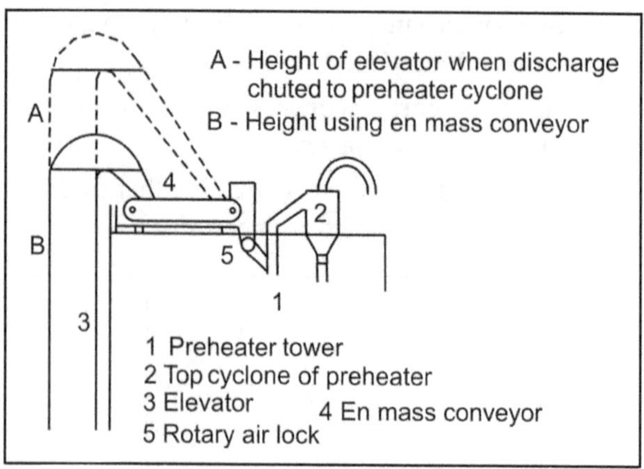

Fig. 6.15.29 Feeding preheater with bucket elevator.

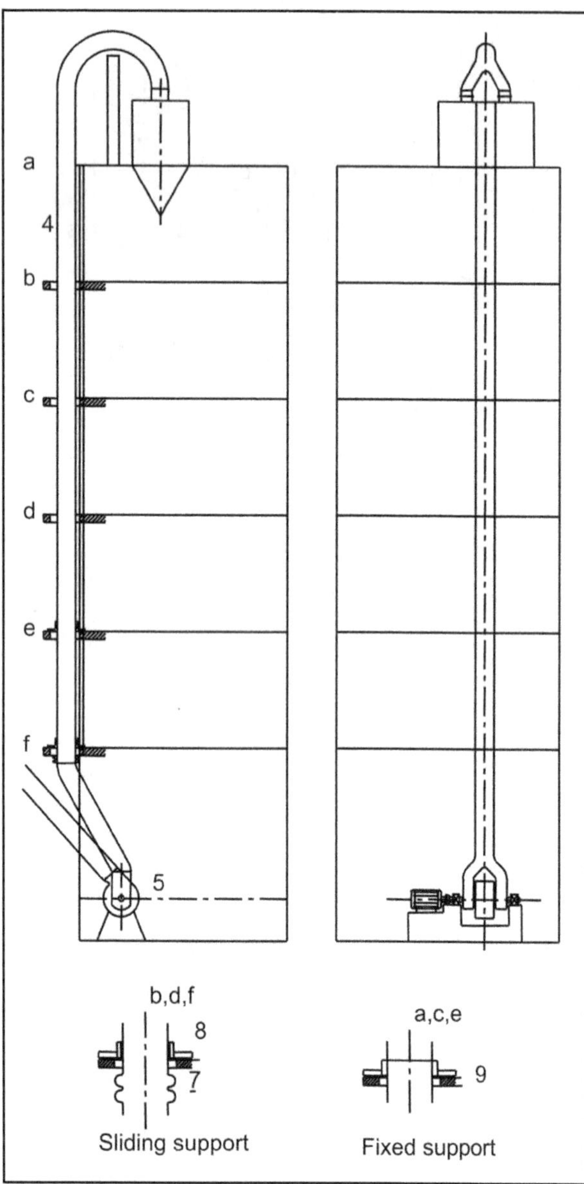

Figs. 6.15.27 & 28 Layout of down comer duct and preheater fan.

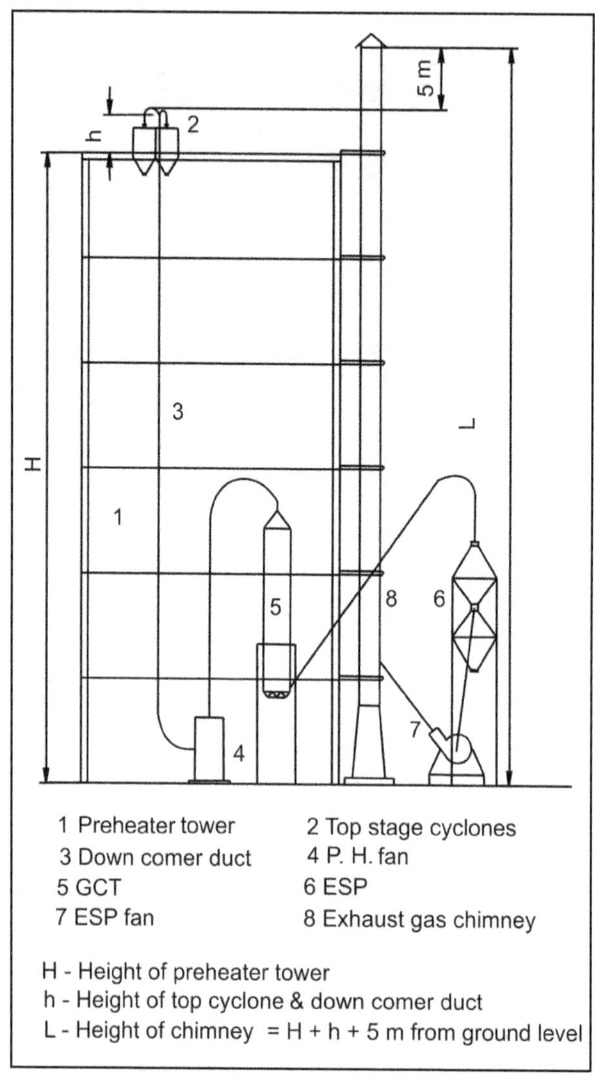

Fig. 6.15.30 Kiln-mill exhaust chimney supported from preheater tower.

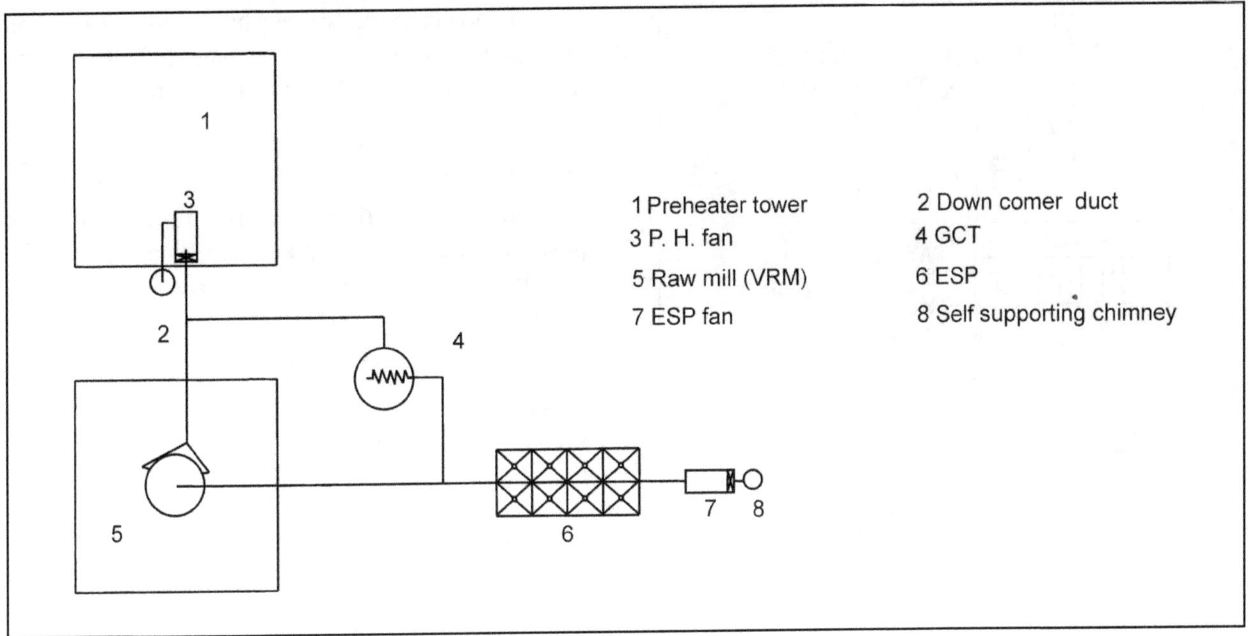

Fig. 6.15.31 Self supporting chimney at a distance from P.H. tower.

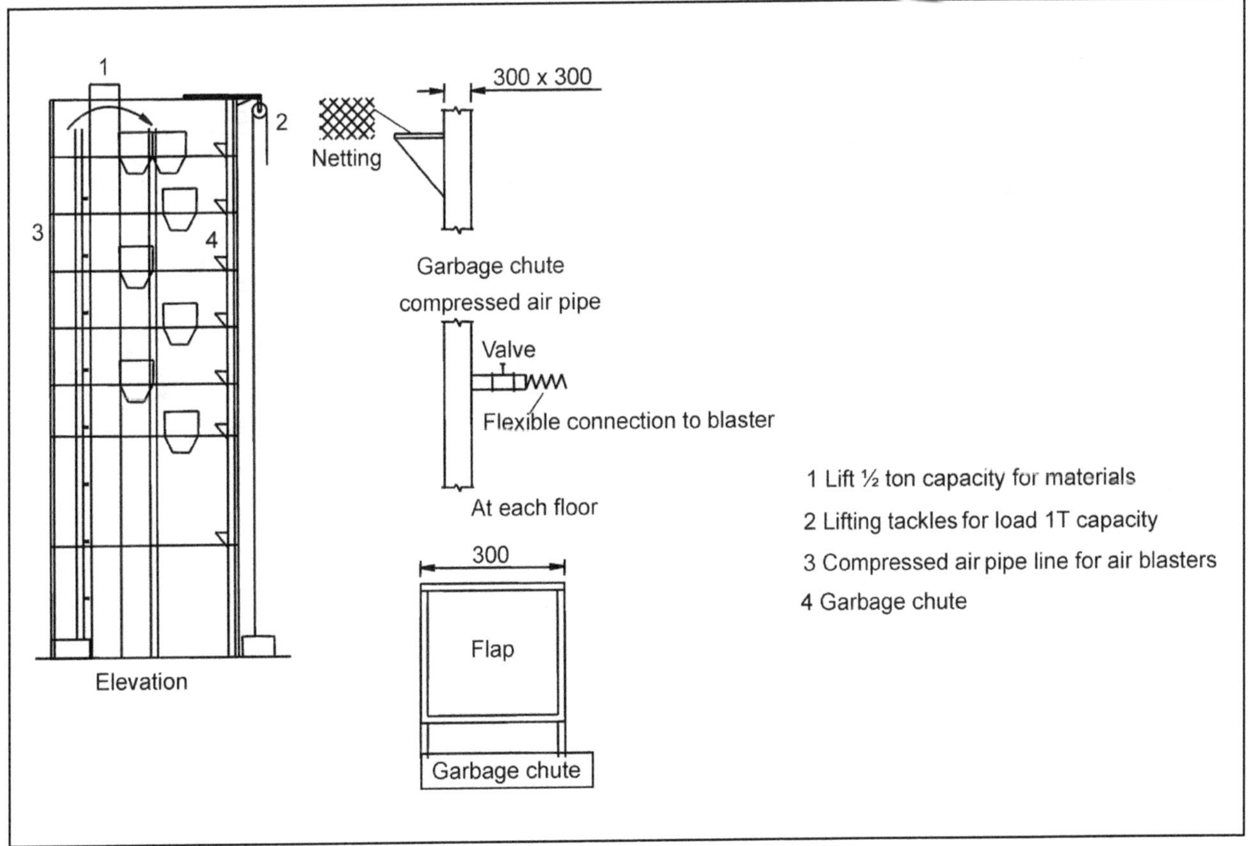

Fig. 6.15.32 Facilities in a preheater tower.

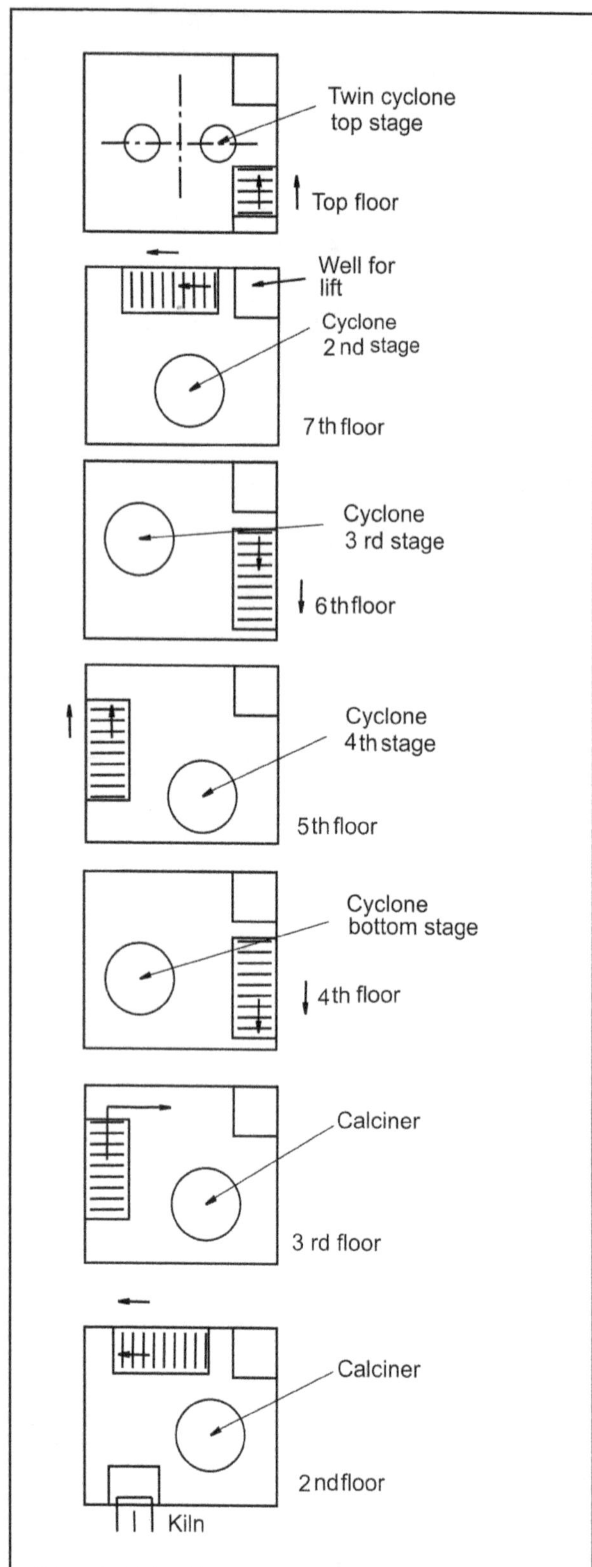

Fig. 6.15.33 Stairs on different floors of a preheater tower to suit positions of cyclones.

preheater tower are determined by the extreme position of cyclones on each floor. They should be such as to leave a clearance of 500-750 mm between cyclone and edge of building.

See Fig. 6.15.33.

It may be possible to keep staircases in same position on all floors by keeping them outside the building. The dimensions of the tower could also be reduced in this arrangement.

See Fig. 6.15.34 and also **Fig. 6.15.52.**

Fig. 6.15.51 shows a tower where stairs are in line but inside the tower; this has resulted in a wider tower.

15.10 Top Cyclones and Ducting

It is best to keep top cyclones in the open and not have a roof over them. Many a time the top cyclones have been replaced by bigger cyclones to increase capacity by 15-20%.

See Figs. 6.15.35 and 6.15.36.

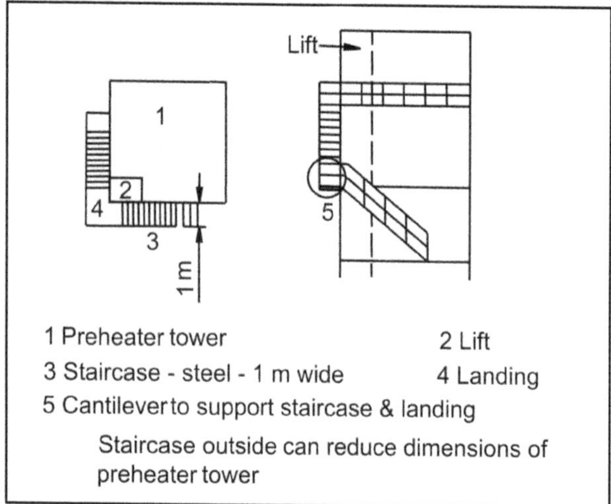

Fig. 6.15.34 Staircase outside the building.

The top ducting needs special attention for supporting it because duct expands in operation and hence supports should be so designed that they always take load of the ducting hot or cold.

See Fig. 6.15.37.

15.11

The Suppliers should furnish layout of preheater tower and calciner.

Consultant should integrate it with raw mill section, coal mill section, blending and dust collector.

As has been mentioned earlier, room must always be kept for expansion.

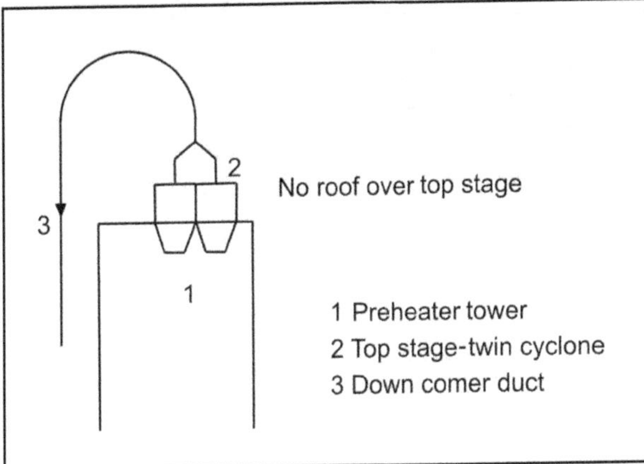

Fig. 6.15.35 Top cyclone – above top of tower.

15.12 Calciner to be Installed Later

When kiln does not have a calciner, room should be kept for a 2nd tower of bigger size; **See Figs. 6.15.38, and 6.15.39** and when first kiln has a calciner, for the 2nd production line. **See Fig. 6.15.40.**

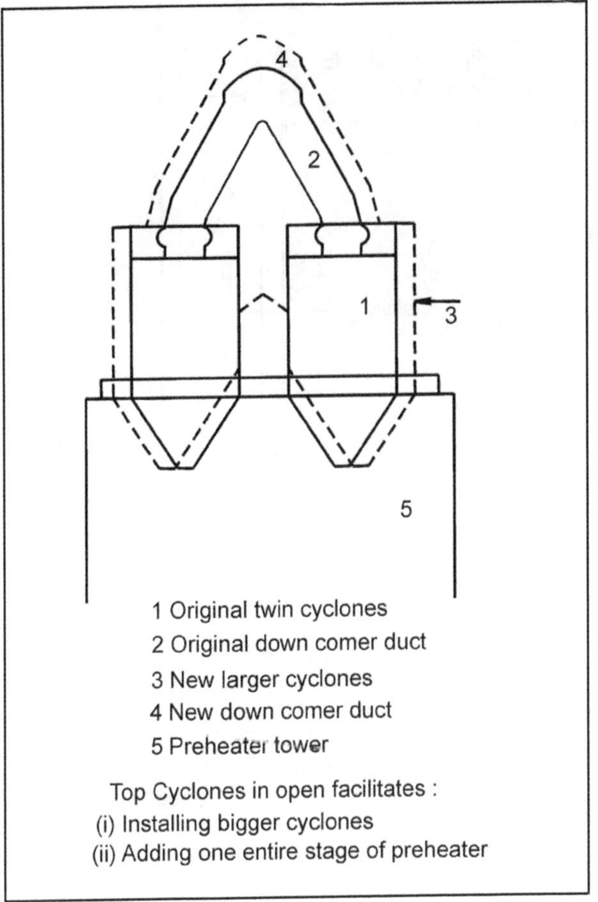

Fig. 6.15.36 Top Cyclones in open.

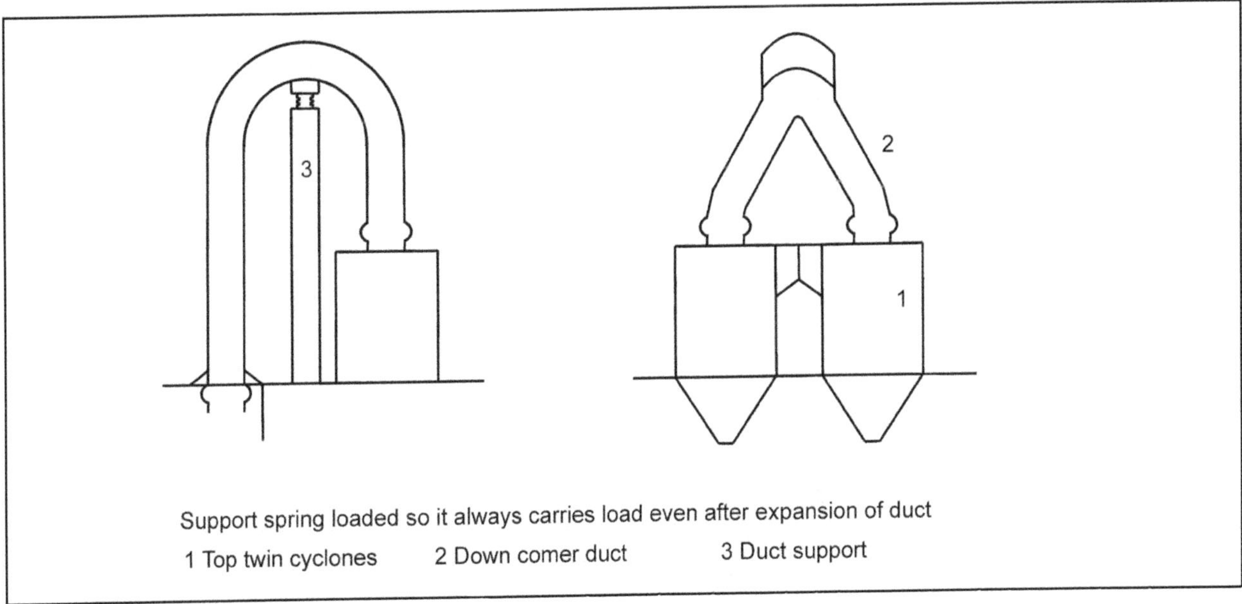

Fig. 6.15.37 Support of down comer duct at top.

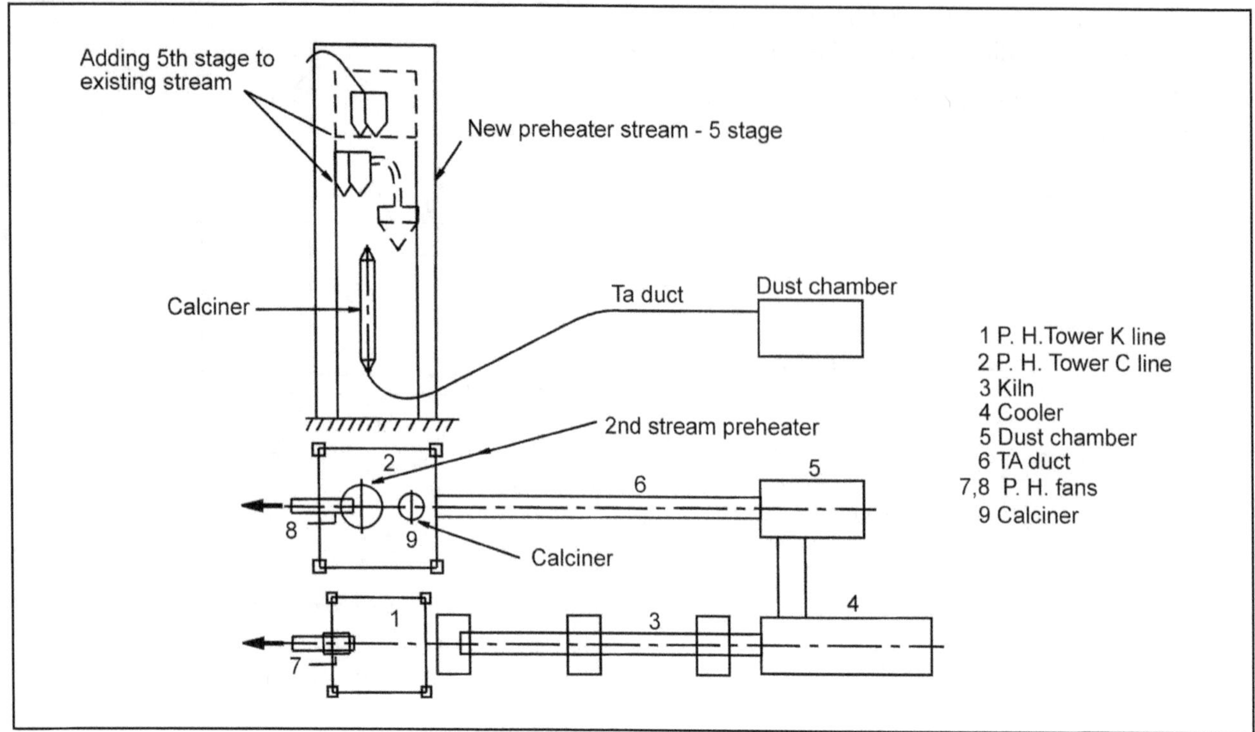

Fig. 6.15.38 Expansion with 2nd stream preheater
Adding one stage to existing preheater.

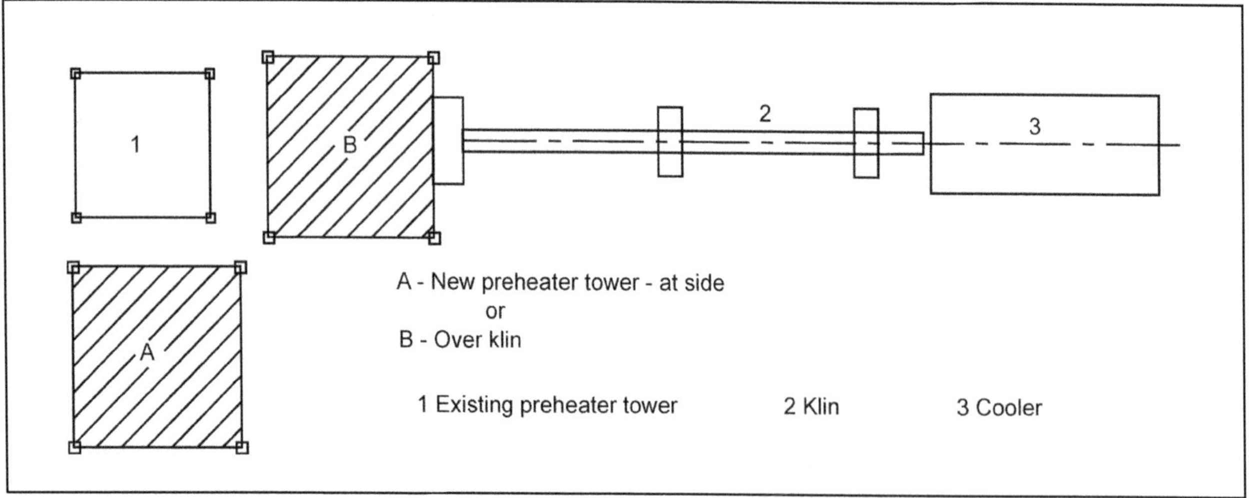

Fig. 6.15.39 Expansion with 2 stream preheater – location of new preheater tower.

Factors which indirectly affect the preheater tower design are :

1. Type of calciner- its dimensions - height etc.
2. How the grate cooler has been installed.

An inline calciner needs bottom cyclone to be at a much greater height above the kiln than an off line calciner.

See Fig. 6.15.41.

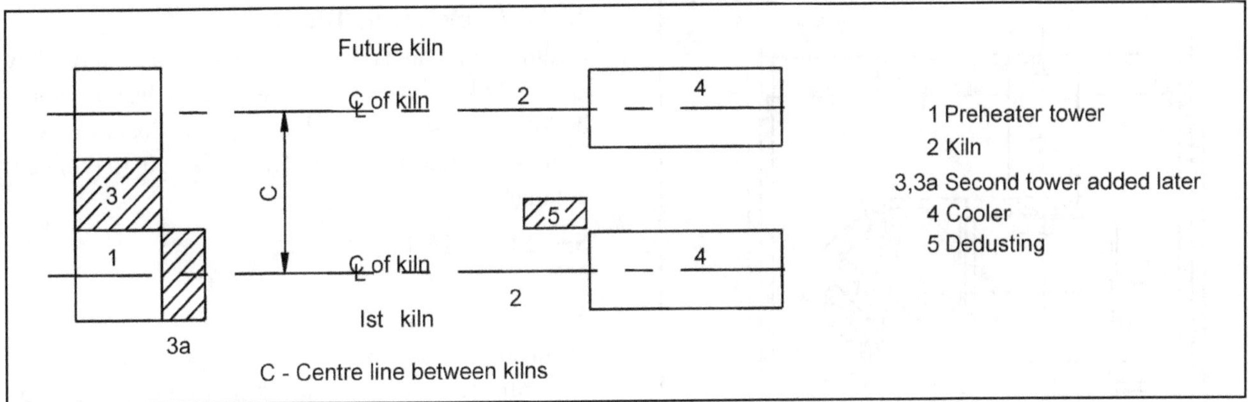

Fig. 6.15.40 Centre line between kilns should allow expansion by 2nd tower of preheater & calciner or 2nd kiln.

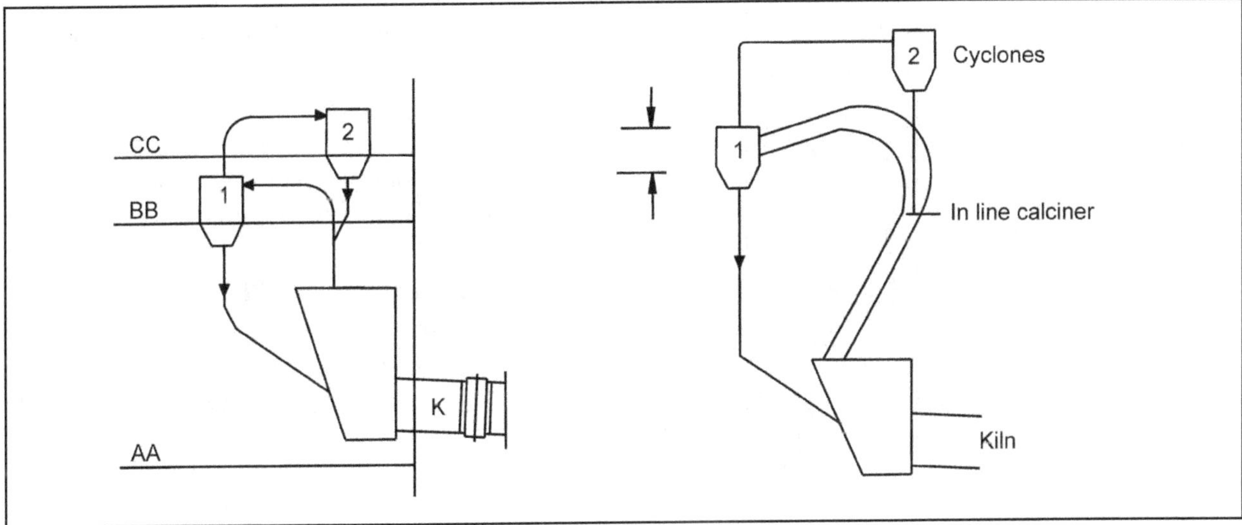

Fig. 6.15.41 In line calciner - impact on height of preheater tower.

In case a calciner is contemplated at a later date, the preheater tower should have room provided for it and beams and columns should be designed to take additional load.

A long riser duct can be replaced by an inline calciner and many small plants have therefore installed bottom cyclone at a considerable height.

See Fig. 6.15.42.

An off line calciner can be installed either within the tower or outside it.

RSP calciner is installed at a higher level because it has a mixing chamber / mixing duct which joins the riser duct from the kiln. Its feed for raw meal and coal are also from top. It is necessarily installed within the preheater tower and hence the tower is to be designed to take load of calciner also.

See Figs. 6.17.11 and 6.17.12 in Chapter 17.

Thus more careful planning of preheater tower design is required if a calciner is to be added later in the same tower.

Because of diverse designs of calciners, it is necessary to decide initially only, which calciner would be installed.

15.13 Grate Cooler and Preheater Tower

Grate cooler with all its auxiliaries , hoppers, drag chain for spillage and deep bucket conveyor under it, takes a considerable amount of head room. Height of tower above ground level depends on how cooler is installed.

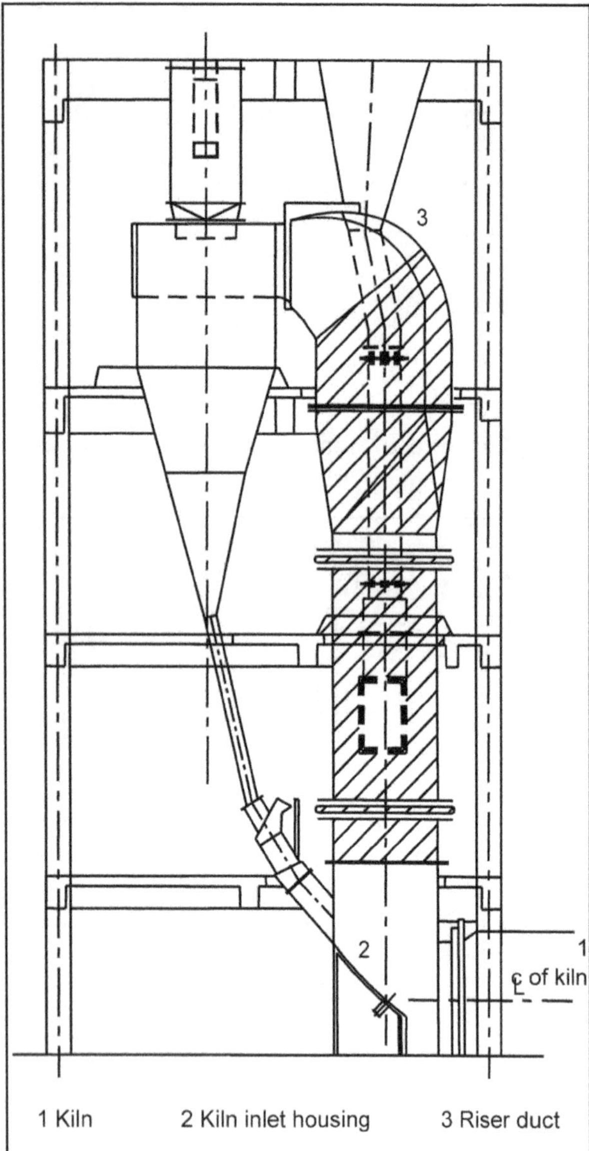

Fig. 6.15.42 Riser duct is replaced by a calciner in case of kilns with in line calciner.

Various options are :

1. Deep bucket conveyor on ground level – height of tower maximum.
2. Spillage drag chain at ground level.
3. Clinker breaker on ground level – height of tower minimum.

From maintenance angle it is desirable to have at least spillage drag chain at ground level, which is a via media solution between two extremes.

See Fig. 6.15.43.

15.14 Cyclones in Tower

Cyclones and ducts rest on beams on each floor, they are not bolted. Auxiliary or secondary beams should be located to support cyclone and duct loads at 3 or 4 points. Loads should get transferred to vertical columns.

See Fig. 6.15.44.

15.15

From point of view of allocating responsibility, it is best to obtain preheater / calciner / kiln from same suppliers though technically it is feasible to have them of different designs.

If this concept is agreed upon, the Supplier may be asked to provide drawings of tower showing calciner in position.

In this way relative heights of floor, beam spacing, their locations and sizes can be arrived at logically and no problems would be faced at a later date when installation of calciner is taken up.

15.16 Calciner in Small Plants

It is thus seen that small plants, which have a potential for upgradation, but can do so only phase wise, need to be planned more carefully to leave room for future additions.

15.17 Increasing Number of Stages

In upgrading old plants – 4 stage preheaters are converted into 5 stages.

Depending on whether the tower is generously designed, the additional load can be taken on the tower but is required to be transferred direct to main beams and columns.

See Fig. 6.15.45.

If the preheater tower design is not known or there is doubt about the margins available in design, it is best to support the load from the ground by a new set of steel columns. **See Fig. 6.15.46**

15.18 New Tower for Increased Capacity

These problems of expansion with calciner can also be solved by scrapping an existing tower and by installing a new tower to house new preheater of the size suited for capacity of kiln with precalciner. The

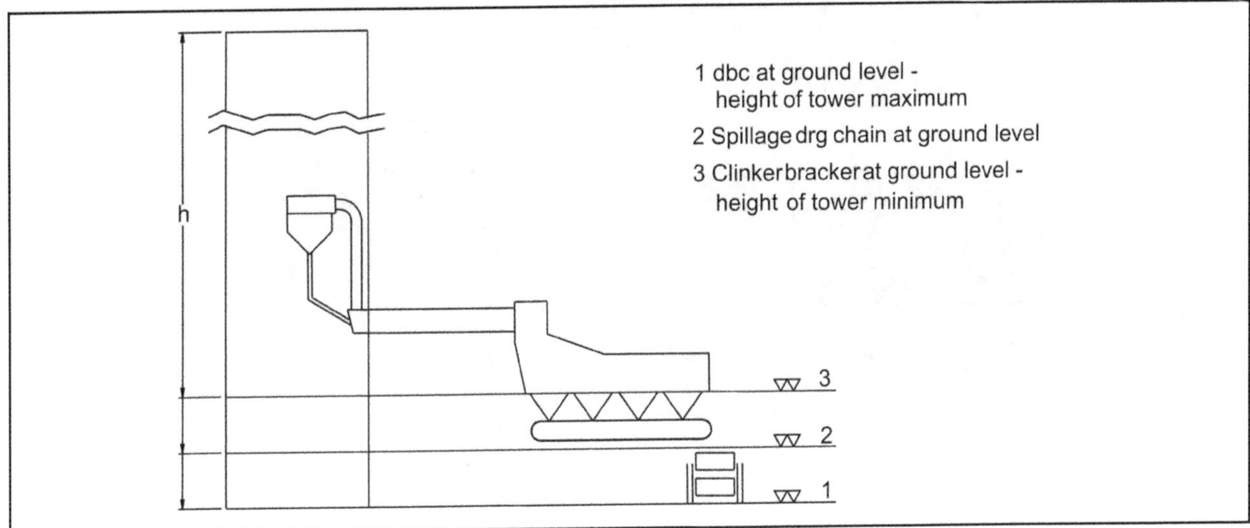

Fig. 6.15.43 Height of preheater tower in relation to cooler layout.

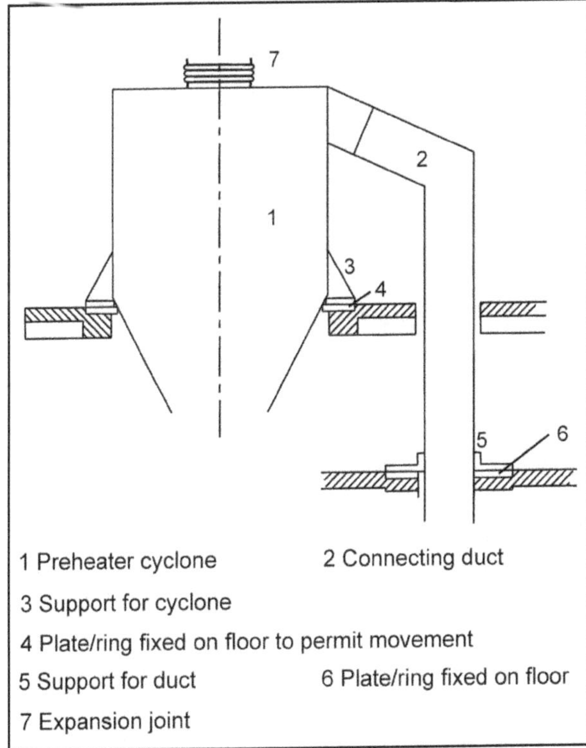

Fig. 6.15.44 Support for connecting duct between preheater cyclones.

existing cyclones are scrapped. This is an expensive solution but more efficient in operation.

See Fig. 6.15.47.

To reduce 'down time' to a minimum, it would be better to house calciner in the new tower. Cyclones would have to be installed in the new tower at a high level for connecting ducting from kiln. This arrangement would need minimum down time but is still expensive.

15.19 Second Stream of Preheater

A more economic solution is to install a 2nd stream of preheater and a calciner. It can be installed by the side or above the kiln as shown. If the design of tower permits it, number of stages in first stream can also be increased side by side.

See Figs. 6.15.26, 6.15.38 and 6.15.39.

Location A is preferred provided there is space on the side of the kiln.

Location B is adopted when A is not possible. It would generally interfere with foundations of columns of the existing preheater tower and also kiln pier.

In this solution, new preheater can be installed with 5 or more stages. Cyclones would be smaller only 1.5 times more in capacity instead of 2.5 times in case of 1 stream preheater.

There is no down time except for hooking up; existing stream does not get disturbed at all. New preheater will have its own fan and own dust collection system.

If first stream can also be upgraded, plant would be as efficient as a new single stream 5 or 6 stage preheater.

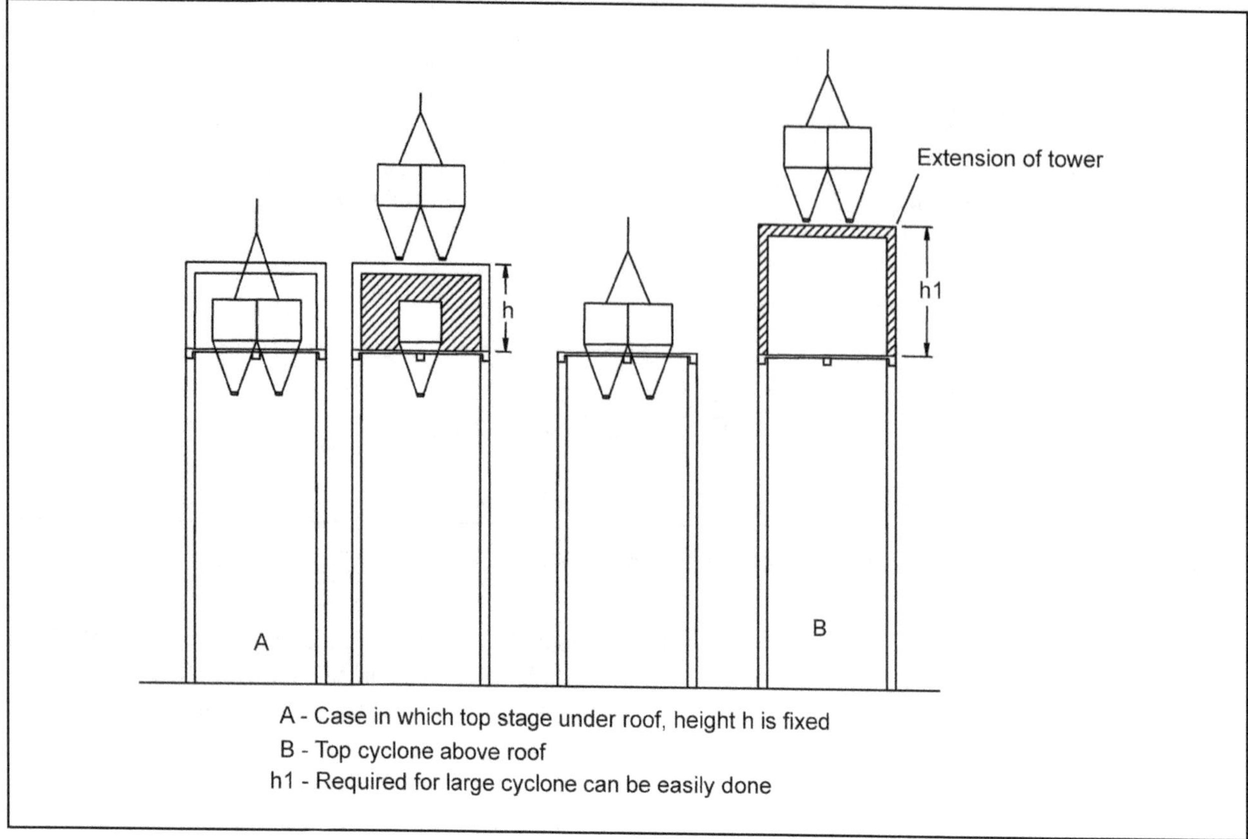

Fig. 6.15.45 Adding a stage - much easier with top cyclone on roof.

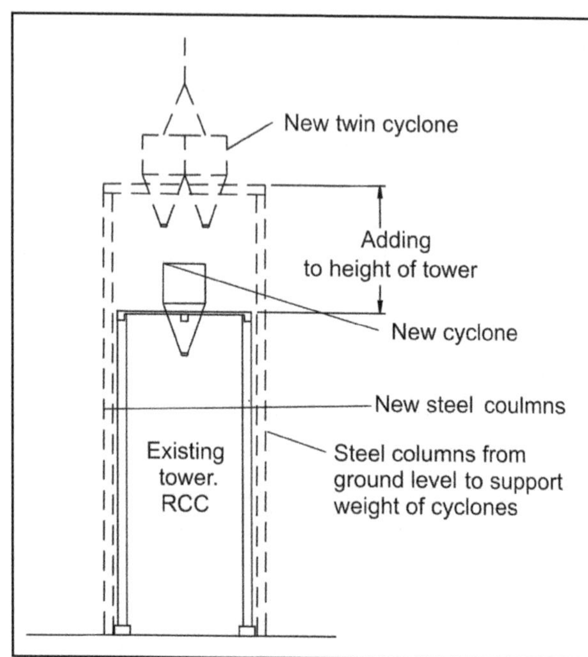

Fig. 6.15.46 Adding of extra stage when existing tower can not take load.

15.20.1 Down draft Cyclone

FLSidth have introduced down draft cyclone in the top stage that eliminates top bend at the beginning of the down comer duct. **See Plate 2.8.4 in Secction 2.**

15.21 Air Blasters

In spite of all the developments in design of cyclones which have been aimed at efficient separation and collection of material from air stream and its smooth discharge to the stage below, it happens many times that thick coatings are formed on the walls of ducts and material gets accumulated in cones of cyclones.

In the bottom two stages these coatings are often due to condensation cycles of alkalies and chlorides. When alkalies and sulphates are not balanced this phenomenon occurs repeatedly.

Therefore to clear the accumulated material in ducts and cones, compressed air is used in periodical blasts through air blasters fixed at locations found to be prone

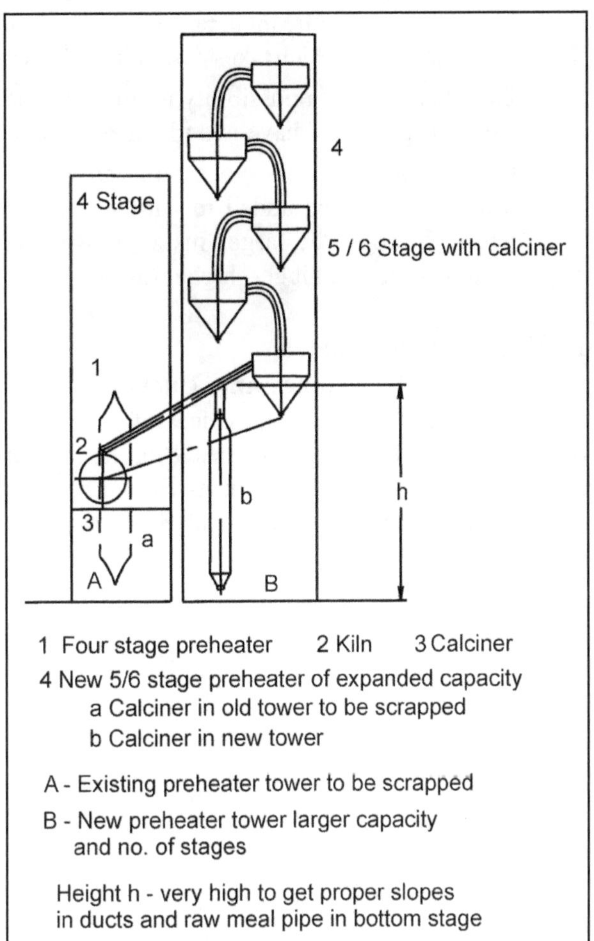

Fig. 6.15.47 Expansion with a new stream preheater and calciner-rejecting existing-system.

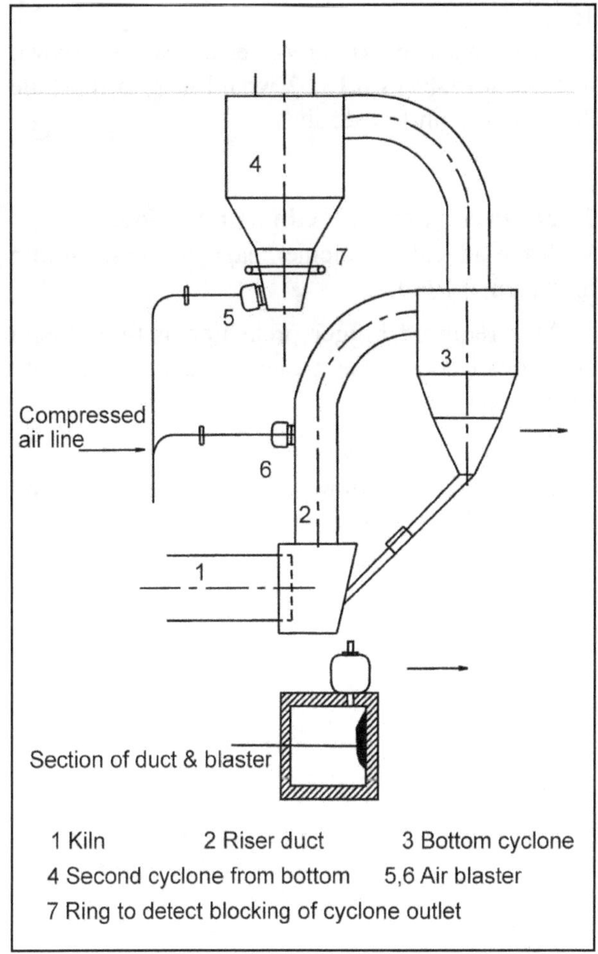

Fig. 6.15.48 Use of air blaster in preheater.

for this trouble. Air blasters are supplied with compressed air from a compressor and a pipeline all along the preheater tower.

Scc Fig. 6.15.48.

When the trouble is severe and frequent, compressed air is used regularly regardless of actual jamming.

It is difficult to predict where jamming could occur. Therefore locations are fixed only after actual operational experience.

15.22 Alkali Bypass

When raw materials have alkalies in excess of 3 % and chlorides in excess of 0.03 %, the cycles of evaporation and condensation resulting in thick and hard coatings mentioned above make operation of the preheater kiln difficult.

Part of the kiln gases are extracted from the riser duct of the kiln to resolve this problem. Gases so extracted are cooled and passed through a dust collector and are re-introduced into the system at a higher level. Some times gases are vented out altogether.

Flow Chart 1.4.1 and 1.4.2 in Chapter 4 of Section 1 show bypass systems mentioned above.

This is the only draw back of a preheater kiln. By pass system increases fuel consumption

In case of claciner kilns, the problem is greatly reduced because only 40 % of gases go through kiln. Cost of installation of a bypass system is greatly reduced and increase in fuel consumption is much less.

15.23

Departmental layouts of preheater and calciner shown in **Figs. 6.15.49 to 6.15.52** would help to illustrate the various points made above.

15.24 Preheater and Kiln with Caciner

With the advent of calciner, capacity of same kiln increased 2.5 times.

This required bigger preheater or two stream occasionally even three stream prehater when capcity exceeded 10000 tpd.

There are many possibilities

1. large capacity from beginning- two stream with equal sizes of cyclones
2. two stream – ones tream to take gases thru kiln and other larger which takes gases thru calciner.

   Even here thre is a possibility to mix kiln and calciner gases and have two identical streams
3. Calciner added later. – A second stream of prheater would be added to handle calciner gaess. It would be larger insize and would require a second, bigger- higher tower.

15.25 Computer Assisted Design of Cyclones and Ducts

Presntly it is possible to make videos of patterns of flow o gases in ducts and cyclones.Witht he help of computers loss in pressure in dacts can be reduced and made optimum. See **plate no 2.8.5 in Chapter 8 of Section 2.**

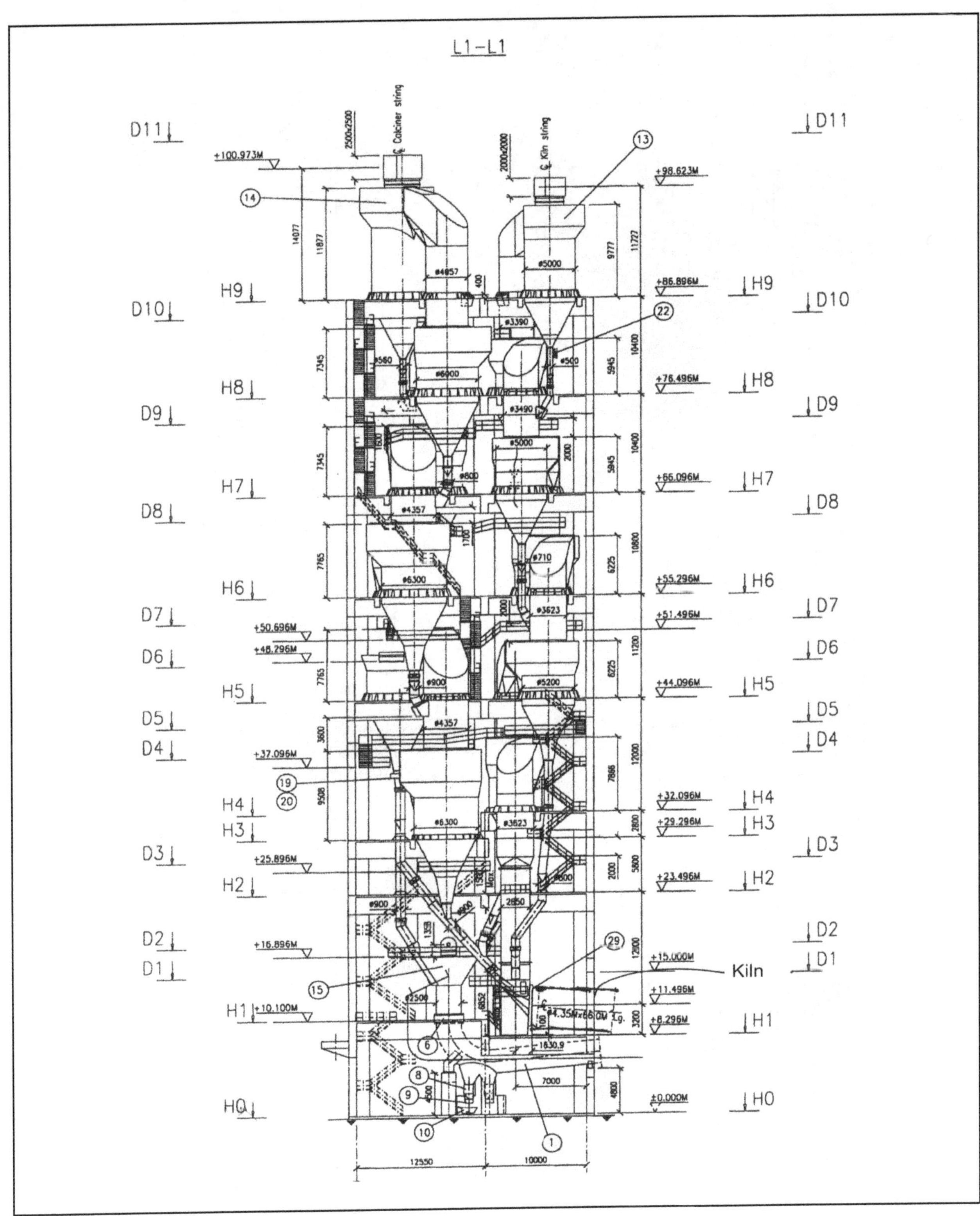

Note : Cyclones in two streams dissimilar in size
Two streams in line with kiln

Fig. 6.15.49 Two stream preheater.

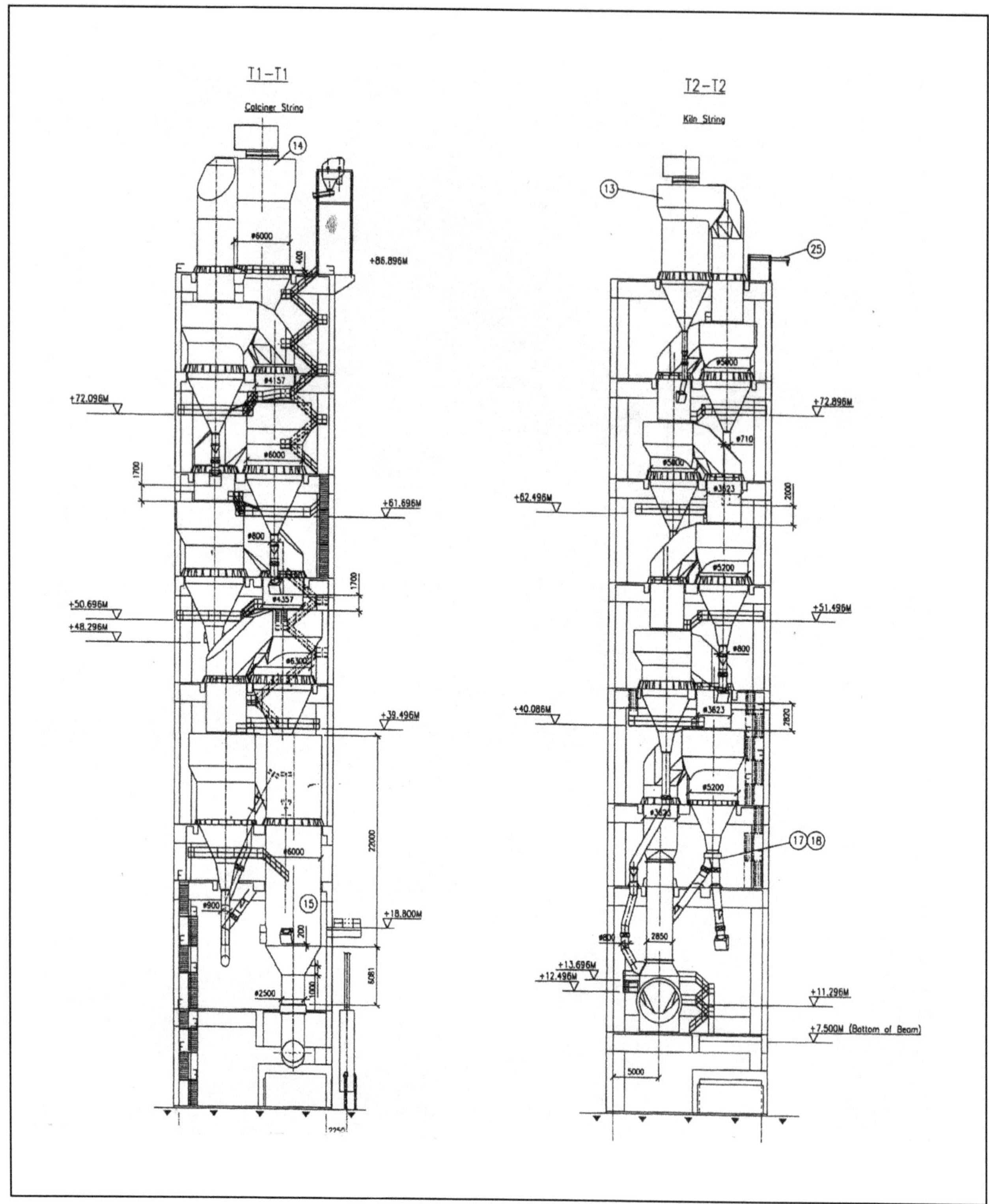

Fig. 6.15.50 Single stream 6 stage preheater with 'off-line' calciner.

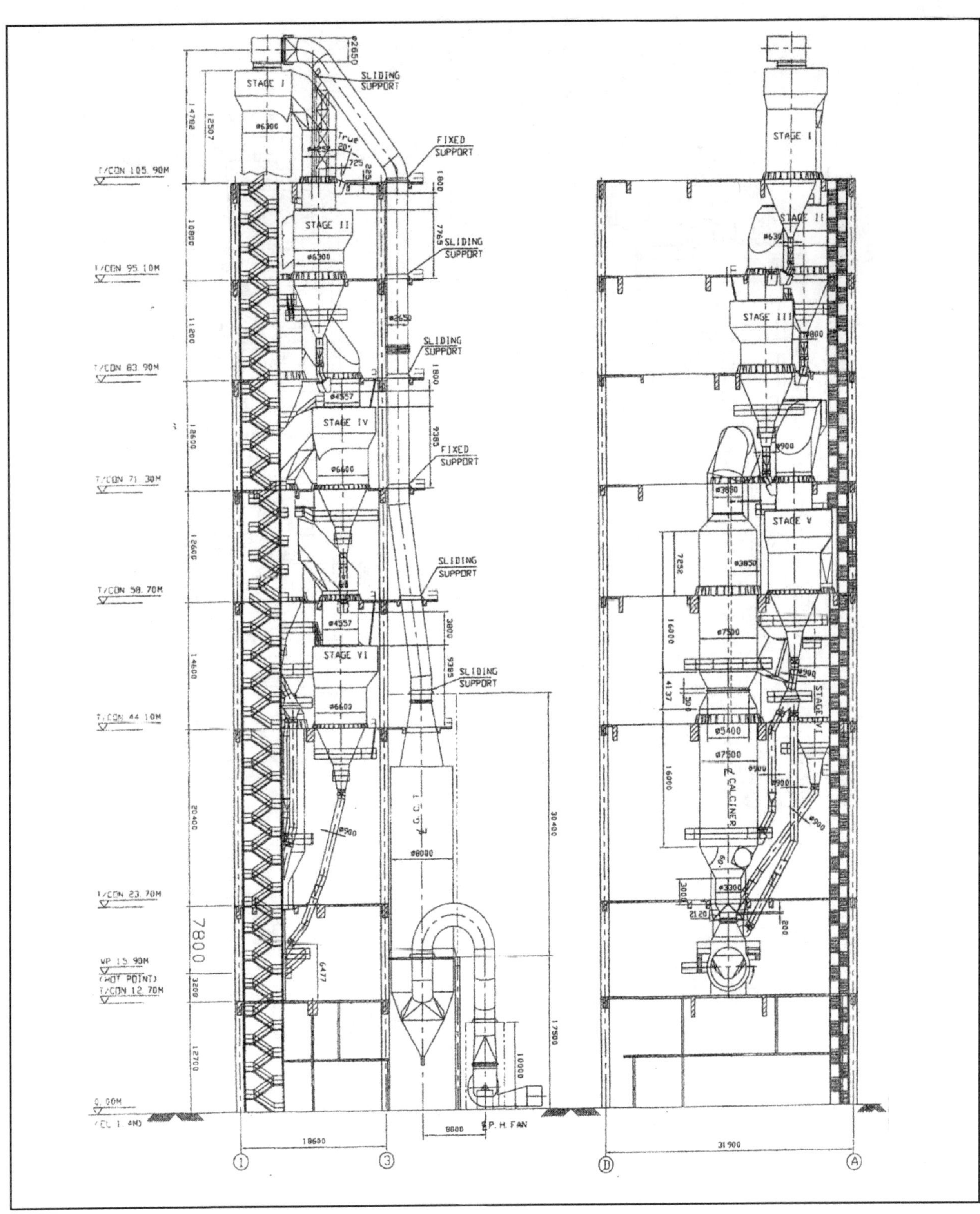

Note : G.C.T. before preheater fan.
Staircases in one line increase dimension of preheater tower.

Fig. 6.15.51 Single stream 6 stage preheater with 'in-line' calciner.

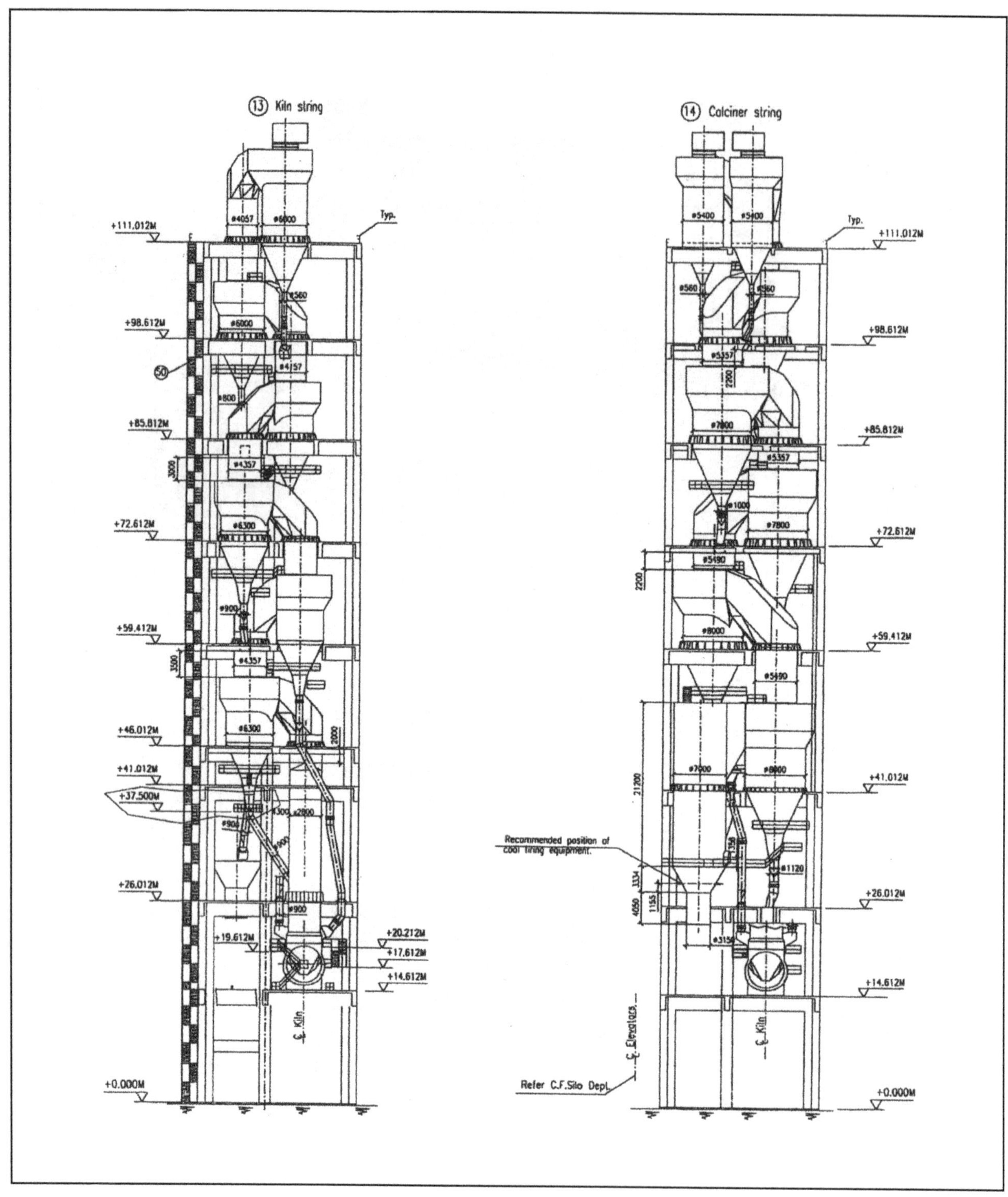

Note : Top twin cyclones are in open
Staircases totally outside the tower; reduces dimensions of tower
Height of tower - 110 meters

Fig. 6.15.52 6 stage preheater and off line calciner.

CHAPTER 16

PREHEATER FAN

16.1 Fans for Preheater

Preheater fan draws gases through the entire kiln preheater / calciner system.

It should be sized to handle the exhaust gases which consist of :

1. Products of combustion including excess air and leakage.
2. CO_2 released during calcining of raw mix.
3. Conveying air when pneumatic conveying like air lift is used.

16.1.1 Sizing of Preheater Fan

It should be sized for the design capacity and should have sufficient margin above it – say 10 to 15%.

If the margin in fan volumetric capacity is 15% then the margin for pressure should be $(1.15)^{1.8} = 1.29$

This is because, resistance in a given system is proportional velocity head which is expressed by density × $v^2/2g$ where v is velocity of gases flowing through the system which is proportional to volume through it which in turn is proportional to the capacity or rate of production.

According to fan laws,

V volume is proportional to speed 'n' in r.p.m.

P static pressure is proportional to n^2.

That is if volume of a fan is increased by increasing its speed, Static pressure developed by it would automatically be appropriate to increased system resistance for the increased volume.

16.2 Dust in Preheater Gases

Preheater fan has to handle dust escaping from preheater. Dust content is determined by cyclone efficiency.

Efficiency of top cyclones usually lies between 92-94 %.

If raw meal / clinker conversion ratio is 1.5 then to produce 1 kg clinker raw meal to be fed to preheater for efficiency of 93 % would be 1.5/0.93 = 1.61 kg / kg clinker.

If . sp. gas volume is 1.5 nm^3, dust concentration would be 0.11/1.5 = 0.07 kg/nm^3. Thus dust concentration fan would have to handle depends on sp. gas volume also.

16.3 Sp. Gas Volume

Sp. gas volume is expressed as gas volume leaving preheater in nm^3/kg clinker. It comprises of all the elements mentioned in para 16.1 above.

CO_2 released is 0.3 nm^3/kg clinker and conveying air is 0.1 nm^3/kg clinker if airlift is used for feeding raw mix to preheater. It is nil if bucket elevator is used.

Products of combustion are directly related to sp. fuel consumption and calorific value of coal. Total leakage in kiln preheater system is taken as 25 % of air required for combustion. It includes 10 % excess air deliberately admitted in the system to ensure complete combustion when fuel is coal.

16.3.1 Sp. Gas Volumes for Different Fuel Consumptions

Sp. gas volumes for different fuel consumptions using coal of 4500 kcal/kg and using airlift are shown in **Table 6.16.1.**

Table 6.16.1 Fuel Consumption KCal/Kg.

	Fuel Consumption Kcal/kg			
	700	750	800	900
Sp. Gas. Vol. in nm^3/kg clinker	1.4	1.5	1.6	1.7

At 93 % efficiency of cyclone –

dust escaping = 1.61 – 1.5 = 0.11 Kg/Kg Clinker.

∴ Dust loss in gm/nm^3 for above sp. gas volumes would be

gm/nm^3 79 73 69 65

Gas temperature at exit would depend on number of stages of preheater

No. of stages

temperature	4	5	6
°C	330 – 360	300 - 330	270 - 300

16.4 Pressure Drop Across Cyclones

Similarly pressure drops across the stages in the preheater would be (as an example for a preheater of same design) : See Table below

Table 6.16.2 Pressure drop across 4 to 6 stage Prheaters with cyclones of modern design.

	4 stage	5 stage	6 stage
		mm wg	
draft at kiln inlet	-20	-20	-20
pr.drop bottom cyclone	70	70	70
2nd stage	70	70	70
3rd stage	70	70	70
4th stage		70	70
5th stage			70
top cyclone	90	90	90
pr drop. Down across preheater	320	390	460
pr.edrop down comer duct	20	30	40
total upto fan	340	420	500

However, with development of low pressure design cyclones, presently, pressure drop in a 6 stage preheater is the same as that in a 4 stage preheater of older design that is between 400- 450 mm.

Considering that the sp. volume is also low with low sp. fuel consumption, and that gaes leave at lower temperatures, sp. power drawn by the preheater fan in kw/ton would be much less.

Drop in down comer duct for design purposes could be taken roughly as 40, 50 and 60 mm respectively for 4, 5 and 6 stage preheaters.

For calciner and t.a. duct, another 80 mm would be added.

Total static pressure for fan (with calciner in ciruit would be)

	4 Stage	5 Stage	6 Stage
old design	530	620	710 mmwg
New design	420	500	580 mmwg

16.4.1 Balancing of Gases

When calciner is off line, to balance flow of gases through kiln and calciner, a throttle is added in riser duct above the kiln. When there are two streams of preheaters with two independent fans, k-line fan would be designed for a static pressures of 450, 540 and 630 mms respectively for 4, 5 and 6 stage preheaters and c-line fan would be designed for static pressures of 530, 620 and 710 mms as mentioned above.

16.5 Design Margins in Preheater Fan

As mentioned earlier appropriate margins are to be added to arrive at fan capacity in terms of volume and static pressure.

In obtaining quotations for fan, the duty requirements should be obtained from designers of preheater for at least 2 sets of operating conditions which would then determine systems resistance curve. **See Fig. 6.16.1.**

Since fan efficiency is different at different operating points the point at which it should be maximum should also be specified.

16.6 Control of Fan Operation

Preheater fans have a louvre dumper, remote controlled, and also a variable speed drive for controlling the flow as required by kiln operation.

16.6.1 Control by Damper

In damper control, fan speed remains the same and hence power drawn remains same. Smaller flow is

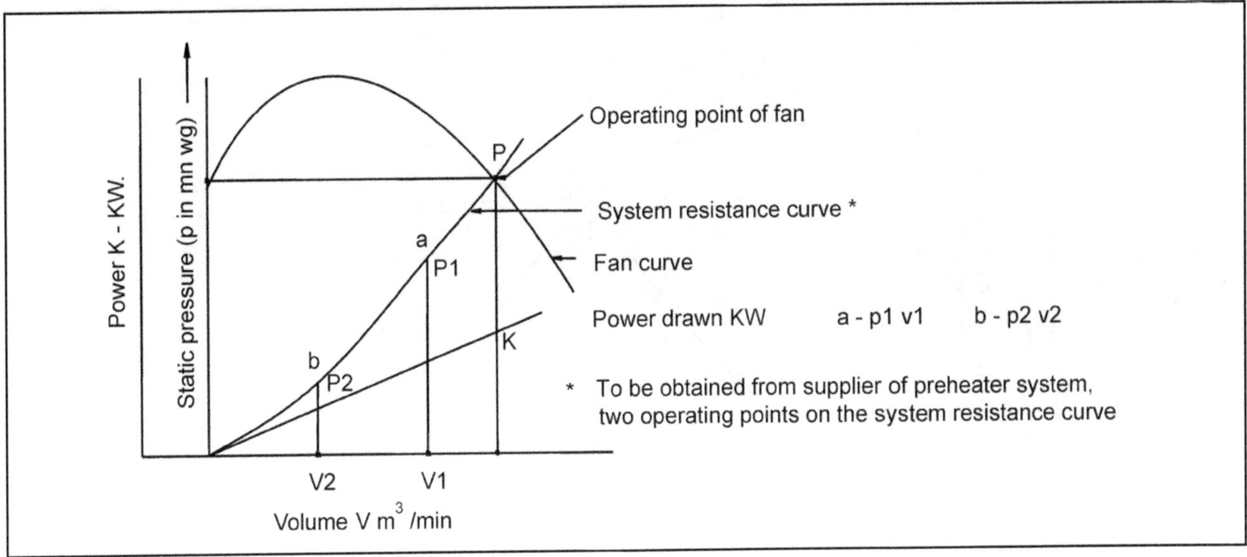

Fig. 6.16.1 Fan curve and system resistance curve.

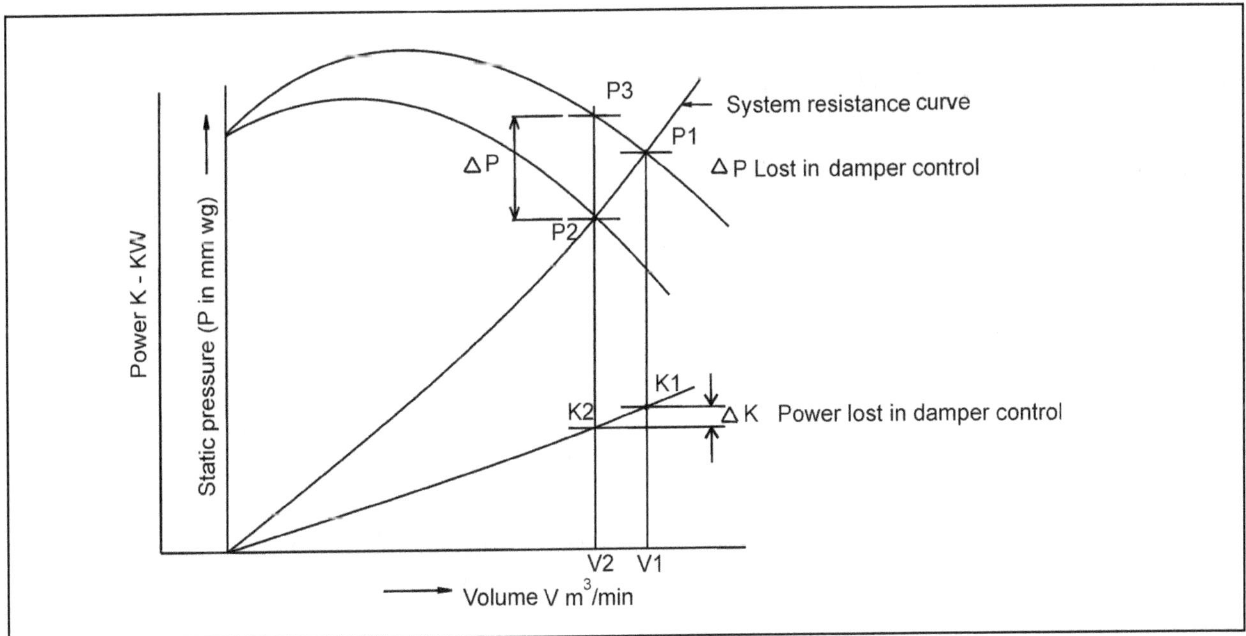

Fig. 6.16.2 Control of preheater fan by damper.

obtained (from V_1 to V_2) by controlling the damper, speed remaining the same. At this volume, fan should develop pressure P_3 but system resistance would drop to P_2. The difference P_3–P_2 is made up by the resistance of the damper.

See Fig. 6.16.2.

16.6.2 Control by Speed

Alternatively to obtain reduced flow, speed of fan could be reduced. Fan will develop reduced pressure appropriate to new speed which will be close to the system resistance at volume V_2.

Thus energy consumption wise, changing speed is a better method of control.

See Fig. 6.16.3.

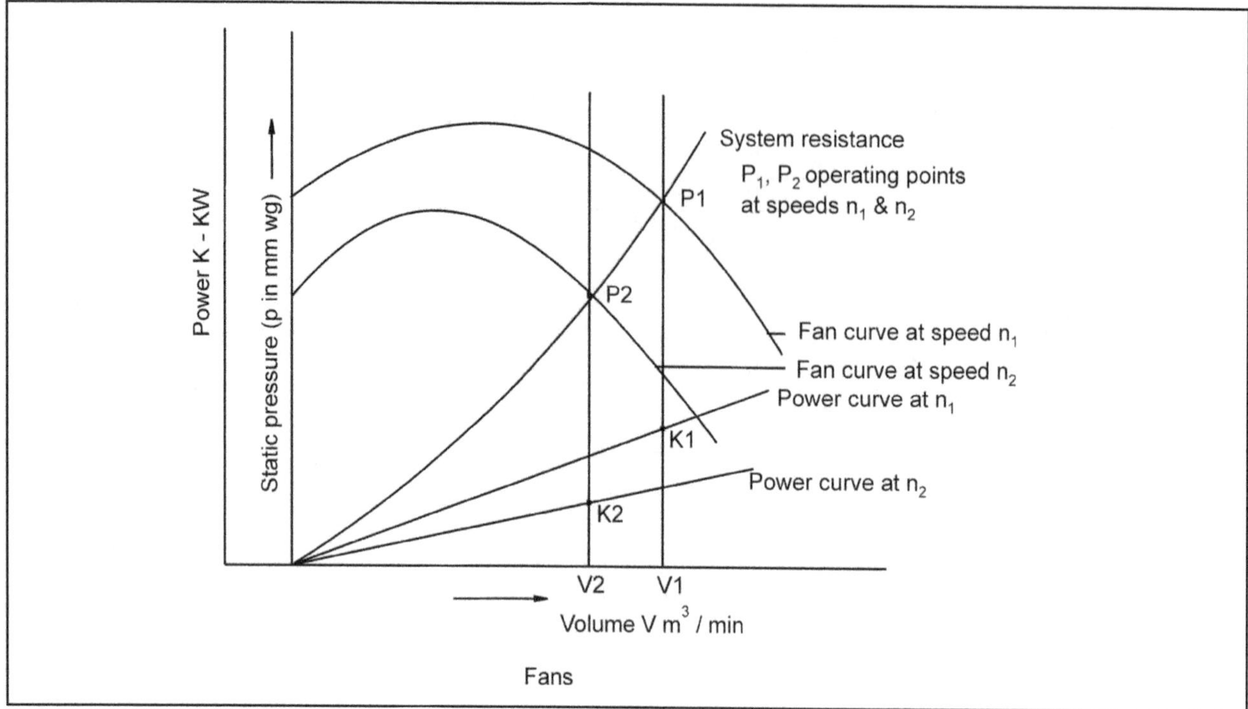

Fig. 6.16.3 Control by speed - no loss in power.

Damper is used to start fan on no load, with damper closed.

In actual practice, control may be using a combination of damper control and speed control.

16.7 Variable Speed Drives for Preheater Fan

1. Slip ring induction motor with rotor resistance.
2. Variable frequency – AC motor.
3. Variable slip - AC motor.
4. DC motor.

Slip ring induction motor is now seldom used for large plants, as it is wasteful of energy and because it gives only a stepped speed variation.

The other three give 'stepless' and smooth speed variation. 'variable slip' type drive is useful when fan normally runs at a much lower speed than the synchronous speed. Choice between AC variable frequency and DC would be from economic considerations.

16.8 Retrofitting of Fans

Because of dust content, earlier, radial bladed impellers were used for P.H. fans, which had poor efficiency.

Now backward curved bladed impellers are used and preheater fans have efficiencies of 70-77%. This considerably reduces power consumption.

It is possible to 'retrofit' fans of older design of poor efficiency by installing impellers of backward curved bladed design of higher efficiency in the existing casing. Since for same volume, power consumption would be less, in most cases the shaft and bearings of the existing fan could be used and also the casing.

16.9 Number of Preheater Fans

1. Single stream preheater will have one fan.
2. Twin stream preheater with identical cyclones can also have 1 fan.
3. With 2 streams preheater one for kiln and other for calciner, there would be 2 fans of different capacity and pressure drops in accordance with sizes and stages of 2 preheaters.

Two preheater fans could be installed under respective towers to save space and also to keep motors under cover.

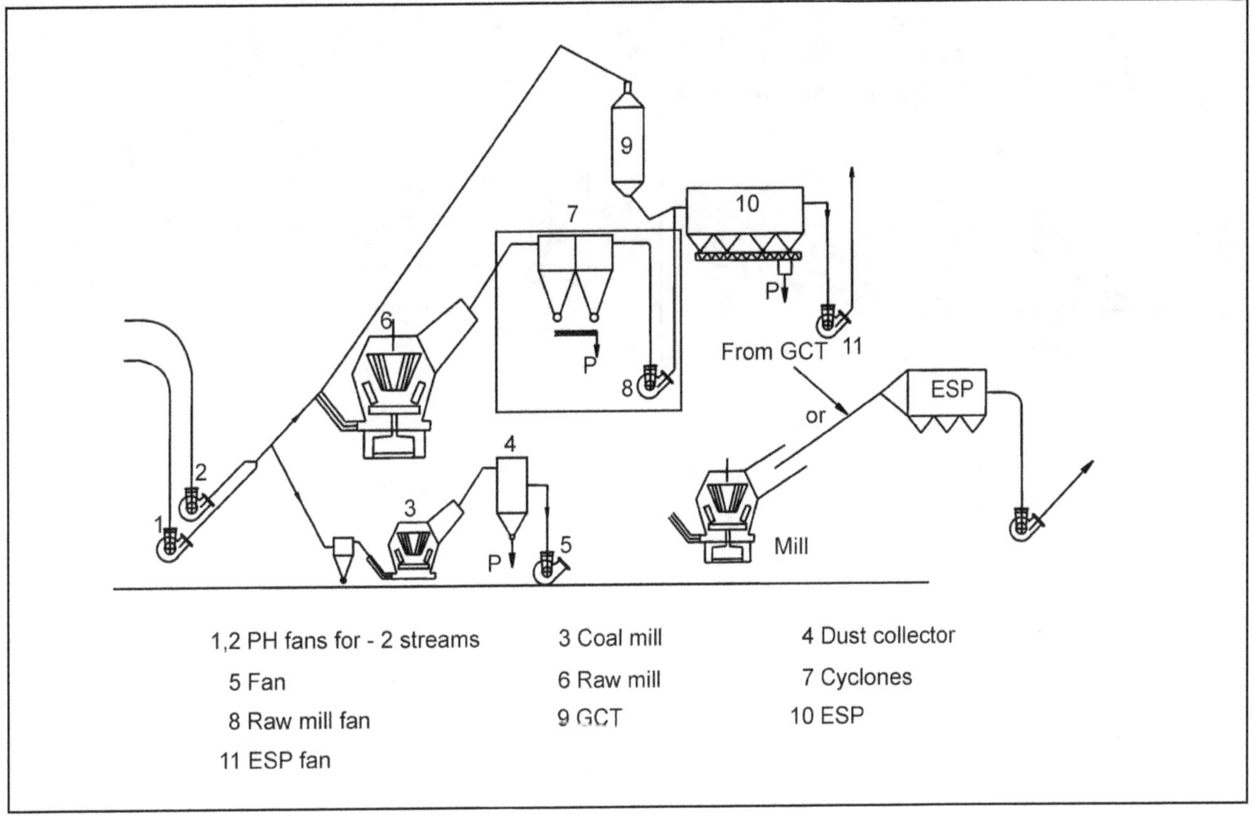

Fig. 6.16.4 Two stream preheater supplying hot gases to raw mill and coal mill.

Between them they would supply hot gases to raw mill and coal mill for drying now that coal mill is located at preheater end.

See Fig. 6.16.4.

16.10 Static Pressure at exit

Pressure at exit of fans should be –15 to –20 mm and draught in conditioning tower should be negative.

Total pressure drop in CT and ESP and ducting would be of the order of 80 mm.

If a cyclone and a mill fan is not included in mill system, the fan after ESP would have to be designed for a very high pressure drop.

See Fig. 6.16.4 and Table 6.16.3.

Table 6.16.3

Case	Pressure drop in mmwg in			
	GCT & ESP	Mill	Duct	Total
Direct operation	100	-	50	150
Combined operation	25	600	50	675

ESP casing also would have to be designed to withstand this pressure drop.

When only kiln is running (direct operation), then the system pressure loss between preheater fan and ESP fan being only 150-200 mm a fan designed for a pressure drop of 700 mm would be ill suited.

It should be possible to run the fan with reduced speed. So this ESP fan should also have a variable speed drive.

See Fig. 6.16.5.

16.11 Construction

Construction of preheater fan would depend on its size. It can be :

1. Overhung impeller for small sizes.

   See Fig. 6.16.6 a and b.

2. Impeller supported on both sides in bearings.

   See Figs. 6.16.7 and 6.16.8.

3. Single or double entry.

   See Figs. 6.16.7 and 6.16.9.

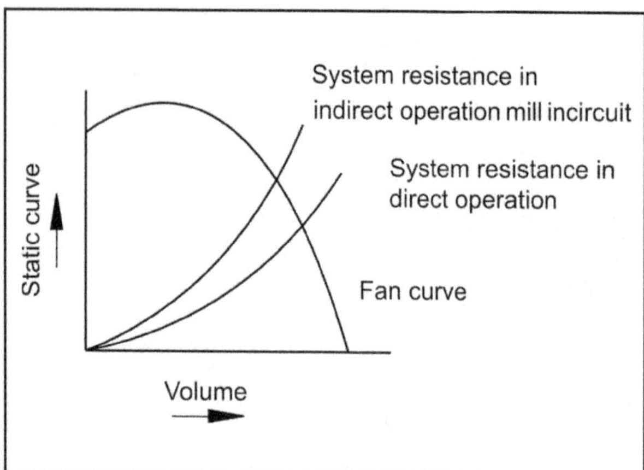

Fig. 6.16.5 Fan operating points to suit direct and indirect operation of kiln and mill system.

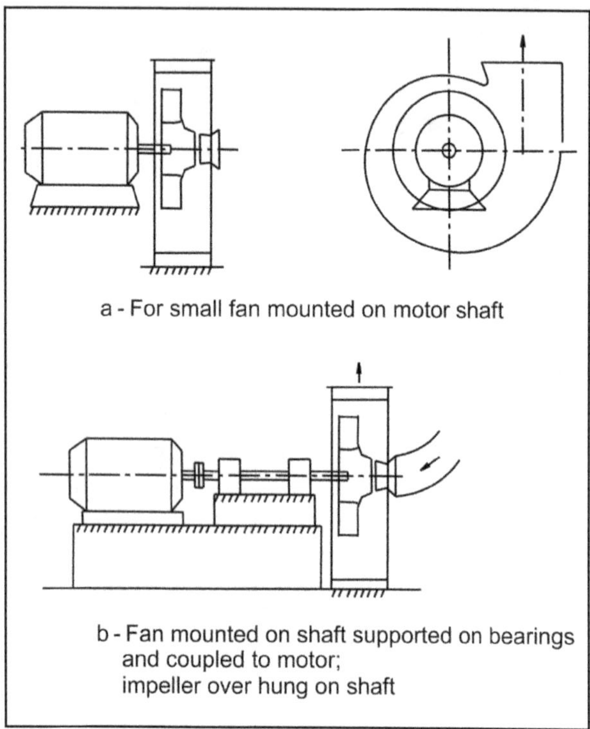

Fig. 6.16.6 Preheater fans with overhung impelller.

16.12 Vibrations of Preheater Fan

Preheater fan can be subject to vibrations because of dust content in gases.

It is possible that dust gets built upon impeller and causes vibrations. To avoid this compressed air may be injected on the impeller to blow away build ups of dust.

See Fig. 6.16.10.

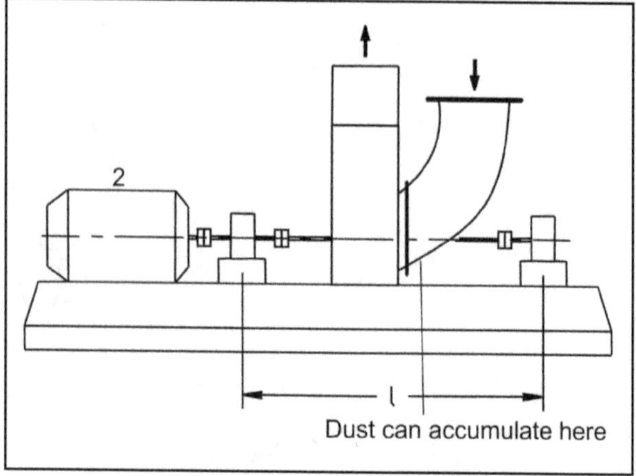

Fig. 6.16.7 In case of large fans – span between bearings increases; dust can accumulate at inlet.

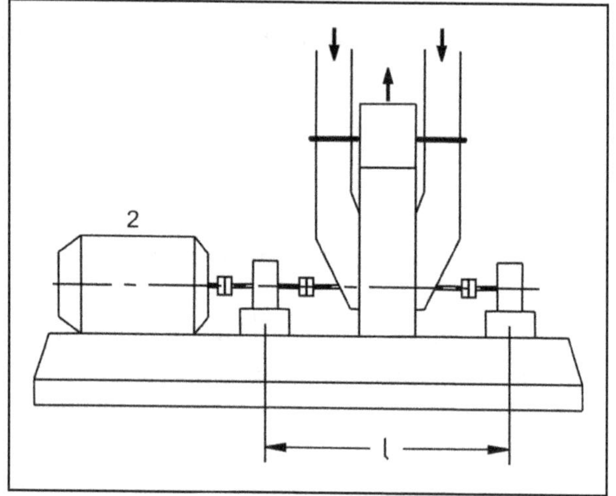

Fig. 6.16.8 Double entry reduces span, prevents accumulation of dust.

16.12.1 Isolators for Vibrations

The fan should be isolated on foundations by vibration dampers so that vibrations are not transmitted to buildings. Suppliers recommendations should be obtained in this respect.

See Figs. 6.16.9 and 6.16.11.

16.13 Ducting Inlets and Outlets for Fan

Another important point in layout of preheater fans is the ducting. The slope of the ducting entering and leaving preheater fan should be minimum 60° with horizontal to prevent accumulation of dust.

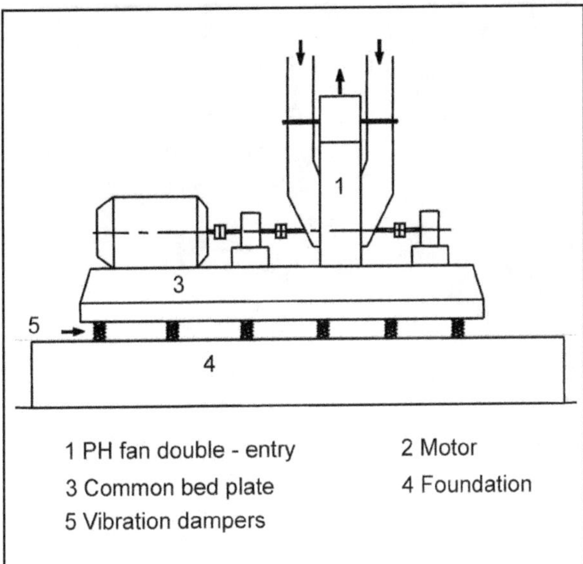

Fig. 6.16.9 Preheater fan – vibration dampers – common base frame – for motor and fan.

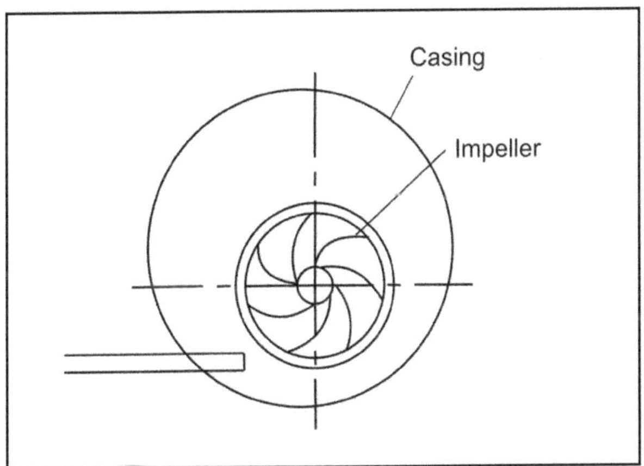

Fig. 6.16.10 Cleaning of impeller blades by compressed air.

Accumulated dust may suddenly flush down the duct and load the impeller. Bends should also be avoided.

For large fans when impeller needs to be supported on both sides, bends could be avoided by using a double entry fan.

See Figs. 6.16.12 and 6.16.8.

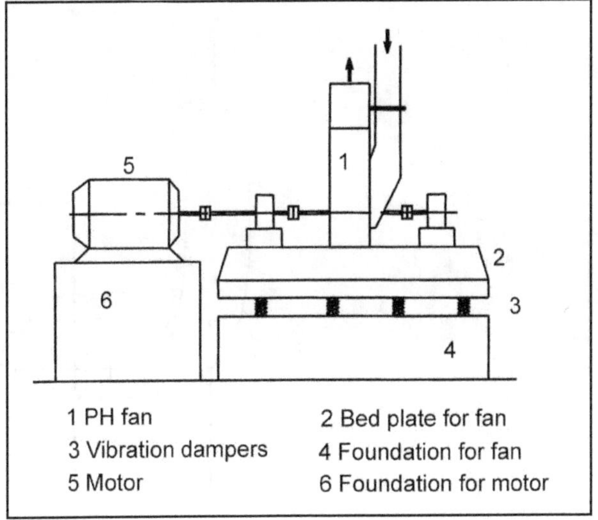

Fig. 6.16.11 Preheater fan and motor on separate foundations.

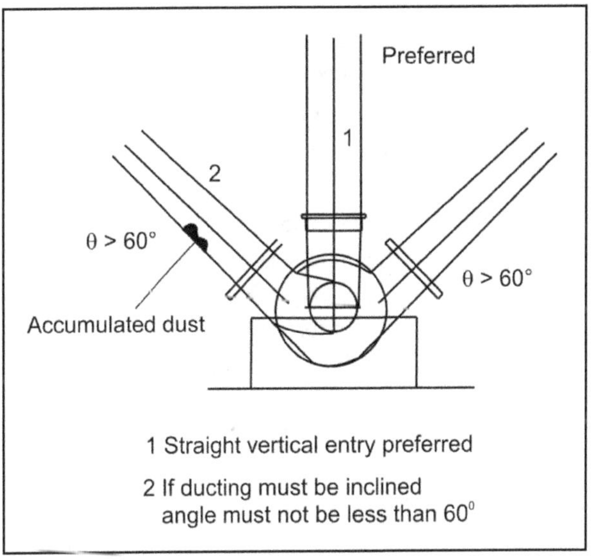

Fig. 6.16.12 Inlet and outlet of PH fan.

16.14 Expansion Joints

Fan should be isolated from the long incoming down comer duct by an expansion joint. Load of ducting should not come on expansion joint. Expansion should preferably in one plane.

Expansion joint should also be installed in the outlet ducting preferably close to the fan and ducting should be supported in such a way that will permit expansion without transferring any load on the joint.

See Fig. 6.16.13.

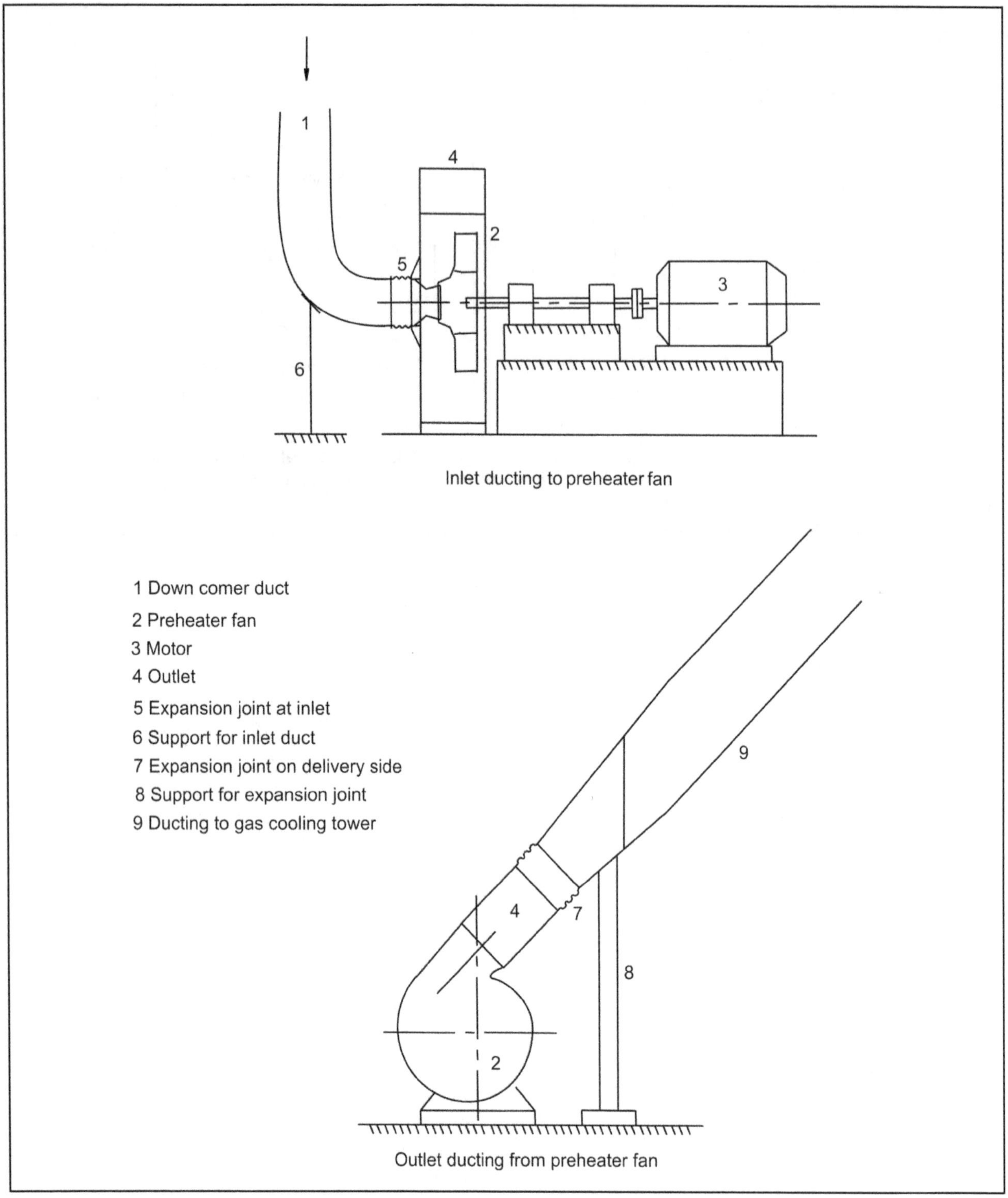

Inlet ducting to preheater fan

1 Down comer duct
2 Preheater fan
3 Motor
4 Outlet
5 Expansion joint at inlet
6 Support for inlet duct
7 Expansion joint on delivery side
8 Support for expansion joint
9 Ducting to gas cooling tower

Outlet ducting from preheater fan

Fig. 6.16.13 Inlet and outlet ductings of preheater fan.

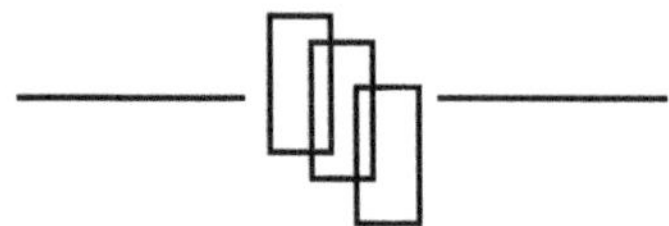

CHAPTER 17

CALCINERS

17.1 Calcination in and Outside Kiln

In a preheater kiln, raw meal enters it, 28-30% calcined. This is because only the bottom 2 cyclones have temperatures suitable for calcination. Cyclones above do the job of preheating the raw meal.

See Fig. 6.17.1.

In kilns with calciners calcinations is almost wholly completed before raw meal enters the kiln.

Calcination is an endothermic process.

$$CaCO_3 = CaO + CO_2 - 420 \text{ kcal/kg}$$

This chemical reaction takes place at constant temperature. Fuel fired in the calciner generates temperatures of 1000-1200 °C and the heat released is immediately absorbed by raw meal and temperature comes down to 850-900 °C.

See Fig. 6.17.2.

Longer the retention time in calciner more complete is the combustion of coal fired and more complete is the process of calcination and lower the temperature of gases leaving calciner.

Fuel fired in the calciner should burn off completely in the calciner itself so that there is no CO in gas leaving the calciner.

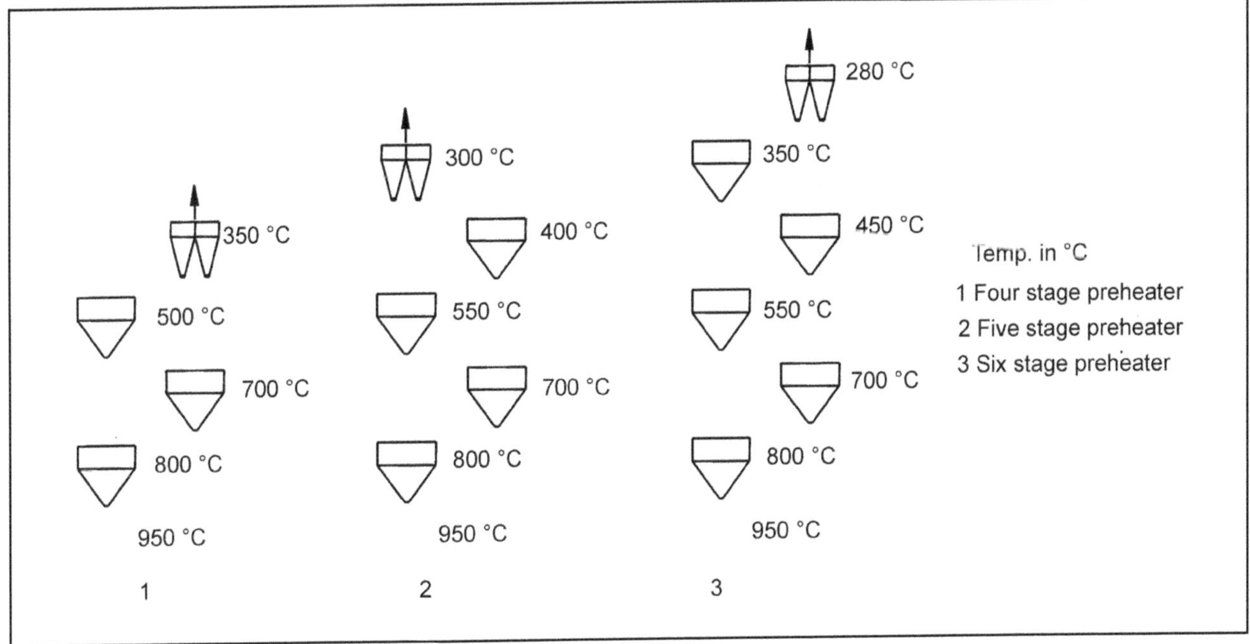

Fig. 6.17.1 Temparature profile in preheaters of various number of stages.

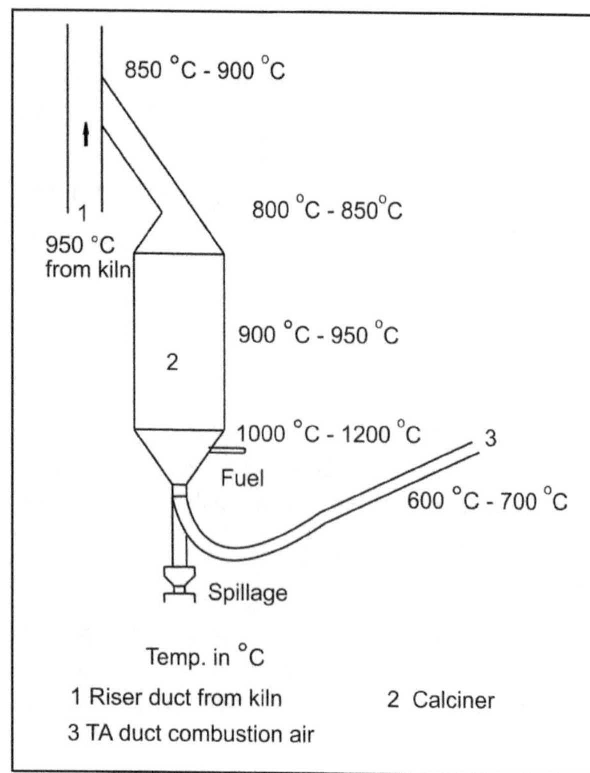

Fig. 6.17.2 Temperature profile in pre calciner.

17.2 Designs of Calciners

There are a great many designs of calciners developed by different designers to suit specific conditions.

The designs have been classified in different groups according to their features. Major designs have been shown schematically in **Figs. 2.9.2 to 2.9.5 in Chapter 9 of Section 2.**

Initially calciners were developed for firing oil as fuel. However, now all designs can fire coal and other fuels and even combustible wastes.

17.3 Secondary Firing

The simplest calciner is an enlarged riser duct. Fuel is fired in the riser duct. Air for combustion can come through the kiln itself, as would be the case for kilns with planetary coolers calcining is partial; it is also called 'secondary firing'. Capacity of kiln will be increased in proportion to fuel fired in rising duct.

Degree of calcinations could be increased up to 50 %. Small plants found this to be a very economical way of increasing kiln capacity till they could install a full calciner.

See Figs. 6.17.3 and 6.17.4.

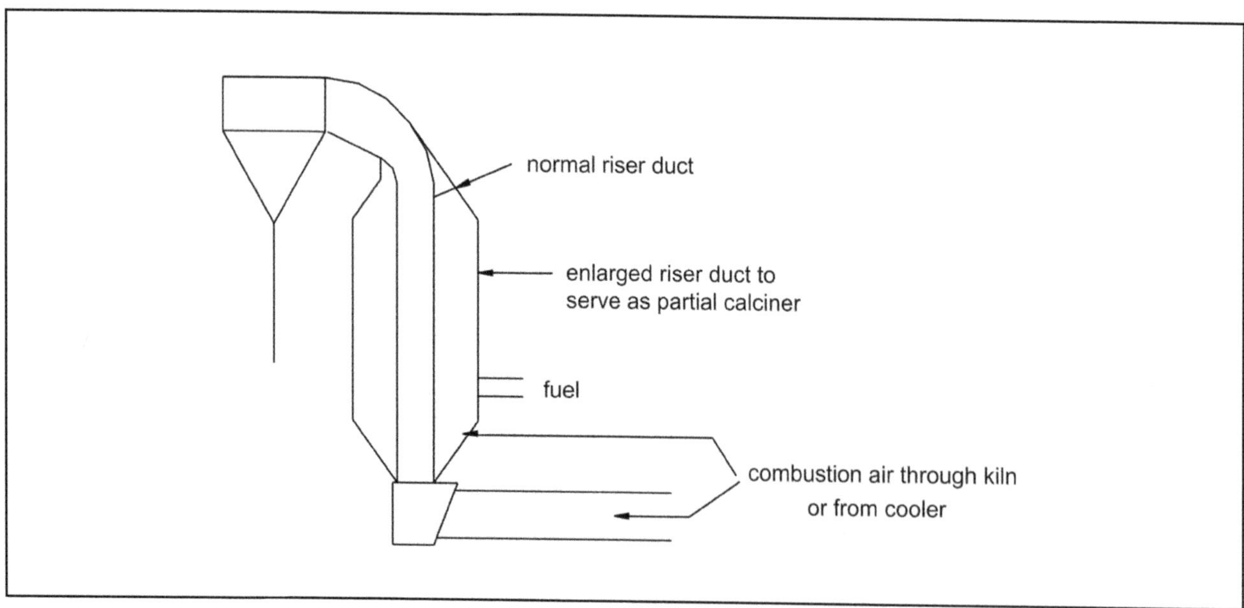

Fig. 6.17.3 Partial calcining or secondary firing.

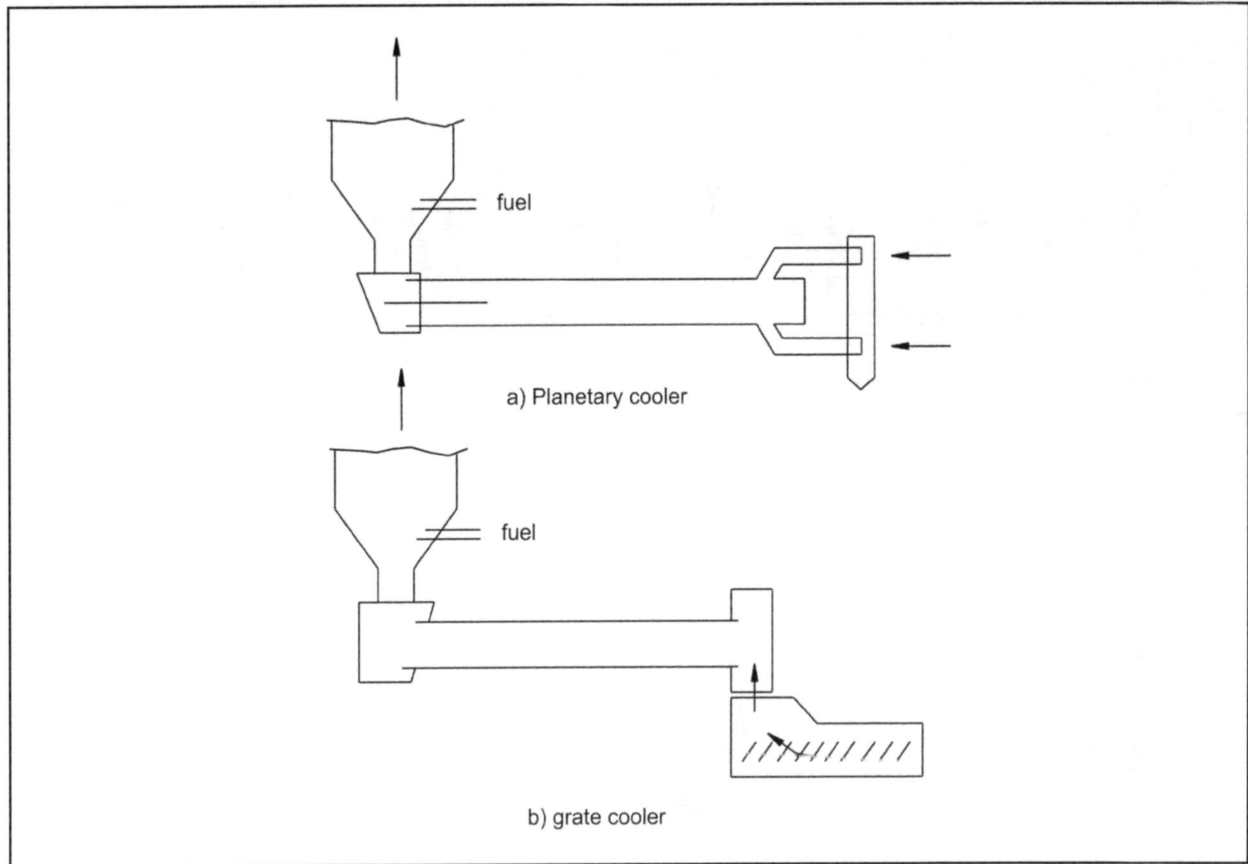

Fig. 6.17.4 Partial calcining; air for combustion through kiln.

When fuel fired in the riser duct was between 10 and 20 %, air for combustion could be drawn through kiln. Therefore secondary firing was also suitable for kilns with planeteary coolers.

See Fig. 6.17.4.

17.4 Classification of Calciners

In **Chapter 9** of **Section 2,** various ways of classifying claciners have been explained.

17.5 In Line Calciners

In line calciners replace the riser duct – calciner can be in the form of a cyclonic vessel or a duct with constrictions or a duct with a bend to increase time of retention.

See Fig. 6.17.5.

The height of calciner is such that the retention time is at least 4-6 seconds.

The velocity of gases in a calciner is about 4-6 m/sec. Therefore, height of a calciner would be 16-30 meters.

In some designs, to save space and heights the duct is bent.

See Fig. 6.17.5.

17.5.1 Flow of Gases and Raw Meal

Since all calcined raw meal must enter the kiln, bottom cyclone is used as a separating cyclone that receives material from both streams in case of a 2 streams preheater. Thus usually, calciner will receive raw meal from last but one stage of cyclone of the streams whether one or two.

In case of a calciner with two stream preheaters, calciner receives raw meal from cyclone No. 2 (from bottom) of each stream, calciner outlet is connected

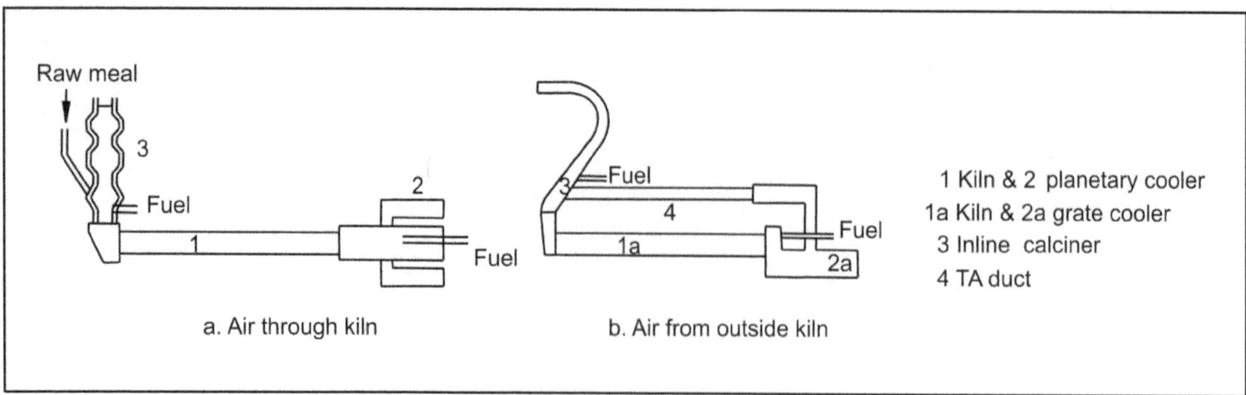

Fig. 6.17.5 In line calciners

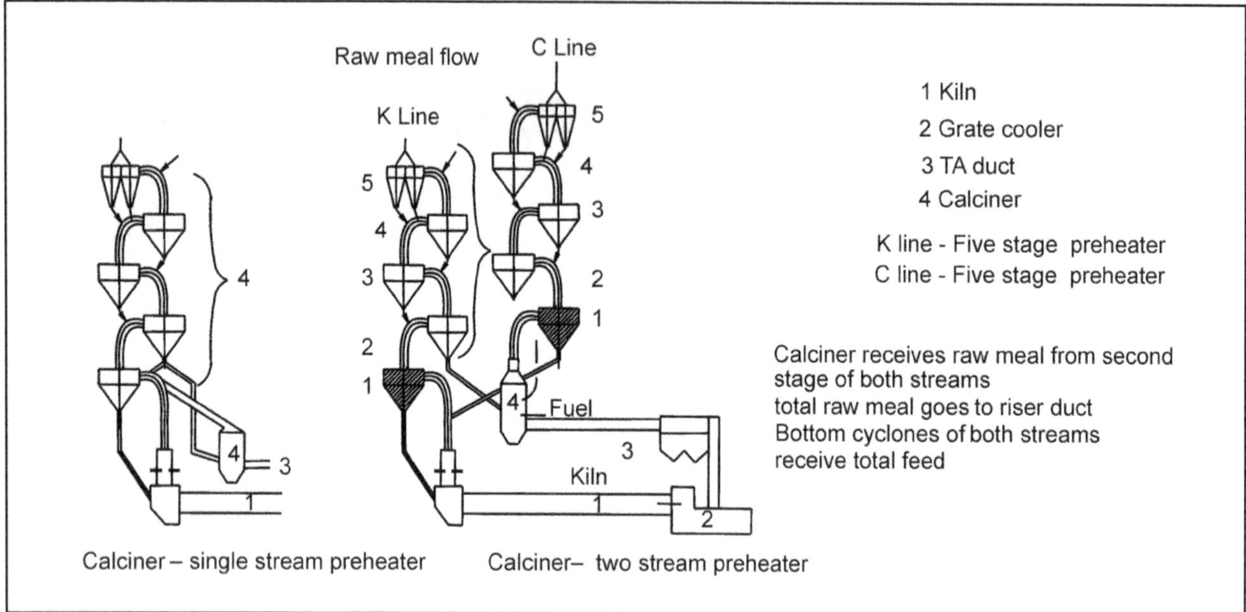

Fig. 6.17.6 Flow of raw meal into and from calciner in single and two stream preheaters for off line calciners.

to inlet of cyclone 1 (from bottom) of C Stream. Raw meal (a + b) enters riser duct of kiln and enters cyclone 1 of K stream. From there it enters kiln.

See Fig. 6.17.6.

There are a great many variations of this basic concept all aimed at improving the efficiency of calcination.

17.6 Off-Line Calciners

In this case, calciner vessel is not a part of the riser duct. The calciner outlet is connected to the riser duct and the kiln and calciner gases enter the bottom (or 1st) cyclone, in case of a single stream preheater and bottom cyclone of C-Line preheater in case of two stream preheaters.

See Fig. 6.17.6 to 6.17.8 and 6.7.13.

Off-line calciners have several advantages over in line calciners in that in most cases calciners can be supported from ground level reducing over all height of preheater.

It is the best solution for a running plant where a calciner is added afterwards. In most such cases preheater tower would neither have room for housing the calciner nor would it be designed civil design wise

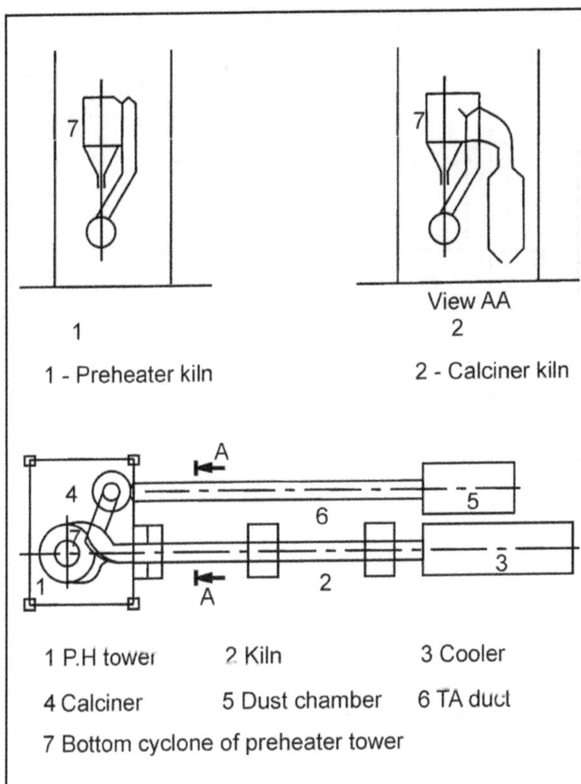

Fig. 6.17.7 Off line calciner to be added later – inside preheater tower – would require bigger tower.

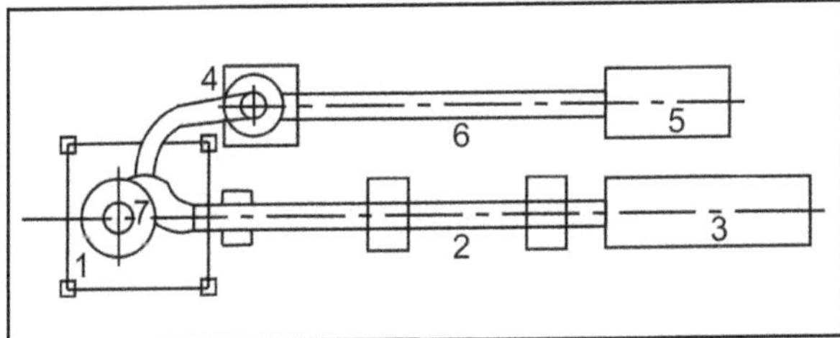

Fig. 6.17.8 Off line calciner out side the preheater tower of normal dimensions.

to support its weight. An off-line calciner can be conveniently housed either inside or outside the tower and connected to the kiln.

See Figs. 6.17.7, 6.17.8 and 6.17.13.

17.6.1 Spillage in Off Line Calciner

In an in line calciner all that material which is not carried away with gases, enters kiln and there is thus no problem of spillage.

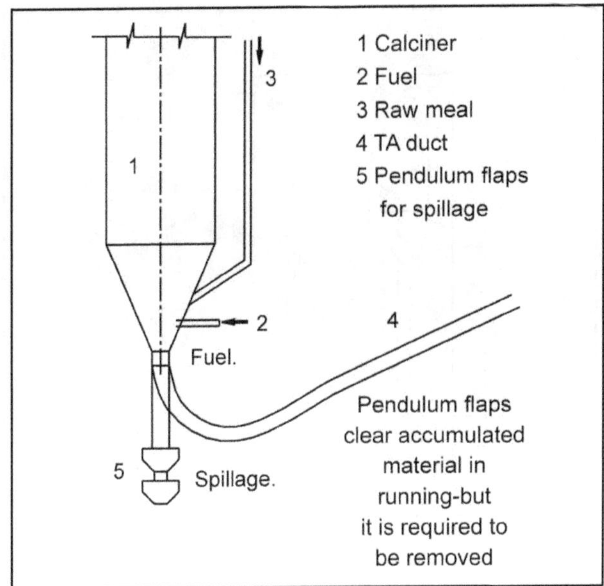

Fig. 6.17.9 Spillage in FLS off line calciner.

In off-line calciners, the material so collected has to be disposed of. This material is very hot at 900 °C and handling it is difficult and therefore hazardous.

In one design, the air for combustion enters from the bottom and carries raw meal with it in the spouted zone. There is a possibility of raw meal dropping out of air stream; unless this accumulated material is periodically removed, it can block entry of air for combustion.

See Fig. 6.17.9.

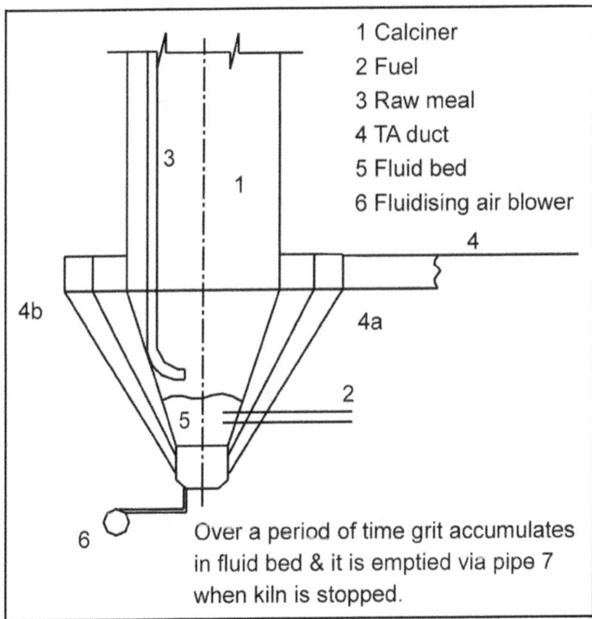

Fig. 6.17.10 Spillage in off line fluid bed calciner.

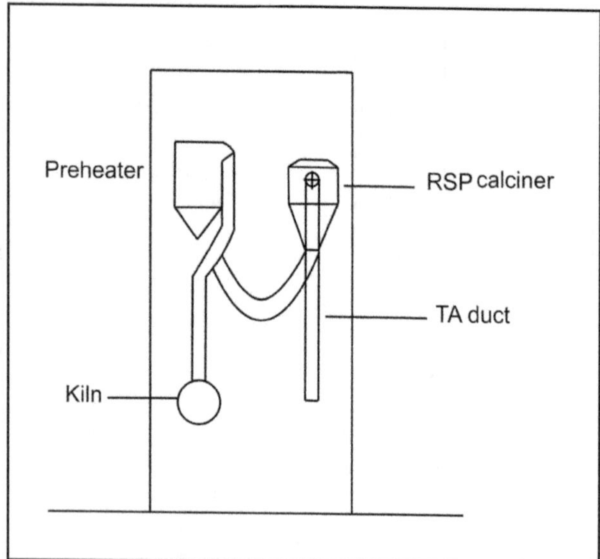

Fig. 6.17.11 RSP calciner supported inside tower; hence tower to be designed to take its weight.

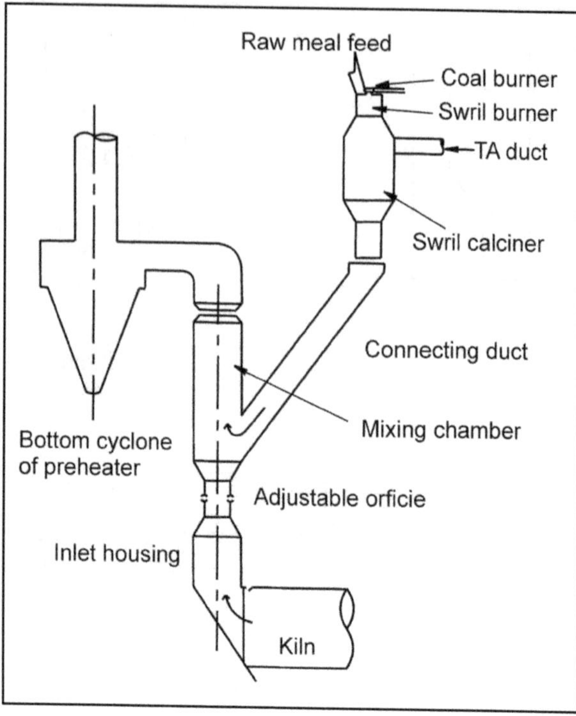

Fig. 6.17.12 Reinforced suspension preheater calciner.

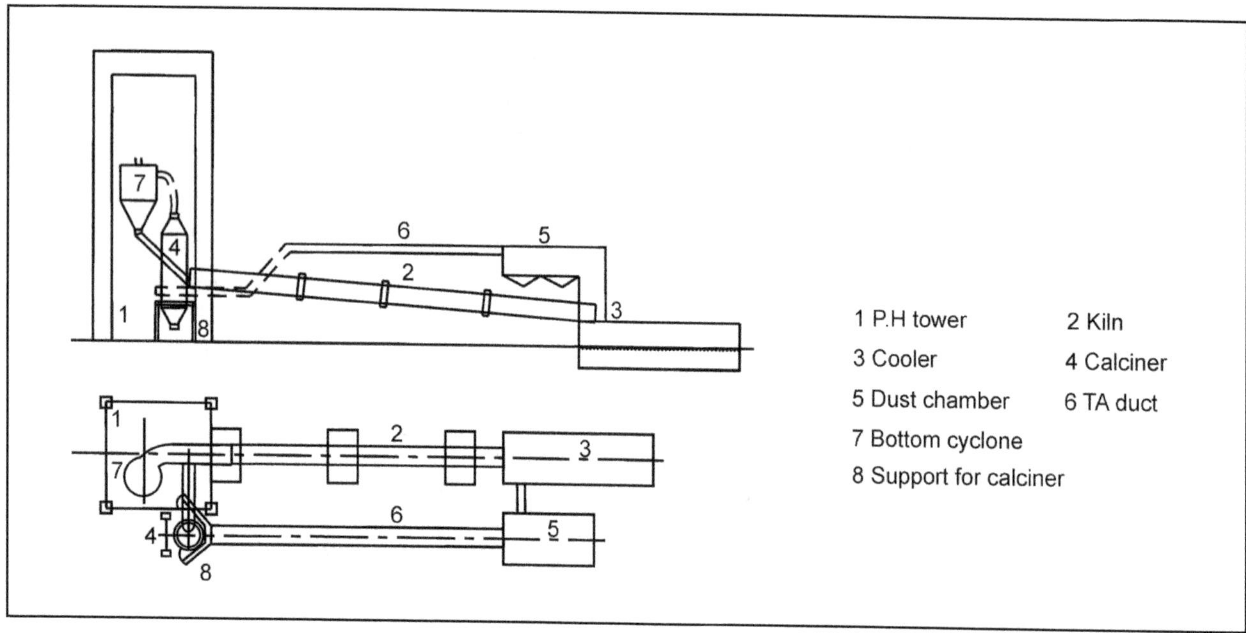

Fig. 6.17.13 Mitsubishi fluid bed calciner supported from ground.

In fluid bed design; there is no spillage but continuous accumulation of material in it can affect efficiency of the fluid bed.

See Fig. 6.17.10.

In RSP design raw meal enters the calciner at top with fuel. It enters mixing chamber and is lifted into preheater cyclone. Therefore there is no problem of spillage.

See Figs. 6.17.11 and 6.17.12.

17.7

Integral parts of a calciner system are:

1. calciner vessel,
2. entry of raw meal,
3. entry of coal,
4. entry of tertiary air for combustion,
5. tertiary air duct and dust chamber bringing hot air from grate cooler,
6. connection of calciner to preheater.

   See Fig. 6.17.14.

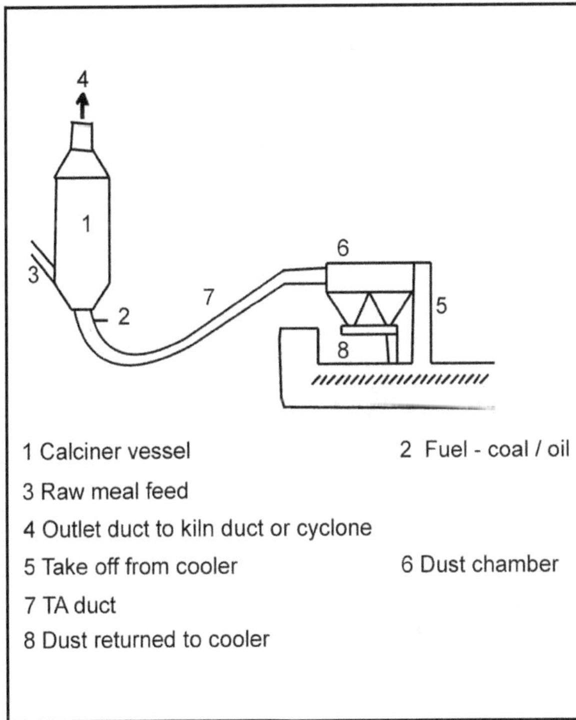

Fig. 6.17.14 Various parts of a calciner system.

Configuration of calciner and connecting duct and connections for air for combustion and fuel and raw meal are shown in for some well known calciners in **Fig. 6.17.15.**

17.8 Division of Combustion Air

In a kiln with a grate cooler and calciner, preheated air from the 1st two compartments of the cooler is divided into secondary air for kiln and tertiary air for calciner, in proportion to quantities of fuel fired in kiln and calciner respectively.

Normally, this proportion is 40 : 60. Thus if 0.8 nm^3/Kg air is drawn through cooler for combustion, 0.32 nm^3 would go into kiln and 0.48 nm^3 through TA duct to calciner.

See Fig. 6.17.16.

Tertiary air can be drawn from cooler or from kiln hood.

ee Figs. 6.17.17 and 6.17.18.

In **Fig. 17.18,** the gases are divided at 'a' as shown.

	a - b - g - f	-	through kiln
and	c - d - e - f	-	through calciner

Since kiln is a long cylinder with large diameter, pressure drop across a - b - g - f is much less than through the calciner leg. i.e., c - d - e - f.

To maintain gas flows in proportions to coal fired, an artificial resistance is introduced at 'g' by means of an adjustable throttle, in case of single stream preheater or in case of a 2 stream preheater with one fan.

In case of 2 streams preheaters however, this is not required. The independent preheater fans draw air as required through two separate routes as mentioned in **para 16.4.1** of **Chapter 16.**

But, Preheater fan in calciner line would have to be designed for a higher pressure drop through calciner and TA duct.

See Fig. 6.17.19.

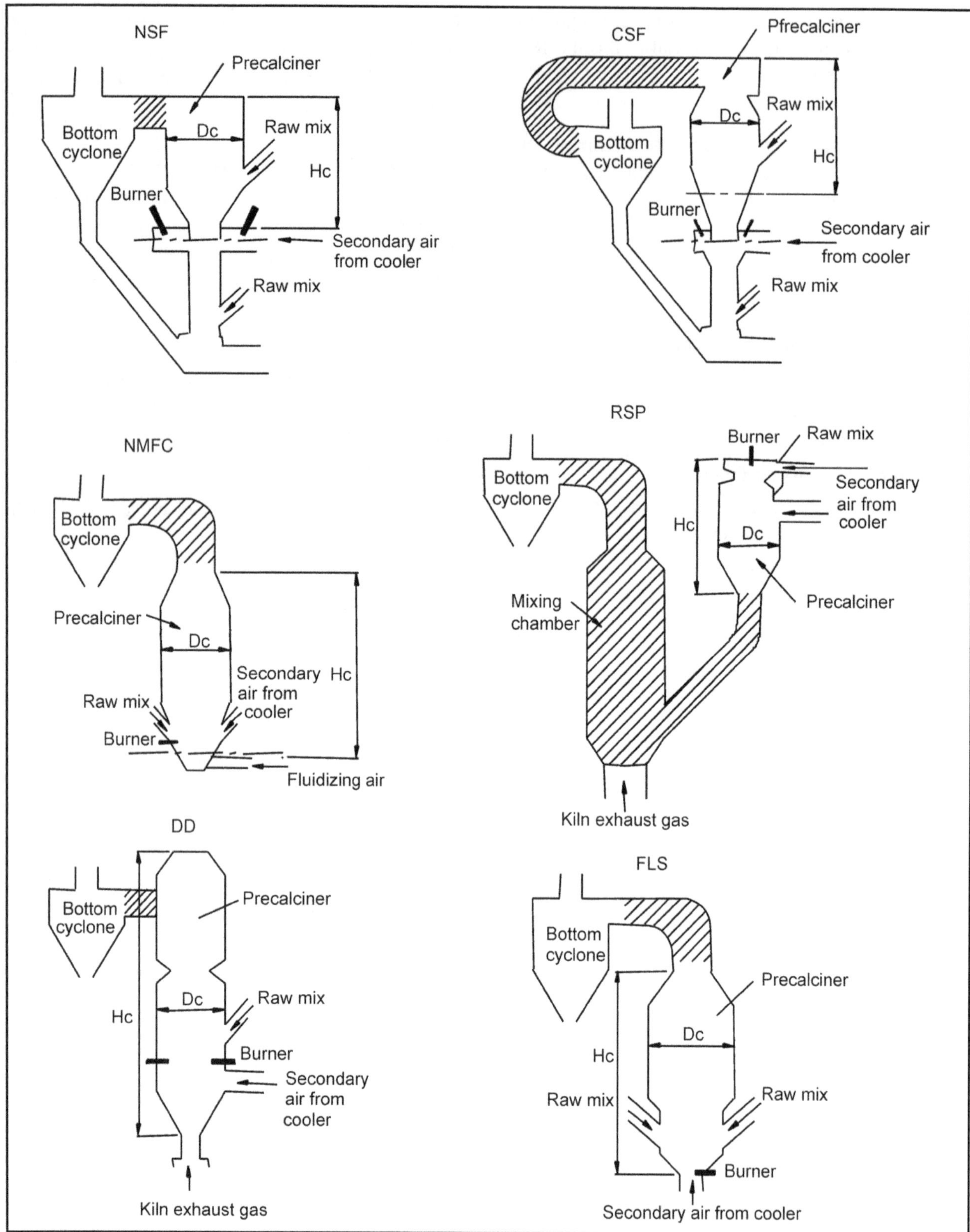

Fig. 6.17.15 Calciners and connecting ducts; rough relative proportions for different designs; also note relative locations of entry of raw meal and fuel.

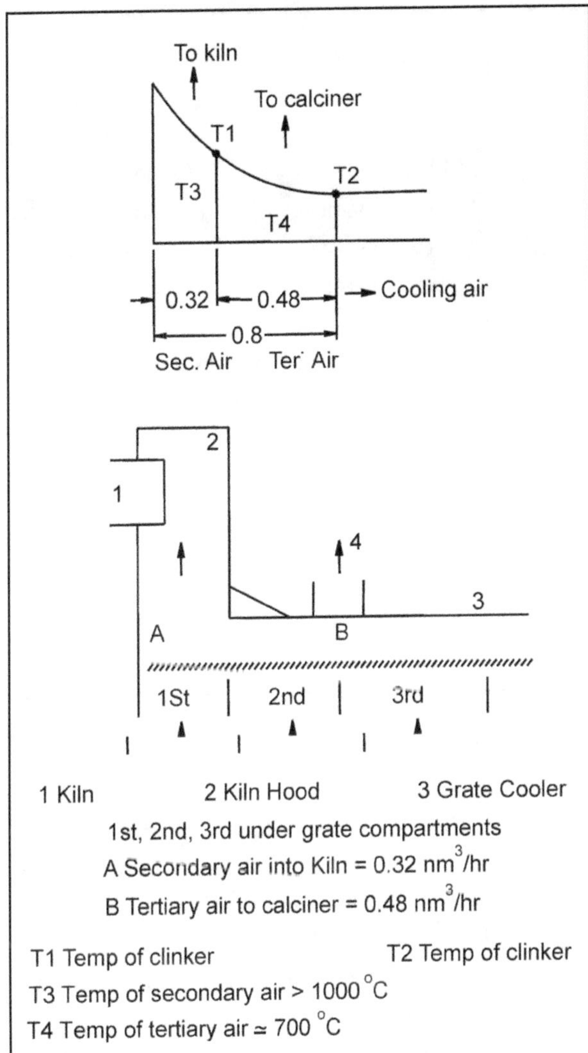

Fig. 6.17.16 Take off point for teritary air for combustion in calciner.

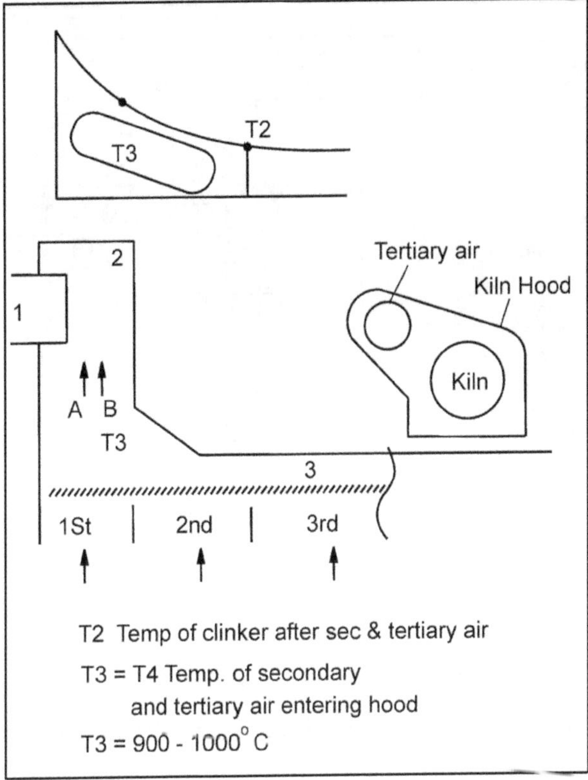

Fig. 6.17.17 Tertiary air taken from hood Low sec. air temp, high t.a. temp.

17.9 Entry of Raw Meal

Raw meal from the last but one cyclone/s of single/ two stream preheaters is admitted into the calciner. Point of entry is such that meal gets maximum time in the calciner for calcination. It is admitted into the calciner through a refractory lined raw meal pipe.

Unlike distribution boxes of preheater cyclones, meal just enters the calciner. In spouted bed designs, raw meal tends to drop down and is picked up by air for combustion and combustion gases. There is re circulation here. In fluid bed design, raw meal is admitted just above the fluid bed. It remains in fluid bed for about 30 seconds. In Reinforced Suspension Calciner, raw meal enters the swirl calciner from top.

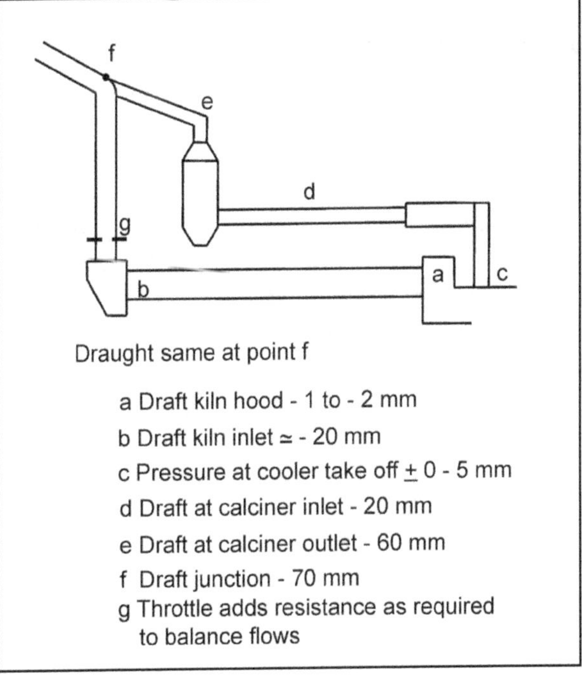

Fig. 6.17.18 Balancing gas flows in kiln and calciner.

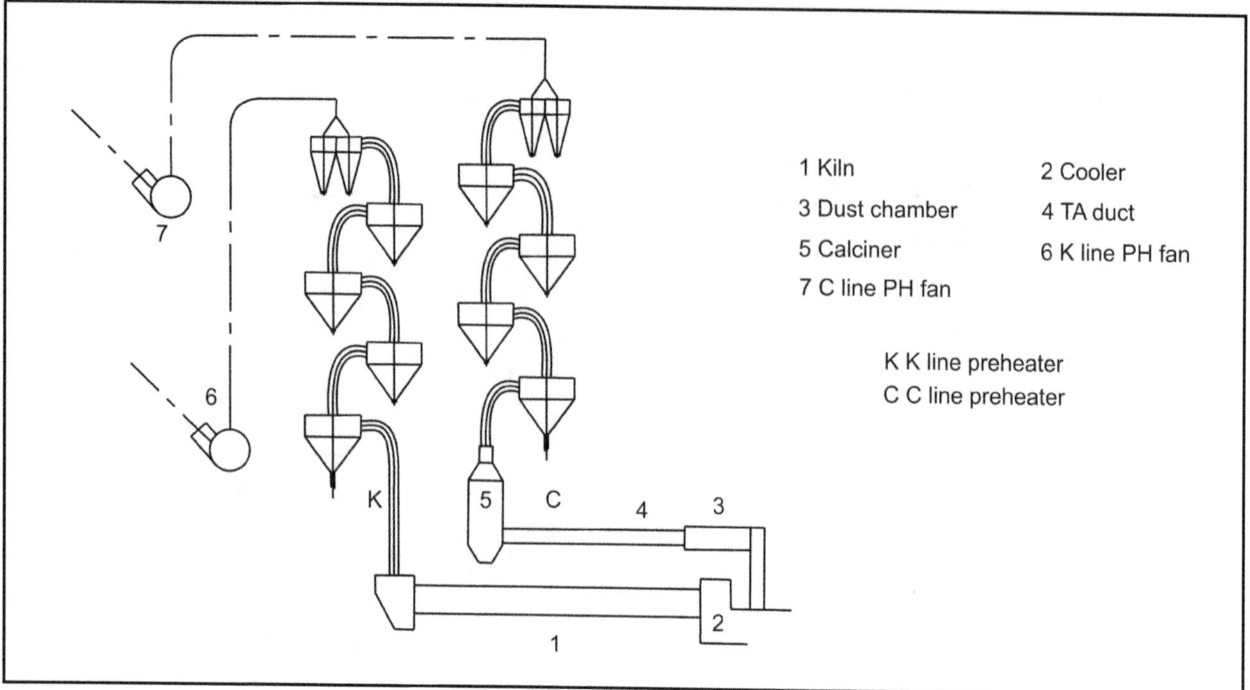

Fig. 6.17.19 Independent streams with separate fans do not need throttle.

Coal is admitted close to the entry of raw meal just under it. In RSP, coal is also fired from the top.

See Figs. 6.17.9, 6.17.10 and 6.17.12.

17.10 Firing Fuel in Calciner

Pulverised fuel is normally fired through pneumatic conveying systems. As temperatures in calciner need not be > 1000 °C, flame is not necessary. It is necessary that fuel is burnt and heat released is absorbed immediately. For this reason, fuel entry point is generally just above entry of air for combustion and just below entry of raw meal. RSP is the exception where both raw meal and coal are admitted from top in swirl calciner.

See Figs. 6.17.9, 6.17.10 and 6.17.12.

Relative positions of air, raw meal and fuel for different calciners are shown in **Fig. 6.17.15.**

Coal is admitted through short nozzles at velocities of about 20 m/sec. FK Pump systems are often used for this purpose. It is also possible to use dense / lean phase conveying system using ejector.

Metering system for coal would be similar to that for the rotary kiln.

When quantity of coal fired is more than 3-5 tons per hour, two burners would be used. Coal carrying pipe line is split and branched into two so that coal enters at two points- sometimes at diametrically opposite points in the calciner vessel. **See Fig. 6.17.20.**

In low NO_x calciners fuel is fired at two levels in the calciner. **See Fig. 6.17.21 and Plate 6.17.1.**

It is also possible to use waste fuel in calciner. If retention time is long enough, coal crushed to – 5 mm size or rice husk or shredded tyres could also be introduced into calciner through a feed pipe for burning.

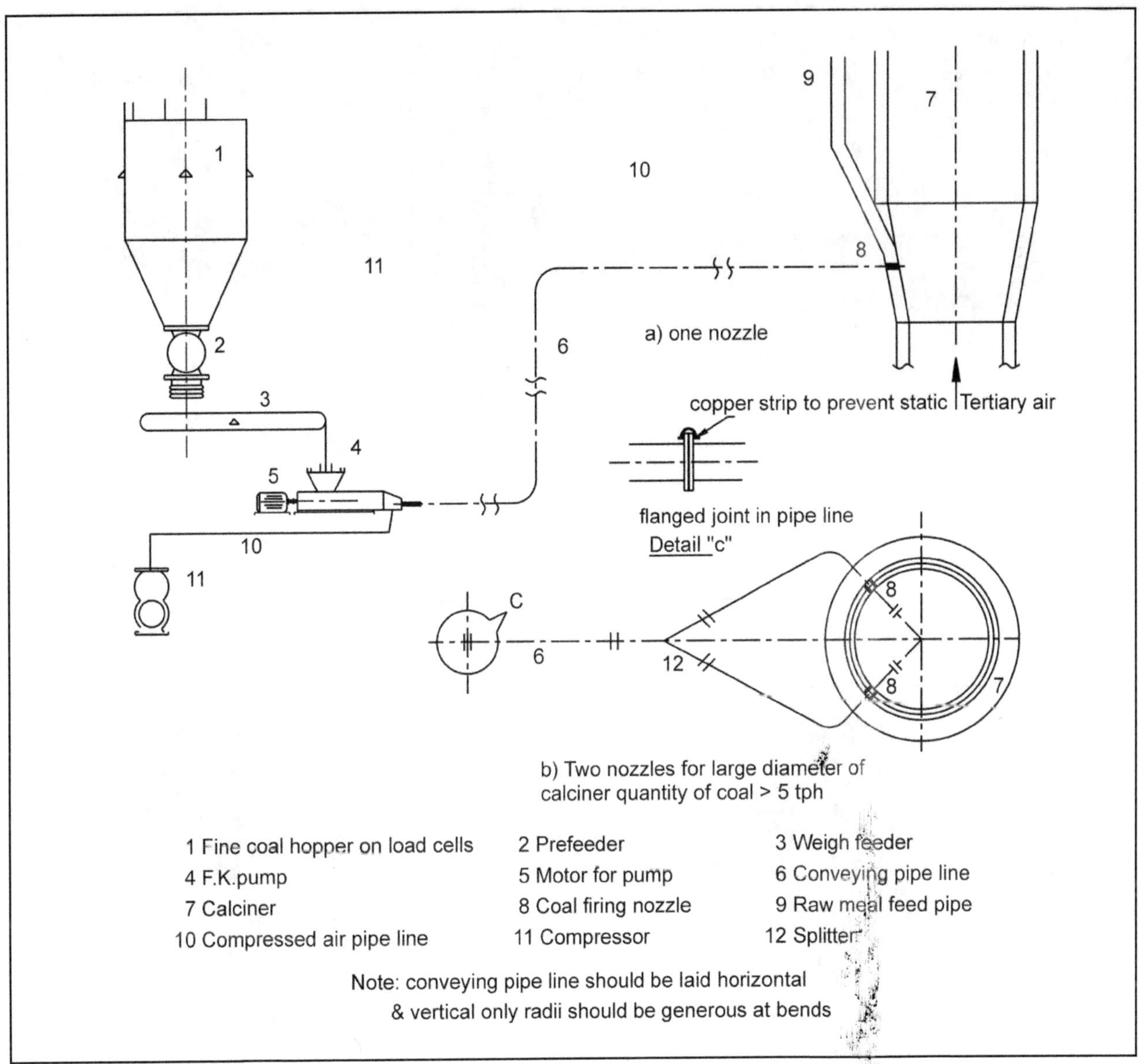

Fig. 6.17.20 Weighing and firing coal in calciners.

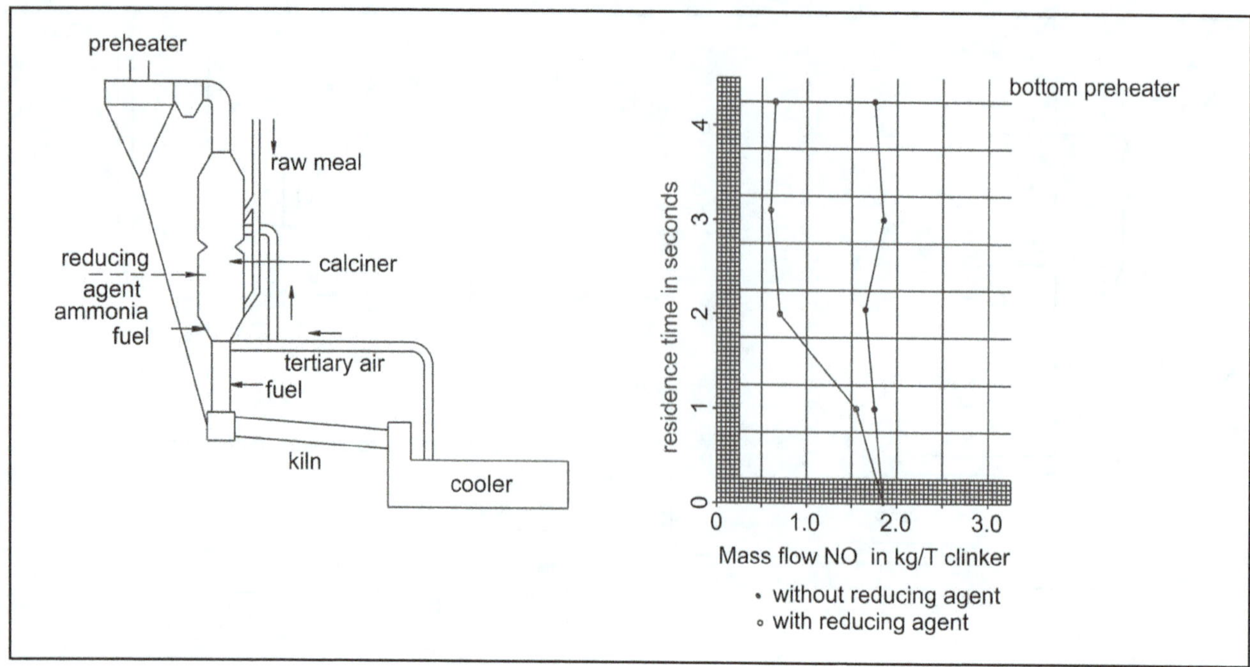

Fig. 6.17.21

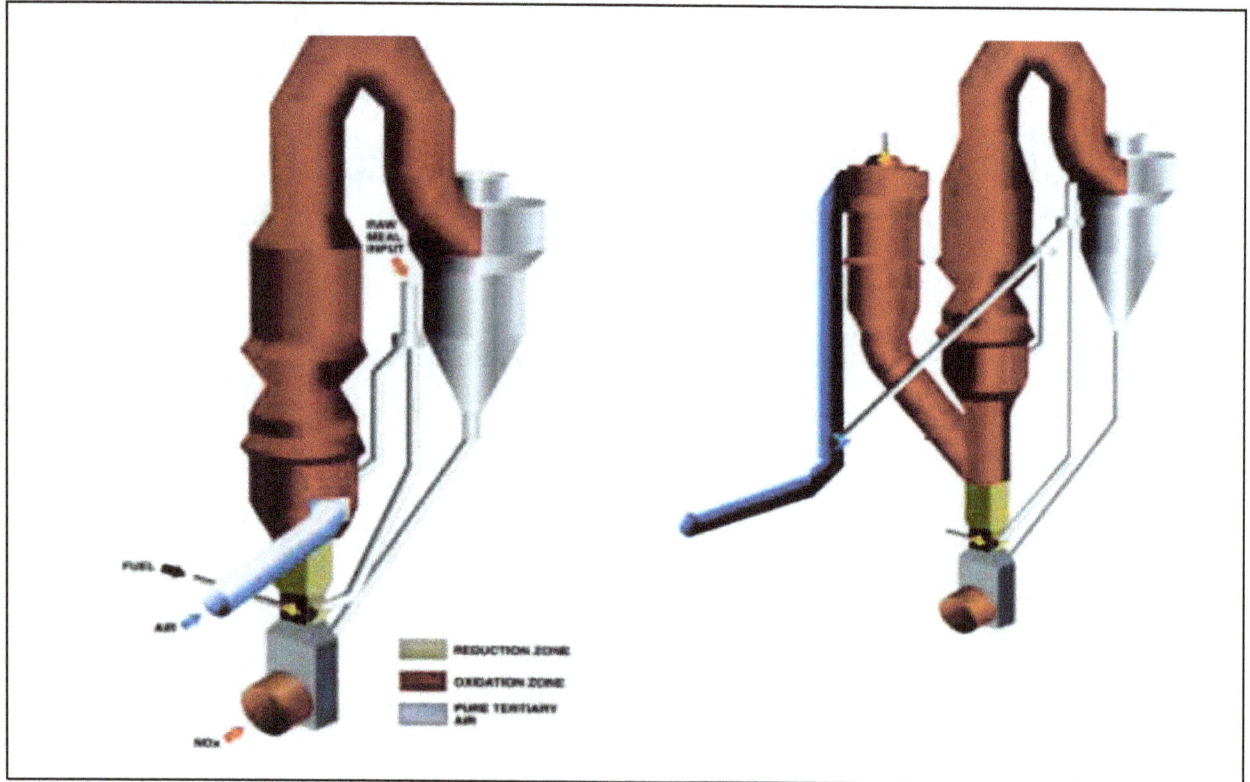

Plate 6.17.1

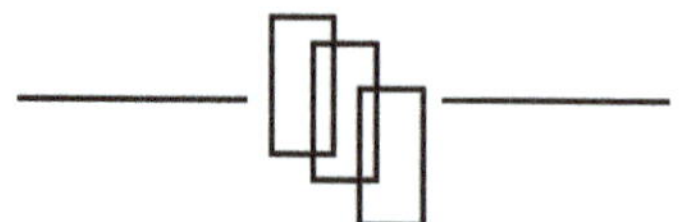

CHAPTER 18

TERTIARY AIR DUCT, DUST CHAMBER AND CONNECTING DUCT

18.1 Tertiary Air

Air drawn through cooler for combustion in kiln is commonly called **'secondary air'** and air through cooler for combustion for fuel fired in calciner, is commonly called **'tertiary air'**.

18.2 Air for Combustion in Calciner

Air drawn into calciner consists of :

1. Primary air – i.e., air used to convey coal pneumatically into the calciner.
2. Fluidizing air (applicable to fluid bed calciners). This air is used to fluidize bed of raw meal and is used for combustion.
3. Tertiary air – air drawn from cooler (in case of planetary cooler, the air is drawn through kiln and hence could also be called secondary air).

As mentioned earlier, calcination begins at 600 °C and is completed at 800-850 °C. Temperatures in calciners are maintained at 800-900 °C. Lower temperatures at the exit of calciner are an indication of full use of heat released in calciner in calcining raw meal.

For design purposes, leakage air at kiln hood is taken as 3% of total air for combustion air for preheater kiln and 1 % for calciner kiln.

Since the draft at kiln hood is between 0 to – 2 mm, this air quantity is small; nevertheless, to ensure that the maximum air is drawn through cooler, seals should be of good design and should be maintained.

'Primary air' is the air used to fire pulverized coal into kiln. In conventional burner, primary air would be between 15-20%.

In multi channel burners it would be about 12.5% or less.

Therefore total air drawn for combustion into kiln and calciner from cooler would be :

Total air for combustion + design excess air – (which is generally 10%) – 5% leakage – primary air (which would be 15% – 20% for conventional burners and 12.5% for multi channel burners) when it is drawn from atmosphere.

It would be combustion air + excess air – 5% leakage – 2 ½% conveying air when primary hot air is drawn at 300 °C through cooler.

In majority of cases, tertiary air is drawn from cooler. **See Figs. 6.17.7, 6.17.8 and 6.17.13 in Chapter 17.** In a few designs it is drawn from kiln hood. **See Fig. 6.17.17**

18.3 Proportions of Secondary and Tertiary Air

The proportions of secondary and tertiary air are the same as proportions of fuel fired in kiln and calciner as explained in **para 17.8** of **Chapter 17.** However they would be at different temperatures as shown in **Fig. 6.17.16.**

When only secondary air is drawn from the kiln its temperature would be high.

If all of air, secondary and tertiary, is drawn through hood, the temperature of secondary air into kiln would drop and that for tertiary air to calciner would increase as shown in above figures.

18.3.1 Air Drawn from Cooler

In calciner where temperature of 850-1000 °C are sufficient, there is no point in supplying tertiary air at temperature of + 800 °C. Where as in kiln where temperatures in burning zone are high at 1400 °C, a higher secondary air temperature of + (1000-1100 °C) helps in increasing fuel efficiency.

From these considerations, it is best to draw tertiary air from cooler. It also avoids dusty conditions and flame can be seen more clearly.

18.4 Sizing of Tertiary Air Duct

Assume that air to TA duct is 0.48 nm^3/kg of clinker **See Fig. 6.17.16.**

Kiln – capacity = 10000 TPD

Therefore preheater kiln = 40000 TPD

For sp. fuel consumption = 700 kcal/kg clinker

Calorific value of coal = 4500 kca/kg

Clinker / hr = 416700 kg.

Fuel in kiln = 40 % ; in calciner = 60 %

Total air for combustion including 10 % excess air = 5.45 nm/kg coal

Coal inkg/kg clinker = 700/4500 = 0.16 kg

Air for combustion/kg clinker = 0.87 nm3/kg @ 60% fuel in calciner, tertiary air = 0.6*0.87 = 0.522 nm^3/kg clinker.

Let tertiary air temp. be 700° c Actual vol of tertiary air = 1.86 m^3 /kg clinker.

At rated capacity of 10000 tpd = 774300 m^3/hr = 215 m^5/sec.

Let velocity in tertiary duct be 20 msec. Area of TA Duct = 10.75 m^2.

Therefore clear dia. of t.a.duct = 3.7 metres.With a refractory lining of 150 mm thick (refractory + insulating blocks), the duct diameter would be 4 metres.

With its length of 60 – 70 Meters, radiation loss on its account is considerable - only slightly less than half as much as that from kiln.

18.5 Layouts of TA Duct

There are different layouts of TA duct for different designs of precalciners. TA duct can be laid :

1. Horizontal, or
2. Inclined

See Figs. 6.18.1 and 6.18.2.

The long length of ducting is supported on trestles of concrete or steel. Expansion joints are provided and fixed and sliding supports are provided alternatively. **See Fig. 6.18.3.**

TA duct is generally located as shown to minimize dust pickup and to provide clear view of the burning zone. It is thus better to install it on the inner side to provide full and clear access for cranes to kiln.

See Figs. 6.18.4 and 6.18.5.

18.5.1 Twin stream preheater

A 10000 TPD kiln is more lkely have a two stream prheater – twin i.e two identical streams or differently sized. In that case it may be convenient to have two tertiary ducts one on either side of the kiln.

See Fig. 6.18.8.

18.6

To provide clear access to two kilns installed side by side, would require kilns rotating in opposite direction and perhaps preheater cyclones of opposite handing for two kilns This of course increases spacing between kilns and preheater.**See Figs. 6.18.6 and 6.18.7**

Generally, however kilns would be rotating in the same direction and cyclones would have same handing. While one kiln is easily approachable, the other is not because of T.A. duct. It would be best to provide a wide enough road for crane. **See Figs. 6.18.8 and 6.18.9.**

18.7 Dedusting Chamber and Connecting Duct

Cooling air emerging through clinker bed has clinker dust in it. Since this is an inert dust, tertiary air taken to calciner should be cleaned of it. A dust chamber is therefore introduced in the TA duct immediately after the cooler, which collects as much of clinker dust as possible and returns it to cooler either inside the cooler or outside on drag chain.

See Fig. 6.18.10.

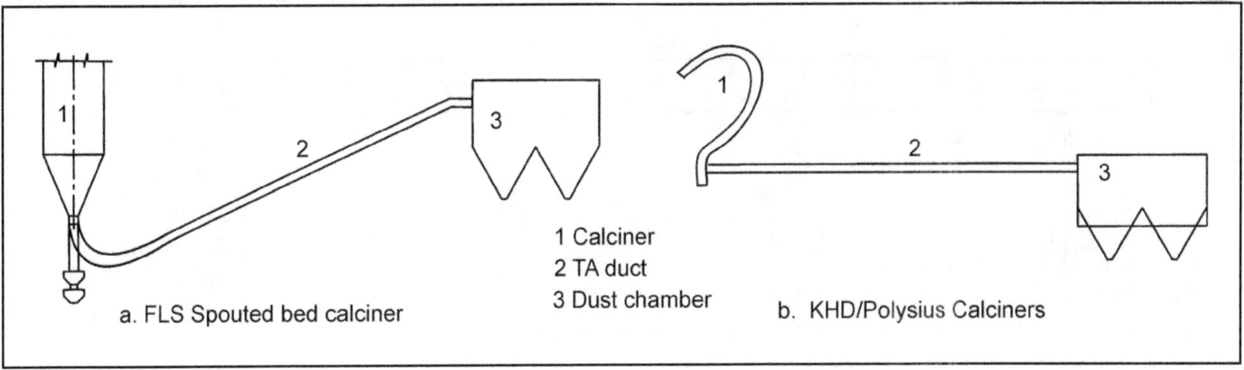

Fig. 6.18.1 Different configuration of TA duct for different calciners.

1
2
3
Plan
a. Mistsubshi fluid bed Calciner
Calciner
Mixing chamber
1 Calciner 2 TA duct 3 Dust chamber
b. RSP calciner

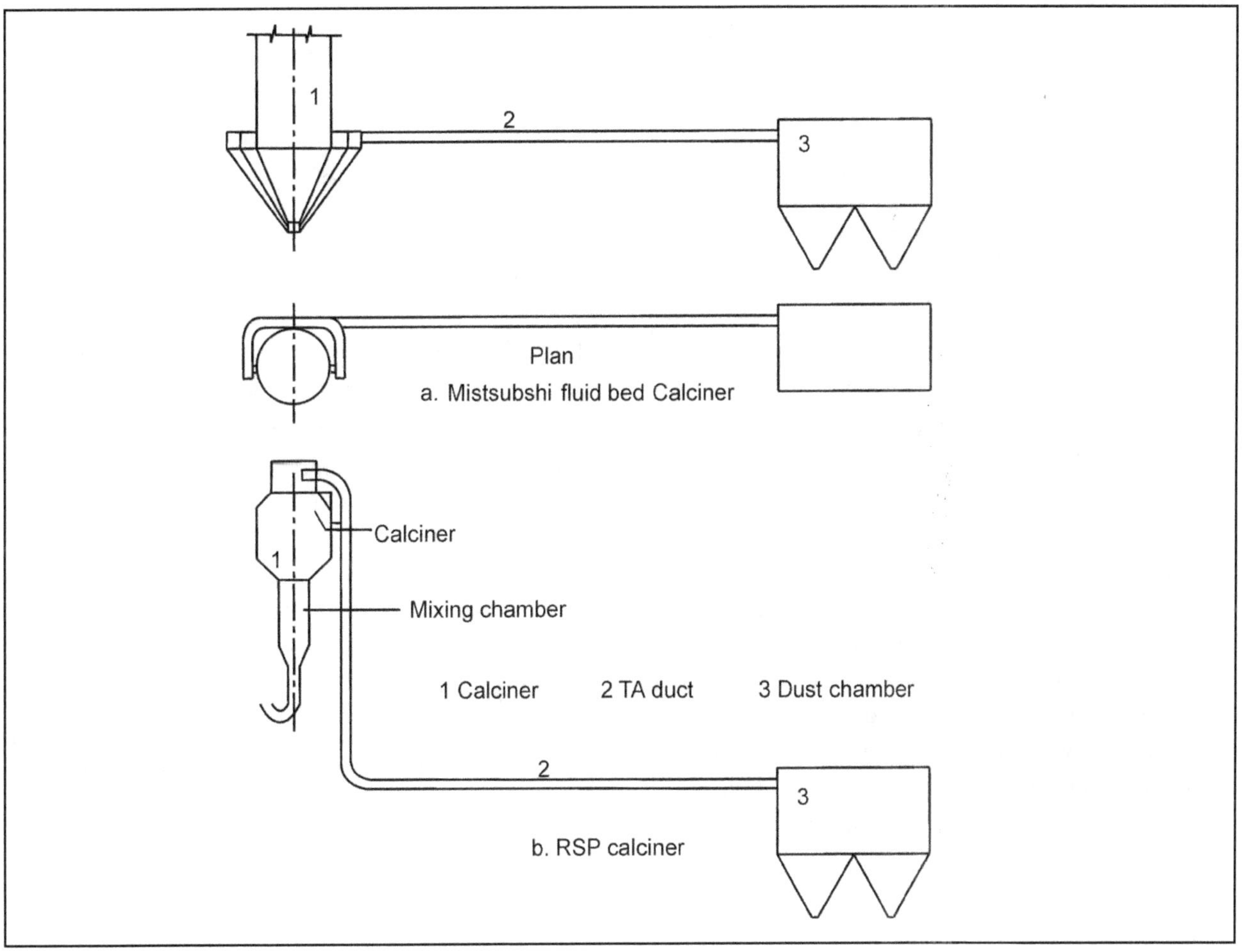

Fig. 6.18.2 Different configuration of TA duct for different calciners.

Dust chamber consists of a brick lined chamber with a sufficiently large cross section so as to reduce air velocity to 4-6 m/sec. The chamber is so designed that air flow through it has to change direction in horizontal and vertical planes to facilitate precipitation and collection of clinker dust.

See Fig. 6.18.10.

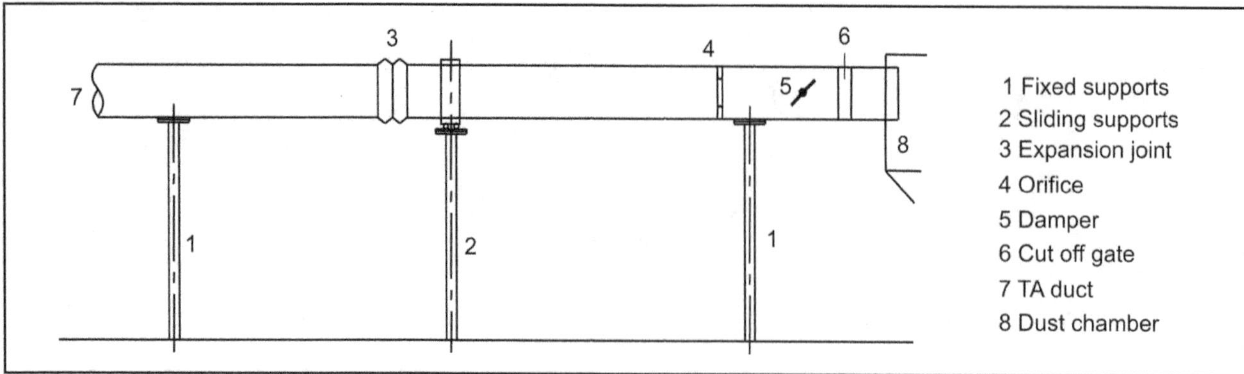

Fig. 6.18.3 TA duct layout – supports.

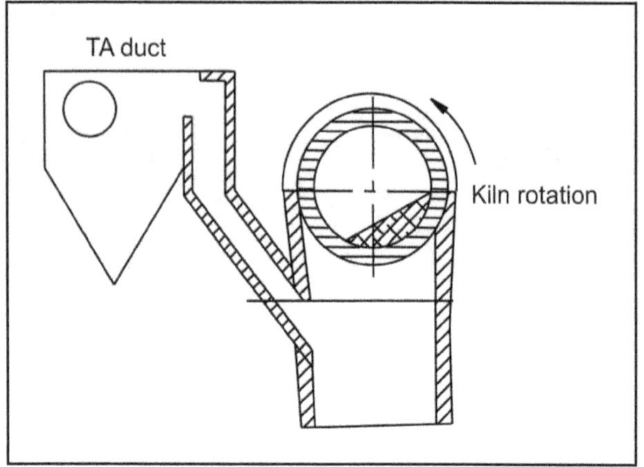

Fig. 6.18.4 Location of TA duct vis-a-vis rotation of kiln.

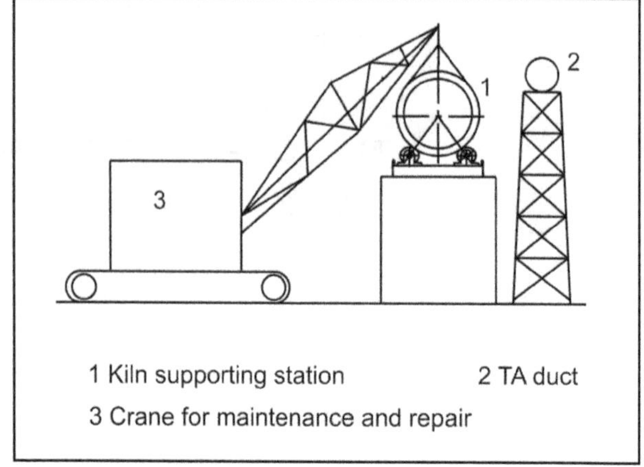

Fig. 6.18.5 TA duct and access to kiln for maintenace.

Dust chamber is connected with the cooler by a refractory lined connecting duct. Its location is so fixed that requisite quantity of tertiary is picked from the cooler at as high a temperature as possible. Its cross section is designed to be large to reduce velocity of tertiary air in it so that dust picked up is minimum before it enters the dust chamber for further cleaning.

The connecting duct may be drawn from the side of the cooler as shown or it may be drawn from the top.

See Figs. 6.18.11 and 6.18.12.

18.7.1 Connecting Duct on Burners Platform

In the latter case, connecting duct emerges on the burners' platform. Taking it from sides has the advantage that the space immediately in front of kiln hood is clear. If the hood is of movable type, it can be moved without requiring to dismantle TA duct.

When duct is taken from the top of the cooler, kiln hood cannot be moved. It should have large movable doors, so that bricks can be taken into the kiln.

See Fig. 6.18.13.

Even then, it is better to have area immediately in front of the hood clear so that refractories can be easily taken in and out of kiln either manually or by mechanical means.

18.8 Location of Take off Point

Temperatures of tertiary air range between 600-800 °C depending on exact location of the take off point of the connecting duct. Nearer this take off point is towards the kiln the better. However, there are practical limitations to this such as dimensions of hood, spacing of beams of the building supporting the kiln hood, column spacing etc.

See Fig. 6.18.14.

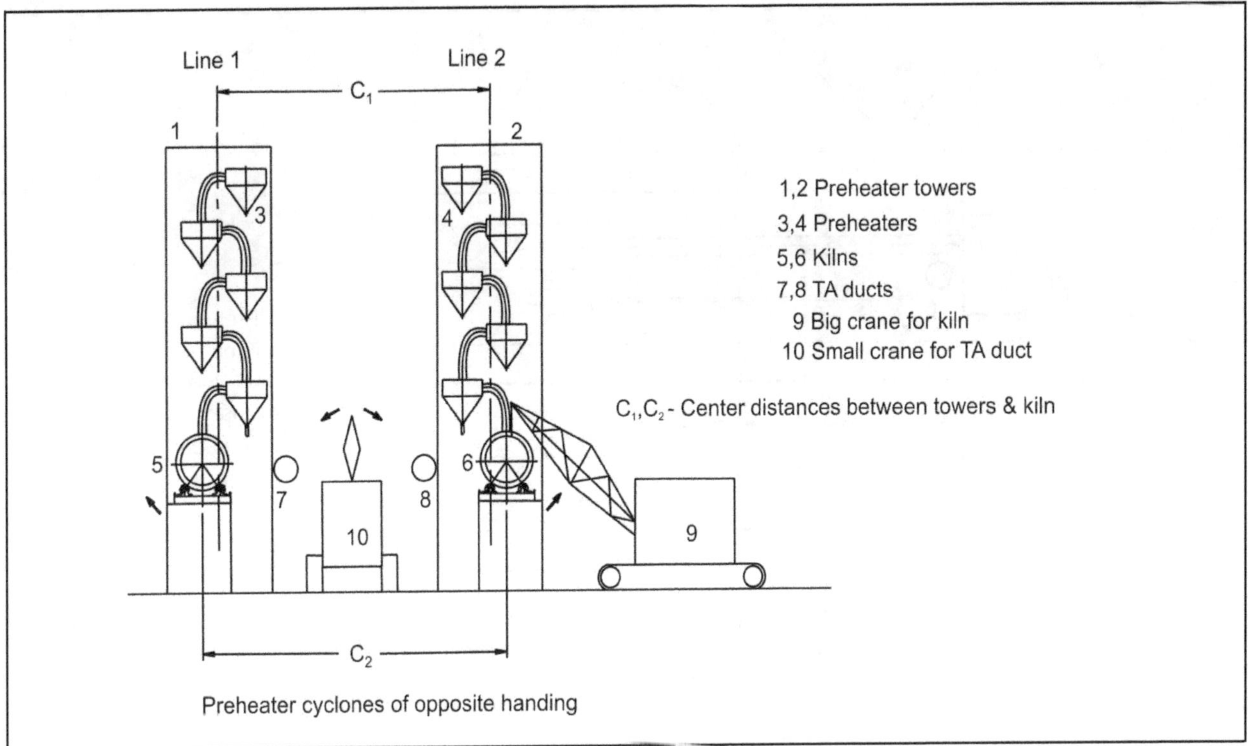

Fig. 6.18.6 Kilns turn in opposite direction; preheater cyclones of opposite handing.

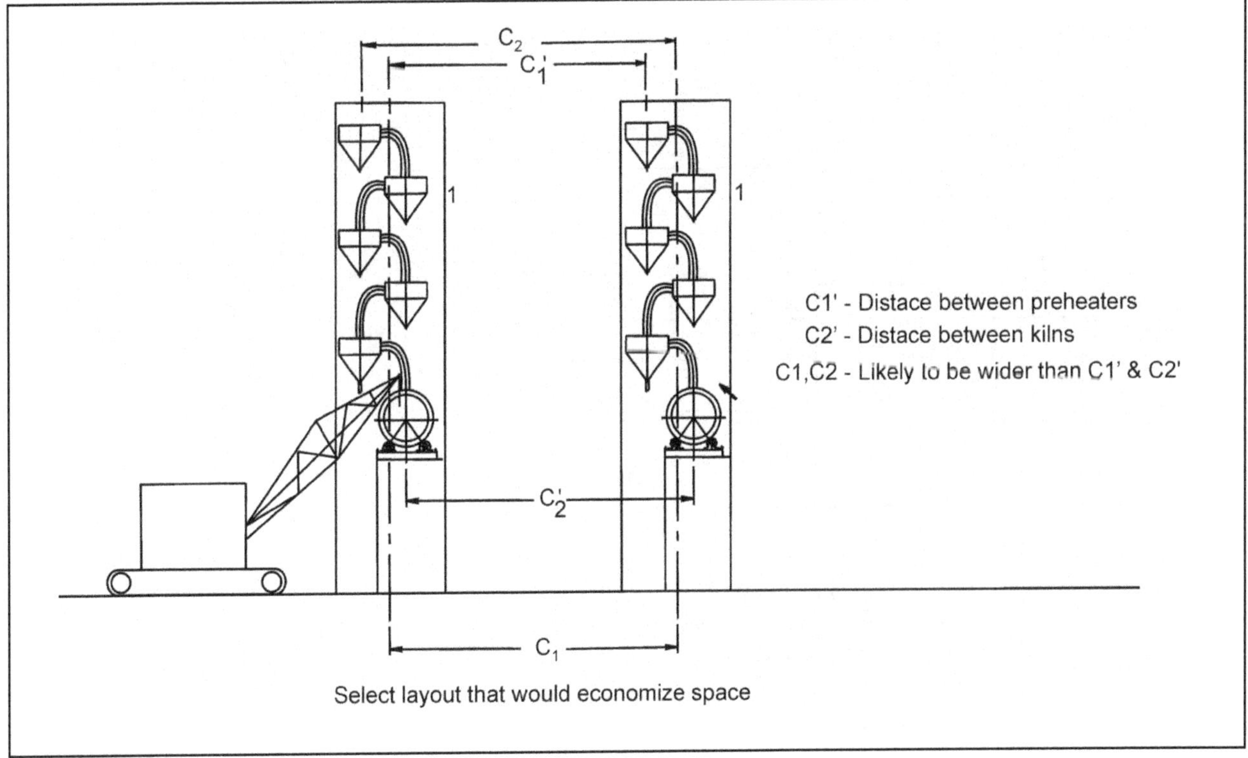

Fig. 6.18.7 Kilns rotate in same direction; preheaters of same handing.

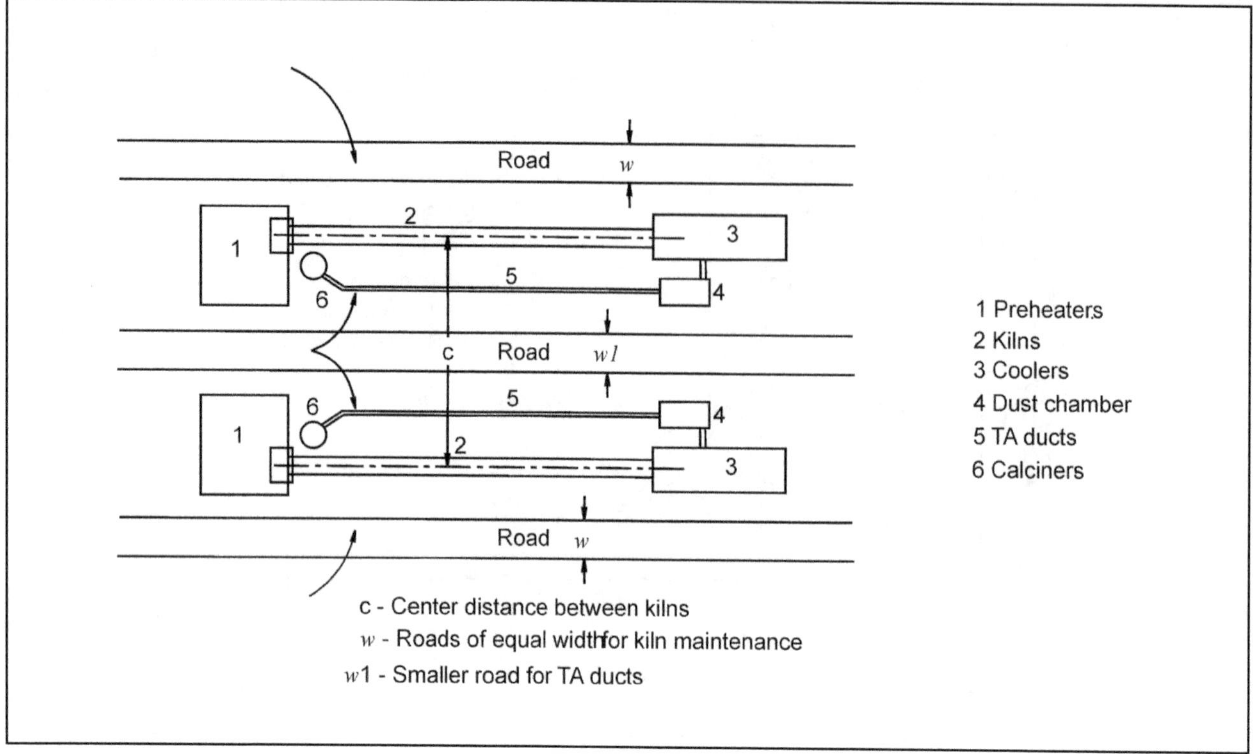

Fig. 6.18.8 Access for TA duct and kiln for maintenance; kilns turn in opposite direction.

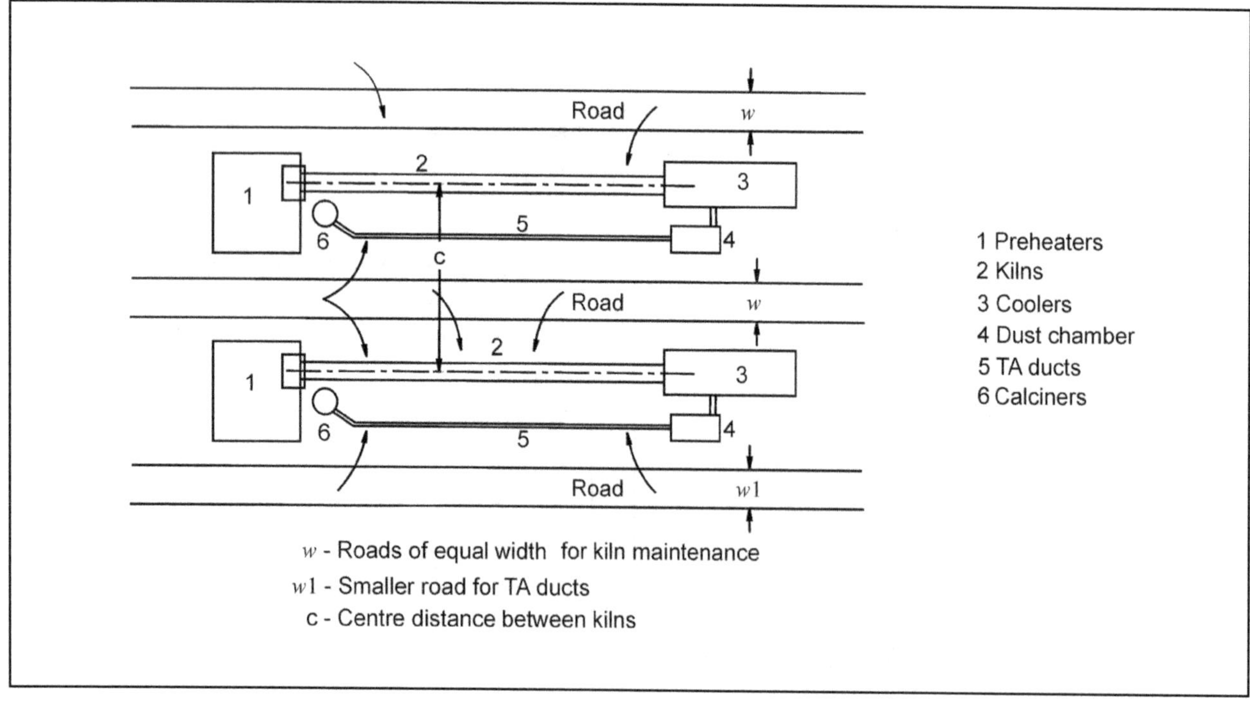

Fig. 6.18.9 Access for TA duct and kilns for maintenance; kilns turn in same direction.

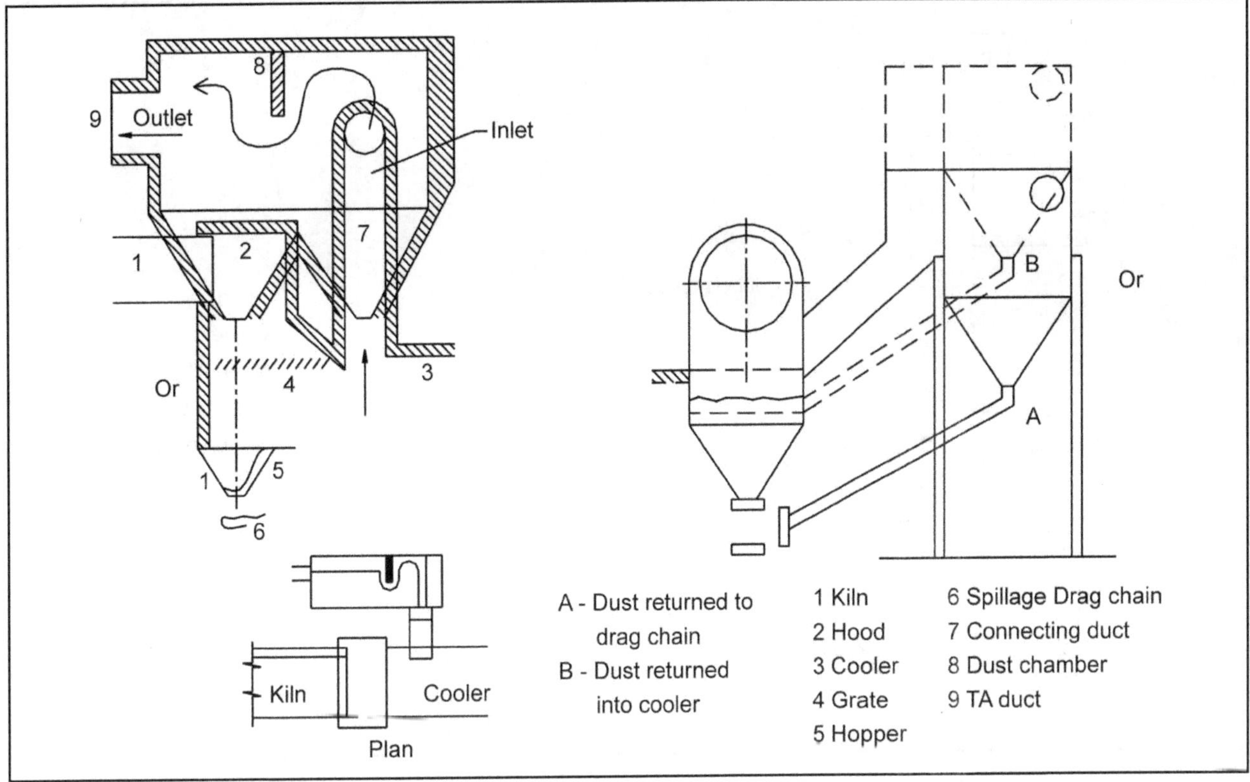

Fig. 6.18.10 Dedusting chamber TA duct taken from cooler.

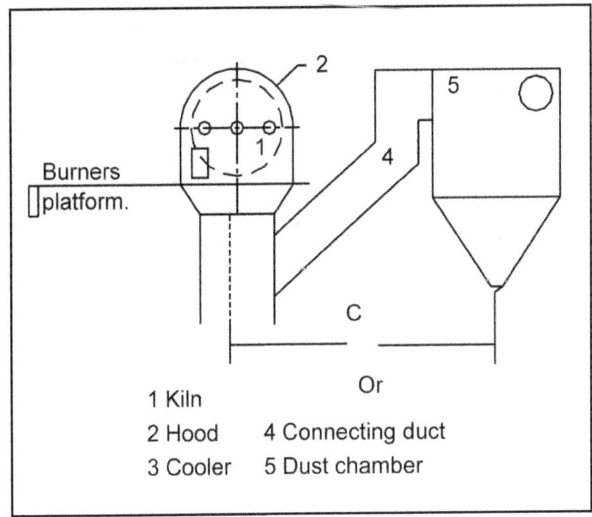

Fig. 6.18.11 TA duct take off from side of cooler.

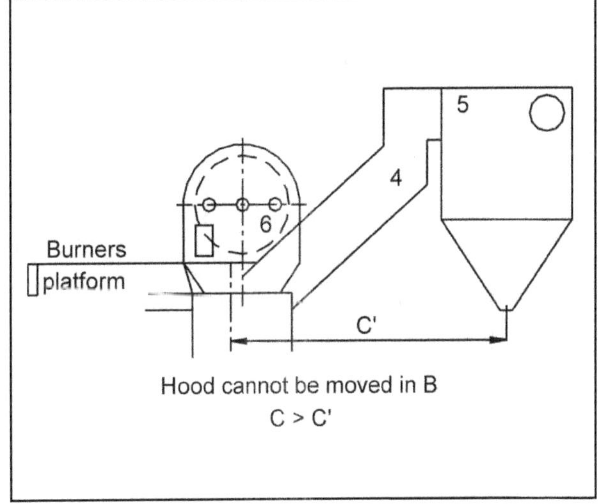

Fig. 6.18.12 TA duct take off from top of cooler.

18.9 Layout of Connecting Duct

Though short, connecting duct is quite heavy because of its size and refractory lining. It needs to be appropriately supported and should have expansion joints.

See Figs. 6.18.15 and 6.18.16.

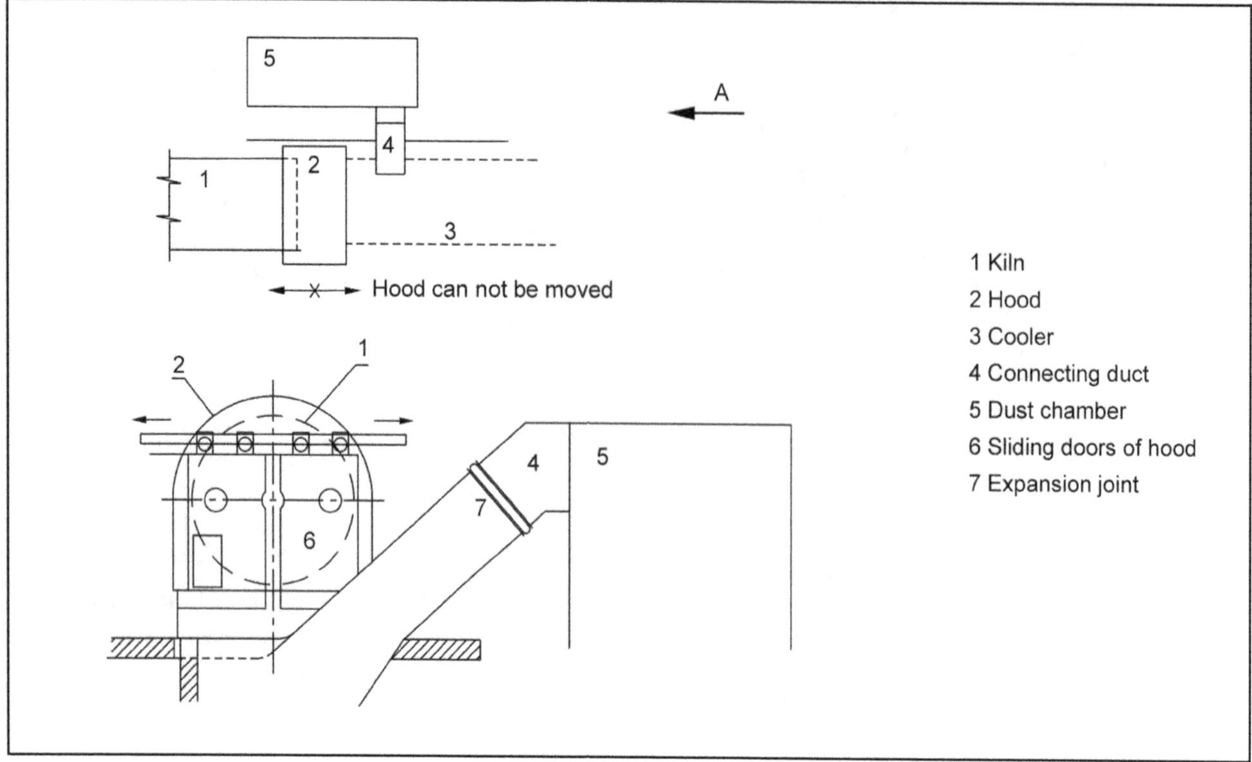

Fig. 6.18.13 Large doors of kiln hood moved side ways for access in to kiln.

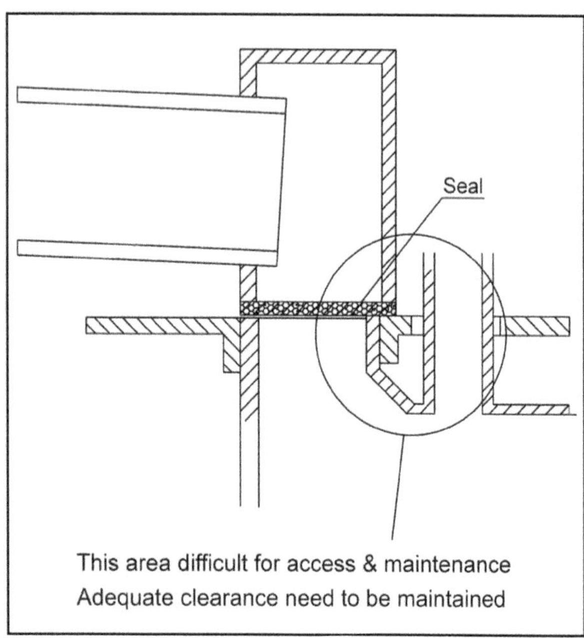

Fig. 6.18.14 Take off point for TA duct.

All problems of locating TA duct vis-à-vis hood, and to some extent precipitation of dust are eliminated if TA duct is drawn from the hood.

See Fig. 6.17.17.

Such an arrangement may not need a dust chamber. However, quantity of air drawn through hood is increased and this may impair visibility of flame and hence measurement of temperatures of burning zone and also 'Video monitoring' of flame and burning zone conditions.

As mentioned earlier temperature of Secondary air would be lower and that of tertiary air unnecessarily higher.

18.9 Return of Clinker Dust

Hot clinker dust collected in the dust chamber is returned either to cooler on grates or to spillage drag chain by pipes made of either heat resisting material or pipes lined with refractories / castable.

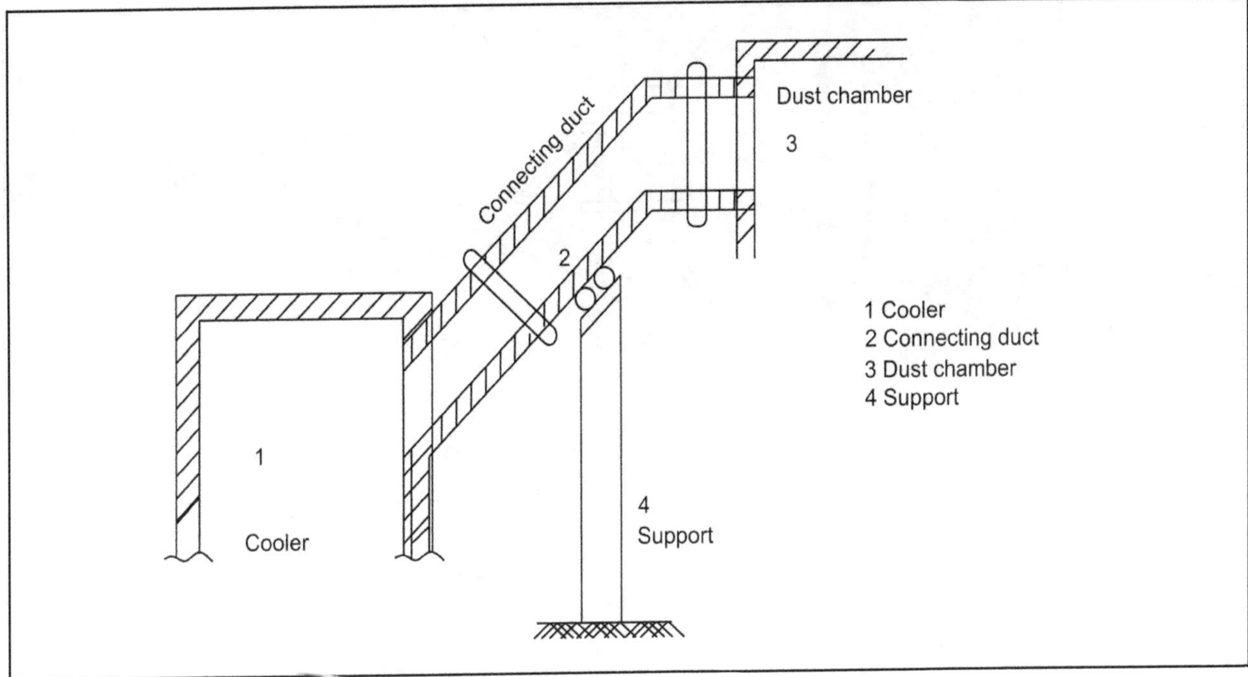

Fig. 6.18.15 Connecting duct between cooler, dust chamber and support.

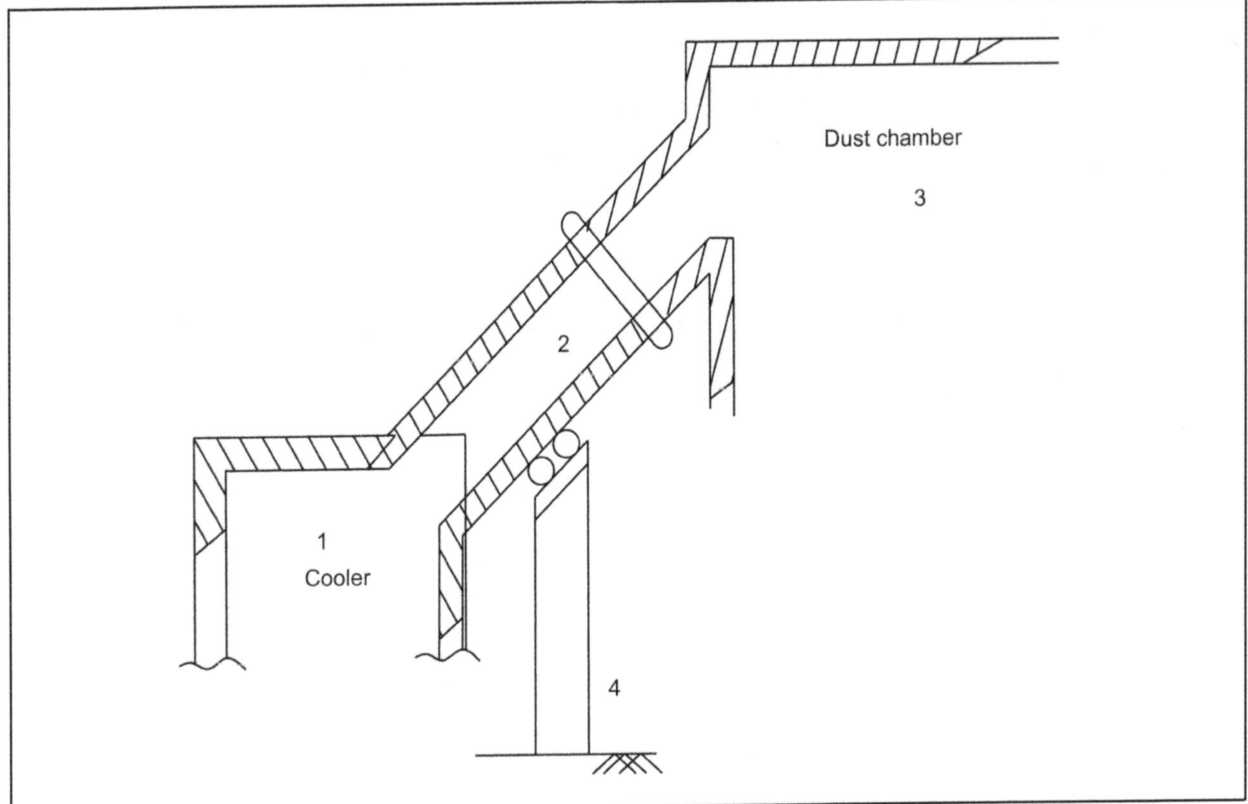

Fig. 6.18.16 Connecting duct between cooler, dust chamber and support.

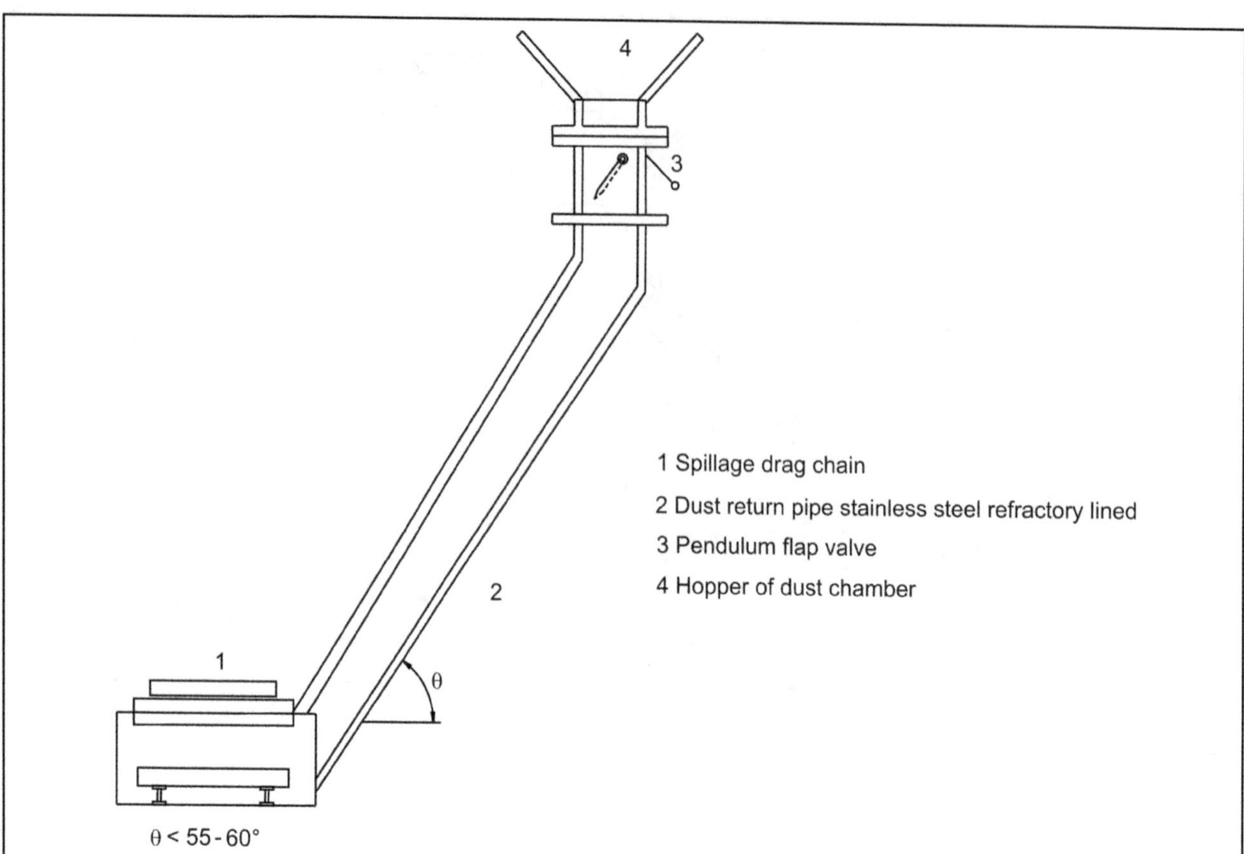

Fig. 6.18.17 Dust returned from dust chamber cooler drag chain.

At temperatures of 600 to 800 °C, clinker dust is not free flowing. Therefore these pipes from hoppers of dust chamber to cooler need be laid at slopes of not less than 55-60° to horizontal.

In case dust is returned inside the cooler, the dust chamber is required to be raised quite high to maintain these slopes as against when dust is returned to the spillage drag chain. It is therefore more common to find dust being returned to the drag chain.

See Figs. 6.18.10 and 6.18.17.

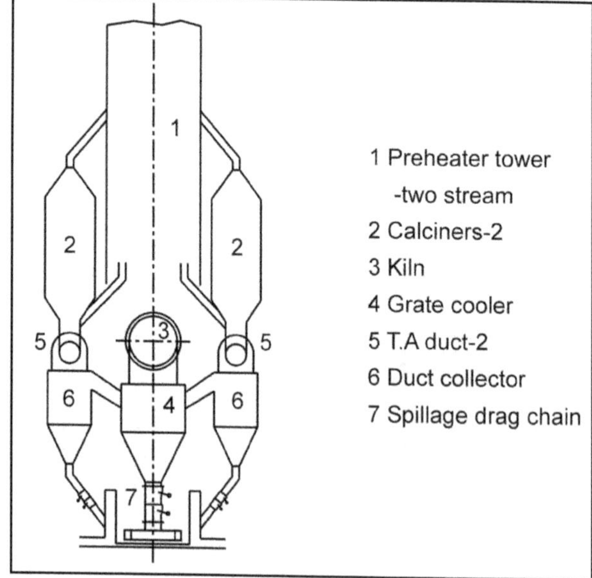

Fig. 6.18.18 End view of 2 stream preheater, 2 Calciners, Kiln & Grate cooler in a large plant.

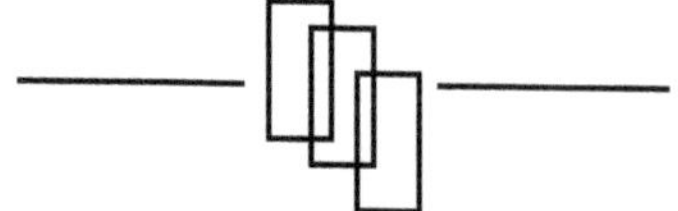

CHAPTER 19

ROTARY KILN

19.1 Kiln

It is the heart of the cement plant. It converts raw meal fed to it into clinker.

According to the processes used kiln is required to carry out various operations inside it till it is converted into clinker as explained in **Section 1** and repeated in **Fig. 6.19.1.**

Thus in wet process where maximum number of steps in the process of manufacture of clinker took place inside the kiln, kilns were long with length to diameter ratios (l/d) of 35-40 / 1.

See Fig. 6.19.1.1.

Dry process kilns with calciner which perform only 10% calcining and balance sintering, can have l/d ratios of 10-12 / 1.

See Figs. 6.19.1.6 and 6.19.2.

Dry preheater kilns have l/d ratio of 14-16 / 1. The reduction in length of the kiln has also reduced the no. of supports which have come down from 6/7 in wet kilns to 2 (max.3) for dry kilns.

See Figs. 6.19.1.5 and 19.1.6.

Kilns with planetary cooler have to support over hung weight at discharge end therefore the last pier is heavier. Shell thickness, tyre size (dia and thickness and width); and roller size (dia and width); bearings; all are heavier and larger for the end section of the kiln and for the last supporting section.

See Figs. 6.19.3 and 6.19.4.

19.2 Physical Features of a Rotary Kiln

The spacing of tyres or kiln supports is calculated so that load is as evenly distributed on all tyres as possible and that sag between spans is within permissible limits.

The overhangs at each end are also calculated from point of view of thermal and mechanical stresses and deflections.

Discharge end is hotter therefore overhang at discharge end is limited to $1 \times d$ where d is diameter of kiln. Over hang at inlet end could be 1.5-2 d, smaller the better.

See Fig. 6.19.5.

19.2.1 Shell Thickness

Shell thickness is calculated to keep ovality within specified limits. It is also decided by deflection or 'sag' mentioned above, and which should be within permissible limits.

Kiln shell becomes oval under its own weight and each section of shell under goes flexing as kiln rotates. **See Fig. 6.19.6.**

This flexing action becomes more pronounced as diameter of shell increases. It has a direct effect on the stability of bricks. Calciners were developed to obtain higher outputs of 7000-10000 tpd from kilns keeping their diameters corresponding to kilns of 2800 to 4000 tpd capacity.

19.3 Components of a Rotary Kiln

Every kiln has the following components :

1. kiln inlet housing and inlet seal
2. kiln shell
3. tyres mounted on shell
4. corresponding supporting stations each station comprising of :
   2 rollers
   4 supporting roller bearings
   1 bedplate

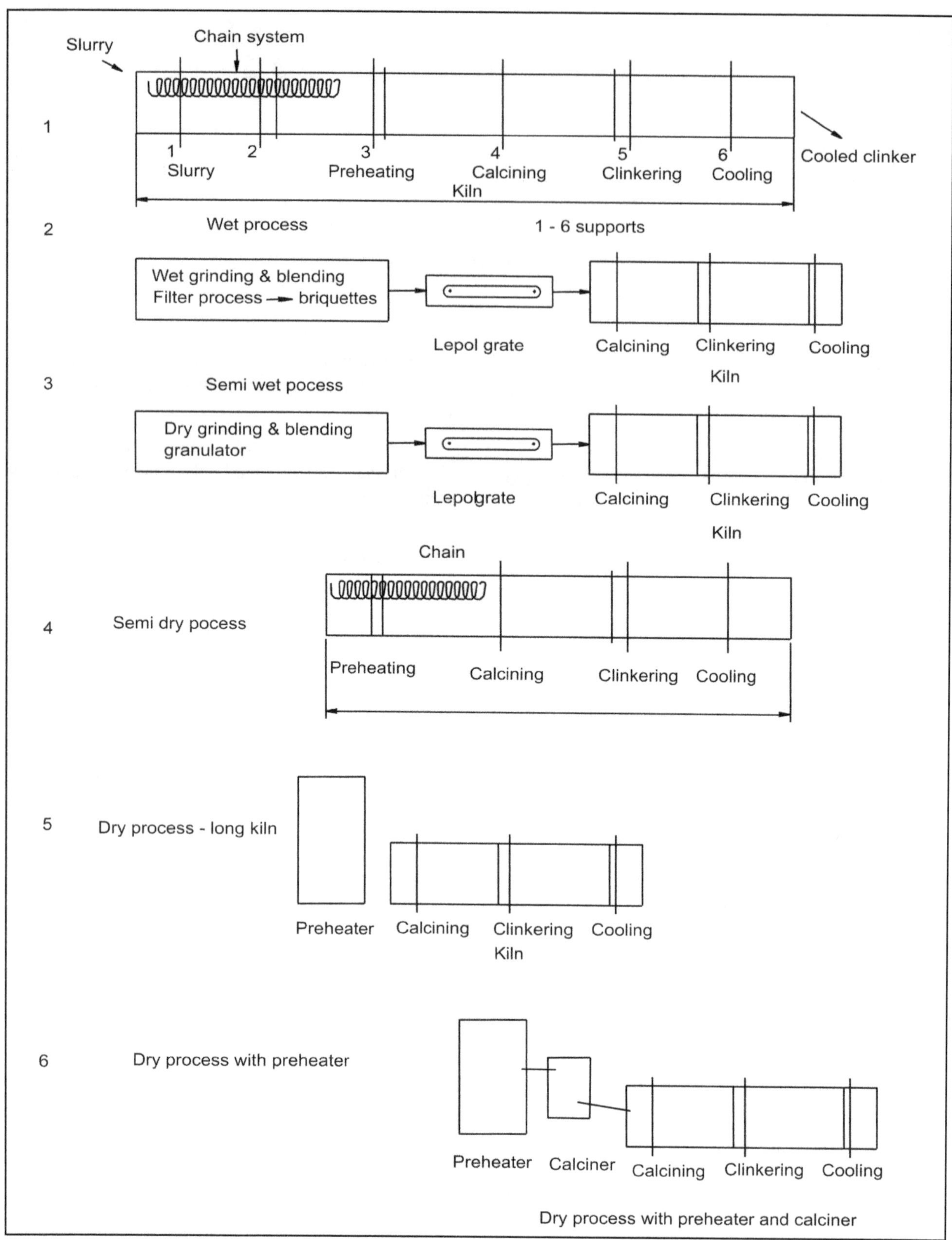

Fig. 6.19.1 Processes in rotary kilns - producing cement clinker.

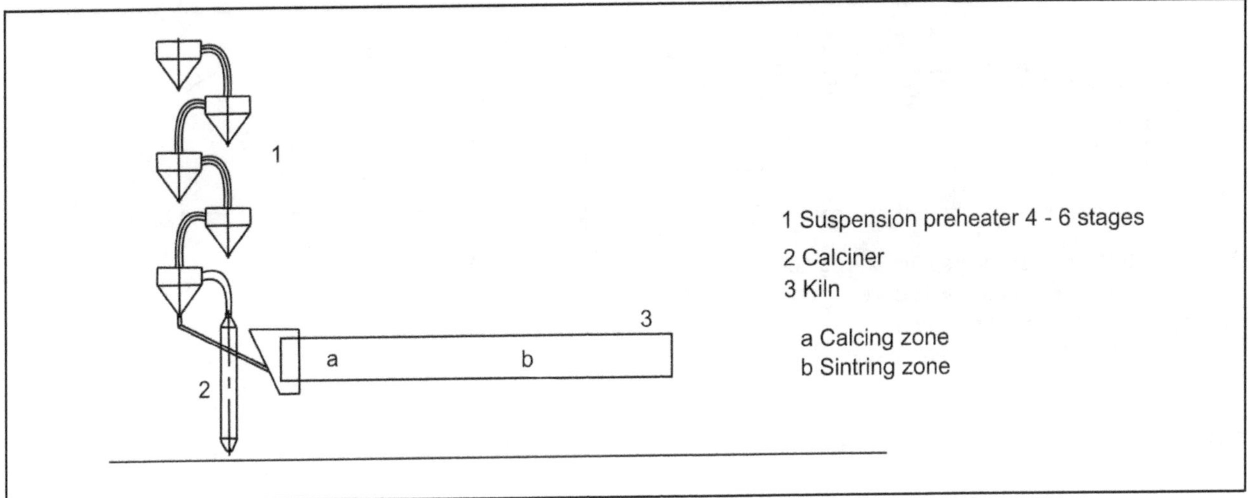

Fig. 6.19.2 Kiln-preheater and precalciner.

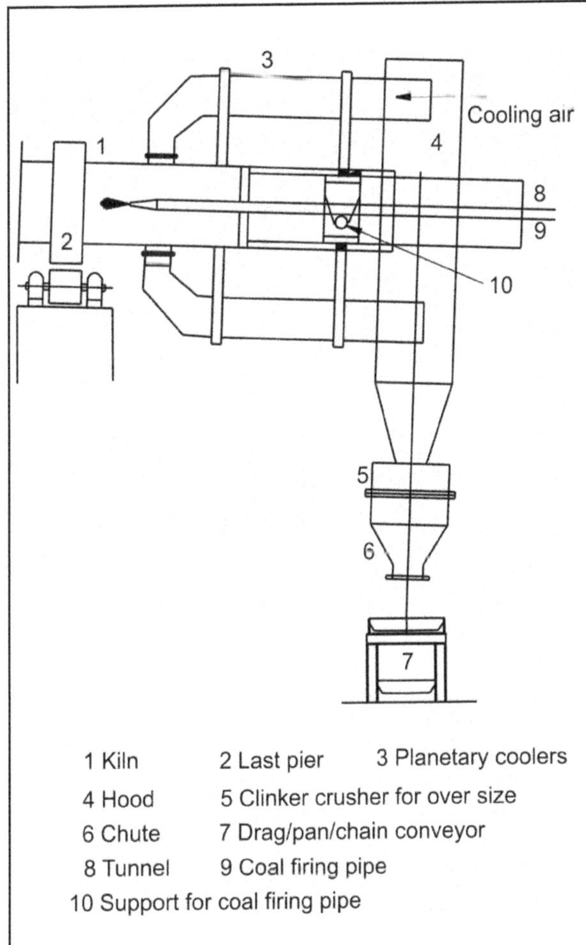

Fig. 6.19.3 kiln with planetary coolers. Cooler tubes 10-12 no.s l/d = < 5-6 : 1

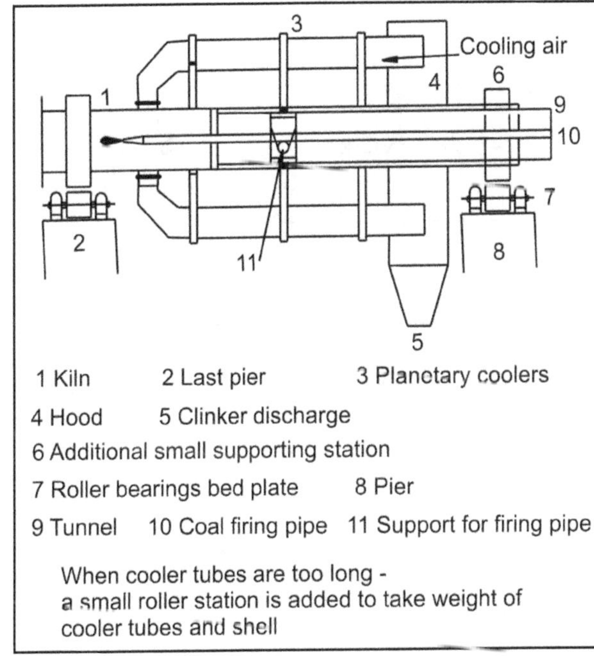

Fig. 6.19.4 Kiln with planetary coolers - long cooler tubes - l/d = > 10 : 1.

5. girth gear, pinion, pinion shaft and bearings
6. torsion shaft, gear box and drive motor
7. kiln hood at discharge end with outlet seal.

See Figs. 6.19.7 and 6.19.8.

Kiln inlet housing, kiln shell and discharge hood are all refractory lined because of temperatures inside the kiln.

Kiln may have a single or a dual drive.

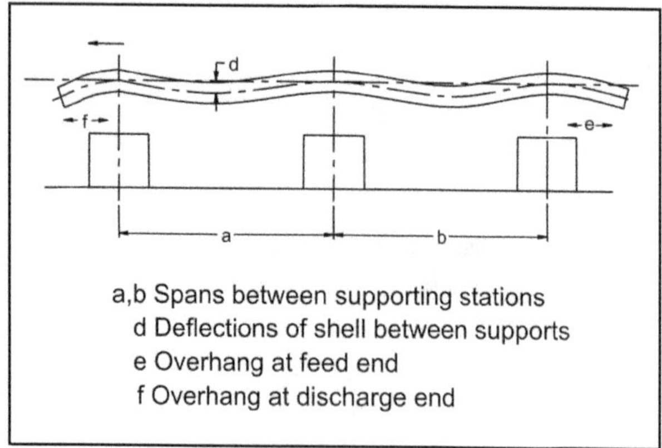

Fig. 6.19.5 Spans between supports.

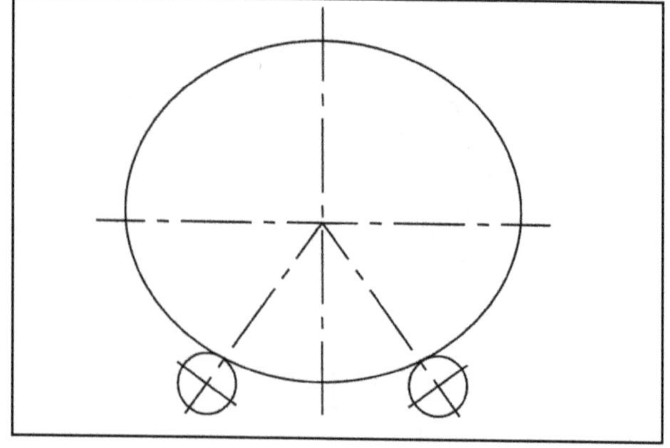

Fig. 6.19.6 Kiln - deformation of shell riding ring at supporting station.

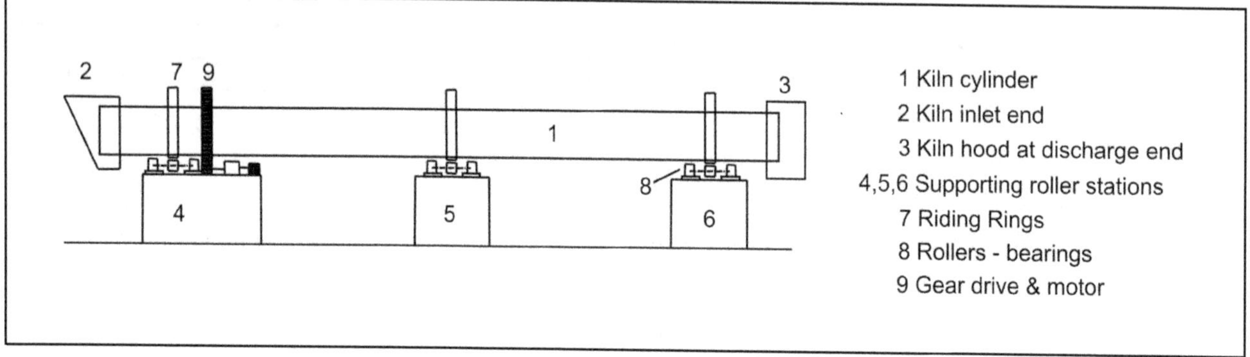

Fig. 6.19.7 Components of rotary kiln.

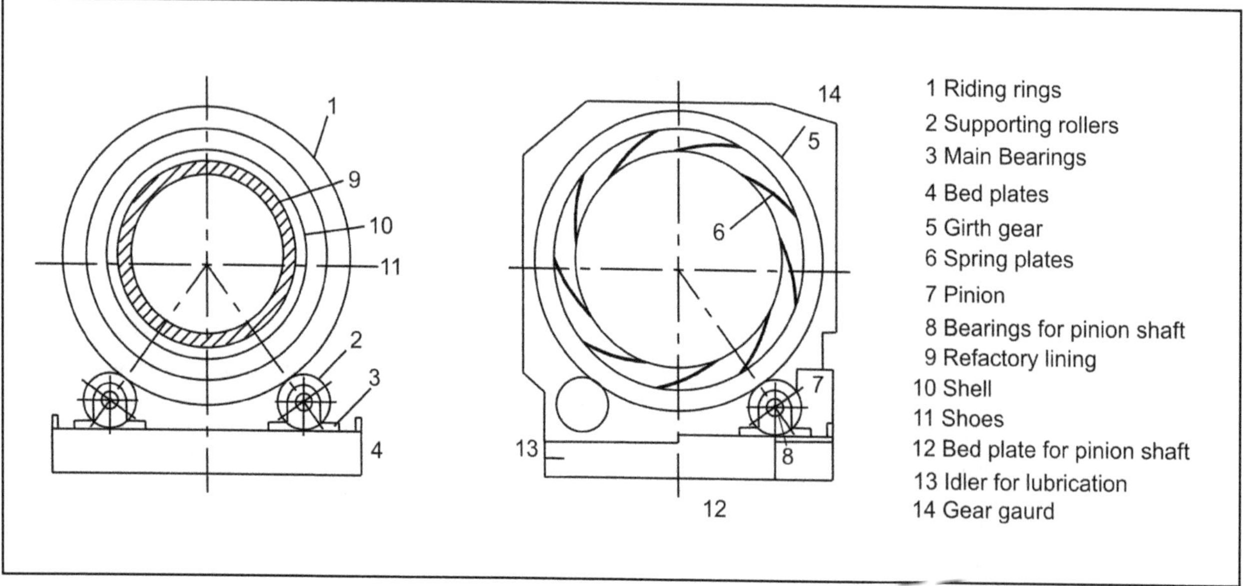

Fig. 6.19.8 Components of rotary kiln.

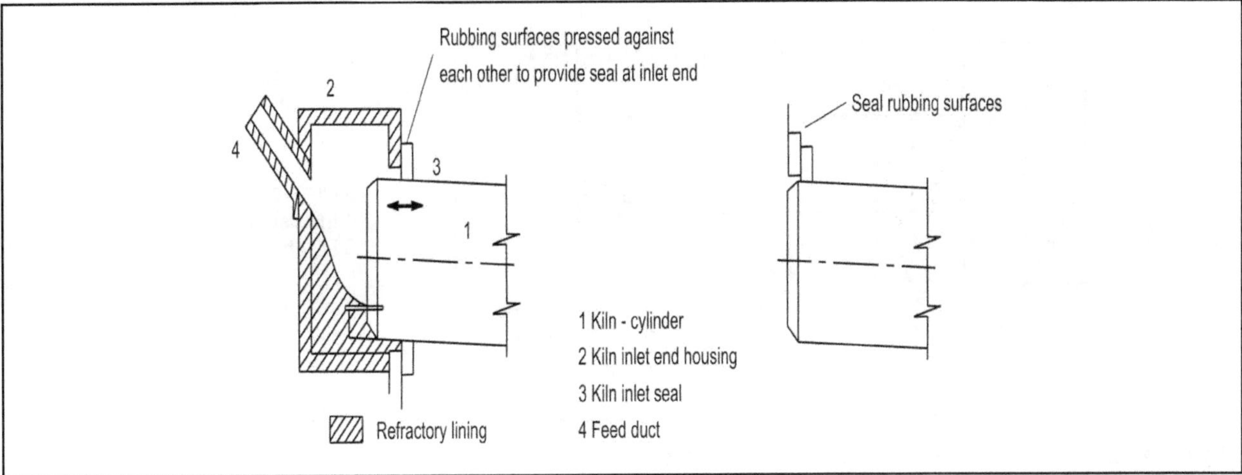

Fig. 6.19.9 Seal - kiln inlet end.

19.4 Seals at Kiln Inlet and Outlet

Kiln inlet seal consists of two parts; one part is fitted onto the stationary kiln inlet housing and the other on the rotating kiln shell.

See Fig. 6.19.9.

19.4.1 Inlet Seals

There are many designs of inlet seals. Commonly used types are:

(i) Hydraulic thrusters or spring loaded thrusters which keep sealing surfaces in contact at all times.

See Fig. 6.19.10.

(ii) Set of springs which keep sections of contacting surfaces on shell pressed against the contacting surface of kiln inlet.

See Fig. 6.19.11.

(iii) Set of spring leaves which cover rorating cylinder and prevent inlcakage of false air.

See Fig. 6.19.12.

Hot gases go out of kiln through a riser duct which connects kiln to the bottom cyclone. Riser duct is also refractory lined and has expansion joints. It is supported on floor/s in preheater tower.

Kiln also recives raw meal for bottom cyclone/s through a pipe/pipes which have pendulum flap valve/s.

See Fig. 6.19.13.

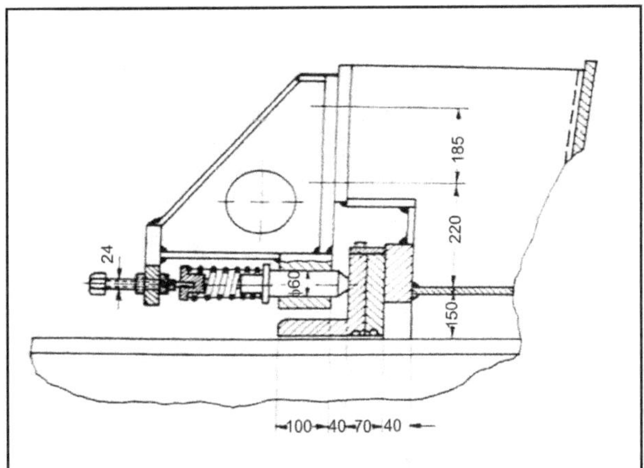

Fig. 6.19.10 Thrustor type seal at kiln inlet end. Thrust is applied either by springs as shown or by hydraulic cylinders.

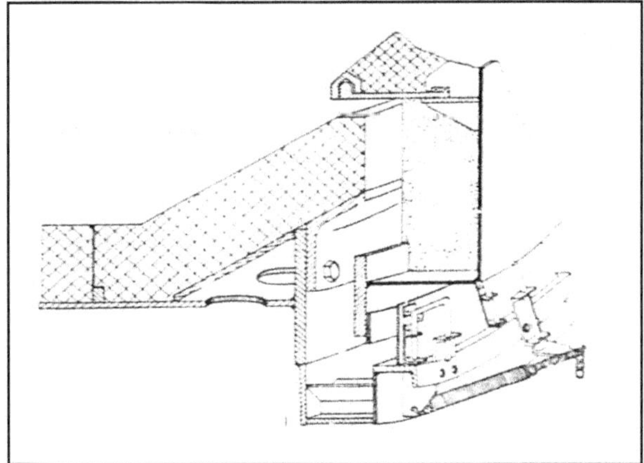

Fig. 6.19.11 Kiln seal at inlet end. Rubbing surfaces kept in contact by a number of springs.

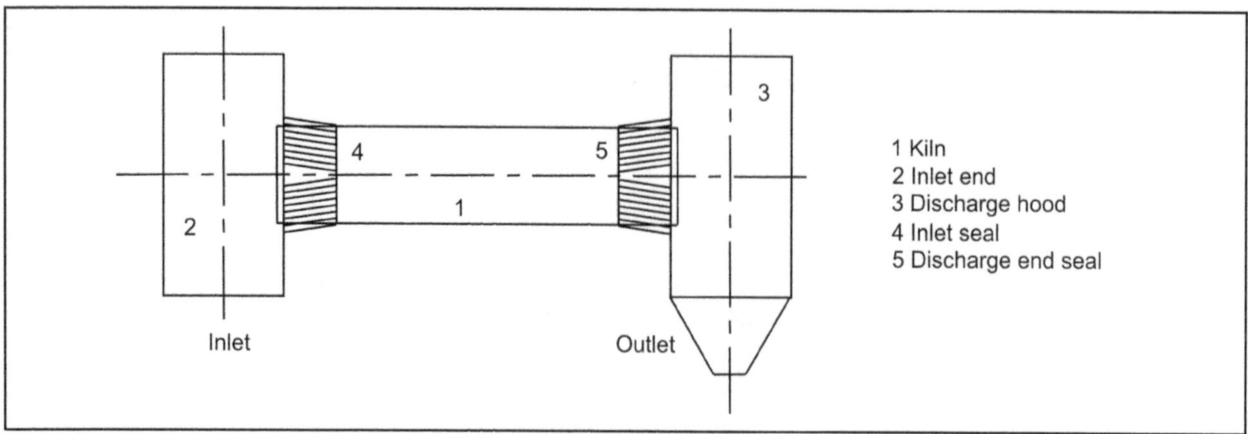

Fig. 6.19.12 Spring plate seals at inlet and discharge ends of kiln.

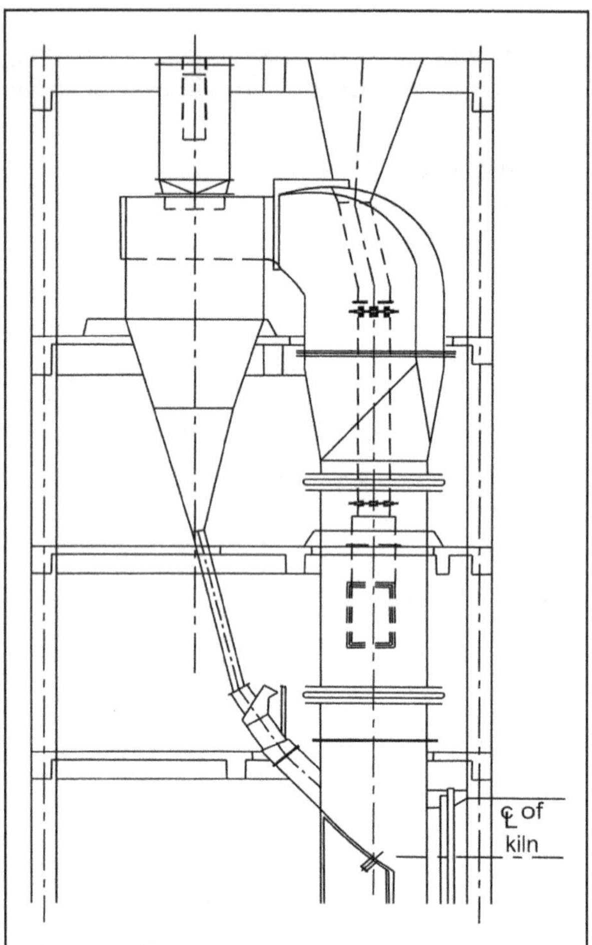

Fig. 6.19.13 3 - Riser duct of kiln and connection to bottom cyclone.
It will be modified to introduce secondary firing ; or replaced by inline calciner.

A throttle will be installed in the riser duct when balancing of pressure drop is required between calciner leg and kiln.

See Fig. 6.19.14.

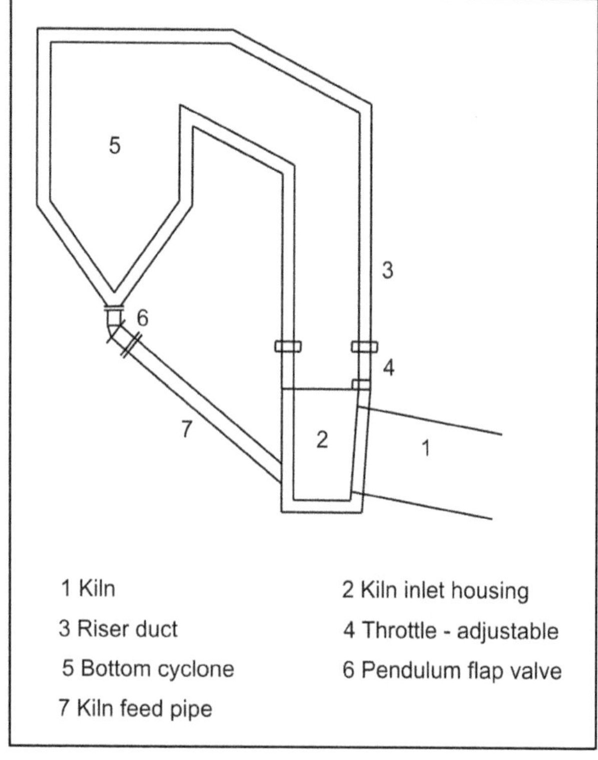

Fig. 6.19.14 In a kiln with calciner and single stream preheater, a throttle is inserted in the riser duct to divide gas flow through kiln and calciner in accordance with fuel fired.

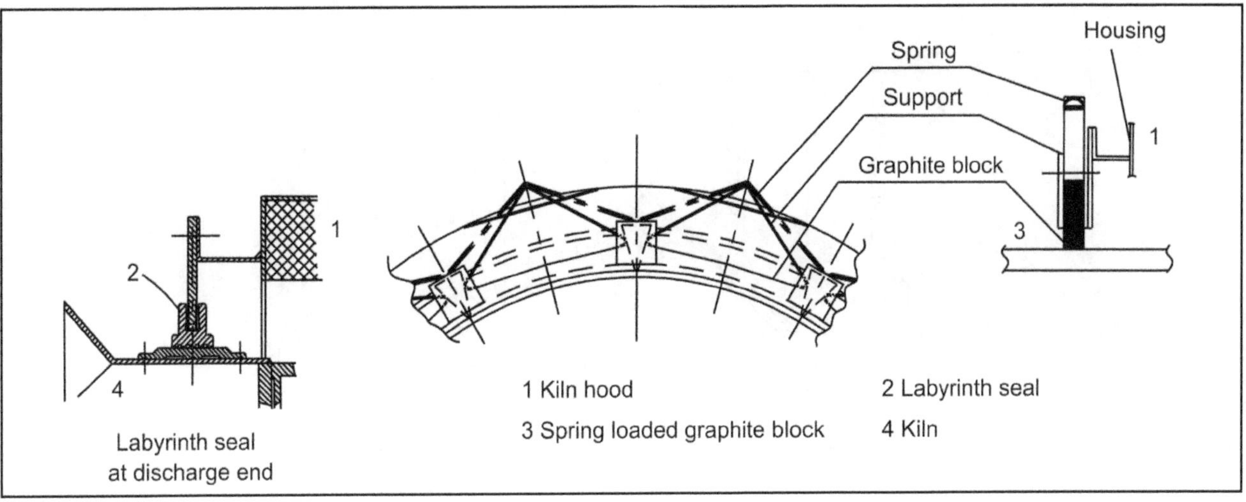

Fig. 6.19.15 Labyrinth and graphite block seals at discharge end of kiln.

19.4.2 Seal at Discharge End

Kiln also has a seal at discharge end. Commonly used types are;

(i) labyrinth seal,

(ii) spring loaded graphite, seal,

(iii) spring leaf

See Figs. 6.19.12 and 6.19.15.

Whatever type of seal, rubbing surfaces, must maintain contact with each other as kiln rotates and even when kiln expands and floats up or down.

Seal should also be able to adjust to the eccentricity if any of kiln shell on diameter and on face.

19.4.3 Materials used for Seals

At the cold end of the kiln, temperature inside are of the order of 900-950 °C and shell temperatures exposed to atmosphere could be 200-250 °C. At the hot end, burning zone temperatures are 1400 °C and shell temperatures would be + 300 °C. Therefore when springs or spring leaves are used, they are likely to lose their tension rendering seal ineffective. Therefore material selected for making springs should be appropriate to the temperature it has to withstand.

19.4.4

Draught at kiln inlet would be between 10 to 20 mms as compared to 1-2 mm at discharge end; damage due to leakage will therefore be more severe at inlet end. Hence this seal should always be maintained in good condition.

19.5 Entry of Raw Meal in to Kiln

Raw meal flowing through raw meal chute of bottom cyclones slides over a crescent shaped heat resisting cast iron / cast steel piece into kiln; and kiln rotation carries it forward into kiln. There should be sufficient overlap between kiln end and the lip even when kiln is cold and floating down.

See Fig. 6.19.16.

19.6 Expansion of Kiln

Expansion of the kiln is calculated taking into account the shell temperature and the length of the kiln.

It is assumed that the kiln expands both ways with girth gear and pinion as the fixed point.

See Fig. 6.19.17.

The supporting rollers and bed plates are so aligned that their center line coincides with the center line of tyres in hot condition.

Thus when kiln is 'cold', center lines of tyre and rollers will not match.

See Fig. 6.19.18.

19.7 Movement of Kiln

Kiln is an inclined rotating cylinder and hence tends to slide down by its own weight like a screw.

It has to be prevented from sliding down by a thrust applied either mechanically or hydraulically1. For small kilns a pair of thrust rollers were installed to take the

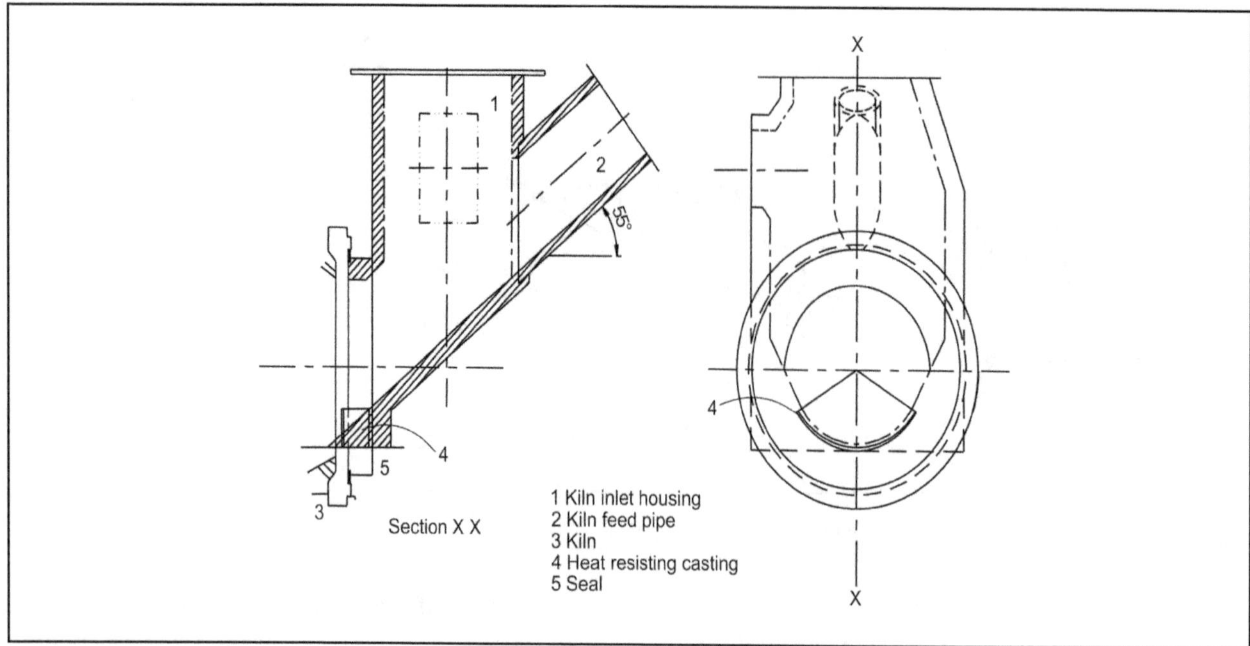

Fig. 6.19.16 Kiln inlet end.

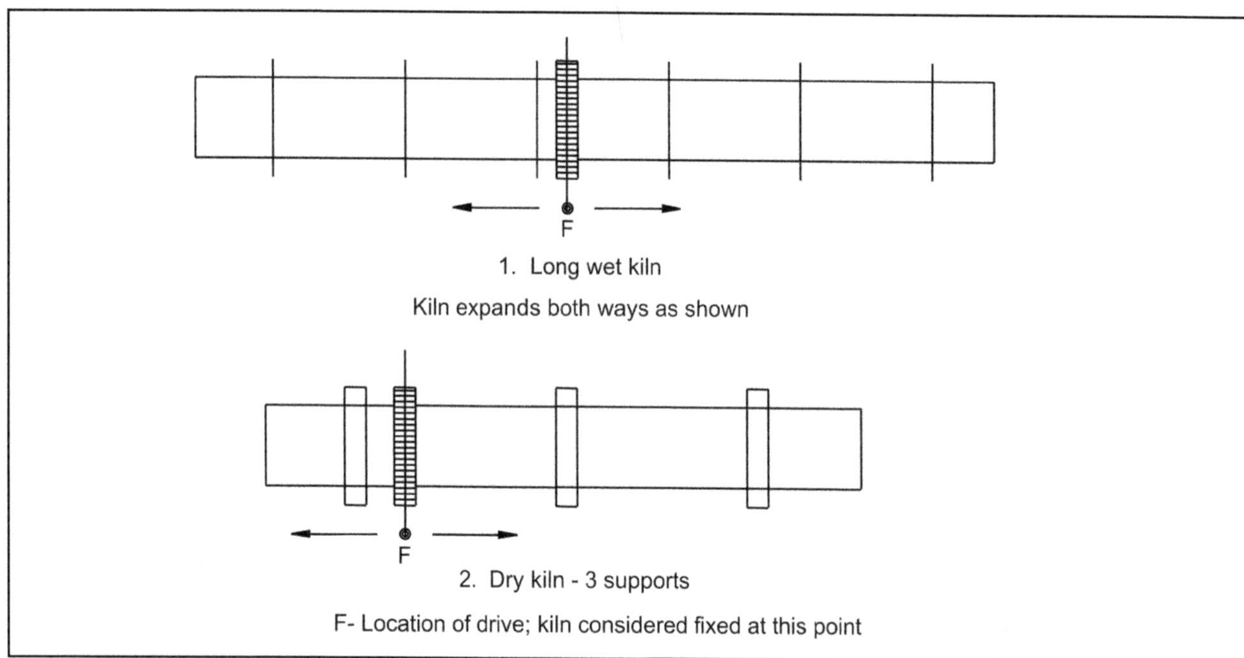

Fig. 6.19.17 Location of kiln drive and expansion of kiln.

thrust of the kiln tending to slide down. For large kilns hydraulic thrusters are used. For long wet kilns thrust rollers were installed on 2 stations.

See Figs. 6.19.19 and 6.19.22.

In case of small kilns without hydraulic thruster, rollers are 'skewed' to counter the tendency of the kiln to slide down. In skewing, axes of rollers are kept at a slight angle to axis of kiln.

See Fig. 6.19.20.

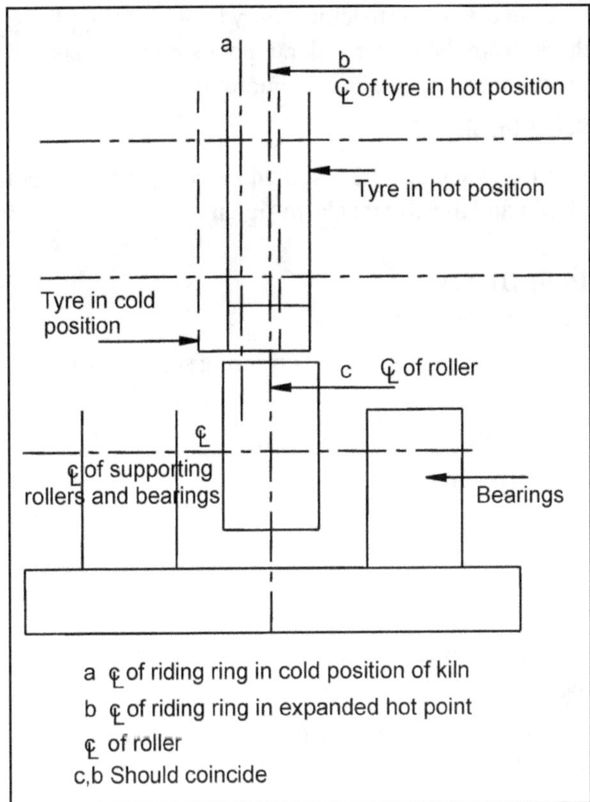

Fig. 6.19.18 Sopporting station so located that center lines of riding rings and rollers come in line in expanded position.

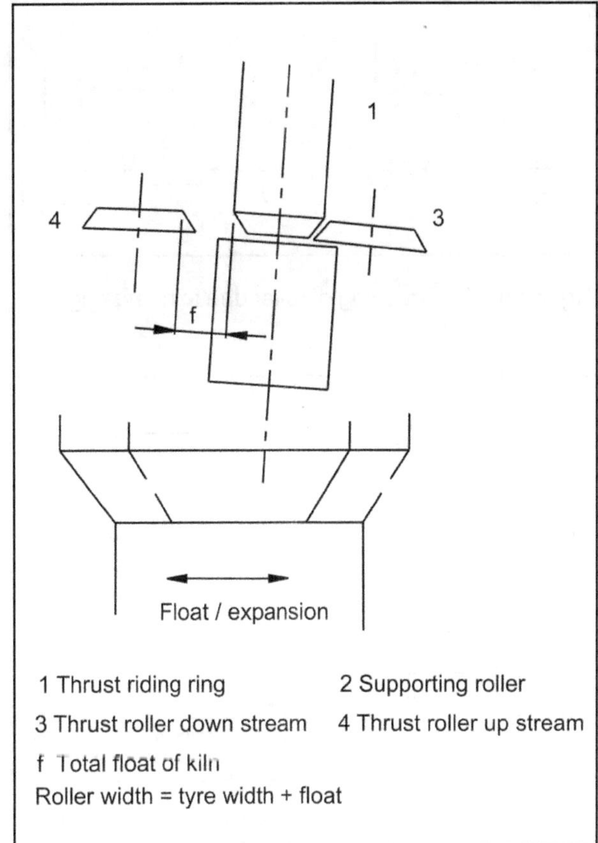

Fig. 6.19.19 Float of kiln.

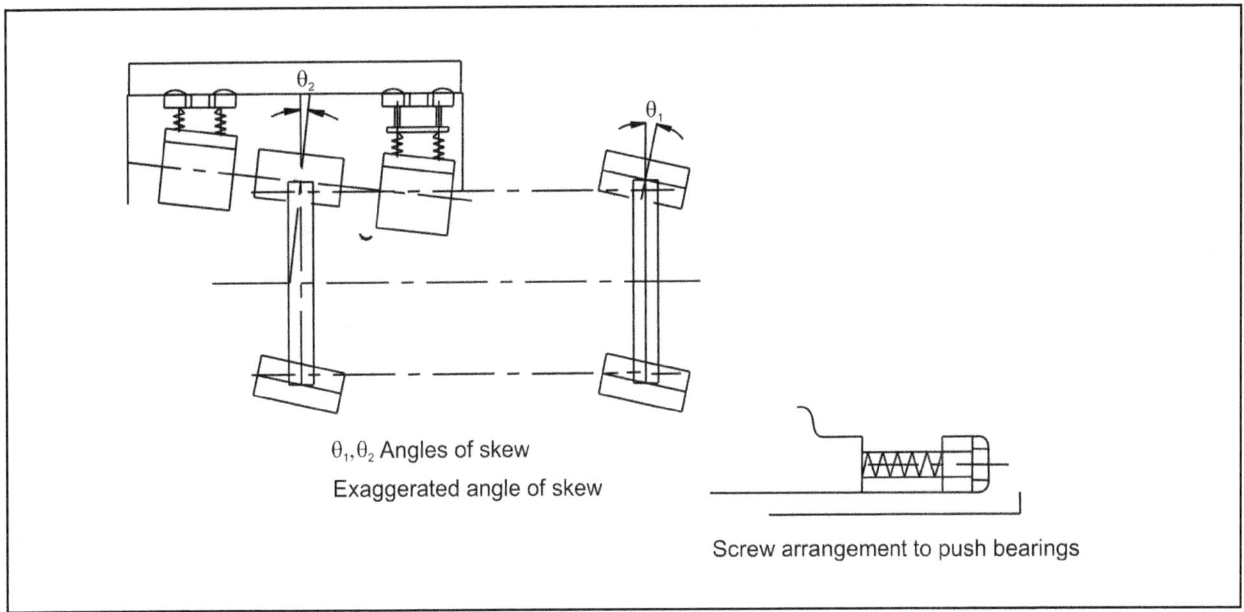

Fig. 6.19.20 Skewing of rollers to produce upward thrust.

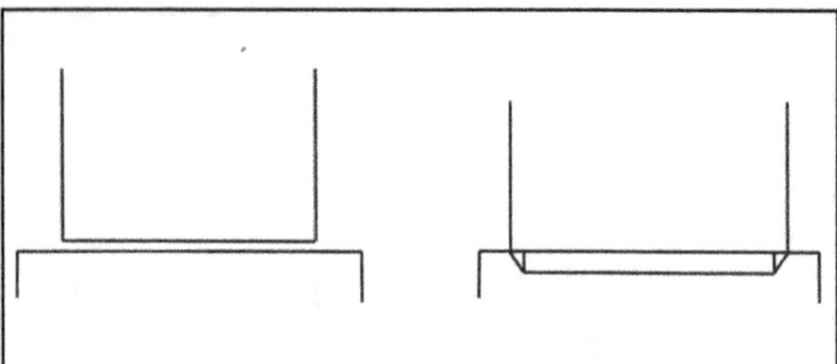

Fig. 6.19.21 Grooving of roller due to riding ring.

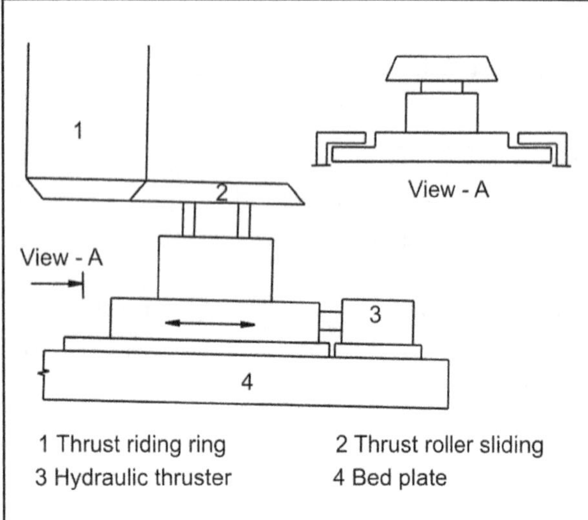

Fig. 6.19.22 Hydraulic thruster.

Amount of skewing could be calculated / established by trial and error. Important point is to maintain uniformity in skewing.

In case of hydraulic thruster where kiln is literally pushed up. Skewing is not necessary.

19.8 Floating

If kiln stayed in one position, always, it would cut grooves in the rollers. The kiln is therefore taken up and down a small distance to prevent formation of such grooves. So that the tyres will rest entirely on the rollers at all times, roller width is kept more than the width of the tyres.

See Fig. 6.19.21.

Kiln must be floated regularly every day in a predetermined manner.

In hydraulic thruster it is easy to do so by applying thrust from the lower roller. Upper side roller is used to prevent accidental over shooting.

See Fig. 6.19.22.

Limit switches are installed to monitor movement of kiln and to activate alarm signals

19.9 Drives

Kiln drive consists of :

1. Girth Gear mounted on spring supports attached to kiln shell.
2. Pinion shaft and bearings.
3. Torsion shaft.
4. Gear box.
5. Motor.

See Figs. 6.19.23 and 6.19.24.

For long wet kilns, drive used to be more or less in the middle.

In dry kilns, it is near the support at the cold end. **See Fig. 6.19.17.**

19.9.1 Mounting of Girth Gear

Spring plates prevent expansion of shell getting transferred to the gear whose axis remains the same and hence clearances between gear and pinion do not get affected in hot running conditions.

See Fig. 6.19.23.

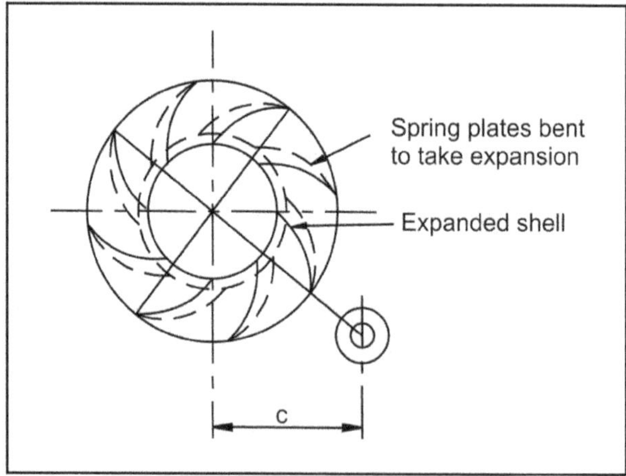

Fig. 6.19.23 Spring plates permit expansion of shell without distrubing center distance between gear and pinion.

Different manufacturers would have different arrangements to take this expansion but use of spring plates is the most common.

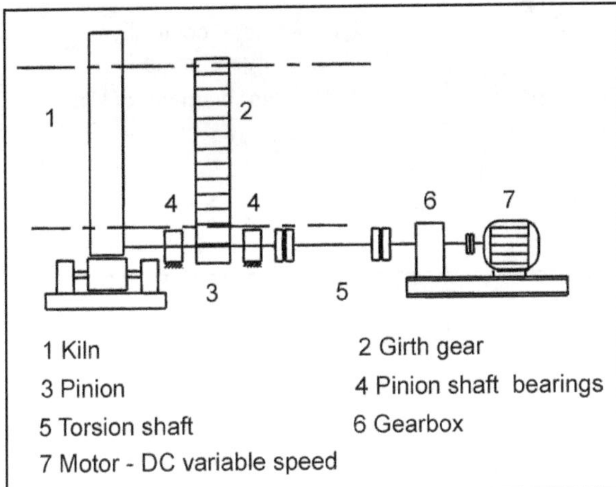

Fig. 6.19.24 Kiln drive - standard arrangement.

19.9.2 Gear and Pinion

Gear and Pinion can be helical or spur. For large kilns helical gears would be preferred as they have higher efficiency compared to spur gears.

See Figs. 6.19.25 and 6.19.26.

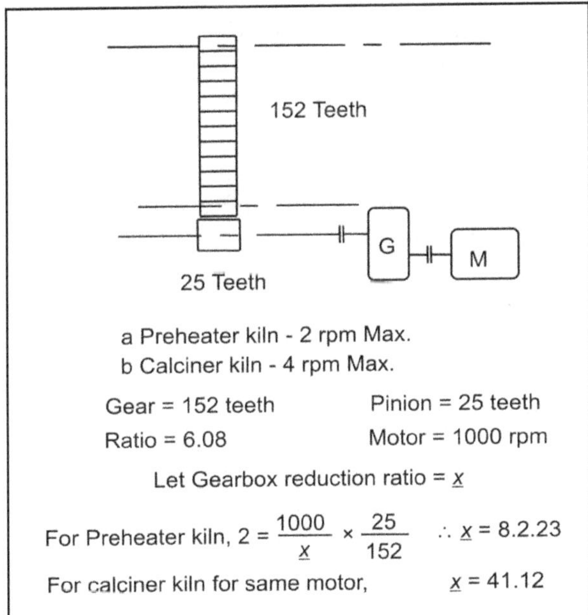

Fig. 6.19.25 If motor speed increased to 1500 rpm, for same gearbox–kiln speed = 3 rpm, same gearbox could be used for increased power.

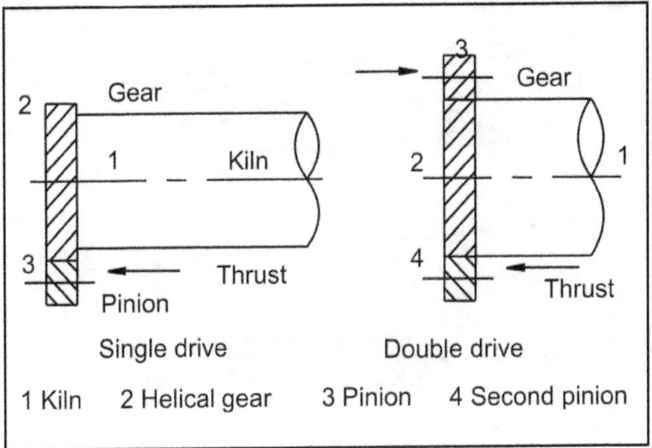

Fig. 6.19.26 Helical gear kiln drive, dual drive neutralizes axial thrust.

19.9.3 Single / Dual Drive

Kiln can have a single or a dual drive. For large kilns, dual drive is preferred. It keeps width of gear down. Dual drive also neutralizes axial thrusts of helical gear.

See Figs. 19.26 and 19.27.

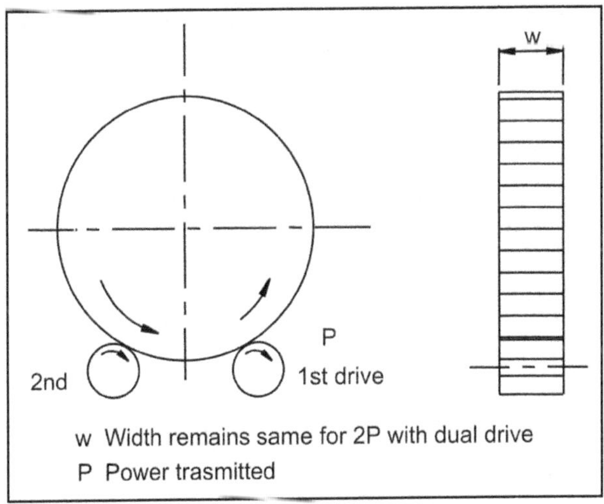

Fig. 6.19.27 Dual drive of kiln- same gear transmits twice the power.

It is also possible to install the 2nd drive when it is required to increase power input of the kiln.

See Fig. 6.19.27.

19.10 Retention Time, Speed and Percentage Filling

Slope and Speed of the kiln determine the 'retention time' of material inside the kiln.

See Fig. 6.19.28.

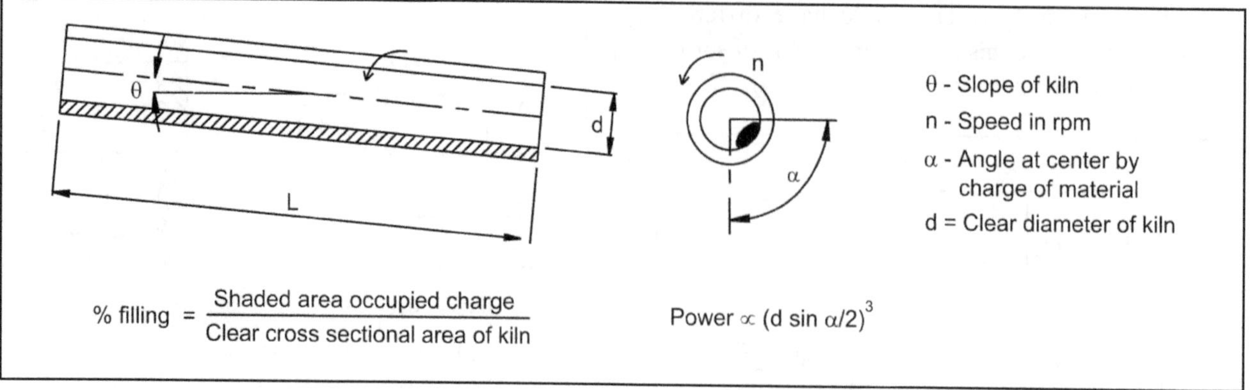

Fig. 6.19.28 Charge in kiln and power of kiln.

The required retention time differs greatly according to the process.

Wet process kilns have the longest retention time of about 120 minutes.

Retention time comes down as the steps of processes carried out inside kiln are reduced. For preheater kiln where material enters kiln, 30% calcined, retention time could be 50-60 minutes. For kilns with calciners, retention time could be as short as 27-30 minutes.

19.10.1 Degree of Filling

The speed and retention times decide the 'percentage filling' of material inside the kiln.

Heat transfer inside kiln takes place by way of :

(i) Direct contact between hot bricks and raw meal / clinker.

(ii) By way of radiation between brick surface and surface of raw meal exposed to it.

(iii) By way of convection between hot gases passing over charge.

To keep the difference between temperature at surfaces of material and its core, to a minimum it is desirable to keep degree of filling between 4-6 %.

See Fig. 6.19.29.

Thus speed, slope and retention time should be so adjusted as to give a degree of filling in this range at design capacity and yet produce clinker of good quality.

19.10.2 Slope of Kiln

Slope is fixed at design stage. It is generally 2° or 3.5 %. For calciner kilns, it can go up to 4 % to reduce speed.

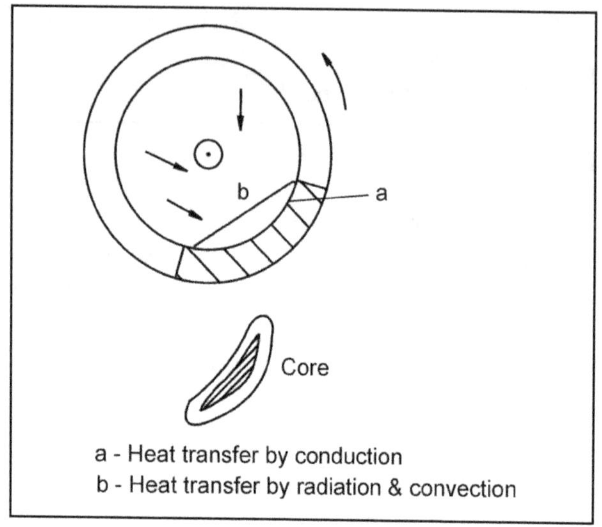

Fig. 6.19.29 Heat transfer to charge inside kiln.

19.10.3 Calculation of Retention Time

Retention time is obtained from the formula

$$T \text{ in minutes} = \frac{1.77 \times Q^{0.5} \times L \times F}{S \times D \times n}$$

where,

Q = Angle of repose, usually 35°

L = Length of kiln / dryer in metres

F = Multiplying factor

for kiln without internal fittings, F = 1

for dryers and coolers, F = 2

S = Slope in degrees

D = Diameter in metres

n = Speed in rpm

19.10.4 Calculation of Degree of Filling

Degree of filling is obtained from formula

$$\% \text{ filling} = \frac{1.667 \times T \times Q \times a}{\text{internal volume of Kiln} \times b}$$

where,

T = Retention time in minutes

Q = Capacity of clinker in tph

a = Ratio raw meal feed / product

b = Average bulk density of material inside kiln

a = Will vary according to % calcination of material entering kiln. It will be lowest for calciner kilns

b for Raw meal = 0.8 t/m^3.

for clinker = 1.2 t/m^3

∴ average b = 1.0 t/m^3

19.10.5 Peripheral Speeds

Earlier peripheral speeds of kilns were limited to 26-35 cm/sec. Linear speed of kiln of 4.0 m dia running at 2 rpm would be 3.14 × 4 × 2 × 100/60 = 41.8 cm/sec.

In calciner kilns, output increases 2.5 times that of preheater kiln.

To keep same or about same degree of filling, retention time is reduced by increasing kiln speed. A kiln that was earlier rotating at 1.67 rpm with a peripheral speed of 35 cm /sec will now be rotating at 4.2 rpm or a perpheral speed of 87.5 cm/sec.

19.11 Kiln Power

Kiln power is a total of :

1. Friction power (P1) and
2. Load power (P2)

19.11.1 Friction Power

Friction power is directly proportional to speed, the total weight of rotating parts and charge inside kiln.

P1 friction power in kw
= 0.595 × (W × Dt × Db × n × *u*) / Dr

where

W = Total weight of rotating parts including weight of refractory and charge in kiln in tons

Dt, Db and Dr are diameters of tyre, bearing and roller respectively in meters

n = speed in rpm

u = friction factor, usually 0.018 for kilns

19.11.2 Load Power

P2 load power in kw = 86.43 × (D Sin θ)^3 × L × K × n

where,

D Sin θ, D is diameter of kiln inside lining,

θ is half angle at center, formed by the charge.

L is length of kiln in metres

K is a constant, usually 0.00076 for kilns

n is speed in rpm

For same degree of filling, (D Sin q)^3 would remain same hence load power would also increase in proportion to speed.

Broadly speaking friction power = 30 to 33 % of load power

Thus a preheater kiln when made into a calciner kiln would run at 2-2.5 times faster and would require about that much more power.

19.11.3

In the process of speeding up, power increases according to speed, torque remaining same. Therefore, existing components of the kiln like gear, tyre, rollers, etc., can transmit the increased power. It is not necessary to make any mechanical changes to these components.

Only wear rate could go high as linear speeds are increased 2.5 times.

19.12 Drive Motor and Gearbox

Motor for a kiln is always a variable speed drive. Slip ring induction motors and commutator 'Schrage' motors have been used in the past.

Now almost always DC variable speed motors are used.

Gearbox is a triple or double reduction helical gear box.

When a preheater kiln is converted into a calciner kiln it is sometimes possible to achieve new power and new speed by changing motor only.

Gear box rating increases according to input speed. A gear box of 100 KW rating at 1000 / 20 rpm ratio would have a rating of 150 KW if input speed is changed to 1500 rpm; output speed changing to 30 rpm.

As the kiln is sloping, the drive is also installed at a slope.

A calciner kiln will require a much smaller gearbox. **See Fig. 6.19.25.**

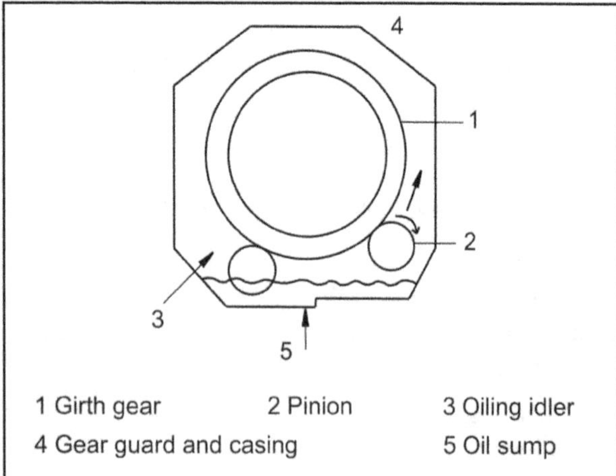

Fig. 6.19.30 Kiln - gear lubrication by oiling idler.

19.13 Gear and Pinion and their Lubrication

Girth gear / pinion ratio is between 6 to 8 / 1. Girth gear and pinion are encased in a guard cum casing, which also holds sump for oil used for lubrication. A lubricating idler is used to keep gear face coated with oil.

See Fig. 6.19.30.

Since kiln is a slow moving machine, oil used must stand high pressure on film and the film should stay on surfaces of teeth till they come in contact again.

Drive can now also be a friction drive through rollers. **See plate no. 2.10.7 in Section 2.**

19.13.1 Torsion Shaft

A torsion shaft is almost always used between pinion and gear box :

(i) To prevent over load,

(ii) to take misalignment.

Kiln being a large cylinder heated from inside and exposed to atmospheric conditions like heat, rain, etc., does not always remain 'true' and hence misalignments are more common and larger in magnitude than in case of other machines like mills during running. A torsion shaft is thus very useful.

See Fig. 6.19.24.

19.14 Measurement of Torque

In Instrumentation and Control schemes, torque of kiln, which is a measure of charge inside it is measured. So both speed and torque would be shown and recorded on kiln control panel.

19.15 Development in Design of Components of Kiln

There have been many developments in design of components of the kiln like splined tyre, self aligining rollers, and friction drive through roller. These have been brought out in Chapter 10 of Section 2.

19.16 Riding Rings

Riding rings or tyres are generally solid in one piece. They are made of cast steel. Split tyres were tried out but were not successful. Tyres can be fixed or floating. Floating tyres move with respect to kiln shell. This was the most commonly used type. Relative movement between tyre and roller is known as 'creep'. In floating tyres there is a gap between tyre and shoes welded to shell even in hot condition. It is maximum at top and nil at points of contact with rollers. This gap should be such that creep is about 1 revolution in 24 hours. Higher gaps cause more slip and more wear. Surface between tyre and shell is lubricated to reduce wear.

See Figs. 6.19.31 and 6.19.32.

19.16.1 Splined Riding Rings

Presently splined tyres have come into use. In this design, splines of the tyre engage with those in shoes and hence there is no relative movement of the tye with respect to the kin. **See Plate 2.10.6 in Chapter 6 of Section 2.**

19.17 Supporting Rollers

Supporting rollers are also made of cast steel. They are generally hollow to keep down weight. Rollers are shrunk fit on forged steel shafts. Shafts may have a small step to locate the roller and roller should be correctly installed on the foundations with respect to the step **See Fig. 6.19.33.**

19.17.1 Roller station

A three support kiln would have:

3 tyres

6 rollers

12 bearings.

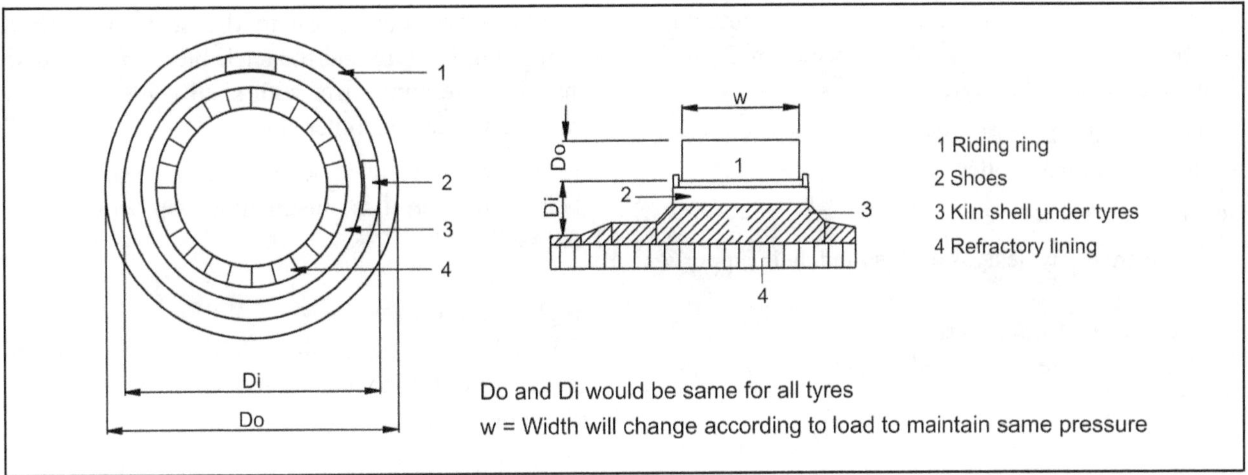

Fig. 6.19.31 Mounting of riding ring on kiln shell.

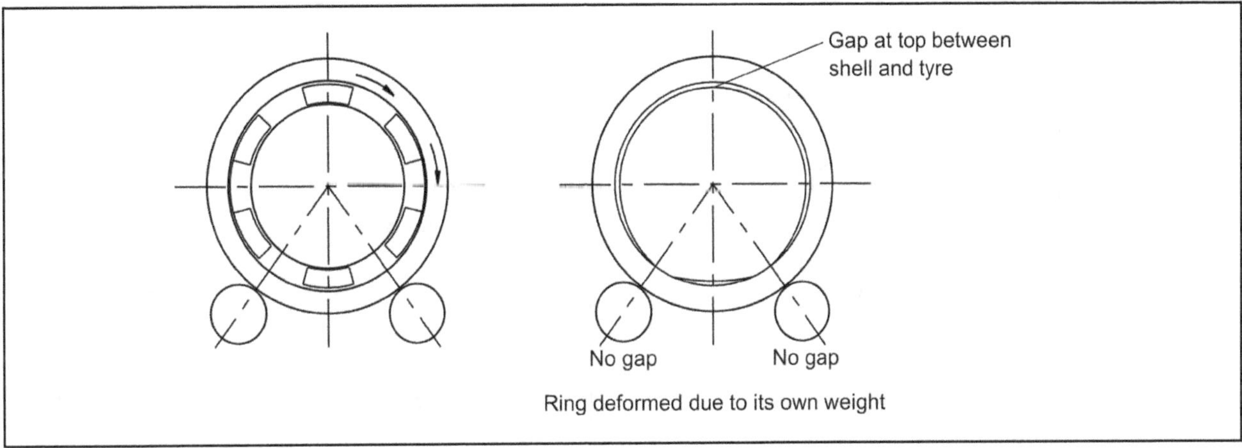

Fig. 6.19.32 Creep is relative movement between kiln and tyre – 1rev. in 24 hours is acceptable.

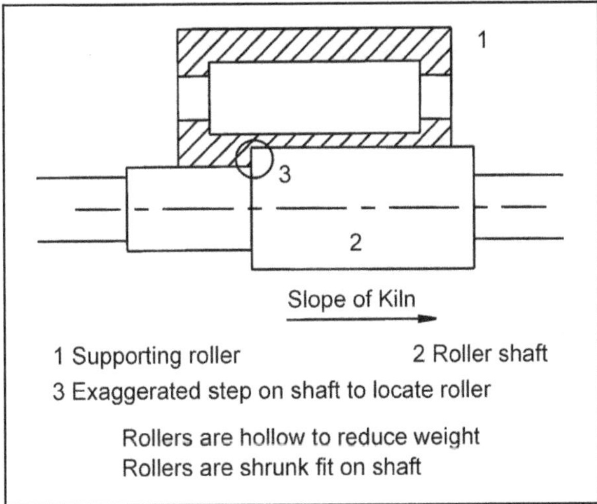

Fig. 6.19.33 Roller and shaft of rotary kiln.

As mentioned earlier, the spacing between tyres is adjusted to distribute load equally on 3 supports to the extent possible.

In case of planetary coolers this is not possible because of over hung load and therefore the last tyre would be considerably heavier compared to the other two. **See Figs. 6.19.3**

19.17.2 Dimensions of Tyres and Rollers

From practical considerations, Do and Di would be kept same for all tyres changing width from load considerations, Hertz pressure remaining same.

See Fig. 6.19.31.

It would be ideal if all rollers also had same diameter and all bearings also of same diameter and width for all roller stations. But it may not always possible to achieve it.

If the tyre width is different, then roller width will also be different to suit as roller is wider by 25 mm than the corresponding tyre.

The tyre taking thrust is wider – as its face is tapered to rub against thrust roller.

See Fig. 6.19.22.

There are many designs of tyres and their supports.

19.17.3 Shell Under Tyre

Shell under the tyre is increased in thickness to maintain ovality.

This may be done in 1 or 2 steps. The edges are beveled and welded to minimize thermal stresses.

See Fig. 6.19.31.

Shoes are welded on to the section of shell supporting the tyre and the shell piece machined so that they are concentric with shell center. Retainer bars are welded on either side.

In some designs shoes are in shape of wedges. This design permits adjustment of gap between tyre and shell.

19.18 Bed Plates

Kiln bed plates are generally of fabricated construction in one piece with machined surfaces to hold 4 bearings of a roller station.

Bed plates should have center lines in both directions and also in a vertical plane on sides.

See Figs. 6.19.34 and 6.19.35.

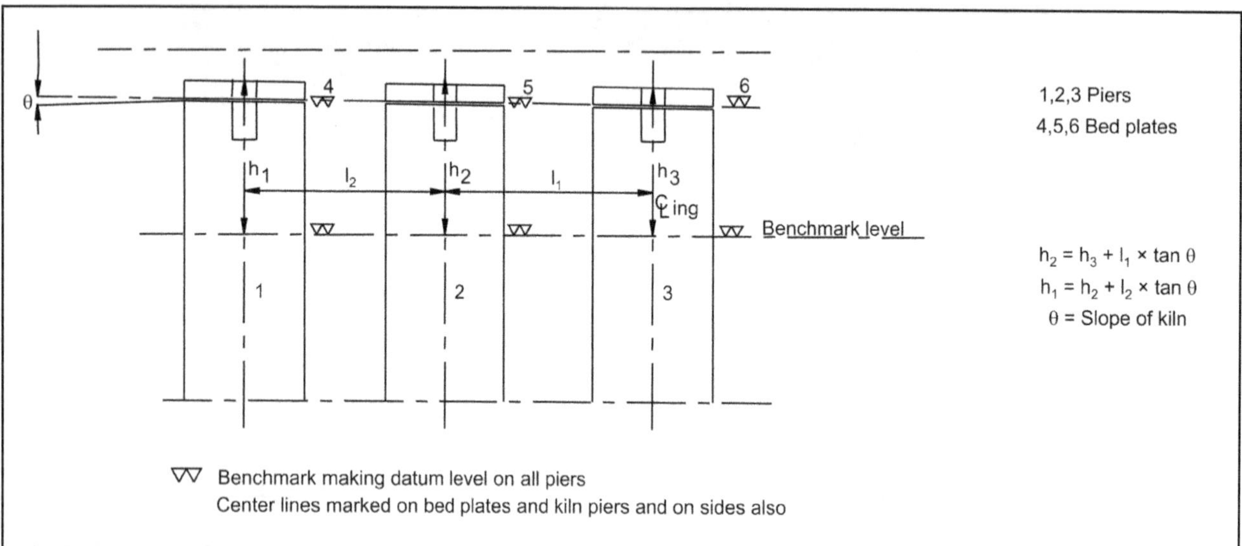

Fig. 6.19.34 Benchmark marked on each pier so that bed plates can be aligned w.r.t. it.

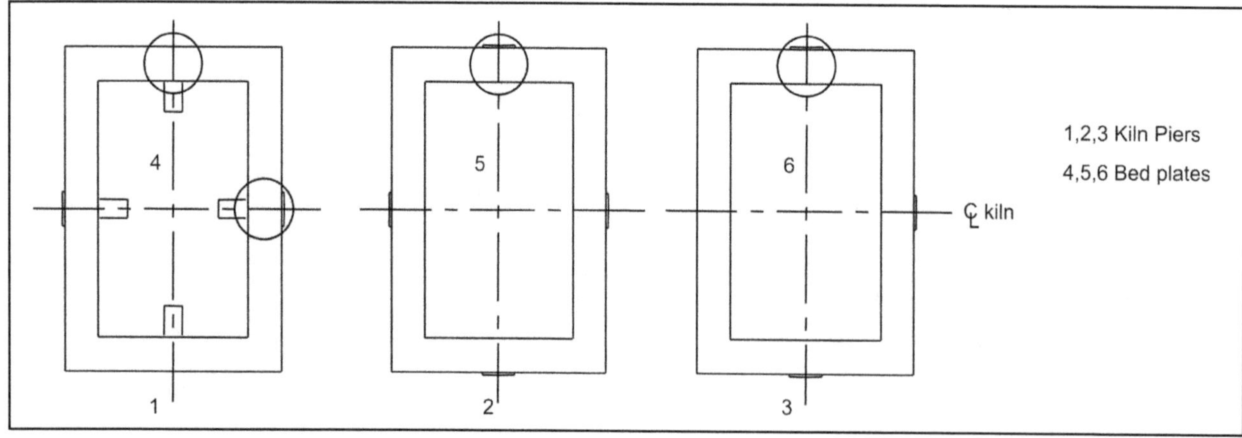

Fig. 6.19.35 Center lines marked on piers as well as bed plates.

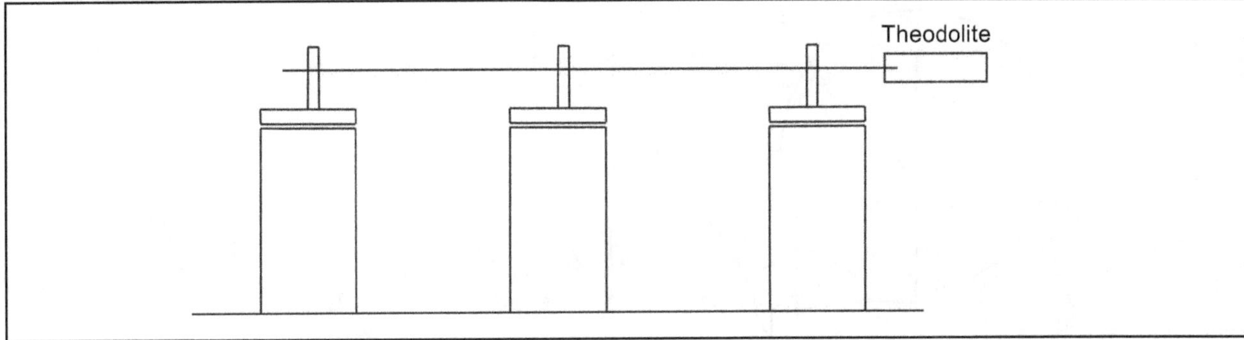

Fig. 6.19.36 Levelling bed plates - using theodolite, or piano wire, laser beam.

Bed plates are aligned and centred and leveled and half grouted before any load is put on them. This is an important job. Centre lines should also be marked on steel plates embedded in the faces of concrete foundations.

See Figs. 6.19.34 and 6.19.35.

Care taken in aligning bed plates can never be enough for future smooth running of kiln for years.

Earlier theodolite was used to align bed plates. Piano wire has also been used to center the bed plates and kiln. Now laser beams are used to do the alignment. **See Fig. 6.19.36.**

Bed plates must have provision for skewing of rollers if conventional thrust rollers are used.

See Fig. 6.19.20.

Many bed plates now have hollows for rollers to dip into water filled in them. When kiln is to be floated oil / grease is put on contacting surfaces.

See Fig. 6.19.38

Bed plates / foundations should have provision for bringing in cooling water and taking it out.

Rollers should be assembled on bedplates according to erection drawings. If bearings and rollers and tyres are made according to drawings and dimensions are within the tolerances, kiln shell put on them should be centered needing only minor adjustments, which can be done by slight shifting or placing shims etc.

See Fig. 6.19.39.

19.19 Kiln Bearings

Bearings are brass / bronze bushes. Bearing covers have thrust pads to take the thrust. Discs with buckets are attached on the shafts. Buckets dip in oil and spread it over the width of the bearings. Because of slow speed kiln bearings are seldom forced lubricated. **See Fig. 6.19.40.**

Bearings are not bolted to the bed plates. They remain in position by the weight of kiln. Their position can be changed/ adjusted with the help of 'skewing' screws provided on bed plates.

See Fig. 6.19.37.

19.20 Hydraulic Thruster

Hydraulic thruster equipment can be installed under the kiln itself so that piping is short in length, when the pier is hollow.

19.21 Kiln piers

Kiln piers support entire assemblies of bed plates, supporting rollers and their bearings. They have to support heavy weight of kiln assembly.Piers are dimensioned on one side by dimensions of bed plates and on the other by height of kiln centers above datum level. Thir own weight can be considerable and it has to be taken into account. Often kiln piers are solid blocks. But to save on weight they could also be made hollow. **See Fig. 6.19.42**

Fig. 6.19.42 Hollow kiln piers.

19.22 Erection of Kiln

Erection of kiln is a specialized job. Kiln is the largest single piece of machinery not only in cement plants but in other industries also.

Kiln shell is received in pieces. They are shipped in specially made wagons or in heavy duty specially designed trailers. Care should be taken to put stays as shown so that shell does not get distorted in transit. Machined surfaces like shoes are protected with wooden slats.

See Fig. 6.19.43.

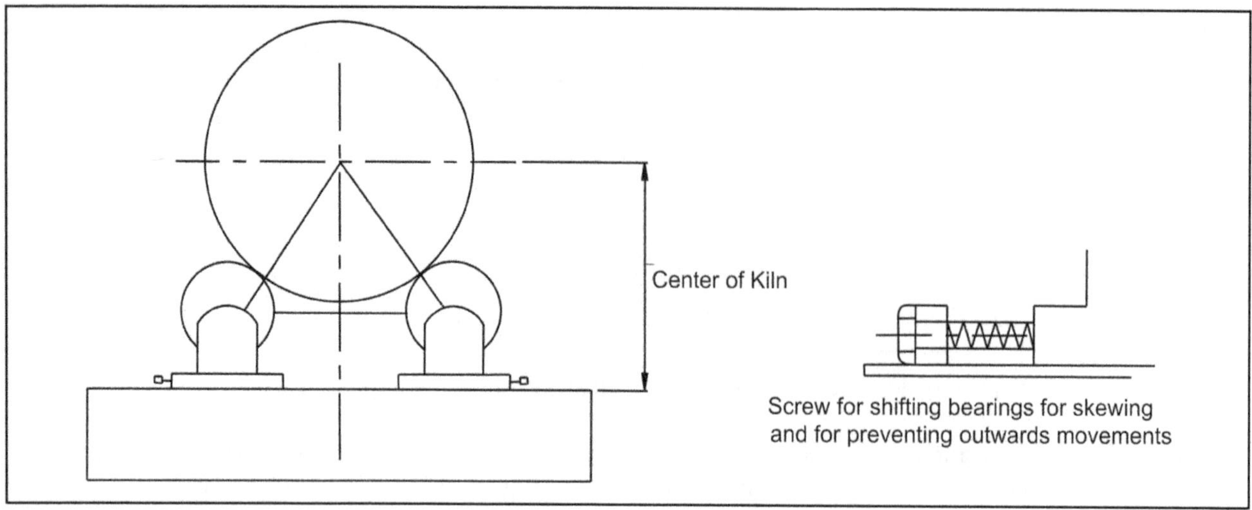

Fig. 6.19.37 Kiln bed plates
Bearings are not bolted on bed plates

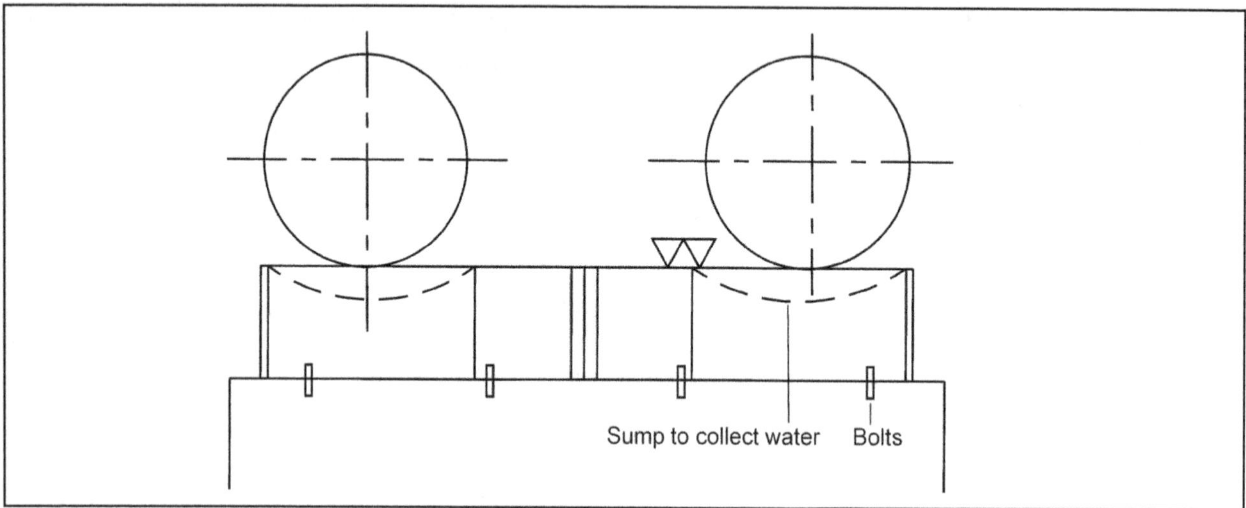

Fig. 6.19.38 Kiln bed plate - fabricated and stress relieved with machined face to receive bearings.

Fig. 6.19.39 'a' to 'c' shows stages in aligning shell sections and their welding.

Previously an erection crane was used for erecting a kiln.

See Fig. 6.19.44.

Kilns are now erected by using traveling boom cranes, which lift kiln shells, put them on tiers, and help in alignment.

See Fig. 6.19.45.

19.23 Maintenance Platform

Kiln piers are distant from one another. A platform is therefore provided to go from one end of the kiln to the other.

It should be wide enough at least 750 mm for workers to carry tools tackles, etc.

See Fig. 6.19.46.

19.24 Cooling of Kiln Shell

Kiln shells are cooled in burning zone to prolong life of refractory.

Fixtures are required to support the axial fans and shift their position angle etc. They are supported from the same walkway.

See Fig. 6.19.47.

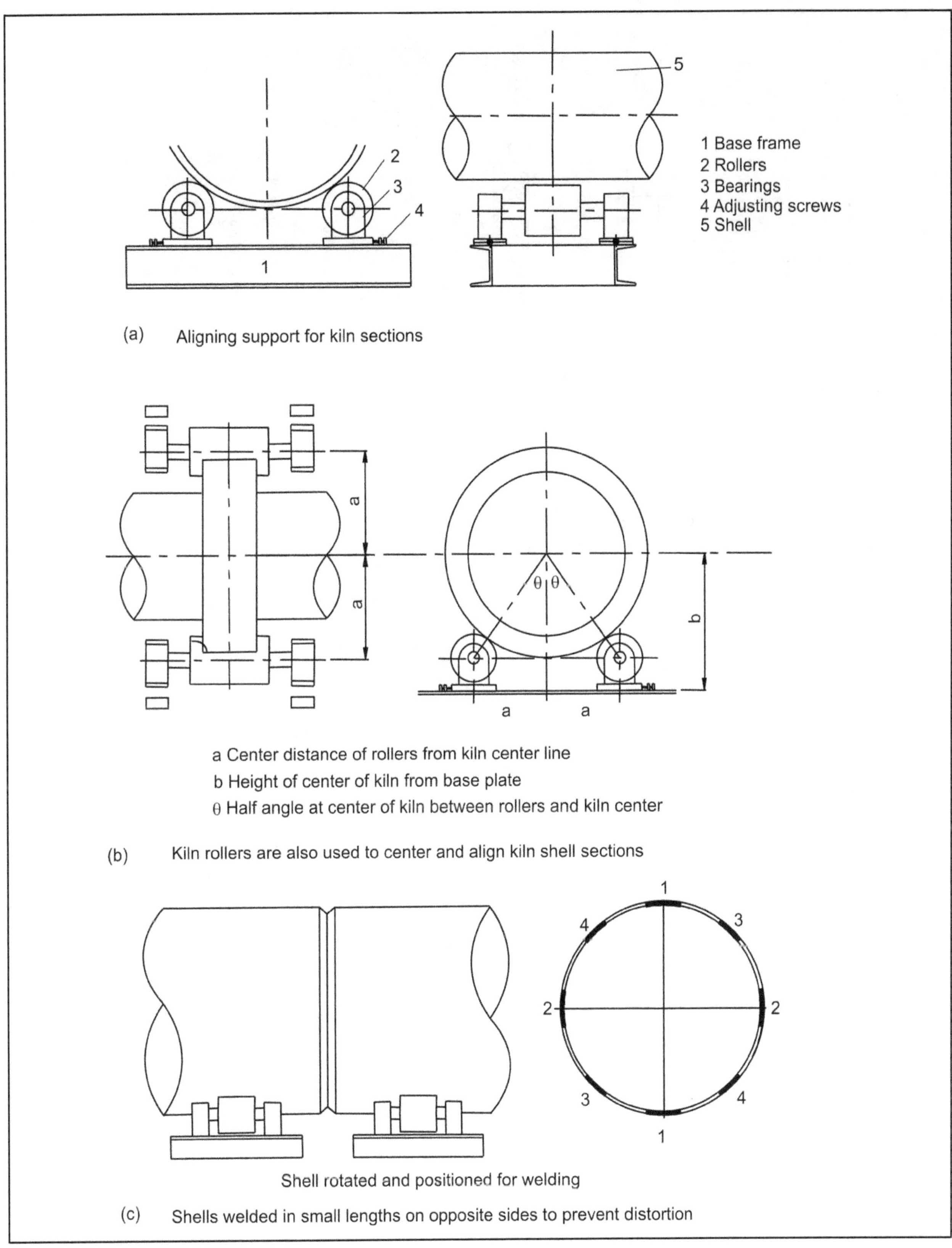

Fig. 6.19.39 Assembly and aligning of kiln shell.

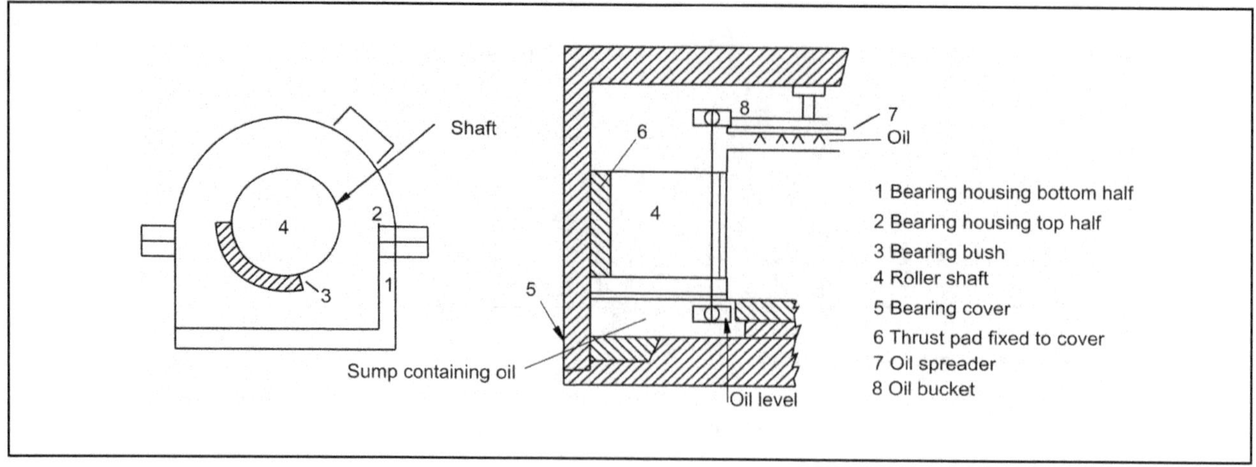

Fig. 6.19.40 Kiln bearing bushes are oil lubricated with buckets and oil spreader ; they are water cooled.

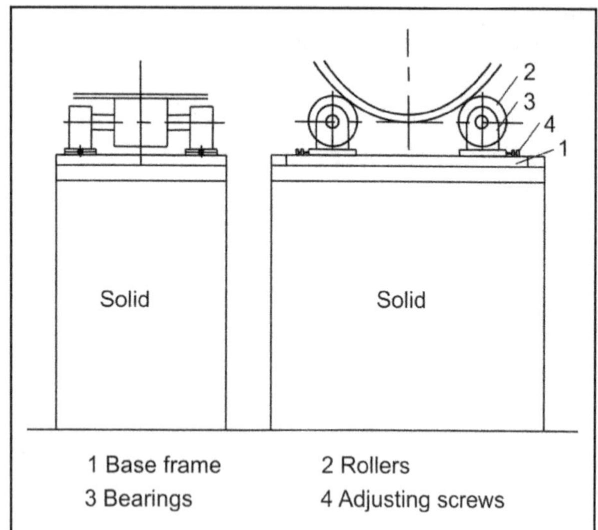

Fig. 6.19.41 Solid kiln piers.

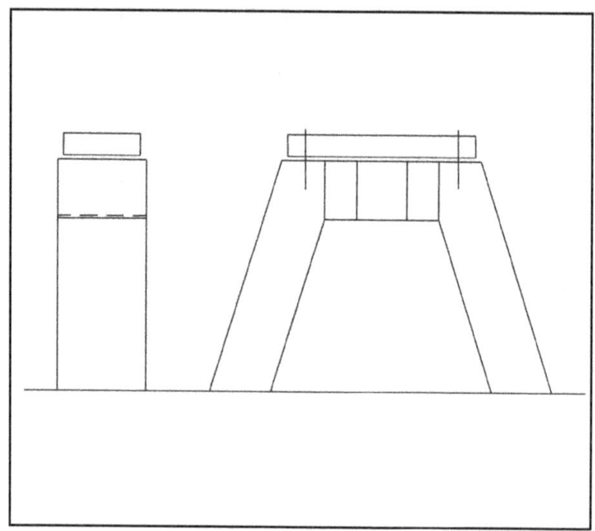

Fig. 6.19.42 Hollow kiln piers.

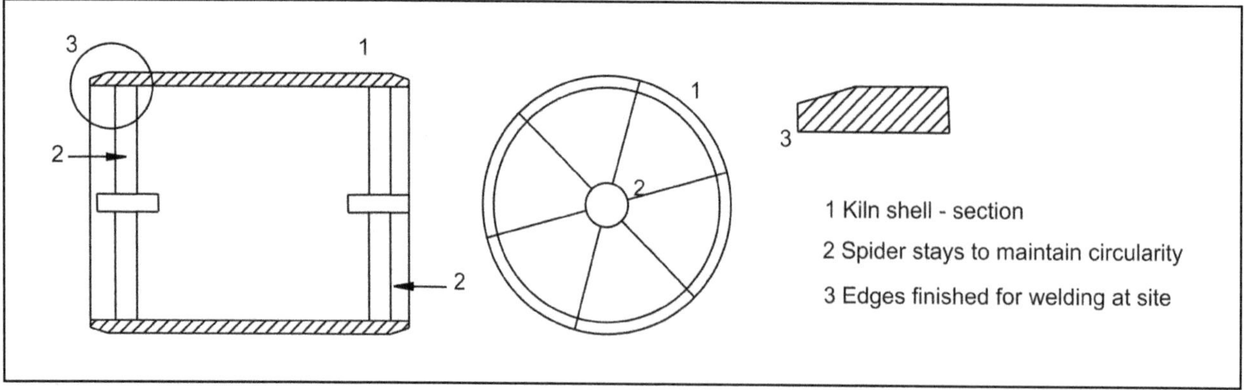

Fig. 6.19.43 Kiln sections for transport.

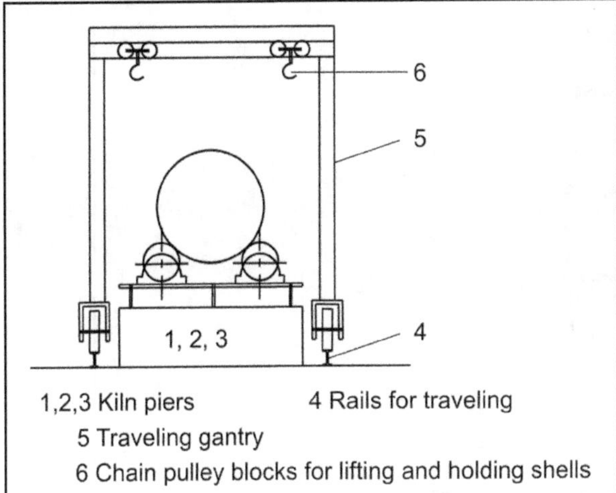

Fig. 6.19.44 Travelling erection crane gantry for erection fo kiln.

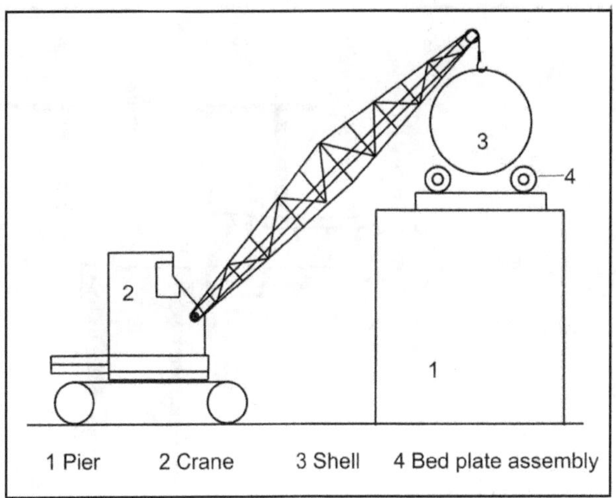

Fig. 6.19.45 Kiln erected with the help of a crane.

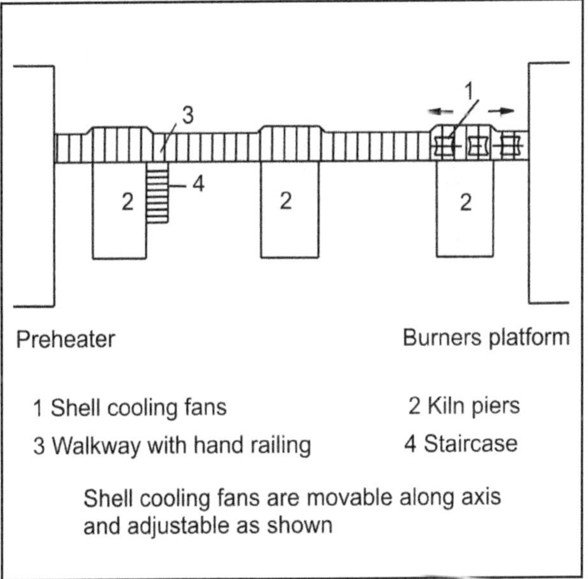

Fig. 6.19.46 Walkway connecting kiln piers.

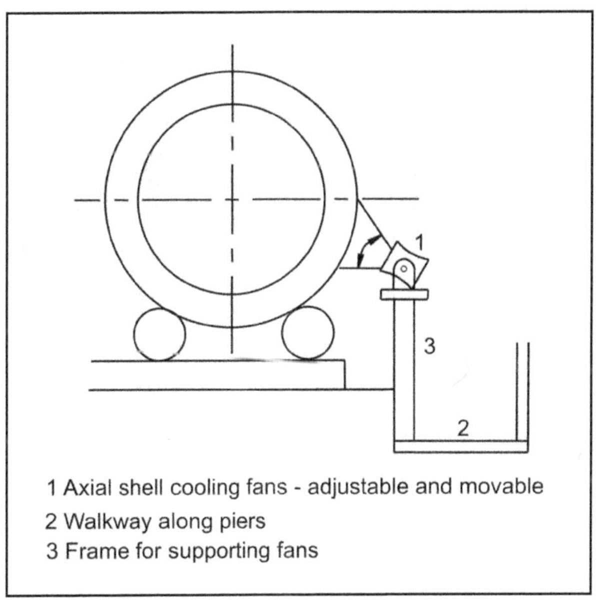

Fig. 6.19.47 Axial shell cooling fans in burning zone.

19.25 Brick Lining

Entire kiln is lined with refractory.

See Fig. 6.19.48.

Brick Lining is also a specialized job and is entrusted to experienced and skilled brick layers and masons. Like in welding, kiln is required to be rotated slowly and positioned during the operation of brick laying. Bricks are jacked to hold them in position. **See Fig. 6.19.49.**

Also **see Chapter 55** on **Refractories.**

Different types of bricks require different procedures, mortars, wedging plates, etc. Some bricks require a cardboard piece or an iron shim to be inserted between them, which burns off.

Brick retaining rings are installed at crucial places and in particular towards end so that bricks do not shift or get dislodged.

See Fig. 6.19.50.

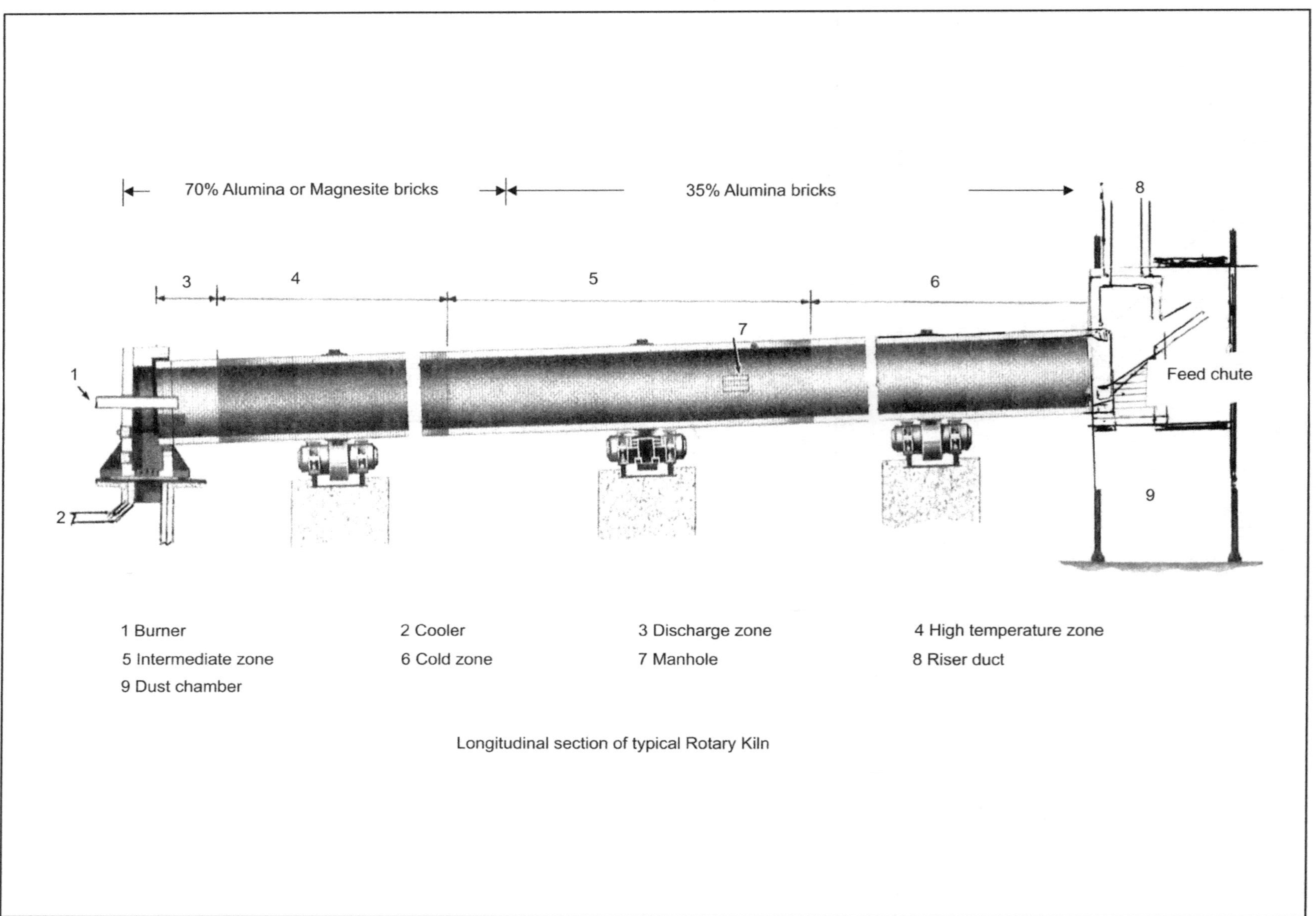

Fig. 6.19.48 Refractory lining of a cement kiln.

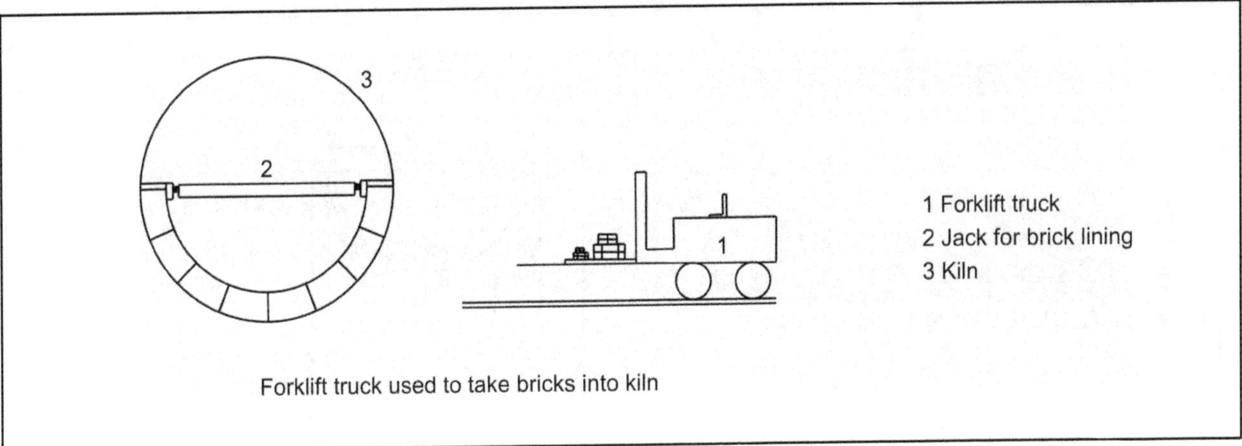

Fig. 6.19.49 Jack for brick lining of kiln.

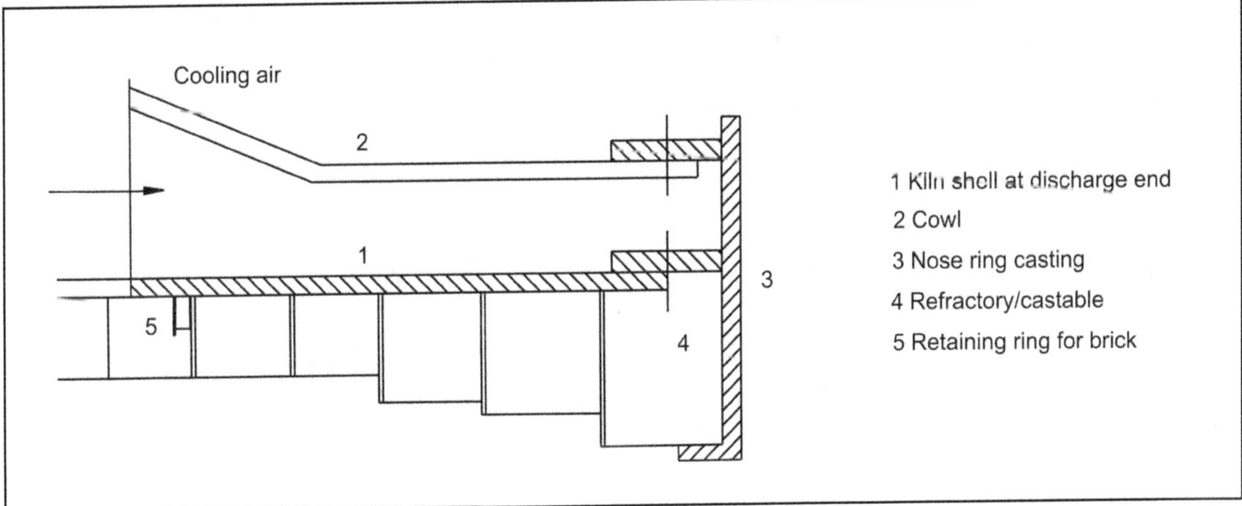

Fig. 6.19.50 Discharge end of kiln shell.

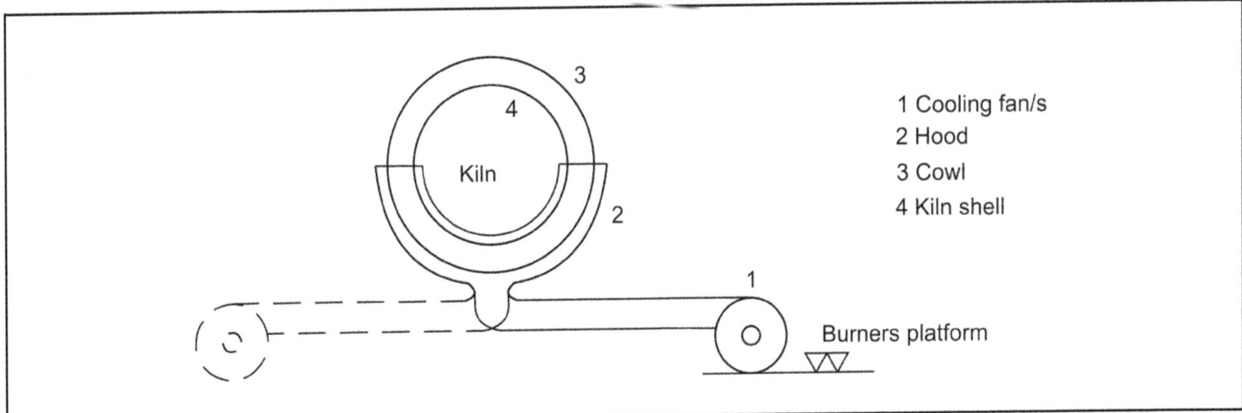

Fig. 6.19.51 Nose ring cooling fans - one or two.

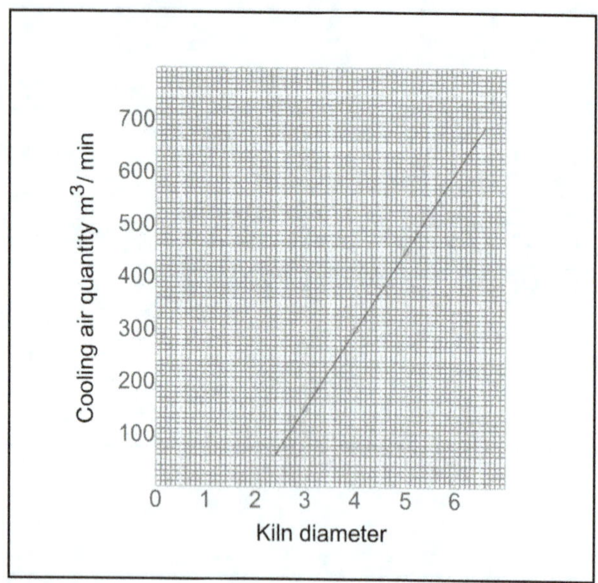

Fig. 6.19.52 Cooling air for nose ring castings.

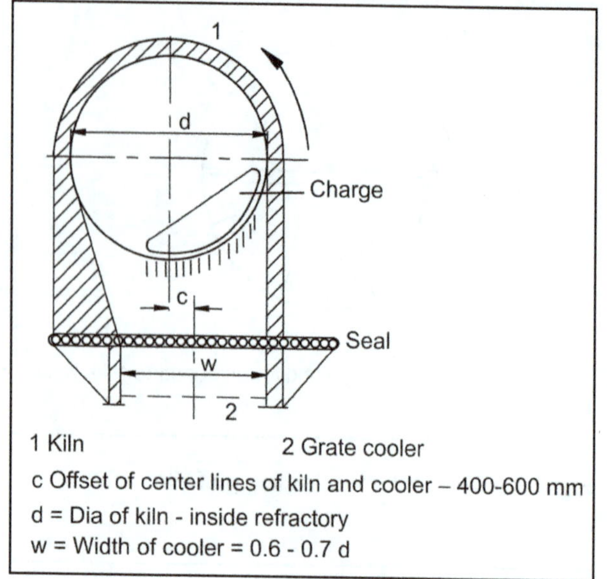

Fig. 6.19.54 Offset centre lines of kiln and cooler.

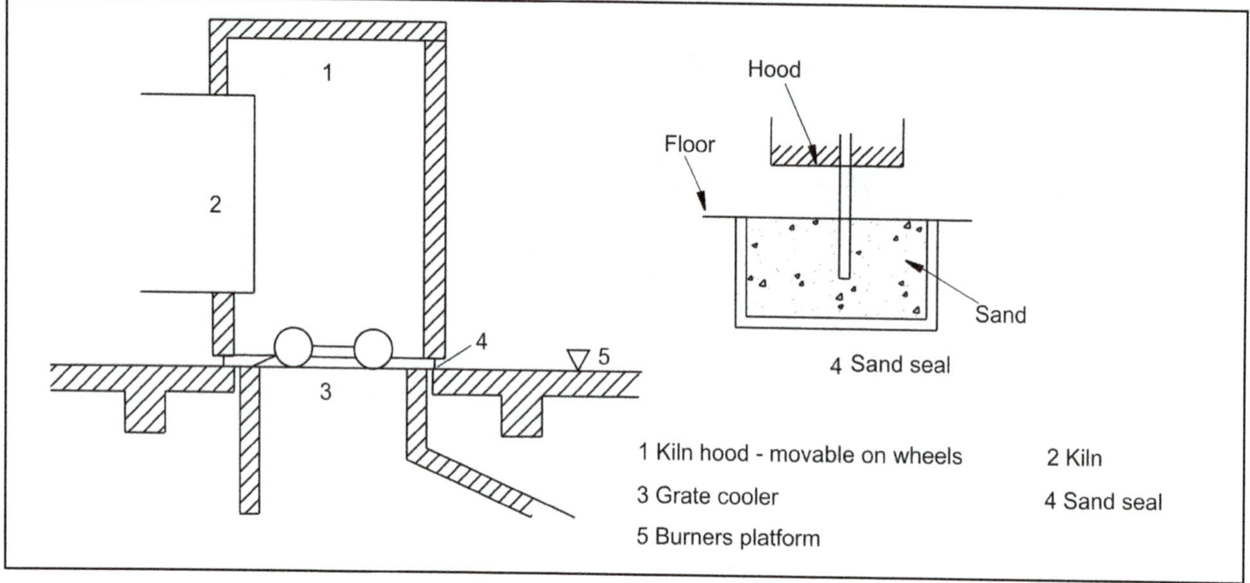

Fig. 6.19.53 Kiln and cooler - kiln hood and seal.

19.25.1 Nose Ring Castings

Kiln shell and bricks are protected at discharge end by installing end or nose ring castings, which are heat resisting cast steel sections bolted to shell. Each manufacturer has its own design of the castings. Sometimes they are in two pieces.

This is a 'maintenance heavy' part of the kiln, requiring stoppages, now and then.

See Fig. 6.19.50.

19.25.2 Nose Ring Cooling Fans

Castings and shell are cooled by nose ring cooling fans. Depending on the size of kiln, cooling blowers can be one or two and they are accommodated on the burners' platform.

See Fig. 6.19.51.

Fig. 6.19.52 provides data on cooling air required for kilns of different diameters.

Fig. 6.19.55 Taking refractories into kiln.

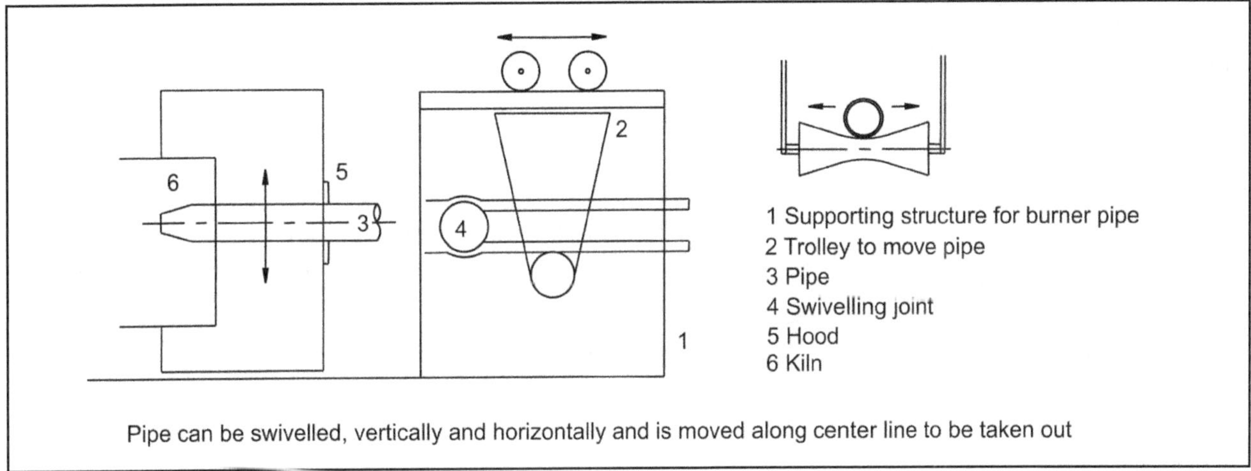

Fig. 6.19.56 Handling and manipulation of burner pipe.

19.26 Kiln Hood

Kiln hood is the connecting piece between rotating kiln and grate cooler. It used to be movable on rails to provide access to kiln.

Center lines of kiln and cooler are offset with respect to one another so that clinker falling from kiln falls in the center of the cooler.

See Figs. 6.19.53 and 6.19.54.

With calciner kilns, the TA duct is likely to come in the way of shifting the kiln hood. Therefore now hoods are stationary with large doors at front through which a fork lift truck can pass for carrying refractories into kiln. **See Figs. 6.19.55**

See Chapter 18 of this Section.

19.27 Burner Pipe

Another important point in layout is the burner pipe.

The burner pipe needs a structure to :

Support it.

To move it backwards and forward.

To position it with respect to kiln center line including tilting through a small angle.

See Fig. 6.19.56.

19.27.1 Location of Burner Pipe

There are different views as to where the burner pipe should be located with respect to kiln center of kiln and its angle.

One school believes that burner pipe should be at the same angle as kiln i.e., 2°. Others believe that it should be horizontal.

Some believe it should be in the center – others prefer to position it in sector b – nearer to the surface of the charge.

See Figs. 6.19.57 and 6.19.58.

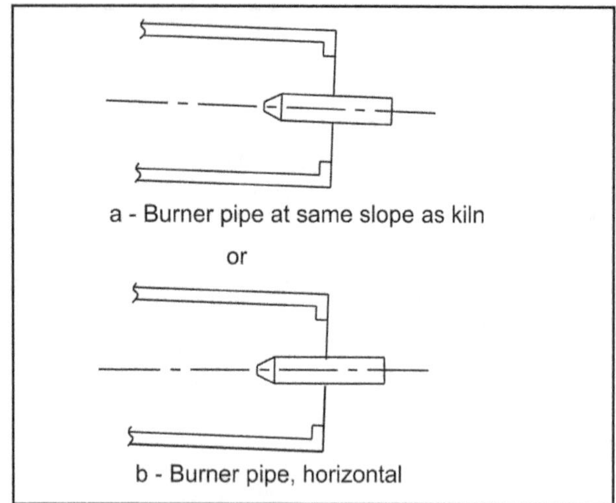

Fig. 6.19.57 Alignment of burner pipe.

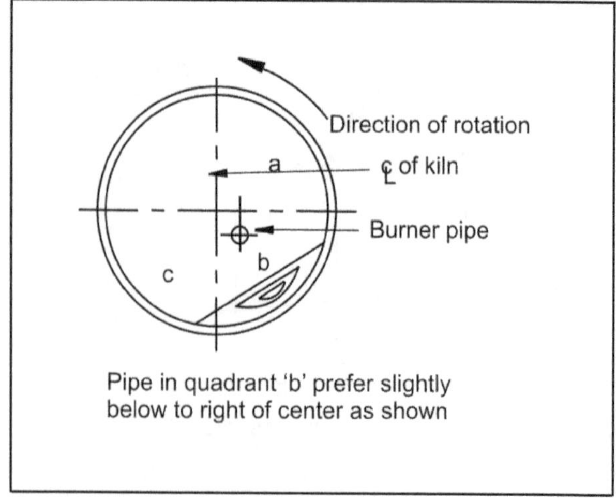

Fig. 6.19.58 Position of burner pipe - kiln section.

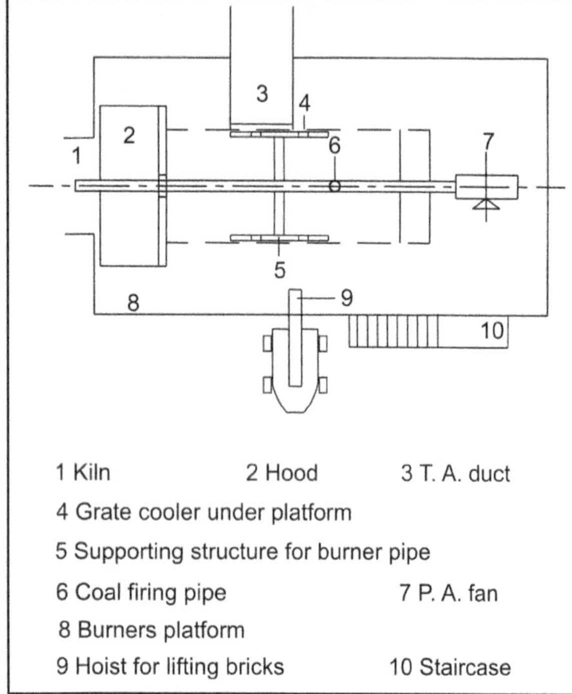

Fig. 6.19.59 Burners platform - layout.

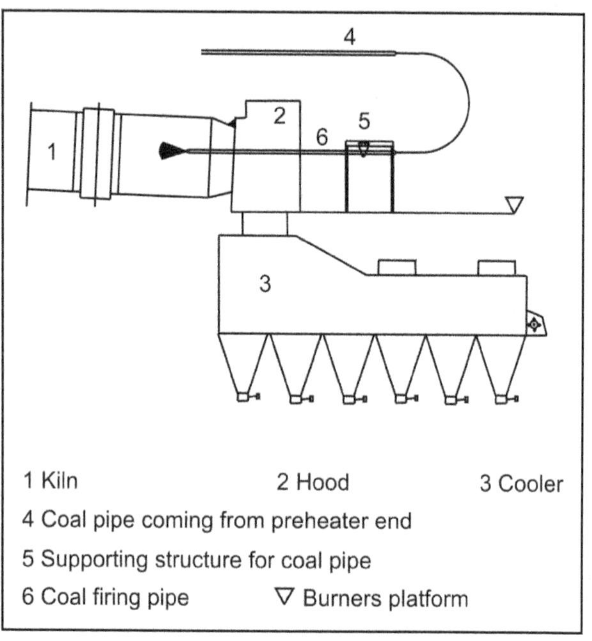

Fig. 6.19.60 Burners platform when coal mill at preheater end coal metering at preheater end.

As mentioned above and shown in **Fig. 6.19.59,** layout must have provision to support the burner pipe and its manipulation.

19.28 Burners' Platform

This has come to be so called because in early days burners in shift used to stay on this floor at the outlet of the kiln and watch conditions in the burning zone almost continuously. The kiln control panel was also located on the burners' platform – so that they could walk over and make adjustments to feeds, speeds and gas flows.

Burner, Chief Burner and Master Burner was the hierarchy amongst burners then. They were drawn from ranks with school level education and learnt the operation under the supervision of senior burners.

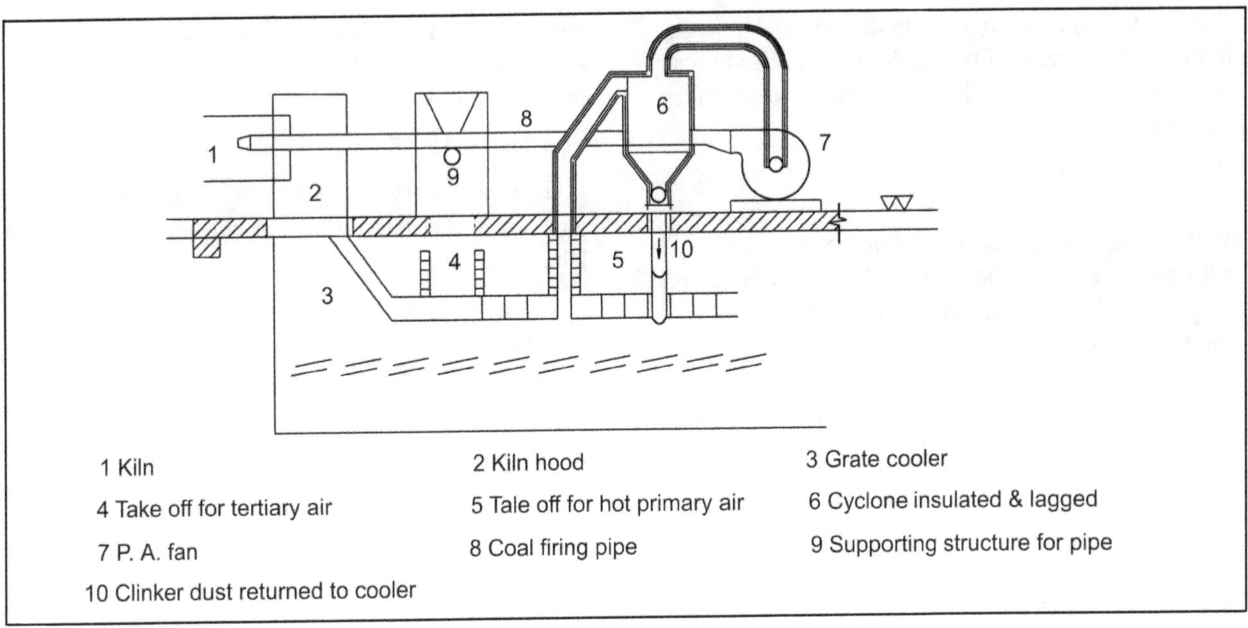

Fig. 6.19.61 Hot primary air from grate cooler.

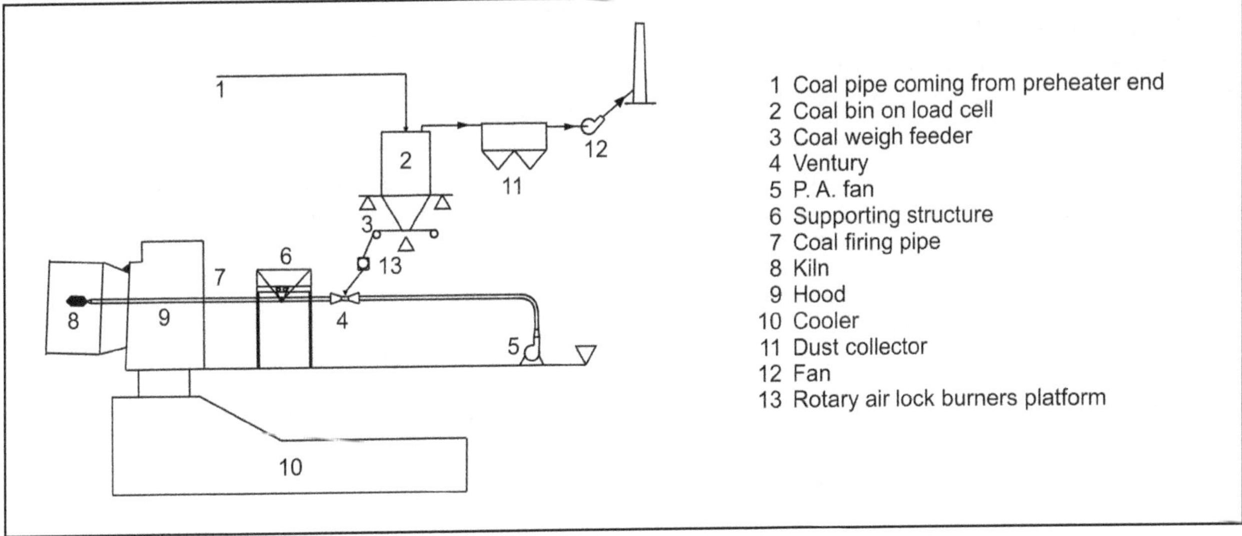

Fig. 6.19.62 Alternative arrangement - fine coal bin and coal feeder on burners' platform - coal mill at preheater end.

They depended on, and carried out instructions received from the Laboratory to maintain quality of clinker.

Primary Air fan was installed at the back end of burners' platform. It conveyed pulverized coal into the kiln through a coal firing pipe. Grate cooler was installed under the Burners' platform.

See Figs. 6.19.59 and 6.19.60.

With advent of calciner and developments in instrumentation and process control – microprocessors etc., the layout has changed.

Opening has to be allowed in the platform for Tertiary Air duct emerging from the cooler.

Control panels are no longer installed on burners' platform. They are housed separately in a building called Central Control Room.

When hot air is drawn from grate cooler as 'primary air' the cyclone and hot Primary Air fan and also blower for conveying coal in a multi channel burner are installed on burners' platform.

See Fig. 6.19.61.

19.28.1 Layout of Burners' Platform

In the past, burners' platform and coal mill buildings merged. The burners' platform providing a sort of roof over the coal mill and also over the grate cooler. Coal mills used to be installed then by the side of the grate cooler. The departmental layout of kiln section beginning with preheater ended with cooler and coal mill and coal firing system.

It ends with cooler when the coal mill is located at the preheater end. **See Fig. 6.19.60**

Alternately a small bin can be installed on burners' platform and coal metered out and fired into kiln.

See Figs. 6.19.62

19.28.2 Layout for Planetary Cooler

With planetary coolers, though it is history now, a tunnel is required to approach the kiln for observations.

In new design of planetary coolers length to diameter ratio of cooler tube is 10 : 1. It is too long a length for overhang. Therefore kiln shell is supported on a small roller station at the discharge end.

See Figs. 6.19.3 and 6.19.4.

Because of lengths of cooler tubes, length of burner pipe is longer. Hence longer traveling space has to be provided to take it out of kiln.

Facilities to be provided for the burner pipe like its handling, withdrawal, positioning are the same as for kiln with grate cooler.

The layout must also provide for installation of a crusher to break oversize lumps before feeding to clinker conveyor.

See Fig. 6.19.3.

19.29 Handling of Bricks

Kilns need to be lined with refractories through its length and also discharge hood and inlet housing.

Bricks are taken into kiln from burners' platform. For small kilns, hood is moved and kiln is made accessible. In case of large calciner kilns, hood doors are pushed apart and kiln is made accessible. For large kilns bricks can be taken inside kilns by small fork lift trucks or portable belt conveyors.

See Fig. 6.19.55.

It is necessary to receive bricks on the burners' platform and also to remove bricks taken out from the kiln.

Burners' platform must therefore have an area to hold bricks and also a hoist to lift bricks brought up to it in truck loads or on trailers.

The same hoist will remove burnt bricks and load them on trucks.

Burners' platform must therefore be easily accessible from ground level for this purpose.

See Fig. 6.19.59.

19.30 Cooler Dust Collector

When coal mill is at preheater end, and coal is pumped from coal mill house, then burners' platform is practically empty. In case of large kilns grate cooler will have 2 or more grates and are long enough. Burners' platform is thus merely a roof over the cooler. It is convenient to install the dust collector for cooler mostly ESP on the burners' platform itself. It is convenient to discharge the clinker dust collected in it on to the deep bucket conveyor.

Figs. 6.24.5, 6.24.8 and **6.24.11** in **Chapter 24 of this Section** show installation of an esp and a heat exchanger respectively on burner's platform.

See Fig. 6.19.63 for a layout of actual installation of a rotary kiln.

19.31 Development in Burners

Burners have also seen considerable development. They use less primary air. Quite a few are multi fuel chnnel designs to be able to fire coal/oil/gas simultaneously. With advent of alternate Fuels, burners are required to receive fue in all forms. Briquettes. Special burners particularly for firing alternate fuels have been designed. **Section 8** of this book deals with these developments.

Annexe 6.19.1 Shows various possibilities of kiln, prheater and calciners for different capacities of kiln.

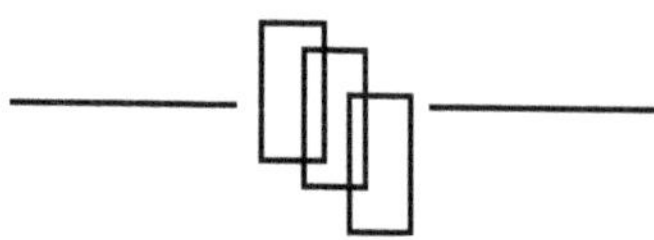

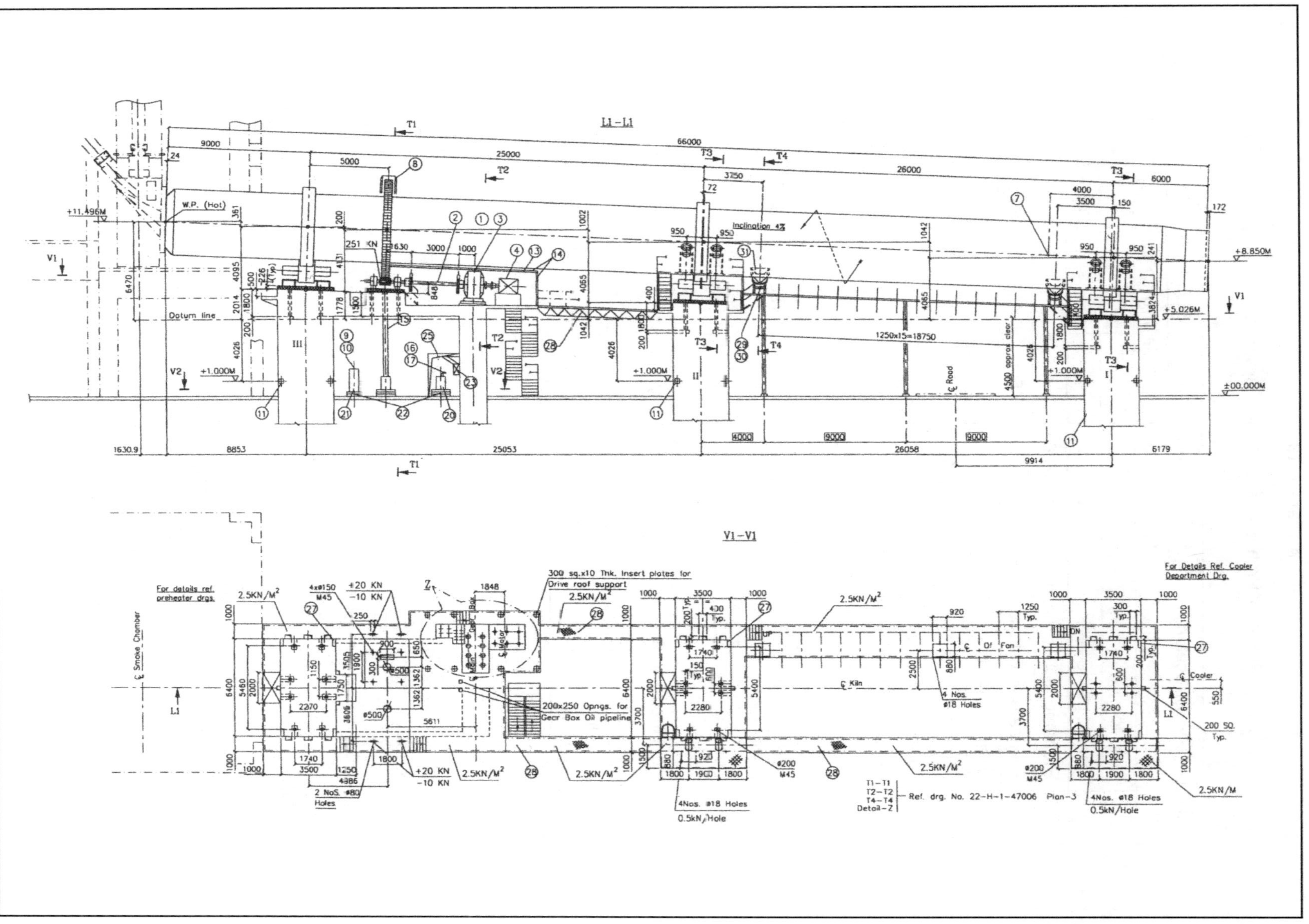

Fig. 6.19.63 Layout of a three support dry process rotary kiln.

Annexe 6.19.1

FLS Kiln, Preheater and Caciner systems.

Sr no	Pyroprocesing system	SP	ILC-E	ILC	SLC-S	SLC
		Air thru kiln		Air thru t.a.duct		
	Max. for					
1	Normal preheater	3000				3000
2	Preheater & Sec. firing		3700			
3	Preheater & calciner			4500	4500	4500
4	Normal layout					
	Preheater streams	1	1	1	1	2
	calciner		1[a]	1	1	1
	Special layouts					
	Preheater streams	(2)	(2)	2 2	2 2	3
	calciner	-	-	1 2	1 2	2
5	Total cap. Normal layout tpd	3000	3700	4500	4500	12000
6	Specific output Tpd/m^3 of kiln	1.8	2.5	4.8	4.8	4.8
7	Coal fired calciner	-	-	yes	yes	yes
8	Low no$_x$ calciner	-	-	yes	yes	yes
9	Max. firing in sec. Firing %	15	25			
	In calciner %			55-65	55-65	55-65
10	Cooler grate	Yes	Yes	Yes	Yes	yes
	planetary	yes	yes	no	no	no
11	Max. bypass for Alkali chlorides %	30	25	100	60	100

Notes : 1 special layouts () used for expansion of existing plants

[a] riser duct modified for sec. Firing

SP preheater kiln

ILC in line calciner

SLC separate line calciner

CHAPTER 19a

COAL AND OIL METERING AND FIRING SYSTEMS

19a.1 Firing Fuels in Kilns

Just as kiln needs to be fed with raw meal of uniform quality at a steady rate it also needs to be fired with fuel of uniform quality at a steady rate to supply heat required for converting that raw meal into clinker of good quality.

In case of calciner kilns heat is supplied to kiln and calciner in predetermined proportions.

Systems for blending raw meal and for feeding it into kiln have been dealt within Chapters 10 to 12. of this Section

19a.1.1 Fuels Suitable for Firing

Kiln systems can be fired with all types of fuels like solids – coal, lignite, petcoke; oil – furnace oil; gas – natural gas. Of these oils and gases are more or less uniform in quality and have no ash to speak of. Calorific value is also uniform and does not fluctuate. They do not need much preparation as such for firing into kiln and calciner. It is therefore easy to use oil and gas as fuel in kiln systems.

Coals from different sources have widely different calorific values and ash content; these fluctuate even in coal coming from same source. Therefore to supply to a kiln system heat at a desired steady rate is more difficult.

Only about two decades back cement plants in developing countries preferred to use oil as fuel and also gas when it was available. However ever since oil prices shot up – and are still rising – cement industry has had to use coal as fuel all over the world. In specific cases like plants in Bangla Desh for example, gas is profusely and conveniently available and hence cement kilns there use gas as fuel.

19a.2 Coal as Fuel

Solid fuel like coal needs to be pulverized for firing in rotary kilns.

It is very necessary to preblend coals (just as limestone is 'pre-blended' in stacker reclaimer systems).

It is then dried and ground in coal mill systems to desired fineness varying between 10-12 % residue on 90 micron sieve and moisture is reduced to 1-2 %.

19a.2.1 Uniform Quality of Coal Fired

Maintaining uniform quality is very important in case of coal because ash in coal gets absorbed in clinker and thereby affects composition of clinker. To maintain uniform quality of clinker it becomes necessary to change composition of raw meal to compensate for the effect of coal ash. This is not very practical in case of large plants.

Ash in coals decides the fineness to which it has to be ground to complete combustion. Thus if ash content fluctuates, fineness needs to be adjusted to suit. This is not very difficult in case of vertical mills but adds another dimension to the operation.

19a.3 Coal Firing Systems

Coal firing systems have evolved over the years. In early days of wet process 'direct firing' was common. In it coal ground in mill was fired straight into kiln.

See Fig. 6.19a.1.

Major disadvantage of direct firing was that it used up too much primary air. With reduced specific fuel consumption the difference became more pronounced. Secondly, if the fan in the system was after the mill, wear of impeller due to high ash coal was excessive.

If it was located before the mill, mill was under pressure and coal leaked from joints. This system is no longer selected.

In 'indirect system', grinding system and coal firing systems are separate and each can be designed according to its need. For instance when multi channel burners are used, quantity of primary air is very small. Hence this is the system now prevalent.

See Fig. 6.19a.2.

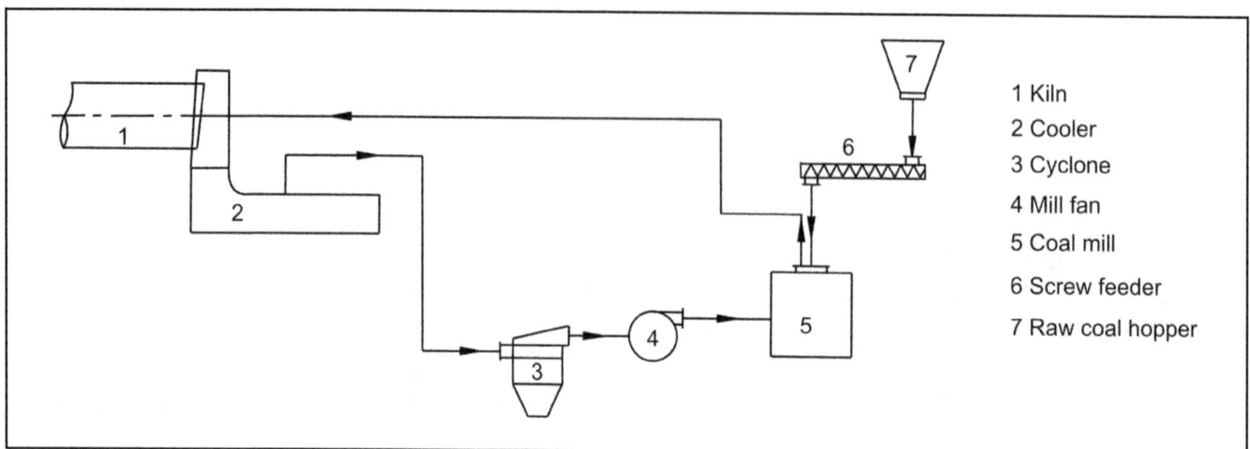

Fig. 6.19a.1 Direct firing system for coal for cement kiln.

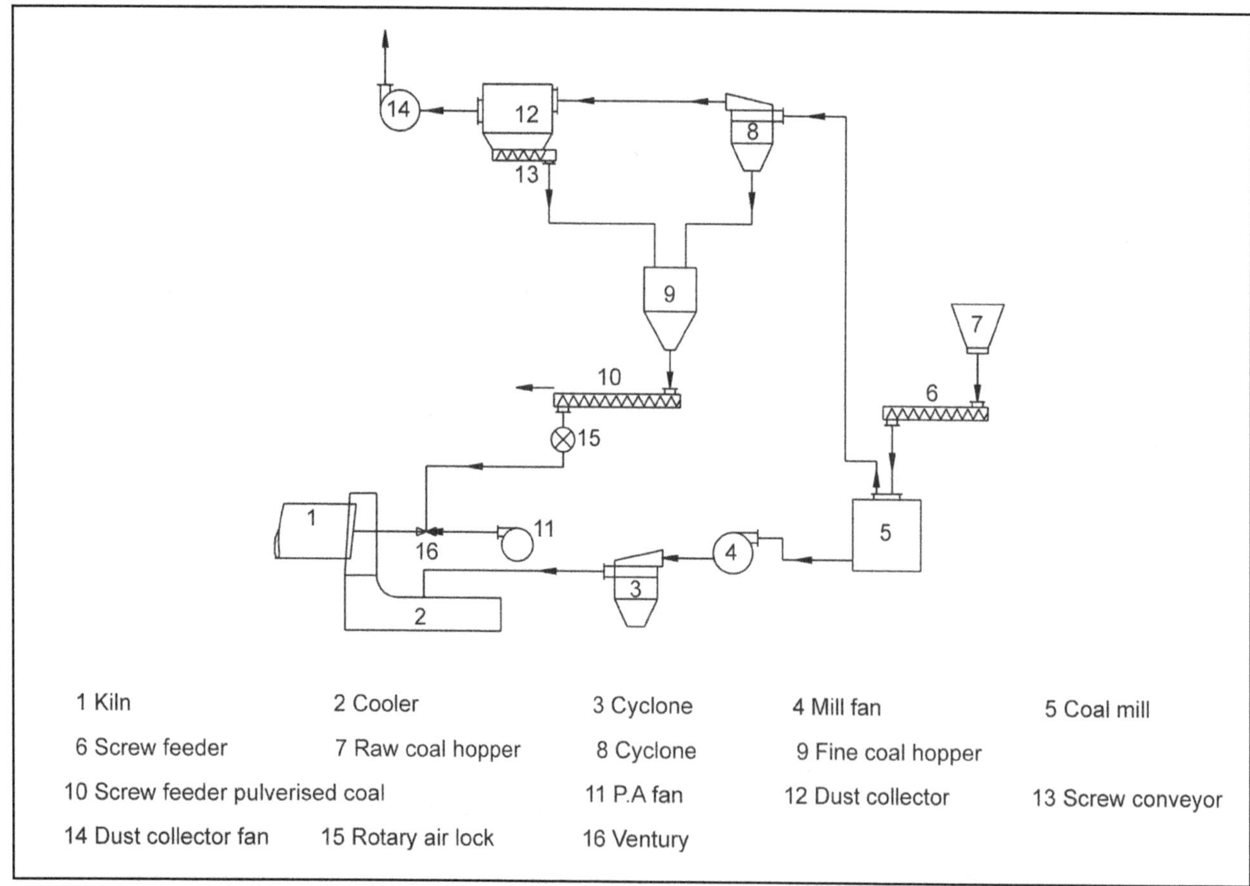

Fig. 6.19a.2 Indirect firing system for coal for cement kiln.

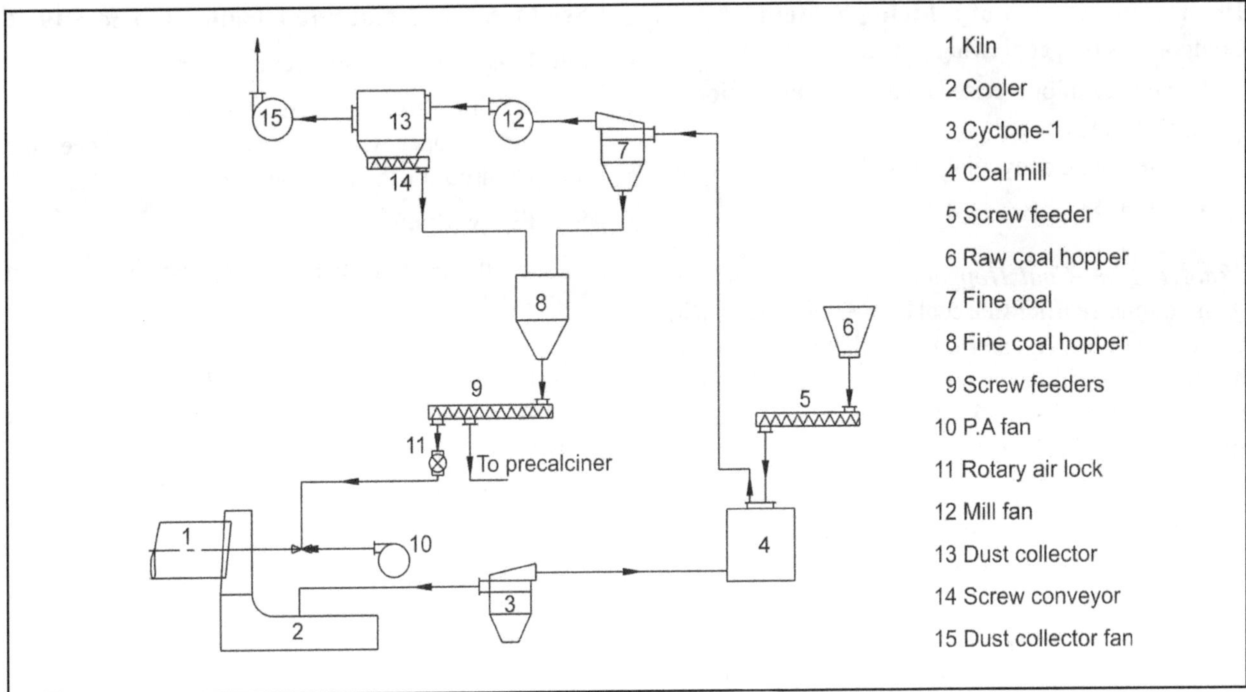

Fig. 6.19a.3 Coal grinding and firing system with vertical mill at cooler end.

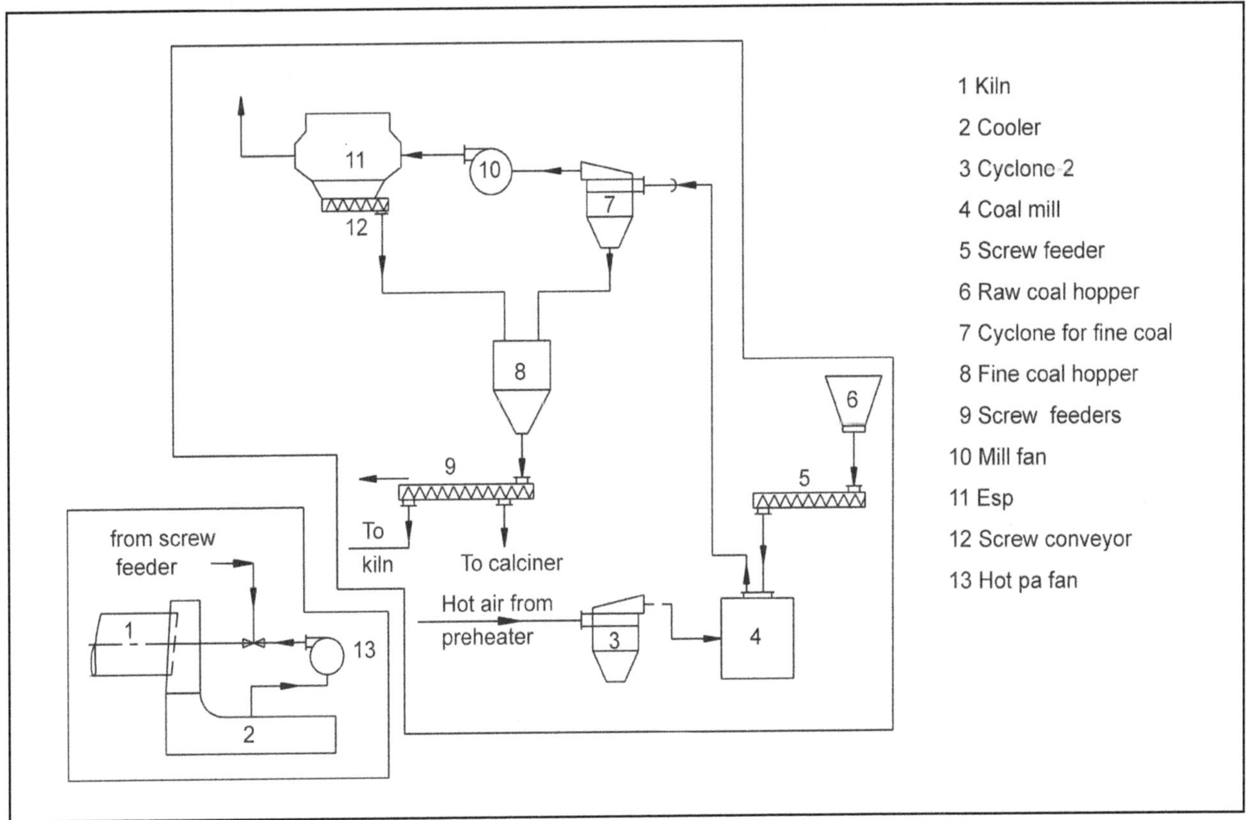

Fig. 6.19a.4 Coal grinding and firing system with vertical mill at preheater end.

19a.4 Components of a Firing System

Components of a coal firing system are:

1. Storage of pulverized coal and its extraction,
2. metering of coal,
3. pneumatic conveying of coal,
4. burners.

19a.4.1 Fine Coal Hoppers

As mentioned earlier fine coal hopper can be one with two outlets for a calciner kiln. Or there could be two hoppers each with its own metering system. In such a case coal of different fineness could be stored and fired in kiln and calciner respectively.

Coal is not always easy to flow because of its moisture. Hoppers have steep slope and they are fitted with stirrers to keep coal in agitation.

19a.4.2 Metering of Coal Fired

Coal is metered gravimetrically either by weigh feeder – pre-feeder system as in case of raw meal in klin feed system or volumetrically by using twin screw feeders with variable speed drive mounted directly under fine coal bin. Such screws have very close clearances between screw and casing and pitch decreases towards discharge end to prevent flushing.

See Figs. 6.12.3 and b in Chapter 12 and 6.19a.2

See Figs. 6.19a.3 and 6.19a.4.

"Loss in weight' systems which monitor rate of emptying hopper with the help of micro processors are also used for the purpose.

See Fig. 6.19a.5.

Yet another coal metering system is shown in **Fig. 6.19a.6.**

19a.5 Conveying of Pulverized Coal

Pulverized coal is conveyed pneumatically to kiln and calciner.

Conveying air could be only part of primary air or whole of it. Though pneumatic conveying consumes more power, impact is small because of low consumption of coal.

Either a ventury is used to introduce coal into conveying pipe or an ejector.

Distances over which coal is to be conveyed depends on location of coal mill. When it is at cooler end, distance to kiln is short and to calciner long; its vice versa when mill is at preheater end.

See Figs. 6.19a.3 and 6.19a.4.

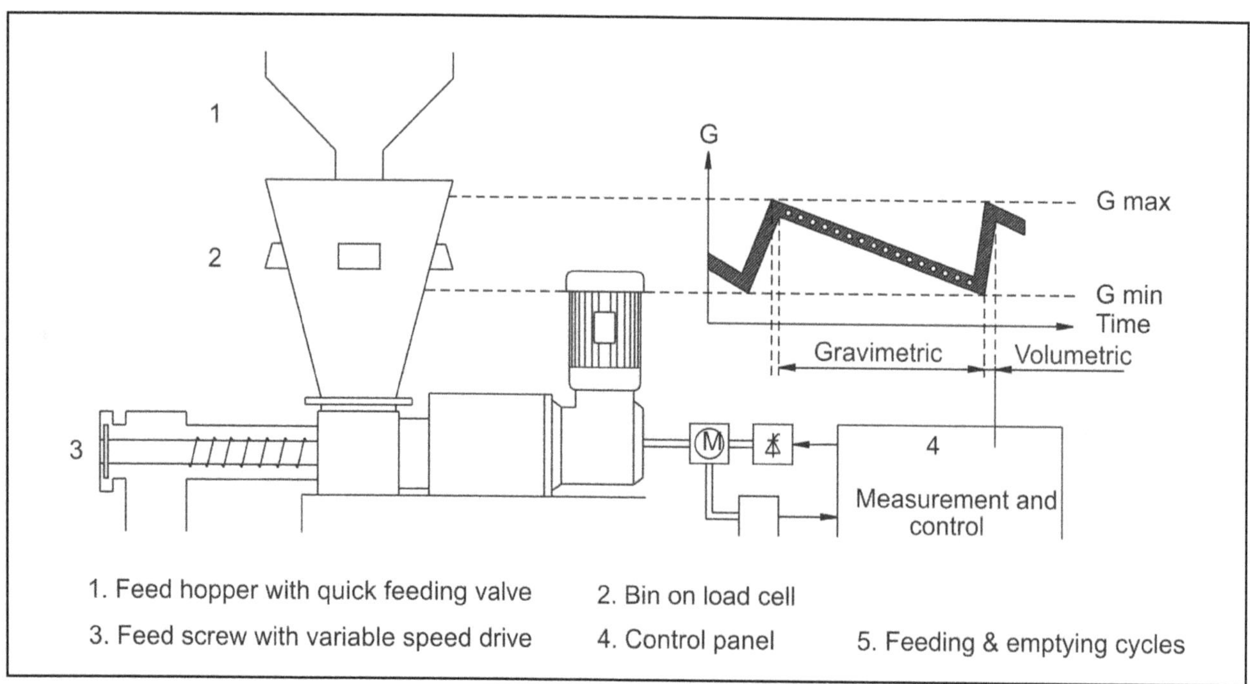

Fig. 6.19a.5 Loss in weight system for pulverised coal feed.

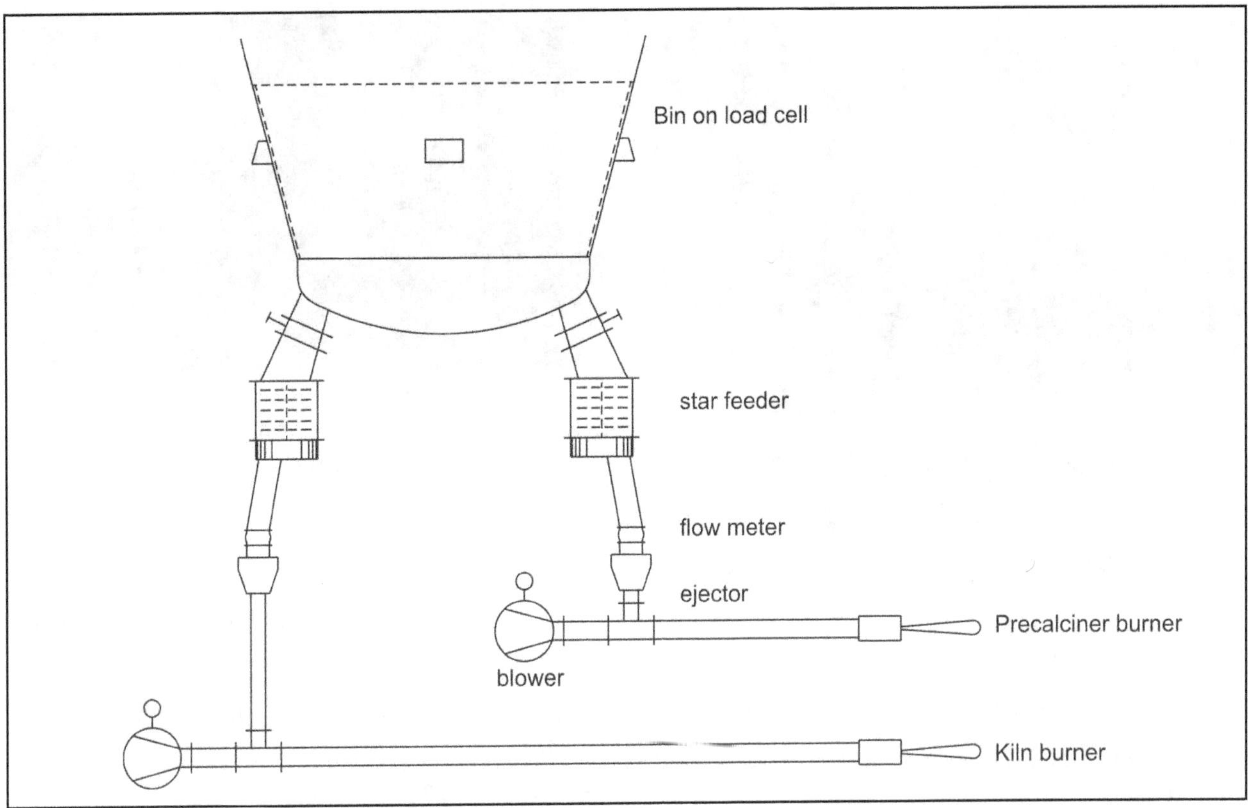

Fig. 6.19a.6 Schematic arrangement of coal feed system for kiln and calciner.

19a.5.1

It is now preferred to locate the coal mill at prehheater end to use inert gases from prheater to dry coal. Coal mill systems have either bag filters or ESPs to vent the exhaust gases. In grinding coals with high volatile content it is possible that combustion takes place and temperature of gases increases and there is CO in exhaust gases. Thus if filter is bag filter,bags burn; if filter is ESP, sparking can cause explosion. Using inert gases from preheater avoids these dangers.

19a.6 Burners

Conventional burner for kiln is a refractory lined pipe with a nozzle. Diameter of nozzle is designed to obtain a velocity of 70 m/sec at the tip. Velocity depends on the diameter of the kiln. It cannot be adjusted in running.

See Figs. 6.19a.14.

Figure 6.19a.14 shows relation between clear diameter of kiln and velocity at tip of nozzle in conventional burners shown in **Fig. 6.19a.7.**

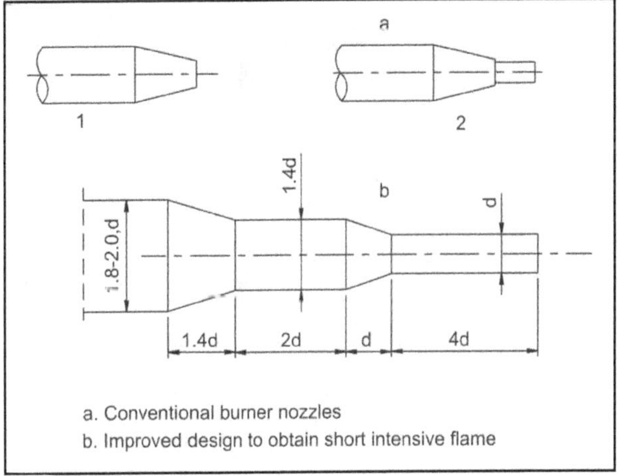

Fig. 6.19a.7 Nozzles for coal burners for rotating kilns.

Burners have seen lot of development . There are burners which can fire more than one fuel- say coal and alternate fuel. Burners are now designed to use minimum primary air.

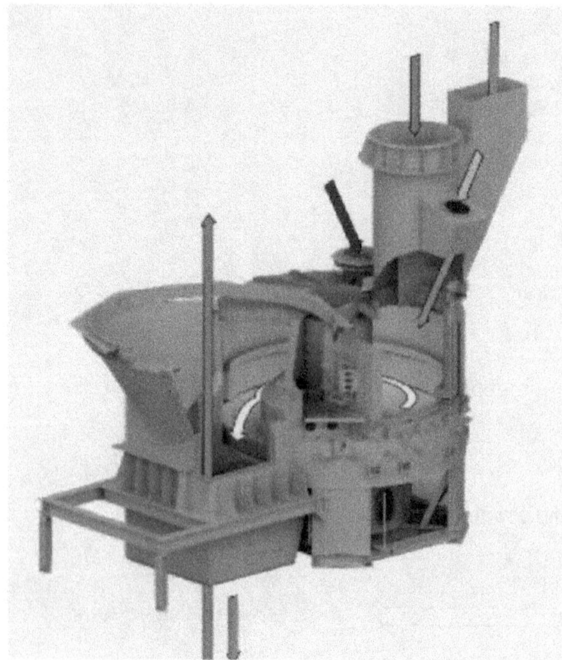

Source: FLS Brochure

Plate 6.19a.1 Hot Disc for feeding Whole Tyres and other Solids.

Source: FLS Brochure

Plate 6.19a.2 'Gunnax' Whole Tyre Firing System.

Gunnax firing waste canister

Source: FLS Review 148

Environmental constraints

Plate 6.19a.3

Now multi channel burners are used to fire coal. They can also be used to fire oil during start up and for pre-heating refractory. A good turn down ratio 6 : 1 is necessary for this purpose. Multi channel burners require minimum primary air. About 2.5 % of primary air is used as conveying air for coal.

See Figs. 6.19a.6 and 6.19a.7.

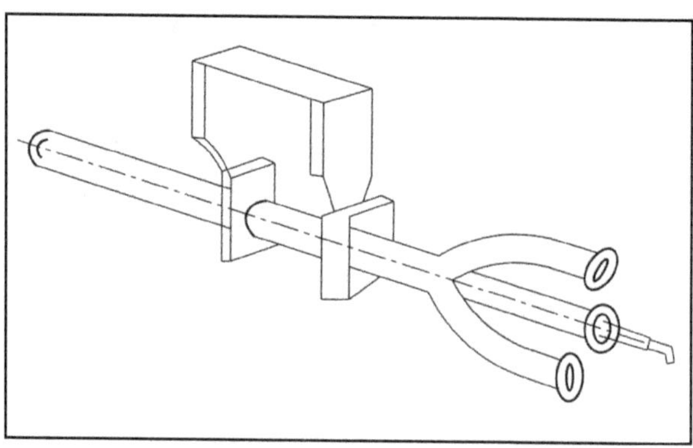

Fig. 6.19a.8 Multi channel swiril flame burner.

As explained earlier, it is possible to use hot air from cooler as primary air. Multi channel burners mix air and coal more thoroughly and give the air coal mix a swirling motion.

Burners should be designed to obtain a short flame and to complete combustion. Flame control is important for producing good quality clinker.

19a.6.1 Oil and Gas Burners

Figs. 6.19a.10 and 6.19a.11 show oil burners and gas burners.

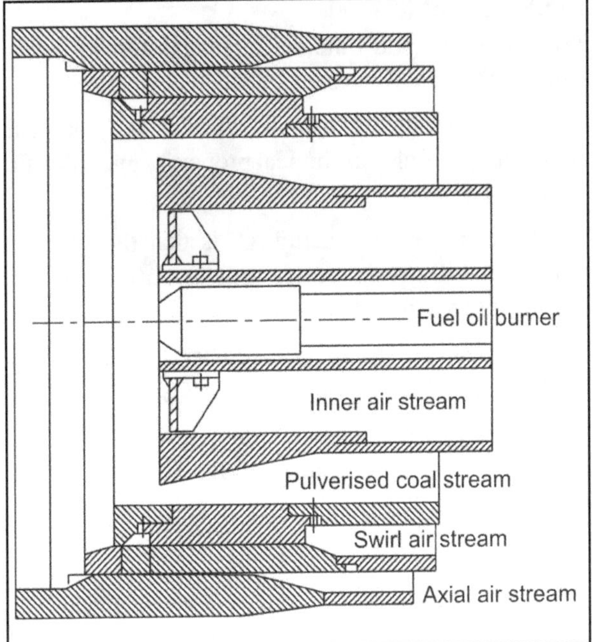

Fig. 6.19a.9 Section of burner nozzle.

Oil firing needs preparation by way of heating oil to temperatures that would facilitate its atomization. Atomization can be by using gear or screw pumps or by using compressed air.

Oil firing units are self contained twin heating and pumping units installed on the burners' platform.

See Fig. 6.19a.12.

Oil is stored in day service tanks located near the burners' platform. Main storage will be at a distance. Its storage and handling has been shown schematically in **Fig. 6.19a.13.**

Gas burners are the simplest as gas does not need any preparation for firing. Metering oil and gas are easy and accurate also. **See Fig. 6.19a.11.**

19.6.2

There are a host of branded burners on the market which incorporate above mentioned features.

They are;

FLS

Polysius

Humboldt

Fives

19a.7 Firing Fuel in Calciner

Metering system to meter and regulate rate of fuel to be fired are the same for Calciner as for kiln. Since temperatures in calciner are much lower than those required to be maintained in the burning zone, velocities at which coal is fed in calciners has been described in **Chapter 17.**

See Fig. 6.17.20.

Fig. 6.8.12 in Chapter 8 show installations of coal grinding, coal storage and metering.

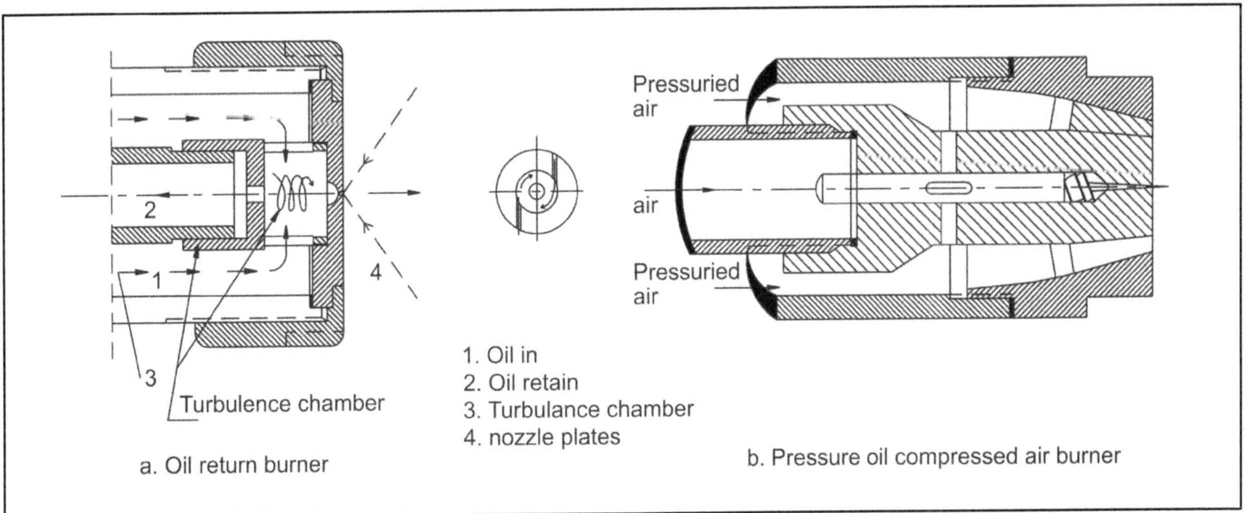

Fig. 6.19a.10 Oil burner nozzles.

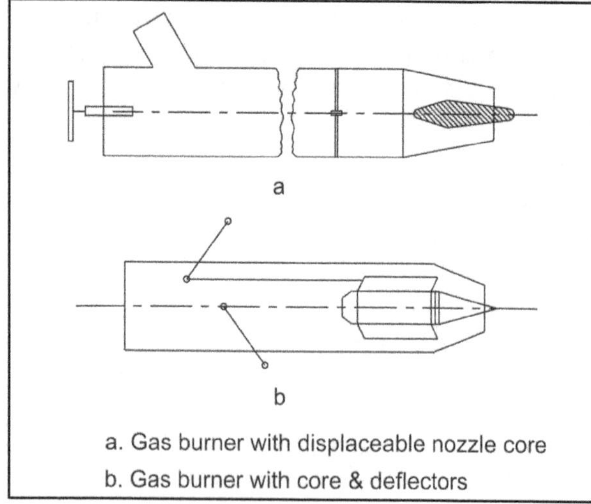

a. Gas burner with displaceable nozzle core
b. Gas burner with core & deflectors

Fig. 6.19a.11 Gas burners for rotary kilns.

19a.8 Alternate Fuels

Now both Kiln and Calciner receive Alternate Fuels which are often Wastes.

They differ greatly from fossil fuels like Coal, Oil annsistency and also in chmical properties and calorific values.

Motivation behind using AF is that they reduce emssion of green house gases and conserve exhaustible fossil fuels.

European countries have replaced fossil fuel by AF by as much as 30 %.

Kilns with their long retention time are able to burn all inds of fossil fuels in various forms including whole tyres.

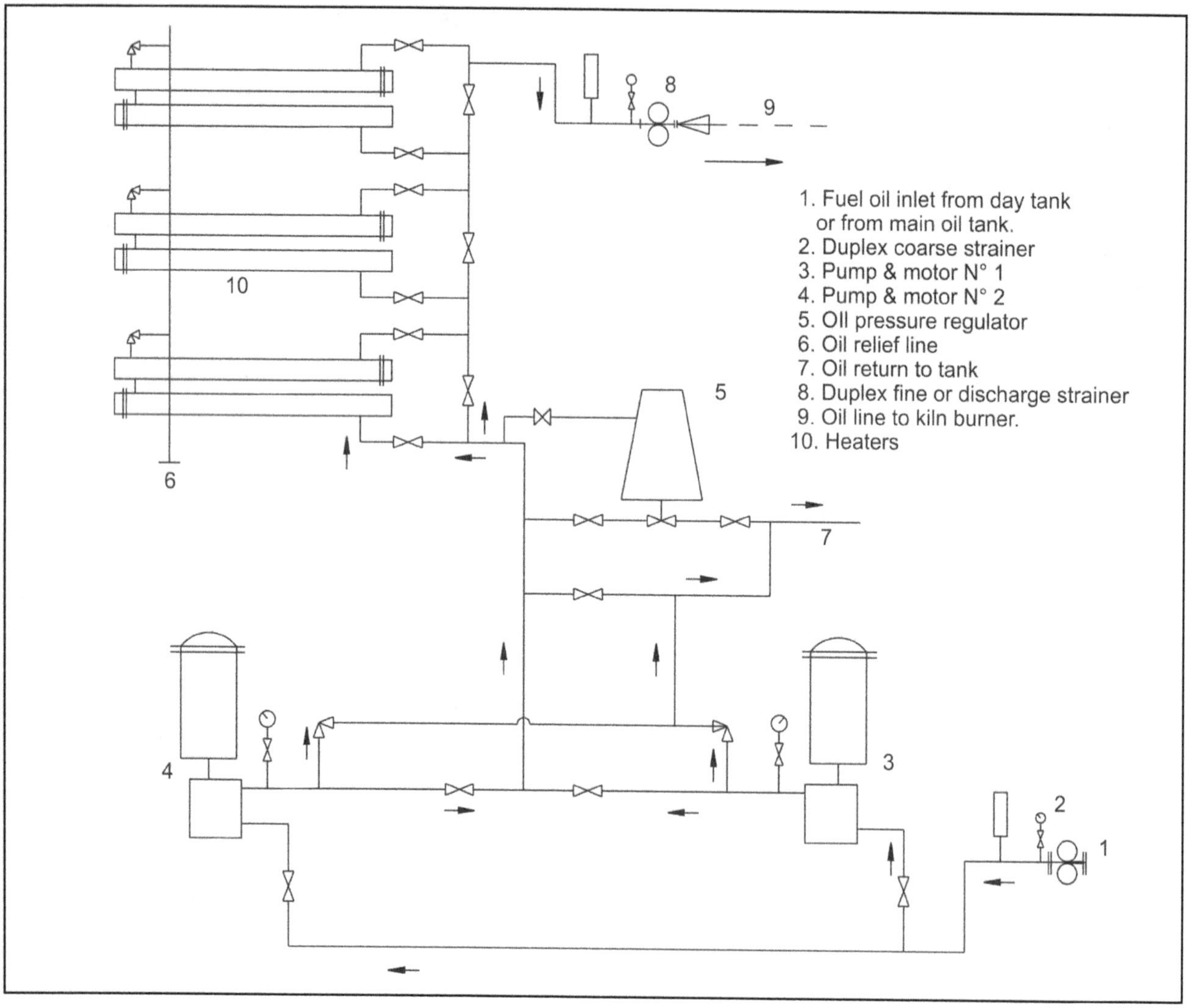

1. Fuel oil inlet from day tank or from main oil tank.
2. Duplex coarse strainer
3. Pump & motor N° 1
4. Pump & motor N° 2
5. OIl pressure regulator
6. Oil relief line
7. Oil return to tank
8. Duplex fine or discharge strainer
9. Oil line to kiln burner.
10. Heaters

Fig. 6.19a.12 Typical pump and heater set for oil burner for rotary kiln.

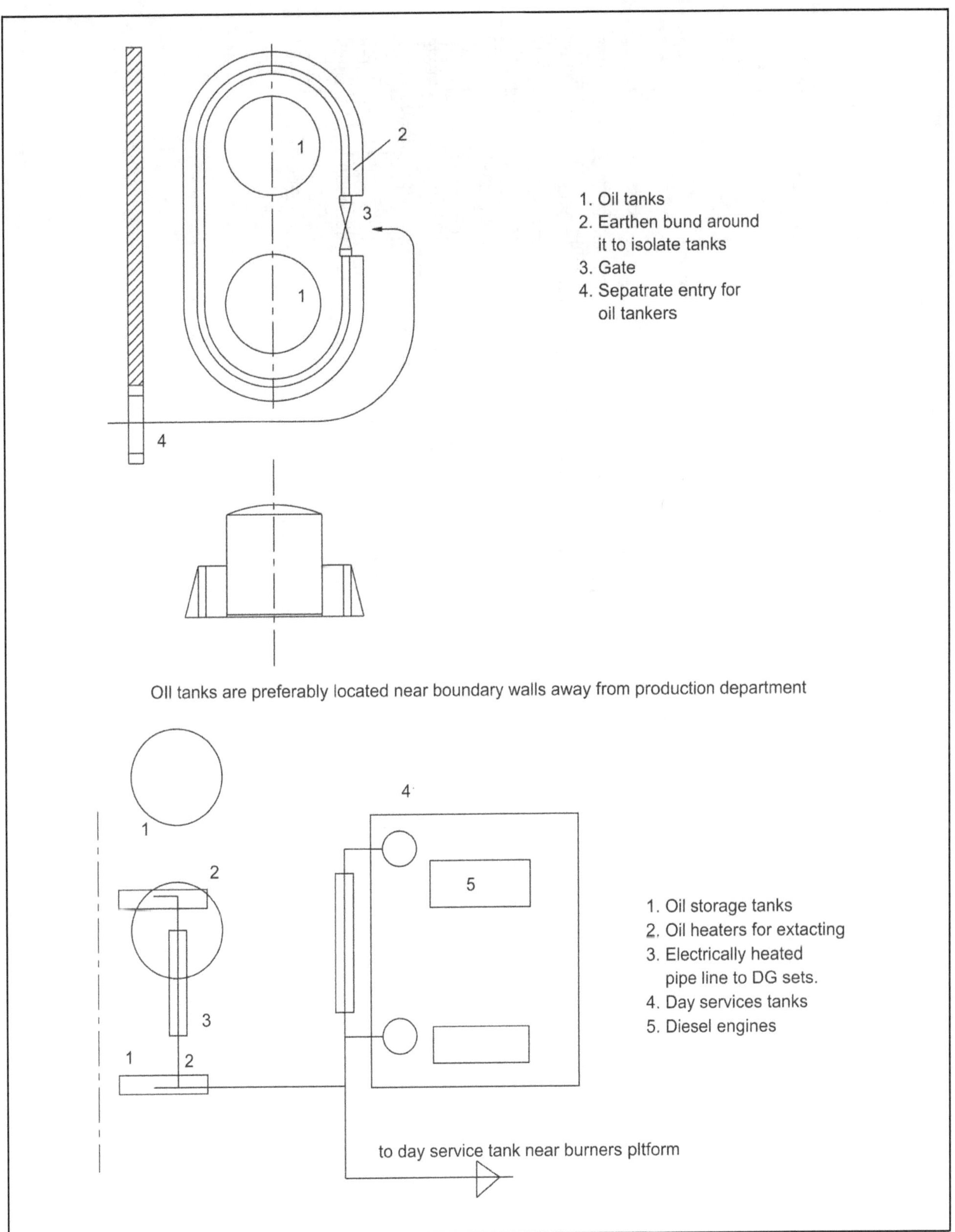

Fig. 6.19a.13 Taking oil to DG sets and for firing it in kilns.

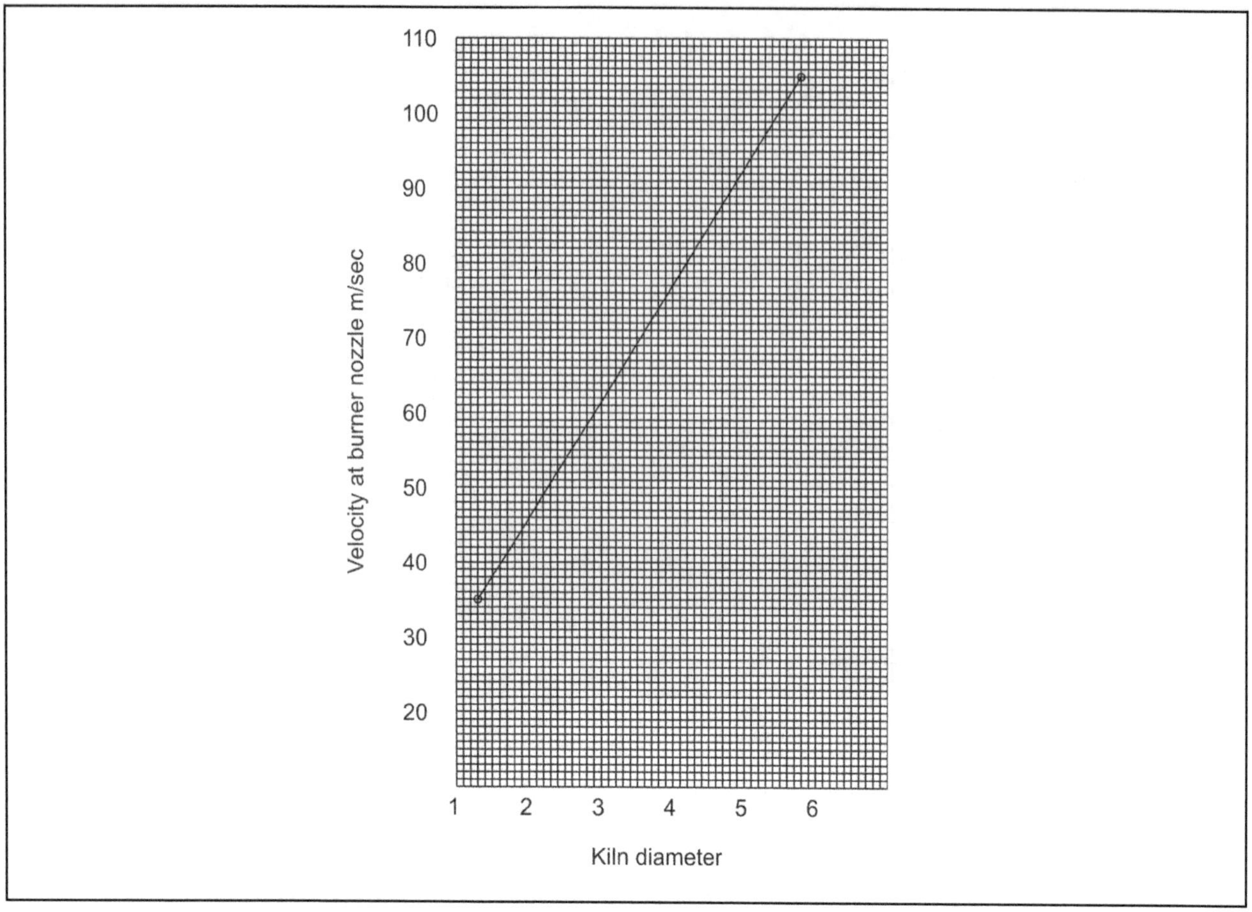

Velocity at burner nozzle m/sec

110
100
90
80
70
60
50
40
30
20

1 2 3 4 5 6

Kiln diameter

Fig. 6.19a.14 Velocity in coal burner nozzle.

Calciners have been designed to be able to receive and burn AF. Feeders like Hot Disc has been developed for firing shredded AF in calciners.

In **Section 8** of this Book there is a Chapter on Alternate Fuels.

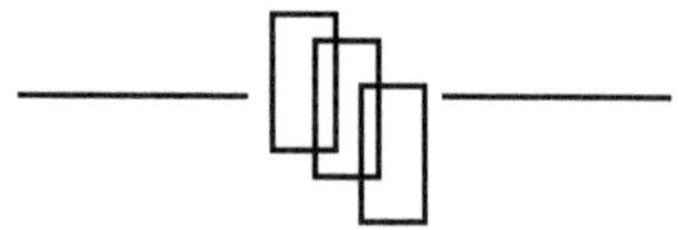

CHAPTER 20

CLINKER COOLERS
PLANETARY COOLERS

20.1 Cooling of Clinker

Clinker as it emerges from the kiln has a temperature of about 1350 °C. It cannot be safely handled at this temperature. It has to be, therefore, cooled to temperatures at which it could be handled by common conveying equipment like belt conveyors, chain conveyors, bucket conveyors.

Air is the media for cooling. It is also the media for combustion. Therefore it followed that air that was itself preheated in cooling clinker should be used in kiln (and also in calciner when there is one) for combustion of fuel.

Clinker coolers thus have a dual function; one to cool clinker to desired temperatures; and to preheat air for combustion in kiln and calciner.

There are 3 types of coolers for cooling clinker, but the prevalent cooler is the reciprocating grate cooler and its variants.

20.2 Rotary Cooler

It is the oldest form of clinker cooler and was used in the days of wet kilns. Rotary coolers up to 2000 tpd capacity were once in operation.

See Fig. 6.20.1.

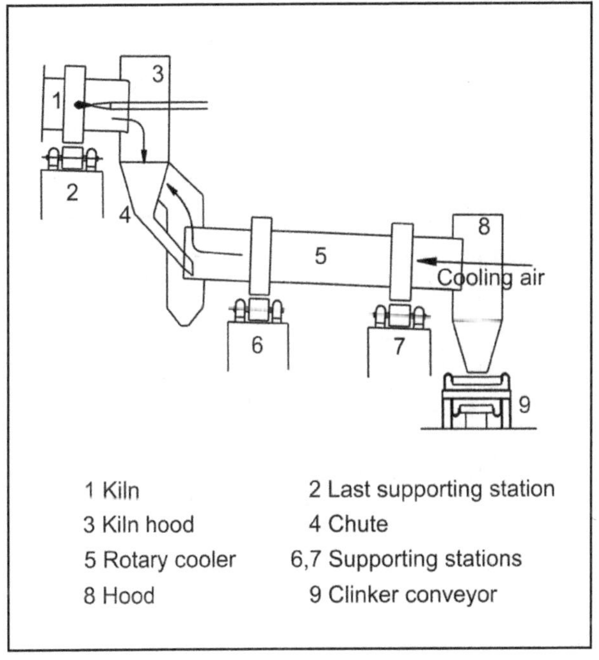

Fig. 6.20.1 Rotary cooler.

In construction it is like a rotary dryer. It has about one third of its length lined with refractory at feed end. Next one third is lined with heat resisting liners and lifters. And the balance is lined with wear resisting liners and lifters.

Kiln I.D. Fan pulls cooling air through the cooler; it passes into kiln as combustion air.

In principle air can be drawn from the rotary cooler for firing in a calciner.

20.3 Planetary Coolers

Rotary cooler was replaced by 'satellite' or 'planetary coolers' which are 10 / 12 tubes mounted on the kiln shell and rotating with it. Even they have become history now.

See Figs. 6.20.2 and 6.20.3 and 6.19.4 in Chapter 19.

Main reasons for their going out are:

1. Cooling air is equal to air sent into kiln as secondary air. It is far less than that required to cool clinker to ~ 100 °C Even grate cooler requires about 2.2- 2.5 nm^3 of air per kg clinker.

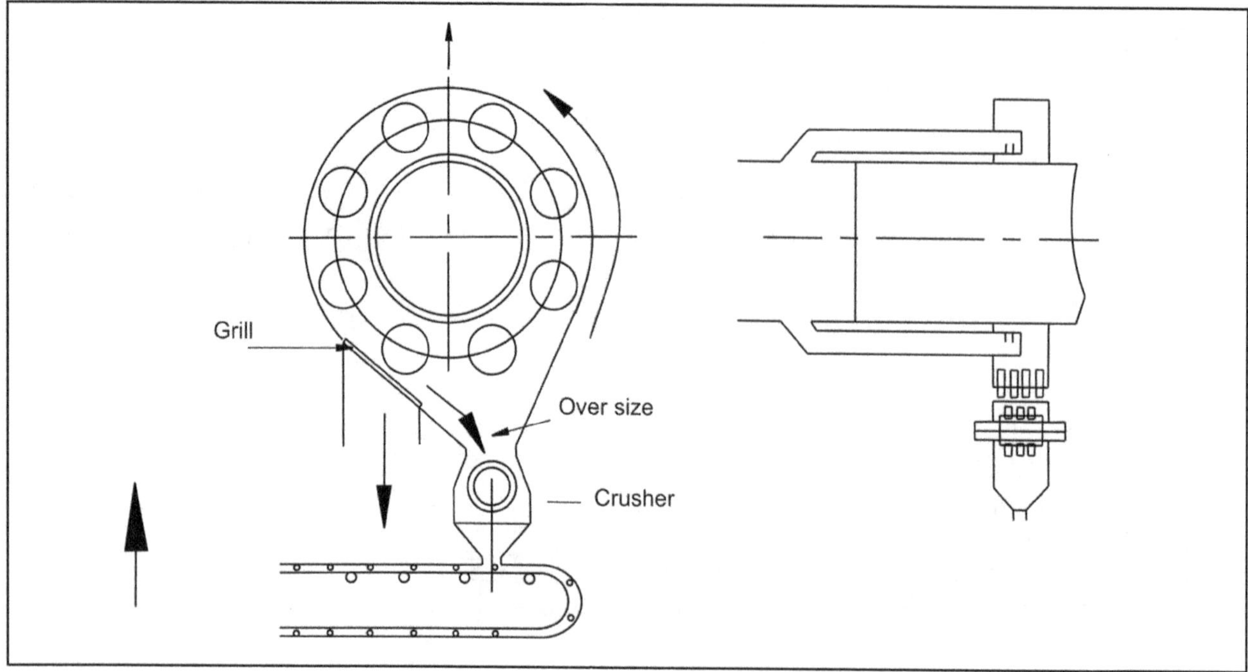

Fig. 6.20.2 Planetary cooler discharge end.

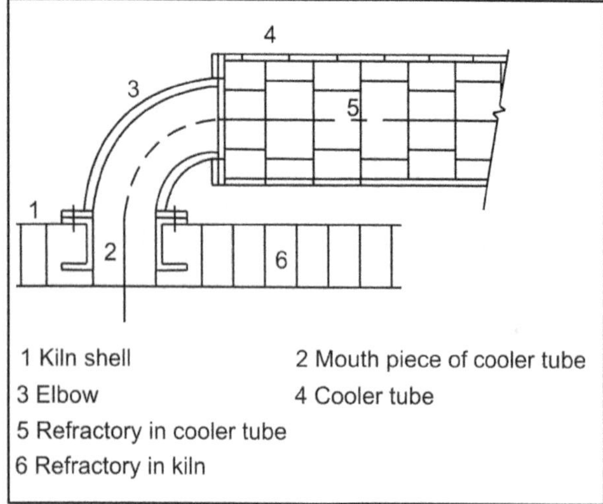

Fig. 6.20.3 Planetary cooler tubes – mouth piece and below.

2. In case of calciner kilns, 60 % fuel is fired in calciner. It was not needed to pass correspondoing amount through the kiln where only 40 % of the fuel was fired..Advent of calciners made planetary coolers irrelevant.
3. Kiln output increased 2.5 times with calciner. It was not possible to accommodate the required number of tubes of correct size on small kilns.
4. Specific fuel consumption of < 700 kcal/kg could not be achieved with planetary coolers.
5. Capacityof a kiln could not be increased in a kiln with planetary coolers as it can be with grate coolers by extending /adding a grate

Many small plants that had planetary coolers, replaced them with grate and consequently could increase the capacity of the kiln by adding a second stream of preheater and a calciner

For these reasons, details furnished about them in the last edition have been omitted in this edition.

20.4 Replacing Planetary Cooler by Grate Cooler

Many old plants have replaced planetary coolers on their kilns by Reciprocating Grate coolers in the course of expansion of their Plants. **See Fig. 6.20.4.**

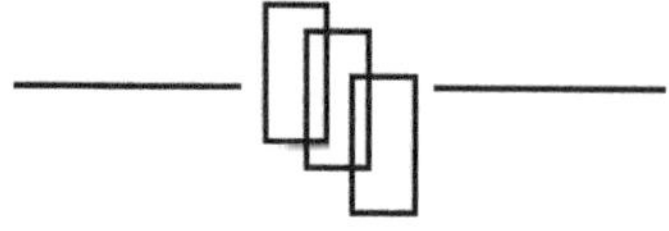

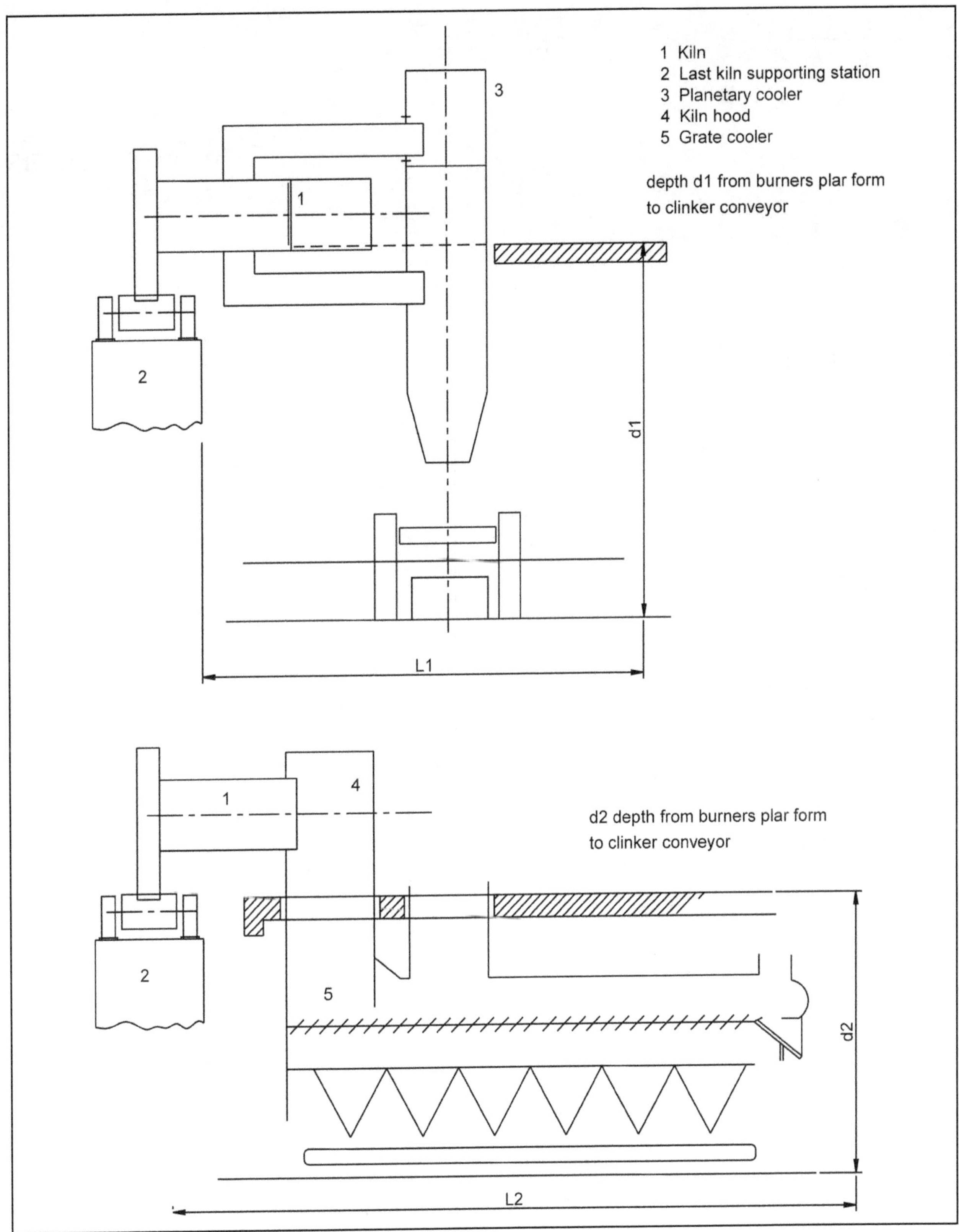

Fig. 6.20.4 Replacing planetary cooler by grate cooler.
depth d2 much larger than depth d1- length wise also grate cooler would need more space.

CHAPTER 21

GRATE COOLERS

21.1 Grate Coolers

The predominantly used coolers to cool clinker are the reciprocating grate coolers and their variations. Clinker coolers have seen a great many developments to cope with large 10000 tpd and above capacity kilns.

Reciprocating grate coolers, have seen considerable development in the last 2 decades and have come a long way from the proto types used in 50s. Major developments are :

1. Combi cooler – with inclined and horizontal grates.
2. Multiple grate coolers.
3. Static grates KIDS / controlled flow grates cooler.
4. Indirect after Cooler – G Cooler
5. Cross Bar cooler
6. Walking Floor Cooler

See Figs. 6.21.1 to 6.21.3.

21.1.1 Static Grate

In **Section 2**, dealing with machinery used to make cement, developments in grate coolers have been described.

Presently most designs have the first compartment with a static grate of about 8-10 rows sloping as an angle of 8-10°. Thickness of clinker bed in this compartment is therefore higher than in other compartments having reciprocating grates. This compartment has therefore its own fan located at the back of the cooler.

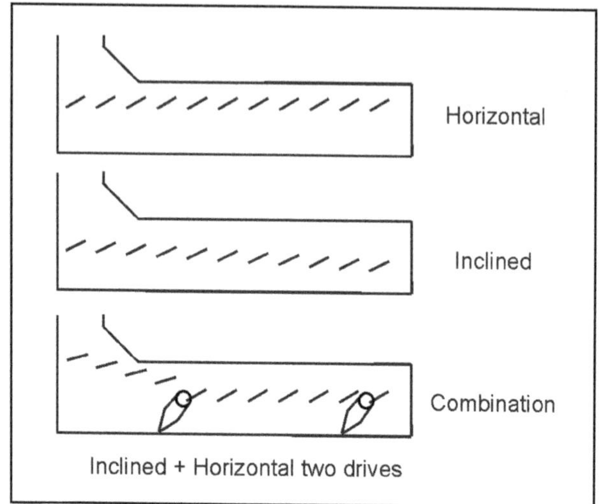

Fig. 6.21.1 Different arrangements of grate coolers.

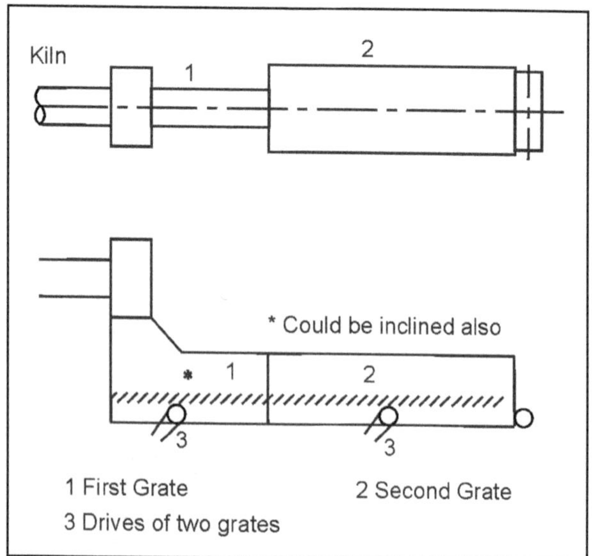

Fig. 6.21.2 Two width cooler.

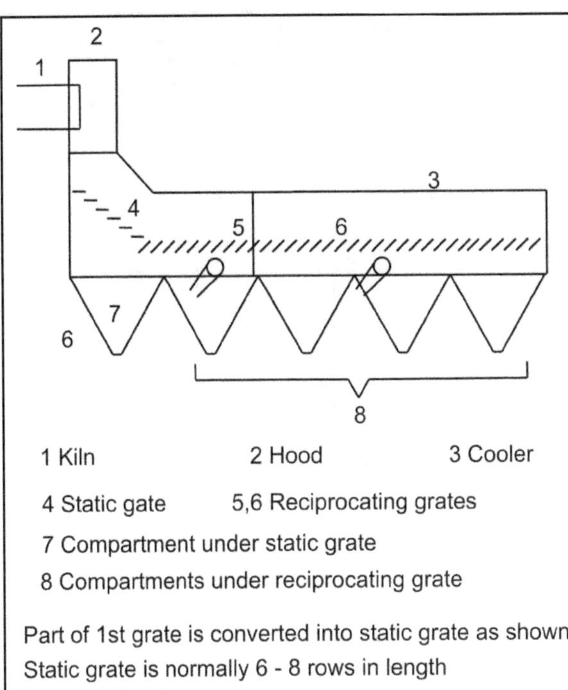

Fig. 6.21.3 Cooler with static grate.

21.1.2 Various Modes of Moving Clinker along the length of the Cooler

New modes of moving clinker along the length of the cooler in the process of cooling have been developed like: Cross Bar, Walking Floor, Poly track.

Cross Bar Cooler:

In Cross bar coolers, a system of cross bars moves inside a casing does the work of reciprocating grates.

Walking floor

Cooler floor is made up of a number of strips. Strips move together in forward direction but return at different speeds

Ploy track cooler

In these designs, spillage – a feature of old design of reciprocating grates has been eliminated.

21.2 Increase in Size and Capacity of Grate Coolers

Coolers have increased in size / area to match with the increase in the capacity of the kiln.Introduction of calciner, increased the capacity of a kiln of a given diameter by 2.5 times.

There being a relation between cooler width and diameter of kiln for ensuring even spread of clinker on the grate plates, coolers of smaller widths can now cope with much higher capacities.

Concepts in cooler design and use of cooling air have also undergone considerable changes even in conventional coolers.

21.3 Reciprocating Grate Cooler

A reciprocating grate cooler as the name suggests consists of a moving grate reciprocating over a stationary grate and thereby pushes clinker over it. The extent of movement is called the stroke. Stroke is also the extent of eccentricity. Clinker falling from the kiln is pushed in small steps grate after grate from inlet end to discharge end.

See Figs. 6.21.4 and 6.21.5.

21.3.1 Recuperation of Heat

As mentioned earlier, a cooler has not only to cool clinker but also to recuperate heat and thereby bring down fuel consumption. Grate coolers have been found to be the best in this respect, reaching efficiencies of 70% and above. **Figs. 6.21.6** shows cooling of clinker,

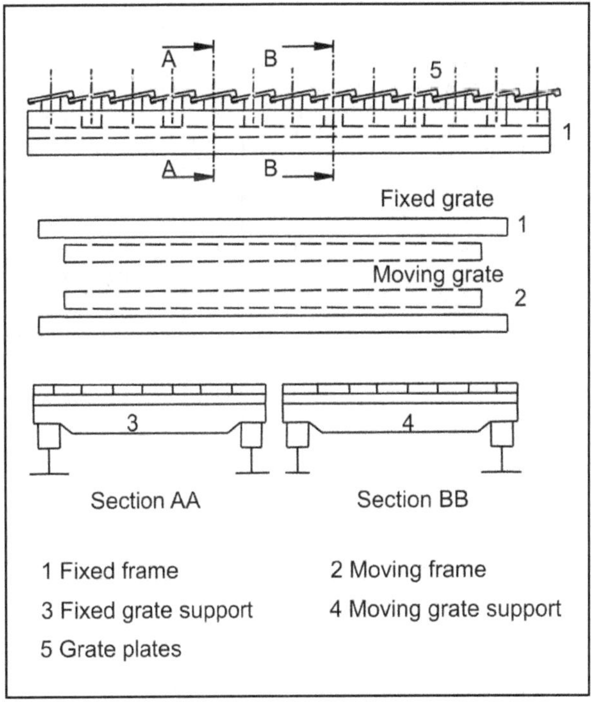

Fig. 6.21.4 Reciprocating grate cooler basic features.

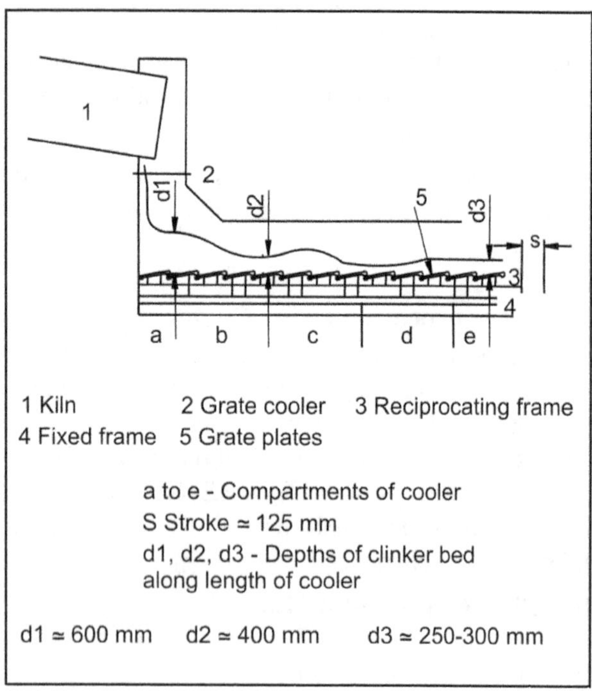

Fig. 6.21.5 Bed thicknesses of clinker in various sections of cooler.

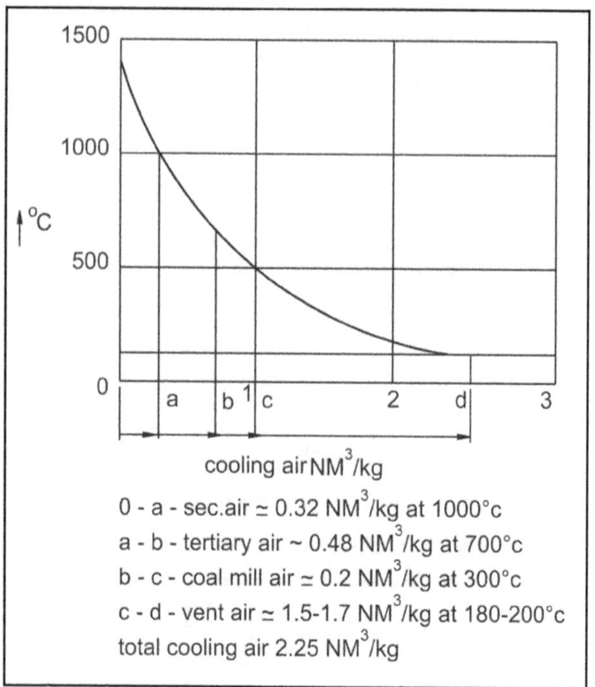

Fig. 6.21.6 Clinker cooling curve.

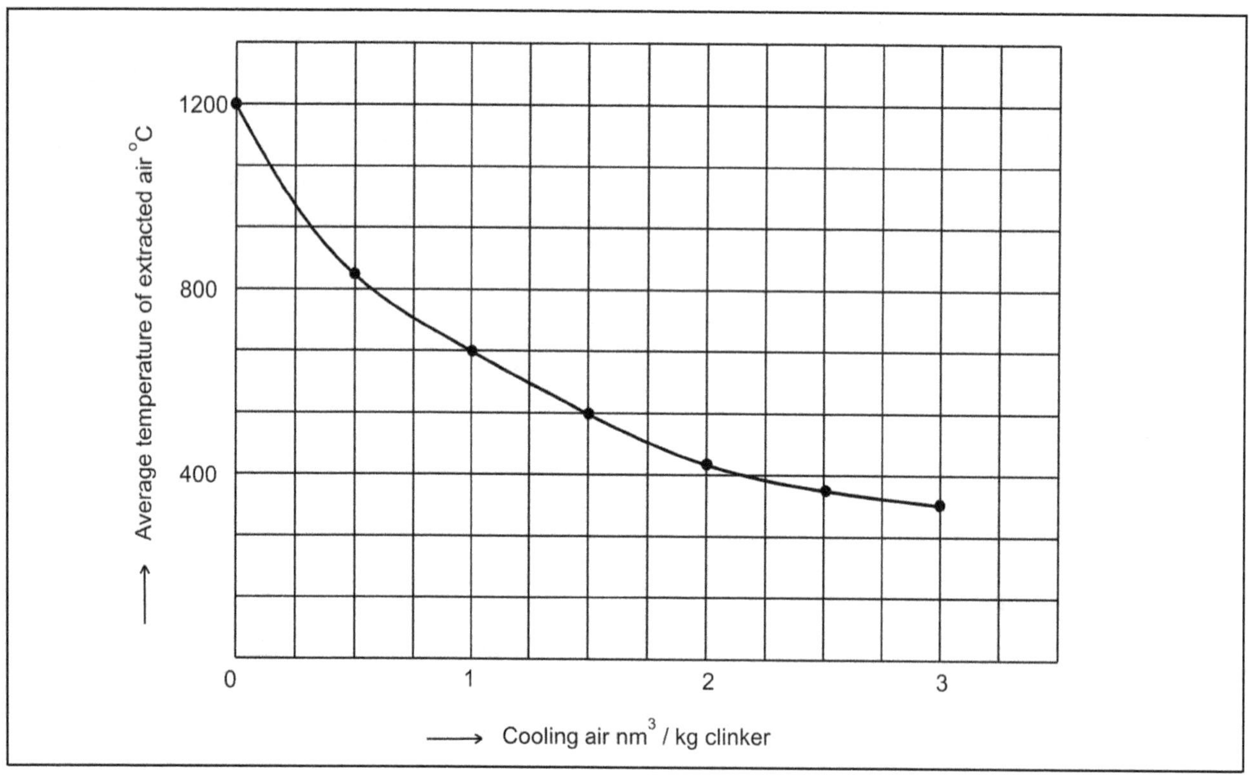

Fig. 6.21.7 Average temperature of air extracted from cooler.

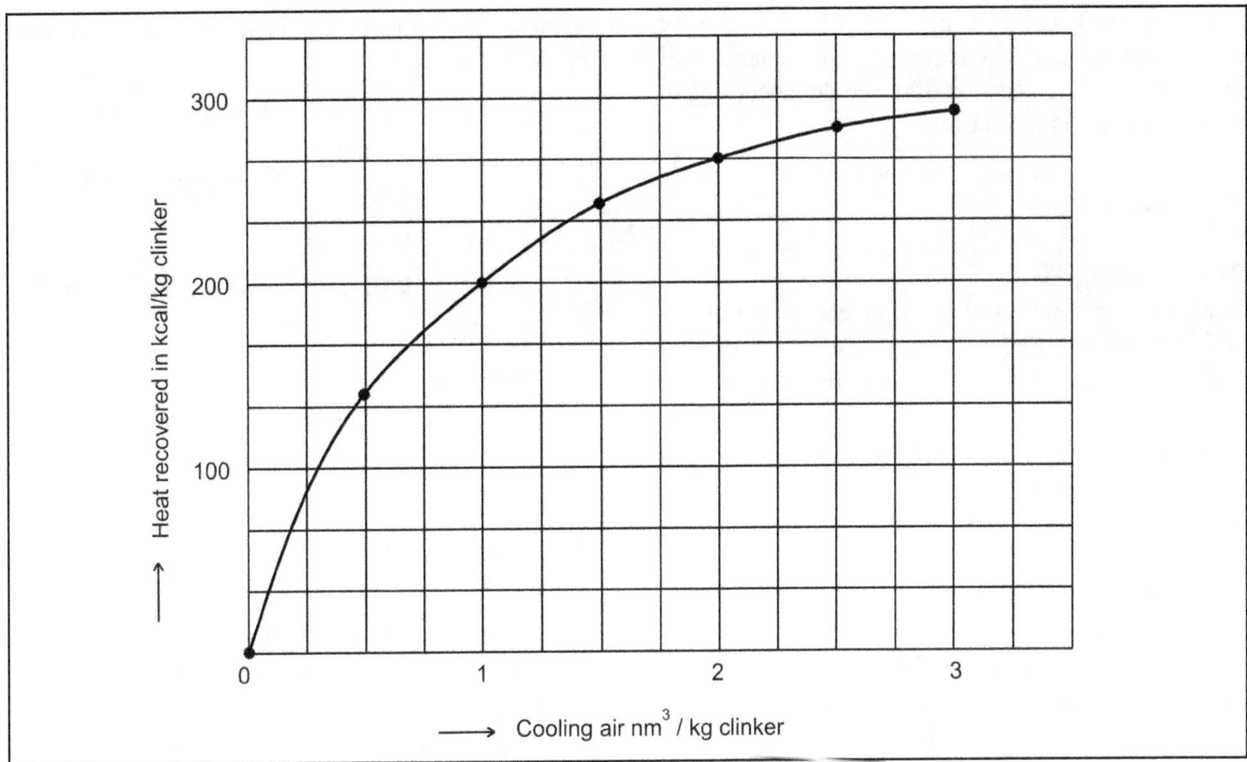

Fig. 6.21.8 Heat recovered by extracted air.

Fig. 6.21.7 increase in temperature of air Admitted for cooling and hence recuperation. **Fig. 6.21.7** shows heat recuperated and **Fig.6.21.8** shows heat recuperated in kcal/kg clinker.

21.4 Variable Speed Drives

The speed of reciprocation or strokes per minute is adjustable and this for a given grate, determines thickness of clinker bed over it for a given output. Coolers have variable speed drives,mostly DC motors.

21.4.1 Individual Drives for Grates

Each grate has its own drive located as for as possible at center of grate to ensure smooth movement of grate without distortion, jerks, misalignment, etc.

All grates can be run either at the same speed or at different speeds. It is customary to run grates slightly faster in sequence. **Fig. 6.21.5 and 6.21.9** show thicknesses of bed in various grates of a multi grate cooler inrelation to their respective speeds.

See Fig. 6.21.9

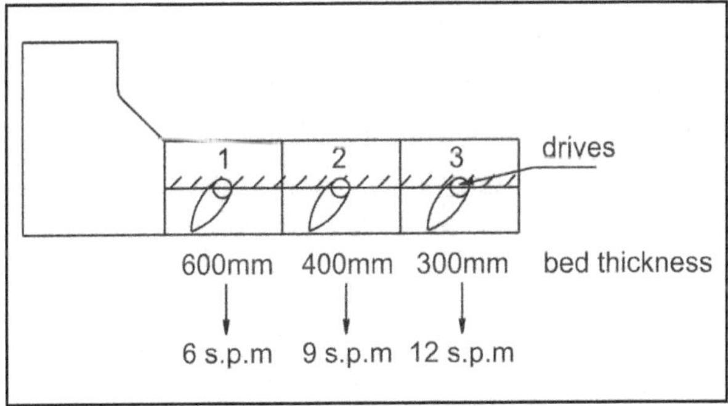

Fig. 6.21.9 Three grates driven at different speeds electrically coupled to maintain ratios and bed thicknesses.
Three grates in series same width.

21.4.2 Electrical Coupling of Drives

Grates could be electrically coupled so that relative speeds between grates and hence bed thicknesses could be maintained at different outputs.

21.5 Thicknesses of Clinker Bed

In older days bed thicknesses were around 150-200 mms because of then prevailing widths of cooles in relation to kiln outputs. Now bed thicknesses are between 300 to 800 mm. Cooler is highly compatmentalised to ensure requisite quantity of cooling air for each section of the cooler according to its needs. **See Fig. 6.21.6.**

In some designs, air supply and its regulation is done individually to grates.

21.6 Cooling Air

In all compartments cooling air enters clinker bed at ambient temperature (except in case of re-circulation) and emerges at different temperatures from clinker bed.

As mentioned earlier, quantities, corresponding to secondary air and tertiary air are drawn into kiln and calciner respectively and the balance is vented out.

21.6.1 Utilising Cooling Air

Generally speaking the 1st and 2nd compartments of the cooler supply air for combustion. The balance of cooling air admitted into cooler can be and is used :

(i) For drying in coal mill.

(ii) As primary air for kiln.

(iii) For drying slag / fly ash in cement mill.

The rest is vented out.

However, when coal mill is located at preheater end,it will not be us d for drying coal.

Presently part of vent air is recirculated as cooling air to through two / there fans in'Duotherm' System.

See Fig. 6.24.11 in Chapter 24.

It is also used in cogeneration of power through'Waste Heat Recovery Systems'.

21.7 Under grate Compartments and Distibution of Air in them

Cooler is highly compartmentalized to ensure requisite quantity of cooling air for each section of the cooler according to its needs.

In some designs, air supply and its regulation is done individually to grates.

21.7.1 Arrangement of Compartments

Different manufacturers of reciprocating grate coolers used different lengths of compartments. Some used a very short 1st compartment; others a fairly long one. Commonly found lengths of cooler compartment expressed in numbers of grate plates in **conventional** coolers were :

3,	6,	9			One grate cooler
3,	6,	9 and			Two grate cooler
3,	6,	9/10			
3,	6,	9			
3,	6,	9/10			Three grate cooler
10	12				

Others use

8*,	10,	17	and		
8*,	10,	10,	15	and	
8*,	10,	17,	8,	10,	17

*effective area could be smaller because of 'horse shoe' shape at the beginning of the cooler.

Above arrangements pertain to conventional grate coolers with reciprocating grates along the entire length.

No authentic data as regards present practice of compartment lengths and air loading in each is available.

21.7.2 Static Grate

'Static grate' is a common feature in several designs of cooler now. The first compartment has an inclined static grate, 6 to 8 rows long. The angle of inclination is ~ 8° Bed is much thicker on the static grate. It has its own fan. **See Fig. 6.21.3.**

21.8 Number of Grates

Whether cooler should have one or more grates is decided by the grate area and the feasibility of having a single drive for it. With calciner kilns cooler width necessarly is small (narrow), requiring coolers of longer lengths.

21.8.1 Width of Cooler

A good width for a cooler is 0.65-0.70 times the clear diameter of a kiln.

	Kiln Dia. m	Cooler Width m
Earlier 1200 tpd Kiln	4.0	2.6
Now 3000 tpd with calciner	4.0	2.6

21.9 Specific Output of a Cooler

Like 'specific output' of a kiln, coolers also have a yardstick of 'specific output' which is tpd / m^2 of grate area. This has continuously increased with increase in thickness of bed.

See Table 6.21.1.

Earlier specific output used to be = 25 tpd/m^2

Now for conventional cooler = 40 tpd/m^2

For cooler with static grate = 50-55 tpd/m^2

Table 6.21.1

		Sp. Output tpd /m^2 grate area		
		25	40	55
Grate area for 4000 tpd cap. m^2		160	100	73
Size of preheater kiln for 4000 tpd = 5.6 m dia inside shell *85 m long				
Dia of kiln inside refractory 5m				
Width of cooler @ 0.7 × 5 = 3.5 m (inside of refractory)				
Grate length	m	46	28.7	21
Required No. of rows		152	95	70
Each plate – 305 mm (effective)				
Therefore each 1 Mts ≃ 3.28 rows				
Kiln capacity raised to 10000 TPD capacity with calciner; cooler width still remains 3.5 m				
For 10000 tpd				
Grate area	m^2		250	182
Grate length	m		72	52
Required No. of rows	-		238	172

Thus if cooler was originally sized for 25 tpd/m^2, it would have 60 rows and can be converted into a 3000 tpd by adding a second grate of 35 rows at 40 tpd / m2 loading. If it was sized for 40 tpd/m^2, the grate area would have to be increased by 95-38 = 57 rows at same specific loading which would mean 2 additional grates.

However, if this cooler is converted into static grate, the increase would be 69-38 = 31 rows i.e., only one additional grate would suffice.

However, the mechanical design of the components like moving and stationary frames, shafts would have to taken into account as bed thickness and speeds increase.

21.10 Two Width Cooler

It is also possible to install a 2 width cooler i.e., width of 1st grate remains 2.6 metres to suit diameter of kiln; subsequqnt grates can be wider by two grate plates or 2.6 + 0.6 = 3.2 metres. This will reduce over all length of cooler for same out put.

See Fig. 6.21.2.

In cases of a large number of small plants increase in capacity has been obtained by :

1. Removing planetary cooler and replacing it by a grate cooler.

   See Fig. 6.20.4 in Chapter 20.

2. Extending length of the existing grate cooler
   (i) Same width – drive of higher rating.
   (ii) 2nd grate – same width – new drive.
   (iii) 2nd grate – wider and new drive.

   See Figs. 6.21.2 and 6.21.10.

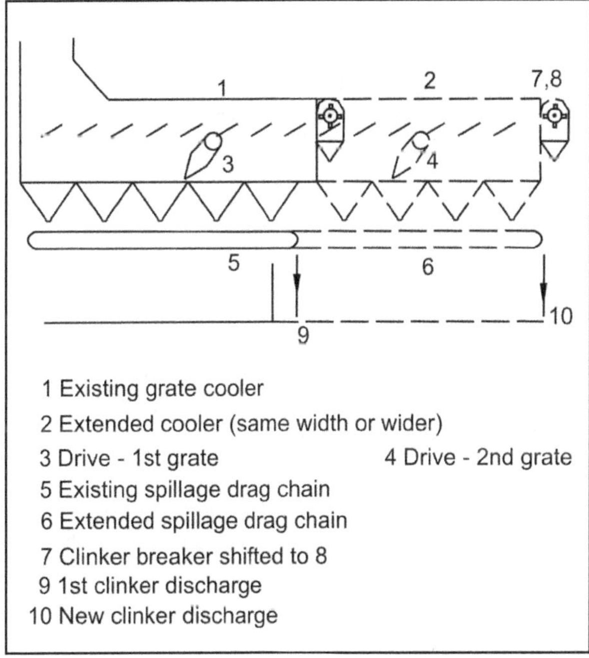

Fig. 6.21.10 Grate cooler - enhancement of capacity by increasing grate area.

21.11 Cooler at Right Angles to Kiln

In some exceptional cases where layout did not permit installation of cooler in line with the kiln, cooler has been installed at right angles but this option should remain as an exception only.

See Fig. 6.21.11.

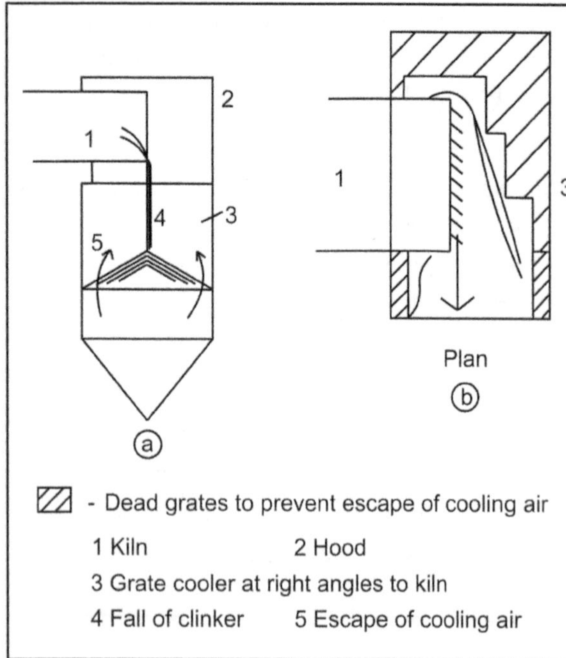

Fig. 6.21.11 Cooler at right angles to kiln; this arrangement permits escape of cooling air.

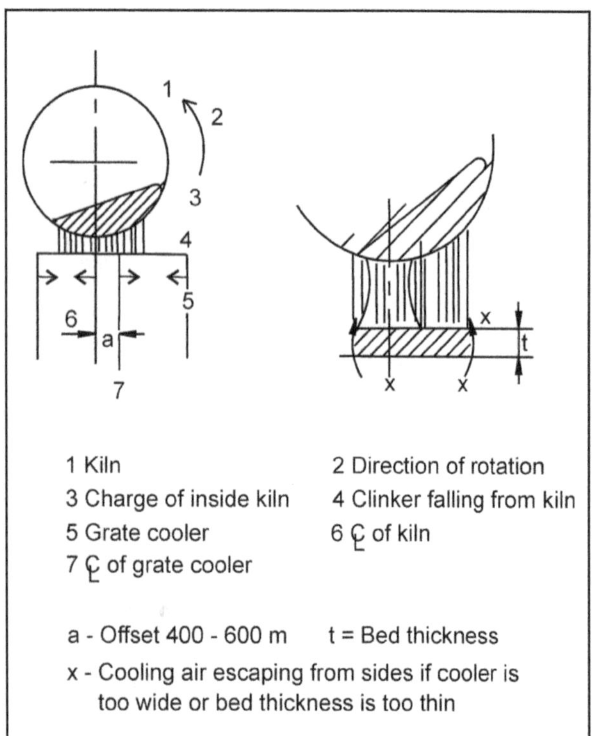

Fig. 6.21.12 Discharge from kiln into cooler need for offset.

Clinker falls from the kiln as shown off center from kiln center because it is raised inside the kiln due to rotation of the kiln. Therefore center lines of kiln and cooler are offset so that clinker falls evenly across the width of the cooler.

See Fig. 6.21.12.

Cooler width should be such that clinker should be spread evenly over it. Otherwise, cooling air would escape from any uncovered area and would be in effective. Hence the convention that cooler width should be 0.65 to 0.7 times diameter of kiln.

An in line cooler has a far greater chance of even distribution than a cooler at right angles.

21.12 Forward and Reverse Flow Coolers

Coolers have mostly been installed in line and in forward flow arrangements.

In some old layouts, coolers have been installed in reverse flow arrangements i.e., under the kiln piers.

See Fig. 6.21.13.

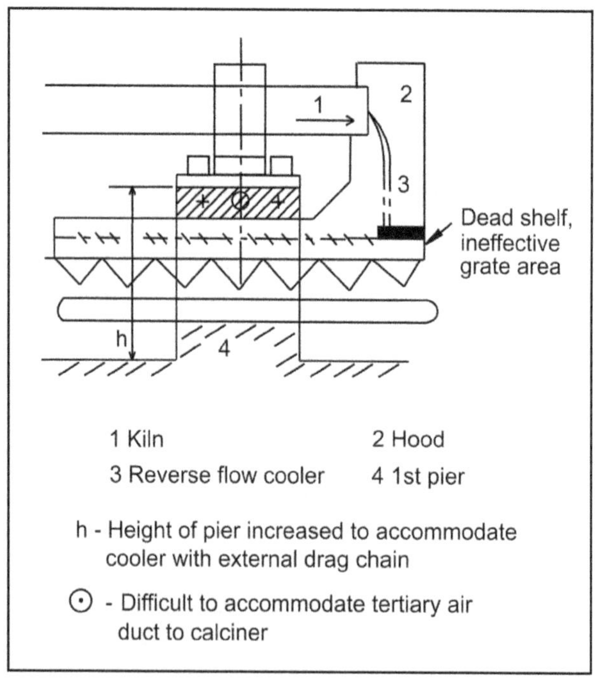

Fig. 6.21.13 Reverse flow cooler.

In case of kilns with calciner, this arrangement is not possible as it will hinder drawing tertiary air from cooler.

21.13 Layout of Cooler with respect to Kiln

In the cooler layout, it is optional whether cooler hoppers and drag chain and deep bucket conveyor should be at ground level or below ground level. Present trend is to install spillage drag chain at ground level and only deep bucket conveyor is installed below ground level.

It raises height of kiln piers and also that of preheater tower but there are no other restrictions. In this arrangement, spillage drag chain is freely accessible. There is no long tunnel and its venting is not required.

See Figs. 6.21.14 and 6.21.15 and 15.43 in Chapter 15.

Cooling fans have to be installed however, on pedestal type foundations and a platform is required along the length of cooler.

21.13.1 Sloping Ground

A sloping ground can thus be used with great advantage in reducing civil costs for preheater kiln and cooler.

See Fig. 6.21.16.

21.14 Cooler Lubrication

Most coolers would be equipped with central grease lubrication systems to lubricate bearings continuously.

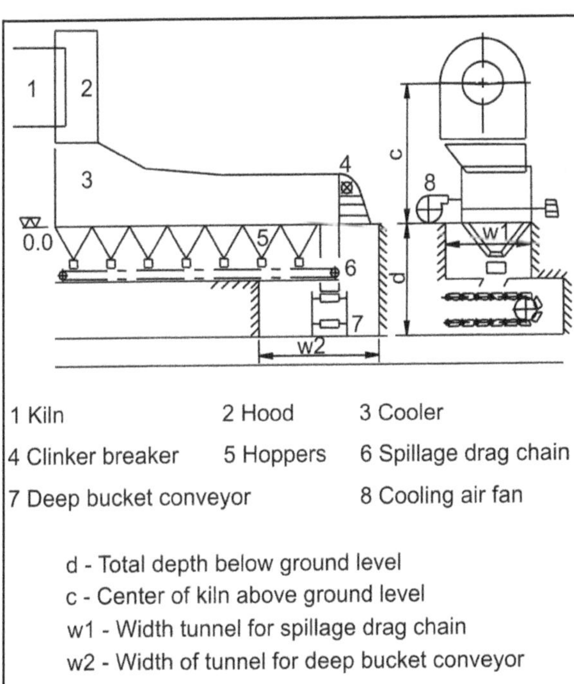

Fig. 6.21.14 Clinker breaker at ground level.

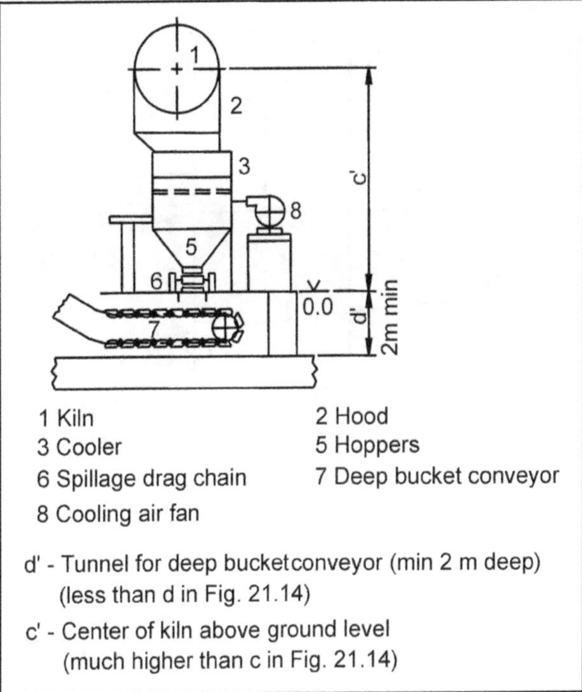

Fig. 6.21.15 Spillage drag chain at ground level; kiln at raised height - fans at higher level.

It is convenient to have bearings outside from point of maintenance and inspection. However in some designs bearings are inside to avoid leakage problems.

When shafts project outside the casing, seals need to be provided to prevent cooling air escaping. This is particularly true of wheel shafts which have a reciprocating motion.

21.15 Drives

Cooler grates are driven by variable speed drives, mostly DC. The drive consists of a d.c. motor, a gear box and roller chain and sprockets. The rotating motion is converted into reciprocating motion by connecting rods and eccentrics. The eccentricity decides the 'stroke' or travel of moving grate over the stationery grate.

Though drive is on one side, to prevent skewing connecting rods are on both sides of the cooler.

See Fig. 6.21.17.

21.15.1 Hydraulic Drive

Presently hydraulic drives have been introduced which can impart reciprocating movement directly. Drive is located centrally under the grate. This is possible because in the present system of supply of cooling air to grate plates, spillage is minimum.

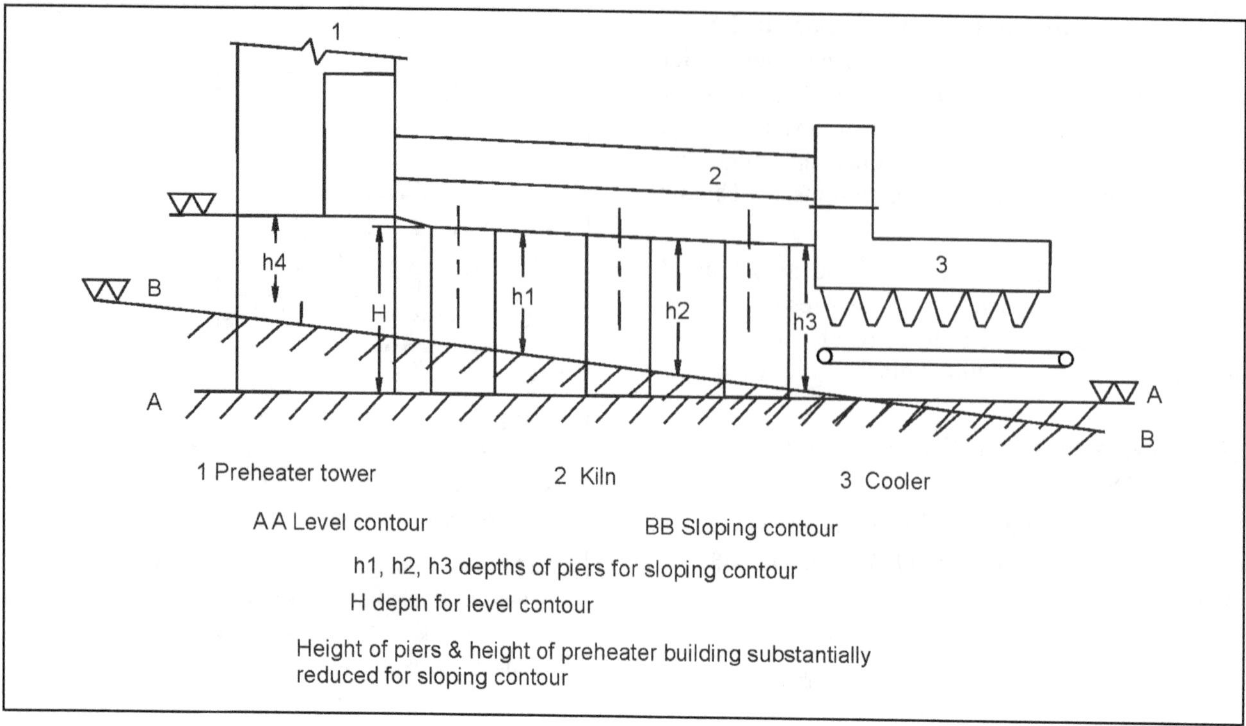

Fig. 6.21.16 Sloping Contour.

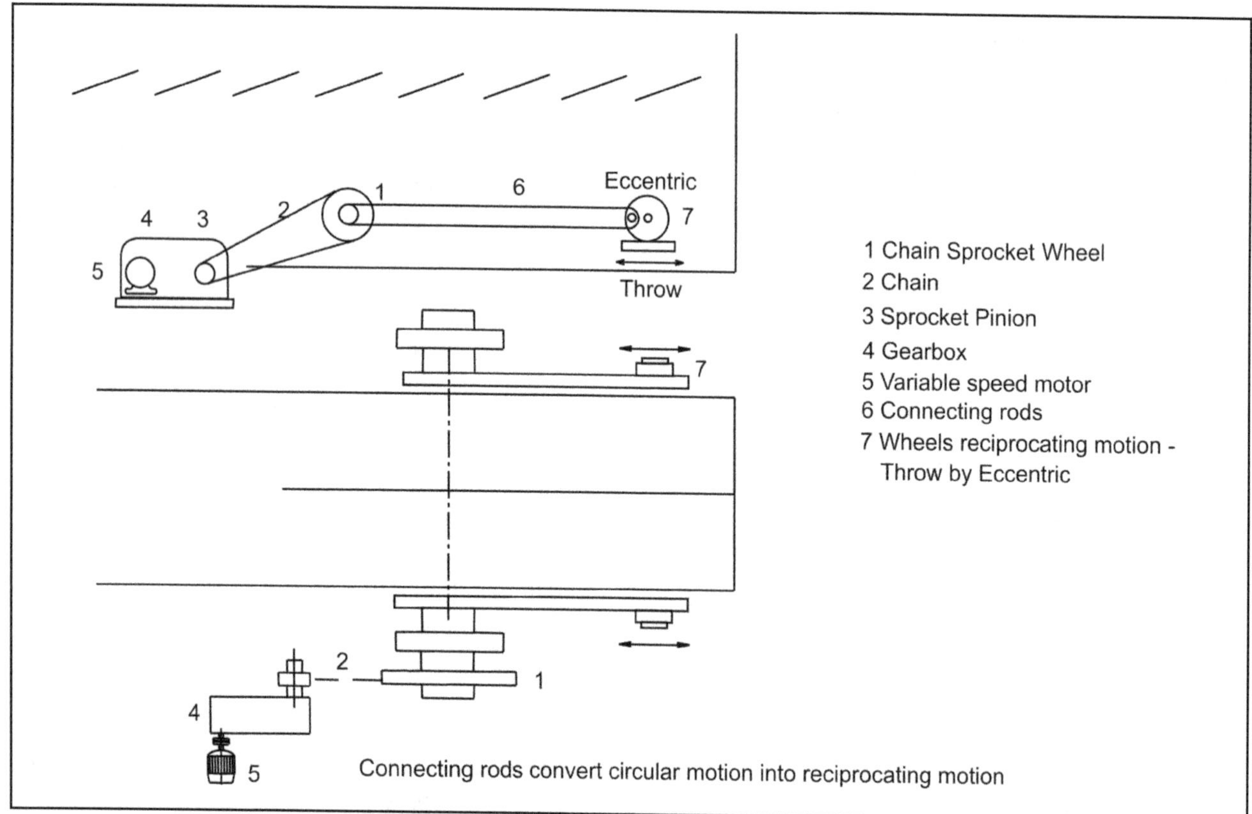

Fig. 6.21.17 Cooler drive - variable speed
(Rotating motion converted to reciprocating motion).

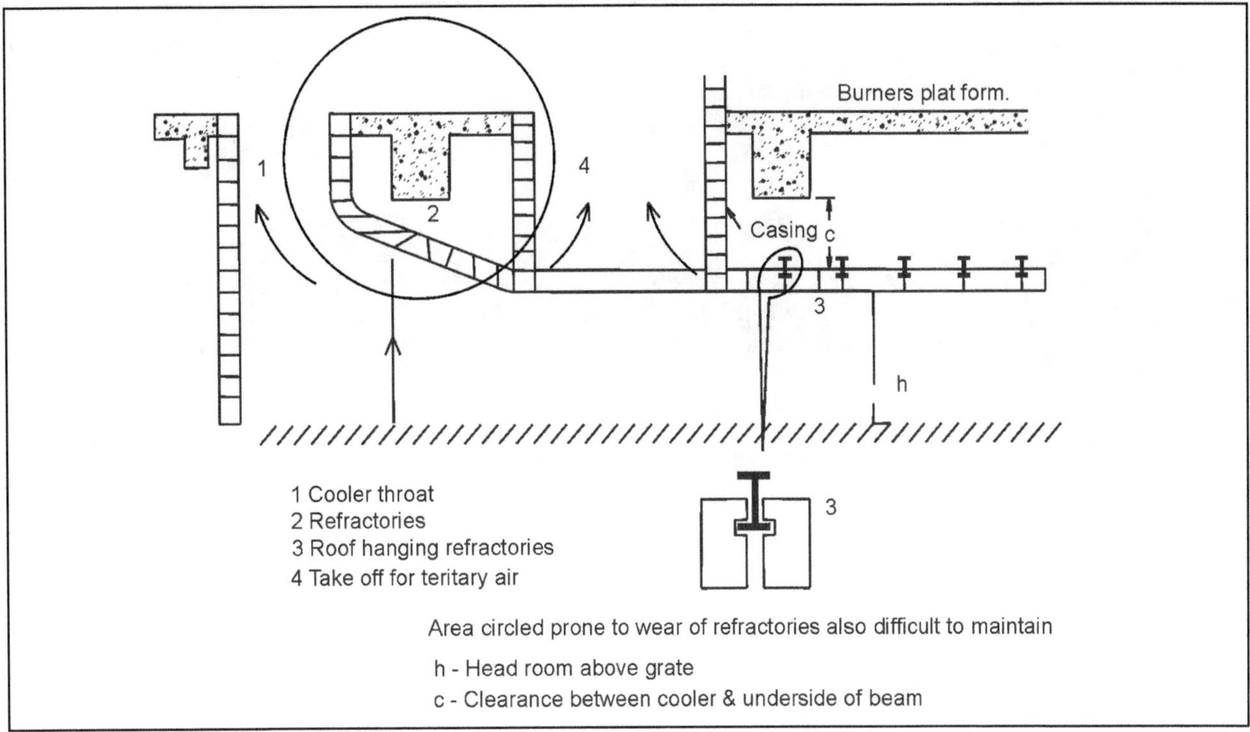

Fig. 6.21.18 Refractories in grate cooler.

21.16 Refractories

Clinker falls into cooler from the kiln at a temperature of about 1370 °C. Therefore cooler inlet, and entire cooler chamber above the reciprocating grates, including roof is lined with refractories and insulating bricks. Roof has hanging refractories.

Hanging refractories are suspended from beams in the roof in the same manner as hanging bricks of roofs of preheater cyclones.

To minimise radiation losses, insulating bricks or hysil blocks are fixed next to the wall. Castables with suitable anchors are also used.

See Figs. 6.21.18 and 6.21.19.

Radiation losses from grate cooler are quite small – only about 5 kcal/kg clinker.

21.16.1 Throat Area of Cooler

The throat area at inlet of cooler is vulnerable to wear. The t.a. duct makes it more difficult to maintain / replace refractories in this area. There should be enough head room inside the cooler to reduce velocity in this zone to 4-6 m/sec.

There should also be enough clearance marked 'c' in **Fig. 6.21.18** between top of cooler and beam of burners' platform to facilitate access and maintenance.

21.17 Typical Layout of a Grate Cooler

Typical layout of a cooler with static grate has been shown in **Fig 6.21.20.** and **Fig. 6.21.21** shows layout of a folax cooler where hot air for calciner is drawn from kiln hood.

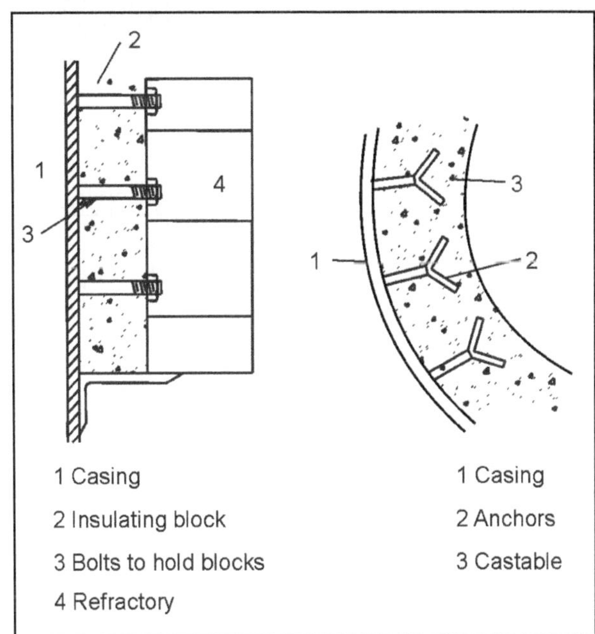

Fig. 6.21.19 Refractory lining in cooler.

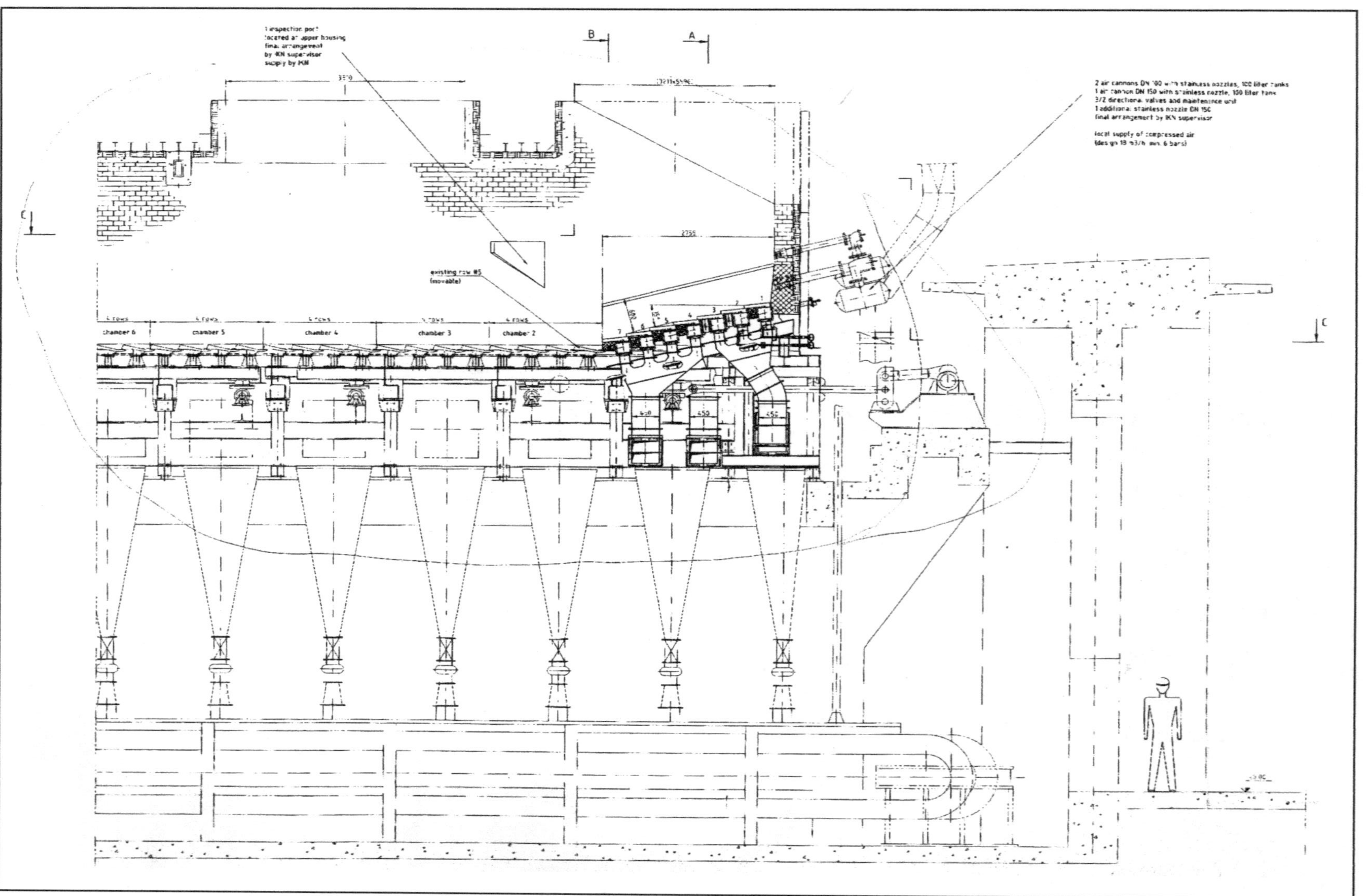

Note : Right and left halves of KIDS get separate air supplies which can be controlled individually
There are clinker pushers in the back wall of the cooler

Fig. 6.21.20 Fitting static grate (KIDS) in an existing grate cooler.

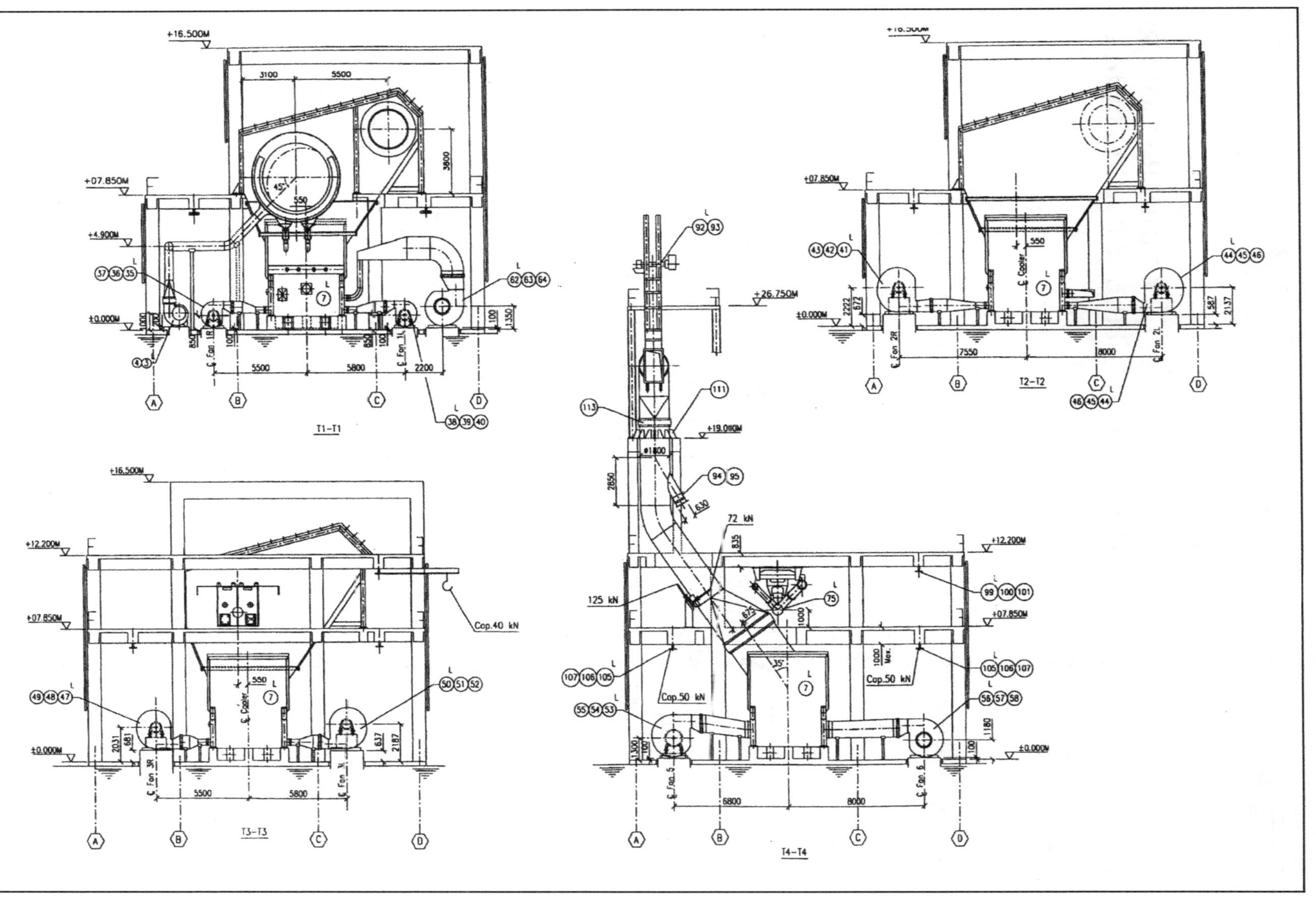

Note : Take off for Tertiary air from Kiln hood

Fig. 6.21.21 Layout of a folax grate cooler.

CHAPTER 22

CLINKER COOLERS, COLLECTING SPILLAGE AND PRODUCT

22.1 Spillage in Conventional Grate Coolers

This Chapter applies mainly to conventional grate coolers. In them, grate plates have small holes of about 6 mm diameter admitting cooling air into clinker bed. Fines from clinker pass through these holes and drop in the air compartment below.

Quantum of fines so collected depends on granulometry of clinker produced. Some kilns produce more fines than the others. In general, in kilns with calciners, percentage of fines is much more than in case of straight preheater kilns.

Dust collected has a temperature of $\simeq$ 1000 °C in the first compartment to $\simeq$ 100 °C in the last compartment.

22.2 Collection and Transport of Dust

This dust has to be collected and removed on a continuing basis for continuous running of the cooler.

It is collected in hopper / hoppers under each air chamber and discharged into a clinker conveyor underneath.

The conveyor underneath can either be only for spillage or also for the total capacity of the kiln. In straight layout of kiln and cooler, it needs to be only for spillage.

22.2.1 Drag Chain Conveyor

The most commonly used conveyor for this application is a drag chain conveyor because of high temperatures and abrasive nature of clinker.

Drag chain conveyor consists of chain links made of manganese steel, which are hard faced. The bottom strand of the conveyor pulls / drags clinker on a bed of clinker at slow speeds.

See Fig. 6.22.1.

The trough in which bottom strand moves is made of concrete lined with steel plates. On the bottom, rails are embedded on which the strand moves at speeds of 1-1.5 m/sec.

22.3 Capacity of Spillage Conveyor

The drag chain conveyor, because of its slow speed, has a small capacity. Since it is required to handle spillage most of the times, its capacity needs to be only about 10-15% for preheater kilns and about 30% for calciner kilns.

Thus for a 10000 tpd kiln with calciner the capacity of spillage drag chain needs to convey at the rate of $10000/24 \times 0.30 = 125$ tph at rated capacity.

22.3.1 Upset Condition of Kiln

Kilns undergo upsets now and then. During upset conditions, under burnt clinker, which is very fine rushes through kiln and cooler at very high rates.

Therefore, the drag chain conveyor for spillage is also required to be sized for coping with upset conditions.

Upset conditions are taken at 1.3 times normal conditions. Therefore drag chain conveyor is sized 20-30% above upset condition.

Kiln rated capacity	-	3000 tpd	10000 tpd
Design capacity	-	3300 tpd	11000 tpd
Output under upset condition			
at design capacity	-	4300 tpd / ~ 180 tph	14300 tpd 600tph
spillage at 30 %	-	54 tph	180 tph
Design capacity for drag chain	-	65-70 tph	~ 240tph

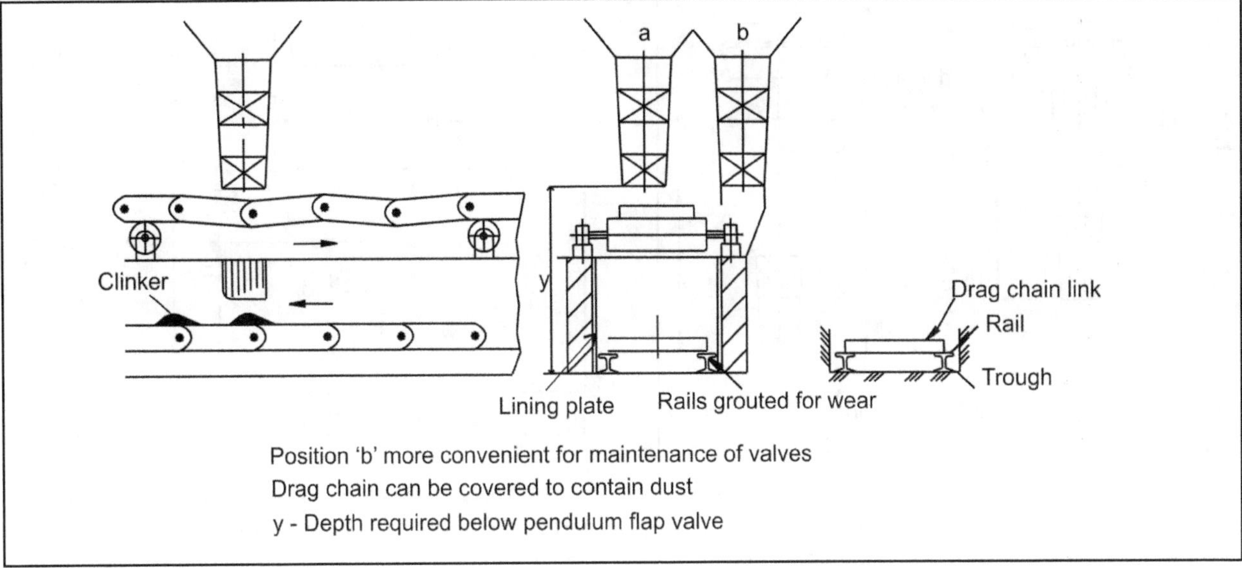

Fig. 6.22.1 Spillage drag chain conveyor under cooler.

Main as well as spillage conveyors for clinker are sized, 60-70 % above their respective capacities.

22.4

Normally drag chain conveyor would be installed horizontally. They can also be installed at angles up to 10-12 degrees but then their capacity drops appreciably. This factor has to be taken into account in selecting the width of the conveyor.

22.5 Product Conveyor

If the conveyor under cooler hoppers is required to handle total product, this conveyor would be most likely, a deep bucket conveyor; hence would require much more height under the pendulum flaps.

See Fig. 6.22.2.

'x' would be same, but 'z' in **Fig. 6.22.2** would be much greater than 'y' in **Fig. 6.22.1**. The overall depth below ground level will be less because drag chain has been eliminated.

The deep bucket conveyor would have to be longer corresponding to length of the cooler and its capacity for a 10000 tpd kiln with usual design margins would be ~ 700 tph. This is a rather costly solution. Best to have a separate drag chain conveyor under cooler for spillage.

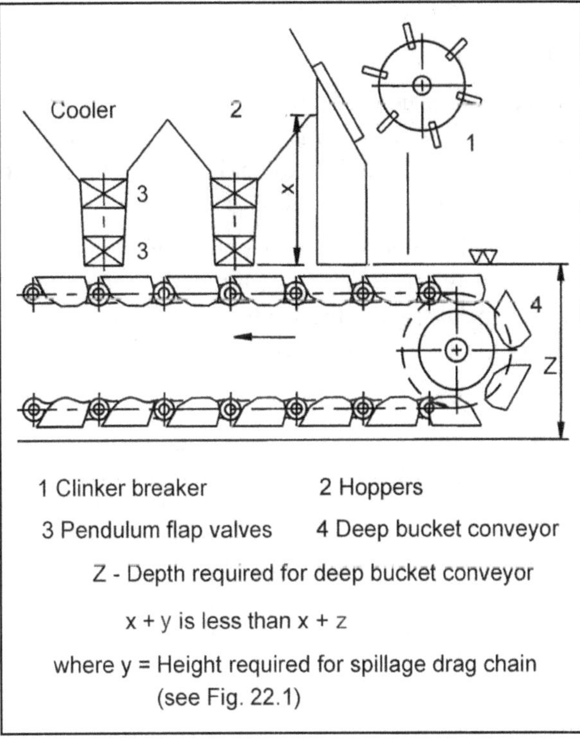

Fig. 6.22.2 Conveyor under cooler handles total product. This reduces total depth.

22.6 Layout of Hoppers and Spillage Conveyor

The air chamber under the grates, is divided into compartments. These are of different lengths, according

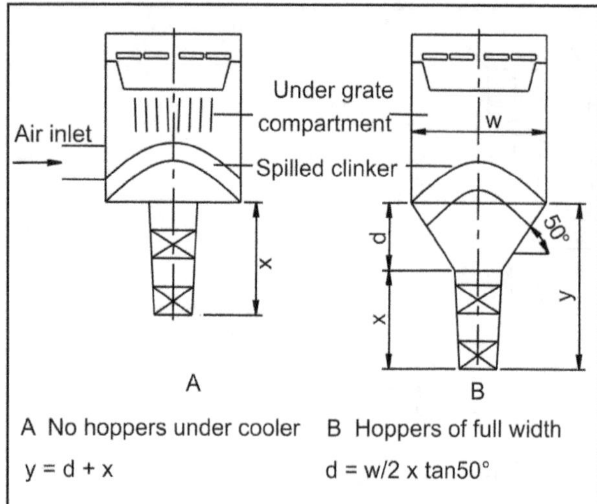

Fig. 6.22.3 Alternative arrangements for collecting spillage in under grate compartments.

design concepts of respective manufacturers as explained in previous chapter.

See Fig. 6.22.5.

If clinker is not collected on regular basis from air chambers, it would form heaps in the chamber and would soon obstruct flow of cooling air and may also rub against the under sides of grates.

Generally, therefore air chambers would have full or partial width hoppers underneath them. Partial hoppers would reduce the depth but would allow bigger clinker heaps in air chamber.

See Fig. 6.22.4.

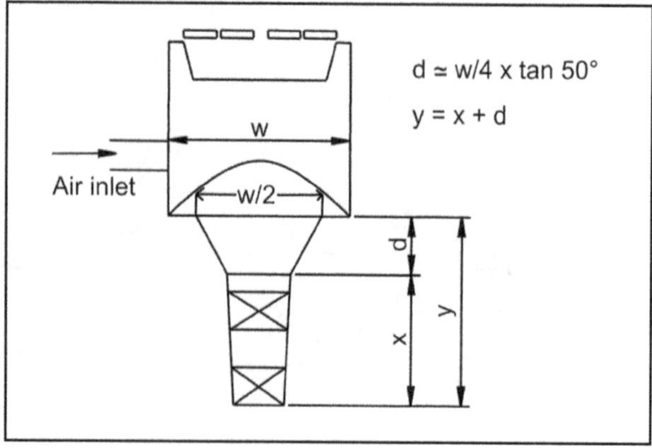

Fig. 6.22.4 Partial hoppers to reduce depth.

In exceptional cases, where it is necessary to keep depth below the cooler to a minimum, hoppers under the air chamber may be eliminated altogether. In this case, spillage drag chain should be oversized to prevent any accumulation of clinker heaps in air chambers.

See Fig. 6.22.3.

Hopper slopes are kept higher than the angle of repose of clinker i.e., 35°.

For short compartments i.e., of 3 or 6 rows, there would be one hopper lengthwise under the compartment. For longer compartments there would be more than one hoppers for one compartment so that, depth is reduced and is uniform for all compartments. 'y' should be same for all compartments throughout the length of the cooler for practical reasons.

See Fig. 6.22.5.

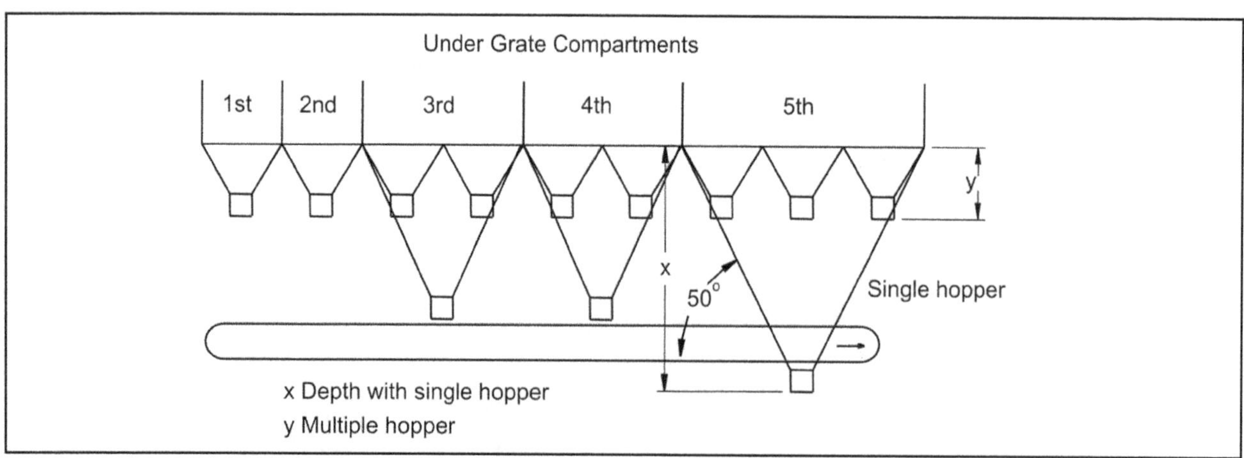

Fig. 6.22.5 Hoppers and compartment lengths. More hoppers per compartment according to width and length of the compartment.

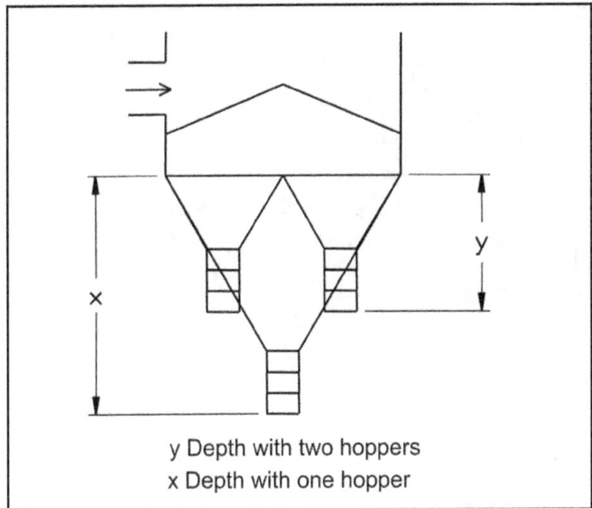

Fig. 6.22.6 If cooler is very wide, then two hoppers abreast across width will reduce depth.

If the cooler width is very large 3.6 m or more, there may be 2 hoppers width wise also requiring 2 drag chains of half the capacity each.

See Fig. 6.22.6.

22.7 Pendulum Flaps under Hoppers

Air chambers receive cooling air at pressures ranging from 700 mm wg to 150 / 200 mm wg. This air would tend to rush out when spillage collected is discharged. Therefore it is most important to provide seals to prevent the escape of air. This seal is provided by, maintaining a small layer of clinker in the hopper and by, using pendulum flap valves singly but preferably in pairs.

Rotary air locks would not be suitable because of abrasive nature of clinker.

Pendulum flap valves can be gravity operated or motor operated with timer, cam and springs or even pneumatically operated.

See Fig. 6.22.7.

A single flap valve is the simplest arrangement but would still permit escape of air whenever valve opens. To avoid this, the valves are normally installed in pairs.

A spacer piece is provided to enable first valve to be closed completely before the lower valve opens.

22.7.1 Drives for Flap Valves

There could be one motor for each valve or one motor common for two valves when the valves would be connected by linkages.

Profiled cams would be used to open the valve, the period for which it would remain open being decided by the profile, and the spring would shut it tight instantly.

The valves may be made to operate sequentially along the length of the cooler to 'even' the load on the drag chain.

See Fig. 6.22.8.

Some manufacturers prefer hydraulically or pneumatically operated thrusters to open and to close the valves. However, they need to be of proven design to work in dusty and hot conditions.

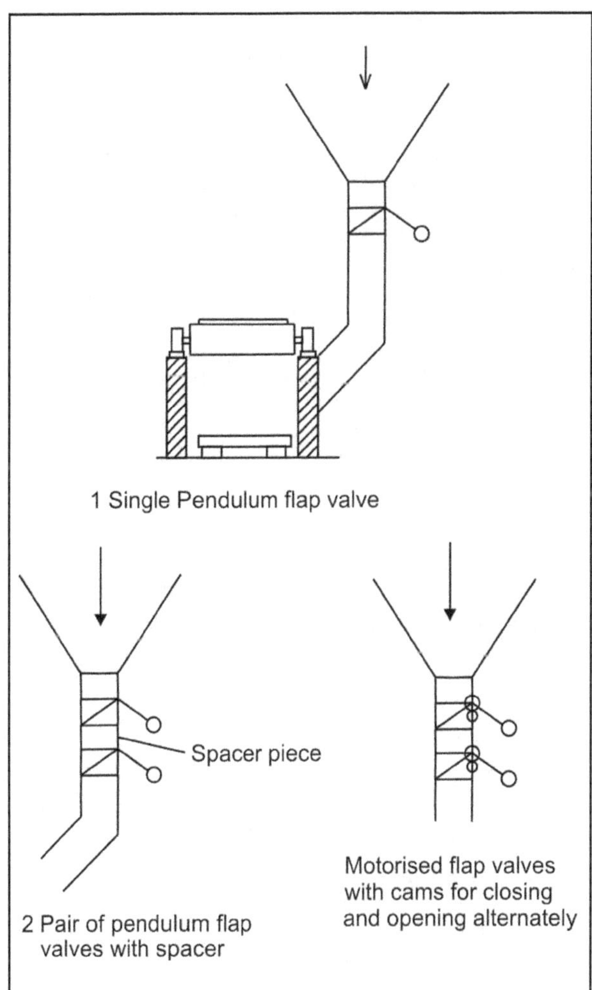

Fig. 6.22.7 Single/double pendulum flap valves for discharge of clinker.

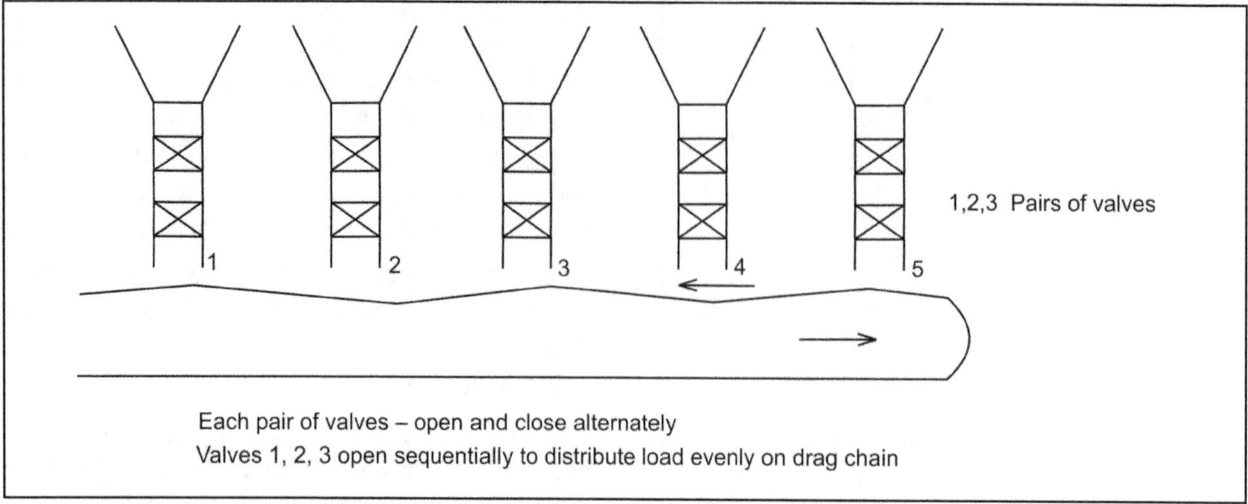

Fig. 6.22.8 Sequential operation of valves.

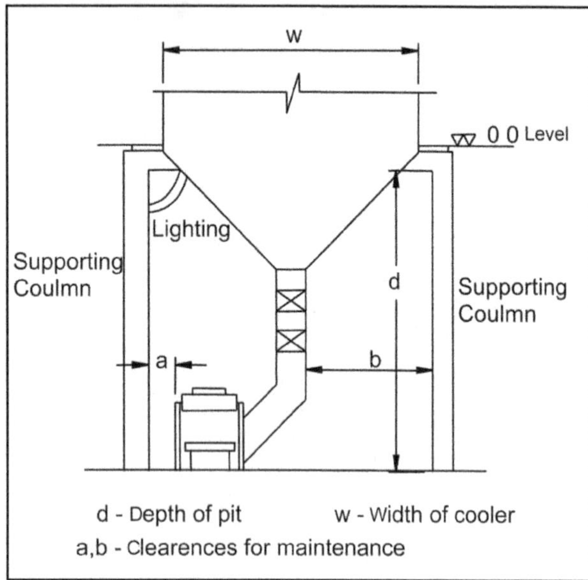

Fig. 6.22.9 Dimension of Tunnel for Hoppers and spillage drag chain; when it is under ground.

22.8 Layout for Maintenance

Easy access to the valves for maintenance and repairs is very necessary. Therefore, sufficient width is provided on both sides and also height for this purpose.

See Fig. 6.22.9.

It is for this reason that it is now preferred to install drag chain at ground level.

To contain dust nuisance, which is very likely when hoppers are empty the drag chain can be totally covered in a M.S. casing.

It is vented out from one or both ends.

See Fig. 6.22.10.

For large kilns these tunnels under the cooler can be very long and hence need to be vented and lighted. The air drawn from the tunnel could be supplied to one of the cooling fans eliminating need for a separate dust collector.

See Fig. 6.22.11.

In one design, clinker collected in hoppers is emptied continuously by using compressed air.

See Fig. 6.22.20.

22.9 Spillage in New Designs

With new types of grate plates that have box like construction with nozzles and where air is supplied individually to each grate plate, spillage is very much reduced. Hence in such cases all that has been described in **paras 22.7 and 22.8** will hardly apply. Elimination of hoppers and spillage conveyor will reduce overall height of cooler.

In a **Cross Bar Cooler**, there are no reciprocating grates as such. Such designs do not need hoppers underneath and spillage collection system and its layout in such cases will be simplified.

In coolers with ' **walking floor'** also there would not be spillage'.

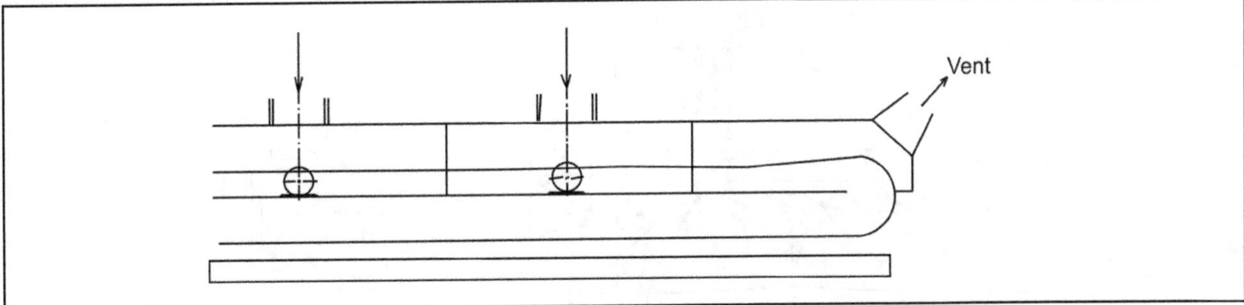

Fig. 6.22.10 Cover on drag chain for containing and venting dust; to keep tunnel dust free.

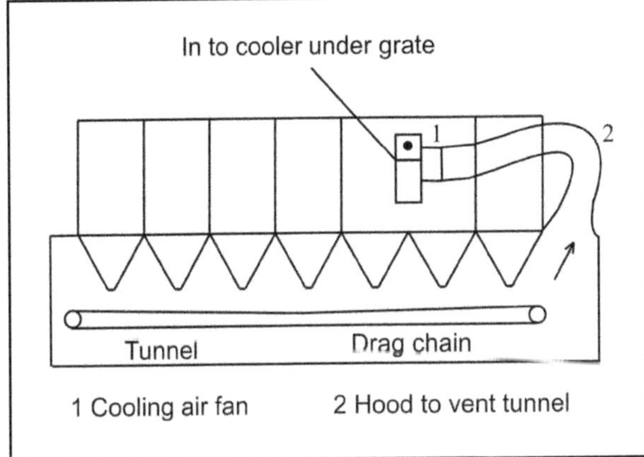

Fig. 6.22.11 Venting of tunnel sucked by cooling air fan and admited into cooler.

22.10 Breaking Lumps of Clinker

In a conventional grate cooler, clinker is discharged through a grate consisting of grate bars spaced at 20-25 mm and installed at an angle of 45-50°.

The oversize lumps are thrown back into the cooler by a hammer type clinker breaker installed at the end. The breaker covers the full width of cooler and consists of heavy manganese steel hammers. A clinker breaker is not a clinker crusher as such. The lumps are not broken inside the crusher against breaker plates as in regular crushers. They are thrown against curtains of chains and get broken. The breaking exposes red hot centers of lumps to cooling air before they are discharged through the grate bars.

See Fig. 6.22.12.

Chain curtains consist of heavy oval link chains or simply pieces of rails linked together. They may be one or more rows of curtain chains.

22.10.1 Width of Clinker Breaker

Clinker breaker has to be as wide as the cooler. A full width clinker breaker can normally cope with the increase of the capacity of the kiln after installing a calciner.

Initial preheater kiln – 1200 tpd or 50 tph

Kiln size	4.0 m dia × 60 m long	
Width of cooler 0.7 × 4.0	-	2.8 m
Therefore clinker breaker installed	-	2.8 m wide.
Capacity of kiln after calciner	-	3000 tpd or 125 tph

Clinker breaker of 2.8 metres width would still be adequate for breaking lumps at 2.5 times the original rate.

See Fig. 6.22.13.

22.10.2 Location of Clinker Breaker

When grate/s is/are added on an existing cooler, the clinker breaker is required to be shifted to the new location **(Fig 6.22.14)**. If the width of the added grate is different then a new clinker breaker corresponding to the width of new grate would have to be added.

22.11

In many small plants where a calciner was installed, coolers had to be extended by adding a second grate. The existing layout in such cases can pose a serious problem in accommodating the grate.

The clinker discharge point has to be shifted. If it is not possible to do so then the drag chain under the new grate will have to be reversed to bring the clinker to original point of discharge. The new drag chain will handle total product plus spillage under the extended grate.

See Fig. 6.22.15.

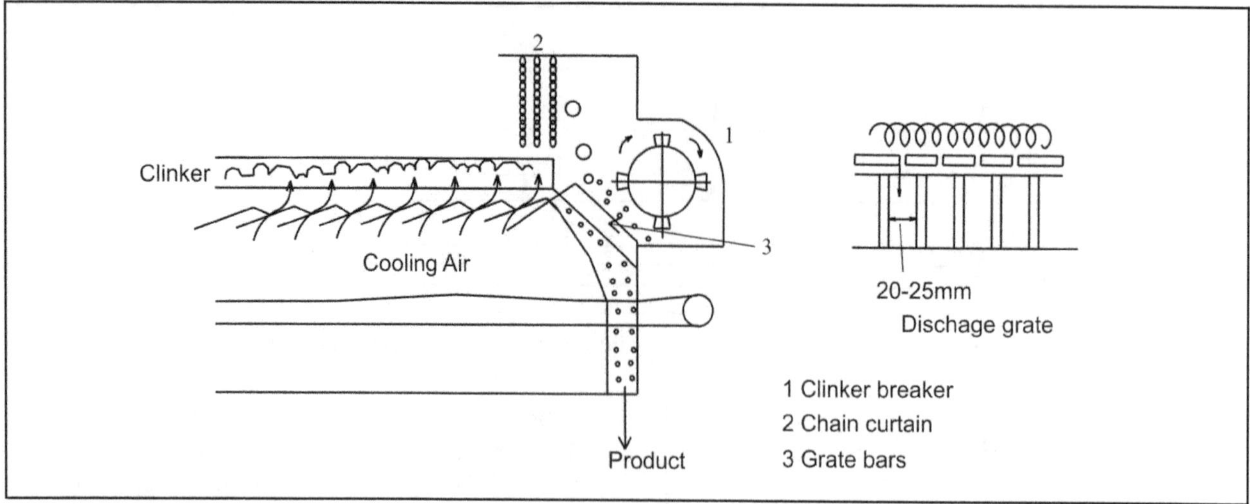

Fig. 6.22.12 Clinker breaker and chain curtain to break lumps.

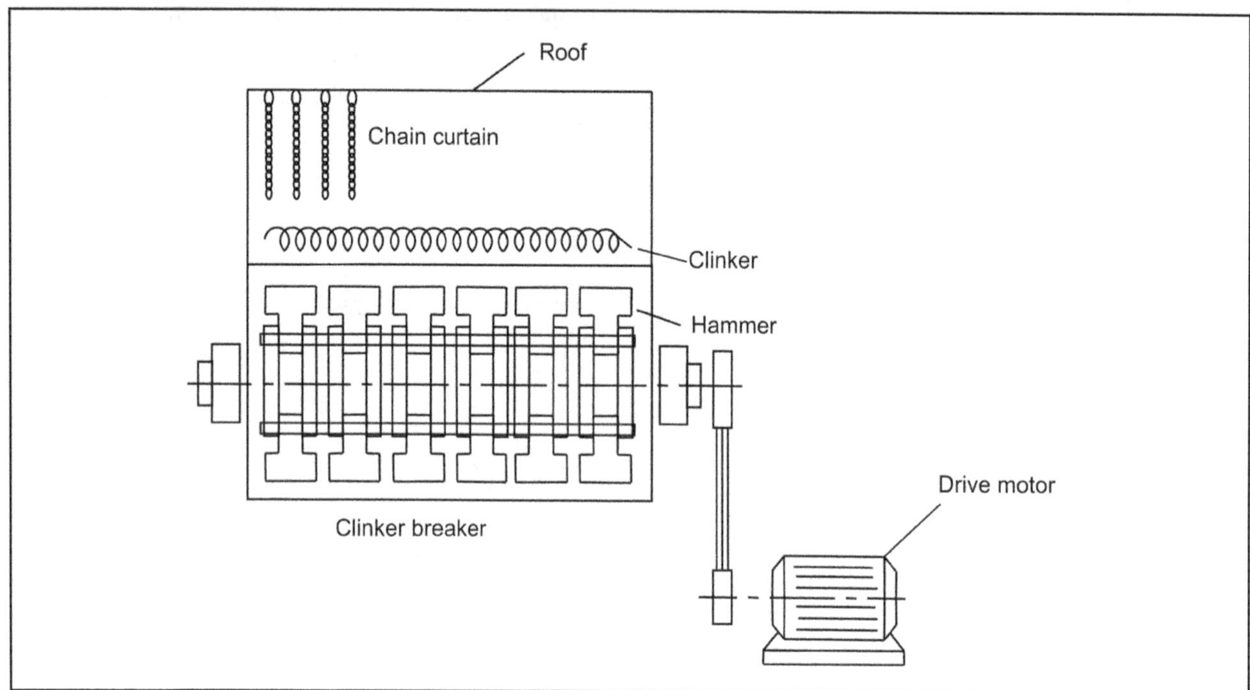

Fig. 6.22.13 Full width clinker breaker is a must for a grate cooler.

22.12 Roll Crusher

In new designs a full width roll crusher is used to break clinker into fines. This crusher unlike breaker described earlier takes the total clinker produced and not just lumps. Therefore it must match the capacity of kiln.

See Fig. 6.22.16.

When there are more than one grate then roll crusher can be installed in the middle and fines produced will be cooled in the next grate.

See Fig. 6.22.17.

Even in old designs clinker breaker was left in its place when a new grate was added. The new grate was installed at a lower level to receive clinker.

See Fig. 6.22.18.

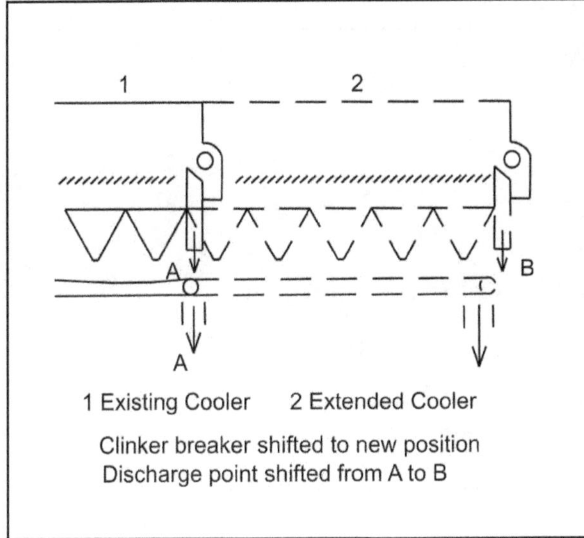

Fig. 6.22.14 Extending length of cooler to increase capacity.

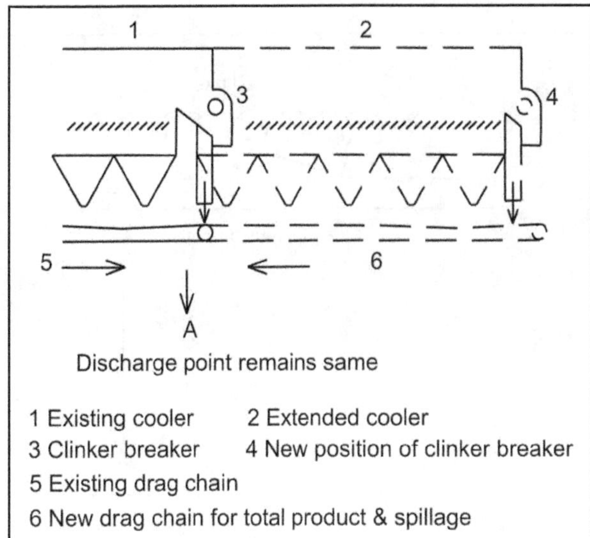

Fig. 6.22.15 Extending length of cooler to increase capacity.

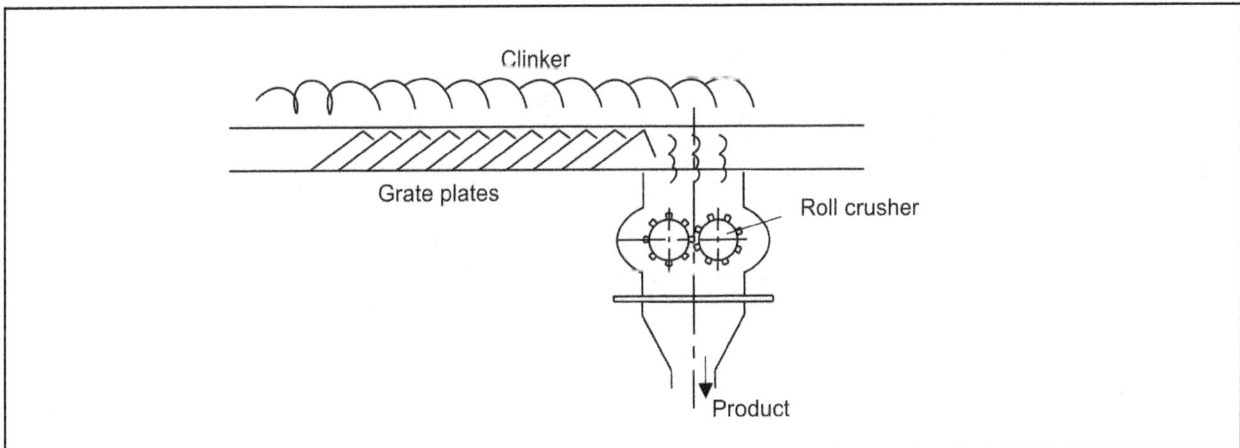

Fig. 6.22.16 Roll crusher to break clinker.

Some times grates were installed at different levels to break layers for better cooling.

See Fig. 6.22.19.

22.13 Continuous Pneumatic Conveying of Spillage

One of the many developments is continuous pneumatic conveying of clinker dust in hoppers, eliminating pendulum flap valves and spillage drag chain. Not sure about its adoption on a large scale because of coarse grain size and abrasive nature of clinker dust

See Fig. 6.22.20.

22.14 Layout of Grate Cooler

Layout of cooler will have to be fashioned to suit the design of main cooler and auxiliaries. There have been very many developments in coolers and it would be difficult to include all of them. As in case of preheater and calciner it would be best to obtain recommendations of Cooler layout from Suppliers.

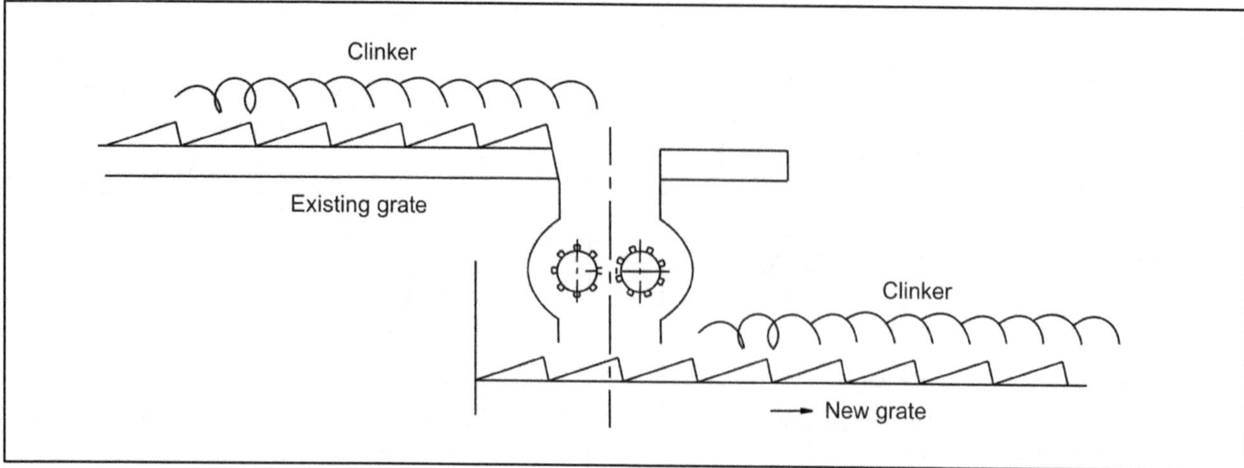

Fig. 6.22.17 Roll crusher in middle.

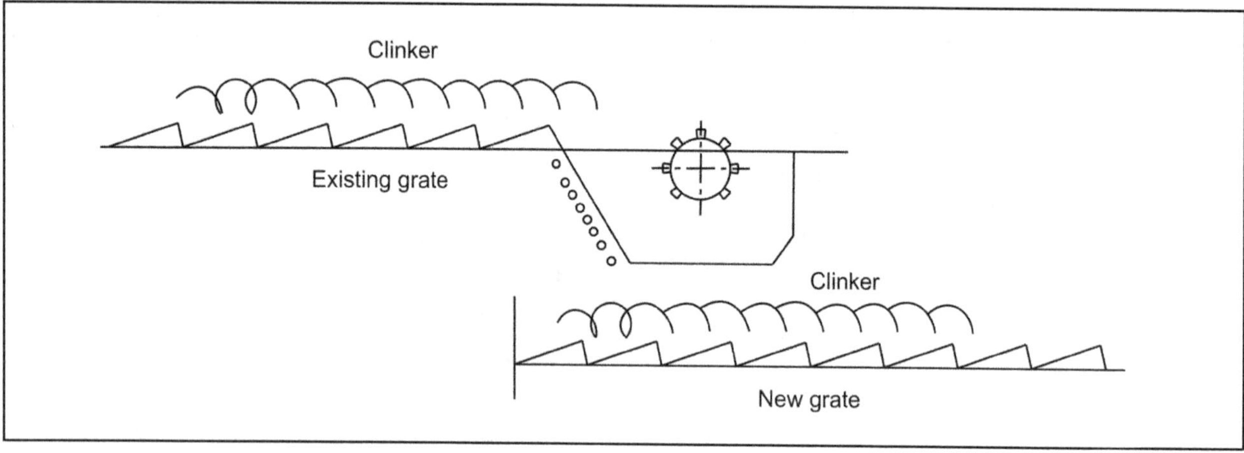

Fig. 6.22.18 Conventional cooler adding second grate.

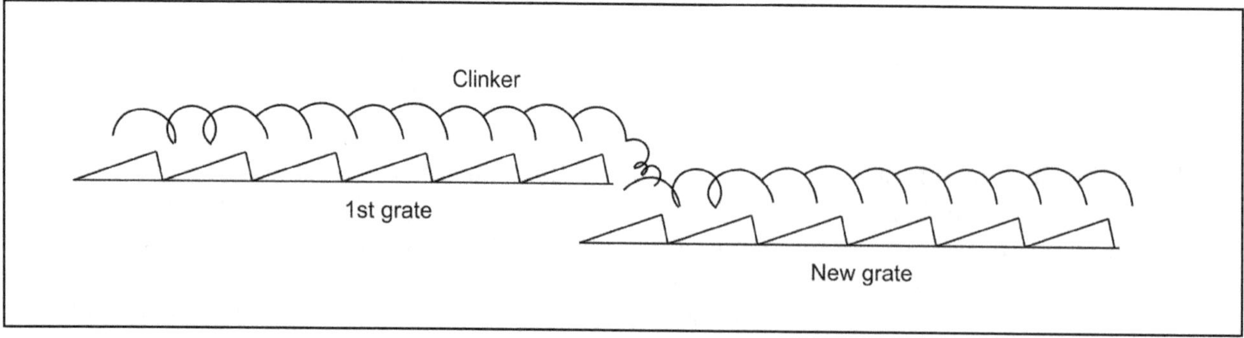

Fig. 6.22.19 Conventional cooler - adding new grate at lower level to turn clinker.

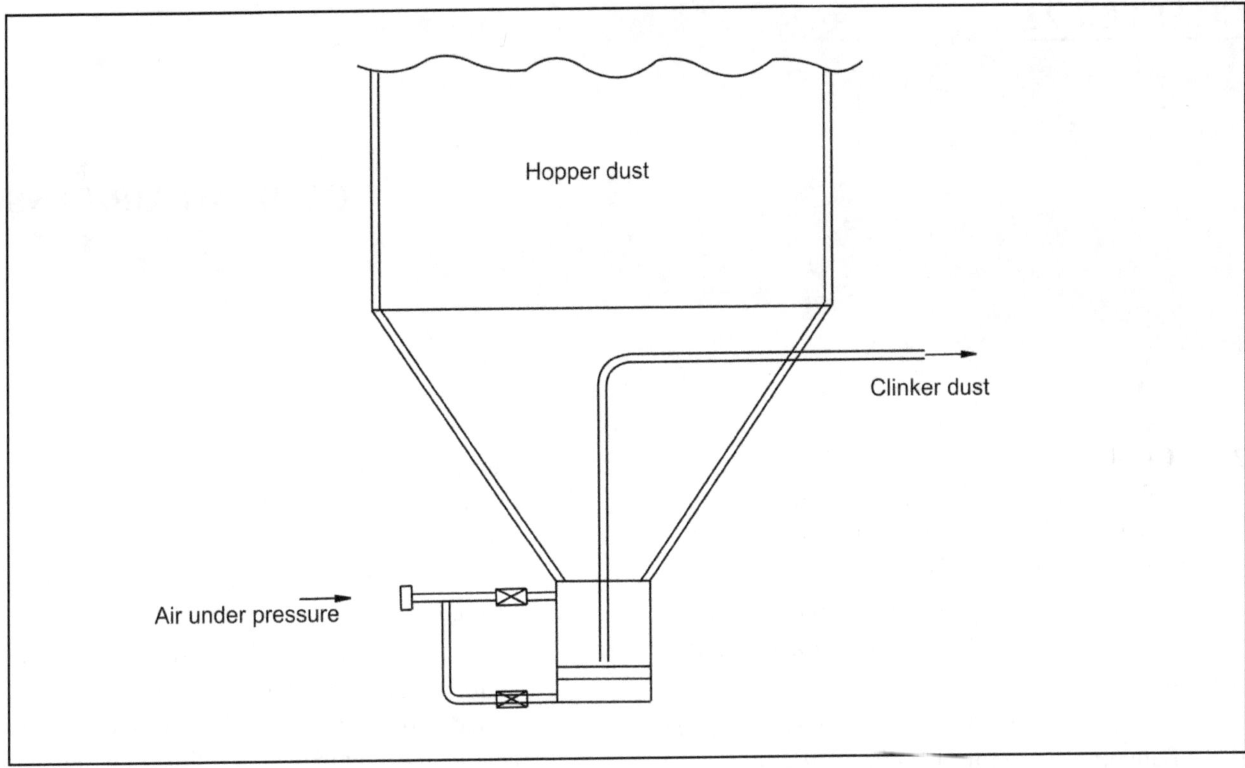

Fig. 6.22.20 Continuous pneumatic conveying of clinker dust to main clinker conveyor.

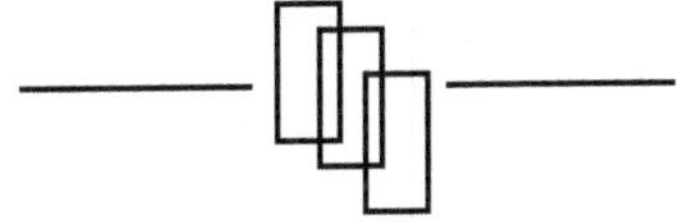

CHAPTER 23

COOLING AIR FANS

23.1 Cooling Air

The total cooling air admitted into cooler ranges between 2.0 nm^3/kg for coolers with static grates to 2.7 nm^3/kg for conventional coolers.

The quantity also depends on the temperature to which clinker is desired to be cooled and ambient temperature.

From the clinker cooling curve, it can be seen that it is a flattening curve towards dischsrge end; for the same quantum of cooling air 'q', drop in temperature of clinker would be much higher at hot end than at cold end.

See Fig. 6.23.1.

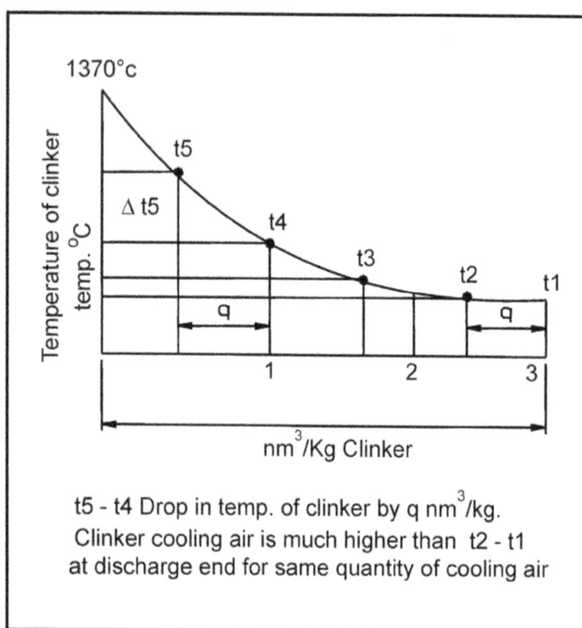

Fig. 6.23.1 Cooling curve.

23.2 Temperature of Clinker at Discharge

The temperature to which clinker should be cooled depends on :

1. Conveyors used to handle clinker after it emerges from cooler.
2. The period between its coming out of cooler and being used in cement mills for grinding. If this period is long enough for clinker to get cooled to near ambient temperatures, clinker coming out of cooler can be comparatively hotter.

Earlier, when belt conveyors were used to handle clinker, it was necessary to cool clinker to less than 100 °C.

With deep bucket conveyors now being used universally, in large and small plants, clinker needs to be cooled down to about 150 °C, assuming that by the time it reaches cement mills it would have cooled down to less than 100 °C.

The static grate helps to bring down the requirement of total cooling air from 2.5 nm^3 to 2.0-2.2 nm^3/kg but clinker neds to be cooled to ~ 120 to 150 oc only as mentioned above.

Less cooling air also means less air to be vented.

23.3 Fans for Cooling Air

It is universal practice to use independent cooling air fans for each compartment to ensure that each section of clinker bed receives adequate quantity of cooling air at required pressure i.e., enough for it to penetrate through the clinker bed above it.

Thus a 5 compartmental grate cooler would have 5 fans; a 6 compartmental grate cooler would have 6 fans and so on.

Perhaps only the last two compartments of the last grate will have a common fan.

23.3.1 Fan for Statice Grate

The static grate has its own cooling air fan requiring a much higher pressure as bed thickness is much higher in this section of the cooler.

Thus a cooler may have anywhere between 3 to 10-12 fans depending on capacity and hence number of compartments

See Fig. 6.23.2.

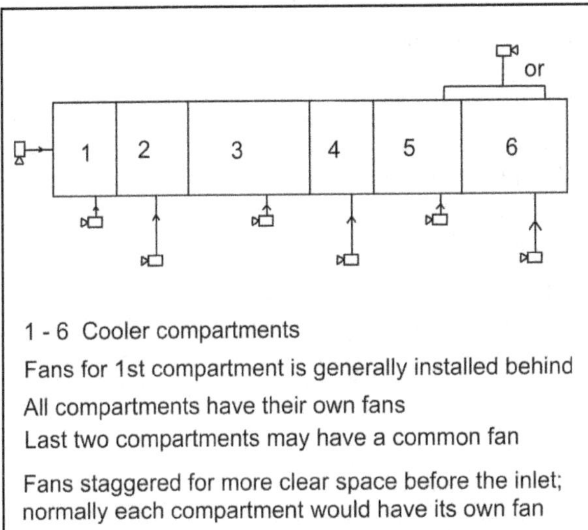

Fig. 6.23.2 Layout of cooling air fans.

23.4 Layout of Fans

The fans could be arranged either all on one side or on both sides of the cooler as shown, which provides enough spacing between fans, and access to cooler is easier. It however requires wider space on both sides of the cooler to accommodate them. **See Figs. 6.23.2 and 6.23.3.**

Fans may be installed at an angle to the cooler to provide sufficient space for a horizontal length of pipe at inlet for measuring quantity of cooling air.

See Fig. 6.23.4.

The cooling air fan for the 1st compartment can also be installed behind the cooler under the kiln.

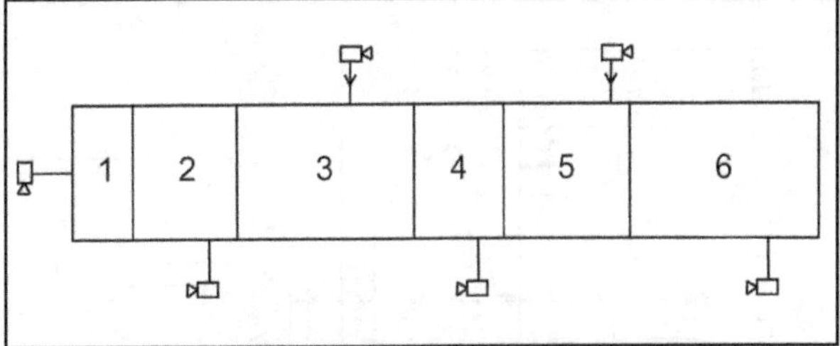

Fig. 6.23.3 Fans installed on opposite sides in uncluttered layout.

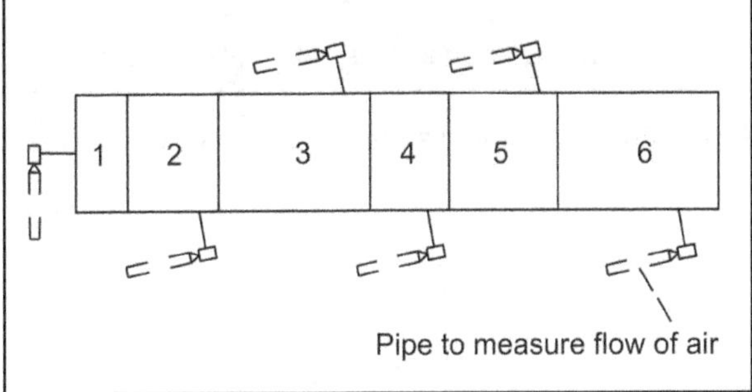

Fig. 6.23.4 Fans may also be staggered or installed - 'skew' to install length of pipes at inlet to measure air flow of cooling air.

23.4.1 Tertiary Air Duct and Dust Chamber

A point to be considered in arranging fans is the location of tertiary air duct and dust chamber.

The connecting duct and dust chamber need to be supported though they themselves would not obstruct installation of cooling fans for the 2nd and 3rd chambers.

See Fig. 6.23.5.

If the fans are laid out staggered on either side of cooler, then the fans would clear the supports of t.a. duct and dust chamber and discharge chutes from dust chamber to cooler or drag chain.

See Figs. 6.23.4 and 6.23.5.

The fans should be so arranged as to have a short straight duct connection to cooler air chamber. Preferably the air inlet should be at center of chamber and its height from the bottom should be such that even if clinker accumulates inside the chamber it would not block the air opening.

See Fig. 6.23.6.

A typical layout of cooling air fans on both sides of a cooler is shown in **Fig. 6.23.10.**

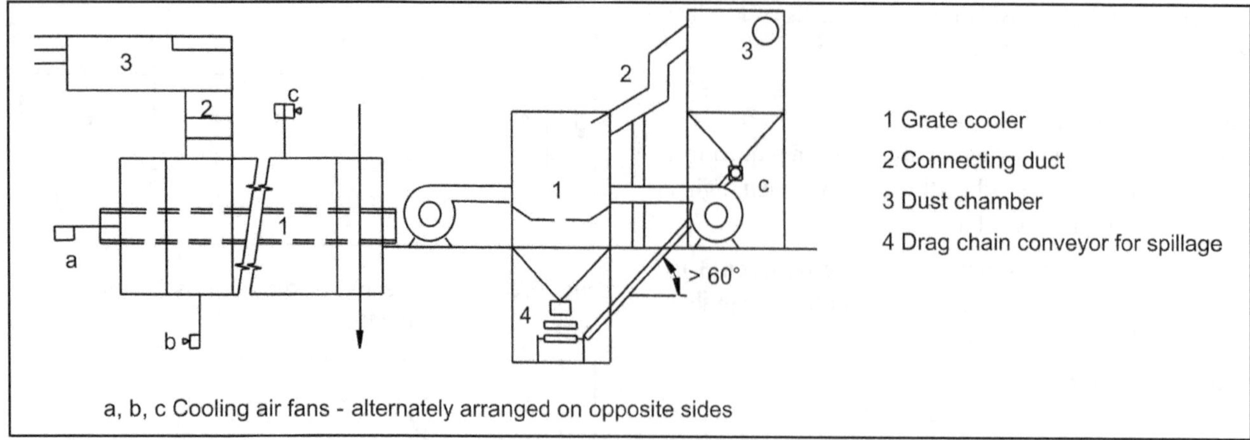

Fig. 6.23.5 Cooling air fans so arranged as to facilitate support of connecting duct and dust chamber.

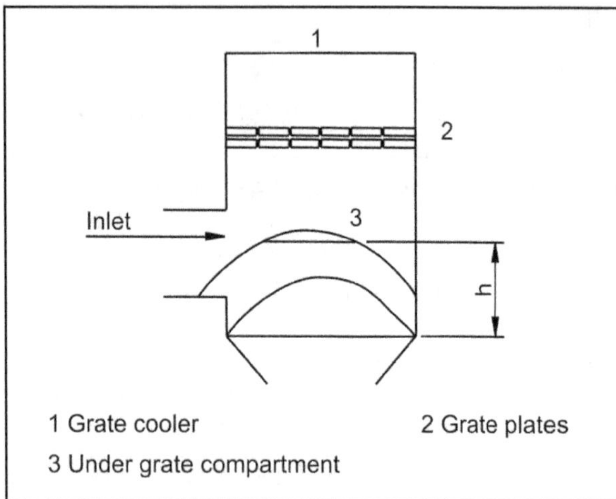

Fig. 6.23.6 Cooling air inlet should not get blocked by heap of clinker.

23.5 Facilites for Measurement of Cooling air

Another important factor in the layout is to provide for the facility to measure air flow and to adjust the damper / speed to maintain the air flow. This is particularly so for 1st, 2nd and 3rd compartments of the first grate. Any diminution of air flow in these compartments due to increase in bed thickness / variations in granulometry etc., would result in reduction in quantity of air for combustion and also in hot clinker at discharge end.

It is therefore common to install piezzometer rings at inlets of fans when it is desired to measure quantity of air delivered by them. It also requires a long straight duct at air inlet for measurements to be accurate and dependable.

The fans should therefore be so laid, at an angle as mentioned above, so that it is possible to have long straight lengths of inlet duct. If this is not possible, the inlet ducts are arranged vertically.

See Fig. 6.23.7.

It is now common to find that not only air flows are measured but they are recorded also.

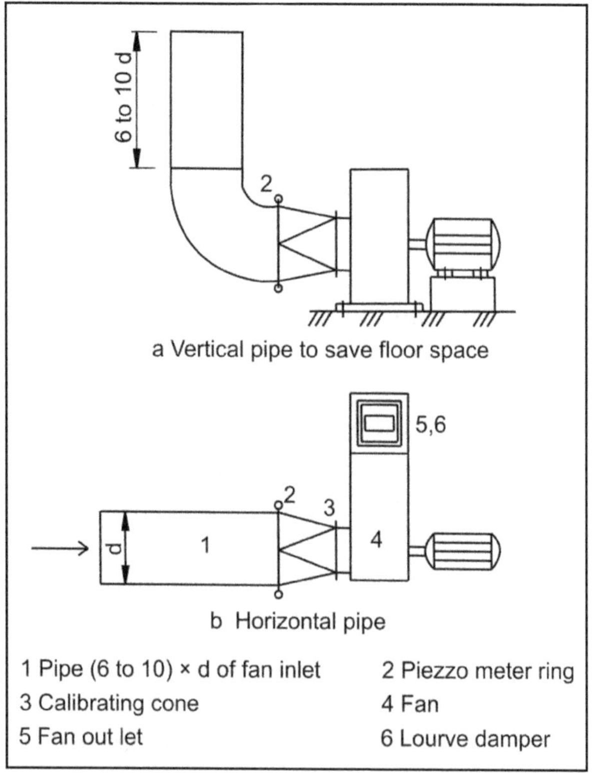

Fig. 6.23.7 Measurement of air flow with piezzo meter ring and calibrated cone.

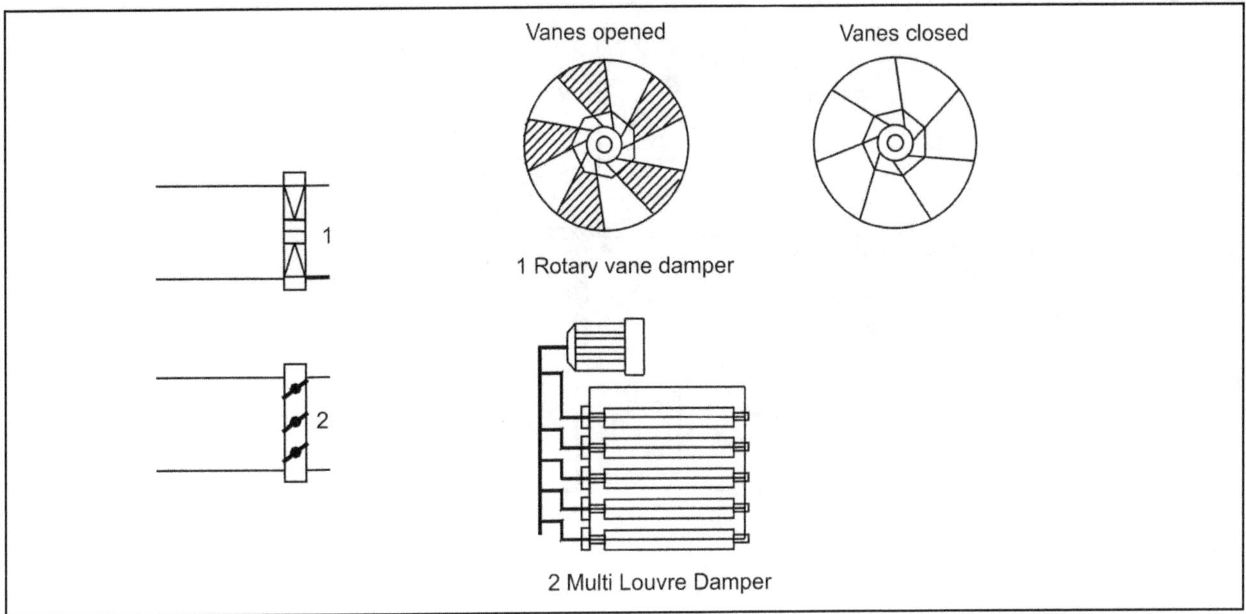

Fig. 6.23.8 Rotary vane and multi Louvre dampers for flow control.

23.6 Dampers for Regulation

Fans are fitted with multi louvre or multi vane radial dampers for regulating air flow. The dampers can be either on inlet or on outlet side of fan.

See Fig. 6.23.8.

The designs that cause least resistance to flow and regulate air flow evenly are preferred. All dampers are remote operated and controlled.

The dampers are connected to air flow measurements and are opened or closed to maintain desired quantity of cooling air.

The bed thicknesses and hence chamber pressures are maintained by regulating grate speeds.

23.7 Cooler Level and Fans

When the cooler is so laid out that air chambers are at ground level, fans would be installed on ground floor.

But if cooler is so arranged that the spillage drag chain is at ground level, fans would either be installed on foundations at a high level with maintenance platforms around them.

Alternatively fans could be installed at ground level and ducts taken to air chamber.

See Fig. 6.23.9.

23.8 Selection of Fans

Fans are selected according to the air loading and pressure for each compartment. Different manufacturers have their own norms for air loading.

23.8.1 Air Loading and Static Pressure of Fans

Air loading, expressed in m^3/ min/m^2 gross area of the compartment, is maximum in the first compartment where clinker temperature is also maximum and reduces progressively as clinker temperature drops along the length of the cooler. The air loadings of the first two compartments are such that they supply the air required for combustion in the kiln and in the calciner.

Static pressure required to be developed by fans is related to the thickness of bed of clinker and temperature of clinker. Pressures of fans decrease in conventional coolers from 800 mm to 250 mm for the last compartment.

Designers and Manufacturers of Coolers have their own norms in this respect and it would be best to go by their advice even if they did not supply the fans

23.8.2 Typical Air Loadings

Typical air loadings for **conventional** coolers using 2.5 to 2.7 nm^3/kg cooling air are shown in **Table 6.23.1** Present designs of coolers would have very different values.

23.8.3 Typical Static Pressures of Fans

Typical air chamber pressures for different air loadings and bed thickneesses would be as shown in **Table 6.23.2.**

These are for guidance only in the absence of authentic data from designers and manufacturers.

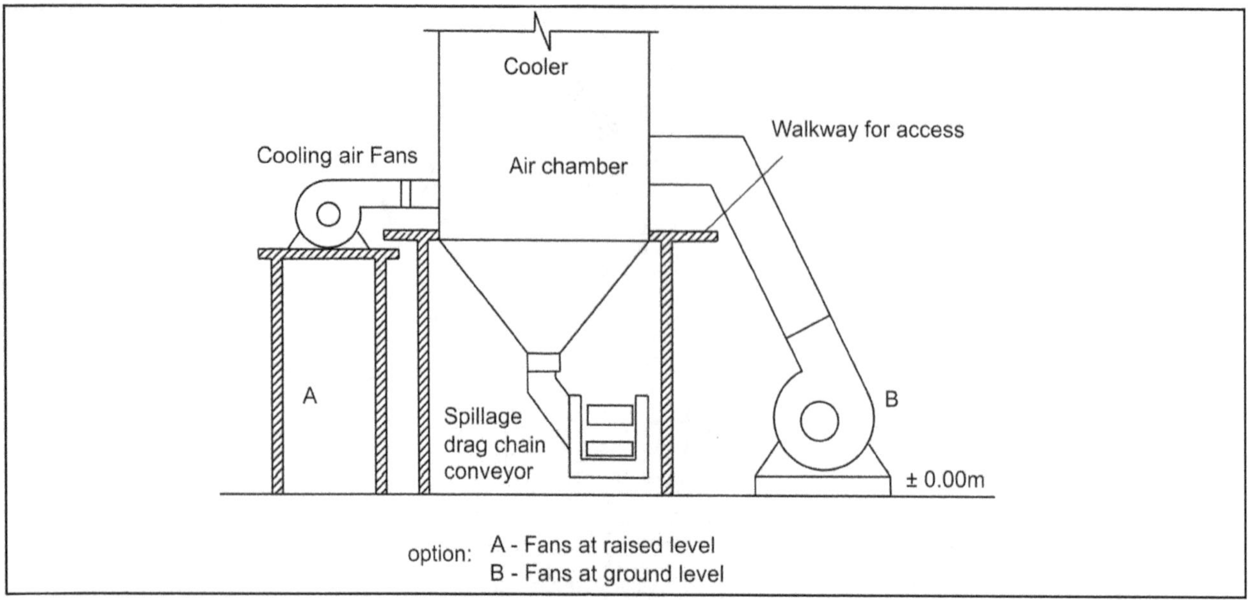

Fig. 6.23.9 Layout of cooling air fans clinker cooler at high level above ground.

Table 6.23.1

	Air loading in m³/min/ m²									
Compt.	1	2	3	4	5	6	7	8	9	10
1 grate	120	100	80	65	45					
2 grates	120	110	95	80	70	55	45			
3 grates	120	110	100	95	80	70	55	45	35	30

Table 6.23.2 Air chamber pressure in mmwg.

Bed thickness	Air loading in m³ /min / m²			
mm	80-120	60-80	50-60	< 50
200	200	200	150	150
300	300	250	200	200
400	400	300	250	200
500	500	400	300	250
600	600	500	400	350
700	700	600	500	450
800	800	700	600	550
900	900	800	700	650

(Reference : data on air loading and static pressures of fans CPAG Manual on clinker coolers for conventional coolers for guidance only).

23.9 Margins in Selection of Fans

In selecting fans margins of 10-15 % on volume and 20-30 % on pressure would be allowed for as in case of other fans.

Since these fans handle clean air at ambient temperature, backward curved bladed fans which have a high efficiency ranging between 75 to 80 % are ideally suited.

More than 60 % of Sp. power consumption for grate cooler is contributed by cooling and vent air fans. Therefore care should be taken in selection and in operation.

23.10 Values for New Designs of Coolers

Not much is known about the configuration of the various compartments that is their lengths, Air loading in m^3/min/m^2 as also design bed thicknesses in each compartment.

Therefore ratings of cooling air fan in terms volumetric capacity and pressure could not be spelled out on a general basis. They have to be obtained from the designers of the cooler.

23.11 Fans for New designs

Specifications of fans for coolers of new design would be somewhat different than for fans for conventional coolers described above.

Presently air is supplied individually to each grate plate and its supply is also controlled individually. Separate fans supply air to right and left sides of the grate.

These developments have to be taken into account in selection of fans and their layout in consultation with the Supplier and the Consultant.

23.12

A typical layout of cooling air fans is shown **in Fig. 6.23.10.**

23.13 Duotherm Circuit

Duotherm circuit is shown in Fig. 6.23.11.

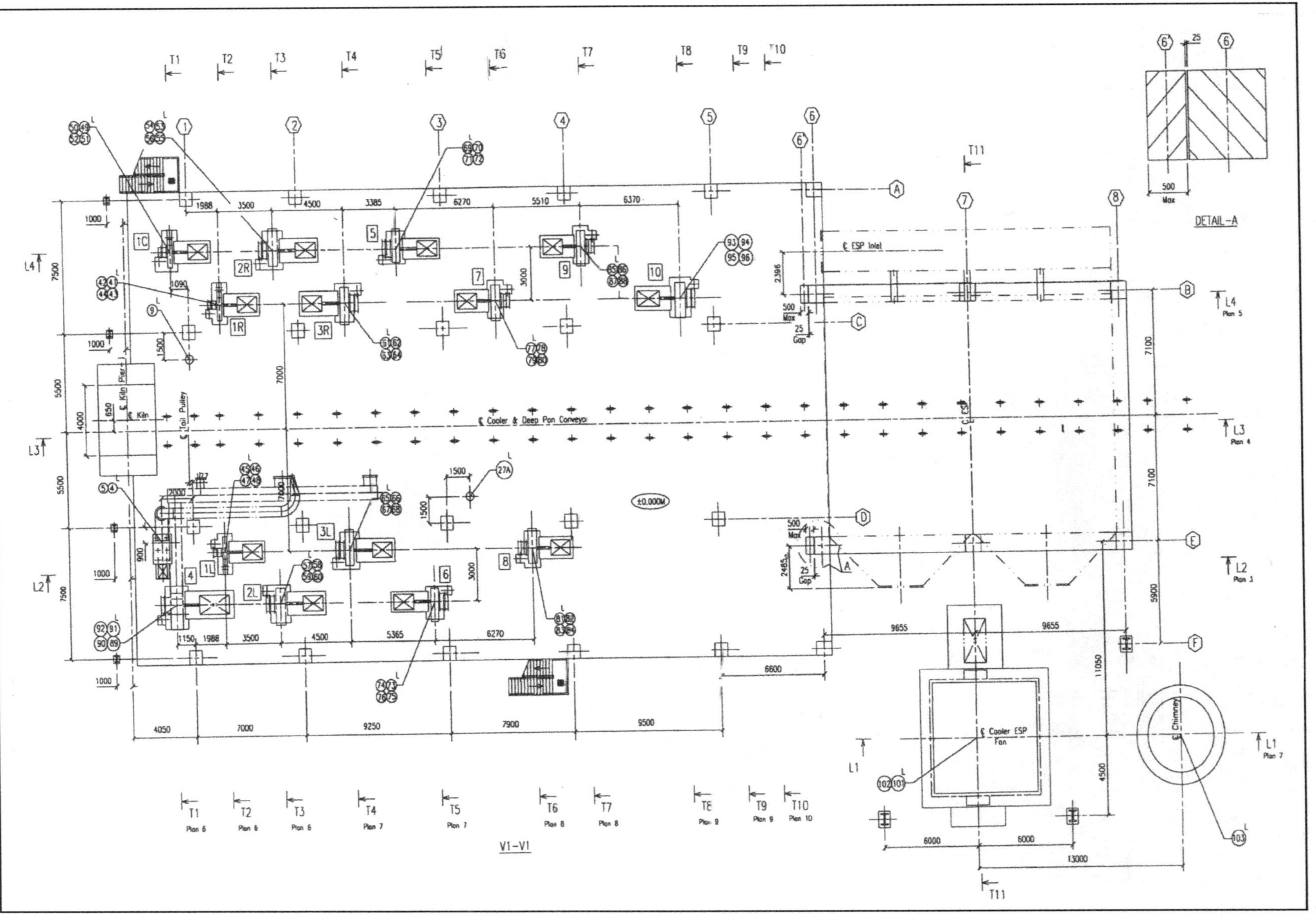

Fig. 6.23.10 Layout of cooling air fans along the length of a grate cooler.

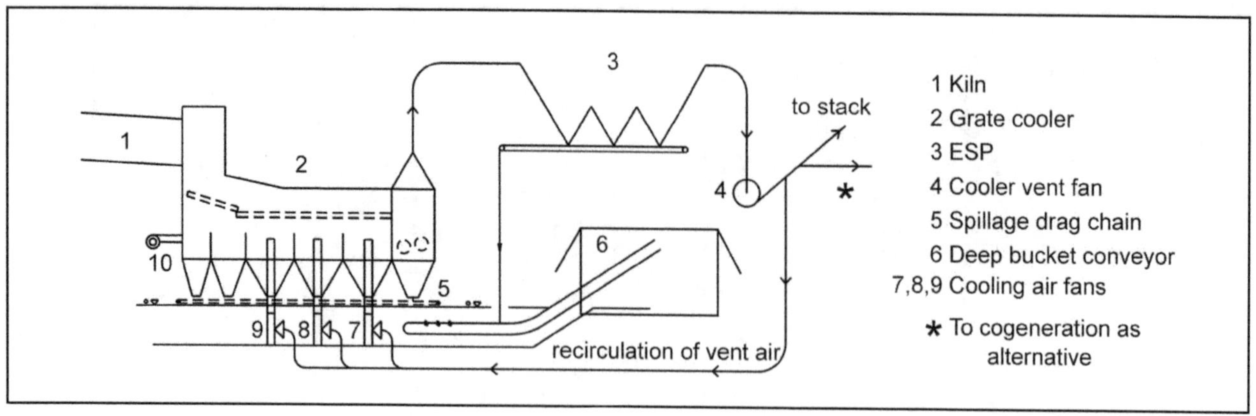

Fig. 6.23.11 Duotherm circuit to recirculate vent air to cooling air fans.

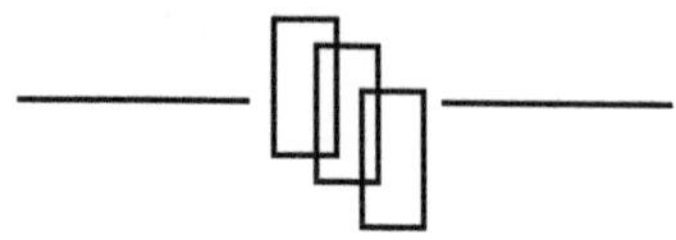

CHAPTER 24

CLINKER COOLERS
COOLER VENT AND DUST COLLECTORS

24.1 Venting of Excess Cooling Air

With improved fuel efficiencies, less and less air is required for combustion, between 0.85 to 0.90 nm^3/kg of clinker where as cooling air would be between 2.2 to 2.7 nm^3/kg in conventional coolers and would be between 2 to 2.2 nm^3/kg for coolers with static grates. The balance is therefore required to be vented. Quantity of vent air would be between 1.3 to 1.7 nm^3/kg clinker for conventional coolers and 1.1 to 1.3 nm^3/kg for coolers with static grates and other new designs.

Thus vent air is only slightly less than sp. gas volumes of preheater which range between 1.4 to 1.6 nm^3/kg of clinker requiring large fans and large dust collectors. Vent air temperatures are between 150 to 220 °C as against 150 °C in earlier times.

Earlier when belt conveyors were used to convey clinker, clinker was designed to be cooled to 65 °C above ambient. Now with all metal deep bucket conveyor as the principal conveyor, clinker is cooled to 120 °C.

24.1.1 Venting Under Upset Condition

Reference has already been made to the upset condition of the kiln. Allowance is required to be made for such conditions of kiln. Vent fan and dust collector are therefore required to be sized 1.3 times more i.e., between 1.7 to 2.2 nm^3/kg clinker. Temperature of vent air under upset condition of kiln could go up to as high as 270 to 300 °C.

24.2 Dust Collectors for Venting

Clinker dust being coarse, till recently multiclone dust collectors were used to clean the vent gases. Dust burden on raw side ranges between 10-15 gm/nm^3 and with 92-95% efficiency, dust content in cleaned gases would be between 500 to 1200 mg/nm^3. The emission of clinker dust from cooler vent stacks was not specifically under the pollution control norms till recently as it settled within a short distance of the factory.

But now even this emission is under the pollution control norms and the stack emission has to be $<$ 110 mg/nm^3 or as low as 55 mg/nm^3 after expansion.

24.2.1 Electrostatic Precipitator

Therefore multiclones are no longer satisfactory. The coolers are now fitted mostly with ESPS for cleaning vent gases. Clinker has poor resistivity and hence ESPS are designed for low migration velocities resulting in dust collectors larger than that for kiln exhaust. Therefore, considerable space is required to install the dust collector at the discharge end of the cooler and for returning the dust to clinker conveyor.

See Figs. 6.24.1 and 6.24.2.

24.3 Layout of Dust Collector

If the clinker storage is in line with the kiln then ESP can be installed by the side.

See Fig. 6.24.3.

See Fig. 6.24.4 for two identical kilns, coolers and dust collectors.

If however coal mill is at preheater end, esp could be installed on the burners' platform itself.

See Fig. 6.24.5.

Also **See Fig. 6.24.11** of actual installation.

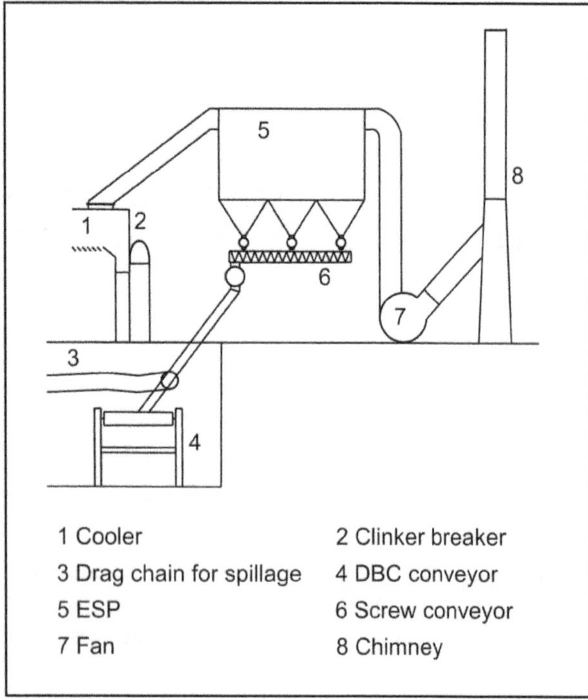

Fig. 6.24.1 Dust collector to vent cooler exhaust gases.

24.3.1 Chimneyys

It is possible to have a common chimney for exhausts of two coolers of two production lines and the chimney can be designed on that basis.

See Fig. 6.24.6.

Cooler chimneys will be self supporting. If the formula recommended by the Pollution Control Board is used to work out the height of the chimneys, then chimneys of 25 to 30 m, height should suffice. But it is better to install taller chimneys so that even if dust emission exceeds 115 mg/nm^3 the quality of ambient air surrounding the plant would not be affected.

The layout of whole kiln section beginning with continuous blending and ending with clinker storage should be worked out not only for the 1st kiln but also for subsequent expansion; firstly, with a view to provide adequate space for expansion and secondly to economise on storages and material handling facilities and utilities etc. **See Figs. 6.24.5 and 6.24.6.**

24.4 Heat Exchanger and Bag Filter

Bag filters can also be used for venting cooler if it is ensured that the temperature is brought down and maintained at < 100-120 °C. This is done by passing vent gases through an air to air heat exchanger. Because of abrasive nature of clinker dust, glass bags are not suitable even if they can withstand temperatures of 240 °C. **See Fig. 6.24.7.**

When kiln is upset, temperature of vent gases can go up by as much as 100 °C. Heat exchanger should therefore be designed for upset conditions also. It would mean that it should be possible to supply cooling air in heat exchanger in larger quantities and higher temperd in during upset conditions. Thus the system has pressure drops to be overcome in heat exchanger and also in bag filter. Two together would be higher than the pressure drop in an esp.

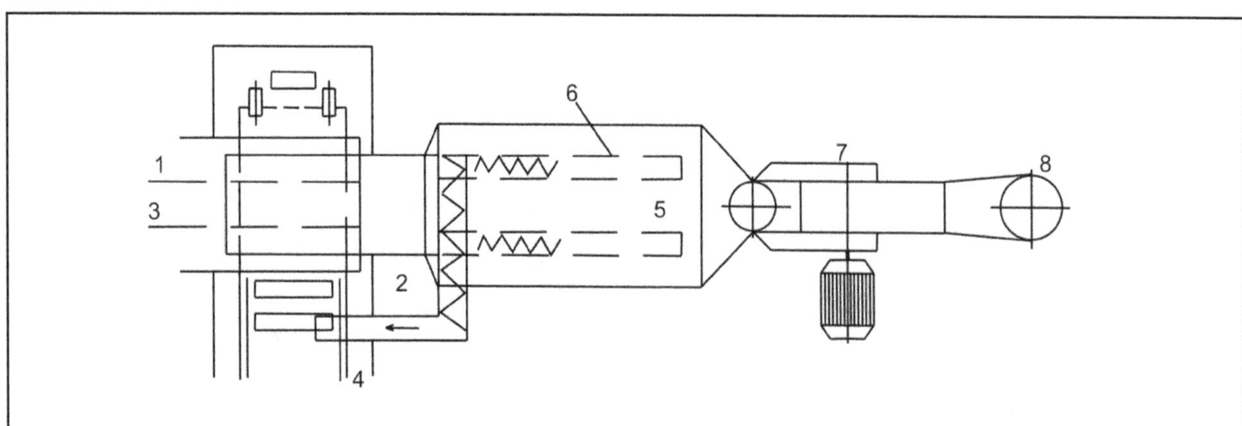

Fig. 6.24.2 Plan of dust collector layout.

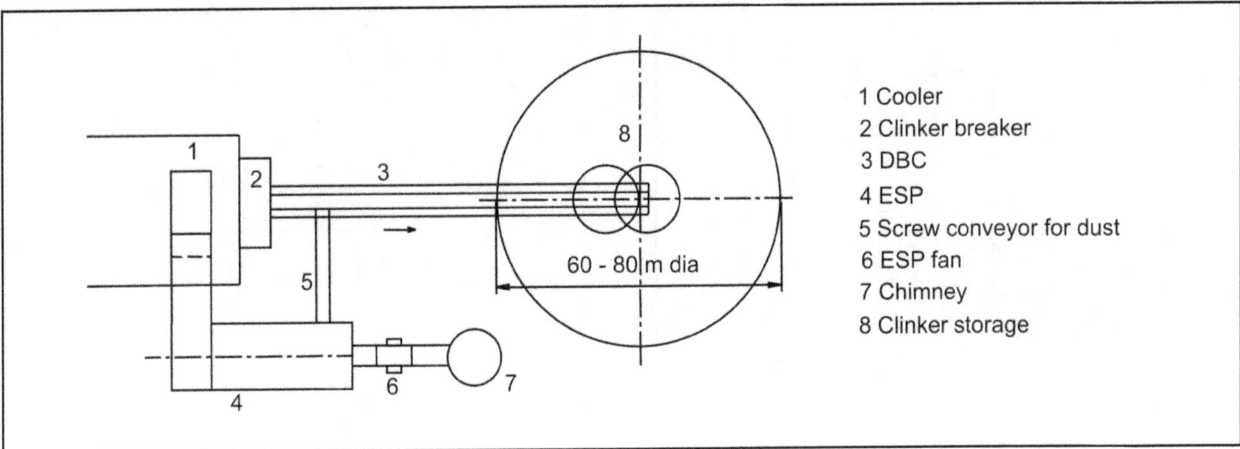

Fig. 6.24.3 When clinker storage is in line with kiln.

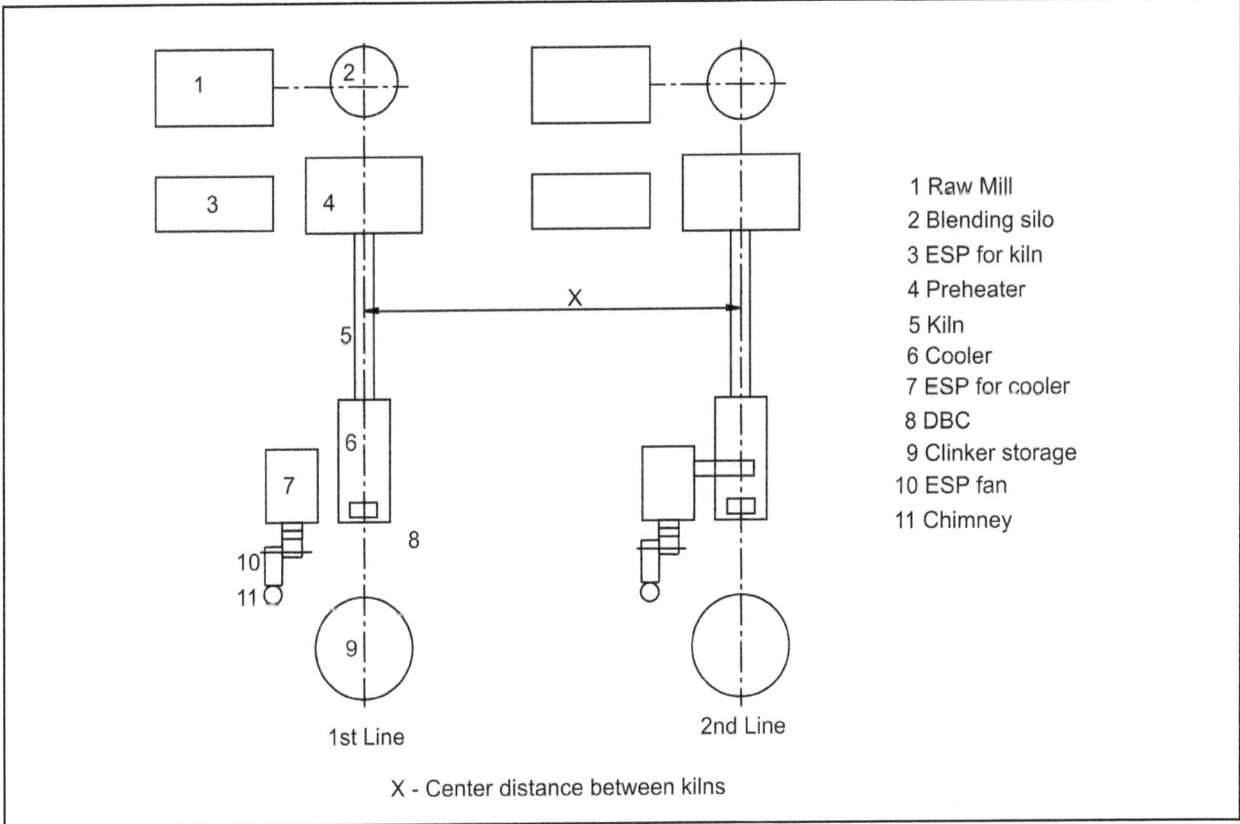

Fig. 6.24.4 Layout with two kilns – identically arranged.

The vent air fan has to be designed for higher volume and pressure.

See Fig. 6.24.8 of an actual installation where heat exchanger is on the burner's platform and bag filter at ground level.

24.5 Gravel Bed Filters

Before ESPs were found to be feasible for cooler vent gases, Gravel bed fitters were developed for cleaning them. Gravel bed filters consisted of a bank of cyclones with gravel beds in two levels. Usually the banks are

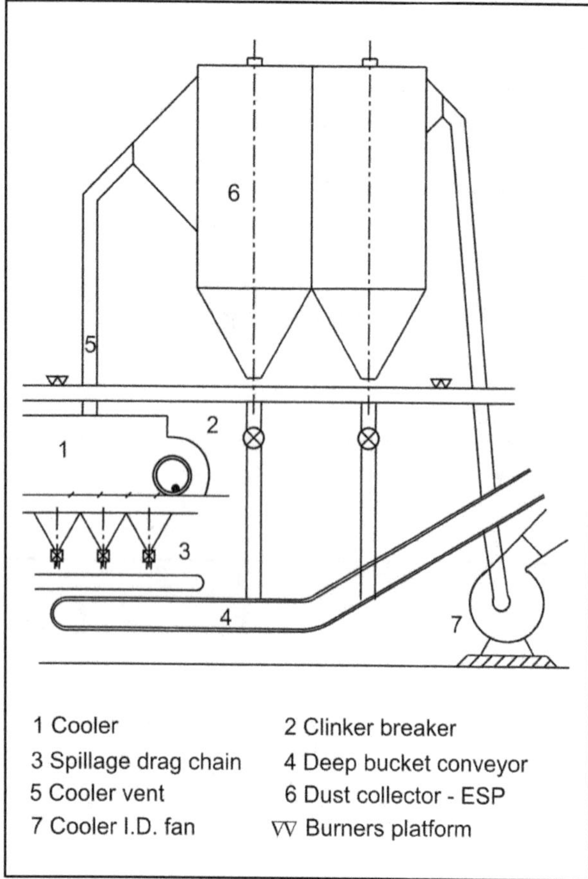

Fig. 6.24.5 Dust collector for cooler vent on burners platform.

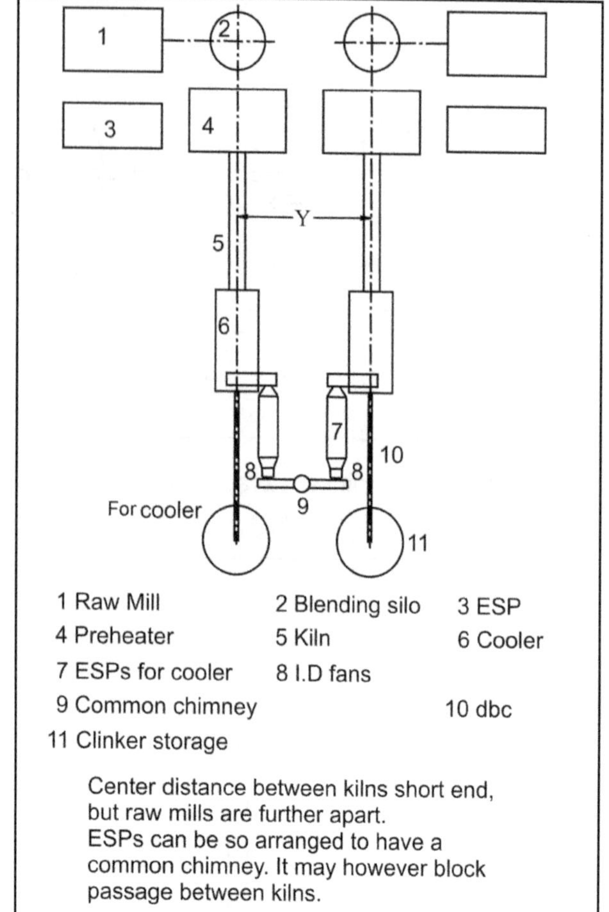

Fig. 6.24.6 Two kilns lay out with mirror image.

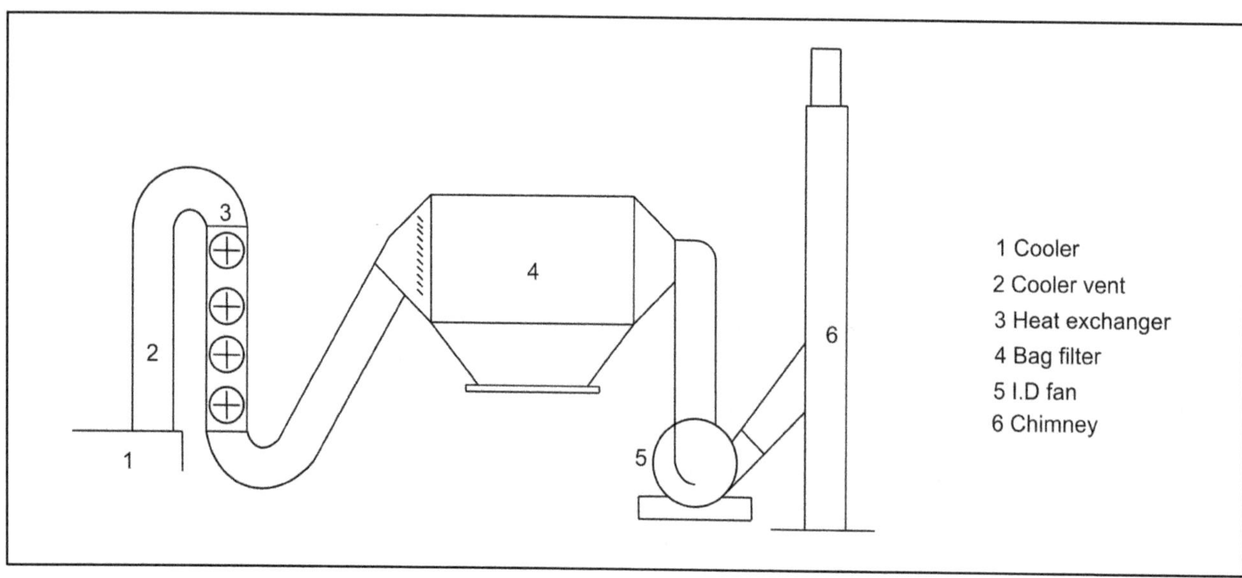

Fig. 6.24.7 Heat exchanger and bag filter for venting cooler gases.

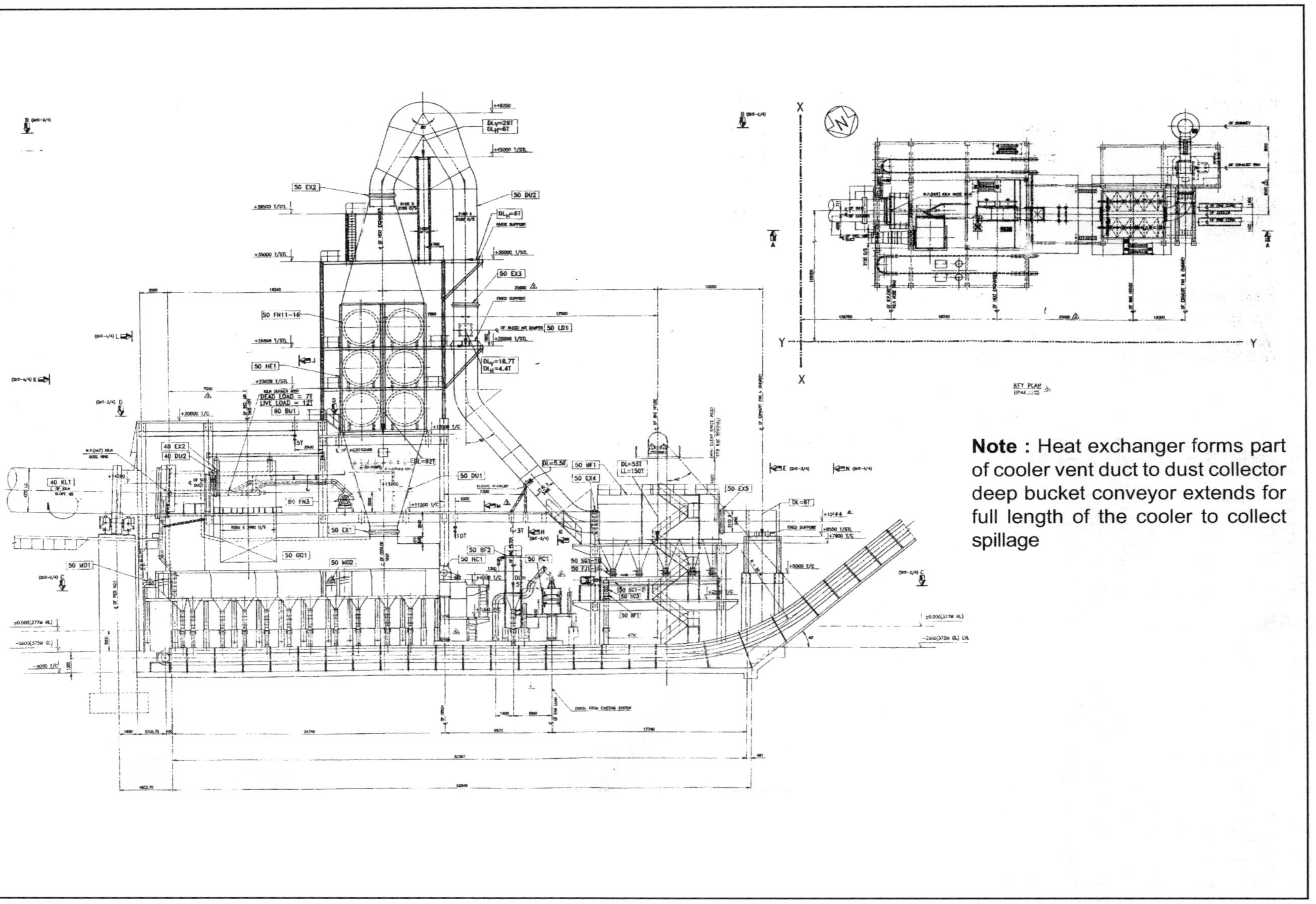

Fig. 6.24.8 Layout of grate cooler, heat exchanger and bag filter to vent cooler exhaust gases.

arranged symmetrically around inlet and outlet gas ducts. Vent gases are precleaned in cyclones before entering the cyclones of gravel bed filter.

Raw gas enters cyclones which has gravel beds in two levels. In passing through gravel beds gases get cleaned and are drawn by a fan and exhausted. Clean scavenging air flows through the system in opposite direction cyclically very much like reverse jets in bag filters. To assist cleaning gravel bed mechanical rakes are used.

But the system pressure drop is high hence this design did not become popular.

Fig. 6.24.9 shows the schematic of the gravel bed filter system.

Plate 2.12.6 in **Chapter 12** of **Section 2** shows system of gravel bed filter.for clinker cooler.

24.6 Indirect Cooling G Cooler

In another design of cooler called 'G-coolers', cooling air did not come into contact with clinker and hence cooler vent was always clean and did not need any dust collectors at all. However, these coolers also did not become popular.

See Fig. 6.25.2 in next **Chapter.**

24.7

Figs. 6.24.9 and 6.24.10 show actual installation of esp for cooler. In **Fig. 6.24.10**, esp is installed on burners platform as mentioned earlier.

24.8 Duo Therm Circuit

In this circuit part of vent gases leaving the dust collector are led back to cooling air fans of the first grate of the cooler to raise temperature of secondary and tertiary airs taken to kiln and calciner respectively. It reduces volume of air to be vented to atmosphere. **See Fig. 6.24.11.**

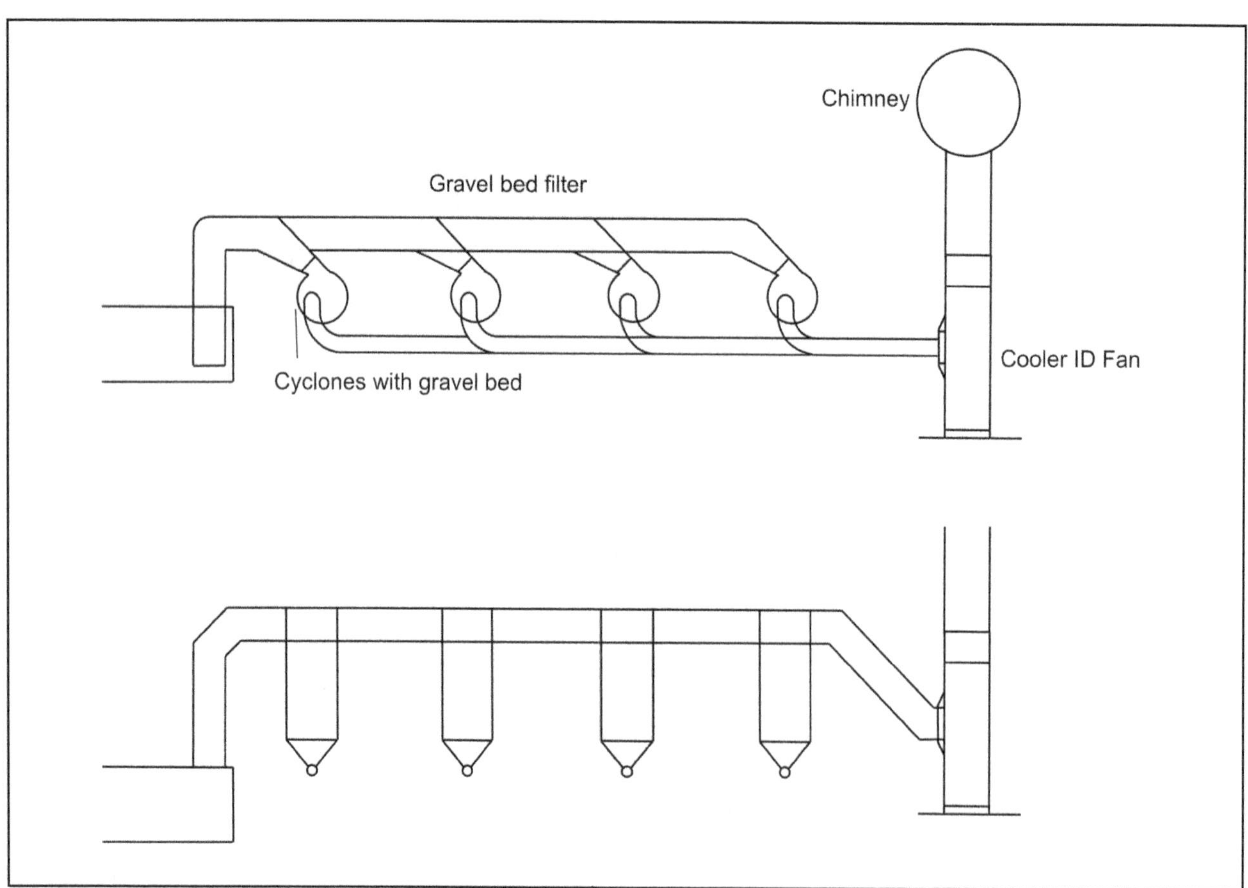

Fig. 6.24.9 Gravel bed filter.

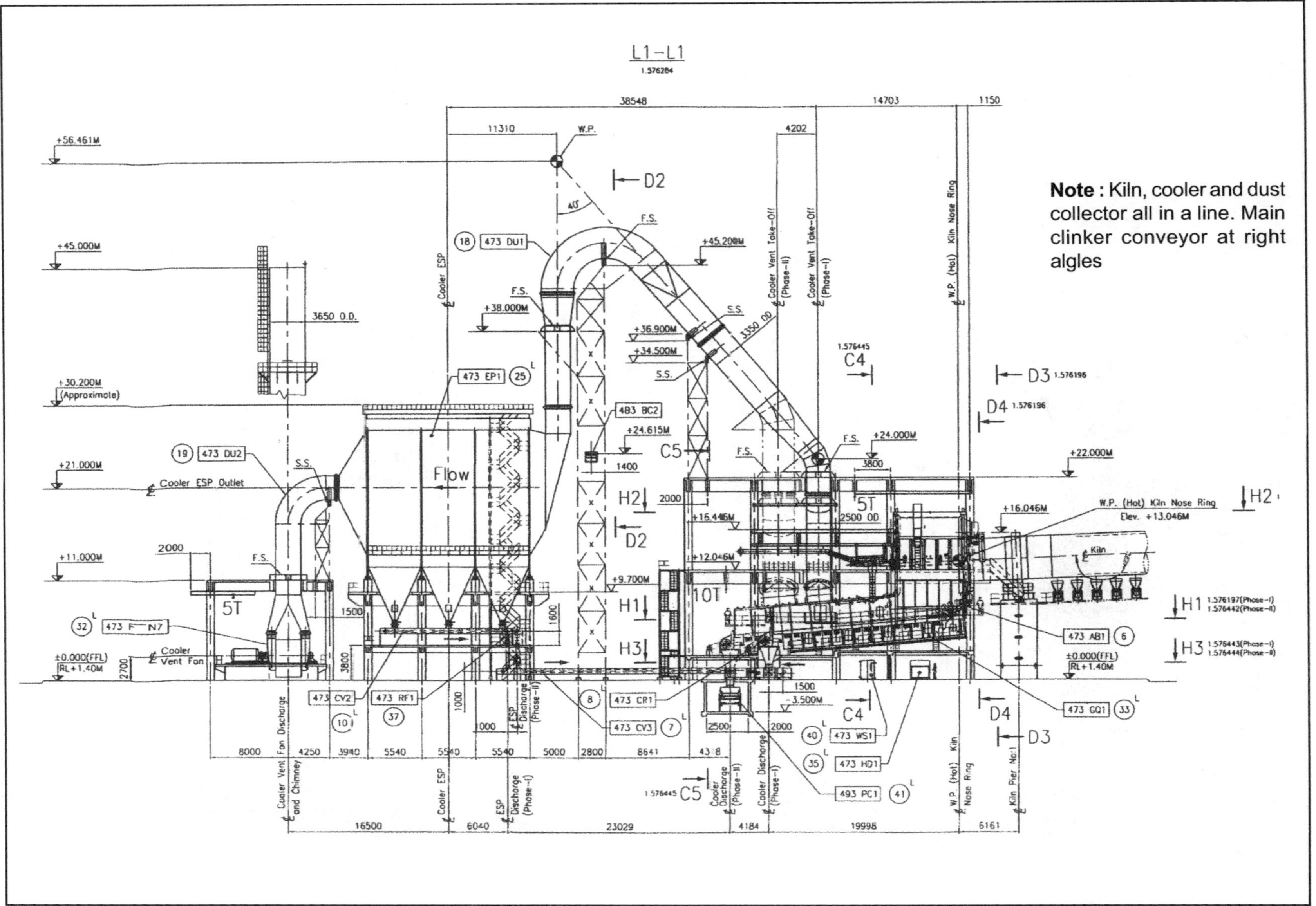

Fig. 6.24.10 Layout of grate cooler and its dust collector.

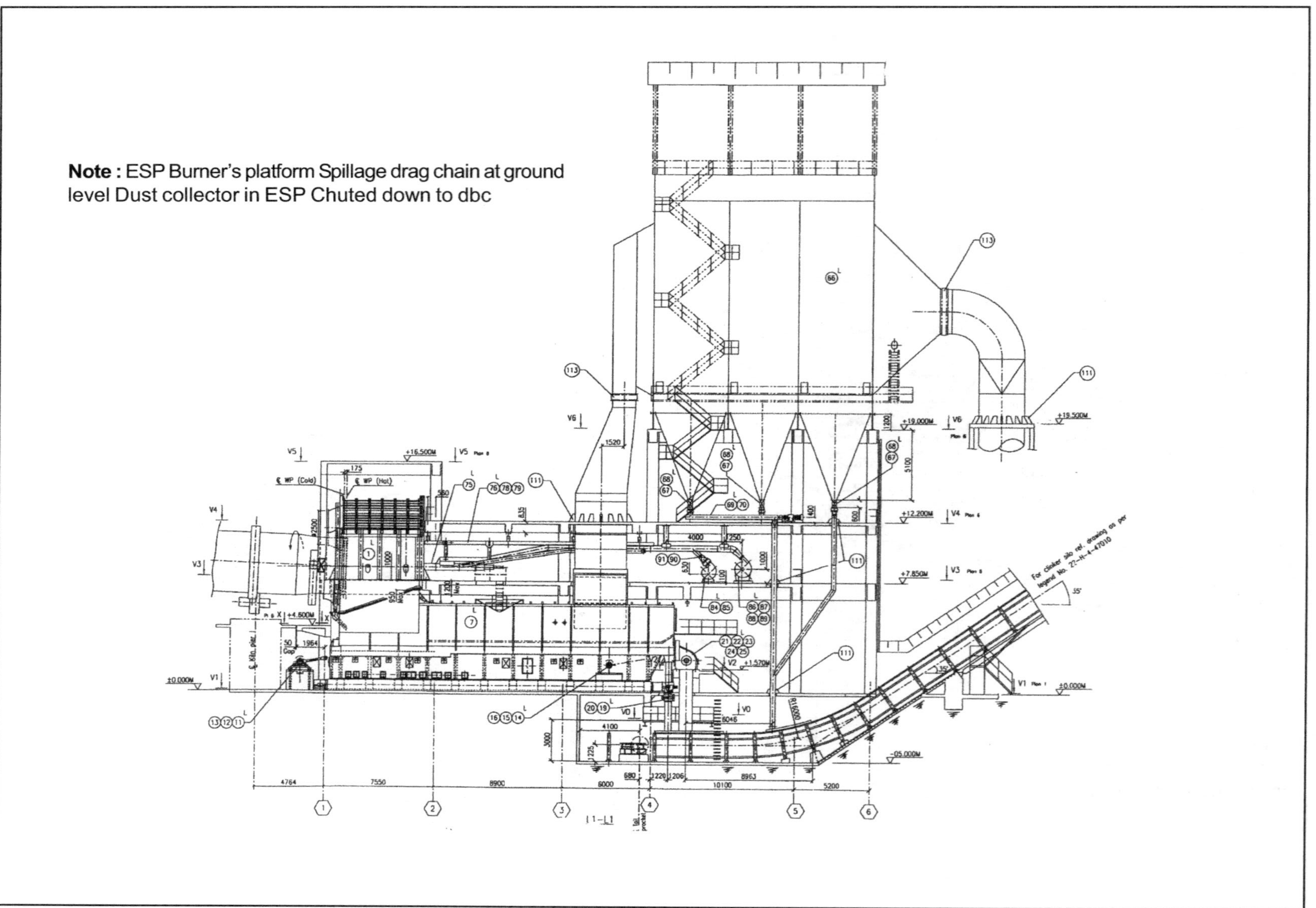

Fig. 6.24.11 Electrostatic precipitator (ESP) for venting cooler exhaust gases.

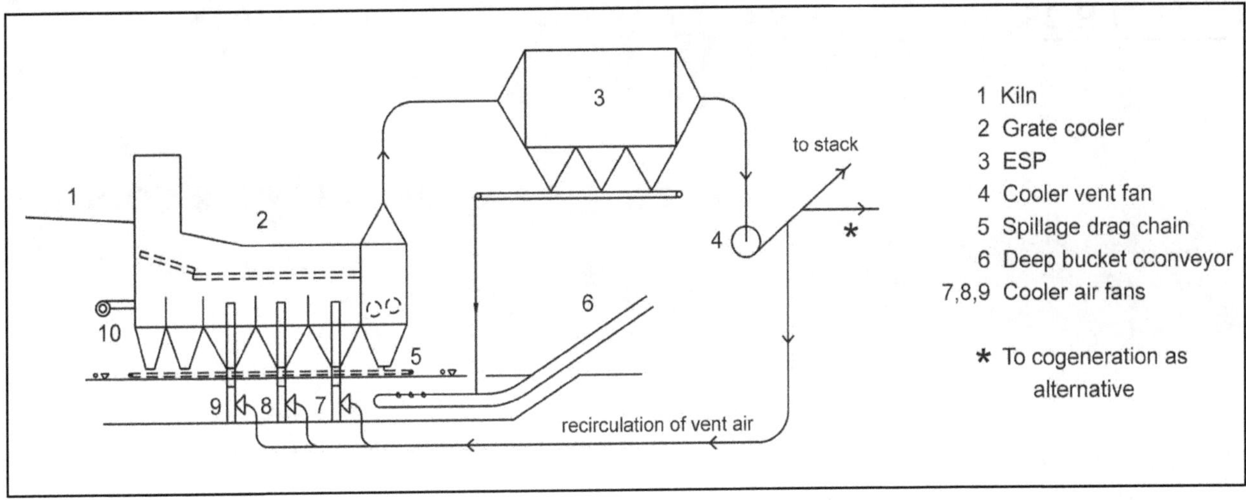

Fig. 6.24.12 Duotherm circuit to recirculate vent air to cooling air fans.

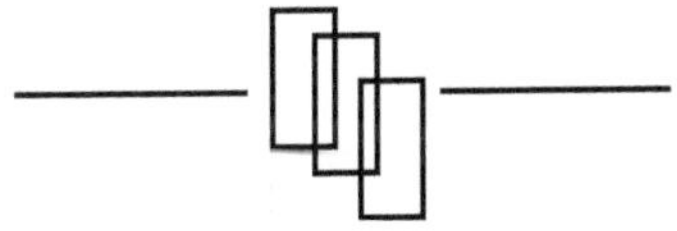

CHAPTER 25

OTHER CLINKER COOLERS

25.1 Shaft Cooler

It is like a shaft kiln and is installed directly under the kiln. Hot clinker falling from kiln moves slowly downwards and is cooled by cooling air admitted from the bottom. Rate of discharge is controlled to regulate temperature of clinker. Like rotary and planetary coolers cooling air is also combustion air. However longer retention in the shaft permits better heat exchange.

Shaft coolers suffer from same limitations as do planetary cooler in that their capacity cannot be increased. It should be possible to take air for combustion from the hood to a calciner through a t.a. duct.

Serious disadvantage is the height required under the kiln. It would push up heights of kiln piers and preheater tower very much.

See Fig. 6.25.1.

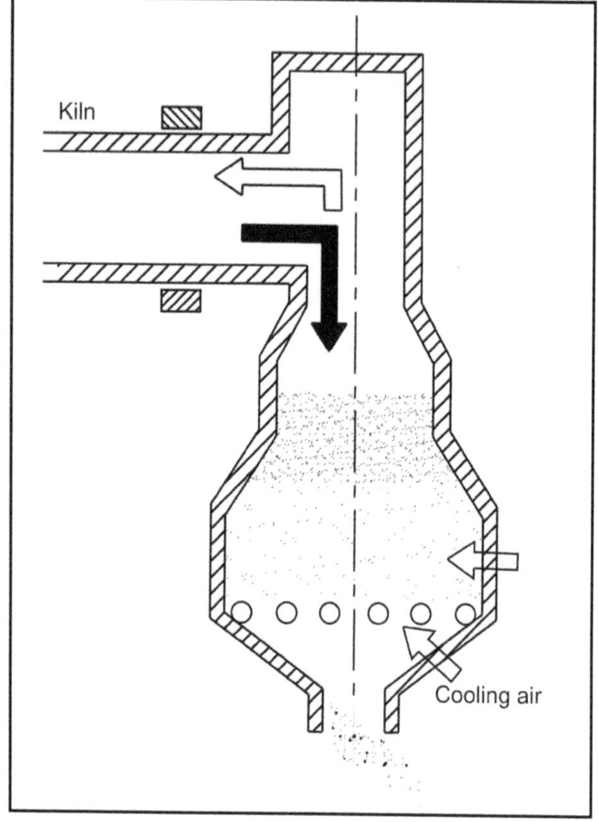

Fig. 6.25.1 Direct - shaft cooler - under kiln.

25.2 Recupol Cooler

It has been mentioned that Lepol Grate was used as a traveling grate cooler. Such Recupol coolers were installed at a few places along with Lepol grate and even with wet process kilns.

Such coolers also require considerable height to be installed under a kiln.

The traveling chain with grates must be made of heat resisting materials as it comes in contact with hot clinker falling from kiln. In reciprocating grate, only grate plates of first few compartments come in contact with hot clinker. Sealing difficulties and high maintenance were other adverse factors.

See Fig. 6.14.1.

25.3 G Cooler

In seventies a direct-indirect cooler called G-Cooler was developed in Europe. A short reciprocating grate cooler sized just adequate to supply air for combustion in kiln (and calciner) was used as the direct cooler. At this point, temperature of clinker was reduced to about 450-500 °C. Clinker was passed through a lump breaker and taken out of the cooler and to the top of

G-cooler. It is spread over the G-cooler by a pivoted pan conveyer. G-cooler is fitted with elliptical passages for cooling air. Air did not come in contact with clinker and could be let out without requiring any dust collector. Flow of clinker in the box was by gravity and was controlled by adjustable gates at the bottom.

Clinker could be cooled to any desired temperature. **See Fig. 6.25.2. See also Plates 2.11.7 and 2.11.8 in Chapter 11 of Section 2 of the book.**

Main difficulty with this cooler was that it needed to handle very hot clinker and required reliable conveyors moving almost vertically and also horizontally. Horizontal pivoting pan conveyor was required to distribute clinker evenly along the length of the G-Cooler.

However such an indirect cooler can be conveniently used along with original reciprocating grate cooler when capacity of kiln was increased by installing a calciner. Temperature of clinker that would be handled would be about 300 °C. The indirect cooler can be installed in line with the main cooler. It can be located at a distance from it according to availability of space. Else it could also be installed at right angles at a convenient place.

See Fig. 6.25.3.

Deep bucket conveyor could be used to raise clinker to the top of the G-cooler. A chain conveyor with hinged plates on its bottom or a pivoting pan conveyor would spread clinker evenly in the box.

Cooling air can be as much as required. Power required for fans would be much less than that required in case of grate cooler. Further there is no capital expenditure for dust collector nor any operational costs.

It is worthwhile to explore the possibility of incorporating a G cooler in plants planning to expand.

There are many variations of Grate Cooler itself which have been mentioned in **Section 2.**

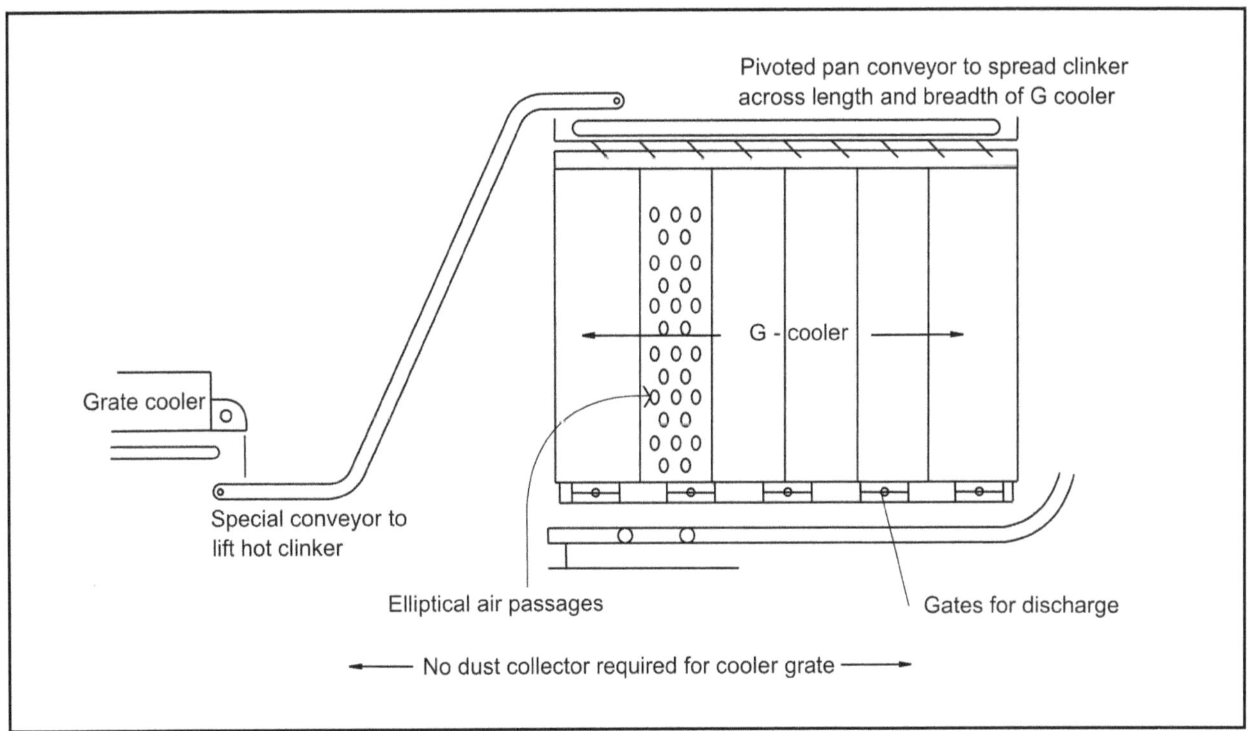

Fig. 6.25.2 G-Cooler.

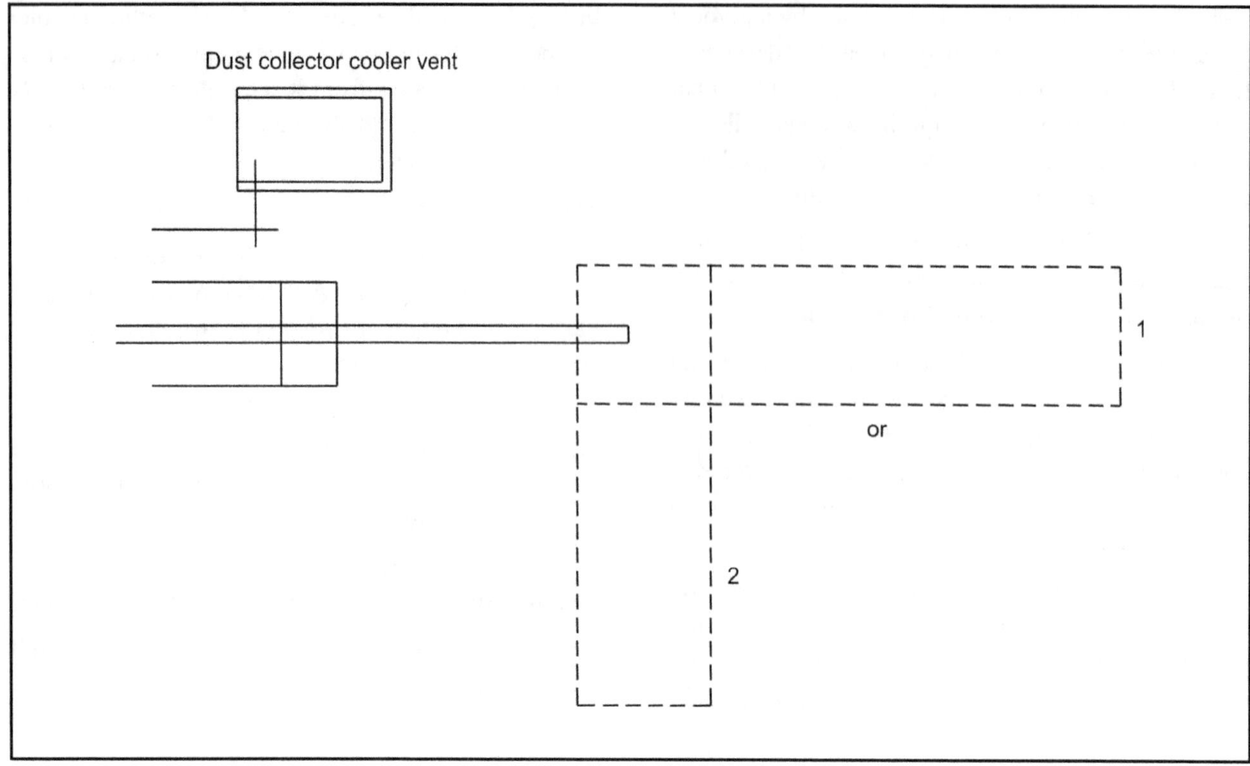

Fig. 6.25.3 Incorporating G cooler in expanding capacity of kiln.

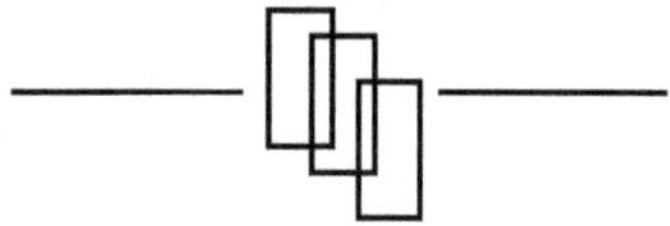

CHAPTER 26

CLINKER CONVEYING AND STORAGE

26.1 Conveying and Storing Clinker

Clinker is conveyed by :

1. Drag chain conveyor,
2. belt Conveyor,
3. deep bucket conveyor,
4. bucket Elevator, or
5. A combination of above according to layout.

26.2 Storage in Crane Gantry

As mentioned earlier, till sixties, the cement plant layout was around a central crane gantry, which served as a common storage for all materials. It received processed raw materials like crushed lime stone and clinker / unprocessed raw materials like clay, laterite, gypsum and coal. It fed the stored materials for further processing to :

1. Raw mills.
2. Coal mills.
3. Cement mills.

See Figs. 6.26.1 and 6.26.2.

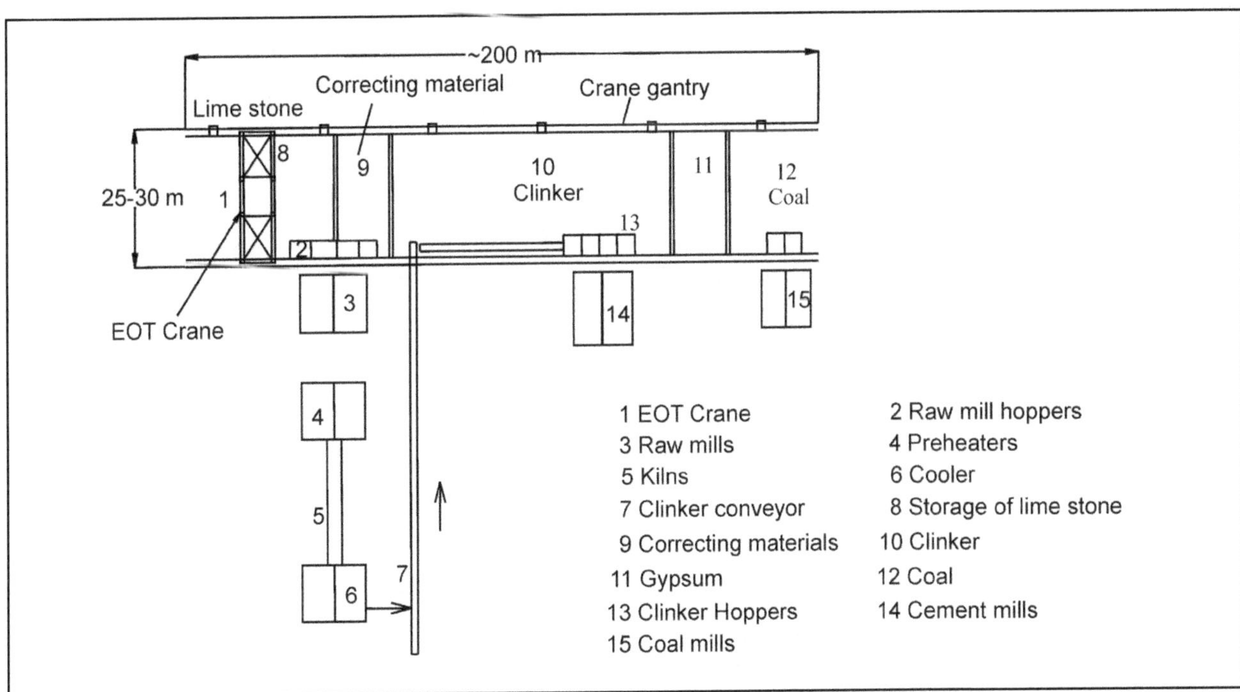

Fig. 6.26.1 Clinker storage of the past in a crane gantry.

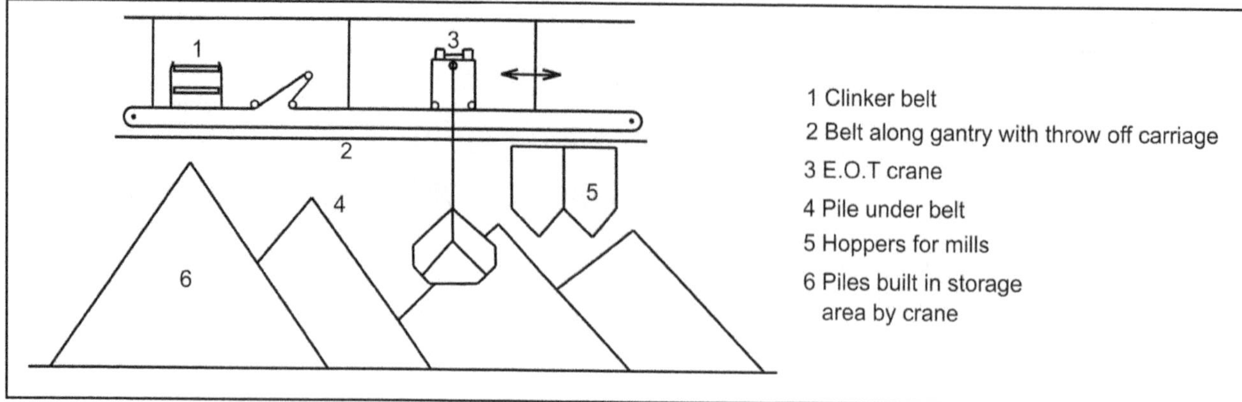

Fig. 6.26.2 Shifting of clinker piles for spreading clinker from storage and feeding it to hoppers.

With increase in capacities of the plant, the stocks to be held would have required very long crane gantries, with 2 / 3 cranes running in it.

26.2.1 Handling in a Crane Gantry

Every material stored in it had to be handled twice :

(i) while receiving, the heaps had to be cleared and shifted to spread material evenly in the allotted bays of the crane gantry,

(ii) while feeding, to hoppers of mills according to their requirements.

Thus cranes had to work round the clock. The cranes also required considerable maintenance of :

(a) alignment of rails along the length of the gantry,

(b) ropes, wheels, gears, etc.

26.2.2 Limitations of Crane Gantry

A stopped crane could cause stoppage of operation of all sections though not immediately. Even if a 2nd crane was available, it had to handle materials for all the plant and for all sections and hence output suffered. Thus when plant sizes reached to about 2000 tpd capacity, crane gantries were not found convenient either as means of storage or for material handling and feeding.

26.3 Scattered Storages

In the dry process plants evolved in seventies and later, storages were scattered as the layout did not have to center around the central storage but could be arranged at a place convenient both for receiving from previous process and for feeding to the next process.

See Fig. 6.1.8 in Chapter 1 in this Section

26.4 Deep Bucket Conveyor

Amongst the various types of conveyors mentioned above, presently the most commonly used conveyor to convey clinker from cooler to clinker storage is the deep bucket conveyor (dbc).

Its advantages are that :

1. it can handle hot clinker as it is an all metal conveyor;
2. clinker does not come into contact with chain moving the buckets. Conveyor is slow moving. Life of conveyor is long and its dependability to work 24 hours a day, day in and day out is very good;
3. it can be installed at steep angles of 35 to 40° to horizontal as compared with drag chains. Therefore length of conveyor required to reach a given height is much shorter.

A sketch of a typical dbc is shown. It is generally installed horizontally under the cooler and then at an angle till it reaches the discharge point.

See Fig. 6.26.3.

26.4.1 Standby Conveyor

Because of its arduous duty, it is very desirable that the dbc should have a standby so that kiln production does not come to a standstill.

See Fig. 6.26.4.

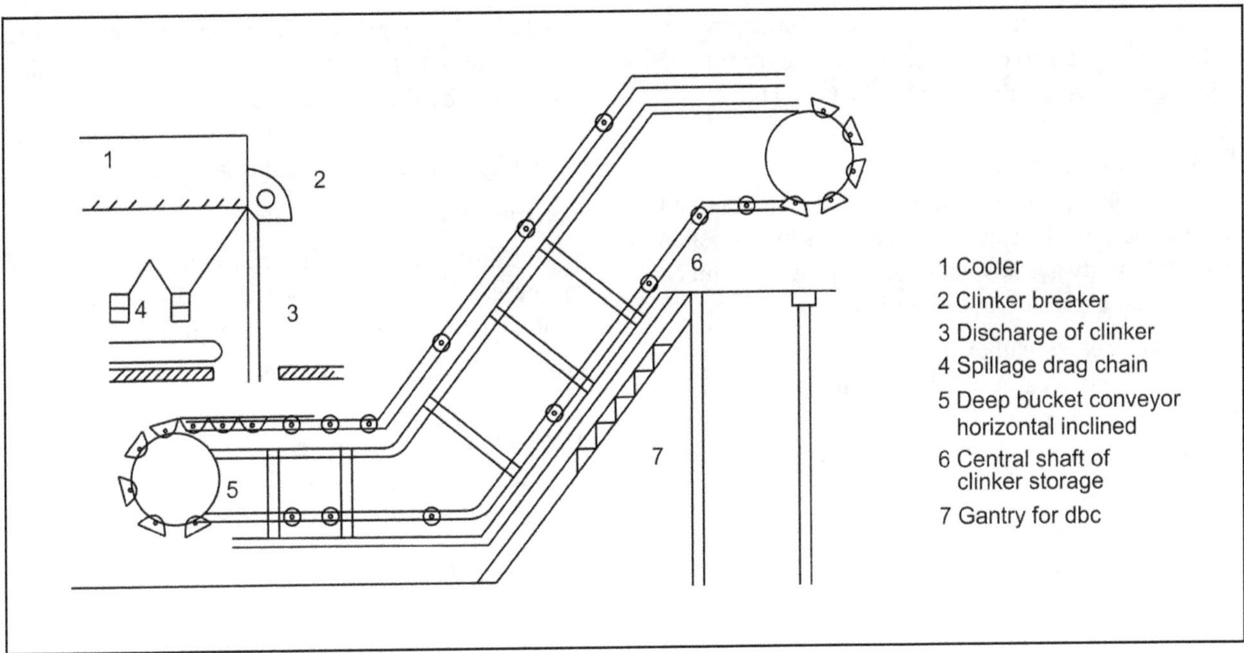

Fig. 6.26.3 Typical dbc conveying clinker from grate cooler to clinker storage.

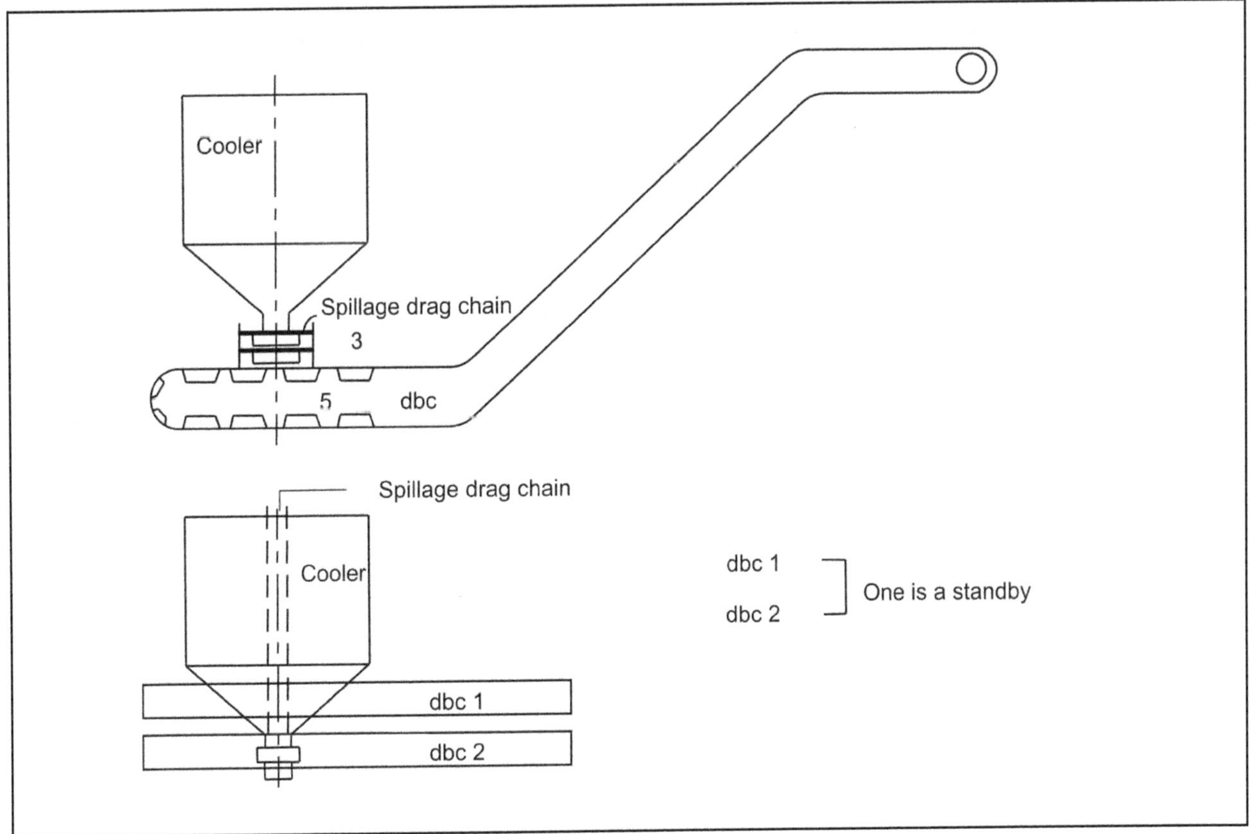

Fig. 6.26.4 dbc conveying clinker to clinker storage at right angles to kiln.

Impact of installing a standby dbc on depth and width of pit and also on dimensions of gantry for dbc has been shown in **Figs. 6.26.9 to 6.26.11.**

26.5 Clinker Stocks

Conventionally, clinker stocks of 14 days' production are maintained. But clinker being a semi finished product, it is always advantageous to provide as much storage place for clinker as possible.

Clinker gets 'weathered' in storage. It is good to supply cooled weathered clinker to cement mills. Therefore, clinker would not be fed to cement mills directly (except only when stocks are nil).

Many a time auxiliary storage or emergency storage of clinker is to be planned for, and also for storage of under burnt clinker. This is a practical necessity as under burnt clinker can be mixed in small proportions of 5 to10% with good clinker to obtain cement that would satisfy all the norms for good cement.

26.6 Storing Facilities for Clinker

Presently, clinker is stored mostly in conical sheds of large diameters of 80-100 meters or more.

See Figs. 6.26.5 to 6.26.7.

Clinker is brought in the deep bucket conveyor to a central shaft with slots at various levels. Clinker discharged by the dbc fills the shaft and spills over into storage area. Heaps around the shaft gradually reach higher and higher heights as lower slots close.

See Fig. 6.26.5.

The dust nuisance is thus confined to the shaft and can be easily contained by passing the escaping air through a dust collector located on top of the shaft.

See Figs. 6.26.5 and 6.26.8.

26.7 Conical Storage Shed

Working out capacity and dimension of this type of storage have been shown in **Table 6.26.1 and in nomogram Fig. 6.26.30**

The entire conical shed is covered to protect clinker from rain.

A retaining wall of 2-3 meters all around periphery considerably increases clinker storage and also prevents over flowing of clinker on all sides and keeps the area clean.

See Figs. 6.26.5 and 6.26.6.

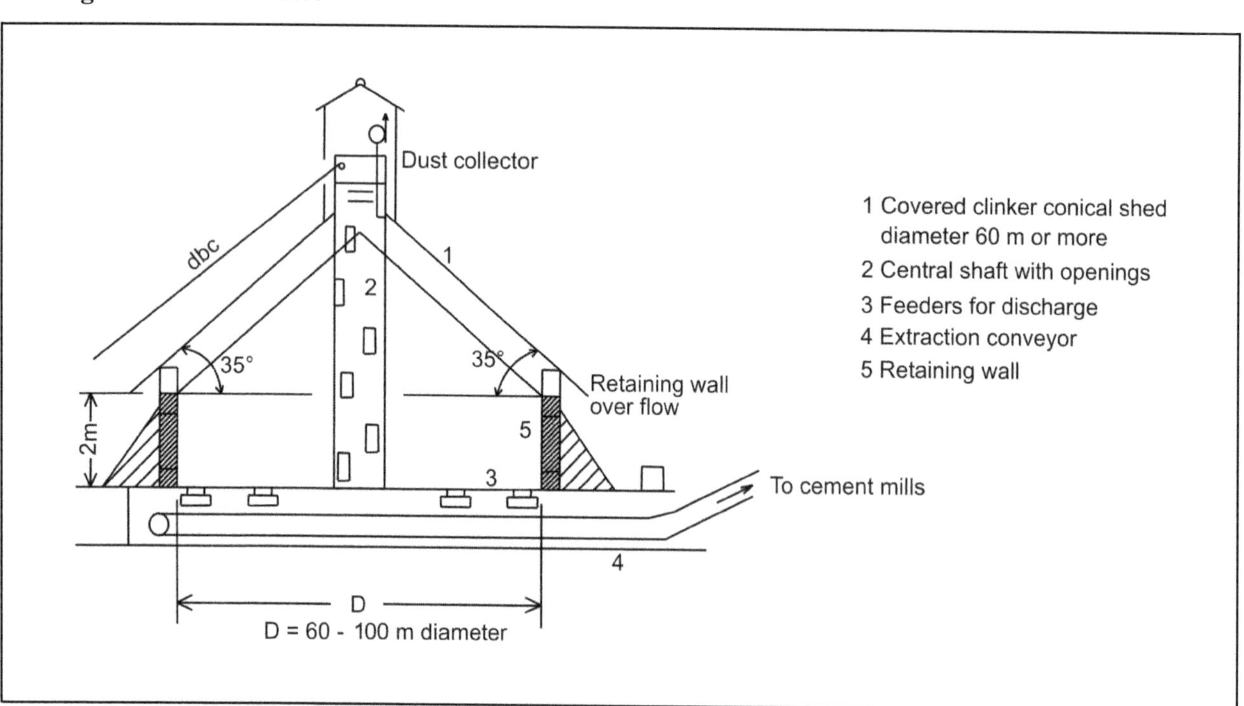

Fig. 6.26.5 Most common clinker storage system.

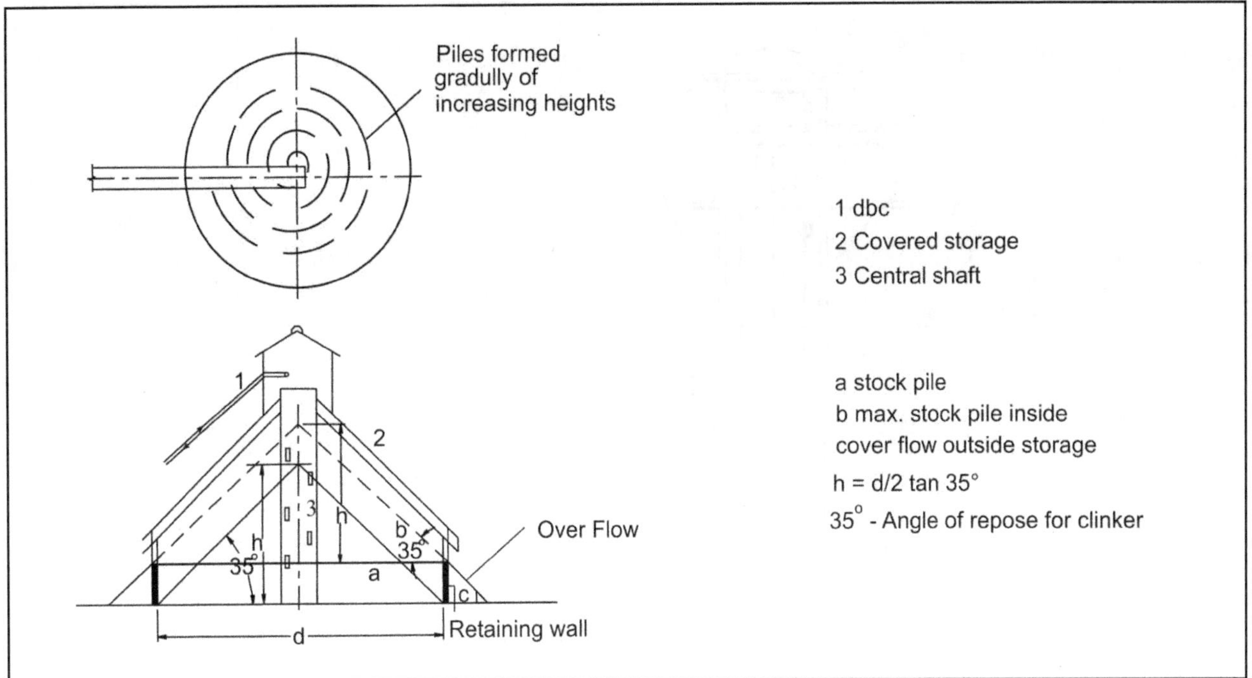

Fig. 6.26.6 Volume and capacity of storage of clinker.

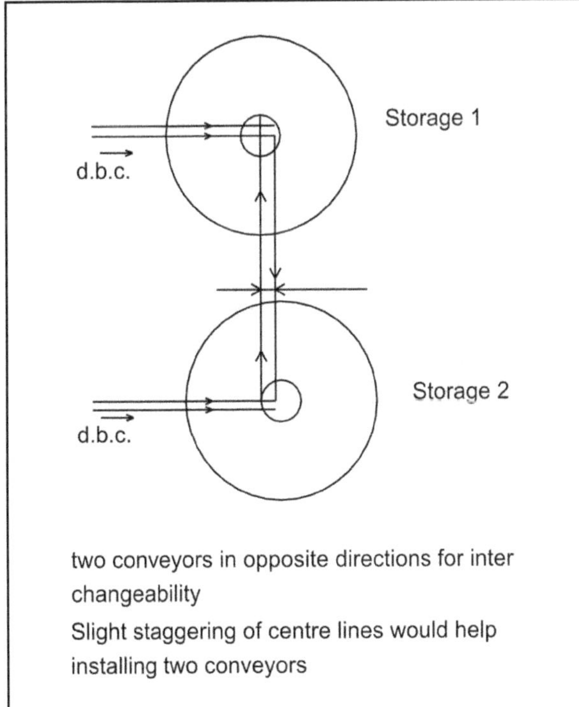

Fig. 6.26.7 Storage of clinker for two kilns with inter changeabillity.

The central shaft should be so sized that it will accommodate two dbcs (including standby) and their drives and also the possibility of taking / receiving clinker to and from storage of the 2nd kiln.

See Figs. 6.26.7 and 6.26.8.

26.8 Interconnection of Two Storages

It is advantageous to interconnect storages of two kilns so that if one over flows the other could be utilized.

By staggering stoppages of kilns, over flowing and consequent double handling of clinker could be avoided.

The conveyor for interchanging can be a reversible belt conveyor if clinker temperature is restricted to 100-120 °C. Since at any time movement of clinker will be moving in one direction only, one reversible belt conveyor should suffice.

It will have to be a pan conveyor if temperatures of cinker would be > 120 °C. But if a pan conveyor cannot be reversible, then there would have to be 2 pan conveyors moving in opposite direction.

See Fig. 6.26.7.

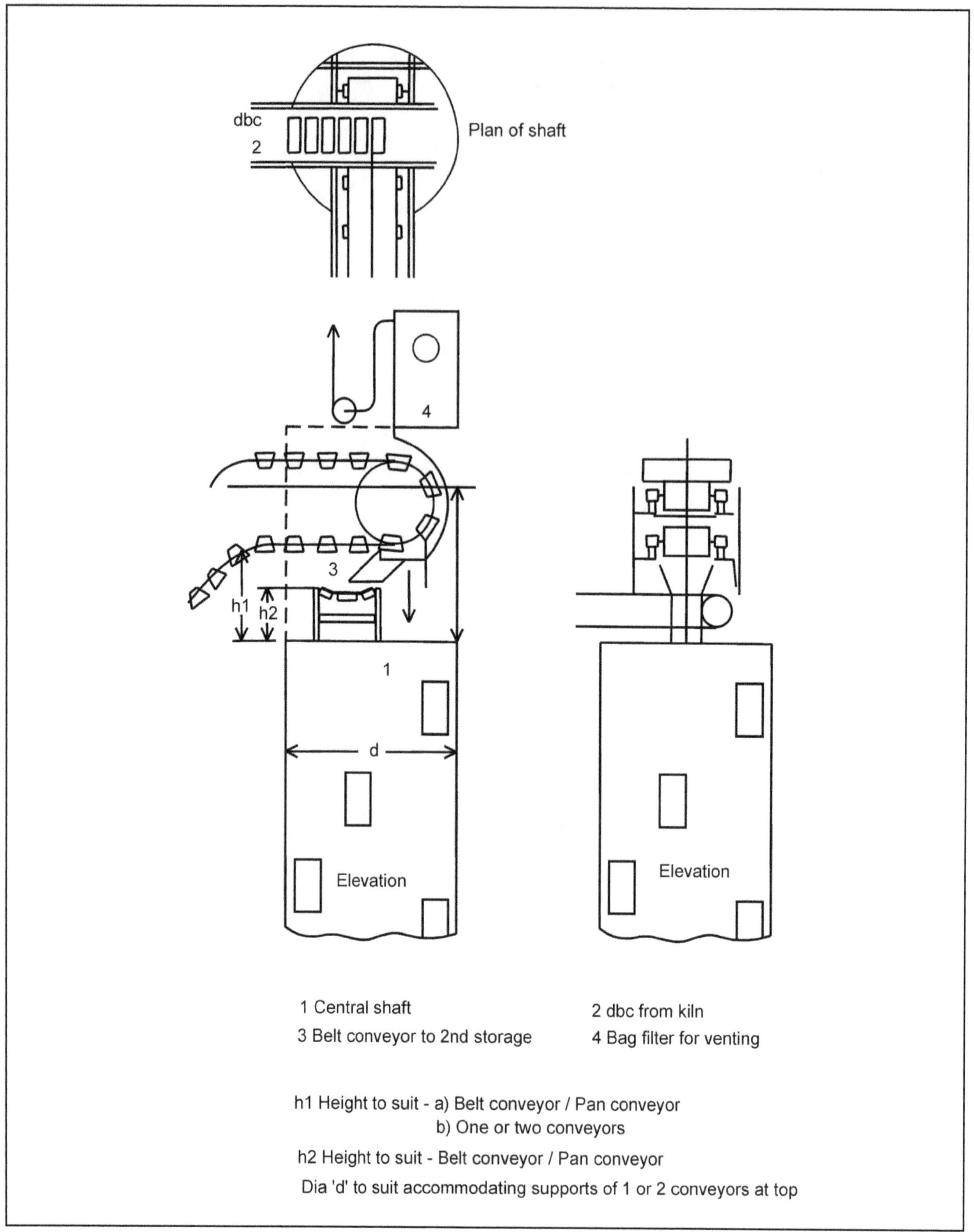

Fig. 6.26.8 Clinker storage - feeding arrangements.

Table 6.26.1

	part 1		calculation of dimensions of clinker storage dia of storage in metres				
			60	70	80	90	100
Sr. no.	**item**	**unit**					
			60	**70**	80	90	100
	area	m^2	2826	3847	5024	6359	7850
	angle of repose				35°		
	height of cone	m	21	24.5	28	31.5	35
	volume of cone	m^3	19782	31413	46891	66764	91583
	height				volume of cylinder		
		height metre					
	volume of	0	0	0	0	0	0
		1	2826	3847	5024	6359	7850
	cylinder	3	8478	11540	15072	19076	23550
		5	14130	**19233**	25120	31793	39250
		10	28260	38465	50240	63585	78500
	total volume	0	19782	31413	46891	66764	91583
		1	22608	35260	51915	73123	
		3	28260	42953	61963	85840	115133
		5	33912	**50646**	72011	98557	130833
		10	48042	69878	97131	130349	170083

Table 6.26.1 ***Contd....***

part 2

			clinkering capacity tpd		
			5000	7500	10000
design factor			1.1	1.1	1.1
no of days			14	14	14
storage	tons		77000	115500	154000
bulk density	t/m^3		1.2	1.2	1.2
volume	m^3		64167	96250	128333
		say	64200	96300	128300
alt. 1					
let dia d	m				
dia meter d =			80	90	100
cyl. Height			3.45	5	5
		say	3.5		
alt 2					
let dia			70		
			8.5		

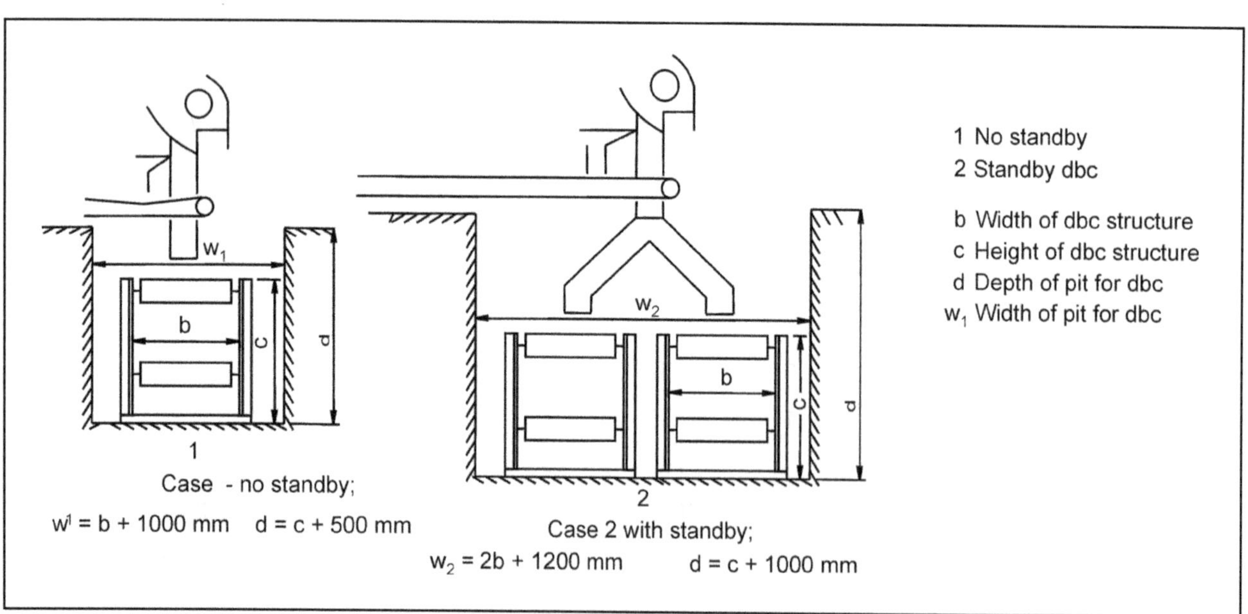

Fig. 6.26.9 Standby for d.b.c conveyor and its impact on dimensions of pit for it.

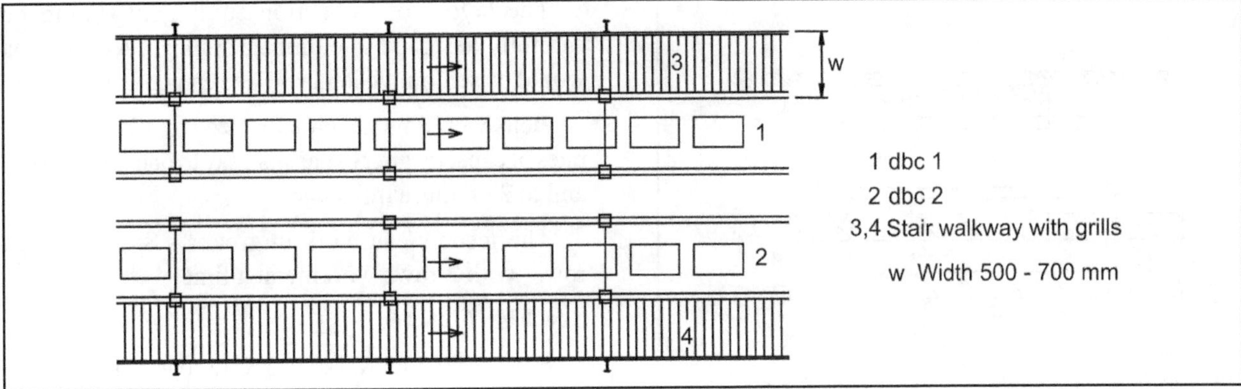

Fig. 6.26.10 Gantry for dbcs.

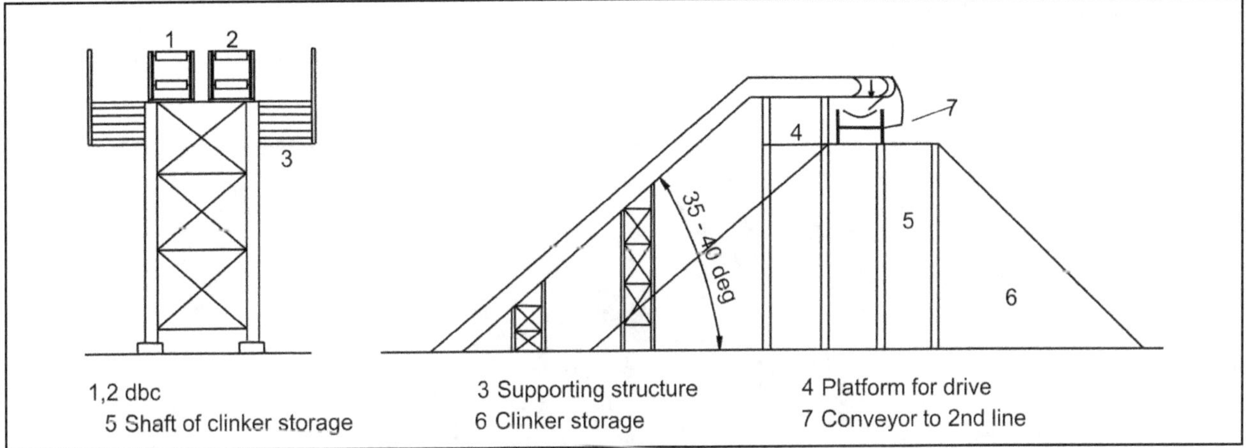

Fig. 6.26.11 Gantry for dbcs.

26.9 Extraction

Extraction of clinker from the storage is by feeders and belt conveyors arranged below ground level. There would be a large number of openings fitted with rod gates and feeders vibro or belt type or short pan conveyors.

See Figs. 6.26.12 and 6.26.13

The more the number of openings, less the dead stock of clinker inside. However, the increase in number also increases the cost of civil construction and equipment.

See Figs. 6.26.14 and 6.26.15.

26.9.1 Extraction Eqipment

The tunnels have to be waterproofed to prevent seepage and should also be vented and dedusted. Good maintenance space around belt conveyors and also good lighting in the tunnel has to be provided for.

See Figs. 6.26.16 and 6.26.17.

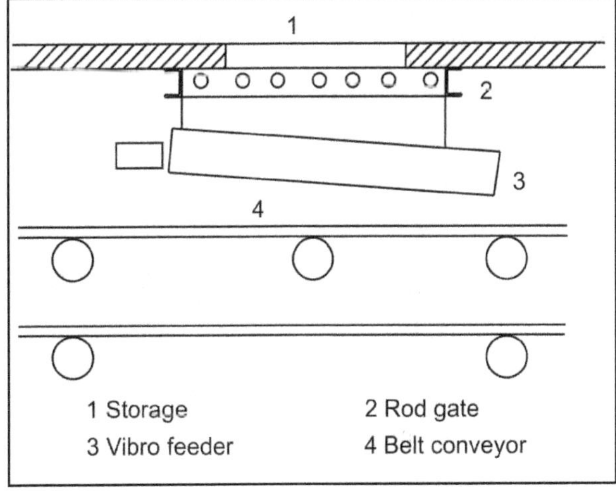

Fig. 6.26.12 Extraction from storage.

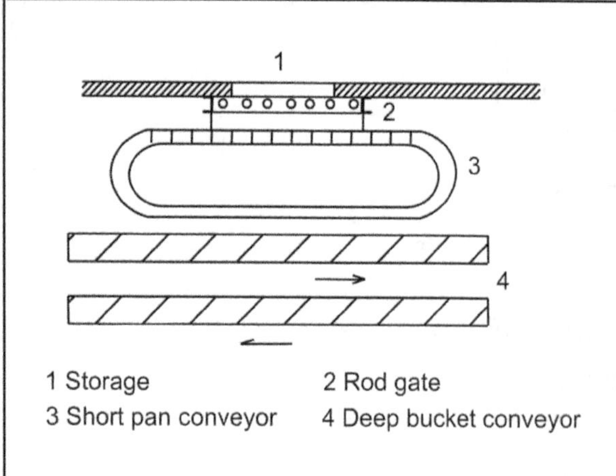

Fig. 6.26.13 Extraction from storage if clinker temperature is likely to be > 120 °C.

The layout of extraction equipment should be carefully designed to eliminate spillage. As this area would be seldom visited, spillage can go out of hand.

Belts would be generously sized to allow for feed rates of capacity after expansion; say to one mill initially and to 2 or more mills later.

The feeders can work in groups - Say 1 in each row – with 2 rows working at a time.

See Fig. 6.26.15.

Say 1 & 11 or 2 & 14 or 3 & 13 together. Capacity of each feeder can also be thereby reduced.

Clinker could also be extracted by installing a 'scraper –reclaimer' system (similar tocircular stacker reclaimer) as shown in **plate 6.26.1**. Such a system would not require under ground network of feeders and belt conveyors.

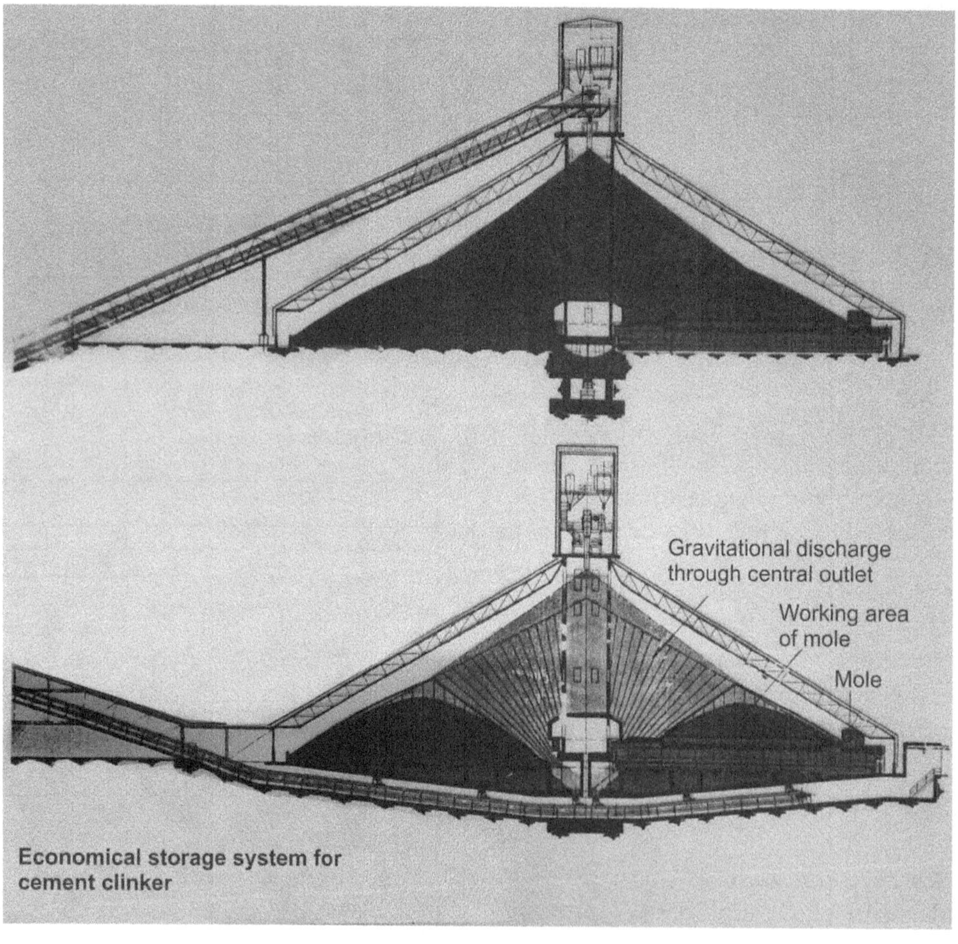

Plate 6.26.1

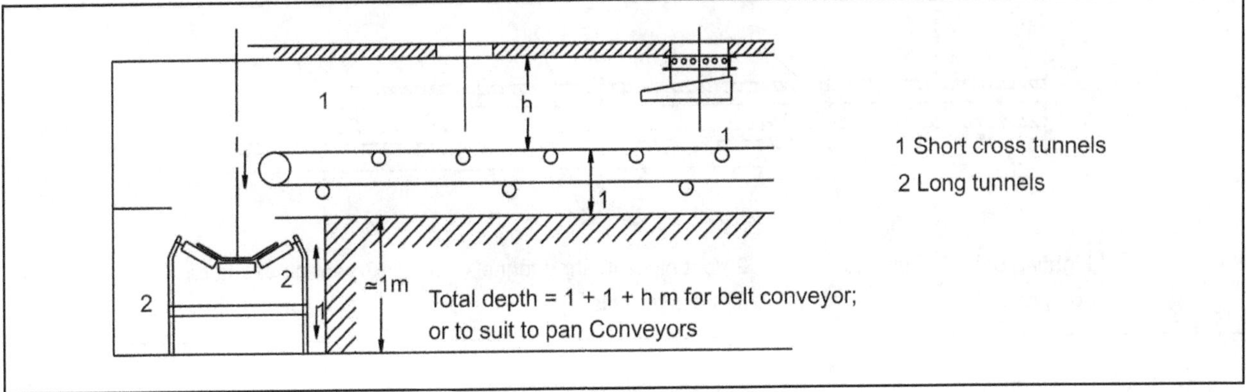

Fig. 6.26.14 Extraction from bottom cross and long tunnels.

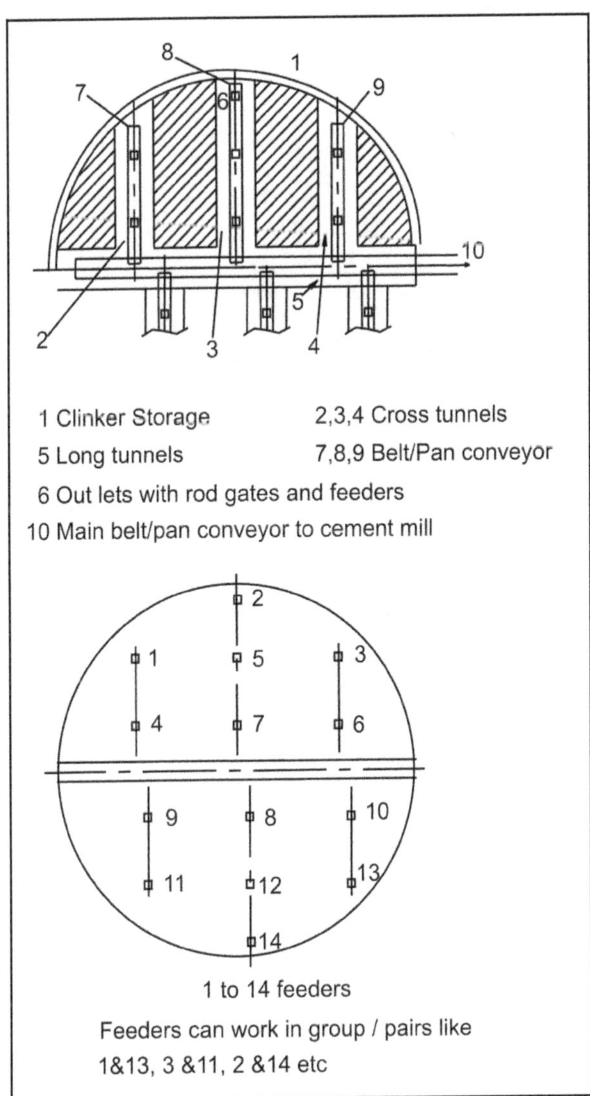

Fig. 6.26.15 Feeders can work in groups/pairs for even extraction.

Since the storage spreads over an area of 4000-5000 m^2, great care needs to be taken from civil engineering point of view taking into account :

(i) Load bearing capacity of soil.

(ii) Underground water table.

(iii) Drainage.

26.10 Shifting of Clinker

Access is also required to be provided for bulldozers and similar equipment to enter into the storage area to shift clinker heaps.

The plinth level should be such that rainwater would not go inside.

For entrances provided for entry of bulldozers, a sloping roof cover or awning could be provided. Floor would be sloped to keep rain water outside.

See Fig. 6.26.18.

26.11 Storage of Under Burnt Clinker

It is best to store under burnt clinker separately and extract it in a metered fashion. Only a small capacity is required to be provided, may be 6 hours' production-maximum 12 hours'.

Thus a capacity of 1000-1200 tons should be sufficient. It is best to store it in a silo.

One way of providing such a storage would be to take clinker from discharge point of d.b.c. to the proposed storage silo by a belt or a pan conveyor. Ideal location for such a silo would be in line with the conveyor for feeding Cement mills.

See Figs. 6.26.19 and 6.26.20.

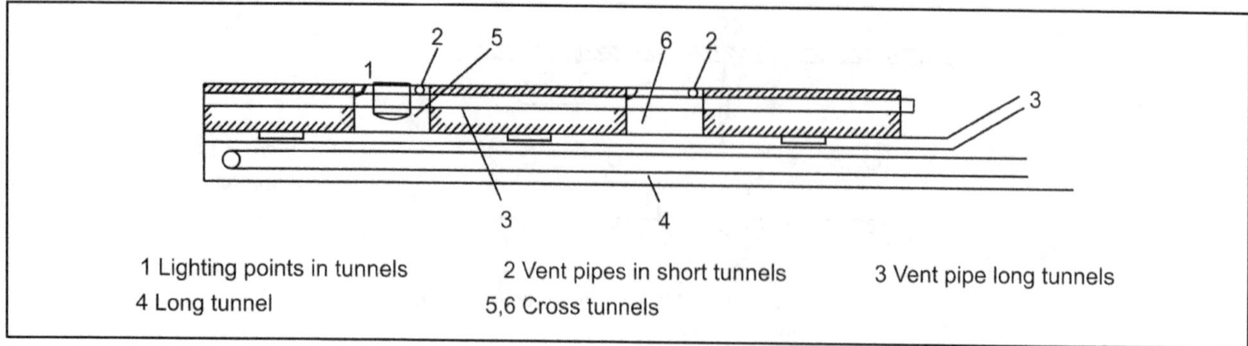

Fig. 6.26.16 Ventilation and lighting in tunnels under clinker storage.

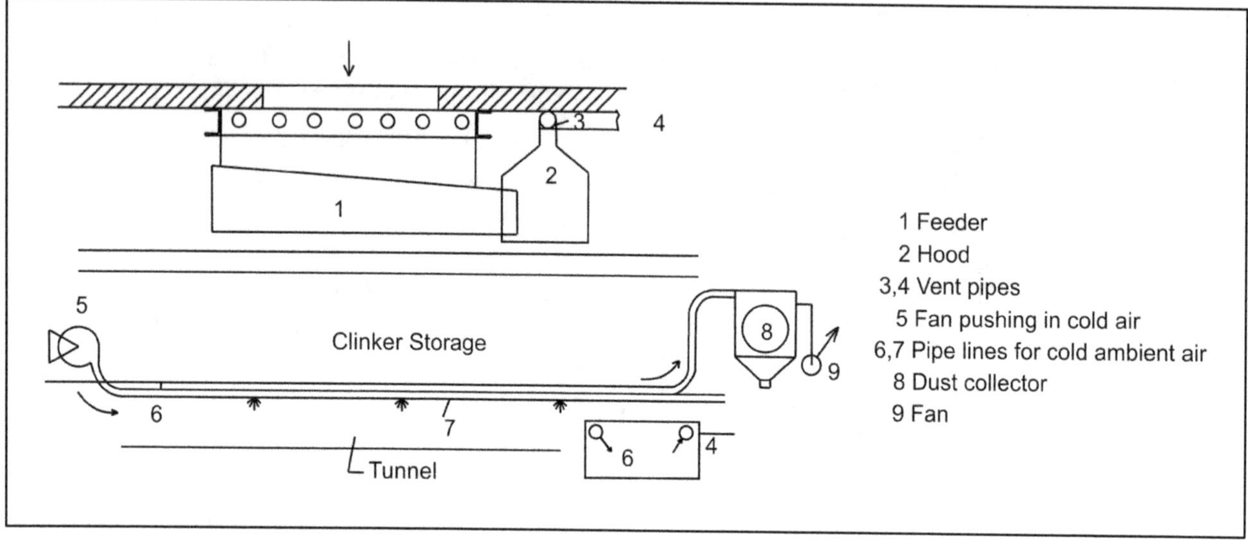

Fig. 6.26.17 Ventilation of tunnels.

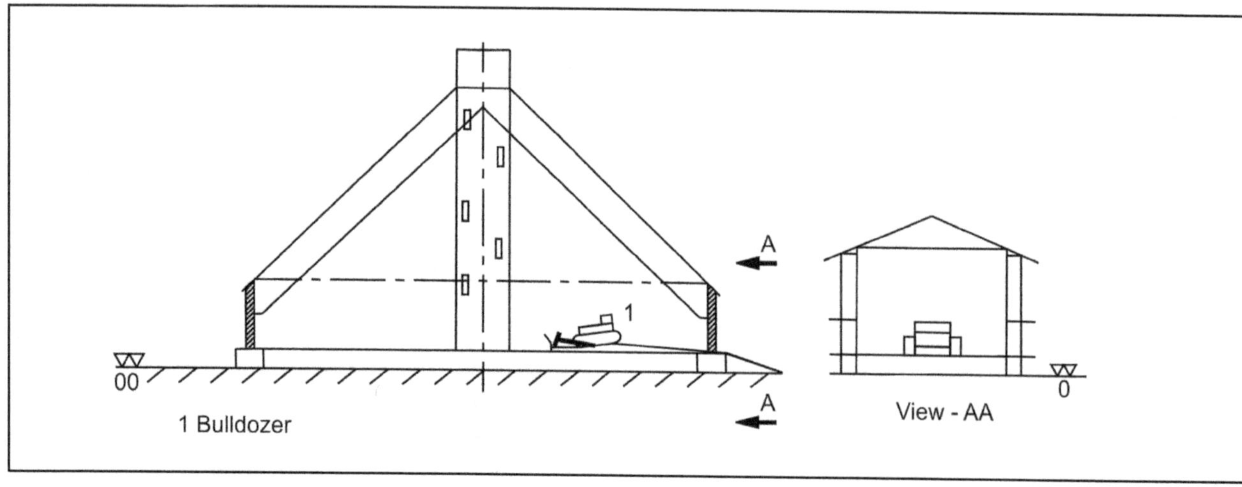

Fig. 6.26.18 Access for bulldozers to push clinker over clinker out lets.

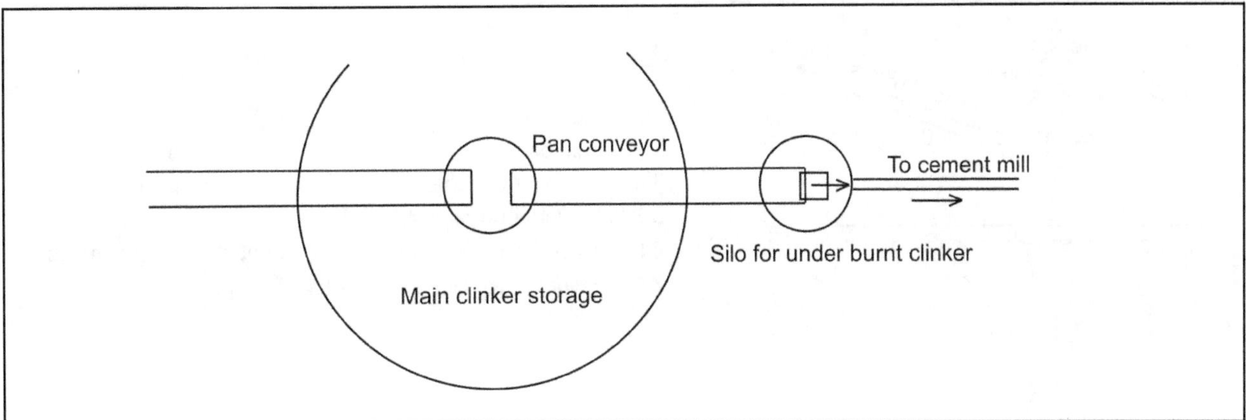

Fig. 6.26.19 Storage of under burnt clinker.

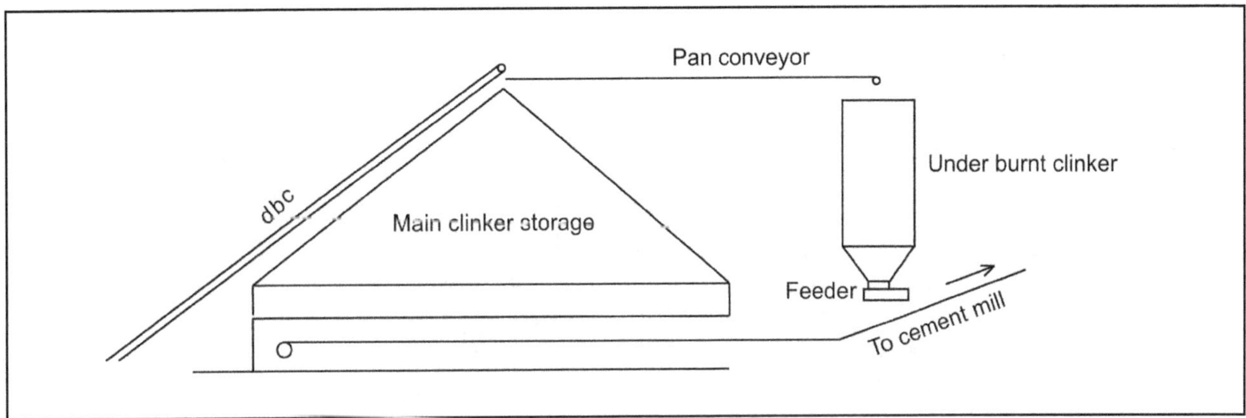

Fig. 6.26.20 Metering arrangement for extraction of under burnt clinker.

For a storage capacity of 1200 t or 1000 m^3, a silo of 8 metre diameter and 20 metre high would suffice.

It is also possible to divert under burnt clinker before it reaches clinker storage by splitting the d.b.c.

It could be fed back to d.b.c. as shown.

See Fig. 6.26.21.

26.11.1 Facility for Drop Test

When such a separate storage and extraction facility is created it could also be used to take 'drop tests' to measure the output of the kiln.

It could also be used to take clinker to a supplementary storage facility, when the existing storage is full but kiln has to be kept running.

26.12 Excess Clinker

Many a plant spends substantial amounts in building up a clinker stock pile, for excess clinker and for withdrawing it when required.

The piles are in the open and hence are covered with plastic sheets / tarpaulins.

This contingency, which can arise at not too distant intervals, can be avoided if a planned storage for excess clinker is fitted in the general plant layout itself as described.

See Fig. 6.26.22.

26.13 Storing Clinker in Silos

Clinker can also be stored in silos of steel or concrete.

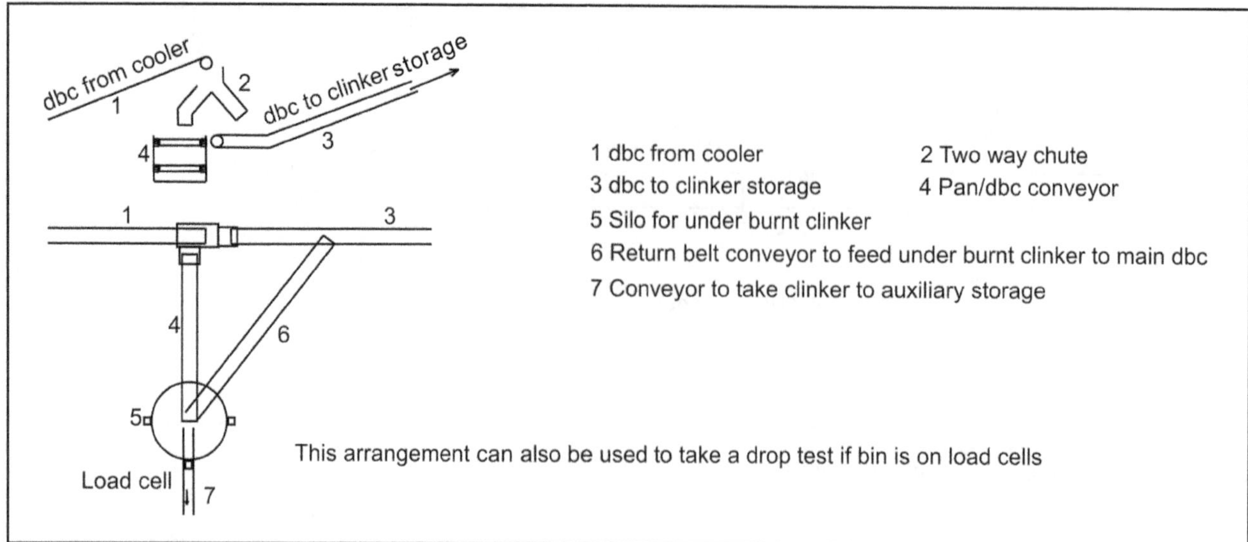

Fig. 6.26.21 Silo for under burnt clinker before main storage.

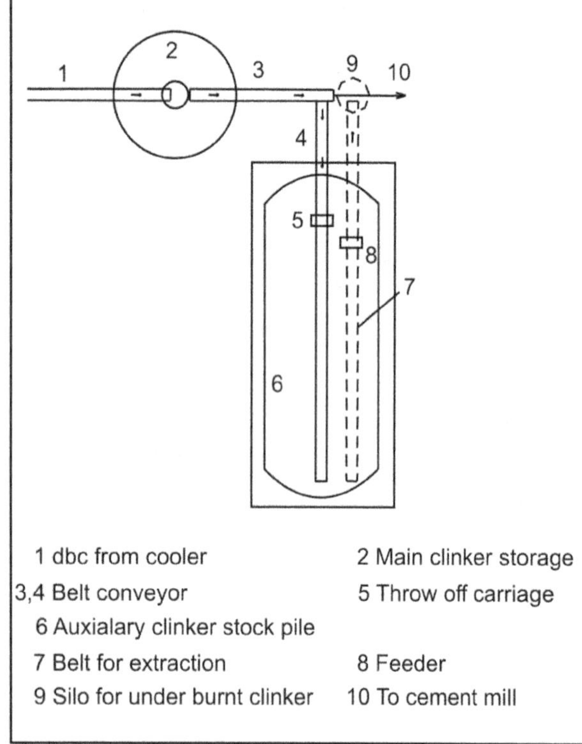

Fig. 6.26.22 Auxiliary storage of clinker when main storage is full.

Taking the example of a 10000 tpd kiln and 14 days' stock, i.e., 154000 Tons or 10000 m^3.

Let the silo be a flat bottom silo with height to diameter ratio of 3:1.

To store 154000 tons of clinker would require a very large silo. Hence number of silos would be increased. **See Table 6.26.2.**

This solution is ideal where floor space is at a premium and from point of view of pollution control. However, it suffers greatly in that there is no facility for over flowing and excess clinker must be conveyed and stored in auxiliary storage.

See Figs. 6.26.22, 6.26.23

Conveyors - mostly dbcs required to reach top of silo would be very long requiring massive supporting structure.

See Fig. 6.26.24.

26.13.1 Extraction From Silo

Extraction arrangements at bottom will be similar to that described earlier but on a smaller scale. Silo, if concrete, should be water proofed so that clinker dust does not harden. If lumps are formed extraction would be very difficult.

Table 6.26.2 Clinker stored in silos working out dimensions of silos.

			clinkering capacity tpd		
Sr. no.	**item**	**unit**	**5000**	**7500**	**10000**
	clinker stock 14days	tons	77000	115500	154000
	bulk density	t/m^3	1.2	1.2	1.2
	clinker stock volume	m^3	**64167**	**96250**	**128333**
	no . of silos		2	3	4
	clinker stock per silo	m^3	32083	32083	32083
	h/d ratio		3	3	3
	dia of silo	M	23.1212478169	23.1212478169	23.1212478169
	rounding off		**23**	**23**	**23**
	height	m	**69**	**69**	**69**
	height to top	m	**79**	**79**	**79**
	no of silos		1	2	3
	vol.per silo		64167	48125	42778
	h/d ratio		3	3	3
	dia of silo	m	**29**	**26**	**25**
	rounding off height	m	**87**	**78**	**75**
	height to top	m	**98**	**88**	**85**

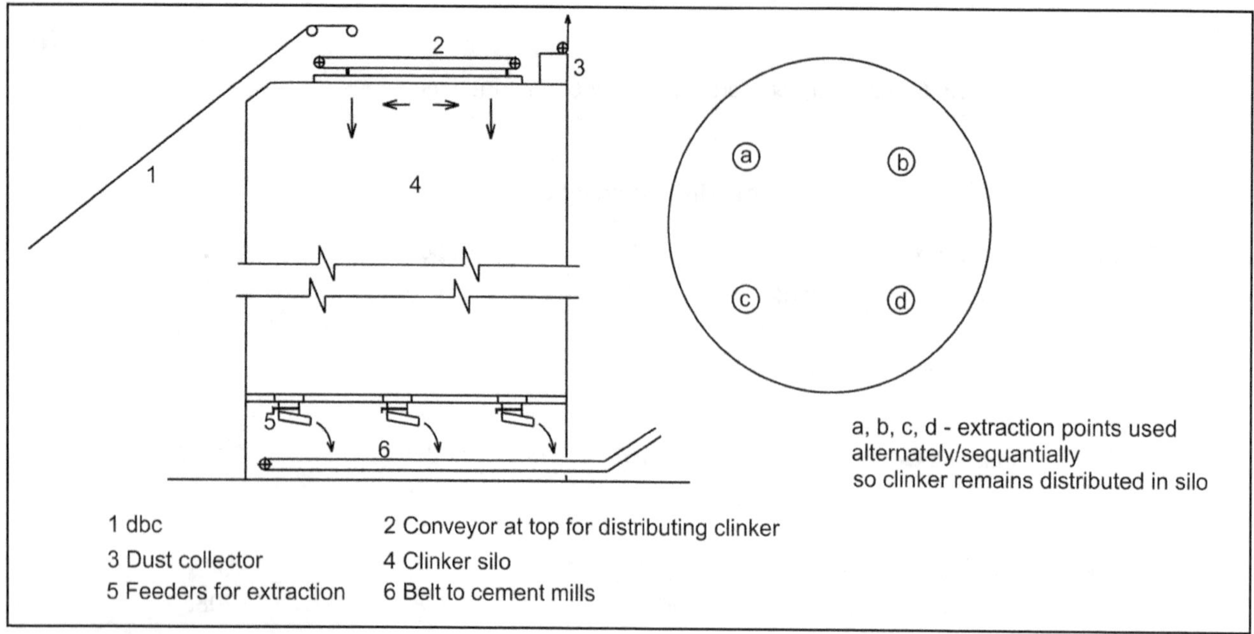

Fig. 6.26.23 Clinker stored in silos.

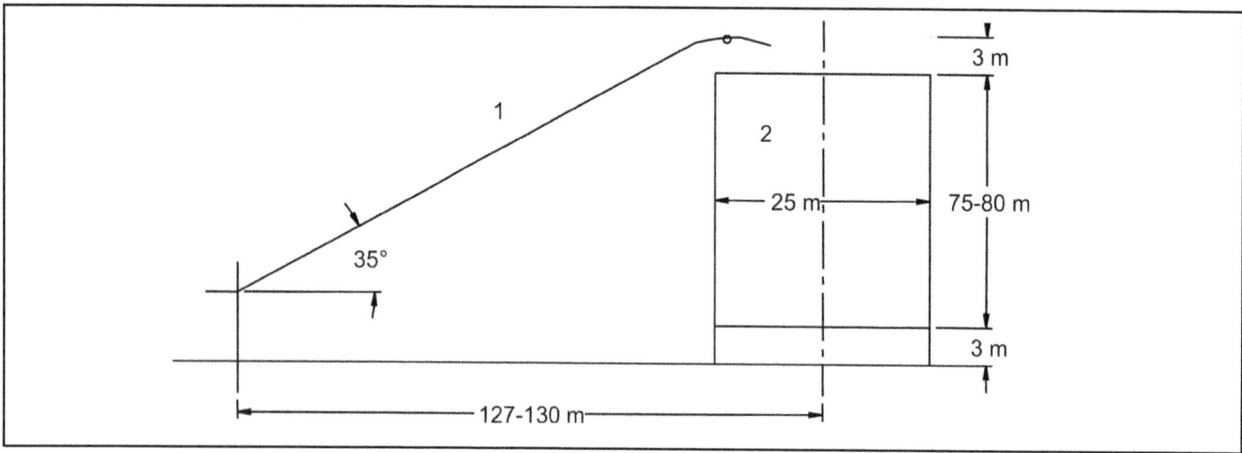

Fig. 6.26.24 Length of dbc required to reach top of clinker silo.

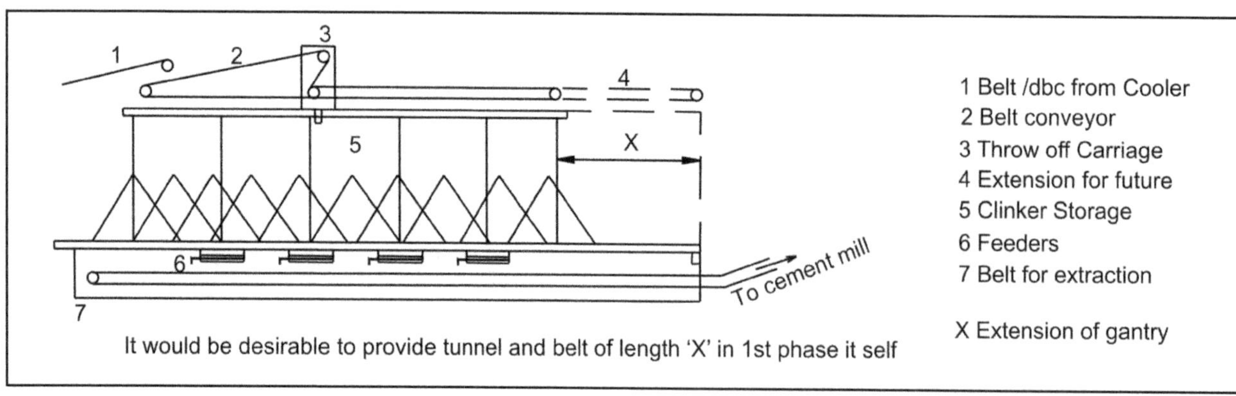

Fig. 6.26.25 Clinker storage for small plants.

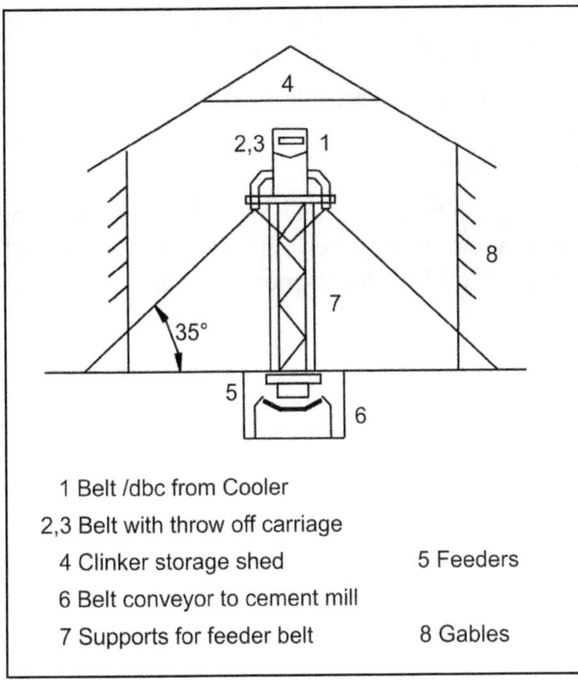

Fig. 6.26.26 Clinker storage for small plants.

In extracting from the silo care should be taken that, clinker is extracted evenly from four quadrants so that load does not come on one section of the wall of the silo.

In a silo, hot clinker once stored, does not get cooled easily. This aspect should be looked into when taking a decision about storing in a silo.

All in all, storing clinker in silos has not been preferred very much at least in India. **See Fig. 6.26.29.**

26.14 Covering Clinker Stored

Clinker sheds would be covered to a great extent from sides also. Access would be provided for a bulldozer to shift clinker piles.

See Fig. 6.26.26.

In between the extraction points would be dead stock, which reduces effective capacity of stockpile. Hence in designing the pile allowance should be made for the dead stock also.

See Fig. 6.5.1 in Chapter 5 of this Section

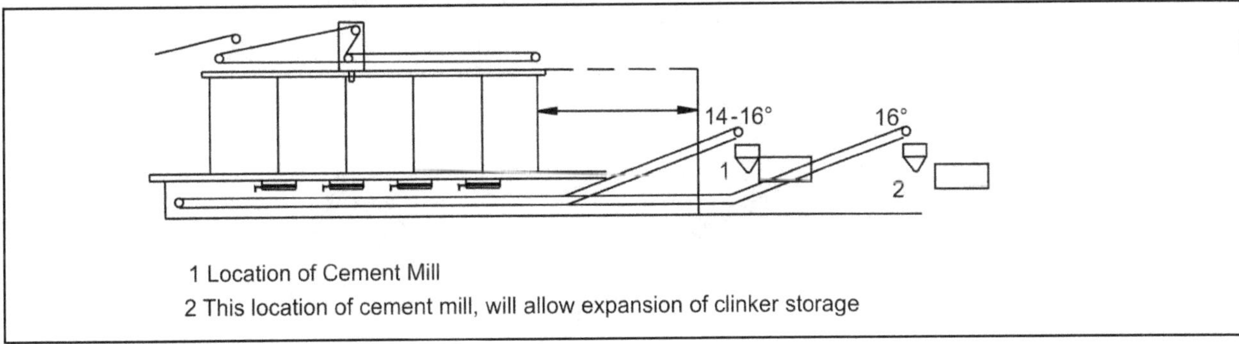

Fig. 6.26.27 To leave room for expansion cement mill should be located further.

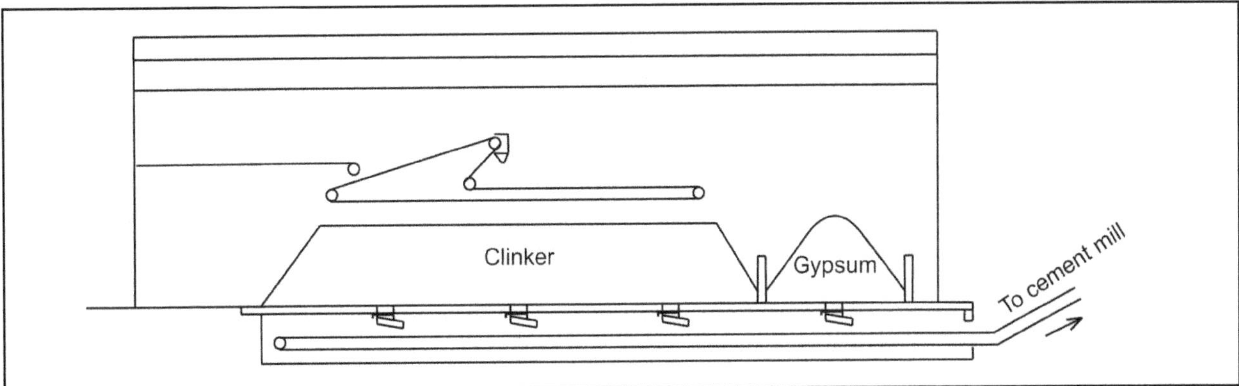

Fig. 6.26.28 Storage of gypsum in the same shed.

26.15 Storing Gypsum

In most cases gypsum would also be stored in a section of the stockpile. Gypsum would be brought in truck loads and unloaded manually for small plants.

See Fig. 6.26.28.

Even for a large plant of 10000 tpd capacity, daily requirement Gypsum at design capacity would be between 400 to 500 tons depending on type and quality of gypsum used.

A month's stock would be between 15000 Tons and 20000 Tons.

Gypsum can be brought by trucks or by wagons. Handling gypsum in rake loads would pose a great many handling problems.

In large plants with circular clinker storage, gypsum would be stored separately under a shed. Handling of gypsum is dealt within the next Chapter.

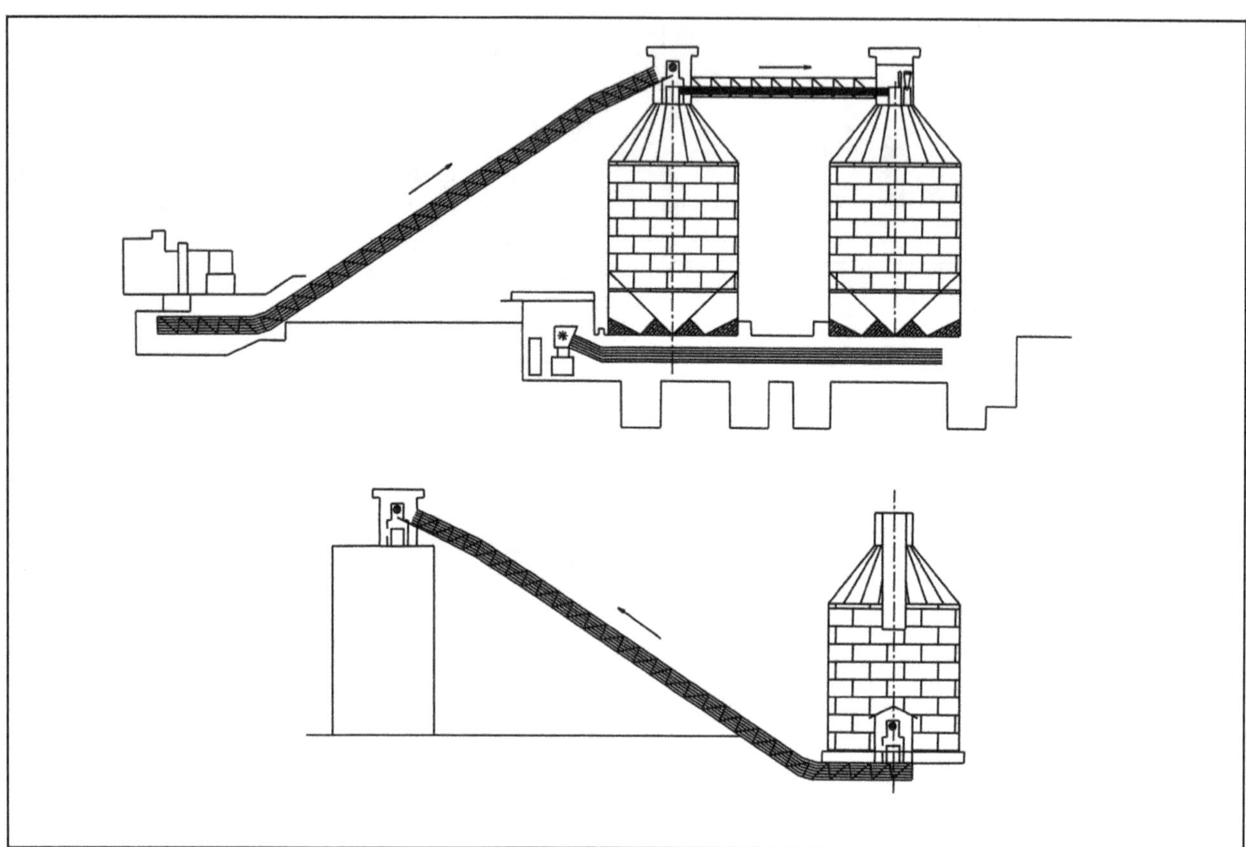

Fig. 6.26.29 Storage of clinker in silos and extraction therefrom to feed cement mills.

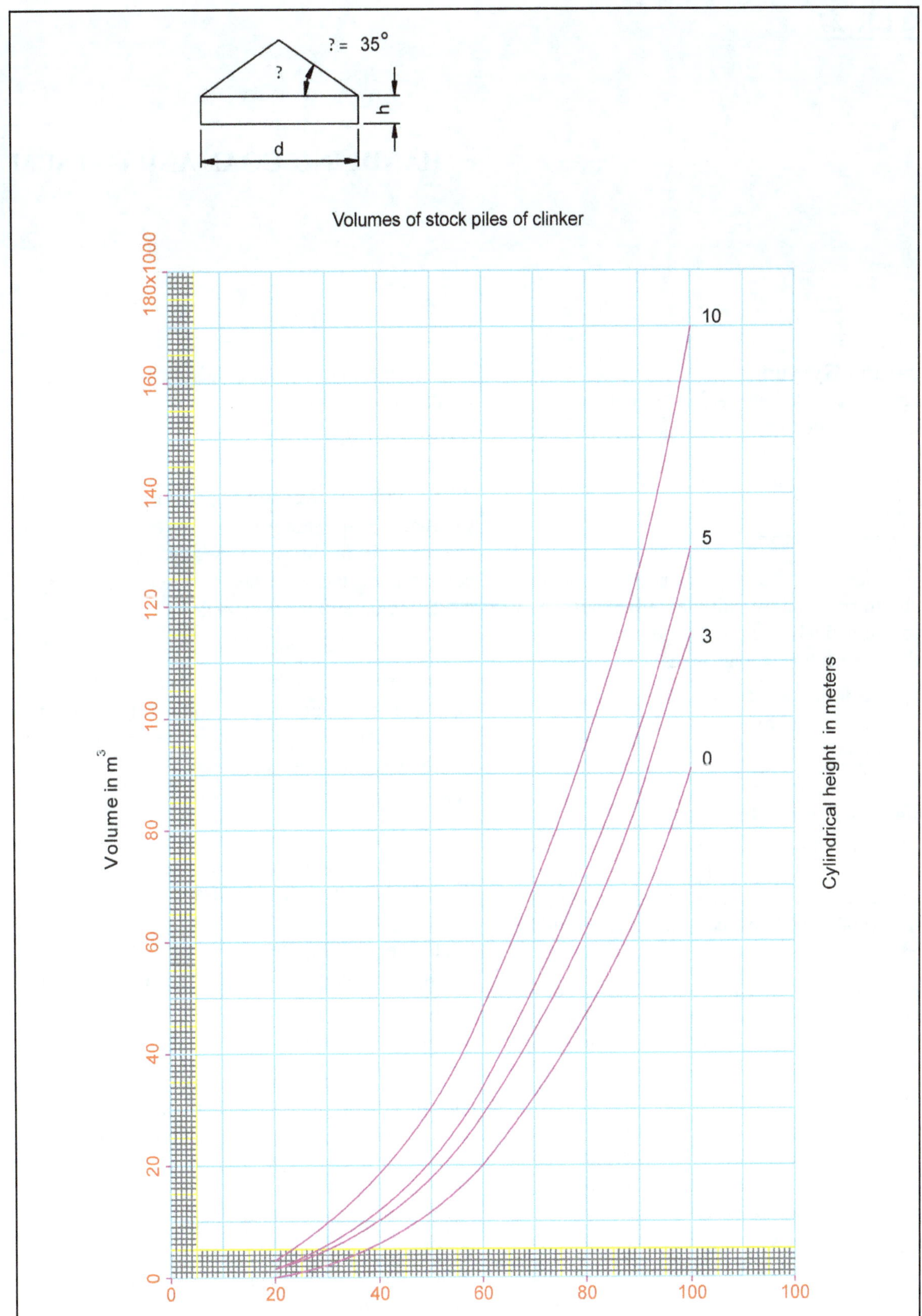

? = 35°
?
h
d
Volumes of stock piles of clinker
Volume in m³
Cylindrical height in meters
10
5
3
0
180x1000
160
140
120
100
80
60
40
20
0
0
20
40
60
80
100
100

Fig. 6.26.30

CHAPTER 27

HANDLING COAL AND GYPSUM

27.1 Coal and Gypsum

Coal and gypsum could be received in truck loads or in wagons. Wet materials like coal and gypsum are not free flowing and hence vibro feeders would not suit. Reciprocating feeder or Pan feeders would be more suitable.

See Figs. 6.27.1 to 6.27.3.

For convenience, coal and gypsum handling systems when both are received by rail, would be common, being used alternatively to unload wagons and convey them to their respective storage areas.

Large plants install wagon tipplers to unload coal. The same could be used for gypsum also.

See Fig. 6.27.4.

27.2 Quantities to be Handled

If fuel consumption is 750 Kcal/kg clinker and coal has a calorific value of 4500 kcal/kg, then consumption of coal would be 16.6 % that is ~1900 tpd at design capacity of 11000 tpd. On as received basis it would be ~ 2100 tpd

Therefore if a rake load is 2200 tons, the plant would require a rake load every day. Thus coal handling would require large enough facilities to unload coal wagons. Hence investment in a 'wagons tippler' could be justified.

Consumption of gypsum on the other hand at 5% would be about 660 tpd on as received basis, assuming 15 % moisture. A rake load of gypsum would be required in approximately 3 days.

Gypsum would come from Rajasthan area if natural gypsum is used and from nearest chemical and fertilizers plants if synthetic gypsum is used. Normally about a month's stock would be maintained.

Thus the coal unloading facility could be used to handle gypsum also.

See Fig. 6.27.4.

27.3 Types of Storage for Coal and Gypsum

Coal and gypsum storages would be covered. For coal, pre blending facilities are installed for large plants consisting of stacker reclaimer systems consisting of either circular or linear stockpiles.

See Figs. 6.27.5 and 6.27.6.

Coal would generally require crushing. Hence coal unloaded by a wagon tippler would be first taken to a coal crusher. The capacity of crusher would have to match that of the tippler.

See Fig. 6.27.4.

It is better to stockpile crushed coal because the voids are minimum and hence supply of oxygen is limited. This minimizes fire hazard.

27.4 Wagon Tippler

To unload a rake load of 2200 Tons in 5 to 6 hours, capacity wagon tippler has to be

2200 / 5-6 = 440-370 tph.

If there is no intermediate stockpile of uncrushed coal, then the coal crusher would have to match the capacity of the wagon tippler that is 450 tph.

27.5

Gypsum if natural, would also need crushing. Synthetic gypsum would be very wet with 15-20% moisture. It is fine, not requiring crushing but it would be very difficult to flow.

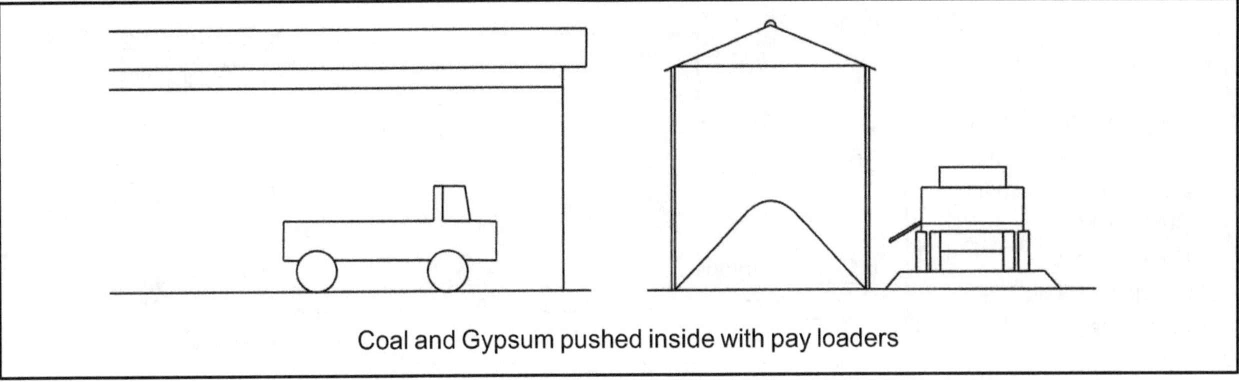

Fig. 6.27.1 Receiving coal and gypsum in truck loads.

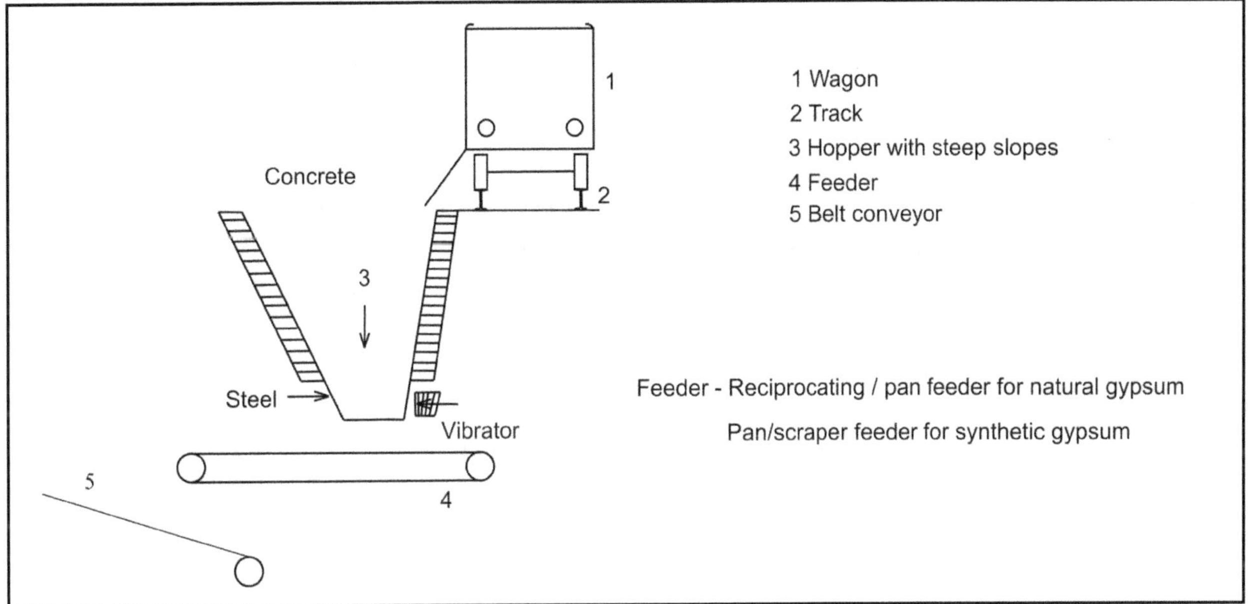

Fig. 6.27.2 Gypsum received in wagons.

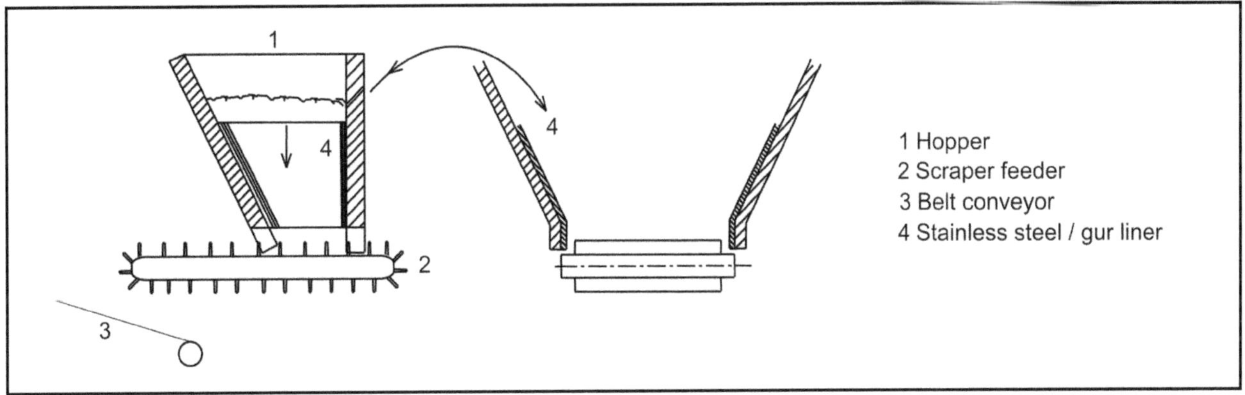

Fig. 6.27.3 Scraper feeder for gypsum.

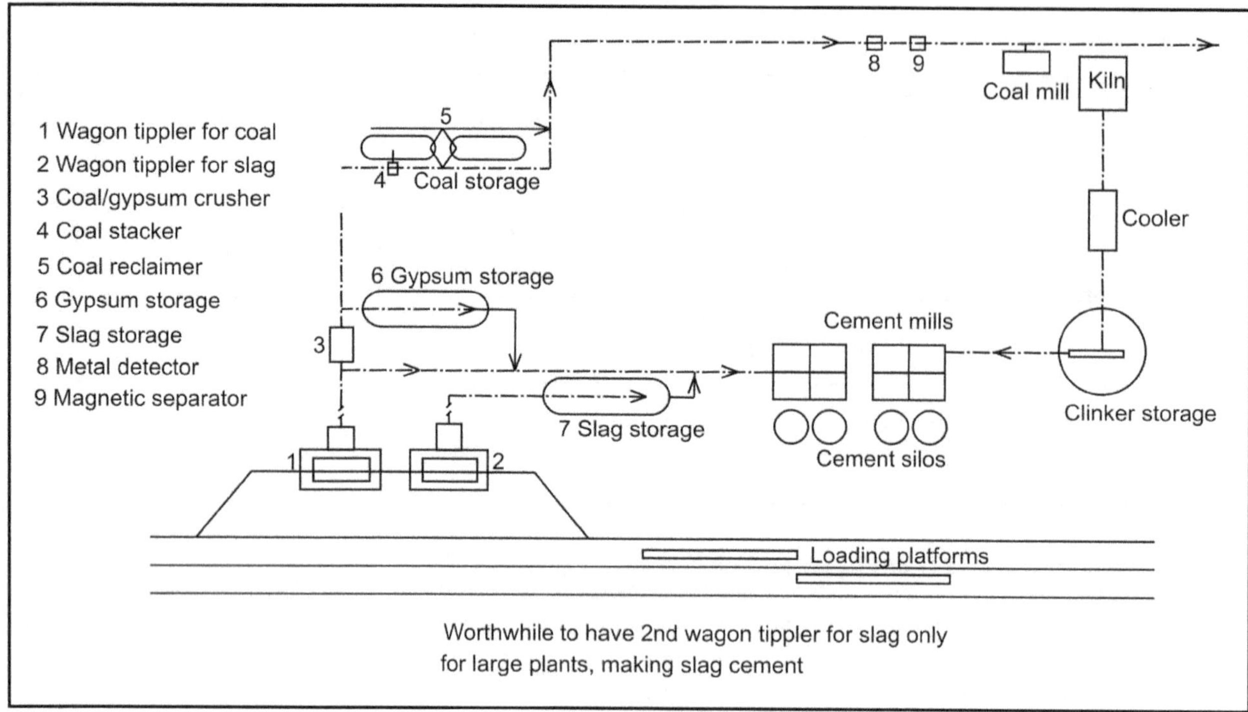

Fig. 6.27.4 Accommodating coal and gypsum handling and storage facilities in general plant layout.

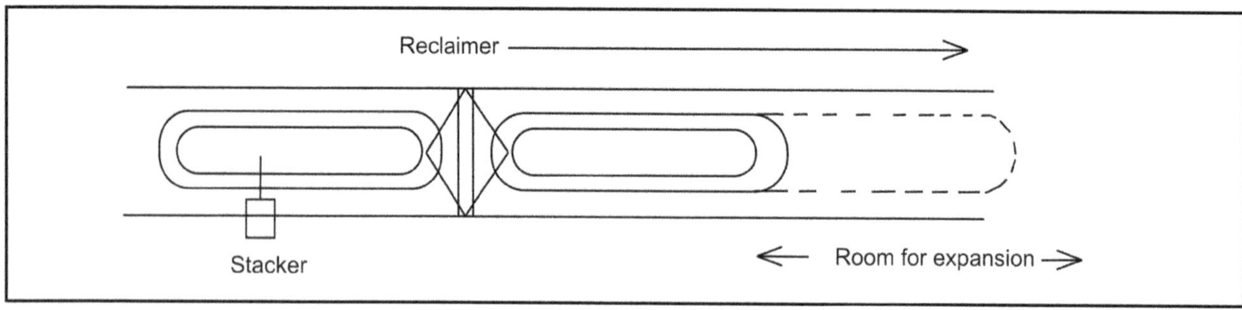

Fig. 6.27.5 Stacker reclaimer system for coal.

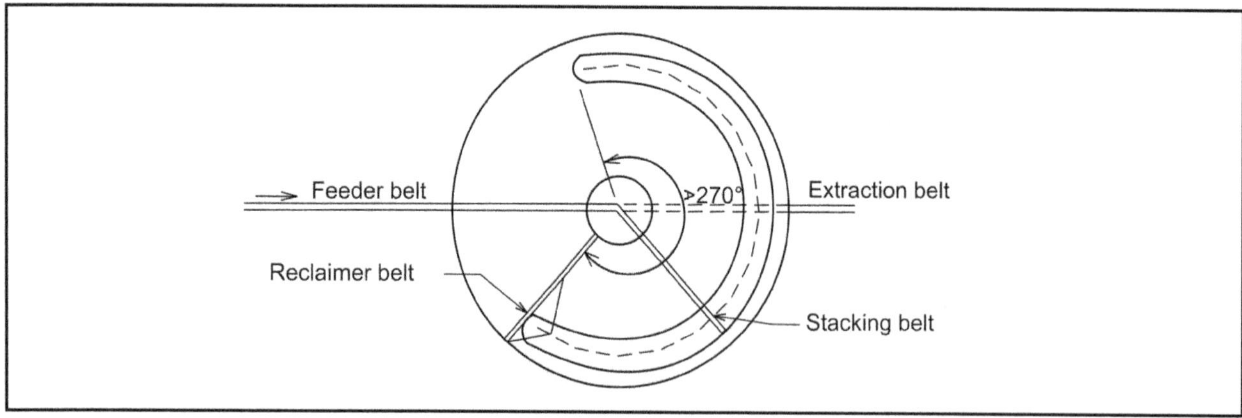

Fig. 6.27.6 Circular stacker reclaimer system for coal.

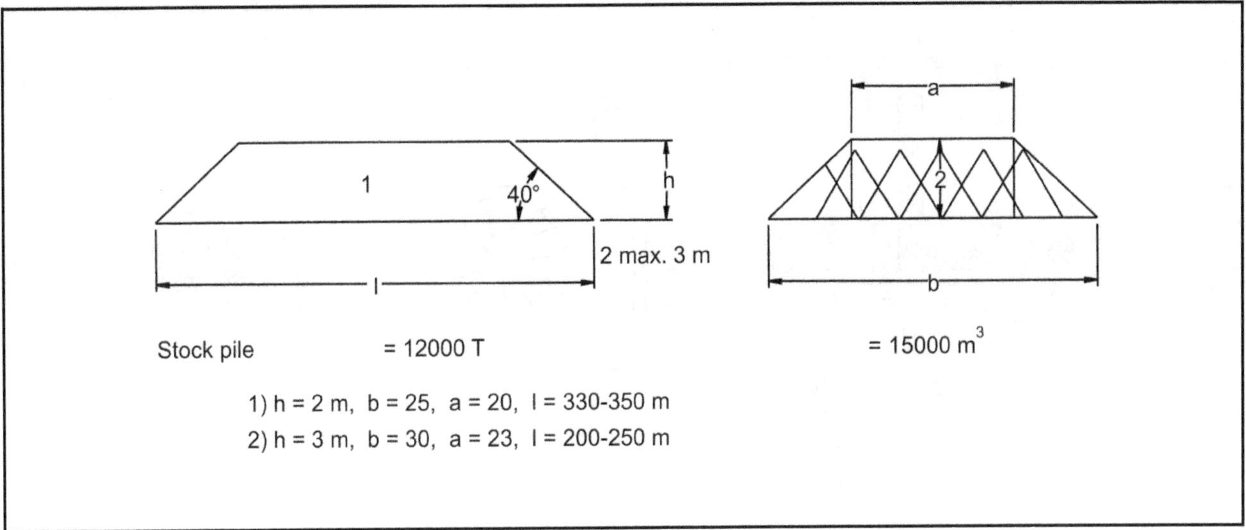

Fig. 6.27.7 Stock piles for coal.

27.6 Stockpile for Coal

To keep 3 week's stock of coal, the stockpile capacity would be 2100 × 14 = 29000 Tons. If linear, 2 stock piles would be required one for building up and one for reclaiming when a stacker-reclaimer system is used to stock coal.

In case of coal, pile height is restricted to 2, max. 3 m on account of fire hazards.

See Figs. 6.27.7

Let height of pile h = 2 m and width of pile be 40 m at bottom and 35 m at top.

Cross sectional area = 75 m^2

Bulk density crushed coal = 0.8 T / m^3

∴ volume of 29000 Tons stockpile = ~ 36000 m^3

∴ length of pile = 36000 / 75 = 480 m if ends are ignored.

If height of pile is kept as 3 metres,

length will reduce to 380 metres.

It is possible to reduce the stock to be maintained, depending on location of the plant vis-à-vis collieries and continuity of supply therefrom.

The linear piles can be extended when plant capacity is expanded. Layout must provide room for such an expansion.

27.6.1 Circular Stockpile

In a circular stockpile, it is not possible to expand later. Therefore diameter of pile has to arrived at taking into account future expansion.

See Fig. 6.27.6.

The circular stock pile should not be more than 270° arc of the circle. Pile is formed at one end and extracted from the other end.

Assuming a pile height of 3 metres, the length of the pile for double the capacity (20000 tpd) would be 760 metres. Mean diameter of the circular stock pile would be about 100 metres.

27.7 Number of Days's Storage

To reduce investment in stockpiles, and storages it is not unusual to reduce the days' storage for expansion.

	Plant Capacity	
	1st unit	**after expansion**
Tons per day	10000	20000
Coal stocks number of days	14	10
Stocks in tons	29000	21000

The layout for stockpile can be designed on this basis i.e., instead of doubling the capacity, it is increased 1.5 times when plant capacity is doubled. Experience

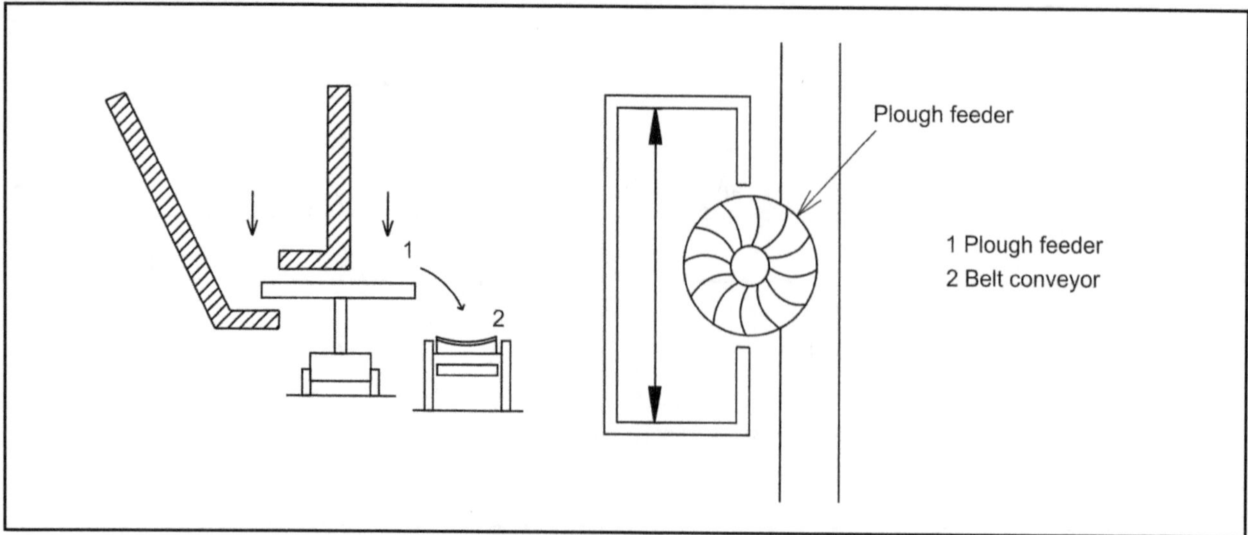

Fig. 6.27.8 Plough feeder for extracting wet gypsum.

gained in operating the 1st unit should indicate to what extent stocks could be reduced on practical basis.

27.8 Precautions Against Fire

It is necessary to provide facility for water spraying near coal stockpiles to extinguish fires that may burn in hot and dry months of the year. The coal stockpiles should be regularly consolidated by bulldozers to remove air pockets.

27.9 Coal Blending

Coal blending is equally or even more important than blending of raw materials as it would lead to kiln operating with reduced fluctuations in calorific value and ash content of coal fired. Stacker Reclaime Systems would be installed as shown in **Figs. 6.27.5 and 6.27.6.**

Piles would be formed in a different manner as compared to piles for limestone because height of pile is restricted.

See Fig. 6.27.7.

Coal will be withdrawn by the reclaimer system and taken to coal mill hoppers. This belt should be provided with a metal detector and a magnetic separator to remove magnetic and non-magnetic metallic pieces before they reach the mill.

See Fig. 6.27.4.

Belt should also have a system to spray water to douse red hot lumps of burning coals if need be.

Layout wise, it is convenient to feed more than one mill if belt is at right angles to the mills.

See Fig. 6.27.4.

27.10 Gypsum

Gypsum being wet, flow through a hopper would be sluggish. Arching can take place. Hence hopper should have steep slopes. Valleys and corners should be rounded. Inner face of hopper may be lined with either stainless steel liners or special plastic-gur-liners for smooth flow. Alternately, bottom 1 metre or so of the hopper could be of steel for installing vibrators for actuating flow of gypsum.

See Figs. 6.27.2 and 6.27.3.

Wet gypsum could also be withdrawn from a hopper by a plough feeder as shown.

See Fig. 6.27.8.

27.10.1 Storage Shed for Gypsum

Gypsum will also be covered under a covered shed. Size of shed will take into account expansion of the plant.

In dry months, gypsum could be dried in the sun. The height of gypsum piles would also be restricted to 3-4 m.

Sometimes coal and gypsum stockpiles are covered partly only. Stocks of 2 weeks would be put under cover and balance would be in the open.

These arrangements would dry out coal and gypsum and hence in dry months, coal / gypsum would be drawn from open piles and in wet months from covered piles. **See Fig. 6.27.9.**

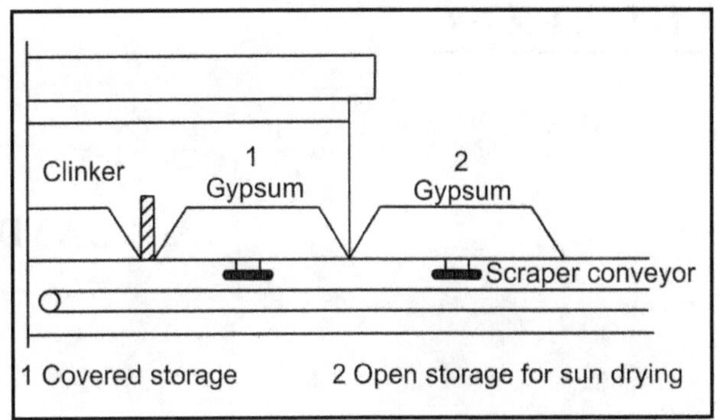

Fig. 6.27.9 Gypsum storage and recovery.

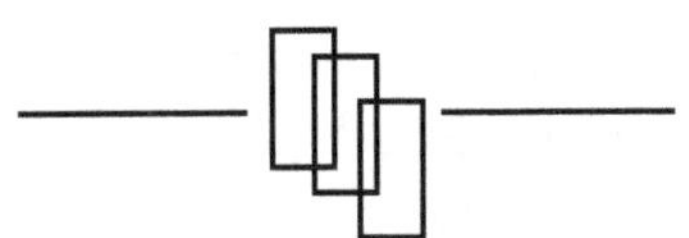

CHAPTER 28

HANDLING SLAG TO MAKE SLAG CEMENT

28.1 Blended Cements

If the plant is proposing to make blended cements, then provisions are required to be made to store and extract pozzolana and or blast furnace slag used as blending materials.

The two materials are quite different from each other though both are suitable as blending materials. They require very different arrangements for receiving, storing and extracting them.

28.2 Granulated Blast Furnace Slag

Slag is granulated by jets of water directed at molten slag. Granulation is done in the steel plant. Slag received from steel plant is therefore granulated but wet. Slag is very hard and brittle and is also very abrasive. It is like glass. Depending on facilities installed in the steel plant to drain water after granulation, slag as received could contain 10-15 % moisture.

Cement Plants making blended cements, enter into contracts with Steel Plants to buy slag on a regular basis. If the quantity is substantial, cement manufacturers may decide to install a slag dryer in the premises of the Steel Plant itself and transport dried slag to their cement plants.

If the movement is in the right direction from market considerations, they may even transport clinker to the Steel Plant and make slag cement in its vicinity.

See Fig. 6.28.1.

28.3 Drying of Slag

In a large number of cases however, slag received would have to be dried in the cement plant before sending it to mills for grinding.

Moisture of 10-15 % in slag and a clinker / slag proportion of 40 : 60 or 50 : 50 in most cases would

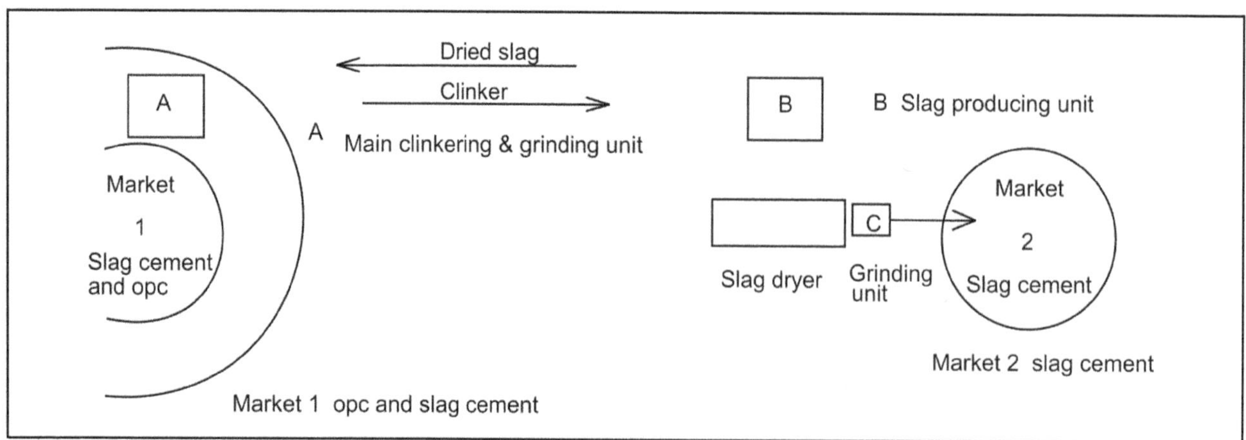

Fig. 6.28.1 Blended cements - movements of clinker and slag.

require a substantial quantity of moisture content to be dried.

	Proportions %		
	Clinker	Slag	Gypsum
	35	60	5
Moisture %	--	15%	20%

resultant moisture = 10 %

It will be difficult to dry this amount of moisture in a drying and grinding system even in a mill with a separate drying chamber. Therefore it becomes necessary to install a dryer for drying slag.

28.4 Slag Dryer

Installing a slag dryer requires double handling of slag.

1. Receiving wet slag and storing it.
2. Extraction and feeding to dryer.
3. Returning dried slag to storage.
4. Extracting and feeding dried slag to feed to C. Mill hoppers.

Therefore, the total storage area for slag would be divided into bays of wet slag and bays of dried slag.

See Fig. 6.28.2.

Slag may be brought in either Wagons (rake loads) or in trucks.

Wagon tippler installed for unloading coal wagons could be used for slag wagons.

The tonnage of slag handled would be much more than that of coal and hence wagon tippler would be primarily installed to handle slag and used for coal and gypsum also.

Depending on the size of the plant separate wagon tipplers could also be installed for unloading slag and coal wagons.

See Fig. 6.27.4.

28.5 Tonnages of Slag to be Handled

A 10000 tpd capacity clinkering unit, making slag cement would produce ~ 28000 tpd Slag Cement when

Slag to the extent of 60 % is added.

Slag to be received per day = ~ 18- 20000 tons on wet basis depending on moisture content.

Thus each day ~ 8-9 rake load of 2200 Tons each, or 5 rake loads of 4000 Tons each, would have to be handled. These quantities are sufficient to warrant investment in a wagon tippler/s

It would be extremely difficult to handle these volumes in truck loads.

To unload trucks which do not have self tipping facilities, truck unloaders which are like wagon tipplers would have to be installed.

See Fig. 6.28.4

Trucks will be locked on the tilting platform. Truck unloader can be designed for end tilting or side tilting.

28.6 Stocks to be Maintained

Even a day's receipts would be 20000 tons. Thus a minimum stock corresponding to 2 days' consumption or ~ 40000 tons is required to be provided for in layout.

A stock of at least half a day's consumption should be maintained for dried slag also.

All in all the gantry would need to stock

40000 + 10000 = 50000 tons.

A crane gantry with overhead traveling crane may be best suited for this purpose.

See Figs. 6.28.3 and 6.28.4.

Bucket elevators are not suited for handling slag, which is very abrasive. Belt conveyors are better suited but would require longer space. Dried slag can be taken either directly to mills or to gantry.

See Fig. 6.28.5.

Hoppers for wet slag need to have steep sides and large openings to prevent arching. The feeders earlier described for wet gypsum could be used for wet slag.

28.7 Slag Dryer

Dryer for slag can be either 'rotary' or fluid bed' with hourly capacity to match daily requirements. 20000 tons of wet slag with 10-15% moisture would correspond to 17000 – 18000 tons of dry slag.

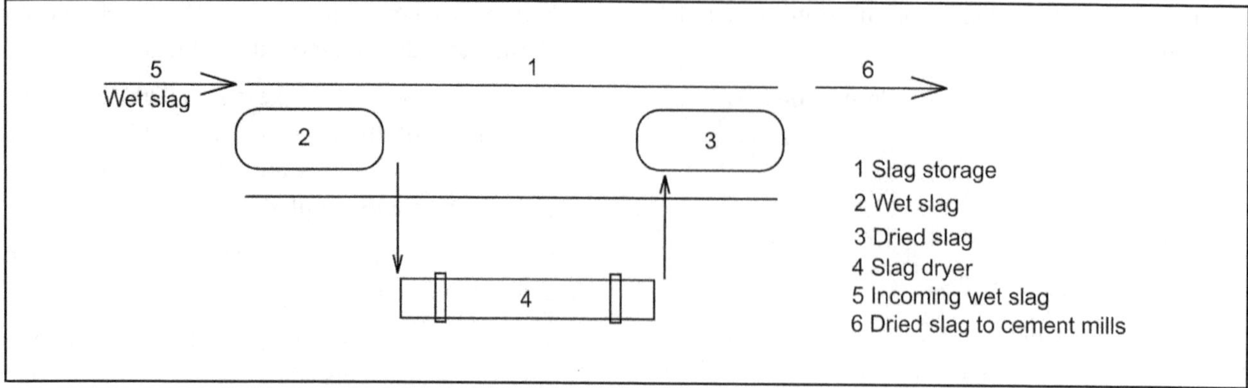

Fig. 6.28.2 Using granulated slag for slag cement.

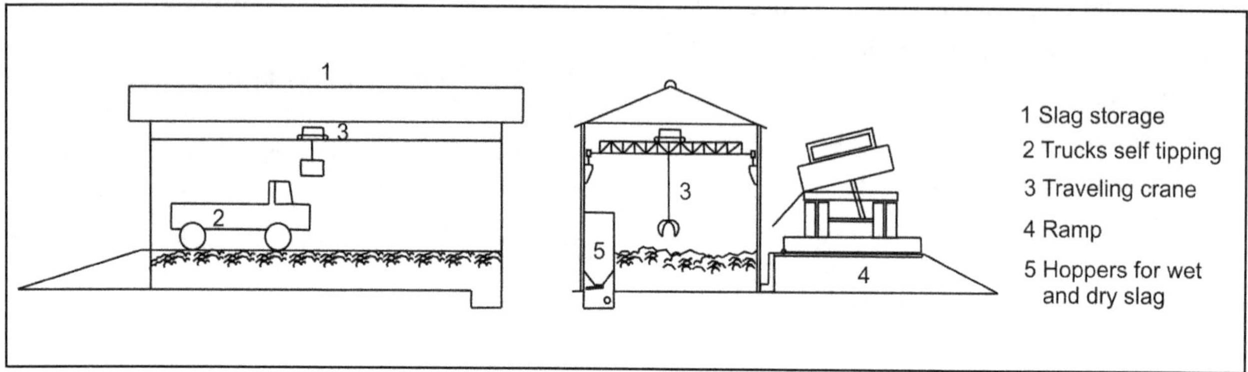

Fig. 6.28.3 Wet slag brought in tipping trucks.

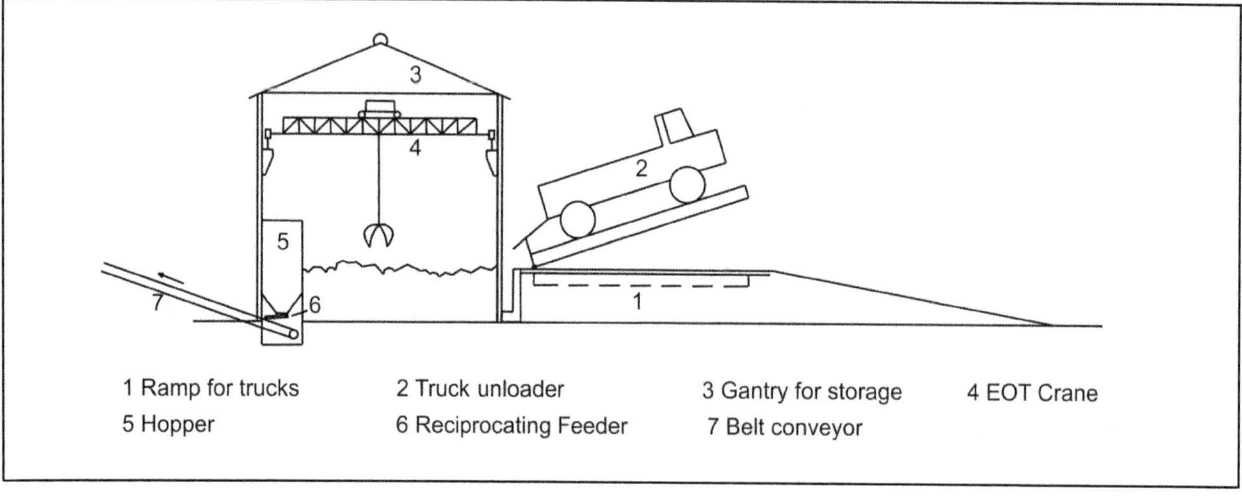

Fig. 6.28.4 Crane gantry for storage.

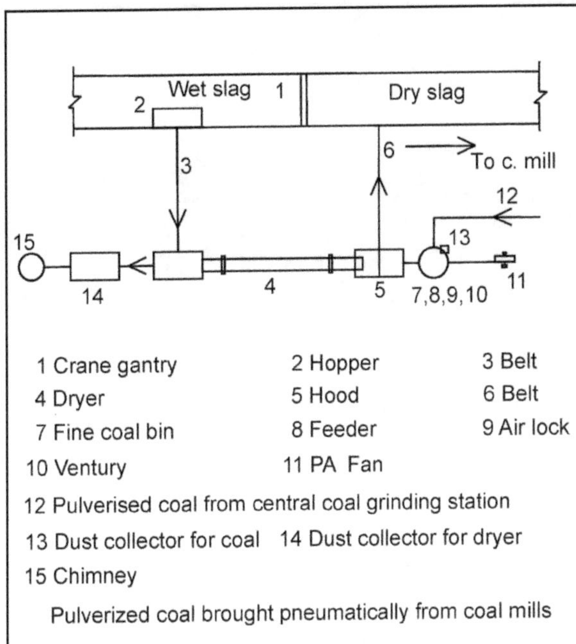

Fig. 6.28.5 Pulversied coal as fuel for burning in dryer.

Therefore, hourly capacity of dryer would have to be ~ 700-750 tons

28.8 Rotary Dryer

A rotary dryer will generally be a 2 support, coal fired dryer. Hence, in the slag dryer installation pulverized coal would be either brought in from main coal mills (in kiln section) or a separate coal mill will be installed to grind coal in the dryer building itself.

See Figs. 6.28.6 and 6.28.7.

A rotary dryer can be counter current or co-current. In a co-current dryer hot gases meet wet slag and dry it up fast. But towards discharge there is not enough heat in cooled gases to dry out residual moisture, which takes much, more heat for drying in kcal/kg water. **See Fig. 6.28.8 a.**

Therefore co-current dryers would be used when a residual moisture of 1-1.5% would be permissible. For drying to bone dry conditions, counter current dryers would be used.

See Fig. 6.28.8 b.

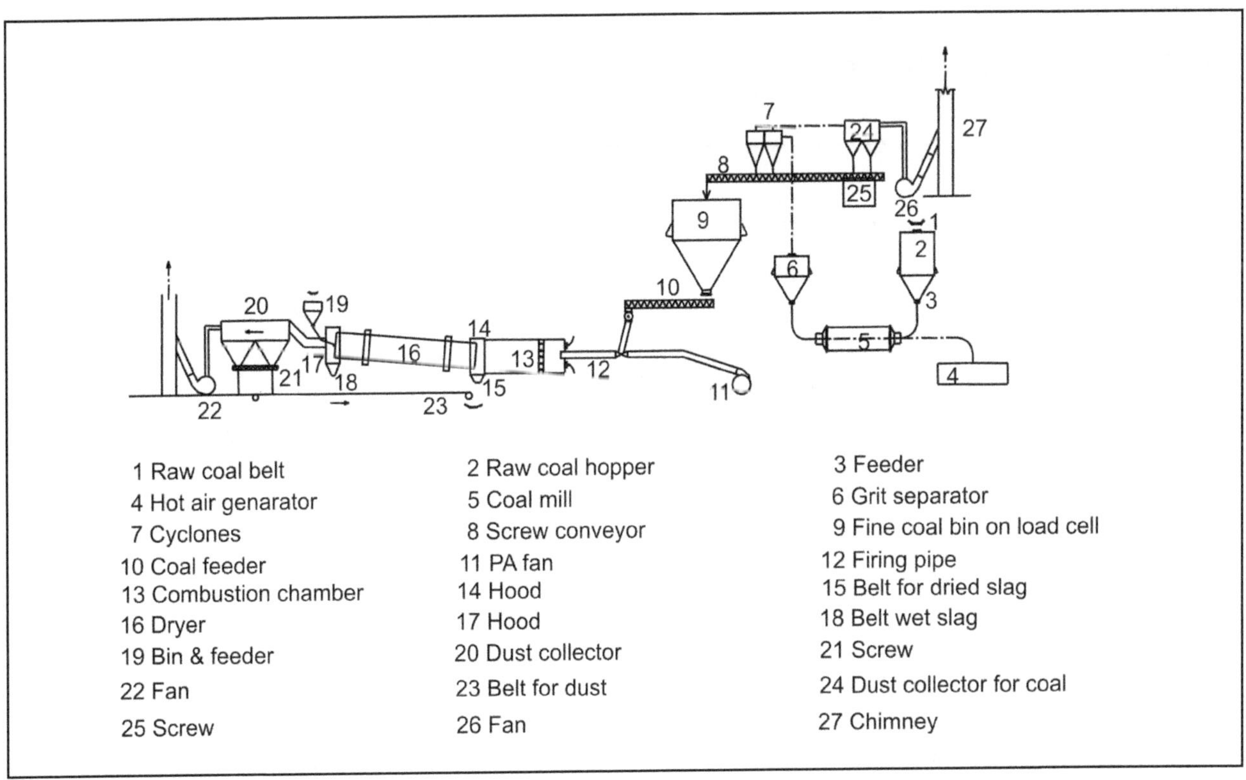

Fig. 6.28.6 Self contained dryer and coal firing system.

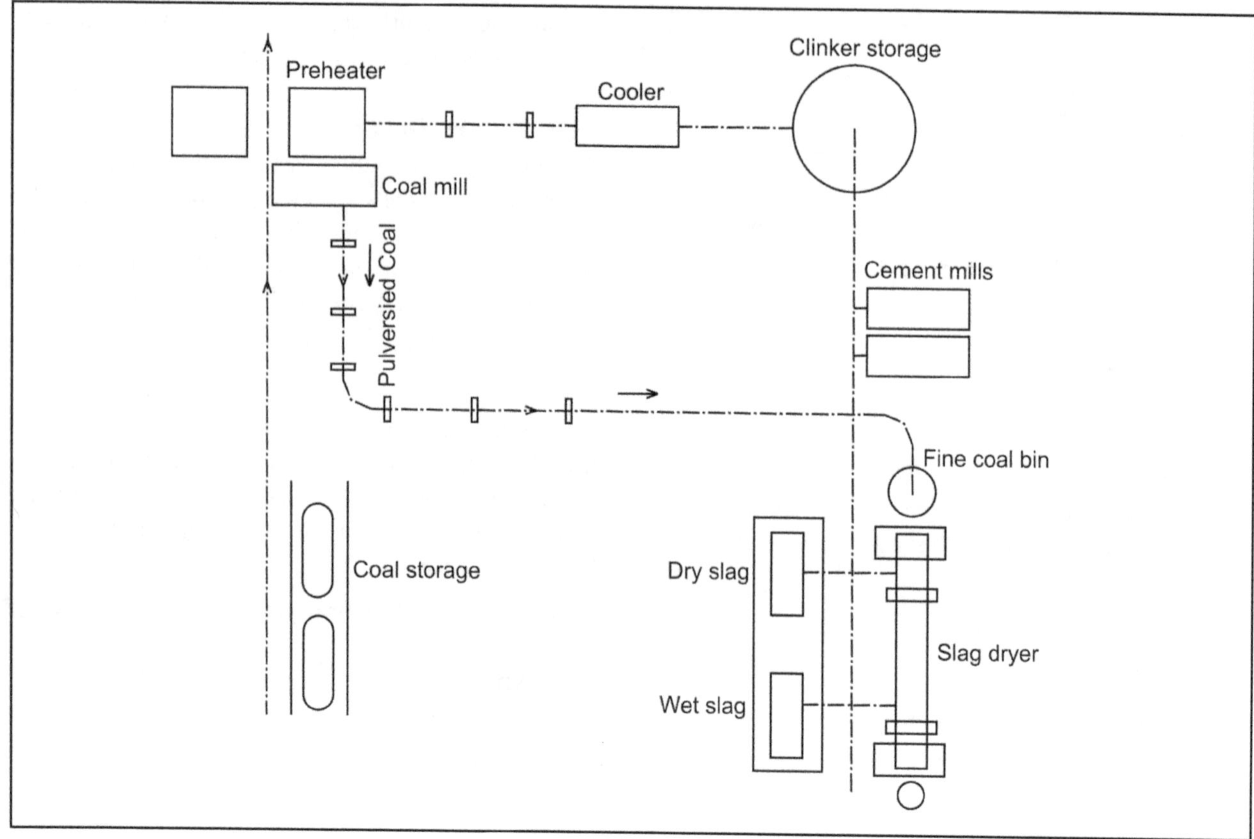

Fig. 6.28.7 Bringing pulversied coal pneumatically from main coal mills.

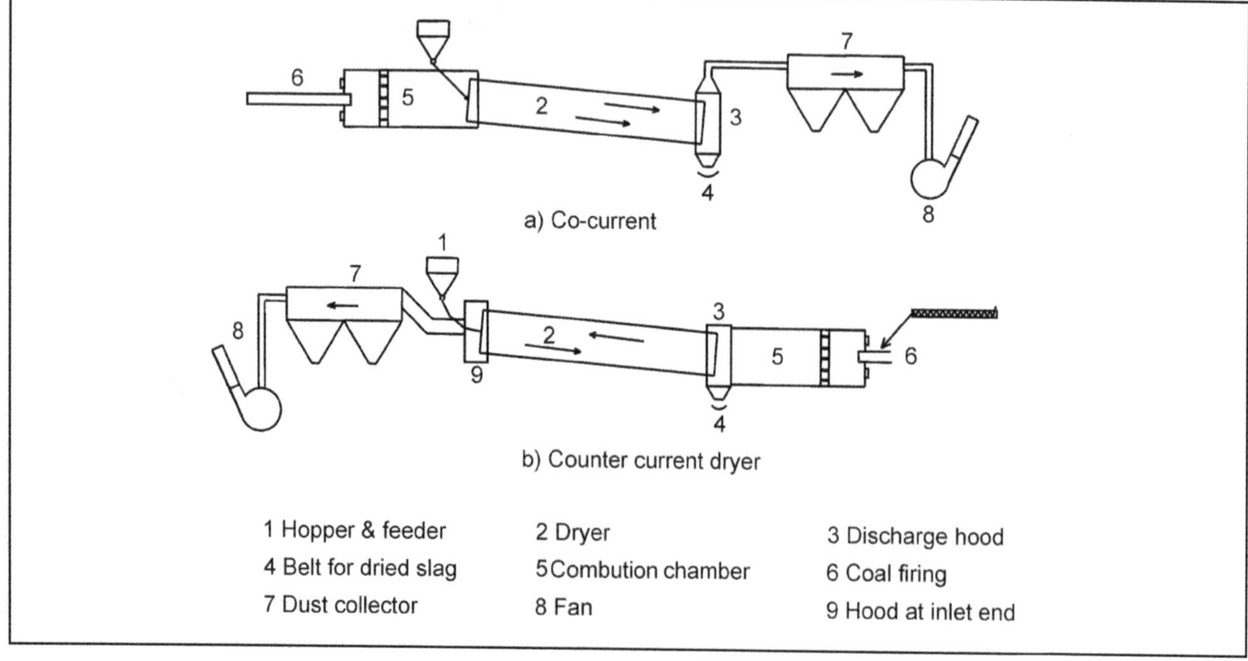

Fig. 6.28.8 Co-current and counter current rotary dryers.

The type of dryer chosen fixes the location of fuel firing system at feed end or discharge end and also that of dust collector on the opposite side.

See Fig. 6.28.8.

Hot gases used for drying would be vented through a bag filter. The gases would contain moisture and hence should leave the dryer at temperatures of 100-110 °C so that the dew point is not too close (50-55 °C) causing condensation in bags.

As in case of all dust collectors, handling wet gases (raw mill, coal mill), the dust collector would be insulated and lagged to prevent condensation.

Temperature of hot gases used would be limited to around 800 °C.

A complete dryer installation, for a counter current dryer, would be as shown in **Fig. 6.28.6.**

28.9 Fuel to be Fired

The coal to be ground in the coal mill specially installed for the dryer would be required to be dried and would require an auxiliary furnace.

See Fig. 6.28.6.

If dry pulverized coal can be pumped from main coal mills, then, this installation would not be required. But capacity of main coal mill installation should allow for requirement of coal for drying slag.

See Fig. 6.28.7.

Let capacity of dryer be 750 tph.

Water to be evaporated - drying from 15% to 0.5% = 108 tph.

Heat required to be supplied per hour at the rate of 1200 kcal/ kg of water evaporated would be ~130 million kcal/hr

If useful calorific value of coal available is 4500 kcal/kg then, coal required per hour would be ~ 30ctph

This is a fairly large quantity and the coal mill installation would be sizeable corresponding to requirement of coal for a 4500 tpd cement plant.

If the kiln is 10000 tpd and specific fuel consumption is 17%, coal mill capacity would be ~ 100 tph. To include requirements of coal for slag dryer, the capacity of a common coal mill to be installed would be 130 tph. It may be advisable to install a dedicated coal mill for the slag dryer.

28.10 Source of Hot Gas for Drying Coal

If the main coal mill is installed near the preheater end as is the practice now, heat in cooler vent would be available for drying slag in the slag dryer.

Presently, new large plants install 'Waste Heat Recovery' System to generate power from exhaust gases of kiln (that is available after drying raw materials and coal) and from cooler vent gases.

In such a case,it would be necessary to install a hot air generator to dry slag.

See Fig. 6.28.9

28.10 Hot Gas from Preheater

A similar exercise can be carried out to see if kiln exhaust gases can be used for drying slag. However in their case the first priority is to dry raw materials and now coal.

It would therefore be best to have an independent source by way of a Hot Gas Generator to dry slag **See Fig. 6.28.6.**

28.11 Fluid bed dryer

Slag dryer could also be a fluid bed dryer as shown in **Fig. 6.28.10.** Feed to dryer would be wet slag and crushed coal of - 6 mm size. Thus when using fluid bed dryer it would not be necessary to install a coal mill of higher capacity. It would be necessary to crush coal required for the dryer in a roll crusher of suitable capacity

28.12 Drying Slag in Small Plants

In case of very small plants, slag would be dried in cement mill during grinding. However it may be required to restrict the quantity of slag to be added. **See Fig. 6.28.11.**

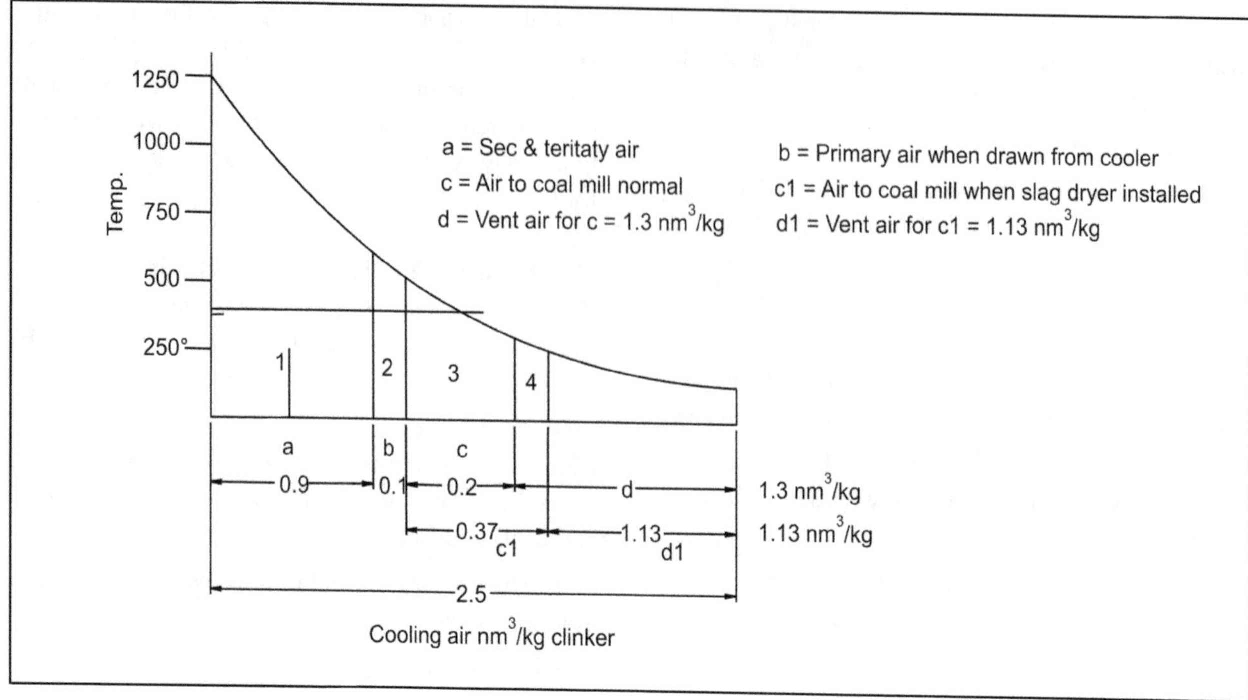

Fig. 6.28.9 Utilisation of hot air from cooler.

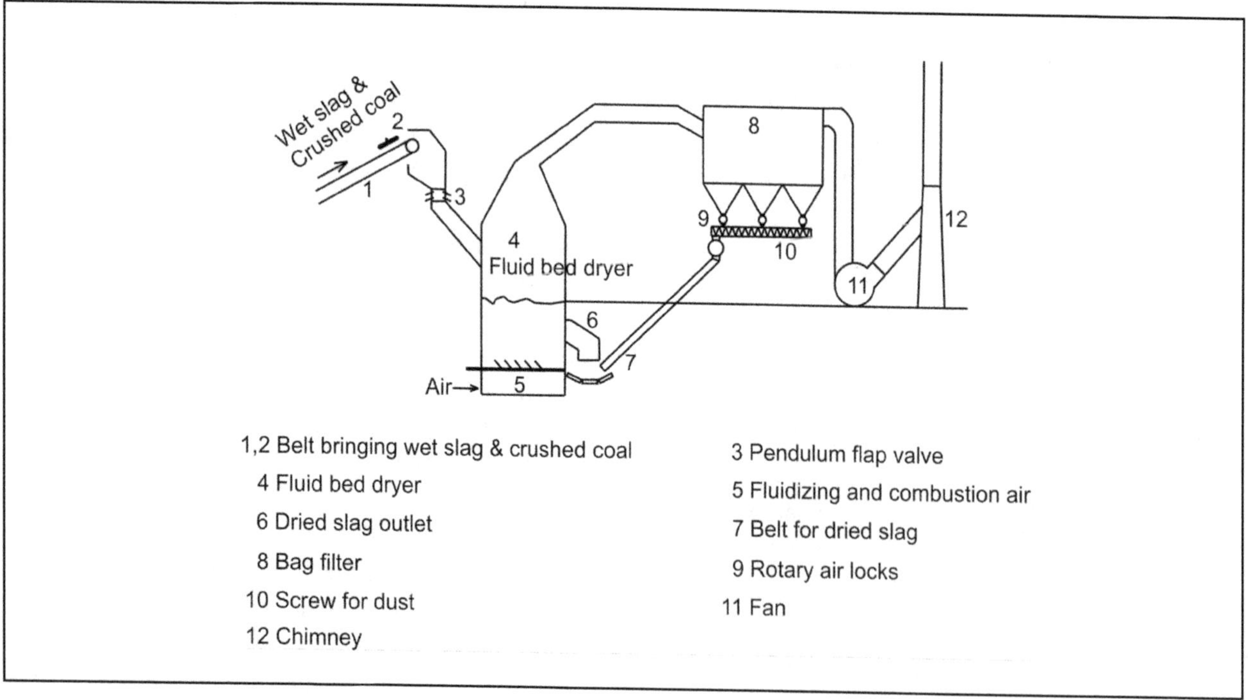

Fig. 6.28.10 Fluid dryer for slag.

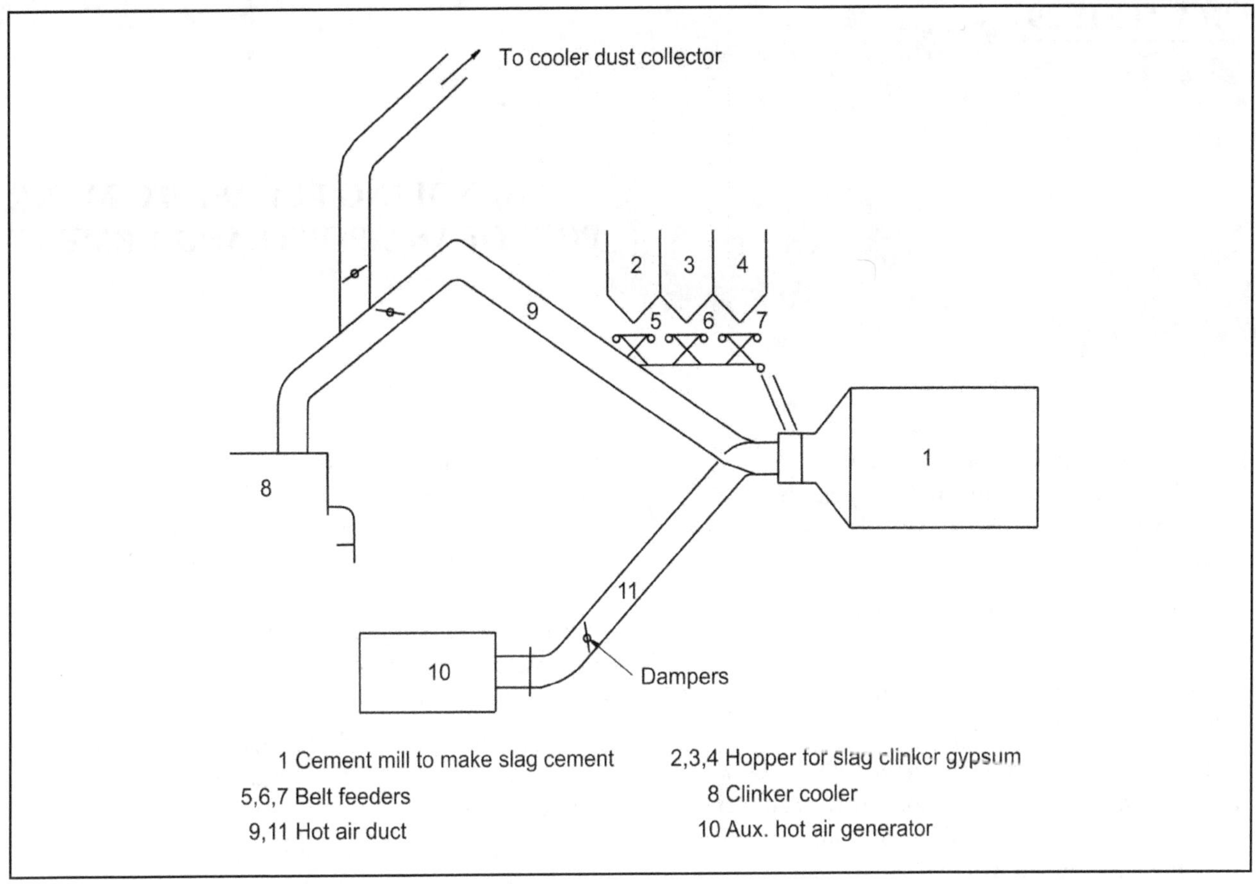

Fig. 6.28.11 Drying slag in cement mill in a small plant using cooler vent gases.

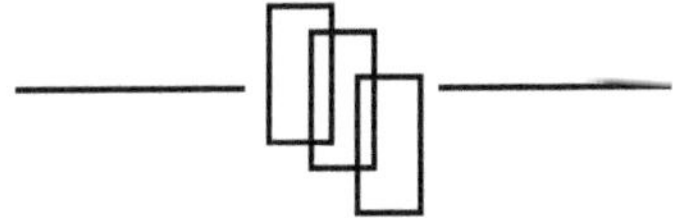

CHAPTER 29

HANDLING FLY ASH TO MAKE POZZOLANA PORTLAND CEMENT

29.1 Pozzolana Portland Cement

Pozzolana Portland Cements (PPC) can be made by blending clinker with 25-30% fly ash; more commonly in proportions, say

Clinker	Fly ash	Gypsum
65%	30%	5%

Fly ash is a very fine ash collected in the dust collectors of power plants.

It is generated dry but is available either wet or dry. In the thermal power stations, fly ash collected in settling chamber and in ESPs / bag filters is collected dry but is then mixed with water and pumped out as a slurry to settling ponds for disposal. Over a period of time water is drained out and / or pumped back for using again in the Thermal plant. Fly ash is then left to dry in the Sun.

Fly ash so collected would have a moisture ranging between 10 to 15%.

Wet fly ash is easy to recover from ponds and transport to the plant (including water) but water would have to be evaporated before it can be fed to Cement mills as in case of wet slag. Hence using wet fly ash would need a dryer to dry moisture to less than 3 to 4%.

Because of increasing making of blended cements, there is growing demand for flyash for making PPC. Cement companies are happy to lift dry fly ash from power plants. Collecting wet fly ash and drying it in cement plant is no longer relevant.

29.2 Blended Cements –PPC

Presently, Blended Cements like Portland Pozzolana Cement (PPC) and Blast Furnace Slag Cement (BFSC) have received universal acceptance and in a great number of countries including India, more than 50 % of total production is of blended cements.

29.2.1 Demand for Fly Ash and Slag

Therefore there is a steady and growing demand for fly ash produced as waste in Thermal Power Plants and also for sl,ag also a waste for Steel Plants.

29.2.2 Availability of Dry Fly ash

Thermal plants are happy to supply fly ash as it they do not have to worry about its disposal.

Fly ash is collected dry. Cement Companies are happy to receive dry fly ash as it can go straight to grinding mills.

Therefore details of processing wet fly ash have been omitted in this edition

29.3 Dry Fly Ash

Collecting fly ash in dry form is advantageous in many ways. The only disadvantage is its transport.

Dry fly ash is very fine and hence its handling generates fine dust. It cannot be transported in open trucks (if done, trucks would have to be covered with plastic sheets / tarpaulins).

At unloading point, manual unloading of ordinary or tipping trucks would cause immense dust nuisance.

It is best stored in silos. If stored in sheds, extracting could again cause dust nuisance while it is being conveyed to mill hoppers.

Dry fly ash is best conveyed over long distance in closed, self unloading bulk carriers. Such bulk carriers are filled from a fly ash storage silo / silos at the thermal power plant.

In such a case fly ash, which is generated dry, is conveyed from ESP hoppers and conveyed pneumatically / mechanically to fly ash storage silo/s. Closed pneumatic systems are best suited as dust can

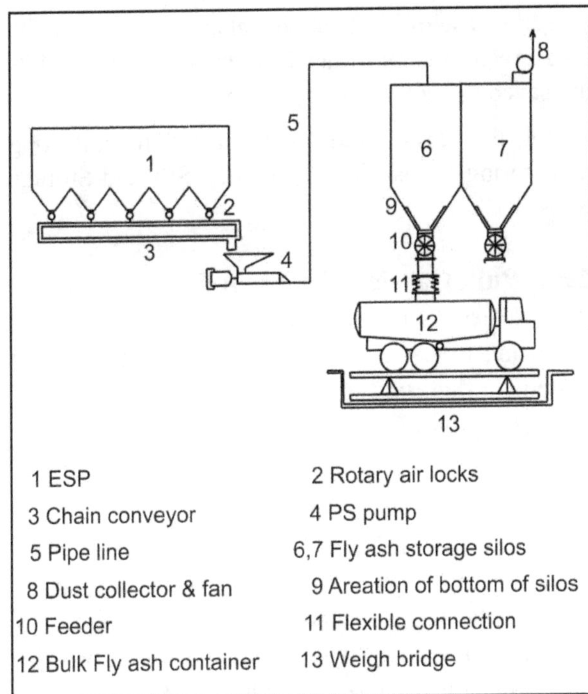

Fig. 6.29.1 Collection of dry fly ash at thermal power station.

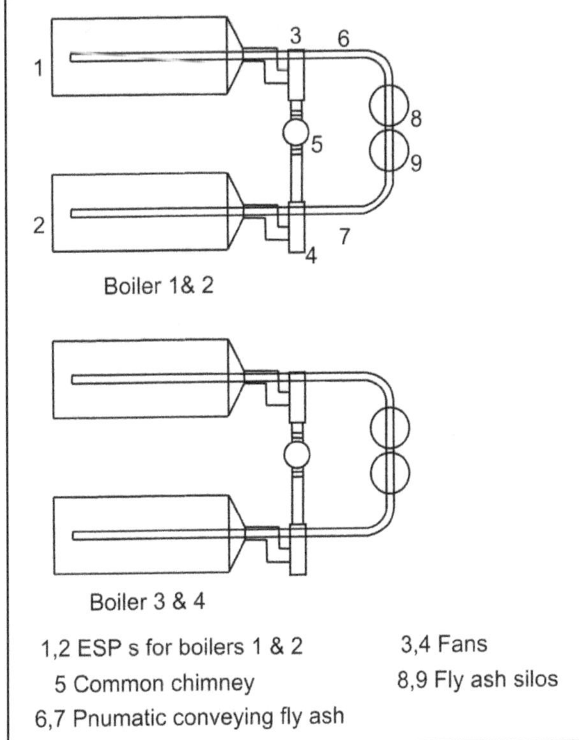

Fig. 6.29.2 Multiple boiler-ESP thermal power station.

be collected from air used for conveying easily in dust collectors located on top of silos.

Fly ash being very fine, it is difficult to convey it in air slides. Chain conveyors are preferred under esps.

See Fig. 6.29.1.

29.4 Collecting Dry Fly Ash

In large thermal plants there would be several boilers, each boiler may have ESPs with 6-7 fields. The first 2 fields collect the maximum amount i.e., about 70-80% of the total ash which is rather coarse. Subsequent fields collect correspondingly less amounts but ash collected is much finer.

See Fig. 6.29.2.

Concentrations at inlets and outlets of successive fields would be approximately as shown in **Table 6.29.1,** assuming original dust burden as 100 gm/nm^3.

Table 6.29.1

	Concentration at Inlet gm/nm³	Concentration at Outlet gm/nm³	Fineness of ash collected in microns
1st Field	100	20	+40
2nd Field	20	4	+10
3rd Field	4	0.8	+2.5
4th Field	0.8	0.16	+1
5th Field	0.16	0.032	-1

Thus fly ash collected for 1st field would need grinding; that collected in other fields needs only mixing with cement. But the quantity being small it is not convenient to collect and store and transport fly ash separately from each field of esp.

But this disadvantage can be overcome at grinding end by introducing metered fly ash first to separator and feed only the coarse fraction from it to the mill for grinding.

See Fig. 6.29.3.

29.5 Layout in Thermal Power Plant (TPP)

A thermal power station would have 2 or more boilers depending on its capacity. Fly ash can also be used for other purposes like making bricks, road surfaces, etc.

Thermal Power Plants (TPPs) now provide facilities for collecting fly ash to their consumers who enter into contract with the TPP for buying it on a regular basis as in case of slag.

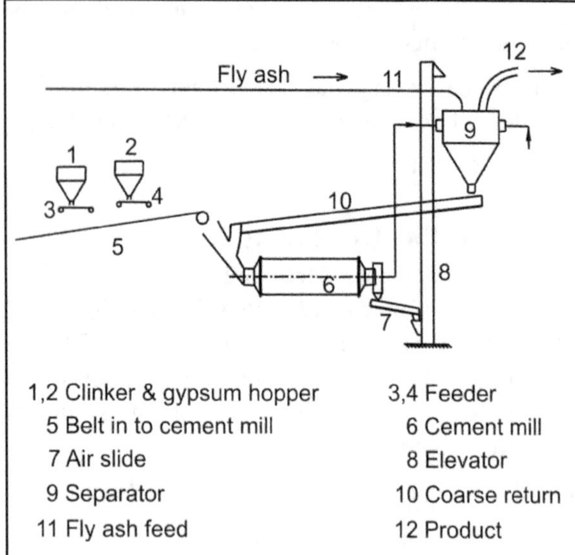

Fig. 6.29.3 Fly ash first fed to air separator.

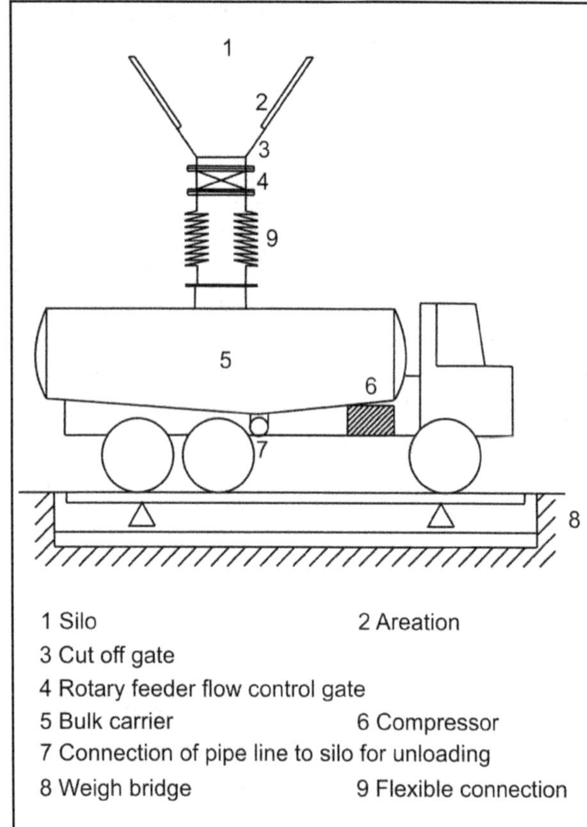

Fig. 6.29.4 Loadings bulk carriers.

As in case of blast furnace slag, if fly ash can be sold directly and immediately after its generation TPPs are saved the bother of disposing it off.

Fig. 6.29.2 shows the scheme in TPP for collecting and storing fly ash for groups of ESPs and Storage Silos.

29.6 Bulk Carriers for Fly Ash

A bulk carrier is a round or an oval tank, installed at an angle on a truck chassis. They have open air slides at bottom for fluidising and for initiating flow.

See Fig. 6.29.4.

29.7 Loading Bulk Carriers

Bulk carriers are filled with the desired quantity of fly ash according to tank's capacity by installing a weigh bridge under the loading point. Tare or weight of empty carrier is noted and weight to which it should be filled is known. Towards closing or when tank is about to be filled feeding can be slowed down or cut off sharply.

See Fig. 6.29.1

The flexible hose from the feeder has a venting pipe inside it to vent the air that is being displaced while filling the tank.

See Fig. 6.29.5.

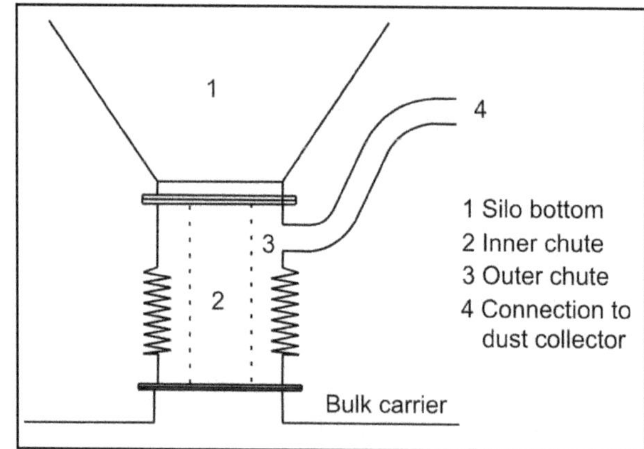

Fig. 6.29.5 Venting of flexible connection.

Alternatively, a small bin holding a measured quantity of fly ash equivalent to the capacity of the container can be kept filled up. The container can then be emptied into the carrier. In this arrangement there is no need to install a weigh bridge at loading station.

See Fig. 6.29.6.

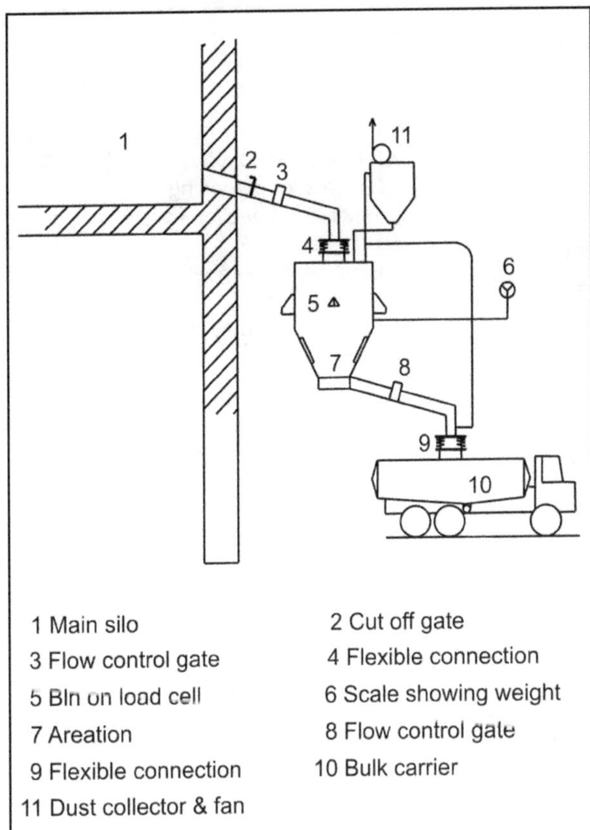

Fig. 6.29.6 Alternative method for loading.

Venting is very important because fly ash is very fine. Bag filters are best suited for this application. Even these dust collectors should be insulated and lagged.

See Fig. 6.29.6.

29.8 Unloading Bulk Carriers

At the unloading end also, dry fly ash would be stored in a silo / silos. Bulk carriers would be parked near the silo and their discharge would be hooked to the pipe conveying it to silo through a flexible connection.

See Fig. 6.29.7.

Small truck mounted compressor would be ideal but it means a lot of investment per truck.

This can be avoided by installing blower/s for unloading under the silo and connecting them to the tanker bottom by flexible hoses. A large silo can have two or more unloading points. It would be best to provide as many blowers for connecting to carriers as the number of unloading points so that a carrier does not have to wait at site for aeration connection.

Layout should provide for smooth movement of carriers, their reversing and exit. Ideally carriers should not be required to wait for unloading.

See Fig. 6.29.8.

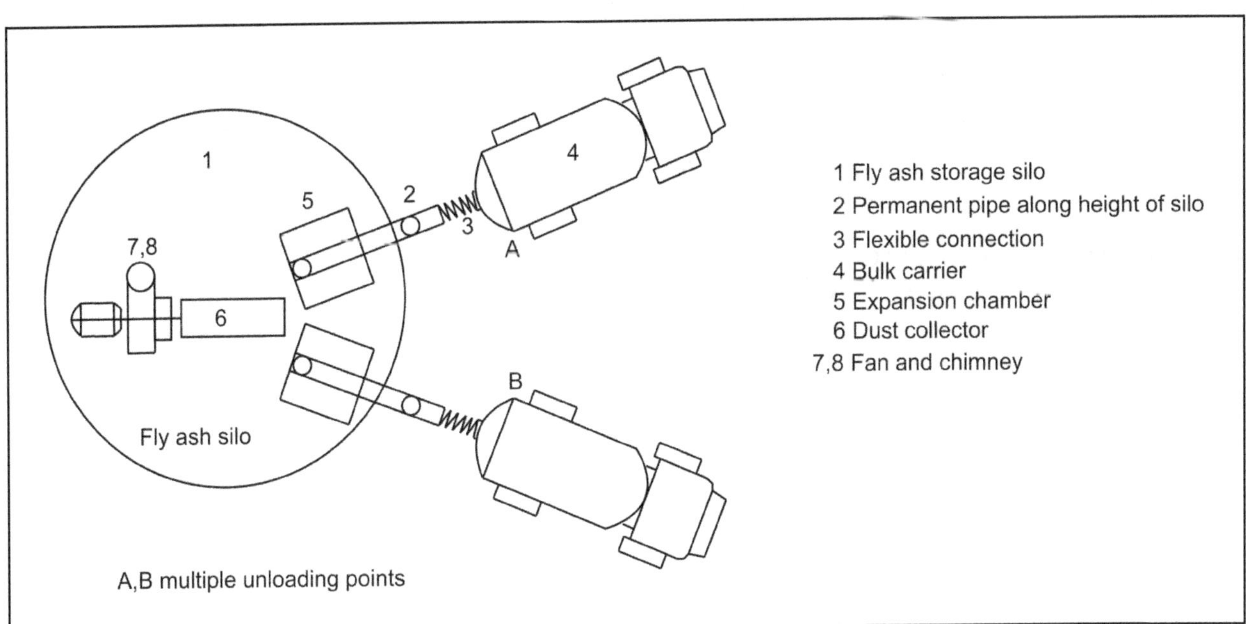

Fig. 6.29.7 Unloading fly ash in storage silo;
bulk carries with self unloading compressor.

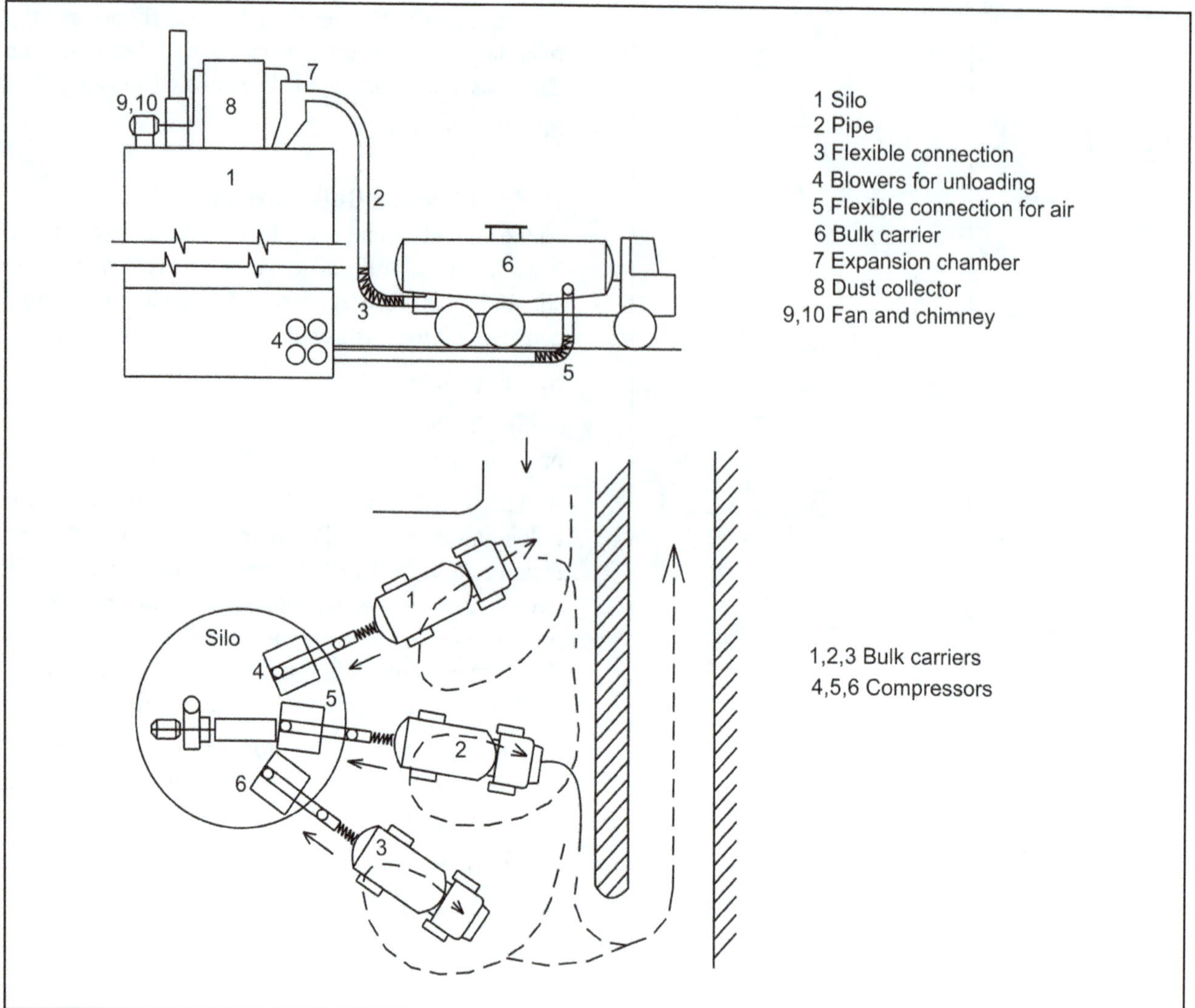

Fig. 6.29.8 Layout arround silo to facilitate handling a number of bulk carriers simultaneously.

29.9 Capacities of Bulk Carriers

Size and capacity of the bulk carrier is decided by the width of roads and terrains through which roads pass, sharp bends, inclination, tunnels and so on.

Fly ash having a low bulk density (aerated fly ash would have even less) the size of tank would be much bigger than normal.

The tanker should be carefully designed and mounted on the truck, the center of gravity remaining within the wheel base so that the tanker does not over turn while negotiating bends or slopes.

29.10

Fly ash can also be handled in plain ordinary self tipping trucks by closing the truck with plastic / tarpaulin or by spraying some water on its surface to make it wet.

But this system though economical in capital costs, is labour intensive and prone to severe dust nuisance both at loading and unloading ends and hence should be avoided.

29.11 Extraction of Fly Ash

Fly ash would be extracted from the silo pneumatically. The bottom would be fitted with open air slides to fluidize fly ash and it can then flow into discharge

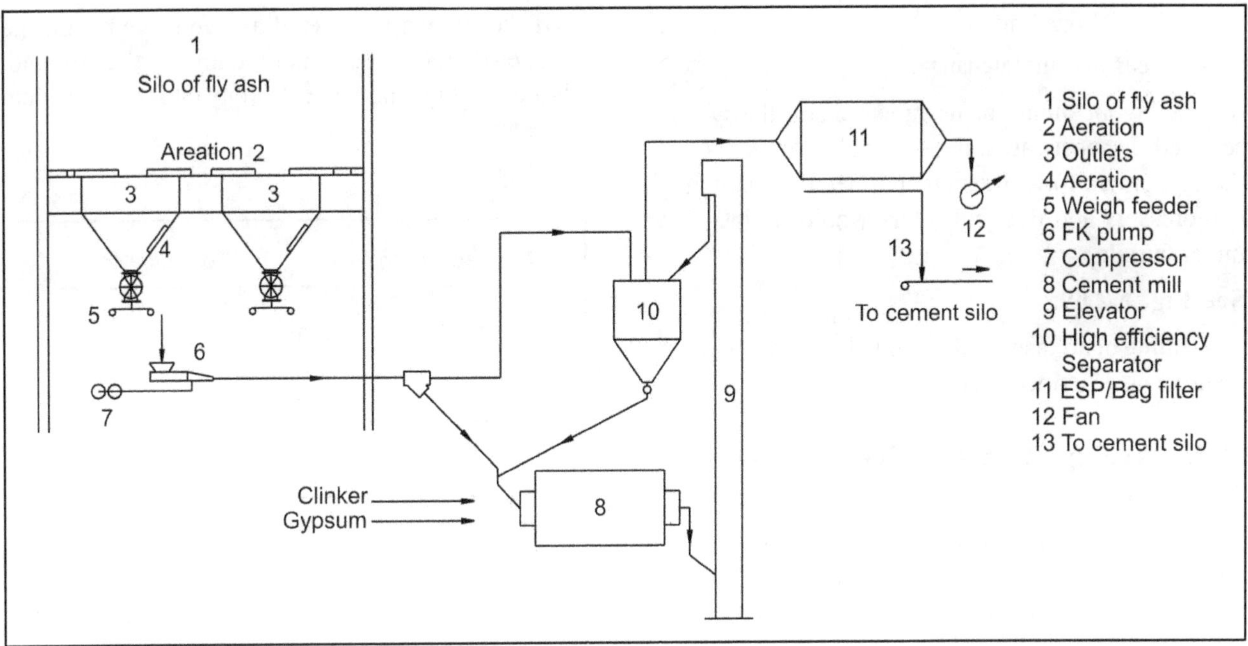

Fig. 6.29.9 Unloading fly ash from storage silo and feeding it to cement mills for grinding.

points. The area so covered would range between 8-10% to 15-20% of the silo area depending on the construction of the bottom.

See Fig. 6.29.9

The number of discharge points would depend on size of silo and also number of mills to which that silo has to supply fly ash.

One silo with one or more discharge points can supply fly ash to one or more Cement Mills.

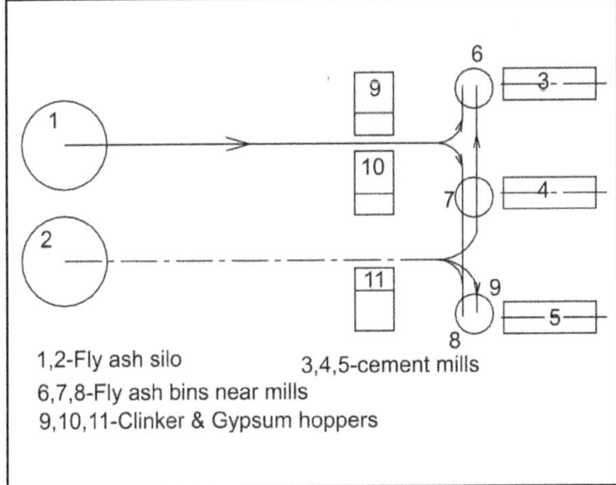

Fig. 6.29.10 Flexibility and interchangeability with multiple silos and cement mills.

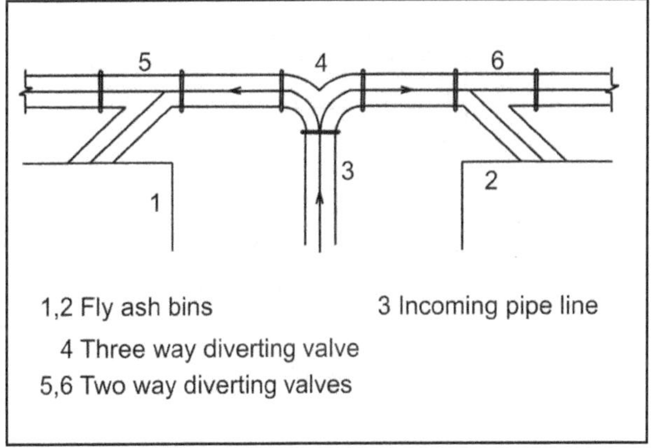

Fig. 6.29.11 Pnumatic conveying multiple discharge.

Interchangeability and flexibility can be achieved with minimum capital investment in pneumatic conveying.

See Figs. 6.29.10 and 6.29.11.

29.12 Metering Systems

Several options exist for metering systems to be selected and for pneumatic conveying.

Main considerations in selection would be :

1. Accuracy desired.
2. Dependability and availability.

3. Power consumption.
4. Wear and maintenance.

The layout should be designed to suit the system selected. In most cases, the silo will be on stilts and the equipment for extraction and metering including compressors and dust collectors would be installed under the silo.

See Fig. 6.29.9.

Venting of air displaced in unloading would have to be done as described earlier.

29.13 Feeding Fly Ash to Cement Mills

The fly ash can be fed at inlet of mill or in separator.

In option 1, fly ash is extracted from silo and conveyed pneumatically to a bin near the mill. A feeder under the bin, volumetric or weighing will extract fly ash from the bin and feed it to mill inlet chute.

See Fig. 6.29.12.

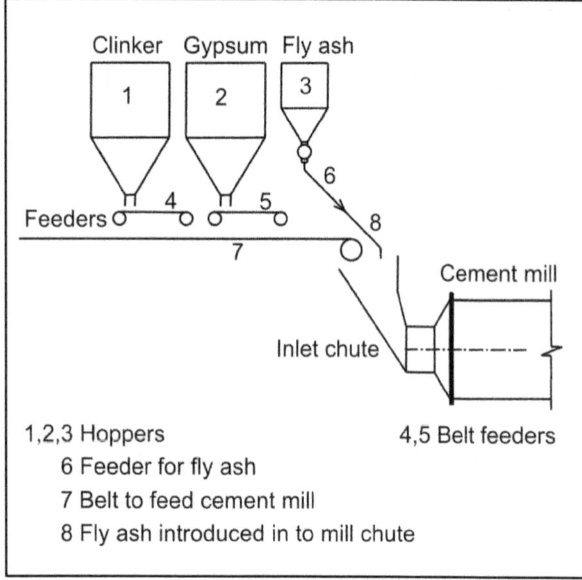

Fig. 6.29.12 Introducing fly ash in mill.

If the fly ash is metered and conveyed from the storage silo then a bin is not required at the mill end. The conveying pipe can just empty into the feed chute of the mill.

See Fig. 6.29.13.

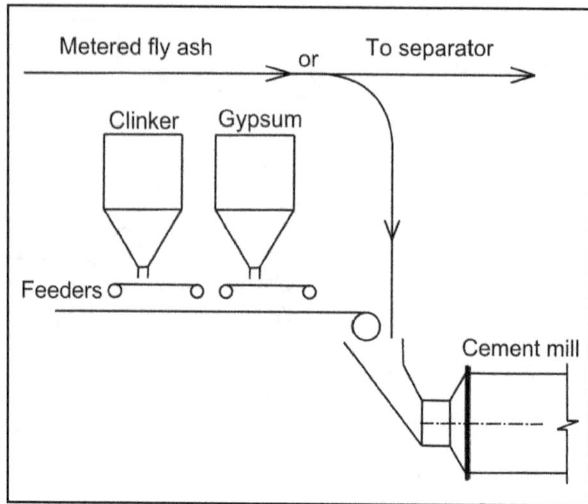

Fig. 6.29.13 Fly ash metered under main storage silo.

If fly ash is sent first to a separator then only coarse fraction would be sent to mill for grinding. Generally 70 to 80 % of fly ash need not really be ground. Therefore there would be saving in energy as well as increase in output of the mill.

29.14 Separate Grinding of Fly Ash

A third option is also available in that fly ash could be ground separately to desired fineness and blended with OPC in a metered manner.

See Fig. 6.29.14 a, b and c.

This alternative of grinding, blending material separately is also available for blast furnace slag cement and will save power on a much larger scale because clinker to slag are in ratio of 40 : 60.

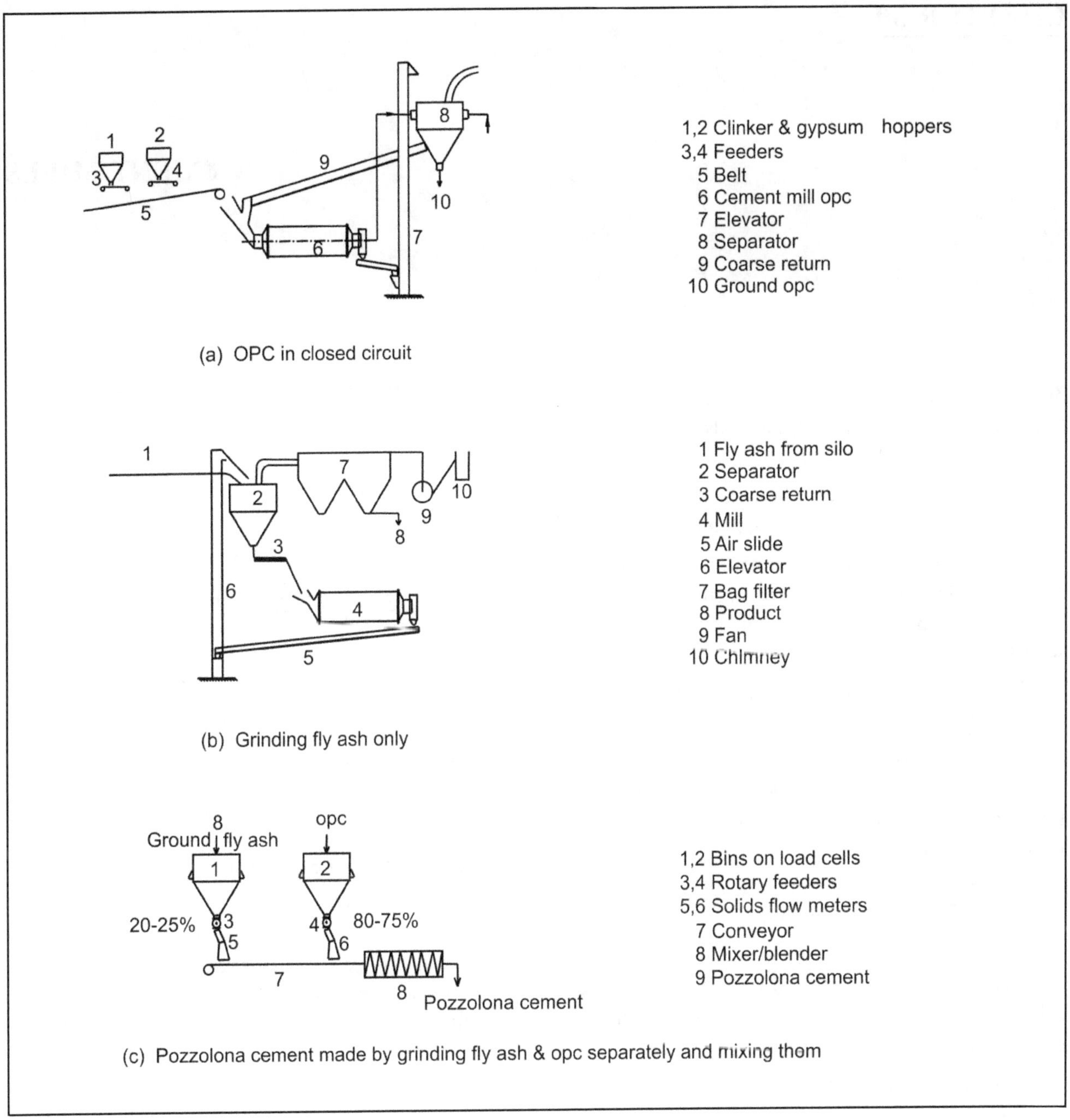

1,2 Clinker & gypsum hoppers
3,4 Feeders
5 Belt
6 Cement mill opc
7 Elevator
8 Separator
9 Coarse return
10 Ground opc

(a) OPC in closed circuit

1 Fly ash from silo
2 Separator
3 Coarse return
4 Mill
5 Air slide
6 Elevator
7 Bag filter
8 Product
9 Fan
10 Chimney

(b) Grinding fly ash only

8 Ground fly ash
opc
20-25%
80-75%
Pozzolona cement

1,2 Bins on load cells
3,4 Rotary feeders
5,6 Solids flow meters
7 Conveyor
8 Mixer/blender
9 Pozzolona cement

(c) Pozzolona cement made by grinding fly ash & opc separately and mixing them

Fig. 6.29.14 Separate grinding of fly ash.

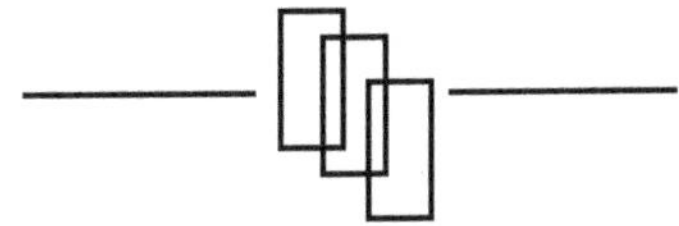

CHAPTER 30

CEMENT MILLS

30.1 Cement Mills

Cement mills are dry grinding ball mills working either in close circuit or open circuit.

Cement mills can also be vertical mills which are closed circuit mills with built in separators.

As in case of raw mills, combinations of roller press and ball mill are also used to make cement. They will be operating in closed circuit with high efficiency separators.

Vertical roller mills are increasingly coming into use to grind OPC and also Slag Cement.

Slag being very abrasive, wear on rollers and table liners would be very high and allowance has to be made for more frequent replacement and more down time for this purpose.

Vertical rollers mills would be heavy in capital cost but would save considerable power in operations.

Separators come in two designs, one conventional also with variable speed dual drive and two, high efficiency separators which always have variable speed drives.

New plants will, have only closed circuit mills because of ease with which different types and grades of cements which need grinding to different finenesses can be made. High efficiency separators after initial troubles are now well established.

30.2 Roller Press and Ball Mill

This combination is more suited for cement grinding because of hardness of clinker and slag.

Clinker when passed through roller press under goes structural change and its grindability is reduced by 1/3rd.

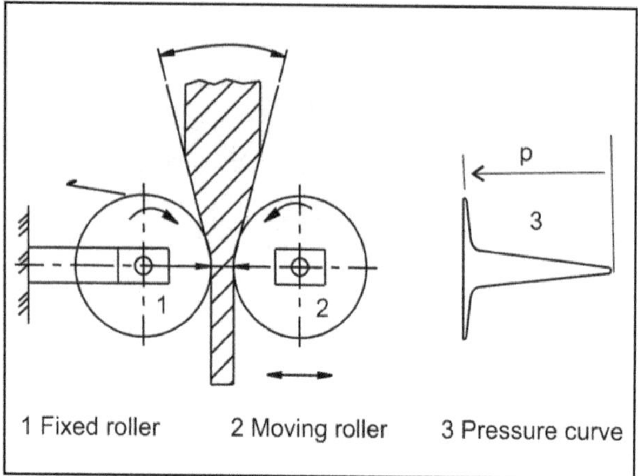

Fig. 6.30.1 Action of roller press.

Action of roller press is shown in **Fig. 6.30.1.**

Thus clinker with a BWI of 16 after passing through the roller press will have a BWI of ~ 11. In passing through the roller press, approximately 1/3rd product is reduced to the desired fineness. Thus the power required to be put in for producing cement is reduced by about 45 % for mill only. Total system power required would be equal to :

'Power of roller press system = power for press + 0.55 × ball mill power + power for auxiliaries'.

30.3 Saving in Power

Power in auxiliaries in terms of kwh/t would be more or less same for ball mill and vertical roller mill circuits. It may be slightly more in case of roller press and ball mill circuit because roller press also works in closed circuit and number of auxiliaries are more.

Thus if a ball mill circuit required 35 units / ton consisting of 28 kwh/t for mill + 7 kwh/t for auxiliaries, with roller press it will be reduced to 2 Kwh/t for roller press + 15.5 kwh for mill + 8.5 kwh for auxiliaries

= 26 kwh / t i.e., a saving of 26 %.

30.4 Layout of a Cement Mill

The layout of a cement mill department is similar to that of raw mill being governed by the same principles.

The ball mills will be longer for the same capacity because of differences in fineness to which raw meal and cement are ground and also because of differences in grindabilities of raw materials and clinker. The differences are compensated to some extent by the fact that 1.5 ton of raw material is required for making 1ton of clinker.

Thus	raw materials	clinker
	BWI	BWI
	10	14
Fineness of	14-16% residue	3000 Blaine
Product	on 90 microns	
Feed size	- 20 mm	- 20 mm
Power kwh /ton		
of material ground.	16	
Per ton of Clinker		28
Kwh/ton of cement	23	27

For a10000 tpd capacity plant with a design margin of 10 % would be

Raw mill	cement mill
825 tph	550 tph
Mill power	
13200 kw	15000 kw

Thus in all probabilities motors and gear boxes would be selected of higher ratings i.e. 15000 kw.

Ball mill can have a dual drive.

However as explained earlier, cement mill will be sized to have additional margin of 10 %. Hence its capacity would be 600 tph and mill power 16500 kw. In this case also motor and gear box of higher rating would be selected both for raw mill and cement mill.

If Roller Press is used in conjunction with Ball Mill, a much smaller Mill of 11-12000 kw would be required as seen above.

30.5 Cement Mill Sizing and Power

Since length / diameter ratios for raw mill and cement mill differ because of finenesses to which they are to be ground. l/d would be

For raw mill, = 1.8 - 2

For cement mill = 3 - 3.5

Ball mill power is a function of diameter, speed and % loading. Speeds of ball mills are generally between 70 to 75 % of cirtical speed and % loadings for same type of mill are similar for raw and cement mills.

Power = $0.515 \times d \times n \times F \times$ Constant

F = Grinding media load in tons - both mills will be loaded to about the same percentage say 30 %. It is calculated from loaded volume and average bulk density of grinding media.

n = mill speed in rpm

n would normally be $31.73/d^{0.5}$

d = diameter of mill inside liners

Constant K is a function of type of grinding media and lining plates and is also related to torque arm which is related to % loading. It could be different for raw mills and cement mills when cement mill is in open circuit.

Briefly, power drawn =

$K1 \times d^{3.5}$ for raw mill

$K2 \times d^{3.5}$ for cement mill

Taking above example, sizes of raw and cement mills would be worked out as follows.

See Table 6.30.1.

Thus though motors will be selected corresponding to power of cement mill, diameters will not be the same. However, if motors and gear boxes can be the same and interchangeable, the spares / standby can be reduced.

If mills were of different diameters their speeds would be different. To keep the same gearbox, gear pinion ratios would be different for two mills.

All these points have an influence on layout and have been brought out for that reason.

Table 6.30.1

Sr. No.	item	unit	Raw mill	Cement mill
1.	loading	%	30	30
2.	l/d ratio		2	3.5
3.	b.density grinding media	t/m^3	4.5	4.7 *
4.	speed	r.p.m.	$31.73/d^{0.5}$	
5.	Clear diameter	m	d	d
6.	Grinding load	tons	$2.12 \times d^3$	$3.87 \times d^3$
7.	constant	0.44	0.44	0.44
8.	Mill power	kw	$15.24 \times d^{3.5}$	$27.83 \times d^{3.5}$
9.	capacity	tph	825	550
10.	Power at shaft	kw	13200	15000
11.	Diameter d	m	7.0	6.1
12.	Grinding length	m	14	21.5
13.	Dia. inside shell	m	7.1	6.2
14.	Total length	m	14.7	22
Assuming same constant				

* bulk densities of grinding media would differ for raw and cement mills because in case of cement mills a large % of balls would be between 25 and 15 mms in diameter.

It may be desirable to have 2 mills for a 10000 tpd plant from the point of availability of motors and gear boxes. Also in case of break down production will be maintained at 50 %.

Thus though motors will be selected corresponding to power of cement mill, diameters will not be the same. However, if motors and gear boxes can be the same and interchangeable, the spares / standby can be reduced.

If mills were of different diameters their speeds would be different. To keep the same gearbox, gear pinion ratios would be different for two mills.

All these points have an influence on layout and have been brought out for that reason

30.6 Drives

Mill can have a single or a dual drive. Single drive will be used to suit maximum capacities of motors and gearboxes available indigenously. Dual drive will be used so as to remain within the range of capacities thus available.

Dual drive also has the advantage of being able to use gear and pinion of the same width i.e., if width of gear for 2500 kws is 750 mm, it would be the same for 5000 kws with dual drive.

See Fig. 6.30.2.

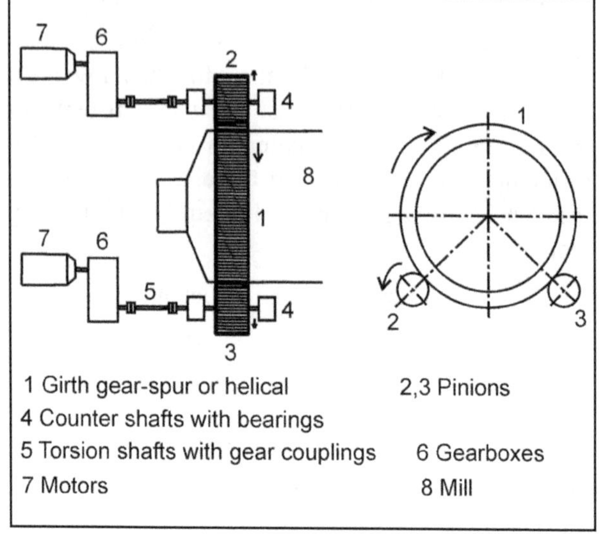

Fig. 6.30.2 Dual drive for ball mill.

Dual drive can be avoided if number of mills are increased. This would however be a very expensive proposition.

The two motors should share load equally. But to be on the safe side a margin of 10% is added to motor rating when using dual drive.

This example has become rather irrelevant, as ball mill would be seldom used for grinding raw materials. Gearbox of a vertical mill is totally different than that of ball mill. Hence, gear boxes would seldom be same.

Perhaps motor ratings of two mills could match.

If it is proposed to select a ball mill only for grinding cement, the size and rating of mill drive puts a limit on capacity of a single mill.

For example cement mills of plants of 5000 to 10000 tpd capacities would be as shown in **Table 6.30.2 below.**

Table 6.30.2

Plant capacity t.p.d.	5000	7500	10000
With Design margin and			
Gypsum - 5%	5800	8700	11600
Mill capacity tph	~300	~440	~600
Mill motor in kw@			
30 kw/ton	9000	~14000	18000
Motor kw	9000	14000	18000
The most probable selection for mills and drives would be			
	Single Mill	Single mill	2 Mills
	Single drive	Dual drive	Single drive

30.7 Blended Cements

Mills are also to be sized according to the cements that they are required to grind. This is particularly so when a plant makes blended cements regularly.

Situation is very different for the two types of blended cements viz. PPC and Slag cement.

In case of PPC, using fly ash, output of a mill will increase when making PPC by about 17 to 18 %. Hence mills should be sized for OPC and auxiliaries for PPC.

In case of Slag cement, situation is very much opposite.

Granulated slag is generally –6 mm in size and needs grinding. There is considerable difference in granulometries of clinker and slag and also in their grindabilities. Slag can either be inter ground with clinker to make slag cement or the two can be ground separately and blended to make slag cement.

If the two are ground separately their respective mills could be sized and designed specifically to suit their properties and hence would give better results than in inter grinding

30.7.1 Power for Slag Cement Mill

Mill required to grind slag would be much bigger than that required to grind OPC because slag is difficult to grind and because it is required to be ground to a greater fineness.

OPC	-	Blaine	3000
Slag cement	-	Blaine	3400

Let Grindability of clnker and slag measured in Bond's Work Index be 16 and 20 respectively. When inter ground, in proportions of 40 slag and 60 clinker, resultant B.W.I. will be 18.4.

If the mill required 26 kwh/ton when grinding OPC of 3000 blaine, it will require 36.6 kwh /ton when grinding slag only to 3400 blaine an increase of 45 %.

When slag cement is interground, to a blaine of 3400, power required would be 37.3 kwh/ton.

For the same capacity, mill required to grind only slag or to make slag cement by intergrinding, will be much larger.

Cement mill capacities when making blended cements for a 10000 tpd plant would be

OPC 640 tph; PPC(30% ash) 930 tph;
BFSC (60% slag) 1730 tph

The number of mills and power required to drive them would have to be worked out taking into account availability of motors and gear boxes on the lines described above.

If the plant starts with making OPC only and blended cements are added later, sizing selection would have to be done taking into account sizes and capacities of the first unit to reduce inventory of spares.

30.8 Capacities of Silos

Capacities of intermediate silos for ground slag and OPC would be minimum – 4 hours' and maximum-12 hours' mill capacities.

Let Clinkering capacity be	-	10000 t.p.d.
Design capacity	-	11000 t.p.d.
Clinker % 35	-	11000 t.p.d.
Slag % 60	-	~ 19000 t.p.d.
Gypsum % 5	-	~ 1600 t.p.d.
Total rounding off	-	31600 Tons Slag Cement
Or	-	11600 Tons OPC

The layout must be arranged to facilitate feed of clinker, slag and gypsum to multiple mills.

See Fig. 6.30.3 which is shown for intergrinding.

30.9 Storage of Slag and Gypsum

If slag and gypsum are stored in a common shed or in adjoining sheds, same conveyor could be used to feed gypsum and slag alternately.

See Fig. 6.30.4.

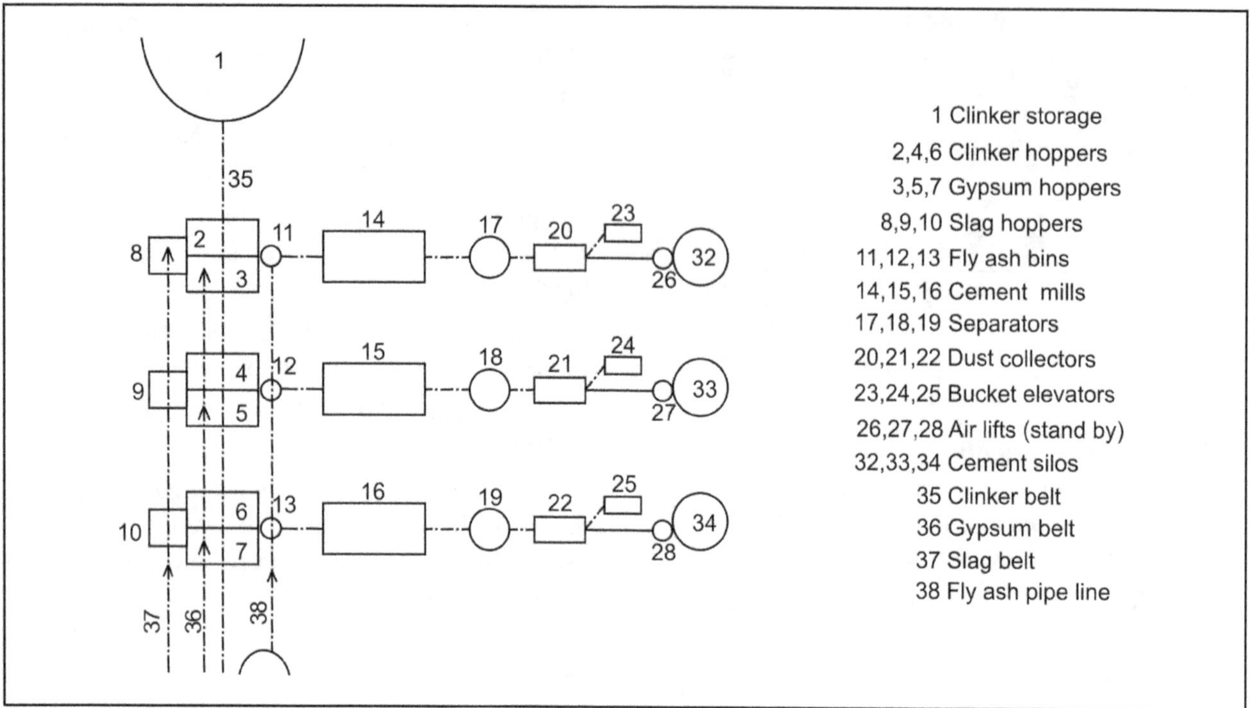

Fig. 6.30.3 Layout - to make slag/pozzolona cement.

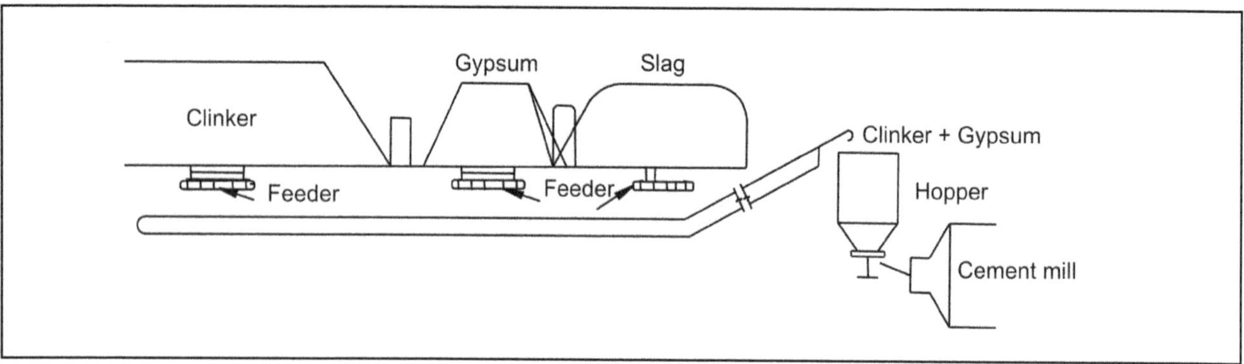

Fig. 6.30.4 Making slag cement in a small plant.

For a large installation however, there could be different storages and either a common belt or separate belts for, slag and gypsum.

See Fig. 6.27.4 in Chapter 27.

30.10 Hoppers for Clinker and Slag

Clinker and slag hoppers would be of same size though proportions of clinker to slag would be 40 : 60.

Gypsum at 4% would require much small hoppers. These gypsum bins could also be made of steel with conical bottom of steep angle and vibrators for ease of extraction of gypsum.

30.11 Making Slag Cement in a Small Plant

In small plants, wet slag (in the absence of dryer for slag) is proportioned and fed on to clinker belt. Same would be the case of gypsum. Mixing of gypsum with clinker helps in flow of gypsum and clinker in the hopper. It would be better though to feed mill with dried slag.

See Figs. 6.30.4 and 6.30.5.

If slag is to be handled and ground wet, it would require hot gases to be made available for cement Mill to dry slag and gypsum. However, in this way a total moisture of 5-6 % only can be dried. Therefore it may not be possible to add full quantity of slag.

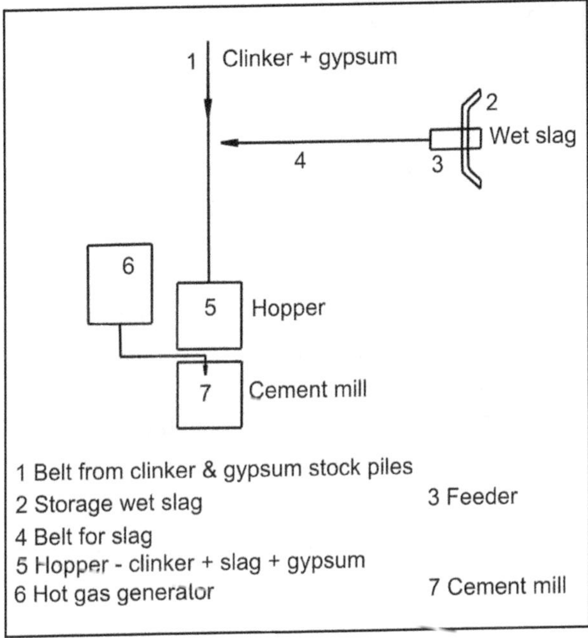

Fig. 6.30.5 Slag cement in a small plant receiving wet salg; drying of slag done in cement mill.

These gases can come from cooler vent at 200 °C. But then the mills would have to be installed close to coolers. This may be possible for small plants but unlikely for large plants.

See Fig. 6.28.11 in Chapter 28.

It would be best to take gases for drying after the cooler dust collector otherwise it would carry clinker dust. This is not a problem as such except that the dust would tend to settle in ducts.

Another alternative would be to install a hot air generator, coal fired, to provide hot gases for drying.

See Fig. 6.28.11.

Which system to be adopted has to be decided in advance with the help of Consultants. This will facilitate firming of flow charts, departmental and plant general layouts.

30.12 Conveying Cement

It has been mentioned earlier that pneumatic conveying systems have been replaced by mechanical conveying systems like bucket elevators and belt conveyors.

There should be a 'standby' for system conveying cement to cement silos so that production is not interrupted.

For example :

Two mills – 100 tph capacity can have 2 bucket elevators of 100 tph each + 1 elevator of 100 tph as standby. Alternatively, there can be two elevators of 200 tph each. One will work and other would be a standby.

See Fig. 6.30.6.

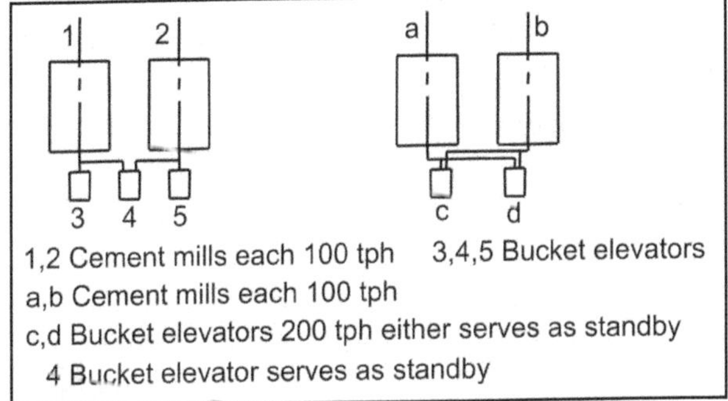

Fig. 6.30.6 Standby for cement elevator in cement mill.

In many cases, existing air lifts would be discontinued and bucket elevators used in their place for saving power. In such a case existing air lifts would be retained to serve as standby.

30.13 Layouts with Roller Press and Ball Mill

Flow chart 1.7.8 in **Chapter 7** of **Section 1** show the various ways in which Roller Press and ball mill could be used together to grind cement. As shown therein, there are three basic ways :

1. Use roller press in open circuit as a precrusher only.
2. Hybrid grinding in which both press and mill work in closed circuit.
3. Use only roller press with a disagglomerator.

Roller press operation necessitates that pressed cake at edges is separated from the cake in the middle and is returned to press for recrushing.

See Fig. 6.30.7.

This introduces more auxiliaries in the circuit.

Therefore as compared to V.R. Mills (even one with an external circuit), roller press –ball mill system will have more number of auxiliaries and would require more space also.

Hoppers feeding roller press would be housed in a separate structure as would be the case with a V.R. Mills.

Roller press itself would be housed in a separate building with facilities for removing rolls and for building them up 'in situ' when worn.

Ball mill and air separator and their auxiliaries and also dust collector would be housed in yet another building.

See Figs. 6.30.8 and 6.30.9.

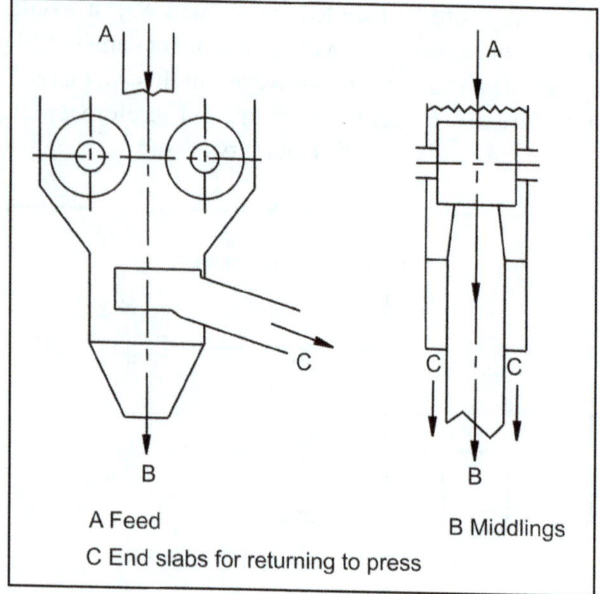

Fig. 6.30.7 Roller press - cake at edges collected seperately for recirculation.

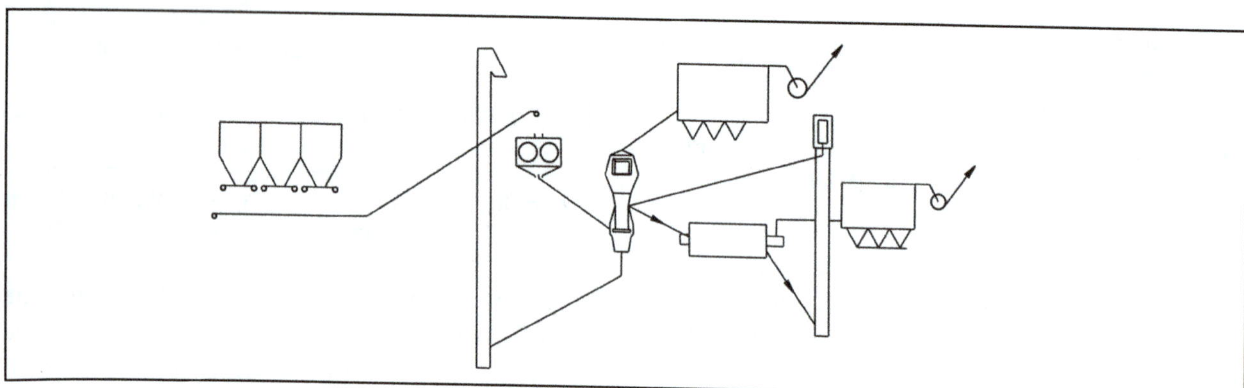

Fig. 6.30.8 Flow chart of roller press and ball mill for cement grinding.

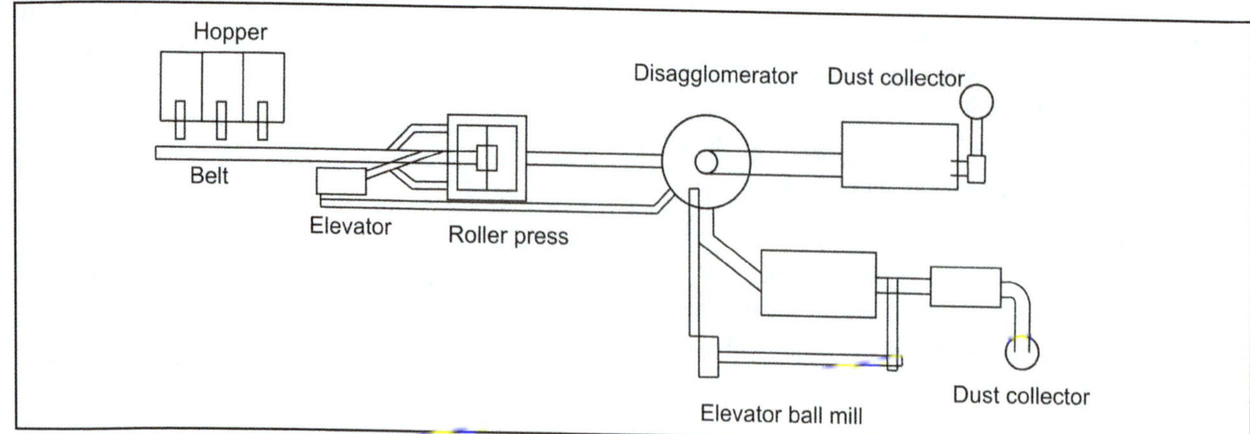

Fig. 6.30.9 Layout of roller press, ball mill and disagglomerator.

Disagglomerator is used to break cake and circuit of roller press and disagglomerator has been shown in the system. Disagglomerator is part of high efficiency separator in Sepax design.

See Fig. 6.30.9.

This design is also used in circuits of roller press where ball mill is eliminated altogether.

30.14 Layouts with V.R. Mills

VRMills are now commonly used to make cements-OPC or Blended, mostly with external circuit.

Layouts with V.R.Mills will be very similar to those for VRM used to grind raw materials except that when making only OPC hot gas circuit will not be required.

When mill has an external circuit and has a high efficiency separator, then also circuits and lay outs will be comparable.

Auxiliaries will be selected taking into account the fact that cement is more abrasive than raw meal. When bag filter used to remove product, (ground cement) from air circuit, it will be sized by selecting a lower air/cloth ratio. It is also possible that a cyclone or at least a precollector would be installed before the bag filter to ensure longer life of bags.

See Figs. 6.30.11 and 6.30.12. which show esp in circuit.

30.15 Pregrinding in Vertical mill

Vertical mill can be used without air circuit to pregrind clinker before feeding it to ball mills for finish grinding. Such an installation would be practical where one pregrinding v. r. mill is used to feed two or more finish mills which will be much smaller in size.

See Plate 2.13.2 in Chapter 13 of Section 2.

See Fig. 6.30.13.

In one system, a fluid bed separator follows a pregrinding mill.

See Figs. 6.30.14 and 6.30.15.

Vertical shaft impact crushers have also been developed as precrushers to reduce size of mill and system power.

See Plate 2.13.1 in Chapter 13 of Section 2.

See Fig. 6.30.16.

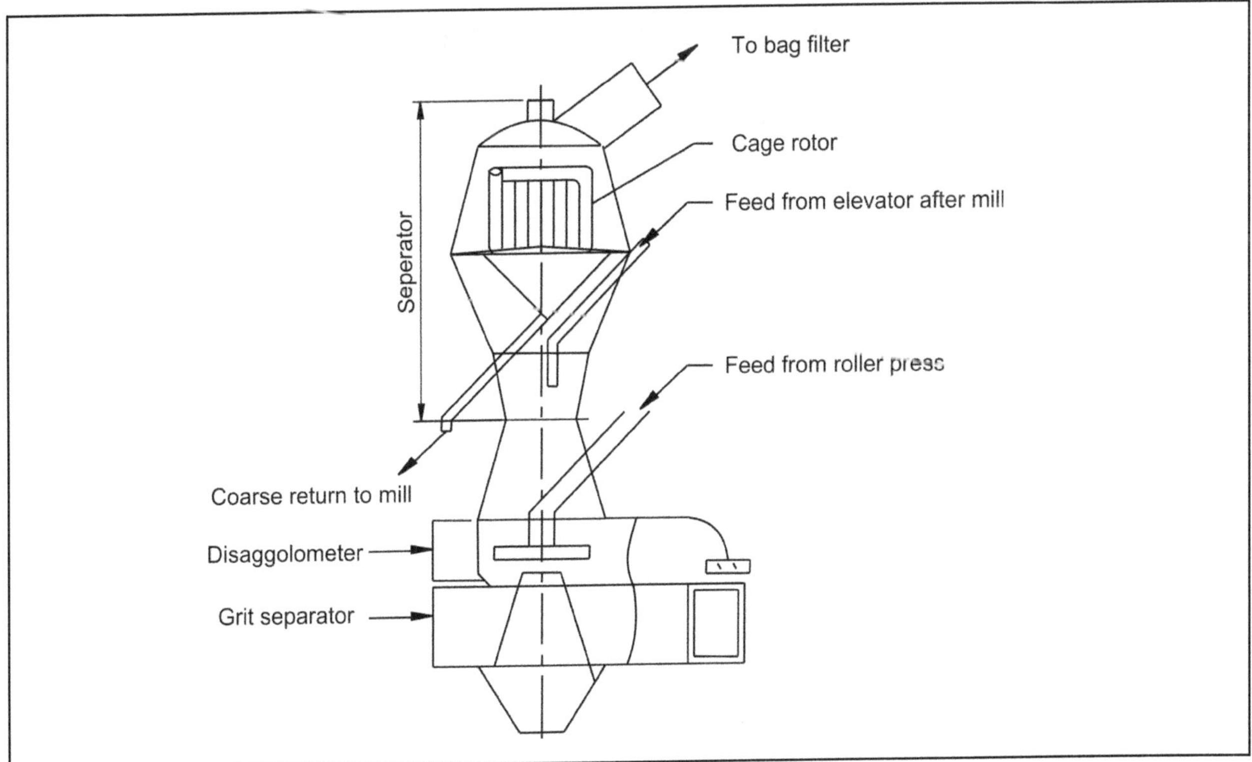

Fig. 6.30.10 Sepax seperator cum disagglomerator and grit separator.

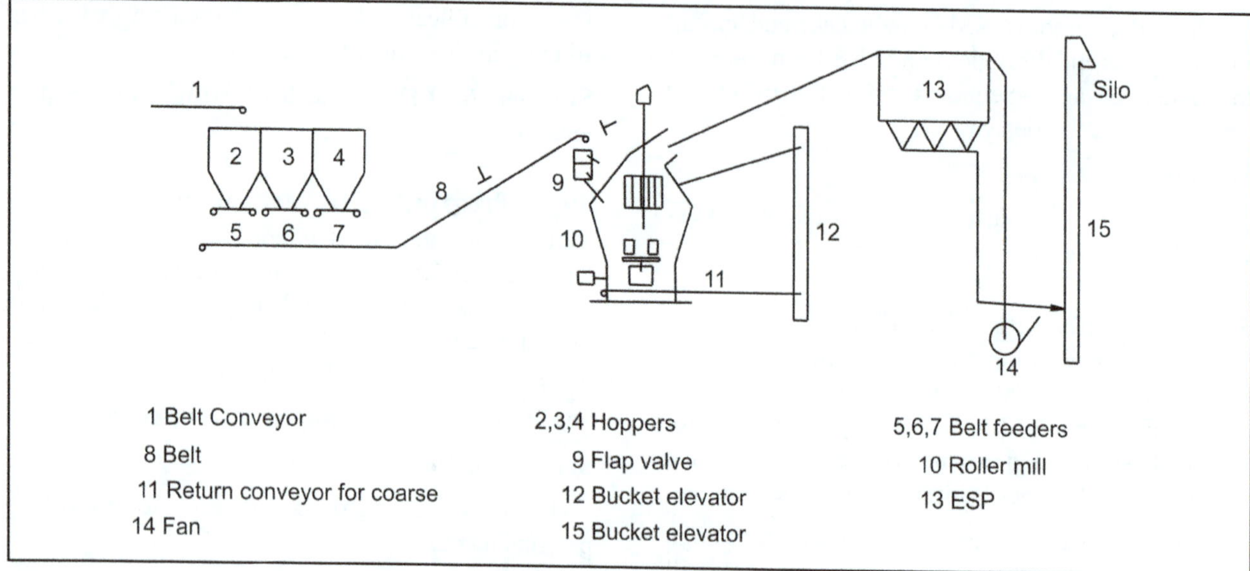

Fig. 6.30.11 Flow chart of V.R. Mill to grind cement/slag cement.

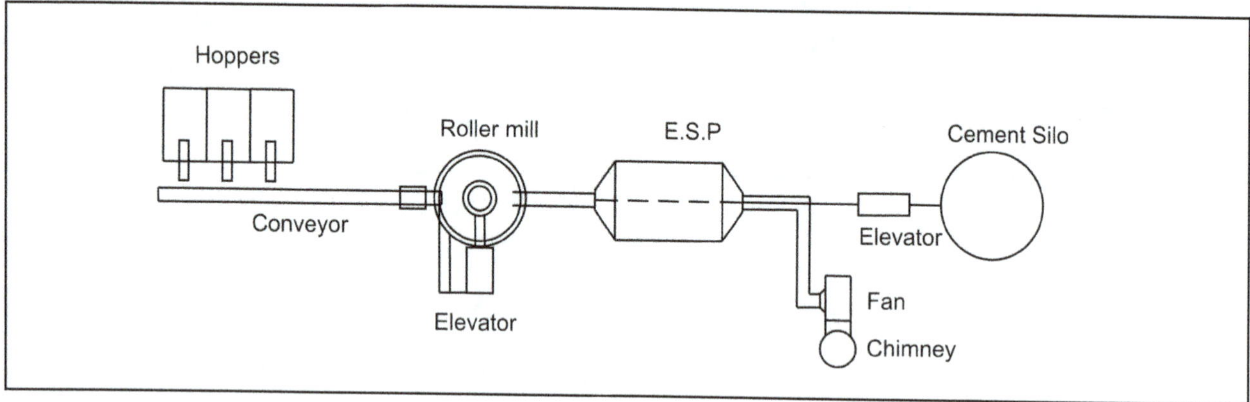

Fig. 6.30.12 Layout of vertical mill for grinding cement/slag.

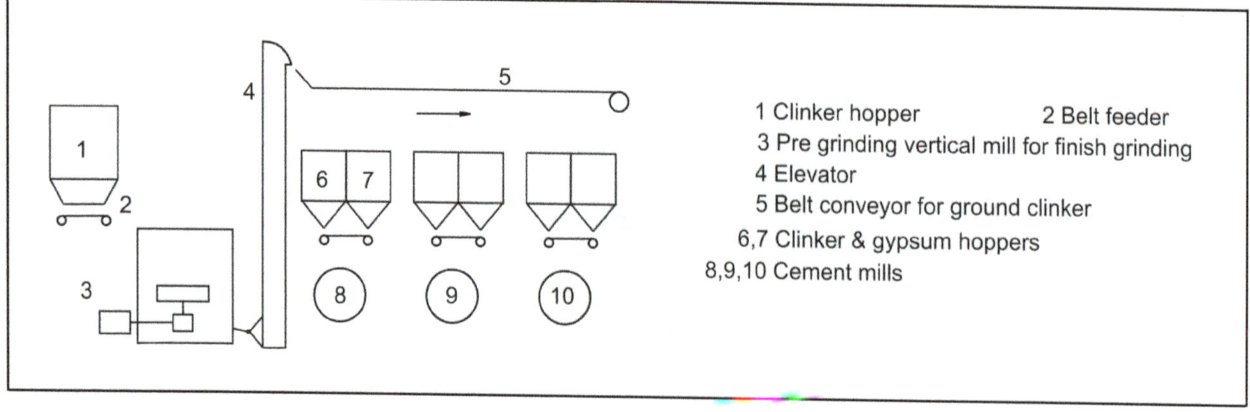

Fig. 6.30.13 Using pregrinding vertical mill for a batch of finish mills.

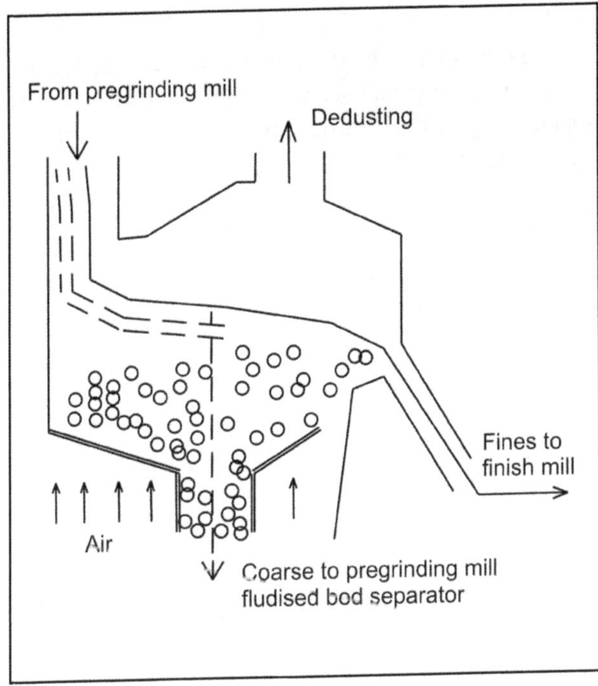

Fig. 6.30.14 Fluidised bed separator in conjunction with pregrinding mill.

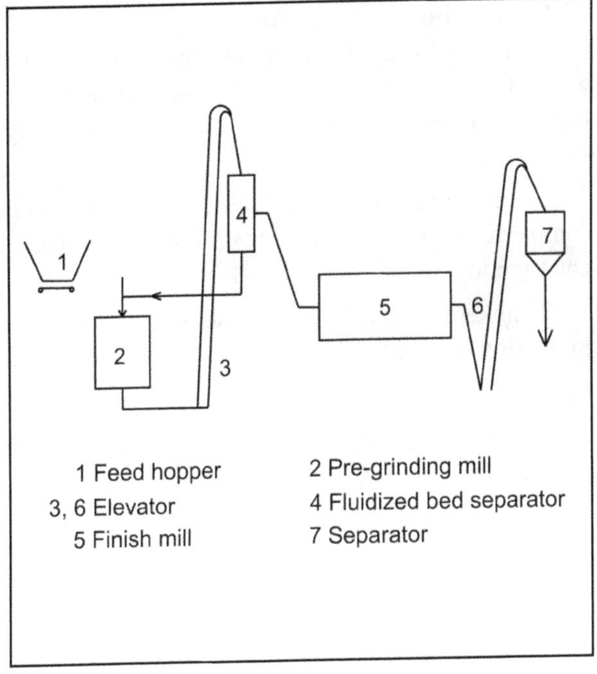

Fig. 6.30.15 Mill circuit with pregrinding mill fluidized bed separator and finish mill.

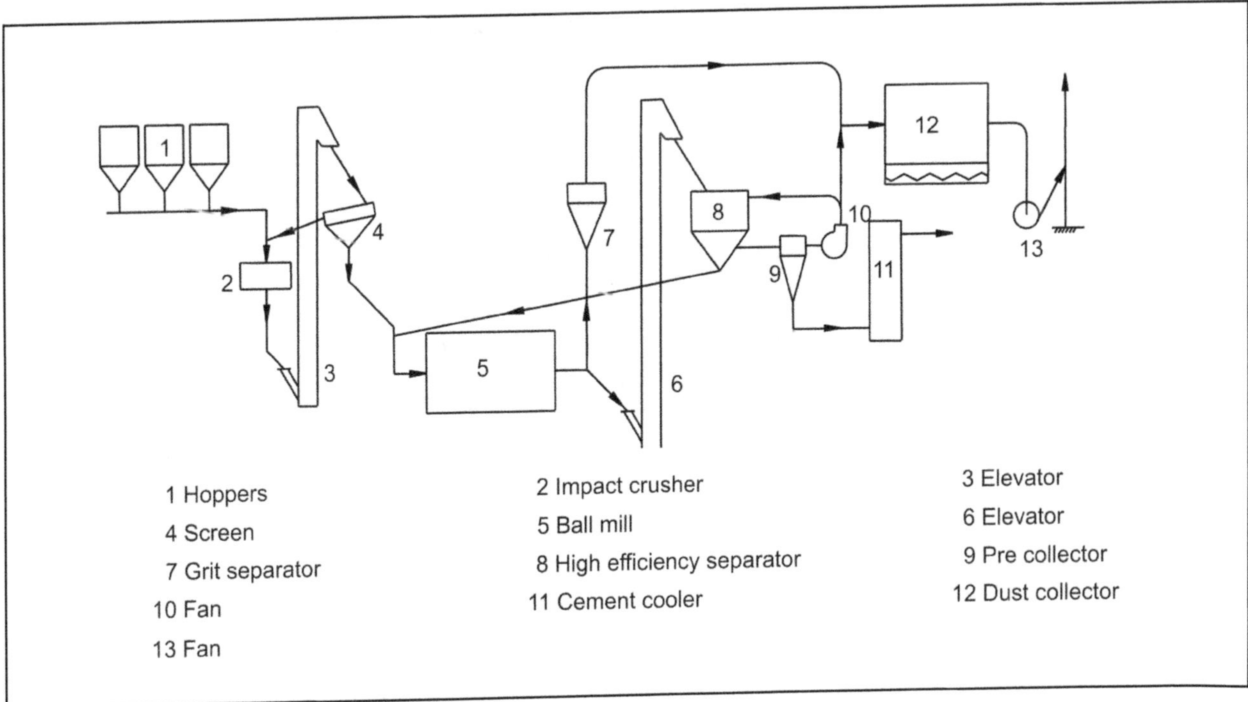

Fig. 6.30.16 Grinding circuit with precrusher and external cement cooler.

30.16 Cooling of Cement

It is necessary to keep temperature of cement produced within 100 to 110 °C to prevent dehydration of gypsum. In the past cement mills were cooled externally by spraying water on the shells.

Later water was also sprayed in the last compartment by inserting a pipe through the discharge end trunnion.

If clinker was very hot then water was also sprayed in the first compartment.

See Fig. 6.30.17.

Cement coolers were also developed where cement was cooled indirectly by flow of water.

An external cooler has been shown in a mill circuit.

See Fig. 6.30.16.

Its construction is shown in **Fig. 6.30.18.**

These practices were prevalent when cement mills used to be open circuit mills.

30.16.1 Cooling in Separators

Cooling could also be done in mechanical separators in mills working in closed circuit - reverse of drying in separators practiced earlier - by admitting cooling air, approximately 0.2 to 0.3 kg/kg of cement.

In case of high efficiency separator, air is used for classification and its quantum is between 1-2 m^3/kg of finish product. Thus cooling is easily and conveniently done in closed circuit mills with high efficiecy separators.

With most cement mills being operated in closed circuit, internal water sprays are hardly required though they could be easily introduced in the circuits of ball mills as well as v.r. mills.

30.17

Fig. 6.30.19 shows a roller press preceding ball mill in a cement grinding circuit.

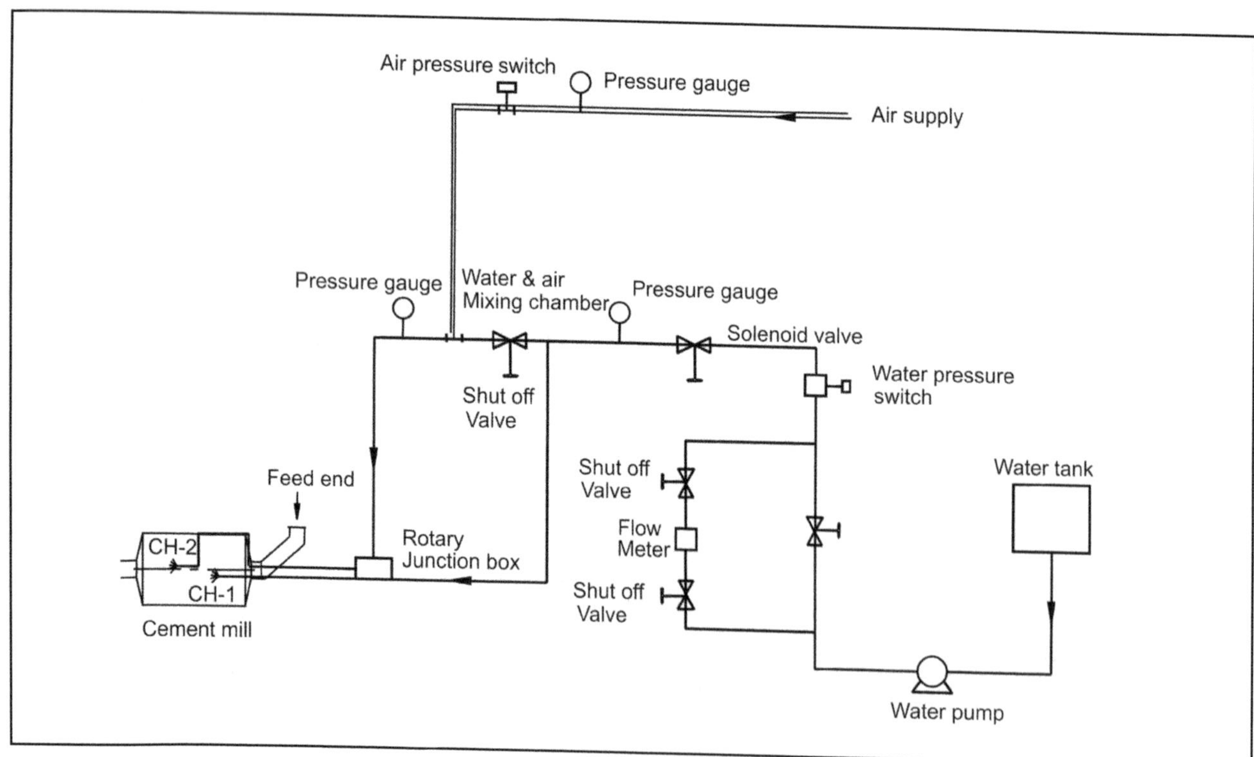

Fig. 6.30.17 Internal water cooling system for a cement mill.

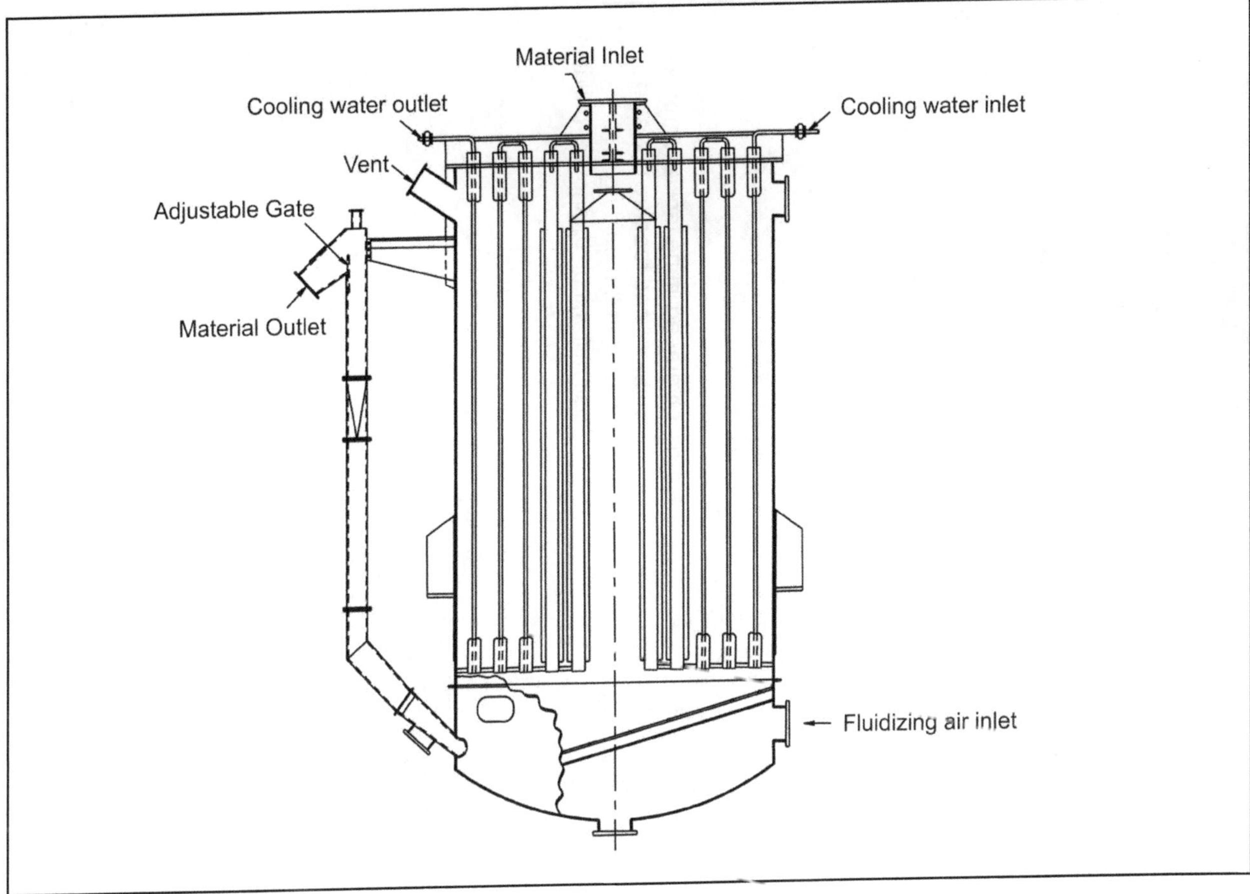

Fig. 6.30.18 External cement cooler.

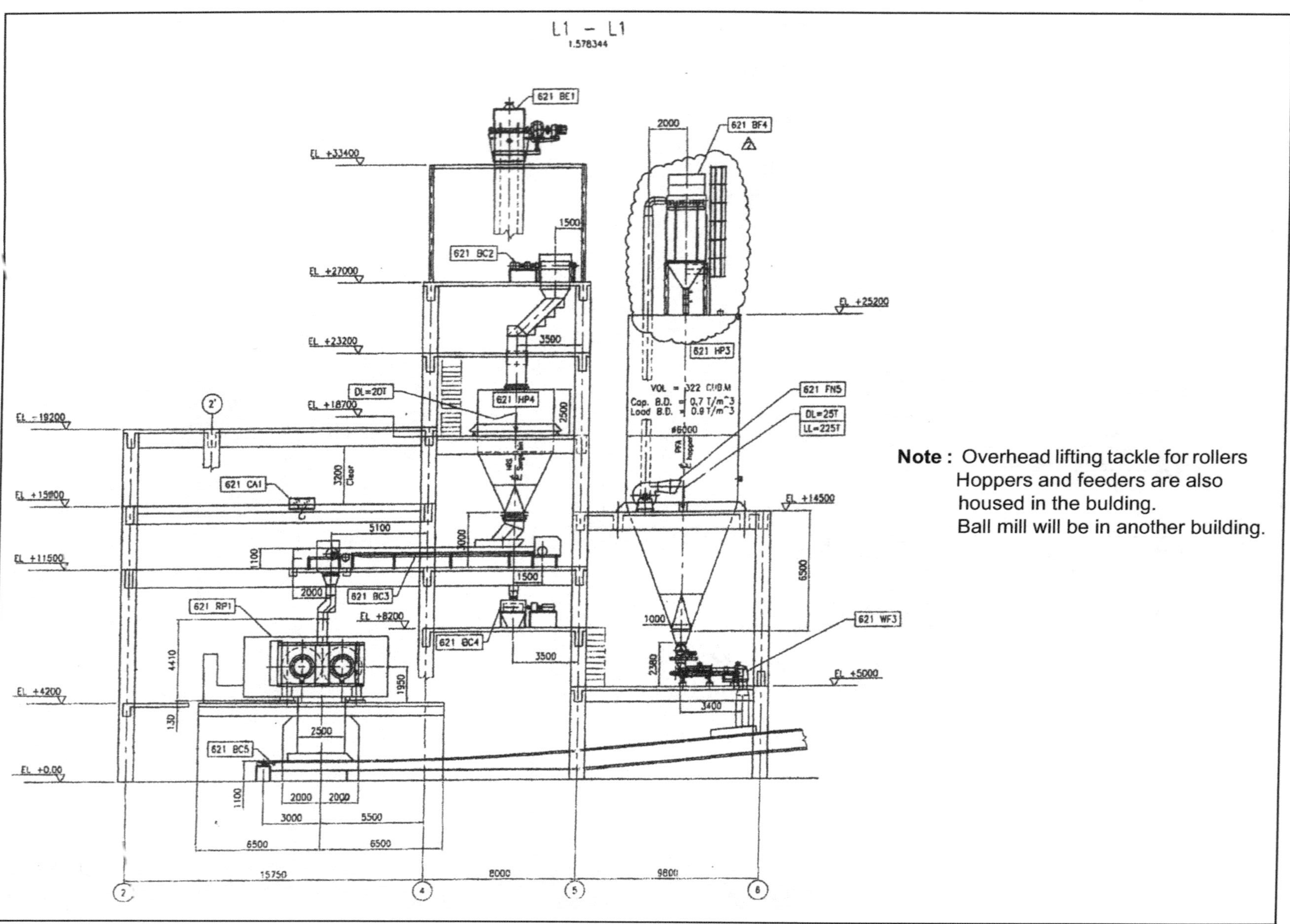

Fig. 6.30.19 Layout of a Roller Press in cement mill circuit.

CHAPTER 31

CEMENT STORAGE

31.1 Storing Cement

Conventionally provision is made to store 7 days' production of cement. However, it could be as little as 4 days' also.

When different types of cements are made they will be stored in separate silos unless the different types are mutually exclusive.

See Fig. 6.31.1.

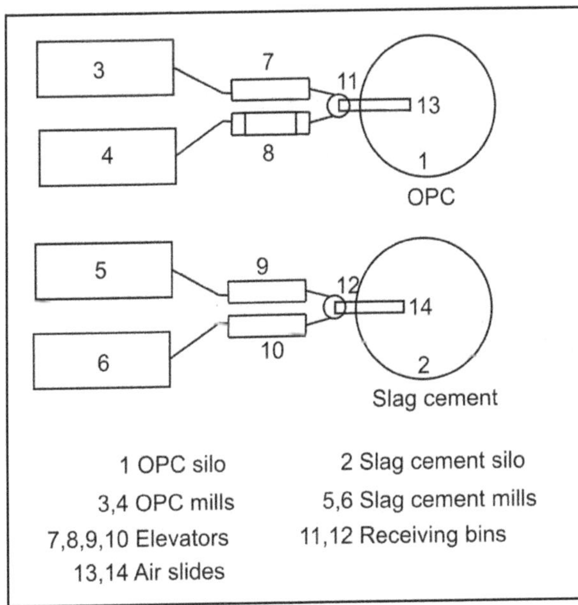

Fig. 6.31.1 Separate storage silos for different types of cement.

31.2 Quantity to be Stored

In deciding the storage capacity of cement silos, it is necessary to take into account :

1. Types of cement to be made.
2. Schedule of production – quantities of each type.
3. Minimum quantity to be stored for each type

   This will be dictated market conditions.

31.2.1

Though the same mills could be used to make different types, the outputs would differ according to the cement made.

For example:

A 3000 kw cement mill may produce :

100 tph - O P Cement

120 tph - P P Cement

60 tph - Slag Cement

Daily production in each case would be 2000, 2400 and 1200 tons respectively.

It is not often that a plant would be making both Slag Cement and PPC but the possibility cannot be ruled out for locations that are close to steel plants and also to thermal power stations.

31.3 Separate Grinding Mills and Storage Silos

If the plant is required to supply OPC and Slag cement and or PPC simultaneously, it would have to use some mills to produce OPC and others to produce Slag and Pozzolana cements.

However, it would be desirable that all mills should be able to produce all types of cement. Therefore feeding systems to mills should be common and similar.

If however some mills are earmarked for OPC only, and others for blended cements only then the respective silos can also be earmarked to hold these types of cement. Some plants make special cements like oil well cement, cement for railway sleepers and such special cements in small quantities; hence may require smaller silos / packing facilities for such special cements.

See Fig. 6.31.2.

Civil construction wise it would be desirable to construct all silos of same diameter; heights differing according to capacities. This would also make it possible to use identical equipments for aeration and extraction. This may not however always be possible.

31.4 Silos and Dispatches

Silos are a link between production units and dispatches.

Dispatches can be in bulk and or in bags. The layout must also provide for both according to the market for each.

Let total annual production be 1,000,000 Tons with break up of types of cements made; of bagged and bulk cement and also of dispatches by road and rail. **See Fig. 6.31.3.**

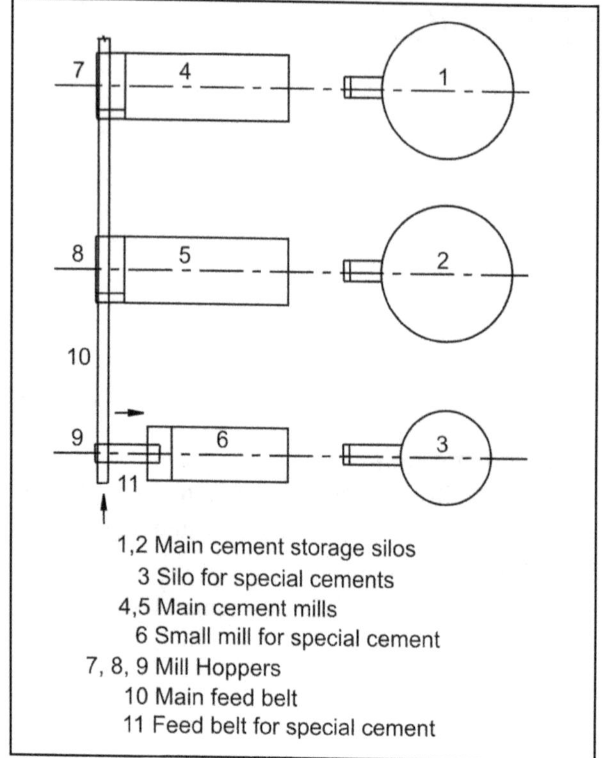

Fig. 6.31.2 Grinding and storage facility for special cements.

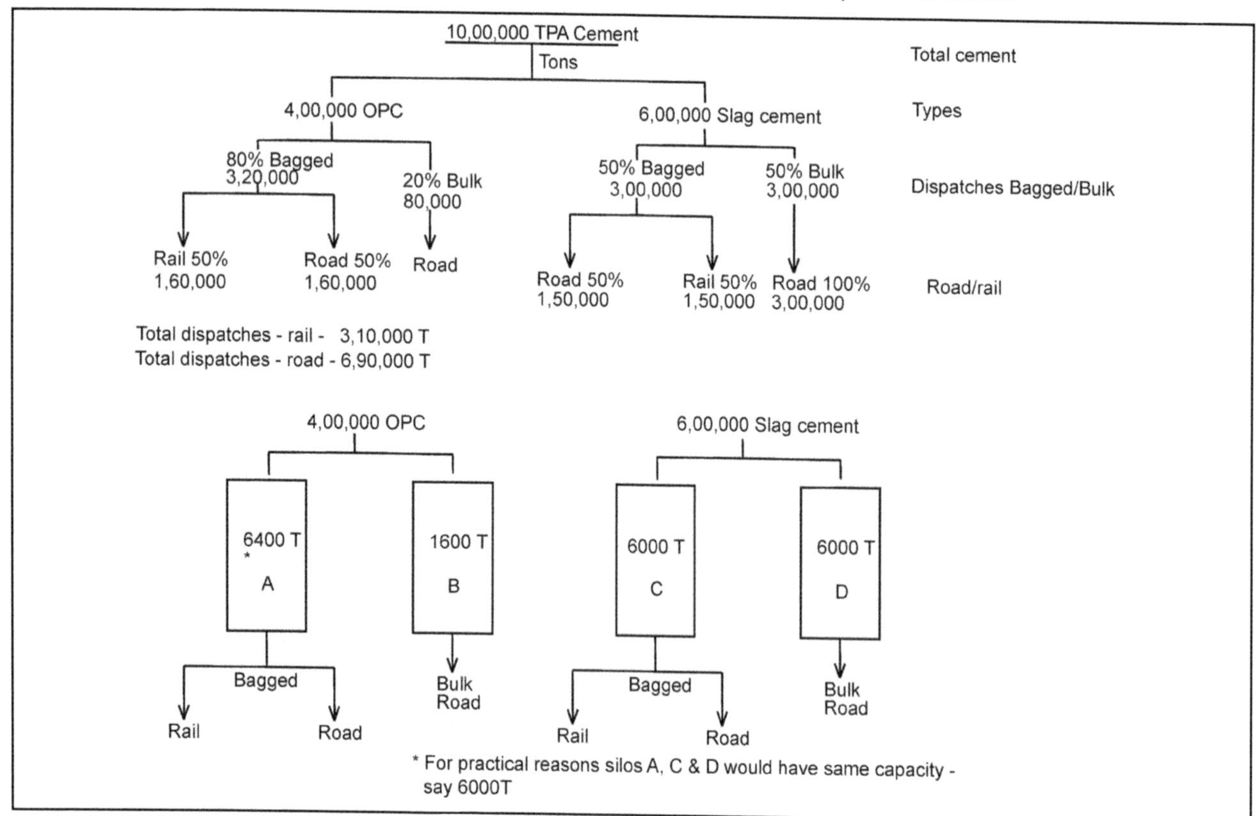

Fig. 6.31.3 Storages provided according to quantities produced and mode of despatches.

In effect it means storage capacities would be as follows :

OPC	4,00,000 / 360	1110 TPD × 7 days = 7800, say 8000 tons
Slag cement	6,00,000 / 360	1670 TPD × 7 days = 12,000 Tons

total = 20,000 Tons

Scheme for storage can be as shown in **Fig. 6.31.3** or as shown in **Fig. 6.31.4.**

31.4.1 Grouping of Silos

Whereas silos used for rail and road dispatches of bagged cement would be linked together along with packing machines, silos used for bulk dispatches could be grouped together as in this instance bulk dispatch is by road only. However, it is possible that bulk dispatches can also be by rail since there is a railhead.

31.4.2

If the demand for special cements is small but regular, it may be worth while to install a separate small mill to grind it together with a small storage silo and bagging facilities. **See Fig. 6.31.2**

Therefore silo for bulk cement = 1600 Tons (7 days' stock)

Thus, 8,000 Tons of OPC storage to be divided into 6400 t for bagged cement and 1600 t for bulk cement.

See Fig. 6.31.3.

Storage capacities of cement silos thus have to be arrived at carefully.

At one end is the market the plant is serving and its share in it.

At the other end are the pro duction units – clinkering and grinding – which produce cements.

31.5 Share of Market and Fluctuations in Demand

A storage silo is a buffer, which maintains continuity of supply. The size of the buffer is determined by the frequency and periods of interruptions.

If the share of the market, a cement plant has, is steady and regular the off take from the cement plant will also be steady and regular.

But markets are anything but regular. There are seasonal fluctuations of demand which can however be anticipated.

Supposing the market a plant serves, covers 3 States with a total demand of 4.0 × 106 tons / annum. (For simplicity we deal with only opc). The plant expects a 40% share of the market i.e., 1.6 m.t.p.a.

Total demand is distributed season wise as shown in **Table 6.31.1.**

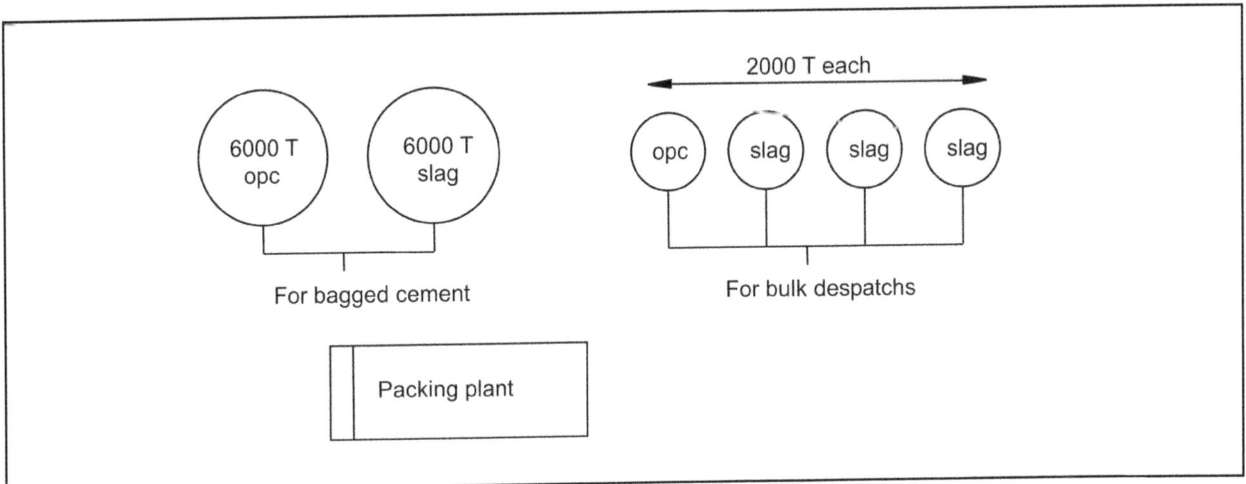

Fig. 6.31.4 Alternative pattern for storage of opc and slag cement for despatches of bagged and bulk cement.

Table 6.31.1

	Summer	Rainy	Winter	Spring
Demand	20%	20%	40%	20%
In million tons	0.8	0.8	1.6	0.8
Plant's share mt	0.32	0.32	0.64	0.32
Monthly demand tons	106700	106700	213300	106700
Daily dispatches tons	3560	3560	7120	3560
Hourly dispatches tons	240	240	480	240

Installed capacity to produce 1.6 million tpa of cement annually with 7200 running hours for mills and 10 % design margin and 10 % margin for cement grinding capacity would be 1.6 × 1.1 × 1.1 1.94 say 2.0 mtpa.

Cement grinding capacity daily

= 2,000,000 / 7200 $\simeq$ 280 t.p.h. = 5600 t.p.d.

Rated Clinkering capacity of the plant would be

= 4850 t.p.d.

Or rounding off 5000 tpd.

Kiln in operation = 5000 t.p.d. Clinker

C.Mills in operation = 5600 t.p.d. Cement

Despatches = 3600 t.p.d. for 9 months

= 7200 t.p.d. for 3 months

Hourly grinding capacity = 5600/20 = 280 tph.

The plant may install 2 mills of 140 tph or 3 mills of 300 tph each.

It will not do to reduce grinding capacity. But for such a pattern of seasonal demand, storage capacity could be reduced to 3-4 days' storage instead of 7 days'. i.e., 22000 tons.

Since grinding capacity is adequate for season's maximum demand, even if one mill is down, stocks of 20000 tons would be adequate to tide over.

31.6 Availability of Trucks for Transport

There are other factors, which can affect 'off take' from the plant.

If the off take is dependent entirely on truck transport, the trucks may not be available due to factors like strike, or their diversion to seasonal application like transport of fruit in appropriate seasons.

In such cases, for days on end, off take would dwindle to zero.

If however, plant has enough storage capacity, silos could be filled up so that when truck supply resumes, dispatches can be made in full swing.

3.17

The plant has to take them also into account to work out storage capacity that would be adequate yet economical.

The product mix decides the number of silos and storage capacity of each type of cement as explained above.

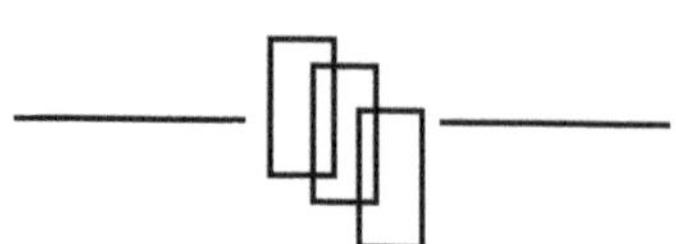

CHAPTER 32

CEMENT STORAGE AND PACKING

32.1 Layout of Packing Plant

The layout of 'cement storage and packing' has 3 sections :

1. Extraction from storage silos and feeding to packing machines.
2. Packing machines with feeding arrangements for bags.
3. Conveyor systems leading to truck and or wagon loading.

See Fig. 6.32.1.

32.2 Silos

Cement storage silos, except for very small capacities, will have flat bottoms. Bottoms have fluidizing facilities consisting of open air slides to fluidize cement and facilitate its flow through openings or discharge points same way as in case of silos for raw meal.

See Fig. 6.32.2.

32.2.1 Aeration of Silo Bottom

The aeration area may be 10-12% or more of cross section of the silo. Aeration air should be free from

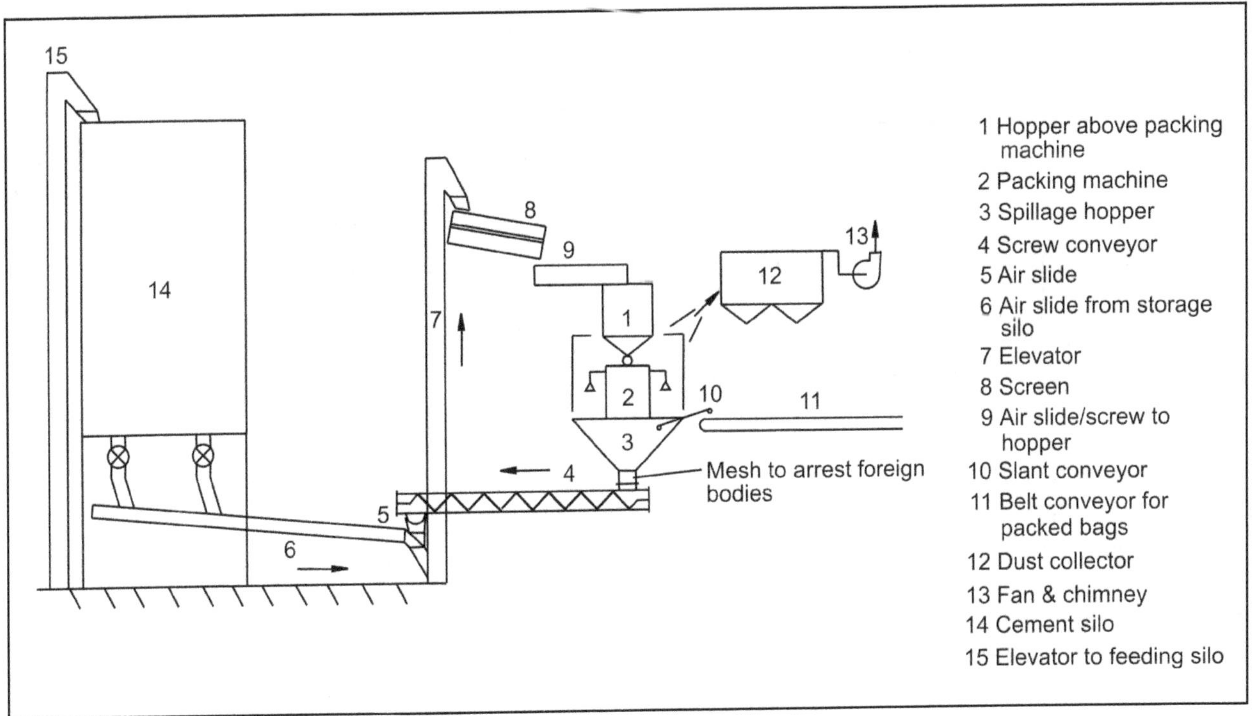

Fig. 6.32.1 Layout of packing plant including recovery and return of spillage.

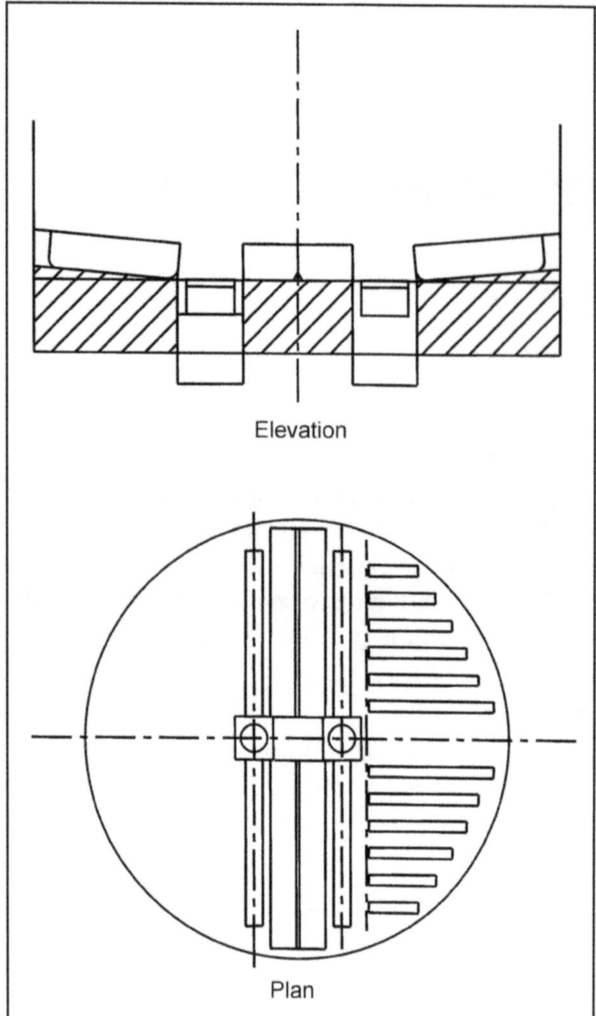

Fig. 6.32.2 Aeration of flat bottom silo.

moisture as cement is apt to harden when it comes in contact with water.

Formation of lumps inside silo is not unusual and periodic inspection and cleaning of lumps should be done.

32.3 Sectionalized Silos

A sectionalized silo can also be used to store cements of different grades / types. This is a ring silo. The central small silo can be used to store special grade cements, which is required now and then. The outer annular area will have several outlet points. It can store different types of cements produced in small quantities

See Fig. 6.32.3.

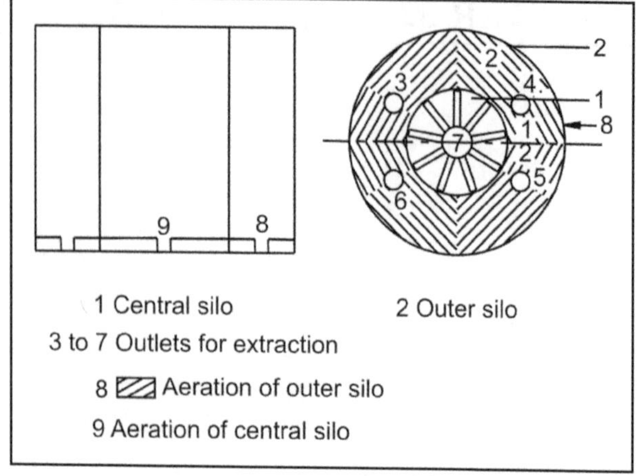

Fig. 6.32.3 Cement silo to hold different types of cements – ring silo.

Another commonly used sectionalization is to have 4 / 6 sectors with a central shaft. Aeration of its bottom is also shown.

See Fig. 6.32.4.

It can store say 5-7 grades of cement. But can be used to store at least 2 /3 grades.

The principal advantage of sectionalization would be that the storage is divided into smaller capacities and can be emptied out for cleaning and operations, cyclically.

32.4 Extraction Arrangements

The extraction arrangements would consist of rotary valves, flow control gates and air slides to bring cement to the bucket elevator/s.

See Figs. 6.32.5 and 6.32.6.

32.4.1 Extraction at Different Levels

Extraction can also be at different levels in sectionalized silos of large diameter and capacity storing more than one type of cement. This is advantageous for simultaneous loading of cement in trucks and wagons.

See Plate 6.32.1.

Two sets of transport and extracting systems would be needed if two types of cement were to be bagged simultaneously. They could also serve as standby so that continuity is maintained.

See Figs. 6.32.6 and 6.32.7.

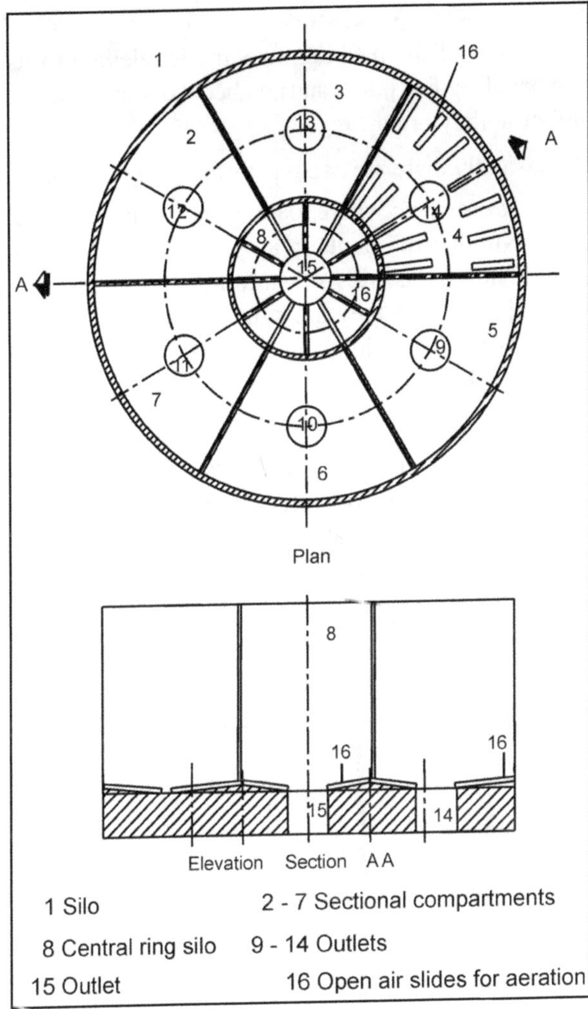

Fig. 6.32.4 Aeration of sectionalised silo for storing cements.

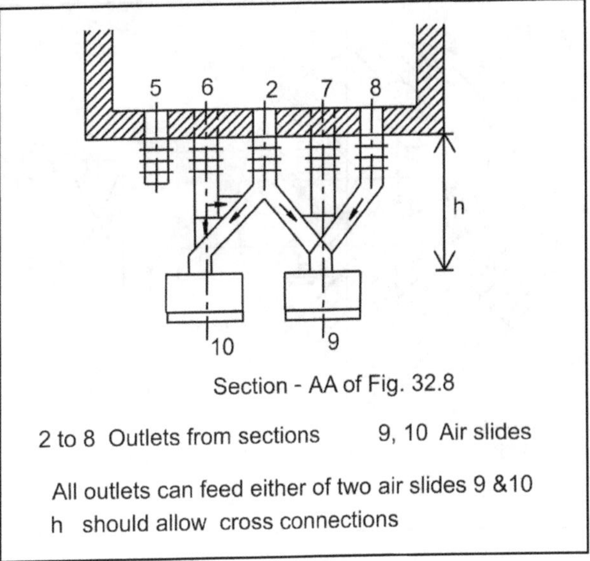

Fig. 6.32.6 Extraction from multi compartment silo to allow flexibility.

32.5 Elevators for Delivering Cement

The elevators can feed either of the two packing machines. If one elevator is feeding two packing machines, its capacity shall be the capacity of 2 machines so that one elevator always remains as a standby.

Heights of elevators should permit cross connection to ensure interchangeability.

See Fig. 6.32.8.

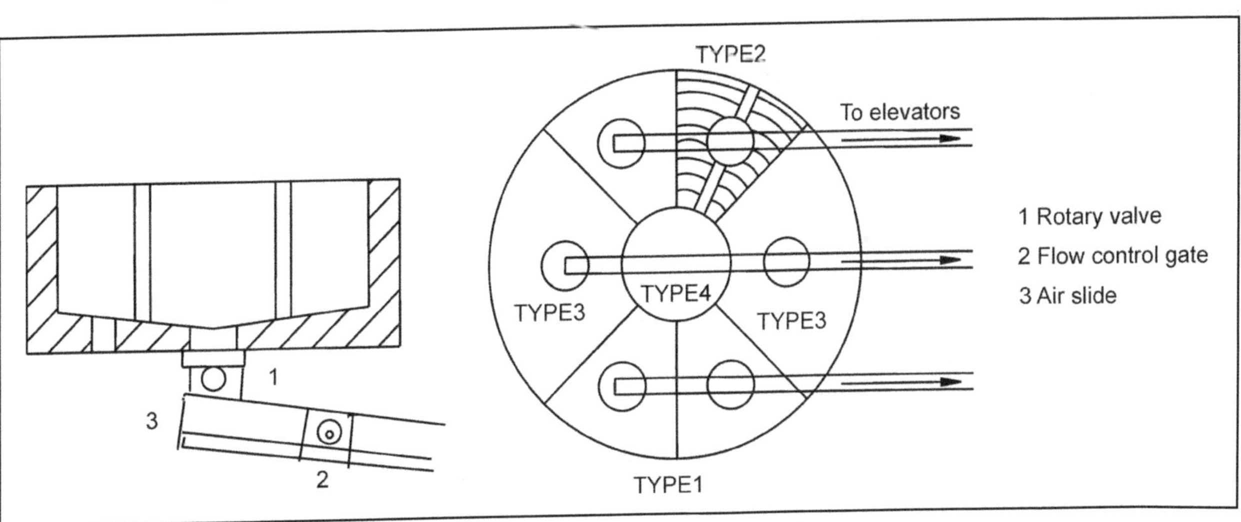

Fig. 6.32.5 Extraction arrangements for a flat bottom silo to handle more than one type of cement at a time.

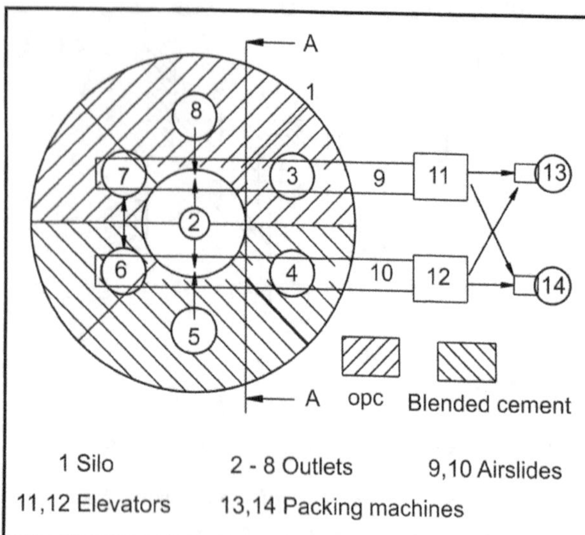

Fig. 6.32.7 Extraction arrangements for a flat bottom silo to handle more than one type cement at a time.

32.6 Packing Machines

The layout of the packing machine is relevant to its number of loading points and number of discharge points and how they are arranged.

Basically there are two types of packing machines.

1. *Stationary* : where loading spouts are in a line. Number of spouts in a line are limited to 4-6. Thus capacities of stationary machines are limited and they are suitable for small plants. Typical weighing mechanism of a stationary machine is shown

   See Fig. 6.32.9.

2. *Rotary:* where filling spouts rotate, bags get filled during rotation and are discharged at a fixed pnt. Typical weighing arrangements of a rotary packer with conventional weighing and with electronic weighing are shown.

   See Figs. 6.32.10 and 6.32.11.

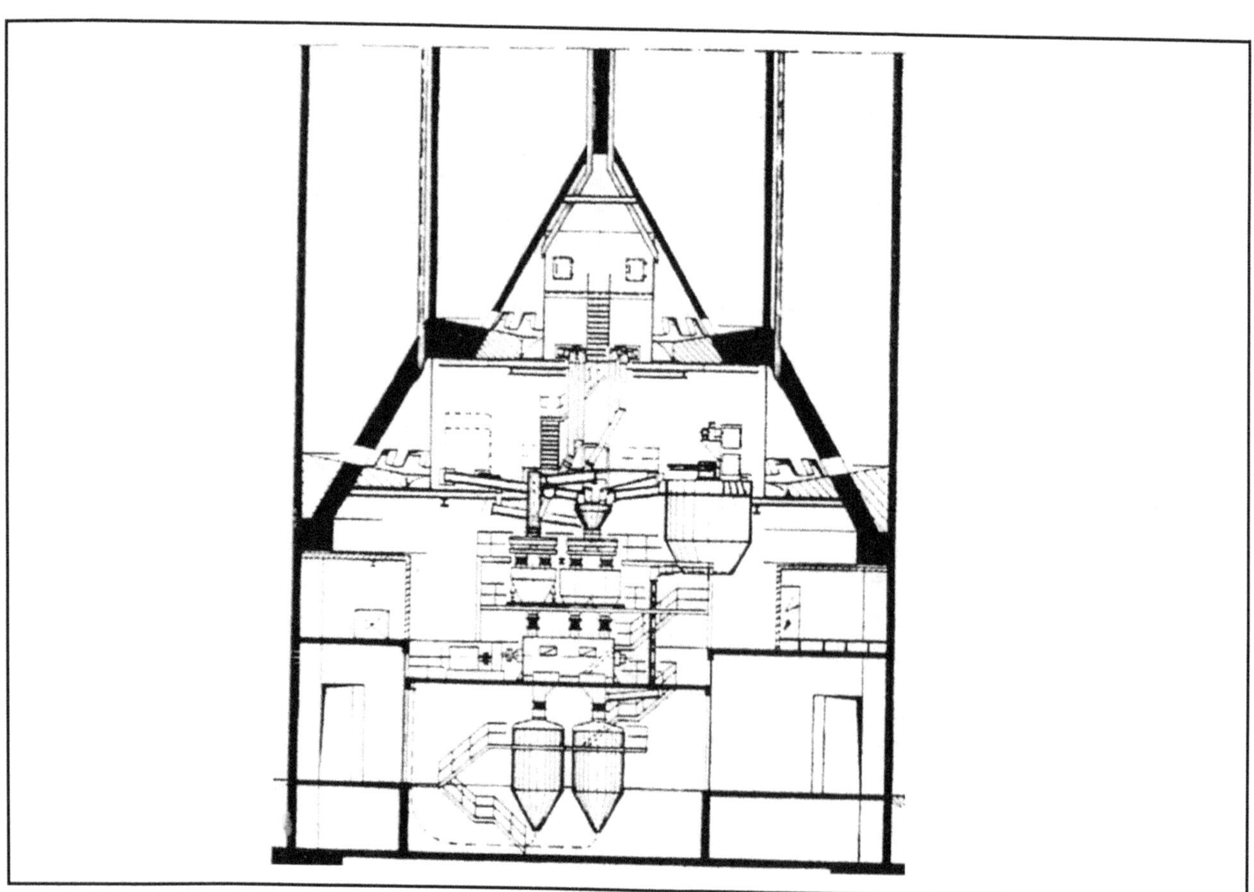

Source : From an article by H.U. Schalkhauser in ZKG

Plate. 6.32.1 Extraction from different levels in multi compartment silos for loading and blending of cement.

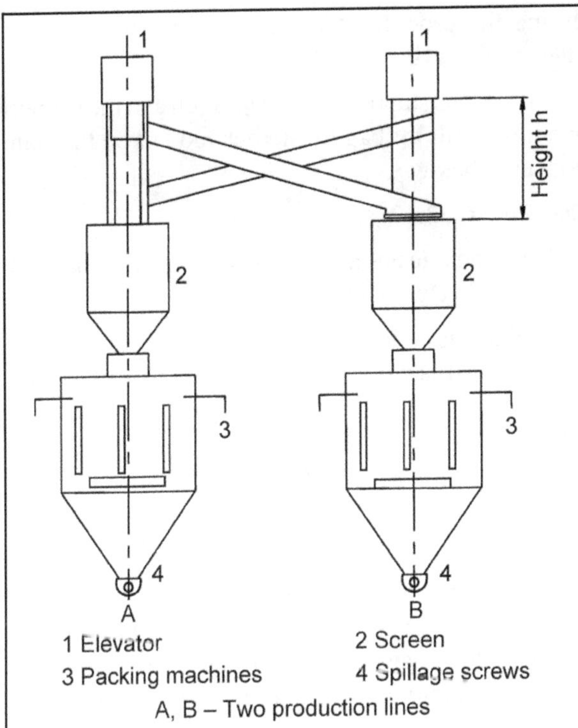

Fig. 6.32.8 Height of elevators should permits interchangeability.

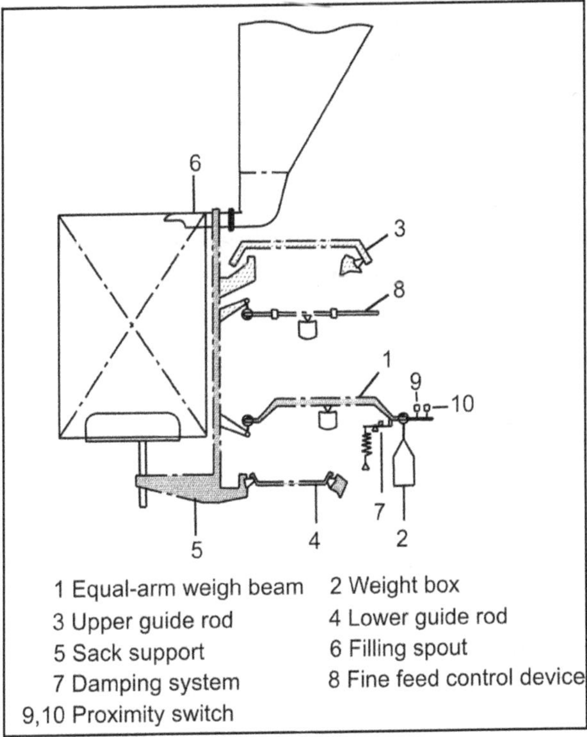

Fig. 6.32.9 Stationary packing machine equal-arm weighing mechanism.

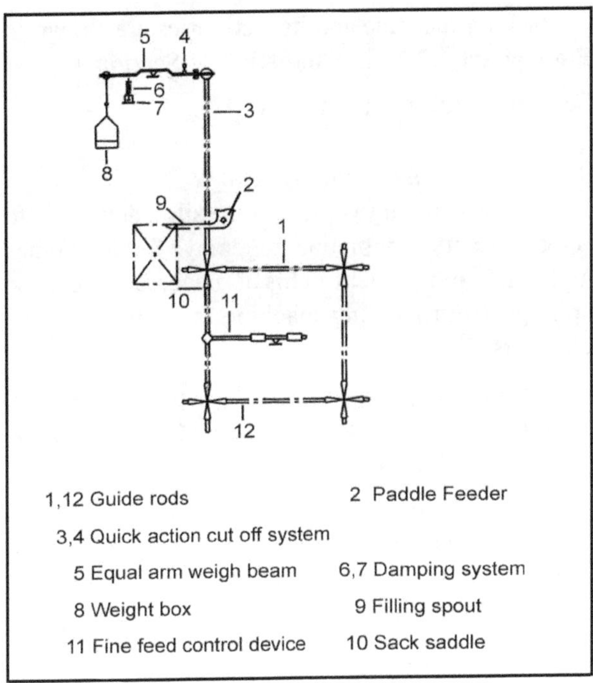

Fig. 6.32.10 Rotary packing machine equal-arm weighing mechanism.

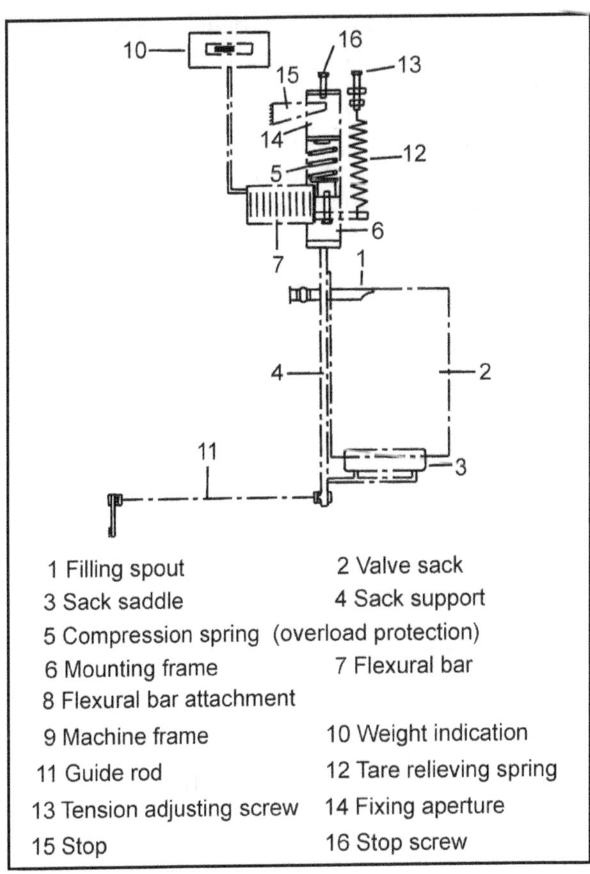

Fig. 6.32.11 Electronic weighing system.

Packing machine and its auxiliaries are shown in **Flow chart 1.7.12** in **Chapter 7** of **Section 1.**

See also Fig. 6.32.1.

32.6.1 Screens in the System

Elevators receiving cement from silos deliver it to screens, rotary or vibrating, to remove foreign bodies and grit from cement. This is necessary because spillage from packing machine is returned to the elevators.

In the layout provision has to be kept for expansion. It would be logical to assume a second production line with a second machine.

As mentioned in paragraph **32.4.1**, there should be interchangeability between two production lines. Thus elevators should be able to deliver cement to either packing machine.

See Fig. 6.32.8.

32.6.2 Components of a Packing Machine

A rotary packing machine consists of a bin with level indicators. This in turn feeds the built in bin with outlets equivalent to number of spouts. The speed of rotation can be changed.

Each bag has a self closing valve. Bag is slipped over the nozzle or spout. It gets filled as the machine rotates. The weight of the bag is measured all along during the process of filling with weights or with electronic load cells.

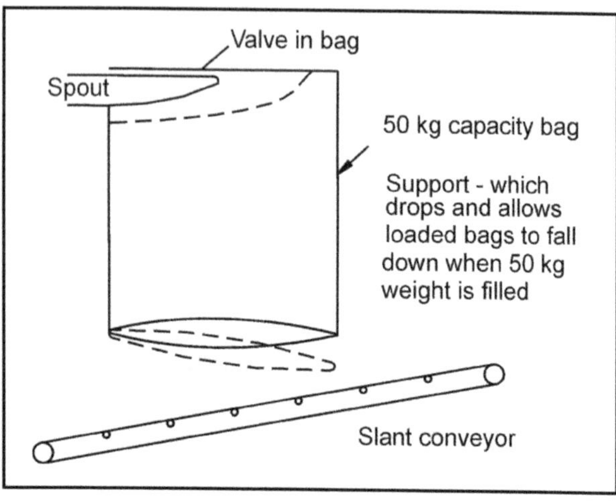

Fig. 6.32.12 Filling of cement in bags - automatic discharge of bags.

The feed is stopped when bag is full and the support and nozzle tilt for bag to slip out and fall on the slant conveyor below.

See Fig. 6.32.12.

The machine has a hood surrounding the spouts to collect and vent dusty air.

Spillage is collected in a hopper under the packing machine and is conveyed by a screw conveyor back to elevator.

32.7 Feeding Empty Bags

The layout must also provide for feeding empty bags to the loading point of the packing machine.

There are twenty bags to a ton. Therefore, a packing machine of 100 tph capacity would need to be fed with 2000 bags / hour or 10,000 bags / shift of 5 hours or 30,000 bags in one day plus a small % of surplus.

Bags are received 'branded' showing type of cement, BIS certification, name and brand of Company, name of unit, etc. Bags are obtained and stored in bags godowns. Bags whether jute or polythene are inflammable and hence bags godown should, for safety's sake be a separate building fitted with fire fighting (extinguishers) equipment.

See Figs. 6.33.1 and 6.33.2 in the next chapter.

32.8 Venting of Packing Machine and Auxiliaries

The packing machine, elevator, screen, air slides and other auxiliaries, need to be vented. A bag filter is installed to vent the displaced air.

Dust collected in the bag filter is also returned to the screw conveying spillage. Bag filter is located at a suitable place in the packing house along with its fan and chimney.

See Fig. 6.32.1.

Suppliers of packing machine should furnish data on vent volumes to be provided for packing machine and auxiliaries. Dust collector and fan would be sized and procured accordingly.

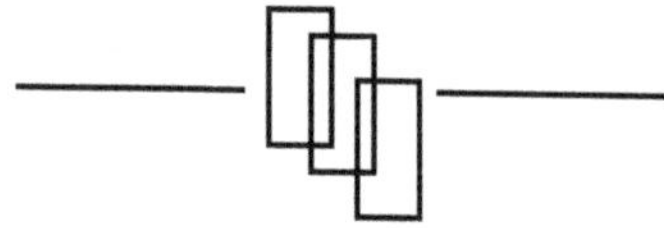

CHAPTER 33

BAGS GODOWN

33.1 Bag Godown for Empty Bags

Bales of empty bags will be received in wagons or in trucks. They have to be unloaded from trucks and wagons as the case may be and shifted to a bags godown. The bags godown must therefore have mechanized facility to handle bales of bags from wagons and trucks.

See Figs. 6.33.1 and 6.33.2.

33.2 Quantity of Bags to be Stored

Number of bags to be stored would depend on the delivery period from the date of placing the order. The buffer stock to be maintained should tide over any delay in receipts so that packing operation does not come to a stand still for want of bags.

A million tons plant will require 20 million bags / year or 1.78 Million bags/month or 60,000 bags per day.

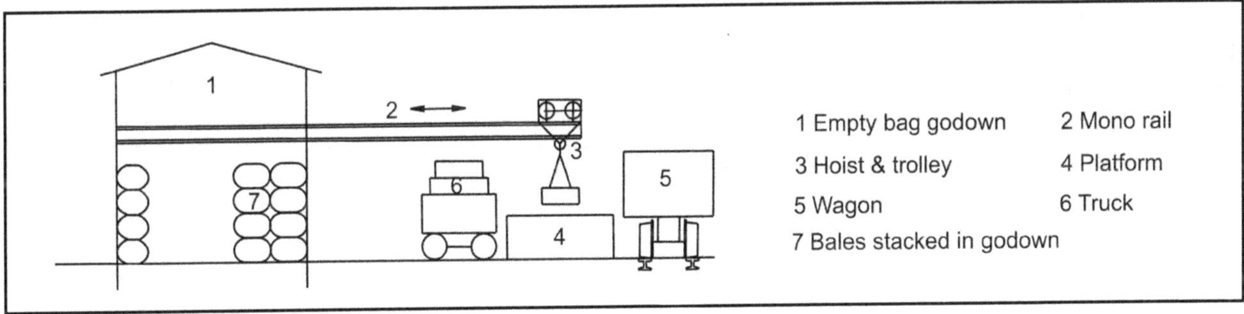

Fig. 6.33.1 Empty bags godown - receipt of bags from railway wagons or trucks.

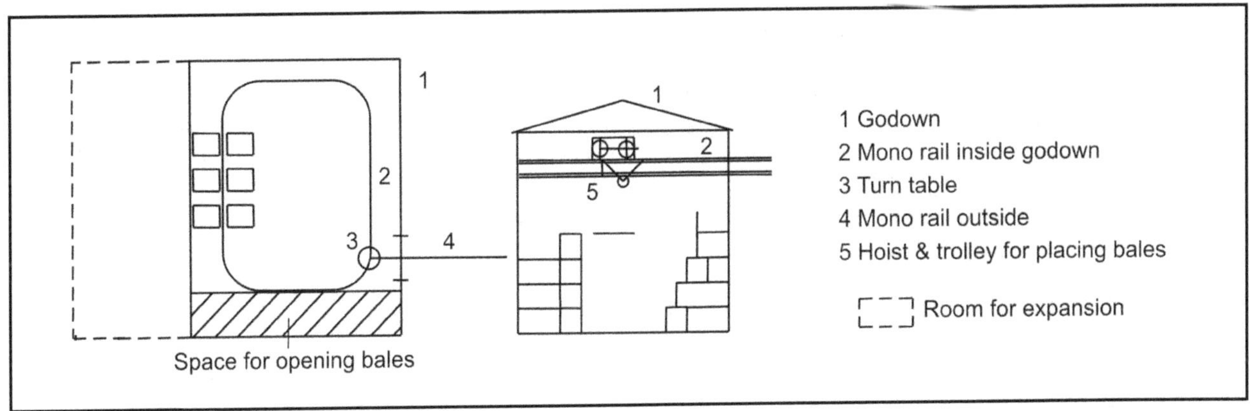

Fig. 6.33.2 Bags godown layout for handling bags inside godown.

Thus even a 15 days' stock would require 9,00,000 bags to be stored.

33.3 Sizing the Godown

The size of bales and number of bags in it and its weight should be ascertained and godown should be sized to hold these bags plus 20-25% more and room for maneuverability. Space should also be provided for expansion.

See Fig. 6.33.2.

The bags are to be fed to the packing machines for bagging.

Approximately 20-25% area of the godown is left vacant to unload bales and load empty bags on to the empty bags conveyor, which will take them to the packing machines.

33.4 Empty Bags Conveyor

The empty bags conveyor runs in all three planes- horizontally and vertically also- as bags would have to be brought from ground level to the loading platform of the packing machines. It consists of pans as shown to hold empty bags. The chain carrying pans moves in different planes as shown. The bags will be loaded in bags godown and unloaded at loading platform. A small stock should be maintained on loading platform so that in case the conveyor stops for some reasons, the packing operation does not stop immediately. The feeding rate of conveying should match the loading rates of the machines; it should be a little higher.

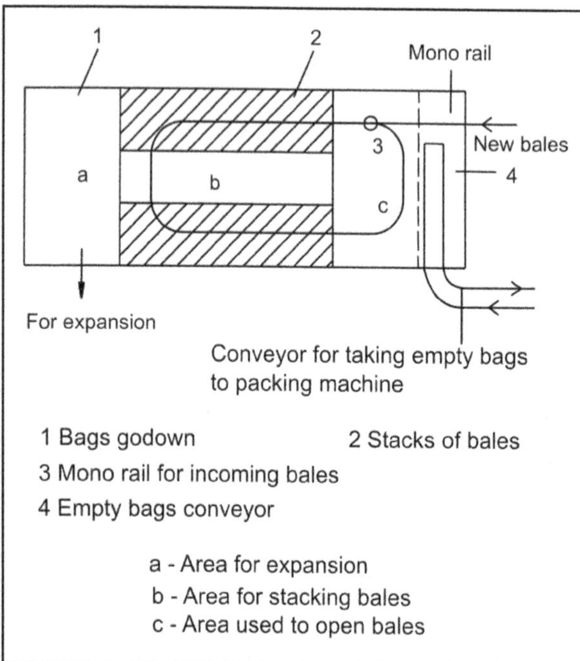

Fig. 6.33.3 Layout of bags godown for empty bags.

See Figs. 6.33.3 to 6.33.6.

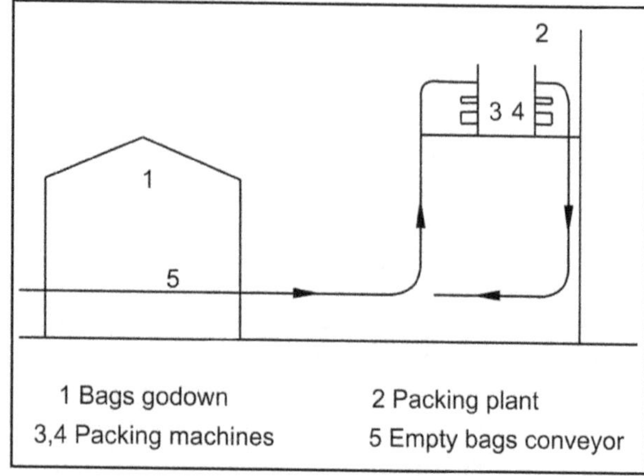

Fig. 6.33.4 Conveying empty bags from godown to packing platform.

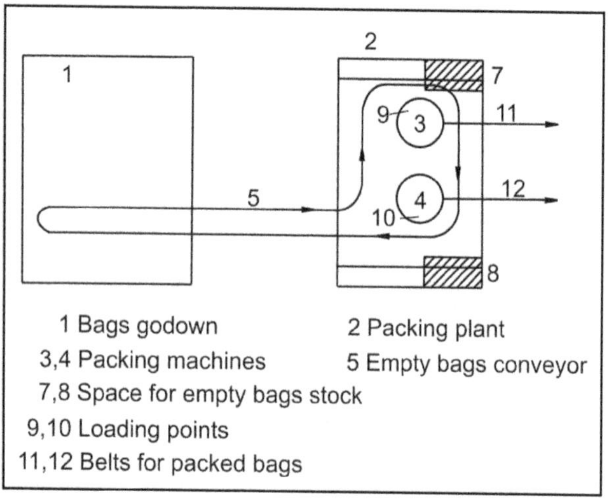

Fig. 6.33.5 Layout of conveyor for empty bags.

33.4.1 Machines with Two Loading Points

For matching with 2 loading points the conveyor would go round the machine for supplying the bags to the second point.

See Fig. 6.33.5.

33.4.2

Another alternative to supply bags on loading platform is by means of a lift.

See Fig. 6.33.7.

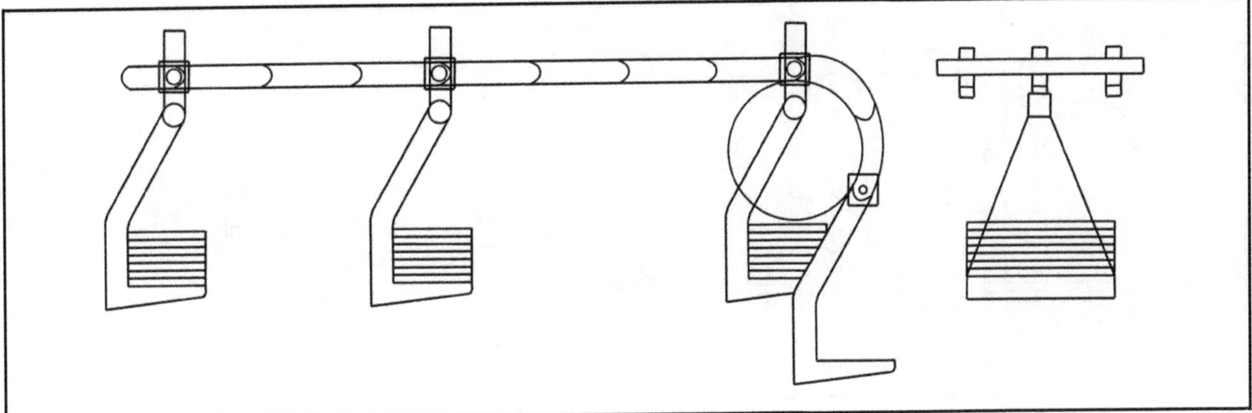

Fig. 6.33.6 Multi level and direction conveyor for empty bags.

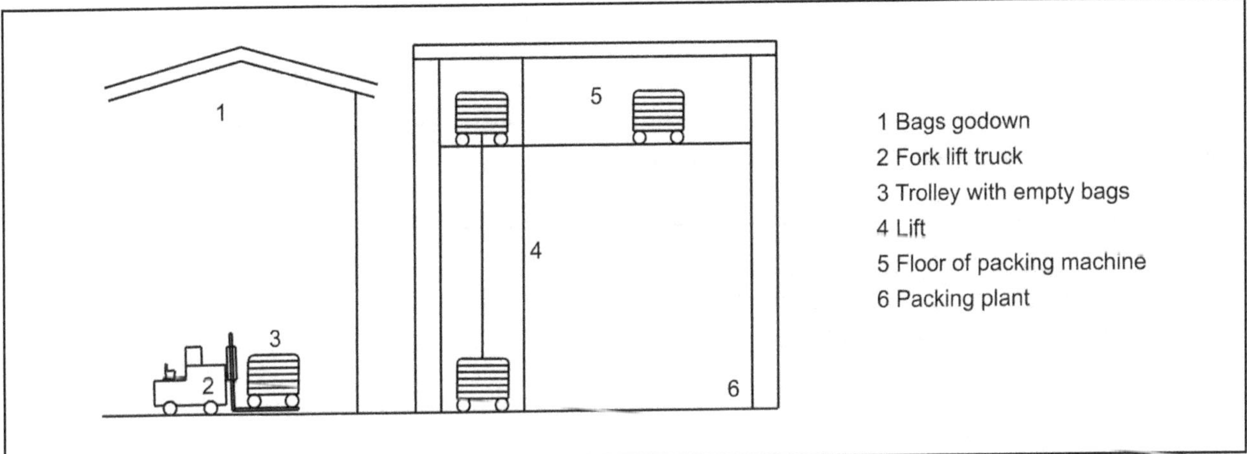

Fig. 6.33.7 Lift and forklift truck for conveying empty bags to packing machine.

33.5 Location of Bags Godown

Of necessity, bags godown would be located near the packing plant.

See Fig. 6.33.8.

33.6 Safety - From Fire Hazard

Bags whether made of paper or jute or polythene are a fire hazard. Bags godown should therefore be declared a 'NON SMOKING ZONE'.

Further it should be provided with fire extinguishers-manual or sprinkler type and also stands with buckets filled with sand.

It should be mandatory to test that the equipment installed is in working order periodically.

See Fig. 6.33.9.

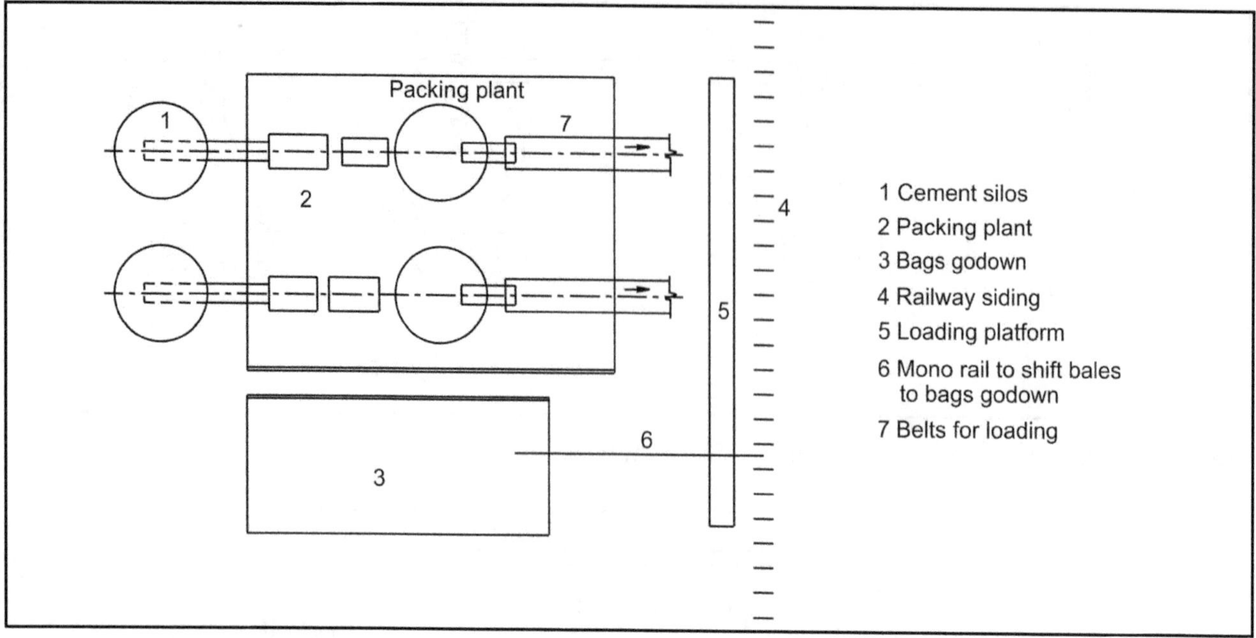

Fig. 6.33.8 Relative positions of bags godown and packing plant.

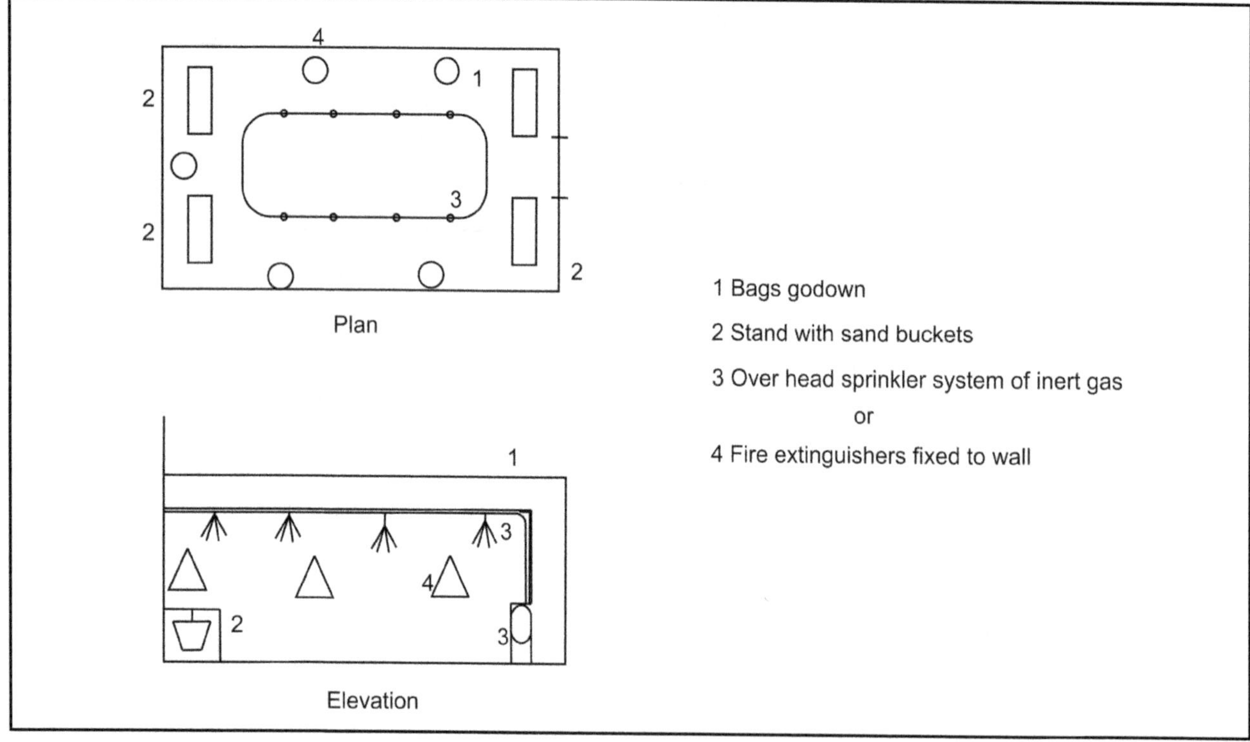

Fig. 6.33.9 Preventing fire in bags godown.

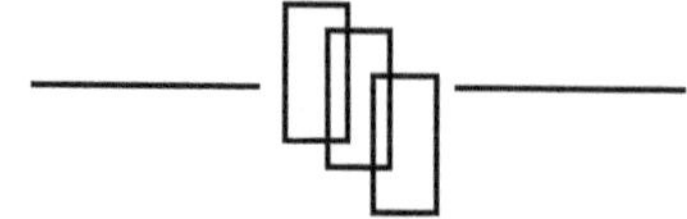

CHAPTER 34

LAYOUT OF PACKING PLANT AND DESPATCHES

34.1 Basic Considerations

The plant must know its market, its size and its fluctuations and how that market is served i.e., the demand pattern and how it is to be satisfied.

As has been pointed out earlier it must know inter alia :

1. Proportions of types of cement produced / cements dispatched.
2. Proportions of loose and bagged cement.
3. Proportions of road and rail dispatches also sub-divided into loose and bagged.

The plant can then arrive at quantities of total cement bagged and total sent in bulk / loose.

Having arrived at these quantities, the bagging capacity can be arrived at.

Say out of 1,000,000 tpa Cement,

60% is bagged = 6,00,000 tpa

40% is sent loose = 4,00,000 tpa

6,00,000 tons are to be bagged per annum. Packing plants generally work in all 3 shifts of the day. Effective working hours / shift are taken as 5. This is to allow for placing of wagons / trucks, change of loading gangs, diverting bags from one loading point to other etc.

Thus maximum number of effective hours per year

$= 15 \times 360$

$= 5400$ (360 working days assumed in 1 year)

Therefore hourly packing capacity required

$= 6,00,000 / 5400$

$= 111$ tph

After allowing a margin of $\simeq 20$ %, packing capacity to be installed will be $\simeq 135$ tph.

See Chapters 9 to 11 in Section 1.

However, Company may opt to install bagging capacity for total production of 1000,000 tons; in which case, capacity of bagging machines to be installed will be 225 tph.

34.2 Selection of Packing Machines

Packing machines come in standard sizes and machine capacity to be selected would be from them.

If standard sizes of machines available are in capacities of 100, 150 and 200 t.p.h. options for installing packing machines would be :

1. 2×100 tph.
2. 1×150 tph.

Option with 2 machines would be preferred as packing operation and dispatches would not come to a stand still even if there is a breakdown of one machine.

Depending on the size of the machine it would have 1 or 2 loading points. Unloading points when bag becomes full would also be 1 or 2.

See Fig. 6.34.1.

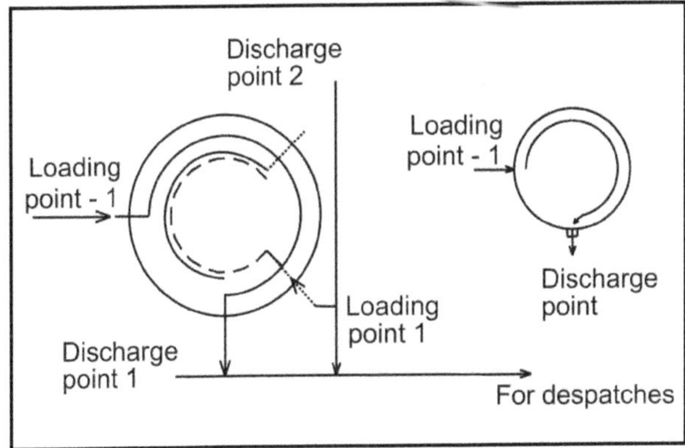

Fig. 6.34.1 Bagging of cement.

Packing machines can be stationary with stationary spouts numbering between 1 to 4. In stationary machines, spouts are arranged side by side. This limits the maximum number of spouts and hence capacity of a stationary machine. Each spout loads roughly at the rate of 15 tph. Loading rate increases for paper bags as compared to jute bags.

Cement is packed in 50 kgs bags jute / polyethylene for Indian markets. Paper bags would be used for export markets.

Thus a 100 tph machine has to receive and fill 2000 bags / hr. The number of loading spouts would depend on the capacity of the machine. A filled bag would fall off the machine every 1.8 seconds.

While ordering packing machines, the supplier should be specifically asked the capacities with each type of bag. The same machine would have a much higher capacity with paper bags as compared to capacity with jute bags.

Thus, for a 100 tph capacity number of spouts would be 100 / 15 = 7

Machines come in standard number of spouts like, 6, 8, 10, 12 etc.

Thus the layout around the packing plant would have to be designed to suit the packing machine selected.

Bags automatically drop off when they are filled up within the tolerance + 0.5 kg for a 50 kg bag.

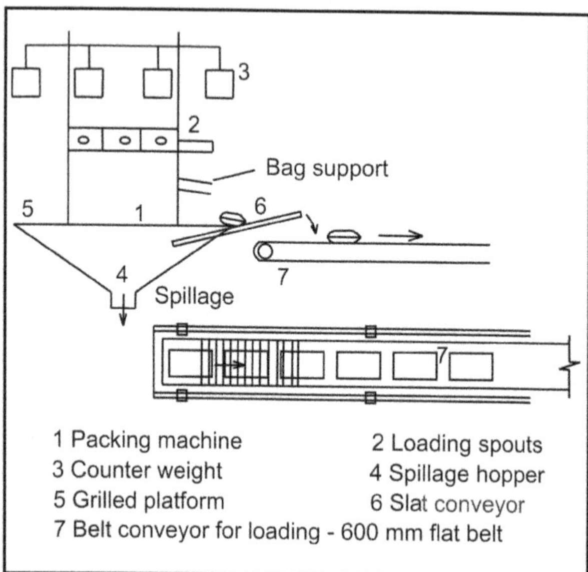

Fig. 6.34.2 Packing of cement in bags.

A 'slat conveyor' takes them to the loading belt. This conveyor allows for collecting spillage if any during the fall in a hopper under the packing machine so that it does not gat carried with bags.

See Fig. 6.34.2.

34.3 Loading Belts

The loading belt is a flat rubber belt on which bags travels with their width across the belt as shown. The common size for a 50 kg bag is 500 × 700 mm.

Therefore generally speaking a 600 mm flat belt is adequate when bags are carried on to the belt correctly.

The speed of the belt must match the rate at which bag would fall on it.

100 tph = 2000 bags per hour

No. of bags / second = 0.55

Let centre to center distance between two bags be 900 mm.

Speed of belt should be = 0.90 × 2000 / 3600
= 0.5 metres/second.

When a machine has two loading points, capacity of the belt carrying total output of the machine will be 200 tph and its speed would 1.0 m/sec.

See Fig. 6.34.3.

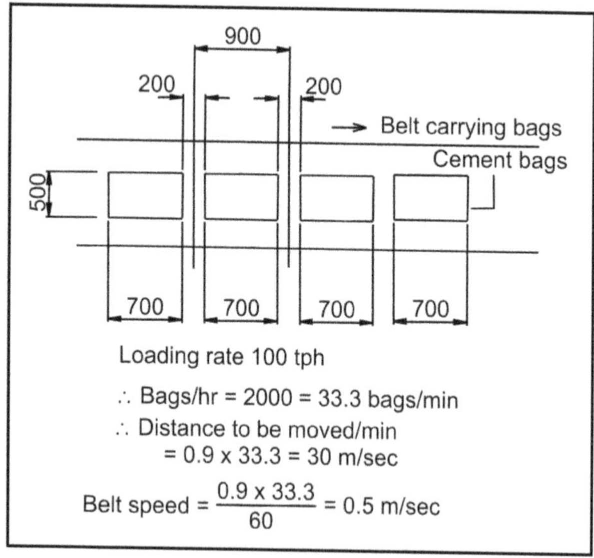

Fig. 6.34.3 Belt speed to suit loading rate.

For keeping margin in belt capacity, speeds of 0.6 m/sec and 1.2 m/sec respectively, could be selected.

34.4 Dispactches

The dispatches can be:

1. By Road only.
2. By Rail only.
3. By Road and Rail simultaneously.
4. By Road or Rail.

If the plant has no rail link whatsoever nor is it likely to have any in future the layout would have to suit dispatches of bagged and bulk cement by road only.

In this case, the layout would be designed to suit truck loading, manual or mechanized and bulk loading.

The bagged cement may be conveyed in standard trucks or on long trailers.

See Fig. 6.34.4.

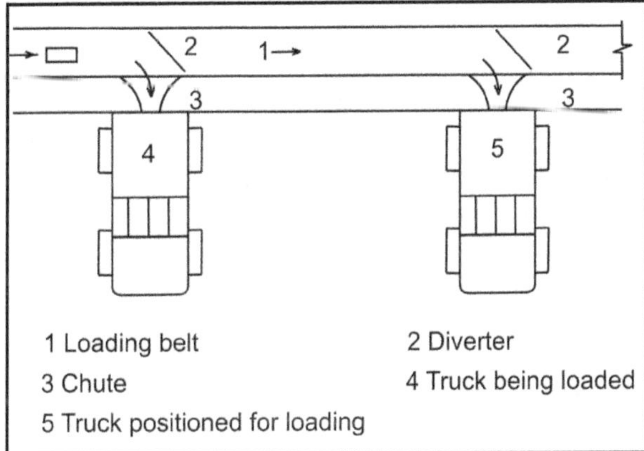

Fig. 6.34.4 Loadings trucks.

The manual loading is becoming exceptional and mechanized loading is becoming popular even in small plants.

34.5 Truck Loading

Truck loading machines are made to suit:

1. Standard trucks.
2. Long trailers.

The plant may thus select the loading machines to suit its needs. If cement is conveyed in standard trucks as well as in trailers, it would be wise to install at least one jumbo loading machine.

Alternatively if the plant is small, it may install a jumbo loading machine and use it for loading both standard trucks and trailers.

There will be minimum of two loading points with a diverter shifting flow of bags from one to the other when one truck / trailer is full. **See Figs 6.34.4 and 6.34.5.**

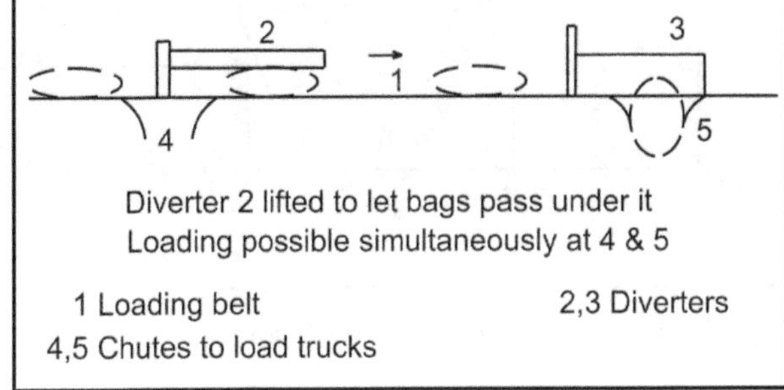

Fig. 6.34.5 Diverters permitting simultaneous loading at two points.

Bag diverters thus play an important part in the loading layout. They are pneumatically or electrically operated.

A diverter can also direct bags alternatively to 2 loading points by getting lifted off the belt to allow a bag to pass underneath it.

See Fig. 6.34.5.

This may be necessary if trucks are manually loaded and hence loading rate is lower than the bagging rate.

If the bagging rate is 100 tph, it means it has to load 10 No.s trucks of 10 t capacity.

Each truck would get loaded in 6 minutes. If there is only one loading point, with manual loading, it is equivalent of shifting and placing and stacking 200 bags in 6 minutes or 1 bag every 1.8 seconds.

In manual loading, there are 2 gangs with 5 to 6 loaders in each. The two gangs load alternately. One gang loads while the other gang rests. Thus man power requirement is virtually double.

See Fig. 6.34.6.

34.6 Mechanised Truck Loading

In mechanized loading, the loading machine has 3 movements.

1. Backward and forward to take bags into the truck. The length and movement of belt would correspond to dimensions of the truck.
2. Tilting – Vertically to facilitate stacking so that it is not necessary to lift a bag while stacking.

Fig. 6.34.6 Manual loading of trucks.

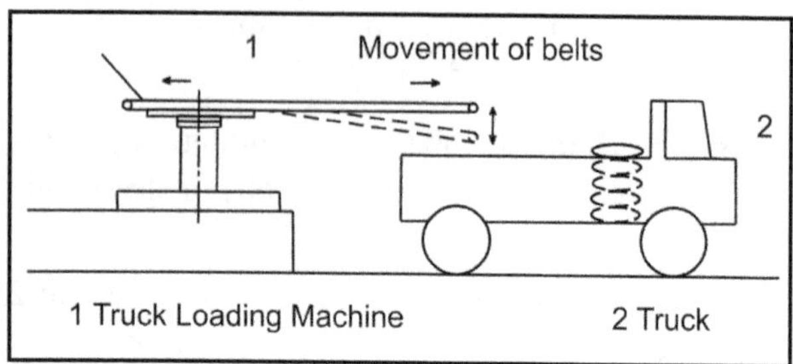

Fig. 6.34.7 Mechanised loadings of trucks .

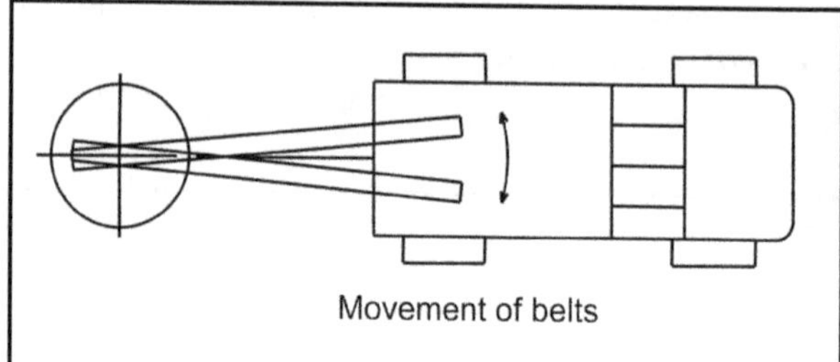

Fig. 6.34.8 Mechanised loadings of trucks.

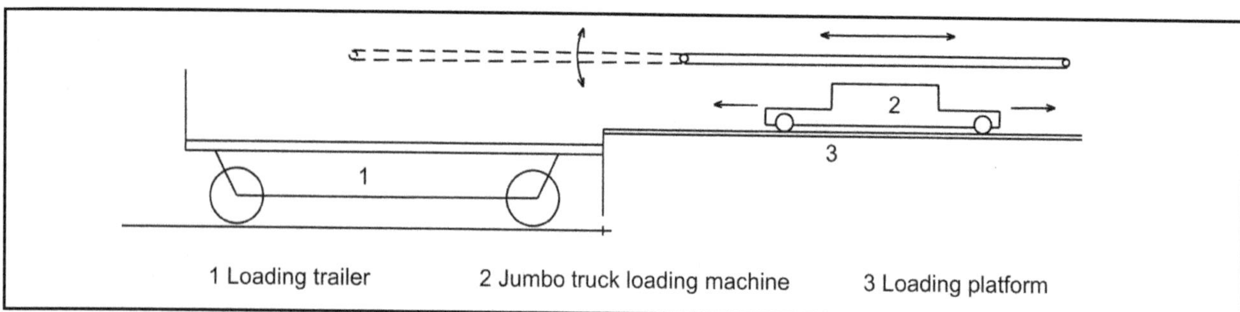

Fig. 6.34.9 Jumbo truck loading machine.

3. Angular motion to reach corners of truck to eliminate shifting of bags.

   See Figs. 6.34.7 and 6.34.8.

The number of men engaged can thus be reduced to 2 or maximum to 3.

Since the manual work and efforts are very much reduced, there need not be 2 gangs (one loading and 1 taking rest) even if it were so, the total number of men employ will be reduced from 10 in case of manual loading to 6 in mechanized loading ; a reduction of 40%.

To accommodate loading of trailers a long platform is required for the belt to go inside the trailer and out. **See Fig. 6.34.9.**

Therefore this platform for loading a trailer would generally be installed at the end of the loading platform. **See Fig. 6.34.10.**

34.7 Mechanized Wagon Loading

Mechanized wagon loaders are also now available to suit door dimensions of wagons of Indian Railways.

See Fig. 6.34.11.

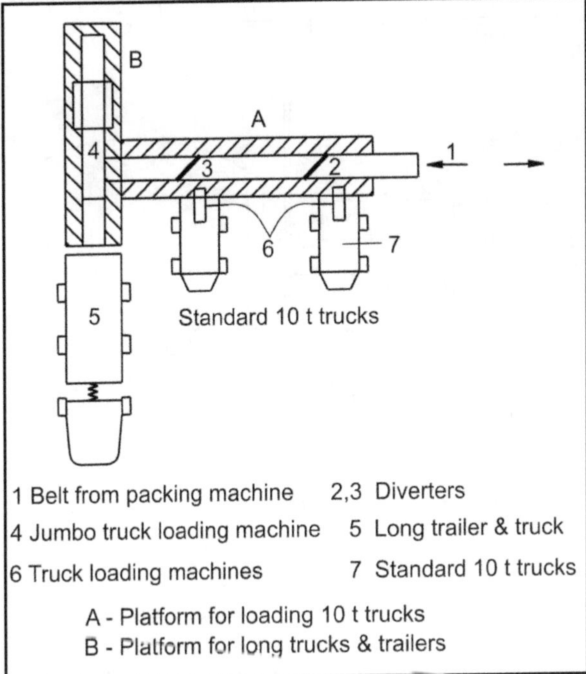

Fig. 6.34.10 Loading trucks and trailers with loading machines.

Wagons loaders have same elements as truck loaders:

1. Enter wagon and be maneuvered to reach corners of wagon.
2. Backward and Forward movement.
3. Tilting movement.
4. Swivel motion of the last belt.

Wagon loader comprises of three belts with independent drives.

When jute bags are used there is more spillage and dust nuisance.

The drives and moving mechanisms of belts need to be protected against dust.

Some kind of 'venting' may also be provided.

34.8 Platforms for Loading Wagons

Platforms for wagon, loading would be different than for truck loading.

It should be wide enough so that wagons can be loaded on both sides in case of manual loading and wide enough to move the wagon-loading machine along the platform and to shift it from one wagon to the other.

See Figs. 6.34.12 and 6.34.14.

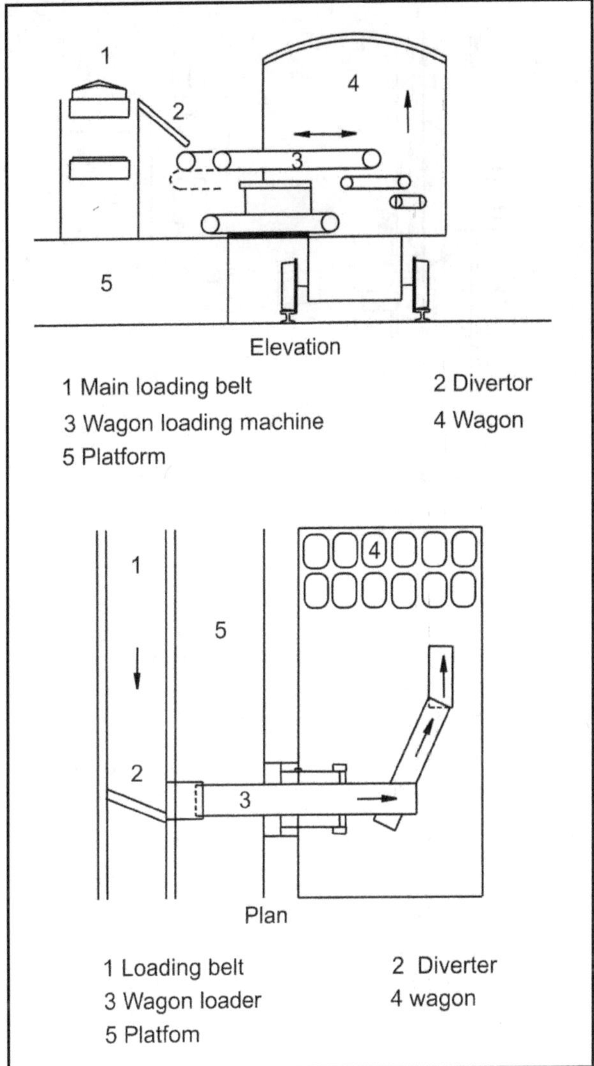

Fig. 6.34.11 Loading wagons by wagon loading machine.

Main loading belt may also be supported from trusses of roof over the loading platform. The wagon loading machine can then go under it and be moved along the platform.

See Fig. 6.34.13.

It can also happen that if the plant has its own locomotives, wagons can be shifted instead of wagon loading machines.

In wagon loading, it is customary to place half rakes on either side of the platform to avoid very long lengths of the loading platforms.

See Fig. 6.34.15.

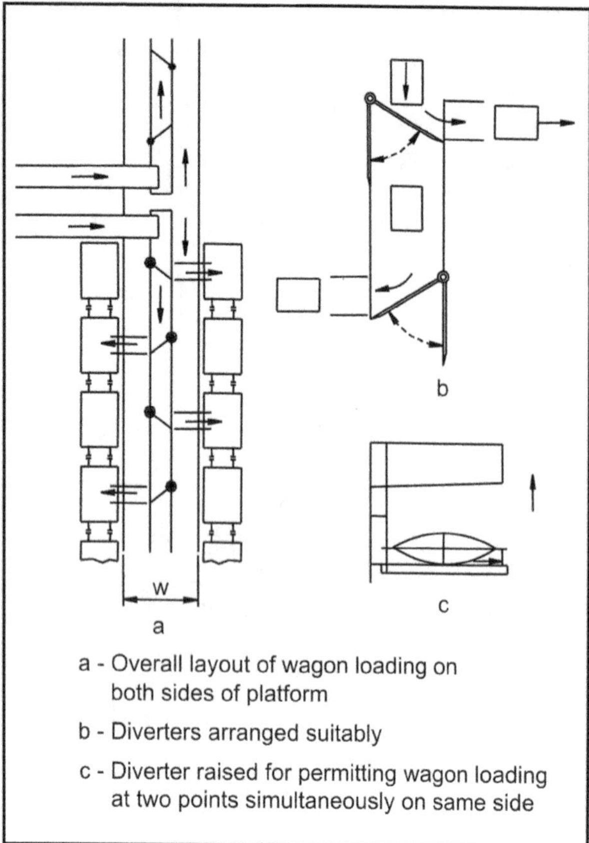

Fig. 6.34.12 Manual wagon loading on both sides of platform.

It also helps loading operations to be close to the packing machines and belt conveyor systems. Supervision is facilitated there by.

The long belt on the platform may be reversible if feeding point is in the middle. But a long reversible belt can easily go out of alignment and hence it would be best to load the belt at one end and it is unidirectional.

For loading the wagons on either side, diverters are installed on either side of the belt as shown. This will save time in shifting diverter. The same diverter is raised and lowered to allow bags to pass under it to be able to load more than one wagon simultaneously.

See Fig. 6.34.12.

As in case of truck loading, it should be possible to have more than one loading points for loading wagons simultaneously.

See Fig. 6.34.12.

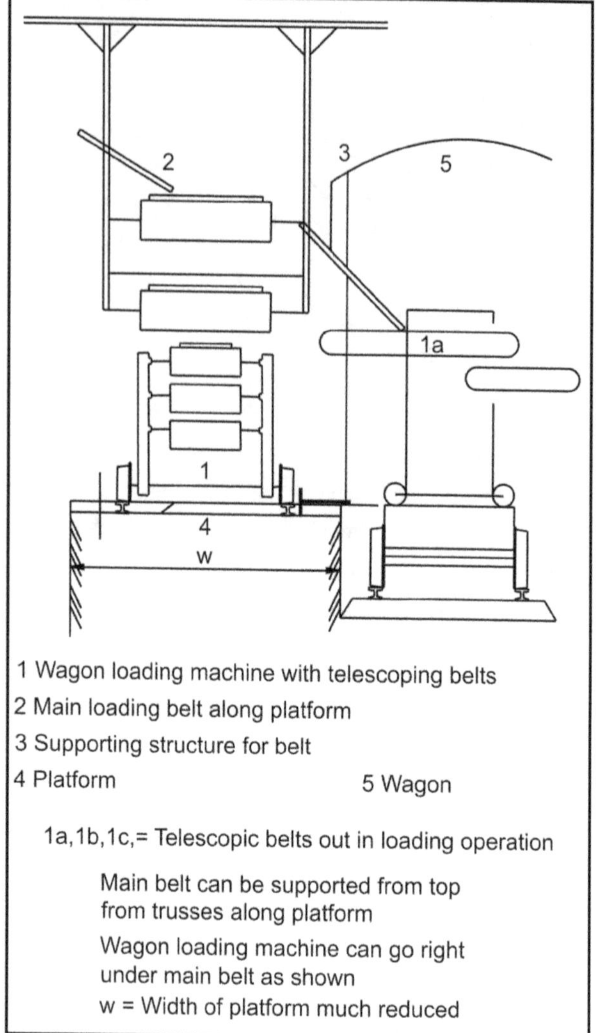

Fig. 6.34.13 Loading wagons with loading machines.

34.9 Maneuverability of Loading Machines

Maneuverability of wagon loading machine is of utmost importance.

The belts of the loading machine telescope – the minimum and maximum lengths should be known. A G.A. drawing of wagon loader with minimum space required to maneuver it, should be obtained from Suppliers.

34.10 Data on Types of Wagons to be Loaded

Rates of loading and hence number of loading points and also capacity and number of feeding points have to be calculated. Data should be collected on types of wagons to be loaded, their capacities and dimensions.

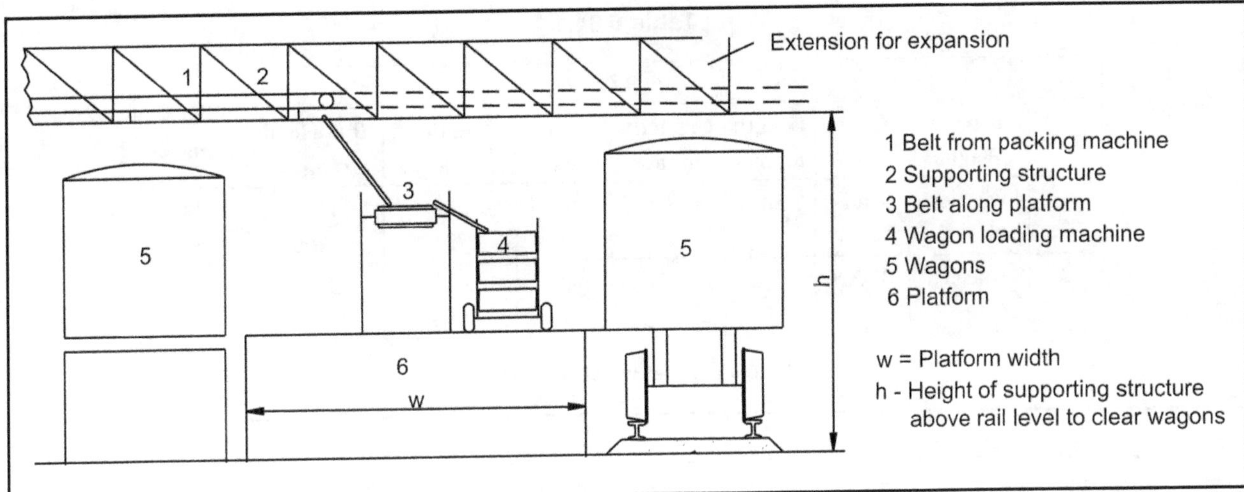

Fig. 6.34.14 Another arrangement for wagon loading machines.

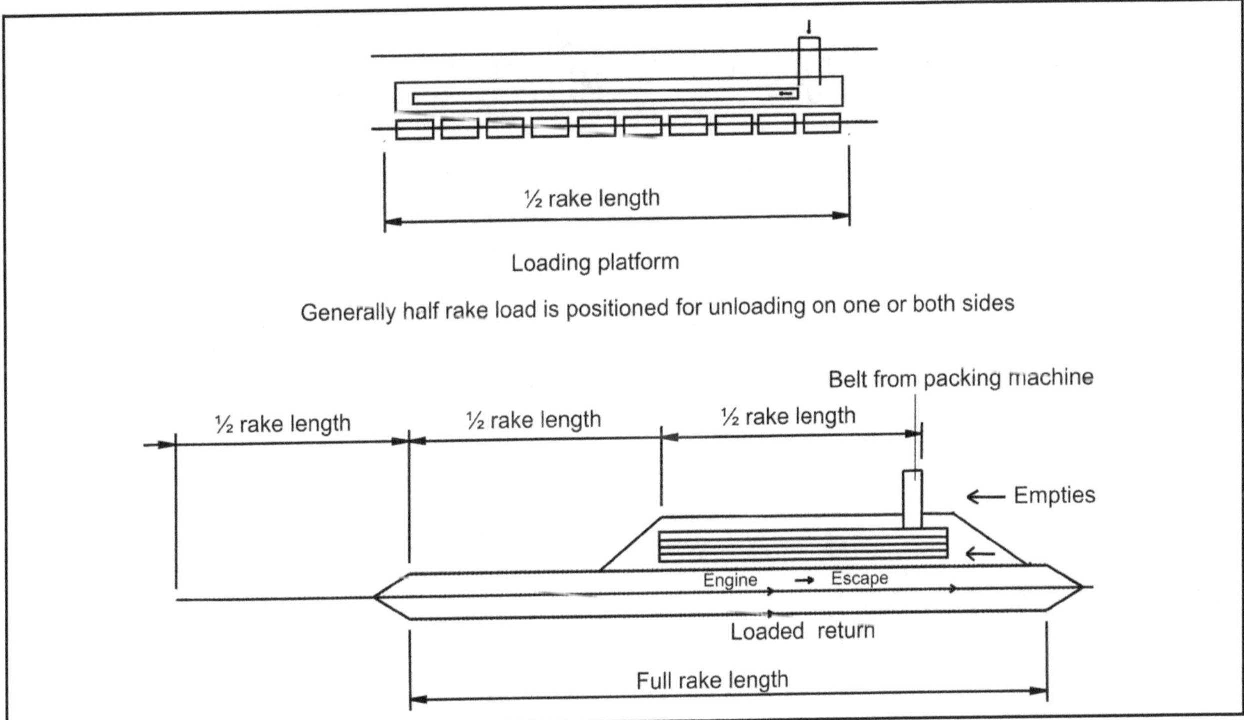

Fig. 6.34.15 Layout of a railway siding.

Types of wagons :

normally closed wagons would be used.

8 wheeler wagons with 40 T capacity.

4 wheeler wagons with 23 T capacity.

Capacities of above wagons are for broad gauge. Similarly capacities for metre gauge wagons should also be obtained.

34.10.1 Rake Loads

Data on rake loads of wagons of different types and gauges should also be collected. It is useful in design of railway siding.

See Table 6.34.1.

34.10.2

Data on wagons should be available from Railways' R.D.S.O. Lucknow. Such details should be procured

Table 6.34.1

B.G.								
Sr No	Type of Wagon	Code no	Length metres	Capacity tonnes	Nos/ rake	Length of rake	Rake load tonnes	remarks
1	4 wheeler	CRT	8.8	27.5	40 80	362 724	1100 2200	Closed wagon
2	8 wheeler	BCXN	15.75	52.7	40 60	645 960	2100 3160	Closed wagon

Length is buffer to buffer in metres.

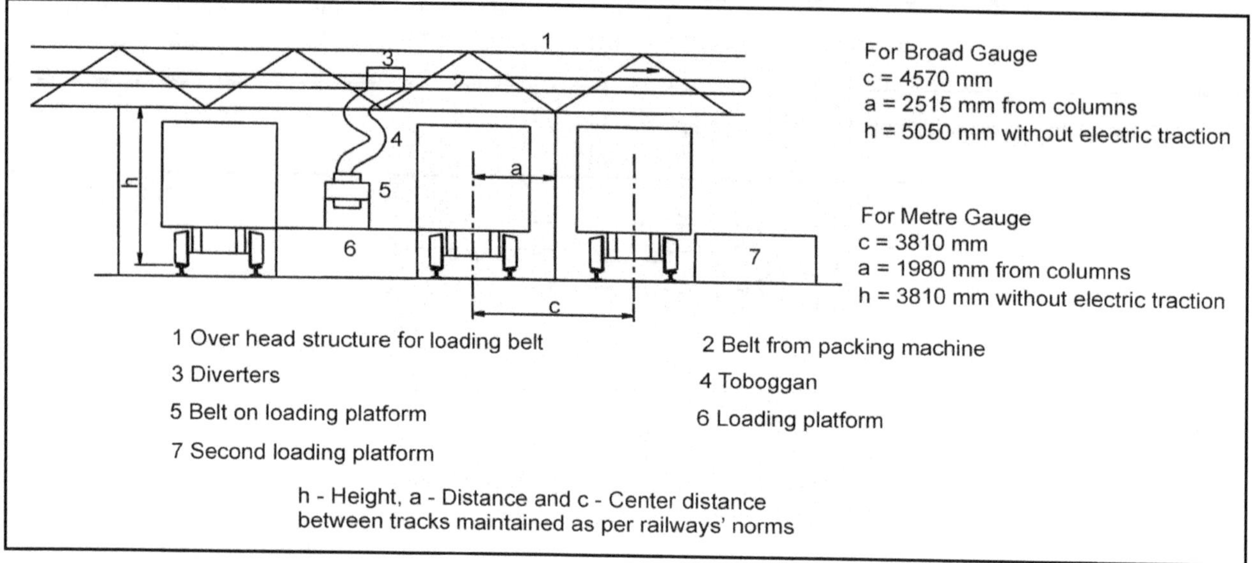

Fig. 6.34.16 Loading platforms for wagons.

as they are useful for preliminary design of railway siding and loading platforms.

34.11 Layout of Loading Belts

When a belt is crossing a railway line, it will be at a considerable height and bags must drop from it on to the loading belt on the platform. This fall may be several meters and bags can be damaged.

See Figs. 6.34.14 and 6.34.16.

Toboggans which are spiral chutes can be used to guide bags from upper belt on to lower belt safely. Toboggan can also change direction.

See Fig. 6.34.17.

If the fall is too much a serrated belt with a slope can be used to feed the bags gradually on to the loading belt.

See Fig. 6.34.18.

In this situation belt will be too high for truck loading also. Hence a luffing belt will have to be introduced between main belt and trucks.

See Fig. 6.34.19.

34.12 Spillage

In use of jute and polyethylene or h.d.p.e bags spillage at transfer points is inevitable. The seams of bags get stuck occasionally at transfer points. Hooks are used to pull bags clear off the obstructions. This also causes spillage. At the end of the day, considerable spillage accumulates on loading platforms, under loading belts etc.

This is now cleaned up by mobile industrial vacuum cleaners and sent back to cement storage.

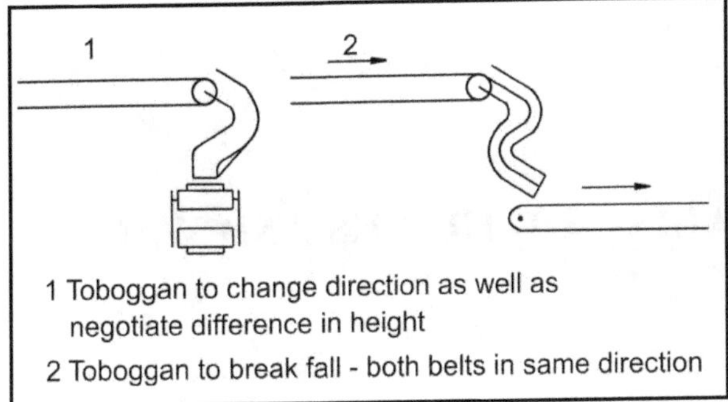

Fig. 6.34.17 Toboggan to break fall from height and change direction.

A system should be installed to return cement either to storage silo or to elevator feeding packing machines. In doing so, the collected material should screened so that items like cotton waste, nails, iron pieces, welding rods, etc., do not find their way into the silo.

The best place to return spillage would to return it to spillage collected under the packing machines.

Also see layouts of packing plants with truck and wagon loading facilities in **Chapter 14 of Section 2**.

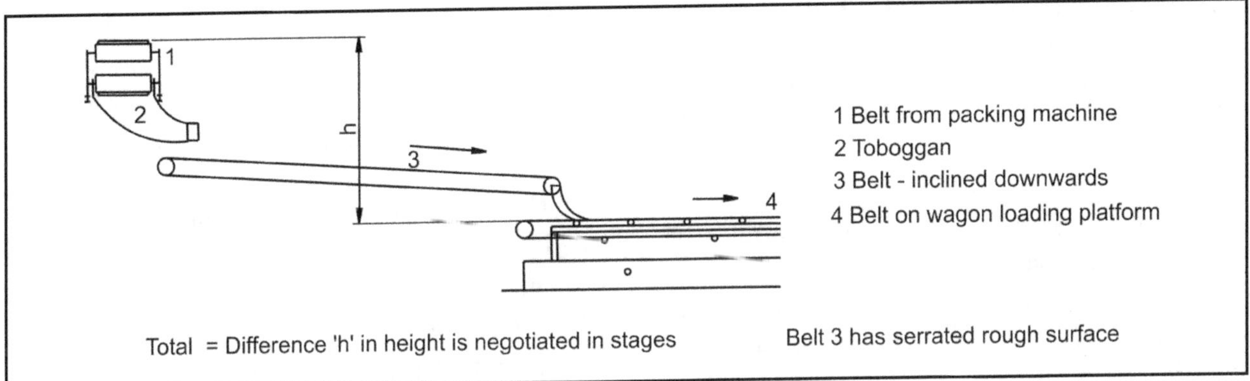

Fig. 6.34.18 Breaking fall between belt from packing machine and loading belt on loading platform.

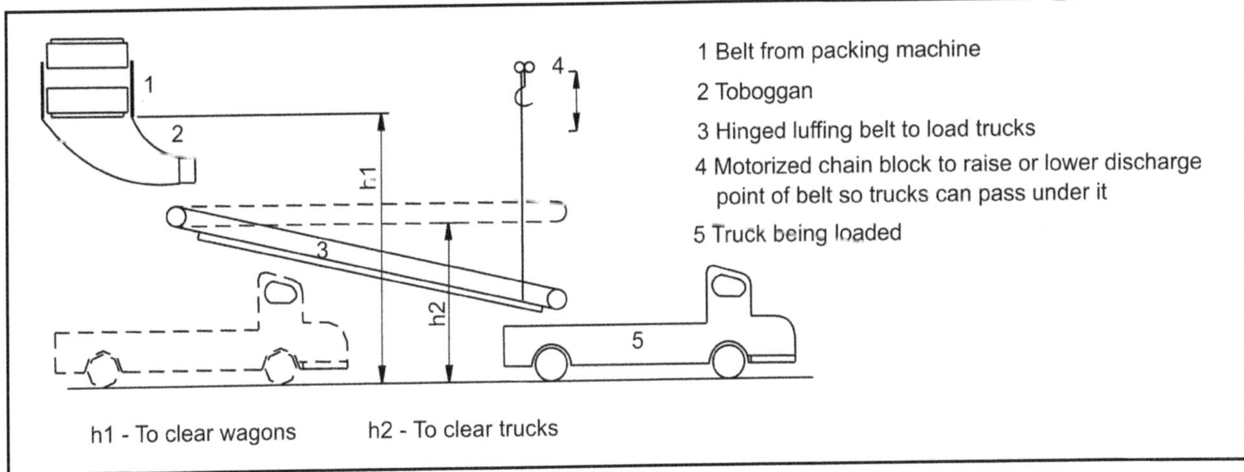

Fig. 6.34.19 Loading of trucks with luffing belts which can change their inclination.

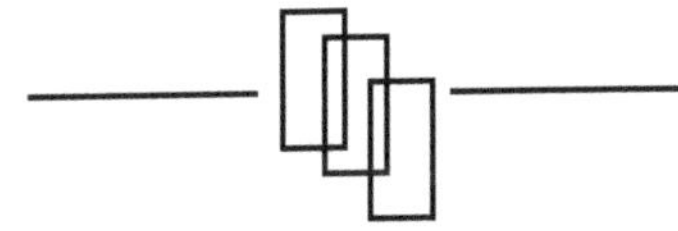

CHAPTER 35

SIMULTANEOUS LOADING OF TRUCKS AND WAGONS

35.1 Dispatches by Road and Rail

About 70 % of cement produced is now sent by road. The plant has therefore to provide facilities for truck loading as well as wagon loading. Often these operations are simultaneous and rarely mutually exclusive.

The layout must therefore provide for facilities for truck and wagon loading both.

Since railway siding is long and has several tracks, it is best to keep it at the extreme boundary of the plant in the general layout. The truck loading facilities are therefore fitted between packing machines and wagon loading. Belts for wagon loading go over truck loading facilities.

See Figs. 6.35.1 and 6.35.2.

As mentioned in previous Chapter, luffing belts with rough surfaces are used as intermediate belts.

See Figs. 6.35.3, 6.35.4 and 6.35.5.

Trucks are positioned and moved backwards / forwards for filling and placing bags.

It is also possible to use standard truck loading machines.

See Fig. 6.35.5.

35.2

Whatever the layout, it should satisfy the following norms :

1. The paths of trucks should not cross.
2. There should be room for backing, reversing and making 'U' turns.
3. Some parking place should be provided in case all loading points are busy.
4. In flow of trucks into the plant premises should be regulated at the gate.
5. Railway lines and truck routes should not cross

   See Figs. 6.35.1, 6.35.2, 6.35.6 and 6.35.7.

35.3 Working Out Requirements

If a million tpa plant dispatches 1 million tons of cement / annum by road only, it would have to dispatch 3000 tpd of cement.

It would thus have to handle 300 trucks (10 tons capacity) in a day of 15 effective hours or 20 trucks / hour.

Though loading takes place in all 3 shifts, trucks would not be coming in evenly.

At an even rate of loading, each truck would have to be loaded in 3 minutes.

But trucks may come say :

20 trucks/hour between 7 a.m. to 9 a.m.;

30 trucks / hour between 9 a.m. to 11 a.m.,

40 trucks / hour between 11 a.m. to 2 p.m.

Bagging rate is already fixed according to capacity of packing machines, say 200 tph.

Thus trucks would have to wait for their turn to come for loading. Therefore, a well designed parking space large enough to hold 50 to 80 trucks should be provided. It should have facilities for drivers' needs like toilet / resting / canteen, etc.

It is best to provide this parking space outside the boundary wall of the plant.

See Figs. 6.35.6 and 6.35.7.

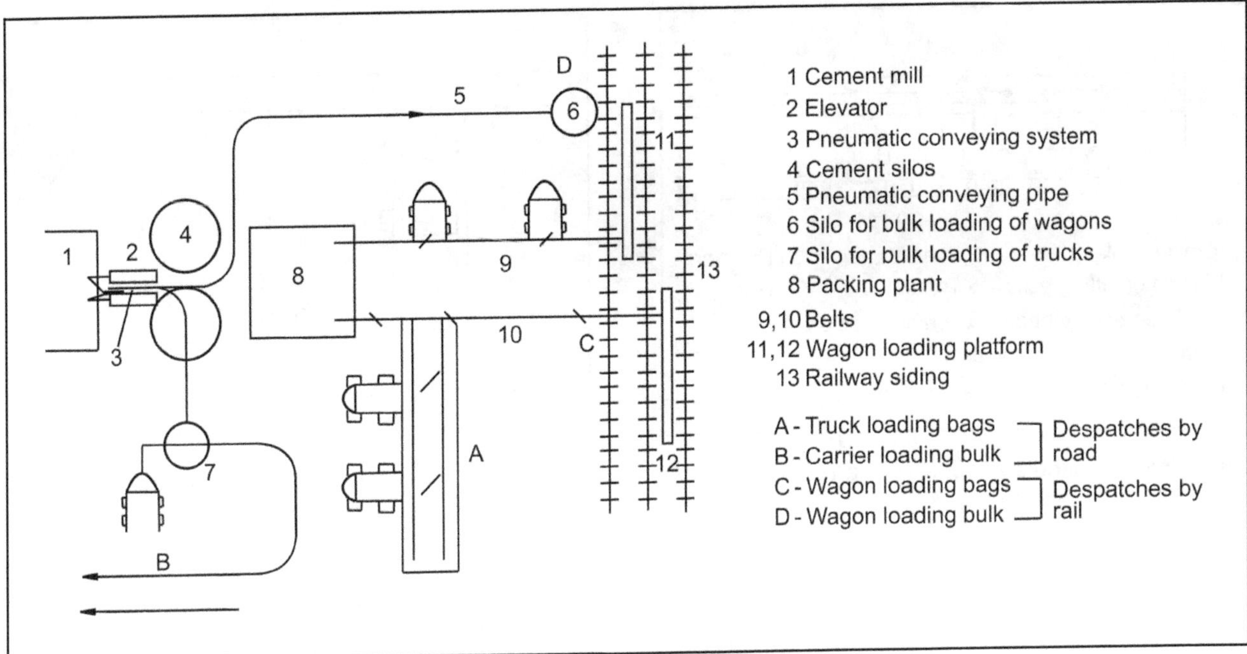

Fig. 6.35.1 Facilities for simultaneous truck and wagon loading of bagged and bulk cement.

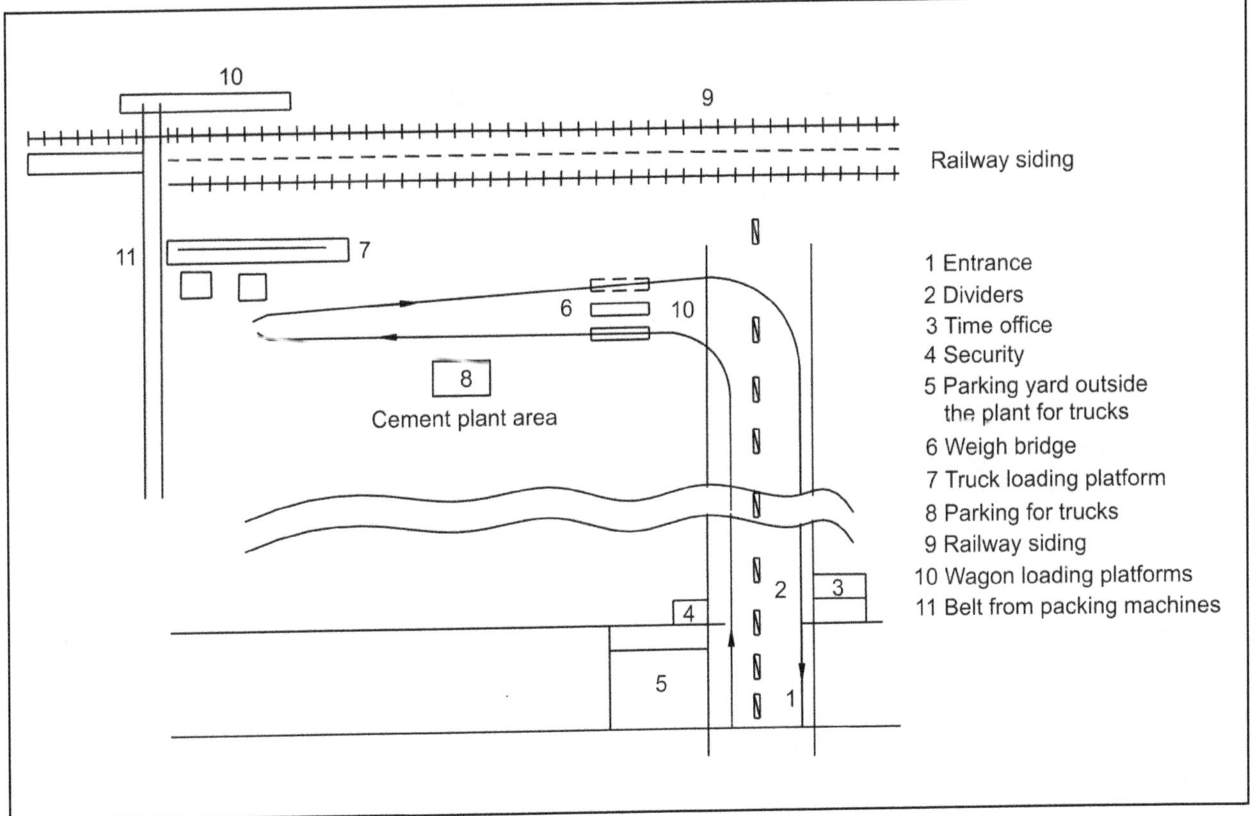

Fig. 6.35.2 Simultaneous truck and wagon loading.

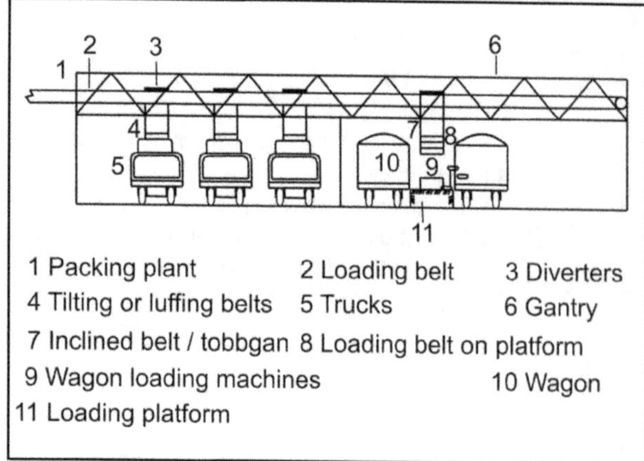

Fig. 6.35.3 Simulataneous truck and wagon loading of bags.

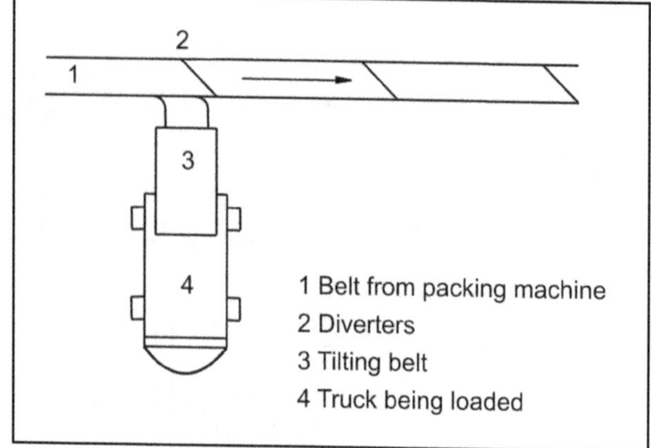

Fig. 6.35.4 Loading trucks with overhead luffing belts.

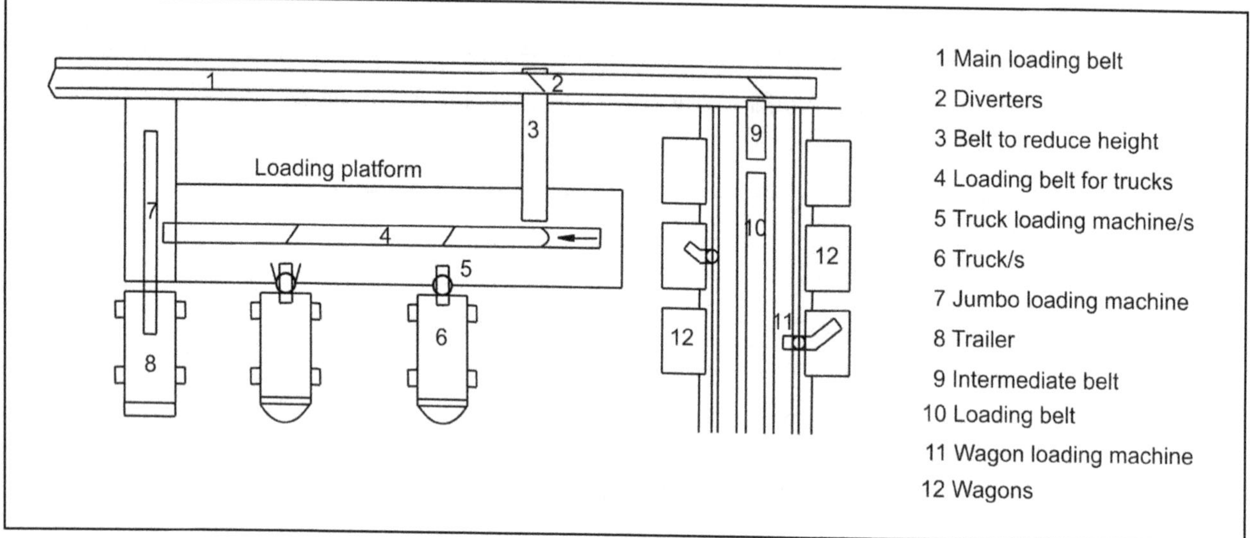

Fig. 6.35.5 Using truck loading machines along with machanised loading in wagons.

35.4 Other Traffic by Trucks

When slag / fly ash etc., is also received in truck loads, it may be better to provide a separate entrance and exit for these trucks out of the way of main traffic for cement.

See Fig. 6.35.8.

It would also be desirable to keep the truck traffic out of the way of traffic of the plant itself like workers and staff coming and going in, cars, cycles, scooters, trucks, etc. This will greatly reduce risk of accidents. Separate gates may be provided for factory traffic and for loaded / unloaded trucks.

See Fig. 6.35.7.

35.5 Railway Siding for Loading Wagons

Truck traffic should not have to cross railway lines. There would be too many hold ups in addition to risks of accidents.

See Figs. 6.35.1 and 6.35.2.

Railways lines should be far away from factory traffic so that workers going from colony to works and back do not have to cross one / more railway lines.

Wagons are shunted and moved all along and at all hours along the railway line. Accidents can happen and should be avoided at the layout stage itself.

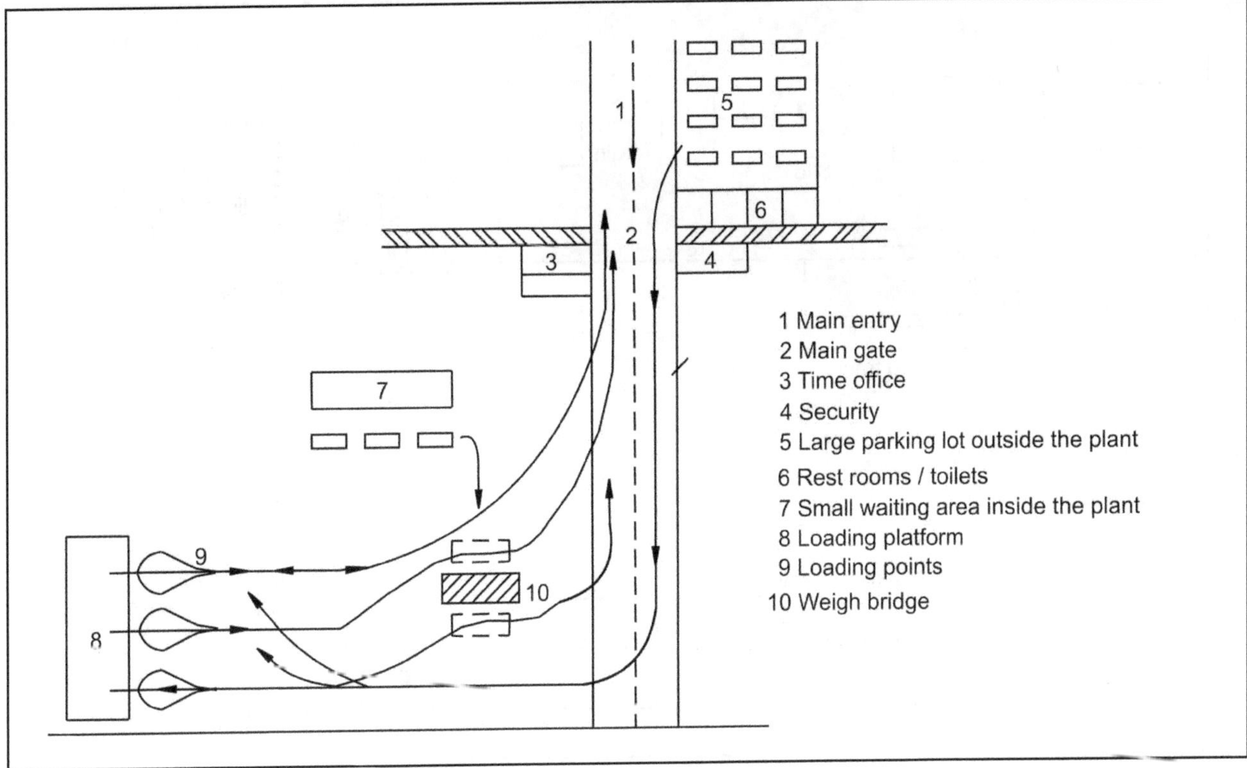

Fig. 6.35.6 Truck loading traffic - waiting and parking arrangements.

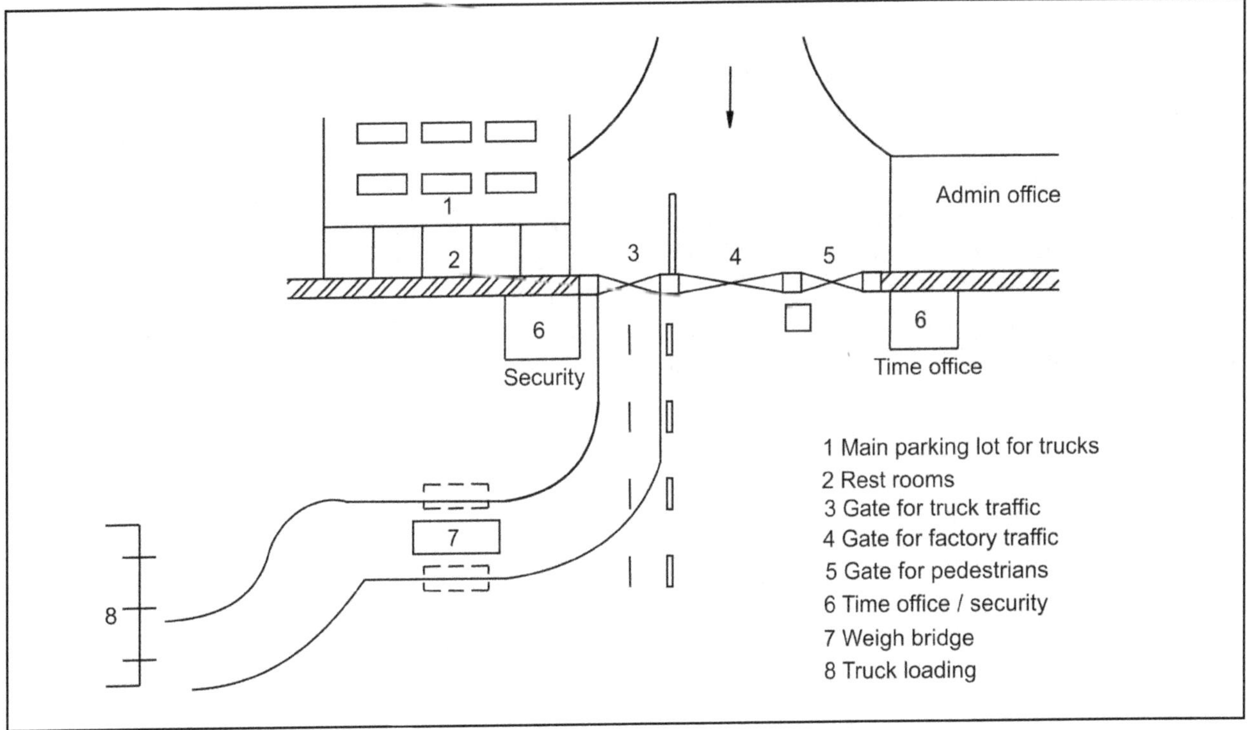

Fig. 6.35.7 Separate entrances for trucks and other factory traffic.

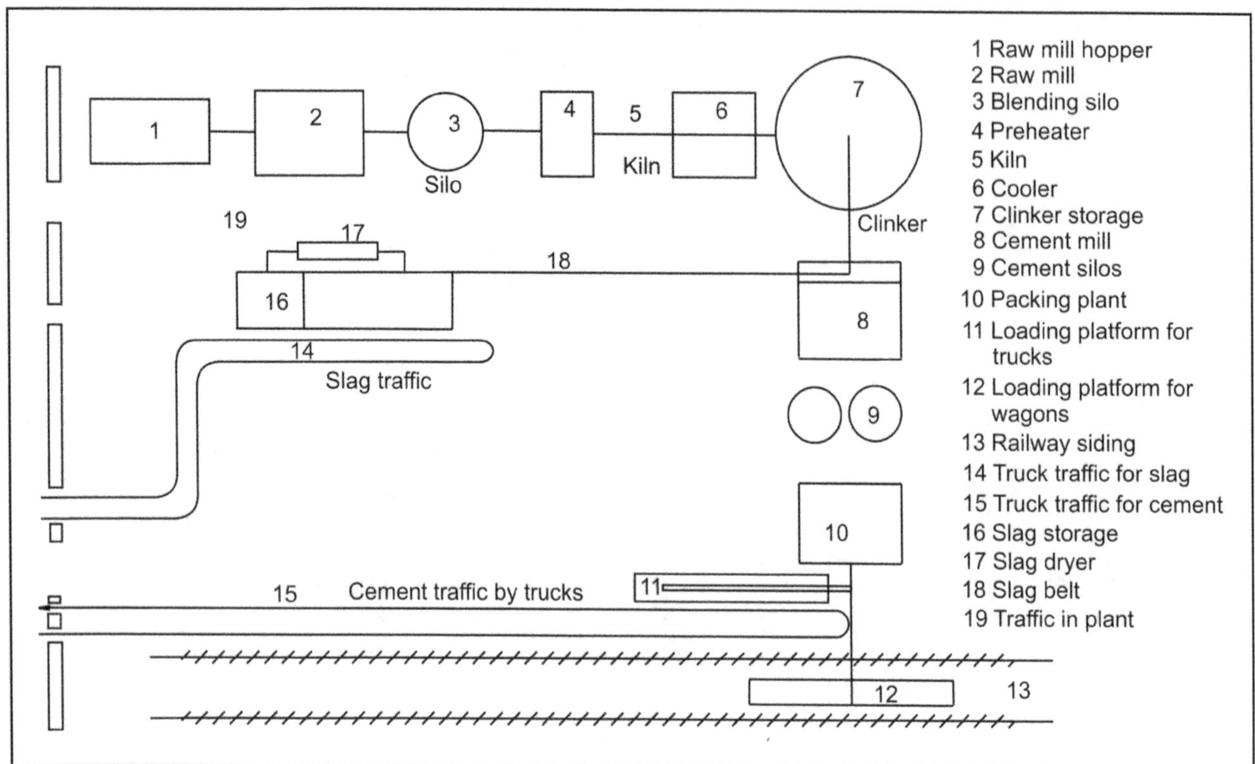

Fig. 6.35.8 Layout showing relative positions of railway siding and clinker and slag storages; separate entries for traffic of slag and cement.

If it is not possible to do so then manned crossings should be provided to avoid accidents.

35.6 Automated Truck and Wagon Loading

It is possible to automate packing and loading operations when paper bags are used. It has already been done.

Automation is not confined to the machines used like packing machines, truck and wagon loaders but covers total operation beginning with receiving orders, placing of trucks and wagons for loading and also raising invoices. Systems of automation are shown schematically in **Figs. 6.35.9 and 6.35.10.**

When operations are thus automated, provisions for parking lots for trucks may sound outdated or unnecessary. Still there are many a plant and many a country which use jute and h.d.p.e bags where full automation is difficult. Provisions suggested above would apply to such situations.

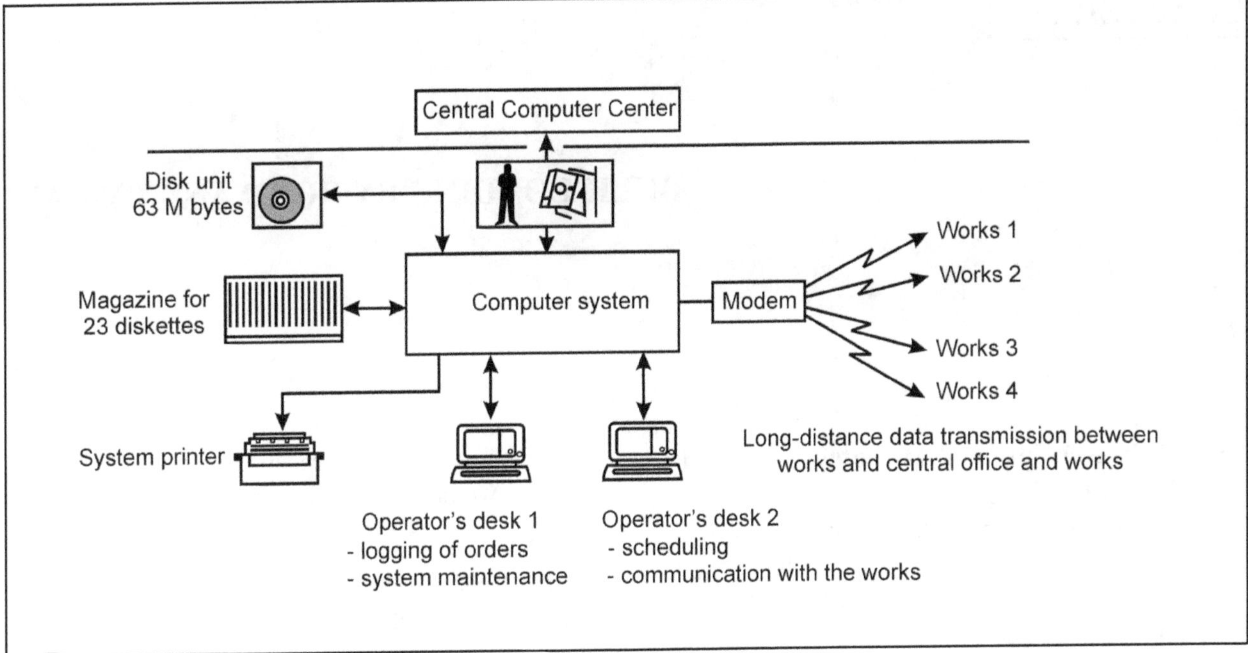

Fig. 6.35.9 Central system configuration.

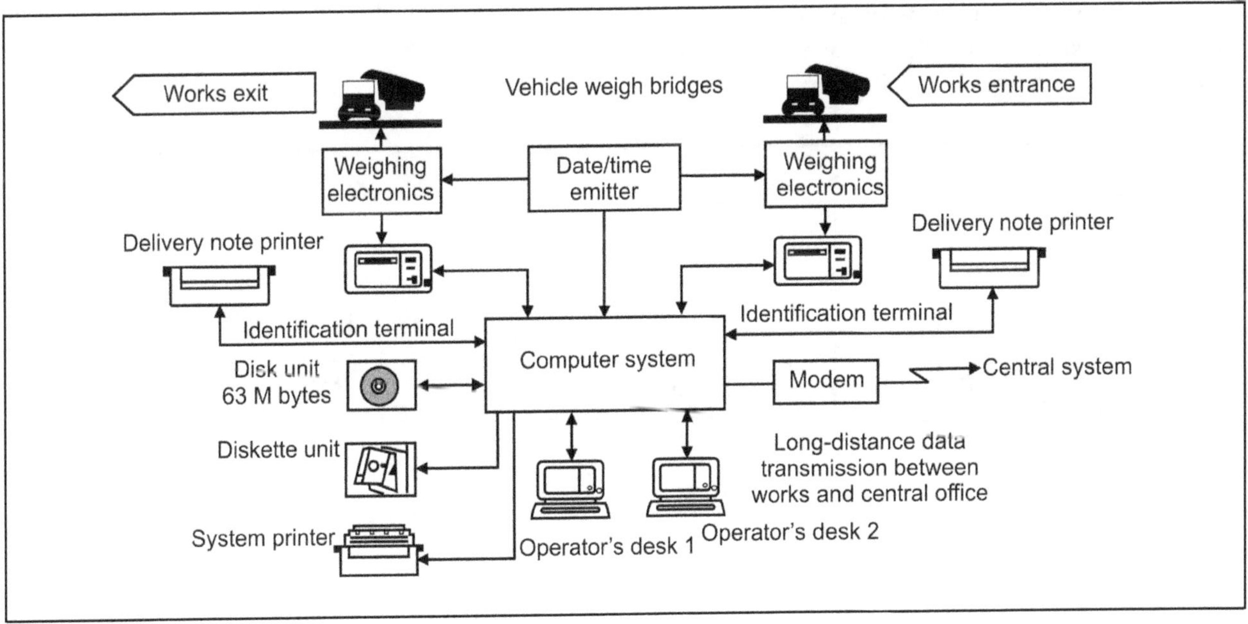

Fig. 6.35.10 Works system configuration.

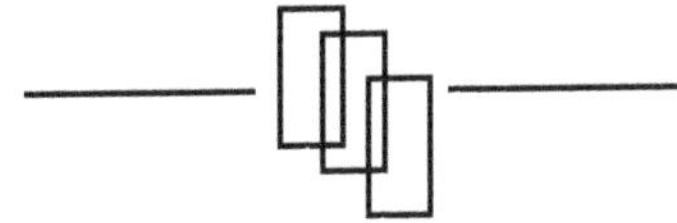

CHAPTER 36

BULK LOADING BY ROAD AND BY RAIL

36.1 Bulk Receipts and Dispatches by Rail / Unloading Wagons

Bulk dispatches of cement by rail would normally require special self-unloading wagons. However, on occasions, loose cement has also been transported in common closed or even open wagons.

For common wagons without built in unloading facilities, wagon un-loaders are used which convey loose cement in the wagon pneumatically to storage silo. Else, cement can be unloaded manually into trough hoppers by the side of rail tracks and cement conveyed pneumatically therefrom by FK Pumps.

However sending or receiving cement in common wagons would be rarely resorted to now.

See Fig. 6.36.1.

Bulk cement received in special wagons can be unloaded pneumatically.

See Fig. 6.36.2.

36.2 Loading Bulk Wagons

Wagons can stand under the silo or silo can be by the side of the track.

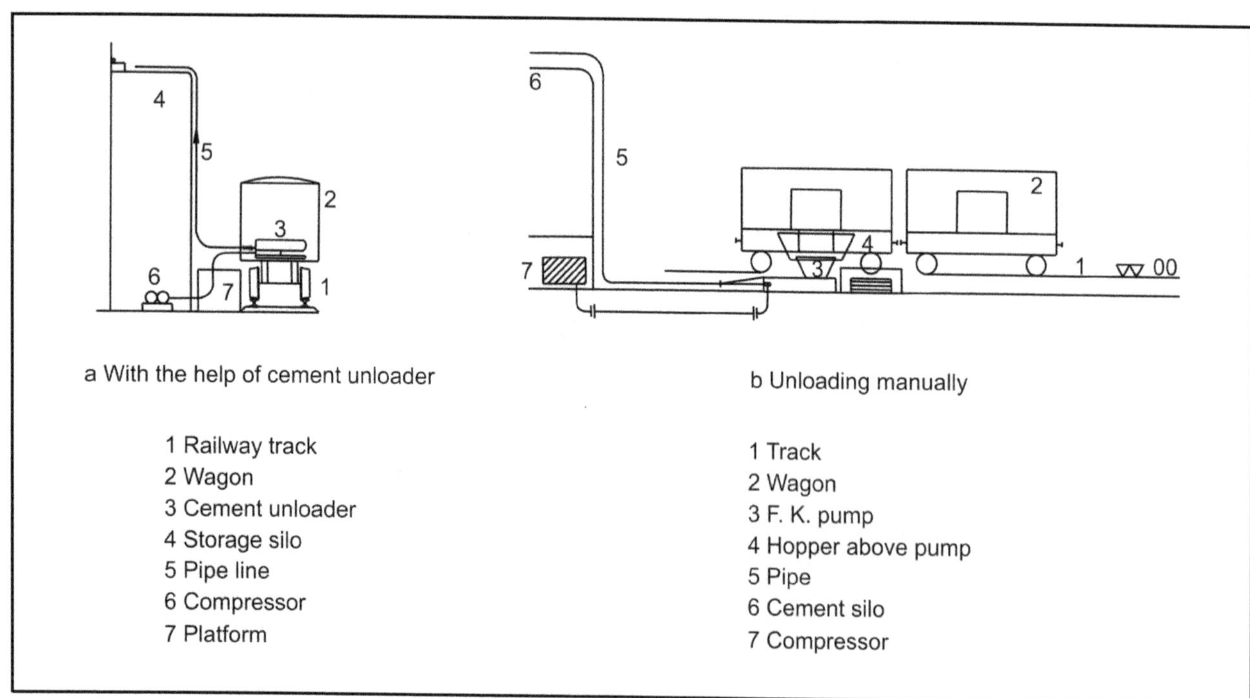

Fig. 6.36.1 Cement / fly ash unloading from wagons.

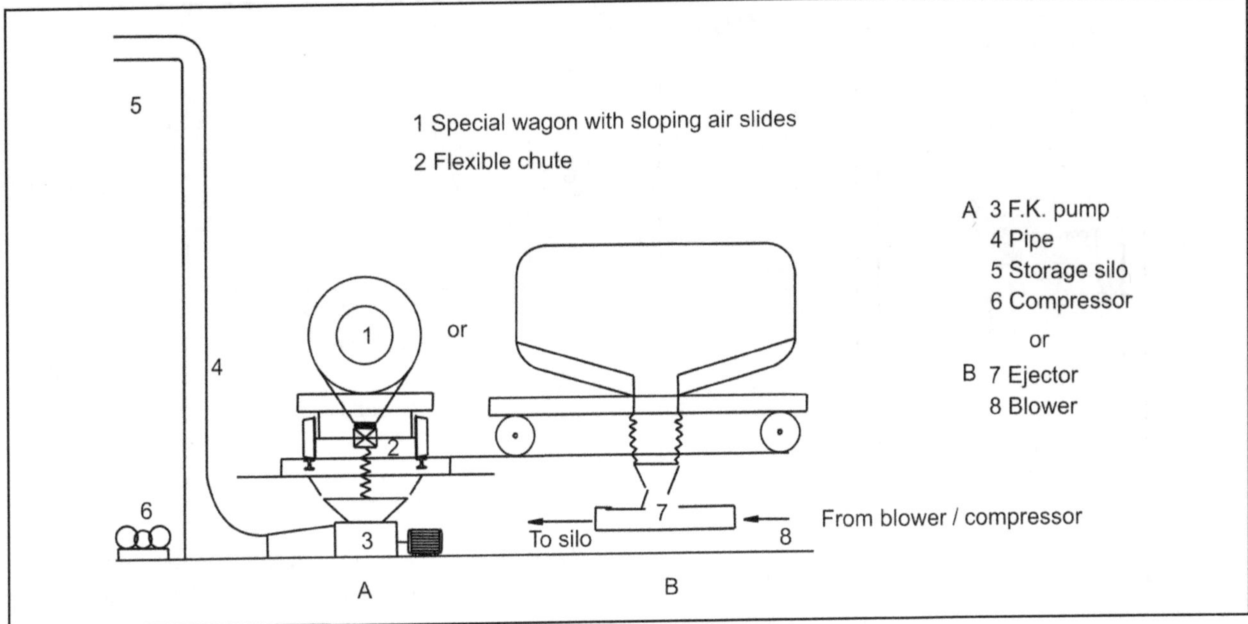

Fig. 6.36.2 Pneumatic unloading of cement from special wagons.

In this case, a weigh bridge of suitable capacity (weight of loaded wagon) would be installed under the track.

See Fig. 6.36.3.

Loading rate should be adjustable reducing quickly and with a sharp cut off as wagon is getting filled up **See Fig. 6.36.4.**

It may also be possible to install silo in such away that truck and wagon loading could be possible simultaneously.

See Fig. 6.36.5.

36.2.1 Bins on Load Cell

Installing weigh bridges can be avoided if a measured quantity can be held in an intermediate bin for unloading into truck / wagons.

See Fig. 6.36.5.

Various possibilities of bulk dispatches by road and rail have been shown in photos attached to **Chapter 15 of Section 2.**

36.3 Bulk Dispatches by Road

Dispatch of dry fly ash in bulk carriers has already been dealt with in **Chapter 29.**

Cement is dispatched by road in similar bulk carriers in a similar fashion. Scale would be much larger.

See Figs. 6.36.3 and 6.36.5

If the storage silo is large enough, instead of a flat bottom, it can have an inverted cone with several outlets. More than one bulk carrier can stand under the silo and can get filled. Dust collector for venting can also be installed under the silo.

See Fig. 6.36.6.

36.4 Venting

Filling operation of cement would need venting. The flexible pipe used to connect carrier with loading point of silo will have a facility to suck vent air. It will be passed through a dust filter.

See Figs. 6.36.3, 6.36.6

36.5 Separate Terminal

If the dispatches in bulk are regular it is best to provide a separate silo with bulk loading facilities either common for road and rail dispatches or separate silos for the two.

See Figs. 6.36.3, 6.36.5 and also Fig. 6.35.1. Chapter 35.

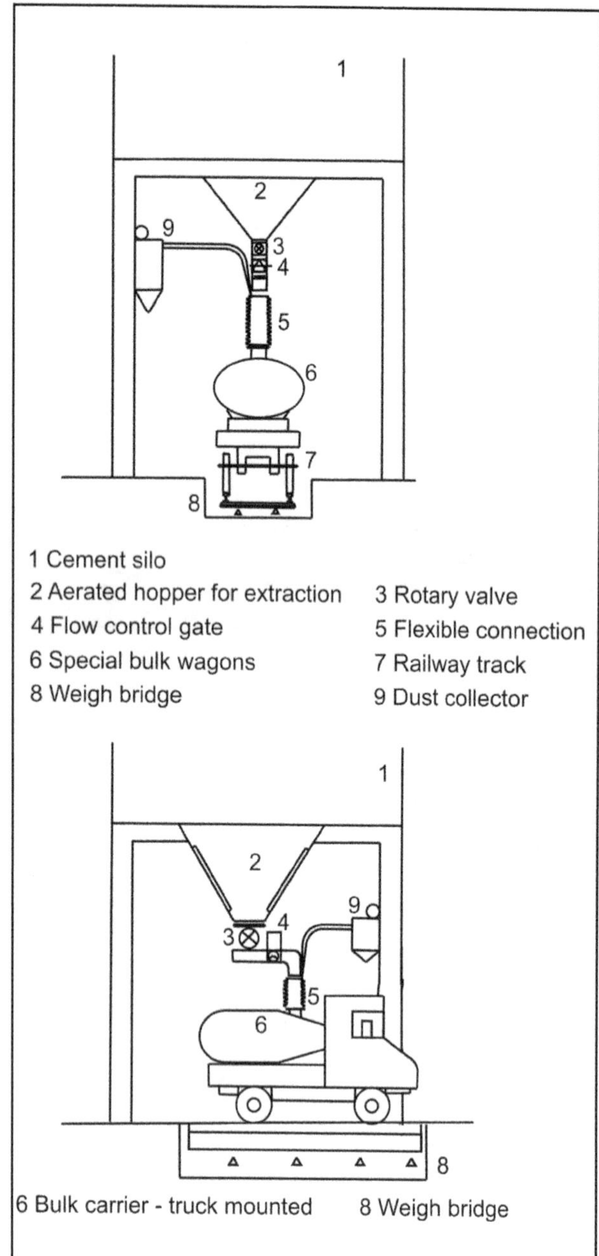

Fig. 6.36.3 Bulk loading of wagon and truck directly from cement silo.

36.6 Automation

Automation is much easier for bulk loading and has been adopted almost everywhere in different degrees. **See figs 6.35.9, 6.35.10 and paragraph 35.6 in Chapter 35.**

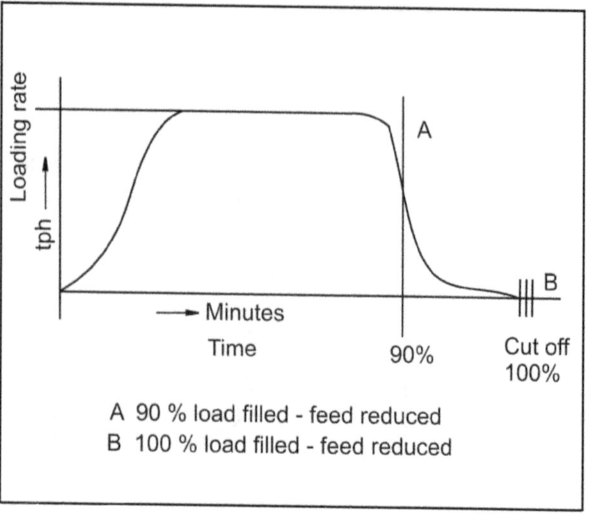

Fig. 6.36.4 Rate of filling reduced to fill exact weight in wagons, bulk carries.

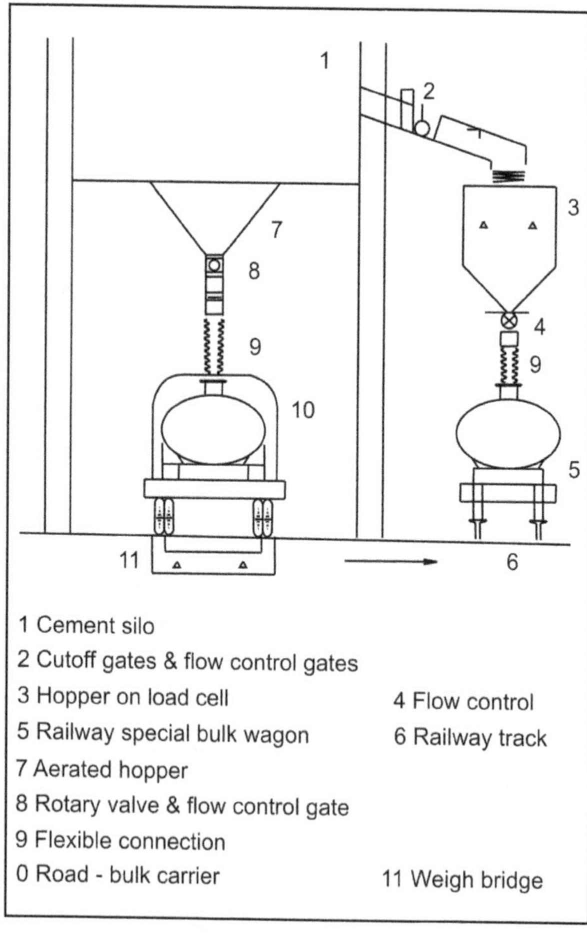

Fig. 6.36.5 Simultaneous bulk loading in wagons and trucks.

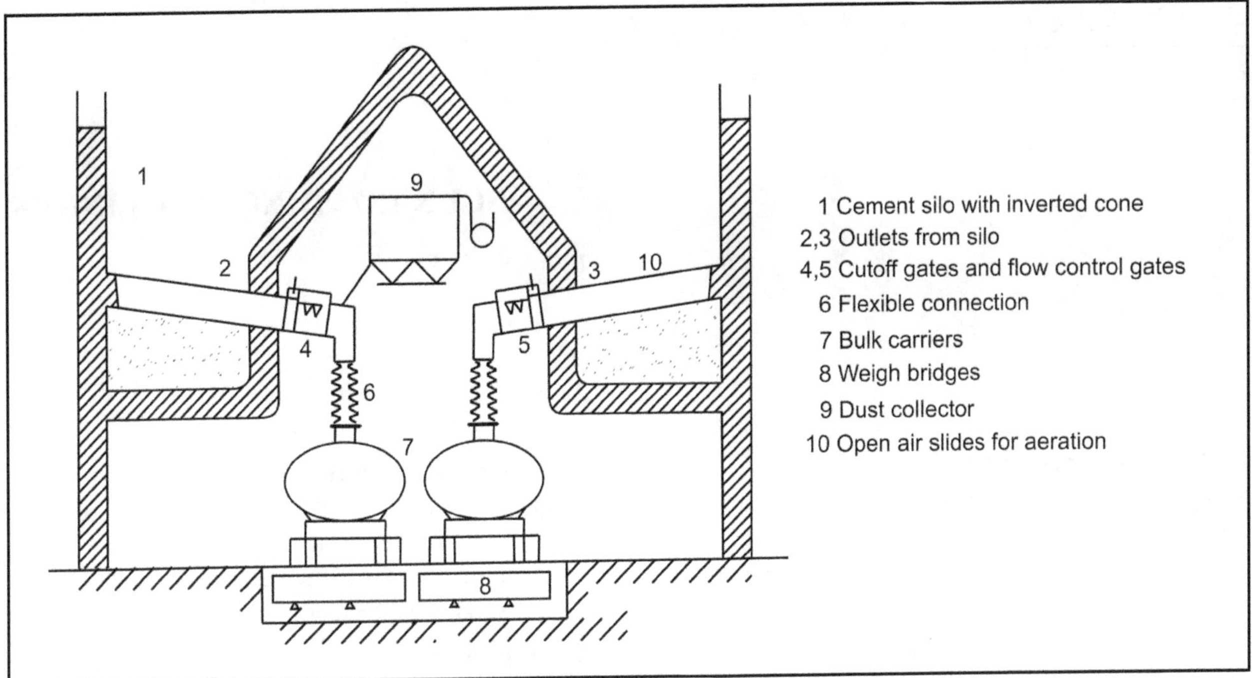

Fig. 6.36.6 Bulk loading in carriers under a silo with an Inverted cone.

36.7 Extraction from Different Levels

In **Chapter 32**, while dealing with sectionalized silos it was mentioned that it was possible to extract different cements simultaneously from different levels.

See plate 6.32.1.

Such an arrangement would be very pertinent in case of simultaneous bulk loading described above.

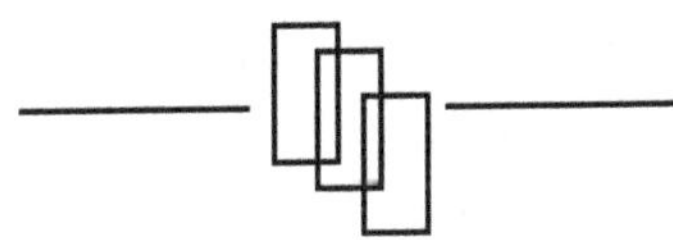

CHAPTER 37

BULK LOADING OF CLINKER

37.1 Dispatches of Clinker

Many plants now sell clinker to small grinding units; others have split locations and hence transport clinker in large quantities to the location where grinding unit is situated.

There are also plants that have a two-way traffic of bringing in slag and sending out clinker.

Some plants export clinker. In all such cases provision has to be made to load clinker in bulk in trucks and or in wagons because clinker is sent loose.

37.2 Loading Clinker in Trucks

For loading clinker in trucks, the possibilities are to use :

1. shovel or pay loaders,
2. grab cranes,
3. belt conveyors,

1 and 2 would cause considerable dust nuisance.

In a small plant which store clinker in a shed with belt conveyor, it may be possible to extend the main conveyors for loading trucks with either fresh or stored clinker.

See Figs. 6.37.1 and 6.37.2.

Clinker goes to the stockpile when trucks are not loaded or during the change over time when loaded truck is replaced by an empty truck. Loading hot fresh clinker would generally be avoided. The transfer point will consist of a two way chute operated manually or electrically.

Loaded trucks will pass over a weigh bridge (having noted weight of empty truck as they come in) and weight of clinker loaded can be found and recorded.

37.3 Separate Bin for Loading Clinker

Frequent shifting of clinker route can be avoided if a small bin with a feeder is installed. If the bin is on a load cell, so much the better because the clinker loaded can be weighed while loading.

See Fig. 6.37.2.

The splitting of belt for installing bin or a take off point would require a head room of 2 to 3 meters which should be provided in the layout. Alternately the loading point may be shifted close to the clinker hoppers.

See Fig. 6.37.3.

This arrangement would avoid splitting but would bring loading point close to mill building blocking access to it.

An alternative to this would be to install a belt conveyor at top of hopper to convey clinker to a loading point at a convenient distance from the cement mill.

See Fig. 6.37.4.

In this arrangement dust nuisance could be controlled by installing a dust collector on the bin.

In arrangements shown of loading without a bin, dust nuisance during loading can not be avoided.

The structure to hold bin and feeder above the track would have to be designed according to norms laid down by Railways regarding heights and clearances.
See Fig. 6.37.6.

37.4 Loading Wagons

If clinker is to be sent in wagon loads for plants with split locations then arrangements are required be made near the rail tracks for loading wagons.

See Fig. 6.37.5.

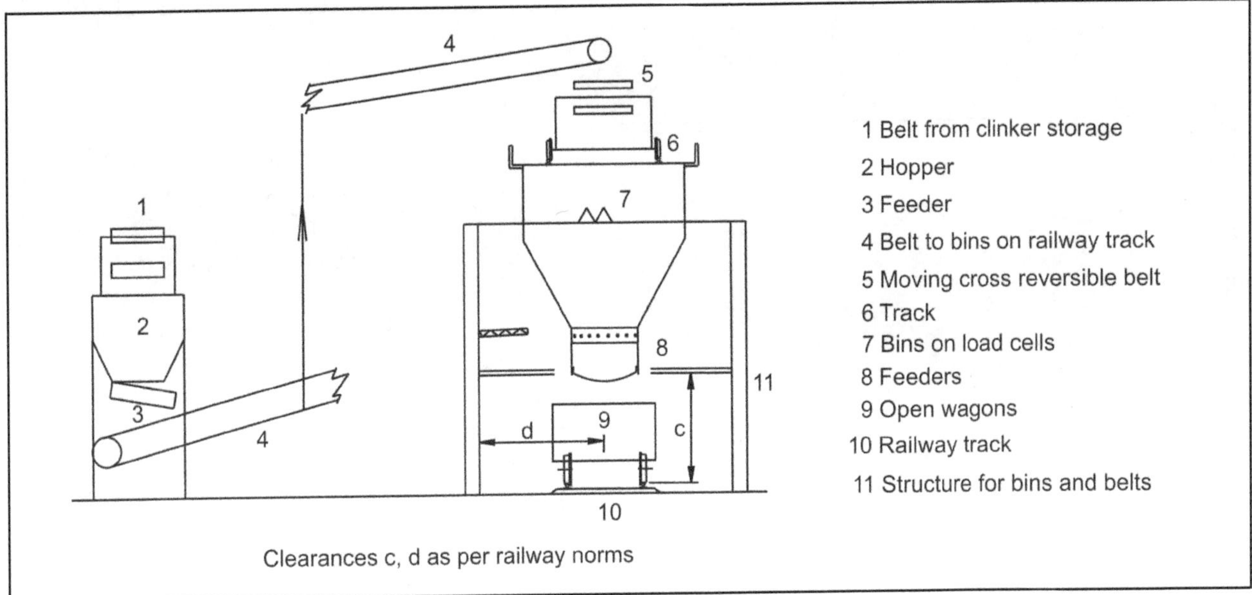

Fig. 6.37.6 Clinkers loading for despatches in wagons for large plants.

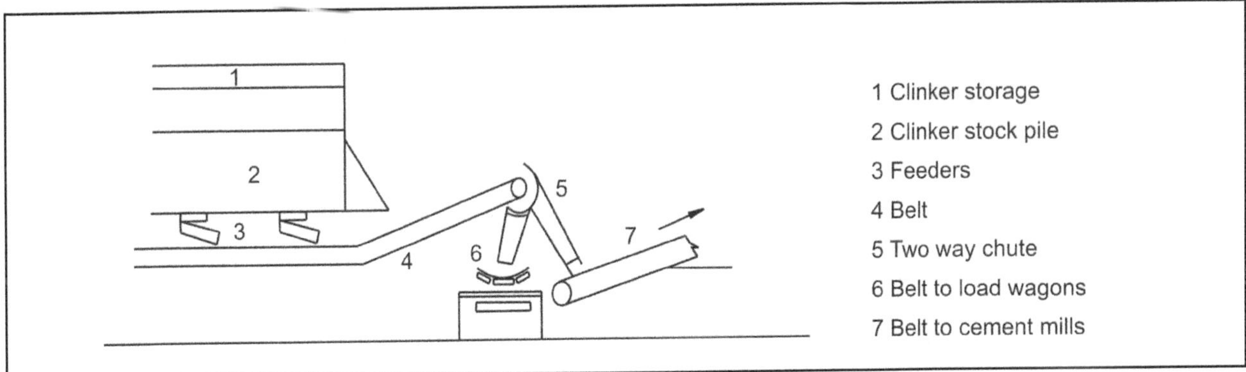

Fig. 6.37.7 Clinker loading in wagons exclusive to filling cement mill hoppers.

If the clinkering unit is located near the coast, double handling of clinker could be avoided. Alternatively, an intermediate storage and extraction facility could be created at the port.

Clinker would be best conveyed by belt conveyors because it is granular and abrasive.

If the barges are owned by the Company, loading time can be adjusted to suit

See Fig. 6.37.8.

37.7 Tides and Coast Line

Sea has tides and hence water levels fluctuate. The jetty should be designed to be able to load barges at low tide.

The difference in high and low tide levels and the distance by which sea goes out at low tide depends on the contours of coast line.

In a gradually sloping coastline, even a small difference vertically between high and low tides would mean considerable difference horizontally.

See Fig. 6.37.9.

In case of steeply sloping coastline, the horizontal difference would be small.

See Fig. 6.37.10.

Draft available would decide size and capacities of barges. For larger barges, jetty would have to go further into sea.

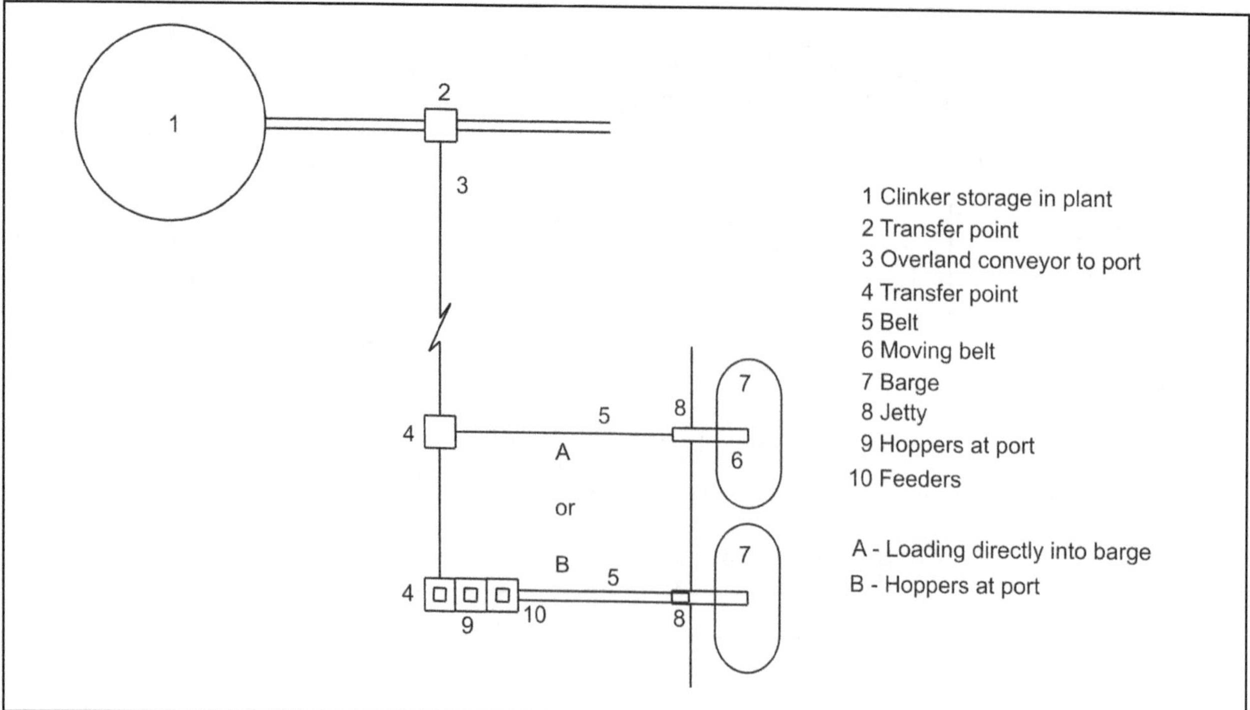

Fig. 6.37.8 Loading clinker in barges.

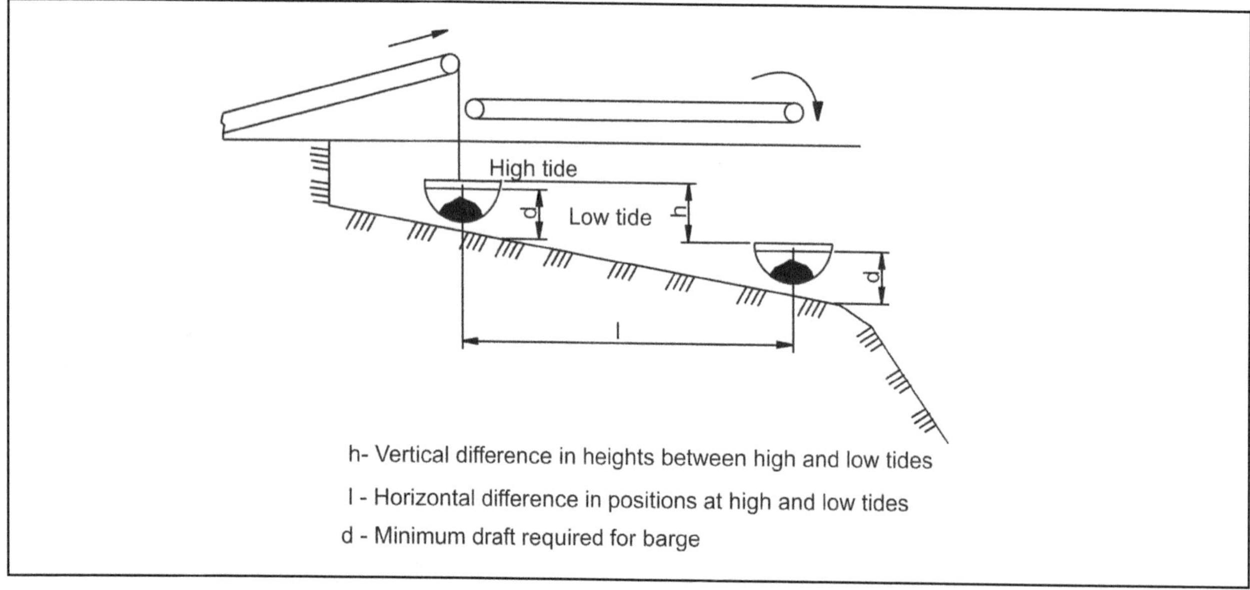

Fig. 6.37.9 High and low tides charge positions of barges ; shallow slope of coast line.

All these factors would have to be considered in locating the jetty and its projection inside the sea so that largest barges can be docked without getting grounded at low tide and can be loaded even then.

The tons of clinker loaded into the barge should be metered or a barge loaded till it sinks up to a line marked on its side.

See Fig. 6.37.11.

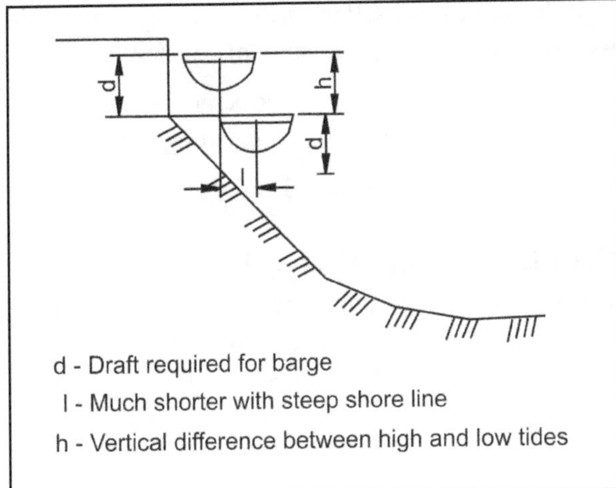

Fig. 6.37.10 Steep slope of coast line.

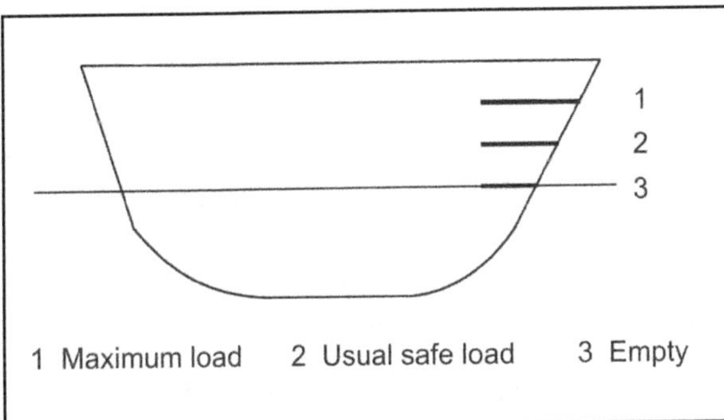

Fig. 6.37.11 Lines marked on sides of barge can indicate tonnage loaded.

37.8 Unloading Ships and Barges

Barges need to be unloaded at the destination port. Portable belt bucket conveyors can be used for the purpose.

The barge holds should have a built in bucket elevator / dbc to facilitate unloading.

For a granular and abrasive material like clinker it would be difficult to use pneumatic systems like airveyor / airstream. Hence, mechanical unloading systems like one described above would be suitable.

See Fig. 6.37.12.

37.8.1 Grab Bucket Crane

Another alternative would be to use grab bucket cranes for emptying the barges.

See Fig. 6.37.13.

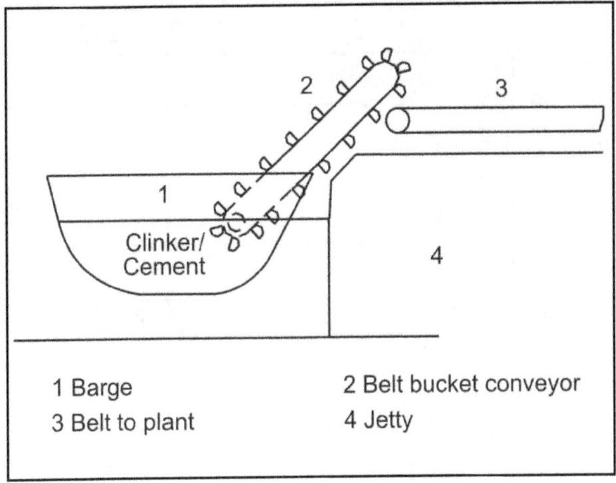

Fig. 6.37.12 Unloading from barge at desitination.

The capacity of crane would depend on the capacity of the barge and time available to unload it. Unloading operations are also affected by tides and the unloading facilities should serve during high and low tides both. **See Fig. 6.37.14.**

Capacity of Crane

Time for one trip consists of

Travel to load empty bucket	$\simeq$ 30 seconds
Filling & Lifting bucket	$\simeq$ 30 seconds
To come to unloading position	$\simeq$ 30 seconds
Unloading	$\simeq$ 10 seconds
	100 seconds
	say 2 minutes

Number of trips / hour = 30

Time available to unload a 10,000 Tons barge – say 24 hours

Unloading rate in tph = 415 say 420 tph

effective bucket capacity = 420 / 30 = 14 tons

say 15 tons

37.9 Barges Available for Dispatches

If the company owns barges, they can all be of uniform size and capacity. Hence, the position of jetty to dock barges and the EOT crane overhead can be designed with certainty.

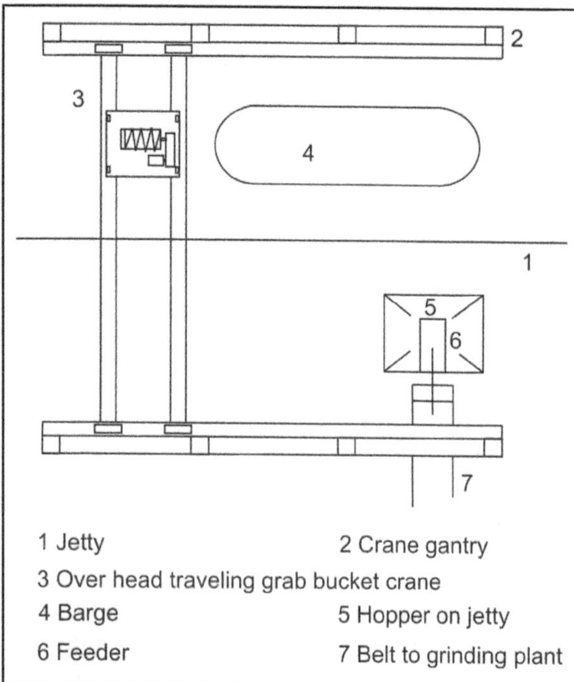

Fig. 6.37.13 Unloading barges with overhead grab bucket crane – plan.

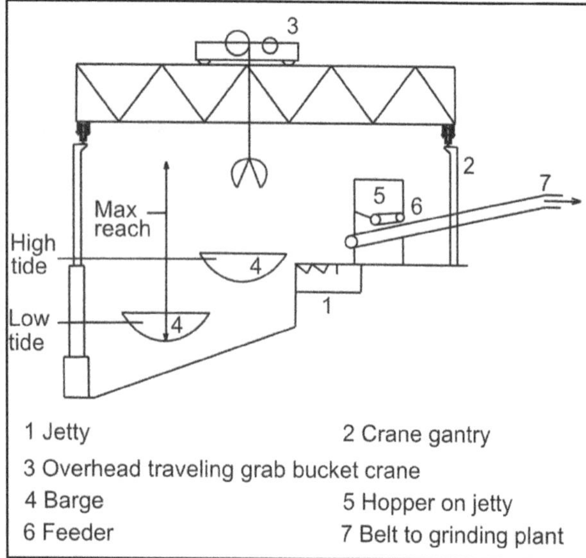

Fig. 6.37.14 Unloading barges with overhead grab bucket crane – elevation.

But if barges happen to be of different capacities and size, the unloading arrangements would have to be designed for the largest barge.

If the barge has a self unloading conveyor, it would also have to be of high capacity of 450-500 tph

The clinker unloaded from the barge should go direct to clinker storage facility of the grinding unit.

See Figs. 6.37.13 to 6.37.15.

37.10 Location of Grinding Unit

It would thus be advantageous to locate the grinding unit near the port. If this is not possible, an intermediate storage capacity has to be created at port and clinker conveyed there from to the plant by truck loads. This is double handling at both ends.

See Figs. 6.37.16 and 6.37.17.

In planning a split location for grinding at port. These aspects should be taken into account to minimize capital costs and operational costs.

37.11 Exporting Clinker

When clinker is exported, the loading arrangements would have to be made on a much larger scale as the ocean going cargo ships would have a much higher holding capacity. Loading time available would be limited and detention would attract demurrages.

Hence the capacity of the ship, its size and draft should be known to design the jetty. This jetty would have to go far into the sea so that the ship can be docked against the jetty at low tides also.

The lifting arrangements or super structures at jetty should be suitable for heights reached by empty ship at high tide.

The reach of the lifting arrangements should be suitable to reach bottom of the hold at low tide when ship is empty.

However, in most case ships that can transport bulk granular materials like clinker / grain would have their own loading facilities. The exporter would have to bring clinker up to this point.

See Figs. 6.37.18 and 6.37.19.

Various possibilities of bulk loading can be seen in photos in **Chapter 15** of **Section 2.**

Loading rates would have to be high to complete loading within the time available.

The exporter would have to obtain all this information about ships, their sizes, capacities, loading and unloading arrangements that are built in, minimum time that would be available for loading. As mentioned earlier, tides and the level difference in high and low tides should also be taken into account and the distance by which ship will have to stand further from the coast at low tide without grounding should also be ascertained.

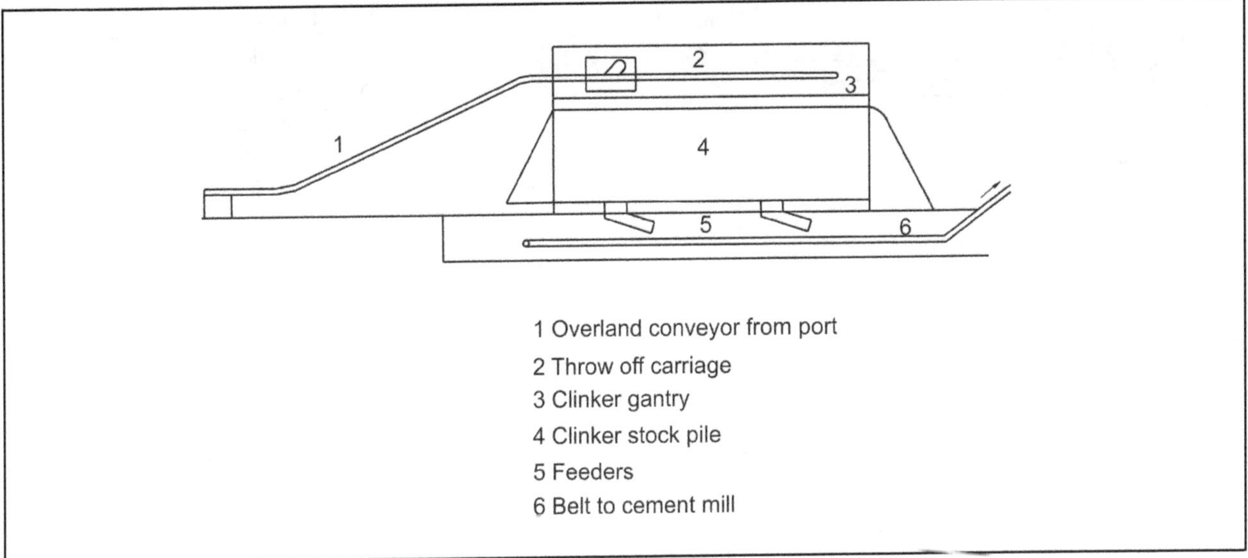

Fig. 6.37.15 Clinker brought in from jetty to clinker storage at works.

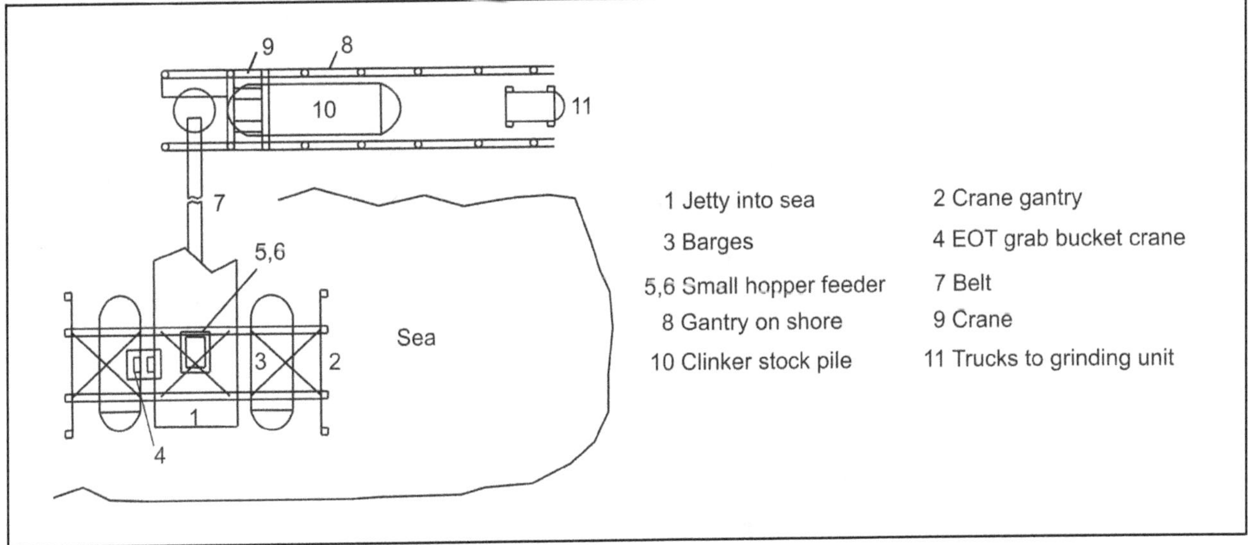

Fig. 6.37.16 Conveying clinker from port to works by trucks.

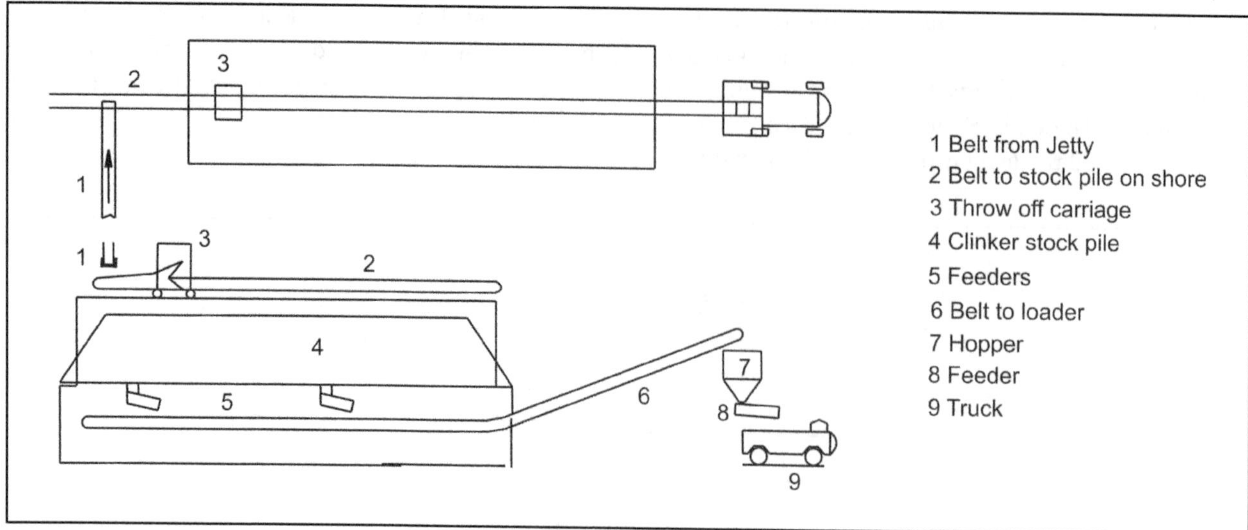

Fig. 6.37.17 Gantry can be replaced by a stock pile and belt conveyor.

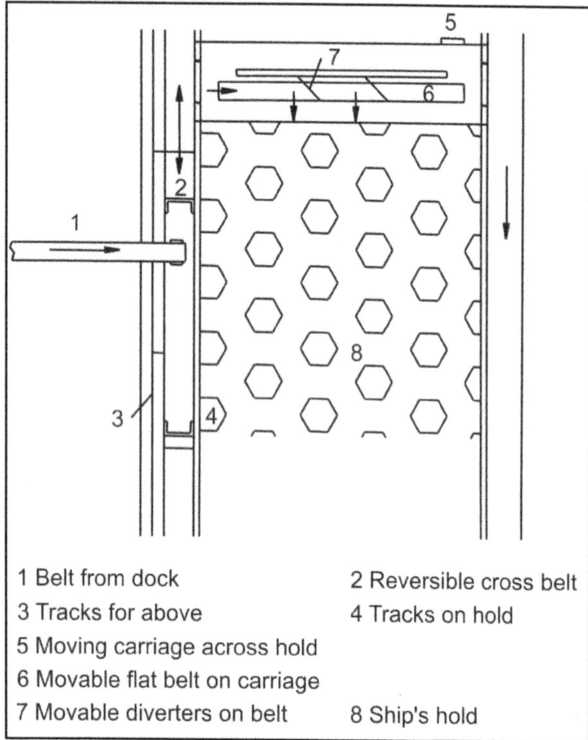

Fig. 6.37.18 Loading clinker in ship's hold - plan.

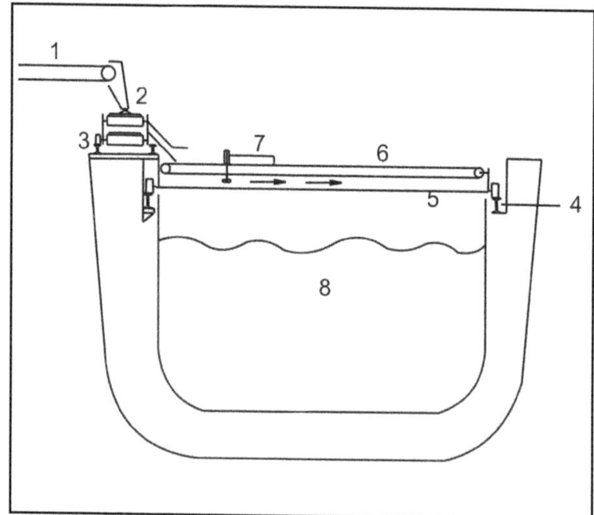

Fig. 6.37.19 Loading clinker in ship's hold - elevation.

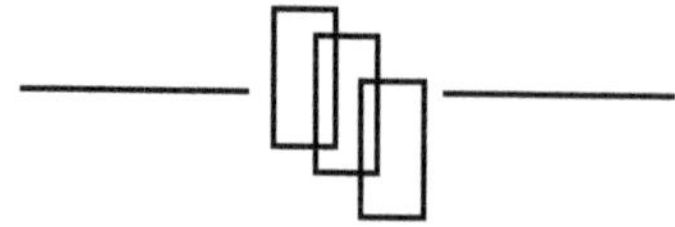

CHAPTER 38

EXPORTING CEMENT

38.1 Export of Cement

If a plant exports cement to take advantage of its location near the port (or because it has to export when domestic market is saturated), then it has to provide for dispatching cement by ships. Export market can be for bagged cement but mostly it would be for loose cement.

Cement can be sent to countries in immediate neighborhood in trucks or by rail or along the waterways.

38.2 Export of Bagged Cement

Bagged cement would be sent in paper bags. Paper bags are easy to handle and can be fed automatically to the packing machine. Thus if a plant is exporting bagged cement regularly it would be worthwhile to install mechanized feeders to feed packing machines.

Feeders are of two types. In the first type, pneumatic suction in cups lift one bag at a time and feed it to the spout of the machine. Movements are synchronized so that the cycle of device returning to the stack and bringing back one empty bag coincides with the movement of the spouts of the packing machine.

In another type, bags are received in a roll, and removed one by one and fed to the packing machine.

See Figs. 6.38.1 and 6.38.2.

Filled bags can be loaded into trucks by truck loading machines.

However, for export purpose, it is convenient to handle a number of bags together say, 10 or 20 (½ t, 1 t).

This is possible by stacking bags on wooden pallets. Pallets are handled by fork lift trucks.

See Fig. 6.38.3.

This way, handling is faster. Loaded pallets can wait on a platform and can be loaded into trucks when trucks are available.

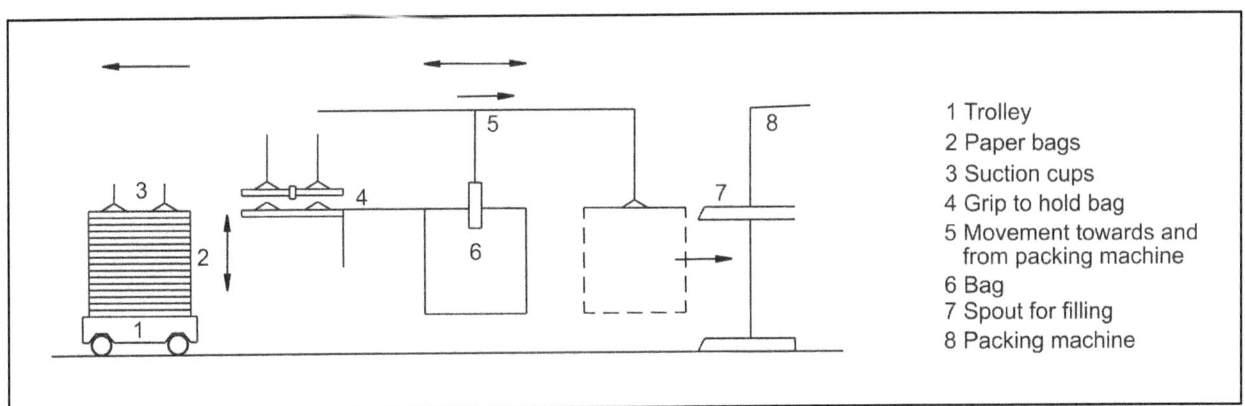

Fig. 6.38.1 Loading paper bags automatically into packing machine.

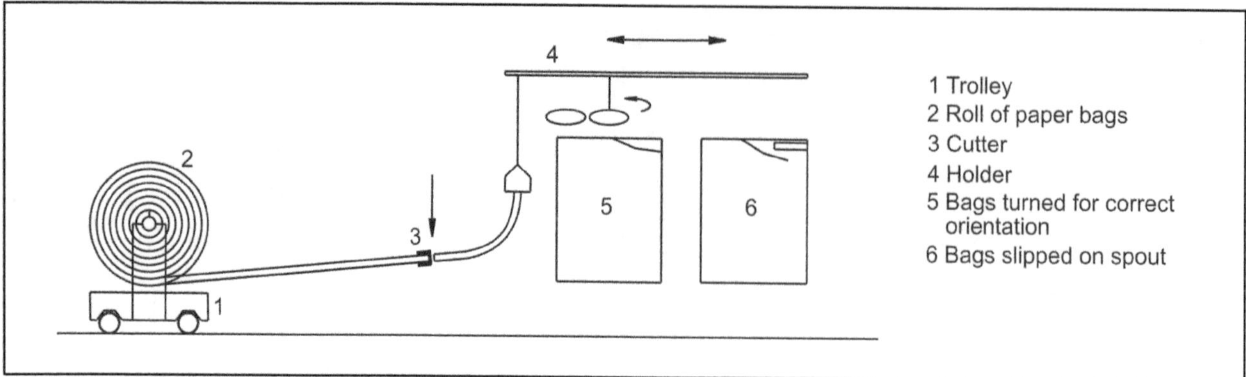

Fig. 6.38.2 When bags are received in rolls.

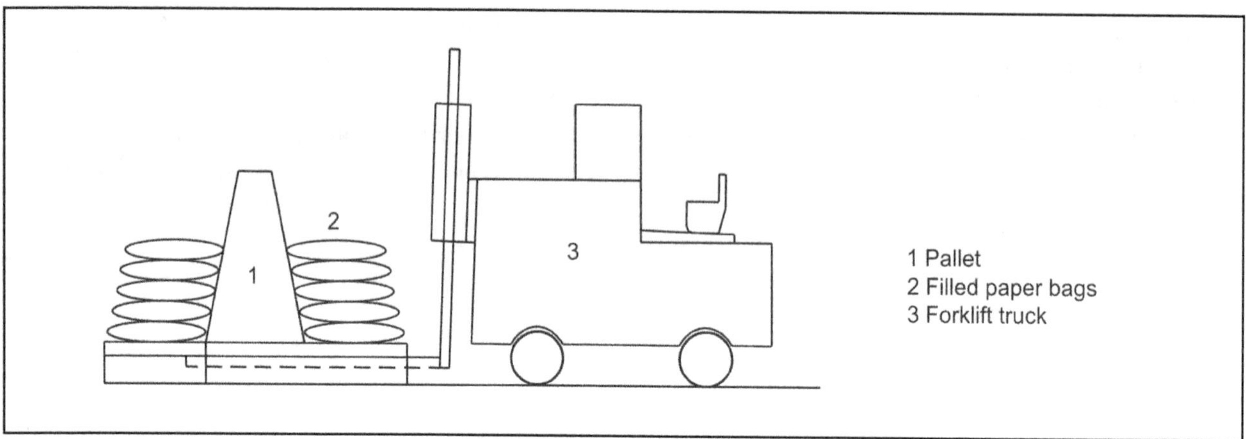

Fig. 6.38.3 Handling of packed bags on pallets.

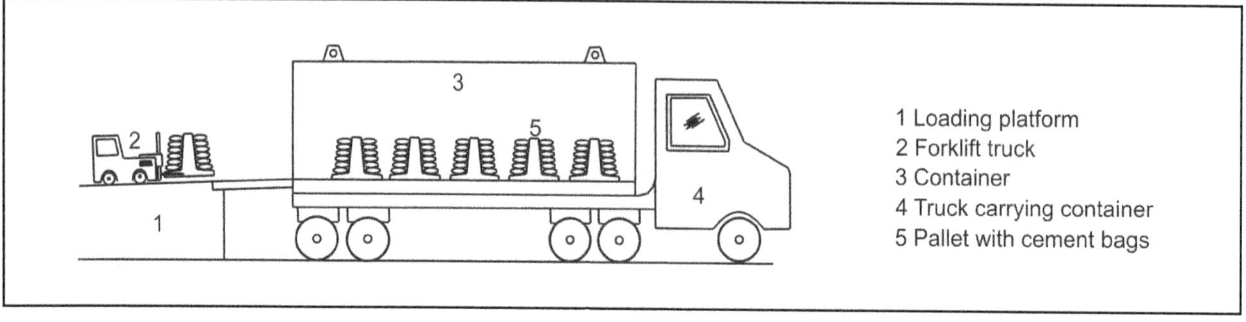

Fig. 6.38.4 Loading pallets in containers.

38.2.1 Pallets and Containers

For export purpose trucks loaded with pallets can reach port; pallets would be unloaded by fork lift trucks from them and loaded into containers and containers are loaded into ships.

See Fig. 6.38.4.

If the export volume is sizeable, containers can be brought into the plant itself and bags loaded into them with the help of fork lift trucks.

At port containers are simply lifted from the trailer and lowered into ship's hold.

See Fig. 6.38.5.

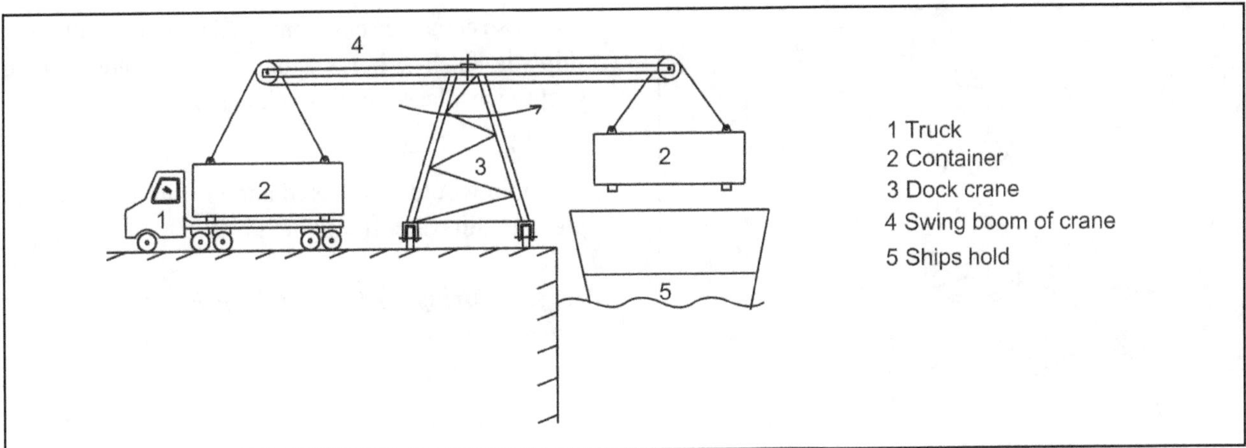

Fig. 6.38.5 Loading containers on ships.

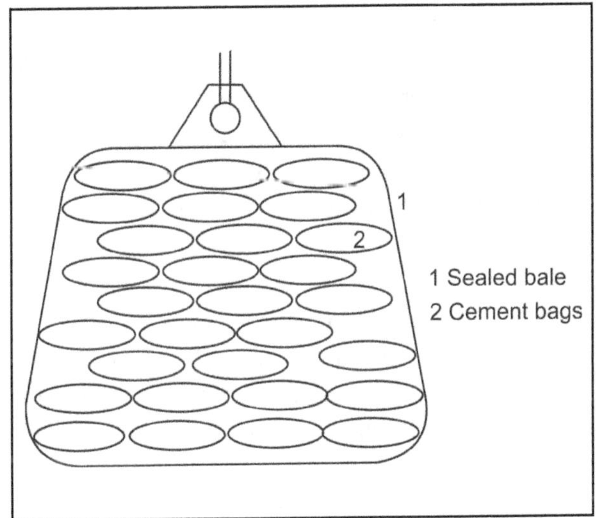

Fig. 6.38.6 Sealed bale of packed paper bags 1 - 1½ or 2 t capacity.

Pallets are a recurring expense and they may or may not be 'returnable'.

38.2.2 Shrink Wrapping, Jumbo Bags

In another method, bags are shrink wrapped into a plastic sheet and sealed and the bale so formed is lifted either by crane or by fork lift trucks.

See Figs. 6.38.6 and 6.38.7.

Cement can also be sent loose in jumbo bags of ¼, ½ and 1 ton capacities.

See Fig. 6.38.8.

38.3 Export of Loose Cement

90% of export business though would be in bulk i.e., loose cement is sent in ships.

As in case of clinker, detailed information should be available as regards ships, capacity, size and time available for loading and interval between arrivals of ships.

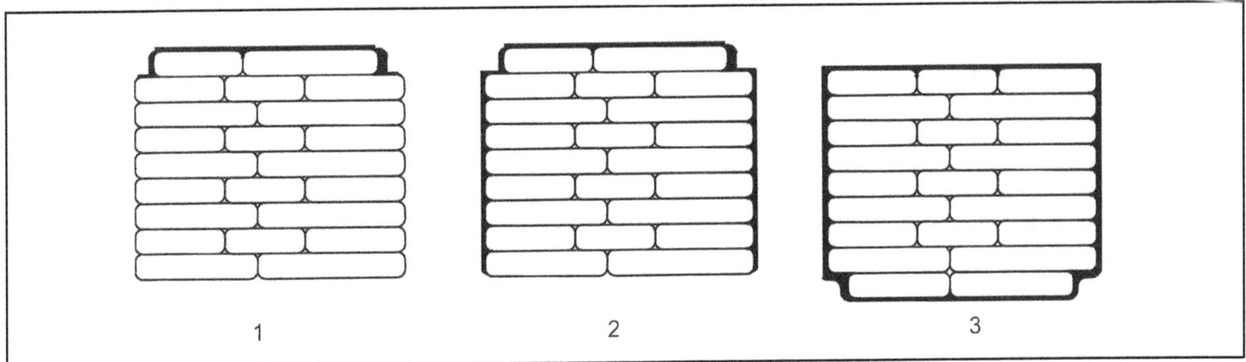

Fig. 6.38.7 Shrink wrapping for palletless despatches.

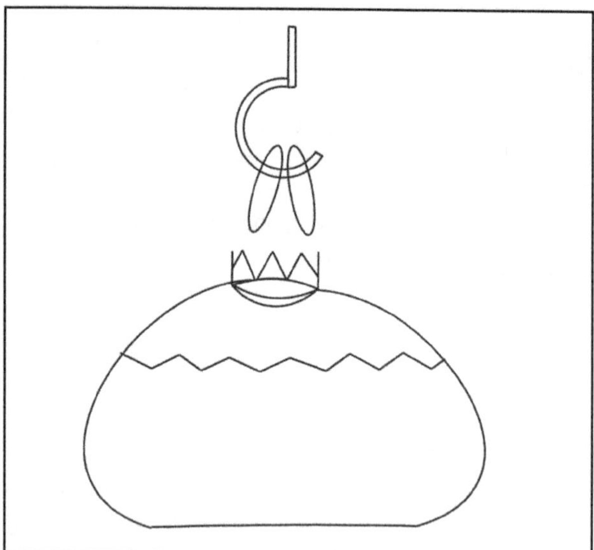

Fig. 6.38.8 Jumbo bags ½, 1 T capacity to ship loose cement.

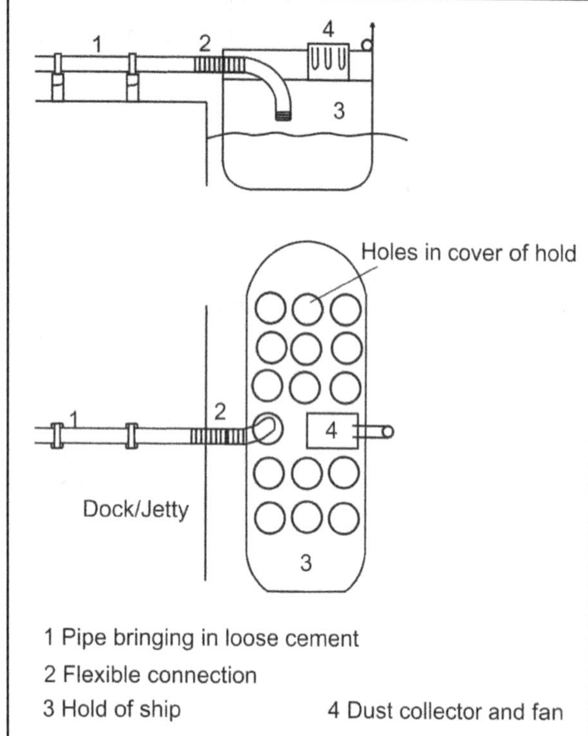

Fig. 6.38.9 Loading loose cement in ships.

It would be necessary to hold or store cement in a silo at the port. The silo capacities should be such that it will hold enough cement, so that loading of ship would not be interrupted when begun.

Loose cement is best conveyed pneumatically into ship's holds. The holds would have arrangements to vent the displaced air through dust collectors.

See Fig. 6.38.9.

The hold would be covered and precautions taken to prevent rain water from going into it.

38.3.1 Bringing Cement from Plant

Cement will be brought in bulk from the plant to be stored in the silo/s at the port.

Bulk carriers would bring in cement and these would be unloaded in the silo.

System would be the same as for bringing in fly ash described in **Figs. 6.29.7 and 6.29.8** in **Chapter 29 of this Section.**

It would be extracted from the silo, which would have one or more outlets (silo would be aerated at bottom to facilitate flow of cement) and brought by bulk carriers to silo at port. From the silo it would be conveyed pneumatically to ship's hold.

See Fig. 6.38.10.

38.3.2 Equipment for Loading

FK Pumps are quite suitable for this purpose. Other pneumatic conveying systems could also be used if the conveying distance and capacity is within their respective ranges.

A standby must be provided so that loading operation is not interrupted while ship is docked.

This operation is however, discontinuous – intermittent by nature – in that this installation works when the ship has docked.

However, it would be good to know when the ship is going to dock so that the silo/s can be kept filled in anticipation.

Suppose a plant exports 2,00,000 tons of cement in a year.

Ship has a capacity of 20,000 Tons

i.e., 10 ships in a year or say one every month.

To erect a 20,000 tons silo at port would be expensive space wise and cost wise. It would be best to install a 10,000 tons silos or 2 × 5000 tons silos.

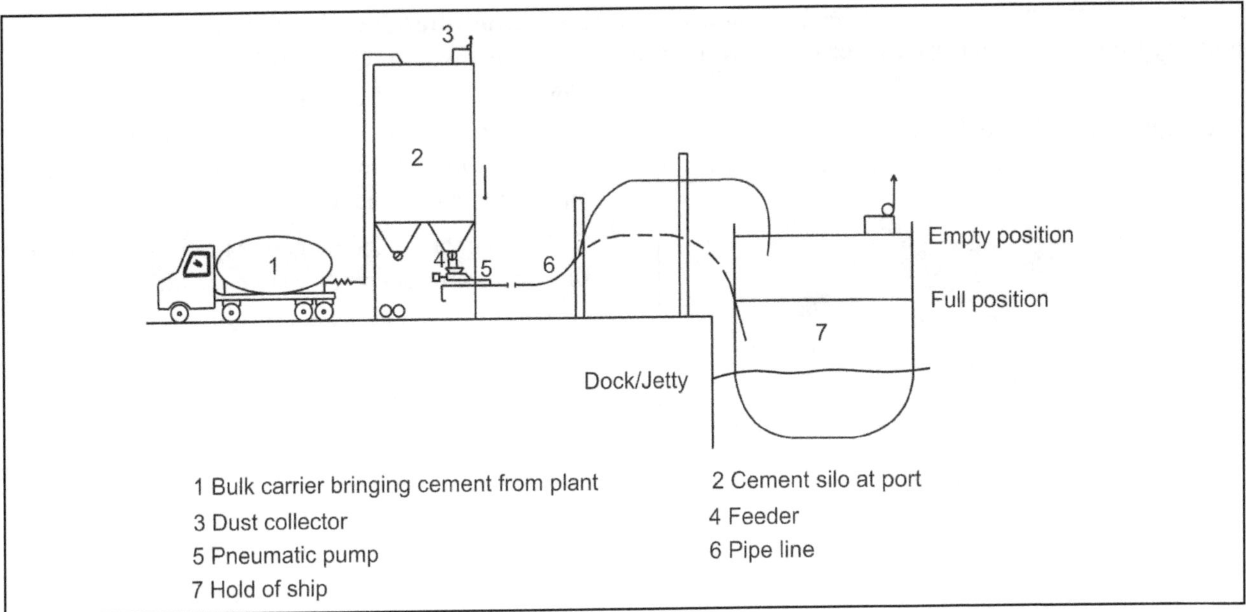

Fig. 6.38.10 Cement silo at docks for loading into ships.

Thus when a ship is docked and say has to be loaded in say 48 hours, loading rate would be 20,000 Tons / 48 = 416 tph requiring a design capacity of 500 tph.

10,000 tons silos would be unloaded in 10,000/500 = 20 hours.

It should therefore receive cement from works also at a similar rate of 500 tph.

If capacity of bulk carriers is 15 tons, then number of bulk carriers to be received and unloaded round the clock = 500 / 15 = 33 Nos per hour.

The plant must thus have bulk loading facility to load bulk carriers at this rate i.e., 500 tph.

See Fig. 6.38.11.

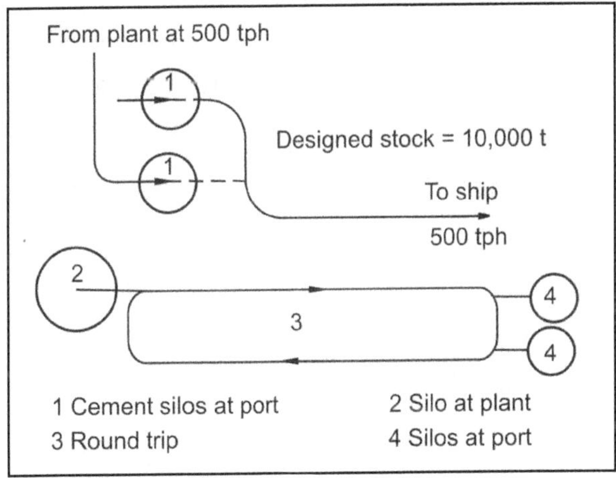

Fig. 6.38.11 Fleet of bulk carriers.

38.4 Development of Scheme

When export is considered the whole scheme is to be integrated from silos and bulk loading facilities at plant to the extraction and unloading facilities at port.

If practicable, silos at port should be emptied after loading the ship as they would be required only next month. If the arrival of ship is known, silos can be filled up but time taken for it can be longer – 4 days or even a week

Assuming 4 days of filling round the clock, loading rate at works need be only half that is 250 tph and would require only about 17 bulk carriers per hour.

The costs of loading equipment for 500 tph or 250 tph would not differ much. Therefore, equipment corresponding to loading rates of 500 tph can be installed.

38.4.1 Fleet of Bulk Carriers

The size of fleet of bulk carriers required for transportation of cement to port would depend on the time taken for the round trip.

If the cement plant is located hinter land and still plans to export cement, it is best for 2 or 3 such plants or companies to join hands and create a common bulk handling facility at port.

Even then cement would have to be brought to the port city from hinterland in rake loads in bulk carriers (wagons).

See Figs. 6.38.12 and 6.38.13.

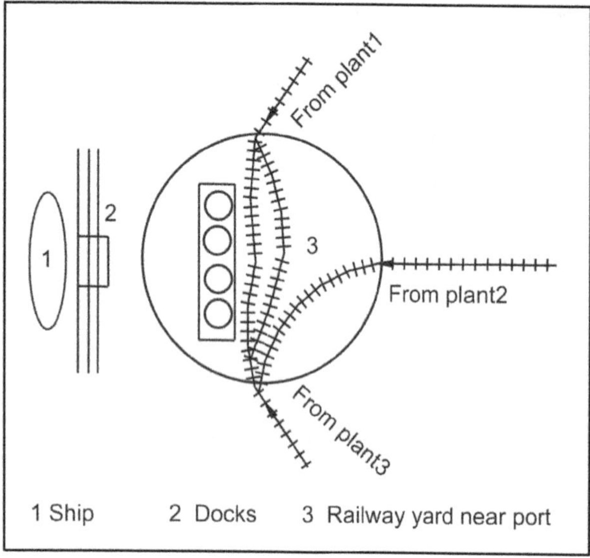

Fig. 6.38.12 Cement brought from hinterland for export to port city.

38.5 Common Facilities for a Number of Companies

Companies joining in this project would have to co-ordinate dispatches of their respective rake loads so that they can be unloaded in turn into their respective silos without bottlenecks.

For maintaining identity of exporters cement from different sources would be stored in different silos.

It would be difficult if not impossible to load cement from more than one source simultaneously in a ship.

See Fig. 6.38.13.

38.5.1

Please see **plates 2.15.3 to 2.15.6** in **Chapter 15** of **Section 2** showing actual bulk loading installations.

Fig. 6.38.14 shows the layout of a handling facility for ships at port.

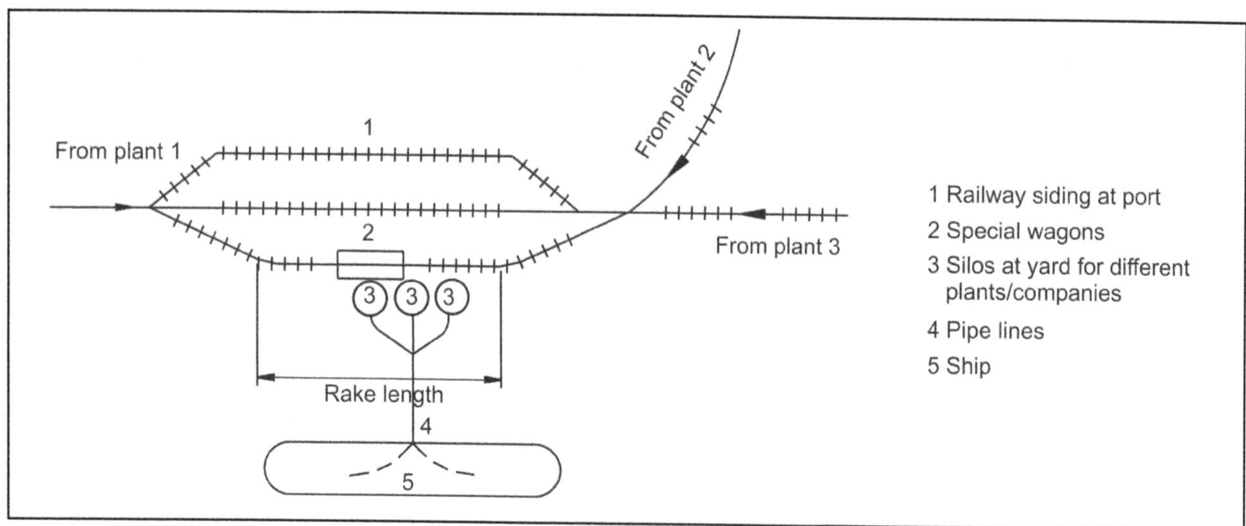

Fig. 6.38.13 Bulk loading of cement in ships at port.

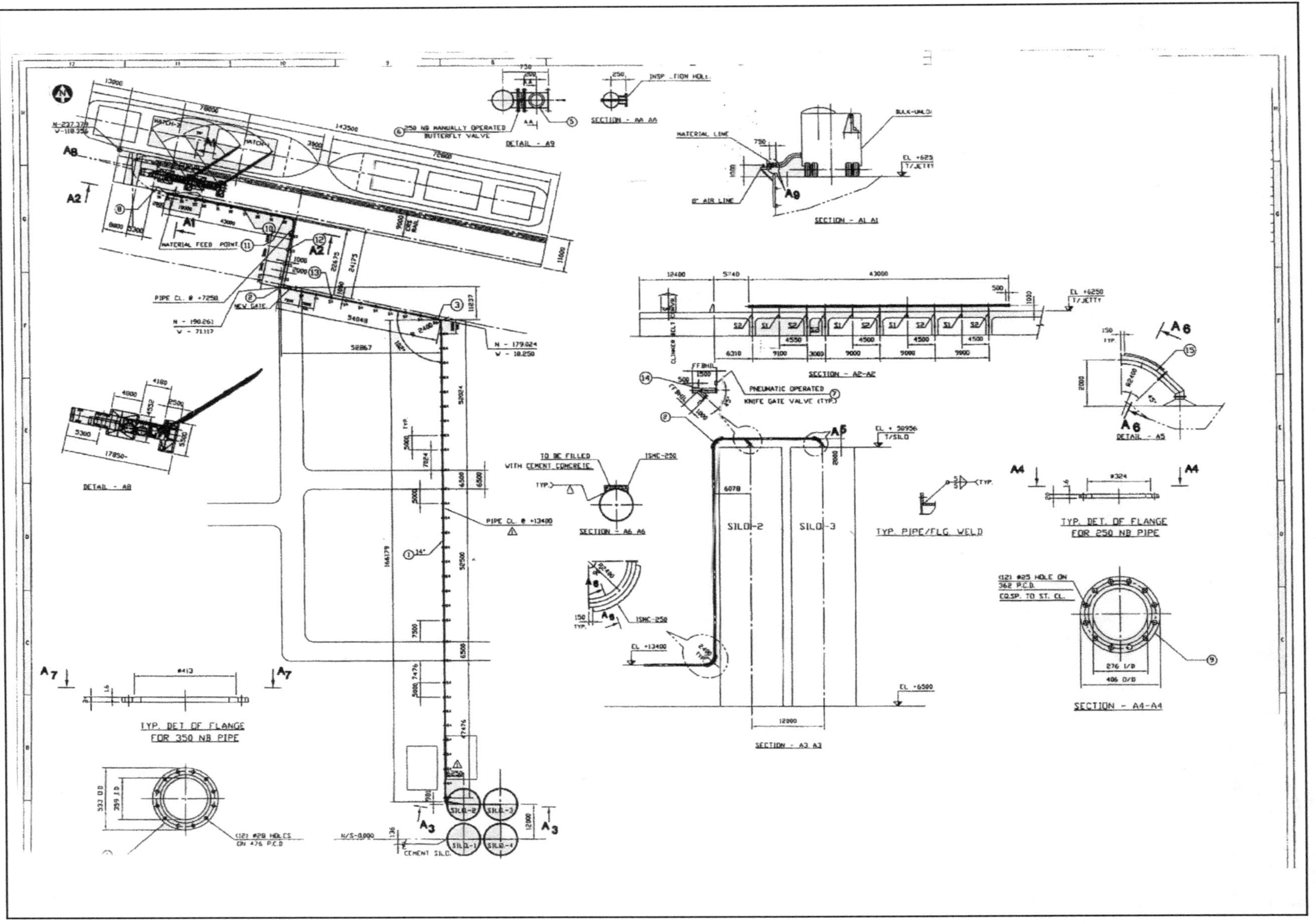

Fig. 6.38.14 Layout of a loading and unloading installation from ships at port.

CHAPTER 39

IMPORT OF COAL

39.1 Quality of Indian Coals

India has large reserves of coal and is very self-sufficient in this respect. On the whole Indian coals, except for collieries in Bengal and Behar / Zarkhand, have high ash and low calorific values. Ash content ranges from 25% to 40% and useful calorific values range between 3500 to 4500 kcal/kg. Coals of calorific values of 4800 / 5000 kcal/kg or above are seldom available to cement plants in India.

Volatiles in coals range between 25-30%.

Moisture ranges between 8-15%.

Fixed carbon ranges between 48-50%.

Except for coals found in Assam, Sulphur is low, less than 0.3 %.

Coals have been classified into A to E categories according to calorific value and ash.

Power plants get worst coals 35-40% ash and calorific values 3500-4000 kcal/kg. Next come cement plants.

While ash content is high, coals are not difficult to burn because of high volatiles.

39.2 Problems with Indigenous Coal

The problems of coals received from collieries are mainly their inconsistency and fluctuations in quality in terms of ash content and calorific value and moisture.

Supplies of coals are often erratic, marked by stoppages and curtailment of production. Many cement plants suffer from shortages of coal for long periods every now and then.

Proposals were on board for installing common 'coal washeries'. However, they did not really take off.

39.3 Import of Coal

After liberalization, it has been possible to import coal from countries like Australia, South Africa and China. Coal is received at ports like Chennai, Visakhapatnam and others and plants take it from there according to their requirements. Imported coal may be taken in rake loads by large plants and in truck loads by small plants.

Even small cement plants began to use imported coal when its cost in rupees per kilocalorie was comparable to indigenous coals.

Many plants used imported coal to blend it with locally available coal to bring about stability in quality.

39.4 Facilities at Ports

At ports stock piles are built up very much like stacker reclaimer systems. Coals would get blended while being reclaimed for dispatches.

Such facilities if used on a 'co-operative basis' i.e., a number of cement companies joining together to import coal, become viable propositions. However as these matters are outside the scope of the 'design of cement plants' they have not been dealt here.

Same kind of facilities are required to be created to receive and store imported coal as are required for indigenous coal. Cement plants may choose to 'blend' imported coal with indigenous coal.

Stock piles of imported and indigenous coals may be kept separately. Coal from them could be drawn in measured quantities for 'blending' purposes. It would need double handling. Therefore plants may also choose not to segregate two coals.

Handling of coal has already been dealt with in **Chapter 27.**

CHAPTER 40

RAILWAY SIDING

40.1 Railway Siding

Design of railway siding is a specialized job and expert advice should be taken to design it.

There are minimum three lines.

1. Incoming.
2. Outgoing.
3. Engine escape.

See Fig. 6.40.1.

It is also required to carry out shunting and positioning operations after the rake has come in and the Railways' engine returns.

A diesel engine / engines, is/are used to shift wagons inside the plant break and form rakes again so that idle time is minimized.

Railways give a specific time for loading wagons after rake has been delivered.

The exact position with respect to starting of time allotted and its end should be known.

If this time is exceeded, railways charge demurrage. It is quite heavy and hence should be avoided as far as possible.

40.2 Rakes for Loading

As has been mentioned earlier, a rake would be broken into two for loading wagons.

See Fig. 6.40.2.

The lengths of the 3 main tracks should of course be long enough (and more) to hold complete rakes.

See Fig. 6.40.1.

Additional lines are required when there are more than one loading platforms.

See Fig. 6.40.2.

40.3 Coal, Slag, Fly ash and Other Materials

When coal and other commodities also come by rail then additional tracks should be provided to place wagon rake loads for these materials.

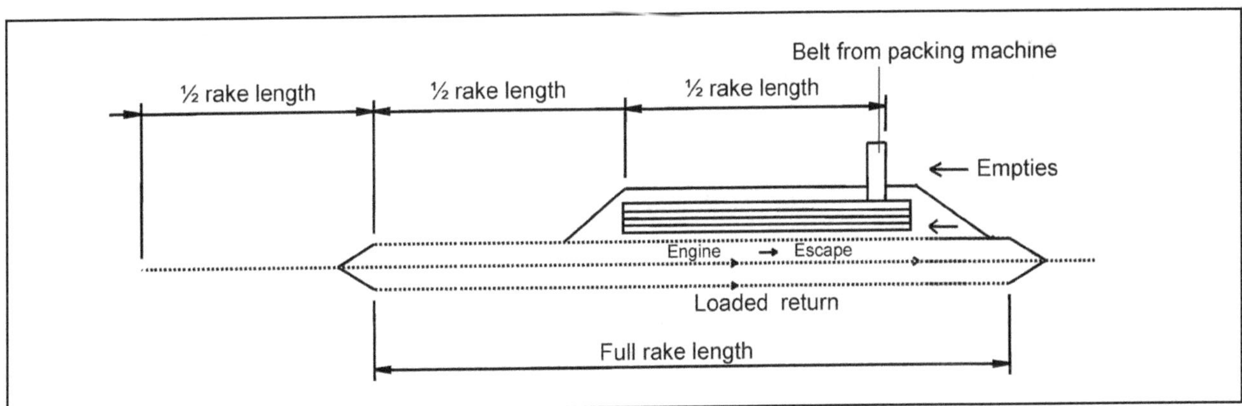

Fig. 6.40.1 Simplified layout of a railway siding.

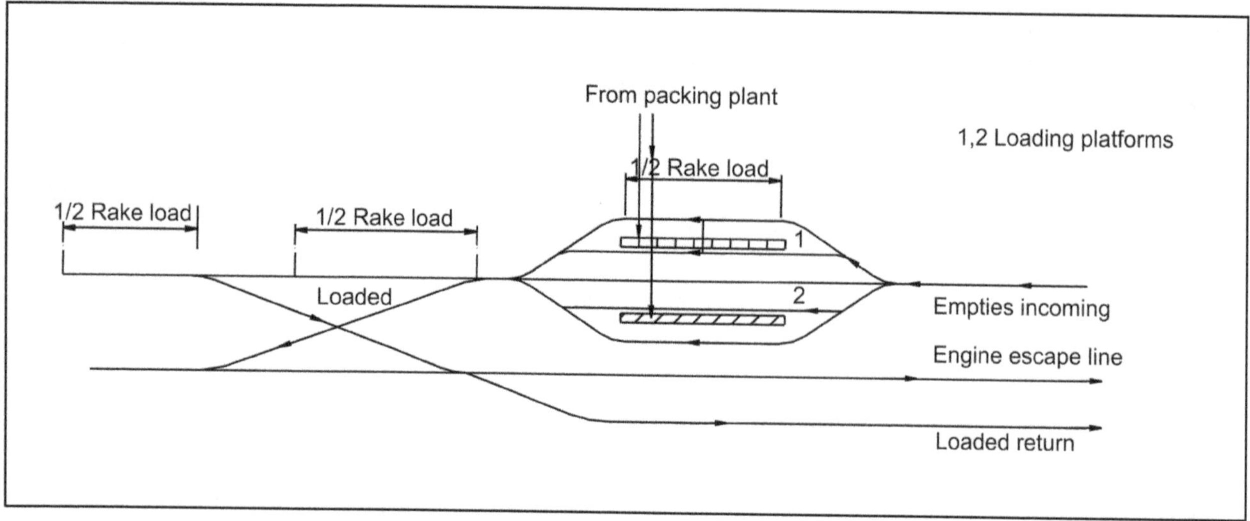

Fig. 6.40.2 Railway siding for wagon loading on more than one loading platform.

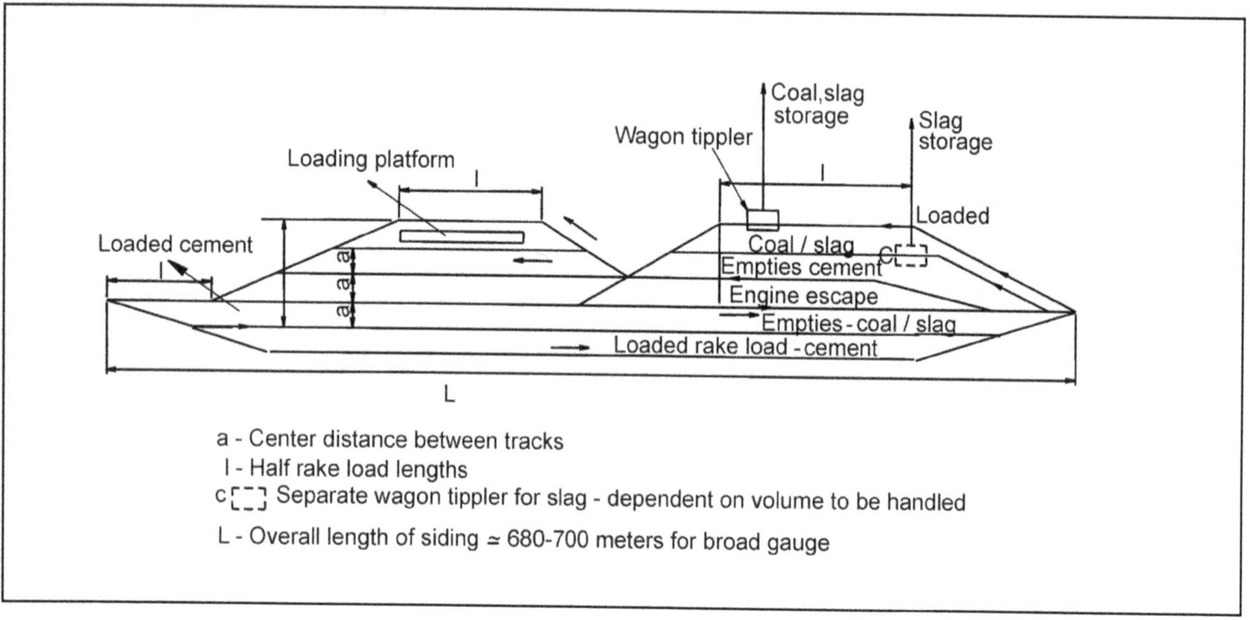

Fig. 6.40.3 Railway siding for loading and unloading coal and slag - by wagon tipplers for large cement plants.

Typical Railway sidings have been shown for dispatches of cement and receipt of coal.

See Fig. 6.40.3.

Since railway siding is capital intensive, the lengths of tracks and their number should be kept as short and few as possible but not so short or few as would become bottlenecks.

It should also be possible to add number of tracks when the plant expands.

40.4 Design of Railway Siding

Spacing between the tracks should be as per norms laid down by the Railways. The siding should be straight and level. If a gradient cannot be avoided, then it should be minimum again as per Railways' norms so that wagons do not start rolling on their own.

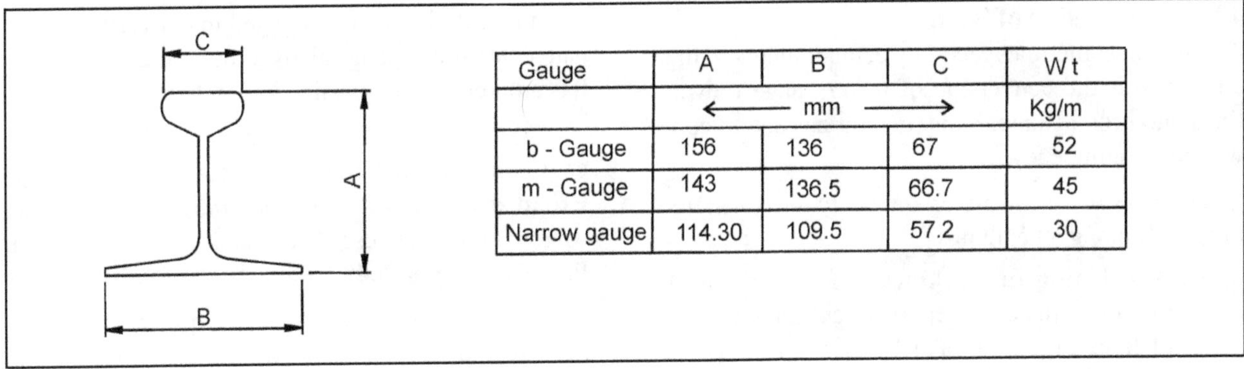

Gauge	A	B	C	Wt
	←	mm	→	Kg/m
b - Gauge	156	136	67	52
m - Gauge	143	136.5	66.7	45
Narrow gauge	114.30	109.5	57.2	30

Fig. 6.40.4 Dimensions of rails for railway tracks of different gauges.

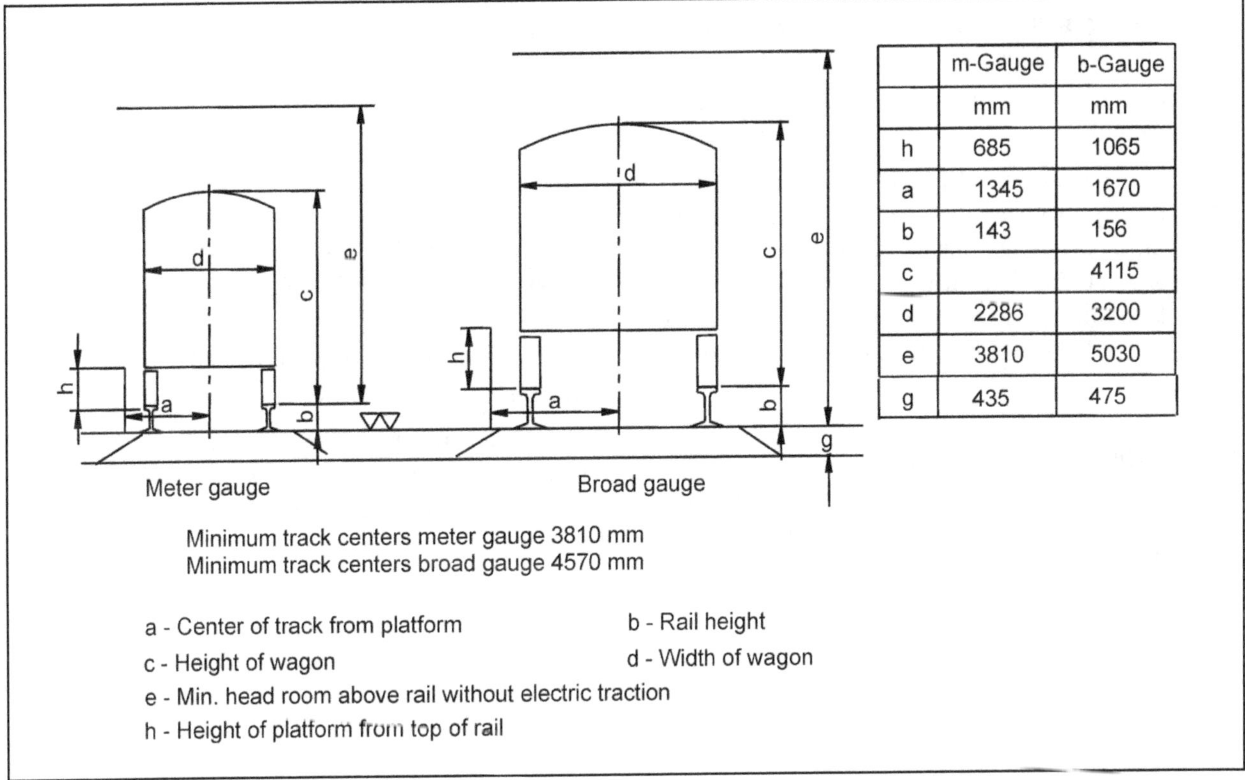

	m-Gauge	b-Gauge
	mm	mm
h	685	1065
a	1345	1670
b	143	156
c		4115
d	2286	3200
e	3810	5030
g	435	475

Fig. 6.40.5 Converson of metre gauge to broad gauge.

Allowance for expansion would require more loading platforms.

Belts would be crossing railway lines. Norms laid down by Railways should be followed as regards overhead clearances for supporting structures of belts. Minimum clearances on sides should also be maintained.

See Fig. 6.34.16 in Chapter 34.

40.5 Data on Railway Wagons and Tracks

Annexure 1 furnishes broad guidelines for railway tracks and wagons. It also furnishes dimensional data and capacities for more commonly used wagons. Broad dimensions of rails are furnished in **Fig. 6.40.4.**

Dctails of center distance between tracks, minimum overhead clearance, and from columns and sides have been furnished in **Fig. 6.40.5. & Fig. 6.34.16.**

40.6 Conversion of Gauge

A situation sometimes faced by cement plants in this country was the conversion of meter gauge railway lines into broad gauge. Many a plant has had to cope with this contingency.

Conversion from meter gauge to broad gauge involved among other things :

1. Broadening of tracks (sleepers and rails); minimum distance between tracks and lengths of lines to handle rake loads of broad gauge wagons.
2. Heights and widths of loading platforms and their lengths to suit broad gauge wagons.
3. Re-alignment of belts with respect to their heights and lengths.
4. Wagon loading machines to suit broad gauge wagons.

If the plant does not have a railway line to start with, it has to provide for truck loading facilities initially. However, if the railways have plans for providing rail links, for commercial traffic in that area, then the plant should take note of such plans and provide for a railway siding in the layout. It would be difficult to acquire land at a later date.

40.7 Assistance from Railways

Divisional offices of Railways can furnish information on their plans for new lines and projected dates of undertaking the rail link, and the completion thereof.

Indian Railways are engaged in converting metre gauge into broad gauge all over the country. Therefore it is unlikely that new rail links would be metre gauge.

The problem is for the plants who have a meter gauge and are required to convert it into broad gauge. Broad differences between broad gauge closed wagons commonly used to dispatch cement have been furnished in **Fig. 6.40.5.**

Thus, for the same capacity of plant, the size of railways siding even for same number of tracks would be considerably longer for a broad gauge line.

The conversion has to be undertaken without disrupting traffic on existing meter gauge line.

The 'siding' outside the plant between Railways', take off point on the main line and the plant also has to be changed.

40.7.1 Railways Expertise

Railways have expertise to design and even install the siding. They may take a contract to do the changes or install a new siding including broadening of culverts, bridges, etc.

It is best to entrust this task to the experts that is Railways themselves – of course at a price.

Indian Railways' R.D.S.O. (Railway Design and Research Organization) is a good source of information on all matters related to railway track and siding.

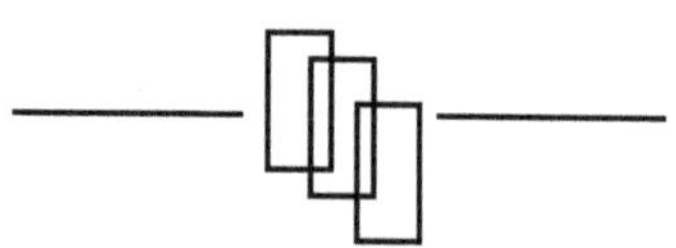

Annexe 6.40.1

GUIDE LINES ABOUT RAILWAY WAGONS AND TRACKS

1 **Railway standard gauges**

Broad gauge 5' 6"

Metre gauge 1 metre

Narrow gauge 2' 6"

Metre gauge is being phased out except in hilly areas

2 **Commonly used railway wagons**

4 wheelers - open and closed

8 wheelers - open and closed

special sink type

special for container

special for bulk dispatch of cement

3 **Data on wagons**

Broad gauge

Type of wagon	Length between buffers	Carrying length	width	height	tare	Pay load	Gross load	No.of wagons Per rake	Rake load
	metres	metres	metres	metres	tons	tons	tons	number	tons
BCN covered	15.43	14.5			25.51	57.7	81.3	40	2250
BCXN covered	15.78	14.85			28.5	52.8	81.3	35	1850
CRT Covered	8.82	7.89			13.1	27.5	40.6	65	1800
Box N Open	10.7	9.78	3.2	3.22	22.47	58.8	81.3	58	3350
Box T open	14.1	13.15			26.3	55	81.3	40	2200
Box C open	13.73	12.8	3.14	3.16	25.1	56.2	81.3	40	2250

4 **rake lengths – broad gauge**

a Normal rake with single engine 40 wagons (BCN,BoxT,Box C)

b Jumbo rake with double engine 58 wagons (Box N)

Rake length in metres

Rake	wagon	nos.	length buffer to buffer	engine	guard	extra van	total length
normal	BCN	40	631	20	6	10	667
Jumbo	BOX N	58	622	40	6	10	678

5 **Railway line requirements and minimum lengths**

A incoming line 680 metres

B inspection line 680 metres

C dispatch line 680 metres

D engine escape line 600 metres

6 **Speeds**

Main line 100-140 kmph

Main line 60-100 kmph

Inside plant 30 kmph max.

7 **Radius of curves**

Main line 250 metres

Main line 220 metres

Inside plant 200 metres

8 **Gradient in yards**

Broad gauge 1 in 400 max.

Metre gauge 1 in 400 max.

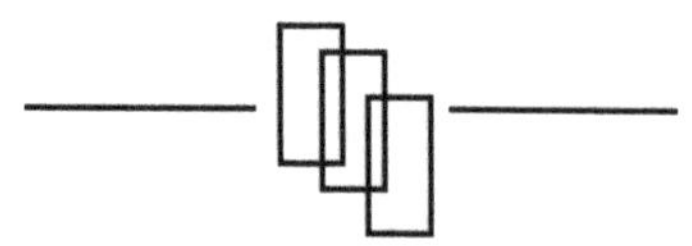

CHAPTER 41

VENTING

41.1 Venting

Every cement plant has to cope with process gases like kiln and cooler exhausts, and system gases like those in grinding systems.

But there are also substantial volumes of gases to be handled and cleaned of dust known as 'vent gases'.

'Vent gases' arise when materials are conveyed from one point to another and displace air in doing so; like when conveyor for crushed limestone discharges it in a bin or a hopper or even on to another conveyor. An equivalent amount of air is displaced. When pulverized raw meal, coal and cement are conveyed pneumatically, conveying air needs to be 'vented' when the material is discharged into bins.

Venting does not mean only provision of outlet for such airs but also providing facility for cleaning the dust carried away with them, collecting it and returning it to the system.

Technically vent air is that which arises because of displacement. Others may be treated as 'process gases'.

Except for cooler vent, vent air temperatures would generally be at ambient or at temperatures of 60-70 °C. In case of cement mills they could go up to 110 °C They can therefore be treated in bag filters.

Venting is thus a must for all material conveying systems and equipments. Some equipment like rotors of crushers , buckets of elevators act like fans and displace air when in motion. Venting is thus necessary in all departments of a cement plant of any size and process.

41.2 Vent volumes

Amounts of 'vent volumes' and their dust content vary in each case like:

1. when bins are filled,
2. when materials are conveyed pneumatically,
3. for different types of equipment and their sizes.

It is necessary to provide venting points at suitable locations to collect displaced air and take it to a dust collector.

When a conveyor empties into a bin, air is displaced and is required to be vented. Vent volume is dependent on the rate of filling, which in turn is related to volume the material occupies (bulk density). Corresponding volume of air is displaced.

See Fig. 6.41.1.

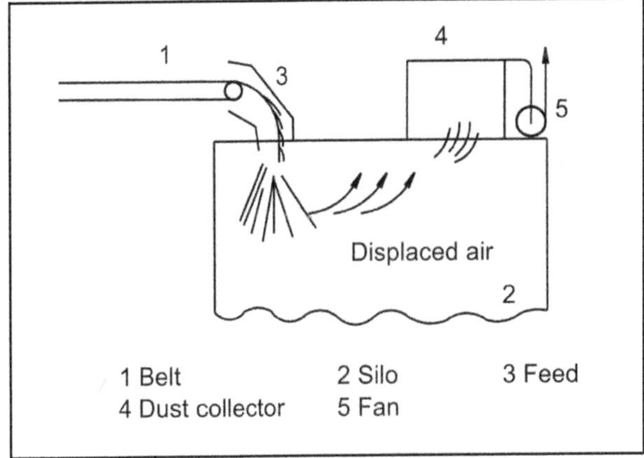

Fig. 6.41.1 Why is venting necessary.

However, it is difficult to be precise in such matters and a 'multiplying factor' is used for designing vent systems.

Alternatively, vent volume is related to the volume of the bin.

Multiplying factor is maximum for small rates of discharge and minimum for large rates.

41.3 Dust Burden in Vent Gases

The dust content, measured in gm/m^3 (or nm^3) varies from case to case being typical of type equipment and the granulometry of material carried. Finer the material, greater would be dust burden in air to be vented. The particle size distribution also varies from case to case.

41.4 Process Gases

Gases passing through mills could be treated as 'process' gases because they are used for drying as in raw mills and coal mills and for cooling in cement mills.

Air swept mills use air / gas to carry half finished product out of the mill.

Even if mills do not do drying / cooling or air lifting like in air swept mills, venting is necessary to keep slots of diaphragms clean and to lift very fine dusts out of the mill.

Rotors of hammer and impact crushers act like fans and displace volume of air at every revolution.

Kiln process gases are exhausted, vented through kiln dust collector in direct operation.

Cooler process gases which are surplus of cooling air over combustion air are exhausted / vented through dust collectors..

41.5 Design of a Venting System

To design a venting system, it is necessary to know:

1. Flow chart of the system.
2. Sizes and capacities of major machinery and auxiliaries.
3. Vent volumes recommended by suppliers for their respective machinery.
4. Dust burden in gases to be treated.
5. Temperature of gases.
6. Moisture content.
7. Sieve analysis of dust.

It is better to obtain this information directly from the respective Suppliers.

In absence of such data forthcoming from suppliers, vent volumes and dust burdens can be estimated from size and type of machinery from reference tables and charts – available in reference books like Otto Labahn, Duda and others. Reference to web pages **W1.96 and W1.97 in Web page1 in Section 9.**

See also Annexes 6.41.1 and 6.41.2.

The system flow chart will, together with departmental drawings, help in selecting :

(i) Location of vent points.

(ii) Clubbing or grouping of a number of vent points to be taken to the dust collector/s.

41.6.1 Separate Dust Collector for Venting

It is advisable to use separate bag filters for venting auxiliaries. Process gases and main machinery should be treated separately in the main dust collector.

See Fig. 6.41.2.

Vent points are best located near feed and discharge points of the machinery when material is transferred from one machine to another.

See Figs. 6.41.8 to 6.41.10

In circuit shown in **Fig. 6.41.2** hot gases are taken to mill for drying. These could be treated as process gases. Volume wise they would be substantial as compared to other vent volumes.

This system thus could have :

Bag filter BF1 - for venting feed belt and hoppers – V_1

Bag filter BF2 - for venting mill and separator – V_6, V_7 and V_8

Bag filter B3 - or venting other auxiliaries - V_2, V_3, V_4, V_5

Bag filters 1, 2 and 3 would be sized to suit respective vent volumes. They would be located close to the points to be vented to reduce lengths of ductings.

Since a Bag filter would have only one inlet and outlet, ducts carrying volumes V2 to V5, will be joined and led to bag filter.

See Fig. 6.41.3.

41.6.2 Venting through Separate Dust Collector

In closed circuit mills with conventional separators, auxiliaries will be vented through a separate bag filter (BF2 in **Fig. 6.41.2**). In mills with high efficiency separators, vent volumes of auxiliaries and mill are

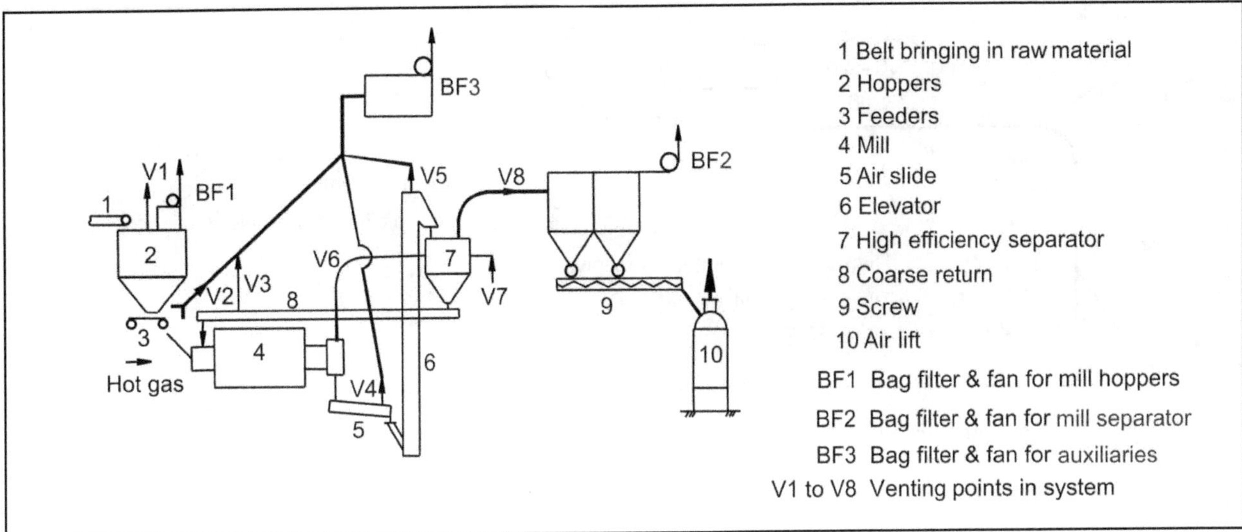

Fig. 6.41.2 Venting systems in grinding circuit.

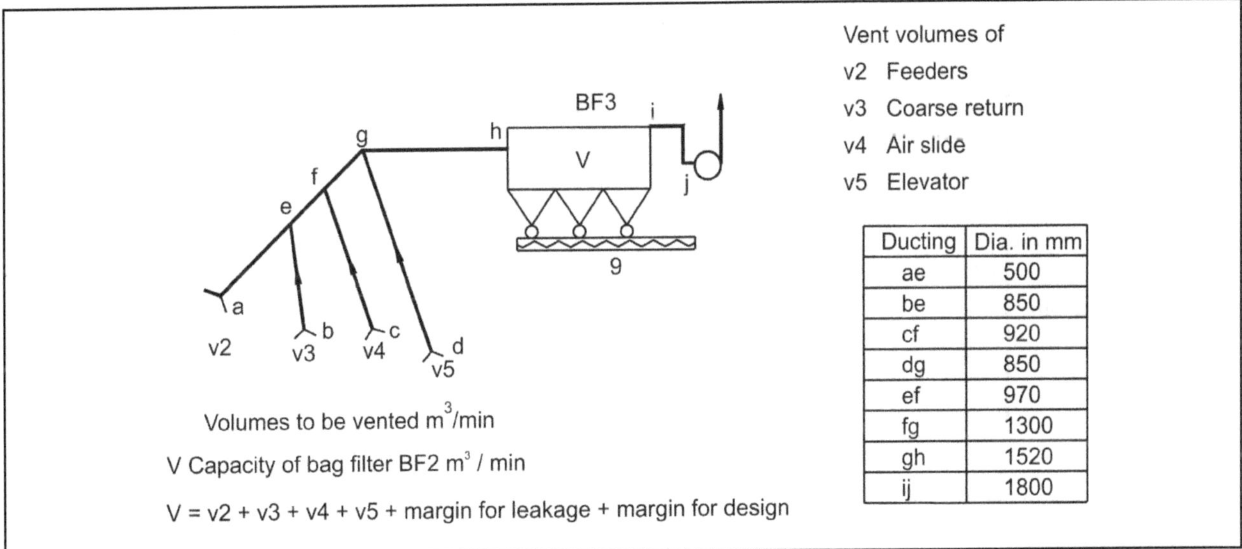

Ducting	Dia. in mm
ae	500
be	850
cf	920
dg	850
ef	970
fg	1300
gh	1520
ij	1800

Fig. 6.41.3 Sizing bag filter for venting.

taken to separator as primary and secondary airs. They emerge out of the separators as V8 to enter bag filter.

See Fig. 6.41.2.

41.6.3 Velocities in Ducts

Velocities V1, V2, V3, etc, or for that matter for all ducts carrying dusty vent air would have to be maintained between 18-20 m/sec.

If dust is coarse and dust concentration high, velocity may have to be as much 20-22 m/sec.

Ducts will be sized accordingly.

See Fig. 6.41.4

Location of the bag filter in the layout fixes lengths of ductings, number of bends, and their angles. This in turn determines the pressure drop in the respective ductings.

See Figs. 6.41.2 and 6.41.5

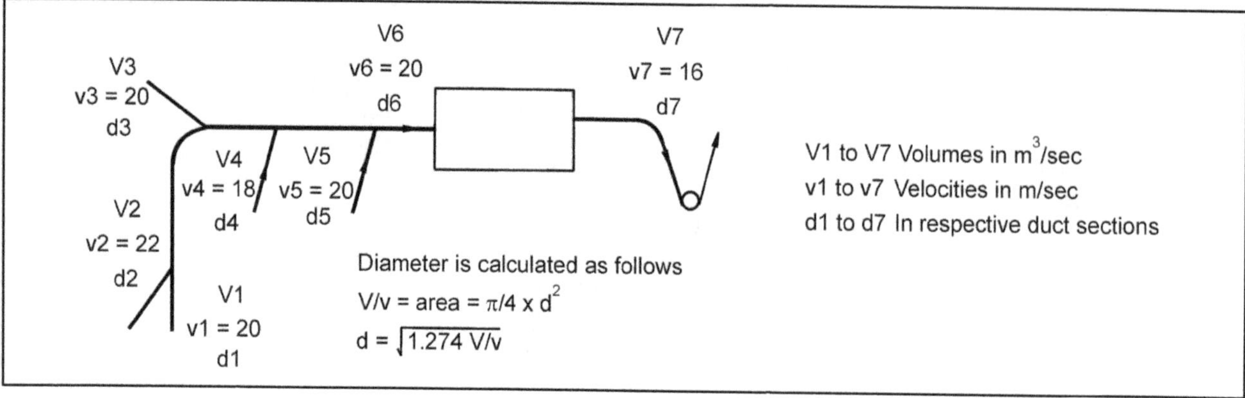

Fig. 6.41.4 Duct sizing according to volumes and velocities.

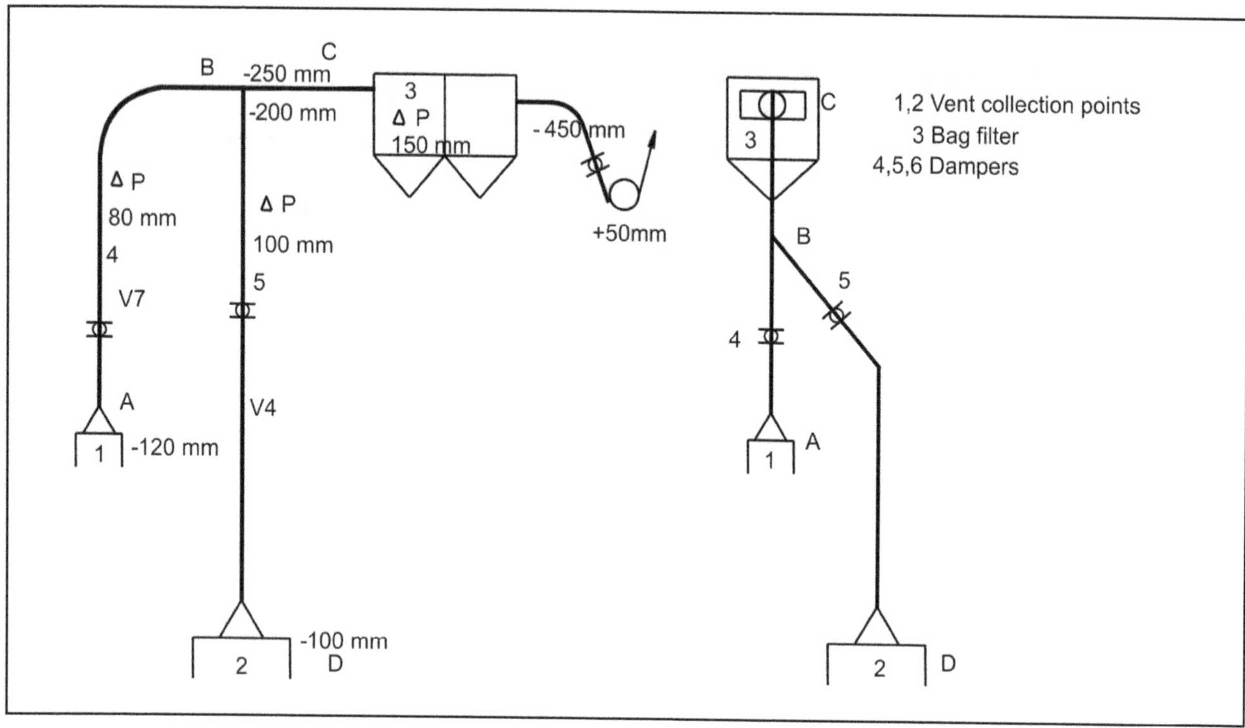

Fig. 6.41.5 Layout of vent ductings (dampers are used to balance pressure drops).

Length of duct AB	=	25 metres.
No. of bends	=	1-90 °C
Radius of bend	=	1 m.
Temperature	=	90°
Dust concentration 200 gm/m³		
Volume in V_7	=	500 m³/min.
	=	8.03 m/sec.
Velocity	=	18 m/sec.
Therefore Duct size	=	770 mm

It would be necessary to calculate pressure drop in ducting from A to B.

Vel. pressure = vel. head × density; and

Vel. head $= (v^2/2g)$

Density of air at 90 ° c and conveying 200 gm / m³ of dust would have to be worked out.

V_4 = 1000 m³/min. at 90 °C, dust concentration – 400 gm/m³

Table 6.41.1
(Fig. 41.3)

Sr.No.	Length of duct	Volume m^3/min		Velocity M/sec	Dia. mm
1	ae	V2	200	18	500
2	be	V3	600	18	850
3	cf	V4	800	20	920
4	dg	V5	600	18	850
5	ef	V2 + V3	800	18	970
6	fg	V2 to V4	1600	20	1300
7	gh	V2 to V5	2200	20	1520
8	ij	V2 to V5 + leakage	2400	16	1800

Length of duct BD = 15 meters.

Number of bends = 1-120°

Velocity in duct B = 20 m/sec.

Size of duct = 1050 mm.

Pressure drop in this leg would have to be worked out since draught at B is common.

Say draught at A = – 120 mm

Pressure drop in duct AB = 80 mm

∴ draught B = – 200 mm

Thus if draught at D = – 100 mm,

pressure drop in duct DB should be

200 – 100 = 100 mm

To draw desired volumes through the legs of the circuit, this condition should be satisfied.

For balancing purposes, resistances are introduced in the circuit through dampers in both legs.

This method has to be followed for bag filter BF3 in **Fig. 6.41.3** also.

See Fig. 6.41.3.

Duct sizes would be as shown in **Table 6.41.1** below.

Pressure drop would be calculated in each length by taking into account length and number of bends, their angle and velocity pressure as explained above.

41.7 Fan in Venting System

The fan of bag filter would be designed for pressure drop in bag filter plus maximum pressure drop in ducting plus margin. Most bag filters would have an operating pressure drop of 150 mm; design pressure drop may be taken as 200 mm.

For ducts handling clean air velocities can be low of about 15-16 m/sec. This will keep pressure drops low.

41.8 Design of Hoods

An important design aspect in venting systems is design of 'hoods' or intake facilities provided to collect vent gases to lead them to the bag filter.

The vent duct itself is very small because the velocity in it is 18-22 m/sec but the area from which vent gases are to be collected is large. The intake or hood should cover the cross sectional area of a discharge chute, so as to collect dust evenly from across the width.

See Fig. 6.41.6.

For example, for a belt conveyor of 1000 mm width, discharge hood would be say 1200 mm wide. The in take should be designed to cover the width of the belt. Assuming vent volume to be 3000 m^3/hr/m belt width, the vent volume is 3000 m^3/hr or 0.833m^3/sec.

If velocity in duct = 18 m/sec,

size of vent duct would be 250 mm dia.

If velocity at in take is to be is say 6 m/sec,

cross section at intake would be 0.139 m^2

For 1 m width (width of belt), breadth of intake would be 140 mm.

The intake should be tapered gradually to reduce pressure loss while increasing its velocity from 6m/sec to 18 /sec.

See Figs. 6.41.6 and 6.41.7.

41.9 Vent Collection Points

The vent collection points are generally at feed / discharge points of conveyors.

They are combined in that discharge point of one is feed point of another.

Point A should serve as vent point for belt 1 (discharge end) and belt 2 for feed end.

See Fig. 6.41.8.

For pneumatic conveyors like air slides, where air is continuously admitted in air chamber throughout its length, it is good to provide vent collection at intermediate points in the length of the conveyor in case of long air slides.

See Fig. 6.41.9.

In designing vents of air slides, air admitted for fluidizing is also to be added to the air displaced by feed.

When elevators are tall say more than 20 metres, it is desirable to provide vent points at hood and at boot both.

41.10 Multiplying Factors for Venting Pneumatic Conveying Systems

When raw meal / coal / cement are conveyed pneumatically to bins, by either FK pumps / fluxo / dense phase system, the air to be vented is not only the air used to convey the material. The quantity of

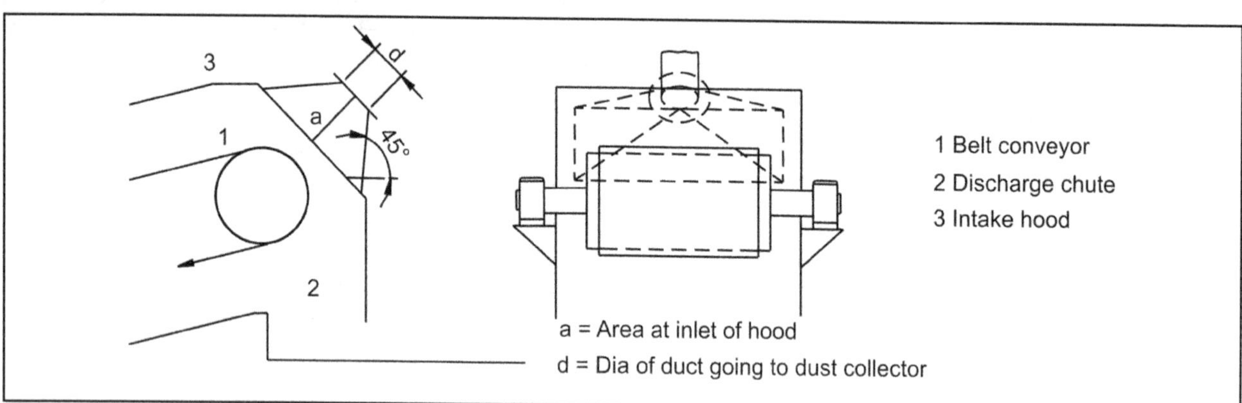

Fig. 6.41.6 Intake hoods to collect vent volumes.

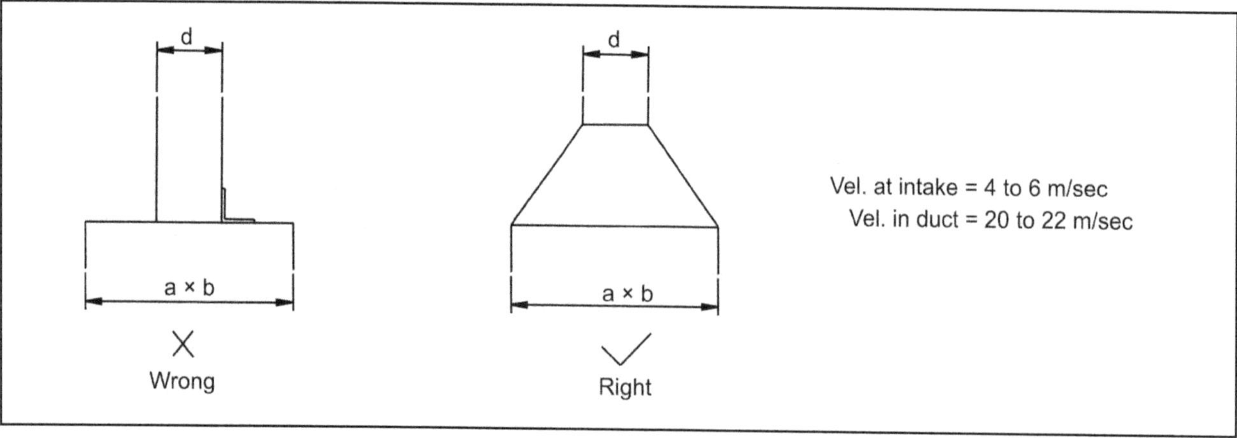

Fig. 6.41.7 Avoid excessive pressure drop at intake points of hoods.

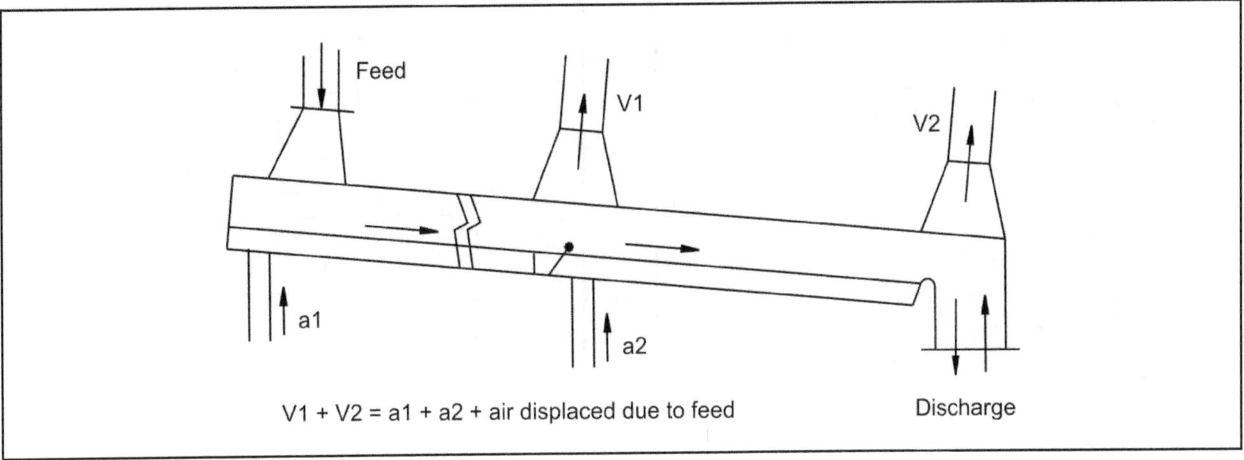

Fig. 6.41.9 Venting airslides - (intermediate vent point if airslides are too long).

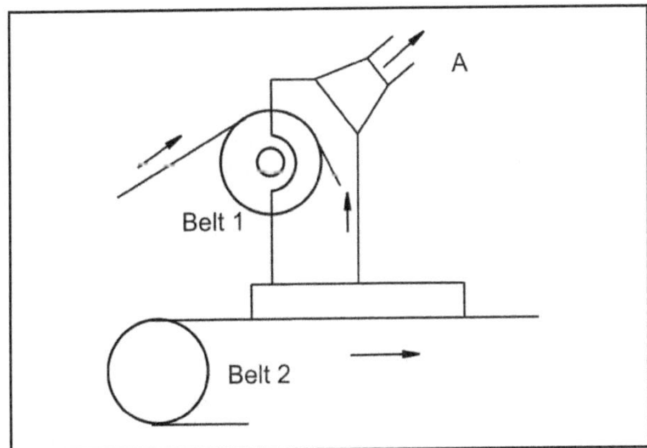

Fig. 6.41.8 Hood designed to collect vent gases from both belts.

this air is usually expressed in 'Free air delivered' i.e., at 20 °C and sea level; whereas when introduced to conveying system, it would be at pressures varying between 7 kg/cm^2 for 'fluxo' pumps to 1- 1.5 kg/cm^2 for air lifts for example. It would be also at high temperatures.

During conveying, the air would tend to expand as pressure is reduced to near atmosphere pressure at discharge points. Therefore, an allowance or multiplying factor is used to convert FAD into actual volume at discharge end.

Instead of working this out from first principles it is customary to use multiplying factors ranging between 1.5 to 2 to FAD depending on distance over which material is conveyed.

While venting silos and bins the conveying air mentioned above plus aeration air introduced in the silo for blending / and or aeration should also be included.

Assume a blending silo 10 m dia – area 78.5 m^2

Blending air at 1.5m^3/min/m^2 = 7065 m^3/hr.

Conveying air to bring raw meal = 3000 m^3/hr.

Use a multiplying factor of 1.5 for conveying air,

Therefore,

total air to be vented from silo = 1.5 × 3000 + 7065
= 11565 m^3/hr + air for aeration to air slides

See Figs. 6.41.10 and 6.41.11.

41.11 Expansion Chambers

Expansion chambers are provided at end of conveying pipe lines, which separate raw feed / coal / cement / conveyed from conveying air. There are many designs of expansion chamber.

See Figs. 6.41.10 and 6.41.12.

41.12

Dust collector and fan would be installed on silo. Many a time dust collector is mounted flush on the silo to eliminate screw conveyor and rotary air lock.

See Fig. 6.41.11.

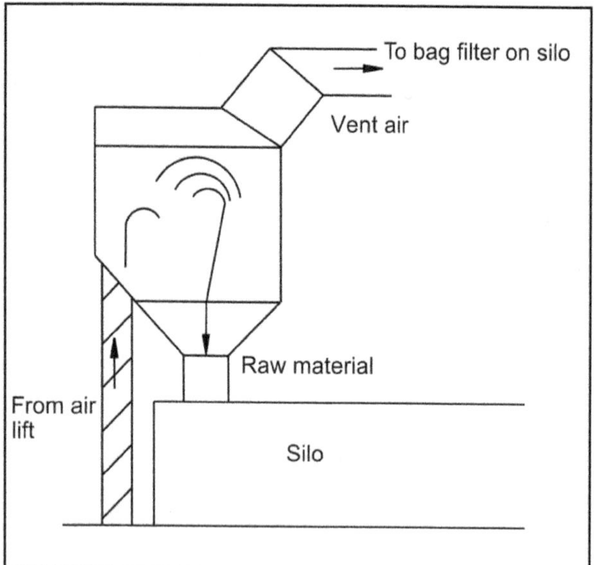

Fig. 6.41.10 Venting of air lift pump at discharge point.

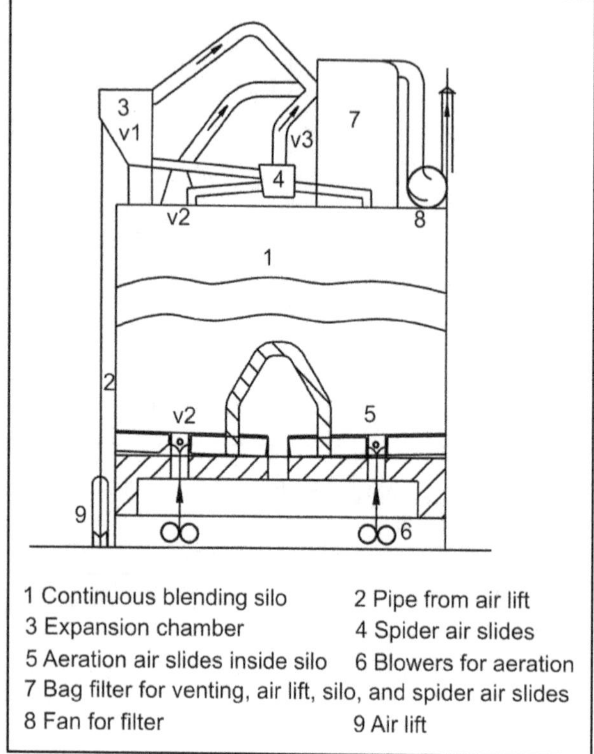

1 Continuous blending silo 2 Pipe from air lift
3 Expansion chamber 4 Spider air slides
5 Aeration air slides inside silo 6 Blowers for aeration
7 Bag filter for venting, air lift, silo, and spider air slides
8 Fan for filter 9 Air lift

Fig. 6.41.11 Venting of continuous blending silo and air lift.

41.13

It is comparatively simple to design hoods or in take pipes for bins, conveyors etc. It is more difficult when the equipment is large like feeders, crushers, crusher hoppers, etc.

41.14

Vent volumes would normally form a small proportion of total gases, (process and system gases plus vent gases). Therefore, often there is temptation to use a common bag filter for system gases and vent gases. However, as explained earlier vent system is to be designed and balanced to collect vent gases from a number of points corresponding to different types of machines. A common dust collector often neglects vent gases. It is therefore better to install small dedicated dust collectors for auxiliaries.

A venting system has to be properly designed to be effective.

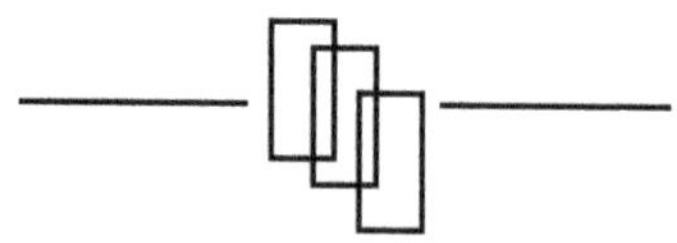

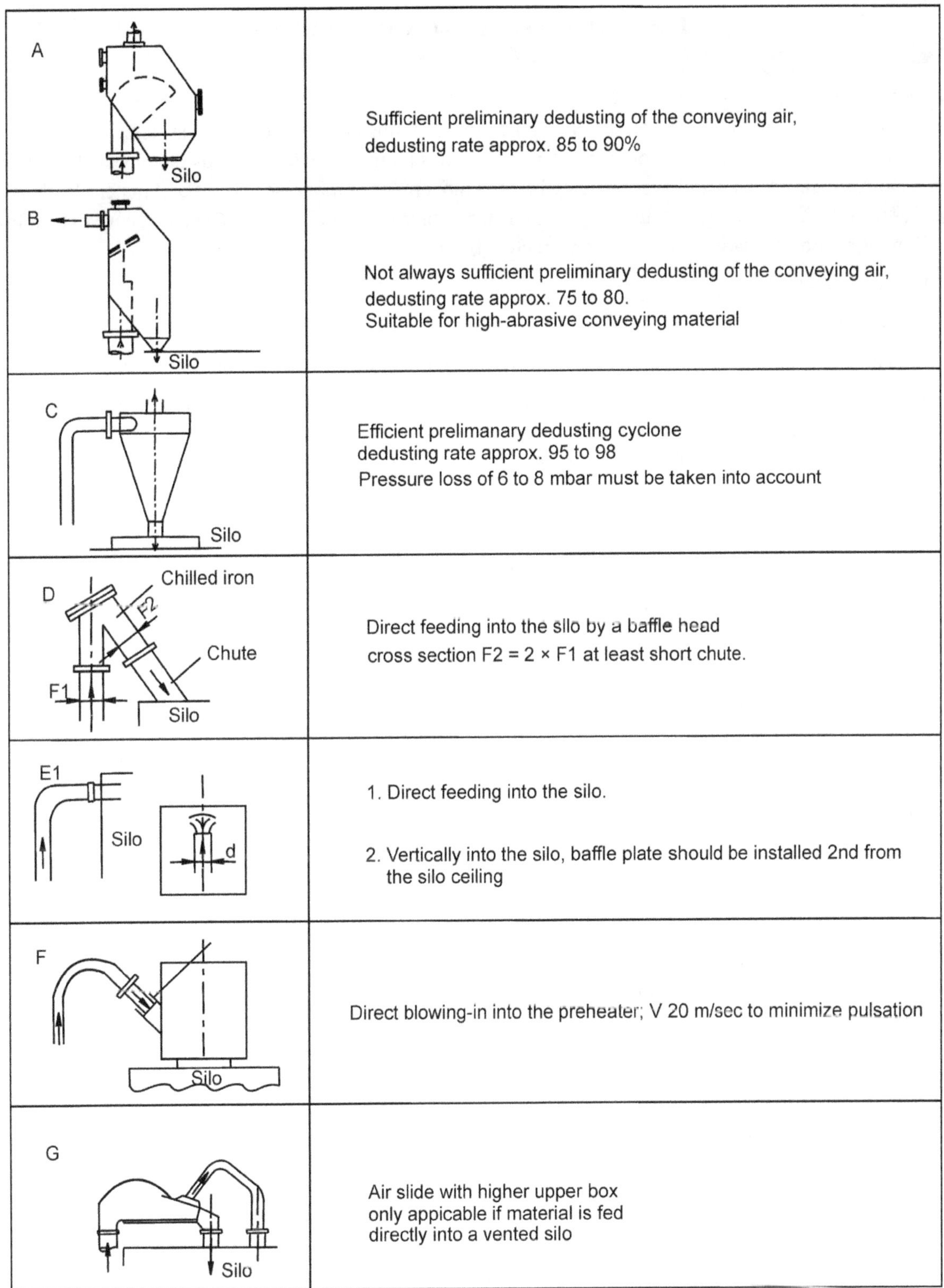

A (Silo)	Sufficient preliminary dedusting of the conveying air, dedusting rate approx. 85 to 90%
B (Silo)	Not always sufficient preliminary dedusting of the conveying air, dedusting rate approx. 75 to 80. Suitable for high-abrasive conveying material
C (Silo)	Efficient prelimanary dedusting cyclone dedusting rate approx. 95 to 98 Pressure loss of 6 to 8 mbar must be taken into account
D (Chilled iron, F2, Chute, F1, Silo)	Direct feeding into the silo by a baffle head cross section F2 = 2 × F1 at least short chute.
E1 (Silo, d)	1. Direct feeding into the silo. 2. Vertically into the silo, baffle plate should be installed 2nd from the silo ceiling
F (Silo)	Direct blowing-in into the preheater; V 20 m/sec to minimize pulsation
G (Silo)	Air slide with higher upper box only appicable if material is fed directly into a vented silo

Fig. 6.41.12 Various arrangements to separate material from conveying air.

Annexe 6.41.1

ESTIMATION OF VENT AIR VOLUMES

1.0

Estimation of vent air volume is the first step in engineering a venting and dust collection system. The air volume to be vented from any machinery hood or transfer point depends on many factors such as the physical nature of the material handled, air currents and surges created by the movement of machinery parts and materials, the type of enclosures provided. Therefore, certain accepted standards which give rule-of-thumb methods have to be followed. The following recommendations have been extracted from :

1. Industrial Ventilation –

   Published by American Conference of Government

   Industrial Hygienists, U.S.A. and could be used as a guideline in estimation of air volumes.

It may be noted here, for guidance, that when the venting volume is expressed in cubic meters, this refers to actual operating conditions at site under consideration. However, if the volume is expressed in nm^3, then this volume is to be corrected for actual operating temperature and altitude above mean sea level.

1. Closed top bins

   Volume = K x volume of entering material.

   The value of K ranges from 1 (for large bins,

   Low feed rates, coarse feed) to 50 (small bins, High feed rate, fine material).

   The minimum rate is 3600 m^3/hr/ m^2 of bin opening.

2. Bucket elevators

   1800 m^3/ hr./ m^2 of elevator casing section.

   For elevators over 10 m height, connect exhaust from

   To both top and bottom

3. Belt Conveyors

   1950 m^3/hr./meter belt width for belt speeds under 1 m/sec.,

   2800 m^3/hr./metre belt width for belt speeds over 1 m/sec.

   Additional volume for greater than 1 meter fall and dusty material:

   1200 m^3/hr. for belt width 300 mm to 1000 mm:

   1700 m^3/hr. for belt width above 1000mm.

4. Flat Deck Screen

   3600 m^3/hr./m^2 through hood openings, but minimum

   900 m^3 /hr./m^2 of enclosure opening.

5. Cylindrical Screen

   1800 m^3/hr/ m^2 circular cross section of screen, but minimum

   7200 m^3/hr/m^2 of enclosure opening.

6. Airveyor and Pneumatic Screw Pump Systems

   Multiply free air volume of the system as follows :-

   For conveying distances upto 150 m .. 1.5 times.

   For conveying distances over 150 m .. 1.75 times.

7. Pneumatic Gravity Conveyors and Blending Systems 1.0 times free air volume of the system.
8. Screw Conveyors

   If dust tight, ventilate feed point only. Use 1100 m^3 /hr. per meter of nominal conveyor diameter corrected for height of fall of feed.
9. Crushers

   At feed end: Minimum of 2700 $m^3/hr/m^2$ of feed hood opening.

   At discharge: Measure air current surges and allow 50% excess.
10. Hammer crushers / Impactors

    These act as fans. Restrict ingress of air at feed point.

    Measure air quantity at discharge and add 50%. (Crusher

    Manufacturers recommendations to be called for).
11. Grinding mills

    Use the larger of the following:

    A. 0.5 m/sec. to 0.6 m/sec. through the cross section area of the mill.

    B. 1m/sec. through each aperture.

    C. 8.5 m^3/hr./h.p. of Raw Mill motor rating.

    D. 13 m^3/hr/h.p. of Cement Mill motor rating.

    (N.B. These figures are for operating temperatures of the mills)
12. Mechanical Air separators

    Where cooling or drying is carried out in Separator, use outlet volumes from detailed calculations. Maximum air volumes at 88 °C are indicated below for Air Separators:

Separator Size. diameter in feet	*Volume at Sea Level. m^3/hr
14	31,300.
16	44,000
18	59,500
20	74,000
22	83,300
24	92,700

For any other temperature or altitude correct the volume suitably.Velocity across the separator is ~1.6-2 m/sec.

High efficiency separators

In high efficiency separators, velocities acrsoo the separator are 3.5-4 m/sec. In one case, vent volumes wer

Size	vent vol. M^3/hr
310	253000
410	440000
520	709000
610	975000

They correspond to a velocity of ~ 3.5m/sec.

13. Grate openings
    A. (Floor or bench dumping – not enclosed)
    2700 to 3600 m^3/hr./m^2 of grate area.
    B. (Enclosed three sides and top)
    2700 m^3/hr/m^2 of net face area.
14. 60 t/hr Rotary Packer
    (i) Packer Guard .. 100 m^3/min
    (ii) Top of Packer .. 15 m^3/min
    (iii) Control Screen .. 20 m^3/min

Annexe 6.41.2

VENTING

DUST BURDENS IN VENT AIRS IN VARIOUS SYSTEMS

Sl.No	Location of Dust Collector	Dust Burden gm/m^3 Max.	Min.	Remarks
	A. Crushing Plant			
1.	Primary & Secondary Crushers.	45	25	Fine Limestone dust
	B. Raw Mill Department			
2.	Cyclone Dust Collector after the Air Separator.	280	150	-do-
3.	Bag type Dust Collector for auxiliaries.	30	20	-do-
	C. Blending and Storage Silos			
4.	Flush bottom Dust Collector Blending Silos	60	35	-do-
	D. Kiln Feed, Kiln, Preheater and Cooler			
5.	Hopper bottom Bag-type Dust Collector for Kiln Feed Section	60	35	-do-
6.	Electrostatic Precipitator or Bag type Dust Collector for Preheater exhaust gases	70	45	When Preheater gases are taken directly to ESP via conditioning Tower
7.	-do-	150	90	Fine limestone powder, when the preheater gases are used for drying raw Material in raw Mill circuit.
8.	Bag or ESP Dust Collector for Grate Cooler vent gases.	15	10	coarse clinker dust.
	E. Coal Mill Department			

9.	Cyclone Dust Collector in hot air duct between Grate Cooler and Coal Mill inlet.	15	10	coarse clinker dust.
10.	Bag filter or ESP for venting Coal Mill circuit with cyclon before dust collector	50	30	Fine Coal Dust.
11.	Bag filter or ESP without cyclone	500	250	Fine coal dust
	F. Cement Mill Department			
12.	Electrostatic Precipitator or Bag-type Dust Collector for open circuit mill.	250	150	Fine Cement Dust.
13.	Electrostatic Precipitator or Bag-type Dust Collector for closed circuit mill.	400	200	Fine Cement dust.
	G. Cement Storage and Packing Plant			
14.	Flush bottom Bag-type Dust Collector above Cement Storage Silos	60	35	Fine Cement dust
15.	Hopper bottom Bag-type Dust Collector for Packing Machine and auxiliaries.	30	20	Fine Cement dust.

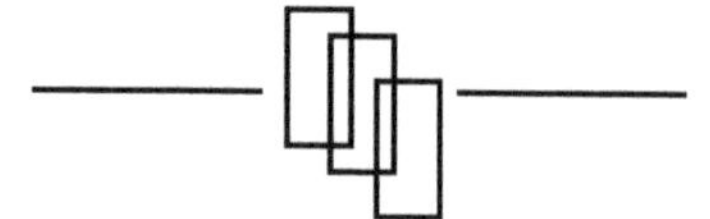

CHAPTER 42

CHUTES

42.1 Chutes

Chutes are connecting pieces that carry solids as against ducts which carry gases.

Chutes are used to convey materials - crushed, granular, or pulverized - so that they convey desired quantity without spillage. Chutes are comparatively short in length. Commonest examples are:

1. From apron feeder to crusher.
2. From feeders to mill inlets.
3. From Mill discharge to conveyors.
4. From Belt conveyors / screw conveyors / chain conveyors to other similar equipment or hoppers and bins.

Where as ducts would normally be round, chutes are normally found to be rectangular / square.

There are also many instances of round chutes. Inlet chutes to ball mills would be round to fit the circular inlets of the mills.

In Preheater, raw meal is conveyed from one stage to the stage below by a round pipe. Technically this is also a chute. Similarly, clinker dust collected in dust chamber before the tertiary air duct is conveyed to cooler in round chutes.

Chutes from air separators in closed circuit grinding systems conveying coarse and fine fractions are also round.

It can therefore be said that chutes are square, rectangular or round according to the shape of opening of the hopper, bin, cyclone from which they originate.

42.2 Sizing of Chutes

As chutes convey materials they should be:

1. sized to suit tonnage to be delivered.
2. installed at an angle greater than the angle of repose of material so that smooth flow is ensured. But the angle should not be too steep as otherwise the velocity at discharge would be too high.
3. dimensioned so as to prevent arching, if material contains moisture.
4. facilitating venting of air that is displaced.

42.3 Design of Chutes

Design of chutes is governed by properties of material being conveyed like its:

42.3.1 Grain size

Chutes should be so sized that largest pieces do not form a bridge and hinder flow. If material is wet, 'arching' can take place at outlets of hoppers and at exit of chutes.

42.3.2 Moisture

Wet material does not flow easily. Hence chutes conveying it should have much steeper slopes not only on surfaces but also in valleys. Valley angles should be grater than angle of repose.

42.3.3 Abrasiveness

Abrasiveness causes wear. Therefore chute should have wearing surfaces when carrying abrasive materials

42.3.4 Flow Characteristics

Some materials are naturally sticky. Others like clay, gypsum become sticky when wet. In case of sticky materials slopes should be steep as in case of wet materials but carrying surfaces should be smooth.

42.3.5

Chutes should also take into account size and dimensions of inlets and outlets provided on connecting machines.

Chutes are used at transfer points to change direction of flow, wholly or partially like between belts, between elevator and belt and screw conveyor.

In majority of application of chutes connecting conveying machinery, chutes would have wider inlets like small hoppers. Outlets would be dimensioned to suit the dimensions of the receiving machine.

Chutes can be combined to form discharge hoods.

Slopes of tapering section of the chute should not be less than the angle of repose of the material.

See Figs. 6.42.1 to 6.42.4.

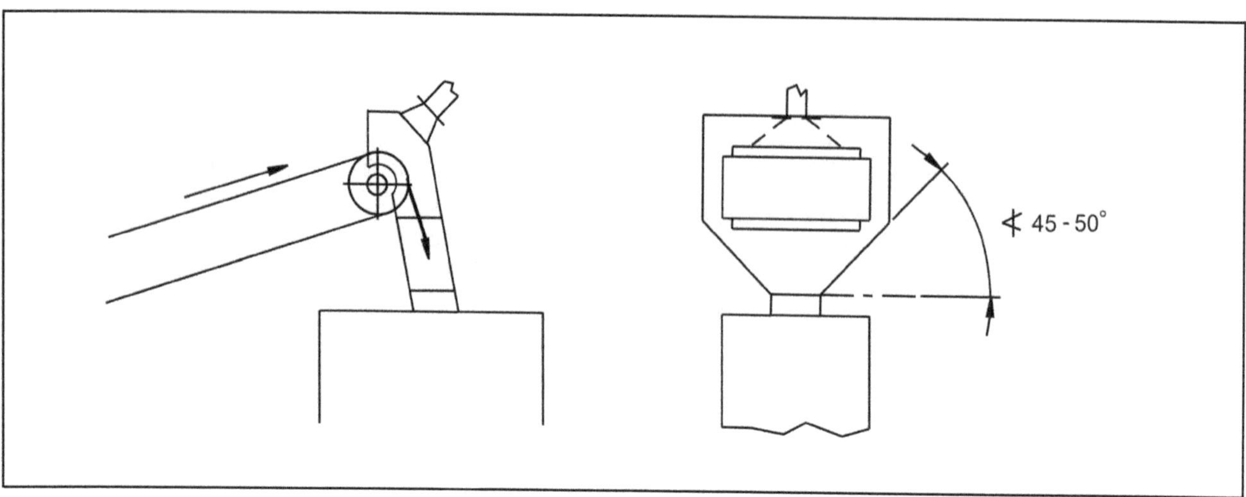

Fig. 6.42.1 Belt to bin/hopper.

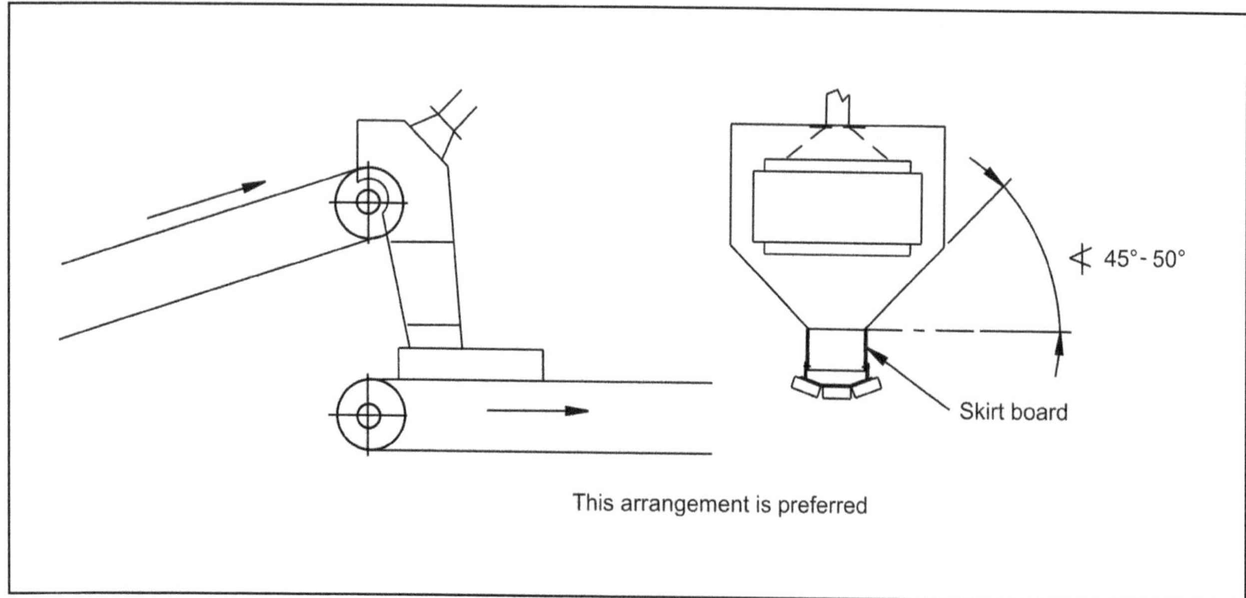

Fig. 6.42.2 Bel t to belt same direction.

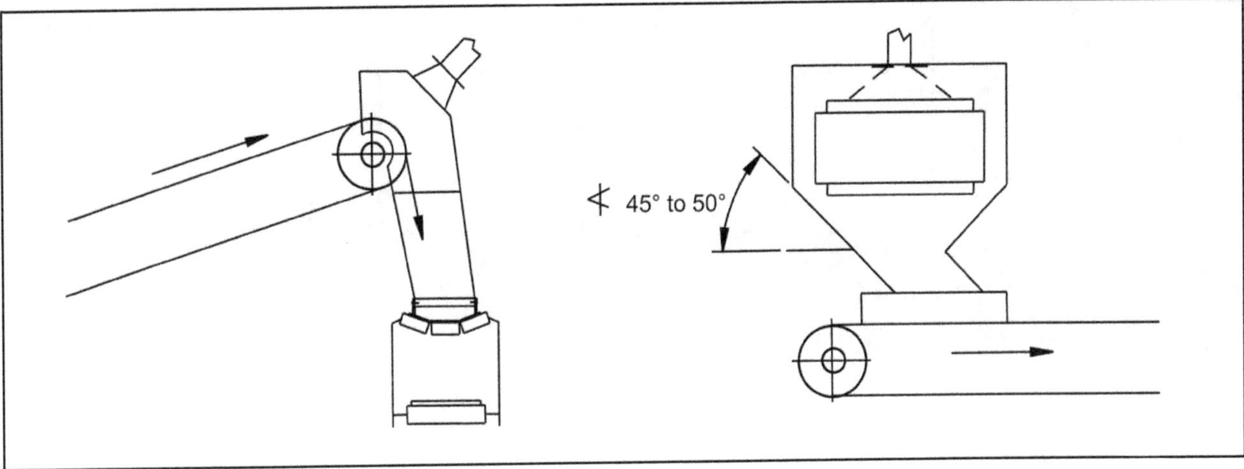

Fig. 6.42.3 Belt to belt with change of direction.

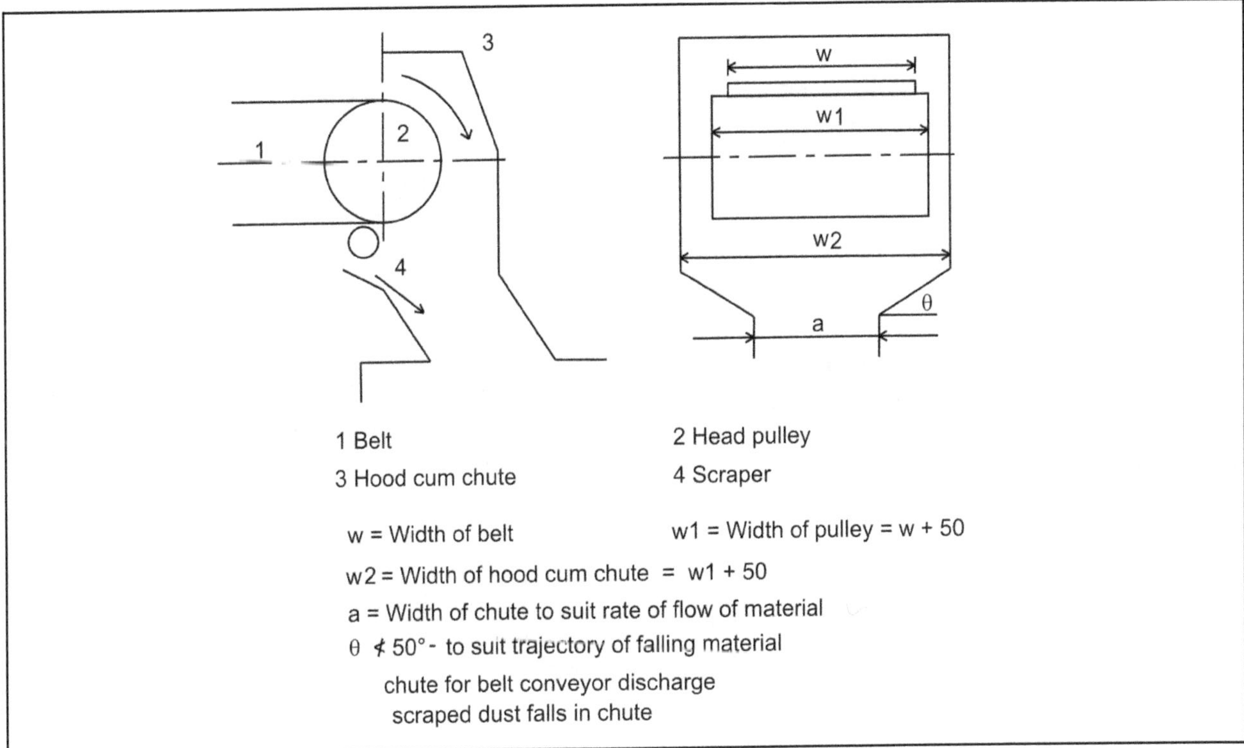

Fig. 6.42.4 Chute cum hood at discharge of belt conveyor.

42.4 'Y' Pieces

Chute can be used to divert the route of the material intermittently.

This is done by introducing a 'Y' piece in the chute. A 'Y' piece also helps to divide the flow between two routes when the hinged Plate is kept in an intermediate position. A scale can be fixed suitably on the face to indicate approximate division ratio of between two routes.The dividing damper can be locked in any desired position. Division will not be very accurate.

See Fig. 6.42.5.

If the flow is to be divided equally then a 'Y' piece with partitions could be installed.

See Fig. 6.42.6.

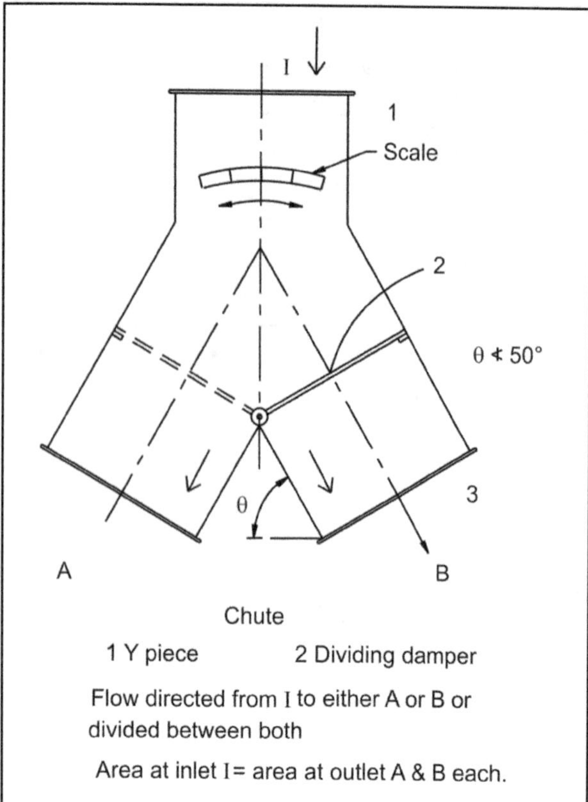

Fig. 6.42.5 Dividing "Y" piece.

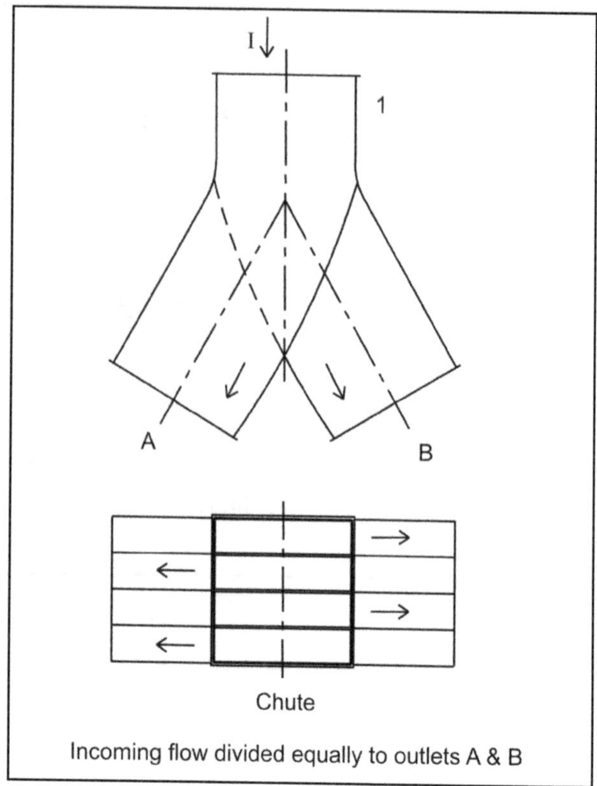

Fig. 6.42.6 Dividing "Y" piece for equal divisions.

42.5 Chutes on Conveying System

Chutes used in conveying systems are to be designed to suit characteristics of flow imparted to the material by the conveyor.

For example, material is thrown off a belt at discharge point in a trajectory; so is the case at the discharge chute of centrifugal elevators.

Chutes should be designed so that material smoothly flows in the chute. They should be angled according to trajectory of flow.

See Fig. 6.42.7.

In a wrong design, material will hit the opposite plate and wear it out.

42.6 Wear in Chutes

Flow of material on the surface of the chute causes wear. Some materials like clinker, slag are naturally abrasive and hence wear is heavy.

The wearing surface of a chute should be identified. That face would be either made of a thicker plate or it would have replaceable liners. Replaceable lines are good in concept but difficult to maintain in practice. Hence a thicker bottom or wear plate is a better option. **See Figs. 6.42.8 and 6.42.9.**

If the impact of material hits the opposite surface of chute that surface should also have a thicker plate. **See Fig. 6.42.7 b.**

When abrasive materials are conveyed in chutes, a simple way to protect it from wear is to weld small angles on carrying surface, so that falling material slides over the accumulated material, rather than the steel plate.

This is also true of hoppers where rings of angles are welded on the inner surface to protect the steel of hopper.

See Fig. 6.42.10.

42.7 Angles of Chutes

In layout, chutes connecting hopper or bin and a machine or connecting two machines, should be straight and direct between connecting points. Bends and corners promote hold up of material and dust.

See Figs. 6.42.11 and 6.42.12.

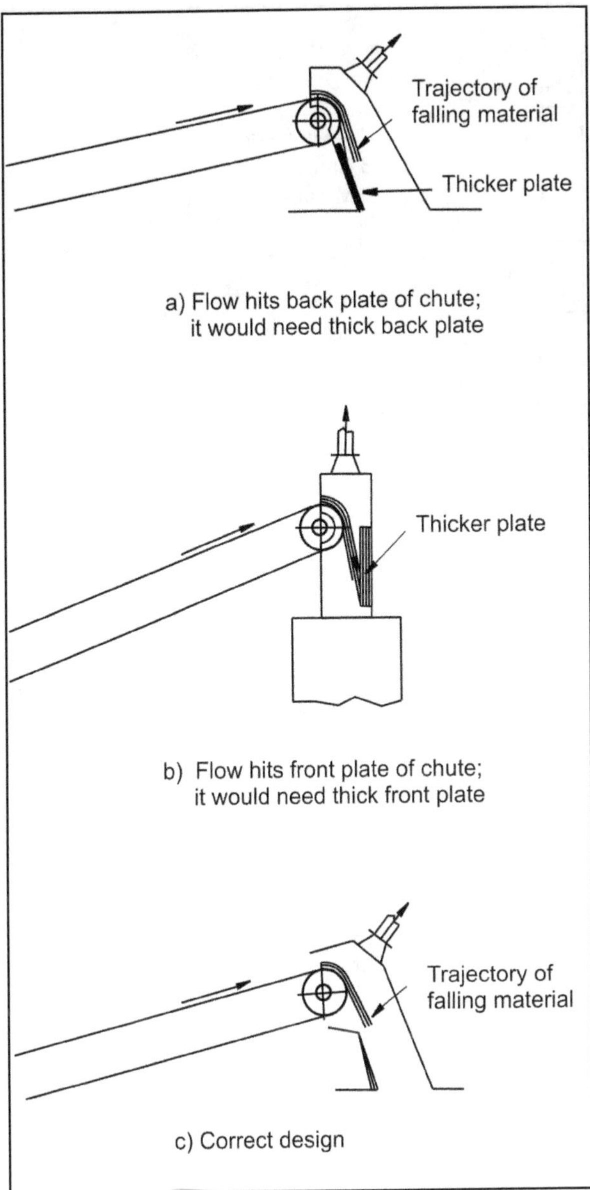

Fig. 6.42.7 Chutes of belt conveyor should be angled according to trajectory of flow.

If bends cannot be avoided, they should be at least in one plane only as real angle is very much reduced hampering flow. Ends should be vertical.

See Fig 6.42.13.

42.8 Basic Guidelines

Basic principles in sizing and layout of chutes are illustrated in **Figs. 6.42.11 to 6.42.16** for machines like a ball mill, a belt conveyor, a screw conveyor and an elevator.

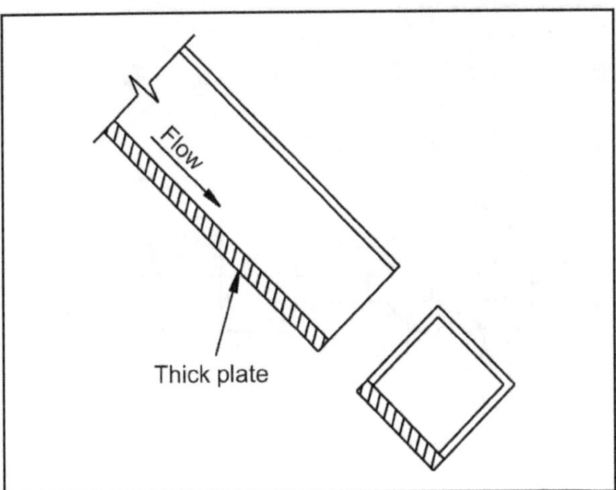

Fig. 6.42.8 Bottom plate thicker than other sides for wear.

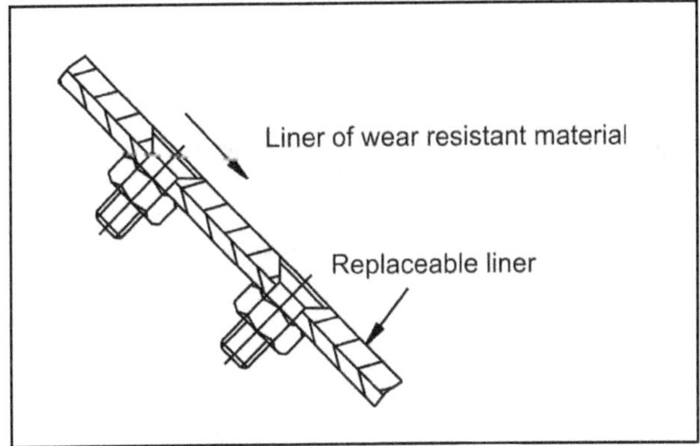

Fig. 6.42.9 Replaceable liner for taking wear.

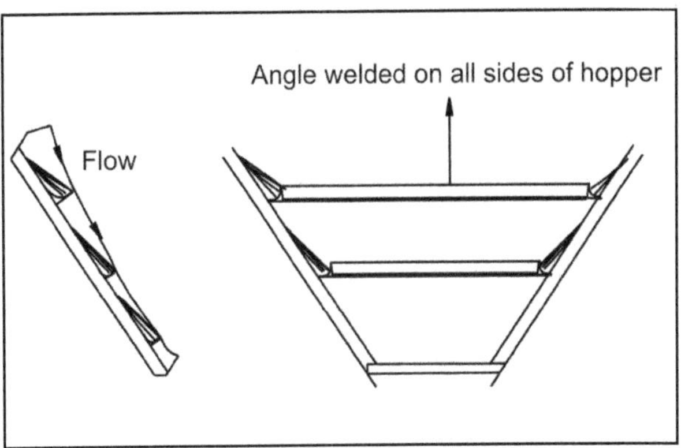

Fig. 6.42.10 Angles welded to chute/hopper material slides over material protection against wear.

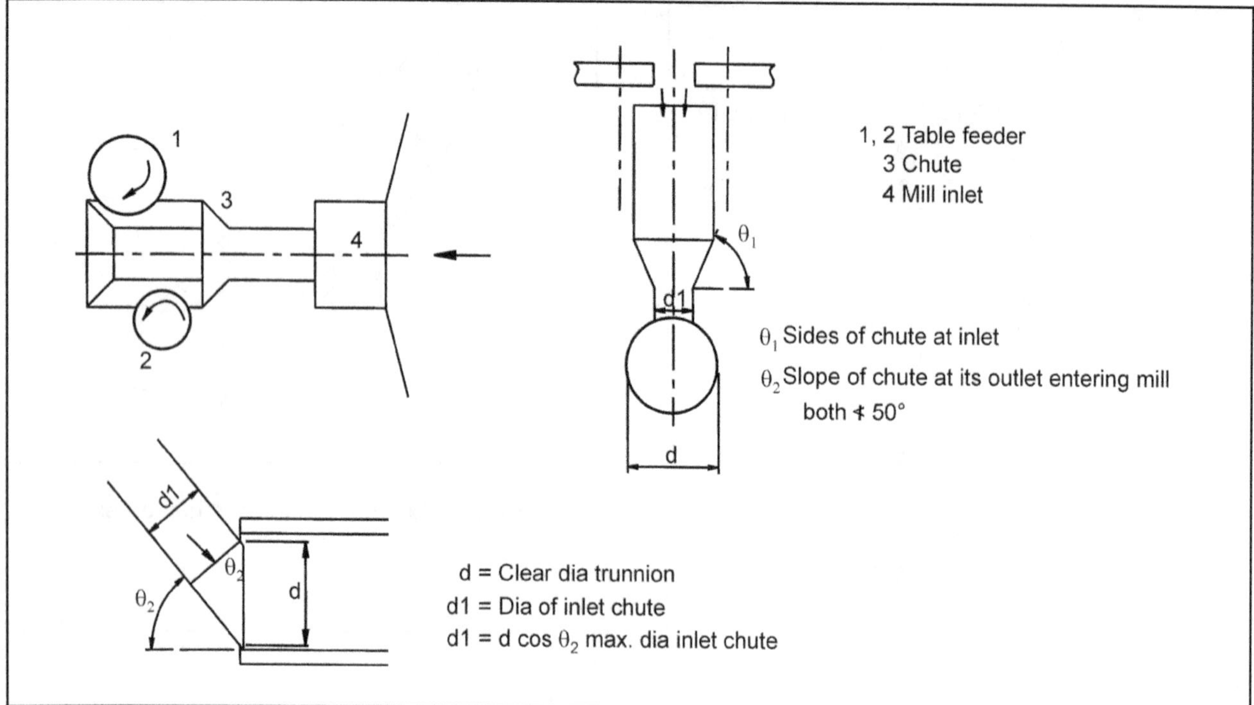

Fig. 6.42.11 Chutes from table feeders to mill.

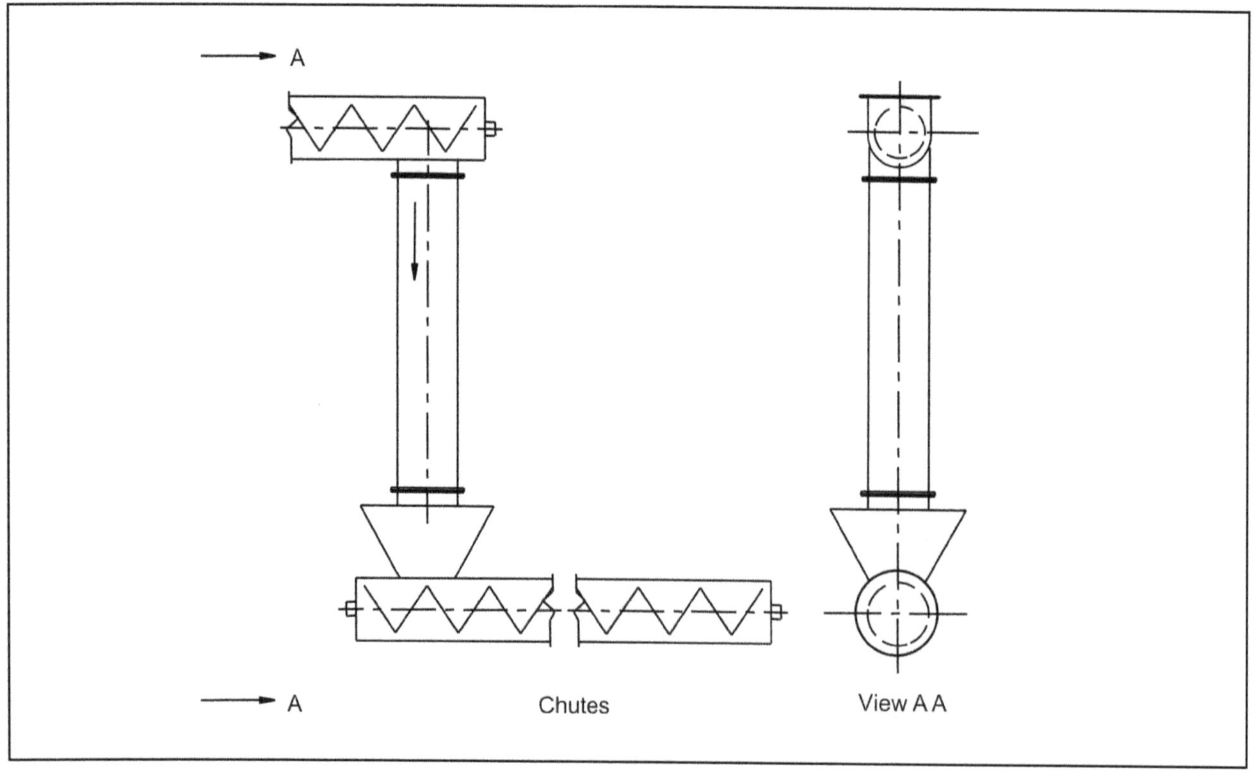

Fig. 6.42.12 Chutes direct connections without bends preferred.

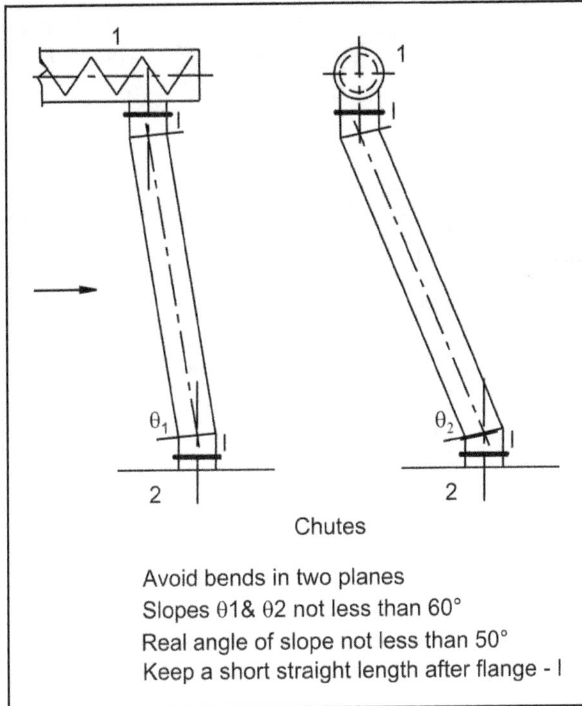

Fig. 6.42.13 Connection between machines.

Avoiding spillage is an important aspect of chute design.

Skirt boards are used to guide material on the receiving conveyor to eliminate spillage.

See Fig 6.42.14.

The chute feeding a ball mill will be circular of necessity to fit in the round mill inlet. Because of the angle, the largest diameter of chute that can be installed will be much smaller than the clear inlet opening.

See Fig 6.42.11.

When chutes connect machines like screw conveyors they are sized to suit the diameters of screws.

See Fig. 6.42.16.

Chute feeding an elevator should be smaller than the width of the bucket and should project into the casing to the extent practical so as to fill buckets without spillage in the casing.

See Fig.6.42.15.

42.9 Support for Chutes

Chutes need to be supported. They can be supported from the hopper or bin they are feeding. A short chute becomes its own support.

See Fig. 6.42.17.

Long chutes would be made in sections. Supports should be so designed that assembling and maintenance is easy.

It is a good idea to have inspection doors on top face of a chute to check flow and to facilitate cleaning. **See Fig 6.42.18.**

42.10 Pendulum Flaps in Chutes

In some chutes pendulum flaps, single or double are inserted to prevent leakage of air or gases. For example Chutes inserted between hoppers of grate cooler and spillage drag chain conveyor are fitted with pendulum flap valves to prevent leakage of cooling air.

See Fig. 6.22.7 in Chapter 22 of this Section.

Pendulum flaps are also inserted in raw meal pipes between two stages of cyclone preheater.

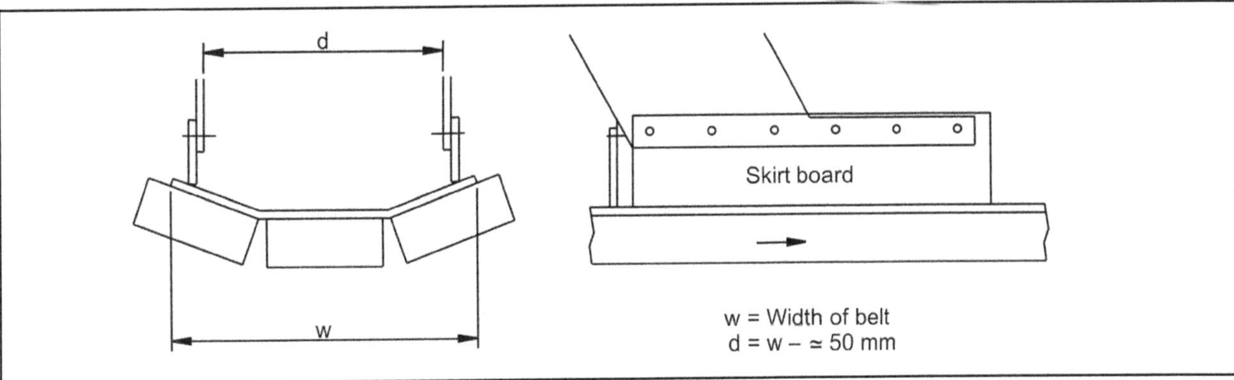

Fig. 6.42.14 Inlet chute belt conveyor.

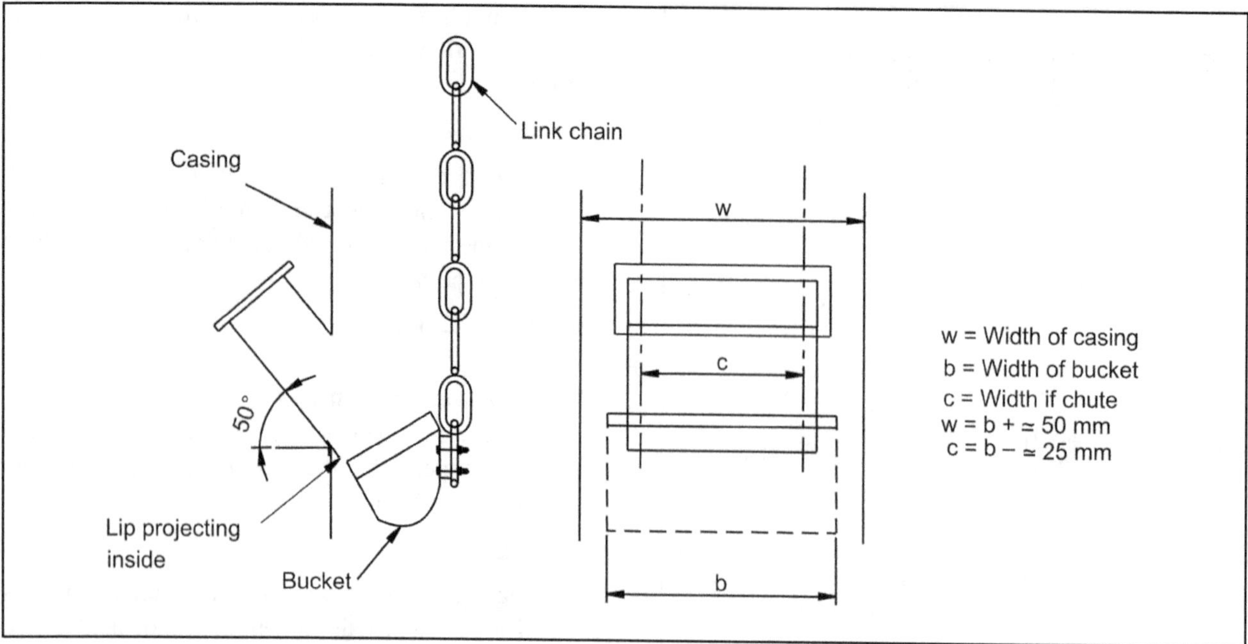

Fig. 6.42.15 Inlet chute – bucket elevator.

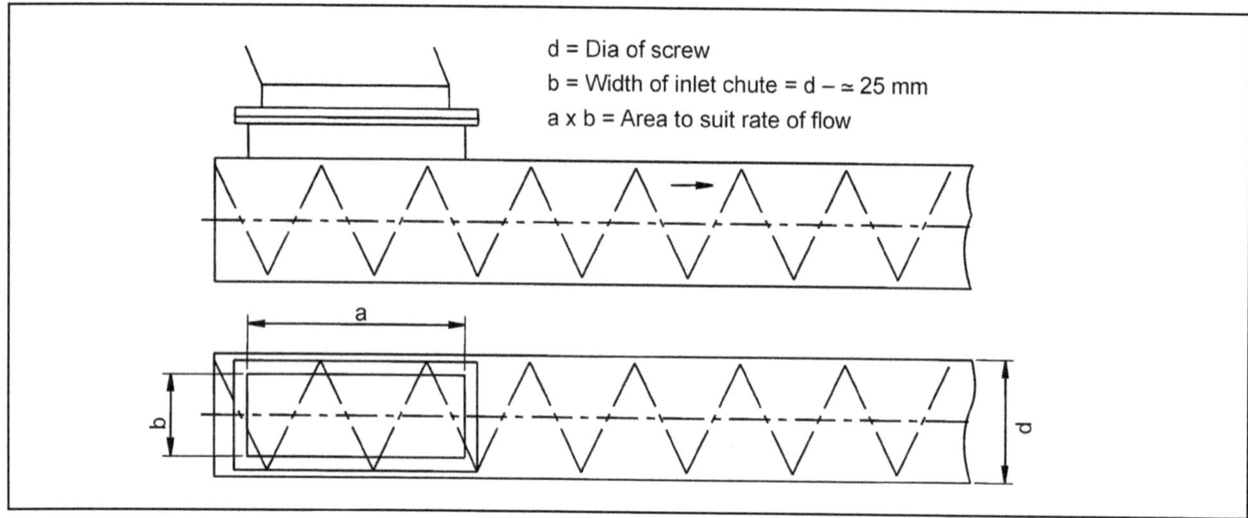

Fig. 6.42.16 Inlet chute – screw conveyor.

42.11 Venting through Chutes

Chutes should facilitate venting. Long chutes and chutes carrying large quantities of material should be vented at more than one point.

42.12 Telescopic Chutes

When chutes discharge material conveyed to form piles like in stock piles for clinker, limestone, etc., then the height of fall is maximum at the beginning of formation of pile; reducing as the pile is built up. This arrangement causes a great amount of dust nuisance.

One way to over come this nuisance is to have a telescopic chute which extends or contracts according to height of fall. It will be motorized so that length of chute can be increased or decreased from a distance. **See Fig. 6.42.19.**

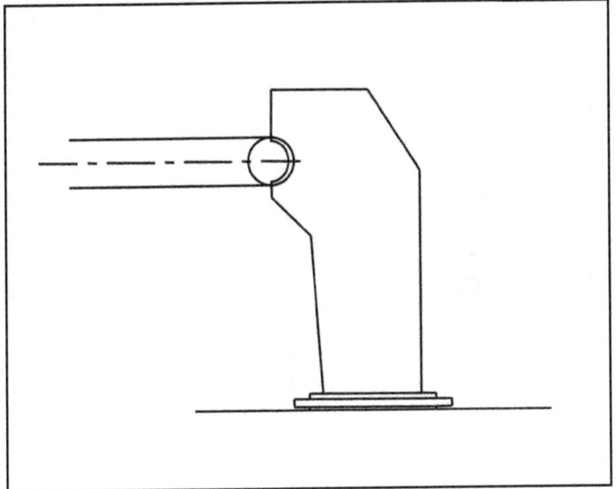

Fig. 6.42.17 Short chute itself acts as a support.

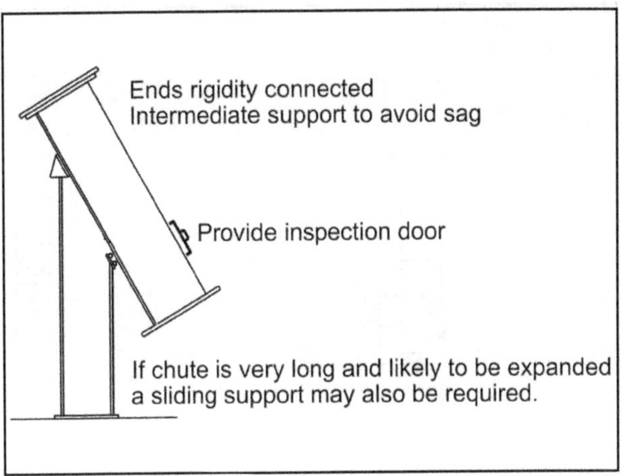

Fig. 6.42.18 Long chutes – require intermediate supports.

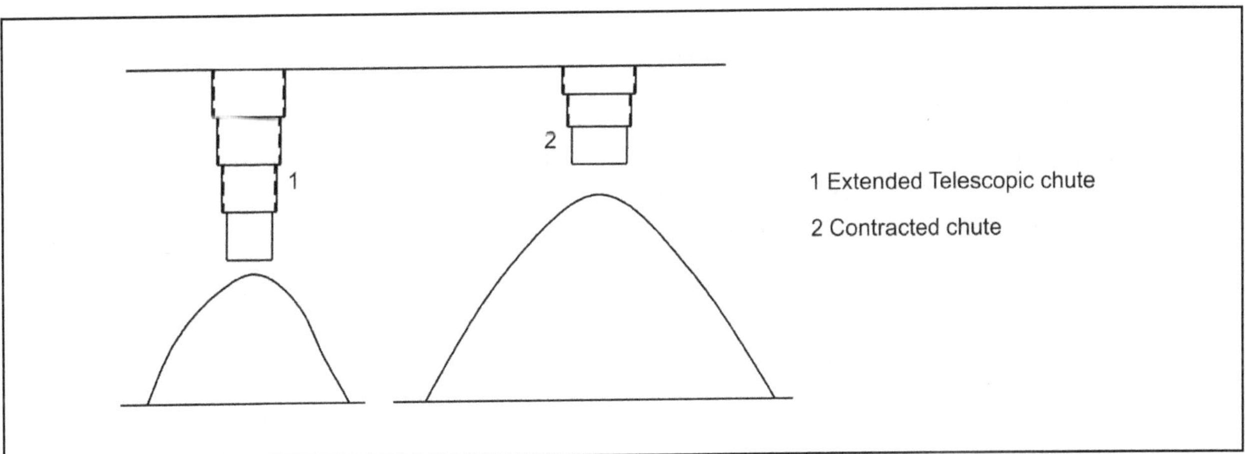

Fig. 6.42.19 Telescopic chute to avoid dust nuisance.

However, telescopic chutes are not easy to maintain because of dusty conditions. Surfaces cannot be lubricated except by using dry lubricants, which would not last for long as they are exposed to extreme weather conditions.

42.13 Flexible Connections

Flexible chutes are used when a machine or a tanker or bulk carrier is to be positioned under a hopper or a bin for discharge. This is particularly applicable for bulk handling.

Flexible chutes – they are made of corrugated pipe. One end would be clamped to the bin. The other end would be clamped to the receiving mouth of the tanker after positioning it under the loading point. These chutes also have facility for a vent pipe so that air displaced can be vented through a dust collector.

See Figs. 6.42.20 and 6.29.5 in Chapter 29 of this Section

Flexible connections are also used when a machine hopper is to be isolated for measuring weight.

When a bin on load cell is introduced between a machine and a feeder, it should not be solidly connected to either hopper above it or feeder below it. Connections are short and made of canvas.

See Fig. 6.42.20.

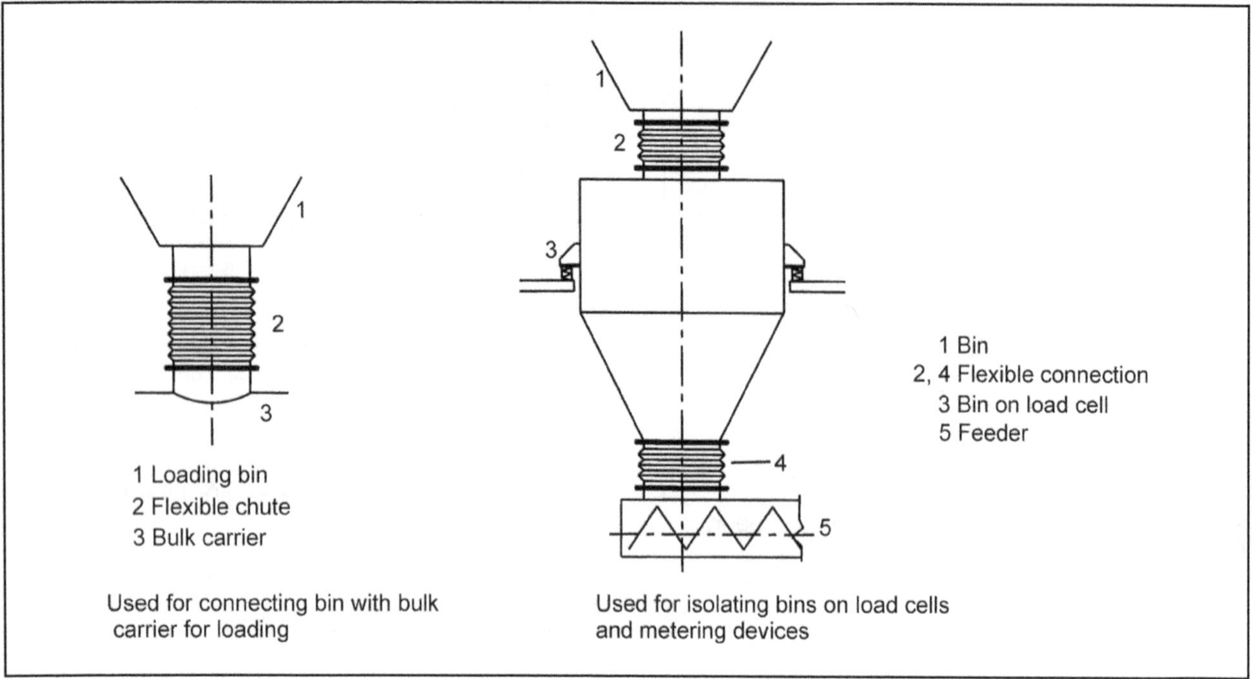

Fig. 6.42.20 Flexible chutes.

42.14 Fabrication of Chutes

Chutes are connecting pieces. They are fabricated according to drawings of chutes and ducts showing true views. Even then they cannot always be fitted as made. It is a good idea to tack weld a chute and finish weld it at site.

Flanges are made in pairs. One flange is welded and the other is supplied loose. This is particularly true of round chutes and pipes.

Flanges should be welded properly at right angle or at the angle shown in the drawing.

To avoid leakage, asbestos rope is inserted between flanges.

See Fig. 6.43.38 in Chapter 43 of this Section

42.15 Raw Meal Pipes

In special category would fall raw meal pipes connecting various stages of preheater and calciner and preheater to kiln. Technically they are 'chutes'. But they are round and carry very hot material and hence are usually called pipes.

Raw meal pipes carry raw meal at various temperatures starting from 60 °C at top stage to 800 °C when entering kiln. Therefore pipes will be m.s. for top stage and stainless steels of suitable grades to suit raw meal temperature for lower stages.

If diameter of pipe is large enough then m.s pipes lined with castable refractory could also be used.

Because of temperature, these pipes would have expansion joints. Generally located just below cyclones. Expansion joints of bellowed construction would also be made from stainless steels to withstand high temperatures.

Pipes also have single / double pendulum flaps so as not to disturb gas circuit in each stage. Pendulum flap valves are also made of heat resisting materials. Separate bearings are installed so as not to overheat them.

Pipes would be installed at an angle of not less than 55° and awkward bends which could hinder smooth flow of raw meal would be avoided.

See Fig. 6.15.16 in Chapter 15 on Preheaters in this Section.

42.16 Spillage at Calciner

Spillage from calciners is also raw meal at vey high temperatures. To handle this spillage pipes of heat resisting material are used.

See Fig. 6.17.9 in Chapter 17 and Fig. 6.53.8 in Chapter 53 on Safety in this Section.

42.17 Spillage at Kiln Inlet from Seal

Sometimes positive pressure develops at kiln inlet and raw meal tends to come out of seal provided. This material is also very hot and should not be allowed to be accumulated on floor of preheater tower. It needs to be collected as shown.

See Fig. 6.53.7 in Chapter 53 of this Section

42.18 Pipes / Chutes for Clinker Dust

Clinker dust collected in the dust chamber between cooler and tertiary air duct is also at very high temperatures of ~ 800 °C. At this temperature it becomes sluggish in flow. Dust is conveyed either to the drag chain for spillage or into the cooler itself by pipes made of heat resisting material from refractory lined hoppers of the dust chamber or in m.s. pipes lined with refractory material.

The pipes should be installed at an angle of not less than 55° to horizontal. Pipes should be suitably supported allowing for expansion. Pendulum flaps would be installed to prevent false air entering into system.

See Fig. 6.18.17 in Chapter 18 of this Section.

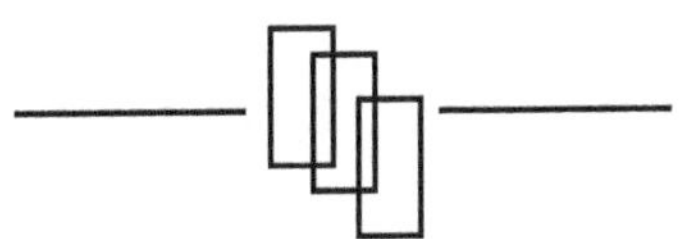

CHAPTER 43

DUCTS

43.1 Ducting

Ducts are used to convey gases such as exhaust gases, flue gases, process gases and vent gases to complete the air/gas circuit of a system usually with the help of a fan or fans. Gases could be air or products of combustion; clean or dust laden.

In **Chapter 41** on 'venting' we have dealt with systems of collecting and exhausting vent gases. Such gases are also conveyed through ducts.

43.2 Aspects of Duct Design

Design and engineering of a ducting system has following aspects:

1. Sizing to cope with volumes of gases to be conveyed.
2. Layout to take into account slopes, bends, allowances for expansion, supports.
3. Construction and fabrication.
4. Refractory lining or insulation and lagging according to specific conditions and uses of gases.
5. Control and regulation like dampers, orifices to be incorporated in the system.

43.3 System Gases

In mill systems:

1. Hot gases are drawn from kiln system to mill systems for drying,
2. from mill to separator,
3. from separator to bag filter / esp,
4. from esp / bag filter to fan,
5. from fan to chimney.

Temperature of gases drawn from kiln system will be between 280 and 350 °C. They will be vented out at about of 90 to 100 °C.

Ducts in system gases would generally be insulated and lagged to conserve temperature.

43.4 Process Gases

In Kiln, preheater calciner and clinker cooler systems ducting includes:

1. Cyclones of preheater are connected with each other by ducts,
2. down comer duct from preheater to preheater fan,
3. from fan to gas conditioning tower (gct) and from gct to esp,
4. from esp to esp fan,

   or to and from bag filter when it is used together with diluting air,
5. tertiary air duct from cooler to calciner,
6. ducts between cooling air fans and cooler,
7. ducts between dust collector and cooler for cooler vent gases and between dust collector and fan.

43.4.1 Conveying Primary Air

An important duct or pipe is from primary air fan to coal burner which conveys primary air and coal to be fired in it. Since concentration of coal is quite high, velocities in it are also very high. If hot air is drawn from cooler to serve as primary air, it would also be conveyed in ducts.

43.4.2 Ducts Carrying Hot Air/Gases

Ducts in kiln, preheater and calciner systems xwould be lined with refractory because of high temperatures of gases.

43.5 Induced and Positive Draught Systems

When fan 'drawing' gases in the system is at the end of the system, the system,will be under negative pressure; when it is at the beginning, it will be under positive pressure.

There are systems that are a combination of negative and positive pressures with an intermediate fan.

43.6 Duct Sizes and Lengths

Systems handing process and system gases generally handle large volumes of gases at different temperatures, dust contents pressures and draughts. Ducts connecting one department to another can be long in lengths. Duct from preheater to preheater fan is also quite long as much as 80-90 metres in case of 6 stage preheaters.

43.7 Sizing

Ducts are to be sized to suit specific conditions of gas flow in each case.

Velocity should be such as to cause minimum pressure drop and yet should not be too low so that material suspended in gases gets separated and tends to settle in ducts.

Dust laden gases would be passed through ducts with velocities of 18-20 m/sec; high velocities are also used (30 m/sec) when dust concentrations are high like in coal firing pipe of kiln.

For handling clean gases or air, velocities can be low so that pressure drop in the duct would be low. However too low a velocity would require a large duct.

In most cases, except in ducts carrying gases after bag filters / esps, gases would be dust laden; dust burden in gases varies between 500-600 gm/nm^3 in mill systems to 80-100 gm/nm3 in kilns.

In case of all such ducts it would be best to maintain sufficiently high velocity, which will enable dust to be carried with gases. Else dust would settle down. This is particularly so far horizontal ducts.

Normally, the velocities in such situations would be 20-22 m/sec. Higher velocities would be used should particles be coarser and dust concentration be high.

43.7.1 Impact of Margins on Velocities in Ducts

In sizing the plant, a design margin is added – about 10%. Major equipment and auxiliaries are selected on the assumption that design margin would be used in acual opaeration. Auxiliaries like fans have another 10-15% margin in them so that they do not become bottlenecks at any time or in the event higher outputs can be reached.

A ducting sized for volumes corresponding to design capacity for a velocity of 20 m/sec and a pressure drop 'p' mmwg, would have when operating at rated capacity and fan capacity respectively, pressure drops as follows:

	Velocity	Pressure Drop
a. at rated capacity	18 m/sec	say 'p' mm

b. at fan capacity 22 m/sec would be 1.5 'p' mm because pressure drop is proportional to square of velocity.

The velocity should not be too low at operating levels nor too high beyond the design static pressure of fan.

Therefore, as explained earlier, if a margin of 10% is added on fan volume, a margin of 20% must be added on its static pressure.

43.7.2 Pressure Drop in a Duct Depends on

(i) Velocity head.

(ii) Material concentration.

(iii) Length of duct.

(iv) Geometry of duct including numbers and type of bends; changes in cross sections.

Losses are also due to lifting of material in it like in case of air lifts and due to introduction of material in gases and due to friction.

43.8 Design of System

System of ducting has to be carefully worked out taking various factors into account to calculate pressure drop in each duct and in the total system.

The fan pushing or pulling gases through the system has to be designed for the maximum pressure drop in the system.

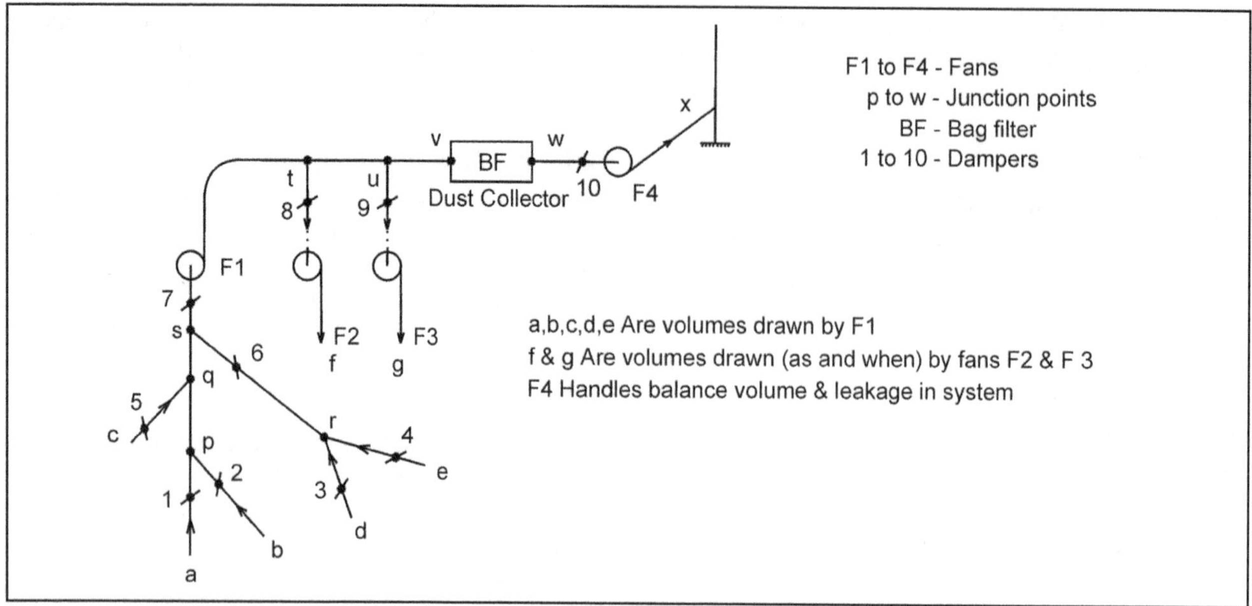

Fig. 6.43.1 A typical gas flow system with fans and dampers.

In a system like kiln or grinding mill or venting, gases are drawn from different points in different quantities and are released also at different points.

See Fig. 6.43.1.

In this circuit gases are drawn by Fan F1 from points, a,b,c,d,e. Quantities and temperatures at each of these points would vary.

Voumes at junction points p, q, r and s would be : (a+b), (a+b+c), (d+e) and (a+b+c+d+e) respectively.

In each of these legs there would be loss in pressure.

Let us say draught at p = - 20 mm

a = - 5 mm

b = 0 mm

Therefore drop in a -p = (20 – 5) = 15 mm

In b- p = -20 – 0 = 20 mm

Let velocity in a p = va

In b p = vb

The duct sizes should be so selected for these volumes that velocity heads of gases flowing in legs ab and bp would cause above respective pressure drops when flowing through the lengths of the ducts.

Therefore velocity pressure in

$$ap = (va^2/2g) \times d1 = Vap$$

$$\text{in } bp = (vb^2/2g) \times d2 = Vbp$$

where *d1* and *d2* are densities of gases in the two legs.

Assuming for the sake of simplicity that the ducts are straight without bends, dampers or orifices

Pressure drop in duct ap = K × Vap × l1 = 15 mm

bp = K × Vbp × l2 = 20 mm

Where l1 and l2 are lengths of ducts ap and bp in metres

If there are bends or dampers in the ducts, allowance has to be made for additional pressure drop or loss of pressure in them.

There are many cases where different volumes of gases are drawn in parallel circuits. In **Fig. 6.43.2,** points A and C are common for both the legs. Therefore pressure drop in the two legs is the same. To divide gases at different rates between kiln and calciner, (to suit fuels fired in kiln and calciner) a resistance R is added in the circuit at point B in riser duct by way of an adjustable throttle. This is adjusted according to rates of fuel fired between kiln and calciner.

See also Fig. 6.43.3 as an example of balancing in kiln esp and raw mill circuit

43.9 Procedure for Design

Flow of gases through ducts follows 'Bernoulli's Theorem' which is based on conservation of energy.

See Fig. 6.43.4.

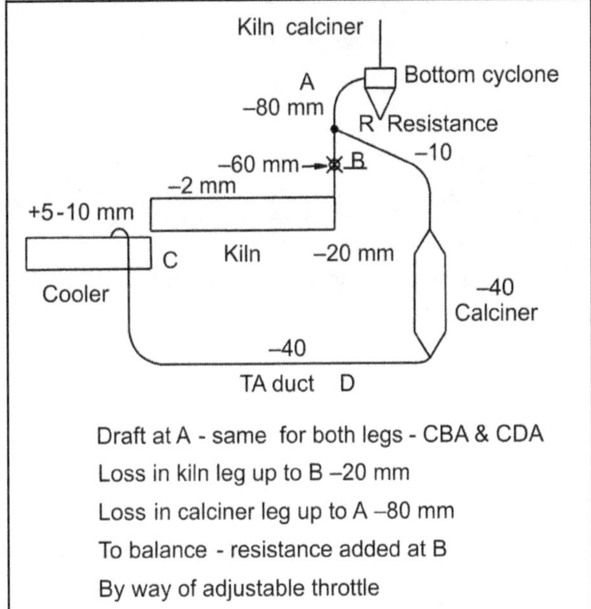

Fig. 6.43.2 Balancing gas circuit - kiln - calciner - cooler.

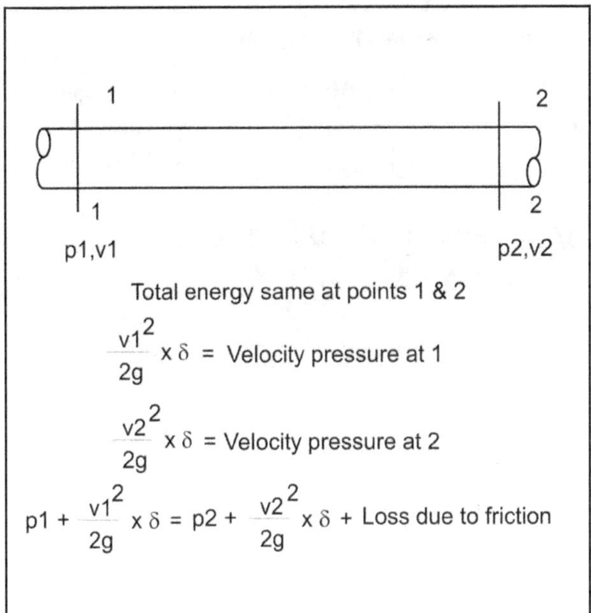

Fig. 6.43.4 Straight flow - uniform diameter.

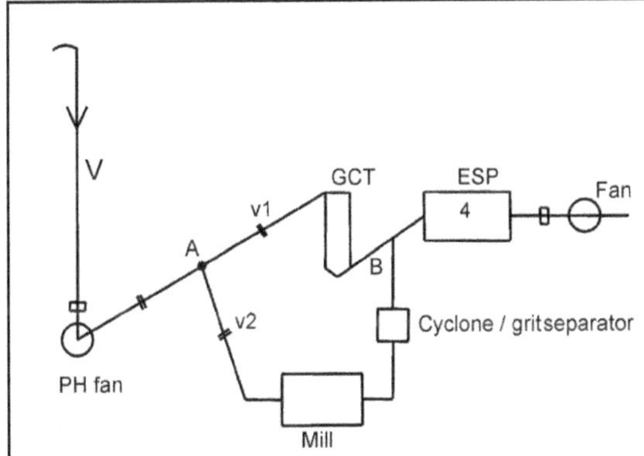

In direct operation, gases entering esp

v1 = V + water spray in gct + leakages

In combined operation,

gases entering esp = v1 + water in gct + leakage + v2 + water dried + leakage in mill

A & B - Junction points of parallel circuit

pr. drop across A – GCT B = pr. drop across A – Mill B

Flow divided according to requirements in two legs using dampers

Ductings sized according to velocity to be maintained

Fig. 6.43.3 Balancing gas circuit - raw mill - GCT - ESP.

Total energy at 1 = Total energy at 2

Let kinetic energy = E and potential energy be P. Kinetic energies at 1 and 2 = E1 and E2 respectively

Potential energies at 1 and 2 = P1 and P2 respectively,

By Bernoulli's theorem,

$$E1 + P1 = E2 + P2 + \text{loss in 1-2}$$

Therefore if $E2 > E1$, then $P2 < P1$

Thus in a converging pipe, kinetic energy expressed in velocity head increases and pressure drops.

This theory is used to introduce material in the pipe like coal in coal firing pipe by adding a ventury

See Figs. 6.43.5 to 6.43.7.

However, if the same velocity is maintained afterwards, pressure loss would be very high. Therefore, velocity head is converted back into pressure head by reducing velocity by adding a divergent piece. Velocity is maintained at a level where air can convey the material with it. This convergent divergent piece is called a 'ventury meter' and is commonly used to introduce coal in coal firing pipes in kilns.

See Figs. 6.43.6.

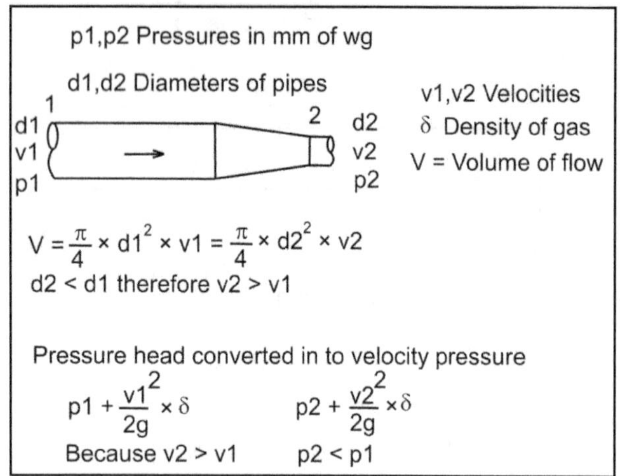

Fig. 6.43.5 Converging section.

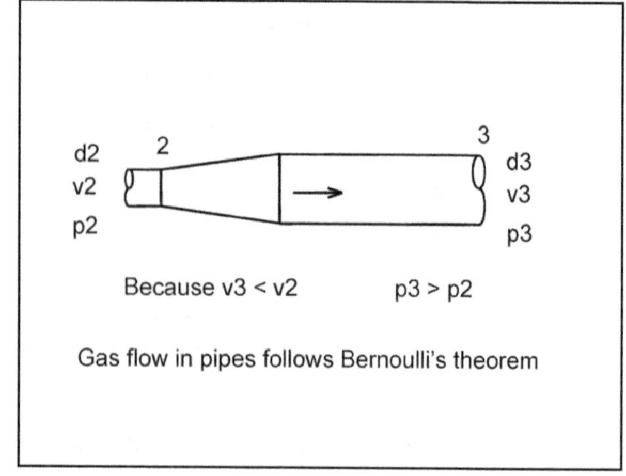

Fig. 6.43.6 Diverging section.

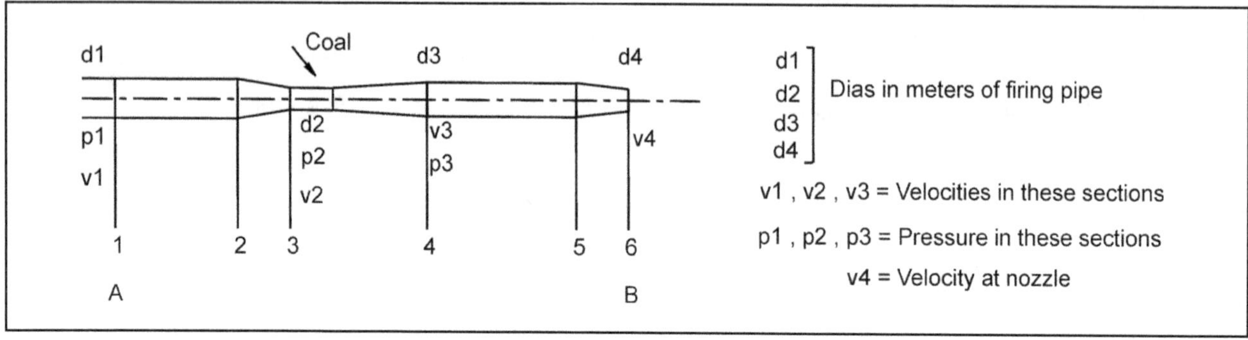

Fig. 6.43.7 Coal firing pipe for kiln with ventury.

Velocity is again increased at the end of the coal firing pipe by decreasing the cross section and shaping it like a nozzle.

See Figs. 6.43.7 and 6.43.8.

Standard procedures are available for calculating pressure drops in a ducting system that takes into account all the factors like, temperature, density, physical configuration like cross sections, bends and length of ducting.

Annexes 6.43.1 to 6.43.3 of this Chapter, outline such a procedure.

43.10 Layout

But first it is the layout, which should be so arranged that the ducts are short and direct and are so laid out that dust if any does not get deposited in transit.

The ducting should be direct i.e., center lines of mill and separator should coincide. **See Fig. 6.43.9.**

See also Fig. 6.43.10.

When bends cannot be avoided because of locations, they should be smooth.

See Fig. 6.43.11.

Ducts should be vertical or at a steep enough slope so that settled dust does not accumulate in one place and does not block the flow.

Thus a duct conveying dust laden air or gas should be laid at an angle of 45° or more to horizontal.

The angle of inclination of ducting would be greater than angle of repose of material so that any settled material would slide down the walls of the duct and not accumulate on the walls of the duct.

Where they must be horizontal like in a tertiary duct, the velocity should be high enough, between 20 / 22 m/sec. so that dust carried does not settle down in the pipe.

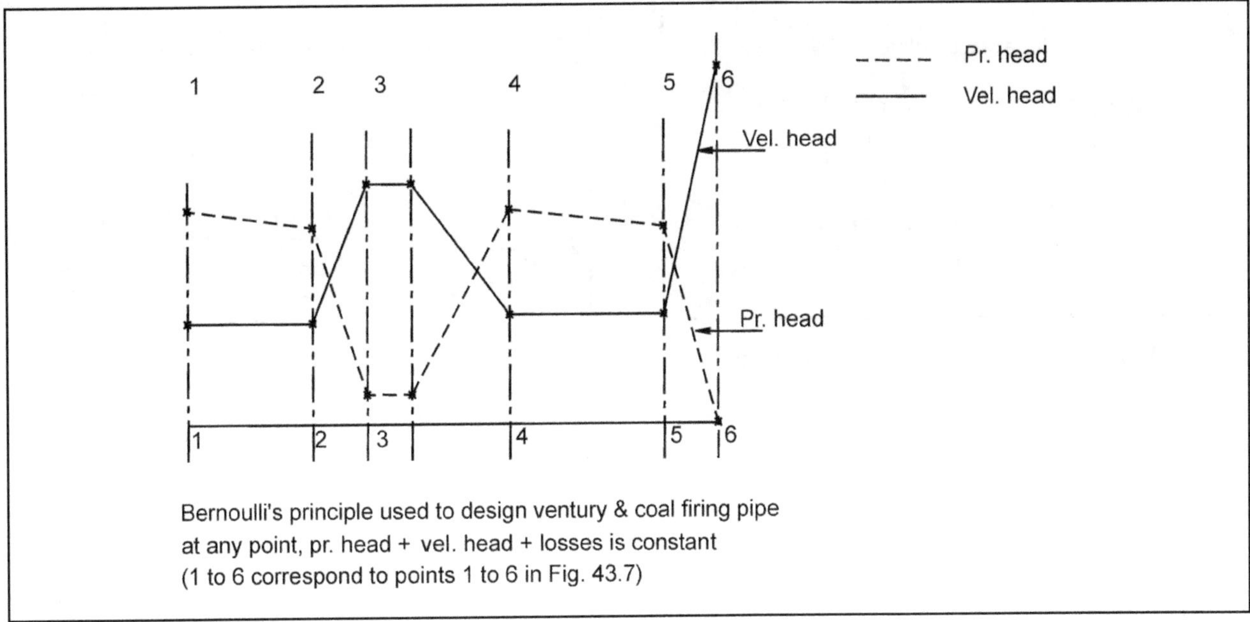

Fig. 6.43.8 Velocity and pressure heads at various points in the pipe.

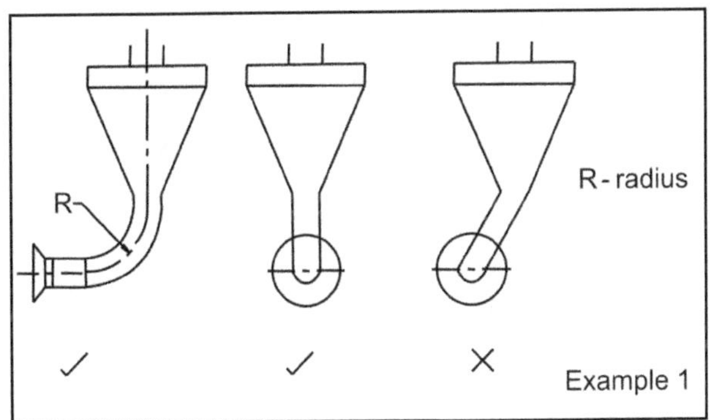

Fig. 6.43.9 Avoid bends in two planes.

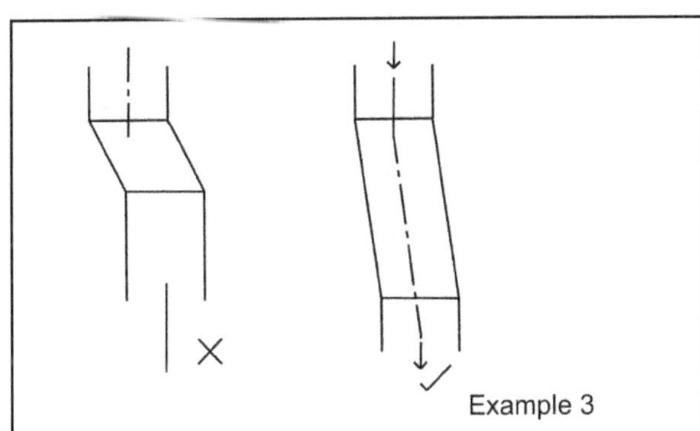

Fig. 6.43.11 Avoid sharp bends.

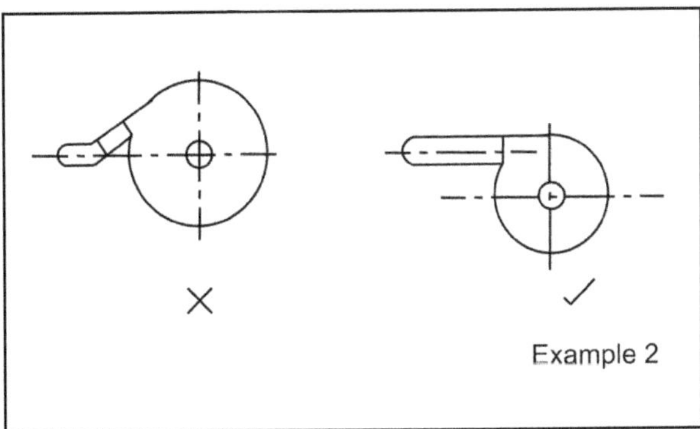

Fig. 6.43.10 Avoid bends.
Layout of ducts.

43.10.1 Correct and Incorrect Arrangements

Examples of right and wrong arrangements of ducting have been shown in **Figs. 6.43.9 to 6.43.14** including a situation where cross section is changed from round to rectangular and vice versa.

When a number of ducts are joined and then are taken to dust collector (for example) the cross section at each stage should correspond to the volume carried in it.

Thus ducting gradually increases in size; the joints should be smooth and new duct should join main duct in direction of flow at a smooth angle helping the flow.

See Fig. 6.43.15.

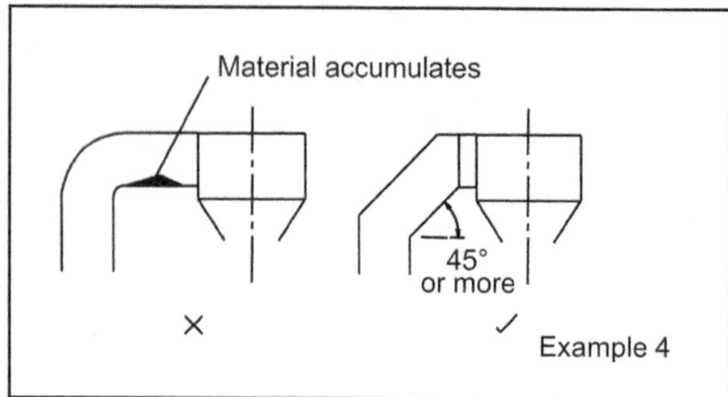

Fig. 6.43.12 Avoid horizontal lengths near bends.

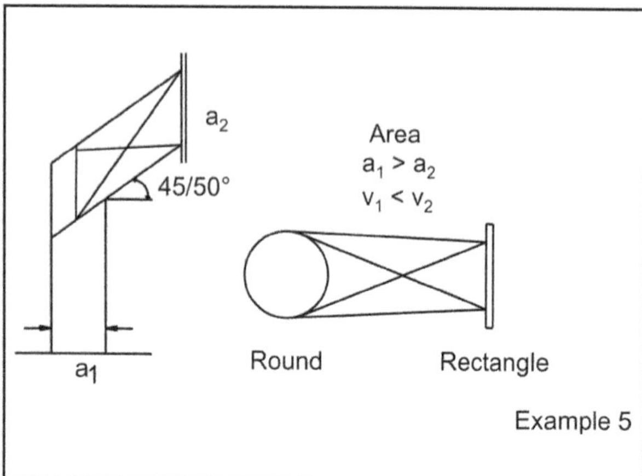

Fig. 6.43.13 Conversion round to rectangle and rectangle to round

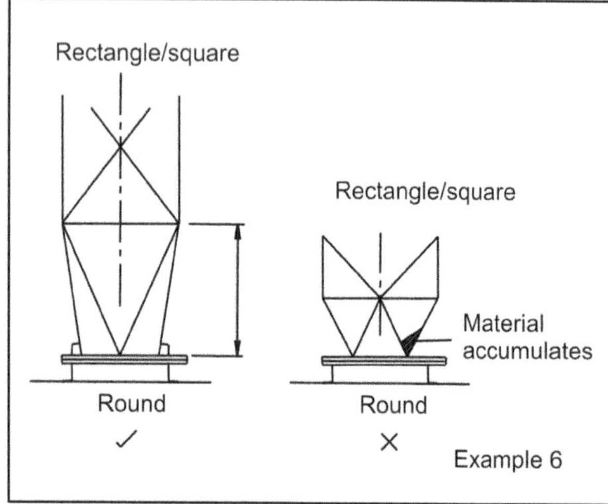

Fig. 6.43.14 Avoid sharp ledges when converting from round to rectangler/square shape. Layout of ducts.

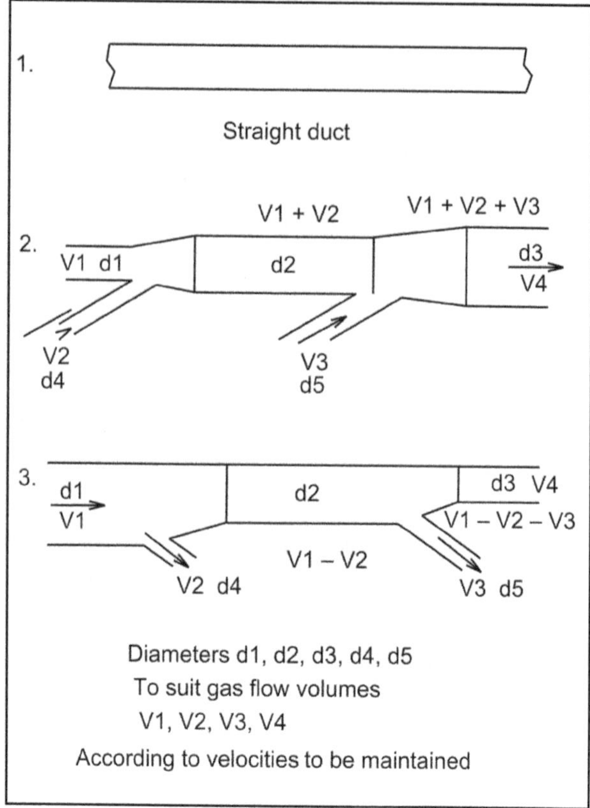

Fig. 6.43.15 Duct cross sections vary as per gas flow through them.

Reverse is also true when cross sections are required to be reduced.

Different types and angles of bends cause different pressure drops. Similarly, radii of bends also affect the pressure drop.

See Figs. 6.43.16 to 6.43.18 and Annexe 6.43.3

Acute and sharp bends cause higher drop than rounded bends with generous radii. Greater the radius R in relation to diameter d of the duct, lower the pressure drop.

If bends are 180°, the radius should be large to reduce pressure drop. In such ducts vanes are fixed in the bend to distribute gas flow evenly and thereby reduce pressure drop.

See Fig. 6.43.16.

A still better way is to enlarge cross section of duct at bend and use camel or hump back bend.

See Fig. 6.43.18.

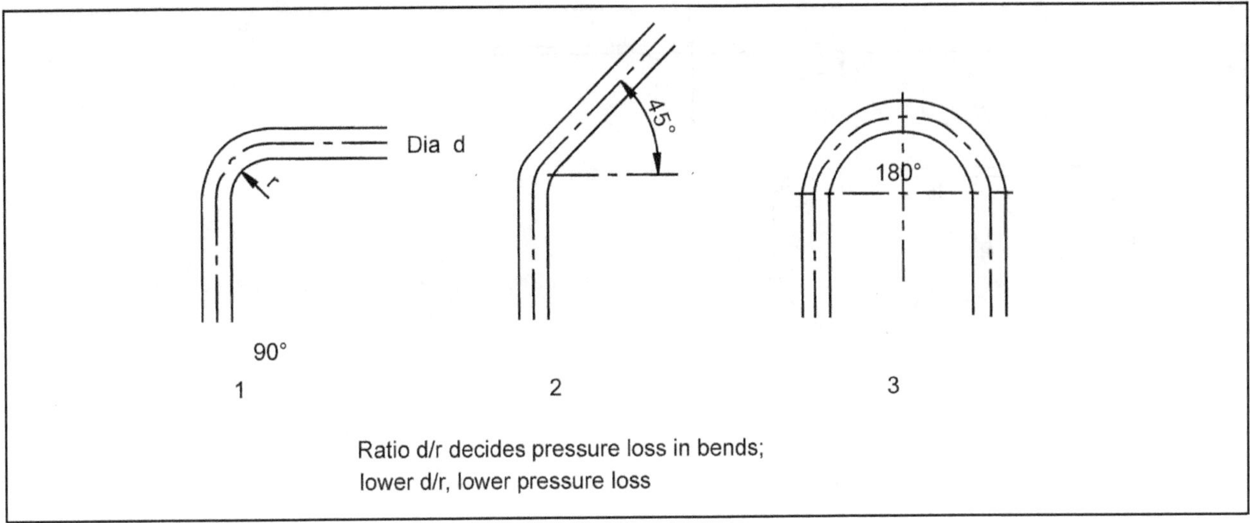

Fig. 6.43.16 Angles of bends.

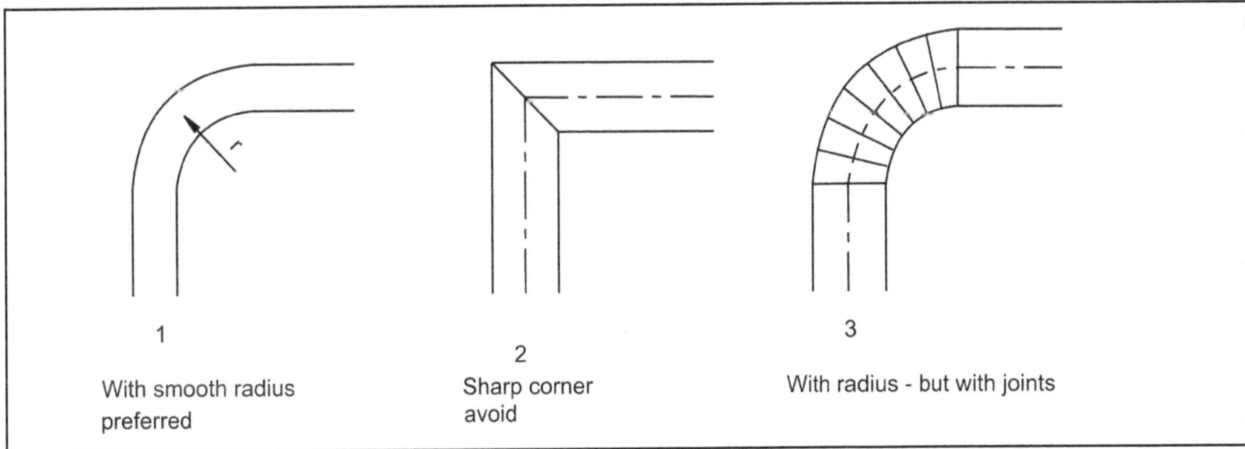

Fig. 6.43.17 Construction of bends.

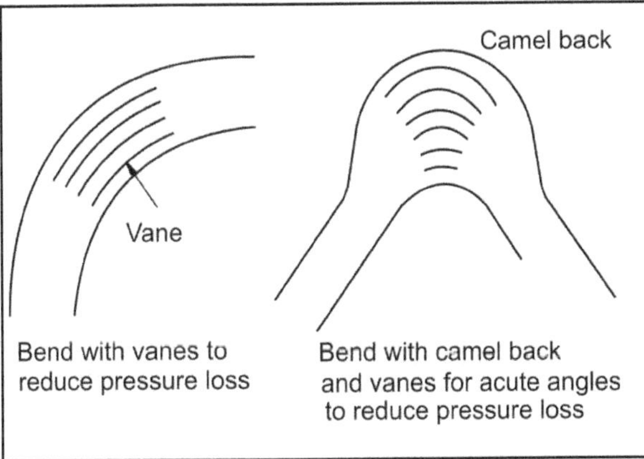

Fig. 6.43.18 Bends.

Conversely, to increase loss of pressure in a ducting, dampers or orifices can be introduced.

See Figs. 6.43.19 and 6.43.20

43.10.2

In a ducting system bends are unavoidable because machines are at different levels and at different locations. But the ducting layout should be stream lined following principles outlined above.

A typical ducting is hot gas drawn from preheater fan outlet to raw mill inlet.

(i) Distance is considerable.

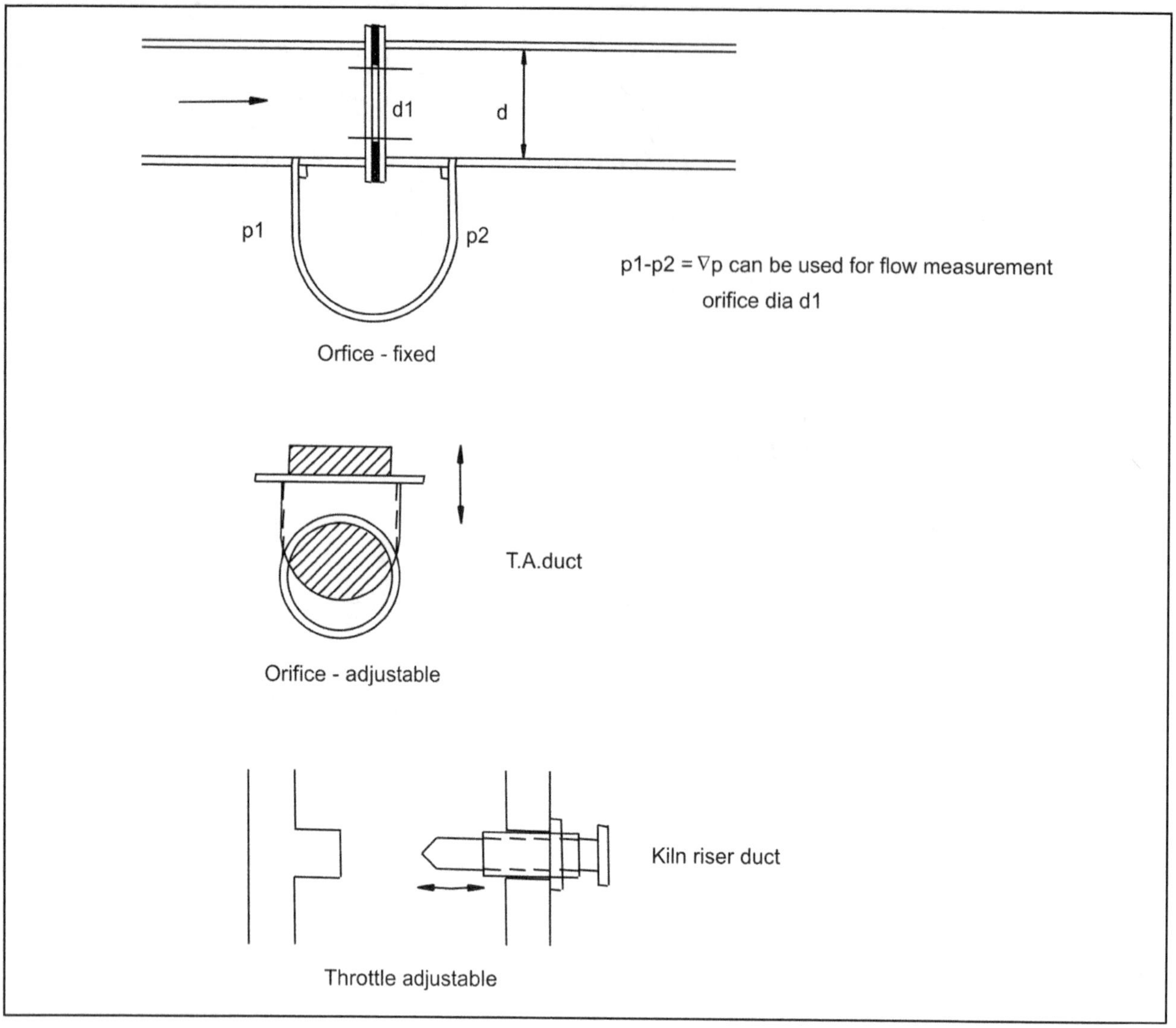

d1
d
p1
p2
p1-p2 = ∇p can be used for flow measurement
orifice dia d1
Orfice - fixed
T.A.duct
Orifice - adjustable
Kiln riser duct
Throttle adjustable

Fig. 6.43.19 Orifices and throttles.

a) Butterfly
b) Louvre
Operated locally or motorzied,
used for introducing resistance
limited use for flow control

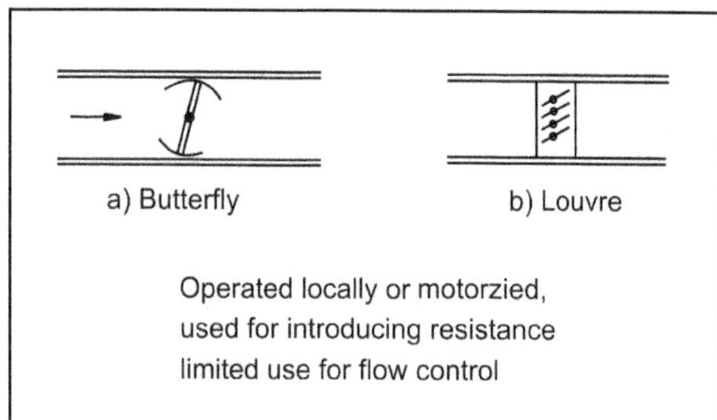

Fig. 6.43.20 Resistances and dampers introduced in ducts for flow control / balancing.

(ii) Centre lines are offset.

(iii) Change of direction to enter mill.

The bend is thus not in one plane. Such bends should be carefully planned. **See Fig. 6.43.21,** which illustrates the point. Detailed ducting drawings need to be made to show actual angles of inclination and actual lengths.

43.11 Duct Supports

Some ducts are so large and long and heavy with lining of refractory and some with insulation and cladding, that, they need specially designed structures to support them.

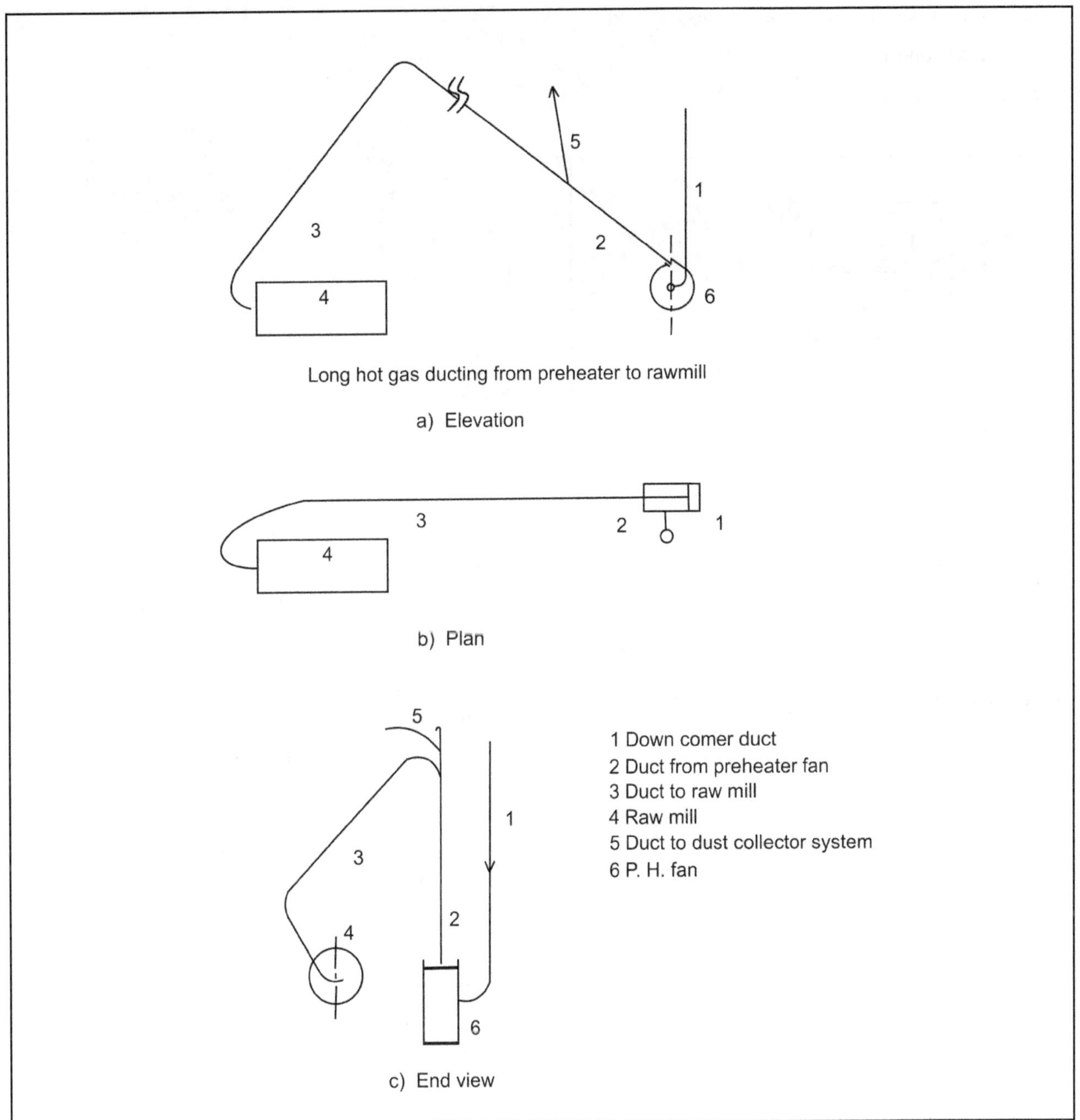

Fig. 6.43.21 Long ducting from PH fan to raw mill inlet.

Duct sizes being large, supports themselves become massive structures with large dimensions and heights. A common support at branching off point can be convenient.

See Fig. 6.43.22.

The ducting is often required to go up and down and hence needs to be supported at several points. They need adequately designed supporting structures. In large plants these supporting structures themselves call for careful engineering and design to support heavy loads and to permit expansions in different directions.

Supporting structures should be so designed that they do not hinder expansion of ducts due to heat in operation.

Thus supports of two kinds are necessary viz;

1. Fixed support.
2. Sliding support.

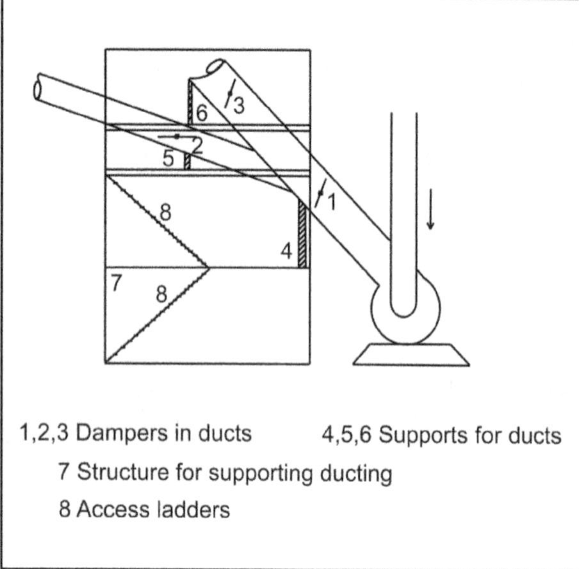

1,2,3 Dampers in ducts 4,5,6 Supports for ducts
7 Structure for supporting ducting
8 Access ladders

Fig. 6.43.22 Large and heavy supports of ductings.

43.11.1 Fixed and Sliding Supports

They would be arranged alternately in long ductings. Supports should be arranged close to an expansion joint so that load of ducting does not come on it. Typical fixed and sliding supports for horizontal and inclined ducts are shown.

See Fig. 6.43.23 and 6.43.24.

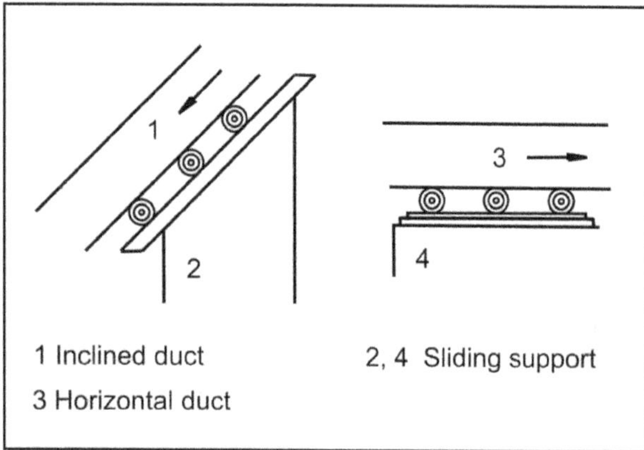

1 Inclined duct 2, 4 Sliding support
3 Horizontal duct

Fig. 6.43.23 Sliding supports for horizontal or inclined ducts.

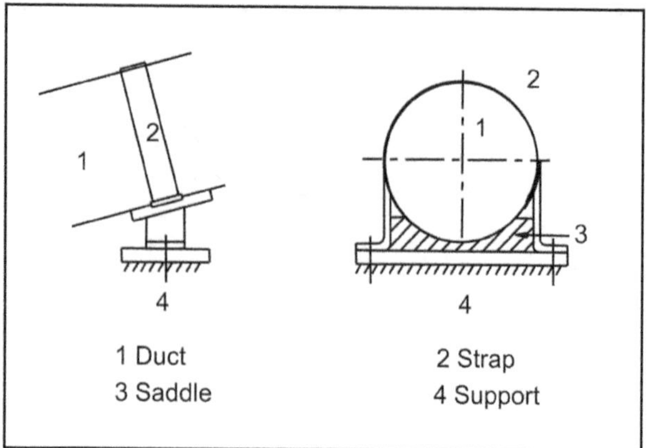

1 Duct 2 Strap
3 Saddle 4 Support

Fig. 6.43.24 Fixed support for horizontal and inclined ducts.

43.11.2 Design of Supports

Supports have to be so designed that they take load at all times. That is pipe must maintain contact with the support.

In the system of preheater cyclones all loads are supported on floors, beams, but are not bolted down. Interconnecting ductings are also supported similarly. It helps if a metal ring is grouted on the slabs to facilitate any relative displacement

See Fig. 6.15.44 in Chapter 15.

The down comer ducts begins with a bend at top from twin cyclones. The support for this bend should be so designed that it takes the load of ducting even in heated condition. It may be mounted on springs to take load of duct at all times.

See Fig. 6.15.37 in Chapter 15 of this Section

Down comer duct is usually supported on brackets from various floors of preheater tower. Supports are fixed and sliding alternately to permit expansion.

See Figs. 6.15.27 and 6.15.28 in Chapter 15 of this Section

Similarly supports for T.A. Duct have been shown in **Fig. 6.18.3 of Chapter 18 of this Section**

43.12 Expansion Joints

Expansion joints are an integral and vital part of a duct system.

Any scheme of ducting carrying hot gases must have expansion joints at regular intervals. Otherwise, the ducting would buckle.

See Fig. 6.43.25.

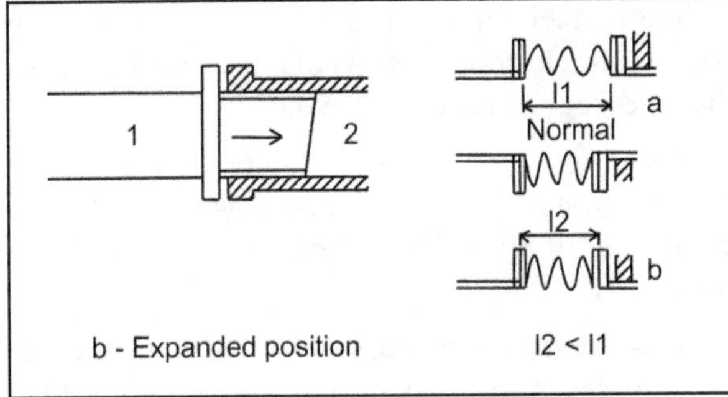

Fig. 6.43.25 Common expansion joints.

Expansion joints come in many types and designs. Selection would depend on:

1. No. of planes in which expansion is to be allowed.
2. Expansion to be allowed in centimetres.
3. Temperature and nature of gases.

As far as practicable expansion joints are so located that they are required to allow expansion in one plane only as shown.

See Fig. 6.43.26 and 6.43.29.

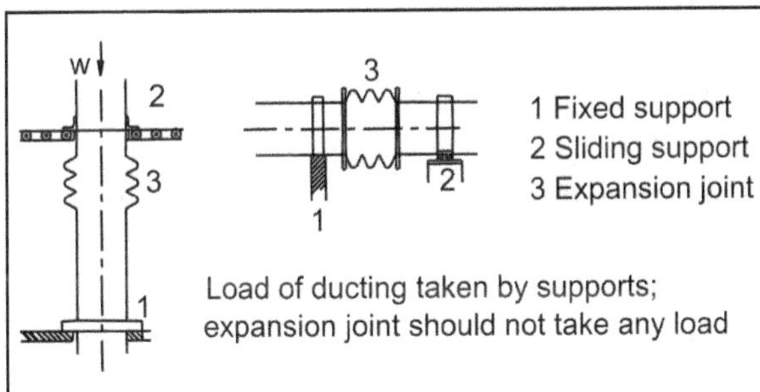

Fig. 6.43.26 Support of ducting at expansion joints.

The number of joints to be included can be worked out using the equation

Expansion = k × l × (T2-T1)

'k' is coefficient of expansion /metre /degree centigrade difference. It can be obtained from reference tables.

'l' is length of ducting and T2 and T1 are temperatures of ducting and abmbient respectively.

The total expansion / expansion per joint = Number of joints in the pipe line.

See Fig. 6.43.27.

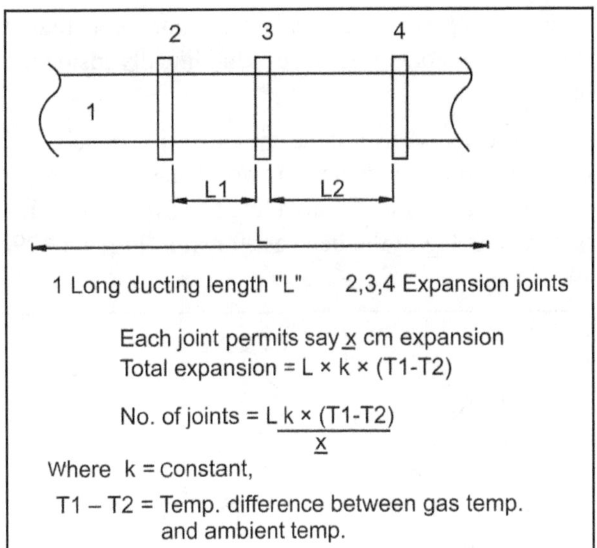

Fig. 6.43.27 Number of joints to be calculated.

As expansion joints can become a sources of leakage, they should be kept to a minimum.

43.12.1 Design of Expansion Joints

Expansion joints are made of thin plate in 1 or more bellows as shown in **Figs. 6.43.26 and 6.43.28**. Plate is 1-1.5 mm thick. This plate therefore cannot with stand load. Therefore, no load should come on an expansion joint

Thus expansion joints would be located close to the supports.

See Fig. 6.43.26.

Most designs permit expansion in one plane.

See Fig. 6.43.25.

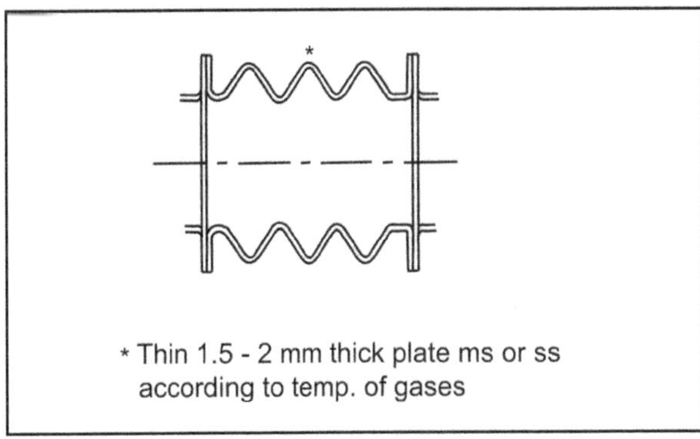

Fig. 6.43.28 Number of bellows according to length of expansion to be allowed for.

In joints permitting expansion in one plane like in sketch shown one pipe can expand lineally inside the other pipe.

Where duct is at an angle, the expansion joint would have to permit expansion in two planes. It is best to avoid such a situation by installing two expansion joints one in each leg of the bend as shown in **Fig. 6.43.29.**

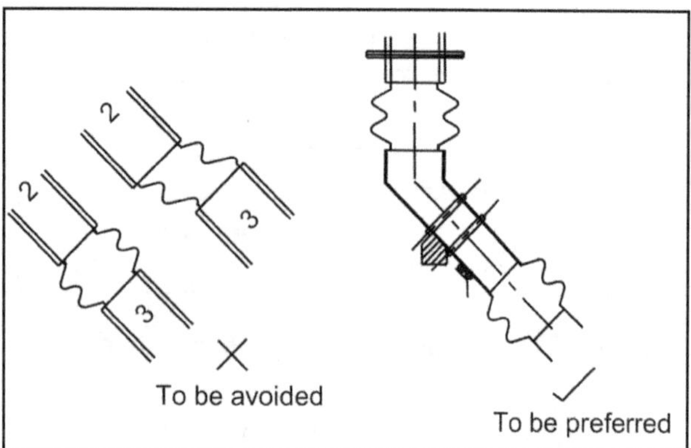

Fig. 6.43.29 Expansion joint will have a short life if it is required to take expansion in two planes. To be avoided.

Material of construction will be m.s. for temperatures up to 300 °C and stainless steel for higher temperatures.

43.12.2 Expansion Joints in Preheater, Kiln Cooler System

The preheater cyclones and the ducts connected to them contain at every stage an expansion joint, just above the cyclone. These joints are large in diameter corresponding to size of preheater (A 6m dia cyclone would have an outlet duct of 3 m dia).

Further, because of change in direction of ducting at every stage the expansion is in two directions, vertical and horizontal.

See Fig. 6.15.44 in Chapter 15 of this Section

Since neither ducting nor cyclone is bolted down, this does not cause a problem.

The expansion joints in preheater system starting at outlet of kiln have to withstand high temperature – 1000 to 1100 °C at kiln inlet and reducing progressively stage by stage to 800, 650, 500, 350 and 280 °C. The construction would be governed by temperature; stainless steel for lower stages and M.S. for top 2 stages. Rock wool, glass wool is used for sealing so that inleakage of false air is avoided.

Expansion joints are also required in the ducting from preheater to preheater fan. This in some cases would be 70-80 m long. Temperatures would be between 350-250 °C.

This ducting is generally supported along the side of preheater tower. Expansion joints are provided at calculated intervals. In most cases, they would have to allow expansion in one plane only. The bend at the top allows for expansion.

See Figs. 6.15.27 and 6.15.28 in Chapter 15 of this Section

On the cooler side, tertiary duct brings air for combustion from cooler to calciner. This is also a fairly long (70 m or so) duct and may be in horizontal or inclined according to the design of the calciner.

Same principles of fixed and sliding supports and expansion joints apply to this ducting and also at take off point from cooler and dust chamber.

This ducting would be lined with refractory from inside and hence would be fairly heavy.

Duct for vent air for cooler, connecting cooler to a dust collector, (now, esp in most cases) should contain expansion joints suitable for temperatures between 180-300 °C.

Preheater fan would be installed at ground level and ducting is taken from it to gct (in case of esp), which is a tall cylinder 5-7 metres in Diameter, and 16-25 metres tall. Ducting is also taken from preheater fan to Raw Mill for drying. It would be installed at an angle so that dust would not settle in it.

All these large and long ducting must have expansion joints.

43.12.3 Connections with Fans

Ducting connection with fan should have an expansion joint at inlet in particular. This serves two purposes; it does not transmit expansion of ducting to fan casing and cause stresses. Secondly it does not transmit vibrations of fan to the ducting.

See Fig. 6.43.30.

This joint may experience expansion in 2 planes therefore, it is made of asbestos cloth.

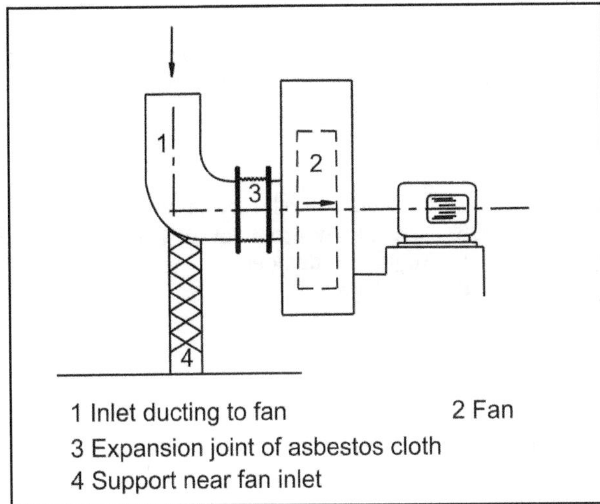

Fig. 6.43.30 Expansion joint at fan inlet and its support.

However, its condition should be inspected regularly as it could be a serious source of false air leakage into the system.

Final exit of exhaust gases is through chimneys. Ducts connect fans to chimneys. These are not very long but temperatures could be high. Therefore this ducting should also have an expansion joint.

43.13 Dampers

Dampers are also an integral part of any system of ducting. They serve:

1. To start fan an no load,
2. to regulate flow,
3. to totally shut off one leg of circuit from the other,
4. to adjust resistance between two legs of circuit.

For example a kiln would be started as a preheater kiln and calciner would be brought into circuit later. Initially therefore it would not be necessary to draw tertiary air from cooler.

When kiln is stabilized and secondary air temperatures reach desired values, fuel would be gradually introduced in calciner, the throttle mentioned above in kiln riser duct together with damper in tertiary air duct would be adjusted to achieve desired flows of gases between kiln and calciner for combustion.

Generally, TA duct would have an orifice to measure the gas flow passing through it.

See Fig. 6.43.31.

43.13.1 Totally Cut off Damper

A totally cut off damper is airtight and permits 'zero' flow past it. Such a damper is used to isolate a piece of equipment like a bag filter either to protect it or to attend to its maintenance by providing an alternate route for the gases.

See Fig. 6.43.32.

43.13.2 Regulation of Flow by Dampers

Dampers are often used to regulate the flow or to divide flow between branches of a circuit. Loss of head in a damper depends on the type of damper – butterfly, louvre, radial vane etc. Multivane or louvre dampers cause low pressure drops. Therefore, if main purpose of introducing a damper is to regulate the flow a louvre damper would be preferred.

See Fig. 6.43.20.

Butterfly dampers do not regulate flow beyond a certain limit and are not very effective.

It has already been mentioned that dampers which are like resistances in electrical circuit cause pressure loss in the circuit of say between 20 to 50 mms. When flow is required to be regulated frequently and altered conditions can continue for long times like in case of kiln operations, variable speed motors are used rather than dampers to regulate flow.

Though fans generally come with dampers their main purpose would be to start fans on no load.

Dampers could also be used to divide flows. However to ensure desired flows, in two legs of system, it would be advisable to have a fan, in each leg.

43.13.3 Orifices

Like dampers, orifices are used to increase resistance. Orifices add considerable resistance and are mainly used to measure the flow. But orifices are generally fixed and if resistance is required to be changed, It has to be replaced by one with a larger or smaller opening.

Sliding plates (with seals) are added to vary the effect of orifices.

See Fig. 6.43.19.

Selection and location of dampers requires considerable thought.

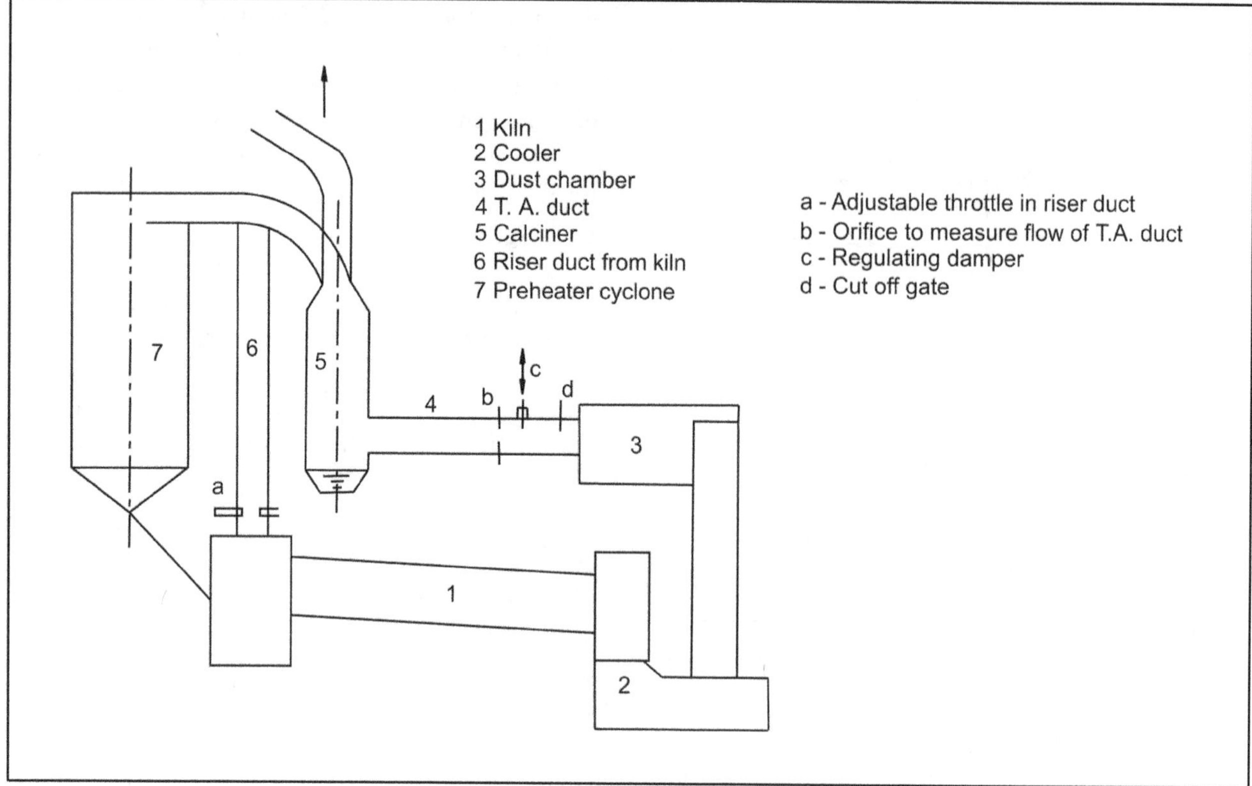

Fig. 6.43.31 Dampers and throttles in T.A. duct and riser duct.

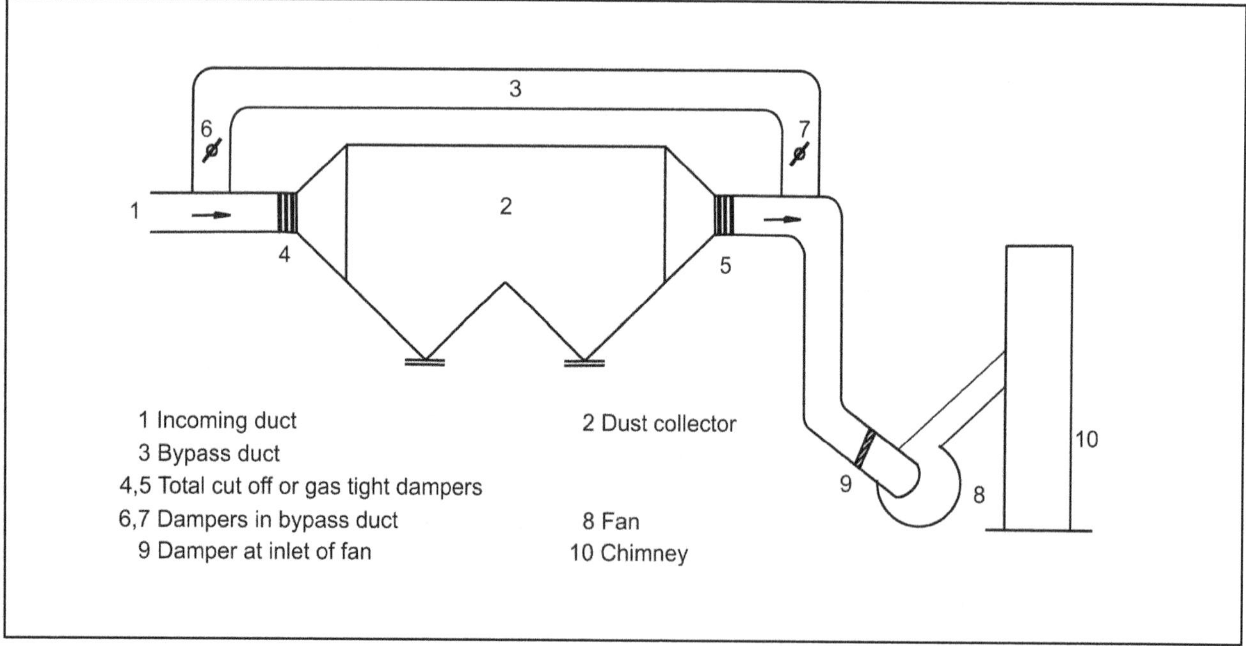

Fig. 6.43.32 Cut off dampers to isolate dust collector.

43.14 Fans in the System

As mentioned earlier fans are used to draw or push gases in a system of ductings. Static pressure of a fan is a sum total of pressure losses in the system. It is calculated systematically as explained in **Annexes 6.43.1 and 6.43.2.**

A preheater fan for example has to deliver a given volume so drawn to say:

1. Raw mill system for drying,
2. coal mill system for drying,
3. bag filter/gct and esp for cleaning.

The total static pressure for which the fan to be designed is the sum total of loss of pressure on suction side and on delivery side.

Suction side loss is say 400 mm

Delivery side loss is say 25 mm

Fan will be designed for a static pressure of 425 mm.

So that a negative pressure of 400 mm is converted into a positive of 25 mm at delivery point which becomes zero at a predetermined point in the circuit so that g.c.t. is under negative pressure.

See Fig. 6.43.33.

According to fan laws, fan would deliver a total pressure of 425 mm, but if system resistance is only 300 mm wg then on delivery side the fan will show a positive pressure of 125 mm and the zero point would then be shifted much further into subsequent system that is in gct and esp which is not at all desirable.

Fans should therefore be selected properly – neither oversized nor undersized.

As mentioned above it is better to achieve regulation by a variable speed motor.

Sometimes, a fan is avoided under the impression that it would save same power. But this is not necessarily so.

See Figs. 6.43.34 and 6.43.35.

The circuit is the same in both cases 'a' & 'b'.

In case 'a' fan F1 is designed for static pressure R1 and fan F2 for static Pressure R2. In case 'b', fan F3 is designed for static pressure R1+R2.

Since power required is proportional to static pressure it is obvious that total power of Fans F1 and F2, in case of 'a' is equal to power of Fan F3 in case 'b'. One disadvantage of having two fans **(Fig. 6.43.34)** would be that fan F1 would have to handle dust laden air.

43.15 Fabrication Drawings of Ductings

Separate drawings should be made of ducting which would show **actual lengths and slopes** and not what appears in standard elevation and plan.

Drawings for ducts should be drawn to scale showing actual views. They should also show development drawings for bends, cones etc., so that these can be made easily.

Development drawings should be made when cross section is changed from round to rectangle and vice versa.

43.16 Fabrication of Ductings

Fabricating straight lengths of ducts is not a problem. Ducts are rolled on plate bending machine to required diameters using standard plate widths and plate lengths. Straight lengths will be made quantity wise, slightly more than the required lengths to allow for any deviations at site in actual construction.

See Fig. 6.43.36.

Flanges would be made in pairs, one supplied fixed other loose to be matched and welded at site. Ducts will be subject to negative or positive pressure according to the process requirements. The duct carrying preheater gases for example would be subject to –500 to –600 mm draft. This duct therefore needs to be reinforced at regular intervals by welding rings of angles around it to prevent collapse by 'implosion' which is reverse of explosion.

See Fig. 6.43.37.

Flanged joints should be minimum consistent with problems of handling / transport.

It is best to machine flanges at least on sides facing each other so that joints can be made leak proof. Flanged joints are made leak proof by using asbestos rope 6 mm thick or more according to diameter of the pipe. Rope should be arranged as shown so that holes are inside.

See Fig. 6.43.38.

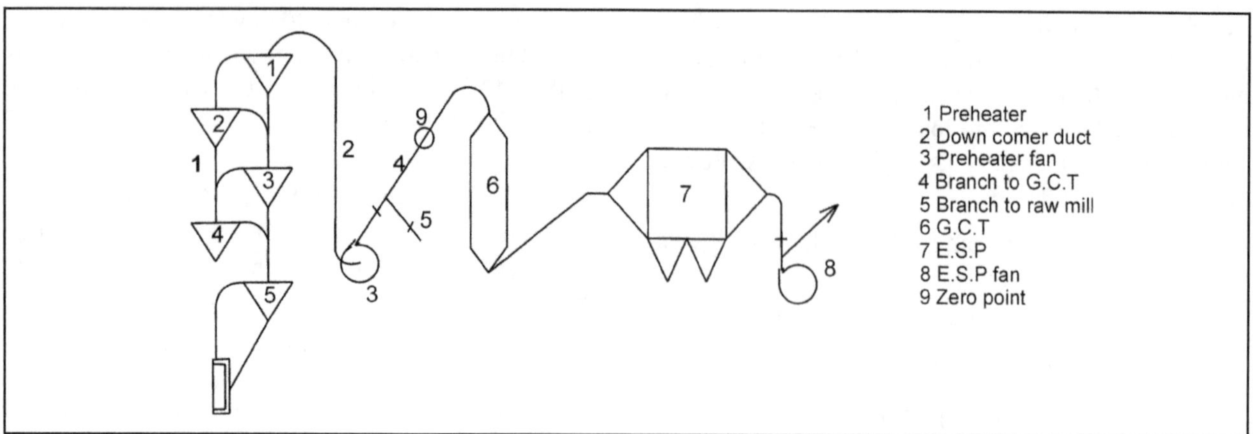

Fig. 6.43.33 Ducting in kiln and ESP circuit zero point in system between P.H fan and ESP fan.

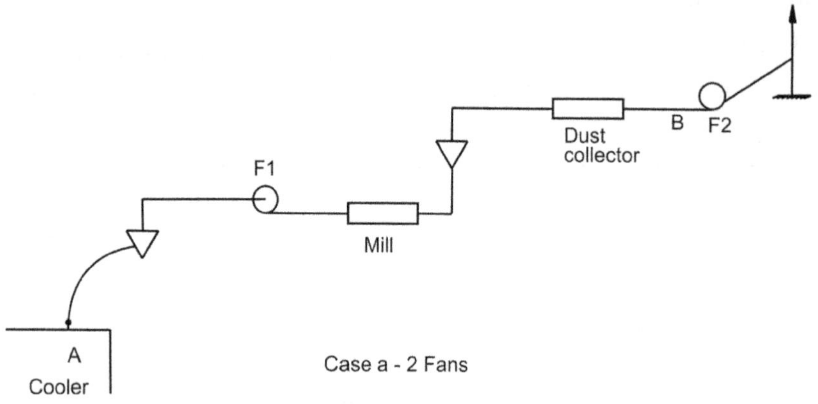

F1
Dust collector
B F2
Mill
A
Cooler
Case a - 2 Fans

Fig. 6.43.34

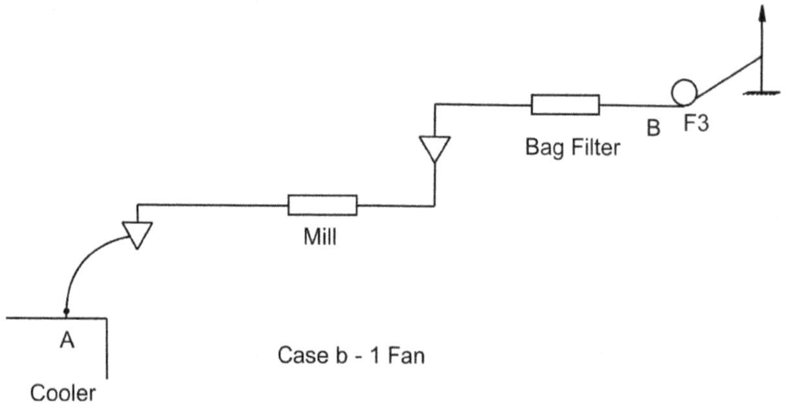

Bag Filter
B F3
Mill
A
Cooler
Case b - 1 Fan

Fig. 6.43.35

Total pressure drops in system remains same in cases a and b
static pressure of fan F3 = sum of static pressure of fans F1 and F2
Power consumption same except for leakage

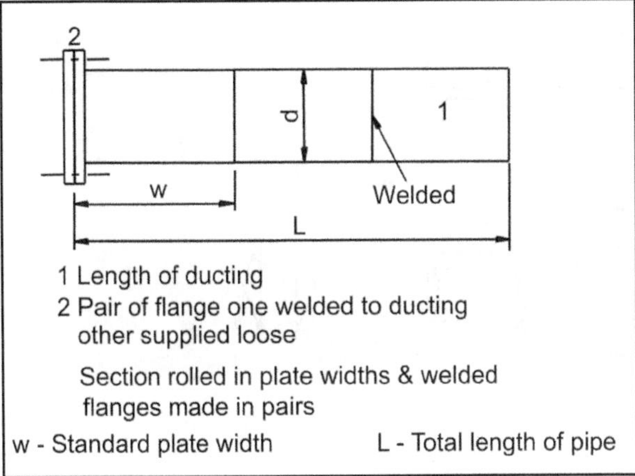

Fig. 6.43.36 Fabricated ducts and pipes.

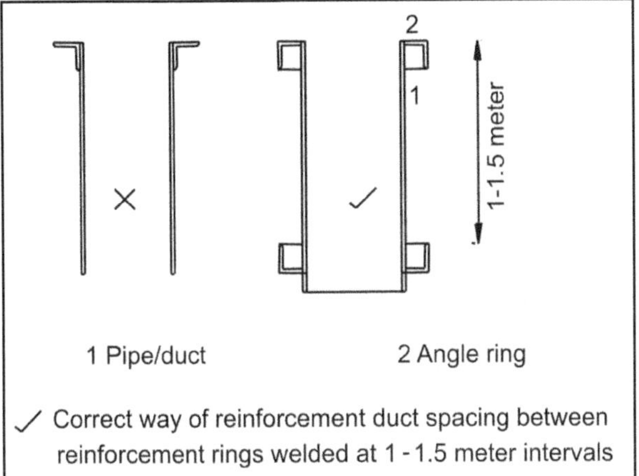

Fig. 6.43.37 Reinforcement of ducting against implosion-explosion.

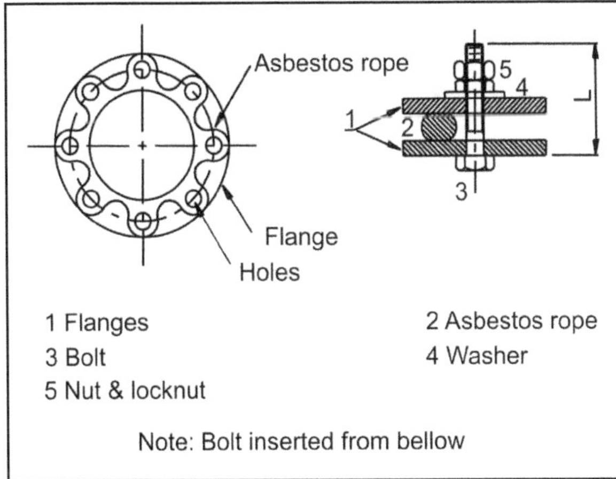

Fig. 6.43.38 Sealing at flanges to prevent in leakage with asbestos rope.

It is practical and time saving to weld one flange and supply the matching flange of the pair loose. This can be welded at site after erection or assembly. This saves considerable time. It is particularly useful when duct changes direction.

When a duct changes direction it is better to keep a small straight portion for joining flanges.

See Fig. 6.43.39.

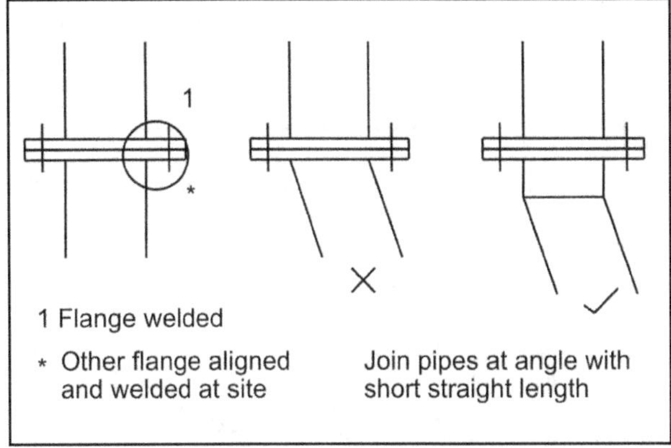

Fig. 6.43.39 Joining of ducts.

Holes should always be drilled and not gas cut. Bolt lengths should allow for flange thickness plus asbestos packing plus 2 nuts and washers.

In assembling bolts are fitted as shown with nuts at top so that if a nut drops off, bolt also drops off.

See Fig. 6.43.38.

When flanges are drilled on machine as pairs, there should not be any difficulty in holes matching. Even then there should be a practical tolerance between bolts and bolt hole as per norms laid down.

See Fig. 6.43.38.

It is difficult to change or modify a bend at site. Hence bends and cones should be made carefully.

It may be a good idea to tack weld the bends, fit up the ducting and then complete the welding.

As has been mentioned earlier, cross sections must change according to volumes to conveyed.

Round ducts are easy to make and are mechanically strong. Rectangular ducts have more welding work and need to be reinforced.

See Figs. 6.43.40 and 6.43.41.

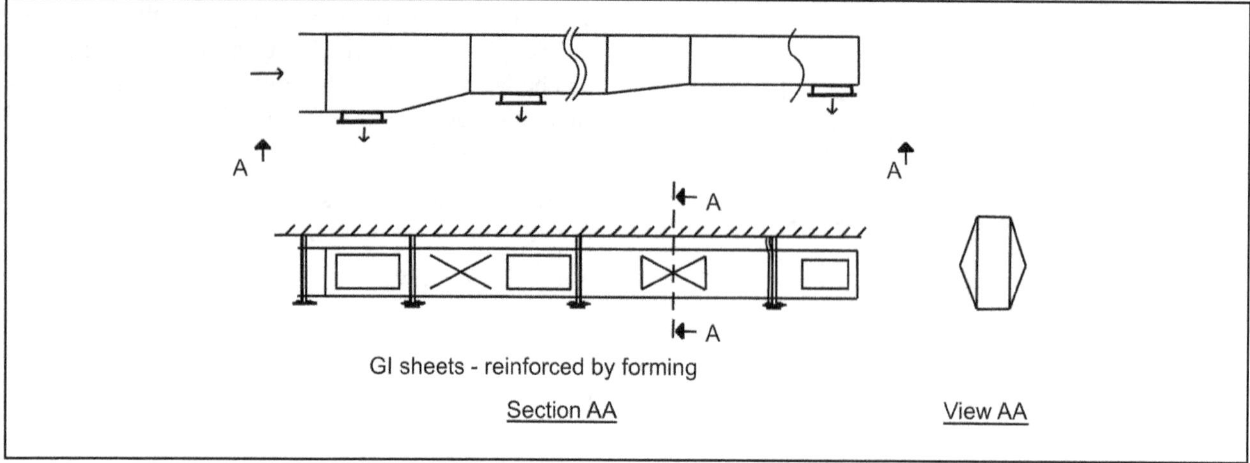

Fig. 6.43.40 Rectangular ducting - with varying cross sections - say for air conditioning.

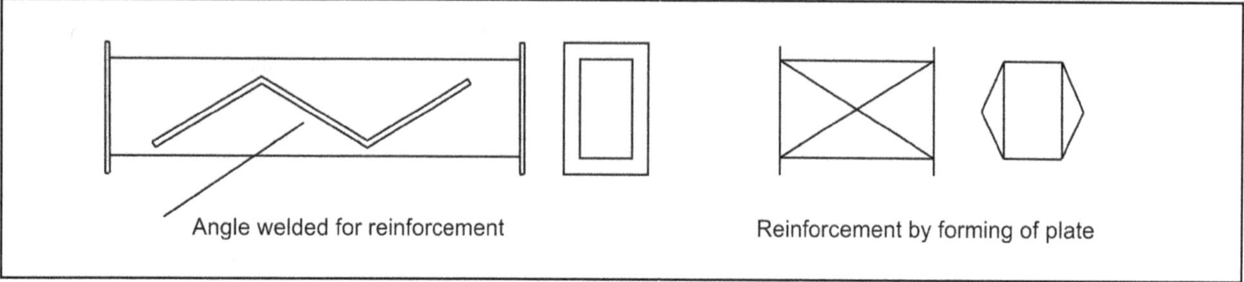

Fig. 6.43.41 Rectangular ducting reinforced.

43.17 Insulation and Lagging

Ducts carrying hot gases at temperatures up to 350 °C are made in m.s. It is necessary to prevent drop in temperature when gases are to be used for drying. Therefore, such ducts are insulated and lagged to conserve heat. When ducts / pipes are required to be insulated and lagged, to conserve heat and temperature, they need to be welded with lugs to hold insulating mats and for fixing aluminum cladding.

The details should be obtained from agencies doing insulation and should be shown in drawing of ducting though actual welding of lugs would be carried out at site.

See Fig. 6.43.42.

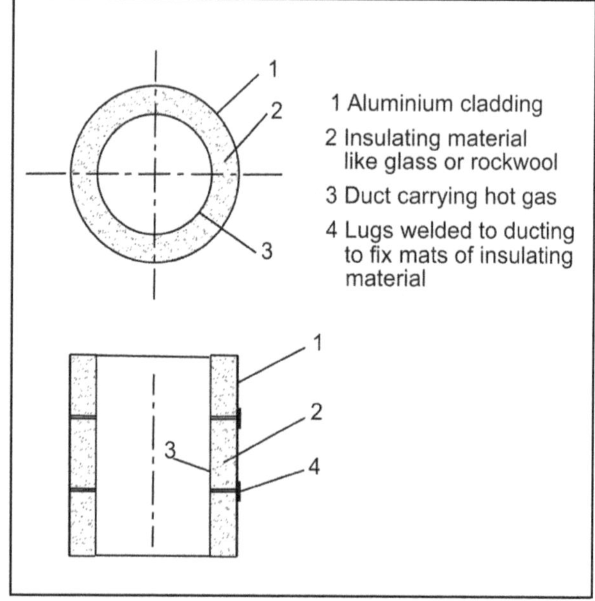

Fig. 6.43.42 Insulating and cladding of ducting.

43.18 Refractory Lining

When temperature of gases is so high that m.s. cannot stand it (> 400 °C) the ducting is lined with refractory lining. Examples are ducts between preheater cyclones,

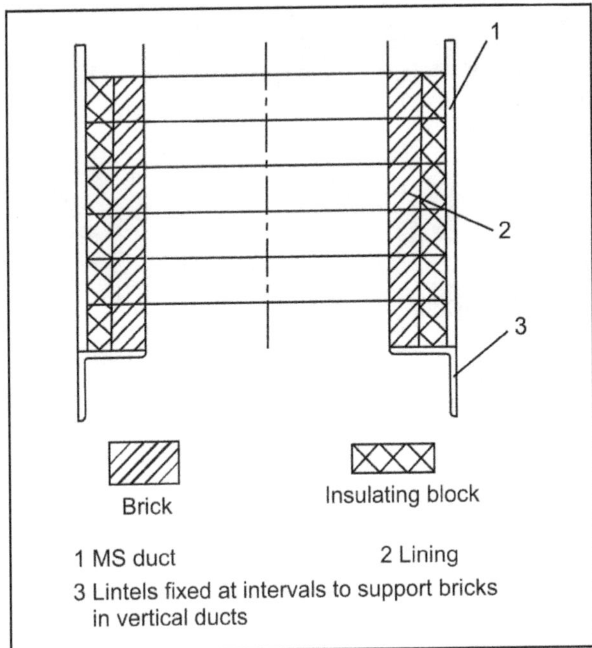

Fig. 6.43.43 Refractory lining of a duct.

hot air duct between cooler and calciner, calciner itself, which is also a kind of duct between kiln and cyclones. Kiln is also a duct in a way.

This refractory can be brick or castable or combination of both. It can be refractory with a backing of insulating bricks or blocks.

See Fig. 6.43.43.

When ducts are so lined, reference is always to clear diameter / dimensions inside of lining. Fabrication drawings of ducts should take note of this. M.S. ducts will be so much bigger in size.

Weight of lining considerably increases weight of ducting which should be taken into account in designing supports.

43.19 Explosion Flaps

In coal mill circuit, there is a possibility of explosion. In these ducts provision will be kept for 'explosion flaps'; should pressure build up, flaps will open to release pressure and protect the duct.

See Fig. 6.53.18 in Chapter 53 of this Section on Safety.

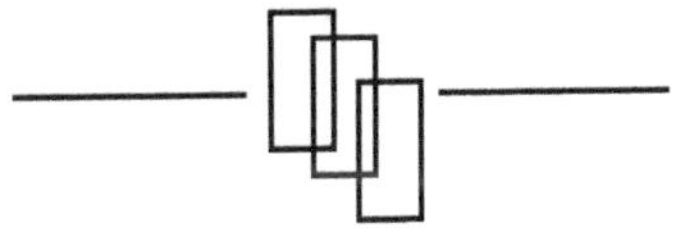

Annexe 6.43.1

Calculating pressure loss in ductings
Calculating Velocity Pressure

1. Definitions

   Velocity head = $V^2/2g$ = V_h

   where

   V = Velocity of gas in a duct in m/ sec.

   g = gravitational acceleration = 9.82 m/sec^2

2. If Q = Volume of gas flowing in metres per second, a duct of diameter 'D' metres,

   A area of duct in sq. ms = 0.785 × D^2

   V Velocity = Q/A metres per second

3. Velocity Pressure = V_p = Velocity head × density in mmwg.

   = $V_h \times d$

   where d = density of gas /air in kg / m^3

4. It is therefore necessary to find out actual density of gas flowing through the duct.

   Density d is inversely proportional to temp. and barometric pressure

   Density d at temp. t °C = d_n × 273 / (273 + t); d_n = density at normal temp. and pressure. (NTP = 0 °C and barometric pressure at sea level.)

   Density d at 500 m altitude = $d_n \times p_a$ at 500 m altitude in mmhg or wg / 760 hg or 10335 mmwg

   Barometric pressure at sea level is 760 mmhg or 10335 mmwg

5. When both temp. and actual pressure (draft) differ from normal conditions, actual density is obtained by multiplying d_n by above mentioned factors on account of temp. and pressure. Thus for temp. t and barometric pr. p_a,

   $d = d_n \times (273/[273 + t]) \times p_a/10335$ when p_a is in mms.

   barometrc pressure $p_a = 10335 \times (1\text{-}0.0000226H)^{5.255}$ in mmwg where H = altitude of place in metres.

6. When system is under suction, then actual pressure is less than atmospheric

   If suction or draft is 400 mm, and barometric pressure is 9900 mmwg, then actual pressure for applying correction factor = (9900-400)=9500 and correction factor is 9500/10335 = 0.92

   If volume through the system is expressed in normal m^3 then it would have to be converted into actual volume by applying similar correction factors for temp. and pressure (actually reciprocals of above factors).

   Q_n = volume in nm^3 / sec at 300 °C and draft is 400 mm and barometric pressure is 9900 mm wg, then correcting factors together would be

   Q = Volume under actual conditions

   $= Q_n \times \{(273 + 300)/273\} \times \{10335/(9900\text{-}400)\}$

   $= Q_n \times 2.098 \times 1.087 = Q_n \times 2.28$

7. If gas flowing is air, its density under ntp conditions is 1.29 kg/m^3. Under actual conditions it would be 1.29 × 1/2.28 = 0.565

8. If velocity through duct is 18 m/sec,

   velocity head $= 18^2/19.63 = 16.51$

   and velocity pressure in mm wg $= 16.51 \times 0.565 = 9.328$

9. For speeding up calculations of Velocity pressures, following graphs are attached which help direct computation

   Figs. 1 & 2 barometric pressures at different altitudes

   Figs. 3 & 4 density correction factors for altitude

   Fig. 5 density of air at different temperatures

   Figs. 7 & 8 velocity pressure for different velocities

   when density is 1 kg/m^3

   For other gases their densities under NTP conditions would have to be taken and

   Expressed in terms of density of air under NTP viz 1.29

   For example if density of exhaust gas is 1.34 kg/nm^3 then to work out its density at different temparatures a multiplier would have to be applied to corresponding densities of air.

   $1.34/1.29 = 1.038$. Density at sea level

   and at 100 °C would be $0.945 \times 1.038 = 0.981$

10. To read **Figs. 7. & 8** actual density would have to be used as these graphs are for a density of 1kg/m^3.

11. **Fig. 8** shows V_p for a velocity of 20 m/sec as 20.4. If density is 0.565,

    then velocity pressure would be $20.4 \times 0.565 = 11.53$ mmwg.

12. Calculation of pressure loss once Velocity pressure has been worked out has been shown in **Anexe 6.43.2.**

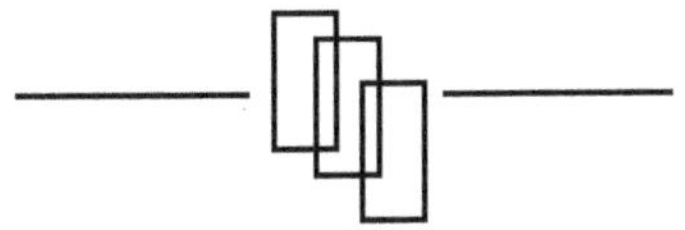

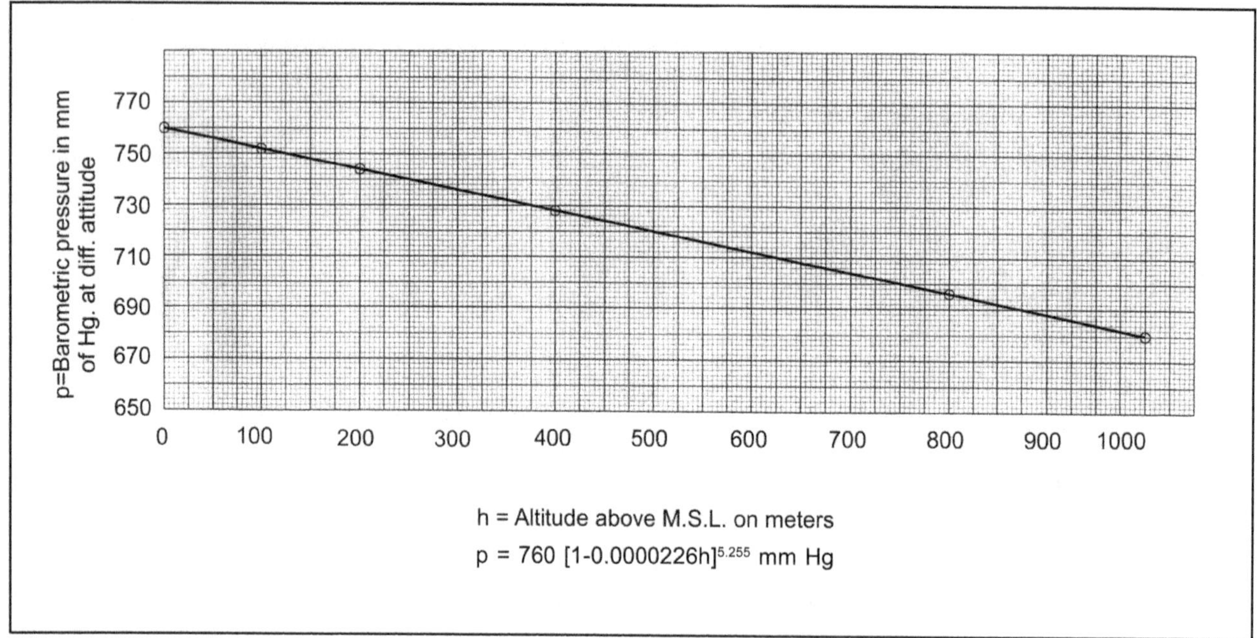

Fig. 1

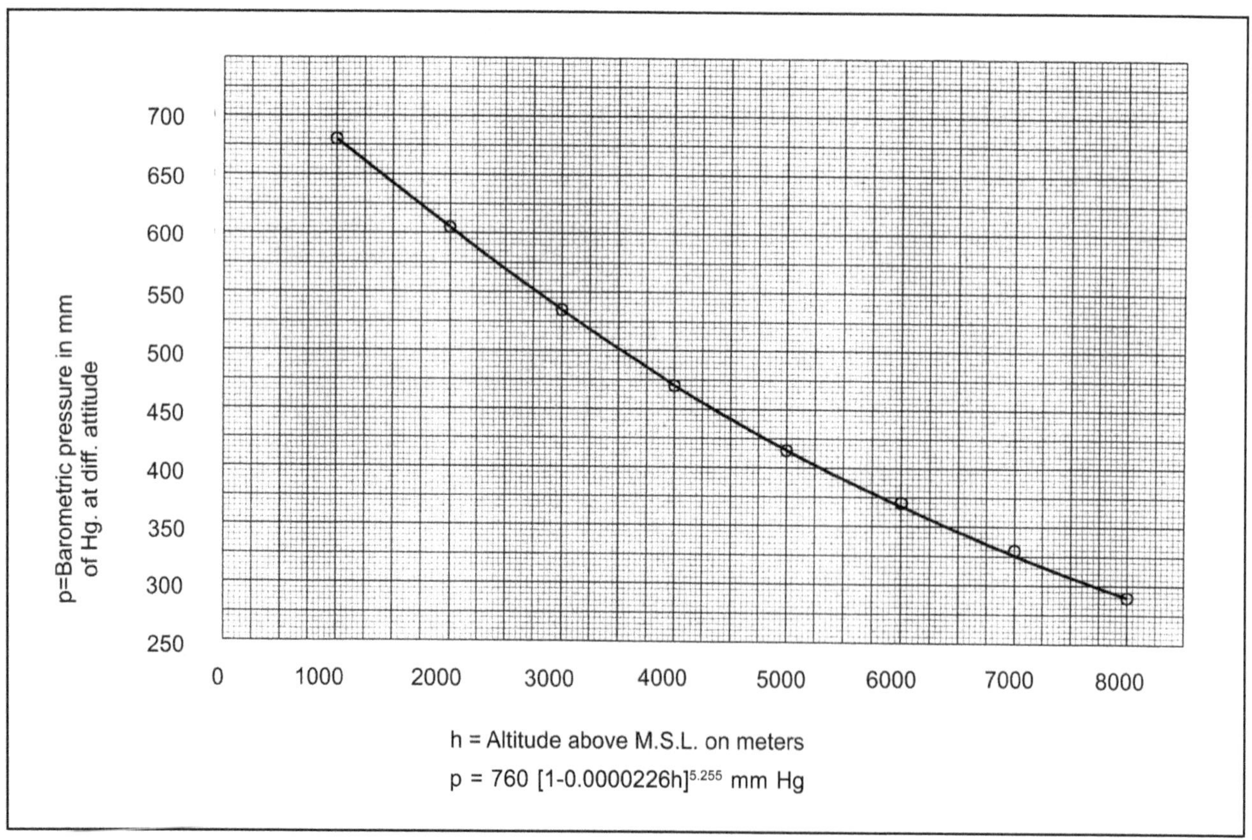

Fig. 2

Annexe 6.43.1

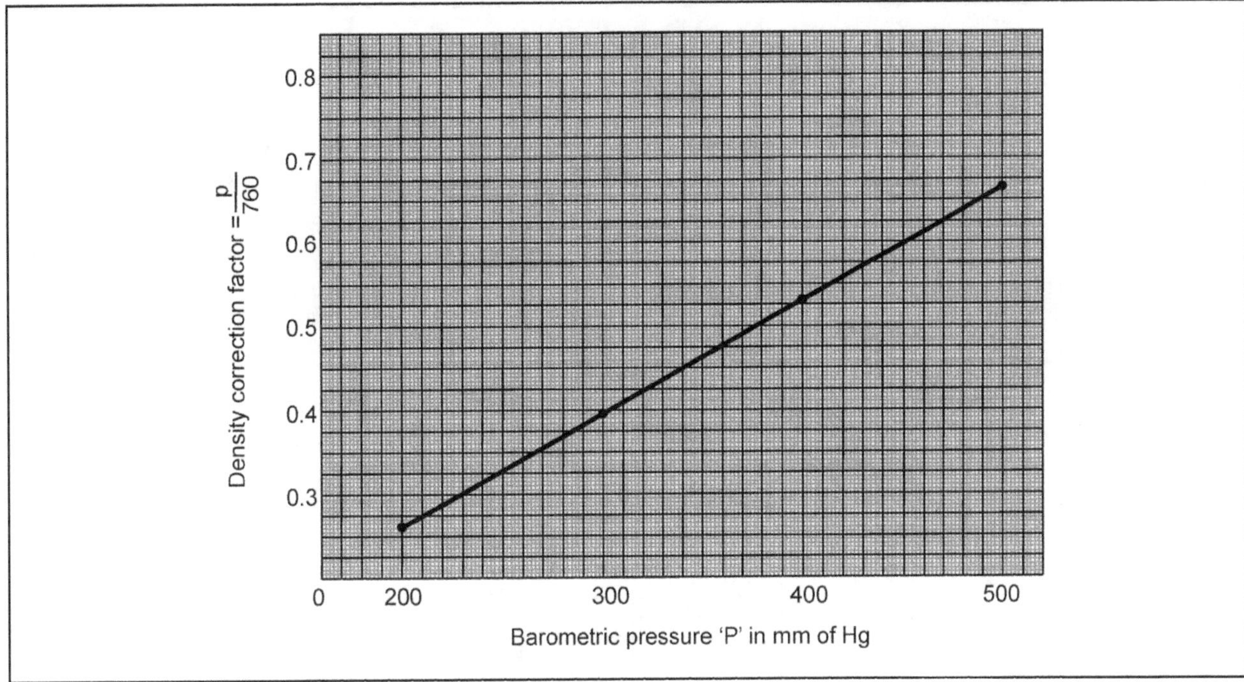

Fig. 3

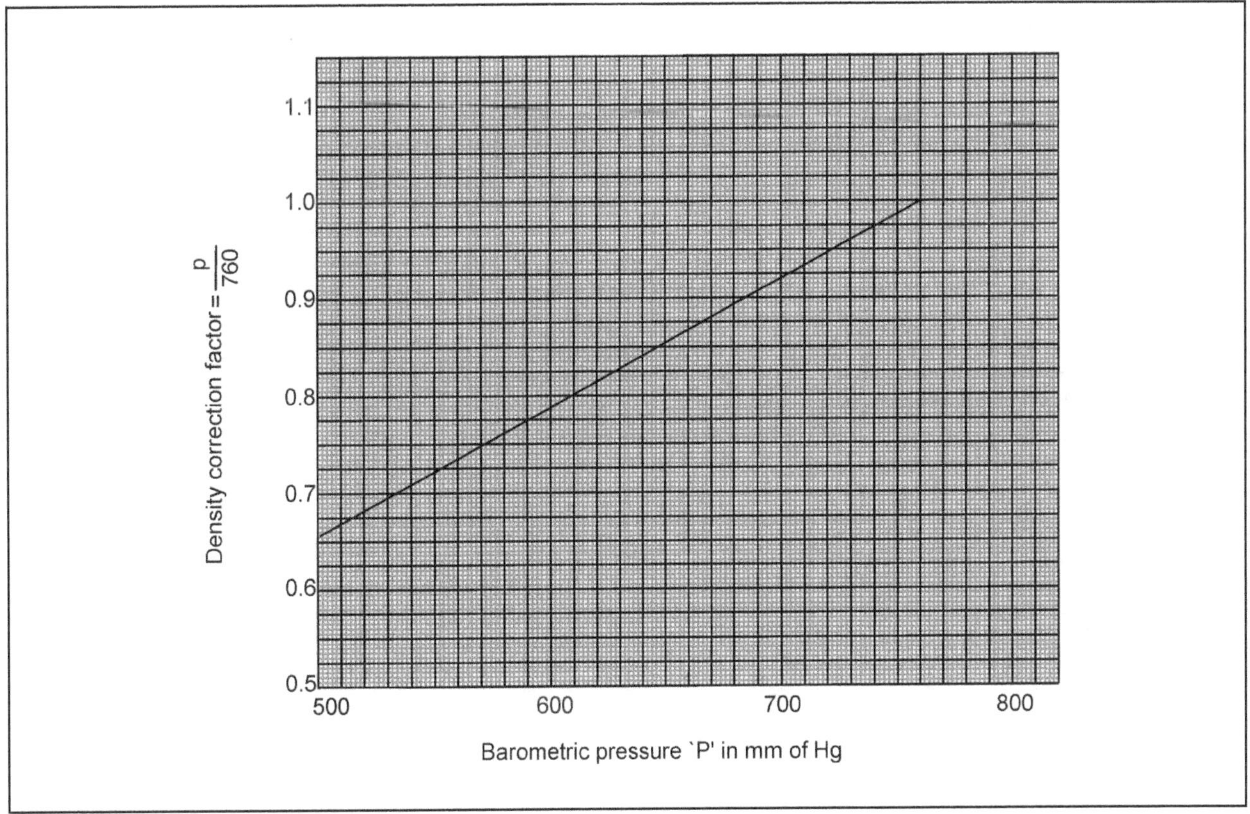

Fig. 4

Annexe 6.43.1

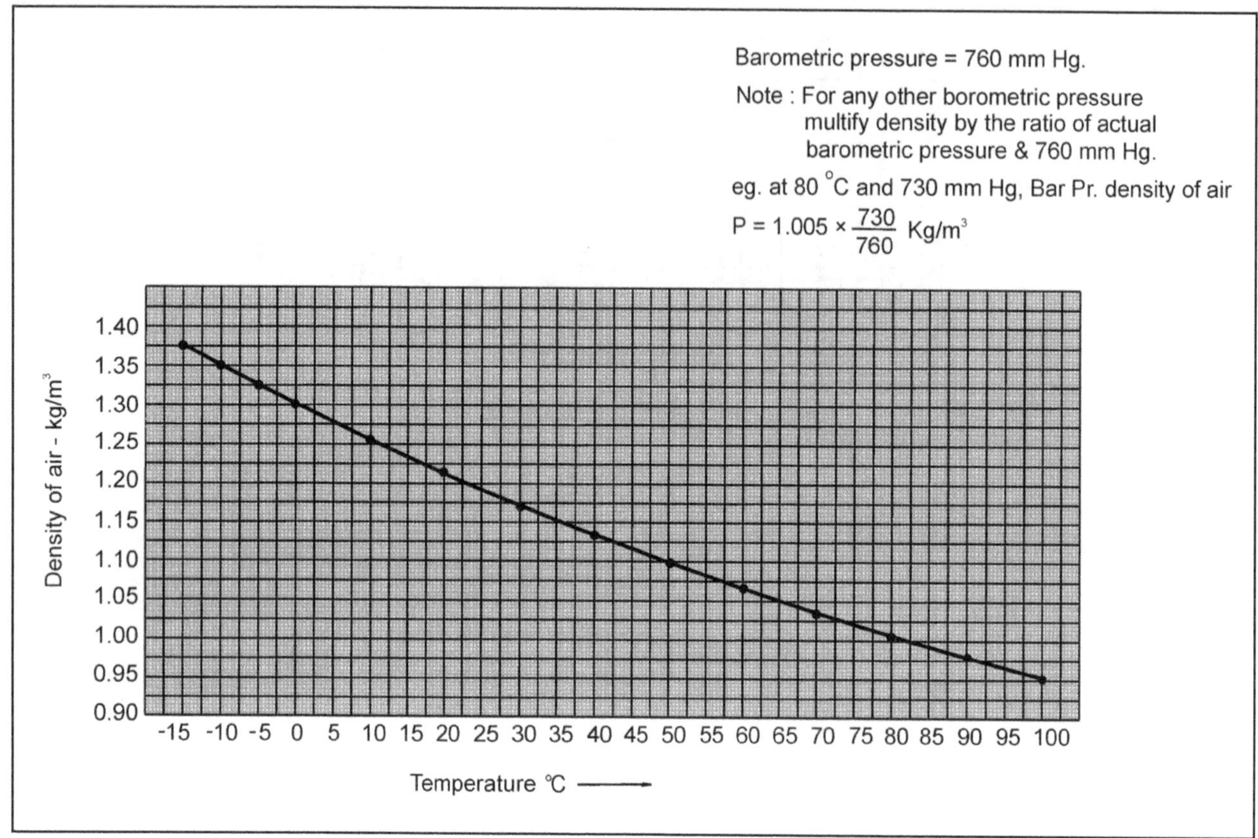

Fig. 5

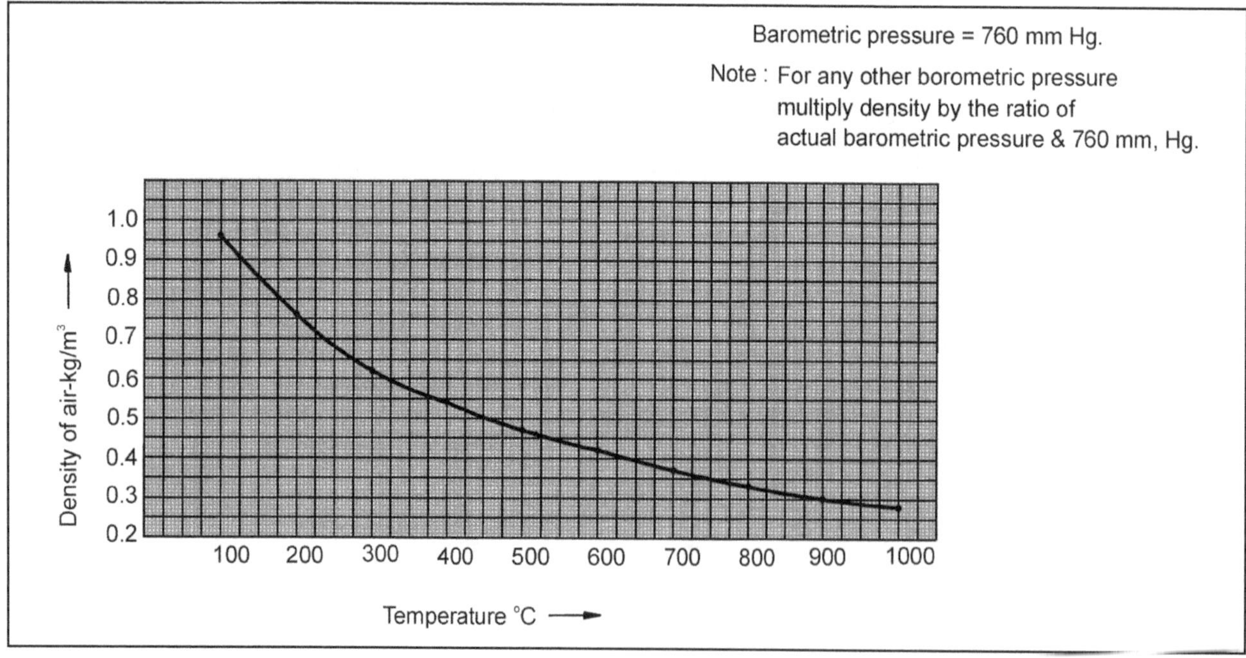

Fig. 6

Annexe 6.43.1

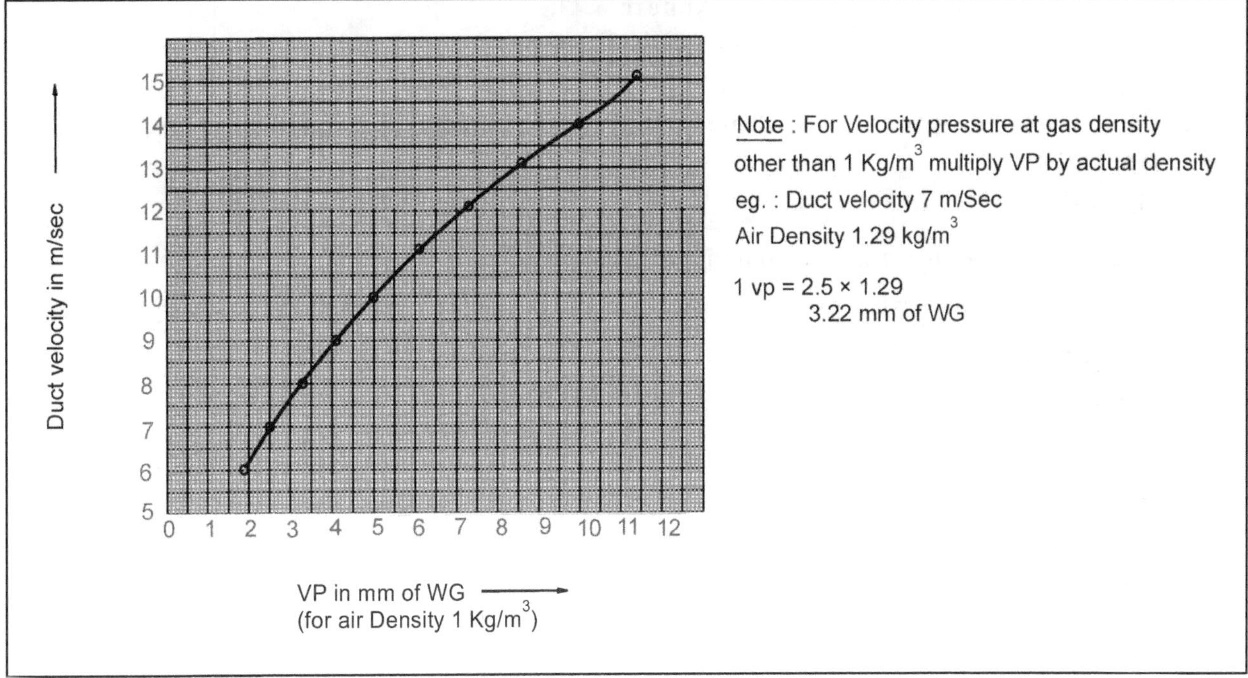

Fig. 7

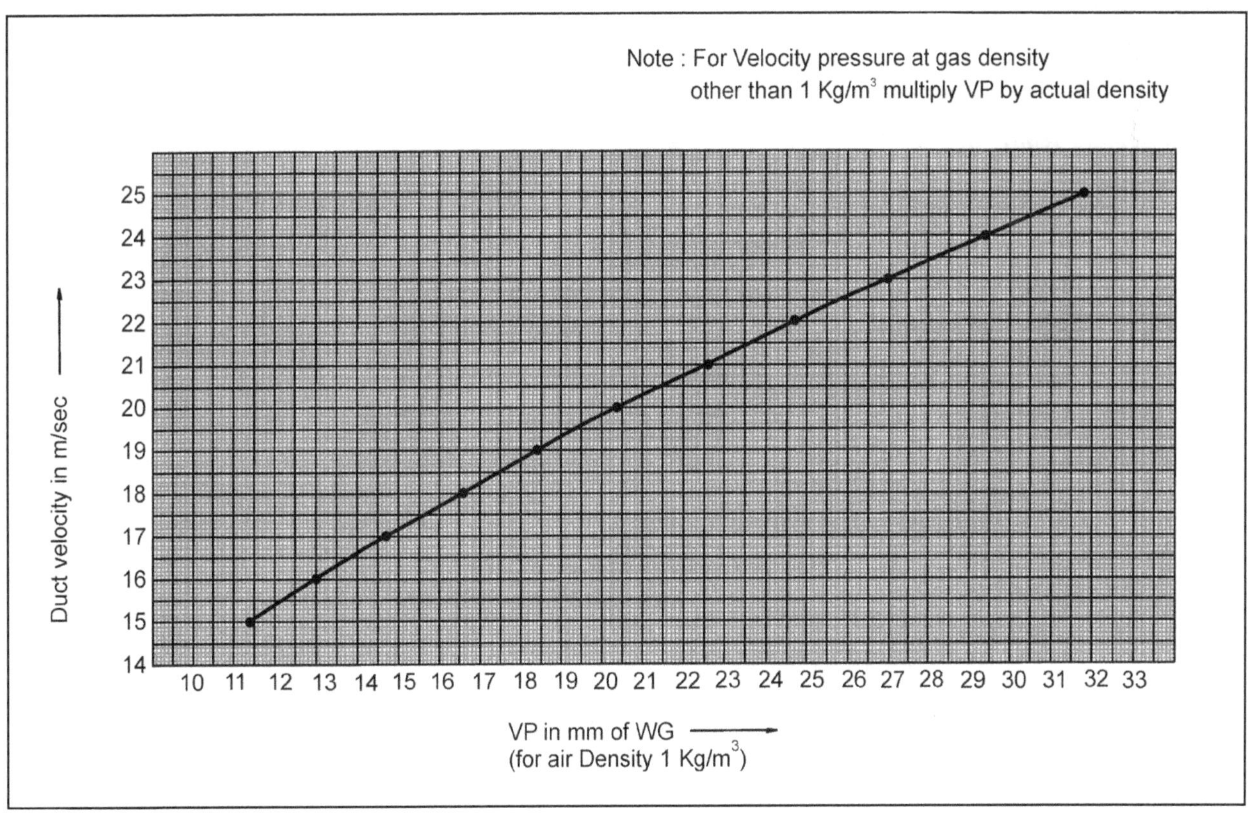

Fig. 8

Annexe 6.43.1

Annexe 6.43.2
Losses in Ductings

1. There are three types of losses when gases (also dust laden) flow through a system of ducts and pipes.

They are:

1 Lift loss – when material is lifted from one level to another.

2 Acceleration loss – when material is introduced in a ducting and attains velocity of carrying gases.

3 Friction loss due to flow of gases (also dust laden) in ducts.

2. Lift loss

Let vertical lift = L metres.

W_m = Weight of material being lifted

W_a = Weight of air (gas) lifting the material

K = a constant = density of water/ density of air under operating conditions.

Density of water =1

Density of air at NTP = 1.29 kg/m^3

$\therefore$ K at NTP = 1/1.29 = 0.0.775

Lift loss in mmwg = $(W_m/W_a) \times L / K$

Let $W_m/W_a = R$

Then, lift loss = $R \times L / K$

3. Acceleration loss

This happens when material enters a pipe like in case of a coal firing pipe. Material is to be accelerated to the velocity of air. In horizontal runs, velocities of material and air would finally be the same.

By law of momentum,

m1v1+m2v2 = MV

m1 and m2 are masses of material and air per unit time.

v1 and v2 are their respective initial velocities.

M = m1+m2 and

V = final velocity of air and material

g = acceleration due to gravity.

W_m = wt. of material = m1 × g

W_a = wt. of air = m2 × g

Energy required to accelerate material = $W_m \times V^2/2g$

Energy lost per unit weight of air = $(W_m/W_a) \times V^2/2g$

This acceleration loss expressed in mmwg = $R \times V^2 / 2g = R \times V_{PA}$

where $R = W_m/W_a$ and V_{PA} = Velocity pressure of air

4. Friction loss

Pressure loss in straight duct (round) expressed in mms. wg = $13.866 \times V^2 \times L \times f / (d \times v \times T)$

where,

V= velocity in duct in m / sec.

f = coeff. of friction for steel which can be taken roughly as 0.018

L = length of duct in metres including developed bends

d = diameter of duct in metres

v = sp. volume of air in m^3/kg, corrected for altitude and temperatures.

T = absolute temp. of air in Kelvin

For rectangular ducts diameter of equivalent round duct is found

If a and b are two sides of a rectangular duct, then dia d of equivalent round duct = $2a \times b/ (a + b)$

Besides losses in straight portions losses also occur in bends which very much depend on type of bend, construction of bend, ratio of radius of bend to diameter of duct. This will be dealt with in **annex 3**

Examples

*1 **Lift loss***

material carried 50 tons per hour

raw meal, bulk density

lift 60 metres

air / material ratio 50: 1

quantity of air : 3000 m^3 /hr

velocity in conveying pipe ≃ 18 m /sec.

dia meter of pipe ≃ 250 mm

altitude : sea level

temperature of conveying air 70 °C

Weight of air W_a = 3078 kg/hr

Weight of material = W_m = 50000 kg/hr

R = 50000/3078 =16.24

Density of air at 70 °C and sea level = 1.026 kg/m^3

K = 1/1.026 = 0.974

Therefore lift loss in mm wg = 16.24 × 60/ 0.974 = 1000 mmwg.

*2 **Acceleration loss***

material coal ; calorific value 4300 kcal/kg.

rate of feed 5 tph = 5000 kg /hr

air for combustion for this coal including 10 % excess air = 5.32 nm^3/kg

primary air at 15 % = 0.8 nm^3 kg coal = 4000 nm^3 /hr

let temp.of primary air be 35 °C and altitude at site sea level

density of air at this temp. = 1.14 kg/m^3

weight of air /hr = 5160 kg / hr.

volume of this air = 4512 m^3/hr.

dia. of coal firing pipe from p.a. fan = 300 mm ; velocity of air in it =16 m/sec.

Let dia, of pipe where coal is introduced in ventury = 150 mm

Velocity of air in this section = 71 m/sec.

Let initial vel. of coal = 0.

Applying principle of momentum,

5000 × 0 + 5160 × 71 = 10160 × vr

therefore vr = 36 m/sec.

W_m/W_a = 5000/5160 = 0.97

Velocity head at 36 m/sec = 66

Density of coal dust laden air at 35 °C = (5000+5160)/(4512+ 62) = 2.22

Velocity pressure = 66 × 2.22 =146.54

Pressure loss = (W_m/W_a) × 16.54 = 0.97 × 146.54 = 142 mm

3 ***Loss due to friction***

length of pipe = 100 m in terms of straight equivalent length

volume of air flowing through it = 300 m^3/min.

let velocity through pipe = 20 m/ sec.

diameter of pipe = 0.564 m.

Temp of air. = 100 °C. ; altitude 300 m

Sp. Volume of air at 100 °C and altitude = 1.097 m^3 /kg

Coeff. of friction for this velocity and diameter of pipe = 0.018

Putting the values in the equation for friction loss

Loss in mms of water = 13.86 × $(20)^2$ × 100 × 0.018 / (0.564 × 1.097 × 373) = 43 mms

friction loss could also be obtained using a chart or nomogram. **See chart attached**

See Fig. 6.43 Anexe 2.1 for metric units and 6.43 Anexe 2.2 for ftlb units and 6.43 Anexe 4

Velocity head at 20 m/sec = 20.38

Density at 100 °C and altitude 300 m = 0.912 kg/m^3

∴ velocity pressure = 18.59

Chart furnishes multiplying factor for these flow conditions = 0.85

Pressure loss in a pipe of 100 m length = 18.59 × 0.85 × 3.28 = 51.85 mms

As compared to the pressure drop of 43 mms calculated from formula above.

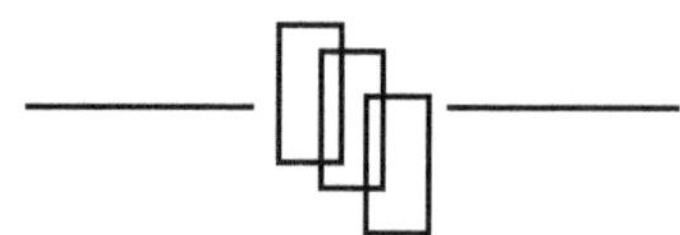

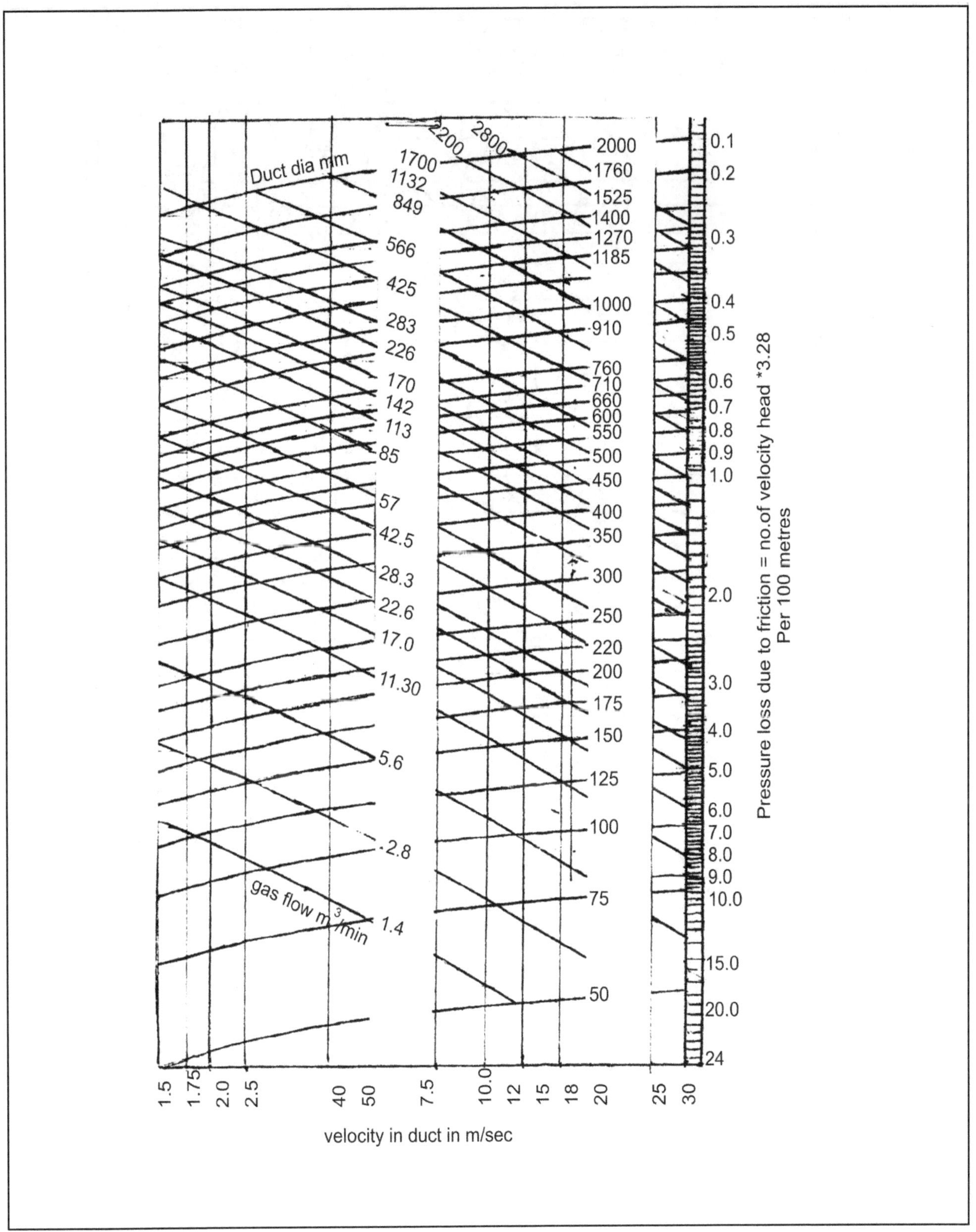

Fig. 6.43 Anex 2.1 Nomogram for calculating pressure drop in straight ducts.

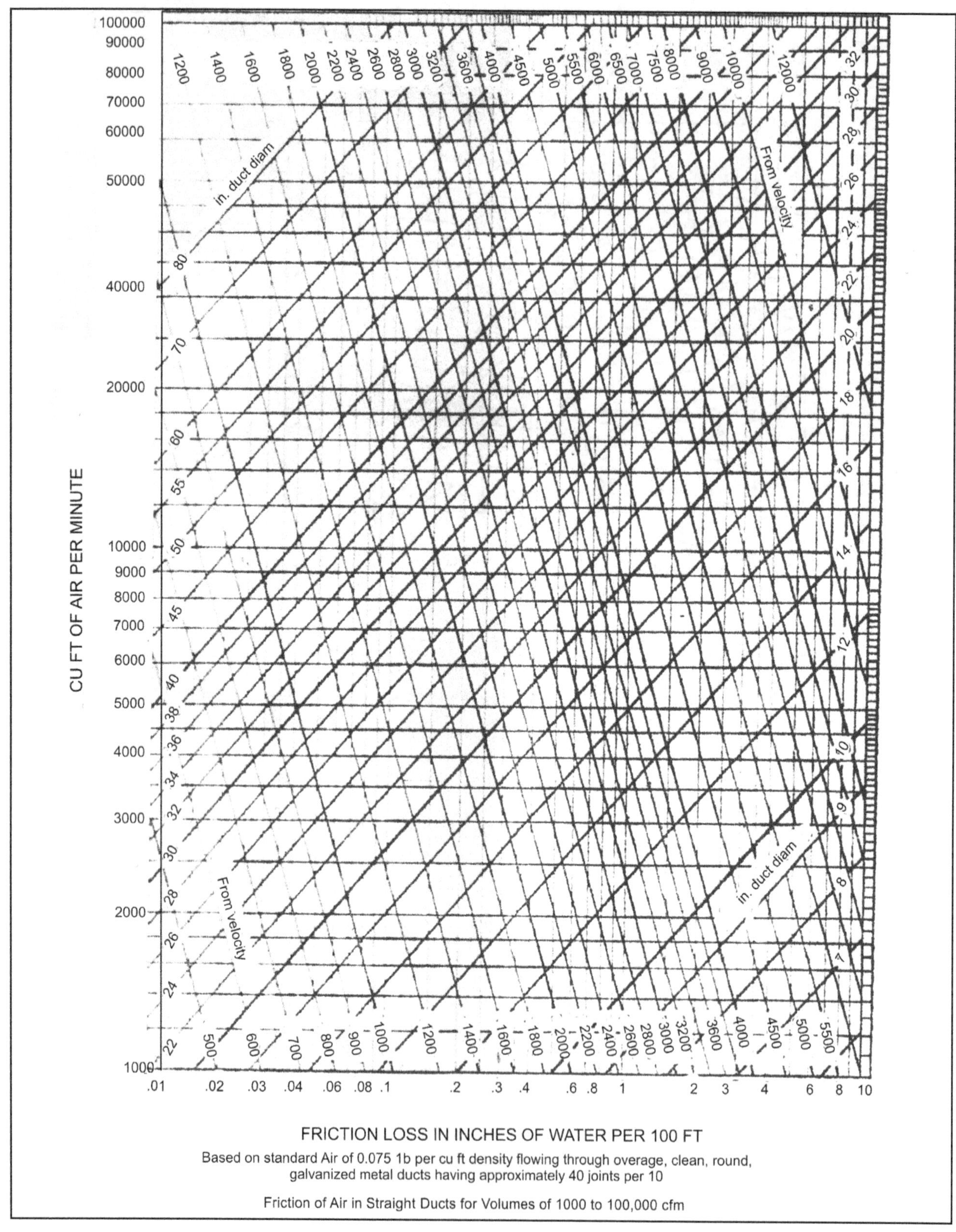

Fig. 6.43 Anex 2.2 Nomogram to work out pressure drops in ducts (ft lb. units).

Annexe 6.43.3

'K' F actors to calculate Pressure loss in Ducts and Bends of different configurations

1. A system of ducts often consists of bends, branches and changes of Cross sections-

   sudden or gradual which cause formation of 'eddies' that result in loss of pressure more than what would have been in a straight duct of uniform cross section.

2. A convenient way to approximate the pressure loss is to use multiplying factors to the velocity pressure. Such multipliers are known as 'k' factors.

   In **Figs. 43 Anex 3.1 to 3.23** 'k' factors for various types of bends and for changes in cross sections etc., have been shown.

3. Following example illustrates how to use the k factors.

   A round duct is carrying air with a velocity pressure of 30 mmwg.

   The system has a bend as shown in **Fig. 6.43 Anex 3.9**

   Let angle of bend be 30°

   Pressure drop in the bend will be K × VP

   K × 30 mmwg

   For 30° bend of circular cross section, K = 0.218

   ∴ pr. Loss in bend = 6.54 mm.

4. This procedure is used to work out total pr. loss in the main duct and branches. The maximum pressure loss thus arrived at is used to arrive at static pressure of the fan in the system.

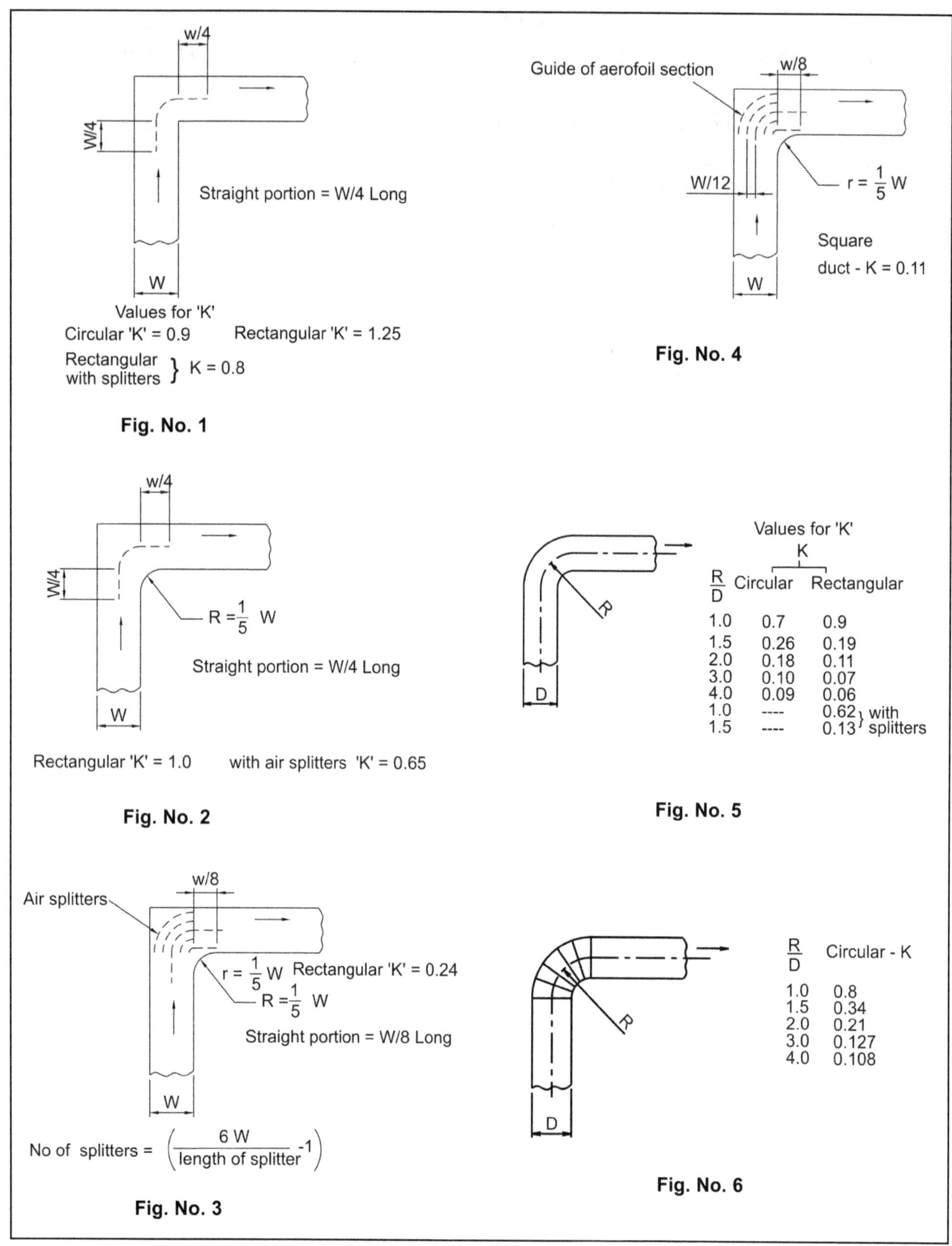

Fig. No. 1

Fig. No. 2

Fig. No. 3

Fig. No. 4

Fig. No. 5

Fig. No. 6

Annexe 6.43.3 'K' factor for various bends, reducers and junctions.

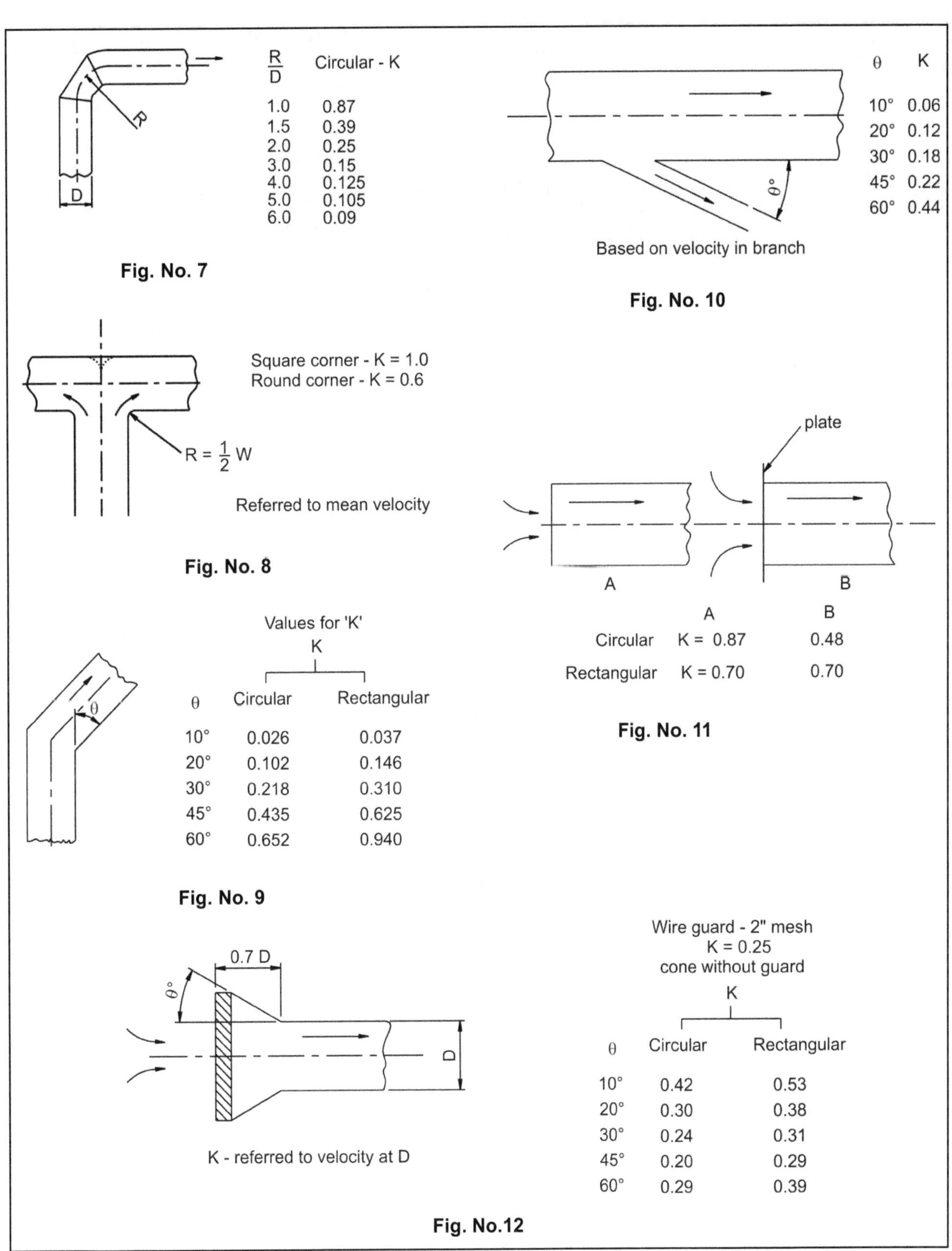

Annexe 6.43.3 'K' factor for various bends, reducers and junctions.

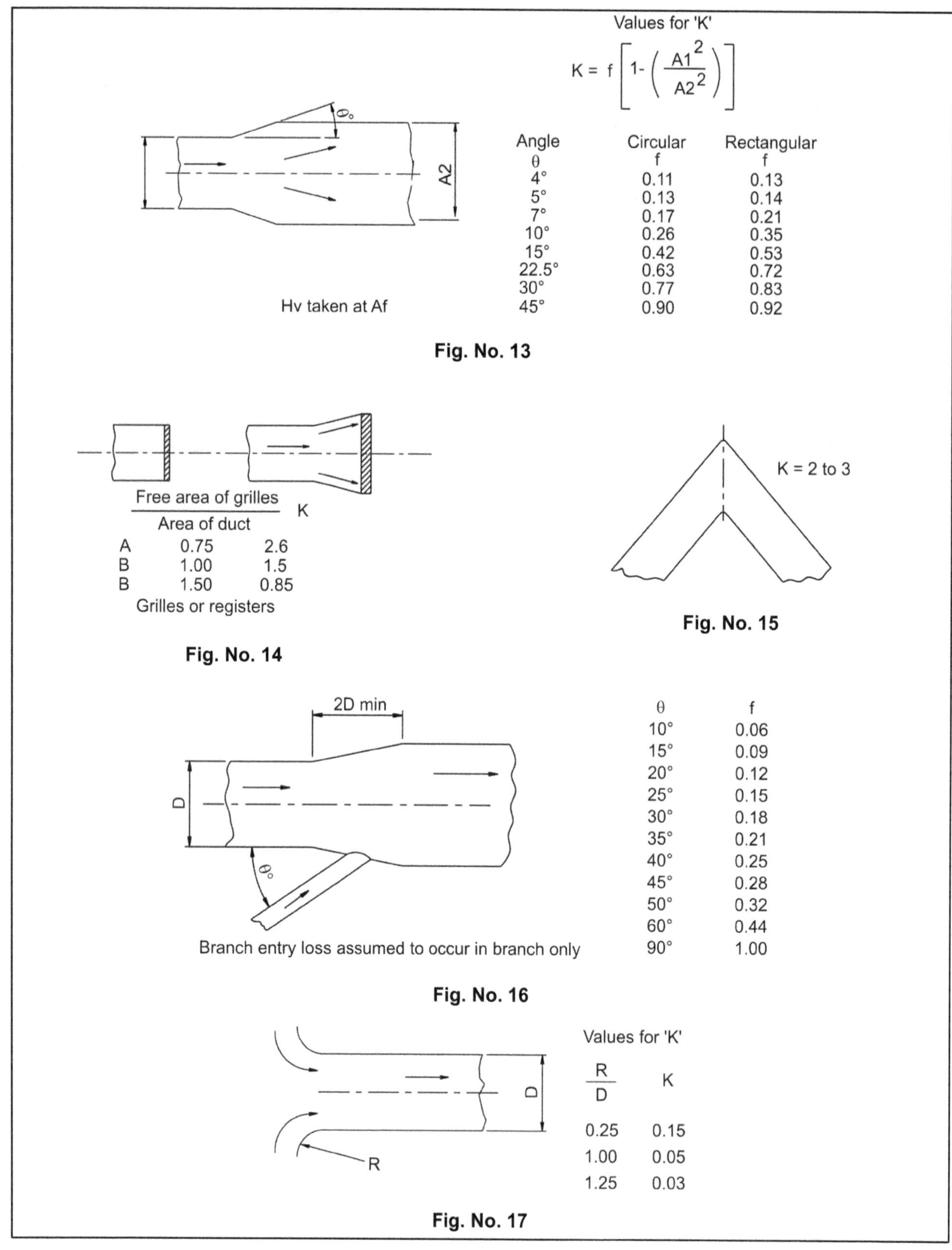

Annexe 6.43.3 'K' factor for various bends, reducers and junctions.

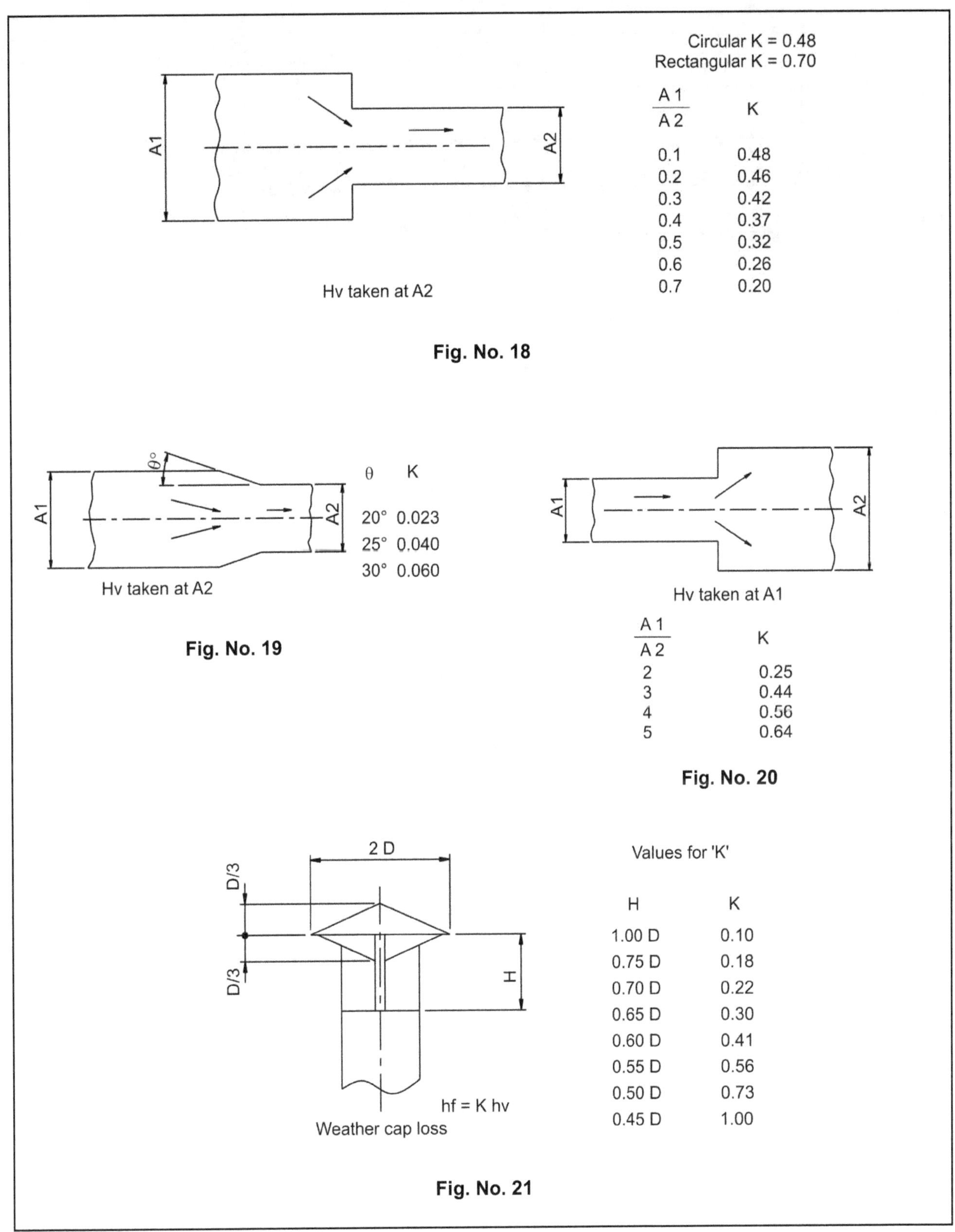

Annexe 6.43.3 'K' factor for various bends, reducers and junctions.

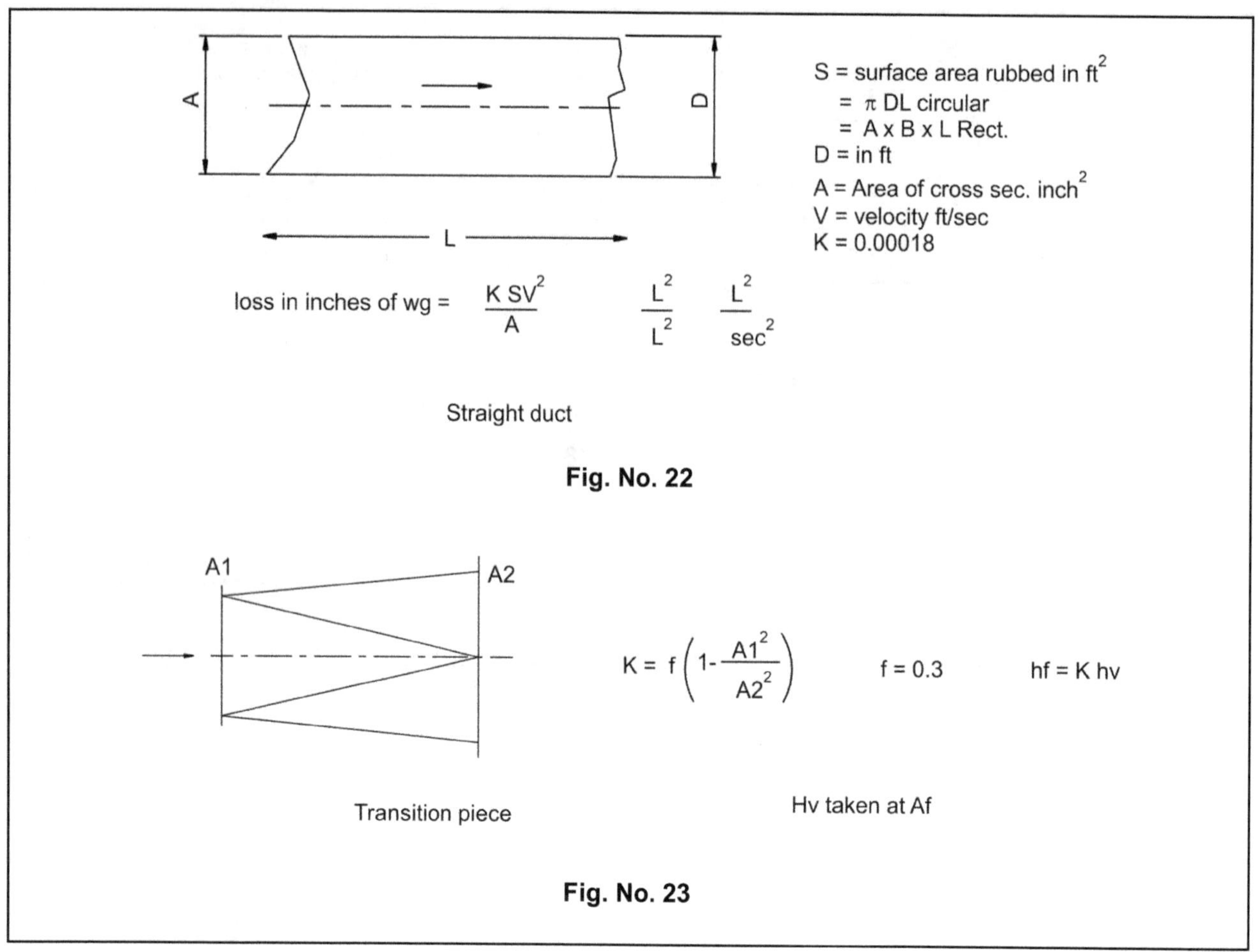

Annexe 6.43.3 'K' factor for various bends, reducers and junctions.

Annexe 6.43.4

Chart for multiplying factor to calculate Pressure loss in a duct due to friction

Sr. No.	Velocity m/sec	Gas flow m^3 /min								
		20	40	60	80	100	400	1000	2000	3000
1	10	10.168	6.89	5.904		4.264	1.673	0.954	0.62	0.46
2	15	12.464	8.528	6.89		4.92	2.034	1.181	0.754	0.56
3	18	14.104	9.184	7.144		5.576	2.23	1.246	0.813	0.59
4	20	14.43	9.512	8.036		5.904	2.296	1.312	0.886	0.64
5	25	16.4	10.824	8.856		6.232	2.624	1.509	1.017	0.722
6	30	18.7	12.136	9.512		6.89	2.257	1.64	1.115	0.787

Pr. loss in mmwg thus arrived at is for a length of 100 metres. Actual length of ducting should be substituted for pressure drop in it.

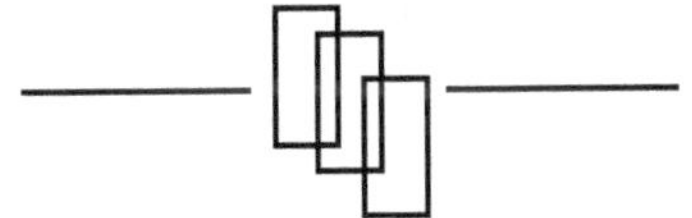

CHAPTER 44

CHIMNEYS

44.1 Chimneys

Final exit of exhaust gases to atmosphere is through chimneys.

Each dust collector, bag filter or cyclone or esp would normally have its own chimney.

One common chimney could be used to vent gases from two streams or two production lines. It should be sized accordingly.

Main function of a chimney is to let out exhaust with the dust they carry, into atmosphere.

Height of a chimney has assumed special significance from pollution control angle. Pollution Control Boards have devised a formula to calculate height of a chimney according to the dust content in the exhaust gases.

44.2 Natural Draught Chimneys

Natural draught chimneys are now rare but used to be common some decades back even on large boilers or grate coolers.

Height of chimney and temperature differential between ambient and that of gases passing through them determine the draught in the chimney that pulled gases through the system and exhausted them out. Taller the chimney higher the draught.

Natural draught chimneys used to be made in brick and were self-supporting.

44.3 Forced Draught Chimneys

Now most chimneys would be forced draught. In cement plants chimneys are made in steel. They could be made in concrete also.

44.4 Height of a Chimney

Effective height of a chimney is measured from point of entry of exhaust gases to the top.

Height H of a chimney is calculated by the following formula furnished by Pollution Control Boards for Cement Industry (in India)

$$H = 74 \times (Q)^{0.27}$$

where H = minimum height of chimney in metres
and Q = particulate emissions in tph.

Minimum height recommended is 30 metres.

Quantity of dust emitted is calculated by multiplying gas volume by dust burden in it for any specified period.

For example : In case of a 10000 tpd plant,

With a sp. gas volume of 1.5 nm^3/kg clinker, exhaust gases in direct operation of kiln section

= 625000 nm^3/hr.

Let permissible dust burden in cleaned gases be 110 mgm/nm^3.

Dust emitted per hour

$$Q = 110 \times 212500 \text{ mg/hr.}$$

$$= 0.023 \text{ t.p.h.}$$

Therefore using above formula, H = 36 metres in round figure

With bag filters and electro static filters that are commonly used for cleaning exhaust and vent gases, heights of chimneys even in case of major exhausts like kiln/raw mill and cooler, heights will work out to less than 30 metres using this formula.

Taller the chimney greater the area of dispersion of the emitted dust over the surroundings. Thus it helps to maintain clean ambient air standards.

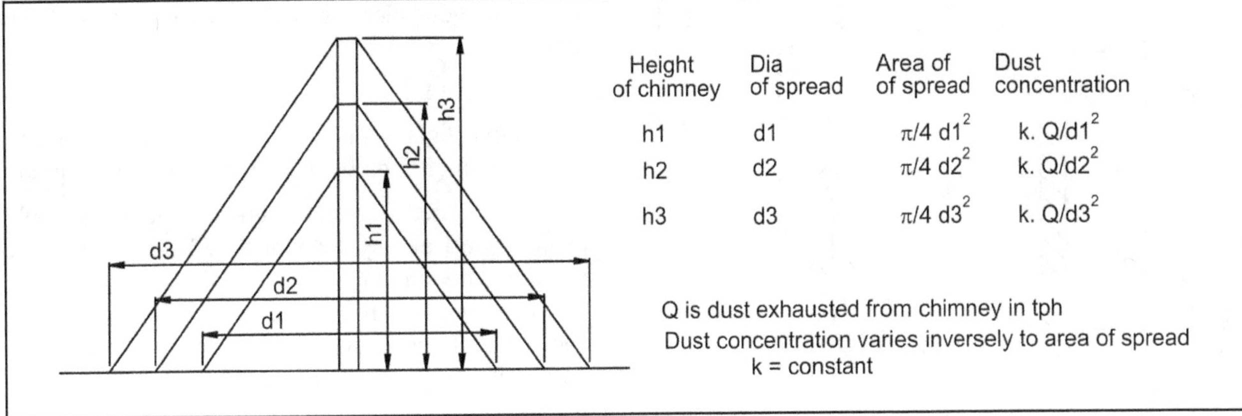

Fig. 6.44.1 Height of chimney helps disperse dust particles over wide areas.

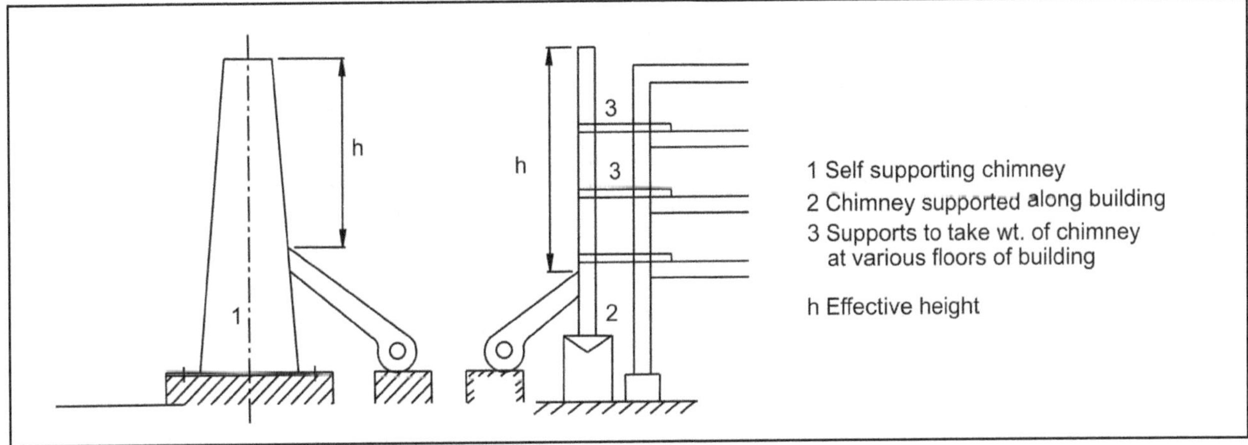

Fig. 6.44.2 Chimneys – self supporting and supported along bulding.

Therefore it is advantageous to install tall chimneys - at least 50 m for kiln and cooler gases regardless of heights worked out by formulae.

See Fig. 6.44.1.

In a cement plant, large volumes of exhaust gases are in kiln, cooler and raw mill sections. These chimneys are large in diameter. Others are comparatively small.

44.5 Location

Whenever possible a chimney should be located near tall buildings like a preheater tower. It can then be supported from floors of the buildings on brackets. **See Fig. 6.44.2.**

When this is not possible, a chimney must be supported on its foundations. They are self supporting when standing alone far away from a building structure **See Fig 6.44.2.**

For small volumes of gases chimneys could be small and can be installed either at ground level or on a floor in the building wherever the fan would be located. **See Fig. 6.44.3.**

Very small chimneys can be mounted on fan outlet itself.

Vibrations of fans would be transferred to chimneys where ducts from fans join chimneys. Expansion joint should be installed in these connecting ducts to avoid this from happening.

44.6 Design of Chimney and its Foundation

Tall self supporting chimneys and their foundations have to be designed carefully. They should withstand wind loads, which are highest at the tallest point; seismic factors are also to be considered.

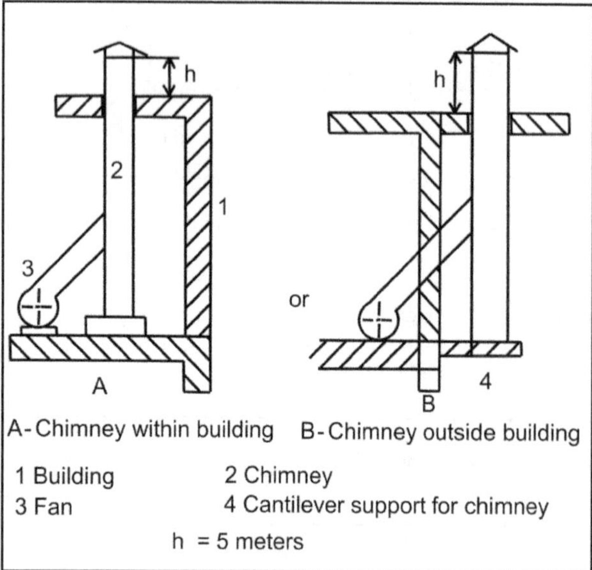

Fig. 6.44.3 Small chimneys starting at higher floor levels.

It is customary to weld a spiral strip along the outer surface of chimney for about one third of its height from top to break wind.

See Fig. 6.44.4.

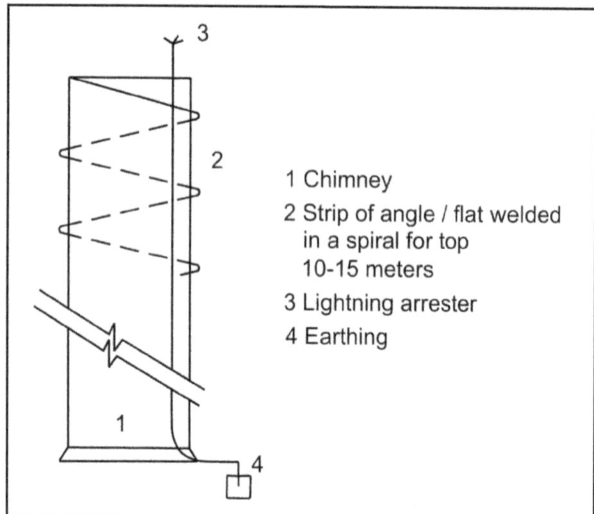

Fig. 6.44.4 Chimney arrangement to break wind velocity and lightning arresters.

Chimneys will tend to sway in winds. The design should keep this swaying within limits. Taller the chimney greater the swaying. The plumb must fall within narrow limits of the base at all times.

If it is the tallest structure in the plant, then a chimney should be fitted with a lightning arrester (It would be a good idea to install a lightning arrester even otherwise).

See Fig. 6.44.4.

Blinking lights should also be fitted on a tall chimney or for that matter on the tallest structure of the plant.

When located alongside a building, chimneys should project well above the working floor, above the machine / ducting in the vicinity. This is particularly so when temperature of gases is high 80-100 °C and above. "h" should not be less than 5 m. This precaution will prevent injury to workers.

See Fig. 6.44.5.

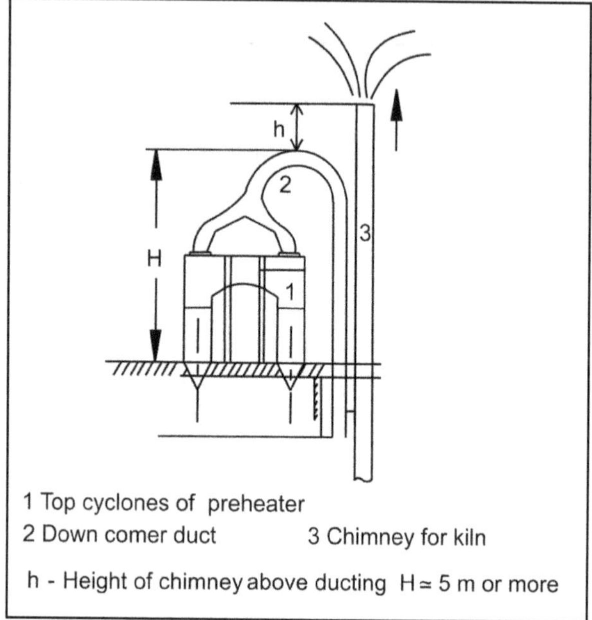

Fig. 6.44.5 Height of chimney above cyclone/ ducts for safety reasons.

This is particularly so for very hot gases like kiln preheater exhausts (100 °C in combined and 250 to 350 °C in direct operation) and Cooler exhaust gases (150-300 °C).

44.6.1 Lightning Arrester

Providing a lightning arrester for tall chimneys as shown in **Figs.6.44.4 and 6.44.9**

44.7 Measurement of Dust Emission

For major chimneys like those for kiln, cooler, coal mill, cement mill and boilers provision should be made to measure dust content in exhaust gases.

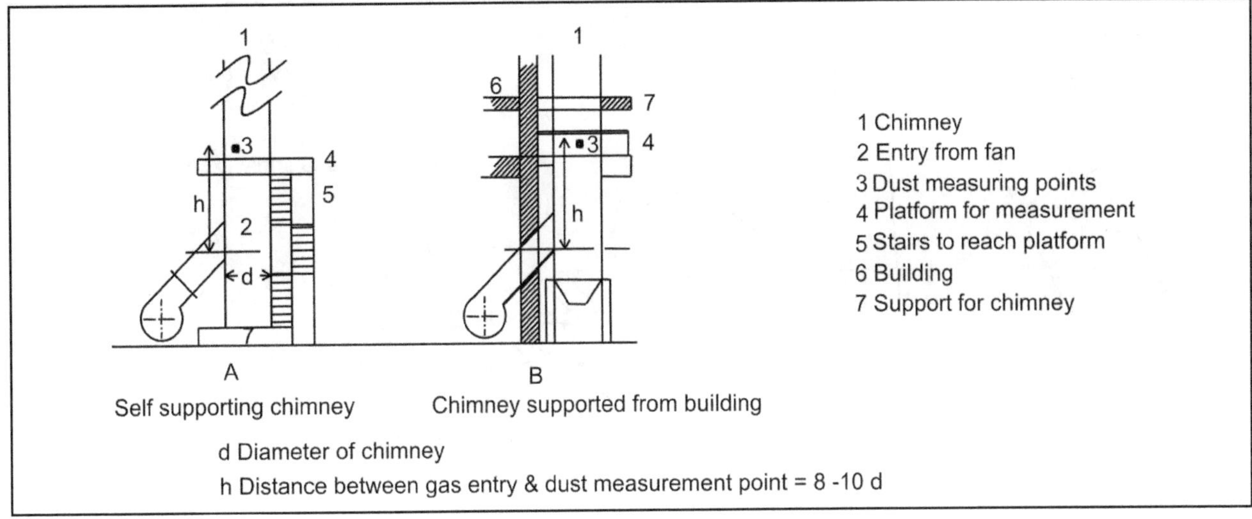

Fig. 6.44.6 Provision for measuring and monitoring dust emission.

Earlier, dust measurements were taken occasionally as and when required. A measuring point to draw out gases and the dust measuring apparatus (which measured gas volume, precipitated and measured dust in it) was used. Power point is also required to be provided. A working platform was necessary and also a service ladder to reach it.

If the chimney were near a building the access to the platform could be from one of the floors.

See Fig. 6.44.6.

Now dust emission is monitored continuously with opacity meters and the dust content in exhaust measured in mg/nm3 can be read at all times / recorded on control panel.

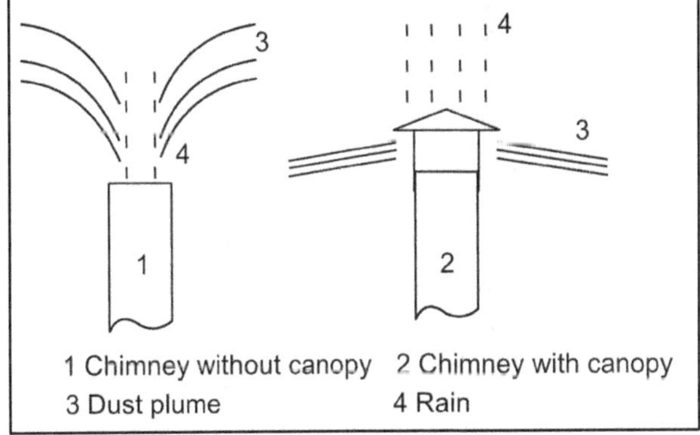

Fig. 6.44.7 Canopies on chimneys.

44.8 Construction

Chimneys can be fitted with canopies to spread out dust gases and to prevent rain water entering them.

However, canopies could result in precipitation in a smaller radius.

Hence they would be fitted on small chimneys only.

See Fig. 6.44.7.

There should be a manhole / access door at bottom to clean out dust collected inside.

See Fig. 6.44.8.

As most of the chimneys would be out exposed to sun, wind and rain, and gases passing through them contain moisture, the ducting and chimney should be

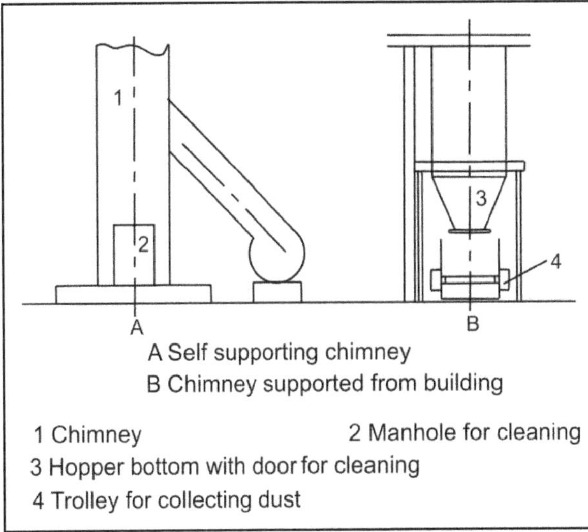

Fig. 6.44.8 Cleaning dust settled at bottom of chimney.

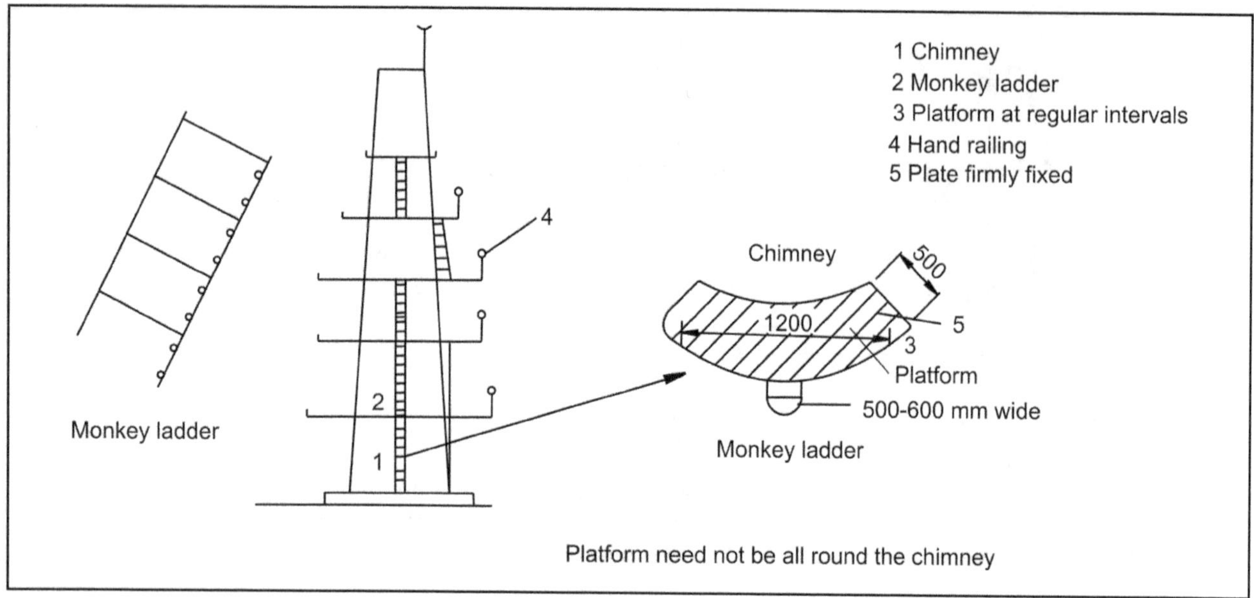

Fig. 6.44.9 Reaching top of stand alone chimney.

protected against corrosion with great care compared to other ducting. Base coat can be red oxide with one or more surface coats. Ideal would be a coat of aluminum paint. Protection with paint is important as once erected chimney is not easily accessible to maintain, or to paint again.

May be tall stand alone chimneys should have a monkey ladder with platforms at suitable levels for reaching to the top.

This certainly would add to weight and cost of chimney but would be a great convenience even though it would be required to be used once in a long while.

The monkey ladder must have a protective loops on outside for safety. Platforms must have strong hand railings and chequered plates securely fixed.

See Fig. 6.44.9.

Chimneys near building could be reached from several floors of the building.

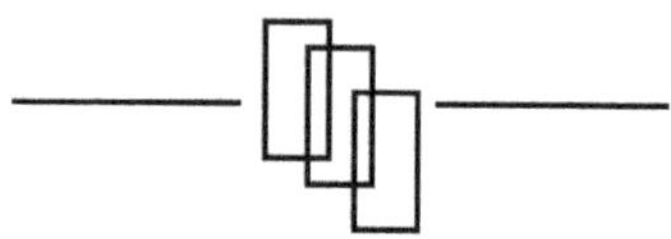

CHAPTER 45

PIPELINES AND SUPPORTS FOR COMPRESSED AIR AND PNEUMATIC CONVEYING PIPE LINES

45.1 Pipelines

Pipes are used to convey compressed air, pulverized materials like raw meal, coal, cement and fly ash pneumatically. They are also used to carry cooling water and lubricating oils for lubricating machinery.

45.2 Pipe Lines for Compressed Air and Pulverized Materials

Pulverised materials like raw meal, coal, cement and fly ash are conveyed pneumatically in pipe lines. There are many types of pneumatic conveying systems. Prominent among them are:

F.K. Pumps.

Air lifts.

Other dense / lean phase conveying systems.

All pneumatic conveying systems need compressors or blowers to supply compressed air and also pipe lines to convey the material pneumatically.

Pipe lines for compressed air and pneumatic conveying will be either seamless steel pipes or ERW pipes capable of withstanding pressures ranging from 0.5 kg/cm^2 to 7 kg/cm^2.

Requirements of compressed air for any particular application is dependent on:

(i) system used,

(ii) material conveyed and its fineness,

(iii) distance conveyed including horizontal and vertical lengths,

(iv) bends their number, radii,

(v) fixtures like diverting valves in the pipe line for multi points of discharge.

Suppliers or designers of pneumatic conveying system would furnish requirement of conveying air which is expressed in fad (free air delivered) and also line pressure. They would also recommend size of pipeline appropriate to conveying velocity.

Some systems like FK Pumps and Fluxo Pumps convey over long distances both horizontally and vertically. Systems like Air lifts convey only vertically.

45.3 Use of Compressed Air

Compressed air supply for different systems has been shown schematically in **Fig. 6.45.1.** It is based on using reciprocating or opposed piston compressor. Hence a receiver has been shown in the line. When rotary compressor or roots blowers are used, receiver may not be necessary.

Compressors and blowers can be individual–dedicated hence located near point of usage or there can be a bank located in a conveniently central place.

In the first case, pipelines are short and direct. In the second they are long.

45.4 Layout of Pipe Lines

Pipes come in standard sizes and lengths. They may have flanges at ends or may be threaded for joining several lengths together.

See Fig. 6.45.2.

45.4.1 Bends

Pipe lines would have bends to change direction.

Bends should have generous radii. Minimum radius should be 10 d where 'd' is diameter of the pipeline. **See Fig. 6.45.3.**

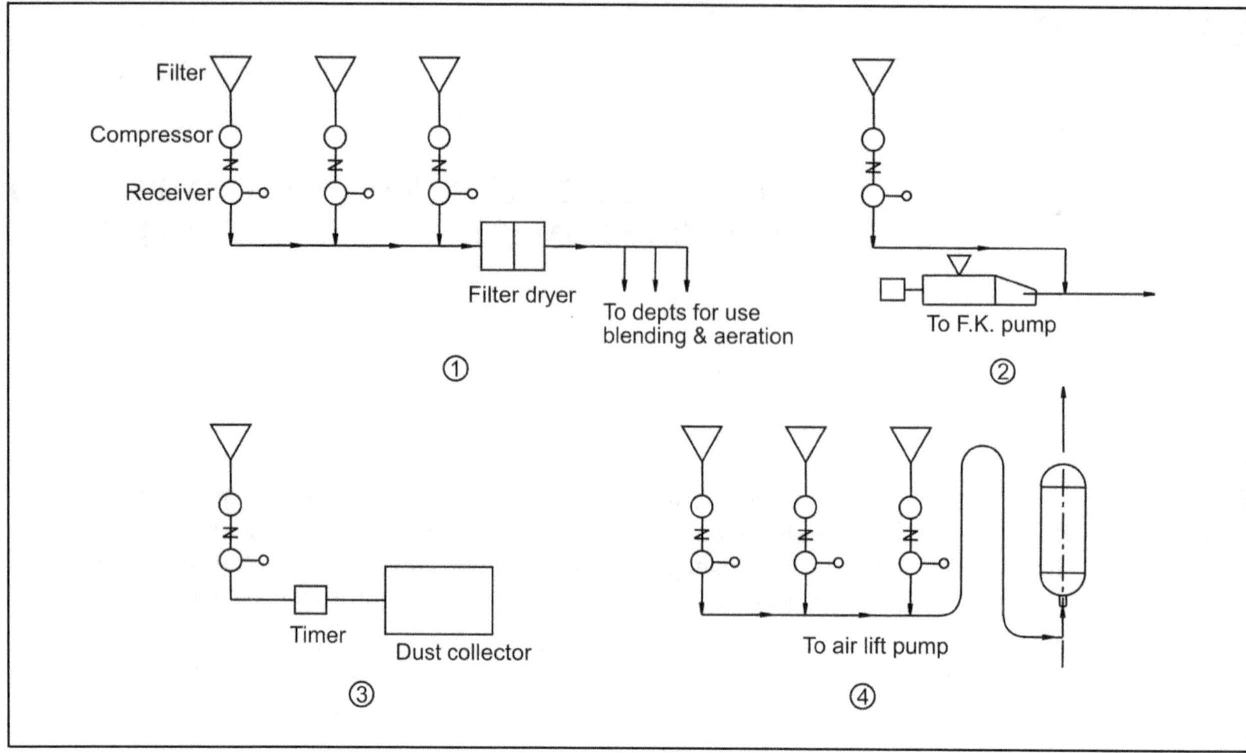

Fig. 6.45.1 Compressed air for different applications.

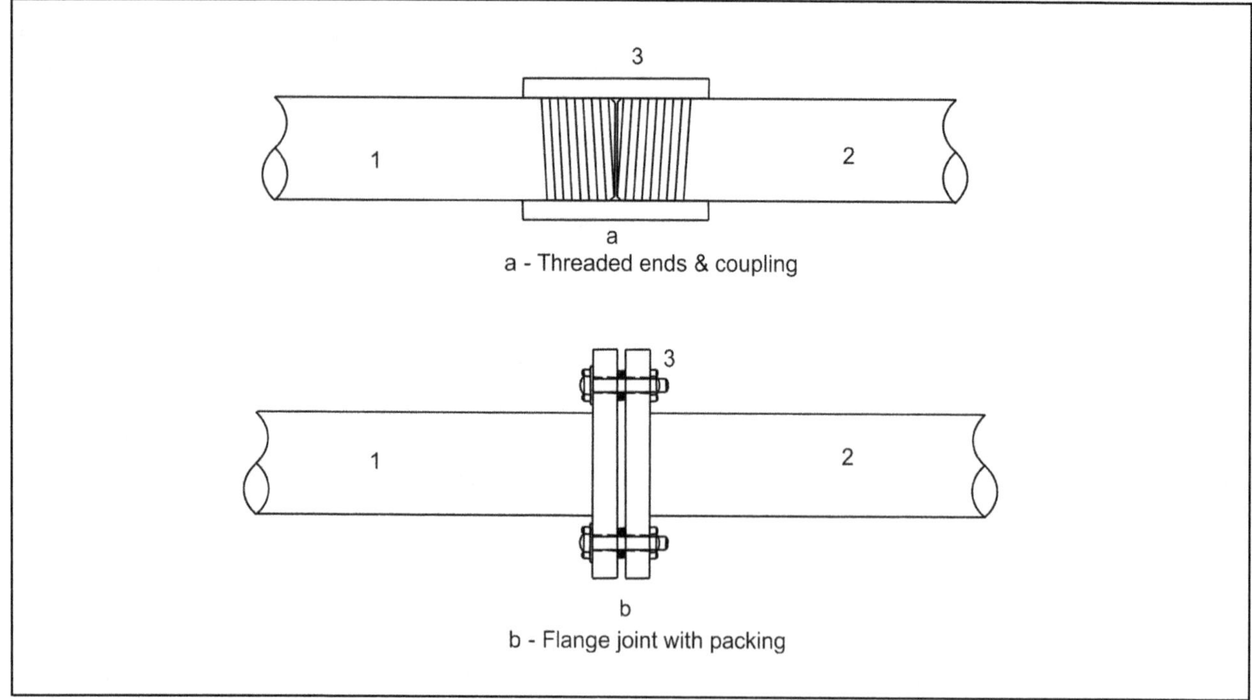

Fig. 6.45.2 Joining pipe lengths.

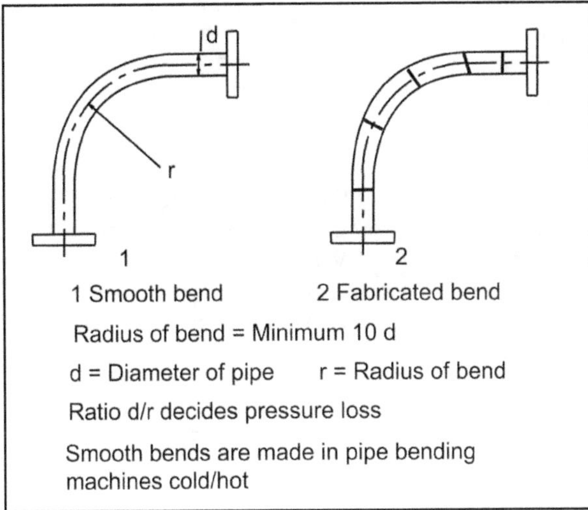

Fig. 6.45.3 Bends of pipe lines.

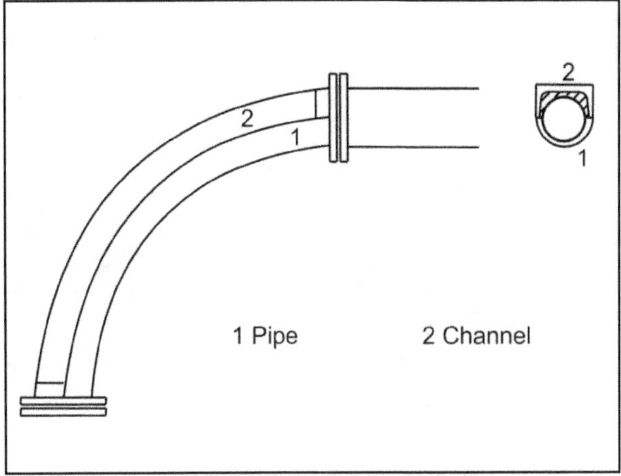

Fig. 6.45.4 Wear back bends for pipe line carrying pulverised coal/raw material / cement.

Where required wear back bends should be used. In the simplest form they would consist of channels welded to the pipe line bends and filled with concrete. **See Fig. 6.45.4.**

If there are a number of bends, then total or longest conveying or pumping distance is arrived at by adding to straight length equivalent lengths of bends. For example one right angle bend could be equivalent to 10 × diameter of the pipe line. Similarly diverting valves if installed are also converted into equivalent straight pipes.

45.4.2 Expansion Joints

For long pipe lines, there should be expansion joints which would permit expansion without damaging the pipe line. Where there are expansion joints in the pipe line there should be a working platform around the pipe line with an access to it for servicing and maintenance.

Dresser couplings are commonly used for this purpose. These should be inserted at suitable intervals.

For long lengths expansion can be provided by U bends. Advantage can be takenfor draining any collected water by providing a tap. **See Figs 6.45.5 and 6.45.12**

Pipe lines carrying material can choke if they are not cleaned out after stopping the flow of material and are started up after a long interval.

Moisture in the material conveyed can also cause choking. The compressor or blower used to supply conveying air is sized to develop pressure 1-1.2 kg/cm2 above the working pressure so that it is able to force air against higher resistance due to choking.

45.4.3 Removal of Chokes

If choking is too frequent, it may be desirable to lay a compressed air pipe line along the conveying pipe line with nozzles at intervals to inject compressed air to clear the blockage.

See Fig. 6.45.5.

45.4.4 Horizontal and Vertical Pipe lines

Vertical pipe lines would be supported along the column of a building or wall of the silo or on the support of a gantry.

See Fig. 6.45.6.

When laid horizontally at ground level it should be so laid that it does not get submerged in water. It would be better to install it in separate troughs, duly covered with slabs, which can be removed for maintenance. These slabs should take the weight of heavy vehicles passing over them.

See Fig. 6.45.7.

Alternatively cable trenches could be used to install the pipe lines.

See Fig. 6.45.8.

45.5 Gantries for Pipe Lines

Pipe lines could be installed along a multipurpose gantry i.e., a gantry which carries cables also. It will save

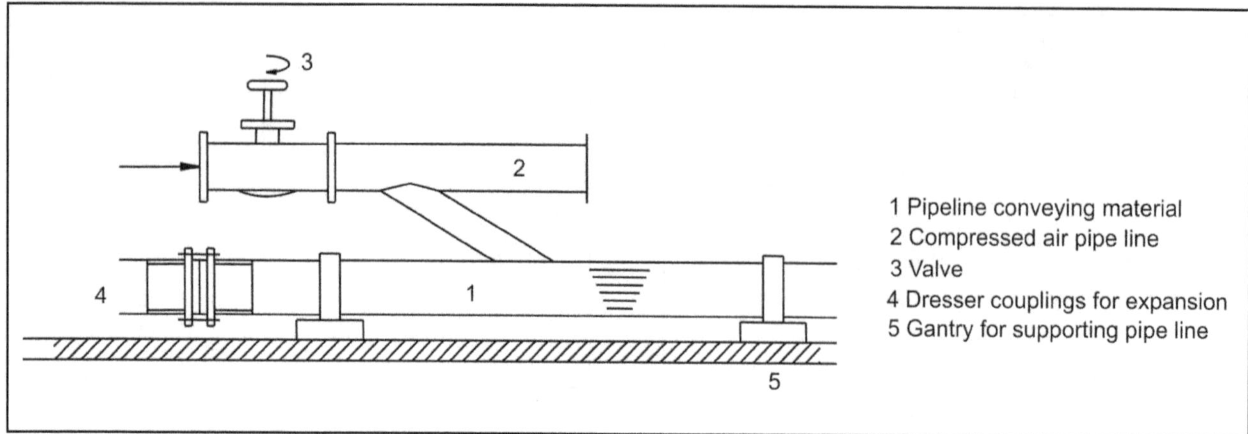

Fig. 6.45.5 Provisions to clear jams in pipe line conveying material.

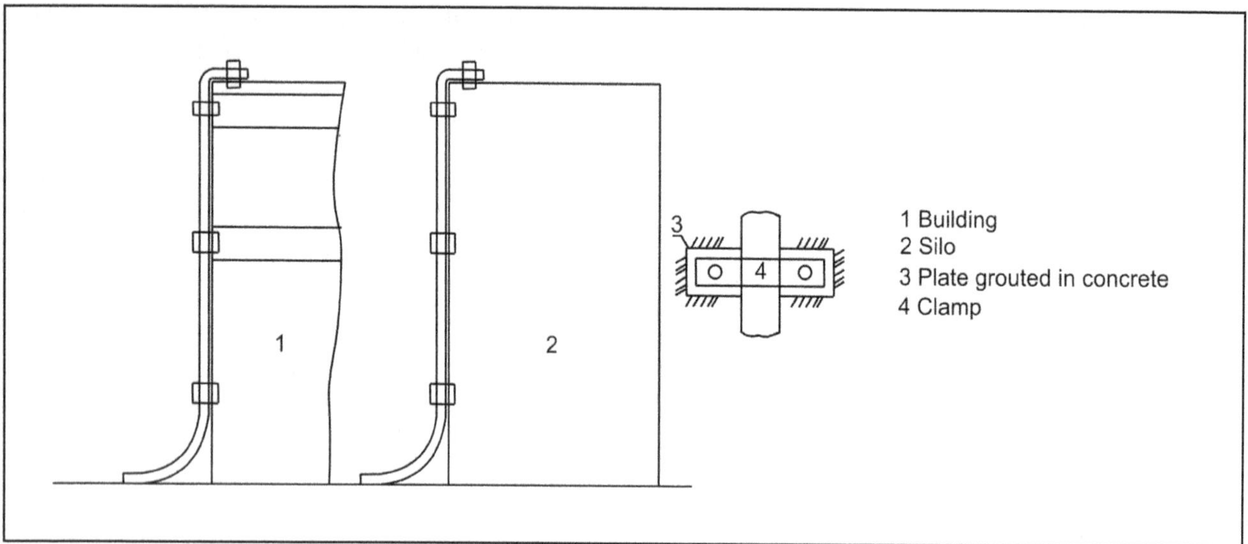

Fig. 6.45.6 Pipe lines along buidings and silos.

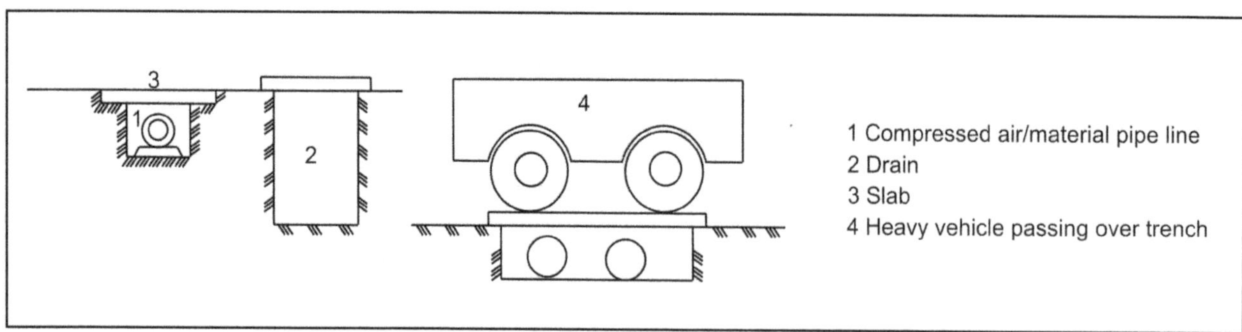

Fig. 6.45.7 Pipe lines at ground level.

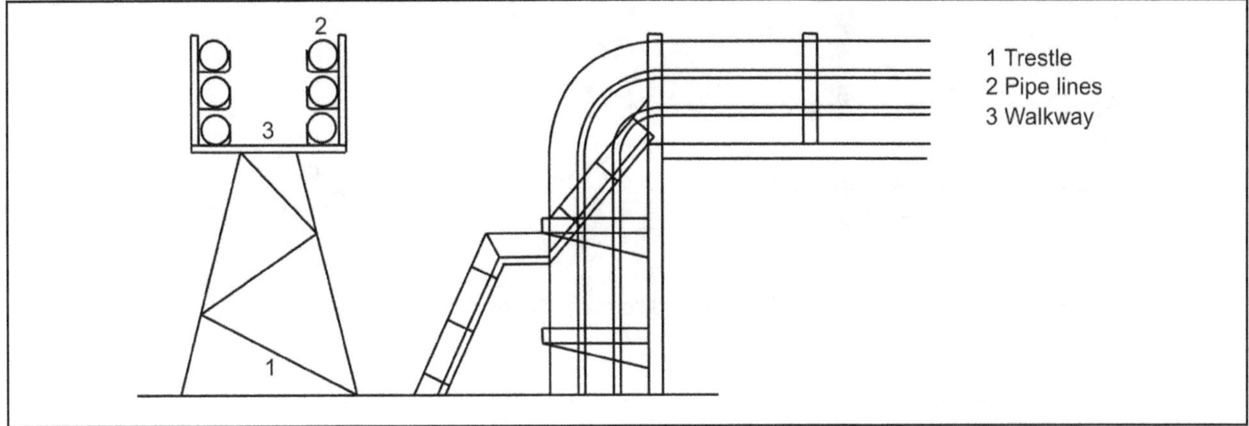

Fig. 6.45.9 Gantries for pipelines of compressed air and conveying raw material/cement/coal.

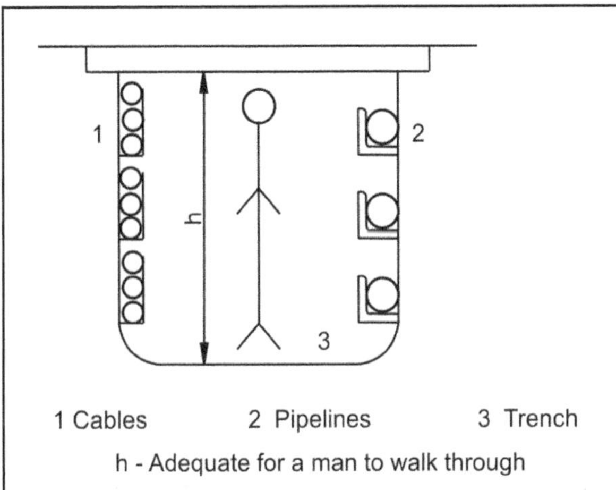

Fig. 6.45.8 Pipeline in trench carrying cables.

expenditure on a separate gantry. However, the guiding principle in the layout of a pipe line is to keep it short, direct and with as few bends as possible.

See Figs. 6.45.9 and 6.45.10.

If a separate gantry cannot be avoided, the gantry should have a walkway for maintenance. It should be designed to take future pipe lines after expansion also.

See Fig. 6.45.9.

Pipe lines should be laid horizontally or vertically but not at an angle.

See Figs. 6.45.11.

45.6 Compressor House and Compressed Air Pipe Lines

Many a company prefers to install a central compressor house and convey compressed air to points of usage. The pipe lines for compressed air should also follow above principles of layout.

The expansion of the pipe line can be provided by U bends. Advantage can be taken to discharge collected water by providing a blow out tap at the bottom of U bend.

See Fig. 6.45.12.

In case of long compressed air pipe lines arrangements should be made at point of usage also to remove water and oil through appropriate filters, and drain taps etc.

See Fig. 6.45.13.

45.7

To enable Suppliers of pneumatic conveying equipment, to recommend the requirements of compressed air correctly, proposed routing of the pipeline should be furnished to them including drawings.

Long pipe lines act as receivers and allowance needs to be made for it by enhancing capacity of compressor to be recommended.

45.8 Receivers

Reciprocating compressors have one working stroke in four. Opposed piston compressors have one in two. Therefore if compressor is connected directly to the pneumatic conveying equipment like FK Pumps etc. then there would be fluctuation in pressure cyclically. To avoid this a receiver of suitable size and volumetric capacity is added in the system. Each compressor has its own receiver. **See Fig. 6.45.1.**

Suppliers of compressors should furnish recommendations for size and capacity of air receivers.

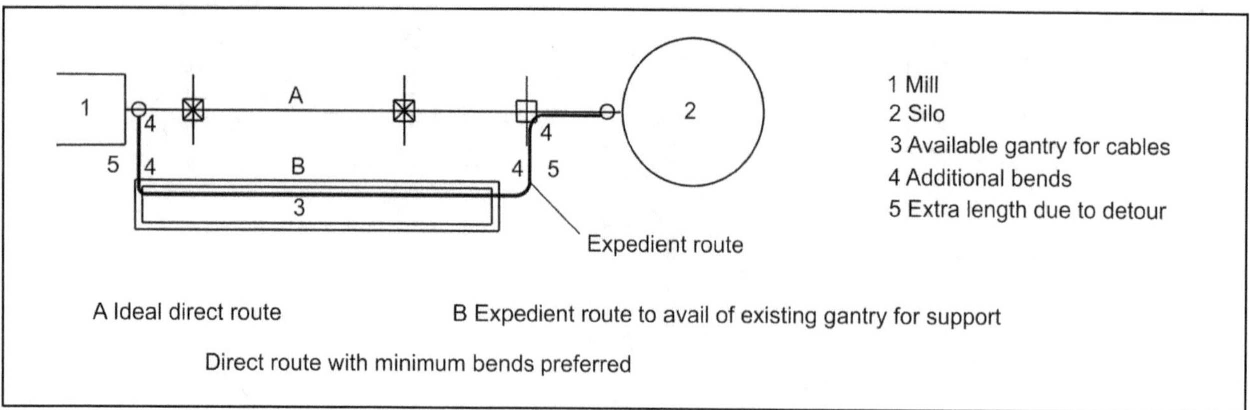

Fig. 6.45.10 Routing of pipe line.

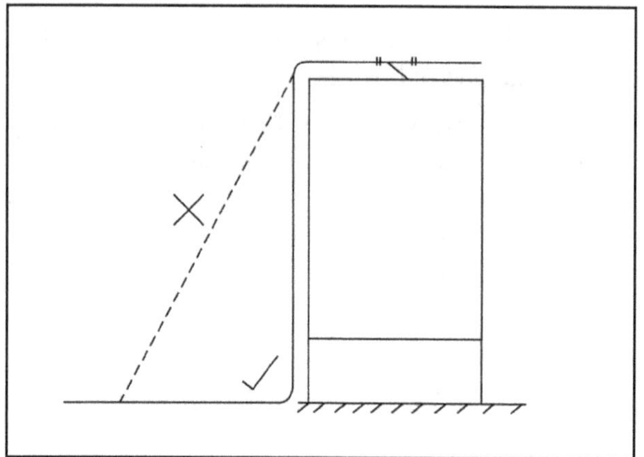

Fig. 6.45.11 Pipelines for pneumatic conveying should be laid horizontally and vertically; not inclined.

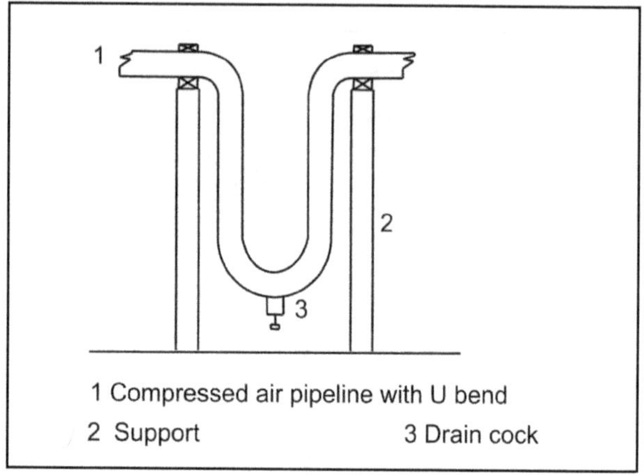

Fig. 6.45.12 U-bends for expansion and draining of condensed water in pipeline for compressed air.

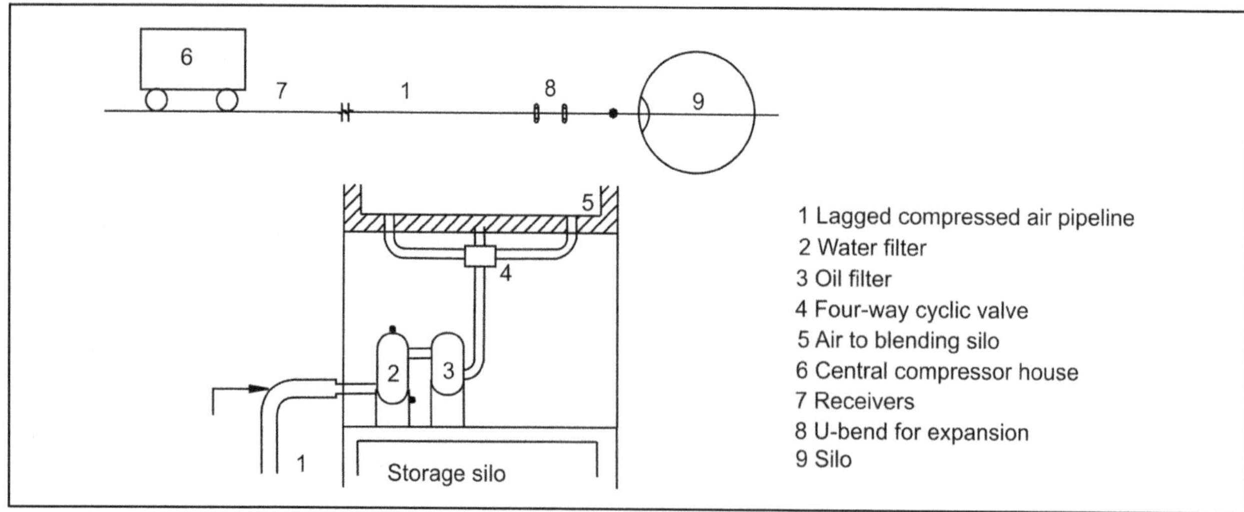

Fig. 6.45.13 Cleaning compressed air of oil and water at point of use.

CHAPTER 46

OIL LUBRICATION AND COOLING WATER SYSTEMS

46.1 Oil Lubrication

Most bearings would have oil sumps and ring lubrication. Rings fixed to shafts dip in oil sumps. Oil carried with them is scraped off and spread on journals.

When the bearings are very large like in case of ball mills or when speeds are low, oil rings are fitted with buckets which lift oil and drop it into a tray with holes that spreads oil on journals.

See Figs. 6.46.1, 6.46.2 and 6.46.3.

Oil in the bearings is water cooled.

See Fig. 6.46.4.

Oil circulation is required in case of bearings which are forced lubricated. That is oil is pumped on to bearings and withdrawn from the bearings, taken to oil coolers and is filtered and pumped back.

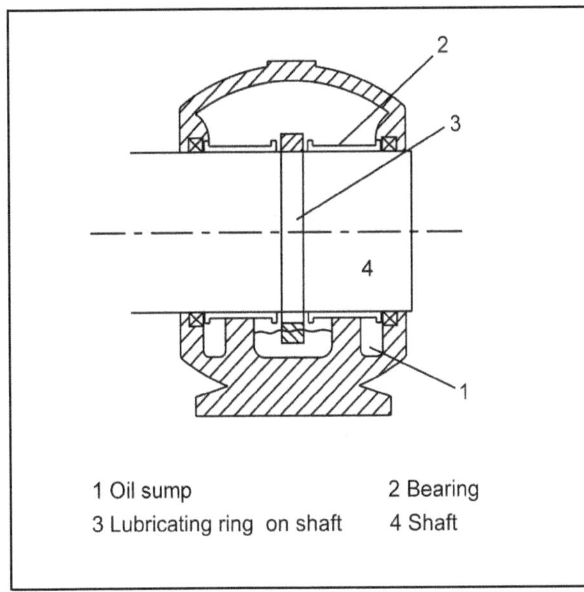

Fig. 6.46.1 Ring oil lubrication conunter shaft bush bearings.

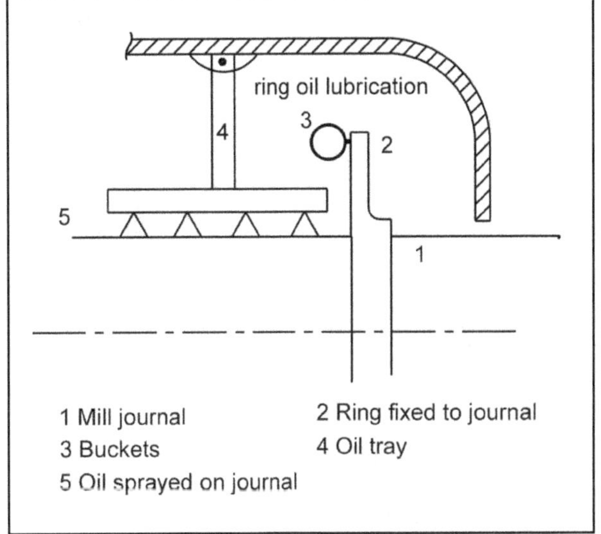

Fig. 6.46.2 Ring lubrication of main bearings of ball mill

See Fig. 6.46.5.

There would be a common circulation system for two bearings of a ball mill.

Oil pipelines would be short. Oil filters, cooler and pumps together would be a self-contained unit. It should be located so that it does not come in the way. Cooling water pipeline to it can be taken from the pipeline supplying cooling water to bearings, compressors etc.

See Fig. 6.46.6.

Girth gear and pinion of mill and kilns are lubricated either by lubricating idlers or by forced lubrication system.

See Figs. 6.46.7 and 6.46.8.

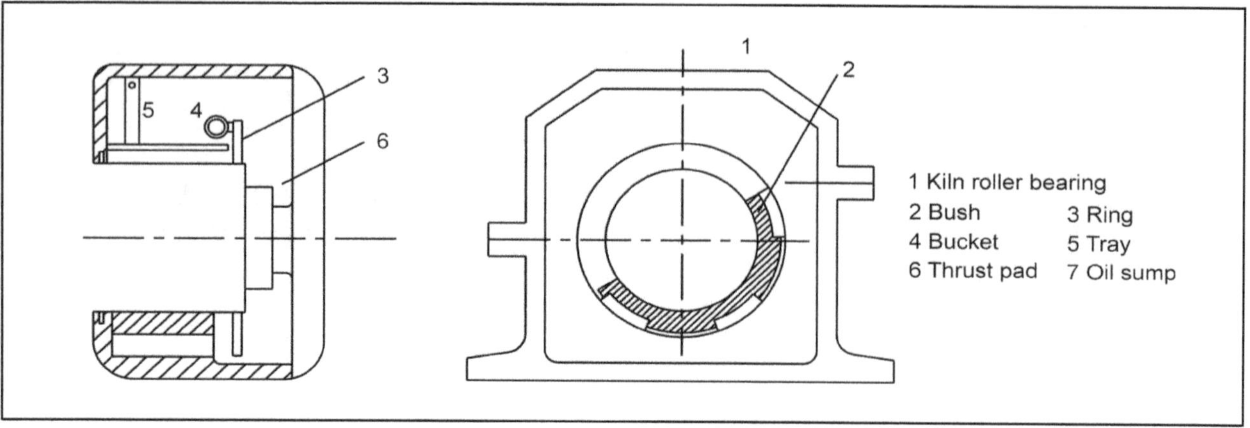

Fig. 6.46.3 Lubrication of kiln bearings; oil lubrication of bronze / bush bearing.

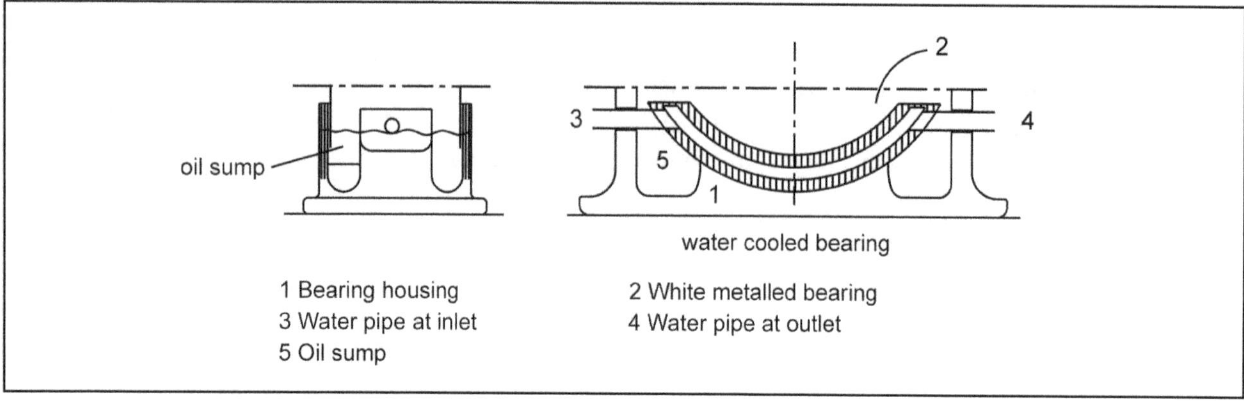

Fig. 6.46.4 Water cooled bearings of a ball mill.

46.2 Specifications of Lubricating System

Machinery Suppliers would furnish recommendations for lubricants to be used and for forced lubrication system. Often they would supply the lubricating system also.

There are specialist suppliers of oil and grease lubrication systems.

When a number of bearings are lubricated from one point it is called a centralized lubrication system.

Oil lubrication systems will be 'closed' in that oil will be used back after filtering and cooling it.

Grease lubrication systems on the other hand would be 'open'.

Presently lubrication is hardly ever done manually.

Details of lubricating system and water cooling system for it should be obtained from Suppliers of machinery together with drawings.

A separate layout showing pipe lines for oil and cooling water should be drawn for each department and the routes fixed and pipe line laid out as per this layout drawing.

When pipe lines are planned in this way, they would be properly designed and laid out. Bills of materials could be prepared in details and pipes and fittings can be procured with some measure of accuracy; they can also be installed speedily.

All pipe lines should be tested for leakages and leaks rectified.

46.3 Cooling Water Systems

Generally Cooling water systems are 'closed' systems.

Cooling water will enter bearings, after coolers of compressors, blowers, etc., at a pressure recommended by the respective suppliers. They would leave bearings or compressors at atmospheric pressure.

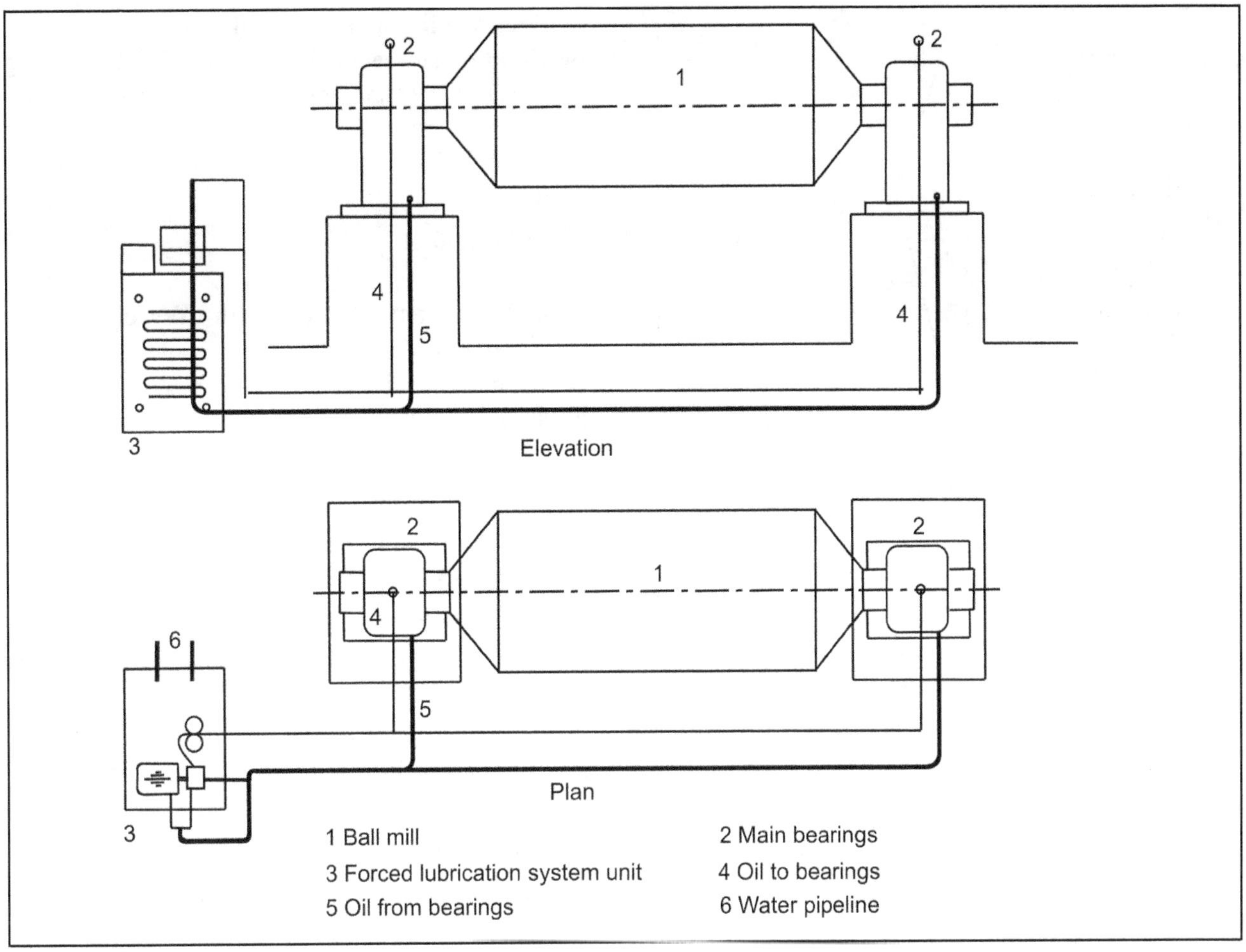

Elevation

Plan

1 Ball mill
2 Main bearings
3 Forced lubrication system unit
4 Oil to bearings
5 Oil from bearings
6 Water pipeline

Fig. 6.46.5 Forced lubrication system for main bearings of a ball mill (girth gear and pinion not shown).

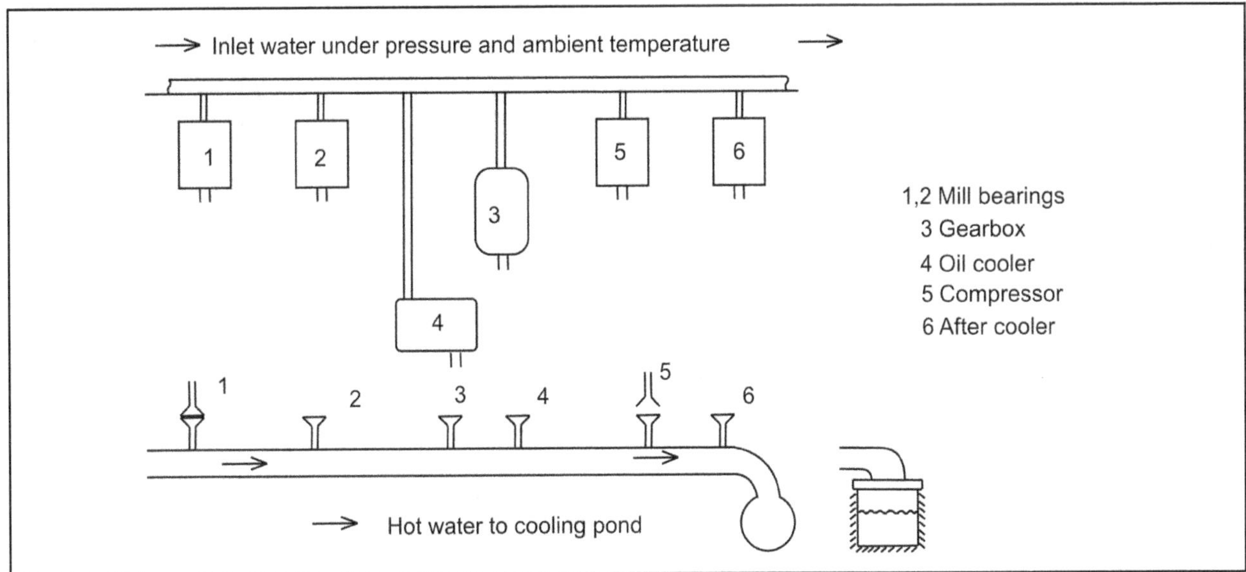

Fig. 6.46.6 Layout of pipe lines for cooling water in a department.

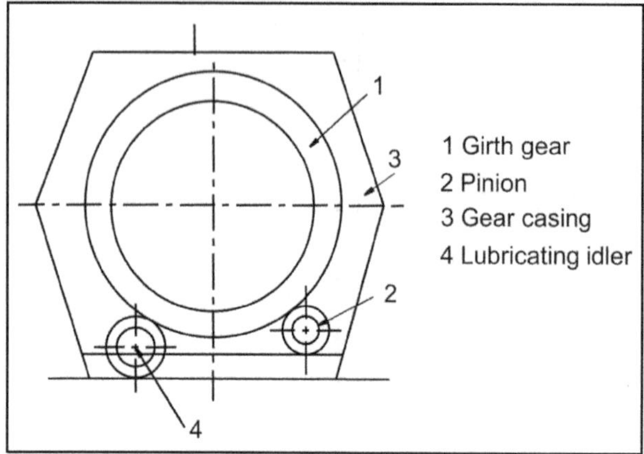

Fig. 6.46.7 Idler lubrication of girth gear on ball mill - similar for kiln also.

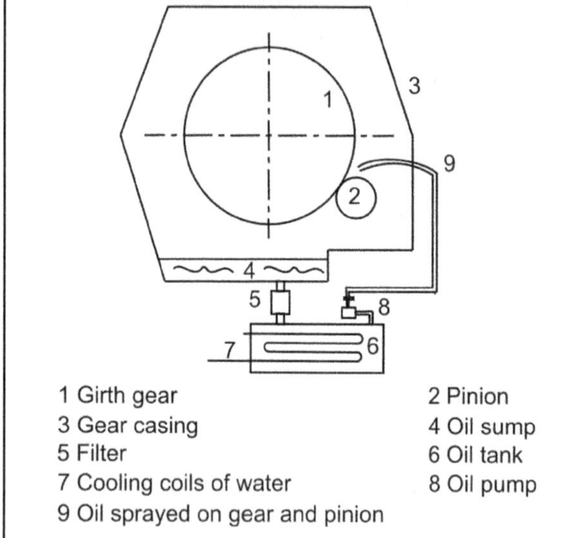

Fig. 6.46.8 Forced lubrication of girth gear of ball mill.

Cooling water pipelines should show at a convenient place in the system layout that water is actually flowing. This can be done simply by breaking the circuit or by inserting a glass window in the pipe line.

See Fig. 6.46.9.

Cooling water returned from different points in a department is collected through pipe lines and brought out in a bigger channel or pipe to be taken to the cooling pond.

See Fig. 6.46.10.

The return pipelines are therefore laid at suitable slopes to facilitate flow of water to the cooling pond.

From cooling pond, cooled water is sent to an overhead tank for recirculation. About 20 % of cooling water is lost by evaporation in the cooling pond and is made up by fresh water. The consumption of water on account of cooling is thus only 20 % of water in the system.

See Fig 6.46.11

46.4 Requirements of Cooling Water

Requirements of cooing water for each machine in a department can be collected to size feeding and discharge pipe lines.

Requirements of various departments are collected to arrive at total needs of the plant.

Water circulation system can be designed from the overhead tank to individual departments step by step after ascertaining flow at every step.

Return pipeline will be from the cooling pond to the over head tank.

Make up water needs to replenish evaporated and lost water are arrived at and it is added in the cooling pond by a separate pipe line.

See Figs. 6.46.10 and 6.46.11.

Overall requirements of water are worked out by adding to make up water, water needed for other purposes.

DG Sets would have their own cooling towers and fresh water would be supplied for make up purposes. **See Fig. 6.46.12.**

When a plant has a 'Thermal Power Station' to supply captive power it will need substantial quantities of water for generation of steam. After passing through turbines, steam condenses into water.

There is a recirculation system here also.

46.5 Water Supply System for the Total Plant

Overall water supply system for the total plant has been shown in **Fig. 4.6.1** in **Chapter 6 of Section 4.**

Pipe lines layouts should be worked out in consultation with departments like Civil Engineering and Electrical Engineering so that efforts are not duplicated, nor do they work against one another.

Layout should be marked in the General Layout of the Plant.

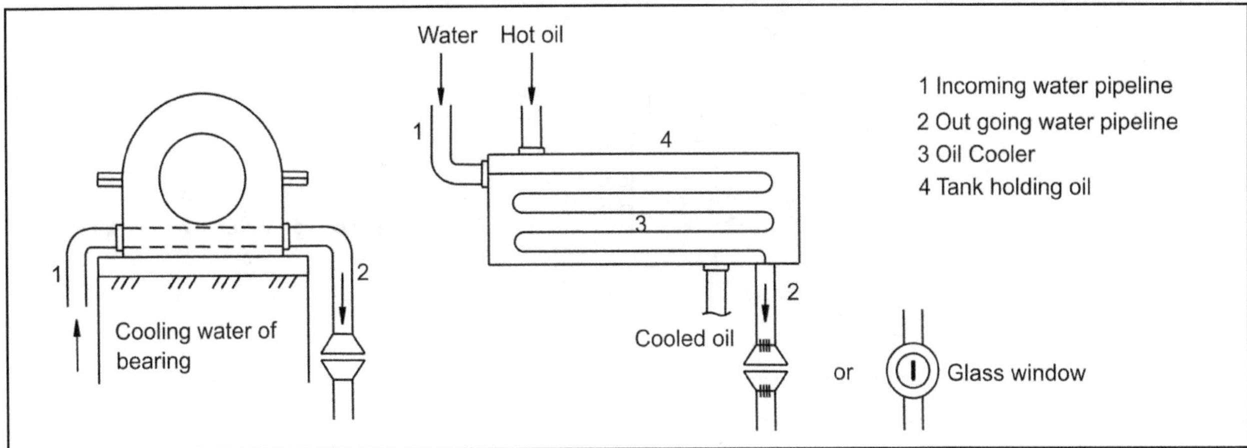

Fig. 6.46.9 Cooling water pipe lines - provision to see continuity of water flow.

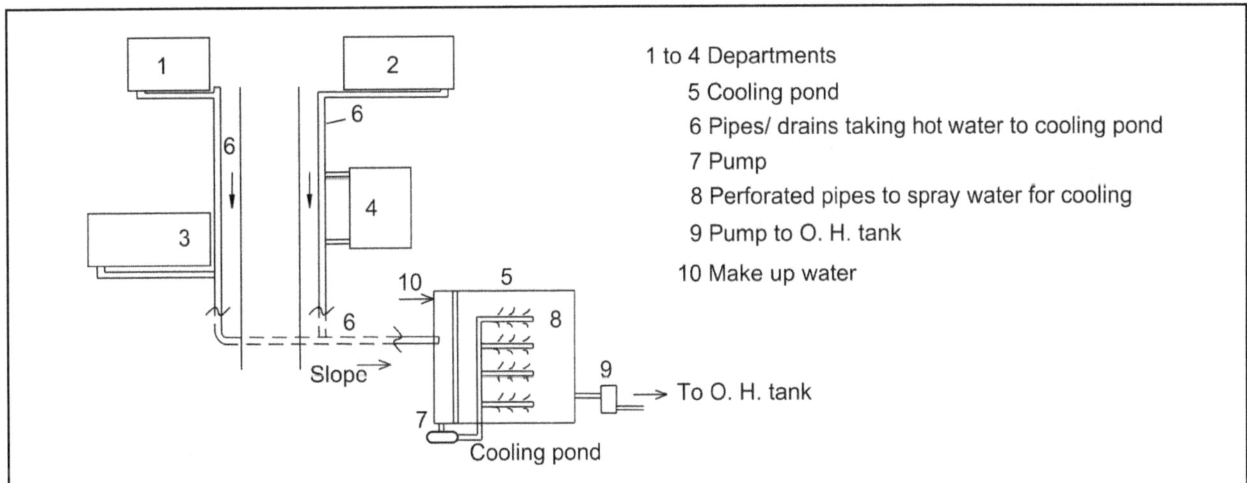

Fig. 6.46.10 Layout of hot water lines from various departments and cooling pond.

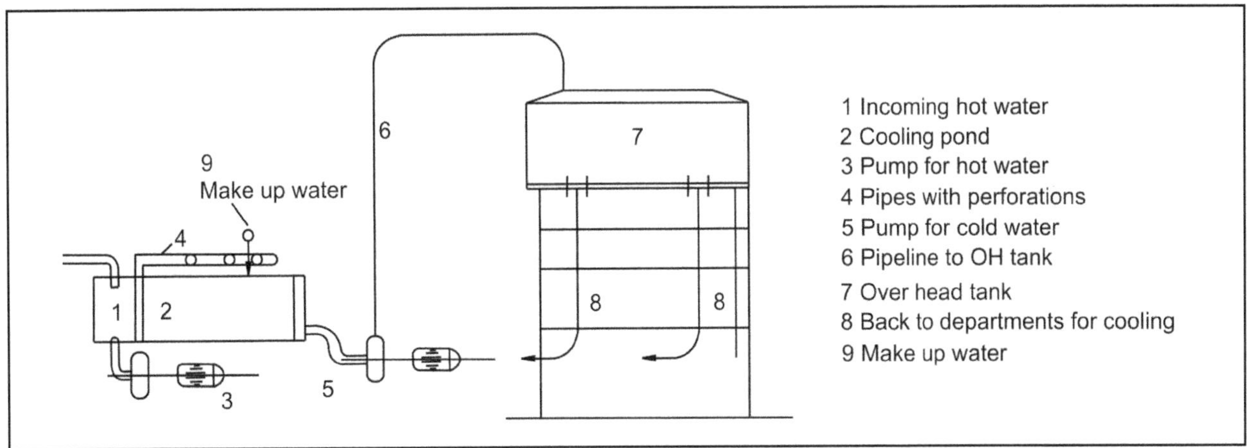

Fig. 6.46.11 Layout of pipe line from cooling pond to over head tank.

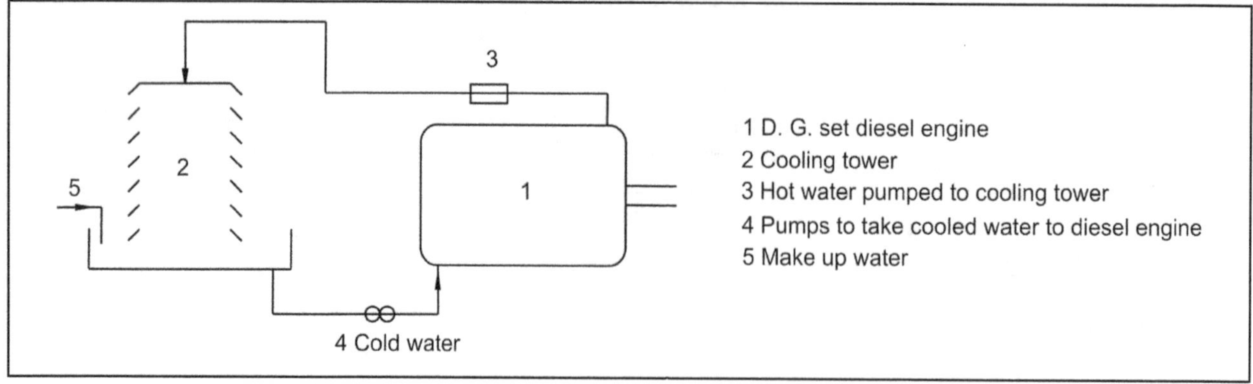

Fig. 6.46.12 Cooling water circuit for D.G. set.

46.5.1 Pipe Lines for Main Water Supply

Pipe lines will be installed to bring in water from main pumping station in river bed or from bore wells to the plant for processing like water treatment. Water will be first taken to a water tank at ground level. It will be chemically treated for softening it for spraying it in gas conditioning towers and for cooling purposes. **See Fig. 4.6.1 in Chapter 6 of Section 4.**

There could be overhead tanks in the plant and also in crusher, in quarries and in colony.

Water would be conveyed in large diameter pipe lines made of CI Spun pipes or ms pipes with concrete coating or concrete pipes.

Such large and heavy pipes will be laid at ground level either above or below ground suitably supported at suitable intervals.

See Figs. 6.46.13 to 6.46.15.

G.I. Pipes were used for water pipelines. They are being replaced by PVC Pipes which are much lighter and easier to install besides being rustproof.

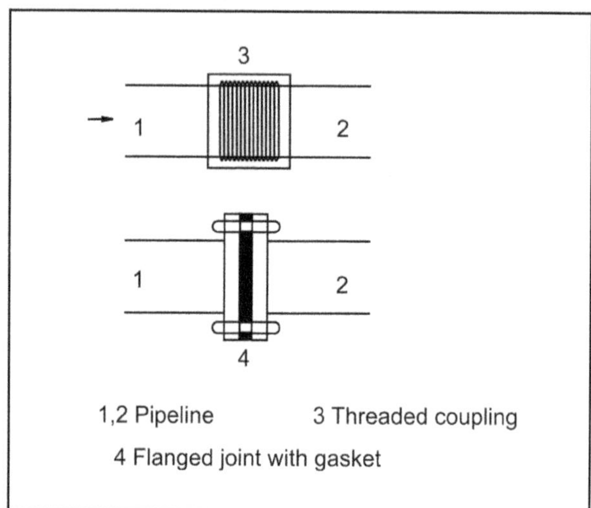

Fig. 6.46.13 Water pipelines.

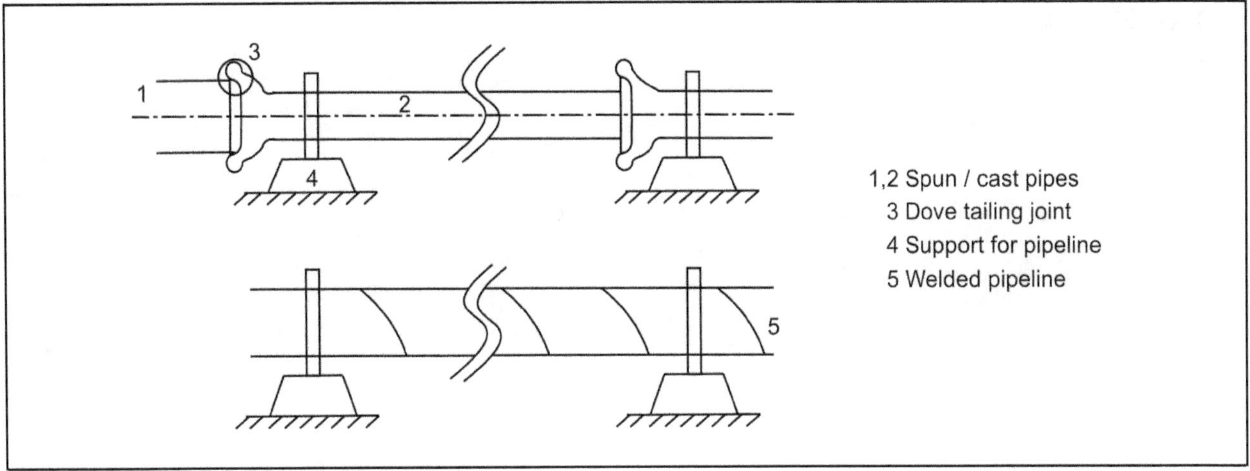

Fig. 6.46.14 Large cast / spun and welded pipelines for water.

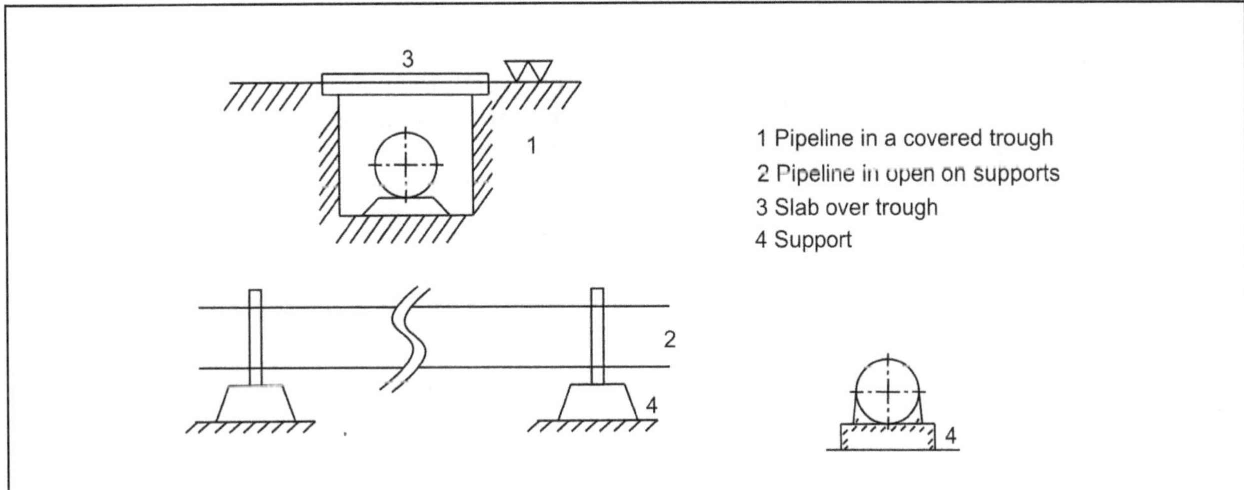

Fig. 6.46.15 Pipelines in trough or in open.

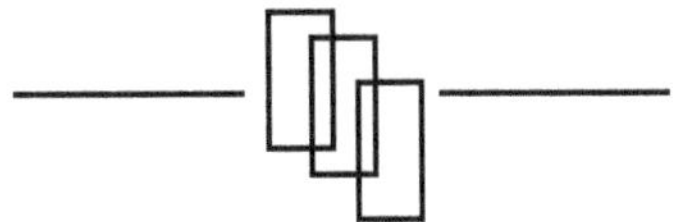

CHAPTER 47

BED PLATES

47.1 Bed Plates

Bed Plates are frames on which machines, motors and gear boxes etc., are fixed. In turn bed plates are installed and grouted on foundations.

Sometimes of course a machine can be installed straight on foundations without a bed plate.

Alignment of a machine is greatly facilitated by using bed plates. Some bed plates have machined surfaces for installing the machines. Others have raised pads which are machined.

47.1.1 Kiln Bed Plates

Kiln bed plates on which are fixed roller bearings, are large in size and support heavy weights. They are therefore to be designed to sustain heavy weights and forces arising out of operation of the kiln.

Each kiln bed plate supports 4 rollers bearings. One of the bed plates will also support thrust roller bearings.

All these would have to be installed at an angle corresponding to the slope of the kiln. A common bed plate to support all these with machined surface greatly helps alignment of the kiln.

47.2

A bed plate is a link between foundation and machine. It must be strong and rigid enough to hold the machine and should not itself buckle or distort when bolts are tightened. It should be suitably reinforced to take the weight without bending when machine is tightened to it.

Some bed plates have facility to lift or adjust their level.

See Figs. 6.47.1 and 6.47.2.

47.3 Design of Bed Plates

Bed plates will be designed and made according to drawings taking into account hole centers of machine /

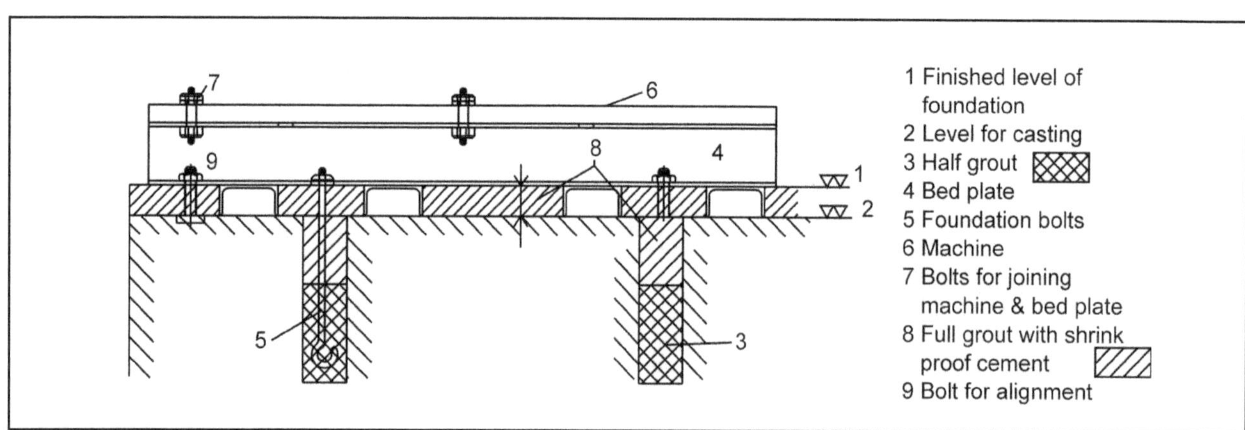

Fig. 6.47.1 Bed plate design.

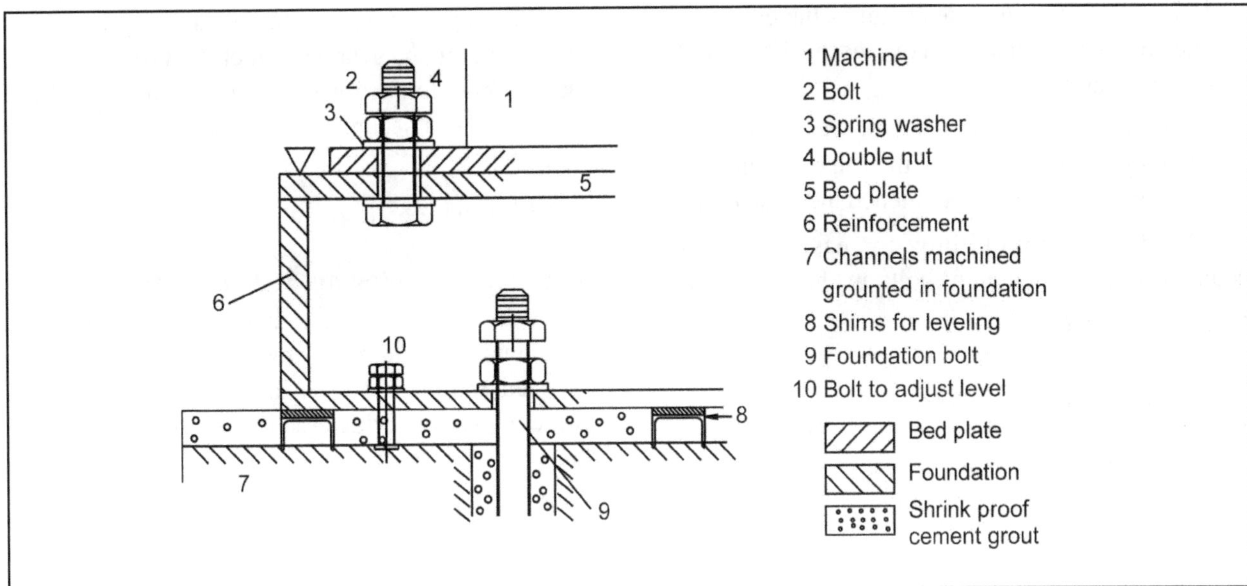

Fig. 6.47.2 Bed plate design.

bearings to be supported and fixed on it and hole centers for foundation bolts to be grouted in the foundation.

For all major machines, the Supplier would furnish drawings of foundations with center lines and levels to be maintained. Often bed plates are included in their scope of supply. This is particularly so for large machines like kilns and mills.

When Supplier is not supplying bed plates he would furnish drawings thereof to enable the Client to take up fabrication.

Bed plates should have facility for locating the machine or bearing back in its original position if removed for maintenance. Dowel pins fixed diagonally, are very useful for this purpose.

See Fig. 6.47.3.

Brass shims, machined plates and wedges are used to level bed plates.

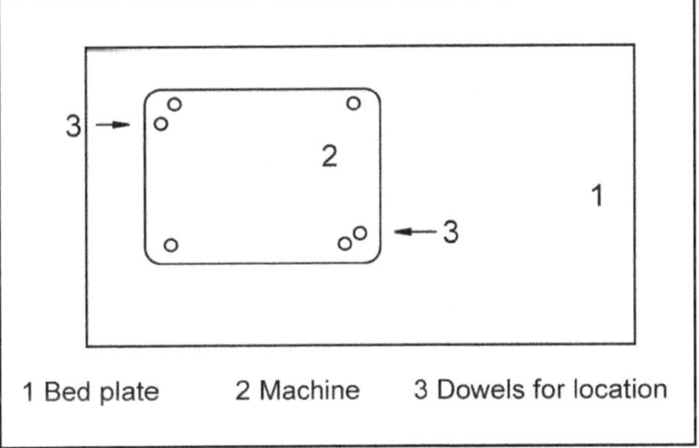

Fig. 6.47.3 Locating machine on bed plate with dowels.

47.3.1 Centring and Leveling of Bed Plates

Normally bed plates would be centered and leveled and half grouted first according to the erection drawings, keeping a small margin for final adjustment.

All bed plates (as also all foundations) should have center lines marked on them (in both directions). Center lines should be punched on the top surface of the bed plates surface and also on sides.

See Fig. 6.47.4.

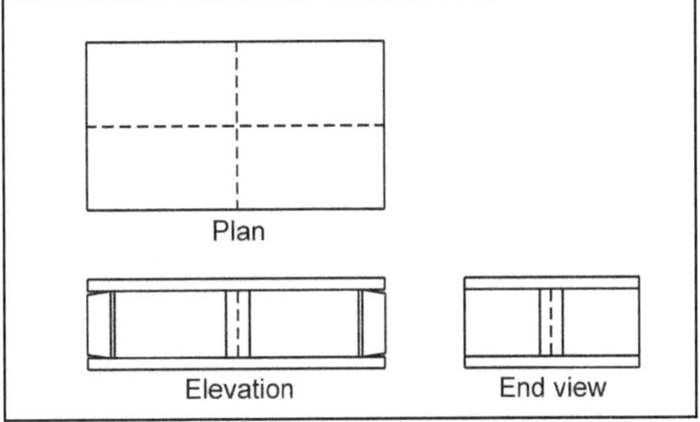

Fig. 6.47.4 Centre lines - on large bed plates for positioning them.

Centre lines on sides are useful as the center line on the top surface may be covered by base of the machine / bearing.

Holes for machine and foundation should be offset so that they do not interfere with each other and bolts can be easily removed and tightened independently.

The design should allow for easy access for spanners used for tightening bolts and holding them.

See Fig. 6.47.2.

47.4 Bed Plate for Drive

When there is a common bed plate for motor and gear box, the bed plate should also take into account the difference in center lines of motor and gear box from their respective bases.

See Fig. 6.47.5.

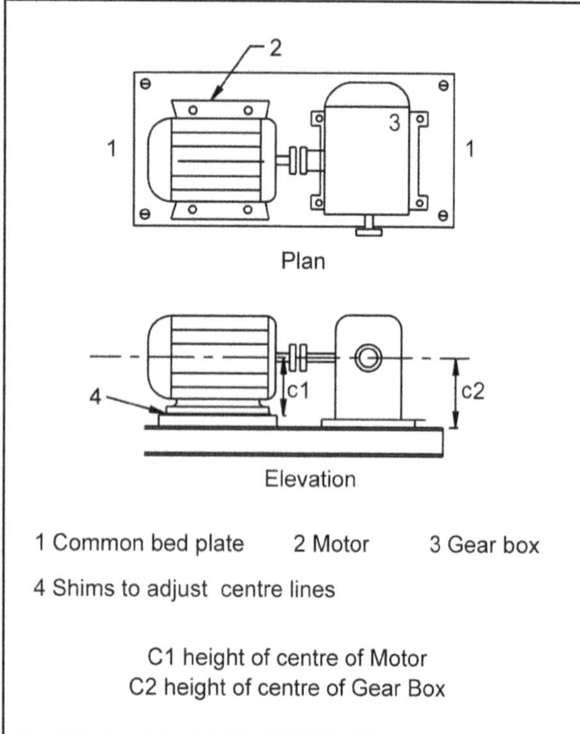

Fig. 6.47.5 Common bed plate for motor and gearbox.

For bringing their respective center lines in line, heights of foundations F1, F2, F3 would be according to centers C1, C2, C3 of bearing, gear box, and motor respectively.

See Fig. 6.47.6.

When a gear box has more than 2 stages, centers lines of motor and gear box outlet shaft would be very much offset. In such cases it may be better to keep two bed plates separate rather than have a common base plate.

See Fig. 6.47.6.

47.5 Stress Relieving of Bed Plates

Large fabricated bed plates like kiln bedplates where considerable welding is involved should be stress relieved and machined so that these stresses do not come in picture during erection and while tightening bolts.

Bed plates must provide adequate machined surface to place a spirit level for leveling it.

47.6 Alignment of Kiln Bed Plates

Kiln bed plates on 2/3 piers would be leveled and aligned with respect to one another with the help of laser beams, theodolite or water level or piano wire. It is absolutely essential that the kiln is perfectly aligned when resting on 2 or more bed plates which are separated by the spans of supporting stations.

See Fig. 6.47.7.

47.6.1 Mill Bed Plates

Same is true of main bearings for ball mills. Mill bed plates can also be made in C.I. They have facility for centering bearings. This can be done with wedges.

See Figs. 6.47.8 and 6.47.9.

The holes for bolts for bearings and foundation bolts should be large so that centering can be facilitated. They are sometimes oval to that movement in another direction is facilitated.

Sometimes during construction, foundations and centre lines are not as per drawings. In such cases this facility is useful.

47.7 Grouting of Bed Plates

Large foundations are generally finished about100 mm below final level. Foundation bolts are inserted in pockets and bed plates are placed on foundations and lowered centered and leveled and bolts grouted up to 100 mm from the surface of foundation. This is called 'half grouting'.

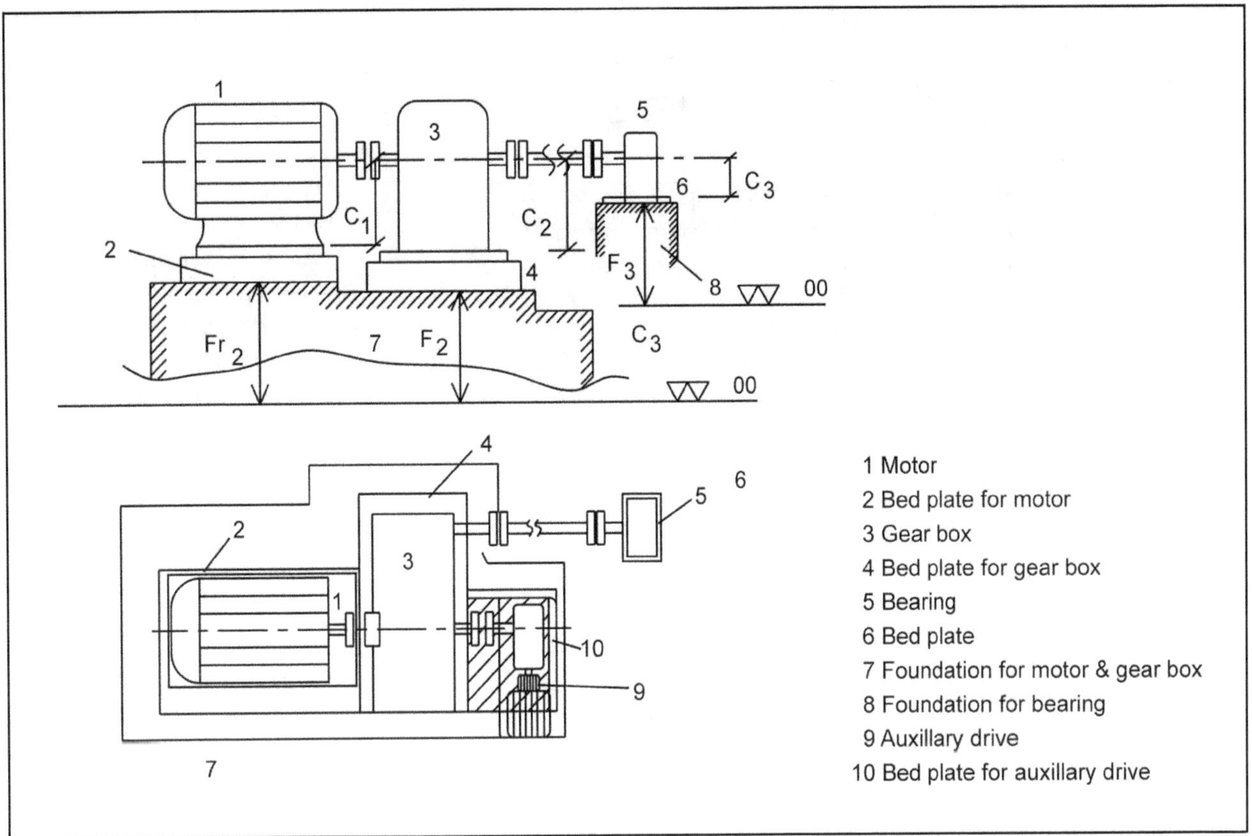

Fig. 6.47.6 Separate bed plates for motor and gearbox for large machines.

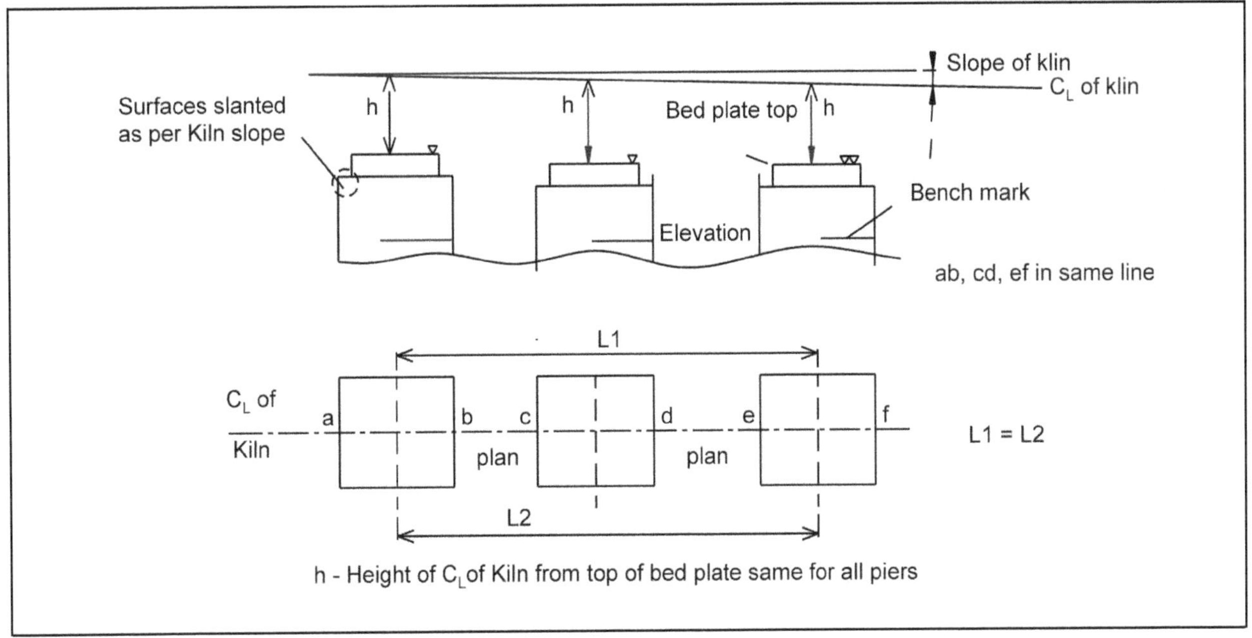

Fig. 6.47.7 Kiln bed plates and centering them.

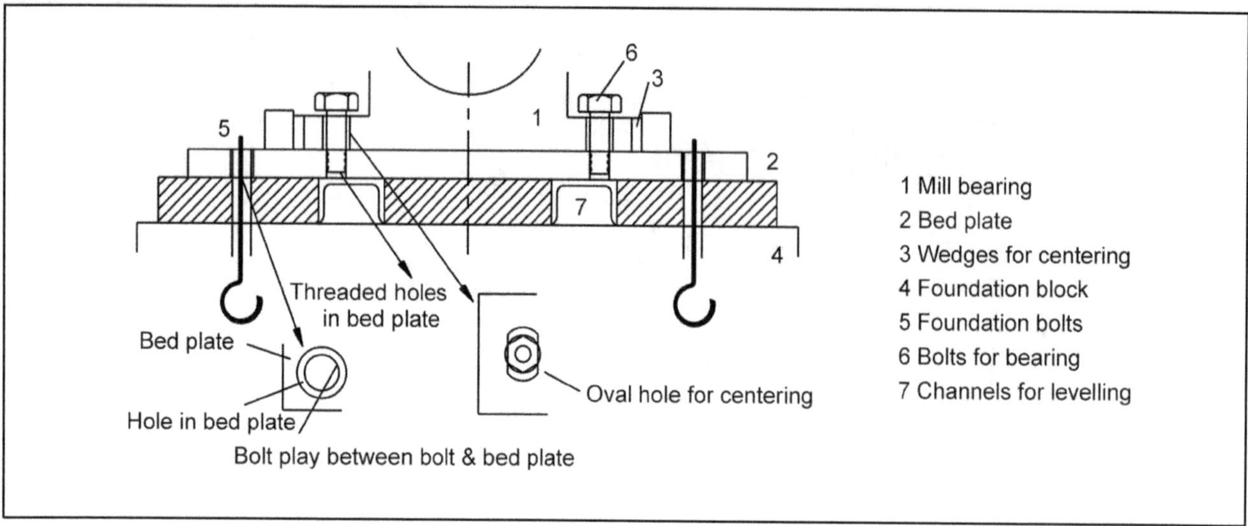

Fig. 6.47.8 Mill bearing - bed plate and fixing on foundation.

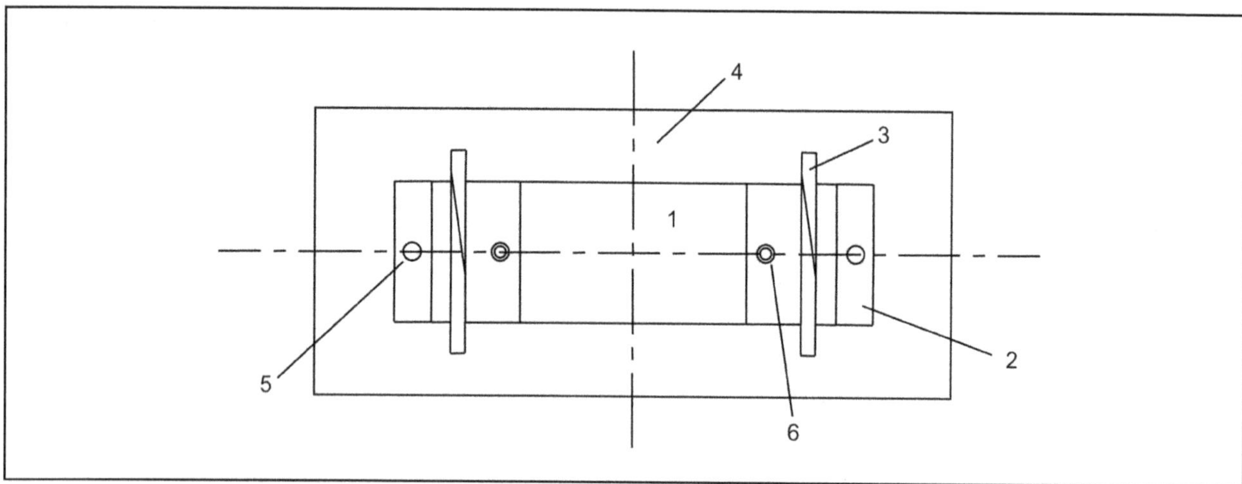

Fig. 6.47.9 Mill bearing - bed plate and fixing on foundation (plan).

Grouting is completed using special shrink proof cement after machine is fully aligned and bolts of bed plate and machine are fully tightened.

See Figs. 6.47.1 and 6.48.7 in Chapter 48.

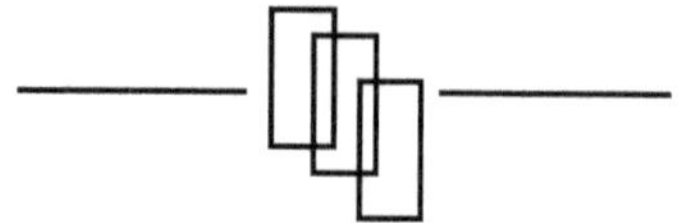

CHAPTER 48

FOUNDATIONS AND LOAD DATA

48.1 Design of Foundations

Foundation drawings and Load Data are supplied by the manufacturers of machines. In doing so, they would have taken into account the bed plate when there are bed plates between machine and foundation because the pockets to be provided in the foundations would correspond to bed plate designed.

48.2 Foundation Drawings

Should Clearly Show

1. Dead weight to be supported,
2. loads coming on to foundations and bolts due to motion of machinery and their directions,
3. in case of kiln additional loads come into play because of the slope of the kiln and its axial up and down movement in running,
4. axial thrusts if any say due to helical girth gears,
5. vertical thrust loads in case of roller mills and axial loads in case of roller presses due to pressure applied between rollers.

   See Figs. 6.48.1, 6.48.2 and 6.48.3.

These details must be furnished in full when there are more than one support like kiln, mill, etc. Load coming on each foundation must be shown separately. They are better presented in a tabular form.

48.3 Loads on Foundations

Dead load should include besides dead weight, loads due to charge, refractory, coating that forms during running and also where applicable, weight of material, hoppers of preheater cyclones and dust collectors can hold, in abnormal conditions.

For example in case of Kiln – dead weight would consist of weights of shell, tyres, girth gear, refractory, lining, coating and also weight of charge in the kiln.

In hoppers of preheater cyclones and hoppers of esps and bag filters dead load would include the weight of material collected in the hopper (when discharge is not working).

Some loads are evenly distributed and some are concentrated point loads.

Building floors would generally be designed to take a load expressed in t/m^2 basis.

Preheater cyclones are supported on beams. In some designs, cyclones are supported at 3 points in others at 4. Cyclones are not bolted neither are ducts.

See Fig. 6.15.44 in Chapter 15 of this **Section**

Not only loads but their points of application are very important to locate and size beams on each floor. Civil designer would so arrange net work of beams that loads get transferred to columns.

It is necessary that these details are received as early as possible so that they could be included in departmental drawings.

Otherwise the departmental drawings will include machine g.a. drawing and foundation details would have to be shown separately in another drawing.

It is necessary to follow Machinery Suppliers' instructions shown on foundation drawings supplied by them.

48.4 Foundations for Kilns, Mills, Roller Presses

Vertical Roller mills and kiln piers have large foundations blocks. Their own weight has to be taken

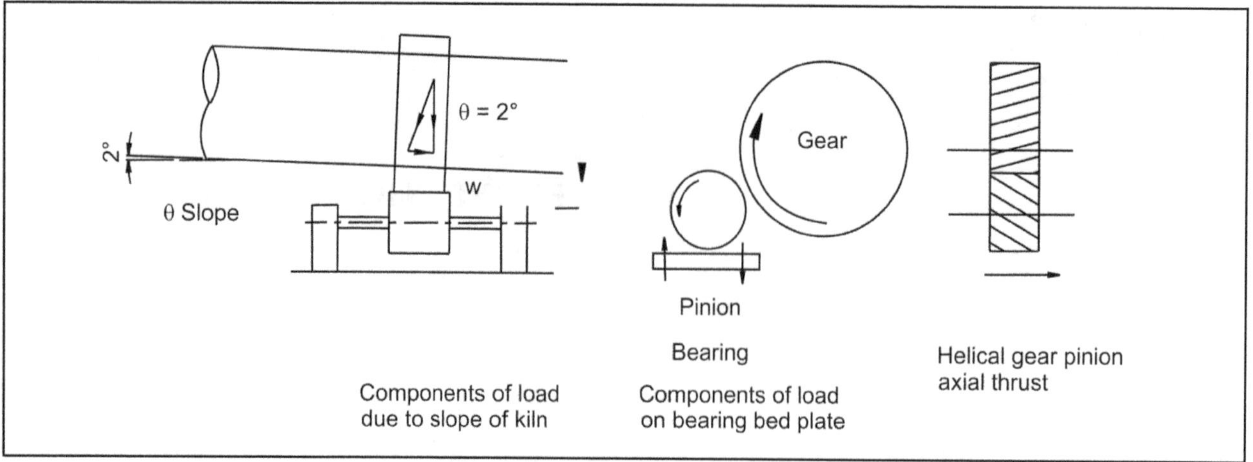

Fig. 6.48.1 Components of load on foundation of kiln.

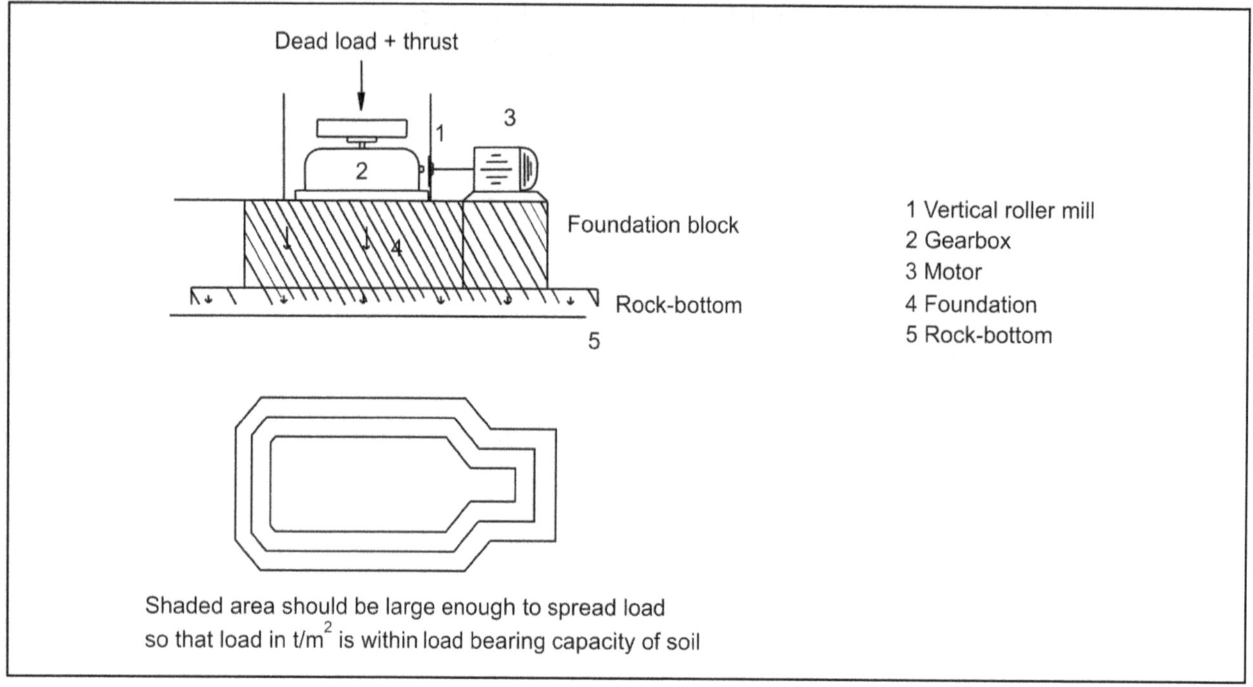

Fig. 6.48.2 Loads in case of vertical roller mill.

into account for spreading the load on the soil according to its load bearing capacity.

Vertical rollers mills have very high vertical thrust loads and need heavy foundation blocks which could weigh as much or more than the mill itself.

The area of foundation block should be such that load distribution is within the load bearing capacity of soil.

See Fig. 6.48.2.

Assuming that load bearing capacity is 20 t/m^2 and load is 800 Tons. Therefore base area will be minimum 40 m^2 or as dictated by the dimensions of the mill whichever is higher.

If the soil is soft or solid rock is too far below the surface, pile foundations are made. In pile foundations, first the number and spacing of piles is required to be settled. For that the foundation drawing should show locations and magnitude of loads clearly.

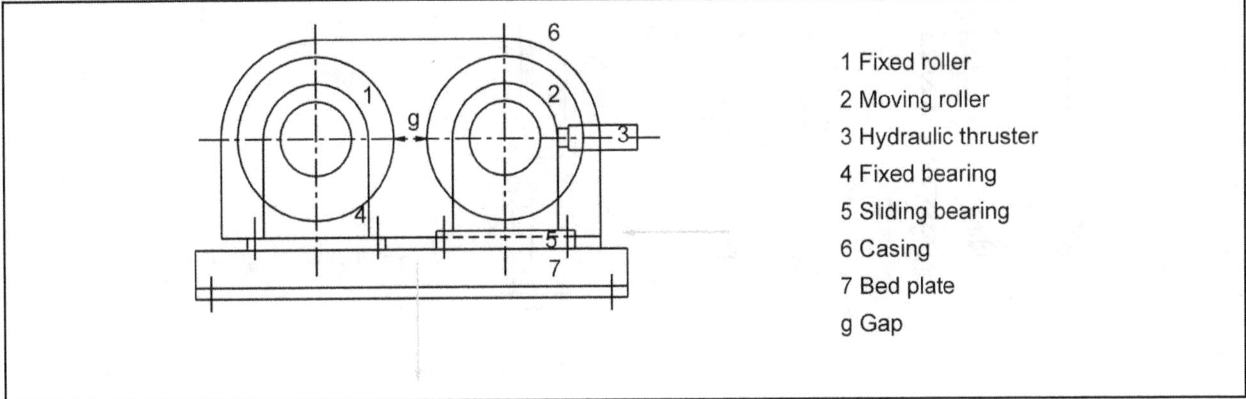

Fig. 6.48.3 Loads in case of roller press.

Foundation drawings should contain enough details for civil designer to take up the design thereof.

48.5 Foundation Bolts and Pockets

Machine designer will also design bolts for foundations viz their diameter, length and number.

There are of course thumb rules for sizing of foundation bolts. It would be better to err on the safer side so that there is sufficient grip between bolts, foundations block and machine.

See Fig. 6.48.4.

Pockets are left in foundations for bolts. Pockets can be straight, rectangular (or round) or taper as shown. The wooden planks used for pockets are removed when machine is to be installed. For easy removal, the planks are smeared with grease.

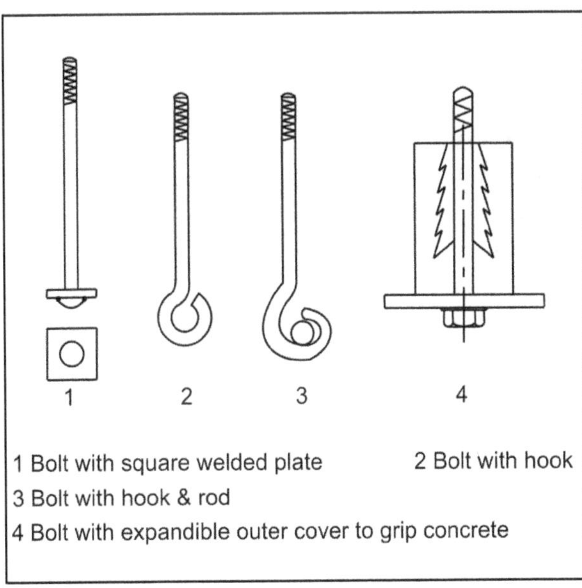

Fig. 6.48.4 Foundation bolts – different types.

See Fig. 6.48.5.

Pockets should be deep enough so that bolt can be lowered into the pocket and pulled to pass the hole in the bed plates. Threaded length h, should also be sufficient for nut, lock nut and spring / taper washer.

See Figs. 6.48.5 and 6.48.6.

Foundations bolts come in different designs. Whichever the design, the pocket size should be such that concrete used for grouting passes around the bottom plate into the pocket so that bolt is fully submerged in concrete on all sides.

See Figs. 6.48.4 and 6.48.6.

First bolts are half grouted; after the erection is complete and bolts are fully tightened, full grouting is done with shrink proof cement.

See Fig. 6.48.7.

Smaller machines like screw conveyors, elevators etc., also need bolts to fix them to the floor and also supports for belt conveyors.

Load of hood of elevator can be taken on beams. Load of boot is taken on a block, which may be deeper, than the rest of the slab or floor to suit length of foundation bolts.

See Fig. 6.48.8.

For light machines, rawl plugs are used to fix machine / bed plate in the floor slab.

See Fig. 6.48.4 item 4.

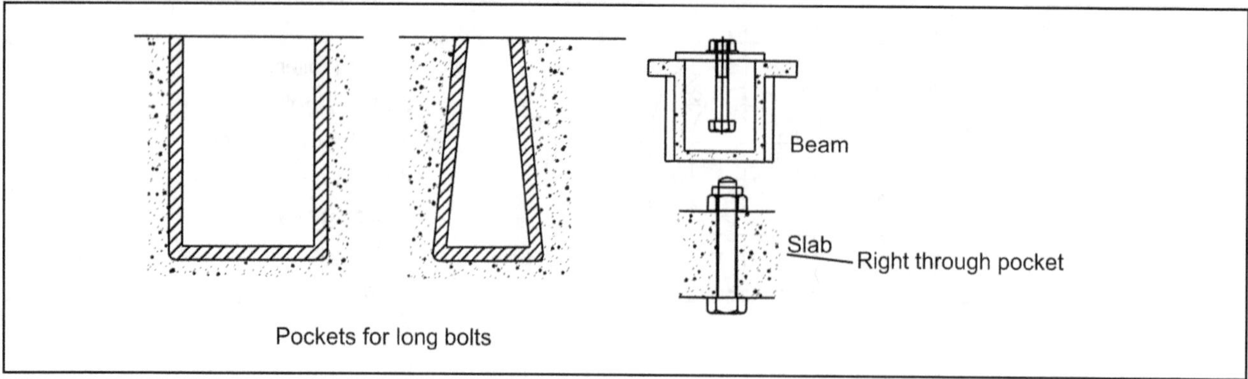

Fig. 6.48.5 Pockets for foundation bolts.

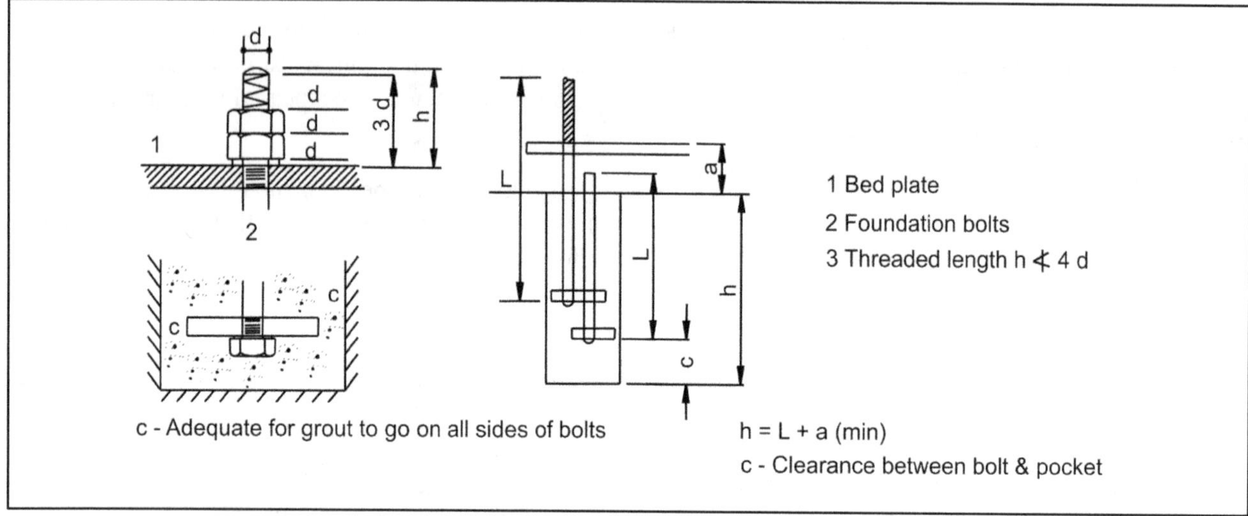

Fig. 6.48.6 Threaded length for tightening to be adequate in length ; pocket should be deep enough for inserting bolt.

In other cases, holes / pockets are left in the slab. Whichever way, the details should be shown on drawing so that execution is done during construction.

When a conveyor is supported on trestles, trestles are supported on footings.

See Fig. 6.48.9.

48.6 Foundation Load Data for a Kiln

Typical Foundation load data for a rotary kiln is shown in **Figs. 6.48.10 and 6.48.11** for a three support rotary kiln by way of illustration.

Loads vary considerably due to bending of kiln shell and misalignment. Such data should be obtained from Suppliers for all major machines.

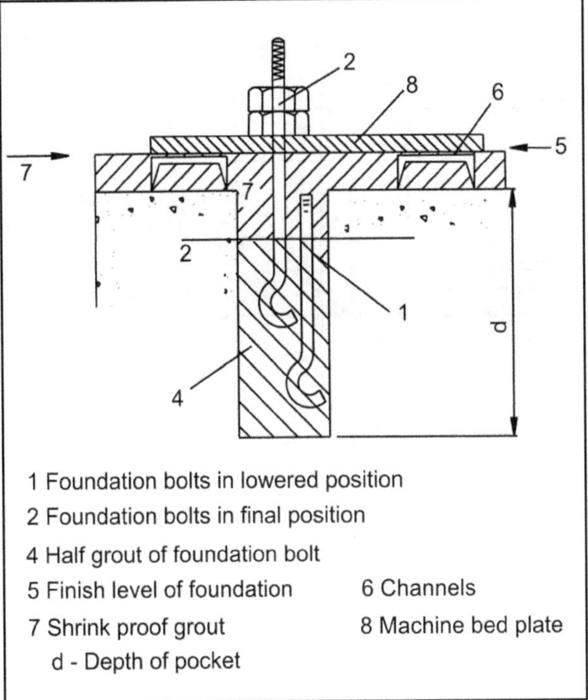

Fig. 6.48.7 Depth of pocket for foundation bolt.

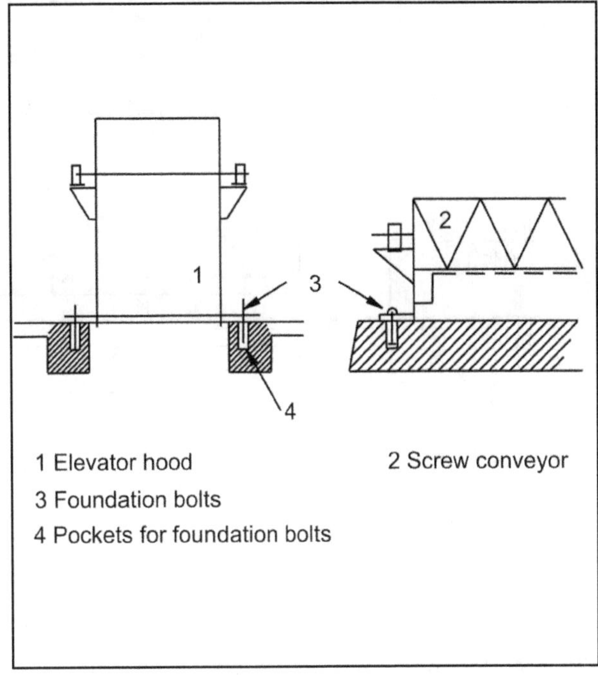

Fig. 6.48.8 Small foundations – elevator hood, screw conveyor etc.

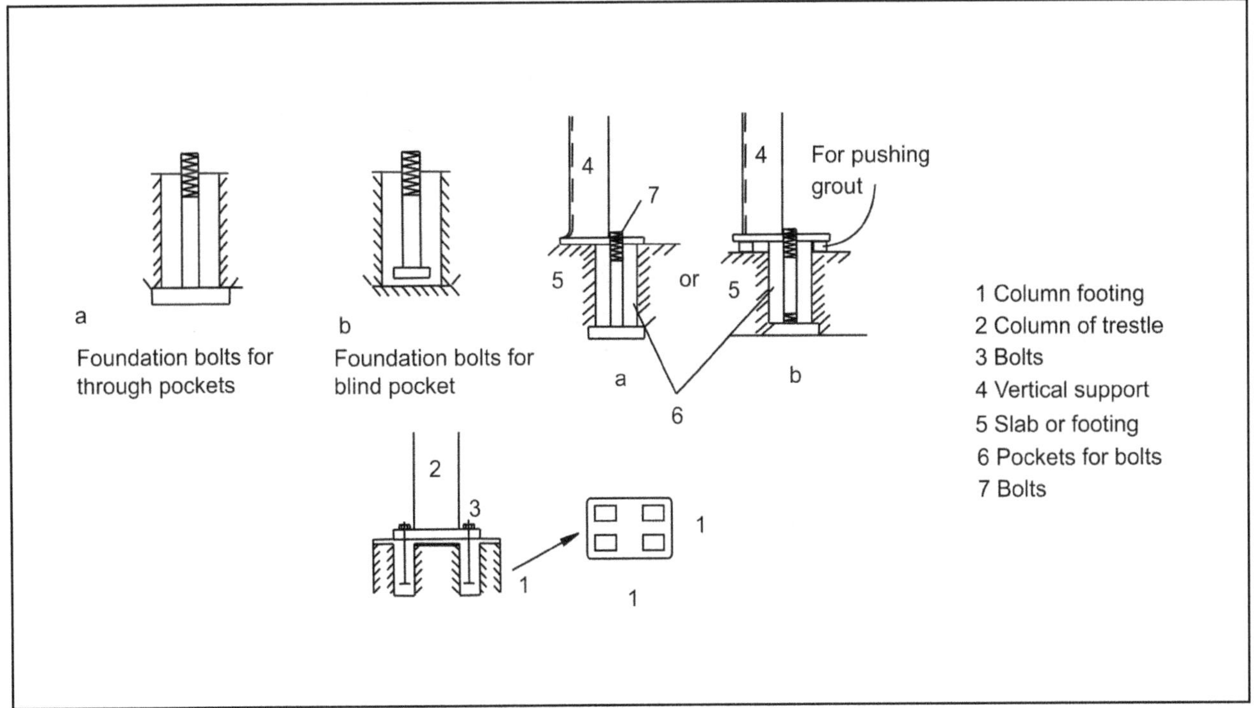

Fig. 6.48.9 Foundations for conveyor and pedestal supports.

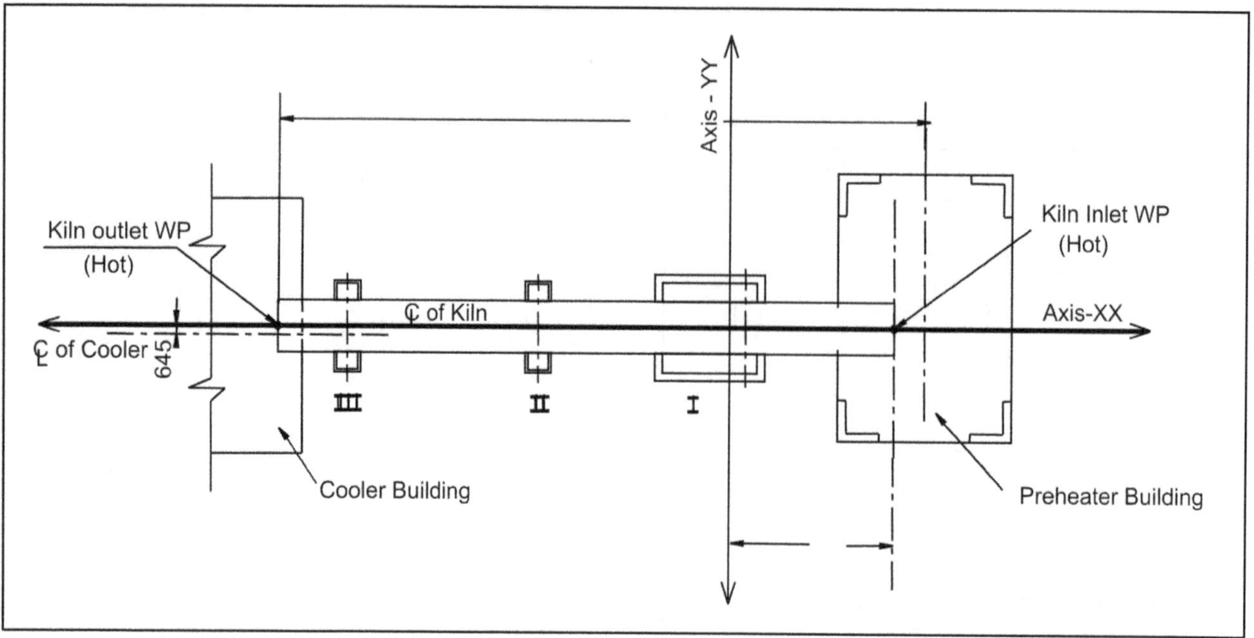

Fig. 6.48.10 Key plan of a 3 support rotary kiln

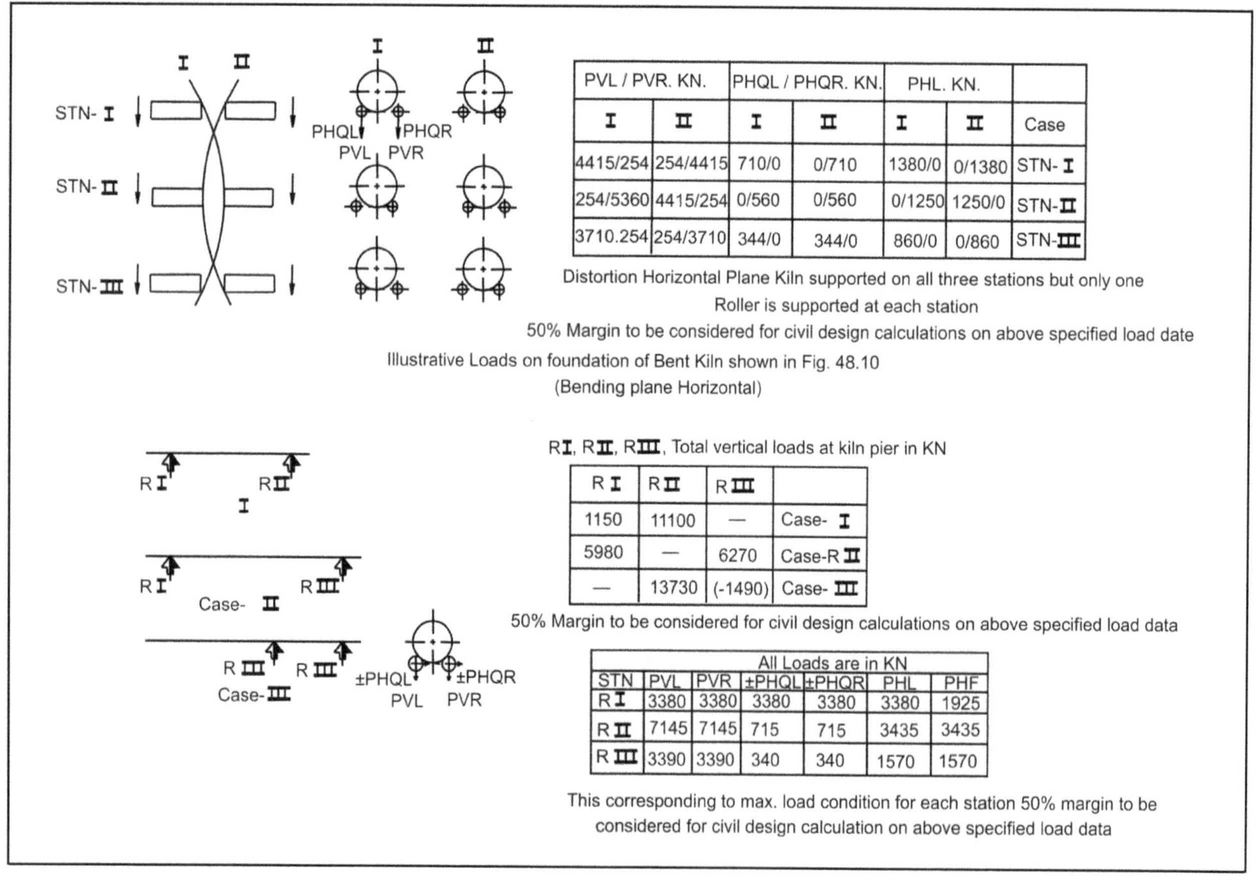

PVL / PVR. KN.		PHQL / PHQR. KN.		PHL. KN.		
I	II	I	II	I	II	Case
4415/254	254/4415	710/0	0/710	1380/0	0/1380	STN- I
254/5360	4415/254	0/560	0/560	0/1250	1250/0	STN-II
3710.254	254/3710	344/0	344/0	860/0	0/860	STN-III

Distortion Horizontal Plane Kiln supported on all three stations but only one Roller is supported at each station

50% Margin to be considered for civil design calculations on above specified load date

Illustrative Loads on foundation of Bent Kiln shown in Fig. 48.10 (Bending plane Horizontal)

RI, RII, RIII, Total vertical loads at kiln pier in KN

R I	R II	R III	
1150	11100	—	Case- I
5980	—	6270	Case-R II
—	13730	(-1490)	Case- III

50% Margin to be considered for civil design calculations on above specified load data

All Loads are in KN

STN	PVL	PVR	±PHQL	±PHQR	PHL	PHF
R I	3380	3380	3380	3380	3380	1925
R II	7145	7145	715	715	3435	3435
R III	3390	3390	340	340	1570	1570

This corresponding to max. load condition for each station 50% margin to be considered for civil design calculation on above specified load data

Fig. 6.48.11 Foundation Load data for a three support rotary kiln.

CHAPTER 49

CIVIL DRAWINGS AND DESIGN OF DEPARTMENTAL BUILDINGS

49.1 Civil Design and drawings

All major loads should be taken on beams and transferred to the main columns of the building. In case of preheater tower, each floor would have different configuration of auxiliary beams on each floor because cyclones are located at different places on each floor.

Heights between floors should be so arrived that the clearances are sufficient for heights of machines and chutes and ducting after deducting depths of beams. **See Fig. 6.49.1.**

49.2 Dimensions of Beams and Columns

At design stage, the dimensions of beams and columns would not be known precisely. Use is made of conventions, thumb rules and past experience and drawings made earlier. Thumb rule : span 'x' meters; depth of beam-'y' cms.

However, when beams and columns are finally designed by civil designers and their dimensions are known, the departmental drawings should be checked for clearances and adjustments made where necessary with the help of Consultants. Civil designers should bring to the attention of Consultants making the departmental drawings any problems faced by them.

49.3 Critical Dimensions for Design of Buildings

The critical dimensions for slopes of chutes and clearances necessary should be shown on departmental drawings before passing them on to civil designer so that the same can be taken care of.

49.4 Taking up of Civil Design

A civil designer begins his design work when he receives departmental drawings showing overall dimensions and loads and foundation details.

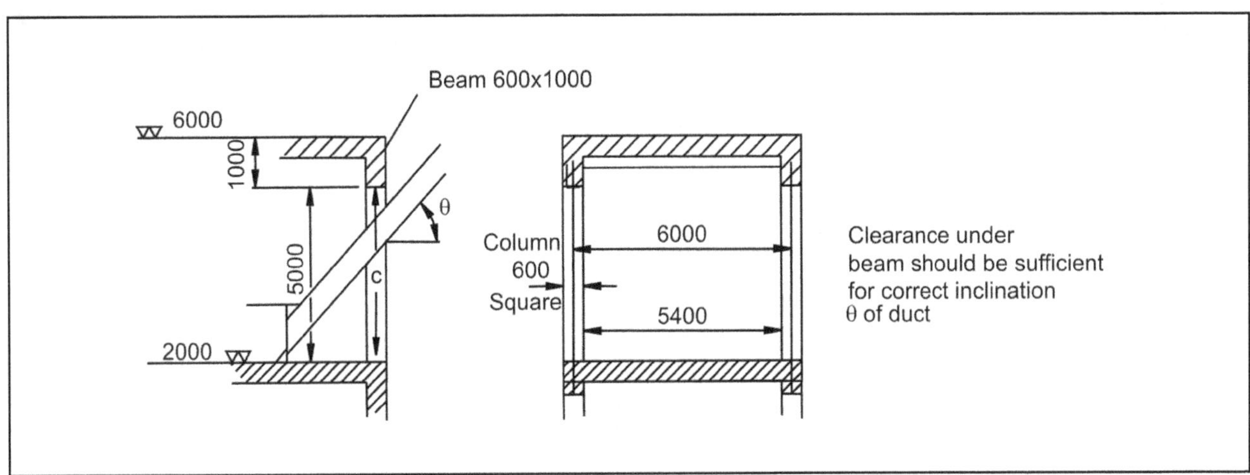

Fig. 6.49.1 Clearances for machinery and ductings - in buildings.

A close co-ordination is thus necessary between machinery manufacturer, consultant, civil designer and other suppliers and also electrical contractors.

49.5 Preparation of Departmental Drawings

If major machinery are brought from one vendor he is in a position to furnish a more complete departmental drawing.

In any case he should furnish a drawing showing overall dimensions of building and floor levels. This he may prepare with some assumptions making use of past experience.

Systems Designers / Consultants furnish specifications for auxiliaries and bought outs.

When orders are finalized, the respective suppliers furnish ga and foundations drawings of their machinery which need to be integrated in the main departmental drawings.

A departmental drawing thus cannot be completed in one go but is completed in stages as information and drawings are received from respective suppliers.

49.6 Data for Departmental Drawings

Finalizing orders is a long drawn out process and if the Company wants to go 'shopping' to save on capital costs, it could buy the same type of machinery from different suppliers.

Sometimes, one manufacturer does not make the entire range required in a plant as happens in case of fans, motors, etc.

Thus there are a host of parties involved who have to furnish ga drawings and foundations drawings to the Company and their Consultants.

At times the main Vendor takes responsibility to make departmental drawings in which case all drawing received from other vendors have to be passed on to him.

49.6.1 Drawings for Drives

Drives are finalized after machineries are finalized. Therefore details of gear boxes and motors are received last.

Fortunately, motors and gear boxes are made to International Standards. For example, motors are made as per International Frame Sizes so that once frame size is known, pertinent details like bolt hole spacing can be obtained from published catalogues.

However, it is best to obtain certified drawings from respective suppliers of motors and gear boxes so that errors due to catalogues being outdated do not arise.

49.7 Co-ordination of Concerned Parties

Cement Entrepreneurs and Consultants have thus to do a great deal of planning and co-ordination so that information and drawings are received in the sequence and in time allotted so that work on drawings proceeds smoothly.

Considerable amount of time can be saved in completing civil design and construction drawings if departmental and load drawings are complete and comprehensive and data is furnished in correct sequence.

49.8

This Chapter should be read together with **Section 4** wherein pertinent aspects of civil design have been dealt with.

49.9 Norms for Green Buildings

Extracts from norms for Green Buildings for factories and homes have already been given in **Chapter 1** of **Section 4** of the book.

They should be kept in mind in designing buildings to conserve energy and water and to provide facilities for disabled people.

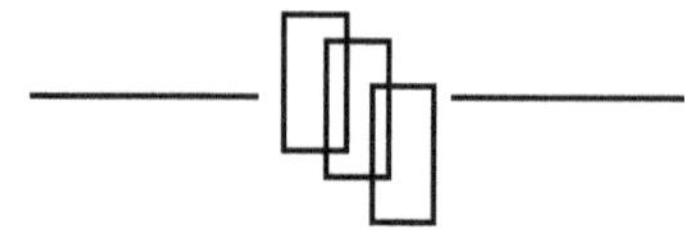

CHAPTER 50

WORKING PLATFORMS

50.1 Working Platform for Machines

Machines need to be attended to in operation. They also need maintenance on regular basis and repairs / replacement in a planned manner or during breakdowns.

To judge the operation and measure operational parameters like temperature, draughts, gas flow, oxygen content etc., field sensors are installed either on permanent basis or on 'as and when' basis at specific points on the machine or the system. They need access. Measurements are also needed to be taken to calibrate the field instruments. Measurements may be required at new locations now and then.

It is not always possible to reach all such points from main floors of the buildings.

It is therefore necessary to reach the points located at a higher level and to have a working space around them.

For these reasons working platforms are constructed around a machine. In machines like bag filters and esps the working platforms are built on them by machinery manufacturers themselves.

See Fig. 6.50.1.

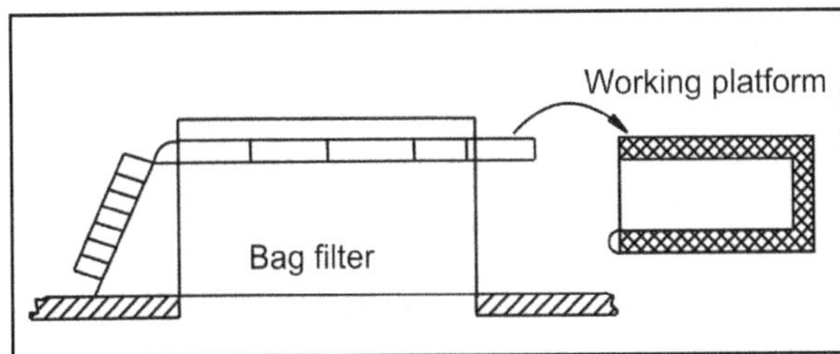

Fig. 6.50.1 Working platforms around machines dust collectors - ESPs - bag filters.

50.1.1 Additional Floors for Working Plarforms

Sometimes such platforms could be in the form of an additional floor in the building.

But if the building is large in dimensions, it is too expensive to provide an additional floor in concrete.

50.2 Fabricated Working Platforms

Working and maintenance platforms would therefore be mostly in steel and with access ladders from nearest floor level and the area would be limited to specific requirements.

50.3 Preheater Cyclones

It is necessary to design and provide working platforms around preheater cyclones. Some manufacturers of preheater supply such platforms along with preheater cyclones.

See Fig. 6.50.2.

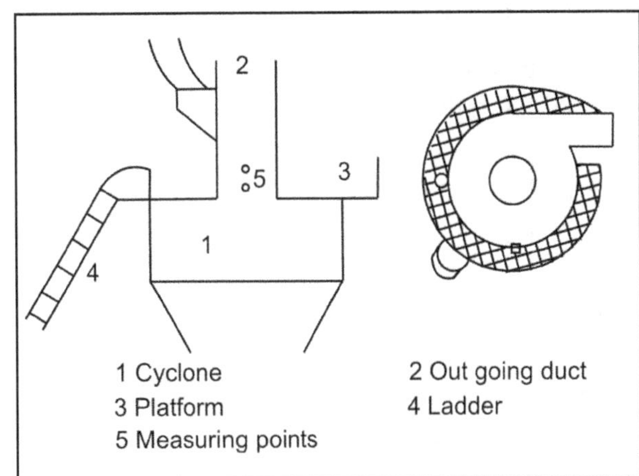

Fig. 6.50.2 Working platforms around preheater cyclones.

Access to working platform can be taken from floor level immediately below or from the landings of the staircases of the building.

See Fig. 6.50.3.

Monkey ladders should be avoided so that heavy equipment could be carried up and down easily.

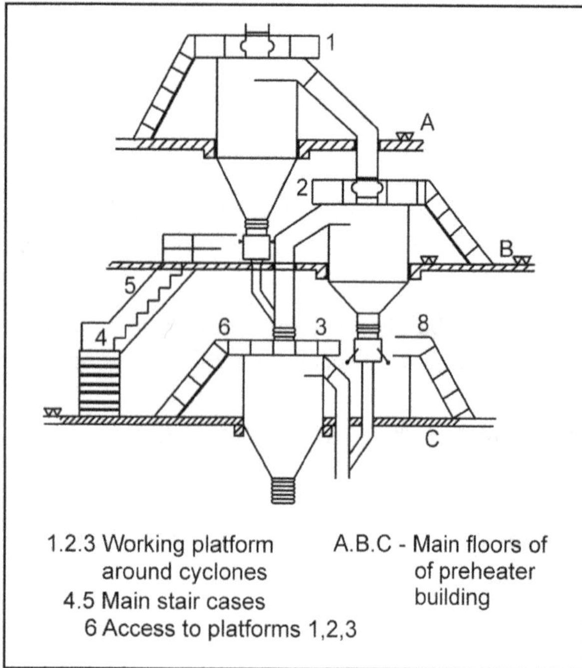

Fig. 6.50.3 Preheater cyclones, working platforms.

The raw meal pipes for preheater and calciners have pendulum flap valves. These need to be accessible and are best located near floor levels.

See Fig. 6.50.3.

The expansion joints in them cannot always be located near floor.

In designing a building the needs for access for such purposes should be ascertained.

The best party to do this is the manufacturer of the machine itself.

He can either include attached working platforms in his scope of supply or can show them in the g.a. and departmental drawing housing the machinery.

The consultant has to integrate them in the final departmental drawing. In consultation with Civil Designers it may be decided whether these could be in RCC as additional floors or should be fabricated and erected at site.

50.3.1 Working Platforms and Staircases and Landings

Working platforms should be planned along with the staircase and ladders of the building so that they do not interfere. Landings can provide ready and easy access to the platforms.

50.4 Platforms for Apron Feeder

Working platform and platform for supporting drive of apron feeder for crusher can be in concrete but should be so constructed that they do not obstruct maintenance of the crusher.

See Fig. 6.50.4.

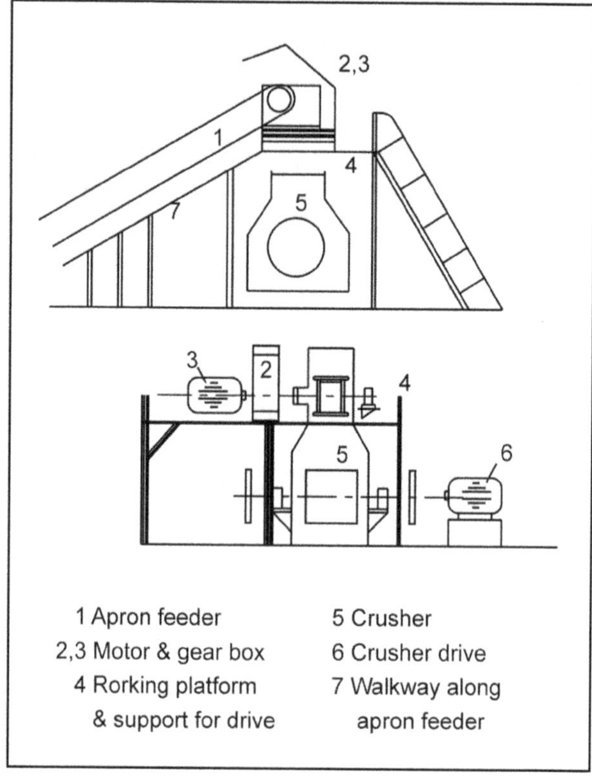

Fig. 6.50.4 Working platform for feeder of crusher.

50.5 Platforms for Belt Conveyors

Long / short belt conveyors have a gantry along them which also serve as access and working platforms. At drive or discharge end of conveyor, working platform is more conveniently made in steel and supported from the supporting structure of the belt itself.

See Fig. 6.50.5.

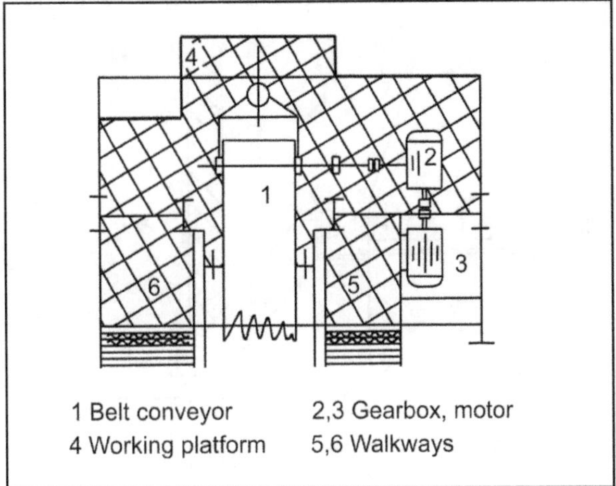

Fig. 6.50.5 Working platform around conveyor drive.

50.6 Platforms Around Kiln Piers

Kiln piers may be made a little wider at top or can have a concrete platform around them. It can be in steel also. Piers will be connected with one another by steel walkways.

See Fig. 6.50.6.

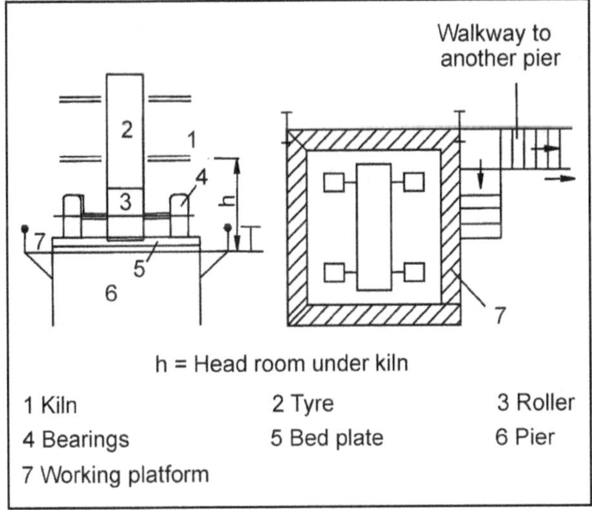

Fig. 6.50.6 Kiln piers.

50.7 Platform for Inspection Doors

Working platforms should be provided around inspection doors and manholes for access and for handling materials like refractories.

When access doors and poking holes are provided, they should be accessible from a platform.

See Fig. 6.50.7.

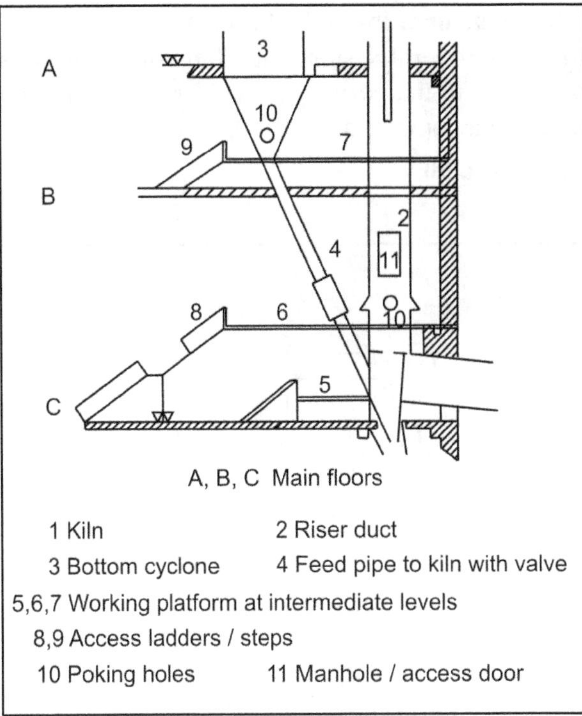

Fig. 6.50.7 Working platforms at kiln inlet end.

50.8 Platforms for Grate Cooler

Working platform / access is provided for grate coolers on both sides throughout the length of cooler. It makes maintenance easy and convenient. Platform can be in concrete or steel.

See Fig. 6.50.8.

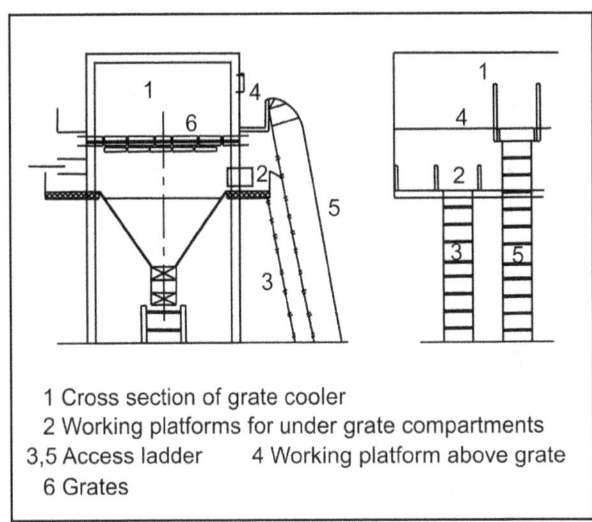

Fig. 6.50.8 Working platforms along grate cooler.

50.9 Platform for Grits Separators

Machines like grit separators need adjustment of vanes from time to time. An access ladder and a platform around it is convenient.

See Fig. 6.50.9.

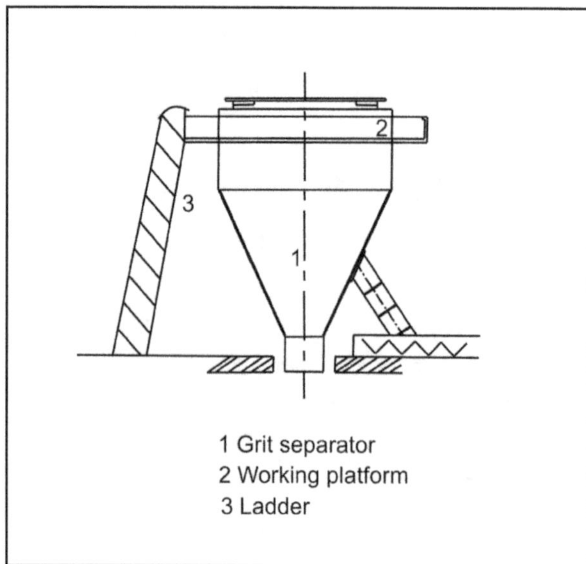

Fig. 6.50.9 Working platform for grit separator in coal mill circuit.

50.10 Platforms for Hoppers

Hopper outlet needs to be poked to remove jams. A working platform is necessary for this purpose. But it should be so located that man stands above manhole and material does not fall on him when jam is suddenly released.

See Fig. 6.6.25 in Chapter 6 of this section

50.11 Platform for Gas Conditioning Tower

Working platform is needed at the level of nozzles fixed on the conditioning tower to attend to them. It will be in steel. Access to it will be provided from the nearest building.

See Fig. 6.50.10.

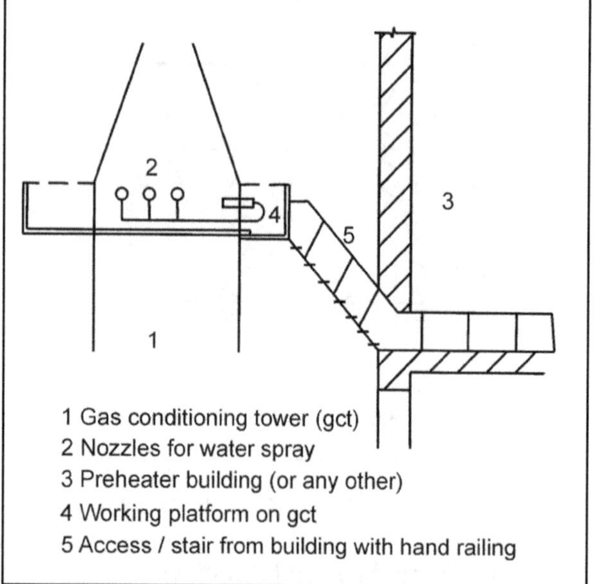

Fig. 6.50.10 Working platform for gct.

50.12 Platforms for Esps and Bag Filters

As mentioned above bag filters will come with access ladders and working platform fixed around them at appropriate level. In bag filters bags are removed and replaced from the top. Therefore working platform will be near the top. On bag filters where explosion flaps are provided, working platform will be so provided that in the event of a blow out no one is injured.

See Fig. 6.53.17 in Chapter 53.

In case of esps working platforms will be needed at two levels. One near bottom where shaking mechanisms for collecting electrodes are provided. And one at top to attend to shaking mechanisms of discharge electrodes and to transformer rectifier sets. Collecting electrodes are removed from the top. Hence a platform at top will facilitate it.

See Fig. 6.52.8 in Chapter 52.

Primarily of course a building should be so designed as to have floors so arranged as to minimize need for separate working platforms and to have a short and direct access to them.

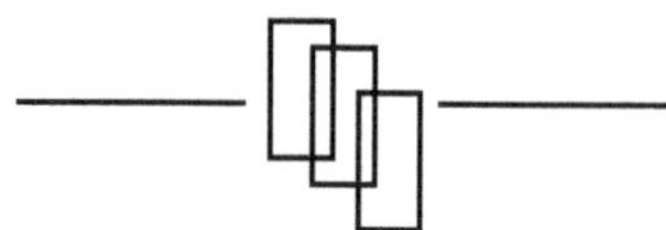

CHAPTER 51

CONVEYOR GANTRIES

51.1 Conveyor Gantries

When not installed at ground / floor levels, most conveyors need structures to support them. Even when installed at ground/floor level, they require structures to support them but of smaller heights to maintain safe clearance of pulleys above floor level. They also need walkways along their length for inspection and maintenance when installed at higher levels.

See Figs. 6.51.1 and 6.51.2.

51.2 Belt Conveyors

A belt conveyor would need a supporting frame along its length to support carrying and return idlers. If the top of conveyor is at a considerable height over the

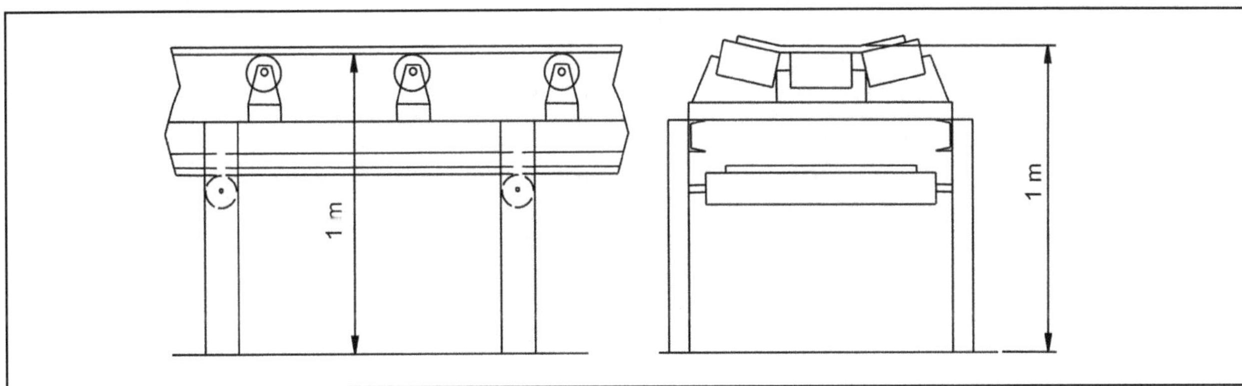

Fig. 6.51.1 Belt conveyor supporting structure when conveyor is level.

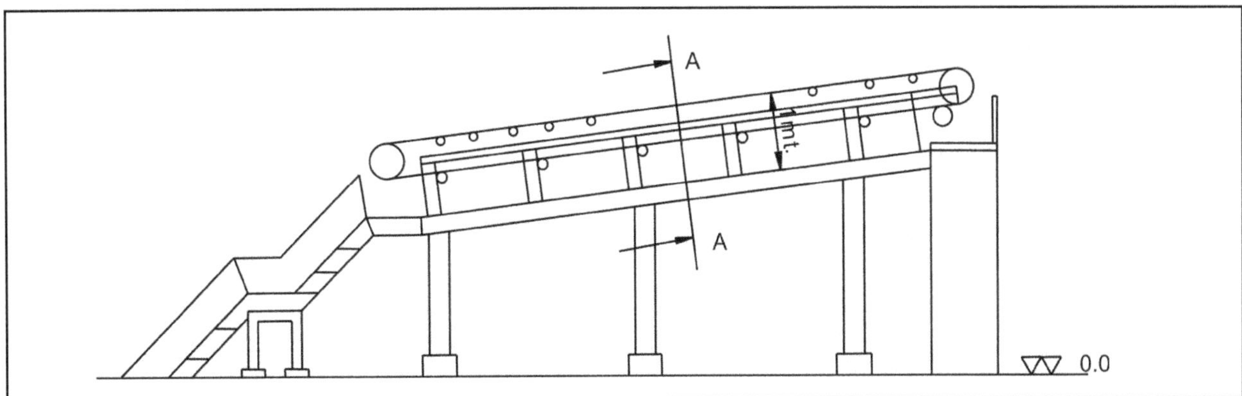

Fig. 6.51.2 Supporting structure for inclined conveyor.

floor level, it would need access to it for maintenance and inspection.

See Fig. 6.51.2.

Structure used to support belt is also used to support walk ways or gantries on one side or more commonly on both sides.

See Fig. 6.51.3.

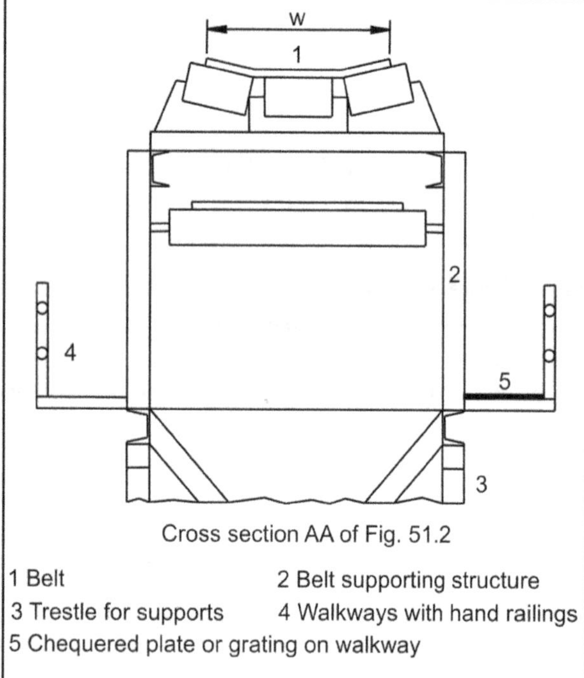

Fig. 6.51.3 Supporting structure with walkways on either side for maintanance - for belt conveyor.

Heights of trestles increase when conveyors are at an angle and can be quite tall if conveyors are unloading into silos.

Spacing between trestles would depend on size of conveyors and their capacities. Number of trestles can be reduced at the expense of cross sections of supporting frame as spans would increase.

One of the trestles would also be used to support belt tensioning device as explained in **Chapter 3, Fig. 6.3.17.**

51.2.1 Covering of Belt

According to its location and the material it is carrying, a belt may be fully covered (say like when conveying cement).

Such covers are made of thin gauge plates / aluminium or plastic sheets. It should be possible to lift the covers, also close and lock them quickly. Therefore they would be hinged on one side and with clamps on the other side.

Arrangement should also be made to support opened cover so that belt can be attended to.

See Fig. 6.51.4.

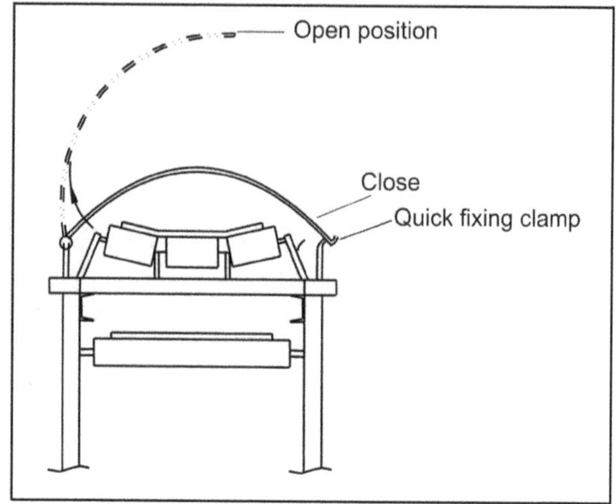

Fig. 6.51.4 Covered belt for conveying cement.

When a cover is provided on the belt, it is not necessary to cover the gantry. Exception would be when belt is used to convey cement. It must then be adequately protected against entry of water.

See Fig. 6.51.5.

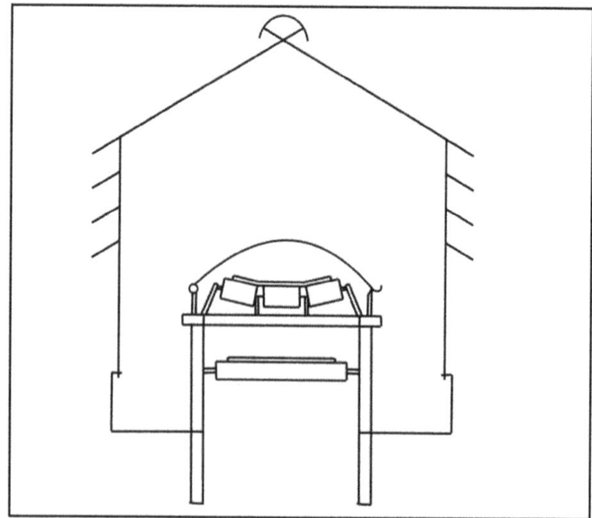

Fig. 6.51.5 For cement additional protection by installing a covered gantry with side cladding.

Such covers are also provided when high winds are likely to lift the belt off the idlers. It is particularly so for conveyors installed at a height.

51.2.2 Head Room Under Conveyor

Long belt conveyors are installed at a height sufficient to clear road, railway lines and vehicles on them when required to cross them.

See Fig. 6.51.6.

However, it is quite expensive to install the belt all through at such a height. Therefore, long belts are installed 'overland' that is at height just sufficient to provide enough clearance from ground and therefore belt follows the contour of the route.

Conveyor height needs to be raised only at crossing of roads / railway etc.

See Fig. 6.51.7.

51.2.3 Overland Conveyors

Overland conveyors are easy to inspect and maintain. But require a strip of land to be acquired for installation and also for a road along side it for maintenance. Such conveyors cause other problems like crossings for cattle if conveyors pass through cultivated land and grazing fields. Hence fencing would also be required along its length.

See Fig. 6.51.8.

Conveyors at elevated height would require only permission for footings to be constructed in the land.

It is of course necessary to protect people, cattle against injury by installing nets under the belt to receive spillage where necessary.

See Fig. 6.51.6.

51.2.4 Trestles and Gantries

Trestles and gantries can be made in steel or concrete or combination of steel and concrete. Concrete trestles or gantries would be cheap. Steel gantries can be fabricated separately in bulk and erected fast.

51.3 Drag Chain Conveyors

Drag chain conveyors are mainly used to convey clinker discharged from Clinker cooler at temperatures ranging between 100 to 120 °C. In small kilns Drag chain was used to convey not only spillage but also the main product.

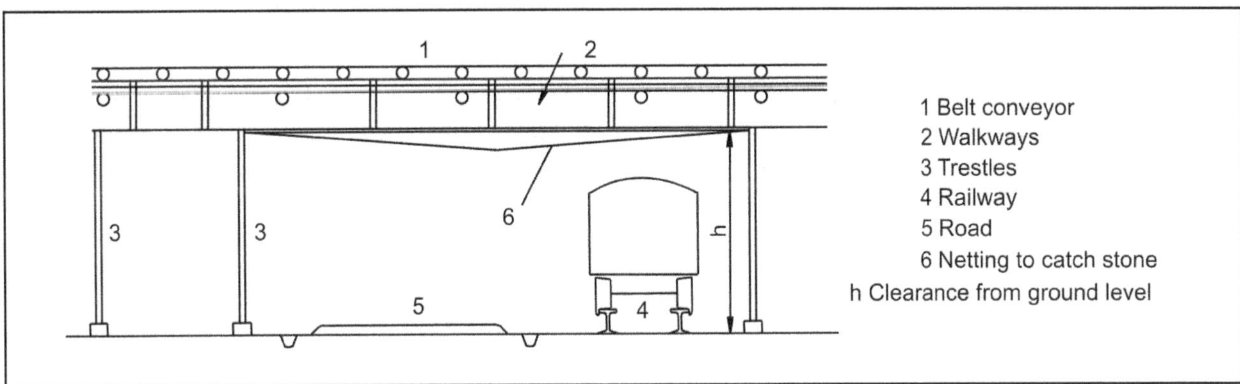

Fig. 6.51.6 Laying belt conveyor overland at crossing of road, rail track etc.

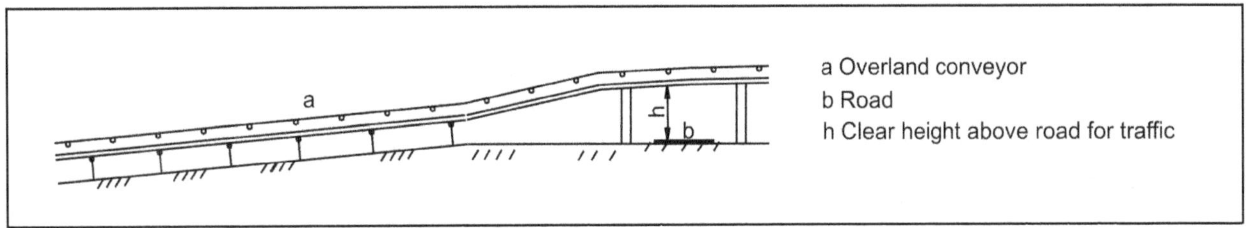

Fig. 6.51.7 Overland belt conveyor and road crossing.

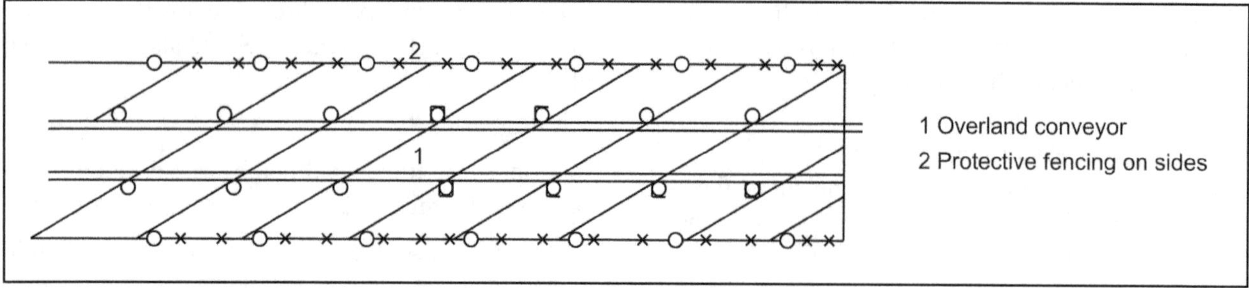

Fig. 6.51.8 Overland conveyor requires protective fencing and acquisition of land throughout its length.

Drag chains would normally be laid in a concrete trough lined with steel plates at sides; rails would be embedded at bottom on which the load carrying strand would slide. Top strand would be supported on pulleys from walls of the trough.

See Figs. 6.22.1, 6.22.9 and 6.22.10 in Chapter 22 of this Section

While a drag chain conveyor thus laid out does not require a gantry, sufficient space must be maintained on either side for maintenance purposes.

Presently drag chain conveyor would be installed at ground level and not in a tunnel. Therefore providing space around it is not difficult.

See Fig. 6.22.9.

When handling spillage from under grate compartments of cooler, the drag chain has to handle very hot clinker dust with temperatures as high as 800 to 900 °C to cooled clinker dust at 100 °C at the discharge end. Cooling air can sometimes escape through pendulum flap valves under the Compartments and creates considerable dust nuisance. Therefore often the Conveyor is covered. Vent points are provided to remove the dust.

See Fig. 6.22.10.

Drag chain conveyors are laid in troughs and would have concrete walkways in steps on either side if they were installed at an angle.

As a thumb rule walkways around conveyors would be in concrete inside the buildings and in steel outside the building.

51.4 Deep Bucket Conveyors

Deep bucket conveyors (DBC) are now used to convey clinker at high rates suitable for kilns of 10000 tpd and above capacities that is above 600 tph capacities. They are slow moving conveyors and if properly fed, clinker does not come in contact with moving chain and wheels.

DBCs start from under the cooler generally at depths of –5 to –6 metres. Initially they are horizontal but then they have inclinations of 35 to 40° to reach top of conical clinker storages.

Since there are no structures or buildings in between, dbcs must necessarily have their own supporting structure and also gantries along it for maintenance.

Till the conveyor emerges from the tunnel at ground level, walkways around it will be in concrete. Above ground level it will be in steel.

See Fig. 6.51.9.

A standby is very desirable for DBCs to maintain continuity of uninterrupted operation of the kiln. The gantry should be designed for a standby conveyor to be added if not installed for the first line. A common standby would suffice for two production lines.

See Figs. 6.26.9 to 6.26.11 in Chapter 26 of this Section.

Since the supporting columns will be inside the clinker storage and hence submerged in clinker, it will be prudent to construct stub columns for future DBC and gantries along it above the conical shed. This way second DBC can be added without any difficulty.

See Fig. 6.26.11 in Chapter 26.

When a second kiln is installed clinker storages of two kilns will be joined by a cross conveyor. This conveyor will be at a high level and hence would need supporting structure and walkways around it.

See Fig. 6.51.10.

2
35°
1
Steel
Steel
Concrete
1 Deep bucket conveyor
2 Supports

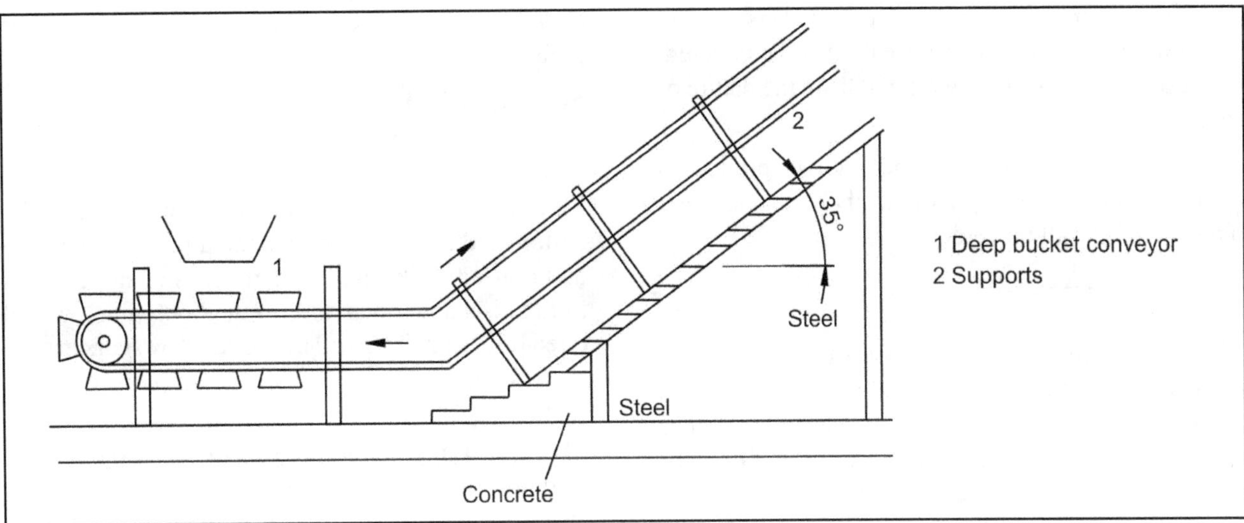

Fig. 6.51.9 Deep bucket conveyors - walkways and supporting structure.

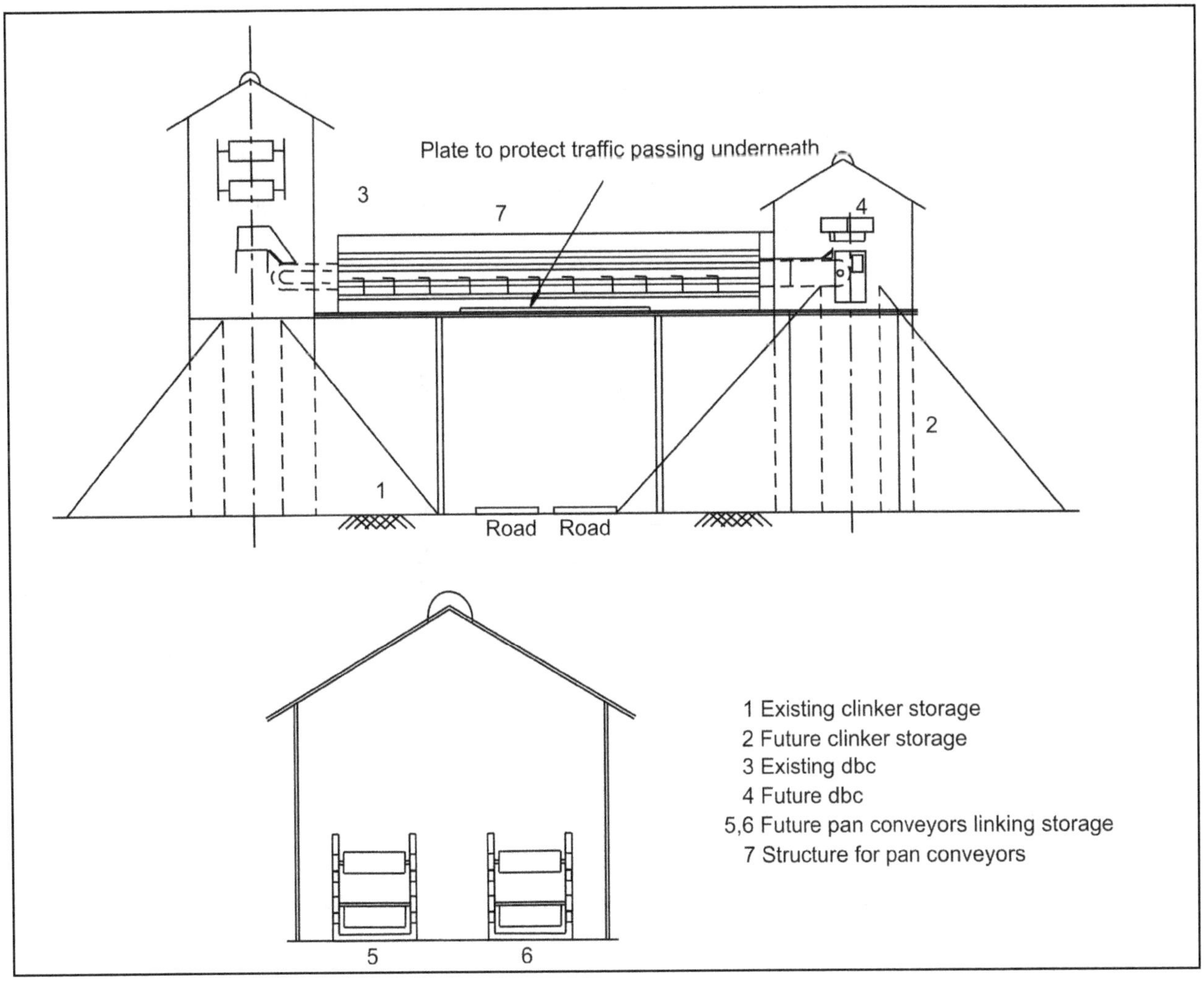

Plate to protect traffic passing underneath
3
7
4
2
1
Road Road
5
6
1 Existing clinker storage
2 Future clinker storage
3 Existing dbc
4 Future dbc
5,6 Future pan conveyors linking storage
7 Structure for pan conveyors

Fig. 6.51.10 Walkways-grantries for clinker conveyors connecting stockpiles of two production lines.

Clinker conveyors will be covered as a rule. Therefore walkways around it will also be covered. Because of its slope walkways will be in the nature of steps.

If a road passes under the dbc it will be prudent to fix a plate under that length of the dbc so that spillage if any does not fall on vehicles.

See Fig. 6.51.10.

51.5 Screw and Chain Conveyors

Dust is collected from hoppers of dust collectors like E.S.P.s and bag filters by screw conveyors or Redler chain conveyors. Sometimes the conveyors are attached to the dust collectors themselves.

Dust collectors if small may be inside a building. In this case the conveyor can be installed at ground level to eliminate a supporting structure and walkway for maintenance for it.

Dust collectors for kiln, raw mill and for cooler vent will be very large and hence would be out in the open. Conveyors under these dust collector will need gantries and walk ways albeit of short heights to be able to attend to the conveyors. The walkways will often be grills with railings.

See Fig. 6.51.11.

51.6 Elevators

Elevators rise vertically and have a feed point at the bottom and discharge point at the top. They therefore do not need walkways or gantries. They need access to the discharge hood and to the boot which would often be in a pit. This will be provided by stairs in steel or in concrete.

51.7 Design of Walkways and Railings

Widths of walkways can be standardized. Design can also be standardized – whether grills or chequered plates so that procurement can be planned and be economical. A good width for a walkway will be between 500 to 750 mm.

While grills would be suitable in most places, it would be better to provide plates in sections where tools or material can accidentally fall on passersby underneath.

It is also necessary to provide hand railings on sides. Their design can also be standardized.

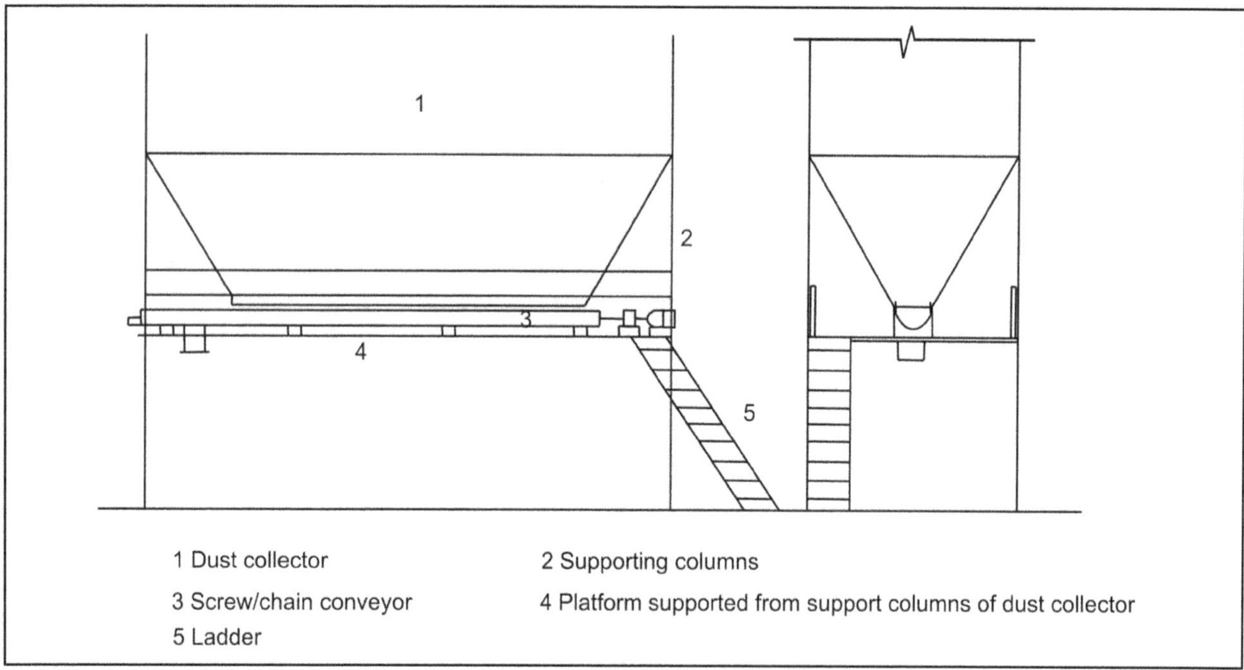

Fig. 6.51.11 Support platform for screw conveyor/chain conveyor under dust collector.

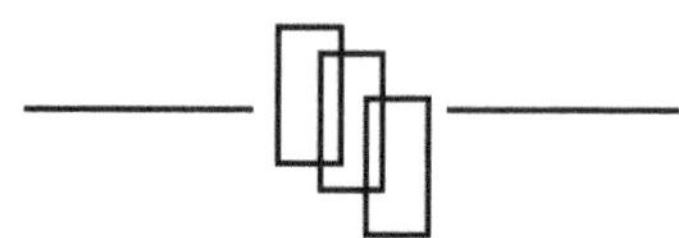

CHAPTER 52

LIFTS, HOISTS AND LIFTING TACKLES

52.1 Lifts for Buildings and Silos

In case of 5/6 stage preheaters of large capacity plants, towers reach heights of 80-90 metres, i.e., equivalent of 30 storied buildings or taller.

It is impossible and impracticable to reach higher floor levels without lifts. Therefore, tall buildings should be provided with lifts, which can carry materials and workers. 1 to 2 tons capacity lifts should be planned at design stage and installed before start up in a preheater tower.

See Fig. 6.15.32 in Chapter 15 of this Section and also Fig. 6.52.1.

52.2 Lifts in Various Buildings of Cement Plant

Lift may be with attendant or automatic. It must have a communication system besides the usual controls for operation. The doors should be wide and should admit large machinery parts, refractories, etc.

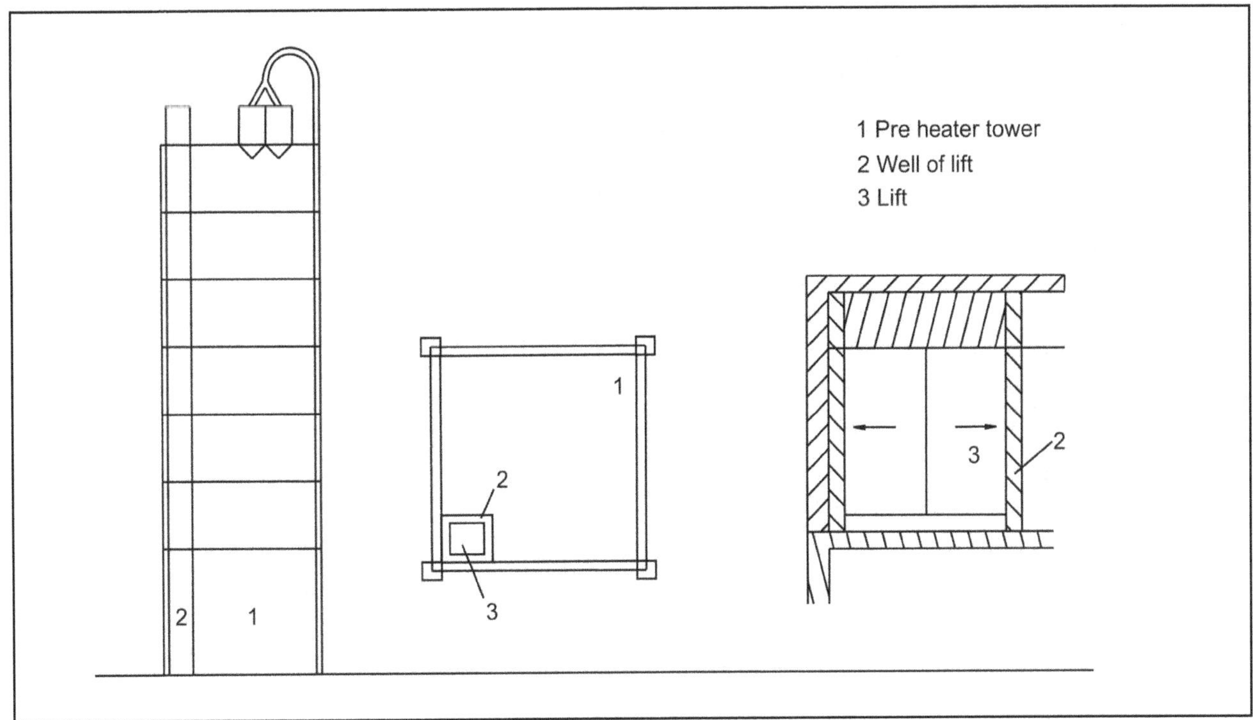

Fig. 6.52.1 Lift in preheater tower; well will be walled up for safety's sake - leaving opening on each floor for exit and entry.

Blending silos, double decked for batch blending systems and continuous blending silos, reach considerable heights so do cement silos because of large storage capacities. Reaching tops of silos, is often necessary to attend to machinery installed there. Attempt is therefore made to locate silos near buildings. For example blending silos can be located near preheater tower and cement silos near packing plant. Access to silos can thus be from the nearest buildings.

See Fig. 6.52.2.

As mentioned above preheater tower will be taller than blending silo and will be fitted with a lift. This lift could be used to go to top of blending silo from appropriate floor level.

Packing plant will not be a very tall building and cement silos will be taller than it. Staircases will be arranged from top of packing plant to go to cement silo.

See Fig. 6.52.2.

It is of course possible to provide 'bubble lifts' from walls of the silo. Care would have to be taken that installing such a lift does not affect waterproofing of the walls of silo.

See Fig. 6.52.3.

52.3 Hoists and Lifting Tackles for Machinery

Hoists and lifting tackles of various capacities and lifts are needed in almost all sections for maintenance and repairs of heavy machinery; large quantities of refractories are required to be handled for lining of preheater, calciner, kiln and cooler.

These requirements have already been dealt with in respective Chapters.

Well planned adequately designed hoists and lifting tackles need to be provided at correct locations for this purpose. Suppliers of respective machinery indicate needs for such tackles and would either supply them or would furnish detailed specifications for procurement.

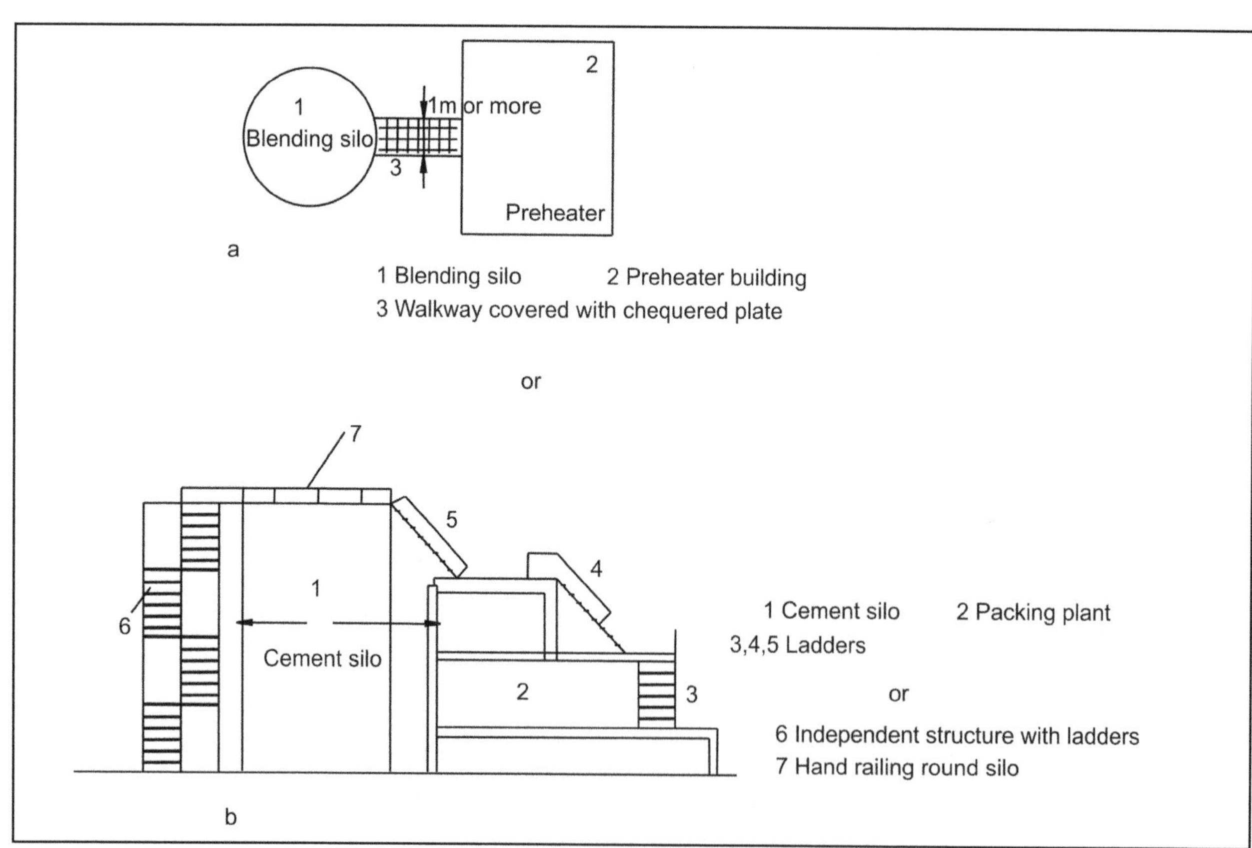

Fig. 6.52.2 Reaching tops of silos - Access stair cases and ladders.

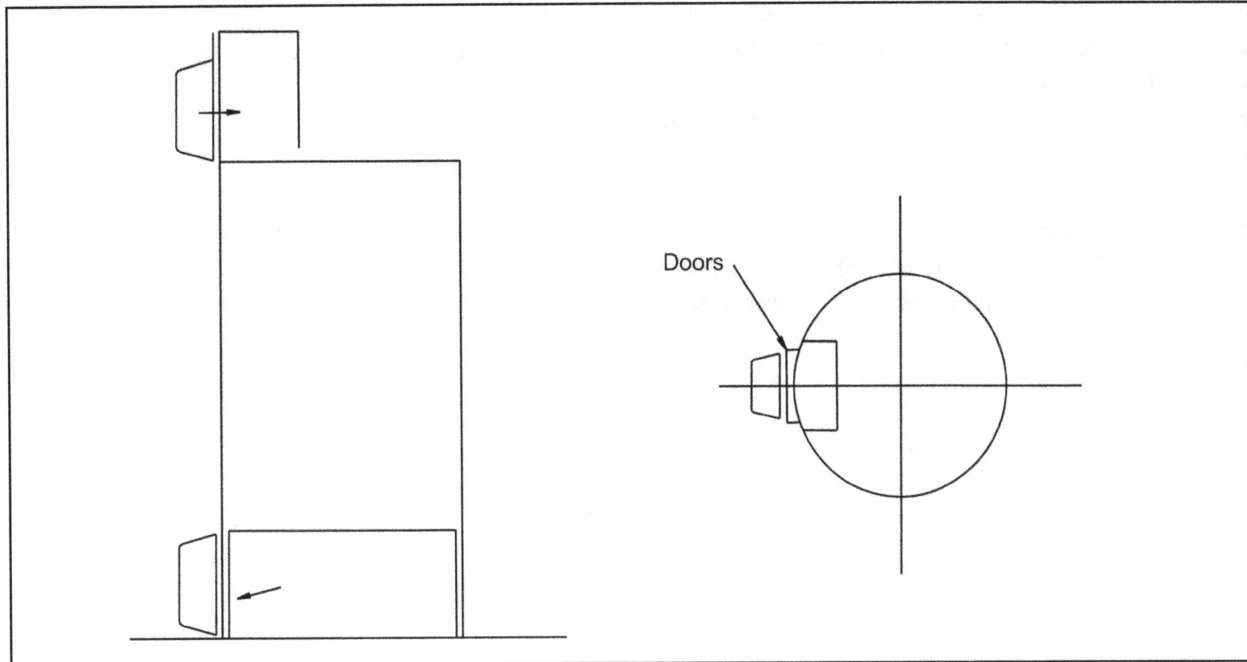

Fig. 6.52.3 Bubble lift for silos.

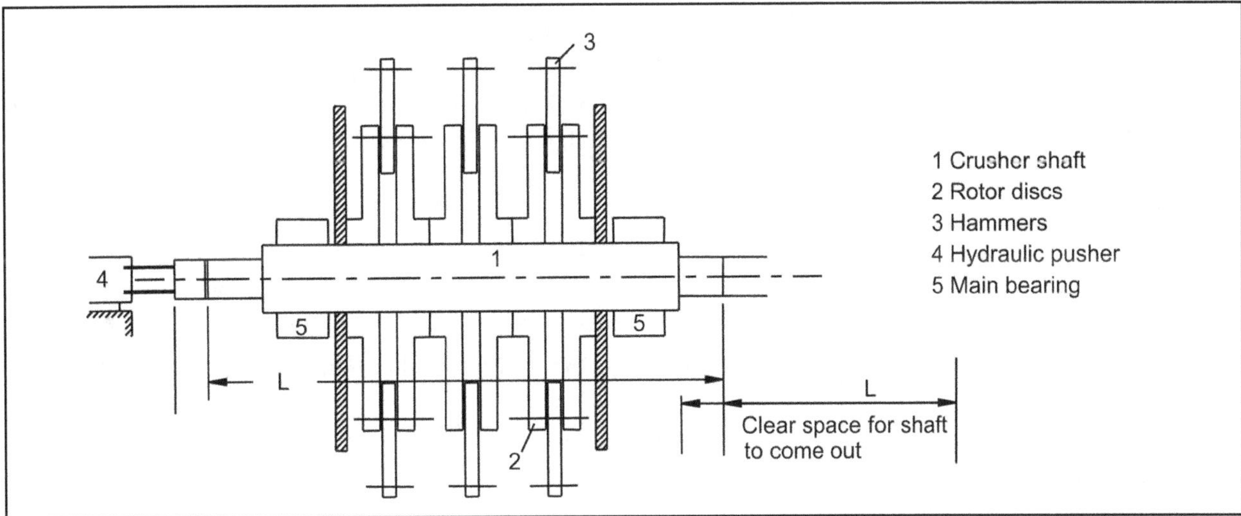

Fig. 6.52.4 Taking out shaft of hammer crusher/impactor.

As a matter of fact it would help greatly if lifting tackle is installed before the erection of the machinery. It will speed up erection.

52.3.1 Crushing Section

In crushing section hammers /impact bars are replaced regularly.

Shaft has to be taken out. An overhead traveling crane is provided in the crusher department for this purpose and also a hydraulic pusher.

See Fig. 6.52.4.

52.3.2 Ball Mills

For ball mills an overhead mono rail with electric hoist with remote control is installed to remove internal fittings from mills. Hydraulic jacks are provided to lift mill off the bearings. In layout slabs near bearings are made thicker to take load of the mill.

See Figs. 6.7.1 to 6.7.6 in Chapter 7 dealing with maintenance.

52.3.3 V. R. Mills

In case of V.R.M.s rollers are required to be handled on regular basis to change liners. Specialised lifting tackles are provided by Suppliers and incorporated in the layout.

Some mills come with built in hoisting beams inside the mill which will have a large access door. In case of others rollers can be swung out.

See Figs. 6.52.5 and 6.52.6.

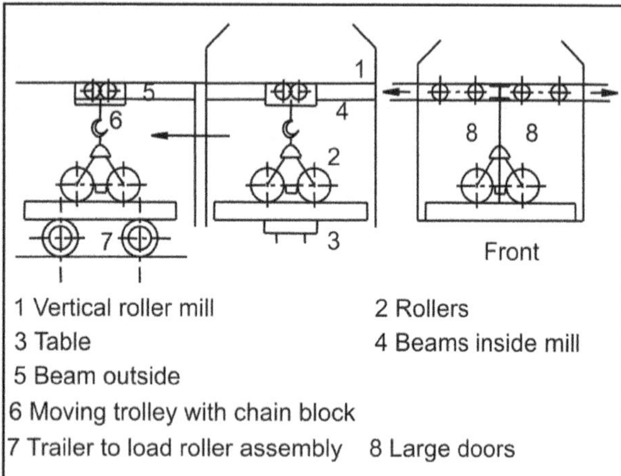

Fig. 6.52.5 Handling facilities for vertical roller mills - 1.

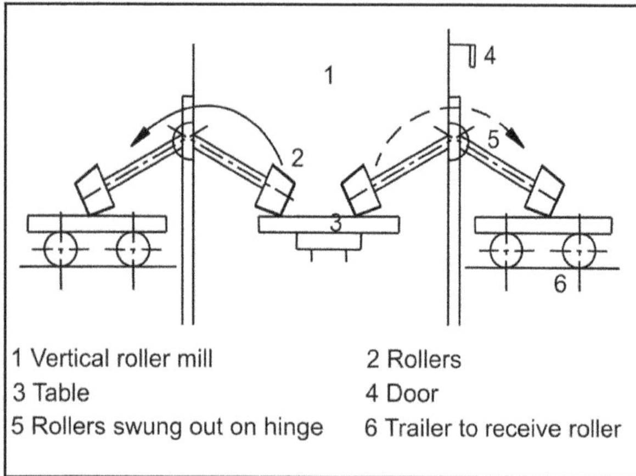

Fig. 6.52.6 Handling facilities for vertical roller mills - 2.

The gearbox which is directly under the mill table. To remove the gearbox the load is lifted off it by jacking.

See Fig. 6.52.7.

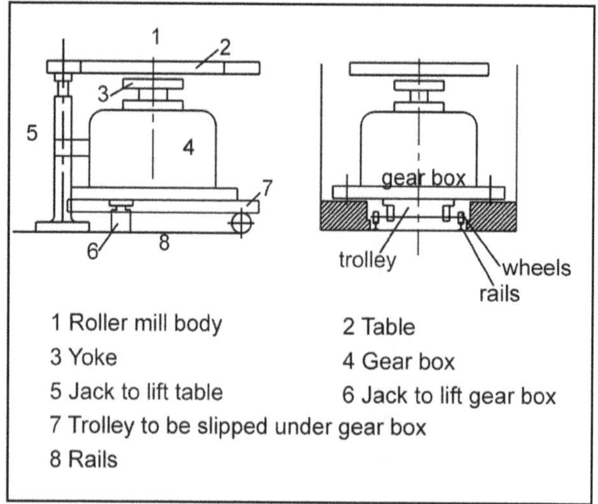

Fig. 6.52.7 Taking gearbox out from under vertical roller mill.

In large 'multi drive' mills this is not necessary.

All vertical mills and roller presses thus come with full specifications of hoisting gear and its layout. It should be incorporated in the departmental layout.

52.3.4 Roller Presses

In case of roller presses rollers are required to be taken out. Hoisting gear is provided to take the roller out of the press and load it on trailer for shifting to workshop.

See Fig 6.7.13 in Chapter 7 of this Section

52.3.5 ESPs

A light hoist is provided on esp top to lift collecting electrodes out of the esp.

See Fig. 6.52.8.

52.3.6 Kilns

Kiln rollers and roller bearings would in many cases be lifted with the help of a crane. Jacks would be required to take load of the kiln and their capacity and specifications should be obtained from Suppliers.

See Figs. 6.19.44 and 6.19.45 in Chapter 19 of this Section.

52.4 Provision for Lifting Arrangements

For infrequent requirements of lifting, provision may be made in layout and construction to be able to install chain pulley blocks when required; like providing beams with hooks above hood of bucket elevators.

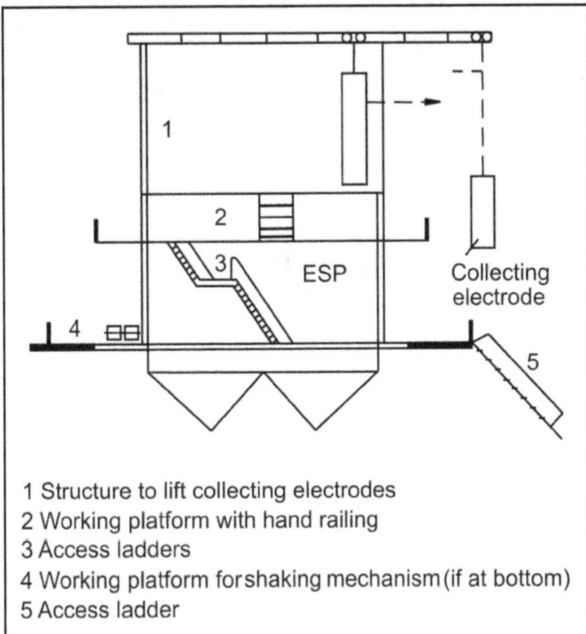

Fig. 6.52.8 Platform for maintanance of ESPs.

A well designed permanent handling system is a part of the machinery layout and machinery supplier provides drawings and specifications of equipment to be installed and integrated into the departmental drawings.

52.5 Collect Data on Specifications of Tackles

In fact, Consultants should ask vendors the needs of maintenance, location and specifications of tackle and space to be provided in layout for accommodating / installing them.

This would avoid requiring tomake makeshift arrangements when a break down happens or even during erection.

Beams of building are used to fix 'I' beams / channels to hang chain pulley blocks.

These blocks may be manually operated or motorized; overhead traveling cranes are used when a large number of drives are arranged in a row like in a motor house.

52.5.1 Details to be sought for

The Consultant should make a list of all handling and hoisting equipment like :

(i) Traveling crane – span – capacity – length of travel and lift.

(ii) Mono rail and hoist – capacity and lift.

(iii) Lifting cranes for rollers of v.r.ms.

(iv) For gear boxes for mills and kilns.

(v) For rollers, motors and gear boxes of Roll press.

(vi) Shafts and rotors of crushers.

(vii) Kiln rollers and bearings.

(viii) Jacks for supporting the kiln.

(ix) Hydraulic jacks for removing bearings / shaft, etc.

These should be incorporated in machinery list / spares and also incorporated in layouts as required.

52.6

The equipment meant for handling the heavy components should be installed first and be readily available for installing the machinery itself.

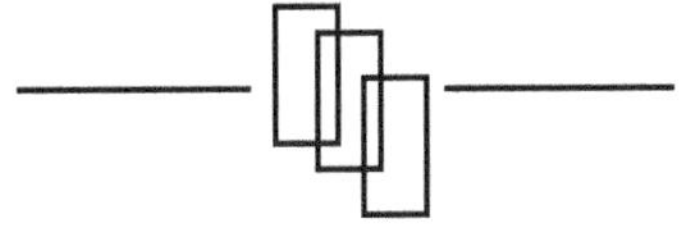

CHAPTER 53

SAFETY

53.1 Safety of Plant and Personnel

Safety of plant and personnel is of paramount importance. Attention paid to ensure safety can never be enough.

'Safety' of course can only be ensured with the help of workers. Without their active co-operation, safety measures even of the highest kind would fail. Therefore there has to be a continuous effort on part of Managements to educate workers in safe ways of working and to see to it they are being followed.

It is necessary to install safety devices at all places, which could cause accidents and be a source of danger to personnel.

53.2 Identify Danger Zones

One of the first steps to be taken would be to identify possible 'danger zones' like:

1. Open pits, floor openings, edges of buildings at higher levels, steep slopes, and ramps.
2. Improperly designed steps, ladders and staircases.
3. Moving machinery, reciprocating and rotating machinery; exposed V belts and chains; Pulleys and couplings; open gears.
4. Quarry operations of drilling and blasting.
5. Hot ductings, shells likely to come in contact with workers.
6. Hazards of getting submerged under falling hot material, pulverized and granular.
7. Getting suffocated / drowned in bins, hoppers and silos.
8. Catching fire due to inflammable substances.
9. Explosions and implosions.

53.3 Safety Measures Built in Machinery and Operation

Safety measures should first be built in the design of the machinery itself, so that accidents and mishaps are minimized. Others are to be installed but these need to be maintained and made use of.

Safety measures are also to be built in operational procedures and practices.

In respective Chapters need for safety and provisions to be made have already been dealt with. However safety is so important that it is worth repeating them here.

53.4 Safety Measures in Buildings and Structures

1. Hand railings should be provided at all levels of the building along edges. They should be provided around floor openings also to prevent accidental falls; along staircases and ladders; monkey ladders if used, should have a protective cage round it to prevent falls.

   See Figs. 6.53.1 to 6.53.3

   Railings can be standardized in design. In some cases, where machinery is to be hauled in, railing could be removed and fixed back.
2. Smaller openings in floors should have grill / chequered plate covers. They should be removed only in emergency but replaced without fail.
3. Edges of building floors and openings should have 'kerbs' to prevent things falling down. **See Fig. 6.53.1.**

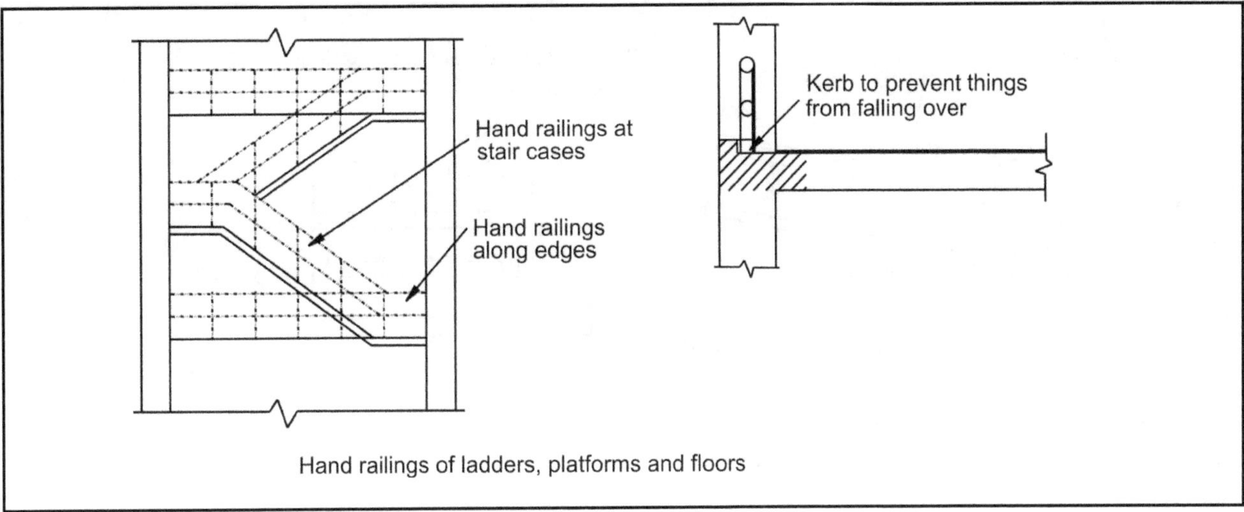

Fig. 6.53.1 Safety measures in building.

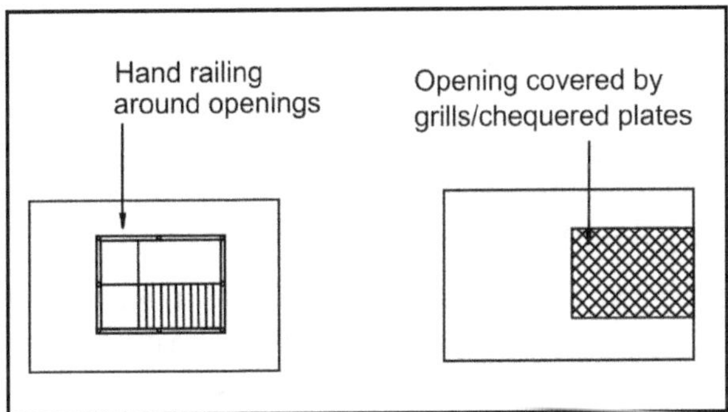

Fig. 6.53.2 Hand railings round openings in floors.

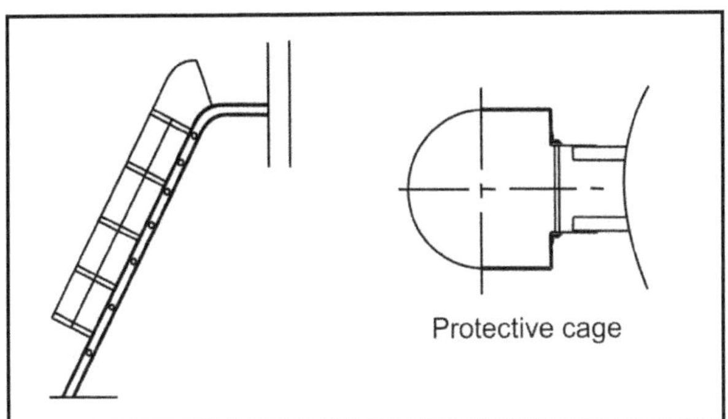

Fig. 6.53.3 Monkey ladders Safety measures.

4. Walkways along conveyors should have hand railings and grilled floors, which cannot be easily removed. In lengths of conveyors that are directly over roads, chequered plates should be installed instead of grills. Such plates should be installed under the belt also to prevent spillage falling on heads of passersby.

   See Fig. 6.53.4.

5. Entry into hoppers, bins and silos can be dangerous and must not be done singly. There should always be two persons one watching the other and raising alarms. All hoppers and bins must have rungs along to wall to go down and climb up. Persons entering hopper must wear a safety belt tied to their waists and a rope by which they can be hauled up.

   See Fig. 6.53.5.

   A red flag may be hoisted on the hopper / silo to indicate work is going on in the hopper and it should not be filled at that time.

6. While unloading a ball mill, area around it should be cordoned off.

   See Fig. 6.53.6.

53.5 Weights Being Raised and Lowered

Care has to be taken while hoisting loads or lowering them. Precautions are to be taken to keep persons away from the area. Area should be cordoned off during the hoisting operation.

53.6 Guards Around Rotating Machinery

Guards for V belts and couplings, projecting parts of machinery that move.

These are standard items and are installed along with machinery. They are designed to be removed quickly for attending to V belts, chains and couplings

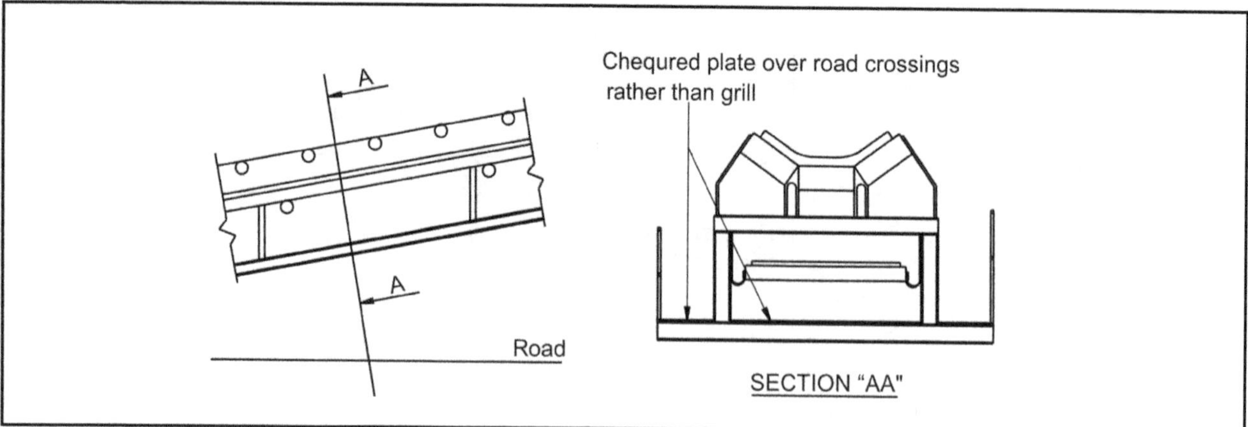

Fig. 6.53.4 Safety measures – belt conveyor.

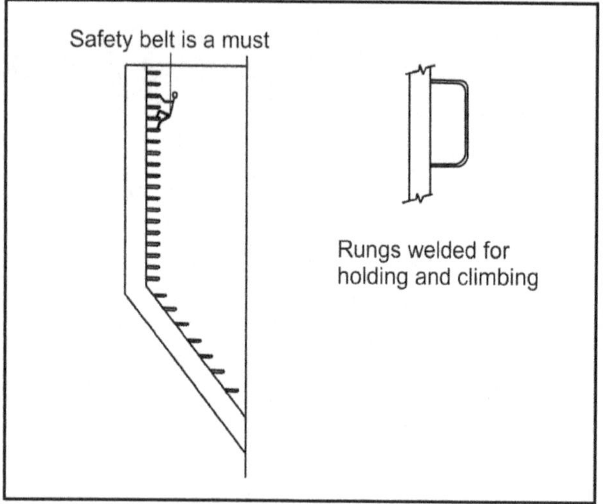

Fig. 6.53.5 Safety measures in hoppers and silos.

and must be fitted back every time when the work is over. Same is true for projecting shafts. They should be guarded.

The guards removed should be replaceable quickly and in exactly same position. Some kind of location device should be provided

53.7 Hot Ductings

1. Ducting carrying hot gases are insulated and lagged to prevent loss of heat and also to prevent injury to workers. Others that are brick lined from inside can still be very hot and should be isolated so that they are not touched accidentally.
2. Kilns would have walkway along their length to attend to supporting rollers and bearings and gear etc. Kiln shell is hot and radiates heat.

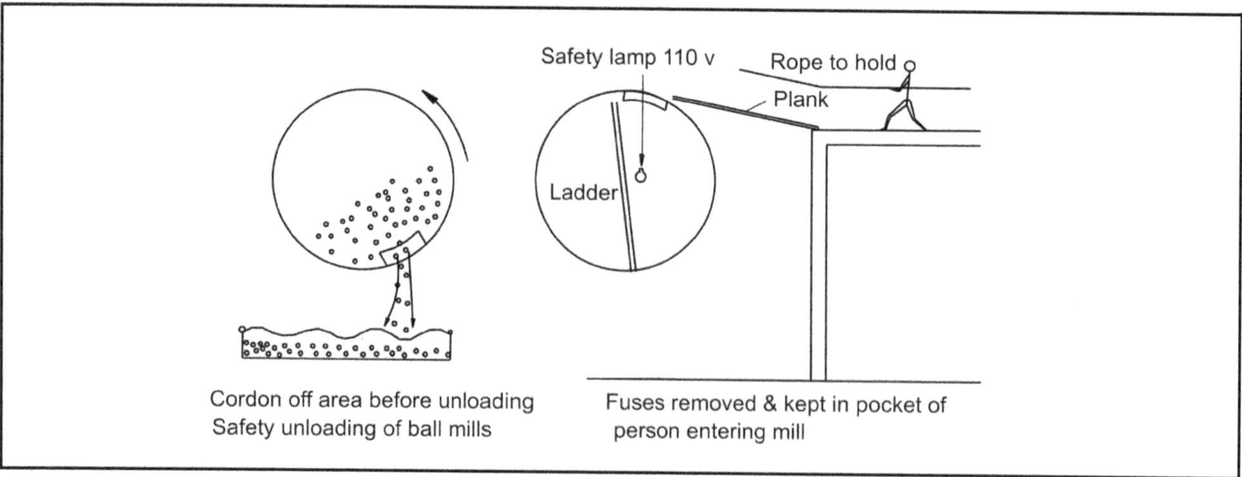

Fig. 6.53.6 Safety - while entering ball mill.

3. Kiln inlet end though normally under negative pressure, may become a source of occasional leakage of hot raw meal and should be isolated. Hot material should be collected in a hopper with a drum / bin underneath which should also be isolated. This is also true for some calciners where material gets accumulated and is discharged from the bottom through pendulum flap valves; this material is also very hot (800-900 °C).

See Figs. 6.53.7 and 6.53.8.

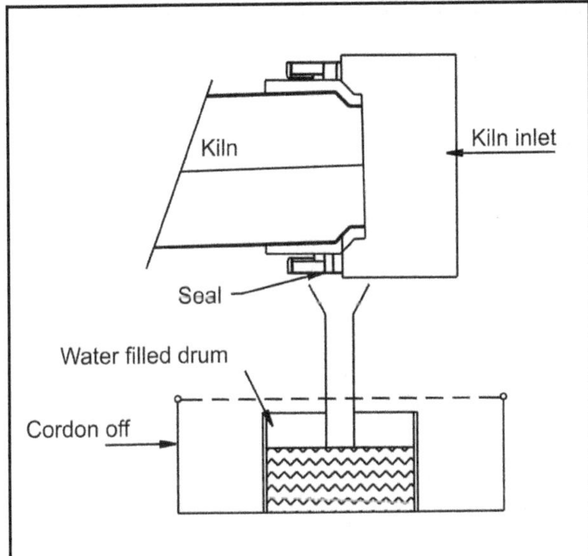

Fig. 6.53.7 Hot spillage at kiln inlet.

53.8 Emptying of Hoppers and Bins

Sometimes, bins and hoppers containing granular or pulverized materials need to be emptied out. At other times, they need to be poked to remove jams. Occasionally, flow may stop by 'arching' but suddenly gush down once the arch is broken. Such a flow can smother workers engaged in poking, if they are standing directly under it. Serious and even fatal accidents can take place. To avoid this, platforms for poking and for men to stand should be above the level of the manholes or openings in hoppers.

See Fig. 6.6.25 in Chapter 6 of this Section.

53.9 Walkways Crossing Conveyors/Screws etc

It is best to avoid crossing conveyors by going around them. But if that is too much of a detour, and workers are found to be taking short cuts, it would be good to provide a walkway over it.

See Fig. 6.53.9.

This situation can arise in packing plant where there can be many belts to cross.

53.10 Safety in Quarries

A major operation in quarrying is drilling holes in rock, filling them with explosives and blasting them. A number of holes drilled to various depths and spaced to suit the rock structure are blasted either simultaneously or sequentially.

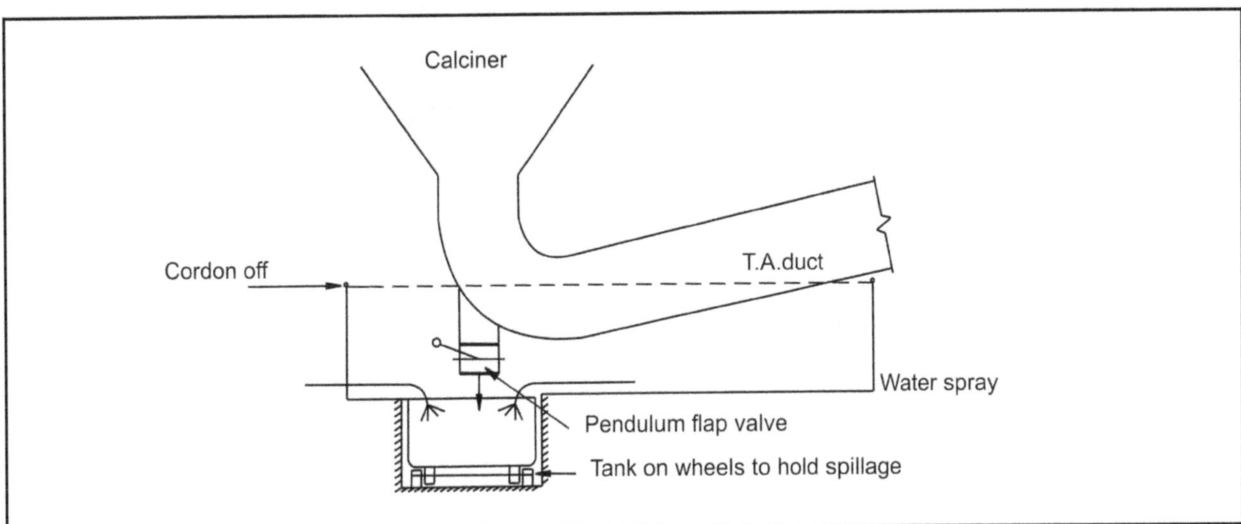

Fig. 6.53.8 Hot spillage from calciner.

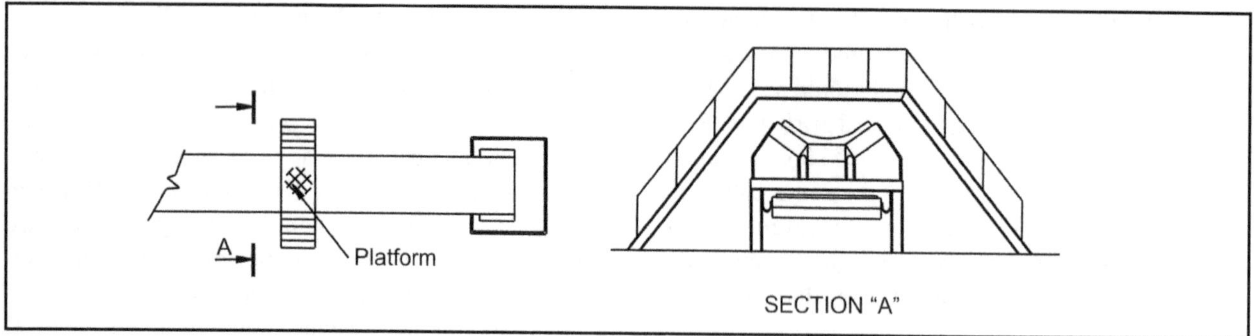

Fig. 6.53.9 Safe crossing over conveyor.

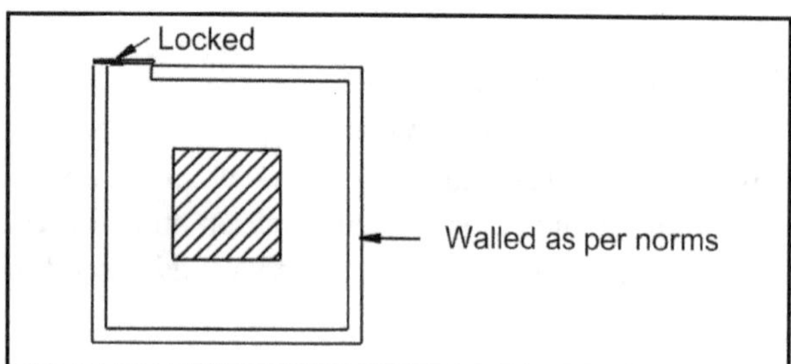

Fig. 6.53.10 Explosive magazine - isolated as per norms.

Explosives required for this purpose are stored in a magazine in a remote isolated place.

See Fig. 6.53.10.

The area in which blasting is planned should be cordoned off and unauthorized entry prevented in it during blasting operations. Red flags or alarms should mark off the area in which blasting is planned. Rock pieces can be thrown over a wide area and hence cordoned off area should be much larger than area of actual operations.

See Fig. 6.53.11.

Shovels, dumpers and drills etc., should be removed from the working place so that they do not get damaged by falling rocks or by subsiding ground levels.

Quarry should be well lighted in the night so that neither vehicles nor men would fall off the vertical faces or in pits.

See Fig. 6.53.12.

In unloading dumpers in hoppers care has to be taken to see that men are not working in the hopper. To prevent damage to dumpers / building it is good to line the beam with tyres / sleepers as shown.

See Fig. 6.53.13.

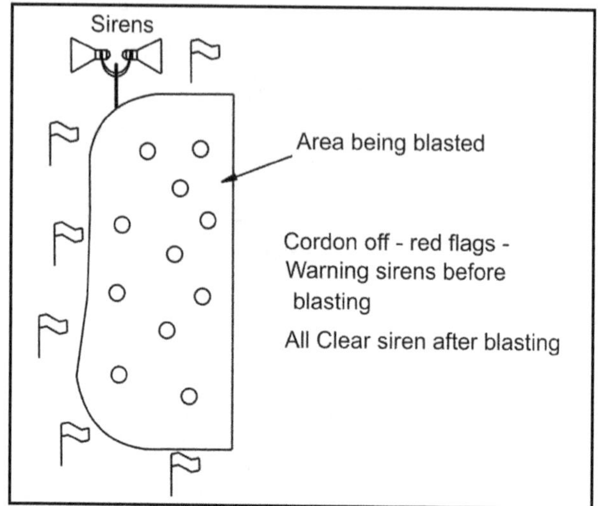

Fig. 6.53.11 Blasting operation.

53.11 Safety of Personnel Entering Machines

Adequate precautions should be taken to safeguard approach and entry into machines like crushers, ball mills and roller mills.

See Fig. 6.53.6.

Safety of personnel working inside machines is largely a matter of following correct procedures. In all sections where men enter machines like crushers for changing crusher hammers; in mills for changing lining plates, grate plates etc., **the most fundamental safety precaution to be taken is to see that the machine cannot be accidentally started when men are inside the machines.**

Strict procedure should be laid down or a protocol created and followed to ensure safety of men.

Such a protocol has assumed great importance in the present times of remote and group starting of motors with the help of PLCs.

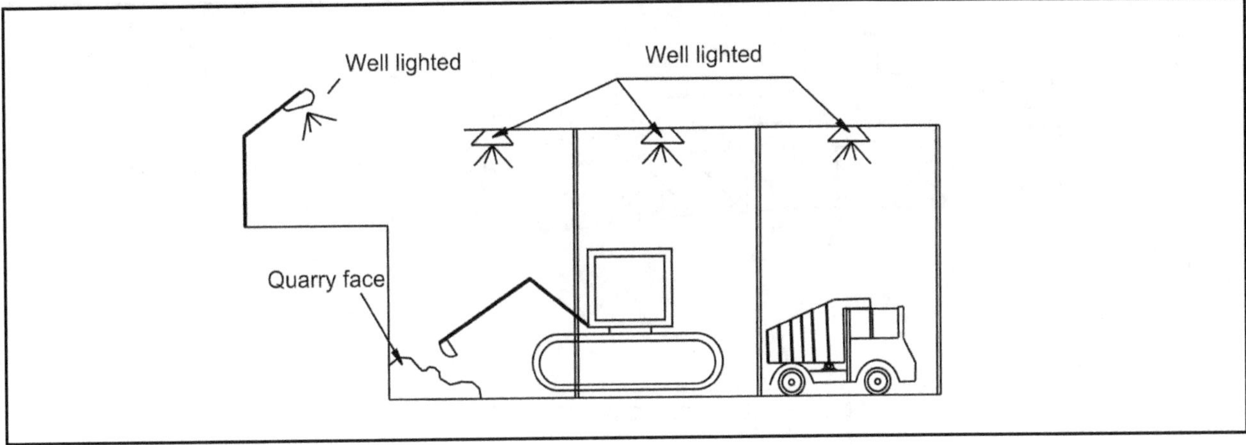

Fig. 6.53.12 Lighted quarry faces in operation - safety measures in quarries.

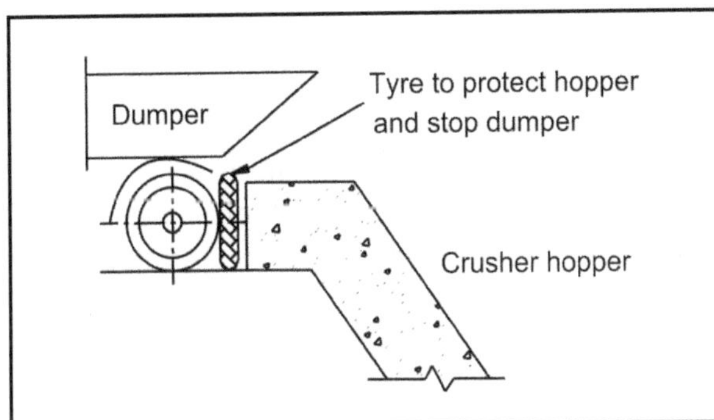

Fig. 6.53.13 Safety measures - kerb at crusher hopper.

Central control room staff should be aware of the maintenance programme and should be informed when work is completed and work can be started again.

In this period, remote starting should be switched off.

All machines being made of steel, the lights taken inside must be low voltage 110 V safety lamps. Special 110 V sockets plugs must be provided near mills, crushers, roller mills etc., so that safety lamps can always be taken inside.

Personnel entering machines must wear safety helmets.

53.12 Conveying Systems

All inspection doors and man holes provided for maintenance must be in place and securely clamped.

Elevator Belt or chain must have a ratchet, which will prevent roll back of buckets, which can hurt maintenance workers.

In case of long overland conveyors, going cross country, grazing cattle can come near the belt and get hurt and can also damage the belt. This should be avoided.

See Fig. 6.53.14.

This is also true of ropeways in which loaded buckets move overhead at great speed. A safety net should be fixed over roads & rail lines which the ropeway may cross.

See Fig. 6.53.15.

Inadvertently even experienced workers put hands inside running screws. This should be prevented by providing netting on inspection covers.

See Fig. 6.53.16.

53.13 Spillage

Sometimes belt conveyors go out of alignment causing belt to run on one side and material spills over. This can hurt workers / vehicles passing underneath when belt is at a height. It would be very expensive to provide a plate under the entire length of the belt. But when belts are crossing roads such a plate should be provided to collect spillage so that nobody is hurt.

See Fig. 6.53.4.

The first precaution to minimize spillage is of course to attend to belt scrapers and keep them in good condition and in contact with the belt.

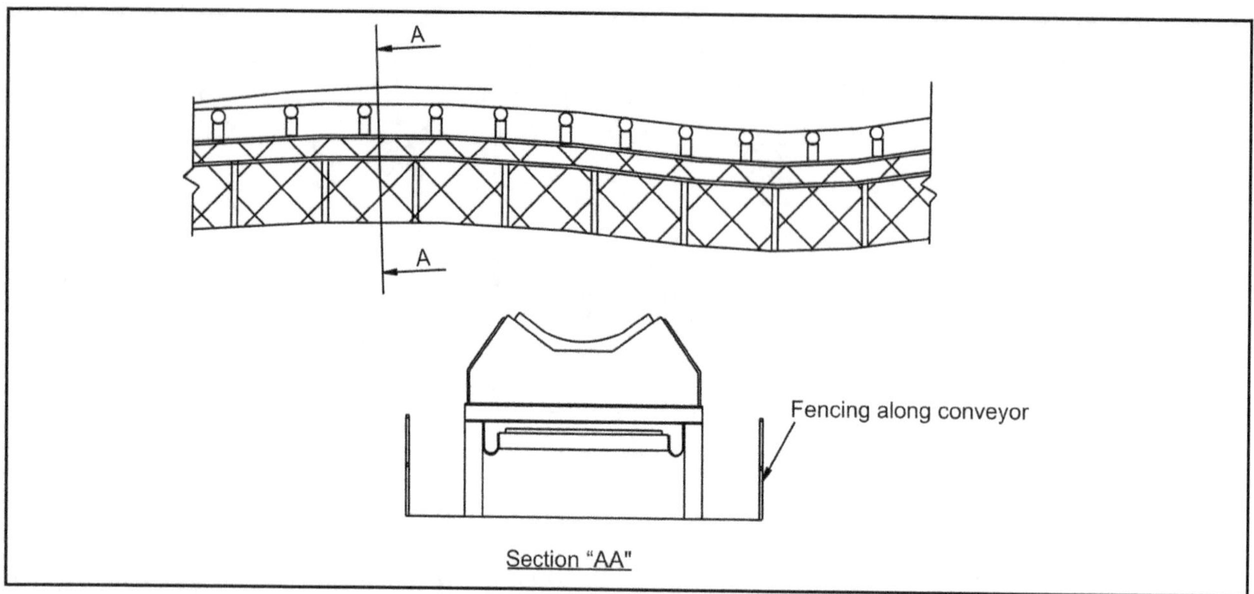

Fig. 6.53.14 Safety measures - isolate overland conveyor to keep out cattle etc.

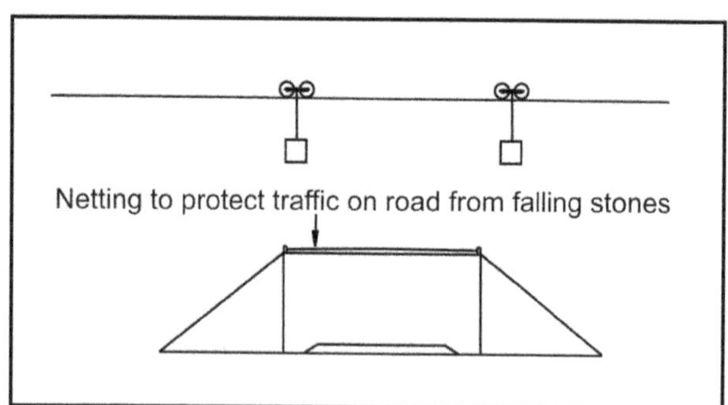

Fig. 6.53.15 Ropeway crossing road.

53.14 Safety Against Explosions

Dust collectors subject to have explosions are fitted with rupturable foils or self-closing spring-loaded explosions doors.

Areas around these safety devices should not be accessible to workers, as it is never known when explosions would take place.

That is why rupturable foils are placed on the roof of the dust collectors - ESP / bag filters. Explosions flaps when fitted on sides should be well above height of man.

See Figs. 6.53.17 to 6.53.20.

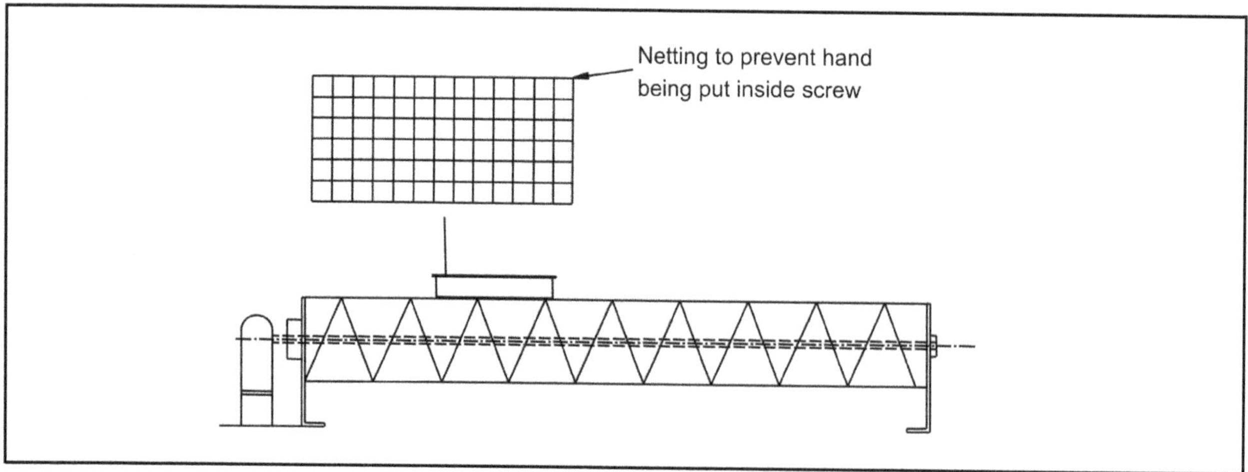

Fig. 6.53.16 Screw conveyor.

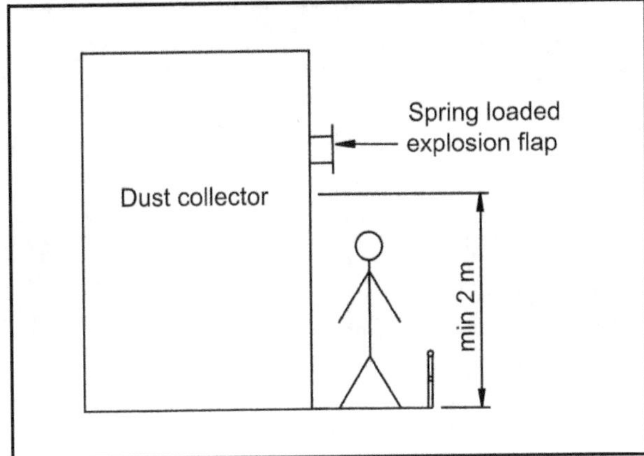

Fig. 6.53.17 Dust collector casing.

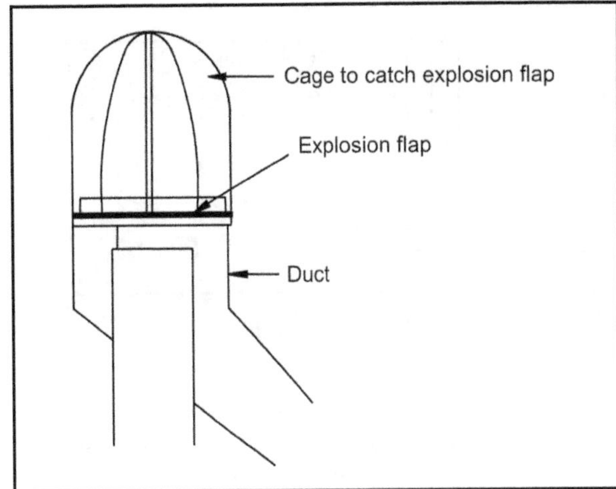

Fig. 6.53.18 Explosion flap on ducting.

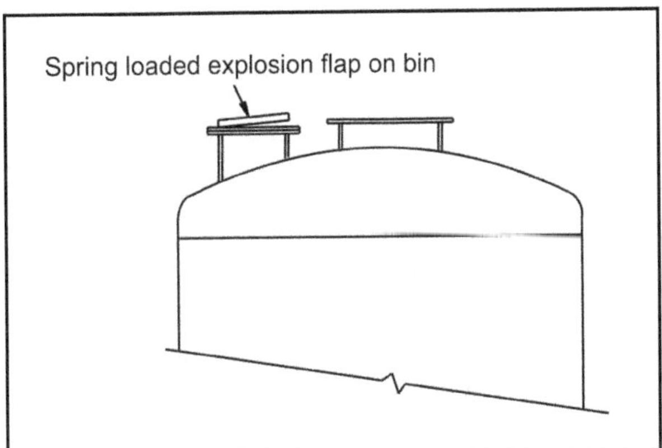

Fig. 6.53.19 Rupture foils on roof of bin.

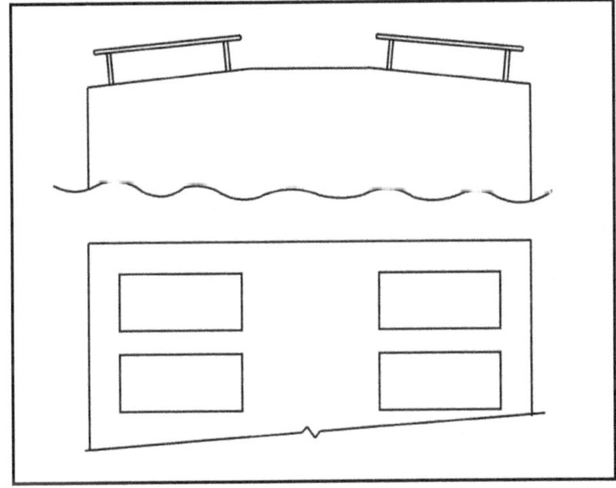

Fig. 6.53.20 Rupture foils on roof of ESP.

53.15 Fighting Fire

Fighting fire and extinguishing it are also an important safety aspects.

Bag filters handling coal dust / ESPs are equipped with CO_2 injection / spray systems consisting of a battery of CO_2 which can be emptied into the filters to smother fire; inert gas generators are also installed to smother fire. If the problem is more severe, ESP can be sprayed with water even.

Using inert gases from kiln in coal mill section can substantially reduce the danger.

Another vulnerable areas subject to fire hazards are :

1. Bags godowns where jute / polythene bags are stored.
2. Where they are temporarily stacked for feeding packing machines.

Packing plant and bags godowns must have fire extinguishers of the right type at various points to fight fire if it breaks out.

A good old measure used to be placing sand filled buckets in different corners of a building to smother fire.

Overhead sprinkling system to introduce inert gas in the offices, godowns etc., can also be installed.

See Fig. 6.53.21.

53.16 Storage of Inflammable Materials

Storage of lubricating oils and greases, kerosene in the General Stores should have similar precautions against fire.

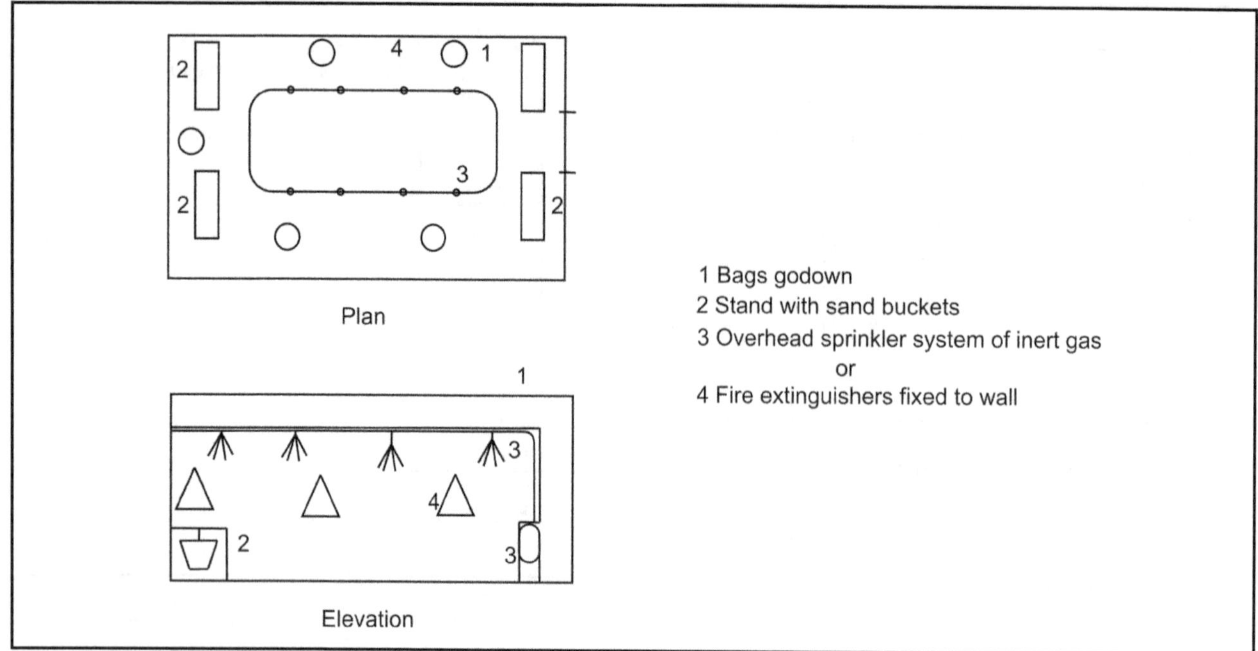

Fig. 6.53.21 Preventing fire in a bags godown.

Diesel or LSHS oils used to genrate power in D.G. Sets must be stored in isolated place with safety measures provided as per Factory Acts.

53.17 Electrical Short Circuits

A major reason for breakout of fire can be electrical short circuits.

Electrical control rooms, central control rooms, MCC, HT, LT substations, etc., need to be provided with readily useable fire fighting equipment as per norms laid down including spray of water or inert gas from roof.

53.18 Coal Stock Piles

Stock piles of coal can catch fire during hot months of the year due to volatiles in coal getting ignited. Coal can continue to burn slowly if it gets oxygen.

Stock piles if rammed, reduce pockets of air and there by minimize risk of slow burn or fire. If under cover, they will not be exposed to extreme heat. From this angle stock piles of crushed coal are less vulnerable to catch fire as voids are less and can be compacted.

In any case, arrangements to spray water over piles should be made. Piles should not be more than 2 metres in height.

Hot burning coal can find its way into mills and can cause fire in dust collector.

As explained in Chapter on coal mills provision is made to inject CO_2 in the dust collector or an inert gas generator is installed to supply instantly inert gas to put out fire. Cut off quick closing dampers are used to isolate dust collector and to protect it.

See Fig. 6.8.8 in Chapter 8 and Fig. 6.43.32 in Chapter 43.

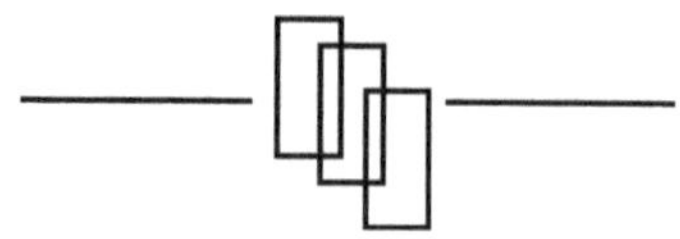

CHAPTER 54

SAMPLING

54.1 Need for Sampling

Tests of different kinds are to be conducted on various materials and results recorded at various stages of cement production.

Some tests are 'statutory', laid down by Bureau of Indian Standards (or equivalent Standards Institutes of each country) and are to be recorded and results submitted to BIS for approval so that cement sold in market can be stated to be BIS approved or as per specific grade of BIS.

Tests start with basic raw material limestone and sampling starts with crushed limestone. Many plants have automatic sampling stations for crushed limestone.

54.2 Samplers and Sampling Station

A sampler moving across the width of the belt backwards and forwards collects small quantities of crushed limestone continuously as sample.

See Fig. 6.54.1.

Sample collected, then undergoes further processing like further crushing and grinding. It can then be conveyed either manually or pneumatically to the central sample preparation plant in the laboratory, which would have an 'X-ray analyzer'. **Fig. 6.54.2** shows flow diagram of a sampling station.

Ground limestone would be formed into pellets of standard dimensions and finish by passing it through a press. Pellets are sent to 'X-ray analyser'.

The X-ray analyzer can indicate the chemical composition almost instantaneously

A sampler if installed on belt of reclaimer transporting crushed limestone from stock pile to raw mill will help in arriving at efficiency of 'preblending'.

54.3 Raw Mix - Automatic Samplers

At raw mill, limestone is proportioned with other correcting materials on a continuous basis to arrive at a raw mix, which would be close to the desired raw mix composition to produce clinker of desired quality.

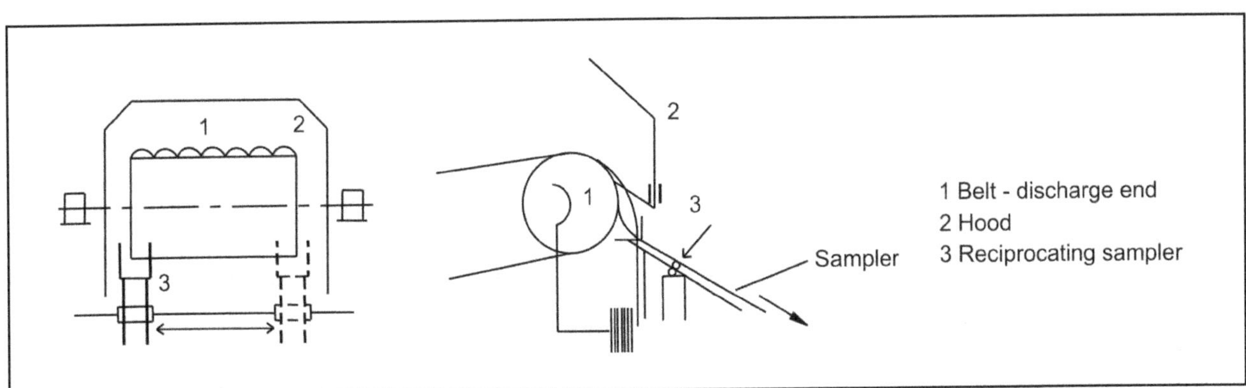

Fig. 6.54.1 Collecting samples of crushed stone from belt conveyor.

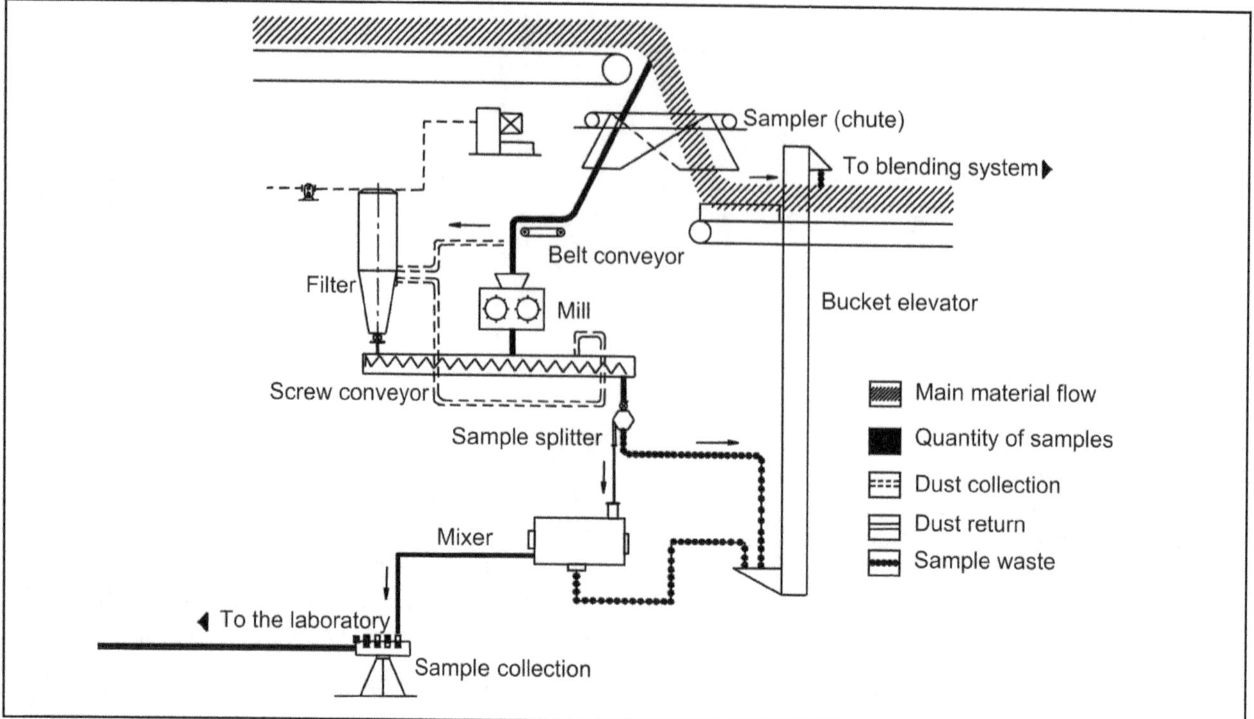

Fig. 6.54.2 Flow diagram of sampling station.

54.3.1 X-ray Analyzer

In large plants 'X-ray analyzer' carries out tests fast and sends signals to the weigh feeders of raw mill section to change the proportions of limestone and additives. To be able to do that, samples of ground raw meal are collected regularly before and after the blending silo and compared with the desired composition of raw mix.

54.3.2 Automatic Samplers

Samples of pulverized raw meal are collected by automatic samplers installed at strategic points. Samplers are small motorized screws which run for predetermined times at intervals and collect specified quantity of material and send it pneumatically to sample preparation room near the x-ray analyzer.

See Fig. 6.54.3.

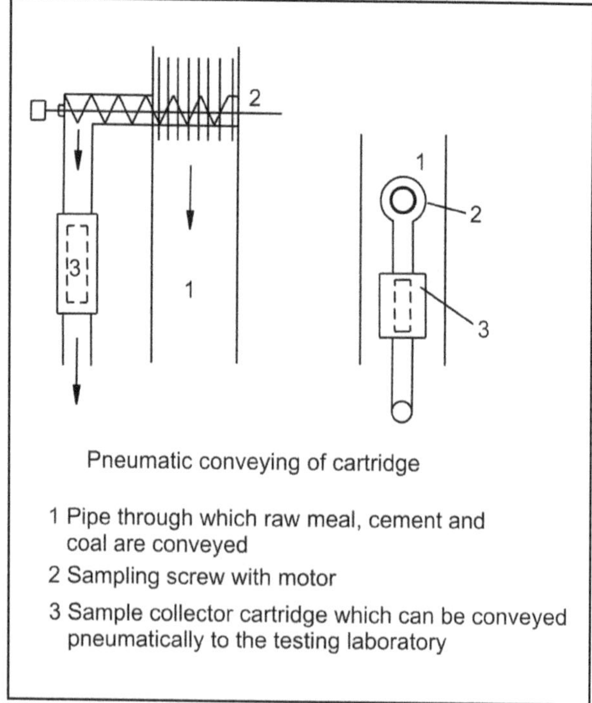

Fig. 6.54.3 Collecting samples of pulverised materials like – raw meal, coal, cement.

54.4 Automation in Sampling and Testing

Today sampling and testing is fully automated so much so that these operations are 'untouched by hand'.

However, intermediate solutions are also possible such as collecting samples in automatic samplers but conveying them manually to the laboratory.

54.5 Testing Cement

Tests on cement are statutory tests and must be carried out at specified intervals. Samples of cement are collected in a similar manner and conveyed to X-ray analyzer for chemical analysis.

Cement samples are to be taken from mill end and also from cement silo before dispatch.

54.6 Coal

It is also useful to carry out tests to determine calorific value, moisture and ash in coal as received and as fired on a regular basis.

The testing time of X-ray analyzer per pellet is so short that same X-ray analyzer can cope with testing of raw meal, raw mix and cement and clinker samples.

54.7 Clinker

Clinker samples are also taken and tested regularly to ensure that poor quality clinker if any is segregated and stored separately.

54.8 Blending Efficiency

Tests should also be carried to check blending efficiencies of stacker reclaimer systems for limestone and coal and of continuous blending system. If found below par action should be taken to set things right.

54.9 Design of Sampling System

Locations of sampling points should be carefully selected in consultation with Suppliers of X-ray analyzer and sampling equipment and Plant's executives responsible for quality.

Suppliers would furnish specifications of automatic samplers and installation drawings; they will also furnish sizes of pipe lines and such details as minimum radii of bends; specifications of blowers to be procured for pneumatic conveying.

54.9.1 Layout of Plant

Suppliers should be furnished with general layout and departmental drawings of pertinent sections of the plant and location of X-Ray analyzer and sample preparation room.

54.9.2 Sample Preparation Room

Lay out of sample preparation room where samples received in cartridges will be converted into pellets will also be furnished by Suppliers.

54.10 X-ray Analyzer

Since same X-ray analyzer is used to test limestone, raw meal, clinker and cement, synchronization of receipt, preparation and feeding of samples from various sources is very important. It may be a good idea to automate partially to start with and fully when systems have been run in.

In any case it would be a good idea to cross check results of X-Ray analyser by conventional methods now and then.

54.10.1 Integration of Sampling, X-Ray Analyser and Quality Control Laboratory.

With growing emphasis on Continuous Monitoring and Automation, these three wings should be integrated and located in the same building and be under one Executive.

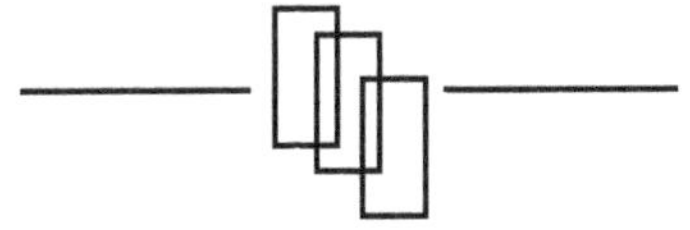

CHAPTER 55

REFRACTORIES

55.1 Refractories

Refractories protect steel plates. They also reduce radiation losses. Therefore furnaces, kilns, preheaters, clinker coolers, calciners and the ducting which carry hot gases are lined with refractory.

Refractory are of three kinds:

1. Brick.
2. Insulating brick or blocks.
3. Castables.

All are used in cement plants.

Bricks are used in kilns and prheaters and calciners and coolers and t.a.duct and dust chamber.

Grade of refractory is selected according to the application and the temperature it has to withstand.

Castables are used to line irregular profiles and contours like elbows of planetary coolers, raw meal pipes where brick lining would be difficult.

Insulating bricks are used to save on total thickness of refractory lining, to reduce weight and to reduce radiation losses. They are now extensively used in preheater cyclones, calciners, grate cooler walls and ducts

55.2 Properties of Refractories

Refractories need to have following properties:

1. resistance to high temperature,
2. resistance to spalling due to changes in temperature. Spalling results in development of cracks in brick,
3. resistance to chemical attack – dust and alkalies in kiln gases can adhere to refractory in burning zone. Brick that does not possess this slag resistance do not last long,
4. abrasion resistance – in a kiln, charge within the kiln, continuously slides over the brick surface; dust in kiln gases also abrades brick. The resistance to abrasion is very important for bricks in zones where there is no formation of coating,
5. coating properties – this is a very important property for bricks to be used in burning zones of kilns. Coating protects refractory in the most severe duty conditions.

55.3 Testing of Refractory

Manufacturers of refractory furnish conventional test results like:

1. compressive and tensile strengths,
2. chemical analysis,
3. density and porosity,
4. thermal conductivity,
5. permeability.

In addition there are tests that are now quite common. They are:

(i) *Melting point* – it is the temperature at which refractory begins to sag and lose its shape.

(ii) *Hot load* – refractory when under load undergo softening at a temperature much below their melting point. Load can be their own weight or weight of charge in the kiln.

(iii) *Linear expansion or shrinkage* – refractory under prolonged service will expand (or shrink). Test determines percent linear change at a specified temperature.

(iv) *Panel spalling* – Frequent heating and cooling can develop stresses in refractory which result in development of cracks. This test determines ability of a refractory to withstand repeated changes in temperature.

(v) Thermal expansion test measures reversible expansion, that is to what extent brick will expand when heated and return to original dimension after cooling.

55.4 Types of refractory

1. *Alumina –Silica Group*

   The two most important components in this group of refractory are Alumina and Silica. Within some limits increase in Alumina results in higher refractoriness.

   Higher alumina also improves spalling resistance and resistance to chemical attack. High alumina bricks have higher reversible thermal expansion.

2. *Basic Group*

   Basic refractory is manufactured mainly from periclase, dead burnt magnesite and chrome ore. In rotary kilns magnesite chrome bricks arc commonly used.

   Basic bricks have a greater resistance to chemical attack but have low spalling resistance compared to alumina bricks.

   They are preferred in burning zonc of the rotary kiln because they take on coating more rapidly and hold it longer.

3. *Dolomite bricks*

   They are mainly composed of CaO and MgO. They are used in burning zone of the rotary kiln – preferably the center of burning zone.

4. *Spinell bonded brick*

   These are bricks with 10-15 % alumina and 80-85 % MgO.

   Life of these bricks is much longer than high MgO-Cr bricks.

   See Fig. 6.55.8.

55.5 Properties and Application of Refractories

Table 6.55.1 shows application of refractories, insulating blocks and castables in cement plants. It also furnishes brief properties of the same.

Table 6.55.2 shows commonly found thicknesses of refractory brick, castable and insulating blocks at different locations.

Tables 6.55.3 and **6.55.4** show properties of different grades of alumina bricks and castables respectively.

55.6 Refractory Shapes

Blocks, arches and wedges are the most commonly used shapes for bricks in kilns. Where metric system is followed, VDZ and ISO shapes are used.

Graph 6.55.1 shows quantity required in m^3 /m length of ISO bricks for kilns and round vessels and ducts.

Tables 6.55.5 and 6.55.6 show combinations in which these bricks could be used. Usually two shapes are used in making a ring or round of a kiln.**See fig. 6.55.2.**

55.7 Estimation of Weight of Refractory for Kiln

Kiln is a cylinder. Refractory lining is 150, 200 and 250 mm thick according diameter of the kiln. Quantity of bricks can be roughly estimated

D = diameter inside shell

d = diameter inside refractory

L = length of kiln

Volume of bricks = $0.785\ (D^2 - d^2)\ L\ m^3$

If bulk density is $1.9\ t/m^3$,

Weight of bricks = $\sim 1.5(D^2 - d^2)\ L$ tons.

This is also true of all round vessels like calciners, ducts like t.a. ducts.

Ready reckoner tables or graphs can be built up to estimate weight of refractory for kilns of different diameter.

See graph 6.55.1.

Using above formula, for a 4.0 m dia × 60 m long kiln with a refractory thick ness of 200 mm, weight of refractory would be

$1.5 \times 60\ (4^2 - 3.6^2) = 274$ tons.

55.8 Estimation of Weight of Refractory in a Preheater

It is possible to estimate roughly weight of refractory of preheater cyclones also.

Table 6.55.1 Properties of refractories commonly used in cement plants.

Sr no	Specification	Hysil blocks	Fireclay Brick 40 % alumina	High alumuna 70 % alumina	Castable Firecrete special	Castable White heat 'A'	Castable White heat 'C'
1.	Location	Preheater Cooler calciner	Preheater kiln canciner cooler	Kiln burning zone	Preheater Cooler calciner	kiln	Preheaeter kiln
2.	Service temp. max °C	950	1400	1600	1350	1750	1500
3.	Refractoriness °C min		1700	1820	1450	1800	1650
4.	Bulk density Min. kg/m³	260	2300	2450	2250	2800	2100
5.	Linear change Max %	2	$\simeq$ 0.8 at 1400 °C for 3 hrs	$\simeq$ 2 at 1500 °C for 2 hrs	$\simeq$ 0.8 at 1350 °C for 2 hrs	$\simeq$ 1.0 at 1550 °C for 3 hrs	$\simeq$ 1.0 at 1550 °C for 3 hrs
6.	Cold crushing strength kg/cm² at 100 °C	120	300	480	400	550	350
7.	Porosity max %	-	25	23	-	-	-
8.	Refractriness under load min °C	-	1400	1480	-	-	-
9.	Chemical analysis Alumina % min	Calcium slicate blocks	40	70	45	90	50
10.	Fe_2O_3 % max		3	2.5	4	0.8	1.5

Table 6.55.2 Typical thicknesses of refractories for applications in cement plants.

Sr.No.	Item	Location	Total Thickness mm	brick	insulating brick/block mm	Castables mm
1.	Kiln	Burning zone	150 for small kilns	150	-	-
			200-250 for large kilns	200 250	-	-
	Rest of shell Discharge end Kiln inlet		As above			
2.	Riser duct		215	115	100	
3.	Preheater	Cyclones cylindrical part	230	150	80	
		Cyclones conical Part	230	150	80	
		Cyclone roof Hanging bricks	240	-	-	
		Inlet duct	190	115	75	
		Outlet duct	190	115	75	
		Raw meal pipes				*
4.	calciner	Vessel	300	150	150	
		Cone	300	150	150	
		Outlet	300			300
		Mixing chamber	215	115	100	
		T. A. duct	215	115	100	
		dust chamber	215	115	100	
		branch duct				180
		take off duct	215	115	100	*
5.	Grate cooler	Throat		*	-	*
		Roof hanging bricks	240	240	-	-
		Side walls	215	115	100	

Table 6.55.3 Typical specifications of high alumina bricks.

Grade	Alumina min	Fe_2O_3 max	B.D. gm/cc	A.P. % max	C.C.S. Kg/cm^2 min	P.C.E. ^oC min	RULTa min ^oC	PLC % at 1500 ^oC per 2 hrs
1.	70	2.25	2.65	22-23	400	38-1835	1600	+ 2.5
2.	70	1.5	2.65	22	500	37-1820	1500	+ 0.5
3.	70	3.5	2.65	23	500	36-1684	1450	+ 2
4.	60	2.5	2.5	21	500	36-1804 37-1820	1550	+ 0.5
5.	60	2	2.5	21	500	36-1804 37-1820	1550	+ 0.5
6.	60	3.5	2.55	24	500	35-1785	1450	+ 1
7.	50	3.5	2.4	25	400	33-1743	1400	+ 1
8.	45	3.5	2.4	25	350	32-1717	1400	+ 1
9.	40	3	2.3	25	300	32-1717	1400	+ 1
10.	80	3.35	2.7	24	500	37-1820	1450	+ 1.5

Data mainly relevant for machine moulded bricks
Size tolerance + 1.5 % or + 2 mm whichever is greater

Table 6.55.4 Typical specifications of castables.

Property		Grade of Castable		
	1	2	3	4
Max. service temp. ^oC	1400	1350	1450	1500
Refractoriness ^oC min	1580	1450	1680	1680
Dry density Kg/m^3	2100	2250	2500	2650
Linear change % max.	+ 1 1400 ^oC/2 hrs	+ 0.8 1350 ^oC/2 hrs	+ 1 1400 ^oC/3 hrs	+ 1 1400 ^oC/ 3 hrs
CCS Kgs/cm^2 min at 110 ^oC 1350 ^oC 1450 ^oC	 250 225 -	 400 300 -	 350 - 450	 500 - 600
Al_2O_3 min % Fe_2O_3 max %	45 4	45 4	70 5	70 4
Max.grain size mm	5	5	5	5

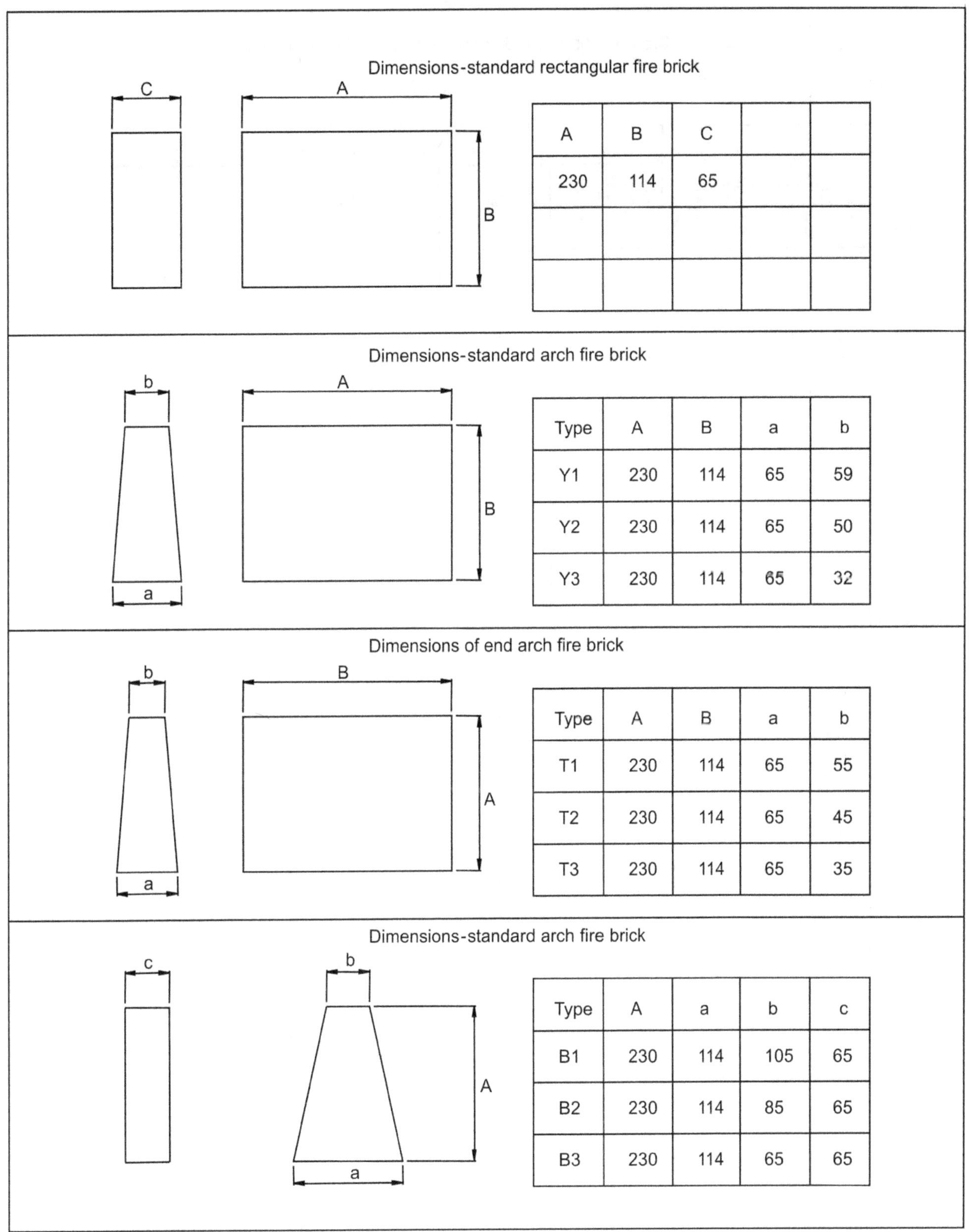

Dimensions-standard rectangular fire brick

A	B	C		
230	114	65		

Dimensions-standard arch fire brick

Type	A	B	a	b
Y1	230	114	65	59
Y2	230	114	65	50
Y3	230	114	65	32

Dimensions of end arch fire brick

Type	A	B	a	b
T1	230	114	65	55
T2	230	114	65	45
T3	230	114	65	35

Dimensions-standard arch fire brick

Type	A	a	b	c
B1	230	114	105	65
B2	230	114	85	65
B3	230	114	65	65

Fig. 6.55.1 Refractories : standard dimensions – ISO Bricks.

Table 6.55.5 Combination of standard ISO key brick for round ducts.

Brick inside diameter of duct mm	number of bricks per round				
	B3	B2	B1	Rectangular Brick	Total
600	30				30
800	23	11			34
1000	15	24			39
1200	8	37			45
1400		50			50
2000		43	23		66
2500		37	43		80
3000		31	62		93
3200		28	70		98
3400		26	78		104
4000		20	100		120
4400		13	117		130
5000		6	141		147
5400			157		157
6000			157		157

Table 6.55.6 Combination of standard ISO arch and rectangular bricks for round ducts.

Brick inside diameter of duct mm	number of bricks per round				
	Y3	Y2	Y1	Rectangular	Total
600	8	30			38
800		48			48
1000		42	15		57
1200		35	31		66
1400		29	46		75
2000		10	93		103
2500			121	5	126
3000			121	28	149
3200			121	37	158
3400			121	47	168
4000			121	76	197
4400			121	95	216
5000			121	122	243
5400			121	141	262
6000			121	168	289

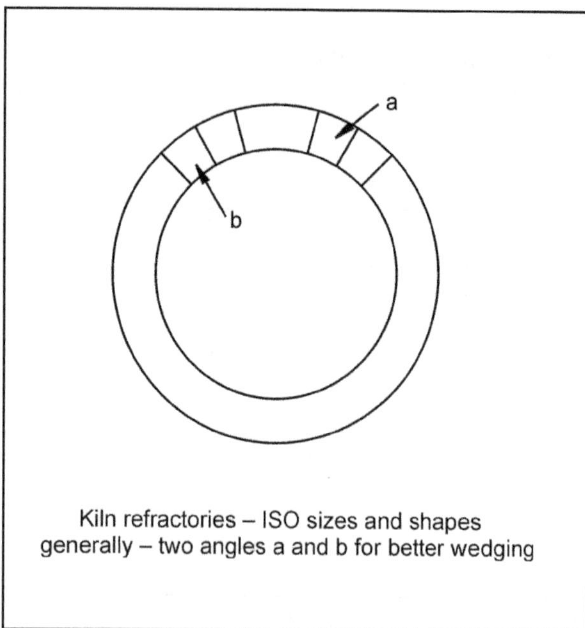

Fig. 6.55.2 Refractory lining of kiln.

If D1 is diameter of bottom cyclone inside shell in mms, then for a 5 stage preheater weight of refractory could roughly be :

1. Firebrick = 0.134 × D1 – 490 tons
2. insulating brick = 0.0457 × D1 – 164 tons
3. castables = 0.0721 × D1 – 272 tons

Thus if diameter of bottom cyclone inside shell is 8500 mm- corresponding to a single stream preheater of 3000 tpd capacity.,

Refractory required would be :

(i) brick = 650 tons
(ii) insulating brick = 225 topns
(iii) castables = 340 tons
a total of 1215 tons

To this are to be added refractory for calciner, ta duct, dust chamber and interconnecting ducting.

It can be seen that very large quantities of refractory are required to be handled and installed.

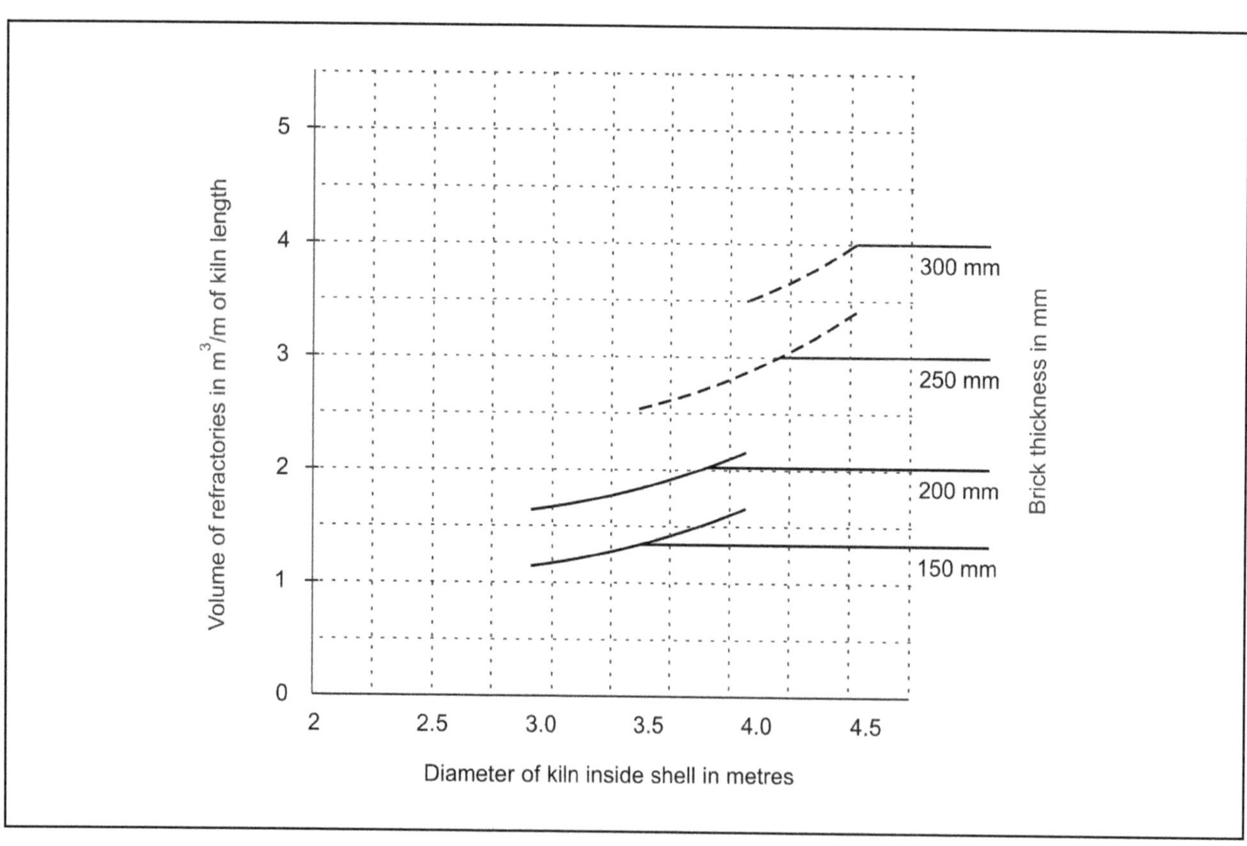

Graph. 6.55.1 Refractories required in round ducts.

55.9 Laying and Replacing Bricks

Replacement is largely in burning zone which is about one third the length of the preheater kiln and about 40- 45 % for calciner kilns.

Various methods to support bricks during lining are used:

1. to 'glue' a number of rows spread evenly around the circle;
2. to use wooden or metal centers;
3. to use props of various kinds or screw jacks.

Large kilns use 'glueing' method. When jack is used, a spreader jack is inserted in the final gap to compress bricks in a ring.

In cyclone cones, each ring has a different radius and hence requires a different size of brick.

See Fig. 6.55.3.

55.9.1 Laying Insulating Bricks

Insulating bricks or block are now commonly used in cyclones to reduce total thickness and weight of refractory lining and to reduce radiation losses.

55.9.2 Refractory in Vertical Walls

In vertical walls, retaining rings are welded at about one metre spacing so that brick walls are properly supported.

See Fig. 6.55.4.

55.9.3 Refractory for roof

Roofs generally have 'hanging bricks' which are slid on beams, welded to under surface of the roof. Intermediate space is filled either by castable or by glass wool to seal them and to prevent leakages. **See Fig. 6.55.5.**

55.10 Permissible Heat Load in Refractories

Refractories in burning zone would either be + 70% alumina or basic bricks. Basic refractories give long life and can sustain high thermal loads of 4.4-4.5 million Kcal/hr/m^2 as compared to alumina refractories. But they cannot sustain severe thermal shocks.

See Fig. 6.55.6 and 6.55.7.

Generally, therefore 1st lining would be done with alumina refractories and basic bricks would be installed when the kiln is stabilized and runs continuously 24 hours of the day.

55.11 Refractory Lining in Calciner

Refractory lining in a duct is shown in **Fig. 6.43.43 in Chapter 43.**

Fig. 6.55.9 shows cross section of refractory drawing of a calciner vessel. For laying insulating blocks bolts are required to hold them in place. Castables are secured with anchors. Bolts and anchors are often in stainless steel to suit temperatures in the vessel and duct.

Fig. 6.55.4 shows typical designs of bolts for insulating brick and **Fig. 6.55.10** shows anchors for castables.

Fig. 6.55.11 shows typical anchors for tubes of planetary coolers.

Laying castables also requires curing with cold water and should be done by experts or Suppliers themselves.

55.12 Storing Refractory

It is desirable to store refractory, castables and insulating blocks inside a fully covered shed. Since most of the refractory materials are required in pyroprocessing section, this godown is preferably located near it.

Quantities to be stored will depend on rates of consumption and also on shelf life –which is limited for castables. Most frequent and regular consumption is that of refractory lining in the burning zone of the kiln and in throat of cooler. Temperature scanners for shell of kiln greatly help in ascertaining actual condition of refractory lining; probable date of stoppage for brick lining can thus be estimated. Bricks can be ordered and stored to be available by that time.

55.13 Handling Refractory

In respective sections arrangements needed to be made to handle large quantities of refractory have already been indicated.

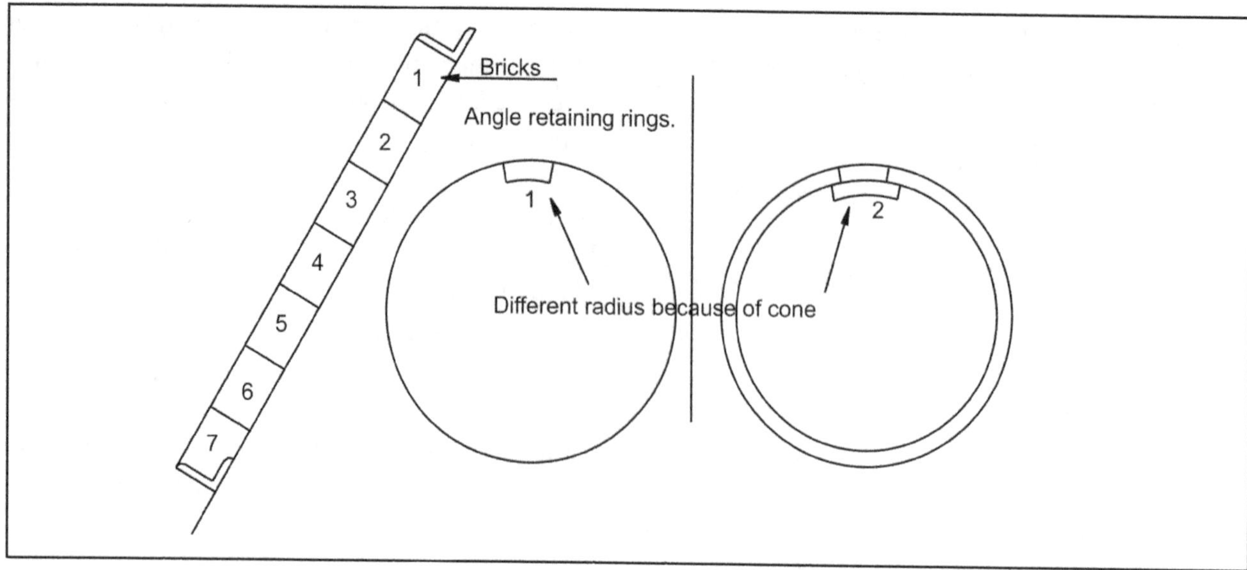

Fig. 6.55.3 Cyclone of preheater.

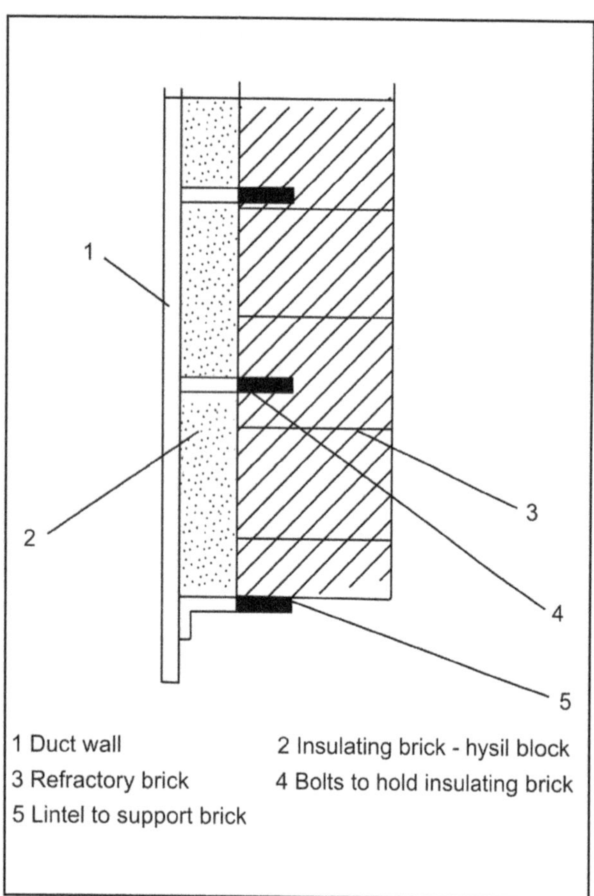

Fig. 6.55.4 Backing of insulating bricks.

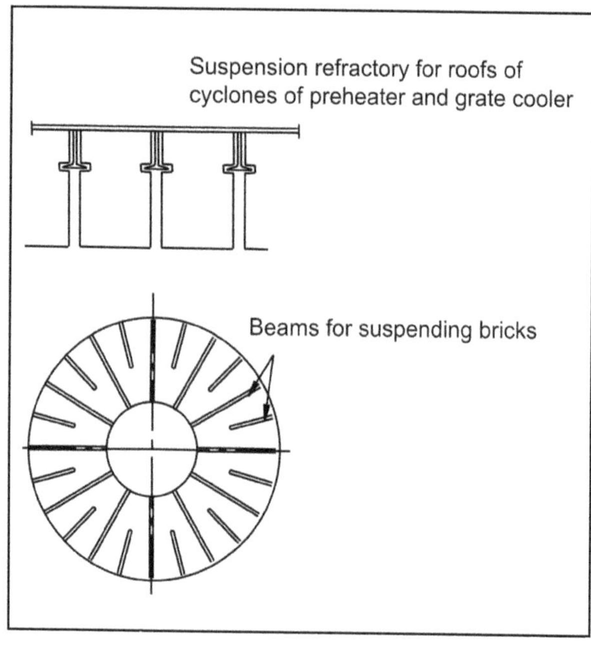

Fig. 6.55.5 Roofs of cyclones of preheater and grate cooler.

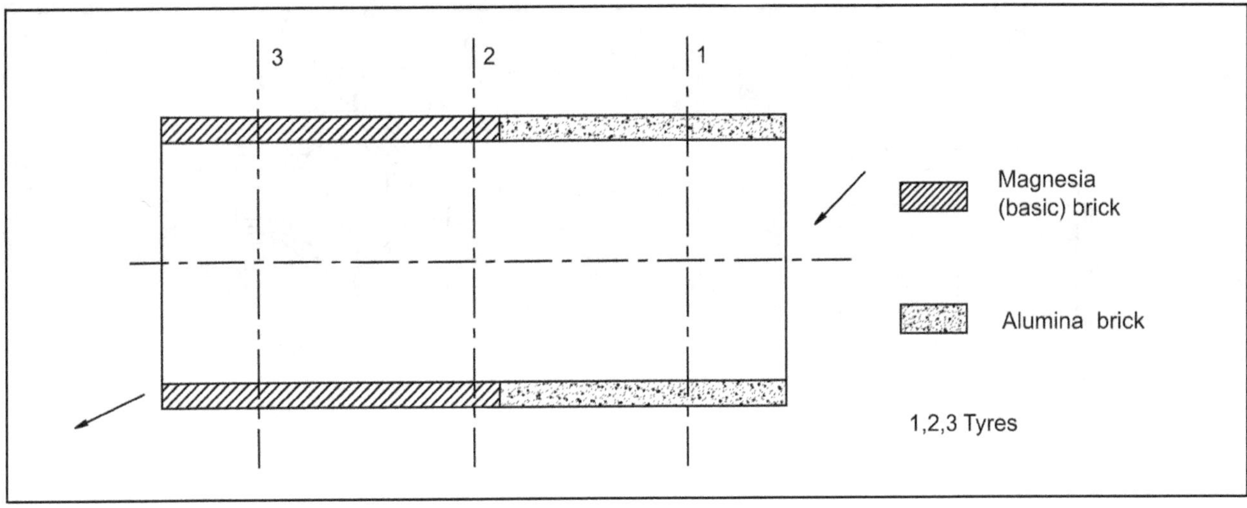

Fig. 6.55.6 Typical refractory lining of a precalciner kiln.

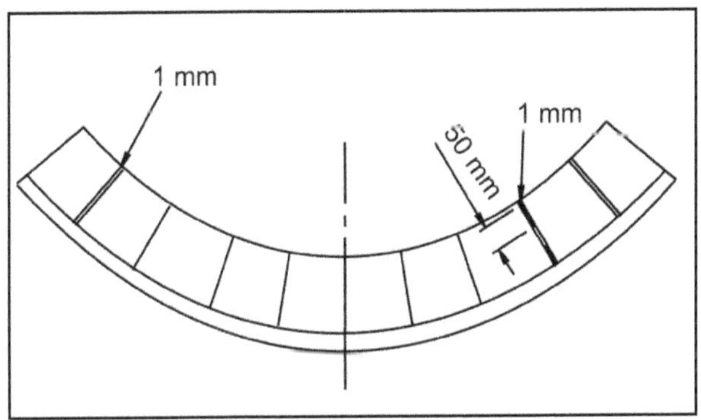

Fig. 6.55.7 Compensation for thermal expansion across the kiln section.

55.14 Consumption of Refractory

Like sp. fuel consumption and sp. power consumption there is also a yardstick of specific consumption of refractory. When plant starts operation, watch should be kept of consumption and 'generic' causes to be looked into if it is too high.

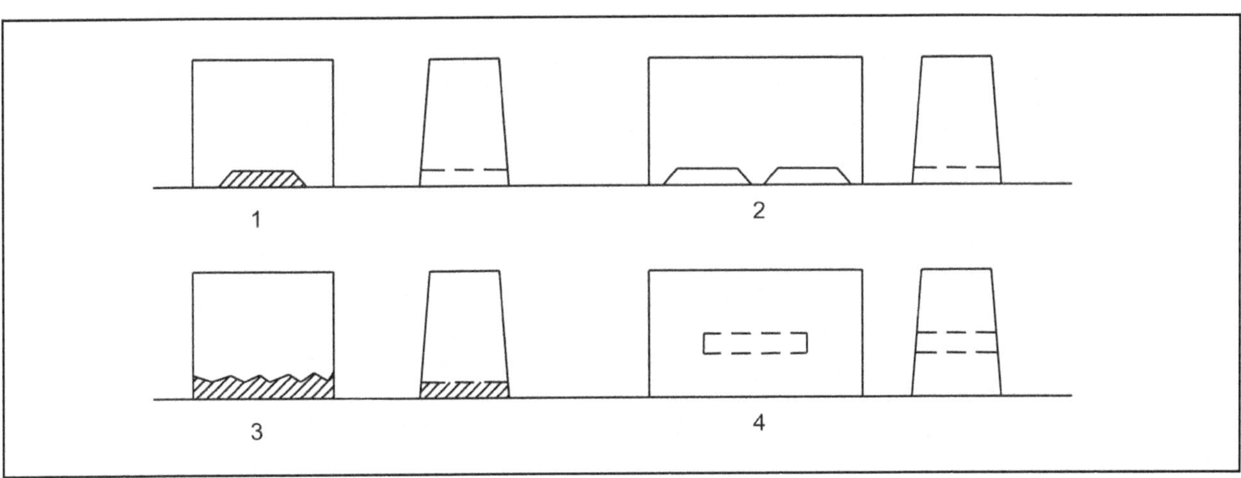

Fig. 6.55.8 Heat insulating magnesite chromite MC or periclase spinel bricks MS.

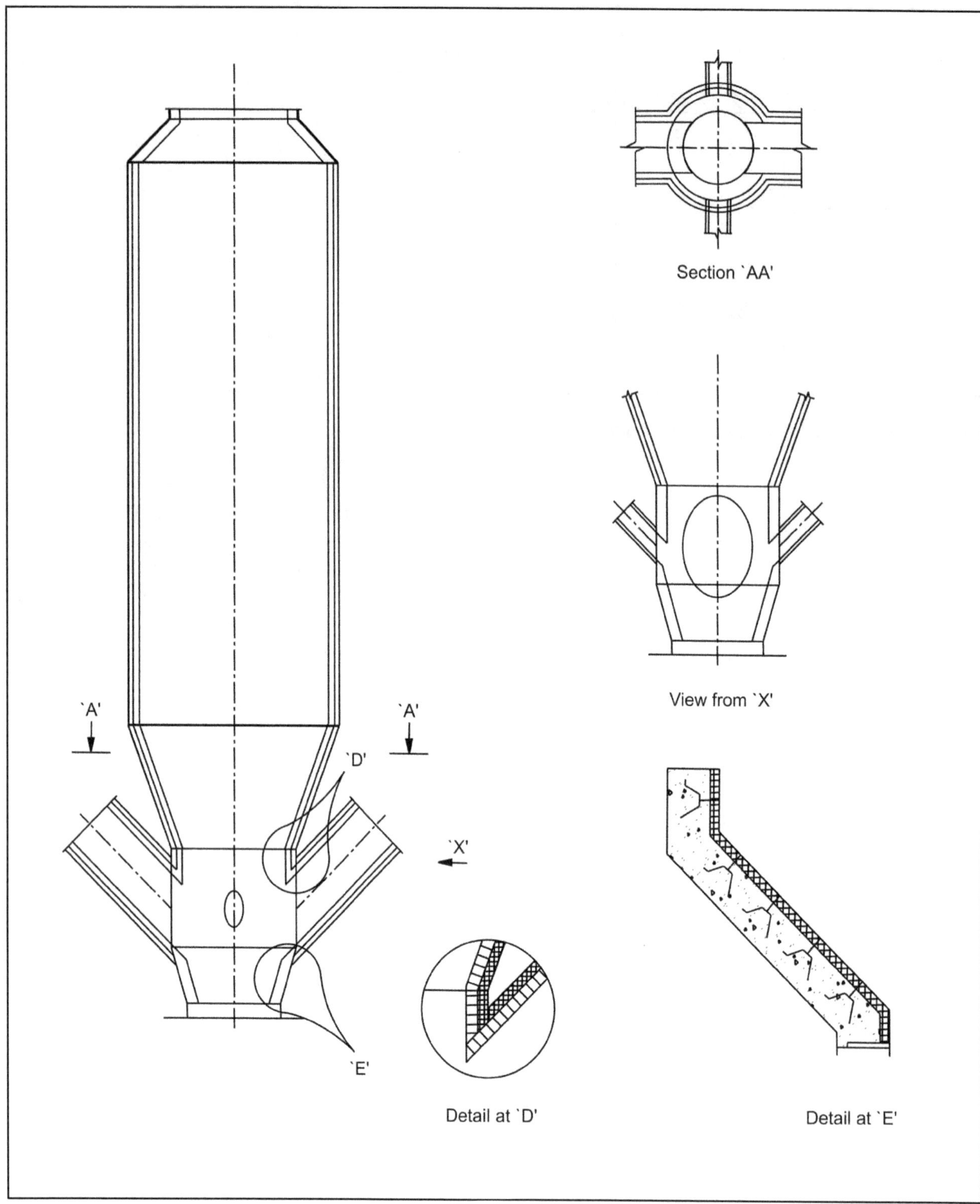

Fig. 6.55.9 Refractory lining of a precalciner showing anchoring of castables and laying of brick and insulation blocks.

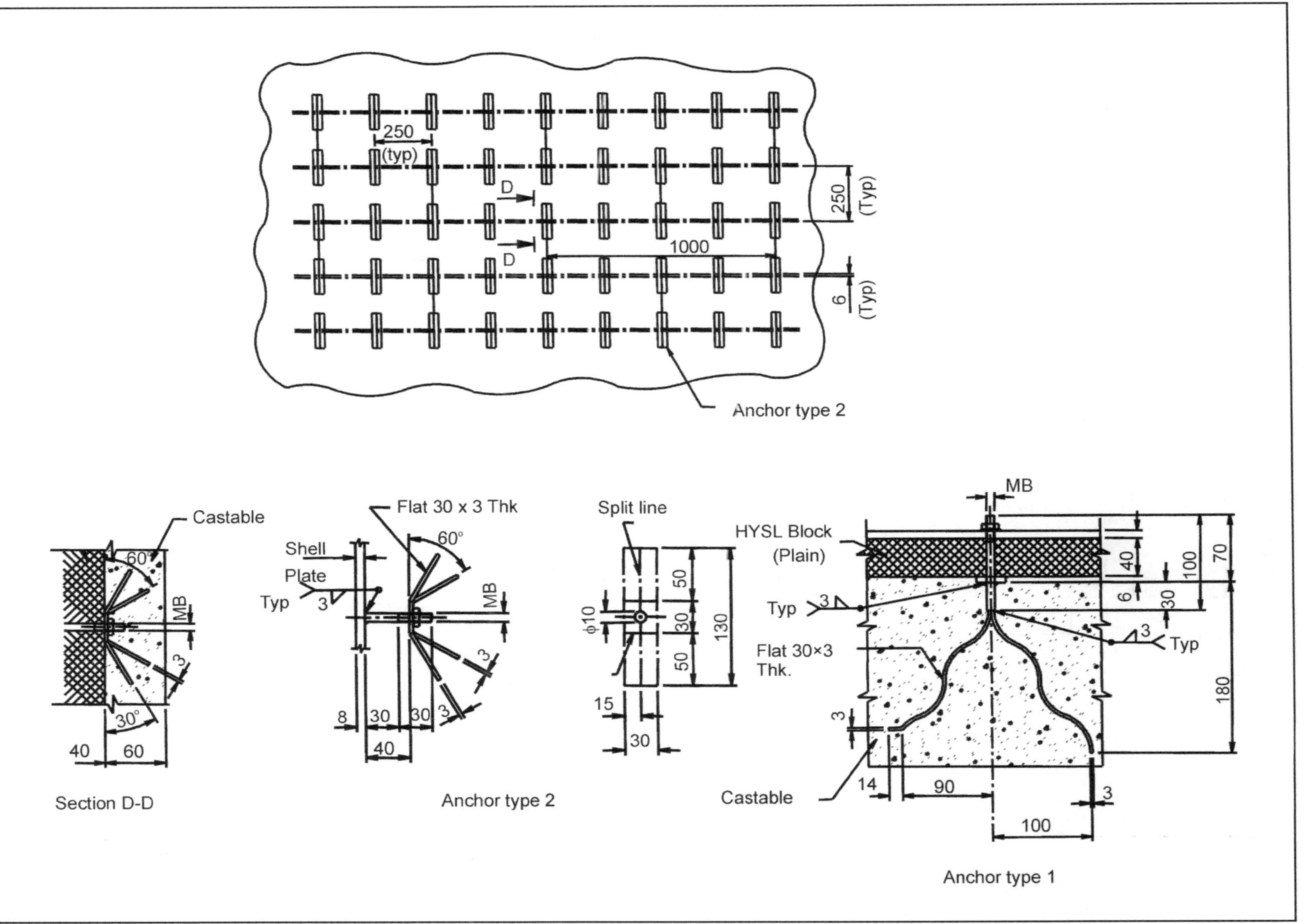

Fig. 6.55.10 Anchors and fixing bolts for hysil blocks and castable.

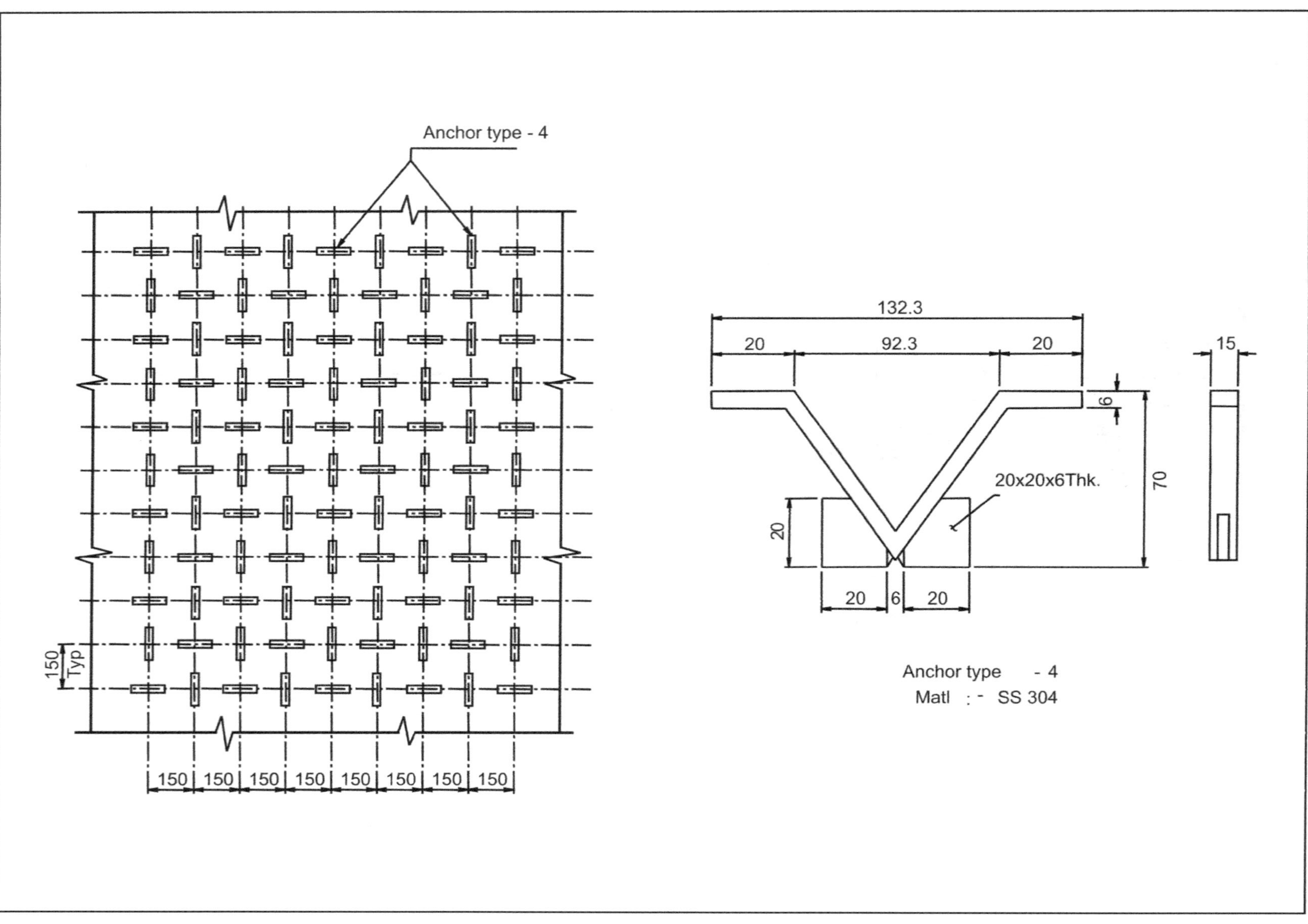

Fig. 6.55.11 Typical anchor arrangement in the elbow and cooler tube.

CHAPTER 56

INSULATION AND LAGGING

56.1 Insulation and Lagging

In ducts carrying hot gases heat would be lost to atmosphere by convection and radiation. Temperature of gases would fall in the process. When gases are used for drying, it is necessary that care is taken that this does not happen. For this purpose, hot ducts are covered with insulating materials like glass wool, rock wool or mineral wool.

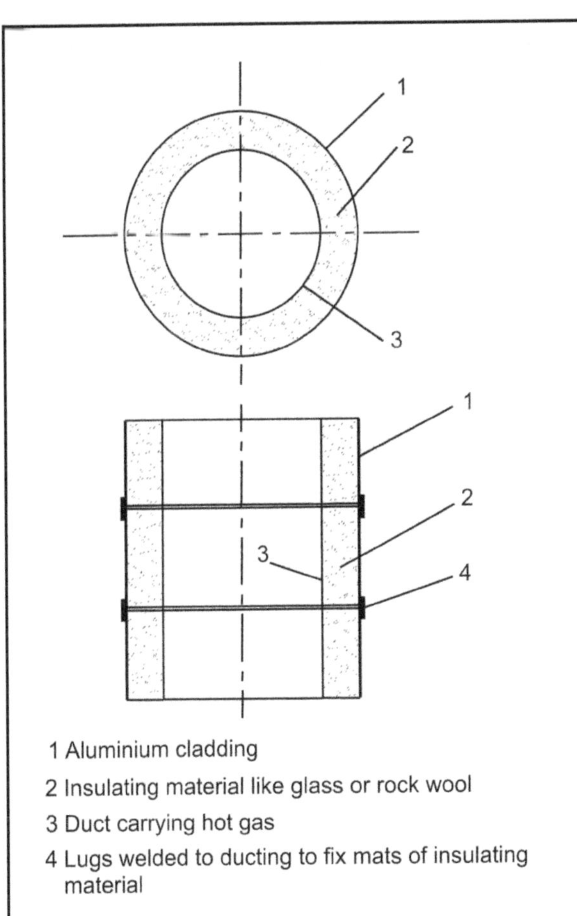

Fig. 6.56.1 Insulating and clading of ducting.

See Fig. 6.56.1.

Should the temperature of gases be more than 350-400 °C on a continuous basis, mild steel would not stand it and in such a case, duct would be lined from inside with refractory as is done in case of duct carrying tertiary air from grate cooler to calciner.

In case of bag filters and esps, where temperaturc of gases being treated is close to dew point, it is necessary to insulate the casing and hoppers to prevent condensation of water on the walls of the casings of dust collectors. This can readily happen because of difference between day and night temperatures. Insulation maintains constant temperature of walls also.

56.2 Insulation of Conditioning Tower

Gas conditioning tower (gct) is used to condition gases. Water is sprayed in the tower to improve humidity. Quantity of water to be sprayed is controlled by either temperature or humidity of gases leaving gct. This control would be ineffective if temperature of casing of the tower varied by exposure to atmosphere. For this reason, casing of gct is insulated and lagged.

56.3 Application of Insulating Materials

Insulating materials are rock, mineral or glass wool. Insulating mats of these materials are available in standard thicknesses. Wire netting and stitching maintain thicknesses of mats. Mats are wrapped around the ducting and casing. Lugs are welded to ducting at suitable intervals and aluminium sheets are wrapped over the matting to hold it in position and to protect it from wind and rain etc. This covering is called 'cladding'. Aluminium cladding also reflects heat. It also indicates that ducting is likely to be hot and should not be touched.

Where duct shapes are irregular and matting cannot be done, loose material would be used. Such material is also used in expansion joints to prevent leakage.

56.4 Properties of Insulating Materials

Insulating material has to be light in weight so as not to add to the weight of the ducting. Another important property is thermal conductivity or 'k factor'.

Properties of insulating materials are covered in standards taken out by Bureau of Indian Standards.

See Tables 6.56.1 and 6.56.2.

Table 6.56.1 Apparent density under load.

Sp. load kg/cm^2	Glass wool	Rock wool
	Max. apparent density kg/m^3	
0.01	130	115
0.02	145	130
0.05	180	165
0.07	210	185
0.10	240	200

Table 6.56.2 Thermal conductivity 'k value" in mW/cm deg

Temp. °C	Glass wool	Rock wool
	at 100 kg/m^3	
	K value in mW/cm deg	
50	0.43	0.50
100	0.52	0.56
150	0.64	0.68
200	0.78	0.80

56.5 Volume of Insulating and Lagging Costs in a Cement Plant

In a large plant ductings carrying hot gases, bag filters and esps requiring insulation are large. Total job of insulating and cladding is substantial in volume and costs. It can be undertaken after the ducting has been installed and after dust collectors have been erected. The job would normally be given on contract. Recommendations of quality of material and thickness of insulation could be obtained from manufacturers of insulating materials and suppliers of dust collectors.

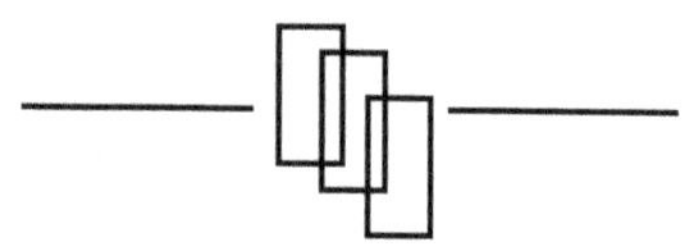

CHAPTER 57

FANS

57.1 Fans

In **Chapter 17 of Section 2**, the various applications of fans in a cement plant have been dealt with.

How they are sized has also been explained application wise.

Fans range from small size required for air slides and for cooling nose ring castings to very large fans for preheater, kiln dust collectors and cooler vent.

Fans must be selected according to the specific duties they have to perform; taking into account volume, static pressure, temperature and dust burden of gases.

Aim will be to select the most energy efficient fan consistent with the duty requirements.

57.2 Requirements of Layout

Requirements of layout in locating fans in different departments have also been dealt with in detail. Separate chapters have been devoted to fans – like **Chapter 16** for preheater fans, **Chapter 23** for cooling air fans and **Chapter 24** for cooler vent in this Section.

Therefore there is no need to repeat them.

57.3 Common Principles

Some common principles though may be stated :

1. For applications like supplying air to closed air slides and open air slides for aeration inside silos, fans must have dry washable filters at inlet.
2. Materials of construction of impellers and casings of fans required to handle dust laden gases must be carefully chosen. If required, impellers should be fitted with wear resistant liners and also casing.
3. Choice of type of impeller is also important for such applications.

   Though backward curved bladed impellers are most efficient and are ideal for fans that handle clean air, they are prone to accumulation of dust when handling dust laden gases like preheater fans.
4. Layout of ducting between machine and fan and another machine or chimney should be short and direct. Bends should be minimum and with generous radii.
5. Fans should not transmit their vibrations to ductings. In the first place they should be statically and dynamically balanced.
6. For large fans moment of inertia should be known and motors should be selected taking it into account so that they do not take too long to start.
7. It is best to use variable speed drives when system warrants them.
8. For large fans using hot gases, heat slingers should be used to protect bearings from heat. When required bearings could be water cooled.
9. Slow speed fans would be bigger for the same capacity but would have longer life other things being the same.
10. Dampers should be selected primarily for starting fans on no load; however if they must remain in circuit in operation, then such types as will cause least pressure drop should be selected.
11. When quantity of air delivered is to be measured and regulated on a continuing basis, then provision

must be made in the layout for the same at the inlet of the fan.

12. Large fans should have their impellers supported on both sides.

    Double entry fans are better from this point. They also eliminate bends at inlet where dust can accumulate.

13. While fans should not become bottlenecks in a system, they should not be oversized either as efficiency at operating point would not be maximum.

57.4

Annexure 1 furnishes data on dust burdens in air/gases to be handled by fans for different applications.

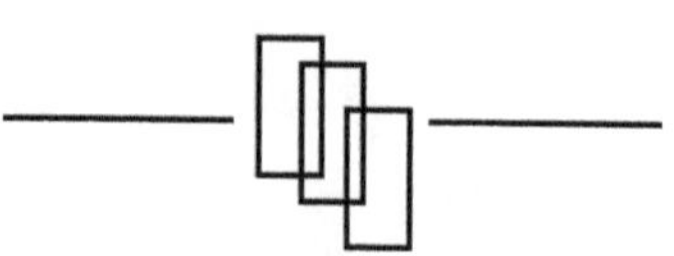

Annexe 6.57.1

Dust Loads in various Exhaust and Vent Gases to be handled by Fans

Sl.No.	Fan required for	Type of Dust Collector	Dust Burden gm/m³ Max.	Min.	Remarks
1.	Crushing plant-Primary & Secondary	Poly-Cyclones	15	9	
2.	Raw Mill Circuit	Bag Type	Negligible		
3.	Pre-Collector for Booster fan	Poly-Cyclones	100	70	
4.	Raw Meal Blending & Storage Silos	Bag Type	Negligible		
5.	Kiln Feed	Bag Type	Negligible		
6.	Waste Gas Fan after Preheater	Nil	30	12	
7.	Exhaust Fan after Poly-Cyclones	Poly-Cyclones &Multi-Cyclones	20	9	
8.	Reciprocating Grate Cooler Vent	Multi-Cyclones	12	9	
9.	Circulating Air Fan for Coal Mill	Poly-Cyclones	100	—	
10.	Primary air Fan		100	30	
11.	Cement Mill Circuit	Bag Type or E.S.P.	Negligible		
12.	Cement Storage Silos	Bag Type	Negligible		
13.	Packing Machine	Bag Type	Negligible		
14.	E.S.P Fan after E.S.P	E. S. P	Negligible		

Notes

1. Polyclones are rarely used as final dust collectors.They are used as primary dust collectors followedby bag filters and or esps.Fans in the circuit do not have to handle high dust loads shown in in the Annexure.
2. Whereas the dust burden to be handled by the exhaust fans after Bag Type Dust Collectors has been indicated to be negligible, there would be occasions when torn bags in any compartment would cause dust to escape with the exhaust gases. Thus, for some period of time the exhaust fan would have to handle approximately 5-10 gm/m3 dust burden.
3. Classification of Dust :

Limestone Dust	...	
Crushed or Pulverised		Mildly Abrasive
Clinker Dust	...	Very Abrasive
Coal Dust	...	Mildly Abrasive
Cement Dust	...	Abrasive

4. All cooling air fans for Grate Cooler and fans for Pneumatic Gravity Conveyors would handle ambient air. The suction of fans and blowers for Pneumatic Gravity Conveyors should have a suitable filter. However Grate Cooler Cooling Fan need not have suction filters.

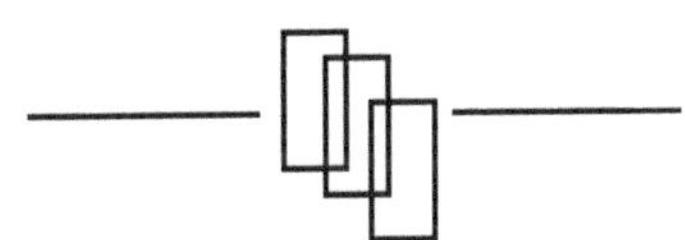

CHAPTER 58

COMPRESSORS AND BLOWERS

58.1 Compressors and Blowers

Fans either draw gases / air through a system overcoming its resistance or push air through the system against this resistance.

Fans would generally develop a maximum of 1000 mm wg pressure.

Blowers (generally meaning Roots Blowers) and Compressors convert air at atmospheric pressure to a higher pressure corresponding to the resistance of the system.

Blowers develop pressures up to 2 kg/cm^2 in one stage. There are multi stage applications like multi stage water pumps.

Compressors develop much higher pressures. Some applications require pressures as high as 7 kg/cm^2.

Most common blowers are 'Roots' or positive displacement blowers. Common types of compressors are rotary, reciprocating and screw compressors. In reciprocating compressor 'opposed piston' compressors have been developed in which there are two working strokes per cycle.

In **Chapter 18** of **Section 2**, various applications of blowers and compressors have been dealt with and need not be repeated.

58.2 Central Compressor House

There is no such thing as a 'central fan house'. But there is a 'central compressor house' in cement plants where compressors required in two or more departments are located.

In theory each application should have its own dedicated compressor or blower. Even if there is a central compressor house it only means a central location.

Supplying compressed air to a group of similar machines through a header receiving it from a battery of compressors and receivers is feasible but not advisable. This is because air will not go in quantities required to each machine always but would tend to go through path of least resistance. Therefore even if requirements of compressed air for dust collectors for purposes of cleaning bags are small and it is tempting to supply a number of dust collectors from one or two compressors working in parallel, it is not advisable to do so.

Locating compressors under one roof has its advantages particularly when they are water cooled.

Air entering compressors can be better filtered (in addition to compressor's own individual filter) by installing filtering media on doors and windows of the compressor house as shown in **Figs. 6.9.11 and 6.9.12 in Chapter 9 of this Section.**

58.3 Filtering and Purifying Air

Specific applications like supplying air to aeration media of blending or storage silos require that air should be free of oil and water. Hence great care needs to be taken to ensure this by installing air and oil filters in pipe lines for these applications close to the point of its entry into silos.

In **Chapter 45** guidelines to be followed in laying out pipe lines have been explained. Layout should permit easy access for inspection and maintenance.

58.4 Recommendations for Compressors

Recommendations for compressors and blowers should be obtained from Suppliers of machines using the compressed air. Suppliers should be told if Company intended to install compressors in one or more compressor houses. They should be furnished with the plant layout showing respective locations of compressors and machines so that in making the recommendations for compressed air Suppliers would take the length of pipe lines into account. If this is not done, machine would get air at lower pressure and hence performance of machine could be affected.

Designers of pneumatic conveying machinery should also indicate safety valve settings of compressors so that Compressors can overcome momentary chokes in the pipe line.

58.5 Layout of Compressed Air and Water Pipe Lines

Layout should integrate water and compressed air pipe lines. They should be laid systematically without cluttering or crossing one another. It is useful to draw the pipe line layouts on drawing board before undertaking installation.

See Figs. 6.9.11 and **6.9.12** in **Chapter 9** and **Figs. 6.45.6 and 6.45.9 and 6.45.10** in **Chapter 45 of this section.**

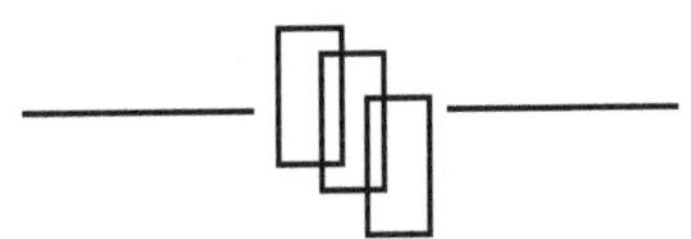

CHAPTER 59

GOOD HOUSEKEEPING

59.1 Good Housekeeping

Good 'Housekeeping' is all about maintaining things neat and tidy and dust free at all times during operation.

As the name implies, good housekeeping is very much like keeping the 'home' neat and tidy which a house wife does day after day with love and care.

It is not a 'one off' job even when all care has been taken in its design and engineering.

In a cement plant we deal with thousands of tons of bulk materials and with thousands of cubic metres of dust laden gases. Even a minute percentage of dust or spillage permitted to accumulate can soon assume ugly proportions and things can begin to go out of hand.

59.2 Continuous Watch Required

Therefore it is necessary to be as watchful as a house wife in spotting dust and dirt and in getting rid of it.

The well known adage that 'a thing not in its place is dirt' very well defines the scope and work involved in housekeeping.

59.3 What is Housekeeping about

In the simplest form housekeeping will seem to be about:

keeping floors and passages clean and washed,

keeping buildings free of dust, dirt and cobwebs, keeping roads swept regularly,

providing facilities like dust bins to deposit waste materials and their regular disposal,

prevent leaking pipelines.

However it is much more than that. It is a self discipline and a way of life that does not tolerate irregularity and uncleanliness and which does not stop at mere tidying up but goes to the bottom of things to find out why some spots are more prone than others in collecting dust and spillage.

59.4 Chronic Problems

Any housekeeping problem that repeats itself needs a solution that removes the root cause of the problem.

For example, it is quite common to come across spillages under belt conveyors. The heaps become higher and higher. Mere removal does not solve anything. Heaps will continue to build up till root cause of spillage itself is studied and removed.

It could be that belt has been running on one side and material spills over; it could be that scraper is not functioning or is not the right choice for the situation.

Therefore those who are in charge of housekeeping should be always on the look out for the weak spots and take assistance of engineering and operations to find lasting solutions.

59.5 Departments that Need Intensive Care

Easily the departments which collect large quantities of dust are packing and dispatches. Jute and hdpe bags that are used in large quantities in this country for packing cement have seams that get stuck in fixtures on belts and truck and wagon loaders. In the process of packing and dispatching operations generation of dust is unavoidable. Hence close and continuous attention needs to be given to collect the spillage on floors and platforms and in returning it to the stream.

In packing plants, dust is returned to elevators which discharge it into silo or to screen above packing machines.

Screening dust to remove foreign bodies is very important.

59.6 Equipment Available for Housekeeping

Fortunately mobile industrial vacuum cleaners are now available to cover long distances and large areas and to deal with large volumes also.

Buildings and offices in particular can be swept by vacuum cleaners. Paper shredding machines can be used to dispose of unwanted waste paper created in the course of the day in offices, drawing offices.

Such equipment should be included in the capital costs of the project at the planning stage itself.

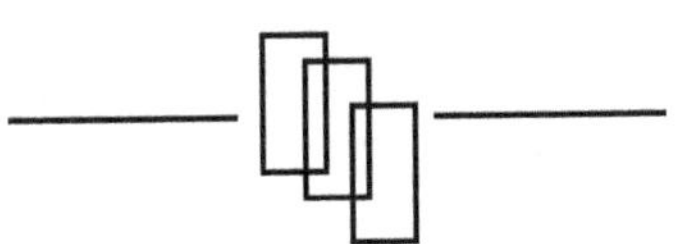

CHAPTER 60

CO-GENERATION

60.1 Co-Generation

Heat in waste gases could be converted into electricity. Waste heat could also be used to generate steam or to heat rooms as is done in central heating.
Heat in waste gases can also be used to generate electricity. This is known as 'co-generation'.

60.2 Waste Heat in Cement Plants

In cement plants waste heat in kiln and cooler exhaust gases has always been used to dry raw materials and coal; some times to dry blending materials also.

This drying naturally gets first priority as it is vitally linked with production of raw mills and coal mills. 'Co generation' can be done if there is still sufficient heat left in exhaust gases.

60.2.1 Heat Content of Waste Gases

Heat content of waste preheater gases has come down appreciably with 6 stage preheaters. Quantity of gases as well as their temperature have come down. Presently sp. gas volumes are about 1.5 nm^3/ kg clinker and temperatures are between 270-280 °C. Further preheater gases are used to dry coal in coal mill also. Vent gases leave cooler at + 200 °C. With new developments in clinker cooler volume of waste gases has come down to 1-1.2 nm^3/kg clinker.

60.3 Feasibility of Cogeneration

In co-generation not only volume but differential in temperature is also important.

With increae in operational efficiencies of the modern cement plants, scope to use conventional 'Rankine cycle' to generate steam for producing power very limited.

60.3.1 Augmenting Capacity of Existing Captive Power Plant

One possibility to use waste heat in kiln and cooler gases is to lead them to the existing Captive Poer Plant and augment its capacity.

60.4 Captive Power Plants

In a great number of countries ' captive power palnts' have become a necessity owing to acute shortage of power available from central grids.

Mostly these are coal base thermal power plants using conventional Rankine Cycle.

60.4.1 Augment capacity of Captive Power Plant (CPP)

Capacity of the proposed CPP can be increased at design stage by making use of available waste heat from kiln and cooler gases. To that extent power to be purchased from grid could be reduced.

60.5 Cogeneration by using New Processes Like 'Organic Rankine Cycle' and 'Kalina Cycle'

The increased size of the modern plant is around 10000 tpd. Therefore cogeneration can be done either by using kiln gases only or by installing a common plant using kiln and cooler gases.

60.6

The subject has been dealt with in detail in Author's book **'Designing Green Cement Plants'.** It also furnishes a work out for calculating ' nett waste heat available for co generation.

It furnishes flow charts of the commonly used systems in today's cement plants that is 1 Organic Rankine Cycle and 2 Kalina Cycle.

60.7

A flow chart of Organic Rankine cycle used to genereate power using kiln and cooler gases is furnished here. **See Fig. 6.60.1**

60.8

Aside advantage of using cogeneration – and not a small advantage either is reduction of 'greenhouse gases' per unit of cement produced. There is also saving in 'transmission losses' due to lesser quantum of power required to be purchase from the grid.

60.9 Co-generation from Exhaust of D.G. Sets

One area where co-generation can still be considered is using waste gases from diesel generating sets. Temperature of these gases is around 300 °C and it is feasible to use them. About 5 % of the rating of the d.g.set could be converted into electricity.

60.10 Other uses of Waste Heat

Use of waste gases for generating steam or for central heating is feasible but these applications are limited in scope in a hot country like India.

But even here they could be considered for plants located in places like Himachal Pradesh and Meghalaya.

60.11 Designng of Co generation Plants

Project Engineering and Design of the power plant to be set up under Co-generation would be done by Consultants and Specialist Vendors. Project engineers would have to furnish data and drawings required by them. In Author's Book, 'Designing Green Cement Plants' a chapter has been devoted to the basics of Co-generation under the title –'Waste Heat Recovery Systems'. It would help to understand the scope of the subject.

60.12

In Section 8 of this book, a brief outline of ' co generation, as a tool for sustainable development has been given with a workout for calculating waste heat available.

A typical scheme for co-generation is shown in **Fig. 6.60.1.**

Please also see **Fig. 4.6.1 in Chapter 6 of Section 4** showing relative locations of Captive Power Plant and Waste Heat Recovery Systems in to water supply. Integration of WHR in power distribution system has been **shown in Figs 5.1.2 to 5.1.4 in Chapter 1 of Section 5.**

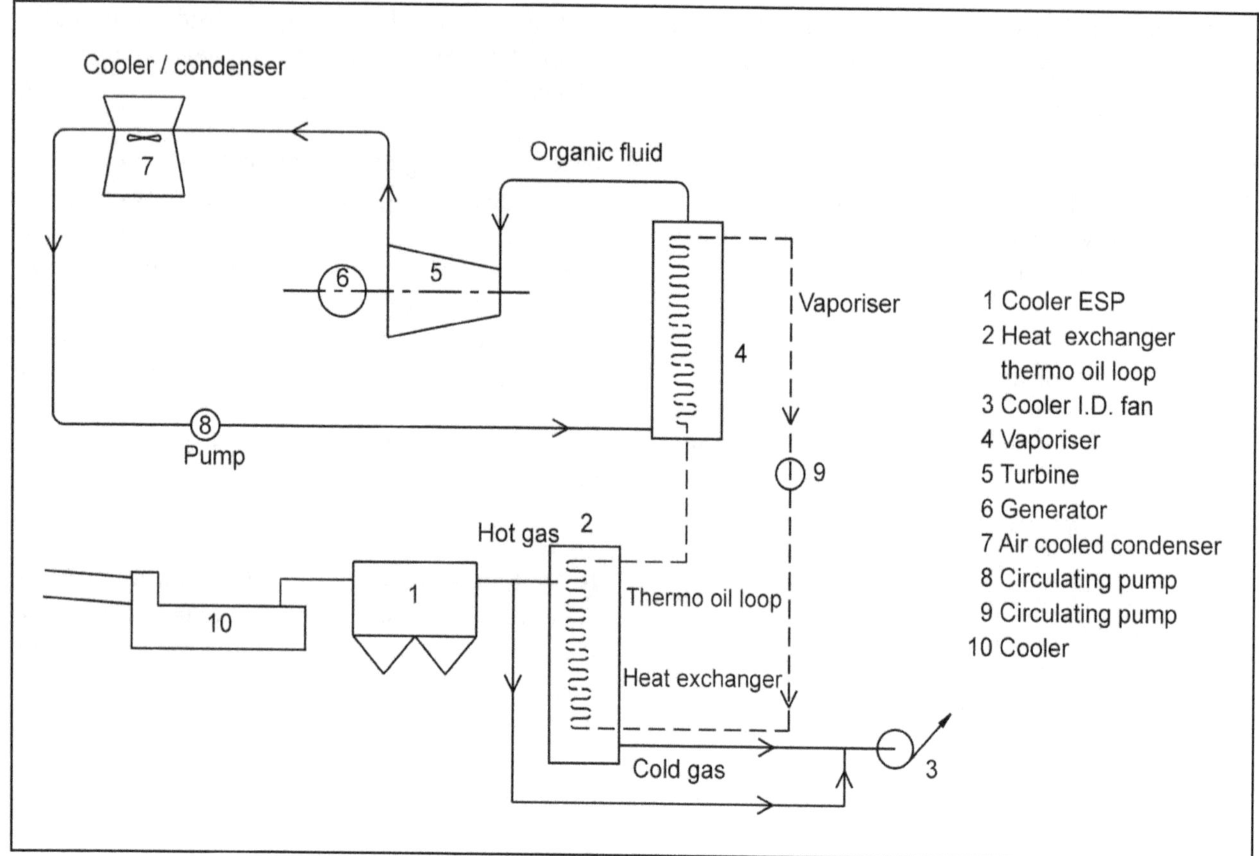

Fig. 6.60.1 Scheme for co-generation in a cement plant.

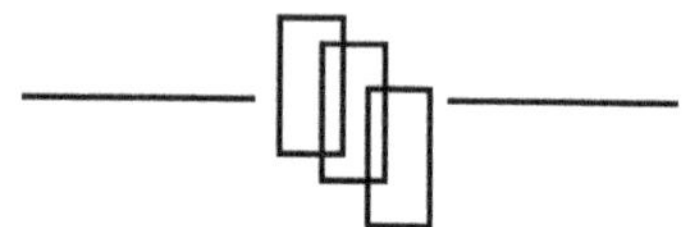

Section - 7

Selecting and Ordering Machinery

Contents

Chapter 1	Ordering Machinery	1-2
Chapter 2	Formulating Enquiries	3-6
	ENQS	7-13
Chapter 3	Technical Data Sheets	14-15
	TDS 1-8	16-29
Chapter 4	Evaluation	30-32
Chapter 5	Crtical Operational and Design Parameters	33-35
Chapter 6	Time Factor	36-36

CHAPTER 1

ORDERING MACHINERY

1.1 Selecting Right Plant and Machinery

An important aspect of setting up a cement plant is to choose the right type of plant and machinery to suit raw materials and fuel proposed to be used to make cement.

In **Section 1** we have seen the processes that are available and in **Section 2** various machinery that is available from a large number of designers of cement plants.

It is not an easy task to select right type of main machinery and auxiliaries from among the several designs available. For instance in crushing there are several options like hammer mill or impactor; one or two rotor crusher; mobile or semi mobile crusher as against conventional stationary crusher.

There are several well known designers and manufacturers of each of these types of crushers. Truthfully their designs are proved in operation.

Which one to use from the several options that are available – choice is often difficult.

Ordering all the machines and auxiliaries and electricals is in itself a big and long drawn out job even for medium sized plants. Stakes are much higher for large plants.

1.2 Indigenous or Imported Machinery

One of the first considerations is whether machinery would be bought from indigenous manufacturers or imported.

Indigenous machinery would be cheaper. What then needs to be looked into is the experience of the manufacturer and the plants that he has already delivered and his track record.

All over the world there are only a handful of original designers and makers of cement machinery. They have over the years 'licensees' in several countries. If the indigenous machinery manufacturer has (or had for sometime in the past) technical collaboration with one or more of this 'elite' group so much the better. Pedigree is thus established.

Some of the recently developed machinery like roller press for instance are not included in the technical collaborations entered into by the 'elite' group with indigenous machinery manufacturers. In such cases it is necessary to import the machinery wholly or partly. For instance casings can be made to drawings furnished and rollers and bearings and drive would be imported. Till recently vertical roller mill used to be ordered in this fashion.

Licensee would make casing and small parts like separator, flap valves. Rollers, table and yoke and gearbox would be supplied by the Principal.

Such options are also acceptable because of the 'pedigree' mentioned above.

One of the first steps therefore in procuring the machinery is to classify it according to source– indigenous and imported.

1.3 Shortlisting of Suppliers

To list out machinery and their reputed and established manufacturers. For example :

Large crushers – Hammermills and Impactors

Foreign designers and manufacturers of imported crushers would be :

Hammermills of U.S.A.

Krupp of Germany.

Hazemag of Germany.

If indigenous procurement is considerd, do these Companies have licensees in India; who are they and what is their track record.

Krupp India and FLSmidth India are licensees of Krupp and Hazemag. If their respective technical collaboration agreements include crushers then it would be possible to obtain indigenously Krupp and Hazemag crushers made by their respective licensees.

This kind of exercise narrows down the number of Vendors who should be approached to invite offers and it should be under taken for all major machinery and auxiliaries in the cement plant.

Next the capabilities of likely indigenous machinery manufacturers should be looked into closely-as regards sizes already made; whether they have facilities to make larger sizes; their set up for ensuring quality in manufacture. ISO Certification may be an indication of their infrastructural set up.

Consistent with the size of the plant and sizes of machinery required, only selected manufacturers should be approached for inviting offers. This would greatly reduce work load and save time in reaching decisions.

1.4 Work Before Floating Enquiries

A lot of preliminary and preparatory work is thus necessary before offers are actually invited. The 'basics' elaborated in **Section 1** should be taken into account and also latest designs of machinery available as explained in **Section 2** to formulate and crystallize the types of machinery and features they must have for the proposed cement plant. Pertinent flow charts and system designs should also taken into account.

1.4.1 Performance Norms

There is always more than one way to reach the destination. Thus almost always machinery of more than one manufacturer could be suitable. To be able to choose what would suit one most, it is necessary first to decide upon the performance norms and look them in the context of raw materials available and fuel that would be used. This will further narrow down the field for selection.

1.4.2 Guidelines for Selection

Assistance should be taken of the **Consultants** in this respect. They should clearly state the type of machinery most suitable for optimum performance in each case, for example:

Single stage crushing - hammer mill or impactor in the context of quality of stone and ratio of reduction.

Mobile or semi mobile crusher - to reduce transportation costs vis a vis spread and location of deposits.

Vertical roller mill with external circuit - with conventional or high efficiency separator.

Whether mill should have one or two fans and where there would be cyclones before the dust collector.

Such aspects and their impact on operational parameters like power consumption, overall operational advantages and cost of machinery should be discussed with an open mind with the **Consultant**.

Enquiries for machinery can then be formulated with the help of **Consultants**.

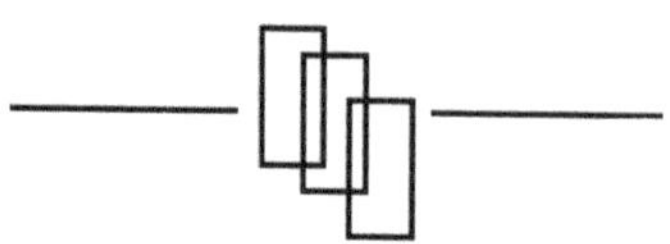

CHAPTER 2

FORMULATING ENQUIRIES

2.1 Formulating Enquiries

To obtain right quotations of machinery it is most desirable that enquiries are spelt out with great care and in detail.

Vendor must be told if he has to quote for a specific machine or the whole section. No manufacturer makes every item of machinery required to make the section complete.

2.2 Buying from one Source

It is a basic policy decision as regards buying machinery from one source or to go 'shopping' for what are commonly expressed as 'bought outs'.

It certainly saves money to buy directly the ' bought outs' rather than buying them through the vendor of main machinery.

In the event of going 'shopping', Vendor may take responsibility for specifications only and not for their performance not even for specific power consumption for the department as a whole.

On the other hand Buyer has to have sufficient organizational strength to do all the work of selecting and buying the 'bought outs'.

2.3 Works Contract

In this category also falls the option whether ordering will be on 'supply' basis or on basis of 'works contract'. In 'works contract', Vendor takes total responsibility for not only supplying but also for erecting and commissioning the plant.

Both modes have some advantages and some disadvantages. Consultant should be in a position to advise the course to be adopted taking into account organizational strengths that would be required by the Purchaser in each case.

2.4 TEFS helps in Formulating Enquiries

If the Buyer has got prepared a Techno-Economic Feasibility Study, with the help of a Consultant, he is already aware of the options available in processes and in equipment. In fact capital costs would have been worked out for processes and equipment that would suit raw materials, fuel, site conditions and type of deposits.

Thus TEFS will be a good starting point for formulating enquiries. System flow charts would be available, capacities of sections and machinery in it would have been worked out from **Section 1**. Sectional machinery schedules could thus be worked out.

2.5 Enquiries Proper

Depending on decisions in matters referred to in **paragraphs 2.2 and 2.3,** enquiries could be section wise or machinery wise under cover of a General Letter inviting quotations.

A typical covering letter is enclosed in **7.2.1.** Quotations are often invited in two parts – 'technical' and 'commercial'. Technical part is made available to technical personnel and Consultants to evaluate offers. Commercial part will be referred to Finance and Accounts departments.

2.5.1 Enquiries Specifications

Besides flow chart and machinery schedule, enquiry should furnish detailed 'duty requirements' of each machine and auxiliary. It will ensure that the Vendor understands buyer's requirements and preferences. He can then prepare his offer as required by the Buyer.

From this point, 'Enquiry Specifications Sheets' (ENQS) are prepared- section wise and machinery wise.

ENQS will spell out site conditions, duty requirements, properties of materials to be processed, fuel to be used, power available, margins to be kept in sizing and performance expected.

ENQS will also define battery limits and scope of supply.

It will also list data and drawings to be submitted with the offer.

Such a procedure would ensure that all Vendors will submit offers which would differ only in specific design features of machinery of respective manufacturers but would satisfy duty requirements fully.

This is particularly important for 'scope of supply'. Considerable time is lost in bringing offers 'on par' in this respect.

Vendor is also clear from ENQS the guarantees and warrantees that he is required to take.

ENQS are like 'standards' which stipulate 'minimum', leaving Vendor free to make 'better' offers.

Typical ENQs are enclosed.

See ENQS 1 and ENQS 2.

2.6 Standardising Offers

Each Vendor has his own way of preparing offers and submitting details. Offers received from different vendors can and almost always do differ in contents and manner of presentation. This makes evaluation difficult and time consuming both for the Buyer and the Consultant.

To make assessment easy and objective, enquiry also contains **'Technical Data Sheets' (TDS)** for major machinery and auxiliaries. More about TDS in the **Next Chapter.**

It is also possible to prepare enquiries by putting together similar machinery in different sections – like material handling equipment, fans, compressors and blowers, bag filters etc.

Enquiries for motors and gear boxes, cables, transformers etc., would generally be sent on this basis.

2.7 Commercial Aspects of Enquiries

Commercial part of the enquiry is equally important and should contain important commercial terms like : Delivery,

terms of payment,

prices – fixed or variable –'for' or delivered at site,

various duties and taxes – what are included and what are not,

performance guarantees and warrantees and their terms,

terms for erection and commissioning,

If commercial terms are received on 'similar basis' and in the same sequence, comparison is easy.

The whole idea in 'formatting' enquiries is to save time in collecting missing data, cross correspondence and in objective assessment.

Normally it is best to have a meeting with Vendors after they have received the enquiry but before they submit the offer so that Buyer and Vendor are on 'same page'.

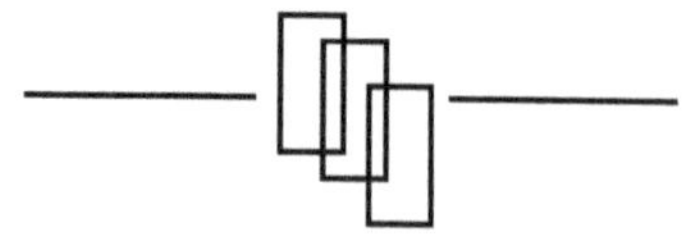

Annexe 7.2.1

GENERAL COVERING LETTER

Ref **Date**

To

Proposed New Cement Plant

At ———————————

Inquiry for

1. Preamble

——————— Company which is already operating — cement plants with an installed capacity of ———mtpa is proposing to install a new cement plant to produce opc / opc and ppc / opc and slag cement with a capacity of——————million tons per annum.

The Company is in the market to buy the plant and machinery required in the various sections of the plant for this expansion.

2. Offers invited for

We are therefore inviting your offers for————————————————————— as per the duty requirements and scope as given in the enclosed Enquiry Specifications (ENQS) Your offers are invited for the complete section that is machinery and auxiliaries that together make the section complete including drives. Where a Vendor is not in a position to include all items of his own manufacture, he should quote for machinery of his manufacture and furnish budgetary prices for other items so that the Company gets a complete picture of the costs of the section.

If you wish to test raw materials/fuel, representative samples can be made available on request

3. Your offer should be in two parts viz. Technical and Commercial.

3a. Technical Aspects

The Technical part should contain:

1. Process flow chart for the section showing battery limits.
2. Material and gas balance, temperature and pressure/draught profile under normal operating conditions at rated capacity.
3. Machinery schedule of the equipment included showing size, capacity, numbers, weight and drive details.
4. Specifications of major machinery and auxiliaries and technical data as per proforma (TDS) enclosed which should be returned duly filled.
5. GA drawings of machinery.
6. Data such as moment of inertia, speed torque curve, pressure volume curve (for fans).

   The performance that you would be prepared to guarantee against penalties.

   In particular in terms of capacity (tph or tpd), quality of product, sp fuel consumption, sp power consumption, degree of accuracy as would be pertinent to the machinery offered.
7. A reference list of plant and machinery similar in size and capacity supplied and in operation.
8. Brochures of major machinery and auxiliaries highlighting their special features.
9. Scope of supply should be clearly defined.

3b. Commercial Aspects

The Commercial part should contain among other things:

1. Prices on 'for' basis for indigenous machinery and fob basis and also 'cif' in US $ for imported machinery; taxes and duties, insurance and freight charges payable in local currency to arrive at price as landed at the plant.
2. Delivery – the shortest that the vendor can offer and guarantee against penalties.
3. Specific performance would have to be guaranteed against penalties. Vendor should indicate his terms for the desired guarantees as indicated in the enclosures.
4. Warranties for design and workmanship and materials of construction would have to be offered for a period of minimum 12 months from date of commissioning. Vendor should indicate his terms for the warranties.
5. Terms of payment – the first advance, progress payments if any and payments against invoices. Company proposes to withhold 5% as retention money, which will be released after satisfactory completion of performance trials. A proper protocol will be drawn with the successful Vendor for this purpose.

Your offer in duplicate should be received by the Company within 4 weeks of this date.

Thanking you,

xxxxxxx

For——— Cement Company

Encl.: General guidelines

Enquiry specifications (ENQs)

Technical data sheets (TDSs)

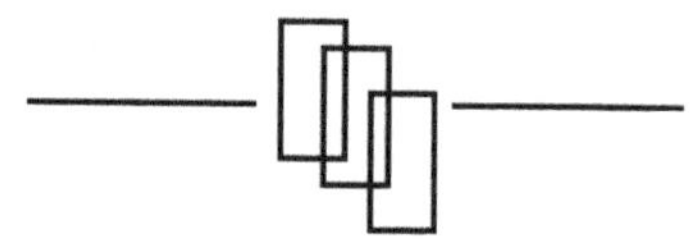

ENQS NO. 1

GENERAL GUIDELINES

1. Site Conditions (illustrative only)

1 altitude	300 m above sea level
2 ambient temperature	35 °C
3 power supply	132 KV, 3 phase, 50 cycles
4 HT motors	6.6 KV, 3 phase, 50 cycles
5 LT motors	415 V, 3 phase, 50 cycles
6 fuel used	coal
calorific value	gross 4800 kcal/kg
	useful 4300 kcal/kg
	ash 35 %
	fixed carbon ≃ 52%
	volatiles ≃ 30 %
	moisture 8-10%

2. Margins for working out capacities and for sizing

1 design margin	10 % over rated capacity
2 in feeders, conveyors, elevators etc.	15 % over design capacity
3 in fans	40 to 80 % respectively for cooler vent fans for upset condition of kiln
4 in clinker conveyors	60 % over design capacity
5 in kiln motor	60 % over b.h.p. at design capacity
6 in other motors	15 % over required bhp at design capacity rounded off to next higherstandard rating and international frame size; high torque motors preferred.

Data on raw material and fuel proposed to be used is furnished in 7.2.2.

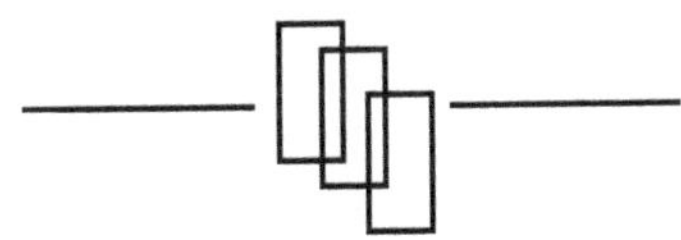

ENQS NO. 2

Enquiry Specifications for

Pyro Processing Section

Beginning with Kiln and Ending with Grate Cooler

1. Scope : The Pyro processing section comprises of
    1. 6 stage cyclone preheater and fan
    2. Dry process rotary kiln with coal firing
    3. on / off line calciner complete with tertiary duct, dust chamber and coal firing equipment and provision for firing AF in it
    4. Reciprocating grate cooler with usual auxiliaries
2. Battery limits: From preheater fan up to the clinker conveyor to take clinker produced to clinker storage
3. Site conditions altitude 300 m above sea level

    Ambient rtemp. 35 °C
4. Duty requirements :

    Data on raw materials and correcting materials and fuel proposed to be used are furnished in **Annexe 1**

    Raw meal as indicated therein will be processed in the pyro processing system to produce well burnt clinker of good quality so as to conform to norms laid down in Standard ref. No IS 269-2013 and will produce when ground with suitable quality and quantity of gypsum O.P.C. of properties laid down in the Standard.

    Rated capacity of the section: 10000 tpd of 24 hours

    Clinker shall be air quenched in a Reciprocating Grate Cooler and shall be cooled to a temperature of not> than 65 °C above ambient. Cooler vent gases shall be passed through an ESP.
5. General conditions

    Please see attached General Guidelines in **ENQS - 1** to arrive at capacities and sizing.
6. Performance guarantees are required to be taken for
    1. capacity in t.p.d.
    2. sp. fuel consumption for the section
    3. sp. power consumption for the section
7. Quotation should be accompanied by
    1. process flow chart
    2. material and gas balances
    3. heat balance of kiln and cooler
    4. temperature and draught profile
    5. machinery schedule for the section
    6. full details- of major machinery included like kiln, preheater, preheater fan, calciner, grate cooler
8. **Technical Data sheets** have been included for kiln, preheater, calciner and grate cooler. These should be completed and returned with the offer.

9. Broad Specifications of auxiliaries not included in scope of supply like belt feeders, esp/bag filter, drives should be furnished.
10. G.A. and cross sectional dimensioned drawings of preheater, kiln, calciner and grate cooler and auxiliaries included should be furnished. Typical departmental drawing for the whole pyro processing section of similar installation supplied earlier should also be supplied.
11. To follow general instructions for sizing and capacities of conveyors, elevators, feeders and fans etc.
12. D.C. variable speed motors will be preferred for preheater fan, kiln, cooling fans and cooler drive/s
13. Special notes on major machinery in the Pyro processing section are given below.

1. Calciner

1.1 Calciner shall be designed to take 60 % of total fuel fired in the system. An off line calciner with fuel burning in air will be preferred.

We shall be firing alternate fuel/s in the calciner. It should designed to take this possibility into account. It should be designed for low NO_x emissions.

Calciner shall be complete with take off point from cooler, dust chamber, tertiary air duct with its supports, pilot and regular burners for coal.

Specifications of refractory lining should be furnished.

1.2 Quotation shall be accompanied by:

Specifications and salient features of calciner

Degrees of calcination achieved at different fuel inputs

Sp. Volume of calciner for a given heat input

2. 6 Stage Preheater

Preheater and Preheater fan will be sized for design capacity with margins as stated in Annexure 1. Cyclones shall be of modern design with low pressure drop. Pressure drop across the 6 stage prheater shall be limited to 400-450 mm at rated capacity and temp. of gases leaving top cyclones shall be 260-280 °C. Eficiency of top cyclone shall be not be < than 93 %.

Preheater shall be complete with cyclones interconnecting ductings, dispersion plates, pendulum flap valves, immersion tubes, expansion joints, down comer duct up to p.h. fan The bottom two cyclones shall have double cones to prevent choking.

Ducting between cyclones and volutes of cyclones should be such that there is no accumulation of dust.

Sp. fuel consumption in the pyroprocessing section should not be more than 700-720 kcal/kg clinker. Sp. gas volume should be commensurate with this; however for sizing preheater an allowance of 5% should be kept on guaranteed fuel consumption.

For the rated capacity of 10000 t.p.d. clearly specify if it would be single stream, two stream with equal/ unequal streams and corresponding fans.

2.1 Quotations shall be accompanied by :

Specifications of preheater with its salient features- pressure drop and efficiencies across each stage, dust circulation at various stages and outlet.

Description of dispersion plates, pendulum flap valves and immersion tubes and materials of construction

Dimensioned drawing of preheater tower showing elevations and floor plans and location of calciner whether inside tower or outside it should be made available.

Details of refractory, insulating brick and castable to be used for lining prehearer cyclones and ducts should also be furnished.

3. Preheater fan/s

Fan will be sized for the gas volume, temperature and pressure drop in the system to meet duty requirements of the 6 stage preheater and calciner as mentioned above.

Preheater fan shall be a high efficiency fan with efficiency not < than 70%.It should withstand a temp. of 400 °C for minimum two hours.. The fan shall have a variable speed drive – d.c. motor. Fan shall be capable of handling a dust burden of 70 –100 gm/nm^3. Impeller should be supported on both sides.

3.2 Quotations shall be accompanied by:

Specifications of prheater fan- construction of impeller, speed variation and drive details. Fan curve appropriate to the fan should be furnished.

G.A. drawing of fan

In case of two dissimilar fans, dissimilar details of two fans should be provided

4. Rotary Kiln

Rated capacity 10000 tpd.

Kiln is a dry process kiln with preheater and calciner. Yearly working days per year would be minimum 330.

In such a kiln, raw meal enters kiln calcined to 85-90 % when fuel fired in calciner 55-60 %.

Properties of raw meal and fuel fired as also site conditions have been furnished.

Kiln will have a variable speed drive and will have all features currently prevalent in modern cement plants

For design purposes a 10 % margin on size and, 5 % on fuel consumption should be provided.

Quotations shall be accompanied by :

Heat balance of kiln section as whole

a g.a. drg of kiln showing spacing between stations, heights; cross sectional view and view across showing centers of rollers and angle at center and height of kiln center above bed plate salient features of the major components of the kiln

Data on refractory with dimensions, quality and weights should be made available

Reference lists of kilns supplied and performance achieved.

Guarantees shall be for:

Capacity in t.p.d . consistent with quality of clinker produced;

Sp. fuel consumption for the pyro processing section as a whole;

Sp. power consumption

5. Grate Cooler

Cooler shall be of the latest design incorporating features currently prevalent in the industry

To receive clinker at~1370-1400 ° c from kiln and air quench it and cool it to a temperature of not more than 65 °C above ambient.

Number of grates would be as per vendor's design parameters. Each grate will have its own variable speed drive.

However details of each compartment in terms of grate area, number of rows,clinker bed thickness in operation shall be furnished.

Details of static grate- its slope, number of rows and bed thickness and details of its fan should be provided.

Details and drawings of special featutres like walking floor, polytrack should be provided.

Details and drawings of grate plates should be provided.

Details of cooling air fans for each compartment should be furnished.

Cooler will be complete with , discharge hoppers with motorized double pendulum flap valves, clinker breaker of full width and spillage drag chain under the full length of the cooler if required.

Tertiary air for calciner will be drawn from cooler

Vent gases from the cooler shall be vented through an ESP.

While cooling air fans shall have the usual margin of 10-15 %, the cooler vent fan shall have a margin of 30-35 % above design capacity. The spillage drag chain conveyor shall have margins of 60-65% over design capacity .

Quotations will be accompanied by

Distribution of cooling air in various compartments as also specific air loading.

Specifications of cooling air and vent air fans

Bed thicknesses in various compartments and speeds of cooler grates to maintain them in strokes per min.

Heat balance of cooler and cooler efficiency

Specifications of materials of construction of grate plates, grate plate supports side castings, refractory and castables used

Sp. power consumption for cooler and auxiliaries with breakdown for grates, fans etc. should be furnished.

Specifications of refractories to be used in the cooler

G.A. and cross sectional drawings with dimensions of grate cooler and hoppers and pendulum flap valves and spillage drag chain; similar drawings for esp should also be submitted

Drive details of cooler grates, clinker breaker, pendulum flap valves, spillage drag chain should furnished in detail

For items not included in scope of supply sufficient details must be given to facilitate procurement

6. Coal Burner for Kiln

Burners should be sized for quantities of coal fired in kiln with a margin of at least 20 % over requirement at design capacity and fuel consumption. Multi channel burners using minimum primary air and which also have facility of oil firing for start up and for preheating of refractory would be preferred for the kiln.

It is proposed to use Alternate Fuels in Kiln and Caciner at a later date.This should be kept in mind in selection of burner offered.

The range of variation possible should be specified. Details of primary air and p.a.fan/ blower necessary should be indicated.

It is proposed to install the coal mill near the preheater. Therefore pulverized coal would be brought to kiln pneumatically by a p.s. pump.

Vendor should furnish his recommendations for the firing system as a whole and not just the burner for calciner.

7. Burner for calciner

Details of burner suitable for fuel fired in calciner should be furnished calciner can fire alternate fuel, details there of should also be furnished.

If it is possible to fire coarse fuel in calciner, the same would be preferred to save on power in pulverizing coal.

Quatation should be accompanied by

specifications of burners and their auxiliaries

catalogues and brochures.

Its capacity in mkcal/hr and turn down ratio

G.A and cross sectional drawing of burner and system layout remembering that 40% fuel will be fired in kiln and 60 % in calciner.

Annexe 7.2.2

Data on Raw Materials and fuel proposed To be used

1. Raw meal feed to kiln representative values are given below

Constituent :	CaO	SiO_2	Al_2O_3	Fe_2O_3	Na_2O	K_2O	Cl_2	L.O.I.
%	40.7	14.4	3.65	1.71	0.97	0.85	0.02	36.01

Other minor constituents would be:	Mn_2O_3	TiO_2	P_2O_5
%	0.03	0.28	0.09

Lime Saturation factor	0.89
Silica Ratio	2.68
Alumina Ratio	2.13
Fineness of raw meal feed	15 % residue on 90 micron sieve
Homogeneity of feed	std. deviation of blended raw meal not more than 0.2%
Fuel used in Kiln and Calciner	coal
Gross calorific value	4800 kcal / kg
Net calorific value	4300 kcal / kg
Volatiles	28-30 %
Moisture	2 %
Ash	30 – 35 %
Fineness coal fired	15 % residue on 90 micron sieve

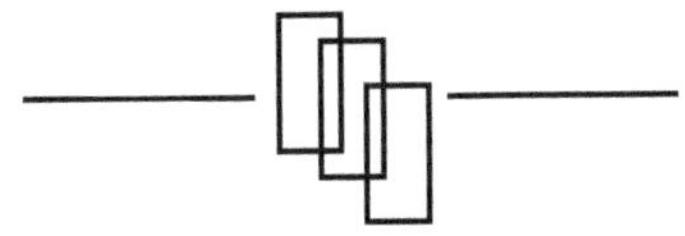

CHAPTER 3

TECHNICAL DATA SHEETS

3.1 Techincal Details of Machinery

To understand the various features of machinery offered, it is useful to receive detailed specifications thereof together with drawings. Each machine has components that are vital and influence its performance and life. There are parts which wear out and which need replacement; hence material used and life expected are important details for them.

From process angle also some parameters are critical, like:

L/d ratio for ball mills, degree of filling, compartment lengths; speed etc.

L/d ratio, slope, retention time, degree of filling, linear speed and specific outputs, thermal load in burning zone for kilns.

3.2 Technical Data Sheets

Vendors must furnish such vital details in their offers. Technical Data Sheets (TDS) are prepared machine wise which help vendors to submit all such details in the format required by the Buyer.

TDSs are made available by the Buyer to Vendors along with the enquiry and Vendors are required to complete them and return them with their offers.

Main advantage of the TDS is that specifications and details are received from participating vendors in a uniform manner that makes evaluation easy and fair.

Buyer should not ask for details that have no bearing on performance and hence should include in the TDS only the essential details. At this stage nobody is interested in 'nuts and bolts' so to say.

It should be 'practical' for the Vendor to complete the TDS in a reasonable time otherwise he would take too long to submit the offer.

3.3 TDSs Help Evaluation

Once completed TDSs are received with offers, Buyer and his Consultant can edit and tabulate them for evaluation.

Even then some homework will have to be done by Buyer and his Consultant to 'derive' essential operating parameters and to cross check if Data 'tallies'.

For example:

does sp. gas volume of gas leaving kiln match with the sp. fuel consumption; does sp. power consumption of mill agree with grindability of material.

Vendors can then be contacted to clarify the doubts. Thus all technical matters can be clarified in just about one meeting with each vendor cutting down time for taking decisions considerably.

It is then Consultant's responsibility to present his recommendations with justification.

3.4 Typical Tachnical Data Sheets

A set of TDSs is enclosed as per list in **Annexe 7.3.1**. They are by way of illustration and do not correspond to any given size of plant.

For the same reason ENQS will not match with TDS for the same machinery enclosed in respective annexes.

Any additional data and drawings over and above what has been mentioned would be welcome.

Actual performance data on plants of similar size in operation would also be most welcome.

Operational data like sp. fuel consumption and sp. power consumption should be furnished.

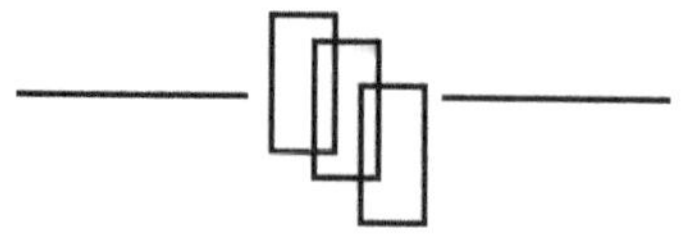

Annexe 7.3.1

List of TDS's Enclosed

TDS No	Description	Page Numbers
1	Rotary Kiln	18-22
2	Preheater	23-29
3	Precalciner	30-34
4	Grate Cooler	35-41

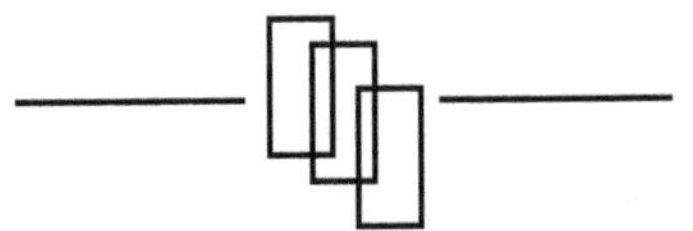

TDS NO. 1

Technical Data Sheet for

ROTARY KILN

Sr.No.	Item	unit	details
Note: Duty requirements for pyro processing section as a whole are furnished separately and hence are not repeated.			
Properties of raw meal to be processed to make clinker and fuel to be fired in the kiln and calciner have also been furnished.			
This TDS is for rotary kiln			
1.	rated capacity	tpd	
2.	design capacity		
3.	process dry process * stage prheater calciner		
4.	fuel coal /oil		
5.	useful calorific value of fuel	kcal/kg	
6.	thermal efficiency of pyro processing section	kcal/kg	
7.	altitude above sea level	m	
8.	ambient temp.	^{o}c	
9.	sp. gas volume at exit of preheater	nm^3 kg clinker	
10.	fuel fired in calciner	%	
11.	degree of calcination achieved at rated capacity and ratio	%	
12.	O_2 at kiln inlet	%	
13.	Kiln		
	Size of kiln dia x length	m	
	1. diameter inside shell	m	
	2. brick thickness	mm	
	3. clear dia. inside brick	m	
	4. length	m	
	5. clear volume of kiln	m^3	
	6. sp. output at rated capacity with calciner	tpd/m^3	
	7. L/D ratio		
	8. Slope	%	
	9. Speed at rated cap.	r.p.m.	
	10. Max. speed	r.p.m.	

Sr.No.	Item	unit	details	
	11. Retention time at rated capacity	minutes		
	12. Degree of filling at Rated capacity	%		
	13. power drawn at rated capacity	kw		
14.	kiln details			
	1. shell thickness	mm		
	2. shell thickness under tyres	mm		
	3. shell thickness of end shell	mm		
	4. Material			
	5. Ovality max. at operating temp.	%		
	6. Number of supports			
	Supports	1st support	2nd support	3rd support
	7. load on supports in tons			
	8. Tyres type : floating/ splined Furnish drawing			
	9. Rollers dimensions in mms			
	Type : self aligning /			
	1st support dia. * width	2nd support dia. * width	3rd support dia* width	
	Furnish drawing			
	10. Main bearings			
	Type			
	Furnish drawing			
	dimensions in mms			
	1st support dia.*width	2nd support dia.*width	3rd support dia.*width	
	11. provision for adjusting rollers / skewing			
	12. Drive single /dual Furnish drawing			
	13. girth gear and pinion spur / helical Material gear and pinion			
	14. method of fixing gear to kiln shell			
	15. lubrication lubricating idler / forced			
15.	floating of kiln			
	total float	mm		
	kiln float controlled by			

Sr.No.	Item	unit	details
16.	downward thrust due to inclination	kn	
	thrust taken by thrust rollers		
	diameter of thrust roller	mm	
17.	hydraulic thruster furnish drawing		
	capacity of thruster		
	description of system		
18.	Drive		
	power at shaft at design capacity and max. speed	kw	
	recommended rating of motor	kw	
	type		
	variable speed		
	d.c.		
	speed range		
	type of speed variation		
	gearbox		
	type		
	installed at slope		
	input / output speeds	r.p.m	
	Max. speed of kiln	r.p.m.	
19.	kiln seals		
	inlet type	furnish drawings	
	outlet type		
20.	Furnish G.A. Drawing of kiln. Furnish details in drawing showing quality and dimensional details in burning zone		
21.	Heat balance Furnish heat balance of kiln at rated capacity		
22.	Main burner Furnish details of burner. It should be capable of firing main fuel as well as alternative fuel.		
23.	Special features like splined tyre, self aligning roller station, friction drive through roller should be highlighted with appropriate drawings.		
24.	Furnish details of refractory lining of the kiln		

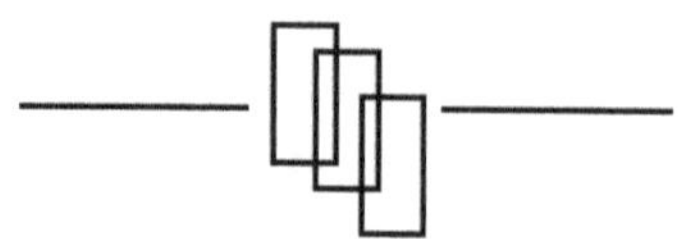

TDS NO. 2

Technical Data Sheet for

6 Stage Ptrheater for a 10000 tpd Dry Process Kiln with Calciner

Sr no.	Item	unit	details
	This TDS deals with Preheater		
1.	rated capacity	tpd	
2.	design capacity	tpd	
3.	fuel :	coal /oil	
4.	useful calorific value of fuel	kcal/kg	
5.	thermal efficiency of pyroprocessing section	kcal/kg clinker	
6.	altitude above sea level	m	
7.	ambient temp.	^{o} c	
8	sp. gas volume at exit of preheater	nm^3 kg clinker	
9.	fuel fired in calciner	%	
10.	degree of calcination achieved at rated capacity and ratio	%	
11.	O_2 at kiln inlet	%	
12.	Preheater make to the designs of / under license agreement with number.of streams twin / dissimilar number.of stages top stage - twin / single		
13.	Furnish G.A, Drawing of Preheater and Drawings of cyclones in all stages		
14.	Furnish profiles of gas flow, temperature and draft in preheater and calciner At rated capacity		
15.	preheater fan/s (if different fans for two streams furnish details of each)		
	capacity	m^3/min	
	temp.	^{o}c	
	max. design temp. at continuous operation	^{o}c	
	max. temp. for half hour	^{o}c	
	pressure	mmwg	
	fan speed	r.p.m.	
	speed variation stepped / smooth		
	speed variation range	r.p.m./ r.p.m.	

Sr no.	Item	unit	details
	impeller type		
	static efficiency of fan	%	
	furnish g.a. drawing		
	motor rating	kw	
	speed	r.p.m.	
	type		
	range of speed variation		
.	Sp. Power consumption of preheater		
	At rated capacity	kw/t	
	Flow pattern of material and gases in cyclone should be furnished if available- particularly for top stage.		
	Furnish g.a. drawing of preheater		

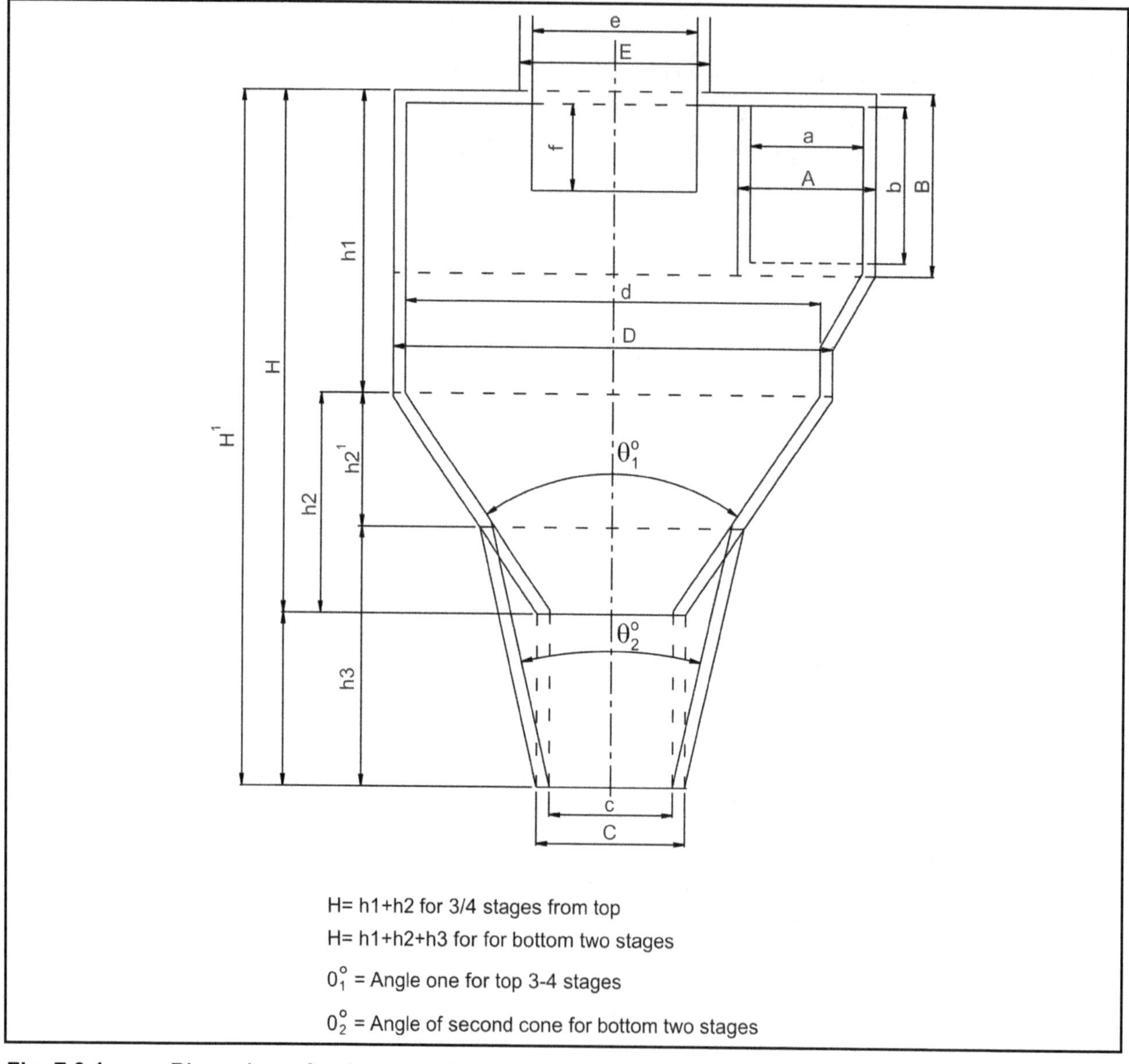

Fig. 7.3.1 Dimensions of preheater cyclone.

TDS NO. 3

Technical Data Sheet for

PRECALCINER

Sr.No.	Item	unit	details

Note: Duty requirements for pyro processing section as a whole are furnished separately and hence are not repeated.

Properties of raw meal to be processed to make clinker and fuel to be fired in the kiln and calciner have also been furnished.

This TDS is for precalciner

Calciner will be fired with alternate fuels in future. It should be designed so that it can be done without major modifications.

Sr.No.	Item	unit	details
1.	rated capacity	tpd	
2.	design capacity	tpd	
3.	fuel	coal /oil	
4.	useful calorific value of fuel	kcal/kg	
5.	thermal efficiency of pyroprocessing section	kcal/kg	
6.	altitude above sea level	m	
7.	ambient temp.	^{0} c	
8.	sp. gas volume at exit of preheater	nm^3 kg clinker	
9.	fuel fired in calciner	%	
10.	degree of calcination achieved at rated capacity and ratio	%	
11.	O_2 at kiln inlet	%	
12.	Precalciner make design		
13.	type in line off line on single stream preheater on separate c line preheater fluid bed spouted bed down draft		

Sr.No.	Item	unit	details
14.	combustion of fuel in air through external duct		
15.	type of cooler grate		
16.	combustion air drawn from cooler from kiln hood		
17.	fuel fired in calciner	% & kg/hr	
18.	tertiary air	nm^3/kg clinker	
19.	temp. of tertiary air	^{0} c	
20.	quantity of gas leaving calciner	 nm^3/kg	
21.	temp. of gas leaving calciner	^{0} c	
22.	velocity of gases in calciner	m/esc.	
23.	retention time in calciner	seconds	
24.	temp. in calciner	^{0} c	
25.	capacity of calciner rated	tpd	
26.	capacity design	tpd	
27.	volume of calciner clear	m^3	
28.	thermal load in calciner	kcal/hr/m^3	
29.	size : clear diameter.* height refractory thickness dimensions inside shell diameter. * height	 m mm m	
30.	connecting duct volume	m^3	
31.	total volume calciner + connecting duct	 m^3	
32.	pr. Loss in calciner	mmwg	
33.	degree of calcinations raw meal entering kiln	fuel % 20 30 calcination % %	40
34.	O_2 in gases leaving calciner	%	
35.	CO in gases leaving calciner		
36.	NO_x in gas leaving calciner		
37.	fuel fired in calciner calorific value fineness % passing 90 micron	 kcal/kg %	
38.	sp. fuel consumption of calciner	kcal/kg	
39.	sp. fuel consumption of prheater		

Sr.No.	Item	unit	details
40.	oveall sp. Fuel consumption		
41.	tertiary air duct		
	quantity of tertiary air	$nm^{3/}kg$	
	quantity	m^3/min	
	temp.	^{o} c	
	Furnish g.a. and cross sectional drawing of calciner and also calciner as installed with t.a.duct, Drawing showing fas flow and temperature and draft profiles in the calciner circuit		
42.	Furnish detailed sprcifications of refractory, insulation and castables recommended		
43.	coal burners		
	coal fired in calciner	tph	
	no. of burners		
	cap. Each burner	tph	
	coal conveyed pneumatically		
	conveying pipe dia.	mm	
	velocity in pipe	m/sec.	
	velocity in burner	m/sec.	
	Layout of coal conveying arrangement		furnish drawing
44.	weights of precalciner, t.a. duct, connecting ducts dust chamber etc.		
45.	Special features of the calciner to reduce NO_X should be mentioned and high lighted with drawings.		
46.	A complete flow chart of kiln, preheater and calciner should be submitted along with the offer. Departmental drawings from actual installations of similarcapacity should also be submitted.		
47.	Possibility of spillage and arrangements provided to handle that should be furnished with drawings.		
48.	Furnish g.a. drawing of the entire pyro processing section showing preheater, calciner, kiln & cooler.		

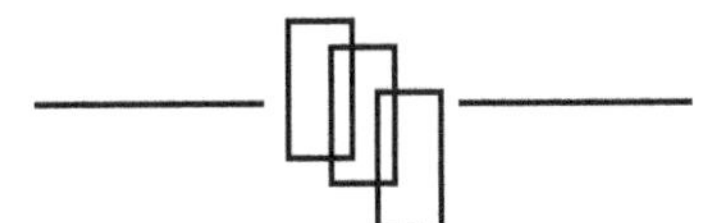

TDS NO. 4

Technical Data Sheet for

GRATE COOLER

Sr.No.	Item	unit	details

Note: Duty requirements for Pyro processing section as a whole are furnished separately and hence are not repeated.

Properties of raw meal to be processed to make clinker and fuel to be fired in the kiln and calciner have also been furnished.

This TDS is for Grate Cooler

Sr.No.	Item	unit	details
1.	rated capacity	tpd	
2.	design capacity	tpd	
3.	thermal efficiency of pyro processing section	kcal/kg	
4.	altitude above sea level	m	
5.	ambient temp.	^{o} c	
6.	temperature of clinker at discharge of cooler	^{o} c	
	temp of clinker discharged under upset conditions	^{o} c	
7.	tertiary air for combustion to calciner will be drawn from in proportion of fuel fired in calciner. calciner		
8.	specific loading at design cap.	Tpd/m^2 grate area	
9.	Size of Cooler		
10.	grate area effective	m^2	
	grate area gross		
11.	Type – cross bar/ pendulum./ walking floor / polytrack horizontal /inclined/combi single width / two width details of static grate		
12.	size of kiln diameter. inside of brick		

Table 1 Details of Grates

Sr no.	Description	Unit	Static	Grate 1	Grate 2	total
1	Width	mm				
	Length	mm				
	No. of rows					
	Gross area	m^2				
	Effective area	m^2				
		mm				
	No. of compartments In each grate No of rows					

Sr.No.	Item	unit	details	
	As per Table 1 above			
13.	Details of Grates			
14.	total cooling air	nm^3/kg		
15.	sp. fuel consumption	kcal/kg		
16.	total air for combustion	nm^3/kg		
17.	excess air	nm^3/kg		
18.	leakage at kiln inlet	%		
19.	fuel in kiln and calciner	%	kiln % :	calciner %
20.	sec. Air into kiln	nm^3/kg		
21.	tertiary air into calciner	nm^3/kg		
22.	temp. of sec. Air	^{0} c		
23.	temp. of tertiary air	^{0} c		
24.	air to be vented cooling air – (sec.+ tertiary air)	nm^3/kg		
25.	temp. of vent air under above conditions			
26.	heat of clinker entering cooler	kcal/kg		
27.	heat recuperated	kcal/kg		

Sr.No.	Item	unit	details
28.	efficiency of cooler	%	
29.	radiation of loss from cooler	kcal/kg	
30.	temp. of vent air normal conditions	^{0} c	
31.	temp. of vent air under upset conditions	^{0} c	
32.	details of cooling air and vent air and fans see **Table 2**		

Table 2

Sr. no	Description	Air loading	Area	Volume	Pr.	Fan capcity	Pr.	motor
		$m^{3}/min/m^{2}$	m^2	m^{3}/min	mmwg	m^3/min	mmwg	kw
1	Static Grate							
2	1St Grate							
	1st compt							
	2 nd compt							
	3 rd compt							
3	2nd Grate							
	1							
	2							
	3							
	4							
	3rd Grate							
	total							
			M^3/min	Temp. ^{o}c	Pr. mmwg			
	Vent air							
	Normal conditions							
	Upset conditions							

33. bed thickness of clinker at rated capacity

grate no	grate area m^2	capcity tpd	grate speed s.p.m.	bed thickness mm
static				
1				
2				
3				

34. drives of grates — motors

grate	area m^2	motor kw	type	grate speed normal r.p.m.	grate speed max. r.p.m.
1					
2					
3					

Sr.No.	Item	unit	details
35.	drives of grates	gear box	
	grate / gear box type / rating kw / speed ratios input/output	chain drive ratio	max. grate spped r.p.m.
	1		
	2		
	3		
36.	hoppers and discharge valves for spillage		
37.	spilllage drag chain if required		
	amount of spillage as % of capacity	%	
	as quantity	tpd	
	capacity of drag chain		
	conveyor	tph	
	size width	mm	
	length c/c	mm	
	linear speed	m/sec.	
	drive rating motor	kw	
	speed	r.p.m.	
	gear box type		
	rating	kw	
	speed ratio		
38.	clinker breaker		
	Type	hammer /roller	
	location		
	(a) hammer type		
	diameter	mm	
	width	mm	
	no. of hammers		
	wt each	kgs	
	material		
	speed	r.p.m.	
	motor	kw	
	speed	r.p.m.	
	grate discharge bars		
	spacing	mm	
	material of grate bars		
	(b) roller crusher type		
	location	middle or at end	
	dia and width of rollers	mms	
	drive	kw	

Sr.No.	Item	unit	details
39.	furnish drawings of cooler assembly and cross sections as also drawings of grate plates		
40.	materials of construction		

Table 3

Sr. no	description	unit			Grate 1	Grate 2	Grate 3	
	Grate plate							
	Grate plate support							

Sr.No.	Item	unit	details
41.	sp. power consumption	kwh/ton	
	grates		
	cooling air fans		
	clinker breaker		
	spillage drag chain		
	vent fan		
	total		
42.	refractories		

location	type	quality	thickness	aprox. Wt
		mm	tons	
throat	brick			
	castsble			
	insulating			
roof	hanging			
sides	brick			
	insulating			
	castable			
furnish drawings				
anchors				
vent duct				

43. reference lists of coolers supplied
feed back on operating coolers performance achieved
literature and brochures
g.a. drawing of cooler
typical layout drawing of cooler, cooling fans, spillage conveyor etc.
Drawings of components like
Grate frames

Sr.No.	Item	unit	details
	Grate plate supports Grate plates Side castings		
44.	note on instrumentation and process control		
45.	main performance parameters like cooler recuperation efficiency sp. power consumption		
46.	cooler heat balance cooler heat balance and process control and instrumentation should be integrated with the pyoprocessing section as a whole including kiln, preheater, and calciner		

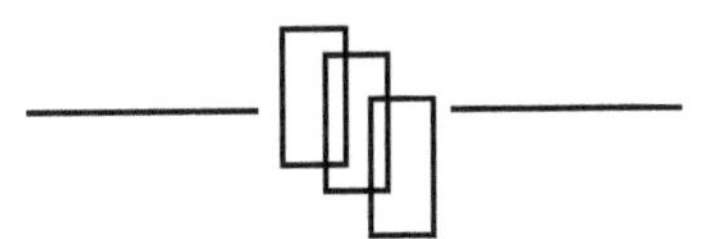

CHAPTER 4

EVALUATION

4.1 Assessment and Evaluation

After TDSs are received from vendors for a machine, and are tabulated (after receiving clarifications if any were required) it is necessary to evaluate and assess them to decide which vendor's machine meets Buyer's requirements best. Some kind of ranking is necessary. However ranking should be on a uniform basis to create 'level playing field conditions'.

4.2 Weightages for Desired Properties

For this purposes it is best to give 'weightages to various properties of the equipment adding up to say 100.

For a ball mill, weightages could be like for example:

	weightage %
for design aspects :	
1. size vis a vis rated capacity (also expressed as sp. output)	10
2. sp. power consumption	10
3. weight of machine (indicative of sturdiness)	5
4. l/d ratio	5
5. drying chamber volume	5
6. % filling	10
7. retention time	10
total	55
for mechanical design features like :	
1. shell thickness	8
2. material of shell	4
3. bearing size; material	7
4. bearing pressure	7
5. deflection	7
6. lining plates: type and material	5
7. gear spur / helical ; materials	7
total	45
Grand Total	100

Above exercise is only for illustrative purposes. Values are not to be taken literally.

For each of these categories a scale has to be established.

4.3 Attributing Weightage

For example if desired sp. output is say 1.7 tpd / m3 volume, and normal sp. outputs range between 1.5 to 2 then scaling and weightage will be done as follows:

Sp. output tpd/m^3	1.5	1.6	1.7	1.8	1.9	2
Weightage	10	10	10	9	8	7

It means those vendors whose machine has a sp. output of 1.7 and less will get full 10 percent points and those with higher sp. outputs will get less as shown.

Once a scale like above is established weightage can be given on a uniform basis.

It means those vendors whose machine has a sp. output of 1.7 and less will get full 10 percent points and those with higher sp. outputs will get less as shown.

Once a scale like above is established weightage can be given on a uniform basis.

4.4 Shortlisting Vendors

Total weightage can be added up to rank vendors. First 3 ranks would normally be considered for final selection

Any vendor earning rank 4 or lower is not 'condemnable'; only that he is not among the best suited in this particular instance.

4.5 Commercial Evaluation

Besides technical evaluation, there is also commercial evaluation involving :

Price.

Delivery.

Terms of payment.

Warranties and guarantees.

Vendors will not have same rankings in both categories.

4.6 Track Record and Service

Over and above all there is the 'track record' of a vendor and his 'after sales service' and his business integrity. They also get some weightage.

A Consultant should therefore 'shortlist' 3 vendors for final negotiations and for finalizing orders. Normally Consultants will excuse themselves from taking part in commercial negotiations.

4.7 Typical Evaluation

A Typical exercise in evaluation of offers for Pyto processing Section is shown in **Annexe 7.4.1.**

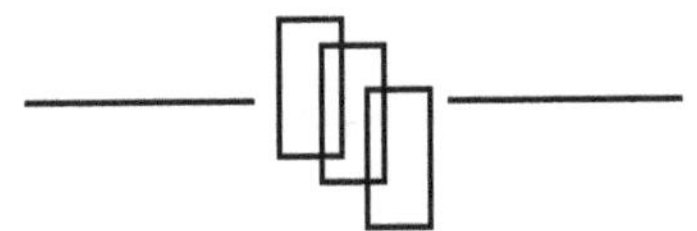

Annexe 7.4.1

Evaluation of offers received for Pyroprocessig section.

Sr. no		Weightage for	Points	Vendors			
				1	**2**	**3**	**4**
1	Kiln	size : diaX length sp.output tpd/m^3 sp.fuel consumption kcal/kg	50	45	48	50	50
2	Preheater	dias of cyclones pressure drop across efficiency top cyclone	50	50	48	45	50
	Calciner	volume m^3 retention time seconds					
3		dgree of calcination at kiln % provision for AF provision to contain nox	50	40	48	50	45
4	grate cooler	grate area m^2 cooling air nm^3/kg cooler efficiency %	50	50	48	50	50
	overall	for the pyro processing section					
5		guarantees offered for output, kcal/kg, sp. power	50	50	45	40	50
6	Overall contd	guarantees offered for delivery	25	20	20	25	20
7	track record	plants of similar capacity suplied and in operation	25	25	20	25	25
8		technical collaboration	25	10	20	20	20
9		import content	25	25	10	20	25
		total	**350**	**315**	**320**	**325**	**335**
		overall ranking for pyroprocessing section		**4**	**3**	**2**	**1**

CHAPTER 5

CRITICAL OPERATIONAL AND DESIGN PARAMETERS

5.1 Critical Operational and Design Parameters

While there are a great many attributes of any machinery that influence its Performance - mechanical and operational, some of them come to be recognized as more critical.

As a matter of fact they are treated as 'yardsticks' of measuring performance. It has become customary to hold a Vendor responsible to fulfill the performance in terms of these yardsticks not just verbally or by way of ' in good faith' or sort of ' gentleman's agreement' but by imposing 'penalties' in the event he fails in this respect.

Such yardsticks are also known as 'Guarantees and Warrantees' of performance.

5.2 Conventional Yardsticks for Cement Machinery

Conventionally in Cement Industry such yardsticks for different machinery are:

Crushers

1. feed opening should be large enough to receive max. size of stone delivered by the shovel without getting overloaded.
2. volume of crusher in relation to output / also ratio of dia. to width.
3. peripheral speeds of hammer, impactor bar tips.
4. **Output in tph.**
5. **Sp. output and sp. power consumption wrt crushing strength / bwi and product size.**
6. Sieve analysis of product from crusher.
7. **Wear rates of hammer / impactor bars in gm/ton of stone crushed.**

Ball mills

1. **Sp. power consumption for grinding stone of a given bwi from a given feed size to a given product fineness,**
2. **output in tph,**
3. drying capacity in drying — % moisture in feed to — % in product,
4. mill dimensions for a given capacity; l/d ratio of mill,
5. % filling of mill,
6. volume of air to be drawn through mill,
7. bearing pressure on main bearings,
8. **sp. power consumption for system or dept,**
9. wear rates of lining plates.

Air separators, conventional and high

1. Circulating load,
2. **efficiency,**
3. % passing or blaine of feed , coarse and product factions,
4. selectivity curves,
5. air /material ratio in feed,
6. pr. drop in separator.

Vertical roller mills - external circuit

1. Mill and fan power
2. **Pr. drop in mill**
3. Air flow in system.
4. **Velocities at throat.**
5. Circulating loads at rollers and in separators.
6. **Wear rates of liners of rollers, table etc.**
7. sp. power for mill and for system as a whole.

Blending systems

(a) Stacker reclaimer systems

1. no. of layers in full stock pile and capacity of each pile,
2. **blending efficiency achieved in terms of ratio of std. deviations of feed and extracted material,**
3. **sp. power for system.**

(b) blending systems batch / continuous

continuous

1. capacity of silo in terms of no. of days' consumption in kiln,
2. **std. deviation of blended raw meal,**
3. blending efficiency achieved in terms of ratio of std. deviations of feed and extracted material,
4. **sp. power consumption of system split into for blending and for extraction.**

Pyroprocessing section comprising

(a) preheater

1. **pr. drop across preheater for — no. of stages,**
2. **efficiency of top cyclones %, or**
   dust burden in gases leaving preheater,
3. **temp. of gases leaving preheater for — — no. of stages,**

(b) preheater fan

1. **efficiency %,**
2. **sp. power consumption.**

(c) kiln

1. output in tpd,
2. **sp. output as preheater kiln and as calciner kiln,**
3. **sp. power for kiln only under two conditions above,**
4. thermal loads in burning zone,
5. degree of filling,
6. retention time,
7. life of refractories in burning zone.

(d) calciner

1. **% fuel in calciner,**
2. degree of calcinations achieved,
3. **retention time in calciner,**
4. temp. at exit of calciner,
5. pr. drop across calciner,
6. CO at outlet of calciner,
7. sp. volume of calciner (including calciner, connecting duct etc.),
8. sp. heat load in calciner (million kcal/hr/m3).
9 NO_X at exit

(e) grate cooler

1. **sp. output in tpd/m2 grate area,**
2. temp. of clinker leaving cooler,
3. **recuperation in cooler / efficiency of cooler,**
4. temperatures of secondary and tertiary airs,
5. temp. of vent air,
6. cooling air in nm3/kg,
7. vent air in nm3 /kg,
8. bed thicknesses of clinker,
9. **sp. power consumption of cooler system as awhole**

(f) pyroprocessing section as whole

1. **sp. fuel consumption in kcal/kg,**
2. sp. gas volume in nm3/kg,
3. **sp. power consumption for total section with break up,**
4. degree of calcination achieved,
5. **output tpd,**
6. quality of clinker produced (analysis, free lime, litre weight etc),
7. heat balance of kiln and cooler.

Coal grinding

1. **output tph at required fineness and moisture,**
2. sp. output / power consumption for mill and for system,
3. **% moisture dried in mill.**

Cement mill

1. **output at given fineness in tph,**
2. **sp. power consumption at agiven fineness for mill,**
3. sp. power consumption for the section as a whole,
4. % of particles between 3-30 microns,

Packing machine

1. **capacity in jute and paper bags,**
2. **accuracy of weight.**

Electrostatic precipitator

1. sp. collection area,
2. migration velocity,
3. **efficiency in % and or clean gas dust burden not to exceed —or—mg /nm3.**

Bag filters

1. total cloth area,
2. air to cloth ratio,
3. **clean gas dust burden,**
4. **pr. drop across bag filter.**

Material Handling Equipment

There are a great many types of material handling equipment which could be broadly classified into mechanical conveyors like screw conveyor, belt conveyor, elevator, chain conveyor etc; and pneumatic conveyors like air slide, screw pump, air lift, lean and dense phase systems etc. Appropriate critical parameters will be selected like say conveying speed in m /sec, inclination, angle of trough, air volume, speed of air borne material in pipe line etc.

Most certainly capacity in tph and sp. power consumption including air power will be selected for guarantees.

fans, blowers and compressors

fans

1. **capacity and static pressure and power drawn,**
2. **static efficiency of fan.**

compressors:

1. **capacity in f.a.d. at stipulated pressure and power drawn.**

Blowers:

1. **Capacity in f.a.d. at stipulated pressure and power drawn.**

5.3 Performance to be Proved

The Buyer, his Consultant and the Vendor should arrive at a clear understanding about operational parameters that would have to be proved by carrying out tests and which would attract penalties if they are not fulfilled.

There should be clear understanding about the testing procedure and about establishing output, power drawn etc.A protocol should be agreed upon and adhered to for this purpose.

Action required to be taken by the Vendor in the event of performance falling short should also be mutually agreed upon.

In the event Buyer cannot fulfill his part of supplying materials and fuel of specifications agreed upon, there should be an agreed procedure of interpretation of performance achieved in relation to guaranteed performance.

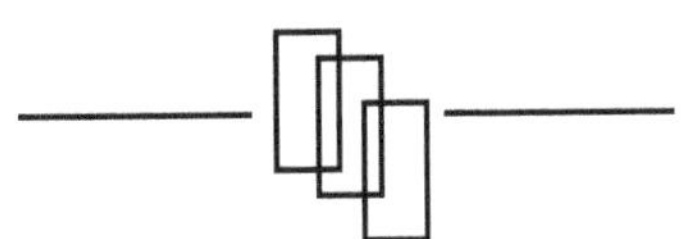

CHAPTER 6

TIME FACTOR

6.1 Time Factor

It would be seen from the foregoing that finalizing orders for machinery is a long drawn out process. Not only there are a large number of machinery to be ordered but also there are a great many Vendors from whom to buy it.

Unless the process is started with careful planning and with great deal of thought given to the objectives, it can go on almost endlessly.

6.2 Factors Governing Time for Ordering Machinery

Geographical factors also intervene such as locations of offices of Buyer, Consultant and Vendors. To bring them together requires planning; hence number of meetings across the table should be kept down to a minimum.

While this would have been very difficult a couple of decades back, developments in IT has made it easy. Today it is not necessary to meet across the table. 'Tele conferences' save time and expenditure. Sending information and drawings by fax, E-mail or CDs is fast and accurate.

Even then having clear objectives and setting a time frame to complete the ordering of machinery is very necessary.

6.3 Organisation for Buying

If a Buyer is clear about whether to :

go in for 'supply basis' or 'works contract' basis,

go 'shopping' or to buy from one source,

undertake fabrication of components of machinery like say casings of dust collectors, cyclones of preheaters etc.

fabricate items like conveyor gantries and working platforms, chutes and ducts and their supports.

he can proceed on the basis of his selected option. He will then strengthen his own organization and arrange to procure materials for items he has under taken to fabricate.

6.4 Consultants' Organisation

Consultant can play an important role in this exercise of selecting and ordering machinery because they have access to the principal vendors in the field. They create a **Data Bank.** They can save a lot of time in preliminaries and in evaluation.

Consultant can also plan his work and organization to fulfill his duties towards the Buyer.

It is not practical nor is it necessary to buy all machinery at the sane time.

Naturally core machinery would come first and then the major auxiliaries.

Items which are available at short notice would be ordered later so as not to lock up capital.

In this context **Chapter on Implementation Schedule** in **Section 3** may be seen. It gives an idea of time required for ordering of machinery.

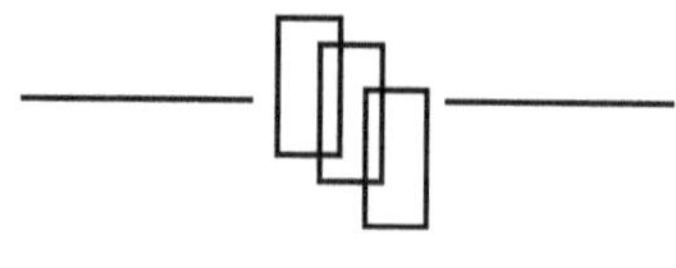

SECTION - 8

Sustainable Development

Contents

Chapter 1	Sustainable Development	1-3
Chapter 2	Blended Cements & Designing Cement Plants to make Blended Cements	4-9
Chapter 3	Alternate Fuels and Raw Materials	10-18
Chapter 4	Co-Generation	19-22

CHAPTER 1

SUSTAINABLE DEVELOPMENT

1.1 Sustainable Development

Over the last few decades, there is a growing awareness universally, that while development is necessary, it should be sustainable. It means that we should use natural resources that are exhaustible judiciously so that they last long. It also meant that our activities should not result in harmful effects like Global Warming.

This has prompted many industries to take a close look in the manner in which they were carrying out their activities.

Cement Industry is one such industry. In the process of production of cement, it was using natural resources like cement grade limestone and fuels like coal and oil that are exhaustible. Moreover in burning fuels and in the process of calcining, considerable quantities of CO_2, a greenhouse gas, were being released into the atmosphere promoting global warming.

Cement Industry therefore has been looking for ways and means to promote sustainable development and has come up with several possibilities. It has been giving effect to them by introducing them into their processes of production of cement.

Principal among them are:

1. **making blended cements** like Pozzolana and Slag Cements whereby quantity of limestone required to make cement is significantly reduced. There is an additional equally significant benefit by way of reduction in emission of CO_2.
2. **using alternate fuels** which are carbon neutral and also wastes that save fossil fuels.
3. **using waste heat** in kiln and cooler exhaust gases to generate power.
4. **Isolate CO_2** released in exhaust gases and put it to other industrial uses. It can be processed to make fuel again.

1.2 Green Cement

Cement so produced is called green cement. Author's book **'Designing Green Cement Plants'** deals with it in detail

Most of the New Cement Plants are going to be 'Green'. It would therefore be appropriate to go into such aspects of Green Cement Plants in this 'Handbook for Designing Cement Plants' as would be and could be dealt with by Cement Companies and their own Designing Personnel.

1.2.1 Blended Cements.

Most Cement Plants are already making 'Blended Cements'- Portland Pozzolana Cement- PPC and or Blast Furnace Slag Cement – BFSC. Cement Companies therefore need to be familiar with design and engineering aspects of making blended cements. In India share of blended cements in total production of **cement is + 75 %.**

1.2.2 Co processing Alternate Fuels and Raw Materials

Another significant step that has already been taken by Cement Companies is to substitute fossil fuels like coal and oil by Alternate fuels- mostly wastes that have calorific value and agricultural or 'bio wastes' that are in addition 'carbon neutral'. Cement Plants have to set up facilities for receiving, processing and firing such Alternate Fuels. Wastes of some Industrial and Chemical plants are useful as substitutes for main raw material – limestone and correcting materials like clay, iron ore. Cement Companies can use them.

1.2.3 Co – Generation

Kiln exhaust gases have heat value albeit to a lesser extent now because of reduced quantum on account of reduced fuel consumption and reduced temperature on account of six stage prehaters.

The first priority is to use them for drying raw materials and coal. In a large number of cases, these can be used to generate power.

Same is the case for Cooler vent air.

A great many plants are therefore installing **Waste Heat Recovery Systems** to generate power from remaining heat of kiln and cooler gases.

While the Power Plant itself will be designed and installed by other specialist companies, Cement Company has to work out the quantum of heat available for this purpose.

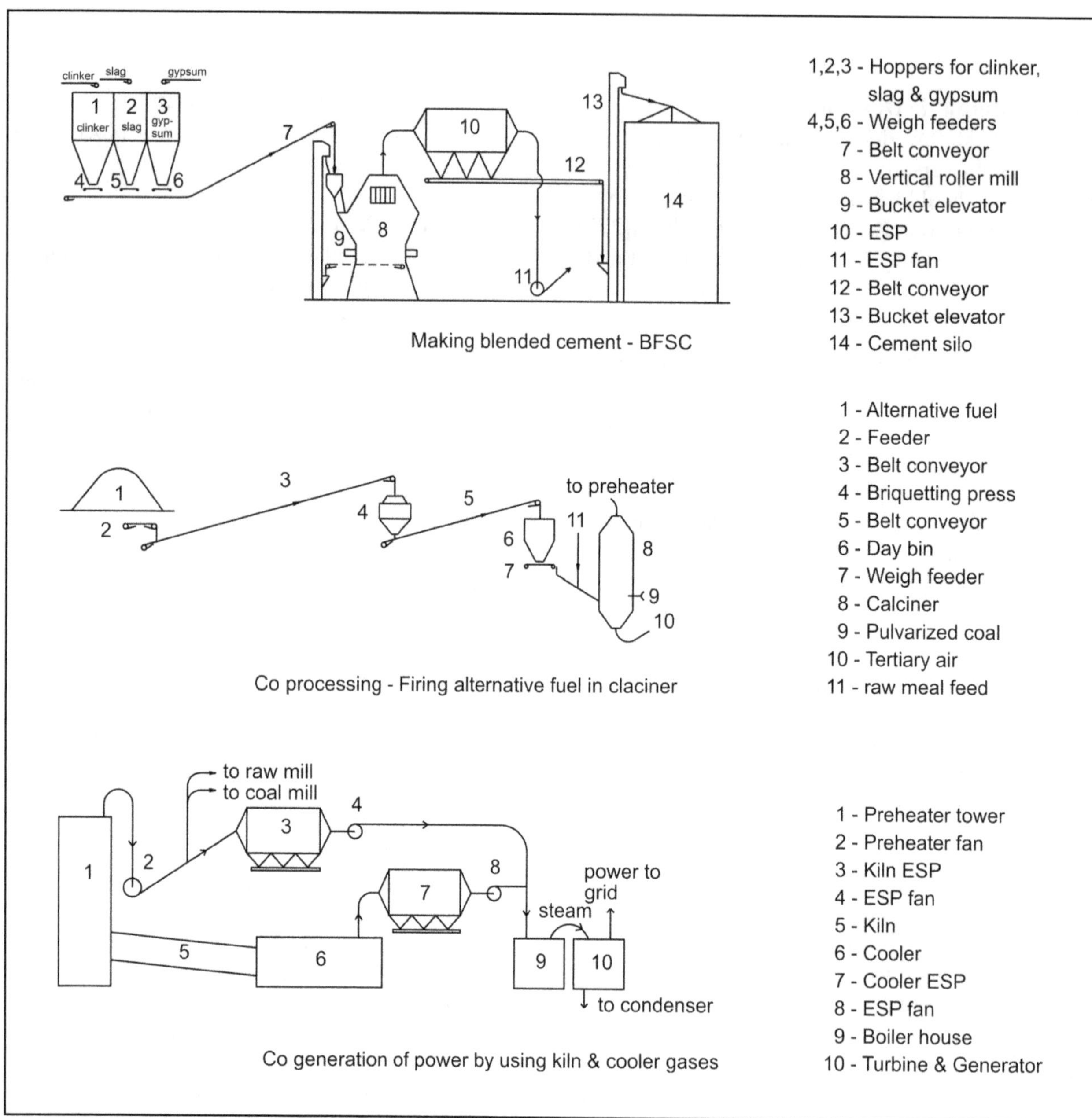

Fig. 8.1.1

Flow charts in **Fig. 8.1.1** show the processes involved in three activities mentioned above.

In this book, only aspects that will be dealt with Cement Company's own personnel are presented.

1.2.4 Green Buildings

CII – Confederation of Indian Industries has evolved a National Code for Bilidings to promote maximum use of light and ventilation in Commercial Industrial Buildings.

It has even evolved a rating system for judging and rewarding this feature.An extract of the code is attached in **Chapter 2 of Section 4 of the book.**

3.0 Carbon Capture and Storage

It is now possible to collect Carbon dioxide from exhaust gases, liquefy them and store them under ground and use it for other Industries. See Author's book – **'Designing Green Cement Plants'**

4.0 Converting CO_2 back into fuel – Methane..

$$2CO_2 + 2H_2O = 2CH_4 + 3O_2$$

Seems It is now possible to convert CO_2 back into Methane on a commercial Scale. This development will not only reduce emissions of CO_2 but will also make fuel and Oxygen available. The Cement Plant itself can install this facility.

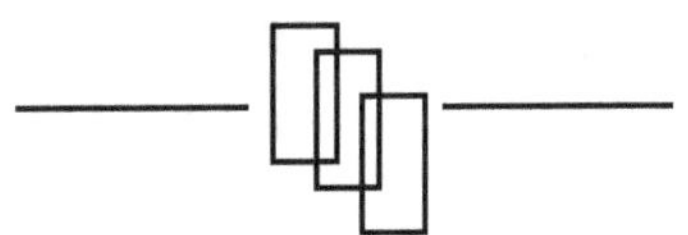

<u>**CHAPTER 2**</u>

BLENDED CEMENTS & DESIGNING CEMENT PLANTS TO MAKE BLENDED CEMENTS

2.1 Requirement of Limestone in making Blended Cements

Assuming 1.5 t of limestone per ton of clinker and 5 % gypsum, limestone required to make 1 ton of cement would be

OPC	PPC (30 % fly ash)	Slag Cement (60% slag)
1.43 ton	**1.0 ton**	**0.53 ton**

It brings out the savings in consumption of limestone in making blended cements.

2.2 Cement/Clinker Ratio and Emission of Greenhouse Gases

Blended Cements increase the cement / clinker ratio. GHG emissions get reduced per ton of cement produced correspondingly, other things remaining the same.

For example:

Type of Cement	**cement/clinker ratio**	**Total ghg emissions Kg CO_2/kg cement**	**% reduction**
O.P.C.	1.05	0.77 median value	-
PPC (30 % fly ash)	1.54	0.52	32
BFSC (60% slag)	2.86	0.28	63

2.3 Win-win Situation as a Result of Making Blended Cements

It would thus be seen that there is very significant saving in consumption of limestone and in emission of greenhouse gases.

The right royal way therefore– and a very effective way towards sustainable development would be to make blended cements.

Since both fly ash and slag are waste products of their respective industries, cost of their procurement is far less than the cost of production of clinker that they would replace.

This is thus a Win-Win situation for all because :.

1. Power and Steel Industries get rid of their respective waste products
2. Cement Industry gets useful material at much less cost
3. Community benefits as the waste products do not have to dumped in landfills, thereby saving large areas of land
4. Significant contribution to the reduction in global warming effect by reduced ghg emissions.

2.4

Table 8.2.1 shows Impact on capacity and size of Kiln to be installed on given annual capacity of OPC, PPC and BFSC.

Table 8.2.1 Cements produced from a given clinkering capacity.

kiln capacity	OPC	PPC produced in MTPA				BFSC produced in MTPA			
TPD	MTPA	fly ash %				slag %			
		20	25	30	35	50	55	60	65
		cement /clinker ratio				cement/clinker ratio			
	1.05	1.33	1.43	1.54	1.67	2.22	2.5	2.86	3.33
5000	1.73	2.2	2.4	2.5	2.8	3.7	4.1	4.7	5.5
10000	3.47	4.4	4.7	5.1	5.5	7.3	8.3	9.4	11.0
15000	5.20	6.6	7.1	7.6	8.3	11.0	12.4	14.2	16.5

2.5 Logistics of Transport of Fly Ash and Slag

Logistics of transport of fly ash and slag to cement plants would play an important part in the rate of growth of blended cements. Cement plants get located near limestone deposits; steel plants near iron ore deposits. Thermal power stations tend to get located near coal deposits.

It is not always possible that the three are located close to one another.

Therefore fly ash and slag have to be transported to cement plants. This puts some limitations and introduces aspects of storage and handling of these materials in cement plants.

When market demands, clinker is transported to steel plant to make BFSC.

2.6 Making of Blended Cements

There is nothing new in making blended cements as both PPC and BFSC are being made the world over for decades.

Processes of making PPC and BFSC are well known so are the machinery to be used.

There are of course small variations like :

1. whether to grind clinker and fly ash / slag together or separately.
2. whether fly ash should be introduced first in the mill or in the separator.
3. selection of grinding system- tube mill, vertical roller mill or roller press and ball mill

They can be examined technically on case to case basis to arrive at an optimum solution for a specific project.

2.7 Differences on Account of Increase in Size of Cement Plants

1. Cement plants have grown much bigger in size. Representative size of plant would be say 3-8 mtpa in terms of cement produced in one place.

   This would require larger quantities of fly ash and slag to be brought in and stored. Slag may have to be dried as well.
2. The storage and handling facilities for these materials have to be designed from this angle.
3. Fly ash would be stored in silos which would be larger in numbers andsize and capacity. Fly ash would be conveyed to mills mostly pneumatically.

   Slag would be stored in covered gantries with facilities for receiving wet slag and also for dried slag. A drier would be a necessity if slag is received wet.

   If slag is ground separately, it can be ground in the VRM.
4. While fly ash would be brought in special self unloading carriers by road in most cases, on occasions it may have to brought in by rail in wagons similar in design to those used to transport cement in bulk.

2.8 Receipt of Slag

Because of its quantity, slag would come in rake loads by wagons. Number of rakes to be handled being quite large, a large railway siding with wagon tipplers would be necessary to cope with the volume of slag to be received and unloaded.

To keep down costs of investment, quantities to be stored in terms of daily consumption could be reduced consistent with maintaining continuity of production.

2.9 Impact on Plant Layout

A corollary of making blended cements is that quantity of OPC made would be reduced correspondingly. More of blended cements would be stored than OPC.

One mill and one silo may be reserved for OPC and the rest for whatever type/s of cements (PPC/ BFSC) are to be made.

If a cement plant is making both PPC and BFSC – which is quite possible- groups of mills would be reserved for each category so as to avoid changes of settings, cleaning of silos to avoid contamination etc.

2.10 Despatches

In a great number of countries cements are mostly dispatched in bulk. Countries like India are also following suit. Layout of cement dispatching section should be designed to meet the requirements of bulk and bagged cement.

Further despatches by road would predominate over despatches by rail.

Layouts of packing plants would be designed to facilitate them.

2.11 Railway siding

Plant capacities at a given place have doubled or even trebled at some places and the volume of traffic has increased. Because of shift to despatches by road.

Railway sidings within the plant are more likely to be designed for receiving coal and slag rather than for handling cement.

2.12 Various aspects of handling fly ash and slag have been dealt with in detail in Chapters 28 & 29 of Section 6 of the Book.

2.13 Choice of Grinding Mill

Presently VRMs are replacing Tube mills for grinding clinker and even slag. Thus there are small differences in layouts of cement mill to allow for use of VRMs and corresponding auxiliaries.

In case of fly ash, there are two possibilities- feed fly ash in the mill or in the separator.

Roller Press and ball mill combination is also being adopted in a variety of ways to make cement.

2.14 System Flow Charts and Layouts

Flow charts **Figs. 8.2.1. and 8.2.2** show grinding circuits for PPC and BFSC based on VRM as the grinding mill.

Figs. 8.2.3 & 8.2.4 show schematic layouts for making PPC & BFSC Respectively using VRMs.

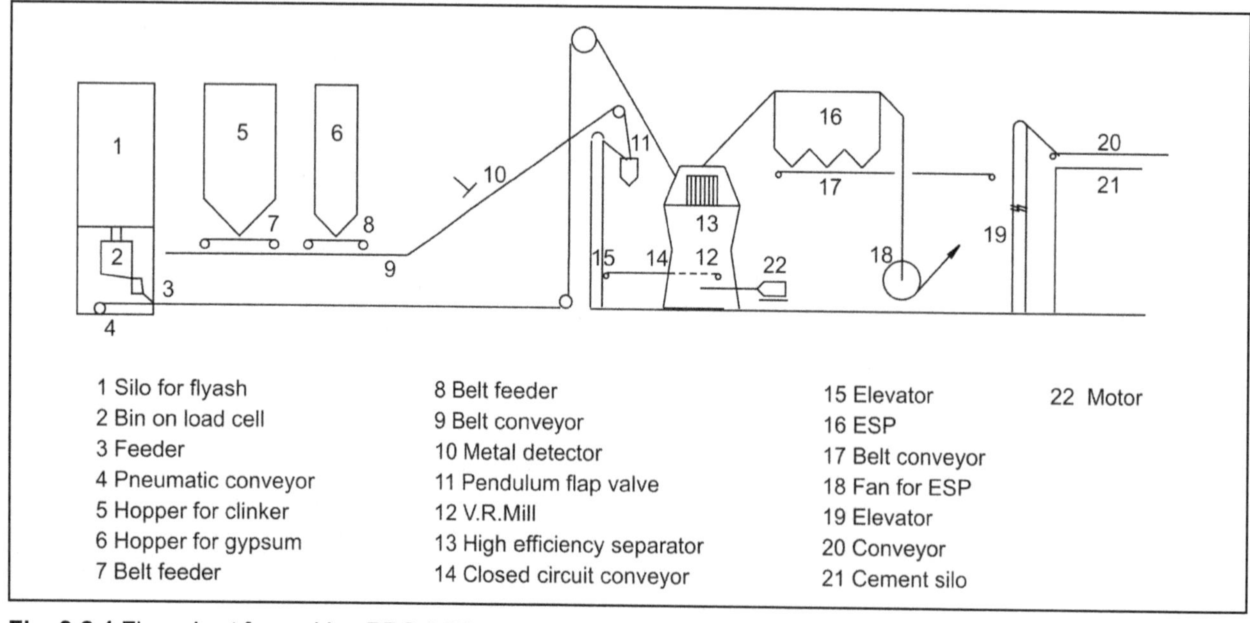

Fig. 8.2.1 Flow chart for making PPC (VRM & High efficiency separator).

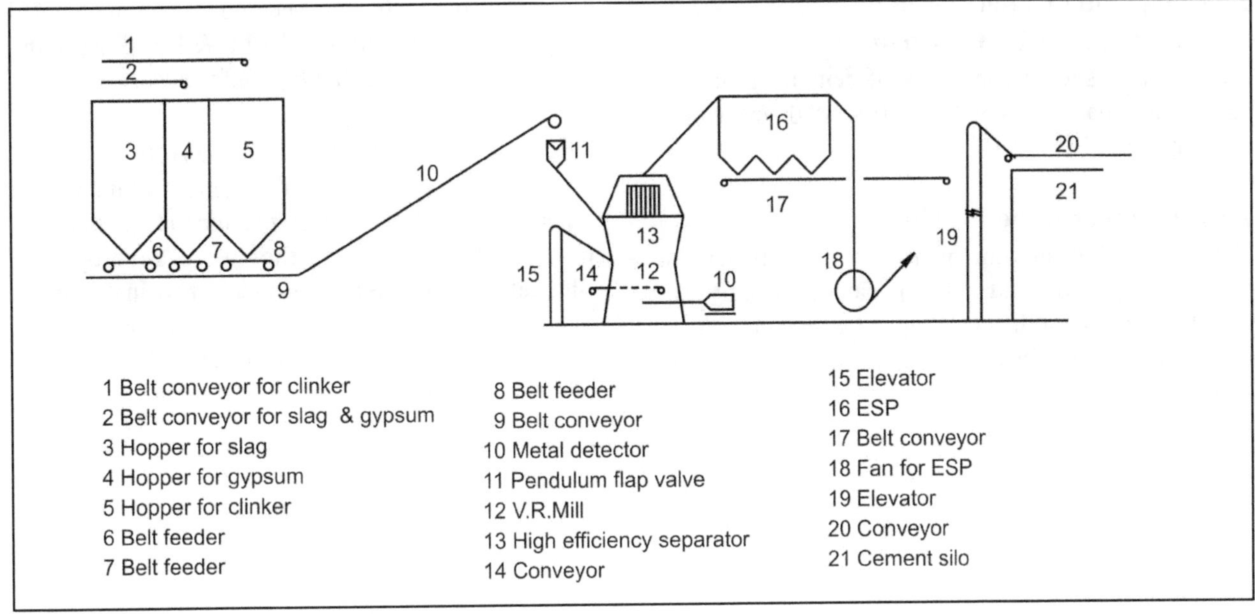

Fig. 8.2.2 Flow chart for making BFSC (slag cement) (VRM with high efficiency separator).

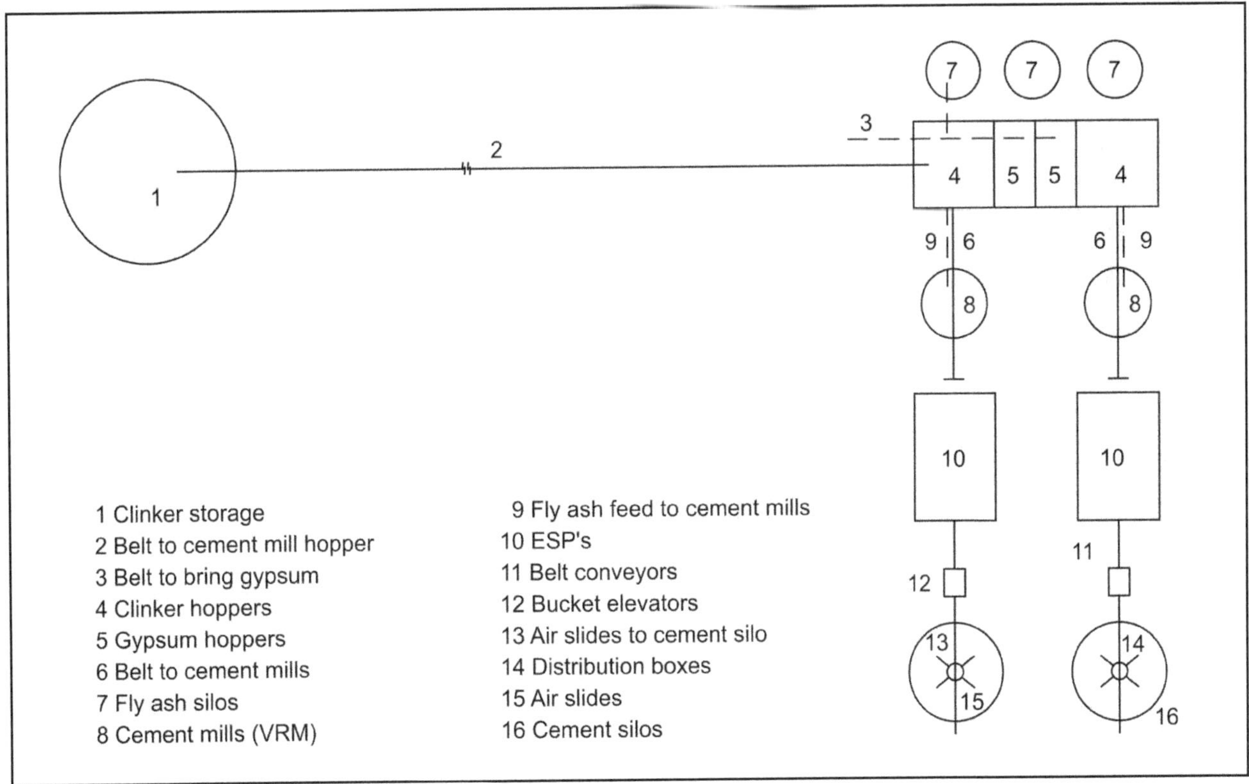

Fig. 8.2.3 Layout for making PPC (pozollana cement).

2.15 Sectional Capacities and Storages to be Provided

Table 8.2.2 shows capacities of cement mills for clinkering capacities of 5000, 7500 & 10000 tpd when making OPC, PPC & BFSC.

2.16 Storages to be Provided

Table 8.2.3 shows requirements of various materials like clinker, gypsum and blending materials like fly ash and slag per day for different capacities and quantities to be stored along with clinker for a 10000 tpd capacity clinkering unit.

It also shows stocks of cements to be maintained (as per convention) for OPC, PPC & BFSC assuming 100 % production of that type in each case.

2.17 General Cement Plant Layout

Typical general layouts of a cement plant with a clinkering capacity of 10000 tpd and making 6.5 mtpa of OPC and BFSC together were developed to illustrate the various aspects mentioned in the design of layout of cement grinding and despatching sections. One such layout is shownin **Fig. 6.1.23 in Chapter 1 of Section 6 of the Book.**

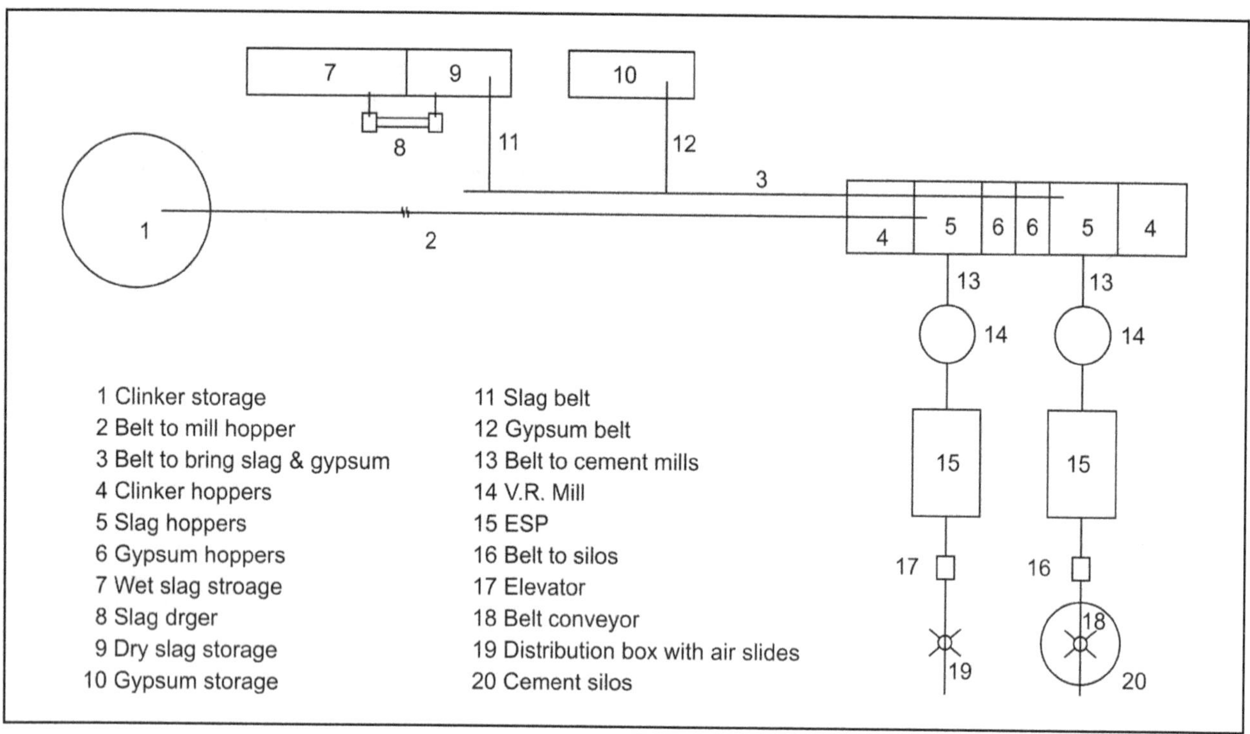

Fig. 8.2.4 Layout for making BFSC (slag cement).

Table 8.2.2 Capacities of Cement mills making OPC and blended cements.

Cement Grinding	cement mill	sizing factor	hours/day	multiplying factor
	OPC	1.27	20	0.064
	PPC 30% ash	1.86	20	0.093
	BFSC 60 %	3.46	20	0.173
	slag			
		kiln capacity tpd		
		5000	7500	10000
	Mill making	Mill capacity in tph		
	100 % OPC	320	480	640
	100 % PPC	465	697.5	930
			700	
	100 % BFSC	865	1298	1730
			1300	

Table 8.2.3 Storages of Clinker, Slag, Flyash and cements in large Cement Plants.

kiln capacity TPD					
Sr. no	**Material**		**unit**		**10000**
1	clinker	factor daily reqmt	1.1	1.1	1.1
		no. of day's	tons		11000
		stock storage	14	14	14
			tons		**154000**
2	gypsum	factor daily reqmt	0.07	0.07	0.07
		no. of day's	tons		700
		stock storage	30	30	30
			tons		21000
3	fly ash	factor daily reqmt	0.4	0.4	0.4
		no. of day's	tons		4000
		stock storage	3	3	3
			tons		12000
4	slag	factor daily reqmt	0.85	0.85	0.85
		no. of day's	tons		8500
		stock storage	2	2	2
			tons		**17000**
5	cement OPC	factor daily reqmt	1.27	1.27	1.27
		no. of day's	tons		12700
		stock storage	7	7	7
			tons		88900
6	cement PPC	factor daily reqmt	1.82	1.86	1.86
		no. of day's	tons		18600
		stock storage	3	3	3
			tons		**55800**
7	cement BFSC	factor daily reqmt	3.18	3.46	3.46
		no. of day's	tons		34600
		stock storage	3	3	3
			tons		**103800**

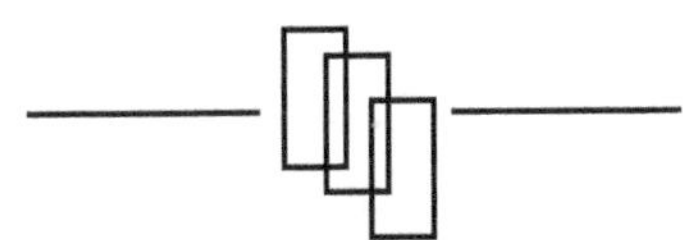

CHAPTER 3

ALTERNATE FUELS AND RAW MATERIALS

3.1 Use of Alternate Fuels and Raw Materials in Cement Plants

Desirability and Feasibility of – Co Processing that is using Alternate Fuels and Raw Mterials has been brought out in Section 4, Chapters 1 and 3 of **Author's Book 'Designing Green Cement Plants'.** Aspects related to designing Co – Proecessing Systems would be dealt with here.

3.2 Desirable Properties in AF to be used

Table 8.3.1 furnishes desirable properties in Wastes to be used as AF. **Annexe 8.3.1** furnishes work out for working out capacity for the Co-processing System for rice husk as an example.

3.3 Flow Charts and Layouts

Fig. 8.3.1 shows layout for receiving, storing, and processing rice husk as AF.

Table 8.3.1 Desirable Properties in AF.

	Parameter		AF –liquid	AF – solid
1.	calorific value	kcal/kg	4000-4400	4000-4400
2.	water	%	< 20	< 20
3.	flash point	^{o}c	> 60	not applicable
4.	chlorides	%	< 1.5	< 1.5
5.	total halogens	%	< 1.5	< 1.5
6.	sulphur	%	< 1.5	< 1.5
7.	PCBPCT	ppm	< 50	< 50
8.	Heavy Metals	%	0.2-1.0	0.2-1.0
9.	Heavy Metalas	ppm		
	TI + CD		< 20	< 20
	V		< 100	< 100
	As		< 60	< 60
	Cr		< 400	< 400
	Hg		< 10	< 10
	CD + TI + Hg		< 100	< 100
	all above + Co +Ni + Se + Te +Sb +Sn +Pb		< 2500	< 25

Table 8.3.1 *Contd....*

	Parameter		AF –liquid	AF – solid
10.	Ph		5-9	5-9
11.	Ash	%	< 5	< 25
12.	odour, toxicity		no strong odour, non toxic	
13.	Contaminations		should be free of inerts like grit, rags, glass Stone, metal etc	

Wastes unsuitable as AF

1. Radio active nuclear wastes
2. infectious wastes
3. entire batteries
4. asbestos containing wastes
5. cyanide wastes
6. mineral acids
7. bio hazardous wastes
8. electronic metal scrap

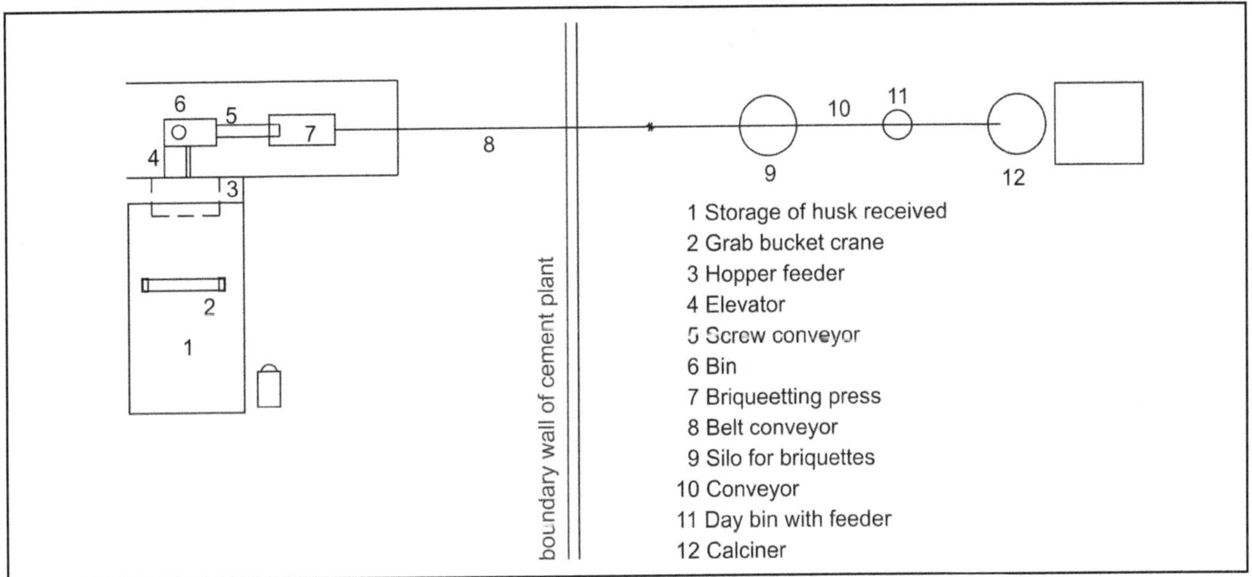

Fig. 8.3.1 Layout showing arrangement for receiving, storing rice husk & for making briquetes thereof for feeding into calciner.

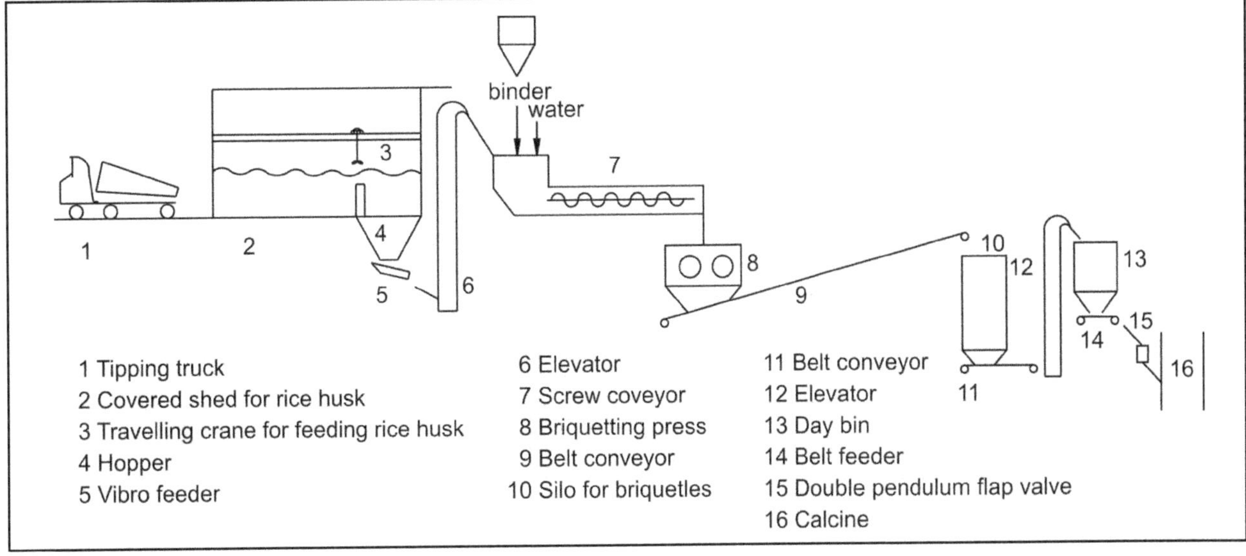

Fig. 8.3.2 Flow chart for rice husk as AF - made in to briquettes before firing.

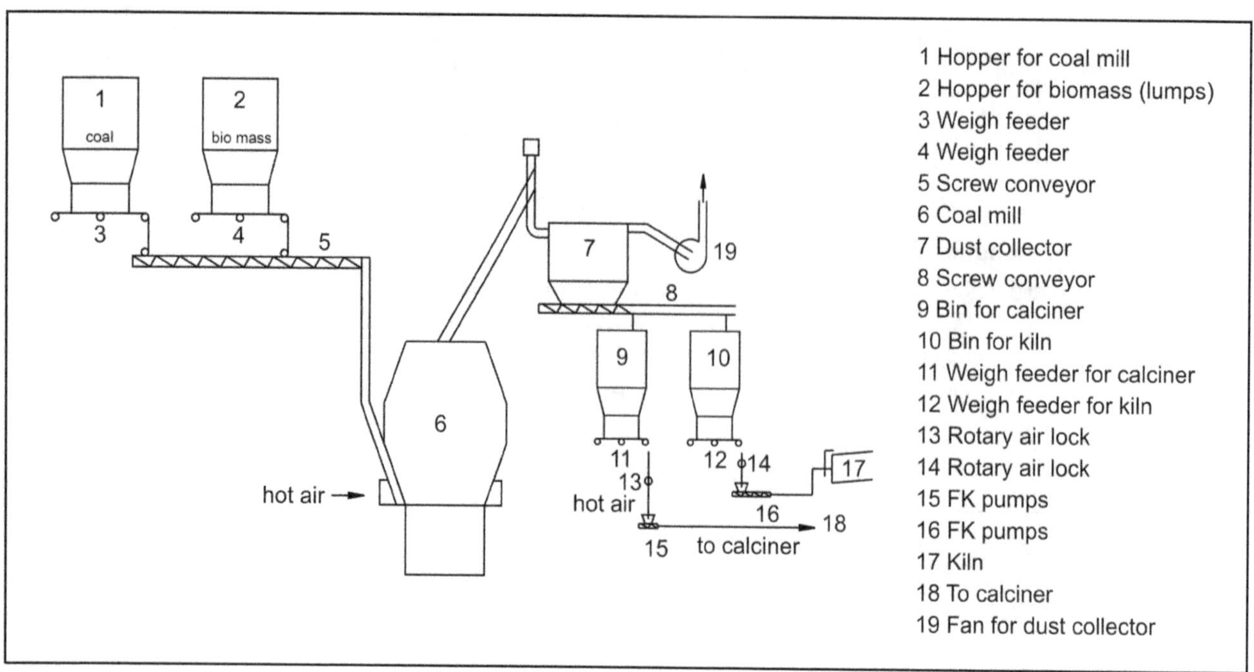

Fig. 8.3.3 Flow chart for grinding bio mass with coal & firing in kiln & calciner.

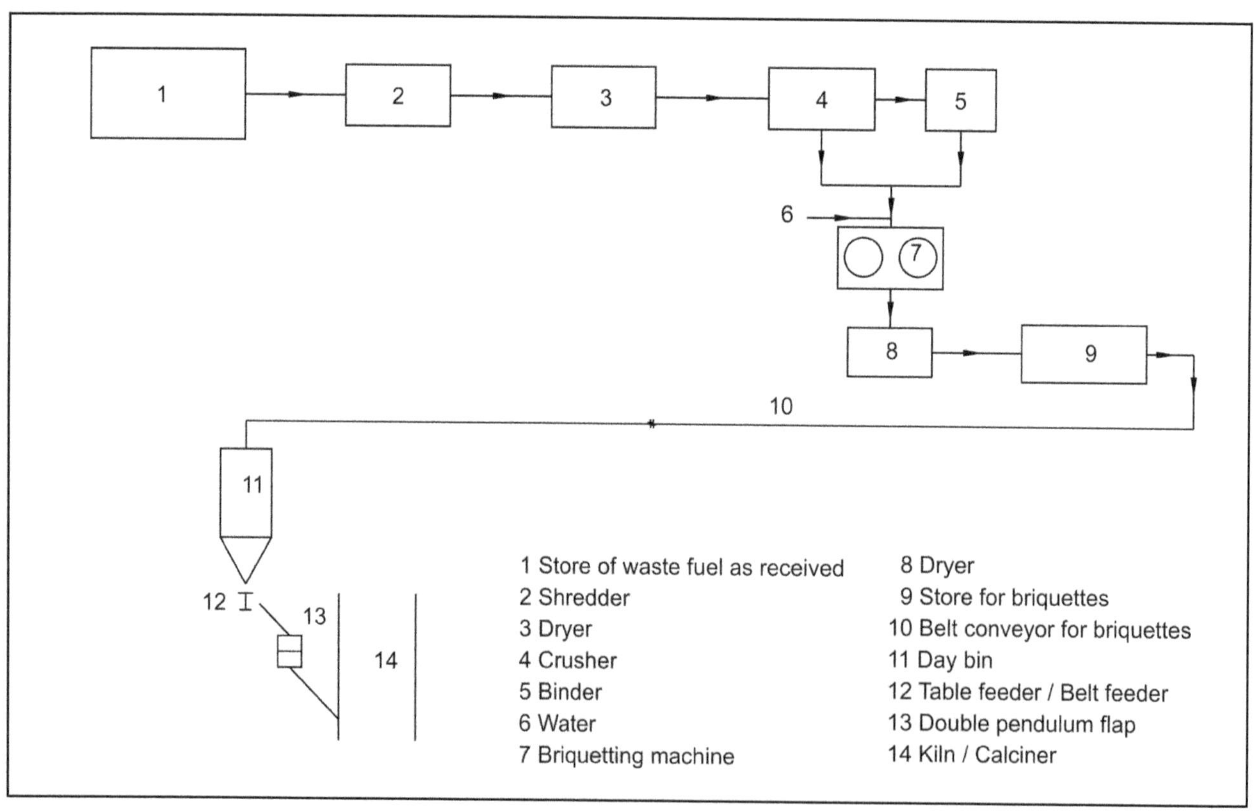

Fig. 8.3.4 Flow chart for making briquettes of an AF for making it suitable for firing.

Figs. 8.3.5 to 8.3.9 show flow charts of various possibilities of firing AFs.

3.4 Design and Engineering of a Specific System

With these details, Project Engineers would be in a position to design and integrate the co processing system suitable for their requirements in their plant.

3.5 Additional Equipment in Laboratories

Because AFs are very different from coals and oils and also vary greatly in amality, it is necessary to equip the Laboratories with additional equipment and testing facilities. They are listed below.

1. Brock fcield viscometer
2. Ion analyzer in quality
3. Calorimeter
4. CHNS analyzer
5. Absorption spectro photo meter

Company should have vans fitted with testing apparatus as it may be necessary to go place for identifying AFs.

3.6 Special Equipment for Firing AFs in Kilns and Calciners

Over the years, several leading Cement Machinery manufacturers have designed special equipment for firing AF in kilns and calciners. Among them are Hot Disc, Gunnax, and Duoflex burners. **See Plates 8.3.1 to 8.3.4.**

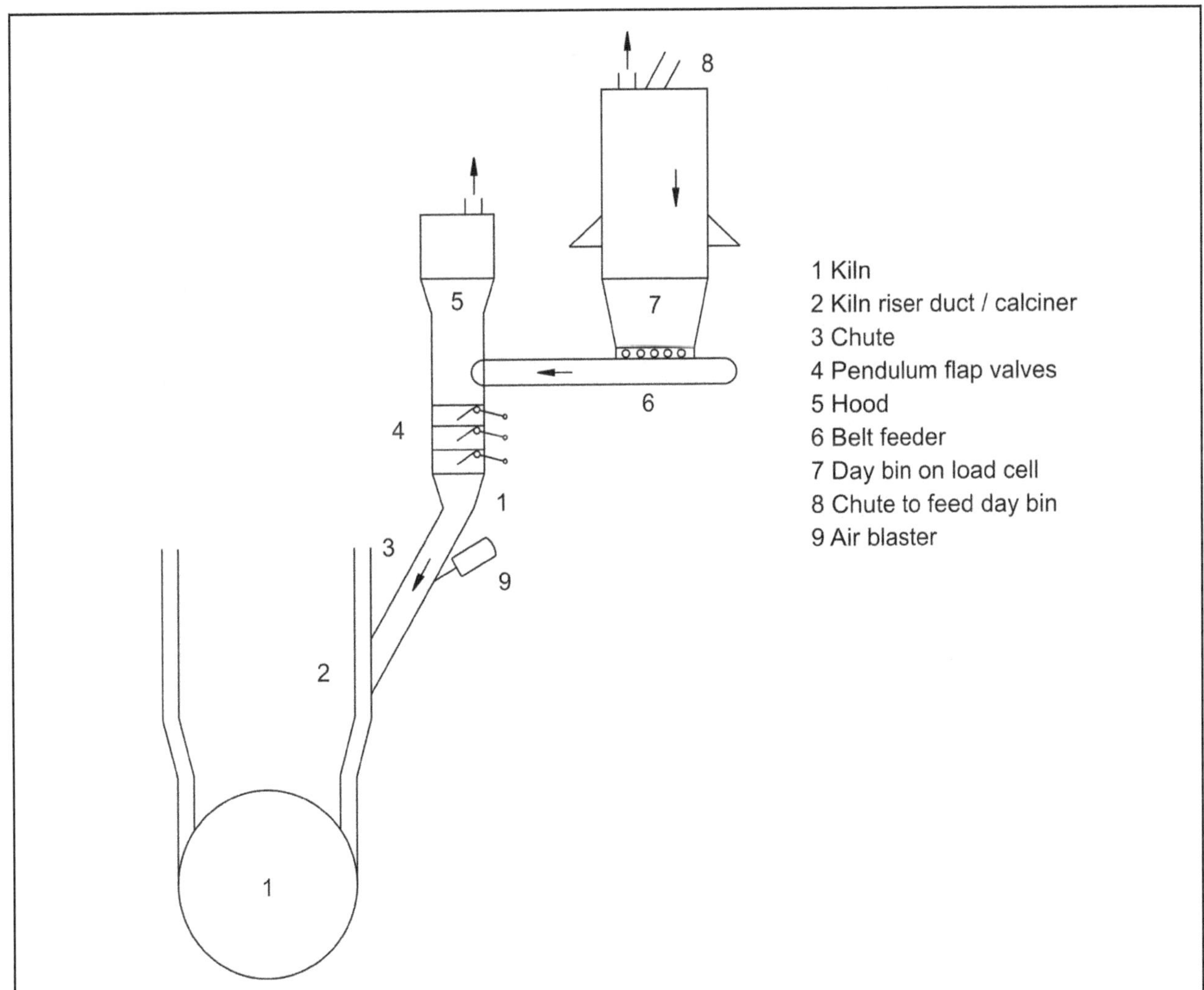

Fig. 8.3.5 Flow chart for feeding shredded AF into calciner / kiln riser duct.

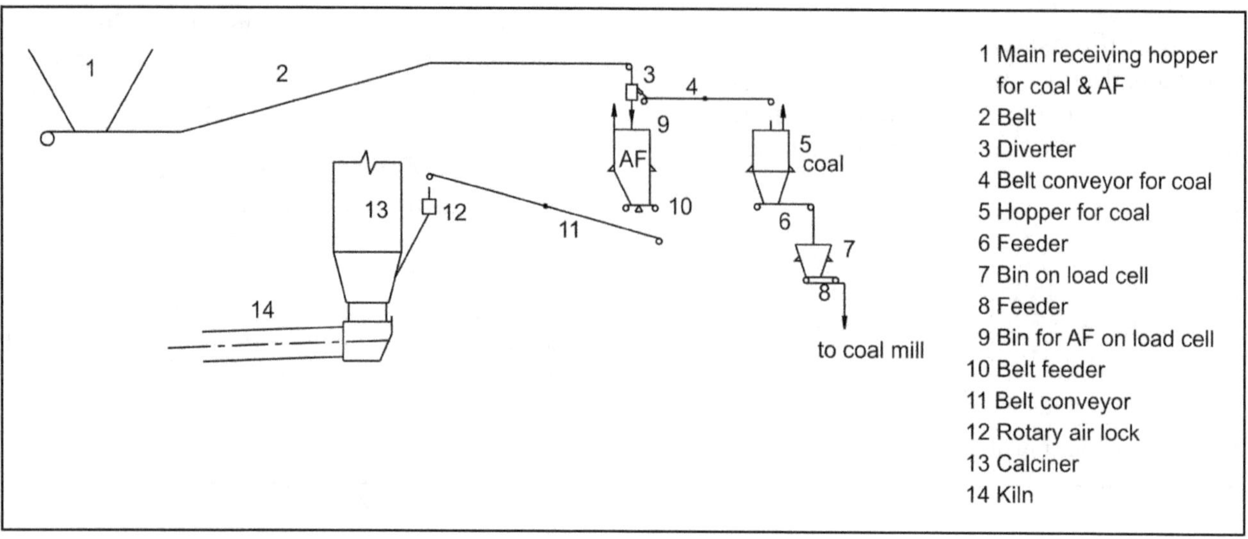

Fig. 8.3.6 Flow chart for a common system to receive coal & AF & feeding AF to calciner.

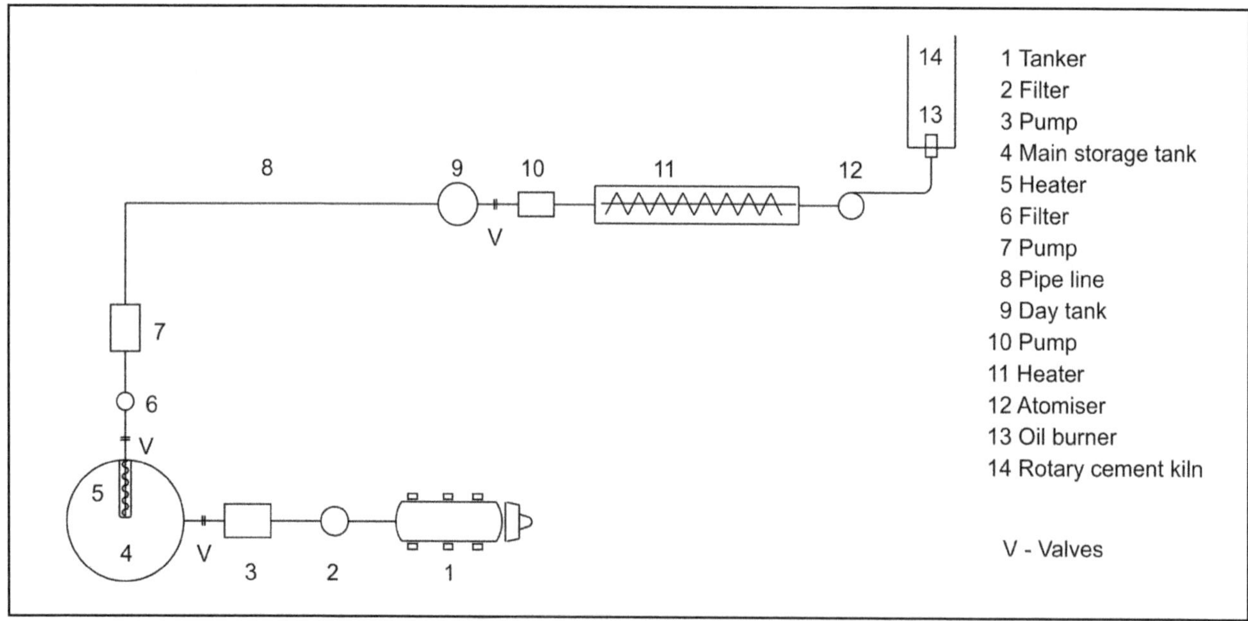

Fig. 8.3.7 Flow chart for firing waste oil as AF in a cement kiln.

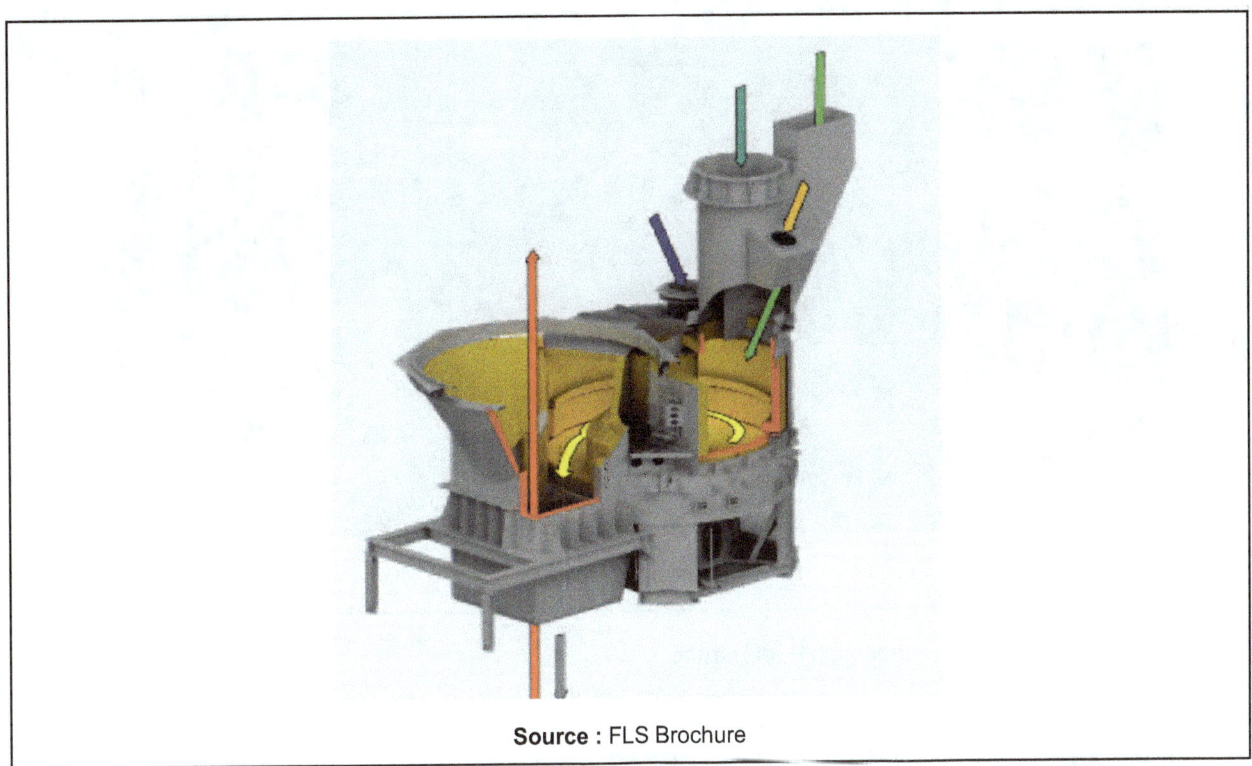

Source : FLS Brochure

Plate 8.3.1 Hot Disc for feeding Whole Tyres and other Solids.

Source : FLS Brochure

Plate 8.3.2 'Gunnax' Whole Tyre Firing System.

Source : FLS Brochure

Plate. 8.3.3 Duoflex Multi Channel Multi Fuel Burner.

Plate. 8.3.4 Gunnax firing waste canister.

Annexe 8.3.1

Workout for arriving at capacities for an AF firing system AF - Rice Husk.

sr no	item	unit	entity
	clinkering capacity	tpd	10000
	design margin	%	10
	sp. Fuel consumpn	kcal/kg	700
	calorific value coal	kcal/kg	4000
	fuel fired in kiln	%	40
	fuel fired in calciner	%	60
	working hours	no	24
	clinkering capacity		
	per hour with design		
	margn	tph	458
	coal in kg/kg clinker	kg/kg	0.175
	total coal per hour	tph	80.21
	coal in kiln	tph	32.1
	coal in calciner	tph	48.1
	AF to be used		rice husk
	to be fired in		calciner
	ueful calorifc value	kcal/kg	3500
	moisture	%	10
	bulk density	kg/m^3	150
	AF @ 20 % of total	%	20
	margin	%	5
	heat supplied	kcal/hr	320833333
	heat supplied		
	to kiln	kcal/hr	128333333
	heat supplied		
	to calciner	kcal/hr	192500000
	when AF fired upto		
	taregted capacity		
	heat to be supplied		
	in calciner	kcal/hr	64166667
	with 5 % margin	kcal/hr	67375000
	rice husk to be fired	kg/hr	19250
		tph	19.25
		say	20
	rice husk to be treated and processed		
	rice husk to be		
	made into briquettes		
	wet rice husk		
	with 10 % moisture	tph	22

Annexe 8.3.1 ***Contd....***

sr no	item	unit	entity
	daily requirement of		
	wet rice husk	t	528
	stock of wet husk		
	no. of days	no	3
	stock to be kept	t	1584
		say	1600
	volume of stock pile	m^3	10000
	dimensions of		
	triangular pile for		
	40° angle of repose	width × length	
		m × m	20 × 110
	at full design capacity for firing rice husk		
	rice husk received in	hours	8
	handling capacity	tph	66
	screen capacity	tph	79.2
		say	80
	briquetting press	tph	80
	belt conveyor	tph	100
	hopper / bin	tons	250
	belt feeder	tph	25
	0-25 tph		
	double pendulum flap		1
	reference flow chart Fig. no. 8.3.2		

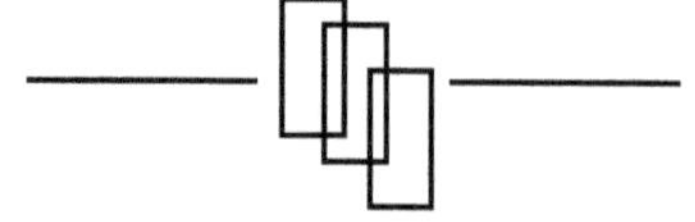

CHAPTER 4

CO-GENERATION

4.1 Scope for Co generation

As mentioned in **Chapter 1**, waste heat in kiln exhaust gases and cooler vent air can be used to generate power. With advanced technology like Organic Rankine Cycle, it is possible to use heat in kiln gases exiting preheater at ~270 °C to generate power.

In large plants, it is possible to combine exhaust gases from kiln and cooler to generate power in a common power plant.

When there is a captive power plant installed in the beginning to maintain continuity of power, its capacity can be augmented by taking kiln and cooler gases to it instead of installing a new power plant.

4.2 Availability of Waste Heat

Heat in waste gases is first used for drying limestone, coal and slag. Balance will be used for power generation.

Therefore it is necessary to work out heat that can be continuously available from waste gases after these requirements are met. **Annexe 8.4.1** furnishes such a workout.

Various aspects of Waste Heat Recovery have been dealt with in **'Designing Green Cement Plants'.**

4.3 Location of Waste Heat Recovery Plant

In the event of WHR System, there would be three sources of power in the cement plant : 1 Grid power, 2 power from captive power plant, 3 power from WHR. It becomes necessary to integrate these sources into one source of continuous availability of power with same, voltage and frequency and involved features of power distribution.

Location of the power plant is also important as it will be using facilities like source of water for generation of steam, cooling and recycling of water. These aspects will be looked after by the Project Engineering Department.

In **Chapter 1 of Section 6,** a General Plant Layout for a 10000 tpd plant has been shown along with its legend. It shows location of Power Plant of Waste Heat Recovery System. **See Fig. 6.1.23 item 36 is WHR.**

4.4 Designing and Engineering of the WHR System

Specialist Vendors would be doing the detailed planning and Engineering of the WHR System based on Data made available by Plant's Project Engineering Department like Heat available (**Annexe 8.4.1**) and details of facilities already available like availability of fresh water, for cooling and recirculation, for receipt and storage of coal and the like. Plant's team would work with the Vendor in installing the WHR System in an integrating manner.

Fig. 8.4.1 shows a flowchart for a cogeneration system based on organic rankine cycle utilising cooler vent.

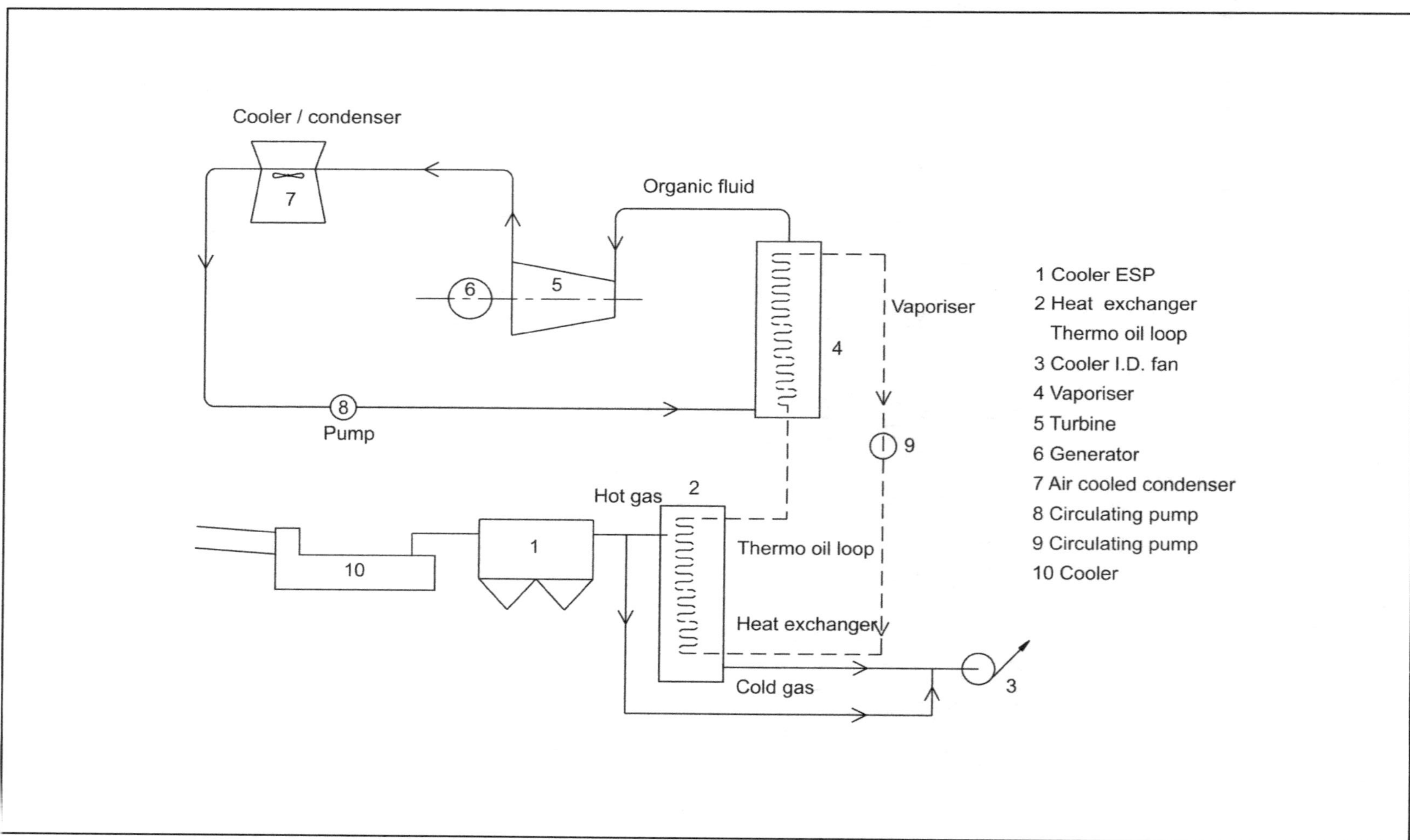

Fig. 8.4.1 General Plant Layout of a 10,000 TPD Clinkering Unit making 6.5 mtpa OPC & BFSC.

Annexe 8.4.1

Example of calculating nett heat available from waste gases for Co generation of power.

	Common WHR for kiln and cooler			
			operating conditions	
	Cement Plant Data		1	2
sr no	**item**	**unit**		
	altitude	metres	sea level	sea level
	ambient temp.	°C	30	30
	fuel		coal	coal
	useful calorific			
	value	kcal/kg	5000	5000
	moisture in wet			
	coal	%	10	10
	to be dried to	%	2	2
	mill running hours	no	20	20
	raw materials to			
	clinker	ratio	1.5	1.5
	moisture in wet			
	raw materials	%	5,3	5,3
	to be dried to	%	1	1
	mill running hours	no	20	20
	preheater stages	no	**6**	**4**
	sp.fuel consumption.	kcal/kg	**700**	**800**
	temp.of gases			
	leaving preheater	°C	270	350
	sp.heat of gases			
	and air	kcal/nm^3	0.3	0.3
	heat required to			
	evaporate water in			
	raw materials	kcal/kg	1500/1800	1500/ 1800
	heat required to			
	evaporate water			
	in coal	kcal/kg	1300	1300
	cooling air	nm^3 /kg	2.2	2.4
	temp. of cooler			
	vent air	°C	250	250
	cooler vent cooled			
	to	°C	90	90
	prheater gases			
	in raw mill and			
	coal mill cooled to	°C	90	90
	CO_2 released per			
	kg clinker	nm^3/kg	0.3	0.3
	sp.gas volume from			
	preheater	nm^3/kg	**1.35**	**1.5**
		clinker		

Annexe 8.4.1 ***Contd....***

sr no	item	unit		
	vent air from cooler	nm^3/kg		
		clinker		
			1.4	**1.5**
	Heat content of	kcal/kg	**72.8**	**117**
	prehater gases	clinker		
	heat content of	kcal/kg	**67.2**	**72**
	cooler vent	clinker		
	gross waste heat	kcal/kg	**140**	**189**
	available	clinker		
		heat required for drying		
	heat required for			
	drying	**case 1 raw material moisture 5 %, coal 10 %**		
	raw material	kcal/kg	113	113
	coal	clinker	19.5	22.3
	total for drying		**132.5**	**135.3**
	nett heat available	**kcal/kg**	**7.5**	**53.7**
	for co-geneartion	**clinker**		
		case 2 raw material moisture 3 %, coal 10 %		
	raw material	kcal/kg	67	67
	coal	clinker	19.5	22
	total heat for drying	kcal/kg	**86.5**	**89**
		clinker		
	nett heat availabe	kcal/kg		
	for co-generation	clinker	**53.5**	**100**
	It would appear that if moisture in raw materials is 5% or more, sufficient heat is not available for co-generation in case of 6 stage preheater If moisture in raw materialsis 3 %, then sufficient heat is available for cogeneration.			
	Quantitty of co-generated power would depend on system installed-rankine or organic rankine or kalina according to their respective efficiencies.			

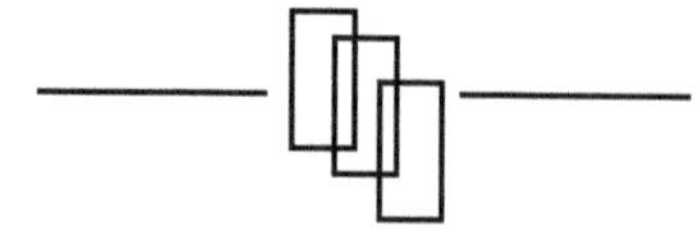

Section - 9

WEB PAGES

Supplementary materials for this Section - 9 is a available in http://www.bspublications.net/9789389354829/

Section - 10

Sources

Contents

	Sources	1-1
1.	Reference Books	2-2
2.	Technical Periodicals	3-3
3.	Reference Manuals	4-4
4.	Sources of Plates	5-8
5.	Sources of Layouts	9-9
6.	Sources of Text and Figures	10-15

SECTION 10

SOURCES

In this Section I have put together Sources of Information and Data collected for the compilation of this Book. It has been classified under some brought categories like

1. Books and Periodicals,
2. Manuals and Brochures
3. Books
4. Plates
5. Layouts
6. Text and Figures

I thank all of them individually and collectively for their valuable help that has greatly contributed to the successful completion of this Project

If I have inadvertently missed out any, I apologize.

1 REFERENCE BOOKS

RB. No	Title	Author/Publisher/ Firm
1	Cement Data Book Vol. 1	W.H.Duda / Bauverlaug
2	Cement Manufacturer's Handbook	Kurt E. Peray / Chemical Publishing Company
3	Cement Engineers' Handbook	Otto Labahn / Bauvelaug
4	Cement Engineers' Handbook	B. Kohlhaas / Bauverlaug
5	Rotary Cement Kiln	K.E.Peray /Edward Arnold
6	Industrial Ventilation	American Conference of Industrial Hygienists
7	Fan Engineering	Buffalo Forge Company
8	Perry's Chemical Engineers Handbook	R.H.Perry / Mcgraw Hill
9	Useful Tables	ACC Babcock Ltd
10	Producers Fact Book	Universal Engineering Corporation
11	Steam –its Generation and Use	Babcock Wilcox & Co.
12	Mathcem	Soft Ideas Pvt L TD
13	Handbook for Designing Cement Plants - 1st Edition	S.P.Deolalkar
14	Designing Green Cement Plants	S.P.Deolalkar
15	Nomograms for Design and Operation of cement plants	S.P.Deolalkar

2 TECHNICAL PERIODICALS

1.	Zement Kalk Gips	Baverlaug Publications, Germany
2.	Cement Industry Review	CMA Publication, India
3.	World Cement	U.K.
4.	Global Cement	U.K.
5.	International Cement Review	U.K.
6.	Powder and Bulk Handling	Technicom, India

3 REFERENCE MANUALS

RM No.	Manual for	By
1	Grinding in Ball Mills	F.L.Smidth A.S. Co.
2	Rotary Kilns. Preheaters and Calciners	Onoda Engineering and Consulting Co Ltd
3	Preheaters and Calciners	Mitsubishi Mining and Cement Co.
4	Grate Coolers, Contiuous Blending, F.K.Pumps, Air Lifts	Claudius Peters A.G.
5	Cement Manufacture Seminar	F.L.Smidth A.S. Co.
6	Engineering Memoranda	ACC Babcock Ltd
7	Manual of Electric Motors	Crompton
8	Gearboxes	Elecon, New Allenbury
9	Crushers	Hammermills

4 SOURCES OF PLATES

SECTION - 2

Chapter no.	Plate no.	Source
1	2.1.1	Bharat Earth Movers Ltd
	2.1.2	do
	2.1.3	do
	2.1.4	do
2	2.2.1	Cement Hand book by Duda
	2.2.2	do
	2.2.3	do
	2.2.4	
	2.2.5	Otto Labahn
	2.2.6	Ottolabahn
	2.2.7	Otto Labahn
	2.2.8	Otto Labahn
	2.2.9	Larsen & Toubro India
3	2.3.1	Krupp Thyssen brochure material handling
	2.3.2	do
	2.3.3	do
	2.3.4	do
4	2.4.1	H Brunding in VGB
	2.4.2 & 2.4.3	from article by Dr. Bapat in Powder & Bulk Materials
	2.4.4	Fives Cail Babcock - Fives
	2.4.5 & 2.4.6	Singhania Systems Technologies
4a	2.4a.1	FL Smidth
	2.4a.2	Krupp Thyssen
	2.4a.3	do
4b	2.4b.1	Cement Data Book by Duda
	2.4b.2	FL Smidth
	2.4b.3	Mitsubishi
	2.4b.4	Pfeiffer
	2.4b.5	Claudius Peters / Fires

Chapter no.	Plate no.	Source
4c	2.4c.1	Thyssen Krupp
	2.4C.2	Humboldt Wedag
	2.4c.3	KHD
5	2.5.1	BHP Engineers Ltd.
	2.5.2	Christian Pfeiffer Ltd
	2.5.3	Krupp Thyssen
	2.5.4	Fuller KCP Ltd
	2.5.5	FL Smidth
	2.5.6	FL Smidth
6	2.6.1	Cement Data Book by Duda
	2.6.2	do
	2.6.3	Thyssen Krupp
	2.6.4	do
	2.6.5	do
	2.6.6	Claudius Peters
7	2.7.1	Jenson Nicholson (schenck)
	2.7.2	do
	2.7.3	FL Smidth
	2.7.4	Jenson Nicholson
8	2.8.1	FL Smidth
	2.8.2	Krupp Thyssen
	2.8.3	do
	2.8.4	FLSmidth
	2.8.5	FLSmidh
9	2.9.1 to 2.9.5	FL Smidth
	2.9.6	Thyssen Krupp
	2.9.7	FLSmidth
	2.9.8	Thyssen Krupp
10	2.10.1	Thyssen Krupp
	2.10.2	do
	2.10.3	Harbison & Walker
	2.10.4	
	2.10.5	FLSmidth
	2.10.6	Thyssen Krupp
	2.10.7	FLSmidth
	2.10.8	FLSmidth
	2.10.9	FLSmidth
	2.10.10	FLSmidth

Chapter no.	Plate no.	Source
11	2.11.1	Walchandnagar Industries Ltd – Thyssen Krupp
	2.11.2	IKN
	2.11.3	FL Smidth
	2.11.4	FL Smidth
	2.11.5	Claudius Peters
	2.11.6	Polysius
	2.11.7	Claudius Peter
	2.11.8	Claudius Peter
	2.11.9	IKN
12	2.12.1	Batliboi (Air Pollution Control Division)
	2.12.2	do
	2.12.3	FL Smidth
	2.12.4	Batliboi
	2.12.5	Batliboi
	2.12.6	Lurgi
13	2.13.1	Enexco Engineering Co.
	2.13.2	ACC Machinery Co. Ltd
	2.13.3	FLSmidth
	2.13.4	article in ZKG
14	2.14.1	Enexco Engineering Co
	2.14.2	BHP Engineers Ltd checksource
	2.14.3	Enexco Engineering Co
	2.14.4	do
15	2.15.1	Enexco Engineering Co
	2.15.2	ACC Machinery Co. Ltd
	2.15.3	H.W.Carlsen
	2.15.4	H.W.Carlsen
	2.15.5	H.W.Carlsen
	2.15.6	Thyssen Krupp
	2.15.7	Thyssen Krupp
	2.15.8	Thyssen Krupp
	2.15.9	Thyssen Krupp
16	2.16.1	ARF Engineering
	2.16.2	BHP Engineers Ltd
	2.16.3	Indiana Engineering Works Ltd
	2.16.4	Aumund Engineering Co
	2.16.5	ACC Babcock Ltd
	2.16.6	Thyssen Krupp

Chapter no.	Plate no.	Source
	2.16.7	Krupp Thysen
	2.16.8	Fuller KCP
	2.16.9	Krupp Thyssen
	2.16.10	Fuller KCP
17	2.17.1 to 2.17.3	Batliboi (Air Pollution Control Division)
	2.17.4	Bachmann
18	2.18.1	Kay International Ltd
	2.18.2	ACC Machinery Co. Ltd
	2.18.3	Kirloskar Pneumatic Co Ltd
	2.18.4	do
20	2.20.1	New Allenberry Works
	2.20.2	do
	2.20.3	GBR Pfeiffer

SECTION - 8

Chapter no.	Plate no.	Source
3	8.3.1 to 8.3.4	FLSmidth

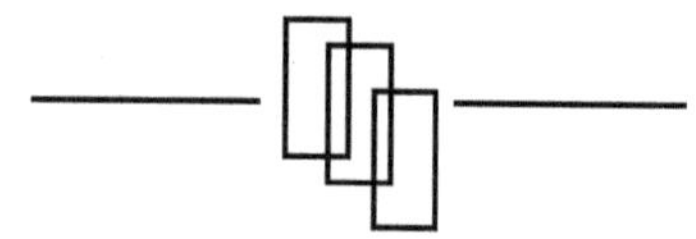

5 SOURCES OF LAYOUTS

SECTION - 5

Chapter no.	Figure no.	Source
1	5.1.7	Wartsila Diesel

SECTION - 6

Chapter no.	Figure no.	Source
1	6.1.28	FLSmidth India
	6.1.29	Layouts of Japanese Cement Plants
	6.1.30	Courtesy ACC
6	6.6.46	Larsen & Toubro India
	6.6.47	Larsen & Toubro India
	6.6.48	FLSmidth India
7	6.7.13	FLSmidth India
8	6.8.12	Larsen & Toubro India
11	6.11.7	FLSmidth India
15	6.15.51	Larsen & Toubro India
	6.15.52	Larsen & Toubro India
	6.15.53	FLSmidth India
	6.15.54	Larsen & Toubro India
19	6.19.63	Larsen & Toubro India
21	6.21.20	IKN
	6.21.21	Larsen & Toubro India
23	6.23.10	Larsen & Toubro India
24	6.24.8	FLSmidth India
	6.24.9	FLSmidth India
	6.24.10	Larsen & Toubro India
30	6.30.19	FLSmidth India
38	6.38.16	FFBH India

SECTION - 8

All flow charts and Layouts are from RB14

6 SOURCES OF TEXT AND FIGURES

SECTION 1

Chapter 2

Table 1.2.2	Respective Standards published by 'Bureau of Indian Standards' as mentioned
Table 1.2.3	Cement Engineers' Handbook by Otto Labahn

Chapter 4

Flow chart no. 1.4.1 and 1.4.2	Chapter 20, RB1

Chapter 5

Para 5.12	RB 1
Para 5.15	RB-1
Para 5.16	Chapter 18,RB1

Chapter 7

Flow charts 1.7.8	Wuestener, Indian Cement Review ZKG, 4/89

Chapter 12

Para 12.3	Onoda/ Mitsubishi manuals See also 'W1.35' and W1.69 In Section 9

SECTION 2

Chapter 1

Table 2.1.1 , 2.1.3 and 2.1.4	Bharat Earth Movers Ltd (BEML)
Table 2.1.2	RB 10

Chapter 2

Table 2.2.1	RB 10

Chapter 3

Fig. no. 2.3.4 & 2.3.5	RB 1, RB 4
Fig. nos. 2.3.8	RB 4

Chapter 4

Fig. no. 2.4.2	Loesche
Fig. nos 2.4.3 & 2.4.4	Thyssen Krupp
Fig. no. 2.4.5 & 2.4.6	FCB catalogue Horomill

Chapter 4a

Fig. no. 2.4a.1	literature from Thyssen Krupp, Slegten, Christian Pfiffer
Fig. no. 2.4a.2	Thyssen Krupp
Fig. no. 2.4a.3 to 2.4a.7	Brussow, Aufebereitungs Technik
Fig. no. 2.4a.8	Haubold and others –ZKG 2/1988

Chapter 4b

Fig, nos 2.4b.2	Vasudeva NCBM
Fig. no. 2.4b.3	H.Wustner Cement Industry Review 1986
Fig. no. 2.4b.4	Mitsubishi literature
Fig. no. 2.4b.5	ZKG 7/1989
Fig. nos. 2.4b.6 & 2.4b.7	Vasudeva – NCBM
Fig. nos. 2.4.b.8 & 2.4b.9	Wuestener- ZKG 1986

Chapter 4c

Figs nos. 2.4c.1 & 2.4c.2	Caldwell ,Kohan – Krupp Polysius
Fig. no 2.4c.3	Kruoo Thyssen literature

Chapter 5

Fig. no. 2.5.7	Christian Pfiffer
Fig. no. 2.5.8	Fuller literature
Fig. nos 2.5.9 & 2.5.10	Krupp Thyssen
Fig. no. 2.5.11	International Cement Production Seminar FLS
Fig. no 2.5.12	Krupp Thyssen literature
Fig. nos 2.5.13 and 2.5.14	Vasudeva – NCB Workshop
Fig. no. 2.5.15	RB 1

Chapter 8

Annexe 2.8.1	RM 2, RM 3 and data on plants for FLS
	KHD data on cyclones of first generation

Chapter 9

Fig. no. 2.9.1	RM 3
Fig. nos. 2.9.2 to 2.9.5	H.M. Garrett, Rock Products, 7/1985
Fig. nos. 2.9.6, 2.9.7 & 2.9.8	Fujisawa and others ZKG 10/1989
Fig. nos. 2.9.9	Champonois ZKG 3/1988
Fig. no. 2.9.10	Matsuoka and others ZKG 7/1980
Table 2.9.2	FLSmidth literature
	Alexander Helm world cement 1994

Chapter 11

Fig. no. 2.11.1	RB 1 & RB 4
Fig. nos. 2.11.10 & 2.11.11	RM 3

Chapter 12

Annexes 2.12.2 & 2.12.3	RB 4
Annexe 2.12.4	RB 4
Annexe 2.12.6	RB 1
Annexe 2.12.7	RB 4

Chapter 13

Fig. 2.13.4	RB 1, RM 6
Table 2.13.1 & 2.13.2	RB 1

SECTION 3

Chapter 2

Annexe 3.2.1 Cement capacity 2011

Fig. no. 3.2.1 as above

Fig. no. 3.2.5 Cement Map of India, Large plants published in Cement Staistics by CMA

Chapter 7

Table no. 3.7.3 Data on Railway Wagons based on publications of R.D.S.O. Indian Railways.

Chapter 12

Table 3.12.8 Based on data from Holtec Consultants

Chapter 13

Table 3.13.16 Based on data from Holtec Consultants

Chapter 14

Table 3.14.1 Based on data from Holtec Consultants

SECTION 4

Chapter 6

Table 4.6.2 Cement Engineers' Handbook, Section L, by B.Kohlass

SECTION 5

Chapter 1

Fig. 5.1.7 Wartsila Diesel

SECTION 6

Chapter 1

Fig. no. 6.1.28	Fuller India / FLSmidth
Fig. no. 6.1.29	Book on Layouts of Cement Plants in Japan
Fig. no. 6.1.30	A.C.C.

Chapter 5

Fig. nos. 6.5.6	RB 4
Figs nos. 6.5.7 and 6.5.8	RB 1
Tables 6.5.2 and 6.5.3	Hilton's Handbook on Belt Conveyors

Chapter 6

Figs. nos 6.6.4 and 6.6.5	FCB group Fives Lille; Division Cimenterie
Figs nos. 6.6.6 to 6.6.8	Wuestener, Indian Cement Review, 1986 also ZKG, 4/89
Fig. no. 6.6.28	RB 4
Fig no. 6.6.46	Larsen and Toubro India
Fig. no. 6.6.47	Larsen and Toubro India
Fig. no. 6.6.48	FLSmidth

Chapter 7

Fig. nos. 6.7.9 and 6.7.10	Christian Pfeiffer-
Plate 6.7.1	Christian Pfeiffer
Fig. 6.7.13	FLSmidth

Chapter 8

Fig. no. 6.8.12	L & T India

Chapter 11

Fig. no. 6.11.4	RM 4
Fig. no. 6.11.7	FLS India

Chapter 15

Fig.no. 6.15.15	O.T.Nielsen, Hasle Refractories
Fig. no. 6.15.51	L&T India
Fig. no. 6.15.52	as above
Fig. no. 6.15	FLSmidth India
Fig. no. 6.15.54	L&T India

Chapter 19

Fig. no. 6.19.10	RB 1
Fig. no. 6.19.11	RB 4
Fig. no. 6.19.15	RB 1
Para 6.19.10	RB 1 and RM 2

Chapter 19a

Fig. no 6.19a.5a	Jenson Nicholson 'Schenck' division
Fig. no. 6.19a.6	RB 1
Fig. no. 6.19a.7 & 6.19a.8	Shakti Textile Engineering co
Fig. no. 6.19a.9 & 6.19a.10	RB 1

Chapter 21

Figs nos. 6.21.8 & 6.21.9	RM3
Fig. no. 6.21.20	courtsey IKN India
Fig. no. 6.21.21	courtsey L & T India

Chapter 22

Fig. no. 6.22.16	IKN India
Fig. no. 6.22.20	IKN India

Chapter 23

Fig. no. 6.23.1	RM 4
Fig. no. 6.23.10	FLSmidth India

Chapter 24

Fig. no. 6.24.8	FLSmidth India
Fig. no. 6.24	FLSmidth India
Fig. no. 6.24.11	L & T India

Chapter 25

Fig. no. 6.25.1	RB 4
Fig. no. 6.25.2	RB 4

Chapter 26

Fig. 6.26.29	Aumund Literature

Chapter 30

Fig. nos. 6.30.6 and 6.30.7	Wuestener, Cement Industry Review 1986
Fig. no. 6.30.14 & 6.30.15	W. Thurst, ZKG, 1989
Fig. no. 6.30.17	RM 5 & RM 6
Fig. no. 6.30.18	Fuller Co.
Fig. no. 6.30.19	FLSmidth India

Chapter 34

Table 6.34.1	Data from R.D.S.O. Indian Railways
Fig. no. 6.34.16	R.D.S.O. Indian Railways

Chapter 35

Fig. no. 6.35.9 & 6.35.10	L. Bauer in ZKG

Chapter 38

Fig. no. 6.38.16	FFE/FLS Bulk handling

Chapter 40	
Fig. no. 6.40.4 & 6.40.5	from Indian Railways
Annexe 6.40.1	Data on tracks etc. from Pawani Trading
Chapter 41	
Fig. no. 6.41.12	RM 4
Annexes 6.41.1 & 6.41.2	RM 6
Chapter 43	
Annexes 6.43.1 to 6.43.4	RM 6
Chaptre 48	
Fig nos. 6.48.10 & 6.48.11	Backau Wolf (Krupp Industries India)
Chapter 54	
Fig. no. 6.54.2	RB 4
Chapter 55	
Fig. no. 6.55.1	RB 2, RM 3
Tables 6.55.5 & 6.55.6	RM 3
Graph 6.55.1	RM 3
Fig. nos. 6.55.7 & 6.55.8	Peter Botha on 'Latest Developments in refractories for cement kilns Kunne in ZKG 10/89
Annexes Tables 6.55.1 to 6.55.4	Gita Refractories and Orient Abrasives Ltd.
Chapter 56	
Table 6.56.1 & 6.56.2	B.I.S.3677/1985 and BIS 3690/1974
Chapter 60	
Fig. No. 6.60.1	Mitsubishi Heavy Industries Ltd

SECTION 7

Chapter 2	
ENQS	by Deolalkar Consultants
TDS	by Deolalkar Consultants

SECTION 8

All Figures, Tables and Annexures are from RB14

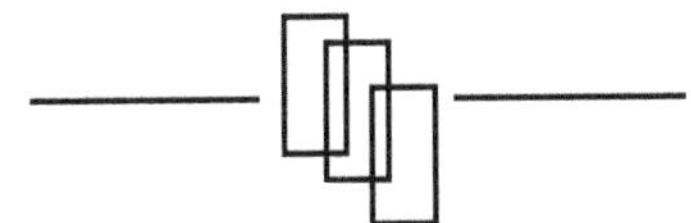

Section - 11

Recommended Reading

Contents

1. Recommended Reading Books/Reports/Articles 1-1

2. Recommended Sources of Data on Cement Production and Capacity 2-2

SECTION 11

Recommended Reading

Books / Reports / Articles

No.	Title	Author / Source
1.	Indian Cement Industry -	Jagdeep Verma - Holtec Consulting
2.	Indian Minereals Year book – 2018 55th Edition – Cement	Govt. of India, Ministry of Mines, Bureau of Mines
3.	Report of Working Group on Cement Industry for 12th 5 year Plan	Department of Industrial Policy and Production- Ministry of Commerce and Industry
4.	Publications of Cement Manufacturers Association	
5.	Publications of National Council for Cement & Building Materials – NCCBM	
6.	Publications of Confederation of Indian Industries - CII & their Green Business Centre	
7.	Designing Green Cement Plants	S.P.Deolalkar
8.	Planning an efficient Power System for a Cement Plant	Mr. P.S.Sharma
9.	Automation in Cement Industry	Mr.P.S.Sharma

Section 11

Recommended Sources of Data on Cement Production and Capacity

Sr. No.	Name of Source	Year	Report on/ of
1.	ICRA	Sept. 2017	Indian Cement Sector
2.	IBEF India Brand Equity Foundation	Jan. 2017	Cement –Advantage India
3.	Holte Consulting		Indian Cement Industry
4.	Dalmia Bharat Group	Dec. 2017	
5.	Care Ratings –Professional Risk Opinion	Jan. 2017	Cement- Research –Perspective & Outlook
6.	Ministry of Mines –Indian Bureau of Mines	Year book 2016	55th Edition Cement -
7.	Department of Industrial Policy & Promotion	2017	Report of Working Group of Cement Industry for 12th 5 Year Plan (2012-17)
8.	Neco Industries Ltd		T.E.F.S. report for a 3 MTPA Cement Plant
9.	Survey of Cement Industry and Directory 2015		Company wise Capacity & Production of Cement Plants in India in 2015-2016
10.	Care Ratings – Professional Risk Opinion		Industrial Research-2017 - Cement Perspective & Outlook
11.	Cement manufacture association		
12.	National council for cement & builder motor		

www.lngramcontent.com/pod-product-compliance
Lightning Source LLC
Chambersburg PA
CBHW080158170726
48282CB00005B/69

* 9 7 8 9 3 8 9 9 7 4 4 2 3 *